Evolutionary Biology

Douglas J. Futuyma

State University of New York at Stony Brook

Evolutionary Biology

Third Edition

Sinauer Associates, Inc. • Publishers
Sunderland, Massachusetts

The Cover

An adult termite of the extinct species *Mastotermes electrodominicus*, preserved in amber dating to the Lower Miocene, more than 20 million years ago. DNA sequences from part of a mitochondrial gene of nine species of insects include the sequence from this species in the second line (De Salle et al. 1992). DNA sequences from this extinct species have helped to determine the phylogenetic relationship of termites to other insects. *(Photograph by Jacky Beckett, courtesy of David Grimaldi, American Museum of Natural History.)*

Evolutionary Biology, Third Edition
© Copyright 1998 by Sinauer Associates, Inc.
All rights reserved.

This book may not be reproduced in whole or in part for any purpose whatever without written permission from the publisher. For information or to order, address: Sinauer Associates, Inc., P.O. Box 407, Sunderland, Massachusetts, 01375-0407 U.S.A. FAX: 413-549-1118.
Internet: publish@sinauer.com; http://www.sinauer.com

Sources of the scientists' photographs appearing in Chapter 2 are gratefully acknowledged:
Photographs of C. Darwin and A. R. Wallace courtesy of The American Philosophical Library
Photograph of R. A. Fisher courtesy of Joan Fisher Box
Photograph of J. B. S. Haldane courtesy of Dr. K. Patau
Photograph of S. Wright courtesy of Doris Marie Provine
Photograph of J. Huxley from the papers of Julian Sorrell Huxley, Woodson Research Center, Rice University Library
Photograph of E. Mayr courtesy of Harvard News Service and E. Mayr
Photograph of G. L. Stebbins, G. G. Simpson, and Th. Dobzhansky courtesy of G. L. Stebbins

Library of Congress Cataloging-in-Publication Data
Futuyma, Douglas J., 1942-
   Evolutionary biology / Douglas J. Futuyma. –3rd ed.
     p.   cm.
   Includes bibliographical references (p.) and index.
       ISBN 0-87893-189-9 (cloth)
       1. Evolution (Biology)   I. Title
QH366.2.F87 1997   97-37947
576.8—dc21   CIP

ISBN 0-87893-185-6 (pbk)

Contents in Brief

Preface   xv
To the Student   xvii

PART I
Background to the Study of Evolution   1

CHAPTER 1
Evolutionary Biology   3

CHAPTER 2
A Short History of Evolutionary Biology   17

CHAPTER 3
Genetics and Development   31

CHAPTER 4
Ecology: The Environmental Context of Evolutionary
   Change   59

PART II
Patterns and History   85

CHAPTER 5
The Tree of Life: Classification and Phylogeny   87

CHAPTER 6
Evolving Lineages in the Fossil Record   127

CHAPTER 7
A History of Life on Earth   165

CHAPTER 8
The Geography of Evolution   201

PART III
**Evolutionary Processes in Populations and
   Species   227**

CHAPTER 9
Variation   231

CHAPTER 10
The Origin of Genetic Variation   267

CHAPTER 11
Population Structure and Genetic Drift   297

CHAPTER 12
Natural Selection and Adaptation   337

CHAPTER 13
The Theory of Natural Selection   365

CHAPTER 14
Multiple Genes and Quantitative Traits   397

CHAPTER 15
Species   447

CHAPTER 16
Speciation   481

PART IV
Character Evolution   517

CHAPTER 17
Form and Function   519

CHAPTER 18
The Evolution of Interactions among Species   539

CHAPTER 19
The Evolution of Life Histories   561

CHAPTER 20
The Evolution of Behavior   579

CHAPTER 21
The Evolution of Genetic Systems   605

CHAPTER 22
Molecular Evolution   625

PART V
**Macroevolution: Evolution above the Species
   Level   647**

CHAPTER 23
Development and Evolution   651

CHAPTER 24
Pattern and Process in Macroevolution   677

CHAPTER 25
The Evolution of Biological Diversity   703

CHAPTER 26
Human Evolution and Variation   727

Epilogue: Evolutionary Biology in the Future   753

Appendix I: Elementary Statistics   755

Appendix II: Contending with Creationism   759

Glossary   G-1

Literature Cited   L-1

Index   I-1

Contents

Preface   *xv*
To the Student   *xvii*

🦋   PART I   🦋
Background to the Study of Evolution   **1**

CHAPTER 1

Evolutionary Biology   3
What Is Evolution?   4
The Evolutionary Perspective   4
Why Is Evolutionary Biology Important?   6
Philosophical Issues   8
Ethics, Religion, and Evolution   8
Hypotheses, Facts, and the Nature of Science   9
Evolution as Fact and Theory   11
The Structure of Evolutionary Biology   12
An Evolving Science   14
Summary   15
Major References   15
Problems and Discussion Topics   16

CHAPTER 2

A Short History of Evolutionary Biology   17
Evolution Before Darwin   17
  Evolutionary Thought Prior to the Nineteenth Century   17
  Lamarck's Contribution   18
Darwin's Contribution   19
  The Life of Charles Darwin   19
  Darwin's Evolutionary Theory   21
Evolution After Darwin   23
  Anti-Darwinian Theories   23
The Evolutionary Synthesis   24
  Major Tenets of the Evolutionary Synthesis   26
  Evolutionary Biology since the Synthesis   28
  Contemporary Evolutionary Studies   28
Summary   29
Major References   29
Problems and Discussion Topics   30

CHAPTER 3

Genetics and Development   31
Chromosomes and Cell Division   31

Meiosis   31
Mendelian Genetics   33
The Relation between Genotype and Phenotype   37
  Dominance   37
  Genotype and Environment   39
  Multiple Loci   41
  Threshold Traits   43
The Genetic Material and Its Functions   43
  Organization of the Genome   45
  Gene Expression   47
Hierarchical Control of Gene Expression in Development   48
Mechanisms of Development   49
  Control of Cell Differentiation   49
  Mechanisms of Morphogenesis   54
Overview of Genetics and Development   55
Summary   56
Major References   56
Problems and Discussion Topics   56
BOX 3.A *Some Variant Patterns of Reproduction and Meiosis in Eukaryotes   34*
BOX 3.B *Elementary Probability   37*
BOX 3.C *Mendelian Ratios   38*
BOX 3.D *A Thumbnail Sketch of Some Important Methods in Molecular Genetic Analyses   50*

CHAPTER 4

Ecology: The Environmental Context of Evolutionary Change   59
Environments   59
Global Climate Patterns   59
Terrestrial Biomes   60
Aquatic Ecosystems   63
  Freshwater Systems   63
  Marine Systems   64
Biogeographic Realms   65
Communities   65
  Ecological Succession   66
  Microhabitats   67
Ecological Niches   68
Responses To Environmental Variation   69
Populations   71
  Population Structure   71

Population Growth   71
Factors Limiting Population Growth   73
Interactions among Species   75
Competition   75
Coexistence of Species at the Same Trophic Level   76
Predation, Parasitism, and Herbivory   77
Mutualism   80
Other Interactions among Species   81
The Unstable Environment   82
Summary   82
Major References   83
Problems and Discussion Topics   83
BOX 4.A *Mathematical Models of Population Growth*   72

🦋 PART II 🦋
Patterns and History   85

CHAPTER 5

The Tree of Life: Classification and Phylogeny   87

The Linnaean Classification System   87
Evolution and Classification   88
Phenetics and Cladistics   92
Phenetics   92
Cladistics   94
Examples of Phylogenetic Analysis   100
Phylogeny of Orders of Insects   100
Relationships of Apes and Humans   103
Evaluating Phylogenetic Hypotheses   105
Phylogenetic Analysis and Extinct Organisms   105
Estimating the History of Character Evolution   107
Some Principles of Evolutionary Change, Inferred from Systematics   108
Principle 1: Homologous Features Are Derived from Common Ancestors   108
Principle 2: Homoplasy Is Common in Evolution   110
Principle 3: Rates of Character Evolution Differ   111
Principle 4: Evolution Is Often Gradual   114
Principle 5: Characteristics Often Owe Their Change in Form to Change in Function   115
Principle 6: Phylogenetic Analysis Documents Evolutionary Trends   116
Principle 7: Most Clades Display Evolutionary Radiation   117
Molecular Data in Phylogenetic Analysis   118
Kinds of Molecular Data Used in Phylogenetic Analysis   119
Molecular Clocks   120
Evidence for Evolution   122
Summary   123
Major References   124
Problems and Discussion Topics   124
BOX 5.A *Taxonomic Practice and Nomenclature*   90
BOX 5.B *Issues in Phylogenetic Analysis*   98

CHAPTER 6

Evolving Lineages in the Fossil Record   127

Some Geological Fundamentals   127
How Good Is the Fossil Record?   129
Taxonomy in the Fossil Record   132
Evolution in the Fossil Record   132
Character Change within Species   133
Speciation   135
Punctuated Equilibrium   137
Evolutionary Transitions among Genera   138
Classes of Tetrapod Vertebrates   143
Correspondence between Phylogeny and the Fossil Record   154
Trends   155
Parallel Trends   156
Rates of Evolution   157
Summary   162
Major References   163
Problems and Discussion Topics   163
BOX 6.A *Correcting for Size in Describing Evolutionary Rates*   159

CHAPTER 7

A History of Life on Earth   165

Before Life Began   165
The Emergence of Life   166
Precambrian Life   169
Prokaryotes   169
Eukaryotes   170
The Tree of Life   171
The Proterozoic Eon   171
Paleozoic Life: The Cambrian Revolution   172
The "Cambrian Explosion" of Animal Phyla   172
Metazoan Phylogeny   175
Paleozoic Life: Ordovician to Devonian   176
Ordovician Diversification   176
Marine Life in the Silurian and Devonian   176
Terrestrial Life: Plants   178
Terrestrial Life: Animals   179
Paleozoic Life: Carboniferous and Permian   180
Terrestrial Life   180
Aquatic Life   182
Mesozoic Life   182
Marine Life   185
Terrestrial Plants and Arthropods   185
Terrestrial Vertebrates   187
The Cenozoic Era   190
Aquatic Life   191
Terrestrial Life   191
The Adaptive Radiation of Mammals   191
Pleistocene Events   197
Summary   199
Major References   199
Problems and Discussion Topics   200

CHAPTER 8

The Geography of Evolution   201

Biogeographic Evidence for Evolution   201
Major Patterns of Distribution   203
Factors Affecting Geographic Distributions   205
Guidelines in Historical Biogeography   207
Competing Explanations for Disjunct Distributions   208
Examples of Historical Biogeographic Analyses   209
The Composition of Regional Biotas   214
Ecological Approaches to Biogeography   216

The Theory of Island Biogeography 216
More about Islands: Interspecific Interactions 218
Effects of Competition on Distributions 219
Community Convergence 220
Gradients in Species Diversity 222
Summary 225
Major References 226
Problems and Discussion Topics 226

PART III
Evolutionary Processes in Populations and Species 227

CHAPTER 9
Variation 231
Definitions 231
Distinguishing Sources of Phenotypic Variation 233
Fundamental Principles of Genetic Variation in Populations 234
Frequencies of Alleles and Genotypes 234
The Hardy-Weinberg Principle 235
The Significance of the Hardy-Weinberg Principle: Factors in Evolution 236
Genetic Variation in Natural Populations 239
Estimating the Proportion of Polymorphic Loci: Protein Electrophoresis 242
Variation at the DNA Level 244
Multiple Loci 245
Variation in Quantitative Traits 247
Variation among Populations 253
Geographic Variation 253
Genetic Distance 256
Patterns of Geographic Variation 257
Kinds of Geographically Variable Characters 259
Species and Higher Taxa 263
Summary 263
Major References 264
Problems and Discussion Topics 264
BOX 9.A *Theories of Evolution, and the Significance of Mendelism* 232
BOX 9.B *Derivation of the Hardy-Weinberg Distribution* 235
BOX 9.C *Mean, Variance, and Standard Deviation* 250
BOX 9.D *Genetic Identity and Distance* 260

CHAPTER 10
The Origin of Genetic Variation 267
Gene Mutations 267
Point Mutations 268
Rates of Mutation 271
Phenotypic Effects of Mutations 276
Effects of Mutations on Fitness 278
Mutation as a Random Process 281
Recombination as a Source of Variation 283
Release of Genetic Variation 283
Some Evidence 284
Erosion of Variation by Recombination 285
Alterations of the Karyotype 286
Polyploidy 286
Chromosome Rearrangements 288

External Sources of Genetic Variation 292
Hybridization 292
Horizontal Gene Transfer 293
Summary 294
Major References 295
Problems and Discussion Topics 295
BOX 10.A *History of the Concept of Mutation* 268
BOX 10.B *Estimating Mutation Rates from Comparisons among Species* 273
BOX 10.C *Accumulation of Recessive Mutations* 275
BOX 10.D *Directed Mutation?* 285

CHAPTER 11
Population Structure and Genetic Drift 297
The Theory of Genetic Drift 297
Evolution by Genetic Drift 300
Genetic Drift in Real Populations 304
Inbreeding 307
Inbreeding in Natural Populations 309
Relationship between Inbreeding and Genetic Drift 314
Gene Flow 314
Models of Gene Flow 315
Gene Flow and Genetic Drift 316
Extinction and Recolonization 316
Estimates of Gene Flow 317
Conclusions about Gene Flow 319
The Neutral Theory of Molecular Evolution 320
Principles of the Neutral Theory 321
Variation within and among Species 323
Rates of Molecular Evolution: Do They Support the Neutral Theory? 323
Population Structure and Gene Trees 327
Haplotype Diversity and Effective Population Size 327
Gene Trees and Population Structure 329
The History of Population Structure 330
Gene Trees and Species Trees 332
Summary 333
Major References 333
Problems and Discussion Topics 334
BOX 11.A *Variation in Allele Frequencies under Genetic Drift* 303
BOX 11.B *The Effect of Inbreeding on Additive Genetic Variance* 309
BOX 11.C *Calculation of Inbreeding Coefficients from Pedigrees* 313
BOX 11.D *Inbreeding, Mutation, and Gene Flow* 315

CHAPTER 12
Natural Selection and Adaptation 337
Examples of Adaptations 337
Explanations of Adaptations 341
What Evolutionary Theory Needs to Explain 342
Experimental Studies of Natural Selection 343
The Nature of Natural Selection 349
Definitions of Natural Selection 349
Levels of Selection 350
Recognizing Adaptations 354
Definitions of Adaptation 354
Nonadaptive Traits 355
Methods for Recognizing Adaptations 356

What Not to Expect of Natural Selection and Adaptation 360
Summary 362
Major References 362
Problems and Discussion Topics 363

CHAPTER 13
The Theory of Natural Selection   365
Evolution by Natural Selection   365
  Fitness   366
  Modes of Selection   366
  Defining Fitness   366
  Relative Fitness and the Rate of Change   367
  Components of Fitness   368
  Components of Fitness: An Example   369
  Dependence of Fitness on Environment   370
Methods of Studying Natural Selection   371
  Correlations across Populations   371
  Deviations from Expected Genotype Frequencies   372
  Temporal Patterns   372
  Response of Populations to Perturbations   372
  Genetic Demography   373
  Functional Studies   373
  Molecular Variation   375
Models of Selection   375
  Directional Selection   375
  Persistence of Deleterious Alleles in Natural Populations   381
  Selection and Gene Flow   381
Polymorphism Maintained by Selection   384
  Heterozygote Advantage   384
  Antagonistic Selection   385
  Varying Selection   385
  Frequency-Dependent Selection   389
Alternative Equilibria   390
  Interaction of Selection and Genetic Drift   392
The Strength of Natural Selection   393
Summary   395
Major Reference   395
Problems and Discussion Topics   395
BOX 13.A *Selection Models with Constant Fitnesses   376*
BOX 13.B *Selection in a Variable Environment   387*
BOX 13.C *The Probability of Fixation when Selection and Genetic Drift Interact   393*

CHAPTER 14
Multiple Genes and Quantitative Traits   397
Evolution at Two Loci   397
  Basic Principles   397
  Directional Selection at Two Loci   398
  Gene Interactions   399
  Linkage Disequilibrium   400
  Multiple Loci in Adaptive Landscapes   402
Quantitative Genetics   409
  Genetic Dissection of Quantitative Traits   409
  Response to Selection   409
  Components of Phenotypic Variation   413
  Estimating Genetic Variance and Heritability   417
Evolution of Quantitative Characters   418

  Responses to Directional Selection   418
  Responses to Artificial Selection   418
❖ *Responses to Stabilizing and Disruptive Selection   421*
  Measuring Natural Selection on Quantitative Characters   422
Selection in Natural Populations   422
  Darwin's Finches   422
  Vestigial Features in Cave Organisms   423
  Evidence of Stabilizing Selection   424
  The Intensity of Natural Selection   424
  Rapid Evolution of Quantitative Traits   425
Correlated Evolution of Quantitative Traits   428
  Correlated Selection   428
  Genetic Correlation   429
  Genetic Constraints on Evolution   436
❖ Special Topics in Phenotypic Evolution   437
❖ *Evolution of Quantitative Characters by Genetic Drift   437*
❖ *What Maintains Genetic Variation in Quantitative Characters?   438*
❖ *Genotype-Environment Interactions   439*
❖❖ *Threshold Traits   440*
❖ *The Epistatic Component of Genetic Variance   442*
Summary   444
Major References   445
Problems and Discussion Topics   445
BOX 14.A *An Example of Genetic Change in a Trait under Directional Selection   399*
BOX 14.B *Genetic Dissection of Quantitative Traits   410*
BOX 14.C *Nonadditive Genetic Variance   415*

CHAPTER 15
Species   447
What Are Species?   447
  A Brief History of Species Concepts   448
  The Biological Species Concept   449
  Alternative Definitions of Species   452
  The Usefulness of the Biological Species Concept   453
  Limitations of the Biological Species Concept   453
Barriers to Gene Flow   457
  Prezygotic Barriers   457
  Postzygotic Barriers   460
Differences among Species   460
  The Temporal Course of Divergence   461
  How Species Are Diagnosed   462
Hybridization   464
  Primary and Secondary Hybrid Zones   464
  Genetic Dynamics in a Hybrid Zone   465
  Other Aspects of Hybrid Zones   467
  The Fate of Hybrid Zones   468
The Genetic Basis of Species Differences   468
  Molecular Differences   468
  Gene Trees and Species Trees   468
  Drosophila Melanogaster *and Its Relatives   469*
  Phenotypic Differences   470
The Genetic Basis of Reproductive Barriers   471
  The Genetics of Postzygotic Isolation   471
  The Genetic Basis of Prezygotic Isolation   476
  The Genetics of Reproductive Isolation: Summary and Significance   477

Summary 478
Major References 479
Problems and Discussion Topics 479
BOX 15.A *Some Terms Encountered in the Literature on Species*
    450
BOX 15.B *Diagnosis of a New Species* *464*

CHAPTER 16
Speciation 481
Modes of Speciation 481
Allopatric Speciation 482
    Kinds of Allopatric Speciation *482*
    Evidence for Allopatric Speciation *484*
    Genetic Models of Allopatric Speciation *485*
    The Role of Genetic Drift *487*
    The Role of Natural Selection *488*
    Ecological Selection and Speciation *488*
    Sexual Selection and Speciation *489*
    Natural Selection for Reproductive Isolation *491*
    Peripatric Speciation *493*
Alternatives to Allopatric Speciation 498
    Parapatric Speciation *498*
    Sympatric Speciation *499*
Polyploidy and Hybrid Speciation 504
    Special Considerations in Plant Speciation *504*
    Polyploidy *504*
    Origin of Species by Hybridization *508*
Rates of Speciation 510
Consequences of Speciation 512
Summary 515
Major References 516
Problems and Discussion Topics 516

PART IV
Character Evolution 517

CHAPTER 17
Form and Function 519
Morphology and Physiology 519
Morphological and Physiological Adaptations 519
    Flight in Birds *520*
    Animals in Hot Deserts *520*
    Plants in Warm, Dry Environments *521*
Body Size 523
    Selection on Size *524*
    Allometry and Isometry *524*
    Allometry and Adaptation *525*
Adaptation and Constraint 528
    Optimal Design and Constraints *528*
    Experimental Analyses of Design *529*
The Evolution of Tolerance 531
    Responses to Stress *531*
    Adaptation to Varying Environments *531*
What Limits the Geographic Range of a Species? 535
Summary 536
Major References 537
Problems and Discussion Topics 537

CHAPTER 18
The Evolution of Interactions among Species 539
Kinds of Interactions 539
Coevolution 539
Phylogenetic Perspectives on Species Associations 541
Coevolution of Enemies and Victims 542
    Theory of Enemy–Victim Coevolution *543*
    Costs of Adaptation *544*
    Empirical Studies of Coevolution between Enemies
        and Victims *545*
Infectious Disease and the Evolution of Parasite Virulence
    548
    Theory of the Evolution of Virulence *549*
Mutualisms 551
    Origins of Mutualisms *552*
    Conflict and Stability *553*
Evolution of Competitive Interactions 554
A Model of the Evolution of Competing Species 554
    Evidence for Coevolution of Competing Species *555*
Evolution and Community Ecology 557
Summary 558
Major References 559
Problems and Discussion Topics 559

CHAPTER 19
The Evolution of Life Histories 561
Life History Phenomena 561
Major Life History Traits and Fitness 561
    Individual Selection and Group Selection *562*
    Life Tables and the Rate of Increase *563*
The Evolution of Demographic Traits 563
    General Principles *563*
    Constraints *563*
    Evidence for a Cost of Reproduction *566*
The Theory of Life History Evolution 566
    Life Span and Senescence *568*
    The Evolution of Age Schedules of Reproduction *568*
    Number and Size of Offspring *571*
    The Evolution of the Rate of Increase *572*
Male Reproductive Success 573
    Effects of Sexual Selection on Life Histories *574*
    Variant Male Life Histories *574*
The Evolution of Dispersal 575
Summary 576
Major References 577
Problems and Discussion Topics 577
BOX 19.A *Life Tables and the Rate of Increase* *565*
BOX 19.B *A Model of the Evolution of Semelparity and Iteroparity*
    570

CHAPTER 20
The Evolution of Behavior 579
Behaviors as Phenotypic Traits 579
    Variation in Behavioral Traits within Species *579*
    Differences among Species *579*
    Phylogenetic Studies of Behavior *580*
Studying the Adaptive Value of Behaviors 581
Optimal Foraging Theory 583

Evolutionarily Stable Strategies   584
Sexual Selection   586
   The Concept of Sexual Selection   586
   Sexual Selection by Contests   588
   Paternity Insurance and Sperm Competition   588
   Sexual Selection by Mate Choice   589
Social Interactions and the Evolution of Cooperation
   594
   Theories of Cooperation and Altruism   594
   Interactions among Related Individuals   595
   Evidence for Evolution by Kin Selection   597
The Role of Behavior in Evolution   601
Summary   603
Major References   603
Problems and Discussion Topics   603
BOX 20.A   *A Model of Optimal Diet Choice*   584
BOX 20.B   *ESS Analysis of Animal Conflict*   585
BOX 20.C   *Calculating Coefficients of Relationship*   600

CHAPTER 21
The Evolution of Genetic Systems   605
Short-Term versus Long-Term Advantage   605
The Evolution of Mutation Rates   605
Sex and Recombination   606
   The Problem with Sex   607
   Can Recombination Rates Evolve?   607
   Hypotheses for the Advantage of Sex and Recombination   608
   Variation and Selection   608
   Evaluating Hypotheses for the Advantage of Sex   611
Haploidy and Diploidy   612
The Evolution of Sexes   612
Sex Ratios, Sex Allocation, and Sex Determination   613
   Sex Ratios in Randomly Mating Populations   613
   The Evolution of Sex Ratios in Structured Populations   614
   Sex Allocation, Hermaphroditism, and Dioecy   615
   The Evolution of Sex-Determining Mechanisms   617
The Evolution of Inbreeding and Outbreeding   619
   Variation in Breeding Systems   619
   Consequences of Inbreeding   620
   Advantages of Inbreeding and Outcrossing   621
   Avoidance of Inbreeding in Animals   622
Summary   623
Major References   624
Problems and Discussion Topics   624

CHAPTER 22
Molecular Evolution   625
Aims and Methods in the Study of Molecular Evolution
   625
   Phylogenetic Insights   626
Evolution of DNA Sequences   626
   Patterns of Variation in DNA Sequences   627
   Interpreting Variation in DNA Sequences: Theory   628
   Evidence on the Causes of Sequence Evolution   629
   Experiments   629
Gene Families and New Gene Functions   635
   Gene Families   635
   Evolution of Novel Functions   637
   Concerted Evolution in Gene Families   637

Transposable Elements   639
   Effects of Transposable Elements   639
   Evolutionary Dynamics of Transposable Elements   640
Highly Repetitive DNA and Genome Size   641
Evolution of Novel Genes and Proteins   642
Summary   644
Major References   645
Problems and Discussion Topics   645
BOX 22.A   *Detecting Natural Selection from DNA Sequence Data*
   633

🦋   PART V   🦋
*Macroevolution: Evolution above the Species
   Level* 647

CHAPTER 23
Development and Evolution   651
Approaches to Studying Development and Evolution
   652
Ontogeny and Phylogeny   652
Developmental Principles of Evolutionary Change   653
   Individualization and Dissociation   654
   Heterochrony   655
   Heterotopy   661
   Changes in Tissue Interactions   662
   Thresholds   662
   Pattern Formation   663
   Developmental Plasticity and Integration   665
Developmental Genetics and Evolution   665
   Hox Genes and Other Regulatory Genes   665
   The Evolution of Hox Genes   666
   The Evolution of Regulatory Gene Function   666
The Problem of Homology   669
Implications of Development for Evolution   670
   Constraints   671
   Nonadaptive Characters   674
   Discontinuity of Evolutionary Change   674
Summary   675
Major References   676
Problems and Discussion Topics   676

CHAPTER 24
Pattern and Process in Macroevolution   677
Gradualism and Saltation   677
   Arguments for Gradualism   678
   Arguments for Saltation   680
Selection and the Evolution of Novelty   681
   Incipient and Complex Features   681
   Functional Integration and Complexity   683
Rates of Evolution   687
   Taxonomic Rates   687
   Rates of Character Change   688
   Genetic Approaches to Evolutionary Rates   688
   Punctuated Equilibrium Revisited   689
Directions, Trends, and Progress   691
   Trends: Kinds and Causes   691
   Discriminating among Kinds of Trends: An Example   694
   Are There Global Trends in Evolutionary History?   694
   The Question of Progress   699

Summary   700
Major References   701
Problems and Discussion Topics   701

CHAPTER 25
The Evolution of Biological Diversity   703
Measuring Diversity   703
Ecological Approaches to Contemporary Patterns of
   Diversity   704
   Factors That Influence Diversity   704
   Are Communities Saturated with Species?   704
   Effects of History on Contemporary Diversity Patterns   706
The History of Diversity and Its Causes   706
   Taxonomic Diversity through the Phanerozoic   707
Patterns of Origination and Extinction   708
   Null Hypotheses   708
   Origination and Extinction Rates through Time   709
   Correlated Rates of Speciation and Extinction   711
Extinction   712
   Causes of Extinction   712
   Mass Extinctions: When and Why?   713
   Selectivity of Extinction   714
   Extinction and the Problem of Adaptation   714
   Tiers of Evolutionary Change   714
Origination and Diversification   715
   Release from Competition   715
   Does Diversity Attain an Equilibrium?   721
Summary   724
Major References   725
Problems and Discussion Topics   725
BOX 25.A *An Enigma: The Evolution of Animal Body Plans   710*
BOX 25.B *The Next Mass Extinction: It's Happening Now   722*

CHAPTER 26
Human Evolution and Variation   727
Controversy and Objectivity   727
Phylogenetic Relationships   728

The Fossil Record   730
   Interpreting the Hominid Fossil Record   730
   Australopithecines   731
   Origin and Evolution of Homo   732
   Homo Sapiens   733
   Causes of Hominid Evolution   734
The Origin of Modern Human Populations   734
   The Transition from Archaic to Modern Homo Sapiens   734
   Migrations   736
   Racial and Ethnic Groups   737
Genetic Variation in Human Populations   738
   Polymorphisms   739
   Rare Deleterious Alleles   739
   Natural Selection and the Evolutionary Future   739
Human Behavior   740
   Culture   741
   The Evolution of Human Behavior   741
   Biological Foundations of Human Behavior   742
   Human Nature   742
   Sociobiology   743
   Evolutionary Psychology   744
Variable Behavioral Traits   744
   Sexual Orientation   745
   Intelligence   747
Evolution and Society   749
Summary   750
Major References   750
Problems and Discussion Topics   751

Epilogue: Evolutionary Biology in the Future 753
Appendix I: Elementary Statistics 755
Appendix II: Contending with Creationism 759
Glossary   G-1
Literature Cited   L-1
Index   I-1

Preface

In the preface to the second edition of Evolutionary Biology (1986), I remarked that during the seven-year interval since its predecessor, the field of evolutionary biology had become even more exuberant, more commanding of the full sweep of biology, than ever. Eleven years later, the same may be said with even greater force. Subjects such as evolutionary developmental biology and evolutionary physiology are growing with new vigor, topics such as the evolution of genetic systems are being plumbed more deeply, and evolutionary biologists are increasingly aware of the contributions they can make to applied fields such as health science, human genetics, agriculture, and conservation. New conceptual approaches have come to the fore in population genetics, phylogenetic methods are now applied to a great range of questions in a new integration of evolutionary history with evolutionary processes, and molecular methods have revolutionized almost every subdiscipline of evolutionary science. Outside of evolutionary biology as such, within such fields as molecular and developmental biology, it is increasingly recognized that the methods, concepts, and perspective of evolutionary biology can provide illumination—that "nothing in biology makes sense except in the light of evolution" (Dobzhansky 1973).

Every teacher knows how hard it is to frame a course for a diverse class of students who differ in background and motivation. The writer of a textbook faces a similar difficulty. The predecessors of this book have been used in advanced undergraduate and beginning graduate courses, and have found use by graduate students and by biologists in other fields as an entry into the evolutionary literature. I have tried again to serve these audiences, and again have emphasized that evolutionary biology, like every other science, is not a static collection of facts and verities, but instead is a dynamic enterprise in which new questions arise, old questions remain only partially answered, and almost all conclusions are incomplete and at least somewhat tentative. It is as important, then, to emphasize not only the accomplishments of a science, but also its procedures—how hypotheses are framed and tested, how evidence is interpreted, how the history of a question engenders the questions we ask today.

The previous editions of this book have been criticized for slighting some important topics, and for too densely written a presentation to serve the needs of some students. I have tried to correct these faults by adding chapters (17–21) on topics that were omitted or treated only briefly in previous editions, and by more expansive, even relaxed expositions that I hope will make the subject more accessible, especially for undergraduates. Hence, I have provided simpler but often longer explanations, especially of basic principles; have presented extended examples of observations and experiments that should serve not only to incarnate concepts but also to provide detailed evidence and to illustrate the methods of evolutionary science; and have indulged in redundancy where it seemed necessary. The cost of these changes has been a book that, at least in size, is something of a hopeful monster. Still, the product, though perhaps excessive, is no longer than some textbooks of biochemistry, molecular biology, and genetics, which, however broad in scope, are narrower than a field that embraces and illuminates all the life sciences.

Probably few instructors will assign the whole book for a one-semester course. They may feel that the core of the subject resides in Chapters 1, 2, 5-7, 9-13, 14 in part, 15, 16, and 23-25. Chapters 3 and 4 may provide needed background for some students, and many instructors will select additional material from Chapters 8, 17-21, and 26. Chapter 22, on molecular evolution, treats one of the most dynamic topics in contemporary evolutionary biology, but molecular methods and some aspects of molecular evolution are also integrated into many other chapters.

In the 1986 edition, I acceded to the practice of many instructors by beginning (after the prefatory background chapters) with genetic mechanisms of evolutionary change and proceeding to phylogeny, macroevolution, and evolutionary history. I have become convinced that, at least for undergraduate courses, the sequence adopted in the 1979 edition is more effective, and I have returned to it. Thus, the exposition in this edition begins with phylogeny and patterns of historical evolution (Part II), turns then to evolutionary processes (Part III), and after treating character evolution (Part IV), returns to macroevolution considered in the light of the theory of processes (Part V). I have readopted this sequence for both pedagogical and intellectual reasons. Many students want to learn about evolution because of curiosity about biological diversity, or dinosaurs, or how humans fit into the grand history of life. Their interest is better captured or held by beginning with the history of life and how we can come to know it than with the abstract concepts and reductions of population genetics—an acquired taste that most students must be taught to appreciate. Phylogeny, diversity, and evolutionary history, moreover, are the foundation and raison d'être of evolutionary biology. These are the subjects that provide evidence for evolution, and that have provided both fundamentally important concepts (e.g., homology, cladogenesis) and the phenomena that evolutionary theory is designed to explain. Evolutionary biology, finally, is an inescapably historical subject—even population genetics is largely devoted to inferring the historical events responsible for the genetic composition of populations—and it is well to emphasize the historical perspective from the outset. Almost every subject in evolutionary biology is informed by history, phylogeny, and "tree thinking," from gene trees to patterns of species diversity.

Acknowledgments

In preparing this book, I have been indebted to many colleagues, too numerous to name, who have answered questions, guided me to literature, and provided illustrations. For reviewing chapter manuscripts and saving me from many errors, I am most grateful to Michael Arnold, Steven M. Carr, Robert Dorit, Douglas Erwin, Steven Frank, Theodore Garland, James Hanken, Jody Hey, Jerry Hilbish, Kent Holsinger, Raymond Huey, David Jablonski, Mark Kirkpatrick, Allan Larson, Rudolf Raff, David Reznick, Wolfgang Stephan, Donald Stratton, Randall Susman, John N. Thompson, Jonathan Waage, Günter Wagner, Bruce Walsh, Kenneth Weiss, and especially to Richard Harrison, David Houle, and Amy McCune, who shouldered a particularly heavy burden. I apologize for not having followed all their suggestions, and bear responsibility for the errors and misinterpretations that remain despite their immensely helpful efforts. Leo Shapiro contributed indispensably by finding illustrations and commenting on much of the text. I am grateful to Siana LaForest for help with obtaining permissions, and to Mary Ann Stewart for typing the manuscript from mountains of handwritten copy. I gratefully acknowledge the John Simon Guggenheim Foundation for support and the Department of Entomology at the Smithsonian Institution for hospitality during preparation of part of the book. With great pleasure I thank Norma Roche for superb copy-editing and Andy Sinauer, Kerry Falvey, and their colleagues at Sinauer Associates for aid, encouragement, and incomparable standards.

Douglas J. Futuyma
Stony Brook, New York
October, 1997

To the Student

All the biological sciences rest on two central principles. One is that all life processes have an entirely physical and chemical (i.e., material) basis. The other is that all organisms and their characteristics are products of evolution. Hence, evolution provides a framework for understanding all the features of living things, and illuminates all the biological disciplines, from molecular biology and biochemistry to physiology, behavior, and ecology. Evolution has important social applications in the health sciences, agriculture, conservation, and other human endeavors. Moreover, evolution has profound implications for anthropology, sociology, and philosophy—in short, for how we view ourselves. Everyone should know something about evolution, and for anyone who envisions a career in the life sciences, an understanding of evolution is indispensable.

The core of evolutionary biology consists of describing and analyzing the history of evolution, and of analyzing the causes and mechanisms of evolution. The study of evolutionary history draws chiefly on the fields of paleontology, systematics, and genetics. Many disciplines, especially genetics, molecular biology, and ecology, contribute to the study of evolutionary mechanisms. Parts of this book (especially Chapters 5, 6, 9–16, and 23–25) treat the core of evolutionary biology, and are likely to be assigned by your instructor. Other parts (Chapters 7, 8, 17–22, 26) treat special aspects of evolutionary history and the evolutionary aspects of various biological disciplines such as physiology, ecology, behavior, molecular biology, and human biology. Your instructor may choose not to assign some of these chapters, but a broadly trained evolutionary biologist should have some acquaintance with all of these topics.

The scope of evolutionary biology is vast: all organisms, and all the features of organisms, are grist for its mill. Therefore, a course on evolution is likely to differ substantially from courses on most other, more narrowly focused, biological subjects. In contrast to fields such as physiology and biochemistry, in which many specific, detailed facts are an essential complement to the field's broad principles, evolutionary biology encompasses so much, and evolutionary biologists engage in such diverse studies, that the number of specific facts that all persons trained in evolution should know is rather limited. (But there are some: for instance, everyone should know the sequence of geological periods and the times of some major evolutionary events.)

Another important difference between evolutionary biology and most other biological disciplines is the role of history. Everything that biologists study, from protein structure to the species composition of ecological communities, is the product of evolutionary history. Many biological disciplines (e.g., biochemistry) focus mostly on describing biological systems and how they function now; they devote rather little attention to how these systems came to be the way they are. (However, some scientists in these disciplines recognize that evolution can be critically important for fully understanding their subject matter.) In contrast, much of evolutionary biology is concerned with understanding the historical pathways and processes that led to the characteristics of organisms, and with understanding why their features are not other than what they are. (Thus an evolutionary biologist might ask why most species have two sexes rather than another number, why they have fixed rather than potentially infinite life spans, or why their genomes include noncoding rather than just coding sequences of DNA.) Consequently, much of evolutionary biology consists of drawing inferences about historical events and processes, such as the probable history of natural selection that led to the evolution of a certain feature or the characteristics of a species' remote ancestors that may have left an imprint on its features today. Such historical inferences may be based on experiments, on fragmentary historical evidence (e.g., from the fossil record), or on comparisons between observed patterns of variation and those predicted by various hypotheses. Evolutionary inference therefore depends on careful reasoning and a strong command of concepts, and differs from the process of inference in most other biological disciplines by having to peer into the past.

Training in evolutionary biology usually emphasizes concepts, general principles, broad generalizations, and the methods and reasoning that underlie evolutionary analyses. For instance, it is critical to understand natural selection and the various ways in which it operates, it is useful to know how high rates of mutation usually are, and it is essential to understand the logical basis for inferring phylogenetic relationships among species or for deducing mutation rates from differences among DNA sequences. It is less important to know how natural selection affects the evolution of coloration in a particular species of butterfly. An example of this kind illustrates the concept of natural selection, and more importantly, shows how natural selection can be studied, but it is likely to be only one of many comparable studies that could have served these purposes.

Much of evolutionary biology, then, deals with concepts. Many of these, such as allele frequencies, population sizes, and natural selection, are rather abstract. Learning about evolutionary processes, therefore, is closer to learning the principles of physics than to learning anatomical structures or biochemical pathways. Like those of physics, in fact, the concepts and principles of evolution have been extensively developed in the form of mathematical models. (These are kept to a bare minimum in this book, but a few are necessary.) Many students at first find the study of evolution difficult because it includes more abstract thinking and reasoning than they have encountered in most other biology courses.

How, then, can you best learn what your instructor probably expects you to know?

First, concentrate on fully understanding the concepts of evolutionary biology, the several evolutionary hypotheses that might account for some feature of organisms, and the methods by which researchers have tested these hypotheses. Very often, in treating a particular topic, I have first described the theory, and have then provided one or more examples of studies that have been done to test the theory. The examples are important for illustrating general methods of evolutionary research and for showing how the theory has been validated. However, remembering the details of the experiments is generally not as important as understanding the ideas that they address.

Second, recognize that evolutionary theory is presented cumulatively in this book. Later chapters build on concepts and principles introduced in earlier ones. You will not understand Chapter 16 (Speciation), for example, unless you are thoroughly familiar with the concepts developed in Chapters 9 through 14, and almost all chapters from 6 through 26 refer to concepts introduced in Chapter 5 (The Tree of Life). You should assume that every paragraph—indeed, every sentence—is there for a reason (because it is!), and you should not proceed if you are not quite sure you understand it.

Third, bear in mind that in evolutionary biology, as in all sciences, words often have specific, technical meanings that differ from their quotidian (everyday) meanings. (Fitness, population, and frequency are examples.) If you do not understand a word, look it up before continuing. Check the glossary or the index first. If it is not there, it may be a word that is broadly used in biology (e.g., dorsal or chloroplast) and can be found in a general biology textbook. Or it may be a nontechnical word (e.g., quotidian) that you happen not to know, in which case you should use your dictionary.

Fourth, I strongly recommend that you work through some of the simpler problems at the end of some chapters and answer some of the simpler questions. (Your instructor may provide better ones.)

Fifth, do not expect to find many pat, dogmatic answers or simple declarations of fact in this book. Very often, the exposition of a topic builds slowly and carefully toward a conclusion, and sometimes the conclusion is that we do not know which of several hypotheses best accounts for our observations. In evolutionary biology, as in every science, achievement of understanding is a continual, perhaps never-ending, process in which old ideas are refined, enlarged, and sometimes entirely supplanted by new ones. In every biological subject, including evolution, the amount we do not know greatly exceeds what we do know. (That is why almost every topic in biology is a subject of continuing research by college professors and other biologists.) It is as important to learn the process by which we achieve partial answers to our questions as it is to learn what those answers are—for the answers will probably be slightly different a few years from now. It is as important to recognize what we do not know as to understand what we think we know, for this divide is where further understanding begins.

I have tried to ease the task of reading the text by highlighting important principles and conclusions in italics. Important terms are highlighted in two ways in the text: those of primary importance are found in boldfaced type; other terms are found in small capitals. To facilitate understanding, I have also presented sometimes extended examples of evolutionary research in the hope that they will provide abstract ideas with more corporeal reality. Some students may at first be distracted by the references inserted throughout the text (e.g., Darwin 1859). You will quickly learn to read past them. They are an integral part of all scientific literature, and in this text serve multiple purposes: crediting the authors of studies, enabling the reader to find the study, and providing references that offer an entry into the scientific literature on the topic.

I have attempted to make this book useful as a foundation for students who expect to draw on evolutionary biology in their professional careers. However, I also have done my best to make it accessible and informative for students who do not expect to become biologists. I hope I have succeeded, and that some readers will find evolution so rich a subject, so intellectually challenging, fertile in insight, and deep in implications that they will turn to the study of evolution for a most rewarding career. *Felix qui potuit rerum cognoscere causas*, wrote Virgil: happy is the person who has been able to learn the causes of things.

PART I

Background to the Study of Evolution

The four chapters in Part I provide an important foundation for studying evolution.

Chapter 1 introduces the field of evolutionary biology. It describes the kinds of questions that evolutionary biology strives to answer, the subjects studied in the subdisciplines of evolutionary biology, and the relevance of evolutionary studies to society and our lives. It also describes the ways in which centuries-old ways of thought were challenged and altered by Darwin's evolutionary theory.

One of the important lessons of evolutionary biology is that the characteristics of organisms cannot be fully understood except in the light of their history. The same applies to any field of human endeavor, including the study of evolution. Chapter 2 sketches the history of evolutionary thought. It describes the earlier beliefs against which Darwin and his successors had to contend, alternative evolutionary theories that we now believe are erroneous, the formulation of modern evolutionary theories, and some contemporary currents in evolutionary research. It provides a useful summary of the principles of evolutionary theory that were formulated about 50 years ago, principles that form the foundation of most contemporary evolutionary theory and research.

All of biology converges in evolutionary biology. Knowledge from all the biological disciplines is relevant to understanding evolution; conversely, evolutionary analyses can illuminate all the fields of biology. Thus some background in all the biological disciplines can be useful in studying evolution. For reasons of space, I have been forced to provide background, in Chapters 3 and 4, only on genetics, development, and ecology, for some knowledge of these fields is not only useful, but indispensable, to the study of evolution. Some of the material in these chapters will arise early in subsequent chapters, and some only much later. The student may wish to skim these chapters just to see what they cover, and refer back to them when necessary.

Evolutionary Biology

If you have never before done so, take an hour on a warm day and look—really look—at the life in a field or woodlot, or even along a roadside, in a back yard, or in an abandoned city lot. You will notice first the familiar creatures—perhaps oak trees, grasses, squirrels, sparrows. Then look more carefully; you will probably see dozens of kinds of plants and insects, perhaps a few more species of birds; if you dig into the soil or break open a rotting log, you will find yet more kinds of insects, and perhaps spiders, earthworms, and millipedes. With a magnifying glass you will see mites and fungi; with a microscope, you would find bacteria, nematodes, and many other minute forms of life. Wherever you look, the world teems with astonishingly diverse living things. Even your skin, your mouth, and your intestine are home to diverse communities of bacteria. More than a million living species have been named, and at least another million—possibly 10 or 20 times that many—have yet to be described.

Despite this extraordinary diversity, organisms have many points in common. You, the squirrel, and the sparrow have fundamentally the same anatomy, and were even more similar as embryos. Even the most different-looking organisms—fungi, plants, insects, vertebrates—are built of cells that are similar in their internal structure. And all the life around you, from bacteria to people, has the same hereditary material, the same genetic code, the same protein-synthesizing machinery, and proteins composed of the same amino acids.

Look closely at any one organism and you may marvel at features that were seemingly designed to enable it to survive and reproduce. The snapdragon in your garden, for instance, has brightly colored petals that attract pollinating bees, and stamens and pistil so placed within the closed flower that they contact the bee's head as she forces her way in (Figure 1.1A). The visiting bee has a thick coat of finely branched hairs that trap pollen grains and an elaborate basket of hairs on her hind legs, into which she will comb the pollen and carry it to the hive to feed the colony's larvae (Figure 1.1B). In humans we marvel at the eye, the immune system, the brain with its apparently endless capabilities. But there are anomalies, too—features in some organisms that do not seem optimally designed. The dandelions in the lawn have bright yellow flowers that, like the snapdragon's petals, seem designed to attract pollinating insects; yet the dandelion's pollen is sterile, and its seeds develop without fertilization. Nor is the dandelion the only species with odd, useless, or even poorly constructed features. Why do humans have an appendix? Why do men have nipples? Why do the human digestive and respiratory tracts cross, so that we occasionally choke on food?

The diversity of life, the differences and similarities among organisms, and the characteristics of organisms, both adaptive and nonadaptive—these are the great themes of the science of evolutionary biology. Of all the biological sciences, evolutionary biology is the most sweeping and comprehensive, for it aims to explain everything about the living world: Why are there are so many kinds of organisms? Why do they share some features but differ in others? Why do these species—from bacterium to human—have the features they do? From these grand questions spring thousands of others that draw on and illuminate every biological discipline: Why is there a universal genetic code? Why is there genetic crossing-over? How did the diverse enzymes in a cell come into being? Why do different species have different life spans? Why do they differ in what they can learn? Why are some species found only in Australia and others only in South America? How did humans come to walk upright, to be almost hairless, and to have such extraordinary mental abilities?

The struggle to answer such questions has continually deepened our understanding of the living world. It has provided a perspective that gives deeper meaning to information in every biological discipline, from molecular biology to ecology. Indeed, *evolution is the unifying theory of biology*. "Nothing in biology makes sense," said the ge-

FIGURE 1.1 Adaptations of plants and animals. (A) The flower of a snapdragon (*Antirrhinum majus*), in external view and in sagittal section, with the terminal parts of the petals deleted. The petals form a closed chamber, so that a bee must force down the lower lip in order to enter and take nectar from the base of the flower. The stigma and anthers are situated at the top of the passageway so that they contact the head of the bee as she enters. (B) A worker honeybee (*Apis mellifera*), and a close-up of the exterior of the hind leg. The bee scrapes pollen from her body hairs and compacts it, using the press, into pellets that are carried on the concave surface of the leg, which is fringed with a basket (corbicula) of long hairs. (A after Lawrence 1951; B after Winston 1987).

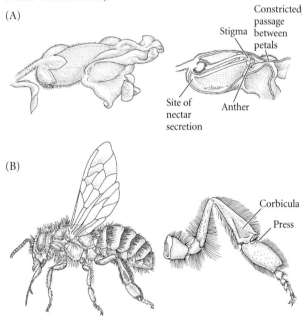

(A)

Stigma

Constricted passage between petals

Site of nectar secretion

Anther

(B)

Corbicula

Press

neticist Theodosius Dobzhansky, "except in the light of evolution."

What Is Evolution?

The word *evolution* comes from the Latin *evolvere*, "to unfold or unroll"—to reveal or manifest hidden potentialities. Like that of most English words, the meaning of *evolve* has changed—evolved—greatly, for it no longer implies that an evolving system merely reveals what it has always contained. We do not believe that the very first forms of life already had packed into them the characteristics of snapdragons or humans.

In its broadest sense, *evolution* simply means "change." For example, the changes in individual stars, from their "birth" to their "death," are sometimes referred to as "stellar evolution." Usually, though, we don't apply the term to the changes an individual entity undergoes during its lifetime. Rather, an evolving system is ordinarily one in which there is descent of entities, one generation from another, over time, and in which characteristics of the entities differ across generations. Thus, **evolution** in a broad sense is *descent with modification*, and often *with diversification*. Many

kinds of systems are evolutionary, even cultural phenomena such as languages, automobiles, cuisines, and computer programs. In all such systems there are **populations**, or groups, of entities; there is **variation** in one or more characteristics among the members of a population; there is HEREDITARY SIMILARITY between parent and offspring entities; and over the course of generations there may be *changes in the proportions* of individuals with different characteristics within populations. This process constitutes **descent with modification**. Populations may become subdivided so that several populations are derived from a COMMON ANCESTRAL POPULATION. If different changes in the proportions of variant individuals transpire in the several populations, the populations DIVERGE, or DIVERSIFY.

The modern theory of evolution holds that the hereditary variation underlying this process in biological systems ultimately originates by some process of **mutation**, in which one reasonably stable, inherited state of a characteristic is somehow transformed into a different reasonably stable, inherited state. It holds also that changes in the proportions of different states are due to some process of **sorting** among the variants, so that some variants survive and reproduce more than others, causing a shift in the representation of the different variants in subsequent generations.

All these properties of an evolutionary process pertain to populations of organisms, in which there is hereditary transmission of characteristics (based on genes, composed of DNA or, in a few cases, RNA), variation owing to mutation, and sorting of variation by several kinds of processes. Chief among these sorting processes are CHANCE (random variation in the survival or reproduction of different variants), and **natural selection** (consistent, nonrandom differences among variants in their rates of survival or reproduction). It is natural selection that causes **adaptation**—improvement in function. Thus **biological** (or **organic**) **evolution** is *change in the properties of populations of organisms, or groups of such populations, over the course of generations*. The development, or **ontogeny**, of an individual organism is *not* considered evolution: individual organisms do not evolve. The changes in populations that are considered evolutionary are those that are "heritable" via the genetic material from one generation to the next. Biological evolution may be slight or substantial; it embraces everything from slight changes in the proportions of different forms of a gene within a population, such as the alleles that determine the different human blood types, to the alterations that led from the earliest organism to dinosaurs, bees, snapdragons, and humans.

The Evolutionary Perspective

As we will see in Chapter 2, the idea of evolution was "in the air" in the early nineteenth century, but it did not become a coherent, influential scientific theory until Charles Darwin published *The Origin of Species* in 1859. Darwin so effectively marshaled evidence for evolution that within about 15 years, a majority of scientists had accepted that diverse organisms have descended with modification from common ancestors. Some decades later, scientists agreed that Darwin had correctly identified one of the major *causes* of evolution: natural selection.

Darwin's theory of biological evolution is one of the most revolutionary ideas in Western thought, perhaps rivaled only by Newton's theory of physics. It profoundly challenged the world view that had long prevailed.

First, the prevailing view had been one of a static world, identical in all essentials to the Creator's perfect creation. Living species, it was believed, had been individually created in their present form (the doctrine of special creation). In fact, it had long been believed that no species had ever become extinct, a belief that crumbled in the century before Darwin as fossils such as dinosaurs and mammoths were discovered. Geologists had come to understand that the earth had had a long history of dynamic change, but it was Darwin who *extended to living things, and to the human species itself, the conclusion that change, not stasis, is the natural order.*

Second, people had long sought the causes of phenomena in purposes: the will of God, or the FINAL CAUSES (the purposes for which events occur) that Aristotle contrasted with EFFICIENT CAUSES (the mechanisms that cause events to occur). Newton revolutionized Western thought by providing purely MECHANISTIC explanations for physical phenomena. Thereafter, physicists would exclude from their theories any reliance on purpose (final causes), divine design, or the operation of any supernatural forces in the day-to-day workings of the physical world. Darwin's immeasurably important contribution to science was to show *how mechanistic causes could also explain all biological phenomena,* despite their apparent evidence of design and purpose. By coupling undirected, purposeless variation to the blind, uncaring process of natural selection, Darwin made theological or spiritual explanations of the life processes superfluous. In the decades that followed, physiology, embryology, biochemistry, and finally molecular biology would complete this revolution by providing entirely mechanistic explanations, relying on chemistry and physics, for biological phenomena. But it was Darwin's theory of evolution, followed by Marx's materialistic (even if inadequate or wrong) theory of history and society and Freud's attribution of human behavior to influences over which we have little control, that provided a crucial plank to the platform of mechanism and materialism—in short, of much of science—that has since been the stage of most Western thought.

Once the framework of evolution rather than special creation is adopted, it follows that *the characteristics of living things can be fully understood only in light of their history.* Because the characteristics of organisms have been modified from antecedent characteristics, the earlier state dictates that only some of the many imaginable later states can be realized. Thus the crossing of the respiratory and digestive tracts in mammals is explicable not by any functional advantage of this arrangement, but rather by a characteristic of the group of fishes that became modified for life on land to which mammals ultimately trace their ancestry (Figure 1.2). Thus there is an important element of HISTORICAL CONTINGENCY in evolution: the condition of a living system, or of its environment, at a certain time may determine which of several paths of change the system will

FIGURE 1.2 A highly diagrammatic view of the respiratory and anterior digestive systems of fishes and mammals (including humans). (A) In most bony fishes, the external nares (nostrils) lead to a blind olfactory sac. An air bladder, used to regulate the animal's depth in the water, opens from the ventral wall of the esophagus. Gases are exchanged as water enters the mouth and is forced out the gill slits. (B) In lungfishes and coelacanths, living members of the primitive group of fishes that gave rise to terrestrial vertebrates, the olfactory sac opens into the pharynx via internal nares, but the flow of water or air is used only for olfaction, not for respiration. The air bladder is used as a lung in lungfishes, but gases reach it as swallowed air via the mouth. (C) In terrestrial vertebrates, gas exchange occurs via the external nares, internal nares, pharynx, trachea, and lungs. The respiratory pathway crosses the path that food follows from pharynx to esophagus. This is not an optimal design; it makes sense only in light of a history in which the nares of primitive fishes were recruited for breathing.

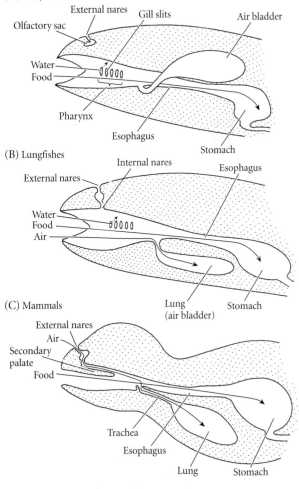

(A) Bony fishes

(B) Lungfishes

(C) Mammals

follow. Historical explanation of the properties of living systems is one of the most important contributions of evolutionary science to biology.

In positing an evolutionary process wherein natural selection sorts among hereditary variations, *Darwin identified variation as a centrally important fact of biological systems.* In doing so, he broke with a 2000-year-old tradition that had dominated Western thought. The tradition stemmed from Plato, whose philosophy was built on the concept of the "eidos," the "form" or "idea," a transcendent ideal form imperfectly imitated by its earthly representations. In his famous metaphor of the cave in *The Republic*, Plato likened earthly objects, such as the triangles or horses we are familiar with, to the shadows cast on the wall of a cave by objects that pass by the entrance. Like people within the cave, bound so that they face the wall, we see only the shadows, the imperfect representations, of reality. Likewise, the reality—the ESSENCE—of the true equilateral triangle is only imperfectly captured by the triangles we draw or construct, all of which are imperfect, and vary from the true, essential triangle. And so it is with horses, or any other species: each has an eternal, immutable essence, but each individual has imperfections. In this philosophy of ESSENTIALISM, variation is accidental imperfection; only essences matter.

Plato's philosophy of essentialism became incorporated into Western philosophy. Its central tenet was that however much the objects in a class might accidentally vary, the class still had a defining essence that could not change. Thus each species—horse, zebra, or ass, for instance—has an essence, and one cannot be changed into another any more than a triangle can vary enough to become a rectangle.

Darwin was bold enough to claim that species do not have essences. *Every* characteristic of a species can vary, and if natural selection or some other sorting process alters the frequencies of variants, *every* characteristic of a species can be altered radically, given enough time. From this evolutionary perspective, then, average characteristics are abstractions; only the pattern of variation is real. To this day, one of the most important contributions of evolutionary biology to our world view is its unceasing emphasis on the reality and importance of variation.

Why Is Evolutionary Biology Important?

Why should scientists study evolution? The foremost reason is that, like every other science, evolutionary biology informs us and helps us to understand the world around us, our place in it, and ourselves. In a more romantic age, John Keats ended his famous "Ode on a Grecian Urn" with the lines:

"Beauty is truth, truth beauty"—that is
All ye know on earth, and all ye need to know.

There is beauty and wonder in knowing how life has changed in the course of more than 3 billion years, how the astonishing diversity of life came to be, and how organisms acquired their marvelous adaptations, just as there is in knowing what a star is, what matter is made of, or how a single cell develops into a human being. Most evolutionary biologists, like other scientists, are consumed with curiosity; for them, learning about how life has evolved is its own reward.

Evolutionary principles and methods make important contributions to research in other biological disciplines, from molecular biology to ecology. Molecular biologists often compare the DNA sequences of genes from different species in the framework of the evolutionary relationships among them, and infer from such comparisons which parts of a gene are likely to encode the functionally most important parts of the gene's product (see Chapter 22). Ecologists must take into account genetic variation and dynamic genetic change in the characteristics of species when they study the factors that determine the distribution and abundance of species (see Chapter 18). Some statistical techniques, such as the analysis of variance, that were developed by evolutionary biologists who studied genetic variation are now used throughout the biological sciences and beyond.

Health sciences Scientific research should not require utilitarian justification, but much basic research nevertheless does have direct or indirect applications to human affairs and social needs. Evolution is directly relevant to our daily lives in many ways. In the health sciences, for example, many organisms that cause infectious diseases, such as malaria, gonorrhea, and tuberculosis, have rapidly evolved resistance to antibiotics. Medical science is challenged to find new drugs or other methods of control and treatment—methods that will succeed only if evolutionary principles are taken into account. Evolutionary biologists studying the human immunodeficiency virus (HIV) that causes AIDS have used phylogenetic methods to trace its origin and spread, and have collaborated with medical scientists to study the evolutionary changes in drug resistance that the virus undergoes within individual carriers. Many pathogens are carried by mosquitoes or other vectors, and are best controlled by controlling the vector. Evolutionary systematics is essential in identifying vector species so that their ecology can be studied and exploited in control efforts (which have often proved short-lived when the vectors evolved resistance to insecticides). Inherited diseases, such as sickle-cell anemia, that are caused by mutant genes are understood in part by applying the principles of evolutionary genetics.

Agriculture The applications of evolutionary biology to food production are legion. Varieties of crops and domesticated animals have traditionally been developed by selective breeding, or artificial selection—which is simply evolution directed by humans. Much of the research in this area has been carried out by individuals who have contributed equally to agricultural genetics and evolutionary genetics, applying the same principles to both. An important extension of selective breeding is the backcrossing of genes for desirable characteristics in wild plant species, such as resistance to insects or fungi, into a related crop species. This technique relies on both evolutionary systematics, which identifies relationships among species, and evolutionary ecology, one branch of which studies the adaptations of plants to their natural enemies. Understanding these adaptations will be increasingly important in the future, when techniques of genetic engineering will be used to transfer desirable genes even among distantly related species. Studies of adaptation will be helpful in identifying plants with useful characteristics that might be genetically transferred. In the same context, evolutionary biologists have been active participants in the debate over the possible risks of genetic engineering of crops, or of genetically altering certain bacteria that can be sprayed on crops to provide benefits such as frost resistance. The concern is that the altered genes might be transferred, by interbreeding, to wild bacteria or plants, which might then become more vigorous and be transformed into pests. The factors that influence the spread of genes are among the major topics of evolutionary genetics.

To mention just one more of the many applications of evolutionary biology to agriculture, resistance to insecticides has evolved in more than 500 species of insects and mites—many of them crop pests—in the last 50 years, causing an enormous increase in the cost of crop production.* Entomologists and agricultural scientists trained in evolutionary biology are working actively on methods of slowing down the evolution of pesticide resistance, and on alternative methods of pest management, such as resistant crop strains, that rely less on pesticides.

Natural products Many hundreds of chemical compounds derived from diverse organisms have proved useful in medicine, industry, and biological research. Penicillin (from a fungus), aspirin (salicylic acid, from willow trees), and quinine (a compound from the cinchona tree, used to treat malaria) are a few of the natural products long used in medicine. More recently, taxol, extracted from yew trees, has been found useful in treating breast cancer, and the clinical use of two chemicals produced by the rosy periwinkle has been shown to reduce death rates from childhood leukemia from 90 percent to 5 percent. Hundreds of natural compounds are used as scents in the cosmetics industry and as emulsifiers, stabilizers, or flavorings in the food industry. Neurobiologists use the neurotoxins in snake and spider venoms to study mechanisms of nerve action. Much of modern molecular biology, including its applications in biotechnology, relies on the polymerase chain reaction (PCR), in which a minuscule amount of DNA is greatly amplified by a DNA polymerase enzyme that can function at high temperatures. The enzyme is found in bacteria that inhabit hot springs.

Many of these compounds were discovered by accident, by trial-and-error screening of thousands of natural compounds, or by testing plants used as folk medicines by peoples throughout the world. But some—such as the polymerase used in PCR—were discovered by using evolutionary principles of adaptation as a guide to where the desired product might be found. The millions of living species constitute a vast "genetic library," hardly yet explored, of many millions of compounds. For guidance in this exploration, evolutionary biology will be indispensable. Evolutionary systematics provides the foundation by describing species and determining their relationships to one another, thereby pointing toward related species that are likely to display variations on a chemical theme. Analysis of adaptations points the way toward species that may have "solved a problem" that we, too, wish to solve. If you want a heat-stable enzyme, it makes sense to look for it in hot-springs bacteria; if you want a compound that protects crops against insects, perhaps a naturally well-defended plant will provide it.

Conservation and environmental management Human activities such as overfishing, uncontrolled hunting, contamination of environments, and (especially) destruction of habitats have already extinguished hundreds of species and threaten to drive thousands of others to extinction within the next century or sooner (Wilson 1992). An all-out effort to conserve biodiversity should be among the highest priorities of the world community, for every extinction is an irretrievable loss from the "genetic library." The arguments for conservation are both aesthetic and utilitarian. Think, for example, of what you or your children would lose if you could not see redwoods in California, bison on a visit to Yellowstone National Park, or gorillas or elephants on a trip to Africa. And think of the opportunities lost if the many species that might be useful as pest control agents, or sources of medicines, or genetic resources for crop improvement are lost even before we can study them.

The rapidly developing field of conservation biology makes extensive use of evolutionary principles and knowledge (Meffe and Carroll 1994). Evolutionary principles tell us how to reduce the likelihood of genetic deterioration (inbreeding depression) in small, endangered populations. They tell us how to measure genetic diversity and how to distinguish species—i.e., what there is to consider saving. They provide guidance in identifying the regions that harbor the greatest diversity of species and the species that are most likely to suffer from habitat destruction or overexploitation—thus helping to set priorities in our conservation efforts. Conservation requires far more than evolutionary biology to be effective, but some understanding of evolution is indispensable.

Understanding ourselves Darwin was well aware of how controversial the idea of human evolution would be. He devoted only a single sentence to it in *The Origin of Species*: "Light will be thrown on the origin of man and his history." His evasion didn't work; human evolution has been controversial ever since *The Origin of Species* was published. Many unsupported speculations about human evo-

---

*It has been estimated that in the United States alone, the evolution of pest resistance by insects and other pests costs $1.4 *billion* per year, over and above what the costs of pesticide use would be, had resistance not evolved (Pimentel et al. 1992).

lution and its implications for "human nature" have been advanced—we will examine some of them in Chapter 26—but evolutionary studies have also provided us with abundant, well-documented evidence on the relationship of humans to other species, on the steps by which some human characteristics evolved, and on genetic variation within and among human populations. Evolutionary biologists have also been among those who have most critically evaluated the evidence on issues such as genetic variation in human behaviors and intelligence and differences among "races" and economic classes. The passionate controversy that surrounds all such issues is testimony to the importance people place on them—and thus to the importance of objective scientific efforts to understand them.

Philosophical Issues

A century after *The Origin of Species* was published, philosophers could still affirm that "there are no living sciences, human attitudes, or institutional powers that remain unaffected by the ideas that were catalytically released by Darwin's work" (Collins 1959). A contemporary philosopher writes "The Darwinian Revolution is both a scientific and a philosophical revolution, and neither revolution could have occurred without the other. . . . If I were to give an award for the single best idea anyone has ever had, I'd give it to Darwin, ahead of Newton and Einstein and everyone else. In a single stroke, the idea of evolution by natural selection unifies the realm of life, meaning, purpose, with the realm of space and time, cause and effect, mechanism and physical law" (Dennett 1995).

With testimonials like these, we obviously cannot do justice in a few pages to the philosophical and social implications of evolution, on which thousands of pages have been written. What, in brief, might be its implications? Some we encountered earlier, in the section on "The Evolutionary Perspective." Darwin undid the essentialism that Western philosophy had inherited from Plato and Aristotle, and put variation in its place. He helped to replace a static conception of the world with the vision of a world of ceaseless change. Above all, his theory of random, purposeless variation acted on by blind, purposeless natural selection provided a revolutionary new kind of answer to almost all questions that begin with "Why?"

It cannot be sufficiently emphasized that before Darwin, both philosophers and people in general answered "Why?" questions by citing *purpose*. Only an intelligent mind, one with the capacity for forethought, can have purpose. Thus questions like "Why do plants have flowers?" or "Why are there apple trees?"—or plagues, or storms—were answered by imagining the possible purpose that God could have had in creating them. The answers might be cast in terms of God's beneficence to humans (He provided apples to feed us) or His chastisement for our misdeeds (plagues and other disasters were created to punish humankind for Adam and Eve's original sin—and for every sin thereafter). Or, God may have created some organisms and their characteristics to complete His scheme, lest there be gaps in an otherwise perfect creation. In Darwin's time, eminent botanists, asked why the female flowers of certain plants should have sterile stamens, answered that they complete the orderly scheme of nature—like unused place settings on a dinner table, as Darwin remarked.

The entire tradition of philosophical explanation by the purposes of things, with its theological foundation, was made completely superfluous by Darwin's theory of natural selection. The adaptations of organisms—long cited as the most conspicuous evidence of intelligent design in the universe—could now be explained by purely mechanistic causes. For evolutionary biologists, the flower of a violet has a *function*, but not a purpose. It was not designed in order to propagate the species, much less to delight us with its beauty, but instead came into existence because violets with brightly colored flowers functioned better—reproduced more prolifically—than violets with less brightly colored flowers. Lice and plague bacteria exist not to punish us, but because their ancestors varied in such a way that some could obtain nutrition from humans and other animals, and these variants survived and reproduced better than others. The profound, and deeply unsettling, implication of this purely mechanical, material explanation for the existence and characteristics of diverse organisms is that *we need not invoke, nor can we find any evidence for, any design, goal, or purpose anywhere in the natural world*, except in human behavior.

It must be emphasized that *all* of science has come to adopt the way of thought that Darwin applied to biology. Astronomers do not seek the purpose of comets or supernovas, nor chemists the purpose of hydrogen bonds, nor molecular biologists the purpose (as opposed to the function) of RNA. Nor, in the light of Darwinian evolutionary theory and the subsequent history of science, do philosophers now seek purpose in the objects around us. Purposes and goals, as far as any philosopher or scientist knows, reside in human minds, and nowhere else.

The question then arises, what about religion and ethics? Do evolutionary biology, and science in general, necessarily render life, morality, and religious belief meaningless?

Ethics, Religion, and Evolution

In the world of science, the reality of evolution has not been in doubt for more than 100 years, but evolution remains an exceedingly controversial subject in the United States and some other countries. Nearly half of Americans believe in special creation rather than evolution, at least of the human species; a vigorous "creationist" movement opposes teaching evolution in public schools, or at least demands "equal time" for creationist beliefs; and many biology curricula and textbooks minimize coverage of evolution, or omit it altogether, to avoid controversy (and to increase sales of textbooks). The opposition arises almost entirely from those who view evolution as a denial of religious belief. They have two main fears: that evolutionary science denies the existence of God; and, consequently, that it denies any basis for rules of moral or ethical conduct, which they believe have a divine source.

Without question, our knowledge of the history and mechanisms of evolution is completely incompatible with a *literal* reading of the creation stories in the Bible's Book of Genesis, as it is incompatible with the hundreds of other creation myths that peoples throughout the world have devised. A literal reading of some passages in the Bible is also incompatible with physics, as when "the sun stood still in the midst of heaven, and hasted not to go down about a whole day" (Joshua 10:13), and with geology, which attests that there could not have been a worldwide flood upon which Noah might have piloted an ark. Such passages must be read as the traditions of a prescientific, pastoral people, or as parables—allegories that tell spiritual truths but not literal scientific facts. The story of how Adam and Eve knew shame and sin when they ate the fruit of the tree of knowledge of good and evil (Genesis 3) symbolically tells a truth, for there can be neither good nor evil, nor sin, unless there is knowledge, consciousness, self-reflection. But it is a symbolic truth, not a history of literal events.

Evolution, and all the rest of science, cannot be reconciled with a literal interpretation of such biblical passages—but does that deny the existence of a supernatural power or powers, of spiritual reality, of God and a human soul? On these questions, science, including evolutionary biology, is silent. By its very nature, science can entertain and investigate only hypotheses about material causes that operate with at least probabilistic regularity. It cannot test hypotheses of supernatural intervention—miracles—nor of the existence of immaterial beings. What science *can* do, and has done, is to posit and document material, natural causes for innumerable phenomena that were once ascribed to the direct actions of supernatural agents. In providing natural, material causes for the diversification and adaptation of species, evolutionary biology has done no more than the physical sciences did when they explained earthquakes and eclipses. The steady expansion of the sciences, to be sure, has left less and less to be explained by a supernatural Creator, but science neither can deny, nor affirm, such a being.

As for their personal beliefs, evolutionary biologists include a small minority of assertive atheists, probably a majority with agnostic or vague spiritual beliefs, and a considerable number who are devout members of all the major religions. Some, in fact, are priests or ministers who teach courses and do research on evolution. Likewise, many nonscientists hold both religious beliefs and belief in evolution. Many, for example, hold the THEIST position that God established the natural laws of physics and let the world develop on its own according to those laws. Science cannot say that they are right or wrong.

Wherever ethical and moral principles are to be found, it is probably not in science, and surely not in evolutionary biology. Opponents of evolution have charged that evolu-

tion by natural selection justifies the principle that "might makes right," and more than one dictator or imperialist has invoked the "law" of natural selection to justify atrocities. But evolutionary theory can offer neither this nor any other precept for behavior. Like any other science, it describes how the world *is*, not how it *should* be. The supposition that what is "natural" is "good" is called by philosophers the NATURALISTIC FALLACY, whereby "natural laws" are taken not merely as regularities in nature, but as morally binding principles that "offer a cosmic backing for the transition from *is* to *ought*" (Collins 1959).

Among animals, behaviors have evolved that by analogy (usually a poor analogy) to human behavior are called cooperation, monogamy, competition, infanticide, rape, slavery, and cannibalism. These phenomena, like hurricanes and friction, *are;* whether they *ought* to be is not a meaningful scientific question. The natural world is amoral—it lacks morality altogether. Despite this, the concepts of natural selection and evolutionary progress have often been taken as a morally proper "law of nature." They were used by Marx to justify class struggle, by the Social Darwinists of the late eighteenth and early nineteenth centuries to justify economic competition and imperialism, by the anarchist Peter Kropotkin to justify cooperative economic institutions, and by the biologist Julian Huxley to support an "evolutionary ethics" leading to higher consciousness and humanitarianism (Hofstadter 1955; Williams 1988). All these ideas, whether we find them appealing or repellent, are philosophically indefensible instances of the naturalistic fallacy.

Hypotheses, Facts, and the Nature of Science

"Oh, no!" groans the reader. "Not another lecture on the scientific method! I learned that in high school!"

Perhaps there is more to learn about it. How, for example, can you be sure that DNA is the genetic material? What if the scientists who "proved" it made a mistake? Has anything really been proved absolutely true? Is science merely one way—the dominant Western way—of perceiving the world, no more or less valid than other perceptions of reality? Is evolution a fact or a theory? Or is it just an opinion I'm entitled to hold, just as creationists are entitled to their opposite opinion?

Consider a hypothetical example. You are assigned to determine why sheep are dying of an unknown disease. You take tissue samples from 50 healthy and 50 sick sheep, and discover a certain protozoan in the liver of 20 of the sick animals, but only 10 of the healthy ones. Is this difference great enough to reject the NULL HYPOTHESIS: that the two groups of sheep do not really differ in the incidence of protozoans? To answer this question, you do a statistical test to see whether the difference between these numbers is too great to have arisen merely by chance. You calculate the chi-square (χ^2) statistic (it is 4.76), look it up in a statistical table of chi-square values, and find that "$0.025 < p < 0.05$." What does this expression, which you will find the like of in almost all analyses of scientific data, mean? It means that (assuming you had a random sample of sick sheep and healthy sheep) the probability is less than 0.05, but more than 0.025, that the difference you found could have been due to chance alone and that there is no real difference in protozoan infection rates of sick and healthy sheep, at large.

Every experiment or observation in science is based on samples from the larger universe of possible observations (all sheep, in this case), and in every case, there is some chance that the data misrepresent the reality of this larger universe. That is, it is always possible to mistakenly reject the null hypothesis—the hypothesis that there is no difference between groups of sheep, that there is no effect of an experimental manipulation, or that there is no correlation between certain variables. In some cases, happily, the probability of rejecting a true null hypothesis, and of accepting as true a false alternative hypothesis, may be 0.00001 or less—in which case you would feel confident that you can reject the null hypothesis, but not absolutely certain.

So the study of 100 sheep *supports* the hypothesis that sick sheep are more likely to have protozoans—but only *weakly*. You suspect that the protozoans might be the cause of death, but you are worried by the imperfect correlation. So you expand your sample to 1000 sheep, take liver biopsies and examine them more carefully for protozoans (revealing cases that you might have missed in your first study, in which the protozoans are present, but at low density), and record which sheep die within the following year. To your great satisfaction, only 5 percent of the sheep in which you did not find protozoans die; 95 percent of the infected sheep die, and when all the survivors are slaughtered at the end of the year, you find that the apparently healthy sheep still show no sign of infection. You triumphantly report to your advisor that the protozoan is the cause of the disease. Right?

Wrong, says she. You haven't eliminated other hypotheses. Maybe the disease is caused by a virus that incidentally also lowers the animals' resistance to a relatively harmless protozoan. Maybe some sheep have a gene that shortens their life and also lowers their resistance to infection. What you must do, she says, is an experiment: inject some sheep, at random, with the protozoan and others with a liquid that is the same except that it lacks the organism. You do so, and after several failed experiments—it turns out that the infection doesn't take unless the sheep consume the protozoan orally—you are delighted to report that 90 of the 100 experimentally infected sheep died within 3 months, and 95 of the 100 "control" sheep lived through the 1-year duration of the experiment. The chi-square test shows that p < 0.0001: there is an exceedingly low probability that your results are due to chance.

At this point, you may have considerable confidence that the protozoan causes disease and death. But you still haven't absolutely *proved* it. Is it possible that you isolated and fed to the sheep not only protozoans, but an unseen virus? Are you sure you infected sheep at random, or might you subconsciously have chosen weaker-looking animals to infect? What do you suppose explains the 15 animals that didn't fit the hypothesis? And even if $p < 0.0001$, there's still a chance, isn't there, that you had a bad "luck of the draw"?

We need not belabor the example longer, but it provides several lessons. First, *data in themselves tell us nothing*: they have to be *interpreted* in the light of theory and prior knowledge. In this example, we need (among other things) probability theory (which underlies statistics such as the chi-square test), the theory of experimental design, and the knowledge that viruses exist and might confound our conclusions. The history of science is full of examples of conclusions that had to be modified or rejected in the light of new theory and information. Until the late 1950s, for instance, almost all geologists believed in the fixed position of the continents; now all believe in plate tectonics and continental drift, and many geological phenomena have had to be reinterpreted in this light. Second, our hypothetical research experience shows us that arriving at a confident conclusion *takes a lot of work*. It is easy to overlook that every sentence in a textbook purporting to state a fact is based on research that required immense effort, usually at least a few years of at least one person's lifetime. For this reason, scientists usually defend their conclusions with considerable vigor—a point to which we will soon return. Third, and most important, research, no matter how carefully and painstakingly conceived and executed, *approaches proof, but never fully attains it*. There is always some chance, although it may seem almost nonexistent, that the hypothesis you have come to accept will someday be modified or rejected in the light of utterly new theories or data that we cannot now imagine. Consequently, almost every scientific paper couches its conclusions in terms that leave some room for doubt. In a paper on *Drosophila* genetics that happened just now to be within reach, I read the conclusion: the experiment "suggests that different mechanisms mediate the two components of sperm displacement" (Clark et al. 1995). The data are, in fact, exquisite, the experiment carefully designed, the statistical analyses exemplary—but the authors do not claim to have *proved* their point. Scientists often have immense *confidence* in their conclusions, but not *certainty*. Accepting uncertainty as a fact of life is essential to a good scientist's world view.

Any statement in science, then, should be understood as a HYPOTHESIS—a statement of what might be true. Some hypotheses are poorly supported. Others, such as the hypothesis that the earth revolves around the sun, or that DNA is the genetic material, are so well supported that we consider them to be *facts*. It is a mistake to think of a fact as something that we absolutely know, with complete certainty, to be true, for we do not know this of anything. (According to some philosophers, we cannot even be certain that anything exists, including ourselves; how could we prove that the world is not a self-consistent dream in the mind of God?) Rather, a fact is a hypothesis that is so firmly supported by evidence that we assume it is true, and act as if it were true.

Why should we share scientists' confidence in the statements they propound as well-supported hypotheses or as facts? Because of the social dynamics of science. A single scientist may well be mistaken (and, very rarely, a scientist may deliberately falsify data). But if the issue is important, if the progress of the field depends on it (as, for example, all of molecular biology depends on the structure and function of DNA), then other scientists will skeptically question the report. Some may deliberately try to replicate the experiment; others will pursue research based on the assumption that the hypothesis is true, and will find discrepancies if in fact it is false. In other words, researchers in the field will test for error, because their own work and their own careers are at stake. Moreover, scientists are motivated not only by intellectual curiosity, but also by a desire for recognition or fame (although they seldom can hope for fortune), and disproving a widely accepted hypothesis is a ticket to professional recognition. Anyone who could show that heredity is *not* based on DNA, or that AIDS is *not* caused by the human immunodeficiency virus, would be a scientific celebrity. Of course, those who originally propounded the hypothesis have a lot at stake—a great investment of effort, and even their reputations—so they typically defend their view passionately, even sometimes in the face of damning evidence. The result of this process is that every scientific discipline is full of controversies and intellectual battles between proponents of opposing hypotheses. There is competition—a kind of natural selection—among ideas, with the outcome decided by more evidence and ever-more rigorous analysis, until even the most intransigent skeptics are won over to a consensus view (or until they die off).

Evolution as Fact and Theory

Is evolution a fact, a theory, or a hypothesis? In science, words are often used with precise meanings and connotations that differ from those in everyday life. This is an exceedingly important point, and we will encounter many examples in this book (e.g., fitness, random, correlation). Among such words are hypothesis and theory. People often speak of a "mere" hypothesis (as in "it is merely a hypothesis that smoking causes cancer") as if it were an opinion unsupported by evidence. In science, however, a hypothesis is an informed statement of what might be true. It may be poorly supported, especially at first, but as we have seen, it can gain support to the point at which it is effectively a fact. For Copernicus, the revolution of the earth around the sun was a hypothesis with modest support; for us, it is a hypothesis with strong support.

Likewise, a THEORY in science is not an unsupported speculation. Rather, it is a mature, coherent *body of interconnected statements*, based on reasoning and evidence, that explains a variety of observations. Or, to quote the *Oxford English Dictionary*, a theory is "a scheme or system of ideas and statements held as an explanation or account of a group of facts or phenomena; a hypothesis that has been confirmed or established by observation or experiment, and is propounded or accepted as accounting for the known facts; a statement of what are known to be the general laws, principles, or causes of something known or observed." Thus atomic theory, quantum theory, and the theory of plate tectonics are not mere speculations or opinions, nor are they even well-supported hypotheses (such as the hypothesis that smoking causes cancer). Each is an elaborate scheme of interconnected ideas, strongly supported by evidence, that accounts for a great variety of phenomena.

Because a theory is a complex of statements, it usually does not stand or fall on the basis of a single critical test (as simple hypotheses often do). Rather, theories evolve as they are confronted with new phenomena or observations; parts of the theory are discarded, modified, added. The theory of heredity, for instance, consisted at first of Mendel's laws of particulate inheritance, dominance, and independent segregation of the "factors" (genes) that affect different characteristics. Exceptions to dominance and independent segregation were soon found, but the core principle of particulate inheritance remained. Building on and adding to this core throughout the twentieth century, geneticists have developed a theory of heredity far more complex and detailed than Mendel could have conceived. Parts of the theory are exceedingly well established, other parts are still tentative, and we may expect many additions and changes as the mechanisms of heredity and development are plumbed further.

In light of the preceding discussion, evolution is a scientific fact. But it is explained by evolutionary theory.

In *The Origin of Species*, Darwin propounded two large hypotheses. One was *descent, with modification, from common ancestors*, or, for simplicity, the hypothesis of descent with modification. I will also refer to this as the "historical reality of evolution." The other large hypothesis was Darwin's proposed *cause* for descent with modification: that *natural selection sorts among hereditary variations*.

Darwin provided abundant evidence for the historical reality of evolution—for descent, with modification, from common ancestors. Even in 1859, this idea had considerable support. Within about 15 years, all biological scientists except for a few diehards had accepted this hypothesis. Since then, hundreds of thousands of observations, from paleontology, biogeography, comparative anatomy, embryology, genetics, biochemistry, and molecular biology, have confirmed it. Like the heliocentric hypothesis of Copernicus, the hypothesis of descent with modification from common ancestors has long held the status of a scientific fact. No biologist today would think of publishing a paper on "new evidence for evolution," any more than a chemist would try to publish a demonstration that water is composed of hydrogen and oxygen. It simply hasn't been an issue in scientific circles for more than a century.

Darwin hypothesized that the *cause* of evolution is natural selection acting on hereditary variation. His argument was based on logic and on interpretation of many kinds of circumstantial evidence, but he had no direct evidence. More than 70 years would pass before an understanding of heredity and the evidence for natural selection would fully vindicate his hypothesis. Moreover, we now know that there are more causes of evolution than Darwin realized, and that natural selection and hereditary variation themselves are more complex than he imagined. Much of this book will be concerned with the complex body of ideas—about mutation, recombination, gene flow, isolation, random genetic drift, the many forms of natural selection, and other factors—that together constitute our current understanding of the causes of evolution. *This complex of interrelated ideas about the causes of evolution is the theory of evolution,* or "evolutionary theory." It is not a "mere speculation," for all the ideas are supported by evidence. It is not *a* hypothesis, but a *body of hypotheses,* most of which are well supported. It is a theory in the sense defined in the preceding section. Like all theories in science, it is incomplete, for we do not yet know the causes of all of evolution, and some details may turn out to be wrong. But the main tenets of evolutionary theory are so well supported that most biologists accept them with confidence.

The Structure of Evolutionary Biology

The two grand questions in evolutionary biology, from which thousands of other questions grow, are:

1. *What has been the history of life?* Can we describe the history of the origin and extinction of the millions of species, and the steps by which their features evolved?
2. *What are the causes of evolution?* Can we understand the causes of the history of life—the origin of species and their characteristics, the differences in rates of evolution, the reasons for extinction?

Like biology in general, evolutionary science can be divided in two ways: taxonomically, by major groups of organisms, and conceptually, according to the kinds of questions asked and the methods used to answer them.

Very important contributions to evolutionary biology are made by scientists who are experts on individual groups of organisms (TAXA), who call themselves botanists, microbiologists, ornithologists, herpetologists, and the like. Through their knowledge of the taxonomy, phylogeny, and biological characteristics of particular groups of organisms, such scientists often contribute to our understanding of both the history of evolution and its causes. Their knowledge is especially critical because each group of organisms poses special questions or is suitable for studying certain problems. Thus, although "model" organisms such as fruit flies, maize (corn), yeast, and the bacterium *Escherichia coli* are enormously important in studying evolution (and other biological problems), we must turn to various parasites to study the evolution of parasite-host interactions, to birds and fishes to study peculiarities of mating behavior, and to bees and primates to study the evolution of cooperative behavior. In all such studies, the knowledge of taxon-oriented biologists is indispensable.

A taxon of special interest, of course, is the family Hominidae, including its single living species, *Homo sapiens.* The closely allied fields of human evolution and human genetics draw on principles and methods developed by evolutionary biologists and geneticists who work with other organisms.

Evolutionary biology may be *conceptually* divided into a number of subdisciplines, some of which are intimately related to other disciplines that do not take an explicitly evolutionary approach (Figure 1.3).

The *history* of evolution is studied primarily in the fields of systematics and paleobiology. The tasks of SYSTEMATICS are (1) to catalogue and name the diversity of life, (2) to discover the phylogenetic (genealogical) relationships among the forms of life (e.g., species), and (3) to classify species into more inclusive groups (HIGHER TAXA). Most systematists agree that such classifications should reflect phylogenetic relationships. Distinguishing species requires analyzing patterns of variation, and thus contributes to our understanding of the causes of evolution. Inferring phylogenetic relationships is neither more nor less than describing the history of evolution of lineages from common ancestors. Studying phylogenetic relationships also provides information on the history of evolutionary changes in the characteristics of organisms (Chapter 5). Such historical information contributes to our understanding of evolutionary processes. Conversely, understanding the causes of evolution can help in inferring relationships among organisms. In recent years. for example, theories of molecular evolution derived from population genetics have contributed to phylogenetic analyses.

PALEOBIOLOGY uses the fossil record to describe the history of organisms and their environments. Among its main tasks are (1) to trace changes in the characteristics of evolving lineages, as well as the relationships among lineages, (2) to provide absolute *times* for evolutionary events, and thus to document rates of evolution, and (3) to analyze changes in biological diversity over time owing to the origination and extinction of taxa. In all these tasks, paleobiologists rely on systematics, as well as on geological information about past environments and earth history. Only paleobiology can provide an absolute time frame for the history of evolution, information on past environments, and direct evidence of the former existence of organisms that are now extinct.

Both systematics and paleobiology make important contributions to our understanding of the *causes*—the processes—of evolution. However, understanding the details of evolutionary processes, and performing experiments to test hypotheses about causes, is mainly the realm of subdisciplines that use living organisms as their subjects. Chief among these are evolutionary genetics (including population genetics), evolutionary developmental biology, and evolutionary ecology. Evolutionary studies of physiology, morphology, behavior, biochemistry, and molecular biology also make important contributions.

POPULATION GENETICS uses mathematical theory and empirical studies to understand genetic variation and the

FIGURE 1.3 The structure of evolutionary biology and its relationship to other biological disciplines. The history and causes of evolution (center) are the subject of the various subdisciplines of evolutionary biology (inner ring), which grade into one another and, as shown by the arrows, are related both to one another and to other disciplines (outer ring). The areas of the segments in the diagram convey an impression of the historical contributions of the subdisciplines to the literature of evolutionary biology.

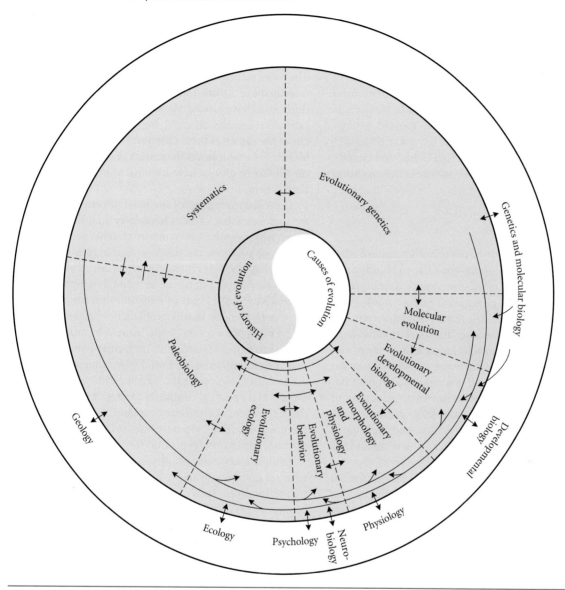

rates and dynamics of genetic changes within and among populations, including those that lead to adaptation and speciation. Its subject matter includes the effects of mutation, random changes, natural selection, and isolation. EVOLUTIONARY GENETICS includes population genetics as well as studies of the origin of genetic variation by mutation and recombination and of the evolution of genomes and breeding systems.

Variation in genes, as well as in environmental conditions, is translated by processes of development into variation in phenotypic characters, such as morphological features. EVOLUTIONARY DEVELOPMENTAL BIOLOGY, now entering into a vigorous new phase because of advances in the understanding of developmental mechanisms, seeks to under-stand the changes in development that are manifested as differences in phenotype among species.

ECOLOGY is the study of interactions among organisms and their environment, including other organisms. ECO-LOGICAL GENETICS studies how the genetic constitution of populations is affected by environmental factors, including those that act as sources of natural selection. It may be included within EVOLUTIONARY ECOLOGY, which also includes analyses of how the ecological characteristics of species, such as their life histories and diets, evolve, how interactions among species evolve, and how evolution affects the composition of assemblages of species (communities).

Each class of characteristics possessed by organisms poses special and interesting questions. In the last few decades, whole subdisciplines of evolutionary biology have arisen that are devoted to applying evolutionary principles in order to understand particular kinds of characteristics. MOLECULAR EVOLUTION has developed hand in hand with advances in molecular biology. This field describes and analyzes variation and evolution in the number, structure, arrangement, and nucleotide sequences of genes. It also provides molecular tools for investigating many questions about the evolution of organisms. EVOLUTIONARY PHYSIOLOGY and EVOLUTIONARY MORPHOLOGY study how the biochemical, physiological, and anatomical features of organisms provide adaptation to their environments, and how these features have evolved. EVOLUTIONARY BEHAVIOR (also known as BEHAVIORAL ECOLOGY) investigates why and how the various behavioral characteristics of animals have evolved, as well as the evolution of the mechanisms of behavior.

An Evolving Science

The rest of this book is devoted to explaining some of what we know so far about evolution—and also to pointing out a very small part of what we don't know. Some people imagine that the study of evolution is a closed book—that we know it all. Nothing could be further from the truth. Yes, it has been thoroughly established that evolution has happened, and many events in evolutionary history have been well documented. Likewise, the major causes of evolution have been well established, and it is unlikely that new discoveries will greatly alter the fundamental principles of evolutionary theory. However, many long-standing questions have not yet been adequately answered, new questions are being posed all the time, and new discoveries prompt still other new questions. Evolutionary biology has probably become a more dynamic, exciting field in the last decade or so than it had been for several decades before. The rate of publication of truly exciting new information and ideas has increased enormously, syntheses between formerly disparate subdisciplines are being developed, and evolutionary principles are illuminating fields such as developmental biology and behavior to a greater extent than ever before. The new excitement in evolutionary science is partly due to an efflorescence of new ideas and partly to technological advances, especially in molecular methods and in computing. The molecular revolution in biology is providing tools for analyzing problems in evolutionary genetics and development in ever-greater depth and detail; the advances in computing enable researchers to analyze large, complex data sets and to perform simulations of evolutionary processes.

To what problems is this ferment of research devoted? Throughout this book, we will encounter questions to which we have only begun to sketch answers. Here are just a few.

In systematics, most phylogenetic relationships among species are only imperfectly understood, or are not known at all. That is, we have only begun to draw the phylogenetic tree of life. The methods for inferring relationships are still a subject of active research. Moreover, we have almost no idea how many species there are on earth, whether 2 million or 20 million. And for many large groups of organisms, such as bacteria, fungi, and nematodes, almost nothing is known about their adaptations, or about how evolutionary processes operate on them.

In paleobiology, new fossils are discovered every year that add important information on the history of evolution. The fossil record continues to yield direct evidence of the intermediate forms that most evolutionary biologists think must have existed. We still have much to learn about how diversity has changed over time, and much more about the causes of these changes. We have only a little evidence, for example, on the causes of mass extinctions, or on why some groups have become so much more diverse than others.

In evolutionary genetics and development, we need a better understanding of how frequently adaptive mutations arise, why populations contain genetic variation, how rapidly populations can adapt to changing environments, and how new species form. An important, little answered question is whether or not genetic and developmental processes channel the kinds of hereditary variations that can arise. A particularly thorny problem is whether or not evolution often proceeds by "leaps," caused by mutations with large, discontinuous effects. In molecular evolution, we are only beginning to learn how DNA sequences and new genes evolve. Molecular biology, together with molecular evolutionary studies, continues to reveal new phenomena that cry out for evolutionary explanation. Why, for example, does more than 90 percent of the DNA in most species not code for any useful products?

Evolutionary ecology, physiology, and behavior likewise are grappling with problems that are only beginning to find solutions. How can plants and animals—including humans—persist if the microbes and insects that attack them have much shorter generation times and should therefore be able to evolve more rapidly? Why do most species reproduce sexually rather than asexually? Why does each species have only a limited range of tolerance of factors such as temperature? Why do some species form pair bonds during reproduction, while others mate promiscuously? What accounts for the prodigious capacities of the human brain?

In an earlier section, we noted that information, concepts, and methods from evolutionary biology are being applied to social needs in areas such as health sciences, agriculture, and conservation. Such applications have only begun. Increasingly, evolutionary biologists are turning their attention to problems such as the origin of new or insurgent diseases (e.g., AIDS), genetic improvement of crops, methods for discovering useful natural products, and strategies for conserving endangered species. As our knowledge of organisms and their evolution grows, it cannot fail to be useful to humanity and to the other species with which we share the earth.

Summary

1. Evolution is the unifying theory of the biological sciences. It aims to discover the history of life and the causes of the diversity and characteristics—all characteristics—of organisms.

2. Biological evolution is change, over the course of generations, in the properties of populations of organisms, or groups of populations. Thus it consists of descent with modification, and often includes diversification from common ancestors.

3. The causes of evolution include the origin of hereditary variation by mutation and recombination, and changes in the proportions of variants due to sorting processes such as random changes in proportions and nonrandom differences in survival and reproduction. The latter process, natural selection, is the cause of adaptations.

4. Darwin's evolutionary theory, published in *The Origin of Species* in 1859, consisted of the hypotheses that (*a*) all organisms have descended, with modification, from common ancestral forms of life, and (*b*) a chief agent of modification is natural selection. The implications of this theory, which revolutionized Western thought, include (*a*) change, rather than stasis, is the natural order; (*b*) biological phenomena, including those seemingly designed, can be explained by purely material causes, rather than by divine creation; (*c*) no evidence of purpose or goals can be found in the living world, other than in human actions; (*d*) the characteristics of organisms can be fully understood only in the light of history; (*e*) variation is fundamentally important; species do not have invariant, "essential" properties.

5. Evolutionary biology makes important contributions to other biological disciplines and to social concerns in areas such as health sciences, agriculture, use of natural products, conservation, and our understanding of ourselves.

6. Like other sciences, evolutionary biology cannot be used to justify beliefs about ethics or morality. Nor can it prove or disprove theological issues such as the existence of a deity. Many people hold that, although evolution is incompatible with a literal interpretation of some passages in the Bible, it is compatible with religious belief.

7. Descent with modification from common ancestors is a scientific *fact*, that is, a hypothesis so well supported by evidence that we take it to be true. The *theory* of evolution, on the other hand, is a complex body of statements, well supported but still incomplete, about the causes of evolution.

8. Evolutionary biology includes many subdisciplines. The history of evolution is studied chiefly in systematics and paleobiology. The causes of evolution are the subject of evolutionary genetics, evolutionary developmental biology, and evolutionary ecology, with important contributions also from evolutionary studies in morphology, physiology, molecular biology, and behavior. All of the biological disciplines both contribute to and are illuminated by the study of evolution.

9. As in most sciences, much remains to be learned in evolutionary biology. Aided by new ideas and technological advances, the study of evolution is a dynamic field of active inquiry.

Major References

The references at the end of each chapter include major works that provide a comprehensive treatment and an entry into the professional literature. Most of the references cited within the text of each chapter will also serve this function.

Darwin, C. 1859. *The origin of species by means of natural selection, or the preservation of favoured races in the struggle for life.* Sixth edition, 1872, reprinted by Modern Library, New York. No one should fail to read at least part of this astonishing book. After some adjustment to its Victorian prose, you will be enthralled by the craft, detail, completeness, and insight in Darwin's arguments.

Hull, D. L. 1973. *Darwin and his critics.* Harvard University Press, Cambridge, MA. This book analyzes and presents excerpts from the objections that Darwin's contemporaries raised against his theory. Most of them are the same as those of today's anti-evolutionists. *Plus ça change, plus c'est la même chose.*

Futuyma, D. J. 1995. *Science on trial: The case for evolution.* Sinauer Associates, Sunderland, MA. A nontechnical exposition, for a general audience, of evolutionary theory, the evidence for evolution, the fallacies in creationist arguments, and the relationship of creationism to broader social issues.

Kitcher, P. 1982. *Abusing science: The case against creationism.* MIT Press, Cambridge, MA. A devastating analysis of so-called scientific creationism by a philosopher of science.

Sober, E. 1993. *Philosophy of biology.* Westview Press, Boulder, CO. Analysis of logical and epistemological problems in biology, especially in evolutionary biology, and some considerations of their relevance to larger philosophical questions. Many of these issues are also treated in A. Rosenberg, *The structure of biological science* (Cambridge University Press, 1985).

Mayr, E. 1988. *Toward a new philosophy of biology.* Harvard University Press, Cambridge, MA. Essays on evolution and its philosophical implications by a leading evolutionary biologist.

Problems and Discussion Topics

1. How does evolution unify the biological sciences? What other principles might do so?

2. Discuss how a creationist and an evolutionary biologist might explain some human characteristics, and the implications of their differences. Sample characteristics: eyes; wisdom teeth; individually unique fingerprints; five digits rather than some other number; susceptibility to infections; fever when infected; variation in sexual orientation; limited life span.

3. Debate the proposition that scorpions do not exist for any purpose. Debate the proposition that humans do not exist for any purpose. Does the second proposition imply that our individual lives lack "meaning"?

4. Should evolution be taught in public schools? Should both evolution and creationism be taught as alternative theories in science classes?

5. Is there anything you absolutely know to be true? How do you know?

6. It is stated in this chapter that "accepting uncertainty as a fact of life is essential to a good scientist's world view." Do you agree? Is it essential for everyone's world view? Does everyone accept uncertainty as a fact of life? What are the implications of doing so versus not doing so?

7. Based on sources available in a good library, discuss how the "Darwinian revolution" affected one of the following fields: philosophy, literature, psychology, anthropology.

8. Debate the validity of the following passage from "Objections to Mr. Darwin's Theory of the Origin of Species," published by the geologist Adam Sedgwick in 1860. Sedgwick affirms his belief that species must have come into being by creation, by which he means "a power I cannot imitate or comprehend; but in which I can believe, by a legitimate combination of sound reason drawn from the laws and harmony of Nature. For I can see in all around me a design and purpose, and a mutual adaptation of parts which I *can* comprehend,—and which prove that there is exterior to, and above, the mere phenomena of Nature a great prescient and designing cause."

CHAPTER 2

A Short History of Evolutionary Biology

The history of thought is not a steady progress toward ever clearer truth. It is intimately bound to social and economic conditions; it ebbs and flows with political ideologies; it sees glimmers of understanding arise and then disappear. In this chapter we can only touch on the high points in the history of evolutionary thought, naming only a few of those who have played important roles. This approach may make it appear as if there has been linear progress toward greater understanding, but in reality there have been false starts and errors along the way.

Evolution Before Darwin

Evolutionary Thought Prior to the Nineteenth Century

It is possible to find some analogy to many modern ideas among those of the ancient Greeks, and so it is with evolution. For example, Anaximander, in the sixth century B.C., held that living creatures were formed from water and that humans and other animals were descended from fishes. Empedocles, in the fifth century B.C., proposed an evolutionary myth whereby heads, limbs, and other organs were joined in random combinations, such as multilimbed creatures with human heads and bovine bodies; only some of such combinations were fit for survival. However, we cannot seriously view these myths as precursors of a scientific theory of evolution.

As we saw in Chapter 1, Plato's philosophy established the notion of fixed essences, eternal ideas in the mind of God. Aristotle later developed this concept into the notion that species have fixed properties. Christian thought elaborated on Platonic and Aristotelian philosophy, arguing that since existence is good and God's benevolence is complete, He must have bestowed existence on every creature, each with a distinct essence, of which He could conceive. Since order is superior to disorder, God's creation must follow a plan: specifically, a gradation from inanimate objects and barely animate forms of life, through plants and invertebrates, up through ever "higher" forms. This **great chain of being**, or *Scala Naturae* (the scale, or ladder, of nature) must be complete and without gaps (or else God would arbitrarily have denied existence to one of the links in the chain); it must be permanent and unchanging, since change would imply that there had been imperfection in the original creation. It was therefore impossible that new forms of life had arisen since creation, or that any form of life had become extinct. Humankind, being both physical and spiritual in nature, formed the link between animals and angels. The great chain of being represented the divinely designed harmony of nature, with an externally fixed role for each being, including humankind—as Alexander Pope tells us in *The Rape of the Lock* (1714):

> Vast chain of being! which from God began,
> Natures aethereal, human, angel, man,
> Beast, bird, fish, insect, what no eye can see,
> No glass can reach; from Infinite to thee,
> From thee to nothing.—On superior pow'rs
> Were we to press, inferior might on ours;
> Or in the full creation leave a void,
> Where, one step broken, the great scale's destroy'd;
> From Nature's chain whatever link you strike,
> Tenth, or ten thousandth, breaks the chain alike.

As late as the eighteenth century, the role of natural science was to catalogue and make manifest the plan of creation so that we might appreciate God's wisdom. For example, John Ray's "The Wisdom of God Manifested in the Works of the Creation" (1701) claimed that the Creator's beneficence is reflected in the adaptations of living things. Carolus Linnaeus (1707–1778), who established the framework of modern classification in his *Systema Naturae* (1735), won worldwide fame for his exhaustive classification of plants and animals, undertaken in the hope of discovering the pattern of God's creation. Linnaeus classified "related" species into genera, "related" genera into orders, and so on. "Relatedness," however, meant not genealogical connectedness, as it does today, but propinquity in the Creator's design.

Well into the nineteenth century, most people believed in the literal truth of the biblical story of creation. But with the rise of modern science, old beliefs gave way. In the late seventeenth century, a more materialist view was forced by Newton's explanations of physical phenomena. The increasingly scientific view that became characteristic of European society in the seventeenth century first fostered skepticism and then, in the eighteenth century—the "Age of Enlightenment"—faith in progress and the power of reason. It was an atmosphere that encouraged the questioning of old beliefs and suggested that the natural world, like the human condition, could change. In the late 1700s, the astronomer Laplace and the philosopher Kant suggested theories of the origin of stars and the solar system. The philosopher Rousseau fancied that the human condition had changed, from "noble savagery" to debasement in modern civilization. Geologists amassed evidence that the earth had undergone profound changes, that it had been populated by many creatures now extinct, and that it was very old. The Scottish geologist James Hutton (viewed by some as the father of geology) argued that the observations of geology should be explicable by processes that we can observe in the present era (a principle called UNIFORMITARIANISM, or more properly, ACTUALISM); and that on this basis, the earth must be so old that we can perceive "no vestige of a beginning—no prospect of an end."

Eighteenth-century concepts of biological change were very different from Darwinian evolution. For example, the Swiss naturalist Charles Bonnet argued in the 1760s that the chain of being was not static, but that God had created a complete series of "germs" in the beginning, each of which came to life at its appointed time in history, the lowest links first and the highest last. Linnaeus held that although each *species* has an eternal, fixed essence, *varieties* within species might be formed as a result of different environmental conditions. Later in life, he suggested that some species within a genus might arise from natural causes, such as hybridization between pre-existing species. Several French materialists of the eighteenth century—Maupertuis, La Mettrie, Diderot, and others—argued that species must have arisen by natural forces; they invoked the common belief in spontaneous generation, even of higher organisms, from non-living matter. Perhaps the nearest approach to an evolutionary theory was that of the great French naturalist Georges Louis Leclerc, known best by his title, Comte de Buffon. Starting in 1749, Buffon published his *Histoire Naturelle* in many volumes. Each species, he said, has an unchanging "internal mold" (*moule interieur*) that organizes the particles of which the organism is made into the species' typical form. In 1753, he raised the possibility that closely related species, such as the horse and the ass, had developed from a common ancestor. By 1766, he affirmed that all species within a genus shared the same internal mold, had arisen from a common ancestor, and had been modified differently by the climatic conditions they had experienced. The original molds arise, he said, by spontaneous generation. They take on their characteristic forms because of intrinsic properties, in the same way that chemical interactions can yield only certain compounds.

Lamarck's Contribution

The most significant pre-Darwinian hypothesis, representing in a sense a culmination of eighteenth-century evolutionary thought, was proposed by Jean Baptiste Pierre Antoine de Monet, chevalier de Lamarck (1744–1829). Lamarck first proposed his hypothesis in 1802, but his definitive work is the *Philosophie Zoologique* of 1809. He accepted the concept of the chain of being, although he recognized two hierarchical chains, animal and plant. Every species we see, he said, originated individually by spontaneous generation at the very bottom of the chain. A "nervous fluid" acts within each species, causing it to progress up the chain over time, along a single predetermined path that every species is destined to follow. Species are continually arising by spontaneous generation and following this course, so even now we see the complete hierarchy, simply because species differ in age (Figure 2.1A). No extinction has occurred: fossil species are still with us, but have been transformed.

Lamarck recognized that not all species of animals fit into a single linear chain, so he argued that some species deviate because of adaptation. Species in different environments have different needs, and so use certain of their organs and appendages more than others. The more strongly exercised organs attract more of the "nervous fluid," which enlarges them; conversely, organs used less become smaller. These alterations, acquired during an individual's lifetime through its activities, are inherited; like everyone in his day, Lamarck believed in such INHERITANCE OF ACQUIRED CHARACTERISTICS. The most famous example of Lamarck's theory is the giraffe: according to Lamarck, giraffes need long necks to reach the foliage above them; because they are constantly stretching upward, the necks grow longer; these longer necks are inherited; and over the course of generations the necks of giraffes get longer and longer. It is im-

JEAN-BAPTISTE PIERRE ANTOINE DE MONET, CHEVALIER DE LAMARCK

FIGURE 2.1 (A) Lamarck's theory of organic progression. Over time, species originate by spontaneous generation, and each evolves up the scale of organization, establishing a Scala naturae, or great chain of being, that ranges from newly originated, simple forms of life to older, more complex forms. In Lamarck's scheme, species do not originate from common ancestors. (B) Darwin's theory of descent with modification, represented by a phylogenetic tree. Lineages (species) descend from common ancestors, undergoing various modifications (marked by horizontal ticks) over the course of time. Some (such as the left-most lineages) may undergo less modification from the ancestral condition than others (the right-most lineages). Branches that do not reach the top represent extinct species. (A after Bowler 1989.)

(A)

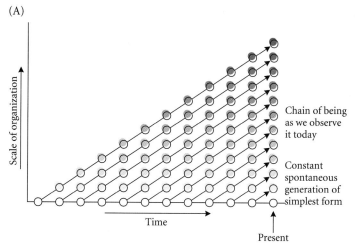

(B)

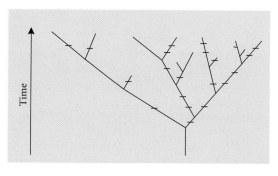

portant to note that for Lamarck, this could happen to *any* and *all* individual giraffes: the entire species grows longer necks because it is made up of individual organisms that are changing during their lifetimes.

Most people who have heard of Lamarck associate him with the theory, now thoroughly disproven, of the inheritance of acquired characteristics. But this theory was not original with Lamarck, and it was only a supplement to his theory of "organic progression" up the scale of nature. Various "neo-Lamarckian" theories of evolution, some based on the idea that alterations caused directly by the environment might be inherited, were popular in the late nineteenth and early twentieth centuries, and have a few advocates even today. Under such theories, for example, individuals who are chronically suntanned would have darker-skinned children than individuals who are not. But Lamarck had little impact during his lifetime, partly because the philosophical ideas of the Enlightenment had gone out of fashion, and partly because his ideas were ridiculed by the most respected zoologist and paleontologist of his day, Georges Cuvier. Cuvier was a superb anatomist who believed firmly that life forms are so complex and so intricately constructed that no organism could function if any part of it were altered. Moreover, most fossil organisms are as complex as living ones. Thus modern species, Cuvier said, could not have evolved from more ancient forms, which had, indeed, become extinct.

In the first half of the nineteenth century, the materialist rationalism of the Enlightenment yielded to romanticism and "idealistic" philosophy: the belief that nature has a harmonious, transcendent order. For some, especially on the European continent, this order had mystical, but not necessarily theological, meaning. For others, especially in Eng-land, idealism joined theology in the growth of "natural theology": proving the attributes of God from those of His creation. In *Natural Theology* (1802), the Reverend William Paley argued that the intricate adaptations of organisms, so obviously designed for their benefit, proclaim a Designer, just as the intricacy of a watch implies a watchmaker. The Earl of Bridgewater commissioned the *Bridgewater Treatises*, a series of books written in the 1830s by leading scientists, to illustrate the "Power, Wisdom and Goodness of God in the Works of Creation." In 1844, however, Robert Chambers anonymously published *Vestiges of the Natural History of Creation*, arguing that in fulfillment of a divine plan, all living things had evolved from simple forms of life, which in turn arose from inanimate matter. His book caused a scandal; it was condemned by nonscientists as an assault on religion and the social order, and by scientists as a collection of errors (which it was, to a considerable extent). The climate in England was not particularly hospitable to the idea of evolution.

Darwin's Contribution

The Life of Charles Darwin

Charles Robert Darwin (February 12, 1809–April 19, 1882) was the son of an English physician and the grandson of Erasmus Darwin, also a physician, who had published speculative essays on the possibility of evolution by Lamarckian mechanisms. (There is little evidence that Erasmus's evolutionary speculations influenced either his grandson or anyone else very much.) Charles briefly studied medicine at Edinburgh, but was so nauseated by observing operations (anesthetics not yet having been de-

CHARLES ROBERT DARWIN

veloped) that he gave up medicine and began studying for a career in the clergy at Christ's College, Cambridge University. As a young man he apparently believed in the literal truth of the Bible. Like most of his fellow students, he spent less time in class than in shooting and other recreations; but, unlike most of his fellows, he was passionately interested in natural history (especially in collecting beetles) and became a companion of the natural scientists on the faculty, especially the botanist John Henslow and the geologist Adam Sedgwick. His life was forever changed in 1831 when he received an invitation to serve as a naturalist on the H.M.S. *Beagle*, a ship the British Navy was sending on an expedition to chart the waters of South America.

With his father's very reluctant agreement, Darwin accepted the position. The *Beagle* left England on December 27, 1831 and spent several years along the coasts of South America, where Darwin was able, on several inland journeys, to observe the natural history of the Brazilian rain forest and the Argentine pampas. After stopping, with what would be momentous consequences, in the Galápagos Islands (on the equator off the coast of Ecuador), the ship headed across the Pacific to Tahiti, the Indian Ocean, southern Africa, back to Brazil, and then homeward, returning to England on October 2, 1836. In the course of those five years, Darwin collected specimens of and made observations on everything geological or biological that he encountered, became an accomplished naturalist, conceived a new (and correct) theory of the formation of coral atolls, and made the observations that would lead him to doubt the fixity of species. He apparently first came to the conclusion that species might have developed from common ancestors shortly after his return, when, in March 1837, the ornithologist John Gould pointed out that Darwin's specimens of mockingbirds (contrary to legend, they were not finches) from the Galápagos were so different from one island to another that they represented different species. Darwin recalled that the giant tortoises, too, differed from one island to the next. It was ap-

parently these facts, and the similarities between fossil and living mammals that he had found in South America, that triggered his belief in evolution.

Shortly after his return, Darwin married his cousin; her family's wealth and the rather comfortable finances of the Darwin family enabled him to devote the rest of his life exclusively to his biological work (although he was chronically ill for most of his life following his voyage). He set about amassing evidence of evolution and trying to conceive of its causes. On September 28, 1838, when he was 29 years old, the great moment arrived. According to his autobiography,

> I happened to read for amusement Malthus on *Population*, and being well prepared to appreciate the struggle for existence which everywhere goes on from long-continued observation of the habits of animals and plants, it at once struck me that under these circumstances favourable variations would tend to be preserved and unfavourable ones to be destroyed.

Malthus, an economist, had argued in his *Essay on the Principle of Population* (1798) that the rate of human population growth is greater than the rate of increase in the food supply, so that unchecked growth must lead to famine. Unlike earlier thinkers who had realized that some *species* might win over other species in the struggle for existence, Darwin saw that if *individuals* of a species with superior characteristics survived and reproduced more successfully than individuals with inferior features, and if those characteristics were inherited, the average character of the species would be altered.

Twenty years passed between this memorable event and Darwin's first publication on the subject. Mindful, perhaps, of the scorn and outrage that had greeted the *Vestiges of the Natural History of Creation*, Darwin occupied himself with amassing evidence for his ideas—and with other researches as well, including an 8-year study that made him the world's authority on the taxonomy of barnacles. In 1844, he wrote an essay outlining his ideas, to be published in the event of his death. In 1856 he finally began what was intended to be his "big book," as he called it. But this book was never completed, for in June 1858 he received a manuscript entitled "On the Tendency of Varieties to Depart Indefinitely from the Original Type" from a young naturalist, Alfred Russel Wallace (1823–1913). Wallace, who was in the Malay Archipelago collecting specimens (which he did for a livelihood), had independently conceived of natural selection. (In numerous later writings, Wallace contributed to evolutionary theory, and especially to the study of the causes of the geographic distribution of species.) At the urging of his friends, Darwin had extracts from his 1844 essay presented orally, along with Wallace's manuscript, at a meeting of the major scientific society in London, and set about writing what he called an "abstract" of the book he had originally intended. The 490-page "abstract," under the title *On the Origin of Species by Means of Natural Selection, or The Preservation of*

ALFRED RUSSEL WALLACE

Favoured Races in the Struggle for Life, was published on November 24, 1859.

For the rest of his life, Darwin continued to read and correspond on an immense range of subjects, to revise *The Origin of Species* (it had six editions), to perform experiments of all sorts (especially on plants), and to publish many more articles and books. Some of these, such as *The Descent of Man, and Selection in Relation to Sex, The Variation of Plants and Animals Under Domestication*, and *The Expression of the Emotions in Man and Animals*, deal directly with major themes of *The Origin of Species*. Others, such as *The Various Contrivances by Which Orchids are Fertilised by Insects, Insectivorous Plants*, and *The Formation of Vegetable Mould, Through the Action of Worms*, seem less related to evolution, but nevertheless explore some of the themes that run through *The Origin of Species*. They reveal Darwin as an irrepressibly inquisitive man, fascinated by all of biology, creative in devising hypotheses and in bringing evidence to bear upon them, and profoundly aware that every fact of biology, no matter how seemingly trivial, must fit into a coherent, unified understanding of the world.

Darwin's Evolutionary Theory

The Origin of Species ("*On*" was dropped from the title of later editions) has two major theses. First, all species, living and extinct, have descended without interruption, from one or a few original forms of life. Living things have not been created repeatedly by a supernatural power, nor have they arisen repeatedly (as in Lamarck's view) by spontaneous generation. Closely similar species have diverged from a relatively recent ancestor by the accumulation of slight differences; more dissimilar groups, such as orders and classes, have arisen from a more ancient common ancestor, and have accumulated more differences in the greater span of time. Darwin's conception of the course of evolution is profoundly different from Lamarck's, in which the concept of common ancestry plays almost no role. Thus, although Darwin was by no

means the first to postulate a history of evolution, his vision of this history's pattern differed greatly from the ideas of his predecessors.

The second, and major, theme of *The Origin of Species* is Darwin's truly original contribution: his theory of the causal agents of evolutionary change. This was his *theory of natural selection:* "if variations useful to any organic being ever occur, assuredly individuals thus characterised will have the best chance of being preserved in the struggle for life; and from the strong principle of inheritance, these will tend to produce offspring similarly characterised. This principle of preservation, or the survival of the fittest, I have called natural selection." This is a *variational theory* of change: the species is altered because the individuals in any generation are descended from, and have inherited the properties of, only some of the individuals in the preceding generation—specifically, those that differed from other individuals in some advantageous respect. This hypothesis is profoundly different from Lamarck's *transformational theory*, in which change is programmed into every member of the species (Figure 2.2). (An example of transformational "evolution" is the history of stars, which all become transformed from red giants to white dwarfs and finally die out as their energy is dissipated.)

Ernst Mayr (1982) has persuasively argued that what is often called "Darwin's theory of evolution" actually includes *five* theories, which are independent enough of each other that early evolutionists differed among themselves as to which ones they accepted and which they rejected. Mayr refers to these theories as (1) evolution as such; (2) common descent; (3) gradualness; (4) populational speciation; and (5) natural selection.

EVOLUTION AS SUCH is the simple proposition that the characteristics of lineages of organisms change over time. This was not an original idea with Darwin; Lamarck and others had proposed it. Nevertheless, before Darwin, most people did not believe in any but the most trivial change, and it was Darwin who so convincingly marshaled the evidence for evolution that most biologists soon accepted that it has indeed occurred.

COMMON DESCENT is a radically different view of evolution from the scheme Lamarck proposed. Darwin was the first to argue that species had diverged from common ancestors, and that all of life could be portrayed as one great family tree (Figure 2.1B). A most unsettling implication of this view was that the human species is but one of millions of twigs, sharing common ancestors with other species.

GRADUALNESS, or gradualism, is Darwin's proposition that the differences between organisms—even radically different organisms—evolve by innumerable small steps through intermediate forms. This belief is contrary to the hypothesis that large differences evolve by leaps, or SALTATIONS, without intermediates between ancestors and descendants. Because intermediates were, and still are, unknown between some living organisms and between many fossil lineages, Darwin had to boldly postulate that the intermediates have been erased by extinction, and that the fos-

FIGURE 2.2 A diagrammatic contrast between transformational (above) and variational (below) theories of evolutionary change, shown across three generations. Within each generation, individuals are represented early and late in their lives. The individuals in the left column in each generation are the offspring of those in the right column of the preceding genera-tion. In transformational evolution, individuals are altered during their lifetimes, and their progeny are born with these alterations. In variational evolution, hereditarily different forms at the beginning of the history are not transformed, but instead differ in survival and reproductive rate, so that their proportions change from one generation to another.

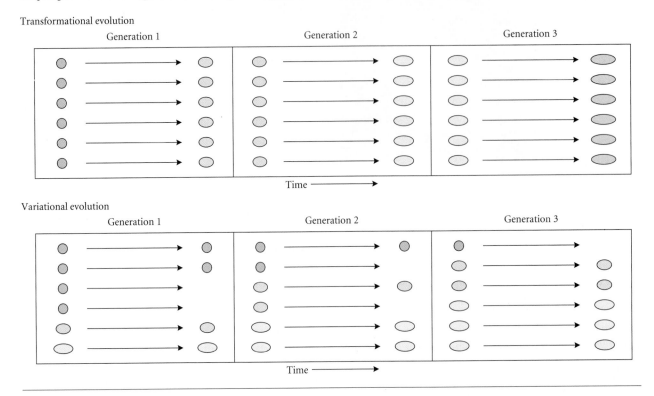

Transformational evolution

Variational evolution

sil record is extremely incomplete. Darwin's gradualism was controversial when it was first proposed, and to some extent it still is.

POPULATIONAL SPECIATION is closely allied to gradualism. By populational, or variational, change, we mean that evolution occurs by changes in the proportions of individuals within a population that differ in one or more hereditary characteristics. This concept was a completely original idea, which Darwin applied both to the evolution of characteristics within a single species and to the formation of new species from a common ancestor. Populational evolution may be contrasted both with the idea of sudden origin of new species by saltational change and with Lamarck's account of evolutionary change by transformation of individuals.

NATURAL SELECTION is Darwin's brilliantly original hypothesis, independently conceived by Wallace, about what causes the proportions of variant individuals within a species to change. Others had previously recognized that the "struggle for life" could cause whole species to become extinct while others flourished; Darwin and Wallace had the insight that this process, operating within species, would result in the evolution of adaptations—features that appear "designed" to fit organisms to their environment. The concept of natural selection, misunderstood and rejected by many for decades after *The Origin* was published, revolutionized not only biology, but all of Western thought.

Darwin believed that no matter how extensively a species has diverged from its ancestor, new hereditary variations continue to arise, so (given enough time) there is no evident limit to the amount of divergence that can occur. Each step in this divergence is small, for Darwin believed that *Natura non facit saltum:* "nature does not take leaps." An individual that differed drastically from others in any one feature would generally not have an adaptive advantage, for its other features would not function harmoniously in concert with the radically altered characteristic; and it is most unlikely that drastic but harmonious alterations in many features would arise *de novo* in a single individual. As to why an ancestral species should give rise to a multitude of descendant species, each with different features, Darwin noted that different variations would be adaptive under different "conditions of life"—different habitats or habits. Moreover, the pressure of competition among species would favor the use of different foods or habitats by different species.

The origins of the hereditary variations that are the necessary raw material for natural selection were quite unknown to Darwin. Gregor Mendel's paper establishing the foundations of modern genetics was published in 1865—but it was ignored by everyone until 1900. The concept of "mutation" developed only after 1900 and was not clarified until considerably later. Darwin knew only that virtually all characteristics of organisms are variable, and that much of that variation is inherited. He did not know the causes of variation. He speculated that a change of environment might stimulate the origin of new variation, but that variations so arising were *not necessarily advantageous* in that environment. The theory that acquired characteristics caused by the direct effects of the environment might be inherited had not yet been disproved; especially in later editions of *The Origin of Species*, Darwin allowed that this might be a subsidiary mechanism of adaptive evolution, although he maintained that natural selection is primary.

These are the principal ideas in *The Origin of Species*, but it is a far richer book than this brief summary can indicate. To appreciate the subtlety of Darwin's thought, the variety of biological phenomena he felt compelled to explain, the ways in which he anticipated modern ideas, and the breadth of knowledge he brought to bear on his themes, it is necessary to read the book itself.

Evolution After Darwin

The Origin of Species engendered enormous controversy, especially among clergy and lay people who viewed it as an attack on religion, but also among many scientists. Some felt that their religious beliefs had been violated; others found it hard to accept that a purely mechanical process of natural selection acting on genetic variation could yield "progress" and adaptive "design." By the 1870s, however, most scientists had accepted the historical reality of evolution by common descent. There ensued, in the late nineteenth and early twentieth centuries, a "golden age" of paleontology and comparative morphology and embryology, during which a great deal of information on evolution in the fossil record and on relationships among organisms was amassed. The evidence collected was so overwhelming that the historical reality of evolution has long since attained the status of fact (see Chapter 1). But this consensus did not extend to Darwin's theory of the *cause* of evolution, natural selection. For about 60 years after the publication of *The Origin of Species*, all but a few faithful Darwinians rejected natural selection, and proposed numerous theories in its stead. By 1900, and for several decades thereafter, scientific thought on the mechanisms of evolution was in a state of amazing disarray.

Anti-Darwinian Theories

The many alternatives to natural selection proposed in the decades around 1900 included neo-Lamarckian, orthogenetic, and mutationist theories. The first and second categories overlap, and both were motivated by the distaste that many scientists felt for the purely materialistic theory of natural selection. They sought philosophically more appealing theories that would allow "life itself to be seen as purposeful and creative. Living things are in charge of their own evolution . . . life becomes an active force in nature, no longer merely responding in a passive manner to environmental pressures" (Bowler 1989, 258).

Neo-Lamarckism Several of the theories proposed were based on the old idea of inheritance of modifications acquired during an organism's lifetime. The modifications may have been consequences of using a feature (as in Lamarck's example of the giraffe stretching its neck to reach foliage) or the direct effects of the environment on development (as in plants that develop thicker leaves when grown in a hot, dry environment). All such NEO-LAMARCKIAN THEORIES were fiercely opposed by the outstanding German biologist August Weismann, an ardent proponent of natural selection. Weismann proposed in 1893 that the "germ plasm" (the reproductive cells, and particularly their chromosomes) is separate from and entirely immune to any influence from the "soma" (the rest of the body), which merely transmits the germ plasm across the generations. In a famous experiment, Weismann cut off the tails of mice for many generations, and showed that this had no effect on the tail length of their descendants. Weismann made many important contributions to evolutionary biology, but his aggressive, uncompromising defense of natural selection (which he called "neo-Darwinism") struck many naturalists and paleontologists as dogmatism, and actually drove some of them further into the neo-Lamarckian camp, even though there was no conclusive evidence of the inheritance of acquired characteristics.

To be sure, some investigators claimed that they had such evidence; but in all such cases, the experiments proved to be flawed. It is insufficient to show, for example, that the offspring of parents exposed to some stress inherit altered features, for this observation does not exclude the possibility that the parents differed genetically from others of their generation, and that their offspring are a selected sample of the previous generation's genes. Similarly, it is not sufficient to show that modified features are retained for several generations by plants grown in a stressful environment: the modified plants must then be grown in the normal environment for several generations to see whether the modifications are inherited or induced directly by the stress. When geneticists performed the proper experiments early in this century, they rejected the principle of inheritance of acquired characteristics.* Thus, as far as we know, neo-Lamarckian ideas are simply wrong. This is one of many examples demonstrating that, in science, it is fallacious to suppose (as many people tend to do) that truth must reside in a compromise, or middle ground, among opposing views.

Orthogenesis Another cluster of theories, espoused especially by paleontologists impressed by the trends they perceived in the fossil record, were ORTHOGENETIC

---

*Evidence for quasi-Lamarckian phenomena is still occasionally reported (see, for example, Chapter 10), but no experiments have yet conclusively demonstrated that the environment can act through the soma to direct hereditary changes in an adaptive fashion.

(straight-line) theories, which held that the variation that arises is directed toward fixed goals, and that a species evolves in a predetermined direction without the aid of selection. Some paleontologists held that such trends need not be adaptive, and could even drive species toward extinction. Alpheus Hyatt proposed, by analogy with the life of an individual organism, that evolutionary lineages undergo "racial senility," degenerating from a peak of greatest development and dwindling perforce to extinction. As an example, the extinct "Irish elk" (*Megaceros giganteus*), a deer with huge antlers, was supposed to have become extinct because it could not stop evolving ever-larger and more cumbersome antlers (see Figure 31 in Chapter 6). None of the proponents of orthogenesis ever proposed a mechanism for it. Orthogenetic theories died out when the participants in the Evolutionary Synthesis (see below), especially the paleontologist George Gaylord Simpson, showed that natural selection, genetics, and a more careful reading of the fossil record could readily account for all the phenomena that orthogenesis was meant to explain.

Mutationism MUTATIONIST THEORIES were based on the principles of Mendelian genetics. Darwin argued that the slight differences among individuals that make up the more or less CONTINUOUS VARIATION we see in features such as body size constitute the important variation on which natural selection acts to change species gradually. As biologists struggled to understand heredity in the late nineteenth century, Darwin's position was maintained by the "biometrical school," initiated by Darwin's cousin, Francis Galton, who hoped to understand heredity by measuring similarities among relatives. The biometricians got off on the wrong track, and they lost prestige after 1900, when Gregor Mendel's 1865 paper was rediscovered. The "Mendelians" followed Mendel in emphasizing DISCRETE VARIATIONS, inherited as single factors, and triumphed after a vituperative battle with the biometricians (Provine 1971). The Mendelians went so far as to maintain (mistakenly) that continuous variation had no genetic basis, so that only discrete variations could play a role in evolution.

How do such variations arise? The Dutch botanist Hugo De Vries, one of the discoverers of Mendel's paper, coined the term *mutation* for variants of evening primroses (*Oenothera*) that appeared spontaneously and differed substantially from their parents. He considered these to be not merely new hereditary variants, but new species. For De Vries, and especially for Thomas Hunt Morgan, the highly influential founder of *Drosophila* genetics, the process of mutation was sufficient to account for the origin of species; natural selection was superfluous. Morgan saw that mutations are not necessarily adaptive, but he denied the importance of both selection and adaptation, arguing that any mutated form that could survive and reproduce would persist (Morgan 1903). He led a groundswell of opinion among geneticists that natural selection and adaptation exemplified old-fashioned speculation, unworthy to be called science, and

that only experimental biology, probing mechanisms in the laboratory, was a properly scientific approach to understanding evolution. In its early stages, Mendelian genetics did evolutionary biology great service by disproving Lamarckian and blending inheritance theories[*] and by laying the foundations for a genetic understanding of evolution; but by supposing that mutation and natural selection are mutually exclusive rather than complementary, it delayed a synthesis that might have occurred far sooner than it did.

One historically important figure who espoused a kind of mutationism was Richard Goldschmidt (1940). He was an accomplished geneticist who rightly stressed that understanding hereditary changes in development is crucial for understanding the evolution of morphology. However, he argued that evolutionary change within species is entirely different in kind from the processes responsible for the origin of new species and higher taxa. These, he said, originate by sudden, drastic changes that reorganize the whole genome. Although most such reorganizations are deleterious, a few of these "hopeful monsters" would be the progenitors of new groups. The genetic concepts underlying Goldschmidt's theory of saltations have been entirely repudiated by modern geneticists, but evolutionary biologists still argue about whether large changes in individual characteristics may have evolved through mutations that have large, discrete effects.

The Evolutionary Synthesis

The resolution of these conflicts occurred in the 1930s and 1940s, in the **evolutionary synthesis** or **modern synthesis**, forged from the contributions of geneticists, systematists, and paleontologists and reconciling Darwin's theory with the facts of genetics (Mayr and Provine 1980; Smocovitis 1996). Ronald A. Fisher (1890–1962) and John B. S. Haldane (1892–1964) in England and Sewall Wright (1889–1988) in the United States, building on the data of genetics, developed a mathematical theory of population genetics, which showed that it is the *conjunction of mutation and natural selection* (among other things) that causes adaptive evolution: mutation is not an alternative to natural selection, but is rather its raw material. The study of genes in wild populations of *Drosophila* (fruit flies) was pioneered in Russia by Sergei Chetverikov in the 1920s, and was continued by Theodosius Dobzhansky (1900–1975) after he moved from Russia to the United States in 1927. These researchers showed that natural populations are not uniform; they harbor genetic variants, including the same kinds of variants that had been observed to arise by mutation in laboratory stocks. This finding showed that laboratory studies of genetics were indeed relevant to evolution in nature. In his influential book *Genetics and the Origin of Species* (1937),

---

[*]The theory of blending inheritance had proposed that heredity was due to fluids, like two colors of paint that are blended in the offspring. A remnant of the concept persists today, when people speak of having, for example, English "blood."

RONALD A. FISHER

J. B. S. HALDANE

SEWALL WRIGHT

Dobzhansky masterfully synthesized the theory of population genetics and data on genetic variation and the genetic differences between species. Dobzhansky's book conveyed the ideas of the mathematical population geneticists to other biologists and influenced their appreciation of the genetic basis of evolution.

The evolution of species—especially animal species—was clarified in *Systematics and the Origin of Species* (1942) by Ernst Mayr (b. 1904), which built on both the knowledge of taxonomists and genetic principles; later, G. Ledyard Stebbins (b. 1906) did the same for plants in *Variation and Evolution in Plants* (1950). These books, especially Mayr's, provided strong evidence against saltational theories such as Goldschmidt's. The vertebrate paleontologist George Gaylord Simpson (1902–1984), in *Tempo and Mode in Evolution* (1944) and its successor, *The Major Features of Evolution* (1953), drew on population genetics to show that paleontological data were fully consistent with the new "synthetic" theory, and that orthogenetic and neo-Lamarckian theories were neither necessary to explain nor supported by the fossil record. The German zoologist Bernhard Rensch (1900–1990), in *Evolution Above the Species Level* (1959), and the English zoologist Julian Huxley, in *Evolution: The Modern Synthesis* (1942), both used genetic principles to explain the major patterns of evolution that had been described by systematists and paleontologists.

The major achievements of these and other scientists were to integrate fully genetics and Darwinian evolutionary theory and to argue persuasively (at least, many people were persuaded) that mutation, recombination, natural selection, and other processes operating within species account for the major, long-term features of evolution (which Dobzhansky termed "macroevolution"). Equally importantly, the Evolutionary Synthesis purged evolutionary biology of anti-Darwinian theories. Orthogenetic, neo-Lamarckian, and simple mutationist theories were refuted, and saltationism was rejected in favor of gradual change. The Synthesis shows that rejection of false hypotheses is an important part of progress in science.

ERNST MAYR

JULIAN HUXLEY

G. LEDYARD STEBBINS, GEORGE GAYLORD SIMPSON, AND THEODOSIUS DOBZHANSKY

Major Tenets of the Evolutionary Synthesis

The principal claims of the Evolutionary Synthesis are the foundations of modern evolutionary biology. They are known collectively as the Synthetic Theory, and serve as a synopsis of much of contemporary evolutionary theory. Many of these points have been extended, exemplified, clarified, or modified since the 1940s. Although some authors have challenged or even rejected some of these principles, the vast majority of evolutionary biologists today accept them as valid and use them as a foundation for evolutionary research. Subsequent chapters of this book will present evidence bearing on these points.

1. The **phenotype** (observed physical characteristics) is different from the **genotype** (the set of genes carried by an individual), and the phenotypic differences among individual organisms can be due partly to genetic differences and partly to direct effects of the environment.

2. Environmental effects on an individual's phenotype do not affect the genes passed on to its offspring. That is, acquired characteristics are not inherited. However, the environment may affect the *expression* of an organism's genes.

3. Hereditary variations are based on particles—genes—that retain their identity as they pass through the generations; genes do not blend with other genes. This is true not only of those genes that have discrete effects on the phenotype (e.g., brown vs. blue eyes), but also of those that contribute to continuously varying traits (e.g., body size, intensity of pigmentation). Variation in continuously varying traits is largely based on several or many discrete genes, each of which affects the trait slightly (polygenic inheritance).

4. Genes mutate, usually at a fairly low rate, to alternative forms (**alleles**). The phenotypic effects of such muta-
tions can range all the way from undetectable to very great. The variation that arises by mutation is amplified by **recombination** among alleles at different loci.

5. Environmental factors (e.g., chemicals, radiation) may affect the rate of mutation, but they do not preferentially direct the production of mutations that would be favorable in the organism's specific environment.

Points 1–5 were important early contributions to the Synthetic Theory from laboratory genetics.

6. Evolutionary change is a populational process: it entails, in its most basic form, a change in the relative abundances (proportions) of individual organisms with different genotypes (and hence, often, with different phenotypes) within a population (see Figure 2.2). Over the course of generations, the proportion of one genotype may gradually increase, and it may eventually entirely replace the other type. This process may occur within only certain populations, or in all the populations that make up a species (see point 11).

7. The rate of mutation is too low for mutation by itself to shift an entire population from one genotype to another. Instead, the change in genotype proportions within a population can occur by either of two principal processes: random fluctuations in proportions (**random genetic drift**) or nonrandom changes due to the superior survival and/or reproduction of some genotypes compared to others (**natural selection**). Natural selection and random genetic drift can operate simultaneously.

8. Even a slight intensity of natural selection can (under certain circumstances) bring about substantial evolutionary change in a relatively short time. Very slight differences between organisms can confer slight differences in survival or reproduction; hence natural selec-

tion can account for slight differences among species, and for the earliest stages of evolution of new traits.

Points 6–8 were among the major contributions of the mathematical theory of population genetics.

9. Selection can alter populations beyond the original range of variation by increasing the proportion of alleles that, by recombination with other genes that affect the same trait, give rise to new phenotypes. (This point is a contribution from genetic studies of agriculturally based plant and animal breeding.)

10. Natural populations are genetically variable: the individuals within populations differ genetically and include natural genetic variants of the kind that arise by mutation in laboratory stocks.

11. Populations of a species in different geographic regions differ in characteristics that have a genetic basis. The genetic differences among populations are often of the same kind that distinguish individuals within populations. A genotype that is rare in one population may be predominant in another.

12. Experimental crosses between different species, and between different populations of the same species, show that most of the differences between them have a genetic basis. The difference in each trait is often based on differences in several or many genes (i.e., it is **polygenic**), each of which has a small phenotypic effect. This finding provides evidence that the differences between species evolve by small steps rather than by single mutations with large phenotypic effects.

13. Natural selection occurs in natural populations at the present time, often with considerable intensity.

Points 9–13 were contributions from those geneticists, most of whom had a background in natural history, who studied natural populations.

14. Differences among geographic populations of a species are often adaptive (hence, are the consequence of natural selection), because they are frequently correlated with relevant environmental factors.

15. Organisms are not necessarily different species just because they differ in one or more phenotypic characteristics; phenotypically different genotypes often are members of a single interbreeding population. Rather, different **species** represent distinct **gene pools**, which are groups of interbreeding or potentially interbreeding individuals that do not exchange genes with other such groups. This **reproductive isolation** of species is based on certain genetically determined differences between them. (This is one version of the **biological species concept**.) Hence, even a mutation that causes substantial change in some phenotypic feature does not necessarily represent the origin of a new species.

16. Nevertheless, there is a continuum in degree of differentiation of populations, with respect to both phenotypic difference and degree of reproductive isolation, from barely differentiated populations to fully distinct species. This observation provides evidence that an ancestral species differentiates into two or more different species by the gradual accumulation of small differences rather than by a single mutational step.

17. **Speciation**—the origin of two or more species from a single common ancestor—usually occurs through the genetic differentiation of geographically segregated populations. Geographic segregation is required so that interbreeding does not prevent incipient genetic differences from developing.

18. Among living organisms, there are many gradations in phenotypic characteristics among species assigned to the same genus, to different genera, and to different families or other higher taxa. This observation is interpreted as evidence that higher taxa arise through the prolonged, sequential accumulation of small differences, rather than through the sudden mutational origin of drastically new "types."

Points 14–18 were contributed chiefly by systematists and naturalists who studied particular taxonomic groups.

19. The fossil record includes many gaps among quite different kinds of organisms, as well as gaps between possible ancestors and descendants. Such gaps can be explained by the incompleteness of the fossil record. But the fossil record also includes examples of gradations from apparently ancestral organisms to quite different descendants. Together with point 18, this leads to the conclusion that the evolution of large differences proceeds by many small steps (such as those that lead to the differentiation of geographic populations and closely related species). Hence we can extrapolate from the genesis of small differences to the evolution of large differences among higher taxa, and can explain the latter by the same principles that explain the evolution of populations and species.

20. Consequently, all observations of the fossil record are consistent with the foregoing principles of evolutionary change (although they do not prove that these mechanisms provide a necessary and sufficient explanation). There is no need to invoke, and in some instances there is evidence against, non-Darwinian hypotheses such as Lamarckian mechanisms, orthogenetic evolution, vitalism ("inner drives"), or abrupt origins by major mutations.

Points 19 and 20 were among the contributions of paleontologists.

In the aftermath of the Evolutionary Synthesis, these twenty points were widely accepted and became the orthodox modern theory of evolution. Some points have been challenged, then and since (Burian 1988). In ensuing chapters, we shall examine evidence pertaining to these principles. At this time, it is important to learn them, not because they are all necessarily correct in every instance, but because they constitute the framework of modern evolutionary theory.

Evolutionary Biology since the Synthesis

In the immediate aftermath of the Synthesis, a great deal of research was devoted to elaborating the basic theory (e.g., the mathematical theory of population genetics), to testing and obtaining further evidence (e.g., studying the nature of species; describing the kinds and amounts of genetic variation in nature; studying the operation of natural selection), and to documenting the *relative prevalence* of different theoretically plausible processes (e.g., the relative importance of random genetic drift versus natural selection, or speciation with versus without geographic segregation of populations).

Important areas of research were opened or stimulated by advances in genetics, especially at the molecular level, beginning in the 1950s and continuing with increasing emphasis up to the present time. Many of the above principles are stated in what may seem to be antiquated language, because the Synthesis occurred before the structure of DNA had been determined (by J. D. Watson and F. H. C. Crick in 1953); in fact, the Synthesis had mostly occurred before there was any evidence that DNA is the genetic material. (Such evidence was first provided by O. T. Avery, C. M. MacLeod, and M. McCarthy in 1944.) Our steadily increasing knowledge of the structure and function of the genetic material has enabled an equally steady increase in our knowledge of topics such as mutation, genetic variation and its causes, species differences, development, and the phylogenetic history of life.

Another development, especially since the mid-1960s, has been the expansion of evolutionary theory into areas such as ecology, animal behavior, and reproductive biology. The Synthesis painted an evolutionary theory in broad brushstrokes, a theory that applied to organisms and their characteristics in general. More specific, detailed theories, intended to explain the evolution of particular kinds of characteristics (e.g., life span, ecological distribution, social behavior) have become an important focus of study.

A major focus of evolutionary research for about two decades after the Synthesis was the further elucidation of mechanisms (especially genetic mechanisms) at the level of populations and species. Relatively little attention was devoted to questions about "macroevolution," such as the origin of novel characteristics or the history and causes of evolutionary patterns at higher taxonomic levels over geological time spans. Since the 1970s, there has been renewed attention to such topics, resulting in, among other things, new methods for inferring phylogenetic relationships, provocative interpretations (e.g., punctuated equilibrium) of the fossil record, and an interest in using developmental biology to explain the evolution of morphology.

Contemporary Evolutionary Studies

Evolutionary biology is more vigorous now than at any time since the 1940s, both because of its expansion and because new concepts and techniques offer promise of resolving many of the questions that the Modern Synthesis answered only tentatively or not at all. The resurgence of debate about some fundamental issues does not reveal a weakness in the field, but rather, is a sign of its intellectual vigor and excitement. In every scientific field, controversies exist, new theories are proposed, old ideas are modified or rejected. Therein lies much of the joy and excitement of science.

Evolutionary biologists today do not concern themselves with trying to demonstrate the reality of evolution. That is simply no longer an issue, and hasn't been for more than a century. Rather, because every step toward understanding leads to new, more detailed questions, and because new techniques and theories enable us to probe new areas, biologists today seek a deeper understanding of the mechanisms and history of evolution. Here are a few of the questions on which research today is centered:

Regarding the evolution of adaptations: How can we explain differences among species in such features as longevity, reproductive rate, body size, migratory habits, sex ratio, sexual vs. asexual reproduction? Why are some species hermaphroditic while others have distinct sexes? Why do some animals change sex during their lives? How has social behavior evolved, including, in its most extreme form, the sterile worker castes of social insects? Why are the sexes in some species almost identical, while in other species they are strikingly different? Why in some species is it the female, and in others the male, that is courted and cares for the young? Why are some species, such as insects that feed on only a single kind of plant, extraordinarily specialized, and others not? Why does a species have ecological and geographic limits to its range, instead of adapting to the terrain beyond? What limits the extent of further adaptation?

Regarding mechanisms of evolution: How do the mechanisms of mutation and recombination limit the range of variations that are available to natural selection? Why is there as much genetic variability within species as there is? Do levels of variability set limits on the rate or direction of a species' evolution? How intense and consistent is natural selection? How does natural selection produce complex adaptations that involve the interplay of numerous features? Does gene exchange among populations limit the rate of evolution? How much of evolution is nonadaptive, occurring by chance rather than by natural selection? What ecological factors and genetic changes cause populations to become new species? How fast do new species form? Are they formed by rapid genetic changes in small populations, or by the slow divergence of large ones? Is geographic segregation always required for speciation? Are all evolutionary changes gradual?

Regarding molecular insights into evolution: Why does the eukaryotic genome carry so much "useless" DNA? Does it affect organisms? How much evolution at the DNA level is caused by natural selection, and how much is nonadaptive? What constitutes an advantageous mutation at the DNA level? Are transposable elements important in evolution? What is the evolutionary importance of changes in gene regulation? What kinds of genes evolve slowly vs. rapidly? What does variation at the DNA level tell us about the history of populations and species?

Regarding patterns in evolutionary history: Why are there as many species as there are (probably between 3 and 10 million), rather than far fewer or many more? Can we explain the changes in numbers of species that have occurred over the history of life? Why do species become extinct rather than adapting to environmental changes? Why have some phyletic lineages produced more species than others, and why do more new spe-

cies occur at some times than at others? Have differences among groups in "success" versus "failure" (extinction) been due to chance? Why is evolution, as viewed in the fossil record, usually so slow? Do adaptive changes depend on speciation? How important have predation, competition, and other interactions among species been in guiding the course of evolution? How can we best deduce the phylogenetic relationships among living species? Can we document trends in the history of life?

Some of these questions had hardly been thought of 20 years ago but have already been answered to a fair approximation; others are old questions, still refractory and still strongly debated; still others have arisen in only the past few years, as molecular biology has revealed new phenomena. For every fact of biology, we can pose questions of an evolutionary nature—which makes evolutionary biology a source of inexhaustible challenge and delight.

Summary

1. In the early eighteenth century, Western Europeans, their world view shaped by Platonic and Aristotelian philosophy and by Christian theology, considered the earth to be only a few thousand years old; held that all species had been created as we see them today, with adaptations conferred by a benevolent Designer; considered the earth and living things to be unchanging, static, and organized according to a divine plan, such as a linear "great chain of being;" and affirmed the special status of humankind, the highest link among material beings in the great chain, to whom dominion over the rest of nature had been granted.

2. During the Enlightenment of the eighteenth century, with its emphasis on rationalism, the stage was set for evolutionary thought. Physics and astronomy provided theories of cosmological change, and geology provided evidence of the great age of the earth, of massive changes in its surface, and of extinction of early forms of life. By the mid-eighteenth century, Buffon and several other French thinkers began to entertain ideas of a limited evolution of species. The culmination of this trend was Lamarck's (1809) proposition that new forms of life arise continually by spontaneous generation, each destined to evolve up the chain of being and achieve adaptation to its environment by inheritance of characteristics that are modified by its behavioral responses to its needs (inheritance of acquired characteristics).

3. In *The Origin of Species* (1859), Charles Darwin was the first to propose an uncompromising, fully materialistic evolutionary theory. He proposed not only that evolution had occurred, but that all living things had descended from one or a few original common ancestors, with species multiplying to give rise to new species. Evolutionary change, he proposed, is mostly gradual and is caused primarily by natural selection operating

on random hereditary variations. Alfred Russel Wallace also conceived of natural selection, but Darwin explored its consequences in depth, and marshaled the most comprehensive evidence for evolution as such and common descent in particular.

4. Although Darwin's works rapidly convinced the scientific world of the reality of evolution and common descent, his hypothesis of natural selection was widely rejected in favor of various neo-Lamarckian, mutationist, or orthogenetic theories. However, these theories were almost entirely rejected during the Evolutionary Synthesis of the 1930s and 1940s, when natural selection, together with the data from genetics, came to be viewed as a sufficient explanation of most of the data of biology and paleontology.

5. The Synthetic Theory that emerged during this period, although not entirely unchallenged, remains the foundation of contemporary evolutionary theory. This chapter includes a summary of that theory.

6. Since the 1940s, evolutionary research has been amassing evidence bearing on the Synthetic Theory, and in recent years it has expanded in many directions. At present, the field is probably more vigorous than ever before.

Major References

Bowler, P. J. 1989. *Evolution: The History of an Idea.* University of California Press, Berkeley. An excellent history of the development of evolutionary thought.

Mayr, E. 1982. *The Growth of Biological Thought: Diversity, Evolution, and Inheritance.* Harvard University Press, Cambridge, MA. A detailed, comprehensive history of systematics, evolutionary biology, and genetics that bears the personal stamp of one of the major figures in the Evolutionary Synthesis. See also Mayr's essays in E. Mayr, *Toward a New Philosophy of Biology* (The Belknap Press of Harvard University Press, 1988).

Ghiselin, M. T. 1969. *The Triumph of the Darwinian Method.* University of California Press, Berkeley. Ghiselin argues that Darwin developed a new scientific method (the hypothetico-deductive method; see Question 7 above), and shows how Darwin applied it in his major works.

Mayr, E., and W. B. Provine (editors). 1980. *The Evolutionary Synthesis: Perspectives on the Unification of Biology.* Harvard University Press, Cambridge, MA. Essays by historians and biologists, including some of the major contributors to the Synthesis. See also V. B. Smocovitis, *Unifying Biology: The Evolutionary Synthesis and Evolutionary Biology* (Princeton University Press, Princeton, NJ, 1996). A historian's account and analysis.

Desmond, A., and J. Moore. 1991. *Darwin.* Warner Books, New York. A colorfully written biography of Charles Darwin, emphasizing the role played by the religious, philosophical, and intellectual climate of nineteenth-century England on the development of his scientific theories. See also P. J. Bowler, *Charles Darwin: The Man and His Influence* (Blackwell, Cambridge, 1990), which emphasizes scientific issues, and J. Bowlby, *Charles Darwin: A Biography* (Norton, NY, 1991), which emphasizes Darwin's personal life.

Problems and Discussion Topics

1. Analyze and evaluate Emerson's couplet,

   > Striving to be man, the worm
   > Mounts through all the spires of form.

   What pre-Darwinian concepts does it express? What fault in it will a Darwinian find?

2. What are the implications of thinking of the human species as one of millions of twigs on the phylogenetic tree of life? What conceptions of our position in nature does it challenge?

3. Imagine yourself as Darwin. At the age of 29, you have the insight that differences in survival among individuals with different characteristics will result in an increase in the proportion that have the more adaptive feature. By the age of 73, you have published explanations of an extraordinary variety of organisms' characteristics (e.g., sterility of worker ants, different forms of flowers in the same species, expression of emotional states in humans, embryological similarities among species, the distribution of plants and animals on islands, the extraordinary "tail" of the peacock); you have also inferred a common origin of all living things, have classified all the species of barnacles, have experimentally shown that competition affects the species composition of plant communities, and have conceived of an explanation (which happens to be wrong) for the mystery of heredity. What would be necessary for you to have achieved these accomplishments, starting with your insight at age 29? Discuss the kinds of knowledge, thought processes, and work habits you would need. What questions and paths of reasoning would have led you to address so great a variety of topics?

4. Debate whether or not it is valid to extrapolate from "microevolutionary" changes within species to the processes by which "macroevolutionary" differences arise among species and higher taxa (genera, families, etc.). Were those who forged the Evolutionary Synthesis justified in this extrapolation? What kinds of data might justify the extrapolation, or show that it was not justified?

5. In learning a scientific subject such as genetics or evolutionary biology, what is the value of learning about hypotheses that were once espoused but have been disproved or rejected? Can you cite examples from other subjects?

6. The distinguished zoologist Richard Owen, a contemporary of Darwin, argued against Darwin's hypothesis of evolution by natural selection. Among his evidence, he cited worker ants. These, he said, "suggest an objection which seems fatal to the theory, since they show extreme peculiarities of structure and instinct in individuals that cannot transmit them, because they are doomed to perpetual sterility." How could (and did) Darwin answer this objection?

7. Many of Darwin's critics criticized the method of his argument, claiming that it was epistemologically invalid. (Epistemology is the analysis of how we can know things.) Darwin posed a hypothesis (e. g., natural selection), deduced predictions of what we should see if it were true or false, and judged its validity by comparing observations against the predictions. This *hypothetico-deductive method* is now widely used in science (see Ghiselin 1969.) The geologist Adam Sedgwick echoed his contemporaries by objecting that "Darwin's theory is not *inductive*,—not based on a series of acknowledged facts pointing to a *general conclusion*,— not a proposition evolved out of facts, logically, and of course including them." Indeed, Darwin had not cited any examples in which a species had been observed to change due to natural selection. Evaluate Sedgwick's objection. What are the pros and cons of using induction versus the hypothetico-deductive method to establish a hypothesis?

CHAPTER 3

Genetics and Development

Many modern studies of the mechanisms of evolution are based on genetics. Studies of development are also becoming increasingly important in evolutionary biology. This chapter reviews those aspects of these topics that are essential for understanding evolution. Genetics and development are reviewed here with little reference to evolution; in later chapters, they are incorporated into discussions of evolutionary theory and evidence. Much of the material in this chapter may be familiar to the reader from introductory biology courses, but some may not be. In subsequent chapters dealing with topics in evolution, I shall expand on the evidence on which our understanding is based, but in most of this chapter, I assume this is unnecessary.

It is absolutely necessary to understand meiosis and Mendelian (transmission) genetics in order to understand the dynamics of genetic variation and change in natural populations. An understanding of linkage and recombination is important for insight into the origin of genetic variation and the joint evolution of different characteristics. The molecular basis of heredity is fundamental to almost all of modern biology, and we will build on this foundation throughout our study of evolution. It is also essential to understand how the phenotypic characteristics that evolve develop under the influence of genes.

Chromosomes and Cell Division

In eukaryotes, most of the genetic material is organized into CHROMOSOMES (Greek, "colored bodies," because of their staining reaction) within the cell nucleus. Each chromosome is composed of a single long, coiled, double-stranded molecule (a double helix) of DNA bound to nucleoproteins. Before the onset of MITOSIS (Figure 3.1), the DNA is *replicated*; each chromosome then consists of two identical SISTER CHROMATIDS joined at a CENTROMERE. The centromere lies almost at one end of an ACROCENTRIC chromosome, but in the interior of a METACENTRIC chromosome, separating its left and right arms.

During PROPHASE of mitosis, the chromatids become condensed (i.e., shorter and thicker) and the nuclear membrane breaks down. During METAPHASE, the centromeres, attached to spindle fibers, lie at the equatorial plane of the cell. The centromeres divide, and during ANAPHASE the chromatids are pulled toward opposite poles of the cell. (Each chromatid, now independent of its sister, is now referred to as a chromosome.) In TELOPHASE, new nuclear membranes form, and the cytoplasm is divided by a new cell membrane. The two resulting "daughter cells" carry an identical set of chromosomes and an identical complement of DNA. With certain exceptions, then, the genetic material carried in the chromosomes (the **nuclear genome**) is the same in all the SOMATIC (nonreproductive) cells of an individual.

In prokaryotes (single-celled organisms without nuclei), the DNA forms a single circular molecule that is referred to as the chromosome but is not enveloped by a nuclear membrane. As in eukaryotic nuclei, daughter cells receive identical genomes during cell division. Mitochondria and chloroplasts within the cells of eukaryotic organisms also have a circular genome, like that of prokaryotes, which is replicated in a similar fashion. The DNA of mitochondria and chloroplasts is often referred to as the **extranuclear genome**.

Meiosis

Were mitosis the only means of cellular replication, the number of chromosomes in sexually reproducing species would double in each generation when cells from two parents united. The reduction division in MEIOSIS re-establishes the characteristic chromosome number. During meiosis, the genes that an individual inherits from its parents are recombined by independent segregation of chro-

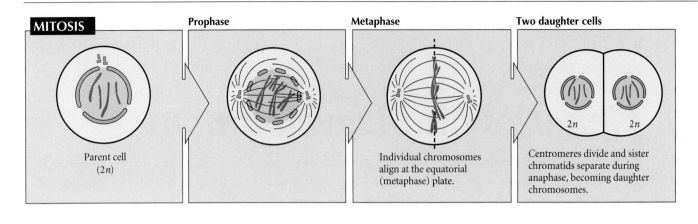

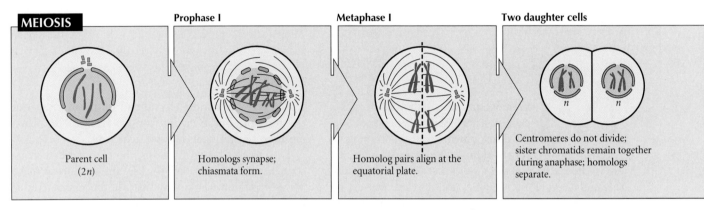

mosomes and by crossing over within chromosomes. RE-COMBINATION is important in evolution because it contributes to genetic variation.

In eukaryotes, sex cells, or GAMETES, are usually formed by meiotic division of a primary oocyte (female)or a primary spermatocyte (male)—known collectively as GERM CELLS—which give rise to egg or sperm cells. We shall describe the most common pattern, as observed in organisms such as humans, fruit flies, and corn plants, but to almost every statement that follows, some organisms provide exceptions. The nature of meiosis and of reproduction varies greatly among organisms (Box 3.A).

Meiosis occurs only in eukaryotes. It consists of two cell divisions, meiosis I and meiosis II (Figure 3.1). During prophase I, each chromosome (consisting of two sister chromatids) pairs with a HOMOLOGOUS chromosome, or HOMOLOGUE. Homologues bear homologous sequences of DNA—they carry much the same genetic information. (Nonhomologous chromosomes, bearing different genetic information, vary in number among species, and often can be distinguished by their size and form.) Humans have 23 pairs of homologues (numbered 1 through 23), and *Drosophila melanogaster* has 4 pairs (the X–Y pair and numbers 2, 3, and 4). During chromosome pairing, or SYNAPSIS, homologous DNA sequences on homologous chromosomes are (usually) precisely aligned with each other.

Because each chromosome consists of two genetically identical chromatids, a pair of chromosomes in synapsis displays four strands. During synapsis, the nonsister chromatids may form cross-shaped structures (CHIASMATA; singular CHIASMA). These are the visible manifestation of **crossing over**, during which there may be breakage and reunion of chromatids, so that the nonsister chromatids swap parts. If crossing over occurs between two sites at which the homologues differ, the recombined chromatids carry new combinations of genes. Thus if one homologue has DNA sequences A and B at two different sites, and the other homologue has the slightly different sequences a and b, crossing over between the sites results in chromatids with the combinations A,b and a,B, whereas the chromatids that did not cross over retain the combinations A,B and a,b (Figure 3.2). Roughly speaking, the chance that crossing over will occur between two sites is proportional to the physical distance between them (although there are many exceptions).

During metaphase I, the synapsed pairs of chromosomes are arranged on the equatorial plane. During anaphase I, the centromeres are pulled to opposite poles, but the centromeres do not divide, so both chromatids are carried together. Therefore each pole (and the resulting daughter cell formed during telophase I) receives one two-stranded chromosome from each homologous pair. Meiosis I is therefore a REDUCTIONAL DIVISION, in which each daughter cell ends up with half the number of chromosomes it had at the start of meiosis.

During metaphase II (metaphase of the second meiotic division), the chromosomes in each daughter cell are

FIGURE 3.1 Mitosis (above) and meiosis (below) in a diploid cell. In mitosis, each of two daughter cells receives one of the sister chromatids of each chromosome. In the first (reductional) division of meiosis, each daughter cell receives one chromosome of each homologous pair. In the second (equational) meiotic division, the sister chromatids are distributed to daughter cells. (After Purves et al. 1998.)

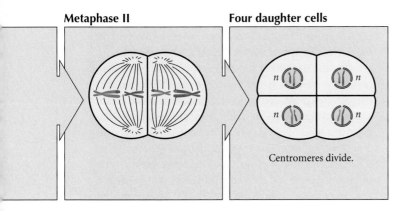

Metaphase II   **Four daughter cells**

Centromeres divide.

arranged on the equatorial plane; the centromeres then divide, and the two chromatids of which each chromosome is composed are pulled to opposite poles. Cell division occurs, and nuclear membranes re-form. Thus the ultimate result of the two stages of meiosis is four cells derived from each germ cell. The four cells derived from a primary spermatocyte differentiate into male gametes (sperm cells in animals, meiospores in plants); only one of the four cells derived from a primary oocyte becomes a functional female gamete, or egg cell.

A cell with a single complement of chromosomes, such as a mammalian gamete, is HAPLOID; one with two complements of homologous chromosomes is DIPLOID. The union of two haploid gametes (produced by meiosis from diploid germ cells) re-establishes diploidy in the ZYGOTE (or fertilized egg), which then contains a MATERNAL complement of chromosomes from the egg and PATERNAL chromosomes from the sperm. All the somatic cells of the developing organism, derived by a succession of mitotic cell divisions from the fertilized egg, likewise contain maternal and paternal chromosomes.

Some organisms are POLYPLOID, possessing multiple complements of chromosomes. Their number is halved during meiosis; for example, a TETRAPLOID organism, with four sets of chromosomes, has diploid gametes.

In most organisms, the movement of the two chromosomes of a pair toward opposite poles during anaphase of meiosis I is independent of that of any other pair (i. e., they segregate independently). Ignoring for now the effect of crossing over, each daughter cell thus receives, at random, either the maternal or the paternal chromosome from each pair, as do the two cells into which this daughter cell divides. Therefore, for each single chromosome site a copy of the maternal representative is transmitted to two of the four cells to which each germ cell gives rise, and the paternal representative is transmitted to the other two.

Mitochondria and chloroplasts, bearing extranuclear genes, may be distributed among the cells that arise by mitosis or meiosis. In many organisms, they are included in eggs, but not in the very small sperm cells. In such organisms, mitochondria and chloroplasts are maternally inherited.

Mendelian Genetics

Many species have both **autosomes**—chromosomes that are the same in both sexes—and **sex chromosomes**. For example, in both *Homo* and *Drosophila*, the male has a pair of morphologically (and genetically) different chromosomes, the X and Y chromosomes, whereas the female has a pair of X chromosomes. Although the X chromosome may carry genes that determine sex, it also carries many genes that are expressed similarly in both sexes; conversely, the expression of some autosomal genes is limited to one sex or the other, as a consequence of a cascade of influences that is ultimately dependent on sex-determining genes carried on the X or Y chromosomes. The following discussion is limited to autosomes in diploid organisms in which recombination occurs in both sexes.

A given gene occupies a particular chromosomal site, or **locus**, and the word *locus* (pl., *loci*) is often used for the gene itself. Different forms of a gene that can be distinguished by some phenotypic effect were originally termed *allelomorphs* (Greek, "different forms"); this term soon was shortened to **alleles**. A HOMOZYGOTE has a double dose of one allele, and a HETEROZYGOTE carries two different alleles. Many different (multiple) alleles can exist at a given locus.

Consider two alleles, say, A_1 and A_2. All F_1 offspring of a mating between unlike homozygotes (A_1A_1 and A_2A_2) will be heterozygous (A_1A_2) at this locus. Because of the orderly segregation that occurs in meiosis, the probability (i.e., "chance") that any given gamete produced by a heterozygote (A_1A_2) will receive A_1 is 1/2 (or 0.5), and likewise for A_2. (In some cases, other factors may intervene to cause SEGREGATION DISTORTION, or MEIOTIC DRIVE. In these cases, the alleles will be unequally represented among a heterozygote's gametes; see Chapter 12). If, then, we consider the BACKCROSS between a heterozygote and a homozygote, such as $A_1A_2 \times A_1A_1$, we expect half the offspring to be A_1A_1 and half to be A_1A_2. Box 3.B reviews the rules of probability, and Box 3.C uses the rules in reviewing the familiar Mendelian ratios.

Using the most elementary rules of probability, one can calculate the proportions of different genotypes expected among the offspring of any cross between two individuals that differ in genotype at one locus. Thus if a corn plant produces 1000 offspring, the cross $Aa \times Aa$ should produce about 250

BOX 3.A

Some Variant Patterns of Reproduction and Meiosis in Eukaryotes

Eukaryotic organisms vary in almost every aspect of reproduction and gene transmission. Some of the variations, such as self-fertilization and the several kinds of asexual reproduction, will be mentioned frequently in subsequent chapters. Others figure in the discussion of the evolution of genetic systems, in Chapter 21.

1. Sexual and asexual reproduction. SEXUAL reproduction occurs when two gametes produced during meiosis join to form a zygote; otherwise, reproduction is termed ASEXUAL. Asexual reproduction has many forms. In VEGETATIVE PROPAGATION, a new "individual" forms from a group of somatic cells of a single parent (e.g., a bud, as in *Hydra*); a subdivision of the parent's body (as in some flatworms); or a stolon (or runner, as in strawberry plants). Especially in plants (and in some animals such as corals), a group of vegetatively produced offspring may be arbitrarily considered separate individuals or one. The whole lineage (perhaps ultimately derived from a zygote) is sometimes called a GENET, and the individual members RAMETS.

   Asexual reproduction via eggs is called PARTHENOGENESIS (adjective: *parthenogenetic*). One form of parthenogenesis is APOMIXIS, in which development proceeds from an unreduced (diploid) egg; that is, no meiosis has occurred. Apomixis is common in plants (e.g., the dandelion *Taraxacum officinale*) and is found sporadically among animals (e.g., in water fleas such as *Daphnia*, in aphids during the asexual part of their life cycle, and in a few vertebrates such as certain species of *Cnemidophorus* lizards). AMPHIMIXIS, in which meiosis occurs, but two of the four nuclei join, is less common (an example is the bagworm moth *Solenobia*). In a few organisms, a haploid egg undergoes mitosis, and the two cells then fuse, forming a diploid "restitution nucleus."

2. Alternation of haploid and diploid generations. In some organisms, such as many algae and fungi, the dominant stage in the life cycle is haploid; reproductive cells formed by mitosis join to form a diploid meiocyte that immediately undergoes meiosis, producing haploid spores that develop into new organisms. In some algae, this diploid cell develops via mitosis into an organism that only later produces haploid cells (by meiosis), which develop into haploid organisms. In the green alga *Ulva* (sea lettuce), for example, the haploid and diploid generations look much alike, and are more or less coequal in the life cycle (Figure 3.A1A).

   Almost all eukaryotes have an alternation of haploid and diploid generations. In mosses, the haploid generation (the GAMETOPHYTE) is the conspicuous, green, vegetative stage, which produces sperm and eggs mitotically in separate antheridia and archegonia, respectively. Union of these gametes results in a (usually nonphotosynthetic) diploid SPOROPHYTE that grows out of the archegonium and ultimately produces, by meiosis, HAPLOSPORES that develop into gametophytes. In ferns and some related plants, the free-living but inconspicuous gametophyte likewise produces sex cells by mitosis; the resulting zygote then gives rise to the conspicuous sporophyte, with leaves bearing sporangia, within which meiosis produces haploid spores, which develop into gametophytes (Figure 3.A1B). In angiosperms, the gametophyte is very greatly reduced in size. The male gametophyte (pollen grain), produced within the anther by meiosis, is a microspore. Its nucleus divides mitotically into two nuclei, one of which divides into two sperm nuclei. The other, the tube nucleus, governs the growth of the pollen tube from the pollen grain. The female gametophyte, one of four haploid megaspores produced by meiosis in a megaspore mother cell, undergoes three mitotic nuclear divisions. One of the eight resulting nuclei is the egg nucleus, which joins with one of the pollen grain's sperm nuclei; two other nuclei of the female gametophyte join with the other sperm nucleus, giving rise to the triploid endosperm that nourishes the developing embryo (Figure 3.A1C). In animals, the haploid generation consists only of the gametes.

3. Differentiation into sexes. In ISOGAMOUS organisms, the haploid cells that join to form a zygote are equal in size. In such organisms, there is no distinction between female and male, although the uniting cells are genetically differentiated into + and – "mating types." Isogamous organisms include green algae such as *Ulva* and fungi such as yeasts and *Neurospora*.

   Most species of eukaryotes are ANISOGAMOUS, producing both large (female) and small (male) gametes. In HERMAPHRODITIC (or COSEXUAL) species, an individual organism produces both female and male gametes. Hermaphroditism occurs among coelenterates, platyhelminths, molluscs, and many other animals, and is extremely common in plants. In cosexual flowering plants, male and female functions may both be performed by the same flower, or by different flowers; there are also species that possess both cosexual flowers and unisexual (female or male) flowers on the same plant. In SEQUENTIAL HERMAPHRODITISM, an individual changes from one sex to the other during its lifetime; this occurs, for example, in some sea basses and other fishes. Organisms such as mammals, flies, and willow trees, which are either male or female throughout life, are termed GONOCHORISTIC (or, in plants, DIOECIOUS).

   An individual's sex may be determined by genetic or environmental factors, depending on the species. In turtles and some other reptiles, for example, temperature during embryonic development determines sex. The stimulus to

FIGURE 3.A1 The life cycles of some green plants. (A) A green alga, the sea lettuce (*Ulva*), in which mitosis of zygotes and zoospores results respectively in diploid sporophytes and haploid gametophytes that are similar in appearance. Green algae are isogamous: the uniting gametes do not differ in size. (B) A fern, in which a haploid meiospore develops into a gametophyte that produces male (sperm) and female gametes (eggs). The sporophyte, which develops from a fertilized egg, grows out of the gametophyte. (C) An angiosperm. The male (pollen grains) and female gametophytes develop by mitosis from microspores and megaspores, which arise by meiosis in the anthers and ovary respectively. Double fertilization gives rise to the triploid endosperm and the diploid zygote, which develops into the sporophyte. (After Campbell 1993.)

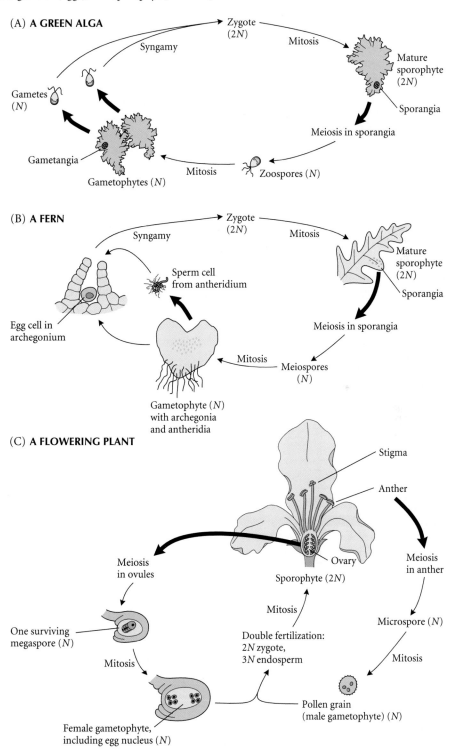

switch sex in sequentially hermaphroditic fishes can be a social one: if the dominant male in a hierarchy ("pecking order") is removed, the largest of the local females may assume his position and become male. Genetic sex determination is often manifested by differentiation in sex chromosomes: the HOMOGAMETIC sex has a pair of like sex chromosomes (XX, for example), and the HETEROGAMETIC sex a pair of unlike chromosomes. In mammals, most insects, and many (but not all) dioecious plants, the male is heterogametic; in birds and Lepidoptera (moths and butterflies), the female is heterogametic. The ways in which sex chromosomes determine sex vary. In mammals, the Y chromosome carries male-determining factors, but in *Drosophila*, it does not: sex is determined by the number of X chromosomes relative to the number of pairs of autosomes. In Hymenoptera (e.g., wasps and bees) and in some mites, sex is determined by ploidy: the female can lay either fertilized (diploid) eggs, which become female, or unfertilized (haploid) eggs, which become male.

4. Self-fertilization and cross-fertilization. Gonochoristic organisms obviously cannot perform self-fertilization ("selfing"), but many hermaphroditic species, especially plants, can and do: pollen from a flower may fertilize the ovules of the same flower or a different flower on the same plant. Many hermaphroditic species, however, engage in self-fertilization to only a limited extent or not at all. The stigmas of a plant, for example, may become receptive to pollen only after the plant's own pollen is likely to have been carried away. Some plants (e.g., violets, *Viola*) have both flowers adapted for cross-fertilization (OUTCROSSING) and flowers adapted for selfing that do not open.

5. Variations in meiosis. In some sexual species, the maternal and paternal chromosomes do not segregate at random; for example, in the Hessian fly (*Mayetiola destructor*), a pest of wheat, meiosis in the female is so organized that the entire paternal chromosome set is relegated to the first polar body, and only the maternally derived chromosomes are included in the egg. In hemipteran insects (e.g., scale insects) and some other organisms, each chromosome has a "diffuse centromere" instead of a single centromere: meiotic spindle fibers attach along the entire length of the chromosome. If, therefore, a chromosome breaks, each fragment can move to one end of the spindle, whereas fragments that lack centromeres are lost in most other organisms. In many male Diptera (true flies, such as *Drosophila*) and female Lepidoptera, crossing over does not occur, even though synapsis does.

These variations on sex and reproduction are only a fraction of those known—and we have not even considered bacteria and viruses. In sex, as in everything else, the living world presents almost infinite variety.

FIGURE 3.2 Crossing over between two loci in a cell undergoing the first meiotic division. Of the four chromatids, two will have new combinations, and two will retain the parental combinations of alleles.

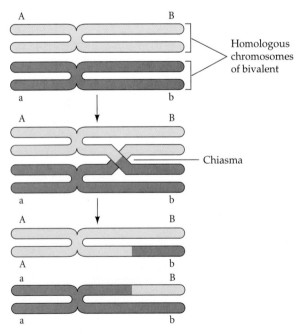

offspring of genotype *AA*, 500 of *Aa*, and 250 of *aa* in the familiar 1:2:1 ratio. If a genetic analysis were carried out on a species that produces only one offspring at a time (say, humans), we would expect the same result among all the offspring resulting from 1000 individual matings of *Aa* females with *Aa* males, but for any individual case, we can say only that the probability that an individual offspring will be *AA* is 0.25 (or 0.5 for *Aa*, 0.25 for *aa*). As Box 3.C shows, the same rules of probability predict the proportions of offspring bearing any particular combination of alleles at two loci (or three, or more) as long as those loci segregate independently in meiosis. They will segregate independently if they are not LINKED (i.e., if they are on different, nonhomologous chromosomes) or if they are so far apart on the same chromosome that crossing over occurs very frequently.

If two (or more) loci do not segregate independently, the expected proportions of different allele combinations in the offspring can be calculated only if the rate of crossing over between the loci has been measured. For such linked loci (i.e., loci located on the same chromosome pair), gametes unlike those that joined to form the parent are produced only from those stem cells in which crossing over occurs during meiosis. For example, if the gametes A_1B_1 and A_2B_2 unite to form a double heterozygote, this individual will produce only these gamete types unless there is crossing over between the loci. If the loci are one

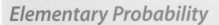 **BOX 3.B**

Elementary Probability

Consider an infinitely large collection of items of types A and B. This collection may exist only in theory: it might be, for example, an infinitely long sequence of coin tosses, in which A and B represent heads and tails. Types A and B occur in proportions P and Q ($= 1 - P$), respectively. If we pick out i items *purely at random*, so that every individual item in the collection has an equal chance of being drawn, we can calculate the chance that we will draw 0 As; 1 A and $i - 1$ Bs; 2 As and $i - 2$ Bs; and so on, up to i As and 0 Bs. If sets of i items are drawn very many times, the *average* number of As per set will ultimately be $P \times i$. Thus, if $P = 0.2$ (20 percent of the items are A) and $i = 100$ in each set, the average (or "expected") number of As per set will be 20. The *probability* that any single item will be A is equal to P, the proportion of A in the collection. Likewise, the probability of drawing a B at random is Q (in this instance, 0.8).

As long as drawing an A does not alter the chance of subsequently drawing an A or a B (i.e., if the events are *independent*), the probability of drawing two As is the *product* of the probabilities associated with the separate events: the probability that both the first *and* the second item will be A is $P \times P = P^2$. When seeking the joint probability of k events, *multiply* their individual probabilities.

There are two compound events, however, that may yield an A,B pair: either A first and B second, or the converse. When seeking the probability that *either* of two (or more) events will occur, *add* their individual probabilities. The probability of A first and B second is $P \times (1 - P)$; the probability of B first and A second is $(1 - P) \times P$. Hence the probability of drawing an A,B pair is $2P(1 - P)$. Thus if $P = 0.2$ and $1 - P$ ($= Q$) = 0.8, the probability of 2 As is $0.2 \times 0.2 = 0.04$, of 2 Bs is $0.8 \times 0.8 = 0.64$, and of an A,B pair is $2 \times 0.2 \times 0.8 = 0.32$. If we draw 1000 pairs of items, we *expect* 40 A-A pairs, 640 B-B pairs, and 320 A-B pairs. By chance, however, we may not get exactly these numbers. Statistical tests are designed to (among other things) indicate whether deviations from such expected numbers can be ascribed to chance alone. If they cannot, factors other than chance are likely to be affecting the outcome.

MAP UNIT apart, 1 percent of the gametes will be recombinant (A_1B_2 or A_2B_1 in this instance). In this case, in which the recombination frequency, r, is 0.01, crossing over occurs in 1 cell in 50 that undergo meiosis (because two nonsister chromatids undergo reciprocal exchange). If two loci are sufficiently far apart on a chromosome, crossing over may occur so frequently that $r = 0.5$: i.e., the loci segregate independently even though they are physically linked.

A GENETIC MAP of the genes along a chromosome can be built up from measurements of recombination frequency. Consider the cross $ABC/abc \times abc/abc$, in which the slash separates the allele combinations that an individual received from its two parents. Considering just loci A and B, suppose the proportions of offspring genotypes are 0.495 $AaBb$, 0.005 $Aabb$, 0.005 $aaBb$, and 0.495 $aabb$. In this case, 1 percent have received new combinations (Ab, aB) from the heterozygous parent, and $r_{AB} = 0.01$. Similarly, if for loci B and C the proportions are 0.490 $BbCc$, 0.010 $Bbcc$, 0.010 $bbCc$, and 0.490 $bbcc$, $r_{BC} = 0.02$. If locus B lies between A and C, then the proportion of the recombinant genotypes $Aacc$ and $aaCc$ should be about 0.03, the sum of r_{AB} and r_{BC}. If this is so, we have established a map of these loci, which lie in the sequence A–B–C along the chromosome, separated by 0.01 and 0.02 map units respectively. Other loci may later be found that lie between them (Figure 3.3).

If many loci have been mapped in this way, it is possible, by suitable crosses, to "construct" genetic stocks of interest. In the fruit fly *Drosophila melanogaster*, by far the genetically best known eukaryote, thousands of gene loci have been mapped (Figure 3.4). Stocks of flies with mutations at various mapped loci, as well as stocks in which parts of chromosomes are missing (deletions), duplicated, inverted 180°, or attached to nonhomologous chromosomes can be used to construct almost any imaginable combination of mutant genes. Such manipulations enable researchers to explore the genes' molecular characteristics and effects on development.

The Relation between Genotype and Phenotype

So far we have calculated only the proportions of genotypes in Mendelian crosses, without saying anything about what the resulting organisms look like. The term **phenotype** refers, in the narrow sense, to the physical manifestation of a **genotype** (gene combination); it can refer to a specific morphological, physiological, biochemical, or behavioral attribute. The term is sometimes used more broadly to refer to a feature (or group of features) of an individual or group of like individuals, without any reference to the genotype.

Dominance

In the simplest case, individuals of a given genotype are phenotypically uniform, and each genotype at a locus is phenotypically distinct from the others. If the phenotype of a heterozygote, A_1A_2, is intermediate between the phenotypes of the homozygotes A_1A_1 and A_2A_2, the locus is said to display INCOMPLETE DOMINANCE; if the heterozygote has a precisely intermediate phenotype, there is no dominance,

Mendelian Ratios

Denote the probability that a sperm carries alleles A_1 or A_2 as $P_m(A_1)$ and $P_m(A_2)$ respectively, and use $P_f(A_1)$ and $P_f(A_2)$ similarly for eggs. In the cross $A_1A_2 \, ♂ \times A_1A_2 \, ♀$, the probability that an offspring will be A_1A_1 is:

$$P(A_1A_1) = P_f(A_1) \times P_m(A_1) = 1/2 \times 1/2 = 1/4 \text{ (or 0.25)}.$$

Similarly,

$$P(A_2A_2) = P_f(A_2) \times P_m(A_2) = 0.25.$$

Heterozygous offspring (A_1A_2), however, can arise either by union of an A_1 egg and an A_2 sperm [$P_f(A_1) \times P_m(A_2) = 1/2 \times 1/2 = 0.25$], or by union of an A_2 egg and an A_1 sperm [$P_f(A_2) \times P_m(A_1) = 1/2 \times 1/2 = 0.25$]. Therefore, the probability that an offspring will be heterozygous is $0.25 + 0.25 = 0.5$. Hence the F_2 generation in a cross between two A_1A_2 heterozygotes will consist of A_1A_1, A_1A_2, and A_2A_2 in the expected proportions 0.25, 0.50, and 0.25 respectively (Figure 3.C1A).

FIGURE 3.C1 Calculation of expected proportions of genotypes among the progeny of crosses, using probability. (A) Cross between two heterozygotes at one locus. (B) Cross between heterozygotes at two unlinked loci. The proportion of each genotype can be found by summing the probabilities of all gametic unions that yield that genotype, or by multiplying the probabilities of obtaining the one-locus genetic constitutions that together compose the two-locus genotype.

(A) $A_1A_2 ♀ \times A_1A_2 ♂$

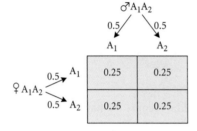

Frequency among progeny

A_1A_1 : 0.25
A_1A_2 : 0.25 + 0.25 = 0.50
A_2A_2 : 0.25

(B) $A_1A_2B_1B_2 ♀ \times A_1A_2B_1B_2 ♂$

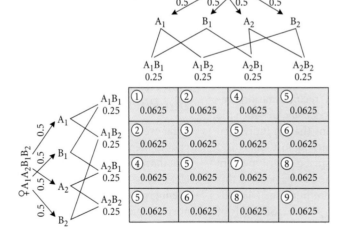

① $A_1A_1B_1B_1$ $1 \times 0.0625 = 0.0625$
② $A_1A_1B_1B_2$ $2 \times 0.0625 = 0.125$
③ $A_1A_1B_2B_2$ $1 \times 0.0625 = 0.0625$
④ $A_1A_2B_1B_1$ $2 \times 0.0625 = 0.125$
⑤ $A_1A_2B_1B_2$ $4 \times 0.0625 = 0.250$
⑥ $A_1A_2B_2B_2$ $2 \times 0.0625 = 0.125$
⑦ $A_2A_2B_1B_1$ $1 \times 0.0625 = 0.0625$
⑧ $A_2A_2B_1B_2$ $2 \times 0.0625 = 0.125$
⑨ $A_2A_2B_2B_2$ $1 \times 0.0625 = 0.0625$

Or, from part A:

Progeny genotype at locus A

	A_1A_1 0.25	A_1A_2 0.50	A_2A_2 0.25
B_1B_1 0.25	① 0.0625	④ 0.125	⑦ 0.0625
B_1B_2 0.50	② 0.125	⑤ 0.250	⑧ 0.125
B_2B_2 0.25	③ 0.0625	⑥ 0.125	⑨ 0.0625

Progeny genotype at locus B

and the alleles are said to have **additive** effects. If the heterozygote has the same phenotype as one of the homozygotes—say, A_1A_1—the A_1 allele is completely DOMINANT, and A_2 is RECESSIVE. Any degree of dominance is a departure from additivity (Figure 3.5). The degree of dominance of an allele is thought to be due to the relation between the amount or activity of gene product, such as an enzyme, and the development of a character that depends on the product. *Dominance* refers to the phenotypic effect of an allele in the heterozygous condition, not to how common the allele is in a population of organisms. For example, Huntington's chorea, a severe nervous disorder in humans, is caused by a dominant allele, but this allele is quite rare in populations.

Genotype and Environment

The phenotype of an individual organism is determined not only by its genes, but also by nutrition, temperature, social conditions, and many other aspects of its environment. Individuals of a given genotype, then, may vary in phenotype because they have experienced different environmental conditions (Figure 3.6). The **norm of reaction** of a genotype is the variety of different phenotypes it can have, expressed in different environments. Different genotypes often differ in their norms of reaction. For example, genotypes at the *Bar* locus in *Drosophila* differ in the effect of temperature on eye size (Figure 3.6B). Note that the *Infra-Bar* and *Ultra-Bar* genotypes have very different phenotypes at high temperature, but do not differ at low temperature. Thus a single genotype can produce different phenotypes under different environmental conditions, and conversely, a given phenotype can be produced by different genotypes.

Because most phenotypic characteristics are influenced by both genes and environment, it is fallacious to say that a characteristic is either "genetic" or "environmental." It is meaningful only to ask whether the differences among individuals are attributable more to genetic differences or to environmental differences, recognizing that both may contribute to the variation. The answer may well depend on the particular individual organisms examined. For example, we might see much the same amount of variation in eye size in *Drosophila* whether we had a mixture of *Infra-Bar* and *Ultra-Bar* flies, all reared at 25°, or a pure

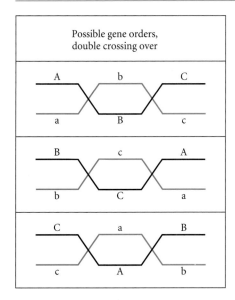
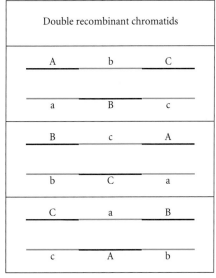

Possible gene orders, double crossing over

Double recombinant chromatids

FIGURE 3.3 Mapping genes by a three-point test cross. The left panels show the three possible orders of three loci, with double crossing over occurring in the heterozygote *ABC/abc*. Only the two chromatids in which exchange occurs are shown. The alleles carried by recombinant chromatids, shown in the right panels, are identified by the phenotypes of the progeny when this individual crossed to an *abc/abc* homozygote. If the combined frequency of *Ab* and *aB* progeny were 0.01, of *Bc* and *bC* progeny 0.02, and of *Ac* and *aC* about 0.03, then only the uppermost gene order would be compatible with the data. (After Suzuki et al. 1989.)

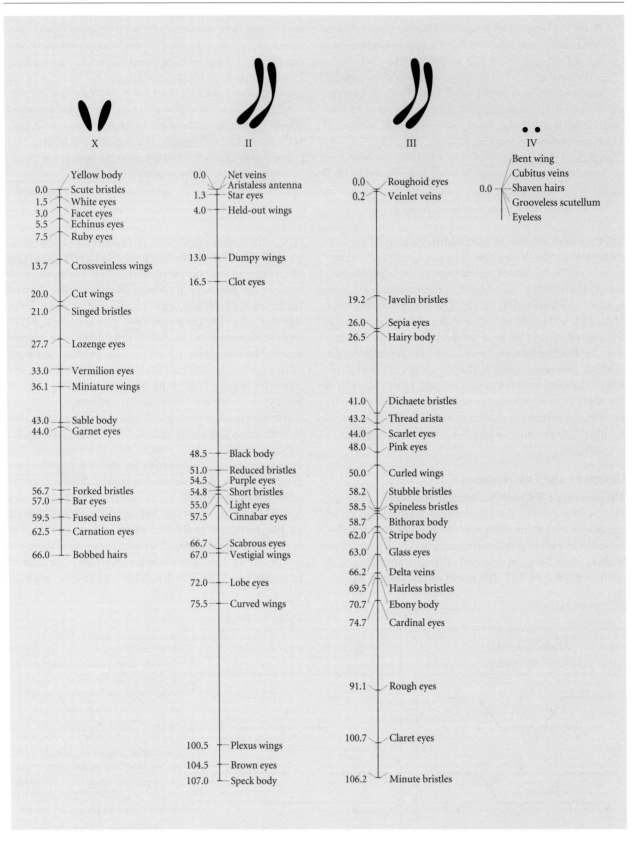

FIGURE 3.4 A partial genetic map of *Drosophila melanogaster*, showing the order of some loci on each of the four pairs of chromosomes. The numbers mark map positions, i.e., recombination distances between successive genes, starting at one end of each chromosome (position 0.00). There is no recombination among genes on the tiny chromosome IV, so all the genes are at position 0.00. (After Sinnott et al. 1958.)

FIGURE 3.5 Two of the possible relationships between phenotype and genotype at a single locus. In curve I, illustrating additive inheritance, replacing A' with A alleles steadily increases the amount of gene product, and the phenotypic character increases accordingly. In curve II, illustrating dominance of A over A', the phenotype of AA' nearly equals that of AA, because the single dose of A in the genotype produces enough gene product for full expression of the character.

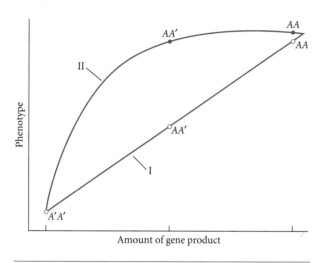

stock of *Ultra-Bar* flies, some of which had developed at 25° and others at 16°.

Multiple Loci

Many characteristics are influenced by several or many loci. Such characters are **polygenic**. In the simplest case, the effects of different loci simply add up to give the phenotype (the loci, then, act *additively*). For example, suppose that in a specific environment, the average height of a plant of genotype $A_2A_2B_2B_2$ is 10 cm, and that replacing A_2 or B_2 with A_1 or B_1 adds 1 cm. (There is no reason for alleles at different loci to add the same amount; the equivalent amounts in this example are chosen for the sake of simplicity.) In this case, $A_1A_2B_1B_2$ would be 12 cm tall, $A_1A_1B_1B_2$ 13 cm tall, and $A_1A_1B_1B_1$ 14 cm tall. In principle, five different heights could occur among the nine possible genotypes.

In any real population, each genotype would vary in size because of environmental effects. (Figure 3.7). Moreover, it is unlikely that only two loci would affect the phenotype; the variation in most characteristics appears to be due to at least five to ten, and often more, loci, many of which have a very slight effect. The consequence of environmental effects and slight genetic effects is a rather smooth continuum of heights, or CONTINUOUS VARIATION. The length of the flower tube (corolla) in the tobacco plant (*Nicotiana longiflora*) is a typical example (Figure 3.8). The mean phenotype of the progeny is the average of the parents' phenotypes, which indicates that the variation is partially genetic and is at least largely additively inherited. That multiple loci are involved is suggested by the continuous distribution of phenotypes in the F_2 generation (rather than a division into two or three discrete classes). The variation in this characteristic is partly due to different genes ("genetic variance," V_G) and partly to environmental effects ("environmental variance," V_E).

In these examples, the effects of alleles at different loci merely add up to give the total phenotype. The loci do not

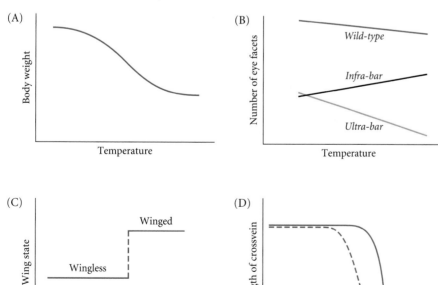

FIGURE 3.6 Four schematic illustrations of real reaction norms. In each case the phenotype developed depends on environmental conditions. (A) The response of adult body weight to temperature during development in *Drosophila* and many other insects. (B) Size of the compound eye, as measured by the number of eye facets, in three genotypes at the *Bar* locus in *Drosophila* reared at different temperatures. (C) A developmental "switch," as found in some aphids, which develop wings if sufficiently crowded at a critical period in development. (D) Differential sensitivity of two *Drosophila melanogaster* genotypes to a heat shock that affects development of the crossvein in the wing. (B after Suzuki et al. 1989.)

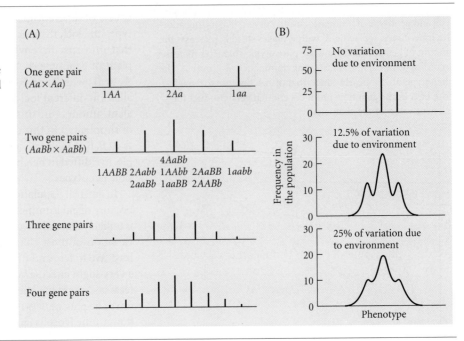

FIGURE 3.7 Continuous variation. The abscissa represents a phenotypic character such as height; the ordinate, the proportions of the phenotypes in a population. (A) All individuals of the same genotype have exactly the same phenotype. As we increase the number of loci involved and decrease the contribution of each locus to the phenotype, the number of phenotypic classes increases. (B) For one gene pair, we superimpose the phenotypic variation that each genotype expresses due to variation in the environment. As environmental effects on variation increase, the genotypes may overlap greatly in phenotype. (A after Strickberger 1968; B after Allard 1960.)

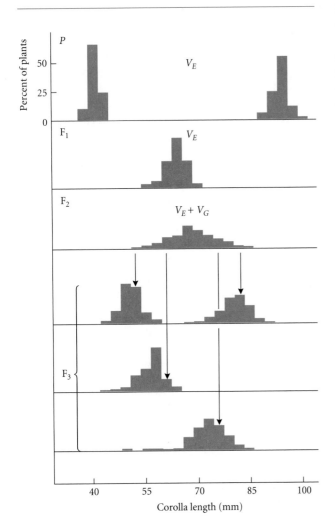

interact; that is, the phenotypic effect of one locus is not influenced by the genotypes at the other loci. Quite often, however, there is such an interaction. For example, the recessive allele *bw* gives brown eyes in homozygous *Drosophila melanogaster* (the normal, or WILD-TYPE, color is red). At another locus, the recessive white allele (*w*) gives white eyes, no matter what the genotype at the brown locus. Thus $w\ w\ +^{bw}+^{bw}$, $w\ w\ +^{bw}\ bw$, and $w\ w\ bw\ bw$ all have white eyes—the latter genotype is not pale brown. (The symbol + represents a wild-type allele; the superscript *bw* is used to indicate that the wild-type allele referred to is at the brown locus. A homozygote for the wild-type allele at both loci might be denoted $+^{w}+^{w}+^{bw}+^{bw}$.) The white allele *w* is said to be EPISTATIC over the brown locus. (The word *epistatic* is etymologically related to the Greek *epistates*, a governor or overseer; the one locus governs the expression of the other.) By extension, the term **epistasis** is used (especially in population genetics) to refer to any interaction among loci that departs from simple additivity. The biochemical basis of this example is quite simple. The red eye color of wild-type

FIGURE 3.8 Inheritance of a continuously varying ("metric") trait, corolla length in the tobacco plant *Nicotiana longiflora*. The two parental (P) strains are two homozygous genotypes; the F_1 is heterozygous but genetically uniform, i.e., all plants have the same genotype. The variation (*V*) in each parental strain and in the F_1 is due to the environment (V_E). In the F_2 and F_3 generations, variation has both genetic (V_G) and environmental (V_E) components. Four F_3 families are shown, from crosses between parents whose means are indicated by arrows. The mean of the offspring of each family is close to that of their parents, indicating that much of the genetic variation is additive. (After Cavalli-Sforza and Bodmer 1971; Mather 1949.)

flies is the result of two pigments, scarlet and brown. A *bw bw* fly lacks the enzyme for producing the scarlet pigment, so only the brown shows. But the white locus governs an earlier step in the biosynthetic pathway leading to both pigments. A *w w* fly does not produce the precursor of the brown and scarlet pigments, and so has white eyes.

Threshold Traits

A special form of epistasis occurs in **threshold traits**. These are characters that may occur in two (or more) discrete states, but in which the phenotype is determined by multiple loci rather than (as one might expect at first glance) a single pair of dominant and recessive alleles. Genetic differences at many loci affect an underlying character, perhaps the amount of some biochemical substance. Different phenotypes develop if the amount of the substance is above or below some critical value (the threshold). For example, in the 1930s, Sewall Wright crossed inbred strains of guinea pigs with three toes (strain C) versus four toes (strain D). The F_1 progeny of C × D have three toes, and in backcross progeny (F_1 × D), half the offspring have three and half have four toes. Moreover, F_1 × F_1 crosses yield a 3:1 ratio of three- and four-toed F_2 offspring. This looks like Mendelian segregation of a dominant allele (*A*) for three toes and a recessive allele (*a*) for four toes. The three- and four-toed backcross progeny (BC_1) should have genotypes *Aa* and *aa* respectively. But the further backcross of three-toed BC_1 (supposedly *Aa*) to strain D (supposedly *aa*) yielded 23 percent three-toed and 77 percent four-toed offspring, instead of the expected 50:50 ratio; and the backcross of four-toed BC_1 × D (both supposedly *aa*) yielded 16 percent three-toed and 84 percent four-toed offspring instead of the expected 0:100 ratio. Clearly the presence or absence of the fourth toe is not determined by a single gene. The results can be explained by postulating polygenic inheritance of a toe-promoting factor; above some threshold, the fourth toe develops (Figure 3.9). There is polygenic variation in the underlying factor among BC_1 animals, so that about half lie above the threshold and half below; further backcrosses to strain D result in increasing numbers of toe-promoting genes, hence in increasing proportions of four-toed progeny. From this model, Wright predicted the phenotype proportions in further backcrosses, and those predictions were confirmed (see Wright 1968).

The Genetic Material and Its Functions

Except in certain viruses in which the genetic material is **RNA** (RIBONUCLEIC ACID), the hereditary information in all organisms is carried by **DNA** (DEOXYRIBONUCLEIC ACID). Each chromosome carries a single long, often tightly coiled, molecule of DNA. The DNA molecule (Figure 3.10) is a double-stranded helix, consisting of a series of nucleotide BASE PAIRS (BP). Each base pair consists of a purine (adenine, A, or guanine, G) and a pyrimidine (thymine, T, or cytosine,

FIGURE 3.9 Model of a threshold trait, based on inheritance of toe number in guinea pigs. The model postulates variation in an underlying character that influences toe development. Below a threshold value in this character, three toes develop; above it, four develop. The underlying character is additively inherited, so that F_1 progeny of the cross between strains C and D lie below the threshold, and have three toes. The F_2 is genetically variable, and some individuals cross the threshold and have four toes. Two generations of back-crossing to strain D (BC_1 and BC_2) have increasing proportions of four-toed progeny. (After Wright 1968.)

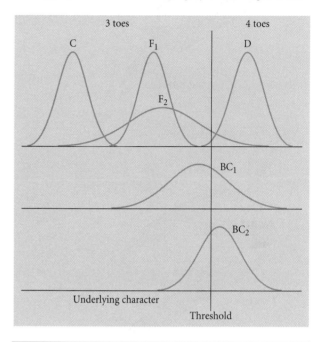

C). A is always paired with T, and C with G. Each of the two strands is so structured that polarity is evident from one end (the 5′ end, usually depicted at the left) to the other (the 3′ end). The polarity of each strand is opposite that of the other strand.

DNA REPLICATION occurs before each cell division. At each of a number of sites along the chromosome, the double helix is unwound, pairs of bases are sequentially separated, and DNA POLYMERASE enzymes bring complementary bases into position along each of the two strands to create a new, complementary strand. Because the DNA molecule is helically twisted, separation of the strands in their entirety would cause the end of the molecule to rotate wildly, but this does not happen because enzymes cut ("nick") one or both strands, enabling short segments to unwind before they are rejoined. Replication produces two DNA molecules (corresponding to the sister chromatids visible in mitosis or meiosis), each with one original and one newly synthesized, complementary strand. Because of the complementarity of A with T and C with G, the two molecules are identical to the original one, unless errors in replication occur.

Many, but not all, sequences of DNA contain information specifying the amino acid sequences of proteins, including enzymes. A *gene* may be thought of as the DNA sequence

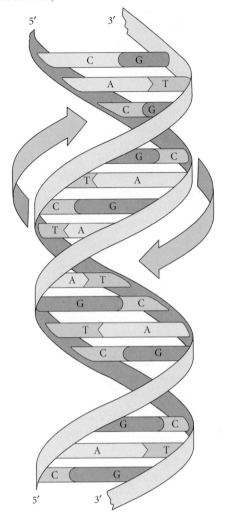

FIGURE 3.10 The DNA double helix. The two strands each consist of sugar-phosphate backbones to which are attached the nucleotide bases. The strands are held together by hydrogen bonds between the pairs of bases, C-G and A-T. (After Purves et al. 1998.)

specifying a single polypeptide, which may be either a complete protein in itself or just one of several polypeptides constituting the mature protein. The full definition of a gene, however, is considerably more ambiguous, because the sequences that together code for a polypeptide are usually interrupted (in eukaryotes) by base pair sequences that are not translated into amino acid sequences. These intervening sequences are called **introns**, and the coding regions **exons**. Moreover, DNA sequences that play regulatory roles are associated with the exons (Figure 3.11). Many such sequences lie upstream (i.e., before the 5′ end) of the sequence that will actually be translated into a polypeptide, but some regulatory regions can lie within or downstream (after the 3′ end) of the coding regions. A **gene**, therefore, is a segment of DNA involved in producing a polypeptide chain, including introns and regions preceding and following the coding region that include sequences affecting its regulation.

TRANSCRIPTION of DNA sequences into RNA is carried out by RNA POLYMERASE enzymes. One strand of a DNA sequence (say, ACGT…) is transcribed into the complementary RNA sequence, a single-stranded molecule, with uracil (U) in place of thymine (UGCA…). The RNA, once synthesized, moves into the cytoplasm. RIBOSOMAL RNA (rRNA) molecules of several sizes become part of the ribosomes, the sites of protein synthesis. TRANSFER RNA molecules (tRNA) fall into 20 sets, each of which transports one of the 20 amino acids that make up proteins. A gene that codes for a protein is transcribed first into a primary transcript; the portions corresponding to the gene's introns (if any) are then spliced out by enzymes. The resulting MESSENGER RNA (mRNA) transcript is then *translated* into a polypeptide or

FIGURE 3.11 Diagram of a eukaryotic gene, its initial transcript (pre-mRNA), and the mature mRNA transcript. Proteins that regulate transcription bind to the enhancer sequences that lie upstream of the coding sequences. Transcription proceeds in the 5′ to 3′ direction. Introns are transcribed, but are excised from the pre-mRNA and discarded. The coding segment of the mature mRNA corresponds to the gene's exons. (After Campbell 1993.)

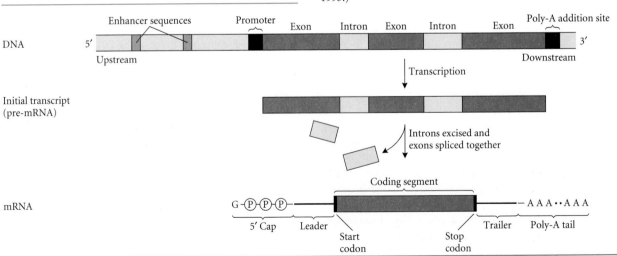

protein on the basis of the **genetic code**. This code resides in triplets of bases. Each triplet is recognized by one of the tRNAs, which, in the process of TRANSLATION, brings its amino acid to join the other amino acids in a growing polypeptide chain. The chain is terminated when one of several "termination" triplets is encountered.

The RNA code, which is complementary to the DNA code, consists of $4^3 = 64$ triplets, or **codons** (Figure 3.12). The code is degenerate: two or more SYNONYMOUS codons code for each of most of the 20 amino acids. The third position in a codon is the most degenerate; for example, all four CC_ codons (CCU, CCC, CCA, CCG) specify proline. The second position is the least degenerate. Thus a substitution of a nucleotide in the second position results in an amino acid substitution in a protein, but a substitution in the third position often does not.

It is a profoundly wonderful fact that the genetic code is almost universal: it is virtually identical in all organisms, from viruses and bacteria to pea plants and mammals. (The few exceptions include differences in the meaning of certain triplets in mitochondria and in some protozoans, compared to the usual meaning of those triplets in nuclear DNA.) Moreover, the machinery of transcription and translation is remarkably uniform among organisms: frogs injected with sea urchin DNA produce RNA transcripts of sea urchin genes, and hemoglobin mRNA from rabbits injected into frog eggs can be translated into rabbit hemoglobin by the frog's translation machinery.

Organization of the Genome

Organisms vary greatly in the amount of DNA they contain (Figure 3.13). Viruses and bacteria have far less DNA than most eukaryotes. Among eukaryotes, however, there is surprisingly little correspondence between the amount of nuclear DNA and our superficial impression of the organisms' complexity. A haploid nuclear genome of *Drosophila* has about 1.5×10^8 base pairs, and that of a human about 3.1×10^9, but some salamanders have more than 100 times as much DNA as humans. Moreover, DNA content varies more than a hundredfold among species of salamanders. If all the DNA constituted functional genes, coding for proteins, the genome of *Drosophila* would consist of about 100,000 genes, and that of an average mammal about 300,000. However, most of the DNA in eukaryote genomes does not encode any functional products, and the number of functional genes is much smaller than the amount of DNA might suggest. For example, in *Drosophila*, each of the approximately 6000 bands visible in the chromosomes of the larval salivary glands (Figure 3.14) appears to carry a single functional gene. The number of functional genes has been estimated at only 30,000–40,000 in mammals, and at fewer than 20,000 in sea urchins. These estimates (which are very inexact, and based on indirect lines of evidence) account for less than 10 percent of the DNA.

Some of the noncoding DNA is in introns within functional genes; some of it consists of **spacers** between genes; and some consists of **repetitive sequences**. These include **highly repetitive sequences** of about 5 to 12 base pairs, sometimes referred to as SATELLITE DNA. In *Drosophila virilis*, 40 percent of the total DNA consists of three satellite sequences with 1.1×10^7, 3.6×10^6, and 3.6×10^6 copies respectively. Much of the satellite DNA occurs in HETEROCHROMATIC REGIONS of the chromosomes: those regions, especially near the centromeres, where the DNA is tightly coiled and is apparently not transcribed.

		Second nucleotide				
		U	C	A	G	
First nucleotide	U	UUU UUC } Phe UUA UUG } Leu	UCU UCC UCA UCG } Ser	UAU UAC } Tyr UAA Chain end UAG Chain end	UGU UGC } Cys UGA Chain end UGG Trp	U C A G
	C	CUU CUC CUA CUG } Leu	CCU CCC CCA CCG } Pro	CAU CAC } His CAA CAG } Gln	CGU CGC CGA CGG } Arg	U C A G
	A	AUU AUC AUA } Ile AUG Met	ACU ACC ACA ACG } Thr	AAU AAC } Asn AAA AAG } Lys	AGU AGC } Ser AGA AGG } Arg	U C A G
	G	GUU GUC GUA GUG } Val	GCU GCC GCA GCG } Ala	GAU GAC } Asp GAA GAG } Glu	GGU GGC GGA GGG } Gly	U C A G

FIGURE 3.12 The genetic code, as expressed in messenger RNA. Three of the 64 codons are "stop" signals ("chain end") that terminate translation. The 20 amino acids are specified by other codons. Note that many codons, especially those differing only in the third nucleotide position, are synonymous, specifying the same amino acid. (The abbreviations for the amino acids are: Ala, alanine; Arg, arginine; Asn, asparagine; Asp, aspartic acid; Cys, cysteine; Gly, glycine; Glu, glutamic acid; Gln, glutamine; His, histidine; Ile, isoleucine; Leu, leucine; Lys, lysine; Met, methionine; Phe, phenylalanine; Pro, proline; Ser, serine; Thr, threonine; Trp, tryptophan; Tyr, tyrosine; Val, valine.)

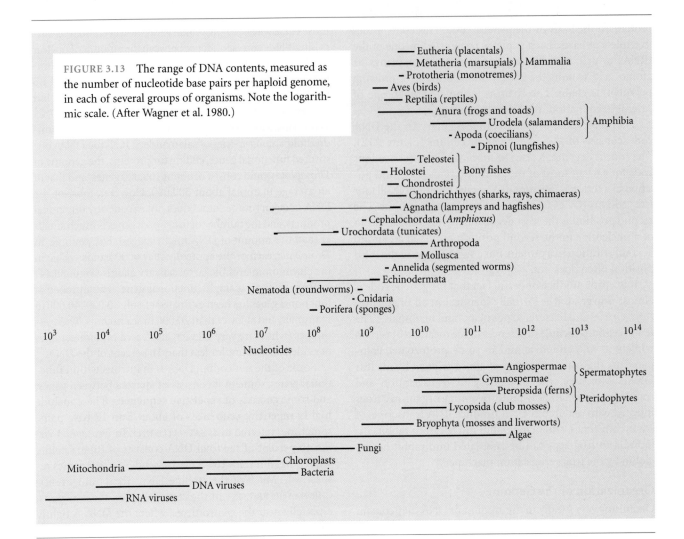

FIGURE 3.13 The range of DNA contents, measured as the number of nucleotide base pairs per haploid genome, in each of several groups of organisms. Note the logarithmic scale. (After Wagner et al. 1980.)

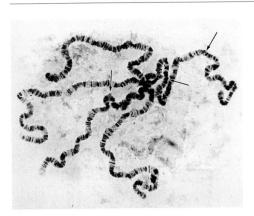

FIGURE 3.14 Polytene chromosomes from the salivary gland of a *Drosophila melanogaster* larva. Each pair of synapsed chromosomes consists of 2048 closely appressed identical strands. The darkly staining bands form patterns that enable experienced researchers to recognize individual chromosome regions. Each of the 6000 or so bands apparently contains one functional gene. The positions of a few such genes, such as the alcohol dehydrogenase locus (ADH), are indicated with arrows. (From Ursprung et al. 1968; courtesy of H. Ursprung.)

Some coding sequences are also repeated. The MODERATELY REPETITIVE fraction of DNA includes several hundred copies of histone genes, as well as hundreds or thousands of copies of the genes that code for ribosomal RNA. Finally, many genes that code for proteins are members of small **gene families**: groups of genes with similar sequences. For example, human adult hemoglobin is made up of α and β polypeptides, the genes for which are similar in sequence. At earlier stages of development, several other α-like and β-like genes are expressed. The α family includes three functional genes on chromosome 11, and the β family includes five such genes on chromosome 16 (Figure 3.15). In addition, each cluster includes one or more **pseudogenes**: sequences that resemble the functional genes with which they are associated, but which differ at a number of base pair sites and are not transcribed because they have internal "stop" codons.

The hemoglobin family shows that functionally related genes may either be linked (i.e., on the same chromosome) or not. With some important exceptions, there is little evidence that the various genes contributing to a particular biochemical pathway, or to the development of a particular structure, are physically associated on the chromosomes.

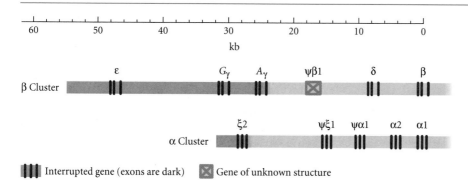

FIGURE 3.15 The α and β families of human globin genes, each located on a different chromosome. Each functional gene is depicted as three dark lines, representing three exons. Pseudogenes are denoted by Ψ. The scale indicates the distance between genes in kilobases (kb). (From Lewin 1985.)

Members of moderately or highly repeated families of sequences are indeed often arranged in tandem, but often such a family is distributed over several or many chromosomes.

Gene Expression

Certain genes (referred to as "housekeeping genes") are transcribed and expressed in most of the cells of an organism throughout its life. These include, for example, the genes that code for the enzymes involved in basic intermediary metabolism (i.e., the release of energy by the metabolism of glucose). Many other genes, however, are expressed only in certain groups of cells, or only at certain stages of development. The hemoglobin genes in vertebrates, for example, are transcribed, and their RNA transcripts translated into hemoglobin, only in prospective red blood cells. In mammals, different genes code for fetal and adult hemoglobin, and they are, accordingly, transcribed at different times in the life cycle. Which genes are transcribed in which cell types is, in fact, largely responsible for the differentiation of the cell types in structure and biochemical function.

Differences among cells in the presence/absence or amount of a particular gene product can arise from several mechanisms, including

1. Gene amplification. Specific genes may be replicated many times in a certain type of cell, while the rest of the genes remain represented by only the original two (in diploid organisms) copies. Amplification, sometimes to extraordinarily high levels, is typical of genes that are called on to produce very large quantities of a gene product, such as chorion (eggshell) protein genes in certain cells of the *Drosophila* ovary or, in the oocytes of many animals, genes that code for ribosomal RNA.
2. Deletion and rearrangement. Deletion is unusual; one example is the complete elimination from somatic cell lines in some gall midges (small flies) of certain whole chromosomes that apparently function only in the germ line. Extraordinarily complex, diverse rearrangements of DNA sequences (including deletion of some intervals) occur in the immunoglobulin

genes of vertebrates. These rearrangements occur in the lymphocytes, which, as a group, are consequently capable of producing an immense variety of immunoglobulin (antibody) proteins.

3. Transcriptional control. A model for the control of gene expression was first developed and verified for bacteria. In 1961, François Jacob and Jacques Monod described the *lac* **operon** in *Escherichia coli*, a cluster of several genes governed by a regulatory sequence (Figure 3.16). These genes include one that codes for β-galactosidase, an enzyme that cleaves lactose into usable sugars, and one encoding a permease that increases the rate at which lactose enters the cell. The cluster includes an operator gene that promotes transcription of these "structural" genes. Another gene, not included in the operon, produces a repressor protein that binds to the operator, inhibiting transcription. When lactose is present, it binds to the repressor protein, dissociating it from the operator and leaving the operon free for transcription. Thus the system can respond to the presence of lactose by producing the enzymes necessary to utilize it.

Although eukaryotes seem not to have operons as such, transcriptional control is by far the most common mode of gene regulation. RNA is transcribed from DNA by RNA polymerase enzymes; the one that transcribes protein-encoding sequences is called RNA polymerase II (POL II). Transcription is initiated upstream (5′) of the transcribed sequence, where a TRANSCRIPTION COMPLEX binds to the DNA. This complex includes pol II (which is not specific to any particular DNA sequence) and several other proteins referred to as TRANSCRIPTION FACTORS. Certain of these transcription factors are specific to particular DNA sequences. The DNA region(s) to which the transcription complex binds is called the PROMOTER region. The promoter often contains a short sequence (e.g., TATA, the "TATA box") that is common to many genes, as well as other sequences that are gene-specific.

In most cases, a gene is transcribed only if it is ACTIVATED by a transcription factor that binds to its specific promoter (the same transcription factor may bind to similar promoter sequences of several genes

FIGURE 3.16 Model of transcriptional control in bacteria. A regulator gene regulates an operon, a cluster of several structural genes and an operator gene. (A) Transcription of the structural genes is repressed when the operator is bound by a repressor protein produced by the regulator gene. (B) An inducing substance from outside the cell binds to the repressor, derepressing the operator, so that transcription of the structural genes can occur. (After Strickberger 1968.)

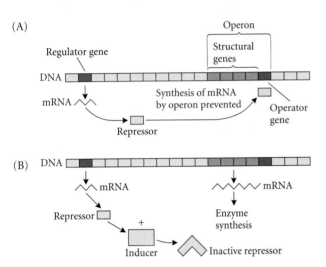

ported from other cells where they are synthesized; both the rate of production and the rate of transport may affect protein levels in the recipient cells.

Hierarchical Control of Gene Expression in Development

The processes that lead from an egg cell to a fully differentiated organism are the subject of developmental biology, at present one of the most dynamic fields of biological research. Because evolutionary changes in morphology result from evolutionary changes in development, a full understanding of evolution requires knowledge of these processes.

One of the most important developmental processes is *differentiation*: the acquisition of different properties by different cells or groups of cells. This process entails, largely, differences among cells in the expression of genes. Among eukaryotes, differentiation is best understood in *Drosophila melanogaster*, owing to the complex genetic manipulations possible in this species. In some of the most exciting research in modern biology, the genetic controls by which the different segments of the insect acquire their identity have been elucidated (DePomerai 1990; Glover and Hames 1989). Developmental biologists have determined much of the "algorithm" that specifies which genes are active in which cells. To do this, they have used two major techniques.

First, mutations that have segment-specific effects show which genes are necessary, in their normal (wild-type) form, for normal development of the affected segments. For example, mutations in the *Bithorax* complex of genes are HOMEOTIC, meaning that they transform structures or segments into structures or segments that are located elsewhere in a normal individual. Certain of these mutations, for instance, transform the third segment of the thorax (the metathorax) so that it resembles the second segment (the mesothorax); the haltere, a metathoracic structure in flies that evolved from the hindwing found in other insects, becomes a wing (Figure 3.17). By using combinations of such mutations, geneticists can infer how the genes interact in directing development.

Second, molecular techniques have enabled biologists to isolate genes, sequence them, and determine the location in the embryo of the genes' products, such as mRNA transcripts (Color Plate 1). These techniques provide direct confirmation of inferences drawn from the phenotypic effects of mutations about when and where the genes are active in the developing insect. Box 3.D provides a brief introduction to some important methods in molecular genetic analysis.

Through such studies, it is now known that the embryo develops largely as a consequence of hierarchical control of gene expression, i.e., control of the expression of certain genes by other genes (Figure 3.18). Consider anterior-to-posterior (A–P) differentiation. What determines which end of an egg will be the head of the embryo, and which parts will differentiate into the various segments that con-

that are regulated in concert). This mechanism is called "positive" regulation. In a few cases, a gene is continuously (CONSTITUTIVELY) transcribed unless its promoter is bound by a protein that prevents, or *represses,* transcription ("negative" regulation). Proteins that bind to promoters often have, as part of their structure, certain similar amino acid sequences, with correspondingly similar base pair sequences in the genes that encode them. These sequences include the HOMEOBOX, coding for about 60 amino acids, found in many genes of *Drosophila* and other eukaryotes. Transcriptional regulation of a particular gene, then, can be based on the production by other genes, in certain cell types and/or at certain times in development, of one or more specific transcription factors that positively or negatively affect transcription. If a gene has two or more promoters, different transcription factors can govern the expression of the gene in different cell types.

4. Posttranscriptional control. In at least some cases, posttranscriptional events influence protein levels in the cell. To mention just two such factors: A primary transcript may be differentially spliced in different cells (e.g., including or excluding certain exons), yielding different mRNAs and hence different polypeptides. Proteins may be present in lower concentrations in some tissues than in others because their transcripts are translated at a lower rate, or because the protein is degraded at a higher rate (or the mRNA may be unstable). Finally, some proteins in certain cells are im-

FIGURE 3.17 A wild-type *Drosophila melanogaster* (A) and a homeotic mutant (B), in which the third thoracic segment has been transformed into another second thoracic segment, bearing wings instead of halteres. The mutant fly was experimental-ly produced by putting together several mutations in the regulatory region of the *Ultrabithorax* (*Ubx*) gene. (Photographs courtesy of E. B. Lewis.)

(A)

(B)

stitute the head, thorax, and abdomen? The answer begins with mRNA transcripts in the egg that were transcribed from the mother's genes. The transcripts of one such MATERNAL EFFECT gene, *bicoid*, decrease in concentration in the egg from anterior to posterior, establishing A–P polarity (directionality). The protein products of these transcripts, also declining posteriorly in concentration, bind to the regulatory regions of, and thereby regulate the spatial expression of, GAP GENES, which divide the embryo into several major regions and confer on them separate identities. The gap genes regulate the expression of PAIR-RULE GENES, which are expressed in transverse bands that specify the borders of future segments. The pair-rule genes govern the spatial expression of SEGMENT POLARITY GENES, which specify the A–P polarity within each segment. In each segment, different combinations of HOMEOTIC SELECTOR GENES are activated, such as the genes in the *Bithorax* complex. For instance, the *Ubx* gene in this complex is expressed in the prospective metathorax and anterior part of the abdomen, and prevents these segments from developing the features of a mesothorax, such as wings. Under the influence of *Ubx*, the abdominal segments would develop the features of a metathorax (such as legs), were it not that two other genes in the *Bithorax* complex are expressed in these segments, and repress the activity of *Ubx*. Thus the spatial expression of these various homeotic selector genes is governed in a complex way by gap genes, pair-rule genes, and interactions among the homeotic selector genes themselves. The homeotic selector genes, governed by a cascade of influences beginning with the maternal effect genes, themselves act as "master switches," activating in each segment numerous other genes that affect the individual features of that segment, such as wings, legs, or genitalia. Thus if a fly carries both a *Ubx* mutation and a mutation that affects the veins of the mesothoracic wings, the *Ubx* mutation transforms the halteres on the metathorax into wings, and those wings express the wing vein mutation.

Mechanisms of Development

The two fundamental processes of development are differentiation and morphogenesis. **Differentiation** is the acquisition of different properties by different cells or groups of cells. These properties include, especially, the expression of different proteins that contribute to the cells' structure and function. **Morphogenesis** is change in the shape of the embryo or any part of it, from cell to tissue to organ. Both processes may be influenced both directly by the genes and indirectly through a series of steps that have been termed **epigenesis**. For example, the enlargement of the eye and brain in vertebrate embryos is a direct mechanical result of hydrostatic pressure within the developing organs; although the mechanics result from growth processes that are ultimately due to the activity of genes, the causal path from genes to the developmental effect is indirect.

Control of Cell Differentiation

As we have seen in *Drosophila*, the organization of maternally derived materials in the cytoplasm of the egg sets the stage for later developmental events by, among other things, establishing anterior-posterior polarity. In the case of *Drosophila*, these gradients in cytoplasmic factors are intercepted by cell divisions, so that cells in different regions acquire different concentrations of these materials. These initial differences appear to set off different chain reactions of gene expression, leading to the differentiation of different groups of cells. Each such group of cells appears

BOX 3.D

A Thumbnail Sketch of Some Important Methods in Molecular Genetic Analyses

Studies of gene structure and function, as well as evolutionary studies that build on this information, use a wide and growing variety of often complex methods. Greatly oversimplified sketches of some of the most important of these methods are presented here. For further details, consult a modern textbook on genetics or molecular biology (e.g., Lewin 1994; Hoelzel 1992).

1. Protein electrophoresis. Some of the amino acid residues in a protein, such as an enzyme, are positively or negatively charged. If a mutation substitutes a different amino acid at such a site, the protein's net charge may be altered. Other mutations may alter the protein's shape or size. When proteins are placed in a gel (such as starch or acrylamide) and an electric field is applied, they migrate through the gel at a rate determined by their net charge and their size. Hence, variant proteins that differ by certain amino acid substitutions can be separated. Their location in the gel is determined by flooding the gel with a substrate for the enzyme and then with other reagents with which the product of the enzymatic reaction reacts, yielding a colored blot. Electrophoresis is an important method for detecting genetic variation in enzymes and other proteins.

2. Restriction enzymes. A number of enzymes have been isolated, mostly from bacteria, that recognize specific, short (e.g., four or six base pairs) DNA sequences ("restriction sites") and cut the DNA within these regions. These restriction enzymes can be used to cut into fragments either a whole (nuclear, mitochondrial, or chloroplast) genome or a specific DNA fragment isolated from such a genome. The resulting DNA fragments can be separated by size by placing the preparation on a gel (e.g., of agarose) and applying an electrical current (electrophoresis). The fragments can be "visualized" by several methods, including AUTORADIOGRAPHY: radioactively labeling the DNA, and then exposing the gel to a radiation-sensitive film. The size of each fragment (in number of base pairs) is measured by its mobility relative to known standards, and the sum of the fragment sizes equals the size of the sequence that has been cut (Figure 3.D1). By comparing the fragments produced by combinations of different restriction enzymes with those produced by each enzyme singly, it is possible (although rather complicated) to map the different restriction sites, i.e., to specify their order and the number of base pairs between them. These restriction sites can then be used as genetic markers for relatively short sequences of DNA, just as mutant alleles are used as genetic markers in traditional genetics.

3. Denaturation, annealing, and "hybridization" of DNA. The two strands of the DNA double helix become separated (denatured) at high temperatures, and reestablish their structure (reanneal) when the temperature is lowered. The heat required to denature a DNA sequence depends on the degree of complementarity of the two strands: the more bases of one strand are mismatched with those of the other (i.e., the greater the number of departures from A-T and G-C pairing), the less heat is required.

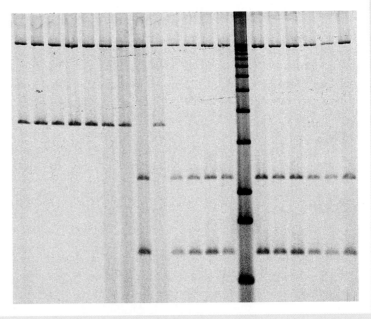

FIGURE 3.D1 Autoradiograph of a gel showing a restriction fragment-length polymorphism (RFLP) in the seaside sparrow (*Ammodramus maritimus*). The mitochondrial DNA of 19 birds has been cut with the restriction enzyme *BclI*. The seventh lane from the right is a molecular weight standard in which the uppermost band is 12 kb, and successively smaller fragments differing by 1 kb are shown by the lower bands. The haplotype in lanes 1–7 and 9 (from the left) has two fragments, so the circular mitochondrial DNA molecule has two restriction sites that are cut by this enzyme. In lanes 8, 10–13, and 15–20, the second fragment from the top is absent and is represented by two smaller (faster) fragments in a second haplotype. Thus this haplotype has a restriction site that the other haplotype lacks. (Courtesy of J. C. Avise.)

A genome that has been fragmented by restriction enzymes can be heated to produce single strands. These DNA fragments are then bound to a filter, and the filter is exposed to another, similarly treated but radioactively labeled, preparation of DNA fragments. The filter-bound DNA might be from one species and the radioactively labeled DNA from another. As the temperature is lowered, the free fragments "hybridize" with the filter-bound fragments to which they are complementary, forming double-stranded (duplex) DNA. The *rapidity* of such "hybridization" depends on how likely a fragment is to "find" a complement; this is more likely for a highly repetitive DNA sequence, so the dynamics of reannealing can measure the proportion of the genome that consists of highly or moderately repetitive sequences. The *stability* of the hybridized duplexes, measured by how much heat is required to separate them again, is a measure of how complementary their base pair sequences are overall. This technique may be used, for instance, to measure the overall sequence similarity of the DNA of two different species.

4. Cloning and amplifying DNA sequences. To isolate and then to produce in quantity a DNA sequence for further study, the first, often difficult, step requires that one have available a PROBE, i.e., a DNA sequence that is similar enough to the "target sequence" to hybridize with it. In one of several methods (Southern blotting), the genome is digested with restriction enzymes, denatured, electrophoresed on a gel, and the single-stranded fragments are then transferred to and immobilized on a nitrocellulose filter. The filter is then washed with a preparation containing the radioactively labeled probe, which hybridizes to the target sequence (and adjacent DNA). The resulting duplex is visualized by autoradiography, and then isolated. The isolated fragment, containing the target sequence, can be introduced by suitable wizardry into a PLASMID, a circular DNA molecule that is taken up by *E. coli* bacteria and is replicated along with the *E. coli* genome. The target sequence is then CLONED by growing the *E. coli* culture, producing enough target DNA to study.

An increasingly important method of producing large quantities of a particular DNA sequence for study is the POLYMERASE CHAIN REACTION (PCR), by which a DNA sequence is multiplied greatly in number. Genomic DNA is placed, together with abundant free nucleotides, DNA polymerase, and two PRIMER sequences, in a solution that is alternately warmed and cooled. Each primer is a short DNA sequence that is sufficiently complementary to a sequence on one of the two strands flanking the target region to hybridize with it. The target duplex is denatured by warming the mixture; the primers then anneal to their complementary sequences when the mixture is cooled. When the mixture is reheated, the DNA polymerase generates two complementary strands, extending from the primers in the 5′ to 3′ direction along each strand (Figure 3.D2). Thus two duplex copies of the original target sequence are produced. The process is repeated again and again, and the number of copies of the target sequence grows exponentially.

4. Sequencing DNA. When sufficient quantities of a DNA sequence have been obtained by cloning or PCR, its nucleotide sequence can be determined. The amplified DNA sequence is labeled at one end by a radioisotope and divided into four subsamples. Each subsample is incubated with a specific primer, DNA polymerase, the four bases, and

FIGURE 3.D2 The general protocol of the polymerase chain reaction for amplifying DNA. In each cycle, double-stranded DNA is denatured by heat, and primers anneal to the single strands as the temperature is lowered. Primer extension, under the action of DNA polymerase, produces about twice as many double-stranded sequences in each cycle. The heat-stable polymerase used in this procedure was obtained from a species of bacterium that inhabits thermal springs. (After Avise 1994.)

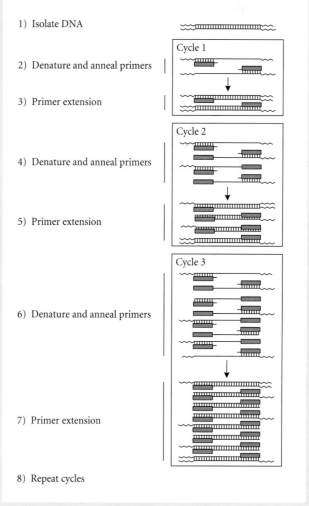

1) Isolate DNA

Cycle 1

2) Denature and anneal primers

3) Primer extension

Cycle 2

4) Denature and anneal primers

5) Primer extension

Cycle 3

6) Denature and anneal primers

7) Primer extension

8) Repeat cycles

a sequenase enzyme, which builds a complementary strand, starting at the radioactively labeled end. Each subsample also includes one of four dideoxy nucleotides, dideoxy-G, -C, -A, or -T. When a dideoxy nucleotide is incorporated into the growing strand, further extension of the strand ceases. Thus, growing strands in the mixture containing dideoxy-G are stopped by insertion of dideoxy-G at each site where C occurs on the complementary strand. Consequently, double-stranded DNA fragments of different lengths are generated, each ending in a C-dideoxy-G pair. The radioactive fragments in the four subsamples ending in dideoxy-G, -C, -A, and -T respectively, are electrophoresed in four parallel lanes of a polyacrylamide gel, and the fragments are sorted by size and visualized by autoradiography. Over the set of four lanes, there is a "ladder" of fragments of increasing size, differing by increments of one base pair (Figure 3.D3). The nucleotide sequence is read across the four lanes, progressing along the ladder.

5. Complementary DNA. COMPLEMENTARY DNA (cDNA) represents only the coding regions of a gene. It can be generated by reverse transcription of a mature mRNA transcript, using the enzyme REVERSE TRANSCRIPTASE from an RNA virus. The cDNA can then be cloned or amplified and used to measure the abundance or detect the existence of different RNA transcripts in various cell types. In IN SITU HYBRIDIZATION, radioactively labeled cDNA can be used as a probe to hybridize to homologous sequences on the chromosomes. The positions of hybridized gene copies can be directly visualized by autoradiography.

6. Transfection. A particular DNA region, marked by a distinctive base pair sequence or mutant phenotypic effect, can be integrated into the chromosome of a recipient organism, where it may express normal or nearly normal activity. In a small but useful minority of individuals, it may be passed on via the germ line. Transfection is one of several techniques used in GENETIC ENGINEERING, in which genes from one species (e.g., a bacterium) are incorporated into the genome of another (e.g., a crop plant). In some instances, incorporation into the host genome is facilitated by enzymatically packaging the sequence in a plasmid (see above), or in a TRANSPOSABLE ELEMENT. In *Drosophila melanogaster*, one such element is the *P* element, a sequence of DNA that can become inserted (together with any donor genes it may carry) into any of many sites in the genome. The sites into which it has been inserted can be visualized by in situ hybridization, and the precise location of insertion can be further studied by sequencing the region. This method can be used to introduce mutant genes from one stock into another genetic background in order to study the effects of different promoters or genetic backgrounds on gene expression.

The many other techniques available include (1) visualizing the presence of a particular protein in different cell types by introducing radioactively or fluorescently labeled antibodies to that protein; (2) the transfer of cytoplasm between embryos of different genotypes by microinjection in order to determine the location and effect of cytoplasmic factors; and (3) transfection with "chimeric" genes, in which the regulatory elements of one gene are inserted into a foreign coding sequence in order to assess factors that influence the activity of the regulatory region.

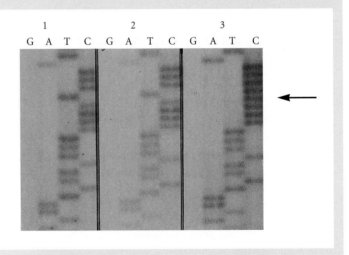

FIGURE 3.D3 DNA sequences as read on a sequencing gel. The figure shows a 22-base pair section from the mitochondrial cytochrome oxidase I gene of two specimens of the leaf beetle *Ophraella bilineata* (panels 1 and 2) and one specimen of the closely related species *O. communa* (panel 3). In each panel, the 4 lanes are G, A, T, and C, and are read from 5′ (top) to 3′ (bottom). This sequence segment includes no Gs. At position 7 (arrow), *O. bilineata* has T and *O. comuna* has C. The first 7 positions in *O. bilineata* are TTACCCT. (Courtesy of L. L. Knowles.)

to develop rather autonomously, influenced little by other groups. Such development is largely MOSAIC. In some organisms with mosaic development, such as tunicates, each of the first few cells derived by mitosis from the fertilized egg (blastomeres) will differentiate into only a partial larva (lacking either right or left side, for example) if the cells are experimentally separated. The development of some other organisms is more REGULATIVE: for instance, if the same experiment is performed on a sea urchin, certain isolated blastomeres will develop into a complete, although miniature, larva. In animals with regulative development, cells soon become "committed" to a particular developmental

FIGURE 3.18 A simplified representation of the hierarchical control of the development of segmentation in the *Drosophila* embryo. Genes in several classes are controlled not only by genes higher in the hierarchy, but also by other genes in the same class. Arrows indicate control of the expression of genes, which are expressed in some body regions, but not others, depending on the regional activity of the genes that control them.

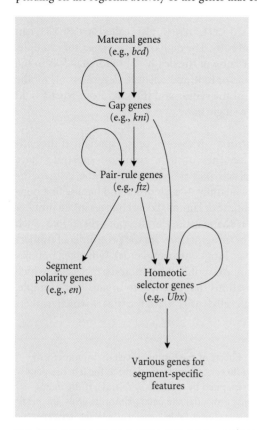

Maternal genes
(e.g., *bcd*)

Gap genes
(e.g., *kni*)

Pair-rule genes
(e.g., *ftz*)

Segment
polarity genes
(e.g., *en*)

Homeotic
selector genes
(e.g., *Ubx*)

Various genes for
segment-specific
features

fate, but this fate depends more on where the cell is located in relation to other cells than on its initial cytoplasmic contents. The differentiation of cells in regulative development depends very largely on *interactions* with other cells. Such interactions characterize much of vertebrate development.

One of the most important such mechanisms of cell differentiation is EMBRYONIC INDUCTION, in which differentiation depends on specific influences (presumably chemical, although the molecules have not been identified) from an adjacent tissue of another cell type. The ability, or COMPETENCE, of cells to respond to such inducers is usually limited to certain parts of the embryo and to specific times in development. (Presumably, competence is determined by changes in the regulation of certain responsive genes.) A commonly cited example is the vertebrate central nervous system, which develops from ectoderm in response to induction by an underlying mesodermal structure, the notochord. Induction is often sequential; for example, the optic cup, an evagination of part of the developing vertebrate brain, induces prospective lens ectoderm to differentiate

into an eye lens vesicle, and the developing lens vesicle in turn induces prospective corneal ectoderm to differentiate into the cornea. Responses to inducers include production of cell-specific proteins (e.g., lens crystalline proteins), alterations of the cytoskeleton within the cell, and changes in the rate of cell division.

Cells that respond to induction can respond, of course, only according to their genetic potentialities. For example, young frog larvae (tadpoles) have suckers near the mouth, and young salamander larvae have balancers. These structures are formed from ectoderm in response to induction by underlying mesoderm in the prospective mouth region. If, at the gastrula stage, flank ectoderm from each species is transplanted to the prospective mouth region of the other, the frog ectoderm is induced to develop suckers and the salamander ectoderm develops balancers, in response to the host's mesoderm (Figure 3.19). Likewise, mammalian hairs and bird feathers are ectodermal structures, induced by underlying mesoderm. Transplanted chicken ectoderm can respond to a mouse mesodermal signal, but it develops feathers, not hair.

Inducers often provide responsive cells with POSITIONAL INFORMATION that "tells" the cells where they are, and

FIGURE 3.19 Genetic specificity of induction. The oral ectoderm develops into balancers in a young salamander (newt), and into suckers in a frog tadpole, in response to induction by underlying mesoderm. When the oral ectoderm of each species is replaced by that of the other, it develops into the feature characteristic of the donor in response to a signal from the mesoderm of the recipient. (After Hamburgh 1970.)

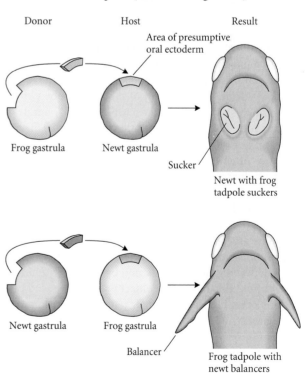

Donor   Host   Result

Area of presumptive
oral ectoderm

Frog gastrula   Newt gastrula

Sucker

Newt with frog
tadpole suckers

Newt gastrula   Frog gastrula

Balancer

Frog tadpole with
newt balancers

consequently, what their fate will be. For example, in the developing limb bud of a bird, a "zone of polarizing activity" (ZPA) at the posterior part of the limb bud base determines the difference between anterior and posterior digits: the farther away from the ZPA cells are, the more anterior the structure into which they will differentiate. This effect is thought to be due to a chemical factor, or MORPHOGEN, that diffuses from the ZPA, forming a gradient of decreasing concentration. The concentration experienced by the cells "tells" them their position and what to differentiate into. Concentration gradients are apparently common developmental signals; another example (above) is the gradient of a maternal gene product that establishes the anterior-posterior polarity of the *Drosophila* embryo.

Mechanisms of Morphogenesis

Changes in the shape of cells and the larger units they compose are responsible for the form of organs (e.g., limbs, lungs) and of cells themselves (e.g., epithelial cells, neurons). These changes result from the activities of both single cells and groups of cells. These activities are mediated, at least in part, by certain of the cell's proteins, and so are dependent on gene expression.

Single cells The shape of a cell is affected by the organization of its cytoskeleton, consisting largely of the proteins actin and tubulin, which form microfilaments and microtubules, respectively. The density and orientation of these structures can result in, for example, the species-specific shape of the head of a spermatozoon. The shapes of many cell types (e.g., columnar epithelial cells) also depend on their adhesion to a substrate, such as the basal lamina that underlies vertebrate epithelia.

Movement of individual cells is important in the development of many animals (but not plants). In vertebrates, for example, the neural crest cells move individually from the periphery of the neural tube (which differentiates into the central nervous system), migrate to various sites, and differentiate into pigment cells, the autonomic nervous system, certain bones of the jaw and face, and other tissues. The paths cells take, and the sites at which they stop, depend on specific cell surface molecules (chiefly proteins) that enable them to "recognize" other (and their own) cell types. Such cell surface properties also determine ADHESION of cells of the same type, which is critically important in conferring form. Migrating cells may be guided along their paths by fibers or junctions between groups of cells, and by contact inhibition, whereby a cell is halted by contact with another cell of a specific type, and takes another path. It is possible that some aspects of form (e.g., the length of a limb bud) are affected by how long the cells remain motile.

The timing, rate, and direction of cell division are important in determining the size and shape of an organ. The relative rates of mitosis parallel to and perpendicular to the axis of a developing limb segment, for instance, affect its length and thickness. Whether the petals of a flower are separate or fused along their edges depends in part on whether the cells between the petal primordia continue to undergo mitosis at the same rate as the cells of the primordia themselves (Figure 3.20). Little is known, at the molecular level, about the factors that govern spatial and temporal variation in cell division rate during development.

Programmed CELL DEATH, also called APOPTOSIS, can play important roles. The digits of vertebrates, for example, are distinct because of the death of interdigital cells in the developing hand or foot, and cell death is responsible for the development of the radius and the ulna from what would otherwise be a single bone.

Cell populations Masses of cells, especially if they adhere to one another or to a common substrate, have developmental activities of their own. Epithelial cells, for example, adhere laterally to each other and basally to a noncellular membrane or lamina. Local changes in cell number (by mitosis) or shape cause the sheet of cells to fold. Such evaginations or invaginations are the primordia of numerous organs (e.g., lungs, salivary glands). Other cell population effects are exemplified by mesodermal cells, which form condensations that give rise to organs. For example, regions of high cell density mark the sites at which feathers

FIGURE 3.20 Effects of the timing of cell division on form. The developing flower, shown from above at the left, has petal primordia (p) separated by intercalary cells (i). If the intercalary cells divide only after the petal primordia grow, the petal lobes of the mature flower (side view, at right) are well separated. If the intercalary cells divide along with those of the petal primordia, the petals are mostly fused, forming a tubular corolla. (After Stebbins 1974.)

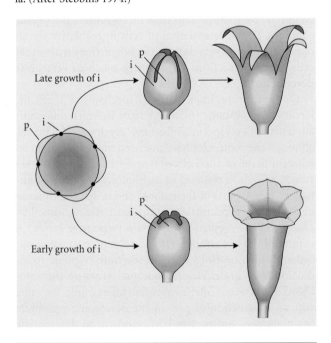

develop. Condensations may form as a result of cell migration, locally increased levels of mitosis, or reduced rates of mitosis in surrounding regions. The density of cell aggregations may affect form; for instance, a mutant in the house mouse has a flat face that is thought to result from the cells' remaining densely aggregated, perhaps due to excessive adhesion of their surfaces.

Overview of Genetics and Development

This chapter has been a condensation of the high points from courses in the two almost inseparable subjects of genetics and developmental biology. When we come to address the mechanisms of evolution, we will find all of this material essential: Mendelian genetics is necessary for understanding genetic variation and the effects of evolutionary factors on the genetic composition of species; molecular genetics provides insight into the origin and nature of new genetic variants; knowledge of gene expression and developmental biology is fundamental to appreciating the evolution of phenotypes.

Our summary of the mechanisms of development leads us to the conclusion that although fundamental developmental mechanisms are rather few in kind, they act in concert in so many ways that the path from genotype to phenotype can be exceedingly complex. For instance, we might expect epistasis to be common if, say, locus *B* affects a property of a tissue (such as part of the nervous system)

that is induced by a group of cells (the vertebrate notochord) that owe some of their characteristics to locus *A*. In such cases a mutation of gene *A* could well affect the expression of gene *B*. Developmental phenomena also explain the great prevalence of **pleiotropy**, that is, multiple effects of a particular gene. Pleiotropy can occur for many reasons: for instance, the gene may be expressed (transcribed and translated) in many different tissues; it may govern a step in a biochemical pathway with more than one product; it may affect a characteristic of a tissue that in turn epigenetically affects the development of other organs. An example of "direct pleiotropy," whereby one gene acts in all affected tissues, is the *achondroplasia* mutant in the rat, which causes an inability to suckle, faulty pulmonary circulation, and arrested development, all because of abnormal development of cartilage. In "relational" pleiotropy, a mutation causes a single primary effect that has a cascade of secondary effects. The mutation of the β-hemoglobin gene that causes sickle-cell anemia in humans is one of many examples of relational pleiotropy (Figure 3.21).

The evolution of a new phenotypic character is an evolutionary change in **ontogeny**, the developmental history of the organism. The potential path of evolution depends on the nature of both the genetic foundation of the trait and the developmental system that translates the genotype into a phenotype. Developmental mechanisms, for the most part little understood, can both provide opportunities for and impose constraints on evolution.

FIGURE 3.21 Pleiotropic effects of the amino acid substitution in the β chain of human hemoglobin that results in sickle-cell anemia. This is an example of relational pleiotropy. (After Raff and Kaufman 1983.)

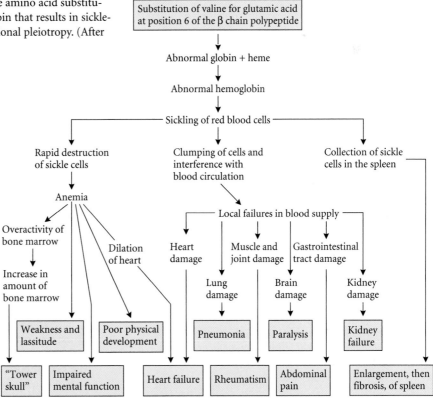

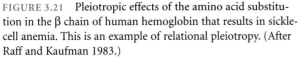

Summary

Among the topics in this chapter that should already be familiar to most readers are mitosis and meiosis; Mendelian ratios; dominance and additivity of gene effects on phenotypes; simultaneous effects of both genetic variation and environmental variation on variation in phenotypic characteristics; the structure, replication, and transcription of DNA; and the genetic code.

Among the important points that may be less familiar are these:

1. A genotype may have different phenotypic expressions in different environments. The set of these possible expressions is called the norm of reaction.

2. Many characters are polygenic, i.e., influenced by several or many loci, which may act additively or may interact with one another (epistasis). Some characteristics (threshold characters) take on discrete alternative states, but are nevertheless polygenically inherited.

3. The genome, especially in eukaryotes, includes a great deal of noncoding, apparently "functionless," DNA, including introns within genes, spacers between genes, highly repetitive DNA, and pseudogenes ("degenerate" genes).

4. All the somatic cells of an organism carry much the same genetic information, but genes are variably expressed in different cell types and at different stages in development. Their expression, as exemplified by the development of segment identity in *Drosophila*, is often hierarchically controlled by other genes that activate or repress them by several molecular mechanisms.

5. The development of an organism, consisting of differentiation of cells and tissues and morphogenesis (acquisition of form), is based on both direct genetic and indirect (epigenetic) effects, together with environmental influences, on the activities of cells and populations of cells.

Major References

Alberts, B. et al. 1995. *Molecular biology of the cell*. Third edition. Garland, New York. Extends molecular biology to cells and development.

Gilbert, S. F. 1997. *Developmental biology*. Fifth edition. Sinauer Associates, Sunderland, MA. A clear and comprehensive treatment of molecular and organismal aspects of development.

Lawrence, P. A. 1992. *The making of a fly: The genetics of animal design*. Blackwell, Cambridge, MA. Includes an evolutionary perspective on the most advanced body of knowledge in developmental genetics. See also D. De-Pomerai, 1990, *From gene to animal: An introduction to the molecular biology of animal development* (Cambridge University Press, Cambridge).

Lewin, B. 1994. *Genes V*. Oxford University Press, Oxford. Detailed treatment of molecular genetics.

Griffiths, A. J. F. et al. 1996. *An introduction to genetic analysis*. Sixth edition. W. H. Freeman, New York. An excellent introductory genetics textbook that treats much of the material in this chapter.

Problems and Discussion Topics

1. *Drosophila melanogaster* has four pairs of chromosomes. Denote these I through IV, and label the two homologous chromosomes of each pair, carried by a single female, a and b. Pretend that crossing over does not occur. Using probability calculations, what proportion of this female's granddaughters will inherit her chromosomes Ia, IIb, IIIb, and IVa ? Chromosome I is the sex, or X, chromosome. Females are XX, males XY. What proportion of her grandsons will carry her chromosomes Xa, IIb, IIIb, and IVa?

2. Using the same notation as in question 1, what is the probability that two of the female's daughters will inherit the same X chromosome (either Xa or Xb)?

3. Suppose we know that the recombination rate between loci A and B is 0.05. We cross $A_1A_1B_2B_2$ females with $A_2A_2B_1B_1$ males, and then cross the F_1 females with $A_2A_2B_2B_2$ males. Among the progeny of this cross, what are the genotypes and the expected proportions?

4. Suppose that the ear length of rabbits is 14 cm in the genotype $A_2A_2B_2B_2C_2C_2$, and that substituting allele 1 for allele 2 at any locus increases length by 0.5 cm. (That is, the effects of allele substitution are additive within and among loci.) How many different genotypes might have an ear length of 15.5 cm, assuming no environmental effects on the characteristic? Can you predict the average ear length of the progeny of a cross between two individuals with 15.5 cm ears? What are the maximal and minimal possible ear lengths among the offspring of such a cross? Show the genotypes of the parents and progeny in answering this question.

5. At any given level of artificial fertilizer use, two genotypes of an apomictic (parthenogenetic) plant differ in height at maturity. Increasing fertilizer use increases the height of both genotypes. What will be the effect of adding fertilizer in one generation on the average height of offspring of a mixture of the two genotypes? What experiment might you do to show that acquired characteristics are not inherited? Suppose now that taller plants produce more seeds. Would fertilizing a field of both genotypes cause the average height in the next generation to change?

6. We observe that in a species of beetle, black-bodied individuals have black legs, and brown-bodied individuals have yellow legs. Assuming you can perform breeding experiments, how can you tell whether or not these differences have a genetic basis? How could you tell whether the correlation between body color and leg color is due to closely linked genes or to pleiotropic effects of a single gene?

7. What accounts for the near universality of the genetic code? What might account for the few cases in which a codon specifies a different amino acid in certain organisms such as yeasts? Would you expect these differences to be peculiar to yeasts, or to be characteristic also of some other fungi?

8. As the text indicates, much is known about differences among organisms in the total amount of DNA, but very little is known about how organisms vary in the number of different functional genes. How might such information be obtained? Do you predict any patterns of variation among major groups of organisms in this respect?

9. What do you suppose we would have to know about genetics and development in order to account for the fact that flies (Diptera) have halteres on the metathorax, in place of the wings borne on this segment by other insects? Read about the effects of mutations in the *Bithorax* complex, and then discuss whether or not it is likely that halteres evolved from wings by a single mutation. How likely is it that a lineage of flies could "re-evolve" wings from halteres?

10. Geneticists sometimes speak of a "genetic program" that "determines" development. Some developmental biologists object to this phraseology. Why? How might you rephrase the relationship between genes and development of the phenotype?

CHAPTER 4

Ecology: The Environmental Context of Evolutionary Change

The biological discipline of ecology (from the Greek *oikos*, "home") is the study of interactions among organisms and their environment. Ecology has important applications in human affairs, and disciplines termed applied ecology or environmental science are concerned with issues such as pollution, global warming, disease, food production, and conservation. However, we are concerned here with ecology as a basic science that contributes importantly to evolutionary biology.

Ecology includes the study of individual organisms, populations, communities, and ecosystems. *Physiological ecology* is part of the study of how individual organisms are affected by environmental factors such as temperature. We will treat some of this subject in Chapter 17. *Population ecology* concerns the factors that influence the abundance, rate of increase, and distribution of populations. These are very important topics for understanding evolutionary concepts such as natural selection (Chapters 12, 13, and 19). *Community ecology* concerns the factors that influence the species composition of assemblages, or communities. Evolutionary biology bears strongly on this subject (Chapters 8, 18, and 25). *Ecosystems*—communities together with their abiotic environment—provide the settings within which ecological and evolutionary dynamics transpire.

Environments

An organism's environment may be thought of as everything other than itself that may affect its development, survival, or reproduction. *Environmental factors include many sources of natural selection* on the characteristics of a population or species. Interactions with environmental factors also affect the *spatial distribution* of organisms, and therefore their population structure: the degree to which the species is divided into separate populations.

The environment includes abiotic and biotic elements. Abiotic features, such as temperature, water, salinity, soil structure, and mineral nutrients, affect organisms and often are affected in turn by their activity (e.g., earthworms affect soil structure; plants deplete soil water). Biotic features include food (other organisms or their products); predators; parasites; mutualists; and competitors, including both conspecifics (individuals of the organism's own species) and other species. Conspecifics are important features of an individual's environment, not only in their capacity as potential competitors, but also as potential mates or social partners.

Which aspects of the environment are important to an organism vary from species to species and depend on the species' evolutionary history. Thus the chemical compounds in the leaves of a tree are among the most important environmental factors for an insect that eats them, but are likely to be less relevant, or entirely irrelevant, to predatory insects that forage on the plant or to birds that nest in it. An important feature of any environmental factor is its pattern and magnitude of variation, but the effect of such variation likewise depends on the organism. Hourly fluctuations in temperature, for example, may be important to a butterfly that requires almost continual solar radiation to maintain the body temperature necessary for flight, but less important to a bumblebee, in which an insulating coat of hairs retains heat generated by its wing muscles.

Global Climate Patterns

Global patterns of temperature and precipitation are among the most important abiotic environmental factors. These patterns have been a major influence on the geographic distributions of species. Some knowledge of the differences among communities on a global scale is essential for understanding biological diversity, a major topic in evolutionary biology.

The annual input of solar energy is greatest near the equator, giving rise to a latitudinal gradient in average temperature. Hence the potential evaporation (or, more exactly, *evapotranspiration*, the water vapor released both by evaporation and by transpiration from plants) is greatest in tropical latitudes. Water vapor condenses and precipitates as the air mass holding it cools. Near the equator, warm air carrying water vapor rises, releasing some of its water as it does so. In the upper atmosphere, the air mass moves poleward, releasing much of its remaining water as it moves, and drops toward the earth's surface in the general regions of 30° north and south latitude—the region of many of the world's deserts. As the now-dry descending air reaches the earth's surface, it spreads out, moving both toward the poles and the equator, setting up south and north winds. This longitudinally circulating air mass, called a Hadley cell, rotates between the equator and about 30° latitude. Other Hadley cells are formed further toward the poles (Figure 4.1).

The longitudinal circulation is deflected by the earth's rotation, so that the winds tend to blow toward the equator from the east in tropical regions (northeast trade winds north of the equator, southeast trade winds south of it) and toward the poles from the west in temperate latitudes (the prevailing westerlies). The winds carry moisture drained by evaporation from land and (especially) sea. If a moisture-laden wind encounters a mountain range, it drops water as it rises and cools; hence the windward slopes of mountain ranges (such as the Cascades in the northwestern United States) are wet, while a dry "rain shadow" lies to the leeward.

The wind patterns set up the surface currents of the oceans, forming clockwise gyres in the North Pacific and North Atlantic and counterclockwise gyres in the southern oceans. These currents affect temperatures on the continents; for example, the warm Gulf Stream runs northeastward toward Europe from the northeast United States, giving Great Britain a far milder climate than that of New-foundland in Canada, which lies at the same latitude. Moreover, the high specific heat of water causes temperature fluctuations in the coastal regions of continents to be less extreme (maritime climates) than in the interior regions (continental climates).

The westward flow of surface water in tropical regions causes an upwelling of deep sea water along the western coasts of the continents; thus a vertical circulation of water moves westward along the sea surface and eastward in the deep sea. This upwelling brings mineral nutrients from the ocean depths to the surface.

The axis of the earth is tilted relative to the plane of the earth's annual revolution about the sun, so that the region of maximal energy input is south of the equator in the southern hemisphere's summer (the northern winter) and vice versa. This tilt creates seasonal changes in temperature at the higher latitudes; near the equator, there is very little seasonal temperature change. However, the distribution of Hadley cells likewise undergoes seasonal shifts toward the south and the north, causing the pattern of wind and rainfall to shift latitudinally. Thus most tropical areas, although experiencing little seasonal change in temperature, have wet and dry seasons, which in some regions are extremely pronounced.

Terrestrial Biomes

Temperature and water interact. Because the rate of evaporation increases with temperature, for a given amount of precipitation, soils are generally drier in a region of high prevailing temperatures than in a cooler region. Many features of terrestrial plants, including their growth form and the size, thickness, and other characteristics of their leaves, are

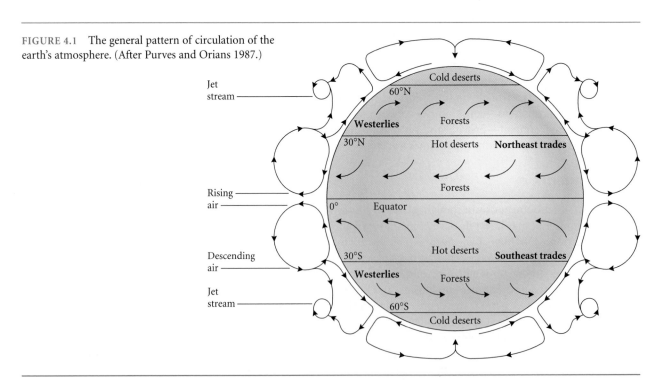

FIGURE 4.1 The general pattern of circulation of the earth's atmosphere. (After Purves and Orians 1987.)

adaptations to temperature and the availability of water. Moreover, competition for limited water can cause plants to be widely spaced (as in deserts). Thus the physiognomy of vegetation—its overall structure and appearance—depends largely on temperature and precipitation (including the seasonal pattern of rainfall). Major physiognomical types of vegetation (as represented in natural, mature plant associations) represent **biomes.** The number of biomes recognized depends on an author's preference for fine or coarse distinctions. Figure 4.2 presents the distribution of nine biomes, which may be briefly described as follows (See Color Plate 2 A–I).

1. Tropical wet forest. This exists where rainfall is high throughout much of the year. The features of the vegetation change with altitude; at low altitudes, where temperatures are high throughout the year, the forest typically has tall, dense canopy trees, with smaller trees and shrubs below them. It is quite dark near the ground (except where there are gaps in the canopy); thus the vegetation near the ground is often sparse, and seedlings grow slowly. Vines and epiphytes (nonparasitic plants that grow on other plants) are usually abundant. The richness (number) of plant species is often extraordinarily high (300 to 400 species of trees in a single square kilometer is not unusual), as is the

species richness of animals. Low- to middle-elevation tropical wet forests have the world's highest terrestrial species diversity, much of which is severely threatened by deforestation. Many of these species have very specific ecological requirements, and many ecological interactions among the species are highly intricate. Predation is intense, and many species have extraordinary defenses against it (Color Plate 3). A high proportion of the plant species depend on animals for pollination and/or seed dispersal. Rates of decomposition of organic matter are very high, and soils often are poor in mineral nutrients.

2. Tropical seasonal forest (Dry forest). These forests experience a pronounced dry season, during which many plants shed their leaves and come into flower. Species that grow near watercourses are often evergreen. Species diversity is very high. Animal-mediated pollination and seed dispersal are common, although quite a few plants have wind-dispersed seeds. Many plants have thorns or spines, perhaps evolved as defenses against herbivorous mammals. Because of deforestation, this is one of the world's most endangered biomes.

3. Tropical scrub forest and savannah. These occur in regions of lower and highly seasonal rainfall. Certainly today, and probably in the past as well, fire has been a

FIGURE 4.2 The distribution of the major terrestrial biomes. (After Begon et al. 1990.)

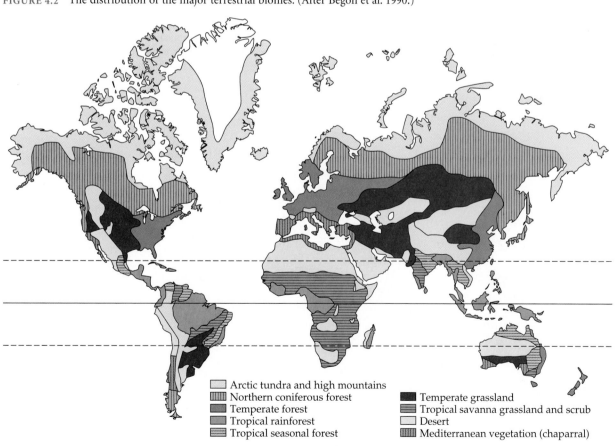

- ☐ Arctic tundra and high mountains
- ☐ Northern coniferous forest
- ☐ Temperate forest
- ☐ Tropical rainforest
- ☐ Tropical seasonal forest
- ☐ Temperate grassland
- ☐ Tropical savanna grassland and scrub
- ☐ Desert
- ☐ Mediterranean vegetation (chaparral)

frequent occurrence, especially in savannahs (grassland with scattered clumps of trees). Many plants have deciduous leaves, thorns, adaptations for dry conditions (e.g., deep roots), and adaptations for episodic fires. Especially in Africa, large social herbivorous mammals are an important component of the ecosystem.

4. Desert. In deserts, the annual precipitation is less than the potential evapotranspiration. Rainfall is low, seasonal, and usually unpredictable. Deserts that receive relatively more rain (e.g., the Sonoran Desert of Arizona and northern Mexico) have a greater density and variety of vegetation than those that receive little (e.g., parts of coastal Peru and Chile). Cold deserts, as in parts of Idaho and Mongolia, have a lower species diversity and lack some of the plant life forms of warm deserts. Adaptations for water shortage are evident in most of the surprisingly many species of desert animals and plants. Desert plants commonly have small leaves or none (Figure 4.3), and often have a growth form (like that of many cacti and cactuslike plants) that minimizes the ratio of surface (over which water loss can occur) to volume. Both animals and plants may undergo ESTIVATION (quiescence in the hot season), and be active for only a limited season (e.g., when rain arrives). Some organisms are dormant for years, awaiting rain.

5. Mediterranean vegetation. Known as chaparral along the west coast of North America, matorral in Chile, and by other names elsewhere, this vegetation type occurs in dry regions that receive rain in winter, before the temperature is high enough for plant growth. It consists largely of dense, woody shrubs and small trees, with small leaves adapted to reduce water loss. Fire is frequent in chaparral, and many plants are adapted accordingly (e.g., with fire-resistant bark).

6. Temperate grassland (Prairie). Relatively low amounts of rainfall occur during the growing season. Fire is rather frequent, and in many areas prevents woody plants from taking over. Native prairie vegetation consists of a rich variety of species of grasses and forbs (herbaceous plants other than grasses). Grazing mammals are important. Natural prairies have been almost entirely eliminated by agriculture; only small patches remain.

7. Temperate broadleaf forest. These occur in regions of moderately high rainfall and strong seasonal variation in temperatures. A (usually) rather high canopy surmounts smaller trees, shrubs, and forest herbs; most plants are deciduous, shedding their leaves in winter, and a thick leaf litter decomposes into an often rich layer of humus. Species diversity of plants and animals is far lower than in tropical wet forests; pollination and seed dispersal are accomplished by animals or wind, depending on the species (Color Plate 4).

FIGURE 4.3 Convergent growth form in desert plants. These leafless succulents, with photosynthetic stems, belong to 3 distantly related families. (A) and (C) American cacti (Cactaceae). (B) An African member of the milkweed family (Asclepiadaceae). (D) An African member of the spurge family (Euphorbiaceae). (A after Benson 1982; B after Heywood 1993; C and D after Ehrlich and Roughgarden 1987.)

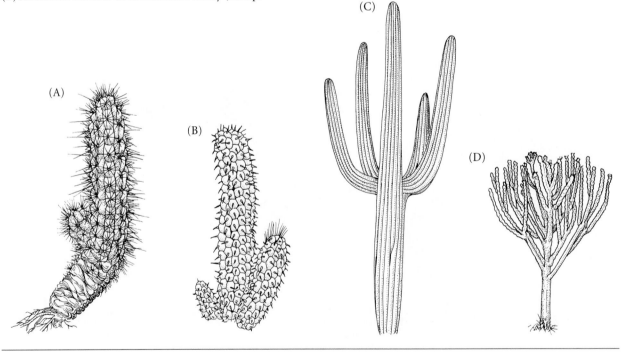

8. Coniferous forest. Although pines are locally a dominant element in parts of the temperate broadleaf forest, the dominance of conifers (especially spruces and firs) farther to the north distinguishes a separate biome, found in cold regions with moderate to high rainfall. Because liquid water is unavailable for much of the year, many of the plants have some dry-adapted features. The plants are evergreen, and the foliage has high quantities of chemical compounds (tannins and aromatic terpenes) that, together with low temperatures, reduce the rate of decomposition of leaf litter. Species diversity is low. Pollination and (usually) seed dispersal of conifers is by wind. Levels of seed production can vary greatly from year to year, causing fluctuations in populations of birds and other animals. Coniferous forests extend toward the equator along mountain ranges.

9. Tundra. Tundra is found beyond the treeline in Arctic regions and on some sub-Antarctic islands. In some such regions, the soil is permanently frozen just below the surface. The growing season is very short, but the extremely long photoperiod (day length) at this time enables rapid growth and reproduction of the few species of plants and animals. The vegetation (Figure 4.4) is extremely low in stature, consisting of lichens, mosses, herbs, and miniaturized "trees" (e.g., willows that form a creeping ground cover a few centimeters tall). Rodents (e.g., lemmings) can be abundant, but undergo pronounced population fluctuations. Most birds migrate to escape the long, almost lightless winter. ALPINE TUNDRA occurs on cold mountain tops, even in tropical latitudes.

Aquatic Ecosystems

Freshwater systems

Freshwater ecosystems can be divided into flowing and standing waters. In rivers and streams, a BENTHIC biota of animals and plants lives on or in the bottom, while various invertebrates (especially insects) and fish may forage in midwater. Especially in rivers, aquatic plants (especially algae) form the basis of the food chain, but organic matter from outside the water, such as dead leaves, is an important source of energy and nutrients, fed on by insects and other invertebrates. Some flowing waters (e.g., the Amazon River) are turbid with silt; turbidity may limit the distribution of some species and is a factor to which some fishes (Figure 4.5) have become adapted (e.g., catfishes with highly developed chemosensory organs; electric fishes that use electrolocation to orient, communicate, and find food). In rapid streams, adaptations for clinging to rocks are prevalent among insects. The oxygen content of rapidly flowing stream water is usually high, and many of the animal species that inhabit these streams are highly sensitive to lower oxygen levels.

Lakes and ponds have a benthic biota and an open-water biota that includes, most prominently, the PLANKTON. Phytoplankton (mostly unicellular algae limited to depths at which there is sufficient light for photosynthesis) are fed on by zooplankton (mostly small crustaceans), which are fed on by many other invertebrates and fishes. The community in temperate zone lakes is strongly affected by the development during summer (and breakdown in autumn) of a THERMOCLINE, an abrupt change in temperature with depth that separates and prevents mixing of warm, oxygen-rich upper water from the cold, nutrient-rich water below. Ponds are generally too shallow to develop a

FIGURE 4.4 A dwarf willow (Salix reticulata) of the Arctic tundra, about 4 centimeters tall. Many species of willows in warmer parts of the world are tall trees. The coin, 18 mm in diameter, is for scale. (Photograph by the author.)

FIGURE 4.5 An African electric fish (family Mormyridae). The lines around the fish represent the weak electric field produced by electric organs. (After Moyle 1993.)

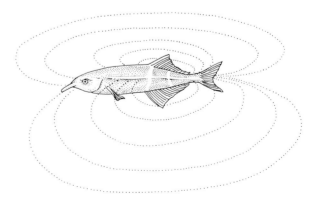

thermocline. Permanent ponds often have different species from temporary ponds that dry out in summer; in particular, the absence of fish in temporary ponds often permits the persistence of amphibians and invertebrates that cannot coexist with predatory fish.

Freshwater organisms tend to lose ions (e.g., Na^+, Ca^{2+}) because the concentration of these is lower in fresh water than in the cells. Such organisms have various physiological mechanisms to maintain ion balance. In many standing waters, oxygen depletion is a factor to which various adaptations have evolved (e.g., hemoglobin in some benthic fly larvae).

FIGURE 4.6 A few of the many peculiar fishes of the deep sea. (A) A whipnose, *Gigantactis*. (B) A gulper, *Saccopharynx*. (C) An angler, *Caulophryne*. *Saccopharynx* is known for its greatly distendable mouth, capable of engulfing fishes as large as the predator itself. (A and B after Fitch and Lawrence 1968; C after Moyle and Cech 1982.)

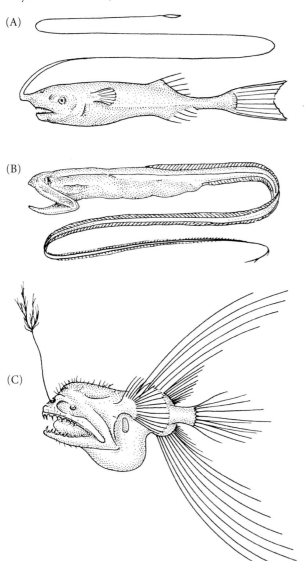

(A)

(B)

(C)

Marine systems

As in fresh water, marine organisms include benthic, planktonic, and actively swimming species. Because light levels are adequate for photosynthesis only down to 100–200 meters, organisms at greater depths either migrate toward the surface to feed (as some fishes do) or are part of a food chain that depends on material descending from the photic zone. Hence, in general, the number of benthic animals present decreases with increasing depth, but (perhaps surprisingly) the number of species present *increases* steadily down to about 2000–3000 meters, beyond which it decreases. Animals at great depth enjoy constant (albeit cold) temperatures and must contend with darkness, great pressure, and scarce food. The adaptations of some deep-sea fishes for obtaining and swallowing food result in some truly bizarre forms (Figure 4.6).

Near the surface in the open ocean, light is adequate for photosynthesis, but mineral nutrients are extremely sparse, except where they are brought from the depths by upwellings. Hence productivity, and consequently the abundance of marine life, is very low in much of the ocean, except in regions such as the coast of Peru and northern Chile, where upwelling promotes high planktonic productivity, and consequently dense populations of fishes and sea birds. Productivity is higher in coastal waters where nutrients are washed in from the land. Estuaries, where fresh and salt water mix, are particularly productive habitats, and are "nurseries" for the young of many marine fishes, including economically important species. In coastal waters, very different communities of organisms inhabit soft sediments and hard bottoms (rocks). Rocky intertidal and subtidal habitats have a rich biota of sessile (attached) algae, barnacles, bryozoans, and other animals that compete for space and provide both habitat and food for many other animal species. The local species composition of this biota is affected by the intensity of wave action, the frequency of disturbances that free surfaces for recolonization, and the activities of predators.

A major marine community is the coral reef. The framework of the reef is formed by colonial coral animals (anthozoan cnidarians) that contain symbiotic algal cells and secrete a calcium carbonate skeleton. Reefs are limited to tropical waters that are shallow enough to support photosynthesis by the symbiotic algae. They are the most species-rich* marine communities, providing a habitat for innumerable invertebrates and fishes.

Although the productivity and BIOMASS (density of living matter) of marine life is generally higher at high than low latitudes, the number of species increases from polar to tropical latitudes for both planktonic and benthic continental-shelf organisms, just as it increases for terrestrial

---

*Many biologists use the word *speciose* to mean "rich in number of species." "Speciose," however, should mean "beautiful," as in scientific names such as *Solidago speciosa*, a "beautiful goldenrod." Orchids, as a group, are both speciose and species-rich (Gill 1989).

species. Most groups of marine organisms, including corals, are more diverse in the Indo-Pacific region than in the Atlantic Ocean.

Biogeographic Realms

A terrestrial biome, which can occur in several parts of the world, is defined by the characteristics of the vegetation. But in most cases, the plants in different regions are not closely related to each other; they have independently evolved similar growth forms and other adaptations. For example, cacti (Cactaceae) are found in deserts only in the New World; African deserts have plants that are similar in form but belong to entirely different families (see Figure 4.3). Such CONVERGENT EVOLUTION (Chapter 5) is often seen in the animals as well (Figure 4.7).

From a taxonomic point of view, the terrestrial biota of the globe can be divided into BIOGEOGRAPHIC PROVINCES that are distributed quite differently from the terrestrial biomes. (Biogeography is the subject of Chapter 8.) Each biogeographic province contains distinctive groups of related species that are largely limited to that region; but within a province, related species may occur in different biomes. For example, the family Bromeliaceae, which includes the pineapple (*Ananas comosus*) and the "Spanish moss" (*Tillandsia usneoides,* which is not from Spain and is not a moss), is a neotropical group, a few species of which extend into the southern United States. Within South and Central America, the various species of bromeliads range from those limited to tropical wet forests to some that are found only in the driest deserts.

The delineation of biogeographic realms depends on the author; the first such classification (Figure 4.8) was by Alfred Russel Wallace (see Chapter 2). Table 4.1 lists a few of the distinctive groups in each realm. Some rather peculiar distributions characterize certain sets of taxonomic groups. For example, South America, Africa, and Australia share such groups as lungfishes, ratite (ostrichlike) birds, and leptodactylid frogs that do not occur elsewhere. Some taxa in eastern North America (but not in the west) are closely related to species in eastern Asia; these include alligators, giant salamanders (Cryptobranchidae), and magnolias. These patterns have historical causes that we will consider in Chapter 8.

Communities

Each biome includes numerous assemblages of species that vary over various spatial scales. Thus within the temperate broadleaf forest biome in New York State, there are marshes, swamps, and various upland forest types such as those dominated by oak and hickory or by beech and maple. The distribution of such assemblages, usually termed **communities,** is governed by factors such as temperature (varying with altitude and slope exposure), water (local variations in rainfall; height of the water table), and soil types (distinctive communities are found on alluvial soils, or on those

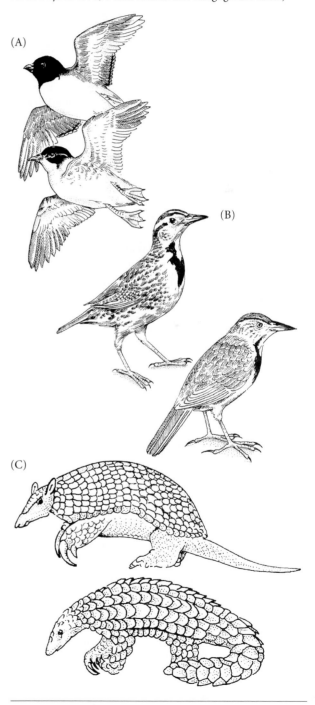

FIGURE 4.7 Examples of convergent evolution in animals. (A) A dovekie (*Alle alle*), family Alcidae (top) and a diving-petrel (*Pelecanoides magellani*), family Pelecanoididae (bottom). The Arctic Alcidae (auks) are related to gulls; the Antarctic diving-petrels are related to albatrosses. Both use their wings below water as they forage for marine invertebrates and fish. (B) A North American meadowlark (*Sturnella magna*, left) and an African longclaw (*Macronyx croceus*, right). Members of distantly related families (Emberizidae and Motacillidae), these grassland birds are similar in coloration and habits. (C) A South American armadillo (order Edentata, top) and an African pangolin (order Pholidota, bottom). Both of these mammals feed on termites and ants, breaking open the nests with strong claws. (A, B after Proctor and Lynch 1993; C after Ehrlich and Roughgarden 1987.)

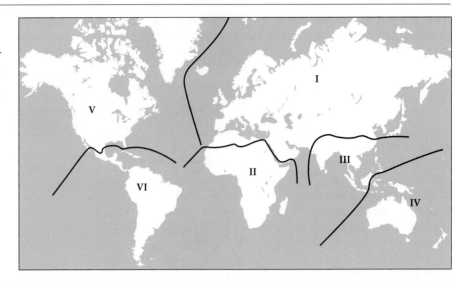

FIGURE 4.8 The zoogeographic regions, or biogeographic realms, recognized by A. R. Wallace: Palearctic (I), Ethiopian (II), Oriental (III), Australian (IV), Nearctic (V), and Neotropical (VI). The distributional limits of most species do not correspond exactly to the borders between regions, which are actually gradients of faunal change.

lying over limestone). Many animals are also found in particular plant communities. Although in some instances, such as the border between a forest and a marsh, a group of species is replaced rather abruptly by another group, most plant species are usually distributed more or less independently of each other, so that no sharp line can be drawn between communities. Hence some ecologists prefer to speak of local assemblages of species rather than of communities, a term that can imply an integrated structure that may not exist. Because of these "individualistic" distributions, a given species will interact with different assemblages of other species in different places. Some plant species have broader ecological distributions than others; for example, in New York forests, red maple (*Acer rubrum*) grows both in swamps and in well-drained upland soils, whereas pitch pine (*Pinus rigida*) is restricted to dry, sandy

soils. With respect to soil type, the maple is an ecological GENERALIST, compared to the SPECIALIST pine. Differences in degree of ecological specialization are an important topic later in this book.

Ecological succession

Ecological succession refers to the change in the composition of species in a habitat (ranging from a fallen log to a piece of landscape) over time. A cleared piece of land undergoes SECONDARY SUCCESSION: in the first few years of succession in a forested region, "pioneer" plants (mostly annual herbs that reproduce and die at the end of their first growing season) are replaced by perennial herbs; over ensuing decades the perennials are replaced by shrubs and certain trees, which in turn are replaced by other trees (CLIMAX SPECIES), in whose shade the earlier plants cannot grow.

Table 4.1 Some characteristic taxa of Wallace's biogeographic realms[a]

| REGION | TAXON | | | | | |
	FLOWERING PLANTS	FISHES	AMPHIBIANS	REPTILES	BIRDS	MAMMALS
I. Palearctic	Butomaceae (flowering rush)	Cyprinidae (carps, in part)	Salamandridae (newts)	—	Alaudidae (larks)	Microtinae (voles)
II. Ethiopian	Melianthaceae	Mormyridae (elephant fishes)	Hyperoliid frogs	Chamaeleontidae (chameleons)	Musophagidae (turacos)	Giraffidae (giraffe, okapi)
III. Oriental	Pandanaceae (screwpines)	Anabantidae (gouramis)	Rhacophorid frogs	Uropeltid snakes	Phasianinae (pheasants)	Tupaiidae (treeshrews)
IV. Australian	Myrtaceae (esp. *Eucalyptus*)	Ceratodidae (lungfish)	—	Pygopodid lizards	Ptilonorhynchidae (bowerbirds)	Macropodidae (kangaroos)
V. Nearctic	Fouquieriaceae (ocotillo)	Centrarchidae (basses)	Plethodontine salamanders	Chelydridae (snapping turtles)	Meleagridinae (turkeys)	Antilocapridae (pronghorn)
VI. Neotropical	Bromeliaceae (bromeliads)	Gymnotidae (knifefishes)	Bolitoglossine salamanders	Teiidae (whiptail lizards, etc.)	Formicariidae (antbirds)	Caviidae (guinea pigs)

[a]These taxa are unique (endemic) to, or most diverse in, the regions indicated.

Because different species occupy different successional stages, the overall species diversity of an area is augmented by moderate levels of disturbance that open sites for recolonization by early-successional species. These are FUGITIVE species, which persist in a region only by colonizing new sites as old ones become untenable. Even a mature ("virgin") forest is a patchwork of small plots of different successional age, caused by treefalls and other local disturbances. The high species diversity of tropical wet forests may be largely a consequence of these GAP DYNAMICS, because light gaps formed by treefalls are necessary for most canopy tree seedlings to grow, and are the major habitat for shade-intolerant tree species as well. Localized disturbances of the soil by badgers, ground squirrels, and other mammals likewise contribute to the persistence of some plant species in prairies.

Microhabitats

A HABITAT is a characterizable place where an organism lives; a RESOURCE is something it uses (food, space, etc.) and which thereby becomes unavailable to other organisms (Begon et al. 1990). In many instances (e.g., for an herbivorous insect living in a tree), habitat and resource are almost the same thing. Large-scale habitats, such as marshes and dry hillsides, are obvious; but each contains innumerable microhabitats that may vary in both abiotic and biotic factors.

For example, if we enter an oak-dominated forest in the Middle Atlantic region of the United States, we will find plant species whose distribution depends on the successional age of each area in the forest. Moreover, within a few meters, differences in soil topography, in the amount of light entering through small gaps in the canopy, and the distribution of logs that retain soil moisture all can affect the distribution of understory herbs and shrubs. Rock outcrops and boulders are likely to be special habitats for lichens, mosses, and ferns. The oak trees themselves are a complex of microhabitats for several hundred species of insects, many of which feed only on oak (Figure 4.9). These include the many species of caterpillars chewing on the leaves; moth larvae that mine within the leaves; and cynipid wasp larvae within galls (growths of leaf or twig tissue formed in reaction to the insect within). The entire larval life (the feeding stage) of a leafminer or gall insect transpires within a single leaf, and its fate is affected by the chemical properties of that leaf and its position on the tree (which affects temperature and levels of proteins and carbohydrates). The leaf gall is also the habitat of other small wasp larvae that feed on the gall or its maker. Within the tree's acorns are larvae of several species of oak-specific weevils; their contents removed by the feeding larvae, the acorn shells may contain colonies of *Leptothorax* ants once they have fallen to the ground. The roots of the tree are food for cicada nymphs and beetle larvae, and are enveloped by fungi that form a symbiotic association, drawing nutrition from the tree but facilitating the tree's uptake of water and nutrients. This fungus-root complex is called a MYCORRHIZA. The bark of the tree is a habitat for lichens, algae, and small insects such as bark-lice (Psocoptera) that feed on them. Beneath the bark and in the wood, there are likely to be larvae of oak-specific cerambycid, buprestid, and scolytid beetles; the distribution of many of these is limited to branches or twigs of a certain diameter and a certain vitality; some prefer dying or dead branches. A hole left by a fallen branch may contain rainwater that is habitat for certain species of mosquito larvae. The dying branch is food

FIGURE 4.9 A few of the insects that may be found on oaks in eastern North America. (A) A leaf-eating moth larva (*Anisota senatoria*). (B) A cicada (*Tibicen sayi*). (C) An acorn weevil (*Curculio* sp.). (D) A leaf gall produced by the larva of a cynipid gall wasp (*Cynips maculipennis*). (E) A minute tree-fungus beetle (Ciidae). (A, B after Lutz 1918; C after Blatchley and Leng 1916; D after Felt 1917; E after White 1983.)

(A) (B) (C) (D) (E)

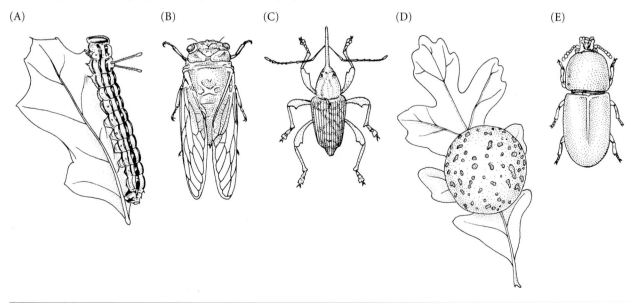

for shelf fungi, which themselves are the habitat for larvae of certain tenebrionid, ciid, and other beetles.

Fallen to the ground, the limb or trunk provides a physical shelter for salamanders, diverse arthropods, and nesting mice. As it decays, the log undergoes a succession of different fungi and of arthropods, including passalid and cucujid beetle larvae that feed on the wood, and an array of springtails, fly larvae, and other creatures feeding chiefly on the fungal mycelia. The decomposing litter on the forest floor has a very diverse community of mites, springtails, isopods, millipedes, and numerous spiders and insects; the soil itself has a similarly diverse community of these groups, as well as annelid worms. All of these microhabitats contain nematodes in high abundance, some parasitic on plants, others feeding on fungi or bacteria in the litter and soil. A dead mouse on the forest floor may be a habitat for fly larvae and other insects such as burying beetles (Silphidae), which lay eggs on the corpse and also transport silphid-specific mites that feed on fly eggs on the dead mouse. The live mice, like most of the vertebrates in the forest, are habitats for external and internal parasites; these might include more or less host-specific fleas, lice, botfly larvae, tapeworms, trematodes, and nematodes. The human visitor to the forest is the habitat for communities of bacteria on her skin, in her mouth, and in her intestine; he may have mites living in the hair follicles of his face, and might be carrying *Phthirius pubis* (crab lice), the sole habitat of which is the human body.

A like complexity, but with mostly different species, could be described for almost any other community; in a tropical wet forest, the variety of microhabitats would be multiplied manyfold.

Ecological Niches

Many factors affect the ability of a species to survive and reproduce sufficiently to persist in a locality. The ecologists Begon, Harper, and Townsend (1990) distinguish *conditions*—abiotic factors such as temperature that are not actually consumed—from resources that are consumed. The set of conditions and resources that a species population requires and/or tolerates are the components of what the eminent ecologist G. Evelyn Hutchinson (1957) defined as the population's **ecological niche**. For example, a clam species might tolerate a certain range of temperatures and feed on plankton in a certain size range. Hutchinson would describe these as the axes of a two-dimensional graph in which a particular point represents an environment with a certain temperature and a certain plankton size. Part of this "space" represents the set of possible environments in which the species can persist (Figure 4.10A). A third critical condition or resource—say salinity—would be represented by a third axis (Figure 4.10B). A portion of this three-dimensional space represents all possible combinations of these three factors that will support a clam population. We can conceive of (and represent mathematically, although not graphically) many more

FIGURE 4.10 (A) The ecological niches of two species, A and B. Perhaps they are estuarine bivalves, each restricted to a range of temperatures and of particle (prey) sizes they can ingest. Each point in the space represents a combination of these two variables. If a locality presents only environments in the vicinity of point 3, both species could exist there, but they will compete. If the locality includes microhabitats represented by points 1 and 2, the species can coexist, because each can use a microhabitat or resource that the other does not. If the only microhabitats are near point 4, neither species can persist. (B) Three dimensions of the niche of a species are represented. The species can persist in a locality if it has combinations of the three variables that lie within the solid figure.

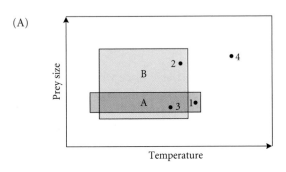

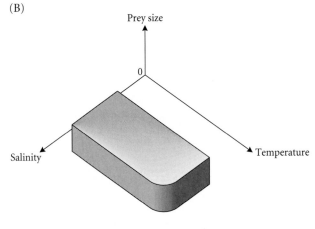

conditions and resources (copper concentration, particle size of the mud in which the clam lives, etc.), each of which is yet another axis, so defining a multidimensional space. A portion of this multidimensional space represents those combinations of factors that are conducive to the species' persistence. This portion of the multi-dimensional space is the species' ecological niche. If the conditions and resources at a particular locality are represented by a point within this niche, the species can persist there (if it can get there in the first place); if not, it cannot persist *unless* it undergoes evolutionary change, adapting to the inimical conditions/resources and thus undergoing an evolution of its niche. The niche of a species, a highly abstract notion, is not the same as its habitat, which is a physical part of the world that is characterized by the niche-defined resources and conditions and can be physically occupied by various species.

The niche concept is used extensively in ecological and evolutionary thought because although a locality may satisfy the niche requirements of a species, that species may be excluded by competing species that use some of the same resources (e.g., plants competing for light). The niches of the competing species are then considered to overlap, at least in part. This has interesting ecological and evolutionary consequences that we shall discuss further.

The tolerance of a species with respect to a particular niche axis may be broad or narrow: the species is relatively generalized or specialized, respectively. For example, larvae of the fall cankerworm (*Alsophila pometaria*) feed on the early foliage of many species of trees in the spring; they cannot develop on mature foliage and have only a single generation per year. In the same localities, larvae of another geometrid moth, *Itame pustularia*, feed only on maple leaves (and, like most host-specific insects, will starve to death rather than eat anything else); however, they can develop on both young and mature foliage, and pass through at least two generations per year, in spring and summer. One species is generalized with respect to plant species, but specialized with respect to leaf age; the other species is the opposite. A question of major ecological and evolutionary interest is why species have such specialized properties.

A closely related question is why species have limited geographical (and altitudinal) distributions. Not counting obvious barriers (e.g., oceans, mountains, cities), the edge of a species' range is set by the environment's failure to meet one or more of the species' niche requirements. For example, the northern and altitudinal limits of the range of the saguaro cactus (*Carnegiea gigantea*), the symbol of Arizona, correspond to localities that do not experience freezing temperatures for more than 36 hours; even an occasional freeze that lasts longer is lethal, and eliminates incipient populations (Hastings and Turner 1965). The distributional limits of many phytophagous insects are set by those of the plants on which they feed. For example, the leaf beetle genus *Monomacra* feeds exclusively on plants of the mostly tropical family Passifloraceae (passionflower); the northernmost species of *Monomacra* extends into the United States only as far as the northernmost species of *Passiflora* does. A puzzling question is why species do not adapt to other resources or to slightly more extreme temperatures, thereby extending their range.

Responses To Environmental Variation

We shall consider adaptations to environmental factors in many contexts, but a few general points are useful. A condition or resource can vary in both space and time. For a mobile organism, spatial variation often becomes temporal variation, if the organism moves among sites with different conditions or resources during its lifetime. Mobile animals often reduce the *effective* variation by **habitat selection**: they move into especially favorable sites. For example, many ectotherms (organisms that do not maintain

a body temperature by their own metabolism) such as lizards and insects move between sunny and shady spots so as to maintain a sufficiently high body temperature. Seasonal migration can have a similar effect. Habitat selection may buffer a species from many apparent environmental changes as well as protecting the species from natural selection that might otherwise lead to evolution of physiological and morphological features. This is one reason why it is insufficient to think of organisms solely as passive victims of the environment: to a considerable degree, they create or define their own environment. Plants have the same capacity, to some degree. For example, roots may grow preferentially into nutritionally favorable microsites, and many plants have seed dormancy that enables seeds to persist in the soil until conditions are favorable for growth.

Spatially and temporally varying factors have several important properties. These include the MEAN (average) and the RANGE (the difference between minimum and maximum). The range of a factor (e.g., temperature) is almost impossible to estimate accurately, because it may on rare occasions take on extreme values. Even though we may not know about these extremes, they may be important, because a single rare event may extinguish a population. However, it is easier, and often more relevant, to measure the VARIANCE, a statistical measure of variation that we will often use (see Chapter 9). If the factor varies regularly, its FREQUENCY or PERIOD of oscillation is important and may be relevant on several scales of measurement (e.g., daily and seasonal changes in temperature). Finally, the PREDICTABILITY of the change can be crucial: seasonal changes in photoperiod are highly predictable, those in temperature less so, and those in rainfall still less.

Organisms usually display adaptations to mean conditions (e.g., adaptations to water stress are more pronounced in desert plants than in trees of tropical wet forests) and to the usual range of fluctuations (as might be characterized by the variance). For example, invertebrates that inhabit estuaries, in which salinity fluctuates, can regulate the ion content of their blood (osmoregulation) so that it remains relatively constant over a range of environmental salinity. Species that live either in the sea or in fresh water, where there is little fluctuation in ion concentration, have little osmoregulatory ability (Figure 4.11). Such examples of HOMEOSTASIS—regulating a feature near a constant value—require that some activities of the organism change (e.g., the rate at which ions are pumped into the blood through cell membranes) in order to keep something else constant.

Alterations in the characteristics of an individual organism have various RESPONSE TIMES. A change in behavior may be almost immediate; changing the levels of an enzyme (e.g., by altering transcription rate) may require several minutes to several days, depending on the system; changes in morphological features are likely to require considerably longer. Which of these kinds of change is the most appropriate or the most feasible depends on the time scale (the period) and rapidity of environmental change (Slo-

FIGURE 4.11 Differences among species of crabs in physiological homeostasis. The osmotic concentration of the blood conforms to that of the environment in a marine crab, *Maja*. *Hemigrapsus* and *Carcinus*, shore crabs that experience greater fluctuations in salinity, regulate their blood osmoconcentration. (After Prosser and Brown 1961.)

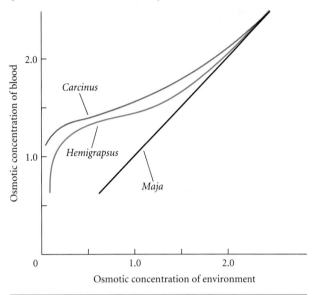

bodkin and Rapoport 1974). Very short-term changes in temperature are likely to be met by behavioral responses, as in lizards that bask in sunlight to increase their body temperatures. Longer term changes in temperature (also salinity and some other factors) are frequently met by PHYSIOLOGICAL ACCLIMATION. Thus fishes moved from cold to warm water perform (in terms of swimming speed, respiration rate, etc.) less well at first than they do after they have been maintained for some days at the new temperature. They undergo changes in enzymes, numbers of mitochondria, and other features that constitute an altered, acclimated, physiological state.

Some features may display adaptive PHENOTYPIC PLASTICITY, the capacity of a genotype to express different phenotypes under different environmental conditions. (The particular variety of phenotypes expressed is the reaction norm; see Chapter 3.) Phenotypic plasticity has two major (not entirely distinct) patterns. Reversible changes are usually adaptations to rather predictable environmental fluctuations with a long period, that the organism is likely to experience during its lifetime. Conspicuous examples are the seasonal changes in the plumage of many birds and the pelage (fur) of some mammals in temperate and polar regions. Less reversible, or irreversible, differences, often initiated early in development, are likely if the organism is likely to experience only one or another state of the environmental factor during its life (see Color Plate 5). For instance, the leaf form of some semiaquatic plants depends on whether the leaf developed below or above water (Figure 4.12). Aphids (in certain of the several generations per year) develop wings if they are crowded (hence likely to run

out of adequate food), but do not if they are not. Such "all-or-none" irreversible differences are often called DEVELOPMENTAL SWITCHES.

Responses to seasonal changes in the environment can include seasonal MIGRATION. Migratory birds such as warblers generally return to a locality very near where they were born; those that have already bred usually return to the same territory they occupied before. Seasonally migrating insects, in contrast, do not make a complete return trip: their offspring do. Monarch butterflies (*Danaus plexippus*) from the northeastern United States fly in autumn to a few specific sites in the mountains of southern Mexico, where they pass the winter without breeding. In early spring they fly to northern Mexico and southern United States, breed, and die. Their offspring fly to and breed in the northern states, and the following generation undertakes the autumnal journey.

In inimical seasons (winter, or dry seasons in arid areas), many organisms "escape" the prevailing conditions by a period of reduced activity, often seeking out protected sites. Plants may drop their leaves or pass the season as a dormant seed or underground stem; some animals hibernate or estivate. Insects and some other arthropods undergo DIAPAUSE, a state of reduced metabolic activity that does not increase even if temperatures should temporarily increase (Tauber et al. 1986). Such responses entail numerous phys-

FIGURE 4.12 Adaptive phenotypic plasticity in response to different environmental conditions in a semiaquatic plant, *Sagittaria sagittifolia*. The form of a leaf depends on whether it developed under terrestrial (left) or aquatic (right) conditions. See also Color Plate 5 for an animal example.

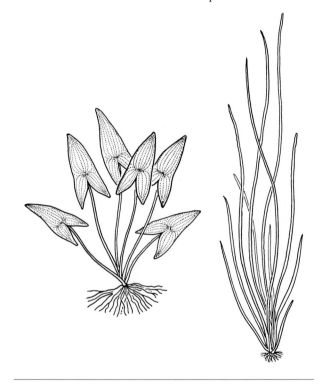

iological and biochemical changes, as does the seasonal cycle of reproduction. In most cases, these physiological changes are set in motion before the environmental change (e.g., low temperature) for which they are adaptive. Likely reasons for this are that the physiological alterations take some time to achieve, and that the exact time of arrival of cold weather is highly unpredictable: an insect that waits for the first frost to hibernate or migrate is likely to become a dead insect. The changing photoperiod (day length), however, is a completely predictable cue that is highly correlated with the average seasonal change in weather, and so it is by far the most commonly employed cue for the physiological and behavioral changes that mark hibernation, diapause, migration, and reproduction.

Populations

Population Structure

Because favorable habitat for a species is discontinuously distributed, almost every species has a discontinuous **population structure**, a fact that has immense importance in evolutionary theory. Most species consist of LOCAL POPULATIONS that may be more or less discrete, and between which individuals may move more or less frequently, depending on the species and on changes in environmental conditions. Especially for fugitive species, some sites become untenable while others become suitable, and are colonized from various source populations. Encounters among individuals from different populations are more frequent in such cases than in species that occupy more stable, but patchy, habitats. In most species, the *ecological* structure of populations, which determines which individuals can interact ecologically (e.g., by competing for food), is similar to the *genetic* structure, which describes which individuals are likely to have the opportunity to mate. But in some instances, the ecological and genetic population structures differ. American eels (*Anguilla rostrata*) are an extreme example: they spend much of their prereproductive lives in rivers and estuaries along the Atlantic coast, and so have a highly discontinuous ecological population structure; but they all swim to the same small region, the Sargasso Sea near Bermuda, to breed, thus forming a single genetic population.

Population Growth

The basic concepts of population growth and limitation are extremely important in evolutionary theory. For example, natural selection consists of differences among genotypes in the rate of change in their numbers. Population size can be calculated as the number of individuals, or more often as the DENSITY of the population—the number per unit area (or volume of water). Individual plants can vary greatly in size; a single genetic individual (a GENET) can consist of numerous stems (RAMETS), distributed over a considerable area by vegetative propagation; plant abundance is often measured, therefore, in other units, such as biomass or stem density.

It is simplest to think in terms of numbers of individuals in a population. A change in number stems from two processes that augment population size (births and immigration from other populations) and two that diminish it (deaths and emigration). For the moment, we shall ignore immigration and emigration and focus only on births and deaths.

Calculation of population increase (Box 4.A) is easiest if all individuals that survive to reproductive age reproduce at the same time and then die. This life history (DISCRETE GENERATIONS) is approximated by many annual plants and insects that have one generation per year. Suppose the number of female eggs (we will count females only, for simplicity) laid by a female grasshopper is $m = 50$ on average, and that we begin at the end of one growing season with 5000 eggs, laid by 100 females. Because of various sources of mortality, only 10% ($L = 0.1$, or 500) survive to adulthood; each adult lays 50 eggs and then dies. The ratio of numbers of eggs from one generation to the next is $25,000/5,000 = 5$, which may be calculated as: (proportion surviving to reproduce) × (number of eggs per survivor) = $L \times m = 0.1 \times 50 = 5$. This is the PER CAPITA RATE OF INCREASE, or the **net reproductive rate**, R. It expresses the ratio of numbers in successive generations, counted at one point in the life cycle. The population size after the passage of one generation, denoted N_1, equals the size in the initial generation N_0, multiplied by R: $N_1 = R N_0$. Thus counting eggs, $N_1 = 5 \times 5000 = 25,000$. If L and m stay the same in the next generation, $N_2 = R N_1 = R^2 N_0$. Or, in general, after t generations, $N_t = R^t N_0$. As long as L and m do not change, and $R > 1$, the population grows *exponentially* (Figure 4.13). If $R < 1$, the population declines exponentially, as it may if the habitat is unsuitable. The *change* in number per generation is

$$\Delta N = N_1 - N_0 = RN_0 - N_0 = N_0\big(R-1\big)$$

When generations overlap, as in human populations, births and deaths are occurring within any almost arbitrarily short time interval. For two successive time points, t and $t + 1$, the number at time $t + 1$ (N_{t+1}) equals the number at time t (N_t), plus the number of births, minus the number of deaths. Again counting only females, the number of births is the number of females (N_t), multiplied by the average number of births (b) per female during this interval, while the number of deaths is N_t multiplied by the probability (δ) that a female dies. Thus $N_{t+1} = N_t + bN_t - \delta N_t$. The *change* in numbers is $N_{t+1} - N_t = bN_t - \delta N_t = N_t(b - \delta)$. The quantity $b - \delta$, the difference between the birth rate and the death rate, *per female*, is denoted r, the **instantaneous rate of increase**. If we denote the change in numbers per change in time as dN/dt, we have a fundamentally important equation for population growth,

$$\frac{dN}{dt} = rN \qquad (4.1)$$

Again, as long as b and δ remain constant, the population grows exponentially if the per capita birth rate exceeds the

BOX 4.A

Mathematical Models of Population Growth

Let x represent the age of an individual, l_x (denoted L in the main text) the probability that a newborn individual (or egg) will survive to age x, and m_x the average number of offspring a female will have at age x (more precisely, within the interval $x - \frac{1}{2}$ to $x + \frac{1}{2}$). Tracing a cohort of newborn ($x = 0$) individuals, which number, say n_0, each of them may lay m_i eggs at age i, m_j eggs at age j, and so on, but only if they survive to those ages. From the original cohort, then, the total production of new offspring is $n_0 \times l_1 \times m_1$, produced at age 1, plus $n_0 \times l_2 \times m_2$ produced at age 2, and so on up to the maximal age (say $x = k$) that any of them attains. These products are the sum $\Sigma (l_x m l_x) = R$, the net reproductive rate. For younger age classes, which do not reproduce, m is zero. In Table 4.A, a hypothetical iteroparous species is shown to have $R = 12.18$ offspring per original individual of age 0. Thus from the original n_0 individuals, $n_0 \times l_x m l_x$ newborns are produced. Suppose that in the table, we set m_3 and m_4 to zero; then the species is semelparous (i.e., reproduces once). Keeping all else equal, R then equals 4.68.

Suppose the ages in the table represent years. Then eventually, since three different age classes reproduce and their descendants overlap in time, the population will consist of all age classes; in fact, it can be proven that if the value of l_x and m_x remain the same over time, a "stable age distribution" is attained, such that the proportions of 1-year-olds, 2-year-olds, etc. remain constant from year to year. There will be a constant average birth rate (b), which is the average of the m_x values, each weighted by the proportion of females of age x. Likewise, the constant average death rate (δ) depends on the proportion in each age class. $b - \delta = r$, the per capita instantaneous growth rate in the equation $dN/dt = rN$. The growth rate r can be very approximately calculated from the l_x and m_x values by the equation $r \approx (\log_e R)/T$, where T is the "average generation time," which is the average time between the birth of an individual and the birth of one of its offspring. An approximate equation for T is $T = \Sigma \, x \, l_x m_x / R$. From the figures in Table 4.A, $T \approx 4.28$ and $r \approx 0.584$.

Table 4.A Hypothetical schedule of age-specific survival and reproduction

AGE (x)	PROPORTION SURVIVING (l_x)	AGE-SPECIFIC FECUNDITY (m_x)	REPRODUCTIVE OUTPUT ($l_x m_x$)
0	$l_0 = 1.00$	$m_0 = 0$	0
1	$l_1 = 0.50$	$m_1 = 0$	0
2	$l_2 = 0.25$	$m_2 = 0$	0
3	$l_3 = 0.125$	$m_3 = 10$	1.25
4	$l_4 = 0.0625$	$m_4 = 100$	6.25
5(=k)	$l_5 = 0.0312$	$m_5 = 150$	4.68
6	$l_6 = 0.000$	—	0
			$R = 12.18$

per capita death rate ($b - \delta > 0$). Under these conditions, the population size at any future time, N_t, may be calculated from its initial size, N_0, by the equation $N_t = N_0 e^{rt}$, where e is the base of natural logarithms (2.718). (Readers who know calculus will recognize that this is found by integrating Equation 4.1.) In fact, of course, numerous factors can alter b and δ, including the density of the species' population. When density is low and resources are superabundant, b is frequently at its maximum, and δ at its minimum, for a particular set of conditions (e.g., temperature). Under these conditions, r is denoted r_m, the **intrinsic rate of natural increase**. However, r_m has different values under different conditions such as temperature, so "intrinsic" is something of a misnomer.

Species with overlapping generations usually reproduce ITEROPAROUSLY, i.e., more than once. Hence in each time interval, offspring are produced by females of several ages, each of which may have a different characteristic fecundity; and of course the probability of death also varies with age (Figure 4.14). This makes the calculation of r a little complicated (See Box 4.A). The rate of increase, r, obviously is greater, the higher survival and fecundity are at each age. Less obviously, the earlier the age at which reproduction be-

FIGURE 4.13 Idealized population growth. Exponential (density-independent) growth is shown by curves E_1 and E_2, representing populations that differ in r either because they are genetically different or because their environment (e.g., temperature) differs. Density-dependent growth of these populations, according to the logistic equation, is shown by curves L_1 and L_2. Both curves level off at a stable equilibrium density, K. Real populations do not show such smooth curves or constant equilibrium densities, even under constant environmental conditions.

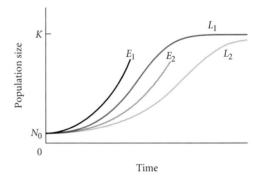

FIGURE 4.14 Age-specific survivorship (A) and fecundity (B) in an Australian fruit fly, *Drosophila serrata*, in the laboratory. The survivorship curves show the fraction of newborns that survive to each age (in days); the fecundity curves show the av- erage egg production by females at different ages. The five curves are data for stocks of this species from different locali- ties, showing that populations differ genetically in these life his- tory characteristics. (After Birch et al. 1963.)

(A)

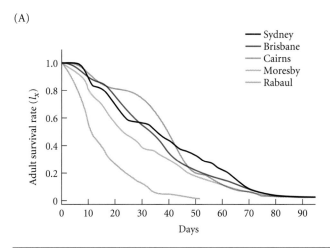

(B)

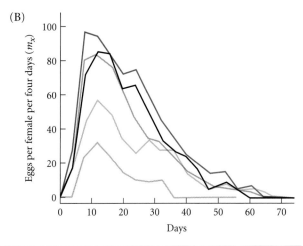

gins, the higher *r* is (all else being equal). An earlier age of first reproduction means a shorter generation length, and the more generations there are per time interval, the greater the increase in numbers. Closely related to this principle, *reproducing at an earlier age contributes more to population growth than reproducing at a later age.* These principles are important in understanding how different life cycles evolve (Chapter 19).

Unchecked population growth, because of its exponential nature, can lead to extraordinary consequences very rapidly. For our hypothetical grasshopper with *m* = 50 and 10% survival, a single female will have 3125 descendants after five generations. Humans have a relatively low rate of increase, but the present population of the world will be doubled in 39 years at its present rate of increase of 1.8% per year (Figure 4.15).

Factors Limiting Population Growth

No populations grow indefinitely. The factors that reduce the rate of growth are frequently divided into DENSITY-IN-DEPENDENT and DENSITY-DEPENDENT LIMITING FACTORS. Changes in weather that increase the mortality rate are often density-independent, in that the chance that an individual dies from harsh weather conditions is likely to be the same whether the population is dense or sparse. Density-dependent factors operate by decreasing the per capita birth rate and/or increasing the probability of an individual's death as population density increases. One obvious potential density-dependent factor is the supply of food or another critical resource: if the supply is fixed, each individual has less as the population increases, so the capacity to survive and reproduce declines. Predators and parasites (including infectious disease organisms) can also act as density-dependent factors if their populations increase in response to population increases in their prey.

It is likely that population numbers of most species are affected by both density-independent and density-dependent factors; some ecologists have argued that the relative importance of density-independent factors such as weather is especially great for many herbivorous insects, whereas density-dependent factors may act more strongly on populations of plants, predators, and decomposer organisms. A population may be limited by different factors at different times. For example, foliage-feeding insects that are thought to be kept at low density by predators may, under especially favorable conditions, undergo spectacular "outbreaks," and are then likely to be limited by shortage of food, which results in equally spectacular population crashes (Figure 4.16).

Populations that are perennially small in number, perhaps because of scarcity of resources, are subject to chance variation in density because of weather, predators, and other factors. Hence a small population is at great risk of extinction, and such extinctions are in fact very frequent (MacArthur and Wilson 1967). Great abundance, however, is no guarantee of persistence. The passenger pigeon *Ectopistes migratorius* was once among the most numerous of American birds; flocks of migrating pigeons are said to have virtually darkened the sky for days at a time. Human hunters slaughtered the breeding colonies, extinguishing the species entirely; the last passenger pigeon died in the Cincinnati Zoo in 1914.

If a population displays no long-term increase or decrease, *R* = 1, meaning that on average, each breeding pair is replaced by one breeding pair in the next generation. (If we use *r* as the rate of population growth per capita, *r* = 0 in a stable population.) Thus on average, *all but two offspring per brood perish before reproductive age*, in species with two sexes. (Read "all but one" for hermaphroditic and asexual species.) The "reproductive excess" is enormous for a high-

FIGURE 4.15 The human population of the world is expected to reach 12 billion by about the year 2100. Horizontal lines show the population levels in 1800, 1960, and 1990. (After Chrispeels and Sadava 1994, based on the Population Reference Bureau, Washington, D.C.)

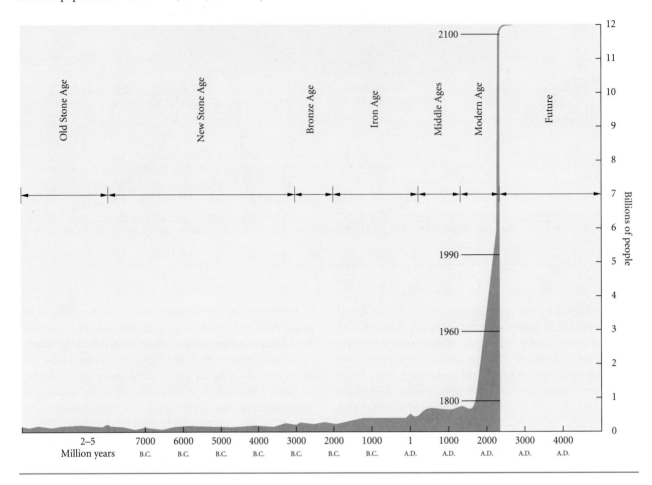

FIGURE 4.16 Fluctuations in the density of the hibernating stage in four species of moths in a pine forest in Germany in 60 consecutive years. Note the logarithmic axis: the population fluctuations were strong. (After Varley 1949.)

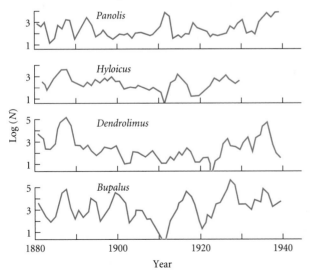

ly fecund species; of all the millions of seeds that an elm tree produces over its lifetime, all but one die before reproductive maturity on average; a particular tree, however, may have many surviving offspring, or none. This is the fact that inspired Darwin to conceive of natural selection (see Chapter 2).

Many populations are so stable in density that it is suspected that they are regulated by density-dependent factors that keep them near an equilibrium. The factor that regulates the population may not be the one responsible for the greatest mortality. For example, the winter moth (*Operophthera brumata*) in England lays eggs in the fall, which hatch in the early spring. The larvae disperse by wind, feed on foliage until early summer, and then pupate in the soil until autumn. In an 18-year-long study, Varley and Gradwell (1968) found that the largest sources of mortality by far (accounting for about 91% of all deaths) were of eggs during winter and, especially, of newly hatched larvae, many of which cannot find food because foliage has not yet appeared. This mortality is density-independent; from year to year, the proportion so dying is not correlated with density. A far smaller proportion of the total mortality (on average, about 2% of all deaths) was caused by predators (beetles and small mammals) feeding on the pupae; but from year to year, the proportion of pupae eaten was positively correlated with the density of pupae. These predators may therefore have been a stabilizing factor.

In a provocative, sweeping proposal, Hairston, Smith, and Slobodkin (1960; also Slobodkin, Smith, and Hairston 1967) argued that as a general rule for terrestrial communities, competition for limiting resources limits the density of plants (competing for space and, with it, light and sometimes water and nutrients), of predators, and of decomposers, whereas herbivorous insects are generally limited by predators. It appears likely that despite some exceptions, these generalizations quite often hold true (Schoener 1983; Connell 1983; Strong et al. 1984; Hairston 1989).

For resource-limited populations, a commonly used equation for population growth (the LOGISTIC EQUATION) supposes that the resource supply can support an equilibrium density of K individuals of the species. K is often called the CARRYING CAPACITY of the environment for this species. If the population number at a given time is N, then $(K - N)/K$ is the proportion of the carrying capacity that is still available for population growth. As N approaches K, this term approaches zero. In the logistic equation, the exponential growth in population size (Equation 4.1) is discounted by the term $(K - N)/K$:

$$\frac{dN}{dt} = r_m N \left(\frac{K - N}{K} \right) \qquad (4.2)$$

This equation says that the increase in numbers, dN/dt, declines toward zero as N grows toward K. The result is a smooth, S-shaped curve that levels off to a stable equilibrium population size at which $N = K$ (see Figure 4.13). Although this equation makes many simplifying assumptions and so does not exactly describe most real populations, it is a useful basis for many very generalized models. The terms r_m and K, in particular, will reappear in later chapters.

Interactions among Species

Species interact with each other in many ways. Of these, competition, predation, parasitism, and mutualism most commonly affect species abundance and distribution, and they all have important evolutionary consequences. Hairston (1989) provides a detailed review of many experiments that have elucidated the ecological effects of such interactions.

Competition

When a resource such as space, food, or nest sites is limiting, individuals of the same species are likely to be competing for it (**intraspecific competition**). Broadly defined (Begon et al. 1990), competition is an interaction between two or more organisms that on average decreases the growth and/or reproductive rate, and/or increases the death rate, of each. This interaction comes about through their common demands on resources. Competition takes

two broad forms. Competition by EXPLOITATION occurs when individuals remove resources so that they are unavailable for other individuals; the competing individuals need never encounter each other. For example, every worm eaten by an early bird is unavailable to later birds. Competition by INTERFERENCE involves direct interactions, with one individual preventing others' access to the resource, or even directly harming others. A taller plant restricts shorter plants' access to light, and some plants release ALLELOPATHIC CHEMICALS that reduce the growth of nearby seedlings. Many breeding birds and other animals defend TERRITORIES against conspecific individuals, behavior which may limit the density of breeding pairs.

Both these forms of competition can also occur between species that, at least in part, use the same resources (**interspecific competition**). Anyone who has weeded a garden knows that plants of different species compete for light and nutrients, and there is abundant evidence that many animal species compete as well. Much of this evidence comes from experiments in which the densities of species are manipulated. For example, Nelson Hairston, Sr. (1980) studied two closely related species of terrestrial salamanders, *Plethodon jordani* and *P. glutinosus*. These amphibians, which feed on invertebrates on the forest floor, occur at higher and lower elevations, respectively, but overlap to some extent in the southern Appalachian Mountains of the United States. The altitudinal zone of overlap is narrow in the Great Smoky Mountains, but broad in the Balsam Mountains. Where the salamanders overlap, *P. jordani* is the more abundant species. In both mountain ranges, Hairston periodically removed *P. jordani* from several large plots, removed *P. glutinosus* from others, and for 6 years monitored the populations both in these plots and in several nonmanipulated control plots. Removal of *P. jordani* resulted in an increase in the density of *P. glutinosus*, relative to their density in the control plots. This response was more pronounced in the Smokies than in the Balsams, suggesting that *P. jordani* imposes more severe competition in the Smokies. Removal of *P. glutinosus* apparently resulted in an increased rate of reproduction in *P. jordani*, because the proportion of juveniles of this species increased.

Species introduced by humans into new areas often reduce the abundance of resident species. Old World grasses, for example, have replaced many of the native grasses and forbs in the hills near San Francisco, California. Another example involves two parasitic wasps, sequentially introduced from different parts of Asia to control the olive scale insect in California: *Aphytis chrysomphali*, the first to be introduced, flourished until *Aphytis lingnanensis* was imported. The second species then entirely replaced the first (DeBach 1966).

The **competitive exclusion principle** holds that two or more competing species that use exactly the same resources cannot coexist indefinitely. Consider two such species, A and B, that have identical net reproductive rates R. The resources can support a total population of K. In each generation, RN_A and RN_B offspring are born; these numbers are reduced, by mortality that falls at random with respect to species, to K adults. We rightly expect that among these, A and B should have the same proportions as in the previous generation. But this is only our *expectation*; in reality, just by chance the proportion of A and B among the survivors will change slightly. These slight random fluctuations will eventually result in either 100% A or 100% B. It is, indeed, more likely that the initially more abundant species will reach 100%. (The random fluctuations in proportions of two equivalent species are conceptually identical to random genetic drift of alleles within species; see Chapter 11.)

Suppose the species are the same, except that A has a higher net reproductive rate than B; then in each generation the proportion of A among the offspring is higher than among the parents, and more A's will be among the K lucky survivors. Species A will replace B. One possible reason for A's higher reproductive rate is that it might more efficiently use some resource than B can, so that A can extract sparse resources from the environment and continue to grow in number when B cannot.

Following Darwin's suggestion in *The Origin of Species*, ecologists often expect closely related species to compete most intensely because, as Darwin pointed out, they are likely to have similar requisites. Studies such as Hairston's analysis of salamanders show that this is often true; in fact, removal of each of these very similar *Plethodon* species affected the other but did not influence the population densities of four other salamander species that are smaller in size and probably have different diets from *P. jordani* and *P. glutinosus*. But in some instances, unrelated species also compete. For example, numerous species of rodents feed on seeds in the deserts of the American southwest. James Munger and James Brown (1981) enclosed some desert plots with fences that had holes large enough to permit passage of small rodents, but too small for the larger kangaroo rats (*Dipodomys*) to pass through. They removed *Dipodomys* from the enclosures. The consequence was that within the plots, populations of several small species of seed-eating rodents, belonging to different genera and, indeed, a different family, increased (Figure 4.17). In anoth-

er experiment, densities of seed-eating ants increased for several years in fenced desert plots from which rodents were removed. After 3 years, however, species of plants with large seeds, which are preferred by rodents, replaced plants with small seeds, the major food of the ants. Hence the eventual effect of rodent removal was a decline in ant abundance, mediated by interspecific competition among plants (Davidson et al. 1984). These experiments, then, illustrate several points: (1) distantly related species may compete; (2) predators can control the abundance of superior competitors, thus enabling inferior competitors to persist; (3) the effects of species on each other's abundance may be either negative or positive, depending on the community structure through which they interact.

Coexistence of Species at the Same Trophic Level

Species at the same trophic level (i. e., level in the food chain) can coexist indefinitely for any of several reasons. (1) They do not compete, because their populations are sparse relative to the supply of resources. That is, their populations may be limited by natural enemies or by weather, rather than by resources. (2) The species compete, but predators or parasites prevent the competitively superior species from achieving numerical predominance. (3) The species compete, but environmental conditions, favoring first one species and then another, fluctuate often enough to prevent one species from excluding the others. (4) The species do not entirely use the same resources, so that each has a resource or a microhabitat in which it is secure from competition, or at least is competitively dominant.

The latter principle—that competing species coexist if their niches do not entirely overlap—is a major theme in ecological and evolutionary biology. This principle is commonly invoked to explain the fact that potential competitors that do in fact coexist usually differ in resource use. (But it must be said that most such evidence is based on studies of animals; it is much less clear that each plant species in a community has a different niche: see Tilman 1988; Harper 1977.) For example, throughout the world, coexisting spe-

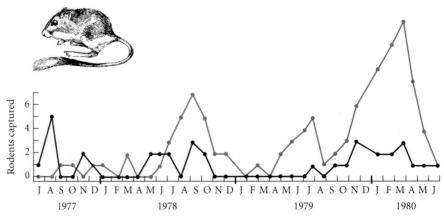

FIGURE 4.17 Population density of small seed-eating rodents in experimental plots in an Arizona desert. The number of captures in plots from which larger kangaroo rats were excluded (colored line) was generally greater than in plots where kangaroo rats were present (black line), implying that competition affects the population density of the small rodents. (After Munger and Brown 1981.)

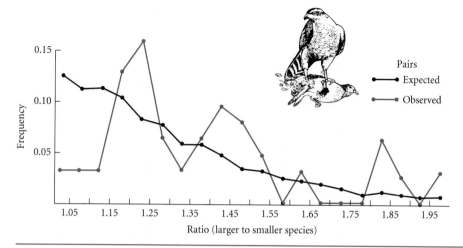

FIGURE 4.18 Coexisting species of bird-eating hawks (*Accipiter*) are more different in body size than expected at random. Hawks of different sizes feed on different species of prey. The data imply either that species can coexist only if they differ in prey, or that coexisting species evolve differences in prey use, to reduce competition. The colored line represents the proportion of pairs of coexisting species, worldwide, with a given ratio of body sizes. The black line is the expected proportion of pairs of species with a given body size ratio, if the world's 47 species of *Accipiter* were paired at random. Few observed pairs of coexisting species have a low ratio (i.e., nearly equal), although many such pairs would be expected at random. (After Schoener 1984.)

cies of bird-eating hawks consistently differ in body size (Figure 4.18), a difference that is correlated with the size of their usual prey items (Schoener 1984). In the Gálapagos Islands, species of Darwin's finches (*Geospiza*) with similar beaks occur on different islands, but species that coexist on any one island differ in beak size (Figure 4.19). The size of the beak is correlated with the size and hardness of the seeds on which the species feeds (Grant 1986).

Reduction of competition by differences in ecological niche has been shown by many experiments. For example, in the southern Appalachian Mountains, the salamander *Plethodon cinereus* occupies forested habitats, while the very similar species *P. shenandoah* is restricted to open, drier, rock-strewn talus slopes. Robert Jaeger (1971) placed both species, separately and together, in enclosures with either forest soil or talus soil. In forest soil, the survival of *P. shenandoah* suffered from the presence of *P. cinereus*, whereas in talus soil, *P. cinereus* survived poorly relative to *P. shenandoah*, which is more resistant to desiccation. Thus *P. shenandoah* appears to persist because talus slopes offer a refuge from competition.

"Niche partitioning" among species of animals sometimes comes about through behavioral responses to the presence of other species. For example, three species of sunfishes (*Lepomis*) compete strongly for food in Michigan ponds. Each species, if placed alone in a pond, forages preferentially for invertebrates in the vegetation. If all three are present, *L. cyanellus* forages in vegetation, but the other species switch to different foraging modes, with *L. gibbosus* feeding mostly on the pond bottom and *L. macrochirus* feeding on open-water plankton (Werner and Hall 1979). Not all species, however, display such behavioral flexibility; for example, two species of crickets (*Allonemobius*) that occupy

moist versus drier vegetation do not alter their habitat range if either species is removed (Howard and Harrison 1984).

Predation, Parasitism, and Herbivory

A PREDATOR usually consumes many individual prey organisms, killing them in the process. A PARASITE obtains nutrition from a single host (or, in some instances, a sequence of hosts); an individual parasite rarely kills its host (although a multitude of parasites may do so). Parasitism takes many forms, and grades into predation; for example, some insects, referred to as PARASITOIDS, feed on a host (usually another arthropod) that lives and behaves rather normally until the parasitoid completes its development, at which time the host dies. Pathogenic microorganisms, such as many viruses, bacteria, and fungi, are properly viewed as microparasites.

Herbivores may eat all or part of a plant. Because a seed, like a seedling, is an individual plant, a seed-eating bird, rodent, or insect is a "seed predator." Grazing mammals or insects, which eat only part of one or many plants, are parasites according to the above definition, and are treated as such by some authors (e.g., Price 1980). Although it takes some imagination to think of a horse as a parasite, the ecological and evolutionary dynamics of many herbivorous insects and mites certainly resembles that of more traditional parasites.

Predation   Most species of predators feed on more than one species of prey, although some are quite specialized in the variety of prey they take. For example, the European honey buzzard (a hawk, *Pernis apivorus*) feeds largely on the larvae of wasps and bees, and many species of snakes are specialists on prey such as crayfish, termites, or other

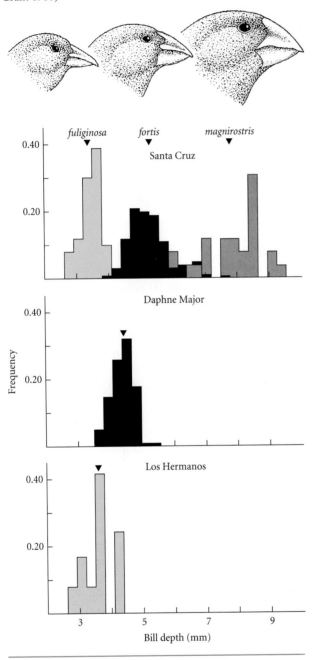

FIGURE 4.19 Bill depth in seed-eating ground finches on three of the Galápagos Islands. The mean beak depths (black pointers) of the three species on Santa Cruz island are equally spaced. Solitary species on different islands, in contrast, may have more similar bill depth, as illustrated by *Geospiza fortis* compared to *G. fuliginosa*. Bill depth is correlated with the size and hardness of the seeds preferred by each population. (After Grant 1986)

snakes. As we noted above, populations of predatory species appear frequently to be food-limited; consequently, their population density can increase if the population density of their prey increases. This is termed the predator's NUMERICAL RESPONSE to prey density. In addition, individuals of many vertebrate predators (and some invertebrate predators as well) change their behavior in response to the relative abundance of different species of prey. This FUNCTIONAL RESPONSE of individual predators is displayed by many birds, which appear to direct their foraging efforts largely to whichever species of prey (e.g., insects) are currently most abundant, switching to other prey species as they in turn become most abundant. It has been hypothesized that the individual forms a "search image" for temporarily abundant prey types, and there is some evidence that this increases the efficiency with which it finds and captures food.

As a consequence of the predator's numerical and functional responses, the intensity of predation suffered by a prey population increases as the prey's population density grows, so that predation can act as a density-dependent factor that contributes to regulation of the prey's population density. Conversely, the abundance of prey controls the density of the predator. According to some oversimplified mathematical models, these mutual influences can cause coupled cycles in the density of both predator and prey. Cycles of prey and predators have been observed in nature, in, for example, snowshoe hare (*Lepus americanus*) and Canada lynx (*Lynx canadensis*) populations (Figure 4.20), although it is not certain that these cycles are caused entirely by the predator-prey interaction. Whether or not cycles occur, predators (and parasites and herbivores) frequently limit prey population densities, as in the examples of winter moths and desert plants cited earlier. Numerous other examples have been demonstrated by experiments (Hairston 1989). Moreover, introduced species have frequently extinguished or greatly reduced the populations of prey species. One of the most spectacular and fascinating examples of biological diversification is that of African cichlid fishes; but they have suffered tragically from human activity. Of the more than 300 species of cichlids that occurred in Lake Victoria in eastern Africa, more than 200 have become extinct or severely threatened with extinction since the early 1980's, due to predation by the Nile perch, which was introduced as a food source for the human population (Figure 4.21).

Parasitism   The definition of a *parasite* is so broad that few generalizations apply to the great variety of parasitic organisms. Some reasonably broad generalizations are:

1. A parasite's generation time is shorter, and the reproductive rate often much greater, than that of the host.
2. The effect of a single parasite may be slight, but the number of parasites per host is often great, and the combined effect is frequently severe.
3. The parasite is often highly specialized with respect to host species, the tissue occupied, and the behavioral and physiological mechanisms by which the parasite attacks the host.
4. Compared to their free-living relatives, parasites often lack many features, such as certain appendages and sense organs. The host constitutes an environment in which such features are not necessary.

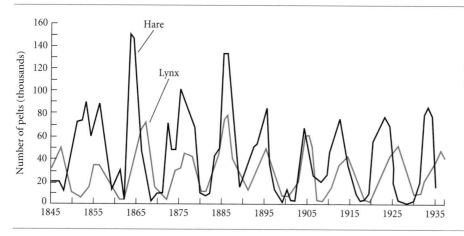

FIGURE 4.20 Fluctuations in abundance of lynx and hares in northern Canada, based on numbers of furs purchased by the Hudson Bay Company. The causes of the coupled cycles are still unclear. (After Purves et al. 1998.)

5. For certain modes of parasitic life, dispersal from one host to another is extremely difficult. Numerous features of the parasite's life history may be viewed as adaptations for dispersal.

6. Hosts have various biochemical, immunological, behavioral, and other defense mechanisms to ward off parasites, and many features of parasites are adaptations for overcoming these defenses.

The life cycles of parasites range from simple to complex, largely in relation to the mode of dispersal among hosts. Lice, such as the human pubic louse *Phthirius pubis*, are transmitted by direct contact between hosts, and have a life cycle much like other insects. The gonorrheal bacterium is one of many microparasites that is adapted for transmission by sexual contact. It is plausible to think that rhinoviruses (common cold viruses) that cause sneezing are manipulating their host in a way that enhances their dispersal. Many parasitic animals have extraordinarily complex life cycles by which they get from host to host. For example, (Figure 4.22), the trematode (fluke) *Dicrocoelium dendriticum* occupies the bile duct of sheep and other grazing mammals (the "definitive hosts"). Eggs are passed with the feces and are eaten by land snails (first intermediate host). Within the snail, two intermediate stages occur, which by asexual reproduction produce many cercaria larvae, which the snail discharges in balls of mucus. These are eaten by ants (second intermediate host). Some of the parasites encyst in the ant's nervous system and cause the ant to climb vegetation and enter a stupor, so that it is readily eaten by a grazing mammal. The larvae then transform into flukes, which travel through the tissues to the bile duct. Life cycles like this, including intermediate hosts and manipulation of the host's behavior, are not unusual among parasites.

The rate at which uninfected members of a host population become infected by a parasite is often greater, the greater the density of the host population (May 1981). Hence parasites, like predators, can have density-dependent controlling effects on their hosts. Examples of parasites that reduce their hosts' populations range from the effect of viruses on gypsy moths and other insects, to the devastat-

ing effect of the bubonic plague bacillus (the Black Death), which killed at least 25% of the human population of Europe in the 14th Century.

Herbivory From the viewpoint of terrestrial plants, the most important herbivores are mammals (especially ungulates, rodents, and rabbits) and insects, of which more than 400,000 herbivorous species have been named and at least that many (perhaps ten times that many) have not yet been named. Plants differ from each other not only in morphological but in chemical features, especially the so-called "secondary compounds." These include an extraordinarily diverse array of phenols, terpenes, acetylenes, alkaloids (such as strychnine and caffeine), and many other classes of organic compounds. Many of these are toxic and/or repellant to various herbivores. Probably because of the chemical and morphological differences among plant species, each herbivore species feeds on only a limited array of plants. This array is quite broad for many mammals and some insects (e.g., many grasshoppers), but most herbivorous insects are highly specialized, many of them being restricted to a few closely related plant species or to only one (Figure 4.23). Moreover, only certain tissues of the plant are eaten, and these often for just a short time of year, in part because of seasonal changes in the plant's features (see reviews in Rosenthal and Berenbaum 1992; Denno and McClure 1983).

Consequently, even if vegetation remains lush throughout a growing season, much of it is inedible for any particular herbivore species. Hence although the population density of many herbivores may be limited by natural enemies (Strong et al. 1984a), some species of herbivores are undoubtedly food-limited. Conversely, many examples are known in which herbivores limit the abundance and distribution of certain species of plants (Crawley 1983). The potential impact of herbivores on plants has been dramatically illustrated by several cases in which a host-specific insect has been introduced to control an alien weed. For example, the European St. John's-wort (*Hypericum perforatum*), introduced into North America, became a serious weed in cattle rangeland. The beetle *Chrysolina quadrigemina*, introduced from Europe to control the plant in parts of the western U.S., quickly eliminated it except for small

FIGURE 4.21 The tragedy of Lake Victoria. (A) The large Nile perch, introduced into Lake Victoria as a food source for humans, has extinguished at least 200 endemic species of small cichlids. (B) Decline of the number of cichlid species captured in trawls at three depths, in samples taken only eight years apart. The cichlids are classified by their feeding habits, and the number of species in all categories has declined. (A from Moyle 1993, B after Witte et al. 1992.)

(A)

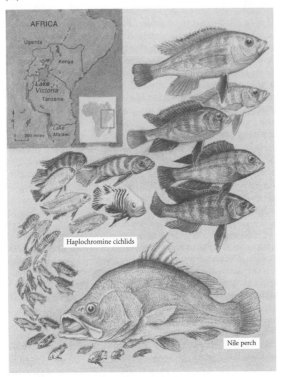

(B)

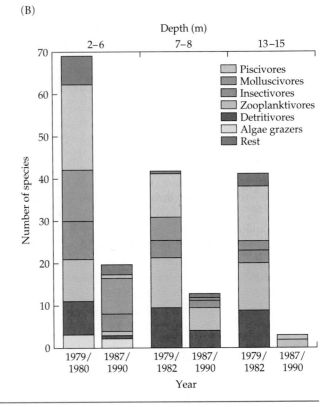

patches that grow in shade, where the beetle does not feed (Huffaker and Kennett 1969).

Mutualism

Interactions between species are termed mutualistic if the interaction enhances the survival and/or reproduction of both species. In some mutualistic interactions, such as those between plants and pollinating animals, each individual of each species may interact with many individuals of the other. Other mutualisms are symbiotic, entailing an intimate, rather long-lasting association of individuals of both species. The term symbiosis, which means "living to-

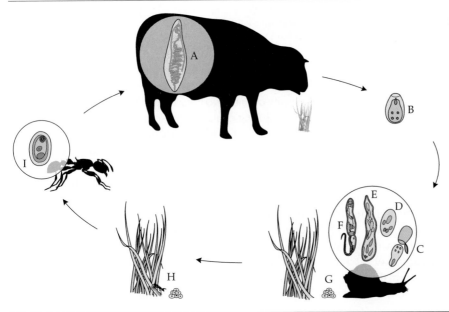

FIGURE 4.22 Simplified life cycle of the fluke *Dicrocoelium dendriticum*. The adult fluke, in a sheep or other grazing mammal, lays eggs (B) which are consumed by snails, within which several distinct larval stages (C–F) reproduce asexually; cercaria larvae (F) are expelled in mucus (G), and are eaten by ants (H), in which the parasites encyst (I). Mammals accidentally ingest the ants, and the parasite completes its life cycle. (After Schmidt and Roberts 1981.)

FIGURE 4.23 An example of diversification and specialization in herbivorous insects. Larvae of species of the gall midge genus *Rhopalomyia* (inset) induce the formation of galls (growths) of the buds, leaves, and stems of the big sagebrush (*Artemisia tridentata*) in western North America. The figure shows an adult gall midge and the characteristic galls (arrows) of 17 species of *Rhopalomyia*, which as far as known feed only on this plant. (From Gagné 1989.)

gether," also includes some parasite/host associations and commensal interactions (in which one species profits from the association but the other is not affected). For example, certain hydroids live only on the tubes of sabellid polychaete worms, which presumably are not affected by the hydroids that depend on them.

Mutualism should probably be called "reciprocal exploitation," because each species interacts with the other so as to obtain some benefit. Plants offer nectar so as to achieve reproduction via the action of pollinators; pollinators visit flowers not to help the plant reproduce, but to obtain food. Some plants "trick" pollinators into providing reproductive service without offering a resource; for example, many nectarless "pseudocopulatory" orchids lure males of specific species of bees or other insects, by a flower pattern that resembles that insect species, as well as by a scent that mimics the female insect's sex pheromone. The male insect pollinates

by "mating" with the flower (Color Plate 6). Conversely, some "nectar-robbing" bees and birds obtain nectar by chewing into the base of flowers, gathering food without offering the plant any benefit.

The balance between mutualism and parasitism is sometimes a fine one. Many vascular plants and fungi form mycorrhizal associations, a root-fungus combination in which the fungus obtains carbohydrates synthesized by the plant and the plant profits from enhanced uptake of water and nutrients; but the same fungus that enhances plant growth in some environments may reduce it in others (Douglas 1994).

Because mutualistic species can theoretically enhance each other's rate of population growth, the interaction theoretically could cause both populations to grow indefinitely large. This does not occur because resources or other factors limit the density of one of the partners (e.g., a plant), so that population growth of the dependent partner (e.g., a bee that depends on pollen and nectar) is likewise limited.

Mutualists vary in their species specificity. For example, many species of mycorrhizal fungi and of pollinating insects are each associated with many species of plants, although some have more specialized associations. Among the most remarkably specialized pollination associations are those between figs (*Ficus*) and fig wasps. With few exceptions, each of the hundreds of species of figs depends for pollination on only a single species of wasp, which pollinates and reproduces in only one species of fig. These wasps, whose larvae feed on the seeds, are both mutualists and predators.

Extreme host specificity is often characteristic of extremely intimate symbiotic mutualisms, especially those in which microorganisms live within an animal or plant host. In many such cases, the two species are so physiologically integrated as to act virtually like a single organism. This is the case with "species" of lichens, each of which is an association of a fungal host with an algal symbiont; of corals and the algal zooxanthellae on which the coral animal is entirely dependent; of sap-sucking insects such as aphids, which depend on intracellular bacteria to synthesize vitamins; and of certain parasitoid wasps, which inherit a virus (polydnavirus) along with their own genome. The wasp inserts into its insect host an egg which is coated with virus particles that prevent the host's immune system from attacking the wasp's egg or larva (Summers and Dib-Hajj 1995). These relationships astonish us by their complexity, but they should also prompt us to ask what prevents the symbiotic microorganism from multiplying so prolifically as to harm the animal or plant in which it resides.

The *ne plus ultra* of symbiotic mutualism is undoubtedly the eukaryotic cell (Chapter 7). There is no doubt that mitochondria and chloroplasts are descended from bacteria that became incorporated, as intracellular symbionts, into a proto-cell more than 3.5 billion years ago.

Other Interactions among Species

Some interactions among species fit the above categories only awkwardly. Of special interest are the several kinds of

FIGURE 4.24 Aggressive mimicry. A grouper attended by a cleaner wrasse (left, and enlarged) and its carnivorous mimic, the saber-toothed blenny (right, and enlarged). (From Owen 1980.)

mimicry (Color Plate 7). These include BATESIAN mimicry, in which a palatable species (the MIMIC) is protected from some predators by its resemblance to an unpalatable or dangerous species (the MODEL) that is avoided by predators, either because they have learned by unpleasant experience with a model or because they avoid it instinctively. The distastefulness of the model is usually "advertised" by bright, readily visible color patterns, such as the black and yellow pattern of many wasps and bees. This "warning coloration" is called an APOSEMATIC pattern. In MÜLLERIAN mimicry, two (or more) unrelated species are both unpalatable and resemble each other; a predator that has had experience with one will avoid both. Especially in tropical forests, there exist "mimicry rings" of species of butterflies that include both Müllerian and Batesian mimics, all with similar color patterns. In AGGRESSIVE mimicry, a predator resembles a nonpredatory species; this is thought to enable it to capture undiscriminating prey. For example, a carnivorous saber-toothed blenny (*Aspidontus taeniatus*) resembles in color pattern and behavior an unrelated fish, the cleaner wrasse (*Labroides dimidiatus*) (Figure 4.24). The wrasse gleans parasites from large fishes such as groupers, which do not eat their helpful consorts. The blenny also escapes being eaten, even though it feeds on the large fish by eating pieces of skin or fin.

The web of interactions among species in a community is usually highly complex (Pimm 1982). It includes not only direct but indirect interactions. We have noted that a predator can help to maintain a competitively weak species in a community by preventing competitively superior species from taking over. Thus the predator that feeds on a competitively dominant species may indirectly benefit another predator that feeds on a competitively inferior species. Likewise, a species may indirectly harm another, as in ants that may harm plants by promoting the population growth of aphids, from which the ants gather honeydew and which they protect against predators. Because of these complex interactions, the addition or extinction of any single species is likely to affect the persistence of at least a few other species, and in some cases can affect many.

The Unstable Environment

Not only are there daily, seasonal, and year-to-year changes in the environment, but the earth's climate has been highly unstable, and the earth itself is in constant flux: the continents themselves have never ceased to move. About 600 years ago, the climate of Europe and North America was rather colder than it is today; during the late Pleistocene, about 12,000 years ago, glaciers covered North America as far south as Wisconsin and New York, walruses occurred in South Carolina, and species were distributed in quite different assemblages than we see today. Fifty million years ago, opossums much like today's species coexisted with condylarths, titanotheres, and many other creatures that have long since become extinct. If biologists seem forever to be ascribing evolutionary changes to hypothetical changes in the environment, it is for very good reason: environments have changed continually, and often drastically, throughout earth's history.

Summary

1. Global patterns of temperature, rainfall, and ocean circulation create major regional differences in climate. On land, these result in different biomes, characterized by differences in vegetation structure. Even in different geographic regions, the species found in a biome have characteristic adaptive features.

2. However, the taxonomic composition of each biome differs among geographic regions, which have been classified into biogeographic realms based on taxonomic patterns. The distributions of species are influenced by barriers to dispersal, by their physiological tolerance of many environmental factors, and by their interactions with other species.

3. The ecological niche of a population or species is the set of possible combinations of environmental variables and resources under which the population can persist. A niche, which is a function of the characteristics of a species, is not the same as a habitat.

4. Because some locations have the features that correspond to a species' niche and others do not, most species are patchily distributed, consisting of local populations.

5. The rate of growth in numbers of a species population depends on the per capita rate of increase (r), the most important components of which are the per capita rates of birth (fecundity) and death. The age at which reproduction begins has an important effect on r. If r remains constant and is greater than zero, the population grows exponentially.

6. On average, populations do not increase; they fluctuate about some average density. Unlimited increase is prevented by density-independent factors such as weather and by density-dependent factors such as predation or competition for limiting resources. Density-dependent factors increase the per capita death rate and/or lower the per capita birth rate, as the population density increases. These factors may establish an equilibrium population density (K).

7. Interactions among species affect population growth and density. Among these interactions is interspecific competition, which can result in exclusion of competitively inferior species. However, several factors can enable species to coexist. One such factor is niche partitioning, i.e., using some different limiting resources. Another is predation, which can alleviate competition.

8. Predators and parasites may limit the population density of their prey or hosts, and in turn can be limited by the availability of these food resources. Parasites, and herbivorous arthropods that are analogous to parasites, often are highly specialized, utilizing only a few species of hosts. The species involved in these interactions have numerous adaptations to each other.

9. In mutualistic interactions, individuals of each of two or more interacting species obtain benefit from the other. Such interactions may affect the density and distribution of the species.

10. The environment–the physical and biological factors that impinge on members of a species–is in a constant state of flux, varying on time scales ranging from hours to millions of years.

Major References

Begon, M., J. L. Harper, and C. R. Townsend. 1996. *Ecology: Individuals, Populations, and Communities*. Blackwell Scientific, Oxford. One of the more comprehensive textbooks on ecology.

Ricklefs, R. E. 1990. *Ecology*. Third edition. W. H. Freeman, New York. Another comprehensive textbook; includes considerable coverage of evolutionary ecology.

Pianka, E. R. 1994. *Evolutionary Ecology*. Fifth edition. HarperCollins, New York. Emphasizes those subjects in ecology for which an evolutionary perspective has been particularly important.

Problems and Discussion Topics

1. Imagine a parthenogenetic moth in which all offspring are female, and females die after laying a clutch of eggs. In moths of type A, 90% of eggs survive to reproductive age and the average clutch size of surviving females is 20 eggs. In type B, only 80% survive, but the average clutch size is 24. What is the value of R for each type of moth? If we begin with 10 adult moths of each type just before egg-laying begins, what proportion of moths, at the same stage, will be type A after 3 generations?

2. For the moths in Question 1, suppose a further difference is that the generation time (egg to egg) of type A is 6 months, whereas that of type B is 12 months. Again beginning with 10 egg-laying adults of each type, what proportion will be type A after 24 months?

3. Other than changes in the environment, what factors might cause the course of population growth to be more irregular than the smooth curve predicted by the logistic equation? What factors might cause fluctuations around an equilibrium, rather than an unchanging equilibrium?

4. Some local populations are not self-perpetuating, but persist only if there is immigration into them from other populations. How might you determine whether this is the case? Why might this be important?

5. Draw on the ecological literature to write an essay on the factors that limit the geographic range of species. Are the borders constant? Do the same factors operate along the different margins of a species' distribution?

6. Suppose a species of herbivore evolves to be more efficient in digesting its food, so that it has a higher birth rate per amount of food consumed. Will its equilibrium population density increase? Why might it not?

7. Many species of weeds and pests, such as the gypsy moth and the kudzu vine in North America, are recent introductions from other parts of the world, where they are relatively uncommon. Discuss possible reasons for this difference.

8. What factors influence the rate at which a parasite population increases? How might these factors differ among parasites that are directly transmitted among hosts (such as the gonorrhea bacterium), parasites that are carried by vectors (such as the malarial protozoan), and those that are dispersed through the environment (such as the influenza virus)?

9. Some species of predators and prey (and of parasites and hosts) coexist, whereas other such interactions are unstable (as in the case of the African cichlids and the Nile perch). What factors determine whether or not stable coexistence occurs? How would evolutionary changes in the species alter the outcome?

10. What factors may determine the outcome of competition between a species with generalized feeding habits and a more specialized feeder, if the diet of the generalist includes that of the specialist? How can you account for the coexistence of specialized and generalized species in communities? What factors might influence species to become more or less specialized?

Patterns and History

Part II describes major patterns of biological diversity, the evolutionary history that has given rise to these patterns, some of the methods and evidence used to ascertain this history, and many sources of evidence for the historical reality of evolution. To a great extent, the subject of these chapters is the history of *macroevolution:* the origin of higher taxa, and of their characteristics and distribution, throughout the long history of life. Understanding the *causal processes* of evolution that have generated these patterns will require that we closely examine genetic and other processes within and among populations and species; these processes are the subject of Part III (Chapters 9–16). Thereafter (in Chapters 23–25), we will return to macroevolution and examine its causes in a new light.

All analyses of biological diversity and its evolutionary genesis rest on *systematics,* the branch of biology that describes patterns of organismal variation and diversity. One task of systematics is to discover, distinguish, and name species and classify them into higher taxa. Another task, the chief subject of Chapter 5, is to infer the genealogy, or *phylogeny,* of species; by determining their relationships to each other, we describe the history of *branching* of evolutionary lineages. In the course of such analyses, it is often possible to infer the stages by which characteristics have changed from ancestors to descendants. Phylogenetic analysis therefore provides insights into both of the major aspects of evolutionary history: *cladogenesis* (branching) and *anagenesis* (evolutionary change within lineages). In the course of their work, systematists have described many of the *major patterns of evolutionary change and diversification.* Their work provides insight into the evolution of many kinds of characteristics, ranging from morphology to DNA sequences (Chapters 17–22), as well as much of the evidence for the historical reality of evolution. Chapter 5 is devoted largely to the methods of phylogenetic study, some patterns of evolution that systematists have documented, and some of the evidence for evolution provided by systematics.

In Chapter 6, we turn to the fossil record, which provides the most nearly direct evidence on evolutionary change in lineages and their characteristics, on rates of change, and on the history of biological diversity. Much of the chapter is devoted to describing patterns in the evolution of morphology within various evolving lineages, to the great differences in rates of evolution that the fossil record reveals, and to evidence that intermediate stages between ancestral and descendant morphology can be found, whether one compares closely related species or higher taxa such as the classes of vertebrates. This chapter presents case studies in much greater detail than most readers will want to remember. The details are intended to provide information for anyone who may have to convince a skeptic that the fossil record does indeed provide abundant evidence for evolution.

Chapter 7 provides an overview of the major events in the history of evolution, set against the background of changes in the earth itself. It includes a description of the origin, fate, and relationships of many of the major groups of organisms, and some portraits of life at various times in the past. This chapter serves two functions: it highlights important events in the history of life that anyone interested in biological diversity should know about, and it also serves as a capsule compendium, or sourcebook, of events in the history of major taxa. For this reason, Chapter 7 also provides much more detail than most readers will need to remember.

Systematics and paleontology are important foundations of the study of biogeography, the geographic distribution of taxa. Geographic patterns, as we shall see in Chapter 8, were among Darwin's most important sources of evidence for evolution. Understanding why taxa have the geographic distributions they do is a challenging task, to which not only systematics and paleontology, but also ecology, contributes. Chapter 8 describes some of the methods of biogeographic study and some of the more interesting patterns of distribution.

CHAPTER 5

The Tree of Life: Classification and Phylogeny

Nobody knows how many species of organisms there are. In fact, no one has even made a catalogue of all the species that have already been named. A rough estimate of the number of described species is 1.4 to 1.8 million (May 1990; Wilson 1992), of which arthropods (especially insects), molluscs, and vascular plants account for more than 80 percent (Figure 5.1). It is not unusual for taxonomists to discover new species of insects even in well-studied regions, such as the environs of New York or London; and it is certain that several million insect species in the tropics remain to be distinguished, or even seen. Only a small fraction of the diverse cryptic organisms (including bacteria, fungi, nematodes, and mites) has been described, and it cannot be doubted that these groups contain hundreds of thousands of species, if not more. Very indirect calculations suggest that at least 3–5 million, and possibly as many as 30 million, species exist (May 1990). Paleontologists agree that the number of living species is a small fraction, probably less than 1 percent, of all the species that have ever lived.

The task of describing, naming, and classifying these myriad species is part of the discipline of *systematics*, which provides a foundation for all biological study. Since Darwin's time, systematists have increasingly sought not only to classify species, but also to discover the evolutionary relationships among them. Systematics is one of the most important and active areas of evolutionary biology, and it underlies many other kinds of studies in evolutionary biology.

The Linnaean Classification System

In the early 1700s, European naturalists, familiar only with European organisms and with specimens brought back from other continents by explorers, had little inkling of how amazingly diverse the living world is. Nonetheless, they saw that a system of classification was needed for the thousands of species they did know of. Moreover, since they believed that God must have created all these species according to some ordered scheme, it was a work of devotion to discover "the plan of creation" by cataloging the works of the Creator and discovering a "natural," true, classification. Many schemes of classification, some highly mystical, were advanced (Mayr 1982), but all were superseded by the work of the Swedish botanist Carolus Linnaeus (1707–1778). In the *Systema Naturae* (1735), Linnaeus listed the known species of his day and provided a classification scheme that made him a celebrity throughout Europe.

First, Linnaeus introduced BINOMIAL NOMENCLATURE, a system of two-part names (genus and species) that replaced the cumbersome polynomials naturalists had used before. For example, the honeybee, previously named *Apis pubescens, thorace subgriseo, abdomine fusco, pedibus posticis glabris utrinque margine ciliatis*, became simply *Apis mellifera*. Second, he replaced earlier systems of grouping species with the HIERARCHICAL CLASSIFICATION (groups nested within groups) that is still used today. (The familiar Linnaean levels of kingdom, phylum, class, order, family, genus, and species are now supplemented by numerous intercalated levels such as suborder, superfamily, subfamily, tribe, subgenus, and subspecies). Each of these levels is today referred to as a taxonomic **category**, whereas a particular group of organisms assigned to a categorical rank is a **taxon**. Thus monkeys in the genus *Macaca*, in the family Cercopithecidae, in the order Primates are taxa that exemplify the categories genus, family, and order. In assigning species to **higher taxa** (those above the species level), Linnaeus ignored characteristics that he considered trivial and used "important" features that he took to represent—what? *Not* evolutionary relationships, for he believed in the separate creation of species, but rather propinquity in God's creative scheme. For example, he defined the order Primates by the features "four parallel upper front [incisor]

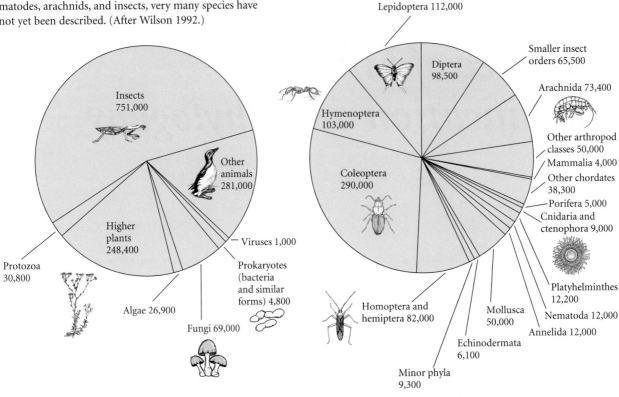

FIGURE 5.1 Estimates of the numbers of described species in the major groups of (A) all organisms and (B) animals. In many groups, such as the prokaryotes, fungi, nematodes, arachnids, and insects, very many species have not yet been described. (After Wilson 1992.)

teeth; two pectoral nipples," and placed in the primates the genera *Homo*, *Simia* (all apes and monkeys), *Lemur*, and *Vespertilio*—the bats! We still retain some of Linnaeus's assignments, but many have been altered; for example, monkeys and apes are now assigned to several families and many genera, and bats are placed in a separate order. (Box 5.A describes some of the rules that have been developed since Linnaeus's time for naming species and higher taxa.)

Evolution and Classification

Classification took on an entirely different significance after *The Origin of Species* was published. Darwin, convinced that species diverge gradually and to an indefinitely great extent from common ancestors, described a hypothetical **phylogenetic tree** (Figure 5.2)—the only illustration in *The Origin*—in a typically flowery Victorian simile:

> The affinities of all the beings of the same class have sometimes been represented by a great tree. I believe this simile largely speaks the truth. The green and budding twigs may represent existing species; and those produced during former years may represent the long succession of extinct species. At each period of growth all the growing twigs have tried to branch out on all sides, and to overtop and kill the surrounding twigs and branches, in the same manner as species and groups of species have at all times

overmastered other species in the great battle for life. The limbs divided into great branches, and these into lesser and lesser branches, were themselves once, when the tree was young, budding twigs; and this connection of the former and present buds by ramifying branches may well represent the classification of all extinct and living species in groups subordinate to groups. Of the many twigs which flourished when the tree was a mere bush, only two or three, now grown into great branches, yet survive and bear the other branches; so with the species which lived during long-past geological periods, very few have left living and modified descendants. From the first growth of the tree, many a limb and branch has decayed and dropped off; and these fallen branches of various sizes may represent those whole orders, families and genera which have now no living representatives, and which are known to us only in a fossil state. As we here and there see a thin, straggling branch springing from a fork low down in a tree, and which by some chance has been favored and is still alive on its summit, so we occasionally see an animal like the Ornithorhynchus or Lepidosiren,* which in some small degree connects

---

Ornithorhynchus, the duck-billed platypus, is a primitive, egg-laying mammal. *Lepidosiren* is a genus of living lungfishes, thought by Darwin to be closely related to the ancestor of tetrapod (four-legged) vertebrates.

FIGURE 5.2 Darwin's representation of hypothetical phyloge- nal lineages become extinct. Distance along the horizontal axis netic relationships, showing how lineages diverge from com- represents degree of divergence (perhaps in body form). Dar- mon ancestors and give rise to both extinct and extant species win recognized that rates of evolution vary greatly, showing this (at top of diagram). Time intervals (between Roman numerals) by different angles in the diagram; for instance, the lineage represent thousands of generations. Darwin omitted the details from ancestor F has survived essentially unchanged. (After of branching for intervals X through XIV. Extant species (at Darwin 1859.) time XIV) can be traced to ancestors A, F, and I; all other origi-

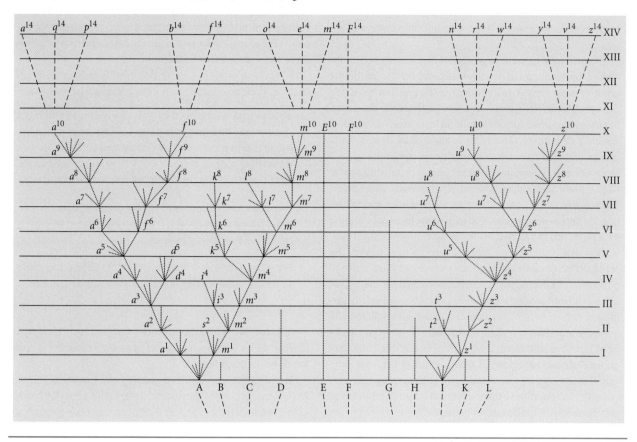

by its affinities two large branches of life, and which has apparently been saved from fatal competition by having inhabited a protected station. As buds give rise by growth to fresh buds, and these, if vigorous, branch out and overtop on all sides many a feebler branch, so by generation I believe it has been with the great Tree of Life, which fills with its dead and broken branches the crust of the earth, and covers the surface with its ever-branching and beautiful ramifications.

Under Darwin's hypothesis of common descent, species in the same genus are similar because they stem from a recent common ancestor; genera in the same family share fewer characteristics because each has departed further from their more remote common ancestor; families in an order stem from a still more remote ancestor and retain still fewer characteristics in common. A hierarchical classification reflects not a mystical ordering of the universe, but a real historical process that has produced organisms with true genealogical relationships, close or distant in varying degrees. Moreover, the "same" structure in different organisms,

such as the tetrapod limb, is the "same" not because it was created according to the same blueprint, but because it was inherited from these organisms' common ancestor (it is **homologous**, in the modern sense of the word). Classification, then, can portray, to some degree, *the real history of evolution*—and this has been a goal of many taxonomists ever since Darwin.

What does it mean for classification to portray or reflect evolution? Under Darwin's hypothesis, evolution has two major features: *branching* of a lineage into two or more descendant lines (= **cladogenesis**), and evolutionary *change* of various characteristics in each of the descendants (= **anagenesis** within each descendant lineage). Some of these changes, such as the evolution of the bird wing from a reptilian forelimb, are particularly striking and adaptively important. Traditional classifications often have been constructed in an attempt to convey both aspects of evolution (insofar as they can be inferred from the available evidence). Let us consider, for example, the taxa skate, shark, whale, baboon, orangutan, chimpanzee, and human; Figure 5.3 portrays the generally understood genealogical (phylo-

Taxonomic Practice and Nomenclature

Taxonomy, the naming and classification of organisms, has evolved rules of procedure, many of which are spelled out legalistically in the International Codes of Botanical Nomenclature and of Zoological Nomenclature. Standardized names for organisms are essential for communication among scientists.

Anyone may name a new species, but most species are named by taxonomists who are experts on the particular group of organisms. A new species may be one that has never been seen before (e.g., organisms dredged from the deep sea), but many unnamed species are sitting in museum collections, awaiting description. (This is especially true of insects, many groups of which have never been studied comprehensively.) Moreover, a single species often proves, on closer study, to be two or more very similar species. A taxonomist who undertakes a REVISION, a comprehensive analysis, of a group frequently names new species. A species name has legal standing if it is published in a journal, or even in a privately produced publication that is publicly available. Numerous rules govern the construction of names (e.g., genus and species must agree in gender: *Rattus norvegicus*, not *Rattus norvegica*, for the brown rat). It is recommended that the name have meaning [e.g., *Vermivora* (worm-eater) *chrysoptera* (golden-winged) for the golden-winged warbler; *Rana warschewitschii*, Warschewitsch's frog], but nonsense words are permissible (e.g., the kingfisher genus *Dacelo*, an anagram of *Alcedo*, which is Latin for kingfisher). It is considered very poor form to name a species after yourself, but very often a taxonomist will honor another biologist by naming a species after him or her.

The name of a species consists of its genus and its specific epithet; both are Latin or latinized words. These are *always* italicized (or underlined), and the genus is always capitalized. (In some names of plants, but not of animals, the specific epithet is also capitalized.) In entomology and certain other fields, it is customary to include the name of the AUTHOR (the person who conferred the specific epithet); for example, *Diabrotica virgifera* LeConte, an agriculturally destructive beetle (corn rootworm).

The first fundamental rule of nomenclature is that no two species of animals, or of plants, can bear the same name. (It is permissible, though, for the same genus name to be applied to a plant and an animal; for example, *Alsophila* is the name of both a fern and a moth.) The second fundamental rule is the rule of PRIORITY: the valid name of a taxon is the oldest available name that has been applied to it. Thus it sometimes happens that two authors independently describe the same species under different names; in this case, the valid name is the older one, and the younger name is a junior SYNONYM. Conversely, it may turn out that two or more species have masqueraded under one name; in this case, the name is applied to the species that the author used in his or her description. To prevent the obvious ambiguity that may thus

arise, it has become standard practice for the author to designate a single specimen (the TYPE SPECIMEN, or HOLOTYPE) as the "name-bearer" so that later workers can determine which of several similar species rightfully bears the name. The holotype, usually accompanied by other specimens (PARATYPES) that exemplify the range of variation, is deposited and carefully preserved in a museum or herbarium.

In revising a genus, a taxonomist usually introduces changes in the taxonomy. Some examples of such changes are:

1. Species that were placed by previous authors in different genera may be brought together in the same genus because they are shown to be closely related. These species retain their specific epithets, but if they are shifted to a different genus, the author's name is written in parentheses.

2. A species may be removed from a genus and placed in a different genus because it is determined not to be closely related to the other members.

3. Forms originally described as different species may prove to be the same species, and be synonymized.

4. New species may be described.

It is sometimes frustrating to non-taxonomists to find that familiar names have been replaced by new ones, but this is usually done for good reason.

These points can be illustrated by the history of North American leaf beetles (Chrysomelidae) in the genus *Ophraella*. In 1764, Müller placed some European species in the genus *Galeruca*. In 1801, Fabricius named several American species, placing them in the same genus: *Galeruca americana* Fabricius, *G. notata* Fabricius, and *G. notulata* Fabricius. Later, several other American species were described, such as *Galeruca integra* LeConte in 1865. In 1873, Crotch divided *Galeruca* into several quite distinct genera, including *Galerucella*, and in 1893 Horn transferred the American species to that genus: thus *Galeruca americana* Fabricius became *Galerucella americana* (Fabricius) (note the parentheses). In 1965, Wilcox decided that the American species that are characterized by stripes, a cocoon, and some other features should be distinguished from other species of *Galerucella*, and named a new genus, *Ophraella*, to hold them. Thus we had *Ophraella americana* (Fabricius), *O. integra* (LeConte), and so on. In 1979, White studied a tropical beetle described as *Trirhabda dilatipennis* by Jacoby in 1886, and placed it in *Ophraella*.

In 1986, LeSage published a revision of *Ophraella*. After studying specimens in many museums, including Fabricius's type specimens in Paris, he introduced several changes: First, *Ophraella dilatipennis* (Jacoby) is not related to other *Ophraella* species, but belongs in *Neolochmaea*, a tropical genus. Second, two species, one with a long stripe and one with a short stripe, had been confused. Everyone had thought that *notulata* Fabricius referred to the short-striped species, but LeSage found that Fabricius's type specimen has the characteristics of the long-striped species, which LeConte had named *integra* in 1865. Thus the long-striped species must be called *Ophraella notulata* (Fabricius), and the name *integra* LeConte is a synonym, no longer to be used. This left the short-striped form (which everyone had been calling *O. notulata*) without a name, so LeSage gave it a new name: *Ophraella communa* LeSage. Third, LeSage discovered museum specimens that differed from the previously described species, and named several new species, such as *Ophraella arctica*. One of his new species, *Ophraella macrovittata* LeSage, resembles *O. sexvittata* (LeConte), but has black rather than striped wing covers. In the course of my studies of *Ophraella*, I found (Futuyma 1990) that female beetles can produce both black and striped offspring in a single brood, and concluded that *O. macrovittata* LeSage is a junior synonym of *O. sexvittata* (LeConte): they are the same species.

The rules for naming higher taxa are not all as strict as those for species and genera. In zoology (and increasingly in botany), names of subfamilies, families, and sometimes orders are formed from the stem of the type genus (the first genus described). Most family names of plants end in -aceae. In zoology, subfamily names end in -inae and family names end in -idae. Thus *Mus* (from the Latin *mus, muris* for mouse), the genus of the house mouse, is the type genus of the family Muridae and the subfamily Murinae; *Rosa* (rose) is the type genus of the family Rosaceae. Endings for categories above family are standardized in some, but not all, groups (e.g., -formes for orders of birds, such as Passeriformes, the perching birds that include *Passer*, the house sparrows). Names of taxa above the genus level are not italicized, but are always capitalized. (Adjectives or colloquial nouns formed from these names are not capitalized: thus we may refer to murids or to murid rodents, without capitalization.)

genetic) relationships among these forms. All except the skate and shark are placed in class Mammalia. Mammalia is a **monophyletic group**, meaning that all its members are believed to stem from a single common ancestor. (As used in this book, a *monophyletic group* includes a given ancestor and *all* its descendants.) This classification therefore reflects the branching history (genealogical relationships) of evolution: all mammals share a more recent common ancestor with one another than they do with sharks; that is, mammal species are more closely related to one another than to sharks. We will shortly consider the kind of evidence used to reach such conclusions.

Now consider the primate species. Orangutan, chimpanzee, and human are placed in the monophyletic superfamily Hominoidea; all share a more recent common ancestor than any does with the baboon. However, orangutan and chimpanzee are traditionally placed in the family Pongidae, whereas humans have traditionally been placed in their own family, Hominidae, even though human and chimp share a more recent common ancestor than chimp and orangutan. Those who favor this classification would argue that, granted the validity of the phylogenetic tree, humans have undergone much more extensive ADAPTIVE DIVERGENCE from the common ancestor of the Hominoidea than has either of the apes, and that this aspect of evolution should be granted recognition by the classification. If the hominids (one living species and several extinct species) are placed in their own family, the family Pongidae is a **paraphyletic group**: a group that is monophyletic except that some descendants (namely, humans) of the common ancestor have been removed. A classification containing such a paraphyletic group does not fully convey information on phylogeny: it does not indicate that humans and chimps are more closely related than chimps and orangutans.

Suppose further that we were to classify whales with sharks as one class, based on their similar shape and aquatic habitats. The class of whales and sharks would have been defined by similar characteristics (e.g., streamlined shape, dorsal "fin") that actually evolved independently (an instance of *convergent evolution*, which we will discuss

FIGURE 5.3 The phylogeny of certain vertebrates, as presently understood. The vertical axis represents time, from past (bottom) to present (top). The diagram is intended to portray only the *relative order of branching* of these lineages from their common ancestors, which are the "nodes" from which lineages diverge.

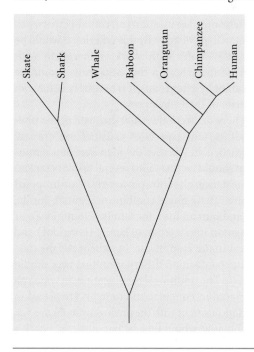

below). Such a taxon, consisting of unrelated lineages, each more closely related to other species not placed in the taxon (e.g., skates, primates), is **polyphyletic**. Many of the higher taxa named by Linnaeus were polyphyletic (e.g., "Vermes," an assortment of unrelated "worms"). Modern taxonomists have generally agreed to eliminate polyphyletic taxa from classifications, and most (but not all) also agree that taxa should not be paraphyletic.

Estimating a phylogeny and constructing a classification of organisms are not the same thing. There is only one true history of life, one true history of divergence of species from common ancestors. A phylogenetic tree is an estimate of part of this history—a hypothesis of what the history has been. Our hypothesis may be true or false; we hope that evidence will lead us closer to the truth. A classification, on the other hand, is a system for conveying or retrieving information and, depending on the uses to which that information is put, different systems of classification might be preferred. A library might classify books by subject matter, by the language in which they are written, or by many other possible criteria. By convention, books are usually classified by subject because this is generally most useful for retrieving information. Likewise, we might classify organisms by their phylogenetic relationships, by their appearance, alphabetically by name, or by many other criteria. Ecologists sometimes find it useful, for example, to classify plants by their growth form (tree, herb, etc.). The question of what criterion to use for classification is not a matter of truth or falsehood, but of what the community of users will find most useful.

Increasingly, biologists have adopted the view that the most useful criterion for classifying organisms is phylogenetic relationship—i.e., common ancestry—because this relationship generally conveys the most information about the characteristics (known and yet to be described) of the members of a taxon. Most systematists therefore agree that the taxa in classifications should be monophyletic groups. Given this criterion, a classification can contain errors, i.e., be inconsistent with phylogeny. By this criterion it would be wrong to classify whales with sharks rather than with mammals. Given the criterion that organisms should be classified in monophyletic groups, it is necessary to have an accurate estimate of their phylogeny.

Even if we have an accurate (true) estimate of the phylogeny, though, certain aspects of classification are still arbitrary: in particular, whether the members of a monophyletic group should be SPLIT into several taxa or LUMPED into one. For instance, given the phylogeny of anthropoids shown in Figure 5.3, we might combine orangutan, gorilla, chimpanzee, and human into one family (Hominidae), or split the orangutan into a separate family (Pongidae) and place the others in the Hominidae. In either case, the classification would conform to the criterion that taxa should be monophyletic. The traditional classification, in which all the apes are placed in Pongidae and humans are placed in Hominidae, is *inconsistent* with the phylogeny in Figure 5.3, because the Pongidae would then be paraphyletic. For this reason, some authors have proposed that the gorilla, chim-

panzee, and human should be placed in a separate family (Hominidae) from the orangutan, the sole member of a redefined Pongidae.

Before the 1950s, taxonomists attempted to construct classifications based on the dual criteria of common ancestry and adaptive divergence. They also proposed phylogenies, but most taxonomists did not articulate very clearly the criteria and logic they used. Some characteristics, they argued, were more "important" than others. "Important" had two meanings: indicative of common ancestry, and indicative of adaptive divergence. The mammary glands and jaw structure of a whale (as well as many other characteristics) were more "important" than its shape in showing that whales are more closely related to other mammals than to fishes. The brain and bipedal posture of humans were "important" aspects of adaptive divergence, which taxonomists felt justified a separate family. Some taxonomists argued that adaptive characteristics were the best indicators of evolutionary relationships. Others said precisely the opposite—that nonadaptive features (or to be more exact, features for which no adaptive function was known) would be best, because they would be less prone to lead us astray by convergent evolution. Determining which characteristics are convergent and which are indicative of common ancestry is not always easy, to say the least—and so taxonomists argued about which characteristics, such as limb muscles or breastbone features in birds, might provide the most reliable guides to relationships, with different authors championing different views of relationships, depending on which characteristics they thought most informative.

Beginning in the 1950s, explicit principles for evaluating evidence for relationships were developed. In reality, many taxonomists had been using these principles all along—they just had not explained what they were doing very well. Systematists comparing the morphologies of living organisms in an attempt to construct evolutionary classifications, together with paleontologists who were doing the same thing with fossils, had uncovered many of the relationships among organisms long before this time, and in doing so they described many important principles of evolution. Most of these principles were already apparent to Darwin but were further documented by systematic research.

Phenetics and Cladistics

Phenetics

In the 1950s, two new approaches to systematics arose. "Numerical taxonomy" was introduced by Charles Michener (an expert on bee systematics) and Robert Sokal (a statistics-minded population biologist), who argued that classification would be more rigorous if it were based not on a few characters that the taxonomist subjectively felt were "important," but rather on the degree of *overall similarity* of species, based on as many features as possible (Michener and Sokal 1957). They used numerical algorithms to create diagrams of overall similarity among spe-

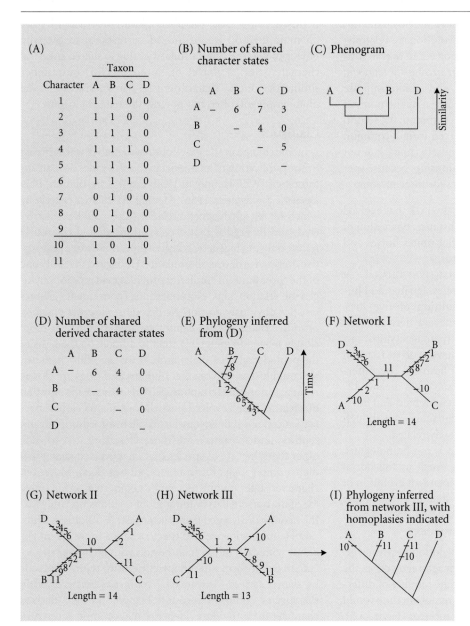

FIGURE 5.4 Illustration, using hypothetical data, of construction of a phenogram and of a phylogenetic tree. (A) Hypothesized data set of 4 taxa and 11 characters, each with states 0 and 1. (B) Matrix of shared character states (i.e., similarity of pairs of taxa) (using characters 1–9 only). (C) A phenogram calculated from the matrix of shared character states. Taxa joined at higher levels are increasingly similar. (D) Matrix of *derived* character states shared by pairs of taxa (using characters 1–9 only), if taxon D is an outgroup. (E) Phylogeny calculated from the matrix in (D), with numbered marks showing where each character evolved a change in state. (F, G, H) Three undirected networks, with locations of changes in all 11 characters in Table A, minimizing the number of character changes. Networks I and II require three characters to change twice, whereas network III requires only characters 10 and 11 to change twice, and thus is the shortest network. (I) The best estimate of the phylogeny, namely, network III, rooted between the outgroup (taxon D) and the common ancestor of the other three taxa. The homoplasious changes in characters 10 and 11 are indicated. Note that the best estimate of the phylogeny, based on shared derived character states, differs from the phenogram (C), which is based on all shared character states.

cies. Such a diagram, or PHENOGRAM (Figure 5.4C), was intended as an objective basis for classification, and their approach came to be known as **phenetics.**

In systematic studies, a **character** is a feature that is thought to vary independently of other features and to be homologous among the taxa of interest (i.e., derived from a corresponding feature of the common ancestor of the taxa). A **character state** is one of the alternative conditions of the character. Thus, "present" and "absent" are two states of the character "thumb" among primates. Likewise, a particular position in a DNA sequence is a character, the possible states of which are A, T, C, and G.

To illustrate phenetic analysis, suppose that we examine nine characters of each of four species (A–D in Figure 5.4A), and that each character is coded as having one of two character states, 0 or 1 (e.g., character 1 of an insect might

be "wings: absent or present," character 2 "mouthparts: chewing vs. sucking"). For each pair of species, we tally how many character states they share (Figure 5.4B). In the most frequently used phenetic algorithm, we then join the pair of species that share the most characters (A and C in our example). We then see which other species is most similar either to this pair or to another unjoined species. In our example, the similarity between B and D is 0, and the similarity between B and the A + C pair is the mean of B/A and B/C, or (6 + 4)/2 = 5. The greater similarity links B to the A + C pair, and the remaining species, D, is then linked to this trio. In a phenogram, the higher the level at which species join, the more similar they are. The pheneticists might then argue that A and C represent, say, a genus, B a different genus in the same subfamily, and D a different family.

It must be emphasized that Michener and Sokal did not intend a phenogram to represent evolutionary relationships, but instead to convey information about overall similarity. Indeed, *a phenogram does not necessarily represent phylogenetic relationships.* Similarity and relationship are two different concepts. The first is defined by the number of character states two species have in common; the second is defined by how recently they diverged from a common ancestor, relative to other species. Similarity, in itself, is not a reliable guide to phylogenetic relationship, because species may share similar character states for three reasons:

1. They share a DERIVED CHARACTER STATE that evolved in their common ancestor. For instance, possession of the placenta is such a feature that unites horses and humans, as compared with lizards.

2. They share an ANCESTRAL (PRIMITIVE) CHARACTER STATE relative to some other species. Iguanas and humans share the primitive condition of five toes, whereas horses, with one toe, have a derived state. With reference to the five-toed (pentadactylous) ancestor of the tetrapods, horses have evolved faster in this feature than have iguanas or humans. This difference in the rate of evolution results in a greater difference (in this character) between humans and horses than between humans and iguanas, even though humans and horses are more closely related.

3. They share a similar feature that is not homologous, but has evolved independently (such as the dorsal "fin" of sharks and whales). Independent evolution of a similar or identical character state in two or more evolutionary lineages is called HOMOPLASY, a general term that includes convergent evolution.

If there were no homoplasy (i.e., if evolution were only divergent, not convergent), and if lineages evolved at the same, constant rate, then the greater the number of differences between two species, the more remote in time would be their divergence from a common ancestor (Figure 5.5).

If this were the case, the phenogram constructed by the phenetic method would indeed correspond to the true phylogenetic tree. But if similarities due to shared ancestral characteristics (2 above) or homoplasy (3) outnumber similarities due to shared derived characteristics (1), the phenogram will often not portray the true phylogeny.

Cladistics

A major change in the practice of systematics was initiated by the publication of an important book by the German entomologist Willi Hennig in 1950 (English translation, *Phylogenetic Systematics*, 1966). Hennig articulated criteria by which he felt phylogenetic relationships could be discovered, and he argued that classifications should rigorously reflect only phylogenetic relationships, not degree of adaptive divergence or overall similarity. He argued that only one of the three kinds of similarity provides evidence for phylogenetic relationships. Homoplasious (convergent) characteristics obviously do not. *Neither do similarities based on primitive characters:* we cannot say that a human is more closely related to an iguana than to a horse on the basis of the pentadactyl (five-digit) limb. If we used this character, we could just as well say that humans and frogs are closely related. Among the tetrapods, pentadactyly means only that this feature hasn't evolved since the organisms' earliest five-toed ancestor—it doesn't mean that all five-toed amphibians, reptiles, and mammals are more closely related to each other than they are to species that *have* evolved changes in digit number (e.g., birds, horses, guinea pigs). Moreover, character states unique to a single taxon, such as the single toe of the horse, tell us nothing about relationships, only that the taxon has diverged from the ancestral condition. Thus, said Hennig, evidence that species share a more recent ancestor with each other than they do with any other species (i.e., that they form a monophyletic group) is provided only by *shared derived* ("advanced") *characters* that evolved in the species' common ancestor. The placenta is a derived character that provides evidence of the common ancestry of

FIGURE 5.5 (A) A hypothetical phylogeny in which evolution occurs at a nearly constant rate. Each tick mark represents the evolution of a new character state. (B) With nearly constant rates of divergence, the relative times of divergence of lineages, and therefore their phylogeny, can be determined from the overall difference (or similarity) between taxa.

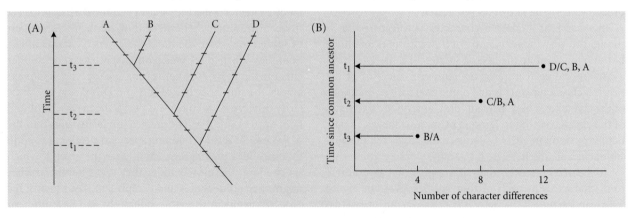

horses, humans, and other eutherian mammals; but the primitive character state (lack of a placenta) does *not* tell us that animals without a placenta (birds, reptiles, fishes—and, for that matter, insects and sponges) are more closely related to each other than they are to mammals (as indeed they aren't). Further, a shared derived character provides evidence that the species are a monophyletic group only if it is *uniquely derived*—i.e., it evolved only once. If it has evolved more than once (homoplasy), we risk mistaking unrelated species for a monophyletic group.

Hennig's principle, therefore, is that monophyletic groups are defined by *shared, uniquely derived character states*. The branching pattern of evolution (the phylogenetic tree) can be found by identifying monophyletic groups nested within more inclusive monophyletic groups; at each level, monophyly is indicated by certain shared derived characters. Thus the amnion is a uniquely derived character marking the monophyletic superclass Amniota (reptiles, birds, and mammals); within amniotes, a lower jaw consisting only of the dentary bone marks the monophyletic Mammalia; within mammals, wings and ever-growing incisors are derived characters uniting certain species into the orders Chiroptera (bats) and Rodentia (rodents), respectively.

If phylogeny can be inferred in this way, said Hennig, it should be reflected in classifications, which should consist only of monophyletic groups. Thus Hennig and his followers would abolish the class Reptilia because it is not monophyletic: it excludes birds (Aves), which are more closely related to crocodilians than crocodilians are to turtles or lizards. That is, birds and crocodilians (as well as dinosaurs and many other extinct "reptiles") share a more recent ancestor with each other than any of them do with turtles (Figure 5.6). Hennig therefore (1) proposed a *method* for inferring the true branching pattern of evolutionary history and (2) presented an *opinion* on criteria for classification. These two very different proposals have together come to be known as **cladistics**, and followers of the Hennigian tradition as CLADISTS. Branching diagrams constructed by cladistic methods are sometimes referred to as CLADOGRAMS, and monophyletic groups are called CLADES. Most systematists accept the method of inferring phylogeny, but not all subscribe to the cladistic philosophy of classification.

Hennig introduced several mouth-watering Greek terms that are now commonly used in systematics:

PLESIOMORPHY (adjective, plesiomorphic): An ancestral (primitive) character state (with reference to another, derived, state).

SYMPLESIOMORPHY: An ancestral character state shared by two or more taxa.

APOMORPHY: A derived ("advanced") character state (with reference to another, ancestral, state).

SYNAPOMORPHY: A derived character state shared by two or more taxa.

AUTAPOMORPHY: A derived character state possessed by only one of the taxa under consideration.

Some other terms commonly used in phylogenetic study are:

COMMON ANCESTOR: A species that at some past time split into two (or more) species, each of which gave rise to one of the clades under discussion. Because all organisms are believed to have descended from one remote common ancestor (see Chapter 7), any two taxa have had a common ancestor at some time in the past. (For example, even though no amniote vaguely resembling either a human or a snake existed 300 million years ago, this is the minimal estimate of the time since these lineages diverged; the synapsid reptiles, from which mammals evolved, were by this time distinct from the diapsid lineage that much later gave rise to snakes.)

NODE: A branch point in a phylogenetic tree. It represents a common ancestor at the time of divergence into two or more lineages.

TERMINAL TAXA: Taxa at the tips of a phylogenetic tree, such as whale and human in Figure 5.3. The term *operational taxonomic unit* (OTU) is approximately synonymous.

SISTER GROUPS (or SISTER TAXA): Two groups with the same immediate common ancestor. (Either group may contain one or more than one species.)

STEM GROUP and CROWN GROUP (used especially in discussing taxa with a fossil record): A stem group is an ancestral group, with relatively primitive characteristics, from which a crown group with relatively "advanced" characteristics has evolved. Often the crown group is extant and the stem group is extinct.

GROUND PLAN: The set of character states typical of relatively unmodified members of a clade. For example, the ground plan of mammals includes teeth differentiated into incisors, canines, premolars, and molars. Some mammals, such as anteaters and whales, have evolved away from this ground plan, having homogeneous teeth or none.

A hypothetical example Consider again the hypothetical data in Figure 5.4. Let us take taxa A, B, and C, and assume that two of them share a most recent common ancestor; the problem is to discover which two (there are three possibilities). Ignore characters 10 and 11 for now. Suppose that we somehow had evidence that for each character, 1–9, state 0 is plesiomorphic (ancestral) and state 1 apomorphic (derived). If for each pair of species we count the number not of shared character states overall (i.e., shared 0's *or* shared 1's, as in the phenetic method), but of *shared derived* character states (shared 1's only), we find that A and B share the greater number—six (Figure 5.4D). Moreover, the four characters shared between each of these and C are a subset of those shared by A and B. Specifically, for characters 3–6, all three species share the derived character state, whereas only A and B share the apomorphic state of characters 1 and 2. Species C has retained the plesiomorphic state of these characters. Hennig's rule leads us to conclude (Figure 5.4E) that A and B are sister taxa (closest

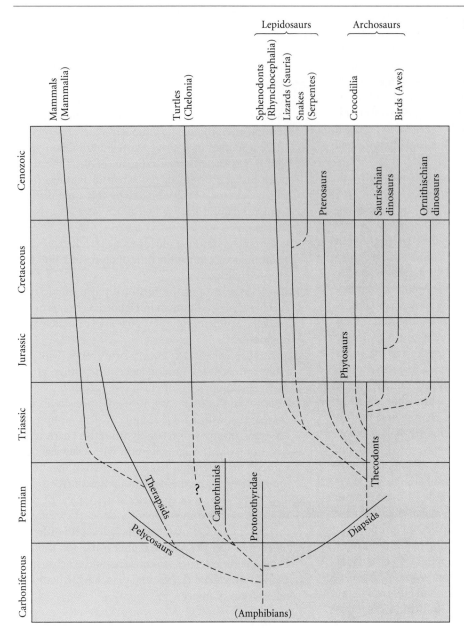

FIGURE 5.6 An oversimplified phylogeny of the amniotes, i.e., the "reptiles" and their derivatives. All the taxa are traditionally placed in the class Reptilia, except for those placed in the class Mammalia (which might include therapsids and pelycosaurs) and the class Aves (birds). Birds are most closely related to dinosaurs and, among living amniotes, to crocodilians. Segregation of birds into the class Aves makes the class Reptilia a paraphyletic taxon. In this diagram, solid lines indicate approximate distribution in the fossil record, and broken lines indicate postulated phylogenetic relationships. Many extinct groups are omitted. (After Romer 1966; Carroll 1982; Benton 1983, 1985).

relatives). This conclusion contrasts with the greater *similarity* of A and C (in the phenogram), which comes about because B has three autapomorphies (characters 7–9), whereas A and C retain the plesiomorphic state of these characters. Remember that neither autapomorphies nor symplesiomorphies provide evidence of immediate common ancestry.

Parsimony and phylogenetic inference Hennig's principle, that uniquely derived character states mark monophyletic groups, poses two difficulties: (1) How can we tell which state of a character is derived? and (2) How can we tell if it is uniquely derived, rather than having evolved more than once? The methods devised by Hennig and his successors to deal with these very difficult problems depend on the concept of PARSIMONY. Parsimony is the principle, dating from

at least the fourteenth century, that an explanation requiring the fewest undocumented assumptions should be preferred over more complicated hypotheses that require more assumptions for which evidence is lacking. This principle was applied to phylogenetic analysis by Hennig (1966), the population geneticists Edwards and Cavalli-Sforza (1964), and the systematic theorists Camin and Sokal (1965) and Farris (1973). (See Sober 1988 for an extensive discussion.) Cladistic parsimony holds that among the various phylogenetic trees that can be imagined for a group of taxa, *the one that is best supported* by evidence is the one that *requires us to postulate the fewest evolutionary changes.* Suppose, for example, we postulate that whales and sharks form one monophyletic group because they have a dorsal fin, and that all the other creatures we call mammals form a different monophyletic group. This would require us to postulate (on the basis of no evidence) that a four-chambered heart, milk, a single aortic arch, and many other features possessed both by whales and by the other creatures we call mammals, each evolved twice (i.e., once in the whales and again in the ancestor of terrestrial mammals). In contrast, if whales are descended from the same ancestor as other mammals, each of these features evolved once, and only the dorsal fin (and a few other features such as body form) evolved twice. The "extra" evolutionary changes we must postulate in a proposed phylogeny are homoplasious changes—those that we must suppose occurred more than once. Thus cladistic parsimony holds that *the best phylogenetic hypothesis is the one that requires the fewest homoplasies*. Whether or not this principle is always a suitable guide is subject to some debate (Felsenstein 1983; Sober 1988). The principle of parsimony in phylogenetic study proposes that we tentatively accept the simplest hypothesis that accounts for the data—a principle that is used throughout science. (Throughout science, as we saw in Chapter 1, hypotheses are always tentatively accepted and may subsequently be rejected in light of new evidence.)

A consequence of the parsimony principle is that we can often infer which character state is plesiomorphic and which is apomorphic (i.e., we can infer the direction, or PO-LARITY, of evolution). This requires some preliminary phylogenetic information. For example, we know that horses and humans are part of a monophyletic group (Mammalia) that excludes frogs and lizards. If we postulated that the single toe of horses represents the primitive state for mammals, then we would have to suppose that either (1) the number was five in amphibians and reptiles, decreased to one in the ancestor of mammals, and reverted back to five in mammals other than horses; or (2) ancestral tetrapods had one digit, and five toes evolved independently in frogs, lizards, and some mammals. Derivation of the one-toed from a many (five)-toed condition, in contrast, implies only one evolutionary change, away from a primitive state found not only in most mammals, but also in more distantly related groups.

This reasoning leads to a general principle of phylogenetic inference: Among the states of a character found in the members of a monophyletic group, *the ancestral character state is the one that is most widely distributed among taxa outside this group* (unless there is contrary evidence). These latter taxa are called **outgroups**, and their features provide invaluable information for inferring relationships among the members of the monophyletic group under study (sometimes called the FOCAL GROUP or INGROUP).

Creating phylogenetic trees

Consider again the hypothetical data in Figure 5.4, this time including characters 10 and 11. We now include taxon D, an outgroup. Our task is to find the relationships among A, B, and C. We might assume that each character state in D is plesiomorphic (especially if D's features are found in other, more remotely related outgroups), but we need not make this assumption. For these four taxa, we can draw three possible "undirected networks" in which each branch point represents an ancestor that gave rise to two lineages (Figure 5.4F–H).

On each network, we plot the minimal number of changes in each character that would result in the observed distribution of character states among the taxa. In network III (Figure 5.4H), for instance, we parsimoniously propose that character 1 changed once, between the pair A + B and the pair C + D, since A and B have state 1 and C and D have state 0.

In contrast to our earlier consideration of just characters 1–9, the addition of characters 10 and 11 results in *conflicts among characters*. Thus, if, as earlier, we take characters 1 and 2 as evidence that A and B form a monophyletic group (network III), this conflicts with character 10, which would lead us to put A and C together (network II). If network III is correct, both characters 10 and 11 must have changed twice (hence a total of 13 changes for 11 characters). Either of the other networks, though, is "longer," requiring us to postulate homoplasious changes in three characters, for a total of 14 character changes. Under the principle of parsimony, network III is the best estimate. Now, since we know that taxon D is an outgroup, we transform the network into a phylogenetic tree by ROOTING it between D and the other taxa (Figure 5.4I). Our conclusion is that A and B share a more recent common ancestor with each other than with C. (In practice, we would want the shortest network to be considerably more than one step shorter than the closest alternative in order to have confidence in our conclusion.)

The number of possible phylogenetic trees to be evaluated grows astronomically with the number of taxa, so the procedures we have described are often carried out by any of several computer programs. Some of the program packages used frequently for this purpose are PAUP, PHYLIP, HENNIG86, and MacClade.

Some limitations of cladistic analysis

Phylogenetic analysis is easy with imaginary data, but in practice it is often difficult to resolve relationships among taxa (see Box 5.B for a more extensive discussion). The difficulties include:

1. Scoring characters. Two serious problems arise, especially in the case of morphological characters. First, how many independent characters are there? If some mammals have, on each side of each jaw, two incisors, one canine, three premolars, and four molars, and if other mammals (e.g., anteaters) have no teeth at all, does this represent a single character difference (loss of teeth), four differences (loss of four kinds of teeth), or ten differences? Second, the systematist relies on anatomical details to decide whether organisms have the same character state or not. This requires extensive knowledge, and is not an easy or trivial task. Even with considerable experience, determining which character states are homologous can be very difficult. Errors in these decisions may appear as extra homoplasies in the phylogenetic tree, and can yield mistakes in estimating the phylogeny. Moreover, the computer programs available deal well with discrete character states (red, white, or blue), but not with measurements that vary continuously, such as size.

Issues in Phylogenetic Analysis

In practice, phylogenetic inference is often much more difficult than the elementary treatment in this chapter indicates. Phylogenetic methods are the subject of a large, sophisticated, and often contentious literature. This box will provide some comprehension of this literature, but interested readers should see the fuller discussions by Swofford et al. (1996) and Maddison and Maddison (1992).

Data

The validity of a phylogenetic analysis depends on the quality (and quantity) of the data, and many decisions must be made at this stage. For morphological (and some molecular) data, *defining characters* may be a problem. For instance, two features might be so closely correlated that they should be considered a single character rather than two. Some characters may be missing for some taxa, either because they cannot be scored (a common problem with fossils) or because the taxa lack the characters (e.g., number of wing veins in wingless insects). Phylogenetic computer programs can handle missing characters, but the more complete the data, the better. Another problem, mentioned in the text, is that most programs require discrete character states. In order to utilize the information provided by continuously varying characters, many authors divide such characters into discrete states (e.g., short, medium, long), but determining how this should be done is a difficult problem.

Problems often arise in determining which characters are homologous among taxa. For instance, if a feature of the sepals varies among some plant taxa, it is necessary to determine which structures are sepals, which is sometimes difficult (e.g., for monocots). Identifying homologous characters among very distantly related taxa can be exceedingly difficult. DNA sequence data can also present serious problems of homology. First, gene duplication may have given rise to a family of PARALOGOUS genes (such as the various hemoglobin genes in a vertebrate genome). To infer relationships among taxa, we need to compare corresponding (ORTHOLOGOUS) genes (e.g., we should compare adult α globins, rather than comparing the adult α globin of one species with the fetal γ globin of another). But identifying orthologous genes may be difficult if the paralogous genes retain similar nucleotide sequences. The second problem is identifying homologous base pair positions in sequences from different taxa–i.e., ALIGNING the sequences correctly. Insertions and deletions in the sequence, differing among taxa, can make this difficult, as can homogeneous repeats (e.g., ATATA …) of variable length. If sequences of different taxa have diverged greatly, they may be so different that they cannot be aligned with confidence. Thus, in order to infer relationships among taxa that diverged very long ago, it is necessary to use slowly evolving genes that preserve considerable sequence similarity. Such genes, however, are likely to be useless for ascertaining relationships among very closely related taxa, because they will be too homogeneous to provide phylogenetic information. For analyzing relationships among very closely related organisms, rapidly evolving sequences are desirable.

A further problem is whether or not to *weight* characters differently. To weight them equally, i.e., to treat them the same, is to assume that they are equally informative. We know, however, that some characters are more likely to be homoplasious than others (e.g., those that evolve more rapidly, or are subject to rapid adaptive adjustment), and these should receive less weight because they are less likely to provide valid phylogenetic information. However, it is difficult to judge, a priori, which characters are most informative and what numerical weights to attach to them. In DNA sequence data, transitions are often found to outnumber transversions by a factor of about three. Because transversions are rarer, a particular transversion is presumably less likely to occur repeatedly in evolution than a particular transition. Thus relative weights of about 3:1 are often applied to transversions vis-à-vis transitions.

Methods for Estimating Phylogenetic Trees

Many algorithms have been proposed for estimating phylogenies (Swofford et al. 1996). Most assume a *branching* history and cannot accommodate reticulation, as when taxa arise by hybridization. The most widely used methods are *parsimony* methods, which hold the *shortest tree* to be the best estimate of the phylogeny. The length of a tree is the minimal number of character state changes, over all characters and all taxa, implied by that tree. The basic method of maximum parsimony is described in the main text. Another, increasingly popular, method is NEIGHBOR-JOINING (Saitou and Nei 1987). This method uses distances between pairs of taxa (e.g., numbers of nucleotide differences) rather than individual characters. It builds the shortest tree by methods too complicated to describe here. A third method, used mostly for sequence data, is known as MAXIMUM LIKELIHOOD (Felsenstein 1981). This method requires a concrete model of the evolution of the characters, involving many parameters such as the rate at which each nucleotide is replaced by each other nucleotide. (With some assumptions, these rates can be estimated from the data.) The method then evaluates the likelihood, for each possible phylogeny (including not only branching orders, but also branch lengths), that the evolutionary model will yield the observed characters (e.g., sequences). The tree with the highest likelihood is the best estimate of the true phylogeny.

Judging from computer simulations, each method is likely to work best for certain evolutionary dynamics. Since we generally do not know much about the dynamics of an evolutionary history, we might have the greatest confidence in our phylogenetic estimate if we were to obtain the same result by several methods (Kim 1993). We should note one important pitfall in the commonly used method of maximum parsimony: long branches attract. If characters have few possible states (e.g., A, T, C, G), high evolutionary rates in two or more distantly related lineages increase the likelihood of convergence. When the parsimony method is used, such lineages will seem more closely related than they actually are. Sometimes this problem can be solved by analyzing more taxa, some of which may "break up" the long branches.

Each method requires some assumptions about evolution. For example, in applying maximum parsimony, one needs to decide whether multiple (>2) character states are ordered or unordered. If the states are ordered, state 1 is assumed to evolve to state 3 via state 2, and at least two evolutionary changes separate taxa with states 1 and 3. If the states are unordered, we assume that any character state can evolve into any other in a single step. The computer program may also be told whether it should allow evolutionary reversals or not ("Dollo parsimony," derived from DOLLO'S LAW, which states that complex characters, once lost in evolution, are unlikely to be regained).

Evaluating Phylogenetic Trees

The number of possible phylogenetic trees for even a few taxa is very large (e.g., 34,459,425 possible trees for 11 taxa!). Programs employing maximum parsimony employ "search routines" that find many or all of the shorter trees. Several indices, called CONSISTENCY and RETENTION indices, measure how short a tree is relative to the shortest possible tree for a given data set. Thus one of the many trees evaluated by the program might have a consistency index (CI) of 0.33, indicating that this tree is three times as long as the shortest possible tree. Such a tree would imply that two-thirds of the character state changes are homoplasious, whereas another tree, with higher CI, would imply less homoplasy. In real data, whether morphological or molecular, the shortest tree often has a CI in the range of 0.6 to 0.7, which declines as the number of taxa studied increases (Sanderson and Donoghue 1989). (Autapomorphies, changes limited to a single taxon in the data set, should not be included in calculations of CI because they are phylogenetically uninformative and inflate CI, giving a false impression of how reliable the tree is.) The greater the level of homoplasy, the less confidence we can have in a phylogenetic estimate; thus CI is sometimes used to evaluate how much confidence we should have.

Another approach is to look at the distribution of the lengths of the various trees evaluated by the program. Frequently, several or many trees have the same minimal length, and several or many are only slightly longer than that. The several trees of equal minimal length can be "merged" to form any of several kinds of CONSENSUS TREES. In a *strict* consensus tree, for example, only those dichotomies are retained that are present in all the trees; taxa whose relationships to each other differ among the trees then form a multifurcation, such as a trichotomy (three taxa attached to a single internal node, or common ancestor). The relationships among those taxa are then said to be UNRESOLVED.

If the shortest tree has *x* steps, but 20 different trees all have *x* + 1 steps, we surely cannot have much confidence that the single shortest tree is the true phylogeny. Several approaches to evaluating phylogenetic estimates have been suggested. A widely used method, introduced by Joe Felsenstein (1985; see also Sanderson 1989), uses a technique developed by statisticians, called BOOTSTRAPPING. Throughout science, we ask whether our data could have been generated by chance; if not, we may tentatively accept one or more alternative hypotheses. If, for instance, a sample of infants consists of 800 males and 200 females, we can use a statistical test (e.g., a chi-square test), and confidently conclude that such a great deviation from the expected numbers (500 of each) could not have arisen by chance. In this case, probability theory (coupled with genetic principles that predict a 1:1 ratio on average) generates an EXPECTED DISTRIBUTION of outcomes, specifying the probability that the genetic process would yield a male:female ratio of 500:500, 501:499, . . . 1000:0.

Now, any data, even random numbers, will yield phylogenetic trees if we feed them into a phylogenetic inference program. We want to know whether our real data are effectively random, or are so organized that the estimated phylogeny could not have arisen by chance. However, no mathematical theories tell us what the distribution of possible trees would be if we had random data. Bootstrapping is an "end run" around this problem (and similar problems in other disciplines). We assume that if we had enough data—enough characters—we could determine the correct answer. We have only a sample of all possible data. Do our data reflect all possible data well enough to give us confidence in our estimate? The bootstrapping method operates by sampling data points (characters) from our data, one at a time, replacing each datum in the data pool, and resampling. This process, repeated, builds up a new data set in which some of the original data points (characters) are represented more than once, and others not at all. We then compute the statistic of interest (in our case, the minimum-length tree). We do this entire procedure repeatedly, perhaps 100 times. If a particular monophyletic group occurs consistently in most (perhaps 95 percent) of the trees, we can conclude with some confidence that our data support the hypothesis that these taxa are indeed a monophyletic group. The evidence for their monophyly is not sensitive to the deletion of some data, so the taxa must share enough synapomorphies to indicate their relationship nevertheless. The same argument is used to evaluate the confidence we can place in our best estimate of the phylogeny as a whole. Figure 5.14 provides an example of a tree with bootstrap values.

An estimate of a phylogeny can be no more reliable than the data that went into it.

2. Homoplasy turns out to be very common. For this reason, a data set may yield several different phylogenetic estimates that are equally good, (i.e., all imply the same number of evolutionary changes) or almost so. It is unwise to rely on a particular phylogenetic estimate if other phylogenetic hypotheses imply just a few extra evolutionary changes. Instead, one should try to get more data (i.e., on other characters).

3. The process of evolution may make it extremely difficult to determine relationships (Figure 5.7). If the taxa under study diverged very long ago, many of their characteristics will have diverged so greatly that homologous characters are extremely difficult to discern. This is a major reason why the relationships among animal phyla are poorly understood. One can only hope to find informative fossils, or to find characteristics that have evolved so slowly that homological similarities have not been erased by evolution. Another difficulty is posed by groups in which all the lineages diverged so rapidly (whether recently or not) that there was little opportunity for the ancestors of each monophyletic group to evolve distinctive synapomorphies. Such "bursts" of divergence of many lineages during a short period, or **evolutionary radiations**, are

often called **adaptive radiations** because the lineages have often acquired different adaptations. Many of the orders of placental mammals, for example, diverged within a rather short time, and their relationships to one another are not well known.

Examples of Phylogenetic Analysis

So far, we have encountered mostly hypothetical examples of phylogenetic analysis. For real examples, I have chosen a morphological phylogeny of higher taxa and a molecular phylogeny of closely related species.

Phylogeny of Orders of Insects

The hexapods (insects and other six-legged arthropods) illustrate both well-established relationships and phylogenetic uncertainties (Kristensen 1991; Figure 5.8). Based on characteristics of the numerous arthropods that are outgroups with respect to hexapods (e.g., trilobites, crustaceans, myriapods such as millipedes), systematists have inferred that the ancestor of hexapods lacked wings and had a pair of jointed appendages on each of most body segments. Several anterior pairs of appendages were modified into mouthparts; these included a pair of mandibles used for chewing, which articulated with the head by a single condyle. The segments bearing the mouthparts were more or less differentiated into a head; the rest of the body (the

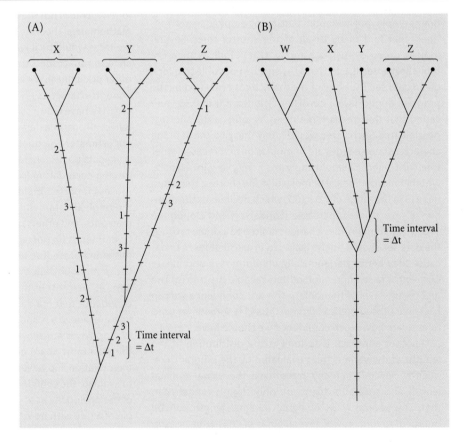

FIGURE 5.7 Two ways in which evolution can make phylogenetic inference difficult. (A) The passage of time. Taxa X, Y, and Z are clearly monophyletic groups, but the relationships among them are hard to determine because derived character states that evolved in the common ancestor of taxa Y and Z have undergone further change in the long time since taxa X, Y, and Z diverged. Tick marks indicate changes in characters 1, 2, and 3, each of which undergoes further changes that erase evidence of relationships. (B) Rapid evolutionary radiation. The relationships among taxa W, X, Y, and Z cannot be determined because they all diverged within such a short interval of time (Δt) that no derived character states evolved in the common ancestors of any subset of these taxa.

trunk) was composed of rather undifferentiated segments. The terminal appendages were not modified. All these features remain little changed in certain living myriapods (Figure 5.9A).

The major synapomorphy of the superclass Hexapoda is the differentiation of the trunk into a thorax of three segments, each with a pair of legs, and an abdomen. The appendages of the abdomen do not serve for walking, but are reduced, lost, or adapted for other functions. The several groups of hexapods include the springtails (Collembola) and the class Insecta. The synapomorphies of the Insecta, in contrast to the Collembola and other non-insect hexapods, include modification of some terminal abdominal appendages into genital structures. These have a primitive form in the order Microcoryphia (bristletails), the basal order of insects, but in "higher" insects they are highly

FIGURE 5.8   A recent estimate of the phylogeny of some of the orders of hexapods, with some synapomorphies indicated. Taxa with unresolved relationships to each other are joined by the same horizontal line (e.g., some orthopteroid orders). The horizontal dashed line indicates that relationships among beetles, wasps, and other holometabolous orders are uncertain. (See also Figure 7.21.) (After Kristensen 1991.)

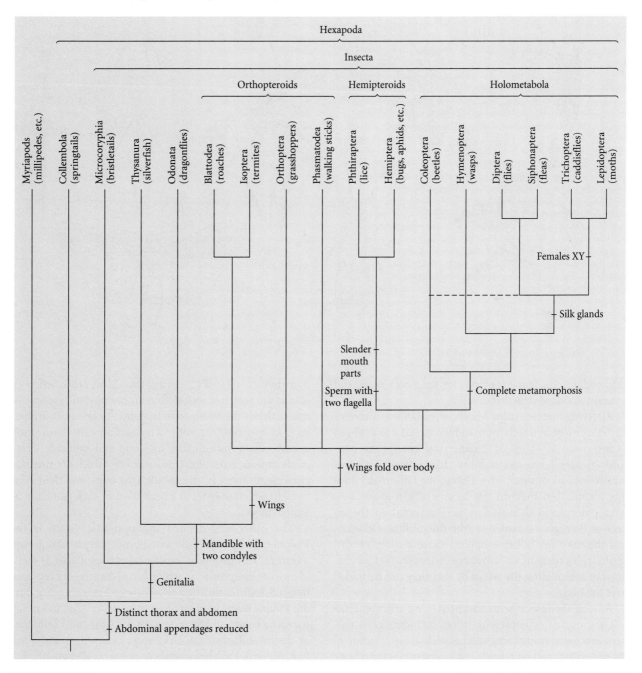

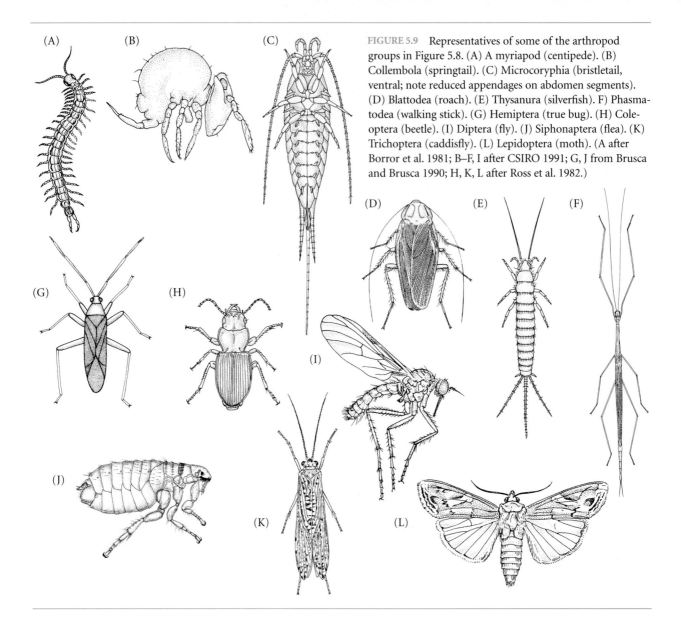

FIGURE 5.9 Representatives of some of the arthropod groups in Figure 5.8. (A) A myriapod (centipede). (B) Collembola (springtail). (C) Microcoryphia (bristletail, ventral; note reduced appendages on abdomen segments). (D) Blattodea (roach). (E) Thysanura (silverfish). F) Phasmatodea (walking stick). (G) Hemiptera (true bug). (H) Coleoptera (beetle). (I) Diptera (fly). (J) Siphonaptera (flea). (K) Trichoptera (caddisfly). (L) Lepidoptera (moth). (A after Borror et al. 1981; B–F, I after CSIRO 1991; G, J from Brusca and Brusca 1990; H, K, L after Ross et al. 1982.)

modified into the female's ovipositor, the male's penis, and several associated structures.

All living insects except Microcoryphia form a monophyletic group, in which the mandible attaches to the head by two condyles—a major "engineering" change that provides greater force and stability (like a door with two hinges instead of one). The Thysanura (silverfish) have this feature, but they do not have wings, a major new synapomorphy of all other insects (Pterygota). All the orders of Pterygota except two (the dragonflies, Odonata, and the mayflies, Ephemeroptera) share a complex synapomorphy, namely, an "advanced" (neopterous) mechanism of articulating the wings so that they can be folded over the body.

Among the insects with advanced wing articulation, a whole group of orthopteroid orders, including grasshoppers, roaches, termites, and walking sticks, have "primitive" characteristics and no reliable synapomorphies to indicate

how they are related to one another. Most other orders of insects fall into two well-defined monophyletic groups. In one of these (the hemipteroids, united by synapomorphies such as spermatozoa with two flagella rather than one), certain of the mouthparts are long and pointed. These *grade in form*, from those of some lice, which are used for chewing, to those of the aphids, true bugs, and their relatives, in which they form a beak used to suck plant sap or animal blood.

The other well-defined monophyletic group is the Holometabola. One of the synapomorphies of this group, a feature with profound ecological consequences, is complete metamorphosis, in which larval features are replaced by adult features during a remodeling that takes place during a pupal stage. In most other insects, the adult has much the same habits and form (except for wings and genitalia) as the juvenile, but the larva (e.g., caterpillar) and adult (e.g., butterfly) of a holometabolous insect differ greatly. The

Holometabola includes some of the largest orders of insects (Coleoptera, beetles; Hymenoptera, wasps and allies; Lepidoptera, moths and butterflies; Diptera, true flies) as well as some smaller orders (e.g., Trichoptera, caddisflies; Siphonaptera, fleas). Many of the relationships among holometabolous insects are unclear; authorities disagree on whether the derived characters that unite certain orders are uniquely derived or convergent. However, a group of five orders is united by labial silk glands (used by the larva to make a cocoon) and certain other features. Within this group, the Trichoptera and Lepidoptera have setae (hairs) on the wings and have "reversed" chromosomal sex determination: females are XY and males are XX.

The members of each order share derived characters that show them to be a monophyletic group. For example, among the Lepidoptera, the wing hairs are modified into small scales, the mouthparts form a coiled sucking proboscis, and the wing veins have a characteristic pattern. Some Trichoptera have wing scales, and the most "primitive" Lepidoptera retain small mandibles and a caddisfly-like pattern of wing veins (Figure 5.10). In this case, *two orders of insects are bridged by intermediate forms.*

Although characteristics such as wing scales clearly show that each order is monophyletic, other characteristics reflect convergent evolution among members of various orders. For example, wings have been lost during the evolution of certain grasshoppers, roaches, true bugs, flies, and many others. Moreover, *each order has some "advanced" and some "primitive" features.* For instance, the Collembola are a basal group (they are not even considered true insects), lacking wings as a primitive condition; but certain of their abdominal appendages are modified into a unique jumping organ, an advanced condition relative to other "primitive" arthropods. Beetles (Coleoptera) are highly "advanced" in many respects (complete metamorphosis, forewings modified into wing covers), but they retain the chewing mouthparts of ancestral insects.

Relationships of Apes and Humans

In traditional classifications, the primate superfamily Hominoidea consists of three families: the Hylobatidae (gibbons), Hominidae (human), and Pongidae (the great apes: the orangutan, *Pongo pygmaeus,* in Southeast Asia; the gorilla, *Gorilla gorilla,* in Africa; and two species of chimpanzees, *Pan,* also in Africa). From anatomical evidence, it has long been accepted that the Hominoidea is a monophyletic group, and that the Hylobatidae are more distantly related to the other species than those species are to one another. Humans are so similar to the great apes that, although Linnaeus placed them in separate genera, he later wrote, "I demand of you, and of the whole world that you show me a generic character . . . by which to distinguish between man and ape. I myself most assuredly know of none. But if I had called man an ape, or vice versa, I would have fallen under the ban of the ecclesiastics. It may be that as a naturalist I ought to have done so."

Physically, *Pongo, Pan,* and *Gorilla* resemble each other more than they resemble *Homo* (see Figure 1 in Chapter 26). Hence the traditional view was that *Homo* branched off first, and that the Pongidae is monophyletic (Figure 5.11A). But in the 1960s, the serum proteins of *Homo, Pan,* and *Gorilla* were shown to be more like each other in immunological characteristics than that of *Pongo,* suggesting that the orangutan diverged from the group before the African apes and *Homo* diverged from each other (Figure 5.11B; Goodman 1963; Sarich and Wilson 1967). This hypothesis has since been supported by many other studies using molecular data. In some of these studies, the base pair sequence of a homologous region of DNA has been determined for each species. Each site that varies among species is a separate character, with four possible states (A, T, C, or G). The best estimate of the phylogeny is the one requiring the fewest base pair substitutions among species. An outgroup is used to root the most parsimonious undirected network.

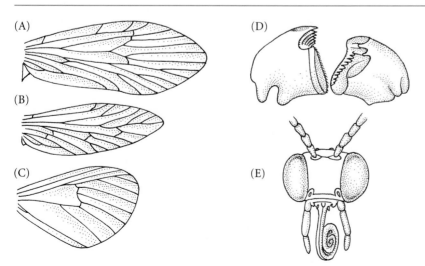

(A)

(B)

(C)

(D)

(E)

FIGURE 5.10 Characteristics that bridge the insect orders Trichoptera (caddisflies) and Lepidoptera (moths). The wing veins of a primitive caddisfly (A) and a primitive moth (B) are very similar, compared with those of an "advanced" moth (C), which has fewer veins. Primitive moths have small mandibles (D) like those of many other orders. These are absent in "advanced" Lepidoptera, in which another mouthpart is modified into a coiled sucking tube (E). (After Imms 1957.)

FIGURE 5.11 Phylogenetic relationships that have been postulated among orangutan (Po), gorilla (G), chimpanzee (P), and human (H). The gibbons (Hy) are an outgroup. (A) The view held before the 1960s: Po, G, and P form a monophyletic group, H having branched off earlier. (B) Po branched off first; relationships among G, P, and H are unresolved. (C) G and P are sister taxa. (D) P and H are sister taxa.

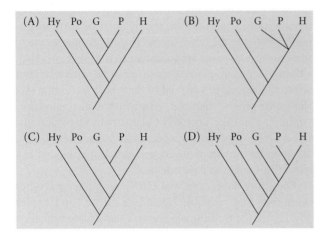

For example, in order to explore relationships among the apes and humans, Morris Goodman and his coworkers (Miyamoto et al. 1987, 1988) sequenced more than 10,000 base pairs making up a segment of DNA that includes a hemoglobin PSEUDOGENE (a silent, nonfunctional DNA sequence derived early in primate evolution by duplication of a hemoglobin gene: see Chapter 3). The outgroups used were the New World spider monkey, *Ateles*, a distant relative of the Hominoidea, and the Old World rhesus monkey, *Macaca*, which belongs to a more closely related family (Cercopithecidae). Many base pairs are the same among all of these species, but the species also differ in base pair substitutions and in short insertions and deletions of bases (Figure 5.12A). At many sites, the orangutan is identical to one or both outgroups, whereas chimpanzee, gorilla, and human share a different base, a synapomorphy providing evidence that they are a monophyletic group. Chimpanzee and human are extraordinarily similar, with more than 98 percent of their sites being identical. Fourteen synapomorphies (including both deletions and base pair substitutions) unite chimpanzee and human as sister groups (Figure 5.12B), whereas only three support the hypothesis that chimpanzee

FIGURE 5.12 Evidence for phylogenetic relationships among primates, based on the Ψη-globin region (Goodman et al. 1989). (A) Portions of the sequence in six primates. *Macaca*, an Old World (cercopithecid) monkey, and *Ateles*, a New World monkey, are successively more distantly related outgroups with reference to the apes and human. Sequences are identical except as indicated. Asterisks represent deletions. Using *Ateles* and *Macaca* as outgroups, positions 3913, 6375, and 8468 exemplify synapomorphies of the other four genera, and position 8230 provides a synapomorphy of *Gorilla*, *Pan*, and *Homo*. Synapomorphies of *Pan* and *Homo* include base pair substitutions at positions 5365, 6367, and 8224, and deletions at 3903-06 and

8469-74. Autapomorphies (nonshared derived states) include 3911 and 3913 (*Macaca*), 8230 (*Pongo*), 6374 (*Gorilla*), 5361 (*Pan*), and 6374 (*Homo*). (B) The most parsimonious (shortest) phylogeny based on the Ψη-globin sequence, using *Ateles* as an outgroup. The minimal number of changes is indicated along each branch. A tree that split up the *Homo-Pan-Gorilla* clade would be 65 steps longer, and one that split *Homo* and *Pan* would be 8 steps longer. Note that from the common ancestor of apes and Old World monkeys (*Macaca*), the number of changes along the branches leading to human (310) or any ape is less than than in the branch leading to *Macaca* (457). The figure also includes the classification proposed by Goodman et al.

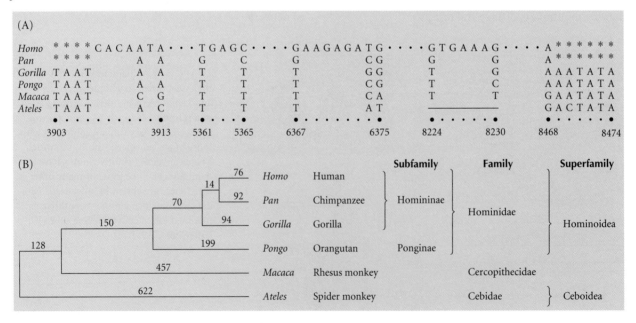

and gorilla are closest relatives (the hypothesis shown in Figure 5.11C). The phylogeny portrayed in Figure 5.11D and detailed in Figure 5.12B, with chimp and human as closest relatives and gorilla as sister group to this pair, is 8 steps (changes) shorter than a tree that separates chimp from human, and 65 steps shorter than any tree that breaks apart the chimp-human-gorilla trio (that is, one would have to postulate 8 or 65 "extra" homoplasious changes).

Many other molecular data sets support this phylogeny better than any other (see Chapter 26). If it is true, it has some interesting implications: (1) The closest relative of the chimpanzee is not the gorilla, but ourselves. (2) The morphological similarity of chimpanzee and gorilla consists largely of shared ancestral characters, features that have evolved little in the chimpanzee, but greatly in the human. Gorillas and chimps both walk on their knuckles and are very hairy. It is likely that these were also characteristics of the common ancestor of the three genera, so we too probably evolved from a hairy, knuckle-walking ancestor. (3) Because the molecular differences among *Gorilla*, *Pan*, and *Homo* are so few, the three lineages probably diverged within a very short time. (4) Under the cladistic philosophy of classification, humans should not be placed in a separate family, because *Homo* is more closely related to certain apes than they are to each other. Some authors have indeed proposed that *Homo*, *Pan*, and *Gorilla* be placed in the same family and subfamily (see Figure 5.12B; see also Chapter 26).

Evaluating Phylogenetic Hypotheses

We do not have a time machine that would enable us to see evolution unfold; nor can we do experiments on evolutionary history. Thus our inferences about evolutionary history are indirect, and are based on the kind of deductive logic that Sherlock Holmes would use to reconstruct the history of a crime. The best estimate of a phylogenetic tree that we obtain from applying our principles of inference to a particular set of data is a phylogenetic *hypothesis* that is *provisionally* accepted (as is any scientific statement). Additional data may lead us to modify or abandon the hypothesis, or may lend further support to it. *The chief way of confirming a phylogenetic hypothesis is its agreement with independent data.* Some such data come from fossils and geographic distributions. Alternatively, we can see whether different kinds of characters yield similar estimates of phylogeny. For example, there is every reason to believe (see Chapter 11) that morphological features and macromolecular sequences provide independent phylogenetic information. With some exceptions, cladistic analyses based on molecular data usually agree substantially with analyses of the same organisms based on morphology (Patterson et al. 1993; Hillis 1987). For instance, a cladogram of vertebrate taxa based on the amino acid sequences of seven proteins conforms in most respects to the relationships postulated by systematists using morphology (Figure 5.13). Likewise, Andrew Smith and colleagues (1992) inferred the phyloge-

ny of 11 species of sea urchins in six families using 81 morphological characters and 380 base pairs of the gene for 28s ribosomal RNA. The relationships among three closely related species could not be confidently determined, but otherwise, the morphological and molecular data yielded exactly the same estimate of the phylogeny (Figure 5.14).

An entirely different way of testing the validity of phylogenetic methods is to apply them to phylogenies that are absolutely known. One approach, taken by many authors, is to simulate evolution on a computer, allowing computer-generated lineages to branch and their characters to change according to various models of the evolutionary process. (For example, characters might change at random at different average rates, or one of two species derived from a common ancestor might evolve faster than the other.) The phylogenetic methods are then evaluated by testing whether they can use the final characteristics of the simulated lineages to portray accurately their history of branching. Perhaps more convincing to the skeptic are a few studies in which phylogenetic methods have been applied to experimental populations of real organisms that have been split into separate lineages by investigators (branching events) and allowed to evolve. In one such study, David Hillis and coworkers (1992) successively subdivided lineages of T7 bacteriophage and exposed them to a mutation-causing chemical so that they would accumulate DNA sequence differences rapidly over the course of about 300 generations. The investigators then scored the nine resulting lineages (Figure 5.15) for differences in restriction sites (see Box D in Chapter 3) and performed a phylogenetic analysis of these data. There are 135,135 possible bifurcating trees for this many populations, but the phylogenetic analysis correctly found the one true tree.

Phylogenetic Analysis and Extinct Organisms

So far, we have discussed how phylogenetic analyses can be used to infer the history of branching that has led to living (extant) organisms. Fossils of the relatives of extant organisms are sometimes available, and they can provide further evidence on relationships (see also Chapters 6 and 7).

In many cases, fossil material confirms inferences based on extant species. Perhaps the most intriguing instances involve studies of DNA from fossils. In all such cases, the DNA sequence of the fossil has been found to resemble that of its living relatives, confirming that the extinct species is indeed related to the modern form. For instance, DNA from 17-million-year-old magnolia leaves was very similar in sequence to that of modern magnolias (Golenberg et al. 1990), and a DNA sequence from a 120-million-year-old weevil, preserved in amber, was more similar to that of living weevils than to those of other beetles (Cano et al. 1993). Such data can also provide evidence on the rate of evolution of DNA sequences. (Incidentally, much of the DNA of even the best such fossils is so degraded that the idea of resurrecting an extinct organism, such as the dinosaurs in the

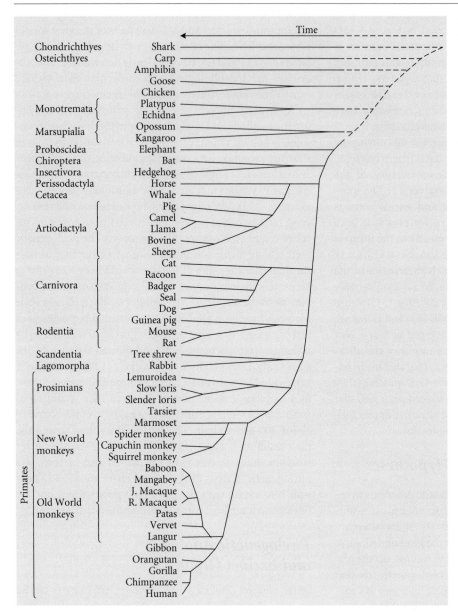

FIGURE 5.13 A phylogeny of some vertebrates, based on amino acid sequences of seven proteins. In most respects, this phylogeny is the same as that previously proposed on the basis of morphology. For example, the relationships among the classes are the same; mammal species fall together into the traditional orders; and most relationships among species within orders of mammals (e.g., Artiodactyla, Primates) conform to morphological evidence. (After Goodman et al. 1982a.)

film and book *Jurassic Park*, must remain only a pleasant fantasy for the foreseeable future.)

Fossilized relatives of the ancestors of extant species can provide evidence on ancestral characteristics that may not be available solely from the extant forms, and which can challenge, confirm, or clarify conclusions based on living species. For example, it has long been evident that the cockroaches, praying mantids, and termites form a monophyletic group, but the relationships among them had been uncertain. Some systematists had postulated that termites are highly modified roaches, most closely related to wood-eating roaches (*Cryptocercus*), which, like many termites, digest cellulose with the aid of certain protozoans that live only in the intestines of these insects (Figure 5.16A). Barbara Thorne and James Carpenter (1992) came to a different conclusion, based on a phylogenetic analysis of morphological characters: they proposed that the termites are the sister group of the roaches and mantids, which together form a monophyletic assemblage (Figure 5.16B). At about the same time, Rob DeSalle and colleagues (1992) analyzed mitochondrial DNA sequences from extant representatives of these taxa as well as from a 25-million-year-old amber-preserved termite. Their phylogenetic analysis agreed with Thorne and Carpenter's: termites are the sister group of roaches and mantids. The DNA sequence of the fossil provided important evidence in this case.

In a considerable number of cases, especially when analyzing relationships among higher taxa such as orders and classes, the taxa have diverged so greatly from their common ancestors that homology and relationships are very difficult to ascertain solely from living representatives. In many such instances, fossils that have not diverged as much provide intermediate character states. Including the extinct forms in the analysis can thus help to resolve ambiguous relationships

(Donoghue et al. 1989). For instance, based solely on anatomical characteristics of extant vertebrates, most authorities maintain that birds are most closely related to crocodilians, but a few have argued that they are most closely related to mammals. Through the inclusion of dinosaurs, mammal-like reptiles, and other extinct reptiles in the analysis, it has become clear that birds are closely related to dinosaurs rather than to the ancestors of mammals—and that their closest living relatives are indeed crocodilians (see Figure 5.6).

Estimating the History of Character Evolution

When a phylogeny has been confidently estimated, it can be used to infer the history of change in the characteristics of taxa. This, in fact, is one of the major ways in which phylogenetic analysis is used to study evolution, and this book provides many examples.

A best estimate of the evolutionary history of a character can be obtained by "mapping" character states on the

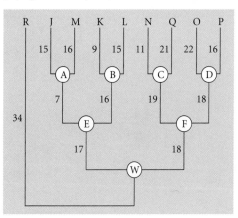

FIGURE 5.15 The true phylogeny of the experimental population of T7 bacteriophage studied by Hillis et al. At each node (indicated by circled letters), an ancestral culture was divided into two cultures. The estimate of phylogeny, based on analyses of restriction site differences among populations at the end of the experiment, was exactly the same as the true phylogeny. The numbers along the branches indicate the number of restriction site changes that evolved along each branch, according to the phylogenetic analysis. (After Hillis et al. 1992.)

phylogeny and inferring the state in each common ancestor, using the principle of parsimony. That is, we assign to ancestors those character states that require us to postulate the fewest "extra" evolutionary changes for which we lack independent evidence. Often there are several equally parsimonious assignments, so that the history of character change in parts of the phylogeny remains uncertain. But in other cases, there is little ambiguity.

For example, take Figure 5.11D as our best estimate of the phylogeny of extant Hominoidea, and consider the evolution of locomotion and body hair. Humans are bipedal and have little body hair; *Pongo*, *Gorilla*, and *Pan* are quadrupedal knuckle-walkers and have dense, long hair. In Figure 5.17, we denote common ancestors A_1, A_2, and A_3 in order of increasing recency, and consider two possible histories of either of these characteristics. Let us assume that A_1 and A_2 were quadrupedal (or hairy) and that A_3, the immediate common ancestor of chimpanzees and humans, was bipedal (or sparsely haired). This would require us to postulate evolution from quadrupedal to bipedal locomotion between A_2 and A_3, and then reversal in the evolution of the chimp. If, however, we assume that A_3 was quadrupedal, like A_1 and A_2, we need to infer only one evolutionary change, namely, the shift to bipedality in the human lineage. This is the more parsimonious hypothesis, so our best estimate is that the common ancestor of humans and chimps was quadrupedal. Likewise, we would infer that this common ancestor was hairier than modern humans. The fossil record shows that hominids indeed underwent a gradual shift toward bipedality (see Chapter 26). Fossils do not provide evidence on hairiness, so our best estimate of the evolution of this characteristic is based entirely on phylogenetic analysis.

FIGURE 5.14 Relationships among genera of sea urchins, estimated by parsimony analysis of (A) 81 morphological characters and (B) a 380 bp sequence of the 28s rRNA gene. The two data sets provide the same estimate of phylogeny (although the molecular data do not resolve relationships within two trios of closely related taxa). The first number on each branch is the number of character state changes estimated along that branch; the second number (in parentheses) is the proportion of bootstrap trees (see Box 5.B) in which the clade occurred. (After Smith et al. 1992.)

(A)

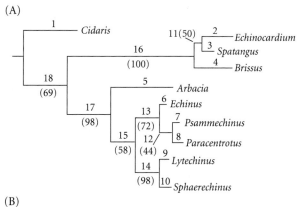

(B)

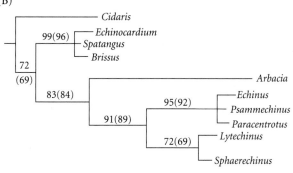

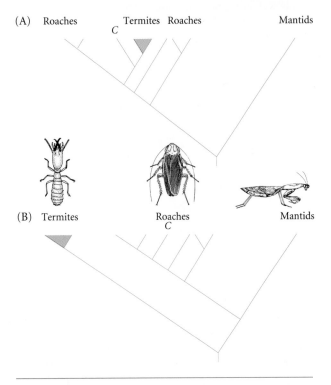

FIGURE 5.16 Two hypotheses about the phylogenetic relationship of termites. (A) The hypothesis that termites are highly modified cockroaches, most closely related to the wood-eating roaches, *Cryptocercus* (labeled *C*). (B) New evidence, including DNA sequence data from a fossilized termite, indicates that termites are the sister group of roaches plus praying mantids.

(A) Roaches   Termites   Roaches   Mantids

(B) Termites   Roaches   Mantids

This is an exceedingly simple, even obvious, example, presented only to illustrate the logic by which character evolution can be inferred. The more interesting cases are more complex. Consider, for example, the insect phylogeny in Figure 5.8, and assume that it is correct, as far as it goes. A characteristic of great interest to students of behavior is eusociality, whereby insects live in colonies of related individuals, some of which are sterile workers. The termites (order Isoptera) and certain members of the order Hymenoptera (namely, ants, social wasps, and social bees) account for most of the eusocial species of insects. It is evident from the phylogeny that eusociality must have evolved at least twice: in the Isoptera (an "orthopteroid" order) and in the Hymenoptera (a holometabolous order). If eusociality had evolved only once—namely, in the common ancestor of these two orders—then it would have persisted throughout the common ancestors of the other insect orders that lie phylogenetically between Isoptera and Hymenoptera. If this were so, we would have to postulate that eusociality was lost independently in the lineages leading to Hemiptera, Coleoptera, and many other orders. This unparsimonious hypothesis would require us to postulate far more than two evolutionary changes of state.

The question of how many times a trait has evolved is exceedingly important when we analyze the adaptive value of characteristics (Chapter 12 et seq.). For instance, since all

of the thousands of species of termites are eusocial, we assume that eusociality evolved once, in their common ancestor, rather than thousands of times. In contrast, the Hymenoptera include many lineages of both social and nonsocial wasps and bees, and detailed phylogenetic studies have shown that eusociality must have evolved independently in at least a dozen distantly related groups. Thus the phylogenetic distribution of eusociality raises the question, what is so special about the Hymenoptera? Why has eusociality evolved so many times in this order, but only once (ignoring a few other minor instances) among the millions of other kinds of insects? (See Chapter 20 if this question intrigues you.)

These examples illustrate how comparisons of the characteristics of organisms within a phylogenetic framework can describe the pattern of evolutionary change. Such systematic studies have provided evidence for some important principles and general patterns of evolution, to which we now turn.

Some Principles of Evolutionary Change, Inferred from Systematics

Because many of the following points will recur in other contexts, we will illustrate them here with only one or two examples, drawn largely from the tradition of comparative morphology in which they were developed. Many of these principles can also be illustrated at the biochemical or molecular level.

Principle 1: Homologous Features Are Derived from Common Ancestors

The term *homology* was coined by the nineteenth-century zoologist Richard Owen (1848) in the context not of evolution (he was not an evolutionist), but of IDEALISTIC MOR-

FIGURE 5.17 Two possible histories of change of a character in the Hominoidea (orangutan, Po; gorilla, G; chimpanzees, P; human, H). Color represents quadrupedal locomotion or dense body hair; white represents bipedal locomotion or sparse hair. (A) If state "white" is hypothesized for the common ancestor (A_3) of chimp and human, two state changes (tick marks) must be postulated. (B) If state "color" is hypothesized for A_3, only one change need be postulated.

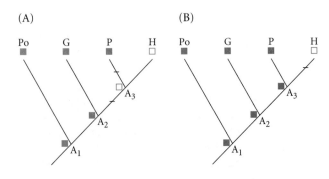

(A)   (B)

PHOLOGY, in which organisms were assumed to have been created according to certain "archetypes" or "plans." Thus Owen thought of homologous structures in two species (e.g., the leg of a lizard and the leg of a mammal) as corresponding parts of the same "plan." In contrast, Darwin gave homology a historical meaning: *organs of two organisms are homologous if they have been inherited (and perhaps modified) from an equivalent organ in the common ancestor.* Although homologous organs in the various descendants of a common ancestor may have undergone different evolutionary modifications, it is presumed that there has been continuity of the genetic "program" that governs the development of the feature. (However, the genetic and developmental underpinnings can sometimes be remarkably different for organs that are thought to be homologous by historical descent; see Chapter 23).

A *character* may be homologous among species (e.g., "toes"), but a given *character state* may not be (e.g., a certain number of toes). The five-toed state is homologous in humans and iguanas (as far as we know, both have an unbroken history of pentadactyly as far back as their common ancestor), but the three-toed state is not homologous in guinea pigs and rhinoceroses, which are thought to have evolved this condition independently from five-toed ancestors. At different levels of description, some character states may be considered either homologous or not. If we call the front limbs of birds and bats "wings," they are not homologous: flight evolved independently in the two groups. If we call them "forelegs," they are indeed homologous to each other (and to the forelegs of all other tetrapods). If we call them "forelimbs," they are homologous not only to the forelegs of other tetrapods, but to the forelimbs (i.e., the pectoral fins) of fishes as well. But in no sense are birds' or bats' wings homologous to the wings of insects, with which they have in common only their name.

Deciding whether or not the characters (or character states) of two organisms are homologous can be relatively easy or extremely difficult. The most common criteria for homology are correspondence of POSITION relative to other parts of the body, and correspondence of STRUCTURE (the parts of which the feature is composed). Correspondence of shape or of function are not useful criteria for homology (consider the forelimbs of a mole and an eagle). In some instances the correspondence may be evident only in early developmental stages. To illustrate: we infer that the hindlimb of a bird is homologous to that of a crocodile, because they correspond in position (both articulate with the acetabulum of the pelvis) and in component structure (the bones of the hindlimb correspond in the two taxa). But the structural correspondence is more evident in the embryo than in the adult: in birds, the tibia, tarsals, and metatarsals are distinct in the embryo (as in adult crocodiles), but later in development the proximal tarsals fuse with the tibia, and the distal tarsals fuse with the metatarsals, so that the leg bones of an adult bird do not appear to correspond with those of a crocodile (Figure 5.18).

Characters are not homologous unless they are INDIVIDUALIZED, i.e., have a distinct developmental determination (Wagner 1989). For example, most reptiles have a large and somewhat variable number of teeth on the maxillary bone, all much the same in shape, and there is no reason to consider any one tooth in one lizard species (or individual) to be homologous to a specific tooth in another lizard. But early in the evolution of mammals, the teeth became reduced in number and differentiated in form, and each acquired a distinct individual identity. Thus homologies can

FIGURE 5.18 The hindlimb of (A) a reptile such as a crocodile and (B) a bird. The differences between them are due to loss of some elements (e.g., digit V) and fusion of others in the bird during development. (C) Embryonic and adult state of the tibia (Tb), fibula (F), tarsals (T), and metatarsals (MT) in a crocodile. (D) The same in a bird, in which the metatarsals fuse together and the several tarsals fuse with both the metatarsals and the tibia. The homologous structure of the limb in birds and other reptiles is more evident in the embryo than in the adult. (A after Romer 1956; C and D after Müller 1989.)

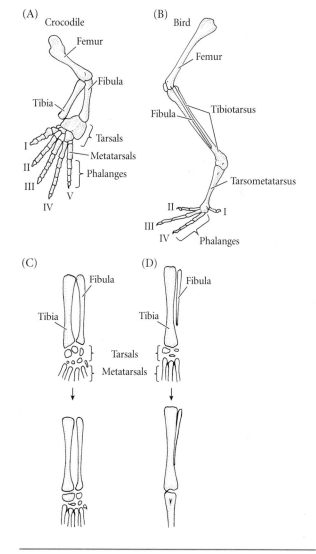

be drawn between individual premolar and molar teeth among most species of mammals.

Homology can be difficult to document for two reasons. First, it is sometimes difficult to tell whether a similar feature of two organisms evolved independently (see the discussion of homoplasy below) or was derived from their common ancestor. This is especially true when the feature is not very complex (e.g., the elongated crest feathers of many birds) or when the character state in question is the absence of a feature (e.g., wings have been lost many times in insect evolution, but we cannot tell that solely from the fact of their absence). Second, a feature that is in fact homologous in two distantly related organisms may have undergone such extensive modification in one or both that correspondence of structure is no longer evident. One would hardly suppose, solely on the basis of structure and position, that the ear ossicles of mammals are homologous to jaw elements of reptiles; we believe this to be true only because of fossils and detailed embryological studies (see Chapter 6). In the flowers of dicotyledonous plants (e.g., magnolias, roses, violets), the distinction between sepals and petals (which together make up the perianth) is usually clear on the basis of their position and vascularization; but in monocotyledonous plants (e.g., lilies, palms), the parts of the perianth are so uniform that it is uncertain whether they are homologous to both the sepals and petals of dicots, or to only one or the other (Figure 5.19).

This discussion has already implied another of the most important principles of evolution: *the features of organisms almost always evolve from pre-existing features of their ancestors; they do not arise* de novo, *from nothing.* Mammalian ear ossicles evolved from jaw bones of fishes. The wings of birds, bats, and pterodactyls are modified forelimbs: they do not arise from the shoulders (as in angels), presumably because the ancestors of these animals had no shoulder structures that could be modified for flight.

Principle 2: Homoplasy Is Common in Evolution

When a similar character (or character state) in two organisms has not been derived from a corresponding char-acter (or state) in their most recent common ancestor, it is said to be HOMOPLASIOUS. An example of a homoplasious character is the superficially similar eye of vertebrates and of cephalopods (squids, octopods). Both have a lens and retina, but their many profound differences indicate that they evolved independently: for example, the axons of the retinal cells arise from the cell bases in cephalopods, but from the cell apices in vertebrates (Figure 5.20). A homoplasious character state is illustrated by the long (as opposed to short) canine tooth that evolved independently in several groups of mammals and mammal-like reptiles (Figure 5.21).

Three more or less arbitrarily distinguished kinds of homoplasy are recognized. In **convergent evolution** (**convergence**), independently evolved features are superficially similar, but arise by different developmental pathways (Lauder 1981). The eyes of vertebrates and cephalopods are an example. **Parallel evolution** (**parallelism**) is thought to involve similar developmental modifications that evolve independently (often in closely related organisms, because they are likely to have similar developmental mechanisms to begin with). For example, similar wing patterns have evolved independently in many moths and butterflies because lepidopterans share a common set of developmental "rules" by which pigment is laid down in different parts of the wing (Nijhout 1991). If a lepidopteran evolves a dark wing band, as many independently have done, these "rules" dictate that it is most likely to develop in certain positions (Figure 5.22). **Evolutionary reversals** constitute a return from an "advanced" character state to a more "primitive," or ancestral, state. For example, almost all frogs lack teeth in the lower jaw, but frogs are descended from ancestors that did have teeth. One genus of frogs, *Amphignathodon*, has "re-evolved" teeth in the lower jaw (Noble 1931). Because the immediate ancestors of *Amphignathodon* lacked teeth, their presence in this genus is a homoplasious reversal to a more remote ancestral condition.

As we have seen, determining whether similar structures in two organisms are homologous or homoplasious can be easy or difficult. It is obvious, on close examination, that the eyes of vertebrates and cephalopods are convergent, but in

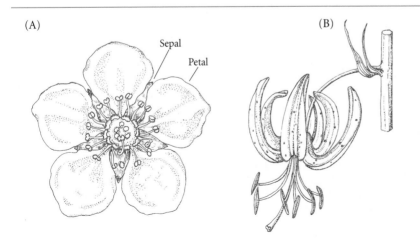

(A)   (B)

Sepal
Petal

FIGURE 5.19 Typical flowers of (A)dicots (*Spiraea*, Rosaceae) and (B) monocots (*Lilium*, Liliaceae). The ground plan of a dicot flower has most elements in sets of five or multiples of five, whereas the monocot ground plan is three-parted. Flowers are highly modified and differ substantially from the ground plan in many taxa. It is not certain whether the six parts of the lily's perianth are homologous to the dicot's sepals, petals, or both. (From Hickey and King 1997.)

FIGURE 5.20 The eye of a vertebrate (A) and of a cephalopod (B), an extraordinary example of convergent evolution. Despite the many similarities, note the several differences, including interruption of the retina by the optic nerve in the vertebrate, but not in the cephalopod. In vertebrates, the axons (nerve fibers) of the retinal cells run over the surface of the retina and converge into the optic nerve, forming a "blind spot." In cephalopods, the axons run directly from the base of the retinal cells into the optic ganglion. (From Brusca and Brusca 1990.)

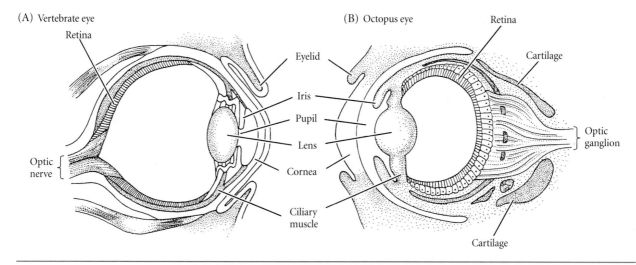

many instances, *we deduce that a character state is homoplasious from its distribution in a phylogeny.* For example, we have seen that the phylogeny of insect orders (Figure 5.8) implies that eusociality evolved independently in termites and Hymenoptera.

Homoplasious features are often (but not always) adaptations by different lineages to similar environmental conditions. In fact, a correlation between a particular homoplasious character in different groups and a feature of those organisms' environment or niche is often the best initial evidence of the feature's adaptive significance. This principle is the basis of the COMPARATIVE METHOD (Chapter 12). For example, if we compare plants that are pollinated by birds with those pollinated by other agents, we find that many unrelated bird-pollinated plants have evolved tubular, often red, flowers (Figure 5.23), whereas the flowers of their relatives have other forms. Likewise, a long, thin beak has evolved in at least four lineages of nectar-feeding birds (Figure 5.24). These correlations are taken to mean that this flower form is an adaptation for bird pollination, and this beak form is an adaptation for feeding on nectar.

Principle 3: Rates of Character Evolution Differ

The phylogenies of the hexapods and of the apes (see above) show that different characters evolve at different rates in different lineages. Another example, using four vertebrate taxa, is given in Figure 5.25. From phylogenetic

FIGURE 5.21 Convergent evolution of the "sabertooth" condition of the canine tooth in four distantly related extinct lineages. (A) A mammal-like reptile, the tapinocephalian *Estemmenosuchus*, from the Permian of Russia. This animal was probably an omnivore that used the canines for fighting, rather than for killing prey as in the other species illustrated. (B) A marsupial, *Thylacosmilus*, from the Miocene of South America. (C) A nimravid carnivore, *Barbourofelis*, from the Miocene of North America. (D) A true cat, *Smilodon*, from the Pleistocene of North America. (A after Cowen 1990; B after Riggs 1934; C after Stearn and Carroll 1989; D after Romer 1966.)

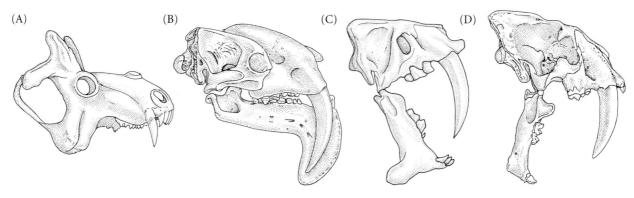

FIGURE 5.22 Parallel evolution of wing pigment patterns in Lepidoptera. Bands of pigment generally conform to the pattern elements labeled in (A), the "ground plan" as seen in brush-footed butterflies (Nymphalidae). Each of these elements has been modified, lost, or regained many times in the Lepidoptera. Similar wing patterns are seen in certain species of (B) inchworm moths (Geometridae), (C) giant silkworm moths (Saturniidae), and (D) pyralid moths (Pyralidae). (After Rensch 1959.)

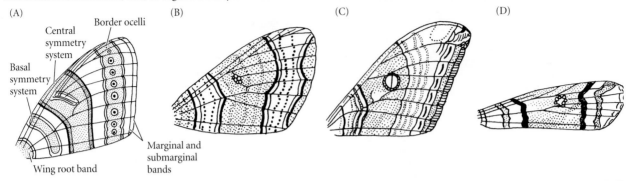

(A)

Central symmetry system

Border ocelli

Basal symmetry system

Wing root band

Marginal and submarginal bands

(B)

(C)

(D)

FIGURE 5.23 Three of the many plant species from different families that are adapted for bird pollination. In each family, the majority of species do not have the long, tubular, often red or orange corolla seen in these bird-pollinated plants. (A) A mint (Lamiaceae: *Salvia henryi*). (B) A legume (Fabaceae: *Erythrina flabelliformis*). (C) A monkey-flower (Scrophulariaceae: *Mimulus cardinalis*). Hummingbirds are depicted with each. (After Johnsgard 1983.)

FIGURE 5.24 Three lineages of birds in which a long, thin bill has evolved in association with a nectar-feeding habit. (A) A hummingbird (Trochilidae, order Apodiformes). This species feeds in flowers with curved corollas. Hummingbirds are restricted to the Americas (see also the species in Figure 5.23). (B) A sunbird (Nectariniidae, order Passeriformes), from the Old World tropics. (C) A Hawaiian honeycreeper (Fringillidae, order Passeriformes). (A after Austin 1985; B and C after Van Tyne and Berger 1976.)

(A)

(B)

(C)

(A)

(B)

(C)

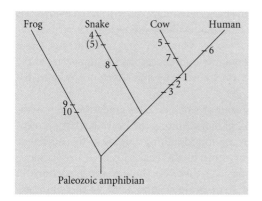

FIGURE 5.25 A simple example of mosaic evolution: primitive (P) and derived (D) states of ten characters in four vertebrate taxa, and the location of character state changes on the phylogeny.

CHARACTER	PRIMITIVE STATE	DERIVED STATE	FROG *(RANA)*	SNAKE *(COLUBER)*	COW *(BOS)*	HUMAN *(HOMO)*
1. Placenta	Absent	Present	P	P	D	D
2. Temperature regulation	Ectothermic	Endothermic	P	P	D	D
3. Aortic arches	2	1	P	P	D	D
4. Legs	Present	Absent	P	D	P	P
5. Toe number	5	<5	P	—	D	P
6. Posture	Quadrupedal	Bipedal	P	—	P	D
7. Ruminant stomach	No	Yes	P	P	D	P
8. Lungs	2	1	P	D	P	P
9. Lower jaw teeth	Present	Absent	D	P	P	P
10. Tail vertebrae form urostyle	No	Yes	D	P	P	P

analysis and from the fossil record (in the case of skeletal features), we can be quite certain what the primitive, or ancestral, condition of each of these features was in the Paleozoic amphibians that were the most recent common ancestor of these living taxa. With reference to the common ancestor's morphology, each taxon has a *mixture* of ancestral and derived character conditions. In frogs, the condition of the teeth and tail vertebrae has evolved faster (i.e., departed further from the ancestral condition) than the aortic arches or legs—different features of a single lineage have evolved at different rates. And a particular character—posture or number of lungs, for example—has evolved faster in some lineages than in others. Some characteristics (often called CONSERVATIVE CHARACTERS) are retained with little or no change over vast periods among the many descendants of an ancestor. For example, all amphibians and reptiles have two aortic arches, and all mammals have only the left arch.

Evolution of different characters at different rates within a lineage is called **mosaic evolution**. It is one of the most important principles of evolution, for it says that an organism (species) evolves not as a whole, but piecemeal: many of its features evolve quasi-independently. (There are important exceptions; for example, features that function together may evolve in concert.) This independence is seen not only in comparisons among distantly related taxa, but even within a species: characteristics usually vary independently among different geographic populations of a species (see Chapter 9). *These observations largely justify the theory of evolutionary mechanisms, in which we analyze evolution not in terms of whole organisms, but in terms of changes in individual features or even individual genes underlying such features.*

Because of mosaic evolution, *it is inaccurate or even wrong to consider one living species more "primitive" or "advanced" than another.* (Darwin reminded himself, in a notebook, "never to say higher or lower.") The amphibian lineage leading to frogs split from the lineage leading to mammals before the mammalian orders diversified (Figure 5.25), so in terms of *order of branching*, frogs are an older branch than cows and humans. In that sense, frogs might be termed more primitive. But each species of frog today has many "advanced" features, and numerous differences among frog species have evolved in the recent past. The members of different families of frogs are at least as different from one another as are the various primates, including humans. Some of them have highly "advanced" modifications; for example, *Eleutherodactylus* has direct development without a tadpole stage, and *Nectophrynoides* gives birth to live young. *Species in "old" groups do not stop evolving.* Although it is inaccurate to describe a living taxon as more primitive than another, it is often useful to speak of a BASAL lineage, one that branches off a phylogenetic tree below others with which it is compared. Frogs, then, are a basal lineage compared to birds and mammals considered together.

Principle 4: Evolution Is Often Gradual

One of the most difficult issues in evolutionary biology, still a matter of contention, is whether or not Darwin was right in arguing that evolution proceeds by small successive changes (GRADUALISM) rather than by large "leaps" (SALTATIONS). The problem is that many higher taxa (e.g., the animal phyla; many orders of insects and of mammals) are very different, and are not bridged by intermediates. The following chapter will treat evidence from the fossil record, and Chapter 24 will discuss this controversy; here we treat only observations on living organisms that have led many biologists to support Darwin's views.

If we contrast higher taxa of living organisms, we frequently find that they differ in *many* characters, *each* of which differs *discretely*. Thus birds differ from living reptiles by having feathers, fused, reduced digits of the forelimb, fusion of tarsal bones to the tibia and the metatarsus, and many other features, none of which is found in an intermediate state in any living species. One of the more extreme saltationists, Otto Schindewolf (1936), declared that "the first bird hatched from a reptilian egg" with its avian features fully formed. Judging from the prevalence of mutant monsters in movies, this seems to be a popular notion of how evolution occurs.

For any particular instance in which intermediates are unknown, we cannot dogmatically assert that they once existed. But innumerable observations on both living and fossil organisms indicate that gradual evolution is common, and

is likely to be the pattern in many cases in which evidence is lacking. Cases of mosaic evolution show that the various features of a higher taxon evolve piecemeal, not all at once. Cases of GRADATION of individual characters among species imply that characters usually evolve by small steps, not by large, discrete jumps.

Numerous examples of character gradation are provided by the beaks of birds. Among the Hawaiian honeycreepers, gradations are found among species in the thickness, length, and curvature of the beak (Color Plate 8). Sandpipers (Scolopacidae) likewise show many gradations in beak length and curvature (Figure 5.26). In these bird examples, there is no evidence that species with intermediate beaks are phylogenetically intermediate, but in many instances there is such evidence. For instance, cladistic analysis of some species of the fly genus *Zygothrica* indicates a phylogenetic progression toward an extraordinarily expanded head (Grimaldi 1987; Figure 5.27).

When different higher taxa are bridged by species with intermediate characteristics, taxonomists often disagree whether the groups should be ranked as distinct higher taxa or "lumped" into one, and indeed, the decision is quite arbitrary. For instance, ornithologists used to divide a large group of birds into the separate families Emberizidae (buntings, cardinals, certain finches and sparrows), Thraupidae (tanagers), Coerebidae (bananaquits), Parulidae (wood warblers), and Icteridae (blackbirds and orioles). It was understood that these groups, differing mostly in bill characters, are closely related. More recently the practice has been to combine all of them into one family, Emberizidae, large-

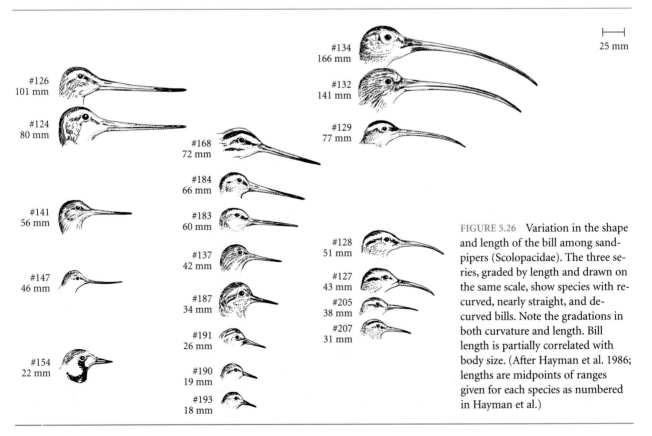

FIGURE 5.26 Variation in the shape and length of the bill among sandpipers (Scolopacidae). The three series, graded by length and drawn on the same scale, show species with recurved, nearly straight, and decurved bills. Note the gradations in both curvature and length. Bill length is partially correlated with body size. (After Hayman et al. 1986; lengths are midpoints of ranges given for each species as numbered in Hayman et al.)

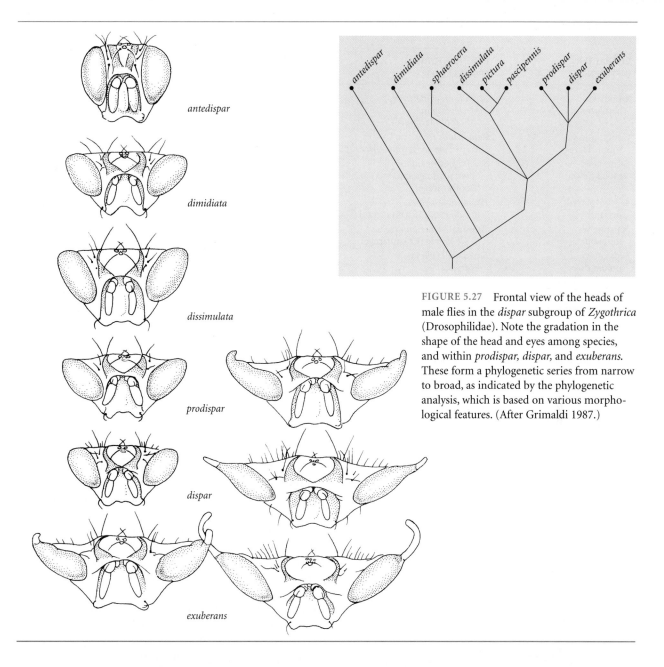

FIGURE 5.27 Frontal view of the heads of male flies in the *dispar* subgroup of *Zygothrica* (Drosophilidae). Note the gradation in the shape of the head and eyes among species, and within *prodispar, dispar,* and *exuberans.* These form a phylogenetic series from narrow to broad, as indicated by the phylogenetic analysis, which is based on various morphological features. (After Grimaldi 1987.)

ly because there are no clear gaps among them. The limits of higher taxa are often arbitrary points along a continuum. Many higher taxa can be defined only because gaps do exist, at least among living species.

Another line of evidence in favor of gradualism is that the same discrete characters that clearly define certain higher taxa frequently vary within or among closely related species in another taxon. For example, one of the major features that divide the flowering plants into two great subclasses is the number of cotyledons, or seedling "leaves": two in the dicots, one in the monocots. But one species of *Claytonia* has only a single cotyledon, and some species of *Pittosporum* have three or four. Both genera belong to clearly dicotyledonous families, and each genus contains species with the usual number of cotyledons. Stebbins (1974), who cites this example, remarks that "within the entire class of angiosperms, every diagnostic character that has been or can be used to separate families or orders may occasionally serve to sepa-

rate different species of the same genus" (Table 5.1). There is nothing intrinsic to a particular character that makes it diagnostic of higher taxa rather than species. Hence, it is reasonable to assume that features that now distinguish higher taxa (such as cotyledon number) arose as variations at the species level. As we will soon see (in Chapter 9), features that distinguish closely related species also vary within species. Thus, at every level of comparison of adjacent taxonomic levels—within vs. among species, species vs. genus, genus vs. family, etc.—many instances of gradation and of mosaic change are known. This pattern implies CONTINUITY of evolutionary change, from slight to great differences.

Principle 5: Characteristics Often Owe Their Change in Form to Change in Function

One of the reasons a homologous character may differ so greatly among taxa is that its form has evolved as its function has changed. Thus two of the ear ossicles of mammals, which

Table 5.1 Some characteristics of plants that vary at each of several taxonomic levels[a]

CHARACTER DIFFERENCE	DIAGNOSTIC AT SPECIES LEVEL	DIAGNOSTIC AT GENUS LEVEL	DIAGNOSTIC AT FAMILY OR ORDER LEVEL
Woody vs. herbaceous	*Mimulus longiflorus* vs. *M. clevelandii* (Scrophulariaceae)	*Zanthorhiza* vs. *Coptis* (Ranunculaceae)	Myrsinaceae vs. Primulaceae
Compound vs. simple leaves	*Ranunculus repens* vs. *R. cymbalaria* (Ranunculaceae)	*Eschscholtzia* vs. *Dendromecon* (Papaveraceae)	Oxalidaceae vs. Linaceae
Bilateral vs. radial flower	*Saxifraga sarmentosa* vs. other *Saxifraga* spp. (Saxifragaceae)	*Tolmiea* vs. *Heuchera* (Saxifragaceae)	Violaceae vs. Cistaceae
4- vs. 5-parted perianth	*Rhamnus crocea* vs. *R. californica* (Rhamnaceae)	*Ludvigia* vs. *Jussiaea* (Onagraceae)	Brassicaceae vs. Moringaceae
Separate vs. united petals	*Crassula Zeyheriana* vs. *C. glomerata* (Crassulaceae)	*Monotropa* vs. *Pterospora* (Monotropaceae)	Pyrolaceae vs. Ericaceae
Separate vs. united carpels	*Saxifraga Lyalii* vs. *S. arguta*	*Delphinium* vs. *Nigella* (Ranunculaceae)	Dilleniaceae vs. Actidiniaceae
Ovary superior vs. inferior	*Saxifraga umbrosa* vs. *S. caespitosa*	*Tetraplasandra* vs. other Araliaceae	Loganiaceae vs. Rubiaceae
Ovules numerous vs. solitary	*Medicago sativa* vs. *M. lupulina* (Fabaceae)	*Spiraea* vs. *Holodiscus* (Rosaceae)	Campanulales vs. Asterales

Source: After Stebbins 1974.

[a]Many of these features are important synapomorphies of certain higher taxa, but nevertheless vary among closely related genera or species within other higher taxa.

FIGURE 5.28 Structures modified for climbing in vines, which demonstrate that structures can become modified for new functions, and that different evolutionary paths to the same functional end may be followed in different groups. (A) Stipules modified into tendrils in Passifloraceae. (B) Terminal leaflets (of tripartite leaves) modified into tendrils and suckers in Bignoniaceae. (C) Leaves modified into tendrils in Ranunculaceae. (D) Inflorescence petioles modified into hooks in Rubiaceae. (From Hutchinson 1969.)

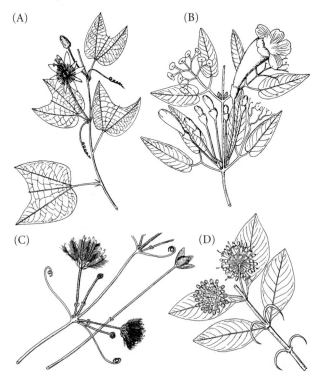

(A)   (B)   (C)   (D)

function in sound transmission, evolved from bones that in reptiles articulate the lower jaw to the skull; in turn, these and associated bones evolved from elements of the gill supports in fishlike ancestors (see Chapter 6). The sting of a wasp or bee is a modification of the ovipositor that other members of the Hymenoptera use to insert eggs into plant or arthropod hosts (that is why only female wasps and bees sting); the structures that make up the ovipositor, in turn, are believed to be homologous to the posterior legs of ancestral arthropods (Snodgrass 1935). In the many groups of plants that have independently evolved a vinelike climbing habit, the structures that have been modified into climbing organs include roots, leaves, leaflets, stipules, and inflorescences (Figure 5.28). The tendrils of a grapevine, arising in place of some of the leaves, illustrate this nicely.

Principle 6: Phylogenetic Analysis Documents Evolutionary Trends

The term EVOLUTIONARY TREND may mean two things. *Within* a lineage, species frequently show a trend, in certain features, away from the ancestral state. For example, among plants in the foxglove family (Scrophulariaceae), there is a trend toward reduction in the number of stamens, from five stamens (the ancestral number, as in mulleins, *Verbascum*) to four fertile stamens plus one sterile, reduced stamen (e.g., beard-tongue, *Penstemon*), to four fertile stamens only (e.g., Indian paintbrush, *Castilleja*), to two stamens (e.g., *Veronica*). Many trends are reversible, but some appear not to be reversed. "Irreversibility" seems to be especially common when structures have been reduced to a small number (e.g., of flower parts) or lost altogether. A descendant of an ancestor with reduced or lost structures may evolve a way of life in which those structures would be advantageous; but in such cases another structure is usually modified to serve their function. Spurges (family Euphorbiaceae, tribe Euphorbieae), for example, have lost their petals, the organs that in most flowering plants serve to attract pollinators. In the poinsettia (*Euphorbia pulcherrima*), the uppermost leaves that surround the flowers are bright red, and serve this function.

PARALLEL TRENDS are those that are independently displayed in many lineages. Among angiosperms, many groups independently display trends from low to high chromosome number and DNA content (largely because of polyploidy; see Chapter 10), from many to few flower parts (e.g., stamens or carpels), from separate to "fused" flower parts, from radial to bilateral symmetry of the flower, from animal to wind pollination, and from woody to herbaceous structure. Among insects, several groups have evolved loss of one or both pairs of wings, loss of the ovipositor, reduced wing venation, reductions in the number of tarsal (foot) segments, greater flight efficiency (in part through similar modifications of the thorax), and elongation of certain mouthparts to serve a sucking rather than a chewing function. A certain kind of progressive *simplification* is one of the most common themes within the evolution of major groups of plants and animals: structures become reduced in number, fused into compound structures, or lost. However, simplification is often accompanied by INDIVIDUATION: the descendant may have fewer repeated structures, but each has a distinctive form and function. Mammals, for example, have fewer, but more highly differentiated, teeth than do reptiles.

Principle 7: Most Clades Display Evolutionary Radiation

Evolutionary radiation is divergent evolution of numerous related lineages within a relatively short time. In most cases, such lineages are modified for different ways of life, and their characteristics are adapted for different environments, diets, or ways of escaping predation. Hence most evolutionary radiations may be called **adaptive radiations**. Because the adaptive paths taken by the different lineages are diverse, few if any characteristics show a progressive trend in any one direction. Because the phylogenetic branching events occurred over a short time, it is often difficult to determine the phylogenetic relationships among such lineages (see Figure 5.7).

Evolutionary radiation, rather than sustained, progressive evolutionary trends, *is probably the most common pattern of long-term evolution.* The term may be used to describe "bursts" of evolution at any taxonomic level. A few examples are:

The orders of eutherian (placental) mammals, most of which differentiated during the Paleocene, a span of about 10 million years (a short time in geological terms).

The Australian marsupials, which in their diverse forms and habits resemble many eutherian mammals outside Australia (see Figure 23 in Chapter 8). Grazing kangaroos, arboreal leaf-eating kangaroos and koala, and marsupial analogues of moles, mice, flying squirrels, rabbits, and dogs are just a few examples of the marsupial radiation.

Several major groups of passeriform (perching) birds. In tropical America, the "suboscines" form an immense, ecologically diverse radiation of tyrant flycatchers, ovenbirds, woodcreepers, antbirds, manakins, cotingas, and others. Muscicapids, chiefly in the Old World, include thrushes, warblers, babblers, and many other groups. The Emberizidae, as noted earlier, includes wood warblers, tanagers, blackbirds, buntings, and grosbeaks. In each of these large groups, size, coloration, feeding behavior, and morphology vary greatly among species, and intermediate species connect very different-looking forms.

Hawaiian honeycreepers and Darwin's finches. The same theme is played out at a lower taxonomic level by two famous groups of birds on island archipelagoes. The Hawaiian honeycreepers (Drepanidini; see Color Plate 8) are diverse in beak form and feeding habits. So are the famous Darwin's finches (Geospizinae: Figure 5.29) of the Galápagos Islands, off the coast of Ecuador (Lack 1947; Grant 1986).

Cichlid fishes. Cichlids are the dominant fishes in the great lakes of the Rift Valley in eastern Africa. Lake Tanganyika has about 140 species, Lake Malawi at least 200 and perhaps more than 500, and Lake Victoria about 200–250. Lakes Tanganyika and Malawi, which are deep, are 1.5–2.0 million years old, and Victoria, which is very shallow, is only 750,000 years old. In all three lakes (most strikingly in Tanganyika and Victoria), cichlid species vary greatly in coloration, body form, and in the form of their teeth and jaws (Figure 5.30), and are correspondingly diverse in feeding habits. There are specialized feeders on insects, detritus, rock-encrusting algae, phytoplankton, zooplankton, macrophytes, molluscs, baby fishes, and large fishes; some species feed on the scales of other fishes, and one has the gruesome habit of plucking out other fishes' eyes. The teeth of closely related species differ more greatly than among some families of fishes. These 600+ species belong to only a few phylogenetic lineages, each of which has speciated extensively (see Echelle and Kornfield 1984 on these and other "species flocks" of fishes.)

The drosophilid flies of the Hawaiian Islands. About a third of the world's species of *Drosophila* and its close relatives occur only in the Hawaiian archipelago, where they form a monophyletic group of more than 800 species. The individual Hawaiian islands range in age from 5.1 million to 500,000 years, and each has its own distinctive species (see Chapter 8). The Hawaiian drosophilids are more diverse than all those in the rest of the world in their morphology and behavior. The species present all sorts of bizarre modifications of the mouthparts, legs, and antennae, asso-

FIGURE 5.29 Adaptive radiation of the 13 species of Darwin's finches in the Galápagos Islands. The phylogeny is tentative because, although there is good evidence that each genus is monophyletic, relationships among and within genera are uncertain. The bills of these species are adapted to their diverse feeding habits. Species of *Geospiza* feed mostly on seeds that differ in size and hardness, depending on the depth of the species' bill. *Platyspiza* feeds on buds and fruits. *Camarhynchus* excavates insects from wood; the species differ in the woody microhabitats they use. The *Cactospiza* species extract arthropods from crevices, using cactus spines or other tools. Species of *Certhidea* feed on nectar and on insects gleaned from vegetation. The blue-black grassquit (*Volatinia jacarina*), found in western South America, may be a close relative of the ancestor of Darwin's finches. (After Purves et al. 1992; Grant 1986.)

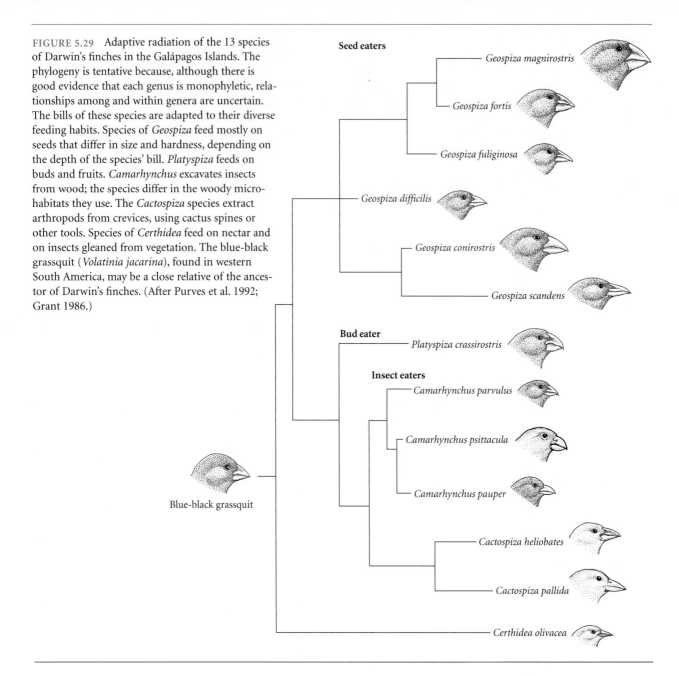

ciated with highly unusual forms of mating behavior (Carson et al. 1970; Carson and Kaneshiro 1976).

The Hawaiian silversword "alliance" consists of 28 species in three closely related genera (*Argyroxiphium, Wilkesia, Dubautia*) of plants in the sunflower family (Asteraceae). They occupy habitats ranging from exposed lava to wet forest, and their growth forms include vines, trees, erect shrubs, unbranched rosettes (all the leaves emerging from the base), branched rosettes, compact "cushion" plants, and herbaceous mats (Figure 5.31). They vary greatly in the form and anatomy of the leaves and in the size, color, and structure of the flowers. In many features their range of variation exceeds that among families of plants, yet almost all of them can be crossed, and the hybrids are often fully fertile. They all appear to have been derived from a single ancestor that colonized the Hawaiian Islands from western America (Baldwin and Robichaux 1995).

Molecular Data in Phylogenetic Analysis

In recent years, molecular data have become very important in inferring phylogenies, for several reasons. Certain kinds of molecular data (especially DNA sequences) can provide many more characters, and hence more information, than morphological data (especially for morphologically similar species). In many instances, molecular characters can be described and coded (e.g., A, T, C, G) much less ambiguously than morphological characters. Certain DNA sequences that evolve rapidly are useful for analyzing relationships

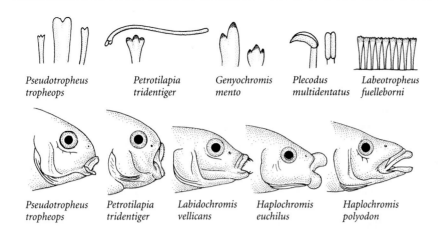

FIGURE 5.30 A sample of the diverse tooth forms and head shapes among the Cichlidae of the African Great Lakes. The differences in morphology are associated with differences in diet and mode of feeding. (After Fryer 1959 and Fryer and Iles 1972.)

Pseudotropheus tropheops *Petrotilapia tridentiger* *Genyochromis mento* *Plecodus multidentatus* *Labeotropheus fuelleborni*

Pseudotropheus tropheops *Petrotilapia tridentiger* *Labidochromis vellicans* *Haplochromis euchilus* *Haplochromis polyodon*

among closely related species, whereas others that evolve very slowly may be useful for assessing relationships among very old groups (see Chapter 11). And in some cases, divergence at the molecular level may provide evidence on the absolute age of a taxon.

Kinds of Molecular Data Used in Phylogenetic Analysis

Phylogenetic studies have used several sorts of molecular data (Hillis et al. 1996b; Avise 1994):

1. Immunological distance. The immunological distance method was one of the first methods used to obtain molecular data for phylogenetic analysis. Antibodies against a protein from a certain species (e.g., human

serum albumin) are obtained from an experimental animal—say, a rabbit. These antibodies are then exposed separately to the same protein (antigen) from several other species. If antigen from one species (e.g., chimpanzee) yields a more intense cross-reaction than that from another (e.g., monkey), that species' protein is presumably more similar to the original antigen (human). The quantified reaction intensity measures immunological similarity (the converse is called "immunological distance"), and has been interpreted as measuring degree of relationship. Such studies, although historically important (e.g., Sarich and Wilson 1967 on hominoid relationships: see above), have been largely superseded by other techniques.

FIGURE 5.31 Some members of the Hawaiian silversword alliance. (A) *Argyroxiphium sandwicense*, a rosette plant that lacks a stem except when flowering. (B) *Wilkesia gymnoxiphium*, a stemmed rosette plant. (C) *Dubautia scabra*, a small shrub. (D) *Dubautia latifolia*, a woody vine. (After Wagner et al. 1990.)

(A) (B) (C) (D)

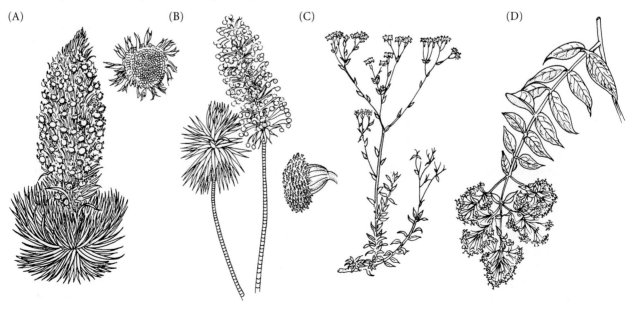

2. DNA-DNA hybridization. This method is explained in Chapter 3, Box D. The thermal stability of heteroduplexes formed by complementary strands of DNA from two species is taken as a measure of the overall similarity of their DNA. The similarity is measured by a *single number* quantifying similarity or its complement, "distance." Many cladists object to the use of "distance values" for inferring phylogeny, partly because one does not have information on the state of each individual character (i.e., each base pair), so one cannot determine how much of the overall similarity is due to synapomorphies, and how much to symplesiomorphies. Advocates of the technique respond that averaged over millions of nucleotide sites, differential rates of evolution are likely to cancel each other out, so that overall similarity should accurately reflect recency of common ancestry. Two groups of investigators have independently used this method to assess relationships among humans and apes (Sibley and Ahlquist 1987; Caccone and Powell 1989), and both arrived at the same phylogeny (Figure 5.12B) that has been derived from DNA sequence data (see below).

3. Protein electrophoresis. This technique, which separates genetically different forms (allozymes) of enzymes or other proteins, has been used mostly to study intraspecific variation, and will be discussed in Chapter 9. Closely related species may share some of the same variant enzymes, so electrophoresis has also been used for phylogenetic studies. Various authors have either treated the enzyme variants as different characters in a typical cladistic analysis, or have computed an index of "genetic similarity" or "genetic distance" (see Chapter 9).

4. Restriction sites. The use of restriction enzymes is described in Chapter 3, Box D. A certain DNA sequence (e.g., mitochondrial DNA) may show either presence or absence of a specific restriction site (i.e., a four- or six-base sequence that is cleaved by a specific enzyme). Each restriction site is treated as a character with two states, and the data can be analyzed by a standard cladistic method.

5. Amino acid sequences. One of the first molecular phylogenetic analyses (Fitch and Margoliash 1970) used the amino acid sequence of cytochrome *c* to describe the phylogeny of numerous vertebrates and other taxa (see Figure 5.13). The characters used are either the amino acid sequences themselves or the nucleotide sequences of the encoding genes, inferred from the genetic code.

6. DNA sequences. A sequence of DNA from minute samples of tissue, sometimes from long-preserved museum specimens or even certain fossils, can be amplified by the polymerase chain reaction (PCR) and then sequenced by the methodology described in Box D of Chapter 3. The most difficult step is extracting the particular gene one wishes to sequence, but once this has been done in one species, copies of the sequence (primers) can be used to extract the sequence from closely, or sometimes distantly, related species. This has become the most popular molecular method in systematics.

Molecular Clocks

All molecular methods indicate that non-interbreeding evolutionary lineages (species and higher taxa) become steadily more different with time. As we will see (in Chapter 11), the theory of population genetics predicts that if natural selection does not favor one molecular variant over others, DNA or protein sequences should diverge, *on average*, at a constant rate. Early in the history of molecular phylogenetic studies, the data suggested that macromolecules may indeed evolve and diverge at a constant rate (which is certainly not true of morphology!). This concept has been dubbed the **molecular clock** (Zuckerkandl and Pauling 1965). If it were true, it would have two important implications for the study of evolutionary history. First, recall from the discussion of phenetics that if the rate of overall divergence is constant, phylogeny can be inferred directly from the *overall similarity* of species (or, conversely, "distance," which includes the autapomorphies that cladists consider anathema). One could then legitimately use phenetic methods to infer phylogeny from "distance measures," such as those yielded by DNA-DNA hybridization, when information is lacking on the ancestral and derived states of individual characters (such as base pairs). Second, if one could calibrate the clock—if one could discover how fast it is "ticking"—one could *estimate the absolute time* since different taxa diverged. The clock might be calibrated by, say, information from the fossil record on the absolute time of divergence of certain taxa, and it could then be used to estimate the divergence times of other taxa that have not left a good fossil record. Bear in mind that the phylogenetic trees we have considered so far portray *relative* times of divergence (the branching *sequences*) of taxa, not absolute times.

First, we need to calculate the number of differences (e.g., in base pairs) that have accrued among pairs of species since their common ancestor. This can be estimated by plotting on our estimated phylogeny where each change took place. For example, Figure 5.12B shows 76 autapomorphic changes between *Homo* and its common ancestor with *Pan* (e.g., site 6374), 14 synapomorphic changes between that common ancestor and the *Gorilla* branch (e.g., site 5365), and 94 autapomorphies along the *Gorilla* branch (e.g., site 6374). Since the common ancestor of all three species, there have been 94 changes leading to *Gorilla*, 76 + 14 = 90 changes leading to *Homo*, and 92 + 14 = 106 leading to *Pan*. From the more remote common ancestor of these species and *Pongo*, there have been 160 changes to *Homo*, 176 to *Pan*, 178 to *Gorilla*, and 199 to *Pongo*.

The *average rate of base pair substitution* in any lineage can be calculated if we have an estimate of the absolute time of divergence. For example, the oldest fossils of cercopithecid monkeys are dated at 25 million years ago (My).

FIGURE 5.32 The relative rate test for constancy of the rate of molecular divergence. Sequences are obtained for living species A and B and for outgroup species E (perhaps also F). C and D represent common ancestors. The genetic distance (e.g., in terms of nucleotide differences) between A and E is $D_{AE} = a + c + d$. That between B and E is $D_{BE} = b + c + d$. If the rate of nucleotide substitution is constant, $a = b$, so $D_{AE} = D_{BE}$. If rate constancy holds throughout the tree, the distance between any pair of species that have D as a common ancestor will equal that between any other such pair of species (e.g., $D_{AE} = D_{BF}$).

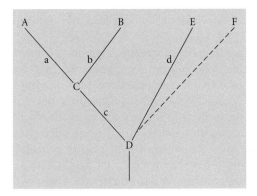

Thus the divergence between *Macaca* and the hominoids (see Figure 5.12) is *at least* this old. This is a *minimal* estimate of divergence time, because it is possible that older cercopithecids (and hominoids) exist that have not yet been found in the fossil record. If, however, we take this date as the age of the cercopithecid/hominoid divergence, we can calculate the average rate of sequence evolution. The number of substitutions per base pair per million years for the cercopithecid lineage (*Macaca*; see Figure 5.12) is 457/10,000 base pairs sequenced/25 My = 0.0457 substitutions per base pair/25 My = 1.83×10^{-3} per My, or 1.83×10^{-9} per year. From the common ancestor to *Homo*, the average rate has been 310/10,000/25 = 1.24×10^{-3} per My, or 1.24×10^{-9} per year.

Relative rates The time that has elapsed from any branch point, or NODE (i.e., common ancestor), on a phylogenetic tree to each of the living species derived from that ancestor is *exactly the same*. If lineages have diverged at a constant rate (i.e., substitutions/time), then the number of changes should be about the same along any path between living descendants and their common ancestor (Figure 5.32). This principle is the basis of the **relative rate test** for equivalence of evolutionary rates (Wilson et al. 1977). In the hominoid example, the number of differences between *Pongo* and *Homo*, *Pan*, or *Gorilla* is about the same (slightly fewer for *Homo*). The number of differences between *Macaca* and the various hominoids ranges from 806 (*Pongo*) to 767 (*Homo*). These data appear to offer some support for a constant rate of divergence, although the human lineage appears to have slowed down somewhat.

The relative rate test has been applied to all kinds of molecular data on various groups of organisms. Some groups have passed the test fairly well, especially when the comparison is made among closely related organisms. Others, especially when comparisons are made among distantly related organisms, fail the test (Britten 1986). For example, the rate of sequence divergence has been about 1.4 times greater in cercopithecid monkeys than in the Hominoidea, and the rate in rodents is at least two or three times greater (Wu and Li 1985). There are several possible explanations for these rate differences, including different generation times. It is also clear that within the same lineage, different DNA sequences can evolve at different rates; for example, mitochondrial DNA appears to evolve five to ten times faster than nuclear DNA in various vertebrate groups.

Calibrated rates Charles Langley and Walter Fitch (1974) were among the first to use data from fossils to examine the validity of the molecular clock hypothesis. From amino acid sequences of seven proteins, determined for mammals representing several orders, families within orders, and genera within families, they estimated (from the genetic code) the number of nucleotide differences between pairs of species. The fossil record provided estimates for the minimal age of divergence of each pair. Langley and Fitch found that the correlation between molecular differences and age is quite strong (Figure 5.33), but is not exact: they calculated that rates of molecular divergence have been about twice as variable as would be expected of a random process (such

FIGURE 5.33 Nucleotide substitutions versus time since divergence, illustrating the approximate constancy of the rate of molecular evolution. Each point represents a pair of living mammal species whose most recent common ancestor, based on fossil evidence, occurred at the time indicated on the abcissa. The ordinate is the number of nucleotide substitutions inferred from the difference between the pair in amino acid sequences of seven proteins. The four open circles represent pairs of primate species, which appear to have diverged more slowly than other groups of mammals. (After Langley and Fitch 1974.)

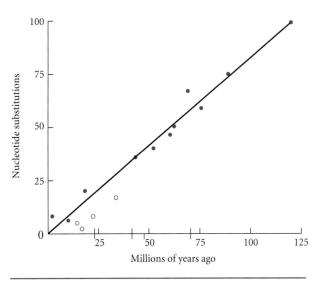

as radioactive decay) with a constant average rate. Their findings suggest that the "clock" is a sloppy one, that might be used for coarse, but not fine, estimates.

Suppose, for example, that for three mammal groups that lack a fossil record, species A and B differ by 41 nucleotide substitutions, and both differ from C by about 60 substitutions. From the relationship shown in Figure 5.33, the best estimate of divergence time would be 50 My for A and B, and 75 My for the divergence of C from the A + B clade. But because of the variation around the line in Figure 5.33, these estimates of age of divergence from the number of nucleotide substitutions are accurate only to within about 15 My or so. Thus this "clock" is not accurate enough to infer phylogenetic relationships among species that diverged within such a time span. The sources of error in estimating divergence times from molecular data include whatever inconstancy there may be in the rate of molecular evolution, as well as the substantial imprecision of the fossil datings used to calibrate the clock. These datings often have a large margin of error, both because fossils provide us with only minimal divergence time, and because the age of a fossil can be determined only approximately (see Chapter 6).

The validity of molecular clocks is the subject of considerable disagreement (reviewed by Avise 1994; Hillis et al. 1996a). It is clear that molecular evolutionary rates differ among genes, among distantly related taxa, and, at least sometimes, among closely related taxa. The effect of differences in generation time on the absolute rate of evolution is still unclear. Whether or not such clocks will prove valid and useful, at least when averaged over enough DNA sequences, remains to be seen.

Evidence for Evolution

In order to classify organisms, systematists compare their characteristics. In doing so, systematists had amassed an immense body of information before *The Origin of Species* was published. This information suddenly made sense in light of Darwin's theory of descent from common ancestors; indeed, Darwin drew on much of this information as evidence for his contention that evolution has occurred. Since Darwin's time, this body of information has increased greatly; it now includes comparative data not only from the traditional realms of morphology and embryology, but also from cell biology, biochemistry, and molecular biology.

All of this information is consistent with Darwin's hypothesis that living organisms have descended from common ancestors. Indeed, innumerable biological observations are hard to reconcile with the alternative hypothesis, that species have been specially created by a supernatural being, unless that being is credited with arbitrariness, whimsy, or a devious intent to make organisms *look* like they have evolved. From the comparative data amassed by systematists, we can identify several patterns that confirm the historical reality of evolution, and which make sense only if evolution has occurred.

1. *The hierarchical organization of life.* Before Linnaeus, there had been many attempts to classify species, but these early systems just didn't work. One author, for instance, tried to classify species in complicated five-parted categories—but organisms simply don't come in groups of five. But organisms do fall "naturally" into the hierarchical system of groups within groups that Linnaeus described. A historical process of branching and divergence will yield objects that can be hierarchically ordered, but few other processes will do so. Thus, languages can be classified hierarchically to some extent, but elements and minerals cannot.

2. *Homology.* Similarity of structure despite differences in function follows from the hypothesis that the characteristics of organisms have been modified from the characteristics of their ancestors, but it is hard to reconcile with the hypothesis of intelligent design. Design does not require that the same bony elements form the frame of the hands of primates, the digging forelimbs of moles, the wings of bats, birds, and pterosaurs, and the flippers of whales and penguins. Modification of preexisting structures, not design, explains why the stings of wasps and bees are modified ovipositors, and why only females possess them. All proteins are composed of "left-handed" (L) amino acids, even though the "right-handed" (D) optical isomers would work just as well *if* proteins were composed only of these. But once the ancestors of all living things adopted L amino acids, their descendants were committed to them; introducing D amino acids would be as disadvantageous as driving on the right in Great Britain or on the left in the United States. Likewise, the universality of an arbitrary genetic code makes sense only as a consequence of common ancestry.

3. *Embryological similarities.* Homologous characters include features that appear during development, but would not be necessary if the development of an organism were not a modification of its ancestors' ontogeny. For example, tooth primordia appear and then are lost in the jaws of fetal anteaters, and we have noted that some terrestrial frogs and salamanders pass through a larval stage within the egg, with the features of typically aquatic larvae, but hatch ready for life on land. Early in development, human embryos are almost indistinguishable from those of fishes, and briefly display gill slits.

4. *Vestigial characters.* The adaptations of organisms have long been, and still are, cited by creationists as evidence of the Creator's wise beneficence, but no such claim can be made for the features, displayed by almost every species, that served a function in the species' ancestors, but do so no longer. The yellow flowers of dandelions (*Taraxacum officinale*) are almost indistinguishable from those of many related plants, in which they attract pollinating insects,

whose visits are necessary for fertilization of the ovules. In the dandelion, however, seeds develop by apomixis, without fertilization. Cave-dwelling fishes and other animals display eyes in every stage of degeneration. Flightless beetles retain rudimentary wings, concealed in some species beneath fused wing covers that would not permit the wings to be spread even if there were reason to do so. In *The Descent of Man*, Darwin listed a dozen vestigial features in the human body, some of which occur only as uncommon variations. They included the appendix, the coccyx (four fused tail vertebrae), rudimentary muscles that enable some people to move their ears or scalp, and the posterior molars, or wisdom teeth, that fail to erupt, or do so aberrantly, in many people. At the molecular level, every eukaryote genome contains numerous nonfunctional DNA sequences, including pseudogenes: silent, nontranscribed sequences that retain some similarity to the functional genes from which they have been derived (see Chapters 3, 22).

5. *Convergence.* We have noted examples, such as the eyes of vertebrates and cephalopods, in which functionally similar features actually differ profoundly in structure. This observation follows logically from the hypothesis that structures are modified from features that differ in different ancestors, but it is inconsistent with the notion that an omnipotent Creator, who should be able to adhere to an optimal design, provided them. Likewise, evolutionary history is a logical explanation, and creation is not, for cases in which different organisms use very different structures for the same function, such as the various modified structures that enable vines to climb.

6. *Suboptimal design.* The "accidents" of evolutionary history explain many features that no intelligent engineer would be expected to design. For example, the paths followed by food and air cross in the pharynx of terrestrial vertebrates (see Figure 2 in Chapter 1). The human eye has a "blind spot," which you can find at about 45° to the right or left of your line of sight. It is

caused by the functionally nonsensical arrangement of the axons of the retinal cells, which run forward into the eye, converging into the optic nerve, which interrupts the retina by extending back through it toward the brain (see Figure 5.20A).

7. *Geographic distributions.* The study of systematics includes the geographic distribution of species and higher taxa. This subject, biogeography, is treated in Chapter 8. Suffice it to say that the distribution of many taxa makes no sense unless they have arisen from common ancestors. For example, many taxa, such as marsupials, are distributed across the southern continents, which is easily understood if they arose from common ancestors that were distributed across the single southern land mass that began to fragment in the Mesozoic.

8. *Intermediate forms.* The hypothesis of evolution by successive small changes predicts the innumerable cases in which characteristics vary by degrees among species and higher taxa. Among living species of birds, we see gradations in beaks; among snakes, some retain a rudimentary pelvic girdle and others have lost it altogether; among primitive moths, some retain small mandibles. At the molecular level, the difference in DNA sequences ranges from almost none among very closely related species through increasing degrees of difference as we compare more remotely related taxa.

For each of these lines of evidence, hundreds or thousands of examples could be cited from studies of living species. Even if there were no fossil record, the evidence from living species would be more than sufficient to demonstrate the historical reality of evolution: all organisms have descended, with modification, from common ancestors. We can be even more confident than Darwin, and assert that all organisms that we know of are descended from a single original form of life. These conclusions, we repeat, are facts (see Chapter 1 on the meaning of "fact"), demonstrable even without a fossil record. But, happily, there is a fossil record—the subject of the next two chapters.

Summary

1. Any of many criteria could be used for classifying organisms. Because genealogical relationships convey a great deal of information, taxonomists have increasingly adopted the position that classifications should reflect phylogenetic (branching) relationships. According to this convention, each taxon in a hierarchical classification (i.e., taxa nested within higher taxa at several levels) should be a monophyletic group, consisting of an ancestor and all its known descendants. Such a classification is based on the best available estimate of the true phylogeny, or evolutionary history of branching.

2. Overall similarity among organisms is not the best indicator of phylogenetic relationships. Two species may be

more similar to each other than to a third because they retain primitive features (whereas the third has diverged), because they independently evolved similar features, or because they share derived ("advanced") features that evolved in their common ancestor. Only unique, shared, derived features are evidence of relationship. Thus monophyletic groups are marked by unique, derived character states shared by the group's members.

3. Phylogenetic relationships can be obscured by homoplasy: the independent evolution of similar character states in different lineages. Several methods are used to estimate phylogenies in the face of these misleading features. Most are based on the criterion of cladistic

parsimony, according to which the best estimate of the phylogeny is the tree that requires one to postulate the smallest number of evolutionary changes to account for the differences among species. A phylogeny obtained by these procedures can be evaluated by its consistency with different data sets, such as those obtained from morphology and from DNA sequences, or from different DNA sequences. Often, the consistency is quite high.

4. Phylogenetic analyses not only describe the branching history of life, but also have many other uses. An important one is providing estimates of the pattern of evolution of interesting characteristics. Thus, systematic studies have yielded information on common patterns of character evolution.

5. Among these common patterns are homoplasy (including convergent evolution, parallel evolution, and reversal); trends in the evolution of certain features; mosaic evolution (in which different characters evolve piecemeal, at different rates); evolution of "new" features from preexisting characters; the gradual evolution of large differences by small steps (common but not necessarily universal); divergence in structure associated with change in a character's function; modification of different structures to serve similar functions in different lineages; and evolutionary (or adaptive) radiation, in which numerous related lineages arise in a relatively short time and evolve in many different directions.

6. The rate of evolution of DNA sequences can be shown in some cases to be fairly constant ("molecular clock"), such that sequences in different lineages diverge at a roughly constant rate. In such cases, degrees of similarity can indicate phylogenetic relationships. The absolute rate of sequence evolution can sometimes be calibrated if fossils of some lineages are known.

7. By comparing characteristics of organisms, systematists have described numerous patterns that provide evidence for evolution, but are inconsistent with special creation.

Major References

Mayr, E., and P. D. Ashlock. 1990. *Principles of systematic zoology.* Second edition. McGraw-Hill, New York. Discusses phylogenetic analysis and, from a largely non-cladistic viewpoint, the practice and philosophy of classification.

Brooks, D. R., and D. A. McLennan. 1991. *Phylogeny, ecology, and behavior.* University of Chicago Press, Chicago. Provides an elementary introduction to cladistic phylogenetic inference, as well as extensive description of applications of phylogenetic analysis to problems in the evolution of behavioral and ecological features of organisms.

Hillis, D. M., C. Moritz, and B. K. Mable (editors). 1996. *Molecular systematics.* Second edition. Sinauer Associates, Sunderland, MA. A "how to" manual of molecular techniques and analyses of molecular data for studies in systematics. Includes extensive description of various methods of phylogenetic inference and their application to biological problems.

Maddison, W. P., and D. R. Maddison. 1992. *MacClade,* version 3.0. Sinauer Associates, Sunderland, MA. A package of computer programs for estimating phylogenies from data, and especially for estimating the evolutionary history of characters from their pattern of variation among taxa. The accompanying manual, available separately, provides a clear introduction to the principles, methods, and uses of phylogenetic analysis.

Mayr, E. 1982. *The growth of biological thought: Diversity, evolution, and inheritance.* Harvard University Press, Cambridge, MA. Includes extensive description of the history of systematics.

Problems and Discussion Topics

1. Suppose we are sure, because of numerous characters, that species 1, 2, and 3 are more closely related to each other than to species 4 (an outgroup). By DNA sequencing, we have found ten nucleotide sites that differ among the species. The sequences of variable nucleotides are:

   (1) GCTGATGAGT; (2) ATCAATGAGT; (3) GTTG-CAACGT; (4) GTCAATGACA.

   Estimate the phylogeny of these taxa by plotting the changes on each of the three possible undirected networks (as in Figure 5.4), determine which network requires the fewest changes, and root that network between species 4 and the common ancestor of the other three species.

2. Suppose the species in question 1 are primates that differ in mating system. Species 1 and 2 are monogamous, and species 3 and 4 are polygamous. We also happen to know that another polygamous species, 5, is more distantly related to 1–4 than these are to each other. Given the best estimate of the phylogeny from question 1, what has been the probable history of evolution of the mating system?

3. Higher taxa defined by shared, uniquely derived character states often include species that lack those definitive character states. For example, the insect suborder Heteroptera (true bugs) has a unique character among all insects: forewings that are horny (sclerotized) at the base but membranous at the tip. Yet some heteropterans, such as bedbugs, have no wings at all. How can the phylogenetic affinity of such species be determined?

4. While comparing DNA sequences among species, students of molecular evolution have discovered a few cases that they interpret as evidence that certain genes have been recently transferred between distantly related organisms, such as cats and primates (see Chapter 22). What phylogenetic information would be necessary to support this hypothesis?

5. In quite a few studies, phylogenies inferred from molecular data, such as DNA sequences, differ in some respects from those inferred from morphological data (Patterson et al. 1993). Do such instances mean that evolution from common ancestors hasn't really occurred? Or that phylogenetic methods really don't work? What should investigators do if faced with this situation?

6. Most evolutionary biologists agree that many of the differences in DNA sequence among species are not adaptive. Other differences among species, both in DNA and in morphology, are adaptive. Do adaptive and nonadaptive variations differ in their utility for phylogenetic inference? Can you think of ways in which knowledge of a character's adaptive function would influence your judgment of whether or not it provides evidence for relationships among taxa?

7. The adaptive significance of a characteristic is often inferred from correlations, among species, between character state and the species' typical environment. The prevalence of leaflessness among desert plants (see Chapter 4), for example, suggests that this trait is adaptive in hot, dry environments (perhaps because it reduces water loss). What role does phylogenetic study play in such analyses of adaptation? Does it matter for such analyses whether or not the hundreds of species of cacti are a monophyletic group? Would it matter if the leafless spurges of African deserts were really members of the Cactaceae rather than the Euphorbiaceae?

8. In order to infer the direction of evolutionary change in a characteristic, some authors have assumed that the primitive character state is the one possessed by the greatest number of species, and that the derived state is shared by fewer species. Why is this an erroneous supposition? Discuss examples (from this chapter and from your own reading) in which this proposition holds and in which it does not. What would be a more accurate statement of the relationship between the direction of a character's evolution and the distribution of character states among taxa?

9. In the absence of a fossil record, how might phylogenetic analysis tell us whether the rate of diversification (increase in number of species) has differed among evolutionary lineages?

CHAPTER 6

Evolving Lineages in the Fossil Record

Although some of the history of evolution can be inferred from living organisms, it is only in the fossil record that we can hope to find reasonably direct evidence of that history. The fossil record tells us of the existence of innumerable creatures that have left no living descendants, of great episodes of extinction and diversification, of the movements of continents and organisms that explain their present distributions. Only from this record can we obtain an absolute time scale for evolutionary events, as well as evidence of the environmental conditions in which they transpired. The fossil record, moreover, provides an independent check—although a very inexact one—on some inferences about some evolutionary processes that we base on studies of living organisms.

In this chapter, we will describe examples in which the fossil record documents evolutionary changes in morphology and the origin of new taxa, ranging from species to classes. Some of these examples are given in greater detail than most readers will need to remember. The purpose of this detail is to provide information for readers who may need to convince skeptics that, indeed, *the fossil record documents evolution.* This chapter also describes important evidence from the fossil record on the *rates and modes of evolution.*

Some Geological Fundamentals

Rock formation Rocks at the earth's surface originate as molten material (magma) that is extruded from deep within the earth. Some of the extrusion occurs via volcanoes, but much rock originates as new crust forms (see Figure 6.1). Rock so formed is called IGNEOUS rock. SEDIMENTARY rock is formed by the deposition and solidification of sediments, which are formed either by the breakdown of older rocks (of all kinds: given long enough, all rocks are lost to erosion) or by precipitation of minerals from water. Under high temperature and pressure, both igneous and sedimentary rocks are altered, forming METAMORPHIC rocks.

Most fossils are found in sedimentary rocks; they are never in igneous rocks, and are usually altered beyond recognition in metamorphic rocks. A few fossils are found in other situations; for example, insects are found in amber (fossilized plant resin), and some mammoths and other mammals have been found in glacial ice caps.

Plate tectonics Although the idea of continental drift was first broached in 1915 (by Alfred Wegener), it was not until the 1960s that both definitive evidence and a theoretical mechanism for continental drift convinced most geologists of its reality. The theory of plate tectonics has revolutionized geology. Like any comprehensive scientific theory (such as evolution), it accounts for a great diversity of observations.

It is now accepted that the lithosphere, the solid outer layer of the earth bearing both the continents and the crust below the oceans, consists of eight major and a number of minor PLATES that move over the denser, more plastic asthenosphere below. The heat of the earth's core sets up convection cells within the asthenosphere. At certain regions, such as the mid-oceanic ridge that runs longitudinally down the floor of the Atlantic Ocean, magma from the asthenosphere rises to the surface, cools, and spreads out to form new crust, pushing the existing plates to either side (Figure 6.1). The plates move at velocities of up to 5–10 centimeters/year. Where two plates come together, the leading edge of one may plunge under the other, rejoining the asthenosphere. The pressure of these collisions is a major cause of mountain building, and is responsible for the major mountain chains of the world. When a plate moves over a "hot spot" where magma is rising from the asthenosphere, volcanoes may be born, or a continent may be rifted apart. The great lakes of eastern Africa lie in such a rift valley; the Hawaiian Islands are a chain of volcanoes that

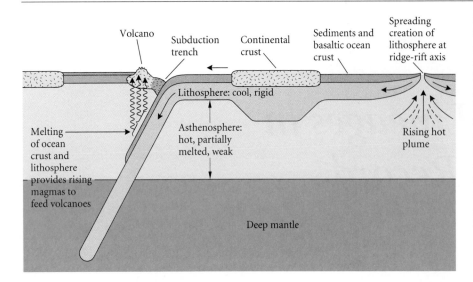

FIGURE 6.1 Diagram of plate tectonic processes in the upper mantle of the earth. A rising hot plume of magma (at right) creates new lithosphere and pushes existing lithosphere, bearing continental land masses, to either side. When such a lithospheric plate encounters another (at left), it plunges under it (at a subduction trench), frequently causing earthquakes and orogeny (mountain building). The heat generated melts crust and lithosphere, causing volcanic activity. (After Press and Siever 1982.)

have been formed by the movement of the Pacific plate over a hot spot(see Chapter 8).

Measuring geological time Astronomers believe that the universe originated in a "big bang" and has been expanding from a central point since then. Light from a receding object is shifted toward longer wavelengths. Thus, from the wavelengths of light emitted by stars and galaxies, astronomers have estimated the age of the universe to be about 14 billion years (one billion = one thousand million). The oldest known rocks on earth are about 3.8 billion years old, but radiometric dating of meteorites and moon rocks, as well as other evidence, consistently indicates that the earth and the rest of the solar system are about 4.6 billion years old. Undoubtedly the earth was so hot for the first billion or so years that its physical, including its geological, properties were extremely different from those of the modern earth.

There is evidence (see Chapter 7) that living things existed by about 3.5 billion years ago. The first evidence of metazoan animal life is about 800 million (0.8 billion) years old. These time spans lie beyond our ready comprehension. As an analogy, if the age of the earth is represented by one year, one day equals 80,000 years, or 40 times the length of the period since Christ. On this scale, the first life appears in late March, the first marine animals make their debut in late October, the dinosaurs become extinct and the mammals begin to radiate on December 26, the human and chimpanzee lineages diverge at about 13 hours before midnight on December 31, and Christ is born at about 13 seconds before midnight. To provide another framework, the time since the American Revolution, 200 years, or about 10 human generations, is 1/22,000 of the time since the earliest hominid of which we have fossil evidence, and 1/32,500 of the time that has elapsed since dinosaurs became extinct. A million years, the unit in which paleontologists commonly measure time, is 5000 times the span since the American Revolution.

Absolute ages of geological events can be determined by the decay of certain radioactive elements in minerals that form in igneous rock. The decay of radioactive parent atoms such as uranium-235 into stable daughter atoms such as lead-207 occurs at a constant rate, with an element-specific half-life. The half-life of U-235, for example, is about 0.7 billion years, meaning that in each 0.7-billion-year period, half the U-235 atoms present at the beginning of the period will decay into Pb-207 (Figure 6.2). The ratio of parent to daughter atoms thus provides an estimate of the rock's age. Several such radiometric isotopes are used (e.g., potassium-40, decaying to argon-40 with a half-life of 8.4 billion years). These radiometric methods can provide cross-checks for the reliability of an age estimate (there are also other meth-

FIGURE 6.2 The loss of parent atoms (e.g., U-235) and the accumulation of daughter atoms (e.g., Pb-207) in the process of radioactive decay. In each time unit (half-life), half the remaining parent atoms decay into daughter atoms. The ratio of the two therefore indicates how many half-lives have elapsed. (After Eicher 1976.)

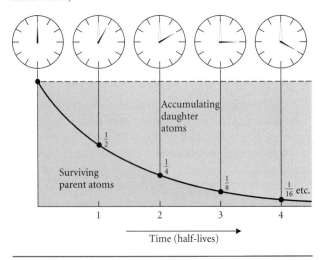

ods of checking reliability). They can be used only on igneous rock (because a sedimentary rock may be formed from sediment particles of many ages), so a fossil-bearing sedimentary rock can be dated only approximately, by bracketing it between younger and older igneous formations. The precision of the age estimate decreases as the age of the rock increases. Dating by radioactive carbon (carbon-14), which has a half-life of only 5730 years, uses rather different principles, and is useful for dating organic materials (plant or animal matter, such as wood) back to 40,000, or at most 70,000, years before the present.

Sedimentary rocks Most sedimentary rocks are formed by deposition of air- or water-borne sediments; because aquatic habitats have a lower elevation than the surrounding terrain, sedimentary deposits most commonly form in aquatic, especially marine, environments. Except in unusual circumstances where rocks have been turned over, younger sediments lie above older ones (the principle of SUPERPOSITION). In many locations, deposition has not been continuous, and sedimentary rocks have eroded; thus in any one area, only a small fraction of geological time is likely to be represented in the STRATIGRAPHIC COLUMN (the local series of STRATA, or sedimentary layers), and those layers that are present are often intermittent. In fact, some geological ages are represented by few sedimentary records worldwide. In general, the older the geological age, the less well it is represented in the paleontological record, because erosion and metamorphosis have had more opportunity to take their toll, and because the oldest rocks are deeply buried, unless canyon formation or other geological processes make them accessible. Although the stratigraphic record is imperfect, there are some places where sedimentation in lakes or oceans has left long, continuous records. Some of these present a long series of distinct annual layers.

Layers deposited at different times have different sedimentary characteristics, and often contain distinctive fossils of species that persisted for a short time and are thus the signatures of the age in which they lived. Contemporaneous strata in different localities can be matched by means of such widespread "index fossils," by widespread patterns of successive sediment types, or by "key beds": distinctive layers deposited over a wide area, such as layers of volcanic ash (Figure 6.3). These few statements constitute a greatly oversimplified caricature of the science of stratigraphy.

The geological time scale Every biologist should memorize the sequence of the eras and periods of the **geological time scale** (Table 6.1 and inner front cover), as well as a few key dates, such as the beginning of the Paleozoic era (and Cambrian period, 543 million years ago, or Mya), the Mesozoic era (and Triassic period, 251 Mya), the Cenozoic era (and Tertiary period, 65 Mya), and the Pleistocene epoch (1.8 Mya). Most of the eras and periods were named and ordered before Darwin, by geologists who did not believe in evolution. They were distinguished, and are still most readily recognized in practice, by distinctive fossil taxa. Great changes

in faunal composition, the result of major extinction events, mark many of the boundaries between them. The absolute time of these boundaries is only approximate, and is subject to slight revision as more information accumulates.

How Good Is the Fossil Record?

> The main cause . . . of innumerable intermediate links not now occurring everywhere throughout nature, depends on the very process of natural selection, through which new varieties continually take the places of and supplant their parent forms. But just in proportion as this process of extermination has acted on an enormous scale, so must the number of intermediate varieties, which have formerly existed, be truly enormous. Why then is not every geological formation and every stratum full of such intermediate links? Geology assuredly does not reveal any such finely graduated organic chain; and this, perhaps, is the most obvious and serious objection which can be urged against the theory. The explanation lies, as I believe, in the extreme imperfection of the geological record.
> Darwin, *The Origin of Species*, Chapter 10

Paleontologists today agree that Darwin was right: the fossil record is dreadfully incomplete (Jablonski et al. 1986). Certain short parts of the record, in some localities, do provide detailed evolutionary histories, as we will see. In general, however, organisms that lack hard skeletal parts, and frail organisms such as birds, are poorly represented or poorly preserved. The best fossil record has been left by groups of marine invertebrates with hard, calcareous skeletons, espe-

FIGURE 6.3 Chronostratigraphic correlation of strata from three localities (1, 2, 3). The sequences of certain distinctive features are common to all three, indicating equivalent times of deposition. These features include distinctive sets of fossils (a, b, e), the polarity of magnetized particles (f), and rock types such as the delta-front sand facies (c). More localized events provide time correlation for fewer localities, but overlapping shared features can correlate all the localities. Sites 2 and 3 share distinctive deposits (channel-sand facies, delta-plain facies, volcanic ash [d]). (After Cooper et al. 1990.)

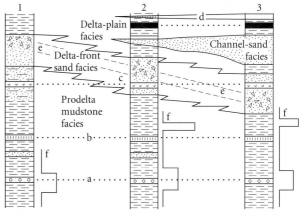

Table 6.1 The geological time scale

ERA	PERIOD	EPOCH	MILLIONS OF YEARS FROM START TO PRESENT	MAJOR EVENTS
CENOZOIC	Quaternary	Recent (Holocene)	0.01	Continents in modern positions; repeated glaciations and lowering of sea level; shifts of geographic distributions; extinctions of large mammals and birds; evolution of *Homo erectus* to *Homo sapiens*; rise of agriculture and civilizations
		Pleistocene	1.8	
	Tertiary	Pliocene	5.2	Continents nearing modern positions; increasingly cool, dry climate; radiation of mammals, birds, snakes, angiosperms, pollinating insects, teleost fishes
		Miocene	23.8	
		Oligocene	33.5	
		Eocene	55.6	
		Paleocene	65.0	
MESOZOIC	Cretaceous		144	Most continents separated; continued radiation of dinosaurs; increasing diversity of angiosperms, mammals, birds; mass extinction at end of period, including last ammonites and dinosaurs
	Jurassic		206	Continents separating; diverse dinosaurs and other "reptiles"; first birds; archaic mammals; "gymnosperms" dominant; evolution of angiosperms; ammonite radiation; "Mesozoic marine revolution"
	Triassic		251	Continents begin to separate; marine diversity increases; "gymnosperms" become dominant; diversification of "reptiles," including first dinosaurs; first mammals
PALEOZOIC	Permian		290	Continents aggregated into Pangaea; glaciations; low sea level; increasing "advanced" fishes; diverse orders of insects; amphibians decline; "reptiles," including mammal-like forms, diversify; major mass extinctions, especially of marine life, at end of period
	Carboniferous		354	Gondwanaland and small northern continents form; extensive forests of early vascular plants, especially lycopsids, sphenopsids, ferns; early orders of winged insects; diverse amphibians; first reptiles
	Devonian		409	Diversification of bony fishes; trilobites diverse; origin of ammonoids, amphibians, insects, ferns, seed plants; mass extinction late in period
	Silurian		439	Diversification of agnathans; origin of jawed fishes (acanthodians, placoderms, Osteichthyes); earliest terrestrial vascular plants, arthropods, insects
	Ordovician		500	Diversification of echinoderms, other invertebrate phyla, agnathan vertebrates; mass extinction at end of period
	Cambrian		543	Marine animals diversify: first appearance of most animal phyla and many classes within relatively short interval; earliest agnathan vertebrates; diverse algae
PROTEROZOIC			2500	Earliest eukaryotes (ca. 1900–1700 Mya); origin of eukaryotic kingdoms; trace fossils of animals (ca. 1000 Mya); multicellular animals from ca. 640 Mya, including possible Cnidaria, Annelida, Arthropoda
ARCHAEAN			3600	Origin of life in remote past; first fossil evidence at ca. 3500 Mya; diversification of prokaryotes (bacteria); photosynthesis generates oxygen, replacing earlier oxygen-poor atmosphere; evolution of aerobic respiration

Source: Dates from Jablonski et al. (1996).

cially those that lived in shallow waters (near shore, or in inland seas that covered major parts of continents). Some groups of planktonic protists with hard shells have left an exceptionally good fossil record. Prominent among these are the protozoan groups Radiolaria and Foraminifera, and single-celled algae called coccolithophores (which make up chalk).

Thus, more often than not, the gradual origins (if indeed there were gradual origins) of species and higher taxa have not been documented: we see the products of speciation and the appearance of new morphologies, but (usually) not the events that led up to them. When you consider what the odds are of finding these origins, however, it is surprising that we have any informative records at all. First, many kinds of organisms rarely become fossilized because they are usually consumed by animals and decomposer organisms. Hence the record of terrestrial organisms, especially of species that occupy forests, where decay is rapid, is generally poor. Second, with some exceptions, such as lake deposits, sediments form in any one locality only very episodically, and so they typically contain as fossils only a small fraction of the species that inhabited the region over time. Third, the fossil-bearing sediments must become solidified into rock, and the rock must persist for millions of years without being eroded, metamorphosed, or subducted below a lithospheric plate. Fourth, the rock must then be exposed and accessible to paleontologists. Finally, there is little reason to suppose that the evolutionary changes we are interested in occurred at the few localities that present strata of that time interval. A lineage that evolved new characteristics elsewhere may appear in a local record fully formed, after having migrated into the area.

The approximately 250,000 described fossil species are a small fraction of those that lived in the past. We *know* that the record is extremely incomplete because, first, geological evidence shows that many time periods are represented by few sedimentary formations worldwide, and that local strata are often separated by gaps of tens of thousands of years or more. Second, many lineages are represented only at very widely separated time intervals; they must have been present in the meantime, but few or no fossils are known. For example, mammals originated in the Jurassic, but their Jurassic and Cretaceous record consists mostly of scattered teeth and skull fragments, and so the Mesozoic history of mammals is only slowly being revealed. Third, many extinct species of large, conspicuous organisms are known from only one or a few specimens. This is the case for many dinosaurs and other terrestrial organisms. Finally, new fossil taxa continue to come to light at a steady rate, indicating that many forms have yet to be discovered. Dinosaurs, for example, are hard to miss, and have surely been collected more assiduously than any other fossils, but at least two dozen new genera of dinosaurs, differing conspicuously from previously known species, have been described in the last 20 years (Figure 6.4).

Even after discovery, the interpretation of fossils poses many difficulties. Their age can be only imprecisely estimated, since they cannot be aged directly by radiometric methods. A sequence of fossils deposited over a short interval is frequently mixed together before the sediment is lithified, so that any given stratum contains a TIME-AVER-AGED sample of specimens that may have lived many thousands of years apart. There is extensive literature on how to tell whether the organism lived at the time and place at which the rock was formed, or was washed in from elsewhere (perhaps from erosion of an older sedimentary rock). Many fossils are crushed or fragmented so that diagnostic characters cannot be studied. For instance, many Paleozoic insects are known only as flattened imprint fossils; almost the entire Mesozoic record of mammals (200 million years or more) consists mostly of isolated pieces of

FIGURE 6.4 New dinosaur species are being found at a steady rate. *Yangchuanosaurus shangyouensis*, a Late Jurassic carnivorous dinosaur, was described in 1978 from two specimens. Its length was nearly 8 meters, and its estimated weight was 1.33 tons. The color pattern has been bestowed by the illustrator's imagination. (From Paul 1988.)

jawbones; many plants are represented by leaves, wood fragments, pollen, and occasional flower impressions, which often cannot be matched together.

Taxonomy in the Fossil Record

Describing and analyzing fossils poses special problems in taxonomy and classification that must be borne in mind when reading paleontological literature.

When discussing contemporaneous forms (such as those living today), most evolutionary biologists use "*species*" to refer to reproductively isolated populations, and "*speciation*" to mean the splitting of one lineage into reproductively isolated gene pools (see Chapter 2). The word *species*, however, is sometimes used simply as a name for a morphologically distinguishable form. This is especially true in paleontology, in which a *single* evolving lineage (gene pool) may be assigned several names for successive, phenotypically different forms. For example, *Homo erectus* and *Homo sapiens* are names for different, distinguishable stages in the same evolving lineage. They are **chronospecies**, rather than separate biological species. The two species names do not imply that speciation (bifurcation into two gene pools) occurred; in fact, it probably didn't in this case. In this book, we shall use "speciation" only in the sense of splitting of lineages (cladogenesis).

In lists of species names for successive geological time periods, a name (e.g., *Homo erectus*) can disappear because the lineage has changed in form and bears a different name. This is "taxonomic extinction," or **pseudoextinction**, and must be distinguished from real extinction, the termination of a lineage that leaves no descendants. Calculations of extinction rates for various geological periods can be inaccurate if pseudoextinctions are included; this problem is less serious with higher taxa than if species names or monospecific genera are counted. Also, many individual species in the fossil record are represented by few specimens and some are ill defined, so paleontologists tend to discuss genera or higher taxonomic categories, for which they have more information. In many groups, living species (especially those that are morphologically almost identical) cannot be distinguished by skeletal features, so a single fossil "species" might well consist of more than one reproductively distinct biological species. In at least some groups, though, living species are skeletally distinct, so the fossil species probably are also (Jackson and Cheetham 1990).

The naming of higher taxa according to a strict formal classification scheme is difficult when the taxa are distributed over time. Suppose that a lineage with a synapomorphic character proliferates in the Cretaceous (Figure 6.5). Call it group A. In the Tertiary, two members of the group evolve different synapomorphies, *b* and *c* respectively, and each proliferates, so that we can recognize groups B and C. These exist at present (the Recent), whereas all members of group A are extinct. Considering only living forms, groups B and C are each monophyletic (they are sister groups), and we might call them families B-aceae and C-aceae. What

shall we call the fossils in group A? In the Cretaceous, they represented a monophyletic group, perhaps the A-aceae. That designation, however, does not reflect the fact that group A is ancestral to the living families; nor would this information be conveyed if we put members of group A in either the B-aceae or the C-aceae. For those who hold a cladistic philosophy of classification, the situation is even more difficult, because cladists do not accept paraphyletic taxa (groups that do not include *all* the descendants of a common ancestor). Whatever we call group A, it is paraphyletic unless it also includes B and C. Every ancestral taxon will be paraphyletic if it is given the same taxonomic rank as one or more of its descendants. There are several conceivable conventions that might be used, but all will meet with some objections. The present controversy on this issue, if resolved, may well result in considerable changes in classification (for example, the class Reptilia would come to include birds, which are probably modified dinosaurs). The issue is one of nomenclature, not truth or falsehood, but it may well affect the presentation and interpretation of evolutionary history.

Evolution in the Fossil Record

The fossil record provides some examples of fairly continuous change in organisms' characteristics, and others in which no continuity is evident. Evolutionary biologists differ on how cases of apparent discontinuity should be explained. In the following sections, we describe some examples of continuous and apparently discontinuous change, as well as speciation (branching of lineages). The

---

FIGURE 6.5   An ancestral clade, A, may give rise to markedly different descendant clades, B and C, with distinctive characters such as *b* and *c*. If, as is often true in traditional classification, all three groups are named as taxa at the same rank (e.g., families), then taxon A is paraphyletic. No scheme of classification conveys both the phylogenetic relationships among the lineages and the phenotypic differences between the "stem group" A and the "crown groups" B and C.

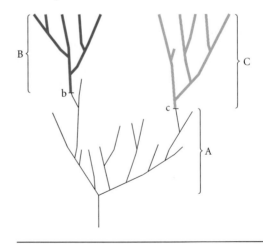

evolution of higher taxa (genera, orders, classes) is given special attention because anti-evolutionists often claim, falsely, that the fossil record does not document the evolution of higher taxa.

Character Change within Species

Our first examples treat changes in characters over geologically short time spans, in cases that present a fairly detailed, continuous record. (Other cases, as well as treatment of subsequent issues, can be found in Levinton 1988; Cope and Skelton 1985; and Hoffman 1989.)

Foraminifera Malmgren, Berggren, and Lohmann (1983) studied a lineage of planktonic foraminiferans (shell-bearing protozoans) that extends over 10 million years (My), from the late Miocene to the Recent. The lineage is called *Globorotalia plesiotumida* during the Miocene and *G. tumida* thereafter. The later form has a larger, thicker shell, with a more massive ridge and a more rapidly increasing height of the whorl (Figure 6.6A). Malmgren et al. analyzed 105 samples through time, concentrating their attention particularly around the Miocene/Pliocene boundary, and commented that "time resolution is very good, between 5×10^3 [5000] yr and 15 $\times 10^3$ yr across the Miocene/Pliocene boundary and about 2×10^5 yr in the remainder of the sequence." (Note that a considerable amount of evolutionary change could occur in 5000 years, which paleontologists would consider a rel-

---

FIGURE 6.6 Evolution of shape in the foraminiferan lineage *Globorotalia*. (A) Scanning electron micrographs, in side and edge views, of specimens from the Late Pleistocene (top), Early Pliocene (middle), and Late Miocene (bottom). (B) A measure of shape, on the horizontal axis, plotted against age in fossil samples from the Miocene (M) to the Pliocene (P) to the Pleistocene (Q). The horizontal bars show the variation around the mean in each sample. Inset shows in detail the rapid change near the Miocene/ Pliocene boundary. (A courtesy of B. Malmgren; B after Malmgren et al. 1983.)

(A) (B)

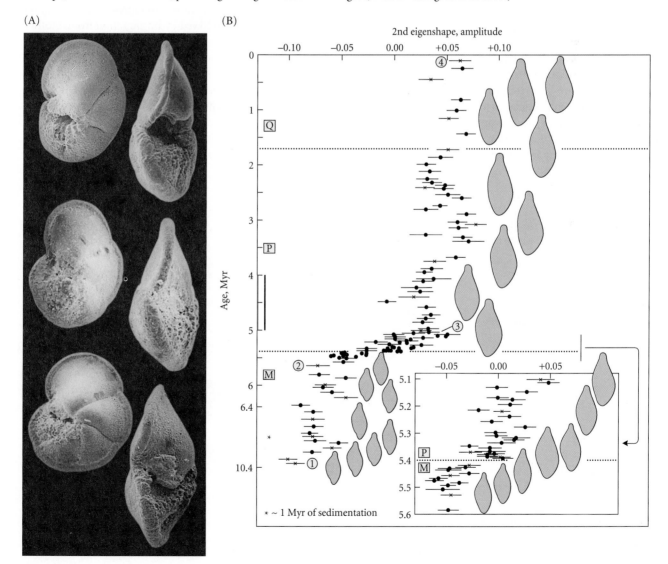

atively detailed record.) Malmgren et al. used a composite measure of shape that combined several of the differences between the early and late forms. Around the Miocene/Pliocene boundary, there was a rapid change in shape, which took about 0.6 My (Figure 6.6B); this was preceded and followed by intervals of about 5 My during which shape fluctuated within a fairly narrow range, without any long-term trend.

This study illustrates several important points that are typical of well-documented fossil lineages:

1. The rate of evolution of this lineage was highly variable over time.
2. The character was seldom absolutely constant; rather, it changed continually, but without any overall direction.
3. The character varied among individual organisms at all times (represented by the horizontal bars in the figure).
4. The species evolved rapidly from one relatively stable phenotype to another.

These points will be relevant when we discuss the topic of "punctuated equilibrium."

Trilobites Each of eight lineages of Ordovician trilobites displays a trend toward an increase in the number of ribs on the pygidium (the rear part of the dorsal region) (Figure 6.7). This trend occurred over a 3-million-year span; the samples are widely spaced in time, but there is no evidence that any of the lineages branched. Within this general trend, fluctuations in rib number occurred. In most of these lineages, the animals at either end of the series had been assigned different species names, and in one case they were assigned to different genera (Sheldon 1987). Thus the origin of higher taxa—genera in this case—is documented by fossils.

Sticklebacks Living sticklebacks, small fishes in the genus *Gasterosteus*, have been the subject of numerous studies of behavior, ecology, and the genetic basis of morphological characteristics. Michael Bell and his colleagues (Bell et al. 1985) studied Miocene fossils of *G. doryssus* in Nevada in an unusually complete series of strata that were laid down annually for 110,000 years. They studied large samples from layers that were on average about 5000 years apart. Three skeletal characters (Figure 6.8) changed more or less independently. All three characters fluctuated, even the two that display an overall trend. The pelvic structure is a particularly interesting feature; the score represents a range from fully developed pelvic bones and fins, through graded levels of reduction, to a vestigial condition. (The abrupt increase in the pelvic structure score is thought to represent the extinction of one population, followed by immigration from another locality.) Absence of the pelvic elements is a major feature of certain families of fishes, but it varies within both the fossil and living species of *Gasterosteus*. It should be noted that if Bell and his colleagues had taken samples spaced at 20,000 years or more—which would be an unusually good degree of resolution in most fossil records—many changes would appear more abrupt and discontinuous than Bell's finer resolution shows them to have been.

FIGURE 6.7 Changes in the mean number of ribs in eight lineages of trilobites over a 3-million-year stratigraphic section in Wales. Irregular but mostly gradual changes are seen in most of the lineages. (After Sheldon 1987.)

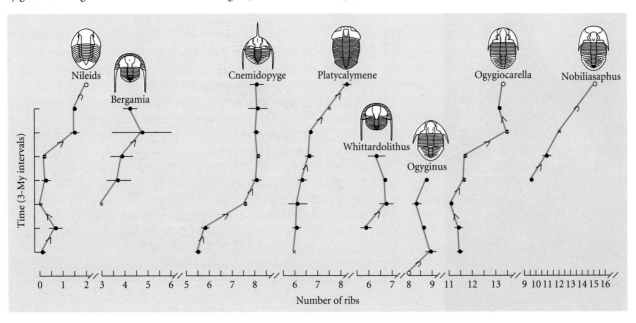

FIGURE 6.8 Changes in the mean values of three characters in the stickleback *Gasterosteus doryssus* over a span of 110,000 years. (After Bell et al. 1985.)

(A) Pelvic structure

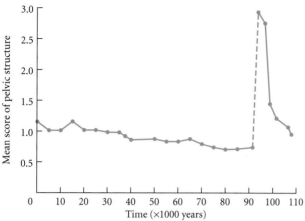

(B) Dorsal spine number

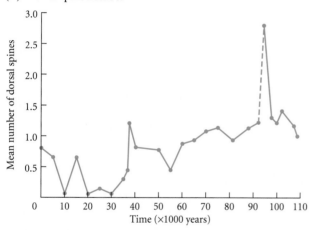

(C) Predorsal pterygiophore number

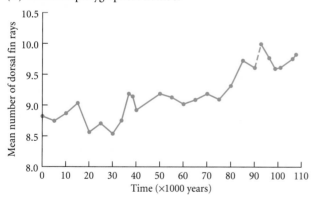

Speciation

As noted in Chapter 2 (and also Chapter 16), most evolutionary biologists hold that speciation events (division into separate, reproductively isolated gene pools) usually occur by genetic divergence of geographically segregated populations. Some authors hold that the populations must be fully **allopatric** (completely geographically isolated) for a considerable time. Therefore, in a geological section from a single locality, we should *not* expect to see a lineage gradually diverge into two or more species; full documentation of the process of speciation would require samples from several or many localities throughout the time span. Such data, especially with the fine time resolution required to show gradual divergence, are very few.

Although fossils are assigned different species names if they are morphologically distinct, we cannot tell whether or not similar forms are different biological species when they are separated in space or time (in fact, the question is meaningless if they are separated in time). This is because different geographic populations of the same species often differ in morphology, yet interbreed where their ranges meet (see Chapter 9). Hence, to show that two distinguishable sets of fossils represent different biological species, they must occur *in the same place at the same geological time* (they must be **sympatric** and **synchronic**).

Speciation is the critically important process that gives rise to the diversity of sexual organisms. Moreover, it plays another role—that of fostering morphological evolution—according to the hypothesis of punctuated equilibrium, as we are about to see.

Radiolarians Radiolarians are another group of planktonic, shelled protozoans. Like foraminiferans, they rain down onto the ocean floor continually, in enormous numbers. The species *Eucyrtidium calvertense* and its larger relative *E. matuyamai* are distinguished most readily by size, but also by shape and the sculpturing of the shell. *E. calvertense* inhabited the North Pacific, south of 40° N latitude, from at least 4 Mya (million years ago) almost until the present. The lineage first appeared north of 40° N about 1.9 Mya, at first being indistinguishable from the southern form, but soon thereafter showing the features of *E. matuyamai*. Sometime between 1.9 and 1.6 Mya, *E. matuyamai* reinvaded southern waters. Davida Kellogg (1975) took samples from a North Pacific sediment core, embracing more than 3 My (Figure 6.9). After *E. matuyamai* became sympatric with its "parent" species, it experienced a rapid, sustained evolution of larger size, whereas *E. calvertense* declined in size. Thus both species displayed a sustained divergence. Interestingly, variation within samples of *E. calvertense* was reduced after the two species became sympatric. Other examples of gradual divergence of radiolarian species from a common ancestor have been described (Lazarus 1983).

***The Devonian trilobite* Phacops rana** A study of a Devonian trilobite lineage by Niles Eldredge (1971) is of considerable historical interest because it is one of the chief examples that Eldredge, together with Stephen Jay Gould, later used to support the hypothesis of punctuated equilibrium (Eldredge and Gould 1972; see below). Eldredge studied trilobites of the *Phacops rana* complex over an 8–10 My span in the Devonian. During this time, trilobites inhabited the eastern marginal sea along the Atlantic coast of North America, as well as a shallower EPICONTINENTAL SEA that covered much of what is now the Ameri-

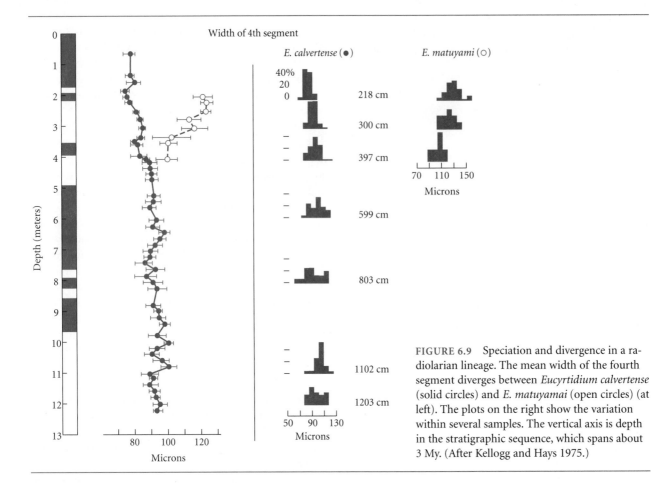

FIGURE 6.9 Speciation and divergence in a radiolarian lineage. The mean width of the fourth segment diverges between *Eucyrtidium calvertense* (solid circles) and *E. matuyamai* (open circles) (at left). The plots on the right show the variation within several samples. The vertical axis is depth in the stratigraphic sequence, which spans about 3 My. (After Kellogg and Hays 1975.)

can Midwest. At several times, the level of the epicontinental sea increased, expanding the sea's area (a TRANSGRESSION) or decreased, lowering its areal extent (a REGRESSION). The trilobites had numerous simple eyes (lenses) arranged in dorso-ventral columns (files); the number of files is the chief character distinguishing the various forms (Figure 6.10A). Both trilobite populations initially had 18 files. According to Eldredge, the eastern marginal population went through a brief stage in which individuals varied in file number, and then reached a stable 17-file condition. About 1–2 My after this event, the 18-file form in the epicontinental sea was abruptly replaced by a 17-file form (Figure 6.10B). Eldredge interprets this to represent extinction of the 18-file form, caused by a regression of sea level, followed by immigration of the 17-file form from the marginal sea during a transgression. Toward the end of the time interval studied, there was again a reduction in file number, from 17 to 15, that occurred at about the same time in both seas.

Eldredge (1971) called attention to the rapidity of these changes, which were followed by an extensive period of no change, or **stasis**. He interpreted the changes as speciation events, involving the origin of new species (twice) in a marginal area, followed by immigration into the central range. Although the change in the marginal population was gradual, the fossil record of the central region shows an abrupt replacement of one form by another. Thus our frequent failure to find intermediate evolutionary changes in the fossil record could be easily explained if most such changes happen when a marginal population, not represented by a fossil record, evolves into a new species with different morphological features. In *Phacops rana*, Eldredge was fortunate to find evidence of change within the marginal population, but paleontologists can seldom hope to be so fortunate.

As we will see (Chapter 16), the same idea had been introduced by Ernst Mayr in 1954. Mayr postulated that most new species arise by evolution in small, peripheral populations, which later expand their range. He pointed out that this would explain many of the gaps in the fossil record. Eldredge suggested that *Phacops rana* exemplified Mayr's theory.

Bryozoans Bryozoans, or "moss animals," are colonial filter-feeding animals in which, as in corals, the colony grows by vegetative propagation of new "individuals" (zooids), each of which secretes a skeletal structure. Alan Cheetham (1987) studied the bryozoan genus *Metrarabdotos* over a 4.5 My stratigraphic series in the Upper and Lower Miocene of the Dominican Republic. He recognized nine "species," some of which are clearly biological species, since they co-occur at the same time. Cheetham studied 46 characters in each lin-

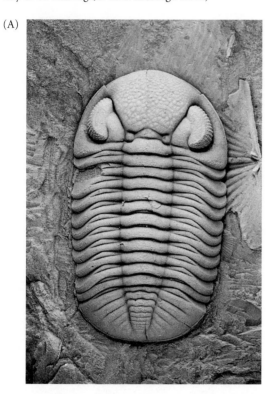

FIGURE 6.10 Evolution in the trilobite lineage *Phacops rana.* (A) Dorsal view, showing the rows of lenses of each eye. (B) Temporal range of forms with different numbers of files in epicontinental and marginal seas. Arrows indicate presumed migration of a form from one region to another; crosses indicate final disappearance. Dashed lines indicate the origin of new forms, interpreted as new species. (A courtesy of N. Eldredge; B after Eldredge 1971.)

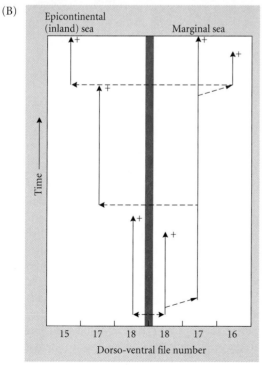

eage, an unusually ambitious effort. These were used both to distinguish species (nine characters were particularly useful in this respect) and to suggest phylogenetic relationships among them (Figure 6.11). Very few of these characters changed significantly over time *within* any of the species that Cheetham recognized, and none of the nine most diagnostic features did so. Rather, *characters changed rapidly, from one stable state to another, as new species originated.* Cheetham concluded that most of the features exhibited a pattern of *extended stasis and occasional rapid change.*

Punctuated Equilibrium

The highly controversial idea of **punctuated equilibrium**, introduced by Eldredge and Gould (1972) and elaborated both by them (e.g., Gould and Eldredge 1977, 1993) and by Steven Stanley (1979), consists of both a *claim about the pattern of change* in the fossil record and a *hypothesis about evolutionary processes* (Figure 6.12). The claim about *pattern* is that many or most phenotypic characters change little over extended spans of geological time (equilibrium, or stasis), but when they do evolve, they change relatively rapidly from one static state to another (that is, the stasis is punctuated by change). The *hypothesis* that Eldredge and Gould introduced is that characters evolve primarily in concert with true speciation (Figure 6.12C). Moreover, if, as Mayr hypothesized, new species (reproductively isolated entities) evolve rapidly in small, isolated populations, the transitional stages in the divergence of these populations will seldom be preserved in the fossil record. Eldredge and Gould, drawing further inspiration from Mayr, suggested that the widespread ancestral form does not change very much because of genetic "constraints" that prevent features from changing, even if natural selection would favor new features. If a newly arisen species becomes widespread and abundant, its new features likewise become stabilized, and little further evolution occurs unless it in turn buds off new "daughter" species. (The nature of the hypothesized genetic constraints is discussed in Chapter 16.)

Eldredge and Gould contrasted their model with what they took to be the traditional view in paleontology—that features evolve slowly, steadily, gradually, without any particular association with speciation. They called this the "phyletic gradualism" model (Figure 6.12B). Another possibility, which Malmgren et al. (1983) called "punctuated gradualism," is that character evolution may not necessarily be associated with speciation, but may nevertheless show rapid transitions between long-stable states (Figure 6.12D).

We will consider in later chapters both the basis of Eldredge and Gould's hypothesis about the evolutionary process and why it is so controversial. Whether or not the hypothesis of punctuated equilibrium is important depends on how common the pattern is. Gould and Eldredge (1993) have concluded that punctuation and stasis is the most common pattern in the fossil record, but some other researchers (e.g., Levinton 1988) disagree.

FIGURE 6.11 The phylogeny and temporal distribution of species of bryozoans ("moss animals") of the genus *Metrarabdotos*. The horizontal distance between species derived from an immediate common ancestor, and between temporal samples of each single species, indicates the degree of morphological difference. The general pattern is one of abrupt shifts to new, rather stable, morphology. (After Cheetham 1987.)

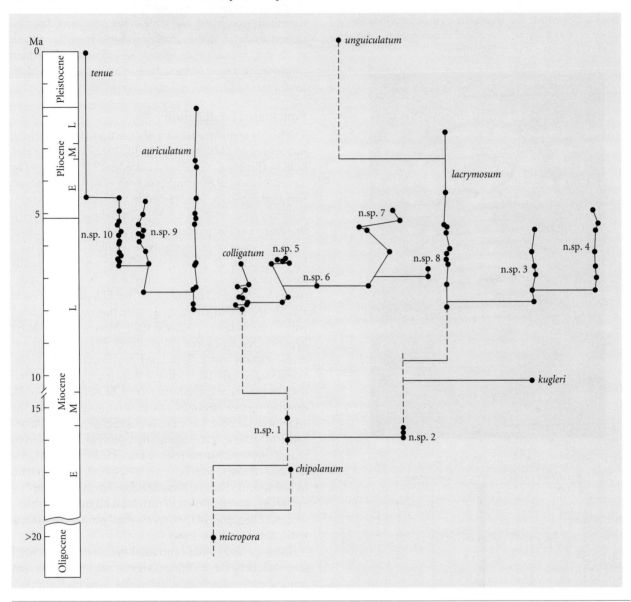

Evolutionary Transitions among Genera

In Chapter 5, we noted that distinctions among genera of living organisms are often arbitrary because of gradations among species in the characters that define genera. The same is often true in the fossil record when an evolving lineage is given different generic names at different time horizons. The following examples begin a series of cases in which the fossil record documents the evolution of higher taxa.

Rotulid echinoids The echinoids, including the sea urchins, are like other echinoderms in having a skeleton composed of small calcareous plates that increase in size as the animal grows. A single lineage, extending from the Miocene to the Recent, is called *Rotuloidea* early in its history and *Heliophora* later (Figure 6.13). In the Miocene, the plates are simple in form, with very slight irregularities appearing in the margin of large plates (i.e., in individuals of greater size and age). In the early Pliocene, small plates (of juvenile specimens) are likewise simple, but large ones have a more scalloped edge. These indentations are more pronounced by the late Pliocene, and in Pleistocene and Recent specimens the indentations are very deep and more numerous. Moreover, they appear earlier in ontogeny: all but the very youngest specimens have pronounced indentations. Thus the *timing* of development of the feature has been advanced to an earlier age during the course of evolution. This is an example of **heterochrony**, an evolutionary change in the

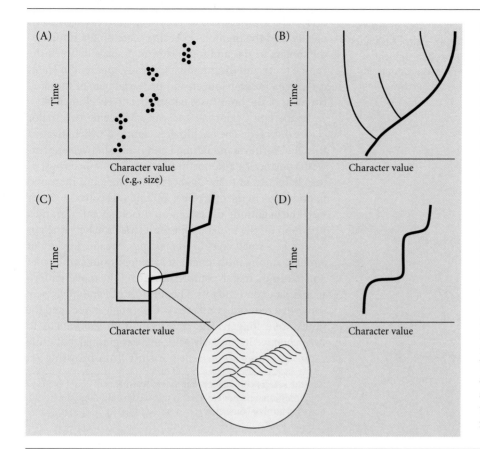

FIGURE 6.12 Three models of evolution, as applied to a hypothetical set of fossils (A). (B) The "phyletic gradualism" model, said by Eldredge and Gould to be the traditional paleontological model. The data in panel A might correspond to the boldly darkened line. Evolutionary change is not associated with speciation. (C) The "punctuated equilibrium" model of Eldredge and Gould, in which morphological change occurs in new species, while the parent species remain unchanged. Morphological evolution, although rapid, is still gradual, as shown in the inset, which illustrates the shift in the mean of the variable character. (D) The "punctuated gradualism" model of Malmgren et al. (1983), in which a lineage rapidly evolves from one equilibrium to another, but speciation need not occur. The absence of intermediates between successive morphologies is attributed to the imperfection of the fossil record.

time of appearance or rate of development of a character during ontogeny. Several forms of heterochrony have been important in evolution (see Chapter 23).

Horses The fossil record of horse evolution is so famous that it is almost overworn with textbook use. It is often presented as if evolution had proceeded in a single direction toward modern horses. In fact, some authors early in the twentieth century believed that evolution inexorably follows ORTHOGENETIC (straight-line) trends, driven by mysterious inner forces. However, the history of horse evolution is much more complicated than its usual textbook portrayal. For this reason, and because the evolution of the horse family (Equidae) illustrates many principles of evolution, we will consider it in some detail. (The following material is drawn from Radinsky 1984; Woodburne and MacFadden 1982; and MacFadden 1986, 1992.)

Horses, together with rhinoceroses, tapirs, and several extinct groups, are members of the order Perissodactyla, characterized by a saddle-shaped facet on the astragalus (a tarsal bone) that facilitates fore-and-aft rotation over the bone below it. The third digit is the central axis of the limb. The perissodactyls diversified about 55 to 45 million years ago, in the Eocene, from an ancestor that was very similar to certain condylarths, a group of mammals that gave rise to various ungulate (hoofed) lineages.

Modern horses (*Equus*) are highly adapted for running in open country and for grazing on grass. The central digit (digit 3) of each foot is greatly enlarged, and digits 2 and 4 are reduced to small remnants of the metapodials, fused to the central metapodial. (In humans, the metapodials include the metacarpals that lie in the palm of our hand and the metatarsals that form our foot between the ankle and the toes.) Only the tip of the enlarged toe, bearing a hoof, contacts the ground. The teeth of horses have important adaptations for feeding on grass, which contains silica and wears teeth rapidly. The premolars and molars are high-crowned (hypsodont) and similar in shape and structure, bearing elevated ridges (lophs) with deep valleys between them that are filled with cement. As the tooth is worn with use, the enamel of the lophs provides an efficient grinding surface. In non-hypsodont teeth, cement is limited to the roots.

Hyracotherium (Lower Eocene of North America and Eurasia) is considered the earliest known equid, although it closely resembles some condylarths (phenacodontids). It weighed about 20–35 kilograms, and browsed on leaves and succulent plants. The molars had incipient lophs, but these varied substantially among individuals. The molars were larger than the premolars. *Hyracotherium* had four digits on the front feet and three on the hind (Figure 6.14). As in some condylarths, the terminal phalanges (finger and toe bones) of the middle digits were slightly enlarged and hooflike, but *Hyracotherium* walked with the full length of the digits appressed to the ground, and had subdigital pads. The legs were rather short, although the metapodials were longer than in phenacodontids.

FIGURE 6.13 Phylogenetic and ontogenetic change in the size and shape of the skeletal plates of rotulid sea urchins. Each horizontal row shows ontogenetic change during development of the species indicated. The adult condition shows pronounced evolution from *Rotuloidea vieirae* to its descendant *Heliophora orbicularis*. (After McNamara 1988.)

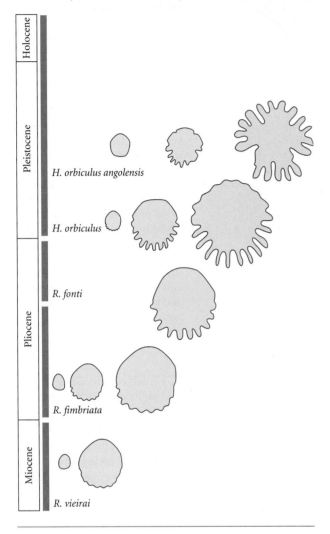

Holocene

Pleistocene

H. orbiculus angolensis

H. orbiculus

Pliocene

R. fonti

R. fimbriata

Miocene

R. vieirai

6.16). The loph pattern of the molars increased in complexity, and the premolars became increasingly similar to the molars in size and loph pattern. A space between the canine tooth and the premolars, already apparent in *Hyracotherium*, became longer, and the facial part of the skull (in front of the braincase) became relatively elongated.

In the late Oligocene and early Miocene, two major clades diverged. The anchitheres, some of which invaded the Old World via the Bering Land Bridge between northwestern America and northeastern Asia, retained most of the characters of their Oligocene ancestors, but increased in body size. In the other line, *Parahippus* evolved a longer face, the beginning of cement on the molar surface, and a new loph on the molars. This loph (the crotchet) first appeared as a small crest in *Mesohippus*, became larger, but variable among individuals, in *Miohippus* and *Parahippus*, and became a major, complex feature of the molar pattern in their later descendants. The side toes of *Parahippus* were somewhat reduced, compared with its predecessors, and the limbs were longer. The transition from *Parahippus* to *Merychippus** in the early Miocene was gradual, but quite rapid, and appears to reflect a shift from browsing on

*For the sake of consistency with most treatments of equid evolution, traditional names are used in this section, although the nomenclature of some taxa is in a state of flux.

FIGURE 6.14 Front (above) and hind (below) feet of *Hyracotherium* (left) and two related "condylarths," *Tetraclaenodon* (center) and *Phenacodus* (right). Although these forms were closely related, *Hyracotherium* shows the elongated metapodials and the reduction in side toes that became accentuated in the later Equidae. (After Radinsky 1966.)

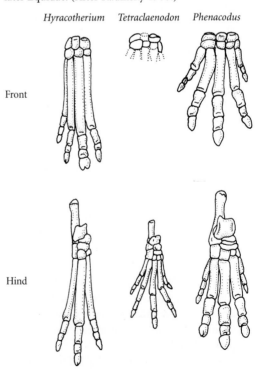

Hyracotherium   *Tetraclaenodon*   *Phenacodus*

Front

Hind

Although *Hyracotherium* was distributed throughout North America and Eurasia, its descendants followed independent evolutionary paths on these continents. *Hyracotherium* gave rise to a short-lived Eurasian lineage (palaeotheres), but most horses evolved in North America, with several American lineages invading Eurasia and South America (Figure 6.15). (*Equus* became extinct in North America, its region of origin, about 10,000 years ago, but persisted in the Old World; the domestic horse was probably derived from the Asian species *Equus przewalskii*, and was brought to America by Spanish explorers.)

In the lineage from *Hyracotherium* to *Mesohippus* (Oligocene), there was a progressive trend toward larger body size and longer limbs. The foot structure and mode of locomotion were much the same as in *Hyracotherium*, but in *Mesohippus* the fourth toe of the front foot was greatly reduced, and is absent in some specimens (Figure

FIGURE 6.15 The evolutionary relationships of the genera of the Equidae. Most of the diversification occurred in North America, but some lineages invaded South America and the Old World (Eurasia, Africa). *Equus*, the only surviving genus, persists only in the Old World. (After MacFadden 1992.)

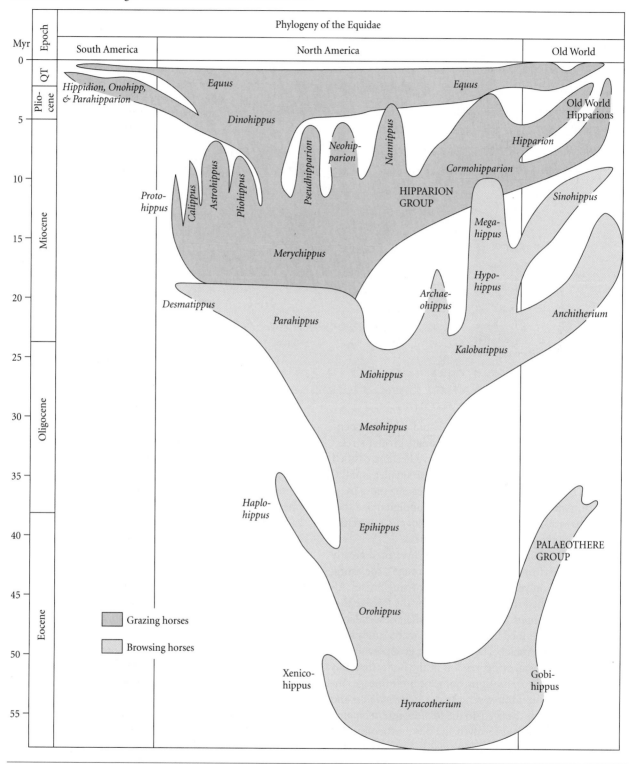

leaves and herbs to grazing on grasses. During the Miocene, the climate of much of North America became drier, and grasslands became widespread. In this context, rapid running was apparently advantageous, and grass-feeding favored a major change in tooth morphology. In *Mery-* *chippus*, the chewing teeth (molars and premolars) became hypsodont, and cement filled the interstices between the highly elevated ridges. Accompanying the increase in tooth height, the tooth row shifted forward, so that the teeth lay anterior to the eye socket. This provided room for the

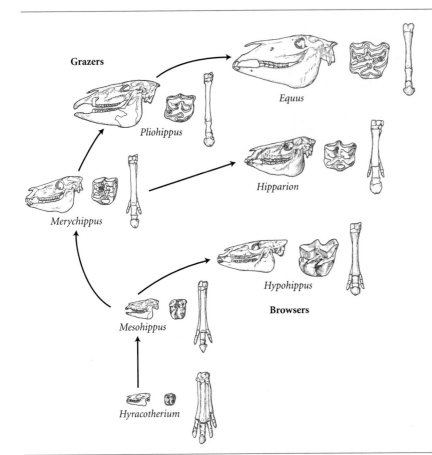

Grazers

Equus

Pliohippus

Hipparion

Merychippus

Hypohippus

Browsers

Mesohippus

Hyracotherium

FIGURE 6.16 Large-scale trends in the evolution of the Equidae, illustrating only a few of the many genera (see Figure 6.15). Heads and teeth are drawn to scale, showing the increase in size in these lineages. Front feet are not to scale. Other trends include: a shift from cusps to complex ridges on the grinding surface of the premolars and molars; elongation of the face and of the space between incisors and cheek teeth; an anterior shift of the cheek teeth in the *Merychippus-Equus* lineage so that they lie forward of the eye; a deeper lower jaw bone; reduction and loss of the side toes; enlargement of the terminal phalanx (hoof); elongation and enlargement of the central metapodial (the longest bone in the foot). (From Stanley 1993, after Simpson 1951.)

elongated teeth, the roots of which would otherwise penetrate the socket. Perhaps to accommodate greater stresses in chewing, a ligament behind the eye became ossified into a bony bar. The lateral digits became smaller and did not reach the ground; the central digit became more massive, terminated in a hoof, and lost the subdigital pad that characterizes all browsing horses. *Merychippus* apparently walked on its toe-tips, as do modern horses.

Grazing descendants of *Merychippus* diversified greatly in the Miocene (from 25 to 8 million years ago), coexisting for several million years with the last browsing horses. All the grazers retained three toes, except for the lineage that gave rise to *Pliohippus, Dinohippus,* and their descendants. Several of the three-toed lineages evolved increasingly hypsodont teeth, in parallel. The *Hipparion* group evolved a large preorbital fossa, or depression in the bone, that may have amplified its vocalizations. Some of these lineages increased in body size, but several (e.g., *Archaeohippus* and *Nannippus*) became dwarfed. In the *Pliohippus/Dinohippus* lineage, from which *Equus* evolved, the side toes became vestigial and were finally reduced to small splints, the central toe and its hoof became enlarged, the ulna and fibula became reduced so that the weight was borne by the radius and tibia, and the loph pattern of the chewing teeth became more complex. In this lineage, body size increased, reaching its maximum in some Pleistocene species of *Equus,* which were somewhat larger than the modern species of this genus, the sole survivors of this eventful history.

The history of the Equidae illustrates several of the themes we have previously emphasized. The differences between *Hyracotherium* and the latest equids, especially the grazers, are many and great, but all these features evolved by many *intermediate stages.* Features such as the crotchet appeared first as slight alterations, *variable among individuals,* and became increasingly accentuated with time. The *rate of evolution varies* among features, from time to time within a lineage, and among lineages, with some lineages retaining ancestral character states while others diverge (see Figure 6.16). Working back in time from the modern horse (*Equus*), one can plot *trends* in many features; but this ignores the *numerous other lineages* in which these features did not follow the *Equus* trend (e.g., browsers), or in which the trend was reversed (e.g., in body size), or which evolved *their own distinctive features* (e.g., the preorbital fossa of the *Hipparion* group). If *Equus* had not been the only surviving equid—if the many other Miocene lineages had not become extinct—it is doubtful that the horses would ever have been presented as an example of straight-line (orthogenetic) evolution. In arguing against authors who believed that evolution is driven by inexorable inner forces toward a goal, George Gaylord Simpson (1953) remarked that "the whole picture is more complex, but also more instructive, than the orthogenetic progression that is still being taught to students as the history of the Equidae. It is a picture of a great group of real animals living their history in nature, not of robots on a one-way road to a predestined end."

Classes of Tetrapod Vertebrates

In general, the fossil record divulges the least information about the modes of origin and relationships of the very highest taxa, such as phyla and classes. Creationists have therefore proclaimed that "the fossils say no!" to evolution (Gish 1974). However, they are simply wrong. The origin of some higher taxa is very well documented. We present the evolution of the tetrapod (four-legged) vertebrate classes as examples.

In reviewing such examples, it is important to realize that we generally cannot point to a particular fossil species as *the* ancestor of any vertebrate class. Rather, we have representatives of the group that included the ancestor. Often, the fossil species in hand are too late in the record, or have specialized features that make it unlikely that they themselves were the ancestors. This is hardly surprising. For example, there are a few species of gliding lizards (*Draco*) in which greatly elongated ribs support a membrane. Who knows but that they may prove, millions of years hence, to have been the ancestors of a major new kind of vertebrate? But if they do, what is the likelihood that among the few fossils that will be left from the 3000 species of lizards living today, those few species will be found? Even if our fossils are only collateral relatives of the actual ancestors of later forms, however, they display many of the novel features (synapomorphies) that characterize the higher taxa.

From Sarcopterygii to Amphibia

Two groups of bony fishes, the ray-finned fishes and the Sarcopterygii, appeared in the early Devonian, about 408 Mya. The Sarcopterygii (Figure 6.17) include lungfishes (Dipnoi), of which three genera survive today; coelacanths, of which only one species survives; and several fossil groups formerly called rhipidistians. Among this last group, the Osteolepiformes are most germane to our story. Many of the features of Osteolepiformes are shared by other Sarcopterygii.

The most relevant features of the skeleton of Devonian Osteolepiformes (Figure 6.18) are the following.* Rays (thin, jointed, bony elements) supported a tail fin. The paired fins were fleshy (hence the name Sarcopterygii), with an axis of several large bones to which short side branches attached (Figure 6.18E). This structure differs markedly from the fins of ray-finned fishes, and displays

---

An editorial opinion: Understanding tetrapod transitions requires attention to anatomical detail and nomenclature, as does all of taxonomy and phylogenetic study based on morphology. There has been an unfortunate recent tendency among biologists to consider morphology (and taxonomy) boring, archaic, and unnecessary. It is not. Morphological structure and nomenclature are as essential for knowing a group of organisms as chemical structure and nomenclature are in biochemistry. Morphology is what organisms do their physiology, behavior, and ecology with, and it is the subject of diverse, innovative, and conceptually rich studies (M. Wake 1992). The special difficulty of morphology and taxonomy is that each major group of organisms has its own nomenclature; but knowledge of at least one group of organisms can provide a rich source of understanding.

FIGURE 6.17 Living lobe-finned fishes. Note the fleshy base of the pectoral (anterior) and pelvic (posterior) paired fins. (A) The Australian lungfish (*Neoceratodus forsteri*), which resembles Devonian lungfishes. (B) The coelacanth (*Latimeria chalumnae*). Coelacanths were thought to have become extinct in the Cretaceous until this species was discovered in the Indian Ocean in 1938. Lungfishes and coelacanths are related to the rhipidistian ancestors of tetrapods. (A after Moyle and Cech 1982; B after Gregory 1951.)

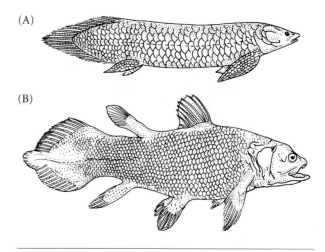

(A)

(B)

pronounced homologies with tetrapod limbs (compare Figure 6.19E). The basal bones of the paired fins articulated to a small posterior pelvic girdle and to a large pectoral girdle that included some of the same bones as in amphibians. The braincase consisted of a movable anterior and a distinct posterior unit. The notochord, which does not persist beyond the embryonic stages in most "higher" vertebrates, ran through a canal below the brain cavity. There was a small opening (the spiracle) between the side of the skull and one of the gill-bearing arches. That arch, called the hyomandibular, formed a brace between the jaw and the otic (ear) region of the skull. (We will trace the evolution of the hyomandibular in the evolution of later vertebrates.)

Surrounding the braincase were dermal bones. In the region of the nostril and the tip of the snout, these were small and variable, but otherwise homologous bones can be identified not only among the various sarcopterygians, but among later vertebrates as well. Their homologies can be traced in part by their positions relative to the external nares (nostrils), eyes, and pineal opening (Figure 6.18B–D), and by the pattern of canals that form part of the lateral line system. We shall make repeated reference to many of these bones, and to those of the lower jaw. We may note at this point that the several bones of the lower jaw include the large *dentary* and the *articular*, which articulated with the *quadrate* bone of the skull. Teeth were borne on several bones of the lower jaw, and on the skull by the premaxilla, maxilla, vomer, palatine, and ectopterygoid. In cross section, the teeth display labyrinthine infoldings of the surface. The external nares led to internal nares, or choanae. De-

FIGURE 6.18 *Eusthenopteron*, a member of the group of lobe-finned fishes (rhipidistians) from which tetrapods arose (cf. Figure 6.19). (A) Skeleton and reconstructed outline. (B–D) Dorsal, ventral, and lateral views of head. (E) Skeleton of pelvic fin and pelvic girdle. The dotted lines in B and D are canals associated with the lateral line system. Note teeth on the premaxilla (pm), maxilla (m), vomer (v), palatine (p), and ectopterygoid (ect). Abbreviations for bones noted in the discussion and in this and other figures of tetrapod evolution are: a, angular; art, *articular*; cor, coronoid; d, *dentary*; ect, ectopterygoid; f, frontal; j, *jugal*; l, lacrimal; m, *maxilla*; n, nasal; p, parietal; pal, palatine; part, prearticular; pm, *premaxilla*; po, postorbital; pop, preopercular; pp, postparietal; prf, prefrontal; ps, parasphenoid; pt, pterygoid; q, *quadrate*; qj, quadratojugal; sa, surangular; sp, splenial; sq, *squamosal*; st, supratemporal; t, tabular; v, *vomer*. The italicized bones are especially important. (A, E after Andrews and Westoll 1970; B–D after Moy-Thomas and Miles 1971.)

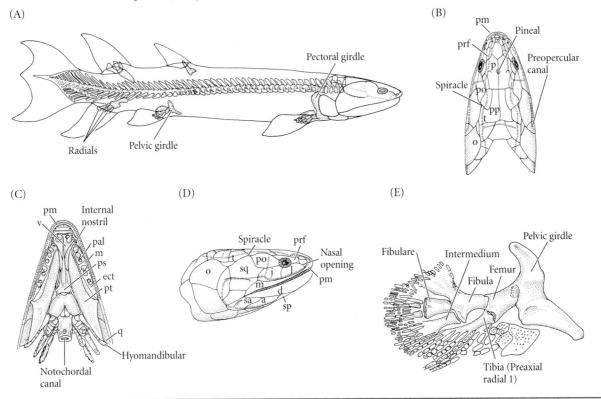

vonian Sarcopterygii, like their living representatives and certain "primitive" ray-finned fishes, had lungs. They were most diverse in fresh water.

The first definitive amphibians are the ichthyostegids (e.g., *Ichthyostega*, Figure 6.19) from the very late Devonian of Greenland. The ichthyostegids closely resemble osteolepiform sarcopterygians in having a tail fin, a notochordal canal in the braincase, a bipartite (but immovable) braincase, a large hyomandibular (called the *stapes* in tetrapods) supporting the skull, and a lateral line system; these features were lost or modified in later amphibians. The dermal bones of the skull and lower jaw are almost all directly homologous to those of osteolepiforms (Figure 6.19B–E); the chief differences are that the opercular and several other bones have been lost, the small dermal bones at the front of the snout are fewer and consistent (they have become "individualized"), and the skull is much longer in front of the eyes and much shorter behind. Thus the skull and many of its component bones have changed in shape. Teeth are borne on the same bones as in osteolepiforms, and have the same labyrinthine structure (hence the name labyrinthodonts, a large group of Paleozoic amphibians that includes the ichthyostegids). As in osteolepiforms, the external nares are connected by a passage to the internal nostrils.

The structure of the limbs and gill arches suggests that ichthyostegids were semiaquatic (in fresh water) and had internal gills as well as lungs (Coates and Clack 1991). However, they had definite legs and highly developed limb girdles. The pelvic girdle is a solid structure, with three bones on each side that are retained in all later tetrapods. The pectoral girdle has the same bones as in sarcopterygians, although modified in form. The proximal limb bones (humerus, radius, ulna, certain carpals of the forelimb; femur, tibia, fibula, certain tarsals of the hindlimb) are directly homologous, and similar in shape, to those of osteolepiforms (compare Figures 6.18E and 6.19E), but there were definitive digits in place of the numerous side branches of the sarcopterygian fin.

Until recently, the distal parts of the limbs of ichthyostegids were not well known. One of the most exciting recent discoveries in vertebrate paleontology has been a well-preserved forelimb of *Acanthostega* and a hindlimb of *Ichthyostega* (Coates and Clack 1990). Both are ichthyostegids from the Upper Devonian of Greenland.

FIGURE 6.19 *Ichthyostega*, a Devonian labyrinthodont amphibian. Although the legs and girdles are well developed, many features, especially of the vertebrae and skull, are very similar to those of rhipidistian fishes (cf. Figure 6.18). (A) Reconstruction of skeleton. (B–D) Dorsal, ventral, and lateral views of skull.

(E) Hindlimb. Compared with *Eusthenopteron*, the snout is longer, and most of the opercular bones (such as o in Figure 6.18) are absent; otherwise the skull is very similar. Note that the limb has more than five digits. (A after Jarvik 1955; B–D after Jarvik 1980; E after Coates and Clark 1990.)

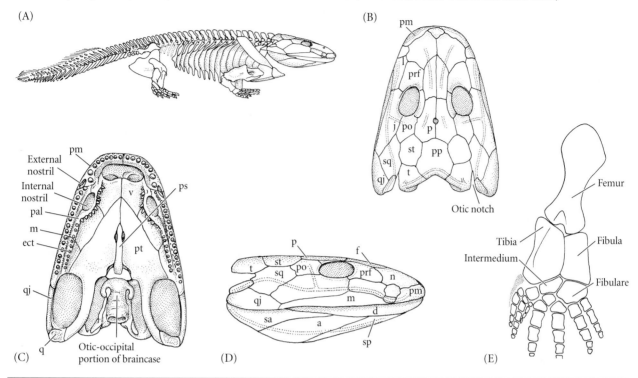

Almost all later tetrapods have five digits or fewer. It had therefore been almost universally assumed that pentadactyly was the ancestral tetrapod condition. The new discoveries, however, show that *Acanthostega* had eight forelimb digits, and *Ichthyostega* had seven hindlimb digits. It therefore appears that pentadactyly in Carboniferous amphibians arose from a polydactylous, possibly variable, state. The limbs of these ichthyostegids seem to have been much less flexible (e.g., at elbow and wrist) than those of later amphibians, which together with other evidence suggests that they were largely aquatic.

The close resemblance of ichthyostegids to osteolepiforms provides an unusually good example of the origin of a major higher taxon (the Amphibia, and for that matter, the Tetrapoda). The transition is marked by MOSAIC EVOLUTION (the limbs evolved faster than the skull, teeth, or vertebrae), and by *gradual change* of individual features (the limbs are intermediate between those of crossopterygians and later amphibians). Certain features that are *variable* in the ancestor (e.g., dermal snout bones, distal fin elements) become *constant* and *individualized*, so that their homologies can be traced thereafter.

From amphibians to amniotes

The amniotes include an enormously diverse array of tetrapods known as reptiles, and the classes Aves and Mammalia, which evolved from them. Because the traditional class Reptilia is paraphyletic, it is not recognized as a taxon by some contemporary systematists, who often use "amniote" rather than "reptile" except when referring to birds and mammals.

The earliest amniotes are similar to, and surely derived from, labyrinthodont amphibians (of which the ichthyostegids were the earliest members). The earliest known amniotes are small, superficially lizardlike animals from the early Pennsylvanian (mid-Carboniferous) of Nova Scotia. They are placed in the Protorothyridae, order Captorhinida (Figure 6.20). (This order will also figure in the origin of mammals.) Although the salient synapomorphy of living amniotes is the amniotic egg, an adaptation for terrestrial reproduction, fossil amniotes can be distinguished from amphibians only by skeletal features. These include two sacral vertebrae supporting the pelvis (rather than one), the fusion of three tarsal bones into one (the astragalus), and an absence of certain features common to many amphibians (e.g., a cleft at the rear of the head that corresponds to the sarcopterygian spiracle). The more highly ossified skeleton and the proportions of the bones of the limb girdle appear to be adaptations for more agile terrestrial locomotion.

The captorhinids are the stem group from which diverse other groups of "reptiles" evolved; many transitional forms are known. The captorhinids are so similar to several groups of what are now considered amphibians that all of them were once classified in a single order (Cotylosauria), show-

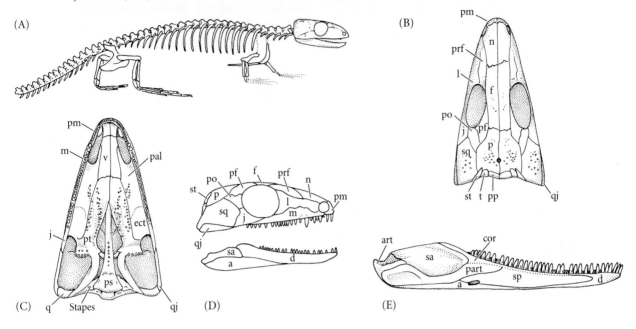

FIGURE 6.20 Early amniotes ("reptiles") in the family Protorothyridae. One of the few differences from amphibians is the lack of a notch at the rear of the skull (cf. Figure 6.19). (A) Skeleton of *Hylonomus*. (B–D) Dorsal, ventral, and lateral views of skull of *Paleothyris*. (E) Medial (inner) view of lower jaw of *Protorothyris*. Compare with amphibians (Figure 6.19) and mammal-like "reptiles." (A–C after Carroll and Baird 1972; D after Carroll 1988; E after Clark and Carroll 1973.)

ing just how difficult it can be to distinguish between amphibians and early "reptiles."

From early amniotes to mammals The origin of mammals, via mammal-like reptiles, from the earliest amniotes is doubtless the most fully, beautifully documented example of the evolution of a major taxon. In no other case are as many steps preserved in the fossil record. The origin of mammals definitively refutes creationists' claims that fossils fail to document evolution. It also shows how characteristics can become highly modified to serve new functions. For these reasons, we present the origin of mammals in some detail.

The living mammals consist of three subclasses, the Prototheria (platypus and echidna), the Metatheria (marsupials), and their sister group, the Eutheria (placental mammals). All are endothermic, have hair, produce milk, and (except for the Prototheria) are viviparous (viviparity has also evolved in some fishes, amphibians, and reptiles). These features are not fossilized, but all mammals share diagnostic skeletal features as well. The primary (and in all except the earliest mammals, the exclusive) *jaw articulation is between the squamosal and the dentary*, rather than the quadrate and articular as in other tetrapods (and crossopterygians). In living mammals, *the dentary is the only bone in the lower jaw; sound vibrations are transmitted by three bones (the auditory ossicles: malleus, incus, stapes) in the middle ear*. The teeth are differentiated into incisors, canines, and premolars, all of which are replaced once in life, and molars, which are not replaced at all.

Other toothed tetrapods have repeated, continual tooth replacement. Other features common to mammals, but which do not define them as such, include (Figure 6.21) legs held vertically below the body; pelvis dominated by a large ileum; several sacral vertebrae; distinct thoracic and lumbar vertebrae; a more mobile pectoral girdle than in reptiles; an enlarged braincase, consisting dorsally mostly of dermal bones; double rather than single occipital condyle; in primitive mammals, an orbit (eye socket) confluent with a large space (temporal fenestra) behind it; a narrow zygomatic arch, consisting of the jugal and squamosal bones, flared out from the skull; a large dorsal projection, the coronoid process, of the dentary bone; cheek teeth with multiple cusps; and a secondary palate, consisting of a medially projecting shelf of the premaxilla, maxilla, and palatine. The jaw muscles are large, originating on the side of the braincase and the medial surface of the zygomatic arch, and inserting on both sides of the dentary; in early amniotes, the smaller jaw muscles ran from the inner surface of the temporal cavity to the inner surface of the jaw.

In mammals, homeothermy may have the advantage of enabling the animal to forage at low temperatures (e.g., at night), but its disadvantage is that it requires a much higher rate of feeding than in ectotherms. Most of the characters listed above can be viewed as adaptations for mobility and speed (the postcranial characters), or for subduing prey by forceful jaw action and for chewing food. The jaw muscle arrangement, for example, enables the upper and lower teeth to occlude and to move in a tri-

FIGURE 6.21   The "ground plan" of the skeleton of modern mammals. Compare with early amniotes ("reptiles"), as in Figure 6.20, and with mammal-like reptiles (Figures 6.23–26). (A) Skeleton of a tree shrew (*Tupaia*). cv, tv, lv, sv: cervical, thoracic, lumbar, and sacral vertebrae; f, femur; fl, fibula; h, humerus; mt, metatarsals; p, pelvic girdle; r, radius; ts, tarsals; u, ulna. (B,C,D) Dorsal, ventral, and lateral views of the skull of a marsupial, the opossum *Didelphis*. See Figure 6.18 for abbreviations. Note: differentiated teeth, the premolars and molars with multiple cusps; frontals (f) and parietals (p) constitute walls of anterior part of braincase; eye socket confluent with a large fenestra, laterally bounded by zygomatic arch consisting of parts of the jugal (j) and squamosal (sq); two occipital condyles; lower jaw of one bone, the dentary (d). (A after Romer 1966; B–D after Carroll 1988.)

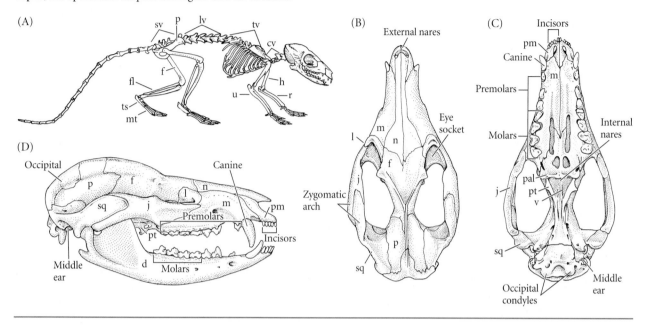

angular pattern with respect to each other, providing more effective chewing.

We shall next trace the evolution of mammals from the earliest amniotes, dwelling on certain steps marked by letters along the phylogeny in Figure 6.22 (Kemp 1982). Most of the genera in this figure are just the best-known members of their particular groups; they are shown as branching off the ancestral lineage leading to mammals, although some of them may have been actual mammal ancestors. We will ignore the great diversity, at almost every stage, of specialized related creatures.

A. The Upper Carboniferous captorhinids, as we have seen, are the earliest amniotes (see Figure 6.20).

B. The several major clades that arose from the captorhinids differ in the number and position of the openings (TEMPORAL FENESTRAE) that developed between certain of the bones behind the orbit (eye socket). The captorhinids lack such openings. In the subclass Synapsida, a fenestra developed between the jugal, below the fenestra, and the postorbital and squamosal bones above it (see Figure 6.23A). The function of the fenestra, which plays a major role in this story, was either to provide space for enlarged jaw muscles to expand into when contracted, or to more securely anchor the jaw muscles along its edges. The synapsids made their first appearance in the mid-Pennsylvanian, almost as early as the first amniotes. Thus the lineage leading to mammals branched off from the other amniotes as the very beginning of amniote history, from a very generalized ancestor. In the early synapsids (order Pelycosauria), the temporal fenestra appears and becomes progressively enlarged (in 6.22 B–F; cf. also Figure 6.23). Other mammal-like features are a broad, flat occipital region of the skull, enlarged canine teeth in some species, and a strong vertebral column.

E, F. The sphenacodontid pelycosaurs (Figure 6.23A) have a coronoid eminence (a dorsal enlargement) on the lower jaw on which large jaw muscles inserted. There are three sacral vertebrae, providing a solid juncture of the pelvic girdle with the vertebral column in these active predators. The jugal and postorbital form a narrow bar, separating the orbit from the large temporal fenestra. The most famous sphenacodontid is *Dimetrodon*, with its huge dorsal sail.

G. The later synapsids, arising from sphenacodontid ancestors (but not *Dimetrodon*), are placed in the order Therapsida; they appear in the mid- to late Permian. *Biarmosuchus* (Figure 6.23B) has the features of a therapsid ancestor. Several skull bones have been lost; the temporal fenestra and the jaw muscles are enlarged; the canines are enlarged; and the vomers (on the palate) are recessed from the roof of the mouth, suggesting partial separation of

FIGURE 6.22 A cladogram illustrating the relationships of the genera most closely related to the lineage leading from basal amniotes (A, B, C) to modern therian mammals (V, marsupials and placentals). Several of these are described in the text. (After Kemp 1982.)

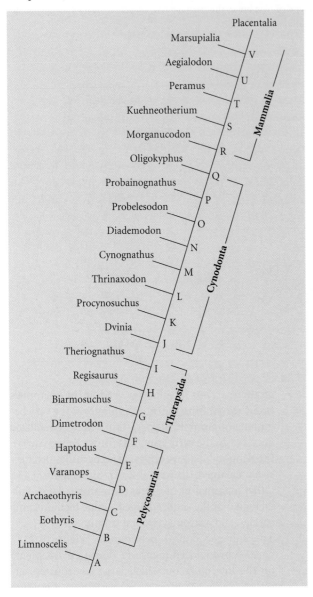

the breathing passage from the mouth cavity. The hind legs could be held fairly erect, and the limb girdles show modifications that also suggest an agile predator.

H, I. The major new features at the next level are a short, anterior secondary palate, a still more enlarged temporal fenestra, and a change in the quadrate, which becomes smaller and movable, rather than fixed to the skull. Recall that the quadrate articulates with the articular bone of the lower jaw.

J, K. The most mammal-like of the therapsids are the suborder Cynodonta, extending from the Upper Permian to the Upper Triassic. They represent many

steps in the approach toward mammals. In primitive cynodonts such as *Dvinia* and *Procynosuchus* (Figure 6.23C), the huge jaw muscles are accommodated by a very large temporal fenestra and by compression of the rear of the skull into a narrow vertical chamber, made up largely of the now almost vertically oriented parietals. The postorbital bar now stands out from the skull, forming a zygomatic arch. This gives the skull a very mammalian aspect. The dentary is enlarged relative to the other bones of the lower jaw. The jaw muscles inserted not only on the medial, but also on the lateral surface of the jaw, and they originated on both the side of the braincase and the inner surface of the zygomatic arch. Hence jaw movements were more complex than in previous groups. The cheek teeth have a row of several cusps. A medial shelf from the premaxilla, maxilla, and palatine forms an incomplete secondary palate, partly separating the vomers from the mouth cavity. The occipital condyle is doubled. The postcranial skeleton of *Procynosuchus* (unknown for *Dvinia*) has partially differentiated thoracic and lumbar vertebrae, and more mammal-like limb girdles.

L, M. The advanced cynodonts *Thrinaxodon* and *Cynognathus* (Figures 6.23D, 6.24), of the Lower Triassic, were medium-sized to rather large predators (skull length 10 cm and 40 cm respectively). The hindlimbs are fully vertical and were probably incapable of a sprawling posture. The lumbar vertebrae have become almost fully distinct (lacking ribs) from the thoracic vertebrae. The secondary palate is complete. The *stapes*, which we first encountered as the hyomandibular bone in sarcopterygians, is small and light. As in earlier tetrapods, it proximally abuts the otic (ear) region of the skull, but distally it contacts the *quadrate*, which is smaller and more mobile than in previous forms and *occupies a socket in the squamosal* (Figure 6.25C). The dentary has a very large coronoid process (continuing the trend), and is by far the largest bone in the lower jaw. The other lower jaw bones are small, and loosely rather than solidly connected to the dentary. The most significant feature is the development of a *secondary articulation between the lower jaw and the skull*: in addition to the old articular/quadrate articulation, the surangular articulates with the squamosal.

P. In *Probainognathus* and related forms from the Middle and Upper Triassic (Figure 6.23E), the greatly bowed zygomatic arch borders an enormous temporal fenestra. The cheek teeth have not only a linear row of cusps, but *a cusp on the inner side of the tooth*. This begins a history of complex cusp patterns in the cheek teeth of mammals, which enable chewing and are modified in different lineages for feeding on different kinds of food. The pineal opening, first seen in sarcopterygians, has been lost. The dentary is very large.

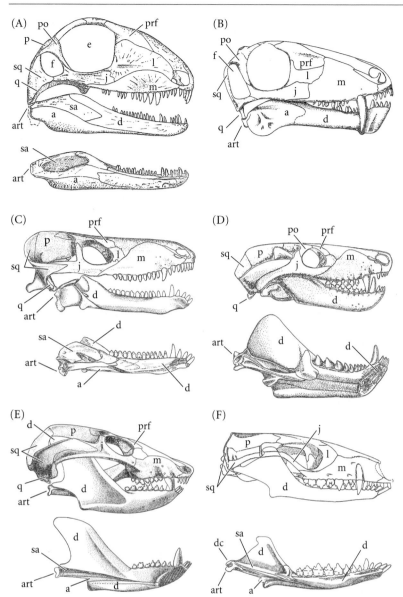

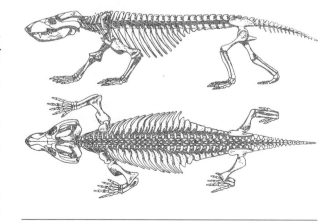

FIGURE 6.23 Skulls and lower jaws (medial view) of some stages in the evolution from early synapsids to the earliest well-known "mammal" (cf. Figure 6.21E, G, K, L, P, R). (A) A pelycosaur, *Haptodus*. Note temporal fenestra (f) bounded by bones j, sq, and po; multiple bones in lower jaw; single-cusped teeth; quadrate/articular (q, art) jaw joint. (B) An early therapsid, *Biarmosuchus*. Note enlarged temporal fenestra. Other changes noted in text are not visible in this view. (C) An early cynodont, *Procynosuchus*. Note greatly enlarged temporal fenestra, vertically oriented side of braincase (p), incipient zygomatic arch (j, sq), enlarged dentary (d). (D) A later cynodont, *Thrinaxodon*, and the jaw of another (*Cynognathus*). Note multiple cusps on rear teeth, large upper and lower canine teeth, fully developed zygomatic arch, and greatly enlarged dentary with a vertical extension (coronoid process) to which powerful jaw muscles were attached. (E) An advanced cynodont, *Probainognathus*. Note cheek teeth with multiple cusps, including an inner cusp not visible here; jaw articulates by the angular (art), surangular (sa), and posterior process of the dentary (d). (F) A morganucodontid (*Morganucodon*), generally considered the first known mammal (Upper Triassic). Note loss of postorbital bar, multicusped cheek teeth (including inner cusps), double articulation of lower jaw, including articulation of a dentary condyle (dc) with the squamosal (sq). (After Futuyma 1995; based on Carroll 1988 and various sources.)

There is a double jaw articulation (articular/quadrate and surangular/squamosal), but most significantly, a posterior process of the dentary also makes contact with the squamosal.

Skeletally, the advanced cynodonts we have just considered are mammalian in almost all respects except for the jaw articulation and the structure and replacement pattern of the teeth. They were rather large, active predators. The secondary palate enabled them to breathe when the mouth was full, and the distinction between thoracic and lumbar vertebrae means that they probably had a diaphragm. These and other lines of evidence imply that they were endothermic. There are also reasons to believe that they had hair (Kemp 1982, 247–251).

R. The Morganucodontidae (*Morganucodon*) of the Upper Triassic and very early Jurassic (Figures

FIGURE 6.24 Skeleton of the cynodont *Thrinaxodon*. Note vertically held limbs, nearly as in mammals (compare with Figure 6.21). (From Brink 1956.)

FIGURE 6.25 Stages in the evolution of the ear apparatus. Posterior view of right side of head: external view in D, cross-sectional view in others. (A) A sarcopterygian fish. The stapes (s) conducts sound from the skull wall to the auditory capsule, within which lie the auditory sac and semicircular canals. The first gill slit forms a passage between the spiracle (sp) and the throat cavity. (B) An early amphibian. The spiracle is closed by a tympanic membrane (tm); the gill slit forms the middle ear cavity (me) and the Eustachian tube (eu); the stapes directly contacts the inner ear. The quadrate (q) and articular (a) form the jaw joint. (C) An early therapsid (mammal-like "reptile"), in which the stapes contacts the quadrate and articular. The articular supports a new tympanum. (D) *Morganucodon*, an early mammal, in external view. A condyle (cond) on the dentary (de) articulates with the squamosal (sq), forming the major jaw joint. A process (ma) on the articular supports a tympanum. The articular contacts the loose quadrate (q), which contacts the stapes (s), which contacts the inner ear as in (C). (E) Close view of the ear of a modern mammal (human), in which the articular [now called the malleus (m) or hammer], the quadrate [now the incus (i) or anvil], and the stapes (now the stirrup) lie within a bony chamber. Sound, entering the outer ear (oe), is transmitted from the eardrum (tympanum, tm) via the three ear bones to the inner ear. (A–C and E after Romer 1966; D after Kermack et al. 1981.)

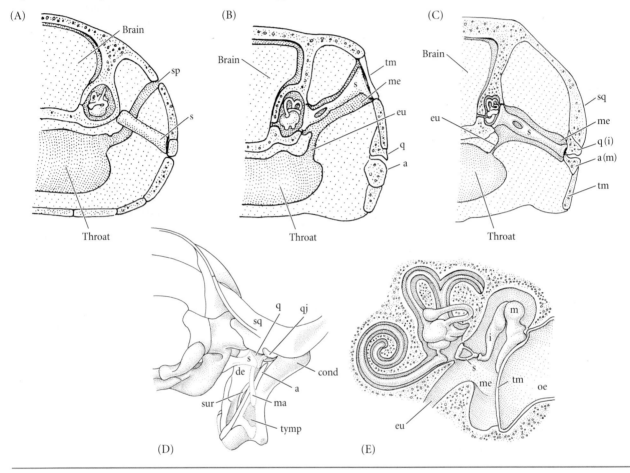

6.23F, 6.25D, 6.26) *are classified as the first, and probably ancestral, Mammalia.* The form and number of teeth are mammalian, and the incisors, canines, and premolars were apparently replaced only once in life, and the molars not at all. Most of the other features are those of advanced cynodonts, including the jaw articulation. The articular/quadrate hinge is still present, although weak, and right next to it is *a fully developed mammalian articulation between the dentary and the squamosal.* The inner ear is large, and *the articular/quadrate/stapes series of bones approaches the condition in modern mammals,* in which these bones, called the hammer, anvil, and stirrup, transmit sound to the inner ear (Figure 6.25D).

S–V. Few alterations need be made in a morganucodont to construct a common ancestor of marsupials and placentals, or therians (V). However, the morganucodonts, unlike many of their predecessors, were very small (skull length 2–3 cm), and so were almost all subsequent Mesozoic mammals (S–V) until the late Cretaceous. Their small size is generally supposed to be a consequence of the competitive pre-eminence of large reptiles. Whether or not this be true, Mesozoic mammals are represented almost exclusively by teeth and fragments of jaws, so little can be said about the evolution of their other features. Therian teeth are known from the Lower Cretaceous (about 140 My after *Morganucodon*), but only in the Upper Cretaceous are more complete fossils

FIGURE 6.26 *Morganucodon*, which links advanced therapsids to mammals and is often considered an early mammal. (A) Reconstructed skeleton. (B,C) Skull in dorsal and ventral view. (D,E) Lower jaw in outer and inner view. See Figure 6.23F for lateral view of skull. Ancestral characters include multiple bones in lower jaw and quadrate/articular articulation. Mammalian features include complete secondary palate (m), multi-cusped cheek teeth, and squamosal/dentary articulation. See also Figure 6.25D. (A after Jenkins and Parrington 1976; B–C after Kermack et al. 1981; D–E after Kermack et al. 1973.)

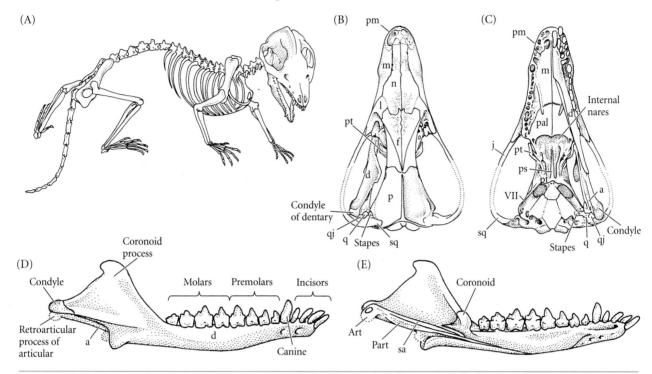

found. These include distinct marsupials and placentals (the latter being rather shrewlike forms). They differ from morganucodontids primarily in the cusp pattern of the cheek teeth, the larger brain, and the greater maneuverability of the front legs, accomplished by changes in the structure of the pectoral girdle (e.g., loss of the interclavicle and coracoids). The salient change, the climax of a long trend, is in the lower jaw and the bones of the middle ear. The lower jaw consists entirely of the dentary, articulated to the squamosal. All other jaw elements have been lost except the very small articular, but this is no longer part of the lower jaw: it resides in the middle ear, in contact with the eardrum (Figure 6.25E). The articular, called the malleus (hammer) in mammals, transmits sound to the greatly reduced quadrate (now called the incus, or anvil), which, following the late cynodont trend, has retreated deeper into the skull. The third middle ear bone, the stapes (stirrup), we have traced ever since the sarcopterygians, in which it braced the skull. It became progressively smaller, lighter, and more capable of transmitting sound, in the cynodonts. These changes, traced in the fossil record, can be seen also in the embryonic development of living mammals (Hopson 1966; Hopson and Crompton 1969). In very young marsupials, the malleus and incus have the same jaw-articulating position as the reptilian articular and quadrate, and later enter the middle ear.

Major points about the origin of mammals   Bearing in mind that we have neglected all the diverse groups of synapsids (now extinct) except those on the mammalian line, we may note that each of many mammalian characters (e.g., posture, limb flexibility, agility, tooth differentiation, skull changes associated with jaw musculature, secondary palate, reduction of the elements that became middle ear bones) evolved gradually, so that in retrospect, a trend can be traced back to the early synapsids or beyond. As in every other origin of a major taxon, evolution has been mosaic, with different characters "advancing" at different rates. At no point in this history has a new bone come on the scene: all the skeletal elements are modified from those of amphibians and even crossopterygians, by fusions or changes in size and shape. Many of the bones of crossopterygians and early amphibians have been sequentially reduced and lost. In many cases, we can discern an adaptive reason for the evolutionary changes, especially in terms of locomotion and feeding, and many of these changes were progressive, associated with more accentuated or efficient function. The functional significance of some changes, when initiated, was almost certainly different from the function that the feature later assumed. For example, the enlargement of the braincase in the cynodonts probably served the major function of increasing the area for attachment of the large jaw muscles; this set the stage for the later enlargement of the brain. It is also likely that the selective advantage of a large brain in early mammals was to provide motor control of the much more complex, agile locomotion (Kemp 1982).

It is tempting to think that therian mammals are "better," or more highly adapted, than other synapsids because most other synapsids are extinct. But this is far from certain. Cynodonts persisted for more than 70 million years, about as long as the two modern groups of therians have. The therapsids were a highly diverse group; we do not know whether modern mammals are any more diverse. Finally, we do not know why the other synapsids became extinct and the Mammalia survived; perhaps mammals survived not because they were more intelligent, or faster, or more efficient, but because they were very small, and thereby avoided competition with (and perhaps predation by) the dominant reptiles of the Mesozoic.

Finally, it is worth repeating that the definition of Mammalia is arbitrary. As more intermediate fossils have been found, the number of suggested definitions has grown (Kemp 1982). As with amphibians and amniotes, the mammals and their predecessors show that higher taxa came into existence by the sequential evolution of small changes.

From dinosaurs to birds Quite a few vertebrate lineages have evolved the ability to glide (e.g., "flying" fishes, frogs, snakes, squirrels, marsupials), but powered, flapping flight has evolved only three times in the vertebrates.* Bats (Chiroptera) evolved in the Paleocene, and are now the second most diverse order of mammals. They possess a wing membrane, extending between greatly elongated fingers. The flight surface of pterosaurs (pterodactyls) was a membrane extending from the greatly elongated fourth finger to the side of the body. Pterosaurs ranged in size from very small to the largest size ever attained by a flying animal: in the late Cretaceous, *Pteranodon* had a wingspan of 7 meters, and *Quetzalcoatlus* of 11–12 meters (broader than a small airplane).

In birds, which presently number about 9000 species, the flight surface is the elongated feathers of the forelimb and tail. The skeleton is highly modified to accommodate the flying habit (Figure 6.27). Many of the bones are hollow. In the

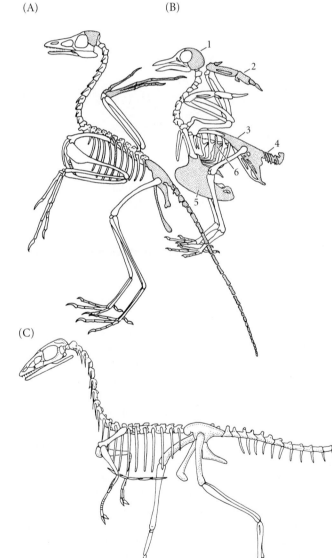

* The South American Gasteropelecidae, known by fish hobbyists as hatchetfishes, have large pectoral fins, greatly enlarged pectoral muscles, and a huge ventral keel, and are apparently capable of powered flight.

FIGURE 6.27 Skeletal features of (A) *Archaeopteryx* and (B) a modern bird (pigeon). Compared with *Archaeopteryx*, the modern bird has (1) an expanded braincase with fused bones, (2) fusion and reduction of the three digits of the hand, (3) fusion of the pelvic bones and several vertebrae into a single structure, (4) fewer tail vertebrae, several of which are fused, (5) a greatly enlarged, keeled sternum (breastbone), and (6) horizontal processes that strengthen the rib cage. (C) The skeleton of a small therapod dinosaur, *Compsognathus*, is very similar to that of *Archaeopteryx* (A, B after Colbert 1955; C after Lucas 1994.)

forelimb, carpals, metacarpals, and digits are fused into a single unit, and the digits lack claws.[*] The tail is short, formed of fused vertebrae (the pygostyle) that support tail feathers. The sacral vertebrae and pelvis form a solid structure. The ribs attach ventrally to a huge breastbone (sternum), with a keel for insertion of the large pectoral muscles that both lower and raise (by a pulleylike arrangement) the wings. All these features provide for flight power and resistance to the stresses of muscular activity. In the hind leg, the fibula is short and nonfunctional, and there is a joint that is not found in any other vertebrates except the dinosaurs: the proximal tarsals (astragalus and calcaneum) are fused with the end of the tibia, and the distal tarsals are fused with the metatarsals, which themselves are fused together. The bones so formed are called the tibiotarsus and tarsus-metatarsus, respectively. In most other tetrapods (including humans), the metatarsals and digits are appressed to the ground, so that two leg segments support the body. In birds, only the digits contact the ground, and the body is supported by three leg segments (tarsusmetatarsus, tibiotarsus, femur). The skull of birds, finally, has extensive fusion of the bones, a large braincase and orbit, and no teeth. Birds are endothermic, and their defining synapomorphy is the feather, which has the same fundamental structure and development as a reptilian scale.

The diapsid reptiles include all living species except turtles, as well as most of the extinct Mesozoic groups. Diapsids, first represented by Upper Pennsylvanian fossils that closely resemble protorothyrids, have two temporal fenestrae rather than the single fenestra of the synapsids (see above). Among them are the archosaurs, which include crocodilians and dinosaurs. The hindlimb of dinosaurs has the same basic structure as that of birds: a joint between the proximal and distal tarsals, elongated metatarsals, and ground contact only by the digits. The hindlimb was held erect below the body. Two great groups of dinosaurs, distinguished by the form of the pelvis, extend from the later Triassic to the end of the Cretaceous. One of these two groups, the Saurischia, includes the bipedal theropods, some of which weighed only about 3 kilograms. In *Compsognathus* (Upper Triassic) and related genera (Figure 6.27C), the skeleton is lightly built, with hollow bones; the forelimb can be rotated and extended forward, the distal tarsals are fused to the metatarsals, the sacral vertebrae form a solid unit, the foot is entirely birdlike, and the orbit and braincase are large. All these features presage those of birds.

Archaeopteryx lithographica (Figure 6.28), the most famous non-missing link of all time, is known from seven pigeon-sized skeletons and an isolated feather from fine Upper Jurassic (140 Mya) limestone deposits in Germany. One specimen, in which the feathers were not recognized, was misidentified as a dinosaur for many years. Most authorities consider *Archaeopteryx* a derivative of the thero-

[*]Claws are present in the young of the hoatzin (*Opisthocomus hoazin*), a probable relative of the cuckoos. These claws have undoubtedly evolved secondarily.

FIGURE 6.28 Cast of a well-preserved specimen of *Archaeopteryx lithographica,* showing the wing feathers and the long tail with feathers on both sides. (Negative #319836, courtesy Department of Library Services, American Museum of Natural History.)

pod dinosaurs (but see Feduccia 1996 for a dissenting view). In almost every respect, except the feathers, *Archaeopteryx* is a dinosaur. It has teeth; the skull bones show little fusion; the tail feathers are borne on a long tail made up of freely articulating vertebrae; the pectoral and pelvic girdles resemble those of several dinosaurs; and it lacks a bony sternum. The forelimbs are very long, and bear long flight feathers, but their skeletal structure is virtually identical to that of the theropods. Each digit ends in a claw. The hind leg is that of a theropod, except that the proximal tarsals are more fully integrated with the tibia, and the distal tarsals with the metatarsals, which are partly fused together. The most birdlike features are those of the hindlimb, and of course the feathers, the form of which shows that *Archaeopteryx* was capable of true flight.

It can hardly be supposed that fully functional flight feathers evolved in one great step from scales, so the question of how a small dinosaur evolved flight has been extensively debated. According to the two major hypotheses, the ancestor of *Archaeopteryx* began as an arboreal glider (Feduccia 1996), or as an active terrestrial predator that leaped into the air to catch insects from a running start (Caple et al. 1983; Hecht et al. 1985; Ruben 1991). A mathematical analysis of the physics of leaping and flying, using the theoretical principles of ballistics and aerodynamics, concluded that from a running start, an animal that leaped into the air could stabilize its movement (and avoid a crash landing) by extending its forelimbs; and that even a very slight extension of the scales on the forelimb could provide enough lift to improve maneuverability and to control landing (Caple et al. 1983). Extended slightly more, the scales would provide further lift, and flapping would extend the animal's trajectory. This ancestor might well have had feathers instead of scales; feathers might first have evolved as insulation. Physiologically, highly active reptiles have a very high power output per unit of muscle mass; if *Archaeopteryx* had reptilian physiology, it would have been capable of sustained flight from a jumping start, even though it lacked the breastbone and highly developed flight muscles of modern birds (Ruben 1991).

Recently, paleontologists have discovered a variety of Upper Jurassic and Cretaceous birds, some of which combine primitive characters, like those of *Archaeopteryx*, with derived characters of modern birds, such as the breastbone and pygostyle (Feduccia 1996). The steps in the evolution of modern birds have not yet been worked out in detail, but even if we had only *Archaeopteryx*, we would have an outstanding example of mosaic evolution—and of the indisputable origin of a major taxon, and a major adaptive change, from a seemingly very different ancestral stock.

Correspondence between Phylogeny and the Fossil Record

In inferring phylogenetic relationships among living taxa, we conclude that certain taxa share more recent common ancestors than others. If such statements are correct, there should be some correspondence between the relative times of origin of taxa, as inferred from phylogenetic analysis, and their relative times of appearance in the fossil record. We can expect the correspondence to be imperfect because of the great imperfection of the fossil record, so that, for example, an ancient group might be recovered only from recent deposits. Moreover, *although a lineage may have branched off early, it may not have acquired its diagnostic characters until much later;* for example, the synapsid clade did not acquire the diagnostic characters of mammals until long after it had diverged from other reptiles.

Nevertheless, there is a strong overall correspondence between phylogenetic branching order and order of appearance in the fossil record. The mammalian orders are more closely related to each other than they are to "reptiles" or amphibians, and appear later in the fossil record. Spore-bearing vascular plants such as ferns and club mosses are phylogenetically more basal than seed plants, and within the seed plants, the angiosperms are believed, on anatomical grounds, to have arisen from the "gymnosperms"; this sequence is mirrored by the sequence of fossils (see Chapter 7). Phylogenetic analysis, entirely independent of fossils, indicates that winged insects (Pterygota) arose from primitively wingless forms, and that within the Pterygota, complete metamorphosis is a derived trait. Correspondingly, primitively wingless insects are known from the Devonian, primitive winged insects (such as roaches and dragonflies) from the Carboniferous, and holometabolous orders from the Permian and later (see Chapter 7).

A striking instance of correspondence is offered by the order Microcoryphia (bristletails, Figure 6.29), which had long been considered "living fossils" because their anatomy was thought to represent the basic "body plan" of primitively wingless insects (Snodgrass 1935). Only a few years ago, a fossil bristletail, corresponding in almost every detail to living species, was discovered in early Devonian deposits (Labandeira et al. 1988). This, the earliest fossil of a true insect, is as old as it *must* have been, given that the Microcoryphia are phylogenetically more basal than the other insect orders, which appear shortly after the Devonian.

Fossils such as the bristletail are assigned to the ancestral stock from which modern species arose not because they occur earlier in time, but because they have features that the ancestor must have had, based on phylogenetic reconstruction of the ancestor. By comparing birds with reptiles, for example, it was possible to say, without reference to fossils, that the ancestor of birds must have had unfused digits, separate tail vertebrae, etc. Thus *Archaeopteryx* could be recognized as an ancestor (or close to the ancestor) of birds because biologists *already knew what to look for.* Other

FIGURE 6.29 Dorsal (A) and ventral (B) views of living bristletails, order Microcoryphia. Among extant orders of insects, bristletails have the most primitive features, such as the reduced legs on the ventral surface of the abdominal segments. (After CSIRO 1991.)

(A)   (B)

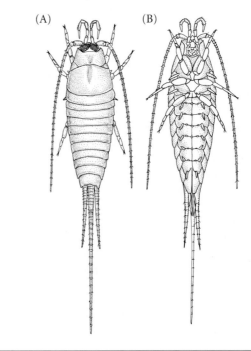

Jurassic fossils, such as pterodactyls, were ruled out because they have characteristics that a bird ancestor surely could not have had (e.g., the lengthened fourth finger). Thus, to some extent, phylogenetic analysis can make *predictions* about fossils not yet discovered. For example, E. O. Wilson, F. M. Carpenter, and W. L. Brown (1967) hypothesized what features the ancestor of ants should have had, based on comparisons between primitive species of living ants and related families of wasps. Subsequently, Cretaceous fossil ants in amber—older than any previously known fossils—came to light, and corresponded to the predicted morphology in all except two features (Figure 6.30).

Phylogenetic relationships can often be clarified by information from extinct species (Donoghue et al. 1989). In any phylogenetic analysis, information on the polarity of changes in phylogenetically informative characteristics is critically important. Much of this information may have been lost through extinction. In particular, the evolution of some features may be hard to trace because they have been lost in many of the extant groups, or so highly modified that it is difficult to trace their evolutionary transformations, or even to determine their homology. Fossils may provide the crucial missing information. For example, the evolution of the articular bone and other features that link mammals with basal reptiles would be almost impossible to discern if we did not have fossil therapsids. The phylogeny of amniotes inferred solely from the morphology of living forms would differ from the phylogeny that incorporates data from the fossils of mammal-like reptiles.

Trends

In Chapter 5, we noted that an evolutionary trend may be described as directional change in a feature within a single lineage, or as a parallel tendency in several, often related, lineages. The fossil record presents many instances of both. More often than not, a trend in one lineage is paralleled, at least in part, in related lineages. We have seen examples in several characters of the Equidae (e.g., body size, high-crowned teeth), and in the evolution of several characters of the Synapsida toward a more mammal-like condition (e.g., enlargement of the dentary, limb position, secondary palate). In these and many other instances, the trend was arrested or reversed in some related lineages. In other cases, evolutionary changes appear never to have been reversed. This is certainly true of the dentary/squamosal jaw articulation of mammals, the modification of hindwings into halteres in Diptera, and the loss of the notochord as an adult feature in all the descendants of early amphibians.

Early in the twentieth century, several competing theories were offered to explain evolutionary trends. Showing that some of them were wrong or superfluous (Simpson 1944, 1953) was an important accomplishment of the Modern Synthesis (see Chapter 2). The major explanations have been:

1. *Neo-Lamarckian theories*, holding that the organism's environment directly evokes the development of

FIGURE 6.30 (A) *Sphecomyrma freyi*, a mid-Cretaceous ant, preserved in amber, that appears to bridge the gap between modern ants and the wasps from which ants are thought to have arisen. (B) Comparison between features of *Sphecomyrma* and those of the previously hypothesized ancestor of ants. (A courtesy of E. O. Wilson; B after Wilson et al. 1967.)

(A)

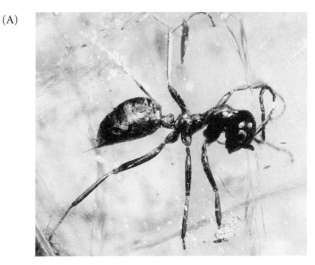

(B)

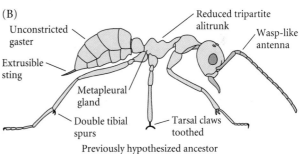

Previously hypothesized ancestor

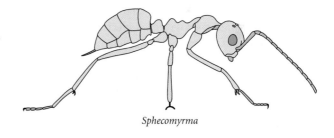

Sphecomyrma

more extreme features. Because of evidence from genetics and developmental biology, neo-Lamarckian ideas have been almost universally rejected.

2. *An inherent drive*, with effectively unstoppable momentum, in a particular direction. Some authors supposed there was an inner drive toward perfection, and that organisms acquired more extreme features before they were actually advantageous. (This is an old, and now abandoned, meaning of the term *preadaptation*.) Others argued that the direction of evolution is not necessarily adaptive, but held that evolutionary "momentum" could carry features to such an extreme that they were disadvantageous. The most famous example of a supposedly disadvanta-

geous, "excessive" form is the Pleistocene "Irish elk," *Megaloceros giganteus* (Figure 6.31). This was the largest of all deer, and had by far the largest antlers, which some authors supposed caused the species' extinction, whether by entanglement in trees or by other effects. Among species of deer, antler size is exponentially related to body size, so that large species have *relatively* larger antlers than small species. The same relationship exists *within* species of deer among individuals of different sizes. (Such a disproportionate, or nonlinear, relationship between two features of an organism is termed **allometry**: see Chapter 23). Gould (1974) showed that the size of the antlers of *Megaloceros* is just what would be expected in a deer of its great size. Whether the huge antlers were a by-product of the evolution of large size, or were advantageous in their own right for antagonistic displays among males competing for mates, they could not have driven *Megaloceros* to extinction, because they persisted for several thousand years before extinction. None of the authors of theories of inherent evolutionary drive ever suggested a biological mechanism for such a drive, and no such mechanism is known.

3. *Directional mutation*, or the theory that mutations occur only or predominantly in one direction (e.g., for larger rather than smaller size). Although it is certainly true that not all imaginable mutations are equally likely (see Chapter 10), the vast majority of characters (especially METRIC characters, such as size or shape, that vary continuously) display mutational and recombinational deviations in both directions from the mean. Moreover, a feature could not evolve by mutation pressure alone if it were opposed by natural selection (see Chapter 13).

4. *Directional natural selection*—that is, progressive selection of variants that more nearly approach an optimum phenotype different from that of the ancestor—is currently by far the most favored explanation of evolutionary trends. In the majority of cases, an adaptive advantage for the progressive modification of directionally evolving features can be readily envisioned or demonstrated. For instance, higher-crowned teeth were clearly advantageous for those horses that fed on grass (but not for those that browsed on other vegetation). The skeletal trends in the synapsid line leading to mammals are associated with the greater speed, jaw strength, and metabolic activity of those predators.

5. *"Irreversibility"* is a term frequently applied to character changes that apparently have never been reversed. Whether or not they are truly incapable of reversal we cannot know for certain. In some cases, it is likely that the feature has not been reversed because the derived state is advantageous in all the descendant lineages that share it, whatever their habitat and habits may be. For instance, almost all mammals retain multicusped cheek teeth; reversal to the single-cusped condition (as in reptiles) has occurred only in the toothed whales. This pattern suggests that reversal is possible, but has not occurred in other orders of mammals because multicusped teeth are advantageous. In some cases, the character is retained because it has become so integrated into the functioning of the organism that many other features would have to be changed if the character were to be lost. For instance, flight is not the only function of feathers, which are retained in ostriches, flightless rails, and many other flightless birds. Feathers provide insulation and are used in sexual and social displays, so much of a bird's physiology and behavior depends on them. The notochord, a primitive feature of chordates, is retained in all vertebrate embryos because it induces the development of the central nervous system. Rupert Riedl (1978) speaks of such a character as carrying a BURDEN, meaning a suite of other features that depend on it for their development or proper function. Another basis for irreversibility is the eventual loss of the developmental foundations of features that have been lost during evolution. The principle that complex characters, once lost, are not usually regained is sometimes termed DOLLO'S LAW.

Parallel Trends

In contrast to a clade, which is a monophyletic group, a **grade** is a group of species that have attained a certain level of structural organization (Figure 6.32). For example, primitively wingless arthropods such as springtails, bristletails, and silverfish form a grade, once classified as the subclass Apterygota. They have attained the hexapod

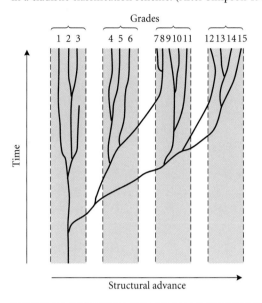

FIGURE 6.32 Grades and clades. A group of species (e.g., 1–3) that share a common ancestor to the exclusion of other species forms a clade. A group of species with the same level of structural organization (e.g., 4–6 or 7–11) forms a grade. If the members of a grade are named as a taxon, the taxon is paraphyletic (e.g., 4–6) or polyphyletic (7–11), and is inadmissible in a cladistic classification scheme. (After Simpson 1961.)

Grades

1 2 3   4 5 6   7 8 9 10 11   12 13 14 15

Time

Structural advance

(six-legged) condition, but have not evolved wings. The Apterygota, like many grade-level taxa, is a paraphyletic group. Grades often are polyphyletic as well, because the same structural features frequently evolve in parallel in related lineages. The fossil record offers many examples. For instance, the "Holostei" are ray-finned fishes that evolved a tail structure intermediate between that of more ancestral fishes (in which the tail is asymmetrical, with the caudal fin arising from the lower side of the vertebrae) and that of the more "advanced" teleost fishes (which have a symmetrical caudal fin attached to highly modified, aggregated vertebrae). The holostean condition evolved in several Mesozoic lineages, and is retained only in the bowfin (*Amia*) and the gars (*Lepisosteus*) among living fishes (Figure 6.33).

Some parallel trends are very common. COPE'S RULE describes the trend toward larger body size in many groups of animals. Another kind of trend is **iterative evolution**, in which the same trend is repeated in several sequential evolutionary radiations. For example, Cretaceous planktonic Foraminifera were represented by several morphological types, the simplest being the globigerines. At the end of the Cretaceous period, all except a few globigerine species became extinct, and these gave rise to species with the same forms as had existed in the Cretaceous. The same thing happened in the Miocene, after all except globigerine types became extinct at the end of the Oligocene.

The adaptive significance of a parallel trend can be tested either by using the comparative method (see Chapter 17), in which the incidence of the change is compared among lineages that differ in their susceptibility to a postulated agent of natural selection, or by direct functional

analysis. Functional analysis of morphological features in extinct organisms is easier if living analogues exist. For example, balanomorph barnacles, which first appeared in the Cretaceous and are still diverse, are enclosed by a cone-shaped skeleton, firmly attached to a rock or other substrate. The wall of the cone is made up of a number of parietal plates that slightly overlap at the sutures between them. Ancestrally, there were eight plates, but in numerous lineages, loss or fusion of plates has reduced the number to six, four, or even one (Figure 6.34). The chief predators of these barnacles are muricacean snails, which drill a hole through the skeleton. The Muricacea also appeared first in the Cretaceous, and have diversified in parallel with the balanomorphs. In experiments with living species of snails and barnacles, Richard Palmer (1982) showed that some muricaceans preferentially drill at the sutures. Drilling through the skeleton can take many hours, and the snails are far more successful when they drill at a suture than when they drill through the middle of a plate. A lower number of sutures would therefore be advantageous to the barnacle. Palmer notes that the primitive number of plates is retained in a barnacle genus that lives on marine turtles, where it is free from snail predation.

Rates of Evolution

Following G. G. Simpson (1953), paleontologists use several measures of evolutionary rates. The TAXONOMIC FREQUENCY RATE is the rate at which new taxa (e.g., genera) replace previous ones. It therefore depends on the rate of

FIGURE 6.33 Three grades of bony fishes, based on tail structure. (A) A chondrostean (sturgeon). The caudal skeleton extends into the upper lobe of the caudal fin. (B) A holostean (the bowfin, *Amia*). The caudal skeleton is somewhat asymmetrical. (C) A teleost (herring). The caudal fin rays are attached to a symmetrical plate at the end of the vertebral column. Holosteans form a grade of organization, not a monophyletic clade. (After Romer and Parsons 1986.)

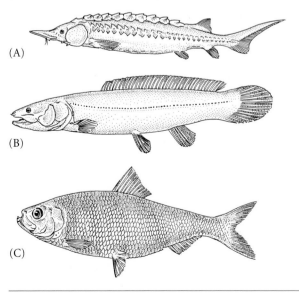

(A)

(B)

(C)

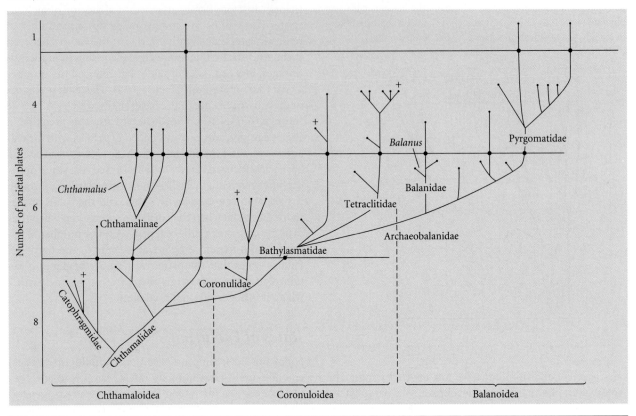

FIGURE 6.34 A phylogeny of the barnacle suborder Balanomorpha, showing parallel reduction in the number of parietal plates from the ancestral number (8) to 6 (5 times), 4 (10 times), and 1 (3 times). The vertical axis is not time, but grade of organization (plate number). Terminal taxa are genera or groups of genera. Names of superfamilies, families, and a few genera are indicated. (After Palmer 1982.)

origin of taxa and the rate of extinction (including taxonomic "extinction"). We defer discussion of taxonomic rates to Chapter 25. The PHYLOGENETIC RATE (or PHYLETIC RATE, another term used by some authors) is the rate of change of single characters, or complexes of characters, within a lineage. For example, Westoll (1949) and Schaeffer (1952) plotted the number of characters that changed per million years in lungfishes and coelacanths, and which distinguished later forms from the earliest Devonian members of these groups (Figure 6.35). After a high initial rate of

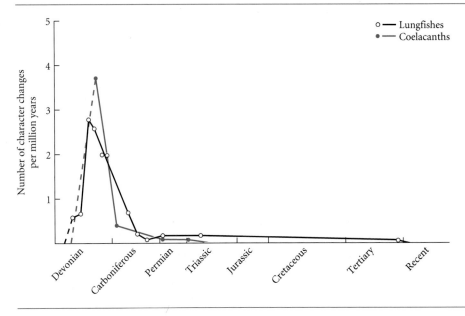

FIGURE 6.35 Change in the rate of morphological evolution during the history of lungfishes and coelacanths. The number of morphological character changes per million years in the various lineages of each group was greatest soon after the groups originated, and became very low thereafter. (After Schaeffer 1952.)

Correcting for Size in Describing Evolutionary Rates

In studying variation in morphology, it is often necessary to correct for size, because the significance of an absolute difference in a measurement depends on how large the organism or characteristic is. Adding 1 millimeter of length to a 3-meter-long snake is a trivial difference, but adding 1 millimeter to a 1-centimeter-long beetle is a substantial 10 percent change. It is more meaningful to say that a feature differs by, say, 10 percent relative to its magnitude in some reference organism. Two rather more complicated ways of expressing relative differences are commonly used in the evolutionary literature.

1. *The standard deviation.* For each of n organisms in a sample (of one species or population), let y_i be the measurement on the ith individual, and $\bar{y}$ the arithmetic mean. The VARIANCE is $s^2 = (\sum[y_i - \bar{y}]^2)/n - 1)$, that is, the average squared deviation of each observation from the mean. The STANDARD DEVIATION is $s = \sqrt{s^2}$. If the frequency distribution approximates a bell-shaped (normal) curve, 99.7 percent of the individuals should lie within three standard deviations above and below the mean (see Appendix A). Suppose, for instance, that $\bar{y} = 10.0$ mm and that $s = 1.0$ mm. Then 99.7 percent of a sample from this population should fall between 7.0 and 13.0 mm. The standard deviation is a convenient measure of how variable the population is. (If s were 0.5 mm in this case, the population would be less variable: the 99.7 percent limits would be 8.5 and 11.5 mm.) A useful measure of the amount of evolutionary change over some interval is, therefore, how greatly the mean has changed relative to the amount of within-pop-

ulation variation, measured in standard deviations. If $s = 1.0$ and the mean changes from 10.0 to 13.0, it has evolved three standard deviations. The COEFFICIENT OF VARIATION, $CV = 100 \ (s/\bar{y})$, is usually independent of size, and is often about 10 percent. Thus if $\bar{y} = 100$ mm and $s = 10.0$, the CV is the same as in the above example, and the same relative amount of evolutionary change would be measured if the mean increased from 100 to 130 mm. To convert this into a rate, the change in standard deviations is divided by the time interval.

2. *The darwin.* If a series of measurements increases (or decreases) by some proportion (e.g., 10, 20, 40, each being twice the preceding value), they form a linear series if transformed to logarithms (e.g., $\log_{10}$ of these values yields 1.000, 1.301, 1.602, each being 0.301 greater than the preceding value). The *increments* on the log scale are independent of the mean: for example, $\log_{10}$ of the series 1, 2, 4 yields 0.000, 0.301, 0.602. If each element in the series is a factor of $e = 2.718$ greater than the preceding element (e.g., 1.000, 2.718, 7.388), then taking the natural log ($\log_e$) yields increments of 1 (e.g., values 0, 1, 2). J. B. S. Haldane (1949) proposed as a rate of morphological evolution the DARWIN, which he defined as a change by a factor of e per million years. On a $\log_e$ scale, therefore, this is a change of one unit (in mm, or whatever the unit of measurement is) per million years. A MILLIDARWIN is a change by a factor of 0.002718 per My.

change, these lineages evolved very slowly; their living representatives, indeed, are often considered "living fossils." This pattern is very common in the fossil record: *the major new features of a higher taxon are established rapidly* (so rapidly that the early steps in the origin of a group often are not documented by fossils), and the overall body plan is then "stabilized" and changes little thereafter.

Although single characters do not convey an adequate sense of the evolution of the organism as a whole, they are easier to describe in quantitative terms. Because of the mosaic nature of evolution, this approach often suffices to describe the rate of evolution of taxonomically or adaptively important features. Thus one may describe the change in, say, the height of a horse's tooth in millimeters per million years. Evolutionary rates are usually described in relative terms. Two measures are often used (Box 6.A). The number of *standard deviations* by which the mean of the character changes per unit of time scales the change to the amount of variation within the population (Figure 6.36); thus if the mean changes by three standard deviations, it has shifted to a phenotypic value that formerly characterized less than 1 percent of the population. Another measure of evolutionary rate is the *darwin*, which is a change

by a factor of 2.718 (the base of natural logarithms) per million years.

The paleontological literature on rates of evolution does not necessarily represent a random sample of evolutionary rates, because paleontologists often focus on characters that display evolutionary change and ignore characters that do not distinguish taxa (i.e., CONSERVATIVE CHARACTERS, those with a very low rate of evolution).

When rates of character evolution have been measured for ancestor-descendant series of dated fossils, the most striking result is that *average rates of evolution are usually extremely low.* For example, Figure 6.37 shows the rate of evolution of the height of one of the molar teeth in a lineage of horses, one of the most important adaptive changes in the evolution of the grazing lineages. The maximum rate was 94 millidarwins (about 7 percent per million years). Expressed another way, the maximum rate was 1.7 standard deviations per million years, and the average rate was 2.8 standard deviations per million generations, meaning that it took this long for the average tooth height to shift less than one-third of the span between the smallest and the largest specimens that the population harbored at any one time. Another example, again from the Equidae, is provided in

FIGURE 6.36 Using the standard deviation of a character to measure evolution. A normally distributed character (e.g., body length) has a standard deviation (*s*) equal to 1.0 in *A* and to 0.5 in *B*. For a normally distributed character, 99.7 percent of the population falls within three standard deviations below and above the mean. Thus at time 1 (curves at left), the mean $\bar{z}$ = 4.0 in both A and B, and 99.7 percent of the population lies between $z = 1$ and $z = 7$ in A, and between 2.5 and 5.5 in B. If $\bar{z}$ evolves, and increases by three standard deviations in a given period of time (curve 2), $\bar{z} = 7.0$ in case A and $\bar{z} = 5.5$ in case B. The *absolute* amount (or rate) of change is greater in case A, but the *relative* change (in units of *s*) is equal in the two cases. If $\bar{z}$ evolves to equal 17.0 (curve 3) in the same time interval in both cases, the absolute change (or rate) is equal in the two cases, but B has evolved *relatively* faster than A: the mean has shifted by 26 standard deviations in B, but only 13 standard deviations in A.

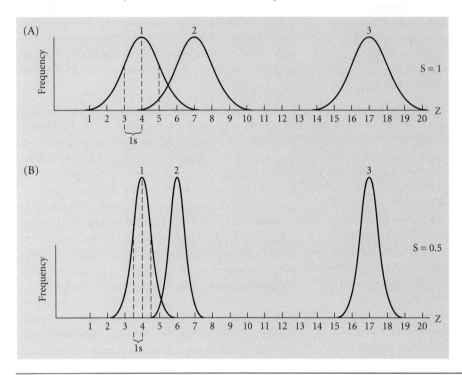

Figure 6.38, which presents the estimated body weights of fossil equids and the rate of change in body weight for some ancestor-descendant series. Note that the rate of evolution of this feature was very low until the early Miocene (25–15 Mya), when the horses began an adaptive radiation into diverse browsing and grazing lineages, some of which increased in size rather rapidly. Even the most rapid rates are quite low, however: 280 millidarwins, or about a 21 percent increase in size per million years on a logarithmic scale. (Note also that some lineages decreased in size, at various rates.)

These are *average* rates, calculated from the difference between the mean values at the beginning and at the end of a time interval, divided by the interval $[(\bar{y}_1 - \bar{y}_0)/t]$. In all cases the time interval is on the order of millions of years. One can readily see that very rapid evolutionary rates would be masked if within this interval the rate fluctuates (e.g., long periods of stasis alternate with bursts of very rapid change) or the direction of evolution changes (e.g., the mean increases and then declines to its original value, yielding no net change). Such changes in rate and direction are indeed evident when successive fossil deposits are closely spaced in time, as the data from sticklebacks and planktonic protozoans illustrate (see Figures 6.6–6.8). In almost all instances in which data of this kind have been analyzed (Charlesworth 1984), the variation in the rate of change between successive samples is much greater than the average rate of change over the whole sequence, largely because of rapid fluctuations in the mean.

Over very short intervals, rates of change may be very rapid, in the tens or even hundreds of darwins (Figure 6.39). Differences between Pleistocene species (less than a million years old) and their living descendants represent some very rapid rates of change, and even faster changes have been documented in populations that have been introduced by humans into new geographic regions. For example, house sparrows (*Passer domesticus*), introduced into North America from Europe about 100 years ago, have become differentiated into geographic races that are adapted in size and coloration to different North American environments (Johnston and Selander 1964). Some of their skeletal dimensions have diverged from those of European populations at rates of 50 to 300 darwins. In *absolute* terms, many of the changes in introduced or post-Pleistocene populations have been slight, and would not be noticed without measuring them; but they have transpired over such a short time that the rate of change is very great. The overall messages of the data in Figure 6.39 are that in dis-

FIGURE 6.37 Analysis of the rate of evolution of a morphological character in the fossil record. Changes in the height of a molar tooth in the horse lineage *Hyracotherium* to *Neohipparion* are plotted arithmetically (black circles) and logarithmically (colored circles). The average standard deviation ($s = 0.055$) on the log scale is shown by horizontal bars. For each interval in this sequence, several calculations are shown: (1) the estimated number of generations (g) between successive time points; (2) the change in mean tooth height (ΔP), measured in standard deviations on the log scale; and (3) the average fraction (f) of the population dying, per generation, if natural selection were responsible for the change and if the change occurred at a constant rate during the time interval. Over this history, the average rate of change was 2.8 mm per million generations (on an arithmetic scale), or 2.6 standard deviations per million generations (on a log scale). The average fraction of mortality that would account for this rate of change, if natural selection were responsible, is 1.0×10^{-6} of the population per generation (see Chapter 14). (After analyses by Lande 1976b.)

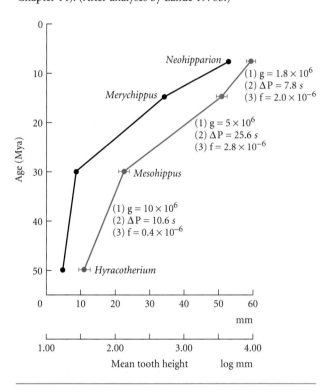

Both proponents and opponents of the hypothesis of punctuated equilibrium have claimed that data on evolutionary rates support their views. "Punctuationists" expect the average evolutionary rate to be low, because they expect brief intervals of rapid change to punctuate much longer periods of stasis. The *Globorotalia* lineage (see Figure 6.6), in which there was a rapid shift from one relatively stable phenotype to another, is an example of the pattern they predict. Opponents of punctuated equilibrium hold that the rapid fluctuations in these characters, even when there is no overall directional trend, provide strong evidence

against punctuated equilibrium. These features, they argue, are not constrained from evolving, as the punctuationists propose; rather, the characters are continually evolving, but simply are not going in any particular direction (perhaps because environmental agents of natural se-

FIGURE 6.38 Evolution of estimated body mass in fossil horses. (A) The body masses of 40 species, plotted against geological time. Notice that although the mean increases toward the present (an instance of Cope's rule), this is largely because of an increase in variation; small species existed throughout much of the history of the Equidae. (B) Rates of evolution of body mass (in darwins) between various pairs of ancestral and descendant taxa, plotted against the midpoint between their geological ages. Both increases (solid circles) and decreases (open circles) in size occurred. The most rapid changes occurred late in equid history, during the evolution of large grazers. Even the highest rate of evolution is much lower than some rates calculated over shorter intervals (see text). (After MacFadden 1986.)

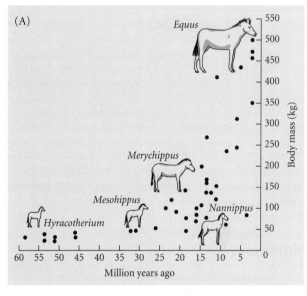

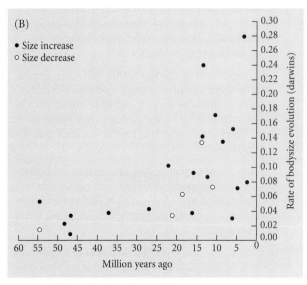

cussing evolutionary rates, it is important to specify the time interval involved, and that *a high rate of evolution is not sustained in one direction for very long.*

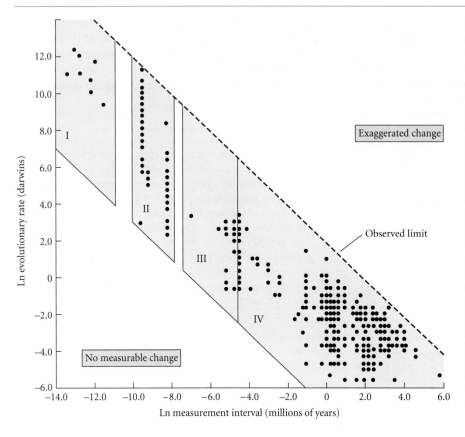

FIGURE 6.39 The inverse relationship between evolutionary rates and the time interval over which the changes are measured. The data are changes in morphological characters of presumed ancestor-descendant pairs of diverse animals, calculated by P. D. Gingerich from data compiled by many authors. The data fall into four classes: I, laboratory selection experiments; II, colonizations of new regions within recent recorded history; III, comparisons of extant populations with post-Pleistocene fossils; IV, more ancient fossil lineages. Many data, not recorded, would fall in the region marked "no measurable change." Lineages that changed so greatly that they would fall in the region marked "exaggerated change" would probably differ so greatly that they would not be compared, because their relationships would not be recognized. Both the evolutionary rate and the time interval are plotted on a log scale. (After Gingerich 1983.)

lection are fluctuating without any long-term trend). Moreover, they argue, detailed fossil records show a history of slight, successive changes, not abrupt jumps between distinct phenotypes.

What factors account for differences among rates of evolution? This is an extremely complex topic, to which almost all of evolutionary biology pertains. Hence, we defer full discussion of it to Chapter 24.

Summary

1. From both geological and biological evidence, it is clear that the fossil record is extremely incomplete, and so cannot be expected to provide documentation of the full course of evolution of most groups of organisms. Nevertheless, quite a few evolutionary histories are rather well known.

2. A detailed record of changes within individual species exists for some exceptional cases, and shows that characters often fluctuate rapidly with little overall change, occasionally shifting rather quickly to new quasi-stable states. The formation of new species has been described in some cases.

3. The hypothesis of punctuated equilibrium holds that most changes in morphology occur in association with the evolution of new species; this hypothesis is highly controversial, and evidence both for and against it has been described.

4. In at least some instances, the evolution of higher taxa (genera, orders, classes) is gradual, entailing both mosaic evolution and gradual change in individual features. For many intermediate fossils, the decision of whether to classify them in one taxon or another is quite arbitrary. Many, if not most, of the evolutionary changes in the origin of higher taxa appear to be adaptive. These points can be illustrated by detailed descriptions of the origin of tetrapod vertebrate classes.

5. Evolutionary trends, which can often be attributed to natural selection, are evident in the fossil record, but such trends are frequently arrested or reversed in different related lineages.

6. Rates of evolution are enormously variable. The inception of higher taxa is frequently marked by the rapid evolution of many characteristics, after which the rate of morphological evolution is frequently much lower. Individual features typically display a very low rate of evolution, averaged over long periods of time, but more detailed fossil records show that very rapid, small fluctuations in characteristics are common. Such changes will not be discerned in more spotty fossil records.

Major References

Stanley, S. M. 1993. *Earth and life through time.* Second edition. W. H. Freeman, New York. A thorough introduction to geological processes, earth history, and major events in the history of life, from a paleontologist's perspective.

Briggs, D. E. G., and P. R. Crowther (editors). 1990. *Paleobiology: A synthesis.* Blackwell Scientific, Oxford. Brief, authoritative essays on numerous topics in paleontology by leading authorities.

Carroll, R. L. 1988. *Vertebrate paleontology and evolution.* W. H. Freeman, New York. The most comprehensive recent treatment of the subject, with extensive coverage of major evolutionary events. Abundantly illustrated.

Kemp, J. S. 1982. *Mammal-like reptiles and the origin of mammals.* Academic Press, London. A detailed, technical history of the evolution of synapsids and early mammals.

Simpson, G. G. 1953. *The major features of evolution.* Columbia University Press, New York. This book and its predecessor (*Tempo and mode in evolution*, 1944) laid the foundation for interpreting paleontological data in the framework of modern evolutionary theory. Well-written and well worth reading, its major conclusions have been little altered by subsequent research.

Gould, S. J., and N. Eldredge. 1977. Punctuated equilibria: The tempo and mode of evolution reconsidered. *Paleobiology* 3: 115–151. One of the more detailed explications of the hypothesis of punctuated equilibrium. This subject is also treated at length by S. M. Stanley in *Macroevolution: Pattern and process* (W. H. Freeman, San Francisco, 1979). See also S. J. Gould and N. Eldredge, Punctuated equilibrium comes of age, *Nature* 366: 223–227 (1993). For extended criticism of the hypothesis of punctuated equilibrium, see J. S. Levinton, *Genetics, paleontology, and macroevolution* (Cambridge University Press, Cambridge, 1988).

Problems and Discussion Topics

1. Suppose you have samples of a single lineage from different times (i.e., different levels in a stratigraphic column). If each sample is a time-averaged set of organisms that lived over a span of tens of thousands of years, what will be the effect on the variation in a characteristic, compared with the variation that existed at any one time? Might it appear that there were two species instead of one? Refer to Bell's study of sticklebacks (Figure 6.8). What would the data look like if he had collapsed samples spanning 20,000 years into single samples, instead of analyzing separate samples at 5000-year intervals?

2. The rate of evolution of DNA sequences (and other features) is often calibrated by the age of fossil members of the taxa to which the living species belong (see Figure 33 in Chapter 5). How do imperfections in the fossil record affect the estimates of evolutionary rates obtained in this way? Is there any way of setting limits to the range of possible rates?

3. An ideal fossil record would enable researchers to distinguish the patterns of phyletic gradualism, punctuated equilibria, and punctuated gradualism (see Figure 6.12). How would you do so? How do imperfections of the fossil record make it difficult to distinguish these patterns?

4. Figure 6.15 presents a phylogenetic tree of the horse family, in which the temporal distribution of taxa is portrayed. Some systematists distinguish phylogenetic trees from cladograms. A cladogram, by definition, portrays branching relationships among taxa, but not their position in geological time. Draw a cladogram for the taxa in Figure 6.15.

5. Using the biological species concept (see Chapter 2), debate whether or not the morphologically different forms in the *Phacops rana* complex of trilobites (Figure 6.10) are different species.

6. Creationists may acknowledge that very similar species can evolve from a common ancestor, but they deny that new "kinds" of organisms can evolve. One creationist (Gish 1974) has defined a "kind" as "including all of those variants which have been derived from a single stock." This stock, they hold, gives rise to variants, but not to different "kinds." Gish (1974) notes that "we cannot always be sure . . . what constitutes a separate kind," but cites fishes, amphibians, reptiles, birds, mammals, and within the mammals dogs, cats, monkeys, chimpanzees, gorillas, and humans as "different basic kinds." What is the validity of the concept of a "kind"? Evaluate the creationist claim, in light of the material in Chapters 5 and 6.

7. Creationists deny that the fossil record provides intermediate forms that demonstrate the origin of higher taxa. Of *Archaeopteryx*, they say that because it had feathers and flew, it was a bird, not an intermediate. Evaluate this argument.

8. Regarding the reptile/mammal transition, Gish (1974) says that reptiles have one auditory ossicle, mammals three, and that no transitional fossils show an intermediate number of auditory ossicles. Nor, he says, has anyone explained how an intermediate could chew or hear while the postulated transition occurred. Refute these arguments on the basis of material in this chapter and the references cited.

9. One of the most frequently cited trends in evolution is the supposed trend toward larger body sizes ("Cope's rule"). Examine the data on the horse family in Figure 6.38A. Note that average body size increases, but not all horses become large. Contrast this situation with a hypothetical case in which all later species are larger than all earlier species. What might you infer about possible reasons for the evolution of size in each of the two cases?

10. Consider the *hypothesis* that Eldredge and Gould advanced to explain the *pattern* they called punctuated equilibria. What would be the implications for evolution if the hypothesis were true versus false?

CHAPTER 7

A History of Life on Earth

This chapter presents a descriptive survey of the major events in the history of life, especially the origin, diversification, and extinction of major groups of organisms. Most of this material treats geological and paleontological evidence, but phylogenetic studies of living organisms have also revealed critically important information. This chapter includes far more details than most readers will want to memorize, but events that many people consider particularly important or interesting are highlighted in italic type.

For the most part, we defer generalizations and theoretical interpretations of evolutionary history to other chapters (especially Chapter 25). However, this chapter provides instances of the following general patterns:

1. The disposition of oceans and land masses has changed over time, as have climates; these changes have affected the geographic distributions of organisms.
2. The taxonomic composition of the biota has changed continually, owing to origination of new forms and extinctions.
3. At several times, extinction rates have been particularly high (so-called "mass extinctions").
4. Evolutionary radiation of higher taxa has sometimes been quite rapid (on the order of tens of millions of years); this has often been the case following mass extinctions.
5. The diversification of higher taxa has included increases both in the number of species and in the variety ("disparity") of their form and ecological habits. There has been an overall increase in the variety of niches occupied.
6. Higher taxa have often been replaced by other, ecologically similar, higher taxa. Not only in time but also in space, members of different higher taxa have sometimes radiated into a variety of "ecological equivalents."
7. The ancestral members of related higher taxa are often ecologically generalized, morphologically simi-

lar forms; the higher taxa diverge with the passage of time.
8. Of the variety of forms in a higher taxon that were present at any time in the remote past, usually only a few have persisted in the long term.
9. The geographic distribution of many taxa has changed greatly.
10. Over time, the composition of the biota increasingly resembles that of the present, as a consequence both of extinction and of the evolution of surviving lineages.

Before Life Began

Several lines of evidence have led most physicists to agree that the current *universe came into existence about 14 billion years ago*, through an explosion (the "big bang") from an infinitely dense point (see Chapter 6). Nothing can be said (at least at present) about what there was "before" this event. (The concept of time "before" the big bang may have no meaning.) Elementary particles formed hydrogen shortly after the big bang, and hydrogen ultimately gave rise to the other elements. The collapse of a cloud of "dust" and gas formed our galaxy less than 10 billion years ago. Throughout the history of the universe, material has been expelled into interstellar space, especially during stellar explosions (supernovas), and has condensed into second- and third-generation stars, of which the Sun is one. *Our solar system was formed about 4.6 billion years ago*, based on radiometric dating of meteorites and moon rocks. The earth is the same age, but because of geological processes such as subduction of the crust, the oldest known rocks on earth are younger, dating from 3.8 billion years ago. The earth was probably formed by the collision and aggregation of many smaller bodies, the impact of which contributed enormous heat.

The early earth was molten, but formed a solid crust as it cooled, releasing gases that formed an atmosphere that *included water vapor but very little oxygen*. As the earth cooled, oceans of liquid water formed, probably by 4.5 billion years ago, and quickly achieved the salinity of modern oceans. By 4 billion years ago, there were probably many small protocontinents, but no large land masses; these may have come into existence toward the end of Archaean time (the span from 3.6 to 2.5 billion years ago).

The Emergence of Life

The simplest things that might be called living must have developed as complex aggregations of molecules. These, of course, would leave no fossil record, so it is only through chemical and mathematical theory, laboratory experimentation, and extrapolation from the simplest living forms that we can hope to develop models of the emergence of life. Nor has anyone yet "synthesized life in a test tube." These points are often used by anti-evolutionists to argue that evolutionary theory is rotten at its core. This criticism, however, misses at least two important points. First, no field of science claims to have explained fully every observation that falls within its realm. Moreover, there is no justification for thinking that because something *has not* been fully explained, it *cannot* be explained or demonstrated. Our extraordinary progress in many areas of science shows how foolish and arrogant such a claim is. Second, whether or not we understand how living things first came into existence is irrelevant to questions about how living things subsequently evolved. Most of this book is devoted to understanding the evolution of "life as we know it."

"Life" is difficult to define at the border of the nonliving. (In isolation from their hosts, for example, viruses lack many of the features of more "typical" life forms.) Undoubtedly the salient feature of living things is their ability to replicate themselves, thereby increasing in number and producing copies of variant forms. All living things, moreover, contain proteins, which catalyze reactions and contribute other phenotypic characteristics. Replication requires energy, which living things obtain either by breaking chemical bonds in complex organic molecules obtained from the environment (heterotrophs) or by building organic compounds through the capture of the energy in light or in simple inorganic compounds such as H_2S (autotrophs). The energy is typically stored as ATP. The several components needed for replication and energy capture are held together in a compartment (such as a cell).

Although it is quite possible that "life" originated more than once, *there is good reason to believe that all organisms we know of stem from a single common ancestor*, because they all share certain features. Among these are the synthesis and utilization of only L isomers of amino acids as building blocks of proteins; L and D optical isomers are equally likely to be formed in abiotic synthesis, but a functional protein can be made only of one type or the other. D isomers could have worked just as well. The universality (with a few minor exceptions; see Chapter 3) of the presumably rather arbitrary genetic code and of the machinery of protein synthesis are among the other features that imply a monophyletic origin of all organisms.

To account for the emergence of systems of molecules that could have given rise to life forms, it is necessary to envision, at the very least, the origin of (1) *simple monomeric compounds* like amino acids and nucleotide bases; (2) *assembly* of nucleotide bases into nucleic acids and of amino acids into polypeptides; (3) *replication* of nucleic acids; (4) *"compartmentation,"* i.e., a mechanism to protect interacting molecules from dilution; and (5) production of an increasing variety of polypeptides or proteins, with reasonably consistent amino acid sequences, to *catalyze the processes of energy capture and replication*. The most difficult problem is that, at least in known living systems, only nucleic acids replicate, but replication requires the action of proteins that are encoded by the nucleic acids. It is a chicken/egg problem—which came first?

Although many formidable obstacles to a fully sufficient theory remain, experiments and theory have led to considerable progress in understanding some of the likely steps in the origin of life (Orgel 1994; Maynard Smith and Szathmáry 1995). The most important advances include these:

1. *Simple organic molecules can be produced by purely abiotic chemical reactions.* These compounds are the building blocks of complex organic molecules. Interstellar space contains gaseous organic molecules (up to at least 11 carbons), including alcohols, aldehydes, and ketones. In a famous experiment, Stanley Miller (1953) found that electrical discharges in a simulated early earth atmosphere of methane (CH_4), ammonia (NH_3), and water (H_2O) yield amino acids and compounds such as hydrogen cyanide (HCN) and formaldehyde (H_2CO), which undergo further reactions to yield sugars (including the components of nucleotide bases), amino acids, purines, and pyrimidines (Figure 7.1). The concentrations of these compounds in the early oceans could have been very high, since there was nothing to consume them. Polymerization (e.g., of amino acids) may have been facilitated by adsorption to clay particles, or by concentration due to evaporation.

2. *Natural selection and evolution can occur in nonliving systems of replicating molecules.* This was first demonstrated by Sol Spiegelman (1970), who placed RNAs, RNA polymerase (from Qβ phage), and nucleotide bases in a cell-free medium. Different RNA sequences were replicated by the polymerase at different rates, so that their proportions changed. In particular, shorter sequences increased in frequency because they were replicated more rapidly. Because a replicase enzyme from a living system was provided, these experiments do not bear on the very first steps in the origin of life, but they do show that systems of replicating macromolecules can evolve.

3. *RNA has catalytic properties, including self-replication.* Because RNA has a complex three-dimensional structure, some RNA sequences can bind smaller molecules and facilitate reactions between them, thus acting like an enzyme. Such RNAs (or ribozymes) can cut, splice, and elongate other oligonucleotides. Moreover, it has

FIGURE 7.1 The apparatus Miller used to synthesize organic molecules under simulated early earth conditions. (After Maynard Smith and Szathmáry 1995.)

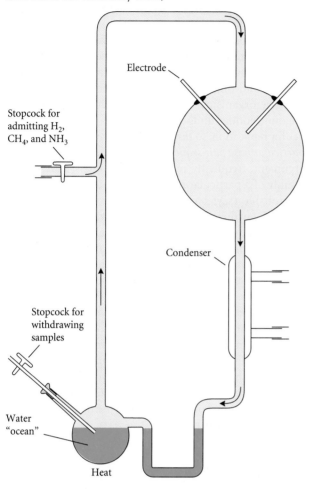

Electrode

Stopcock for admitting H_2, CH_4, and NH_3

Condenser

Stopcock for withdrawing samples

Water "ocean"

Heat

been experimentally demonstrated that in the absence of protein enzymes, *short RNA template sequences can self-catalyze the formation of complementary sequences* from free nucleotides (Figure 7.2). Thus, for example, poly(C) directs the assembly of poly(G), although the error rate (mutation rate) is quite high. Because of the high error rate, specific RNA sequences cannot increase greatly in number, but instead generate families of molecules with various related sequences. (Such an assemblage of variant sequences is called a "quasi-species.") In the absence of a replicating enzyme, the error rate of self-replicating RNAs would be so high that self-replication would be limited to short sequences of less than 100 nucleotide bases. Manfred Eigen (1971) has postulated that longer sequences might have evolved if two (or more) RNA sequences were coupled in a "hypercycle," each sequence catalyzing the replication of the other in a positive feedback loop. This would require that the interacting RNAs in the hypercycle be coupled together into "individuals," rather than encountering each other at random in solution. These proto-individuals, consisting of two or more simple RNA "genes," might be formed by adhesion to clay or other surfaces, or by envelopment in a molecular membrane. Some of these steps are theoretically plausible, but have not yet been experimentally demonstrated.

4. *Lipid membranes can form spontaneously.* As we have just seen, the evolution of anything more complex than short oligonucleotides requires that interacting molecules be *compartmentalized* into "individuals." Within these compartments, molecules can be held in sufficient proximity and concentration to undergo reactions that otherwise would be rare. Lipid membranes may have played this role in the evolution of life, as they do in cells today. Because of their hydrophobic tails, straight-chain fatty acids aggregate and exclude water, spontaneously forming double layers that can take on spherical form. Such spheres divide under some conditions. They are permeable to certain ions and small organic molecules, which would steadily diffuse into the globule if it enclosed RNAs that assembled the diffusing molecules into replicate sequences, thus removing

FIGURE 7.2 A scenario for the origin of RNA replication. Reactive nucleotides, consisting of bases joined to ribose and phosphate (A), form RNA polymers, some of which have catalytic properties (B), which enable them to form complementary strands from free nucleotides (C,D). Steps C and D have occurred in the laboratory. A further requirement for the evolution of populations of replicating molecules is that the complementary strands in (D) separate and each repeat the replication process. This has not yet been achieved in the absence of polymerase enzymes. (After Orgel 1994.)

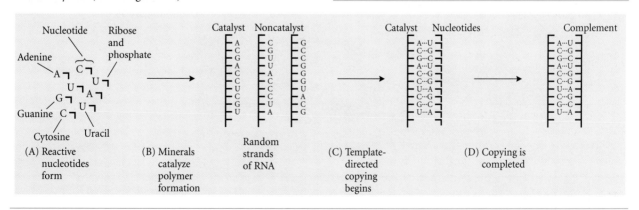

(A) Reactive nucleotides form — Nucleotide, Adenine, Ribose and phosphate, Guanine, Cytosine, Uracil

(B) Minerals catalyze polymer formation — Random strands of RNA — Catalyst, Noncatalyst

(C) Template-directed copying begins — Catalyst, Nucleotides

(D) Copying is completed — Complement

them from solution. The empirical stumbling block at this point is that, although fatty acids do form under simulated prebiotic conditions such as those in Miller's experiment, most of them are branched fatty acids, which do not form bilayered membranes.

5. *The evolution of protein enzymes is perhaps the greatest unsolved problem.* In the living systems we know, replication of long RNA (or DNA) sequences, like many other processes, is based on the catalytic activities of proteins rather than ribozymes. Eörs Szathmáry (1993) has suggested that the evolution of proteins began when cofactors, consisting of an amino acid joined to a short oligonucleotide sequence, aided RNA ribozymes in self-replication (Figure 7.3). He notes that many contemporary coenzymes have nucleotide components. The next step may have been the stringing together of several such amino acid-nucleotide cofactors. Ultimately, the ribozyme evolved into the ribosome, the oligonucleotide component of the cofactor into transfer RNA, and the strings of amino acids into catalytic proteins. This hypothesis promises to explain the origin of proteins, the specificity of tRNAs for amino acids, and the genetic code itself.

Many of the preceding points are speculative but testable hypotheses that point the way toward understanding the origin of some of the fundamental properties of life. Maynard Smith and Szathmáry (1995) provide a useful introduction to ideas about the origin of some other properties, such as the evolution of the metabolic capacity to capture energy for growth and replication, the linkage of RNA sequences (genes) into chromosomes, and the evolution of DNA as the prevalent hereditary macromolecule.

We close this passage on the origin of life with a few reflections. Does the origin of life fall within the realm of realistic probability? However self-replicating systems may have been formed, it is likely that replication was slow and inexact, and only much later acquired the fidelity that modern organisms display. Moreover, many different oligonucleotides undoubtedly had the necessary properties for replication, so *the first "genes" need not have had any particular base pair sequence.* Thus there is no force to the argument, frequently made by creationists, that the assembly of a particular nucleic acid sequence is extremely improbable (say, 1 chance in 4^{50} for a 50-bp-long sequence). Furthermore, even an event with very low probability is almost certain to happen many times over the course of a billion years or so in a worldwide reaction system containing perhaps 10^{10} oligonucleotides per liter.

Our sketchy understanding of the origin of life is due, in part, to the youth of scientific inquiry into the subject. The first substantive ideas about the mechanisms of the origin of life, proposed by British evolutionary theorist J. B. S. Haldane in 1928 and Russian biochemist A. I. Oparin in 1934, are only about 60 years old; the first experiments on abiotic synthesis of organic compounds date only from 1953. Substantial progress has been made since then, so it would be a counsel of despair to suppose that we will never be able to complete the story.

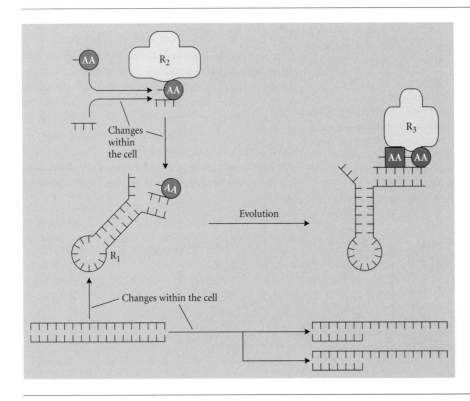

FIGURE 7.3 A hypothesis for the origin of polypeptide synthesis and the genetic code. An RNA ribozyme (R_1), ancestor of mRNA, binds to a cofactor consisting of an amino acid (AA) and a short oligonucleotide, which have been joined by another ribozyme (R_2) that joins specific oligonucleotides to amino acids according to a primitive code. This system evolves to one in which ribozyme R_3 links amino acids together, and is ancestral to the ribosome. (After Maynard Smith and Szathmáry 1995.)

Could life originate de novo today? Ever since Louis Pasteur's experiments, scientists have denied the possibility of spontaneous generation—under current conditions. The oxidizing atmosphere of the modern world prevents the formation of the compounds that would form under anoxic conditions—and even if primitive proteins or nucleic acids were to form spontaneously, they would soon be eaten by living organisms.

Precambrian Life

Prokaryotes

The Archaean Eon, from 3.6 billion to 2.5 billion years ago, and the Proterozoic Eon, from 2.5 billion to 543 million years ago, are together referred to as PRECAMBRIAN TIME. The oldest known rocks (3.8 billion years) contain carbon deposits that may indicate the existence of life. *The first fairly definite evidence of life dates from 3.5 billion years ago*, in the form of bacteria-like microfossils and layered mounds (STROMATOLITES) with the same structure as those formed today along the edges of warm seas by cyanobacteria (blue-green bacteria, or blue-green "algae") (Figure 7.4). Thus cellular life was apparently well established within a billion years after the formation of the earth.

At the time of the earliest life, the atmosphere virtually lacked oxygen. The earliest cells were surely *anaerobic prokaryotes* (bacteria), possibly photosynthetic or chemosynthetic. When *photosynthesis* evolved in cyanobacteria and other bacteria, it *introduced oxygen*, which at first became bound in iron-rich rock, but later built up in the atmosphere. Moderate atmospheric levels of O_2 had already been attained by about 2.3 billion years ago; there was probably a marked increase about 700 million years ago (Mya), and an approach toward modern levels during the Cambrian period (543–505 Mya). As atmospheric O_2 increased, so did the ozone (O_3) layer. Ozone reduces the flux of ultraviolet light, which causes mutations in nucleic acids. As oxygen built up, living organisms evolved the capacity for *aerobic respiration*, as well as mechanisms, such as the enzyme superoxide dismutase, that protect the cell against oxidation.

The bacteria fall into two groups that are as different from each other in many aspects of ultrastructure, molecular design, and DNA sequence as they are from eukaryotes (Figure 7.5). These two groups have been classified by Woese (1987) as the kingdoms Archaebacteria and Eubacteria (also called Archaea and Bacteria). The Archaebacteria include extremely thermophilic ("heat-loving") forms that live at temperatures near or above the boiling point of water (in hot springs), as well as several other groups, such as the extreme halophiles ("salt-lovers"), that probably evolved from thermophilic ancestors. None is completely photosynthetic, but some are partly so. Most are anaerobic. It is likely that they arose from an anaerobic, sulfur-reducing, extremely thermophilic ancestor within the first billion years of the earth's history, when temperatures were high and oxygen was lacking. The Eubacteria are extremely diverse in their metabolic capacities. Many are photosynthetic; it is likely that photosynthesis arose once within the Eubacteria and was independently lost in many lineages (Woese 1987). Thermophilia (although not as extreme as in Archaebacteria) is not uncommon among Eubacteria, and may have characterized their common ancestor.

FIGURE 7.4 Stromatolites formed by (A) living cyanobacteria in Australia and (B) Proterozoic photosynthetic bacteria. A © F. Gohier, The National Audubon Society Collection/Photo Researchers; B © K. and B. Collins/Visuals Unlimited.

(A)

(B)

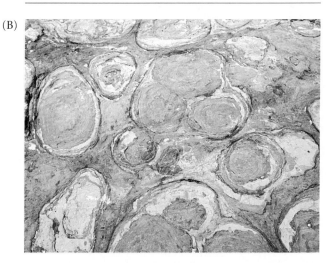

(A)

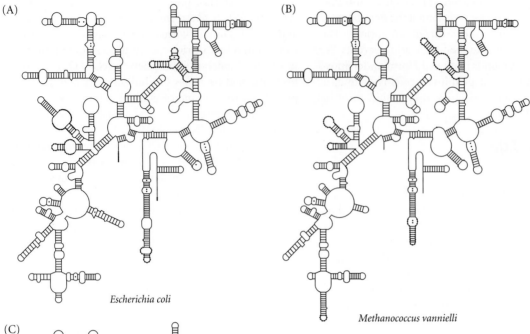

Escherichia coli

(B)

Methanococcus vannielli

(C)

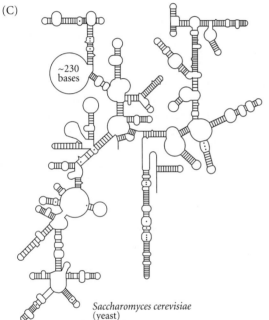

~230 bases

Saccharomyces cerevisiae
(yeast)

FIGURE 7.5   The structure of small-subunit ribosomal RNA in representative Eubacteria (A), Archaebacteria (B), and Eukaryota (C). Cross-linked stems are regions of complementary base pairing; open loops are regions without pairing. Note the overall similarity of structure throughout the three major divisions of life. (After Woese 1987.)

Eukaryotes

A major event in the history of life was the *origin*, from prokaryotes, *of eukaryotes*, which are distinguished by a nucleus with distinct chromosomes, a mitotic spindle, and meiosis, the basis of highly organized recombination and sexual reproduction (see Chapter 20 on the evolution of recombination). Most of the evolutionary stages from prokaryote to eukaryote are unknown, but more than 1000 species of eukaryotic protists that primitively lack mitochondria, and thus represent an intermediate stage, are known (Cavalier-Smith 1987).

Almost all eukaryotes have mitochondria, and many have chloroplasts. It is now quite clear that *these organelles are de-*

scended from bacteria that at first were probably eaten by heterotrophic prokaryotes and later became *endosymbionts* (Margulis 1993; Maynard Smith and Szathmáry 1995). These organelles are similar to bacteria not only in their ultrastructure, but also in their DNA sequences (Schwartz and Dayhoff 1978). Mitochondria are derived from the group of purple bacteria (which includes *E. coli*), and chloroplasts are derived from cyanobacteria. Many eukaryotes today have symbiotic bacteria or algae in their cells (e.g., the zooxanthellae of corals, the bacteria in mycetocytes of certain insects), and at least one protozoan (*Pelomyxa*) that lacks mitochondria relies on three types of endosymbiotic bacteria for respiration (Parker 1982, I, 514). It is quite likely that chloroplasts originated more than once, in different lineages of algae (Howe et al. 1992). Many of the genes that code for cytochromes and other constituents of mitochondria, and which must have been part of the original mitochondrial genome, have long since been transferred to the nuclear genome—a process that has been observed in experimental populations of yeast and other organisms (Gray 1989).

The *earliest probable eukaryotes* in the fossil record, *single-celled algae, are about 1.4 billion years old;* unquestionable eukaryotes date from about 0.9 billion years. Perhaps as many as 2 billion years passed between the emergence of prokaryotes and that of eukaryotes. The eukaryotes include many lineages of algae and protozoans (collectively sometimes called Protoctista), as well as the kingdoms Fungi, Plantae, and Animalia. Because most protoctistan groups are distinguished by biochemical and ultrastructural features that fossilize poorly, the fossil record offers only sketchy evidence of their relationships and times of origin (see Knoll 1992). The first multicellular algal fossils date from 0.9 billion years ago (there are possible fossils at 1.3 billion years), and the first definite protozoans from 0.8 billion years.

The Tree of Life

DNA sequences show that Archaebacteria, Eubacteria, and Eukaryota are monophyletic "primary kingdoms" (Woese 1987; Iwabe et al. 1991). But how are these three groups related to each other? Finding a "root" for an unrooted minimal-length tree usually requires the use of an outgroup, but there is no outgroup for a tree that includes all of life! A clever solution for rooting the "tree of life," suggested by Margaret Dayhoff long before information became available to act on, is to use the DNA sequences of genes that were duplicated in the ancestor of all living things (Figure 7.6). Two copies of a gene (call them *A* and *B*) retained in the same genome (often adjacent to each other) are termed PARALOGOUS genes, and will diverge in base pair sequence over time. After an ancestral lineage diverges into two, both the *A* and the *B* sequences will also diverge between the lineages. Synapomorphies in the *A* sequences of two sister groups will be recognized because they do not appear in the *B* sequences (and therefore are derived characters), and likewise for synapomorphies in the *B* sequence.

Naoyuki Iwabe and collaborators (1989) employed this approach, using sequences of proteins coded by pairs of duplicated loci. For example, organisms in all three primary kingdoms have two loci that code for the two subunits of the ATPase enzyme. The sequence similarity between the two loci implies that they were derived from duplication of a single locus in the common ancestor of the three kingdoms. A phylogeny of the sequences of both loci in an archaebacterium, several eubacteria, and several eukaryotes shows, as expected, that the ATPase-α genes of all these organisms are genealogically more closely related to each other than they are to any of the ATPase-β genes (and vice versa) (Figure 7.7). Moreover, the archaebacterium (*Methanococcus*) is most closely related to the eukaryotes. Another pair of paralogous genes gave the same result. Thus, despite the extraordinary characteristics of Archaebacteria, they appear to have diverged from the common ancestor with eukaryotes after the eubacterial line had already diverged. [Note also in Figure 7.7 that mitochondrial (yeast mt) and chloroplast (*Euglena* chl) sequences are closely related to those of the Eubacteria.]

Phylogenetic studies based on rRNA gene sequences indicate that groups such as microsporidians, flagellates, and slime molds represent early branches on the eukaryotic tree, and that other eukaryotic groups, including ciliates,

FIGURE 7.6 Determining the phylogeny of the three primary kingdoms from duplicated genes. (A) The real history of branching of kingdoms 1, 2, and 3. An ancestral gene, *A*, gives rise to genes *A* and *B* by duplication; both duplicate genes are carried by all three descendant lineages. Mutations *a* and *b* distinguish genes *A* and *B*. In the common ancestor of lineages 2 and 3, mutations *c* and *d* occur in genes *A* and *B* respectively. (B) A table of the mutations (synapomorphies) shared by the duplicate genes in the three lineages. Lineages 2 and 3 share more mutations (i.e., *c* and *d*) than lineage 1 shares with either of them. Thus we can infer from these mutations that lineages 2 and 3 are sister taxa.

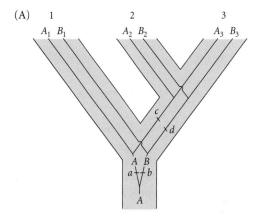

(B)

	A_1	A_2	A_3	B_1	B_2	B_3
A_1	–	*a*	*a*	0	0	0
A_2		–	*a, c*	0	0	0
A_3			–	0	0	0
B_1				–	*b*	*b*
B_2					–	*b, d*
B_3						–

fungi, plants, and animals, diverged nearly simultaneously (Figure 7.8). Animals and fungi appear to be more closely related to each other than to other major lineages of eukaryotes.

The Proterozoic Eon

The assembly of land masses during the Proterozoic formed the continents of Laurentia (including the present Greenland and parts of North America, Great Britain, and perhaps Siberia) and Gondwanaland (including South America, Africa, India, Antarctica, and Australia). (The land masses that later formed Eurasia were separate at this time.) The temperature at the earth's surface declined so greatly that there were widespread glaciers about 2 billion years ago and again about 800 million years ago.

For most of the Proterozoic, the only evidence of life is of prokaryotes and eukaryotic algae. A diverse group of

A History of Life on Earth 171

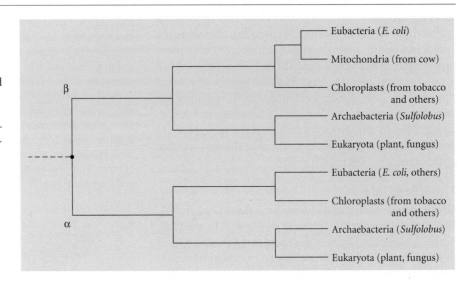

FIGURE 7.7 Phylogeny of the duplicate genes encoding subunits α and β of an ATPase enzyme in Eubacteria, Archaebacteria, and eukaryotes, as well as chloroplasts and mitochondria. Archaebacteria are more closely related to eukaryotes than to Eubacteria. Note that chloroplasts and mitochondria, from eukaryotic cells, are related to the Eubacteria, from which they have been derived. (A mitochondrial sequence for one of the subunits was not available for this study.) (After Iwabe et al. 1989.)

protoctistans, the acritarchs, are thought to have evolved from protozoan ancestors, and are recorded as far back as 1.4 billion years ago. The *first evidence of animal life is less than a billion years old*, in the form of TRACE FOSSILS, i.e., burrows and other disturbances of sediments. Before this time, compacted layers of fine sediments show no evidence of burrows. The *oldest fossils of multicellular animals are about 640 My old*. The best-known Precambrian animals are the EDIACARAN FAUNA, which extends from about 600 Mya to the beginning of the Phanerozoic, 543 Mya (Grotzinger et al. 1995). The Ediacaran animals are soft-bodied, lacking skeletons, and most appear to have been flat creatures that crept or stood on the sea floor. There is no evidence that any were predatory. Although some have been tentatively placed in the Cnidaria (jellyfishes), Annelida (segmented worms), and possibly the Arthropoda, the majority of the Ediacaran creatures do not obviously fit into any of the later phyla. Authorities differ on whether these animals are related to any of the various modern phyla or are an extinct sister group of the modern Metazoa.

Paleozoic Life: The Cambrian Revolution

Phanerozoic time, divided into the Paleozoic, Mesozoic, and Cenozoic eras, embraces all of earth's history since the end of the Proterozoic. The Phanerozoic eras, and many of their periods, were defined by pre-Darwinian geologists on the basis of their distinct fossil faunas, and so are marked by major episodes of extinction or by the diversification of major new groups of animals.

The "Cambrian Explosion" of Animal Phyla

The beginning of the *Cambrian period* has been dated at about 543 million years ago (Bowring et al. 1993). A few representatives of modern animal phyla (e.g., Cnidaria, and possibly Annelida, Arthropoda, and Echinodermata) are known from the late Precambrian and from the first 14

million years of the Cambrian. Fossils of a "small shelly fauna" of minute shelled animals whose relationships are difficult to determine have also been found. The abundance and diversity of trace fossils—fossilized tracks and burrows—from this time period is very low.

During the later part of the early Cambrian, starting at about 530 Mya, *almost all the modern phyla and classes of skeletonized marine animals, as well as many groups that may represent extinct phyla and classes, suddenly appear in the fossil record*. Recent geological studies indicate that this "**Cambrian explosion**" may have occurred *within about 30, and perhaps only 5 to 10, million years* (Bowring et al. 1993). This interval marks the first appearance of brachiopods (Figure 7.9), trilobites (Figure 7.10) and other classes of arthropods, various classes of molluscs, echinoderms, and many others. Many peculiar animals from this time period may represent extinct classes or phyla. The best-known of these are included in the remarkably well preserved fauna of the BURGESS SHALE of British Columbia (Whittington 1985). Among them (Figure 7.11) are *Opabinia*, with five eyes and a proboscis ending in a claw, *Wiwaxia*, with bladelike spines projecting from a coat of scales, and the half-meter-long arthropod *Anomalocaris*, with peculiar feeding appendages and a circular mouth rimmed by tooth-bearing plates. The Burgess Shale also includes an early nonvertebrate chordate, *Pikaia* (Figure 7.11D). Most of the fundamentally different body plans, or BAUPLÄNE, as they are sometimes called, known in animals had evolved by the end of the Cambrian, and the diversity of body plans of Cambrian animals equals, or perhaps exceeds, that at any later time in evolutionary history (Gould 1989; Wills et al. 1994). (Differences among organisms in their degree of morphological or ecological difference are also referred to as DISPARITY.)

The extraordinarily rapid diversification of higher taxa in the Cambrian fossil record is one of the most stunning events in the history of evolution, one that has prompted strong debate (see reviews in Bengtson 1995; Erwin 1991; Fortey et al. 1996; Lipps and Signor 1992; Valentine et al.

FIGURE 7.8 A recent estimate of the phylogeny of eukaryotes, inferred from DNA sequences of 16s rRNA genes. Taxa to the right of the dashed line (M) have mitochondria, and are presumed to stem from a mitochondria-bearing ancestor; taxa to the left lack mitochondria. Note the great diversity of distantly related unicellular lineages (those to the left of the Rhodophytes), which include many organisms often referred to as protozoans, or protoctistans. Other unicellular forms (e.g., alveolates, choanoflagellates) are more closely related to multicellular taxa. (After Schlegel 1994.)

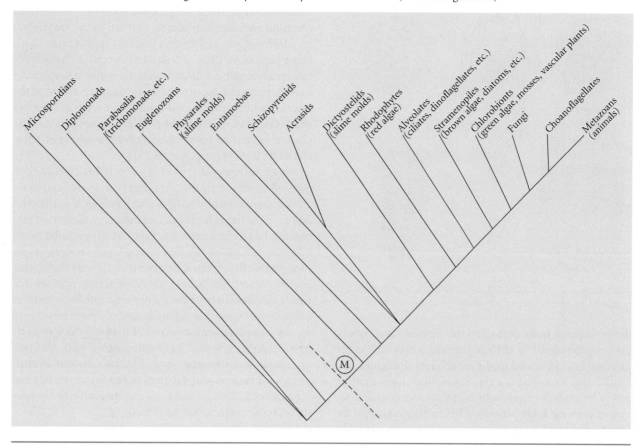

1991; Wray et al. 1996). It raises two major questions. First, why did fundamentally different body plans evolve in such great profusion early in evolutionary history, but hardly at all thereafter? We consider this question in Chapter 25. Second, how and why did so many great changes evolve in such a short time—or did they?

Researchers have offered two answers to this question; which answer they support depends on whether or not they take the fossil record at face value. Some paleobiologists believe that the fossil record provides a fairly accurate portrait of early animal evolution, and that not only the fossilizable skeletons, but also the other major features of

FIGURE 7.9 (A) A Silurian brachiopod. (B) A Cambrian archaeocyathid. Although brachiopods have clamlike shells, they are not related to true bivalves, which are in a different phylum (Mollusca). A few brachiopod species survive today, but archaeocyathids occurred only in the Cambrian. (A courtesy of P. Copper; B after Boardman et al. 1987.)

(A)

(B)

FIGURE 7.10   A sample of the great diversity of trilobites. (A) *Olenellus* (Early Cambrian); (B) *Cryptolithus* (Ordovician); (C) *Eoharpes* (Ordovician); (D) *Scutellum* (Silurian); (E) *Phillipsia*, one of the last surviving genera (Permian). (After Stearn and Carroll 1989.)

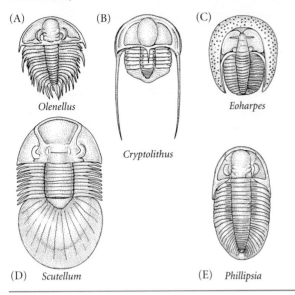

(A)   (B)   (C)

Olenellus

Cryptolithus

Eoharpes

(D)   *Scutellum*

(E)   *Phillipsia*

Cambrian in parallel with the diversity of body fossils, implying that animal diversity really did increase greatly at that time. Their opponents suggest that the degree of molecular sequence divergence among living members of certain phyla is so great, relative to the divergence among classes within phyla, that the phyla must have originated from common ancestors considerably earlier, in the Precambrian. Moreover, some of the taxa that first appear in the Cambrian explosion, such as trilobites, occupy a cladistically rather advanced position within their phyla, suggesting that the common ancestor of the phyla (such as that of the diverse Cambrian arthropods) must have been considerably older (Fortey et al. 1996). The problem of the Cambrian explosion has not yet been solved, but it is a subject of active research (see Box A in Chapter 25).

We are naturally interested in our vertebrate ancestry, so it is well worth noting that the Cambrian diversification included early vertebrates. Like many Paleozoic vertebrates and even the living hagfishes and lampreys, Cambrian vertebrates lacked vertebrae. But they did have cellular bone, a feature unique to the Vertebrata. Until recently, it was thought that the earliest known vertebrates were Ordovician jawless "fishes" (see below). However, it now appears that fossils long known from the Cambrian, toothlike structures called CONODONTS, belong to the *earliest known vertebrates*, for the body form and position of the teeth have recently been described (Figure 7.12). Moreover, the teeth are made of cellular bone (Sansom et al. 1992). Conodont animals were small and eel-shaped; they lacked vertebrae, but had a notochord. In these respects, they resemble living hagfishes, to which they may be related.

The end of the Cambrian (500 Mya) *was marked by mass extinction.* The trilobites, of which there had been more than 90 Cambrian families, were greatly reduced, and several classes of echinoderms became extinct. As Gould (1989) has emphasized, if the early chordates had also succumbed, we would not be here today. The same may be said about every point in subsequent time: had our ancestral lineage been among the enormous number of lineages that

diverse animal body plans, indeed evolved very rapidly, perhaps stimulated by environmental changes such as the increase in oxygen and in the availability of calcium carbonate that occurred at this time. Others hold that the major lineages diverged from common ancestors and acquired different body plans long before they appear in the Cambrian record. In this view, animals were already highly diverse, but either lacked carbonate skeletons or were so small that their Precambrian remains have not been found (Fortey et al. 1996). The diversity of animal classes and phyla became "visible" when many lineages either acquired skeletons or evolved larger size.

Those who argue for rapid radiation point out that the abundance and diversity of trace fossils increases in the

FIGURE 7.11   Artist's reconstruction of several of the peculiar animals of the Cambrian Burgess Shale. (A) *Opabinia*; (B) *Wiwaxia*; (C) *Anomalocaris*, a large predatory arthropod; (D) *Pikaia*, an early chordate, showing the notochord and muscle bands. (After Marianne Collins in Gould 1989.)

(A)   (B)   (C)   (D)

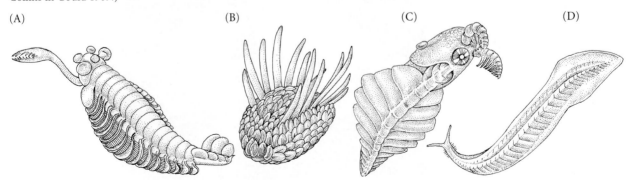

FIGURE 7.12 (A) A Cambrian conodont, showing the position of the bony toothlike structures that formed its feeding apparatus. (B) A hagfish (Myxinoidea), the most basal lineage of living vertebrates. Conodonts may have been related to these agnathans. (A after Purnell et al. 1992; B by Charles H. Coles, negative 318382, courtesy of the Department of Library Services, American Museum of Natural History.)

(A)

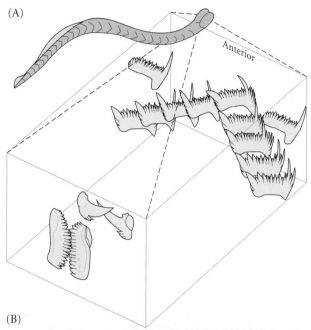

(B)

became extinct, humans would not have evolved, and perhaps no other form of life like us would have, either.

Metazoan Phylogeny

With a few exceptions, phylogenetic relationships among the animal phyla are very poorly understood. Most of these groups differ so greatly from each other that rather few morphological characters unite any of them, and for many of these characters, it is arguable which state is ancestral and which is derived, and whether or not they are homoplasious. Based on morphological characters, there has been modest agreement on a few points (Figure 7.13).

It is widely believed that multicellular animals arose from colonial flagellate protozoans. Sponges (Porifera) very possibly arose independently from the lineage that gave rise to the other phyla (together classified as the subkingdom Metazoa). The coelenterates—the Cnidaria (jellyfishes, corals) and Ctenophora—are radially symmetrical and have only two embryonic germ layers; they are thought to have diverged early from the lineage that gave rise to the other, bilaterally symmetrical, phyla. Of these, the Platyhelminthes (flatworms) have long been considered the most primitive, since they lack a coelom (body cavity); however, some authors have suggested that this is a secondarily reduced condition, and that flatworms may have been derived from coelomate ancestors (possibly Annelida).

Traditionally, the other phyla fall into several groups that, according to some authors, may all have originated independently from flatworm-like ancestors. One group of about nine phyla, of which the best known are the Nematoda (roundworms) and Rotifera (rotifers), may or may not be a monophyletic assemblage, united by their possession of a pseudocoel (a body cavity that surrounds the digestive tract, and is derived from the embryonic blastocoel). One group that is generally agreed to be monophyletic is the nine or so protostome phyla, the largest of which are the Mollusca, Annelida (segmented worms), and Arthropoda. In protostomes, the blastopore (the entry to the archenteron, or primitive gut, that is formed during gastrulation) becomes the mouth, embryonic cleavage is spiral, and a coelom (body cavity) develops from cavities that form within masses of mesodermal tissue. The annelids and the arthropods share many features, including a ventral nerve cord and a truly segmented (metameric) body plan; traditionally, it has been believed that the ancestors of arthropods were related to polychaete annelids. The embryology of molluscs is very similar to that of polychaetes, and they have similar (trochophore) larvae. One group of molluscs (Monoplacophora) has serially repeated structures; there has been great debate about whether or not this indicates that the ancestral mollusc was metameric (and thus, whether molluscs branched off before or after the annelids evolved segmentation).

The other monophyletic group of phyla is the deuterostomes, in which the blastopore becomes the anus, embryonic cleavage is radial, and the coelom develops from a series of evaginations (pouches) of the primordial gut. Of the four deuterostome phyla, the largest are the Echinodermata (starfishes and relatives) and the Chordata (including the vertebrates, as well as the subphyla Tunicata and Cephalochordata).

In addition to these groups of phyla, there are several "lophophorate" phyla that share a similar feeding apparatus (the lophophore), but may not be a monophyletic group. These phyla include the Bryozoa (moss animals),

FIGURE 7.13 Two possible phylogenies for some animal phyla. (A) A phylogeny based on maximum parsimony analysis of morphological and other phenotypic characters. (B) A phylogeny based on several recent studies of DNA sequences. Note that relationships among some phyla are unresolved in this portrayal. Ongoing molecular phylogenetic studies will probably result in revisions of the phylogeny of animal phyla in the future. (A after Brusca and Brusca 1990; B after Erwin et al. 1997.)

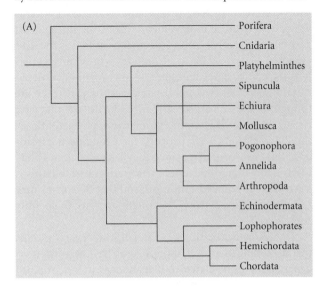

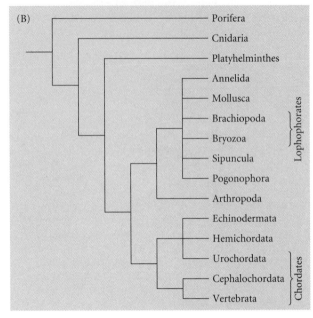

which are believed to be related to the protostomes, and the Brachiopoda (lamp shells), which are possibly closer to the deuterostomes.

Traditional estimates of the relationships among animal phyla are being challenged by new information from DNA sequences of slowly evolving genes, such as those encoding ribosomal RNA. Many relationships have not yet been resolved, for sequence differences among many of the phyla are roughly equal, probably due to the multiple changes that have erased the synapomorphies that related phyla possessed earlier in their history. Figure 7.13B illustrates one of several phylogenies obtained from DNA sequence data, but it should not be considered definitive.

Paleozoic Life: Ordovician to Devonian

Ordovician Diversification

Following the mass extinction at the end of the Cambrian, many of the phyla diversified greatly in the Ordovician (500–439 Mya), giving rise to many *new classes and orders*. Moreover, many of the predominant Cambrian groups did not recover their earlier diversity, so the Ordovician fauna had a very different character from that of the Cambrian. Among the Ordovician groups were as many as 21 classes of echinoderms, which differed greatly from each other and from the five classes that survive today (Figure 7.14). New groups appeared among the trilobites (which reached their greatest diversity in this period), brachiopods, bryozoans, gastropods (snails), and bivalves. Fishlike vertebrates also appeared at this time (see below). All of these animals were aquatic, and *most were epifaunal* (i.e., living on the surface

of the sea floor), although some bivalves evolved an infaunal (burrowing) habit. The major large predators were starfishes and nautiloids (shelled cephalopods, that is, molluscs related to squids). Archaeocyathid reefs were replaced by reefs built by two groups of corals (tabulate and rugose), with contributions from stromatoporoid sponges, bryozoans, and stromatolite-building cyanobacteria. By the end of the Ordovician, the fossil record contains about *400 families of marine animals*, which is *about the level maintained throughout the rest of the Paleozoic*, except when mass extinctions occurred. One of these extinctions closed the Ordovician; in terms of the proportion of taxa that became extinct, this *mass extinction* may have been the *second largest of all time*. It may have been caused by a drop in temperature and a drop in sea level, for there were glaciers at this time in the polar regions of the continents.

Marine Life in the Silurian and Devonian

Most of the groups that suffered in the end-Ordovician mass extinction increased in diversity during the Silurian (439–408 Mya). One conspicuous group was the eurypterids: large, predatory "sea scorpions" that had originated in the Ordovician, but diversified thereafter, and became extinct in the Permian. The *nautiloids gave rise to the ammonoids in the Devonian* (408–354 Mya). The ammonoids are among the most diverse (and extensively studied) groups of extinct animals, more than 5000 species having arisen during successive adaptive radiations, which we shall note in due course (see Figure 7.23).

From our egocentric point of view, perhaps the most interesting event during this time was the diversification of early fishlike vertebrates. Among living chordates, the am-

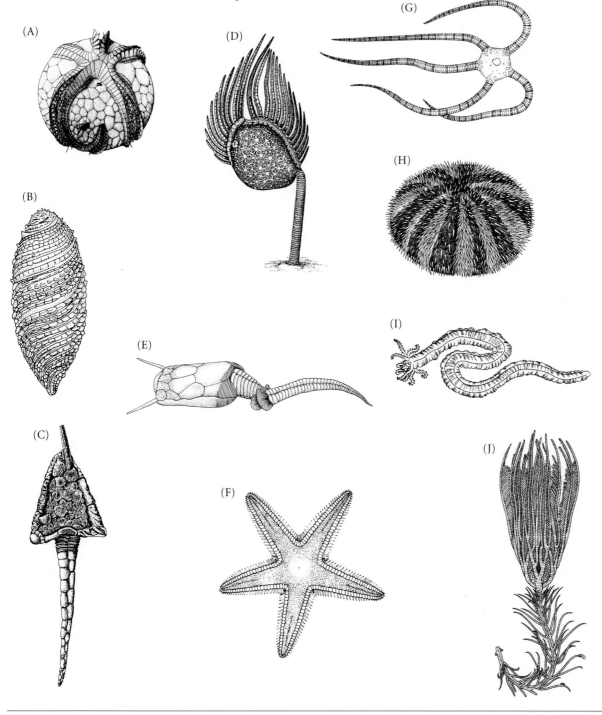

FIGURE 7.14 Extinct and living classes of echinoderms. Classes that became extinct before the end of the Paleozoic include (A) Edrioasteroidea, (B) Helicoplacoidea, (C) Homoiostelea, (D) Paracrinoidea, and (E) Stylophora. The living classes of echinoderms are (F) Asteroidea (starfish), (G) Ophiuroidea (brittle stars), (H) Echinoidea (sea urchins), (I) Holothuroidea (sea cucumbers), and (J) Crinoidea (sea lilies). (A,C,D from Broadhead and Waters 1980; B after Moore 1966; E after Parsley 1988; F–J from Rowe and Gates 1995.)

phioxus (Figure 7.15) (*Branchiostoma*), in the subphylum Cephalochordata, is thought to represent best the ancestral "ground plan" from which vertebrates evolved. It is an infaunal filter feeder with the notochord, dorsal hollow nerve cord, and pharyngeal slits that characterize the Chordata, and it is segmented (segmentation evolved independently in the chordates and the protostomes). It is not a verte-

brate. However, in many details, it is similar to the larvae of lampreys, one of the two living groups of agnathan (jawless) vertebrates.

The first known vertebrates are the jawless fishes (agnathans) known as *Heterostraci*. Pieces of probable heterostracan armor are known from the late Cambrian; fossils of a modest variety of heterostracans occur in the

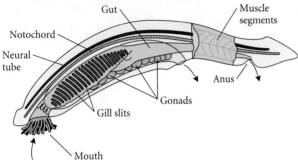

FIGURE 7.15 The amphioxus, an invertebrate chordate (subphylum Cephalochordata). The notochord, neural tube, gill slits, and muscle segments mark the basic body plan of the chordates. (After Purves and Orians 1998.)

Ordovician, and the group was diverse in the Silurian and Devonian (Figure 7.16A). Heterostracans and other early jawless vertebrates had a dermal armor composed of bone, a substance unique to vertebrates and profoundly important because of the many functions that the internal skeleton of later vertebrates serves. Most heterostracans were bottom dwellers that probably fed on detritus and invertebrates. Some jawless fishes had paired pectoral fin–like structures, which, however, lacked an internal skeleton and are not homologous to the paired appendages of jawed vertebrates. In another group of early jawless vertebrates, the Cephalaspida, the structure of the brain and cranial nerves, deduced from grooves on the interior of the skull, is very similar to that of lampreys (living agnathans).

Gnathostome (jawed) vertebrates may well have diverged from agnathans in the Cambrian, but the first known fossil (an acanthodian) is from the early Silurian. Gnathostomes

FIGURE 7.16 Extinct Paleozoic classes of vertebrates. (A) An agnathan, class Heterostraci (*Pteraspis*, Devonian); (B) a gnathostome, class Acanthodii (*Climatius*, Devonian); (C) a placoderm (*Bothriolepis*, Devonian). (A,C after Romer 1966; B after Romer and Parsons 1986.)

(A)

(B)

(C)

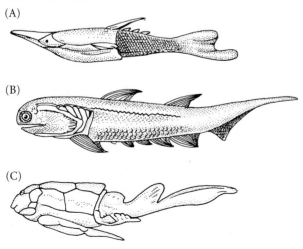

have jaws and paired limbs. By the late Silurian, three classes of gnathostome fishes had evolved. The Acanthodii (Figure 7.16B) became extinct in the early Permian, and the Placodermi (Figure 7.16C) survived only into the earliest Carboniferous. The *Osteichthyes* (bony fishes) include two subclasses, both of which are known first from the *late Devonian*. We have already (in Chapter 6) mentioned the subclass Sarcopterygii (lobe-fins), which gave rise to the amphibians. Among the sarcopterygians, rhipidistians and lungfishes were most diverse in the Devonian and Carboniferous; only a few lungfishes and a single coelacanth represent the Sarcopterygii today. The subclass Actinopterygii (ray-finned fishes) in the late Silurian and the Devonian had a chondrostean grade of anatomical organization (see Figure 33 in Chapter 6), seen today in sturgeons.

From the Ordovician through the Devonian (which is often called the "Age of Fishes"), *marine life increased* not merely in number of taxa, but *in diversity of ways of life*. To the Cambrian fauna, consisting mostly of bottom dwellers subject to few large predators, were added infaunal filter feeders, actively swimming organisms, and eurypterids, cephalopods, and fishes with predatory lifestyles that the world had never before seen.

Terrestrial Life: Plants

The Devonian provides the first evidence of terrestrial life. Terrestrial plants, derived from green algal ancestors, constitute two divisions. The Bryophyta (mosses, liverworts, hornworts) lack vascular tissues; this is generally (but not universally) thought to represent a primitive condition. The Tracheophyta, or vascular plants, have vascular conducting tissues (phloem and xylem).

The life cycle of a plant consists of haploid gametophyte and diploid sporophyte stages (see Box A in Chapter 3.) The green stage of most bryophytes is the gametophyte; in tracheophytes, photosynthesis may occur in both stages, or only in the sporophyte (as in seed plants). The tracheophytes include several groups, such as ferns, that reproduce by one-celled spores (as do bryophytes), and several that reproduce by seeds. In bryophytes and spore-bearing tracheophytes, specialized structures of the gametophyte produce eggs and swimming sperm, which travel through water to the eggs. The fertilized egg (zygote) develops into the sporophyte, which grows out of the gametophyte. The sporophyte bears, often on the leaves, sporangia within which meiosis occurs. The haploid products of meiosis (meiospores) disperse and grow into gametophytes. Many plants are homosporous, producing only one kind of meiospore, but in heterosporous plants, different sporangia produce either megaspores or microspores, which give rise, respectively, to "female" megagametophytes that produce eggs and "male" microgametophytes that produce sperm.

In seed plants, the gametophytes are not free-living, but remain attached to the sporophyte. Male and female gametophytes develop within separate sporangia, which are located on the upper surfaces of structures (sporophylls) that are homologous to leaves. The microgametophyte consists of a few nuclei within a pollen grain, which has a desiccation-resistant wall. There are no free-swimming sperm. The multicellular megagametophyte provides nutrition to the developing embryo and is surrounded by at least one integument (the seed coat).

We may now turn to the fossil record of plants, which is complicated by two facts. First, some of these diagnostic life cycle characteristics (e.g., chromosome number) do not fossilize well. Second, the several structures of a fossilized plant are often dissociated, so that it is sometimes hard to know which vegetative structures go with which reproductive structures or pollen grains.

The earliest known bryophyte is from the late Devonian, but bryophytes do not fossilize well, and they probably evolved considerably earlier. *The first definite tracheophytes are found in the late Silurian.* The group to which they are assigned, Rhyniopsida, extends through the middle Devonian, and may have given rise to two living genera (the psilopsids, or whisk ferns) with similar features. These ancient *rhyniopsids* (Figure 7.17A) were short and *simple:* about 5–10 cm high, with *sporangia at the ends of leafless, dichotomously branching, erect stalks* arising from a prostrate axis *lacking true roots.* The rhyniopsids are thought to have given rise to the Lycopsida, which appear first in the early Devonian (Figure 7.17B). The stem of lycopsids bears simple leaves, on the upper side of which the sporangia are located. In many lycopsids, these sporophylls are densely packed at the ends of the branches, forming a strobilus, or cone. Early lycopsids were small, but by the *late Devonian,* some were large (5 meters) *trees.* Some late Devonian lycopsids were the first known heterosporous plants. The only living lycopsids are club mosses (*Lycopodium*) and several other small plants.

The trimerophytes, known only from the Devonian, are also thought to have originated from the rhyniophytes. They had a *more complex vascular system,* and had a main stalk with lateral branches and paired terminal sporangia (Figure 7.17C). It is from the trimerophytes that the three remaining clades of vascular plants—the horsetails, ferns, and seed plants—may have independently arisen. The Sphenopsida (horsetails) have whorls of leaves and short branches on erect stems with terminal whorls of sporangia-bearing structures (Figure 7.17D). They originated in the middle Devonian, and by the late Devonian, some had evolved to be large trees (20 meters or more). Only one genus (*Equisetum*) persists today. The earliest "pre-ferns" appeared in the middle Devonian, but only in the late Devonian did true ferns (Filicopsida) diversify. The sporangia of ferns are borne on the lower surface of the leaves. We will consider the seed plants below. The temporal distribution and estimated phylogeny of vascular plants are portrayed in Figure 7.18.

Terrestrial Life: Animals

Most of the earliest terrestrial animals were arthropods. These fall into two major groups, the *chelicerates* and

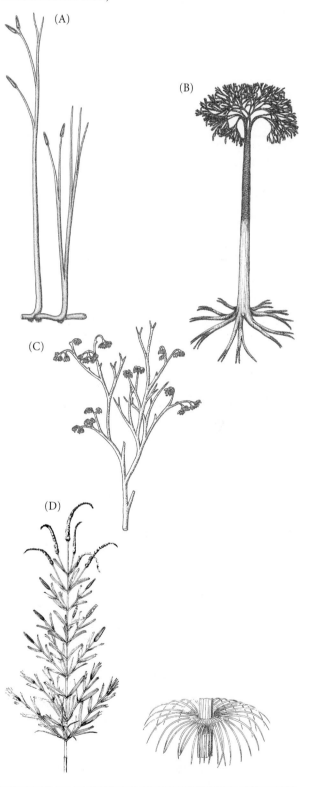

FIGURE 7.17 Paleozoic spore-bearing vascular plants, portrayed at different scales. (A) A Devonian rhyniopsid, the most primitive known group of vascular plants. (B) A Carboniferous lycopsid tree, *Lepidodendron.* (C) A Devonian trimerophyte. (D) A sphenopsid, with a close-up of part of the stem of a Permian species. (A from Kidston and Lang 1921; B from Stewart 1983; C from reconstruction by Kasper, Andrews, and Forbes 1974; D from Boureau 1964.)

FIGURE 7.18 The temporal distribution in the fossil record of some groups of vascular plants, and their possible phylogenetic relationships. The width of the envelopes expresses changes in diversity of a group over time. Solid lines portray generally ac-cepted relationships among groups; dashed lines, less confi-dently established relationships; dotted lines, probable temporal extensions not documented in the fossil record. (Data from Stewart 1983; Doyle and Donoghue 1986; Donoghue 1994.)

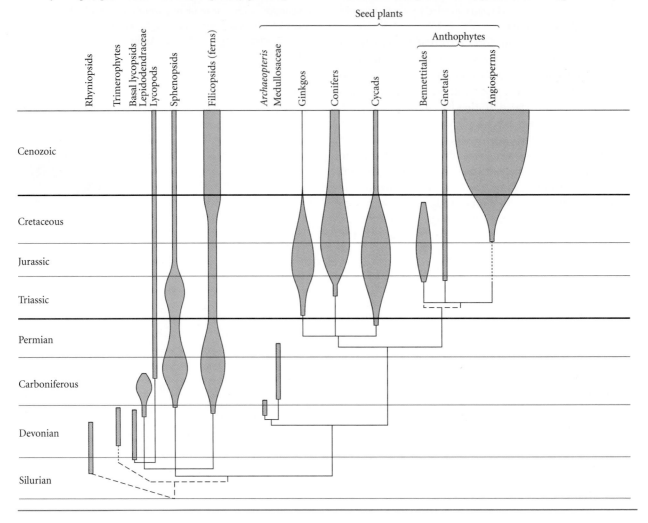

mandibulates, both of which are *first recorded on land in the early Devonian*, about 400 to 380 Mya, but which have ma-rine antecedents. The Devonian chelicerates, most or all of which were predators, include spiders, mites, scorpions, and several other groups. The terrestrial scorpions were pre-ceded by marine forms, some of which were very large. The earliest mandibulates include myriapods, such as millipedes, from the early Devonian; these fed on plant material, in-cluding wood and detritus. Centipedes, which are predato-ry myriapods, are recorded from the middle Devonian. The first known hexapods are springtails (Collembola) in the early Devonian, and a bristletail (Microcoryphia), the *first true insect, in the middle Devonian*. As we noted in Chapter 6, these forms are cladistically more primitive than the winged insects, which are not recorded before the Carboniferous.

Terrestrial vertebrates, as we have seen (in Chapter 6), *arose late in the Devonian*. These ichthyostegid amphibians presumably fed on arthropods such as scorpions and large myriapods, some of which had defensive spines.

Paleozoic Life: Carboniferous and Permian

Terrestrial Life

The *Carboniferous* (354–290 Mya) is sometimes divided into the Mississippian and the Pennsylvanian periods. At this time, land masses were aggregated into the supercontinent Gondwanaland in the Southern Hemisphere and several smaller continents in the Northern Hemisphere. Although glaciation occurred in polar regions, widespread tropical cli-mates favored the development of extensive swamp forests, which were preserved as coal beds. For much of the Car-boniferous, and extending into the Permian (290–251 Mya), these forests were dominated by sphenopsid and ly-copsid trees; among the latter, *Lepidodendron* (see Figure 7.17B) was especially conspicuous. Ferns, including tree ferns, were common.

The *origin of the seed plants* can be traced to the late De-vonian, in which several "progymnosperms" are recorded.

(A)

(B)

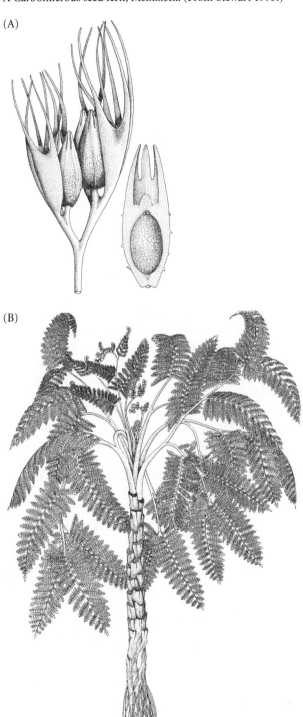

These plants, although bearing spores on the upper leaf surfaces, had a wood structure like that of later "gymnosperms." [The term "gymnosperms" refers to all the seed plants except the angiosperms. The gymnosperms are not a monophyletic group. Living gymnosperms include conifers, the ginkgo, Mormon tea (*Ephedra*), and a few

other taxa.] The progymnosperms are believed to stem from the trimerophytes, and by the end of the Devonian gave rise to early seed plants such as *Archaeosperma* (Figure 7.19A), in which a multicellular megagametophyte was enclosed in a cupule, a structure that is thought to be the antecedent of the integument that surrounds true seeds. *Archaeosperma* and its relatives are sometimes included among the *seed ferns* (pteridosperms), which were diverse in the Carboniferous and became extinct in the Permian. Like the later seed plants, the seed ferns were heterosporous, producing *both seeds and pollen*. The evolution of pollen freed the seed plants from dependence on water for transmission of gametes; the evolution of the seed, and the suppression of the gametophyte as a free-living stage, provided protection against desiccation, as well as a store of nutrients that enabled the young sporophyte to grow rapidly and overcome adverse conditions. Bear in mind that among all these diverse plants, none had flowers.

Because terrestrial deposits from the early Carboniferous are sparse, the arthropod record is best for the late Carboniferous. Several new orders of arachnids appear (e.g., harvestmen, Phalangida) with much the same morphology that they have today. *Winged insects* make their first appearance (about 325 Mya) in considerable variety. Those in several extinct orders were of the *paleopterous* type (see Chapter 5), in which the wings could not be folded over the back and were usually held outstretched (as in dragonflies). It is especially interesting that the wings developed gradually through several immature stages, a condition seen today, in modified form, only in the paleopterous mayflies. Moreover, some Carboniferous insects had wings on all three thoracic segments, whereas living insects lack wings on the prothorax. Many of these Carboniferous forms (e.g., Palaeodictyoptera) had long sucking mouthparts (Figure 7.20) and

FIGURE 7.20 Reconstruction of a member of the extinct order Palaeodictyoptera, shown feeding on cones with its long sucking mouthparts. The wing pattern is evident in fossils. (After Shear and Kukalová-Peck 1990.)

probably fed on the sap, pollen, and seeds of pteridosperms. Some of the Permian paleopterans were the largest known insects; they included primitive dragonflies with wingspans exceeding 700 mm (28 inches).

Also in the Carboniferous and Permian, many groups of *neopterous insects*, which fold the wings over the back, appeared. In the Carboniferous, there were roaches, grasshoppers, and many related (orthopteroid) types, as well as hemipteroid insects (leafhoppers and allies). The sap-sucking hemipteroids diversified greatly in the Permian, in concert with the decline of the ecologically similar Palaeodictyoptera. In the Permian, a great profusion of neopterous insect groups with *complete metamorphosis* made their first appearance. These included beetles, scorpionflies, and the stem lineage of the closely related orders Trichoptera (caddisflies) and Lepidoptera (moths and butterflies). Most of these insects had characteristics that are judged to be more primitive than those of the majority of living representatives of these orders. Many of the insect orders that were most diverse in the late Paleozoic became extinct at the end of the Permian. Figure 7.21 portrays the fossil distribution of the orders of insects, together with an estimate of their phylogenetic relationships, many of which are poorly understood (see Chapter 5).

Amphibians were diverse in the Carboniferous, but only a few of the many groups of amphibians persisted through the Permian. They included both fully aquatic and highly terrestrial forms, some of which were considerably larger (about 2 m) than most modern amphibians. As far as we know, all were predatory. In the order Anthracosauria, *several groups in the Pennsylvanian and early Permian acquired many reptilelike skeletal features*, and have been variously classified as amphibians or reptiles. It is surely from among these anthracosaurs that the *first known amniotes*, the protorothyrid captorhinomorphs of the *early and middle Pennsylvanian*, arose (see Figure 20 in Chapter 6). By the late Pennsylvanian, these primitive amniotes gave rise to the pelycosaurs and other *synapsids*, which included the *ancestors of mammals* (see Chapter 6), and to the *diapsids*, a major reptilian stock characterized by two temporal openings in the skull. The first diapsids were similar to the protorothyrids in most of their features.

Both the synapsids and the diapsids diversified in the Permian. Among the synapsids, the *therapsids increasingly acquired mammal-like features*, especially the cynodonts, which extended from the late Permian to the middle of the Mesozoic era (see Chapter 6). The late Permian diapsid radiation included lepidosauromorphs, which gave rise to lizards and (much later) snakes, and archosauromorphs, whose Mesozoic history is particularly interesting (see below).

Aquatic Life

During the Carboniferous, many groups of marine organisms, such as ammonoids, crinoids, brachiopods, gastropods, and bivalves, recovered from the late Devonian extinction, increasing again in diversity. The diversity of a few groups, such as tabulate and stromatoporoid corals, declined, and the placoderm "fishes" became extinct. Lung-

FIGURE 7.21 Temporal distribution in the fossil record of orders and some suborders of insects and their close relatives, and their possible phylogenetic relationships. Thick bars represent known temporal distributions; dotted lines, tentative relationships; solid lines, postulated relationships. (Data from Kristensen 1991 and Kukalová-Peck 1991.)

fishes reached their greatest diversity, in both salt and fresh water, and sharks diversified. Among bony fishes, more modern features, associated with more efficient swimming, evolved in several groups (known informally as holosteans) in the late Permian.

During the Permian, the continents approached one another, and by its end *formed a single world continent, Pangaea* (Figure 7.22A). The sea level dropped to its lowest point in history, and climates were greatly altered by the arrangement of land and sea. Equatorial regions became hot and dry, the great swamp forests contracted, and the polar seas became warm. Although its exact causes are debated, these changes may have been instrumental in bringing about one of the most significant events in the history of life, the *end-Permian mass extinction* (Erwin 1993a). It is estimated that in this, *the most massive extinction event of all time (so far)*, at least 52 percent of the families, and more than 90 percent of all species, of skeleton-bearing marine invertebrates became extinct over a span of 5 to 8 million years. Groups such as anthozoan corals, ammonites, stalked echinoderms (e.g., crinoids), brachiopods, and bryozoans declined greatly, and major taxa such as trilobites, fusulinid foraminiferans, and tabulate and rugose corals became extinct. With the exception of the loss of many families of amphibians and therapsids and of some orders of insects, extinctions on land were, curiously, much less pronounced.

Mesozoic Life

The Mesozoic era, divided into the Triassic (254–206 Mya), Jurassic (206–144 Mya), and Cretaceous (144–66 Mya) periods, is often called the Age of Reptiles. By its end, the earth's flora and fauna were acquiring a rather modern cast, but this was preceded by the evolution of some of the most extraordinary creatures of all time. Pangaea began to break up during the Triassic, first with the formation of the Tethyan Seaway between Asia and Africa, then (in the Jurassic) with the full separation of a northern from a southern continent (see Figure 7.22B). The northern continent, LAURASIA, consisted of North America and Eurasia; these began to separate late in the Jurassic, but northeastern North America, Greenland, and Western Europe remained connected until well into the Cretaceous. The southern continent, GONDWANALAND, consisted of Africa, South America, India, Australia, New Zealand, and Antarctica. These slowly separated in the late Jurassic and the Cretaceous, but even then the South Atlantic formed only a narrow seaway between Africa and South America (see Figure 7.22C). Throughout the Mesozoic, the sea level rose, and many continental regions were covered by shallow epicon-

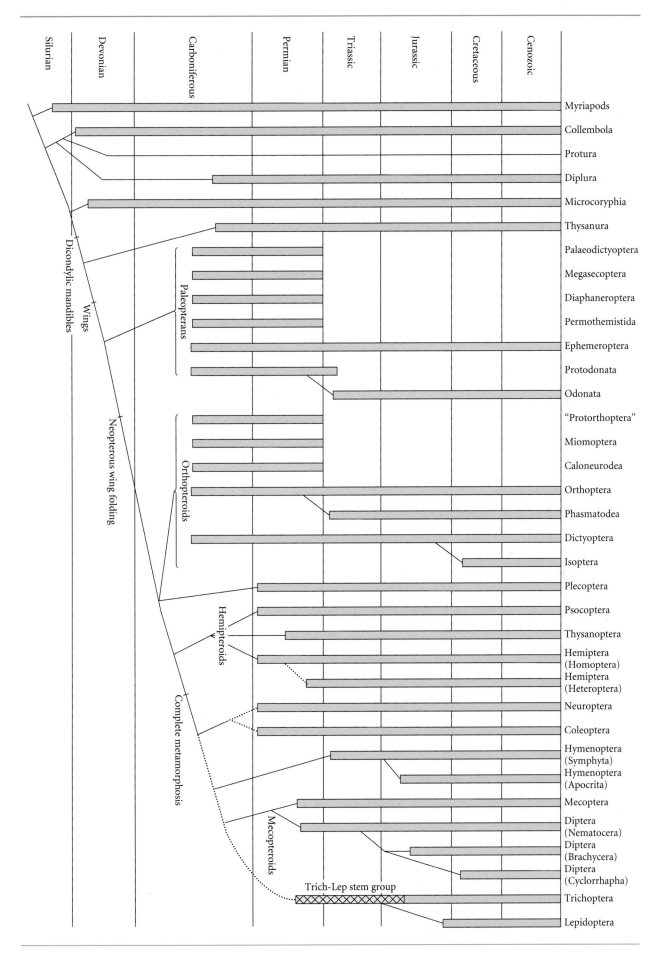

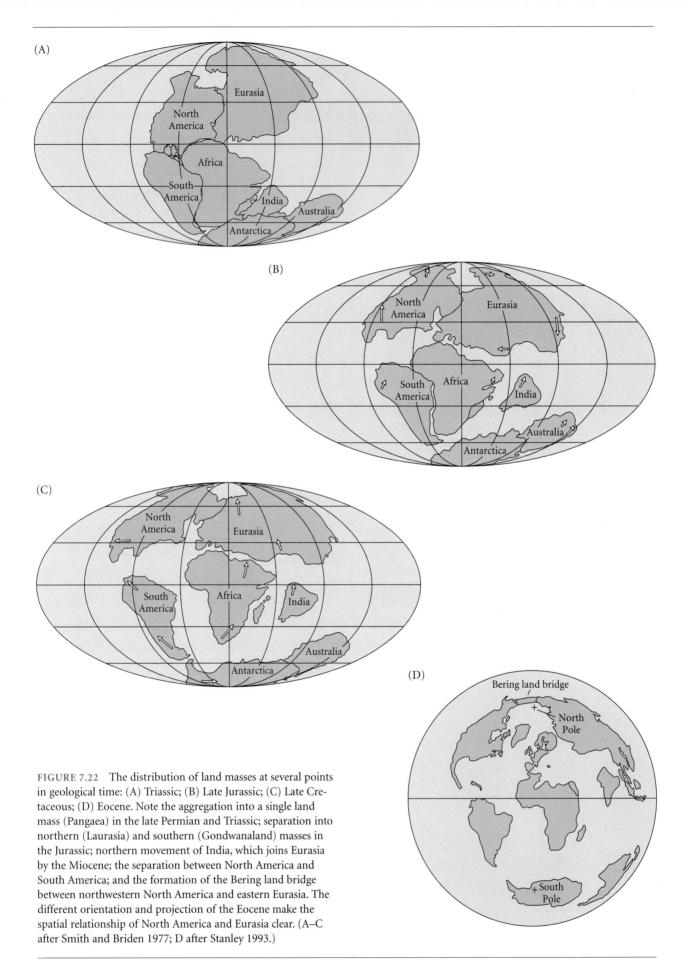

FIGURE 7.22 The distribution of land masses at several points in geological time: (A) Triassic; (B) Late Jurassic; (C) Late Cretaceous; (D) Eocene. Note the aggregation into a single land mass (Pangaea) in the late Permian and Triassic; separation into northern (Laurasia) and southern (Gondwanaland) masses in the Jurassic; northern movement of India, which joins Eurasia by the Miocene; the separation between North America and South America; and the formation of the Bering land bridge between northwestern North America and eastern Eurasia. The different orientation and projection of the Eocene make the spatial relationship of North America and Eurasia clear. (A–C after Smith and Briden 1977; D after Stanley 1993.)

tinental seas. Although the polar regions were cool, most of the earth enjoyed warm climates, with global temperatures reaching an all-time high in the mid-Cretaceous, after which substantial cooling occurred.

Marine Life

During the Triassic, most (but not all) of the *marine groups* that had been decimated during the end-Permian extinction *again diversified*; for example, ammonoids increased from two genera to more than a hundred by the middle Triassic (Figure 7.23). Planktonic Foraminifera and modern corals evolved, and bony fishes continued to radiate. Another *mass extinction* occurred at the *end of the Triassic*, again decimating groups such as ammonoids and bivalves. These groups underwent yet another adaptive radiation during the Jurassic, which marked the beginning of the "*Mesozoic marine revolution*," so called because of the coevolution of shelled molluscs and their predators. New kinds of predators evolved, including crabs, bony fishes with the ability to crush mollusc shells, and carnivorous gastropods. For their part, many of the prey, especially among the bivalves and gastropods, evolved elaborate protective mechanisms, such as the thick shells and spines that characterize many molluscs today (Vermeij 1987; see Chapter 18). A radiation of sharks began early in the Jurassic that has continued to the present. *Modern bony fishes, the teleosts, evolved* from "holostean" ancestors; by the late Jurassic, several of the more primitive groups of teleosts, such as the herringlike fishes, began to diversify.

The *ammonoids* experienced another *major extinction* at the end of the Jurassic, and then, in the Cretaceous, underwent their *last (fifth) evolutionary radiation*. Modern groups of gastropods, bivalves, and bryozoans rose to dominance, and a peculiar group of gigantic sessile bivalves, the rudists, formed reefs. Late in the Cretaceous, the *marine plankton became dominated by groups that retain their dominance today*, such as diatoms and planktonic foraminiferans, as well as coccolithophores, algae whose calcareous skeletons form the chalk deposits of the late Cretaceous in England and elsewhere. ("*Cretaceous*" comes from the Latin for "chalk.") From the upper Triassic through the Cretaceous, the seas also harbored several groups of aquatic reptiles, which we shall mention below.

The *end of the Cretaceous* is marked by what is surely the best-known *mass extinction*, as the last of the dinosaurs became extinct at this time. Many groups of planktonic organisms (e.g., coccolithophores) became extinct rather suddenly; a more gradual decline led to the *total extinction* of ammonoids, rudists, marine reptiles, and many families of invertebrates. Few hypotheses about evolutionary history have so captured public attention as the possibility that this extinction was caused by the impact of an asteroid or some other extraterrestrial body—a hypothesis for which there is considerable evidence (Chapter 24).

Terrestrial Plants and Arthropods

The *flora of the Mesozoic was dominated by "gymnosperms*," especially by the cycads (Figure 7.24A) and by the conifers and their relatives. The cycads, of which only a few persist today in tropical and subtropical regions, have superficially fernlike leaves, cones bearing seeds and pollen, and a thick trunk. The conifers appeared first in the mid-Triassic; modern families (such as pines) differentiated in the Jurassic. The closely related Ginkgoales evolved early in the Tri-

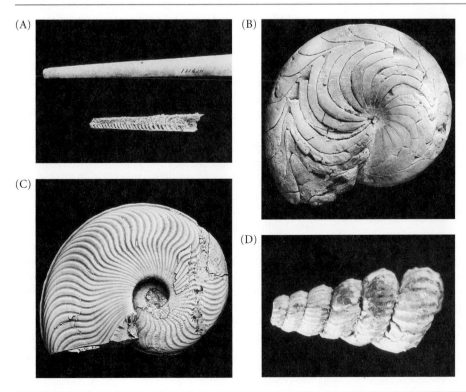

(A)

(B)

(C)

(D)

FIGURE 7.23 A few of the diverse forms of ammonoids. The shells housed the squidlike body of these cephalopod molluscs. (A) *Mooreoceras* (Permian); (B) *Imitoceras* (Carboniferous); (C) *Harpoceras* (Jurassic); (D) *Turrilites* (Cretaceous). (Photographs courtesy of the National Museum of Natural History/ The Smithsonian Institution.)

FIGURE 7.24 Seed plants. (A) A living cycad, a member of a once diverse group. (B) A leaf of the living ginkgo (*Ginkgo biloba*) next to a fossilized *Ginkgo* leaf from the Paleocene. (A, © J. Lepore, The National Audubon Society Collection/Photo Researchers; B, © B. P. Kent/Earth Scenes.)

(A)

(B)

assic. Fossils of the genus *Ginkgo* appear in upper Triassic deposits; *Ginkgo* is represented today by a single species, *G. biloba*, a "living fossil" that is extinct in the wild, but was cultivated in the imperial gardens of China and today is planted extensively along city streets in the United States and elsewhere (Figure 7.24B).

The *flowering plants* (angiosperms) probably originated in the late Triassic, because the extinct Bennettitales and the extant Gnetales, which according to cladistic analysis have an ancestor in common with the angiosperms, already existed at this time. However, definitive fossils of angiosperms do not appear until the *early Cretaceous*. Angiosperms are characterized by several features that do not fossilize (e.g., double fertilization, triploid endosperm) and several that are at least occasionally preserved in fossils (e.g., xylem vessels, two integuments surrounding the seed, features of the pollen, modification of the sporophyll into a carpel that encloses the ovules). Several of these features are shared by related "gymnosperms," and *almost all of those features that fossilize*, including showy flowers that undoubtedly attracted insects, *had evolved individually in various Jurassic groups of "gymnosperms."* Angiosperm fossils are rare until the middle Cretaceous, when they become rather abundant and quite diverse, and include both dicotyledons and monocotyledons (which, according to cladistic analysis, evolved from dicotyledons). Late Cretaceous fossils of leaves and flowers (Figure 7.25A) show that many modern families had evolved by this time (e.g., sycamores, magnolias, poplars, water lilies). During the Cretaceous, the *angiosperms achieved ecological dominance over the "gymnosperms,"* beginning an adaptive radiation that has continued since then; there are now about 250,000 species of flowering plants in more than 300 families.

The anatomically most *"advanced" orders of insects made their appearance in the Mesozoic.* There are fossil Diptera in the early Jurassic and perhaps in the early Triassic; during the Mesozoic, successively more "advanced" groups of flies appeared and radiated in sequence. Lepidoptera (moths and butterflies) appear first in the Jurassic, although they probably evolved considerably earlier. The earliest Hymenoptera (early Triassic) are related to today's herbivorous sawflies; this group is anatomically more primitive than the parasitic and predatory wasplike hymenopterans, which appear in the Jurassic and early Cretaceous. Among these are the sphecoid solitary wasps, from which bees evolved. The earliest fossil bee (Figure 7.25B), from late Cretaceous amber of the northeastern United States, belongs to a genus of social bees (*Trigona*) that today is entirely tropical. In many respects, this is one of the most advanced groups of bees, so it is likely that bees evolved in the early Cretaceous (Michener and Grimaldi 1988). The late Cretaceous record also contains fossils of the earliest ants, which retained features of their wasplike ancestors (see Chapter 6). In the Cenozoic, ants became one of the most species-rich and ecologically dominant groups of terrestrial organisms.

Throughout the Cretaceous and the Cenozoic, *insects and angiosperms affected each other's evolution* and may have augmented each other's diversity (see Chapter 18). Early angiosperm flowers, like those of some Mesozoic "gymnosperms," were probably pollinated mostly by beetles, and later by dipteran flies. During the late Cretaceous and Cenozoic, hymenopterans (especially bees) and lepidopterans became important pollinators, and adaptive modification of flowers to suit different pollinators gave rise to the great floral diversity of modern plants. It is also likely (see Chapters 18, 25) that herbivorous insects and angiosperms exerted reciprocal selection pressures that increased each other's diversity (Ehrlich and Raven 1964). *It is largely because of the spectacular increase of angiosperms and insects that terrestrial diversity is greater today than ever before.*

Fossil evidence of the timing of diversification of insect-pollinated angiosperms and their pollinators. (A) *Protomimosoidea*, a Paleocene/Eocene member of the insect-pollinated family of legumes that includes mimosas and acacias. (B) *Trigona prisca*, a Late Cretaceous stingless bee preserved in amber. This genus of flower-visiting social bees is diverse in the tropics today. Parts of this fossil were missing or difficult to see, so were not drawn. (A, photograph courtesy of W. L. Crepet; B after Michener and Grimaldi 1988.)

(A)

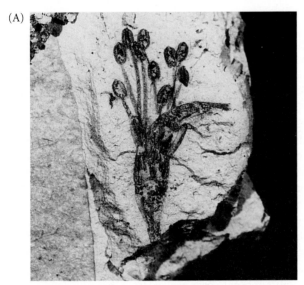

(B)

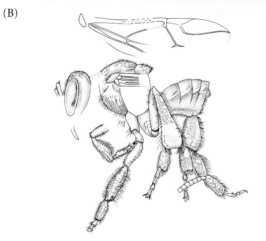

Terrestrial Vertebrates

Mesozoic vertebrate history is so rich that it is best to trace each of the major lineages separately. Turning first to the amphibians, a few archaic groups persisted through the Triassic but became extinct at its end, yielding dominance of the land to the reptiles. No fossils connect Triassic and pre-Triassic amphibians to modern amphibians. Frogs are recorded early in the Jurassic, with modern frog families appearing sequentially throughout the Cretaceous and Cenozoic; a few salamanders are known from as far back as the middle Jurassic.

The major groups of amniotes, as we have seen, are distinguished by the presence or absence of temporal fenestrae, openings in the temporal region of the skull (at least in the stem members of each lineage). The earliest (Carboniferous) amniotes, the *captorhinomorphs*, lacked temporal fenestrae. All such reptiles became extinct by the end of the Triassic, except for turtles, which appear near the end of the period without intermediate predecessors. An unusual pattern of temporal fenestrae distinguishes the plesiosaurs and ichthyosaurs, marine reptiles that flourished from the late Triassic to the end of the Cretaceous. Ichthyosaurs were astonishingly porpoiselike in form, and gave birth to live young (Figure 7.26).

The *diapsids*, with two temporal fenestrae, became the most diverse group of reptiles. One major diapsid lineage, the lepidosauromorphs, diversified in the late Permian and Triassic into various stocks, including *lizards* and sphenodontids. Lizards became differentiated into modern suborders in the late Jurassic, and into modern families in the late Cretaceous. Among the late Cretaceous lizards, a notable extinct family is the mosasaurs, fishlike marine predators that reached more than 10 meters in length. Probably during the Jurassic, one group of lizards (most likely the one that includes the modern monitors) gave rise to snakes. The sparse fossil record of snakes begins only in the early Cretaceous, however, and by the late Cretaceous it includes the family of boas and pythons (anatomically among the most primitive living snakes). It is likely that the snakes evolved from a burrowing ancestor that resembled the several groups of lizards that have become legless.

The *archosauromorph* diapsids were the most spectacular and diverse of the Mesozoic tetrapods; it is largely on their account that the Mesozoic is often termed the *Age of Reptiles*. From the great number of archosaur groups, we can single out only a few for particular mention. Most of the late Permian and Triassic archosaurs were fairly generalized quadrupedal predators a meter or so in length (Figure 7.27); some of them were lightly built, rapidly running forms, of which a few may have been facultatively bipedal. From this generalized body plan, numerous specialized forms evolved:

FIGURE 7.26 Extinct marine reptiles. (A) A plesiosaur; (B) an ichthyosaur, preserved with the outline of its skin. The dorsal and tail fins are superficially similar to those of sharks and porpoises, although the tail fin of porpoises is horizontal. (A after Saint-Seine 1955; B from the Field Museum, negative 72728, Chicago.)

(A)

(B)

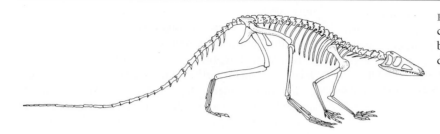

FIGURE 7.27 *Lagosuchus*, a Triassic thecodont archosaur, showing the generalized body form of the stem group from which dinosaurs evolved. (After Bonaparte 1978.)

Phytosaurs, in the late Triassic, were aquatic predators that closely resembled modern crocodiles. They are not closely related to crocodiles, however, and differ from them in several anatomical features, including the position of the nostrils (Figure 7.28).

Pterosaurs, ranging from the end of the Triassic to the end of the Cretaceous, are one of the three vertebrate groups that evolved true powered flight. Their ancestry among the archosaurs is unknown. Some were as small as sparrows, but by the end of the Cretaceous they included the largest flying vertebrate known (*Quetzalcoatlus*, with a wingspan estimated at 12 meters and a weight of 65 kilograms!). Some, it is thought, fed on fish, and others on insects. It is likely that they had a high metabolic rate. The wing consisted of a membrane extending to the body from the rear edge of a greatly elongated fourth finger (Figure 7.29).

Crocodilians, extending back to the mid-Triassic, are the only surviving archosaurs other than birds. The order Crocodilia, distinguished by skeletal synapomorphies, began as a Triassic–Jurassic radiation of agile terrestrial animals with long, slender legs. One of these stocks radiated from the late Jurassic to the early Cenozoic, and included terrestrial, semiaquatic, and fully marine forms. Modern crocodilians are descended from a third adaptive radiation, which began in the late Cretaceous; almost forty genera are known from Cenozoic deposits, of which only eight survive.

Dinosaurs, technically, are not simply any old large, extinct reptiles. The term "*dinosaur*" refers specifically to members of the orders Saurischia and Ornithischia. These are distinguished by the form of the pelvis and certain other features, but they share a unique birdlike joint between the proximal and distal tarsals. In effect, this joint provided an extra leg segment. The legs were held vertically below the body. It is likely that the quadrupedal dinosaurs were descended from bipedal ancestors. Not all dinosaurs were large: the smallest weighed less than 4 kilograms. Dinosaurs appear to have evolved from thecodont archosaurs related to the middle Triassic lagosuchids (see Figure 7.27); the first saurischians are recorded from the middle Triassic, and the first ornithischians (which probably evolved from saurischians) from the late Triassic. Neither order became diverse until the Jurassic.

In all, about 39 families of dinosaurs are recognized. The *Saurischia* (Figure 7.30) included the carnivorous, bipedal theropods (middle Triassic—end of Cretaceous) and the herbivorous, quadrupedal sauropods (late Jurassic—late Cretaceous). Among the noteworthy theropods are *Compsognathus* (see Figure 27 in Chapter 6), an agile animal less than 4 kilograms in weight; *Deinonychus*, a fierce late Cretaceous predator with a huge, sharp claw that it probably used to disembowel prey; and the enormous predatory carnosaurs, the largest of which was the renowned *Tyrannosaurus rex* (late Cretaceous), which stood 15 meters high and weighed about 7000 kilograms. The sauropods, herbivores with small heads and long necks, include the largest animals that have ever lived on land, such as *Apatosaurus* (= *Brontosaurus*), *Brachiosaurus*, which weighed more than 80,000 kilograms, and *Diplodocus*, which reached about 30 meters in length.

The *Ornithischia*, which likewise included both bipedal and quadrupedal forms, were herbivorous, with specialized, sometimes very numerous, teeth. Among the bipeds were the small (less than 5 meters), agile hypsilophodontids and the larger hadrosaurs, or duck-billed dinosaurs, which flourished in the late Cretaceous. The skulls of many hadrosaurs bore curious hollow crests (Figure 7.31A), which may have served as resonating devices for vocaliza-

FIGURE 7.28 Skulls of (A) a phytosaur, *Machaeroprosopus*, and (B) a crocodilian, *Geosaurus*. These groups converged in habits and morphology, although many differences are evident, such as the position of the nostrils (en). Crocodilians became the ecological replacements of phytosaurs after the latter became extinct. (After Romer 1960.)

(A)

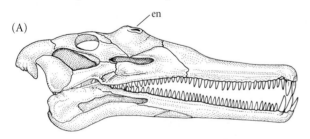

en

(B)

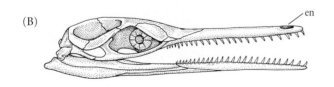

en

FIGURE 7.29 A Jurassic pterosaur, *Rhamphorhynchus*, showing the membrane that extended from the fourth finger to the body, the large sternum on which flight muscles inserted, and the terminal tail membrane in this genus, possibly used for steering. (After Williston 1925.)

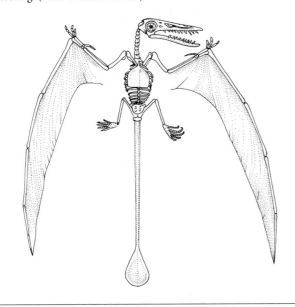

tions used in social behavior. The quadrupedal ornithischians included the well-known stegosaurs (mid-Jurassic to late Cretaceous), with dorsal plates that probably served for thermoregulation; the late Cretaceous armored dinosaurs (ankylosaurs, Figure 7.31B); and the ceratopsians (horned dinosaurs, Figure 7.31C), of which *Triceratops* is the best known. The distinctive horns and bony neck frill evolved only in the later ceratopsians, which are limited to North America. *The ceratopsians are among the last surviving dinosaurs; they became extinct at the end of the Cretaceous,* or possibly slightly later. Contrary to popular belief, only a few groups of dinosaurs became extinct at the end of the Cretaceous; as a group, dinosaurs had been declining in diversity throughout the later Cretaceous, and many families became extinct before its end.

Birds are the dominant group of living archosaurs. *Archaeopteryx*, from the late Jurassic, closely resembles a small theropod dinosaur in its skeletal features (see Figures 27 and 28 in Chapter 6). The Mesozoic record of birds is sparse, although several forms with mixtures of ancestral and derived characters have recently been discovered. A few Cretaceous fossils represent extinct groups that retained teeth but otherwise have the features of modern birds. Among extant orders of birds, the Procellariiformes (albatrosses and relatives) and the Charadriiformes (gulls, sandpipers, and relatives) are known from the late Cretaceous. The fossil record of birds is so fragmentary that it provides few clues to the origin or relationships of the major groups.

We turn now from the diapsid clade of amniotes to the descendants of the late Paleozoic *synapsids,* which, as Chap-

ter 6 recounts, *gave rise to the mammals.* The synapsids included many groups that became extinct by the end of the Permian, but two groups flourished in the Triassic: the herbivorous dicynodonts, which had reduced teeth and a turtlelike beak, and the very mammal-like cynodonts (see Figure 23 in Chapter 6). Except for the lineage leading to mammals (which diverged in the Triassic), the last cynodonts became extinct in the middle Jurassic.

As we noted in Chapter 6, the first animals that are usually considered mammals, the morganucodonts (see Figure 26 in Chapter 6), were small predators of the *late Triassic and early Jurassic.* Several extinct lineages of mammals, most of which are known only from teeth and jaws, have been identified from later Mesozoic deposits. The most diverse and most fully documented of these were the multituberculates, which had rodentlike teeth and habits (Figure 7.32). Their fossil record extends from the late Jurassic to the early Cenozoic (Oligocene), when they became extinct. Another group, the therians, were very generalized mammals extending from the late Triassic into the Cretaceous. This is believed to be the stock from which the two major subclasses of modern mammals, the Metatheria (marsupials) and the Eutheria (placental mammals), evolved. In the late Cretaceous, some therians had characteristics that would be expected in the common ancestor of metatherians and eutherians, but it is quite possible that these subclasses had arisen considerably earlier. Then, after a 20-million-year gap in the fossil record, *defini-*

FIGURE 7.30 Some saurischian dinosaurs. (A) *Deinonychus* (Late Cretaceous). (B) *Tyrannosaurus* (Late Cretaceous). (C) A sauropod, *Camarosaurus* (Late Jurassic). (A after Carroll 1988; B after Lucas 1994; C after Osborn and Mook 1928.)

(A)

(B)

(C)

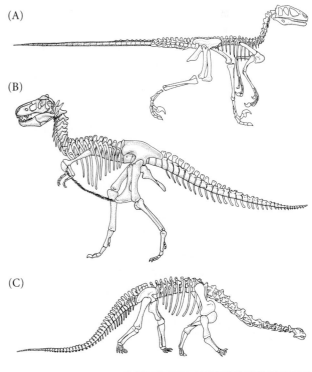

FIGURE 7.31 Some ornithischian dinosaurs. (A) A bipedal hadrosaur, *Parasaurolophus* (Late Cretaceous). (B) An ankylosaur, *Polacanthus*, with spikes and dermal armor (shown only in part). (C) A ceratopsian, *Triceratops* (Late Cretaceous), one of the last surviving dinosaurs. (After Lucas 1994.)

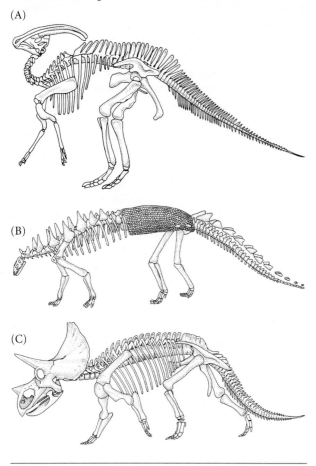

(A)

(B)

(C)

tive marsupials and eutherians appear in the late Cretaceous, about 20 My before the end of the Mesozoic.

The late Cretaceous marsupials are members of the Didelphidae (and related families), the group that includes the living American opossum (*Didelphis*), which they resembled. Curiously, these records are North American, even though modern marsupials are almost entirely limit-

ed to Australia, New Guinea, and South America. Marsupials became extinct in North America in the Cenozoic (Miocene), and opossums reinvaded Central and North America from South America less than 3 million years ago.

The *late Cretaceous eutherians* resemble tree shrews and insectivores in many of their characteristics. These features, many of which also typify opossums, represent the primitive *"ground plan" of a generalized mammal* (Figure 21 in Chapter 6). Primitive eutherian characters include a complete zygomatic (cheek) arch and many other details of the skull; a dental formula of three incisors, one canine, four premolars, and three molars (on each side, in each jaw); triangular molars with three cusps; small, separate carpals and tarsals; and separate metacarpals and metatarsals that were at least partly appressed to the ground. The limbs were generally held in a flexed, crouching posture, as in rats and many other modern mammals. The later orders of mammals are distinguished by various departures from the "ground plan," especially in features of the limbs and teeth. Among the generalized insectivore-like eutherians of the latest Cretaceous, a few have features that mark them as early members of the orders Insectivora (represented today by shrews, hedgehogs, and others) and Primates (known only from teeth). The great adaptive radiation of the eutherians occurred in the Cenozoic era, to which we now turn.

The Cenozoic Era

The Cenozoic era embraces six epochs (Paleocene through Pleistocene). We are really still in the Pleistocene, but the last 10,000 years are often distinguished as a seventh epoch (the Holocene, or Recent). Traditionally, the first five epochs (from 65 Mya to 1.8 Mya) are referred to as the Tertiary period, and the Pleistocene and Recent (1.8 Mya–present) as the Quaternary period. Some paleontologists divide the era into Paleogene (65–24 Mya) and Neogene (24 Mya–present) periods.

Early in the Cenozoic, North America moved so far to the west that it became separated from Europe in the east, but joined northeastern Asia in the west, forming the broad *Bering land bridge* that extended between Alaska and Siberia throughout most of the era (Figure 7.22D).

(A)

(B)

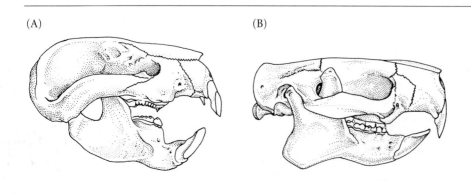

FIGURE 7.32 Convergent evolution in mammals: (A) A Paleocene multituberculate (*Taeniolabis*) and (B) an Eocene rodent (*Paramys*). Multituberculates, extending from the Cretaceous to the Oligocene, were nonplacental mammals, ecologically similar to squirrels and other rodents. Competition with rodents may have driven them to extinction (see Chapter 25). (After Romer 1966.)

Gondwanaland broke up into the separate island continents of South America, Africa, India, and, far to the south, Antarctica plus Australia (which separated in the Eocene). About 18 to 14 Mya, during the Miocene, Africa made contact with southwestern Asia, India collided with Asia (forming the Himalayas), and Australia moved northward, approaching southeastern Asia. *During the Pliocene,* about 3.5 Mya, the *Isthmus of Panama arose,* forming a bridge between North and South America for the first time.

This reconfiguration of continents and oceans contributed to major climatic changes. In the late Eocene and Oligocene, there was *global cooling and drying;* extensive savannahs (sparsely forested grasslands) formed for the first time, and Antarctica acquired glaciers. Sea level fluctuated, dropping drastically in the late Oligocene (about 25 Mya). During the Pliocene, temperatures increased to some extent, but toward its end, *temperatures dropped, and a series of glaciations began* that have persisted throughout the Pleistocene.

Aquatic Life

Most of the marine groups that survived the end-Cretaceous mass extinction proliferated early in the Cenozoic, and a few new taxa evolved, such as the burrowing echinoids (sea urchins) known as sand dollars. The taxonomic composition of marine Cenozoic communities is quite similar to that of modern ones. The Great Barrier Reef of Australia formed as the continent moved northward in the Oligocene. Substantial extinctions, followed by rediversification, occurred in the late Eocene and especially in the late Oligocene, as well as during the Pleistocene glaciations, when temperatures fell and so much water was sequestered in glaciers that the sea level dropped as much as 100 meters below its present level. As many as 70 percent of the mollusc species along the Atlantic coast of North America became extinct at this time.

Teleost fishes continued to diversify throughout the Cenozoic, becoming by far the most diverse aquatic vertebrates. By the Eocene, minnows, characins, and highly "advanced" teleosts such as Tetraodontiformes (puffers) and Pleuronectiformes (flatfishes) evolved, and many modern genera became differentiated. Except for a few turtles, an iguana, and the quite young family of sea snakes, reptiles are absent from Cenozoic seas. Whales and sea lions made their first fossil appearances in the Eocene and Oligocene, respectively.

Terrestrial Life

The Paleogene record indicates that *most of the modern families of angiosperms and insects had become differentiated by the Eocene,* if not earlier. Angiosperms were pollinated by wind, beetles, flies, bees, and lepidopterans. Many fossil insects of the late Eocene and Oligocene belong to genera that still survive, and some of the species are strikingly similar to living ones. In the Oligocene, forests were replaced in many areas by *savannahs.* The dominant plants in these habitats were grasses, which underwent a major adaptive radiation at this time, and herbaceous plants, in many families, most of which evolved from woody ancestors. Among the most important of these families is the Asteraceae (Compositae), which includes sunflowers, daisies, ragweeds, and many others. It is one of the two largest plant families today.

Among vertebrates, we may note the appearance in the Paleocene (65.0–55.6 Mya) of modern families of frogs and some modern orders of birds. Other orders of birds, as well as many modern families, are recorded from the Eocene (55.6–33.5 Mya) and Oligocene (33.5–23.8 Mya). By the Miocene (23.8–5.2), many modern genera of birds had differentiated, and the largest order of birds, the *perching birds* (Passeriformes), began its radiation early in this epoch. Several giant flightless predators convergently evolved, including the Eurasian and North American diatrymids of the Paleocene and Eocene, and the (chiefly) South American phorusrhacids of the Oligocene through Pliocene (Figure 7.33). Standing 2–3 meters tall and possessing huge skulls, these birds flourished when there were few large predatory mammals, but are now extinct.

In addition to birds, two other groups of terrestrial vertebrates radiated in the Cenozoic. One is the *colubroid snakes,* which began an *exponential increase in diversity in the Oligocene,* and by the Miocene had given rise to the venomous elapids (cobras and relatives) and vipers (including pit vipers such as rattlesnakes). Colubroid snakes today feed on a great variety of animal prey, from worms and termites to bird eggs and wild pigs, and they range in habits from burrowing to arboreal forms, some of which can glide.

The other group that evolved great diversity and ecological dominance in the Cenozoic is the Mammalia.

The Adaptive Radiation of Mammals

As we have seen, most eutherians of the late Cretaceous were generalized insectivorous or omnivorous creatures of modest size and a primitive ground plan. In overall aspect and habits, moreover, the marsupials of the time resembled the generalized eutherians. Few of the modern eutherian orders had become differentiated. *Much of the differentiation of the modern orders occurred during the Paleocene,* and most of them can be recognized in Eocene fossils. Unfortunately, the 10- to 15-million-year period in which this differentiation occurred has about the worst fossil record of the entire Cenozoic. Thus we lack most of the critical record that might elucidate the phylogenetic relationships among the orders of mammals. Moreover, we have seen (Chapter 5) that the reconstruction of phylogenetic relationships among lineages is most difficult when they diverge within a short span of time. It is little wonder, then, that despite intensive study, the phylogenetic relationships among mammalian orders, and among major clades within many orders, are quite controversial (Figure 7.34).

The difficulty of determining phylogenetic relationships among the orders of living mammals, even from DNA sequences, suggests that the orders emerged almost

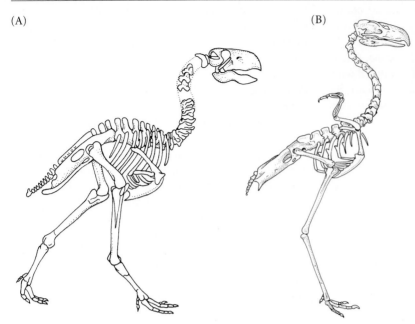

(A)

(B)

FIGURE 7.33 Convergent evolution of giant flightless predatory birds that stood 1.5–2 meters tall. (A) The Diatrymidae occurred in Eurasia and North America during the early Tertiary. (B) The Phorusrhacidae were distributed during the Tertiary in South America, Florida, and Europe. (A after Andrews 1901; B after Matthew and Granger 1917.)

simultaneously from within a group of closely related species (see Figure 7B in Chapter 5). This is confirmed by the fossil record in two ways: (1) the short time between the late Cretaceous and the Eocene appearance of the orders; and (2) the fact that except for the particular synapomorphies that characterize each order, the earliest fossil representatives of many of the orders are very similar to one another and to the generalized ancestral ground plan. For example, *Hyracotherium* is recognized as a member of the Perissodactyla by its characteristic perissodactyl astragalus (see Chapter 6), and *Diacodexis* is considered an early (Eocene) artiodactyl chiefly because of its artiodactyl astragalus. In most other respects, these animals resembled each other, as well as many other generalized herbivorous mammals, known as condylarths, which in turn are differentiated from other early eutherians only by the form of their teeth (adapted for an herbivorous diet). Thus, the earliest members of these orders did not have the full panoply of features that typify the later members; they had only those few critical features that enable us to recognize them as members of those later lineages.

We now briefly consider the history of some groups of mammals, focusing especially on some extant orders. There were also as many as 15 extinct orders, of which we will mention only a few.

Marsupials Marsupials are recorded from the late Cretaceous of both North America and South America (which were positioned fairly close to each other at that time); they may have migrated from North America into Eurasia (Eocene to Miocene fossils) and from South America to Australia via Antarctica (where an Eocene fossil has been found). *In South America, they experienced a great adaptive radiation;* some resembled kangaroo rats, some were bear-sized carnivores, and one group, the thylacosmilids of the

Miocene and Pliocene, were convergent with saber-toothed cats (see Figure 21 in Chapter 5). Except for various opossum-like creatures, South American marsupials became extinct by the end of the Pliocene, perhaps because of competition with placental mammals, which entered South America from North America at that time. The first Australian marsupial fossils are from the late Oligocene; by the Miocene, many of the families of the well-known Australian adaptive radiation had become distinct.

Eutherians Only a few departures from the basal eutherian ground plan distinguish the living tree shrews (Scandentia), elephant shrews (Macroscelidea), and members of the order Insectivora. Among the latter, the shrew and mole families are known from the Eocene. The bats (Chiroptera), today the second most diverse order of mammals, may have been derived from an insectivore-like stock; the earliest known bat skeleton, from the early Eocene, has modern features. Both morphological and molecular evidence suggests that the edentates (Xenarthra) diversified very early in mammalian history. Most of their radiation occurred in South America. The oldest edentate fossil, an armadillo, dates from the mid-Paleocene. Later edentates include the anteaters, as well as the armored glyptodonts and giant ground sloths (Figure 7.35), both of which entered North America and survived until the late Pleistocene.

Among eutherian orders, one of the earliest and, in many ways, structurally most primitive is the *Primates*. In the late Paleocene and Eocene, three now extinct primate groups were quite diverse; one of these, the plesiadapiforms, is known from late Cretaceous teeth. These animals are so *similar to basal eutherians* that it is rather arbitrary whether they be called primates or not, but they

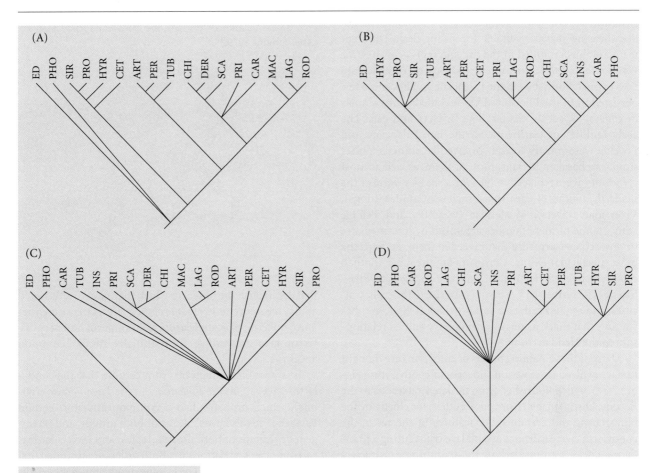

FIGURE 7.34 Four different hypotheses for the phylogenetic relationships among the orders of eutherian mammals, showing that the relationships among many orders are not well established. (A) An analysis of morphological characters by Shoshani in 1986. (B) An analysis of protein sequences by Miyamoto and Goodman in 1986. (C) An analysis of morphological characters by Novacek et al. in 1988. (D) A composite of several molecular analyses, showing only those relationships on which all the studies agreed. (A–C after Benton 1988b; D after Honeycutt and Adkins 1993.)

have primatelike dental features, and some indications of an arboreal habit (Figure 7.36). Adaptations for life in the trees are the chief distinguishing features of later primates, including an Eocene group that was quite similar to the later lemurs. The ancestral origin of the Anthropoidea (monkeys and apes), known first from the Oligocene, is not documented in the fossil record. The first recorded apes (superfamily Hominoidea) are baboon-sized animals from the Miocene, about 22 Mya; *Ramapithecus*, which is considered closely related to the orangutan (*Pongo*), dates from 17 Mya. There is no fossil

record of gorillas or chimpanzees. The *earliest member of the Hominidae, Ardipithecus ramidus*, is from African deposits dated at 4.4 Mya in the early Pliocene. It is followed by several species of *Australopithecus*, which evolved (from about 2.4 to 1.8 Mya) into *Homo*. Fossils from about 0.3 Mya (300,000 years before present) onward are called *Homo sapiens*. Stone tools were in use by at least 2 million years ago. We will consider more of the details of human evolution in Chapter 26.

The early carnivorous mammals are very similar to basal insectivorous eutherians, differing chiefly in the structure of certain of the cheek teeth, which have blade-like shearing surfaces suitable for slicing flesh. Different teeth are thus modified in the *two orders of carnivores, the Creodonta and the Carnivora.* Creodonts, all of which are now extinct, flourished in the Paleocene and Eocene; a few persisted until the Pliocene. They included forms similar in general aspect to weasels, dogs, hyenas, and cats. The early families of Carnivora, in the late Paleocene and Eocene, were mostly weasel- or civetlike animals, rather similar to basal eutherians; most of them, as well as most creodonts, were replaced in the *Oligocene* by several of the *modern families of carnivores,* such as Canidae (dogs), Viverridae (civets), Mustelidae (weasels), and Felidae (cats). Some of these families are linked by intermediates to the archaic Carnivora that preceded them. Perhaps the most spectacular extinct Carnivora are the saber-toothed cats, which lived from the Oligocene to the late Pleistocene. The terrestrial Carnivora gave rise to the seals, walruses, and sea lions, the earliest of which (Oligocene) has features that mark its terrestrial ancestry and its relationship to the Ursidae (bears).

Many of the remaining orders of eutherians are thought to have evolved from various lineages of condylarths (Figure 7.37), which flourished from the late Cretaceous to the Eocene. Distinguished from basal eutherians chiefly by the blunt cusps and certain other features of the teeth, the condylarths were a diverse assemblage constituting a grade (a paraphyletic group) lacking the apomorphic characters

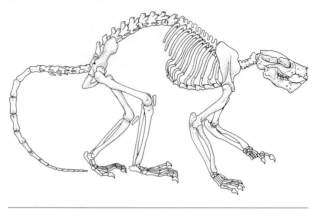

FIGURE 7.36 An early primate, the Paleocene *Plesiadapsis.* (After Simons 1979.)

that distinguish the orders to which they gave rise. As noted above, *the stem groups of some of those orders were closely related, rather similar condylarths,* an example of how very different taxa can evolve from initially similar common ancestors.

It is possible, though far from certain, that the condylarths gave rise to the *Rodentia* and the *Lagomorpha* (rabbits), which probably had a common ancestor: certain Paleocene fossils have features of both groups, and phylogenetic analysis of both morphological and molecular features indicates a relationship. Both orders are recorded first from the *late Paleocene.* The lagomorphs have never been diverse, but the *rodents became the most diverse order of mammals,* among which 50 Cenozoic families and more than 1700 living species are counted. Many of the modern families are known from the Eocene and Oligocene. Two evolutionary radiations of rodents are noteworthy. In South America, the caviomorph rodents began to diversify in the Oligocene, giving rise to 16 families, including porcupines, guinea pigs, chinchillas, capybaras, and others. The rats and mice (superfamily Muroidea) appeared in the late Eocene and experienced a series of radiations, the most prolific of which occurred within the last 10 million years in Eurasia, Africa, and North America; within the last 3 million years, muroids invaded and proliferated in South America and Australia as well. They now make up about two-thirds of the species of rodents.

The many orders of *ungulates* (hoofed mammals) and their relatives are doubtless *descended from condylarths,* probably from several different condylarth groups. One almost surely monophyletic cluster includes the orders Proboscidea (the elephants), Hyracoidea (the small, rather rodentlike hyraxes of Africa), and Sirenia (the aquatic manatees), all of which are first recorded in the Eocene. Of these, the greatest diversification by far was that of the Proboscidea, which over time differentiated into at least 40 genera (of which 2 survive). The earliest proboscideans were relatively small (about a meter tall), and lacked the tusks and most of the other peculiarities of later forms (some of which were very peculiar indeed: Figure 7.38).

FIGURE 7.35 Two extinct groups of Pleistocene edentates (Xenarthra). (A) Reconstruction of a glyptodont (*Glyptotherium*), with a shell of bony armor. (B) The huge ground sloth *Nothrotherium.* (A after Gillette and Ray 1981; B after Stock 1925.)

(A)

(B)

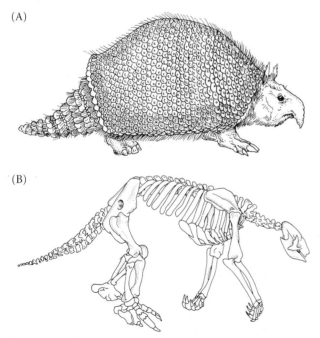

FIGURE 7.37 Condylarths, exemplifying the ground plan from which ungulate lineages evolved. (A) *Ectoconus* (Paleocene). (B) Skull of *Phenacodus* (Eocene). (A after Gregory 1951; B after Romer 1966.)

(A)

(B)

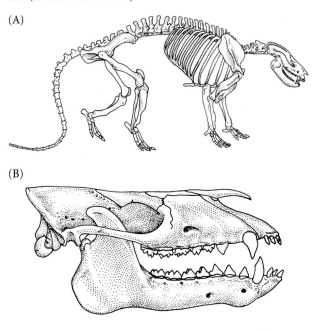

FIGURE 7.38 Skulls of some extinct Proboscidea. (A) An early, generalized proboscidean, *Moeritherium* (late Eocene–Early Oligocene). (B) *Phiomia* (early Oligocene). (C) *Gomphotherium* (Miocene). (D) *Deinotherium* (Miocene). (E) *Mammuthus*, the woolly mammoth (Pleistocene). (After Romer 1966.)

Originating in Africa, the proboscideans spread in the Miocene into Eurasia, thence to North America, and into South America in the Pleistocene.

South America was an island throughout the Tertiary, but before it separated from the rest of Gondwanaland, it harbored a group of condylarths that gave rise to a *major radiation of South American ungulates* that are assigned to at least six orders. Various of these orders were convergently similar, in many ways, to rodents, sheep, rhinoceroses, camels, and elephants. *Diadiaphorus*, in the order Litopterna, was astonishingly similar to modern horses in limb structure, with one digit per foot. The endemic South American ungulates declined in diversity after the connection to North America was established in the late Pliocene; the last of them became extinct in the late Pleistocene. Their extinction has been attributed to the ecological impact of North American immigrants, including, very possibly, *Homo sapiens*.

The *Perissodactyla*, or odd-toed ungulates, also evolved from condylarths. The major groups of perissodactyls, which are distinguished chiefly by dental features, were distinct by the *early Eocene*. The earliest members of these groups (e.g., early horses, tapirs, and rhinoceroses) were very similar to one another in most respects; only later did they diverge appreciably. The diversification of horses was described in Chapter 6. Rhinoceroses, in several families, were very diverse (about 80 genera) in Africa, Eurasia, and North America from the late Eocene to the Miocene. The earliest rhinoceroses were small, stocky animals rather like the living tapirs, but later they evolved into many different forms, including slender, fleet running rhinoceroses and the gi-

(A)   (B)   (C)

(D)   (E)

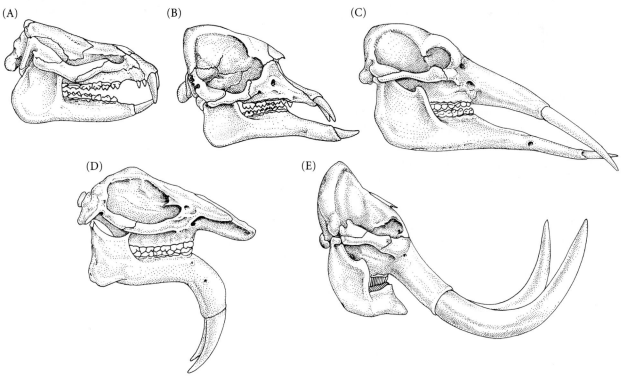

gantic *Indricotherium* (= *Baluchitherium*), which stood 5 meters tall and was the largest terrestrial mammal known. Only five species of rhinoceroses survive, all facing imminent extinction. The several extinct families of perissodactyls include the titanotheres, which display gradual evolution from small, hornless forms to large species with a bony nasal horn, and the peculiar chalicotheres, some of which had retractable, split claws.

The earliest fossil assigned to the Artiodactyla (even-toed ungulates) is *Diacodexis* (early Eocene), a fleet, rabbit-sized animal that has the diagnostic limb characters of the Artiodactyla, but is otherwise similar to arctocyonid condylarths. By the late Eocene, artiodactyls were quite diverse, including early pigs, camels, and ruminants. *In the Miocene, the ruminants began a sustained radiation,* mostly in the Old World, that is correlated with the increasing prevalence of grasslands, to which other ungulates, such as horses, also became adapted. Among the several families that proliferated are the Cervidae (deer), Giraffidae (giraffes and relatives), and Bovidae (the diverse family of antelopes, sheep, goats, and cattle). The few species of deer and bovids in the New World are descended from species that moved across the Bering land bridge late in the Cenozoic.

Among the Paleocene condylarths was a family, the *Mesonychidae*, that was exceptional in being carnivorous (Figure 7.39). Based on features of the teeth and ear bones, mesonychids include the *ancestors of whales (Cetacea)*. Recently, discoveries of spectacular mid-Eocene (52–46 Mya) *fossils have documented some of the anatomical changes that this lineage underwent* in its adaptation to marine life. Modern cetaceans have lost the hindlimbs (except for vestigial bones). They propel themselves by vertical flexure of the tail, which bears horizontal flukes that lack an internal skeleton. The skull is highly modified, with dorsally situated nostrils. The ear bones are highly modified for hearing under water.

Among the important recent discoveries (Thewissen and Hussain 1993; Thewissen et al. 1994; Gingerich et al. 1994) is *Pakicetus*, known from a skull found in a riverine deposit along with terrestrial animals. It is very similar to a mesonychid, but the ear bones are intermediate between those of mesonychids and those of cetaceans. *Ambulocetus*, a slightly younger form, is an exquisite intermediate. The fingers were separate and terminated in small hooves like those of mesonychids, but the structure of the lower back vertebrae, the joint between the hind leg and the body, and the large hind feet show that it swam by undulating the rear part of the body, as whales do. This form of locomotion was accentuated in the somewhat younger *Rodhocetus*. Younger still, by about 10 My, is *Basilosaurus*, which has long been recognized as a primitive cetacean. A specimen of *Basilosaurus* with complete hind legs has recently been found. The legs are complete in structure, but are far too small to have supported the weight of this large (25 feet), fully aquatic mammal. Thus the evolutionary origin of this highly modified order of mammals is becoming increasingly well documented.

FIGURE 7.39 Reconstructions, based on skeletal anatomy, of stages in the evolution of the Cetacea. (A) Mesonychid; (B) *Ambulocetus*; (C) *Rodhocetus*; (D) *Basilosaurus*. (From Futuyma 1995.)

(A)

(B)

(C)

(D)

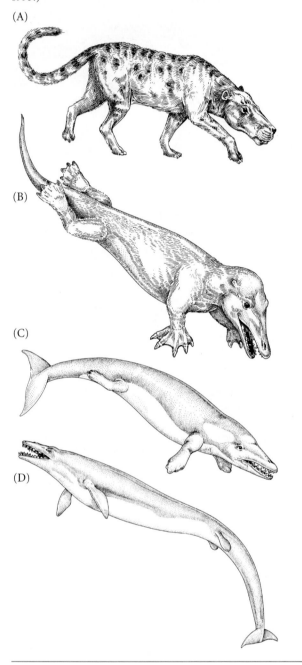

We have noted that, according to anatomical evidence, both the Cetacea and the Artiodactyla evolved from condylarth ancestors. Recent studies of DNA sequences show that, indeed, Cetacea are more closely related to Artiodactyla than to any other order of mammals. Some DNA data, in fact, suggest that cetaceans may be more closely related to ruminant artiodactyls (deer, antelopes, etc.) than to nonruminants (pigs, camels). That is, in a phylogenetic sense, cetaceans may *be* artiodactyls, not just related to artiodactyls (Graur and Higgins 1994).

Pleistocene Events

We have almost completed our survey of the history of evolution over more than 3 billion years, during which the biota has become more and more like that of the present day. The last Cenozoic epoch, the *Pleistocene*, embraces a mere 1.8 million years, but because of its recency and the dramatic events that have transpired, it is *critically important for understanding today's organisms.*

By the beginning of the Pleistocene, the continents were situated as they are now (although they continue to move, slowly, as in the past). North America was connected in the northwest to eastern Asia by the Bering land bridge, in the region where Alaska and Siberia almost meet today. North and South America were (and are) connected by the Isthmus of Panama. Except for those species that have become extinct, Pleistocene species are very similar to, or indistinguishable from, the living species that descended from them.

Global temperatures began to drop about 3 million years ago, and then, in the Pleistocene itself, underwent violent fluctuations with a period of about 100,000 years. When temperatures cooled, continental glaciers as thick as 2 kilometers formed at high latitudes, receding during the warmer intervals. *At least four major glacial advances,* and many minor ones, occurred. During the most recent glacial stage, termed the Wisconsin in North America and the Riss-Würm in Europe, an ice sheet extended across northern Asia, Europe, Greenland, and North America, where it advanced as far south as Long Island (New York), northern Ohio, and Montana. *It reached its maximum about 18,000 years ago, and melted back 15,000 to 8000 years ago.* Ice also covered southern South America, parts of southern Africa and Australia, and the tops of many mountain ranges, although certain mountaintops stood ice-free above the surrounding continental glaciers. During glacial episodes, the great mass of ice withdrew water from the *oceans,* which *dropped* as much as 100 meters below the present sea level at certain times. This exposed parts of the continental shelves, extending many continental margins beyond their present boundaries and *connecting many islands to nearby land masses.* (Japan, for example, was a peninsula of Asia, New Guinea was connected to Australia, and the East Indies were an extension of Southeast Asia.) Temperatures in equatorial regions were apparently about as high as they are today, so the latitudinal temperature gradient was much steeper than at present. Rainfall patterns were much altered. Although regions near the glacial margins, as well as certain other places such as the Sahara and parts of Australia, were wetter than they are now, *global climate during glacial episodes was generally drier.* Thus *mesic and wet forests became restricted* to relatively small favorable areas, and grasslands expanded. For example, the Amazonian rain forest appears to have retracted into large "islands" surrounded by savannah. During interglacial periods, such as the present time, the climate became warmer and generally wetter. (There is no reason to doubt that we are still in the Pleistocene—that more glacial episodes are yet to come.)

These events had profound effects on the distributions of organisms (see Chapter 8). The lower sea level during the glacials enabled many terrestrial species to move between land masses that are now isolated; for example, the Bering land bridge, although dry and cool, was broad and unglaciated, and served as a route from Asia to North America for species such as woolly mammoths, bison, and humans. The *distributions* of many, probably most, species *shifted* toward lower latitudes during glacial episodes and toward higher latitudes during interglacials, when tropical species extended far beyond their present limits. Fossils of elephants, hippopotamuses, and lions have been taken from interglacial deposits in England, whereas arctic species such as spruce and musk ox inhabited the southern United States during glacials. The vicissitudes of climate extinguished many species over broad areas; for instance, beetle species that have been found as Pleistocene fossils in England are now restricted to such far-flung areas as northern Africa and eastern Siberia (Coope 1979). An extremely important effect of these climatic changes was that many species that had been broadly, rather uniformly distributed became isolated in separate areas (REFUGES or REFUGIA) where favorable conditions persisted. *Some such isolated populations diverged* genetically and phenotypically, in some instances becoming different species (see Chapter 16). In some cases, populations have remained in their glacial refuge areas, isolated from the major range of their species; for example, mountaintops in the southern Appalachians have populations of cold-adapted plants that are otherwise found only much farther north (see Figure 1 in Chapter 8). On the other hand, many species have rapidly spread over broad areas from one or a few local refugia, and have achieved their present distributions only in the last 8000 years or so. (See Pielou 1991 for more information on postglacial history.)

It should not be supposed that in response to Pleistocene climatic changes, animals migrated by abandoning their former homes and moving en masse to more favorable areas. Climatic changes occurred slowly enough for suitable habitats to develop at the edge of a species' range—to the south, say, during a Northern Hemisphere glaciation—and to be colonized by individuals dispersing as they always do; farther to the north, formerly favorable sites would become unfavorable, and the growth of populations would become negative due to both emigration and mortality. Thus a species' distribution shifted rather gradually over the landscape. From fossil pollen of extant plant species, it has become clear that some of the species that are typically associated in communities today moved rather independently, at different rates, over the land; thus *the species composition of communities changed kaleidoscopically* (Davis 1976; Figure 7.40). For some species, new favorable areas were inaccessible, and the species became extinct, at least regionally. For example, the Alps prevented the southward movement of many European species, such as magnolias, and the biota of Europe is rather poor in species as a consequence.

It is a most interesting fact that although some Pleistocene species are distinguishable from their present-day descendants, the great environmental changes of the time seem not to have stimulated particularly rapid evolution; in fact, many species have not changed discernibly in morphology. For example, almost all Pleistocene and late Pliocene fossil beetles are indistinguishable from living species (Coope 1979). The divergence of populations, as in different Pleistocene refuges, is a process that occurs at all times, and seems not to have been particularly rapid in the Pleistocene. Aside from changes in species' geographic distributions, the most conspicuous events in the Pleistocene were extinctions. *A substantial fraction of shallow-water marine invertebrate species became extinct, especially among tropical species*, which may have been poorly equipped to withstand even modest cooling. No major taxa became entirely extinct, however. On land, the story was different. Although small vertebrates seem not to have suffered, a *very high proportion of large-bodied mammals and birds became extinct.* These included mammoths, saber-toothed cats, giant bison, giant beavers, giant wolves, ground sloths, and all the endemic South American ungulates. There is con-

siderable argument, still unresolved, over whether these animals became extinct because of changes in climate and habitat or were hunted to extinction by weapon-wielding humans.

Whether one is studying the genetic structure of populations, the adaptations of species, or the structure of ecological communities, it is well to bear in mind that history, especially recent dynamics such as those of the Pleistocene, casts its shadow on the present. The vicissitudes of the Pleistocene must have destabilized many genetic and ecological equilibria, and the glaciers had hardly retreated when major new disruptions began. The advent of human agriculture about 12,000 years ago began yet another reshaping of the terrestrial environment. For the last several thousand years, deserts have expanded under the impact of overgrazing, forests have succumbed to fire and cutting, and climates have changed as vegetation has been modified or destroyed. At present, under the impact of an exponentially growing human population and its modern technology, tropical forests, with their richness of species, face

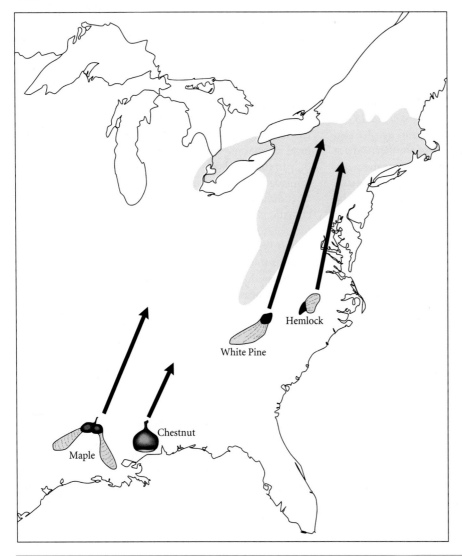

FIGURE 7.40 Different rates of spread of four North American tree species from refugia after the most recent glacial episode. All four species occupy the shaded area today. After the glacial, maple and chestnut spread from the Gulf region, and white pine and hemlock from the mid-Atlantic coastal plain. The lengths of the arrows are proportional to the rate of advance. (After Pielou 1991.)

almost complete annihilation, temperate zone forests and prairies have been eliminated in much of the world, and even marine communities suffer pollution and overexploitation. In the twenty-first century and beyond, one of the greatest mass extinctions of all time will come to pass—unless we act now to prevent it (Wilson 1992).

Summary

1. The first fossil evidence of life dates from about 3.5 billion years ago, about a billion years after the formation of the earth. Some progress has been made in understanding the origin of life, but a great deal remains unknown. The earliest life forms of which we have evidence were prokaryotes. Evidence from living organisms indicates that all living things are descended from a single common ancestor.

2. Among the most significant events in the subsequent history of life are (*a*) the evolution of eukaryotes, including the symbiotic origin of mitochondria and chloroplasts from bacteria, at least 1.4 billion years ago (*b*) the "explosive" diversification of animal phyla near the beginning of the Cambrian period, about 543 million years ago (*c*) the invasion of the land by plants and arthropods in the late Silurian and early Devonian (about 400 million years ago) and by amphibians in the late Devonian (*d*) the most devastating mass extinction of all time at the end of the Permian (about 251 million years ago), which profoundly altered the taxonomic composition of the earth's biota (*e*) the diversification and ecological dominance of seed plants and reptiles during the Mesozoic era (251–65 million years ago) (*f*) the substantial extinction at the end of the Mesozoic era, which included the extinction of the last dinosaurs (*g*) the steady, spectacular increase in the diversity of flowering plants and plant-associated insects from the middle of the Cretaceous (about 100 million years ago) onward (*h*) the rapid differentiation of the orders of mammals in the early Tertiary (about 65–50 million years ago), and their subsequent evolutionary radiation (*i*) climatic changes during the Cenozoic era, especially a drier climate since the Oligocene, which favored the development of grasslands and the evolution of herbaceous plants and grassland-adapted animals (*j*) the series of Pleistocene glacial and interglacial intervals over the last 1.8 million years, during which some extinctions occurred and the distributions of species were greatly altered.

3. In many instances, the sequence in which the major groups within a higher taxon appear in the fossil record matches the sequence in which they evolved, as inferred from phylogenetic analyses (chiefly of extant organisms) that do not take the fossil sequence into account. That is, independent lines of evidence often agree on the relative times of origin of taxa and their distinctive characteristics.

4. Many higher taxa appear "suddenly" in the fossil record, and their ancestry is not recorded. In some cases, they probably evolved very rapidly. In many other cases, the ancestry of a higher taxon is documented by the close relationship of a "stem group," often with a generalized morphology, to the stem groups of other taxa.

5. Throughout evolutionary time, taxa have originated and become extinct. The rates of these processes have varied, being accentuated in a number of "mass extinctions" and in numerous episodes of evolutionary radiation, some of which followed mass extinctions.

6. With the passage of time, the composition of the earth's biota has become increasingly similar to its composition today.

Major References

General references:

Maynard Smith, J., and E. Szathmáry. 1995. *The major transitions in evolution*. W. H. Freeman, San Francisco. History and theoretical interpretations, ranging from the origin of life to the origins of societies and languages, by leading evolutionary theoreticians.

Allen, K. C., and D. E. G. Briggs (editors). 1989. *Evolution and the fossil record*. Smithsonian Institution Press, Washington, D.C. An excellent summary of evolutionary history in the fossil record.

On the evolution of invertebrate groups:

House, M. R. (editor). 1979. *The origin of major invertebrate groups*. Academic Press, New York.

Clarkson, E. N. K. 1993. *Invertebrate paleontology and evolution*. Chapman and Hall, London.

Brusca, R. C., and G. J. Brusca. 1990. *Invertebrates*. Sinauer Associates, Sunderland, MA. A textbook of invertebrate zoology, treating the biology and phylogeny of living forms.

On the fossil record and phylogeny of insects:

Kristensen, N. P. 1991. Phylogeny of extant hexapods. In CSIRO, *The Insects of Australia*, pp. 125–140. Melbourne University Press, Melbourne, and Cornell University Press, Ithaca, NY.

Kukalová-Peck, J. 1991. Fossil history and the evolution of hexapod structures. In CSIRO, *The Insects of Australia*, pp. 141–179. See also W. A. Shear and J. Kukalová-Peck, *Canadian Journal of Zoology* 68:1807–1834 (1990).

Wootton, R. J. 1990. Major insect radiations. In P. D. Taylor and G. P. Larwood (editors), *Major evolutionary radiations*, pp. 187–208. Clarendon, Oxford.

On the fossil history and phylogeny of land plants:

Stewart, W. N. 1993. *Paleobotany and the evolution of plants.* Second edition. Cambridge University Press, Cambridge. Well illustrated, rather technical survey, especially of morphological evolution. Includes a short treatment of nonvascular plants.

On the fossil history of vertebrates:

Carroll, R. L. 1988. *Vertebrate paleontology and evolution.* W. H. Freeman, New York. A comprehensive, well-illustrated treatment of the subject.

Benton, M. J. (editor). 1988. *The phylogeny and classification of the tetrapods.* Clarendon, Oxford. Quite technical essays by various authors on controversial aspects of tetrapod phylogeny.

Benton, M. J. 1997. *Vertebrate palaeontology.* Second edition. Chapman and Hall, New York. Vertebrate fossil history and research methods.

Weishampel, D. B., P. Dodson, and H. Osmolska. 1990. *The dinosauria.* University of California Press, Berkeley. Comprehensive, fairly technical treatment of dinosaurs.

Problems and Discussion Topics

1. Why, in the evolution of ancestral eukaryotes, might it have been advantageous for separate organisms to become united into a single organism? Can you describe analogous, more recently evolved, examples of intimate symbioses that function as single integrated organisms?

2. Early in the origin of life, as it is presently conceived, there was no distinction between genotype and phenotype. What characterizes this distinction, and at what stage of organization may it be said to have come into being?

3. If we employ the biological species concept (see Chapter 2), when did species first exist? What were organisms before then, if not species? What might the consequences of the emergence of species be for processes of adaptation and diversification?

4. How would you determine whether the disparity, or morphological diversity, of animals has increased, decreased, or remained the same since the Cambrian?

5. Discuss the kinds of evidence that might solve the problem of the Cambrian "explosion."

6. Evolutionary radiations occur when many lineages proliferate from a common ancestor within a relatively short time. They are called *adaptive radiations* if the different lineages acquire different adaptations. Compare five of the evolutionary radiations mentioned in this chapter (e.g., ammonites, flowering plants, holometabolous insects, orders of mammals, dinosaurs, rodents, passerine birds), and discuss whether or not they should be considered adaptive radiations.

7. Compare terrestrial communities in the Devonian and in the Cretaceous, and discuss what may account for the difference between them in the diversity of plants and animals.

8. Referring to the sources cited at the end of this chapter, trace the evolution of reproductive systems and of architecture (physical form) in vascular plants. Which feature(s) may account for the increase in diversity? How could you rigorously test your hypothesis?

9. To what extent is there a correlation between the sequence of origin of higher taxa, as inferred from phylogenetic analysis, and the sequence in which they appear in the fossil record?

10. Discuss the implications of the spread and change of species distributions after the retreat of the Pleistocene glaciers for evolutionary changes in species and for the species composition of ecological communities (cf. Chapters 16, 25).

CHAPTER 8

The Geography of Evolution

Biogeography, sometimes divided into phytogeography and zoogeography, is the study of the geographic distributions of organisms. It attempts to explain why species and higher taxa are distributed as they are, and why the diversity and taxonomic composition of the biota vary from one region to another.

Biogeography is intimately related to geology, paleontology, systematics, and ecology. The history of the distribution of land masses and climates, a product of geological study, often sheds light on the causes of organisms' distributions. Conversely, organisms' distributions have sometimes provided evidence for geological events. For example, the Great Lakes of North America presently drain into the Atlantic Ocean via the Saint Lawrence River, but their fish fauna is similar to that of the Mississippi River—an indication that until recently, the lakes drained via the Mississippi into the Gulf of Mexico. The geographic distributions of organisms were used by some scientists as evidence for continental drift long before geologists agreed that it really happens.

The role of paleontology in biogeography is obvious, since present distributions often can be interpreted in light of past distributions as indicated by the fossil record. Systematics is relevant in many ways. Biogeography is completely reliant on an accurate understanding of phylogenetic relationships. For example, a peculiar lizard (*Lanthanotus*) restricted to the East Indian island of Borneo was once considered a close relative of the Gila monster (*Heloderma*) of the southwestern United States and Mexico. It was hard to explain how close relatives could have such a peculiarly disjunct distribution. It turned out, however, that an explanation was unnecessary: further study showed that *Lanthanotus* is really related to the monitor lizards (Varanidae), which are broadly distributed throughout the Old World tropics. ("Old World" refers to lands other than North and South America, which are called the "New World.") Other aspects of the intimate relationship between biogeography and systematics will be described below.

In some instances, the geographic distribution of a taxon may best be explained by historical circumstances; in other instances, ecological factors operating at the present time may provide the best explanation. In fact, the field of biogeography may be roughly subdivided into HISTORICAL BIOGEOGRAPHY and ECOLOGICAL BIOGEOGRAPHY. Systematists and paleontologists tend to emphasize past events (such as continental drift) in explaining distributions, whereas ecologists tend to invoke contemporary or recent factors (such as the distribution of habitats and competition among species). Biogeographers are coming increasingly to recognize that both kinds of explanations are important (Brown and Gibson 1983; Myers and Giller 1988; Ricklefs and Schluter 1993b).

Among the kinds of questions that biogeography addresses are these: Why do some higher taxa have very broad distributions, whereas others are restricted (**endemic**) to a single region, such as a single land mass or part thereof? How can we account for the disjunct distributions of taxa that are found in widely separated localities, but not in between? Is there any correspondence between the phylogeny of taxa and the history of geological events? Why is a taxon richer in species in some regions than in others? Why is the biota of some regions and habitats more diverse than in others?

Biogeographic Evidence for Evolution

Before we consider modern studies, we should take note of the great importance of biogeography to both Darwin and Wallace. Wallace devoted much of his later career to the subject, but here we will note only the role it played in providing inspiration and evidence for his and Darwin's conviction that evolution had occurred. To us, today, the reasons for certain facts of biogeography seem so obvious that they hardly bear mentioning. If someone asks us why there are no elephants in the Hawaiian Islands, we will naturally answer that they couldn't get there. However, this answer assumes that elephants originated somewhere else, namely, on a continent. But in a pre-evolutionary world view, the view of special divine creation that Darwin and Wallace were combating, such an answer would not hold: the Creator could have placed each species anywhere, or in many places at the same time. In fact, it would have been reasonable to expect the Creator to place a species wherever its habitat, such as rain forest, occurred.

Darwin devoted two chapters of *The Origin of Species* to showing that many biogeographic facts that make little sense under the hypothesis of special creation make a great deal of sense if a species (1) has a definite site or region of origin, (2) achieves a broader distribution by migration (movement), and (3) becomes modified and gives rise to descendant species in the various regions to which it migrates. (Darwin emphasized migration as the chief explanation of distributions because in his day there was little inkling that continents might have moved. Today, the movement of land masses also serves to explain certain patterns of distribution.) We shall briefly examine some of Darwin's major points.

Darwin's "*first great fact*" is that "neither the similarity nor the dissimilarity of the inhabitants of various regions can be wholly accounted for by climatal and other physical conditions." Similar climates and habitats, such as deserts, occur in both the Old and the New World, yet the organisms inhabiting them are unrelated. For example, the cacti are restricted to the New World, whereas cactuslike plants in Old World deserts are chiefly members of the family Euphorbiaceae (see Figure 3 in Chapter 4). Conversely, within a geographic region, extremely different habitats may be occupied by closely related species. In tropical America, for example, different members of the pineapple family, Bromeliaceae, occupy rain forests, cold mountaintops, and the driest of the world's deserts.

Darwin's "*second great fact*" is that "barriers of any kind, or obstacles to free migration, are related in a close and important manner to the differences between the productions [organisms] of various regions." As examples, Darwin cites the great difference between the marine species on the eastern and western coasts of South America, and between western South America and the islands of the western Pacific: both land and wide expanses of deep water, he suggests, are barriers to dispersal. He notes, however, that there is considerable similarity between the marine species on either side of the Isthmus of Panama, and that "this fact has led naturalists to believe that the isthmus was formerly open." We now know that, indeed, it was open until the late Pliocene, about 3.5 million years ago.

A "*third great fact*," which Darwin notes is related to the preceding, is "the affinity of the productions of the same continent or of the same sea, though the species themselves are distinct at different points and stations." Drawing on his experience in South America, he gives the example of a rodent, the vizcacha, on the plains of Argentina that has an alpine relative on the lofty peaks of the Cordillera. Furthermore, the aquatic rodents of South America, the coypu and capybara, are related to the vizcacha, not to the northern beaver or muskrat.

"We see in these facts," says Darwin, "some deep organic bond, throughout space and time, over the same areas of land and water, independently of physical conditions . . . . The bond is simply inheritance, that cause which alone, as far as we positively know, produces organisms quite like each other."

The geographic proximity of related species likewise led Wallace to conclude (as Darwin quotes him) that "every species has come into existence closely coincident both in space and time with a pre-existing closely allied species." Thus species are "related" not in the sense that they were created along the same design, but in the sense that they descend from common ancestors.

It was important to Darwin to show that a species had not been created in different places, but had a single region of origin. He drew particularly compelling evidence from the inhabitants of islands. First, distant *oceanic islands generally have precisely those kinds of organisms that have a capacity for long-distance dispersal*, and lack those that do not. For example, the only native mammals on many islands are bats. Such islands, moreover, conspicuously lack amphibians, which cannot survive salt water, but they do have certain kinds of lizards (e.g., geckos), the resistant eggs of which can be transported in floating vegetation. Second, *the number of indigenous species on oceanic islands is far less than in comparable continental areas*, yet numerous continental species of plants and animals have flourished on islands to which humans have transported them. Thus, says Darwin, "he who admits the doctrine of the creation of each separate species, will have to admit that a sufficient number of the best adapted plants and animals were not created for oceanic islands." Third, most of the species on islands, although distinctly different from those elsewhere, are clearly *related to species on the nearest mainland*, implying that that was their source. This is the case, as Darwin says, for almost all the birds and plants of the Galápagos Islands. Fourth, on many islands the *proportion of endemic species* is very high, but it *is particularly high when the opportunity for dispersal to the island is low*. (If dispersal is frequent, interbreeding between island populations and mainland immigrants will slow down the differentiation of the island populations; see Chapter 11.) For example, Darwin notes that in the Galápagos Islands, almost all the land birds, although related to continental species, are distinct from them, whereas the water birds, which have much greater powers of over-water dispersal, are identical to mainland populations. In contrast to the Galápagos, to which land birds rarely disperse, there are no endemic species in Bermuda, to which many American birds disperse yearly despite its equivalent distance from the mainland. Fifth, *island species often bear marks of their continental ancestry*. For example, Darwin notes, hooks on seeds are an adaptation for dispersal by mammals, yet on oceanic islands that lack mammals, many endemic plants nevertheless have hooked seeds.

In arguing that each species has dispersed to a greater or lesser extent from a single, restricted region of origin, Darwin had to explain how some taxa have achieved broad, sometimes peculiarly disjunct, distributions. He invokes two kinds of explanations. First, he cites evidence from geology that the spatial distribution of climatic zones has changed over time, and argues that we should expect that organisms will follow the habitats to which they are adapted. During the Pleistocene glaciations, in particular, arctic species would move southward, and as mountains became covered with ice, "their former Alpine inhabitants would descend to the plains." These species would then become broadly distributed in the lowlands, but "as the warmth returned, the arctic forms would retreat northward" as well as upward along mountain slopes, so that ultimately the same species would be found both in the Arctic "and on many isolated mountain summits far distant from each other." Darwin thereby explains how the same species, or slightly different forms, can be found in Scandinavia, Scotland, and the Pyrenees Mountains along the border of France and Spain, just as certain species occur in both the Appalachian Mountains and northern Canada (Figure 8.1). His explanation holds good today.

Darwin's second explanation for unusual distributions was organisms' means of dispersal, to which he devoted a great deal of study. He describes how the seeds of plants can be transported in the digestive tracts of birds, or in mud adhering to their feet, and how freshwater animals likewise can be carried to new bodies of water. In one of many passages that show what an acute observer Darwin was, and how he could see the importance of the seemingly most trivial facts, he tells us that a freshwater beetle landed on the *Beagle* when it was 45 miles from land; small molluscs, he says, have been found adhering to these flying insects. He did experiments that showed that many kinds of seeds could survive a prolonged sojourn in salt water. Since Darwin's time, it has become known, further, that large masses of vegetation from flooded rivers may float for hundreds of miles, carrying both plant and animal life. We also know that some species of plants and animals can expand their range very rapidly. Many species of plants have expanded across most of North America from New York and New England, to which they were brought from Europe by humans within the last 200 years, and some birds, such as the starling (*Sturnus vulgaris*) and the house sparrow (*Passer domesticus*), have done the same within a century (Figure 8.2).

Major Patterns of Distribution

The geographic distribution of almost every species is limited to some extent, and many higher taxa are likewise restricted (*endemic*) to a particular geographic region. The level of endemism usually depends on taxonomic rank. For example, except for one European genus, the lungless salamanders (family Plethodontidae) are found only in the New World; the genus *Plethodon* itself is limited to North America; and the range of some species of *Plethodon*, such as *Plethodon caddoensis*, which occupies only the Caddo Mountains of western Arkansas, is very narrow. From this example, it is immediately evident that any meaningful discussion of biogeography depends on accurate, phylogenetically informative taxonomy: the previous sentence would be meaningless if Plethodontidae and *Plethodon* were not each unified by common ancestry. The distributions of some higher taxa are almost cosmopolitan (e.g., the pigeon family, Columbidae), whereas others are narrowly endemic [e.g., the Sphenodontidae, consisting of two species of tuataras, which are restricted to islets along the coast of New Zealand (Figure 8.3)].

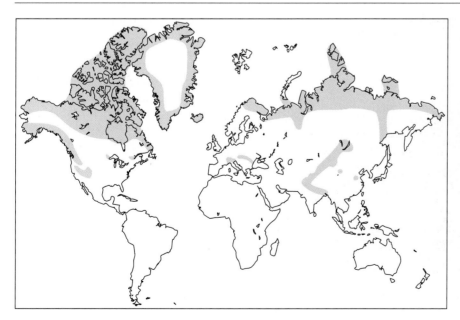

FIGURE 8.1 The disjunct distribution of the plant *Saxifraga cernua* (Saxifragaceae) in northern and mountainous regions of the Northern Hemisphere. Relict populations persist at high elevations, following the species' retreat from the southern region that it occupied during glacial periods. (After Brown and Gibson 1983.)

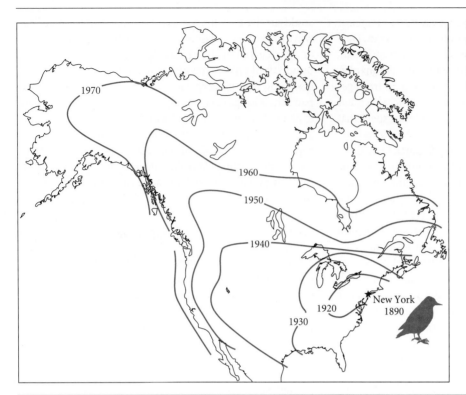

FIGURE 8.2 The history of range expansion of the European starling (*Sturnus vulgaris*) following its introduction into New York City in the 1890's. (After Cabe 1993.)

Wallace and other early biogeographers recognized that many higher taxa have roughly similar distributions, and that taxonomic composition is more uniform within certain regions than between them. Wallace designated several BIOGEOGRAPHIC REALMS for terrestrial and freshwater organisms (see Figure 8 in Chapter 4) that are still widely recognized today (although some workers would divide them more finely). These are the Palearctic (temperate and tropical Eurasia and northern Africa), the Ethiopian (sub-Saharan Africa), the Oriental (India and Southeast Asia), the Australian (Australia, New Guinea, New Zealand, and nearby islands), the Nearctic (North America), and the Neotropical (South and Central America). Each of these realms has many higher taxa that are much more diverse there than elsewhere, or are restricted to that region. For example, the Neotropical realm is characterized by endemic taxa (Figure 8.4) such as edentates (anteaters and allies), platyrrhine primates (such as spider monkeys and marmosets), hummingbirds, a large assemblage of suboscine birds such as flycatchers and antbirds, many families of catfishes, and plant families such as the pineapple family (Bromeliaceae). Within each biogeographic realm, individual species have more restricted distributions, and regions that differ markedly in habitat, or which are separated by mountain ranges or other barriers, have rather different sets of species. Thus a biogeographic realm can often be divided into faunal and floral provinces.

FIGURE 8.3 The tuatara *Sphenodon punctatus*, one of two closely related species that are the sole surviving members of the ancient family Sphenodontidae. This family has a relictual distribution, limited to small islands along the coast of New Zealand. (From Bellairs 1970.)

FIGURE 8.4 Examples of taxa endemic to the Neotropical biogeographic realm. (A) An armadillo (order Edentata). (B) An anteater (order Edentata). (C) A platyrrhine primate, the howler monkey (*Alouatta*). (D) An antshrike (Formicariidae), representing a huge evolutionary radiation of suboscine birds found throughout the Neotropics. (E) An armored catfish (Callichthyidae), one of many families of freshwater catfishes restricted to South America. (A,B after Emmons 1990; C after Hershkovitz 1977; D after Haverschmidt 1968; E after Moyle and Cech 1982.)

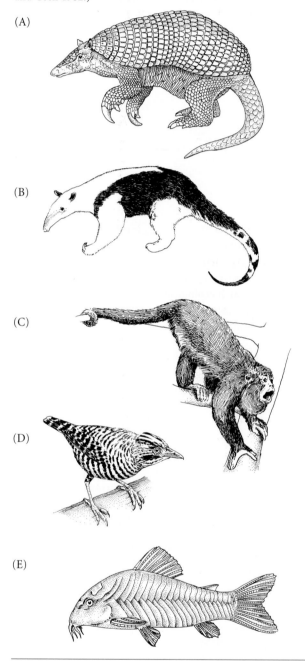

(A)

(B)

(C)

(D)

(E)

The borders between biogeographic realms (or provinces) cannot be sharply drawn because some taxa infiltrate neighboring realms to varying degrees. In the Nearctic realm (North America), for instance, some taxa are related to Palearctic taxa: examples include bison, deer, chickadees,

trout, and birches. Some other North American species are related to, and have been derived from, Neotropical stocks: examples include the armadillo, the opossum, hummingbirds, whiptail lizards, and bromeliads (the "Spanish moss," *Tillandsia usneoides*, that festoons southern trees is a bromeliad, a flowering plant). We shall explain below how one can deduce the ways in which the biota of a region has been assembled.

Some taxa have **disjunct distributions**; that is, their distributions have gaps. Some such taxa are found in more than one biogeographic realm. Typically, such higher taxa have different species or genera in each area. For example, the mostly flightless birds known as ratites (Figure 8.5) include the ostrich in Africa, rheas and tinamous in the Neotropics, the emu and cassowaries of Australia and New Guinea, and the kiwis and the recently extinct moas of New Zealand. Because it was a challenge to explain how flightless birds could have crossed oceans, it was long debated whether the ratites were actually related to one another or had independently evolved, by convergent evolution from flying ancestors, in each area. Both morphological and DNA studies affirm that the ratites are indeed monophyletic (Cracraft 1974; Sibley and Ahlquist 1990)—and we can now attribute their distributions to continental drift.

Many other taxa are distributed, like the ratites, across the southern land masses (Goldblatt 1993). Africa, Australia, and South America share lungfishes, leptodactylid frogs, and quite a few other groups; Australia and South America share marsupials, chelydid turtles, and southern beeches (*Nothofagus*); Africa and South America share cichlid and characiform fishes. Another common disjunct pattern is illustrated by alligators (*Alligator*), skunk cabbages (*Symplocarpus*), and tulip trees (*Liriodendron*), which are among the many genera that are found both in eastern North America and temperate eastern Asia. Finally, some disjunct distributions do not fit any common pattern: the Camelidae are represented by South American llamas and central Asian camels, and the lizard family Iguanidae is diverse in the New World, but has a few far-flung representatives in Madagascar and on the western Pacific islands of Fiji and Tonga (Figure 8.6). One of the tasks of biogeographic study is to seek explanations for such peculiar distributions.

Factors Affecting Geographic Distributions

The many factors that can affect the distribution of a species may be roughly classified as biological or abiotic. On the biological side, processes that might *increase* a species' range are

1. *Adaptation* to conditions, such as temperature or competing species, that previously limited the range of the species.
2. *Range expansion*, not necessarily entailing genetic change. This term is used for the normal dispersal of individuals across expanses of more or less continuously favorable habitat.

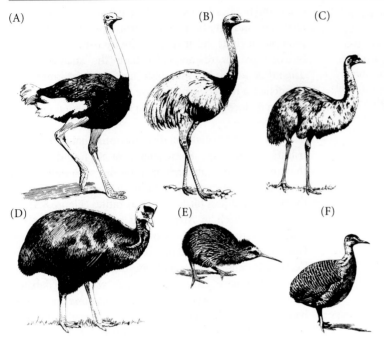

(A) (B) (C)

(D) (E) (F)

FIGURE 8.5 The living families of ratite birds. (A) Struthionidae (ostrich), Africa; (B) Rheidae (rhea), South America; (C) Dromiceiidae (emu), Australia; (D) Casuariidae (cassowary), Australia and New Guinea; (E) Apterygidae (kiwi), New Zealand; (F) Tinamidae (tinamou), tropical America. Despite their disjunct distribution, these birds, which except for the tinamou are flightless, are a monophyletic group. They diverged from a common ancestral stock that inhabited Gondwanaland before it split into the southern land masses. (From Van Tyne and Berger 1959.)

3. *Jump dispersal* (Myers and Giller 1988), a term used to describe colonization across a substantial barrier. For terrestrial organisms, such barriers include bodies of water, mountain ranges, and expanses of unfavorable habitat. The effectiveness of a barrier in preventing dispersal depends on the characteristics of a species. Some birds of the understory of tropical rain forests, for example, seldom fly across even narrow clearings, and can be restricted by rivers.

ESTABLISHMENT of new populations, by either short-range or jump dispersal, can be prevented by numerous in-

imical ecological factors, such as intolerable physical conditions, scarcity of suitable food, or other species such as predators or competitors. The range of a species may *decrease* if certain of its populations become extinct because of changes in such ecological factors.

The abiotic factors that can alter a species' range are chiefly climatic and geological.

1. A change in CLIMATIC REGIME can be imposed on a region (as during the Pleistocene glaciations), or a land mass can be carried by plate tectonic events into a different climatic zone (e.g., the northward move-

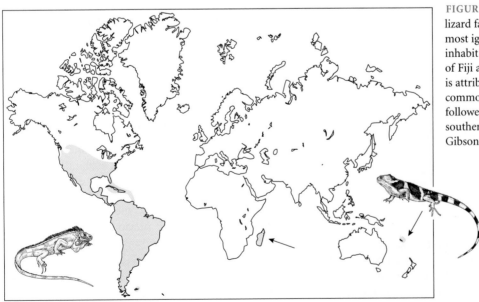

FIGURE 8.6 The distribution of the lizard family Iguanidae. Although most iguanids are American, a few inhabit Madagascar and the islands of Fiji and Tonga. Their distribution is attributed to divergence from a common Gondwanaland ancestor followed by extinction in most other southern regions. (After Brown and Gibson 1983.)

ment of India in the early Cenozoic). Changes in climate, together with associated changes in biota, may result in a SHIFT in the range of a species, as it expands along one front and becomes extinct along another. Climatic changes can also sunder distributions if, for example, expanses of forest become separated by grasslands.

2. EUSTATIC events, i.e., changes in sea level, can disconnect bodies of land or water, creating barriers, or connect them, providing bridges for range expansion.

3. TECTONIC (large-scale geological) events include the formation of mountain ranges, as well as movements of plates that separate and join land masses (see Figure 22 in Chapter 7).

The distribution of a clade, or higher taxon, is affected by these same factors, as well as by DIVERGENCE and SPECIATION. Populations that are separated by a barrier may diverge over time, becoming different species (see Chapter 16), which in turn may give rise to other species in each of the occupied regions. Thus two or more regions may each, over time, come to harbor monophyletic subgroups of a major clade (Figure 8.7). A simple pattern like this may be altered, however, by extinction of the clade in certain regions, or by subsequent movements of species from one area to another.

It is obvious that the factors responsible for a taxon's distribution, to say nothing of those responsible for the overall composition of a region's biota, are many and complex.

Any or all of them may have played a role in any particular case, and so inferring the history and causes of distributions is among the most difficult problems in evolutionary and ecological biology. In many cases, it is difficult or apparently impossible to obtain definitive evidence for one hypothesis over another.

Guidelines in Historical Biogeography

Among the major questions posed in historical biogeography are: Did the taxa in a particular region originate there or move there from elsewhere? If they moved there, where was the source? Can we find the causes of disjunct distributions of taxa? Are there historical reasons (e.g., barriers to movement) for the absence of taxa from certain areas?

Biogeographers have used a variety of guidelines in inferring the history of distributions. Some of these guidelines are well founded. For example, the distribution of a taxon cannot be explained by an event that occurred before the taxon originated: a genus that originated in the Miocene cannot have achieved its distribution by continental movements that occurred in the Cretaceous. Some other guidelines are more debatable. Some authors in the past assumed that a taxon originated in the region where it is presently most diverse. But this need not be so. For example, members of the horse family (Equidae) are today native only to Africa and Asia, yet the fossil record shows that they are descended from North American ancestors, now extinct (see Chapter 6). Another suggestion has been that the phylogenetically

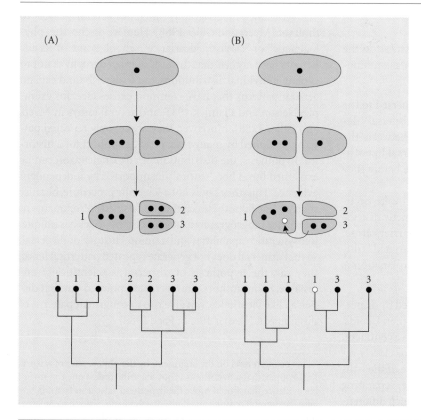

FIGURE 8.7 (A) A simple model of the distribution of a clade endemic to three regions. The phylogeny of the group reflects the order of separation of the areas the different taxa occupy. The dots represent species, which increase in number due to speciation. The cladogram, whose tips are labeled with the areas occupied by each taxon, is referred to as an *area cladogram*. (B) Realistic complications: After separation and divergence, a lineage colonizes area 1 from area 3, and the clade becomes extinct in area 2. The cladogram provides evidence of the dispersal, but it cannot tell us whether or not the clade ever occupied area 2.

most basal ("primitive") members of a clade occupy the region of its origin. Other biogeographers, however, have held precisely the opposite: that primitive forms survive only in peripheral areas, and have been excluded from the region of origin by more recently evolved groups. Probable examples of each hypothesis have been cited; suffice it to say that neither position is a hard-and-fast rule.

The value of the fossil record in tracing biogeographic history is a matter of considerable debate. Certainly the *presence* of fossils in a region that a taxon currently does not occupy must be taken into account. Also, fossils provide a MINIMUM AGE for a group, as well as for its sister group, which, having originated from the same common ancestor, is by definition equal to it in age. (This assumes, of course, that the phylogeny has been reliably estimated.) The *absence* of fossils is more problematic. A group with a sparse fossil record may be older than the known fossils indicate, and it may have inhabited regions from which fossils have not been recorded. If, however, a group is abundantly represented in the fossil record, and if it has not been recorded from a region that offers a rich fossil record of organisms with which it is elsewhere ecologically associated (indicating that inhabitants of its habitat are well represented), one may strongly suspect that the group did not occur there (although this suspicion may be proven false by new fossils). For example, we can be confident that the South American orders of Tertiary ungulates did not occur in North America, where there is a rich fossil record of other ungulates, but no representatives of the South American groups.

Competing Explanations for Disjunct Distributions

If a taxon occurs in two or more disjunct areas (as do the ratite birds, for example), two explanations (or a combination of the two) may be offered.

1. The taxon originated in one area and dispersed to the other, by range expansion or by jump dispersal (depending on the nature of the barrier between the areas). This explanation, called the **dispersal hypothesis,** assumes that the areas were separate before they became occupied.
2. The areas were formerly contiguous, and were occupied by the ancestor of the present, disjunctly distributed species, which differentiated after barriers arose. In this case the taxon is said to have "vicariated" (differentiated into different species); this explanation is known as the **vicariance hypothesis**.

A simple example of a distribution explained by dispersal is the occurrence of *Drosophila* in the Hawaiian Islands, which, apparently, have never been connected to a continent. An example of vicariance is the distribution of closely related species of fish (e.g., gobies) on either side of the Isthmus of Panama, which arose in the Pliocene, separating eastern Pacific and Caribbean populations, which differentiated into different species.

Dispersal, the more traditional hypothesis, has often been inferred from several kinds of evidence:

1. A fossil record of early distribution in one area, with the taxon appearing in the other only later. In some instances, geology provides evidence of the appearance or disappearance of barriers. For example, fossil armadillos are limited to South America throughout the Tertiary, and are found in North American deposits only from the Pliocene and Pleistocene, after the Isthmus of Panama joined the continents.
2. An area is assumed to have been colonized by dispersal if it has a highly "unbalanced" biota—i.e., if it lacks a great many taxa that it would be expected to have if it had been joined to other areas. This assumption has been applied especially to islands that lack forms such as amphibians and nonflying mammals.
3. Dispersal is often inferred if the species in one area are phylogenetically derived from lineages that have more "primitive" (phylogenetically basal) members in another area, which is inferred to be the source area.

This last point illustrates that, at least in principle, a well-supported phylogeny can provide evidence for dispersal (Figure 8.8A). In Chapter 5, we noted that the history of evolution of a character may be inferred from its distribution in a phylogeny. If one monophyletic subgroup has character state 1, and several sequentially more remotely related subgroups have state 0, it is most parsimonious to infer that the character evolved from state 0 to state 1. The same kind of inference may be made if the character states are areas in which taxa are found.

"Vicariance biogeography" includes several schools of thought (Myers and Giller 1988). Here we discuss the "phylogenetic" or "cladistic vicariance" school. Some of its adherents have argued that the task of biogeography is not to explain individual distributions, but to describe and explain general patterns that have common causes (see, for example, Nelson and Platnick 1981, and several essays in Myers and Giller 1988). Moreover, they argue, only repeated patterns, exhibited by many taxa, provide support for a historical hypothesis; the distribution of any single taxon can be explained by ad hoc "stories" unsupported by independent evidence. This, they say, is not a scientific procedure, because such ad hoc hypotheses cannot be falsified.* Proponents of vicariance biogeography argue that each species has unique, idiosyncratic capabilities and a unique history of dispersal, so that dispersal does not generate repeated patterns. Hence, they say, the hypothesis of dispersal is scientifically untestable. Some authors, however, have pointed out that dispersal, like vicariance, *can* give rise to congruent patterns of distribution (Endler 1983b; Page 1988).

---

* This position rests on the argument, by the philosopher of science Karl Popper, that a hypothesis is not scientific unless one can imagine evidence that would prove the hypothesis false if in fact it is false. For several reasons, most philosophers of science have rejected this argument (cf. Cohen and Laudan 1983; Kitcher 1984).

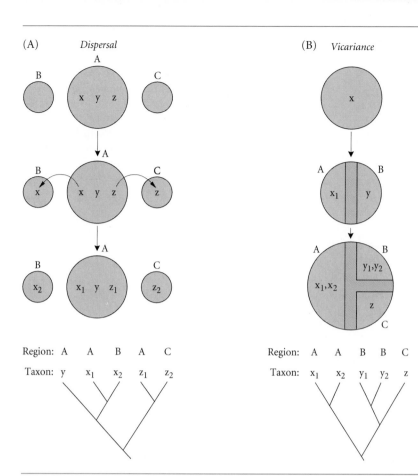

(A) *Dispersal*

(B) *Vicariance*

Region: A A B A C
Taxon: y x_1 x_2 z_1 z_2

Region: A A B B C
Taxon: x_1 x_2 y_1 y_2 z

FIGURE 8.8 Phylogenetic relationships may indicate that species were formed by dispersal or vicariance. (A) Areas B and C are colonized from area A by species x and z. Evolutionary divergence gives rise to pairs of related species, x_1, x_2 and z_1, z_2. The phylogeny shows the relationships among the species in the source area and those that originated by dispersal. (B) A region becomes successively subdivided into regions A, B, and C. An ancestral species, x, evolves into sister species x_1 and y; after areas B and C are formed, the descendant of y in area C evolves into species z. Speciation may also occur within areas, yielding species pairs x_1, x_2 and y_1, y_2. The phylogenetic relationships among species mirror the history of formation of the areas they inhabit.

Vicariance biogeographers argue that *a history of vicariance should generate repeated patterns* because the fragmentation of one area into several disjunct areas will affect all the taxa that were spread across the original area. This hypothesis predicts a COMMON PHYLOGENETIC PATTERN: each taxon that is distributed across three or more disjunct areas should display the same pattern of phylogenetic relationship as other taxa that occupy the areas, a pattern that should mirror the historical sequence in which the original area was broken up (Figure 8.8B). Thus, if land areas B and C were formed by the division of a land mass that was previously separated from area A, the inhabitants of areas B and C should be phylogenetically more closely related to each other than to the inhabitants of area A—and this should be true for every group that occupies the three areas.

Just as the stages in the evolution of a morphological character in a single clade can be inferred by phylogenetic study (see Chapter 5), the history of dispersal or vicariance of a single clade should be detectable. In practice, inferring the biogeographic history of taxa can be challenging, because present-day distributions have often been affected by both vicariance and dispersal, as well as by extinction. Page (1994) describes phylogenetic methods for testing biogeographic hypotheses in the face of some of these complications.

The several examples of historical biogeographic analysis that follow illustrate how both dispersal and vicariance have given rise to the distributions of contemporary organisms. They are based primarily on phylogenetic evidence, supported by geological evidence. Following these cases, we show that some distributions are not as simply explained, and continue to challenge biogeographers.

Examples of Historical Biogeographic Analyses

Central American fishes "Primary" freshwater fishes are those that are thought to be incapable of survival in salt water even briefly; "secondary" freshwater fishes have some tolerance of salt water. Freshwater fishes, especially primary groups, are very useful in historical biogeography because they presumably require land connections (in which they move from one stream to another) to extend their distributions. North America has several families, most of which also occur in Eurasia, that do not occur in South America. There are fossils of these families throughout the Tertiary in North America and Eurasia, but not in South America. South America harbors numerous other families. The North American families extend southward at most to Nicaragua (Figure 8.9). Only a few of the many South American families extend into Central America, where they are not very diverse; a single species from each of two families that are extremely diverse in South America reaches the southernmost United States.

The simplest interpretation of these patterns is that North America was connected to Eurasia, but that the North and South American fish faunas diversified in iso-

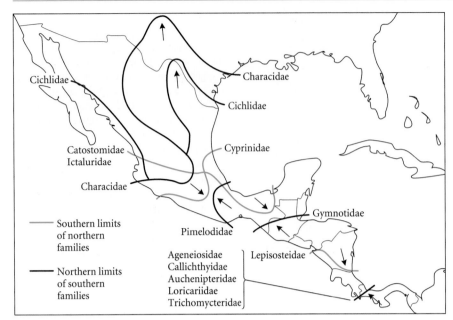

FIGURE 8.9 The distributional limits of some families of freshwater fishes in Central America. Only a few South American families enter Central America, and only two extend as far north as the southern United States. Only a few North American families have spread into Central America, and none has reached South America. The recency of the continuous land bridge in the Central American region explains this pattern. (After Brown and Gibson 1983.)

lation from each other, and could not extend their ranges into the other continent until recently. The geological record supports this interpretation. In the Cretaceous, North and South America were isolated from each other (see Figure 22 in Chapter 7). Between them lay a series of islands, which in the late Cretaceous moved northeastward, forming the Antilles. Present-day Central America formed first as another series of islands in the early Tertiary, but a solid connection between North and South America was not formed until the Isthmus of Panama arose in the Pliocene, a mere 3.5–4 million years ago. Thus geological information supports the inference, from simple patterns of distribution, that taxa that diversified on each continent extended their ranges into Central America only after a barrier broke down in the recent past. Mammals likewise provide evidence of a post-isthmian interchange between North and South America (Stehli and Webb 1985; see Chapter 7).

Tropical American snakes Our next example also concerns dispersal out of South America, but adds phylogenetic evidence for both range expansion and jump dispersal. The snake family Colubridae, which includes most snakes except the boas, cobras, and vipers, is distributed throughout much of the world. One group of colubrid snakes, referred to as xenodontines, is found in tropical America. There are two major clades of xenodontines, which are largely restricted to Central America and South America, respectively (Figure 8.10). Assuming that the protein differences between these two clades conform to a "molecular clock" (see Chapter 5), John Cadle (1985) estimated that the two lineages diverged 40 to 50 Mya. Only 9 of the 23 genera in the Central American clade extend into South America, and only 4 of the 45 genera in the South American clade extend into Central America. Thus the phylogenetic relationships of the genera indicate that there

has been asymmetrical dispersal (range expansion) between the two regions.

All except 1 of the 10 xenodontine genera that occur in the Greater and Lesser Antilles are members of the South American clade. Several of these genera also occur in South America, implying that they must have occupied the Antilles only recently. This pattern implies that they dispersed (across water) to these islands (see also Williams 1989). The alternative (vicariance) hypothesis, that they occupied the Antilles while these land masses were connected to (or very close to) the mainland, is less plausible. Recall from the discussion of freshwater fishes that the Antilles moved in the late Cretaceous (about 65 Mya) into the Caribbean from their former position, which is now occupied by Central America. According to the vicariance hypothesis, one would expect the Antilles to have carried members of the Central American, not the South American, clade of xenodontines. But the protein data suggest that the Central American and South American clades did not become distinct until after the Cretaceous; that is, after the Antilles had moved into the Caribbean. Moreover, the close relationship between Antillean and South American members of the same genera implies that the Antillean snakes diverged from their South American relatives much more recently than the Cretaceous.

Drosophila and other organisms in the Hawaiian Islands The Hawaiian Islands, in the middle of the Pacific Ocean, have been formed as a crustal plate has moved northwestward, like a conveyor belt, over a "hot spot," which has caused the sequential formation of volcanic cones. This process has been going on for tens of millions of years, and a string of submerged volcanoes that once projected above the ocean surface lies to the northwest of the present islands. Of the current islands, Kauai, at the northwestern end of the archipelago, is about 5.1 My old; the southeasternmost island, the "Big Island" of Hawaii, is the youngest, and is less than

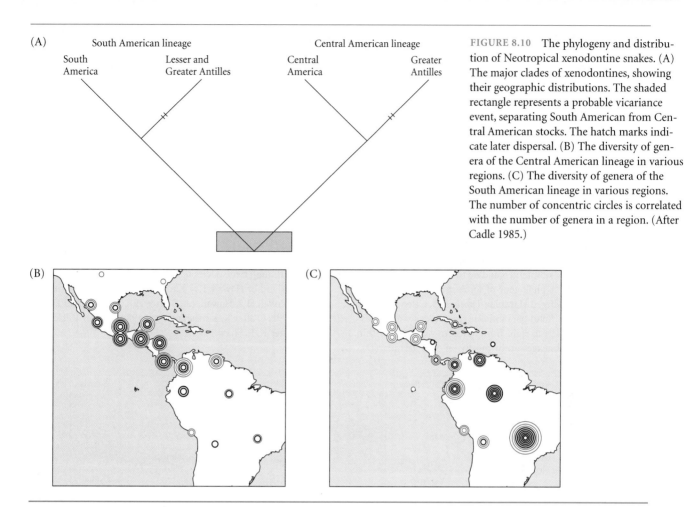

(A) South American lineage

South America / Lesser and Greater Antilles

Central American lineage

Central America / Greater Antilles

(B)

(C)

FIGURE 8.10 The phylogeny and distribution of Neotropical xenodontine snakes. (A) The major clades of xenodontines, showing their geographic distributions. The shaded rectangle represents a probable vicariance event, separating South American from Central American stocks. The hatch marks indicate later dispersal. (B) The diversity of genera of the Central American lineage in various regions. (C) The diversity of genera of the South American lineage in various regions. The number of concentric circles is correlated with the number of genera in a region. (After Cadle 1985.)

500,000 years old (Figure 8.11A,B). Given the geological history of the archipelago, the simplest phylogeny expected of a group of Hawaiian species would be a "comb," in which the most basal lineages occupy the oldest island (Kauai) and the youngest lineages occupy the youngest island (Hawaii). This would occur if species successively dispersed to new islands as they were formed, did not disperse from younger to older islands, and did not suffer extinction.

The Hawaiian archipelago has clearly been colonized by over-water dispersal. Few lineages of plants and animals have colonized it, but many of those that arrived have diversified greatly. For example, most of the land birds belong to one lineage, the Hawaiian honeycreepers, which has adaptively radiated into many niches (see Color Plate 8). The most spectacular radiation is that of the drosophilid flies. These fall into two major clades, comprising more than 800 species, that have diversified greatly in sexual behavior, reproductive morphology, and ecological habits.

As we noted in Chapter 3, the chromosomes of the salivary glands of *Drosophila* larvae have bands that enable a practiced worker to identify small segments of chromosomes (see Figure 14 in Chapter 3). Very occasionally, a chromosome segment becomes inverted 180° by breaking and rejoining at two points. The banding patterns enable researchers to specify the breakpoint of each inversion precisely. Because it is unlikely that exactly the same pair of breaks would occur twice, each chromosome inversion is

probably unique, and so serves as a uniquely derived character that unites all the species that possess it. These inversions provide an unusually firm basis for inferring the phylogeny of species.

In a heroic series of studies, Hampton Carson and his colleagues (Carson 1983) developed a phylogeny for more than 100 species in a single group of these flies (the "picture-winged" group) that consists of several subgroups. Almost every species is confined to a single island, or to a group of islands that were connected during the Pleistocene. A phylogenetic analysis based on chromosome inversions (Figure 8.11C) shows that within most of the species subgroups, basal lineages are found primarily on the older islands (Kauai and Oahu), and derived lineages are found mostly on the younger islands (Maui and Hawaii) (Kaneshiro et al. 1995). Using mitochondrial DNA, Rob DeSalle (1995) found the same pattern (Figure 8.11D). In several other groups of Hawaiian organisms, such as crickets, basal lineages likewise occupy older islands than younger lineages, although there is also evidence of back-colonization from younger to older islands (Figure 8.11E).

In a more comprehensive molecular analysis of the Hawaiian *Drosophila*, DeSalle (1995) concluded that the three major lineages form a monophyletic group that is most closely related to a cosmopolitan subgenus that is known from 30-million-year-old fossils. The remarkable inference is that the Hawaiian species are all descended from an ancestor that colonized older, now submerged, Hawaiian islands long before the oldest current island, Kauai, was formed 5.1 Mya.

Vicariance biogeography of midges In Chapter 7, we saw that the world's continents became aggregated in the late Permian into a single land mass, Pangaea, which remained more or less intact until the early Jurassic (see Figure 22 in Chapter 7). After Pangaea became divided into the northern and southern land masses of Laurasia and Gondwanaland, respectively, a land mass consisting of Australia and Antarctica (which then had a temperate climate) separated from South America in the early Cretaceous, and Africa separated from South America later in the Cretaceous.

FIGURE 8.11 The Hawaiian Islands and the phylogeny of some Hawaiian organisms. (A) The Hawaiian archipelago at present. The Big Island, Hawaii, lies just to the northwest of the "hot spot," shown by the broken circle, where islands have been successively formed. (B) The archipelago 5 million years ago, when Kauai was emerging over the hot spot, and several islands existed that are now submerged and lie to the northwest of the present islands. (C) A phylogeny of some species in several subgroups of picture-winged Hawaiian *Drosophila*, based on chromosome inversions. Solid bars indicate the presence of species on islands, arrayed left to right from oldest to youngest. Within most groups, basal lineages occur mostly on older islands. (D) An area cladogram, showing the distribution of five species in the *antopocerus* group of *Drosophila*, based on mitochondrial DNA. Compared with the outgroup (OG), successively younger clades are found on the successively younger islands Molokai, Maui, and Hawaii. (E) A similar diagram for species of crickets (*Laupala*). Successively younger groups are found on younger islands, although two species have colonized Maui from the younger island of Hawaii. K = Kauai, O = Oahu, M = Maui, Mo = Molokai, H = Hawaii. (A, B after H. L. Carson and D. A. Clague 1995; C after K. Kaneshiro et al. 1995; D after R. DeSalle 1995; E after K. L. Shaw 1995)

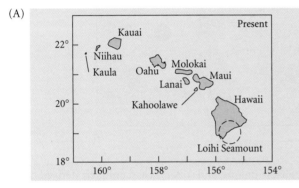

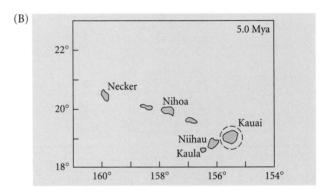

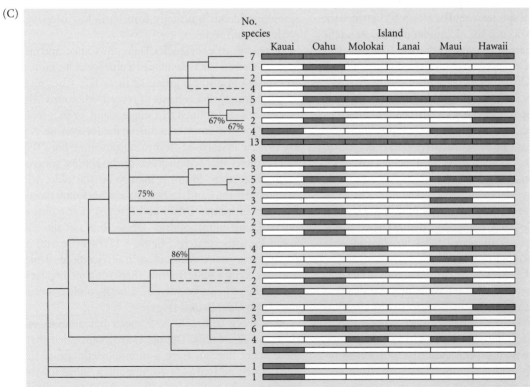

One of the first examples of vicariance biogeography was proposed by Lars Brundin (1965; also 1988), who studied the phylogeny of a group of midges (Chironomidae), flies whose larvae are found in cool streams. Almost all the members of two closely related subfamilies are restricted to the southern continents. If one examines Brundin's phylogeny (Figure 8.12) in detail, one observes numerous examples of pairs of sister lineages that are split between South America and Australia or New Zealand (e.g., numbers 1–3 vs. 6, 36 vs. 37, 44 vs. 45). Brundin interprets these patterns as divergence from common ancestors that were isolated when South America and Australia became separated. The several genera of Chironomidae in Laurasia (the northern continents), Brundin suggested, may represent the early separation of Laurasia from the southern continents.

Some difficult problems

Not all distributions are so easily explained. To illustrate an unsolved puzzle, we turn again to primary freshwater fishes of Africa and South America (Lundberg 1993).

Lungfishes arose in the Devonian, and diversified in many regions. Today, however, only three genera of lungfishes survive, all strictly limited to fresh water: the Australian *Neoceratodus*, in the family Ceratodontidae, and the African *Protopterus* and South American *Lepidosiren*, both in the family Lepidosirenidae. The two lepidosirenids share a more recent common ancestor with each other than with

Neoceratodus (Figure 8.13). Their disjunct distribution suggests that the three genera became isolated by the breakup of Gondwanaland (i.e., vicariance). Their story, however, may be more complex. Fossils of each *genus* are known only from the continent it now inhabits: *Neoceratodus* from the early Cretaceous and both of the other genera from the late Cretaceous. However, the *family* Ceratodontidae occurs in Triassic deposits on all continents except South America and Antarctica, and the Lepidosirenidae are known from the late Paleozoic of Europe and North America (Carroll 1988). Thus the ancestral stock of both families may have ranged widely over Pangaea. Perhaps, then, the three living genera did not differentiate because they became isolated by the fragmentation of Gondwanaland; rather, each genus may be located where it is because that is the only place in which it did not become extinct. Certainly the divergence of *Neoceratodus* from the lepidosirenids was not caused by continental drift, because these two lineages were already distinct before the continents separated.

The characiform fishes are a huge group of strictly freshwater forms that includes piranhas and many popular aquarium fishes. They are diverse in Africa and especially in South America, are not known from fossils elsewhere (except a few in Europe that are not relevant), and are commonly cited as an example of vicariance caused by continental drift. However, although the phylogeny of the characiforms is not fully understood, it seems not to con-

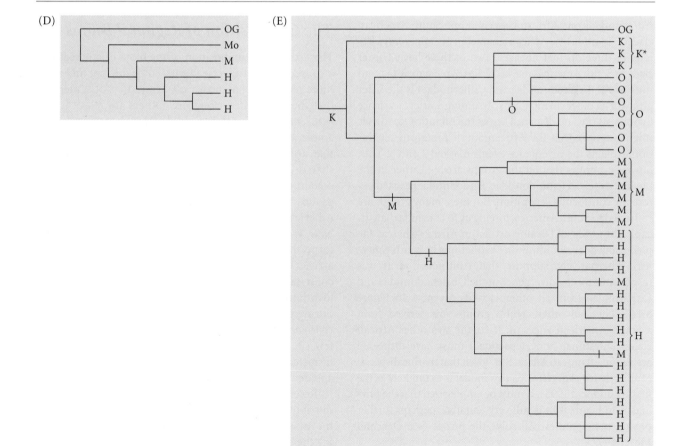

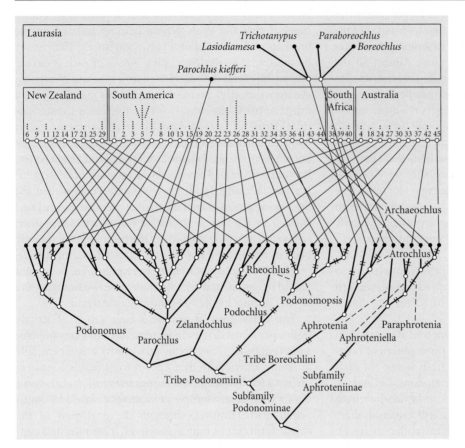

FIGURE 8.12 The phylogeny of a lineage of chironomid midges, based on an analysis of morphological characters by Brundin. Many sister groups occupy different land masses in the Southern Hemisphere. Especially notable are the many pairs of taxa shared by South America and Australia, which during the Mesozoic were connected by Antarctica, where the climate was temperate. Brundin interpreted these distributions of closely related taxa as a consequence of continental drift. In the boxes denoting land masses, the numbers refer to species groups, and the dots indicate the number of species in each group. (After Brundin 1965.)

sist simply of one African clade and one South American clade (Figure 8.14). African characiforms apparently belong to three distinct lineages. Two of these have a sister clade in South America (marked 2 and 3 in Figure 8.14), and the third (Distichodontidae and Citharinidae) is the sister group of all other characiforms. Assuming a strict vicariance hypothesis (i.e., no dispersal across the Atlantic Ocean), we might suppose that the African/South American separation explains the divergence events marked 2 and 3. This, however, implies that all the groups arising earlier in the phylogeny (e.g., Hemiodontidae, Parodontidae) existed before the continental separation. But these many groups are restricted to South America, forcing us to conclude that, inexplicably, they all became extinct in Africa (whereas the phylogeny does not imply any extinctions in South America). Alternatively, suppose that continental drift was responsible for the divergence of the Distichodontidae and Citharinidae from all other cichlids (event 1 in Figure 8.14). Then all other cichlid groups are derived from a South American ancestor. If so, the two other African clades (Alestiinae and Hepsetidae) must have dispersed across the Atlantic to Africa. But "given that marine dispersal is unlikely for characiforms this model is as unlikely as the first" (Lundberg 1993). It must be emphasized that the phylogeny in Figure 8.14 is still very tentative; perhaps further phylogenetic studies will solve the puzzle (see Ortí and Meyer 1997).

The Composition of Regional Biotas

The taxonomic composition of the biota of any region is a consequence of diverse events, some ancient and some more recent. Certain taxa are ALLOCHTHONOUS, meaning that they originated elsewhere. Others are AUTOCHTHONOUS, meaning that they evolved within the region. For example, the biota of South America has (1) some elements that are remnants of the Gondwanaland biota and are shared with other southern continents (e.g., lungfishes, certain midges, some plant families); (2) autochthonous groups that became distinct and diversified in South America after it became isolated by continental drift (e.g., New World monkeys, caviomorph rodents); (3) some forms that diversified from allochthonous progenitors that arrived from North America in the mid-Tertiary [e.g., raccoon family, cricetid (ratlike) rodents]; (4) some allochthonous species that entered from North America during the Pleistocene (e.g., the mountain lion, *Panthera concolor*, which also occurs in North America); and (5) a few species that have colonized South America within historical time (e.g., the cattle egret, *Bubulcus ibis*, which apparently arrived from Africa in the 1930s: see Chapter 4).

Recent events have left their stamp on the diversity and distributions of species. Almost all the large mammals of the Americas became extinct during the Pleistocene, including the diverse endemic orders of South American hoofed

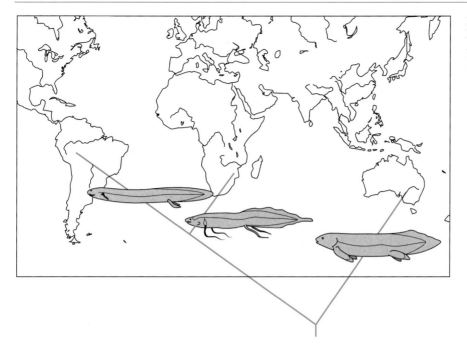

FIGURE 8.13 The phylogenetic relationships among the three genera of living lungfishes. (From *Cladistics* 10(3), 1994 [cover illustration]).

mammals (see Chapter 7). Some have postulated that they were exterminated by human hunting (Martin and Klein 1984). The succession of Pleistocene glacial and interglacial events both extinguished species and altered distributions (see Figure 8.1). In North America, for example, many taxa apparently survived the glacial periods in southeastern and southwestern refugia (refuge areas). Some of these differentiated into different subspecies or species, such as the eastern and western diamondback rattlesnakes (Figure 8.15).

The composition of a local assemblage of species, such as the plants of an Appalachian forest, is likely to be the consequence of both historical and contemporary ecological, factors. As the paucity of freshwater fishes in Central America shows, a locality may lack some species simply because they have not been able to get there yet. Others may have occurred there in the past, but may now be locally extinct. Still others may be absent because of unsuitable habitat, exclusion by predators or competitors, or the absence of spe-

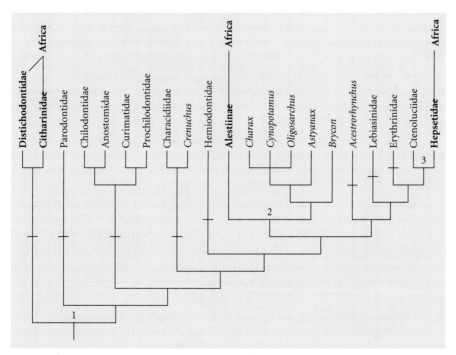

FIGURE 8.14 A provisional phylogeny of the families of characiform fishes. All are South American except the four families indicated as African. Separation of African from South American clades is marked by 1, 2, and 3. A single disjunction of an African from a South American stock cannot account for all three African/South American pairs of clades. (After Lundberg 1993.)

FIGURE 8.15 The western and eastern diamondback rattlesnakes, *Crotalus atrox* and *C. adamanteus*, closely related species thought to have differentiated in southwestern and southeastern refugia during the Pleistocene. (Photographs from Klauber 1972; map after Conant 1958.)

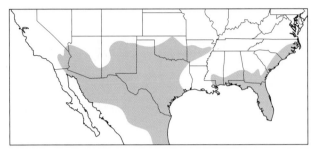

cies on which they depend. Thus historical biogeography becomes inseparable from ecological biogeography.

Ecological Approaches to Biogeography

Whereas systematists often look first to evolutionary history in order to understand the reasons for a taxon's distribution, ecologists tend to look to factors operating now or in the very recent past. To some extent, this difference in viewpoint arises from differences in the particular questions posed, and from differences in the spatial scale of the distributions under study. For example, phylogenetic history is likely to explain why cacti (Cactaceae) are native only to the Americas, but to explain why the saguaro cactus (*Carnegiea gigantea*) is restricted to certain parts of the Sonoran Desert, it is necessary to look toward ecological factors, such as the species' tolerance for rainfall and temperature, or perhaps the effects of competitors, herbivores, or pathogens. Such factors may explain, say, the latitudinal limits of the species' distribution, and within these, the variety of habitats it occupies. The ecologist is likely to assume that except for limits set by impassable barriers (such as oceans), the species has been able to spread as far as its physiological tolerances (for temperature, pathogens, etc.) allow. This is tantamount to assuming that the distribution of the species has reached an *equilibrium*: that it is not in the process of changing.

It would be more exact, in this instance, to say that ECOLOGICAL EQUILIBRIUM is assumed: given the properties of the species today, and the current distribution of ecological factors such as climate, the species' range is assumed to be constant. Over a longer time scale, we can expect its distribution to change, because climates change, and because the species' physiological tolerances may evolve. On an evolutionary or geological time scale, we cannot assume equilibrium. Ecologists are likely to justify the idea of ecological equilibrium by assuming that the ecological processes that affect a species' distribution—such as its rate of dispersal into suitable unoccupied areas—are much more rapid than either changes in the environment or genetic changes in the species' physiological properties. This assumption may not always be warranted.

In contrast, a NONEQUILIBRIUM HYPOTHESIS holds that the current state of any system is temporary, even within the existing framework (e.g., of climate). There might exist a theoretical equilibrium, but the system hasn't gotten there yet, perhaps because it is recovering from a perturbation. The change in the system, however, might be too slow for us to observe directly.

As we will see, both equilibrium and nonequilibrium hypotheses best explain particular instances. For some biogeographic questions, moreover, a combination of historical and ecological factors may provide answers.

The Theory of Island Biogeography

What determines the number of species on an island? This question provides a simple example of the difference between equilibrium and nonequilibrium hypotheses. An early, simple answer, prompted by the observation that an island usually has fewer species than a patch of the same size on the nearest large land mass, is that because of the low probability that each continental species will colonize the island, most of the continental species haven't gotten there—yet. The "yet" implies that given enough time, the island will achieve the continental number of species; it has not yet reached its equilibrium. Robert MacArthur and Edward O. Wilson (1967) pointed out that this explanation ignores extinction. The number of species on an island is increased by new colonizations, but decreased by extinctions. As long as the rate of new colonizations exceeds the rate of extinctions, the number of species grows, but when the rates become equal, the number no longer changes; it is at equilibrium. MacArthur and Wilson suggested that smaller islands will have greater extinction rates because smaller populations are more likely to suffer extinction. This theory (Figure 8.16A) appears to explain the consistent relationship that is often observed between island area and the number of indigenous species (Figure 8.16B). This regularity, or consistent pattern, in itself suggests to many ecologists that the factors affecting species number are strong enough to have brought the system near equilibrium, because the state of a system that is far from equilibrium is often less predictable.

Not all islands, however, have the number of species that would be predicted from the overall relationship between

FIGURE 8.16 (A) The theory of equilibrium species diversity on islands. The rates of immigration of new species and extinction of resident species are plotted against the number of species on an island at a given time. The equilibrium number of species, *S*, is found by dropping a line from the intersection of the immigration and extinction rate curves (at which immigration balances extinction). Differences in rates of immigration and extinction, which may depend respectively on distance from a source of colonists and on island size, determine different equilibria. (B) The number of species of amphibians and reptiles on West Indian islands, plotted against island area on a log-log plot. (After MacArthur and Wilson 1967.)

FIGURE 8.17 Relationship between area and number of species of high-altitude mammals on isolated mountaintop "islands" in the western United States (solid circles). The slope of the relationship is steeper than for samples of equivalent area in a source region, the Sierra Nevada (open triangles), suggesting that the "islands" have less than the expected equilibrium number of species. (After Brown 1971a.)

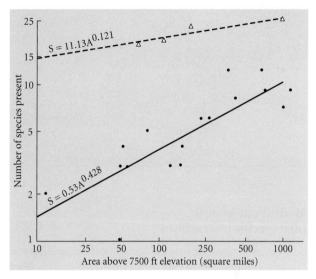

(A)

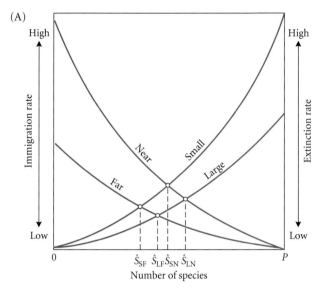

(B)

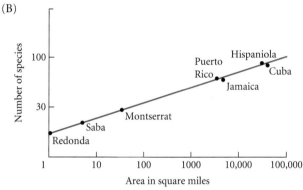

the number of species depends on the rates of extinction and of ORIGINATION (by speciation) rather than colonization. Equilibrium is reached if the speciation rate and the extinction rate are equal (Figure 8.18).

FIGURE 8.18 In any area, such as a continent, both the rate of origination of new species and the rate of extinction (per existing species) may increase with an increase in the number of species. An equilibrium number of species would be due to equality of these rates. The equilibrium number (*S*) is higher if the extinction rate is lower (compare e_1 and e_2) or if the speciation rate is higher. (After Rosenzweig 1975.)

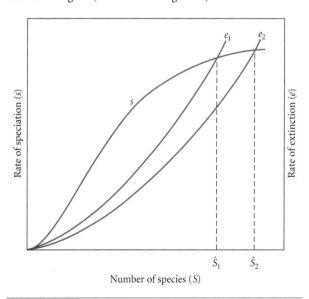

species number and area. Mountaintops, for example, can be considered islands for species that do not inhabit lower elevations. James Brown (1971a) found that the number of species of mammals on certain mountain peaks in western North America is considerably lower than expected from the species equilibrium theory (Figure 8.17). He proposed that these peaks, which were uninhabitable during Pleistocene glaciations, have not been colonized by certain species that should be able to survive there. That is, they haven't gotten there—yet. The fauna of these peaks, he proposed, is not at equilibrium.

We shall have reason to note below that MacArthur and Wilson's simple model can be modified to describe the number of species in a taxon on a continent. In this case,

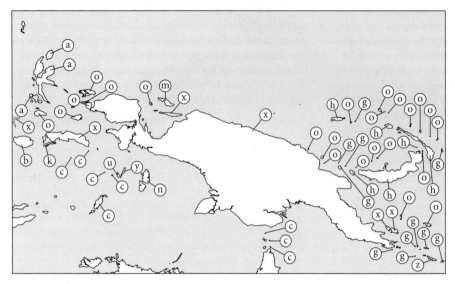

FIGURE 8.19 A "checkerboard distribution" of twelve species of white-eyes (*Zosterops*) on islands in the New Guinea region. Almost no islands have more than one species, even though many of the species, each denoted by a letter, are widely distributed. (After Diamond 1975.)

More about Islands: Interspecific Interactions

Small islands may have fewer species simply because chance events are more likely to extinguish small populations than large populations. Interactions among species—specifically, predation and competition—can lower population sizes and thus contribute to extinction. For example, a smaller area offers fewer microhabitats and fewer "pockets" in which a prey species can escape predators. There are, indeed, numerous examples in which introduced predators such as rats, cats, and mongooses have extinguished island species of birds, snails, and other creatures. A particularly dramatic and sad story is the extermination of the diverse land snails (genus *Partula*) of the island of Moorea in the South Pacific. These snails were extirpated by a predatory snail (*Euglandina rosea*) that was introduced to control another

introduced land snail that had become a pest. *Euglandina* brought to an end an almost century-long series of studies on the evolution of *Partula* (Murray et al. 1988).

Interspecific competition can also cause extinction, or prevent colonizing species from becoming established. If the variety of microhabitats and other resources is lower on a small island, fewer species can be expected to coexist; and if the total amount of resources for a group of competing species is less on a small than on a large island, the species' small population sizes will enhance the likelihood of extinction. The distributions of species among real islands or "habitat islands" often fit this hypothesis. Figures 8.19 and 8.20 provide two examples from the birds of New Guinea and nearby islands. No two of the twelve species of white-eyes (small, warblerlike birds) occur on the same island, even though three of them (labeled a, c, and g) are dis-

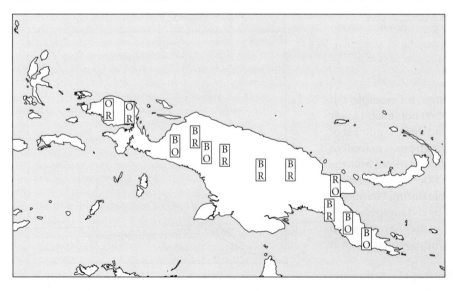

FIGURE 8.20 Another "checkerboard" distribution." Among the various mountain ranges in New Guinea, three species of honeyeaters (*Melidectes*), denoted by letters O, R, and B, are distributed in pairs. Each pair has mutually exclusive altitudinal ranges, as shown by the stacked letters. (After Diamond 1975.)

tributed over many islands. If these species were randomly distributed, they would be expected to coexist on at least a few islands, yet they have a mutually exclusive "checkerboard" pattern. Likewise, three species of nectar-feeding honeyeaters occur in the mountains of New Guinea, but each mountain range has only two species (with mutually exclusive altitudinal distributions). Which species is missing from a mountain range appears to be a matter of chance.

Effects of Competition on Distributions

In some instances it is rather clear that species can coexist because they do not use entirely the same resources, and that other species are missing because they would be subjected to excessive competition. Peter Grant and his colleagues (Grant 1986) have provided extensive evidence of such a pattern from their studies of Darwin's finches in the Galápagos Islands. Each of the 17 islands has from two to five of the six species of ground finches (*Geospiza*). Although each species is widely distributed through the islands, the species do not occur together at random; for example, we would expect that, by chance, six islands would each have five species of finches, but only one island does. Grant and his colleagues have shown that the bird populations are often limited by food supply, and that the species that are most similar in beak size have the most similar diets, and so are the most likely to compete (Figure 8.21). Species that coexist on the same island almost always differ in beak size to a greater extent than expected if sets of two, three, or four species were composed at random.

Based on extensive ecological data, Dolph Schluter and Peter Grant (1984) developed a predictive model of what beak size to expect finches to have on each of 15 islands. First, they collected data on the kinds of seeds eaten by each type of finch: finches with larger bills feed on larger, harder seeds. Second, they measured the abundance of each type of seed on each island. This enabled them to calculate how much food is available on each island for a population

of finches with a given mean beak size. Third, from bird censuses, they determined how dense a population of finches can be maintained by different levels of food. Schluter and Grant used all this information to ask, "If we had a single hypothetical finch species, with a beak depth of *x* millimeters, how dense a population could this island (say, Daphne) support, given the known abundance of the various seed types on this island?" They made such a calculation of expected population density for each hypothetical (but realistic) beak size (*x*).

Figure 8.22 shows their results. For example, on the small island of Daphne, the most abundant kinds of seeds are those that could support a hypothetical population of finches with a (log) beak depth of slightly over 2.2 mm at a greater density than if the finch had any other beak size. (In other words, this is the "optimal" beak size for maximizing the density of a finch population.) The small square in the panel shows that the single finch species on this island, *Geospiza fortis*, has almost exactly the theoretically optimal beak size!

On some islands (e.g., Marchena), several plant species with quite different seeds are so abundant that there are several peaks—i.e., a finch would reach high density if it had a beak of about 1.9 *or* 2.3 *or* 2.9 mm. On Marchena, and on most other such islands, several finch species coexist, with approximately the "expected" beak sizes. Moreover, in *no* case does an island have two species with the same beak size, even though different species can have nearly identical beaks if they occupy different islands (e.g., *G. difficilis* on Wolf and *G. fuliginosa* on Gardner and elsewhere).

In some cases, (e.g., Fernandina), the expected population density is quite high for finches with any of a range of beak sizes. Such islands have several finch species, with beak sizes rather widely spaced. Using computer simulations that would take too much space to describe here, Schluter and Grant found that in such cases, the beak sizes of coexisting species are more different than would be expected if the finches' beaks were adapted only to the availability of certain sizes of seeds. Rather, they are spaced apart as would be expected if competition among species depleted the food supply, making it advantageous to feed on the seed types least utilized by competing species.

The thrust of this example is that in at least some instances, (1) the distributions of species are affected by those of competing species, and (2) the numbers and distributions of species in a locality are predictable from ecological information. Such predictability has led some ecologists to think that the species composition of communities is often near equilibrium.

Among the Galápagos ground finches, species with similar beak dimensions and diets do not coexist, and coexisting species differ more than would be expected at random. This pattern could be caused by one or both of two processes: First, a species that is excessively similar to another cannot invade an island, or is driven to extinction by competition. This is a purely ecological explanation of the

FIGURE 8.21 The greater the difference in beak depth between the members of various pairs of coexisting species of Galápagos finches, the lower the overlap in their diet. (After Grant 1986.)

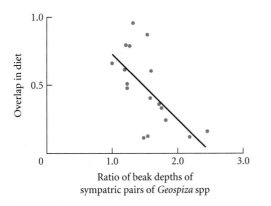

"assembly" of a community. Second, coexisting species evolve to become more different from each other, thereby reducing competition. The following section presents further examples suggesting that evolutionary responses to competition may affect the structure of communities. We will consider this topic in more detail in Chapter 18, under the heading of COEVOLUTION.

Community Convergence

CONVERGENT EVOLUTION (a form of HOMOPLASY; see Chapter 5) is well known. For example, desert plants such as cacti and Euphorbiaceae have independently evolved similar morphological features in many parts of the world (see Figure 3 in Chapter 4). Hummingbirds in the New World, sunbirds (Nectariniidae) in Africa and Asia, honeyeaters in the Australian region, and honeycreepers (Drepanididae) in Hawaii have independently evolved features suitable for feeding on nectar, such as a long, slender bill (see Figure 24 in Chapter 5; Color Plate 8). The question arises, are these individual instances part of a larger pattern of convergence of whole communities? If two regions present a similar array of habitats and resources, will species evolve to utilize and partition them in the same way? Are the same "niches" (see Chapter 4) predictably occupied by phylogenetically independent groups of organisms?

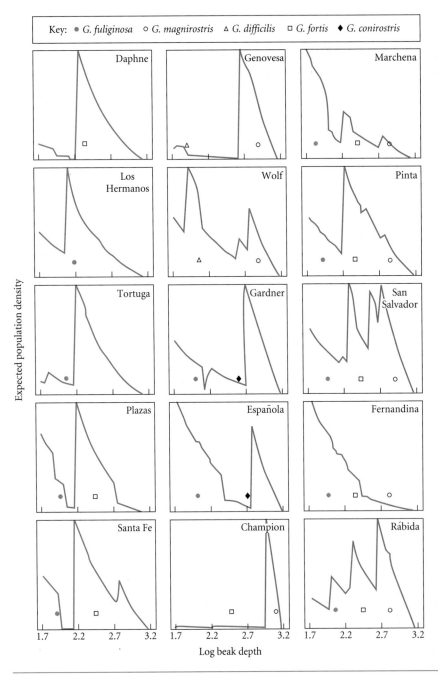

Key: ● *G. fuliginosa* ○ *G. magnirostris* △ *G. difficilis* □ *G. fortis* ◆ *G. conirostris*

Daphne · Genovesa · Marchena · Los Hermanos · Wolf · Pinta · Tortuga · Gardner · San Salvador · Plazas · Española · Fernandina · Santa Fe · Champion · Rábida

Expected population density

Log beak depth

FIGURE 8.22 A plot, for each of 15 islands in the Galápagos archipelago, of the relative population density of hypothetical ground finches (*Geospiza*) with different beak depths that the island could support. These plots were calculated from the abundance of different kinds of seeds, in relation to the observed capacity of finches with different beak depths to utilize such seeds. The symbols (see key) show the mean beak depths of the *Geospiza* species that actually inhabit each island. (After Schluter and Grant 1984.)

Placentals   Marsupials

Ocelot
(*Felis*)

Native "cat"
(*Dasyurus*)

Anteater
(*Myrmecophaga*)

"Anteater"
(*Myrmecobius*)

Mouse
(*Mus*)

"Mouse"
(*Dasycercus*)

Flying squirrel
(*Glaucomys*)

Flying
phalanger
(*Petaurus*)

Wolf
(*Canis*)

Tasmanian "wolf"
(*Thylacinus*)

FIGURE 8.23 Convergent evolution of placental (left) and Australian marsupial (right) mammals from other continents. Each pair is similar in form and in ecological habits. Such examples of parallel evolutionary radiations have led evolutionary ecologists to postulate that ecological communities in similar environments should have similar structures at equilibrium. (From Luria et al. 1981.)

At least superficially, such convergence seems evident in cases such as the radiation of Australian marsupials, among which many analogies to placental mammals can be drawn (Figure 8.23). With careful study, even more exact, though subtle, parallels can sometimes be discerned, as illustrated by anoline lizards.

The anoles are a species-rich group of insectivorous, mostly arboreal Neotropical lizards (Color Plate 9). The more than 70 species in the West Indies range in size (of adult males) from 43 to 180 millimeters (snout to vent). The several major phylogenetic lineages are placed in different genera by some authors, but others refer to all of them as *Anolis*. The systematics and ecology of West Indian anoles have been extensively studied by Ernest Williams and many of his students and colleagues (see, e.g., Williams 1972, 1983; Schoener 1970; Roughgarden 1995; Losos 1990 a,b; 1992).

In the West Indies, anoles compete for food, and competition has influenced the structure of anole communities. This structure has different forms in different groups of islands. In the Lesser Antilles, each of the small islands has either a single (solitary) species or two species. Solitary species are generally moderate in size, whereas pairs of small and large species coexist on two-species islands. These species can coexist because they take insect prey of different sizes and also differ in microhabitat. The small species of the various islands are a monophyletic group, and so are the large species. Thus, it appears that coexisting pairs of species are due to the assembly of species from the small-sized and the large-sized clades (Losos 1992).

The large islands of the Greater Antilles harbor greater numbers of species (up to 37 in Cuba). These occupy certain "structural habitat categories," such as tree crown,

The Geography of Evolution   221

twig, trunk, and grass and bushes (Figure 8.24). Each island has different species, yet most of these categories are filled, although several habitats are occupied only on the larger islands. When several species on an island require the same habitat category (e.g., grass-bush anoles in Hispaniola), they usually occupy distinct habitats (moist versus arid) or are allopatric, restricted to different parts of the island.

The occupants of different structural habitats have consistent, characteristic morphologies; Williams has termed these "ecomorphs." For example, the grass-bush ecomorph is small and has a slender body, a long head, and enlarged mid-dorsal scales; the trunk ecomorph is also small, with a short body and head and uniform scales. These features affect locomotory abilities (size, body proportions) and temperature regulation (scales) (Losos 1990 a,b). The important point is that these ecomorphs have evolved repeatedly: there are numerous cases of convergent evolution within the genus. The anoles of Jamaica, for example, are a monophyletic group that has diversified into several ecomorphs within the island; conversely, species representing a given ecomorph on different islands have evolved repeatedly in different phylogenetic lineages (see also Chapter 18).

The anoles and other striking examples of "community convergence" suggest that there are predictable ways in which species can divide resources, and that species will evolve to take advantage of the resources that are available. But this conclusion does not always hold. For example, the structure of communities of desert lizards is very different in North American, African, and Australian deserts: the number of species is about twice as great in Australian deserts as on the other continents, but the species' overlap in resource use is the lowest (Pianka 1986). Several studies of other organisms have also failed to find evidence of community convergence (Orians and Paine 1983; Schoener 1988). And some niches appear conspicuously empty: there are blood-drinking vampire bats in the tropics of the New World, but not the Old World, and sea snakes in the Indian and Pacific, but not the Atlantic, oceans. The cattle egret (*Bubulcus ibis*, Figure 8.25), which throughout the Old World feeds on insects stirred up by grazing ungulates, now thrives in North and South America, after having colonized the Americas within the past century. It seems not to have had a negative effect on native species of birds, and fills what apparently had been an empty niche. Thus new species do not necessarily evolve just because ecological opportunity awaits them: nature does not so abhor a vacuum that it creates new species just because they fit in. Consequently, although instances of community convergence suggest that certain habitats may be "saturated" with certain kinds of organisms (i.e., are at equilibrium), not all communities are necessarily saturated with species (i.e., they may not be at equilibrium). This theme is explored further in Chapter 25.

FIGURE 8.24 Diagrams of the habitats occupied by anoles in the lowlands of (A) Hispaniola and (B) Puerto Rico. The name and maximum size (in mm) of each species is printed next to its typical microhabitat. Anoles in similar microhabitats on the two islands have convergent morphological characteristics. (After Williams 1983.)

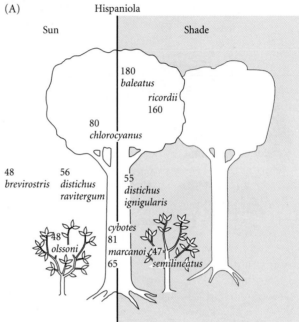

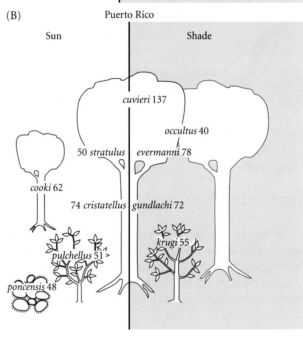

Gradients in Species Diversity

The preceding sections bear on one of the most extensively discussed issues in biogeography and community ecology: gradients in species diversity (Brown and Gibson 1983; Brown 1988; Huston 1994). Among the more conspicuous gradients, most terrestrial taxa increase in diversity with decreasing altitude and increasing moisture. Relatively rare environments that we think of as harsh, such as caves and thermal springs, have relatively few species.

By far the most conspicuous of these patterns is the LATITUDINAL GRADIENT, and it is this gradient that we shall consider here. In both marine and terrestrial environments, many major taxa (e.g., birds, amphibians, insects, plants, gastropods) are far more diverse at tropical than at higher latitudes. Within North America, for example, the number of breeding land bird species declines steadily from 600 at 8° north latitude to 50 at 66° N—even though the land mass at 8° N (in Central America) is far smaller than in the Arctic. Similar patterns hold for most lower-level taxa, such as families of insects.

Many explanations for this pattern have been proposed. All invoke evolution, but in rather different ways. "Equilibrium" explanations, relying heavily on ecological theory, assume that species rapidly evolve adaptations to climate, to one another, and to available resources, so that communities at all latitudes are nearly saturated with species and are near equilibrium. Authors inclined toward this view are impressed by phenomena such as checkerboard distributions, niche segregation of potential competitors, and community convergence—the topics of the preceding sections—which suggest that competition and evolution have produced an optimal match between species and resources. Other authors, taking a "nonequilibrium" stance, explain latitudinal diversity gradients in terms of history: species have not yet evolved to stock the higher latitudes as fully as the tropics. For the sake of brevity, we shall treat this high-ly complex topic only by listing a few of the hypotheses that have been suggested, with one or two comments on each.

The first of several more ecological ("equilibrium") hypotheses we shall consider involves *degree of specialization*. Because of the supposedly more constant climate in the tropics, it is possible for species to evolve to specialize on particular foods or microhabitats to a greater degree than in nontropical areas, and their special resources are less likely to be eliminated by environmental changes. Hence a given set of resources (e.g., types of insects for insectivorous animals) can be divided more finely (Figure 8.26). The problems with this idea are that tropical environments are not more constant than climates elsewhere, that it is hard to measure a species' degree of resource specialization, and that the few available data do not provide evidence that tropical species are more specialized in resource use (Beaver 1979).

Greater primary productivity and a greater variety of resources have both been proposed to enable more species to coexist. (Primary productivity is the rate of growth of plant biomass by fixation of carbon.) The problems with the productivity hypothesis (Ricklefs and Schluter 1993b) are that tropical wet forests, despite their year-round growth, do not have higher productivity per day of growing season than high-latitude forests; that highly productive communities generally have lower species diversity than moderately productive communities; and that some ecological theories predict that increasing productivity should *decrease* the number of coexisting species by giving some of them a competitive edge over others (Huston 1994). A greater variety of resources (see Figure 8.26) may well help to explain the diversity of many groups of tropical animals, since these depend, directly or indirectly, on the great diversity of tropical plants, but this hypothesis seems unable to explain why the number of plant species in tropical forests is so high (or, to shift our emphasis, why temperate zone forests have relatively few species).

Another hypothesis involves the *physiological tolerances* of organisms. The range of temperatures over the course of a year increases with latitude. We might therefore expect tropical species to be adapted to a narrower range of temperatures than species at higher latitudes. A plausible expectation, then, is that as we move away from the equator, individual species will have broader distributions with respect to both latitude and altitude, because they will be less sensitive to average differences in temperature among different locations. Data from several groups of organisms agree with both of these predictions (Huey 1978; Stevens 1989).

Tropical ecologist Daniel Janzen (1967) proposed a hypothesis for the diversity gradient based on the idea of physiological sensitivity. As in MacArthur and Wilson's theory of island biogeography, the equilibrium number of species in a region depends on the relative rates of origination and extinction of species (see Figure 8.18). All else being equal, a region will have more species if its inhabitants speciate more rapidly. Janzen assumed that most species

FIGURE 8.25 A cattle egret accompanying a cow in Alabama. This heron feeds on insects stirred up by grazing ungulates in both the Old World and the New World, which it invaded about 60 years ago. Photograph © A. Morris/Visuals Unlimited.

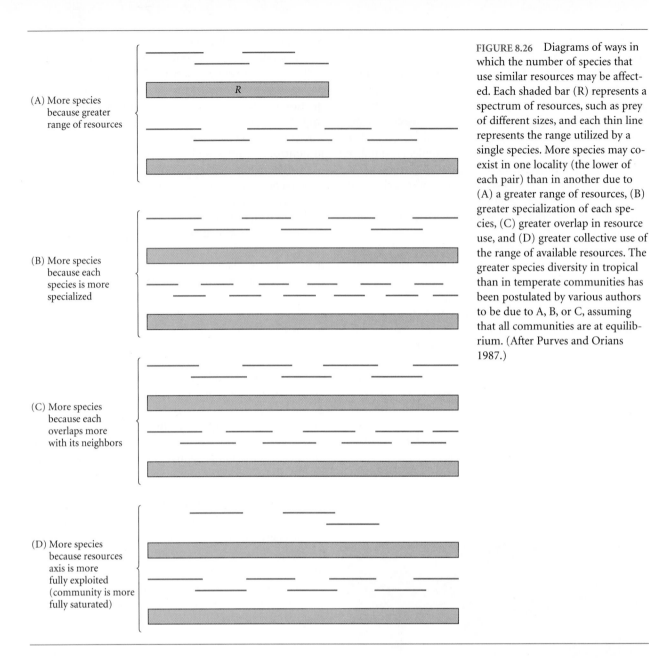

(A) More species because greater range of resources

R

(B) More species because each species is more specialized

(C) More species because each overlaps more with its neighbors

(D) More species because resources axis is more fully exploited (community is more fully saturated)

FIGURE 8.26 Diagrams of ways in which the number of species that use similar resources may be affected. Each shaded bar (R) represents a spectrum of resources, such as prey of different sizes, and each thin line represents the range utilized by a single species. More species may co-exist in one locality (the lower of each pair) than in another due to (A) a greater range of resources, (B) greater specialization of each species, (C) greater overlap in resource use, and (D) greater collective use of the range of available resources. The greater species diversity in tropical than in temperate communities has been postulated by various authors to be due to A, B, or C, assuming that all communities are at equilibrium. (After Purves and Orians 1987.)

originate when topographic barriers reduce genetic exchange among populations, causing them to diverge. If tropical organisms are less tolerant of temperature variation than temperate zone species, low temperatures at high altitudes will present a more effective barrier to dispersal (and hence a greater barrier to gene exchange) for lowland tropical species than for lowland high-latitude species. From the organism's point of view, "mountains are higher in the tropics." The narrow altitudinal ranges of tropical species are consistent with Janzen's proposal.

In contrast to these equilibrium hypotheses, some authors have looked to evolutionary history to explain latitudinal diversity gradients. It may be useful, in this context, to shift our viewpoint. Instead of asking why tropical communities have so many species, as if this were extraordinary, perhaps we should consider the relative paucity of species at higher latitudes to be the peculiarity that requires explanation. From this point of view, we may entertain the pos-

sibility that the number of high-latitude species is still increasing toward an equilibrium that has not yet been attained. At least two such hypotheses may be considered.

First, some authors have supposed that the vicissitudes of the Pleistocene extinguished many high-latitude species. It has also been proposed that during the dry glacial periods, tropical rainforests (e.g., in South America) were restricted to a few isolated pockets, in which isolated populations of many organisms evolved into new species. Thus tropical diversity is attributed, in part, to a high rate of speciation in the recent past. Whether or not there were such refugia, and whether or not new species developed in them, is controversial (Haffer 1969; Endler 1982; Lynch 1988). However, Pleistocene speciation cannot possibly explain latitudinal diversity gradients, for tropical regions are much richer than temperate regions not only in species, but in numbers of genera and even families, which evolved in the mid-Tertiary or earlier. Thus the diversity gradient would still exist even if no new species had originated during the Pleistocene (Ricklefs 1989). Nor is there any paleontological evidence that species diversity at higher latitudes equaled or even approached tropical levels before the onset of the Pleistocene glaciations.

The second historical hypothesis begins with the observation that moist tropical environments seem more favorable, less harsh, for terrestrial organisms than do colder (or drier) environments, which require special adaptations (especially in seasonally cold environments). To be sure, an arctic caribou or a snowy owl is no more adapted to a tropical environment than a spider monkey or toucan is to an arctic climate: whether an environment is harsh or benign depends on the properties of the organism in question, *which have evolved at some prior time in its phylogenetic history*. Perhaps, given a long enough time, any tropical lineage of plant or animal could adapt to survive cold winters, but obviously most of them have not done so—at least not yet.

Evolving a complex set of adaptations to a drastically different environment takes *time* and *opportunity*. Few species encounter, or are forced to adapt to, rare environments, such as very arid deserts and hot springs, so it is not surprising that the diversity of species in such environments is low (Brown 1988). The few phylogenetic lineages that have acquired adaptations to such habitats may be rich in species—some genera of insects are very diverse in arid deserts, for example—but the adaptive "breakthrough" has occurred in rather few lineages.

For almost the entire history of terrestrial (and marine) life—until the mid-Tertiary, a mere 30–40 million years ago—most of the earth had a tropical or subtropical climate (see Chapter 7). Cold environments, whether at high latitudes or altitudes, were restricted to much smaller areas than they are now. Thus most living things evolved and diversified under tropical or semitropical conditions, over a far longer time than has been available for them to encounter and adapt to cold environments. Some phyletic lineages, to be sure, have probably had a long history of evolution in cold regions, and are diverse at high latitudes: examples include microtine rodents (voles), sandpipers, and the plant family Ericaceae (heather, blueberries, azaleas). But in a great many lineages that are primarily tropical, only a few species (or none) have acquired adaptations to temperate zone conditions. In the United States, for example, skunk cabbage (*Symplocarpus*), pawpaw (*Asimina*), the monarch butterfly (*Danaus*), whiptail lizards (*Cnemidophorus*), a few hummingbirds, and the porcupine (*Erethizon*) are northern outliers of primarily tropical groups. Other groups, such as palms, begonias, heliconiine butterflies, and antbirds, are diverse in the tropics, but extend only into the southernmost United States, or not even that far.

It is plausible, therefore, that many primitively tropical groups have not acquired the adaptations necessary to persist and diversify in seasonal habitats, which until comparatively recently were geographically very restricted. The distributions of many groups of animals may be restricted to the tropics not only by the necessity of evolving adaptations to pass the winter, but also by the failure of many groups of tropical plants, on which the animals depend, to do so. Insects provide a striking example (Farrell and Mitter 1993). Rove beetles (Staphylinidae) and leaf beetles (Chrysomelidae) are huge groups. Rove beetles are generalized predators, whereas most species of leaf beetles are highly specialized, each feeding on perhaps a single genus of plants. Moreover, in many groups of leaf beetles, host associations evolve slowly: related species (e.g., the members of a genus) often feed on closely related plants. Both families of beetles are far more diverse in the tropics than in the temperate zone, but the disparity in species richness is far greater for leaf beetles. Each family is represented by about 300 species in the *state* of Indiana, but at a *single locality* in a Peruvian rainforest, about 310 species of rove beetles have been collected, and about 750 species of leaf beetles! Relative to rove beetles, leaf beetles have undergone less diversification in the temperate zone, perhaps not only because they would have to adapt to cold temperatures (like the rove beetles), but also because they would have to adapt to new plants, since many of the tropical plant groups on which they depend are not found in the north. Because of the long tropical history of most organisms, invasion of the temperate zone requires new adaptations, which, in the relatively brief time available, may have been acquired by rather few groups.

In summary, the biogeographic problem of latitudinal variation in species diversity has not been solved. Several ecological and historical hypotheses (including some not discussed here) could account, at least in part, for the pattern, and it is quite possible, even likely, that many or all of them contribute to it in part. This problem will challenge the ingenuity of ecologists and evolutionary biologists for some time to come.

Summary

1. The geographic distributions of organisms provided Darwin and Wallace with some of their strongest evidence for the reality of evolution.

2. Biogeography, the study of organisms' geographic distributions, has both historical and ecological components. Certain distributions are the consequence of long-term evolutionary history, and others of contemporary ecological factors.

3. The historical processes that affect the distribution of a higher taxon include dispersal, vicariance (fragmentation of a continuous distribution by the emergence of a barrier), adaptation, speciation, extinction, and climatic change.

4. Disjunct distributions are attributable in some instances to vicariance and in others to dispersal. Either of these hypotheses may be favored by evidence from

phylogenetic relationships, paleontology, geology, or the taxonomic composition of biotas.

5. The local distribution of species is affected by ecological factors, including both abiotic aspects of the environment and biotic features such as competitors and predators.

6. The diversity of species in a local region may or may not be at an equilibrium. Accounting for consistent geographic patterns in species diversity is difficult because of the difficulty in choosing among competing hypotheses. This problem is illustrated by the latitudinal gradient of species diversity, for which both equilibrium and nonequilibrium hypotheses have been advanced.

7. Similar environments in different parts of the world often harbor ecologically similar species that have evolved by convergent evolution. In some instances (but not others), sets of species have independently evolved to partition resources in similar ways.

Major References

Brown, J. H., and A. C. Gibson. 1983. *Biogeography*. Mosby, St. Louis. A comprehensive textbook of biogeography, with excellent coverage of both historical and ecological aspects of animal and plant distributions.

Myers, A. A., and P. S. Giller (editors). 1988. *Analytical biogeography*. Chapman and Hall, London. Essays on both ecological and historical biogeography, emphasizing concepts, analytical methods, and recent controversies.

Ricklefs, R. E., and D. Schluter (editors). 1993. *Species diversity in ecological communities: Historical and geographical perspectives*. University of Chicago Press, Chicago. Includes both ecological and historical approaches to understanding species diversity.

Problems and Discussion Topics

1. Until recently, the plant family Dipterocarpaceae was thought to be restricted to tropical Asia, where many species are ecologically dominant trees. Recently, a new species of tree in this family was discovered in the rain forest of Colombia, in northern South America [*Science* 269: 1049 (1995)]. What hypotheses can account for its presence in South America, and how could you test those hypotheses?

2. Provide and suggest a test for a hypothesis to explain the distribution of taxa, such as alligators, that are restricted to temperate eastern Asia and eastern North America.

3. The number of species in a taxon is frequently greatest in one region and declines away from that region, as in the case of xenodontine snakes (see Figure 8.10). Provide a historical hypothesis and an ecological hypothesis to account for this pattern. How might you distinguish between these hypotheses? Are they mutually exclusive?

4. Vicariance biogeographers often postulate that clades have diverged in concert with the separation of the land masses occupied by their common ancestors. Thus the phylogeny of the organisms should mirror the sequence of separation of the land masses. Under this hypothesis, the *absence* of taxa from a region often presents a puzzle. For example, if we postulate that African and South American characiform fishes diverged from ancestors that occupied Gondwanaland, the absence of characiforms in Australia is puzzling, because Australia and South America separated after Africa became separated from them (see Figure 22 in Chapter 7). How might you explain this pattern and test your hypothesis?

5. Some taxa have amphitropical distributions: they occur in high latitudes of both the Northern and Southern Hemispheres, but not in between. For instance, creosote bush (*Larrea*) occurs in the deserts of western North America and southern South America. By what mechanisms could such distributions have arisen?

6. Except for birds and bats, there are almost no native land vertebrates in New Zealand. The fauna includes one frog, a few lizards, *Sphenodon*, and several flightless birds (kiwis and moas, the latter recently extinct). But there are no snakes, freshwater fishes, or terrestrial mammals. Account for this peculiar situation.

7. Suppose we wish to decide whether the biota of a land mass such as New Zealand was derived primarily by dispersal from another region (such as Australia) or by vicariance (i.e., from ancestors on a formerly contiguous land mass, such as Australia plus New Zealand). What is the significance of the missing elements (such as freshwater fishes)? (See question 6.)

8. In some cases it can be shown that species are physiologically incapable of surviving temperatures that prevail beyond the borders of their range. Do such observations prove that cold regions have low diversity because of their harsh physical conditions?

9. How might we decide whether or not the biota of a region is at equilibrium? Does equilibrium mean that it is "saturated" with species? Is it possible to determine whether there are "empty niches?"

10. Describe mechanisms by which dispersal could give rise to congruent distributions of many different taxa.

Evolutionary Processes in Populations and Species

Many, perhaps most, of the phenotypic characteristics of organisms—their morphology, physiology, and behavior—have evolved because they are adaptive. Certainly the marvelous adaptations of organisms are the features that, as Darwin said, "excite our admiration." Adaptations are features that have evolved by natural selection because they serve some function that enhances the survival or reproduction of the organisms that bear them. Thus, natural selection in its manifold guises must be the centerpiece of our study of the causes of evolution.

We will indeed examine natural selection at some length. But before we can appreciate it fully, we will need some foundations. Adaptive evolution is caused by natural selection acting on genetic variation. Hence, we must first (in Chapter 9) understand how genetic variation is described and measured. We need to know whether or not there is genetic variation within natural populations of species, and how much there is. We will find that this kind of information has resulted in a substantial difference between Darwin's view of the course of evolution and the modern view. We will also need to understand concepts such as allele frequencies, genotype frequencies, genetic variance, and heritability in order to describe evolution in modern terms.

Next (in Chapter 10), we need to understand where this genetic variation comes from. This requires a close look at mutation and recombination, the processes on which all evolution ultimately depends.

We cannot assume, a priori, that all of evolution is adaptive; we need to test our observations against the alternative hypothesis that it is not adaptive. In Chapter 11, we will see that there is another possible mechanism of evolution besides natural selection, one that, unlike natural selection, is purely a matter of chance. This process, random genetic drift, provides a kind of "null hypothesis:" it tells us how evolution proceeds if mutations are *not* acted on by natural selection. We will find that some characters, especially at the molecular level, appear to have evolved by genetic drift, and therefore are not adaptations. In the course of analyzing genetic drift, we will also encounter concepts having to do with the structure of populations that affect our understanding of adaptation as well as speciation.

With this knowledge, we will be prepared to examine natural selection, a subject that embraces so many aspects of evolution that we will devote three chapters to it. In Chapter 12, we will examine the meaning of natural selection and its various modes and levels of operation. In Chapters 13 and 14, we will put natural selection together with genetics and population structure to shape a comprehensive theory of evolution within species. This theory, in turn, provides the necessary background for understanding the origin of new species (in Chapters 15 and 16), the evolution of adaptations (in Part IV), and evolution above the species level (in Part V).

I close these prefatory remarks with a few words about *models*. We have already encountered models of evolution—for example, in our discussion of phylogeny (Chapter 5)—but now we meet models head-on.

All sciences use models. A model is a deliberately oversimplified representation of some aspect of the complex real world. It is used to gain insight into how the system that is modeled will "behave" or change as various factors are altered. We gain understanding by omitting (at least initially) as many aspects of reality as possible—namely, those that we suppose to have a trivial effect, or those that may be important only under certain circumstances. We do this in order to simplify for the sake of understanding. For example, in simple physics such as ballistics, the effect of the moon's gravity on a projectile is omitted—even though it is real—because it is negligible compared with that of the earth's gravity. The model may also omit the atmosphere's friction in order to calculate an ideal trajectory. The model will provide a description of the behavior of some objects that is sufficient for practical purposes. For other objects, the model will suffice only if we add more aspects of reality, such as friction. But if we begin by trying to incorporate every aspect of reality that could conceivably be relevant, the model will be as confusing as reality (it will be not a model, but a mimic), and it will fail to enlighten us.

A model may serve several purposes. It may serve as a *generalization* that enables us to describe the common features of many systems that all differ in detail, and none of which match the model exactly. (The simple models of population growth in Chapter 4 were intended to serve this purpose. They show that populations of all kinds of species tend to grow exponentially, but that the growth rate declines toward zero as the population increases.) A model may tell us *how the outcome depends on the relative magnitude* of each of several factors—which can often be difficult to intuit. (In ballistics, for example, we might want to consider the simultaneous effects of gravity, friction, and the force applied to a projectile.) Closely related to this function, a model may tell us that some factors can be safely *ignored* because their effects, relative to those of other factors, are minimal; or it may tell us that although quite a few outcomes are possible, *certain outcomes are very unlikely or impossible*. This can be very useful to know because it may limit the possible explanations for a phenomenon to a more manageable number. Another important function of models is to formulate hypotheses with explicit assumptions, and thereby to specify predictions of what data should be observed if the hypotheses are true. We will encounter many instances of models that are used for this purpose in evolutionary theory. Most predictions in evolutionary biology, in fact, are statements about the patterns that new data will reveal—not predictions about how evolution will proceed in the future. Finally, a model may be used to make a precise numerical *prediction* of future events; this usually requires a complex model that takes all important variables into account. (And to make a real prediction, these variables have to be measured with sufficient accuracy.)

We will encounter examples of all these kinds of models in evolutionary biology, although rather few of the kind that make precise numerical predictions. This is simply because evolution is influenced by a great many factors (including both genetic and environmental factors) that can be very hard to specify, much less measure. However, a "precise" prediction need not be quantitative—it can be qualitative. In some circumstances, we can successfully predict what the end state of an evolutionary process will be, even though we may not be able to say how long it will take to get there.

There are several kinds of models. Some are simply verbal, such as the statement that a warm body radiates heat faster if it has a relatively large surface area. To achieve greater usefulness, we can go from a verbal model to a physical model (e.g., we could make warm objects of various sizes and measure their heat radiation). Or we can go from a verbal to a graphical model (cf. the model of island biogeography in Chapter 8) or a mathematical model with equations. Mathematical models are often the most powerful, and are favored in all the sciences, including evolutionary biology. In this textbook, mathematical models are kept to a bare minimum, but some are necessary. Some mathematical models are too complex to solve analytically; in that case, a computer model that simulates the system under study may be used. Computer models are often used in evolutionary studies, and we will encounter a few examples.

Mathematical models have the virtue that (as long as the math has been done right) the conclusions are *true within the framework of the model's assumptions*. This is a powerful statement, because *if* we can be confident that a real system obeys, or nearly obeys, the assumptions, then we may be sure that the system will follow the predictions of the model. Consequently, there are many examples, in sciences ranging from evolutionary genetics to cosmology, in which models' predictions have been validated by data long after the models were conceived. Models also, therefore, tell us *what to look for:* science becomes not merely the accumulation of random facts, but an enterprise directed toward critical observations or experiments that will help us to accept or reject hypotheses—hypotheses that often have grown out of models.

I have raised the subject of models because some students (and some professional biologists) think they are so abstract that they can't have much to do with real organisms. But every science faces the dilemma of finding a balance between generality and precision, between the complexity of the natural world and understanding its essential features. Whenever we make a verbal statement about how we think the world works, we are presenting a model—an oversimplified generalization. Mathematical models force us to be precise in our arguments, they often bring us to unexpected conclusions, and they often show us that our verbal arguments are wrong.

CHAPTER 9

Variation

Darwin proposed, and modern evolutionary biology affirms, that evolution is based on variation in the characteristics of organisms: differences among individuals within populations and among populations and species. Evolution is a two-step process, in which (1) variations arise among individuals and (2) the proportions of variant types (their relative numbers) change from generation to generation. However, changes in the proportions of variants in one generation are not carried over to subsequent generations unless the variation is at least partly inherited. Thus, genetic variation is the foundation, the sine qua non, of evolution. Mendel's discovery that inheritance is based on particulate genes, and the disproof of the theories of Lamarckian transformation and blending inheritance (Box 9.A), made possible the modern theory of evolutionary processes.

The modern study of evolutionary processes is based on understanding the amount and nature of genetic variation, the processes by which it originates, and the factors that alter the pattern of variation. In this chapter, we describe variation within and among populations. In subsequent chapters, we will treat the origin of variation and the factors that, by altering the proportions of variants, constitute the causes of evolution within populations and species.

Definitions

Repeating in part material from Chapter 3, let us first define some important terms.

PHENOTYPE: A morphological, physiological, biochemical, or behavioral characteristic of an individual organism, or of a group of like individuals that differ in this respect from other individuals. The term sometimes refers to more than one characteristic.

GENOTYPE: The genetic constitution of an individual organism, or of a group of organisms alike in this respect, at one or more loci singled out for discussion. Different genotypes often, but not always, differ in phenotype. Conversely, a given genotype may manifest more than one phenotype.

LOCUS (plural: *loci*): A site on a chromosome, or more usually, the gene that occupies a site.

GENE: The definition of this term often depends on the context in which it is used. For our purposes, we will generally think of a gene as a nucleic acid sequence (DNA or, in some viruses, RNA) that encodes a product (RNA or polypeptide) that, alone or in combination with other products, has a distinct function within an organism. Variation in the base pair sequence of the gene may or may not be reflected in the sequence or functional properties of the product. (On occasion, we will refer to "multigene families," meaning several or many sites in the genome that are occupied by similar or identical DNA sequences with functionally similar products.)

ALLELE: A particular form of a gene. The criteria used to distinguish alleles depend on the circumstances. Often alleles are distinguished by their effects on the phenotype. Sometimes alleles are distinguished (in theory or in practice) by differences in base pair sequence. Often, few or none of the base pair differences affect the phenotypic characteristics studied. In that case, each of the alleles distinguished by phenotypic effects may include several DNA sequences.

HAPLOTYPE: One of the various DNA sequences of a given gene that can be distinguished by molecular methods such as DNA sequencing. Haplotypes are equivalent to alleles *if* alleles are distinguished by base pair sequence.

MUTATION: An alteration of genetic material, usually of a DNA sequence, that gives rise to a new allele (however distinguished) or haplotype. The new allele or haplotype itself may be referred to as a mutation, especially if it has arisen recently. The term "*mutation*" also refers to the process by which genes are altered.

231

BOX 9.A

Theories of Evolution, and the Significance of Mendelism

Variational and Transformational Theories of Evolution

Until late in the nineteenth century, and even into the twentieth, it was widely believed that alterations of an individual organism's features that were acquired during its lifetime could be transmitted to its offspring. For example, children of laborers were expected to inherit greater muscular development than children of courtiers, and individuals who had dark skin because of lifelong exposure to sun should, by this theory, have heavily pigmented offspring. According to this theory, characteristics could be acquired either through the effects of *use* (e.g., muscle size) or through the direct effects of the *environment* (e.g., pigmentation). Lamarck used this principle of the INHERITANCE OF ACQUIRED CHARACTERISTICS, or SOFT INHERITANCE, as it is sometimes called, as part of his theory of the mechanisms of evolution (see Chapter 2). Under this theory, if all members of a species engaged in similar activities (e.g., if all giraffes stretched their necks upward to browse on foliage), or if all experienced the same environment, then all offspring might deviate from their parents in the same way. Thus the entire species could be altered from one generation to the next. This is a TRANSFORMATIONAL THEORY of evolutionary change (Lewontin 1983).

Lamarckian theories are radically different from Darwin and Wallace's theory, in which *some* individuals of a species fortuitously have inborn differences from others (variations), and thus survive or reproduce to a greater extent. If their variant feature is inherited, the proportion of individuals in the next generation bearing this feature will be greater than in previous generations. Repetition of this process will result in a gradual increase in the proportion of the variant phenotype, which may ultimately replace the ancestral phenotype altogether. Evolution occurs not by the transformation of individuals of the species, but by the replacement of one type of individual by another. This is a VARIATIONAL THEORY of evolutionary change. *Transformational and variational theories are the only fundamentally different scientific theories of evolution that have yet been conceived.*

Among biologists of the nineteenth century, the German biologist August Weismann (1834–1914) has been called second only to Darwin in his impact on evolutionary theory (Mayr 1988a). Weismann absolutely rejected any kind of inheritance of acquired characteristics. From his own and others' cytological studies, he concluded that the germ cells that give rise to gametes are segregated early in development from the somatic tissues, and cannot be influenced by alterations of those tissues due to environment or experience. He argued, further, that Darwinian selection can explain anything that Lamarckism can explain, and moreover, that Lamarckism cannot explain a great many characteristics. For example, the external skeleton (cuticle and its features) of a butterfly is not altered dur-

ing the butterfly's life, yet it develops in the pupa—before the butterfly emerges. Moreover, he said, acquired modifications are simply not inherited. Despite centuries of circumcision, Jewish boys are still born with a foreskin. Weismann himself bred mice for many generations, amputating their tails every time, yet the tails did not evolve to be shorter. Weismann consequently became an even more ardent selectionist than Darwin, and set in motion the rejection of belief in inheritance of acquired characteristics. By the 1930s and 1940s, many other scientists had done experiments that led to a total rejection of Lamarckian inheritance.[*] By this time, Mendelian genetics had developed, providing a foundation for the variational principle that Darwin's theory required. Subsequently, molecular genetics demonstrated that changes in DNA are translated into differences in proteins. No mechanism is known whereby changes in proteins (or in other constituents of the body, other than nucleic acids) can alter the information encoded in the DNA.

The Significance of Mendelism

Because offspring are often intermediate between their parents in characteristics such as size, a popular theory of heredity in the nineteenth century was BLENDING INHERITANCE. The factors responsible for a trait were imagined to blend together much as a mixture of red and white paint yields some shade of pink. Based on this idea, the engineer Fleeming Jenkin posed a serious challenge to Darwin in 1867. Just as red and white colors cannot be reconstituted from pink paint, so intermediate offspring would not give rise to the original parental phenotypes. Thus as interbreeding among different individuals proceeded, a population would become homogeneous. The variation that Darwin's theory requires would rapidly be lost.

Gregor Mendel's experiments, published in 1866 but not noticed until 1900, showed that heredity is based on particles that, rather than losing their identity when combined with other hereditary factors, can segregate generation after generation. Heterozygous pink-flowered plants (A_1A_2) can indeed have red- and white-flowered offspring (A_1A_1, A_2A_2). Thus variation persists. Moreover, characteristics that vary continuously (such as body size or human skin color) are also inherited as

---

[*]"Soft inheritance" had been rejected by geneticists and most other experimental biologists by this time, but it remained popular among many paleontologists and naturalists, probably because Lamarckism imputed some purposeful, creative element in life, which was philosophically more appealing than the random mutations and the slings and arrows of outrageous fortune inherent in Darwinism (Bowler 1989). "Neo-Lamarckism" did not die out in Western biology until the Evolutionary Synthesis of the 1930s and 1940s. In the meantime, Lamarckism became part of the ideology of the Soviet Union, where Trofim Lysenko, untrained in genetics, gained control over agricultural and genetic research, with Stalin's support. From the 1930s until he was deposed in 1965, Lysenko purged the Soviet Union of Mendelian geneticists (many of whom were sent to labor camps or to their deaths) and reshaped plant breeding on Lamarckian principles. The disastrous effects on Soviet agricultural production and on Soviet biological research were felt for decades.

GENE COPY: We will use this term when counting the number of representatives of a gene in a sample or population. In a diploid population (e.g., humans), each individual carries two copies of each autosomal gene, and each gamete carries one copy. A sample of 100 individuals represents 200 gene copies. The term "*gene copy*" is used without distinguishing allelic or sequence differences among the copies. For this purpose, we will sometimes refer to "ALLELE COPIES." Thus, a heterozygous individual, A_1A_2, has one copy of allele A_1 and one copy of allele A_2. If a sample includes 25 A_1A_1, 50 A_1A_2, and 25 A_1A_2 individuals, it represents 200 gene copies, of which 100 are copies of A_1 and 100 are copies of A_2.

Distinguishing Sources of Phenotypic Variation

We noted in Chapter 3 that individuals may differ in phenotype because of genetic differences, environmental differences, or both. We need to understand the causes of these differences, and how they can be distinguished.

The major sources of variation in phenotype are these:

1. *Differences in genotype*, i.e., in the DNA sequence at one or more loci. Although most genetic variations can be transmitted through either eggs or sperm, some are strictly maternally or paternally inherited. For example, mitochondrial genes are transmitted mostly or only via eggs in most animals, and chloroplast genes are transmitted through male gametes in conifers.

2. *Differences in environment.* Some features, such as many physiological and behavioral traits, are affected by immediate or very recent environmental conditions, and may change repeatedly throughout life. At the other extreme, differences that persist through part or all of an individual's lifetime may be caused by environmental differences experienced very early in development (e.g., before birth in humans), or even in the egg.

3. *Maternal effects.* This term refers to characteristics of the offspring of a mother that are due not to the genes they inherit from her, but rather to nongenetic influences, such as the amount or composition of yolk in her eggs, the amount and kind of maternal care she provides, or her physiological condition while carrying eggs or embryos. Differences among mothers may be due either to their genotypes or to nutrition or other environmental factors. Nongenet-

ic "paternal effects" have also been described in a few organisms.

With this background, we can see that CONGENITAL differences among individuals—those present at birth—are not necessarily *genetic* differences. Congenital differences may be caused not by genes, but by nongenetic maternal effects or by environmental factors that act on an embryo before birth or hatching. For example, consumption of alcohol, tobacco, and many other drugs by pregnant women increases the risk of nongenetic birth defects in their babies.

We must again emphasize that phenotypic variation often arises from *both* genetic and environmental sources. To determine whether it is at least partly due to genetic variation, several methods may be used. (Textbooks of genetics treat these methods in greater depth.)

1. Phenotypes can be experimentally crossed to produce F_1, F_2, and backcross progeny. Any of several standard Mendelian ratios among the phenotypes of the progeny (e.g., 3:1 or 1:2:1) are taken as evidence of simple genetic control.

2. Simple Mendelian ratios are not expected if the phenotypic variation is due to many variable genes, but resemblance between relatives is. Correlation between the average phenotype of offspring and that of their parents, or greater resemblance among siblings than among unrelated individuals, suggests that genetic variation contributes to phenotypic variation. However, we must take maternal effects into account, and must also be sure that siblings (or relatives in general) do not share more similar environments than nonrelatives. A correlation between the phenotype of offspring and that of their fathers excludes the possibility of maternal effects, and is usually taken as evidence of genetic variation. To reduce the likelihood that siblings share a common environment as well as genes, an investigator may distribute eggs, seeds, or hatchlings from different parents at random within a growth chamber or garden (or may use more complex experimental designs to ensure that relatives do not grow in correlated environments). In a variation on this theme, researchers who study birds and other species that provide parental care sometimes distribute eggs or offspring among foster parents (cross-fostering) so that genetic differences among parents are not correlated with environmental differences. Likewise, human geneticists rely strongly on studies of adopted children to determine whether behavioral or other similarities

among siblings are due to shared genes rather than shared environments.

3. The offspring of phenotypically different parents can be reared in a uniform environment. Differences among offspring from different parents that persist in such circumstances are likely to have a genetic basis. To ensure that maternal effects due to environmental differences among mothers are not responsible for variation among offspring, it is advisable to propagate the sample in the uniform environment for at least two generations, on the assumption that maternal effects last for only one generation. This COMMON-GARDEN METHOD has often been used in studying plants.

Fundamental Principles of Genetic Variation in Populations

We embark now on our study of genetic variation and the factors that influence it: the factors that cause evolution within species. We begin with some simple theory. *The definitions, concepts, and principles introduced here are absolutely essential for understanding evolutionary theory.*

Frequencies of Alleles and Genotypes

The scarlet tiger moth, *Panaxia dominula*, has one generation per year. It has three color forms in England. These forms differ in the amount of white spotting on the black forewings and in the amount of black on the largely red hindwings (Figure 9.1). From experimental crosses, it is known that both pigmentation differences are caused by a difference at one locus. (This single gene difference with two phenotypic effects is an example of **pleiotropy**.) The heterozygote is intermediate in both respects, and so can be distinguished from both homozygotes. We shall refer to the genotype with extensive white spotting as A_1A_1, to the intermediate as A_1A_2, and to the dark form as A_2A_2.

Each year from 1939 to 1970, the geneticist E. B. Ford collected specimens at a certain locality. The total numbers collected were:

A_1A_1	A_1A_2	A_2A_2	Total
17,062	1,295	28	18,385

We will use this sample to establish a few basic concepts. First, a **genotype frequency** is the proportion of a population that has a certain genotype. For genotype A_1A_1, this is estimated as 17,062/18,385 = 0.928. The frequency of A_1A_2 is likewise calculated as 0.070, and of A_2A_2 as 0.002. These frequencies, or proportions, must sum to 1, and they do.

The frequency of an allele (**allele frequency**, often called **gene frequency**) is the proportion of gene copies in the population that are of that allelic type. Each moth has two gene copies, so the sample represents 18,385 × 2 = 36,770 copies. A particular allele—say, A_1—is carried both by homozygotes and by heterozygotes. Because homozygotes each have two copies and heterozygotes have one, the number of A_1

copies in the sample is (17,062 × 2) + (1,295 × 1) = 35,419. The frequency (proportion) of A_1 is this number divided by the total number of gene copies in the sample: 35,419/36,770 = 0.963. Likewise, the number of copies of allele A_2 is (1,295 × 1) + (28 × 2) = 1,351, and the frequency of A_2 is 1,351/36,770 = 0.037. The frequencies of the alleles must sum to 1, and they do.

These figures are *estimates* of the true genotype frequencies and allele frequencies in the population because they are based on a sample rather than a complete census. The larger the sample, the more confident we can be that we are obtaining accurate estimates of the true values. (This assumes that the sample is a *random* one, i.e., that our likelihood of collecting a particular type is equal to the true frequency of that type in the population.)

We will now generalize from this example and introduce some algebraic notation. The sample consists of N moths (N = 18,385 in this case) and $2N$ gene copies ($2N$ = 36,770). The number of individuals of each genotype might be denoted n_D, n_H, and n_R for genotypes A_1A_1, A_1A_2, and A_2A_2 respectively. The genotype frequencies D, H, and R are therefore n_D/N, n_H/N, and n_R/N. If this is a full list of the genotypes in the sample, $D + H + R = 1$.

The *number* of A_1 copies is $2n_D + n_H$, and the number of A_2 copies is $n_H + 2n_R$. For simplicity, let us call these quantities n_{A1} and n_{A2}, the number of each of the two alleles. Then $p = n_{A1}/2N$ is the frequency of allele A_1, and $q = n_{A2}/2N$ is the frequency of allele A_2. We will often use the symbols p and q for allele frequencies. Notice that $p + q =$

FIGURE 9.1 Variation due to two alleles at a single locus in the moth *Panaxia dominula*. Top: the most common genotype, A_1A_1; middle: the heterozygote, A_1A_2; bottom: the rare homozygote, A_2A_2. In A_2A_2, the central white spot on the forewing is reduced or absent and the amount of black on the hindwing is less than in A_1A_1. The heterozygote is intermediate. (After Ford 1971.)

BOX 9.B

Derivation of the Hardy-Weinberg Distribution

Let the frequencies of genotypes A_1A_1, A_1A_2, and A_2A_2 be D, H, and R respectively, and let the frequencies of the alleles A_1 and A_2 be p and q. The genotype frequencies sum to 1, as do the allele frequencies. The probability of a mating between a female of any one genotype and a male of any one genotype equals the product of the genotype frequencies. For reciprocal crosses between two different genotypes, we take the sum of the probabilities. The mating frequencies and offspring produced are:

Mating	Probability of Mating	Offspring		
		A_1A_1	A_1A_2	A_2A_2
$A_1A_1 \times A_1A_1$	D^2	D^2		
$A_1A_1 \times A_1A_2$	$2DH$	DH	DH	
$A_1A_1 \times A_2A_2$	$2DR$		$2DR$	
$A_1A_2 \times A_1A_2$	H^2	$H^2/4$	$H^2/2$	$H^2/4$
$A_1A_2 \times A_2A_2$	$2HR$		HR	HR
$A_2A_2 \times A_2A_2$	R^2			R^2

Recalling that $p = D + H/2$ and $q = H/2 + R$, the frequency of each genotype among the offspring is:

A_1A_1: $D^2 + DH + H^2/4 = (D + H/2)^2 = p^2$

A_1A_2: $DH + 2DR + H^2/2 + HR = 2[(D + H/2)(H/2 + R)] = 2pq$

A_2A_2: $H^2/4 + HR + R^2 = (H/2 + R)^2 = q^2$

This result may also be obtained by recognizing that if genotypes mate at random, gametes, and therefore genes, also unite at random to form zygotes. Because the probability that an egg carries allele A_1 is p, and the same is true for sperm, the probability of an A_1A_1 offspring is p^2. A "Punnett square" shows the probability of each gametic union:

		Sperm	
		A_1 (p)	A_2 (q)
Eggs	A_1 (p)	A_1A_1 (p^2)	A_1A_2 (pq)
	A_2 (q)	A_1A_2 (pq)	A_2A_2 (q^2)

Whenever mating is random, therefore, the frequency of homozygous progeny of genotype A_iA_i is the square of the frequency of allele i, and the frequency of heterozygotes of genotype A_iA_j is twice the product of the frequencies of alleles i and j (because two kinds of unions can produce a heterozygote). The subscripts i and j refer to any two numbers that denote alleles (say, A_1 and A_2). We can extend this principle to multiple alleles. For alleles $A_1, A_2, A_3, \ldots A_k$, denote their frequencies, $p_1, p_2, p_3, \ldots p_k$. These frequencies sum to 1. After one generation of random mating, the frequency of, say, A_3A_3 will be p_3^2, that of A_2A_3 will be $2p_2p_3$, and similarly for all other genotypes. These genotype frequencies, again, sum to 1. In passing, notice that the frequency of all homozygous genotypes taken together is

$$\sum_{i=1}^{k} p_i^2$$

and that of all heterozygous genotypes is

$$1 - \sum_{i=1}^{k} p_i^2$$

The notation

$$\sum_{i=1}^{k}$$

means "add all the values of p_i^2, from $i = 1$ to $i = k$" where k is the last in the list of alleles.

1 (and that $p = 1 - q$). Note also that $p = D + H/2$ and $q = H/2 + R$.

In our example from *Panaxia dominula*, $D = 0.928$, $H = 0.070$, and $R = 0.002$. The frequencies of alleles A_1 and A_2 are, respectively, $p = 0.963$ and $q = 0.037$.

The Hardy-Weinberg Principle

Let us work with some easier numbers, and imagine that a population of *Panaxia dominula* has 1000 individuals, of which 400 are A_1A_1, 400 are A_1A_2, and 200 are A_2A_2. Assume that each genotype is equally represented by females and males, and that *individuals mate entirely at random*. (This is an entirely different situation from the crosses encountered in elementary genetics exercises, in which familiar Mendelian ratios are produced by crossing females of *one* genotype with males of *one* other genotype.) At the population level, there are *nine* possible matings in our example because females of three genotypes can pair with males of three genotypes. Box 9.B shows how, for any array of genotypes among the parents,

we can construct a table of matings and determine the frequency of each genotype among the progeny.

Let us consider what fraction of progeny will be A_2A_2 in our example. The genotype frequencies among the parents are $D = 0.4$ $(= 400/1000)$, $H = 0.4$, $R = 0.2$. The allele frequencies of A_1 and A_2 respectively are $p = D + H/2 = 0.6$ and $q = H/2 + R = 0.4$. The frequency of A_2A_2 among the progeny can be found by summing the proportions of A_2A_2 among the offspring of each type of mating and weighting each proportion by the probability of that kind of mating. Four types of matings yield A_2A_2 progeny. For example, half of the progeny of the mating $♀A_1A_2 \times ♂A_2A_2$ will be A_2A_2. The probability of this mating is $0.4 \times 0.2 = 0.08$ (i.e., the chance that a female is A_1A_2 is 0.4, and that a male is A_2A_2 is 0.2, because these are the frequencies of these genotypes). Thus $(0.5)(0.08) = 0.04$ of the next generation's individuals will be A_2A_2 moths produced by this type of mating. Following this procedure for all matings that yield A_2A_2, we may write:

Mating	Probabilitiy of mating	Fraction of A_2A_2 offspring	Production of A_2A_2
♀$A_1A_2 \times$ ♂A_1A_2	$H_2 = 0.4 \times 0.4 = 0.16$	$\times \quad 0.25 =$	0.04
♀$A_1A_2 \times$ ♂A_2A_2	$HR = 0.4 \times 0.2 = 0.08$	$\times \quad 0.50 =$	0.04
♀$A_2A_2 \times$ ♂A_1A_2	$RH = 0.2 \times 0.4 = 0.08$	$\times \quad 0.50 =$	0.04
♀$A_2A_2 \times$ ♂A_2A_2	$R_2 = 0.2 \times 0.2 = 0.04$	$\times \quad 1.00 =$	0.04
		Sum =	0.16

Thus the frequency of A_2A_2 among the offspring is 0.16. If we similarly calculate the expected frequencies of the other two genotypes, we find that the genotype frequencies, the new values of D, H, and R (Figure 9.2), are

A_1A_1	A_1A_2	A_2A_2
0.36	0.48	0.16

Now let us calculate the allele frequencies. These are $p = D + H/2 = 0.6$ and $q = H/2 + R = 0.4$. *The allele frequencies have not changed from one generation to the next*, although the alleles have become distributed among the three genotypes in new proportions. Now note that the frequency of A_1A_1 is $0.36 = (0.6)^2 = p^2$, the frequency of A_1A_2 is $0.48 = 2 \times 0.6 \times 0.4 = 2pq$, and the frequency of A_2A_2 is $0.16 = (0.4)^2 = q^2$. This is the **Hardy-Weinberg distribution** of genotype frequencies, named after G. H. Hardy and W. Weinberg,* who independently calculated this result in 1908. If this generation again mates at random, the genotype frequencies in the following generation will again be p^2, $2pq$, and q^2. The HARDY-WEINBERG PRINCIPLE, broadly stated, holds that whatever the initial genotype frequencies for two autosomal alleles may be, *after one generation of random mating, the genotype frequencies will be $p^2:2pq:q^2$.*

---

*Because Weinberg was German, his name should be pronounced "vine-berg," rather than "wine-berg."

Moreover, both these genotype frequencies and the allele frequencies *will remain constant in succeeding generations* unless factors not yet considered should change them. When genotypes at a locus have the frequencies predicted by the Hardy-Weinberg principle, the locus is said to be at HARDY-WEINBERG EQUILIBRIUM.

The Hardy-Weinberg genotype frequencies can be calculated only because of the orderly segregation of alleles in meiosis, which enables us to predict the frequency of each offspring genotype from each type of mating. The regularity of meiosis is therefore the foundation of the entire theory of genetic change in sexually reproducing populations. The Hardy-Weinberg principle can be extended to multiple alleles, as well as to more complicated patterns of inheritance.[†]

The Significance of the Hardy-Weinberg Principle: Factors in Evolution

The Hardy-Weinberg principle is the foundation on which almost all of the theory of population genetics of sexually reproducing organisms—which is to say, most of the genetic theory of evolution—rests. Its importance cannot be overemphasized. We will encounter it repeatedly in the theory of natural selection and other causes of evolution. In some circumstances, moreover, deviations of genotype frequencies in natural populations from their ideal Hardy-Weinberg values provide evidence that factors such as natural selection are operating.

An important consequence of the Hardy-Weinberg principle is that no matter what the past history of a population may have been, a single generation of random mating yields the Hardy-Weinberg genotype frequencies. If, for

---

[†]Some readers will recognize that the genotype frequencies are given by the binomial expansion $(p + q)^2$. For k alleles, the genotype frequencies are given by $(p_1 + p_2 + \dots p_k)^2$.

FIGURE 9.2 The hypothetical example used in the text to illustrate attainment of Hardy-Weinberg genotype frequencies after one generation of random mating. (A) Genotype frequencies among parents, not in Hardy-Weinberg equilibrium. (B) Allele frequencies in the population of parents. (C) Genotype frequencies among offspring, if all assumptions of the Hardy-Weinberg principle hold.

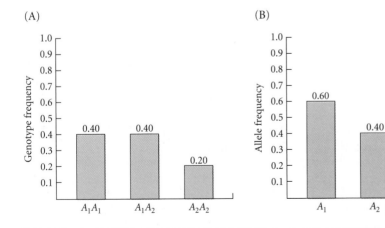

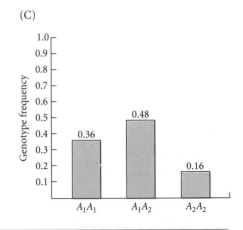

example, a new population is founded by 3000 A_1A_1 and 1000 A_2A_2 individuals, with each genotype equally represented by both sexes, $p = 0.75$, $q = 0.25$, and one generation of random mating yields 0.5625 A_1A_1, 0.375 A_1A_2, and 0.0625 A_2A_2. The same result would be obtained from any mixture in which $p = 0.75$ and $q = 0.25$, no matter what history gave rise to that mixture.

In most populations, the genotype frequencies at most loci fit the Hardy-Weinberg distribution very well. (We will illustrate this shortly.) However, allele frequencies and genotype frequencies often change from one generation to another. Thus, the assumptions of the Hardy-Weinberg formulation do not always hold. Therefore *the study of genetic evolution consists of asking what happens when one or more of the assumptions are relaxed.*

The most important assumptions of the Hardy-Weinberg principle are:

1. *Mating is random.* If members of the population do not mate at random, there may be changes in the genotype frequencies so that these depart from the ratios $p^2:2pq:q^2$. Certain kinds of nonrandom mating also alter allele frequencies, but these are usually classified as a form of selection.

2. *The population is infinitely large* (or so large that it can be treated as if it were infinite). The calculations are made in terms of probabilities. If the number of events is finite, the actual outcome is likely to deviate, purely by chance, from the predicted outcome. If we toss an infinite number of unbiased coins, probability theory says that half will come up heads, but if we toss only 100 coins, we are likely not to obtain exactly 50 heads, purely by chance. Similarly, among a finite number of offspring, both the genotype frequencies and the allele frequencies may change from those in the previous generation, *purely by chance.* Such random changes are called **random genetic drift**.

3. *Genes are not added from outside the population.* Immigrants from other populations may carry different frequencies of A_1 and A_2; if they interbreed with residents, this will alter allele frequencies and, consequently, genotype frequencies. Mating among individuals from different populations is termed **gene flow** or **migration**. We may restate this assumption as: There is no gene flow.

4. *Genes do not change from one allelic state to another.* Such alteration is termed *mutation,* and it clearly would change allele frequencies. Hence the Hardy-Weinberg principle assumes no mutation.

5. *All individuals have equal probabilities of survival and of reproduction.* If these probabilities differ among genotypes (i.e., if there is a consistent difference in the genotypes' rates of survival or reproduction), the frequencies of alleles and/or of genotypes *may* be altered from one generation to the next. Thus the Hardy-Weinberg principle assumes that there is *no natural selection* affecting the locus.

Inasmuch as nonrandom mating, chance, gene flow, mutation, and selection can alter the frequencies of alleles and genotypes, these are the major factors of evolutionary change within populations.

Certain subsidiary assumptions of the Hardy-Weinberg principle can be important in some contexts. As presented, the principle applies to autosomal loci; it can also be modified for sex-linked loci (which have two copies in one sex and one in the other), which arrive at stable genotype frequencies only after a number of generations. Also, the principle assumes that alleles segregate into a heterozygote's gametes in a 1:1 ratio; in some cases, however, one allele is carried by more than half of the gametes of the heterozygote (**segregation distortion**, sometimes called **meiotic drive**). Segregation distortion changes allele frequencies, and is usually viewed as a peculiar form of selection.

For simplicity, we have considered only two alleles. Often, however, multiple alleles of a gene can be distinguished. For instance, the human blood groups A, B, O, and AB are determined by three alleles, denoted A, B, and O. At Hardy-Weinberg equilibrium, the frequency of homozygotes for the ith allele (A_iA_i) is p_i^2, and the frequency of heterozygotes for any two alleles i and j (A_iA_j) is $2p_ip_j$, where p_i and p_j denote the frequencies of the two alleles in question. For instance, the frequencies of blood type alleles A, B, and O might be denoted p_1, p_2, and p_3 (which sum to 1), and the frequency of the AB heterozygote would then be $2p_1p_2$. The Hardy-Weinberg principle thus applies as well to multiple alleles as to two (see Box 9.B).

In the hypothetical case with which we introduced the Hardy-Weinberg principle, we began with 400 A_1A_1, 400 A_1A_2, and 200 A_2A_2 individuals—i.e., with genotype frequencies 0.4 : 0.4 : 0.2. In this case, the locus was not in Hardy-Weinberg equilibrium (which, as we have seen, requires the genotype frequencies 0.36 : 0.48 : 0.16, established after one generation of random mating). This demonstrates that genotype frequencies can deviate from Hardy-Weinberg proportions. We will encounter real examples of such deviations below.

If the assumptions we have listed hold true in a particular case, a locus will display Hardy-Weinberg genotype frequencies. But if we observe that a locus fits the Hardy-Weinberg frequency distribution, we cannot conclude that the assumptions hold true! For example, mutation or selection may be occurring, but at such a low rate that we cannot detect a deviation of the genotype frequencies from expected values. Also, under some forms of natural selection, we might observe deviations from Hardy-Weinberg equilibrium if we measure genotype frequencies at one stage in the life history, but not at other stages.

A real example From our hypothetical example, we have seen that an array of genotype frequencies (D, H, R) may or may not be in the Hardy-Weinberg proportions $p^2:2pq:q^2$. The principle's assumptions, such as infinite population size and no mutation, are manifestly unrealistic, so we might imagine that real populations would never fit its predictions. But as we noted above, the forces acting on a real locus may be so weak, or may balance each other in such a way, that reality may conform closely to the Hardy-Weinberg predictions. In fact, genotype frequencies in human and other populations very often fit Hardy-Weinberg proportions.

Let us return to E. B. Ford's collection, made over 32 years, of *Panaxia dominula*. The sample consists of 17,062 A_1A_1, 1295 A_1A_2, and 28 A_2A_2 moths. We calculated $p = 0.963$ and $q = 0.037$. From our estimates of p and q, we can calculate the expected frequencies p^2, $2pq$, and q^2; thus the expected *frequency* of A_1A_1 is $0.963^2 = 0.9274$. The expected *numbers* are calculated by multiplying the number of moths (N) by the expected frequencies; thus the expected number of A_1A_1 is $0.9274 \times 18,385 = 17,050$. Continuing these calculations, we obtain the following results:

Genotype:	A_1A_1 p^2	A_1A_2 $2pq$	A_2A_2 q^2
Expected frequency	0.9274	0.0713	0.0013
Observed frequency	0.9280	0.0704	0.0015
Expected number	17,050	1,311	24
Observed number	17,062	1,295	28

The difference between the observed numbers of each genotype and the theoretically expected numbers is so slight that it can be readily attributed to accidents of sampling (SAMPLING ERROR). Thus, averaged over 32 years, this locus appears to be in Hardy-Weinberg equilibrium.

In the years 1955 and 1956, Ford collected the following numbers of moths, from which we calculate as above, for each year, the observed genotype frequencies, the allele frequencies, the expected genotype frequencies, and the expected numbers.

	1955			1956		
Genotype:	A_1A_1	A_1A_2	A_2A_2	A_1A_1	A_1A_2	A_2A_2
Observed numbers	308	7	0	1,231	76	1
Observed genotype frequencies	0.978	0.022	0.000	0.941	0.058	0.001
Allele frequencies	$p = 0.989, q = 0.011$			$p = 0.970, q = 0.030$		
Expected genotype frequencies	0.978	0.022	0.0001	0.941	0.058	0.001
Expected numbers	308.1	6.9	0.03	1,230.8	75.9	1.31

Within each year, the frequency and number of each genotype fits the Hardy-Weinberg expectation extremely well. From 1955 to 1956, however, the frequency of the A_1 allele decreased slightly (and A_2 necessarily increased, since $q = 1 - p$). This change in allele frequency was apparently real, because a statistical test can show that a dif-

ference of this magnitude, estimated from a sample of this size, is very unlikely due to accidents of sampling alone. Thus the population *evolved slightly* from 1955 to 1956. The amount of evolutionary change is unimpressive and may seem trivial, but *if* the allele frequencies were to change in the same direction, to the same degree, each year, it would take only 52 years (generations) for the frequency of A_1 to reach zero and the frequency of A_2 to reach 1.0, and the population would have evolved from almost entirely white-spotted moths (carrying A_1) to the almost spotless phenotype (A_2A_2). In this population, though, the decline in p between 1955 and 1956 was only a temporary reversal of an overall trend, in which p increased from about 0.91 in 1939 to 0.97 in 1970.

Notice that the rare allele, A_2, is carried almost entirely by heterozygotes; q^2, the frequency of A_2A_2, is nearly zero. The ratio of heterozygotes to A_2A_2 homozygotes is $2pq:q^2$, or $2p:q$; thus the fraction of A_2 carriers that are heterozygous is $2p/(2p + q)$, which is nearly 1 if q is very small. It will be important in several contexts to realize that *a rare allele exists mostly in the heterozygous state*. The frequency of heterozygotes is greatest when the alleles have equal frequencies (Figure 9.3).

Dominance When, as in the example of *Panaxia*, the heterozygote can be distinguished from both homozygotes, neither allele is fully dominant. A dominant allele masks the expression of a recessive allele in heterozygotes. If the dominant and recessive alleles A and a, with frequencies p and q, are in Hardy-Weinberg equilibrium, the dominant phe-

---

FIGURE 9.3 The Hardy-Weinberg genotype frequencies at a locus with two alleles as a function of the frequency (p) of allele A_1. Note that the frequency of heterozygotes ($2pq$) is greatest when $p = q = 0.5$, and that heterozygotes are the most common genotype in the population if the allele frequencies are between 1/3 and 2/3. (After Hartl and Clark 1989.)

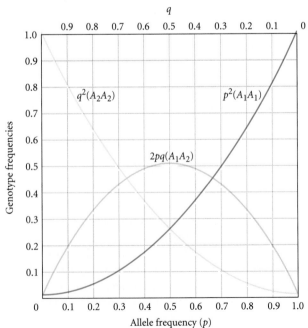

notype makes up $p^2 + 2pq$ of the population, and the recessive phenotype, q^2. Thus if a is very rare, say, $q = 0.01$, only $(0.01)^2 = 0.0001$ of the population displays this recessive allele in its phenotype. Almost all the carriers of a are heterozygotes ($2p/(2p + q) = 1.98/1.99 = 99.5$ percent in this case), which are not distinguishable from dominant homozygotes. If we know only the proportions of *phenotypes* in the population, we can estimate allele frequencies by test-crossing many individuals to determine (from their progeny) whether they are homozygous or heterozygous. If this is impractical, we can estimate allele frequencies from phenotype frequencies, but *only if we assume that the locus is in Hardy-Weinberg equilibrium.*

For example, in another British moth, *Cleora repandata*, black (melanic) coloration is inherited as a dominant allele, and "normal" gray coloration is recessive. In a certain forest, 10 percent of the moths were found to be black (Ford 1971). The *phenotype* frequencies are therefore 0.1 black, 0.9 gray. *If* we assume Hardy-Weinberg equilibrium, $q^2 = 0.9$, so $q = 0.95$ and $p = 0.05$. This would imply that $(0.05)^2 = 0.0025$ of the population are dominant homozygotes and $2(0.05)(0.95) = 0.095$ are heterozygotes.

This example, incidentally, highlights an important point: *a dominant allele may well be less common than a recessive allele.* "DOMINANCE" refers to an allele's phenotypic effect in the heterozygous condition, not to its numerical prevalence. According to the Hardy-Weinberg principle, the frequencies of alleles and phenotypes remain the same from generation to generation, whether or not one allele is dominant over another. If factors such as genetic drift or natural selection cause allele frequencies to change, they may cause either a dominant or a recessive allele to increase in frequency.

In summary, when the assumptions of the Hardy-Weinberg principle hold, the frequencies of alleles and of genotypes in a population remain constant from one generation to another. For two alleles with frequencies p and q ($= 1 - p$), a single generation of random mating establishes the Hardy-Weinberg genotype frequencies p^2, $2pq$, and q^2 for genotypes A_1A_1, A_1A_2, and A_2A_2 respectively. For multiple alleles, the Hardy-Weinberg frequencies are p_i^2 for homozygote A_iA_i and $2p_ip_j$ for heterozygote A_iA_j. If allele frequencies change from p, q to p', q', genotype frequencies p'^2, $2p'q'$, and q'^2 are established after one generation of random mating. The Hardy-Weinberg principle assumes random mating, effectively infinite population size, no mutation, no gene flow, and no natural selection. When these assumptions do not hold, changes in the frequencies of alleles and/or genotypes—evolution—may occur.

Genetic Variation in Natural Populations

Genetic polymorphism is the presence of two or more genetically determined, more or less discretely different phenotypes within a single population of interbreeding individuals. By extension, this term also refers to the presence of two or more alleles in a population, irrespective of

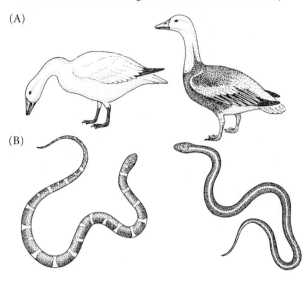

FIGURE 9.4 Single-locus polymorphisms for color pattern. (A) Different color forms of the snow goose *Chen caerulescens*. (B) The two patterns of the king snake *Lampropeltis getulus* found in California. In both cases, the two forms are often found in the same litter. (A after Pough 1951; B after Stebbins 1954.)

their phenotypic effects; thus we often speak of a "POLYMORPHIC LOCUS." The term *polymorphism* is usually further qualified to mean that the rarer allele has a frequency greater than 0.01. A locus or character that is not polymorphic is MONOMORPHIC.

Polymorphisms caused by allelic differences at a single locus have been described in many species of plants and animals. In animals, single-locus polymorphisms have been described not only for color pattern (Figure 9.4), but also for features of anatomy, life cycle, and behavior. For example, in the fruit fly *Drosophila melanogaster*, different genotypes at the *period* locus differ in whether their daily activity rhythm follows a 24-hour cycle or a cycle of slightly different length. Some polymorphic variation is due to multiple alleles. For example, at least 15 alleles at one locus are responsible for the great variation in the color pattern of the ladybird beetle *Harmonia axyridis* (Figure 9.5; Tan 1946).

Beginning with work by the Russian geneticist Sergei Chetverikov in 1926, geneticists found that the same mutations of *Drosophila* that had been studied in laboratory populations—mutations affecting coloration, bristles, wing shape, and the like—also existed in wild populations. In order to study these recessive alleles, it is necessary, using special genetic crosses (Figure 9.6), to create flies that are homozygous for a particular chromosome (e.g., chromosome 2).* Through a series of crosses between a wild fly population and a laboratory stock that carries a dominant marker allele and an inversion that suppresses crossing over (see

---

*The four pairs of homologous chromosomes of *Drosophila melanogaster* are denoted X, 2, 3, and 4. Reference to "hundreds of second chromosomes" means hundreds of copies of the second chromosome, originally carried by wild flies. The symbol "+" refers to a "wild-type" allele at a locus, or to a chromosome that is not known, prior to an experiment, to carry any specific mutation.

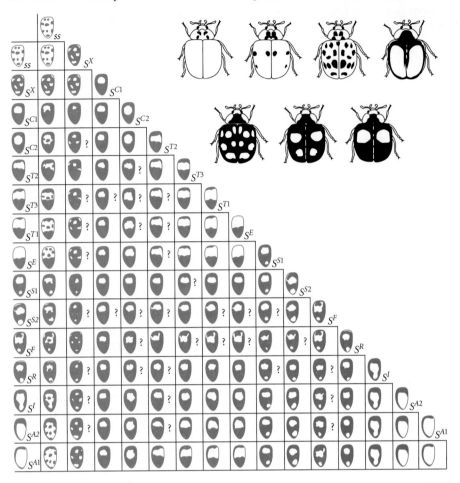

FIGURE 9.5 Variation in the color pattern of the ladybird beetle *Harmonia axyridis* due to 15 alleles at one locus. Each cell shows the pattern on the left wing cover (elytron) of offspring of a cross between the form at the left end of the row and the form at the top of the column. Cells with a question mark show the predicted phenotype of crosses that had not yet been made when this study was published. Letters such as S^x designate alleles; *ss* is the fully recessive homozygote. A sample of beetle phenotypes is shown below. (After Tan 1946 and Dobzhansky 1937.)

Chapter 10), it is possible to "extract" a single chromosome from the wild fly, i.e., to bring it into homozygous condition. The phenotypic effects of recessive alleles are therefore revealed. In addition to recessive alleles that affect morphological features, those that affect traits such as viability (survival from egg to adult) can be detected. The crosses produce a family of flies of which one-fourth are expected to be homozygous for the wild chromosome. The degree of deviation from that expected proportion is a measure of the mortality of such flies within this family. For instance, if no wild-type offspring appear, the wild chromosome must carry at least one recessive LETHAL ALLELE, i.e., one that causes death before adulthood. Performing such crosses with many different wild flies makes it easy to determine what fraction of wild chromosomes cause complete lethality or reduce the likelihood of survival.

Theodosius Dobzhansky,* a towering figure in evolutionary biology who worked first in Russia and then, for most of his life, in the United States, performed many such studies together with his students and other collaborators. They focused on viability, and their results are striking (Figure 9.7). In *Drosophila pseudoobscura*, about 10 percent of the hundreds of second chromosomes sampled from wild flies are lethal when made homozygous. About half of those chromosomes that do not cause death of almost all their bearers (lethal and semilethal) reduce survival to at least some extent (subvital). Moreover, the percentage is almost as high for both the third and fourth chromosomes. Almost every wild fly carries at least one chromosome that, if homozygous, substantially reduces the likelihood of survival. Moreover, many chromosomes cause sterility as well. Many other species of *Drosophila* have been examined, with similar results. Using a very different analytical method, Morton, Crow, and Muller (1956) concluded, from the mortality of children from marriages between relatives, that humans are just like flies: "the average person carries heterozygously the equivalent of 3–5 recessive lethals [lethal alleles] acting between late fetal and early adult stages."

Such data pointed to a surprising, indeed staggering, incidence of life-threatening genetic defects. They implied,

---

*Pronounced Dob-zhan-ski, the *zh* pronounced like *z* in *azure*.

FIGURE 9.6 Crossing technique for "extracting" a chromosome from a male *Drosophila melanogaster* and making it homozygous in order to detect recessive alleles. The "+" denotes a wild-type chromosome that does not carry the mutations (*Cy*, *L*, *Pm*) used as markers. The extraction process is shown for two wild-type males (with chromosomes $+_1, +_2$, and $+_3, +_4$), each of which is crossed to a laboratory stock that has dominant mutant markers *Cy* (curly wing) and *L* (lobed wing) on one chromosome and *Pm* (plum eye color) on the other. Each of these chromosomes also has inversions (see Chapter 10) that prevent crossing over. Consequently, one of a male's wild-type chromosomes, such as $+_2$, is transmitted intact. The third offspring generation consists, in principle, of equal numbers of four genotypes, including a +/+ homozygote and the *CyL*/+ and *Pm*/+ heterozygotes. The viability of these genotypes is measured by their proportion in the family, relative to the expected 1:1:1:1 ratio. The lowermost family illustrates how flies heterozygous for two wild-type chromosomes ($+_2$ and $+_3$) can be produced. Their viability is also measured by deviation from the 1:1:1:1 ratio expected in this cross. (After Dobzhansky 1970.)

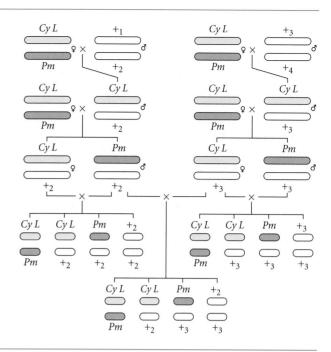

moreover, that *natural populations carry an enormous amount of concealed genetic variation,* that becomes manifested only when individuals are homozygous. However, when genetic crosses are made in *Drosophila* so that two different lethal chromosomes (i.e., derived from two wild flies) are brought together into heterozygous condition, the heterozygous progeny almost always have normal viability. This finding implies that the two chromosomes carry recessive lethal alleles at different loci. Thus one lethal homozygote is *aaBB*, the other is *AAbb*, and the heterozygote, *AaBb*, has normal viability because each recessive lethal is masked by a dominant "normal" allele. From such data, it can be determined that the

lethal allele at any one locus is very rare ($q < 0.01$ or so), and that the high proportion of lethal *chromosomes* is caused by the summation of rare lethal alleles at *many loci*.

The revelation of this variation was surprising in the 1930s, when most people assumed that populations were genetically quite uniform. It led Dobzhansky and others to believe that species are genetically highly variable, and that there is no such thing as a wild-type, or normal, genotype. Rather, Dobzhansky felt, the norm *is* diversity. These studies initiated attempts to estimate levels of genetic variation more quantitatively—an enterprise that took on new meaning in 1966.

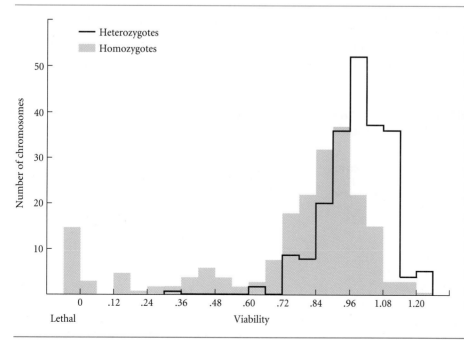

FIGURE 9.7 The distribution of relative viabilities of chromosomes extracted from a wild population of *Drosophila pseudoobscura* by the method illustrated in Figure 9.6. A viability value of 1.00 would indicate conformity to the expected ratios of laboratory and wild-type genotypes, as explained in Figure 9.6. The shaded distribution shows the viability (relative survival from egg to adult) of homozygotes for about 195 wild chromosomes. The great majority lower viability, indicating that they carry deleterious recessive alleles. Heterozygotes for various wild chromosomes, shown by the solid line, have higher average viability. (After Lewontin 1974a.)

Estimating the Proportion of Polymorphic Loci: Protein Electrophoresis

Knowing how much genetic variation a population carries requires that we know what fraction of the loci are polymorphic, how many alleles are present at each locus, and what their frequencies are. To know this, we need to count both the monomorphic (invariant) and polymorphic loci in a random sample of loci. Ordinary phenotypic characters cannot provide this information, because we cannot count how many genes contribute to phenotypically uniform traits.

In 1966, two landmark papers addressed this question. Richard Lewontin and John Hubby (1966), working with *Drosophila pseudoobscura*, and Harry Harris (1966), studying human populations, reasoned that because most loci code for proteins (including enzymes), an invariant enzyme should signal a monomorphic locus, and a variable enzyme a polymorphic locus. Biochemists had already devised techniques for visualizing certain proteins. In ELECTROPHORESIS, a tissue extract (or a homogenate of a whole animal such as *Drosophila*) is placed in a starch gel or some other medium through which proteins can slowly move. An electrical current is applied to the gel, and the proteins move at a rate that depends on the molecules' size and net charge. Certain amino acid substitutions alter the net charge, so some variants of the same protein, encoded by different alleles, can be distinguished by their mobility. The position of a particular enzyme in the gel can be visualized by letting the enzyme react with a substrate, the product being visible as a colored blot when subjected to further reactions. If the locus is monomorphic, all individuals display the same electrophoretic mobility; if it is polymorphic, mobility varies, and the various homozygotes and heterozygotes are distinguishable (Figure 9.8). A number of different enzymes and other proteins, each representing a different locus, can be investigated, thus providing an estimate of the fraction of polymorphic loci. The amount of genetic variation is often underestimated by this technique, because not all amino acid substitutions alter electrophoretic mobility under a particular set of conditions (or under any conditions, for that matter). Thus a particular electrophoretic variant is often called an ELECTROMORPH, and may represent more than

FIGURE 9.8   An electrophoretic gel, showing genetic variation in the enzyme phosphoglucomutase among 18 individual killifishes (*Fundulus zebrinus*). Five electromorphs (alleles) can be distinguished by differences in mobility. The fastest, at top, is allele 1; the slowest, at bottom, is allele 5. The enzyme is a monomer, so homozygotes (such as "2/2") display a single band and heterozygotes (such as "2/5") display two bands. From left to right, the genotypes are 2/2, 2/2, 2/2, 2/5, 2/5, 2/3, 3/3, 2/4, 1/2, 2/5, 2/2, 2/2, 1/2, 2/2, 2/2, 1/2, 2/2, 2/2. (Courtesy of J. B. Mitton.)

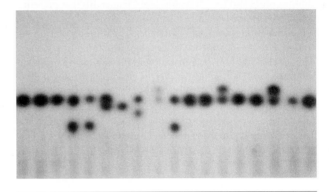

one allele (Coyne 1976). Electrophoretically distinguishable forms of an enzyme that are encoded by different alleles are called ALLOZYMES.*

Lewontin and Hubby examined 18 such loci in samples of *Drosophila pseudoobscura* from five locations in the western United States. In every locality, about a third of the loci were polymorphic, represented by two to six different alleles (Table 9.1). Moreover, unlike the rare lethal alleles studied by Dobzhansky, these alleles had high frequencies. The proportion of heterozygotes is a good measure of how nearly equal in frequency the alleles are (see above). If we assume Hardy-Weinberg equilibrium, the frequency of

---

*Some researchers incorrectly refer to these as *isozymes*, which properly are variant forms of an enzyme encoded by different loci. An individual can be homozygous for two isozymes, but must be heterozygous for two allozymes, which represent segregating alleles at a single locus.

Table 9.1   **Polymorphism and heterozygosity at 18 enzyme loci in five populations of *Drosophila pseudoobscura***

POPULATION	NUMBER OF LOCI POLYMORPHIC	PROPORTION OF LOCI POLYMORPHIC	PROPORTION OF GENOME HETEROZYGOUS PER INDIVIDUAL
Strawberry Canyon	6	0.33	0.148
Wildrose	5	0.28	0.106
Cimarron	5	0.28	0.099
Mather	6	0.33	0.143
Flagstaff	5	0.28	0.081
Average	—	0.30	0.115

Source: After Lewontin and Hubby (1966).

Table 9.2   Genetic variation at allozyme loci in animals and plants

	NUMBER OF SPECIES EXAMINED	AVERAGE NUMBER OF LOCI PER SPECIES	AVERAGE PROPORTION OF LOCI	
			POLYMORPHIC PER POPULATION	HETEROZYGOUS PER INDIVIDUAL
Insects				
Drosophila	28	24	0.529	0.150
Others	4	18	0.531	0.151
Haplodiploid wasps[a]	6	15	0.243	0.062
Marine invertebrates	9	26	0.587	0.147
Marine snails	5	17	0.175	0.083
Land snails	5	18	0.437	0.150
Fishes	14	21	0.306	0.078
Amphibians	11	22	0.336	0.082
Reptiles	9	21	0.231	0.047
Birds	4	19	0.145	0.042
Rodents	26	26	0.202	0.054
Large mammals[b]	4	40	0.233	0.037
Plants[c]	8	8	0.464	0.170

Source: After Selander (1976).

[a]Females are diploid, males haploid.

[b]Human, chimpanzee, pigtailed macaque, and southern elephant seal.

[c]Predominantly outcrossing species.

heterozygotes at a locus is $1 - \sum p_i^2$, where p_i is the frequency of the ith allele and p_i^2 is the frequency of homozygote A_iA_i. This value, averaged over all 18 loci (including the monomorphic ones) was about 0.12 in each population. This is equivalent to saying that an average individual is heterozygous at 12 percent of its loci. (This calculation is called the AVERAGE HETEROZYGOSITY, $\bar{H}$.) If *Drosophila* has 6000 loci (a conservative estimate), this would represent more than 700 heterozygous loci per individual, and about 2000 loci would be polymorphic in a population. Remarkably, Harris found almost exactly the same proportion of polymorphic loci, and the same average heterozygosity, in

human populations. Since these pioneering studies, other investigators have examined hundreds of other species, most of which also proved to have high levels of genetic variation (Figure 9.9, Table 9.2). These other species include "living fossils" such as the horseshoe crab (*Limulus polyphemus*), which have undergone little morphological evolution in hundreds of millions of years, implying that low rates of evolution are not necessarily caused by an insufficiency of genetic variation.

Lewontin and Hubby's paper (see also Lewontin 1974a) had a great impact on evolutionary biology, not only because their data were so striking, but also because they explained

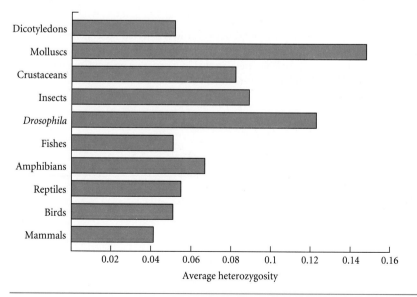

FIGURE 9.9   Average heterozygosity, measured by electrophoresis for protein-encoding loci, averaged across sexually reproducing species within several higher taxa. The abundant data on many *Drosophila* species are plotted separately from those on other insects. Average heterozygosity is generally lower in vertebrates than in invertebrates, although still considerable. (After Gillespie 1991.)

the important questions it raised. These data, and Harris's, confirmed that *almost every individual in a sexually reproducing species is genetically unique* (even with only two alleles, 3000 polymorphic loci, the estimate for humans, could generate $3^{3000} = 10^{1431}$ genotypes, an unimaginably large number). Populations are far more genetically diverse than almost anyone had previously imagined. Lewontin and Hubby asked what factors could possibly be responsible for the existence of so much variation, found most existing hypotheses inadequate to account for it, and set in motion a research agenda that kept population geneticists busy for more than 20 years. Their central question was, "Do forces of natural selection maintain these polymorphisms, or are they neutral, subject only to the operation of random genetic drift?" We will consider this question in Chapters 13 and 22.

Electrophoresis, moreover, provided a tool for investigating many other questions. All of population genetic theory is cast in terms of the frequencies of discrete, identifiable alleles. Before 1966, genes in nature could be studied only with simple phenotypic polymorphisms like the coloration of *Panaxia dominula*—but relatively few species have such polymorphisms, their Mendelian basis often could not be documented in species that cannot be bred in captivity, and it is frequently difficult to distinguish heterozygotes from homozygotes. Proteins provided, in almost every species, abundant polymorphisms with a clear genetic basis. These polymorphisms could be studied in their own right (e.g., to study natural selection), or they could be used simply as GENETIC MARKERS to determine, for example, which individuals mate with each other. They were widely used to determine how much related species or populations differ from each other genetically. And they were a major step toward the ever-increasing use of molecular information in evolutionary biology.

Variation at the DNA Level

For many years, especially after enzyme electrophoresis revealed such abundant genetic variation, evolutionary geneticists hoped for the time when it would be possible to measure variation at the ultimate level—in the nucleotide sequences of genes. That time arrived in the early 1980s, when technical advances made DNA sequencing relatively easy. The technique of using sequencing gels was briefly described in Chapter 3, as was the polymerase chain reaction (PCR), which is used to amplify single DNA strands from small quantities of tissue in order to obtain sufficient copies for sequencing. For analyzing large samples of genes, as is required for estimating population variation, another technique, whereby restriction enzymes are used to reveal RESTRICTION FRAGMENT LENGTH POLYMORPHISMS (RFLPs), is often useful. This method does not yield a complete nucleotide sequence, but instead reveals whether or not genes differ in the presence or absence, at various sites, of specific 4- or 6-base-pair sequences that are cleaved by specific restriction enzymes (see Chapter 3).

The first study of variation by complete DNA sequencing was carried out by Martin Kreitman (1983), who sequenced 11 copies of a 2721-base-pair (bp) region in *Drosophila melanogaster* that includes the locus coding for the enzyme alcohol dehydrogenase (ADH). It was already known that throughout the world, populations of this species are polymorphic for two common electrophoretic alleles, "fast" (Adh^F) and "slow" (Adh^S). In his first study, Kreitman sequenced only 11 gene copies (from three Old World and three U.S. regions) because DNA sequencing at that time was still extremely laborious. Subsequently, he and his colleagues have examined larger samples (87 gene copies) for RFLPs, and they have also examined *Adh* sequences in related species.

The region sequenced includes the four exons and three introns of the *Adh* gene, as well as noncoding flanking regions on either side (Figure 9.10). Among the 11 gene copies, there were 18 variable sites out of 755 sites in the introns (2.4 percent), 8 of 1102 in other nontranslated regions (0.7 percent), and 14 of 765 in the exons, or coding regions (1.8 percent).

FIGURE 9.10 Nucleotide variation at the *Adh* locus in *Drosophila melanogaster*. Four exons (rectangles) are separated by introns (solid lines). The shaded regions are the translated parts of the sequence. The numbers above the diagram of the locus show the positions of 43 variable base pairs and 6 insertions or deletions (indicated by arrows) detected by sequencing 11 copies of this gene. The numbers below show the positions of 27 variable sites detected by restriction enzymes among 87 gene copies. (After Kreitman 1987.)

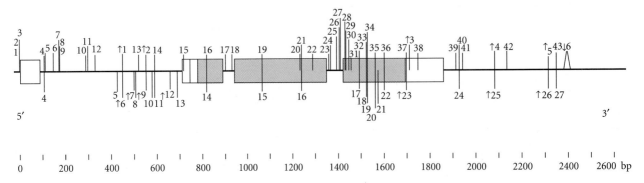

In addition, 6 sites in noncoding regions varied in the presence or absence of short runs of base pairs (INSERTION/DELETION POLYMORPHISMS). The 6 copies of the "slow" allele (Adh^S) were more variable, differing on average at 15.5 sites, than the 5 copies of the "fast" allele (Adh^F), which differed at 8.1 sites on average. The average difference between copies of Adh^S and Adh^F was greater than that among copies of either type, even if these came from different continents.

The most striking discovery was that 13 of the 14 variations in the coding region were SYNONYMOUS SUBSTITUTIONS that would not alter the amino acid sequence of the enzyme. A single REPLACEMENT SUBSTITUTION, substituting threonine for lysine in the protein, differentiates the Adh^S and Adh^F alleles. Thus only one of the 2721 nucleotide sites varies so as to alter the gene product and potentially affect the organism's biochemistry and physiology. [Note also that DNA sequencing did not reveal any replacement substitutions that had not already been detected by electrophoresis. The same result has since been found for several other enzyme loci, but some loci that are highly polymorphic for allozymes have "extra" replacement substitutions that are detected by sequencing, but not by electrophoresis (Riley et al. 1992).]

The other striking feature of Kreitman's sequence data, supported also by the RFLP data, is that the density of silent (synonymous) polymorphisms is greater in exon 4 than elsewhere. They appear to be clustered around the single site responsible for the Adh^S/Adh^F amino acid polymorphism.

In later chapters, we will consider why most of the variations in this gene are silent polymorphisms, and why they should be clustered near the replacement polymorphism. At this point, it is sufficient to emphasize that populations are genetically very variable at the molecular level.

Multiple Loci

Genes do not exist in isolation. Each is embedded in a genome containing thousands of other loci with diverse functions. Two or more genes may affect a single character or different characters. Moreover, each gene is linked to certain other genes, meaning that they are *physically associated* on the same chromosome. (In genetics and evolutionary biology, *linkage* refers *only* to such a physical association between loci, not to the functional relationship or any other relationship among genes.) We must therefore ask whether our descriptions of genetic variation must take multiple loci into account. It is simplest, and sufficient for our present purposes, to consider just two loci (A and B), each with two alleles (A_1, A_2 and B_1, B_2).

Suppose that for some reason allele A_1 became associated exclusively with B_1, and A_2 with B_2, at some point in a population's history. That is, the only genotypes in the population are A_1B_1/A_1B_1, A_1B_1/A_2B_2, and A_2B_2/A_2B_2, where the slash separates the allele combinations an individual received from its two parents. Such an association between specific alleles at different loci is referred to as **linkage disequilibrium**. This common term is an unfortunate choice, because alleles can be associated for various reasons even if

they are not linked (i.e., not located on the same chromosome). Moreover, two loci may be physically linked, yet not be in a state of linkage disequilibrium (in which case the loci are in **linkage equilibrium**).

The strength of linkage disequilibrium—the degree of association between alleles at different loci—varies. Recombination during meiosis reduces linkage disequilibrium and brings the loci toward linkage equilibrium. If there were no recombination in our example, gametes would carry only the allele combinations A_1B_1 and A_2B_2, which, when united, could generate only the same three genotypes as before. Recombination in the double heterozygote, however, gives rise not only to A_1B_1 and A_2B_2 gametes, but also to A_1B_2 and A_2B_1 gametes, which by uniting with A_1B_1 or A_2B_2 gametes produce genotypes such as $A_1A_1B_1B_2$ (A_1B_1/A_1B_2). If the rate of recombination is low, the loci are tightly linked (by definition), and the numerically deficient allele combinations (A_1B_2, A_2B_1) and genotypes increase in frequency slowly from generation to generation (Figure 9.11). If linkage is loose, or if the loci are not linked, the degree of association between alleles at the two loci breaks down (decays) more rapidly. If the decay of the association is not opposed by any other factor, the association ultimately breaks down entirely, and the loci achieve a state of linkage equilibrium, even if they are closely situated on a chromosome.

When the loci arrive at linkage equilibrium, if we were to examine the alleles carried by an egg or sperm taken at random from the population, knowing which A allele it carried would not help us predict its B allele. (In our original hypothetical population, in contrast, we could be sure that a gamete with A_1 also carried B_1.) Therefore, at linkage equilibrium, the probability (frequency) with which a gamete carries any one of the four possible allele combinations is the product of the probabilities of the alleles considered separately. These probabilities are the allele frequencies. If we denote the frequencies of alleles A_1 and A_2 as p_A and q_A ($p_A + q_A = 1$), and of B_1 and B_2 as p_B and q_B ($p_B + q_B = 1$), the frequency of the A_1B_1 combination among the eggs (or sperm) produced by the population as a whole is then $p_A p_B$, and the frequency (proportion) of genotype $A_1A_1B_1B_1$ (A_1B_1/A_1B_1) in the next generation is $(p_A p_B)^2$, or $p_A^2 p_B^2$. Thus, if alleles at two loci are in linkage equilibrium, we can predict the frequency of each combination of alleles, whether in gametes or in diploid organisms, if we know the allele frequencies at each locus. Conversely, if all the genotypes conform to the expected frequencies thus calculated, we can conclude that the loci are in a state of linkage equilibrium.

Whether loci are in linkage equilibrium or linkage disequilibrium, the genotype frequencies at each locus viewed individually conform to Hardy-Weinberg frequencies (unless the fundamental assumptions of the Hardy-Weinberg principle, described earlier, are violated). However, linkage disequilibrium has many important consequences that will be described in subsequent chapters. For example, suppose we drastically change genotype frequencies by eliminating all individuals except

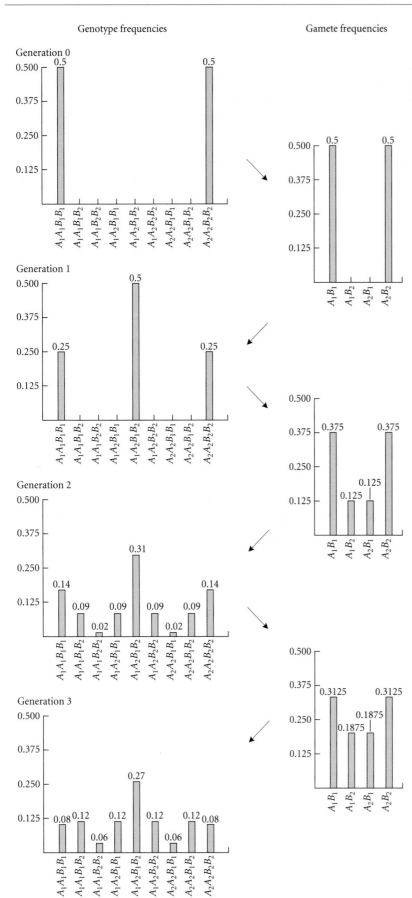

FIGURE 9.11 The decay of linkage disequilibrium between two unlinked loci (recombination rate = 0.5) in a population initiated with equal numbers of two genotypes ($A_1A_1B_1B_1$ and $A_2A_2B_2B_2$) in both sexes. Beginning in the first offspring generation, the frequency of numerically deficient allele combinations in the gametes increases, and that of numerically excessive combinations decreases, due to recombination in the double heterozygotes ($A_1A_2B_1B_2$). Therefore, the frequency distribution of genotypes among the zygotes changes. Throughout, the allele frequencies are $p = q = 0.5$ at both loci.

those that are A_1A_1 (regardless of their genotype at the B locus) from the population. If the loci are in linkage equilibrium, A_1 allele copies are not preferentially associated with either B_1 or B_2, so within the A_1A_1 class, the frequencies of B_1B_1, B_1B_2, and B_2B_2 are p_B^2, $2p_Bq_B$, and q_B^2 respectively. Thus, the elimination of A_2-bearing individuals will not alter the frequencies of alleles or genotypes at the B locus. If, however, A_1 tends to be associated with one of the B alleles, say, B_1, changing the frequency of A_1 will cause a correlated change in the frequency of B_1. In our hypothetical population of parents, in which A_1 and B_1 were perfectly associated, eliminating all except A_1A_1 individuals would eliminate all copies of B_2 along with A_2.

In outbreeding, sexually reproducing populations, pairs of polymorphic loci usually are found to be in linkage equilibrium, or nearly so. For example, Brian Charlesworth and Deborah Charlesworth (1973) used electrophoresis to study five polymorphic enzyme-encoding loci in *Drosophila melanogaster*. Although some weak associations of alleles at different loci were found, even closely linked loci were generally at or near linkage equilibrium.

In certain cases, however, linkage disequilibrium has been found in natural populations. For example, the European primrose *Primula vulgaris* is *heterostylous*, meaning that plants within a population differ in the length of stamens and style (pistil). Almost all plants have either the "pin" phenotype, with long style and short stamens, or the "thrum" phenotype, with short style and long stamens (Figure 9.12). In most experimental crosses, this difference is inherited as if it were due to a single pair of alleles, with thrum dominant over pin. Rarely, however, "homostylous" progeny are produced, in which the female and male structures are equal in length (either short or long). Thus style and stamen length are actually determined by separate, closely linked loci: alleles G and g determine short and long style respectively, and alleles A and a determine long and short stamens respectively. Thrum plants have genotype GA/ga, and pin plants are ga/ga.

Although recombination gives rise to short-homostyled (Ga/ga) and long-homostyled (gA/ga) progeny, these are exceedingly rare in natural populations (Ford 1971). Rather, the allele combinations GA and ga greatly predominate over Ga and gA. The reasons are several. Thrum and pin phenotypes are apparently most successful in cross-pollination, since pollen from one is placed on an insect in a position that corresponds to the stigmatic surface of the other. Moreover, this complex polymorphism actually includes not two, but seven closely linked loci, all in linkage disequilibrium. One of these loci governs an incompatibility reaction, such that pollen placed on the stigma of the same type of flower produces few seeds. Thus the alleles at the several loci appear to be packaged in adaptive combinations that are maintained by natural selection. (For more on heterostyly, see Barrett 1992a and Chapter 21.)

Linkage disequilibrium is common among polymorphic sites within genes (Chapter 22) and in asexual popu-

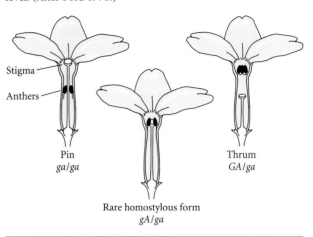

FIGURE 9.12   Heterostyly in the primrose *Primula vulgaris*. Natural populations consist almost entirely of "pin" and "thrum" plants. Rarely, crossing over produces homostylous forms, in which the stigma and anthers are situated at the same level. (After Ford 1971.)

Pin
ga/ga

Rare homostylous form
gA/ga

Thrum
GA/ga

lations (Chapter 21), because recombination rates are very low in these instances.

Variation in Quantitative Traits

Sources of variation   Discrete genetic polymorphisms in phenotypic traits are much less common than slight differences among individuals, such as differences in weight, hair color, and nose shape among humans. In fact, it is hard to think of a phenotypically variable human trait that varies discretely, rather than continuously. Continuous characters such as weight are also called METRIC characters. MERISTIC (countable) characters, such as the number of bristles on a certain part of a fly, vary discretely, but only because the number of bristles must be an integer. In Chapter 3, we saw that the genetic component of such variation is often POLYGENIC: it is due to variation at several or many loci, each with a rather slight contribution to the variation in phenotype. A simple model of the relation between genotype and phenotype, in which we envision only two variable loci, might be:

		A_1A_1	A_1A_2	A_2A_2
		3	4	5
B_1B_1	2	5	6	7
B_1B_2	4	7	8	9
B_2B_2	6	9	10	11

In this example, relative to the genotype $A_1A_1B_1B_1$, each A_2 allele adds one unit, on average, and each B_2 allele adds two units to the phenotype. This is a model of purely ADDITIVE allele effects. (Departures from additivity can be caused by dominance within loci, or by epistatic interactions between loci: see Chapter 3.) A notable feature of this additive model is that the mean phenotype of offspring equals the mean phenotype of their parents. (For example, the mean phenotype of $A_1A_2B_1B_1 \times A_2A_2B_1B_2$ is 7.5, and so is the mean of the four equally frequent genotypes of their progeny.) Another important feature of such inheritance is that the same phenotype can correspond to several different genotypes. In the rest of this section, we will assume that traits are additively inherited.

In addition to this genetic variation, individuals with identical genotypes usually vary at least slightly in phenotype. This nongenetic variation is caused by environmental factors as well as by "developmental noise." For example, the number of scales along the side of a fish may depend not only on the animal's genotype, but also on the temperature during its development, so that variation in temperature contributes to phenotypic variation. But even fish of the same genotype reared at exactly the same temperature will vary slightly in scale number because of ineradicable variations in developmental processes. Such "developmental noise" is often evident in asymmetry between the two sides of the same animal, which obviously have shared both the same genotype and the same environment.

If a character is variable for both genetic and environmental reasons, two individuals may differ because they differ in genotype, because they have had different environmental experiences, or both. We can either perform genetic tests (e.g., examining their offspring) to determine whether they differ in genotype, or focus our attention not on these individuals, but on the population as a whole, to determine the relative importance of genetic versus nongenetic variation.

In Chapter 3, we encountered the concept of a NORM OF REACTION, the variety of different phenotypic states that can be produced by a single genotype under different environmental conditions. An idealized example is shown in Figure 9.13A, and a real example in Figure 9.13B. In the latter case, Gupta and Lewontin (1982) measured the mean number of abdominal bristles on each of ten genotypes of *Drosophila pseudoobscura* raised at three different temperatures. They found a GENOTYPE × ENVIRONMENT INTERACTION, meaning that the effect of temperature on phenotype differed among genotypes. Figures 9.13C and D show that in a hypothetical population consisting of just two of these genotypes, most of the phenotypic variation would be between genotypes if the temperature varied between 17° and 20°, but there would be no substantial effect of genetic variation on phenotypic variation if the temperature varied between 23° and 25°. This example shows that variation in a feature may stem from *both* genetic and environmental causes (so that it is useless to ask whether a characteristic is "genetic" *or* "environmental"). Moreover, the same population may display either a greater genetic component of variation or a greater environmental component, depending on the circumstances.

Estimating components of variation Because each locus contributing to quantitative variation has a small effect on the phenotype relative to the effects of other loci plus nongenetic sources of variation, it is difficult, and often impossible, to identify alleles and measure their frequencies at the various loci. The description and analysis of quantitative variation therefore is based on statistical measures. This "BIOMETRIC" approach to genetic variation is conceptually founded on the theory of allele frequencies, but it does not refer to them explicitly because they cannot be measured in practice. The biometric description of genetic variation was developed largely in the context of breeding domesticated plants and animals, and has been an important tool in developing improved varieties.

If, for simplicity, we ignore the variation that arises from genotype × environment interactions, the variation in any phenotypic characteristic is due to genetic differences and environmental differences among individuals. A statistical measure of variation is the **variance**, which quantifies the spread of individual values around the mean value. (To be precise, the variance is the average squared deviation of observations from the mean: see Box 9.C.) In simple cases, the variance in a phenotypic character (V_P) is the sum of **genetic variance** (V_G) and **environmental variance** (V_E): $V_P = V_G + V_E$. Oversimplifying, we imagine that each genotype in a population has an average phenotypic value (of, say, body length), but that individuals with this genotype vary due to environmental effects or developmental noise. The amount of such variation is the environmental variance, V_E. The rest of the phenotypic variation, which is due to the presence of various genotypes, is the genetic variance, V_G. The magnitude of V_G depends on how many genotypes there are, how greatly the average phenotype of each genotype differs from that of other genotypes, and on the genotype frequencies.

Suppose, for example, (Figure 9.14) the mean length of beetles of genotypes A_1A_1, A_1A_2, and A_2A_2 is 11, 10, and 9 mm respectively. If p and q are the frequencies of A_1 and A_2, and the locus is in Hardy-Weinberg equilibrium, then the genotype frequencies are 0.25, 0.50, and 0.25 if $p = q = 0.5$. The mean length for the population is $(0.25)(11) + (0.50)(10) + (0.25)(9) = 10$, and fully half the population (the A_1A_1 and A_2A_2 individuals) deviates from the mean by 1 mm. If, however, $p = 0.9$ and $q = 0.1$, the genotype frequencies are 0.81, 0.18, and 0.01, the mean for the population is $(0.81)(11) + (0.18)(10) + (0.01)(9) = 10.8$, and fully 81 percent of the population (the A_1A_1 individuals) deviates from the mean by only 0.2 mm. This is a less variable population.

The proportion of the phenotypic variance that is genetic variance is the **heritability** of a trait, denoted h^2. Thus, $h^2 = V_G/(V_G + V_E)$.

One way of detecting a genetic component of variation is to measure correlations[*] between parents and offspring, or between other relatives. In fact, the values of V_G and h^2 can be estimated from such correlations. For example, suppose that in the medium ground finch (*Geospiza fortis*) of the Galápagos Islands, the mean bill depth of the members

---

[*]More properly, the *regression* of offspring mean on the mean of the two parents. The regression coefficient measures the slope of the relationship, and is conceptually related to the correlation (see Appendix I).

FIGURE 9.13 Effects of genotype and environment on phenotype, with and without genotype × environment interaction. (A) An idealized, hypothetical case in which there is no interaction between genotype and environment. The two genotypes, although different in phenotype, respond similarly to differences in temperature. (B) The mean number of bristles on the abdomen of male *Drosophila pseudoobscura* of ten genotypes, reared at three temperatures. The genotypes respond differently to temperature, so there is an interaction between genotype and environment. (C) If flies of genotypes 5 and 10 in B developed at various temperatures between about 17° and 20°C, the distribution of phenotypes would be bimodal, and most of the variation would be genetic, because at these temperatures the genotypes differ greatly in phenotype. (D) A unimodal distribution of phenotypes would be observed if the temperature varied between about 26° and 28°C. Most of the phenotypic variation would be environmental because of the genotypes' similar responses to these temperatures. (After Gupta and Lewontin 1982.)

(A)

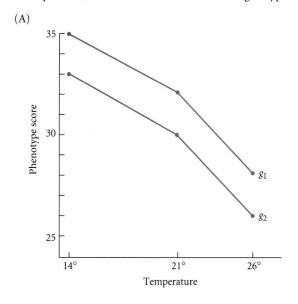

(C)

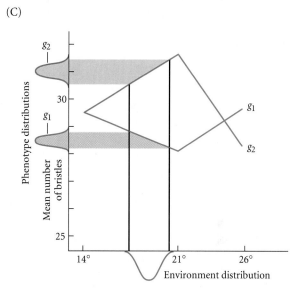

(B)

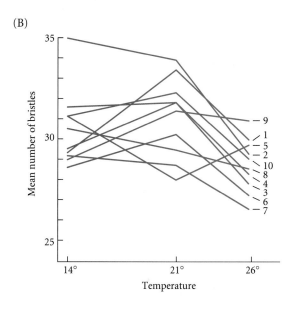

(D)

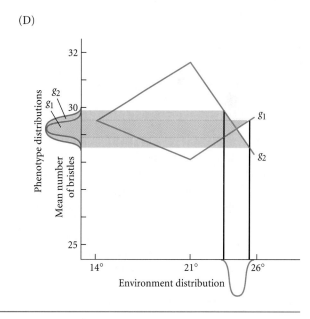

of each brood of offspring were exactly equal to the mean bill depth of their parents (Figure 9.15). So perfect a correlation clearly would imply a strong genetic basis for the trait. In fact, in this instance, V_G/V_P (i.e., the heritability, h^2) would equal 1: all the phenotypic variation would be accounted for by genetic variation. If the correlation were lower, with some of the phenotypic variation perhaps due to environmental variation, the heritability would be lower.

In a real-life example, Peter Boag (1983) kept track of mated pairs of *Geospiza fortis* and their offspring by banding them so that they could be individually recognized, measured the phenotypic variance of bill depth and several other features, and correlated the phenotypes of parents and their offspring. He estimated that the heritability of variation in bill depth was 0.90; that is, about 90 percent of the phenotypic variation is attributable to genetic differ-

Mean, Variance, and Standard Deviation

Suppose we have measured a characteristic in a number of specimens. The character may vary continuously, such as body length, or discontinuously, such as the number of fin rays in a certain fin of a fish. (These are sometimes referred to as metric and meristic characters, respectively.) Let X_i be the value of the variable in the ith specimen (e.g., $X_3 = 10$ cm in fish no. 3). If we have measured n specimens, the sum of the values is $X_1 + X_2 + \ldots X_n$, or

$$\sum_{i=1}^{k} X_i$$

(or simply $\sum X_i$). The **arithmetic mean** (or AVERAGE) is

$$\bar{x} = \frac{\sum X_i}{n}$$

If the variable is discontinuous (e.g., fin rays), we may have n_1 individuals with value X_1, n_2 with value X_2, and so on for k different values. The arithmetic mean is

$$\bar{x} = \frac{n_1 X_1 + n_2 X_2 + \ldots n_k X_k}{n_1 + n_2 + \ldots n_k}$$

The sum of n_i equals n, so we may write this as

$$\bar{x} = \frac{n_1 X_1}{n} + \frac{n_2 X_2}{n} + \ldots + \frac{n_k X_k}{n}$$

If we denote $n_i/n = f_i$, the FREQUENCY of individuals with value X_i, this becomes

$$\bar{x} = \sum_{i=1} (f_i X_i)$$

For example, in a sample of $n = 100$ fish, we may have $n_1 = 16$ with 9 fin rays ($X_1 = 9$), $n_2 = 48$ with 10 rays ($X_2 = 10$), and $n_3 = 36$ with 11 rays ($X_3 = 11$). There are three phenotypic classes ($k = 3$). The mean is

$$\bar{x} = \sum_{i=1} (f_i X_i) = (0.16)(9) + (0.48)(10) + (0.36)(11) = 10.2$$

Because we have only a sample from the fish population, this sample mean is an *estimate* of the true (parametric) mean, which we can know only by measuring every fish in the population. (See Appendix I, which repeats and expands on this material, for a more extensive discussion.)

How shall we measure the amount of variation? We might measure the RANGE (the difference between the two most extreme values), but this measure is very sensitive to sample size. A larger sample might reveal, for example, rare individual fish with 5 or 15 fin rays. These rare individuals do not contribute to our impression of the degree of variability. For this and other reasons, the most commonly used measures of variation are the **variance** and its close relative, the **standard deviation**. The true (parametric) variance is estimated by the mean value of the square of an observation's deviation from the arithmetic mean:

$$V = \frac{(X_1 - \bar{x})^2 + (X_2 - \bar{x})^2 + \ldots (X_n - \bar{x})^2}{n - 1} = \frac{1}{n-1} \sum_{i=1} n_i (X_i - \bar{x})^2$$

For statistical reasons, the denominator of a sample variance is $n - 1$ rather than n. In our hypothetical data on fin ray counts,

$$V = \frac{16(9 - 10.2)^2 + 48(10 - 10.2)^2 + 36(11 - 10.2)^2}{99} = 0.485$$

The variance is statistically very useful, but it is hard to visualize because it is expressed in squared units. It is easier to visualize its square root, the standard deviation:

$$S = \sqrt{V}$$

For our hypothetical data, $S = \sqrt{0.485} = 0.696$. The meaning of this figure can perhaps be best understood by contrast with a sample of 1 fish with 9 rays, 18 with 10 rays, and 81 with 11 rays—an intuitively less variable sample. Then $\bar{x} = 10.8$, $V = 0.149$, and $S = 0.387$. V and S are smaller in this than in the previous sample because more of the individuals are closer to the mean.

A continuously distributed variable often has a bell-shaped, or NORMAL, frequency distribution (Figure 9.C1). In its mathematically idealized form (which many real samples approximate quite well), about 68 percent of the observations fall within one standard deviation on either side of the mean, 96 percent within two standard deviations, and 99.7 percent within three. If, for example, body lengths in a sample of fish are nor-

FIGURE 9.C1   The normal distribution curve, with the mean taken as a zero-reference point, and showing how the variable represented on the *x*-axis can be measured in standard deviations (σ). The bracketed areas show the fraction of the area under the curve (that is, the proportion of observations) embraced by one, two, and three standard deviations on either side of the mean. The true value of the standard deviation is denoted σ; the estimate of σ based on a sample is denoted *S* in this book.

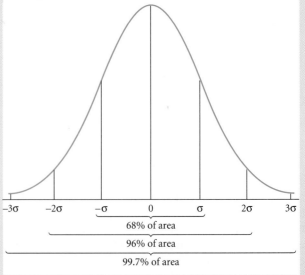

mally distributed, then if $\bar{x} = 100$ cm and $S = 5$ cm (hence $V = 25$ cm^2), 68 percent of the fish are expected to range between 95 and 105 cm, and 96 percent to range between 90 and 110 cm. If the standard deviation is greater, say, $S = 10$ cm ($V = 100$ cm^2), then for the same mean, the range limits embracing 68 percent of the sample are broader: 90 cm and 100 cm.

The variance is often due to several factors that contribute to variation. An important property of variances (but not of standard deviations) is that the variances attributable to the several factors may be added to give the total variance. Table 9.C1 provides a simple example in which the total variance in a phenotypic character is the sum of the variance *among* genotypes and the variance *within* genotypes (which might be the consequence of an environmental variable that affects individuals' development).

The variance among group means is $[(90 - 100)^2 + (100 - 100)^2 + (110 - 100)^2]/3 = 66.67$ (for these calculations, we divide by *n*). Within each group, the variance is 6.00: for example, $[(87 - 90)^2 + (90 - 90)^2 + (93 - 90)^2]/3 = 6.00$. The total variance

is found by summing the squared deviations of all nine values from the grand mean: $[(87 - 100)^2 + (90 - 100)^2 + \ldots + (113 - 100)^2]/9 = 72.67$. In this example, because the sample sizes are equal for all groups, the total variance is the simple sum of among- and within-group variances: $72.67 = 66.67 + 6.00$. The proportion of the variance "explained" by genetic variation is $66.67/72.67 = 0.92$. This is a highly contrived example; when groups differ in size, the sample sizes enter into the calculation of the variances.

Table 9.C1 Additivity of variances

GENOTYPE:	A_1A_1	A_1A_2	A_2A_2
Individual values:	87, 90, 93	97, 100, 103	107, 110, 113
Genotype means:	90	100	110
Within-genotype variances:	6.00	6.00	6.00
Grand mean:		100	

ences, and 10 percent to environmental differences, among individuals.

More commonly, genetic variance and the heritability of various traits have been estimated from organisms reared in a greenhouse or laboratory, in which it is easier to set up controlled matings and to keep track of progeny. The laboratory environment is usually less variable than a natural environment, so estimates of heritability are often higher than if the measurements had been made in wild populations. The examples in Table 9.3 are only a small sample of those measurements that have been made. They show that almost all characteristics, in almost all species, are genetically variable to some extent.

Responses to artificial selection A character can be altered by natural selection only if it is genetically variable.

Therefore, we can infer that if a character has been altered by selection, some of the phenotypic variation in that character must have a genetic basis. This principle underlies the use of **artificial selection** to examine genetic variation. Under artificial selection, an investigator (or an animal or plant breeder) breeds only those individuals that possess a particular trait (or combination of traits) of interest. A common mode of artificial selection is TRUNCATION SELECTION, in which only those individuals that exceed some threshold are allowed to breed: that is, those beyond some fixed character value or, more commonly, a certain percentage at either the upper or lower end of the distribution of phenotypes (Figure 9.16). Artificial selection may grade into natural selection, but the conceptual difference is that the reproductive success of individuals is determined largely by a single characteristic chosen by the investigator,

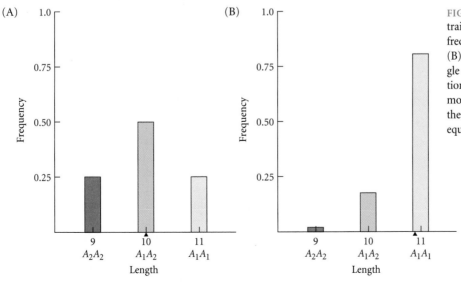

FIGURE 9.14 Variation in a metric trait such as length due to alleles with frequencies (A) $p = 0.5$, $q = 0.5$ or (B) $p = 0.9$, $q = 0.1$. The black triangle denotes the mean. The distribution of lengths is more even, hence more variable, when the alleles have the same frequency. Hardy-Weinberg equilibrium is assumed in both cases.

FIGURE 9.15 The relationship between the phenotypes of offspring and parents. Each point represents the mean of a family of offspring, plotted against the mean of their two parents (midparent mean). (A) A hypothetical case in which offspring means are nearly identical to midparent means. The heritability is nearly 1.00. (B) A hypothetical case in which offspring and parent means are not correlated. The slope of the relationship and the heritability are approximately 0.00. (C) Bill depth in the ground finch *Geospiza fortis* in 1976 (open circles) and in 1978 (solid circles). Although offspring were larger in 1978, the slope of the relationship, shown by the line of best fit to the points, was nearly the same in both years. The heritability, estimated from the slope, was 0.90. (C after Grant 1986, based on Boag 1983.)

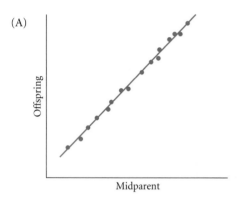

(A)

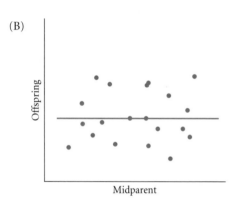

(B)

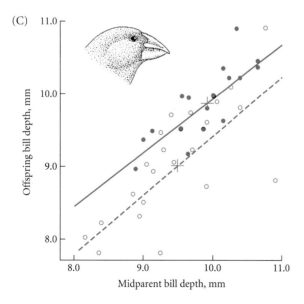

(C)

Table 9.3 Heritability for some traits of animals

TRAIT	HERITABILITY ($h^2 = V_G/V_P$)
Cattle	
Amount of white spotting in Friesian breed	0.95
Milk yield	0.3
Conception rate	0.01
Pigs	
Body length	0.5
Litter size	0.15
Sheep	
Length of wool	0.55
Body weight	0.35
Chickens (White Leghorn)	
Egg weight	0.6
Age at first laying	0.5
Egg production	0.2
Viability	0.1
Mice	
Tail length	0.6
Litter size	0.15
Drosophila melanogaster	
Abdominal bristle number	0.52
Length of thorax	0.43
Ovary size	0.30
Egg production in 4 days	0.18

*These are "narrow-sense heritability" values, V_A/V_P; see Chapter 14.
Source: After Falconer (1981); from various sources.

rather than by their overall capacity (based on all characteristics) for survival and reproduction.

From among hundreds of such experiments, we choose as an example artificial selection on two behavioral traits in the fruit fly *Drosophila pseudoobscura*. Theodosius Dobzhansky and Boris Spassky (1969) crossbred 20 strains of flies (from 20 inseminated females collected at a single locality in California) to form a base population, and from this drew flies to establish eight selected populations, each maintained in a large cage with food. Two of these populations were assigned to each of four selection treatments: positive and negative selection on phototaxis and on geotaxis, i.e., positive and negative responses to light or gravity. To select for phototaxis, the investigators introduced flies (males and virgin females) into a maze in which they had to make 15 successive choices between light and dark pathways, ending up in one of 16 tubes (Figure 9.17A). Flies that made 15 turns toward light arrived at tube 1, those that made 15 turns toward dark arrived at tube 16, and those that made equal numbers of turns toward light and dark ended up in tubes 8 and 9. Flies selected for geotaxis were run through a similar maze, going up or down at each choice point. The mean and variance of phototactic or geotactic scores were estimated from the number of flies in each of the 16 tubes. In each generation, 300 flies of each sex, from each population, were released into the maze, and the 25 flies of each sex that had the most extreme high score (in the positively selected populations) or low score (in the negatively selected populations) were saved to initiate the next

FIGURE 9.16 Diagram of truncation selection, a common method of artificial selection. The solid bell-shaped curve represents the frequency distribution of a trait in a population. In selecting for increases in the trait, an experimenter breeds only the individuals in the shaded region above the truncation point. The dashed curve shows the trait distribution in the next generation, when selection may be repeated, breeding only individuals in the darkened region.

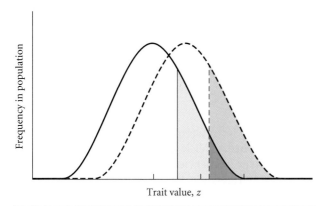

generation. This procedure was repeated for 20 generations.

The data (Figure 9.17B,C; Table 9.4) show that initially, the flies were, on average, neutral for both phototaxis and geotaxis: the mean scores were 8 to 9. Very quickly, however, both the positively and negatively selected populations diverged, in opposite directions, from the initial mean. By generation 15, the range of variation within positively and negatively selected populations hardly overlapped at all.

We may therefore infer that variation among flies in their responses to light and gravity is partly hereditary. It is well understood that the greater the extent to which variation is inherited, the faster the mean will change when subjected to selection. It is therefore easy to estimate heritability from the rapidity of change in an artificial selection experiment of this kind (see Chapter 14). On this basis, Dobzhansky and Spassky calculated that the heritability of phototaxis is about 0.09, and of geotaxis is about 0.03. Thus most of the variation in these behavioral traits (more than 90 percent) is not heritable. Nevertheless, a small genetic component exists, and is sufficient for rapid, rather extensive evolution when selection is strong.

Experiments of this kind have shown that *Drosophila* species are genetically variable for almost every imaginable trait, including features of behavior (e.g., mating speed), morphology, life history (e.g., longevity), physiology (e.g., resistance to insecticides), and even features of the genetic system (e.g., the rate of crossing over). Artificial selection has been the major tool of breeders who have produced agricultural varieties of corn, tomatoes, pigs, chickens, and every other domesticated species, which often differ extremely in numerous characteristics. Such experiences reinforce the conclusion that species contain genetic variation that can serve as the foundation for evolution of almost all of their characteristics.

Genetic variation in uniform traits   In Chapter 3, we encountered the concept of a THRESHOLD TRAIT, a polygenic trait that shifts from one phenotypic state to another only if the "gene dosage" (i.e., of + versus − alleles) exceeds a certain threshold. Such a trait may be phenotypically uniform, yet genetically variable. For example, *Drosophila melanogaster* almost always has 4 precisely situated chaetae (bristles) on the scutellum, a certain part of the thorax. Essentially no variation exists among wild flies. In the laboratory, a mutation, *scute* (*sc*), has been found that lowers the average number of chaetae to about 2.5. The interesting point is that *sc* flies vary. It is possible to make up a population of homozygous *scute* flies by crossing *sc* mutants to wild-type (+) flies and saving only the homozygous *sc* progeny, which vary in chaeta number. James Rendel (1967) did this, and then artificially selected for higher numbers of scutellar bristles. The selection was successful in raising the mean number of bristles. Because the flies were uniform at the *scute* locus (they carried only the *sc* allele), this response to artificial selection indicated that wild flies vary at other gene loci that can affect the number of scutellar chaetae—but that this genetic variation is not expressed in flies that carry the wild-type allele at the *scute* locus. Thus, some characters have *unexpressed genetic variation*, which becomes manifested only under certain circumstances. Such characters have a latent potential for evolution.

Variation among Populations

Geographic Variation

Differences among different geographic populations of the same species have been extensively studied, in part because they are critical for distinguishing species. Studies of **geographic variation** have provided many insights into the mechanisms of evolution, so we shall frequently refer to geographic variation in subsequent chapters. At this point, we will simply illustrate the phenomenon with a few examples.

A classic example is the variation in color pattern of the widespread ladybird beetle *Harmonia axyridis*, studied by Dobzhansky (whose interest in evolutionary genetics grew out of his background as a beetle taxonomist). This beetle displays numerous patterns of black, white, yellow, and red (see Figure 9.5), some of which carry informal varietal names. Experimental crosses led to the conclusion that the color patterns are determined by a large number of alleles at a single locus. The frequencies of some of the phenotypes at various localities from western Siberia to the Pacific rim of Asia are given in Table 9.5. Western populations (e.g., Altai Mountains) are almost uniformly of the *axyridis* type, which is replaced in the east by numerous other "morphs" that vary in frequency from one region to another with no particular pattern.

A pattern is evident, though, in the polymorphic enzyme alcohol dehydrogenase (ADH) in *Drosophila melanogaster*, which we encountered earlier. All three genotypes for the

(A)

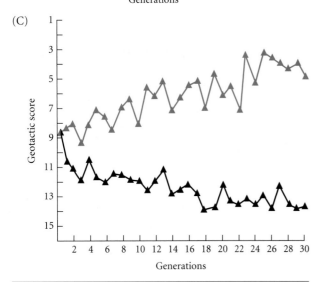

(B)

(C)

FIGURE 9.17 Selection for phototaxis and geotaxis in *Drosophila pseudoobscura*. (A) Diagram of part of a maze used to select for phototaxis (movement in response to light). The maze is oriented vertically below a light source, and flies introduced at the bottom move upward, choosing light or dark at each intersection. Lateral movement across intersections is prevented by barriers. The diagram shows the first 3 of the 15 choices made by the flies in the selection experiment. To select for geotaxis (movement in response to gravity), the maze was rotated 90°, and illumination was uniform. (B) The mean phototactic scores of flies in the first 20 generations of selection in two populations selected for positive (colored triangles) and negative (black triangles) phototaxis. (C) The mean geotactic scores in the first 30 generations of selection. (After Dobzhansky and Spassky 1967, 1969.)

fast and slow electrophoretic alleles (Adh^F and Adh^S) can be distinguished, so allele frequencies are easily calculated from genotype frequencies. The frequency of Adh^F among local populations gradually increases from low to high latitudes *on three continents* (Australia, North America, Eurasia: Figure 9.18). In each area, the locus exhibits a **cline**, which is a gradual change in an allele frequency or character along a geographic transect. By statistical analysis, John Oakeshott and his colleagues (1982) showed that the latitudinal trend is best correlated not with temperature, but with rainfall. The consistency of this pattern strongly suggests that it is adaptive: that the locus is affected by some source of natural selection that varies with latitude. What this selective agent may be, or what it has to do with rainfall, is not yet known.

Our third example draws on an extensive pioneering study of variation by the botanists Jens Clausen, David Keck, and William Hiesey (1940). *Potentilla glandulosa*, the sticky cinquefoil, is an herbaceous perennial in the rose family, found throughout much of western North America from sea level to above the timberline. Populations from different habitats differ in many characteristics, including plant size, the color and size of flowers, flowering time, and many details of leaf morphology, yet even the most extremely different forms produce fertile offspring when crossed, and intermediates are found wherever different forms meet. Certain of these morphological types are **ecotypes**, genetically distinct forms that are consistently found in certain habitats.

Clausen and his collaborators set out to determine whether the differences among ecotypes are genetically

Table 9.4 Responses to selection for phototaxis and geotaxis in *Drosophila pseudoobscura*[a]

| | PHOTOTAXIS | | | | GEOTAXIS | | | |
| | POSITIVE | | NEGATIVE | | POSITIVE | | NEGATIVE | |
GENERATION	MEAN	S.D.	MEAN	S.D.	MEAN	S.D.	MEAN	S.D.
0 (base)	8.70	2.74	8.70	2.74	8.24	4.25	8.24	4.25
1	7.78	2.75	6.77	2.39	10.75	3.55	7.89	4.27
15	13.43	1.81	2.37	1.48	12.14	3.41	4.71	3.17

Source: Dobzhansky and Spassky (1967).

[a]Mean and standard deviation of scores for females at several generations over the course of selection for positive and negative responses

Table 9.5 Frequencies of color patterns (in percentages) in the ladybird beetle *Harmonia axyridis* from different regions of Asia

REGION	SUCCINEA, FRIGIDA, 19-SIGNATA	AULICA	AXYRIDIS	SPECTABILIS	CONSPICUA	OTHERS	NUMBER EXAMINED
Altai Mountains	0.05	—	99.95	—	—	—	4,103
Yeniseisk Province	0.9	—	99.1	—	—	—	116
Irkutsk Province	15.1	—	84.9	—	—	—	73
West Transbaikalia	50.8	—	49.2	—	—	—	61
Amur Province	100.0	—	—	—	—	—	41
Khabarovsk	74.5	0.3	0.2	13.4	10.7	—	597
Vladivostok	85.6	0.8	0.8	6.0	6.8	0.1	765
Korea	81.3	—	—	6.2	12.5	—	64
Manchuria	79.7	0.5	—	11.2	8.6	—	232
North China	83.0	0.4	—	8.8	7.3	0.5	9,676
West China	42.6	2.9	0.01	28.8	25.1	0.8	1,074
East China	66.6	0.6	—	16.5	16.1	0.2	6,231
Japan	17.2	—	11.0	14.3	47.4	—	154

Source: From Dobzhansky (1937).

based or are directly caused by differences in the environment. To do this, they divided, or cloned, each of a number of plants from populations of several different ecotypes, or "races," from sites that differed greatly in altitude, but were at most a few hundred miles apart. Plants from each site were grown in a common garden at Stanford, California (eleva-

(A)

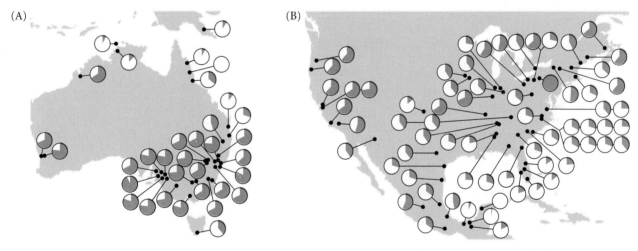

(B)

(C)

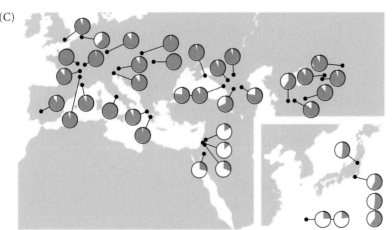

FIGURE 9.18 Clines in the frequency of the *Adh^F* allele at the alcohol dehydrogenase locus of *Drosophila melanogaster* in (A) Australia, (B) North America, and (C) Eurasia. The colored area of each "pie" diagram represents the frequency of *Adh^F*, which decreases at lower latitudes in all three regions. (After Oakeshott et al. 1982.)

tion 30 meters), at Mather (in the foothills at 1400 m), and at Timberline (above the tree line at 3050 m). For several years, the investigators measured the characteristics of the resulting plants.

Many of the differentiating features, such as flower color and leaf morphology, remained distinctive even when the different populations were grown together in all three common gardens. Clausen et al. concluded that these features differ genetically, and are not substantially affected by the environment. Other features, such as plant height and flowering time, varied among the three gardens, showing that they were affected by the environment, but the ecotypes nevertheless differed one from another in each garden, implying that they differ genetically in these features. Figure 9.19 presents data for the ecotypes *typica*, which naturally grows at low altitudes, *hanseni*, from middle elevations (1200–1800 m), and alpine *nevadensis*, from high elevations (above 2700 m). In nature, plant height, from tallest to shortest, is *hanseni-typica-nevadensis*. In the common gardens, *hanseni* is always tallest but the growth of *typica* is so stunted at the higher sites that it is shorter than *nevadensis*. In fact, very few *typica* plants survived at Timberline, showing that the populations differ greatly in their physi-

ological adaptation to different environments. For all ecotypes, flowering time is later at higher than at lower elevations, but at each of the three sites, *nevadensis* flowers first and *hanseni* last. Again, this characteristic varies genetically among populations, even though it is also directly affected by the environment. In further work, Clausen et al. concluded, from F_1 and F_2 crosses between mid- and high-elevation ecotypes, that each of the differences between them—height, flowering time, flower color, and the rest—is based on differences at several loci. All of these features are polygenic.

Genetic Distance

It is sometimes useful to quantify the degree of genetic differentiation among two or more populations of the same species, or among different species. Electrophoretic data have been most used for this purpose, as the loci have been assumed to represent a random sample of differences throughout the genome. At each of several loci encoding electrophoretically detectable proteins, we can estimate the frequency (p_{ij}) of allele i in population j. These values can be used to calculate several indices of genetic similarity or genetic difference (**genetic distance**) between a pair of

FIGURE 9.19 Clausen et al.'s common garden study of the cinquefoil *Polentilla glandulosa*. (A) Representative specimens of coastal, mid-altitude, and alpine ecotypes, grown together at Mather, California (elevation 1400 m), shown in early June. (B) The mean height of three ecotypes, *typica*, *hanseni*, and *nevadensis*, grown in common gardens at three elevations. The differences among ecotypes at each elevation reflect genetic differences, and the differences in each ecotype among elevations show the effect of environment. (C) A similar plot for flowering time. (After Clausen et al. 1940.)

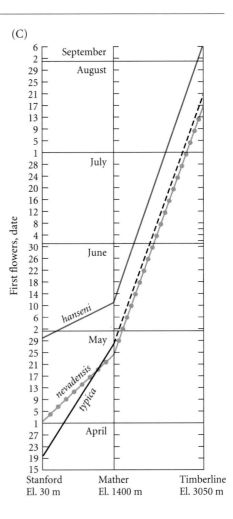

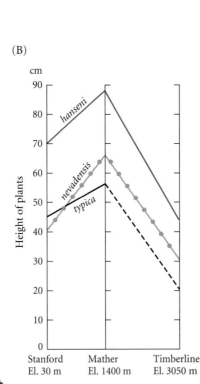

populations. The most frequently used indices were suggested by Masatoshi Nei (1987), and are referred to as "Nei's *I*" (for similarity) and "Nei's *D*" (for distance). If all alleles at all loci have the same frequencies in two populations, the genetic distance is zero, whereas it increases indefinitely as the populations share fewer alleles or differ more in allele frequencies. Box 9.D presents the formula, as well as an example that illustrates the common finding that electrophoretic divergence among populations is correlated with other aspects of genetic divergence.

Patterns of Geographic Variation

If distinct forms or populations have overlapping geographic distributions, such that they occupy the same area and can frequently encounter each other, they are **sympatric** (Greek *syn*, "together," and *patra*, "fatherland"). Populations with adjacent but nonoverlapping geographic ranges that come into contact are **parapatric** (Greek *para*, "beside"). Populations with separated distributions are **allopatric** (Greek *allos*, "other").

The genetic and phenotypic differences among populations of a species vary from very slight to very pronounced, and may occur over short or long distances. (The factors responsible for different patterns are considered in subsequent chapters.) Often, allele frequencies differ among populations at some loci, but not others, and the same is true of phenotypic characters. Several kinds of patterns of geographic variation are frequently encountered (Figure 9.20).

A **subspecies**, or geographic race, in zoological taxonomy means a recognizably different population, or group of populations, that occupies a different geographic area from other populations of the same species. (In botanical taxonomy, subspecies names are sometimes given to sympatric, interbreeding forms.) Subspecies are distinguished by one or more characters, but no rules specify how different populations must be to be named as subspecies. Hence they are arbitrarily defined. Different characters that have different patterns of geographic variation may distinguish subspecies. For example, in the rat snake *Elaphe obsoleta* (Figure 9.21), coloration distinguishes *Elaphe obsoleta quadrivittata* (yellow) from *E. o. rossalleni* (orange) from *E. o. obsoleta* (black); lengthwise stripes distinguish *E. o. quadrivittata* and *E. o. rossalleni* from the blotched subspecies *E. o. spiloides*, and so on. It is quite possible that other characteristics such as allozymes could be used to break each of these down into still more subspecies. Because the number of distinguishable populations may depend only on how many characteristics are studied, some systematists have argued that subspecies shold not be named in such cases (Wilson and Brown 1953).

In some instances, subspecies differ in a number of features with concordant patterns of geographic variation (Figure 9.20A). For example, in the northern flicker (*Colaptes auratus*), populations in eastern and western North America, subspecies *auratus* and *cafer* respectively, differ in the color of the wing feathers, in crown and mustache marks, in the presence or absence of several plumage marks, and in size (Short 1965; Moore and Price 1993). Despite these differences, the two populations interbreed in the Great Plains (Figure 9.22), forming a wide **hybrid zone** (a region in which genetically distinct parapatric forms interbreed).

When one or more characters differ among allopatric populations, it is difficult to specify whether they should be treated as different species or as geographic variants of a single species. This is the case, for instance, with island forms of drongos (Figure 9.23), crowlike birds of the Old World tropics, which differ in the form of the crest. Unless we were to bring drongos from different islands together, there is no way to know whether they would interbreed or not.

FIGURE 9.20 Highly diagrammatic representations of some common patterns of geographic variation within species. (A) Two classic subspecies that interbreed along a narrow border. Size and color are correlated. (B) Abrupt transition in each of two characters (size, color) that have discordant distributions. (C) Concordant clines in two characters. (D) Discordant clines in two characters. (E) An east-west cline in the frequency of one allele or phenotypic state, represented by the colored portion of each pie diagram, in a polymorphic locus or character. (F) A mosaic distribution of two phenotypes, as might be observed if they were associated with mosaic habitats (e.g., wet and dry).

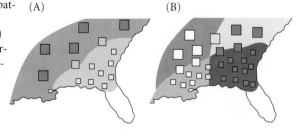

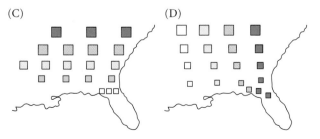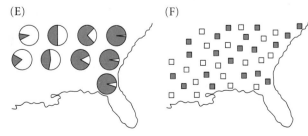

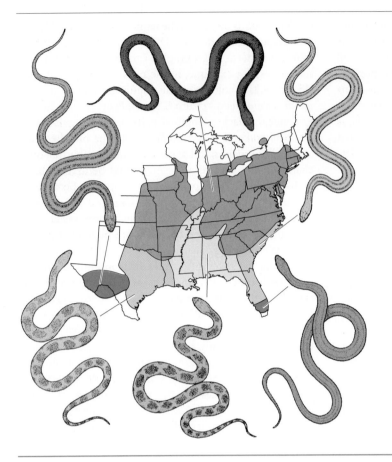

FIGURE 9.21 Classic subspecies: the rat snake *Elaphe obsoleta*. These allopatric geographic races interbreed where their ranges meet. The subspecies differ in pattern (stripes or blotches) and in color (*E. o. obsoleta* is black, *E. o. quadrivittata* yellow, *E. o. rossalleni* orange, *E. o. spiloides* and *E. o. lindheimeri* different shades of brown). (After Conant 1958.)

As we noted above, a gradual change in a character or in allele frequency along a geographic transect is called a cline. Clines may extend over broad areas; for instance, body size in the white-tailed deer (*Odocoileus virginianus*) increases with increasing latitude over much of North America. This relationship between body size and latitude is so common in mammals and birds that it has been named *Bergmann's rule*. Two other such "rules," or generalizations, of geographic variation are that populations of homeotherms in colder climates have shorter appendages (*Allen's rule*) and

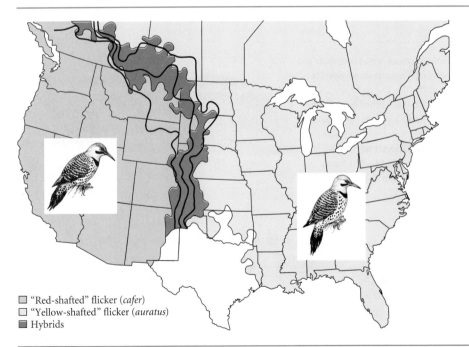

"Red-shafted" flicker (*cafer*)
"Yellow-shafted" flicker (*auratus*)
Hybrids

FIGURE 9.22 A hybrid zone in a common North American woodpecker, the northern flicker (*Colaptes auratus*). The eastern ("yellow-shafted") subspecies (*C. a. auratus*) and the western ("red-shafted") subspecies (*C. a. cafer*) form a broad hybrid zone in the region indicated by the wavy lines. From west to east, these lines indicate regions where the birds are generally more red-shafted, intermediate, and yellow-shafted in their features. (After Moore and Price 1993.)

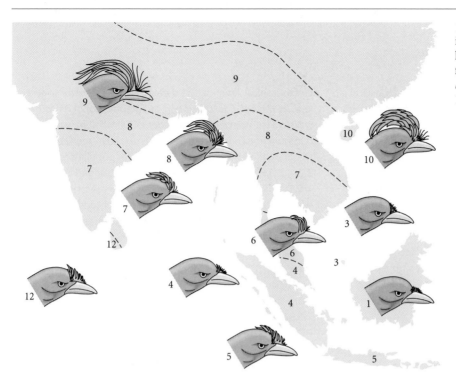

FIGURE 9.23 Geographic variation in the form of the crest among populations of drongos, currently classified as a single species, *Dicrurus paradisaeus*. (After Mayr and Vaurie 1948.)

that populations in more arid environments are paler in color (*Gloger's rule*). These patterns, because they occur in so many species, provide important evidence of ADAPTIVE GEOGRAPHIC VARIATION due to natural selection, since they are too consistent to be attributed to chance. Larger body size and shorter appendages are thought to be advantageous for homeotherms in colder climates because they reduce the surface area, relative to body mass, over which body heat is lost (see Chapter 17). Pale coloration, matching the pale soils and vegetation common in arid regions, probably provides protection against predators.

Clines in different characters are often discordant, i.e., they have different geographic patterns (Figure 9.20D). For instance, the house sparrow (*Passer domesticus*) in North America displays a north-south cline in body size (Bergmann's rule) and a more or less east-west cline in intensity of coloration (Gloger's rule). These clines have evolved since the species was introduced about 100 years ago from Europe to New York, whence it spread throughout the continent (Johnston and Selander 1964).

In some instances, a character may show a more or less mosaic pattern (Figure 9.20F), usually correlated with an environmental factor. For example, the short form of the cinquefoil *Potentilla glandulosa* that Clausen and his collaborators studied is restricted not to a single contiguous geographic area, but to the high elevations of many different mountains. As noted above, such habitat-associated phenotypes are often called *ecotypes*. In some instances, it has been shown that the same ecotypic feature in different areas differs in its genetic basis, implying that it evolved independently. Thus convergent evolution may occur among different populations of the same species (see Chapter 14).

Kinds of Geographically Variable Characters

Any characteristic of a species may vary from one population to another. We have already provided examples of variations in morphology and allozymes. Physiological features also vary; for instance, in the goldenrod *Solidago virgaurea*, populations from exposed habitats display a high photosynthetic rate over a greater range of light intensities than populations from shaded habitats (Björkman and Holmgren 1963). Very often, as in this case, the pattern of geographic variation is what one would expect if populations have become adapted to their local environments. In some cases, the nature of the adaptive variation is at first surprising. For example, in nature, larval development in montane populations of the green frog (*Rana clamitans*) is slower than in lowland populations, simply because of lower temperatures at high altitudes. When reared at the same low temperature in the laboratory, however, montane larvae develop faster than lowland larvae (Berven et al. 1979). Thus the genetic differences among populations compensate, in part, for the effects of temperature in the field, and run counter to the phenotypic differences observed in nature. This pattern is termed **countergradient variation**. This case also exemplifies geographic variation in life history.

The ecological characteristics of species—their ecological niches—also vary among populations. For instance, in coastal California and Oregon, the garter snake *Thamnophis elegans* feeds on slugs, whereas slugs do not occur in drier inland regions where the snake also lives. Newborn, naive snakes from coastal populations display a pronounced feeding response to slugs, whereas those from inland populations do not (Arnold 1981b; see Chapter 14). In some cases, sympatric populations of two species differ more than allopatric

BOX 9.D

Genetic Identity and Distance

At one locus, let p_{i1} and p_{i2} be the frequencies of the ith allele (of k alleles) in populations 1 and 2 respectively. Within population 1, the probability that two randomly chosen genes are both allele 1 is p_{11}^2, and similarly for each other allele (p_{21}^2, p_{31}^2, etc.). Hence the probability of identity of two genes is

$$J_1 = \sum_{i=1} p_{i1}^2$$

within population 1, and similarly

$$J_2 = \sum_{i=1} p_{i2}^2$$

within population 2. If we take one gene copy from each population, the probability that both are allele 1 is $p_{11}p_{12}$, and similarly for each other allele, so the probability of sampling two identical genes is

$$J_{12} = \sum_{i=1} p_{i1}p_{i.}$$

Nei's (1977) index of genetic identity is defined as

$$I = \frac{J_{12}}{\sqrt{J_1 J_2}}$$

and his index of genetic distance is D = $-\log_e(I)$. Given allele frequencies at several loci, I or D is calculated for each locus, including those that display no allelic variation in one or both populations, and the arithmetic mean of these values is then the genetic identity or distance.

For example, Francisco Ayala and coworkers, in a series of papers, estimated allele frequencies at 28 to 30 electrophoretic loci in numerous populations of *Drosophila willistoni* and related species in tropical America. The data include frequencies of six alleles at the *Lap-5* locus for two geographic populations of *D. willistoni* (subspecies *willistoni*), one of *D. willistoni* (subspecies *quechua*), and the related species *D. tropicalis* (see Table 9.D1). Denoting the Santo Domingo and Caracas populations as 1 and 2, $J_1 = 0.432$, $J_2 = 0.350$, $J_{12} = 0.345$, so I = 0.887 and D = $-\log_e(I)$ = 0.119. These populations interbreed freely in the laboratory, whereas the subspecies interbreed less freely, and are also less similar at this locus (I = 0.66, D = 0.42 for Caracas *willistoni* versus *quechua*). The fully reproductively isolated species *D. tropicalis*, although it shares many of the same alleles, is still more dissimilar (I = 0.33, D = 1.12 for Caracas *willistoni* versus *tropicalis*).

These calculations are for only one locus. Averaged over many loci, Ayala et al. calculated the mean genetic identity as I = 0.97 (D = 0.03) among local populations of *D. willistoni* and I = 0.80 (D = 0.23) between its subspecies, whereas between species in the *willistoni* group, the mean identity was 0.35 (D = 1.06). Note that there is no reason to believe that the loci studied cause divergence into different species; they are presumed only to be markers, or indicators, of genetic divergence in general.

Table 9.D1 **Allele frequencies in *Drosophila* and related species**

ALLELE:	1	2	3	4	5	6
D. w. willistoni (Santo Domingo)	0.006	0.057	0.494	0.429	0.015	0.000
D. w. willistoni (Caracas, Venezuela)	0.010	0.110	0.230	0.520	0.120	0.000
D. w. quechua (Lima, Peru)	0.000	0.035	0.719	0.246	0.000	0.000
D. tropicalis	0.000	0.002	0.008	0.120	0.848	0.021

populations in their use of food or other resources, and the difference may be reflected in their morphology. Such a pattern is termed **character displacement**. Several instances are presented by the ground finches of the Galápagos Islands (Grant 1986). For example, *Geospiza fortis* and *G. fuliginosa* are the only ground finches on the islands of Daphne and Hermanos respectively; these populations have almost the same bill size and feed on the same kinds of seeds. On islands on which these same species co-occur, such as Santa Cruz, *G. fortis* has a substantially larger bill than *G. fuliginosa*, and feeds on larger, harder seeds (Figure 9.24).

Geographic populations also vary in the degree to which they are reproductively compatible with each other. Differences among populations may take the form of SEXUAL ISO- LATION, i.e., failure to mate. For example, Stephen Tilley and colleagues (1990) examined sexual isolation among populations of the dusky salamander, *Desmognathus ochrophaeus*, from various localities in the southern Appalachian Mountains of the eastern United States. They scored the fraction of matings that occurred between males and females in various pairings of different populations (heterotypic pairings) and within populations (homotypic pairings). The strength of sexual isolation was measured by an isolation index ranging from zero (when the proportions of heterotypic and homotypic matings were equal) to two (when all homotypic pairs mated, but no heterotypic pairs did so). They found that among the various pairings of populations, the isolation index varied continuously from almost no to almost complete sexual isolation—and that the more distant the populations, the stronger was the strength of

FIGURE 9.24 Character displacement in bill size in ground finches of the Galápagos Islands. The species *Geospiza fortis* and *G. fuliginosa* are more similar where they occur separately (A,B) than where they occur together (C,D). (A) *G. fortis* from the is- land of Daphne Major; (B) *G. fuliginosa* from Hermanos; (C) *G. fortis* and (D) *G. fuliginosa* from Bahía Académia, Santa Cruz. (A by P. R. Grant; B by D. Schluter; C and D by W. Clark. From Grant 1986; photographs courtesy of P. R. Grant.)

(A)

(B)

(C)

(D)

sexual isolation between them (Figure 9.25). Moreover, sexual isolation was correlated with genetic distance (Nei's *D*), measured by allele frequency differences at enzyme-encoding loci. Thus the more genetically differentiated populations were less prone to interbreed.

Reproductive incompatibility among populations of a species can also take the form of varying levels of inviability or infertility (sterility) of hybrids between them. For instance, the fertility of experimentally produced hybrids among Californian populations of the plant *Streptanthus glandulosus* is lower than that of within-population crosses in most cases, and the greater the geographic distance between populations, the lower the fertility of the hybrids (Kruckeberg 1957). The geneticist Richard Goldschmidt (1940), an important figure whose name will resurface in our analysis of macroevolution (Chapter 24), obtained fascinating results from crosses among eastern Asian populations of the gypsy moth (*Lymantria dispar*). These form what Goldschmidt called "sex races" that vary from "weak" to "strong" (Figure 9.26). A cross between females from a "weak" population and males from a "strong" population produces daughters that are phenotypically male (although they have the sex chromosomes of females); crosses between females from "half-weak" populations and males from "strong" populations yield daughters that are sterile intersexes, combining the morphological features of both sexes. Thus certain local populations of gypsy moths would be incapable of exchanging genes if they were to meet.

Sometimes even features that are used to distinguish higher taxa can vary within and among populations of a sin-

FIGURE 9.25 The degree of sexual isolation between various populations of the salamander *Desmognathus ochrophaeus* in experimental pairings in the laboratory. Sexual isolation is equally well correlated with the geographic distance between populations and with Nei's genetic distance, which measures the degree of difference in allele frequencies at several enzyme-encoding loci analyzed by electrophoresis. The more divergent and geographically distant the populations, the less likely they are to interbreed. (After Tilley et al. 1990.)

(A)

(B)
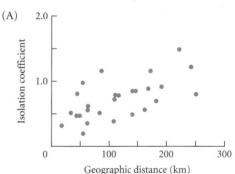

Variation 261

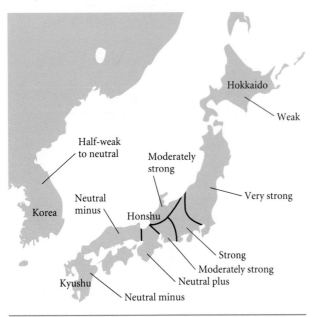

FIGURE 9.26 Richard Goldschmidt's characterization of the "sex races" of the gypsy moth (*Lymantria dispar*) in eastern Asia. (After Goldschmidt 1940.)

gle species. For example, one of the characters that distinguishes the two great classes of flowering plants, the Monocotyledonae (lilies, palms, grasses, etc.) and the Dicotyledonae (roses, magnolias, and in fact the majority of angiosperms), is the number of "seedling leaves," or cotyledons, respectively one or two. But even this fundamental character varies within species of certain shrubs in the genus *Pittosporum*, and it has been possible to select strains of snapdragons (*Antirrhinum*) with three cotyledons (Stebbins 1974). Within the family Asteraceae (sunflowers and relatives), taxonomists traditionally distinguished the tribe Helenieae from the tribe Heliantheae by the lack of ray florets and bracts in the former. Jens Clausen and collaborators (1947) discovered a small population of plants that lacked these features, and classified it as a new genus and species, *Roxira serpentina*, in the Helenieae. However, it proved to be an aberrant population of *Layia glandulosa* in the Heliantheae; crosses with this species showed that just two loci control the presence or absence of ray florets and bracts (Figure 9.27).

Conclusions about variation among populations

The most important implications of these examples are:

1. A species is not genetically uniform over its geographic range. Populations differ in allele and genotype frequencies, often very considerably and in many different characteristics.

2. At least some of these differences appear to be adaptive consequences of occupying different environments.

3. The genetic differences between populations are the same in kind as genetic differences among individuals within populations. Some differences are polygenic,

and are inherited in the same way in between-population crosses as in within-population crosses. In other cases we can examine individual loci, and find that populations differ in the frequencies of the same alleles that segregate within populations. Sometimes, as for *Adh* in *Drosophila*, populations differ greatly in allele frequency, but we can find other populations with intermediate allele frequencies.

These observations lead us to conclude that

4. Differences among populations range from slight to great, and it may be difficult or even arbitrary to determine if populations belong to one or more than one species. Conspecific populations vary in the degree of reproductive isolation, and even in characters that ordinarily distinguish higher taxa.

5. Differences among populations arise by the transformation of genetic variation *within* populations into variation *among* populations, due to changes in allele frequencies that transpire differently from one population to another. This is a conclusion of the highest importance.

6. Variation among conspecific populations forms a continuum with variation among species.

FIGURE 9.27 A plant first described as a new genus, *Roxira* (P_2, at upper right), is actually conspecific with *Layia glandulosa* (P_1). Segregation of phenotypes in the F_2 generation showed that the distinguishing characteristics are determined by only a few gene differences. (From Clausen et al. 1947.)

Species and Higher Taxa

Genetic variation exists among individuals within and among populations of a species—but what is a species? This is a highly controversial question; no single definition of species has ever satisfied all biologists. This book uses the definition of a species that has been most widely adopted in evolutionary biology, the **biological species concept**, as proposed by Ernst Mayr in 1942: "Species are groups of actually or potentially interbreeding populations that are reproductively isolated from other such groups." (Note that a species is a population or a group of populations of individual organisms. A particular organism such as Leonardo da Vinci is a member of a species, but is not in itself a species. The term "species" should be used to refer only to the collectivity of individuals that compose the species, not to the individuals themselves, for we will often contrast species with individuals.)

The reasons for controversy over the meaning of species, and the features and implications of the biological species concept, are described in Chapter 15. For our present purposes, a few points are useful.

1. A biological species is a population, or a group of populations, among which there is interbreeding (gene exchange). Two individual organisms might not be able to interbreed, but they are members of the same species (CONSPECIFIC) if they are part of the same "gene pool" (Chihuahuas and Great Danes are an example).

2. The *degree* of phenotypic difference (e.g., in morphology) is not a criterion, in and of itself, of species. Very different-looking organisms are conspecific if they interbreed; virtually indistinguishable organisms may not interbreed, and so represent different species.

3. The definition applies to sexually reproducing organisms in which the genes of different individuals are combined in offspring. It does not apply to asexually reproducing organisms or to those that reproduce entirely by self-fertilization. (Those organisms, however, may bear species names, such as the mostly asexually reproducing bacterium *Escherichia coli*, which illustrates the confusing fact, discussed in Chapter 15, that the word "species" does double duty. It denotes a bio-

logical entity with an important role in evolution, and it also denotes a taxonomic category, a unit of classification. *Escherichia coli* is the name of a taxon, but it is not a biological species.)

Variation in a gene or in a character may thus be partitioned into variation within populations, variation among populations within species, and variation among species. This is easily seen by considering variation in a feature such as body size. Among North American deer, for example, individuals vary in size within every population; geographic variation in size exists among populations of the white-tailed deer (*Odocoileus virginianus*), which is large in the north and very small in the Florida Keys; and different species of deer (e.g., the white-tailed deer, the elk, *Cervus canadensis*, and the moose, *Alces americana*) differ one from another. At a single gene locus, likewise, there may exist variations at all these levels. Allele frequencies at polymorphic loci often differ among conspecific populations. Related species may have entirely different alleles at a locus, or they may share some or all of the same alleles, usually with different allele frequencies (as in Box 9.D). At some loci, species may share polymorphisms, whereas at other loci, they have different alleles (see Chapter 15).

In sexually reproducing organisms, biological species are usually classified as different taxa with different species names (e.g., the gray squirrel, *Sciurus carolinensis*, and the fox squirrel, *Sciurus niger*). The species, as a unit of classification, is considered by many biologists to be the only taxonomic category that has objective biological reality (defined by the criterion of interbreeding). The other taxonomic categories may also provide a way to express the variation among organisms, but they are usually considered to be arbitrary constructs. Below the species level, subspecies, as we have seen, are arbitrarily defined; they are merely sets of populations that differ in one or more features, and the number of subspecies recognized is likely to depend on how many features are examined and how great a difference a taxonomist considers worthy of naming. Above the species level, genera and other higher taxa may be nonarbitrary insofar as they are monophyletic, but every monophyletic clade is nested within a larger clade, and no criteria exist for determining which branches of a larger phylogenetic tree should be named as higher taxa.

Summary

1. Lamarckian "transformational" theories of evolution required that inherited changes be directly caused by the environment, and guided by the environment in adaptive directions. There is no evidence for, and much evidence against, this notion. Transformational theories have therefore been rejected in favor of the variational theory, proposed by Darwin and Wallace, that evolution occurs by the replacement of some genotypes by others. Hence evolution requires genetic variation.

2. Mendel's discovery that inheritance is particulate rather than blending implies that genetic variation can persist indefinitely.

3. The all-important concepts of allele frequency and genotype frequency are central to the Hardy-Weinberg principle, which states that in the absence of perturbing factors, allele and genotype frequencies remain constant over generations, with genotype frequencies given by the binomial expansion $(p + q)^2 = p^2 + 2pq + q^2$. For k multiple alleles, the binomial expansion $(p_1 + p_2 + \ldots p_k)^2$ gives the genotype frequencies.

4. The potential causes of evolution at a single locus are those factors that can cause deviations from the Hardy-Weinberg equilibrium. These factors are (*a*) finite population size, resulting in random changes in allele fre-

quencies (genetic drift); (*b*) nonrandom mating; (*c*) transformation of alleles (mutation); (*d*) incursion of genes from other populations (gene flow); and (*e*) consistent differences among genes or genotypes in reproductive success (natural selection).

5. Evolution would be very slow if populations were genetically uniform and only occasional mutations arose to replace pre-existing genotypes. However, populations of most species contain a great deal of genetic variation. This variation includes rare alleles at many loci, which usually appear to be deleterious. But it also includes many common alleles, so that many loci—perhaps up to a third of them—are polymorphic. Much of this variation has been revealed by enzyme electrophoresis and, more recently, by analyses of DNA sequence variation.

6. Many phenotypic traits, including morphological, physiological, and behavioral features, exhibit polygenic variation. Although the individual contributing loci cannot usually be distinguished and studied, the magnitude of this variation can be estimated by breeding experiments and by artificial selection.

7. Variation in most phenotypic traits includes both a genetic component and a nongenetic ("environmental") component. The proportion of the phenotypic variance that is due to genetic variation is the heritability of the trait.

8. Alleles at different loci, affecting the same or different traits, sometimes are nonrandomly associated within a population, a condition called linkage disequilibrium (even though the genes are sometimes not physically linked). More commonly, specific alleles at different loci are not associated (linkage equilibrium). Linkage disequilibrium affects the frequencies of genotypes considered as pairs (or greater numbers) of loci, but at each locus, genotype frequencies remain in Hardy-Weinberg equilibrium.

9. Genetic differences exist not only among individuals within populations, but also among different populations of a species. These differences take the form of differences in the frequencies of alleles that may also be polymorphic within populations. Thus variation within populations is convertible into variation among populations. At least some genetic differences among populations appear to be adaptive.

Major References

Dobzhansky, Th. 1970. *Genetics of the evolutionary process.* Columbia University Press, New York. This book and its predecessors, the several editions of *Genetics and the origin of species*, are among the most influential books on evolution from the viewpoint of population genetics. The material on genetic variation is out of date, but has enormous historical importance.

Stebbins, G. L. 1950. *Variation and evolution in plants.* Columbia University Press, New York. A seminally important work that summarizes extensive early information on variation and the nature of species in plants.

Mayr, E. 1963. *Animal Species and Evolution.* Harvard University Press, Cambridge, MA. This classic work on evolution contains a wealth of information on geographic variation and the nature of species in animals, with important interpretations of speciation and other evolutionary phenomena.

Lewontin, R. C. 1974. *The genetic basis of evolutionary change.* Columbia University Press, New York. A description of genetic variation as revealed by classic methods and by electrophoresis, together with a review of the history of the subject and a thorough analysis of the applications of population genetic theory to data.

Hartl, D. L., and A. G. Clark. 1998. *Principles of population genetics.* Third edition. Sinauer Associates, Sunderland, MA. Includes extensive treatment of much of the material in this chapter.

Problems and Discussion Topics

1. In an electrophoretic study of enzyme variation in a species of grasshopper, you find 62 A_1A_1, 49 A_1A_2, and 9 A_2A_2 individuals in a sample of 120. Show that $p = 0.72$ and $q = 0.28$ (where p and q are the frequencies of alleles A_1 and A_2), and that the genotype frequencies of A_1A_1, A_1A_2, and A_2A_2 are approximately 0.51, 0.41, and 0.08 respectively. Demonstrate that the genotype frequencies are in Hardy-Weinberg equilibrium.

2. In a sample from a different population of this species, you find four alleles at this locus. The frequencies of A_1, A_2, A_3, and A_4 are $p_1 = 0.50$, $p_2 = 0.30$, $p_3 = 0.15$, $p_4 = 0.05$. Assuming Hardy-Weinberg equilibrium, calculate the expected proportion of each of the 10 possible genotypes (e.g., that of A_2A_3 should be 0.09). Show that heterozygotes, of all kinds, should constitute 63.9 percent of the population. In a sample of 100 specimens, how many would you expect to be heterozygous for allele A_4? How many would you expect to be homozygous A_4A_4?

3. In the peppered moth (*Biston betularia*), black individuals may be either homozygous (A_1A_1) or heterozygous (A_1A_2), whereas pale gray moths are homozygous (A_2A_2). Suppose that in a sample of 250 moths from one locality, 108 are black and 142 are gray. (*a*) Which allele is dominant? (*b*) Assuming that the locus is in Hardy-Weinberg equilibrium, what are the allele frequencies? (*c*) Under this assumption, what *proportion* of the sample is heterozygous? What is the *number* of heterozygotes? (*d*) Under the same assumption, what proportion of black moths is heterozygous? (Answer: approximately 0.85.) (*e*) Why is it necessary to assume Hardy-Weinberg genotype frequencies in order to answer parts *b–d*? (*f*) For a sample from another area consisting of 287 black and 13 gray moths, answer all the preceding questions.

4. In an experimental population of *Drosophila*, a sample of males and virgin females includes 66 A_1A_1, 86 A_1A_2, and 28 A_2A_2 flies. Each genotype is represented equally in both sexes, and each can be distinguished by eye color. Determine the allele and genotype frequencies,

and whether or not the locus is in Hardy-Weinberg equilibrium. Now suppose you discard half the A_1A_1 flies and breed from the remainder of the sample. Assuming the flies mate at random, what will be the genotype frequencies among their offspring? (Hint: The proportion of A_2A_2 should be approximately 0.23.) Now suppose you discarded half of the A_1A_2 flies instead of A_1A_1. What will be the allele and genotype frequencies in the next generation? Why is the outcome so different in this case?

5. In an electrophoretic study of a species of pine, you can distinguish heterozygotes and both homozygotes for each of two genetically variable enzymes, each with two alleles (A_1, A_2 and B_1, B_2). A sample from a natural population yields the following numbers of each genotype: 8 $A_1A_1B_1B_1$, 19 $A_1A_2B_1B_1$, 10 $A_2A_2B_1B_1$, 42 $A_1A_1B_1B_2$, 83 $A_1A_2B_1B_2$, 44 $A_2A_2B_1B_2$, 48 $A_1A_1B_2B_2$, 97 $A_1A_2B_2B_2$, 49 $A_2A_2B_2B_2$. (*a*) Determine the frequencies of alleles A_1 and A_2 (p_A, q_A) and B_1 and B_2 (p_B, q_B). (*b*) Determine whether locus A is in Hardy-Weinberg equilibrium. Do the same for locus B. (*c*) *Assuming* linkage equilibrium, calculate the expected *frequency* of each of the nine genotypes. (Hint: The expected frequency of $A_1A_1B_1B_2$ is 0.103). (*d*) From the results of part *c*, calculate the expected *number* of each genotype in the sample, and determine whether or not the loci actually are in linkage equilibrium. (*e*) From these calculations, can you say whether or not the loci are linked?

6. Until a few decades ago, most population geneticists believed that populations were genetically uniform, except for rare deleterious mutations. We now know that most populations are genetically very variable. Contrast the implications of these different views for evolutionary processes.

7. The text states that genetic differences among populations arise by transformation of genetic variation within populations into variation among populations, and that "this is a conclusion of the highest importance." Why is it so important?

8. Different characters may vary more or less independently among geographic populations of a species (as in the rat snake example), or may vary concordantly (as in the flicker example). Suggest processes that could produce each pattern.

9. Different populations of white-crowned sparrows (*Zonotrichia leucophrys*) in western North America differ in certain features of the males' song. Songs also vary to some extent among males within populations. Turning to another behavioral trait, IQ ("intelligence quotient") scores vary among people, and differ on average among some socioeconomic and "racial" groups. Suggest hypotheses to account for the variation in each of these cases, and ways in which the hypotheses might be tested.

10. Several researchers have shown that groups of gypsy moths (*Lymantria dispar*) reared on more nutritious foliage lay eggs that hatch into larger larvae than those laid by groups reared on less nutritious foliage. Pose three hypotheses to account for this observation, and describe experiments to determine which is valid.

CHAPTER 10

The Origin of Genetic Variation

Evolution cannot occur unless there is genetic variation. As we have seen (in Chapter 9), there is considerable genetic variation within and among populations of most species. We now turn our attention to the processes by which this genetic variation originates.

We will first treat gene mutations, the alterations of individual genes that are so fundamentally important in evolution. The many aspects of this topic occupy much of this chapter. We will then consider recombination and changes in the structure and number of chromosomes, sometimes referred to as chromosomal mutations. Finally, we will note the significance of genetic variation acquired from other populations and species. (These topics are also treated in Chapters 11 and 22.)

The word "*mutation*" refers both to the process of alteration of a gene or chromosome and to the product, the altered state of a gene or chromosome. It is usually clear from the context which is meant.

Gene Mutations

Mutational changes of individual genes are overwhelmingly important in evolution. Many, perhaps most, evolutionary changes in phenotypic characters are attributable to changes in enzymes or other proteins, and thus to changes in the DNA sequences that encode them. At the molecular level, however, many alterations of DNA sequences occur that have slight or no phenotypic consequences.

In a broad sense, a **gene mutation** is *an alteration of a DNA sequence.* Thus, our modern knowledge of the molecular basis of heredity provides a definition in molecular terms. Before the development of molecular genetics, however, a mutation was identified by its effect on a phenotypic character (Box 10.A describes the history of the concept of mutation). That is, a mutation was a newly arisen change in morphology, survival, behavior, or some other property

that was inherited and could be mapped (at least in principle) to a specific locus on a chromosome. In practice, many mutations are still discovered, characterized, and named by their phenotypic effects. Thus, we will frequently use the term "*mutation*" to refer to an alteration of a gene from one allele to another in which the alleles are distinguished by their phenotypic effects. However, not all alterations of DNA sequences have phenotypic consequences. Hence, in a molecular context, the term "*mutation*" refers to a change in DNA sequence, independent of whatever phenotypic effect it may have.

Mutations have evolutionary consequences only if they are transmitted to succeeding generations. If a mutation occurs in a somatic cell, it is extinguished with the organism's death in the case of many animals; it may, however, be inherited in certain animals and plants in which the reproductive structures arise from somatic meristems. In those animals in which the germ line is segregated from the soma early in development, a mutation is inherited only if it occurs in a germ line cell. The chance that a gamete will carry a new mutation increases with the number of cell divisions that transpire in the germ line between the mutation event and gametogenesis. In humans, more cell divisions have preceded spermatogenesis than oogenesis in individuals of equal age, and the incidence of new mutations appears to be higher in sperm than in eggs (Crow 1993). An individual may produce many gametes with the same new mutation if the mutation occurred early in the germ line's history, or few if it occurred immediately before gametogenesis.

Both during replication and at other times, DNA is frequently damaged by chemical and physical events, and changes in base pair sequence occur. Many such changes are repaired by DNA polymerase and other "proofreading" enzymes, but some are not. These alterations, or mutations, are considered by most evolutionary biologists to be *errors.* That is, *the process of mutation is thought not to be an adap-*

267

History of the Concept of Mutation

The meaning of *"mutation,"* like that of many other words, has evolved. As far back as the seventeenth century, it was used to describe any drastic change in an organism's form, as in the fossil record. Early in the twentieth century, it was given a different meaning by the Dutch botanist Hugo DeVries, who is widely known as a discoverer of Mendel's neglected paper. DeVries was interested in the origin of new species, and thought he had solved this problem when he found discretely different, true-breeding forms among the offspring of the evening primroses (*Oenothera lamarckiana*) in his experimental garden. He termed these "mutations" and concluded that a new species arises by a spontaneous, discrete change in one or more features. To DeVries and his followers, Darwin's theory of natural selection therefore became superfluous, because the mutation process created new species in a single step, in which natural selection and the environment played no role. Moreover, the slight, continuous hereditary variations in characteristics such as size and shape were considered by the "mutationists" to have an entirely different genetic basis from discrete mutations, and to play no role in evolution. (It was later found that the "mutations" or "new species" that DeVries had observed in *Oenothera* were mostly rare recombinations of several genes, produced in a plant with a very unusual system of chromosomes.)

"*Mutation*" underwent a further change in meaning when the pioneering *Drosophila* geneticist Thomas Hunt Morgan, at Columbia University in New York, discovered newly arisen aberrations, such as white-eyed flies, that obeyed Mendelian rules of inheritance. Thus *mutation* came to mean not necessarily the origin of a new species, but a spontaneous alteration of a gene. (Nonetheless, Morgan continued for much of his life to affirm that new species arise by mutation, and that natural selection plays no causal role in evolution.)

If a mutation is an alteration of a single gene, it may be (and usually is) a genetic variant rather than a new species. If the same mutation occurs only rarely, mutation "pressure" will generally not be adequate to transform a species, and something else (such as natural selection) is required to increase the frequency of the mutation in the population. This reasoning, with its emphasis on evolution as a *population-level* process rather than the origin of species as mutant *individuals*, is the foundation of the Evolutionary Synthesis of the 1930s and 1940s (see Chapter 2), in which mutation and natural selection are complementary rather than mutually exclusive ingredients of evolution.

When geneticists came to realize that continuous variation is based on multiple genes that are inherited in the same way as discrete Mendelian factors, it became understood that the mutational process generates both kinds of variation: mutations with small phenotypic effects are the basis of continuous variation, and those with large effects generate discrete variations. Moreover, there is a continuum of effects from very small to quite large. When the molecular nature of the gene was elucidated in the 1950s, mutation could be recognized as an alteration of the base pair sequence of a gene—including base pair changes that have no effect whatever on the phenotype, even on the amino acid sequence of the protein that the gene encodes.

What mutation is *not* is the birth of new organisms utterly unlike their parents. These exist only in science fiction. Dinosaurs or birds do not hatch fully formed from lizard eggs. Some mutations may be monstrous, such as flies with their antennae transformed into legs, but mutations can only alter already immanent developmental processes, and so must have a limited range of possible effects.

---

tation, but a consequence of unrepaired damage. Both the existence of repair enzymes and the theory of the evolution of mutation rates, discussed in Chapter 21, are the basis for this conclusion.

Mutational changes of DNA sequences are of many kinds.

Point Mutations

The simplest mutation is a substitution of one base pair for another (Figure 10.1). In classic genetics, a mutation that maps to a single gene locus is called a **point mutation**; in modern usage, this term is often restricted to single base pair substitutions. A **transition** is a substitution of a purine for a purine (A↔G) or a pyrimidine for a pyrimidine (C↔T). **Transversions**, of eight possible kinds, are substitutions of purines for pyrimidines or vice versa (A or G↔C or T).

Some base pair changes occur in nontranslated DNA, and have no known phenotypic effect. Mutations in genes that encode ribosomal and transfer RNA potentially affect the function of these gene products. Other base pair changes may result in amino acid substitutions in polypeptides or proteins. These may have little or no effect on the functional properties of the polypeptide, and thus no effect on the phenotype, or they may have substantial consequences. For example, the change from the (RNA) triplet GAA to GUA causes the amino acid valine to be incorporated instead of glutamic acid. This is the mutational event that in humans caused the abnormal β-chain in sickle-cell hemoglobin, which in turn has many pleiotropic effects (see Figure 21 in Chapter 3) and is usually lethal in homozygotes.

Because of the redundancy of the genetic code, many substitutions at the third base position in codons, and quite

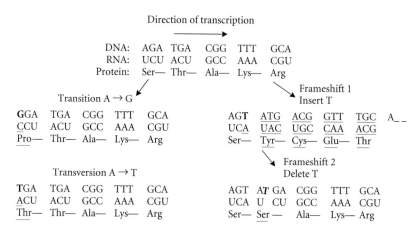

FIGURE 10.1 Examples of kinds of point mutations and their consequences for messenger RNA and amino acid sequences. (Only the transcribed, "sense" strand of the DNA is shown.) At left, transition and transversion mutations at the first base position. At right, a frameshift mutation, caused by the insertion of T between sites 2 and 3, shifts the reading frame so that downstream bases are read in new triplets, altering the amino acid sequence. A second frameshift mutation, a deletion of one base at the fifth site, reestablishes the original reading frame downstream from the site. The encoded amino acids can be found from the code shown in Figure 12 in Chapter 3.

a few at the first base position, are synonymous: they do not alter amino acids. About 24 percent of the possible substitutions in the code are synonymous, but the proportion of synonymous mutations that occur in a species' genome depends on the proportions in which the various codons are represented, as well as any nonrandomness of substitution that may exist.

Three of the triplets in the RNA code are "stop" codons, signaling termination of translation into a polypeptide product. Mutation of an amino acid-encoding triplet into a stop codon will result in an incomplete, usually nonfunctional, gene product. Mutations to termination codons are often found within nonfunctional pseudogenes.

If a single base pair (or more) becomes inserted into or deleted from a DNA sequence, the triplet reading frame is shifted by one nucleotide, so that downstream triplets are translated into different amino acids (Figure 10.1). This is a **frameshift mutation**. The greatly altered gene product is usually nonfunctional.

Sequence changes arising from recombination

When homologous DNA sequences differ at two or more base pairs, **intragenic recombination** between them can generate new DNA sequences. In molecular terms, intragenic recombination is not mutation, but the new haplotypes might be distinguished as alleles or mutations if they have phenotypic effects. Thus recombination between DNA sequences that code for, say, the amino acid sequence Val-Thr-Arg-Leu and Glu-Thr-Arg-Gly could give rise to the new polypeptide product Val-Thr-Arg-Gly. Precisely this kind of polymorphism was described for the amino acid sequence of the enzyme 6-phosphoglycerate dehydrogenase in the Japanese quail (Ohno et al. 1969). Since then, direct DNA sequencing has revealed many examples of variant haplotypes that apparently arose by intragenic recombination. Kreitman described some instances in his study of sequence variation in the alcohol dehydrogenase gene of *Drosophila melanogaster* (see Chapters 9 and 22).

Recombination appears to be the cause of a peculiar mutational phenomenon called **gene conversion**, which has been studied most extensively in fungi. The gametes of a heterozygote should carry the two alleles (A_1, A_2) in a 1:1 ratio. Occasionally, though, they occur in different ratios, such as 1:3. An A_1 allele has been replaced specifically by an A_2 allele rather than by any of the many other alleles to which it might have mutated: it seems to have been converted into A_2. In some cases gene conversion is unbiased (conversion of A_1 to A_2 is as likely as the converse), but cases of biased gene conversion have been described whereby one allele is preferentially converted to the other. The details of the molecular mechanisms thought to underlie this process need not concern us; suffice it to say that it is believed that a damaged DNA strand of one chromosome is repaired by enzymes that insert bases complementary to the sequence on the undamaged homologous chromosome.

Transposable elements and their effects

Until recently, geneticists thought that all genes occupied fixed sites on the chromosomes, except when moved to new positions by inversion or translocation events. This is indeed true of many genes. But in the 1940s, Barbara McClintock described several genes in maize (corn, *Zea mays*) that frequently moved to new sites. Her work was considered a mere curiosity until the 1980s, when it was discovered that apparently all organisms carry in their genomes numerous **transposable elements**: sequences that can move to any of many places in the genome. These DNA sequences carry genes that encode enzymes (transposases) that accomplish the transposition (movement), and sometimes they carry with them other genes near which they had been located. In some cases, the transposable element leaves one "host" gene and becomes inserted elsewhere (conservative transposition). In most cases, a parent element remains in situ, but produces copies that become inserted elsewhere (replicative transposition). The process of insertion generates short (4–12 bp) repeats of the host DNA sequence on either end of the inserted element, called flanking repeats, which are useful for recognizing transposed sequences (Figure 10.2).

The several kinds of transposable elements include

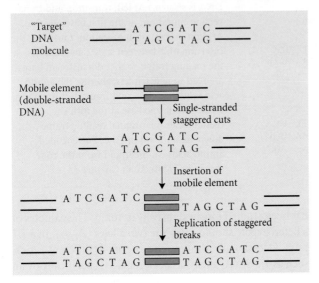

1. *Insertion sequences*, of about 700–2600 bp. Their only functional genes encode transposases, the enzymes that cause transposition. They have been found in bacteria, phage, and maize, among other organisms.
2. *Transposons*, of about 2500–7000 bp. They encode not only transposases, but other functional genes as well. Some plasmids (circular DNA molecules in bacteria that in some cases integrate into the bacterial chromosome and in other cases do not) carry genes that confer resistance to antibiotics and other stresses. Transposons are common in plants, fungi, and animals. There are many kinds; *Drosophila melanogaster* may have as many as 50 to 100. Typically, many copies of a particular kind of transposon are scattered throughout a genome, their locations varying among individuals.
3. *Retroelements.* The traditional view that information flows only from DNA to RNA to protein was changed in the early 1970s by the discovery of **reverse transcription**. The enzyme reverse transcriptase uses RNA as a template for the synthesis of a DNA copy (cDNA). Reverse transcriptase genes are carried by RETROVIRUSES, which are RNA viruses (including the HIV virus that causes AIDS) that invade a cell, make DNA copies of themselves, and insert them into the host genome. These are then transcribed into more RNA virus copies, which infect other cells. RETROPOSONS act similarly except that they do not cross cell boundaries, and spread only by cell division in the host. *Copia* is a retroposon that has been studied extensively in *Drosophila melanogaster.*

Transposable elements have many effects on host genomes, including

1. An increase of total genome size, by replicative transposition.
2. Alteration of expression of host genes. Insertion of a transposable element into the coding region of a host gene can abolish the gene's function. Insertion of a transposable element into the control region of a gene affects its expression; this is the cause of many well-known mutations in *Drosophila*, such as those at the *white* locus, which affect eye color. The promoters carried by a transposable element, which regulate its own transcription, can affect the rate of transcription of nearby host genes as well. The departure of a transposable element is often imprecise, causing the deletion or addition of a few base pairs of the host gene.
3. An increase in the mutation rate of host genes. (We will describe an experiment on mutation rates below.)
4. Chromosome rearrangements in the host genome can result from recombination between two copies of a transposable element located at different sites. This can cause an inversion (a 180° reversal in the orientation of part of a chromosome), or a deletion of the DNA sequence between the transposable elements (Figure 10.3). The deleted material, attached to one copy of the transposable element, can be inserted elsewhere in the genome; the FB transposable elements of *Drosophila* are known to move sequences of hundreds of kilobases. Moreover, the numerous copies of a transposable element promote unequal crossing over, resulting in deletions and duplications of host DNA. For example, a mutation of a human lipoprotein gene, resulting in high cholesterol levels, consists of the deletion of one of the gene's exons, caused by unequal crossing over between repeated sequences distributed throughout the gene's introns (Figure 10.4). (Inversions, deletions, and other chromosome rearrangements are described more fully later in this chapter.)
5. Transposable elements that encode reverse transcriptase sometimes form, and insert into the genome, DNA copies (cDNA) not only of their own RNA, but also of RNA transcripts of the host's genes. A processed RNA transcript lacks the sequences corresponding to the gene's introns, and also lacks the gene's nontranscribed control regions. Therefore a cDNA copy of RNA (a "retrogene") is easily recognized by DNA sequencing: its sequence resembles that of the exons of an ancestral gene located elsewhere in the genome, but it lacks control regions and introns, and its ends correspond precisely to those of the transcribed region of the ancestral gene (Figure 10.5).

Most retrogenes are nonfunctional, partly because they lack control regions. In humans, however, phosphoglycerate kinase is encoded not only by an "ancestral" X-linked gene with introns, but also by an autosomal gene that lacks introns but has a sequence corresponding to the exons of

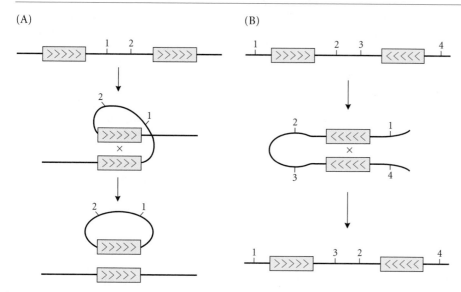

FIGURE 10.3 Recombination between repeated sequences, such as copies of a transposable element, can result in deletions and inversions. The boxes represent repeats, with the polarity of base pair sequence indicated by the arrows within. The numerals represent genetic markers (in different genes or within a single gene). (A) Recombination (×) between two direct repeats (i.e., with the same polarity) excises one repeat and deletes the sequence between the two copies. (B) Recombination between two inverted repeats (with opposite polarity) inverts the sequence between them. (After Lewin 1985.)

the X-linked gene. It is particularly interesting that the X-linked gene is expressed in many tissues, but the autosomal gene is expressed only in the testes, and so has acquired a novel tissue-specific pattern of expression (Li and Graur 1991).

Most retrogenes are **processed pseudogenes** (Figure 10.5), which do not produce functional gene products. They lack sequences corresponding to the ancestral genes' introns, but otherwise show some similarity of sequence. Because they are nonfunctional, however, they accumulate mutations, including termination codons, that are presumably not affected by natural selection, and so their sequence degenerates, diverging from the ancestral gene over time. Mammalian genomes are highly laden with pseudogenes, which may make up as much as 20 percent of the DNA content (Walsh 1985). For example, the hemoglobin gene family includes at least three pseudogenes, and the glyceraldehyde-3-phosphate dehydrogenase sequence is represented by one functional locus and about 20 pseudogenes in humans—and about 200 pseudogenes in the mouse (*Mus musculus*) (Li and Graur 1991). In mammals, a 300-bp sequence called *Alu*, which seems to have been derived by reverse transcription from 7SL RNA, is highly repeated: with more than 500,000 copies, it constitutes about 5 percent of the human genome.

Rates of Mutation

Estimates Estimates of mutation rates depend on the method used to detect mutations. In classic genetics, a mutation was detected by its phenotypic effects, such as a white vs. red eye in *Drosophila*. Such a mutation, however, might be caused by the alteration of any of many sites within the locus; moreover, many base pair changes have no phenotypic effect. Thus phenotypically detected rates of mutation underestimate the total mutation rate at a locus. With molecular methods, mutated sequences can be de-

FIGURE 10.4 A mutated low-density-lipoprotein gene in humans, labeled here as FH 626-a, lacks exon 5. It is believed to have arisen by unequal crossing over between two normal gene copies, due to out-of-register pairing between two of the repeated sequences (*Alu* sequences, shown as dark and light shaded boxes) in the introns. The black boxes represent exons. The other product of unequal crossing over, labeled FH-?, has not been found in human populations. (After Hobbs et al. 1986.)

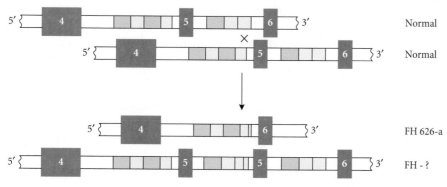

FIGURE 10.5 Comparison of the DNA sequence of the first two exons of the functional Cu/Zn superoxide dismutase gene (*SOD*-1) in humans, and homologous parts of the processed pseudogene (ψ69.1) that arose from *SOD*-1 by reverse transcription from the SOD-1 mRNA. The letters above and below the DNA sequences represent different encoded amino acids, and the dots, dashes, and plus signs between the sequences mark nucleotide substitutions, deletions, and insertions, respectively. Note that the pseudogene precisely lacks the sequence corresponding to the functional gene's intron, and that the deletion of one nucleotide shortly before the intron has generated a chain-termination codon, TGA (marked by asterisks). (After Li and Graur 1991.)

```
          M   A   T   K   A   V   C   V   L   K   G   D   G   P   V
SOD-1    ATG GCG ACG AAG GCC GTG TGC GTG CTG AAG GGC GAC GGC CCA GTG
          ·  ·       ·           ·   ·       ·           ·  ·  ·
ψ69.1    ATA ATG ATG AAG GTC ATG TAC ATG TTG AAG GGC CAG AGC CCG GTG
          I   M   M   K   V   M   Y   M   L   K   G   Q   S   P   V


          Q   G   I   I   N   F   E   Q   K           E   S   N   G
SOD-1    CAG GGC ATC ATC AAT TTC GAC CAG AAG G intron AA AGT AAT GGA
          ·  ·  –       ·       –       ·           ·   – – –   ·
ψ69.1    CAG GCG A C ATC CAT TT  GAG CAG AAG G        AA AAT     GAA
          Q   V   T   S   I   *  * *


          P   V   K   V   W   G   S   I   K   G   L   T   E   G   L
SOD-1    CCA GTG AAG GTG TGG GGA A GC ATT AAA GGA CTG ACT GAA GGC CTG
          ·  ·       ·       –   +       ·       ·           ·  ·
ψ69.1    CCA TTT ATG GTG T C AGA ATGC ATT ACA GGA TTG ACT GAA CGC CAG


          H   G   F   H   V   H   E   F   G   D   N   T   A
SOD-1    CAT GGA TTC CAT GTT CAT GAG TTT GGA GAT AAT ACA GCA intron
          ·  ·                                   –   –   ·  ·
ψ69.1    CAC AGA TTC CAT GTT CAT CAG TTT GGA G  T A  T AAC ACA
```

tected directly. These findings are often expressed as mutation rates per base pair.

Rates of mutation are estimated in several ways. A direct method is to count the number of mutations arising in a laboratory stock (which is usually initially homozygous), scoring mutations by their phenotypic effects or by molecular methods. Another, indirect method (Box 10.B) is based on the number of base pair differences between homologous genes in different species, relative to the number of generations that have elapsed since they diverged from their common ancestor. This method depends on a theory of population genetics that is described in Chapter 11.

Direct measures of mutation rate usually count the number of mutations, usually scored by their phenotypic effects, among the offspring of an initially homozygous stock. For example, a bacterial culture can be grown from a single haploid cell, and the rate of origin of mutations conferring resistance to antibiotics can be measured by the number of cells that yield colonies, out of a known number of cells placed on antibiotic-containing medium. This procedure yields the rate of mutation per cell division. In *Drosophila* and other multicellular organisms, the number of chromosomes bearing a particular mutation is scored among a large number of descendants from a stock initially homozygous for a different allele (perhaps the "wild type"). It may be necessary to use special crosses (perhaps using inverted chromosomes; see Figure 6 in Chapter 9) to make the stock homozygous and to make descendant chromosomes homozygous in order to detect recessive mutations. The autosomal chromosomes of N flies represent the $2N$ gametes that formed them, so the mutation rate is calculated as the number of new mutations per gamete per generation. Subsequent crosses, and perhaps mapping of the mutations, may be necessary to determine whether phenotypically similar mutations all occurred at the same locus.

Measured by phenotypic effects, an average locus mutates at a rate of about *10^{-6} to 10^{-5} mutations per gamete per generation* (Tables 10.1, 10.2), an estimate that has been confirmed by studying individual proteins (Neel 1983). The *average mutation rate per base pair*, based mostly on the indirect method of comparing DNA sequences of different species, has been estimated at about 10^{-9}. There is a great deal of variation around these values: mutation rates vary among genes and even among regions within genes, and they can be elevated by transposable elements. Moreover, some genes are known that, in mutant condition ("mutator alleles"), increase the rate of mutation elsewhere in the genome.

Back mutation is mutation of an allele (a "mutant") back to the allele (usually the "wild type") from which it arose. Back mutations are ordinarily described at a phenotypic level. They usually occur at a much lower rate than "forward" mutations from wild type to mutant, presumably because many more substitutions can impair gene function than can restore it. At the molecular level, most phenotypically detected back mutations are not restorations of the original sequence, but instead result from a second amino acid substitution that restores the function that had been altered by the first substitution (Allen and Yanofsky 1963).

Evolutionary implications of mutation rates An average per-locus mutation rate of, say, 10^{-5} (one in 100,000 gametes) is so low that *the rate of change in the frequency of an allele, due to mutation alone, is very low*. Suppose that of two phenotypically distinguishable alleles, A_1 and A_2, with allele frequencies $p (= 1 - q)$ and q, A_1 mutates to A_2 at rate $u = 10^{-5}$. In each generation, the frequency of A_2 is increased by $u \times p$ (that is, a fraction u of the A_1 alleles mutate). Denoting the change in the frequency of A_2 from one generation to the next by Δq, we have

$$\Delta q = up = u(1 - q).$$

BOX 10.B

Estimating Mutation Rates from Comparisons among Species

In Chapter 11, we will describe the "*neutral theory of molecular evolution*." This theory describes the fate of purely neutral mutations, i.e., those that neither enhance nor lower fitness. One possible fate is that a mutation will become fixed—that is, attain a frequency of 1.0—entirely by chance. The probability that this will occur equals u, the rate at which neutral mutations arise. In each generation, therefore, the probability is u that a mutation that occurred at some time in the past will become fixed. After the passage of t generations, the fraction of mutations that will have become fixed is therefore $x = ut$.

If two species diverged from a common ancestor t generations ago, the expected fraction of fixed mutations in both species is $D = 2ut$, since various mutations have become fixed in both lineages. If the mutations in question are base pair changes, a fraction $D = 2ut$ of the base pairs of a gene should differ between the species, assuming that all base pairs are equally likely to mutate. Thus the average mutation rate per base pair per generation is $u = D/2t$. Thus we can estimate u if we can measure the fraction of base pairs in a gene that differ between two species (D), and if we can estimate the number of generations since they diverged from their common ancestor (t). This requires an estimate of the length of a generation, information from the fossil record on the absolute time at which the common ancestor existed, and an understanding of the phylogenetic relationships among the living and fossilized taxa.

In applying this method to DNA sequence data, it is necessary to assume that most base pair substitutions are neutral, and to correct for the possibility that earlier substitutions at some sites in the gene have been replaced by later substitutions ("multiple hits"). Uncertainty about the time since divergence from the common ancestor is usually the greatest source of error in estimates obtained by this method. Often, the common ancestor has not been identified in the fossil record, and the time estimate is based on the earliest known fossils of either lineage. For example, divergence between a primate and a rodent would date from the late Cretaceous, about 70 Mya, in which the earliest primates are known. Such dates are minimal estimates of divergence time.

The best estimates of mutation rates at the molecular level have been obtained from interspecific comparisons of pseudogenes, other nontranslated sequences, and fourfold degenerate third-base positions (those in which all mutations are synonymous), since these are thought to be least subject to natural selection (although probably not entirely free of it). In comparisons among mammal species, the average rate of nucleotide substitution has been about 3.3–3.5 per nucleotide site per 10^9 years, for a mutation rate of 3.3–3.5×10^{-9} per site per year (Li and Graur 1991). If the average generation time were 2 years during the history of the lineages studied, the average rate of mutation per site would be about 1.7×10^{-9} per generation. Comparison of human and chimpanzee sequences yielded an estimate of 1.3×10^{-9} per site per year, assuming divergence 7 Mya. If the average generation time in these lineages has been 15–20 years, the mutation rate is about 2×10^{-8} per generation. The human diploid genome has 6×10^9 nucleotide pairs, so this implies at least 120 new mutations per genome per generation—an astonishingly high number (Crow 1993).

If, for the sake of argument, $q = 0.5$ (A_2 already accounts for half the gene copies), then in the following generation $q' = q + \Delta q = 0.50000495$. At this rate it will take about 70,000 generations to get to $q = 0.75$, and another 70,000 to get to $q = 0.875$. [In fact, each successive halving of the difference between an initial allele frequency and the final equilibrium frequency takes about 70,000 generations (Crow and Kimura 1970).]

With such a low mutation rate per locus, it might seem that mutations occur so rarely that they cannot be impor-

Table 10.1 **Comparative spontaneous mutation rates**

SPECIES	BASE PAIRS PER GENOME	MUTATION RATE PER BASE PAIR REPLICATION	MUTATION RATE PER GENOME PER GENERATION
Bacteriophage lambda	4.7×10^4	2.4×10^{-8}	0.0001
Bacteriophage T4	1.8×10^5	1.1×10^{-8}	0.0002
Salmonella typhimurium bacteria	3.8×10^6	2.0×10^{-10}	0.0001
Escherichia coli bacteria	3.8×10^6	4.0×10^{-10}	0.0002
Neurospora crassa fungus	4.5×10^7	5.8×10^{-11}	0.0003
Drosophila melanogaster[a]	4.0×10^8	8.4×10^{-11}	0.93

Source: After Drake (1974).

[a]The *Drosophila* values are per diploid genome per generation of flies, not per generation of cells, as in the other species.

Table 10.2 Spontaneous mutation rates of specific genes, detected by phenotypic effects

SPECIES AND LOCUS	MUTATIONS PER 100,000 CELLS OR GAMETES
Escherichia coli	
Streptomycin resistance	0.00004
Resistance to T1 phage	0.003
Arginine independence	0.0004
Salmonella typhimurium	
Tryptophan independence	0.005
Neurospora crassa	
Adenine independence	0.0008–0.029
Drosophila melanogaster	
Yellow body	12
Brown eyes	3
Eyeless	6
Mus musculus (mouse)	
a (coat color)	7.1
c (coat color)	0.97
d (coat color)	1.92
ln (coat color)	1.51
Homo sapiens	
Retinoblastinoma	1.2–2.3
Achondroplasia	4.2–14.3
Huntington's chorea	0.5

Source: After Dobzhansky (1970)

tant. However, the number of functional genes in a genome is quite large: about 10,000 in *Drosophila* and perhaps 150,000 in humans. This implies that *almost every gamete carries a new, phenotypically detectable mutation somewhere in its genome* (10^{-5} mutations per gene × 10^5 genes = 1 mutation per haploid genome in humans). So in a population of 500,000 individuals, about one million new mutations arise every generation. If even a tiny fraction of these were advantageous, the amount of new "raw material" for adaptation would be substantial, especially over the course of thousands or millions of years. These figures may be underestimates, for at the molecular level, each human haploid genome may carry about 200 new nucleotide substitutions (Kondrashov and Crow 1993; see Box 10.B). Recall from Chapter 9, furthermore, that many alleles that arose by mutation in the past have high frequencies in natural populations, so the amount of potentially adaptive genetic variation is very high.

Experiments on *Drosophila* have confirmed that the total mutation rate per gamete is quite high. In a heroically large experiment, Terumi Mukai and colleagues (1972) counted more than 1.7 million flies in order to estimate the rate at which the second chromosome accumulates mutations that affect egg-to-adult survival (viability). They used crosses (see Figure 6 in Chapter 9; Box 10.C) in which copies of a single wild chromosome were carried in a heterozygous condition so that deleterious recessive mutations could persist without being eliminated by natural selection. Every ten generations, they made large numbers of these chromosomes homozygous, and measured the proportion of them that reduced viability. The mean viability declined, and the variation among chromosomes increased steadily (Figure 10.6). From the changes in the mean and variance, Mukai et al. calculated a mutation rate of about 0.15 per second chromosome per gamete. This is the sum, over all loci on the chromosome, of mutations that affect viability. Because the second chromosome carries about a third of the genome, the total mutation rate is about 0.5 per gamete. Thus, almost every zygote carries at least one new mutation that affects viability. In a similar experiment, David Houle and colleagues (1992, 1994) confirmed this conclusion, and found that new mutations substantially affect not only viability, but also other aspects of fitness such as reproductive rate.

Because the genetic basis of most phenotypic traits is polygenic, it is important to know how rapidly mutation supplies genetic variance (see Chapter 9) in such characteristics. In order to facilitate comparisons among traits, the variance attributable to new mutations (V_m) is often expressed as a fraction of the environmental variance (V_E) of the trait. One of several methods for estimating this quantity is to propagate a population that has been made entirely homozygous (e.g., by inbreeding), and after several generations to estimate, by the usual methods of quantitative genetics, the genetic variance that has accumulated (Chapters 9, 14). V_m/V_E can also be estimated by artificially selecting a character in an initially homozygous population, for any response to selection must then be based on new mutations.

FIGURE 10.6 Effects of the accumulation of spontaneous mutations on the viability (egg-to-adult survival) of *Drosophila melanogaster*. The mean viability of flies made homozygous for chromosomes carrying recessive mutations decreases, and the variation (variance) among chromosomes increases. (After Mukai et al. 1972.)

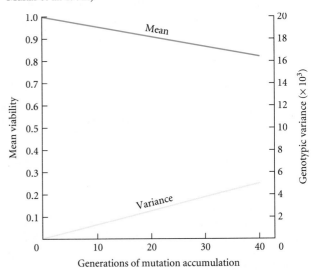

BOX 10.C

Accumulation of Recessive Mutations

In the crossing scheme used by Mukai, a single wild male fly, with second chromosomes that we may denote $+_1$ and $+_2$, is mated with females of a stock Cy/Pm (as in Figure 6 in Chapter 9). Each of the female's two chromosomes has an inversion that prevents crossing over, a recessive lethal gene, and a dominant "marker" gene (*Curly* wing, *Cy*, or *Plum* eye color, *Pm*). A single male F_1 offspring bearing *Pm* (hence $Pm/+_1$ or $Pm/+_2$) is then crossed to Cy/Pm females. Thus all wild (+) chromosomes in the F_2 are identical (either $+_1$ or $+_2$). Each of a large number of $Pm/+$ F_2 males is mated to Cy/Pm females, and this procedure is repeated in subsequent generations. Note that the + chromosomes, all initially identical, are kept in heterozygous condition (so recessive mutations that affect viability are not expressed) and that the + chromosomes remain intact because they cannot recombine with *Cy* or *Pm* chromosomes.

In each generation, the cross $Cy/Pm \times Pm/+$ yields both females and males of each of four genotypes: Cy/Pm, $Cy/+$, Pm/Pm, and $Pm/+$ (but Pm/Pm is lethal and does not appear among the adult progeny). Every ten generations, from each chromosome line, Mukai mated males and females of the $Cy/+$ genotypes, yielding Cy/Cy, $Cy/+$, and $+/+$. These should fit a 1:2:1 ratio, but Cy/Cy is lethal, so we expect a 2:1 ratio of $Cy/+$ to $+/+$. If the ratio of $Cy/+$ to $+/+$ is greater than 2:1, we can assume that relative viability of the + chromosome in that cross has been reduced. The magnitude of the deviation measures how much viability has been reduced by recessive mutations that have accumulated on this chromosome. Since all the fly lineages started with copies of a single + chromosome, the variation among lineages in their degree of departure from a 2:1 ratio can be used to estimate how many different mutations have occurred.

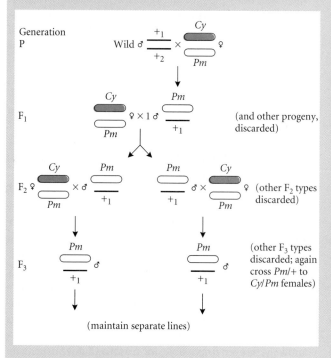

For bristle number in *Drosophila* and pupal weight in flour beetles (*Tribolium*), V_m/V_E is about 0.1 percent per generation; that is, the genetic variance increases by a factor of 0.001 of the environmental variance, per generation, due to new mutations. Thus, it would take about 500 generations for an initially homozygous population to reach a heritability of 0.5, assuming that none of the new genetic variation were eroded by genetic drift or natural selection. The magnitude of this MUTATIONAL VARIANCE varies somewhat among characters and species (Lynch 1988).

The rate of polygenic mutation can be increased substantially by transposable elements. In *Drosophila melanogaster*, transposable elements called *P* elements cause a syndrome called HYBRID DYSGENESIS (Kidwell 1983; Engels 1983). "P" strains of flies carry *P* elements, and "M" strains do not. "P" strains carry a maternally transmitted cytoplasmic factor that represses transposition of *P* elements, whereas "M" strains lack this factor. In F_1 progeny of the cross ♀M × ♂P, *P* elements transpose freely, causing the "hybrid dysgenesis" syndrome, which includes reduced fertility, recombination in males (crossing over does not normally occur in males), and a high incidence of chromosome breakage and gene mutation.

Trudy Mackay and her colleagues have carried out several experiments (e.g., Mackay et al. 1992) in which they estimated the variance due to mutations in dysgenic flies, compared with that in inbred strains without *P* elements. On average, each transposed *P* element (each "insert") reduced viability by about 12 percent in the homozygous condition and by 5.5 percent in the heterozygous condition. Most interestingly, the genetic variance in bristle number increased by about $V_m/V_E = 0.03$ per generation—about 30 times the rate at which mutation adds genetic variance in "normal" flies. Many of the *P*-induced mutations had only slight effects on bristle number, but a few had rather large effects.

In summary, although any given mutation is a rare event, the rate of origination of new genetic variation in the genome as a whole, and for individual polygenic characters, is appreciable, and can sometimes be quite high.

This conclusion has many important implications that we will discuss in subsequent chapters. The high input of mutations may account for much of the genetic variation within populations (see Chapter 14). Because mutations are more likely to interfere with enzyme function and developmental pathways than to improve them, the steady input of mutations is likely to cause characters to degenerate in the long term, unless such mutations are removed by natural selection. In the same vein, the input of deleterious mutations lowers fitness, and would ultimately cause the extinction of populations if it were not counteracted by "purifying" natural selection. Selection for characteristics that mitigate the effects of deleterious mutations is thought by some authors to be an important factor in the evolution of recombination and sex (see Chapter 21).

Phenotypic Effects of Mutations

It is useful to make a somewhat arbitrary distinction between the effects of mutations on phenotypic features such as morphology and physiology, and their effects on components of fitness (viability, fertility, developmental rate). The distinction is arbitrary because all differences in fitness are based on physiological, morphological, or behavioral differences. Nevertheless, we can often describe a phenotypic difference, and then ask whether or not it affects fitness. (However, we are often ignorant of the physiological or morphological basis for manifest differences in fitness.)

Effects on the phenotype   The phenotypic effects of mutational changes in DNA sequence range from none to drastic. At one extreme, synonymous base pair changes are expected to have no evident phenotypic effect, and this is apparently true also of many amino acid substitutions, which seem not to affect protein function. (Even synonymous mutations, however, can affect fitness, probably because of their effects on the rate of translation of RNA into protein; see Chapter 22.) Mutations of slight effect are exemplified by the mutational increase in variation in bristle number in *Drosophila* described above. In this and many other such instances, the effects of the mutations in aggregate can be measured, but many of the mutations that contribute to the variation have so slight an effect that they cannot be isolated individually for study. Identifiable mutations, however, range from subtle to striking; in *Drosophila*, some have minor effects on features such as eye color and wing veins, and others have large effects, such as reduction of the wings to tiny, useless structures. Similarly drastic mutations are known in humans and in domesticated plants and animals.

Many mutations affect behavior. For example, in mutants of the *per* (*period*) locus in *Drosophila melanogaster*, a variety of biological rhythms, including circadian activity cycles and the frequency of the wingbeat in the male's courtship behavior, are aberrant (Hall and Kyriacou 1990; Kyriacou 1990). The *yellow* mutation in this species affects not only body color, but also the rate at which males perform certain courtship behaviors, and such males are less successful in mating than wild-type males. A fascinating example of behavioral mutations is illustrated by the honeybee (*Apis mellifera*), in which larvae reside in individual cells of the comb, and are sometimes killed by a bacterial disease that spreads through the hive unless dead larvae are removed. Workers must uncap the cells, then remove the cadavers. Each of these behaviors is abolished by a single dominant mutation, so the normal behavior depends on the bee's being homozygous recessive at two loci. (See Ehrman and Parsons 1981 for more information on behavioral genetics.)

Perhaps the most fascinating of the drastic mutations that affect morphology are HOMEOTIC MUTATIONS, which redirect the development of one part of the body into another. These mutations occur in the genes that determine the basic "body plan" of an organism, conferring a distinct identity on each part of the developing body and regulating the activity of other genes that affect the features of each such part (see Chapters 3, 23). In *Drosophila*, mutations in the *Antennapedia* and *Bithorax* gene complexes transform the identity of certain body segments, so that, for instance, legs may develop in place of antennae, or wings in place of halteres (structures in the Diptera that have been derived in evolution from the second pair of wings).

The limits of mutation   It cannot be stressed too strongly that even the most drastic mutations cause alterations of one or more *pre-existing* traits. Mutations with phenotypic effects alter developmental processes, but they cannot alter developmental foundations that do not exist. We may conceive of winged horses and angels, but no mutant horses or humans will ever sprout wings from their shoulders, for the developmental foundations are lacking. (J. B. S. Haldane, one of the founders of population genetics, once quipped that humans could never evolve into a race of angels, because they lack the genetic capacity for both wings and the moral sense.) Thus, some morphologies are highly unlikely, or even impossible, for reasons that we usually do not understand because of our ignorance of developmental pathways. For instance, the numerous ankle bones of a salamander can be organized (by fusions and fissions) in many imaginable ways, but some conceivable patterns have never been found, either as intraspecific variants or as species-typical characters. For example, variations in the foot structure of the newt *Taricha granulosa* almost all occur as typical conditions in other species of salamanders (Figure 10.7), whereas other conceivable variations have not been seen either within *Taricha* or among other species (Shubin et al. 1995). Such data suggest that some variations are more likely to arise and to contribute to evolution than others. Similarly, in laboratory stocks of the green alga *Volvox carteri*, new mutations affecting the relationship between the size and number of germ cells correspond to the typical state of these characters in other species of *Volvox* (Koufopanou and Bell 1991). Because mutations can supply only certain kinds of variation, constrained by existing developmental pathways, some imaginable paths of evolution are closed to a species, or at least are less likely than others.

The likelihood of a given evolutionary change may also be affected by the number of different mutations that generate a particular phenotype. When several or many loci affect the trait (polygeny), evolution can be based on many combinations of mutations, so genetic variation is likely to

FIGURE 10.7 The normal arrangement of tarsal bones in some salamanders. The tarsals lie between the metatarsals (mt1–mt5 in part A) and the tibia (T) and fibula (F). (A) The usual arrangement in the newt *Taricha granulosa*. Almost all variations found in this species are the normal condition in certain other species (B–F). For example, fusion of dt4 and dt5 (dt4 + dt5) is found in *Cynops* (B) and some other salaman-

ders. The other species shown, and their tarsal condition relative to that of *T. granulosa*, shown by the shaded bones, are (C) *Liua* (two extra bones); (D) *Bolitoglossa dolfleini* (dt3 + dt4 + dt5); (E) *Necturus* (dt3 + dt4; f + i); and (F) *Thorius* (dt4 + dt5 + c; f + i). Digit V has been lost in *Necturus* and *Thorius*. (After Shubin et al. 1995.)

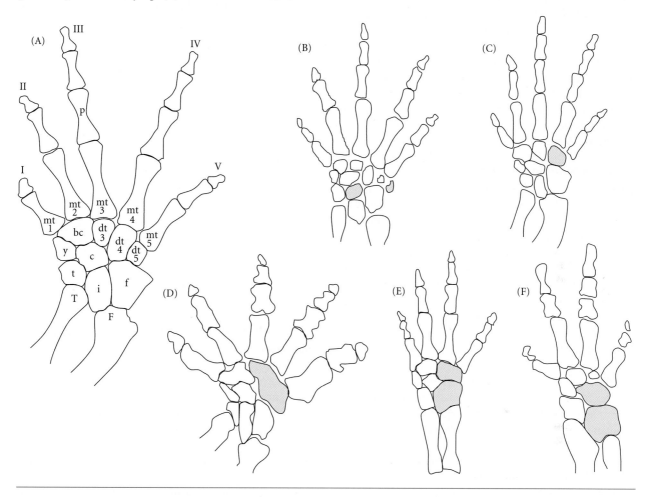

arise. Very often, a similar phenotype can be produced by mutation at several different loci, as we know both from mutation in laboratory stocks (e.g., phenotypically similar eye color mutations in *Drosophila*) and from genetic analyses of natural populations. For example, when different copper-tolerant populations of the monkeyflower *Mimulus guttatus* are crossed, the variation in copper tolerance is greater in the F_2 generation than within either parental population, indicating that the populations differ in the loci that confer tolerance (see Cohan 1984 for this and other examples).

Nevertheless, the number of loci providing variation in some polygenic traits is not all that great—perhaps five to ten, with a few loci responsible for much of the variation. In *Drosophila*, for example, some mutations that have small effects on quantitative variation in bristle number are alleles of well-known mutations with large effects (see Chapter 14). Certain phenotypes can apparently be produced by mutation at a very few loci, or perhaps only one. Insecticide

resistance offers particularly striking examples. For instance, resistance to dieldrin in different populations of *Drosophila melanogaster* is based on repeated occurrences of the same mutation, which, moreover, is thought (because of its map position) to represent the same mutation that confers resistance in flies (*Lucilia cuprina, Musca domestica*) that belong to two other families (ffrench-Constant et al. 1990). A remarkable case is that of organophosphate resistance in the cosmopolitan mosquito *Culex pipiens* (Raymond et al. 1991). Resistance is associated with an electrophoretically distinguishable form of an esterase enzyme that is "overproduced": the locus is repeated (amplified) about 60 times in resistant strains. Unlike other alleles at this locus, which display considerable variation in DNA sequence (based on restriction site mapping), this allele showed identical DNA sequences in resistant mosquitoes from Pakistan, Egypt, Congo, Ivory Coast, California, and Texas. The most reasonable inference is that within the few decades in which organophosphates have been used for mosquito control and have imposed selection for resistance, a single mutation (i.e., amplification of this locus) has spread by migration across three continents from its site of origin. Even though untold millions of mosquitoes throughout these regions have experienced natural selection for resistance, apparently only one mutational event has proved successful. This may mean that very few genes—perhaps only this one—undergo suitable mutation, and that such a mutation is a very rare event.

In such instances, we can conclude that the supply of rare mutations can limit the capacity of species for adaptation. This may help to explain why species have not become adapted to a broader range of environments, or why, in general, species are not more adaptable than they are (Bradshaw 1991).

Effects of Mutations on Fitness

The distribution of the effects of new mutations on fitness extends from very positive to very negative (e.g., lethal or sterile). We should like to know what the form of this distribution is: what fraction of mutations are strongly or slightly advantageous, neutral, or slightly or strongly disadvantageous (Figure 10.8). The precise form of the distribution is not known, but it is surely not fixed, for the fitness consequences of many mutations depend on the population's environment and even on its existing genetic constitution.

Undoubtedly many mutations are neutral or nearly so, having very slight effects on fitness (see Chapter 11). However, many do affect fitness, and their *average*, or net, effect is deleterious. This was shown, for example, by the decline in mean fitness as mutations accumulated in Mukai's (1972) experiment, described above (see Figure 10.6). These were spontaneous mutations; new mutations experimentally induced by the mutagenic (mutation-causing) chemical EMS also have a net deleterious effect. Ellen Wijsman (1984) showed that the offspring of *Drosophila melanogaster* that had been exposed to EMS suffered in competition with *Drosophila simulans* to a greater extent than nonmutagenized controls.

In experiments such as those performed by Mukai, it is evident that mutations that affect fitness have occurred, but nothing further is known about which genes have mutated, or what physiological or developmental features have been altered. Conversely, we can study the fitness effects of mutations of known phenotypic effect. These effects often depend on the environment; for example, mutations that slightly alter color pattern might or might not be advantageous, depending on whether or not some environmental factor were to favor a change in this feature. However, we must also recognize that *most mutations have pleiotropic effects,*—i.e., they affect more than one characteristic (see Chapter 3). We noted above, for example, that the *yellow* mutation in *Drosophila* affects not only body color, but also several components of male courtship behavior. The polygenic mutations that generate variation in bristle number in *Drosophila* often reduce larval viability (Kearsey and

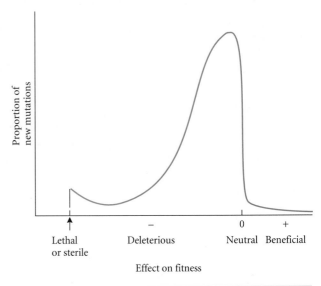

FIGURE 10.8   A possible frequency distribution of the effects on fitness of newly arisen mutations. (The real frequency distribution is unknown.) This figure reflects the widespread belief that the great majority of mutations are deleterious or nearly neutral (i.e., with nearly zero effect), and that only a very small proportion are beneficial. The curve rises at the left end because even the worst "superlethal" or "supersterile" mutations cannot cause more than total lethality or sterility.

Barnes 1970; Mackay et al. 1992). Many of the conspicuous mutations studied by *Drosophila* geneticists (as well as those in other organisms) have strongly negative pleiotropic effects on survival and reproduction. Some such mutations appear to lower fitness unconditionally, but others adversely affect viability only under certain conditions (e.g., of temperature) (Dobzhansky 1937). In some cases, the basis of deleterious pleiotropic effects is understood; for example, some mutations that affect *Drosophila* bristles disrupt the development of the nervous system. The developmental disharmonies caused by many major mutations (i.e., those with large phenotypic effects) are cited by those who argue that major mutations play little role in adaptive evolution, which, they claim, is based mostly on the accumulation of many genetic changes, each with only a slight phenotypic effect (Fisher 1930). This is a controversial subject, however, and some authors have argued that mutations with large effects have been important in adaptive evolution (see Chapter 23).

Not all mutations are deleterious. The following examples illustrate experiments in which advantageous mutations have been shown to occur in bacteria. Because of their short generation times and the ease with which huge populations can be cultured, microorganisms such as bacteria and yeasts are exceptionally useful for studying mutation— as well as other aspects of the evolutionary process (Dykhuizen 1990).

Adaptation to temperature   Experimental populations of bacteria such as *Escherichia coli* can be kept in a state of continual population growth (cell division) in several ways, such as transferring a sample to a new vessel of liquid nutrient broth every day. Because bacteria can be frozen (during which time they undergo no genetic change) and later revived, samples taken at different times from an evolving population can be stored, and their fitness can later be directly compared. The fitness of a genotype can be measured by its rate of increase in numbers, relative to that of another genotype with which it competes in the same culture, but which bears a genetic marker and so can be distinguished from it. Suppose, for example, that a culture is begun with equal numbers of genotypes A and B, and that after 24 hours B is twice as abundant as A. If the numbers have grown at a constant rate of cell division, each initial cell has produced 2^x descendants, where x is the number of generations (cell divisions). Thus if genotypes A and B have grown at the respective rates of 2^5 and 2^6 (i.e., B has one more generation per 24 hours), their relative numbers are 32:64, or 1:2.

The relative fitnesses of the genotypes—i.e., their relative growth rates—are measured by their rates of cell division per day, namely 5:6 or 1.0:1.2. If the genotypes had the same growth rates—e.g., 2^5—both would increase in number, but their fitnesses would be equal.

Albert Bennett, Richard Lenski, and John Mittler (1992) studied adaptation to temperature in *E. coli*, using replicate cultures derived from a single bacterial cell taken from a culture that had been propagated for 2000 generations at 37°C (Figure 10.9). Among the progeny of this cell, they found a mutant (Ara^+) that forms white colonies on nutrient agar plates containing a certain compound, whereas the Ara^- parent forms red colonies. Thus the relative numbers of Ara^+ and Ara^- bacteria in a culture can be measured by plating out a sample (i.e., spreading a dilute suspension of bacterial cells on a nutrient agar, on which each parent cell forms a separate colony of asexually produced descendants). Ara^+ is only a *marker*: it does not affect fitness, be-

FIGURE 10.9 The design of Bennett et al.'s experiment to monitor adaptation to different temperatures in cultures of *Escherichia coli* that were derived from a single cell, and hence initially lacked genetic variation. The fitness of each culture was measured by allowing it to compete with a genetically marked ($Ara+$ or $Ara-$) sample from the ancestral culture.

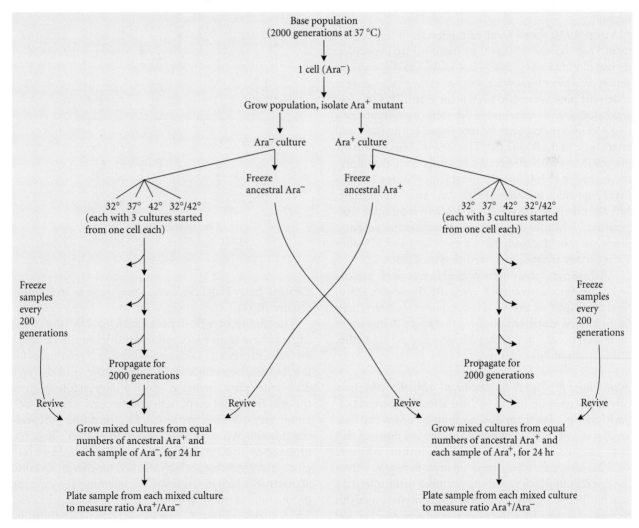

cause when *Ara*⁺ and *Ara*⁻ were grown together, their relative numbers did not change. Except for the marker mutation, the *Ara*⁺ and *Ara*⁻ cells were initially genetically identical, cloned from the same ancestral cell.

Bennett et al. grew sufficient numbers of *Ara*⁺ and *Ara*⁻ cells so that they could freeze a sample of each genotype (the ANCESTORS) and also grow several different populations of each genotype at each of several temperatures: 32°, 37° (the ancestral temperature), 42°, and a regime of daily alternation between 32° and 42°. They grew these populations at these temperatures for 2000 generations. Every 200 generations or so, they removed and froze a sample of each population. They could then determine whether a population had adapted by thawing and competing a sample from, say, the 42° population after 200 generations with a sample from the ancestral population that had not experienced the novel temperature regime. In order to distinguish the two competitors, they grew the 42° *Ara*⁺ cells with the ancestral *Ara*⁻ (as well as the converse, 42° *Ara*⁻ with ancestral *Ara*⁺, to be sure that the *Ara* allele did not affect the course of adaptation—which it did not). In this manner, the fitness of the bacteria that had been propagated at 42° could be measured, relative to the fitness of the ancestral genotype, by the change in the relative number of the two types over 24 hours.

Figure 10.10 shows some of the results of this experiment. Over the course of 2000 generations, each population increased in fitness (growth rate) relative to the ancestral genotype, when fitness was measured at the temperature at which the population had been maintained. This increase was based on a combination of two factors: new mutations that provided the capacity for more rapid cell division, and an increase in the frequency of these new alleles within the population by natural selection. It is interesting that fitness increased not only under the novel temperatures, but also in the population maintained at the "ancestral" temperature, 37°. The gain in fitness was less at this than at the other temperatures, which is to be expected because the ancestral genotype should already have been adapted to 37°—but some further adaptation nevertheless took place.

All these populations were derived from a single haploid cell. Moreover, this strain of *E. coli* is strictly asexual. Therefore, the adaptation to several temperatures must have been based on new mutations, showing that adaptive mutations do indeed occur. (In Chapter 17, we will describe further elaborations of this experiment.)

Mutations for novel biochemical abilities Bacteria can be screened for mutations that affect biochemical capacities by placing them on a medium on which that bacterial strain ordinarily cannot grow, such as a medium that lacks an essential amino acid or other nutrient. Whatever colonies do appear on the medium must have grown from the few cells in which mutations occurred that conferred a new biochemical ability. Several investigators have used this technique to select for the evolution of new biochemical pathways (e.g., Clarke 1974; Hall 1982, 1983). The experi-

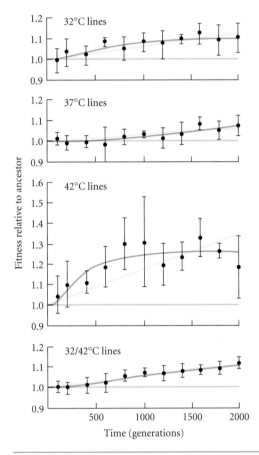

FIGURE 10.10 Adaptation to temperature in experimental populations of *Escherichia coli* studied by Bennett et al. A population's fitness is its growth rate relative to that of the ancestor, which is set at 1.0 (gray line). Solid circles with vertical bars show the mean fitness of replicate populations and a measure of variation (confidence interval) among them. The colored lines are two statistically calculated expressions of the increase in fitness with time. Because the populations initially lacked genetic variation, the increase in adaptation to different temperatures was due to natural selection acting on new advantageous mutations. (After Bennett et al. 1992.)

ments of Barry Hall (1982, 1983) may serve as an example (Figure 10.11).

The ability of wild-type *Escherichia coli* to use the sugar lactose as an energy source resides in the *lac* operon (see Chapter 3). In this operon, the *lacZ* gene codes for β-galactosidase, which hydrolyzes lactose, and the *lacY* gene codes for a permease, which allows lactose to enter the cell. Transcription of both genes occurs when the operator gene (*O*) is de-repressed, i.e., freed from a repressor protein. When lactose leaks into the cell, it is hydrolyzed by small standing levels of β-galactosidase into glucose and galactose, which are used for energy, and into allolactose, which de-represses the operator, allowing the production of more β-galactosidase and of permease, which enables lactose to enter more freely, thus permitting growth of the cell.

FIGURE 10.11 Diagram of mutational changes in populations of *Escherichia coli* selected for the ability to metabolize lactose in experiments by Hall. Lines ending with crossbars indicate repression of one gene's transcription by another gene. (1) Genes in wild-type *E. coli*. The existence of *R* and *ebg* was not known when the experiment began. (2) A strain with a deletion of *lacZ* (encoding the enzyme that normally metabolizes lactose) is used for the experiments. (3) Cells are placed on a medium containing lactose and IPTG, which enables transcription (symbolized by the curlicue) of the *lacY* gene. This gene encodes a permease. A few cells grow; these prove to have a mutation ([II]) in *ebg*. (4) Selection for better growth is successful due to the occurrence of a mutation in *R* that enhances the transcription of *ebg*. (5) Further selection, without IPTG, is successful, based on occurrence of another mutation in *ebg*. The new *ebg* enzyme produces lactulose from lactose; lactulose de-represses the operator (*O*), which promotes greater transcription of *lacY*, resulting in greater flux of lactose into the cells.

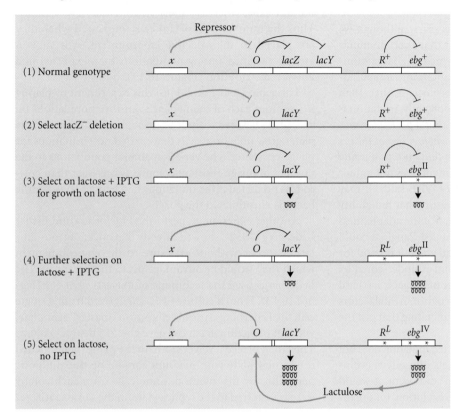

Hall obtained a strain in which most of the *lacZ* gene had been deleted. This strain, which may be denoted $O^+lacZ^-lacY^+$, cannot synthesize β-galactosidase, and so cannot grow on a medium in which lactose is the sole energy source. On such a medium, moreover, the *lacY* gene remains repressed (so that little lactose can enter the cell) because no allolactose is formed. A chemical additive (IPTG), however, will induce the permease.

Hall selected for the ability to use lactose by putting this strain on a lactose medium with IPTG. He obtained mutants that could grow on lactose. These proved to have mutations in a previously unknown gene, located on the chromosome far from the *lac* operon, which he called *ebg* ("evolved β-galactosidase"). These mutants (*ebg*[II]) grew slowly because the *ebg* gene is largely (but not completely) repressed by its regulatory gene (*R*[+]). By selecting for mutants with high growth rates, Hall obtained regulatory mutations (*R*[L]) in which lactose turns off the repression by the regulatory gene. Thus he had obtained a double mutant, $O^+lacZ^-lacY^+ebg^{II}R^L$, which, however, still required IPTG in order for lactose to enter the cell. By growing the double mutant strain on lactose without IPTG, he obtained *ebg*[IV], a second mutation of the *ebg* gene (i.e., *ebg*[II] → *ebg*[IV]). The enzyme produced by *ebg*[IV] metabolizes lactose to lactulose, which induces the permease gene *lacY*[+].

This genotype ($O^+lacZ^-lacY^+ebg^{IV}R^L$), then, represents the evolution of an entire system of lactose utilization, consisting of changes in enzyme structure enabling hydrolysis of the substrate; alteration of a regulatory gene so that the enzyme can be synthesized in response to the presence of its substrate; and evolution of an enzyme reaction that induces the permease needed for the entry of the substrate. One could not wish for a better demonstration that mutation and selection in concert can give rise to complex adaptations.

Mutation as a Random Process

Mutations occur at random. It is extremely important to understand what this statement does and does not mean. It does not mean that all conceivable mutations are equally likely to occur, because, as we have noted, the developmental foundations for some imaginable transformations do not exist. It does not mean that all loci, or regions within a locus, are equally mutable, for geneticists have described differences in mutation rates, at both the phenotypic and molecular levels, among and within loci (Woodruff et al. 1983; Wolfe et al. 1989). It does not mean that environmental factors cannot influence mutation rates: ultraviolet and other radiation, as well as various chemical mutagens and poor nutrition, do indeed increase rates of mutation.

Mutation *is* random in two senses. First, although we may be able to predict the *probability* that a certain mutation will occur, we cannot predict which of a large number of gene copies will undergo the mutation. The spontaneous process of mutation is stochastic rather than deterministic. Second, and more importantly, mutation is random in the sense that *the chance that a particular mutation will occur is not influenced by whether or not the organism is in an environment in which that mutation would be advantageous.* That is, the environment does not induce adaptive mutations. As Dobzhansky (1970) said, "It may seem a deplorable imperfection of nature that mutability is not restricted to changes that enhance the adaptedness of their carriers. However, only a vitalist Pangloss could imagine that the genes know how and when it is good for them to mutate."* Dobzhansky's point is underscored by our knowledge of the molecular nature of the gene: it is hard to conceive of a mechanism whereby environmental factors could direct the mutation process by dictating that just the right base pair changes should occur.

The argument that adaptively directed mutation does not occur is one of the fundamental tenets of modern evolutionary theory. If "directed mutation" did occur, it would introduce a Lamarckian element into evolution, for organisms would then acquire adaptive hereditary characteristics in response to their environment. The death blow to such "neo-Lamarckian" ideas came in the 1940s and 1950s from experiments with bacteria that appeared to demonstrate that adaptation is a consequence of spontaneous, random mutation followed by natural selection, rather than mutation directed by the environment.

Salvador Luria and Max Delbrück (1943) based an experiment on the following reasoning. Suppose that a large number of bacterial populations, all initially genetically identical, are each grown to size N from a single individual and then placed in a selective environment. If the environment induces an adaptive mutation (one that confers resistance to the selective factor) with some low probability p, the average number of resistant cells in each population will be pN. But as in any probabilistic

process, there will be variations among populations in the number of resistant cells. The number of populations with 0, 1, 2 ... k resistant cells will follow a POISSON DISTRIBUTION, in which the variance equals the mean. On the other hand, if the resistance mutations arose spontaneously *before* the populations were placed in the selective environment, any cell that mutated during that time will transmit the mutation to all its descendants, so that some of the populations will include a large number of mutant cells, whereas others will include few or none. Thus the number of resistant cells, measured when they are placed in the selective environment, will show a greater variance among populations than expected in a Poisson distribution (Figure 10.12).

Luria and Delbrück did just this experiment by plating *E. coli* from each of many phage-sensitive populations on agar plates covered with T1 bacteriophage. Colonies on the plates grew only from cells that carried new mutations for phage resistance. The variation among populations in the number of phage-resistant colonies was greater than Poisson, as predicted if the mutations had happened *before* the bacteria encountered the phage.

Another experiment, performed by Joshua and Esther Lederberg (1952), showed directly that advantageous mutations occur without exposure to the environment in which they would be advantageous to the organism. The Lederbergs used the technique of REPLICA PLATING (Figure 10.13). From a culture of *E. coli* derived from a single cell, the Lederbergs spread cells onto a "master" agar plate, without penicillin. Each cell gave rise to a distinct colony. They then placed a velvet cloth on the plate, then touched it to a new plate with medium containing the antibiotic penicillin. By this method, some cells from each colony were transferred to the replica plate, in the same spatial relationships as the colonies from which they had been taken. A few colonies appeared on the replica plate, having grown from penicillin-resistant mutant cells. When all the colonies on the master plate were tested for penicillin resistance, those colonies (and only those colonies) that had been the source of penicillin-resistant cells on the replica plate displayed resistance, showing that the mutations had occurred before the bacteria were exposed to penicillin.

Since these classic experiments, biologists have generally accepted that mutation is adaptively random rather than directed. Recently, however, several investigators have reported results, again with *E. coli*, that at face value seem to suggest that some advantageous mutations might be directed by the environment (Box 10.D). Their interpretations have been challenged by other investigators, who find no compelling evidence for directed mutation in these experiments. The issue is being vigorously debated at this time (Sniegowski and Lenski 1995). It is hard to imagine mechanisms by which mutations for many characteristics, such as morphological features, could be adaptively directed by environmental factors such as natural enemies and a species' social interactions; but a convinc-

---

*Dobzhansky refers to Dr. Pangloss, a character in Voltaire's satire *Candide,* who despite the most tragic contrary evidence taught that "all is for the best in this best of all possible worlds." Vitalism is the belief, rejected by biologists, that processes in living beings are produced by an immaterial vital force, rather than by purely chemical and physical processes.

FIGURE 10.12 Two hypotheses for the origin of mutations for phage resistance in bacteria. Luria and Delbrück's experiment was designed to determine which hypothesis is correct. In both A and B, the pedigrees of bacteria in several cultures, each derived from a single cell, are shown. Bacteria are exposed to phage in the fourth generation. Shaded cells carry a mutation for phage resistance. It is assumed (and experimentally demonstrated) that a small minority of cells are resistant. (A) Under the hypothesis that exposure to phage directly induces muta- tions for resistance, the few resistant cells should be distributed at random among the cultures. (The mathematical description of the expected frequency distribution is called the Poisson distribution). (B) Under the hypothesis that mutations for resis- tance occur at random, before exposure to phage, resistant cells will be "clumped" in certain cultures and absent in all others, because mutations in ancestral cells are inherited by multiple descendants. (After Stent and Calendar 1987.)

(A) Mutation induced by environment

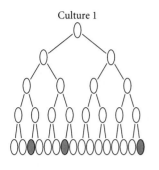
Culture 1

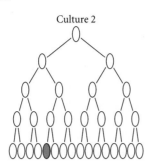
Culture 2

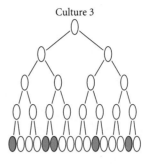
Culture 3

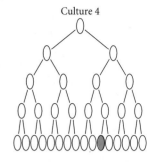
Culture 4

(B) Random mutation

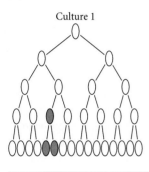
Culture 1

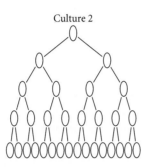
Culture 2

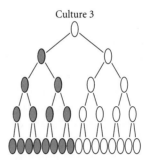
Culture 3

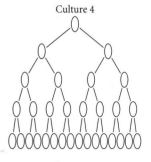
Culture 4

ing demonstration that the environment can direct even some mutations, such as those altering biochemical ca- pacities, would constitute a major change in our under- standing of evolution.

Recombination as a Source of Variation

Release of Genetic Variation

All genetic variation owes its origin ultimately to mutation, but in the short term, a great deal of the genetic variation within populations arises through recombination. Several mechanisms of genetic exchange occur in prokaryotes, but we will restrict our discussion to sexually reproducing eu- karyotes, in which genetic variety arises from two pro- cesses: the union of genetically different gametes, and the formation of gametes with combinations of alleles differ- ent from those that united to form the individual that pro- duces them. The formation, during meiosis, of genetically variable gametes is the consequence of the independent segregation of nonhomologous chromosomes and of crossing over between homologous chromosomes.

The potential genetic variation that can be released by recombination is enormous. If *Drosophila* has about 10,000 functional loci, and if an average individual is het- erozygous at 10 percent of its loci (based on estimates from electrophoresis; see Chapter 9), then an individual could conceivably produce 2^{1000}, or approximately 10^{300}, genetically different kinds of gametes, which is immense- ly greater than Avogadro's number (6.023×10^{23}), the number of molecules per mole. (In actuality, the vast ma- jority of theoretically possible gene combinations will never be formed, because they would require an immense number of crossovers between tightly linked loci.) At a more modest level, if an individual is heterozygous for only one locus on each of five pairs of chromosomes, in- dependent segregation alone generates $2^5 = 32$ allele com- binations among its gametes, and mating between two such individuals can give rise to $3^5 = 243$ genotypes among their progeny. (This assumes that both parents have the same heterozygous genotype, say, A_1A_2, at each locus. If each is heterozygous for a different pair of alleles, A_1A_2 and A_3A_4, their progeny can carry $4^5 = 1024$ different geno- types.)

FIGURE 10.13 The method of replica plating, used to show that mutations for penicillin resistance arise spontaneously before exposure to penicillin, rather than being induced by it. We begin (upper left) with an agar plate with numerous colonies of bacteria, each derived from one cell. Part of each colony is transferred to a velvet cloth (i.e., an "imprint") (step 1), and the plate is saved (2). Each colony is tested for penicillin resistance by growing it in medium with penicillin (3). A clean plate with penicillin is pressed onto the velvet (4). This plate is thus seeded with cells from each original colony (5), but only cells from original colonies shown (in step 3) to be penicillin-resistant grow on the new plate. (After Srb et al. 1965.)

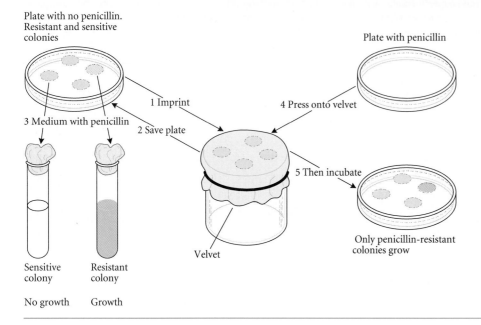

Plate with no penicillin.
Resistant and sensitive colonies

Plate with penicillin

1 Imprint

4 Press onto velvet

3 Medium with penicillin

2 Save plate

5 Then incubate

Only penicillin-resistant colonies grow

Velvet

Sensitive colony

Resistant colony

No growth

Growth

If each locus affects a different feature, this represents a great variety of character combinations. If all five loci have equal and additive effects on a single character, such as size, the range of variation among the offspring can greatly exceed the difference between the parents. For instance, if each substitution of + and – alleles in a genotype adds or subtracts one unit of phenotype, two quintuply heterozygous parents (both +–+–+/–+–+–), both of size 20, could have offspring ranging in size from 15 (–/– at all five loci) to 25 (+/+ at all five loci). Each of the two extreme phenotypes, however, would have very low frequency among the offspring (1/1024, assuming independent segregation). By the same token, if + and – alleles have intermediate frequencies (near 0.5) in a population, the majority of individuals will be heterozygous at many loci, hence intermediate in size. Multiple homozygotes, with extreme phenotypes, will segregate only infrequently. (For example, if the frequencies of the + and – alleles are $p = q = 0.5$ at each of five independently segregating loci, the frequency of individuals homozygous +/+ at all loci is $(p^2)^5 = 0.0009766$.)

Some Evidence

The validity of these theoretical constructs has frequently been demonstrated. For example, Figure 7 in Chapter 9 shows results typical of several studies of *Drosophila pseudoobscura* by Theodosius Dobzhansky and his colleagues, who, using crosses by which wild chromosomes can be "extracted" from natural populations (see Figure 6 in Chapter 9), measured the egg-to-adult viability of homozygotes for numerous chromosomes, as well as flies heterozygous for random pairs of these same chromosomes. The variation among homozygotes is greater than among heterozygotes, which is what we would expect if several or many alleles on the chromosome confer extreme phenotypes when homozygous.

In order to judge how much variation is released by recombination, the Dobzhansky team (Spassky et al. 1958) made all possible crosses between homozygotes for only ten different chromosomes. All these chromosomes conferred almost the same, nearly normal viability when homozygous. From the F_1 female offspring, in which crossing over occurred, they then extracted recombinant chromosomes and measured their effect on viability when homozygous (Figure 10.14). Even though the original ten chromosomes differed little, the variance in viability among the recombinant chromosomes was more than 40 percent of the variance among homozygotes for much larger samples of chromosomes from natural populations (as illustrated in Figure 7 in Chapter 9). Thus a single episode of recombination among just ten chromosomes generates a large fraction of a wild population's variability. Some of the recombinant chromosomes were "synthetic lethals," meaning that recombination between two chromosomes that yield normal viability produced chromosomes that were lethal when they were made homozygous. This implies that each of the original chromosomes carried an allele that did not lower viability on its own, but

BOX 10.D

Directed Mutation?

The classic experiments by Luria and Delbrück and by the Lederbergs show that some mutations arise spontaneously, but do not rule out the possibility that some other mutations, under certain conditions, might be environmentally directed. John Cairns and his colleagues (1988) and Barry Hall (1988) reported results suggesting directed mutation for metabolic capacities. In both cases, the investigators scored mutations that occurred not in rapidly growing populations, but in bacteria that were plated on agar with an energy source that the original strain could not use. Cairns et al. plated *lac*⁻ bacteria, unable to use lactose, on lactose medium, and scored the incidence of *lac*⁺ mutants that could grow on lactose. They reported that these mutations occurred with a lower variance than if they had been spontaneous (cf. the Luria-Delbrück experiment), and that they occurred only after a long delay (implying that the mutations had not already been carried by the bacteria first plated).

Hall plated bacteria on a medium with salicin, which, he said, could be used for energy only if two mutations occurred in a galactosidase operon. The mutants that grew on salicin had to have experienced both a mutation of the regulatory region (R) and excision of a transposable element (IS) from the structural gene, enabling this operon to function. Hall presented evidence that the IS excision mutations (IS⁻) occurred before the R mutations, that the IS⁻ mutations alone could not use salicin,

and that the combination of the IS⁻ and R mutations occurred much more frequently than expected. He suggested that the IS⁻ excision occurred only on medium with salicin, and that it "serves only to create the potential for a secondary selectively advantageous mutation"—which implies not only directed mutation, but also "anticipatory" mutation.

Several authors have provided criticisms and alternative interpretations of these experiments (see Sniegowski and Lenski 1995 and references therein). For example, the delay in the appearance of mutants able to grow on the medium, as well as the low variance, might be due to an undetected first mutation that provided some capacity for growth, resulting in populations large enough for ordinary spontaneous mutations, conferring a high growth rate, to occur. In the same vein, the high incidence of the double mutant in Hall's experiment might be explained if IS⁻ excision is stimulated by starvation (rather than by salicin per se), if the IS⁻ mutants can indeed grow in the presence of salicin, and if spontaneous R mutations then occur in the large resultant populations. Mittler and Lenski (1992) provide evidence for just this hypothesis. So far, it seems that explanations consistent with orthodox neo-Darwinian theory can be found for apparently directed mutation. (For the contrary view, see Thaler 1994.) Most geneticists agree that truly directed mutation should be an explanation of last resort.

---

did cause death when combined with another allele, at another locus, on the other chromosome. This is a striking example of epistasis (nonadditive interaction among genes).

Erosion of Variation by Recombination

Recombination is a double-edged sword: it both generates and destroys genetic variation. In sexually reproducing populations, genes are transmitted to the next generation, but genotypes are not: they end with organisms' deaths, and are reassembled anew in each generation. Thus an unusual, favorable gene combination may occasionally arise through recombination, but if such individuals mate with other members of the population, it will be lost immediately by the same process. For example, if + and − alleles that affect an additive polygenic trait have frequencies $p = 0.9$, $q = 0.1$ at each of three independently segregating loci, a triply homozygous −/− genotype arises with frequency $(q^2)^3 = 10^{-6}$; but because these few individuals mate with other genotypes, their progeny will be heterozygous, and therefore less phenotypically extreme. In the heterozygous progeny, moreover (such as $++-/---$), recombination spawns a variety of gametes bearing a mixture of + and − alleles, which unite to yield offspring with intermediate traits. This is true also if the

loci are linked, although then the proportions of different gamete types depend on the recombination rate (r) between them.

In Chapter 9, we saw that if two or more loci are initially in a state of *linkage disequilibrium* (in the extreme, only the gamete types AB and ab, or $++$ and $--$, are present), then recombination generates the other gene combinations (Ab and aB, or $+-$ and $-+$) until the association between alleles has broken down completely (i.e., *linkage equilibrium* is achieved). Thus, for example, the + and − alleles that affect a polygenic trait will be carried by gametes as various +, − mixtures. Consequently, the phenotype of most individuals will be near the mean, and relatively few will have either extreme phenotype. That is, *recombination reduces phenotypic variance*, compared with the variance that might be achieved in its absence. By the same token, it might happen that in a population that is polymorphic for two loci, both double homozygotes ($AABB$ and $aabb$) are better adapted than any other genotype; but if mating is random, most of the progeny in each generation will be less well adapted. (For example, if $p = q = 0.5$ at each of two independently segregating loci, the frequency of each of the double homozygotes will be only 0.0625.)

The conflict between natural selection, which tends to increase the frequency of favorable gene combinations,

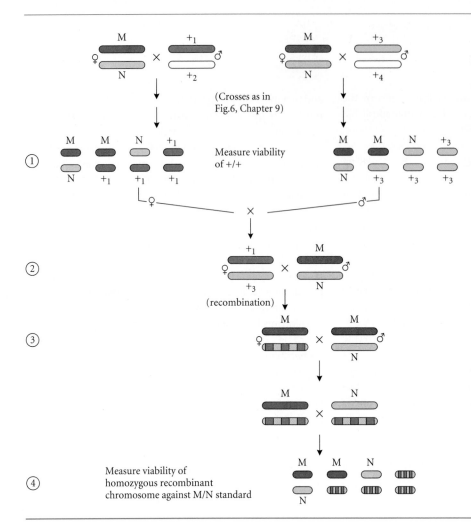

FIGURE 10.14 The method used by Spassky et al. to measure the effect of recombination on variation in egg-to-adult survival (viability) in *Drosophila*. Chromosomes M and N have dominant markers and inversions that prevent crossing over. Two wild chromosomes, say, $+_1$ and $+_3$, are made homozygous (as in Figure 6 in Chapter 9), and their viability is measured by the proportions of wild-type and M/N phenotypes in the progeny (step 1). Heterozygous $+_1/+_3$ females are produced (2). Crossing over between $+_1$ and $+_3$ occurs. Each female is crossed to an M/N male (only one such case is shown), yielding a variety of recombinant chromosomes (shown with dark and light bars) among the progeny. Further crosses (3) bring the recombinant chromosomes into the homozygous state, and the viability of each homozygote is measured (4). Similar crosses are made to obtain recombinants among other wild chromosomes.

and recombination, which tends to destroy them, has many consequences for adaptation, speciation, and the evolution of factors that modify recombination rates, including sexual and asexual reproduction (see Chapters 16 and 21).

Alterations of the Karyotype

An organism's **karyotype** is the description of its complement of chromosomes: their number, size, shape, and internal arrangement. The karyotype may be altered in many ways. In considering these alterations, it is important to bear in mind that the loss of a whole chromosome, or a major part of a chromosome, usually reduces the viability of a gamete or an organism because of the loss of genes. Also, a gamete or organism often is inviable or fails to develop properly if it has an ANEUPLOID, or "unbalanced," chromosome complement—for example, if an otherwise diploid organism has three copies of one of its chromosomes. (For instance, humans with three copies of chromosome 21, a condition known as Down syndrome, have brain and other defects.)

Alterations of the karyotype fall into two major categories: changes in the number of whole sets of chromosomes (*polyploidy*) and *rearrangements* of one or more chromosomes.

Polyploidy

A diploid organism has two entire sets of homologous chromosomes ($2N$); a **polyploid** organism has more than two. (In discussing chromosomes, N refers to the number of different chromosomes in the gametic, or haploid, set, and the numeral refers to the number of representatives of each autosome.) In a diploid species, diploid, or UNREDUCED, gametes are occasionally produced when the reduction division fails to occur in meiosis. (This event can be experimentally induced by inhibitors of cell division such as colchicine.) The union of an unreduced gamete (with $2N$ chromosomes) and a reduced gamete (with N chromosomes) yields a TRIPLOID ($3N$) zygote. Triploids often develop normally, but they are largely sterile because most of their gametes have aneuploid chromosome complements (i.e., at segregation, each daughter cell may receive one copy of certain chromosomes and two of certain others). However, if two unreduced gametes unite, a tetraploid ($4N$) offspring is formed. Other such unions can form hexaploid ($6N$), octoploid ($8N$), or even more highly polyploid genotypes.

In many cases, each set of four homologous chromosomes of a tetraploid is aligned during meiosis into a quartet (quadrivalent), and then may segregate in a balanced (two by two) or unbalanced (one by three) fashion (Figure 10.15). Many ANEUPLOID GAMETES result, so fertility is greatly reduced. In other cases, the four chromosomes align not as a quartet, but as two pairs that segregate normally, resulting in balanced, viable gametes, so that fertility is normal, or nearly so. This would seem to require that the chromosomes be differentiated so that each can recognize and pair with a single homologue rather than with three others.

Many species of plants, and a few species of trout, tree frogs, and other animals, are polyploid (Chapter 16). Polyploidy is common among plants. Many polyploid species appear to have arisen recently from diploid ancestors, and some major groups of plants are believed, because of their high chromosome numbers, to stem from ancient polyploid ancestors. Estimates of the proportion of polyploid angiosperms range from 30 percent (Stebbins 1950) to 50–70 percent (Stace 1989). A substantial minority of recently arisen polyploids has arisen by the union of unreduced gametes of the same species; these are **autopolyploids**. But the majority are **allopolyploids**, which have arisen by hybridization between closely related species.[*] They present a mixture of, or an intermediate condition between, the characteristics of their parents. In allopolyploids, the parental species' chromosomes apparently are different enough for the chromosomes of each parent to recognize and pair with each other, so that meiosis in an allotetraploid involves normal segregation of pairs rather than quartets of chromosomes (Figure 10.15C). In such a case, the karyotype has $2N$ chromosomes, but N may be based on a smaller ancestral haploid genome of x chromosomes. For example (Anderson 1936), a common blue flag, or iris (*Iris versicolor*), has $2N = 108$ ($N = 54$) chromosomes. It appears to have been derived from hybridization between *Iris virginica*, with $2N = 72$ chromosomes ($N = 36$), and *Iris setosa*, with $2N = 36$ ($N = 18$). The base, or ancestral, chromosome number appears to be $x = 18$, relative to which *I. virginica* has four chromosome sets and *I. versicolor* has six. This is one example of a POLYPLOID COMPLEX, some of which are much more complex than the *Iris* case.

---

[*]There is a full spectrum of intermediate cases between auto- and allopolyploidy, and the meiotic behavior of the chromosomes can be very complicated, with some chromosomes forming pairs and others quartets in the same genome. Stebbins (1950) presents an extended discussion of polyploidy.

(A)

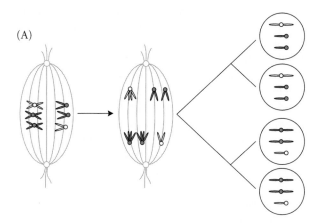

(C)

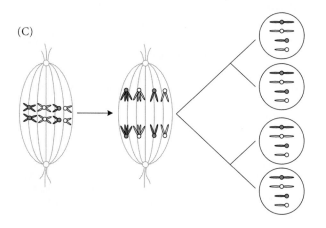

(B)

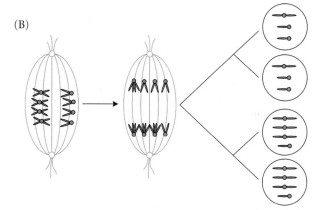

FIGURE 10.15 Segregation of chromosomes in meiosis in various polyploids. Each panel shows synapsis and segregation on the spindle in meiosis I, and the four haploid products of meiosis II. Two sets of homologous chromosomes are shown; each begins meiosis with two identical sister chromatids. In hybrids (parts A and C), all chromosomes with different colors (colored or white) are derived from different species. (A) In a triploid, trios of homologous chromosomes may align, and unequal numbers of chromosomes from each homologous set move to opposite poles. Hence gametes are aneuploid, with unbalanced numbers of chromosomes. (B) In an autotetraploid, quartets of homologues undergo synapsis, but may not segregate two by two. In this figure, they are shown segregating three by one, yielding aneuploid gametes. (C) In an allotetraploid, each chromosome may pair with a single homologue from the same species. Segregation is normal, and the gametes receive euploid (balanced) chromosome complements.

Polyploidy often has many direct effects. Polyploid cells are generally larger and divide more slowly than diploid cells. (This is a specific instance of a very widespread relationship between DNA content and the size and division rate of cells, discussed further in Chapter 22.) Perhaps for this reason, polyploid plants often have thicker leaves, and certain structures (and sometimes the whole plant) are larger. Levels of enzymes, hormones, and other biochemical constituents are often higher, and various physiological functions may differ (Stebbins 1950; Levin 1983b). Some authors have suggested that polyploids may be particularly capable of colonizing stressful environments, possibly because of their physiological vigor.

Many genes in a tetraploid are represented by four, rather than two, copies. If, as in an allotetraploid, they are inherited as two distinct loci, they may undergo independent evolutionary changes and diverge from each other in nucleotide sequence. We will return to this topic in Chapter 22.

Chromosome Rearrangements

Changes in the structure of chromosomes constitute another class of karyotypic alterations. These changes are caused by breaks in chromosomes, followed by rejoining of the pieces in new configurations. Some such changes can affect the pattern of segregation in meiosis, and therefore affect the proportion of viable gametes. Most chromosome rearrangements seem not to have direct effects on morphological or other phenotypic features, but some such effects do occur, because an alteration of gene order may bring certain genes under the influence of the control regions (e.g., the promoters) of other genes, and so alter their expression. Such POSITION EFFECTS have often been described in the laboratory, but it is not certain that they have contributed to evolutionary change.

Individual organisms may be homozygous or heterozygous for a rearranged chromosome, and are referred to as HOMOKARYOTYPES or HETEROKARYOTYPES respectively. It is also useful to distinguish ACROCENTRIC chromosomes, in which the centromere is near one end, from METACENTRIC chromosomes, in which the centromere is somewhere in the middle and separates the chromosome into two arms.

Inversions Consider a segment of a chromosome in which ABCDEFGHIJ denotes a sequence of markers such as genes. If a loop is formed, and breakage and reunion occur at the point of overlap, a new sequence, such as AB<u>I-</u><u>HGFEDC</u>J, may be formed. (The inverted sequence is underlined.) Such an **inversion** is PERICENTRIC if it includes the centromere, and PARACENTRIC if it does not. Inversions rearrange gene order, and may also transform metacentric into acrocentric chromosomes, or vice versa.

Because inversion is a rare event, it alters one of the two homologous chromosomes in a cell; if this is a germ line cell, heterokaryotypic daughter cells undergo meiosis. During meiotic synapsis, alignment of genes on the normal and inverted chromosomes requires the formation of a loop,

which can sometimes be observed under the microscope (Figure 10.16). Now suppose that in a paracentric inversion, crossing over occurs between loci such as F and G (Figure 10.16). Two of the four strands are affected. One strand lacks certain gene regions (A, B), and also lacks a centromere; it will not migrate to either pole, and is lost. The other affected strand not only lacks some genetic material: it also has two centromeres, so the chromosome breaks when these are pulled to opposite poles. The resulting cells lack certain gene regions, and will not form viable gametes. Consequently, in inversion heterokaryotypes, (1) *fertility may be reduced* because many gametes are inviable; and (2) *recombination is effectively suppressed* because gametes carrying the recombinant chromosomes, which lack some genetic material, are inviable. (If you examine Figure 10.16 carefully, you may deduce that if crossing over occurs at *two* sites within the inversion loop, all the meiotic products are balanced and viable. However, recombination may be fully suppressed if one chromosome carries a series of overlapping inversions.) Crossing over and gamete formation proceed normally in homokaryotypes, whatever gene arrangement they may carry.

Related species often differ in their karyotypes as a consequence of inversions. However, it is rather unusual to find two or more inversions in a polymorphic state within a population because of the reduced fertility of heterokaryotypes: one of the arrangements is rapidly eliminated from the population by natural selection (see Chapter 13). The most prominent exceptions are in *Drosophila* and some other flies (Diptera), in which the unbalanced recombinant chromosomes enter the polar bodies during meiosis, so that female fecundity is not reduced.* It is particularly easy to study inversions in *Drosophila* and some other flies because the larval salivary glands contain giant (polytene) chromosomes that remain in a state of permanent synapsis (so that inversion loops are easily seen), and because these chromosomes display bands, each of which apparently corresponds to a single gene. The banding patterns are as distinct as the computer-scanned bar codes on supermarket products, so an experienced investigator can identify different sequences. **Inversion polymorphisms** are common in *Drosophila*—more than 20 different arrangements of the third chromosome have been described for *Drosophila pseudoobscura*, for example—and they have been extensively studied from both population genetic and phylogenetic points of view (Chapters 8, 13).

Translocations By breakage and reunion, two nonhomologous chromosomes may exchange segments, resulting

---

*This is true for heterozygous paracentric inversions. Crossing over within a heterozygous pericentric inversion results in genetically unbalanced products of meiosis, in Diptera as in other organisms, and reduces fertility. Male fertility is not reduced in a heterokaryotype, because crossing over does not normally occur in male Diptera.

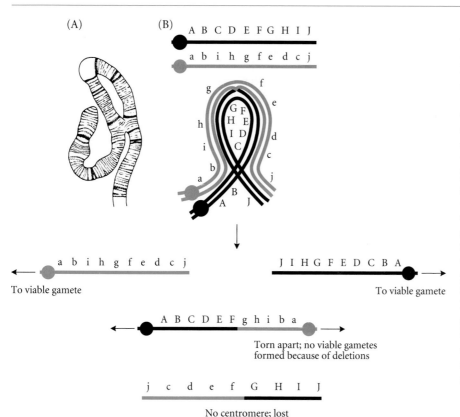

(A)

(B)

A B C D E F G H I J

a b i h g f e d c j

g
f
G
H
I
C
D
E
F
e
h
d
i
c
b
a
j
B
A
J

a b i h g f e d c j

To viable gamete

J I H G F E D C B A

To viable gamete

A B C D E F g h i b a

Torn apart; no viable gametes
formed because of deletions

j c d e f G H I J

No centromere; lost

FIGURE 10.16 Chromosome inversions. (A) Synapsed chromosomes in a salivary gland cell of a larval *Drosophila pseudoobscura* heterozygous for *Standard* and *Arrowhead* arrangements. The two homologous chromosomes are so tightly synapsed that they look like a single chromosome. The "bridge" forms the loop shown diagrammatically in part B. Similar synapsis occurs in germ line cells undergoing meiosis. (B) Two homologous chromosomes differing by an inversion of the region C to I, and their configuration in synapsis. Crossing over between two chromatids (between F and G) yields products that lack a centromere or substantial blocks of genes. Because these do not form viable gametes, crossing over appears to be suppressed. Only cells that receive the two chromatids that do not cross over become viable gametes. (After Strickberger 1968.)

in a **reciprocal translocation**. In a translocation heterokaryotype, synapsis of homologous regions results in contorted configurations such as the cross-shaped arrangement in Figure 10.17. In the simple case illustrated, the four chromosomes may segregate pairwise in three different ways, two of which result in gametes with unbalanced gene complements (some genes are lacking). These gametes, or the resulting progeny, are inviable, so the fertility of translocation heterokaryotypes is often reduced by 50 percent or more. Consequently, polymorphism for translocations is rare in natural populations. Nevertheless, related species sometimes differ by translocations, which have the effect of moving groups of genes from one chromosome to another. For example, the Y chromosome of the male *Drosophila miranda* includes a segment that is homologous to part of one of the autosomes of its close relative *D. pseudoobscura* (Dobzhansky 1970).

Fissions and fusions In the simplest form of chromosome **fusion**, two nonhomologous acrocentric chromosomes, in which the centromeres are nearly terminal, may undergo reciprocal translocation near the centromeres so that they are joined into a metacentric chromosome. Conversely, a metacentric may undergo **fission**, becoming two acrocentrics if it undergoes a reciprocal translocation with a minute "donor" chromosome (Figure 10.18). A simple fission heterokaryotype, then, has a

metacentric, which we may refer to as AB, with arms that are homologous to two acrocentrics A and B. AB, A, and B together synapse as a "trivalent." Viable gametes and zygotes are formed if AB segregates from A and B, but if homologous chromosomes do not segregate to opposite poles (i.e., if **nondisjunction** occurs) aneuploid gametes result, reducing fertility or zygote viability. The frequency of aneuploid gametes is often about 5–25 percent, and sometimes up to 50 percent (Lande 1979a; see also Chapter 16). Chromosome fusions and fissions often distinguish related species or geographic populations of the same species (Figure 10.19), and are sometimes found as polymorphisms within populations.

Changes in chromosome number Summarizing what we have covered so far, the number of chromosomes may be altered by polyploidy (especially in plants) or by translocations and fusions or fissions of chromosomes. These are the mutational foundations for the evolution of chromosome number. For example, the haploid chromosome number varies among mammals from 3 to 42 (Lande 1979a), and among insects from 1 in a species of ant to about 220 in some butterflies (the highest number known in animals). Related species sometimes differ strikingly in karyotype: in one of the most extreme examples, two very similar species of small deer, *Muntiacus reevesii* and *M. muntjac*, have haploid chromosome numbers of 46 and 3

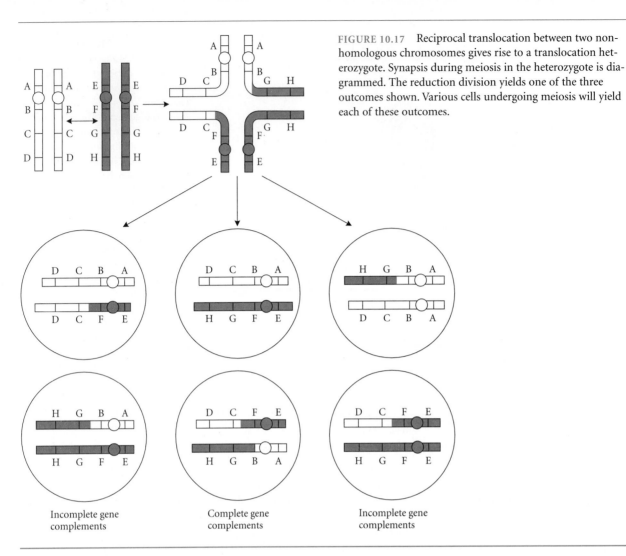

FIGURE 10.17 Reciprocal translocation between two nonhomologous chromosomes gives rise to a translocation heterozygote. Synapsis during meiosis in the heterozygote is diagrammed. The reduction division yields one of the three outcomes shown. Various cells undergoing meiosis will yield each of these outcomes.

Incomplete gene complements

Complete gene complements

Incomplete gene complements

or 4 (in different populations) respectively (Figure 10.20). As for all characteristics, the evolution of the karyotype requires not only mutation, but other processes as well (see Chapters 11, 13).

The spontaneous rate of origin of a given class of chromosome rearrangement (e.g., reciprocal translocation) is quite high: about 10^{-4} to 10^{-3} per gamete (Lande 1979a). However, a rearrangement involving breakage at any particular site(s) rarely arises, and is usually considered to be unique.

Duplications and deletions The rearrangements described above, unlike polyploidy, do not greatly alter the amount of genetic material. Several other processes, however, can change the amount of DNA. At this point we will treat only one of these, **unequal crossing over** (unequal exchange). If crossing over occurs between two chromosomes that are not perfectly aligned, a region will be tandemly duplicated on one recombination product and deleted from the other (Figure 10.21). The length of the affected region (e.g., the number of loci) depends on the amount of displace-

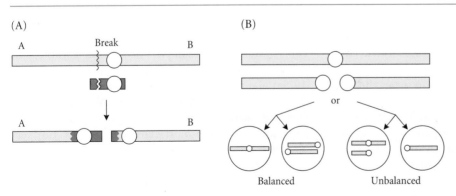

(A)

A   Break   B

A   B

(B)

or

Balanced   Unbalanced

FIGURE 10.18 (A) A simple fission of a metacentric chromosome, with arms A and B, into two acrocentric chromosomes, by translocation with a minute "donor" chromosome. (B) Segregation in meiosis of a fission heterozygote can yield euploid (balanced) or aneuploid (unbalanced) complements of genetic material.

ment of the two misaligned chromosomes. Deletion of a substantial amount of genetic material results in inviable gametes or zygotes, but duplications are sometimes advantageous, perhaps because more gene product is synthesized. Chromosomes with a tandem duplication (ABBC) are even more likely subsequently to engage in unequal crossing over because the duplicate regions can pair out of register,

$$\begin{bmatrix} ABBC... \\ ...ABBC \end{bmatrix}$$

generating further duplications (ABBBC). Because misalignment occurs rarely, each of the repeated genes is generally inherited as a separate locus, undergoing independent mutations, so that their DNA sequences may diverge over time.* As a result, a great many genes are members of **gene families**: sets of genes that have similar DNA sequences, but differ to some degree in sequence and often in function. For example (Figure 22 in Chapter 3), humans have eight loci that code for the several α-like and β-like hemoglobin polypeptides (as well as several related degenerate sequences that are not translated). These loci differ in their time of expression (embryonic, fetal, or adult). Unequal crossing over is probably one of the processes that have generated the extremely high number of copies of nonfunctional sequences (in satellite DNA) that constitute much of the DNA in most eukaryotes. This

---

* There also exist processes that may homogenize repeated DNA sequences, giving rise to the phenomenon of "concerted evolution" (Chapter 22).

FIGURE 10.19 Geographic variation in the karyotype of the pocket mouse *Perognathus goldmani* in northwestern Mexico. (A) The distribution of populations with different karyotypes, designated by Greek letters. (B) The proposed sequence of evolutionary changes in karyotype, involving four fusions (e.g., of chromosomes 3 and 4 in race ε) and two inversions (e.g., the X chromosome in race α). Karyotypes β and ε are most similar to the hypothetical ancestral karyotype, which was inferred by phylogenetic comparison with other species of *Perognathus*. 2N is the diploid number of chromosomes, FN the number of chromosome arms. (After Patton 1969.)

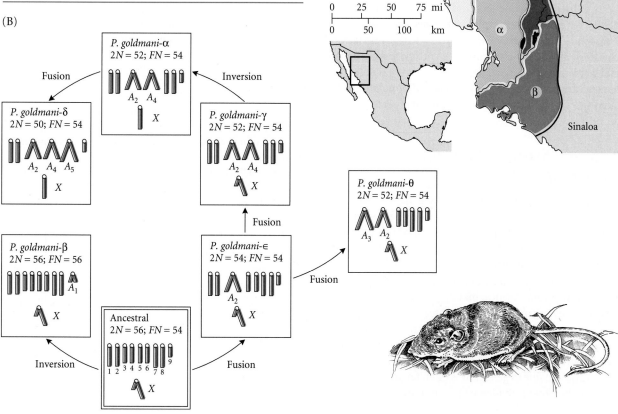

FIGURE 10.20 The diploid chromosome complements, taken from photographs, of two closely related species of muntjac deer. This is one of the most extreme differences in karyotype known among closely related species. Despite the difference in chromosomes, the species are phenotypically very similar. (A) *Muntiacus reevesi* (2N = 46). (B) *Muntiacus muntiacus* (2N = 8). (From White 1978.)

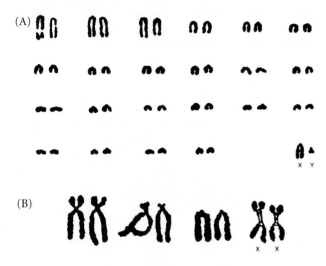

process has also been extremely important in the evolution of greater numbers of genes and of total DNA (see Chapter 22).

Unequal exchange may also occur on a very fine scale, namely, within a single gene. This gives rise to short repeats of base pair sequences, and to variation in the number of repeats. For example, in the *per* (*period*) gene of *Drosophila melanogaster*, which affects various behavioral rhythms, a six-base-pair motif, encoding threonine and glycine, is tandemly repeated 14 to 23 times, with several length variants occurring within natural populations. Related species also differ in the number of repeats (Costa et al. 1991).

External Sources of Genetic Variation

So far, we have discussed the origin of genetic variation by mutation and recombination within a single population of a species. In almost all species, an important source of genetic variation is **gene flow** from other populations of the species: the influx of alleles that have arisen or have reached high frequency in other populations, and are carried by migrants that join the local mating population. Genes may be carried by organisms (including seeds and spores) or by their gametes (e.g., pollen). We will return to this important topic in Chapter 11 and subsequent chapters. Suffice it to say that in the short term, the increase in a population's genetic variation due to gene flow is often far greater, per generation, than that provided by new mutations.

In some cases, genetic variation enters the population from *other species*, either by **hybridization** or by **horizontal gene transfer**.

Hybridization

Closely related species often can hybridize and produce at least some partly or fully fertile offspring, which, by backcrossing, introduce genes from one species into the other. Hybridization (more fully discussed in Chapter 15) may occur at a low level wherever the species meet, or may be limited to certain localities, such as disturbed habitats. Some hybrids can differ considerably from both parent forms (Figure 10.22). In some cases it may be arbitrary whether the hybridizing forms are called species or some other term, such as "semispecies." Hybridization has been documented in many groups of animals, and appears to be especially frequent in plants (Harrison 1990; Stebbins 1950; Grant 1971; Mayr 1963; but see also Mayr 1992).

The botanist Edgar Anderson (1949) applied the term **introgressive hybridization** to the incorporation into a gene pool of genes from other species that nevertheless remain fairly distinct. An example from animals is presented by two morphologically different forms of house mice, *Mus musculus* and *M. domesticus*, that are distributed throughout eastern and western Europe respectively, meeting along a narrow zone of hybridization. Robert Selander and his colleagues (1969) described patterns of enzyme polymorphism from several regions on either side of the hybrid zone in the Jutland Peninsula of Denmark (Figure 10.23). The two forms differ substantially in allele frequencies at 13 of 17 polymorphic loci; at 6 of the loci, the two species are fixed, or nearly so, for different alleles. However, some of these alleles do cross the species boundary—especially alleles typical of *M. domesticus*, such as *Es-1b*, which penetrates into *M. musculus* populations, particularly those located nearest *M. domesticus*.

FIGURE 10.21 Ordinarily, as at left, crossing over results in reciprocal exchange. But in unequal crossing over, at right, a segment of one chromosome, marked by locus B, is transferred to the other chromosome. Thus one chromosome has a deficiency, and the other bears a tandem duplication for one or more loci. Deletions are often deleterious, but duplications are sometimes advantageous.

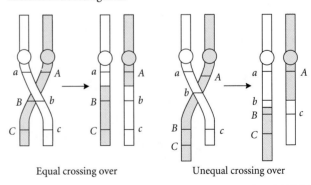

Equal crossing over   Unequal crossing over

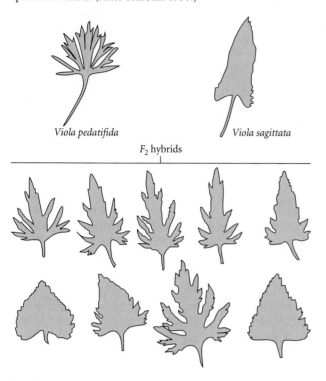

FIGURE 10.22 An example of the phenotypic variation that can arise from hybridization between species. These variations in leaf form have been found in natural hybrids between two species of violets. (After Stebbins 1950.)

Viola pedatifida

Viola sagittata

F_2 hybrids

Horizontal Gene Transfer

Species that have diverged so greatly that they cannot hybridize should not be able to exchange genes. Nevertheless, a few cases have emerged in which transfer of genetic material between very distantly related species is suspected. The evidence, in most cases, consists of a striking similarity of certain base pair sequences that is incongruent with phylogeny. For example, sequences called virogenes are normal constituents of vertebrate genomes, yet are similar to, and surely originally derived from, retroviruses. One virogene is found in Old World monkeys. It is apparently ancient, because the phylogeny of monkey species based on the virogene DNA sequence matches that based on other information. A similar virogene sequence is found in six closely related species of cats, but not in other Felidae or other Carnivora. Its sequence closely matches that of only certain monkeys, namely, baboons (Figure 10.24). This sequence appears, therefore, to have been transferred to the ancestor of these cat species (probably about 5–10 million years ago) from the ancestor of the baboon species. (The reader is invited to suggest why the transfer was not in the opposite direction.)

How might such transfers occur? The most probable route is cross-infection by agents such as retroviruses, which can incorporate host DNA into their genomes. Plasmids play a similar role in transferring genes among unrelated species of bacteria (Heinemann 1991). Because some similarity of DNA sequences may result from convergent evolution, only very similar sequences that are highly in-

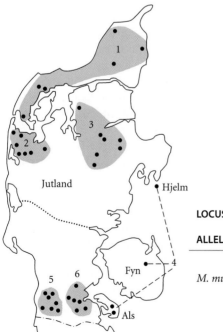

Jutland

Hjelm

Fyn

Als

FIGURE 10.23 Introgressive hybridization between species of house mice in the Jutland Peninsula of Denmark. *Mus musculus* (regions 1–4) and *Mus domesticus* (regions 5 and 6) form a narrow zone of hybridization in the region indicated by the dotted line. The frequencies of allozyme alleles at several loci in each of the six regions show introgression of *domesticus* alleles as far north as region 1. Locus *Idh*-1 reveals introgression of *musculus* alleles into *domesticus* populations as well. (After Selander et al. 1969.)

LOCUS:			*Es*-1		*Hpd*-1			*Idh*-1	
ALLELE:			*a*	*b*	*a*	*b*	*c*	*a*	*b*
	Site								
M. musculus	1		1.00	0.00	0.00	0.00	1.00	0.17	0.83
	2		1.00	0.00	0.00	0.00	1.00	0.23	0.77
	3		1.00	0.00	0.00	0.00	1.00	0.00	1.00
	4		0.95	0.05	0.03	0.03	0.95	0.08	0.92
M. domesticus	5		0.00	1.00	0.50	0.50	0.00	0.93	0.07
	6		0.00	1.00	0.53	0.47	0.00	0.93	0.07

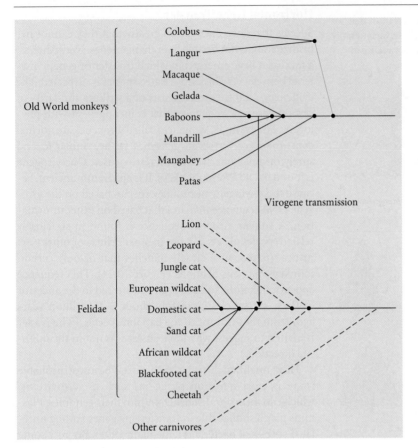

FIGURE 10.24 Horizontal gene transfer, as inferred from the phylogenies of Old World monkeys and cats (Felidae). The solid branches of the phylogenetic trees lead to species whose genomes include similar sequences of a virogene. Broken lines lead to species that lack the virogene. The sequence of the virogene in cats is most similar to that in baboons, suggesting that the gene was transferred from the ancestor of these primates to the ancestor of the small cats, after the lion, cheetah, and leopard lineages had diverged. (After Li and Graur 1991.)

congruent with the phylogeny of organisms can be considered strong evidence for horizontal gene transfer. So far, the number of well-documented cases is small (Li and Graur 1991), and it is unlikely that this phenomenon will prove to be common.

Finally, we should touch on a related subject, the evolution of extremely *intimate associations of unrelated organisms* (see Chapter 18). Lichens, formed by associations of algae with fungi, are the classic example, but many other mutualistic or parasitic associations have formed in which two species effectively become a single organism: intracellular bacteria that provide vitamins and amino acids to aphids, the algal endosymbionts essential for the growth of corals, the many viruses and plasmids that become inte-

grated into the genomes of bacteria and other hosts. Molecular evidence suggests that transfer of genes between hosts and symbionts may occur in such instances (Heinemann 1991), but even without DNA exchange, an endosymbiont that is tightly integrated into its host's life can provide, in effect, a major addition to the host's genome, and an expansion of the host's functional capacities. Antibiotic resistance in bacteria, for example, is usually conferred by genes carried by plasmids. The ne plus ultra of such a functional integration of different genomes is the eukaryotic cell (see Chapter 7), in which mitochondria and chloroplasts, derived from two lineages of bacteria, have joined forces with a proto-eukaryote represented by the cell nucleus—a union that has proved successful in the extreme.

Summary

1. Mutations of chromosomes or genes are alterations that are subsequently replicated. They ordinarily do not constitute new species, but rather variant chromosomes or genes (alleles, haplotypes) within a species.

2. At the molecular level, mutations of genes include base pair substitutions and several other types of changes. Intragenic recombination also gives rise to new DNA sequences (haplotypes).

3. The rate at which any particular mutation arises is quite low: on average about 10^{-6} to 10^{-5} per gamete for mutations detected by their phenotypic effects, and

about 10^{-9} per base pair. The mutation rate, by itself, is too low to cause substantial changes of allele frequencies. However, the total input of genetic variation by mutation, for the genome as a whole or for individual polygenic characters, is appreciable.

4. The magnitude of change in morphological or physiological features caused by a mutation can range from none to drastic. Mutations alter pre-existing biochemical or developmental pathways, so not all conceivable mutational changes are possible. Some adaptive changes may not be possible without just the right mu-

tation of just the right gene. For these reasons, the rate and direction of evolution may in some instances be affected by the availability of mutations.

5. In part because most mutations have pleiotropic effects, the *average* effect of mutations on fitness is deleterious, but there is abundant evidence, including many experiments, that some mutations are advantageous. Mutations with large effects are often deleterious, but some evolutionary biologists believe that such mutations have sometimes been important in evolution.

6. Mutations appear to be random, in the sense that their probability of occurrence is not directed by the environment in favorable directions, and in the sense that specific mutations cannot be predicted. The likelihood that a mutation will occur does not depend on whether or not it would be advantageous.

7. Recombination of alleles can potentially give rise to astronomical numbers of gene combinations, and in sexual organisms generates far more genetic variation per generation than mutation alone. However, recombination also breaks apart favorable gene combinations, and constrains the amount of variation displayed by polygenic characters.

8. Mutations of the karyotype (chromosome complement) include polyploidy (which can give rise to new species) and rearrangements that alter the chromosome number or the arrangement of genes. Many such rearrangements reduce fertility in the heterozygous condition.

9. Unequal crossing over causes deletions and duplications of genes, and is one of the processes responsible for gene families and increases in genome size and gene number.

10. The genetic variation in most populations is augmented by gene flow from conspecific populations. In some cases, genes acquired by hybridization with closely related species add genetic variety. A few examples are known of horizontal gene transfer between very distantly related organisms, but this is probably extremely rare.

Major References

Lewin, B. 1994. *Genes V.* Oxford University Press, New York. Includes molecular aspects of mutation.

Li, W.-H. 1997. *Molecular Evolution.* Sinauer Associates, Sunderland, MA. Treats evolutionary aspects of mutation at the molecular level.

White, M. J. D. 1973. *Animal cytology and evolution.* Third edition. Cambridge University Press, Cambridge. The authoritative work on chromosome evolution in animals.

Stebbins, G. L. 1971. *Chromosomal variation in higher plants.* E. Arnold, London. See also the classic by Stebbins, *Variation and evolution in plants* (Columbia University Press, New York, 1950), on chromosome evolution in plants.

Problems and Discussion Topics

1. Consider two possible studies. (*a*) In one, you capture 3000 wild male *Drosophila melanogaster*, mate each with laboratory females heterozygous for the autosomal recessive allele *vestigial*, *vg* (which causes miniature wings when homozygous), and examine each male's offspring. You find that half the offspring of each of three males have miniature wings and have genotype *vgvg*. (*b*) In another study, you determine the nucleotide sequence of 1000 base pairs for 20 copies of the cytochrome *b* gene, taken from 20 wild mallard ducks. You find that at each of 20 nucleotide sites, one or another gene copy has a different base pair from all others. From these data, can you estimate the rate of mutation from the wild-type to the *vg* allele (case *a*) or from one base pair to another? Why or why not?

2. From a laboratory stock of *Drosophila* that you believe to be homozygous wild type (++) at the *vestigial* locus, you obtain 100,000 offspring, mate each of them with heterozygous +/*vg* flies, and examine a total of 1 million progeny. Two of these are *vgvg*. Estimate the rate of mutation from + to *vg* per gamete. What assumptions must you make?

3. The following DNA sequence represents the beginning of the coding region of the alcohol dehydrogenase (*Adh*) gene of *Drosophila simulans* (Bodmer and Ashburner 1984), arranged into codons:

CCC ACG ACA GAA CAG TAT TTA AGG AGC TGC GAA GGT

(*a*) Find the corresponding mRNA sequence, and use Figure 12 in Chapter 3 to find the amino acid sequence. (*b*) Again using the figure, determine for each site how many possible mutations (nucleotide changes) would cause an amino acid change, and how many would not. For the entire sequence, what proportion of possible mutations are synonymous versus nonsynonymous? What proportion of the mutations at first, second, and third base positions within codons are synonymous? (*c*) What would be the effect on amino acid sequence of inserting a single base, G, between sites 10 and 11 in the DNA sequence? (*d*) What would be the effect of deleting nucleotide 16? (*e*) For the first 15 (or more) sites, classify each possible mutation as a transition or transversion, and determine whether or not the mutation would change the amino acid. Does the proportion of synonymous mutations differ between transitions and transversions?

4. A genus of Antarctic fishes, *Channichthys*, lacks hemoglobin. In its relatives, such as *Trematomus*, hemoglobin serves its usual functions. Assuming that the gene encoding hemoglobin in *Channichthys* has no function, and is not transcribed, how might you expect the nucleotide sequence of this gene to differ between these genera? As a challenge problem, is it possible to estimate how long the gene has been nonfunctional in the *Channichthys* lineage? How might you attempt such an estimate?

5. Many mutations with large phenotypic effects that have been found in laboratory stocks of *Drosophila* are caused by insertions of transposable elements. However, it is not yet clear that transposable elements have caused many of the mutations that have contributed to adaptive evolution in nature. Discuss research strategies that might address this gap in our knowledge, and the possible pitfalls that might be encountered.

6. Researchers have used artificial selection (see Chapter 9) to alter many traits in *Drosophila melanogaster*, such as phototactic behavior and wing length. No one has selected *Drosophila* to be as large as bumblebees (although I'm not sure if anyone has tried). Do you suppose this could be done? How would you attempt it? If your attempts were unsuccessful, what hypotheses could explain your lack of success? What role might mutation play in your experiment?

7. Ultraviolet light (UV) can induce mutations in organisms such as *Drosophila*. Because it damages DNA, and therefore essential physiological functions, it can also reduce survival. Suppose you expose a large number of *Drosophila* to UV, screen their offspring for new mutations, and discover some mutations that increase the amount of black pigment, which protects internal organs from UV. The progeny of an equal number of control flies, not exposed to UV, show fewer or no mutations that increase pigmentation. Can you conclude that the process of mutation responds to organisms' need for adaptation to the environment?

Population Structure and Genetic Drift

This chapter treats two profoundly important facts: natural populations, unlike ideal populations at Hardy-Weinberg equilibrium, are finite in size; and species are structured, by geographic or other factors, so that mating is not entirely random. In finite populations, allele frequencies can fluctuate by chance, a process called **random genetic drift**. Random fluctuations can result in the replacement of old alleles by new ones, resulting in NON-ADAPTIVE evolution. Genetic drift and natural selection are the two important potential causes of allele substitution,—that is, of evolution within populations.

Adaptations do not result from genetic drift, so this process is not responsible for many of the most interesting features of organisms. Genetic drift nevertheless has many important consequences, some of which can influence the course of adaptation. The topic of genetic drift is intimately related to POPULATION STRUCTURE, especially the subdivision of species into local breeding units (Figure 11.1), which exchange genes to a greater or lesser degree. Such structuring has many important consequences.

Because all populations are finite, alleles at all loci are potentially subject to random genetic drift—but all are not necessarily subject to natural selection. For this reason, and because genetic drift constitutes evolution by chance alone, some evolutionary geneticists feel that genetic drift should be the "null hypothesis" used to explain an evolutionary observation unless there is positive evidence for natural selection or some other factor. This perspective is analogous to the "null hypothesis" in statistics: the hypothesis that the data do not depart from those expected on the basis of chance alone.[*]

*For example, if we measure height in several samples of people, the null hypothesis is that the observed means differ only because of random sampling, and that the means of the populations from which the samples were drawn do not differ. A statistical test, such as a *t*-test or analysis of variance, is designed to show whether or not the null hypothesis can be rejected. It will be rejected if the sample means differ more than expected if samples were randomly drawn from a single population.

According to this view, we should not assume that a characteristic, or a difference between populations or species, is adaptive or evolved by natural selection unless there is evidence for this conclusion.

Genetic drift and population structure are the subjects of some of the most highly refined mathematical models in population genetics—or in all of biology, for that matter. Much of the theory was developed by Sewall Wright, starting in the 1930s, and by Motoo Kimura, starting in the 1950s. We will present the material as verbally as possible, but readers who enjoy mathematics can consult the boxes in this chapter for a taste of the models. (See Hartl and Clark 1997 or Crow and Kimura 1970 for more extensive treatments.) For each topic, we will begin with theoretical considerations, and then consider empirical data.

In our discussion of the theory of genetic drift, we will describe random fluctuations in the frequencies (proportions) of two or more kinds of self-reproducing entities that do not differ *on average* (or differ very little) in reproductive success (fitness). For the purposes of this chapter, the entities are alleles. But exactly the same theory applies to any other self-replicating entities, such as chromosomes, asexually reproducing genotypes, or even different species.

The Theory of Genetic Drift

Sampling error   That chance should affect allele frequencies is readily understandable. Imagine, for example, that a single mutation, A_2, appears in a large population that is otherwise A_1. If the population size is stable, each mating pair leaves an average of two progeny that survive to reproductive age. From the single mating $A_1A_1 \times A_1A_2$ (for there is only one copy of A_2), the probability that two surviving progeny are both A_1A_1 is $(1/2)(1/2) = 1/4$, which is the probability that the A_2 gene will be immediately lost. We may assume that pairs vary at random in the number of surviving offspring they leave (0, 1, 2, 3 …), in which case, as

(A)
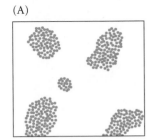

(B)
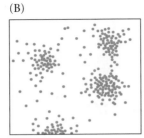

(C)
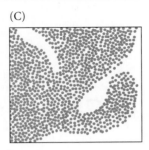

FIGURE 11.1 Some patterns of spatial population structure. Each dot represents an individual. (A) Discrete populations. (B) Perhaps the most common pattern in nature: ill-defined populations between which density is low. (C) A more or less uniform distribution.

the pioneering population geneticist Ronald Fisher (1930) calculated, the probability that A_2 will be lost, averaged over the population, is e^{-1}, or 0.368. In each subsequent generation, there is likewise a probability of loss; Fisher calculated that after 127 generations, the probability is 0.985. These calculations assume that the allele neither increases nor decreases the likelihood of survival of A_1A_2 relative to A_1A_1. However, even if A_1A_2 has a 1 percent advantage, the chance is only 0.027 that A_2 will still be in the population after 127 generations.*

This example illustrates that the frequency of an allele can change (in this instance, to zero from a frequency very near zero) purely by chance: the one or few copies of the A_2 allele might happen not to be included in those gametes that unite into zygotes, or might happen not to be carried by those few newborn individuals that survive to reproductive age. Similarly, if a population carries two (or more) alleles in any frequencies, their frequencies will change from one

generation to the next because of random variation in the proportions that unite to form zygotes, variation in the number of offspring produced by carriers of the different alleles, and variation in the number that survive to reproduce. The gene copies in any generation of adult organisms represent a SAMPLE of the gene copies carried by the gametes of the previous generation; and any sample is subject to random variation, or SAMPLING ERROR.

Coalescence   The concept of genetic drift is so important that we will develop it by two theoretical approaches. Both theoretical perspectives will recur in later chapters. The first approach is coalescent theory.

Figure 11.2 shows a hypothetical, but realistic, history of lineages of reproducing objects. First, imagine the figure as a lineage of individual asexual organisms such as bacteria. We know from our own experience that not all members of our parents' or grandparents' generations had equal numbers of descendants; some had none. Figure 11.2 diagrams this familiar fact. We note that the individuals in generation t are the progeny of only some of those that existed in the previous generation $(t - 1)$: purely by chance, some individuals in generation $t - 1$ failed to leave descendants. Likewise, the population at $t - 1$ stems from only some of those

*Fisher (1930) assumed that the number of surviving offspring per pair has a Poisson distribution with a mean of 2. Here, e is the base of natural logarithms, 2.718. You may wonder why Fisher calculated the probability of loss after such an odd number of generations as 127. So do I. Perhaps, since he didn't have a computer or calculator, he got tired of doing arithmetic and quit.

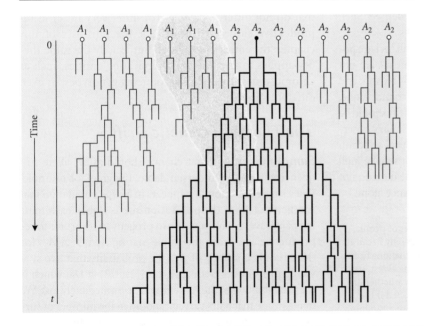

FIGURE 11.2 A possible history of descent of gene copies in a population that begins (at time 0) with 15 copies, representing two alleles. Each gene copy has 0, 1, or 2 descendants in the next generation. Gene copies present at time t are all descended from (coalesce to) a single ancestral copy, which happens to be an A_2 allele (this lineage is shown by the heavier black lines in the figure). Gene lineages descended from all other original gene copies have become extinct. If the failure of gene copies to leave descendants is random, the gene copies at time t could equally likely have descended from any of the other gene copies present at time 0. (After Hartl and Clark 1989.)

individuals that existed in generation $t - 2$, and similarly back to the original population, only one member of which has descendants at time t.

Exactly the same model of descent applies to individual genes within a sexually reproducing population as would apply to individuals in an asexually reproducing haploid population; in other words, we can trace the descendants of a gene just as we would trace the descendants of an individual bacterium. Therefore, if we think of the objects in Figure 11.2 as copies of genes at a locus, in either a sexual or an asexual population, Figure 11.2 shows that as time goes on, more and more of the original gene lineages become extinct, so that the population consists of descendants of fewer and fewer of the original gene copies. This implies that *the average degree of relationship among individuals increases* with the passage of time: more and more of them can be traced back to the same common ancestors. In fact, if we look backward rather than forward in time, *all the gene copies* in the population ultimately *are descended from a single ancestral gene copy*, because given long enough, all other gene lineages become extinct. The genealogy of the genes in the present population is said to COALESCE to a single common ancestor. Because that ancestor represents one of the several original alleles, the population, descended entirely from that ancestor, must eventually become monomorphic: one or the other of the original alleles becomes **fixed** (reaches a frequency of 1.00). In our example, all gene copies have descended from a copy of an A_2 allele, but because this is a random process, A_1 might well have been the "lucky" allele if the sequence of random events had been different. If, in the generation that included the single common ancestor of all of today's gene copies, A_1 and A_2 had been equally frequent ($p = q = 0.5$), then it is equally likely that the ancestral gene copy would have been A_1 or A_2; if A_1 had had a frequency of 0.9 in that generation, then the probability is 0.9 that the ancestral gene would have been an A_1 allele. *Our analysis therefore shows that by chance, a population will eventually become monomorphic for one allele or the other, and that the probability that allele A_1 will be fixed, rather than another allele, equals the initial frequency of A_1.*

How long will this process take? Suppose the population in Figure 11.2 has a constant size of N gene copies (carried by N individual haploid organisms, or $N/2$ diploid organisms). If we were to pick two copies at random from the current population, the chance that the second came from the same parent copy as the first would be $1/N$. This is the probability that two gene copies coalesce to an ancestor in the previous generation. The probability that the two gene copies come from different parent copies is $1 - (1/N)$. By the same reasoning, the probability that those two parents had the same parent is $1/N$, so the probability that the two current copies had the same "grandparent" is $[1 - (1/N)] \times 1/N$. Thus the probability that they coalesce two generations back in time is $(1/N)[1 - (1/N)]^{(2-1)}$. We can similarly calculate the probability that the two gene copies coalesce 3, 4, or in general G generations back as $(1/N)[1 - (1/N)]^{G-1}$. Because N is in the denominator, the smaller the population is, the

larger this expression. For example, if the population size is $N = 5$ (i.e., 5 gene copies), the probabilities of common ancestry 1, 2, or 3 generations ago are 0.200, 0.160, and 0.128, whereas for $N = 10$, these probabilities are 0.100, 0.091, and 0.081. The mean of this distribution, the average time back to the common ancestor of random *pairs* of gene copies, can be shown to equal N generations. The mean time back to common ancestry of *all* gene copies in the population is $2N$ generations.*

Thus the coalescence of all gene copies in the current population back to a single ancestral copy is faster, the smaller the population. Viewed from past to present, a single gene copy at some time in the past becomes the ancestor of all gene copies in the population after $G = 2N$ generations on average—i.e., faster in a smaller than in a larger population. If, for the sake of argument, we suppose that each gene copy at some time t generations ago was a different allele or haplotype, (i.e., a distinguishable DNA sequence), then after $2N$ generations we would expect all gene copies to be the same allele or haplotype (assuming no new mutations have occurred during the $2N$ generations). On the other hand, if the population at time t had, let us say, only two alleles, with m copies of A_1 and n copies of A_2 ($m + n = N$), the probability that a copy of A_1 would become the ancestor of all gene copies is $m/N = p$, the allele frequency at time t.

If this process occurs in a large number of independent populations, each of size N, and if allele A_1 had an initial frequency p in each population, then we would expect a fraction p of the populations to become fixed for A_1, and a fraction $1 - p$ to become fixed for other alleles such as A_2. Thus the genetic composition of populations diverges by chance.

We have arrived at the following important conclusions about evolution by random genetic drift.

1. Allele (or haplotype) frequencies fluctuate at random, but eventually one or another allele becomes fixed.
2. Thus, the population eventually loses its genetic variation.
3. Initially similar populations diverge in allele frequency, and may become fixed for different alleles.
4. The probability, at time t, that an allele will eventually become fixed equals the frequency of the allele at that time.
5. The rate at which these events occur is greater, the smaller the population.

Bear in mind that this model, as developed so far, includes only the effects of random genetic drift. Other evolutionary processes—namely, mutation, gene flow, and natural selection—are assumed not to operate. Thus the model, as developed so far, does not describe the evolution of adaptive traits, those that evolve by natural selection. We will incorporate these other evolutionary factors later.

---

*For a diploid locus, the average time to coalescence for a pair of genes is $2N$ generations, and for all genes in the population is $4N$ generations, which is the expected time to fixation of a newly arisen mutation, i.e., of a single gene copy.

Random fluctuations in allele frequencies Let us take another, more traditional, approach to genetic drift. Assume that the frequencies of alleles A_1 and A_2 are p and q in each of many independent populations, each with N breeding individuals (representing $2N$ gene copies in a diploid species). The populations do not exchange individuals or genes (i.e., there is no gene flow), there is no mutation during the period of time considered, and there is no natural selection (i.e., the genotypes do not differ, *on average*, in survival or reproductive success)—that is, the alleles are NEUTRAL with respect to fitness. (Some of these assumptions will be relaxed later.) Small populations are sometimes called DEMES, and the ensemble of populations may be termed a METAPOPULATION.

In any generation, the large number of newborn zygotes is reduced to N individuals by the time they breed, by mortality that is random with respect to genotype. On average, we expect p (and q) to be the same among the survivors as in the previous generation (say, 0.5), just as we expect half of a number (say, $2N$) of tossed coins to be heads. But it is possible to have as much as 100 percent A_1 among their gene copies, just as it is possible to get all heads. In fact, all proportions, from 0 to 1, are possible, and the probability of each can be calculated from the binomial theorem, generating a PROBABILITY DISTRIBUTION. Among a large number of populations (or demes), the new allele frequencies (p') will vary, by chance, around a mean, namely, the original frequency p. The Hardy-Weinberg principle (see Chapter 9) shows that in an infinite population, allele frequencies do not change. In a finite population of size N, however, allele frequencies do indeed fluctuate by chance.

Now if we trace one of the populations, in which p has changed from 0.5 to, say, 0.47, we see that in the following generation, it will change again from 0.47 to some other value, *either higher or lower with equal probability*. This process of random fluctuation continues. It may seem counterintuitive, but instead of tending to wander back toward 0.5, p will eventually wander (drift) either to 0 or to 1: *the allele is either lost or fixed.* (Once the frequency of an allele has reached either 0 or 1, it cannot change unless another allele is introduced into the population, either by mutation or by gene flow from another population.) The allele frequency describes a "RANDOM WALK," analogous to a New Year's Eve reveler staggering along a very long train platform with a railroad track on either side: if he is so drunk that he doesn't compensate steps toward one side with steps toward the other, he will eventually fall onto one of the two tracks, if the platform is long enough.

The copies of the allele that is fixed, say, A_2, may represent several or many of the gene lineages pictured in Figure 11.2. The coalescence of all of a population's gene copies back to a single ancestral gene copy is merely the logical consequence of a random drift process extended far enough back in time.

Just as an allele's frequency may increase by chance in some populations from one generation to the next, it may decrease in other populations. As a result, allele frequencies vary among the populations. The *variance* in allele frequency among populations continues to increase from generation to generation (Figure 11.3). Eventually some populations reach $p = 0$ or $p = 1$, and can no longer change. Among those in which fixation of one or the other allele has not yet occurred, the allele frequencies continue to spread out, with all frequencies between (but not including) 0 and 1 eventually becoming equally likely (Figure 11.4). Those that approach 0 or 1 tend to "fall over the edge" soon, so the number of populations fixed for one or another allele continues to increase, until all populations in the metapopulation have become fixed. Thus, *initially genetically identical populations evolve by chance to have different genetic constitutions.* (Remember, though, that at this stage in the theory, the alleles have identical effects on fitness.)

Evolution by Genetic Drift

The following points, presented without mathematical derivations, are some of the most important aspects of evolution by genetic drift.

First, probability theory tells us that in repeated tosses of N coins, the variance in the proportion of heads among the several trials will be greater if N is small than if it is large. The same theory applies to the sampling of genes. If each generation of adults represents $2N$ gene copies, the variance in allele frequency after one generation is $V = p(1 - p)/2N$ (Box 11.A). This may be thought of as the breadth of the probability distribution of p for a single population, or as the variance among populations that all began with the same allele frequency p. Because $2N$ is the denominator, this expression tells us that the random deviations from the initial allele frequency are likely to be greater (V larger), the smaller the population size. Therefore, *evolution by genetic drift proceeds faster in small than in large populations.*

Second, a drunk who staggers down a railway platform is clearly more likely to fall off the side to which he is closer. Similarly, a rare allele (p near zero) is more likely to be lost, and a common allele to be fixed. In fact, *the probability, at time t, that an allele will ultimately be fixed equals its frequency (p_t) at that time.* Thus, when $p = 0.7$, the allele's probability of fixation in the future is 0.7. But if, 100 generations later, the allele's frequency has drifted to 0.2, its new probability of fixation is 0.2, and its probability of loss is 0.8. At any time, an allele's probability of fixation versus loss equals its frequency at that time, and is not affected or predicted by its past history of change in frequency.

If an allele has just arisen by mutation, and is represented by only one among the $2N$ gene copies in the population, its frequency is $p_t = 1/(2N)$, and this is its likelihood of reaching $p = 1$. Clearly it is more likely to be fixed in a small than in a large population. Moreover, if the same mutation arises in each of many demes, each of size N, the mutation should eventually be fixed in a proportion $1/(2N)$ of the demes. Similarly, of all the new mutations (at all loci) that arise in a population, a proportion $1/(2N)$ should eventually be fixed. Notice that if a mutation has drifted to high

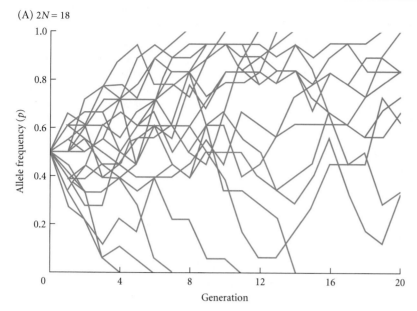

(A) 2N = 18

Allele frequency (p) / Generation

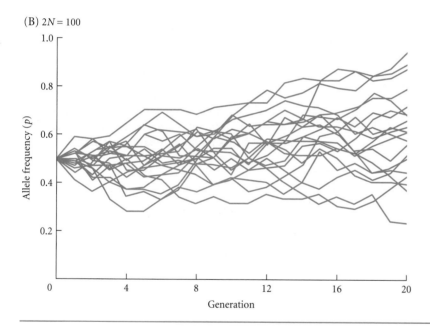

(B) 2N = 100

Allele frequency (p) / Generation

FIGURE 11.3 Computer simulations of random genetic drift in populations of (A) 9 diploid individuals (2N = 18 gene copies) and (B) 50 diploid individuals (2N = 100 gene copies). Each line traces the frequency (p) of one allele for 20 generations. Each panel shows allele frequency change in 20 replicate populations, all of which begin at p = 0.5 (i.e., half the gene copies are A_1 and half A_2). Note that the oscillations are larger in A than in B, and the allele is more rapidly fixed (p = 1) or lost (p = 0), due to the smaller population size. (After Hartl and Clark 1997.)

frequency, say, p = 0.5, its probability of fixation at that time equals that frequency.

If alleles A_1 and A_2 have initial frequencies p and q in each of a number of populations of equal size, then a proportion p of the populations become fixed for A_1 and q for A_2. If there are k populations, the total number of gene copies is 2Nk. The total number of A_1 copies is the number of populations fixed for A_1 (i.e., pk) multiplied by the number of genes per population (2N). Thus the overall frequency of A_1, summed over the entire metapopulation, is 2Npk/2Nk = p, the initial frequency.

Third, a simple consequence of the effect of population size on the rate of genetic drift is that for those alleles that do become fixed, *the average time required to reach fixation (the "fixation time") is shorter, the smaller the population.*

The average time to fixation of a newly arisen neutral allele that does become fixed is 4N generations for a diploid population, which may be a long time. A new mutation that is lost, on the other hand, is lost rapidly, even if the population is large.

Fourth, the frequency of heterozygotes (heterozygosity), H, is maximal when all alleles in a population are equal in frequency (see Chapter 9); for example, when p = q = 0.5 in the case of two alleles. Within any population undergoing genetic drift, an allele *may* drift from low frequency to high frequency or fixation, in which case H increases temporarily and then decreases. But *on average*, rare alleles drift toward p = 0, and in all populations an allele ultimately becomes fixed, so on average, the frequency of heterozygotes declines, and ultimately reaches zero. Thus *genetic drift*, un-

FIGURE 11.4 Changes in the probability that an allele will have various possible frequencies as genetic drift proceeds. (A) The probability distribution of allele frequencies between 0 and 1 when the initial frequency of the allele is 0.5. The probability distribution after $t = N/10$ generations is shown by the uppermost curve. This may be thought of as the distribution of allele frequencies among numerous populations, each of size N, that began with the same allele frequency. With the passage of generations, the curve becomes lower and broader, until at $t = 2N$ generations, all allele frequencies between 0 and 1 are equally likely. This panel does not show the proportion of populations in which the allele is fixed or lost. (B) After $t = 2N$ generations, the proportion of populations in which the allele is lost ($p = 0$) or fixed ($p = 1$) increases at the rate of $1/(4N)$ per generation, and each allele frequency class between 0 and 1 decreases at rate $1/(2N)$ per generation. (A after Kimura 1955; B after Wright 1931.)

(A)

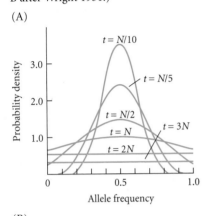

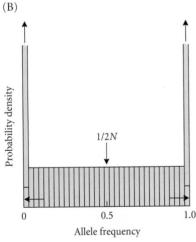

(B)

heterozygotes and homozygotes retain approximate Hardy-Weinberg genotype frequencies *within* each population, but in the metapopulation as a whole there is a deficiency of heterozygotes and an excess of homozygotes relative to Hardy-Weinberg proportions.

Not surprisingly, the *rate of decline in the frequency of heterozygotes in the metapopulation depends on the size of the component populations or demes*. If H_{t-1} and H_t represent the frequency of heterozygotes in two successive generations, the expected change per generation is given by $H_t = [1 - 1/(2N)]H_{t-1}$; given an initial heterozygosity H_0, the frequency of heterozygotes after t generations is $H_t = [1 - 1/(2N)]^t H_0$ (Figure 11.5): it decreases faster, the smaller the population.

Effective population size The theory presented so far assumes highly idealized populations of N breeding adults. In such populations, the rate of change under genetic drift can be described by the rate of decline in the frequency of heterozygotes. If we measure the actual number of adults in real populations, however (such as pondsful of fish or bottlesful of fruit flies), the number we count (the CENSUS SIZE) may be greater than the number that actually contribute genes to the next generation. For example, a few dominant male elephant seals mate with all the females in a population, so the alleles these males happen to carry contribute disproportionately to following generations, and from a genetic point of view, the unsuccessful subdominant males might as well not exist. Thus the rate of genetic drift of allele frequencies, and loss of heterozy-

FIGURE 11.5 The average decrease in heterozygosity in populations of different sizes (N) due to genetic drift. The vertical axis is the proportion of the initial frequency of heterozygotes that is retained. (After Strickberger 1968.)

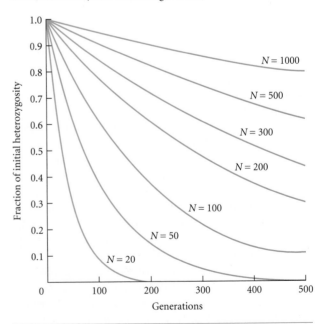

less opposed by other factors, *reduces genetic variation within a* population, and ultimately eliminates it altogether. Within a metapopulation, different alleles drift toward fixation in different demes, each of which declines in heterozygosity. Therefore the total frequency of heterozygotes, measured over the entire metapopulation, decreases steadily, and ultimately reaches zero. As we saw above, both alleles persist in the metapopulation as a whole, and retain the same *average* frequency; however, when complete fixation has occurred, some populations are monomorphic A_1A_1 and others are monomorphic A_2A_2. During this process,

Variation in Allele Frequencies under Genetic Drift

The $2N$ gene copies carried in a population of N adults are a sample of the very large number of gametes produced by the N adults in the previous generation. If the frequencies of alleles A_1 and A_2 are p and q (= $1 - p$) respectively, the probability $\Pr(i)$ that i of the $2N$ gene copies among the offspring will be A_1 can be calculated from the binomial expansion $(p + q)^{2N}$. For example, if $p = q = 0.5$, the probability that all the gene copies in the next generation will be A_1 (i.e., that the new allele frequency p' equals 1) is the first term of this expansion, p^{2N}. (For example, if the population size $N = 10$, $\Pr(20) = 0.5^{20}$.) We may similarly calculate each term $\Pr(i = 19)$, $\Pr(i = 18)$, ..., $\Pr(i = 0)$ from the expression

$$\Pr(i) = \binom{2N}{i} p^i (1-p)^{2N-i}$$

where

$$\binom{2N}{i}$$

means

$$\frac{(2N)!}{i!(2N-1)!}$$

This yields a PROBABILITY DISTRIBUTION with a mean and variance. From probability theory, the mean of this distribution is p, and the variance is $V_p = p(1 - p)/(2N)$. The greater the population size (N), the lower the variance, i.e., the more likely the new allele frequency is to be near the mean. The breadth of the probability distribution is more easily apprehended from the standard deviation, $S = \sqrt{V}$. For a large population, 96 percent of the probability distribution lies within two standard deviations on either side of the mean. For example, if $2N = 20$ and $p = 0.5$ in the first generation, $V_p = (0.5)(0.5)/20 = 0.0125$ in the next generation, $S = \sqrt{0.0125} = 0.1118$, and $2S = 0.2236$. Thus, 96 percent of the time, the allele frequency in the second generation will be 0.5 ± 0.2236, i.e., between 0.2764 and 0.7236. But if $2N = 200$ and $p = 0.5$, $V_p = 0.00125$, $2S = 0.0707$, and 96 percent of the time the new allele frequency will lie between 0.4293 and 0.5707.

---

gosity, will be greater than expected from the population's census size, and will correspond to what we expect of a smaller population. The population is *effectively* smaller than it seems. The **effective size** of an actual population is the number of individuals in an ideal population (in which every adult reproduces) in which the rate of genetic drift (measured by the rate of decline in heterozygosity) would be the same as it is in the actual population. The effective size is usually denoted N_e, and N is used for the census size.[*] For instance, if we count 10,000 adults in a population, but only 1000 of them successfully breed, genetic drift proceeds at the same rate as if the population size were 1000, and this is the effective size, N_e.

Without going into mathematical detail, the effective population size can be smaller than the census size for several reasons:

1. *Variation in the number of progeny* produced by females, males, or both. The elephant seal represents an extreme example. One factor that can cause variation is natural selection; for instance, if the reproductive rate depends on body size, so that larger individuals have more offspring, the rate of genetic drift may be increased at all neutral loci because individuals with certain body sizes contribute fewer gene copies to subsequent generations.

2. *Unequal numbers of females and males.* For autosomal loci in a sexual population, half the gene copies in each generation come from each sex. The sex that is in the minority therefore has more progeny per individual than the majority sex. This is a special case of variation in number of progeny. For example, the effective size (N_e) of a population of 20 males and 80 females ($N = 100$) may be calculated to be 64. The elephant seal again provides an example (if we think of the nonreproducing males as nonexistent). Skewed sex ratios are common in many species, especially in groups such as wasps and mites in which females can regulate the sex ratio of their progeny (see Chapter 21).

3. *Overlapping generations.* This is a very complicated subject, but the thrust of it is that if generations overlap, offspring may mate with their parents, and since these pairs carry identical copies of the same genes, the effective number of genes propagated is reduced.

4. *Fluctuations in population size.* All populations fluctuate in size, sometimes drastically. The rate of genetic drift is increased during periods of small population size; in fact, the effective size in the long run is more strongly affected by the smaller than by the larger size.[†] For example, if the number of breeding adults in five successive generations is 100, 150, 25, 150, and 125, N_e is approximately 70 rather than the arithmetic mean 110. Fluctuation may well be the most pervasive reason for low effective population size—and the most difficult to evaluate.

---

[*] Technically, the effective size has several related but distinct definitions, and may differ slightly depending on whether or not population size fluctuates (see Crow and Kimura 1970; Hartl and Clark 1997).

[†] If a breeding population consists of N_f females and N_m males, $N_e = 4N_f N_m/(N_f + N_m)$. If the population size fluctuates, N_e approximately equals the harmonic mean population size, calculated for t generations.

Founder effects The restrictions in size through which populations may pass are called **bottlenecks**. A particularly interesting bottleneck occurs when a new population is established by a small number of colonists, or founders—sometimes as few as a single mating pair (or a single inseminated female, as in insects in which females store sperm). The random genetic drift that ensues is often called a **founder effect**.

A colony founded by a small number of colonists will suffer some loss of genetic variation: uncommon alleles, in particular, are unlikely to be represented. The average level of heterozygosity, however, is not greatly reduced in the first generation: it is $(1 - 1/2N)H_0$, where N is the number of founders and H_0 is the heterozygosity in the source population. Thus in a colony founded by one mating pair ($N = 2$), the heterozygosity is, on average, reduced by only 25 percent in the first generation. Recalling that the genetic variance of a character is proportional to the population's heterozygosity at loci that affect that character, we see that most of the heterozygosity and genetic variance of a large population are, on average, carried over into a colony founded by few individuals. This is because rare alleles, which are likely to be lost, contribute little to quantitative measures of heterozygosity and genetic variance, whereas moderately common alleles, which contribute the most, are likely to be carried by the founders. If the colony increases very rapidly in numbers, further loss of genetic variation will be slow; however, if it remains small, genetic drift continues to erode genetic variation rapidly. If the colony persists and grows, new mutations eventually restore heterozygosity to higher levels (Figure 11.6).

Summary on genetic drift Genetic drift is random change in the frequencies of alleles or haplotypes due to accidents of sampling caused by random variation in rates of survival or reproduction by different genotypes. The frequency of an allele fluctuates at random until the allele is ultimately fixed ($p = 1$) or lost ($p = 0$). Random changes in frequency are larger, so that genetic drift proceeds more rapidly, the smaller the population. The genetically "effective size" of a population is usually less than the "census size."

Because one allele is eventually fixed and other alleles are lost, the genetic variation in a population (measured by the frequency of heterozygotes) eventually declines to zero due to genetic drift. In any given generation, the probability that an allele will eventually become fixed equals the allele's frequency at that time. Thus even a very rare allele, such as a newly arisen mutation, has some probability of replacing all other alleles. Among an ensemble of populations in which genetic drift proceeds independently, the frequency of an allele is likely to increase in some populations and decrease in others. Hence populations diverge in genetic composition, and may become fixed for different alleles, by chance.

FIGURE 11.6 Effects of a bottleneck in population size, as in a newly founded population beginning with two (black lines) or ten (colored lines) individuals. Genetic variation is measured by heterozygosity, which is 14 percent in the source population. Heterozygosity declines more substantially in the newly founded population if the rate of population increase is low ($r = 0.1$) than if it is high ($r = 1.0$). Eventually, mutation supplies new genetic variation, and heterozygosity increases. (After Nei et al. 1975.)

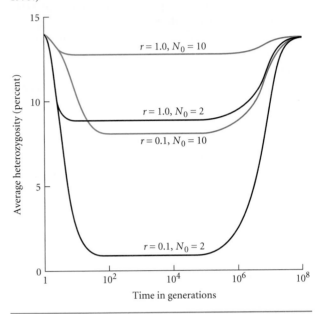

Genetic drift affects allele frequencies at all loci simultaneously. It has the consequences summarized here if no other factors affect allele frequencies, especially natural selection. That is, the dynamics of genetic drift described so far apply to selectively neutral alleles, those that do not differ in their effect on survival or reproduction.

Because evolution at any locus could, in principle, transpire by random genetic drift, many population geneticists hold that genetic drift should be the "null hypothesis" used to explain evolutionary data, unless there is evidence in favor of another hypothesis (such as natural selection).

Genetic Drift in Real Populations

Laboratory experiments Peter Buri (1956), who studied with Sewall Wright, initiated 107 experimental populations of *Drosophila melanogaster*, each with eight males and eight females, all heterozygous for two alleles (bw and bw^{75}) that affect eye color (by which all three genotypes are recognizable). Thus the initial frequency of bw^{75} was 0.5 in all populations. He propagated each population for 19 generations by drawing eight flies of each sex at random and transferring them to a vial of fresh food. The frequency of bw^{75} rapidly spread out among the populations (Figure 11.7); after one generation, the number of bw^{75} copies ranged from 7 ($q = 7/32 = 0.22$) to 22 ($q =$

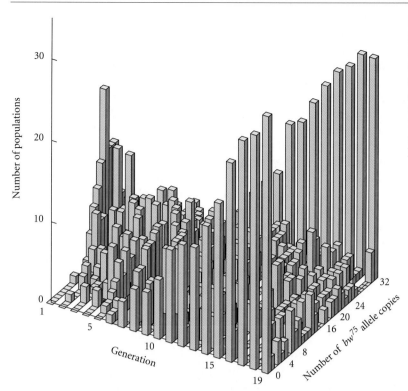

FIGURE 11.7 Random genetic drift in 107 experimental populations of *Drosophila melanogaster*, each founded with 16 bw^{75}/bw heterozygotes, and each propagated by 16 flies per generation. Over the course of 19 generations, allele frequencies (number of bw^{75} alleles) became more evenly distributed between 0 and 1.0 (32 copies), and the bw^{75} allele was lost (0) or fixed (32) in an increasing number of populations. (After Hartl and Clark 1997.)

0.69). By generation 19, 30 populations had lost the bw^{75} allele, and 28 had become fixed for it; among the unfixed populations, intermediate allele frequencies were quite evenly distributed. The results nicely matched those expected from genetic drift theory (cf. Figure 11.4), except that the rate of genetic drift was higher than expected in a population of $N = 16$. In fact, Buri estimated that the average *effective* population size was about 9. In other experiments, Buri could find no evidence of differences in fitness among the genotypes; that is, they did not differ *consistently* in survival or reproductive rate. He therefore ascribed the genetic changes in these small populations entirely to random drift.

In an experiment that we will describe further in Chapter 16, Edwin Bryant and his colleagues established four replicate laboratory populations of houseflies (*Musca domestica*) at each of three bottleneck sizes: 1, 4, and 16 pairs, taken from a natural population. Each population rapidly grew to an equilibrium size of about 1000 flies, and the populations were then again reduced to the same bottleneck sizes. This procedure was repeated as many as five times. Control populations, initiated with 1000 pairs, were maintained at this size throughout the experiment. After each recovery from a bottleneck, the investigators estimated the allele frequencies at four polymorphic enzyme loci for each population, using electrophoresis (see Chapter 9). From these data, they estimated average heterozygosity ($\bar{H}$) and the total number of alleles (initially 13, summed over loci). They expected each bottleneck episode to reduce $\bar{H}$ by $(1 - 1/2N)/\bar{H}_0$, where $\bar{H}_0$ is the value for the control popu-

lations (in which $\bar{H}$ should not decline appreciably in so short a time because of their large size).

The data (Table 11.1) conform very well to the predictions of genetic drift theory. Heterozygosity declined after each bottleneck episode, most appreciably in the 1-pair lines and least in the 16-pair lines. In almost every case, the magnitude of $\bar{H}$ was very close to the predicted value. A statistical analysis showed that, again as predicted, the loss of rare alleles in the bottlenecked population was relatively greater than the loss of heterozygosity.

Table 11.1 **The proportion of average heterozygosity ($\bar{H}$) in bottlenecked populations of houseflies, relative to control populations**[a]

EPISODE	BOTTLENECK SIZE (NUMBER OF PAIRS)		
	1	4	16
1 Observed	0.681[b]	0.910[b]	0.986
Expected	0.750	0.938	0.984
3 Observed	0.386[b]	0.891[b]	0.934
Expected	0.422	0.824	0.954
5 Observed	0.284[b]	1.006	0.921
Expected	0.237	0.724	0.924

Source: McCommas and Bryant (1990).

[a]The values are averaged over 4 replicate populations at each bottleneck size for 1 and 3 bottleneck episodes, and over 2 replicates for 5 episodes.

[b]Represents a statistically significant difference in heterozygosity between bottlenecked and control populations.

Natural populations When we describe the genetic features of natural populations, the data usually are not based on experimental manipulations; nor do we usually have detailed information on the populations' histories. We therefore attempt to *infer* causes (such as genetic drift or natural selection) by interpreting *patterns. Such inferences are possible only on the basis of theories* that tell us what pattern to expect if one or another cause has been most important.

Occasionally, we can check the validity of our inferences using independent information, such as historical data. For example, a survey of electrophoretic variation in the northern elephant seal (*Mirounga angustirostris*) revealed *no* variation at any of 24 enzyme loci (Bonnell and Selander 1974)—a highly unusual observation, since most natural populations are highly polymorphic (see Chapter 9). However, although the population of this species now numbers about 30,000, it was reduced by hunting to about 20 animals in the 1890s. Moreover, the effective size was probably even lower, because less than 20 percent of males succeed in mating. The hypothesis that genetic drift was responsible for the monomorphism, a priori a likely hypothesis, is supported by the historical data.

In the same vein, genetic drift explains the pattern of electrophoretic variation in introduced populations of the common mynah (*Acridotheres tristis*), an abundant Indian bird related to the starling (*Sturnus vulgaris*). In the late nineteenth century, humans released mynahs in Australia, New Zealand, Fiji, Hawaii, and South Africa (at Durban). The sparse historical records suggest that the numbers introduced ranged from perhaps 250 or more in Australia to "several" pairs in South Africa. As is commonly the case with introduced species, some of the colonies remained small for several decades before the birds became abundant. In an electrophoretic survey of birds from these regions and from seven localities throughout India (which presumably represent the original condition), Allan Baker and Abdul Moeed (1987) found that the introduced populations had, on average, a lower number of alleles per locus than the native populations (1.24 versus 1.57), a lower percentage of polymorphic loci (14.3 versus 18.8), and significantly lower heterozygosity (0.05 versus 0.06). Genetic variation was lowest in South Africa (9 percent polymorphic loci, heterozygosity 0.03), and was further reduced in Johannesburg, presumably colonized by birds from Durban (about 300 miles away, as the mynah flies). All the alleles lacking in

introduced populations were rare in Indian populations. Allele frequencies at some loci varied more among the introduced than among the native populations (at one locus, for example, an allele with a frequency of at most 0.01 in India reached frequencies of 0.07 to 0.08 in some introduced populations). Overall, the mean "genetic distance" (see Chapter 9) among the introduced populations (Nei's *D* = 0.006) was greater than among the native populations (0.001); thus alleles both increased and decreased in frequency relative to the source populations. The mean genetic distance among introduced populations was equivalent to that commonly found among subspecies of birds—a measure of the rapidity with which genetic drift can alter the constitution of small populations.

The data for the elephant seal and the mynah conform qualitatively to what we would expect if the loci examined have been evolving by genetic drift. *Quantitative* comparisons of observed and expected patterns, like those Buri applied to his laboratory populations, are sometimes also possible. For example, the cattle in Iceland have been isolated since their introduction from Norway 1000 years ago. From historical records, Kidd and Cavalli-Sforza (1974) estimated sex ratios and fluctuations in population size, and thereby estimated effective population sizes. These estimates provided an expectation of the allele frequency difference that should have accumulated between the populations by genetic drift. Their analysis is too complex to present here, but they found that the genetic differences at several loci controlling blood groups (analogous to the A-B-O and Rh blood group polymorphisms in humans) quantitatively matched the theoretical expectation.

Genetic patterns in natural populations for which historical information is lacking sometimes conform to what we would expect if the loci are affected by genetic drift. From among many examples, we may consider Robert Selander's (1970) study of two loci in house mice (*Mus musculus*) from widely scattered barns in central Texas. Selander considered each barn to harbor a separate population, because mice rather seldom disperse to new barns, and those that do are often excluded by the residents. Having estimated the population size in each barn, Selander found that although small and large populations had much the same *mean* allele frequencies, the *variation* (variance) in allele frequency was much greater among the small populations, as we would expect from random drift (Table 11.2).

Table 11.2 **Frequency of alleles at two loci relative to population size of house mice**

ESTIMATED POPULATION SIZE	NUMBER OF POPULATIONS SAMPLED	MEAN ALLELE FREQUENCY		VARIANCE OF ALLELE FREQUENCY	
		Es-3[b]	Hbb	Es-3[b]	Hbb
Small (median size 10)	29	0.418	0.849	0.0506	0.1883
Large (median size 200)	13	0.372	0.843	0.0125	0.0083

Source: After Selander (1970).

A note on the study of genetic drift   Although genetic drift has been shown to affect morphological and other phenotypic characters, biochemical polymorphisms have provided many more examples of the effects of genetic drift in natural populations, for two major reasons. First, because we are concerned with evaluating a theory of allele frequencies, it is necessary to distinguish and count alleles rather unambiguously. It is especially useful to be able to distinguish heterozygotes from homozygotes. Since the advent of electrophoresis, it is far easier to find single-locus polymorphisms for enzymes than for morphological features, in which variation usually is contributed by more than one locus, and in which genotypes (e.g., heterozygotes) are less easily distinguished because of dominance and environmental sources of variation. Second, in most studies of simple morphological polymorphisms, there is evidence, or at least reason to suspect, that the allele frequencies are affected not only by genetic drift, but also by natural selection—and we are presently concerned with genetic drift only. In contrast, many investigators have argued that many biochemical polymorphisms are affected only slightly or not at all by selection—although this is a very controversial topic (see Chapter 13).

Inbreeding

Allele frequency change by genetic drift is critically affected by the breeding structure of populations: who mates with whom. The breeding structure also has many consequences for the effects of natural selection. Again, we will develop the theory and then confront it with data. In doing so, we will see how an abstruse concept (identity by descent) that cannot be directly measured in the real world gives rise to predictions that can be tested with data.

A PANMICTIC population is one in which individuals mate entirely at random. The common eel (*Anguilla rostrata*) may be panmictic: it is thought that all eels from rivers of both eastern North America and western Europe travel to a single area near Bermuda to breed. And some extremely localized species, such as the Devil's Hole pupfish (*Cyprinodon diabolis*), which is restricted to a single sinkhole near Death Valley, Nevada, are presumably panmictic. But most species are subdivided into several or many separate populations, within which most or all mating takes place. Also, individuals may, for any of several reasons, be more likely to mate with relatives than with nonrelatives, a phenomenon called INBREEDING. We will see that inbreeding and population subdivision are closely related concepts. The theory of inbreeding was pioneered by Sewall Wright.

Imagine that in an ancestral population we could label every gene copy at a locus and trace the descendants of each gene copy through subsequent generations. (In practice, we cannot do this, although it might be possible if every copy had a unique base pair sequence.) Some of the gene copies are allele A_1, and others are A_2. If we label one of the A_1 copies A_1^*, then after one generation, a sister and brother may both inherit A_1^* from a parent (with probabil-

FIGURE 11.8   Two pedigrees showing inbreeding due to (A) sib mating and (B) parent-offspring mating. Squares represent males, circles females. Copies of an A_1 allele, A_1^*, are traced through three generations. Individual I possesses alleles that are identical by descent at this locus. The average inbreeding coefficient of progeny from either of these matings is 0.25, the probability of identity by descent. Note in (A) that I's mother is homozygous for allele A_1, but the two copies are not identical by descent.

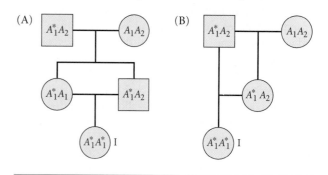

ity $0.5^2 = 0.25$). Among the progeny of a mating between sister and brother, both heterozygous for A_1^*, one fourth will be $A_1^* A_1^*$. These individuals carry two gene copies that are not only the same allele (A_1), but are also **identical by descent** (A_1^*) (Figure 11.8). The $A_1^* A_1^*$ individuals are said to be not only homozygous, but **autozygous. Allozygous** individuals, on the other hand, may be either heterozygous or homozygous (if the two copies of the same allele are not identical by descent). *An inbred population is one in which the probability that an individual is autozygous is greater, as a consequence of mating among relatives, than in a panmictic population.* We define the **inbreeding coefficient**, denoted F, as the probability that an individual taken at random from the population will be autozygous. In a population that is not at all inbred, $F = 0$. In a fully inbred population, $F = 1$: all individuals are autozygous (Figure 11.9).

Now consider a population that is inbred to some extent. F is the fraction of the population that is autozygous, and

FIGURE 11.9   Genotype frequencies at a locus with allele frequencies $p = 0.4$, $q = 0.6$, when mating is random ($F = 0$) and when the population is completely inbred ($F = 1$).

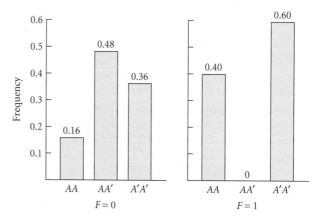

$1 - F$ is the fraction that is allozygous. If there are two alleles, A_1 and A_2, with frequencies p and q, then the probability that an individual is allozygous *and* that it is A_1A_1 is $(1 - F) \times p^2$. Likewise, the fraction of the population that is heterozygous is $(1 - F) \times 2pq$, and all the heterozygotes are allozygous (by definition). The fraction that is allozygous and A_2A_2 is $(1 - F) \times q^2$. Turning our attention now to the fraction, F, of the population that is autozygous, we note that none of these individuals is heterozygous, because a heterozygote's alleles are not identical by descent. If an individual is autozygous, the probability that it is autozygous for A_1 is p, the frequency of A_1. Thus the fraction of the population that is autozygous and A_1A_1 is $F \times p$. Likewise, $F \times q$ is the fraction that is autozygous and A_2A_2.

Thus, taking into account the allozygous and autozygous fractions of the population, the genotype frequencies (see Figure 11.9) are:

	Allozygous	Autozygous		Genotype frequency
A_1A_1	$p^2(1-F)$	$+$	pF	$= D$
A_1A_2	$2pq(1-F)$			$= H$
A_2A_2	$q^2(1-F)$	$+$	qF	$= R$

Therefore, the consequence of inbreeding is that the frequencies of homozygous genotypes are higher, and the frequency of heterozygotes is lower, than in a population that is in Hardy-Weinberg equilibrium. [The increase in the frequency of homozygotes is evident if, for example, we note that $D = p^2 (1 - F) + pF = p^2 + F (p - p^2)$. The term $p - p^2$ is positive if $p > 0$, so the entire expression is greater than p^2.]

Note also that H, the frequency of heterozygotes in the inbred population, equals $(1 - F)$ multiplied by the frequency of heterozygotes we expect to find in a randomly mating population $(2pq)$. Denoting $2pq$ as H_0, we have $H = H_0(1 - F)$, or $F = (H_0 - H)/H_0$. Thus we can estimate an utterly abstract parameter, the probability of autozygosity, by two quantities that we can readily measure: the observed frequency of heterozygotes, H, and the "expected" frequency, $2pq$, which we can calculate from data on the allele frequencies p and q. In practice, then, F is a measure of the fractional reduction in heterozygosity in an inbred population, as compared with a panmictic population with the same allele frequencies.

Increase in F with inbreeding

If mating among relatives is a consistent feature of a population, F will increase over generations. How rapidly it does so depends on how closely related the average pair of mates is: regular mating between sisters and brothers (full sib mating) increases F and decreases heterozygosity more rapidly than mating between first cousins (Figure 11.10). Consider a simple, extreme form of inbreeding: self-fertilization, or selfing, which occurs in many species of plants. In selfing, meiosis is normal, producing both female and male gametes in different sexual organs; the gametes of a single individual then unite. [In many plants, such as wheat (*Triticum*), some or

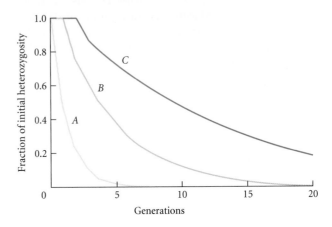

FIGURE 11.10 Decreases in heterozygosity with successive generations of inbreeding due to exclusive (A) self-fertilization, (B) full sib mating, and (C) double first cousin mating. (After Crow and Kimura 1970.)

all flowers are CLEISTOGAMOUS: they do not open, and fertilization occurs within the flower.]

Suppose the genotypes A_1A_1, A_1A_2, and A_2A_2 initially have frequencies D, H, and R, and that reproduction occurs only by self-fertilization. All offspring of A_1A_1 are necessarily A_1A_1, and similarly for A_2A_2. Among the offspring of heterozygotes, assuming random union of A_1 and A_2 gametes within each plant, 1/4 are A_1A_1, 1/4 are A_2A_2, and 1/2 are A_1A_2. Therefore the genotype frequencies after one generation are $D' = D + H/4$, $H' = H/2$, $R' = R + H/4$. The same happens in the next generation. Therefore, the frequency of heterozygotes (H) is halved in each generation. Ultimately, the frequency of heterozygotes reaches zero, so that $F = (2pq - H)/2pq = (2pq - 0)/2pq = 1$, and the population consists entirely of A_1A_1 and A_2A_2 homozygotes (with frequencies p and q respectively. If only a portion of the population engages in self-fertilization (or less extreme forms of inbreeding), and the rest engages in outcrossing, F will reach a stable value between 0 and 1.

Genetic consequences of inbreeding

Among the effects of inbreeding, these are the most important:

1. As our analysis of selfing illustrates, inbreeding redistributes alleles from the heterozygous to the homozygous state. The *genotype frequencies are changed*, relative to Hardy-Weinberg frequencies, but *inbreeding in itself does not alter allele frequencies*.

2. *The genetic variance of a phenotypic character within a population is usually increased by inbreeding*. This may seem surprising to those who have only a slight acquaintance with the concept of inbreeding, but it follows from the definition of variance as the average squared deviation from the mean (see Chapter 9). Consider two alleles that act additively such that, relative to the heterozygote, the average phenotype of

BOX 11.B

The Effect of Inbreeding on Additive Genetic Variance

If the loci contributing to the genetic variance of a metric character (e.g., height) are in linkage equilibrium, the additive genetic variance in the character is the sum of the additive variance contributed by each locus (see Chapter 9). For one such locus, suppose the phenotypic effects of genotypes A_1A_1, A_1A_2, and A_2A_2 are respectively $+a$, 0, and $-a$, and that these genotypes have the frequencies $p^2(1-F) + pF$, $2pq(1-F)$, and $q^2(1-F) + qF$, where F is the inbreeding coefficient. The mean phenotype is

$$\bar{x} = a[p^2(1-F) + pF] + (0)[2pq(1-F)] - a[q^2(1-F) + qF]$$

$$= a(p^2 - q^2)$$

or, since $(p^2 - q^2) = (p + q)(p - q)$ and $p + q = 1$,

$$\bar{x} = a(p - q)$$

The variance is the sum of the squared deviations of phenotypes from the mean, weighted by their frequencies:

$$V_F = [p^2(1-F) + pF](a - \bar{x})^2 + [2pq(1-F)](0 - \bar{x})^2 + [q^2(1-F) + qF](-a - \bar{x})^2$$

When $F = 0$, $V_F = V_0 = 2pqa^2$.

When $F = 1$, $V_F = V_1 = pa^2 + qa^2 - \bar{x}^2$ (from the alternative expression for the variance,

$$V = \sum_i f_i X_i^2 - \bar{x}^2$$

in Appendix I). Substituting $a(p - q)$ for $\bar{x}$, we find that $V_1 = 4pqa^2$. Thus the additive genetic variance in a fully inbred population is twice that in a randomly mating population. For a population with any inbreeding coefficient F, the expression for V_F can be recast as $V_F = (1 - F)V_0 + FV_1$.

A_1A_1 is a units and that of A_2A_2 is $-a$ units. Suppose that $p = q = 0.5$. Then, if the locus is in Hardy-Weinberg equilibrium ($F = 0$), half the population consists of heterozygotes ($2pq = 0.5$), the phenotype of which (i.e., 0) equals the population mean; only the homozygotes deviate from the mean and so contribute to the variance. If the population is inbred, more individuals contribute to the variance, because both homozygotes increase in frequency (Box 11.B). In a panmictic population, the genetic variance is $V_0 = 2pqa^2$, while in an inbred population it is $V_F = 2pqa^2(1 + F)$.

3. **Inbreeding depression** is a reduction in the mean of a phenotypic character in an inbred population due to the increased frequency of homozygous recessive phenotypes. In evolutionary literature, the term *inbreeding depression* refers most frequently to reductions in components of fitness, such as survival and fecundity. We noted in Chapter 9 that in most populations, rare recessive alleles at many loci reduce survivorship or fecundity. These alleles are seldom expressed in a large panmictic population because they are carried mostly in heterozygous condition. If the population becomes subjected to inbreeding, however, the frequency of recessive homozygotes will increase, so that the mean fitness of the population is temporarily reduced. (Because the recessive deleterious alleles are then exposed to natural selection, they eventually will be reduced in frequency or eliminated altogether. This process reduces the amount of genetic variation in the population. Thus although inbreeding *in itself* does not change allele frequencies or reduce genetic variation, inbreeding and natural selection in concert can do so.)

4. *Inbreeding can promote linkage disequilibrium*, i.e., nonrandom associations of alleles at different loci

(see Chapter 9). Recall that with two alleles at each of two loci (*A,a* and *B,b*), linkage disequilibrium exists if certain combinations—say, *AB* and *ab*—are carried by gametes more frequently than expected by chance, so that *Ab* and *aB* are less frequent than expected. Linkage disequilibrium declines because recombination in the double heterozygote (say, *AB/ab*) generates gametes with the deficient allele combinations (*Ab*, *aB*). Because inbreeding reduces the frequency of heterozygotes, the opportunity for recombination is reduced, so the rate at which linkage equilibrium is approached is reduced. If two loci are initially in linkage disequilibrium in a population of plants that reproduce exclusively by selfing, this may become a permanent condition, with most of the individuals having one of only two (say *AB/AB* and *ab/ab*) of the four possible homozygous genotypes.

Inbreeding in Natural Populations

Selfing and outcrossing in plants Self-fertilization, the most extreme pattern of inbreeding, occurs in some hermaphroditic animals (e.g., some snails) and in many plants. In extreme cases, pollination may occur almost exclusively within individual flowers. In other cases, some pollen may be transported both to other flowers on the same plant (which has the same genetic consequences as intra-flower selfing) and to other plants, in which case the mating system of the population is a mixture of selfing and outcrossing.

As theory predicts, genotype frequencies deviate from Hardy-Weinberg expectations in populations that engage in selfing. For example, about 95 to 98 percent of offspring result from self-fertilization in the wild oat *Avena fatua*. S. B. Jain and D. K. Marshall (1967), who studied several local populations in California, estimated genotype frequencies at three loci, each with two alleles. These loci determine the color (B,b) of the lemma (a bract enclosing the flower in members of the grass family), the degree of pubescence of the lemma (H,h), and the pubescence of the leaf sheath (L,l). At each locus, all three genotypes have different phenotypes. Table 11.3, presenting data from one of the populations, shows that at each locus, heterozygotes were much rarer, and both homozygotes much more abundant, than expected in a randomly mating population. Note, moreover, that the inbreeding coefficient, estimated from the relationship $F = 1 - H/2pq$, had almost the same value at each locus—which is just as it should be, because *inbreeding affects all loci in the same way*.

Another member of the grass family, the wild emmer wheat *Triticum dicoccoides*, has an outcrossing rate of only about 0.5 percent. Edward Golenberg and Eviatar Nevo (1987) identified seven polymorphic enzymes by electrophoresis, and another polymorphic locus affecting coloration. Each locus has two alleles, denoted *0* and *1*. The investigators determined the genotype, at all eight loci, of each of about 50 plants in each of five localities, spaced approximately evenly along an 8-kilometer transect in Israel. At the first two localities, almost every plant had the same genotype, the homozygous genotype *00001111/00001111* (written as *00001111* for the sake of simplicity). Likewise, almost all plants at the two sites at the other end of the transect had the "opposite" genotype, homozygous *11110000*. In the middle locality (chosen in order to analyze the contact between the two different genotypes), Golenberg and Nevo analyzed one plant every meter for 50 meters. They found only five recombinant genotypes, one more common than the others, and all localized within 28 meters (Figure 11.11). Only one of the plants was heterozygous.

These data illustrate not only the great excess of homozygotes and deficiency of heterozygotes resulting from strong inbreeding, but also the extreme level of linkage disequilibrium that inbreeding promotes. The loci studied are located on at least three chromosomes, but the nonrandom association of alleles, even of unlinked loci, is almost as complete as it can be.

Several features of the biology of a species may prevent self-fertilization. The most obvious is separation of the sexes into different individuals. Many plant species have "perfect" flowers with both male and female organs, but the male and female functions often are separated in time (e.g., the stigma may not be receptive to pollen until after the flower's pollen has been shed). Often the flower is structured so that self-fertilization is unlikely (Figure 11.12). We encountered in Chapter 9 the polymorphism known as heterostyly, whereby individuals have either short stamens and a long style or the reverse, so that pollen placed on a pollinating insect is likely to be deposited on a stigma of the other morph. An especially common mechanism that promotes or enforces outcrossing in plants is **self-incompatibility**, which is based on a genetic polymorphism of self-sterility alleles ($S_1, S_2, S_3, \dots$). In the more common of two types of self-incompatibility, a pollen grain carrying any one of these alleles (S_i) fails to grow down the stigma of a plant whose genotype (S_iS_j) includes that allele. Thus self-fertilization is not possible. At least three alleles must segregate in such a population (S_1 pollen can fertilize S_2S_3 plants, S_2 can fertilize S_1S_3, S_3 can fertilize S_1S_2), and most self-incompatible plant species have many more.[*]

---

[*]This is a description of gametophytic self-incompatibility, in which the haploid genotype of the pollen determines which stigmas are incompatible. In sporophytic incompatibility, pollen carrying either allele of its parent plant is incompatible with a stigma carrying either of these same alleles, even though only one of these alleles is actually included in the pollen grain.

Table 11.3 Genotype frequencies and inbreeding coefficients for three loci in the self-fertilizing wild oat *Avena fatua*

GENOTYPE	OBSERVED FREQUENCY	q^a	EXPECTED FREQUENCY[b]	INBREEDING COEFFICIENT[c]
BB	0.548		0.340	
Bb	0.071	0.417	0.486	0.854
bb	0.381		0.174	
HH	0.312		0.118	
Hh	0.062	0.657	0.451	0.862
hh	0.626		0.432	
LL	0.571		0.366	
Ll	0.071	0.395	0.478	0.851
ll	0.358		0.156	

Source: After Jain and Marshall (1967).

[a]Frequency of lowercase allele.

[b]Genotype frequencies expected under Hardy-Weinberg equilibrium.

[c]F, estimated from the expression $F = (H_0 - H)/H_0$, where H and H_0 are observed and expected frequencies of heterozygotes, respectively.

FIGURE 11.11   The spatial distribution of genotypes of the wild wheat *Triticum dicoccoides* along a 50-meter transect (locations 1–50). The genotype at eight loci, homozygous in all but one case, is shown for each of the 50 plants sampled. The loci are listed vertically for eight plants in two patches where recombinants were found. Asterisks indicate heterozygous loci. This selfing species is genetically variable, but displays complete homozygosity and strong linkage disequilibrium. (After Golenberg and Nevo 1987.)

```
                                         1 1 1 0              1 1 1 1
                                         1 1 1 0              1 1 1 0
                                         0 1 1 0              1 1 1 *
1---------14  15---------28   1 0 0 0 33-----39 1 0 0 0 44-------50
     (00001111)       (11100100)    0 0 0 1 (11100100)   0 0 0 0     (00001111)
                                         1 1 1 1              0 1 1 1
                                         0 0 0 1              0 0 0 *
                                         0 0 0 1              0 0 0 *
```

We would expect genotype frequencies to depart further from Hardy-Weinberg proportions in self-compatible than in self-incompatible species, and this is indeed the case. For example, Donald Levin (1978) scored genotypes at each of 20 enzyme loci in numerous populations of three species of *Phlox* in Texas. Two of the species (*P. drummondii* and *P. roemariana*) are self-incompatible. *P. cuspidata* is self-compatible, but it also outcrosses, at a rate of about 22 percent. (The outcrossing rate can be established by comparing the genotypes of individual plants—the female parents—and their offspring, if one has estimates of population allele frequencies. If an A_1A_1 plant has some A_1A_2 offspring, we know that it outcrossed; but how many of its A_1A_1 offspring resulted from self-pollination versus outcrossed A_1 pollen can be estimated only by calculating the probability of receiving A_1 pollen from other plants.)

Levin found that the proportion of polymorphic loci was the same in all three species. Within most populations of *P. cuspidata*, however, the deficiency of heterozygotes, relative to Hardy-Weinberg expectations, was greater at each locus than in the other two species. For example, at the *Adh* locus, the value of *F*, averaged over populations, was 0.67 in *P. cuspidata*, but only 0.38 in *P. drummondii* and 0.36 in *P. roemariana*.

Inbreeding in animals   Inbreeding consists not only of self-fertilization, but also of matings among relatives. In organisms with separate sexes, like most animals, this is most likely to occur if individuals do not disperse far, and thus reproduce close to their site of birth. Extreme cases are represented by some small wasplike hymenopteran parasites of insects, in which several offspring of a single female emerge from a single host and immediately mate with each other (Godfray 1994).

The average degree of relationship among individuals in a small, isolated population is higher than that in a large population, or one that receives many immigrants. For example, Arie van Noordwijk and Wim Scharloo (1981) analyzed data from a 20-year population study of the great tit (*Parus major*). This small songbird (related to North American chickadees) nests naturally in treeholes, but readily uses nestboxes provided by the investigators. For more than a decade, van Noordwijk and Scharloo studied a population on a small island in Holland, marking adult birds and their offspring (readily accessible in the nestboxes) with colored leg bands that enabled them to recognize each individual. Consequently, they could determine which birds formed mating pairs, and could trace the ancestry of many birds for as many as four generations. That is, they could construct

FIGURE 11.12   One of many mechanisms that promote outcrossing in plants. In these protandrous species of *Clerodendrum* (Verbenaceae), the flower functions first as male (left), with anthers presented to approaching pollinators and the style recurved. In the later female stage (right), the stigma intercepts pollinators, and the anthers droop (in *C. speciosissimum*, top), or are coiled (in *C. thomsonae*, bottom). The name *speciosissimum*, incidentally, means "most beautiful." (After Endress 1994.)

FIGURE 11.13 A pedigree for an individual great tit (marked "?") in a population in Holland. All individuals shown are related to its parents, B and A. The inbreeding coefficient of individual "?" is 0.145, and is calculated (see Box 11.C) by tracing all ten pathways of relationships between B and A through their common ancestors (italicized in the pathways as follows): BC*F*A, B*HE*A, BCDI*KHE*A, BCDI*J*HEA, BHK*ML*GFA, BHK*N*LGFA, BCDIK*N*LGFA, BCDIK*ML*GFA, BCFGL*MK*HEA, BCFGL*NK*HEA. (After van Noordwijk and Scharloo 1981.)

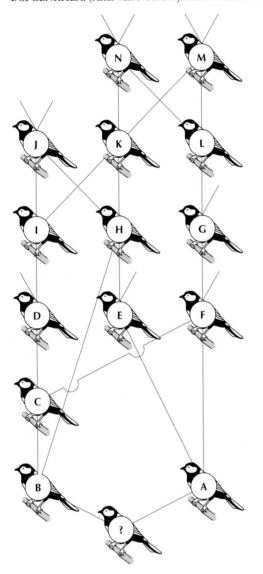

a **pedigree** for many of the birds in the population (Figure 11.13). A pedigree enables one to compute the inbreeding coefficient (*F*) of an individual—that is, the probability that it has received genes identical by descent through two or more related ancestors. (Box 11.C provides a simple example of the calculation of an inbreeding coefficient from a pedigree.) The average inbreeding coefficient ($\overline{F}$) of the population is then the average of the individual inbreeding coefficients.

The average size of the great tit population was 50 breeding pairs. For 1706 broods of progeny, both parents were known in 76 percent of the cases, although some of these parents were not banded and may have been immigrants from outside the population. In 33 percent of the cases in which both parents were known, all the grandparents were also known. In the latter category, the parents of 44 percent of the broods had a common ancestor, and the average inbreeding coefficient of progeny was $\overline{F}$ = 0.036. For all clutches for which both parents were known, $\overline{F}$ = 0.015, assuming that unmarked birds were not related to their mates. If the birds had mated entirely at random throughout the island, the inbreeding coefficient expected from the simple fact that a population of 50 pairs should contain many relatives would be only 0.005. The higher value of $\overline{F}$ calculated from the pedigree (at least 0.015), van Noordwijk and Scharloo suggested, is caused by some ancestors' leaving a greater number of descendants than others.

The value of $\overline{F}$ for this population might be underestimated if clutches from related parents were disproportionately likely to fail, because such clutches would not be included in the data. On the other hand, van Noordwijk and Scharloo assumed that the birds that raised each brood were the parents of those nestlings; if they were not, then $\overline{F}$ might be overestimated. Recently, DNA fingerprinting and other molecular markers have been used to determine parentage in birds (much as they are used in legal cases in which paternity is questioned). These studies have shown that in many species of supposedly monogamous birds, the frequency of "extra-pair paternity"—offspring fathered by males other than the female's apparent mate—is as high as 65 percent (Birkhead and Møller 1992). Inbreeding coefficients are likely to be overestimated if such behavior is not taken into account.

Inbreeding depression As we saw in Chapter 9, populations carry deleterious recessive alleles at low frequency at many loci. Because such alleles are rare, they are carried mostly in heterozygous condition, and have little effect on their carriers' fitness. Inbreeding increases the frequency of homozygotes for these alleles, as it does for all alleles, and thus causes inbreeding depression—a decline in components of fitness. Inbreeding depression has long been known in domesticated and laboratory populations (Figures 11.14, 11.15). In human populations, the consequences of inbreeding include a higher incidence of mortality, mental retardation, albinism, and other physical abnormalities (Stern 1973). It was from pedigree studies of inbreeding that, as mentioned in Chapter 9, Morton et al. (1956) estimated that each of us, on average, carries the equivalent of three to five lethal recessive alleles in heterozygous condition. Nestling mortality is up to 70 percent greater among the offspring of related than unrelated birds in natural populations (Greenwood et al. 1978), and in species of plants that engage in both selfing and outcrossing, fitness components in offspring produced by selfing are often depressed by as much as 50 percent as compared with outcrossed progeny (Schemske 1983; see also Chapter 21).

Calculation of Inbreeding Coefficients from Pedigrees

Suppose individuals I and J are the progeny of a sister and brother, C and D, whose parents (A and B) were not inbred. The pedigree, using squares for males and circles for females, is shown at the left, and is recast at the right to show the transmission of genes in gametes denoted by lowercase letters. For a diploid species, the probability of transmission of a given gene copy is 1/2 along each segment in the pathway.

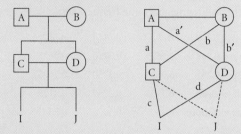

Given a particular gene copy in individual I, what is the probability that the other copy in I is identical by descent? I could be autozygous for a gene from either of its grandparents (A, B). The probability that it is autozygous for one of A's genes is the probability $[\Pr(c = d)]$ that gametes c and d carry the same gene descended from A. This requires that $c = a$ in terms of gene identity, that $a = a'$, and $d = a'$. Each step has a probability of 1/2, so $\Pr(c = d) = (1/2)^3 = 1/8$. Alternatively, gametes c and d could carry the same gene descended from B, with probability $\Pr(c = d) = \Pr(c = b) \times \Pr(b = b') \times \Pr(d = b') = (1/2)^3 = 1/8$. So the inbreeding coefficient of I, the likelihood that I is autozygous for a gene carried by either A or B, is $1/8 + 1/8 = 1/4$.

The "chain-counting" technique for determining F is therefore the following: Construct a transmission diagram like the one on the right, above. Trace a path from I up through each common ancestor and back to I. (In this case there are two paths, through A and through B.) Count the number of individuals (n) in each path, exclusive of I. The inbreeding due to the common ancestor in that path is $f = (1/2)^n$. Sum the values of f over the several paths to find F.

For example, in the complicated pedigree for a great tit in Figure 11.13, there are ten paths. Italicizing the common ancestor in each, and denoting the path length in parentheses, these are: BC*FA*(4), B*HE*A(4), BCDIK*HE*A(8), BCDI*J*HEA(8), BHK*ML*GFA(8), BHK*NL*GFA(8), BCDIK*NL*GFA(10), BCDIK*ML*-GFA(10), BCFGL*MK*HEA(10), BCFGL*NK*HEA(10). $F = (2)(1/2)^4 + (4)(1/2)^8 + (4)(1/2)^{10} = 0.145$.

As outlined, this procedure assumes that the individuals in the first generation are not inbred, and that the locus is autosomally inherited in a diploid population. The calculations may be modified for cases that do not meet these assumptions.

A pedigree also enables us to calculate the **coefficient of relationship** (r) between two individuals. This coefficient equals twice the inbreeding coefficient (F) of their offspring, were the individuals to mate. In our example of brother-sister mating, $r = 2F = 1/2$. Two sisters can mate in the abstract world of mathematical theory, even if not in reality, so their coefficient of relationship is also 1/2. This concept will prove useful when we consider the evolution of certain kinds of social behavior (see Chapter 20).

FIGURE 11.14 Effects of outcrossing and inbreeding on stature in corn. The two plants at the left are both inbred, homozygous strains. The third plant from the left is the F₁ offspring of these two strains, which displays heterosis ("hybrid vigor"). Successive self-fertilized generations from the F₁ are shown to the right. They display inbreeding depression in their stature, as well as in fitness components such as seed production. (From Jones 1924.)

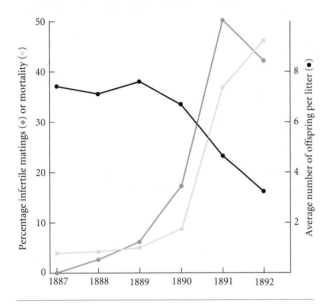

FIGURE 11.15 An early study of inbreeding depression in a laboratory population of rats maintained for about 30 generations of parent-offspring and sib matings. The proportion of infertile matings and the rate of mortality of newborns in the first month of life increased over time. The average number of offspring per litter decreased. (Data from Lerner 1954.)

Relationship between Inbreeding and Genetic Drift

Genetic drift causes changes in allele frequencies, whereas inbreeding alters the proportions of genotypes, but does not alter allele frequencies. Despite this difference, inbreeding and genetic drift are closely related, as we can see from several perspectives. First, imagine that a large, randomly mating population becomes divided into demes, and that mating occurs only within these demes. Within each deme, the probability that two randomly chosen gene copies are identical by descent increases with the passage of generations, as we saw in our discussion of the coalescence of gene lineages (see Figure 11.2), simply because the gene copies in each generation are descended from a smaller number of copies in previous generations. The probability of autozygosity (F) increases faster, the smaller the population. It can be shown (Box 11.D) that after t generations,

$$F_t = 1 - \left(1 - \frac{1}{2n}\right)^t$$

From the relationship between F and H, $H_t = (1 - 1/2N)^t H_0$. The quantity $(1 - 1/2N)^t$ approaches zero as t becomes large, so ultimately F equals 1 and H equals 0: the population has become completely inbred.

Even if the individuals within each deme mate at random, mates are more closely related to each other than they would be if mating occurred at random throughout the entire metapopulation. Thus, a group of isolated demes is conceptually similar to a population consisting of inbred lines propagated by, say, mating between siblings. In both cases,

the population *as a whole* exhibits an excess of all homozygous genotypes and a deficiency of heterozygotes, relative to the Hardy-Weinberg proportions expected if mating occurred at random throughout the population.

Suppose, for example, that the frequencies p and q of alleles A_1 and A_2 are initially 0.5 in a large population that is then subdivided into demes. If A_1 and A_2 each drift to fixation in half of the demes, the overall (or mean) allele frequencies are still 0.5, but there are no heterozygotes: $F = (H_0 - H)/H_0 = (0.5 - 0.0)/0.5 = 1.0$. Less extremely, suppose p has drifted to 0.75 in half of the demes and to 0.25 in the others. Then, if there is random mating *within* demes, we have the following genotype frequencies:

	A_1A_1	A_1A_2	A_2A_2
where $p = 0.75$:	0.5625	0.3750	0.0625
where $p = 0.25$:	0.0625	0.3750	0.5625
Overall (mean):	0.3125	0.3750	0.3125
H-W expectation:	0.25	0.50	0.25 (because $p = 0.5$)

Because p in the population as a whole is $(0.75 + 0.25)/2 = 0.5$, the frequency of heterozygotes *expected* if all individuals mated at random (H_0) would be $2(0.5)(0.5) = 0.5$. Since the observed frequency of heterozygotes is $H = 0.375$, $F = (0.5 - 0.375)/0.5 = 0.25$. In this context, F represents *the probability that two gene copies within a deme are the same, relative to gene copies taken at random from the entire metapopulation*. This value of F, which stems from divergence among demes (or populations of any size) by random genetic drift, was denoted F_{ST} by Sewall Wright, and is sometimes called the **fixation index**. If each deme has an effective size N, then after t generations, $F_{ST} = 1 - (1 - 1/2N)^t$, as we saw earlier.

F_{ST} *is often used as a measure of the observed variation in allele frequencies among populations* (regardless of how the variation may have arisen), and in fact may be calculated from the variance in allele frequencies as $F_{ST} = V_q/[\overline{q}(1 - \overline{q})]$. For example, applying this calculation to the data on the frequency of the $Es-3^b$ allele in house mice in Table 11.2 yields $F_{ST} = 0.0506/(0.418)(0.582) = 0.208$ for small populations and $F_{ST} = 0.0125/(0.372)(0.628) = 0.054$ for large populations.

Gene Flow

Natural populations of a species typically are not completely isolated, but instead exchange genes with one another to a greater or lesser extent. This process is called **gene flow**. *Gene flow, if unopposed by other factors, homogenizes the populations of a species,*—that is, it brings them to the same allele frequencies. Thus conspecific populations differ genetically only if gene flow is sufficiently counterbalanced by the divergent forces of genetic drift or natural selection.

Models of gene flow treat organisms as if they formed either discrete populations (e.g., on islands or in ponds) or continuously distributed populations. In models of dis-

BOX 11.D

Inbreeding, Mutation, and Gene Flow

The probability of autozygosity, F, may be calculated for a single population of effective size N in which inbreeding occurs (denoted F_{IS}) or for a population subdivided into subpopulations, each of effective size N (denoted F_{ST}). In either case, we first note that the rate at which F approaches 1 depends on population size. F in the current generation is denoted F_t, in the immediately previous generation F_{t-1}, and in the original population F_0. Choosing a gene copy at random from the current generation (t), the probability that a second copy is identical by descent (ibd) from the parental generation—i.e., that both are immediately descended from a single parental gene—is $1/(2N)$, because there were $2N$ genes in generation $t-1$. The probability that this is not the case is $1 - 1/(2N)$. But the two different genes may nevertheless be ibd because of descent from the same gene in generation $t-2$; the probability of this is, by definition, F_{t-1}. Therefore, the probability of autozygosity in generation t is

$$F_t = \frac{1}{2N} + \left(1 - \frac{1}{2N}\right) F_{t-1}$$

Multiply both sides by -1, add 1 to both sides, and after a little algebra,

$$1 - F_t = \left(1 - \frac{1}{2N}\right)\left(1 - F_{t-1}\right)$$

a recursion with the general form

$$1 - F_t = \left(1 - \frac{1}{2N}\right)^t \left(1 - F_0\right)$$

If $F_0 = 0$, this may be rearranged as

$$F_t = 1 - \left(1 - \frac{1}{2N}\right)^t$$

which approaches 1 as t becomes large, more rapidly for smaller N.

Now if in each generation (say, generation $t-1$), a proportion u of genes have mutated, or a proportion m have entered from other populations by gene flow, any individual in generation t that inherits one such gene cannot be autozygous. (We assume that u or m is small, so inheritance of two such genes is improbable.) The probability that neither of two uniting gametes carries a mutant or immigrant gene is $(1-u)^2$ or $(1-m)^2$. Thus the probability of autozygosity is, in the case of mutation,

$$F_t = \frac{1}{2N} + \left(1 - \frac{1}{2N}\right) F_{t-1}(1-u)^2$$

At some point, the increase in autozygosity is balanced by the rate of mutation; then $F_t = F_{t-1} = F$. Entering F in the above equation, some algebra yields

$$F = (1-u)^2 / [2N - (2N-1)(1-u)^2]$$

Proceeding further and dropping $2u$ and terms in u^2 because they are small,

$$F = 1/(4Nu + 1)$$

Likewise for gene flow,

$$F = 1/(4Nm + 1)$$

---

crete populations, individuals may move from any population to any other with equal probability (ISLAND MODELS, as in Figure 11.1A), or, more realistically, they may move only to neighboring populations (STEPPING-STONE MODELS). Continuously distributed species, such as trees in a forest, are treated in ISOLATION BY DISTANCE MODELS (as in Figure 11.1C), in which individuals or their gametes move from their site of origin to other sites with a probability that declines with increasing distance. In an isolation by distance model, each individual is the center of a NEIGHBORHOOD, within which the probability of mating declines with distance from the center. The population as a whole consists of overlapping neighborhoods. Most real organisms probably fall in between these idealized extremes.

Gene flow is the movement of genes from one population into the gene pool of another. Immigrants that do not succeed in reproducing within their new population do not contribute to gene flow, so ecological studies of the movement of organisms may overestimate gene flow. Seasonal migrations do not contribute to gene flow if individuals return to their natal population to breed, as appears frequently to be the case in migratory birds (Greenwood and Harvey 1982). From a genetic point of view, it does not matter whether genes are carried by moving individuals (such as most animals, as well as seeds and spores) or by moving gametes (such as pollen and the gametes of many marine animals and plants). In some species, both individuals and gametes move: gene flow in plants, for example, is accomplished by dispersal of both seeds and pollen.

Models of Gene Flow

Consider first a large group of populations, each so large that the rate of divergence due to genetic drift is negligible compared with the effect of gene flow. Suppose that migrants are equally likely to disperse among all the populations, all of which are equal in size. The RATE OF GENE FLOW, m, is the proportion of the gene copies of the breeding individuals in a population that have been carried into that population in that generation by immigrants from other populations. The frequency of an allele, say, A_1, varies among the populations. One of these populations, say population i, has allele frequency p_i (or simply p), and the average allele frequency in the other populations is $\bar{p}$. Within population i, a proportion m of the gene copies enter, per generation, from other populations, and the frequency of A_1 among these gene copies is $\bar{p}$. A proportion $1 - m$ of the gene copies are nonimmigrants, and among these the frequency of A_1 is p. After one generation, therefore, the population's new allele frequency, p', is

$$p' = p(1 - m) + \bar{p}m$$
$$= p - pm + \bar{p}m$$

The *change* in frequency, denoted Δp, is $p' - p$, so

$$\Delta p = p' - p = m\,(\bar{p} - p)$$

The EQUILIBRIUM FREQUENCY within the population, the value of p when it has ceased to change, is found by setting Δp at zero and solving for p:

$$\Delta p = 0 = m(\bar{p} - p)$$
$$p = \bar{p}$$

Therefore, each population will ultimately attain the same allele frequency, namely, the average allele frequency among the ensemble of populations that exchange genes. If the populations differ in size, $\bar{p}$ is a weighted average.

Gene Flow and Genetic Drift

Now we turn to the *joint effects of gene flow and genetic drift.* In the absence of gene flow, populations tend to diverge from each other by genetic drift. The extent of divergence, as you will recall, is measured by F_{ST}. As we saw earlier, after t generations, $F_{ST} = 1 - (1 - 1/2N)^t$, which equals 1 after t has become large. If, however, new gene copies enter the populations in each generation by gene flow at rate m, some fraction of the offspring cannot be autozygous. Therefore, F_{ST} must reach an equilibrium between gene flow and the tendency toward autozygosity caused by random drift. It can be shown (see Box 11.D) that the equilibrium value of F_{ST}, denoted $\hat{F}$, is approximately

$$\hat{F} = \frac{1}{4Nm + 1}$$

The quantity Nm, which is encountered frequently in population genetic theory, is the *number* of immigrants per generation (whereas m is the *fraction* of immigrants in the population). As Nm increases, $\hat{F}$ rapidly decreases. For example, if $m = 1/(N)$, (i.e., only one breeding individual per deme is an immigrant, per generation), $Nm = 1$, and $\hat{F} = 0.20$. That is, the demes are only 20 percent more inbred, with even this low rate of gene flow, than if they constituted a single randomly mating population. Hence, a moderate level of gene flow keeps all the demes fairly similar in allele frequency, and heterozygosity remains high. Note that $\hat{F}$ will be low if either individual populations are large or the rate of gene exchange is high.

In isolation by distance models, one can imagine that the birthplace of each reproductive individual is the center of a mating neighborhood, the size of which depends on the distance moved by individuals (or their gametes) between birth and reproduction. If the neighborhood is small, each individual has few potential mates. Because this situation is conceptually similar to a small effective population size, theory predicts that genetic drift will give rise to more pronounced local variation in allele frequencies than if the neighborhood is large (Figure 11.16).

Extinction and Recolonization

The models described above envision what has sometimes been called "trickle gene flow," in which there is a fairly constant rate of migration among established populations. In some cases, another kind of gene flow may well be more important. If demes at some sites become extinct, and those sites are then colonized by individuals from several other demes, the allele frequencies in the new colonies may be a more nearly even mixture of those among the source populations than when a few migrants contribute their genes to a large, established population. The mathematical models involved are quite complex, but in general, if the number of colonists founding new colonies is greater than twice the average number of immigrants (Nm) in the island model of "trickle gene flow," the extinction-colonization process may be more effective in reducing the differentia-

---

FIGURE 11.16   Effects of neighborhood size on spatial variation in allele frequencies in the isolation by distance model. We are looking down from an angle at an area populated by a 100 × 100 array of equally spaced individuals. Changes in allele frequencies were simulated by a computer. Local allele frequency differences are shown by different heights above the horizontal plane after 0, 50, and 110 generations of mating. In series A, each individual had an equal probability of mating with any other. In series B, each mated equiprobably with the nine nearest individuals. Genetic differentiation is more pronounced in series B. (After Rohlf and Schnell 1971.)

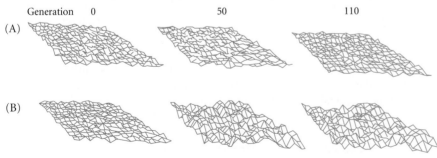

Generation   0   50   110

(A)

(B)

tion of demes by random genetic drift. If, however, each new colony is founded by individuals from a single source population, the extinction-colonization process increases the genetic differentiation among demes (Slatkin 1977; McCauley 1993).

Estimates of Gene Flow

Rates of gene flow among natural populations may be estimated by direct methods, such as following the movement of individuals or their gametes, or by indirect methods, whereby the mathematical models described above are used to infer migration rates from genetic data. Many geneticists prefer the indirect methods, because direct observations are usually insufficient to detect processes such as long-distance migration, rare episodes of massive gene flow, and the perhaps rare but nevertheless important processes of extinction and recolonization (Slatkin 1985a).

Direct estimates   Estimates of gene flow often conform reasonably well to what we would expect from the biology of species; for example, animals such as land snails, salamanders, and wingless insects move little, and are divided into smaller, genetically more distinct populations than are more mobile species. However, even seemingly mobile species often display remarkably restricted dispersal.

The most common direct method of estimating gene flow is the MARK-RECAPTURE METHOD used by population ecologists. For example, Frank Blair (1960) marked rusty lizards (*Sceloporus olivaceus*) so that individuals were recognizable; by recapturing them repeatedly, he traced their lifetime movements. On average, the lizards moved only 78 meters from their hatching site to their final home, and females wandered about 29 meters to lay eggs. The average individual's neighborhood was about 10 hectares, encompassing 225–270 individuals—small enough that considerable spatial variation in allele frequencies might be expected to develop due to genetic drift (Kerster 1964). As in all such studies, individuals that dispersed beyond the study area were not recorded, so the possibility of long-distance gene flow could not be evaluated.

Dispersal is sometimes studied by releasing individuals bearing a *genetic marker* that distinguishes them from others. One of the first such studies (Dobzhansky and Wright 1943) used about 4800 *Drosophila pseudoobscura* flies, which carried an orange eye mutation to distinguish them from the native population. The flies were released in a California woodland, at the center of a cross-shaped array of banana-baited traps set every 20 meters for a distance of 300 meters in each direction from the release point. From the distances at which the orange-eyed flies were recaptured over the next several days (the average interval between emergence and mating), Sewall Wright calculated the neighborhood radius at about 250 meters. The natural density of flies at the time of the study, in early summer, was about $50/100{,}000 \text{ m}^2$, although it may well be lower in early spring. Hence an average fly might encounter 500–1000 other flies, a panmictic unit that the authors felt was "so large that but little perma-

nent differentiation can be expected in a continuous population of this species owing to the genetic drift alone." A later study of the same species in Colorado (Crumpacker and Williams 1973) found dispersal distances 50 percent greater than Dobzhansky and Wright had found, and flies released at an oasis in the hostile desert environment of Death Valley, California, were recaptured at other oases almost 15 kilometers away (Coyne et al. 1982). A low level of long-distance gene flow may provide considerable genetic integration of populations, even if most dispersal is more local.

The seeds of many plant species fall near the parent plant, thus contributing rather little to gene flow. In other species, seeds are distributed by fruit-eating animals such as birds and bats, and may be transported considerable distances. Few measurements have been made, however, because of the difficulty of tracing the movements of the animals and determining where they defecate. The extent of seed transport is sufficient for many species of plants to have extended their range a thousand miles or more since the last Pleistocene glacier withdrew from North America about 10,000 years ago, but that does not mean that the average level of gene flow among established populations is great enough to overcome genetic drift.

The movement of pollen may contribute more to gene flow than does seed dispersal, and is more easily studied. Pollen dispersal in species that are pollinated by insects or other animals has been inferred by tracing the movements of pollinators. Pollinators generally move among neighboring plants, so that most gene flow is very localized (Levin and Kerster 1974). This is the case in the species of *Phlox* studied by Donald Levin, and probably accounts for the evidence of inbreeding that he found even in self-incompatible species (see above). However, pollinators occasionally fly for long distances. Moreover, offspring number and viability are often greater in the offspring of plants from moderately distant populations than in the offspring of close neighbors, which may enhance the effectiveness of long-distance gene flow (Levin 1983a).

Gene flow by pollen dispersal in wind-pollinated plants has been studied most by tracing the movement of genetic markers. For example, Bateman (1947) studied gene flow in maize (= corn, *Zea mays*) by counting the number of heterozygous progeny of homozygous recessive plants ("seed parents") situated at various distances from a stand of homozygous dominant plants ("pollen parents"). At only 40 to 50 feet away, less than 1 percent of the progeny were fathered by the dominant plants (Figure 11.17). Similar studies on a large number of crop species have shown that fields separated by a kilometer or more are effectively isolated in most cases (Levin 1984).

Few data provide a basis for estimating levels of gene flow owing to extinction and recolonization. We do know that these processes occur at rather high frequencies in many species (Hanski and Gilpin 1991). For example, David McCauley (1993) found that local populations of a beetle (*Tetraopes tetraophthalmus*) that feeds and reproduces only on the milkweed *Asclepias syriaca* became extinct at a rate

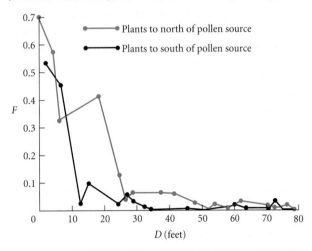

FIGURE 11.17   Gene flow in corn, a wind-pollinated plant. The vertical axis, *F*, gives the proportion of offspring of recessive plants, grown at different distances from a dominant strain, that were fathered by the dominant strain. The curves for plants situated north and south of the pollen source differ because the direction of prevailing winds affects the dispersal of pollen. (After Bateman 1947.)

of about 2 percent per generation, owing chiefly to human destruction of stands of its host plant. Newly planted patches of milkweed were almost all colonized by small numbers of beetles each year, from populations at least several kilometers away. The genetic consequences of these population dynamics are unclear because it is not known whether the colonists at a site are drawn from several source populations (which would tend to reduce genetic differentiation) or from one (which would enhance it).

Indirect estimates   An alternative to tracing the movement of genes directly is to infer the level of gene flow from the differences in allele frequency among populations. F_{ST} is one measure of such differences. By rearranging the equation $F_{ST} = 1/[4Nm + 1]$, we might estimate the average number of immigrants into each population per generation as $Nm = (1/F_{ST} - 1)/4$. In doing so, we infer that the more similar the allele frequencies among populations, the higher the rate of gene flow; conversely, strong divergence among populations is taken to indicate that the balance between genetic drift and gene flow is tipped toward genetic drift.

This inference is valid only if two assumptions hold true. First, the alleles for which we calculate F_{ST} must be selectively neutral. F_{ST} would underestimate gene flow if natural selection favored different alleles in different areas, and would overestimate gene flow if selection favored the same allele everywhere. This assumption can be evaluated by the degree of consistency among different loci used to calculate F_{ST}. Genetic drift and gene flow affect all loci the same way, whereas natural selection affects different loci more or less independently. Therefore, if each of a number of polymorphic loci yields about the same value of F_{ST}, it is likely that

selection is not strong. Second, allele frequencies must have reached an equilibrium between gene flow and genetic drift. This might not be the case if, for example, the various sites have only recently been colonized, and the populations have not yet had time to differentiate by genetic drift. Their genetic similarity would then lead us to overestimate the rate of gene flow. This assumption can be difficult to evaluate.

The black-tailed prairie dog (*Cynomys ludovicianus*), a social ground squirrel of North American prairies, has a complex population structure. A group of coteries, each with one or two breeding males and several females, makes up a ward; a number of wards, each occupying a patch of suitable habitat, constitute a deme. Ecological studies have shown that males disperse farther than females, and that they move more frequently among coteries within wards than among wards. Ronald Chesser (1983) used electrophoresis to estimate allele frequencies at seven polymorphic enzyme loci in 21 populations of prairie dogs, scattered widely throughout four regions of eastern New Mexico. Most, although not all, of the loci showed fairly similar levels of gene frequency differentiation (F_{ST}), as illustrated in Table 11.4. From the mean F_{ST} values of the several loci, *Nm* was estimated at about 1 animal moving into each coterie from other coteries in the same ward per generation. At higher hierarchical levels, about 4.4 enter each ward, 4.9 enter each deme, and 2.2 enter each population region from other regions. Thus the level of migration even among coteries is low, probably because these social groups tend to exclude outsiders. Gene flow among the larger population units is also low, but nonetheless seems high enough to prevent complete divergence by genetic drift.

Even more extreme subdivision is evident in the pocket gopher *Thomomys bottae* (Patton 1972; Patton and Yang 1977). This burrowing rodent seldom emerges from the soil, and the maximal dispersal distance of marked individuals is only about 900 feet. This species is famous for its localized variation in coloration and other morphological features, which has led taxonomists to name more than 150 subspecies. Moreover, local populations differ more in chromosome configuration than in any other known species of mammal. A study of 21 polymorphic enzyme loci in 825 specimens from 50 localities in the southwestern United States and Mexico also showed extreme geographic differentiation, with different loci displaying different patterns of variation (Figure 11.18). F_{ST}, averaged over these loci, was extraordinarily high at 0.412 across all 50 populations (which might imply $Nm = 0.36$). It was likewise high within smaller regions (e.g., $F_{ST} = 0.198$, $Nm = 1.01$, among 17 localities in Arizona). Patton and Yang found that populations were genetically most different when they were geographically distant and/or segregated by expanses of unsuitable habitat—both factors that would reduce gene flow.

Generally, the magnitude of gene flow estimated from genetic data corresponds fairly well with what we might ex-

Table 11.4 Genetic differentiation (F_{ST}) and estimated rate of gene flow (Nm) among population subunits of the black-tailed prairie dog

LEVEL OF COMPARISON	LOCUS	F_{ST}	Nm
Among 12 populations in one region	Ada	0.0168	14.63
	Gdh	0.0588	4.00
	Got-2	0.0809	2.84
	Np	0.0689	3.38
	6-Pgd	0.0632	3.71
	Pgm-2	0.0538	4.40
Mean		0.0489	4.86
Among 21 populations in all four regions		0.1031	2.17
Among four wards in one population		0.541	4.37
Among coteries within one ward		0.1830	1.12

Source: Data from Chesser (1983).

pect from the natural history of species. For example, the blue mussel (*Mytilus edulis*) is a marine bivalve with planktonic larvae that can be transported for long distances by currents. From allele frequencies in this species, Slatkin (1985b) estimated *Nm* as 42.0, one of the highest rates of gene flow described. At the other extreme, two species of wholly terrestrial plethodontid salamanders had very low values of *Nm* (0.10 and 0.16). Plethodontid sala-

manders had already been considered highly sedentary on the basis of ecological studies and patterns of geographic variation.

Conclusions about Gene Flow

The major conclusion to be drawn from many studies such as these is that although some species display substantial gene flow over fairly long distances, in a great many species

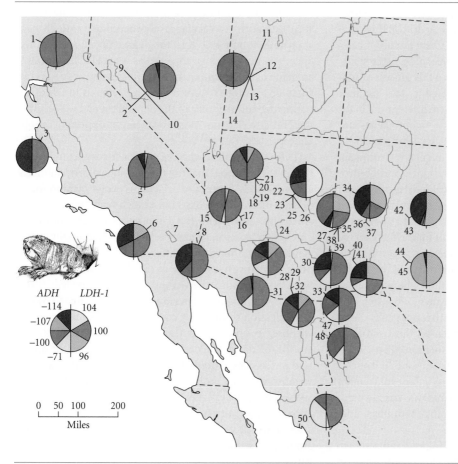

FIGURE 11.18 Spatial variation in allele frequencies at two electrophoretic loci in the pocket gopher *Thomomys bottae*. The left half of each circle shows frequencies of up to four alleles of *Adh*, and the right half up to three alleles of *LDH-1*, at various numbered localities. The allele frequencies at these loci vary greatly even among geographically close localities, as expected if gene flow is low. (After Patton and Yang 1977.)

the average level of gene flow is greatly restricted, even over surprisingly small distances. This conclusion has extremely important implications for evolution. It implies that local populations of many species can diverge substantially by genetic drift, and also, as we shall see in later chapters, that there is potentially great opportunity for populations of a species to adapt to local conditions. The level of gene flow also has important consequences for theories of speciation and for the mechanisms and course of adaptation on a macroevolutionary time scale.

The Neutral Theory of Molecular Evolution

Our discussion so far has ignored any new mutations that may arise during the course of genetic drift. A new mutation that eventually becomes fixed takes, on average, $4N_e$ generations to do so; if N_e, the effective population size, is large, the time taken for a new mutation to approach fixation may be so great (4000 generations if $N_e = 1000$) that other new mutations are likely to arise in the interim. For example, with a mutation rate of 10^{-9} per base pair per generation, a gene with 10^3 base pairs will have 10^{-6} mutations per gamete per generation; a population of $N_e = 1000$ ($2N_e = 2000$ gene copies) should experience $10^{-6} \times 2000 \times 4000 = 8$ new mutations in that gene over the course of 4000 generations. Thus there is a continual influx of new mutations that augment variation, while at the same time variation is eroded by genetic drift. If all new mutations are selectively neutral, some will increase in frequency by genetic drift, at least temporarily; *in the long run*, because each new mutation has a probability of fixation of $1/(2N_e)$, *there will be a steady substitution of one allele for another*. During this process of substitution, the locus will be polymorphic. We are here assuming that each mutation occurs at a different site in the gene, so that each mutation generates a unique allele (haplotype). (This is often called the "infinite alleles" model.)

The theory of evolution by random genetic drift of selectively neutral alleles became important—and controversial—in the 1960s because of two kinds of molecular data. At that time, most evolutionary biologists believed that almost all alleles, including those found as polymorphisms within populations, differed in their effects on organisms' fitness, so that their frequencies were affected chiefly by natural selection. In the aftermath of the Evolutionary Synthesis, so much evidence of the ubiquity of natural selection was accrued that Ernst Mayr (1963: 162, 211) declared that "it is altogether unlikely that two genes would have identical selective values under all the conditions under which they may coexist in a population." Thus, "cases of neutral polymorphism do not exist," and "it appears probable that random fixation is of negligible evolutionary importance." Mayr's belief, widely shared, was based on numerous studies of genes with morphological or physiological effects.

In 1966, however, Lewontin and Hubby, having shown that a high proportion of enzyme loci are polymorphic in

Drosophila populations, argued that natural selection could not actively maintain so much genetic variation, and suggested that much of it might be selectively neutral. At about the same time, the Japanese geneticist Motoo Kimura (1968) analyzed published data on the amino acid sequences of proteins in several species, and calculated their rates of evolution from the number of amino acid differences between species relative to the time since they diverged from their common ancestor (based on the fossil record). He calculated that a given protein evolved at a similar rate in different lineages, and concluded that such constancy would not be expected to result from natural selection, but would be expected if most evolutionary changes at the molecular level are caused by mutation and genetic drift. These authors, as well as J. L. King and T. H. Jukes (1969), who advanced similar arguments, initiated a major controversy, the "neutralist-selectionist debate," which has not yet been resolved.

The NEUTRAL THEORY OF MOLECULAR EVOLUTION holds that although a small minority of mutations in DNA or protein sequences are advantageous and are fixed by natural selection, and although some are disadvantageous and are eliminated by "purifying" natural selection, the great majority of mutations that are fixed are effectively neutral with respect to fitness, and are fixed by genetic drift. According to this theory, most genetic variation at the molecular level—whether revealed by DNA sequencing or by enzyme electrophoresis—is selectively neutral and lacks adaptive significance. Moreover, evolutionary substitutions at the molecular level proceed at a roughly constant rate, so much so that the degree of sequence difference between species can serve as a MOLECULAR CLOCK, enabling us to determine the divergence time of species (see Chapter 5). Motoo Kimura, who developed much of the mathematical structure of the neutral theory, was its most forceful advocate; its most vocal recent opponent is John Gillespie (1991).

Before proceeding with the neutral theory, let us be clear that it does *not* hold that the morphological, physiological, and behavioral features of organisms evolve by random genetic drift. Many, perhaps most, such features may evolve chiefly by natural selection. Differences in such features are encoded in base pair differences in DNA, and usually are engendered by amino acid differences in the enzymes and other proteins that partake in the development of organisms. Furthermore, the neutral theory acknowledges that many mutations are deleterious, and are eliminated by natural selection so that they contribute little to the variation we observe. Thus the neutral theory does not deny the operation of natural selection on *some* base pair or amino acid differences. It holds, though, that *most* of the variation we observe at the molecular level, both within and among species, has little effect on fitness, either because the differences in base pair sequence are not translated into differences at the protein level, or because most variations in the amino acid sequence of a protein have little effect on the organism's physiology.

Principles of the Neutral Theory

In the discussion that follows, we retain the term *mutation* for an alteration of a single copy of a gene during DNA replication, and use the term SUBSTITUTION for the replacement of one DNA or protein sequence by another in a population or species. "*Substitution*," therefore, is synonymous with the fixation of a new allele (haplotype). A SILENT SUBSTITUTION is a change at the DNA level (e.g., of a single base pair) that does not alter the amino acid sequence of a protein, either because the DNA is not translated or because it has mutated to a synonymous codon (e.g., AAA to AAG, both coding for phenylalanine). A REPLACEMENT SUBSTITUTION is a change in DNA sequence that alters an amino acid in the encoded protein (e.g., AAA to AGA, phenylalanine to serine). Proteins that vary in electrophoretic mobility or other properties differ at at least one amino acid position.

Assume that mutations occur in a gene at a constant rate of u_T per gamete per generation, and that because of the great number of mutable sites, every mutation constitutes a new DNA sequence (or allele, or haplotype). Of all such mutations, some fraction (f_0) are effectively neutral, so the NEUTRAL MUTATION RATE, $u_{0 = f_0 u_T}$, is less than the total mutation rate, u_T. By "effectively neutral," we mean that the mutant allele is so similar to other alleles in its effect on survival and reproduction (i.e., fitness) that changes in its frequency are governed by genetic drift alone, not by natural selection. It is of course possible that the mutation does affect fitness to some slight extent, say, s (see Chapter 13). Then natural selection and genetic drift operate simultaneously, but because random changes in allele frequency are stronger in small than in large populations, the changes in the mutant allele's frequency will be governed almost entirely by genetic drift if the population is small enough. It can be shown that the effect of natural selection is negligible, and that genetic drift is the predominant factor, if $s \ll 1/(2N_e)$. Thus if the difference in relative survival rate caused by two alleles is $s = 0.001$ (i.e., 0.1 percent), the allele frequencies will fluctuate by random genetic drift if the effective population size (N_e) is much less than 500. However, if $s = 0.001$ and $N_e = 5000$, natural selection will be more important than genetic drift. Therefore a particular allele may be EFFECTIVELY NEUTRAL, relative to another allele, when the population is small, but not when the population is large.

The rate of origin of effectively neutral alleles, u_0, can depend on the gene's function. If many of the amino acids in the protein it encodes cannot be altered without seriously affecting an important function—perhaps because they affect the shape of a protein that binds to DNA or to other proteins—then a majority of mutations in the gene will be deleterious rather than neutral (i.e., s will be rather large), and u_0 will be much lower than the total mutation rate u_T. Such a locus is said to have many functional CONSTRAINTS. On the other hand, if the protein can function well despite any of many amino acid changes (i.e., it is less constrained), the mutation rate to effectively neutral alleles will be higher. Within regions of DNA that code for proteins, we would

expect the neutral mutation rate to be highest at third-base-pair positions in codons[*] and lowest at second-base-pair positions, because the genetic code indicates that these have the highest and lowest redundancy respectively (see Figure 12, Chapter 3). (For example, the amino acid leucine is specified by six RNA triplets: CUU, CUC, CUA, CUG, UUA, and UUG.) We expect constraints to be least, or even nonexistent, and the neutral mutation rate to be greatest, in DNA sequences that are not transcribed and have no known function, such as introns and pseudogenes.

Now consider a population of effective size N_e in which the rate of neutral mutation at a locus is u_0 per gamete per generation (Figure 11.19). The *number* of new mutations is, on average, $u_0 \times 2N_e$, since $2N_e$ gene copies might mutate. From genetic drift theory, we have learned that the probability that a mutation will be fixed by genetic drift is its frequency p, which equals $1/(2N_e)$ for a newly arisen mutation. Therefore the number of neutral mutations that arise

---

[*]This assumes that base pair changes can affect fitness only by changing amino acid sequences. They may affect fitness in other ways, however, such as by influencing the rate of translation of mRNA to protein. See the discussion of codon bias in Chapter 22.

FIGURE 11.19 The neutral theory of evolution by genetic drift. Each graph plots the number of copies of an allele in a population of N individuals ($2N$ gene copies) against time. Mutations to new neutral alleles at a locus arise at an average interval of $1/u$. Panel A shows that most of these are lost soon after they arise; these are not shown in panels B and C. The populations in panels A and B are smaller than that in C; B and C have a longer time scale. Occasionally, an allele increases toward fixation by genetic drift. The time required for such a fixation, $\bar{t}$, is longer in the large population (C). Hence at any time there will be more neutral alleles in the large population. (After Crow and Kimura 1970.)

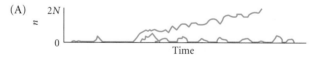

(A)

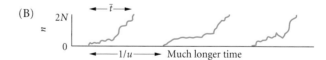

(B)

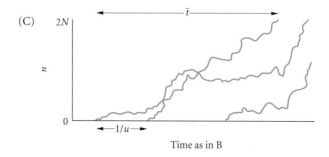

(C)

in any generation *and will someday be fixed* is $2N_e u_0 \times 1/(2N_e) = u_0$. Since on average, it will take $4N_e$ generations for such mutations to reach fixation, about the same number of neutral mutations should be fixed in every generation: *the rate of fixation of mutations is theoretically constant, and equals the neutral mutation rate.* This is the theoretical basis of the molecular clock (see Chapter 5). Notice that, surprisingly, the rate of substitution does not depend on the population size: each mutation drifts toward fixation more slowly if the population is large, but this is compensated by the greater number of mutations that arise.

The average number of mutations fixed per locus per generation is much less than one, because u_0 is a small fraction. If at a given locus, u_0 were, say, 10^{-7}, a mutation would be fixed every 10^7 (10 million) generations, on average, because the average time between events is the reciprocal of the rate at which the events occur per unit of time. Suppose each mutant allele differs from others at one nucleotide site. Then, using the hypothetical neutral mutation rate of 10^{-7}, we would expect that after 40 million generations, the currently prevalent allele would differ from the original one at four base pair positions in the DNA sequence. That is, the number of substitutions after t generations is $X = u_0 t$. We would detect a greater number of substitutions if we examined many more sites, perhaps by combining data from many genes, if they were all subject to the same neutral mutation rate.

If two species diverged from their common ancestor t generations ago, and if each species has experienced u_0 substitutions per generation (relative to the allele in the common ancestor), then the number of base pair differences (D) between the two species should be $D = 2 u_0 t$, because each of the two lineages has accumulated $u_0 t$ substitutions. Therefore the neutral mutation rate can be estimated as $u_0 = D/2t$, if we have an estimate of the number of generations that have passed (cf. Box B in Chapter 10). This formula requires qualification, however. Over a sufficiently long time, some sites experience repeated substitutions: a particular site may undergo substitution from, say, A to C and then from C to T or even back to A. Thus the observed number of differences between species will be less than the number of substitutions that have transpired. As the time since divergence becomes greater, the number of differences begins to plateau, as is evident from the number of differences per base pair between the mitochondrial DNA of pairs of mammal species (Figure 11.20). In this figure, each point represents a pair of taxa for which the age of their common ancestor has been estimated from the fossil record. The number of base pair differences increases linearly for about 5–10 million years and then begins to level off; after about 40 million years, little further divergence is evident because of multiple substitutions. From the linear part of the curve, the mutation rate can be readily calculated, assuming that all the base pair differences represent neutral substitutions (in Figure 11.20, it is about 0.01 mutations per base pair, per lineage, per million years, or about 10^{-8} per year), but as the curve begins to level off, the rate can be estimated only by making corrections for multiple substitutions (Li 1997). Data from taxa on the plateau cannot be used to estimate the substitution rate.

As we noted above, "neutralists" such as Kimura argue that much of the molecular polymorphism in populations, revealed by techniques such as electrophoresis, is selectively neutral. At a locus that evolves purely by genetic drift, genetic variation is ultimately lost, and all individuals become autozygous: $F = 1$. But an individual must be allozygous (so that $F < 1$) if one of its genes is a new mutation. Thus some steady-state level of genetic variation is ultimately attained, at which the rate of loss of variation by genetic drift is balanced by the rate of gain of variation by mutation (see Box

FIGURE 11.20 The number of base pair differences per site between mitochondrial DNA of pairs of mammal species is plotted against the estimated time since their most recent common ancestor. Sequence difference was estimated from restriction enzyme analysis, divergence time from paleontological data and protein differences. The black line estimates the rate of substitution by linear extrapolation from the initial slope of the curve. Multiple substitutions at the same sites after 10–15 My cause the curve to level off. (1 = variation among individual humans; 2 = goats and sheep; 3–9 = pairs of primates; 10–12 = pairs of rodents; 13–20 = various rodent-primate species pairs.) The dotted line represents data available in 1979 on divergence in single-copy nuclear DNA, which evolves at a lower rate than mtDNA. (After Brown et al. 1979.)

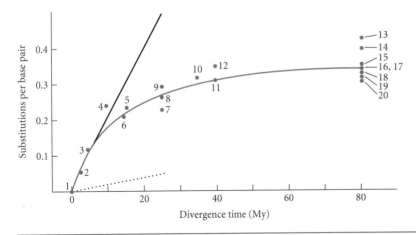

11.D). At this steady state, $F = 1/(4Nu + 1)$, and from the relation between F and the frequency of heterozygotes, it follows that $H = 4Nu_0/(4Nu_0 + 1)$. H is the frequency of heterozygotes when the population is in a STEADY STATE. In this state (Figure 11.21), new neutral alleles arise continually by mutation; many are immediately lost by genetic drift, but others drift to higher frequency, and very occasionally (about every $4N$ generations) one approaches fixation. In the meantime, since all allele frequencies sum to 1, previously common alleles have drifted to low frequency and are ultimately lost. Thus there is a steady turnover, or flux, of alleles. Although the identity of the several or many alleles present in the population changes over time, the level of variation, represented by H, remains about the same, and is higher in a large population than in a small one (Figure 11.21). For example, suppose the mutation rate for allozymes is 10^{-6} per gamete, as observed (Voelker et al. 1980). Then in a population of effective size $N_e = 1000$, the equilibrium frequency of heterozygotes would be 0.004, whereas it would be 0.50 in a population of $N_e = 250,000$ (for which $4N_e u_0 = 1$).

Finally, we can relate the two major tenets of the neutral theory—the steady rate of allele substitutions over time and the equilibrium level of heterozygosity—by noting that both of these measures are proportional to the neutral mutation rate u_0. This fact suggests that if, because of differences in constraint or other factors, various DNA sequences or kinds of base pair sites differ in their rate of neutral mutation, those sequences or sites that differ more *between* related species should also display greater levels of variation *within* species. That is, *there should be a positive correlation between the heterozygosity at a locus and its rate of evolution.*

Variation within and among Species

The expectation, from the neutral theory, that polymorphism within species should be correlated with the rate of divergence between species has been examined in two ways. In one test, the level of allozyme heterozygosity at each of several loci proved to be strongly correlated with the degree of divergence of that locus among species of mammals, in terms of both differences in allele frequency and the number of amino acid differences (Skibinski and Ward 1982). These data are compatible with the neutral theory. A different conclusion was reached by John McDonald and Martin Kreitman (1991), who sequenced 6 to 12 copies of the coding region of the *Adh* (alcohol dehydrogenase) gene in each of three closely related species of *Drosophila*. Polymorphic sites (differences within species) and substitutions (differences between species) were classified as either synonymous or amino acid-replacing. From the neutral theory, we would expect that if the rate of effectively neutral mutation is u_R for replacement changes and u_S for synonymous changes, then the ratio of replacement to synonymous differences should be the same—$u_R:u_S$—for both polymorphisms and substitutions, if indeed the replacement changes are subject only to genetic drift. The data (Table 11.5) show, however, that only 5 percent of the polymorphisms, but fully 29 percent of the substitutions that distinguish species, are replacement

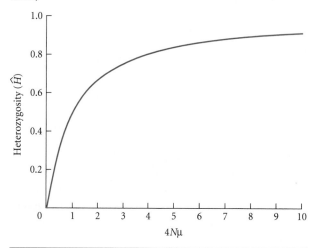

FIGURE 11.21 The theoretical "steady-state" level of heterozygosity at a locus. The equilibrium level of heterozygosity increases as a function of the product of the effective population size and the neutral mutation rate. (After Hartl and Clark 1989.)

changes. McDonald and Kreitman considered these data as evidence that the evolution of amino acid-replacing substitutions is an adaptive process governed by natural selection. If most replacement substitutions are advantageous rather than neutral, they will increase in frequency and be fixed more rapidly than by genetic drift alone. They will therefore spend less time in a polymorphic state than selectively neutral synonymous changes, and will contribute less to polymorphic variation within species. In Chapter 13, we will return to the problem of whether the presence of genetic variation within populations is better explained by genetic drift or by natural selection.

Rates of Molecular Evolution: Do They Support the Neutral Theory?

For about 25 years, the neutralists and selectionists have debated two major questions: Is the genetic variation revealed by electrophoresis selectively neutral or not? Is most evolution at the molecular level attributable to genetic drift, and if so, is its rate constant? These questions are fundamental, for they address the extent to which the major mechanism of evolution advocated by Darwin and by the neo-Darwinians, natural selection, extends even to the molecular

Table 11.5 **Replacement and synonymous substitutions within and among three *Drosophila* species[a]**

	POLYMORPHIC	FIXED
Replacement	2	7
Synonymous	42	17
% Replacement	4.5	29.2

Source: Data from McDonald and Kreitman (1991).

[a] *D. melanogaster, D. simulans,* and *D. yakuba.*

level. We will defer a discussion of the argument about allozyme variation to Chapter 13, and will focus at this point on rates of molecular evolution.

Recall from our discussion of the uses of molecular data in phylogenetic inference (see Chapter 5) that the rate of nucleotide or amino acid substitution can be estimated from comparisons among species in two ways. First, as we have seen, the number of substitutions per year can be calculated (perhaps with a correction factor for multiple substitutions) from the number of sites in a DNA or protein sequence that differ between species and from fossil evidence of the time since they diverged from a common ancestor. As we noted in Chapter 5, this method, which estimates *absolute* rates of evolution, suffers from the imprecision and incompleteness of the fossil record.

The second method is the RELATIVE RATE TEST, which determines, from a phylogenetic tree, how many differences have accumulated in each member of a monophyletic group, relative to an outgroup (Figure 11.22A). Because the absolute time of divergence may not be known, this procedure enables us only to test the constancy of rate by asking whether the number of substitutions in each lineage is approximately the same. A variant of this method is to find a "star phylogeny" (Figure 11.22B): a set of lineages that all diverged from a common ancestor at about the same time. If the rate of evolution has been constant, all pairs of lineages should differ by about the same number of substitutions. Some authors have treated the orders of placental mammals as a star phylogeny because they apparently originated within a short period in the late Cretaceous and Paleocene.

We must here interject that the neutral theory does not predict an exactly equal number of substitutions in each time period in each lineage. Rather, the theory postulates that the *probability* of a substitution is the same throughout, and that it is *small* (u_0, after all, is a small number). These assumptions conform to the Poisson distribution in probability theory. Suppose, for example, that in a light snowfall, you repeatedly hold out a saucer just long enough to catch an av-erage of 0.5 snowflakes (i.e., the mean, $\bar{x}$, equals 0.5: a low number, implying a low probability). In a large number of such trials, some saucers will catch 0, 1, 2, 3 … snowflakes. The *variance* (V) in the number per saucer will equal the mean, if we have enough trials to estimate both V and $\bar{x}$ accurately. The ratio $V/\bar{x} = 1$ characterizes a Poisson process—that is, one in which the rate is constant and each rare event occurs independently. If in 200 trials you obtained 0, 1, 9, and 10 snowflakes, each 50 times, you would suspect, from the ratio $V/\bar{x} = 20.5/5.0 = 4.1$, that the rate was not constant (or that snowflakes fall in clusters, not independently). The saucer in our analogy corresponds to the genes in a population or species, exposed for a generation at a time to a low probability of a substitution (a snowflake).

From the first suggestion of a molecular clock, by Emil Zuckerkandl and Linus Pauling (1962), until recently, most calculations of rates of molecular evolution were based on differences among species in the amino acid sequences of proteins; using the genetic code, the number of nucleotide differences in the corresponding genes could be estimated. Figure 33 in Chapter 5 presents one such analysis, based on seven proteins in a variety of mammals. The variance, or scatter around the line, is almost twice the mean. In particular, the rate of substitution in the hominoid primates (humans and apes) is lower than in most other mammals. These and other such data impress advocates of the molecular clock by the relatively high constancy of rate, and impress skeptics by the differences in rate among some lineages.

Recently, direct sequencing of DNA has provided a wealth of data on rates of molecular evolution. Moreover, the genetic code can be used to specify whether each nucleotide substitution in a protein-coding sequence is a replacement substitution or a synonymous substitution. First, we may note that the rate of synonymous substitutions is greater than the rate of replacement substitutions, as in various genes of humans versus rodents (Table 11.6). That is, substitutions occur most frequently at third-base positions in codons and least frequently in second-base positions. Second,

FIGURE 11.22 The basis for using phylogeny to test the neutral theory of molecular evolution. Each tick mark represents fixation of a base pair mutation in a DNA sequence. Under the neutral theory, the mean number of substitutions is nearly the same in any two lineages since their common ancestor. (A) The relative rate test. The average number of differences is the same between either species A or B and either species C or D. (B) A modification of the relative rate test, in which taxa have an unresolved ("star") phylogeny because they diverged within a short time. The expected number of differences between any pair of taxa is the same.

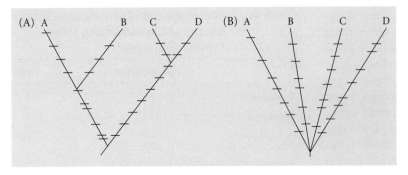

Table 11.6 Rates of synonymous and replacement substitutions in several protein-coding genes, calculated from the divergence between humans and several rodent species[a]

GENE	NUMBER OF bp COMPARED	REPLACEMENT SUBSTITUTION RATE ($\times 10^9$)	SYNONYMOUS SUBSTITUTION RATE ($\times 10^9$)
Histone 3	135	0.00 ± 0.00	6.38 ± 1.19
Histone 4	101	0.00 ± 0.00	6.12 ± 1.32
Insulin	51	0.13 ± 0.13	4.02 ± 2.29
Insulin C-peptide	35	0.91 ± 0.30	6.77 ± 3.49
α-hemoglobin	141	0.55 ± 0.11	5.14 ± 0.90
β-hemoglobin	144	0.80 ± 0.13	3.05 ± 0.56
Glyceraldehyde-3-phosphate dehydrogenase	331	0.20 ± 0.05	2.84 ± 0.37
Lactate dehydrogenase A	331	0.20 ± 0.04	5.03 ± 0.61
Albumin	590	0.91 ± 0.07	6.63 ± 0.61
Average of 36 genes		0.85	4.61

Source: From Li and Graur (1991).

[a]The rate is the number of substitutions per base pair per 10^9 years. A divergence time of 80 million (8×10^7) years is assumed.

rates of substitutions are higher in introns than in coding regions of the same gene, and even higher in pseudogenes, the nonfunctional genes related in sequence to functional genes (Figure 11.23). Third, some genes, such as histone genes, evolve much more slowly than others (Table 11.6). The genes that evolve most slowly are those thought to be strongly constrained by their precise function. A striking example is that of the peptide hormone insulin, which is formed by the linking of two segments of a proinsulin chain, the third segment of which (the C peptide) is removed, apparently playing no role other than in the formation of the mature insulin chain. Among mammals, the average rate of amino acid substitution has been six times greater in the C peptide than in the other portions of the protein(Kimura 1983).

These classes of evidence all indicate that the *rate of evolution is greatest at DNA positions that when altered are least likely to affect function*, and therefore least likely to alter the organism's fitness. This conclusion provides strong support for the neutral theory. No other hypothesis yet devised can explain why pseudogenes should evolve so rapidly. (Rates of

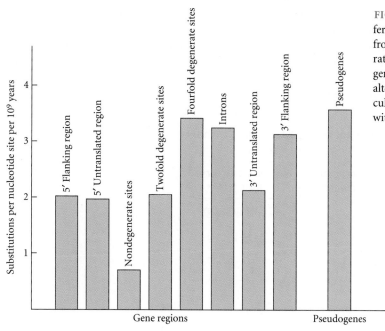

FIGURE 11.23 Average rates of substitution, in different parts of genes and in pseudogenes, estimated from comparisons between humans and rodents. The rate is highest in pseudogenes, and lowest at nondegenerate nucleotide positions, in which any mutation alters the amino acid. Differences in the rate of molecular evolution among these classes are consistent with the neutral theory. (After Li 1997.)

synonymous substitution also vary among genes, and also among evolutionary lineages. We consider reasons for this in Chapter 22.)

The rate of evolution depends on the DNA sequence (and codon positions) examined; but is this rate constant over time and among phyletic lineages?

Kimura (1983) cited several examples, from amino acid sequence data, of apparently equal rates of substitution in each of several orders of mammals. More recently, Simon Easteal (1990, 1991) used relative rate tests for DNA sequences among orders of mammals and among primates. For synonymous substitutions, he concluded, the rates are uniform in both contexts. For example, using various outgroups to compare the rate of evolution of as many as 35 genes in the human lineage with that in various other primates, Easteal found statistically significant evidence for a difference in only 1 of 73 comparisons.

Kimura and Easteal (among others) conclude that substitution rates are approximately constant per unit of *time*. This conclusion is surprising, because the neutral theory is built on the supposition of a constant rate per *generation* (the fixation rate equals the rate of neutral mutation per generation). But generation length presumably differs among orders of mammals (e.g., shorter in rodents than in artiodactyls or primates) and among primates (longer in humans than in monkeys), so the number of neutral substitutions per unit of time should vary. Several sources of evidence (Li and Graur 1991) indicate that mutations occur during DNA replication (i.e., during cell division), so mutation rates should be uniform per generation (if the number of cell divisions between zygote and gamete production is roughly the same in various mammals, as is thought to be the case). However, the phylogenetic evidence of rate uniformity per unit of time has led some authors to suggest that mutations may accumulate at a constant rate in nondividing cells. The conflict between these views has not yet been resolved.

The resolution of this conflict may reside in more data, because while some studies have found evidence of rate uniformity, others have not. Rates of molecular evolution do vary among lineages of organisms (Britten 1986; Li and Graur 1991; Gillespie 1991). For example, Chung-I Wu and Wen-Hsiung Li (1985) used a relative rate test to compare rates of substitution in 11 DNA sequences in humans and rodents, using artiodactyls (cattle, etc.) and carnivores as outgroups. They concluded that the rate of replacement substitution in the rodent lineage has been 1.3 times as fast, per year, as in the lineage leading to humans, and that the rate of synonymous substitution has been 2.0 times as fast. Using paleontological data to calculate rates, Li and his colleagues (1987) found evidence of a greater rate of divergence, per year, between rodent species than between primate species (Table 11.7). They attributed the difference to the shorter generation times of rodents, thus to more generations' having elapsed per unit of time.

John Gillespie (1991) has argued that generation length is not sufficient to explain all the variance in substitution rates among, say, different orders of mammals. This variance is the sum of two components: the variance in the average rates of different orders and the "residual" variance. The average rate for each order, attributable to generation times and perhaps other lineage-specific influences such as differences in mutation rate, was calculated from a number of genes. For each of these genes, Gillespie then calculated how much of the variance among mammal orders is accounted for by the variance in average rates, and then asked whether the remaining (residual) variance is small enough to be explained by a Poisson process. Comparing the rates of evolution of 20 genes among primates, rodents, and artiodactyls, Gillespie concluded that the residual variance, on average, is perhaps no greater than the neutral theory predicts for silent substitutions, but is 7.8 times greater than expected for replacement substitutions. He argued, therefore, that replacement substitutions do not support a molecular clock (at least when we compare major lineages such as orders), and that many or most amino acid substitutions have evolved by natural selection rather than by genetic drift.

Many other examples of substantial differences in the rate of molecular evolution among higher taxa have come to light. For example, mitochondrial DNA sequences have evolved more slowly in turtles than in other vertebrates (Avise et al. 1992). Among more closely related species, rates are often more homogeneous, but nevertheless vary considerably in some cases (e.g., Smith et al. 1992; Martin et al. 1992). Thus, it is not yet possible to make a general statement about the accuracy of the molecular clock and the

Table 11.7 Rates of synonymous substitution per base pair per year within the orders Primates and Rodentia, from several genes

SPECIES COMPARED	ESTIMATED TIME (Myr)[a]	NUMBER OF bp	PERCENT DIVERGENCE	SUBSTITUTION RATE (×10^9)[a]
Human vs. chimpanzee	7 (5–10)	921	1.9	1.3 (0.9–1.9)
Human vs. Old World monkeys	25 (20–30)	998	11.0	2.2 (1.8–2.8)
Mouse vs. rat	15 (10–30)	3,886	23.7	7.9 (3.9–11.8)

Source: From Li and Graur (1991).

[a]The lower and upper estimates of divergence times, and of corresponding substitution rates, are given in parentheses.

applicability of the neutral theory—which will undoubtedly be controversial for some time to come.

Population Structure and Gene Trees

In the genetic theory we have encountered so far, as in most of the literature on evolutionary genetics, we have considered changes in simple counts of the numbers and frequencies of alleles. However, the alleles at a locus have a genealogical history, which is the subject of a fairly new and rapidly growing theory, **coalescent theory**, which can be used to derive important inferences about the size, structure, and history of species populations. The mathematical theory, too complex for us to treat in detail here, is reviewed by Hudson (1990).

Figure 11.2, with which we began our exploration of genetic drift, shows how the gene copies in a population are derived from common ancestors. The gene copies in the diagram are not necessarily different alleles. To recapitulate some of that earlier discussion, recall that if we take pairs of copies at random from the present population (at time $t = 0$), some of these pairs have a common ancestor in the previous generation (i.e., they are copies, in two siblings, of one copy in a single parent). Such a pair of gene copies is said to *coalesce* one generation back. In other cases the common ancestor of two gene copies occurred several generations in the past. If we go far enough back in time, we find an ancestral gene copy from which all copies in the present population have been derived, because all the gene lineages derived from other copies that existed at that remote time have become extinct by genetic drift. The genealogy of the present gene copies forms a **gene tree**, fully analogous to a phylogenetic tree of species. To use the language of phylogenetic systematics (see Chapter 5), these gene copies form a monophyletic lineage, derived ultimately from a single common ancestor.

Recall that in a haploid population of constant size N, the average time back to the common ancestor of random pairs of gene copies, which we will denote t_{CA}, is $t_{CA} = N$ generations, and that the expected time back to the common ancestry of all gene copies in the population is $2N$ generations. In a diploid population of N individuals, $t_{CA} = 2N$ generations, and coalescence of all gene copies takes $4N$ generations on average. Thus the larger the population, the further back in time one must go to find the ancestor of all the gene copies in the population. That is, the divergences in the gene tree will be "deeper," or older, in a large than in a small population (Figure 11.24).

Figure 11.25 reproduces the gene history in Figure 11.2, but adds another consideration. Neutral mutations—single nucleotide changes—occur in this gene tree at an average rate of u per nucleotide site per gene lineage per generation. We assume that each mutation occurs at a different site, thus constituting a unique haplotype. Each mutation is inherited by all descendant gene copies. We therefore are able to (1) calculate the number of base pair differences that we would expect to find among a random pair of copies, and (2) infer a phylogeny of many of the gene copies in the population—that is, to reconstruct the gene tree. The first calculation will enable us to estimate the population's effective size.

Haplotype Diversity and Effective Population Size

If two gene copies had a common ancestor t_{CA} generations ago, and each lineage experiences on average u mutations per generation, then each will have accumulated $u \times t_{CA}$ mutations since the common ancestor, and the number of base pair differences between them will be $S = 2ut_{CA}$, because there are two gene lineages. For two randomly chosen gene copies, the average time to coalescence is $t_{CA} = N$ generations, so $S = 2uN$. That is, the larger the population, the more slowly gene lineages are lost by genetic drift. Therefore the genes present in a large population have an older genealogy than those in a small population, and have had more time to accumulate differences by mutation. We therefore *expect the average number of base pair differences between gene copies to be greater in large than in small populations*. In fact, if we have an estimate of the mutation rate per base pair (u), and measure the average proportion of sites that differ between random pairs of gene copies (S), *we can estimate the effective population size as $N = S/2u$.*

This theory has been applied, with interesting results, by John Avise and his collaborators, who have studied nucleotide diversity and gene phylogenies in the mitochondrial DNA (mtDNA) of many species (Avise 1989, 1994; Avise et al. 1987). Mitochondrial DNA is a single circular molecule that includes about 37 genes in most vertebrates. In most animals, it is apparently inherited only through females and does not undergo recombination; it lacks complicating features such as introns and transposable elements; and in vertebrates, it evolves as much as ten times faster than most nuclear genes, perhaps because of a higher mutation rate owing to less effective DNA repair.

For the most part, Avise and colleagues have not used direct sequencing, but instead have used a large number of restriction enzymes (see Chapter 3), each of which recognizes and cuts a specific short sequence (a *restriction site*) of four to six base pairs. A change in any of these base pairs can be detected because the enzyme then cannot cut the DNA. The relative position within the genome of each restriction site is mapped, so that the mtDNA of any two specimens can be compared with respect to both the number of sites at which they differ and the positions of the differences.

In one study (Avise et al. 1988), these investigators analyzed mtDNA from 127 red-winged blackbirds (*Agelaius phoeniceus*) from 19 widely scattered localities across the United States. Restriction enzyme analysis revealed 34 haplotypes among the 127 birds. The red-winged blackbird is one of the most abundant and widespread North American birds; based on breeding bird censuses in various regions, the investigators calculated that a very conservative estimate of the number of breeding females would be 20 million. The neutral mutation rate per base pair of mtDNA in vertebrates has been estimated at about 10^{-8} per

FIGURE 11.24 Coalescence time in relation to population size. The average time back to coalescence (common ancestry) of all gene copies in a population is $2N$ generations for a population of N gene copies. (A) $N = 6$, $t = 12$ generations. (B) $N = 12$, $t = 24$ generations. Note that the gene tree is "deeper," extending back further in time, in the larger population.

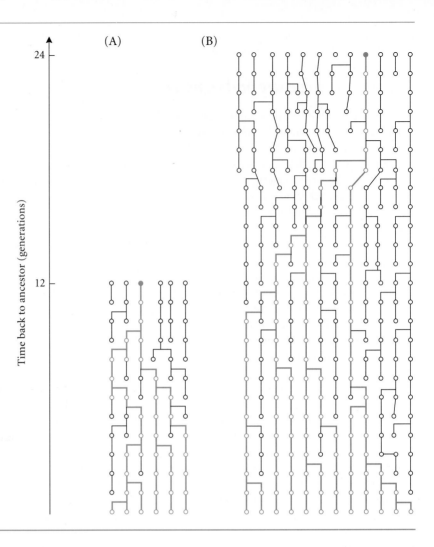

Time back to ancestor (generations)

(A)   (B)

gamete, based on the rate of divergence among species whose time of common ancestry has been estimated from the fossil record (as in Figure 11.20). The expected proportion of site differences between pairs of mtDNAs among the 127 birds was therefore $S = 2uN = (2 \times 10^{-8})(20 \times 10^6) = 0.4$. However, the observed average proportion of site differences was at most 0.008. What accounts for the discrepancy?

FIGURE 11.25 Gene trees in a population, as in Figure 11.2 except that mutations (indicated by tick marks) have been added, so that the gene copies represent variant sequences (haplotypes). The number of sequence differences between any two gene copies at time t can be found by counting the mutations along the paths between these gene copies and their common ancestor. By analyzing these differences by the phylogenetic methods described in Chapter 5, an investigator can estimate the genealogy of haplotypes. Such estimates of gene trees cannot determine the pattern of ancestry of three or more gene copies that do not differ in sequence.

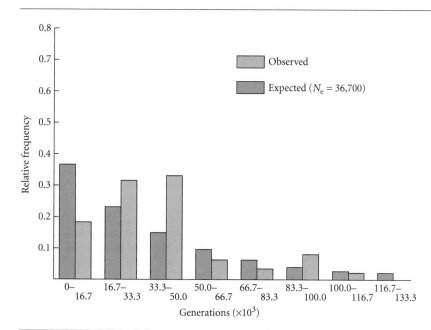

FIGURE 11.26 Frequency distribution of times to common ancestry of mitochondrial DNA haplotypes in the red-winged blackbird. For each pair of haplotypes, the time since their common ancestor was estimated by sequence differences. The distribution of these "observed" times (see key) is fairly close to the distribution expected if the effective population size were 36,700, according to coalescence theory. If the effective population size equaled the estimated current population of this species, the average time to common ancestry would be much greater than observed, and indeed would lie off the scale. (After Avise et al. 1988.)

Avise et al. considered and rejected the possibility that the mutation rate u is much lower than 10^{-8}: it would have to be far lower than any evidence warrants to yield the observed value of S. It is much more likely that the effective population size has been much lower in the recent past than it is now. In fact, the data conform fairly well to what would be expected if N_e were about 36,700 (Figure 11.26), a far cry from 20 million! The most likely explanation is that during the Pleistocene glaciations, as recently as 10,000 years ago, blackbirds were much lower in number, and were probably confined to one or a few small areas, from which they have recently expanded over North America. (We will see shortly that they may well have expanded from only one area.)

Gene Trees and Population Structure

In Figure 11.25, we depicted the occurrence of mutations in a genealogy of gene copies. Assuming that back mutation does not occur, all the descendants of a mutated copy carry the same base pair change. In a phylogenetic framework, each mutation is a synapomorphy (see Chapter 5), a derived character shared by descendant gene copies. The methods of phylogenetic inference can therefore be used to infer the phylogeny of haplotypes. These phylogenetic methods assume that the sequences do not undergo recombination, so they are particularly suitable for mtDNA analysis. In a small population, most gene copies will have a recent common ancestor (Figure 11.24A), whereas some copies in a large population will have much more remote common ancestors, giving rise to "deep" branches in the gene phylogeny (Figure 11.24B).

Now suppose (Figure 11.27) that a large population becomes divided into two populations that do not exchange genes. At first, the gene copies in each population will be genealogically as closely related to those in the other population as they are to other copies in their own population. But now the process of genetic drift, proceeding independently

in the two populations, increases the degree of genealogical relationship among the copies in each population independently. If there is no gene flow between the populations, then ultimately all the gene copies in each will be descendants of *one* of the copies that was included in each population at the time of isolation. That is, each population will have a "monophyletic" gene tree. If mutations have accumulated along the branches of each gene tree, then a phylogenetic analysis should reveal that the gene copies in each population form a monophyletic clade. Depending on the population dynamics, it will take $2N$ to $4N$ generations from the time of isolation for monophyly of the genes in each population to be attained (where N is the effective size of each haploid population).

Eldredge Bermingham and John Avise (1986) sampled four species of freshwater fish in drainages of the southeastern United States, from South Carolina to Louisiana. In the sunfish *Lepomis punctatus*, they found 17 mtDNA haplotypes among 79 specimens. A phylogenetic analysis revealed two clades (Figure 11.28). These clades are geographically segregated: one is distributed from South Carolina to the Suwannee River drainage in western Florida, and is replaced to the west by the other clade, from the Apalachicola drainage in western Florida across to Louisiana. Almost the same disjunction of clades was found in the other three fish species (see Figure 4 in Chapter 16). The simplest interpretation of this pattern is that at some time in the past, these fishes were segregated into two separate populations long enough to evolve monophyletic gene trees, and have since expanded from these "refuges" to meet in western Florida. This conclusion is supported not only by the mtDNA data, but also by the geographic ranges of more than 200 fish species, many of which reach their limits in western Florida. If the mutation rate of mtDNA in fishes is the same as that in mammals, calculations from the theory we have outlined imply that the two sunfish populations became isolated in the early Pliocene (3

FIGURE 11.27 A gene tree for two populations, A and B, derived from a common ancestor. When the two populations are isolated (indicated by a vertical colored bar), each inherits several gene lineages from the common ancestor. The gene copies in each population are initially polyphyletic, as at time T_1, when population A has gene copies descended from ancestral copies a and b, and population B has copies descended from b and c. As gene lineages are lost by genetic drift, the gene copies in each population eventually form monophyletic gene trees, as at time T_3, when all gene copies are descended from ancestor e in both populations. In between, such as at time T_2, one population (A) may have a paraphyletic gene tree with respect to the other (B). That is, at time T_2, some gene copies in A are more closely related to some of B's gene copies, both having descended from ancestor e, than to other gene copies in A (those descended from d). The duration required from the time of isolation to attainment of a monophyletic gene tree is shown to be t_1 for population B and t_2 for population A. The duration t_2 would be greater than t_1 if the population size of species A were greater. (After Avise 1994.)

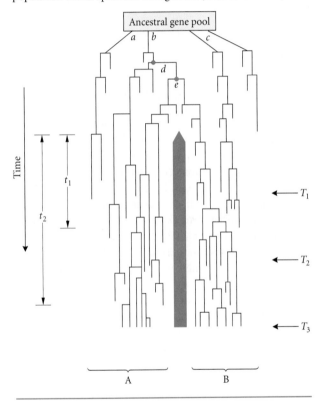

to 4.5 Mya), at which time the Gulf coast was covered by the sea, and freshwater organisms had refuges in peninsular Florida, as well as north and west of the Gulf coast.

That all the mtDNA haplotypes in *Lepomis punctatus* are restricted to either the east or the west suggests that there has been no recent gene flow across western Florida. In contrast, in a phylogenetic analysis of haplotypes in the red-winged blackbird, Avise et al. found little geographic segregation: closely related haplotypes were distributed throughout North America. This might imply that the level of gene flow, even over great distances, is high; but this is doubtful, because even migratory birds like the red-winged blackbird generally breed near their site of birth. It is more

likely that the blackbird has recently expanded over North America from a single small region, and that the populations in different regions today have not been established long enough to have developed different monophyletic gene trees.

Montgomery Slatkin and Wayne Maddison (1989) have developed a method for estimating the rate of gene flow (Nm) between populations, based on the discordance between geography and gene phylogenies. Consider their analysis of the nine mtDNA haplotypes that Bermingham and Avise found in the western clade of *Lepomis punctatus* (Figure 11.28). Using the phylogenetic methods described in Chapter 5, the localities are treated as states of a character that is parsimoniously mapped onto the gene phylogeny. (For example, haplotypes 9, 10, and 11 are all found in locality j, so it is most parsimonious to suppose that their common ancestor occurred there.) Some localities are distributed discordantly over the phylogeny, like a homoplasious character state. (For example, haplotypes 9, 12, 13, and 16, which are not genealogically close to each other, are found in locality l.) Each such discordance implies movement of a haplotype from one locality to another. In the *Lepomis* example, there have been at least five migration events among the western localities, from which Slatkin and Maddison estimate (using computer analysis) that Nm (the average number of immigrants in a population per generation) is about 0.2. They caution, however, that the current level of gene flow might be lower than this if the distribution of haplotypes is a consequence of the recent spread of populations.

The History of Population Structure

On the basis of mitochondrial gene trees, John Avise (1989) has classified species according to two criteria: the extent of sequence divergence among haplotypes, and the extent to which clades of haplotypes are geographically localized (Figure 11.29). Category I includes species such as the sunfish *Lepomis punctatus*, in which clades differ substantially in sequence and are highly localized. In such species, isolation of populations in different areas (from which they may recently have expanded) must have persisted long enough for substantial divergence to occur. Species in category III, such as the slider turtle *Trachemys scripta*, have geographically localized clades, but they differ little in sequence, indicating that the clades are relatively young. This pattern is expected if the species form small populations, with little gene flow, that diverge rapidly by genetic drift, but which have not been isolated for a long time. This category grades, via species such as the red-winged blackbird and *Homo sapiens* (see Chapter 26), into category IV, which includes species such as the American eel (*Anguilla rostrata*). In these species, clades differ relatively little in sequence and are not geographically localized. Such species must have had, at least in recent history, a rather small effective population size, accounting for the low sequence divergence, and either form a nearly panmictic population (as in the eel, as mentioned above) or have recently expanded from an effectively panmictic population (as is probable for the blackbird).

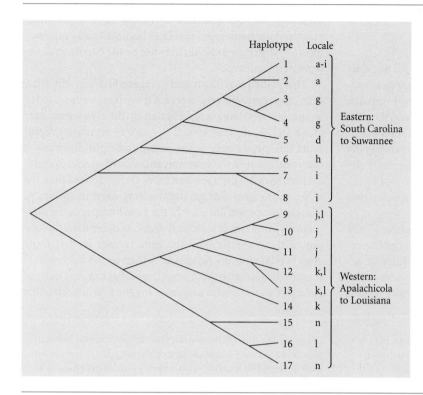

FIGURE 11.28 A phylogeny (gene tree) of 17 mitochondrial haplotypes in the sunfish *Lepomis punctatus*, with the localities in which each was found. Eastern and western groups of sites have separate clades of haplotypes, interpreted to mean that these regions were populated from two isolated ancestral populations. In the western region, the three haplotypes found at site l do not form a monophyletic group, an indication of gene flow between sites. (After Bermingham and Avise 1986.)

To date, as Avise points out, no species have been found in category II: those with highly divergent mtDNA clades that are *not* geographically localized. The category II pattern would be expected of a species with a combination of high effective population size and high gene flow. Apparently, few species have these characteristics, even though many (such as humans and eels) *appear* to have. The implication is that effective population sizes are, or have been, far lower than

they seem today, probably because of episodes of great reduction in numbers. In species that do display highly divergent, localized clades (category I), the haplotypes within each broad geographic region are generally similar in sequence, implying that local populations within each region have not been strongly isolated for very long.

Such data, then, imply that *many species have spread over large geographic areas quite recently from relatively small, lo-*

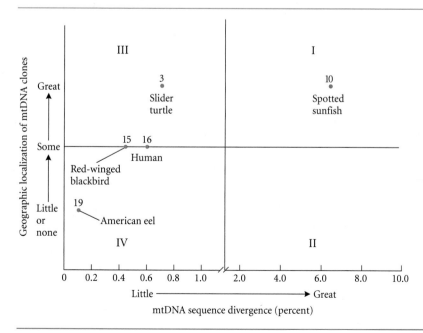

FIGURE 11.29 Observed relationship between the magnitude of sequence divergence among mitochondrial DNA haplotypes within species and the extent of geographic partitioning among mtDNA clades in 5 species of vertebrates. These variables can be used to divide species into four categories of presumed phylogenetic histories. The species shown are 3, *Trachemys scripta* (slider turtle); 10, *Lepomis punctatus* (spotted sunfish); 15, *Agelaius phoeniceus* (red-winged blackbird); 16, *Homo sapiens* (human); 19, *Anguilla rostrata* (American eel). (After Avise 1989.)

calized populations, and that their *population demographics have been highly dynamic over time and space.*

Gene Trees and Species Trees

Coalescent theory has important implications for our ability to infer phylogenetic relationships among species. Suppose we obtain the DNA sequence of a gene copy from each of three species (A, B, C). By using an outgroup and the phylogenetic method described in Chapter 5, we obtain a confident estimate of the phylogenetic relationships among the sequences. Coalescent theory tells us that under some circumstances, *this gene tree*, even if correct, *may not be the same as the species tree*, i.e., the phylogeny of the species from which the gene copies were taken.

The reason becomes evident if we again examine Figure 11.27, which shows that if a species is initially polymorphic—if it has multiple gene lineages that differ in sequence—it takes $2N_e$ to $4N_e$ generations for the gene copies in the population to become monophyletic under genetic drift. If, during this interval, two successive speciation events occur, each daughter species is initially polymorphic, and the haplotypes that later become fixed in the two most closely related species may not be the two most closely related haplotypes.

This principle is illustrated in Figure 11.30, in which the thin lines represent gene trees for three haplotypes, carried in species populations represented by the enveloping gray areas. In the left diagrams, the gene tree accurately represents the species tree. In the center and right diagrams, it does not, because the common ancestor of species B and C carries two gene lineages, one more closely related than the other to the gene lineage that became fixed in species A. Thus the common ancestor of the gene lineages in the populations ancestral to species B and C is older than the origin of these species, and the gene lineage in B is more closely related to the lineage in A than to the lineage in C (center diagrams), or vice versa (right diagrams). Thus, the gene tree in the center diagrams suggests, incorrectly, that

FIGURE 11.30   A gene tree (solid black lines) may or may not reflect the phylogeny of the species (outer envelopes). Species A, B, and C have arisen by two successive speciation events. If the common ancestor of B and C has a monophyletic gene tree, the gene tree will mirror the species tree (left column). If it contains two gene lineages, one more closely related than the other to the gene copies in species A, the gene tree will not reflect the species tree, and will give an incorrect estimate of the species phylogeny (center and right columns). The probability of one or the other outcome depends on the interval between speciation events relative to the population size (N) of the lineage ancestral to B and C. This interval is 0.5N generations in the upper row, 2N generations in the middle row, and 4N generations in the lower row. The number associated with each diagram is the probability that the gene tree will be congruent with the species tree (left column), or that it will be incongruent in each of two possible ways (center and right columns). A diploid organism is assumed. (After Maddison 1995.)

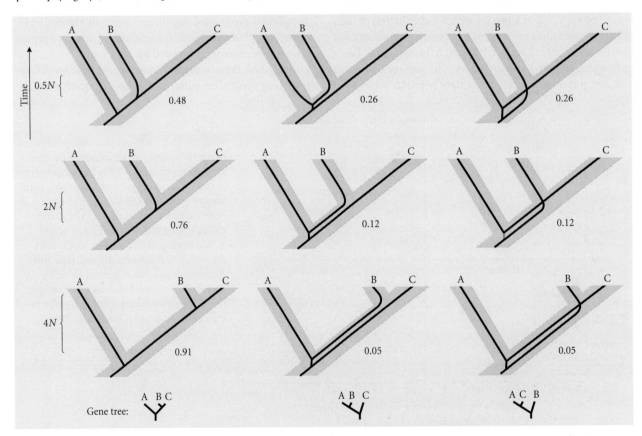

species A and B are most closely related, and in the diagrams at right it suggests, again incorrectly, that species A and C are most closely related.

The probability of each of these outcomes depends on the effective population size (N_e) of the common ancestor of species B and C, relative to the time (t) that elapses between the successive speciation events (Pamilo and Nei 1988; Takahata 1989). If N_e is small or t is large [if $t \gg 1/(2N_e)$ generations], the ancestor is likely to become monomorphic for haplotype (i.e., its gene copies coalesce to a single ancestor) before new mutations arise that are inherited by species B and C and thus mark their common ancestry. In this case the gene tree will reflect the species tree. If, however, N_e is large or t is small, the ancestor of B and C is likely to be polymorphic at the time of speciation, both daughter species may carry both haplotypes, and each species may become fixed for either haplotype, irrespective of the haplotype's genealogical relationship to the haplotype in species A.

Thus, a gene tree is most likely to provide accurate information about the phylogeny of species if populations have been small or the time between successive speciation events has been great. It is least likely to provide an accurate estimate of phylogeny if the populations have been large or if multiple speciation events have occurred in rapid succession (Figure 7 in Chapter 5). This is probably why, as we saw in Chapter 5, the phylogenetic relationships among some taxa, such as the orders of placental mammals, have been difficult to determine. Perhaps the best hope in such cases is offered by phylogenies estimated from the sequences of many different gene loci, since in all but the worst cases, multiple gene trees should, on average, correspond to species trees.

Summary

1. The frequencies of alleles that differ little or not at all in their effect on organisms' fitness (neutral alleles) fluctuate at random. This process, random genetic drift, reduces genetic variation (as measured by the frequency of heterozygotes) and leads eventually to the random fixation of one allele and the loss of others, unless it is countered by other processes such as gene flow or mutation. Different alleles are fixed by chance in different populations, which thereby diverge in genetic composition.

2. Random genetic drift operates more rapidly the smaller the effective size of the population. The effective population size is often much smaller than the apparent population size for many reasons, including temporary reductions in numbers (bottlenecks). Patterns of allele frequencies at some loci in both experimental and natural populations conform to predictions from the theory of genetic drift.

3. Inbreeding (mating among relatives) reduces the frequency of heterozygotes and increases the frequency of homozygotes within a population. Its effects on genotype frequencies are especially evident in plants that practice self-fertilization and in animals with limited dispersal. Divergence among small populations by genetic drift is formally equivalent to inbreeding within a single population: across an ensemble of populations, genetic drift leads to a deficiency of heterozygotes and an excess of homozygotes.

4. Gene flow among populations counteracts genetic drift, by increasing the genetic similarity of different populations. For organisms with more or less continuous distributions, the distance over which gene flow occurs determines the effective population size. Gene flow can be measured either directly, by tracing the movement of organisms or gametes, or indirectly, from the distribution of allele frequencies. Rates of gene flow and effective population sizes vary greatly among species.

5. The theory of genetic drift has been applied especially to variation at the molecular (DNA, protein) level. The "neutral theory" of molecular evolution holds that although many mutations are deleterious and a few are advantageous, most molecular variation within and among species is selectively neutral. The theory predicts that over long spans of time, alleles are substituted (fixed) at an approximately constant rate for a given gene ("molecular clock"). As the theory predicts, synonymous mutations and mutations in less "constrained" genes are fixed more rapidly than those that are more likely to affect function. The rate of molecular evolution, as measured by differences among species, appears to be more nearly constant for noncoding sequences than for coding sequences; the reality and universality of "molecular clocks" are still debated.

6. DNA sequences or similar data can be used to estimate the genealogy of variant genes (haplotypes). Coalescent theory can be applied to such data to estimate effective population sizes, rates of gene flow, and the history of isolation and geographic expansion of populations. It appears from such studies that population sizes and geographic distributions are highly dynamic. The contemporary population structure of many species has arisen only recently.

7. The average time required for the gene copies in a population to form a monophyletic gene tree (i.e., with all copies descended from one original copy) is $2N$ generations, where N is the number of gene copies in the population. The shorter the interval between successive speciation events compared with this time span, the more likely it is that the phylogeny of gene copies from the several species will not reflect the phylogeny of those species.

Major References

Crow, J. F., and M. Kimura. 1970. *An introduction to population genetics theory.* Harper & Row, New York. This exposition of mathematical population genetics includes extensive treatment of inbreeding and genetic drift.

Hartl, D. L., and A. G. Clark, 1997. *Principles of population genetics.* Third edition. Sinauer Associates, Sunderland, MA. Includes a clear treatment of the theory covered in this chapter.

Wright, S. 1968–1978. *Evolution and the genetics of populations.* 4 volumes. University of Chicago Press, Chicago. A technical treatise on population genetics by one of its founders; includes mathematical theory and interpretations of data, especially from classic studies.

Slatkin, M. 1985. Gene flow in natural populations. *Annual Review of Ecology and Systematics* 16: 393–430. Reviews the theory and empirical studies of gene flow and its interaction with natural selection. See also M. Slatkin (1987), Gene flow and the geographic structure of natural populations (*Science* 236: 787–793).

Li, W.-H. 1997. *Molecular evolution.* Sinauer Associates, Sunderland, MA. Includes the topics discussed in this chapter and many others as well.

Kimura, M. 1983. *The neutral theory of molecular evolution.* Cambridge University Press, Cambridge. A comprehensive discussion of the neutral theory by its foremost architect.

Gillespie, J. H. 1991. *The causes of molecular evolution.* Oxford University Press, New York. A provocative argument for the role of natural selection in molecular evolution.

Problems and Discussion Topics

1. For a diploid species, assume one set of 100 demes, each with a constant size of 50 individuals, and another set of 100 demes, each with 100 individuals. (*a*) If in each deme the frequencies of neutral alleles A_1 and A_2 are 0.4 and 0.6 respectively, what is the expected variance in the frequency of A_1 after one generation, among the demes in each set? Take the square root of this value to find the standard deviation. What does this tell you about the allele frequencies of 66 percent of the demes in each set? (*b*) What fraction of demes in each set are likely to become fixed for allele A_1 versus A_2? (*c*) Assume that a neutral mutation arises in each deme. Calculate the probability that it will become fixed in a population of each size. In what *number* of demes do you expect it to become fixed? (*d*) If a fixation occurs, how many generations do you expect it to take?

2. In a population of 50 diploid individuals, the frequencies of neutral alleles A_1 and A_2 are 0.7 and 0.3 respectively. Assuming Hardy-Weinberg equilibrium, calculate the frequency of heterozygotes. What is the expected frequency of heterozygotes after five generations? Repeat the exercise for a population of 20 individuals.

3. In a sample of 150 specimens, 11 are A_1A_1, 44 are A_1A_2, and 95 are A_2A_2. Compare the observed genotype frequencies to those expected under Hardy-Weinberg equilibrium. Calculate F. Provide two possible explanations for the discrepancy from Hardy-Weinberg expectations, based on mating system or population structure.

4. Assuming that the average rate of neutral mutation is 10^{-9} per base pair per gamete, how many generations would it take for 20 base pair substitutions to be fixed in a gene with 2000 base pairs? Suppose that the number of base pair differences in this gene between species A and B is 92, between A and C is 49, and between B and C is 91. Assuming that no repeated replacements have occurred at any site in any lineage, draw the phylogenetic tree, estimate the number of fixations that have occurred along each branch, and estimate the number of generations since each of the two speciation events.

5. Consider a focal population of N diploid individuals in which the frequencies of alleles A_1 and A_2 are 0.7 and 0.3 respectively. It is surrounded by many other populations in which the frequencies of A_1 and A_2 are 0.1 and 0.9 respectively. Calculate the expected frequency of A_2 in the focal population at equilibrium if (*a*) $N = 49$ and it receives one immigrant individual per generation; (*b*) $N = 99$ and it receives one immigrant; (*c*) $N = 99$ and it receives ten immigrants.

6. Consider a population, A, of 100 individuals that mate randomly, and another population, B, of 100 individuals in which most matings are between sisters and brothers. In both cases, the probability that an individual's two gene copies are identical by descent increases as generations pass. Explain the causes of this in the two cases, and the difference between them. How do the populations differ, if at all, in the pattern of change of allele frequencies and genotype frequencies?

7. Suppose that a sample of 100 specimens of a species from each of two locations revealed that for two electrophoretic alleles, A_1 and A_2, the genotype frequencies were 0.36, 0.48, and 0.16 in one sample and 0.16, 0.48, and 0.36 in the other sample, for genotypes A_1A_1, A_1A_2, and A_2A_2 respectively. Calculate the genotype frequencies you would expect if the samples were pooled and were imagined to represent a single panmictic population. From these calculations, estimate F_{ST}. (Answer: 0.04.) What are the possible explanations for the observation that F_{ST} is greater than zero? Now suppose that a similar analysis of nine other loci yielded approximately the same estimate (0.04) for eight of them, but at one locus F_{ST} was calculated as 0.20. What is the most likely explanation for the data?

8. We have noted that some species of plants typically outcross, and others commonly self-fertilize. How would you expect them to differ in the degree of inbreeding depression if you were to examine the survival or seed production of offspring from matings between siblings?

9. Suppose you sequence about ten gene copies from each of three species (and an outgroup), and estimate the gene genealogy. Most, but not all, of the haplotypes in each species form a monophyletic branch, and from this you estimate the phylogenetic relationships among the species. Each species, however, has a few copies of haplotypes that fall within the gene trees of the other two species. Two hypotheses for the incongruence be-

tween the gene tree and the species tree occur to you: (1) There has been incomplete sorting of ancestral polymorphisms. (2) Some introgressive hybridization (see Chapter 9) between species has occurred. Can you think of ways to distinguish these hypotheses?

10. Some evolutionary biologists have argued that the neutral theory should be taken as the null hypothesis to explain genetic variation within species or populations and genetic differences among them. In this view, adaptation and natural selection should be the preferred explanation only if genetic drift cannot explain the data. Others might argue that since there is so much evidence that natural selection has shaped species' characteristics, selection should be the explanation of choice, and that the burden of proof should fall on advocates of the neutral theory. Why might one of these points of view be more convincing than the other?

CHAPTER 12

Natural Selection and Adaptation

The two pervasively important features of living things, which form the two great themes of *On the Origin of Species* and of most evolutionary science since Darwin, are their similarities and their adaptations. Darwin accounted for the unity of life, the similarities among organisms, by a history of common descent. To account for the adaptations of organisms, those innumerable features that so wonderfully equip them for survival and reproduction, he proposed, as did Alfred Russel Wallace independently, the central theory of evolutionary process: natural selection. Natural selection, probably the most important theory in biology, is still an inexhaustible fount of insight into why organisms are what they are. In order to understand what adaptations are, and what the theory of natural selection aims to explain, we begin with some examples of features that, as Darwin said, "excite our admiration." It is also a pleasure simply to reflect on some of the truly wonderful features of organisms.

Examples of Adaptations

Aroids Among the most common household ornamental plants are *Philodendron*, *Monstera*, and other members of the family Araceae (aroids), a large, primarily tropical family with a few temperate zone members such as skunk cabbage (*Symplocarpus*). Many tropical aroids are vines. In some of these, the young plant, having germinated in soil, grows toward dark regions in its vicinity (rather than toward light, as do most plants' growing tips). This strategy often brings it to a tree trunk, up which it grows. In some species, the leaves produced along the trunk-clinging stem, called "shingle leaves," are flat and oval, and are closely appressed to the trunk; but upon reaching lighter regions high above the ground, the stem produces adult leaves, entirely different in shape and borne on long petioles (Figure 12.1). The tiny flowers, borne on a fleshy stem surrounded by a large bract, are visited by flies and beetles attracted to their fetid odor, which is intensified by an exothermic reaction that heats the in-

florescence above ambient temperatures and volatilizes the odorous compounds. (The same reaction enables skunk cabbage to emerge through semifrozen mud in the early spring in temperate regions.) The notable features, then, are the plant's "habitat selection behavior," its capacity to produce leaves suitable for different environments, and its remarkable method of enhancing visits by pollinating insects.

Orchids The Orchidaceae, with 18,000 to 25,000 species, is the second largest (possibly the largest) family of plants. Descended from monocots (such as lilies) with three petals, six stamens, and a three-parted stigma, the orchids have evolved extraordinarily modified flowers. In most species, one of the three petals, called the labellum, is greatly modified into any of a wide variety of forms. The sexual organs are united into a single structure, the column. This includes, usually, only one anther, the pollen of which forms two or more aggregated masses (pollinia). Each pollinium is released as a unit, rather than as separate pollen grains as in almost all other plants. The stigmatic surface lies on the column just below the anther. These profound modifications of structure are the basis for an astonishing variety of pollination mechanisms (Dressler 1990), of which we can mention only a few examples.

Extraordinary, but still simple, is the phenomenon of pseudocopulation (Nilsson 1992). Many orchids throughout the world attract male insects of certain species, not by offering food in the form of nectar or pollen, but by a modification of the labellum to look at least vaguely like a female insect, together with a scent that mimics the attractive sex pheromone (scent) of the female (Color Plate 6). As the insect—variously a bee, fly, or a tiphiid wasp—"mates" with the flower, a pollinium is attached to it by a sticky disc. As in most orchids, the pollinium is attached precisely on that part of the insect's body that will contact the stigma of the next flower visited, where the entire pollen mass will be deposited.

337

FIGURE 12.1 Different forms of leaves on the vine *Monstera tenuis* (Araceae) in tropical American forests. The small "shingle leaves," appressed to a tree trunk, give way via transitional leaves to "adult leaves" far above ground. The adult leaves are much larger, are deeply lobed, and are borne on long petioles. (After Lee and Richards 1991.)

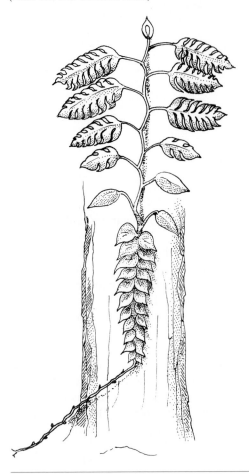

Darwin wrote a long book called *The Various Contrivances by which Orchids are Fertilised by Insects* (1877b), in which he used orchids to show how adaptations result from the modification of pre-existing features. He described both his own and others' painstaking observations on many species, among them *Coryanthes speciosa* (Figure 12.2). The labellum of this orchid is modified into a huge bucket, into which a watery secretion drips copiously from appendages above. A small space separates the spout of the bucket from the column, which lies above it. Bees scramble about and fall into the bucket, and can crawl out only via the spout, so that the bee's dorsal surface rubs first against the stigmatic surface, leaving on it any pollinia that the bee may be carrying, and then against the flower's pollinium, which is attached to the bee's back as it exits.

Darwin described a still more intricate mechanism in *Catasetum saccatum*, which has separate male and female flowers (Figure 12.3). In the male flower, the pollinia are held within an upper chamber on the column, and are connected by a broad pedicel to a sticky disc that is turned inward within a lower chamber in the column, and so is held under

pressure. The disc is connected to two long, asymmetrical "antennae" that overhang the labellum, which has the form of a narrow trough. Euglossine orchid bees, the males of which collect orchid scent to mark mating arenas that attract females, chew material from the trough. If a bee touches the "antenna," the stimulus is transmitted to a membrane that ruptures, releasing the sticky disc. The pollinia are shot out of their chamber, disc first, onto the bee's thorax, with such force that they may be impelled up to three feet if not intercepted. The female flower is so different from the male in form that it was originally placed in a different genus. It is oriented upside down, so that a bee sits on the labellum back down. When it enters the flower, the pollinia on its back are inserted into the stigmatic chamber (which in the male flower holds the disc of the pollinia), and are removed when the bee backs out. Various of the male flower's structures are present, but greatly reduced, in the female flower, including rudimentary pollinia.*

---

*Darwin remarks that "every detail of structure which characterises the male pollen-masses is represented in the female plant in a useless condition . . . At a period not far distant, naturalists will hear with surprise, perhaps with derision, that grave and learned men formerly maintained that such useless organs were not remnants retained by inheritance, but were specially created and arranged in their proper places like dishes on a table (this is the simile of a distinguished botanist) by an Omnipotent hand 'to complete the scheme of nature.'"

FIGURE 12.2 The remarkable flower of the orchid *Coryanthes speciosa*. The right part of the flower, portrayed in lateral view, is the labellum (L), part of which forms a bucket (B), into which liquid drips from the structure marked H. Bees crawl out of the bucket via its spout (P), above which lie the stigma and anther. (From Darwin 1877b.)

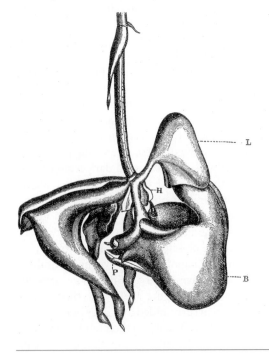

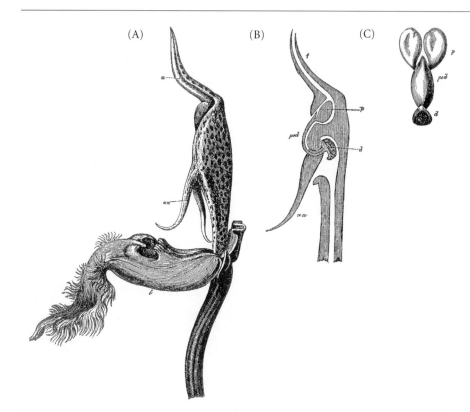

FIGURE 12.3 The orchid *Catasetum saccatum*, as depicted by Darwin. (A) Lateral view of the male flower, showing the labellum (l), the "antennae" (an), and part of the anther (a). (B) Sagittal section of part of the flower, showing how the anther consists of a pollinium (P) connected by a pedicel (ped) to a sticky disc (d) held under pressure in a chamber. (C) A detached pollinium, with its pedicel and disc. (From Darwin 1877b.)

Snake skulls Unlike the majority of terrestrial vertebrates, in which most bones of the skull are rather rigidly attached to each other, snakes have a kinetic skull that is a marvel of engineering. Most snakes can swallow prey items much larger than their heads, and without benefit of hands can manipulate their prey with astonishing versatility. They accomplish this by drawing the prey into the gullet with recurved teeth mounted on a number of freely moving bones that act as levers and fulcra, operated by complex muscles. The lower jawbones (mandibles) articulate to a long, movable quadrate that can be rotated downward so that the mandibles drop away from the skull (Figure 12.4). The front ends of the mandibles are joined not by a fusion of the bones, but by skin that stretches widely. Each mandible is independently moved forward to engage the prey, then pulled back to bring it farther into the throat. A similar action is exerted by the maxillary bones and the palatine-pterygoid series, the outer and inner tooth-bearing bones suspended from the bottom of the skull. These are brought forward (independently on each side) by the action of several muscles, which pull forward the pterygoid and therefore the maxilla. Movement of the maxilla, both fore and aft and in a vertical rotation relative to the skull, is facilitated by rotation of the prefrontal, which forms a hinge for the maxilla.

This system is elaborated still further in vipers (such as rattlesnakes), in which the maxilla is short and bears only a long, hollow fang, to which a duct leads from the massive poison gland (a modified salivary gland). The fang lies against the roof of the mouth when the mouth is closed. When the snake opens its mouth, the same lever system that moves the maxilla in nonvenomous snakes rotates the maxilla 90° (Figure 12.4E,F), so that the fang is fully erected and protrudes well beyond the margin of the mouth. Rattlesnakes and their close relatives have another exquisite adaptation in the form of a highly innervated pit on each side of the head, between the nostril and the eye, that is sensitive to infrared radiation and can detect temperature differences of about 0.2°C. The receptive fields of the two pits form cones extending and overlapping in front of the head, providing precise information on the direction of a heat source (as our ears do for sound). These organs are used primarily to locate mammals and birds, the snakes' major prey.

Parasites and hosts Many species of parasitic animals disperse from one primary host (in which sexual reproduction occurs) to another by complicated means, including secondary hosts. The trematode *Dicrocoelium dendriticum*, for example, reproduces sexually in ungulates and other herbivorous mammals, and passes through asexual stages first in land snails and later in ants, which are induced to climb vegetation and enter a stupor so that they are readily swallowed by grazing mammals (see Figure 22 in Chapter 4).

Even more impressive than the life cycles of parasites are their many defenses against the antibodies of vertebrate hosts, the proteins that recognize and bind to foreign proteins and other molecules. Some parasitic worms coat themselves with the host's own proteins, which do not elicit attack; others secrete a coat of proteins that closely mimic the host's. The most remarkable defense is found in the try-

FIGURE 12.4 The kinetic skull of snakes. The movable bones of the upper jaw are stippled. (A) and (B) show a nonvenomous colubrid snake with jaws closed and opened; (C) and (D) show the same for a viper. (E) and (F) show the movement of the bones that erect the fang of a viper into a striking position. Depression of the quadrate (q) advances the pterygoid (pt), ectopterygoid (ec), and palatine (pal), which rotate the maxilla (mx) and prefrontal (prf), thus erecting the fang (f). In most tetrapod vertebrates, these bones are fixed in position. (After Porter 1972.)

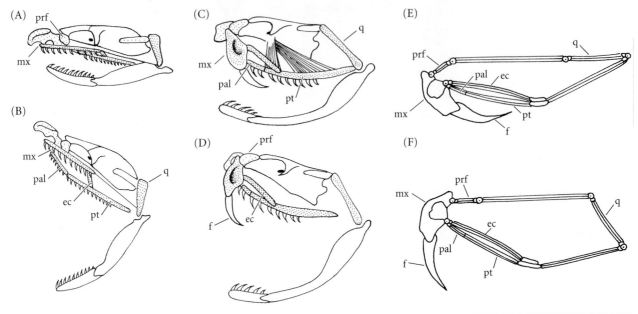

panosome protozoans that cause sleeping sickness. The cell is coated with a variant-specific surface glycoprotein (VSG), which may be any one of about 1000 variants. The active VSG-coding gene can be replaced by any of about 10^3 different VSG genes that have been silent. The silent gene can be activated by any of several mechanisms, particularly duplicative transposition (movement) to one of several sites near the ends of the chromosomes, where the gene is turned on by a local promoter (Borst and Greaves 1987). Individual trypanosomes with new coat proteins escape attack by antibodies that the host has produced in response to previously prevalent trypanosome types. A similar mechanism operates in some parasitic bacteria, such as the species that causes gonorrhea.

The marvels of the trypanosome hardly compare with those of the vertebrate immune system, which would take many pages to describe fully (cf. Tizard 1988). An antibody (immunoglobulin) protein consists of several polypeptides, certain of which are translated from the spliced mRNA transcripts of several genes. In one class of immunoglobulins, two of these kinds of genes (V and J) are themselves spliced together imprecisely so that the nucleotide sequence at the junction is variable. Moreover, the splicing occurs between one of about five J genes and one of 90 to 300 V genes, so there is enormous potential variety. These events transpire differently in each stem cell that gives rise to antibody-producing cells (such as B cells), so an immense variety of different clonal lineages of cells develops. These cells produce antibody that coats the cell surface. A foreign protein (antigen), such as the surface protein of a parasite, that happens to fit a particular cell's antibody binds to the cell surface. This binding sets in motion chemical events that stimulate cell division, so that the clone of cells proliferates, sending cells into the blood to attack the invader. This process of clonal selection provides potential defense against many thousands of different foreign molecules.

Social insects Many species of animals engage in cooperative behavior, but in none does it reach such extremes as in some social insects. An ant colony includes one or more inseminated queens and a number of sterile females, the workers. The workers of many species are differentiated—by chemical stimuli during development rather than by genetic differences—into castes that differ in behavior, size, and often in other morphological traits. Among the many examples of fascinating social behavior in ants (Hölldobler and Wilson 1990; Wilson 1991), we may mention *Zacryptocerus varians* (Figure 12.5), a twig-nesting species in which the largest workers (majors) have a huge saucer-shaped head, with which they block the nest entrance against intruders. In the honeypot ants (*Myrmecocystus*) of American deserts, the largest workers serve as storage vessels for food and water. They hang from the roof of the underground nest chamber, imbibing nectar regurgitated by foraging workers and storing it in their enormously distended abdomens. They regurgitate it to other members of the colony during seasons of food shortage.

The leaf-cutter ants (e.g., *Atta*, *Acromyrmex*) of tropical America form colonies of up to 8 million workers (Figure 12.6), which excavate a nest that may extend six meters into the soil and have a thousand entrances. Columns of workers radiate daily from the nest, along paths worn into the soil

FIGURE 12.5 The ant *Zacrypto-cerus varians* in a dead stem that houses the colony. A major worker, at right, blocks the entrance hole with her saucer-shaped head. The queen is at left, and behind her, another major worker is being fed by regurgitation by a minor worker. (From Hölldobler and Wilson 1990; drawing by T. Hölldobler-Forsyth.)

by the daily tramp, year after year, of millions of little feet. They ascend trees, cut fragments of leaves, and return them to the nest, all the while guarded by huge, aggressive majors ("soldiers"). Within the nest, very small workers chop the leaves into small fragments, chew them and coat them with digestive enzymes, and implant them in a huge mass of fungus, of a species that grows nowhere else. This garden of fungus, nurtured on leaf material and carefully tended, is the chief food of the colony.

Perhaps the most complex behavior is displayed by the arboreal weaver ants (*Oecophylla*) of the Old World tropics. The nest, formed of living leaves, is constructed by the intricately coordinated action of numerous workers, lines of which draw together the edges of leaves by grasping one leaf in their mandibles while clinging to another (Figure 12.7). Sometimes several ants form a chain to collectively draw together distant leaf edges. The leaves are attached to each other by the action of other workers carrying larvae, which, when stimulated by the workers' antennae, emit silk from their labial glands. (Adult ants cannot produce silk.) The workers move the larvae back and forth between the leaf edges, forming silk strands that hold the leaves together. In contrast to the larvae of all other ants, which spin a silk cocoon in which to pupate, *Oecophylla* larvae produce silk only when used by the workers in this fashion.

Explanations of Adaptations

The adaptations described above are marked by two conspicuous features. Most are *complex*, involving, in even the simplest cases, complicated underlying biochemical processes, control by hormones or nervous systems, or morphogenetic developmental pathways. Each of the features we have described, whether it be the shape of an aroid's leaf or the behavior of a weaver ant, could easily take many other forms, or fail to develop altogether, if the underlying processes were altered. Second, all these features have the appearance of *design*; they are constructed or arranged so as to accomplish some *function*, such as growth, defense, or dispersal, that appears likely to promote reproduction or survival. In inanimate nature, we see nothing comparable—we would not be inclined to think of erosion as a process de-

FIGURE 12.6 Worker castes of the leaf-cutter ant *Atta laevigata*, which perform different tasks. (From Oster and Wilson 1978; drawing by T. Hölldobler-Forsyth.)

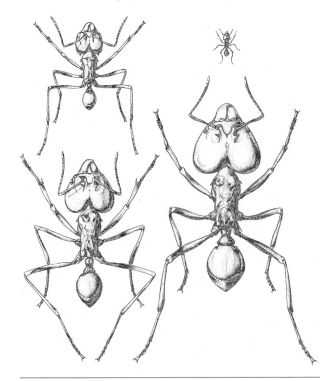

FIGURE 12.7 Weaver ants (*Oecophylla smaragdina*) constructing part of their nest. While one group of workers holds two leaves together, other workers move silk-spinning larvae back and forth across the gap. (From Wheeler 1910, after F. Doflein.)

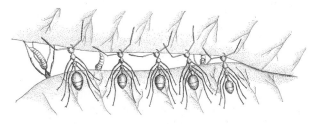

signed to shape mountains, or wind patterns as serving the function of delivering rain.

The complexity and evident function of organisms' adaptations cannot conceivably arise from the random action of physical forces. (In fact, we know what happens to an organism subjected only to such action—namely, a dead organism: it falls apart into its constituent molecules and atoms.) For hundreds of years, it seemed that adaptive design could have only one explanation: an intelligent designer. In fact, the "argument from design" was one of the strongest proofs of the existence of God. Among biologists, the best-known presentation of this argument was the Reverend William Paley's *Natural Theology—or Evidences of the Existence and Attributes of the Deity Collected from Appearances of Nature* (1802), with its famous analogy of a watch and its maker. If I find a stone in the wilderness, Paley wrote, I should be less inclined to wonder how it happened to be there than if I find a watch, for the intricacy of the watch implies that it had a maker "who formed it for the purpose which we find it actually to answer; who comprehended its construction, and designed its use." In every aspect of living nature, such as the human eye, there exists "every indication of contrivance, every manifestation of design, which exists in the watch," except "greater or more, and . . . in a degree which exceeds all computation." Nature must have had a Designer, then, as the watch has had a maker.

Supernatural processes cannot be the subject of science; and so when Darwin offered a purely natural, materialistic alternative to the argument from design, he not only shook the foundations of theology and philosophy, but brought every aspect of the study of life into the realm of science. His alternative to intelligent design was natural selection, according to which organisms possessing variations that enhance survival or reproduction replace those less suitably endowed, which therefore survive or reproduce in lesser degree. Thus the adaptations of organisms have indeed been "designed," but by a *completely mindless* process. This process cannot have a goal, any more than erosion has the goal of forming canyons, for *the future cannot cause material events in the present*. Thus Aristotle's notion of a "final cause," the end or purpose for which events occur, has no place in biology, nor in any other of the natural sciences. According to Darwin, and contemporary evolutionary theory, the weaver ant's behavior has the appearance of design because among many random genetic variations (mutations) in the behavior of an ancestral ant species, those displayed by *Oecophylla* enhanced survival and reproduction under its particular ecological circumstances; other genetic variations, which did not do so as effectively, no longer exist in this species. (Or, if they do come into existence by mutation, they are eliminated because they impair reproductive success.)

Adaptive biological processes *appear* to have goals: weaver ants act as if they have the goal of constructing a nest; migrating birds fly to the tropics "in order to" escape winter; an aroid's leaf develops toward a suitable shape and stops developing when this is attained. Much philosophical discussion has been devoted to how such goal-directed phenomena should be described (e.g., Rosenberg 1985). TELEOLOGICAL statements are those that invoke goals, or ends (Greek *teleos*, "end"), as causes (e.g., "He went to the store *in order* to get milk."). But evolutionary theory does not admit anticipation of the future (i.e., conscious forethought), either in the process of evolution of an adaptive characteristic or in the development or behavior of an individual organism. No conscious anticipation of the future resides in the gene rearrangements that generate antibodies, nor in the organization of cell division underlying the shape of an aroid's leaf, nor, as far as we can tell, in the behavior of weaver ants or migrating birds. In order to distinguish these processes from truly teleological behavior (such as a great deal of human behavior), some authors call the "goal-directed" processes of organisms TELEONOMIC. Ernst Mayr (1988a) defines a teleonomic process or behavior as one that "owes its goal-directedness to the operation of a program," which is "coded or prearranged information that controls a process"—such as the information residing in DNA. A program likewise resides in the arrangement of cogs in a watch, but whereas these have been shaped by an intelligent designer to accomplish some end, the information in DNA has been shaped by a historical process of natural selection, meaning that this information has survived and multiplied to a greater extent than DNA sequences that contain different, or no, information. The information in DNA sequences may not be the only program that directs processes in organisms; for example, the central nervous system carries programs for behavior that may frequently be repatterned (as in learning). But whatever the program, modern biology views an organism as a machine, operating according to programmed instructions, as does a clock or a computer.

What Evolutionary Theory Needs to Explain

A naive interpretation of natural selection is "the survival of the fittest," whereby adaptations evolve that ensure the survival of the species. Impressed by the complexity not only of individual organisms but also of communities of species, many people also perceive a harmony that has evolved to preserve the "balance of nature." On the other hand, some people see natural selection as "Nature red in tooth and claw," a ceaseless, amoral battle that offers no opportunity for cooperation or harmony. Although most biologists would incline more toward the last interpretation, all these views are flawed by inexact metaphor, and are naive and insufficient.

The theory of evolution by natural selection must account for the origin of the complex adaptations that promote the survival and reproduction of the organisms that bear them; but it must also account for features that seem not to benefit individual organisms, such as genetic recombination, the sex ratio of a species, and programmed life spans in which senescence is followed by death. It must account for features that enhance the likelihood of survival of the species, as well as features that do not: for example, a recessive allele that causes sterility in house mice and is transmitted to more than 90 percent of a heterozygous male's sperm, and so can increase in frequency to the point of threatening a population's survival. The theory must explain the origin of cooperation within species, as in ants, and between species, as among plants and their pollinators; but it must also explain antagonistic relationships, both among members of a species, as in langur monkeys in which males kill the offspring of other males, and between species, as in parasites that castrate their hosts, with negative consequences both for the host population and potentially for their own. A full theory, finally, must explain failures of adaptation—the great majority of species that have ever existed are extinct—and instances of apparent maladaptation. We should be able to explain why a honeybee has a barbed sting that is ripped from her body, causing her death, when she stings a mammal, and why humans can be so hyperallergic as to suffer from anaphylactic shock in response to bee stings.

To gain insight into the nature and consequences of natural selection, it is perhaps best to introduce some specific examples.

Experimental Studies of Natural Selection

Microorganisms We have seen that evolution can be studied in laboratory cultures of microorganisms. Chapter 10 provided an example of adaptation in bacteria based on favorable new mutations. Several other examples may be used to develop our understanding of natural selection.

Populations of *Escherichia coli* have often been used to study selection at individual loci (Dykhuizen 1990). For example, Dean et al. (1986) cultured a wild-type strain with several different mutants of the gene that codes for β-galactosidase, the enzyme that breaks down lactose. For each such combination, they did a pair of experiments in which either the wild-type or the mutant strain was marked by another mutation, for resistance to T5 phage, that enabled them to distinguish the strains and thus to estimate their frequencies among samples taken at several times during the growth of a culture. Dean et al. showed that the phage-resistance marker did not influence growth rate. The competing strains were then made genetically identical except at the marker locus and the β-galactosidase locus, and placed in vessels with lactose as their sole source of energy. For several combinations, the ratio of mutant to wild type did not change over the course of several days; apparently these mutations were selectively neutral (see Chapter 11). One mutant strain (TD10.3), however, decreased in frequency (Figure 12.8), and so had lower fitness than the wild type; another (TD10.4) increased in frequency, displaying a greater growth rate (by 0.2 percent per hour) than the wild type. The enzyme activity* of the TD10.3 strain was 25 percent lower than that of the wild type, whereas that of strain TD10.4 was 36 percent higher.

---

*V_{max}/K_{m}, for readers familiar with biochemistry.

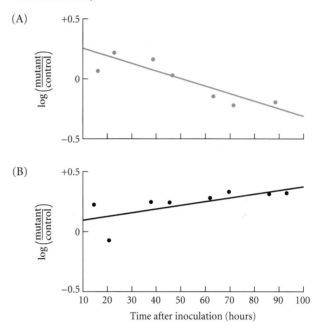

FIGURE 12.8 Natural selection on mutations in the β-galactosidase gene of *Escherichia coli* in laboratory populations maintained on lactose. In each case, a strain bearing a mutation competed with a control strain bearing the wild-type allele. Populations were initiated with equal numbers of cells of each genotype, i.e., with log (mutant/control) initially equal to zero. Without selection, no change in the log ratio would be expected. (A) Mutation TD10.3 decreased in frequency, showing a selective disadvantage. (B) Mutation TD10.4 increased in frequency, demonstrating its selective (adaptive) advantage. (After Dean et al. 1986.)

In another experiment, Atwood et al. (1951) monitored the frequency of selectively neutral alleles in laboratory populations of *E. coli*. The wild-type allele, *his*+, codes for an enzyme that synthesizes histidine, an essential amino acid, whereas *his*− alleles are nonfunctional. However, *his*+ and *his*− are selectively neutral if histidine is supplied so that the mutant can grow. Starting with cultures of *his*−, Atwood et al. observed, to their surprise, that every few hundred generations the allele frequencies changed rapidly and drastically (Figure 12.9). The experimenters showed that a genotype (say, *his*+) occasionally increases drastically in frequency because of linkage to an advantageous mutation at another locus. Recombination is so rare in *E. coli* that alleles at virtually all loci can increase in frequency, even if they themselves are not advantageous, by **hitchhiking** with an advantageous mutation in this manner. Subsequently, an alternative allele (such as *his*−) may increase because it is linked to a new advantageous mutation at another locus altogether. This phenomenon, which Atwood et al. termed PERIODIC SELECTION, has frequently been observed in bacterial cultures: a neutral allele changes in frequency because it is in linkage disequilibrium (see Chapter 9) with an advantageous allele at another locus. In asexual organisms, a genotype that differs from others at many loci may replace

FIGURE 12.9 Periodic selection in laboratory populations of *Escherichia coli*. The vertical axis represents the frequency of a selectively neutral mutant marker allele compared with that of the wild-type allele. Nevertheless, the frequency of the neutral marker fluctuates. It increases if a cell bearing the marker experiences an advantageous mutation at another locus, and then decreases when a different, more advantageous mutation occurs in a cell that lacks the marker allele. In organisms such as bacteria that have little or no recombination, neutral alleles thus change in frequency by hitchhiking with advantageous mutations. A shows a population that lacks, and B one that carries, a "mutator" allele that increases the mutation rate throughout the genome. Advantageous mutations occur more frequently in the presence of the mutator allele, so that genetic changes happen more rapidly. (Note the difference in scale on the axes.) (After Nestmann and Hill 1973.)

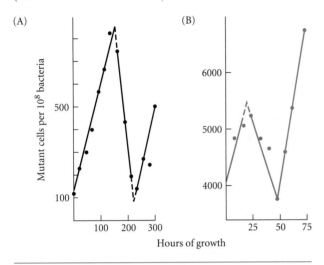

other genotypes because of an advantageous difference at just one locus. (The same process occurs, but usually to a lesser degree, in sexually reproducing species.) Thus selection occurs not only among alleles at a locus, but also among whole genotypes, or among groups of genes, that do not recombine.

These experiments illustrate two points. First, adaptation, such as the evolution of greater enzyme activity in the experiment by Dean et al., results from a difference in the rates of reproduction of different genotypes caused by a difference in some property of those genotypes (such as enzyme activity in this case). This experiment conveys the essence of natural selection, a mindless process without forethought or goal. Second, natural selection may increase the frequency of a characteristic, such as histidine synthesis in Atwood et al.'s experiment, not because the characteristic enhances survival and reproduction, but because it is correlated with another characteristic that does so.

Drosophila Natural populations of *Drosophila pseudoobscura* are highly polymorphic for inversions on the third chromosome (see Chapter 10). In his early studies of these inversions, Theodosius Dobzhansky had assumed that they were adaptively equivalent (selectively neutral), for it

seemed improbable that the arrangement of a chromosome should affect fitness. He changed his mind, however, when he observed that the frequencies of several such arrangements within natural populations displayed a regular seasonal cycle (Figure 12.10A), which implied changes in their relative fitnesses as a consequence of environmental changes, perhaps in temperature (Dobzhansky 1943). He followed these observations with several experiments using POPULATION CAGES: boxes in which populations of several thousand flies are maintained by periodically providing cups of food in which larvae develop and removing exhausted food cups.

In one such experiment, Dobzhansky (1948) used flies with Standard (ST) and Arrowhead (AR) chromosome inversions, the latter named for the locality in which it had first been found. These, like other arrangements, can be distinguished under the microscope by their banding patterns. One cage was initiated with 1119 ST and 485 AR chromosome copies, i.e., frequencies of 0.70 and 0.30. A second cage was initiated with ST and AR frequencies of 0.19 and 0.81

FIGURE 12.10 Changes in the frequencies of chromosome inversions in *Drosophila pseudoobscura*. (A) Seasonal fluctuations in the frequencies of the chromosome arrangements ST and CH in a natural population. The consistency of such cycles suggested to Dobzhansky that the inversions affect fitness. (B) Changes in the frequency of ST chromosomes in experimental populations carrying both ST and AR inversions. Their convergence toward the same frequency, irrespective of starting frequencies, shows that these inversions affect fitness, and that natural selection maintains both in a population in a stable, or balanced, polymorphism. (A after Dobzhansky 1970; B after Dobzhansky 1948.)

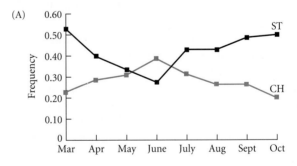

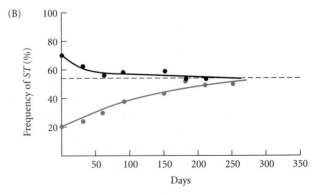

respectively. Within seven months (about 15 generations), the frequency of ST had dropped to and leveled off at about 0.54 in the first cage, and it had risen to almost the same frequency (0.50) in the second (Figure 12.10B). Later experiments showed that such changes occurred consistently in replicate populations. The critical implication of this experiment is that the chromosome frequencies approach a **stable equilibrium**, no matter what the initial frequencies are. This can only be due to natural selection, for genetic drift would not show such consistency. Moreover, natural selection must be acting in such a way as to *maintain variation* (polymorphism); it does not necessarily cause fixation of a single best genotype.

Male reproductive success The *white* mutation (*w*) in *Drosophila melanogaster* changes eye color from red to white. It is located on the X chromosome; hence females have two copies of the locus and males have one. If a population is initiated with both wild-type (+) and mutant (*w*) alleles and is propagated by moving progeny to new bottles of food each generation, the frequency of the *white* allele rapidly approaches zero (Figure 12.11). Reed and Reed (1950) found that *white* and wild-type flies had equal rates of survival from egg to adult. However, when both genotypes of virgin females (+*w* and *ww*) were confined with both male genotypes (+*Y* and *wY*), the wild-type males achieved about 25 percent more matings than did the *white* males. The difference in male reproductive success accounted for the elimination of *w* in the experimental populations.

The courting males of many species of animals have elaborate morphological features and engage in conspicuous displays; roosters provide a familiar example. Some such features appear to have evolved through FEMALE CHOICE of males with conspicuous features, which therefore enjoy higher reproductive success than less elaborate males. For example, in the long-tailed widowbird (*Euplectes progne*), an African finch that does not form permanent pair-bonds, the male has extremely long tail feathers (Fig-

ure 12.12). Malte Andersson (1982) shortened the tail feathers of some wild males and attached the clippings to the tail feathers of others, thus elongating them well beyond the natural length. He then observed that males with shortened tails mated with fewer females than did normal males, and males with elongated tails mated with more (Figure 12.12).

The calls of male frogs attract females. Males of the Central American túngara frog (*Physalaemus pustulosus*) emit a complex call that may or may not end in a "chuck" sound. Females respond preferentially to males that produce the "chuck," the frequency of which corresponds to the greatest frequency sensitivity of the female's auditory system. However, males that emit the "chuck" are also at greater risk of predation by a frog-eating bat (*Trachops cirrhosus*) that homes in on this element of the call (Figure 12.13; Ryan 1985, 1990). Thus the call is subjected to CONFLICTING SELECTION PRESSURES.

The evolutionary consequences of conflicting selection pressures are evident in the guppy (*Poecilia reticulata*), a South American fish that is popular in the aquarium trade because of the male's pattern of colorful, and very variable, spots. In Trinidad, males have smaller, less variable, and less contrasting spots in streams inhabited by their major predator, the fish *Crenicichla*, than in streams that lack this predator. John Endler (1980) moved a sample of 200 guppies from a *Crenicichla*-inhabited stream to a site that lacked the predator. About two years (15 generations) later, he found that the size of the males' spots and the diversity of color patterns had increased, so that the population now resembled those that naturally inhabit *Crenicichla*-free streams. He also set up populations in large artificial ponds in a greenhouse, with either fine or coarse gravel of several colors. After six months of population growth, he introduced *Crenicichla* into four ponds, released a less dangerous predatory fish (*Rivulus*) into four others, and maintained two control populations free of predators. In censuses after 4 and 10 generations, the number and brightness of spots per fish had increased in the ponds

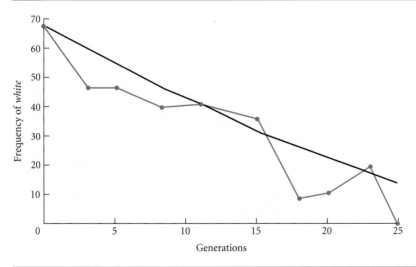

FIGURE 12.11 Elimination of the mutant allele *white* from a laboratory population of *Drosophila melanogaster*. The black line shows the decrease expected as a result of the lower mating success of white-eyed than wild-type males. (After Wallace 1968.)

FIGURE 12.12 Effects of experimental alterations of tail length on mating success in male long-tailed widowbirds, measured by the number of nests (each with one female) found in the male's territory. Nine birds were chosen for each of four treatments: shortening or elongating the tail feathers, or leaving them unaltered (in controls of two types: I, cut and repasted; II, not manipulated). A shows the number of nests per male before treatment; B shows the number of new nests built in the male's territory after treatment. Because the males retained their territories, which did not change in size, territorial interactions among males were probably not responsible for the greater number of mates acquired by longer-tailed males. Their greater success was attributed to female choice. (After Andersson 1982.)

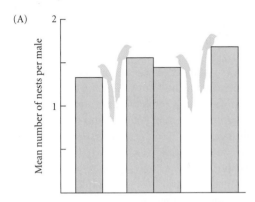

(A)

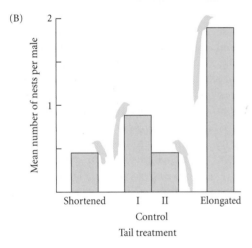

(B)

rather than adaptations for survival. We note also from these experiments that a feature may be subjected to conflicting selection pressures.

Evolution of population size in flour beetles

The small beetle *Tribolium castaneum* breeds in stored grains and can be reared in containers of flour. Larvae and adults feed on flour, but also eat (cannibalize) eggs and pupae. Michael Wade (1977, 1979b) created a genetically variable stock population (S) by crossing four inbred strains; from this stock he drew founders for populations subjected to three experimental treatments. Forty-eight populations were subjected to each treatment, each of which was propagat-

FIGURE 12.13 (A) Oscillograms of four calls of male túngara frogs, *Physalaemus pustulosus*, showing calls with zero (top) to three (bottom) "chucks." The oscillograms display the amplitude against time. Female frogs are most attracted to calls with "chucks." (B) This frog-eating bat (*Trachops cirrhosus*) preys more often on male túngara frogs that emit the chuck than on those that do not. (A after Ryan 1985; B by M. D. Tuttle, Bat Conservation International.)

(A)

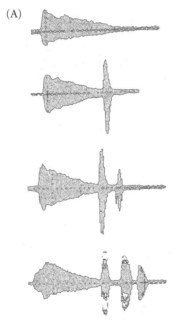

(B)

without *Crenicichla* and had declined in those with it; there were also differences in the amount of variation (Figure 12.14). In the ponds with predators, the guppies had larger spots if the gravel was coarse than if it was fine: this match to the background would make them less conspicuous. But in the absence of predators, the guppies had larger spots if on a fine-grained background and smaller spots if on a coarse-grained background, making them more conspicuous. This evolution toward more conspicuous patterns was the result of greater mating success.

These experiments show that *natural selection may sometimes lie only in differences in reproductive rate*, not survival. Differences in mating success, which Darwin called **sexual selection**, result in adaptations for obtaining mates,

346   Chapter 12

FIGURE 12.14 Evolution of male color pattern in experimental populations of guppies. The number of spots increased in control populations without predators (K) and in populations (R) to which a weak predator (*Rivulus*, which preys mostly on very young guppies) was added, but declined in populations (C) to which a predator that feeds on adult guppies (*Crenicichla*) was added. The guppy populations were initiated at time F, predators were added at time S, and coloration was measured at times I and II. The vertical bars measure the variation among males. The increase in mean spot number in K and R populations was due to female choice. (After Endler 1980.)

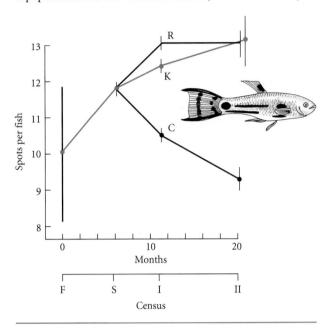

liferate based on a phenotypic characteristic of each *group*—namely, its size. This process, called **group selection** or **interdemic selection**, operates *in addition to* individual selection among genotypes within populations. We must distinguish selection *within* populations from selection *among* populations.

The first question these data raise is why population size declined in the control (C) populations, a pattern that seems like the very antithesis of adaptation. Wade was able to rule out inbreeding depression (fixation of deleterious alleles by genetic drift: see Chapter 11) as the cause, and so concluded that natural selection was somehow responsible. For populations in each treatment (including the original stock, S), Wade measured various characteristics, of which the most relevant turned out to be fecundity (number of eggs laid), the development time of each life stage, and the propensity of larvae and adults to cannibalize eggs and pupae. Compared with the foundation stock (S), the C populations had evolved in several ways: adults were more likely to cannibalize pupae; development time was more variable, which made more individuals susceptible to cannibalism because they were pupae while others were in the feeding stages; and females were prone to lay fewer eggs when confined with other beetles. Each of these features could be advantageous to an individual beetle: for example, cannibalism is a way of obtaining protein, and it may be advantageous for a female not to lay eggs if she perceives the presence of other beetles that may eat them. But although these features are advantageous to the individual, they are disadvantageous, in a sense, to the population, which declines in reproductive rate.

By selecting groups for low population size (treatment B), Wade reinforced these same tendencies. In the A treatment, on the other hand, selection at the group level opposed the consequences of individual selection within populations. Wade found that, compared with the C and B populations, beetles from treatment A had higher fecundity, even in the presence of other beetles; they developed more rapidly and were thus less susceptible to cannibalism; and they had evolved a lower propensity for adults to cannibalize eggs and pupae. Thus selection among groups had affected the course of evolution.

These experiments show that natural selection among individual organisms need not enhance the adaptedness of populations, if adaptedness is measured by population size or growth rate. They also introduce the concept of selection at two levels: among individuals and among populations.

Selfish genetic elements In many species of animals and plants, there exist "selfish" genetic elements, which are transmitted at a higher rate than the rest of an individual's genome and are detrimental (or at least neutral) to the organism (Werren et al. 1988). Many of these exhibit segregation distortion, or meiotic drive, meaning that the element is carried by more than half of the gametes of a heterozygote (see Chapter 9). For example, the *t* locus of the house mouse (*Mus musculus*), actually a complex of tight-

ed from 16 adult beetles each generation (allowing 37 days per generation). The control (C) populations were propagated simply by moving beetles to a new vial of flour: each population in one generation gave rise to one population in the next. In treatment A, Wade deliberately selected for high population size by initiating each generation's 48 populations with sets of 16 beetles taken only from those few populations (out of the 48) in which the greatest number of beetles had developed. In treatment B, low population size was selected in the same way, by propagating beetles only from the smallest populations (Figure 12.15A).

Over the course of nine generations, the populations changed markedly and surprisingly (Figure 12.15B). The average population size declined in all three treatments, most markedly in treatment B and least in treatment A. The net reproductive rate (*R*; cf. Chapter 4) also declined. In treatment C, these declines must have been due to evolution *within* each population, and if genetic change was a consequence of natural selection, it must have been due to genetic differences in survival and reproduction among *individual* beetles within each population. This process is **individual selection**, of precisely the same kind we have assumed to operate on, say, the color patterns of guppies. But in treatments A and B, Wade imposed another **level of selection** by allowing some *populations*, or groups, but not others, to pro-

Natural Selection and Adaptation   347

FIGURE 12.15 Effects of individual selection and group selection on population size in the flour beetle *Tribolium castaneum*. (A) The experimental design. In each generation, only the most (treatment A) or least (treatment B) productive populations were propagated (group selection). In treatment C, all populations were propagated, so any changes were due to natural se- lection among individuals within populations. (B) Changes in the mean number of adult beetles in the three treatments. The decline of population size due to individual selection (C) was enhanced (B) or counteracted (A) by group selection. (After Wade 1977.)

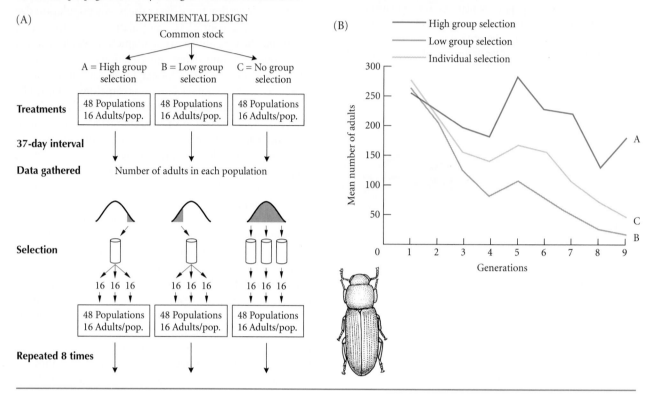

ly linked loci, has several alleles that, in a heterozygous male, are carried by more than 90 percent of the sperm. In homozygous condition, certain of these *t* alleles are lethal, and others cause males to be sterile. Despite these disad- vantages to the individual, calculations show that the "transmission advantage" of these alleles is so great that they should reach a frequency of about 0.7, and in some populations they do indeed reach a frequency of about 0.4. In seeking to explain why these alleles are not as common as they "ought" to be, Richard Lewontin (1962) calculated that the small populations into which mice are divided could easily become extinct when the alleles are "driven" to high frequency by meiotic drive;* the extinction of whole populations with *t* alleles (selection among groups) would lower their frequency in the species as a whole.

In *Drosophila pseudoobscura*, an element on the X chro- mosome causes the male to father only daughters because the Y-bearing products of meiosis fail to develop, so that only X-bearing sperm are transmitted. This element should result in all-female populations and consequently extinc- tion; so far, it is not known what prevents it from becom- ing common enough to cause this. In the parasitic wasp *Nasonia vitripennis*, a small chromosome called "paternal sex ratio," or *psr*, is transmitted mostly through sperm rather than eggs; when an egg is fertilized, this genetic ele- ment causes the destruction of all the other paternal chro- mosomes. In *Nasonia*, as in all Hymenoptera, diploid eggs become females and haploid eggs become male. The *psr* el- ement thus converts female into male eggs, thus ensuring its own future propagation through sperm. Perhaps the most extreme such "selfish" effect is exerted by the *Medea* element that is common in the flour beetle *Tribolium cas- taneum* (Beeman et al. 1992). This transposable element (see Chapter 3) can occupy many different positions on the chromosomes. If a mother carries a *Medea* element, all her offspring die upon hatching, unless they have inherited a copy of *Medea* at the same chromosome site.[†] (This is an ex- ample of a maternal effect.) This obviously has the effect of increasing the frequency of *Medea* in the population.

---

*There is no direct evidence that this occurs, but it is known that mice discriminate against +/*t* individuals when choosing mates (Lennington and Egid 1985). Thus it is not clear that Lewontin's explanation is correct, but his theory is sound, and shows how in- crease of a selfish genetic element could, in principle, extinguish a population.

[†]In Greek mythology, Medea, spurned by Jason, killed their chil- dren in revenge.

If natural selection is defined by differences in survival and reproduction, then selfish genetic elements provide another example of different levels of selection: in these cases, selection among *genes* acts in opposition to selection among *individual organisms*. We note further that selection among genes may not only be harmful to individual organisms, but might also cause the extinction of populations or species.

The Nature of Natural Selection

Definitions of Natural Selection

The examples presented above show that selection can take many forms. Consequently, many definitions of natural selection have been proposed (Endler 1986); it is for this reason that we have not yet provided a definition in this text. Most authors agree that the definition must include the following concepts: some attribute or trait must *vary* among biological entities, and *there must be a consistent relationship*, within a defined context, *between the trait and one or more components of reproductive success*, where "reproductive success" includes both survival (a prerequisite for reproduction) and the reproductive processes themselves. An entity's reproductive success is its **fitness**, defined as the average per capita rate of increase. If the entities are different classes of individual organisms or genes (such as genotypes or alleles), then fitness is usually measured as the mean number of descendants, per individual organism or gene copy, counted as newly produced offspring (fertilized egg or newborns) after one generation. (Thus fitness is usually measured by the entity's rate of increase, R or r, as defined in Chapter 4; see also Chapter 13. Occasionally, it is useful to measure fitness by counting descendants after two generations rather than one; see Chapter 21.)

Some authors treat sexual selection as a process distinct from natural selection, and restrict natural selection to differences in survival and fertility. More commonly, sexual selection is considered a kind of natural selection, and will be so considered in this book.

For selection to exist, there must be average differences in reproductive success among different classes of entities. Evolutionary biologists differ on whether or not the definition of selection should require that the classes differ genetically. Some authors, such as Russell Lande and Stevan Arnold (1983), define selection as a consistent difference in fitness among *phenotypes*, acting within a single generation. Whether or not it alters the frequencies of phenotypes in the next generation depends on whether and how the phenotypic differences are inherited. The change in the population from one generation to another is termed the **response to selection**. Authors who advocate this phenotypic definition distinguish the response, which is solely a matter of inheritance, from differences in survival and reproduction, which constitute selection itself. Thus the experiment on widowbird phenotypes demonstrated selection on tail length, but provided no information on the response to selection, because nothing was discovered about the inheritance of this fea-

ture. This definition emphasizes that *selection acts on phenotypes, but may change allele and genotype frequencies if the phenotypes differ in genotype*. Genetically identical members of an asexual clone may differ in phenotype, perhaps because of environmental influences, and in reproductive rate, but the phenotype frequencies in the next generation will not be altered unless the distribution of environmental effects has changed. According to Lande and Arnold, there is selection in this case, but no response to selection, and no evolution.

Biologists have more commonly included the genetic response to selection in the definition. For example, one of the founders of population genetics, Sewall Wright (1969), defined selection as "any process in a population that alters gene frequency in a directed fashion without change of the genetic material (mutation) or introduction from without (immigration)." John Endler (1986) defined natural selection as a process in which (to paraphrase Endler), if a population exhibits (*a*) variation in a trait, (*b*) a consistent relationship between the trait and fitness, and (*c*) inheritance of the trait, then the frequency distribution of the variations (1) will differ among age classes and (2) may differ between generations. Parts *a* and *b* of this definition describe differences among phenotypes; parts 1 and 2 describe their consequences (response to selection), mediated by inheritance (*c*).

Some authors include under natural selection only selection at the level of genes, genotypes, and individual organisms, excluding selection among groups (such as populations or species). However, many authors (e.g., Endler 1986; Sober 1984) include all these entities within natural selection, which then may operate at a variety of levels, ranging from the gene to the species. We shall follow this convention.

For our purposes, we will define natural selection as *any consistent difference in fitness* (i.e., survival and reproduction) *among phenotypically different biological entities*. The entities may be individual genes (which must have some phenotypically variable property if they differ consistently in fitness), groups of genes, individual organisms, populations, or taxa such as species. (Although we have adopted a phenotypic definition, we will almost always discuss the fitness of phenotypes that are inherited to at least some degree, because selection has no evolutionary effect unless there is inheritance.)

Natural selection and chance A critical feature of our definition is that selection operates only if phenotypes differ *consistently* in fitness. If one neutral allele replaces another in a population by genetic drift (see Chapter 11), the bearers of this allele in this population have had a greater rate of increase than the bearers of the other, but natural selection has not occurred, and the genotypes are not considered different in fitness, because the allele does not consistently confer higher rates of survival or reproduction: in any generation, one of the alleles will increase in frequency in about half of a number of replicate populations, and the other allele will increase in the other half. There is no *average* difference between them, *no bias* toward increase of one relative to the other. Fitness differences, in contrast, are *average* differences, *biases*, differences in the *probability* of reproductive success. This does not mean, of course, that every individual of a fitter genotype (or phenotype) survives and reproduces prolifically, while every individual of an inferior genotype perishes; a great deal of mortality and variation in reproductive rate occurs independent of—that is, at random with respect to—phenotypic differences. Thus the difference in fitness among phenotypes is the difference that is *not* due to chance, but is *caused* by some characteristic difference between them. Therefore natural selection is the difference in rates of increase among biological entities that is *not* due to chance. *Natural selection is the antithesis of chance.*

If fitness and natural selection are defined by consistent, or average, differences, then we cannot tell whether a difference in reproductive success between two *individuals* is due to chance or to a difference in fitness. We cannot say that one identical twin had lower fitness than the other because she was struck by lightning (Sober 1984), or that the genotype of Pyotr Tchaikovsky, who had no children, was less fit than the genotype of Johann Sebastian Bach, who had many. Fitness cannot be measured for an individual gene, organism, or population, but only as the average of some number of like genes, organisms, or populations. The biologists who performed the experiments described above could ascribe genetic changes to natural selection rather than genetic drift because they observed consistent changes in replicate populations, or measured numerous individuals of each phenotype and found an average difference in survival or mating success.

Selection of and selection for   The philosopher of science Elliot Sober (1984) uses a child's "selection toy" (Figure 12.16) to make some useful conceptual points. Balls of several sizes, placed in the top compartment, fall through holes in partitions, the holes in each partition being smaller than in the one above. The toy thus selects small balls. Natural selection may similarly be considered a sieve that selects organisms with a certain body size, mating behavior, or other feature. If the small balls in the toy are green, and the larger ones have other colors, the toy will select small, green balls. Sober emphasizes, however, that we should distinguish *selection of objects* from *selection for properties*. The objects selected are the balls, but they are selected *for* the property of small size; i.e., *because of* their small size. They are not selected for their color, or because of their color, but nonetheless there is selection *of* green balls. A biological analogy is provided by the hitchhiking effect in the experiment by Atwood et al.: in the populations of *E. coli*, there was selection *of* the capacity for histidine synthesis, but there was selection *for* linked advantageous mutations at other loci, not for histidine synthesis itself. The property of histidine synthesis was not the cause of natural selection of genotypes with this property. Thus the evolution of histidine synthesis was an **effect** of natural selection, but not its cause, just as uniform green color is

FIGURE 12.16   A child's toy that selects small balls, which drop through smaller and smaller holes from top to bottom. In this case there is selection *of* darkly shaded balls, which happen to be the smallest, but selection *for* small size. (After Sober 1984.)

an effect of selection in the toy, but not a cause of the prevalence of green balls.

These semantic points are more important than they may seem. When we speak of the FUNCTION of a feature, we imply that there has been natural selection *of* organisms with the feature and *of* genes that program it, but *for* the feature itself. We suppose that the feature *caused* its bearers to have higher fitness. The feature may, however, have other *effects*, or consequences, that were not its function, and *for* which there was no selection. For instance, there was selection for an opposable thumb and digital dexterity in early hominids, with the effect, millions of years later, that we can play the piano. There was selection for a large brain and a great capacity for learning, for reasons on which we can only speculate, with the effect that we can do calculus and invent computers. There is selection for more cryptic coloration in a population of guppies exposed to predation, and an *effect* of this might well be a lessening of the likelihood that the population will become extinct, but *avoidance of extinction is not a cause of evolution* in the guppy population. *There has not been selection for avoidance of extinction;* avoidance of extinction is *not the function* of the guppies' coloration, nor, probably, of any feature of any organism.

Levels of Selection

Selection of organisms and groups   It is common to read, in student essays and even in professional biological literature, statements to the effect that clams have a high reproductive rate "to ensure the survival of the species," or that antelopes with sharp horns engage in ritual displays rather than physical combat because combat would lead to the species' extinction. These naive statements betray a misunderstanding of natural selection.

In 1962, the ecologist V. C. Wynne-Edwards placed a provocative interpretation on many social behaviors of animals, such as the flocking behavior of starlings and choruses of male frogs. He proposed that these behaviors had evolved as mechanisms of population control to prevent populations from exhausting food or other resources. These aggregations form, he suggested, so that individuals can assess the population's density and reduce their reproductive rates accordingly. Thus populations are self-regulatory as a consequence of individual reproductive restraint.

Wynne-Edwards realized that neither reproductive restraint nor behaviors promoting it could evolve by natural selection as Darwin conceived it—that is, by selection among individual organisms. A genotype practicing reproductive restraint, amid genotypes that did not, would necessarily decline in frequency, simply because it would leave fewer offspring, per capita, than the others. Likewise, if a population were to consist of genotypes that practiced reproductive restraint, a mutant that did not—a "cheater"—would increase to fixation, even if the consequent high population growth were later to cause exhaustion of resources and extinction of the population. Thus an ALTRUISTIC TRAIT, a feature that reduces the fecundity or increases the likelihood of mortality of an individual for the benefit of the population or species, *cannot evolve by individual (organismal) selection.*

Wynne-Edwards therefore proposed that reproductive restraint had evolved by group selection. Populations made up of genotypes that do not practice restraint should have a higher extinction rate than populations made up of genotypes that do. Therefore the species as a whole should evolve restraint through the differential survival of groups with lower average reproductive rates, even though individual selection would act in the opposite direction (Figure 12.17A).

Wynne-Edwards's ideas elicited critiques from several authors. The most influential of these was *Adaptation and Natural Selection* by George Williams (1966). Williams considered not only the behaviors described by Wynne-Edwards, but also many other features of organisms that other authors, implicitly or explicitly, had supposed benefited the species at a cost to the individual: for example, senescence and fixed life spans, which were supposed to alleviate population density and make resources available for the next generation. Williams called characteristics that supposedly benefit the whole population or species BIOTIC ADAPTATIONS, and those that benefit the individual organism ORGANIC ADAPTATIONS. He argued forcefully that biotic adaptations simply do not exist: either the feature in question is not an adaptation at all, or it can be plausibly explained by individual selection and is really an organic adaptation. For example, females of many species do indeed lay fewer eggs when population density is high, but this is not restraint for the good of the species. At high densities, when food is scarce, a female simply cannot form as many eggs, so the reduction in fecundity is a physiological neces-

FIGURE 12.17 Conflict in the evolution of traits advantageous to the individual versus the population. Each circle represents a population of a species, traced through four time periods. New populations are founded by colonists from established populations, and some populations become extinct. The dark and light areas in each circle represent the proportions of an "altruistic" genotype (dark) and a "selfish" (light) genotype, the latter having a higher reproductive rate (individual fitness). Lateral arrows indicate gene flow between populations. *Within* each population, the selfish genotype increases in frequency by individual selection, but extinction of a population is more likely, the higher the frequency of the selfish genotype. (A) Wynne-Edwards assumed that altruistic behavior, such as reproductive restraint, would evolve because group selection by differential extinction vs. survival of populations is strong. (B) Williams argued that genotype frequencies within populations would change much faster than rates of extinction and founding of whole populations, so that the selfish genotype would become fixed.

(A)   (B)

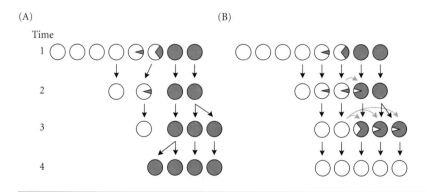

sity, not an adaptation. Alternatively, an individual female may indeed be more fit if she forms fewer eggs in these circumstances and either disperses to a richer area or waits until food becomes more abundant, so that her subsequent offspring will have a greater chance of survival. The net reproductive rate of a genotype that reproduces only when food is plentiful may be higher than that of a genotype that reproduces even when food is sparse, when most offspring will starve. Such individual selectionist hypotheses, proposed by Williams (1966), William Hamilton (1964), and others, gave rise to a major new field of evolutionary study, the analysis of ecological, behavioral, and reproductive adaptations (see Chapters 19, 20, and 21).

Williams based his opposition to group selection on a simple argument. Individual organisms are much more numerous than the populations into which they are aggregated, and they turn over—are born and die—much more rapidly than populations, which are formed (by colonization) and become extinct at relatively low rates. Selection at either level requires differences—among individuals or among populations—in rates of birth or death. The rate of replacement of less fit by more fit individuals is potentially much greater than the rate of replacement of less fit by more fit populations. (Here, the fitness of a population may be thought of as its probability of survival per unit of time.) Therefore, if individual selection is opposed to group selection, individual selection will prevail (Figure 12.17B). By analogy, if I walk down an up escalator (or moving stairway), I will reach the bottom only if I walk faster than the escalator moves. No one denies that group selection, if sufficiently intense, might prevail—Wade (1977) showed this by imposing strong group selection on populations of *Tribolium*—but many, perhaps most, evolutionary biologists believe that it is only rarely an important force of evolution. Hence the majority view is that *few characteristics have evolved because they benefit the population or species.* We shall, however, encounter a few examples of traits that may have evolved by group selection, and some evolutionary biologists argue that group selection is more important than is generally believed (Wilson 1983).

Species selection Consider two groups of species—they might be two related clades—that differ in several characters. For example, the orchids (Orchidaceae) generally have floral characters (scents, flower forms) that promote pollination by specialized insects that differ among orchid species, and the stalk of the flower is twisted. Their relatives, the lilies (Liliaceae), have straight stalks and are usually pollinated by less specialized insects. Suppose, as is likely to be the case, that evolutionary changes in flower form that induce pollination by different species of insects are more likely in the orchids than in the lilies, so that old and new flower types are more likely to be reproductively isolated. The rate of speciation might therefore be greater in orchids than in lilies, and the number of species of orchids might grow more rapidly. Considering both families together, we would expect, over evolutionary time, a

change in the proportion of species with modified flowers (e.g., a labellum: see above) and twisted stalks. The average state of species in the clade as a whole would change because of this difference in the "birth rate" (speciation rate) of new species with one or another feature, analogously to a change in the proportions of different phenotypes within a population that differ in birth rate (Figure 12.18). Differences in extinction rates among sets of species with different characteristics similarly could change average phenotypes, as does mortality-based selection within species. Group selection in which each group is a species has been called **species selection** by Steven Stanley (1975, 1979). It has been invoked especially by some paleontologists, particularly advocates of punctuated equilibrium, to explain features of the fossil record. Williams (1992b) prefers to call it "taxon selection," because the units that proliferate at different rates could be taxa at any level.

In the orchid/lily example, the proportion of species with specialized pollination increases because specialized pollination causes a higher speciation rate. Selection among species selects *for* specialized pollination. The proportion of species with twisted stalks increases, but only as an inci-

FIGURE 12.18 Two bases for species selection (or lineage selection), i.e., differential proliferation of species with different character states. The horizontal axis represents a character, such as body size. (A) Differential speciation: Lineages toward the right of the phylogeny have higher rates of speciation, analogous to higher birth rates of individual organisms, than those toward the left. (B) Differential extinction and survival: Lineages toward the right have longer survival times than those toward the left, analogous to higher survival rates of individual organisms. In both cases, the character value, averaged across species, is greater at time B (upper dashed line) than at time A (lower dashed line). If the character—say, body size—causally affects speciation or extinction rates, many authors would call this an illustration of species selection for large size. (After Gould 1982a.)

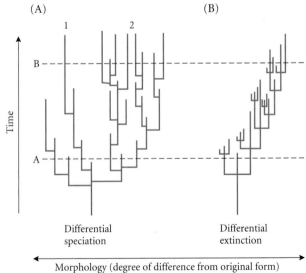

dental consequence of the correlation between stalk structure and mode of pollination. There has been selection *of*, but not selection *for*, twisted stalks. The increasing incidence of twisted stalks among these plant species is an *effect* of a fortuitous association with speciation rate.

The problem of causation

Suppose, said Williams (1966), that a species of deer is divided into herds (groups), some composed of fleet individuals and some composed of slow individuals. Predators such as wolves extinguish groups of slow deer at a higher rate, so the average degree of fleetness in the deer species as a whole increases. Is this group selection, because slow herds suffer a higher extinction rate? Or is it individual selection, because a slow herd is merely a herd of slow deer? Williams opted to call it individual selection, because the fate of the herd arises simply from the summation of the effects of the characteristics of the individuals that compose it. Similarly, one might argue that rapid proliferation of plant species with specialized pollination is caused not by any property of those species, but by the floral features of individual plants. In each case, events at a higher level (group, species) can be REDUCED to, or explained by, properties at a lower level (individual organisms).

A voluminous philosophical literature has emerged on the question of how to describe and identify selection at different levels (e.g., Vrba 1984; Vrba and Eldredge 1984; Maynard Smith 1976; Sober 1984; Williams 1992a). Some authors (Williams 1992b) say that selection operates at a given level if entities are sorted at that level by differences in the rate of increase of entities at that level. Thus the sieve of speciation rate sorts species with specialized pollination from those without it, tends to accumulate the former, and represents an instance of selection at the level of species. The other authors listed above argue that selection operates at a given level only if differences in proliferation rates are *caused by* characteristics specific to entities at that level; if the properties of higher-level entities can be reduced to properties at a lower level, selection should be said to operate at the lower level. For example, Elisabeth Vrba (1980) proposed that ecologically specialized species may speciate, and so grow in number, faster than generalized species. She argued, however, that this is not species selection for specialized habits, but rather a by-product—an effect—of the habits of the individual organisms. She proposed EFFECT HYPOTHESIS as a name for this concept of nonrandom differences in species diversification caused by characteristics of individual organisms.

Features of groups or species that are "emergent" at that level, and are not simple summations of individual organisms' features, are few. They might include the abundance, sex ratio, or genetic variability of a population, since none of these can be described for an individual organism—but even these are arguably organismal features in aggregate. Those who adopt the more stringent definitions of group selection and species selection—nonrandom proliferation caused by differences in group-level and species-level properties—consider these concepts to have very narrow domains of application. The elimination by predators of slow herds of deer but not fleet herds would not constitute group selection, because the speed of the individual deer, rather than their membership in one herd or another, determines their fate. But a greater extinction rate of species with little genetic variation than of more genetically variable species might qualify as species selection.

Reductionism and the selfish gene

Similar philosophical questions of causation and definition apply to selection at the levels of gene and organism. An increase in the frequency of a "selfish" genetic element, such as a *t* allele in the mouse or *Medea* in flour beetles, is evidently due to **genic selection**: selection of an allele because of its capacity for distorting segregation, irrespective of its effect on the organism that bears it. But if an allele increases in frequency because it enhances the organism's survival or fecundity, shall we call that genic selection or individual selection? Again, opinions differ.

Consider an instance of heterozygous advantage (see Chapter 13), such as the case of human hemoglobins (Sober 1984). In parts of Africa, homozygotes for the "normal" hemoglobin allele (*AA*) suffer high mortality from malaria. Heterozygotes for the sickle-cell allele (*AS*) have slight anemia, but have a higher rate of survival because they are resistant to malaria. Homozygotes for the sickle-cell allele (*SS*) have sickle-cell disease, and die at a very high rate before reproduction. Thus the fitnesses of the genotypes are $AS > AA > SS$. These fitnesses result from the physiological effects of these *combinations* of alleles—combinations that exist only in organisms. It is mathematically possible to calculate, as an abstract quantity, the average fitness of each of the alleles *A* and *S*. This value is the average of the fitness values of the two genotypes in which the allele occurs, weighted by the frequency with which the allele occurs in each genotype. Some authors view evolution by natural selection as a competition among alleles for representation in the gene pool, and would use these abstract allele fitnesses to predict the course of evolution. For these authors, the gene is the unit of selection (Dawkins 1989). Other authors hold that in cases such as these hemoglobins, evolution is most readily understood and predicted as a consequence of the fitness of individual organisms, or more precisely, of genotypes: combinations of alleles at one or more loci. They distinguish genic selection from individual selection, and attribute most adaptation to individual selection (Sober 1984).

Authors who view the gene as the unit of selection emphasize that the outcome of natural selection depends on which allele has the higher rate of survival and reproduction, *averaged* over all the genetic contexts (and ecological circumstances) in which it occurs. Richard Dawkins (1989) uses the analogy of a coach who selects a team for a crew race by repeatedly shifting oarsmen among several boats; after several trials, the winning boat will tend to have the same oarsmen. A member of the crew finally chosen will have been in the company of good oarsmen at some times and of inferior ones at others, but on average his performance has contributed more to winning the trials than that of an individual who is not chosen. Likewise, says Dawkins, natural selection within populations can be understood simply as competition among alleles, the winner being the one that confers some characteristic on organisms that provides that allele with the highest rate of survival and reproduction, averaged over all the gene combinations in which the allele occurs. There is selection of genes for their ability to create organisms that enhance the genes' survival. In this view, just as a chicken is said to be an egg's way of making more eggs, organisms are the genes' "survival mechanisms," within which the genes "swarm in huge colonies, safe inside gigantic lumbering robots" (Dawkins 1989). All genes are "selfish," manipulating their bearers in ways that enhance the genes' survival.

This perspective can provide insight, but it is criticized as too reductionist by those who believe that selection at the level of organisms is a better way to describe natural selection (Sober and Lewontin 1982; Sober 1984). Genes interact so that the properties of organisms—genotypes—are not the simple sum of their genes' effects. The fitness of the sickle-cell heterozygote cannot be predicted, they say, solely from knowledge of the individual genes' activities. The allele frequencies change *because of the phenotypic effects of genotypes*, so it is at this level that natural selection operates. It is possible to calculate the fitnesses of individual alleles, but only as a mathematical abstraction from measurements of the fitnesses of organisms—and these abstract quantities play no causal role. There is selection *of* genes, but only as an effect of selection *for* resistance to malaria or other phenotypic properties of organisms. There is selection *of* crew oarsmen, but the coach selects *for* winning races—a property of the team.

If, then, our concept of levels of selection includes causality, natural selection can act at the level of gene (as in meiotic drive), organism, and at least in principle, population and species.

Recognizing Adaptations

Definitions of Adaptation

The word *adaptation* has three meanings in biology. Physiologists sometimes use it to describe an individual organism's phenotypic adjustments to the environment, such as physiological acclimation to temperature. We will not employ this usage. In evolutionary biology, the word can refer either to the *process* of becoming adapted (as in "adaptation cannot occur without genetic variation") or to the *features* of organisms that enhance reproductive success (including survival) relative to other possible features. Even in this last sense, however, adaptation is a complex concept, with several possible definitions (Krimbas 1984; Coddington 1988; Brandon 1990; West-Eberhard 1992; Reeve and Sherman 1993). All authors agree that an adaptive trait enhances fitness compared with at least some alternative traits. How-

ever, some authors include a historical perspective in their definition, and others do not.

Perhaps the broadest ahistorical definition is provided by Reeve and Sherman: "An adaptation is a phenotypic variant that results in the highest fitness among a specified set of variants in a given environment." This definition refers only to the current effects of the trait on reproductive success, compared with those of variant traits that might be found in the species (or in related species), or which might be devised by experimental manipulation, or which might only be imagined. (Imagine what the fitness of a toothless horse might be, and you may conclude that teeth are an adaptation for feeding in horses.)

At the other extreme, Harvey and Pagel (1991) hold that "for a character to be regarded as an adaptation, it must be a derived character that evolved in response to a specific selective agent." This definition explicitly requires an inference about history, namely, that the feature evolved by natural selection for a specific function. Furthermore, we can conclude that the feature is an adaptation only by comparing its effects on fitness with those of a specific variant, namely, the ancestral character state from which the currently prevalent character state evolved. This comparison requires knowing what the ancestral state was.

One reason for this emphasis on history is that the condition of a trait may be a consequence of phylogenetic history rather than an adaptation. Winglessness in fleas, which had winged ancestors, may be an adaptation, but winglessness in bristletails (Microcoryphia; cf. Chapter 5) is not: they are primitively wingless, none of their ancestors having ever had wings. Darwin saw clearly that a feature might be beneficial, yet not have evolved for the function it serves today, or for any function at all: "The sutures in the skulls of young mammals have been advanced as a beautiful adaptation for aiding parturition [birth], and no doubt they facilitate, or may be indispensable for this act; but as sutures occur in the skulls of young birds and reptiles, which have only to escape from a broken egg, we may infer that this structure has arisen from the laws of growth, and has been taken advantage of in the parturition of the higher animals" (*On the Origin of Species*, Chapter 6). The lack of petals in the poinsettia, in which leaves are modified to attract pollinators (see Chapter 5), is not an adaptive feature of the poinsettia, but a primitive characteristic of its genus, *Euphorbia*, in which, to be sure, the original loss of petals might have been adaptive, but which is perhaps now constrained from re-evolving petals even if they would be useful. Thus before declaring that a feature is an adaptation for its present function, it is wise to ask from what ancestral state it evolved, and whether its evolution from that state was caused by natural selection for the function it now serves (Coddington 1988).

Because traits evolve from pre-existing traits, the features a species can evolve depend strongly on its phylogenetic position and history, and these potentialities differ strongly among lineages. Knowing that beak length varies considerably within and among species of Galápagos finches, it makes sense to ask whether there is an adaptive reason for *Geospiza fortis* to have an average beak length of 11.5 mm, rather than, say, 8.1 mm as in *Geospiza fuliginosa* or 14.5 mm as in *Geospiza scandens*, which occupy the same island (Grant 1986). But it is not sensible to ask whether it is adaptive for *Geospiza* to have four toes rather than five, like the rats and iguanas on the island, because the ancestor of birds lost the fifth toe and it has never been regained in any bird since. Five toes are probably not an option for birds because of genetic developmental constraints. Thus if we ask why a species has one feature rather than another, the answer may be adaptation, or it may be phylogenetic history.

A **preadaptation** is a feature that fortuitously serves a new function. For instance, parrots have strong, sharp beaks, used for feeding on fruits and seeds. When domesticated sheep were introduced into New Zealand, some were attacked by an indigenous parrot, the kea (*Nestor notabilis*), which pierced the skin and fed on the sheep's fat. The kea's beak was fortuitously suitable for a new function, and may be viewed as a preadaptation for slicing skin. It did not evolve in anticipation of the new function—no foresight was involved—but it was constructed in such a way that this bird was more capable of adopting a carnivorous habit than, say, a hummingbird, the beak of which could not possibly penetrate skin. The coelurosaur dinosaurs to which *Archaeopteryx* is related had hollow bones and fused clavicles (the "wishbone"). These features provided the skeleton with lightness and structural strength without which the evolution of flight would have been less likely (Ostrom 1976), and may be considered preadaptations for flight.

Stephen Jay Gould and Elisabeth Vrba (1982) suggest that if an adaptation is a feature evolved by natural selection for its current function, a different term is required for features that, like the hollow bones of birds or the sutures of a young mammal's skull, did not evolve because of the use to which they are now put. They suggest that such characters that evolved for other functions, or for no function at all, but which have been co-opted for a new use be called **exaptations**. The same feature that is now an exaptation is a preadaptation, say Gould and Vrba, when in the hindsight of history we view the feature in the ancestor before it was co-opted for a new function. Thus hollow bones in coelurosaurs are a preadaptation, but also a potential exaptation, for flight.* Useful features therefore include both adaptations and exaptations, which together constitute "aptations": features that are fit (Latin *aptus*) for their current role. An exaptation may be further modified by selection, so that the modifications are adaptations for the feature's new function; for example, the tail feathers of birds doubtless originally evolved as an adaptation for flight, but in widowbirds may be viewed as an exaptation for sexual display, for which they have been modified.

If we take the concept of exaptation to an extreme, a feature that has persisted unchanged from a remote ancestor

to each of many descendant species might not be considered an adaptation in the descendants, but rather a consequence of their phylogenetic heritage. In doing so, we would be led to deny that many functional characteristics, such as the feathers and hollow bones of birds, the teeth of horses, or the eyes of vertebrates, are adaptations. Advocates of a less historical definition of adaptation argue that in most such cases, the character is retained in each descendant species only because natural selection maintains it by ridding populations of mutations that impair its function. Whatever the origin of the vertebrate eye may have been, they would say, we can show that it is an adaptation in humans and birds, not by comparing these taxa with a hypothetical eyeless ancestor of the vertebrates, but by comparing the functional effects of "normal" eyes with those of variant and "degenerate" eyes, such as those of certain cave fishes.

Perhaps no definition of adaptation will satisfy everyone, but for our purposes, we will conclude that *a feature is an adaptation for some function if it has become prevalent or is maintained in a population (or species, or clade) because of natural selection for that function.*

Nonadaptive Traits

Given this definition of adaptation, it is evident that a given trait might *not* be an adaptation, for any of several reasons. In some cases we must be very cautious in defining the trait in question.

1. The trait may be a necessary consequence of physics or chemistry. Williams (1966) gives the example of flying fishes (Exocoetidae; see Figure 9 in Chapter 23), which leap into the air to escape predators and glide (not fly) some distance before falling back to the water. A flying fish could not live indefinitely in air, but its return to the water is not an adaptation for respiration; it is a simple effect of gravity. However, the behavior of leaving the water, and the large size of the pectoral fins that enable it to glide, are probably adaptations for escape from predators. In the same vein, hemoglobin gives blood a red color, but there is no reason to think that redness is an adaptation: it is a by-product of the structure of hemoglobin.

2. The trait may have evolved by random genetic drift rather than by natural selection. For example, an allele coding for a β-galactosidase enzyme in *E. coli* might be substituted by genetic drift for a functionally equivalent allele that differs by one amino acid. The new allele is not an adaptation relative to the old allele. But β-galactosidase, irrespective of which amino acid occupies that one position, is undoubtedly an adaptation, shaped by selection from some antecedent protein for the function of metabolizing lactose. Likewise, the complex cryptic color pattern of grouse chicks (Figure 12.19) is probably an adaptation for evading predation; but the patterns of different species, perhaps all equally cryptic, might have di-

---

*For etymological reasons, Gould and Vrba suggest that the term "preadaptation" be replaced by "preaptation."

verged by genetic drift, so the particular pattern of, say, the sage grouse might not be an adaptation relative to the ancestral pattern from which it evolved. Winglessness in fleas and other flightless insects that evolved from winged ancestors might be an adaptation for reallocating energy from unused wings to, say, egg production; or it might not be an adaptation, if winged and wingless variants in the ancestors of these insects did not differ in fitness, and alleles causing winglessness were fixed by genetic drift.

3. The feature may have evolved not because it conferred an adaptive advantage, but because it was correlated with another feature that did. Genetic hitchhiking, exemplified in the bacterial experiment by Atwood et al. described above, is one cause of such correlation. Pleiotropy, the phenotypic effect of a gene on multiple characters, is another. It is often expressed as correlations in the development of different features or dimensions of an organism. If small species of deer have smaller antlers than large species, this might merely be due to organism-wide responses to hormones and other growth-promoting factors, not to an adaptive advantage of small antlers per se.

4. As we have discussed, the condition of a trait in a species may be a consequence of its phylogenetic history (e.g., winglessness in primitively wingless insects such as bristletails). A related possibility is that the feature has not yet become adaptively altered in response to a recently changed environment. For example, the large fruits of many tropical trees seem to be adapted for dispersal by large mammals that became extinct in the Pleistocene (Janzen and Martin 1982).

Methods for Recognizing Adaptations

Recognizing the many factors other than adaptation that may account for an organism's features, Williams (1966) wrote that although "adaptation is often recognized in purely fortuitous effects, and natural selection is invoked to resolve problems that do not exist," we should consider adaptation "an onerous concept that should be used only where it is really necessary." That is, we should not suppose that a feature is an adaptation unless the evidence favors this interpretation. This is not to deny that a great many of an organism's features, perhaps the majority, evolved as adaptations, either in that species or in its ancestors.

Several methods are used to infer that a feature is an adaptation for some particular function. We shall note these only briefly and incompletely at this point, exemplifying them more extensively in later chapters.

Complexity Even if we cannot immediately guess the function of a feature, *we often suspect it has an adaptive function if it is complex*, for complexity cannot evolve except by natural selection. The histological complexity of the lateral line system of fishes, and its uniformity within many clades, implied to physiologists that it had an adaptive function, which after extensive study proved to be sensation of differences in water pressure. When a biochemist finds a new kind of protein, she assumes from its complexity that it has a function, even though individual variants of the protein might not differ adaptively. A peculiar, highly vascularized structure called a pecten projects in front of the retina in the eyes of birds (Figure 12.20). Although its function is uncertain, it is generally assumed to play some important functional role, both because of its complexity and because it is ubiquitous among bird species.

Design The function of a morphological structure is often inferred from its *correspondence with the design* an engineer might use to accomplish some goal, such as locomotion or the dissipation or retention of heat; indeed, many structures are analogous in design to human implements. The fields of physiology and functional morphology are devoted to analyzing design and experimentally testing for function (see Chapter 17). For instance, in hot environments, plants with large leaves often have leaves that are finely divided into leaflets, or torn along fracture lines. This structure reduces the leaf's temperature because the thin, hot "boundary layer" of air at the leaf surface is more readily dissipated by wind passing over a small than over a large surface. Among related species or conspecific populations of mammals and birds, those in cold environments often are larger in body size (Bergmann's rule) and have shorter ears and legs (Allen's rule), both of which, by

FIGURE 12.19 Plumage patterns of newly hatched grouse and ptarmigans (Tetraoninae). These down patterns are replaced within a few days by a juvenile plumage. The patterns doubtless provide cryptic protection, but whether or not the differences among species are adaptive is not known. (A) Sage grouse (*Centrocercus urophasianus*); (B,b) blue grouse (*Dendragapus obscurus*); (C,c) sharp-winged grouse (*Dendragapus falcipennis*); (D) spruce grouse (*Dendragapus canadensis*); (E,e) rock ptarmigan (*Lagopus mutus*); (F,f) willow ptarmigan (*Lagopus lagopus*); (G,g) white-tailed ptarmigan (*Lagopus leucurus*). (Reprinted from *Grouse of the World* by P. A. Johnsgard by permission of the University of Nebraska Press. Copyright 1983 by the University of Nebraska Press.)

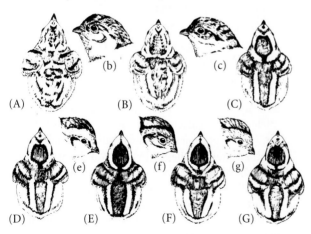

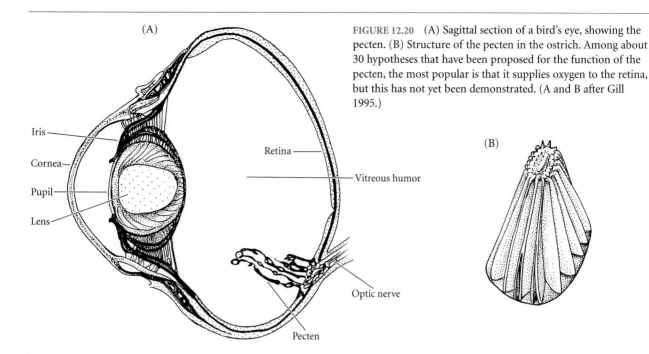

(A)

Iris

Cornea

Pupil

Lens

Retina

Vitreous humor

Optic nerve

Pecten

(B)

FIGURE 12.20 (A) Sagittal section of a bird's eye, showing the pecten. (B) Structure of the pecten in the ostrich. Among about 30 hypotheses that have been proposed for the function of the pecten, the most popular is that it supplies oxygen to the retina, but this has not yet been demonstrated. (A and B after Gill 1995.)

decreasing the ratio of surface area to body mass, reduce the rate of loss of body heat (see Chapter 9).

Very frequently, MODELS are devised to describe the kinds of features organisms might be expected to have, to achieve a specified function in a particular context. Such models are often used to *predict* features of organisms that have not yet been examined in detail, or to *explain* why a feature has one form rather than another. For example, mathematical models, based on principles of engineering, explain why, if fleetness is advantageous, horses should have evolved the one-toed condition (a single bone has greater resistance to bending stresses than several bones of the same cross-sectional area and weight). Both mathematical and verbal models are used to predict or explain adaptive features of all kinds, ranging from biochemistry to behavior.

In all such instances, however, we should ask whether the feature evolved as an adaptation in the species we are examining, or is an ancestral feature that might have evolved for a different reason. For example, *Artemisia carruthii*, a sage that grows in hot regions of the American Southwest, has finely dissected leaves, whereas closely related species in both America and Eurasia have broad, less dissected leaves (Figure 12.21A,B). It is likely that *A. carruthii* evolved from a broad-leaved ancestor (a hypothesis that might be tested by inferring the phylogenetic relationships among *Artemisia* species). If so, we may plausibly infer that dissected leaves evolved in *A. carruthii*, perhaps as an adaptation to heat. Now if we examine a species of *Acacia* from tropical Africa and a species of *Desmanthus* ("mimosa") from temperate North America (Figure 12.22A,B), we find dissected compound leaves in both cases; but we cannot safely call these an adaptation in either case, because almost all members of this family (Mimosaceae) have compound leaves,

wherever they grow. A phylogenetic perspective is important in analyzing adaptation.

Experiments Experiments may show that a feature enhances survival or reproduction, or enhances performance (e.g., locomotion or defense) in a way that is likely to in-

FIGURE 12.21 *Artemisia carruthii* (A) has narrower, more dissected leaves than its close relative *A. ludoviciana* (B), perhaps as an adaptation to the hotter environment in which *A. carruthii* generally grows. (After Gleason 1952.)

(A)   (B)

FIGURE 12.22 Finely dissected compound leaves in Mimosaceae from very different environments. (A) *Acacia persiciflora*, a tree from Kenya in Africa. (B) *Desmanthus illinoiensis*, an herb from temperate North America. (A after Coe and Beentje 1991; B after Gleason 1952.)

(A)

(B)

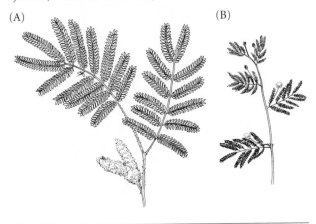

crease fitness, relative to individuals in which the feature is modified or absent. In order to infer the ADAPTIVE SIGNIFICANCE, or function, of the feature, the experiment should be done in the context of a postulated selective factor. The individuals whose performance or fitness is compared may differ either genetically or by experimental alteration. For example, H. B. D. Kettlewell (1955) suspected that the melanic (dark) genotype of the peppered moth (*Biston betularia*) had increased in frequency in Britain following the Industrial Revolution because, compared with the light form, it more closely resembled the color of tree trunks. Before the Industrial Revolution, many trees were clothed with pale lichens, but air pollution killed many of these lichens, leaving bare, dark trunks. Kettlewell showed that when marked moths of both colors were released in a polluted woodland, a higher proportion of melanic than of typ-

ical moths were recaptured in subsequent days, whereas the reverse occurred when he released moths in a nonpolluted wood. He confirmed that differential predation occurs by pinning moth specimens of both color forms on both bare and lichen-covered trees and recording the rates at which birds attacked them (Figure 12.23). Typical moths suffered less predation than melanic ones on lichen-covered trees, whereas melanic moths had an advantage on dark lichen-free boles. (However, in both *Biston* and many other insects that display industrial melanism, predation is probably not the only selective factor that affects gene frequencies: Lees 1981). Andersson's (1982) alteration of the tail length of male widowbirds, described earlier in this chapter, illustrates how artificially created variation may be used to demonstrate a feature's adaptive function, in this case its role in mating success.

The interpretation of such experiments can be beset by many pitfalls (Endler 1986). If the performance of different genotypes is compared, there is always a risk that the superiority of one over the other is due not to the particular character examined, or to the function it seems to serve, but rather to another, unknown, correlated difference caused by linked genes or by pleiotropic effects of the genes that code for the character. (This problem is often less severe if the feature is experimentally altered, because then correlated differences are less likely.) If the feature is shown to affect some aspect of performance, such as defense, it may be difficult, but necessary, to show that the difference in performance actually affects survival or reproduction in nature. For example, many experiments have shown that the various terpenes, alkaloids, and other secondary compounds of plants deter or poison insects and other herbivores. However, few experiments have shown that natural levels of herbivory are high enough to substantially affect the survival and reproduction of plants, or that variation in the kind or amount of sec-

(A)

(B)

FIGURE 12.23 Wild-type (gray) and melanic (dark) peppered moths (*Biston betularia*) on (A) a lichen-covered tree trunk and (B) a trunk lacking lichens. The gray form is more cryptic in A, and the dark form is more cryptic in B. In one of Kettlewell's experiments, fresh specimens of both forms were pinned to both kinds of trunks. The more conspicuous form in each situation was removed more rapidly by birds. (Photographs by H. B. D. Kettlewell.)

ondary compounds affects plants' fitness in nature (Feeny 1992; Futuyma and Keese 1992). In other words, differences in performance must be shown to result in natural selection. But even this may not be enough. The feature may be shown to have a positive *effect* on fitness, but, recalling the concept of exaptation, this need not be the *function* for which it was selected. For example, some authors argue that the secondary compounds of plants evolved as a mechanism of storing toxic waste products so as to make them harmless to the plant (plants cannot excrete toxins), and that the repulsion of herbivores is only an incidental, albeit beneficial, effect (see Chapter 18).

The comparative method   A powerful means of inferring the adaptive significance of a feature is the **comparative method**, which takes advantage of "natural evolutionary experiments" provided by convergent evolution. If a feature evolves independently in many lineages because of a similar selection pressure, it is correlated, among lineages, with that selection pressure. Hence we can often infer the function of a feature by determining the ecological or other selective factor with which it is correlated, or conversely, predict the correlation by postulating, perhaps on the basis of a model, the adaptive features we would expect to evolve in response to a given selective factor. *The comparative method may be defined as the use of comparison of sets of species to pose or test hypotheses on adaptation and other evolutionary phenomena.* For example, the independent evolution of large body size in northern populations of many mammal species implies that this is an adaptation to low temperatures; the way in which it confers such an advantage is deduced from physiological and physical principles.

An instance of a prediction confirmed by the comparative method was occasioned by the observation that the testes of the chimpanzee are larger than those of the gorilla, even though the gorilla is a considerably larger animal. However, a female gorilla in estrus mates with only one male, whereas female chimpanzees often mate with several males in rapid succession. These observations suggest the hypothesis that male chimpanzees compete to fertilize the female's egg, and that a male could therefore increase his probability of success by producing large numbers of sperm, which would require large testes. Thus large testes should be expected to provide a greater reproductive advantage in polygamous than in monogamous species. Paul Harvey and collaborators compiled data from numerous primates, and confirmed that, as predicted, the weight of the testes, relative to body weight, is significantly higher among polygamous than monogamous taxa (Figure 12.24). The hypothesis that testis size evolves *at least partly* as an adaptation for sperm competition was therefore supported. Later studies showed a similar difference in testis size between polygamous and monogamous species in other taxa, such as deer and birds.

This example raises several important points. First, Harvey and his collaborators gathered data on both testis size and mating system in primate species from numerous re-

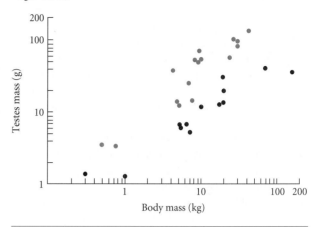

FIGURE 12.24   The relationship between weight of the testes and body weight among primate taxa. Matched for body weight, polygamous taxa (colored circles) have larger testes than monogamous taxa (black circles). (After Harvey and Pagel 1991.)

ports in the literature by various authors who had described these features for other reasons. Although each individual datum was already known (e.g., testis size in the rhesus monkey), the relationship between the two variables was not known until Harvey and collaborators compiled the data. No one had previously compiled these data to look for a relationship, because no one had had any reason to do so *until an adaptive hypothesis had been formulated*. Hypotheses about adaptation can be fruitful because, whether or not the data ultimately support them, they suggest investigations that would not otherwise occur to us.

Second, because the consistent relationship between testis size and mating system was not known a priori, the hypothesis generated a PREDICTION about testis sizes in various primates, of which Harvey et al. were ignorant when they set out to compile data. The predictions made by evolutionary theory, and in particular by adaptive hypotheses, are most often predictions of what we will find when we collect data on existing (or extinct) organisms— what character state a species will have, what relationship we will find among variables such as morphology and mating system, or what the results of an experiment (such as Endler's experiment with guppies) will be. Prediction in evolutionary theory does *not* usually mean that we predict the future course of evolution of a species. (To be sure, under some circumstances and to a limited degree, we can predict future evolutionary events, but not with great accuracy, and we may not be here to test our predictions.) Predictions of data, deduced as the expected consequences of hypothesized selective (or other) causes, constitute the **hypothetico-deductive method**, of which Darwin was one of the first effective exponents (Ghiselin 1969; Ruse 1979).

Third, the data support the hypothesis in the form of a CORRELATION between two variables. This point raises two issues. The correlation with mating system does not account for all the variation in testis size (e.g., all polygamous species do not have an identical testes weight/body weight ratio), so it may be presumed that unknown factors other than mating system have also affected the evolution of this feature. And, most importantly, a *correlation* of variable Y with variable X *does not necessarily mean that X caused Y*. Conceivably, Y caused X, or both X and Y are independent consequences of a causal factor Z. It is at least conceivable that some characteristic of a species' diet, for example, imposes selection on both testis size and mating system independently. But inasmuch as no such factor has yet been identified, or even conceived, whereas a plausible causal hypothesis has made a successful prediction, we are justified in *tentatively* accepting the hypothesis about adaptation to sperm competition. But *every hypothesis in science is accepted tentatively, and is potentially subject to modification or rejection in the light of new data or theories.*

Fourth, we cannot say that polygamous and monogamous taxa differ in mean testis size unless we demonstrate that the average values of the two groups show a STATISTICALLY SIGNIFICANT difference. To do this, it is necessary to have a sufficient number of data points—i.e., a large enough sample size (16 polygamous and 13 monogamous taxa in Figure 12.24). For a statistical test to be valid, each data point must be INDEPENDENT of all others. For example, it would not be valid to conclude that college professors are more likely to vote for progressive candidates, and bank officials to vote for conservatives, if we recorded the votes of the same three individuals in each group for 20 years and pretended that we had 60 data points for each group, because an individual's behavior is correlated—it is not independent—across years.

Harvey et al. could have had a larger sample size if they had included, say, 20 species of marmosets and tamarins (Callitrichidae), all of which are monogamous. But the uniformity of monogamy in this monophyletic family suggests that monogamy (or testis size) evolved not 20 or more times, but only once, and has been retained in all descendants of the ancestral species. Perhaps it has been retained because of some internal constraint in development, rather than because of natural selection; or perhaps the callitrichids have originated so recently from their common ancestor that there has not been enough time for their mating systems or testes to diverge. That is, the condition of these features in the various callitrichid species may be a result of history, rather than adaptation in each species. Because our hypothesis is that testis size *evolves* in response to the mating system, we must suspect that the Callitrichidae represent only one evolutionary event, and so provide only one data point (Figure 12.25). Many authors argue that in order to use convergent evolution—i.e., the comparative method—to test hypotheses of adaptation, we must count the *number of convergent evolutionary events* in which a character state (such as testis size) evolved under one ecological circumstance (e.g., mating system) versus another (Ridley 1983; Felsenstein 1985; Harvey and Pagel 1991). In their view, phylogenetic information is essential for proper use of the comparative method. However, some practitioners of the

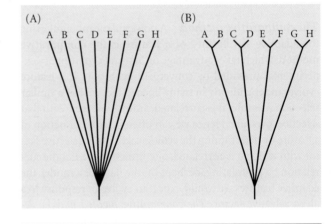

FIGURE 12.25 The problem of phylogenetic correlation in employing the comparative method. Suppose we test a hypothesis about adaptation by calculating the correlation between two traits, such as testis size and mating system, in eight species (A–H). If the species are related as in A, each has evolved independently of the others from a common ancestor, and we have a sample of eight. If the species are related as in B, both traits may be similar in each pair of closely related species due to their common ancestry rather than to independent adaptive evolution. Some authors maintain that the two species in each pair are not independent tests of the hypothesis; we would have a sample size of four in this case. (After Felsenstein 1985.)

comparative method argue that when there is reason to think that the characters are genetically variable, and hence potentially capable of adaptive change, this phylogenetic approach may not be necessary (Reeve and Sherman 1993; Westoby et al. 1995).

What Not to Expect of Natural Selection and Adaptation

We conclude this discussion of the general properties of natural selection and adaptation by considering a few common misconceptions of, and misguided inferences from, the theory of adaptive evolution.

The necessity of adaptation It is naive to think that if a species' environment changes, the species must adapt or else become extinct. That is true when some factor changes beyond the tolerance limits of the majority of the population, but many environmental changes, even if they reduce the species' abundance, do not press these limits. Some changes enhance the abundance of certain species; the conversion of farmland to suburbs, for example, increases the abundance of some bird species while reducing others. An environmental change that does not threaten extinction may nonetheless set up selection for change in some characteristics. Thus white fur in polar bears may be advantageous, but not necessary for survival (Williams 1966). Just as a changed environment need not set in motion selection for new adaptations, conversely, new adaptations may evolve in an unchanging environment. For example, new mutations may arise that are superior to any pre-existing genetic variations. We have already stressed that the probability of extinction of a population or species does not in itself constitute selection on individual organisms, and so cannot cause the evolution of adaptations.

Perfection   This is not "the best of all possible worlds," as Dr. Pangloss in Voltaire's *Candide* taught, and we should not expect organisms' adaptations to conform to the best possible design. Darwin noted that "natural selection tends only to make each organic being as perfect as, or slightly more perfect than the other inhabitants of the same country with which it comes into competition," so that "natural selection will not produce absolute perfection, nor do we always meet, as far as we can judge, with this high standard in nature" (*On the Origin of Species*, Chapter 6). Selection may fix only those genetic variants with a higher fitness than other genetic variants in that population at that time. It cannot fix the best of all conceivable variants if they do not arise, or have not yet arisen, and the best possible variants often fall short of perfection because of various kinds of constraints. (For example, with a fixed amount of available energy or nutrients, a plant might evolve higher seed numbers, but only by reducing the size of its seeds or some other part of its structure.)

Progress   Whether or not evolution is "progressive" is a vexed question (Nitecki 1988; Ruse 1996), to which we shall return in Chapter 24. The word "progress" has the connotation of a goal, and as we have seen, evolution does not have goals. But even if we strip away this connotation and hold only that progress means "betterment," the possible criteria for "better" are many. Innumerable criteria must be very specific to individual taxa, such as better echolocation for bats, better drought resistance for cacti, better ability to detoxify potato alkaloids for Colorado potato beetles. Better learning ability or greater brain complexity, frequently taken as a criterion of progress by those who wish to view humans as the pinnacle of evolution, has no more evident adaptive advantage for most animals—e.g., rattlesnakes—than a capacity for sensing the direction of a heat source, or an effective poison delivery system, would have for humans. Measurements of "improvement" or "efficiency" must be relevant to each species' special niche or task. To be sure, for specifically defined tasks, we have encountered (see Chapter 6), and will again encounter, examples of adaptive trends, each of which might be viewed as progressive within its special context. We shall later consider whether there exists evidence of other types of improvement, such as the possibility that the average lifetime of species, before extinction, may have increased in the last 500 million years (see Chapter 25).

Harmony and the balance of nature   As we have seen, selection at the level of genes and individual organisms is inherently "selfish": the gene or genotype with the highest rate of increase "wins" (unless countervailing forces establish stable polymorphism) at the expense of other individuals. Thus we do not expect organisms to be truly "altruistic" to other members of their species. The variety of selfish behaviors organisms inflict on conspecific individuals, ranging from territory defense to parasitism and infanticide, is truly stunning. Indeed, those cases in which organisms are cooperative require special explanations. For example, a parent that forages for food for her offspring, at the risk of exposure to predators, is cooperative, but for an obvious reason: her own genes, including those coding for this parental behavior, are carried by the offspring, and the genes of individuals that do not forage for their offspring are less likely to survive than the genes of individuals that do. This is an example of KIN SELECTION, an important basis for the evolution of cooperation within species (see Chapter 20).

Because this principle of kin selection cannot operate across species, "natural selection cannot possibly produce any modification in a species exclusively for the good of another species, though throughout nature one species incessantly takes advantage of and profits by the structures of others … If it could be proved that any part of the structure of any one species had been formed for the exclusive good of another species, it would annihilate my theory, for such could not have been produced through natural selection" (Darwin, *On the Origin of Species*, Chapter 6). If, other than incidentally (as when the growth of plants is increased by nutrients in the dung of grazing animals), a species exhibits behavior that enhances the fitness of another species, either the behavior is profitable to the individuals performing it (as in bees that obtain food from the flowers they pollinate), or they have been duped or manipulated by the species that profits (as are insects that pseudocopulate with orchids and receive no benefit). Most mutualistic interactions between species, then, consist of reciprocal exploitation (see Chapter 18).

If mutualistic relationships among species are founded on individual self-interest, other aspects of the equilibrium we may observe in communities—the so-called "balance of nature"—reflect even less any striving for harmony. We observe coexistence, with greater or lesser stability of numbers, of predators and prey, but this is not because of any restraint on the part of the predators. It is because prey species have defenses sufficient for persistence, or because the abundance of predators is limited by some factor other than food supply. Predator-prey systems that happen not to have been stabilized by such factors are not seen, simply because, being unstable, they have become extinct.

Similar explanations apply to characteristics of ecosystems. Nitrogen and mineral nutrients are rapidly and "efficiently" recycled within tropical wet forests not because ecosystems are selected for or aim for efficiency, but because under competition for sparse nutrients, microorganisms have evolved to decompose litter rapidly, and plants have similarly evolved to capture nutrients released by decomposition. Selection of individual organisms for their ability to capture nutrients has the *effect*, in aggregate, of a dynamic that we measure as ecosystem "efficiency."

Well-meaning, idealistic environmentalists sometimes take the view that ecosystems are harmonious, integrated "superorganisms" designed to foster the living things that compose them. (A recent version of this idea is the "Gaia hypothesis," which propounds a superorganismic interpretation of the earth as a whole, with both its living and nonliving constituents adjusted to form a self-regulating support system.) It is certainly true that most perturbations of ecosystems, including even the removal or addition of a single species, have numerous consequences, often including the extinction of some species. But there is neither any evidence nor any scientific foundation for these often mystical notions of a beneficent force guiding ecosystems toward harmony and maintaining balance (Williams 1992a).

Morality and ethics   The theory of natural selection, of incessant competition and struggle, of individual self-interest as the motive force of adaptive evolution, paints a dark picture of "Nature red in tooth and claw," as dark and aesthetically unappealing, perhaps, as untrammeled capitalism. It portrays nature as utterly amoral—not immoral, for that implies conscious violation of ethical norms, but simply lacking in any moral or ethical qualities whatever. For this reason, evolutionary theory, since its inception, has met with vigorous, emotional opposition: some find it merely aesthetically distasteful, some feel that it assaults their need to find meaning in the universe, others fear that it will be used to justify immorality in human society. Indeed, many have misused evolutionary theory in just this way. Darwin expressed distress over an article "showing that I have proved 'might is right,' and therefore that Napoleon is right, and every cheating tradesman is also right." The Nazis followed many precedents to their ultimate horrible conclusion in justifying racism by invoking evolution. But neither evolutionary theory nor any other field of science can speak of or find evidence of morality and immorality. These do not exist in nonhuman nature, and science describes only what *is*, not what *ought* to be. The NATURALISTIC FALLACY, the supposition that what is "natural" is "good," has no philosophical foundation. Neither natural selection nor any other theory of natural science can provide a code of ethics.

Summary

1. A feature is an adaptation for a particular function if it has evolved or is maintained by natural selection for that function by enhancing the relative rate of increase—i.e., the fitness—of entities with that feature. Natural selection is a consistent (biased, or nonrandom) difference in fitness among phenotypically different biological entities, and is the antithesis of chance. Selection is a mindless process of differential survival and reproduction, without forethought or goal. It is important to distinguish between the function of a feature—the reason for which the feature was selected—and the effects of a feature—incidental consequences that may or may not include survival of the species as a whole.

2. The entities that may be selected can be described at several levels of organization, including genes, individual organisms, and groups such as populations, species, or other taxa. If selection pressures conflict at different levels, the evolutionary outcome is likely to be determined by selection at the level of genes or organisms because the numbers and turnover rates of these entities are greater than those of populations or species. Therefore, most current theory holds that the majority of features are unlikely to have evolved by group selection, the one form of selection that could in theory promote the evolution of features that benefit the species even though disadvantageous to the individual organism. On the other hand, selection may occur at the gene level that is detrimental to organisms. It can be conceptually and practically difficult to define the level at which selection is acting in certain instances.

3. Although a great many features are adaptations, criteria for adaptations are necessary, because a trait may not be an adaptation for its apparent function, or may not be an adaptation at all, for any of many reasons. Methods for identifying and elucidating adaptations include studies of function and design, experimental studies of the correspondence between fitness and variation within species, and comparisons of traits among species that establish correlations with environmental or other features (the comparative method). Phylogenetic information may be necessary for proper use of the comparative method.

4. Natural selection does not necessarily produce anything that we can justly call evolutionary progress. It need not promote harmony or balance in nature, and, utterly lacking any moral content, it provides no foundation for morality or ethics in human behavior.

Major References

Williams, G. C. 1966. *Adaptation and natural selection.* Princeton University Press, Princeton, NJ. A clear, insightful, and influential essay on the nature of individual and group selection. A classic.

Williams, G. C. 1992. *Natural selection: Domains, levels, and challenges.* Oxford University Press, New York. Further analysis of natural selection.

Sober, E. 1984. *The nature of selection: Evolutionary theory in philosophical focus.* MIT Press, Cambridge, MA. Evolution from the perspective of a philosopher of science.

Dawkins, R. 1989. *The selfish gene.* New edition. Oxford University Press, Oxford. See also *The extended phenotype* (W. H. Freeman, Oxford and San Francisco, 1982) and *The blind watchmaker* (Norton, New York, 1986), by the same author. Written in a lively style for both lay and professional audiences, these books explore the nature of natural selection in depth, as well as many other topics. *The blind watchmaker* is devoted to the evolution of complex adaptations and to rebutting criticisms of adaptationist thinking and of evolutionary theory generally. Dawkins's books have drawn the criticism that they advocate excessive reductionism and adaptationism.

Gould, S. J., and R. C. Lewontin. 1979. The spandrels of San Marco and the Panglossian paradigm: A critique of the adaptationist programme. *Proceedings of the Royal Society of London B* 105: 581–598. A well-argued but controversial critique of unwarranted adaptationist speculation.

Harvey, P. H., and M. D. Pagel. 1991. *The comparative method in evolutionary biology.* Oxford University Press, Oxford. On the use and phylogenetic foundations of the comparative method.

Problems and Discussion Topics

1. In the flour beetle experiment described in this chapter, population size declined partly because beetles evolved an increased propensity for cannibalism. Is this propensity an adaptation? Generalizing from this example, need the process of adaptation result in an increase in the abundance or growth rate of a population or species?

2. Discuss criteria or measurements by which you might conclude that a population is better adapted after a certain evolutionary change than before.

3. Consider the first copy of an allele for insecticide resistance that arises by mutation in a population of insects exposed to an insecticide. Is this mutation an adaptation? If, after some generations, we find that most of the population is resistant, is the resistance an adaptation? If we discover genetic variation for insecticide resistance in a population that has had no experience of insecticides, is the variability an adaptation? If an insect population is polymorphic for two alleles, each of which confers resistance against one of two pesticides that are alternately applied, is the variation an adaptation? Or is each of the two resistance traits an adaptation?

4. Adaptations are features that have evolved because they enhance the fitness of their carriers. It has sometimes been claimed that fitness is a tautological and hence meaningless concept. According to this argument, adaptation arises from the "survival of the fittest," and the fittest are recognized as those that survive; consequently there is no independent measure of fitness or adaptiveness. Evaluate this claim. (See Sober 1984.)

5. Elisabeth Vrba and Niles Eldredge (1984) have argued that when selection ("sorting") can occur at two or more hierarchical levels, either "upward" or "downward" causation can determine the pattern of variation that results. For example, selection among alleles or individual organisms can affect the characteristics of species and higher taxa ("upward causation" of properties at a higher level by selection at a lower level), whereas differences among clades in rates of speciation or extinction can affect the prevalence of alternative character states of individual organisms ("downward causation"). Discuss the usefulness and implications of this distinction, and the conditions under which each form of causation would have a greater effect.

6. Consider a characteristic for which it is debatable whether or not its observed beneficial *effect* is also its *function*, i.e., the reason for which it evolved and is maintained by natural selection. For instance, plant compounds that are repellent or toxic to herbivores (such as nicotine in tobacco plants) might be waste products, and only incidentally have this beneficial effect. What kind of evidence might help to resolve debate in such cases?

7. It is often proposed that a characteristic that is advantageous to individual organisms is the reason for the great number of species in certain clades. For example, wings have been postulated to be a cause of the great diversity of winged insects compared with the few species of primitively wingless insects. How could an individually advantageous feature cause greater species diversity? How can one test a hypothesis that a certain feature has caused the great diversity of certain groups of organisms?

8. Provide an adaptive and a nonadaptive hypothesis for the evolutionary loss of useless organs, such as eyes in many cave-dwelling animals. How might these hypotheses be tested?

9. List the possible criteria by which evolution by natural selection might be supposed to result in "progress," and search the biological literature for evidence bearing on one or more of these criteria.

10. There is great controversy about whether or not various common behavioral traits of humans, such as aggressiveness and the "traditional" sex roles of women and men, are traits that evolved by natural selection. What evidence would be required to answer the question for one such trait?

CHAPTER 13

The Theory of Natural Selection

As an elementary concept, natural selection—differential reproductive success—is very simple, but its modes and consequences are exceedingly diverse. The great complexity of the subject arises from the many ways in which ecological factors and other agents of selection can operate, and from the diverse relationships between variation in phenotypes, which survive and reproduce in varying degree, and genotypes, which transmit the effects of selection to subsequent generations.

The factors of evolution that we examined in Chapter 11—genetic drift, inbreeding, and gene flow—act at the same rate on all loci.* Selection is profoundly different, for in a sexually reproducing species, *allele frequency changes proceed independently at different loci* (except for some complications discussed in Chapter 14). We have seen (in Chapters 5 and 6) that different characteristics of a species evolve at different rates (mosaic evolution), as we would expect if natural selection brings about changes in certain features while holding others constant. Thus we are justified in beginning our analysis of natural selection with a single variable locus that alters a phenotypic character. After examining selection at a single locus, we will consider (in Chapter 14) the more complicated cases in which several or many loci, affecting one or more than one trait, are subject to selection.

Evolution by Natural Selection

Before delving into the theory of natural selection, we should recall several important points.

1. *Natural selection is not the same as evolution.* Evolution is a two-step process: the origin of genetic variation by mutation or recombination, followed by a change in the pattern of variation, such as replacement of some genotypes by others. Natural selection is one agent of change in the pattern of variation; genetic drift is another. Both can be responsible for the spread of traits through populations, but neither natural selection nor genetic drift accounts for the origin of variation.

2. *Natural selection is different from evolution by natural selection.* Just as evolution can occur without natural selection (e.g., by genetic drift), natural selection can occur without evolution. In some instances that we will soon encounter, genotypes differ in each generation in survival or fecundity, yet the proportions of genotypes and alleles stay the same from one generation to another.

3. Although natural selection may be said to exist whenever different phenotypes vary in average reproductive success, *natural selection can have no evolutionary effect unless phenotypes differ in genotype.* For instance, selection among genetically identical members of a clone, even though they differ in phenotype, can have no evolutionary consequences.

4. Natural selection *is* variation in average reproductive success (including survival) among phenotypes. Without such variation, there can be no evolution by natural selection. Thus a feature that merely increases an individual's "comfort" cannot evolve by natural selection; it must make a positive contribution to reproduction or survival. The long-haired tail of a horse, used as a fly-switch, could not have evolved by selection for this function unless it increased reproductive success, perhaps by lowering mortality caused by fly-borne diseases.

Hence, the consequences of natural selection depend on (1) the relationship between phenotype and fitness, and (2) the relationship between phenotype and genotype. These, then, yield (3) a relationship between fitness and genotype, which determines (4) whether or not evolutionary change occurs (Figure 13.1).

---

*The exceptions are loci that differ in copy number from the rest of the genome, such as sex-linked loci in a diploid species.

FIGURE 13.1 The fitnesses of different genotypes depend on both the relationship between phenotype and fitness and the relationship between genotype and phenotype. For a phenotypic character such as size, dominance, additive inheritance (no dominance), and overdominance are illustrated. Because of the linear relationship between phenotype and fitness, the fitnesses of the genotypes also show dominance, no dominance, and overdominance (heterozygous advantage; see below), respectively.

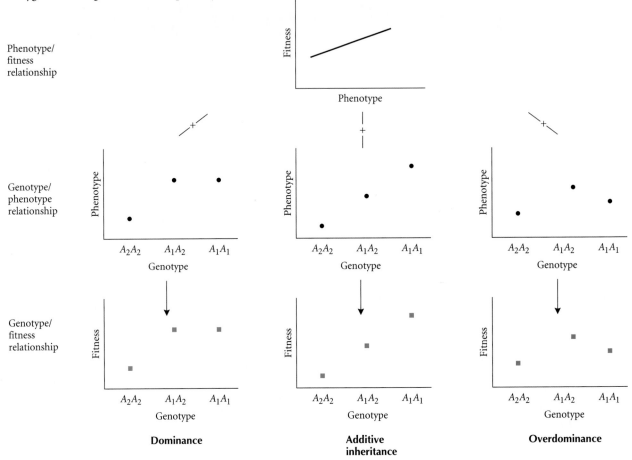

Fitness

Modes of Selection

The relationship between phenotype and fitness can often be described as one of three modes of selection (Figure 13.2). For some variable trait, such as size, selection is **directional** if one extreme phenotype is fittest, **stabilizing** (*normalizing*) if an intermediate phenotype is fittest, or **disruptive** (*diversifying*) if two or more phenotypes are fitter than the intermediates between them. Which *genotype* has the highest fitness under a given selection regime depends on the relationship ("mapping") between phenotype and genotype. For example, under directional selection for large size, genotype A_1A_1 is most fit if it is largest, but A_1A_2 is favored if it is larger than either homozygote.

The selection regime—i.e., the fitness/phenotype relationship—in a real population can depend on the environment (including, as we have noted earlier, the "genetic environment" in which the variation is embedded). It also depends on how the variation is distributed relative to the fitness/phenotype relationship. Thus if the mean body size is below the optimum, it will be directionally selected until

it corresponds to the optimum (at least approximately); after that, it is subject to stabilizing selection (Figure 13.2).

Defining Fitness

Because we are concerned with evolutionary effects of selection that depend on inheritance, we will use models in which an average fitness value is assigned to each genotype. A genotype is likely to have different phenotypic expressions as a result of environmental influences in development (norms of reaction; see Chapter 3), so the fitness of a genotype is the average of the fitnesses of all individuals of that genotype, whatever their phenotypes may be. For example, the number of eye facets in the *Bar* mutant of *Drosophila melanogaster* depends on the temperature at which the fly develops (see Figure 6, Chapter 3), so if fitness depended on facet number, the fitness of the *Bar* genotype in a particular population would depend on the average number of facets among *Bar* flies, which in turn would depend on the proportions of flies that developed at each temperature.

The fitness of a genotype is the average per capita lifetime contribution of individuals of that genotype to the population after one or more generations, measured at the same stage in

FIGURE 13.2 Modes of selection on (A) a heritable quantitative (continuously varying) trait and (B) a polymorphism inherited as two alleles at one locus. In both cases, the phenotype is assumed to be additively inherited (i.e., heterozygotes are intermediate between homozygotes; there is no interaction among loci that contribute variation to the quantitative trait). The vertical axis is the proportion of the population with each phenotype. The upper rows of figures in both A and B show the distribution of phenotypes in one generation, before selection occurs. The shaded portions represent individuals with a relative disadvantage (lower reproductive success). The lower rows of figures in both A and B show the distribution of phenotypes in the following generation, after selection among the parents has occurred. X marks the mean of the quantitative trait before selection. Left: Directional selection increases the proportion of genotypes with higher values of the trait. Center: Stabilizing selection, if symmetrical about the character mean, does not alter the mean, but may reduce the variance (variation). Right: Disruptive (or diversifying) selection is unlikely to be exactly symmetrical, and thus usually shifts the mean. (After Endler 1986.)

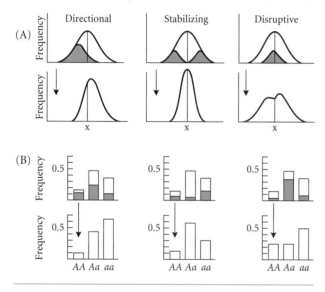

viving) × (average fecundity) = 0.05 × 60 = 3. This is the per capita replacement rate, or population growth rate, of this genotype, which in Chapter 4 we denoted R (Figure 13.3). Likewise, genotype B might have a survival rate of 0.10 and an average fecundity of 40, yielding a fitness of 4. If the frequencies of genotypes A and B are p and q (p + q = 1), then the growth rate of the population as a whole is $\bar{R} = pR_A + qR_B$. Thus if 20 percent of the population were A, the growth rate per generation would be $\bar{R} = (0.2)(3) + (0.8)(4) = 3.8$.

The per capita growth rate, R_i, of each genotype i is that genotype's **absolute fitness**. $\bar{R}$ may be termed the average fitness of the population—its actual rate of increase. The **relative fitness** of a genotype, W_i, is its value of R relative to that of some reference genotype. By convention, the reference genotype, often the one with highest R, is assigned a relative fitness of 1.0. Thus in our example, $W_A = 3/4 = 0.75$ and $W_B = 1.0$. The AVERAGE FITNESS, $\bar{w}$, is then *the average fitness of individuals in the population, relative to the fittest genotype*. In our example, if $p = 0.2$, the average fitness is $\bar{w} = (0.2)(0.75) + (0.8)(1.0) = 0.95$. The mean fitness, $\bar{w}$, does not indicate whether or not the population is growing, because it is only a relative measure.

Relative Fitness and the Rate of Change

The rate of genetic change under selection depends on the relative, not the absolute, *fitnesses of genotypes.* We see this if we calculate the change in genotype frequencies per generation. Let the frequencies of asexual genotypes A and B be $p = N_A/N$ and $q = N_B/N$, where N is the population size at the beginning of a generation (e.g., in the egg stage). After

the life history. (Usually it suffices to define fitness as the contribution to the population after one generation, but we will encounter circumstances in which fitness must be evaluated by the number of descendants after two generations.) Thus we may think of fitness as the average number of eggs or offspring one generation hence that are descended from the average egg or offspring born. A general term for this average number is REPRODUCTIVE SUCCESS, which includes not simply the average number of offspring produced by the reproductive process, but the average number that survive from birth to reproductive age, since survival is prerequisite to reproduction. Fitness is most easily conceptualized for an asexually reproducing organism in which all adults reproduce only once, all at the same time (nonoverlapping generations), and then die, as in some parthenogenetic weevils and other insects that live for a single growing season. Suppose that in a population of such an organism, the proportion of eggs of genotype A that survive to reproductive age is 0.05, and that each reproductive adult lays an average of 60 eggs. Then the fitness of A is (fraction sur-

FIGURE 13.3 The growth of two asexually reproducing genotypes in a population with discrete generations. The proportion of the more prolific genotype (R = 4) rapidly becomes far larger than that of the less fecund genotype (R = 3). The differential rate of population growth of the genotypes is an instance of natural selection.

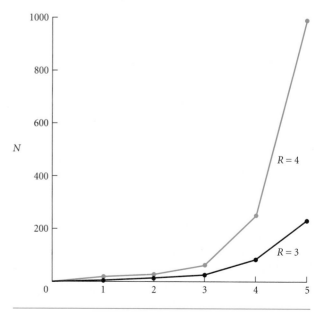

one generation, the numbers of A and B individuals are $N_A R_A$ and $N_B R_B$ respectively. The new frequency of A, denoted p', is

$$p' = \frac{N_A R_A}{N_A R_A + N_B R_B} = \frac{pNR_A}{pNR_A + qNR_B} = \frac{pR_A}{pR_A + qR_B}$$

The change in the frequency of A is $\Delta p = p' - p$, or

$$\Delta p = \frac{pR_A}{pR_A + qR_B} - p = \frac{pR_A - p(pR_A + qR_B)}{pR_A + qR_B}$$

After some algebra (using the relation $1 - p = q$), this becomes

$$\Delta p = \frac{pq(R_A - R_B)}{pR_A + qR_B} \tag{13.1}$$

If, for example, $p = 0.2$ and $q = 0.8$, and R_A and R_B are 3 and 4 respectively, $\Delta p = -0.042$. The same result is found if R_A and R_B are 6 and 8, or 9 and 12, or 300 and 400. Thus the relative fitnesses W_A and W_B, which in this instance are 0.75:1.0, determine the rate of change.

Let us set the largest fitness value (W_B in this example) at 1.0, and denote the fitness of each other genotype, i, as $1 - s_i$. The term s_i is the **coefficient of selection**, and measures the intensity of selection against the less fit genotype, or the SELECTIVE ADVANTAGE of the fitter genotype. (In our example, $W_A = 0.75$, so $s = 0.25$.) If in Equation (13.1) we substitute relative for absolute fitnesses, and enter the values $1 - s$ and 1 for W_A and W_B respectively, we have

$$\Delta p = \frac{pq([1-s]-1)}{p(1-s) + q(1)} = \frac{-spq}{1 - sp} \tag{13.2}$$

Note several consequences of this equation. First, Δp, the change in the frequency of genotype A from one generation to the next, is negative as long as p, q, and s are positive: the genotype A declines in frequency because its fitness is lower. Second, the magnitude of Δp is *directly proportional* to the numerator, i.e., *to the coefficient of selection*, and *also to the product pq*. That is, the rate of change is greater when both genotypes are common than when they have extreme frequencies (in fact, pq is maximized when $p = q = 0.5$). Consequently, the rate at which genotype A is reduced in frequency becomes lower, the lower its frequency becomes. Also, this means that the potential rate of evolution is greater if there are two or more common genotypes than if the variation is contributed only by rare genotypes. Third, Δp is *inversely proportional* to the denominator, which in the form $p(1 - s) + q(1)$ is readily seen to be the *mean* (average) *fitness* of individuals in the population. Thus as p approaches zero and more of the individuals are of the fittest genotype, the rate of evolution declines. Fourth, the population ceases to evolve ($\Delta p = 0$) only when $p = 0$. This is a stable equilibrium, for if p should again exceed zero (perhaps because of mutation), it will return to zero due to selection.

Many environmental variables, as well as genetic variables, cause variation among individuals in survival and reproductive success. Of these, only some, perhaps only one, may differentially affect genotypes at a particular locus, or genotypes that differ in a particular character. Thus much, perhaps most, of the mortality suffered by a population may be random with respect to this locus or character. These NONSELECTIVE DEATHS may be contrasted with SELECTIVE DEATHS, those that contribute to the difference in fitness between genotypes. Even if most mortality is nonselective, the selective deaths that do occur can be a potent source of natural selection. For instance, genetic differences in swimming speed in a small planktonic crustacean might well not affect the likelihood of being eaten by baleen whales, which might be the major source of mortality. But if swimming speed affects escape from another predator species, even one that accounts for only 1 percent of the deaths, there will be an average difference in fitness, and swimming speed may evolve by natural selection.

Components of Fitness

We have so far considered a simple case in which fitness has two **components**: (1) the probability of a genotype's survival from birth to reproduction and (2) its average fecundity. The components of fitness are more complex if a species reproduces sexually and if it reproduces repeatedly during the individual's lifetime. When generations overlap, as in humans and many other species that reproduce repeatedly, the absolute fitness of a genotype may be measured in large part by its instantaneous per capita rate of increase, r (see Chapters 4 and 19). In this case, fitness depends not only on the number of offspring females produce, but also on whether they produce them early or late in life. If a female's average offspring is born when she is 6 months old in the case of genotype A, but when she is 12 months old in the case of genotype B, then if all else is equal, the rate of increase (the fitness) of A is about twice that of B, for the simple reason that A will have a full complement of grandchildren (i.e., two generations of descendants) in the time B needs to produce one generation of descendants (see Chapter 19). Moreover, differences among males in reproductive success may also contribute to differences in fitness.

In sexually reproducing species, genotypes do not make copies of themselves; instead, they transmit haploid gametes. Therefore genotype frequencies depend on the allele frequencies among uniting gametes. These allele frequencies are affected by several components of fitness at the organismal stage of the life cycle (sometimes called "zygotic selection") and sometimes by fitness at the gametic (haploid) stage as well (Figure 13.4; Christiansen 1990). Table 13.1 summarizes the components of selection in a sexual species.

Evolution by natural selection is described by the way in which changes in allele frequencies are determined by the components of fitness of each zygotic and each gametic genotype. These components of fitness are combined (usually by multiplying them) into the overall fitness of each genotype. For instance, the overall fitness of genotypes in the simple example above was found by multiplying the survival and fecundity of each genotype. A genotype may be superior to another in certain components and inferior in others, but it is its *overall*, or *total fitness*, that *determines the outcome of natural selection*.

FIGURE 13.4 An oversimplified portrait of the components of natural selection that may affect the fitness of a sexually reproducing organism over the life cycle. Beginning with newly formed zygotes, (1) genotypes may differ in survival to adulthood; (2) they may differ in the numbers of mates they obtain, especially males; (3) those that become parents may differ in fecundity (number of gametes produced, especially eggs); (4) selection may occur among the haploid genotypes of gametes, as in differential viability or meiotic drive; and (5) unions of some combinations of gametic genotypes may be more compatible than others. (After Christiansen 1984.)

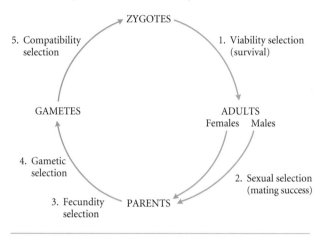

Components of Fitness: An Example

In order to analyze selection in a natural population, one should measure the total fitness of each genotype of interest. In practice this is extremely difficult, and most such studies are incomplete in that investigators usually measure only one or a few components of fitness (Endler 1986). One

of the few thorough studies of components of fitness was performed by the Danish biologist Freddy Christiansen and his colleagues (Christiansen et al. 1977), who studied a polymorphism of an esterase enzyme in the eelpout (*Zoarces viviparus*), a small marine fish. The females bear live young, and the genotype (A_1A_1, A_1A_2, or A_2A_2) of each adult and of each offspring carried by a pregnant female can be determined by electrophoresis. Christiansen et al. determined the genotypes in large samples of adult males, nonpregnant and pregnant females, and offspring carried by the latter, as well as in samples from the age class represented by these offspring when they reached maturity two years later. Some of the data from this study are presented in Table 13.2.

First, half the progeny of heterozygous females should be heterozygous if *meiotic segregation is normal* and if A_1 and A_2 *eggs have equal viability*.* This proved to be the case ($1360/2696 = 0.504$ of the offspring of A_1A_2 females were heterozygous).

Second, almost all mother/offspring combinations provide information on the allele carried by the sperm. For example, A_1A_1 offspring carried by A_1A_2 mothers represent A_1-bearing sperm. The inferred frequency of A_1 and A_2 among the sperm, i.e., p and q, should be the same among offspring of all three maternal genotypes if mating has been random, and it was. Moreover, the values of p and q inferred for sperm from the mother/offspring combinations were the

---

*With 1:1 segregation, the probability that an egg is A_1 is 1/2, and likewise for A_2. If the frequencies of A_1 and A_2 among sperm are p and q, the probabilities of the four egg/sperm unions are $P(A_1, A_1) = p/2$, $P(A_1, A_2) = q/2$, $P(A_2, A_1) = p/2$, and $P(A_2, A_2) = q/2$. The expected frequency of A_1A_2 progeny is therefore $q/2 + p/2 = 1/2$, since $p + q = 1$.

Table 13.1 Components of selection in sexually reproducing organisms

I. *Zygotic selection*

1. *Viability.* The probability of survival of the genotype through each of the ages at which reproduction can occur. The length or probability of survival beyond the last age of reproduction does not usually affect the genotype's contribution to subsequent generations, and so does not usually affect fitness.

2. *Mating success.* The number of mates obtained by an individual. Mating success is a component of fitness if the number of mates affects the individual's number of progeny, as is often the case for males, but less often for females, all of whose eggs may be fertilized by a single male. Variation in mating success is the basis of sexual selection.

3. *Fecundity.* The average number of gametes per individual, but usually measured as the number of viable offspring produced. In species with repeated reproduction, the contribution of each offspring to fitness depends on the age at which it is produced (see Chapter 19). The number of offspring resulting from a mating may depend only on the maternal genotype (e.g., number of eggs or ovules), or it may depend on the genotypes of both mates (e.g., if they display some reproductive incompatibility).

II. *Gametic selection*

4. *Segregation advantage.* An allele has an advantage if it segregates into more than half the gametes of a heterozygote (*meiotic drive* or *segregation distortion*). This component may be difficult to distinguish from the next two.

5. *Gamete viability.* Dependence of a gamete's viability on the allele it carries.

6. *Fertilization success.* An allele may affect the gamete's ability to fertilize an ovum, e.g., if there is variation in the rate at which a pollen tube grows down through a style. Dependence of fertility on the mate's genotype may be based on this component.

Table 13.2. Selection analysis of an esterase polymorphism in *Zoarces viviparus*

A. Mother-offspring combinations (number of progeny of each genotype)

MOTHERS' GENOTYPES	OFFSPRING GENOTYPES			
	A_1A_1	A_1A_2	A_2A_2	SUM
A_1A_1	305	516[a]		821[a]
A_1A_2	459	1360[b]	877	2696[b]
A_2A_2		877	1541[a]	2418[a]
Sum	764	2753	2418	5935
Offspring genotype frequencies	0.129	0.464	0.407	

B. Genotype numbers and frequencies, and allele frequencies, among adults

GENOTYPE	FEMALES				MALES	
	PREGNANT		NONPREGNANT			
	NO.	FREQ.[c]	NO.	FREQ.[c]	NO.	FREQ.[c]
A_1A_1	821	0.138	43	0.118	693	0.133
A_1A_2	2696	0.454	161	0.442	2332	0.446
A_2A_2	2418	0.407	160	0.440	2201	0.421
Sum	5935		364		5226	
p		0.365		0.339		0.356[d]
q		0.635		0.661		0.644[d]

Source: Data from Christiansen (1990).

[a]Among progeny of A_1A_1 females, 516/821 = 0.629 represent A_2-bearing sperm; among progeny of A_2A_2 females, 1541/2418 = 0.637 represent A_2-bearing sperm. A similar, but more complicated, calculation may be made for many progeny of A_1A_2 females. Random mating is implied.

[b]Half the progeny of heterozygous mothers are heterozygous (1360/2696 = 0.504).

[c]Genotype frequencies among pregnant females are nearly the same as in other segments of the population, implying no differences in female mating success.

[d]Allele frequencies in males are nearly the same as among sperms that fertilized eggs (see note b); hence there is no evidence for gametic selection or genotypic differences in male mating success.

same as in the population of adult males. This finding implies that GAMETIC SELECTION was not occurring in sperm, and also implies that genotypes did not differ in MALE MATING SUCCESS.

Third, the genotype frequencies among pregnant females were almost the same as in the adult population at large, and the average number of offspring was the same for all genotypes of females. Thus the genotypes did not differ in FEMALE MATING SUCCESS or FECUNDITY.

Fourth, VIABILITY SELECTION was examined by comparing the genotype frequencies of offspring in the same cohort before birth and after two years, when they were reproductive adults. The actual genotype frequencies at each age were compared against those expected from the Hardy-Weinberg distribution (see Chapter 9). Before birth, the frequencies of the three genotypes were at Hardy-Weinberg equilibrium, but not when this cohort had become adults:

		A_1A_1	A_1A_2	A_2A_2	
Before birth:	Observed	0.129	0.464	0.407	$p = 0.36$, $q = 0.64$
	Expected	0.130	0.461	0.408	
Adults:	Observed	0.135	0.450	0.415	$p = 0.36$, $q = 0.64$
	Expected	0.129	0.461	0.410	

Although allele frequencies had not changed, heterozygotes had declined in frequency relative to homozygotes, which Christiansen et al. attributed to differential mortality. From the changes in genotype frequencies, they estimated the relative viabilities of the genotypes to be 1.065:1.000:1.037, where the viability of A_1A_2 is set at 1.0. Because viability differences appear to be the only component of selection acting in this case, these are also estimates of the relative fitnesses of the genotypes.

Dependence of Fitness on Environment

Equations such as those presented above use fitness values to predict the magnitude of allele frequency changes; under some circumstances, they may be rearranged in order to use observed frequency changes to estimate fitnesses. For example, Dobzhansky observed rapid changes in the frequencies of two chromosome inversions, ST and CH, in laboratory populations of *Drosophila pseudoobscura* maintained at 25°C. From these data, Sewall Wright calculated that the rates of change conformed to those expected if the relative fitnesses of ST/ST, ST/CH, and CH/CH were respectively 0.89, 1.00, and 0.41. In populations maintained at 16°C, however, the chromosomes did not change in frequency, and the genotypes appeared not to differ in fitness (Dobzhansky 1970). In Chapter 12, we encountered a case in which the relative advantage of different sizes of spots in male guppies depended on whether contrast with the background made the males conspicuous to predators.

Methods of Studying Natural Selection

The simplest question we can ask about the state of a characteristic or an allele frequency in a population is (1) whether it is determined by natural selection rather than by one of several other possible causes: genetic drift, gene flow, or historical events that no longer operate. As we noted in Chapter 11, many population geneticists hold that chance, or genetic drift, should be assumed to explain the data unless there is evidence that natural selection or other factors have played a role. If the trait is subject to selection, then we wish to determine (2) the relationship between the trait value and fitness, i.e., which phenotypes have higher versus lower fitness under one or several environmental conditions. We would then like to know (3) the causal relationship between the trait and fitness—namely, which fitness components differ and what the ecological, mechanical, physiological, or biochemical reasons for these differences may be. To judge the evolutionary consequences of fitness differences among phenotypes, we need to know (4) the correspondence between phenotype and genotype. Finally, if we can (5) quantify the differences in fitness, we can estimate selection coefficients and employ these in models that predict the dynamics of evolution.

Many methods are used to address these questions. We will briefly survey some of these methods, drawing on Endler's (1986) comprehensive treatment.

Correlations across Populations

Geographic variation within species has probably provided more evidence for the existence of natural selection than any other source of information. A selectively neutral locus or trait should show a random pattern of variation among populations, whereas nonrandom patterns are likely to imply selection. In Chapter 12, we described a latitudinal cline in the frequency of two alleles at the alcohol dehydrogenase locus in *Drosophila melanogaster*. If this had been observed only in, say, North America, it could possibly be explained by gene flow if, for example, separate colonies of this species with different allele frequencies had first been established in the north and south, and had expanded and interbred. But the cline is also observed on two other continents. A *consistently repeated pattern* of this kind is unlikely to arise from random events: it must result from selection. Similarly, parallel clines among different species provide a consistent pattern that implies selection. For example, Richard Harrison (1977) found the same clinal pattern of allele frequencies at the phosphoglucose isomerase locus in two closely related species of field crickets (*Gryllus*) in eastern North America. Among populations ranging from North Carolina to Connecticut, the two most common alleles vary substantially in frequency, but are nearly the same in samples from the two species taken from any one locality (Figure 13.5).

In these cases, it is not clear what selective factor is responsible for the correlations: we infer the existence of selection, but not its cause. Geographic variation is most informative when a trait is statistically correlated with an environmental factor. For example, male guppies (*Poecilia reticulata*) have less conspicuous coloration in streams with predators than in streams that lack them, implying that predation may be a selective factor (see Chapter 12). The European land snail *Cepaea nemoralis* has a complex genetic polymorphism for both shell color (brown, pink, or yellow) and the number of dark bands on the shell (0 to 5). In samples from numerous wooded English localities, there is a high proportion of unbanded, brown or pink snails, whereas in open habitats such as fields, banded yellow snails are more abundant (Figure 13.6). Selection must be responsible for this pattern; likely factors might be predation, if the coloration matches the background, or temperature, if shell color affects heat exchange.

In such comparisons among populations, it is important to determine that the phenotype is not directly induced by the local environment. For some features, such as allozyme mobility, this is unlikely; for others, such as body size, environmental effects may be assessed by rearing samples in a common environment (see Chapter 9).

FIGURE 13.5 The frequency of the PGI^{100} allele at the phosphoglucose isomerase locus in two species of field crickets, *Gryllus pennsylvanicus* and *G. veletis*, sampled from 11 localities, from New York and Connecticut in the north to North Carolina in the south. Along the dashed line, the allele frequencies are equal in the two species. This strong correlation suggests that some geographically variable agent of natural selection affects fitnesses of genotypes at this locus similarly in both species. (After Harrison 1977.)

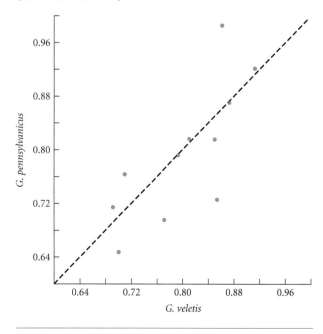

FIGURE 13.6 Microgeographic variation in the frequencies of shell color and banding variants in the land snail *Cepaea nemoralis* in the vicinity of Oxford, England. Each symbol represents a sample from the habitat indicated. In wooded habitats, there is a high proportion of unbanded, brown or pink shells (note that most circles are toward the lower right). In more open habitats (triangles and squares, toward upper left), more of the snails have yellow, banded shells. (After Cain and Sheppard 1954.)

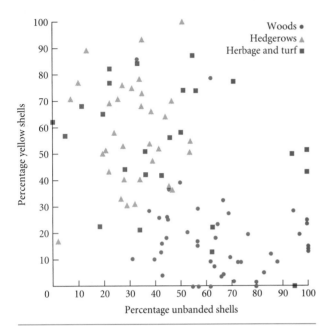

Deviations from Expected Genotype Frequencies

Deviations from Hardy-Weinberg genotype frequencies may sometimes indicate the existence of selection. For example, Dobzhansky and Levene (1948) found a slight but significant excess of individuals heterozygous for chromosome inversions, compared with the expected frequency $2pq$, in a natural population of *Drosophila pseudoobscura*, as well as in caged laboratory populations in which there was intense competition for food. They inferred that heterozygotes had higher egg-to-adult survival. Generally, though, the fit of genotype frequencies to the Hardy-Weinberg distribution is a very weak test for selection, because selection may be operating and yet not cause a detectable deviation from Hardy-Weinberg frequencies. Moreover, other factors can cause deviations: for example, inbreeding or population subdivision can cause heterozygote deficiencies (see Chapter 11).

Selection may also be implicated if two or more loci display strong linkage disequilibrium despite the potential for recombination to break down such gene combinations. For example, the several genes that together produce either the pin or the thrum phenotype in heterostylous plants (see Chapter 9) would not show such strong and consistent associations unless selection eliminated recombinants.

Finally, because gene flow among populations homogenizes their allele frequencies (see Chapter 11), large differences between neighboring populations imply that selection counteracts gene flow by favoring different genotypes in each. For example, despite the high rate of dispersal in *Drosophila melanogaster*, the frequency of the *F* allele of alcohol dehydrogenase is about 0.90 in the vicinity of an African brewery, but only about 0.03 at countryside sites about a kilometer away (Vouidibio et al. 1989). Only selection at this or a linked locus could maintain such a difference in the face of gene flow.

Temporal Patterns

A selectively neutral polymorphism will give way, under genetic drift, to fixation of one allele at a rate that depends on population size (see Chapter 11). Pleistocene fossils of the snail *Cepaea nemoralis* display the same polymorphism in bands on the shell as contemporary populations (Diver 1929); the apparent stability of the polymorphism implies that selection has prevented fixation by genetic drift. Likewise, some molecular polymorphisms detected by DNA sequencing are shared by species that diverged more than 10 million years ago, and must therefore be maintained by some form of selection (see Chapter 22). Conversely, if the genetic constitution changes too fast to be accounted for by drift, selection is likely to be responsible. The most conspicuous such example is the famous case of "industrial melanism" in the peppered moth (*Biston betularia*). Museum collections dating since the mid-nineteenth century in England, less than a century after the onset of the Industrial Revolution, show a decline of the "typical" pale gray form and a rapid, rather steady increase of the black (melanic) form, from about 1 percent to more than 90 percent in some areas (see Figure 23 in Chapter 12). The rate of change is so great that it implies a very substantial 50 percent selective advantage for the melanic form (Haldane 1932). In this instance, we know further that the frequency of the melanic allele tends to be highest in heavily industrialized areas, where air pollution has killed off the pale lichens that invest tree trunks in nonpolluted areas.

Response of Populations to Perturbations

Selection is often inferred from genetic changes in populations exposed to an altered environment. Such "natural experiments" occur when a population's environment changes because of human activity—industrial melanism in *Biston* and other insects is a case in point—or more natural events. For instance, during a drought in the Galápagos Islands, the mean beak and body size of the finch *Geospiza fortis* increased substantially within a single generation, apparently because of the higher survival of larger birds, which could feed on larger seeds, and so could use a resource that smaller birds could not. Similar evidence comes from changes in populations accidentally or deliberately introduced into new environments, and in populations subjected, in either the field or the laboratory, to experimental manipulation of environmental variables.

Genetic Demography

A variety of approaches to the study of natural selection compare one or more components of fitness among different phenotypes or genotypes. Occasionally selection on a trait is inferred by artificially creating phenotypic variants; an example is Andersson's experimental manipulation of the tail feathers of male widowbirds, which demonstrated a correlation between tail length and mating success (see Chapter 12). More often, measurements are made on naturally occurring phenotypes within a population. The example of fitness component analysis in the previous section, on an esterase polymorphism in the eelpout *Zoarces*, is unusually complete; most such studies address only one or a few components of fitness.

Many researchers have compared the phenotypes of survivors versus nonsurvivors, or breeding versus nonbreeding individuals. For instance, in a study on the banding polymorphism of the land snail *Cepaea nemoralis*, Cain and Sheppard (1954) collected the broken shells of snails that had been eaten by thrushes and compared the proportions of phenotypes among these shells with those among samples of live snails from the same vicinity. In one locality, banded shells made up 47 percent of live snails, but 56 percent of dead ones, implying that the banded phenotype was more susceptible to predation at this site. In general, predation by thrushes was higher on shell types that the investigators judged to be more conspicuous in the local habitat (woods versus fields). Quantitative studies of this kind can yield not only evidence of selection, but estimates of its intensity and sometimes information on its causes.

Functional Studies

Many of the methods described above may serve to detect and measure selection, but often they do not tell us how a difference in phenotype causes a difference in fitness. Often, experimental studies of function and performance are necessary for this purpose. These studies do not in themselves tell us that the phenotypes differ in fitness, but may be used in conjunction with methods that do, such as genetic demography.

An example of such a multifaceted analysis is a study by Ward Watt and his colleagues (Watt 1977, 1983; Watt et al. 1983, 1985) on the enzyme phosphoglucose isomerase (PGI) in several species of sulfur butterflies (*Colias*). This glycolytic enzyme plays several key roles in allocating carbohydrates among biochemical pathways, particularly in the flux of glucose from glycogen storage to central metabolism. The enzyme is a dimer, composed of two polypeptide chains both translated from the same locus; thus, for example, an individual heterozygous for alleles *1* and *2* (denoted genotype *1/2*) has three kinds of enzyme: the homodimers *1-1* and *2-2* and the heterodimer *1-2*. The functional properties of a heterodimer may differ from the average of the two homodimers.

Several electrophoretically distinguishable alleles constitute a polymorphism in each species of *Colias*. These species include *C. meadii*, which occurs at high altitudes, and *C. philodice* and *C. eurytheme*, both of which have lower altitudinal distributions in western North America. These butterflies can fly only when their body temperature is high (often nearly 40°C). They cannot maintain flight temperature early in the morning, or—especially at high altitudes—when clouds block the sunlight. Watt argued that selection should favor PGI variants with high catalytic efficiency, especially at low temperatures, because such enzymes would maximize glucose metabolism and so enable the animal to fly rapidly and escape predators and sudden storms. Perhaps more importantly, males that fly over the greatest time span per day would have an advantage in finding females, especially in the early morning, when young females emerge from the pupal state. At excessively high temperatures, however, thermal instability of these enzymes may lower the animal's ability to fly.

Alleles *2* and *3* are most common in high-altitude *C. meadii*, whereas alleles *3* and *4* are most common in *C. philodice* and *C. eurytheme*. Using standard biochemical methods, Watt measured the enzyme activity* of each genotype at 10°, 30°, and 40°, as well as the enzymes' thermal stability at 50°. At low temperatures, genotypes *2/2* and *2/3* have higher enzyme activity than most other genotypes (e.g., *3/3* and *4/4*), but they lose activity more rapidly at high temperatures (Figure 13.7A). These in vitro results fit nicely with the high frequency of allele *2* in the high-altitude *C. meadii*. Even more interesting was the observation that in several instances, heterozygotes have higher enzyme activity than either homozygote, rather than being intermediate. Genotype *3/4* is particularly striking in this respect (Figure 13.7B). Perhaps the heterodimer in such heterozygotes has unusually high activity.

These biochemical studies show that the allozymes differ in functional properties, and the differences correspond to the prevalence of different alleles in species that live in different environments. But these observations do not prove that the enzyme variants affect fitness. Such evidence comes from studies of genotype frequencies and genetic demography.

The first evidence of selection on PGI in *Colias* was the observation that in samples from several species, heterozygotes were more frequent, and homozygotes less frequent, than expected under the Hardy-Weinberg principle. Moreover, heterozygote excess was observed among old, but not young, butterflies. (These can be distinguished by the tattering and other signs of wear on the wings of old butter-

---

*Watt measured V_{max}, the maximal velocity of the enzyme-mediated reaction at a given temperature, and the Michaelis constant K_m, the substrate binding affinity. A lower value of K_m indicates greater binding affinity. The ratio V_{max}/K_m is a "coupling ratio" that measures the rapidity of response of the enzyme to changes in substrate flux from preceding steps in the biochemical pathway. Watt argues that selection in *Colias* should favor low values of K_m and high values of V_{max}/K_m. The text reference to "enzyme activity" refers to this coupling ratio.

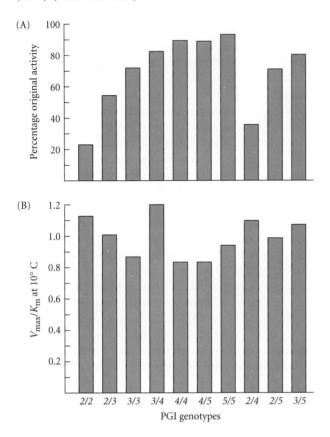

Perhaps the most dramatic differences in fitness components were revealed by comparing allele frequencies among males caught in flight with those among males that had mated. If a female mates with several males, the sperm from the most recent mate fertilizes all the eggs laid within a short time thereafter. This fact was demonstrated by an experiment in which females, their offspring, and two males with which each female had sequentially mated were electrophoretically typed. (For example, if a *3/3* female mated first with a *3/3* and then with a *4/4* male, all her offspring were *3/4*.) Knowing that all a female's offspring have the same father, one can capture wild females, rear and determine the genotype of several offspring from each, and deduce the genotype of each father. For example, if seven progeny of a *3/3* female are all *3/3*, we can have more than 99 percent confidence that the father was *3/3* also, because the probability that he was heterozygous for a second allele that was transmitted to none of seven offspring is only $(0.5)^7 = 1/128 = 0.008$. Similar but more complicated calculations can be used to determine the father's genotype when the mother is heterozygous.

In one of several analyses of male mating success, Watt and his colleagues captured 78 flying males and 80 females of *Colias eurytheme* in a field in California on September 24, 1983. From the electrophoretic patterns among the progeny of the 80 females, the genotype frequencies among the mating males were estimated. In this sample, the heterozygous genotypes, such as *3/4*, that on

flies.) In sequential samples of a population of *C. philodice* taken throughout the summer, the excess of heterozygotes grew steadily as the butterflies aged, and then dropped to zero when this generation of insects was replaced by a new brood of young butterflies (Figure 13.8). Superior survival of heterozygotes is the most plausible explanation for this pattern.

Second, Watt and his colleagues predicted from the biochemical data on enzyme activity and thermal stability that heterozygotes, especially genotype *3/4*, should fly more frequently than other genotypes in the cool morning, and should fly over a greater span of time during the day. They tested this prediction six times, in large populations of *C. philodice* and *C. eurytheme*, by collecting samples of flying butterflies at four or five times throughout the day. Their predictions were upheld in each case (Figure 13.9). They predicted also that genotypes carrying alleles *4* and *5* should survive better than other genotypes at high temperatures, because these allozymes are most heat-stable. During an unusual heat wave in Colorado, this proved to be the case. The combined frequency of these alleles increased from 0.109 to 0.177 within the same cohort of *C. philodice* as it aged.

FIGURE 13.8 Changes in the excess of PGI heterozygotes over Hardy-Weinberg expected frequencies in a population of the butterfly *Colias philodice eriphyle* in 1975. The expected frequency of heterozygotes was calculated from the allele frequencies in each sample; the vertical axis plots the difference between this value and the observed frequency of heterozygotes. The excess of heterozygotes increased as the first brood of butterflies aged in July. In September, newly emerged butterflies of the second brood showed no excess. A sample from the end of the first brood of 1974, plotted for comparison, suggests that the increase in heterozygotes with age is a consistent pattern. The change in the frequency of heterozygotes within a cohort is probably due to superior survival of heterozygotes. (After Watt 1977.)

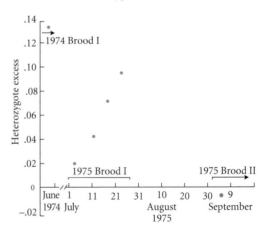

FIGURE 13.9 Changes over a single day in the density of actively flying individuals of different PGI genotypes in the butterfly *Colias eurytheme*. The vertical axis measures the number of flying individuals relative to the greatest total number observed at any time during the day. All genotypes are highly active in early afternoon, but genotype *3/4* was relatively more active in the morning than either of the homozygotes *3/3* or *4/4*. The dots indicate the environmental temperature (T_a) at each time. (After Watt 1983.)

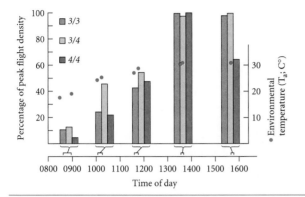

biochemical grounds were predicted to have the greatest flight span made up only 43.6 percent of the flying males, but 68.8 percent of the mating males. The frequency of the biochemically least favored genotypes was 17.6 percent among the flying males, but only 9.3 percent among males that had mated. The fitness advantage of the heterozygotes, predicted from a combination of biochemical data and knowledge of the behavioral ecology of the species, is remarkably high.

Molecular Variation

Ever since electrophoretic analyses of protein variation in the 1960s revealed abundant polymorphism, there has been debate about how to interpret variation at the molecular level within and among species. The same controversy surrounds variation in DNA sequences. "Neutralists" hold that most mutations are either deleterious, and are eliminated or kept at very low frequencies by natural selection, or selectively neutral. In their view, neutral mutations, changing in frequency by genetic drift, account for most of the molecular variation within and among species (see Chapter 11). "Selectionists," in contrast, hold that natural selection maintains much of the molecular variation within species and is the cause of many of the molecular differences among species.

Many models have been proposed to resolve this "neutralist-selectionist controversy." Most of them are meant to detect the existence of selection by deviations in the pattern of variation from the pattern predicted by the neutral theory. In Chapter 11, for instance, we discussed the work of McDonald and Kreitman (1991), who showed that the ratio of nonsynonymous to synonymous substitutions was lower for intraspecific polymorphism than for divergence between species of *Drosophila*, whereas the neutral theory predicts that the ratio should be the same in both. Several

other tests for natural selection, designed specifically for molecular data, are described in Chapter 22.

Models of Selection

In the following simple models, we make the following assumptions: the population is very large, so genetic drift may be ignored; mating occurs at random; mutation and gene flow do not occur; and selection at other loci either does not exist or does not affect the locus we are considering. We will later consider the consequences of changing these unrealistic assumptions. We also assume, for the sake of simplicity, that selection acts by differential viability among genotypes in a species with discrete generations. The principles are much the same for other components of selection and for species with overlapping generations, although these factors introduce complications when data from real populations are analyzed. In the first few models, we assume that within a given environment, the relative fitness of each genotype can be specified by a single number (a *constant*).

In all models of sexual populations, the frequency of an allele—say, A_1—in the next generation—say, p'—is the proportion of all transmitted gene copies that are of that allele. These are all the gene copies transmitted in gametes by A_1A_1 homozygotes (p^2, weighted by their fitness W_{11}) plus half those transmitted by heterozygotes ($1/2 \times 2pq$, weighted by W_{12}, the fitness of heterozygotes). Of all the gene copies present in the population before selection acts, the total proportion transmitted is the sum of each genotype's frequency weighted by its fitness, or $p^2W_{11} + 2pqW_{12} + q^2W_{22} = \bar{w}$. Thus

$$p' = \frac{p^2W_{11} + pqW_{12}}{p^2W_{11} + 2pqW_{12} + q^2W_{22}} \quad (13.3)$$

Assuming that gametes unite at random, the genotype frequencies in the next generation will have changed from p^2, $2pq$, q^2 to p'^2, $2p'q'$, q'^2, where $q' = 1 - p'$. Building on this foundation, we now consider several models of selection. Box 13.A provides a mathematical framework for these models.

Directional Selection

Theory The replacement of relatively disadvantageous alleles by advantageous alleles is the fundamental basis of adaptive evolution. This occurs when the homozygote for an advantageous allele has a fitness equal to or greater than that of the heterozygote or of any other genotype in the population. An advantageous allele may initially be very rare if it is a newly arisen mutation or if it was formerly disadvantageous, but the environment has changed so that it is now favorable. Or it may initially be fairly common if under previous environmental circumstances it was selectively neutral or was maintained by one of several forms of balancing selection (see below). An advantageous allele that increases from a very low frequency is often said to INVADE a population, and it SPREADS through a population as it increases in frequency.

Selection Models with Constant Fitnesses

We first present a general model of selection at one locus,[*] and then modify it for specific cases. Suppose three genotypes at a locus differ in relative fitness due to differences in survival:

	A_1A_1	A_1A_2	A_2A_2
Frequency at birth	p^2	$2pq$	q^2
Relative fitness	w_{11}	w_{12}	w_{22}

The ratio of $A_1A_1 : A_1A_2 : A_2A_2$ among surviving adults is

$$p^2w_{11}:2pqw_{12}:q^2w_{22}$$

and the ratio of the alleles ($A_1 : A_2$) among their gametes is

$$[p^2w_{11} + 1/2\,(2pqw_{12})]:[1/2\,(2pqw_{12}) + q^2w_{22}]$$

which simplifies to

$$p(pw_{11} + qw_{12}):q(pw_{12} + qw_{22})$$

The GAMETE FREQUENCIES, which are the allele frequencies among the next generation of offspring, are found by dividing each term by the sum of the gametes, which is

$$p(pw_{11} + qw_{12}) + q(pw_{12} + qw_{22})$$
$$= p^2w_{11} + 2pqw_{12} + q^2w_{22}$$
$$= \bar{w}$$

Thus the allele frequencies after selection (p', q') are the gamete frequencies, or

$$p' = p(pw_{11} + qw_{12})/\bar{w}$$
$$q' = q(pw_{12} + qw_{22})/\bar{w}$$

The *change* in allele frequency between generations is $\Delta p = p' - p$, or

$$\Delta p = \frac{p(pw_{11} + qw_{12}) - p\bar{w}}{\bar{w}}$$

Substituting for $\bar{w}$ and doing the algebra yields

$$\Delta p = \frac{pq[p(w_{11} - w_{12}) + q(w_{12} - w_{22)}]}{\bar{w}} \tag{A.1}$$

We can analyze various cases of selection by entering explicit fitness values for the w's. A few important cases are the following:

1. Advantageous dominant allele, disadvantageous recessive allele ($w_{11} = w_{12} > w_{22}$).

For w_{11}, w_{12}, and w_{22}, substitute 1, 1, and $1 - s$ respectively in equation (A.1). The mean fitness is $p^2(1) + 2pq(1) + q^2(1 - s) =$

$1 - sq^2$ (bearing in mind that $p^2 + 2pq + q^2 = 1$). The equation for allele frequency change is

$$\Delta p = \frac{spq^2}{1 - sq^2}$$

or, equivalently,

$$\Delta q = \frac{-spq^2}{1 - sq^2} \tag{A.2}$$

2. Advantageous allele partially dominant, disadvantageous allele partially recessive ($w_{11} > w_{12} > w_{22}$).

Let h, lying between 0 and 1, measure the degree of dominance for fitness, and substitute 1, $1 - hs$, and $1 - s$ for w_{11}, w_{12}, and w_{22}. (If $h = 0$, allele A_2 is fully recessive.) After sufficient algebra, we find

$$\Delta p = \frac{spq[h(1 - 2q) + sq]}{1 - 2pqhs - sq^2} \tag{A.3}$$

which is positive for all $q > 0$, so allele A_1 increases to fixation. If $h = 1/2$, equation (A3) reduces to $\Delta p = spq/[2(1 - sq)]$.

3. Fitness of heterozygote is greater than that of either homozygote ($w_{11} < w_{12} > w_{22}$).

Using s and t as selection coefficients, let the fitnesses of A_1A_1, A_1A_2, and A_2A_2 be $1 - s$, 1, and $1 - t$ respectively. Substituting these in equation (A1), we obtain

$$\Delta p = \frac{pq(-sp + tq)}{1 - sp^2 - tq^2} \tag{A.4}$$

There is a stable "internal equilibrium" that can be found by setting $\Delta p = 0$; then $sp = tq$. Substituting $1 - p$ for q, the equilibrium frequency p is $t/(s + t)$. Thus the frequency of A_1 is proportional to the relative strength of selection against A_2A_2.

4. Fitness of heterozygote is less than that of either homozygote ($w_{11} > w_{12} < w_{22}$).

As this is the reverse of the preceding case, let $1 + s$, 1, and $1 + t$ be the fitnesses of A_1A_1, A_1A_2, and A_2A_2. The equation for allele frequency change is

$$\Delta p = \frac{pq(sp - tq)}{1 + sp^2 + tq^2} \tag{A.5}$$

Δp is positive if $sp > tq$, and negative if $sp < tq$. Setting $\Delta p = 0$ and solving for p, we find an internal equilibrium at $p = t/(s + t)$, but this is an *unstable* equilibrium. For example, if $s = t$, the unstable equilibrium is $p = 0.5$, but then Δp is positive if $p > q$ (i.e., if $p > 0.5$), and negative if $p < q$.

---

[*]The derivation of equation (A1) follows Hartl and Clark 1989.

The same equations that describe the increase of an advantageous allele describe the fate of a disadvantageous allele. (If A_1 and A_2 are advantageous and disadvantageous alleles with frequencies p and q, and if $p + q = 1$, then $\Delta p = -\Delta q$.) For instance, if A_2 is a deleterious mutation, selection that reduces its frequency or eliminates it is referred to as **purifying selection**, which is simply directional selection in favor of a prevalent homozygous genotype, such as A_1A_1.

If the fitness of the heterozygote is precisely intermediate between that of the two homozygotes, neither allele is dominant with respect to fitness. We can denote the fitnesses W_{11}, W_{12}, and W_{22} as 1, $1 - (s/2)$, and $1 - s$ respectively:

Genotype	A_1A_1	A_1A_2	A_2A_2
Frequency	p^2	$2pq$	q^2
Fitness	1	$1 - (s/2)$	$1 - s$

The coefficient of selection against A_1A_2 ($s/2$) is half that against A_2A_2. From Box 13.A, we find that the advantageous allele A_1 increases in frequency, per generation, by the amount

$$\Delta p = \frac{(1/2)spq}{(1-sq)} \quad (13.4)$$

where $(1 - sq)$ equals the average fitness, $\bar{w}$. Because $q = 1 - p$, $\Delta q = -\Delta p$, so the rate of decline of the disadvantageous allele is given by the negative form of the same expression ($\Delta q = -spq/2\bar{w}$). This expression describes the rate at which purifying selection reduces the frequency of the deleterious allele.

Equation (13.4), like equation (13.2), which describes selection in two asexual genotypes, tells us that allele A_1 increases to fixation ($p = 1$), and that $p = 1$ is a stable equilibrium. The rate of increase is proportional to both the coefficient of selection s and the allele frequencies p and q, which appear in the numerator. Therefore Δp, the rate of evolutionary change, increases as the variation at the locus increases. (When selection is weak, Δp is approximately proportional to the product pq, and is maximal when $p = q = 0.5$.) Thus Δp is also proportional to $2pq$, the frequency of heterozygotes.

Another important aspect of equation (13.4) is that Δp is positive as long as s is greater than zero, even if it is very small. Therefore, as long as no other evolutionary factors intervene, *a characteristic with even a minuscule advantage will be fixed by natural selection.* We could seldom hope to measure a difference in fitness among genotypes of, say, 0.0001, but a feature providing such a small advantage would nevertheless evolve by selection—although it would take a very long time to be fixed. Therefore, even very slight differences among species, in seemingly trivial characteristics such as the distribution of hairs on a fly or veins on a leaf, could conceivably have evolved as adaptations. *This principle explains the extraordinary apparent "perfection" of some characteristics.* Some katydids, for example, resemble dead leaves to an astonishing degree, with transparent "windows" in the wings that resemble holes and blotches that resemble spots of fungi or algae (Color Plate 3B). One might suppose that a less detailed resemblance would provide sufficient protection against predators, and some species are indeed less elaborately cryptic; but if an extra blotch increases the likelihood of survival by even the slightest amount, it may be fixed by selection (providing, we repeat, that no other factors intervene).

Dominance and recessiveness influence the rate of allele frequency change, but not its qualitative outcome. For example, if allele A_1 is fully dominant with respect to fitness, we may denote the fitnesses as

Genotype	A_1A_1	A_1A_2	A_2A_2
Frequency	p^2	$2pq$	q^2
Fitness	1	1	$1 - s$

and the equation for the rate of change (Box 13.A) is

$$\Delta p = \frac{spq^2}{1-sq^2} \quad (13.5)$$

Allele A_1 again increases to fixation, but the maximal rate of change occurs when $p = 1/3$ and $q = 2/3$, rather than when $p = q = 1/2$.

How many generations are required for an advantageous allele to replace one that is disadvantageous? This depends on the initial allele frequencies, the selection coefficient, and the degree of dominance (Figures 13.10, 13.11). For example, if the fitnesses of A_1A_1, A_1A_2, and A_2A_2 are 1.00, 0.99, and 0.98 ($s = 0.02$, no dominance), and the initial frequency of A_1 is 0.75, it requires 110 generations to reach $p = 0.90$, and another 240 generations to reach $p = 0.99$. But if A_1 is dominant, the recessive allele A_2 is eliminated very slowly, *because a rare recessive allele occurs mostly in heterozygous form, and is thus shielded from selection.* Therefore, if the fitnesses of A_1A_1, A_1A_2, and A_2A_2 are 1.0, 1.0, and 0.99 (complete dominance), A_1 increases from $p = 0.75$ to $p = 0.90$ in 710 generations, and another 9,240 generations are required to attain $p = 0.99$. It is to be expected, then, that *deleterious recessive alleles should exist at very low frequencies at many loci* because it takes so long for them to be eliminated entirely—and this is indeed the case (see Chapter 9). (Recurrent mutation also introduces deleterious alleles into populations, as we will see below.)

By the same token, a new advantageous mutation, being initially very rare, increases far more rapidly if it is expressed in the heterozygous state (i.e., if it is partially or completely dominant) than if it is recessive. Until it reaches a fairly high frequency, it is carried almost entirely by heterozygotes, so selection initially has little opportunity to increase its frequency if it is advantageous only when homozygous. Thus, if in a population of A_1A_1 homozygotes, both a dominant mutation A_2 and a recessive mutation A_3 causing the same advantageous phenotype were to arise at the same time, the dominant would increase more rapidly at first, and the recessive, being selectively neutral with

FIGURE 13.10 Rates of increase of an advantageous allele (A_1) from initial frequencies of (A) $p_0 = 0.01$ and (B) $p_0 = 0.10$. For the advantageous dominant A_1, the fitnesses of genotypes A_1A_1, A_1A_2, and A_2A_2 are 1.0, 1.0, and 0.8 respectively; for the "intermediate" case (neither allele is dominant), they are 1.0, 0.9, and 0.8; for the advantageous recessive A_1, they are 1.0, 0.8, and 0.8. Note that the recessive mutation increases only very slowly when its initial frequency is low because it is rarely exposed in homozygous form. The final approach to fixation of the recessive is very rapid, however, because a deleterious dominant allele is being eliminated, a rapid process. Conversely, although an advantageous dominant allele increases rapidly at first, its approach to fixation is very slow because elimination of the increasingly rare recessive is a very slow process. Note also that the steepness of each curve is greatest when allele frequencies are intermediate (between about 0.3 and 0.7).

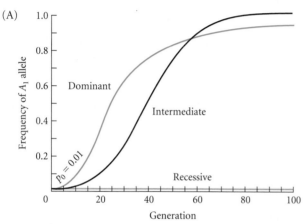

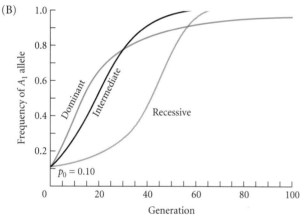

when $\bar{w} = 1 - sq^2$), the mean fitness increases as the frequency (q) of the deleterious allele decreases. *The mean fitness therefore increases as natural selection proceeds*, and reaches its maximum ($\bar{w} = 1$ in this instance) when the population consists entirely of the fittest genotype. In a graphical representation of this relationship (Figure 13.12A), *we may think of the population as climbing up a "hillside" of increasing mean fitness until it arrives at the summit.*

Examples of directional selection　It is difficult to measure all the components of fitness, so it should not be surprising that few investigators have compared observed changes in allele frequencies with those predicted from measurements of fitness. In one such comparison, Peter Dawson (1970) traced the decline in frequency of a recessive lethal mutation ($W_{aa} = 0$) that had been introduced at high frequency ($q = 0.5$) into a laboratory population of the flour beetle *Tribolium castaneum*. He calculated the expected decrease in two ways, assuming either that the mutation was completely recessive ($W_{Aa} = W_{AA} = 1$) or that it lowered the heterozygote's fitness by 10 percent ($W_{Aa} = 0.9$) and so was not completely recessive. The data (Figure 13.13) conformed nicely to those expected if the heterozygote suffered a slight loss of fitness.

If a locus has experienced consistent directional selection for a long time, the advantageous allele should be near equilibrium—that is, near fixation. Thus the dynamics of directional selection are best studied in recently altered environments, such as those altered by human activities. In few, if any, cases is there sufficient information on both genotype fitnesses and the time course of genetic change for us to compare observed with expected changes. In many instances, however, rapid, more or less directional change has been observed.

One such case is the evolution of warfarin resistance in brown rats, *Rattus norvegicus* (Bishop 1981). Warfarin is an

FIGURE 13.11 Rates of increase in the frequency of an advantageous dominant allele (A_1), showing the effect of differences in the magnitude of its fitness advantage s. The fitnesses of A_1A_1, A_1A_2, and A_2A_2 are $1 + s$, $1 + s$, and 1 respectively. (After Hartl and Clark 1989.)

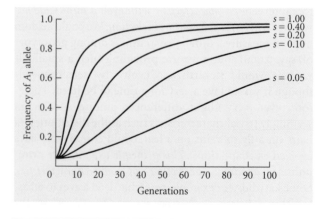

reference to the dominant since they have the same phenotype, would not increase rapidly in frequency. Thus, one might expect adaptive evolution to be based more often on dominant than on recessive mutations. However, an allele that is partly expressed in the heterozygote (i.e., one that is not completely recessive) will increase to fixation nearly as fast as a dominant allele. Much of evolution is based on alleles that are neither completely dominant nor completely recessive.

One more theoretical conclusion can be drawn from equations (13.4) and (13.5). The denominator in both instances is the average relative fitness of individuals in the population, $\bar{w}$. In both cases (i.e., when $\bar{w} = 1 - sq$ and

FIGURE 13.12 Plots of mean fitness ($\bar{w}$) against allele frequency (p) for one locus with two alleles when fitnesses differ due to differences in survival. Each of these plots represents an "adaptive landscape," and may be thought of as a surface, or hillside, over which the population moves. From any given frequency, p, the allele frequency moves in a direction that increases mean fitness ($\bar{w}$). The arrows show the direction of allele frequency change. (A) Directional selection, in which A_1A_1 has the highest fitness. Selection fixes A_1, and the equilibrium ($\hat{p} = 1$) is stable: the allele frequency returns to $p = 1$ if displaced. (B) Directional selection, in which the relative fitnesses are reversed compared with graph A, perhaps because of changed environmental conditions. A_2A_2 is now the favored genotype. (C) An example of overdominance (heterozygous advantage). From any starting point, the population arrives at a stable polymorphic equilibrium ($\hat{p}$). (D) An example of underdominance (heterozygous disadvantage). The interior equilibrium ($\hat{p} = 0.4$ in this example) is unstable because even a slight displacement initiates a change in allele frequency toward one of two stable equilibria: $\hat{p} = 0$ (loss of A_1) or $\hat{p} = 1$ (fixation of A_1). This adaptive landscape has two peaks ($\hat{p} = 0$, $\hat{p} = 1$). (After Hartl and Clark 1989.)

(A) Directional selection
$w_{11} = 1, w_{12} = 0.8, w_{22} = 0.2$

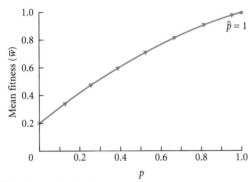

(B) Directional selection
$w_{11} = 0.2, w_{12} = 0.8, w_{22} = 1$

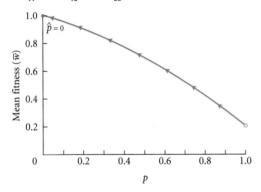

(C) Overdominance
$w_{11} = 0.6, w_{12} = 1, w_{22} = 0.2$

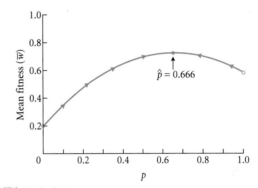

(D) Underdominance
$w_{11} = 1, w_{12} = 0.4, w_{22} = 0.8$

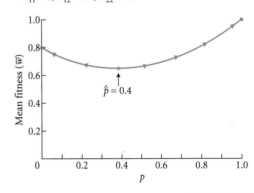

anticoagulant: it inhibits an enzyme responsible for the regeneration of vitamin K, a necessary cofactor in the production of blood-clotting factors. Vitamin K is oxidized when it partakes in this process, and is then reduced to its active state by the enzyme. Susceptible rats poisoned with warfarin often bleed to death from slight wounds. Mutations confer resistance by altering the enzyme to a form that is less sensitive to warfarin, but also less efficient in regenerating vitamin K, so that a higher dietary intake of the vitamin is necessary. It is interesting to note that the mutant alleles are dominant with respect to resistance, but recessive with respect to the animal's vitamin K requirement—which illustrates that dominance and recessiveness are not intrinsic properties of alleles, but rather of the phenotypes that arise from complex biochemical or developmental pathways.

Warfarin has been used as a rat poison in Britain since 1953, and by 1958 resistance was reported in certain rat

FIGURE 13.13 The decrease in the frequency of a recessive lethal allele in a laboratory population of the flour beetle *Tribolium castaneum*. The black line is the expected change in frequency if the allele has no effect on the fitness of the heterozygote; the gray line is the expected change if it lowers the heterozygote's fitness by 10 percent (i.e., is not completely recessive). The triangles represent the actual data. (After Dawson 1970.)

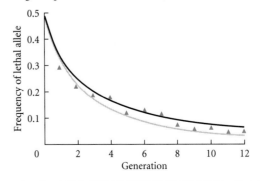

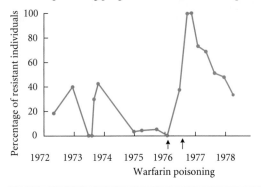

populations. Under exposure to warfarin, resistant rats have a strong survival advantage, and the frequency of the mutation can increase rapidly (Figure 13.14). Resistant rats can suffer a strong disadvantage compared with susceptible rats, however, because of their greater need for vitamin K, and the frequency of the resistance gene drops rapidly if the poison is not administered. Perhaps because of these conflicting selection pressures, the frequency of resistance seldom exceeds 0.5.

Many similar tales can be told of insecticide resistance in insects and mites (Wood 1981; Georghiou 1972; Roush and McKenzie 1987). First reported in a scale insect in 1908, resistance appeared in many species in the 1940s, when synthetic pesticides came into wide use. By 1976,

populations of 364 species were known to be resistant to one or more insecticides (Figure 13.15). Insects that have evolved resistance to one insecticide are often cross-resistant to other insecticides to which they have not been exposed, because the mechanisms of resistance—which include altered detoxification enzymes and reduced sensitivity of the biochemical targets of the poisons—can incidentally confer protection against a variety of chemicals. In other cases, multiple genes confer resistance to multiple chemicals. Some species, such as the Colorado potato beetle (*Leptinotarsa decemlineata*), have evolved resistance to all the major classes of pesticides (Roush and McKenzie 1987). (The evolution of resistance adds immensely to the cost of agriculture and is a major obstacle in the fight against insect-borne diseases such as malaria. For these reasons, as well as the devastating toxic effects of many pesticides on natural ecosystems and on human health, supplementary or alternative methods of pest control are a major topic of research in entomology.)

Insecticide resistance in natural populations of insects is often based primarily on single mutations of large effect (Roush and McKenzie 1987). The resistance allele (R) is usually partially or fully dominant over the allele for susceptibility (S). In populations that have not been exposed to insecticides, R alleles are often so rare as to be undetectable by simply measuring the insects' resistance. However, their presence becomes manifest when they rise rapidly in frequency after a pesticide is applied. Because of the extremely high mortality of susceptible genotypes exposed to strong doses of pesticide (i.e., the selection coefficient s may be nearly 1.0), dominant R alleles often

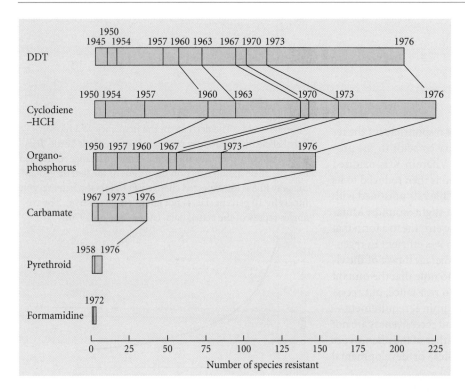

FIGURE 13.15 The cumulative numbers of arthropod pest species known to have evolved resistance to six classes of insecticides by 1976. The insecticides were first used on the leftmost dates in each row. (After Wood 1981.)

increase nearly to fixation within two or three years—usually fewer than 25 generations (Figure 13.16). In the absence of the insecticide, however, the R genotypes usually decline in frequency, because under these circumstances they are likely to be about 5 to 10 percent less fit than S genotypes.

Persistence of Deleterious Alleles in Natural Populations

Although the most advantageous allele at a locus should in theory be fixed by directional selection, deleterious alleles often persist because they are repeatedly reintroduced, either by recurrent mutation or by gene flow from other populations in which they are favored by a different environment. In either case, the frequency of the deleterious allele moves toward a STABLE EQUILIBRIUM that is a *balance between the rate at which it is eliminated by selection and the rate at which it is introduced by mutation or gene flow.*

Suppose, for example, that a deleterious recessive allele A_2, with frequency q, arises at a mutation rate u from other alleles that have a collective frequency of $p = 1 - q$. The increase in frequency of A_2 due to mutation in each generation is up, whereas the decrease in its frequency due to selection (from Equation [13.5]) is $-spq^2/\overline{w}$. At equilibrium, the rate of increase equals the rate of decrease, i.e.,

$$up = spq^2/\overline{w}$$

Assume that A_2 is rare, so that $\overline{w}$ is approximately equal to 1, and solve for the equilibrium frequency, denoted $\hat{q}$. We find that $\hat{q}^2 = u/s$, and

$$q = \sqrt{u/s}$$

Note that $\hat{q}^2$ is the equilibrium frequency of the recessive homozygous genotype.

The equilibrium frequency of a deleterious allele is directly proportional to the mutation rate, and inversely proportional to the strength of selection. Thus if s is much greater than u, the allele will be very rare; for example, $s = 1$

if A_2 is a recessive lethal allele. Then if the mutation rate is 10^{-6}, the equilibrium frequency will be $\hat{q} = 0.001$, but almost all the individuals with this allele will be heterozygous. If the deleterious allele is partly or entirely dominant, its equilibrium frequency will be even lower, because selection then eliminates both homozygous and heterozygous carriers.

We noted in Chapter 9 that a large fraction of chromosomes in *Drosophila* populations are lethal when homozygous because of individually rare lethal mutations at each of a great many loci. (The same is true for humans and almost all other species.) The frequency of each such mutation is, on average, actually lower than would be expected if it lowered fitness only when homozygous—low enough to imply that the average such mutation reduces fitness by about 2 to 5 percent in heterozygous condition (Simmons and Crow 1977; Crow 1993). In human populations, a severe degenerative disorder of the neuromuscular system, called Huntington's disease, is caused by a dominant mutation (*H*). Heterozygotes (*Hh*) usually do not develop the disease until about age 35, by which time they have often reproduced (Hartl and Clark 1989); thus their fitness is not zero. From records of birth rates, the fitness of *Hh* has been calculated to be 0.81 relative to normal homozygotes *hh* (for which *W* = 1). As expected of a deleterious dominant allele, the frequency of *H* is very low—about 0.00005 in Michigan.

Selection and Gene Flow

Different regimes of directional selection often fix different alleles among different populations of a species. Thus in the absence of gene flow, the frequency, q, of an allele A_2 will be 1 in certain populations and 0 in others. Gene flow among populations can introduce each allele into populations in which it is deleterious, and the allele frequency thus arrives at an equilibrium ($\hat{q}$) set by the balance between selection and gene flow (Figure 13.17). If the gene flow rate m is much less than the strength of selection s, $\hat{q}$ is approximately mq_m/s, where q_m is the frequency of the allele among the immigrants. For example, if A_2 is

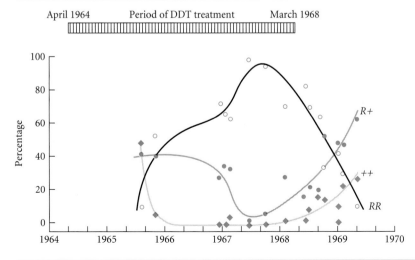

April 1964   Period of DDT treatment   March 1968

FIGURE 13.16 Changes in the proportions of genotypes in a population of mosquitoes (*Aedes aegypti*) in a suburb of Bangkok, Thailand, during and after a period of treatment with DDT. *R* is the resistance allele, + the wild-type (susceptible) allele. The curves fit the changes in frequency of genotypes *RR* (open circles), *R+* (solid circles), and *++* (diamonds). In each sample, the genotype frequencies fit Hardy-Weinberg expectations. The frequency of the resistance allele was high during DDT application. It declined after application ceased, suggesting that the susceptible allele confers higher fitness in the absence of DDT. (After Wood and Bishop 1981.)

FIGURE 13.17 Effects on allele frequencies of gene flow opposed by selection. In a local population (such as an island population), the recessive homozygote A_2A_2 has a selective advantage of 0.20. Gene flow from another (e.g., continental) population in which A_1 is fixed introduces A_1 alleles at a rate m per generation. Selection increases q toward 1.0, so that the change in allele frequency, Δq, is positive unless gene flow, which decreases q, is too great (in which case Δq is negative). For any given value of m, the point at which the Δq curve crosses the horizontal axis (q) is the equilibrium point, the allele frequency at which selection and gene flow are balanced. The lower the influx of A_1 alleles (lower m), the higher the equilibrium frequency of A_2 is. (After Hartl and Clark 1989.)

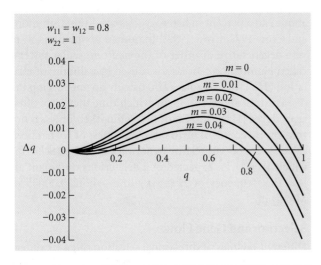

fixed in a source population so that $q_m = 1$, but reduces the fitness of heterozygotes by 10 percent ($s = 0.1$) in a local population of which 1 percent are immigrants from the source population in each generation, then $\hat{q} \approx 0.1$. Thus gene flow can contribute to genetic variation within populations. If gene flow is much greater than selection, a population inhabiting a small patch of a distinctive environment will not become genetically differentiated from surrounding populations.

If a series of local populations is distributed along an environmental gradient over which selection changes, then in the absence of gene flow, we should expect an abrupt shift in allele frequencies (a "step cline") at the point at which selection favors different alleles. This is true even if the environment changes gradually, as long as one homozygote or the other has highest fitness in each population. However, a smooth *cline in allele frequencies may be established if there is gene flow* among populations along the gradient (Figure 13.18). The width of the cline (the distance over which q changes from, say, 0.2 to 0.8) is proportional to V/s, where V measures the distance that genes disperse, and s is the strength of selection against them. Thus a cline will be wide and shallow if gene flow is high or if selection is weak on the scale over which genes disperse (as it may be if the environment changes very gradually). Conversely, if selection is strong relative to gene flow, steep clines in allele frequencies will result, so that populations are strongly differentiated. Several authors have argued that genetic uniformity in a widespread species generally cannot be due to contemporary gene flow, because the distance over which gene flow occurs in most species is very small rela-

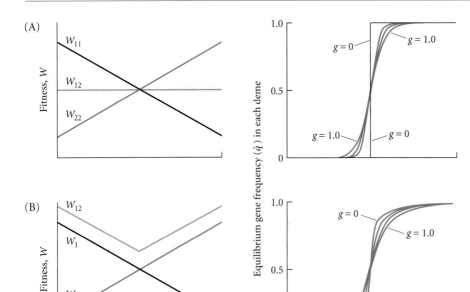

FIGURE 13.18 Interaction between selection and gene flow. At left, genotype fitnesses change gradually along an environmental gradient from west to east. The fitness of A_1A_2 is intermediate in (A) and superior in (B). At right, the frequency of A_2 in an array of demes changes clinally. The steepness of the cline depends on the fitnesses and the level of gene flow (g) between adjacent demes, ranging from 0 to 100 percent ($g=1$). (After Endler 1973.)

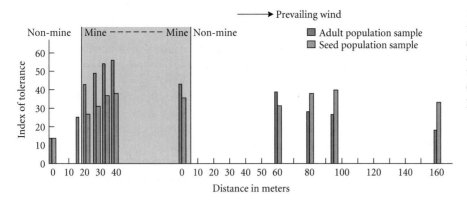

FIGURE 13.19 The average copper tolerance of the grass *Agrostis tenuis* along a transect through the edge of a copper mine is higher downwind than upwind from the mine because of gene flow due to wind-borne pollen. (After Macnair 1981.)

tive to their total geographic range (Ehrlich and Raven 1969; Barton and Clark 1990).

An example of a local cline is provided by copper tolerance in the wind-pollinated grass *Agrostis tenuis* growing at mine sites in Britain. Copper in mine wastes is toxic to nontolerant genotypes, but these are competitively superior to copper-tolerant genotypes on normal soil. The frequency of copper-tolerant genotypes declines rapidly away from copper-laden soil, but it is still moderately high, especially downwind from the copper-tolerant populations, due to the movement of pollen (McNeilly 1968; Figure 13.19).

A longer cline is exhibited by alleles at the aminopeptidase I locus in the blue mussel (*Mytilus edulis*) in Long Island Sound, a body of water about 100 miles long in which salinity increases from the west to the east, where the Sound meets the ocean. The frequency of one of several electrophoretic alleles, ap^{94}, increases from about 0.12 in the west to 0.55 at the oceanic sites (Figure 13.20). The planktonic larvae of these mussels disperse over long distances, so gene flow is high; therefore the cline must be maintained by countervailing selection at this or a closely linked locus. Richard Koehn and Thomas Hilbish (1987) and their col-

leagues found that aminopeptidase I activity is higher in animals carrying the ap^{94} allele than in other genotypes. This enzyme cleaves terminal amino acids from proteins, increasing the intracellular concentration of free amino acids and thereby helping to maintain osmotic balance in saline waters. Because ap^{94} genotypes have higher intracellular amino acid levels than other genotypes, the ap^{94} allele is favored at oceanic salinity. In less saline waters, however, high enzyme activity is less advantageous because a lower concentration of amino acids suffices for osmotic balance; indeed, high aminopeptidase activity is disadvantageous because the breakdown of protein is costly in terms of both energy and nitrogen, and must be compensated by feeding. The investigators found that within Long Island Sound, ap^{94} has a high frequency among recently settled young mussels, but that its frequency decreases as they age, indicating that ap^{94} mussels have a higher mortality rate than others. The decline in frequency is most marked during the autumn, when a combination of low food supply and high temperature makes it difficult to maintain a positive energy budget. The persistence of ap^{94} within the Sound therefore seems to be due to incursions of larvae from the ocean in each generation.

FIGURE 13.20 The frequency of the ap^{94} allele, indicated by the dark portion of each circle, in samples of the mussel *Mytilus edulis* in Long Island Sound and nearby sites. The frequency of the allele drops rapidly over a 30-kilometer distance in the Sound, where salinity decreases from east to west, despite high gene flow. (After Koehn and Hilbish 1987.)

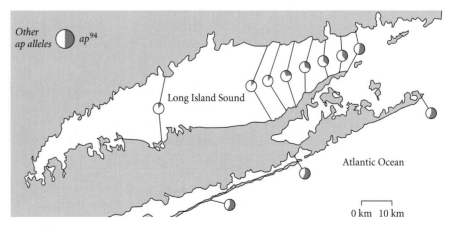

Polymorphism Maintained by Selection

The most prominent subject of study by population geneticists—both those who devise theory and those who do experiments—is polymorphism. Polymorphisms provide valuable material for studying natural selection and the other factors of evolution because they constitute much of the genetic variation that is the foundation of evolutionary change, and accounting for the very existence of genetic variation has been a major challenge. Until the 1930s and beyond, the prevalent, or CLASSIC, view had been that at each locus, a best allele (the "wild type") should be nearly fixed by natural selection, so that the only variation should consist of rare deleterious alleles, recently arisen by mutation and fated to be eliminated by purifying selection. As we have seen (in Chapter 9), studies of natural populations revealed instead a wealth of variation. The factors responsible for this variation might be (1) recurrent mutation producing deleterious alleles, subject to only weak selection; (2) gene flow of locally deleterious alleles from other populations in which they are favored by selection; (3) selective neutrality, i.e., genetic drift; and (4) maintenance of polymorphism by natural selection.

From the 1940s on, studies of many polymorphisms led to the conclusion that they were subject to rather strong selection, so that neither recurrent mutation nor genetic drift could explain all the variation. British ecological geneticists led by E. B. Ford and American population geneticists influenced by Theodosius Dobzhansky, representing the BALANCE school, held that natural selection maintains a great deal of genetic variation. For their part, theoreticians devised many models of how natural selection might maintain two or more alleles in a population. We treat here only the most commonly discussed models of selectively maintained polymorphism.

Heterozygote Advantage

If the heterozygote has higher fitness than either homozygote, both alleles are necessarily propagated in successive generations, in which, of course, union of gametes yields all three genotypes among the zygotes. Heterozygote advantage is also termed OVERDOMINANCE or HETEROSIS for fitness. If the fitnesses of A_1A_1, A_1A_2, and A_2A_2 are $1 - s$, 1, and $1 - t$ respectively, selection will bring the allele frequencies from any initial value to the stable equilibrium $\hat{p} = t/(s + t)$, $\hat{q} = s/(s + t)$, where $\hat{p}$ and $\hat{q}$ are the equilibrium frequencies of A_1 and A_2 respectively. The equilibrium frequencies of the alleles and genotypes thus depend on the balance of fitness of the two homozygotes (see Box 13.A, Figure 13.12C, Figure 13.21A–C).

Genotypes that are heterozygous at several or many loci often *appear* to be fitter than more homozygous genotypes. For example, inbreeding depression is commonly observed when organisms become more homozygous under inbreeding (see Chapter 11); they decline in fertility, viability, and overall vigor. Conversely, crossing inbred strains often results in HYBRID VIGOR, or heterosis (see Figure 14 in Chapter 11). Electrophoretic studies of plants, bivalves, butterflies, and other organisms have often shown that the number of loci at which an individual is heterozygous is correlated with its growth rate or other indicators of "health" or "vigor."

None of these observations, however, gives unequivocal evidence that the heterozygote at any given locus is more fit or vigorous than the homozygotes. Inbreeding depression is best explained by dominance rather than by overdominance, and the same may be true for correlations between allozyme heterozygosity and fitness. Suppose there were linkage disequilibrium between favorable dominant alleles (A, B) and unfavorable recessive alleles (a, b) at two closely linked loci such that the chromosomes Ab and aB were more prevalent than AB and ab. If the loci independently

---

FIGURE 13.21 (A–C) Genotype frequencies among newborn zygotes at a locus with heterozygous advantage (overdominance) for several relative fitnesses of the two homozygotes. The fitnesses of A_1A_1, A_1A_2, and A_2A_2 are $1 - s$, 1, and $1 - t$; these values, and those of s and t, are shown below the genotypes. (D) Expected frequencies for the sickle-cell hemoglobin polymorphism, using fitness values estimated for an African population exposed to malaria. AA is the normal homozygote, AS has the sickle-cell trait, and SS expresses sickle-cell anemia.

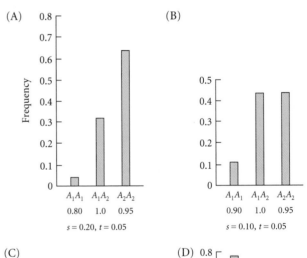

(A)

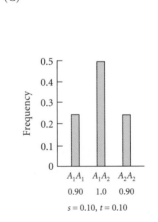

(B)

(C)

(D)

affect the probability of survival, the fitness of a two-locus genotype equals the product of the fitness values at each of the loci. Then, if the fitness of each recessive homozygote is $1 - s$ and that of the dominant homozygotes and heterozygotes is 1, the three most prevalent genotypes are Ab/Ab with fitness $1 - s$, Ab/aB with fitness 1, and aB/aB with fitness $1 - s$. If we observe only the marker locus A, a (perhaps by electrophoresis), the heterozygote will appear fittest, but only because of its association with undetected alleles at the B locus. Less pronounced linkage disequilibrium yields the same, but a less pronounced, effect. The apparent but spurious superiority of the heterozygotes at the marker locus is called ASSOCIATIVE OVERDOMINANCE.

The possibility of linkage disequilibrium with recessive alleles means that most of the many cases of apparent heterozygous advantage are equivocal. For example, Dobzhansky and his colleagues (see Dobzhansky 1970) have described several ways in which heterozygotes for chromosome inversions in *Drosophila pseudoobscura* are superior to chromosome homozygotes in laboratory populations (cf. Chapter 12). At some temperatures, caged populations initiated with different frequencies of inversions such as ST and AR converge toward an intermediate equilibrium (see Figure 10B in Chapter 12). Chromosome heterozygotes develop faster, mate faster, and appear to survive better than homozygotes, and polymorphic populations have a higher growth rate and produce more flies than monomorphic populations. Such results might be expected if each chromosome type carried, among the many loci embraced by the inversion, deleterious recessive alleles that are masked by complementary dominant alleles in the other chromosome type. Dobzhansky, however, believed that each chromosome carried a COADAPTED set of alleles at several or many loci, which together confer greatest fitness when heterozygous. That is, ST chromosomes might carry a combination ABC, and AR chromosomes a combination $A'B'C'$, which confer higher fitness in the $ABC/A'B'C'$ heterozygote than in any other combination (see Chapter 14).

Single-locus heterozygote advantage has been documented in a few cases, including Watt's study of PGI in *Colias* butterflies (see above). The best-known case is the β-hemoglobin locus in some African and Mediterranean human populations (Cavalli-Sforza and Bodmer 1971). One allele at this locus, sickle-cell hemoglobin (S), is distinguished by a single amino acid substitution from normal hemoglobin (A). At low oxygen concentrations, S hemoglobin forms elongate crystals, which carry oxygen less effectively and cause the red blood cells to adopt a sickle shape and to be broken down more rapidly. Heterozygotes (AS) suffer slight anemia; homozygotes suffer severe anemia (sickle-cell disease) and usually die before reproducing. However, if the red blood cells of heterozygotes are infected by the sporozoan protozoan (*Plasmodium falciparum*) that causes malaria, the cells are broken down more rapidly than in "normal" homozygotes (AA), so the growth of the protozoan is curtailed. In parts of Africa with a high incidence of falciparum malaria, the frequency of S is quite high (about $q = 0.13$ in some regions) because heterozygotes survive at a higher rate than either homozygote (Figure 13.21D). (The relative fitnesses have been estimated as $W_{AA} = 0.89$, $W_{AS} = 1.0$, $W_{SS} = 0.20$.) The heterozygote advantage therefore arises from a balance of OPPOSING SELECTIVE FACTORS: anemia and malaria. In the absence of malaria, balancing selection yields to directional selection, because then the AA genotype has the highest fitness. In the African-American population, the frequency of S is about 0.05, and is presumably declining due to mortality.

Several other hemoglobin mutations similarly provide heterozygous resistance to malaria, such as those responsible for thalassemia anemia, most prevalent in Mediterranean regions. The various mutations tend to have different geographic distributions within malarial regions, because if one such mutation has attained high frequency in a given area, a different mutation that provides similar protection against malaria cannot increase in frequency if it is rare (Hartl and Clark 1989). Only under exceptional circumstances can heterozygous advantage maintain three or more alleles as a stable polymorphism.

Antagonistic Selection

The opposing forces acting on the sickle-cell polymorphism are an example of ANTAGONISTIC SELECTION, which in this instance maintains polymorphism, but *only* because the heterozygote happens to have the highest fitness. Unless this is the case, antagonistic selection usually does not maintain polymorphism (Curtsinger et al. 1994). Suppose, for example, that the survival rates of an insect in the larval stage are 0.5 for genotypes A_1A_1 and A_1A_2 and 0.4 for A_2A_2, whereas the proportions of surviving larvae that then survive through the pupal stage are 0.6 and 0.9 respectively, so that the recessive homozygote is first disfavored and then favored. For genotypes A_1A_1 and A_1A_2, the proportion surviving to reproductive adulthood is $(0.5)(0.6) = 0.30$, and for A_2A_2 it is $(0.4)(0.9) = 0.36$: A_2A_2 has a net selective advantage, and allele A_2 will be fixed. The same is true if antagonistic selection favors different genotypes in terms of survival versus fecundity: polymorphism is maintained only if the net fitness of the heterozygote, found by multiplying the fitness components, exceeds that of the homozygotes. In theory, however, directional selection for different genotypes in the two sexes can sometimes maintain polymorphism. Also, polymorphism can result, under some circumstances, if antagonistic selection acts on the gametic and zygotic stages. For example, t alleles in the house mouse are maintained by segregation distortion, even though they cause lethality or sterility (see Chapter 12).

Varying Selection

Within a single breeding population, a fluctuating environment may favor different genotypes at different times; or different genotypes may be fittest in different microhabitats that provide spatial variation. Intuitively, one would expect that such variable selection would promote genetic variation. But this is an instance in which intuition is misleading. Variable selection is a very complex subject (Box 13.B; see also reviews by Felsenstein 1976; Hedrick et al. 1976; Hedrick 1986), and according to the mathematical theory, *a variable environment does not necessarily maintain genetic variation*; it does so only under special circumstances. TEMPORAL FLUCTUATION in the environment may slow down the rate at which one or another allele approaches fixation, but *usually it does not preserve both of two alleles* indefinitely unless the fitness of the heterozygote, averaged over the varying conditions, exceeds that of both homozygotes.

SPATIAL VARIATION, in which a mosaic of resources or microhabitats exists within the area occupied by a single population, is more likely to maintain polymorphism (sometimes called **multiple-niche polymorphism**). However, the likelihood of polymorphism depends on whether the variation is coarse-grained or fine-grained, and whether selection is soft or hard. Variation in the environment is COARSE-GRAINED if each individual experiences only one or another state of the environment during its lifetime, as does a plant growing in a patch of one kind of soil, or a parasite that develops in one species of host. The same environmental states may be FINE-GRAINED for an organism that samples more than one of them, perhaps by moving about or by eating different food types.

HARD SELECTION (Figure 13.22A) occurs when the likelihood of survival of an individual depends solely on how well its genotype equips it for the microenvironment in which it settles: for example, grass seeds that land on copper-contaminated soil have either a very low or a high probability of survival to adulthood, depending on whether they are of a susceptible or a resistant genotype. The number of survivors then varies depending on what fraction of the colonizing seeds have the resistant genotype. In contrast, SOFT SELECTION (Figure 13.22B) occurs when the number of survivors is determined by competition for a limiting factor such as space or food; then selection among the colonists does not determine the number of surviving adults, but only their genotypes. A pure culture of any of several genotypes might attain the same density, but in competition, inferior genotypes suffer higher mortality. Which genotypes these are depends on the local environmental conditions.

Mathematical models have shown that *coarse-grained* environmental variation and *soft* selection are *more likely to maintain genetic polymorphism* than fine-grained, hard selection. Even so, only rather special combinations of selection coefficients and proportions of different microenvironments will maintain genetic variation (Figure 13.23). The conditions for polymorphism are broader (i.e., easier), however, if individuals have a better than random chance of colonizing the kind of microhabitat to which their genotype best suits them. In plants, this could occur if seeds fall near their parent, in the same microhabitat to which their mother is presumably well adapted (since she survived there). Animals can accomplish this correlation between

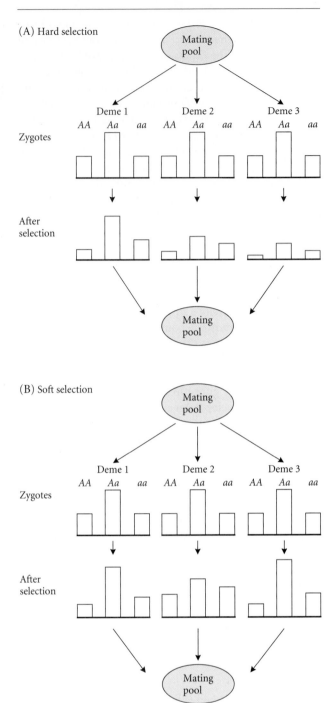

FIGURE 13.22 Hard selection and soft selection. The height of each bar represents the number (not the proportion) of individuals of each genotype. Zygotes produced by a pool of randomly mating parents disperse at random into several patches, forming demes. The relative fitnesses of the genotypes differ among the demes because of environmental differences. (A) In hard selection, mortality affects the total number of individuals that survive to adulthood, so the size of the adult population depends on the genotypes' fitnesses. (B) In soft selection, mortality of some genotypes is compensated by greater survival of others, so the size of the adult population is not affected by natural selection. (After Hartl and Clark 1989.)

BOX 13.B

Selection in a Variable Environment

When different populations of a species, exchanging genes at a low rate, experience selection for different homozygotes, selection maintains genetic variation in the metapopulation as a whole. But when the environment varies within the area of a single breeding population, variable selection may or may not maintain polymorphism.

In the case of fine-grained variation between two environmental conditions, we imagine that the fitness of an individual is the weighted arithmetic average (mean) of the fitness experienced under each condition. Thus if the fitness of genotype i in environmental state j is w_{ij}, and environmental states 1 and 2 occur with relative frequencies c_1 and c_2 ($c_1 + c_2 = 1$), the fitness of genotype i is $W_i = c_1W_{i1} + c_2W_{i2}$. A stable polymorphism is maintained only if the mean fitness of the heterozygote is highest. For example, if the fitnesses of A_1A_1, A_1A_2, and A_2A_2 are 1.0, 0.95, and 0.85 under warm conditions, but 0.87, 0.95, 1.0 under cool conditions, then if warm and cool conditions are equally frequent ($c_1 = c_2 = 0.5$), the genotypes have mean fitnesses of 0.935, 0.950, 0.925. Polymorphism is maintained in this case by MARGINAL OVERDOMINANCE, even though under any single condition the heterozygote is not overdominant. If warm conditions were more common, say, if $c_1 = 0.9$, then the genotypes' fitnesses would be 0.987, 0.950, 0.865; despite the environmental variation, A_1A_1 would have the highest average fitness, and allele A_1 would become fixed.

Now consider coarse-grained temporal variation. Suppose that sequential, discrete generations occur under warm and cool conditions, and that the genotypic fitnesses under the two conditions are as in the previous example. There is directional selection for A_1A_1 in warm generations, and for A_2A_2 in cool generations. But because the disadvantage of A_1A_1 under cool conditions is not as great as the disadvantage of A_2A_2 under warm conditions, the net trend over a sequence of generations is a decline in the frequency of A_2. For example, if the frequency of A_1 is $p = 0.5$ at the beginning of a warm generation, equation (A1) in Box A can be applied to show that after this generation, $p' = p + \Delta p = 0.5 + 0.08 = 0.58$, and that after the next, cool, generation it is $p'' = p' + \Delta p = 0.58 - 0.17 = 0.563$. Thus there is a net increase in the frequency of A_1 over the course of a two-generation cycle. Haldane and Jayakar (1963) showed that only under special conditions will variable selection among generations maintain polymorphism indefinitely.

For coarse-grained spatial variation, we imagine that the environment consists of a mosaic of patches of types 1 and 2, that adults emerge from these patches and mate at random, and that irrespective of genotype, a fraction c_1 of offspring settle at random in type 1 patches and a fraction c_2 in type 2 patches. The mathematical treatment (Levene 1953; Dempster 1955) then depends on whether selection is hard or soft. Under hard selection, the average fitness of a genotype is the weighted arithmetic mean ($c_1W_{i1} + c_2W_{i2}$). As in fine-grained spatial variation, polymorphism is maintained only if the heterozygote has higher average fitness than either homozygote (marginal overdominance).

Under soft selection, the fitness of a genotype in a patch depends on the frequencies of the several genotypes in that patch. Howard Levene (1953) showed that the measure of a genotype's average fitness that represents its contribution to the next generation in these circumstances is not the arithmetic mean, but the harmonic mean, $1/[c_1(1/W_{i1}) + c_2(1/W_{i2})]$, and that polymorphism is stable if the heterozygote has the highest harmonic mean fitness. An example of fitness values satisfying this condition when $c_1 = c_2 = 0.5$ is:

	A_1A_1	A_1A_2	A_2A_2
Fitness in patch type 1:	1.05	1	1.07
Fitness in patch type 2:	0.7	1	1.5

In this case, the heterozygote does not have the highest arithmetic mean fitness, but the polymorphism is stable. If, however, the patch types differ substantially in abundance (say, if $c_1 = 0.3$, $c_2 = 0.7$), the polymorphism is not stable (A_2 would be fixed in this case). This example shows that the conditions for polymorphism are broader under soft than under hard selection.

It may be intuitively evident that if offspring, instead of settling at random, preferentially select the patch type to which they are best adapted, polymorphism is more likely to persist, for each patch type serves as a "refuge" that preserves one or the other homozygote (Maynard Smith and Hoekstra 1980; Wilson and Turelli 1986). In fact, a genetic polymorphism for habitat preference may persist in this way, even if the genotypes differ only in habitat preference and not in their physiological adaptation to different patch types (Rausher 1984).

---

genotype and environment by HABITAT SELECTION (resource selection), i.e., choosing their environment.

The topic of multiple-niche polymorphism is one of many in evolutionary genetics in which mathematical theory has advanced farther than data analysis. Many cases have been described, such as the enzyme polymorphisms studied by Watt in butterflies and by Koehn and Hilbish in mussels, in which different environmental conditions favor

different genotypes. But in many of these cases, the variation within local populations may be due to migration among populations in different environments. Only a few cases have been sufficiently well studied to show that the genotypes' fitnesses conform to the conditions under which a polymorphism would be stable in the absence of gene flow. Thomas Smith (1993) has studied an African finch, the black-bellied seedcracker (*Pyrenestes ostrinus*), in

FIGURE 13.23 The conditions under which stable polymorphism is maintained by environmental heterogeneity in space (multiple-niche polymorphism) and in time (temporal variation in relative fitnesses). Two environmental states are either mixed spatially or occur at different times. The relative fitnesses of A_1A_1, A_1A_2, and A_2A_2 are 1, 1, and x_1 in environment 1, and 1, 1, and x_2 in environment 2. Combinations of x_1 and x_2 lying within the shaded regions yield stable polymorphism—for example, if the fitness of A_2A_2 is 2.0 in environment 1 and 0.5 in environment 2. The set of combinations of fitness values that yield polymorphism is more restricted when the environment varies temporally than spatially. (After Hedrick et al. 1976.)

FIGURE 13.24 Multiple-niche polymorphism in the black-bellied seedcracker. (A) Probability of survival to adulthood of juvenile birds in relation to their lower mandible width, a measure of bill size. These curves are based on the data in (B), which shows the number of banded juveniles that survived (solid bars) and that did not survive (open bars). The distribution of lower mandible width among adults (C) is bimodal. The peak centered at 12.8 mm corresponds to recessive homozygotes, the other peak to heterozygotes and dominant homozygotes. (After Smith 1993.)

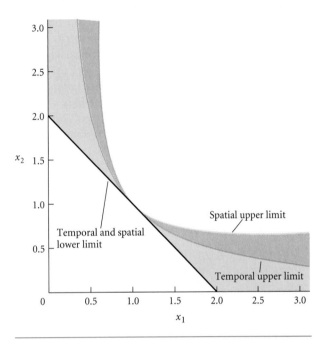

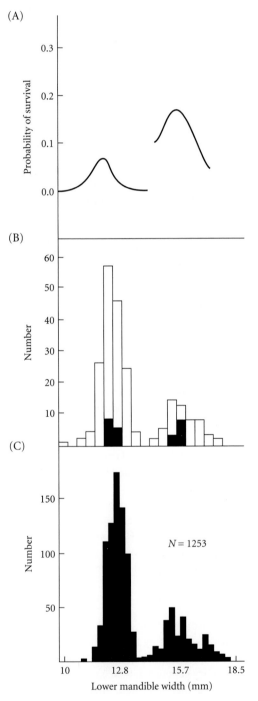

which populations have a bimodal distribution of bill width (Figure 13.24). The inheritance of this difference suggests that it is due to two alleles, with wide bill dominant over narrow. The two morphs differ in the efficiency with which they process seeds of different species of sedges, their major food; wide-billed birds process hard seeds, and narrow-billed birds soft seeds, more efficiently. Smith banded more than 2700 juvenile birds, and found that survival to adulthood was lower for birds with intermediate bills than for either wide- or narrow-billed birds. Thus, disruptive selection, arising from the superior fitness of different genotypes on different resources, appears to maintain the polymorphism.

A more surprising example is provided by the freshwater crustacean *Asellus aquaticus*, in which two allelic forms of the enzyme amylase, which breaks down starch into glucose, are distinguishable by electrophoresis. Bent Christensen (1977) found that in a small pond bordered at one end by willows and at the other by a beech forest, samples of this leaf-eating isopod taken from beech litter were dominated by genotype A_1A_1, whereas A_2A_2 was the prevalent genotype in willow litter. He put 1472 isopods into an aquarium with beech and willow litter at opposite sides, and after two weeks found that the genotype frequencies among individuals recovered from the two microenvironments were quite different:

Genotype	% in beech	% in willow
A_1A_1	42.9	57.1
A_1A_2	44.2	55.8
A_2A_2	27.6	72.4

Thus the genotypes differ in the degree of their preference for willow. A sufficiently great difference in preference could in itself maintain such a polymorphism, even if the genotypes do not otherwise differ in fitness (Rausher 1984; Wilson and Turelli 1986).

This example, incidentally, raises an interesting question: What causes the correlation between an individual's enzyme genotype and its behavior? This question arises whenever we find a polymorphism in which individuals choose a resource to which their particular features best suit them (see Chapters 14 and 20).

Rather few examples have been described in which the models of environmental variation that we have discussed so far explain the persistence of genetic variation (Futuyma and Peterson 1985; Hedrick 1986). Instead, selection in heterogeneous environments is often a bit more complicated, taking the form of frequency-dependent selection.

Frequency-Dependent Selection

In the models we have considered so far, the fitness of each genotype is assumed to be constant within a given environment. In many cases of selection, however, *the fitness of a genotype depends on the genotype frequencies in the population.* This principle bears on many aspects of evolution, especially the evolution of many kinds of behavior (see Chapter 20).

In **inverse frequency-dependent selection**, the rarer a phenotype is in the population, the greater its fitness (Figure 13.25A). For example, the per capita rate of survival and reproduction of a dominant phenotype may be greatest when it is very rare, and may decrease as it becomes more common; and likewise for the recessive phenotype. (We might model this by writing $W_{11} = W_{12} = 1 - s(1 - q^2)$ and $W_{22} = 1 - sq^2$, for genotypes A_1A_1, A_1A_2, and A_2A_2 respectively.) Then when A_2 is at high frequency, it declines because A_2A_2 has lower fitness than A_1A_1 and A_1A_2, and likewise for A_1. Whatever the initial allele frequencies may be, they change under selection toward a stable equilibrium value, which in this case occurs when the frequencies of the two phenotypes are equal (i.e., when $q^2 = 0.5$). At this point the average fitnesses of both phenotypes are the same: neither has an advantage over the other.

Inverse frequency-dependent selection acts on many kinds of characteristics, and can maintain many alleles if each is advantageous when rare. Such is the case with self-incompatibility alleles in many plants (see Chapters 11, 21),

in which pollen carrying allele S_i can effectively grow only on stigmas with genotypes that do not include S_i. Thus for three alleles, S_1, S_2, and S_3, pollen of type S_1 can grow on stigmas of genotype S_2S_3, but not on S_1S_2 or S_1S_3. (Because of this self-incompatibility, no plants are homozygous for any of the alleles.) If a new rare allele S_4 arises by mutation, S_4 pollen can fertilize almost any flower on which it lands, whereas pollen carrying S_1 (or S_2 or S_3) can fertilize only about one-third of the flowers in the population. Thus S_4 will increase until its frequency reaches about 0.25, at which point it loses its greater probability of fertilization. A fifth such allele may arise and increase similarly. We should expect to find a large number of self-incompatibility alleles, all

FIGURE 13.25 Two forms of frequency-dependent selection, in which the fitness of each genotype depends on its frequency in a population. (A) Inverse frequency-dependent selection, in which a genotype's fitness declines with its frequency, which depends on allele frequency (p, the frequency of A_1). The fitnesses of A_1A_1, A_1A_2, and A_2A_2 are $W_{11} = 1 - p^2$, $W_{12} = 2pq$, and $W_{22} = 1 - q^2$ in this model; the curves are calculated for $t = 1$. A stable equilibrium exists at $p = q = 0.5$. (B) Positive frequency-dependent selection, in which a genotype's fitness increases with its frequency. The fitnesses of A_1A_1, A_1A_2, and A_2A_2 are $1 + tp^2$, $1 + 2tpq$, and $1 + tq^2$. The curves are calculated by setting $t = 1$ and subtracting 1 from each fitness value. (A from Hartl and Clark 1989.)

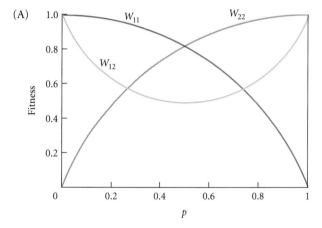

(A)

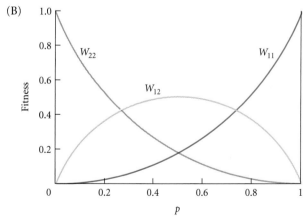

(B)

about equal in frequency, at equilibrium. In some plant species, indeed, hundreds of such alleles exist.

Many biological phenomena give rise to inverse frequency-dependent selection; here we will mention only predation, mating preferences, and competition.

Many animals act as if they have a "search image" when foraging for food: an individual temporarily "focuses attention" on a certain food type, more or less ignoring others that it might accept at other times. For example, many songbirds, such as tits, prey selectively on one species of caterpillar at a given time—namely, the species that is most abundant at that time. (The ethologist Niko Tinbergen coined the term "search image" based on his studies of tits.) A similar preference for the most abundant prey type is often shown by predators when selecting among different phenotypes of a single species of prey. Figure 13.26 shows an example in which fish in aquaria, presented with different proportions of color morphs of an aquatic bug, consumed a greater percentage of each morph when it was relatively abundant than when it was relatively rare.

The mating preferences of females apply inverse frequency-dependent selection in some animal species. In experimental cultures of *Drosophila* species, a female is more likely to mate with the rarer genotype in a mixture of males; for example, a male carrying any of several chromosome inversions in *D. pseudoobscura* has greater mating success if he belongs to a minority genotype than to a majority genotype (Spiess 1968; Petit and Ehrman 1969). This pattern would tend to maintain polymorphism.

Selection is often frequency-dependent when genotypes compete for limiting resources. Returning to the model of soft selection in a heterogeneous environment (see above), suppose that two phenotypes A_1 and A_2 can survive on either resource 1 or resource 2, but that A_1 is a superior competitor for resource 1 and A_2 is a superior competitor for resource 2. If A_1 is rare, each A_1 individual will compete for resource 1 chiefly with inferior competitors, and so will have a higher per capita rate of increase than A_2 individuals; but as A_1 increases in frequency, each A_1 individual competes with more individuals of the same genotype, so its per capita advantage, relative to A_2, declines. The same pattern of rarity advantage, lost as phenotype frequency increases, applies to phenotype A_2.

This effect is illustrated in an experiment by Norman Ellstrand and Janis Antonovics (1984). Although neither genotypes nor resources were specified, the experiment shows how a model can be tested by matching observations to predictions. The grass *Anthoxanthum odoratum* can be propagated either sexually by seed or asexually by vegetative cuttings. In a natural environment, Ellstrand and Antonovics planted small "focal" cuttings and surrounded each with competitors. These were either the same genotype as the focal individual, taken as vegetative shoots from the same parent plant, or different genotypes, obtained from sexually produced seed. After a season's growth, focal individuals on average were larger and produced more seed (i.e., had higher fitness) if surrounded by different genotypes than if surrounded by the same genotype. This is just what we would expect if different genotypes use somewhat different resources (perhaps certain mineral nutrients are partitioned differently by adjacent plants that differ in genotype). The more intense competition among individuals of the same genotype would then impose frequency-dependent selection. In the same vein, agricultural researchers have sometimes found that a mixture of crop varieties yields more than fields of a single variety.

Alternative Equilibria

One of the most important principles in evolution is that initial genetic conditions often determine which of several paths, or trajectories, of genetic change a population will follow. Thus *the evolution of a population often depends on its previous evolutionary history.* We have already encountered one example in the different hemoglobin mutants that confer resistance to malaria among human populations: a high frequency of one such mutation can prevent another such mutation from increasing in frequency. Positive frequency-dependent selection and heterozygote disadvantage are two other important factors that can give rise to MULTIPLE STABLE EQUILIBRIA.

Positive frequency-dependent selection Positive frequency-dependent selection is the mirror image of inverse frequency-dependent selection. If the fitness of a genotype

FIGURE 13.26 An example of inverse frequency-dependent selection. In a series of trials, three color morphs of an aquatic bug, the water boatman *Sigara distincta*, were offered, in different proportions, to fish in aquaria. If the percentage of each morph eaten were the same as the percentage offered, the data point would lie on the colored line. Mortality due to predation (percentage eaten) was disproportionately great for the more common form, irrespective of which color morph it was. (After Clarke 1962, based on data of Popham 1942.)

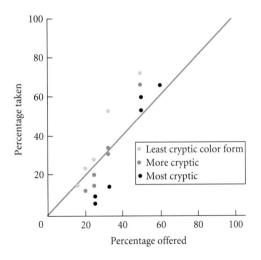

is greater the more frequent it is in a population, then *whichever allele is initially more frequent will be fixed* (Figure 13.25B).

For example, the unpalatable tropical butterfly *Heliconius erato* has many different geographic races of distinct color forms, with black or blue wings marked with red, yellow, or white. Each race is monomorphic. Adjacent geographic races interbreed at zones only a few kilometers wide. The geographic variation in color pattern in this species is closely paralleled by that in another unpalatable species of *Heliconius*—a spectacular example of Müllerian mimicry (see Color Plate 7). Because both species are unpalatable, predators avoid both if they have had unpleasant experience of either. Thus gene flow from one geographic race to another is countered by positive frequency-dependent selection: immigrant butterflies that deviate from the locally prevalent color pattern are selected against.

On either side of a contact zone between two geographic races in Peru (Figure 13.27), James Mallet and Nicholas Barton (1989) released *H. erato* of the other race, and, as a control, butterflies of the same race from a different locality. The butterflies, marked so they could be recognized, were repeatedly recaptured for some time thereafter. Compared with control butterflies, with the same color pattern as the population into which they were released, far fewer of those with the other color pattern were captured. Based on bill marks left on the wings of butterflies that had escaped from birds, the authors concluded that the missing butterflies were lost to bird predation, and calculated an average selection coefficient of 0.52 against the "wrong" color pattern in either population. This amounts to a selection coefficient of about $s = 0.17$ at each of the three major loci that control the differences in color pattern between the races—very strong selection indeed.

Heterozygote disadvantage HETEROZYGOTE DISADVANTAGE, also called UNDERDOMINANCE, the case in which the heterozygote has lower fitness than either homozygote, is also one in which monomorphism for either A_1A_1 or A_2A_2 is a stable equilibrium, and the initially more frequent allele is fixed by selection. Suppose genotypes A_1A_1, A_1A_2, and A_2A_2 have the constant fitness values $1 + s$, 1, and $1 + s$. If a population is initially monomorphic for A_1A_1, and A_2 then enters at low frequency by mutation or gene flow, almost all A_2 alleles are carried by heterozygotes (A_1A_2); since their fitness is lower than that of A_1A_1, selection reduces the

FIGURE 13.27 The proportions of two color forms of the butterfly *Heliconius erato* (A) at various sites in a small region of Peru, and of the almost identical color forms of *H. melpomene* (B) at the same sites. These species are Müllerian mimics. In both species, almost complete replacement of one color form by the other occurs over a very short distance. The four named localities are those at which Mallet and Barton obtained *H. erato* for the transfer experiment described in the text. (After Mallet and Barton 1989.)

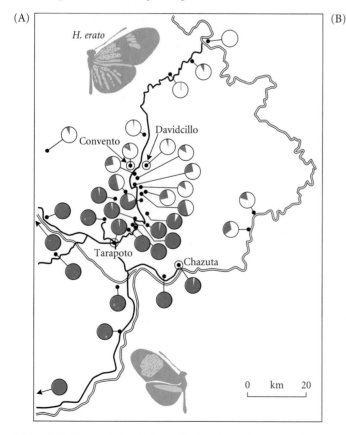

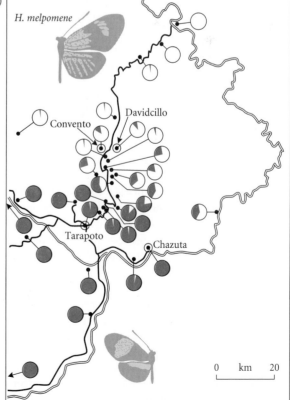

frequency (q) of A_2 to zero. Likewise, A_1 is eliminated if it enters a monomorphic population of A_2A_2, which also has a greater fitness than A_1A_1. Therefore $p = 1$, $q = 0$ and $p = 0$, $q = 1$ are alternative stable equilibria. In this case selection fixes A_1 whenever $p > 0.5$, and fixes A_2 whenever $q > 0.5$. The allele frequency $p = q = 0.5$ is also an equilibrium (i.e., $\Delta p = 0$), but this equilibrium is UNSTABLE: if q diverges even slightly from 0.5 by chance, A_2 will be either lost or fixed.

If the homozygotes' fitnesses are different, but both are greater than that of the heterozygote, there will still be both stable equilibria at $p = 1$ and $q = 1$ and an unstable "internal" equilibrium (i.e., an equilibrium frequency other than 0 or 1). In this case, the mean fitness in a population fixed for the less fit homozygote is less than if it were fixed for the other homozygote, but *selection cannot move the population from the less fit to the more fit condition*. Thus *a population is not necessarily driven by natural selection to the most adaptive possible genetic constitution*.

Adaptive landscapes The reader will recall, from our treatment of directional selection, that we can calculate the mean fitness ($\bar{w}$) of individuals in a population with any conceivable allele frequency and plot $\bar{w}$ as a function of p (see Figure 13.12A,B). When fitnesses are constant, natural selection changes allele frequencies in such a way that $\bar{w}$ increases, so the population moves up the slope of the $\bar{w}$ versus p curve from wherever it is. The current location of the population on this slope is then a simple guide to how allele frequencies will change under selection: simply see which direction of allele frequency change will increase $\bar{w}$. For an underdominant locus, the curve dips in the middle and slopes upward to $p = 0$ and $p = 1$ (Figure 13.12D). Thus p decreases or increases, depending on whether a population begins to the left or the right of the minimum of the $\bar{w}$ curve.

The curves in Figure 13.12 are often referred to as **adaptive landscapes**, or ADAPTIVE TOPOGRAPHIES. The curve in Figure 13.12D represents two ADAPTIVE PEAKS separated by an ADAPTIVE VALLEY. This metaphor, introduced by Sewall Wright, is widely used in evolutionary biology. Because we shall encounter more complex examples, it is important to be clear on just what it represents. Each point on the curve (the adaptive landscape) represents the *average* fitness of individuals in a hypothetical *population* made up of (in this case) three genotypes with frequencies p^2, $2pq$, and q^2. Each possible value of p—each possible hypothetical population—yields a different point on the landscape. When, as in Figure 13.12D, the relationship of $\bar{w}$ to p is not monotonic, two genetically different populations (e.g., with $p = 0$ or $p = 1$ in this case) can (but need not) have the same average fitness ($\bar{w}$) *under the same environmental conditions*. Different environments, which might alter the fitnesses of the genotypes relative to each other, are represented not by different points on any one landscape, but rather by *different landscapes*—different relationships between $\bar{w}$ and p.

Interaction of Selection and Genetic Drift

In developing the theory of selection so far, we have assumed an effectively infinite population size, in which the population arrives at the allele frequency equilibrium specified by the relative fitnesses of the genotypes. However, *in a finite population, allele frequencies are simultaneously affected by both selection and chance*. As the movement of an airborne dust particle is affected both by the deterministic force of gravity and by random collisions with gas molecules (Brownian movement), so the effective size (N_e) of a population and the strength of selection (s) both affect changes in allele frequencies. The effect of genetic drift is negligible if selection on a locus is strong relative to the population size—if s is much greater than $1/(4N_e)$. Conversely, if s is much less than $1/(4N_e)$, selection is so weak that the allele frequencies change mostly by genetic drift: the alleles are *nearly neutral*. For instance, if the genotypes AA, AA', and $A'A'$ have fitnesses 1.0, 0.995, and 0.990 (where $0.990 = 1 - s$, so $s = 0.01$), selection is overwhelmingly important if the effective population size is greater than about 250; but if it is less than about 10, genetic drift is more important.

The effect of population size on the efficacy of selection has several important consequences. For one, a population may not attain exactly the equilibrium frequency predicted from the genotypes' fitnesses; instead, it is likely to wander by genetic drift in the vicinity of the equilibrium frequency, according to a probability distribution given by s and N_e (Figure 13.28). Second, a slightly advantageous mutation is less likely to be fixed by selection if the population is small than if it is large, because it is more likely to be lost simply by chance. In fact, a deleterious mutation can become fixed by genetic drift despite natural selection, especially if selection is weak and the population is small (Box 13.C).

Another important consequence is that if a population fluctuates in size, *bottlenecks of low size provide temporary conditions under which genetic drift may sufficiently counteract selection so that a deleterious allele may increase in frequency*. For example, slightly deleterious mutations could be fixed, contributing to divergence among populations at the molecular level. This principle may be especially important if heterozygotes are inferior in fitness, so that the adaptive landscape has two peaks (see Figure 13.12D). Selection alone cannot move a population from one peak to another, even from a lower to a higher peak. But during episodes of very low population size, allele frequencies may fluctuate so far by genetic drift that they cross the adaptive valley—after which selection can move the population "uphill" to the other peak (Figure 13.29). The probability that such a **peak shift** will occur (Barton and Charlesworth 1984) depends on the mean fitness at the valley ($\bar{w}_v$) relative to the mean fitness at the initially occupied peak ($\bar{w}_p$); it is proportional to $(\bar{w}_v/\bar{w}_p)^{2N_e}$. Suppose, for example, that A_1A_1, A_1A_2, and A_2A_2 have fitnesses of 1.0, 0.9, and 1.0 respectively; then $\bar{w}$ is at its minimum when $p = q = 0.5$, and $\bar{w}_v = 0.95$. If $N_e = 10$ during a bottleneck, the probability of a peak shift is proportional to $(0.95/1.0)^{20}$, or 0.358.

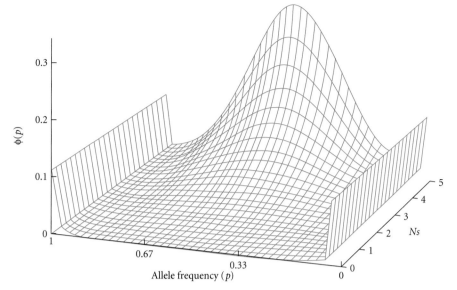

FIGURE 13.28 The probability distribution of p [$\phi(p)$] as a function of Ns, for a locus at which the fitness of A_1A_1, A_1A_2, and A_2A_2 are $1 - s$, 1, and $1 - s$. N is the effective population size. The probability that p is near the equilibrium determined by selection alone (0.50) increases with Ns. (After Hartl and Clark 1989.)

Thus when there are multiple stable equilibria, *genetic drift and selection may act in concert to accomplish what selection alone cannot*, moving a population from one adaptive peak to another. This theory explains how populations come to differ in chromosome rearrangements, such as translocations and pericentric inversions. Many chromosome rearrangements may conform to the model of underdominance, because heterozygotes are thought to have lower fertility than either homozygote (see Chapter 10). Selection alone cannot cause a newly arisen rearrangement to increase in frequency, but in a sufficiently small population, it can increase by genetic drift to the critical frequency (about 0.5) above which selection will carry it to fixation. For example, local populations of the Australian grasshopper *Vandiemenella viatica* are monomorphic for various chromosome fusions and pericentric inversions,

heterozygotes for which have many aneuploid gametes. Any such chromosome, introduced by gene flow into a population monomorphic for a different arrangement, is reduced in frequency by natural selection, so no two "chromosome races" are sympatric; instead, they meet in "tension zones" only 200–300 meters wide (White 1978). Because these grasshoppers are flightless and quite sedentary, local populations are effectively small, providing the opportunity for genetic drift to occasionally initiate a peak shift whereby a new chromosome arrangement is fixed.

The Strength of Natural Selection

Until the 1930s, most evolutionary biologists followed Darwin in assuming that the intensity of natural selection is usually very slight. Perhaps for this reason, few attempted

---

BOX 13.C

The Probability of Fixation when Selection and Genetic Drift Interact

Following Kimura (1983), let the fitnesses of genotypes A_1A_1, A_1A_2, and A_2A_2 be 1, $1 + s$, and $1 + 2s$ respectively, where s is positive if A_2 is advantageous and negative if it is deleterious. If the initial frequency of A_2 is q, the probability of fixation of A_2 is

$$P = \frac{1 - e^{-4Nsq}}{1 - e^{-4Ns}} \tag{C1}$$

where N is the effective population size and e is the base of natural logarithms. If A_2 is a new mutation represented by a single copy in the population, $q = 1/(2N)$, and the fixation probability becomes $P = 1/(2N)$ if $s = 0$ (i.e., if A_2 is selectively neutral). If $s > 0$,

$$P = \frac{1 - e^{-2s}}{1 - e^{-4Ns}} \tag{C2}$$

which becomes $P \approx 2s/(1 - e^{-4Ns})$ if s is small, and $P \approx 2s$ if N is large and s is positive.

To illustrate this numerically (Li and Graur 1991), if A_2 arises in a population of $N = 1000$ individuals and its selective advantage is 0.01, equation (C2) tells us that its probability of eventual fixation is 0.02. If A_2 has a selective advantage of $s = 0.01$ and its initial frequency is $q = 0.01$, its fixation probability (from equation [C1]) is approximately 0.04. If A_2 is deleterious, and $s = -0.001$, its probability of fixation is 0.00004. Thus because of genetic drift, an advantageous mutation does not always become fixed in a population, and a deleterious mutation can be fixed despite selection. However, even a very slightly deleterious mutation has only a small probability of fixation.

FIGURE 13.29 A peak shift due to the joint action of genetic drift and natural selection. For two alleles or chromosome rearrangements that lower the fitness of the heterozygote, the fitnesses of A_1A_1, A_1A_2, and A_2A_2 are 1, $1 - s$, and 1 respectively. The curve shows the mean fitness ($\bar{w}$) for each value of p, the frequency of A_1 (colored dots), and may be considered an adaptive landscape with two peaks (at $p = 0$ and $p = 1$). An unstable equilibrium exists at $p = 0.5$, indicated by the dotted line (cf. Figure 13.12D). The frequency of A_1, initially near 0 (A), may increase by genetic drift (B), but returns toward 0 (C) if Ns is large enough (i.e., if selection is strong and/or the population is large). With some low probability, however, p may increase beyond the unstable equilibrium frequency 0.5 due to genetic drift (D). The probability that this will occur is enhanced by a reduction in population size. If it does occur, selection increases the frequency of A_1 toward fixation (E,F).

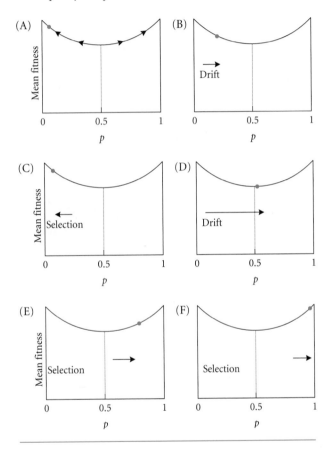

er sample—nearly 10,000—to show that a difference in survival of 1 percent really exists. Needless to say, few investigators are able to monitor the lives of such huge numbers of animals or plants.

By the 1930s, however, examples such as industrial melanism in the peppered moth (*Biston betularia*) began to come to light, with the surprising implication that selection might at times be very intense (s may be as high as 0.5 in *Biston*). Since then, ecological geneticists have quantified differences in components of fitness for traits in natural populations of many species by measuring variables such as the survivorship, fecundity, and mating success of different genotypes. Much of this information, collated by John Endler (1986), is summarized in Figure 13.30, which presents a distribution of selection coefficients for numerous polymorphic morphological and physiological traits found in various studies. Endler notes that because investigators often do not report instances that do not provide evidence of selection, there is an underrepresentation of cases in which s is close to zero. Moreover, the figure does not include studies of allozymes and other molecular polymorphisms that may well be usually neutral. Nonetheless, the conclusion we must draw is that the intensity of selection acting on polymorphic loci ranges from negligible to extremely strong. In some instances, s is nearly 1.0 (one or more geno-

FIGURE 13.30 A compilation of selection coefficients (s) reported in the literature for discrete, genetically polymorphic traits in natural populations of various species. The total height of each bar represents the percentage of all reported values in each interval, and the solid portion represents the percentage shown by statistical analysis to be significantly different from zero. N is the number of values, based on one or more traits from each of a number of species. A and B contrast results from undisturbed populations (A; 36 species) with results from populations in field cages, stressful environments, or perturbation experiments (B; 12 species). C and D contrast selection based on mortality (C; 34 species) with fecundity, fertility, and sexual selection (D; 13 species). Figure 19 in Chapter 14 presents comparable data for selection on quantitative characters. (After Endler 1986.)

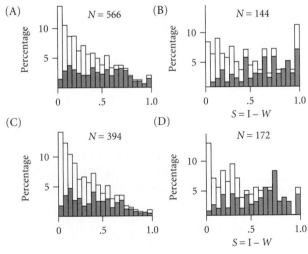

to measure selection in natural populations, for it would have been exceedingly difficult to detect. Suppose, for example, that one of two genotypes has a 1 percent disadvantage in survival ($s = 0.01$) relative to the other. You follow the survival of 1000 marked individuals of each genotype, and find that 500 of the superior genotype survive. If $s = 0.01$, then the number of survivors of the inferior genotype is 495 (0.500:0.495 = 1.00:0.99). Despite the labor you put into tracing the fate of 2000 animals or plants, this difference of 5 individuals cannot be shown by any statistical test to be greater than chance alone could yield, so you have no evidence that selection is operating (i.e., that s differs from zero). You would need a much larg-

types are effectively lethal or sterile), and differences of 50 percent or more are not uncommon. Selection acting through differences in fecundity or mating success is, if anything, stronger on average than selection acting through differential survival. The conclusion is inescapable that, at least for morphological and physiological traits, natural selection is often a very powerful factor of evolution.

Although the neutralist-selectionist debate over molecular evolution is far from resolution, evidence is mounting that selection is a major cause of divergence in DNA sequences among species and variation within species (Kreit-man and Akashi 1995). Many of the methods used to detect evidence of selection rely on comparing DNA sequences among species that diverged millions of years ago. Fitness differences far too small to detect by short-term experiments with small populations can, over the course of millions of generations, have effects that are manifested in comparisons among species. For example, even synonymous substitutions that do not alter amino acid sequences are influenced by weak natural selection. Evidence bearing on the neutralist-selectionist controversy is described in Chapter 22.

Summary

1. Even at a single locus, the diverse genetic effects of natural selection cannot be summarized by the slogan "survival of the fittest." Selection may indeed fix the fittest genotype, or it may maintain a population in a state of stable polymorphism, in which inferior genotypes persist.

2. The absolute fitness of a genotype is measured by its rate of increase, the major components of which are survival, female and male mating success, and fecundity. In sexual species, differences among gametic (haploid) genotypes may also contribute to selection among alleles. Rates of change in the frequencies of alleles and genotypes are determined by differences in their relative fitness, and are also affected by genotype frequencies and the degree of dominance at a locus.

3. Much of adaptive evolution by natural selection consists of replacement of previously prevalent genotypes by a superior homozygote (directional selection). However, genetic variation at a locus often persists in a stable equilibrium condition, owing to a balance between selection and recurrent mutation, between selection and gene flow, or because of any of several forms of balancing selection.

4. The kinds of balancing selection that maintain polymorphism include heterozygote superiority, inverse frequency-dependent selection, and variable selection arising from variation in the environment. Variable selection, however, maintains polymorphism only under special circumstances.

5. Examples of all these forms of selection have been described, but we do not know which are most important for explaining the abundant genetic variation found in most populations, nor whether selection, rather than genetic drift, is the most important determinant of evolution at the molecular level.

6. Often the final equilibrium state to which selection brings a population depends on its initial genetic constitution: there may be multiple possible outcomes, even under the same environmental conditions. This is especially likely if the genotypes' fitnesses depend on their frequencies, or if two homozygotes both have higher fitness than the heterozygote.

7. When genotypes differ in fitness, selection determines the outcome if the population is large; in a sufficiently small population, however, genetic drift is more powerful than selection. When the heterozygote is less fit than either homozygote, genetic drift is necessary to initiate a shift from one homozygous equilibrium state to the other.

8. Studies of variable loci in natural populations show that the strength of natural selection varies greatly, but that selection is often strong, and thus is a powerful force of evolution. Many cases of rapid evolution due to directional selection in recently altered environments have been observed.

Major Reference

Endler, J. A. 1986. *Natural selection in the wild.* Princeton University Press, Princeton, NJ. An analysis of methods of detecting and measuring natural selection, and a review of studies of selection in natural populations.

Problems and Discussion Topics

1. If a recessive lethal allele has a frequency of 0.050 in newly formed zygotes in one generation, and the locus is in Hardy-Weinberg equilibrium, what will be the allele frequency and the genotype frequencies at this locus at the beginning of the next generation? (Answer: $q = 0.048$; $p^2 = 0.9063$, $2pq = 0.0914$; $q^2 = 0.023$.) Calculate these values for the succeeding generation. If the lethal allele arises by mutation at a rate of 10^{-6} per gamete, what will be its frequency at equilibrium?

2. Suppose the egg-to-adult survival of A_1A_1 is 80 percent as great as that of A_1A_2, and the survival of A_2A_2 is 95 percent as great. What is the frequency (p) of A_1 at equilibrium? What are the genotype frequencies among zygotes at equilibrium? Now suppose the environment changes so that the relative survival rates of A_1A_1, A_1A_2, and A_2A_2 become 1.0, 0.95, and 0.90. What will be the frequency of A_1 be after two generations in the new environment? (Answer: 0.208.) Now suppose the relative survival rates of A_1A_1, A_1A_2, and A_2A_2 had originally been 0.90, 1.0, and 0.90. Repeat the exercise, assuming

the same new fitnesses in the new environment. What is the magnitude of the change in p in the first generation in the new environment, in each of the two cases?

3. If the egg-to-adult survival rates of genotypes A_1A_1, A_1A_2, and A_2A_2 are 90, 85, and 75 percent respectively, and their fecundity values are 50, 55, and 70 eggs per female, what are the approximate absolute fitnesses (R) and relative fitnesses of these genotypes? What are the allele frequencies at equilibrium? Suppose the species has two generations per year, that the genotypes do not differ in survival, and that the fecundity values are 50, 55, and 70 in the spring generation and 70, 65, and 55 in the fall generation. Will polymorphism persist, or will one allele become fixed? What if the fecundity values are 55, 65, 75 in the spring and 75, 65, 55 in the fall?

4. Let A and B represent two chromosome arrangements, such as a pericentric inversion of one with respect to the other, and suppose the fecundity of the heterozygote is reduced by 10 percent relative to the homozygotes, which are equal in fecundity. Suppose that populations 1 and 2 are fixed for A and B respectively. How do you expect the chromosome frequencies to change if the populations expand their ranges and meet at a border, if the populations are large and individuals disperse only a short distance relative to the geographic extent of each population? Now suppose instead that population 2 consists of small demes, each of 100 breeding individuals. Considering one such deme near the border with population 1, what will be the approximate equilibrium frequency of chromosome A in this deme if in each generation two of the mating individuals are immigrants from population 1? (Answer: 0.2.) As a challenge, use equation (C1) in Box 13.C to calculate the probability that chromosome A will become fixed in this deme if it starts from the frequency you have calculated above.

5. In pines, mussels, and other organisms, investigators have often found that components of fitness such as growth rate and survival are positively correlated with the number of allozyme loci at which individuals are heterozygous rather than homozygous (Mitton and Grant 1984; Zouros 1987). The interpretation of such data has been controversial (see references in Houle 1994). Provide two hypotheses to explain the data, and discuss how they might be distinguished.

6. Considering the principles of mutation, genetic drift, and natural selection discussed so far in this and preceding chapters, do you expect *adaptive* evolution to occur more rapidly in small or in large populations? Why? Answer the same questions with respect to *non-adaptive* (neutral) evolution.

7. The presence in a population of genotypes with inferior fitness reduces the average fitness of a population, compared with its maximal fitness (when no inferior genotypes are present). The reduction in mean fitness is called GENETIC LOAD, and is expressed as $L = 1 - \bar{w}$, where the fittest genotype has a relative fitness set at 1, and $\bar{w}$ is mean fitness. (*a*) If a deleterious recessive allele arises recurrently with a mutation rate u, calculate $\bar{w}$ and L at equilibrium between mutation and selection. (*b*) What would be the effect of such mutations at many, say n, loci on the load? (*c*) Under what conditions would you expect the load to lower the absolute size of a population? Could it cause extinction of a population? (*d*) What factors other than recurrent mutation could contribute to a population's genetic load? (For a review of the complex, controversial topic of genetic load, see Crow 1993.)

8. Bearing in mind that evolution by natural selection requires genetic variation, and that abundant genetic variation has been revealed by electrophoresis and other molecular techniques, discuss the implications of the neutralist-selectionist controversy for our expectation of how rapidly populations may adapt to changes in the environment.

9. The process of adaptation is generally considered to be the replacement, by natural selection, of inferior genotypes by genotypes with higher fitness in a population's environment. Now that we have analyzed the process of natural selection, discuss whether or not selection would be expected to (*a*) increase the abundance (population size) of populations or species; (*b*) increase the rate at which new species evolve from ancestral species, thus increasing the number of species.

10. Criticize the following statements, commonly expressed by people with an imperfect understanding of natural selection. (*a*) Natural selection is the survival of the fittest. (*b*) Evolution by natural selection is a matter of chance. (*c*) Natural selection is a tautology, because it is survival of the fittest, and we identify the fittest by seeing which survive.

CHAPTER 14

Multiple Genes and Quantitative Traits

Chapters 9 through 13 introduced the single-locus concepts that are the indispensable foundation of all evolutionary genetics. Very often the evolution of a single gene, or of a character strongly affected by a single allele substitution, can be quite well understood without reference to other loci. However, the function of any character often depends on the condition of other characters of the organism, which are likely to be independently inherited; and most individual characters are themselves *polygenic*, varying as a consequence of allelic variation at several or many loci. Moreover, pleiotropy is exceedingly common: most genes affect more than one feature. Thus it is necessary to consider the evolution of polygenic traits and multiple characteristics. The evolution of phenotypic traits is the subject of this chapter.

It would be virtually impossible to model, or understand, evolution of organisms as a whole, immensely complex as they are, so we will consider only elementary, oversimplified models. Even these constitute a far more complex subject than we will venture into. Mathematical treatments of these subjects include some of the most complex of biological models. This chapter will not delve into the mathematical intricacies, although a few simple equations are unavoidable. Even so, for many readers this may be the most difficult chapter in the book, requiring careful attention to the explanation of concepts. However, this chapter includes many important points about the dynamics of evolution, highlighted in italicized, purely verbal statements. A few of the topics covered are fairly specialized; the headings for these sections are marked with the symbol (❖), and those that are both specialized and difficult are marked with two (❖❖).

The evolution of multiple genes and characters is modeled and conceptualized in two seemingly very different ways. Explicit ALLELE FREQUENCY MODELS are extensions of the one-locus models we have already encountered. This explicit approach provides theoretical understanding, but it

is often very difficult to apply the theory to real data because measuring the variables can be a hopeless task. It is usually difficult, for example, even to know how many loci affect a trait, much less the allele frequencies and genotype fitnesses at each locus. Nevertheless, allele frequency models provide an indispensable foundation for understanding the evolution of polygenic traits.

The second approach is the biometric, or statistical, approach usually known as QUANTITATIVE GENETICS. It was developed largely as a tool for plant and animal breeders, so it uses concepts and variables that can be readily measured, such as statistical means, variances, and correlations. The theory of quantitative genetics is founded on explicit allele frequency models, but has been developed in such a way that allele frequencies and the other variables in those models do not appear (see Falconer 1981). Because variation in most characters is polygenic, biologists who study the evolution of morphology, life history characteristics, behavior, and other phenotypic traits use quantitative genetics extensively in their work.

In this chapter, we will first treat allele frequency models rather briefly, and then, referring to these when necessary, we will consider quantitative genetics at greater length.

Evolution at Two Loci

Basic Principles

We can understand many important concepts and principles of evolution at multiple loci by considering two loci, with alleles A_1, A_2 and B_1, B_2. These loci may be related to the phenotype (e.g., morphology) in several ways:

1. Each locus may affect a *different character*. In this case, *each evolves independently* of the other, according to the principles in the previous chapters, *unless* the loci are *linked* or the characters have a *functional relationship* to each other. If the loci are linked, allele frequency changes at one locus may (or may not) cause

correlated changes at the other locus. If the characters are functionally related, the fitness of a genotype (or phenotype) with respect to one locus may be affected by the genotype (or phenotype) at the other locus.

2. The two loci may contribute to variation in a *single character*. (Variation in a character due to two or more loci is called POLYGENIC VARIATION.) The phenotype, and perhaps also the fitness, then depends on the allelic constitution at both loci.

3. Either or both loci may affect two (or more) characters. This occurs when the loci are *pleiotropic*. In the simplest case, both loci affect the same two features (perhaps the dimensions of two body parts). Then the evolution of the two features is likely to be COR-RELATED. Moreover, evolutionary change at each locus is affected by the action of natural selection on both characters.

If there are two alleles at each locus, there are nine possible genotypes (three at locus *A* × three at locus *B*). Evolution by natural selection depends on the relative fitnesses of these genotypes, which depend on the phenotype—of one character or of two, as the case may be. It may also depend on the strength of linkage between the loci.

The events from one generation to the next, assuming that fitness depends on survival differences, are:

(1)	**(2)**	**(3)**	**(4)**
	survival		
zygotes →	(fitness) → adults	→ gametes	→ zygotes
$A_1A_1B_1B_1$	$A_1A_1B_1B_1$		$A_1A_1B_1B_1$
$A_1A_1B_1B_2$	$A_1A_1B_1B_2$	A_1B_1	$A_1A_1B_1B_2$
$A_1A_2B_1B_2$	$A_1A_2B_1B_2$	A_1B_2	$A_1A_2B_1B_2$
.	.	A_2B_1	.
.	.	A_2B_2	.
.	.		.

The zygotes formed in one generation (1), with some set of genotype frequencies and allele frequencies, differ in fitness (2) because of differences in, let us say, survival. Their gametes (3), which we assume unite at random, determine the genotype frequencies, and hence the allele frequencies, of the next generation (4). In order to predict the change in allele frequencies and genotype frequencies from one generation to the next, then, we need to know the genotype frequencies of the parents, their fitnesses, and, as we shall see, the frequency of each of the four types of gametes.

Directional Selection at Two Loci

The relative fitnesses of the nine genotypes at two loci might vary in many possible ways. A simple condition might be that the two loci contribute to variation in one trait, that they have ADDITIVE effects on the trait (Figure 14.1A), and that fitness is directly (linearly) related to the trait value. Suppose, for example, that replacing either A_2 with A_1 or B_2 with B_1 increases body size and fitness in this

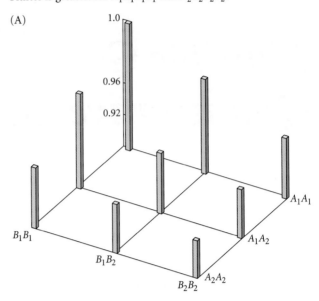

FIGURE 14.1 A pictorial representation of the fitnesses of nine genotypes. (A) Fitnesses are additive both within and between loci (case 1). (B) An instance of strong epistasis (case 2). Fitness is greatest for $A_1A_1B_1B_1$ and $A_2A_2B_2B_2$.

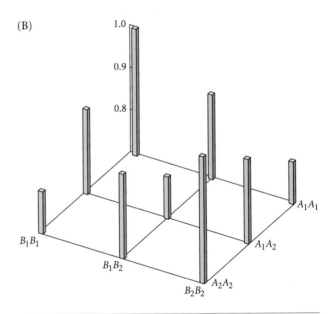

simple, additive fashion, where each allele substitution adds 0.05 to size and 0.02 to fitness. Each element in the tables below is the average phenotype (or fitness) of individuals with the *A*-locus genotype of that row and the *B*-locus genotype of that column.

Case 1

	Phenotype				Fitness		
	B_1B_1	B_1B_2	B_2B_2		B_1B_1	B_1B_2	B_2B_2
A_1A_1	1.00	0.95	0.90	A_1A_1	1.00	0.98	0.96
A_1A_2	0.95	0.90	0.85	A_1A_2	0.98	0.96	0.94
A_2A_2	0.90	0.85	0.80	A_2A_2	0.96	0.94	0.92

BOX 14.A

An Example of Genetic Change in a Trait under Directional Selection

Suppose substitution of A_1 for A_2 or B_1 for B_2 in an individual's genotype increases a character (perhaps the number of bristles on a part of a fly) by 1, and that the frequencies of both A_1 and B_1 (p_A and p_B) are 0.2. Assuming Hardy-Weinberg frequencies and no linkage disequilibrium, the nine genotypes will have these phenotypes and frequencies. Note that the frequency of $A_1A_1B_1B_1$, with the highest bristle number, is only 0.16 percent.

Phenotype				Frequency		
	B_1B_1	B_1B_2	B_2B_2		B_1B_1	B_1B_2	B_2B_2
					0.04	0.32	0.64
A_1A_1	4	3	2	A_1A_1 0.04	0.0016	0.0128	0.0252
A_1A_2	3	2	1	A_1A_2 0.32	0.0128	0.1024	0.2048
A_2A_2	2	1	0	A_2A_2 0.64	0.0256	0.2048	0.4096

Now suppose that of 1000 individuals, we select 50 (5 percent) with the highest bristle number, and breed them. The 50 "top" flies will include approximately (because of rounding) 2 $A_1A_1B_1B_1$ (A_1B_1/A_1B_1) individuals, 13 $A_1A_1B_1B_2$ (A_1B_1/A_1B_2) individuals, and 12 $A_1A_2B_1B_1$ (A_1B_1/A_2B_1) individuals (the gametes produced by these individuals are given in parentheses). This accounts for 28 flies. The remaining 22 of the 50 flies will be taken at random from the approximately 26 $A_1A_1B_2B_2$ (A_1B_2/A_1B_2), the 102 $A_1A_2B_1B_2$ (A_1B_1/A_2B_2 and A_1B_2/A_2B_1, in equal numbers), and the 26 $A_2A_2B_1B_1$ (A_2B_1/A_2B_1), which all have the same phenotype. The number selected, of each of these genotypes, is approximately as listed in the table below,

which gives the number (out of $2N = 100$) of each type of gamete that these $N = 50$ individuals represent.

Geno-type	Pheno-type	Number of individuals	Number of gametes of type			
			A_1B_1	A_1B_2	A_2B_1	A_2B_2
A_1B_1/A_1B_1	4	2	4			
A_1B_1/A_1B_2	3	13	13	13		
A_1B_1/A_2B_1	3	13	13		13	
A_1B_2/A_1B_2	2	4		8		
A_1B_1/A_2B_2	2	7	7			7
A_1B_2/A_2B_1	2	8		8	8	
A_2B_1/A_2B_1	2	3			6	
		Sum	37	29	27	7
		Frequency	0.37	0.29	0.27	0.07

Among the offspring (F_1) of these individuals, the frequency of A_1 is given by the sum of the frequencies of the A_1B_1 and A_1B_2 gametes; it has risen to $p_A = 0.37 + 0.29 = 0.66$ from $p_A = 0.20$ in the previous generation. Likewise, p_B has risen to 0.64. If the gametes unite at random, the frequency of $A_1A_1B_1B_1$ offspring (with the highest bristle number), due to union of A_1B_1 gametes, will be $0.37^2 = 0.137$, far higher than in the previous generation (0.0016). (As a challenge exercise, the reader may wish to use the *gamete* frequencies to calculate the frequency of each genotype in the F_1 generation and to show that the mean bristle number has increased from 0.80 in the parental generation to 2.53 in the F_1.)

In this case, there is directional selection for larger size. The size-enhancing alleles at each locus (A_1 and B_1) will increase in frequency and will ultimately be fixed. These alleles may have been so rare initially that the largest genotypes (such as $A_1A_1B_1B_1$) were so infrequent as seldom or never to be seen. (Perhaps, for example, A_1 and B_1 recently arose by mutation.) If $p_A = p_B = 0.05$, the expected frequency of $A_1A_1B_1B_1$ is 0.00000625. The most frequent genotypes are $A_2A_2B_2B_2$ (with frequency 0.8145) and $A_1A_2B_2B_2$ and $A_2A_2B_1B_2$ (each with frequency 0.0857). The higher fitness of the latter two genotypes (because of their greater size) increases the frequency of alleles A_1 and B_1, so that the still larger genotypes increase in frequency. *Thus selection acts as a sieve, increasing the "concentration" of favorable alleles in the population, and formerly rare gene combinations appear with increasing frequency. Selection, then, is the "creative" factor in evolution, making improbable phenotypes probable.* A more extended example is given in Box 14.A.

Gene Interactions

The previous hypothetical example (case 1) represents the familiar ADDITIVE MODEL. Because of its simplicity and mathematical tractability, and because many morphologi-

cal and other phenotypic traits do seem to be inherited in this way (see Chapter 3), much of the theory of multiple loci (and of quantitative genetics) uses this additive model. However, alleles often interact. One kind of departure from additivity is *dominance*, which can be thought of as an interaction between alleles at a locus so that the heterozygote is not precisely intermediate between the homozygotes. (This concept includes overdominance and underdominance, in which the heterozygote lies outside the range of the homozygotes; see Chapter 13.)

Interactions between alleles *at different loci* also cause the inheritance of the phenotype (or fitness) to be nonadditive. Nonadditivity caused by interactions between loci is called *epistasis*, and it can take many forms. An extreme case might be this hypothetical example:

Case 2

	B_1B_1	B_1B_2	B_2B_2
A_1A_1	1.0	0.9	0.8
A_1A_2	0.9	0.8	0.9
A_2A_2	0.8	0.9	1.0

In this case, the fitness of the organism's genotype at each locus depends on its genotype at the other locus: A_1A_1

Multiple Genes and Quantitative Traits 399

is advantageous only if with B_1B_1, and B_2B_2 is advantageous only if with A_2A_2 (Figure 14.1B). Because of the functional relationships among an organism's characters, epistatic fitness patterns of this kind may be very common, especially if the two loci affect different features. In Chapter 9, for example, we described primroses and other heterostylous plants that have either short styles and long stamens or long styles and short stamens; these combinations of features are advantageous because they promote cross-pollination. Another example can be found among herbivorous insects that feed on plants that may contain toxic compounds. Females often vary genetically in which plants they prefer to lay eggs on, and their offspring vary in their ability to tolerate toxins. These traits generally seem to be inherited independently (Thompson 1988). If alleles A_1, A_2 conferred tolerance or intolerance to a toxin, and B_1, B_2 determined acceptance or avoidance of a plant, the combinations A_1, B_1 and A_2, B_2 would confer higher fitness than A_2, B_1 and probably A_1, B_2.

For a final example, consider two loci that contribute additively to a trait such as body size, but in which fitness is not linearly correlated with the trait. In this case fitness will show epistasis, even though the trait does not. In one model of stabilizing selection (Wright 1968), the intermediate phenotype is optimal, and the fitness of each phenotype declines as the square of its deviation from the optimum (Figure 14.2). Thus the fitness of phenotype i might be $W_i = 1 - c(z_i - z_O)^2$, where z_i is the phenotypic value, z_O is the optimum phenotype, and c expresses the strength of selection. If $c = 0.1$ and the optimal phenotype is $z_O = 2$, the phenotypes and fitnesses of the nine genotypes might be:

Case 3

	Phenotype				Fitness		
	B_1B_1	B_1B_2	B_2B_2		B_1B_1	B_1B_2	B_2B_2
A_1A_1	4	3	2	A_1A_1	0.6	0.9	1.0
A_1A_2	3	2	1	A_1A_2	0.9	1.0	0.9
A_2A_2	2	1	0	A_2A_2	1.0	0.9	0.6

Notice that in this case, three different genotypes have the same optimal phenotype ($z = 2$): the double heterozygote ($A_1A_2B_1B_2$) and the two homozygotes $A_1A_1B_2B_2$ and $A_2A_2B_1B_1$.

Linkage Disequilibrium

In order to describe evolution at two or more loci, it is often important to stipulate the frequencies of the types of gametes that unite to form the next generation, because the gamete frequencies most simply express the degree of association between alleles at the several loci. Let g_{11}, g_{12}, g_{21}, and g_{22} be the frequencies of the gametes A_1B_1, A_1B_2, A_2B_1, and A_2B_2. Then the frequency of $A_1A_1B_1B_1$ zygotes is $(g_{11})^2$, the frequency of $A_1A_2B_1B_1$ is $2(g_{11})(g_{21})$, and so on.

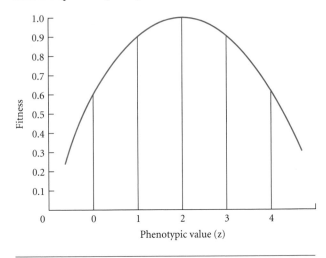

FIGURE 14.2 An example of stabilizing selection, in which the fitness of a phenotype declines as the square of its deviation from the optimum (case 3).

Gametes with like alleles (A_1 and B_1, A_2 and B_2 in our example) are referred to as COUPLING GAMETES, and those with unlike alleles (A_1 and B_2, A_2 and B_1) as REPULSION GAMETES. "Like" and "unlike" may be defined by any of a variety of criteria, such as whether the alleles have similar or opposite effects on a phenotypic character or fitness. The frequency of a gamete type, say, g_{11}, is determined in part by the frequency, among the reproducing adults, of each of the genotypes that produce that type of gamete. For instance, $A_1A_1B_1B_1$, $A_1A_2B_1B_1$, $A_1A_1B_1B_2$, and $A_1A_2B_1B_2$ all produce A_1B_1 gametes. But there is one complication: the double heterozygotes $A_1A_2B_1B_2$ are of two types, which may be written as the gametes that united to form them: A_1B_1/A_2B_2 and A_1B_2/A_2B_1. If the loci were so tightly linked that there were no recombination, A_1B_1/A_2B_2 would yield only coupling gametes (A_1B_1, A_2B_2) and A_1B_2/A_2B_1 would yield only repulsion gametes (A_1B_2, A_2B_1). The frequency with which A_1B_1/A_2B_2 generates repulsion gametes and A_1B_2/A_2B_1 generates coupling gametes depends on the rate of recombination. Therefore the gamete frequencies (g_{11}, etc.) depend not only on the frequencies of the nine genotypes among the parents, but also on the recombination rate (or, conversely, on the tightness of linkage.)

In Chapter 9, we briefly considered the concepts of *linkage equilibrium* and *linkage disequilibrium*. Let p_A and q_A be the frequencies of alleles A_1 and A_2 ($p_A + q_A = 1$), and p_B and q_B be the frequencies of alleles B_1 and B_2 ($p_B + q_B = 1$). If the alleles at the two loci were inherited independently, each gamete frequency would be simply the product of the frequencies of the alleles it carries (i.e., multiply independent probabilities). Thus, for example, g_{11} (the frequency of gametes carrying alleles A_1 and B_1) would equal $p_A \times p_B$. For instance, if $p_A = 0.4$, $q_A = 0.6$ and $p_B = 0.3$, $q_B = 0.7$, g_{11} should be 0.12. Thus with random mating, $AABB$ offspring should have a frequency of $g_{11} \times$

$g_{11} = (0.12)^2 = 0.0144$. The frequency of each of the other eight genotypes can be found similarly. (As an exercise, show that the frequency of $A_1A_2B_1B_2$ offspring is 0.2016 in this case.) When the frequency of each gamete type equals the product of the allele frequencies, the loci are in linkage equilibrium.

With the allele frequencies in our hypothetical example, we expect, if there is linkage equilibrium, that $g_{11} = 0.12$, $g_{12} = 0.28$, $g_{21} = 0.18$, and $g_{22} = 0.42$. We might instead find in a sample from a population (perhaps by scoring the offspring of many individuals crossed with an $A_2A_2B_2B_2$ stock) that $g_{11} = 0.22$, $g_{12} = 0.18$, $g_{21} = 0.08$, and $g_{22} = 0.52$: an excess of coupling and a deficiency of repulsion gametes. The loci would then be in linkage disequilibrium. A COEFFICIENT OF LINKAGE DISEQUILIBRIUM may be defined as $D = (g_{11} \times g_{22}) - (g_{12} \times g_{21})$, i.e., the product of coupling gamete frequencies minus the product of repulsion gamete frequencies. In our example, $D = 0.10$. Compared with a population in linkage equilibrium, genotypes formed from the union of coupling gametes will be more frequent, and certain other genotypes will be less frequent. In our example, the genotypes $A_1A_1B_1B_1$ and $A_1A_2B_1B_1$ would, if there were linkage equilibrium, have the respective frequencies $(g_{11})^2 = (0.12)^2 = 0.0144$ and $2g_{11}g_{21} = 2(0.12)(0.18) = 0.0432$. But when $D = 0.10$, their frequencies are $(0.22)^2 = 0.0484$ and $2(0.22)(0.08) = 0.0352$. *Linkage disequilibrium therefore alters the frequency distribution of genotypes and phenotypes.*

With the passage of generations, *the level of linkage disequilibrium declines due to recombination in the double heterozygotes* (Figure 14.3). For example, if coupling gametes are in excess, there are more A_1B_1/A_2B_2 than A_1B_2/A_2B_1 double heterozygotes, so by recombination, the coupling heterozygotes A_1B_1/A_2B_2 give rise to more repulsion gametes (A_1B_2, A_2B_1) than the repulsion heterozygotes A_1B_2/A_2B_1 do coupling gametes (A_1B_1, A_2B_2). Linkage disequilibrium "decays" at a rate proportional to the recombination rate (R) between the loci: After t generations, $D = D_0 (1 - R)^t$, where D_0 is the original level (see Chapter 9). Eventually, D declines to zero unless other factors intervene.

Causes of linkage disequilibrium

Linkage disequilibrium is an important concept, because when loci are in a state of linkage disequilibrium, allele frequency changes at one locus are not independent of changes at another. Linkage disequilibrium is relevant to topics such as speciation (see Chapter 16), sexual selection (see Chapter 20), the evolution of sex and mating systems (see Chapter 21), and molecular evolution (see Chapter 22).

Why might two loci, or two sites within a gene, be in linkage disequilibrium? There are several possible reasons.

1. The sample of organisms that display linkage disequilibrium actually includes two different species that differ in allele frequencies at both loci. More generally, nonrandom mating can maintain linkage disequilibrium, as discussed in Chapter 11.

2. When a new mutation arises, the single copy is necessarily associated with specific alleles at other loci on the chromosome, and therefore is in linkage disequilibrium with these alleles. The copies of this mutation in subsequent generations will retain this association until it is broken down by recombination.

3. The population may have been formed recently by the union of two populations with different allele frequencies, and linkage disequilibrium has not yet decayed.

4. Recombination is very low or nonexistent. Chromosome inversions have this effect (see Chapter 10), as does parthenogenesis.

5. Linkage disequilibrium may be caused by genetic drift. The simplest way to think about this is to picture a very low recombination rate. Then the four gamete types may be thought of as if they were four alleles at one locus. One of these "alleles" may drift to high frequency by chance, creating an excess of that combination relative to others.

6. Natural selection may cause linkage disequilibrium, as described in subsequent sections.

Linkage disequilibrium can affect the rate of evolution of a trait (Felsenstein 1965). Suppose that two loci contribute to variation in body size (as in Case 1 above). If selection favors large size, and coupling gametes are in excess ($D > 0$), then the trait will evolve faster than if the loci were in linkage equilibrium. This is because both $A_1A_1B_1B_1$ and

---

FIGURE 14.3 The decrease in linkage disequilibrium (D), relative to its initial value (D_O), over generations. The decay is shown for pairs of loci with different recombination rates (R); $R = 0.50$ represents unlinked loci. Random mating is assumed. Note that if the recombination rate is fairly high, the association between alleles at different loci is rapidly reduced to zero. (After Hartl and Clark 1989.)

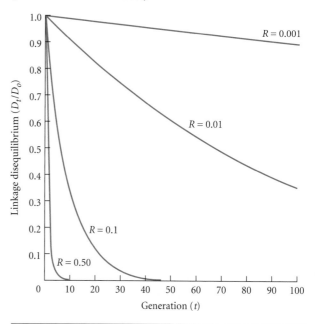

FIGURE 14.4 The frequency distribution of phenotypes (e.g., size) for an additively inherited trait when two alleles at each of two loci have equal effects (case 1). The allele frequencies are $p = q = 0.5$ at each locus. The height of each bar is found by summing the frequencies of all genotypes with the same phenotypic value. The open bars are the phenotype frequencies when the loci are in linkage equilibrium ($D = 0$); the colored bars, when $D = 0.1$ (alleles with like effects are partly associated); the black bars, when $D = -0.1$ (alleles with unlike effects are partly associated). Note that positive linkage disequilibrium increases the variance by increasing the frequency of the extreme phenotypes and decreasing the frequency of the intermediate phenotype. Negative linkage disequilibrium has the opposite effect.

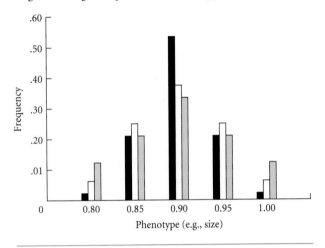

$A_2A_2B_2B_2$ will have higher frequency than when $D = 0$. Because these are the genotypes that differ most in fitness, selection can eliminate A_2 and B_2, and increase A_1 and B_1, more rapidly. Another way of putting this is to note that the *variance* in both body size and fitness is greater when $D > 0$ than when $D = 0$ (Figure 14.4). We will shortly see that the rate of evolution under selection is proportional to the variance in fitness. Conversely, if repulsion gametes (A_1B_2, A_2B_1) are in excess, $D < 0$. In this case a greater proportion of the population is clustered near the mean, the extreme phenotypes $A_1A_1B_1B_1$ and $A_2A_2B_2B_2$ are less common, the variance in fitness is reduced, and the response to selection is initially retarded. Thus if, initially, $D < 0$, but selection favors the coupling homozygote, the character changes little at first; then, when selection alters gamete frequencies so that $D > 0$, evolution of the feature accelerates. Such a LAG in response is sometimes seen in artificial selection experiments (see below).

Multiple Loci in Adaptive Landscapes

Recall from Chapter 13 that when the fitnesses at a single locus are specified, average fitness values ($\bar{w}$) can be calculated for populations with all possible allele frequencies and plotted against allele frequency (say, q) (Figure 12 in Chapter 13). Furthermore, since natural selection increases $\bar{w}$, a population can be envisioned as traveling up the $\bar{w}$ curve to a peak as selection changes the allele frequency. The $\bar{w}$-versus-q curve describes a "hillside," or

adaptive landscape, which may have more than one peak (as in Figure 12D in Chapter 13).

Sewall Wright, who developed the metaphor of the adaptive landscape, noted that quite a complex landscape can occur if fitness is affected by the interaction of two or more loci. We consider only two loci here, and must understand that our general conclusions hold for greater numbers as well. Beginning first with a simple example, suppose that at locus A, the fitnesses are additively determined, that of A_1A_1 being 1, of A_1A_2 $1 - s/2$, and of A_2A_2 $1 - s$. Then $\bar{w}$ declines linearly with q_A, the frequency of A_2 (Figure 14.5A). Let locus B be overdominant for fitness, with B_1B_1, B_1B_2, and B_2B_2 having the fitnesses $1 - t$, 1, and $1 - t$ respectively (Figure 14.5B). We know from Chapter 13 that the equilibrium frequency of A_2 is 0, and that of B_2 is 0.5. (Refer back to Chapter 13 if this is not obvious.)

Now for simplicity, suppose that the fitness of each two-locus genotype is the simple sum of the contribution of each locus (e.g., the fitness of $A_1A_1B_1B_2$ is 2)—that is, there is no epistasis. We can calculate the mean fitness $\bar{w}$ for every possible population–that is, for every possible combination of q_A and q_B, the frequencies of A_2 and B_2. For example, if $s = 0.1$ and $t = 0.2$, the mean fitness of a population with $q_A = 0.2$ and $q_B = 0.2$ is the sum of the nine genotypes' fitnesses, each weighted by the frequency of the genotype, or $\bar{w} = 1.824$. (Suggestion: Obtain this conclusion as an exercise in understanding what $\bar{w}$ is.) This value may be plotted as a point in a three-dimensional space for which the y-axis is $\bar{w}$ and the x- and z-axes are q_A and q_B, i.e., a space in which we plot average fitness as a function of the allele frequencies at both loci. With enough work on a pocket calculator, or better still a computer, we can calculate $\bar{w}$ for each of a great many pairs of allele frequencies, and thus calculate an ADAPTIVE SURFACE, the height of which represents $\bar{w}$ for all possible allele frequency combinations. In our example, this surface (Figure 14.5C) is concave downward with respect to the q_B axis, because as q_B approaches 0.5, the frequency of the superior B_1B_2 heterozygote increases. It slopes upward with respect to the q_A axis, from $q_A = 1$ toward $q_A = 0$ (when the superior A_1A_1 homozygote is fixed). Recalling now that $\bar{w}$ increases as natural selection increases the frequency of fitter genotypes, *evolution may be envisioned as the movement of a point* (representing the population) *on the $\bar{w}$ surface. The point moves steadily upslope until it arrives at the peak—the allele frequency equilibrium.*

It is important to understand that a point on this landscape represents not the fitness of a genotype, but the mean fitness of the individuals in a possible population consisting of nine genotypes (unless a locus is fixed), and that any real population, with its current allele frequencies at two loci, can be represented by a single point on this surface.

Because three-dimensional figures are hard for many people to draw, population geneticists, following the lead of cartographers, usually represent an adaptive surface as a "topographic map," in which a line of elevation connects

FIGURE 14.5 Simple adaptive landscapes. (A) Mean fitness, $\bar{w}$, declines as the frequency of A_2 increases, at a locus at which A_1A_1, A_1A_2, and A_2A_2 have fitnesses 1, $1 - s/2$, and $1 - s$. Each point on the line is the average of the fitnesses of the three genotypes, weighted by their frequencies, which are determined by the allele frequencies. (B) Another such plot, for a locus with heterozygous advantage, when B_1B_1, B_1B_2, and B_2B_2 have fitnesses 1 $- t$, 1, and $1 - t$ respectively. (C) A fitness surface, or adaptive landscape, for the two loci in (A) and (B) considered together, if there is no epistatic interaction. Each point on the floor of the three-dimensional space represents a population with particular allele frequencies at loci A and B. The height of the fitness surface above that point is the mean fitness of the nine genotypes, weighted by their frequencies. (D) The fitness surface in (C) represented as a "topographic map" in which all points on a contour line represent possible populations with different allele frequencies, but the same mean fitness ($\bar{w}$). The peak is at point H; minimal values of $\bar{w}$ are at points L.

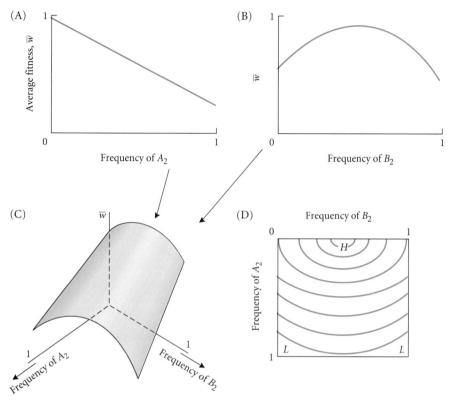

populations (allele frequency combinations) of equal mean fitness. Figure 14.5D illustrates such a map, with H marking the population of highest fitness (a PEAK) and L marking those with lowest fitness (VALLEYS).

An adaptive landscape provides a visualization of how a population will evolve in a particular environment. If the environment changes so that the genotypes' relative fitnesses are altered, the new fitnesses determine a new adaptive landscape—and what was a valley may become a peak. For instance, if $A_2A_2B_2B_2$ in our example were to become the fittest genotype because of an environmental change, the new peak would be at the lower right corner of Figure 14.5D, and the contour lines would decline away from it. Thus in a continually fluctuating environment, an ocean surface might be a better metaphor than a landscape.

The reason for developing this subject at such length is that the adaptive landscape is probably the most common metaphor in evolutionary genetic discourse, providing a shorthand vocabulary for describing evolutionary hypotheses (although like any metaphor, it can mislead the un-

wary).* It is particularly useful for discussing cases in which genes interact in their effect on fitness.

Multiple equilibria In Chapter 13, we learned that when the heterozygote is less fit than either homozygote (underdominance), allele frequencies may arrive at either of two stable equilibria, depending on the initial frequencies. This pattern shows that the past history of a population (whatever it was that determined the "initial frequencies") can affect the direction of its subsequent evo-

---

*The hill-climbing metaphor is widely used in the evolutionary literature, and we will employ it to convey some useful distinctions among evolutionary hypotheses. Strictly speaking, however, the metaphor is imperfect, or even wrong, if fitnesses are frequency-dependent, if selection consists of differences in reproductive rate rather than survival, or if selection at two or more loci is strong and recombination is weak. In such cases, $\bar{w}$ may not increase (i.e., the population may not climb to a peak). The adaptive landscape is a useful metaphor, then, if fitnesses within a given environment are constant and are due to viability differences, and if selection is weak and the recombination rate is high.

lution. The same effect often arises from interactions among loci. Returning to our hypothetical case 2, for example, it should be evident that the population with highest mean fitness ($\bar{w} = 1.0$) would be *either* monomorphic for $A_1A_1B_1B_1$ *or* monomorphic for $A_2A_2B_2B_2$, the superior homozygotes. All other possible (i.e., polymorphic) populations would have lower $\bar{w}$, so we would expect the population to evolve to either of the two equilibrium states at which $\bar{w} = 1.0$. The population moves only upslope under natural selection, so it arrives at the peak on whose slope it is initially situated. Following through on the example in which locus *A* determines tolerance to toxins by an herbivorous insect and locus *B* determines egg-laying choice, the population may evolve *either* to avoid the toxic plant *or* to use it and become adapted to the toxin, depending on the initial allele frequencies at both loci (Futuyma 1983; Castillo-Chávez et al. 1988; Rausher 1993).

The first adaptive landscape constructed from real data was created by the population geneticist Richard Lewontin and Michael White, an authority on animal chromosomes (1960). The Australian grasshopper *Keyacris* (*Moraba*) *scurra* is polymorphic for inversions A_1, A_2 and B_1, B_2 on two different chromosomes. In a wild population, the frequencies of the nine genotypes differed from those expected under Hardy-Weinberg equilibrium and linkage equilibrium. Lewontin and White assumed that this disparity between observed and expected genotype frequencies was caused by natural selection, and from the magnitude of the disparity estimated the fitnesses as

	A_1A_1	A_1A_2	A_2A_2
B_1B_1	0.79	0.67	0.66
B_1B_2	1.000	1.006	0.66
B_2B_2	0.83	0.90	1.07

From these fitnesses, they calculated an adaptive landscape (Figure 14.6). The pronounced epistasis among the estimated fitness values gives rise to a complex landscape with two peaks, or stable equilibria (at $q_A = 1$, $q_B = 1$ and $q_A = 0.55$, $q_B = 0$), separated by a "saddle" (S in the figure), which represents an unstable equilibrium. Depending on the initial frequencies of the two inversions, a population would evolve to either of the two adaptive peaks, as the "trajectories" in Figure 14.6 illustrate.

An experimental example is provided by Cavener and Clegg's (1987) study of changes in the frequency of alleles at the alcohol dehydrogenase (*Adh*) and α-glycerol phosphate dehydrogenase (α-*Gpdh*) loci in laboratory populations of *Drosophila melanogaster*. The allele frequencies at both loci changed over the course of 50 generations in populations given larval food to which ethanol was added, but not in control populations (Figure 14.7A,B). The authors estimated the fitness of each genotype from the disparity between the observed genotype frequencies and those expected under Hardy-Weinberg equilibrium and linkage equilibrium. These estimates (Table 14.1) showed strong epistasis in both environments. For example, in the

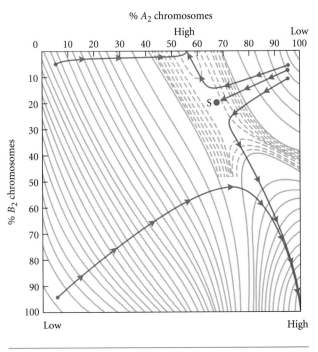

FIGURE 14.6 An adaptive landscape for two polymorphic chromosomes in the Australian grasshopper *Keyacris scurra*. Genotype fitnesses, estimated from deviations between observed and expected genotype frequencies, were used to calculate $\bar{w}$ for populations with various possible genotype frequencies. Each contour line represents populations with the same mean fitness. The dashed lines indicate finer distinctions than the solid contours. There are two peaks (High) and a saddle point, or unstable equilibrium (S). The trajectories represent theoretical changes in genetic composition that a population would follow from five initial states. (After Lewontin and White 1960.)

ethanol environment, the fitnesses of α-*Gpdh* genotypes were *SS* < *SF* > *FF* if on a homozygous *Adh-SS* background, but *SS* > *SF* < *FF* if on a homozygous *Adh-FF* background. In the "normal" (control) environment, the fitnesses were altered; for example, the α-*Gpdh* fitnesses on an *Adh-FF* background are *SS* < *SF* > *FF*. When Cavener and Clegg used these fitnesses to calculate (by computer simulations) the expected changes in allele frequencies in both the ethanol-treated and the control populations, their predicted changes matched the observed changes quite well (compare graph C with graphs A and B in Figure 14.7). Notice that in the ethanol-treated populations, the *S* allele of α-*Gpdh* declined at first, because the *Adh-SS* genotype had a high initial frequency, and α-*Gpdh-SS* was then inferior to α-*Gpdh-SF* and *FF*. But as the *S* allele of *Adh* declined, the *S* allele of α-*Gpdh* reversed its course and increased, because the genotype *SS* at this locus had high fitness when associated with the *Adh* genotypes *SF* or *FF*. This experiment demonstrates that epistatic interactions between loci can reverse the course of allele frequency change: *allele frequencies respond not only to the external environment, but to the "genetic environment" as well.*

FIGURE 14.7 Changes in the frequency of (A) the *Adh^S* allele and (B) the α–*Gpdh^S* allele in experimental populations of *Drosophila melanogaster* reared in medium with (E₁, E₂) and without (C₁, C₂) ethanol. Two populations were maintained in each environment. (C) Computer simulations of expected changes in the allele frequencies, based on estimates of the fitness of each of the nine genotypes in each environment. The correspondence with the data in panels A and B is quite high. (After Cavener and Clegg 1981.)

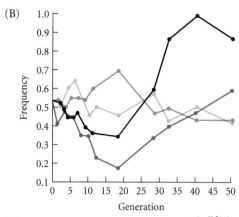

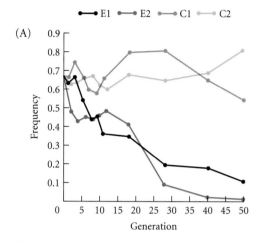

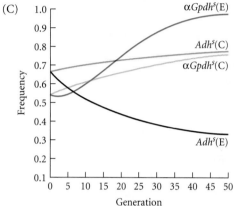

The importance of multiple peaks cannot be overestimated: because of gene interactions, "redundancy" among loci, and alternative characters that may provide adaptation to an environmental factor, *different populations* exposed to the same adaptive "problem" frequently *arrive at different genetic "solutions."* For example, when copper-tolerant populations of the monkeyflower *Mimulus guttatus* from different copper mines are crossed, the F₂ generation is more variable in copper tolerance than either parent population, indicating that the populations differ in the loci that confer copper tolerance. (If uppercase letters denote tolerance alleles, two parent populations might be *AAbb* and *aaBB*; the F₂ will then be variable, whereas it would not be if both populations were simply *AA*.) Different natural populations of houseflies (*Musca domestica*) have achieved resistance to DDT by several different physiolog-

Table 14.1 **Fitnesses of two-locus genotypes of *Drosophila melanogaster* in two experimental environments**[a]

ETHANOL PRESENT

		Adh GENOTYPE		
		SS	SF	FF
α–*Gpdh* GENOTYPE	SS	0.596	1.288	0.932
	SF	0.964	1.000	0.836
	FF	0.909	0.968	0.864

ETHANOL ABSENT

		Adh GENOTYPE		
		SS	SF	FF
α–*Gpdh* GENOTYPE	SS	0.992	1.059	0.863
	SF	1.080	1.000	0.935
	FF	0.765	1.164	0.750

Source: Modified from Cavener and Clegg (1981).

[a]The fitness of the double heterozygote, taken as a standard, has been set at 1.000 in each environment. Fitness was measured as the proportion of larvae surviving to adulthood.

ical and biochemical mechanisms. Some populations of spider mites (*Tetranychus urticae*) have adapted to organophosphate pesticides by detoxifying the poison, whereas others have decreased the sensitivity of the nervous system to the toxin. These and many other examples are summarized by Cohan (1984).

Adaptive landscapes and quantitative traits In case 3 above, we presented a model in which two loci contribute additively to a trait (such as body size or number of bristles), and in which there is stabilizing selection for an intermediate phenotype (a phenotype of 2 in the example). What will the allele frequencies be at equilibrium in this example? Figure 14.8 shows approximately the adaptive landscape calculated from the hypothetical fitnesses and phenotypes given in case 3. Notice that although $A_1A_2B_1B_2$ has the same, maximal, fitness (1.0) as the two fittest homozygotes, there cannot be a pure population of heterozygotes in a sexual population, so the mean fitness of a genetically variable population will be less than 1.0. For example, $q_A = 0.5$ and $q_B = 0.5$ specifies a population in the center of the figure (S). At each locus, the genotype frequencies are $p^2 = 0.25$, $2pq = 0.50$, $q^2 = 0.25$; assuming linkage equilibrium for simplicity, the frequency of each of the nine two-locus genotypes is readily calculated (for example, the frequency of double heterozygotes $A_1A_2B_1B_2$ is 0.25). Adding the products of the genotypes' frequencies and fitnesses, we find that when $q_A = 0.5$ and $q_B = 0.5$, the mean fitness is $\bar{w} = 0.775$. This polymorphic population in fact represents an *unstable* gene frequency equilibrium, a saddle point on the landscape from which a ridge rises to each of two stable equilibria, at $q_A = 0$, $q_B = 1$ (monomorphic for $A_1A_1B_2B_2$) and $q_A = 1$, $q_B = 0$ (monomorphic for $A_2A_2B_1B_1$). Thus, *the population will evolve to either of two monomorphic equilibria* with the same mean fitness, *representing different genetic constitutions but with the same phenotype*. This example also illustrates the general conclusion that *superiority of an intermediate state of a polygenic character does not maintain genetic variation indefinitely* (Lewontin 1964; Felsenstein 1979).

This example has several significant implications. First, it raises the question of *what does maintain the genetic variation* that most quantitative traits display. Second, it implies that two populations subjected to identical selection for the same character state can *evolve different genetic constitutions, even though they evolve the same phenotype* (Cohan 1984). The case of different genetic bases for copper tolerance in the monkeyflower, cited earlier, is a likely example. Third, if alleles at many loci each contribute only slightly to the phenotype, the fitness valleys between nearby adaptive peaks (different genetic constitutions) are likely to be shallow: mean fitness probably will not be greatly diminished by substituting an allele at one out of 10 or 20 loci. Therefore, *the genetic basis of a characteristic under stabilizing selection is likely to change by genetic drift*; as the frequency of a + allele at one locus drifts toward fixation, the frequency of a + allele at one or more other loci is decreased by selection so that the mean phenotype stays near the optimum (Wright 1935; Kimura 1981). Due to genetic drift, therefore, two populations with the same phenotype can differ in genotype.

The fourth implication of our example is that two populations with the same mean phenotype but differing in genotype may, when crossed, produce a hybrid population with a lower mean and a higher variance. This would be the case if we obtained an F_2 generation from a cross between equal numbers of individuals from the two equilibrial populations ($A_1A_1B_2B_2$ and $A_2A_2B_1B_1$) in our example: the F_2 generation would be highly variable, and its mean fitness would be lower (it would be 0.775, as calculated earlier). Precisely this pattern has been observed in the viability and fecundity of the F_2 offspring of crosses between different geographic populations in several species of *Drosophila* (Table 14.2), as well as in some other organisms. This reduction in fitness, termed F_2 BREAKDOWN, implies that recombination between sets of genes from the two populations has given rise to unfavorable gene combinations. Thus alleles that work well together as a team within a population do not necessarily function harmoniously with alleles from other populations. Theodosius Dobzhansky (1955) used the term **coadapted gene pool** to refer to a system of alleles that interact harmoniously within a population, but which form disharmonious combinations with alleles from other populations. The concept of coadapted gene pools is important in some theories of speciation (see Chapter 16).

FIGURE 14.8 An approximate adaptive landscape for two additively acting loci that affect variation in a single trait, such as size, that is under stabilizing selection for the intermediate phenotype. Genotype frequencies and fitnesses are as given in case 3; linkage equilibrium is assumed. Allele frequencies of 0.5 at both loci are represented by an unstable equilibrium, or saddle point, on the landscape (S). Depending on initial allele frequencies, populations will evolve, as shown by the arrows, toward either of two monomorphic equilibria, $A_1A_1B_2B_2$ or $A_2A_2B_1B_1$ (H). L indicates regions of lowest mean fitness.

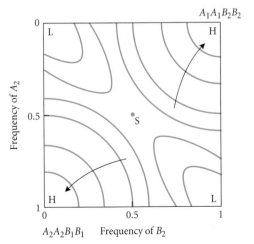

Table 14.2 **Offspring fitness in crosses within and between geographic populations of *Drosophila*[a]**

	F₁	F₂
D. pseudoobscura		
Within	59.6	59.6
Between	70.4	49.2
D. willistoni		
Within	59.9	57.2
Between	68.3	51.5
D. paulistorum		
Within	54.7	53.1
Between	58.0	51.3

Source: After Wallace and Vetukhiv (1955).

[a]The table shows the mean percentage of surviving larvae in F₁ and F₂ crosses of flies from the same population (within) and between widely separated populations (between). The decrease in the F₂ of crosses between populations, especially evident in *D. pseudoobscura*, is termed "F₂ breakdown."

Divergence of populations When two populations occupy different adaptive peaks, so that the fitness of both is higher than that of hybrids between them, we can envision at least three histories by which they might have arrived at these peaks (Figure 14.9). Assume that the common ancestor of the populations was nearly fixed for genotype $A_1A_1B_1B_1$, but had low frequencies of A_2 and B_2.

1. Population 1 has continued to occupy an environment in which monomorphism for $A_1A_1B_1B_1$ represents an adaptive peak. The environment of population 2, however, has changed, so that a different genetic constitution—say, monomorphic $A_2A_2B_2B_2$—is now the adaptive peak, and $A_1A_1B_1B_1$ has low fitness in that environment (Figure 14.9A). Population 2 has evolved simply by adaptation to a new environment. One of the founders of evolutionary genetics, R. A. Fisher, held that this is the most common mode of adaptive evolution.

2. The environment of both populations has been altered in the same way, but the two populations have responded in different ways. The adaptive landscape has been transformed into one with two peaks, representing monomorphism for two different genotypes, $A_1A_1B_2B_2$ and $A_2A_2B_1B_1$ (Figure 14.9B). Copper tolerance in *Mimulus guttatus* may exemplify this history.

3. The environment of neither population has changed, but it is one for which the adaptive landscape has two or more peaks (Figure 14.9C). One population has remained at the ancestral adaptive peak ($A_1A_1B_1B_1$), and

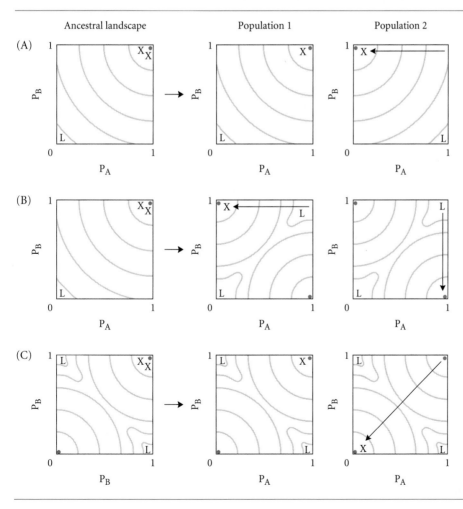

FIGURE 14.9 Some possible histories by which two populations may have evolved to different adaptive peaks. A colored circle marks an adaptive peak; L marks a region of minimal average fitness; X marks the position (allele frequencies) of a population. The diagrams at the left represent the adaptive landscape determined by selective factors acting similarly on the ancestral populations, both of which are located near the peak at the upper right (nearly fixed for $A_1A_1B_1B_1$). The center and right diagrams show the adaptive landscape for each of the contemporary populations determined by current or recent selective factors, the position (X) of each population, and the trajectory of allele frequency change for each population (line with arrow). (A) Change in the environment of population 2 (only). (B) Identical change in the environment of both populations. (C) No change in the environment of either population; divergence of population 2 by a "peak shift."

the other has shifted to the other peak ($A_2A_2B_2B_2$). To do so, however, it would have to cross an adaptive valley, which should result in a temporary decline in fitness that is counter to natural selection. How this might happen was described by Sewall Wright, and constitutes one of the most controversial theories in evolutionary biology.

The shifting balance theory Sewall Wright, who introduced the concept of the adaptive landscape, used it to develop a comprehensive theory of evolution very different from that of R. A. Fisher. Fisher (1930) believed that long-term "progressive" evolution of greater and greater adaptation occurs by the action of natural selection, within large populations, on mutations that enhance fitness more or less additively. Gene interactions play relatively little role in this simple theory of "mass selection." Wright, however, based on his experience with experimental genetics and animal breeding (Provine 1986), was greatly impressed with the importance of gene interactions. This experience inspired him to develop his **shifting balance theory** of evolution (Provine 1986; Wade 1992). Wright (1931, 1982), throughout his long professional life, maintained that this complex and controversial theory described a mode of long-term progressive adaptation that was faster than mass selection, and was the mode by which such evolution had actually occurred. The difference between Fisher's and Wright's points of view persists in population genetics today, and gives rise to considerable controversy about several aspects of evolution.

We learned in Chapter 13 that if fitnesses at one locus are such that there are two or more stable equilibria at which mean fitness ($\bar{w}$) is maximal (i.e., adaptive peaks), *natural selection cannot move a population "downhill" from one peak to the other.* However, *genetic drift can move the genetic composition away from the peak it occupies, and may shift the allele frequencies across the adaptive valley* to the slope of the other peak. Natural selection can then move the population uphill to the new equilibrium (see Figure 28 in Chapter 13). The chance of such a PEAK SHIFT is lower, the stronger selection is (i.e., the steeper the slope of the original peak), but its probability increases, the smaller the effective population size is (for example, if a temporary bottleneck occurs).

Exactly the same principles hold if interactions of two or more loci give rise to a landscape with multiple adaptive peaks. These peaks may differ in height—i.e., in mean fitness, $\bar{w}$—and if, due to genetic drift followed by selection, a population shifts from a lower peak to a higher peak, its adaptedness presumably improves. Moreover, there may be many other fitness peaks (especially if we consider many other loci), some of which are higher still; so that by repeated peak shifts, a population might over time move to higher and higher peaks. (It might also, of course, shift to lower peaks, and become less well adapted.) None of this would happen in a large, panmictic "Fisherian" population because genetic drift would be negligible, so that selection

would confine the population to the first peak it evolved to, however low that peak might be.

Wright argued that the optimal conditions for progressive adaptation occur in a species subdivided into small populations connected by low levels of gene flow. According to his shifting balance theory, all the populations initially are situated on one or another adaptive peak and have the same allele frequencies as a result of natural selection in the past. *Variation among populations* then arises because *random genetic drift* causes the allele frequencies in each population to vary around the theoretical equilibrium frequencies corresponding to the peak. Occasionally the allele frequencies in one or another population drift across an adaptive valley ("phase 1"). *Selection within that population* then shifts the genetic composition up the hillside to a new peak ("phase 2"). If a population thereby shifts to a higher peak, representing higher mean fitness, it should have a higher rate of population growth; therefore, more migrants should disperse from this population than from others, carrying their superior gene combinations into other populations. This gene flow shifts the allele frequencies in the recipient populations toward those of the donor population (see Chapter 11). If they are shifted far enough, the *recipient populations will be shifted across the adaptive valley, and will evolve to the superior adaptive peak* ("phase 3"). This process increases the number of populations that, having the superior genetic constitution, reproduce so prolifically as to flood yet other populations with migrants, until the species as a whole has evolved to the superior adaptive peak. Wright referred to the difference in productivity among populations, which causes phase 3, as INTERDEMIC SELECTION. Thus in Wright's view, adaptation is the consequence of (1) selection within populations, (2) genetic drift, (3) selection among populations (interdemic selection), and (4) gene flow.

Fisher's view of adaptation by natural selection, in which advantageous mutations are fixed by selection in large populations, is simple and is not controversial. Many examples of this process, some of which are described in Chapter 13, have been documented in natural and experimental populations. Wright's theory, being much more complex than Fisher's, is very difficult to test, and whether or not this process has played an important role in evolution is unknown. Population geneticists differ strongly on its plausibility, and many of them adhere to Fisher's simpler theory (e.g., Barton and Clark 1990; Coyne et al. 1997). Mathematical theory shows that genetic drift can indeed initiate peak shifts if selection is weak and populations go through sufficiently small bottlenecks (Barton and Rouhani 1987; Rouhani and Barton 1987; Charlesworth and Rouhani 1988). The likelihood of "phase 3," however, in which gene flow from highly adapted populations "drags" other populations to the superior peak, is controversial, since it depends on the recombination rate, the rate of gene flow, and the distribution of populations (Crow et al. 1990; Coyne et al. 1997). Theoreticians who doubt the importance of

Wright's theory argue that drift-induced peak shifts, although possible, should generally be so rare that they are unlikely to account for much of evolution, compared with the well-documented process of simple Fisherian natural selection within populations.

Most evidence bearing on the validity of the shifting balance theory is very indirect. It seems likely that many species meet at least some of its assumptions. Subdivision into local populations, and genetic differentiation among them, is very common (see Chapter 9), and at least some of this genetic differentiation is likely to be caused by genetic drift (see Chapters 9 and 13). Crosses between populations often provide evidence of epistasis for fitness, i.e., multiple adaptive peaks or different "coadapted gene pools" (Whitlock et al. 1995). But it is not known whether populations have achieved these genetic differences through drift-induced peak shifts or simply through divergent selection in different environments. Nor is it known whether differences in productivity, necessary for interdemic selection, often arise from the genetic differences among populations.

Quantitative Genetics

Variation in most phenotypic characters is *continuous*; such characters are called QUANTITATIVE TRAITS. That component of the variation that is genetically based is *polygenic*, i.e., it arises from allelic variation at several or many loci. (Some of the variation also arises from nongenetic effects of the environment). Most of the principles of the evolution of polygenic traits can be understood from two-locus models of the kind we have already considered. However, because many of the loci involved affect the phenotype only slightly, the effects of individual loci cannot usually be distinguished.

Genetic Dissection of Quantitative Traits

The loci that produce polygenic characters are often called POLYGENES or QUANTITATIVE TRAIT LOCI (QTLs). Various authors have suggested that the number of loci controlling typical polygenic characters might be rather low (less than about ten), or might be up to several hundred. An approximately continuous distribution of phenotypic variation can arise from the action of only three or four loci, partly because of environmental variation, and partly because at each locus, a population may harbor multiple alleles that vary in their effect. Alleles with "major" and "minor" effects may arise from different mutations of the same locus. The number of loci that contribute to the development of a character may be considerably greater than the number that are polymorphic and contribute to its variation. For example, more than 70 loci are known (through the study of mutations in the laboratory) to affect eye color in *Drosophila melanogaster*, but few of them vary in natural populations. Box 14.B discusses in detail the analysis of polygenic characters in *Drosophila melanogaster*.

Response to Selection

We will use a simple model of selection on a quantitative character to introduce some important concepts. We assume that the character, z—say, the tail length of rats—has a *normal* (bell-shaped) FREQUENCY DISTRIBUTION in a population.* An experimenter imposes *truncation selection* (see Figure 16 in Chapter 9) for greater tail length by breeding only the 16 percent of rats in the population with the longest tails (Figure 14.10A). The mean tail length of these selected parents ($\bar{z}_s$) differs from that of the population from which they were taken ($\bar{z}$) by an amount S, the **selection differential**. If the average tail length of each brood of offspring precisely equaled the average tail length of its two parents, the mean of all the offspring ($\bar{z}'$) would exceed that of the preceding generation (before selection—namely, $\bar{z}$) by exactly S (Figure 14.10B, right-hand graph). The change in the population mean ($\bar{z}' - \bar{z}$) is the **response to selection**, denoted R, and is a measure of how much the character has evolved in one generation.

If the 16 percent of rats with the longest tails were selected as parents, but the population from which they were taken was less variable, S would be smaller (Figure 14.10C). If the relationship between the phenotypes of these parents and their offspring is again one of equality, R will again equal S. Thus we have shown that, all else being equal, *the response to selection is proportional to the amount of variation in the character*. Recall from Chapter 9 that the amount of variation in a phenotypic trait is measured by the *phenotypic variance* (V_P).

Now consider Figure 14.10D, in which the character is just as variable as in Figure 14.10B and the selection differential is the same, but there is considerable variation among broods of offspring from a given value in their parents. This variation may be due to the environment or to other causes that we will soon describe. According to the theory of statistics, a variable y (offspring phenotype) can best be predicted from x (parent phenotype) by a line representing a REGRESSION EQUATION ($y = a + bx$), in which the y-intercept (a) and the slope (b) are calculated so as to minimize the sum of the deviations of the y-values from the line. (Actually, it minimizes the squared deviations.) In Figure 14.10D, the slope of this line is less than 1.0, as it is whenever y is not perfectly predicted by x. Therefore, R is less than S, because the resemblance between parents and offspring is imperfect—the trait is not completely *heritable*. In Chapter 9, we defined the *heritability* (h^2) of a trait as the proportion of phenotypic variance attributable to genetic variation, or more precisely, to the *genetic variance*, V_G—that

---

*Some characteristics are not normally distributed on the original scale of measurement (such as millimeters), but will assume a normal distribution if "transformed" by some mathematical operation, such as taking the logarithm of each measurement. The data are then analyzed in the usual way. The scale of measurement—whether arithmetic or logarithmic, for example—is entirely arbitrary. It is only by convention that we commonly measure length on an arithmetic scale of meters, and sound intensity on a logarithmic scale of bels.

BOX 14.B

Genetic Dissection of Quantitative Traits

The number of loci contributing to quantitative variation in a character can best be estimated by QTL MAPPING, in which each of a number of chromosomes or chromosome regions is distinguished by a marker (perhaps a mutation), and the phenotypic effect of each marked region is assessed by recombining it into an otherwise uniform genetic background. Perhaps the most extensively studied quantitative characters are the numbers of bristles (chaetae) on various parts of the body of *Drosophila melanogaster*. These bristles are mechanoreceptors that are sensitive to air movement or contact with objects, and are important components of the nervous system. The molecular and developmental genetics of mutations that substantially alter bristle patterns have been the subject of extensive study, which has provided considerable understanding of the development of bristle patterns (Held 1991; Campos-Ortega and Jan 1991). Such "drastic" mutations are seldom found in natural populations, but there exists considerable quantitative variation in the number of bristles on various parts of the fly, including the sternopleuron (part of the side of the thorax) and the ventral sclerites of the abdomen.

A. E. Shrimpton and A. Robertson (1988) used "classic" QTL mapping to estimate the number of loci contributing to the difference in the number of sternopleural bristles between two laboratory stocks that had been artificially selected for high and low numbers respectively. Using a "balancer" chromosome, that is, one with inversions and a mutant marker (see Figure 6 in Chapter 9), they extracted a single third chromosome (called "*C*") from the "high"-bristle population and one from the "low"-bristle population (as in Figure 6 of Chapter 9). By recombination with various mutant stocks, they inserted six mutant markers (such as *ru* and *ca*, which affect eye characteristics) into the "low"-bristle chromosome, dividing it into five segments (Figure 14B-1A). Flies homozygous for this low-bristle, marked chromosome (called "*ruseca*") had, on average, 24 fewer bristles than those homozygous for the unaltered *C* chromosome.

Crosses between the *C* and *ruseca* lines (Figure 14.B1A) yielded chromosomes carrying one or more of the segments of the *C* chromosome (i.e., those segments not carrying the mutant markers interspersed with marker-bearing segments of the *ruseca* chromosome). A difference in bristle number between *ruseca* homozygotes and any such genotype indicates that *at least one* locus affecting bristle number lies in this region. For example, flies had, on average, 1.74 more bristles than *ruseca* flies if they were heterozygous for segment 1 of chromosome *C*, and 6.05 more bristles if they were homozygous for this segment. Segments 2 and 3 also had such effects (Figure 14.B1B). Segment 5 had little effect on its own, but increased bristle number if combined with either segment 1 or segment 2. Thus segment 5 carries at least one gene that in-

teracts epistatically with genes in the other segments. Overall, Shrimpton and Robertson found evidence for at least 5 loci on the third chromosome that have substantial effects on bristle number, and in a more detailed analysis found evidence for at least 12 other loci that contribute some additional variation. The third chromosome accounted for about half the difference in bristle number between the high and low selected lines; the other two major chromosomes in the *Drosophila* genome are also known to affect this character.

Shrimpton and Robertson's analysis did not indicate what these genes are or how they function. Some insights into these questions have been provided by Trudy Mackay, Charles Langley, and their colleagues. In one experiment (Long et al. 1995), they selected for high and low abdominal bristle numbers in laboratory populations founded by large numbers of wild flies. After 25 generations of selection, the lines differed substantially not only in the number of abdominal bristles, but also in the number of sternopleural bristles (an example of a CORRELATED RESPONSE to selection.) Using marked balancer chromosomes, they extracted numerous X chromosomes and third chromosomes from the "high"-bristle and "low"-bristle populations. H1; L3, for example, is a derived line homozygous for an X (first chromosome) from a "high"-bristle population and for the third chromosome from a "low"-bristle population; the researchers obtained many such derived lines with different H1 chromosomes. Crosses between H1; L3 and L1; L3 yielded, in the F_2 generation, flies with mixtures of H and L segments of the X chromosomes, due to crossing over. These recombinant X chromosomes were then made homozygous. The researchers then used a DNA probe to determine the location, on each recombinant X chromosome, of a transposable element (called *roo*) that occurs at each of a great many chromosome sites, but is rare at any individual site. If a region of the X chromosome from "high"-bristle and from "low"-bristle stocks differs in the presence or absence of the *roo* element, then a difference in bristle number between the genotypes indicates that a gene near that *roo* element affects bristle number. The *roo* elements located at various sites on the X chromosomes, then, served as markers for chromosome regions, like the mutant markers used by Shrimpton and Robertson. A similar analysis was applied to the third chromosomes.

The *important results* were:

1. Two factors (loci or clusters of tightly linked loci) affecting bristle number were located on the X chromosome, and five were located on the third chromosome. Each of these factors had a large effect (0.5 to 6 bristles each), and these factors collectively accounted for much of the difference in bristle number between the selection lines. The variant alleles at these loci had rather high frequencies in the natural population from which the stocks were derived.

2. The effects of loci in combination often differed substantially from the sum of their individual effects, indicating that a large proportion of the variation is due to gene interaction (epistasis).

3. Most, but not all, of the loci affecting abdominal bristle number also affected sternopleural bristle number, suggesting that the loci are pleiotropic.

4. Several of the loci affected larval survival; the alleles that increased bristle number the most lowered viability to the greatest extent.

5. All the chromosome segments that affected bristle number included genes for which major mutations are known to affect the development of bristles, and of the nervous system of which they are a part.

Mackay, Langley, and their collaborators examined two such loci in detail to determine whether they contribute to natural variation in bristle number (Mackay and Langley 1990; Lai et al. 1994). They extracted 64 X chromosomes in one study, and 47 second chromosomes in another, from natural populations of *D. melanogaster*. They made these chromosomes homozygous on an ISOGENIC genetic background (i.e., one that is uniform for the other chromosomes). Each chromosome line was then scored for abdominal and sternopleural bristle count. Molecular differences among *achaete-scute* gene copies on the X and *scabrous* gene copies on the second chromosome were found by using restriction enzymes that revealed varia-

tion in restriction sites (see Chapter 3, Box D) and length (i.e., insertions and deletions of DNA). Several molecular differences in each gene were correlated with bristle number, especially insertions in the *achaete-scute* gene and nucleotide sequence differences in the *scabrous* gene. The differences in bristle count among the molecular variants were quite large, and variation at these two loci, which are known to affect bristle development, accounted for a considerable fraction of natural variation. Another important result was that at the *scabrous* locus, a few molecularly distinguishable alleles each occurred at rather high frequencies in the population. These alleles had large effects on bristle number.

These and other such studies have important implications for the study of quantitative traits (Mitchell-Olds 1995). The statistical models of quantitative variation generally assume that variation in a trait is due to alleles with individually small effects that segregate at many loci. However, *the number of loci that contribute most of the variation is rather low, and the allelic effects may be large*. Some models of quantitative variation assume that it is due to many slightly deleterious alleles that exist at low frequencies—but at the *scabrous* locus (though not the *achaete-scute* locus), the *allele frequencies are high*. The statistical models have often assumed that variation stems from the sum of additively acting alleles at functionally interchangeable loci. Variation in bristle number, however, is due to loci that are not interchangeable: they have *different developmental roles*, and *interact with one another* in producing the phenotype.

FIGURE 14.B1 Estimation by QTL mapping of the minimal number of loci on the third chromosome of *Drosophila melanogaster* that contribute to variation in the number of sternopleural bristles. (A) Crosses by which segments of a chromosome (*C*) from a selected population are inserted into the "ruseca" chromosome, which bears mutant markers. The effect of each segment on bristle number is found by comparison with "ruseca" homozygotes. (B) The effect of each chromosome segment on bristle number, standardized by the number of chromosome bands in each segment in order to correct for differences in length of the segments. Note that segment 5 affects bristle number only by interacting with other segments, such as segment 2. Segments 2a and 2b were combined in this analysis. (After Shrimpton and Robertson 1988.)

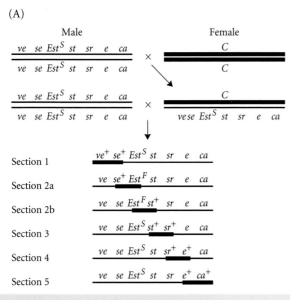

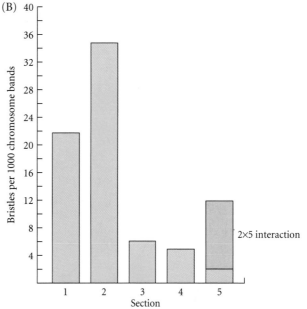

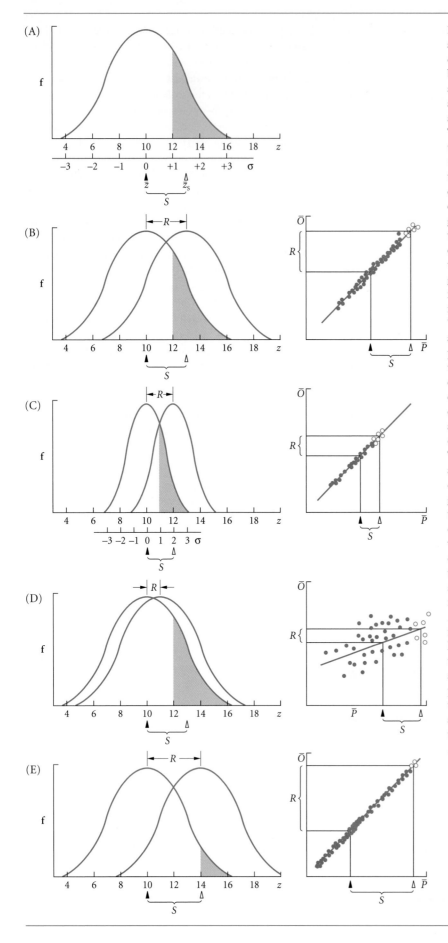

FIGURE 14.10 Factors that affect the response of a quantitative character to truncation selection. (A) The frequency distribution of a normally distributed trait, z, such as tail length. The trait can be measured in units such as centimeters, or in standard deviations (σ). Here, $\sigma = 2$ cm. The approximately 16 percent of the population more than one σ above the mean (i.e., with tails longer than 12 cm) breeds; the rest of the population does not. The difference between the mean of the selected parents, $\bar{z}_S$ (marked by the open triangle), and the mean of the whole population, $\bar{z}$ (solid triangle) is the selection differential, S. (B) The left-hand curve in the graph at left is the same as in A. The graph at right plots the mean phenotype in each brood of offspring ($\bar{O}$) against the mean phenotype of their two parents (the "midparent" value, $\bar{P}$). In this case, they are nearly identical, and the regression line predicting $\bar{O}$ from $\bar{P}$ has a slope of 1. Thus the response to selection (R), the difference in the mean between generations, equals the selection differential, S. (The open circles represent the selected parents and offspring, and the solid circles the rest of the population, were it to have bred.) This is shown by the distribution of z among the offspring of the selected parents, given by the right-hand curve in the graph at the left. Its mean lies R units to the right of the mean of the parental generation. (C) A less variable population, in which $\sigma = 1$ cm. Under the same intensity of selection, in which individuals more than one σ above the mean breed, S is smaller, and so is R. (D) As in B, but with a more variable relationship between $\bar{O}$ and $\bar{P}$. The regression line that best predicts $\bar{O}$ from $\bar{P}$ has a slope of less than 1, so R is less than S. Thus, as shown in the graph at the left, the mean of the next generation is a little greater than that of the selected parents. (E) As in B, but with a greater intensity of selection. The parents are the approximately 2 percent that lie more than two σ above the population mean, $\bar{z}$. S and R are greater than in the other cases.

is, $h^2 = V_G/V_P$. Thus *the response to selection depends on the heritability* of the trait.

Figure 14.10E is the same as Figure 14.10B, except that selection is more intense: the experimenter breeds only the 2 percent of rats with the longest tails. Both S and R are therefore greater. The intensity of selection expresses, in some sense, *the relationship between the phenotype and fitness*. In our example, a hypothetical investigator has dictated the fitness of rats with different tail lengths, but the same principle would hold if natural selection acted on the character.

Thus *the response of a quantitative trait to selection depends on the relationship between phenotype and fitness, the phenotypic variance*, and *the degree to which the trait is heritable*. For a fuller understanding of the evolution of phenotypic characters, however, we must examine these concepts more closely and take into account some important complications.

Components of Phenotypic Variation

Additive genetic variance and dominance variance.

Recall from Chapter 9 that the *variance* (V) is the average of the squared deviations of observations from the arithmetic mean. The square root of the variance is the *standard deviation* (s.d. $= \sqrt{V}$) measured in the same units as the observations. If a variable has a normal (bell-shaped) frequency distribution, about 68 percent of the observations lie within one standard deviation on either side of the mean, 96 percent within two standard deviations, and 99.7 percent within three. Figure 14.10 shows how a phenotypic character, z, can be measured in standard deviation units. In panels B, D, and E, s.d. = 2 cm, whereas s.d. = 1 cm in panel C, which represents a less variable distribution. The 16 percent of the population selected for breeding in panels B and C includes those individuals with character values more than one standard deviation above the mean; the 2 percent selected in panel E are those lying more than two standard deviations above the mean.

Although the amount of variation in a sample can be readily visualized by means of the standard deviation, the variance is a more useful measure for analyzing variation, because the total variance is the sum of the variances contributed by each source of variation. To begin with a simple case, suppose that variation in a phenotypic trait, z, is due to segregation at one locus with two alleles. In Figure 14.11A, each of the three genotypes is phenotypically variable due to environmental effects. Thus the breadth of each curve in the figure is a measure of the *environmental variance*, V_E. The differences among the mean values of the three genotypes give rise to the *genetic variance* in the character, denoted V_G. If we take the midpoint between the two homozygotes' means as a point of reference, the mean phenotype of A_1A_1 individuals deviates by $+a$, and that of A_2A_2 by $-a$. In the example, $a = 2$ (perhaps this is 2 cm, if we are measuring rats' tails). The quantity $+a$ is the *additive effect of an allele*: by replacing two A_2 alleles with two A_1 alleles in the genotype, the phenotype is increased (on average) by $2a$ (or 4 in this instance).

FIGURE 14.11 Visualization of genetic and environmental variances in a character, z, affected by two alleles at one locus. (A) Additive inheritance, in which the heterozygote lies at the midpoint between the homozygotes. The additive effect of substituting an A_1 or an A_2 allele in the genotype is a. The magnitude of a affects that of the additive genetic variance, V_A. The phenotypic variation in each genotype is measured by the environmental variance, V_E. (B) In this case, A_1 is partly dominant. The degree of dominance is d. (C) The distribution of phenotypes for the additive case, when $p = q = 0.5$. The additive genetic variance is high because both homozygotes, which differ the most, are abundant. (D) The distribution of phenotypes when $p = 0.9$, $q = 0.1$. V_A is lower than in C, because A_1A_1 constitutes most (81 percent) of the population. V_E is the same as in the other graphs, and so constitutes a higher proportion of the total phenotypic variance. The solid triangle marks the population mean.

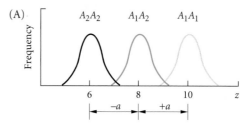

(A)

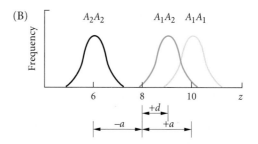

(B)

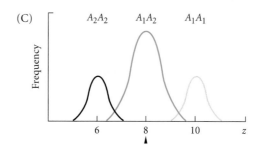

(C)

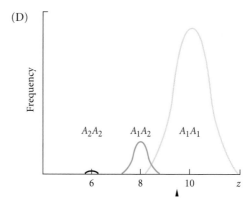

(D)

In Figure 14.11B, *a* again equals 2, but the mean of the heterozygote lies above the midpoint of the means of the two homozygotes at a distance *d* ($d = 1$ in the example). In this case, allele A_1 is partially dominant; the degree of dominance is given by *d*. If A_1 were fully dominant, *d* would equal *a*. When alleles have entirely additive effects, as in Figure 14.11A, $d = 0$.

The amount of phenotypic variation depends on allele frequencies, as Figures 14.11C and D show. A greater fraction of the population deviates further from the population mean when $p = 0.5$ than when $p = 0.9$, so the genetic variance is greater in that case.

The variance attributable to genetic differences among individuals, therefore, depends on both genotype frequencies and the additive effects of alleles on the phenotype. *The additive effects are responsible for a critically important component of genetic variance*, the **additive genetic variance**, denoted V_A. When, as in Figure 14.11A, alleles have purely additive effects, the genetic variance is entirely additive, and $V_G = V_A$. It can be shown (see Falconer 1981 and Hartl and Clark 1989) that when inheritance is entirely additive, V_A at a single locus is

$$V_A = 2pqa^2$$

Assuming Hardy-Weinberg equilibrium, V_A is proportional to the frequency of heterozygotes ($2pq$) and the additive effects of alleles on the phenotype. [For example, $V_A = (0.5)(2^2) = 2.0$ in the example in Figure 14.11C, and $V_A = (0.18)(2^2) = 0.72$ in Figure 14.11D.] When several loci contribute additively to the phenotype, V_A is the sum of the additive genetic variance contributed by each locus.

The *additive genetic variance* plays a key role in evolutionary theory because it *is the basis for response to selection within populations*. This is because *the additive effects of alleles are responsible for the degree of similarity between parents and offspring*—which, as we have seen, determines the magnitude of the response to selection.

To see this, let us contrast the effects of additivity and of dominance on parent-offspring resemblance. Table 14.3 shows, for each possible mating between two genotypes at a diallelic locus, the mean phenotype of the parents and that of their offspring if the alleles lack dominance and if A_1 is dominant. Ignoring environmental effects, the mean of a brood of offspring exactly equals that of their parents if inheritance is purely additive, whereas the correlation between parents and offspring is imperfect if there is dominance. Referring back to Figure 14.11B, we see that there is an additive component of variation among genotypes, measured by *a*, but that there is also a component of variation, measured by *d*, that is due to dominance. Thus the genetic variance in a population includes *both* additive genetic variance (V_A) and the "variance of dominance deviations," or DOMINANCE VARIANCE, V_D. Box 14.C provides the formulas for V_A and V_D when there is dominance. Both V_A and V_D depend on the allele frequencies in the population (Figure 14.12). In fact, it may be said that V_A is converted to V_D, or vice versa, as allele frequencies change.

V_D depends in part on how gene action affects the phenotype, i.e., how closely the heterozygote resembles a homozygote (expressed by *d*). But because V_D and V_A also depend on genotype frequencies in the population, *most of the genetic variance is additive* (unless *p* is close to 1), *even if one allele is fully dominant* (see Figure 14.12).

Because variances can be added,

$$V_P = V_G + V_E$$

where V_P is the phenotypic variance, V_G is the total genetic variance attributable to genetic differences among individuals, and V_E is the variance among individuals due to

Table 14.3 **Examples of the effect of additivity versus dominance on the correlation between average phenotypes of parents and offspring**

A. PHENOTYPES

	A_1A_1	A_1A_2	A_2A_2	*a*	*d*
Additive effects only	10	8	6	2	0
Dominance	10	10	6	2	2

B. CALCULATION OF MEAN PHENOTYPE OF PARENTS ($\bar{P}$) AND OFFSPRING ($\bar{O}$) FOR EACH MATING

MATING	OFFSPRING A_1A_1	A_1A_2	A_2A_2	WITHOUT DOMINANCE $\bar{P}$	$\bar{O}$	WITH DOMINANCE $\bar{P}$	$\bar{O}$
$A_1A_1 \times A_1A_1$	1			10	10	10	10
$A_1A_1 \times A_1A_2$	0.5	0.5		9	9	10	10
$A_1A_1 \times A_2A_2$		1		8	8	8	10
$A_1A_2 \times A_1A_2$	0.25	0.5	0.25	8	8	10	9
$A_1A_2 \times A_2A_2$		0.5	0.5	7	7	8	8
$A_2A_2 \times A_2A_2$			1	6	6	6	6

Nonadditive Genetic Variance

❖ This box contains somewhat advanced material; some readers may simply want to accept the conclusions (in *italics*).

It may be best to introduce the concept of variances arising from interaction between alleles with a nongenetic example (see Appendix A). Suppose we measure the effects of low versus high levels of fertilization with nitrogen (N) and phosphorus (P) on the weight of plants. We treat two plants with each of the four combinations. The effects might reveal interactions between these factors, or they might not. The mean for each treatment is given in parentheses.

	(a) No interaction		**(b) Interaction**	
	Phosphorus			
	Low	High	Low	High
Nitrogen Low	9,11	14,16	9,11	14,16
	(10)	(15)	(10)	(15)
High	19,21	24,26	19,21	29,31
	(20)	(25)	(20)	(30)

In (a), the difference between high and low P treatments, on average, is 5, that between high and low N treatments is 10, and (high P, high N) – (low P, low N) is 15. The effects are additive: there is no interaction. In (b), high P adds 5 if N is low, but 10 if N is high; and high N adds 10 if P is low, 15 if P is high. High P and high N interact synergistically to increase weight beyond the additive expectation.

The analysis of such data (a "two-way analysis of variance" in any textbook of statistics) computes four components of variance: that due to factor A (phosphorus), to factor B (nitrogen), to their interaction (A x B), and to the variation within each treatment. For these hypothetical data, the results of this analysis are:

Source of variation	**(a) No interaction**		**(b) Interaction**	
	Variance	**% of total**	**Variance**	**% of total**
Phosphorus (P)	12.0	18.9	26.6	23.9
Nitrogen (N)	49.5	78.0	77.6	69.6
P x N interaction	0.0	0	5.3	4.7
Within treatment	2.0	3.1	2.0	1.8
Total	63.5		111.5	

The total variance, therefore, is the sum of the contributions of each of the "main effects" (P and N), the interaction between them (if it exists), and an "error" (within-treatment) variance due to unspecified causes.

Dominance at a locus, from a statistical point of view, is an interaction between alleles at a locus. If a is half the difference between the trait values of the two homozygotes and d is the deviation of the heterozygote from the midpoint between them, it can be shown (Falconer 1981) that the *additive genetic variance* is

$$V_A = 2pq \, [a + d \, (q - p)]^2$$

and the *dominance variance* is

$$V_D = (2pqd)^2$$

Suppose, for example, that the genotypes have the following trait values (z) and fitnesses (w). For either of these, we can calculate a and d.

	A_1A_1	A_1A_2	A_2A_2	a	d
Frequency	p^2	$2pq$	q^2		
Trait value (z)	3	2	1	1	0
Fitness (w)	0.5	1.0	0.5	0	0.5

It is instructive to calculate V_A and V_D for two different populations, for example, with $p = 0.6$ and $p = 0.5$. Note that the fitnesses imply that $p = 0.50$ at gene frequency equilibrium (Chapter 13).

$p = 0.6$	$p = 0.5$
For trait z:	
$V_A = (0.48)(1)^2 = 0.48$	$V_A = (0.50)(1)^2 = 0.50$
$V_D = [(0.48)(0)]^2 = 0$	$V_D = [(0.50)(0)]^2 = 0$
For fitness w:	
$V_A = (0.48)[0 + 0.5 \, (0.2)]^2 = 0.0048$	$V_A = (0.50)[0 + 0.5(0)]^2 = 0$
$V_D = [(0.48)(0.5)]^2 = 0.0576$	$V_D = [(0.50)(0.5)]^2 = 0.0625$

The phenotypic trait z is inherited additively; it has no dominance variance ($V_D = 0$) at any allele frequency. V_A is maximized when $p = 0.5$ (which happens to be the equilibrium frequency in this case). Fitness, however, displays strong dominance (overdominance, in fact), and some of the variance in fitness consists of dominance variance, V_D. *The additive genetic variance in fitness is positive when $p = 0.6$, but is zero at allele frequency equilibrium ($p = 0.5$ in this case).* This is a very important conclusion, for it implies that *when allele frequencies have reached the equilibrium determined by natural selection, there is no heritable variation in fitness* (using heritability in the narrow sense of V_A/V_P), even though there is genetic variation in fitness. (This follows from Fisher's "fundamental theorem of natural selection," which states that the rate of change of mean fitness equals the additive genetic variance in fitness). Thus if a population does contain additive genetic variance in fitness, it is because some factor or factors have displaced allele frequencies from the equilibrium that selection alone would specify, and have converted V_D to V_A. Mutation, gene flow, and genetic drift are such factors, as is a change in environment that alters fitnesses, so that the population has yet to evolve to its new equilibrium.

Calculation of the epistatic variance (V_I) arising from interactions among two or more loci is too complicated to present here. As in the fertilizer example, variance might arise from locus A, from locus B, and from the interaction between them.

direct effects of the environment. We have described two components of genetic variance:

$$V_G = V_A + V_D$$

so that

$$V_P = V_A + V_D + V_E$$

Other components of phenotypic variance Phenotypic variation may include several other components, which we will note briefly here and return to at the end of this chapter.

Alleles at different loci may interact so that the effect of one locus on the phenotype of an individual depends on the allelic constitution at one or more other loci, as in Case 2 above. Such gene interaction, or *epistasis*, gives rise to variation in the population, or **epistatic variance**, denoted V_I. The magnitude of V_I (also called INTERACTION VARIANCE) depends not only on the magnitude of gene effects on the phenotype, but also on allele frequencies at each of the loci. Like dominance variance (V_D), V_I is generally small compared with V_A, even if the alleles at different loci interact strongly in determining the phenotype. Just as V_A and V_D are converted into each other as allele frequencies change, so V_A and V_I may be converted into each other (see Box 14.C). V_D and V_I together are often referred to as NONADDITIVE GENETIC VARIANCE.

Another source of phenotypic variation is *genotype × environment interaction*, which occurs when the effect of the environment on the phenotype differs among genotypes. Figure 13 in Chapter 9 illustrates this phenomenon with the effect of temperature on bristle number in different genotypes of *Drosophila*. The resulting variance (V_{GxE}) depends not only on the phenotypic effect of the interaction, but also on the frequencies of genotypes and the frequencies of the different environmental conditions.

A correlation between the genetic differences among individuals and environmental effects can occur if genotypes are distributed nonrandomly among the environmental settings that can affect the phenotype. For instance, if genotypes that are prone to grow large also tend to select nutritious environments (or can more successfully defend good resources), and small genotypes tend to be found in poor environments, this distribution pattern would increase the variance in body size, whereas the opposite correlation between the distribution of genotypes and environments would decrease the variance. This correlation is closely related to a statistical quantity called the COVARIANCE; thus the phenotypic variance may include a positive or negative covariance between genotype and environment, cov(G,E).

In summary, the components of phenotypic variance are

$$V_P = V_A + V_D + V_I + V_E + V_{G \times E} + \text{cov}(G,E)$$

Except when we discuss V_I, $V_{G \times E}$, and cov(G,E) explicitly, we will assume that they are negligible.

Heritability The degree to which a trait is inherited is measured by the resemblance between parents and offspring, which depends specifically on the additive effects of alleles. The proportion of phenotypic variance due to additive effects is the **heritability in the narrow sense**, h^2_N:

FIGURE 14.12 The dependence of the total genetic variance (V_G) on the allele frequency for a locus at which A_1 is dominant and A_2 is recessive ($a = d = 0.0701$). At each allele frequency, $V_G = V_A + V_D$. Note that although V_D exceeds V_A when *p* is near 1, most of the genetic variance is additive at lower allele frequencies. (After Hartl and Clark 1997.)

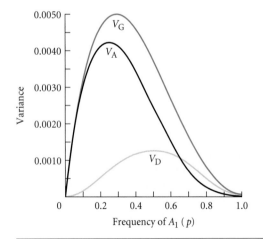

$$h_N^2 = \frac{V_A}{V_P}$$

or

$$h_N^2 = \frac{V_A}{V_A + V_D + V_I + V_E}$$

(In Chapter 9, we defined heritability as $h^2 = V_G/V_P$. This is the "heritability in a broad sense," h_B^2, since V_G may include V_D and/or V_I. We will generally use "heritability" only in the narrow sense, and understand that h^2 refers to h_N^2.)

Because the heritability of a trait is the fraction of *variance* that is due to additive allele effects, it includes the effects of only those loci that vary within the population (not monomorphic loci), and it depends on allele frequencies. These may vary from population to population, or within one population over the course of time. The denominator of the expression $h^2 = V_A/V_P$ includes the environmental variance V_E, which may be greater or less depending on how variable the environmental factors are that affect the development or expression of the character. Therefore, *an estimate of heritability is strictly valid only for the population in which it was measured, and only in the environment in which it was measured.* It cannot necessarily be extrapolated to other populations of the species, much less to different species. Thus *it is a fallacy to think of a characteristic as being "genetic" to a certain extent and "environmental" to another extent, as if these were fixed properties of the feature.* In fact, if the genotype × environment interaction is substantial, it is meaningless to partition variation into "genetic" and "environmental" quantities (Lewontin 1974a).

Because the resemblance between parents and offspring is caused by the additive effects of alleles only, and because $h^2 = V_A/V_P$, it is not surprising that h^2 is related to the correspondence between parent and offspring phenotypes. In fact, if we ignore some possible complications, *the heritability of a trait equals the slope of the regression of offspring phenotypes on parent phenotypes* in a sample from a population, as illustrated in Figure 14.10.

Heritability is an important concept in plant and animal breeding, and it is commonly measured in evolutionary studies. However, the heritability of a trait is not as informative a measure as the additive genetic variance (Houle 1992). The latter indicates how great the amount of genetic variation is, and therefore how great the potential is for rapid evolution [what Houle (1992) calls EVOLVABILITY]. Heritability does not provide this information because it is the ratio $V_A/(V_A + V_D + V_I + V_E)$. Thus even if the additive genetic variance is very small, h^2 may be high if the other terms in the denominator are small. Likewise, even if V_A is high, h^2 may be low if V_E (or other terms) is higher.

Estimating Genetic Variance and Heritability

Statistical methods can be used to estimate additive genetic variance, heritability, and sometimes other components of phenotypic variance. The data required are measurements on relatives—easily obtained if the organisms are bred in a laboratory or garden, where offspring can be produced from known parents, but more difficult in natural populations, where at best one parent is likely to be known.

Moreover, in estimating heritability, it is important that genetic and environmental differences not be confounded, as they are when different families occupy different microenvironments. A common way of dealing with this problem is to randomize the location of each individual (e.g., of plants in a garden) or to divide each family among several randomly selected locations (e.g., when rearing fruit flies in an incubator). Nevertheless, some "common environmental" effects may be virtually unavoidable, especially maternal effects (see Chapter 3) owing to nongenetic differences among females in postnatal or prenatal influences on their offspring (Mousseau and Dingle 1991; Roach and Wulff 1987.) The similarity of offspring of the same mother is more likely to result from effects of nongenetic influences than is the similarity of offspring of the same father. This problem of common environment (i.e., growing up in the same household or neighborhood) makes it especially difficult to estimate the heritability of many human traits, including behavioral ones (see Chapter 26).

The most common methods (Falconer 1981) of estimating heritability are the following:

Parent-offspring regressions   Heritability can be estimated by the correlation (or, more properly, regression) of the phenotypes of offspring with that of one or both parents. Chapter 9 provided an example in which the heritability of bill depth in the ground finch *Geospiza fortis* was estimated by this method as 0.90.

Comparisons of variance among full-sib families
Very often, offspring can be obtained from pregnant or inseminated wild females. It is assumed that all the offspring of a female are full sibs, i.e., have the same father. Similarity of sibs—i.e., greater variation among than within families—may imply inherited variation. For example, in coastal Oregon, the garter snake *Thamnophis sirtalis* often feeds on the brilliantly colored salamander *Taricha granulosa*. This salamander, however, does not occur in Idaho, where the snake also lives. The salamander's skin has a high concentration of a potent neurotoxin, tetrodotoxin (TTX). *Thamnophis ordinoides*, a congeneric snake living in Oregon, does not eat the salamander. Brodie and Brodie (1990) tested resistance to TTX in newborn snakes by measuring each individual's speed on a garter snake racetrack before and after injecting it with a nonlethal dose of TTX. From the Oregon *T. sirtalis* population, which feeds on the salamander, they tested 328 newborns from 23 females captured while pregnant. The variation in resistance among families was greater than that within families, yielding a heritability estimate of 0.715. This finding implies that resistance can evolve, since it is genetically variable. The average resistance of newborn Oregon *T. sirtalis* was much higher than that of Idaho *T. sirtalis* or of *T. ordinoides*. The researchers concluded that the Oregon population of *T. sir*-

talis had evolved an adaptation to the chemical defense of its prey.

Comparing siblings from different mothers is the only possible way of estimating the heritable component of variation in asexually reproducing (parthenogenetic or clonal) organisms. Whether applied to sexually or asexually reproducing organisms, this method has the drawback that the importance of nongenetic maternal effects is difficult or impossible to estimate.

Comparisons among half-sibs Half-sibs share only one parent, full sibs both parents. A common experimental design, with organisms that can be bred in the laboratory or greenhouse, is to mate each of a number of males to several females and measure several offspring from each. Similarity among progeny of the same father but different mothers is generally thought to be due entirely to additive genetic effects. Thus the variance among progeny of different fathers directly estimates the additive genetic variance, which in turn provides an estimate of h^2. For example, Futuyma and coworkers (1995) mated each of several males of a highly host-specific herbivorous beetle (*Ophraella communa*) to several females, and measured the larval feeding response to several plant species in several offspring of each female. The plants used are the hosts of other species in the same beetle genus. The authors were interested in determining whether the evolution of certain host associations during the divergence of these species from their common ancestor might have been influenced by the availability of genetic variation. Substantial additive genetic variance, estimated from the variance among offspring from different fathers, was found in the larval feeding responses to some plants, but not to others. Both this and related species of beetles were most likely to display genetic variation in their responses to plants closely related to their own normal host plants (Futuyma et al. 1995). The authors concluded that paucity of genetic variation in some of the responses might make adaptation to some plants less likely than to others.

Evolution of Quantitative Characters

Responses to Directional Selection

From our analysis of Figure 14.10, we know that when a trait (*z*) is subject to truncation selection, the response to selection (*R*) is related to the selection differential (*S*) by the slope of the offspring-parent regression. We now know that the slope equals the heritability, h^2. Therefore,

$$R = h^2 S \tag{14.1}$$

This equation arose in animal and plant breeding, and is used chiefly in that context, as well as by evolutionary biologists who use artificial selection in their research. For example, since Equation (14.1) can be rearranged as $h^2 = R/S$, heritability can be estimated from a selection experiment in which *S* (which is under the experimenter's control) and *R* are measured. Such an estimate of h^2 is called the REALIZED HERITABILITY. This is how Dobzhansky and Spassky esti-

mated that the heritability of phototaxis in *Drosophila pseudoobscura* was about 0.09, and of geotaxis about 0.03 (see Chapter 9).

Natural selection does not usually operate by truncation selection. Under directional selection, fitness usually increases or decreases gradually in relation to a character's value (see Figure 2 in Chapter 13), and there is variation in reproductive success among individuals with the same character value. For evolutionary studies, an equation introduced by Russell Lande (1976) is useful:

$$\Delta \bar{z} = \frac{V_A}{\bar{w}} \times \frac{d\bar{w}}{d\bar{z}}$$

Assuming constant rather than frequency-dependent fitnesses, the mean fitness in the population ($\bar{w}$) must depend on the mean character value ($\bar{z}$) if the fitness of individuals (*w*) depends on their phenotype (*z*). The steeper the relationship between *w* and *z*, the stronger selection is, and the more $\bar{w}$ will be altered by a change in $\bar{z}$. The slope of the relationship between $\bar{w}$ and $\bar{z}$, $d\bar{w}/d\bar{z}$, thus expresses the strength of selection (Figure 14.13). The character mean will change only if there is heritable—i.e., additive—genetic variation. Thus the change in the character mean, $\Delta \bar{z}$, is proportional to the additive genetic variance (V_A) and the strength of selection, as reflected by $d\bar{w}/d\bar{z}$. This expression is conceptually similar to Equation (14.1) ($R = h^2 S$), but it shows explicitly that for a given strength of selection, the rate of evolution is directly proportional to the additive genetic variance.

As selection proceeds, it increases the frequency of those alleles that make phenotypes closer to the optimum. If favorable alleles initially have low frequencies, the additive genetic variance (and therefore the heritability) is low at first (since at each locus, the contribution to genetic variance is $V_A = 2pqa^2$, proportional to the frequency of heterozygotes). As the favorable alleles increase in frequency due to selection, *the genetic variance and heritability may temporarily increase.* In the absence of complicating factors such as gene flow or antagonistic selection, *prolonged directional selection should ultimately fix all favored alleles,* eliminating genetic variation. *Further response* to selection *would then require new variation,* arising from *mutation.* We noted in Chapter 10 that for many features, the "*mutational variance*", V_m— the infusion of new additive genetic variance by mutation—is on the order of $10^{-3} \times V_E$ per generation, i.e., about one-thousandth the environmental variance. Thus a fully homozygous population could, by mutation, attain $V_A/(V_A + V_E) = h^2 = 0.5$ in about 1000 generations.

Responses to Artificial Selection

Ever since *The Origin of Species*, which Darwin opened with an analysis of variation and selection in domesticated organisms, evolutionary biologists have drawn useful inferences about evolution from artificial selection. Artificial selection differs from natural selection in that selection focuses on one trait rather than on overall fitness due to all the organism's characteristics, and in that it usually entails

FIGURE 14.13 Natural selection on a quantitative character (z). (A) The fitness of individual phenotypes. The optimal phenotype is z_{opt}. (B) For any variable population, the mean phenotype ($\bar{z}$) and mean fitness ($\bar{w}$) can be calculated if the fitnesses (as in A) and phenotype frequencies are known. The maximal mean fitness of a population is lower than that of the single optimal phenotype if the population is variable. The frequency distribution of three possible populations is shown. Selection shifts the distribution of the population, as shown by the dotted curves, and increases $\bar{w}$. (C) The rate of change of $\bar{z}$ depends on the additive genetic variance of the trait, and on the slope of the relation between $\bar{w}$ and $\bar{z}$, as shown for three points. Selection is directional except when $\bar{w}$ is maximal, in which case the slope is zero, and selection is stabilizing.

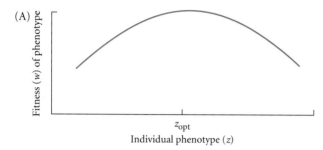

(A)

Fitness (w) of phenotype

z_{opt}

Individual phenotype (z)

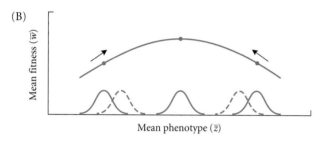

(B)

Mean fitness ($\bar{w}$)

Mean phenotype ($\bar{z}$)

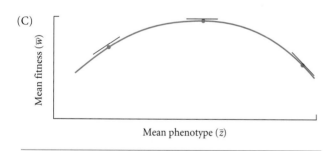

(C)

Mean fitness ($\bar{w}$)

Mean phenotype ($\bar{z}$)

the breeding of only those individuals whose traits exceed a certain value (truncation selection). Nevertheless, natural selection (especially hard selection; see Chapter 13) often operates much like artificial selection, and it would be a mistake to suppose that artificial selection is irrelevant to evolutionary theory. Incidentally, selection in a laboratory environment is natural selection, not artificial selection, if it sorts genotypes on the basis of their fitness under laboratory conditions rather than on the basis of a specific trait chosen by the experimenter. The studies by Bennett et al. of adaptation to temperature in laboratory populations of bacteria, described in Chapter 10, are examples of natural selection.

Responses to artificial selection over just a few generations generally are rather close to those predicted from estimates of heritability by correlations among relatives, such as parents and offspring (Hill and Caballero 1992). However, the heritability estimates seldom serve to predict accurately the change in a trait over many generations of artificial selection (Clayton and Robertson 1957; Hill and Caballero 1992; Barton and Turelli 1989). Among the factors that can cause a disparity between predicted and observed responses are changes in linkage disequilibrium among loci, small population size, changes in genetic variance due to the allele frequency changes caused by selection, input of new genetic variance by mutation, and the action of natural selection, which often opposes artificial selection.

In a massive experiment, B. H. Yoo (1980) selected for increased numbers of abdominal bristles in six lines of *Drosophila melanogaster* taken from a single laboratory population. In each generation for 86 generations, Yoo scored bristle numbers on 250 flies, and bred from the top 50 flies, of each sex. In the base population from which the selection lines were drawn, the mean bristle number was 9.35 in females and 6.95 in males, the phenotypic variances (V_p) were 2.07 and 2.99 respectively, and the heritability was 0.20. (For females, the phenotypic standard deviation was $\sqrt{2.07} = 1.44$: more than 99 percent of females would have been less than three standard deviations above the mean, i.e., with fewer than $9.35 + (3)(1.44) = 13.67$ bristles.) After 86 generations, mean bristle number in the six replicate lines had increased to 35–45, i.e., 20 to 30 bristles over the original value (Figure 14.14). This represents an average increase of 316 percent, or 12 to 19 phenotypic standard deviations. Thus in a very short time, *selection had accomplished an enormous evolutionary change*, at a rate far higher than is usually observed in the fossil record (see Figure 39 in Chapter 6).

This progress was by no means constant, however. The replicate populations *increased at different rates*, and some (e.g., line *CRb*) showed temporary plateaus during which there was little increase, followed by episodes of rapid increase. Several populations (e.g., *Ua*) eventually stopped responding: they reached a **selection plateau**. This *cessation of response to selection was not caused by loss of genetic variation*, which remained present despite the prolonged selection. Its presence is demonstrated by what happened when Yoo terminated ("RELAXED") selection after 86 generations: mean bristle number declined, proving that not all alleles for low bristle number had been replaced by alleles for higher bristle number. Thus the mean would have been increased still further if these "low" alleles had been eliminated by further selection. The persistent presence of genetic variation was apparently due at least partly to *new mutations* that occurred during the course of the experiment (Yoo 1980; Hill 1982; Hill and Caballero 1992). However, in this and other selection experiments, retention of genetic variation cannot be explained entirely by new mutations.

A selection plateau and a decline when selection is relaxed are commonly observed in selection experiments.

They are caused by *natural selection, which opposes artificial selection*: genotypes with extreme values of the selected trait have low fitness. In many other similar experiments, such as a pioneering experiment by Mather and Harrison (1949), viability and fertility have declined so much that the experimental populations have actually died out. At least one reason for this result is *hitchhiking of deleterious alleles* with those that increase the selected trait. A chromosome that happens to carry many "plus" alleles for bristle number increases in frequency so rapidly under strong artificial selection that there is little opportunity for recombination to dissociate the "plus" alleles from deleterious alleles at closely linked loci. Yoo found, for example, that lethal alleles had increased in frequency in the selected populations, and Mather and Harrison found that their lines selected for high abdominal bristle number also showed changes in the number of bristles on other parts of the body as well as in the number and form of spermathecae (the sperm storage organs in the female reproductive tract). These *correlated responses to selection* can be caused not only by *linkage disequilibrium*, but also by *pleiotropy*: effects of the genes that affect the selected trait on other phenotypic traits and on fitness. [Pleiotropic correlations between two bristle traits were also noted in the experiment by Long et al. (1995) described in Box 14.B.] We will return to this point below.

Each of Yoo's experimental populations was bred from 50 pairs of flies per generation. What would be the response to selection in larger populations? Weber and Diggins (1990) studied the effects of population size on responses to selection for resistance to ethyl alcohol vapor in *Drosophila melanogaster*. Weber devised an "inebriometer" in which flies that lose their grip due to ethanol vapor tumble down a se-

ries of baffles into collecting tubes that are replaced each minute. Thus "time to inebriation" could be used as a measure of ethanol resistance, and large numbers of flies could be selected in this apparatus.

From a population founded by only 10 wild fertilized females, Weber established four "small" selection lines, propagated from 160 selected parents per generation, and two "large" lines, propagated from 1600 selected parents. The selected parents were the most resistant 20 percent of those flies exposed to ethanol in each generation. Two unselected control lines were also maintained at 320 parents per generation. Weber also imposed selection on two lines of highly inbred, initially homozygous flies (with 400 parents per generation), and maintained one inbred control line without selection. Selection on the outbred lines was practiced for 60 generations, and on the inbred lines for 29.

Ethanol resistance increased in all of the selected populations; the initial response indicated that the realized heritability was 0.22. *The large* outbred populations *responded faster than the small ones*, and to a greater extent: after 60 generations, the average "knockdown time" was 27.9 minutes for the large lines, 19.4 minutes for the small lines, and 5.0 minutes for the unselected controls (Figure 14.15). Relative to the inbred control lines, the inbred selected lines evolved about 1.1 minutes of improved resistance, which was attributed to new mutations.

The theory of quantitative genetics predicts that for additively inherited polygenic traits, the *response to selection should be greater in larger populations* (Robertson 1960) for several reasons. First, more genetic variance is introduced by mutation in large than in small populations. In Weber's experiment, however, new mutations could account for only

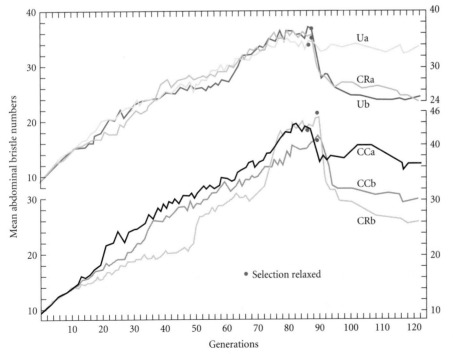

FIGURE 14.14
Responses to artificial selection for increased number of abdominal bristles in six laboratory populations of *Drosophila melanogaster*. Mean bristle number in females is shown for two sets of populations separately, for ease of visualization. After about 86 generations, the means had increased greatly. Selection was terminated at the points indicated by colored circles, and the means declined thereafter. Note that the response slowed or stopped in the later generations of selection. (After Yoo 1980.)

• Selection relaxed

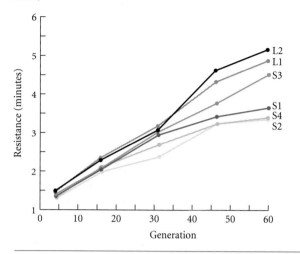

FIGURE 14.15 Response to selection for resistance to ethanol vapor in two large populations (L) and four small populations (S) of *Drosophila melanogaster*. Population S3 shows a spuriously high resistance because the flies evolved the peculiar behavior of clinging to the baffles even unto death. (After Weber and Diggins 1990.)

about 11 percent of the observed difference in response between large and small populations. Second, large populations lose variation by genetic drift more slowly. For the time scale and population sizes in Weber's experiment, this factor was probably not important. Third, when differences in fitness are slight, *selection is more efficient in large populations*. (Recall from Chapter 13 that whether allele frequency change is affected more by selection or by genetic drift depends on the relationship between the selection coefficient and the population size.) When many loci and environmental effects generate variation in a trait, a particular allele with only a small effect on the phenotype contributes so little to the difference between those individuals selected for breeding and those that are not that the selection coefficient at that locus is very small. Thus, a larger sample is required for a mean difference between individuals that carry the allele and those that do not to be manifested. The greater efficiency of selection in larger populations was probably the major reason for Weber's results.

To *summarize* the conclusions of these and other *laboratory studies of artificial selection*:

1. Even in populations initiated with rather small numbers, most traits respond to selection. Much of the response is far too rapid to be explained by new mutations; it is due instead to pre-existing variation.

2. Selection can shift the mean far beyond the original range of variation, at rates far greater than have been usual in evolutionary history.

3. Even under prolonged, intense selection, genetic variation persists.

4. New mutations contribute appreciably to genetic variation throughout the course of selection, and therefore to the sustained response to selection.

5. The contribution of new mutations to responses to selection, even in populations that have diverged greatly from their initial state, implies that mutation may provide the genetic variation required for indefinitely great divergence.

6. As the mean of an intensely selected trait changes, components of fitness such as fecundity and viability often decline; thus natural selection opposes pronounced, rapid change in any one character.

7. This decline in fitness is sometimes due to hitchhiking by deleterious alleles. (Whether or not the altered trait itself, or the genes responsible for it, also reduce fitness is discussed below.)

8. The response to selection is often irregular and differs among replicate populations. Initial differences in allele frequencies, changes in allele frequencies, new mutations, and recombination can all contribute to such irregularities.

9. Selection is more efficient, and generates a greater response, in large than in small populations.

❖ Responses to Stabilizing and Disruptive Selection

In our analysis of selection at two loci, we noted that if alleles at the loci have additive effects on a character, and have no other effects on fitness, then *stabilizing selection* for an optimal intermediate character state *should ultimately eliminate genetic variation*, resulting in a population monomorphic for a homozygous genotype with a mixture of "plus" and "minus" alleles. The same is true for greater numbers of loci (Wright 1935; Lewontin 1964; Felsenstein 1979). For instance, if + and – alleles at each of six loci add or subtract one unit of the character, and a precisely intermediate phenotype is optimal, then the population should eventually become monomorphic for a genotype such as +–+–+–/+–+–+–. Thus, like sustained directional selection, *stabilizing selection reduces genetic variation* in additively inherited traits.

Disruptive (diversifying) selection (see Figure 2 in Chapter 13), as occurs when two phenotypes have higher fitness than the intermediates between them, can increase genetic variance, at least temporarily. With constant fitnesses, however, *such selection is very unlikely to maintain a bimodal frequency distribution* of the trait (Felsenstein 1979). Instead, the distribution is likely to be unimodal (more or less bell-shaped), with the mean lying near one or the other optimum. Unless disruptive selection is frequency-dependent, favoring rare phenotypes, genetic variation will eventually be lost.

In artificial selection experiments with characters such as bristle number in *Drosophila*, stabilizing selection—i.e., breeding from individuals near the mean—indeed tends to reduce phenotypic variation, and disruptive selection to increase it, in accordance with theoretical predictions (Mather 1983). Genetic variance (V_A) and heritability, however, often remain considerably higher in the face of artificial stabilizing selection than expected. For example, Kaufmann et al. (1977) imposed stabilizing selection on pupal weight in the flour beetle *Tribolium castaneum* by breeding from individuals of intermediate weight for 95 generations. The phenotypic variance V_P decreased, but V_A and h^2 hardly decreased at all. The decrease in phenotypic variance was due largely to a reduction of the environmental variance V_E; selection apparently favored genotypes that were better buffered against the effects of the environment.

Measuring Natural Selection on Quantitative Characters

Many studies of selection on quantitative characters in natural populations have been performed, and in some of these, the additive genetic variance or heritability has also been estimated. Recall, however, from Chapter 12 that many researchers define *selection* as variation in reproductive success among different phenotypes, without reference to their genetic basis. Whether or not selection has any effect on the following generations—i.e., whether or not there is a response to selection—depends on additive genetic variance (Lande and Arnold 1983). For our present purposes, we will define "*selection*" in this broad, purely phenotypic sense, although some authors maintain that natural selection cannot occur unless there is inheritance (Endler 1986).

Several indices of the strength of directional selection can be used (Endler 1986). One is the standardized selection differential, or INTENSITY OF SELECTION, $i = (\overline{z}_s - \overline{z})/\sqrt{V_P}$, where $\overline{z}$ is the population mean before selection, $\overline{z}_s$ is the mean after selection, and V_P is the phenotypic variance. Selection is measured in units of standard deviations. Often, $\overline{z}_s$ is the mean of those individuals within a generation that successfully survive and reproduce. Another measure of selection is the slope, b, of the relation between the fitness of a phenotypic class and its phenotype, z. As in studies of selection at single loci (see Chapter 13), total fitness is less frequently measured than certain of its components, such as survival or fecundity.

Often, several characters are correlated with one another to some degree, such as, say, beak length (z_1) and body size (z_2). Lande and Arnold (1983) introduced a widely adopted method of estimating selection on each such feature (say, z_1) while (in a statistical sense) holding the others (say, z_2) constant. If z_1, z_2, and fitness (w) have been measured for each of a number of individuals, the equation $w = a + b_1 z_1 + b_2 z_2$ is used to find the slopes (or "SELECTION GRADIENTS"), b_1 and b_2, of the relationships of fitness to both characters. (In statistical terms, b_1 and b_2 are partial regression coefficients, and a is a constant.) In effect, this enables one to estimate, for instance, how greatly variations in beak length affect fitness among individuals with the same body size.

Not only the mean, but also the variance of a character may be altered as selection proceeds. For example, a lower variance in a cohort of adults than in the same cohort when it was younger may indicate the operation of stabilizing selection. Conversely, an increase in variance may provide evidence of disruptive (diversifying) selection. One measure of

the intensity of such "VARIANCE SELECTION" is $j = (V_a - V_b)/V_b$, where V_b and V_a are the phenotypic variances before and after selection, respectively (Endler 1986). This index is negative if selection is stabilizing, positive if it is disruptive.

Selection in Natural Populations

Many studies of selection on quantitative characters in nature have been performed; we can mention only a few of them here.

Darwin's Finches

Peter and Rosemary Grant (Grant 1986; Grant and Grant 1989) and their colleagues have carried out long-term studies on some of the species of Darwin's finches on certain of the Galápagos Islands. We saw in Chapter 9 that by banding birds so that they were individually recognizable, these workers could estimate the heritability of various traits from offspring-parent regressions. For example, in the ground finch *Geospiza fortis*, h^2 was 0.91 for weight, 0.79 for bill depth, and 0.90 for bill width. The length, depth, and width of the bill are all correlated with one another, and with body weight, to a considerable degree. Extensive behavioral observations showed that birds with larger (especially deeper) bills feed more efficiently on large, hard seeds, whereas there is some evidence that small, soft seeds are more efficiently utilized by birds with smaller bills.

During the Grants' study, the islands suffered a severe drought in 1977. Seeds, especially small ones, became sparse, the finches did not reproduce, and their population sizes declined greatly due to mortality. The survivors in 1978 were larger than the pre-drought population of 1976, having increased by 0.3 standard deviations in body weight and by 4 percent in bill size—a considerable change to have transpired within a single generation (Figure 14.16). From the differences in morphology between the survivors and the pre-drought population, the intensity of selection i and the selection gradient b were calculated:

Trait	i	b
Weight	0.28	0.23
Bill length	0.21	−0.17
Bill depth	0.30	0.43
Bill width	0.24	−0.19

The values of i show that each character increased by about 0.2 to 0.3 standard deviations. The values of b show the strength of the relationship between survival and each character while holding the other characters constant. Selection strongly favored birds with deeper bills because they could more effectively feed on large, hard seeds, virtually the only available food. The negative b values show that selection favored shorter, narrower bills. Nevertheless, bill length and width increased during this period (as indicated by the positive values of i) because all the bill dimensions are positively correlated. Thus *a feature can evolve in a direction opposite to the direction of selection, as a correlated re-*

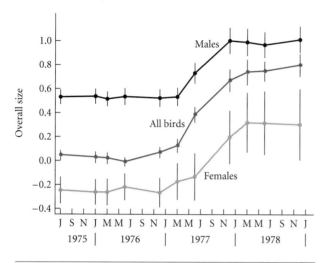

FIGURE 14.16 Changes in the mean overall size of the ground finch *Geospiza fortis* on Daphne Island in the Galápagos Islands due to mortality during a drought in 1977. The increase in 1977 and 1978 is the result of the loss of smaller birds from the population; no reproduction occurred during this period. "Overall size" is a composite of measurements of several characters. The vertical bars are a measure of variation (standard error) in each sample. (After Grant 1986.)

sponse, if it is strongly correlated with another trait that is more strongly selected. This study also shows that *natural selection* on a quantitative trait *can be very strong.*

Why don't these finches evolve ever-larger bills? The Grants found that during normal years, larger males, with larger bills, hold larger territories and are more successful in attracting mates. However, birds with smaller bills survive better in their first year of life, probably because they feed more efficiently on small seeds; also, small females tend to breed earlier in life than large ones. Thus *conflicting selection pressures create stabilizing selection* that on average favors intermediate bill size.

Vestigial Features in Cave Organisms

A long-standing puzzle is why, over the course of evolution, features that have no function become vestigial and are ultimately lost. The human appendix (a vestigial storage sac off the intestine), the vestigial stamens or pistil in many flowers that have only female or male function, and the legs of the ancestor of the snakes are only a few of the innumerable such examples. In many cave-dwelling animals, the eyes and pigmentation have been greatly reduced or completely lost, whereas other sensory organs such as antennae have become exaggerated.

The Lamarckian explanation, whereby organs are maintained or lost as a result of use or disuse, has long since been rejected. Under neo-Darwinian theory, either (1) mutations that cause degeneration of an unused character become fixed by genetic drift because variations in the character are selectively neutral, or (2) selection against an unused organ exists. Possible reasons for such selection in-

clude the following: (*a*) the organ interferes with some important function (e.g., legs might be a hindrance to snakes); (*b*) development of the organ requires energy and materials that could better be used to build other body parts; (*c*) due to pleiotropy, the organ has a negative genetic correlation with another, functional feature (for example, eyes would degenerate if the development of eyes and antennae were negatively correlated and selection favored longer antennae.) These hypotheses are discussed further by Fong et al. (1995).

David Culver and colleagues have explored these hypotheses in their studies of the amphipod crustacean *Gammarus minus*, some populations of which live in surface springs and others in caves. Surface and cave populations are electrophoretically more similar to each other within the same drainage basin than cave populations in different drainage basins are to each other, implying that the cave populations have been derived independently from the surface populations. Nevertheless, the cave populations all have longer antennae than the surface populations, and have reduced compound eyes, consisting of about 3 to 5 facets (ommatidia) compared with about 18 in the surface populations. Measurements of the offspring of pregnant females taken from several cave and surface populations yielded high heritability values (in the broad sense, h^2_B) for both facet number and antenna length. There was a tendency, especially in the surface populations, for families with lower facet numbers to have longer antennae, but these negative genetic correlations were generally weak (Fong 1989). (Genetic correlations are discussed later in this chapter.)

After mating, males of this species remain mounted on a female for a week or two, guarding her against insemination by other males. Jones and Culver (1989) took this pairing as evidence of reproductive success, whereas unpaired individuals give no evidence of reproductive success, and were assumed to have somewhat lower fitness on average. In each of two years, the researchers collected all the paired and unpaired amphipods they found in one cave, and calculated the intensity of selection (*i*) and the selection gradient (*b*) (the partial regressions) for several features, assigning fitness values of 1 and 0 to paired and unpaired individuals respectively. Mating males in both years, and mating females in one year, had longer antennae than unpaired individuals. In both years, in both sexes, the number of eye facets was lower in paired than in unpaired animals; the selection gradient ($b = -0.30$ to -0.35 in males) was so pronounced as to indicate quite strong selection for small eyes. This analysis corrects for any negative correlation between eye size and antenna length, which according to Fong's study is quite weak anyway. Thus, smaller eyes appear to be advantageous in their own right. The authors speculated, on the basis of neurobiological studies by other researchers, that reduction of the unused visual system could enhance the function of other sensory systems, such as the chemosensory antennae, by freeing more of the central nervous system to process non-visual sensory input. Whatever the mechanism, the data imply that *unused organs can be reduced and lost due to natural selection.* In other cases, though, reduction may be due to fixation of mutations by genetic drift (Fong et al. 1995).

Evidence of Stabilizing Selection

Seldom does natural selection favor an ever more extreme state of a characteristic (unless it be complete absence). At some point, antagonistic agents of selection come into play, so that components of selection that favor and disfavor an increase in the trait become balanced, establishing a regime of stabilizing selection for an intermediate optimum (Travis 1989). These antagonistic agents of selection are often called **trade-offs**. We should expect to find evidence of stabilizing rather than directional selection quite often if we study populations in environments they have occupied for a long time, so that many of their characters have evolved to their equilibria. Postnatal mortality in human infants, for example, is appreciably lower for those near the population mean than for lighter or heavier infants (Figure 14.17; Karn and Penrose 1951). The mean body size of lizards of the genus *Aristelliger* seems to be determined by a balance between the advantage larger lizards have in defending territories and their greater susceptibility to predation by owls (Hecht 1952).

In one of the most extensive studies of natural selection, Arthur Weis and colleagues (1992) studied 16 populations of the goldenrod gall fly, *Eurosta solidaginis*, for four years in central Pennsylvania. The egg of this fly hatches in early summer, and the larva induces a globular growth (gall) on the stem of its goldenrod host. It feeds and pupates within the gall, emerging in spring. During its development over the summer, it may be attacked by the larvae of two species of parasitoid wasps, *Eurytoma gigantea* and *E. obtusiventris*, which lay eggs on the fly larvae by inserting their ovipositors through the gall wall. During winter, woodpeckers and chickadees open galls and feed on the fly pupae. Mortality caused by parasitoids and birds can be determined by examining galls in the spring.

The researchers had previously determined that although gall size is affected by plant genotype and by environmental factors that affect plant growth, a considerable amount of variation in gall size is due to genetic variation in the fly. By measuring the diameter of galls in which pupae were still alive or had succumbed to either parasitoid or bird attack, they calculated the intensity, direction, and form of selection in each of 64 population-year combinations. The parasitoid *E. gigantea* very consistently selected for wide galls (i.e., mortality was greatest for fly larvae in narrow galls), because the wasp's ovipositor cannot penetrate thick gall tissue. *Eurytoma obtusiventris* generally selected for intermediate-sized galls, whereas birds most frequently attacked wide galls, selecting for narrower gall diameter (Figure 14.18A). Taken together, the parasitoids generally (in 85 percent of the 64 cases) imposed strong directional selection (at an intensity, *i*, frequently greater than 0.3) for wide galls, whereas selection by birds for narrow galls was weaker, and in many cases (about 60 percent) not statistically significant. The opposing selection pressures created a stabilizing selection component, but the net effect of selection was generally directional toward wider gall diameter, due to the greater effect of parasitoids than of birds. Thus *selection on a trait may simultaneously be both directional and stabilizing* (Figure 14.18B). Although the direction of selection was quite consistent, the *intensity of selection varied greatly* among populations and among years, presumably because the abundance of parasitoids and birds varies in space and time. In some studies of other organisms, the direction of selection has also been found to vary (Endler 1986).

The Intensity of Natural Selection

John Endler (1986) has compiled estimates of the intensity of directional selection (*i*) and variance selection (*j*) from several studies of quantitative traits in natural populations (Figure 14.19), complementing his compilation of selection coefficients for single-locus polymorphisms (see Figure 30 in Chapter 13). In many cases, *directional selection was very intense,* often exceeding half a standard deviation and sometimes extending even to two standard deviations per generation. Selection due to mortality as well as to other components of fitness such as fecundity or mating success can be intense. An unexpected result, shown in 14.19E, was that *selection on variance,* although *most often stabilizing* as expected, *is frequently disruptive.* In some instances, this is because individuals with different trait values are adapted to use different resources. For in-

---

FIGURE 14.17 Stabilizing selection for birth weight in humans. Early mortality, shown by the points and the line fitting them, is lowest near the mean for the population. The histogram shows the distribution of birth weights in the population. (After Cavalli-Sforza and Bodmer 1971.)

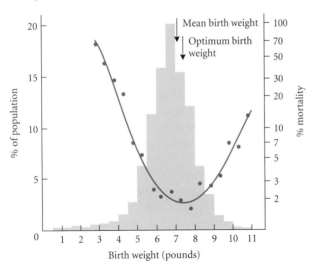

stance, Rosemary Grant (1985) observed that some individuals in a population of *Geospiza conirostris*, one of Darwin's finches, fed mostly by extracting soft seeds from cactus fruits, and others mostly by cracking hard seeds or stripping bark to find insects. These groups of birds differed in their bill dimensions. Moreover, young birds that survived their first dry season, when food is sparse, had a somewhat higher variance in bill size than those that did not survive. In Chapter 13, a similar case was described in which large-billed individuals of an African finch species feed on larger seeds than small-billed birds.

FIGURE 14.18 Selection on the size of galls made by the goldenrod gall fly (*Eurosta solidaginis*). Mortality is greatest for flies in narrow galls, due to parasitism, and in wide galls, due to attack by birds. (A) Survival in relation to gall diameter; data pooled from 16 populations in one season. (B) The frequency distribution of gall diameter for the population as a whole (open bars) and for survivors (colored bars). The positive value of i indicates directional selection for greater size; the negative value of j' indicates stabilizing selection. The %s is the percentage surviving. (After Weis et al. 1992.)

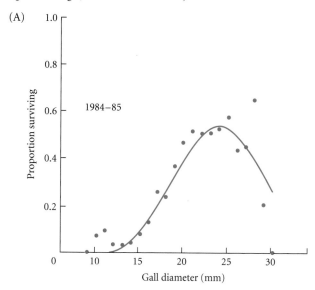

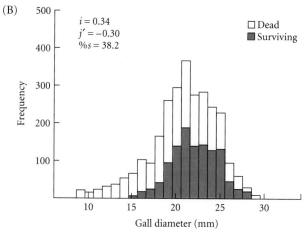

FIGURE 14.19 Intensities of selection on traits in natural populations, compiled from data from numerous studies. The open bars show the proportion of all reported values of i or j; the solid portion of each bar shows the cases in which the intensity of selection was statistically different from zero. N is the number of reported values; more than one value was reported for some species. A–D present values of i, the intensity of directional selection. E shows values of j, selection on the variance. Negative values indicate stabilizing selection; positive values, disruptive selection. (A) Directional selection in undisturbed populations, 25 species. (B) Directional selection in disturbed populations, 5 species. (C) Directional selection based on mortality differences, 17 species. (D) Directional selection based on fitness components other than mortality, such as fecundity; 9 species. (E) Selection on the variance, 32 species. (After Endler 1986.)

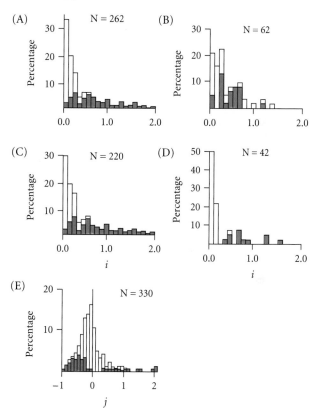

Rapid Evolution of Quantitative Traits

Because of the high additive genetic variance and heritability of many traits, and because selection is often intense, we should expect many evolutionary changes to be rapid, especially if environmental factors change, as when a species is introduced into a new region or when humans alter features of its environment. Indeed, *rapid adaptation has been observed in many features of many species* (Endler 1986; Bishop and Cook 1981; Taylor et al. 1991). *These changes*, in which trait means have changed detectably in usually much less than a century, *have occurred at rates far greater than the average rates of long-term evolution*, as documented by paleontological data or comparisons of species

FIGURE 14.20 Geographic variation in the critical photoperiod for entering diapause in two species of moths at various latitudes. (A) *Chilo suppressalis* in Japan, where it is native. (B) *Cydia pomonella* in North America, where it is an introduced pest that has spread within the last 200 years. Adaptation of local populations of *Cydia* is comparable to that seen in *Chilo*, which has occupied its geographic area far longer. (After Tauber et al. 1986.)

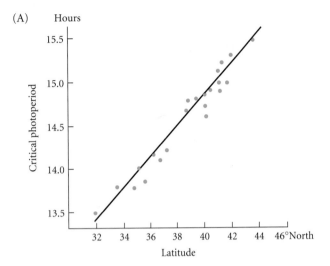

(A)

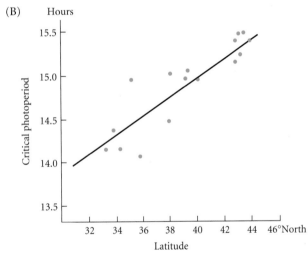

(B)

for which the time since common ancestry has been estimated (see Figure 44 in Chapter 6). We briefly describe only a few examples.

Responses to photoperiod
For many insects, a critical day length, or photoperiod, is the cue to enter diapause, the state of low metabolic activity in which they pass the winter or other harsh seasons. In the north temperate zone, species often show a latitudinal cline in critical photoperiod (Figure 14.20A), with northern populations genetically programmed to enter diapause at longer day lengths than southern populations. This pattern is adaptive because a September day length signals less time before the onset of cold weather in the north than in the south. The codling

moth (*Cydia pomonella*), a major pest of apples, is a European species that was first recorded in North America in New England in 1750, and has since expanded its range over 12 degrees of latitude. Riedl and Croft reported in 1987 that populations tested in the laboratory to determine their critical photoperiod for diapause display the same kind of clinal variation as other north temperate insect species in their native environments (Figure 14.20B). Thus adaptive differentiation of local populations has occurred within two centuries. Tauber et al. (1986) summarize many similar examples of rapid evolution of insect diapause.

Responses to food
The checkerspot butterfly *Euphydryas editha* is highly specialized in its choice of plant species on which females lay eggs and which the larvae subsequently eat. Its natural hosts in California are *Collinsia* and other plants in the family Scrophulariaceae. Michael Singer and colleagues (1993) have discovered local populations that also feed on a weedy plant in a related family, *Plantago lanceolata*, that was introduced into North America from Europe less than 100 years ago. Testing individual butterflies' oviposition responses by alternately presenting them with *Collinsia* and *Plantago*, they found that a considerable fraction of females in one *Plantago*-using population actively preferred *Plantago* over *Collinsia* for egg laying, whereas no females from a population living where *Plantago* is absent preferred this plant. Moreover, the proportion of females that prefer *Plantago* has increased over the course of 8 years of this study (Figure 14.21). Offspring of females from these populations, reared in the laboratory on *Collinsia*, had preferences similar to those of their mothers, with a broad-sense heritability (h_B^2) of about 0.90.

FIGURE 14.21 Rapid changes in the proportion of checkerspot butterflies (*Euphydryas editha*) that preferred a native host plant (*Collinsia parviflora*, black line) versus an introduced host plant (*Plantago lanceolata*, colored line). The percentages do not sum to 100 because some females displayed no preference. (After Singer et al. 1993.)

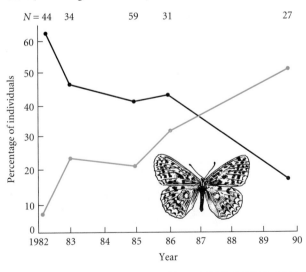

FIGURE 14.22 Genetic differences in the direction of autumn migration in the blackcap (*Sylvia atricapilla*). Each symbol shows the mean compass direction displayed by a single bird in repeated tests at night. The vector shows the mean direction for all the birds. (A) Adults captured in Britain in winter and tested the following autumn. (B) F_1 offspring of the birds shown by triangles in A. (C) Hand-reared offspring of birds captured in Germany at the location where the birds in B were reared. The direction of migration of birds that overwinter in Britain differs from that of the particular population of German birds, and the difference is evidently inherited. This species has consistently overwintered in Britain only since the 1950s. (After Berthold et al. 1992.)

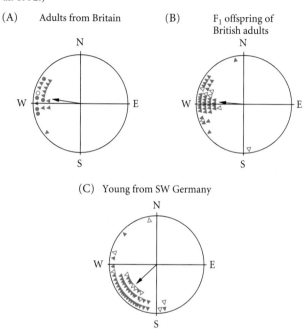

(A)   Adults from Britain

(B) F_1 offspring of British adults

(C)   Young from SW Germany

1500 kilometers north of its traditional wintering area. The authors suggest that the selective advantage lies in improved winter weather and other conditions in Britain, and the earlier spring return from Britain than from the Mediterranean.

Heavy metal tolerance   Heavy metal tolerance refers here not to one's capacity to bear certain excruciating music, but to the ability of some plants to grow under soil concentrations of copper, zinc, and other metals that are normally toxic. In several species of grasses and other plants, metal-tolerant populations have evolved where soils have been contaminated by mine works that range from over 700 to less than 100 years old. In some cases tolerance has evolved within decades on a microgeographic scale, such as in the vicinity of a zinc fence. Tolerance, usually measured by the rate of root growth in a metal solution (Figure 14.23), is based on a variable number of genes, depending on the population, and its genetic basis may vary among populations of the same species. When tolerant and nontolerant genotypes of a species are grown in competition with other plant species in the absence of the metal, the relative fitness of the tolerant genotypes is often much lower than that of the nontolerant genotypes, implying a "cost" of tolerance. It is interesting that rather few plant species have evolved heavy metal tolerance. Large numbers of seedlings taken from populations of several species growing on normal soils were tested for copper tolerance. Small numbers of tolerant seedlings were found in those species that have evolved copper tolerance in other locations, but no tolerant seedlings were found in most of the species that have not formed tolerant populations. Thus the genetic

Migration   Until the 1950s, the blackcap (*Sylvia atricapilla*), a European songbird, migrated only to the western Mediterranean region for the winter. Since that time, an increasing number of blackcaps have overwintered in Britain, migrating northwest from parts of Germany and Austria rather than southwest to the Mediterranean. Peter Berthold and colleagues (1992) took overwintering birds from Britain to a region in Germany where the birds retain their traditional southwestern route, and there reared their offspring, as well as those of some local birds. In autumn, the birds were placed in funnel-shaped cages lined with typewriter correction paper and with a view of the night sky, which provides them with information on compass directions. Scratches on the paper indicated the directions in which they had fluttered during the night in their attempts to migrate (Figure 14.22). Both the overwintering adults from Britain and their offspring oriented WNW, whereas the young German birds oriented SW, a difference of about 50°. A selection experiment, moreover, demonstrated realized heritability for orientation direction. Thus within about 30 years, a substantial proportion of the central European population has evolved a novel migration route ending 1000–

FIGURE 14.23   Variation in root growth of the monkeyflower *Mimulus guttatus* grown in a copper solution. This variation has a genetic basis, which has enabled some populations of this and other plant species to adapt rapidly to high soil concentrations of copper. (From Macnair 1981; courtesy of M. Macnair.)

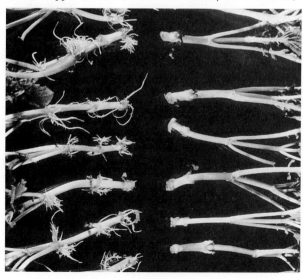

variation required for the evolution of heavy metal tolerance appears to be present in some species, but not in others (Antonovics et al. 1971; Macnair 1981; Bradshaw 1991).

Morphology In the 1860s, house sparrows (*Passer domesticus*) were introduced by European settlers into New Zealand and into New York, whence they rapidly spread throughout North America. Measurements of skeletal dimensions of sparrows from various North American localities (Johnston and Selander 1971) and from various sites in New Zealand (Baker 1980) have revealed, in both regions, significant geographic differentiation among populations in various dimensions, as well as in overall size and in sexual dimorphism. Some of these differences are correlated with climatic differences among localities, such as a cline in body size in North America conforming to Bergmann's rule (larger in colder regions: see Chapter 9). Although genetic data are lacking, these apparently adaptive patterns suggest that the differences among populations have evolved rapidly by natural selection.

Correlated Evolution of Quantitative Traits

When we compare a number of different species or populations, we often observe correlations among various characteristics. For example, many features are correlated with body size (see Chapter 17). This pattern suggests that the characteristics have evolved in a correlated fashion. *Correlated evolution can be caused by* either or both of two factors: *correlated selection* for the traits, or *genetic correlation* between the traits. In CORRELATED (or correlational) *selection*, there is independent genetic variation in the traits, but selection favors some combination of character states over others, usually because the characters are functionally related. **Genetic correlation** means that the genetic variation in one trait is not independent of genetic variation in the other. The joint, correlated evolution of two traits may be enhanced if both correlated selection and genetic correlation exist, if these two factors are oriented in the same direction rather than working at cross-purposes (so to speak).

Correlated Selection

We are already familiar with the idea of an adaptive landscape for two loci in which epistatic interactions affecting fitness can give rise to different combinations of equilibrium allele frequencies—different adaptive peaks (see Figures 14.5 and 14.6). An analogous adaptive landscape can be conceived for two quantitative characters (Figure 14.24), in which the horizontal axes represent the population means of two different traits (denoted $\bar{z}_1, \bar{z}_2$) and the vertical dimension is the mean fitness of a population with a particular combination of trait means. The landscape may be simple, with a single peak, or, if fitness depends on functional interactions between the two characters, it may have multiple peaks and ridges. As for allele frequency adaptive landscapes (see Figure 14.6), the evolution of quantitative characters may be visualized as a hill-climbing process: the population is a point on the adaptive surface that moves upslope, arriving at whichever adaptive peak that slope brings it to, whether the peak be relatively high or low (Lande 1976b).

For some purposes, it is useful to calculate a similar figure in which the horizontal axes represent not the means of the traits ($\bar{z}_1, \bar{z}_2$), but the individual character values (z_1, z_2). The vertical axis then represents the fitness of those indi-

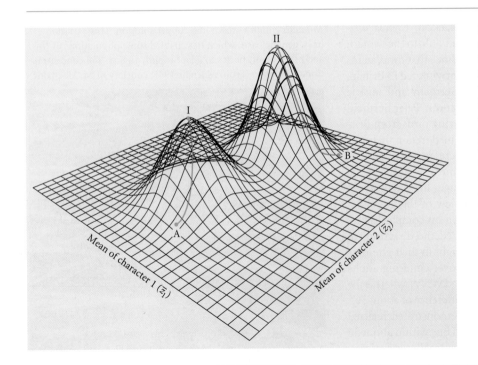

FIGURE 14.24 *An adaptive landscape for two quantitative characters. The vertical axis is the mean fitness of possible populations with various combinations of character means, $\bar{z}_1$ and $\bar{z}_2$. Either or both characters may vary within any such population. This landscape has two peaks (I, II), which will be attained by populations that begin at different starting points, such as A and B. This figure is an extension to two characters of the kind of fitness surface depicted for one character in Figure 14.13B. (After Slatkin 1983.)*

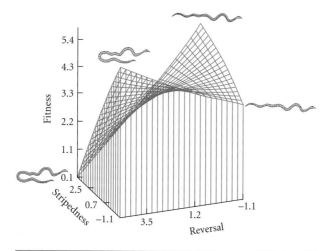

FIGURE 14.25 *A* fitness surface for combinations of two traits in the garter snake *Thamnophis ordinoides*, based on survival in the field. "Fitness" is the relative survival estimated for snakes with various values of stripedness and tendency to reverse course when fleeing (see text). The surface represents fitness of individual phenotypes, not mean fitness of populations. Snakes with highest fitness have a high stripe score and low reversal score *or* a low stripe score and high reversal score. (After Brodie 1992)

viduals in a population that have particular values of z_1 and z_2. This approach was taken by Edmund Brodie (1992) in a study that illustrates correlated selection (Figure 14.25).

In western North America, the garter snake *Thamnophis ordinoides* can have a uniform color, spots or blotches, or lengthwise stripes. Brodie found that the snakes vary in their behavior when chased down an experimental runway: some flee in a straight line until exhausted, while others repeatedly reverse their course. Both coloration and escape behavior are highly heritable. Other investigators had noted that among different species of snakes, striped forms tend to flee rapidly and rectilinearly, while blotched or spotted species have irregular escape patterns or tend to be sedentary, relying on their cryptic patterns or on aggression to avoid predation. It is thought that stripes are advantageous for rapidly moving snakes because visually hunting predators find it difficult to judge the speed and position of a moving stripe.

Brodie obtained 646 newborn snakes born to 126 pregnant females captured at a locality in Oregon, scored their color pattern (z_1), and measured their propensity to reverse course (z_2) when chased down a runway. He marked each snake by clipping individual combinations of ventral scales (a permanent mark that does not injure the animal), released the snakes in a suitable habitat, and periodically sought them thereafter. Snakes that were recaptured after 1 year (101 of them) were given a survival score of 1, and those not recaptured a survival score of 0. Brodie had reason to believe that most of the snakes that were not recaptured were eaten by crows and other predators.

The survival rate was greatest for those that had both strong striping and a low reversal propensity, and for those that had both an unstriped pattern and a high reversal propensity. Other phenotypes, such as striped snakes that reversed course when chased, had lower survival rates. Thus there was correlated selection on color pattern and antipredator behavior, in the direction that had been predicted from comparisons among species of snakes and from the theory of visual perception. By calculating the fitnesses (based on survival rates) of phenotypes with different stripedness and reversal scores, Brodie described the adaptive surface (Figure 14.25) as one with two peaks: two optimal combinations of color pattern and behavior.

Genetic Correlation

Many characteristics of a species are correlated with each other to some degree. The magnitude of a correlation—the degree to which two features vary in concert—is expressed by the *correlation coefficient* (r), which ranges from +1.0 (for a perfect correlation in which both features increase or decrease together) to –1.0 (when one feature decreases exactly in proportion to the other's increase). For uncorrelated characters, $r = 0$.

The **phenotypic correlation**, r_P, between, say, body size and fecundity is simply what we measure in a random sample from a population. But just as the phenotypic variance may have both genetic and environmental components, so too may the phenotypic correlation. Two features of individuals with the same genotype, such as those of an asexual clone, may vary together because both are affected by environmental factors such as nutrition. They display an *environmental correlation*, r_E. Likewise, the correlated variation may be caused by genetic differences that affect both characters, causing a *genetic correlation*, r_G.

Genetic correlations can have two causes (Figure 14.26). Probably the most common and important is *pleiotropy*—i.e., the influence of the same genes on different characters. The other cause is *linkage disequilibrium* among the genes that independently affect each character. Both can change over time, so *that genetic correlations need not be constant, but may evolve* (Turelli 1988). A genetic correlation due to linkage disequilibrium, such as the correlation between pistil length and stamen height in the primrose *Primula* (see Chapter 9), will decay as recombination reduces the linkage disequilibrium, unless selection for the adaptive gene combinations maintains it.

Correlations due to pleiotropy may change more slowly than those due to linkage disequilibrium, but they nevertheless can change for several reasons. Suppose that allele A_1 increases both size and fecundity, and A_2 decreases both. If the locus is polymorphic, it contributes a positive correlation between the traits, but if A_1 becomes fixed, it does not, because it contributes no variation (Figure 14.26B). Thus changes in allele frequencies can alter r_G. If at a second locus, allele B_1 increases size but decreases fecundity while B_2 does the reverse, this locus contributes negative correlation. The net genetic correlation will then

FIGURE 14.26 Causes of correlations between two phenotypic characters, z_1 and z_2. (A) Linkage disequilibrium. The marginal values give the effects of each genotype at locus A and at locus B on the characters (z_1/z_2), and the values of the characters for all nine two-locus genotypes are given in the cells. Additive effects are assumed. Note that locus A affects only z_1, and locus B affects only z_2. The sizes of the circles in the plots of z_2 against z_1 convey the frequencies of the phenotypes when the allele frequencies are $p = q = 0.5$ at each locus. At linkage equilibrium ($D = 0$), the characters are uncorrelated. Partial association of A_1 with B_1 and A_2 with B_2 ($D = 0.15$) yields a positive correlation, and the reverse association ($D = -0.15$) yields a negative correlation. (B) Pleiotropy. Two additively acting alleles at one locus have a positive pleiotropic effect on characters z_1 and z_2, causing a positive genetic correlation in a polymorphic population, as in the left and center graphs. Environmental effects on the two characters can cause environmental correlations, indicated by shaded ovals, that may be either positive (left) or negative (center). If the population is monomorphic for A_1A_1 (graph at right), there is no genetic correlation, but there may be an environmental correlation.

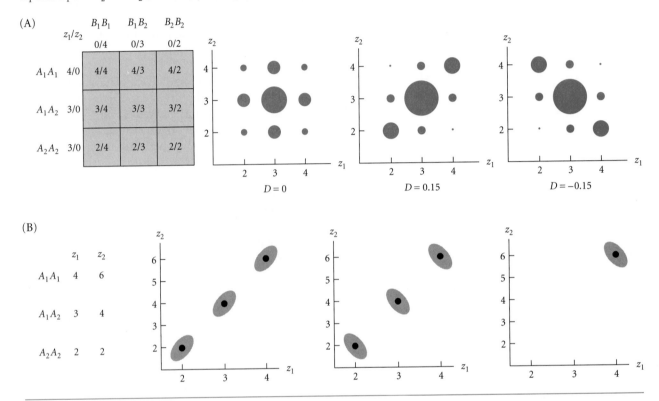

depend on how many A-type and B-type loci there are, as well as on their allele frequencies and the magnitude of the phenotypic effects of the alleles. The net r_G may well be zero (Figure 14.27).

These considerations imply that an increase in the frequency of new mutations or of rare alleles may substantially alter a genetic correlation. For example, natural populations of Austrian blowflies (*Lucilia cuprina*) exposed to the insecticide diazinon rapidly evolved resistance due to an increase in the frequency of a resistance allele, R. At first, resistant flies had reduced viability (when tested in the absence of diazinon) and a high incidence of bilateral asymmetry; these were pleiotropic effects of the R allele. Over the course of several years, viability increased and the incidence of asymmetry decreased. This change was not due to change at the R locus, because the R allele still reduced viability when it was backcrossed from the resistant field population into susceptible laboratory stocks. Rather, natural selection had increased the frequency of so-called *modifier alleles* at other loci, which altered and mitigated the deleterious effects of the R allele. An initially negative genetic correlation between diazinon resistance and viability evolved nearly to zero (McKenzie and Clarke 1988).

The relationship of the phenotypic correlation to the sum of the genetic correlation and the environmental correlation is analogous to the expression $V_P = V_G + V_E$ for the variance in a single trait.[*] The genetic and environmental correlations can differ substantially in magnitude, just as genetic and environmental variances do. Often they have the same sign (Figure 14.26); for instance, in one study, body weight and tail length in mice (*Mus musculus*) were positively correlated genetically ($r_G = 0.29$), environmentally ($r_E = 0.56$), and phenotypically ($r_P = 0.45$) (Falconer 1981). Occasionally they differ in sign, as in a study of weight gain and

---

[*]But is slightly more complicated. For features x and y, $r_P = h_x\, h_y\, r_G + e_x\, e_y\, r_E$, where $h = \sqrt{s^2}$, $e = \sqrt{1 - h^2}$, and $r_G = \mathrm{cov}_{xy}/\sqrt{V_{AX} \cdot V_{AY}}$. V_{AX} and V_{AY} are the additive genetic variances, cov_{XY} is the additive genetic covariance (estimated from correlations among relatives), and h^2 is heritability.

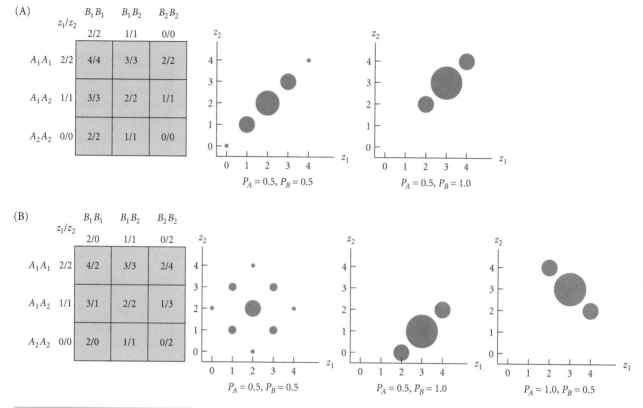

FIGURE 14.27 Effects of multiple pleiotropic loci and changes in allele frequencies on the genetic correlation between characters z_1 and z_2. (A) Two loci have positively correlated pleiotropic effects on characters z_1 and z_2, as indicated by the marginal values. The phenotypes (z_1/z_2) of the nine possible genotypes are shown in the cells. The sizes of the circles in the plots of z_2 against z_1 reflect the frequencies of the phenotypes when allele frequencies equal 0.5 at both loci (left) and when $p_A = q_A = 0.5$ but allele B_1 is fixed ($p_B = 1.0$). (B) Locus A has positively cor-

related pleiotropic effects on characters z_1 and z_2, but locus B has equal, negatively correlated pleiotropic effects on these characters. When the allele frequencies are $p = q = 0.5$ at both loci (left graph), the correlation due to locus A is canceled by that due to locus B, and there is no genetic correlation. When B_1 is fixed, the genetic correlation is positive due to variation at locus A (center graph). When A_1 is fixed, the variable locus B causes a negative genetic correlation between the characters (right graph).

the thickness of fat in pigs (*Sus scrofa*), in which r_G was 0.13, r_E was −0.18, and r_P was nearly zero. For organisms in which kinship among individuals cannot readily be determined by marking them in the field or by experimental breeding, it would be useful to use r_P as an index of r_G; there is some controversy about whether or not this might be possible, and not nearly enough data to tell (Cheverud 1988a; Willis et al. 1991).

Genetic correlation is an exceedingly important subject of study, bearing on topics as diverse as sexual selection, sexual dimorphism, the evolution of life histories, ecological specialization, and rates of morphological evolution. Its importance lies in the fact that *genetic correlations among characters cause correlated evolution*. For example, selection on one character may cause a *correlated response to selection* in another character that itself is selectively neutral. The extent to which this occurs depends on the *magnitude* and the *constancy* of the genetic correlation.

Examples of genetic correlation Genetic correlations between traits can be estimated from correlations between

relatives in the same way that genetic variances can. Often such data are obtained from full sibs born to field-caught females. (In such cases there is some risk that the correlation includes a contribution of nongenetic maternal effects.) Studies on garter snakes (*Thamnophis*) provide some interesting examples of this method.

Stevan Arnold (1981a) tested the responses of naive, newborn *Thamnophis elegans*, born in the laboratory to pregnant females captured in the field, to the odors of a variety of potential prey species, including several species of frogs and salamanders as well as a slug, a leech, and a fish. The response scored, the rate at which the snake flicks its tongue at a cotton swab dipped in an extract of the prey, is correlated with the snake's propensity to eat that kind of prey (Figure 14.28). Arnold studied snakes from a population in coastal California that feeds primarily on terrestrial slugs, and from an inland Californian population that lives where slugs do not occur and feeds mostly on fish and aquatic amphibians.

The mean response to both slugs and leeches was considerably higher in the slug-eating coastal population than the

FIGURE 14.28 (A) A newborn garter snake investigating the odor of a potential prey item on a cotton swab. (B) The positive genetic correlation (r_G) between response to slugs (z_1) and to leeches (z_2). The small points represent individual genotypes; the large point, the population mean for both traits. (C) A possible fitness landscape for a snake population where slugs and leeches both occur. Mean fitness ($\bar{w}$), shown by contours for various mean responses to slugs and leeches, is highest (H) for populations that accept slugs but avoid leeches. L represents a mean phenotype with minimal fitness, and the large dot is the initial mean of an evolving population. (D) The solid-color trajectory shows the course of evolution if selection against eating leeches is stronger than selection for eating slugs. The genetic correlation of the two traits causes a temporary, maladaptive, increased aversion to slugs before the population eventually evolves to the optimum. The shaded-color trajectory shows the course evolution would take if the genetic correlation were zero. (E) If selection for eating slugs is stronger than selection against eating leeches, the propensity to eat leeches may temporarily increase due to the positive genetic correlation. The shaded-color trajectory again shows the course of evolution if $r_G = 0$. (F) If the population encounters slugs, but not leeches, the response to leeches is selectively neutral, and the fitness contours are linear, increasing in value from left to right. As a correlated (pleiotropic) response to selection for feeding on slugs, an enhanced response to leeches evolves. (Photograph courtesy of S. J. Arnold.)

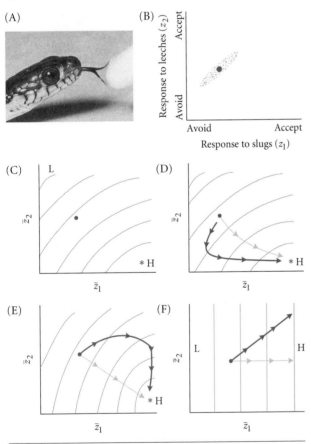

(A)

(B)

(C)

(D)

(E)

(F)

a few of which are listed in Table 14.4, in some instances were very different between the two populations, but in other cases were quite similar. In both populations there were generally high, positive correlations between responses to most of the amphibians (both frogs and salamanders), and to slugs and leeches. Neither population feeds on the poisonous salamander *Taricha* (which figured in Brodie and Brodie's study of another species of garter snake, described earlier), and the response to this prey was poorly correlated with responses to most other items.

Arnold suggested that these patterns of genetic correlation might be due to pleiotropy—specifically, that genetically variable substrate-specific chemoreceptors might be sensitive to similar compounds in the skin of various prey species. Various species of frogs, for example, may generally have similar skin compounds. Different receptors may respond to compounds in the skin of *Taricha* (perhaps to its poison, tetrodotoxin). The slug/leech correlation, possibly also due to similar compounds in these two organisms, is particularly interesting. Inland snakes do not encounter slugs, but do encounter leeches, which may attack the gut lining if swallowed. Natural selection for aversion to leeches might also cause an aversion to slugs (in the laboratory) as a correlated response. Conversely, selection for slug feeding in coastal snakes has probably caused a correlated positive response to leeches, which do not pose a danger to the coastal population because the snakes do not encounter them.

Arnold's study illustrates that (1) *genetic variation and genetic correlation apply to behavior* as well as to other traits; (2) measurements of *genetic correlation can give rise to hypotheses* about physiological and biochemical mechanisms (e.g., classes of chemoreceptors, chemical similarities among certain prey species); (3) *natural selection on one character can cause evolution of other characters* as a correlated response; and (4) *genetic correlation may* in some cases *remain fairly constant* for considerable spans of evolutionary time (these snake populations have probably been separated for hundreds or thousands of generations).

Some other interesting conclusions are illustrated by Brodie's (1989, 1993) analysis of escape behavior and color pattern in *Thamnophis ordinoides*. As described earlier, survival of first-year snakes in the field was highest for those that had high stripedness and low reversal propensity, and for those that had the opposite combination of traits—low stripedness and high reversal propensity.

In all, Brodie scored 1388 snakes, born to 251 mothers taken from four local populations, for color pattern and the number of times the snake reversed course. The greater variance among than within families showed that all these traits were highly heritable. In two populations, there was a statistically significant negative genetic correlation between stripedness and the number of reversals. Striped individuals tended to flee until they were exhausted, whereas blotched individuals generally changed direction after a short distance.

Within these populations, therefore, there appears to be a correlation between inherited morphology and in-

inland population (most snakes from the inland population refuse to eat slugs). There was significant genetic variation in the tongue-flicking response to these and to all other prey items. The genetic correlations of responses to pairs of items,

Table 14.4 **Some genetic correlations between responses to different prey odors in two populations of the garter snake *Thamnophis elegans*[a]**

	Hyla (FROG)	*Batrachoseps* (SALAMANDER)	*Taricha* (SALAMANDER)	FISH	SLUG	LEECH
Hyla (frog)	—	1.10	−0.24	0.18	0.88	1.01
Batrachoseps (salamander)	0.81	—	0.07	1.00	1.34	0.98
Taricha (salamander)	−0.45	0.57	—	0.09	−0.55	−0.88
Fish	0.89	1.27	0.02	—	0.59	0.84
Slug	−0.03	0.56	−0.79	0.19	—	0.89
Leech	0.07	0.77	−0.01	−0.38	0.89	—

Source: Adapted from Arnold (1981).

[a]Genetic correlations above the diagonal are for a coastal population, and below the diagonal are for an inland population. Some values exceed 1.0 due to variation in estimates of the components of V_G.

herited behavior that is adaptively appropriate to that morphology—in other words, the traits are *coadapted*. It seems unlikely that the genes that control pigmentation also control behavior in snakes. Thus the most likely explanation for the genetic correlation is linkage disequilibrium among alleles at the two kinds of loci, which must be maintained by natural selection in opposition to recombination. We will see in the next section that a positive correlation between traits subject to selection pressures in the same direction enhances the rate of response to selection—i.e., the rate of adaptation. Because the color pattern is more advantageous if it is used properly, so to speak, and the behavior is more advantageous if it is coupled with the right color pattern, *the genetic correlation facilitates the rapid evolution of a coadapted complex of functionally interdependent features*, such as the patterns and behaviors of different species of snakes.

How genetic correlation affects evolution

Genetic correlations among characters *can either enhance or retard the rate of adaptive evolution*, depending on the circumstances. In extreme cases, they may severely *constrain* adaptation.

When two characters, with values z_1 and z_2, are genetically correlated, the rate of evolution of the mean of character 1, $\bar{z}_1$, depends on both direct selection on z_1 and the correlated response of z_1 to selection on z_2; $\bar{z}_2$ likewise evolves due to both direct selection and selection on z_1. Character 1 evolves at a rate ($\Delta\bar{z}_1/\Delta t$) that is also dependent on its additive genetic variance (V_{A1}) and on the magnitude of the genetic correlation between z_1 and z_2.

❖ Skip this paragraph if you are uncomfortable with math. Otherwise, recall Lande's equation for the rate of evolution of a character,

$$\Delta\bar{z} = \frac{\overline{V}_A}{\overline{w}} \cdot \frac{d\overline{w}}{d\overline{z}}$$

which can also be written

$$\Delta\bar{z} = \frac{V_A d(\log_e \overline{w})}{d\overline{z}}$$

because $d\overline{w}/\overline{w} = d(\log_e \overline{w})$. The magnitude of the genetic correlation between z_1 and z_2 is taken into account by using the covariance between them, cov_{12}, which is the numerator of the genetic correlation. Thus the two characters evolve at rates

$$\Delta\bar{z}_1 = V_{A1}\frac{d(\log\overline{w})}{d\bar{z}_1} + \mathrm{cov}_{12}\frac{d(\log\overline{w})}{d\bar{z}_2}$$

$$\Delta\bar{z}_2 = V_{A2}\frac{d(\log\overline{w})}{d\bar{z}_2} + \mathrm{cov}_{12}\frac{d(\log\overline{w})}{d\bar{z}_1}$$

In the equation for each character, the first term is the response to direct selection on the character, which depends on the genetic variance and the strength of selection, expressed by $d(\log \overline{w})/d\bar{z}$. The second term is the correlated response due to selection on the other character and the covariance between the characters.

Because the rate of evolution of character z_1 depends on both direct selection and selection on a correlated character, z_2, z_1 may evolve primarily due to direct selection *or* primarily due to selection on z_2, depending on which character experiences stronger selection (and on the genetic variances and genetic correlation coefficient). Suppose, for example, that Arnold's garter snakes, which show a genetic correlation of 0.89 between responses to slugs and responses to leeches (see above), occurred where both slugs and leeches were present. Direct selection on the slug-feeding response might favor the evolution of slug feeding, but because of the genetic correlation, such individuals would also be likely to eat leeches, which could kill them, since snakes swallow prey alive and leeches can attack the digestive tract. If selection against eating leeches were stronger than selection in favor of eating slugs (as it might be if the snakes could turn to frogs as an alternative food), then the population might evolve at least a temporary aversion to slugs as a correlated effect of aversion to leeches (Figure 14.28D). Conversely, if leeches were rare and slugs were the most abundant food, selection for feeding on slugs would be stronger, the population would evolve a slug-feeding habit, and it would also evolve the maladaptive habit of occasionally eating leeches (Figure 14.28E).

The snake example is one in which the characters are positively correlated, but selection can also work in opposite directions. When it does, the two character means may reach their joint optimum only slowly, or possibly not at all. If selection favors the evolution of two positively correlated characters in the same direction, the genetic correlation enhances the rate of evolution toward the optimal state of both characters, as we noted above in discussing the coloration and antipredator behavior of snakes. However, if the characters are negatively correlated, but selection favors the evolution of both in the same direction, a conflict exists, and the more strongly selected character is likely to approach its optimum more rapidly than the less strongly selected character. Figure 14.29 provides some hypothetical examples of such a situation.

As close study of Figure 14.29 reveals, when there is a conflict between selection on two features and the genetic correlation between them, then *one feature may evolve to its optimum and the other may evolve to a permanent suboptimal or maladaptive state.* In this case, pleiotropy, expressed as genetic correlation, can act as a strong constraint on adaptation. However, this occurs only if the genetic correlation is perfect, i.e., if it equals +1.0 or −1.0, and only if the genetic correlation itself does not change. *If the genetic correlation is not complete* (i.e., $r_G \neq |1|$), then *the more weakly selected character may evolve very slowly toward its optimum, and may even temporarily evolve away from its optimum, but eventually attains its optimal state.* When the genetic correlation is less than complete, each character has some "free" genetic variation, perhaps due to loci that affect only that

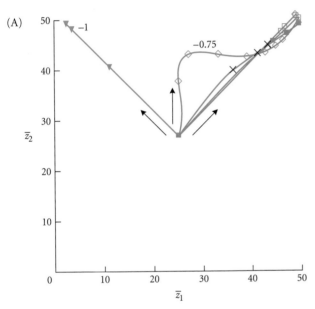

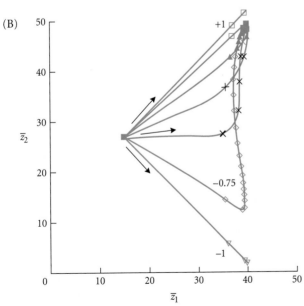

FIGURE 14.29 The effects of genetic correlation on the evolution of two characters, based on computer simulations. In both graphs, the optimal mean phenotype is the point at the upper right, and the population begins at the point from which the trajectories radiate. The trajectories differ for different values of the genetic correlation coefficient: +1 (open squares), 0.75 (open circles), 0.375 (open triangles), 0.0 (plus signs), −0.375 (Xs), −0.75 (diamonds), −1 (upside-down triangles). The symbols along each trajectory are at intervals of 50 generations. The genetic correlations are assumed to remain constant during the course of evolution. (A) Selection is equally strong on both characters. The optimum is $\bar{z}_1 = \bar{z}_2 = 50$, and the population begins at $\bar{z}_1 = 25$, $\bar{z}_2 = 27$. Both characters evolve rather directly and rapidly to the optimum, except when $r_G = -0.75$ or −1. The joint optimum is eventually attained when $r_G = -0.75$; when $r_G = -1$, $\bar{z}_2$ evolves to its optimum, but $\bar{z}_1$ evolves to a maladaptive state. (B) Selection on z_1 is about twice as strong as on z_2. When r_G is positive or weak, $\bar{z}_1$ evolves to its optimum faster than $\bar{z}_2$ does. When $r_G = -0.75$, $\bar{z}_2$ evolves in a maladaptive direction at first, as a correlated response to the strong selection on z_1. When $\bar{z}_1$ reaches its optimum, however, the component of variation in z_2 that is not correlated with z_1 enables $\bar{z}_2$ to respond to selection, and it slowly evolves to its optimum. When $r_G = -1$, the strong selection on z_1 causes correlated evolution of $\bar{z}_2$ to a permanently maladaptive state. (After Via and Lande 1985.)

character and do not have pleiotropic effects on the other character. This "free," or uncorrelated, genetic variation can eventually enable the character to respond to selection independently of the other character.

This conclusion, however, does not ensure that every feature of a species will evolve to its optimum state. First, the feature that evolves mostly as a correlated response may approach its optimum very slowly indeed, and as we have noted, may evolve in a maladaptive direction for a considerable time. Thus, some features that we observe in organisms today might well be in such a transitional suboptimal state. Moreover, evolution to the optimum may be so slow that the environment and the agents of selection change in the interim, creating entirely different optima; or other features altogether, not genetically correlated with the suboptimal feature, may evolve to "remedy" the maladaptive condition (as we saw in the case of mitigation of deleterious effects of insecticide resistance in blowflies above).

Second, we have considered only two genetically correlated characters. Many characters may be intercorrelated, as in the responses of garter snakes to diverse amphibian prey (see Table 14.4), or as might be expected for morphology if many organs covary in response to, say, variation in a growth hormone. As the number of intercorrelated characters increases, it becomes more likely that some of them may be prevented from attaining their selectively optimal state (Via and Lande 1985).

These considerations support the idea that genetic correlations can be important constraints on adaptive evolution. On the other hand, they assume that genetic correlations remain consistent for long periods of evolutionary time—a proposition that has been greatly debated (e.g., Lande 1979b; Turelli 1988; Barton and Turelli 1989; other references in Brodie 1993). If genetic correlations are highly malleable, then they should impose less severe constraints.

Genetic correlations may constrain adaptive evolution, but they can also enhance its rate, as Günter Wagner (1988) has shown in complex models of multiple, functionally interdependent traits. Wagner found that for certain patterns of natural selection, the more characters there are, and the more variation there is in each, the slower will be the response to selection on any one character (Figure 14.30)—quite the opposite of the relationship between genetic variation and response to selection on a single trait. Because genetic correlations reduce the amount of "free" genetic variation in each character, it is possible for the characters to evolve more rapidly if they are genetically correlated, as when they are subject to the same developmental controls. (For instance, evolution of a longer leg would be very slow indeed if the development of bones, muscles, blood vessels, and nerves were not intercorrelated, but if each instead varied independently of the others.)

This conclusion raises the question of whether or not functionally related characters tend to be genetically correlated. Some authors have suggested that the evolution of adaptive "morphological integration" or "developmental integration" should result in coordinated variation in func-

tionally related traits (Olson and Miller 1958; Riedl 1978; Cheverud 1988b). Some evidence for this hypothesis was presented by James Cheverud (1982), who measured 56 dimensions of the skulls of female rhesus macaques (*Macaca mulatta*) and their offspring. He found some evidence, in accord with the theory of morphological integration, that both genetic correlations and phenotypic (r_P) correlations were higher between traits in the same "functional set" (e.g., various dimensions of the jaws) than in different functional sets. Such "adaptive correlations," however, are by no means universal. For example, within species of herbiv-

---

FIGURE 14.30 Excessive uncorrelated variation in functionally related characters may retard the response to selection. (A) A fitness landscape for characters z_1 and z_2. The fitness ($\bar{w}$) surface has the form of a steep-sided ridge because character z_2 must be closely matched to z_1 to function properly. Under directional selection (D), the phenotype should evolve upward and to the right. Stabilizing selection (S) acts orthogonally to the axis of the ridge. If z_1 and z_2 are uncorrelated, as shown by the circle, much of selection is stabilizing. The response of the characters to directional selection is then slower than if the characters are correlated (as shown by the ellipse) and the optimum lies along the axis of correlation. (B) The rate of evolution along the ridge under directional selection as a function of the amount of uncorrelated variation in other, functionally interacting, characters. The greater the number (n) of characters and the amount of uncorrelated variation in each, the slower the rate of evolution. Genetic correlations among characters can reduce this retarding effect. (After Wagner 1988.)

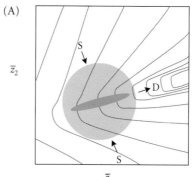

(A)

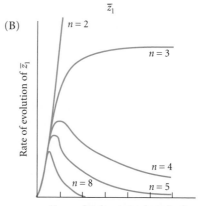

(B)

Rate of evolution of $\bar{z}_1$

$n = 2$
$n = 3$
$n = 4$
$n = 5$
$n = 8$

Uncorrelated variation in each of $n-1$ other characters

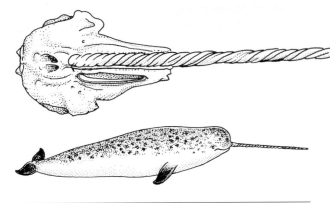

FIGURE 14.31 Although most features seem to show little genetic variation for directional bilateral asymmetry, such asymmetry has evolved in many organisms. A striking example is the male narwhal (*Monodon monoceros*), in which the upper left incisor forms a spirally coiled tusk, used in competition for mates. Part of the skull, viewed from above, has been removed to show the root of the tusk and the small, unerupted right incisor. (After DeBlase and Martin 1981.)

orous insects such as butterflies, in which the rather sedentary larvae feed on the plant on which their mother laid her eggs, there appears to be little genetic correlation between females' oviposition preferences for one plant species or another and the capacity of the larvae for growth and survival on the preferred plant (Thompson 1988).

Genetic Constraints on Evolution

Having alluded several times to genetic constraints, let us now directly address their nature and importance. We will return to this topic in Chapter 23, in considering whether such constraints have strongly influenced the course of macroevolution over geological time. We must realize that genetic constraints may well range from virtually absolute (we can be quite certain, for example, that no vertebrate lineage will ever change its genetic material from DNA to RNA) to quite temporary and local (e.g., lack of genetic variation in a small population, which might later be supplied by mutation or gene flow). We now briefly discuss some of the most likely constraints.

Absence of genetic variation
An absence of genetic variation might be a temporary or historically localized condition, as in an inbred population, or it might be a typical condition for a specific character. However, since we ordinarily cannot examine all members of a species, we can never say that genetic variation is definitely absent—only that we cannot detect any in our samples.

Although the great majority of features display genetic variation in most species, there are some apparent exceptions. Genetic variance for copper tolerance was not detected in large samples of several species of plants (Macnair 1981; see above). Maynard Smith and Sondhi (1960) attempted to select populations of *Drosophila* for absence of either the right or the left ocellus (ocelli are simple eyes on the top of the head). The level of "fluctuating asymmetry" (i.e., asymmetry that is inconsistent in its direction) increased, but they could not obtain consistently right-eyed or left-eyed stocks. Perhaps directional asymmetry is unlikely to evolve—but it has nevertheless done so in a variety of organisms. Fin whales have a white stripe on the right side of the jaw, and the left incisor of the narwhal has evolved into a long, spirally coiled tusk (Figure 14.31). Some of our own internal organs are asymmetrical in form and position, such as the heart.

In some cases, a character is genetically variable, but *the variation is not selectable*, and selection engenders no response. Recall that only the additive genetic variance provides response to selection; other components of genetic variance, due to dominance and epistasis, do not. If allele frequencies change, however, some of this nonadditive genetic variance may be converted into additive genetic variance, and response to selection becomes possible, as described above.

Genetic correlation
Genetic variation in a trait may not respond to selection because of antagonistic selection on genetically correlated traits, as we have seen. How severe and permanent a constraint is genetic correlation likely to be? We noted above that genetic correlation of two characters is not a permanent constraint unless r_G is +1 or −1 (i.e., perfect correlation). However, most estimates of r_G have not been this extreme, and it would be very difficult to demonstrate that a genetic correlation was exactly +1 or −1.

Are genetic correlations stable, or can they change rapidly? We noted earlier that there are theoretical reasons to expect them to change over the course of time (Turelli 1988). In some selection experiments, genetic correlations have remained stable throughout the selection process and uniform among different populations; in other cases, however, selection on one character has altered its genetic correlation with another character (reviewed by Wilkinson et al. 1990). In his study of responses to prey in garter snakes, Arnold (1981a) found that the pattern of genetic correlations was much the same in two genetically divergent populations. On the other hand, genetic correlations among various skull measurements differed between two species of *Peromyscus* mice (Lofsvold 1986), and the correlation between the developmental rate of wood frog (*Rana sylvatica*) tadpoles and the size at which they metamorphosed was dramatically different between lowland (r_G = 0.65) and montane (r_G = −0.86) populations (Berven 1987). On the whole, it does not appear that genetic correlations remain immutable for substantial periods of evolutionary time. Thus they probably more often affect evolution in the short term than in the long term.

How finely can selection shape an organism? Can it hone every detail to a functionally optimal form? When we consider that a wing or a leg or a mandible grows as a whole, under the influence of growth factors that might control much of its structure, we might suppose that the many measurements we could make on such a structure would be so highly intercorrelated that the possible changes in its shape would be quite limited. Some developmental biologists have indeed proposed that developmental "rules" can strongly limit the variety of possible forms (see Chapter 23). Weber (1992), on the other hand, asked, "How small are the smallest selectable domains of form?" and concluded that they are very small indeed. He chose two of the shortest distances identifiable by landmarks on the wing of *Drosophila melanogaster* (Figure 14.32A). One (D_1) was about 0.15 mm (about 35 cell diameters) and the other (D_2) about 0.11 mm

(20 cell diameters). He selected one line for large D_1, relative to D_2, and one for the reverse. After only 11 generations of selection, both populations had diverged from control unselected lines (Figure 14.32B). Thus the shape of this exceedingly small part of the wing was rapidly altered.

Weber argued that if this result is general, then virtually any small domain of the wing could be altered independently by selection, so that it is indeed possible that every minor feature of the wing—its exact shape, the position, thickness, and length of every wing vein, the length of the hairs on its edge—has been molded to fit functional demands. This line of thought tends to justify what Stephen Jay Gould and Richard Lewontin (1979) called the "adaptationist program"—the belief that virtually all characteristics are adaptive and have nearly their optimal form.

The "adaptationist program" is the subject of a major debate (Mayr 1983; Williams 1992b). Some evolutionary biologists hold that genetic and developmental constraints strongly limit the variety of phenotypes that might evolve in a lineage (Gould and Lewontin 1979; Wake 1991), while others minimize their importance (Reeve and Sherman 1993). We will return to this problem repeatedly in subsequent chapters.

❖ Special Topics in Phenotypic Evolution

❖ Evolution of Quantitative Characters by Genetic Drift

If variation in a feature is caused by selectively neutral alleles—i.e., if neither variation in the character nor pleiotropic effects of the alleles affect fitness—then the character can evolve only due to new mutation (which increases variation) and genetic drift (which erodes it). For additively acting alleles, variance due to mutation arises at a rate V_m per generation, which equals $2nua^2$, where n is the number of loci affecting the character, u is the average mutation rate per locus, and a^2 is the average phenotypic effect of a mutation. We have noted previously that for morphological traits, V_m is often about 0.001 V_E (see Chapter 10).

In an extensive theoretical analysis, Michael Lynch and William Hill (1986; see also Lynch 1994) showed that the genetic variance (V_A) in an initially uniform population rises to an equilibrium level at which mutation is balanced by genetic drift. This equilibrium level is approximately $2N_eV_m$, and it takes about $6N_e$ generations to get 95 percent of the way to this level. Because $V_m \approx 0.001 V_E$, the expected heritability of a selectively neutral polygenic character $[V_A/(V_A + V_E)]$ can be calculated. The expected heritability increases steeply as a function of effective population size (N_e), and is very high if N_e is greater than 1000 or so (Figure 14.33). Thus *if a polygenic character is selectively neutral, mutation alone can maintain high levels of genetic variation.*

If a number of isolated populations are derived from an initially uniform ancestral population, *mutation and drift can cause genetic divergence* in a polygenic character, just as they

FIGURE 14.32 Response to selection for the shape of a very small part of the wing of *Drosophila melanogaster*. (A) An enlargement of part of the wing shows the distances D_1 (between the two upper points) and D_2 (between the lower points). The ratio D_1/D_2 was selected upward in population E$^+$ and downward in population E$^-$. (B) Change in D_1/D_2 over the course of 11 generations. The vertical bars measure the variation in each sample. (After Weber 1992.)

(A)

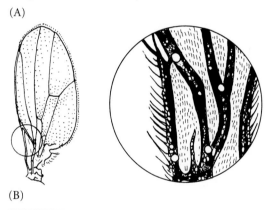

(B)

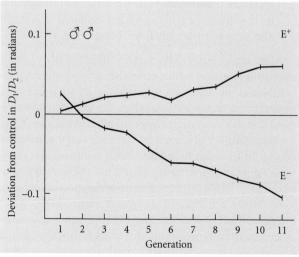

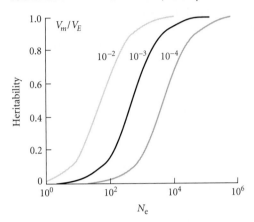

FIGURE 14.33 The theoretical equilibrium level of heritability of a selectively neutral character as a function of effective population size (N_e) for several levels of V_m/V_E. V_m is the genetic variance arising by mutation per generation, and V_E is the environmental variance of the trait. (After Lynch and Hill 1986.)

do at a single locus (see Chapter 11). *The rate at which the means diverge is a constant value,* $2V_m$ *per generation, that does not depend on population size.* (Recall that this is also true of the rate at which populations diverge by substitution of neutral mutations at a single locus; see Chapter 11.) It must be emphasized that this conclusion holds only when selection does not act on the character or on the alleles underlying it. It tells us, however, that the amount of phenotypic divergence among populations or species should be proportional to the time since their common ancestry if the character is neutral. If populations are either less or more divergent than expected, it is likely that natural selection has either had a stabilizing effect or contributed to divergence. In Chapter 24, this theory will be applied to divergence among extant species and to data from the fossil record.

❖ What Maintains Genetic Variation in Quantitative Characters?

Many, if not most, quantitative traits are genetically variable. Accounting for the existence of this variation is as difficult, and as unresolved, a problem as accounting for the variation in proteins and DNA sequences.

If phenotypic variation is selectively neutral, then a balance between the rate at which variation arises by mutation and the rate at which it is lost by genetic drift would result in lower heritability values in small than in large populations. However, although there exist few comparisons of h^2 among closely related populations or species that differ in population size, estimates of h^2 for characters of species that are thought to have small effective population sizes (e.g., Darwin's finches, salamanders, mice) are often just as substantial as those for more populous species (Houle 1989). Furthermore, alleles that contribute to quantitative variation are probably seldom selectively neutral (see below).

Although many factors have been postulated to maintain genetic variation in quantitative traits (Barton and Turelli

1989), little empirical evidence is available to evaluate their importance in natural populations. These possible factors include:

1. *Heterozygote superiority* (overdominance) at the individual contributing loci because of their pleiotropic effects on functionally important features (Gillespie 1984). There is little evidence for overdominance for fitness, other than a few special cases such as human sickle-cell hemoglobin (see Chapter 13).
2. *Antagonistic pleiotropy,* i.e., superiority of some genotypes with respect to some fitness components and of other genotypes with respect to others (Rose 1982). Except under very special circumstances, antagonistic pleiotropy will not maintain variation (Curtsinger et al. 1994).
3. *Gene interaction.* Theoretically, epistatic interactions among the loci with respect to fitness can in some circumstances maintain polymorphism (Gimelfarb 1989). There is not yet sufficient evidence to evaluate this proposition.
4. *Variable selection.* Fluctuations in the optimal phenotype from one generation to another can delay the loss of genetic variation; random variation in the direction of selection, however, acts like genetic drift, and ultimately leads to homozygosity (Kimura 1965). Disruptive selection, as when different phenotypes are adapted to different resources, can maintain polygenic variation if fitness is frequency-dependent, as explained for a single locus in Chapter 13 (Slatkin 1979).
5. *Gene flow.* As for a single locus, gene flow between populations selected for different optimal phenotypes can maintain variation in each population. Even if the optimum is the same in all populations, the turnover in the genetic basis of a polygenic trait (see above) within populations leads to variation in allele frequencies among populations. Thus gene flow among the populations may maintain variation (Goldstein and Holsinger 1992). However, there is not much difference between h^2 estimates from natural populations and from populations that have long been maintained in the laboratory under rather uniform conditions and isolated from gene flow (Bürger et al. 1989).

Many quantitative characters are subject to fairly intense stabilizing selection, which ultimately should eliminate genetic variation. Although some characters seem not to be important to fitness in themselves, the alleles that control them nonetheless may be subject to selection. For example, the exact number of sternopleural bristles on the side of an adult fruit fly is not thought to affect its fitness. But genotypes for intermediate bristle counts have higher survival as larvae (which lack bristles) than do genotypes for higher or lower bristle numbers (Kearsey and Barnes 1970). Adult bristle number is a pleiotropic effect of genes that affect other, functionally important, characters such as the nervous system. Thus even selectively "trivial" characters are probably subject to indirect selection because of the pleiotropic effects of the underlying genes (Dobzhansky 1956).

For these reasons, much attention has been focused on the hypothesis, advanced especially by Russell Lande (1976b), that levels of polygenic variation reflect a balance between the erosion of variation by stabilizing selection and the input of variation by mutation (V_m). This balance depends on the strength of selection, the number of loci (n), the mutation rate per locus (u), and the average effect of an allele (a) on the phenotype. The adequacy of this mutation-selection balance model has been extensively debated (e.g., Turelli 1984; Houle 1989; Bürger et al. 1989) because the mutational variance has been considered inadequate to account for observed levels of additive genetic variance. However, Houle et al. (1996) have presented evidence consistent with mutation-selection balance. Houle et al. reasoned that since fitness is the product of many physiological and developmental processes, far more loci contribute to a fitness-related trait, such as survival or fecundity, than to a single morphological trait. If V_A is maintained in a population by mutation-selection balance, then the input of variance by mutation (V_m) should be higher for fitness-related traits because of the number of loci involved. In fact, V_m appears to be about six times greater for fitness-related traits than for morphological traits, as predicted by the mutation-selection balance hypothesis.

❖ Genotype-Environment Interactions

The *norm of reaction* of a genotype is the set of phenotypes it expresses in different environments (see Chapters 3 and 9). The norm of reaction can be visualized by plotting the genotype's phenotypic value in each of two or more environments, as for the hypothetical data in Figure 14.34 or the real data in Figure 13 in Chapter 9.

When the effect of environmental differences on the phenotype differs from one genotype to another in the population, the phenotypic variance includes a component due to *genotype × environment* (G × E) *interaction* ($V_{G \times E}$) (Figure 14.34B–E). If all the genotypes have parallel reaction norms, so that the difference between the phenotype in environment 1 and in environment 2 is equal for all genotypes (Figure 14.34A), there is no G × E interaction ($V_{G \times E} = 0$).

The magnitude of $V_{G \times E}$ may, but does not necessarily, suggest whether or not an *adaptive norm of reaction* can evolve. Some alterations of the phenotype by the environment are *not adaptive*—stunted growth due to inadequate nutrition, for example. Other environmental alterations of the phenotype are evidently *adaptive responses* to environmental stimuli, and have evolved by *natural selection for those genotypes with norms of reaction that most nearly yield the optimal phenotype for each environment the organism commonly encounters*. Notice that here we are speaking not of genetic polymorphism—different genotypes adapted to different conditions—but of the evolution of a genotype capable of adaptively altering the development or expression of a feature to suit the conditions that the individual organism encounters. This capability is often called **phenotypic plasticity** (Via et al. 1995).

In some cases these adaptive responses are *graded* responses to continuously (rather than discretely) varying

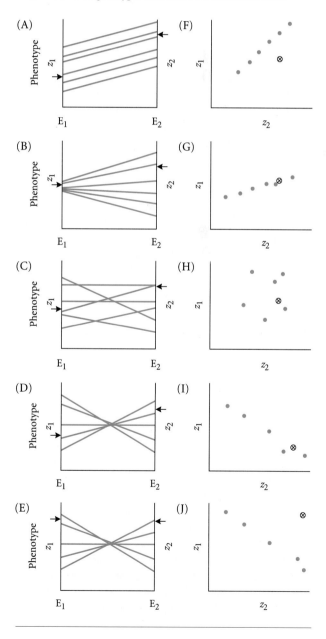

FIGURE 14.34 Genotype × environment interaction and the evolution of adaptive phenotype plasticity. In A–E, each line represents the reaction norm of a genotype—its expression of a phenotypic character in environments E_1 and E_2. These character states are labeled z_1 and z_2. The genotypes differ in the effect of environment on phenotype (G × E interaction) in B–E, but not in A. The arrows indicate the adaptively optimum phenotype in each environment. Panels F–J plot z_1 against z_2 for the genotypes in the reaction norm plots A–E respectively. The phenotypes expressed in environments E_1 and E_2 may be thought of as two characters, displaying a genetic correlation that is positive in F and G, nearly zero in H, and negative in I and J. The adaptively optimal character states (optimal reaction norm) are shown by ⊗, corresponding to the arrows in the reaction norm plots. A genotype with nearly optimal phenotypic plasticity—the capacity to produce a nearly optimal phenotype in each environment—may be fixed by selection in G, H, and I; a slightly suboptimal reaction norm will evolve in F; and the negative genetic correlation will prevent evolution of the optimal reaction norm in J, unless other genotypes than those shown should arise.

environmental factors. The production of an enzyme is often finely regulated in response to the availability of substrate. Behaviors are often graduated in response to graded stimuli, as in the level of aggression territorial males display in response to intruders' persistence. Other adaptive phenotypic responses are more or less discrete "*alternative phenotypes*" (West-Eberhard 1989; see Color Plate 5). In still other cases, the most adaptive norm of reaction may be a *constant* phenotype, buffered against alteration by the environment. It may be advantageous, for example, for an animal to attain a fixed body size at maturity or metamorphosis despite variations in nutrition or temperature that affect the rate of growth.

One way of thinking about the evolution of reaction norms is to view them as genetic correlations (Via 1994; Via and Lande 1985). Considering only two environmental states for the sake of simplicity, we can think of the phenotypes that a genotype expresses in two environments as two characters: z_1 in environment 1 and z_2 in environment 2. Then plots of reaction norms (Figure 14.34A–E) can be transformed into plots of each genotype's phenotype in environment 1 versus environment 2 (Figure 14.34F–J). Selection in each of the two environments favors a joint optimum, a pair of phenotypes that the ideal genotype should be able to express. Whether or not this joint optimum is likely to be attained depends on where it lies relative to the correlated distribution of the two "characters," just as in the analysis of two genetically correlated characters (see above). Thus panels D and E exemplify a case of strong G × E interaction, but recasting these phenotypes as two negatively correlated characters shows that the population might easily evolve toward the joint optimum (panel I) or not (panel J), depending on what the optimal phenotypes are. Notice that an organism capable of developing the optimal phenotype in each of two or more abundant environments might be capable of using or persisting in both—it would be an ecological *generalist*. In contrast, a species that could not evolve optimal phenotypic plasticity might be restricted to one environment or another, and so would be regarded as an ecological *specialist*.

❖❖ Threshold Traits

Threshold traits, introduced in Chapters 3 and 9, are expressed as discrete alternatives, but are controlled by polygenic variation rather than by single loci. Examples include the number of digits on a vertebrate's foot, the number of bristles on certain parts of an insect's body, such as the scutellum of *Drosophila*, and probably the number of flower parts in plants such as lilies. In Chapter 3, we described evidence for the commonly accepted model of threshold traits. An underlying variable, such as the concentration of a growth-promoting factor (or "morphogen") at a particular site in the developing foot or other structure, has a continuous frequency distribution in the population, owing to polygenic variation as well as environmental influences. Whether or not a structure (e.g., a toe or bristle) develops at that site depends on whether or not

the underlying variable exceeds some threshold level, which is determined by some other feature of the developmental system, such as the competence of certain genes to be transcribed in response to the morphogen.

Willem Scharloo (1987), following earlier authors, presents a simple model of threshold traits (Figure 14.35). A "gene-environmental factor/phenotype mapping function" (GEPM) relates the phenotype to the amount of gene product (morphogen) determined by the genes and the environment. This function may be linear (Figure 14.35A) so that the phenotype changes in proportion to the amount of morphogen, as appears to be the case for many traits that

FIGURE 14.35 Effects of the gene-environmental factor/phenotype mapping function (GEPM) on the distribution of phenotypes. Each diagram shows the frequency distribution of a gene product in a population. Panels A and C show the effects of shifting the distribution by altering its mean. For each distribution of gene product, there is a corresponding distribution of a phenotypic character, the development of which depends on the gene product. How the phenotype depends ("maps") on the amount of gene product is determined by the GEPM, represented by the line within the interior graph. The form of this "mapping function" is determined by developmental processes that are generally not well understood. In (A), the GEPM is linear, and a unit of change in gene product yields a unit of change in phenotype. In (B), the GEPM is a threshold function. A unimodal distribution of gene product can yield a bimodal distribution of phenotypes. In (C) and (D), the GEPM has a zone of canalization, where its slope is shallow. This is a region in which a large difference in gene product yields only a slight change in phenotype. Even if, as in D, the amount of gene product is highly variable, the phenotype shows little variation. In all cases, variability or change in the distribution of gene product can be due to genetic variation, environmental variation, or both. (After Scharloo 1987.)

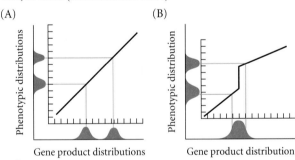

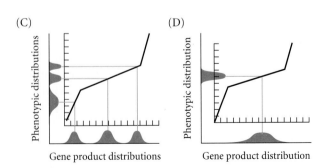

vary continuously, or nearly so. Structures that are highly repeated in the body, such as abdominal bristles (10 or more) in *Drosophila*, often vary in this way. But for other traits, such as the number of scutellar bristles (typically 4), the GEPM is *nonlinear* (Figure 14.35B–D). Within a certain zone, considerable variation in gene product yields little or no variation in phenotype. In other words, the developmental system has the capacity to reduce the effect of perturbing influences on the phenotype (reviewed by Scharloo 1991). The developmental biologist Conrad Waddington (1957), one of the first to integrate developmental and evolutionary biology, referred to this phenomenon as **canalization**. A nonlinear GEPM may have a "zone of canalization" within which the phenotype varies little; the borders of this zone mark the *thresholds* at which change in amount of gene product yields a change in phenotype.

Nonlinear GEPMs may retard evolution by preventing the expression of variation. Lande (1978) has shown that under selection for, say, reduced digit number, little evolution occurs while the distribution of the underlying variable, *z* (the morphogen), lies between thresholds in a canalization zone (Figure 14.36). When, perhaps due to genetic drift or mutation, the distribution of *z* comes to straddle a threshold, phenotypic variation is exposed to selection, and the phenotypic character (digit number) can evolve rapidly to the next state. *A threshold trait, then, is likely to evolve in spurts of rapid change interspersed with long periods of stasis.*

Waddington used the concept of canalization to interpret some curious experimental results (e.g., Waddington 1953). A crossvein in the wing of *Drosophila* sometimes fails to develop if the fly is subjected to heat shock as a pupa. By selecting and propagating flies that developed a crossveinless condition in response to heat shock, Waddington developed a population in which most individuals were crossveinless when treated with heat. But after further selection, a considerable portion of the population was crossveinless even without heat shock, and the crossveinless condition was heritable. *A character that initially developed in response to the environment had become genetically determined*, a phenomenon that Waddington called **genetic assimilation**.

This result is reminiscent of the discredited theory of inheritance of acquired characteristics that is often associated with Lamarck (see Chapter 3). However, it has a simple Mendelian, non-Lamarckian interpretation. Genotypes of flies differ in their susceptibility to the influence of the environment (heat shock)—i.e., they differ in the degree of canalization, so that some are more easily deflected into an aberrant developmental pattern. Selection for this pattern, Waddington suggested, favors alleles that canalize development into the newly favored pathway. As such alleles accumulate, less environmental stimulus is required to produce the new phenotype. The finding that genetic assimilation does not occur in inbred populations that lack genetic variation (Scharloo 1991) supports this interpretation.

There is some evidence that genetic assimilation could be important in evolution (Hall 1992). For example, high temperature during development reduces body size in *Drosophila*. This ontogenetic response is paralleled by genetic differences among geographic populations that experience different temperature regimes. The possibility that these differences evolved by genetic assimilation is given plausibility by an experiment in which laboratory populations of *Drosophila pseudoobscura* from the same initial stock were propagated at different temperatures for 6 years. They diverged so that even when reared at the same temperatures, the warm-adapted flies were smaller than the cold-adapted ones (Anderson 1966).

Are the developmental processes that determine the shape of the GEPM, including the thresholds and the canalization zones, inherent, fundamental properties of developmental mechanics, or can they be altered by natural selection? Waddington and the Russian biologist I. I. Schmalhausen (1949) independently proposed that canalization is an adaptive property, evolved by natural selection to enhance the likelihood that development will proceed normally. Experiments on the scutellar bristles of *Drosophila* (Rendel 1967; Scharloo 1987) imply that this is likely to be the case. Wild-

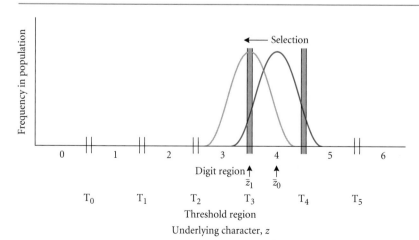

FIGURE 14.36 A model for the evolution of a threshold character, such as number of digits. A character *z*, such as a gene product, is polygenically determined and continuously distributed in the population. If the distribution lies between thresholds, i.e., in a "digit region" or zone of canalization, little or no phenotypic variation is expressed, and the response to selection is slight. If, perhaps due to selection for fewer digits, the mean shifts from $\bar{z}_0$ to $\bar{z}_1$, more phenotypic variation arises, and the character can rapidly evolve to a new state in response to selection. (After Lande 1978.)

type *Drosophila melanogaster* almost always have four precisely situated bristles on the scutellum, but as we noted in Chapter 9, the *scute* mutation destabilizes canalization; stocks homozygous for *scute* have a lower and variable bristle number, expressing the polygenic variation in an underlying morphogen (Figure 14.37). When Rendel imposed stabilizing selection for a mean of two bristles on a stock homozygous for *scute*, he succeeded in obtaining a stock with almost no phenotypic variation. Conversely, Scharloo selected for increased bristle number, and obtained variable stocks with a mean of eight bristles. He then imposed stabilizing selection for eight bristles. Flies with seven or nine bristles soon became less frequent. Both experimenters had selected for developmental processes that canalized bristle development at a lower or a higher number: they had altered the GEPM by selection.

We may therefore assume that developmental constraints on the expression of variation often are genetically controlled characteristics that have evolved and are potentially capable of further evolution. This conclusion is highly relevant to macroevolution, a subject in which developmental constraints are considered important (see Chapter 23). For example, no living species of vertebrate has more than five digits per limb. Polydactylous mutations have been described, but they usually have deleterious pleiotropic effects (as in polydactylous guinea pigs: see Chapter 3). Thus developmental processes may constrain the number of digits. However, the earliest known Devonian amphibians had a variable number of about seven or eight toes (see Chapter 7). Therefore, there is no inescapable rule of foot development that specifies a maximum of five toes. Perhaps developmental processes constrain toe number in most living tetrapod lineages, but if so, this constraint has evolved.

❖ The Epistatic Component of Genetic Variance

Except in our consideration of threshold traits and canalization, our discussion of quantitative genetics has explicitly or implicitly assumed that the relationship between genotype and phenotype is approximately additive. We now return briefly to issues of gene interaction, which we considered in a two-locus model earlier in this chapter.

Although genes undoubtedly interact, in a biochemical sense, in determining every character, the statistical measure of the amount of variation attributable to gene interaction (V_I) may nevertheless be small, depending on the character (Wade 1992). For example, major loci interact in determining the numbers of abdominal and sternopleural bristles in *Drosophila*, yet most of the variance in these traits appears to be additive (see Box 14.B), because V_A and V_I are determined by allele frequencies, not just by gene action. The epistatic variance can also depend on the environment; for example, V_I for several traits of tobacco was greater in both "good" and "bad" environments than in "average" environments (Jinks et al. 1973).

The epistatic variance in a quantitative character is considerably harder to measure than V_A, and is often estimat-

FIGURE 14.37 An interpretation of the response to selection on a canalized character, the number of scutellar bristles in *Drosophila*. (A) A mapping function (GEPM; see Figure 14.35) that relates variation in an underlying variable, such as a gene product, to the phenotype (bristle number). The notations 0/1 and so forth mark the thresholds at which the bristle number is changed by 1. The region between 3/4 and 4/5 is the "canalization zone." (B) The variable distribution of gene product in wild-type (++), homozygous *scute* (*scsc*), and heterozygous flies. Both ++ and +*sc* lie within the canalization zone, and have 4 bristles. However, *scsc* flies vary in bristle number from 0 to 3. (C) Selection for higher bristle number in an *scsc* stock increased the number (as shown by the shift in the *scsc* curve) due to selection on variation at other loci that is ordinarily not expressed in ++ or +*sc* flies. Replacing *sc* with + alleles in the selected stock resulted in an increase (to 5 or 6) in bristles in +*sc* and ++ flies.

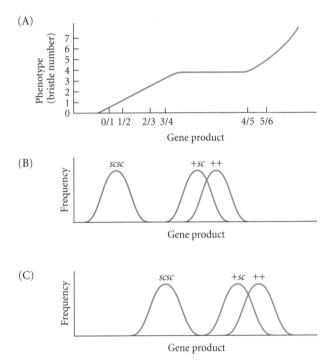

ed by rather complex statistical analyses of the phenotypes of crosses (F_1, F_2, and backcrosses) among inbred lines or separate populations. Few such estimates have been made for natural populations. It is widely believed, although supporting data are few (Houle 1996), that V_I is relatively greater for traits that are components of fitness, such as fecundity, than for traits, such as bristle number, that are thought to be less intimately, or at least less consistently, related to fitness. This pattern would conform to theoretical expectations (Falconer 1981; Wade 1992). Alleles that affect fitness additively—i.e., those that improve fitness regardless of the genetic background they occur in—will be directionally selected toward fixation, and so will contribute little genetic variance at equilibrium. Much of the genetic variance for fitness components will instead be contributed by the residue of alleles that increase fitness when in company with certain alleles at other loci, but decrease it when with certain other alleles—i.e., those that act epistatically.

If epistasis is an important feature of genetic variation in natural populations, then theory predicts several important consequences (Wade 1992; Whitlock et al. 1995). First, the genetic variance contributed by polymorphic loci may be more additive or more nonadditive depending on the allele frequencies. That is, *epistatic variance* (which does not yield response to natural selection) *may be converted into additive variance* (which does) if allele frequencies change (see Box 14.C). Genetic drift, for example, could change allele frequencies and thereby engender additive genetic variance that responds to selection. Thus *a character that does not evolve in a large population may evolve if the population is temporarily reduced in size, or is subdivided into small populations* (Goodnight 1988). Second, due to gene interaction, an allele may be favored in some populations, but not in others, if the genetic background with which it interacts varies from one population to another. Thus, as noted above, *the genetic basis of response to selection may differ among populations*, even if all are exposed to the same ecological regime of selection. These conclusions are among the critical assumptions of Wright's shifting balance theory, and they are highly relevant to some contemporary arguments about the mechanism of speciation (see Chapter 16).

Only a few experiments have been done to evaluate these hypotheses. Michael Wade and his colleagues conducted a series of experiments with laboratory populations of the flour beetle *Tribolium castaneum* to investigate genetic differentiation in fitness, which they measured by the production of offspring from one generation to another. The initial populations displayed very little additive genetic variance for fitness, yet small subpopulations taken from the initial stock rapidly diverged in fitness, at a rate consistent with the hypothesis that allele frequency changes initiated by genetic drift converted epistatic variance to additive genetic variance, enabling evolution by natural selection to occur (Wade 1992).

The same principle may explain the intriguing results of experiments performed by Edwin Bryant, Lisa Meffert, and their colleagues (Bryant et al. 1986; Bryant and Meffert 1988, 1990). From a large collection of wild houseflies (*Musca domestica*), they established a control population maintained at 1000 flies as well as experimental populations that were founded with 1 pair, 8 pairs, or 16 pairs. These bottlenecked populations rapidly grew to a population size of 1000. After six generations, the investigators, using correlations between offspring and parents, estimated the additive genetic variance for eight morphometric traits, such as wing length and head width, in each population. For purely additively inherited traits, one would expect V_A to be less, the smaller the bottleneck in population size, due to loss of genetic variation by genetic drift. For most of the characters, however, V_A proved to be highest in the populations that had been founded with 4 pairs—even higher than in the control population (Figure 14.38). In the bottlenecked populations, moreover, the correlations among the charac-

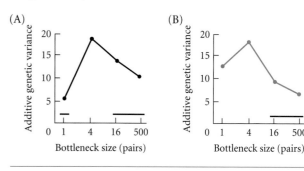

FIGURE 14.38 The additive genetic variance for two measurements [wing length (A) and scutellum length (B)] in experimental populations of houseflies passed through bottlenecks of several sizes and a control population that was maintained at a large size (500). The increase of V_A in the bottlenecked lines reflects conversion of nonadditive to additive genetic variance due to changes in allele frequencies caused by genetic drift. The horizontal lines indicate treatments that did not differ significantly in V_A. (After Bryant et al. 1986.)

ters had changed, so that the populations had diverged in body shape. Thus the genetic correlations, which might be considered constraints on the evolution of form, had been alleviated. The authors concluded that epistatic variance had been converted into additive variance in the bottlenecked lines, and that changes in the frequencies of interacting alleles had altered the genetic relationships among characteristics as well.

The role of epistasis in divergence among natural populations has been studied by Jeffrey Hard, William Bradshaw, and Christine Holzapfel (1992, 1993) in the pitcher-plant mosquito, *Wyeomyia smithii*. Larvae of this mosquito live only in the water-filled tubular leaves of the pitcher plant, an insectivorous plant that grows in bogs. (The larvae are resistant to the plant's enzymes, which digest other insects that drown in the water.) This mosquito is widely distributed in North America, and is believed to have progressively colonized the continent from south to north. As in most temperate-zone insects, winter diapause (a state of low metabolism) is triggered by a certain day length (photoperiod). The critical photoperiod is longer for northern than for southern populations. Northern populations have a single, diapausing, generation per year, whereas southern populations have several generations, only one of which diapauses.

Hard et al. measured the additive genetic variance for critical photoperiod within each of several populations from Florida north to southern Canada. They also crossed Florida stocks with more northern ones, and by comparing the parental populations, F_1, F_2, and backcrosses, estimated the degree to which the difference between populations in mean critical photoperiod was due to additive versus epistatic effects. All crosses between populations indicated that both additive and epistatic effects contributed to the differences in critical photoperiod. Thus interactions among genes that affect photoperiod response have contributed to population divergence.

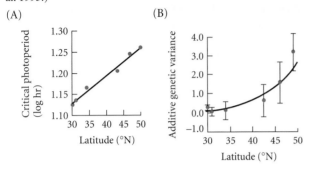

FIGURE 14.39 Variation in the critical photoperiod for diapause in the pitcher-plant mosquito. (A) The mean critical photoperiod varies clinally with latitude. (B) The additive genetic variance for the trait also increases with latitude. (After Hard et al. 1993.)

Hard et al. also found that the additive genetic variance for this trait within populations increased consistently from south to north (Figure 14.39A). Stabilizing selection on

photoperiod response, which should decrease V_A, should be more, not less, intense in the north, where every generation must enter diapause. Thus the trend in V_A does not parallel the strength of selection—but it does parallel the direction of colonization. The investigators suggested that this mosquito has progressively colonized more northern areas in a series of founder events, representing bottlenecks in population size. As new populations were founded, the epistatic component of genetic variation was reorganized, increasing the level of additive variation. The northern populations may have been so recently established that stabilizing selection has not yet reduced the additive variation to its ultimate equilibrium level.

These studies are among the few that have addressed the role of epistasis in evolution. More research will be needed before its importance can be assessed. Evaluating its importance will be necessary before some of the liveliest controversies in evolutionary biology, such as the importance of interdemic selection, the validity of the shifting balance theory, and the mechanisms of speciation, can be resolved.

Summary

1. Insight into the joint evolution of two characters can be obtained a model in which each trait is controlled by one locus. The same model describes the evolution of a single trait affected by two loci (a polygenic trait). When fitness depends on two or more loci, the dynamics of evolution depend on the fitness of each genotype and on the rate of recombination between loci. Linkage disequilibrium can enhance or retard the rate of evolution under natural selection.

2. If there is epistasis for fitness, there may exist multiple genetic equilibria, or gene frequency combinations toward which selection can move a population (multiple adaptive peaks on the adaptive landscape). Thus even in identical environments, different populations may evolve different genetic compositions. In some cases this results in different combinations of characters (alternative adaptations). In other cases, the phenotypes are similar, but their genetic basis differs among populations.

3. Wright's shifting balance theory holds that when there are multiple genetic equilibria (when epistasis yields multiple peaks), subdivided populations can diverge by genetic drift, and some may evolve by a combination of genetic drift and natural selection from an inferior to a superior genetic equilibrium. (This cannot occur in a single large population.) The flow of these superior gene combinations into less fit subpopulations can alter their gene frequencies so that they too evolve to the superior equilibrium. With repetition of this process, progressive adaptive evolution may be faster in subdivided populations than in large, randomly mating populations. Wright's theory is controversial. Many of its assumptions probably hold true, but there is no direct evidence.

4. When two (or more) loci contribute additively to variation in a phenotypic trait, and stabilizing selection favors

an intermediate phenotype, genetic variation is eroded, and the population becomes monomorphic for a genotype that has a mixture of "plus" and "minus" alleles at different loci. When two such populations, even with the same phenotype, interbreed, many of the F_2 and backcross progeny will have inappropriate mixtures of genes because of recombination, and may have lower fitness. The gene combinations in the parent populations are "coadapted," whereas those of the "hybrids" are not.

5. The two-locus theory can be extended to multiple genes (polygenes) that affect a single quantitative trait, but because the individual genes are difficult to detect, the evolution of such traits is described in statistical terms (means, variances, correlations). Special methods for estimating the number of polygenes indicate that at least a dozen or so, and often many more, loci affect variation in most traits, although just a few of these loci contribute much of the variation.

6. Variation (variance) in a phenotypic trait (V_P) may include genetic variance (V_G) and variance due to the environment (V_E), as well as other components. Genetic variance can include both additive genetic variance (V_A) due to the additive effects of alleles and nonadditive genetic variance due to dominance (V_D) and epistasis (V_I). Only the additive variance creates a correlation between parents and offspring (and it can be measured by this correlation). Thus only V_A enables response to directional selection—i.e., a change in the mean character state of one generation as a consequence of selection in the previous generation. The ratio V_A/V_P is the heritability (h^2) of a trait. Heritability is not fixed, but depends on allele frequencies and on the amount of phenotypic variation induced by environmental variation.

7. Most characters show substantial additive genetic variance in natural populations, and can therefore evolve

rapidly if selection pressures change. Many examples of rapid evolution, within a century or less, have been described. The causes of high levels of genetic variation (high V_A and h^2) in natural populations are uncertain, but input by mutation may balance losses due to selection and genetic drift.

8. Artificial selection experiments, which in some ways mimic natural selection, show that traits can often be made to evolve rapidly and dramatically (far beyond the initial range of variation); that the changes are due both to genetic variation in the original populations and to variation arising from new mutations that occur during the experiments; that a greater response can be obtained in larger populations; that characters other than the selected trait can also change (correlated responses) due to pleiotropy or linkage disequilibrium (hitchhiking); and that these correlated responses sometimes impair fitness so greatly that natural selection prevents further response to artificial selection. Similarly, antagonistic agents of natural selection, acting on one character or on genetically correlated traits, can stabilize a character in natural populations.

9. Linkage disequilibrium and especially pleiotropy cause genetic correlations among characters, which, together with correlations caused by environmental factors, give rise to phenotypic correlations. The evolution of a trait is governed both by selection on that trait and by selection on other traits with which it is genetically correlated. The effect of a genetic correlation depends on its strength and degree of permanence. Genetic correlations can enhance the rate of adaptation (if functionally interdependent features are genetically correlated in the right way), can cause one trait to evolve in a maladaptive direction (if selection on a correlated trait is strong enough), or may reduce the rate at which characters evolve toward their optimal states. Genetic correlations are known to change over time, but in some instances appear to have remained fairly constant for considerable spans of evolutionary time (perhaps hundreds or thousands of generations).

❖10. Genotype × environment interactions, which can contribute to phenotypic variance, exist when genotypes vary in the degree to which the phenotype is altered by the environment in which an individual develops. The nature of such G × E interactions affects the likelihood that an optimal norm of reaction—i.e., the capacity to develop the phenotype appropriate to the environment an individual experiences—can evolve. Some characters exhibit adaptive phenotypic plasticity, whereas selection in other cases favors constancy of phenotype despite differences in environment.

❖11. Some characteristics have evolved so as to be buffered against alteration by either environment or genetic variation. Such features may exhibit a nonlinear relationship between the phenotype and genetic variation in underlying characters (such as growth factors) that give rise to the phenotype. Threshold characters do not change in phenotype unless the underlying character is substantially changed (either by environmental influ-

ences or by genetic change). Experimental evidence indicates that some such characters have been selected to be canalized (channeled into a consistent pattern of development). Threshold characters, and canalized characters generally, are likely to evolve in episodes of rapid change interspersed with long periods of constancy.

❖12. Interactions among genes give rise to epistatic or interaction variance (V_I). Epistatic variance does not contribute to response to selection (since it does not cause a correlation between parents and offspring); thus a population may harbor nonselectable genetic variation in some traits. However, if allele frequencies change due to genetic drift or other causes, epistatic variance can be converted into additive variance, allowing evolution by natural selection. Both theory and experimental data suggest that in some cases, episodes of genetic drift (as in populations that experience a bottleneck in size), rather than reducing genetic variance, may thereby increase additive variance. Some differences in traits among natural populations are due to different sets of epistatically interacting genes.

Major References

Falconer, D. S. 1996. *Introduction to quantitative genetics.* Fourth edition. Longman, Essex, England. Perhaps the clearest introduction to the subject (and the most widely read by evolutionary biologists). This book treats methods of quantitative genetics and their application to animal and plant breeding, and provides some information on the evolution of quantitative traits. For the population genetic theory of multiple loci, see Hartl and Clark (1989).

Barton, N. H., and M. Turelli. 1989. Evolutionary quantitative genetics: How little do we know? *Annual Review of Genetics* 23: 337–370. Despite the skeptical title, this essay summarizes much of the available information on the principles of evolution of quantitative characters.

Roff, D. A. 1997. *Evolutionary quantitative genetics.* Chapman and Hall, New York. The most comprehensive reference on the evolution of quantitative traits, treating many of the subjects of this chapter in depth.

Loeschcke, V. (editor). 1987. *Genetic constraints on adaptive evolution.* Springer-Verlag, Berlin. A collection of essays on many of the subjects included in this chapter.

Bishop, J. A., and L. M. Cook (editors). 1981. *Genetic consequences of man-made change.* Academic Press, London. Includes detailed descriptions of many studies of rapid evolution in response to human alterations of the environment. Endler (1986) cites many other examples.

Problems and Discussion Topics

1. Let the array $A_i A_j$ (x, y, z) represent the fitness of genotypes with $A_i A_j$ at the A locus and $B_1 B_1$, $B_1 B_2$, and $B_2 B_2$, respectively, at the B locus. Assume Hardy-Weinberg equilibrium, and let p_A and p_B be the frequencies of al-

leles A_1 and B_1 respectively. The fitnesses are A_1A_1 (0.7, 0.8, 0.9), A_1A_2 (0.8, 0.9, 1.0), A_2A_2 (0.6, 0.7, 0.8). Calculate the mean fitness ($\bar{w}$) of a population in which (A) $p_A = 1.0$, $p_B = 0.3$ (Answer: 0.84); (B) $p_A = 0.5$, $p_B = 0.5$; (C) $p_A = 0.5$, $p_B = 0$. (D) What are the allele frequencies and the frequencies of the nine genotypes at equilibrium?

2. Using the same notation as in problem 1, suppose the fitnesses are A_1A_1 (0.6, 0.8, 1.0), A_1A_2 (0.7, 0.7, 0.7), A_2A_2 (0.8, 0.6, 0.4). Calculate the mean fitness of a population in which (A) $p_A = 0.5$, $p_B = 0.5$ (Answer = 0.7); (B) $p_A = 0.9$, $p_B = 0.1$; (C) $p_A = 0.1$, $p_B = 0.9$. (D) What allele frequencies and genotype frequencies might a population with these fitnesses attain at equilibrium? (E) Try sketching the adaptive surface in three dimensions, using the vertical coordinate for $\bar{w}$.

3. Using the same notation as in problem 1, let the fitnesses be A_1A_1 (0.60, 0.80, 1.0), A_1A_2 (0.55, 0.75, 0.95), A_2A_2 (0.50, 0.70, 0.90). Calculate the additive genetic variance for fitness when the allele frequencies are (A) $p_A = 0.5$, $p_B = 0.5$ (Answer: 0.02125); (B) $p_A = 0.2$, $p_B = 0.2$; (C) $p_A = 0.2$, $p_B = 0.8$. (D) If $V_E = 0.01$, what is the heritability (h^2_N) in each case?

4. Under truncation selection for increased body weight, what will be the response to selection (R), after one generation, for the following values of phenotypic variance (V_P), additive genetic variance (V_A), environmental variance (V_E), and selection differential (S)? (A) $V_P = 2.0$ grams^2, $V_A = 1.25$ g^2, $V_E = 0.75$ g^2, $S = 1.33$ g; (B) $V_P = 2.0$ g^2, $V_A = 0.95$ g^2, $V_E = 1.05$ g^2, $S = 1.33$ g; (C) $V_P = 2.0$ g^2, $V_A = 1.25$ g^2, $V_E = 0.75$ g^2, $S = 2.67$ g. (D) (optional) By how many phenotypic standard deviations will the mean weight increase in each case? (E) If the parameters remain the same for successive generations of selection, and the initial mean weight is 10 grams, what is the expected mean after two generations of selection in each case?

5. In "phase 3" of Wright's shifting balance theory, adaptive combinations of alleles are spread by gene flow from well-adapted to more poorly adapted populations. The consequent change in allele frequencies moves the recipient populations across the adaptive valley to a position at which it can move upslope toward the higher adaptive peak, so that the mean fit-

ness increases. Why might this be unlikely? (See Coyne et al. 1997.)

6. It is generally assumed that character differences among species have a genetic basis—i.e., that they have evolved—even when breeding experiments have not been done, and may not even be possible. For example, the differences among sandpiper species in the length and shape of the bill (see Figure 26 in Chapter 5) are assumed to have a genetic basis. What evidence or reasoning might justify this assumption? When might you suspect that it is not justified?

7. If most quantitative genetic variation within populations were maintained by a balance between the origin of new mutations and selection against them, then most mutations might be eliminated before environments could change and favor them. If the "residence times" of most mutations were short enough, the alleles that distinguish different populations or species would generally not be those found segregating within populations (Houle et al. 1996). Does this mean that it was wrong to state (in Chapter 9) that differences among populations and species arise from the transformation of genetic variation within populations into genetic variation among populations? How might one determine whether the alleles that contribute to among-population differences in the means of quantitative traits segregate within populations?

8. It has been suggested that genetic correlation between the expressions of a trait in the two sexes is responsible for some apparently nonadaptive traits, such as nipples in men and the muted presence in many female birds of the bright colors used by males in their displays (Lande 1980). Suggest ways of testing this hypothesis. What other traits might have evolved because they are genetically correlated with adaptive traits, rather than being adaptive themselves?

9. Debate the proposition that paucity of genetic variation and genetic correlations do not generally constrain the rate or direction of evolution.

10. "Gene-environmental factor/phenotype mapping functions" (GEPMs), such as those that underlie canalization and threshold characters, must have developmental causes. Suggest some developmental mechanisms that could give rise to such GEPMs.

CHAPTER 15

Species

In the six preceding chapters, we have considered the causes of evolution within populations and species. Although such evolutionary changes may be considerable over the course of sufficient time, they are often referred to as MICROEVOLUTION. In Chapters 23 through 25, we will analyze the principles of MACROEVOLUTION, that is, the origin and diversification of higher taxa. Many biologists consider the study of species and speciation to constitute the bridge between microevolution and macroevolution.

There are both methodological and conceptual reasons for this point of view. The processes that cause slight genetic or phenotypic changes within and among populations are often rapid enough for us to observe in natural or laboratory populations, whereas those that give rise to the differences among genera or still higher taxa usually are not. Speciation—the origin of new species—stands at the threshold: some steps toward speciation may occur fast enough for us to study directly, but the full history of the process is usually too prolonged for one generation, or even a few generations, of scientists to observe. Conversely, the process is often too *fast* to be fully documented in the fossil record, which is usually too coarse to trace events on a time scale of less than a few hundred thousand years. Thus the study of speciation, which is seldom amenable to either experimental or paleontological study, is based largely on inferences from living species.

Conceptually, speciation forms the bridge between the evolution of populations and the evolution of taxonomic diversity. The diversity of organisms is the consequence of *cladogenesis*, the branching or multiplication of lineages, each of which then evolves (by *anagenesis*, or evolution within species) along its own path. Each branching point in the great phylogenetic tree of life is a **speciation event**: the origin of two species from one. In speciation lies the origin of diversity.

Many events in the history of evolution are revealed to us only by virtue of speciation. If a single lineage evolves great changes, but does not branch, the record of all steps toward its present form is erased, unless they can be found in the fossil record. But if the lineage branches frequently, and if the ancestral or intermediate stages of a character are retained in some branches that survive to the present, the history of evolution of the feature may be represented, at least in part, among living species (see Figure 27 in Chapter 5). This fact is used routinely to infer phylogenetic relationships among living taxa that collectively display both ancestral and derived character states (see Chapter 5), and to trace the evolution of characteristics.

Some authors, including advocates of the hypothesis of punctuated equilibrium (see Chapters 6), have also suggested that speciation may facilitate the evolution of new morphological and other phenotypic characters—i.e., that a characteristic that would not evolve in a single, unbranched lineage may be able to do so if the lineage branches. We will discuss this controversial hypothesis in Chapters 16 and 24.

What Are Species?

Many definitions of "species" have been proposed (see Table 15.1). It is important to bear in mind that a definition is not true or false: scientific research cannot determine whether the word "carrot" should designate a certain kind of root or a certain kind of plant—or a kind of airplane, for that matter. The definition of a word is a convention. But a definition can be more or less useful, and it can be more or less successful in accurately characterizing a concept or an object of discussion. And if a conventional definition of a word has been established, one can apply it in error. I will be wrong if I call a Boeing 747 a carrot.

Table 15.1 The biological species concept and some recently proposed alternatives

BIOLOGICAL SPECIES CONCEPT A species is a group of individuals fully fertile inter se, but barred from interbreeding with other similar groups by its physiological properties (producing either incompatibility of parents, or sterility of the hybrids, or both). (Dobzhansky 1935)

Species are groups of actually or potentially interbreeding natural populations that are reproductively isolated from other such groups. (Mayr 1942)

EVOLUTIONARY SPECIES CONCEPT A species is a single lineage (an ancestral-descendant sequence) of populations or organisms that maintains its identity from other such lineages and which has its own evolutionary tendencies and historical fate. (Wiley 1978)

PHYLOGENETIC SPECIES CONCEPTS A phylogenetic species is an irreducible (basal) cluster of organisms that is diagnosably distinct from other such clusters, and within which there is a parental pattern of ancestry and descent. (Cracraft 1989)

A species is the smallest monophyletic group of common ancestry. (de Queiroz and Donoghue 1990)

RECOGNITION SPECIES CONCEPT A species is the most inclusive population of individual biparental organisms that share a common fertilization system. (Paterson 1985)

COHESION SPECIES CONCEPT A species is the most inclusive population of individuals having the potential for phenotypic cohesion through intrinsic cohesion mechanisms. (Templeton 1989)

ECOLOGICAL SPECIES CONCEPT A species is a lineage (or a closely related set of lineages) that occupies an adaptive zone minimally different from that of any other lineage in its range and which evolves separately from all lineages outside its range. (Van Valen 1976)

INTERNODAL SPECIES CONCEPT Individual organisms are conspecific by virtue of their common membership in a part of the genealogical network between two permanent splitting events or between a permanent split and an extinction event. (Kornet 1993)

Source: Coyne (1994).

A Brief History of Species Concepts

In biology, species (Latin for "kind") at first were taxonomic units: the named categories to which Linnaeus and other taxonomists of the eighteenth century assigned specimens, largely on the basis of appearance. A bird specimen was a member of the species *Corvus corone*, the carrion crow, if it looked like a carrion crow. A larger bird with shaggy throat feathers and a more pointed tail was a raven (*Corvus corax*) (Figure 15.1).

Linnaeus and other early taxonomists held what Ernst Mayr (1942, 1963) has called a "TYPOLOGICAL" or "ESSENTIALIST" NOTION OF SPECIES. Individuals were members of a given species if they sufficiently conformed to that "type," or ideal, in certain characters that were "essential" fixed properties—a concept descended from Plato's "ideas" (see Chapter 1).

Nevertheless, species are variable. Quite often, the variation among specimens falls into discrete groups, or clusters, as in the case of raven versus crow (Figure 15.2). These clusters, for early taxonomists, were species. Each had certain "essential" defining properties, such as the raven's pointed tail, that varied only slightly. But often gradations between such clusters exist. For instance, carrion crows, which are entirely black, differ from hooded crows (*Corvus cornix*), which have a gray back and belly, except in certain parts of Europe, where some specimens are intermediate, having only a bit of gray. The existence of such cases has led some authors to conclude that species are arbitrary constructs of the human mind imposed on a continuum of variation. Darwin, for one, took this position. In fact, despite the title of his greatest book, Darwin did not solve, and scarcely addressed, the problem of how two different species evolve from a common ancestor.

With the typological concept of species, based on morphological characteristics, there coexisted, even in the eighteenth century, *another criterion: common descent.* Offspring of the same parents were the same species, even if they differed considerably. This concept led to the idea that organisms are the same species (are CONSPECIFIC) if they can produce fer-

FIGURE 15.1 (A) Raven, *Corvus corax.* (B) Carrion crow, *Corvus corone.* (C) Hooded crow, *Corvus cornix.* Ravens are readily distinguished from crows by their larger size, heavier bill, shaggy throat feathers, more pointed tail (when viewed from above or below), and voice. Even though carrion and hooded crows are superficially more different than either is from the raven, hybrids showing various mixtures of their plumage patterns are found in central Europe. (From D. Goodwin 1986.)

(A) (B) (C)

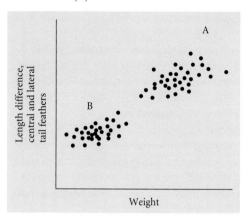

FIGURE 15.2 A hypothetical example of how different species usually form clusters of phenotypes, distinguishable by several characters. Each point represents an individual measured for two characters. These points might represent ravens (A) and carrion crows (B).

tile offspring, and are different species if they cannot. *This criterion of intersterility is no longer considered sufficient for defining species*, because some distinct species can produce fertile hybrids. However, it led to the important recognition that *morphologically different individuals may well be members of the same species*. The striped and banded forms of the California king snake (see Figure 4B in Chapter 9), known to be born to the same mother, represent a genetic polymorphism, not different species. A mutation that causes a fruit fly to have four wings rather than two is just that: a mutation, not a new species.

By the early twentieth century, taxonomists, having studied many groups of organisms, had accumulated a great deal of evidence on several points, which led to the most widely used modern concept of species. These points, which Ernst Mayr documented at length in his influential book *Systematics and the Origin of Species* (1942), included the following:

1. Many, perhaps most, characteristics vary among the members of a single population of interbreeding individuals. (We will consider below how to tell whether they interbreed, even without direct observation.) Sometimes this variation is continuous quantitative variation; sometimes discrete, large differences exist (as in the king snake polymorphism).

2. Populations in different geographic locations usually differ in the mean of one or more characteristics. There exists a spectrum from slight to great difference. Very often, intermediate forms are found where such populations meet, providing evidence that they interbreed (i.e., that there is gene flow between them).

3. What appears at first to be a single, morphologically uniform species often proves to include two or more populations that occupy the same area, but do not interbreed. In entomology, such morphologically cryptic

SIBLING SPECIES are often recognized by slight differences in the form of the genitalia; in other organisms, variations in ecology or behavior were the first indication of sibling species, and slight morphological differences were later found by which they could be distinguished. For example, the ornithologist Erwin Stresemann demonstrated that there were two, not one, species of treecreepers (*Certhia*) in central Europe. These small, brown, tree-climbing birds have very different songs, but differ only slightly in the color of the flanks and the length of the bill and claws (Figure 15.3). The discovery that the European mosquito *Anopheles* "*maculipennis*" is actually a cluster of six sibling species had great practical importance because some transmit human malaria and others do not. (See Box 15.A for the definition of sibling species and some other terms.)

The Biological Species Concept

The abundance of variation within and among populations, together with the adoption of Darwin's view that all characteristics can vary and evolve, led to abandonment of the typological species concept. *There are no immutable, "essential" characters.* Also abandoned was the concept that species could be defined by degree of morphological difference. (That does not mean, however, that morphology cannot be used to distinguish species, as we shall see.) Rather, the view became increasingly widespread among

FIGURE 15.3 Two sibling species that were distinguished only after careful study. The short-toed treecreeper (*Certhia brachydactyla*, left) and the treecreeper (*Certhia familiaris*, right) are morphologically almost identical. The short-toed treecreeper has slightly different wing markings, especially near the tip; a slight brown wash on the rear of the flank; and a slightly shorter claw on the hind toe. However, the two species' vocalizations are very different, and they differ ecologically. The treecreeper is found at higher altitudes and often inhabits denser forests. (After Jonsson 1992.)

BOX 15.A

Some Terms Encountered in the Literature on Species

Some of the following terms are frequently, and others infrequently, used in the literature on species. These definitions conform to usage by adherents to the biological species concept (e.g., Mayr 1963).

Geographic isolation: Reduction or prevention of gene flow between populations by an extrinsic barrier to movement, such as topographic features or unfavorable habitat.

Reproductive isolation: Reduction or prevention of gene flow between populations by genetically determined differences between them.

Allopatric populations: Populations occupying disjunct geographic areas.

Parapatric populations: Populations occupying adjacent geographic areas, meeting at the border.

Sympatric populations: Populations occupying the same geographic area, and capable of encountering each other.

Sibling species: Reproductively isolated species that are difficult to distinguish by morphological characteristics.

Sister species: Species that are thought, on the basis of phylogenetic analysis, to be each other's closest relatives, derived from an immediate common ancestor. Cf. *sister groups* in phylogenetic systematics: two clades that share a more immediate common ancestor than either shares with any other clade.

Subspecies: A taxonomic term for populations of a species that are distinguishable by one or more characteristics, and are given a subspecific name (e.g., the subspecies of the rat snake *Elaphe obsoleta*; see Figure 21 in Chapter 9). In zoology, subspecies have different (allopatric or parapatric) geographic distributions, so are equivalent to "geographic races;" in botany, they may be sympatric forms. No criteria specify how different populations should be to warrant designation as subspecies, so some systematists have argued that the practice of naming subspecies should be abandoned.

Race: A vague, meaningless term, sometimes equivalent to subspecies and sometimes to polymorphic genetic forms within a population.

Ecotype: Used mostly in botany to designate a phenotypic variant of a species that is associated with a particular type of habitat; may be designated a subspecies.

Variety: Vague term for a distinguishable phenotype of a species; used in plant systematics, sometimes in the sense of *ecotype*.

Polytypic species: A geographically variable species, usually divided into subspecies. (Most species are polytypic, whether or not subspecies have been named.) The German term *Rassenkreis* (*Rasse* = "race," *Kreis* = "circle," in the sense of "a circle of friends") is equivalent.

Hybrid: A loosely defined term for the offspring (F_1, F_2, and/or backcross) resulting from interbreeding between genetically different populations, usually so different that they are considered different species or semispecies (q.v.), but sometimes merely differing in one or more heritable characters.

Hybrid zone: A region where genetically distinct populations meet and interbreed to some extent, resulting in some individuals of mixed ancestry; usually the distinct populations are parapatric. "Primary" zones of hybridization are those thought to have developed in situ, whereas "secondary" hybrid zones are formed by interbreeding between formerly allopatric populations.

Introgression: The movement, or incorporation, of genes from one genetically distinct population (usually considered a species or semispecies) into another; "introgressive hybridization" is the process of interbreeding that results in introgression.

Semispecies: Usually, one of two or more parapatric, genetically differentiated groups of populations that are thought to be partially, but not fully, reproductively isolated; nearly, but not quite, different species.

Superspecies: Usually, the aggregate of a group of semispecies. Sometimes designates a group of closely related allopatric or nearly allopatric forms that are designated as different taxonomic species. The equivalent German term, sometimes encountered in English-language literature, is *Artenkreis* (*Art* = "species," *Kreis* = "circle," as in "a circle of friends").

Ring species: A chain of interbreeding "races" or subspecies, the ends of which overlap but do not interbreed; equals "circular overlap," and is not the same as a *Rassenkreis* (see *polytypic species*).

---

systematists (especially ornithologists and entomologists at first) and geneticists that a species is a *population* (or group of populations) of *variable* individuals that *exchange genes by interbreeding*, and do not interbreed with other such populations because of biological differences between them. (Physical isolation of populations due to separate geographic distributions does not mean that they are separate species.) This concept was promulgated among evolutionary geneticists especially by Theodosius Dobzhansky (1937) and Hermann Muller (1942), and among

systematists by the ornithologist Ernst Mayr (1942). Mayr became the most vocal champion of the **biological species concept** (**BSC**), which he defined in 1942: "*Species are groups of actually or potentially interbreeding populations, which are reproductively isolated from other such groups.*"

This definition applies to populations viewed in a narrow interval of time. To describe evolution over extended spans of time, the paleontologist George Gaylord Simpson proposed the concept of an EVOLUTIONARY SPECIES: a lineage that evolves separately from all others. A single, unbranching lineage that evolves changes in morphology may be divided into successive named "species," or **chronospecies**. However, they are not different biological species. When one lineage branches into two reproductively isolated lineages (biological species), it may be arbitrary whether one of them bears the same name as the common ancestral lineage or not (Figure 15.4B).

Mayr's biological species concept has been widely adopted in evolutionary discourse. Several aspects of the definition of the BSC require discussion.

First, Mayr's definition (like all definitions) has a restricted domain of application. It cannot be (and is not meant to be) applied to entirely asexual organisms, for instance. And as we have noted, it does not explicitly state whether or not we should call a contemporary population and its ancestors of 20 million years ago by the same name.

Second, the definition is phrased in terms of populations, not individual organisms. Two individuals might be incapable of interbreeding (two males, for instance, or the familiar example of a Great Dane and a Chihuahua among dogs), but yet be members of a single reproductive community, or gene pool.

Third, the criterion is *interbreeding*, or to be more exact, *gene exchange*, among populations in nature, not fertility or sterility. To be sure, many of the matings that can be induced between different species yield sterile offspring (such as the mule, a hybrid between horse and ass), or no offspring at all. However, there exist many, many instances in which different species, induced or forced to mate in a zoo, garden, or laboratory, yield partly or fully fertile offspring, yet seldom or never hybridize in their natural environment, because they simply do not usually mate with other species in nature (Figure 15.5).

Fourth, Mayr's definition refers to actually or *potentially* interbreeding populations. The word "potentially" causes many difficulties in applying the concept to real popula-

FIGURE 15.4 Evolutionary species and chronospecies. (A) Simpson's evolutionary species. Here the lineages a, b, and c, surrounded by shaded areas, are distinct lineages that persist and evolve through time. At time t_1 there are two, and at time t_2 three, biological species in Mayr's sense. (B) Chronospecies. Morphological differences in a single evolving lineage may lead to different names for the lineage in different geological intervals. If a and b are given different names, they are chronospecies. Chronospecies a ends in "pseudoextinction" by evolving into a differently named form, chronospecies b.

FIGURE 15.5 Despite the conspicuous differences between the mallard (A, *Anas platyrhynchos*) and the northern pintail (B, *Anas acuta*), these species can hybridize. However, they very seldom do so in nature. (After the National Geographic Society 1987.

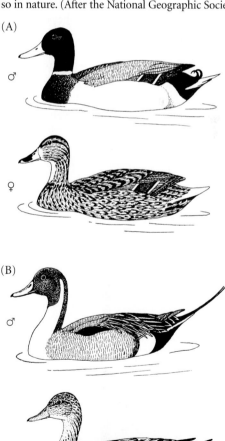

(A)

(B)

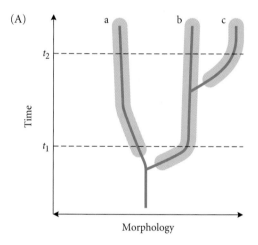

(A)

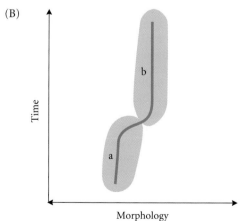

(B)

tions, but it has been felt necessary. The local populations of green frogs (*Rana clamitans*) in various ponds immediately north of New York City no doubt actually interbreed, as the result of movement of frogs among ponds. They are reproductively isolated from bullfrogs (*Rana catesbeiana*) in the same ponds by genetically encoded *biological differences* between the two species (including differences in the male calls to which females respond). The green frogs on Long Island, extending east from New York City, are not perceptibly different from, and undoubtedly *would* interbreed with, those green frogs north of the city, if they could swim through salt water or hop through a metropolis that is a formidable barrier even to most human tourists. The Long Island green frogs are isolated by external barriers, not by biological differences. Probably all biologists would agree that it would be absurd to designate every geographically isolated population as a separate species.

Alternative Definitions of Species

Many definitions of species beside the biological species concept have been proposed (see Table 15.1). Undoubtedly the most popular current contenders with the BSC are the various **phylogenetic species concepts** (PSC), which are gaining acceptance especially among systematists. These concepts emphasize not the present properties of organisms (such as gene exchange) or their hypothetical future (i.e., whether or not they would interbreed if they came into contact), but rather their phylogenetic history (see, e.g., de Queiroz and Donoghue 1988; McKitrick and Zink 1988; Cracraft 1989; Nixon and Wheeler 1990; Baum 1992). Joel Cracraft (1989) has provided one of several definitions of a phylogenetic species: "an irreducible (basal) cluster of organisms diagnosably different from other such clusters, and

within which there is a parental pattern of ancestry and descent." Any character unique to a population or set of populations would diagnose them as a species, even if they interbreed with other species. This definition would presumably apply to both sexual and asexual organisms.

A related concept, proposed by David Baum and Kerry Shaw (1995), is the "genealogical species concept," which draws on coalescence theory (see Chapter 11). It is intended to apply to sexually reproducing organisms, in which members of a population have a reticulate (netlike) pattern of descent, since different ancestor-descendant lineages join by interbreeding. When populations become isolated, the reticulating pattern of the ancestral population gives way to a divergent pattern of ancestry and descent. At first, gene copies in each population will often be genealogically more closely related to some gene copies in the other population than to some gene copies in their own population (see Figure 27 in Chapter 11). Due to genetic drift and natural selection, however, all gene copies will eventually be more closely related to each other within their own population than to those in the other population—that is, each population will acquire a monophyletic gene tree (Figure 15.6). Baum and Kerry proposed that a species be defined as a set of populations for which the gene trees at all loci examined are monophyletic. In some instances, reproductively isolated populations will not yet have achieved monophyletic gene trees, in which case entities distinguished as biological species will differ from those distinguished as genealogical species (Figure 15.7). Avise and Ball (1990) offered a similar perspective, but suggested that reproductive isolation should take precedence over gene genealogy when the two approaches to diagnosing species conflict.

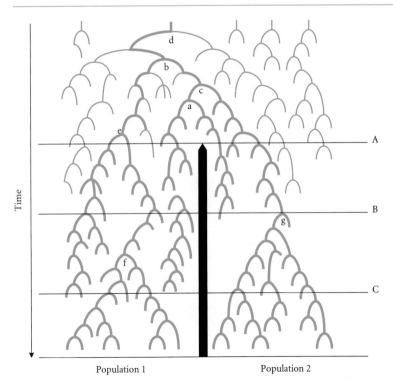

Population 1   Population 2

FIGURE 15.6 The lineage of haplotypes at a single locus in two populations derived from an ancestral population by isolation (black bar) at time A. Due to genetic drift or natural selection, all ancestral gene lineages except one are eventually lost, so that all the gene copies in the contemporary populations (at bottom) coalesce to ancestral gene copy b. Population 1 loses gene lineages more slowly than population 2, perhaps because it is larger. Between times A and B, both populations are polyphyletic for haplotype lineages, since some gene copies in each population, such as those derived from a, are more closely related to some gene copies in the other population than to some other gene copies (e.g., those derived from b) in their own population. Population 2 becomes genetically monophyletic sooner (at time B) than population 1, perhaps because of its smaller size. Between times B and C, population 1 is genetically paraphyletic, since its gene copies derived from ancestor c are more closely related to population 2's gene copies than they are to its own copies derived from ancestor e. At time C, population 1 also becomes monophyletic, with all gene copies derived from f. (After Avise and Ball 1990.)

FIGURE 15.7 Examples of ways in which different species might be diagnosed by different species concepts. (A) Populations 1, 2, and 3 interbreed where their ranges meet, but population 4 is sexually isolated (indicated by a black bar). All four populations are morphologically different, and their phylogenetic relationships are shown. Two biological species (BS) might be treated as four phylogenetic species (PS) under the phylogenetic species concept. (B) Population 3 is sexually isolated from populations 1 and 2, which are not sexually isolated from each other. The gene genealogy for one locus is shown by the thin solid lines within the shaded area, which shows the historical relationship among the populations. Because the gene genealogies of populations 2 and 3 are not monophyletic, these populations might be treated as a single species under the genealogical species concept (GS) of Baum and Shaw. If other genes showed different patterns of monophyly and polyphyly, the genealogical species might be difficult to delimit.

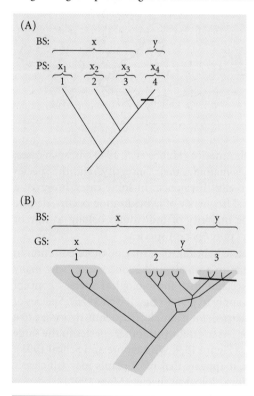

At this time, the biological species concept is more widely used than any alternative definition. Moreover, it plays a key role in evolutionary theory. For example, the entire theory of speciation uses the BSC. For these reasons, this book uses the biological species concept.

The Usefulness of the Biological Species Concept

As Mayr (1963) has stressed, the problem of *how species multiply*—of how the diversity of species has arisen—*could not be solved until it was understood that the evolution of reproductive discontinuity is the key event*. The simple evolution of changes in morphological (or other) features is not at all the same as the evolution of many species from one, so a theory of evolutionary changes in morphology within a population is insufficient to account for diversity.

BSC describes an evolutionary entity—*a unit of evolution*—*that is fundamental to discourse on the mechanisms of evolution*. No matter how different in morphology (or other features) two individuals or populations may be, they partake in, or could partake in, a common evolutionary history if they are actually or potentially members of the same gene pool. In theory at least, an advantageous mutation that arises in a green frog in Connecticut could someday be part of the genome of all green frogs in North America—but it will never be incorporated into the gene pool of the bullfrog. Looking into the past instead of the hypothetical future, any gene carried by all green frogs today, and which is unique to that species, arose in a single frog somewhere in North America, and has been transmitted by gene flow, then fixed by natural selection or genetic drift, in populations throughout the continent. A series of such events defines for the green frog a unique evolutionary history, in which the bullfrog did not partake, but all populations of green frogs did (or at least all populations that have left descendants). This is not to deny that *some* genetic changes are restricted to certain populations of a species; the point is that some can potentially encompass the species as a whole.

An exceedingly important feature of biological species is that *they can*, potentially, *become sympatric*. Geographically segregated populations undergo genetic divergence, and eventually differ at many loci, affecting many characteristics. If two such populations later encounter each other, and if they are *not* reproductively isolated, recombination will break down their separate constellations of characters. It will also break down the difference in any single polygenic trait, which will assume a unimodal rather than a bimodal distribution (see Chapter 14). Therefore two "types" of individuals, differing genetically in a multitude of traits, cannot ordinarily occupy the same region unless they are reproductively isolated.* The buildup of local diversity, the formation of a community of ecologically different organisms, depends largely on the existence of reproductive isolation.

Limitations of the Biological Species Concept

Domain and uses All concepts have limitations. The concept may have a limited *domain of application* (for example, "matter" is ambiguous on a subatomic scale). A concept may *inadequately describe borderline cases*. [For instance, the concept of an "individual organism" is very ambiguous in a grove of quaking aspen trees (*Populus tremuloides*), all of which may have grown, by vegetative propagation, from a single seed.] There may also be *practical limitations*: existing objects may really be instances of a concept, yet we may be barred, for reasons of technology or expense, from determining readily (or perhaps at all) whether or not they are.

---

*This statement does not hold for special cases in which either (1) loci governing different traits are in a state of extreme linkage disequilibrium, forming a "supergene," or are otherwise prevented from free recombination, as by an inversion; or (2) a single polymorphic locus acts as a "switch" gene, triggering differential expression of many other genes.

The domain of the BSC is restricted to sexual, outcrossing populations and to short intervals of time. Thus it does not apply to asexually reproducing bacteria or to populations widely separated in geological time. Bacteria, however, do have names, such as *Escherichia coli* and *Pseudomonas aeruginosa*—i.e., they are classified into "species." And morphologically different temporal segments of a single evolving lineage are given different "species" names by paleontologists, such as the mid-Pleistocene *Homo erectus* and its descendant *Homo sapiens* (Figure 4 in Chapter 26). These instances illustrate that *the word "species" has two overlapping but distinct meanings and uses in biology*. One meaning is *embodied in the BSC*: a species is a set of interbreeding populations that is reproductively isolated from other such groups. The other meaning of *species* is *a category in classification*, just like "genus" and "family." The organisms that bear binomial names (such as *Escherichia coli*) are taxa in the species category, just as *Escherichia* and *Homo* are taxa in the genus category. For many sexual organisms, such as frogs, species-as-biological-entities and species-as-units-of-classification correspond nicely: the green frog, *Rana clamitans*, and the bullfrog, *Rana catesbeiana*, are both "taxonomic species" and "biological species." With bacteria, however, species are taxa only. It is convenient to give *E. coli* and *P. aeruginosa* different names; but if it were convenient to do so, we could as readily give every genetically distinct clone of *E. coli* a different name. Advocates of the BSC hold that in contrast, a species of sexually reproducing organisms is not an arbitrary convenience, because sexual species are not just taxa, but real gene pools, objectively recognized by interbreeding and reproductive isolation. Failure to distinguish between the species as an evolutionary entity and the species as a unit of classification—between "species" as used in evolutionary theory and as used in some areas of taxonomy—has caused no end of confusion.

Restricted gene exchange Like almost everything else in biology, interbreeding versus reproductive isolation is not an either/or, all-or-none distinction. There exist graded levels of gene exchange among adjacent (parapatric) populations, and sometimes between sympatric, more or less distinct populations. The most frequently encountered situations are:

1. *Narrow hybrid zones*, at which genetically distinct populations meet and interbreed to a limited extent, but in which there exist partial barriers to gene exchange (Figure 15.8). The hybridizing entities, neither fully interbreeding nor fully reproductively isolated, are often called **semispecies**, and a collection of semispecies is a **superspecies** (Mayr 1963).

2. *Sympatric hybridization.* Partial, but not completely free, interbreeding sometimes occurs between populations that are rather broadly sympatric. Hybridization between some species is only very occasional: for instance, 5 F_1 hybrids were found in a collection of 2000 *Catostomus catostomus* and *C. commersoni* (fishes in the sucker family) from the Platte River of cen-

FIGURE 15.8 The narrow hybrid zone between the carrion crow (*Corvus corone*), distributed throughout western Europe and southern Britain, and the hooded crow (*Corvus cornix*), found in northern Scotland and eastern Europe, extending into Asia (see Figure 15.1). Some taxonomists classify these forms as subspecies of a single species. (After Meise 1928.)

tral North America (Hubbs et al. 1943). In such cases, the species maintain their "integrity," and the species concept is not threatened. In some cases, however, a substantial frequency of hybridization occurs, despite which the majority of individuals belong to one or the other "pure" parent species.

Sympatric hybridization may be more common in plants than in animals, and is one reason why many botanists claim that the biological species concept does not apply to plants (e.g., Raven 1976). Species of oaks (*Quercus*), for example, frequently hybridize to a greater or lesser extent with other species in the same subgenus (Figure 15.9; Whittemore and Schaal 1991). However, it appears that most plants may fall clearly into distinct biological species (Mayr 1992).

3. *Geographic variation in status.* Under some circumstances, genetically different populations may appear to be conspecific in certain geographic regions, but to be different species elsewhere. An example is Charles Sibley's (1954) analysis of two species of towhees (*Pipilo erythrophthalmus* and *Pipilo ocai*) that differ greatly in plumage (Figure 15.10). Sibley assigned a numerical score (a "hybrid index") of 24 to "pure" *erythrophthalmus* phenotypes, 0 to "pure" *ocai* phenotypes, and intermediate scores to birds with various mixtures of the characteristics of each. Throughout much of Mexico, the mean score in various populations was intermediate, providing evidence of extensive hybridization. In several localities in southern Mexico, however, both "pure" types coexisted with little or no evidence of hybridization. Thus, the towhees behave as if they are the same species in some locali-

FIGURE 15.9 An example of sympatric hybridization. The bear oak (*Quercus ilicifolia*, A) and the blackjack oak (*Q. marilandica*, E) have broadly overlapping ranges along the Atlantic coastal plain of the United States. Hybrids, showing variation in leaf shape and other features (B–D), are found in many places. (After Stebbins et al. 1947.)

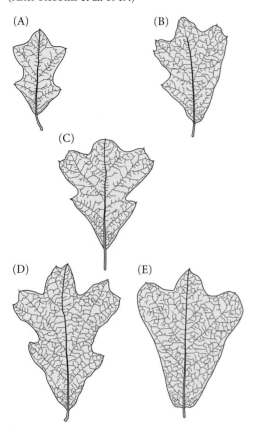

(A)   (B)

(C)

(D)   (E)

FIGURE 15.10 (A) The rufous-sided towhee (*Pipilo erythrophthalmus*, top), the collared towhee (*P. ocai*, second from top), and two of the many plumage patterns found in hybridizing populations. (B) Mountain areas of central Mexico. In some areas, the towhees coexist without hybridizing, while in others (solid color), they hybridize freely. (After Sibley 1950 and 1954.)

(A)

(B)

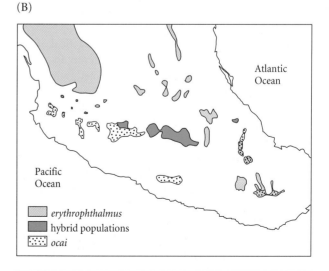

Atlantic Ocean

Pacific Ocean

☐ *erythrophthalmus*
■ hybrid populations
⸬ *ocai*

ties, but different species in others. Sibley suggested that human alteration of the birds' habitats had induced hybridization, a factor that has been implicated in many cases of sympatric hybridization.

Another example of geographic variation in status is the phenomenon of **circular overlap**, or "ring species," in which a chain of interbreeding populations loops around such that the terminal populations coexist without interbreeding. Probably the most fully studied example is the salamander *Ensatina eschscholtzii* (Wake et al. 1989; Moritz et al. 1992), which in California extends from north to south in the Coast Range and the Sierra Nevada, on either side of the Central Valley. A chain of seven subspecies that differ strikingly in coloration (Figure 15.11A) forms a ring around the valley. Intermediate phenotypes and electrophoretic studies indicate that the subspecies interbreed at the northern end of the valley and along both mountain ranges on its sides. Thus the chain of interbreeding races apparently constitutes one species. In southern California, however, the Sierran form *klauberi* meets the Coast Range form *eschscholtzii*, and they coexist with little or no hybridization. Here they appear to be distinct spe-

FIGURE 15.11 The "ring species" *Ensatina eschscholtzii*. (A) The distribution of the seven subspecies. The coastal forms *xanthoptica* and *oregonensis* resemble form A (*eschscholtzii*). Form *xanthoptica* has colonized the Sierras, where it hybridizes with C (*platensis*). In southernmost California, forms E (*klauberi*) and A (*eschscholtzii*) coexist without evident hybridization. The other adjacent pairs of subspecies all interbreed where they meet. (B) An estimate of the phylogeny of mitochondrial DNA haplotypes from several localities in the range of each subspecies. Both the coastal (*xanthoptica*, *eschscholtzii*) and Sierran (*platensis*, *croceater*, *klauberi*) forms appear to have been derived from *oregonensis*. This finding is consistent with the hypothesis that this species spread from the north southward in both mountain ranges and became genetically differentiated to the point that the southernmost forms do not interbreed. (After Stebbins 1954 and Moritz et al. 1992.)

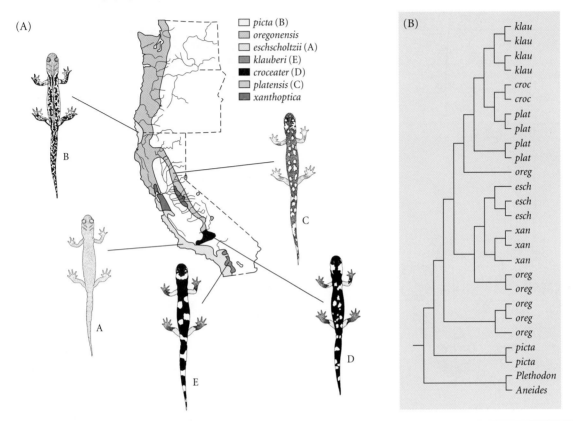

cies. Phylogenetic evidence (Figure 15.11B) suggests that this salamander slowly extended its range from the north down the two mountain chains, becoming increasingly genetically differentiated, so that the southernmost populations of each chain are genetically least compatible.

Allopatric populations The greatest practical, if not conceptual, limitation of the BSC surely lies in determining whether or not geographically segregated (allopatric) populations belong to the same species, for applying the BSC requires us to assess whether or not they would *potentially* interbreed if they should someday encounter each other. In some instances, this is an unrealistic possibility, and it is probably unimportant to determine whether the organisms are the same species or not. Chinese and American frogs, for instance, are unlikely ever to become sympatric. But in many other instances, range extension or colonization could well bring presently disjunct populations into contact. Humans have inadvertently or purposely introduced many species into new areas, some of which have hybridized with native populations (Abbott 1992), and many native species have extended their ranges

considerably over the course of decades. After the Pleistocene glaciations, disjunct populations of a great many species expanded their ranges, met one another, and in many instances now interbreed (see examples in Chapter 16). There is every reason to expect that future changes in climate will similarly affect some currently disjunct populations.

Unfortunately, there is no simple way to judge whether allopatric populations are conspecific or not. Occasionally, we can perform experiments to determine whether or not organisms from allopatric populations will interbreed and produce viable, fertile offspring, but we must bear in mind the possibility that under natural ecological conditions they would not do so. The most common criterion for *classifying* allopatric populations as different species (rather than as subspecies or races) has been the existence of morphological or other phenotypic differences as great as those usually displayed by other species in the same group (e.g., genus) that are sympatric. Avise and Ball (1990) have proposed that if the gene genealogies in allopatric populations are found to be monophyletic, they should be designated different species, but this would not necessarily indicate potential reproductive isolation.

Barriers to Gene Flow

According to the biological species concept, populations or groups of populations are different species if gene flow between them is mostly or entirely prevented by biological differences, which we will term **reproductive barriers** or *barriers to gene flow*. If populations freely exchange genes, they are conspecific, even if they differ strikingly in morphology; conversely, if they are reproductively isolated, they are different species, even if they are indistinguishable by most phenotypic characters. Thus *speciation—the origin of two species from a common ancestral species—consists of the evolution of biological barriers to gene flow*. As we noted above, mere physical isolation by topographic or other barriers does not define populations as different species, although it is considered to be instrumental in the formation of species.

The biological barriers to gene flow are frequently called **isolating mechanisms**, a term coined by Theodosius Dobzhansky (1937). Dobzhansky believed that many of the barriers to gene exchange evolved, by natural selection, specifically for the function of preventing gene exchange. The word "*mechanism*" suggests such a function. Mayr (1963) and others have expressed doubt that "isolating mechanisms" usually evolve in order to isolate, for reasons discussed in Chapter 16. The term "*reproductive barriers*" is more neutral, and is perhaps preferable for that reason.

Table 15.2 provides a classification of reproductive barriers. The most important distinction is between "prezygotic" and "postzygotic" barriers. **Prezygotic barriers** prevent (or reduce the likelihood of) the formation of hybrid zygotes (i.e., the fusion of egg and sperm nuclei). In many cases, these take the form of **positive assortative mating**, i.e., nonrandom mating between individuals of like phenotype or genotype. **Postzygotic barriers** are factors that reduce the fitness of hybrid zygotes (perhaps to zero) by reducing their survival or reproductive rates. Any of these barriers may be incomplete; for example, interspecific mating may occur at a low rate, or the hybrid offspring may have reduced fertility (be "partially sterile"). Often several different barriers reduce, or potentially can reduce, gene flow between species; for example, the hybrid offspring may be sterile, but the sterility may seldom be manifested in nature because the species seldom if ever mate with each other.

Prezygotic Barriers

Temporal isolation (1Ai in Table 15.2) Some species of insects seek mates for only a few hours of the day or night, and related species do so at different times, as in some fireflies (Lloyd 1966). More common is seasonal isolation. For example, two closely related field crickets (*Gryllus pennsylvanicus* and *G. veletis*) reach reproductive age in the fall and spring, respectively, in the northeastern United States (Alexander and Bigelow 1960; Harrison 1979).

Habitat or resource isolation (1Aii) Differences in the habitats occupied or resources used by different species commonly lower the opportunity for cross-mating and so contribute to reproductive isolation. In some cases, this may be the only barrier to gene flow. In Japan, for example, two sympatric species of herbivorous ladybird beetles, *Epilachna nipponica* and *E. yasutomii*, feed on thistles (*Cirsium*) and blue cohosh (*Caulophyllum*) respectively. Moreover, each species mates exclusively on its respective host plant, which is its microhabitat. No other barriers to gene exchange seem to exist, for hybridization experiments yielded no evidence for sexual isolation, gametic incompatibility, or hybrid inviability or sterility (Katakura et al. 1989; Katakura and Hosogai 1994). Habitat isolation is frequently an important barrier to gene flow among species of plants. For example, the wild irises *Iris fulva* and *I. hexagona* hybridize in Louisiana where their habitats (bayous and marshes respectively) come into contact and have been disturbed (Anderson 1949; Nason et al. 1992).

Ethological isolation (1B) In animals, ethological ("behavioral" or "sexual") isolation is surely the most common and important barrier to gene flow among sympatric species, which may frequently encounter each other, but simply do not mate. In animals, the courting individual (usually male) often does not display to members of other species; if he does, the courted individuals (usually female) do not indicate that they are receptive to mating. Ordinarily, then, communication between the sexes entails one or more signals from one sex (e.g., male) and a response by the other sex. [These may be the elements of what Paterson (1985) has called the **specific mate recognition system** of a species.] Both signal and response are ordinarily species-specific, suggesting that these two characteristics have diverged during speciation.

Ethological isolation has been studied in many groups of animals. In some cases, the relevant signals can be readily

---

Table 15.2 A classification of barriers to gene flow ("isolating mechanisms")

1. Prezygotic barriers ("premating" or prezygotic "isolating mechanisms")
   A. Potential mates (although sympatric) do not meet
      i. Temporal isolation (by season or time of day)
      ii. Habitat isolation
   B. Potential mates meet but do not mate ("ethological," "behavioral," or "sexual" isolation)
   C. Copulation occurs but no transfer of male gametes takes place (mechanical isolation)
   D. Gamete transfer occurs, but egg is not fertilized (gametic incompatibility)
2. Postzygotic barriers ("postmating" or postzygotic "isolating mechanisms")
   A. Zygote dies (zygotic mortality soon after fertilization)
   B. F_1 hybrid has reduced viability (hybrid inviability)
   C. F_1 hybrid viable, but has reduced fertility (hybrid sterility)
   D. Reduced viability or fertility in F_2 or backcross generations (F_2 breakdown and related phenomena)

Source: After Mayr (1963) and Templeton (1989).

identified in experiments. For example, female crickets and other orthopterans exposed to tape recordings of the songs of conspecific males and those of another species, played through different loudspeakers, almost always move toward the conspecific male's song (Alexander 1968). Marta Martínez Wells and Charles Henry (1992a,b) studied three morphologically indistinguishable species of green lacewings (*Chrysoperla*) in which a male and a female engage in a duet, initiated by the male, of low-frequency songs produced by vibrating the abdomen. Mating does not occur unless the female sings back to the male. Martínez Wells and Henry found that females sing much more often in response to tape recordings of their own species (Figure 15.12). When the researchers played songs synthesized on a microcomputer, they found that changes in the frequency ("pitch") of the entire song had little effect, but changes in the frequency modulation within the song or in the time interval between units of song reduced the females' responses.

Probably more species of animals use chemical communication in mating than any other modality. The compounds they use as signals are called **sex pheromones**. Sex pheromones have been studied most extensively in insects, especially moths (Löfstedt 1990; Cardé and Baker 1984). Male moths respond to sometimes astonishingly low concentrations of female sex pheromones (usually short-chain acetates). Related species differ in which chemical compounds constitute the pheromone, or simply in the proportions of the same compounds. Placed in wind tunnels, male moths will fly to and attempt to mate with a small sponge or other object impregnated with their own species' pheromone, but respond little or not at all to other species' pheromones or to experimental preparations of, say, a single component of their own species' blend. Moreover, electrophysiological recordings from single olfactory receptor

cells in the male antennae show that different receptors are typically "tuned" to one or another compound, and that the relative numbers of each receptor type correspond to the male's behavioral response. For example, two "strains" (perhaps sibling species) of the European corn borer moth (*Ostrinia nubilalis*) differ in the female pheromone: the E strain produces about a 97:3 ratio of two isomers and the Z strain produces approximately the reverse. The receptors of the males, and their attraction to the two different blends, show a parallel difference.

Visual signals, sometimes accompanied by acoustic or chemical signals, are important elements in the courtship displays of many insects, fishes, birds, and other animals. Among birds, the ducks, pheasants, hummingbirds, and birds of paradise (Figure 15.13) are particularly striking in the diversity of color patterns and ornaments that males display to females. In many of the Hawaiian *Drosophila* species, males have extraordinary modifications of the mouthparts, legs, or head shape that are used in display (Carson et al. 1970; Carson and Kaneshiro 1976). In many animals, such structures and displays are also used in territorial or other aggressive encounters among males (see Chapter 20).

Sexual isolation can be measured even if the mechanism of isolation is not known. For example, numerous investigators, over more than 60 years, have described sexual isolation among various species of *Drosophila*, among semispecies, and among populations within species (e.g., Ehrman 1965). The level of isolation is commonly measured by the frequency of between-species mating compared with that of within-species mating in mate-choice experiments in the laboratory.

In plants, the nearest equivalent of ethological or sexual isolation is pollination of different species by different pollinating animals (Grant 1971). Some, though by no means all, pollinators are quite specific, and transfer little pollen among flowers that differ in color, scent, or structure.

A fairly frequent barrier to gene flow in plants is the evolution of self-fertilization (Stebbins 1957; Barrett 1989). This mechanism fits only awkwardly into the classification of reproductive barriers in Table 15.2 because it engenders inbred lines that do not fit the biological species concept very well. Outcrossing species of many plants have given rise to populations, often morphologically distinct enough to have been named as separate species, in which self-pollination is the norm. For example, outcrossing species of water hyacinths (*Eichhornia*) are "tristylous," a condition resembling the heterostyly described for primroses in Chapter 9, except that there are three rather than two kinds of flowers. The stigmas of a given plant are set at one of three levels (long, medium, or short), with the stamens set at both of the other levels. Spencer Barrett (1989, 1995) has shown that populations consisting exclusively of a predominantly self-fertilizing "medium" morph have arisen in many localities, owing to a combination of genetic drift and natural selection (Figure 15.14). The flowers of self-fertilizing "species" of *Eichhornia* and other plants are smaller and less showy than those of outcrossing species, and differ from them in other ways as well.

FIGURE 15.12 Oscillographs of three song types of green lacewings. Although all these insects are referred to as *Chrysoperla plorabunda*, they appear to be distinct sibling species. (After Martínez Wells and Henry 1992a.)

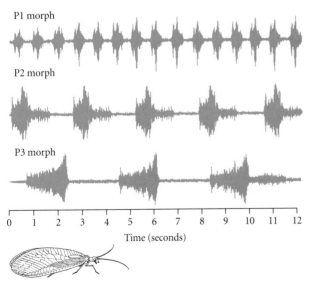

P1 morph

P2 morph

P3 morph

0 1 2 3 4 5 6 7 8 9 10 11 12
Time (seconds)

FIGURE 15.13 Secondary sexual characteristics vary greatly among male hummingbirds. Featured prominently in courtship displays, they undoubtedly contribute to reproductive isolation. Females of different species are much more similar to one another. (Left to right, above: *Sappho sparganura, Ocreatus underwoodii, Lophornis ornata*; below: *Stephanoxis lalandi, Popelairia popelairii, Topaza pella*). (After illustrations by A. B. Singer in Skutch 1973.)

Mechanical isolation (1C) Mechanical isolation, resulting from an imperfect structural fit between the sexes, is seldom a barrier to gene exchange among animals, even in insects, in which the genitalia are often highly complex and differ substantially in structure among related species (Shapiro and Porter 1989; Eberhard 1986). However, the structural fit between flowers and pollinators can affect exchange of pollen among some species of plants.

In a study of two Swedish orchids, for example, Anders Nilsson (1983) found that *Platanthera bifolia* is pollinated mostly by hawkmoths, and *P. chlorantha* mostly by smaller noctuid moths. These insects appear to respond to the different fragrant compounds produced by the two orchid species; moreover, *P. bifolia* has a larger spur, so that its nectar is more readily accessible to hawkmoths. The pollinia (pollen masses) of *P. bifolia*, situated close to each other, are placed on the base of the proboscis of hawkmoth pollinators, whereas those of *P. chlorantha*, more widely spaced, adhere to the eyes of its pollinators (Figure 15.15). Nilsson provided evidence that hybrid orchids, with intermediate flower morphology, have lower success in pollination and seed production, in part because their pollinia are placed at disadvantageous positions on the moths' faces.

Gametic incompatibility (1D) Incompatibility of the gametes of different species is an important reproductive barrier among some externally fertilizing animals, especially those marine invertebrates, such as many molluscs and echinoderms, that release eggs and sperm into the water, where they may easily encounter the gametes of other species. In abalones (large gastropods), the sperm carries a lysin protein that dissolves a hole in the egg's vitelline envelope—

FIGURE 15.14 A model of the evolution of self-fertilization in the water hyacinth *Eichhornia paniculata*. Outcrossing populations are polymorphic for style length (L, long; M, medium; S, short). Each morph has stamens at levels that facilitate pollen transport to the other two morphs (arrows). In small populations, the *S* allele is lost by genetic drift, yielding a dimorphic population. If pollinators are sparse, self-fertilization becomes advantageous, resulting in fixation of the M morph. Selection on other loci alters the lengths of the floral parts of the M morph, facilitating self-fertilization. (After Barrett 1989.)

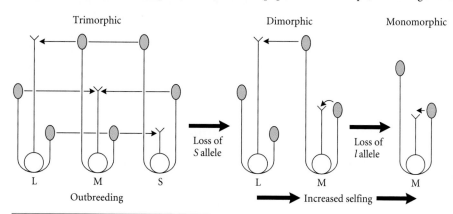

FIGURE 15.15 Pollinia of two species of orchids, adhering to the heads of moths, viewed from the right. (A) Pollinia of *Platanthera chlorantha* attached to the eyes of a noctuid moth. One pollinarium adheres to the right eye; the ends of two others, adhering to the left eye, are visible. (B) Eighteen pollinia of *Platanthera bifolia* attached to the base of the proboscis of a large hawkmoth. (After Nilsson 1983.)

(A) (B)

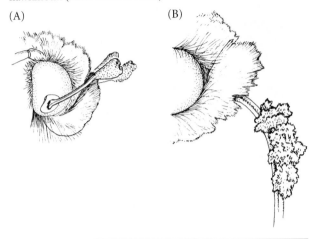

but only if the egg and sperm are of the same species. Vacquier and colleagues (Lee et al. 1995; Swanson and Vacquier 1995) have found that the DNA sequence of genes encoding both the lysin and another sperm protein display far more amino acid-changing nucleotide substitutions than silent substitutions among species, the reverse of what is usually found in comparisons of genes among closely related species.

In flowering plants, pollen from one species is frequently deposited on the stigmas of another, yet often few or no viable hybrid seeds are produced. In some cases, this is because the pollen tube of one species is arrested during its growth through the style of another (Heslop-Harrison 1982). Although the physiological causes of gametic incompatibility between plant species are poorly understood, they include both the factors that cause self-incompatibility within species (Chapters 11, 21) and other factors as well (de Nettancourt 1977, Taylor 1996).

Postzygotic Barriers

Hybrid inviability (2A, 2B) In many cases, though by no means all, hybrids between species have lower survival rates than nonhybrids. Quite often mortality occurs in the embryonic stages; for example, hybrids between the distantly related leopard frog (*Rana pipiens*) and wood frog (*Rana sylvatica*) do not survive beyond the early gastrula stage (Moore 1961). In some cases, the degree of hybrid inferiority depends on the environment. The higher incidence of hybrids in disturbed or intermediate habitats is sometimes a consequence of their higher survival there than in either of the parent species' typical habitats (Anderson 1949; Cruzan and Arnold 1993).

The genetic basis of hybrid inviability has been studied extensively (see below). However, very little is known about the malfunctions in development that cause mortality in hybrids.

Hybrid sterility (2C) The sterility (infertility) of hybrids ranges continuously from none to complete, depending on the hybrid. We noted in Chapter 9 that infertility, like other reproductive barriers, can occur in crosses among populations within species, as well as between species. Hybrid sterility can have several causes, of which the most common appear to be (*a*) the segregation of at least some *aneuploid gametes* during meiosis—i.e., gametes with an unbalanced complement of chromosomes or genes—due to structural differences between the chromosomes (see Chapter 10) or improper pairing of homologous chromosomes; (*b*) *differences between the genes* from the two parents, which interact disharmoniously.

Very frequently only one sex is sterile in the hybrid. This is also true (although less often: Wu and Davis 1993) of hybrid inviability. J. B. S. Haldane (1922), one of the founders of population genetics (see Chapter 2), first noted an important regularity now known as HALDANE'S RULE: when in the F$_1$ offspring of two species or populations, one sex is inviable or sterile, that sex is usually the heterogametic sex. The HETEROGAMETIC sex is the one with two different sex chromosomes, or with only one sex chromosome; the HOMOGAMETIC sex has two sex chromosomes of the same type. Thus in mammals, the female is XX (homogametic) and the male is XY (heterogametic); in some insects the male is XO (with O representing "no chromosome"). However, in birds and lepidopterans, the male is homogametic (XX) and the female heterogametic. Haldane's generalization holds generally, no matter which sex is heterogametic, and appears to be one of the most consistent generalizations that can be made about speciation (Coyne and Orr 1989b). As we shall see, the asymmetry embodied in this rule is found mostly in species that have been recently formed; hybrids between older species are likely to be inviable or infertile in both sexes. Thus the development of sterility or inviability between populations evolves more rapidly in the heterogametic than in the homogametic sex.

Hybrid infertility and/or inviability may also be manifested, in varying degrees, in F$_2$ offspring and backcrosses with the parent populations (2D in Table 15.2). This phenomenon has been observed in crosses both between species and between different geographic populations of the same species. For example, survival of F$_2$ larvae in a cross between *Drosophila pseudoobscura* from California and from Utah was lower than that in either "pure" population (see Table 2 in Chapter 14). This was interpreted to mean that recombination in the F$_1$ generated various combinations of alleles that were "disharmonious." In contrast, alleles at different loci within the same population have presumably been selected to form harmonious, i.e., *coadapted* combinations (Dobzhansky 1955; Wallace and Vetukhiv 1955).

Differences among Species

The various differences among closely related species include *those that are responsible for reproductive isolation* (i.e., those that are responsible for their status as separate species) and *those that are not.* Moreover, we can, at least in principle, distinguish differences that (1) existed between geographically segregated populations *before* reproductive barriers evolved; (2) *evolved during*, and may have caused, speciation; and (3) *evolved after* the evolution of the reproductive barriers.

In addition to features that affect reproductive isolation, related species display presumably nonadaptive ones, such as different synonymous codons and functionally equivalent amino acids at certain positions in various proteins, as well as adaptive differences. For example, they usually differ in ecological features, such as their use of resources or mode of protection from predators, as the examples in Chapters 4 and 9 illustrated. Some of these differences originate in geographic variation before speciation. In some cases, ecological differentiation may actually cause speciation; we will discuss this concept especially in relation to the controversial topic of sympatric speciation (see Chapter 16). Other ecological differences evolve after speciation, and may in some cases be due to selection arising from competition between the newly formed sister species (see Chapter 18).

In some instances, the ecological differences are too slight to prevent COMPETITIVE EXCLUSION, which is probably responsible for some instances of mutually exclusive geographic distributions (see Figure 19 in Chapter 8).

The Temporal Course of Divergence

Assuming that many or most species are formed by the genetic divergence of geographically segregated populations (see Chapter 16), both kinds of differences—those that contribute to reproductive isolation and those that do not—accumulate before, during, and after speciation.

An example has been provided by a study of allozyme differences in the *Drosophila willistoni* group by Francisco Ayala and his colleagues (1974a,b). They scored the frequencies of alleles (electromorphs) at 36 loci in populations that had been designated geographic populations of the same species, subspecies, semispecies (which show partial reproductive isolation in laboratory tests), sibling species (which are fully reproductively isolated, but morphologically almost indistinguishable), and nonsibling species (which are fully reproductively isolated, morphologically distinguishable, and presumed to have speciated longer ago). Allele frequencies were most similar among conspecific populations and least similar among nonsibling species (see Box D in Chapter 9). Fully reproductively isolated species have different alleles at some loci, share some of the same polymorphic alleles at other loci, and are almost fixed for the same allele at yet other loci. Overall genetic similarity declines, the more distantly related the populations are (Figure 15.16). This study illustrates, first, that species are not necessarily divergent at all their loci, and second, that divergence occurs both before and after speciation.

There is substantial evidence, from this and other studies, that "genetic distance," measured by the degree of difference in allozyme allele frequencies, increases fairly linearly with the time since gene flow between populations was curtailed (whether by geographic isolation or by biological reproductive barriers). Thus the genetic distance, *D* (see Box D in Chapter 9), can be used as an approximate

"molecular clock" to estimate the relative times of divergence of various pairs of populations or species. Jerry Coyne and Allen Orr (1989a) used such information to plot the *temporal pattern by which reproductive isolation evolves*. They compiled data on reproductive isolation from reports, published over the previous 60 years, of experiments on 119 combinations of populations or species of *Drosophila*. Some studies reported levels of postzygotic isolation (inviability or sterility of one or both sexes of the hybrid). Others reported levels of prezygotic (usually sexual) isolation, which is commonly measured by the frequency of between-species mating compared with that of within-species mating.

FIGURE 15.16 The percentage of electrophoretic loci that exhibit different levels of similarity in allele frequencies in contrasts among (A) geographic populations, (B) named subspecies, (C) semispecies, (D) sibling species, and (E) nonsibling species of the *Drosophila willistoni* complex. The genetic similarity index (Nei's *I*) ranges from zero to one. (After Ayala et al. 1974b.)

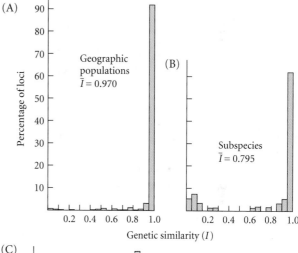

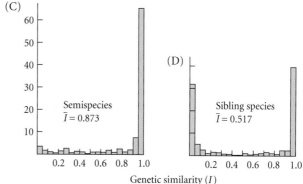

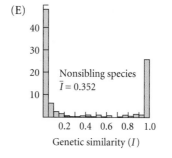

Among the 119 comparisons, many were not phylogenetically independent. If species A and B are more closely related to each other than to C, the degree of isolation between A and C is not independent of that between B and C, because some of the isolation evolved in the common ancestor of A and B. Thus a correction for phylogeny was required (see Chapter 5), which reduced the sample to 42 independent comparisons. Coyne and Orr expressed each form of isolation in each of these pairs by an index ranging from 0 (no measurable reproductive isolation) to 1 (complete isolation).

Coyne and Orr's analysis bears on many questions about speciation. Their results indicate that:

1. The strength of both prezygotic and postzygotic isolation increases gradually with the time since the separation of the populations (as measured by the allozyme genetic distance, D) (Figure 15.17). Thus, as was already known from many other studies, speciation is a gradual process.

2. The time required for full reproductive isolation to evolve is very variable, but on average, it is achieved when D is about 0.30–0.53, which (based on a molecular clock calibrated from the few fossils of *Drosophila*) corresponds to about 1.5–3.5 million years. However, a considerable number of full species appear to have evolved in less than 1 million years.

3. Prezygotic isolation evolves more rapidly than postzygotic isolation in sympatric pairs, but not in allopatric pairs. At low genetic distances ($D < 0.5$)—i.e., in recently diverged populations or species—sexual isolation is, overall, a stronger barrier to gene exchange than hybrid sterility or inviability. However, this effect was entirely due to sympatric taxa. The strength of postzygotic isolation did not differ between sympatric and allopatric taxa at any given value of D, but prezygotic isolation was stronger among sympatric than among allopatric pairs of taxa (Figure 15.18). This result has an important bearing on whether or not sexual isolation evolves because of natural selection to prevent hybridization, a controversy treated in the next chapter.

4. Postzygotic isolation in the early stages of speciation ($D < 0.5$) almost always consists of sterility or inviability of hybrid males only; female sterility or inviability appears only when taxa are older. Thus Haldane's rule holds, and can be extended: genetic divergence yielding postzygotic isolation evolves more rapidly in males than in females.

How Species Are Diagnosed

Biological species are *defined* as reproductively isolated populations, but *diagnosing* species—distinguishing them in practice—is seldom done by directly testing their

FIGURE 15.17 The level of (A) prezygotic (sexual) isolation and (B) postzygotic isolation among pairs of populations and species of *Drosophila*, plotted against genetic distance (D) at enzyme-encoding loci. Reproductive isolation increases gradually with time since divergence, as inferred from genetic distance. (After Coyne and Orr 1989a.)

FIGURE 15.18 Prezygotic isolation in *Drosophila*, plotted against genetic distance separately for (A) allopatric and (B) sympatric pairs. At low genetic distances—interpreted as recent divergence—sexual isolation is greater among sympatric than among allopatric forms. (After Coyne and Orr 1989a.)

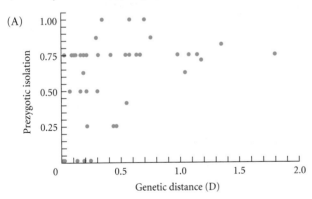

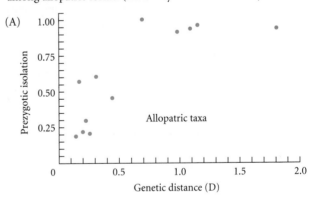

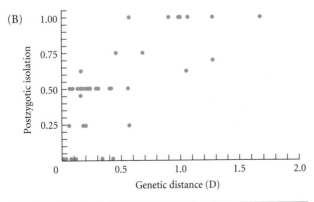

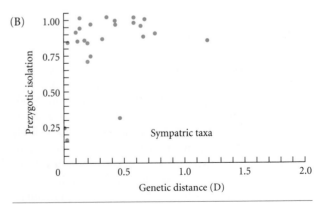

propensity to interbreed or their ability to produce fertile offspring. Species are not defined by their degree of phenotypic difference, yet phenotypic characters are the usual evidence used for diagnosing species. A great many, perhaps most, named taxonomic species are indeed biological species, even though they were diagnosed by differences in morphology (Figure 15.19). This is no contradiction at all, because morphological and other phenotypic characters, judiciously interpreted, can serve as markers for reproductive isolation or community among sympatric populations.

From elementary population genetics (see Chapter 9), we know that a locus in a single population with random mating should conform fairly closely to Hardy-Weinberg genotype frequencies. Two or more loci should be nearly at linkage equilibrium, unless very strong selection and/or suppression of recombination exists. If these loci affect a more or less additively inherited character, its variation will have a unimodal, more or less normal distribution. If they affect different characters, variation in these characters is likely not to be strongly correlated. Conversely, if a sample includes two (or more) reproductively isolated populations with different allele frequencies, then a single locus should show a deficiency of heterozygotes compared with the Hardy-Weinberg expectation; variation in a polygenic character may have a bi-

modal distribution; and variation in different, genetically independent characters may be strongly correlated.

If, therefore, in a sample of, say, mosquitoes from a single locality, we find that a particular feature shows two distinct states, but no intermediates, we may suspect that the sample contains two species, on the supposition that like most features, it is polygenically inherited. Still, this might be a simple single-locus polymorphism of a single species (e.g., two alleles, one dominant). If however, the two morphs differ in a number of other features, so that each feature is bimodally distributed, and the characters are correlated, our suspicion that these are two species grows stronger as the number of such distinguishing features increases. The evidence for two species mounts if we find further differences in such features as ecology, chromosomes, or, especially, markers such as allozymes that ordinarily show clear genetic segregation. Box 15.B provides a real example of such a diagnosis.

Morphological distinctiveness is a good, but not infallible, indicator of separate species. Sibling species may remain undetected unless allozymes or other molecular markers are studied. And in some groups of organisms, developmental "switches" may cause conspecific individuals to develop as discrete morphs that differ in several or many characters (West-Eberhard 1986; see Color Plate 5).

FIGURE 15.19 An example of species distinguished by morphological characters. These seven species of horned lizards (*Phrynosoma*) from western North America can each be distinguished by several differences in the number, size, and arrange-ment of horns and scales, as well as body size and proportions, color pattern, and habitat. (A, *P. cornutum*; B, *P. coronatum*; C, *P. platyrhinos*; D, *P. solare*; E, *P. modestum*; F, *P. m'calli*; G, *P. douglassi*.) (After Stebbins 1954.)

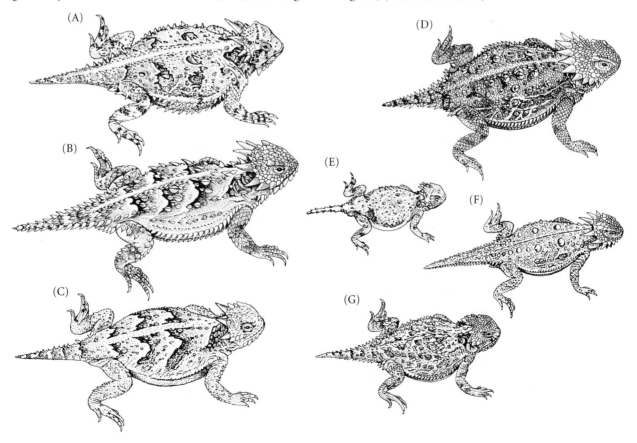

Diagnosis of a New Species

Species in the leaf beetle genus *Ophraella* each feed on, and are found on, at most a few related species of plants. *O. notulata*, for example, has been recorded only from two species of *Iva*, especially the marsh elder *Iva frutescens*, a shrub found in salt marshes of the eastern coast of the United States. The feature that most readily distinguishes this beetle from other species is four dark stripes that run almost the full length of each wing cover (elytron). The number and length of these stripes is different in other species of *Ophraella*.

Some specimens found in Florida closely resembled *O. notulata*, but were collected on ragweed, *Ambrosia artemisiifolia*, a widespread plant found in disturbed habitats. This host association suggested the possibility that these beetles were a different species. In a broader study of the genus, samples of beetles from both plants throughout Florida, including some from both plants in the same localities, were examined by enzyme electrophoresis. Consistent differences were found in allele frequencies between samples from *Iva* and *Ambrosia* at three loci (Futuyma 1991). In the most extreme case, one allele had an overall frequency of 0.968 in *Ambrosia*-derived samples, but was absent in *Iva*-derived samples, in which a different allele had a frequency of 0.989. No specimens, even from localities with both plants, had heterozygous allele profiles that would suggest hybridization. Thus these genetic markers were evidence of two reproductively isolated gene pools.

A careful examination of morphology then revealed average differences between *Ambrosia*- and *Iva*-associated beetles

in the shape of one of the mouthparts, the density of hairs on the border of the thorax, and the ratios of certain body dimensions (for example, *Ambrosia*-associated beetles tend to have a relatively broader thorax and shorter tibiae). None of these morphological differences distinguishes all individuals of one species from all individuals of the other, and no other morphological differences were discovered. Later studies showed that adults and newly hatched larvae strongly prefer their natural host plant (*Ambrosia* or *Iva*) when given a choice, and that the beetles mate preferentially with their own species. In laboratory crosses, viable eggs were obtained by mating female *Ambrosia* beetles with males from *Iva*, but not the reverse. Few of the hybrid larvae survived to adulthood, and none laid viable eggs. The *Ambrosia*-associated form, which I named *Ophraella slobodkini* in honor of the ecologist Lawrence Slobodkin, is clearly a different species (Figure 15.B1).

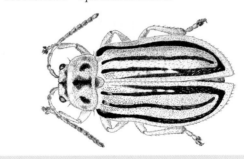

FIGURE 15.B1 *Ophraella slobodkini*

Hybridization

Hybridization occurs when offspring are produced by interbreeding between genetically different populations (Harrison 1990). Hybridization in nature interests evolutionary biologists because it provides them with opportunities to study interactions between selection and gene flow, factors that restrict gene exchange, and the genetics of species differences. In some cases, hybridization may be the source of new adaptations or even of new species (see Chapter 16).

Primary and Secondary Hybrid Zones

A **hybrid zone** is a region where genetically distinct populations meet and mate, resulting in at least some offspring of mixed ancestry (Harrison 1990, 1993). A character or locus that changes across a hybrid zone exhibits a *cline*, often quite steep, in frequency. For example, the all-black, smaller carrion crow (*Corvus corone*) and the gray-cowled, larger hooded crow (*Corvus cornix*) are broadly distributed in western and eastern Europe, respectively, but form a narrow hybrid zone within which all degrees of intermediate birds are found (see Figure 15.8). In cases such as these, the effects of hybridization are quite localized despite the interbreeding, and it is arbitrary whether the interbreeding

forms are called species or "semispecies." Such indeterminate cases are to be expected if speciation is a gradual process.

Hybrid zones are thought to be caused by two processes. PRIMARY HYBRID ZONES originate in situ as natural selection alters allele frequencies in a series of more or less continuously distributed populations. Thus the position of the zone is likely to correspond to a sharp change in one or more environmental factors. **Secondary hybrid zones** are formed when two formerly allopatric populations that have become genetically differentiated expand so that they meet and interbreed (**secondary contact**). On first consideration, we might expect primary zones to occur only at regions of pronounced ecological change and to affect only characters that are subjected to differences in natural selection stemming from the difference in environment (Figure 15.20A). Different loci might be differentially affected by the environmental gradient, so we would expect the clines in each locus to occupy different positions. In contrast, secondary zones might occur wherever expanding populations met, which need not be at an ecological border. Moreover, they should show coincident clines (clines located at the same place) for all features that differentiate the parent populations, including selectively neutral alleles such as, perhaps, some allozymes or other molecular markers (Figure 15.20B).

FIGURE 15.20 Expected patterns of change in the frequency (*p*) of alleles or characters across a hybrid zone originating by (A) divergent selection along an environmental gradient (a primary zone) and (B) secondary contact. "Stepped clines" are shown for four loci. In (A), the loci are assumed to be affected differently by the environmental gradient, so the clines have different geographic positions. Allelic variation at locus d (dotted line) is assumed to be nearly selectively neutral. (B) In a secondary hybrid zone, clines for all loci, including that for the nearly neutral alleles at locus d, are expected to have about the same location. The width of the cline depends on the strength of selection, relative to gene flow, at the locus or at closely linked loci. After a long enough time, the cline at locus d will come to resemble that in (A) due to gene flow. In some circumstances, the two causes of hybrid zones can result in indistinguishable patterns.

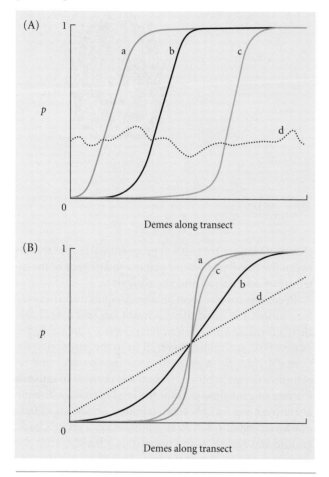

However, it is not all that easy to determine whether a hybrid zone is primary or secondary (Endler 1977). In the case of primary (in situ) differentiation, it is possible, at least in principle, for alleles (*B*, *b*) at a locus that is not affected by environmental selection to form a coincident cline with alleles at a locus (*A*, *a*) that is, if the fitness effects at the first locus depend on the allele frequencies at the second. In other words, coadapted gene systems (*AABB*, *aabb*) may develop on either side of the zone. Conversely, when two populations come into secondary contact, the hybrid zone may move over time to a region where either the population den-

sity is low or the environment changes abruptly, because the genotypes of the two populations may be differentially adapted to the two environments.

Nevertheless, many investigators take secondary contact to be the most likely explanation of hybrid zones when many characters show coincident clines, especially when these include allozymes or other molecular markers. For example, the fire-bellied toad (*Bombina bombina*), widely distributed in central and eastern Europe, and the yellow-bellied toad (*B. variegata*), distributed throughout western and southern Europe, meet in a long hybrid zone that is only about 6 kilometers wide (Szymura 1993). The two species differ in color pattern, in the presence or absence of male vocal sacs, in habitat, and in several other features, including allozyme loci that are diagnostic of each species in areas remote from the hybrid zone (Figure 15.21). Hybrids display various states and combinations of these characters. These species (or semispecies) are thought to have diverged during the Pliocene, and to have spread from different refuges in southeastern and southwestern Europe that they occupied during the Pleistocene glacial periods.

DNA sequences provide strong evidence for secondary contact when the gene genealogies are distinct and monophyletic in each of two broadly distributed forms that interbreed in a hybrid zone. For example, eastern and western populations of the sunfish *Lepomis punctatus* have monophyletic mitochondrial DNA genealogies (see Figure 28 in Chapter 11), the expected consequence of genetic drift in isolated populations.

Genetic Dynamics in a Hybrid Zone

Dispersal, selection, and linkage all affect the distribution of alleles in hybrid zones. Genetic data, interpreted by population genetic theory, can provide insights into these factors (Barton and Hewitt 1981a, 1985; Barton and Gale 1993).

If, in a zone of secondary contact, the F_1 and backcross hybrids are at least partly fertile, then as these individuals disperse, the alleles from each population are carried into the other. Thus if subscripts 1 and 2 designate alleles from the two populations, allele A_1 is carried into population 2 and A_2 into population 1. This *gene flow*, the rate of which depends on the dispersal distance, σ, *broadens the cline in gene frequency. Selection, however, narrows the cline. Selection may be due to a difference in environment*, such that A_1 is favored on one side and A_2 on the other side of the zone; *it may be due to epistatic interactions*, so that A_1 is disfavored if it is combined with the alleles of population 2 at other loci, and likewise for A_2; or the heterozygote A_1A_2 may have reduced fitness, as in the case of chromosome rearrangements that lower the fertility of heterozygotes. *If selection acts entirely by epistatic incompatibility or heterozygote disadvantage, the position of the hybrid zone is not fixed by the environment*, and the hybrid zone (called a TENSION ZONE) may wander. An allele that "leaks" across a tension zone (say, by the movement of an individual of population 1 into population 2) will not increase in frequency if it has low fit-

(A)

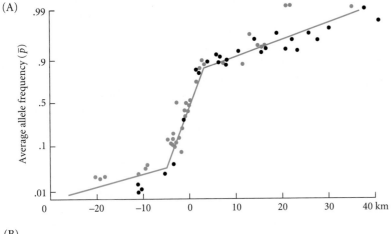

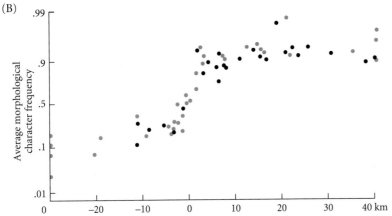

(B)

FIGURE 15.21 The hybrid zone between two species of *Bombina* toads. (A) Average allele frequencies at diagnostic enzyme loci. (B) Average frequencies of seven morphological characters. Colored and black circles represent two different 60-kilometer transects in Poland. The center of the cline (where the average allozyme allele frequency is $p = 0.5$) is denoted by zero on the *x*-axis. The frequencies are on a log scale. (After Szymura 1993.)

ness in hybrid offspring, being embedded in the wrong genetic background.

Suppose the two populations are differentiated at several or many loci, such as A and B. Then if only the parent types and F_1 hybrids occur at the contact zone, there will be complete linkage disequilibrium among these loci: the genotypes will be A_1B_1/A_1B_1, A_2B_2/A_2B_2 (the parental genotypes), and A_1B_1/A_2B_2 (the F_1 hybrid). If the F_1 hybrids are at all fertile and backcross, however, some recombinants, such as A_1B_2/A_1B_1, are formed. These may further backcross, so that with dispersal, an allele such as B_2 is introduced into population 1 farther away from the contact zone. If B_2 happens to be selectively advantageous regardless of environment or genetic background, it will ultimately replace B_1, so that the populations no longer differ at this locus. If the alleles B_1 and B_2 are selectively neutral, then there will at first be a steep cline, with the frequency of B_2 declining in population 1 with increasing distance from the contact zone, and B_1 likewise declining in population 2. Eventually, however, the differences in allele frequencies between the two populations will fade away. Thus the width of the cline for neutral alleles increases with time, at a rate that depends on gene flow and also, as we will shortly note, on the degree of linkage to a selected locus. If, however, the allele B_2 is selected against within population 1 (and similarly for B_1 in population 2), then a stable cline is established. The width of the cline, which may be thought of as the distance over which the allele frequency

changes from, say, 0.2 to 0.8, is proportional to the ratio between the dispersal distance and the square root of the intensity of selection against the allele ($\sigma/\sqrt{s}$).

Recombination reduces linkage disequilibrium between loci, so that the *associations between loci in a hybrid zone should break down. However, this process is opposed by two factors.* First, continued dispersal from the parent populations (A_1B_1/A_1B_1, A_2B_2/A_2B_2) into the contact zone replenishes the association. (In fact, it is possible to estimate the rate of gene flow, σ, from the level of linkage disequilibrium among loci.) Second, selection against an allele at one locus (against, say, A_1 in a population 2 genetic background and against A_2 in a population 1 background) also reduces the frequency of the associated alleles at other loci (such as B_1 or B_2) if the loci are linked. This is true even if the linked alleles are selectively neutral. *Thus a lower fitness of hybrids at one locus acts as a barrier to the flow of alleles—even neutral alleles—between populations. The strength of this barrier depends on how closely linked the loci are and on how strong selection is* (i.e., how low the fitness of hybrids is). Thus if A_1 and A_2 are subject to selection, but B_1 and B_2 are neutral, many B_1 copies will fail to cross the hybrid zone because they are associated with A_1 copies (and likewise for B_2). Some B_1 copies, however, will become dissociated from A_1, and may then spread by gene flow into population 2. The result is that alleles that are nearly neutral display a "stepped cline" that is steep in the middle of the hybrid zone, but shallow further from the zone (see Figure 15.21).

Several important conclusions follow from this theory:

1. The width of a cline of a locus or character in a hybrid zone varies from locus to locus, depending on the strength of selection on the locus or on linked loci.

2. Thus even if the fitness of hybrids is quite low, alleles at some loci may penetrate far beyond the hybrid zone that other loci display. Such penetration is called **introgression**.

3. Neutral alleles will show a steeper cline than expected if they are tightly linked to a locus that is subject to selection, such as the selection that arises from epistatic incompatibility.

An important consequence of the third point is that if we assume that a genetic marker such as an allozyme is selectively neutral, or nearly so, then we can infer that it is closely linked to a selected locus if it shows a steep cline in allele frequency. If we have a number of such markers, located on various chromosomes, then *the proportion of markers with steep clines indicates how many linked loci differ between the parent populations and are subject to selection.* This measurement can be taken as an estimate of how many loci contribute to the reduced fertility or viability of hybrids—i.e., to postzygotic reproductive isolation. Furthermore, the fitness of hybrids—the strength of postzygotic isolation—can be estimated from the steepness of such clines (Barton and Gale 1993).

For example, Jacek Szymura and Nicholas Barton (Szymura 1993; Barton and Gale 1993) have studied the hybrid zone of the *Bombina* toads described above. Several allozyme loci show steep clines in the frequencies of alleles that are diagnostic for each species in areas remote from the zone (see Figure 15.21). The clines at all these loci and in morphological characters are located in the same place. The clines in allele frequencies are "stepped;" i.e., they are steep in the middle of the zone, but have a long "tail" of introgression on either side of it.

In the middle of the zone, linkage disequilibrium, or association among the *bombina* alleles and among the *variegata* alleles at the several loci, is fairly strong. This finding implies a fairly high rate of dispersal of "pure" *bombina* and *variegata* toads into the zone, estimated at about 1 kilometer/generation based on the magnitude of linkage disequilibrium. The stepped form of the clines, as well as their narrowness, implies that hybrids have low fitness, which creates a barrier to gene flow. From the shape of the cline, the fitness of hybrids was estimated as 0.58—i.e., 58 percent of that of nonhybrids. Assuming that the selection against hybrids is the product of n heterozygous loci, each with fitness $1 - s$, then $w = 0.58 = (1 - s)^n$ if the loci were not in linkage disequilibrium. Because the width of the cline is related to the ratio of the dispersal rate (estimated as described above) to the selection coefficient, s, s can be calculated as 0.22. This coefficient applies to each chromosome region carrying an allozyme marker for which cline width was measured, so n is the number of such regions; n was estimated to be about 5. Each such region, carrying a marker, includes a number of loci that are in strong

linkage disequilibrium, so s (the selection coefficient for the region) is the sum of the coefficients of selection on some number of loci linked to the marker, multiplied by their linkage disequilibrium with the marker. Using a mathematical model too complex to describe here, the authors estimated the number of such loci as 11. Hence, at least 55 loci are estimated to cause the reduction in the fitness of hybrids.

The hybrid zone in *Bombina* appears to be a *tension zone, maintained by the low fitness of hybrids* due to genetic incompatibility of the species' genomes, rather than an effect of a steep environmental gradient. The genetic incompatibility is apparently due to many loci. The zone is likely to be *stable*, in the sense that *B. bombina* and *B. variegata* are unlikely to fuse into a single species; but it might well shift in position, and in that sense it is unstable.

Other Aspects of Hybrid Zones

Many hybrid zones involve concordant clines for allozymes and other characters, and therefore probably represent secondary contact (Hewitt 1989). Some of the differences between the hybridizing taxa may confer adaptation to different habitats. In some such cases, the hybrid zone is a complex "mosaic" in which relatively "pure" populations of each parent species are found in patches of the species-typical habitat and hybrids are most frequent where habitat distinctions break down. For example, the crickets *Gryllus firmus*, found in sandy soil, and *G. pennsylvanicus*, found in loamy soil, form a mosaic hybrid zone (Figure 15.22).

Barton and Hewitt (1985) estimated that at least 37 percent of the hybrid zones in Europe and North America have developed since the last glaciation through secondary contact of populations that were isolated during the Pleistocene. Quite often, the hybrid zones between different pairs of taxa are located in more or less the same region, suggesting a common history of isolation and expansion. In North America, for example, such "suture zones" (Remington 1968) include the Great Plains, where eastern and western pairs of birds such as orioles (*Icterus*), flickers (*Colaptes*), and grosbeaks (*Pheucticus*) hybridize to varying extents.

Many hybrid zones provide evidence of a breakdown of coadapted gene complexes and "genomic disruption." For instance, hybrids between the house mice *Mus musculus* and *Mus domesticus* in Europe had a higher incidence of intestinal parasites (tapeworms and nematodes) than nonhybrids (Sage et al. 1986), and some plant hybrids are more heavily infested by insects than the parent species (Strauss 1994). Hybrids often display greater developmental instability and apparently elevated rates of mutation and recombination (Fontdevila 1992). For reasons that are not well understood, both unique chromosome rearrangements and unique allozymes have been found in hybrid zones (Bradley et al. 1993). However, in some cases, hybrids have high fitness (Arnold 1992) and introgression can *sometimes provide advantageous genetic variation and contribute to adaptation* (Rieseberg and Wendel 1993).

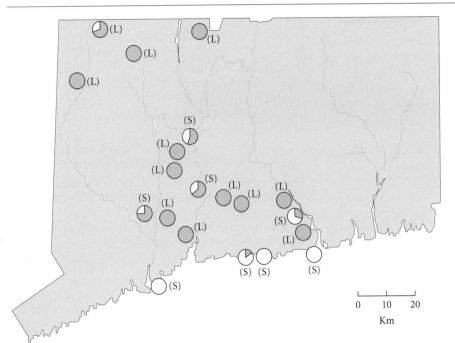

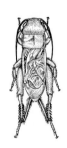

FIGURE 15.22 A mosaic hybrid zone between two species of field crickets in Connecticut. Each circle shows the frequencies of two mitochondrial DNA haplotypes typical of *Gryllus pennsylvanicus* (shaded) and *G. firmus* (white) respectively. L and S designate sites with predominantly loamy or sandy soil, the habitats preferred by *G. pennsylvanicus* and *G. firmus* respectively. (After Rand and Harrison 1989.)

The Fate of Hybrid Zones

As the last glaciers receded about 8000 years ago, many organisms rapidly colonized the newly exposed landscape. Thus, many postglacial hybrid zones, such as that between the European carrion and hooded crows, are undoubtedly several thousand years old (Mayr 1963). Old museum collections show that the positions of some hybrid zones, such as those between geographic races of aposematic *Heliconius* butterflies in tropical America, have been stable for at least 200 years (Turner 1971). Such apparent stability raises the question of whether or not hybridization will continue indefinitely.

Hybrid zones may have several fates:

1. The hybrid zone may persist indefinitely, with selection maintaining steep clines at some loci even while the clines in neutral alleles dissipate over the course of introgression. If the hybrid zone is a tension zone, it may move. It may become lodged in a region of low population density, or may eventually move to the far edge of the range of one of the semispecies, resulting in its extinction. If, however, some of the character differences are favored by different environments, the positions of the clines in these characters will be stable.
2. Natural selection may favor alleles that enhance prezygotic isolation, resulting ultimately in full reproductive isolation.
3. Alleles that improve the fitness of hybrids may increase in frequency. In the extreme, the postzygotic barrier to gene exchange would break down, and the semispecies would merge into one species.
4. In at least part of the hybrid zone, the hybrids themselves may evolve reproductive isolation from the parent forms, and become a third species.

The last three possibilities are discussed in Chapter 16.

The Genetic Basis of Species Differences

The genetic differences among closely related species appear to be the same in kind as the variation found within and among conspecific populations, but often are quantitatively greater.

Molecular Differences

For both allozymes and DNA sequences, the degree of difference among closely related species varies substantially, but generally increases with time since isolation. *The pace of molecular change may be quite different from the pace of morphological change.* For example, in Lake Malawi in Africa, more than 500 species of cichlid fishes have radiated in less than 2 million years, or possibly as little as 200,000 years. They have achieved great ecological and morphological diversity, yet the species surveyed so far display less than 2 percent sequence divergence in their mitochondrial genes (Meyer et al. 1990). In contrast, cichlids of the genus *Tropheus* in Lake Tanganyika, a much older lake, are morphologically almost indistinguishable except by coloration, but display as much as 14 percent sequence divergence (Sturmbauer and Meyer 1992).

Gene Trees and Species Trees

Recall from Chapter 11 that a phylogeny of DNA sequences—a gene tree—can be estimated by phylogenetic methods, just as a phylogeny of populations or species can be. The gene tree for variant sequences within and among species may not match the species tree. The reasons for this can be understood if we recall the COALESCENT THEORY (Chapter 11), which describes how a population that initially includes many lineages of DNA sequence variants eventually contains only one lineage due to the loss of the others by genetic drift. Thus two populations (incipient species) that become isolated from each other at first share many of the same gene lineages (i.e., they share polymorphisms: see Figure 15.6). Each population is at first polyphyletic with respect to gene lineages (i.e., its genes are derived from several ancestral genes). *Thus with respect to a specific gene or set of genes that do not recombine, some individuals in each population are genealogically less closely related to one another than they are to some individuals in the other population* (Figure 15.6, times A–B).

Gene lineages in each population are lost by genetic drift at a rate inversely proportional to the effective population size. At some point, one population (population 2 in Figure 15.6) may become monophyletic for a single gene lineage, while the other population (population 1), especially if it is larger, retains both this and other gene lineages (Figure 15.6, times B–C). At this point, the more genetically diverse population (1) will be paraphyletic with respect to this gene, and some gene copies sampled from population 1 will be more closely related (and may be more similar in sequence) to any of population 2's gene copies than they are to other copies from population 1. *Thus the phylogenetic relationships between genes from several individuals in each population will not correspond to the relationships among the individual organisms or the populations. Eventually, however, each population will become monophyletic for gene lineage* (Figure 15.6, time C). This process of **lineage sorting** requires about $4N_e$ generations for a diploid locus, or N_e generations for a maternally transmitted mitochondrial genome, to achieve monophyly in two populations derived from a common ancestor (Neigel and Avise 1986; Pamilo and Nei 1988). Thus the process is faster if effective population sizes (N_e) are small.

An example of incomplete lineage sorting is described by Paul Moran and Irv Kornfield (1993), who used restriction enzymes to study mitochondrial DNAs from 30 species in 11 genera of cichlid fishes from Lake Malawi. These species are diverse in morphology and ecology, including fishes that specialize on algae, molluscs, the scales and fins of other fish, and other foods. Among species for which multiple individuals were analyzed, seven had haplotypes belonging to both of the major gene lineages (Figure 15.23). This retention of ancestral polymorphisms, as well as the very low level of sequence divergence among species (1.8 percent on average), suggests that speciation in these cichlids has been quite recent. Some of the species are restricted to locations that were dry land until about 25,000, and perhaps as little as 2000, years ago, further supporting the hypothesis of recent evolutionary radiation.

Drosophila melanogaster and Its Relatives

Drosophila melanogaster has spread throughout the world, mostly by human transport, from Africa. Its close relatives include *D. simulans*, which has had a similar history; *D. mauritiana*, which is found only on the island of Mauritius in the Indian Ocean; and *D. sechellia*, which is restricted to the Seychelles Islands in the Indian Ocean and is ecologically the most specialized, since its larvae feed only on the fruits of a single species of tree. Because *D. melanogaster* is

the genetically best known eukaryotic organism, and because some of these species can be hybridized, they have been important subjects in the study of speciation.

Jody Hey and Richard Kliman (1993; Kliman and Hey 1993) sequenced three genes on the X chromosome in six individuals of each of these four species. They found the same ratio of silent (synonymous) to amino acid-altering nucleotide substitutions within and among species. This finding suggests that the variation is selectively neutral and that sequence evolution has been due to genetic drift rather than to selection (see Chapters 11 and 13). Hence sequence divergence among species may have been fairly constant (see Chapter 11).

Second, the level of intraspecific nucleotide polymorphism was used to estimate effective population size (see Chapters 11 and 22). The levels of polymorphism varied among species, and implied that the effective population sizes have the ratios 1.000:1.600:1.296:0.111 for *D. melanogaster: D. simulans: D. mauritiana: D. sechellia*. Thus N_e has been very low for the specialized island endemic *D. sechel-*

FIGURE 15.23 An estimate of the genealogy (gene tree) of mitochondrial DNA haplotypes in 32 species of cichlid fishes (plus two outgroup species, at bottom) from Lake Malawi. Species acronyms (3-letter labels) shown in boldface represent taxa that were polymorphic for haplotypes in the α and β lineages. Many of the species have originated so recently that they share these polymorphic gene lineages. (After Moran and Kornfield 1993.)

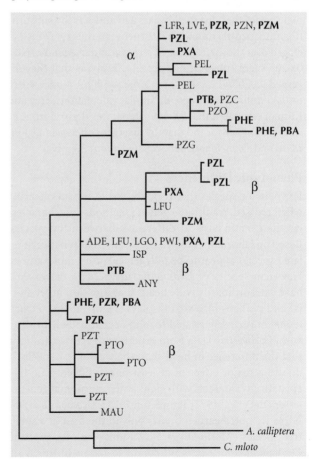

lia, but remarkably high for the island endemic *D. mauritiana*. (These results are relevant to theories of speciation; see Chapter 16.)

Third, all three genes implied the same phylogenetic relationships among the species (Figure 15.24). However, some gene copies of *D. simulans* are more closely related to genes of *D. sechellia* and of *D. mauritiana* than to other *D. simulans* genes, suggesting that both *D. sechellia* and *D. mauritiana* are descended from *D. simulans* stocks that colonized the two islands independently (since *D. sechellia* and *D. mauritiana* are not each other's closest relatives). Thus *D. simulans* is a "paraphyletic species"; its gene lineages include those that have given rise to the gene copies in the two island species. Now note in Figure 15.25 that the *D. simulans* gene copies are traced back to a "deeper" common ancestor than in any other species, and have accumulated more mutational differences from one another. This pattern implies that the gene lineages are old, that gene lineages have been lost by genetic drift at a low rate, and that the overall genetic composition of *D. simulans* has changed less than that of *D. sechellia* or *D. mauritiana* since the time of their divergence—all conclusions that are consistent with the larger effective population size.

A few Oligocene and Miocene fossils of the subgenus to which these species belong suggest that it is about 40 million years old. This estimate has been used to calibrate rates of molecular evolution in the group, yielding an estimated substitution rate of about 12×10^{-9} synonymous substitutions per year within a species. Recall from Chapter 11 that $D = (u)(2t)$, where D is the number of substitutions between two species that diverged t years ago and u is the substitution rate (neutral mutation rate) per year; thus $t = D/2u$. Averaged over the three genes, *D. sechellia* and *D. simulans* have 0.0205 silent substitutions per site, implying that they diverged $0.0205/(12 \times 10^{-9}) = (1.708 \times 10^6)/2 = 0.85 \times 10^6$, or 0.85 million years ago. By similar calculations, Hey and Kliman estimated that *D. mauritiana* was derived from *D. simulans* 0.77 Mya, and that the divergence between *D. simulans* and *D. melanogaster* occurred about 3.4 Mya.

Phenotypic Differences

For pairs of species that produce fertile hybrid offspring when crossed, traditional genetic methods can characterize the genetic basis of differences between the species. Based on laboratory crosses, many differences between related species appear to be polygenic. For example, among Hawaiian species of *Drosophila*, males of *D. heteroneura* have extraordinarily wide heads that function in territorial and perhaps in sexual displays (Figure 15.25). *D. heteroneura* is closely related to, and occasionally hybridizes with, *D. silvestris*. Data from experimental crosses indicate that the difference in head shape between these species is due to a major effect of at least one X-linked gene that epistatically interacts with eight to ten autosomal genes of lesser effect (Val 1977; Templeton 1977). *Drosophila sechellia*, as we noted earlier, breeds only in the fruit of a single tree species (*Morinda citrifolia*, Rubiaceae). Its ancestor, *Drosophila simulans*, is intolerant to toxins in this fruit:

adults and eggs soon die if confined with it. In F_1 and backcross hybrids, tolerance, measured by survival, behaved as a dominant trait, whereas female oviposition preference was inherited additively. Thus the behavioral

---

FIGURE 15.24 Gene trees for two loci in the *Drosophila melanogaster* species group, based on DNA sequences. Six individuals from each species were sampled: ME, *melanogaster*; MA, *mauritiana*; SE, *sechellia*; SI, *simulans*. The branch lengths are proportional to the numbers of nucleotide substitutions. (After Hey and Kliman 1993; Kliman and Hey 1993.)

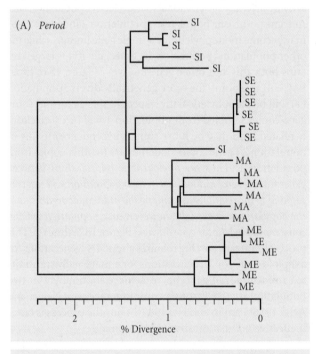

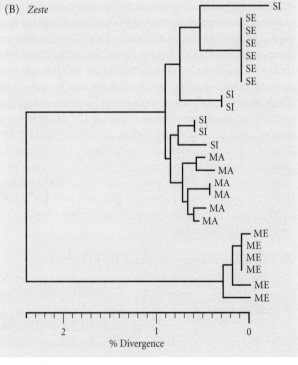

FIGURE 15.25 Males of the sister species *Drosophila silvestris* (left) and *Drosophila heteroneura* (right) in an aggressive display. Note the great difference in head shape, due to differences at eight to ten loci. The species also differ in the position in which they hold their wings during displays. (Photograph courtesy of K. Kaneshiro.)

and physiological adaptations to the plant are independently inherited, and each is controlled by at least one gene (R'Kha et al. 1991).

Although many of the characters that distinguish species are polygenically determined, there are also numerous cases, especially in plants (Gottlieb 1984), in which a difference is due to one or two genes. This is especially true of qualitative differences, such as the presence or absence of a structure, and less true of quantitative characters such as fruit shape. In some of these cases, it is likely that a single mutation causes a character to fail to develop, whereas the original evolution of the structure may have been based on several genes.

The Genetic Basis of Reproductive Barriers

In analyzing those characters that are instrumental in speciation—the "isolating mechanisms" or barriers to gene exchange—we may wish to know whether reproductive isolation is based on few or many genes, and how those genes act. Some answers to these questions were adumbrated in the preceding sections—e.g., in the discussions of gamete incompatibility and of the hybrid zone between *Bombina* toads.

Because some genetic differences that contribute to reproductive isolation may have accrued after speciation, *two species may differ in more incompatibility factors than were responsible for speciation.* Therefore, in order to determine how many and what kinds of genetic differences are *required* for speciation, we need to compare populations that have speciated very recently, or are still in the process of doing so. We shall consider examples of the genetic bases of prezygotic isolation and of three kinds of postzygotic isolation, based on nuclear gene differences, on structural differences in chromosomes, and on cytoplasmic incompatibility (Coyne 1992).

The Genetics of Postzygotic Isolation

Hybrid sterility in Drosophila

By far the most extensive information on the genetics of sterility and other reproductive barriers has been obtained for certain *Drosophila* species because of the availability of the genetic markers that are the sine qua non of any genetic analysis. Theodosius Dobzhansky (1937), one of the most influential figures of the Evolutionary Synthesis, pioneered the use of genetic markers to study hybrid sterility between the sibling species *Drosophila pseudoobscura* and *D. persimilis.* This method, effectively the same as QTL mapping (see Chapter 14), is still being used by modern researchers.

A simple example of this method is Jerry Coyne's (1984b) analysis of hybrid sterility in crosses between *D. simulans* and its island derivative *D. mauritiana* (see above), in which F_1 hybrid males are sterile, but hybrid females are fertile. Both species have two major autosomes (2 and 3), each with two arms, and X and Y sex chromosomes. The Y chromosome, in males, carries few genes. Coyne used a *D. simulans* stock that was homozygous for a recessive visible mutation on the X and on each of the autosomal arms; this stock is designated *f; nt pm; st e* (the semicolons separate chromosomes X, 2, and 3). Female *D. simulans* of this stock were crossed to *D. mauritiana* males (which carried no visible mutations). The fertile F_1 females were backcrossed to *f; nt pm; st e D. simulans* males. The resulting male progeny (with only one X) may display the *f* mutation or not, depending on whether they inherit the *simulans* or *mauritiana* X from their mothers. A male that lacks (for example) only the *pm* mutant phenotype is heterozygous for the right arm of chromosome 2 from *D. mauritiana*, is otherwise homozygous for *simulans* autosomes, and carries the *simulans* X (Figure 15.26). Thus the constitution of each chromosome arm—*simulans* or *mauritiana*—can be inferred from the markers. (In reality, because crossing over between *simulans* and *mauritiana* chromosomes occurs in the F_1 females, the species origin is known only for genes in the vicinity of each marker; a chromosome with the *simulans* marker is likely to include *mauritiana* genes some distance away.)

Coyne scored backcross males for the presence or absence of motile sperm in the testes, an index of fertility that is correlated with their ability to father offspring. Note that genotypes 1 and 2 in Figure 15.27 both have *simulans* autosomes, but genotype 1 has a *simulans* X and genotype 2 has a *mauritiana* X. The latter is completely sterile. Thus *at least one* X-linked gene differs between the species and contributes to sterility. There might be several such X-linked genes, but several X-linked markers would be required to determine this. (Later studies showed that there are at least five such X-linked genes: Coyne and Charlesworth 1986; Wu et al. 1993.) Thus this technique can detect no more gene differences between the species than there are genetic markers.

The X chromosome of *D. mauritiana* does not cause sterility in *D. mauritiana*, of course. Therefore the sterility of genotype 2 in Figure 15.28 must be due to an *interaction* between *mauritiana* allele(s) on the X and *simulans* al-

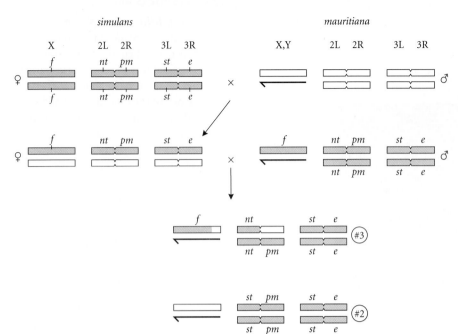

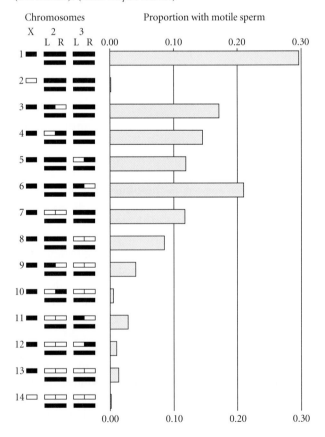

FIGURE 15.26 Crosses for determining the chromosome locations of genetic differences between species such as *Drosophila simulans* and *D. mauritiana*, which produce fertile female hybrids. The recessive genetic markers enable one to determine, for example, which male progeny of the F₁ × *simulans* backcross carry a *mauritiana* X chromosome (genotype 2) or right arm of chromosome 2 (genotype 3). 2L, 2R and 3L, 3R refer to the left and right arms of chromosomes 2 and 3. The arrow-shaped lines represent Y chromosomes. Two of the many possible backcross offspring are shown, with genotypes numbered as in Figure 15.27. Because of crossing over in the F₁ females, a chromosome bearing a *simulans* marker, such as the X in genotype 3, may still include parts of a *mauritiana* chromosome.

lele(s) on one or more autosomes. Thus sterility can be ascribed to *epistasis*, i.e., interaction between loci. *At least two loci must differ between the species to cause hybrid sterility* (Figure 15.28A).

Genotypes 3–6 in Figure 15.27 show that each substitution of a single autosome arm from *D. mauritiana* into the *D. simulans* genome reduces fertility compared with that of the "all-*simulans* control," genotype 1. Therefore at least four autosomal alleles must differ between the species; each such *mauritiana* allele interacts with the *D. simulans* genome to reduce fertility. Therefore, at least five genes in all—the most that can be detected by this experiment—contribute to sterility. *The most conspicuous effect is that of the X chromosome*, which causes complete sterility, overriding the effects of the autosomes.

The greater effect of the X chromosome, documented in almost all of the many studies of hybrid sterility in *Drosophila* is thought to be related to *Haldane's rule*. Most of the several explanations for both generalizations (Coyne and Orr 1989b; Wu and Davis 1993; Turelli and Orr 1995) elaborate on Herman Muller's (1940) suggestion that the heterogametic sex in the hybrid will be sterile or inviable, while the homogametic sex will not, if the X chromosome of one species carries *recessive* alleles that interact unfavorably with autosomal alleles of the other species. The deleterious effect of such a recessive allele will be masked in the homogametic hybrid (the female in *Drosophila*) by the dominant allele on the other X chromosome, whereas it is not masked in the heterogametic (male) hybrid. Autosomal recessive alleles that cause incompatibility will not be expressed in the F₁ hybrid, in which they are heterozygous. However, not all data support Muller's hypothesis, and the causes of Haldane's rule are still uncertain (Hollocher and Wu 1996; Wu et al. 1996).

FIGURE 15.27 The proportion of males with motile sperm in backcross hybrids with various combinations of chromosome arms from *Drosophila simulans* (solid arms) and *D. mauritiana* (open arms). All genotypes have a *D. simulans* Y chromosome (not shown). (After Coyne 1984b.)

In some instances, one or a few genes appear to have a large effect on hybrid sterility—or inviability, which generally shows similar genetic patterns (Orr 1989). For instance, sterility of the male progeny of the cross between female *D. mojavensis* and male *D. arizonensis* appears to be caused by an interaction between the *arizonensis* Y chromosome and a small region—perhaps a single gene—on one of the *mojavensis* autosomes (Zouros et al. 1988). Thus, *the number of allele substitutions required for speciation may be small*, at least in some cases.

In contrast, Chung-I Wu and his coworkers have proposed that sterility is often based on *epistatic interactions among many genes* (Wu and Palopoli 1994; Cabot et al. 1994; see also Naveira 1992). Wu's group has backcrossed short segments of the X chromosomes of *Drosophila mauritiana* and *D. sechellia* into a genome otherwise derived from *D. simulans*, and has mapped several genes affecting fertility by their association with various combinations of visible and DNA markers. The *D. sechellia* X chromosome bears at least two closely linked genes, neither of which reduces fertility alone, but which in combination cause male sterility (Figure 15.28B). The *D. mauritiana* X chromosome has at least two clusters, one of two genes and the other of three, that have similar effects. *These genes individually have little effect, but cause sterility in combination.* By extrapolation from these short segments of chromosomes, Wu and Palopoli (1994) suggested that among these closely related species, as many as 40 allele differences on the X chromosome and 100 in the genome as a whole might cause hybrid male sterility, due to epistatic interaction. Their evidence supports the convictions of earlier workers, such as S. C. Harland (1936) and Ernst Mayr (1963), who argued that species consist of distinct coadapted gene pools, or "interactive systems" of genes, that interact harmoniously within species, but interact disharmoniously if mixed together.

Chromosome differences and postzygotic isolation

Chromosome differences among species include differences in the number of chromosome sets (polyploidy) as well as alterations of chromosome structure that may or may not change the number of chromosomes. Speciation by polyploidy is treated in the next chapter. We focus here on structural rearrangements and their possible effects on gene exchange. Some authors ascribe a major role in speciation to chromosome rearrangements (White 1978; King 1993), while others are skeptical (e.g., Sites and Moritz 1987; Coyne et al. 1993). *The critical question is whether heterozygosity for chromosome rearrangements causes reduced fertility (postzygotic isolation) in hybrids.* Fertility may be reduced if heterozygotes produce a high proportion of aneuploid gametes (Chapter 10). The major relevant classes of chromosome rearrangements are inversions, translocations, and fusions or fissions of chromosomes (White 1978 and King 1993).

Heterozygotes for *inversions* may produce aneuploid gametes due to crossing over within the inversion, which generates chromatids with duplications and deficiencies

of genetic material (see Figure 16 in Chapter 10). In Diptera such as *Drosophila*, but not necessarily in other organisms, *paracentric inversions* (which do not include the centromere) do not reduce fertility because there is no crossing over in males, and the unbalanced chromosomes resulting from crossing over in females are relegated to polar bodies. Paracentric inversions are commonly polymorphic within *Drosophila* populations. Heterozygotes for *pericentric inversions* (which include the centromere) are expected to produce gametes with deficiencies when crossing over occurs, in *Drosophila* as in other organisms, and thus should show reduced fertility. However, the expected reduction of fertility in this and other classes of rearrangements does not always occur.

A *reciprocal translocation* is an exchange between two homologous chromosomes. Suppose, for example, that 1.2 and 3.4 represent two metacentric chromosomes in one population, with 1 and 2 representing the arms of one and 3 and 4 the arms of the other. A second population that is fixed for a translocation might have chromosomes 1.4 and 3.2. The F_1 hybrid has all four chromosome types (1.2, 3.4, 1.4, 3.2). Only if the two parental combinations segregate (1.2 and 3.4 to one pole, 1.4 and 3.2 to the other) are balanced (euploid) gametes (Figure 17 in Chapter 10) formed.

FIGURE 15.28 Two models for gene interactions that cause sterility or inviability in hybrids between species. The X chromosome and one autosome of males of two species are shown above backcross genotypes. (A) The interaction between complementary loci A and C causes sterility. (B) Complex epistasis of the kind described by Wu et al. Neither of the alleles at two loci (A and B) on the X chromosome of one species causes sterility when combined with the autosome of the other species. However, the pair of alleles (A, B) of one species together interact with an autosomal allele (C′) of the other species so as to cause sterility.

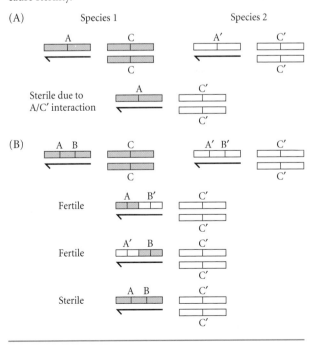

Other patterns of segregation (e.g., 1.2 and 3.2 to one pole, 3.4 and 1.4 to the other) yield aneuploid gametes.

Some species differ by multiple translocations. For example (Dobzhansky 1951), the jimsonweeds *Datura stramonium* and *D. discolor* differ by multiple translocations among 5 of the 12 pairs of chromosomes. If these chromosomes in *D. stramonium* are designated 1.2, 3.4, 5.6, 7.8, and 9.10, those of *D. discolor* represent 1.3, 2.7, 4.10, 5.9, and 6.8. In synapsis, a ring of ten chromosomes is formed, with the two arms of each *stramonium* chromosome aligned with an arm of each of two *discolor* chromosomes (Figure 15.29). The opportunities for aneuploid segregation are numerous.

The several kinds of FUSIONS and FISSIONS of chromosomes include **Robertsonian rearrangements:** either the fusion of two acrocentrics (1, 2) to form a metacentric (1.2) or the fission of a metacentric (1.2) into two acrocentric chromosomes (1, 2). Related populations often differ by several or many such rearrangements. For example, one population may carry the fusion 1.2, while another population may carry the fusion 3.4. In the F$_1$ hybrid, two trios of chromosomes are formed in meiosis (1.2 with 1 and 2, 3.4 with 3 and 4). These may or may not segregate to form euploid gametes (Figure 15.30A). Moreover, two fusions in different populations may display "monobrachial homology," such that one arm of a metacentric chromosome is homologous in the two populations, but the other is not. For example, acrocentrics 1 and 2 may become joined into metacentric 1.2 in one population, whereas acrocentrics 1 and 3 form metacentric 1.3 in the other (Figure 15.30B). Synapsis in the hybrid then entails chains of chromosomes, such as 1.2, 1.3, 2, and 3 in this example. Frequently failure of pairing or nondisjunction occurs, forming aneuploid gametes.

On the whole, related taxa differ less often by rearrangements that are thought to reduce fertility greatly, such as translocations, than by rearrangements with less pronounced effects, such as simple fusions (Lande 1979a). Often a certain kind of rearrangement is typical of a group (White 1978). For example, many species of *Chironomus* midges and of *Oenothera* evening primroses differ by reciprocal translocations, whereas Robertsonian differences predominate in morabine grasshoppers, house mice (*Mus*), and many other groups. The reason for such patterns is not known (King 1993). Clearly related species sometimes differ by many rearrangements. For example, the 23 pairs of chromosomes in the muntjac *Muntiacus reevesi* (a small deer) have been reduced by 20 fusions to 3 pairs in *Muntiacus muntiacus vaginalis* (see Figure 20 in Chapter 10).

Pericentric inversions, reciprocal translocations, and Robertsonian alterations are only infrequently found as polymorphisms within populations. More often, they are *nearly or entirely monomorphic* in different populations, except where these meet in *narrow hybrid zones*. For example, parapatric races of the burrowing mole-rat *Spalax ehrenbergi* in Israel, with 26, 27, 29, and 30 chromosome pairs, meet in zones that range from 2.8 km to only 0.3 km wide (Figure 15.31).

Are differences in chromosome structure important postzygotic barriers to gene exchange? That chromosome differences are seldom polymorphic within populations and that parapatric chromosome races form such narrow hybrid zones both *suggest that chromosomal heterozygotes have lower fitness* than homozygotes (are *underdominant*), perhaps due to reduced fertility. If so, a chromosome introduced by gene flow from one population into another would seldom increase in frequency, because its initial frequency would be low, it would occur mostly in heterozygous condition, and it would probably be eliminated by selection (see Chapter 13). Thus the border between chromosome races would be a *tension zone* (see above). If the sterility of the F$_1$ hybrid is nearly complete due to a high incidence of aneuploidy, there will be almost no gene flow between the populations. If only a single chromosome differs by an underdominant rearrangement, the hybrid may have only incomplete sterility, and sufficient backcrossing may occur to allow substantial introgression, especially of genes located on unaffected chromosomes. Thus *the strength of the barrier to gene flow would depend on the degree to which chromosome differences reduce hybrid fertility*.

Whether chromosome rearrangements reduce fertility and restrict gene flow in nature is uncertain (King 1993; John 1981; Sites and Moritz 1987; Searle 1993). The major problems are that (1) various mechanisms may reduce the incidence of aneuploid gametes or genetically imbalanced offspring, so that fertility is not reduced as much as one would expect; and (2) when aneuploidy or reduced fertility does occur in hybrids, it is very difficult to tell whether it is caused by the *structural differences* between the chro-

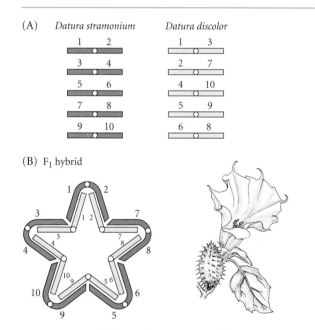

FIGURE 15.29 (A) Five chromosomes of the jimsonweeds *Datura stramonium* and *D. discolor*, which differ by five reciprocal translocations. Homologous chromosome arms are correspondingly numbered. Only one member of each chromosome pair is shown. (B) A diagram of the possible arrangement of these chromosomes in synapsis in an F$_1$ hybrid.

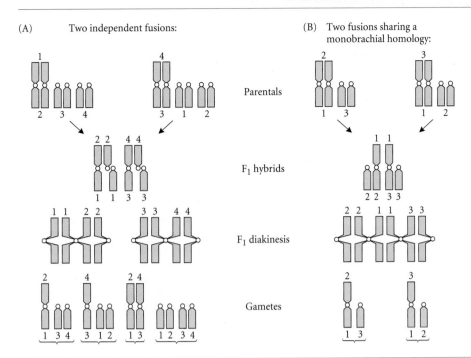

(A) Two independent fusions:

Parentals

F₁ hybrids

F₁ diakinesis

Gametes

(B) Two fusions sharing a monobrachial homology:

Parentals

F₁ hybrids

F₁ diakinesis

Gametes

FIGURE 15.30 Robertsonian fusions. (A) Two parental species have undergone independent fusions of acrocentric chromosomes into metacentrics (1.2 and 3.4). (B) Two parental species have undergone different monobrachial fusions (1 and 2 in one species and 1 and 3 in the other). Meiosis in the F₁ hybrid, with each chromosome consisting of two chromatids, may result in balanced gametes (as shown) or not. (After King 1993.)

mosomes or by differences between the *genes* of the parental populations.

For example, pericentric inversions "should" reduce the fertility of heterozygotes, but this proved not to be the case for about 40 percent of newly arisen inversions in *Drosophila melanogaster* (Coyne et al. 1993). In at least some cases, crossing over was suppressed within the inversion. Spontaneous Robertsonian fusions in laboratory mice (*Mus domesticus*) reduce fertility due to aneuploidy, as do multiple Robertsonian fusions introduced from wild populations into laboratory stocks; but neither females nor males have reduced fertility in crosses between wild populations that differ in chromosome number (Britton-Davidian et al. 1990; Sage et al. 1993). Whether or not rearrangements cause aneuploidy and reduce fertility cannot be reliably inferred without examining each individual case. In a few cases, it has been possible, using special techniques, to show that both the structural differences and gene differences contribute to infertility (Shaw et al. 1993).

The distinction between sterility due to structural rearrangements and to gene differences is important, because if hybrid sterility is due to one or a few chromosome rearrangements, this would provide evidence that genetic drift in small populations contributes to speciation, since it is the most likely mechanism by which underdominant chromosome rearrangements can become fixed (see Chapters 13 and 14). If, however, the sterility is caused by multiple gene differences, these could have accrued gradually, by natural selection, in populations of any size.

Symbiont-induced incompatibility

In certain insects representing at least five orders, hybrids between populations are inviable or sterile owing to *cytoplasmic factors rather than nuclear genes or chromosomes*. The hybrids are viable, however, if the parents are treated with antibiotics.

FIGURE 15.31 The distribution of four chromosome "races" of the mole-rat *Spalax ehrenbergi* in Israel. Pairs of races meet at very narrow hybrid zones, indicated by the dotted lines. (After Nevo 1991.)

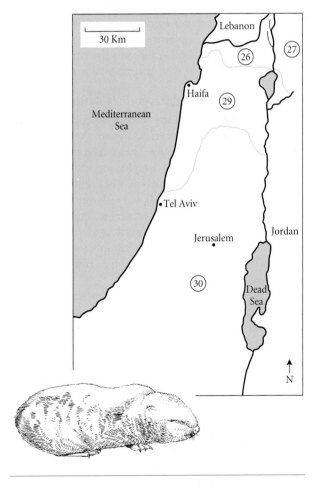

The incompatibility is caused by infectious bacteria (*Wolbachia*) that are transmitted in the cytoplasm.

In at least one case, the incompatibility arises because the bacteria destroy chromosomes (Breeuwer and Werren 1990, 1993). The parasitic wasps *Nasonia vitripennis* and *N. giraulti*, when crossed, produce only male progeny. These are not hybrids, however, because, as in other hymenopterans, males develop from unfertilized eggs, and are therefore haploid (Figure 15.32B). When males of either species were treated with antibiotics and mated to females of the other species, a large proportion of the progeny were fertile hybrid females (Figure 15.32C). Microscopic examination and genetic analysis showed that in incompatible crosses, the paternal set of chromosomes in the diploid fertilized egg fails to condense properly in the first mitotic division, and apparently degenerates. These eggs, having only a single, maternal complement of chromosomes, develop into nonhybrid males. A strain of wasps was constructed that had *giraulti* chromosomes and *vitripennis* cytoplasm (carrying bacteria); such females produced hybrid female offspring when mated to infected *N. vitripennis* males (Figure 15.32D). However, crosses of these females with infected male *N. giraulti* yielded no hybrid offspring (Figure 15.32E).

This finding shows that the incompatibility between normal *N. vitripennis* and *N. giraulti* is due not to incompatibility between chromosomes or genes, but to incompatibility between the bacteria carried by the two species. It therefore appears likely that in *Nasonia* wasps and perhaps some other insects, *speciation might be caused by divergence of symbiotic bacteria, not by divergence of the insects' own genes*. It should be emphasized, though, that in this and some other such cases, other reproductive barriers, such as sexual isolation, may also be instrumental in speciation.

The Genetic Basis of Prezygotic Isolation

Hybrid sterility or inviability often is not the major barrier to gene flow between species; instead, they may simply not produce hybrid progeny. The prezygotic barriers to gene exchange may be quite complex, especially in animals, in which male courtship signals may be entirely behavioral or may consist of behavior coupled with morphological or other display characters. Moreover, mating is a consequence of communication between sender and receiver: between *male signal and female response* (or, sometimes, the converse). Reproductive isolation results from the divergence of both.

Divergence of the morphological characters displayed in courtship has the same kind of genetic bases as that of morphological characters generally. We have already described the polygenic basis of the difference in head shape between the Hawaiian *Drosophila silvestris* and *D. heteroneura*, in which the X chromosome has a major effect. Differences in the shape of the male genitalia of *Drosophila simulans* and *D. mauritiana*, which may cause interspecific matings to terminate prematurely, are affected by at least one gene on each of the three major chromosomes (Coyne 1993; Coyne et al. 1991).

FIGURE 15.32 Evidence that incompatibility between the wasps *Nasonia giraulti* (G) and *N. vitripennis* (V) is caused by cytoplasmic symbionts. Gametes and zygotes are shown with haploid or diploid genomes of either species in the nucleus; letters in the surrounding cytoplasm show the species source of the cytoplasm in crosses D and E. (A) Normal reproduction. Unfertilized eggs develop into haploid males, fertilized eggs into diploid females. (B) In matings between species, only males are produced, because the paternal genome degenerates. These male offspring are not hybrids. (C) If the males in matings between species are treated with antibiotics (*), fertile hybrid female offspring are obtained. Thus a bacterial agent is implicated. (D) The nuclear genome of G was backcrossed into the cytoplasm of V to create a new strain. Such females, mated to V males, produce fertile hybrid female offspring. Thus the nuclear genomes of the two species are compatible. (E) The same females, mated to symbiont-carrying males of their own species, produce only male offspring, due to loss of the paternal genome. The incompatibility is thus due to the differences in cytoplasmic factors—i.e., the symbiotic bacteria.

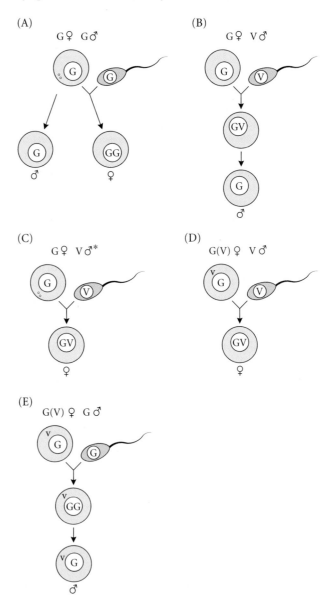

Sexual isolation between several species pairs of *Drosophila* has been studied by inserting genetically marked chromosomes (or chromosome regions) from one species into the genome of another, as described earlier for the study of hybrid sterility. For example, female *D. simulans* readily mate with male *D. mauritiana*, but female *D. mauritiana* only rarely mate with male *D. simulans*. Using *D. mauritiana* that carried a genetic marker on each of the three major chromosomes, Coyne (1989) found that the proportions of backcross females that mated with male *D. simulans* was reduced by each *mauritiana* chromosome that they carried. These results indicate that *D. mauritiana* has *at least one* gene on each chromosome that reduces the propensity of the female to mate with a male *D. simulans*.

The males of these species must also differ, or else the females could not distinguish them. Eleftheriou Zouros (1981) found that in another closely related pair of species, *D. mojavensis* and *D. arizonensis*, the genetic factors that made males of one species unacceptable to females of the other were located on the Y chromosome and one autosome, and that these differed from the genes responsible for the discrimination ability of the females.

Drosophila males engage in an elaborate courtship that includes a "song" produced by rapidly vibrating the wings. This behavior enhances female mating behavior, although that need not mean that the response is species-specific. The songs, consisting of hums punctuated by pulses, have species-specific rhythms. Charalambos Kyriacou, Jeffrey Hall, and their coworkers have come closest to a molecular characterization of a species difference in courtship in their study of this behavior (Wheeler et al. 1991). Mutations of the *period* (*per*) gene shorten, lengthen, or abolish several very different rhythms in the life of *Drosophila melanogaster*: the daily rhythm of locomotion activity, the time of day of eclosion from the pupa, and the rhythmic interval in the courtship song (see Chapter 10). The mutant *per*[01] is arrhythmic. The researchers cloned the *per* gene of *D. simulans* and inserted it into *per*[01] *D. melanogaster* by P-element transfection (see Box D in Chapter 3). The male flies carrying the insertion displayed the song rhythm typical of *D. simulans*, whereas *per*[01] flies carrying an insertion of the normal *melanogaster per* gene displayed the typical rhythm of *D. melanogaster*. Transfection with only part of the *per* gene narrowed the effect down to a region in which the species differ by only four amino acids (inferred from the DNA sequence). Thus the difference between the species' courtship songs is presumably caused by four or fewer amino acid differences in the protein product of a single gene.

Other than in *Drosophila*, one of the most complete analyses of the genetic basis of sexual isolation has been carried out with the European corn borer moth (*Ostrinia nubilalis*) by Wendell Roelofs, Christer Löfstedt, and coworkers (Roelofs et al. 1987; Löfstedt 1990). There are two "pheromone strains"—or species—that interbreed to only a limited extent. Females of both types produce a pheromone blend of two optical isomers (E and Z) of a tetradecenyl acetate. "E"-type females produce about 99%

of the E isomer, whereas the "Z" type produces the opposite ratio. Hybridization experiments demonstrated that the difference is caused by a single autosomal factor, and the heterozygote is intermediate. Other loci contribute a little variation among backcross progeny. Both in the field and in wind tunnels, males of each strain are almost exclusively attracted to the E:Z ratio produced by females of their own strain. This behavioral response is sex-linked. Electrophysiological recordings from male antennae exposed to each of the compounds revealed two kinds of specialized olfactory receptor cells, one of which generates a high-amplitude spike in the recording, and the other a low-amplitude spike. The high spike is evoked by the E isomer in E-strain males and by the Z isomer in Z-strain males. F[1] hybrid males have equal responses to the two compounds. The electrophysiological response is controlled by a single autosomal gene that is not linked to the gene for pheromone ratio in the female. In this moth, the evolution of signal and response, leading to reproductive isolation, appears to be due to very few allele substitutions, and the genetic control of the male response is independent of the genetic control of the difference in the females' chemical signal. In general, *there is little evidence for genetic coupling between the male and female components of communication that cause reproductive isolation* (Butlin and Ritchie 1989).

The Genetics of Reproductive Isolation: Summary and Significance

Both prezygotic and postzygotic reproductive incompatibility are based on at least a few, and in some cases many, gene differences. *Reproductive isolation between species, if based on genic differences rather than differences in chromosome structure, requires that populations diverge by at least two allele substitutions.* Thus an $A_1A_1B_1B_1$ ancestral population may give rise to populations with genotypes $A_1A_1B_2B_2$ and $A_2A_2B_1B_1$, the incompatibility between A_2 and B_2 being the cause of reproductive isolation (Dobzhansky 1937). The data indicate that incompatibility usually arises from more than two such genes. The number of gene differences that confer postzygotic isolation *may be rather small at first, but increases over time.* Thus among older, fully established species, the number of gene differences contributing to postzygotic isolation becomes high, and *reproductive isolation becomes irreversible.*

One of the most important tenets of the theory forged during the Evolutionary Synthesis of the 1930s and 1940s was that "macroevolutionary" differences among organisms—those that distinguish higher taxa—arise from the accumulation of the same kinds of genetic differences that are found within species. Opponents of this point of view believed that "macroevolution" is qualitatively different from "microevolution" within species, and is based on a totally different kind of genetic and developmental repatterning. The iconoclastic geneticist Richard Goldschmidt (1940), who held this opinion, believed that the evolution of species marks the break between "microevolution" and "macroevolution"—that there is a "bridgeless gap" between species that cannot be understood in terms of the genetic variation within species. Genetic studies of species differences have decisively disproved Goldschmidt's claim. *Differences between species* in morphology, behavior, and the processes that underlie reproductive isolation all *have the same genetic properties as variation within species:* they occupy consistent chromosomal positions, they may be polygenic or based on few genes, they may display additive, dominant, or epistatic effects, and they can in some instances be traced to specifiable differences in proteins or DNA nucleotide differences. *The degree of reproductive isolation between populations,* whether prezygotic or postzygotic, *varies from little or none to complete.* Thus, *reproductive isolation, like the divergence of any other character, evolves in most cases by the gradual substitution of alleles in populations.*

Summary

1. Pre-Darwinian concepts of species as "types," distinguished by discrete morphological differences, gave way to a concept of species as reproductively isolated gene pools, or sets of interbreeding populations, due to studies of variation within and among populations and the discovery of reproductively isolated sibling species that are morphologically almost indistinguishable. The biological species concept (species are groups of actually or potentially interbreeding populations that are reproductively isolated from other such groups) is the one most widely used by evolutionary biologists today. Species are defined by reproductive discontinuity, not by phenotypic difference. (Phenotypic differences, however, may be indicators of reproductive discontinuity.) Some systematists and others adhere to other species concepts, such as a phylogenetic species concept, in which species are sets of populations with characters that distinguish them.

2. The biological species concept has a restricted domain; moreover, some populations cannot be readily classified as species or not because speciation, which is the origin of reproductive discontinuity, is a gradual process.

3. Species exist because of the evolution of biological differences that constitute barriers to gene exchange (isolating barriers or "mechanisms"). These are many in kind, the chief distinction being between prezygotic (e.g., ecological or sexual isolation) and postzygotic barriers (hybrid inviability or sterility). Species may be reproductively isolated by both kinds of barriers, or entirely by one or the other.

4. Reproductive barriers evolve gradually, so that any one of them may vary in strength. In *Drosophila,* pre- and postzygotic isolation seem to evolve at similar rates in allopatric populations, but the rate of evolution of prezygotic isolation may be greater when populations are sympatric. Sterility or inviability of hybrid males usually evolves before its manifestation in females (Haldane's rule).

5. Species (or semispecies) sometimes hybridize to some degree, often in "hybrid zones" that in many cases are regions of contact between formerly allopatric populations. Different alleles may flow from one species into the other at different rates, depending on rates of dispersal, the strength of selection, and linkage to selected loci. Hybrids often manifest genetic "disruption" due to deleterious interactions among alleles at different loci (epistasis).

6. Levels of molecular divergence between closely related species vary; in some instances they are so low as to suggest that speciation has occurred within a few thousand years or less, although more commonly speciation may require 1–3 million years. Some species share ancestral molecular polymorphisms, suggesting recent origin. In some cases, a species is more closely related to certain populations of another species than the various populations of the latter are to each other; this suggests that a widespread species, which may have changed little, has given rise to a "daughter" species that acquired reproductive isolation.

7. The morphological and other phenotypic differences between related species include those that contribute to reproductive isolation (i.e., to their status as species) and those that do not. Both kinds of differences vary continuously in degree, from slight (among some conspecific populations) to great (among other conspecific populations and among species). The genetic bases of differences among species are the same as for variation within species. Thus, speciation is a consequence of gradual substitution of alleles among populations.

8. Postzygotic isolation can be due to differences in nuclear genes, structural differences in chromosomes, or incompatibility between cytoplasmic factors such as symbiotic intracellular bacteria. *Genic* differences that yield hybrid sterility or inviability consist, typically, of differences at *at least* two loci that interact disharmoniously in the hybrid (complementary loci). In some cases, alleles at two or more loci of one species must together interact with at least one locus of the other species to yield postzygotic isolation (complex epistasis). Postzygotic isolation between newly formed species may be due to only a few genes, but more such interacting gene differences accumulate after reproductive isolation has evolved.

9. Structural chromosome differences are expected to reduce the fertility of hybrids because aneuploid gametes may be produced. However, various meiotic mechanisms can reduce the incidence of aneuploidy, which may be severe only when the hybridizing populations differ by multiple rearrangements. Populations are often monomorphic for different chromosome arrangements, and form narrow hybrid zones, suggesting that hybrids have reduced fitness and that the chromosome differences restrict gene exchange. However, it is difficult to tell whether reduced fitness is due to the difference in chromosome structure or to differences in the genes on the chromosomes.

10. Among prezygotic barriers to gene exchange, sexual (ethological) isolation is important in animals. It entails

a breakdown in "communication" between the courting and the courted sexes, and therefore, usually, genetic divergence in both the "signal" and the "response" (usually by the male and female, respectively). The male "signal" may be entirely behavioral, or may also entail morphological, chemical, or other features. In some cases the species differences in signal and response are due to only a few, and in some cases to many, gene differences. The differences in male and in female components are often or usually independent in a genetic sense.

Major References

Mayr, E. 1963. *Animal species and evolution.* Harvard University Press, Cambridge, MA. This book and its abridged successor, *Populations, species, and evolution* (Harvard University Press, 1970), are the definitive classic works on the nature of animal species and speciation. Although some of Mayr's views on speciation are controversial, and although information has grown greatly since these books were written, no serious student of species—or of evolution—can afford not to have read at least parts of this work or of Mayr's foundational *Systematics and the origin of species* (Columbia University Press, 1942). For plant species and speciation, an equally foundational work is G. L. Stebbins, *Variation and evolution in plants* (Columbia University Press, 1950), a topic also treated by V. Grant in *Plant speciation* (Columbia University Press, 1971).

Otte, D., and J. A. Endler (editors.) 1989. *Speciation and its consequences.* Sinauer Associates, Sunderland, MA. This unusually diverse and thought-provoking collection of papers from a symposium includes several essays on species concepts.

Coyne, J. A. 1992. Genetics and speciation. *Nature* 355: 511–515; Wu, C. I. 1994. Genetics of postmating reproductive isolation in animals. *Annual Review of Genetics* 28: 283–308. These articles review the genetic bases of reproductive isolation, especially in animals.

Harrison, R. G. (editor.) 1993. *Hybrid zones and the evolutionary process.* Oxford University Press, New York. Essays, including case studies of animals and plants, on this complex topic.

Problems and Discussion Topics

1. The biological species concept was justified in this chapter on the grounds that all populations of a species can potentially partake in a common evolutionary history and that reproductive isolation prevents recombination from breaking apart constellations of coadapted characters. Some degree of genetic exchange occurs in bacteria, which reproduce mostly asexually. What evolutionary factors should be considered in debating whether or not a certain rate of genetic exchange would be enough to justify applying the biological species concept to bacteria?

2. Suppose that the phylogenetic species concept, as defined by Cracraft (Table 15.1), were to be adopted in place of the biological species concept. What would be the implications for (*a*) evolutionary discourse on the mechanisms of speciation; (*b*) studies of species diversity in ecological communities; (*c*) estimates of species diversity on a worldwide basis; (*d*) conservation practices under such legal frameworks as the U.S. Endangered Species Act?

3. Botanists have generally been more reluctant than zoologists to adopt the biological species concept. Read and discuss their arguments in, for example, papers by Raven (1976) and Levin (1979).

4. How would you determine whether an increased incidence of hybrids between two plant species in a disturbed habitat was due to a greater rate of production of hybrid zygotes or to greater survival of the hybrids?

5. The higher rate of amino acid-changing substitutions than of synonymous substitutions in the divergence of genes that encode sperm surface proteins in abalones indicates that natural selection has caused rapid divergence. The agents of this selection are unknown. Suggest some hypotheses and ways in which they might be tested.

6. Studies of hybrid zones have shown that mitochondrial and chloroplast DNA markers frequently introgress farther, and have higher frequencies, than nuclear gene markers (Avise 1994, pp. 284–290). Thus, far from the hybrid zone, individuals that have no phenotypic indications of hybrid ancestry may have mitochondrial or chloroplast genomes of the other species (or semispecies). How would you account for this pattern?

7. The models of hybrid zones described in this chapter envision diffusion of alleles, by short-distance dispersal, from each semispecies into the other, resulting in smooth clines in allele frequencies. Some hybrid zones have been formed by expansion of populations from widely separated "refuges" that they occupied during Pleistocene glacial periods. How might the structure of a hybrid zone be affected if the intervening area were colonized by long-distance migrants? What would be the consequences of repeated cycles of expansion and contraction of populations during repeated glacial and interglacial episodes? (See Hewitt 1989, 1996.)

8. Suppose that two or more related taxa are polyphyletic for gene lineages, as in the cichlid fishes in Figure 15.23. What evidence might enable you to decide whether this is due to incomplete sorting (i.e., lack of coalescence) of ancestral polymorphism or to introgressive hybridization?

9. Knowing that epistatic interaction among two or more divergent loci causes hybrid inviability or sterility does not tell us what the functions of these genes are, nor what processes within an organism cause these effects. Viewing this as a problem in developmental biology, can you find in the developmental literature any evidence bearing on these questions?

10. Based on the kind of evidence discussed in this chapter, how often is reproductive isolation among species in nature due to postzygotic barriers alone, rather than prezygotic barriers?

CHAPTER 16

Speciation

Despite the title of his greatest book, the problem of the origin of species, as we now understand it, is one of the rather few major questions in evolution which Darwin hardly began to solve. As we noted in Chapter 15, Darwin thought of species as merely well-marked "varieties;" that is, as populations defined by their degree of difference in morphological features. Thus for Darwin, as for his contemporaries, explaining the origin of species was the same as explaining the evolution of morphological and other phenotypic characters—which Darwin indeed did, spectacularly well. But although he grappled, not very successfully, with the problem of hybrid sterility, Darwin hardly addressed the problem of how barriers to gene exchange evolve, which we recognize today as the essence of speciation.

Perhaps the most basic of the many questions about the causes of speciation is whether or not it is an adaptive process. The ecologist G. Evelyn Hutchinson (1968) asked "When are species necessary?" and suggested that organisms might form discrete species because the environment presents discrete ecological niches for them to fill. He was not the first to suggest this. Theodosius Dobzhansky (1951) argued that "the living world is not a formless mass of randomly combining genes and traits, but a great array of . . . gene combinations, which are clustered on a large but finite number of adaptive peaks." Thus, he suggested, one species of insect may feed on oak and another on pine, but an insect that required an intermediate food would probably starve. Interbreeding between the harmonious genotypic systems adapted to these different ecological niches would break them down. "The gene combinations whose adaptive value has been vouchsafed by natural selection must be protected from disintegration," Dobzhansky believed, by the barriers to gene exchange that he termed isolating mechanisms, and which he believed evolved to serve just this function.

There is no doubt that different species do usually occupy different niches and have different coadapted gene pools, which are broken down in hybrids and lower their fitness. But these *effects* of interbreeding do not necessarily mean that speciation is an adaptive process, whereby natural se-

lection builds isolating mechanisms because they achieve these results. Perhaps the more prevalent view, expressed by Herman Muller (1940), Ernst Mayr (1963), and many later authors, is that *reproductive isolation evolves as a by-product of genetic changes that occur for other reasons*, so that speciation is an incidental, nonadaptive consequence of divergence of populations. What these genetic changes are and why they occur is a major topic of this chapter. As we deal with these questions, we will also ask what roles natural selection and genetic drift play in speciation, what the geographic setting of speciation is, and what factors enhance or retard the rate of speciation. The chapter closes with the question of what the larger evolutionary consequences of speciation may be.

Modes of Speciation

For a single randomly mating population to give rise to two reproductively isolated populations, it is ordinarily not sufficient for a single mutation to confer reproductive isolation on its bearers. Such a mutation (say, A_2) may arise in one or a few individuals, in heterozygous condition (A_1A_2). Whether it confers postzygotic isolation (e.g., sterility) or prezygotic isolation (e.g., failure to mate with the "normal" type), its reproductive fitness is lowered, so it will ordinarily be eliminated by natural selection.

Chapter 15 provided evidence that reproductive isolation between populations is usually based on more than one locus. However, the problem with a polygenic trait as a cause of reproductive isolation is that recombination generates intermediates. If several loci govern, for example, time of breeding, $A_1A_1B_1B_1C_1C_1$ and $A_2A_2B_2B_2C_2C_2$ might breed early and late in the season, respectively, and so be reproductively isolated; but the many other genotypes, presumably with intermediate breeding seasons, would constitute a "bridge" for the flow of genes between the two extreme genotypes. *The problem of speciation, then, is how two different populations can be formed without intermediates.* This problem holds, whatever the character that confers prezygotic or postzygotic isolation.

The many conceivable solutions to this problem are the "modes of speciation." Two authors' classifications of the modes of speciation that have been suggested are listed in Table 16.1. We shall follow Mayr's classification, although not in the order listed.* We will not consider his mode 2A, speciation due to a single mutation, because it is unlikely for the reasons discussed above, and because no convincing examples of it have been described.

The evolution of reproductive barriers based on several or many allele substitutions is referred to by Mayr as *gradual speciation* in Table 16.1 (Mayr's mode 3). We devote the bulk of this chapter to gradual speciation, considering three geographic settings in which it may occur (Figure 16.1). *Allopatric speciation* is the evolution of reproductive barriers in populations that are prevented by a geographic barrier from exchanging genes at more than a negligible rate. In *parapatric speciation*, neighboring populations, between which there is modest gene flow, diverge and become reproductively isolated. *Sympatric speciation* is the evolution of reproductive barriers between subsets of a single, initially randomly mating population.

Some instances are also known of species that have arisen from *hybrids* (Table 16.1, mode 1). Speciation due to

*This treatment ignores several suggested mechanisms of speciation, entailing factors such as meiotic drive and transposable elements, that are not favored by empirical evidence (Coyne 1992).

a single genetic event (instantaneous speciation; mode 2), rather than by multiple allele substitutions, occurs chiefly by *polyploidy*. Discussion of these modes of speciation closes the chapter.

Allopatric Speciation

Kinds of Allopatric Speciation

Allopatric speciation *is the evolution of genetic reproductive barriers between populations that are geographically separated* by an extrinsic, physical barrier such as topography, water (or land), or unfavorable habitat. The physical barrier reduces gene flow sufficiently for genetic differences between the populations to evolve, by natural selection or genetic drift, which prevent gene exchange if the populations should later come into contact (Figure 16.1). In species that disperse little or are strongly tied to a particular habitat, extrinsic barriers may isolate populations on a "microgeographic" scale, such as among segregated habitats within a lake. Allopatry is defined by a severe reduction of movement of individuals or their gametes, not by geographic distance. The extrinsic barrier need not reduce gene exchange to zero, just to a very low level (see below). Although there is debate about how frequently parapatric and sympatric speciation occur, *all evolutionary biologists agree that allopatric speciation does occur, and many hold that it is the prevalent mode of speciation in animals* (Mayr 1963; Futuyma and Mayer 1980; Coyne 1992).

Table 16.1 Two classifications of potential modes of speciation in sexual organisms[a]

By geography and level (after Mayr 1963)
1. Hybridization (maintenance of reproductively isolated hybrids between two species)
2. Instantaneous speciation (through individuals)
   - A. Genetically (single mutation)
   - B. Cytologically
     - *a.* Chromosome rearrangement
     - *b.* Polyploidy
3. Gradual speciation (through populations)
   - A. Sympatric speciation
   - B. Parapatric speciation
   - C. Allopatric (geographic) speciation
     - *a.* Peripatric speciation (by evolution in an isolated colony)
     - *b.* Vicariant speciation (division of range by an extrinsic barrier or extinction of intervening populations)

By population genetic mode (after Templeton 1982a)
1. Transilience
   - A. Hybrid maintenance (selection for hybrids)
   - B. Hybrid recombination (selection for recombinants following hybridization)
   - C. Chromosomal (fixation of chromosome rearrangements by drift and selection)
   - D. Genetic (founder event in a colony)
2. Divergence
   - A. Habitat (divergent selection without isolation by distance)
   - B. Clinal (selection on a cline with isolation by distance)
   - C. Adaptive (erection of an extrinsic barrier followed by divergent microevolution)

[a] The categories in Mayr's classification correspond approximately to those in Templeton's as follows: Templeton's 1A, 1B = Mayr's 1; 1C = 2Ba; 2A = 3A; 2B = 3B; 1D = 3Ca; 2C = 3Cb. Templeton's classification does not explicitly include speciation by polyploidy or by single mutations; the latter is included by Mayr only for historical reasons. I have reorganized these authors' classifications to facilitate comparison between them.

FIGURE 16.1 Diagrams of successive stages in each of four models of speciation, differing in geographic setting. (A) Allopatric speciation by vicariance. A physical barrier divides a widespread species into two populations, which diverge in features that cause reproductive incompatibility. They become sympatric if the barrier breaks down or if they disperse over it. (B) The peripatric, or founder effect, model of allopatric speciation. A localized colony diverges from a widespread "parent" species, which remains little changed. Range expansion may establish sympatry. (C) Parapatric speciation, in which populations of a widespread species diverge by adaptation to different environments. Divergent selection even at a narrow environmental discontinuity may oppose gene flow and result in reproductive isolation. (D) Sympatric speciation, in which genetic differences (white versus dark symbols) develop gradually among members of an initially randomly mating population, resulting in reproductive isolation.

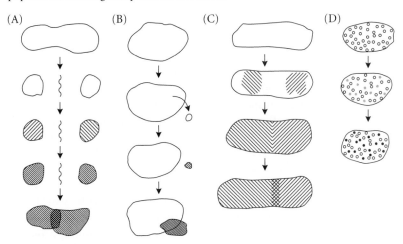

Two major models of allopatric speciation, differing in population structure and genetic dynamics, have been postulated. **Vicariant speciation** (vicariance) occurs when *two rather widespread populations are divided* by the emergence of an extrinsic barrier, the extinction of intervening populations, or migration into a separate region (Mayr's 3C*b* in Table 16.1; Figure 16.1A). The emergence of the Isthmus of Panama in the Pliocene, for example, divided many marine organisms into Pacific and Caribbean populations, some of which have diverged into distinct species (Knowlton et al. 1993). "Vicariant species" of tulip poplar (*Liriodendron*) in eastern Asia and eastern North America are the remnants of a once wider distribution across northern North America and Siberia, where climatic change has made much of the region uninhabitable for these trees. During the colder periods of the late Pliocene and the Pleistocene, many species moved into several disjunct refuges (refugia), where they differentiated into semispecies or species (Hewitt 1996; Figure 15 in Chapter 8). Vicariant speciation is generally supposed to proceed by the operation of natural selection, and perhaps of genetic drift, in each of the separated populations, perhaps to a greater extent in one than the other.

The other major mode of allopatric speciation occurs when a *colony* derived from a more widespread "parent" population *diverges and acquires reproductive isolation* (Figure 16.1B; Table 16.1, Mayr's 3C*a*). Ernst Mayr, who from 1942 onward argued that many species originate as local populations along the periphery of a broad species distri-

bution, has called this **peripatric speciation** (Mayr 1982b). A great deal of controversy surrounds the genetic changes postulated for such populations. Many population geneticists believe that the genetic changes underlying peripatric speciation are the same as those in vicariant speciation. In contrast, Mayr (1954, 1963), Hampton Carson (1975), and Alan Templeton (1980) have posited variant models in which the colony, during its founding or at other times when its effective population size is low, undergoes allele frequency changes at some loci by genetic drift; this change in the "genetic environment" creates selection at other loci or new coadapted gene combinations. In effect, the population experiences a *peak shift* (see Chapter 14), which Templeton (1980) calls a TRANSILIENCE. This model, sometimes called *founder effect speciation*, is very controversial. Mayr (1954) illustrated peripatric speciation with the kingfisher *Tanysiptera galeata*, which shows only slight geographic variation in plumage throughout climatically very different regions of New Guinea, but which has given rise to several markedly different forms on small islands nearby (Figure 16.2). The hundreds of species of *Drosophila* in the Hawaiian Islands are another commonly cited example of peripatric speciation, for most of these species appear to have evolved rapidly following colonization of one island (or part of an island) from another (Carson and Kaneshiro 1976).

We shall treat the controversial genetics of the founder effect model below, in our discussion of peripatric speciation. In most of the present discussion of allopatric speci-

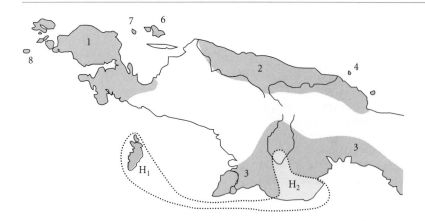

FIGURE 16.2 Subspecies of the kingfisher *Tanysiptera galeata* in New Guinea and on nearby islands. The mainland subspecies (1–3) are much more similar to one another than the local insular subspecies (4–8). The distribution of two other species (H_1, H_2) is shown by the dotted line. (After Mayr 1942.)

ation, we will not distinguish the two models, except as noted, for many of the principles apply to both.

Evidence for Allopatric Speciation

Probably thousands of examples provide evidence for allopatric speciation. Most of these arise from systematists' studies of geographic variation within species and the geographic distributions of semispecies and closely related species (Mayr 1942, 1963). There are several classes of evidence for allopatric speciation.

Geographic patterns In earlier chapters (Chapters 9, 15), the point was made that geographic populations of most species differ genetically. In many instances, the variation is adaptive, reflecting selection for different features in different environments. The differences are often most pronounced when gene flow is greatly reduced by extrinsic barriers, as among islands or mountaintops, especially for organisms, such as land snails and salamanders, that have relatively poor powers of dispersal. Most importantly, geographic populations classified as the same species frequently vary in the strength of prezygotic or postzygotic isolation when tested in the laboratory. For example, disjunct populations of the dusky salamander *Desmognathus ochrophaeus* exhibit weak to strong sexual isolation (see Figure 25 in Chapter 9). Instances of circular overlap, as in the salamander *Ensatina* (see Figure 12 in Chapter 15), dramatically illustrate how a reduction of gene flow with distance enables populations to evolve reproductive isolation.

In many cases, related species replace each other geographically, reflecting their presumably allopatric origins. For example, the leopard frogs of North America, most of which were at one time considered geographic races of *Rana pipiens*, consist of about 27 species in two clades (Hillis 1988). The geographic ranges of the species in each clade overlap only slightly or not at all (Figure 16.3), but these species are actually or potentially isolated by habitat, breeding season, mating call, and/or hybrid inviability.

Correspondence with present or past barriers The greatest differentiation among allopatric populations or species is frequently associated with abundant, strong topographic barriers. For example, freshwater animals show the greatest regional diversity in mountainous regions where there are many isolated river systems, and in many plants and animals that live at high altitudes, dozens of distinguishable populations are likely to occupy different peaks or ranges. On islands, a species that is homogeneous over much of its continental range may diverge spectacularly in appearance, ecology, and behavior (see Figure 23 in Chapter 9).

Instances of "double invasion," many of which were noted by Mayr (1942), provide further evidence for allopatric speciation. For example, the thornbill *Acanthiza pusilla*, a small passerine bird, is widespread on the Australian continent and has a slightly differentiated population on the nearby island of Tasmania, where a markedly differentiated species, *A. ewingi*, also occurs. The latter species doubtless originated from an early colonization of Tasmania, and a more recent invasion established *A. pusilla* on the island. Such multiple invasions explain why each island of an archipelago often has several related species, whereas a remote island of similar size has only one. Most of the islands in the Galápagos archipelago have several species of Darwin's finches, but there is only one species on Cocos Island, an isolated island 600 miles away. The Cocos Island finch has been there long enough to have become morphologically very distinct, but it has not had the opportunity to proliferate into more than one species. Within the archipelago, in contrast, populations can diverge on different islands and recolonize the islands from which they came, having evolved reproductive isolation during their period of allopatry.

CONTACT ZONES between differentiated forms are often best interpreted as a consequence of the expansion of formerly allopatric populations. Many hybrid zones, such as the one between European fire-bellied toads (see Chapter 15), may be examples of this phenomenon. In some cases, geological evidence supports this conclusion. For example, Eldredge Bermingham and John Avise (1986; Avise 1994) analyzed the genealogy of mitochondrial DNA in samples of six fish species from rivers throughout the coastal plain of the southeastern United States. In each of the six species, which belong to three families, eastern and western popu-

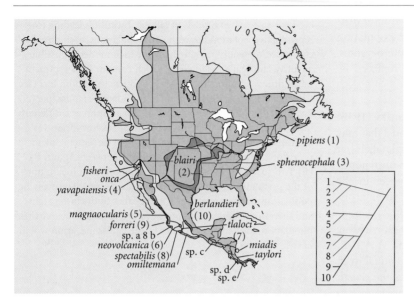

FIGURE 16.3 The distribution of one clade of species in the *Rana pipiens* (leopard frog) complex. Only recently have studies of the overlap zones (dark gray areas) between some of these forms show that they are reproductively isolated species. The largely allopatric or parapatric distributions of these species suggest that physical separation has enabled them to become differentiated. The greater number and narrower distributions of species in Mexico and Central America may be due to the mountainous topography, which affords more opportunities for isolation, and the longer residence time of populations in these areas, which were less affected by Pleistocene glaciations than more northern regions. (After Hillis 1988.)

lations have DNA sequences in two different clades (Figure 16.4). This finding implies that gene flow between east and west was reduced for a considerable time at some time in the past. The eastern and western clades of each species meet in western Florida or eastern Alabama; rivers to the east of this contact zone drain into the Atlantic, and those to the west into the Gulf of Mexico. The same region marks a faunistic break: many other species of fishes are distributed either to the east or to the west of this region. In most of the species studied by Bermingham and Avise, the amount of sequence divergence between the two DNA clades suggests that isolation occurred 3–4 million years ago, during a Pliocene interglacial when sea level stood 50 to 80 meters above the present level. Thus much of the coastal plain was flooded by salt water, which formed a barrier to dispersal by freshwater fishes. These species, which presumably had been broadly distributed before the sea level rose, probably passed the interglacial in two disjunct areas, in the headwaters of the western rivers (in northern Alabama and Georgia) and in peninsular Florida, which was an island at various times in the late Cenozoic.

Laboratory experiments

Many experiments have shown that incipient prezygotic and postzygotic isolation can develop between isolated laboratory populations of *Drosophila* and houseflies (*Musca*) (Rice and Hostert 1993). These experiments will be described at several points in the following pages, where they bear on controversial issues.

Genetic Models of Allopatric Speciation

Three major classes of population genetic models can be envisioned for the evolution of incompatibility (prezygotic or postzygotic isolation) between two populations. We apply them here to allopatric differentiation, but the models hold also, with some elaboration, for parapatric and sympatric speciation. These models are speciation by natural selection,

FIGURE 16.4 Evidence for genetic divergence in isolated "refuges," followed by range expansion and secondary contact. In each of six freshwater fish species, mitochondrial DNA haplotypes fall into two clades, one with a western (colored circles) and one with an eastern (black circles) distribution (see Figure 28 in Chapter 11 for the gene tree of the spotted sunfish). Three families of fishes are represented: Amiidae (bowfin), Centrarchidae (sunfishes), and Poeciliidae (mosquitofish). (After Avise 1994.)

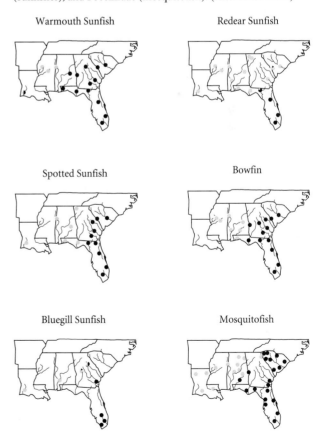

by genetic drift, and by peak shift (a combination of drift and selection). In the first two models, neither population experiences a polymorphism with inferior heterozygotes. Genotypes in one or both populations are replaced by genotypes of equal or greater fitness.

Speciation by natural selection Dobzhansky (1936) and Muller (1939) introduced a two-locus model that may be generalized to multiple loci (or to multiple alleles at one locus). Suppose the ancestral genotype in both populations is $A_2A_2B_2B_2$ (Figure 16.5A). Differences in the regime of natural selection (perhaps owing to environmental differences) favor replacement of A_2 by A_1 in population 1 and of B_2 by B_1 in population 2, yielding populations monomorphic for $A_1A_1B_2B_2$ and $A_2A_2B_1B_1$ respectively. Both A_1A_2 and A_1A_1 have fitness equal to or greater than A_2A_2 in population 1, *as long as the genetic background is B_2B_2*; likewise, B_1B_2 and B_1B_1 are superior to B_2B_2, as long as the genetic background is A_2A_2. However, an epistatic interaction between A_1 and B_1 causes incompatibility, so that either the hybrid $A_1A_2B_1B_2$ has lowered viability or fertility, or $A_1A_1B_2B_2$ and $A_2A_2B_1B_1$ are isolated by sexual behavior or another prezygotic barrier. The important feature of this model is that *neither population has passed through a stage in which inferior heterozygotes existed*: lower fitness is exhibited only when certain alleles that have been fixed at different loci in the two populations are conjoined. In a variant of this model, the successive allele substitutions that cause reproductive isolation may all transpire in one of the two populations (Figure 16.5B). Thus if both populations are ancestrally $A_2A_2B_2B_2$, B_1 may be fixed by selection in one population. If allele A_1 is advantageous on a B_1B_1 background, but not on a B_2B_2 background, A_1 will then be fixed, yielding populations $A_1A_1B_1B_1$ and $A_2A_2B_2B_2$ (the ancestral state), which are reproductively incompatible.

This model is widely favored by population geneticists. It supposes that *reproductive isolation arises as a by-product of adaptive divergence* in the two populations. It does not specify whether divergent selection acts *directly on the trait* that will reduce gene exchange when the populations later meet (e.g., courtship behavior or habitat use) or on other features, the genes for which *cause reproductive isolation as a pleiotropic side effect*. We shall address this question below, as well as some others: Do both ecological selection and sexual selection cause the divergence leading to speciation? Does natural selection ever favor the evolution of isolating mechanisms *because* they prevent gene exchange?

Speciation by genetic drift Precisely the same model may be envisioned, but with genetic drift, rather than natural selection, as the cause of allele substitution (Nei et al. 1983). On a B_2B_2 background, genotypes at the A locus are selectively neutral, as are genotypes at the B locus if the genetic background is A_2A_2. Thus no incompatibility occurs within population 1 as it evolves from $A_2A_2B_2B_2$ to $A_1A_1B_2B_2$, nor within population 2 as it evolves from

FIGURE 16.5 Two models of allele substitution leading to reproductive isolation between two populations, each initially composed of genotype $A_2A_2B_2B_2$. For each model, a possible adaptive landscape is shown, in which contour lines represent mean fitness as a function of allele frequencies at both loci (see Chapter 14). P represents a peak in the adaptive landscape. (A) Each population undergoes an allele substitution at a different locus. The hybrid combination $A_1A_2B_1B_2$ has low fitness (as indicated by the height of the landscape in its center) due to prezygotic or postzygotic incompatibility between A_1 and B_1. (B) Two successive substitutions occur at different loci in one of the populations; the other remains unchanged. The evolving population climbs an adaptive "ridge," first to the lower right and then to the upper right on the adaptive landscape. A hybrid population, located in the center of the landscape, again would have low fitness.

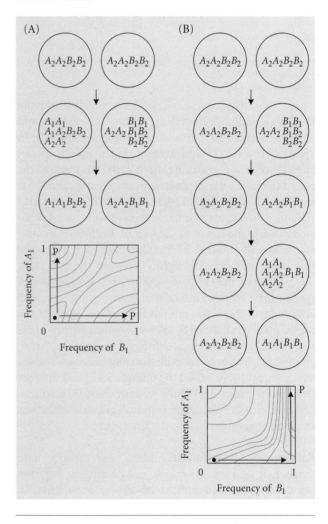

$A_2A_2B_2B_2$ to $A_2A_2B_1B_1$. However, the interaction between A_1 and B_1 yields reproductive incompatibility, as in the previous model.

Speciation by peak shift Also known as peripatric speciation or founder effect speciation, this process involves a combination of genetic drift and natural selection in a colony that experiences a bottleneck of small population size, perhaps because it has been founded by a few individuals. At certain loci, according to this model, rare alleles that lower the fitness of heterozygotes increase in frequency by genetic drift, and are then fixed by natural selection. This may, in turn, create epistatic selection for allele frequency changes at other loci. The new genetic constitution of the colony is therefore incompatible with that of the ancestral species.

To use Wright's metaphor of the adaptive landscape (see Chapter 14), the colony undergoes a *shift to a new adaptive peak* (Figure 16.6B). Suppose the ancestral population of a colony contains, at a low frequency, a chromosome rearrangement that greatly lowers fertility in heterozygous condition. The ancestral homokaryotype, A_1A_1, is therefore strongly favored by natural selection. If, due to genetic drift in a sufficiently small colony, the alternative chromosome arrangement A_2 exceeds some critical frequency, such as 0.5, natural selection will increase its frequency and eventually fix it (see Chapter 13). This may occur if, by chance, A_2 has a high frequency among the few founders of the colony (e.g., its initial frequency would be 0.25 if one member of a single founding pair were heterozygous). The same principle applies to alleles at one or more loci when they lower fitness in heterozygous condition.

There are several variant models of speciation by peak shift. Ernst Mayr (1954, 1963) postulated that most genetic variation would be lost in a colony that experienced a bottleneck during and after its founding, so that selection would favor rare alleles or new mutations that confer high fitness when they function on a highly homozygous genetic background (or "genetic environment") at other loci. Fixation of such alleles would in turn select for new, favorably interacting alleles at many other loci, leading to massive, rapid genetic change. Alan Templeton (1980, 1996) postulated that genetic changes of this kind are limited to a smaller number of loci. In his view, genetic variation in the colony is not depleted very much, but genetic drift, followed by natural selection, fixes new alleles at a few "major" loci that strongly affect fitness. This process creates selection for new alleles at "modifier" loci that interact with the major loci. Thus a new allele combination evolves that affects one or a few important features of the organism. Hampton Carson (1975) emphasized the genetic effects of rapid, exponential growth in a newly formed colony. During exponential growth, selection is weaker than in a stable population (Slatkin 1996), and permits new combinations of alleles to increase in frequency, initiating evolution along a new trajectory to a different adaptive peak.

An important difference between these founder effect models and speciation by natural selection alone is that in these models, selection is not caused by differences in the environments of the colony and the parent populations; the initial and final adaptive peaks are alternative genetic constitutions toward which natural selection can drive populations even if they experience the same environment. In contrast, the model of speciation by divergent selection envisions *different adaptive landscapes* for the two populations, due to differences in ecological or other agents of selection (Figure 16.6A).

We now examine evidence, and some further theory, bearing on these models of allopatric speciation.

The Role of Genetic Drift

Deferring to a later section our discussion of speciation by peak shift, in which genetic drift plays a role, we ask here

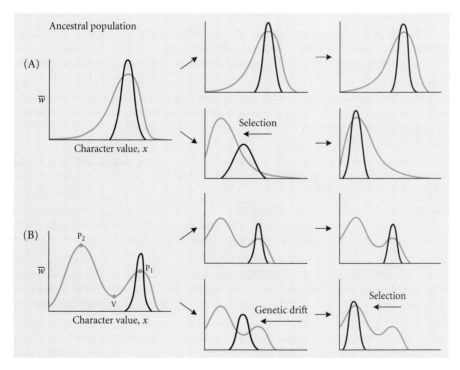

FIGURE 16.6 Models of speciation by (A) adaptive divergence and (B) peak shift. The black curve is the distribution of a character, x, and the colored curve is the relation between $\bar{x}$ and mean fitness, $\bar{w}$. Two populations are derived from an ancestral population, shown at left. (A) One population occupies an altered environment, and $\bar{x}$ evolves to a new (lower) value by natural selection. Divergence in the character confers reproductive isolation. (B) The populations occupy similar nvironments, but in one, $\bar{x}$ evolves by genetic drift past the adaptive valley (V), and then by selection to another adaptive peak (P_2).

whether substitution of *neutral* alleles within populations can give rise to reproductive isolation. There is little or no evidence that reproductive isolation among natural populations has arisen by this mechanism. However, such evidence would be exceedingly difficult to obtain, because it is very difficult to rule out the operation of natural selection. By the same token, genetic drift cannot be ruled out in the absence of evidence for natural selection, and it is philosophically supportable to argue that phenomena (such as speciation) should be ascribed to random processes (a "null hypothesis") unless there is evidence for deterministic processes (such as selection). In a cross between two *Drosophila* species, for example, hybrid sterility arises from epistatic interactions between multiple genes in one species (*D. sechellia*) and genes in the other (*D. simulans*), whereas each of the *D. sechellia* genes individually has little or no measurable effect (see Chapter 15). It is quite possible, but difficult to demonstrate, that these alleles had so little effect in *D. sechellia* that they were fixed by genetic drift.

Many experimenters have sought to determine whether reproductive isolation arises between populations of *Drosophila* (and a few other organisms) in the laboratory. Almost all the experiments designed to determine whether genetic drift causes reproductive isolation have used a design more appropriate for our discussion of the role of founder effects, and will be treated in that section below.

The Role of Natural Selection

The most widely held view of vicariant speciation is that it is caused by *natural selection, which causes the evolution of genetic differences that create prezygotic and/or postzygotic incompatibility.* Some, perhaps most, of the incompatibility evolves while the populations are allopatric, so that a substantial or complete barrier to gene exchange exists when the populations first meet if their ranges expand (Mayr 1963). Thus both pre- and postzygotic isolation are *effects*—by-products—of the divergent selection that occurred during allopatry. There is also the possibility that prezygotic (e.g., sexual) isolation may evolve, or be enhanced, by natural selection when the populations meet if genotypes that do not produce unfit hybrid offspring have greater reproductive success. This controversial idea, known as *reinforcement of isolating mechanisms*, will be treated below. We first discuss whether or not natural selection in spatially segregated populations has the *effect* of engendering reproductive isolation. Both pre- and postzygotic barriers might evolve due to differences in selection stemming from *ecological* factors, both abiotic and biotic. Prezygotic isolation might also evolve due to *sexual selection.*

Ecological Selection and Speciation

Because differences among conspecific populations and among related species often include adaptations to different ecological conditions, many biologists have assumed that reproductive isolation is a side effect of adaptive divergence. This interpretation, although it may well be correct, requires evidence that the genes underlying the

adaptive differences either have pleiotropic effects resulting in reproductive isolation or be closely linked to those responsible for isolation (divergence in which, then, would be ascribed to hitchhiking). There exists remarkably little such evidence from natural populations.

Almost nothing is known of the developmental or physiological function of those genes that cause postzygotic isolation, so it is perhaps not surprising that their divergence can seldom be ascribed to an identifiable ecological factor. One of the few exceptions was found in the monkeyflower *Mimulus guttatus* (Macnair and Christie (1983). Most hybrids between a copper-tolerant population (a recently evolved trait) and an intolerant population were inviable. Genetic analysis revealed that the allele for copper tolerance (or an allele at a very closely linked locus) interacts with another locus in the copper-intolerant plants to cause mortality of hybrid seedlings. It appears, therefore, that postzygotic isolation has arisen as a pleiotropic effect of selection for copper tolerance.

Prezygotic isolation has been convincingly ascribed to ecological selection in several cases in which isolation is a consequence of divergence in the trait itself, rather than a pleiotropic effect of the divergent genes. Verne and Karen Grant (1965; Grant 1993) and others have described instances in which related species or conspecific populations of plants are adapted to different, locally abundant pollinators, and are not strongly isolated by any other barriers. Among columbines (*Aquilegia*) in western North America, for example, species at lower altitudes are pollinated by hummingbirds and those at higher altitudes by hawkmoths (Sphingidae). The two groups differ accordingly in coloration, flower form, and time (day versus night) of nectar and scent production. Within a single species of columbine, geographic variation in flower characteristics is correlated with the relative abundance of hawkmoths and bumblebees (Miller 1981; Hodges and Arnold 1994). Similarly, at higher altitudes in Japan, *Cimicifuga simplex*, in the buttercup family, is pollinated by bumblebees and does not produce fragrance; at lower altitudes, where bumblebees are scarce during the flowering season, the plant releases scent, has a somewhat different inflorescence, and is pollinated mostly by butterflies (Pellmyr 1986).

One of the few examples of reproductive isolation in animals that is ascribable to ecological selection is found among the ground finches (*Geospiza*) of the Galápagos Islands, in which divergence in the size and shape of the bill represents adaptation to different diets (see Figure 21 in Chapter 8). The species are very similar in coloration. Males will court a stuffed female bird (Figure 16.7) more readily if the specimen is of their own species, and they discriminate between heads of different species placed on the same body (Grant 1986). Although other differences probably also contribute to assortative mating, the bill apparently is used in species discrimination.

Another likely example of speciation caused by natural selection is provided by the three-spine stickleback (*Gasterosteus aculeatus*). In western Canada (British Columbia),

FIGURE 16.7 A male ground finch, *Geospiza difficilis*, courting a stuffed female in an experiment on species discrimination. Female specimens with the "wrong" body size or beak size elicited less courtship. (From Grant 1986; photograph courtesy of P. T. Boag.)

a marine ancestor has given rise independently to several freshwater populations, in several rivers, that are smaller in body size (Figure 16.8A). Also, several freshwater lakes harbor two forms, a small open-water "species" and a large bottom-feeding "species," that have evolved from the same marine ancestor, probably by two successive colonizations of each lake (Figure 16.8B). Experiments on mating preferences have shown that in each comparison (marine versus river fish, open-water versus bottom fish), females of each form prefer to mate with males of the size typical of their own species. These body size differences almost certainly evolved by natural selection related to habitat use, since they have consistently evolved by parallel evolution in different localities. (Recall from Chapter 12 that parallel evolution is often good evidence for adaptation.) Thus "parallel speciation"—independent evolution of the "same" reproductively isolated forms in different localities—suggests that ecological selection can cause speciation (Schluter and Nagel 1995).

In the laboratory, using houseflies and several species of *Drosophila*, investigators have tested for reproductive isolation among subpopulations drawn from a single base population and subjected to divergent selection for various morphological, behavioral, or physiological characteristics (summarized by Ringo et al. 1985; Rice and Hostert 1993). In many cases, sexual isolation and/or postzygotic isolation, sometimes quite strong, was found, demonstrating that *substantial progress toward speciation can be observed in the laboratory*, and that it can arise as a by-product of divergent selection. In two populations of houseflies (*Musca domestica*) selected for 16 generations for the propensity to move up or down, the frequency of within-population (homogamic) mating was 60 percent greater than the frequency of between-population (heterogamic) mating (Hurd and Eisenberg 1975). Populations of *D. pseudoobscura* that had been maintained at 16°, 25°, and 27°C for 4 1/2 years showed slight but significant sexual isolation and significant F_2 breakdown in the viability of interpopulation crosses (Ehrman 1964; Kitagawa 1967). The *D. pseudoobscura* experiment shows that divergent *natural* selection can cause in-

cipient reproductive isolation, because the populations were allowed to adapt naturally to their different environments.

Sexual Selection and Speciation

Sexual selection is variation in reproductive success due to variation in the ability to acquire mates (see Chapter 12). It arises from competition among individuals of one sex (usually the male) for access to the other sex, or from the preference ("choice") of one sex (usually the female) for certain characteristics in the other sex (Andersson 1994). The male's secondary sexual characteristics frequently play a role in both contexts (Berglund et al. 1996). We are concerned here mostly with sexual selection by female preference.

Experiments with many species of animals have shown that females choose among conspecific males on the basis of the state of one or more features of the courtship display.

FIGURE 16.8 "Parallel speciation" in the three-spined stickleback. The large marine ancestor has given rise independently to small-bodied forms in several rivers (A), and to pairs of open-water and bottom-foraging forms (represented by solid and hatched environments) within each of several lakes (B). Females mate assortatively with males on the basis of body size. This isolating character is evidently adaptive, since it has repeatedly evolved in the same way. (After Schluter and Nagel 1995.)

(A)

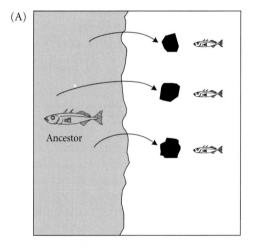

(B)

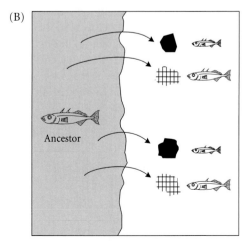

Often, although not always, they prefer males with more stimulating traits—those that are larger, brighter, or more complex (Ryan 1990). For instance, male widowbirds whose tail feathers were artificially lengthened acquired more mates (see Figure 12 in Chapter 12); the females appeared to respond to what ethologists call a SUPERNORMAL STIMULUS.

Why females should have such preferences is a controversial question (Kirkpatrick 1987; Pomiankowski 1988; Andersson 1994). Several hypotheses have been advanced. These include:

1. The preference is the result of a self-reinforcing process, known as RUNAWAY SEXUAL SELECTION, whereby the evolution of male trait and female preference reinforce each other.
2. The preference is adaptive, because the male trait is correlated with other characteristics that will increase the fitness of the female's offspring.
3. Females prefer traits that, inherited by their sons, will enable the sons to compete successfully with other males for territories or access to females.
4. Females may perceive louder, brighter, or otherwise more conspicuous male traits more easily against background "noise" in the environment, and preference for such traits may be advantageous because it enables females to find mates faster. Related to this hypothesis is the possibility that females prefer *novel* or *supernormal* stimuli because this releases them from habituation to commonly encountered stimuli (Hinde 1970; West-Eberhard 1979).
5. Females have sensory biases, residing in the organization of the sense organs and the nervous system, toward certain stimuli (Ryan 1990).

These hypotheses and the evidence bearing on them are discussed further in Chapter 20. The possible roles of novelty and sensory bias in female preferences are particularly interesting because they suggest that sexual selection may favor the evolution of "arbitrary" traits that lack any evident adaptive value with respect to ecological factors. Sensory bias, alone or as an initiator of runaway sexual selection, may therefore cause the evolution of different female preferences and male traits in different populations, should mutations that alter different male characters occur. The result would be divergence in what Paterson (1982) has called the SPECIFIC MATE RECOGNITION SYSTEM, the system of communication between male and female that underlies both

sexual selection and mating within rather than between species (Ryan and Rand 1993b).

Although much of the evidence for the role of sexual selection in speciation is circumstantial, the weight of it strongly implies that prezygotic (sexual) isolation often arises as a by-product of sexual selection. This evidence is of several kinds. First, *sexually selected characters act as barriers to interbreeding.* For example, characteristics of the male call in the túngara frog, *Physalaemus pustulosus*, are sexually selected by female preferences. Females of this species also prefer calls of their own species to calls of other *Physalaemus* species, implying that the call differences act as reproductive barriers (Figure 16.9; Ryan and Rand 1993a). Some species of African lake cichlids are distinguishable only by male color pattern, and females choose mates on this basis (McKaye et al. 1984; Seehausen et al. 1997). Sexual selection, presumably responsible for the diverse color patterns of these cichlids, has been postulated as a major contributor to their extraordinarily high rate of speciation (Dominey 1984).

A second line of evidence is the *high species richness of many taxa in which sexual selection appears to be intense and diverse.* The African lake cichlids are one such example. Others include hummingbirds, (Trochilidae: 341 species; Gill 1995), pheasants (Phasianidae, including grouse and peafowl: 205 species), ducks (147 species), birds of paradise (Paradisaeidae: 42 species, almost all on a single island, New Guinea), and the Hawaiian *Drosophila* (more than 800 species; Kaneshiro 1983). In all these cases, many of the species are genetically very similar (judging from allozymes) and so must have arisen recently and rapidly. The groups of birds mentioned include species that are interfertile, since interspecific hybrids occur rather often (Sibley 1957); thus sexual isolation may be a stronger barrier to gene exchange than postzygotic isolation. In all these groups, females of different species may be almost indistinguishable, but males differ in their extraordinary modifications of a great variety of characteristics. Male hummingbirds, for example, may have brilliant, iridescent feathers on the throat or crown, long crests or whiskers, and long, often peculiarly shaped tail feathers (see Figure 13 in Chapter 15). The birds of paradise present an even more riotous variety of brilliant plumages and ornaments, which they display to the accompaniment of bizarre behaviors and loud vocalizations. In all these groups of birds, the majority of species do not form pair-bonds, which enhances the opportunity for high variation in mating suc-

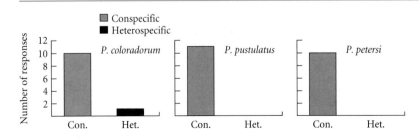

FIGURE 16.9 Responses of female túngara frogs, *Physalaemus pustulosus*, to recordings of calls of conspecific males compared with those of three other species in the same genus. The difference in calls among these species acts as a barrier to interbreeding. (After Ryan and Rand 1993.)

cess—for strong sexual selection—among males. Sexual selection by female choice has been demonstrated in birds of paradise and grouse, in which males vary greatly in breeding success (Gibson and Bradbury 1985; Pruett-Jones and Pruett-Jones 1990).

Acting on the assumption that pronounced sexual dimorphism in birds is due to sexual selection by female choice, Timothy Barraclough et al. (1995) compared the species diversity of sister groups of perching birds, based on phylogenetic hypotheses derived from DNA-DNA hybridization (Sibley and Ahlquist 1990; see Chapter 5). More often than not, the clade with a higher proportion of sexually dimorphic species contained more species than the related clade. This analysis suggests that sexual selection increases the rate of speciation.

Natural Selection for Reproductive Isolation

Our discussion of speciation thus far has assumed that reproductive isolation arises as a *side effect* of genetic divergence that transpires, perhaps due to natural selection, in allopatric populations. However, many authors have supposed that reproductive isolation evolves, at least in part, as an *adaptation to prevent the production of unfit hybrids*. The champion of this viewpoint was Theodosius Dobzhansky, who expressed the hypothesis this way:

> Assume that incipient species, A and B, are in contact in a certain territory. Mutations arise in either or in both species which make their carriers less likely to mate with the other species. The nonmutant individuals of A which cross to B will produce a progeny which is adaptively inferior to the pure species. Since the mutants breed only or mostly within the species, their progeny will be adaptively superior to that of the nonmutants. Consequently, natural selection will favor the spread and establishment of the mutant condition. (Dobzhansky 1951, 208)

To Dobzhansky, who introduced the term "isolating mechanisms," reproductive barriers were indeed mechanisms designed to isolate.

In contrast, Ernst Mayr (1963), among others, held that although natural selection might enhance reproductive isolation, reproductive barriers arise mostly as side effects of allopatric divergence. Mayr cited several theoretical objections to Dobzhansky's view (discussed below) and several lines of evidence for his own: sexual isolation exists among fully allopatric forms; it has not evolved in several hybrid zones that are thought to be thousands of years old; and features that promote sexual isolation between species are usually not limited to regions where they are sympatric and face the "threat" of hybridization. This issue continues to be a topic of debate (Butlin 1989; Howard 1993).

Suppose that two populations that have diverged in allopatry expand their ranges and form a hybrid zone, in which hybrids have lower fitness than the parental genotypes (see Chapter 15). We will call the differentiated forms semispecies, and in each semispecies we distinguish parapatric local populations, in the contact zone, from allopatric populations that are distant from the other semispecies. It is possible that (1) the hybrid zone will persist indefinitely, although its location may move; (2) mutations will arise and spread that improve the fitness of hybrids; or (3) mutations will spread that, as Dobzhansky envisioned, reduce the frequency of hybridization. What determines which of these outcomes will occur?

If mutations that *improve the fitness of hybrids* increase in frequency, this reduces the advantage of genes that accentuate reproductive isolation, leading ultimately to fusion of the semispecies into a single species. A probable example of this phenomenon has been described in the common shrew (*Sorex araneus*) in England, in which populations differ in complex metacentric chromosome fusions that should greatly reduce the fertility of heterozygotes. However, shrews in the contact zones have a high frequency of simple acrocentric chromosomes that do not reduce hybrid fertility very much. These have apparently replaced the chromosomes that cause lower hybrid fitness, and provide a bridge for gene exchange between the semispecies (Searle 1993). In the same vein, the fertility of hybrids between two species of sunflowers (*Helianthus*), measured by the proportion of viable pollen, increased greatly in the 16 years following contact between the two (Stebbins and Daly 1961).

Selection to improve the fitness of hybrids bears on an important point: *natural selection cannot strengthen postzygotic isolation between hybridizing populations* because this would require that alleles that reduce fertility or survival increase in frequency. This would be precisely antithetical to the meaning of natural selection! The only exception is the possibility that in species that provide parental care (such as yolk or endosperm, or postnatal feeding), selection might favor the ability to abort or neglect hybrid offspring if by doing so the parent could allocate resources to nonhybrid offspring whose prospects for high fitness were better (Grant 1966a; Coyne 1974).

The evolution of prezygotic barriers that Dobzhansky envisioned in regions of hybridization or overlap as a response to selection against hybrids is often called **reinforcement of prezygotic isolation**. This process has been cited as the cause of **reproductive character displacement**, meaning a *pattern* whereby characters that reduce mating between populations differ more where the two taxa are sympatric than where they are allopatric (Brown and Wilson 1956; Figure 16.10).[*]

---

[*]Some authors follow Butlin (1989) in using *reproductive character displacement* to mean the process of evolution of accentuated prezygotic barriers to mating between fully developed species—i.e., populations that do not actually exchange genes. They may be fully isolated by postzygotic barriers (e.g., full hybrid sterility), or selection may favor greater mating discrimination or more different courtship signals for other reasons (e.g., to avoid wasting time dallying with members of the other species in unconsummated courtship). Butlin restricts the term *reinforcement* to the evolution of stronger mating barriers between taxa that can exchange genes through hybrids that have low but nonzero fitness. This text uses the traditional definitions (Howard 1993).

FIGURE 16.10 Character displacement is a greater difference in a feature between sympatric than between allopatric populations of two species (1 and 2). The character may function in courtship or other aspects of reproduction (reproductive character displacement) or in an ecological context such as the use of food or other resources (ecological character displacement; cf. Figure 19 in Chapter 4).

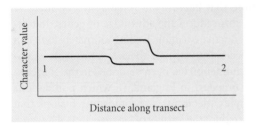

Selection for reinforcement of sexual or other prezygotic barriers arises from the reduced fitness of hybrids, as the quote from Dobzhansky makes clear. At least one locus—say, with alleles A_1 and A_2—must have low fitness when heterozygous. Prezygotic isolation is based on other loci; for example, B_1 carriers might tend to mate with other B_1 carriers, and B_2 with B_2. This might hold if, for example, B_1 and B_2 differed in mating season, or preferred different habitats. Or, conceivably, the B_1 and B_2 alleles might cause differences in both male signals and correlated female responses, but as we saw in Chapter 15, these are usually controlled by separate loci. If prezygotic isolation is based on multiple loci (as most data indicate; see Chapter 15), then a single locus, such as B here, will cause incomplete reproductive isolation.

There are several theoretical reasons why reinforcement of prezygotic isolation may be unlikely (Mayr 1963; Spencer et al. 1986; Sanderson 1989; Butlin 1989; but see Liou and Price 1994). We discuss them here with reference to hybrid zones between semispecies, but some of them apply in other contexts, such as sympatric populations.

1. Selection for isolating alleles (e.g., at the *B* locus in our example) occurs only where the semispecies meet. It is hard to see why the alleles should spread from the contact region into the allopatric populations and come to characterize each species in its entirety (Moore 1957). Yet many pairs of sister species are indeed so characterized.

2. If an isolating allele (say, B_2) has any pleiotropic disadvantage, it certainly will not spread through the allopatric populations. Worse, the allele favored in allopatric populations (say, B_1) will continually enter the contact zone by gene flow, counteracting selection for the B_2 allele. Because B_2 is favored in only a narrow area, but B_1 is favored in a broader area, this influx of genes may have a strong effect (Sanderson 1989).

3. If the fitness of F_1 hybrids is reduced by many heterozygous loci, each with only a slight effect, then backcrossing will produce a relatively wide hybrid zone in which most individuals are various backcross genotypes, perhaps with higher fitness than the F_1 hy-

brids. Thus "pure" genotypes may hybridize only rarely; if they interbreed mostly with relatively fit backcrosses, their progeny will have relatively high fitness, so selection for prezygotic isolation will be weakened (Butlin 1989). This argument exemplifies the importance of knowing the genetic basis for hybrid inviability and infertility (see Chapter 15).

4. If an allele such as B_2 does increase in frequency, so that cross-mating is reduced, but not eliminated, the strength of selection for further reinforcement will become lower because unfit A_1A_2 heterozygotes will arise less often. Thus alleles at other loci (say, C_2) that might perfect prezygotic isolation are unlikely to increase in frequency. Full reproductive isolation is not likely to be achieved.

5. Recombination between the isolating locus B and the fitness-reducing locus A counteracts natural selection for reinforcement (Felsenstein 1981; Sanderson 1989).

The effect of recombination is very important, and requires further explanation. Initially, the two populations in our example have genotypes $A_1A_1B_1B_1$ and $A_2A_2B_1B_1$. They interbreed freely, since they are identical at the "mating locus," B, and produce hybrids (A_1A_2) with reduced fitness. An allele B_2, causing its bearers to mate preferentially with each other, can increase in frequency only if its bearers have higher fitness than B_1 carriers. This will be the case only if B_2 carriers are more likely to be homozygous at the A locus than B_1 carriers. If B_2 copies are almost exclusively associated with A_2 copies (i.e., if there is strong linkage disequilibrium), most progeny of matings between B_2 carriers will be A_2A_2. Because their average fitness is higher than that of B_1 carriers (many of which are less fit A_1A_2 hybrids), and because they inherit B_2 alleles, B_2 increases in frequency, and incipient sexual isolation evolves.

If, however, recombination distributes copies of B_2 equally among A_1A_1 and A_2A_2 individuals (i.e., if linkage disequilibrium breaks down), then B_2 carriers are as likely as B_1 carriers to have unfit A_1A_2 offspring. Thus, the fitness of B_2 will be no higher than that of B_1, B_2 will not increase in frequency, and no sexual isolation will evolve. Hence, an allele for assortative mating (partial sexual isolation) has no selective advantage unless it is associated, in linkage disequilibrium, with alleles that affect hybrid fitness (Maynard Smith 1966; Felsenstein 1981). If linkage disequilibrium exists between a locus that reduces the fitness of hybrids and a locus that reduces the incidence of cross-mating, it will be reduced by recombination in F_1 hybrids (A_1B_1/A_2B_2) if they succeed at all in reproducing. The greater the rate of recombination, and the fitter the hybrids are, the greater the rate at which the allele for assortative mating becomes randomly combined with the alleles that reduce the fitness of hybrids, and hence the lower the intensity of selection for the assortative mating allele. Thus *recombination opposes selection for alleles that confer prezygotic isolation*. The problem becomes even worse if other loci are required to further increase assortative mating beyond the level that the

B locus confers, for then the combinations of alleles at the several loci will remain intact only if all the loci are tightly linked, or if the fitness of the F₁ hybrid is very low indeed (Felsenstein 1981). (If the hybrid does not reproduce—i.e., if the hybridizing populations are reproductively isolated by postzygotic inviability or infertility—then recombination does not occur.)

If both the postzygotic decrement of hybrid fitness and prezygotic isolation are caused by one or a few genes with large effects, reinforcement of sexual isolation may occur if the genes *happen to be tightly linked*. If either pre- or post-zygotic isolation is due to several or many loosely linked genes, then the multiple associations of loci necessary for selection to increase prezygotic isolation break down by re-combination. We have seen (in Chapter 15) that both post- and prezygotic isolation are generally based on several, and in some cases many, genes that are not necessarily (or usu-ally) linked. This fact suggests that *recombination will often, perhaps usually, be a powerful force against the reinforcement of prezygotic barriers to gene exchange,* and thus that *"iso-lating mechanisms" have probably seldom evolved because of their isolating function.* The major exceptions should be cases in which virtually complete postzygotic isolation (hybrid sterility or inviability) evolved in the populations before they came into contact, for in such cases recombi-nation does not dissociate the alleles at the different loci. Thus, *full allopatric speciation*—the evolution of a com-plete barrier to gene exchange—*normally requires that ei-ther complete prezygotic or complete postzygotic isolation evolve before contact*—i.e., before the opportunity for mas-sive hybridization presents itself.

If reinforcement does occur, prezygotic barriers should be accentuated in contact zones relative to populations dis-tant from the contact, because selection for prezygotic iso-lation should occur only where the divergent populations interact. That is, *patterns of reproductive character displace-ment should constitute the evidence for reinforcement.* The displaced characters might be features such as flowering time, male courtship signals, or simply the level of mating isolation observed in laboratory tests. Howard (1993), in a review of such studies, reported that a pattern of character displacement has been found in 33 of 48 cases in which it has been sought. However, in relatively few instances is there evidence that this is a consequence of reinforcement (i.e., selection for prezygotic isolation). In many of these cases, for example, there is little or no evidence that the taxa ever mate with each other.

Although some authors do not feel that any natural ex-amples of reinforcement have been well documented (But-lin 1989), a few examples seem convincing. For example, wing color elicits courtship in two species of damselflies (*Calopteryx*), natural hybrids of which have low fitness. The species differ more in wing color where they are sympatric than where they are allopatric (Figure 16.11; Waage 1979). Moreover, we noted (in Chapter 15) that in Coyne and Orr's (1989a) analysis of laboratory studies of reproductive isolation in *Drosophila*, prezygotic isolation was found to be

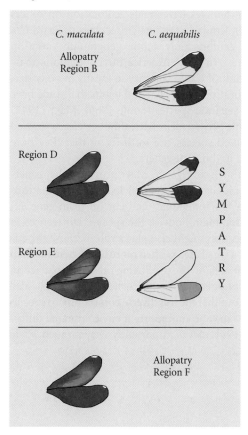

FIGURE 16.11   Reproductive character displacement in female wing color of the damselfly *Calopteryx aequabilis*. *Calopteryx maculata* has dark wings that display little geographic variation. *Calopteryx aequabilis* has smaller wing spots and paler wings in regions where this species is sympatric with *C. maculata* than in more northern regions where it is allopatric. (After Waage 1979.)

more pronounced between sympatric than between al-lopatric pairs of taxa during the early stages of divergence. That is, prezygotic isolation appears to evolve faster be-tween populations that are in contact than between those that are not, even if the degree of postzygotic isolation is the same. This striking pattern suggests that *reinforcement might be common after all.* The importance of reinforcement of reproductive isolation is still a controversial topic.

Peripatric Speciation

The hypothesis   One of Ernst Mayr's most influential and controversial hypotheses was FOUNDER EFFECT SPECI-ATION (1954), which he later termed *peripatric speciation* (1982b). This process may be thought of as speciation due to a *shift, initiated by genetic drift and followed by natural se-lection, between adaptive peaks.* Mayr did not, however, ex-press it in these terms.

Mayr's hypothesis stemmed from the observation, in many birds and other animals, that *isolated populations with restricted distributions,* in locations peripheral to the distribution of a probable "parent" species, *often are highly divergent*, even though their ecological environment appears similar to that of the "parent." The kingfisher described earlier (see Figure 16.2) is one example. Another is the robin *Petroica multicolor*, in which the bright plumage of the male contrasts with the duller plumage of the female throughout eastern Australia. Yet among the various Melanesian islands nearby, males have "female" coloration on some, and females have "male" coloration on others (Figure 16.12). Similarly, the small insectivorous lizard *Uta stansburiana* exhibits only subtle geographic variation throughout western North America, but populations on islands in the Gulf of California vary so greatly in body size, scalation, coloration, and ecological habits that some have been named separate species (Soulé 1966). Mayr (1942) noted that some peripheral isolates, although clearly derived from a mainland species, are so different that they have been assigned to different genera.

To explain such divergence, Mayr proposed that genetic change could be very rapid in localized populations founded by a few individuals and cut off from gene exchange with the main body of the species. Differences in ecological selection might also play a role, because the environment of a small area is often more homogeneous than that of a larger area, so the conflicting pressures that act on a widespread population may be less numerous. (It has also been pointed out that the less diverse community of species in a small area may be an important environmental difference: Turner 1981.) More importantly, Mayr argued, allele frequencies at some loci will differ from those in the parent population because of accidents of sampling—i.e., genetic drift—simply because a few colonists will carry only some of the alleles from the source population, and at altered frequencies. (He termed this initial alteration of allele frequencies the *founder effect*.) *Because epistatic interactions among genes affect fitness, this initial change in allele frequencies at some loci alters the selective value of genotypes at other, interacting loci*—that is, the "genetic environment" of other loci is altered. Another important difference between the colony and the source population, Mayr suggested, is that the colony is isolated from the gene flow among populations that increases the level of genetic variation in any local population of the parent species. Thus, in the colony, the genetic environment that imposes selection on alleles—including new mutations—is less variable, more homozygous, and may favor alleles that confer high fitness in specific combinations with certain alleles at other loci. In contrast, selection in populations of the parent species favors "generalist" alleles that are advantageous in any of many gene combinations.

Mayr (1963) believed that epistatic interactions are so strong and ubiquitous that each genetic change might trigger others, so that massive genetic change (a "genetic revolution") might transpire, yielding reproductive isolation as a by-product. An important implication of this hypothesis, Mayr (1954) pointed out, is that substantial evolution is likely to occur so rapidly and on so localized a geographic scale

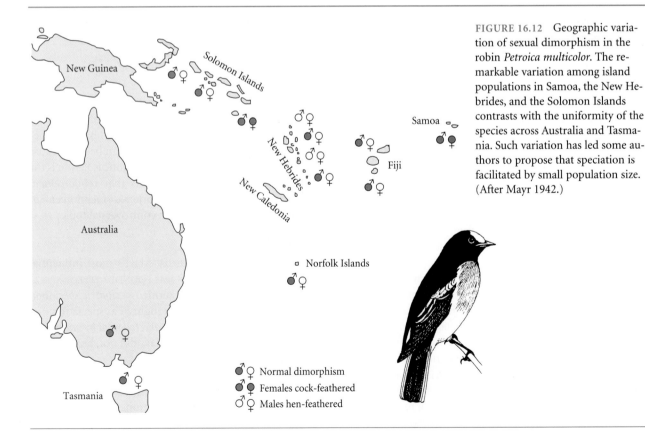

FIGURE 16.12 Geographic variation of sexual dimorphism in the robin *Petroica multicolor*. The remarkable variation among island populations in Samoa, the New Hebrides, and the Solomon Islands contrasts with the uniformity of the species across Australia and Tasmania. Such variation has led some authors to propose that speciation is facilitated by small population size. (After Mayr 1942.)

that it will seldom be documented in the fossil record. If such a new species expanded its range, it would appear suddenly in the fossil record, without evidence of intermediate steps. Thus this hypothesis, he said, may help to explain the rarity of fossilized transitional forms among species and genera. Mayr thus anticipated, and provided the theoretical foundation for, the idea of *punctuated equilibrium* (see Chapters 6, 24) advanced by some paleontologists (Eldredge and Gould 1972).

A somewhat similar theory was advanced by Hampton Carson (1975), who suggested that genetic reorganization may be limited to certain groups of genes among which there are strong epistatic interactions, and that it may be enhanced by repeated fluctuations in population size. He suggested that genes controlling mate recognition and courtship behavior may be among those that are affected (Carson 1982). In a related model, Alan Templeton (1980) emphasized that genetic drift may initiate changes in the frequency of alleles of large effect at only a few loci; polygenic modifier loci are then altered by natural selection to bring about a new coadapted state of a character—such as courtship behavior—that is controlled by the major loci.

All these versions of the founder effect speciation model envision a shift in the colony from one adaptive genetic constitution (that of the parent population) through a less adaptive constitution (an adaptive valley) to a new adaptive equilibrium (see Figure 16.6B and Figure 14.9C). The most elementary such change would be substitution of one allele for another when the heterozygote has lower fitness than either homozygote, e.g., when A_1A_1, A_1A_2, and A_2A_2 have fitnesses of 1, $1 - s$, and 1 (see Chapter 13). The same model applies to a chromosome rearrangement that reduces fertility in the heterozygous state. However, Mayr, Carson, and Templeton envision not a single-locus substitution, but a shift between peaks determined by epistasis among loci, as, for example, when $A_1A_1B_1B_1$ and $A_2A_2B_2B_2$ both have higher fitnesses than any other combinations of alleles, and represent, respectively, the genetic composition of the parent population and the state to which the colony evolves (see Figures 14.5 and 14.6). The new "genetic environment" that these loci constitute in the colony may select for changes at yet other loci.

This model differs importantly from the genetic model of Dobzhansky and Muller, in which each successive allele substitution in one or both populations is caused entirely by natural selection, and neither population ever passes through a stage in which mean fitness is lowered. In the peak-shift models, one or more alleles increase in frequency despite their disadvantage in heterozygous condition, so that mean fitness is temporarily decreased. This stage of allele frequency change requires genetic drift.

Theoretical considerations Many theoretical population geneticists are skeptical about the likelihood that peak shifts engender speciation (Barton and Charlesworth 1984; Barton 1989; Charlesworth and Rouhani 1988; see Chapter 14). The theoretical arguments and counterarguments are mathematically too complex to describe here in depth. One of the important points, however, is that the likelihood that genetic drift will carry allele frequencies across an adaptive valley is inversely proportional to $(\bar{w}_p/\bar{w}_v)^{2N}$, where $\bar{w}_p$ is the population's fitness at the initial adaptive peak, $\bar{w}_v$ is its fitness at the adaptive valley separating peaks, and N is the effective population size of the colony. This expression says, in effect, that *the more intense selection is against genotypes (heterozygotes) that are intermediate between those in the two "optimal" populations, the less likely a peak shift is.* If selection is not very strong, $\bar{w}_v$ is fairly large, so a peak shift is more likely. But reproductive isolation requires that hybrids ($A_1A_2B_1B_2$) have low fitness, either because they fail to mate or because of postzygotic inviability or infertility. *Thus if the adaptive valley is shallow enough for a peak shift to be likely, the genetic difference between the populations will cause little reproductive isolation; if selection is strong and the valley is deep, the populations will be reproductively well isolated, but the shift to the new genetic composition is unlikely to occur.*

This argument (which is related to criticisms of Wright's shifting balance theory; see Chapter 14) is a strong one, but perhaps is not fatal to the hypothesis. Several models show that peak shifts may be much more likely if different assumptions are made (Kirkpatrick 1982a; Price et al. 1993; Wagner et al. 1994; Gavrilets and Hastings 1996). In some models, for instance, certain kinds of epistatic interactions between loci severely reduce the fitness of hybrids, but peak shifts are nevertheless fairly likely to occur (Wagner et al. 1994). Another consideration, not incorporated into either Mayr's hypothesis or the theoretical arguments against it, is that changes in allele frequencies due to genetic drift can convert epistatic genetic variance into additive genetic variance (Goodnight 1988; Whitlock et al. 1993; see Chapter 14). Because response to selection is proportional to the additive genetic variance in a character, *some characteristics may evolve more rapidly in a colony derived from a few founders than in the large population from which they came.*

Evidence from natural populations Evidence is mounting rapidly that *many species do originate*, as Mayr said, *as localized "buds" from a widespread parent species.* This evidence consists of phylogenies of DNA sequences (gene genealogies) taken from various geographic populations of widespread species and closely related species with narrower, often peripheral distributions. The localized species often prove to be more closely related to certain populations of the widespread species than the populations of that species are to one another (Avise 1994). For example, the beach mouse *Peromyscus polionotus*, of the southeastern United States, has long been postulated to be a peripheral derivative of the deer mouse *P. maniculatus*, which occurs throughout much of the rest of North America. The genealogy of mitochondrial DNA indicates that this is indeed the case (Figure 16.13). Similarly, the island endemics *Drosophila mauritiana* and *D. sechellia* appear to be derived from African *D. simulans* (see Chapter 15). Other such cases are steadily being revealed.

FIGURE 16.13 Relationships among mitochondrial DNA haplotypes in the widespread deer mouse, *Peromyscus maniculatus*, and the beach mouse, *P. polionotus*, distributed in the southeastern United States near the periphery of the distribution of *P. maniculatus*. The clades of haplotypes within *P. maniculatus* are localized within different regions throughout the United States. This gene tree implies that *P. polionotus* originated from a local, peripheral population of *P. maniculatus*. (After Avise 1994.)

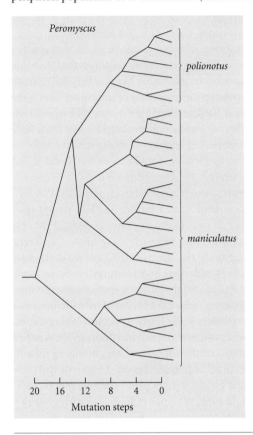

Rapid divergence and speciation in localized peripheral populations, however, does not in itself constitute evidence for speciation by peak shift, because the divergence might have been caused solely by natural selection, and not at all by genetic drift during population bottlenecks. Indeed, many evolutionary geneticists hold that this is the simpler, and therefore preferable, explanation (Barton and Charlesworth 1984; Coyne 1992, 1994). Other sources of evidence are needed.

Kenneth Kaneshiro (1983) and others have found that sexual isolation between closely related species of *Drosophila* is often, although by no means always, asymmetrical, especially among the diverse Hawaiian species in which sexual selection has caused the evolution of extraordinary courtship-related characters. When ancestral and derived pairs of Hawaiian species are tested for sexual isolation, females of the ancestral species generally discriminate against males of the derived species, but females of the derived species mate rather readily with males of the ancestral species. Although their mating behavior has not been analyzed in detail, Kaneshiro suggested that the derived males may have lost

elements of the complex ancestral courtship behavior due to genetic drift during a founder event. Females in the colony may have adapted to the altered, simpler courtship behavior because the initially low population density favored those that mated indiscriminately. Thus the derived females respond to ancestral males, whose courtship includes all the elements of the derived males' behavior, but the ancestral females reject derived males because of their "incomplete" courtship signals. Although plausible, this hypothesis remains to be confirmed.

The high rate of speciation in Hawaiian *Drosophila* has been attributed to bottlenecks during and after colonization among and within islands (Carson and Templeton 1984). A bottleneck in population size sufficient to allow a peak shift at even a few strongly selected loci should reduce genetic variation, so that ancestral and derived populations should display substantial differentiation of allozymes and other molecular markers. Closely related Hawaiian species of *Drosophila*, however, have high heterozygosity and low genetic distances (*D*) (see Barton 1989). This is true of many pairs of closely related species, and is perhaps *the strongest evidence against the founder effect as an important cause of speciation.*

In some cases, though, derived species have undergone bottlenecks in population size. The island species *Drosophila sechellia* has lower levels of DNA polymorphism than its progenitor *D. simulans* (see Chapter 15). Among the nine European species of small ermine moths, *Yponomeuta*, *Y. rorellus* is unique in that it almost lacks genetic variation at about 75 enzyme loci. Unlike other members of the genus, which produce a blend of sex pheromone compounds, the female *Y. rorellus* produces only a single compound. Whereas males of the other species respond only to a blend, male *Y. rorellus* are attracted to either a blend or the single compound, and have antennal receptors for the compounds in the other species' pheromone blends, implying that female production of these compounds was lost during the evolution of *Y. rorellus*. We cannot know whether or not a bottleneck was responsible for the simplified pheromone composition without knowing how this feature is genetically controlled, but the almost complete homozygosity suggests that this species has indeed experienced a population bottleneck (Löfstedt et al. 1986; Menken 1987). In all such cases, however, it is impossible to say whether the bottleneck accompanied speciation or occurred at a later time.

Speciation by chromosome rearrangement In Chapter 15, we noted that closely related species often differ by chromosome rearrangements that might be expected to reduce the fertility of F₁ hybrids, and thus to confer postzygotic isolation. However, the incidence of aneuploid gametes in the hybrids is sometimes not as high as might be expected, and it is difficult to determine the degree to which hybrid infertility is caused by structural chromosome differences or by gene interactions.

Substitutions of rearrangements that lower the fertility of heterozygotes require, in part, the action of genetic drift, and therefore reduced population size. Thus if speciation occurs by this mechanism, it is similar to founder effect speciation. For the reasons described above, rearrangements that lower heterozygotes' fertility slightly are more likely to be fixed than those that lower it greatly, and the more severe the effect on fertility, the smaller the population size or bottleneck required (Wright 1941; Walsh 1982; Lande 1984). Genetic isolation is more likely to be accomplished by successive chromosome substitutions, each of which reduces fertility slightly, than by a single substitution that reduces it severely (Walsh 1982). The average level of allozyme heterozygosity is lower in those animal groups in which the rate of evolution of chromosome structure and number has been high, as expected if small population size favors substitution of rearrangements by genetic drift (Coyne 1984a).

Chromosome differences associated with hybrid sterility are common among plants (Stebbins 1950; Grant 1971), such as the genus *Clarkia* in the evening primrose family (Onagraceae). Harlan Lewis (1962) studied several cases in which very localized species appear to have arisen from more broadly distributed species in California. For example, *Clarkia lingulata* is restricted to a single canyon at the edge of the range of *C. biloba*, which appears to be its parent (Figure 16.14). The two species differ in the shape of their petals, and their chromosomes differ by two translocations and an inversion, which are thought to cause the

greatly reduced fertility of the hybrids. Studies of allozymes have supported the relationship of these species (Gottlieb 1974). Lewis (1962) suggested that this and the other localized species of *Clarkia* arose by what he called CATASTROPHIC SPECIATION when chromosome substitutions occurred in a population greatly reduced by drought. These plants are capable of self-fertilization, which would enhance the likelihood of chromosome substitutions. Lewis's hypothesis is clearly similar to Mayr's hypothesis of peripatric speciation.

Experimental evidence on founder effect speciation

Several investigators have passed laboratory populations through repeated bottlenecks to see whether reproductive isolation can evolve, as in Carson's scenario for speciation. In the largest such study, Agustí Galiana and coworkers (1993, 1996) established two large "control" populations of *Drosophila pseudoobscura*, from two North American localities, and from these set up 45 bottleneck lines. Each line went through cycles of population growth, recovering to a large size after being reduced to 1, 3, 5, 7, or 9 pairs in each of nine bottleneck episodes. Various lines were examined for assortative mating with respect to one another and to their parent (control) populations. The authors found evidence of sexual isolation between bottleneck and "ancestral" lines in 8 of 118 tests, and between various bottleneck lines in 26 of 370 tests. Of the 45 bottleneck lines, 23 displayed sexual isolation from other lines, some to a substantial degree. The sexual isolation between some lines persisted for more than 40 generations. John Ringo and colleagues (1985) performed a similar experiment with *D. simulans*, adding to the design several lines that were selected for a variety of morphological, physiological, and behavioral traits. Sexual isolation between the large control population and the selected lines did not evolve, but 3 of 8 bottleneck lines did show weak, albeit erratic, isolation. Perhaps most interestingly, postzygotic isolation among the bottleneck lines and between these and the control population was appreciable, whereas selection caused only slight reduction of the fitness of hybrids.

In Chapter 14, we described an experiment with houseflies (*Musca domestica*) in which populations that had been bottlenecked at 1, 4, or 16 pairs displayed an increase in additive genetic variance for several morphological traits, as well as changes in the pattern of genetic correlation among those traits. These results suggested that evolution by natural selection might be faster, and might proceed along novel lines, in bottlenecked populations. Lisa Meffert and Edwin Bryant (1991) tested for sexual isolation among these housefly populations and the large base population from which they had been derived; they also filmed courtship behavior and scored several of its elements. Only 1 pair of lines, among 15, showed significant assortative mating. However, there was substantial divergence among bottlenecked lines in the rapidity and frequency of mating, by both sexes, with some bottlenecked lines exceeding the values of the control population. Most

FIGURE 16.14 Distribution in California of several subspecies of *Clarkia biloba*, showing the location (•) of the single colony of *C. lingulata*, derived from *C. biloba* and differing from it by several chromosome rearrangements. (After White 1978.)

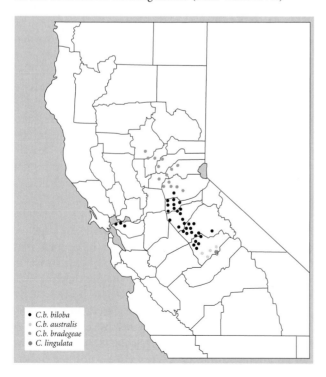

- • *C.b. biloba*
- ◦ *C.b. australis*
- ◦ *C.b. bradegeae*
- • *C. lingulata*

intriguingly, the lines diverged in the frequency with which several elements of courtship behavior were displayed. In some lines, certain elements were accentuated compared with the control population, and in some they were reduced (Figure 16.15). Moreover, changes in the genetic correlations among courtship elements were observed. Thus the *pattern* of courtship behavior was altered.

Whether these results will cheer proponents or opponents of founder effect speciation may depend on whether they consider the glass half full or half empty. To some (Rice and Hostert 1993; Galiana et al. 1993), the relatively low frequency, inconsistency, and weakness of reproductive isolation observed in the experiments inspire skepticism about bottlenecks as an important agent of speciation; to others (Templeton 1996), the experiments support the theory, considering that some instances of incipient isolation, some of them quite appreciable, were revealed in every such experiment. Moreover, it must be borne in mind that the initial genetic changes engendered during and shortly after a bottleneck might only set the stage for further genetic divergence. It is perhaps unlikely that full speciation is often attained immediately by one or a few bottleneck events, but founder events and bottlenecks may still increase the rate of speciation.

In summary, divergence of localized populations from more widespread, slowly evolving parent populations may well prove to be a common *pattern* of speciation. Whether this divergence is more frequently due to natural selection alone or to peak shifts initiated by genetic drift and completed by selection remains to be seen.

FIGURE 16.15 Divergence in courtship behavior among bottlenecked laboratory populations of houseflies. For each of five lines that underwent reduction to 1, 4, or 16 pairs, the percentage of courtships that included the designated behavioral element and terminated in successful copulation are shown, expressed as a deviation from the values in a large (control) population. The data suggest that genetic drift has altered male display elements and female responses, either diminishing or enhancing the likelihood of successful mating. (After Meffert and Bryant 1991.)

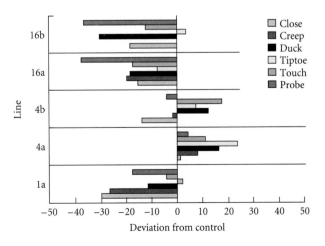

Alternatives to Allopatric Speciation

Parapatric speciation is the evolution of reproductive isolation between populations that are continuously distributed in space, so that there is substantial movement of individuals, and hence gene flow, between them. **Sympatric speciation** is the evolution of reproductive isolation within a randomly mating population. *Allopatric, parapatric, and sympatric speciation form a continuum,* from little to more to much gene exchange between the diverging groups that eventually evolve biological barriers to gene exchange. Even in allopatric speciation, there may be some gene flow between populations, but it is low compared with the divergent action of natural selection and/or genetic drift. Parapatric speciation is the same process, but since the rate of gene flow is higher, the force of selection must be correspondingly stronger to engender genetic differences that create reproductive isolation. Because gene flow into a local population increases its effective size (see Chapter 11), the "more parapatric" the populations are, the less likely genetic drift is to cause speciation. This is also true of selection, but selection is sometimes very strong (and is more effective in large populations).

Parapatric Speciation

All evolutionary biologists agree that allopatric speciation occurs, both because of the evidence from geographic patterns of differentiation and because theory tells us that divergence due to selection or genetic drift is virtually inevitable. The importance of parapatric speciation, however, is a matter of dispute (e.g., Mayr 1982b; Futuyma and Mayer 1980; Barton and Hewitt 1985; Endler 1977), largely because the empirical evidence is usually ambiguous. *The observation that two sister species have parapatric (neighboring) or sympatric distributions does not mean that they speciated parapatrically or sympatrically.* They may have diverged in different locations and later expanded their ranges to become parapatric or sympatric. As we have seen, there is abundant evidence that this has often been the case. When empirical evidence is difficult to interpret, theoretical considerations play a larger role in judging the relative merits of alternative hypotheses. Sometimes we may be fortunate enough to have other sources of evidence that help to decide the case; for example, we may have a historical record of the process of divergence.

The major question to ask about speciation is how the different combinations of genes that cause reproductive isolation can develop in the face of gene exchange that tends to scramble those very combinations. One answer is that divergent selection for the different gene combinations may be stronger than the rate of gene flow (Endler 1977). If gene flow is low, as it might be over a very wide area, then relatively weak selection suffices to establish clines at the relevant loci; if gene flow is appreciable, as between neighboring populations, then a steep cline is formed only if divergent selection is strong. To say that selection is strong, for instance, at a sharp border between different habitats is to say that there exists a barrier to gene exchange, caused by reproductive failure of individuals with the "wrong" genotype or phenotype that migrate across the border. Thus the steepness of a cline caused by divergent selection is a measure of the degree to which gene exchange between the populations is reduced (see Chapter 15). In principle, therefore, strong divergent selection at certain loci, causing a steep cline, acts like a physical barrier, so that genetic divergence occurs at other, more weakly selected loci. Consequently, clines at various loci may tend to develop at the same location, resulting in a hybrid zone that has developed in situ (a primary hybrid zone), but looks like a secondary hybrid zone (Endler 1977; Barton and Hewitt 1985). Moreover, hybridization at the zone eventually ceases, not because of reinforcement of prezygotic isolation, but because of steady genetic divergence. However, it is very difficult indeed to show that a hybrid zone, or the evolution of reproductive isolation between parapatric species or semispecies, developed in this way—although some authors have interpreted particular cases in this light (e.g., Mallet 1993).

We have noted (see Chapters 9 and 15) that in broadly, more or less continuously distributed species, there often exists geographic variation in features that affect pre- or postzygotic compatibility (revealed when distant populations are tested in the laboratory). Thus populations isolated by distance can evolve reproductive incompatibility. The question is whether these divergent features can spread, supplanting ancestral features as they travel, and prevent gene exchange when they eventually meet. Russell Lande (1982) has theorized that prezygotic isolation could arise in this way due to divergent sexual selection. If, for example, a male character, X, and a female preference for that character arose and spread from one location, replacing the ancestral character and preference, while a different male character, Y, and a female preference for it arose and spread from another, distant location, individuals with these two sexual communication systems might not interbreed when they met.

These theories have not yet been shown to be a better explanation than allopatric speciation for any real cases, but they are very real possibilities. They rely on the same principle as allopatric speciation—namely, that divergence in reproductively isolating characters, as in any other characters, can occur if selection is much stronger than gene flow. Possibly the best-documented example of the parapatric origin of reproductive isolation, however, is attributable not to these theories, but to selection for isolation—i.e., reinforcement (which, as we have noted, is generally improbable on theoretical grounds). *Anthoxanthum odoratum* is one of several grasses that have evolved tolerance to heavy metals in the vicinity of mines within the last several centuries (Antonovics et al. 1971; Macnair 1981; see Chapter 14). Several populations, under very strong selection for heavy metal tolerance, have diverged from neighboring nontolerant populations (on uncontaminated soil) not only in tolerance, but in flowering time; moreover, they self-pollinate more frequently, having become more self-compatible (Figure 16.16). Both characteristics provide considerable reproductive isolation from adjacent nontolerant genotypes. In this case, the historical evidence indicates that the divergence is very recent, and the species is so ubiquitously distributed that parapatric divergence cannot be doubted.

Sympatric Speciation

The controversy

Sympatric speciation is one of the most controversial subjects in evolutionary biology. Spe-

FIGURE 16.16 Parapatric evolution of incipient reproductive isolation over a very short distance in the grass species *Anthoxanthum odoratum*. (A) Sample sites and distances along a transect from an area contaminated with lead and zinc (shaded) into an area with uncontaminated soil. Plants growing in the contaminated area are tolerant of the metals, while those in the uncontaminated area are not. (B) Flowering time, in a common garden, of plants derived from different sample sites along the transect. The later flowering time of the plants in the contaminated area reduces gene flow from nontolerant plants in the uncontaminated area. (C) Seed set by plants with flowers bagged to prevent cross-pollination. The capacity to produce seed by self-fertilization was higher in plants from the contaminated area in both years in which the experiment was performed. (After McNeilly and Antonovics 1968 and Antonovics 1968.)

(A)

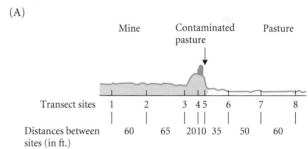

(B)

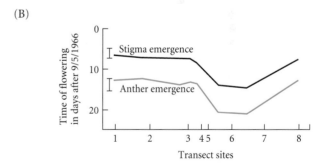

(C)

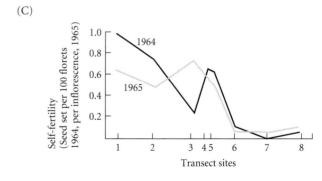

ciation would be sympatric if a biological barrier to gene exchange arose within the confines of a panmictic (randomly mating) population *without any spatial segregation of the incipient species*—that is, if speciation occurred despite high initial gene flow. In the past, many biologists assumed that sympatric species had arisen in situ, a completely unwarranted inference. Moritz Wagner in 1868, Karl Jordan in 1896, and several other authors soon pointed out the difficulties with this assumption and argued that spatial segregation was necessary for speciation. Since 1942, Ernst Mayr (1942, 1947, 1963, 1982) has been the most vigorous and influential opponent of the sympatric speciation hypothesis, demonstrating that many supposed cases are unconvincing and that the hypothesis must overcome severe theoretical difficulties. Under certain special circumstances, however, these difficulties are not all that severe (Maynard Smith 1966; Felsenstein 1981; Kondrashov 1986; Diehl and Bush 1989).

The one mode of sympatric speciation that is not very controversial is speciation by polyploidy in plants (see below), in which a single, instantaneous change reproductively isolates a new polyploid from its ancestor. Except in this case, however, reproductive isolation between sister species is usually based on several or many loci, so speciation is usually gradual. *The difficulty any model of sympatric speciation must overcome is how to reduce the frequency of the intermediate genotypes that would act as a conduit of gene exchange.*

Models of sympatric speciation

Most models of sympatric speciation postulate *disruptive selection*, whereby certain homozygous genotypes have high fitness on one or the other of two resources (or in two microhabitats) and intermediate (heterozygous) phenotypes have lower fitness, perhaps because they are not as well adapted to either resource. Selection then favors alleles at one or more other loci that cause *assortative mating*, reducing the frequency of unfit heterozygous offspring. A common simple model, discussed by Mayr (1963, 468–480), Maynard Smith (1966), Bush (1975), Felsenstein (1981), and other authors, is often applied to host-specific herbivorous insects, but can apply to other organisms as well.

For simplicity, we shall consider two loci in an herbivorous insect, A (for "adaptation") and B (for "behavior"). The ancestral form, with genotype A_1A_1, survives well on plant species 1. A mutation, A_2, confers high survival on plant species 2 if it is homozygous. A_2A_2 is superior in fitness to A_1A_1 on plant 2, but A_1A_1 remains superior on plant 1. The heterozygote has lower fitness on either plant than the better-adapted homozygote. Selection increases the frequency of A_2, creating a multiple-niche polymorphism (see Chapter 13), but because of random mating, unfit A_1A_2 offspring are produced. Another locus, B, controls mate preference (we will shortly consider what happens if it controls host plant preference instead). Both sexes of the ancestral genotype B_1B_1 prefer to mate with each other rather than with B_2B_2 individuals, which also prefer to mate with each other, and which are initially rare. If the B_2 allele is as-

sociated (in linkage disequilibrium) with the A_2 allele, and B_1 with A_1, then A_2B_2 gametes unite more frequently with other A_2B_2 than with A_1B_1 gametes, yielding fit A_2A_2 rather than unfit A_1A_2 offspring. Thus selection favors B_2, it increases in frequency, and a partial barrier to interbreeding between the two host-adapted forms (or "host races") develops. Other loci that further reduce cross-mating may evolve in the same way.

This model should sound familiar, since it is very similar to the model of reinforcement of prezygotic isolation in a hybrid zone. As in the case of reinforcement, the strength of selection for the mating preference allele depends entirely on its degree of association with one or the other of the alleles that are under selection due to differences in survival. Recombination breaks down this association, just as it does in the case of reinforcement of isolation between hybridizing populations (see above). *The antagonism between selection*, which promotes association between alleles for adaptation and alleles for assortative mating, *and recombination*, which destroys this association, *is the greatest barrier to sympatric speciation* (Felsenstein 1981; Rice and Hostert 1993). In the model we have discussed so far, tight linkage and strong selection would be necessary for progress toward speciation. Because the B locus alone does not (according to the hypothesis) confer complete reproductive isolation, further progress toward complete speciation would require alleles at other loci that affect mate preference to be selected because of their association with the A_1B_1 and A_2B_2 combinations. By destroying the $A_1B_1C_1$ ... and $A_2B_2C_2$ ... combinations, recombination would make further progress toward speciation less likely, the greater the number of loci. *Thus, sympatric evolution of sexual isolation by disruptive selection is very unlikely.*

Sympatric evolution is more probable, however, in a variant model, verbally posed by Guy Bush in 1969 and later explored by computer simulation (Diehl and Bush 1989). This model was motivated by Bush's study of true fruit flies (see below) in the family Tephritidae, which, like many other host-specific herbivorous insects, *mate almost exclusively on the host plant.* The female lays her eggs on this plant, and the larvae cannot move to another plant. Thus a genetic difference in host preference (or, more generally, habitat preference), if it affects both sexes, automatically causes assortative mating. *Speciation then occurs by sympatric evolution of ecological isolation* (see Chapter 15, Table 1) rather than by sexual isolation as such.

In this model, as in the previous one, A_1A_1 and A_2A_2 survive best on different hosts, and A_1A_2 has lower fitness than either homozygote. B_1 carriers prefer host 1. A mutation, B_2, carriers of which prefer host 2, increases in frequency for two reasons. B_2 carriers are attracted to the initially underutilized host species, so their offspring suffer less competition and have higher survival rates (until carrying capacity is reached). If, moreover, the B_2 and A_2 alleles are associated, and B_2 carriers mate with each other on the host they prefer, then most of their progeny will be A_2A_2, and will have high fitness on the plant on which they are deposited. The optimal gene combinations are $A_1A_1B_1B_1$ (adapted to and attracted to host 1) and $A_2A_2B_2B_2$ (adapted to and attracted to host 2), so *selection promotes linkage disequilibrium*— it favors the divergent gene combinations. Thus the frequency of the mutation B_2 and of the A_2B_2 gene combination increases. As this occurs, a greater fraction of the population prefers and mates on host 2, lowering the frequency of $A_1A_1B_1B_1 \times A_2A_2B_2B_2$ matings; as fewer $A_1A_2B_1B_2$ heterozygotes are produced, the opportunity for recombination declines. Thus *the antagonism between selection and recombination is lower* than in the model described previously. In computer simulations, the frequencies of alleles such as A_2 and B_2 that confer adaptation to and preference for a new host rapidly increase, as does linkage disequilibrium, so that the population divides into two host-associated, ecologically isolated incipient species (Figure 16.17).

Possible examples of sympatric speciation Some authors believe that sympatric speciation may be quite frequent in certain groups of organisms. Guy Bush (1975) and Catherine and Maurice Tauber (1989), for example, argue that the biological characteristics of many insects may conform to the conditions that make sympatric speciation likely, and cite many possible instances. For example, Thomas Wood, with collaborators, has shown that a "species" of treehopper, *Enchenopa binotata*, actually consists of six sympatric sibling species, each restricted to a different genus of host plant. These species are reproductively isolated not only by their mating on different host plants, but also by sexual isolation (when they are confined together). Moreover, because their eggs hatch at different times, they reach reproductive maturity and mate at somewhat different times during the summer (Wood and Guttman 1982; Figure 16.18). This partial ALLOCHRONIC ISOLATION may be mediated directly by the host plant, because the eggs, laid within twigs, hatch in response to spring sap flow, the timing of which differs among the plant species (Wood and Keese 1990). Wood has suggested that gene flow among the species was initially reduced by differences in the timing of the life history caused directly by the host plants, and that this reduction allowed subsequent genetic divergence in host preference and mate preference.

Wood's hypothesis for *Enchenopa* differs from the models of sympatric speciation considered above in that reproductive isolation is initiated not by a genetic difference, but by an extrinsic barrier—an ecologically imposed asynchrony of life cycles—that is analogous to a physical barrier in allopatric speciation (see also Alexander and Bigelow 1960). An insect that conforms more nearly to the model of habitat isolation driven by genetic divergence is the apple maggot fly, *Rhagoletis pomonella*. Guy Bush (1969), a proponent of sympatric speciation, has proposed that this fly is undergoing that process. Subsequent study has shown this to be one of the few convincing cases of incipient sympatric speciation (Feder et al. 1990, 1993).

Adult apple maggot flies emerge from pupae in July and August and mate on the host plant; the larvae develop in ripe

FIGURE 16.17 Results of computer simulations of sympatric speciation by ecological isolation. (A) The increase of a mutation, B_2, that increases preference for a new habitat, within which mating occurs. The strength of preference by B_2 carriers for habitat 2 is g, ranging from 0 (no preference) to 1 (exclusive preference). (B) The increase of an allele, A_2, that improves fitness (to extent s, ranging from 0 to 1) in habitat 2. Each graph shows simulations that differed in the values of g, s, and m (the rate of gene flow, set at 0.5 for fully sympatric habitats and 0.1 for partially separated habitats). In A, the black ($m = 0.5$, $g = 0.4$, $s = 0$) and gray ($m = 0.5$, $g = 0.2$, $s = 0$) curves show that increasing strength of habitat preference enhances the rate of increase of the B_2 allele, and hence causes partial reproductive isolation, even if fitnesses do not differ between habitats. The contrast between the dark-colored ($m = 0.1$, $g = 0.2$, $s = 0.5$) and light-colored ($m = 0.5$, $g = 0.2$, $s = 0.5$) curves shows the effect of reducing gene flow by spatial separation. In B, the allele that enhances fitness in habitat 2 increases more rapidly if selection is stronger (compare the black curve, $m = 0.5$, $g = 0$, $s = 0.5$, with the gray curve, $m = 0.5$, $g = 0$, $s = 0.1$). Again, spatial separation of habitats enhances the rate of genetic differentiation (compare the dark-colored curve, $m = 0.1$, $g = 0.8$, $s = 0.1$, with the light-colored curve, $m = 0$, $g = 0.8$, $s = 0.1$). (After Diehl and Bush 1989.)

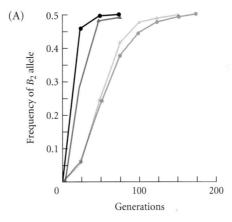

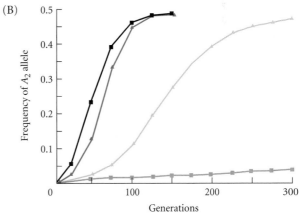

fruits, and in the autumn drop to the ground, spending the winter as pupae. The major ancestral host plants throughout eastern North America were hawthorns (*Crataegus*). About 150 years ago, *R. pomonella* was first recorded in the northeastern United States as a pest of cultivated apples

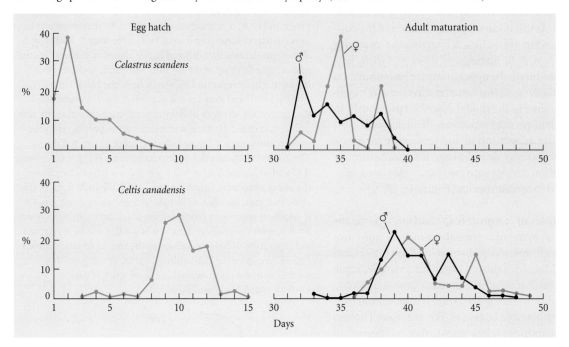

FIGURE 16.18 Partial temporal isolation among *Enchenopa* treehoppers associated with different host plants. Earlier egg hatch on bittersweet (*Celastrus scandens*) than on redbud (*Celtis canadensis*), for example, results in different maturation times for adults, and hence in partially segregated breeding seasons. This separation may have enabled the sibling species to become genetically differentiated in sympatry. (After Wood and Guttman 1982.)

(*Malus*), which are related to hawthorns. Subsequently, infestation of apples spread westward and southward. Although some polymorphic allozymes have similar frequencies in samples from both host species (suggesting persistent gene exchange), several loci differ significantly between apple- and hawthorn-derived flies (Figure 16.19). These differences are found both in larvae and in adults sampled on trees to which they have moved after emergence from the pupal state; this implies that the alleles at these loci are correlated with a difference in host preference. Gene exchange is further reduced by a mean difference of about 3 weeks between mating activity on apple and on hawthorn, which is correlated with a genetically determined difference between apple- and hawthorn-derived flies in the time of emergence from the pupal state. These differences in host preference and mating time are far from complete, but together may reduce the rate of gene flow between these "host races" to less than 10 percent (compared with 50 percent for a panmictic population). The authors suggest that because apple fruits ripen and are suitable for larval development earlier than hawthorn fruits, selection for divergent timing of emergence and reproduction may be responsible for the genetic differences. This species may well be in the initial stage of sympatric speciation.

It has often been suggested that the enormous diversity of cichlid fishes in the African Great Lakes (see above) arose by sympatric speciation. However, there is plentiful opportunity for allopatric speciation within each lake, because most of these species are sedentary and are restricted to one or another of several distinct habitats (rocky shore, soft bottom, etc.) that are discontinuously distributed along the lake periphery. Thus populations could readily be isolated (Fryer and Iles 1972). However, the likelihood of such "microallopatric" speciation seems remote in the case of eleven and nine species, respectively, that are confined to two small lakes in volcanic craters in eastern Africa (Schliewen et al.

FIGURE 16.19 Genetic differentiation between *Rhagoletis pomonella* flies collected from hawthorn and apple at several sites over a latitudinal transect. The average allele frequencies at several enzyme loci, scored by electrophoresis, are plotted. The differentiation between these forms during the past century may have occurred by sympatric divergence in host use. (After Feder et al. 1990.)

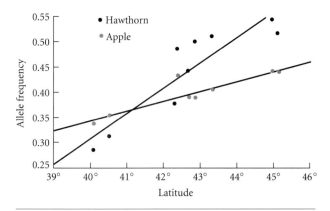

1994). Mitochondrial DNA sequence data indicate that the cichlids in each lake are related to a species in a nearby river, but appear to be monophyletic, implying that speciation has occurred within the crater lakes. The lakes lie in simple conical basins lacking the habitat heterogeneity that might otherwise provide opportunities for spatial isolation. The factor responsible for speciation—possibly assortative mating based on coloration or other differences—is not known, but if the phylogenies are correct, these fishes provide perhaps the most convincing example of completed speciation without spatial separation.

Experimental studies William Rice and Ellen Hostert (1993) have reviewed the many experiments, mostly with *Drosophila*, in which laboratory populations subjected to disruptive selection have been tested for prezygotic isolation. They distinguish between "double-variation" and "single-variation" experiments, which correspond to the sexual isolation and habitat isolation models discussed above. In "double-variation" experiments, the investigator imposes disruptive selection on a character such as bristle number that is genetically independent of variation in mate preference. Sexual isolation evolves if linkage disequilibrium develops between genes for the two characters. Because this process is opposed by recombination, most such experiments have yielded little or no reproductive isolation, except when the offspring of matings between the divergently selected phenotypes have been entirely prevented from breeding. In "single-variation" experiments, the investigator breeds from both extremes of the distribution of a character, such as habitat preference, that would automatically cause assortative mating as a correlated effect. The antagonistic influence of recombination is lower in this case, so such experiments have been more successful in yielding reproductive isolation. For example, Rice and Salt (1990) disruptively selected a laboratory population of *Drosophila melanogaster* by running the flies through a maze in which they were sorted by their positive or negative responses to light, gravity, and choice of two chemical vapors. They were also disruptively selected for adult emergence time. Those that differed most in all four properties were allowed to mate among themselves, and their progeny were treated in the same way. Within fewer than 30 generations, two subpopulations of flies had developed, strongly isolated by development time and to a lesser extent by habitat preference (Figure 16.20). Rice and Hostert emphasize that strong selection, on several characters simultaneously, is most likely to cause sympatric reproductive isolation. Although these experiments show that sympatric speciation is possible in principle, such intense disruptive selection may not be common in natural populations.

FIGURE 16.20 Sympatric divergence in habitat preference and development time in experimental populations of *Drosophila melanogaster*. (A) control (nonselected) populations; (B) experimental (disruptively selected) populations. The experimental population was disruptively selected for three habitat choices and for development time, and flies with similar phenotypes were mated with one another. In the control population, flies mated at random, regardless of habitat choice or development time. The lines in each figure show the mean phenotypes of progeny of females that differed in all four characteristics. In the selected population, divergence occurred in flies' choice to move up rather than down (left graphs) and especially in time to maturation (right graphs). These differences divide the population into partially reproductively isolated groups. (After Rice and Salt 1990.)

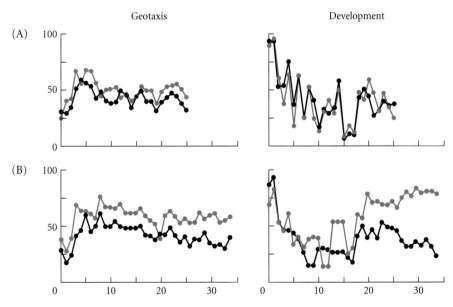

Polyploidy and Hybrid Speciation

Special Considerations in Plant Speciation

Several phenomena play a considerably greater role in diversification and speciation in plants than in animals (Stebbins 1950; Grant 1971). As a result, "taxonomic species" described by morphological differences often correspond less well in plants than in animals to "biological species" defined by reproductive isolation. These phenomena include:

1. *Asexual reproduction*, either by vegetative propagation or by *apomixis* (parthenogenesis, or development from an unfertilized egg). Asexual lines neither fit nor vitiate the biological species concept, but they may enhance ecological and morphological diversity beyond that achieved by sexual populations. Some plant genera, such as *Rubus* (blackberries and raspberries), *Crataegus* (hawthorns), *Taraxacum* (dandelions), and *Hieracium* (hawkweeds), include not only sexual species but also diverse apomicts. In the latter two genera, several hundred apomictic "species," differing in morphological and ecological features, have been named in Europe. Some sexually sterile hybrids are abundant because they reproduce by apomixis.

2. *Self-fertilization*. Strongly self-fertilizing plants form inbred lines that are reproductively isolated from others. Moreover, genetic or chromosomal differences that might lower the fitness of outcrossing genotypes (by producing heterozygous offspring with low fertility) do not have this disadvantage if the plants inbreed. Some cases of incipient or complete speciation by evolution of self-fertilization were described above and in Chapter 15.

3. *Ecological isolation and adaptation*. To a far greater extent in plants than in animals, closely related populations have been described that differ in many morphological and physiological respects and are associated with different habitats, but are highly interfertile and appear to retain their distinctness chiefly because of divergent selection (see Chapter 15). Prezygotic isolation in these cases appears to be weak. Such forms may be called ecotypes or subspecies; some have been classified as species.

4. *Hybridization*. Botanists believe that hybridization is very frequent among closely related plant species. In some cases, *interspecific hybrids become distinct species*, as we will see below.

5. *Polyploidy*. A POLYPLOID is an organism with more than two complements of chromosomes. Polyploid species are much more frequent among plants than animals; in fact, *polyploidy is a major mode of speciation in plants*. In the majority of cases, the polyploid species stems from hybridization between genetically differentiated ancestors.

In the following pages, we discuss speciation by polyploidy and by hybridization that does not entail polyploidy. When a hybrid becomes a new species, the *phylogeny is no longer strictly bifurcating, but rather is reticulate* (netlike) (Figure 16.21).

Polyploidy

The nature of polyploidy   Polyploids are reproductively isolated by hybrid sterility from their diploid (or other) progenitors, and are therefore distinct biological species. Speciation by polyploidy has two distinctions: it is the only known mode of *instantaneous speciation* by a *single genetic event*, and it is the only mode of *sympatric speciation* that all authorities acknowledge is widespread.

Among animals, polyploidy is quite rare. Sexually reproducing polyploids are known in a few groups of fishes (especially the trout and sucker families) and frogs, but

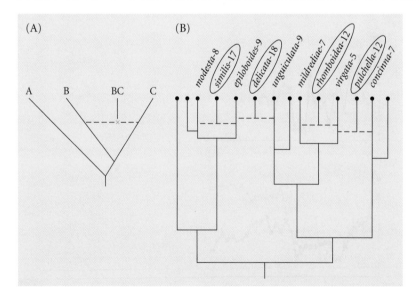

FIGURE 16.21   (A) When a new species (BC) arises by hybridization, the phylogeny of the taxon is not fully divergent, but reticulate (netlike). (B) Part of a large, complex phylogeny postulated for the plant genus *Clarkia*, in which hybridization has frequently given rise to species (circled) with chromosomes derived from both parent species. The gametic (haploid) chromosome number is given after each species' name. Unlabeled branches indicate the positions of other clades. (After Lewis and Lewis 1955.)

polyploid animals are more often parthenogenetic, as are various weevils, grasshoppers, salamanders, and lizards (see Lewis 1979a). The reasons for the rarity of polyploid animals are not fully understood. Polyploidy is very common in plants, and has been studied extensively by systematists, evolutionary biologists, and plant breeders (Stebbins 1950, 1971; Grant 1971; Lewis 1979a).

Polyploidy is usually revealed by the observation that the chromosome numbers in related plants are multiples of some basic number (Figure 16.22). For example, in *Chrysanthemum*, different species have the somatic (diploid) chromosome numbers 18, 36, 54, 72, and 90. In other words, the basic number 9 is multiplied by 2, 4, 6, 8, and 10. The *observed* number of somatic chromosomes is counted as $2n$ ($2n = 18, 36, \ldots 90$ in this case), and the observed number of chromosomes in a gamete (the gametic number) as n ($n = 9, 18, \ldots 45$).

Polyploidy usually occurs because of a failure of the reduction division in meiosis, yielding a $2n$ gamete. Union of two such gametes yields a tetraploid ($4n$) zygote, but this event is most unlikely, because unreduced gametes are rare. Probably more often, union of a $2n$ gamete with an n (normal) gamete yields a triploid ($3n$) plant; unreduced ($3n$) gametes from the triploid then unite with reduced (n) gametes to produce the tetraploid. Similar events can result in hexaploids ($4n + 2n$) and other higher levels of ploidy.

Natural polyploids span a continuum between two extremes called autopolyploidy and allopolyploidy. An **autopolyploid** is formed by the union of gametes from genetically and chromosomally compatible individuals that may be thought of as belonging to the *same species*. For instance, the cultivated potato (*Solanum tuberosum*) is thought to be an autotetraploid of a South American diploid species. An **allopolyploid** is a polyploid derivative of a diploid hybrid between two species—or, to be more exact, two ancestors that form hybrids that are partly sterile due to genetic or chromosomal incompatibility. Because the degree of such incompatibility varies from one case to another, auto- and allopolyploidy are the extremes of a spectrum.

Ideally, auto- and allopolyploids can be distinguished by the behavior of their chromosomes in meiosis (Figure 16.23). An autotetraploid, for example, has four homologues of each chromosome (rather than the two in a diploid). These often tend to synapse in QUADRIVALENTS (or more generally, multivalents)—i.e., sets of four synapsed chromosomes. Such a quadrivalent may segregate three chromosomes to one pole and one to the other, rather than two and two. Because the different quadrivalents segregate independently, many *aneuploid gametes*, with low viability, are formed. Thus autopolyploids often have *reduced fertility*. In contrast, the chromosomes of allopolyploids form BIVALENTS (synapsed *pairs*) in meiosis, segregate balanced, viable gametes, and *generally have nearly normal fertility*. This happens because the corresponding ("homoeologous") chromosomes (say, A_1 and A_2) of the two ancestral diploid species are differentiated so that they fail to pair properly in the diploid hybrid (A_1A_2); but in the tetraploid hybrid ($A_1A_1A_2A_2$), each of the homoeologous chromosomes has a nearly identical partner with which to pair (A_1 with A_1, A_2 with A_2).

This ideal distinction between meiosis in auto- and in allopolyploids often does not hold, for there are many intermediate cases in which a mixture of quadrivalents and bivalents is formed. Moreover, chromosomes of autopolyploids often show a remarkable tendency to form bivalents, apparently because genetic factors can rapidly alter organization or recognition among chromosomes so that they can pair normally. For example, a tetraploid strain of corn (*Zea mays*) in which multivalents were initially predominant evolved in just ten generations, by natural selection for fer-

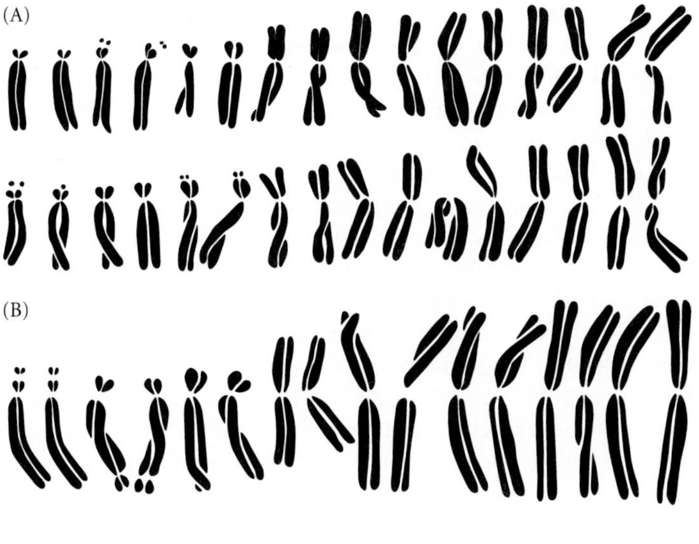

(A)

(B)

FIGURE 16.22 Drawings of metaphase chromosomes in two species of anemones (*Anemone*, in the buttercup family, Ranunculaceae). The somatic (diploid) number is 32 in *A. quinquefolia* (A) and 16 in *A. rivularis* (B), as well as in several other species. This pattern suggests that *A. quinquefolia* arose by polyploidy. (From Heimburger 1959.)

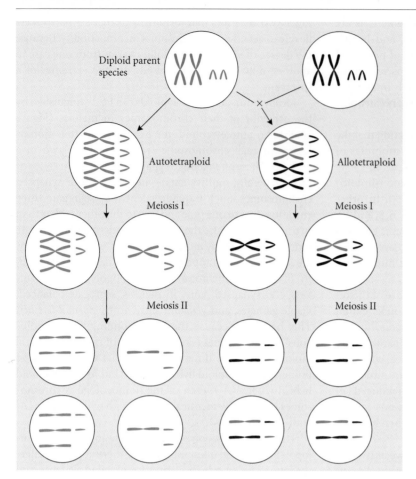

Diploid parent species

Autotetraploid

Allotetraploid

Meiosis I

Meiosis I

Meiosis II

Meiosis II

FIGURE 16.23 A highly schematic diagram of chromosome constitution and meiosis in an autotetraploid (left) and an allotetraploid (right). Segregation in the autotetraploid may be unbalanced, resulting in aneuploid gametes (after the second meiotic division) with unbalanced chromosome complements. In the allotetraploid, homologous chromosomes from each of the parent species (black or colored) may pair, resulting in orderly segregation and balanced gametic complements.

tility, to a high level of bivalent formation, and its fertility increased correspondingly (Gilles and Randolph 1951).

Speciation by polyploidy Plants with odd ploidy (e.g., triploids, $3n$, and pentaploids, $5n$) are generally nearly sterile. The sets of three chromosomes in a triploid, for example, segregate two and one to opposite poles, so most gametes are aneuploid and inviable. (There are naturally occurring odd polyploids that reproduce asexually, as in the genus *Crepis* of the Asteraceae, which contains $3n$, $5n$, and $7n$ populations.) *Because the triploid hybrid between a tetraploid and its diploid ancestor is largely sterile, the tetraploid is reproductively isolated*, and is therefore a distinct biological species (and the same is true at higher ploidy levels). Because a change in ploidy is a single genetic event, *speciation is instantaneous*, in contrast to the gradual process of allele substitution that characterizes all other modes of speciation.

A milestone in the study of speciation was the *experimental reproduction of a natural polyploid species* by Arne Müntzing in 1930. Müntzing suspected that the mint *Galeopsis tetrahit*, with $2n = 32$ chromosomes, might be an allotetraploid derived from the diploid ($2n = 16$) ancestors *G. pubescens* and *G. speciosa*. He crossed the two diploid species, and from the F_1 diploid hybrids, which are about 8 percent fertile, he obtained one triploid offspring. The triploid, backcrossed to *G. pubescens*, yielded a single

tetraploid offspring, which Müntzing propagated. The tetraploids closely resembled *G. tetrahit* in morphology, were highly fertile, and were reproductively isolated from the diploids, but yielded fertile progeny when crossed with wild *G. tetrahit*.

A conspicuous feature noted in this and some other experimental studies of allopolyploids is that the diploid hybrids between species are mostly sterile and form few bivalents in meiosis, whereas the tetraploid hybrids are highly fertile and have normal, bivalent chromosome pairing. In these cases, *the sterility of the diploid hybrid is not due to interactions between the genes of the two parent species, but instead to whatever mechanisms inhibit chromosome pairing.* The diploid and tetraploid hybrids have the same genes in the same proportions, so genic differences cannot account for the sterility of the one but not the other (Darlington 1939; Stebbins 1950).

The evidence that a species arose by allopolyploidy lies not only in chromosomes, but also in morphological and, more recently, molecular characters. For example, three diploid European species of goatsbeards (Asteraceae), *Tragopogon dubius*, *T. porrifolius*, and *T. pratensis*, have become broadly distributed in North America. They frequently form F_1 hybrids that are highly sterile. In 1950, Marion Ownbey described two fertile tetraploid goatsbeard species from eastern Washington State that had apparently arisen within the previous decade, for they had not been found earlier. On the basis of their chromosomes and morphological features, he concluded that *T. mirus* was a tetraploid hybrid of *T. dubius* and *T. porrifolius*, and that *T. micellus* was likewise derived from *T. dubius* × *T. pratensis* (Figure 16.24A). He proposed, further, that *T. mirus* had arisen independently in several localities. Ownbey's conclusions have been confirmed by studies of allozymes (Roose and Gottlieb 1976) and DNA markers (D. Soltis and P. Soltis 1989; P. Soltis and D. Soltis 1991) in these plants. The alleles of the diploid species at several enzyme loci are readily distinguished by electrophoresis, and the tetraploid species have exactly the combinations of alleles predicted by Ownbey's hypothesis. *Tragopogon mirus* is a *fixed heterozygote* for the alleles of *T. dubius* and *T. porrifolius*, as is *T. micellus* for the alleles of *T. dubius* and *T. pratensis*. In other words, if A_d and A_p designate alleles at an enzyme locus in *T. dubius* and *T. pratensis* respectively, the tetraploid *T. mirus* carries both, since it has both the *dubius* and *porrifolius* chromosomes (hence the term "fixed heterozygote"). Moreover, since some of the enzymes are multimers (aggregates of two or more polypeptide chains), the tetraploids have enzymes that differ from those of either diploid parent, and may well confer different physiological properties. For example, if the enzyme is a dimer, it has the combination A_dA_d in *T. dubius*, whereas the tetraploid *T. mirus* carries the dimers A_dA_d, A_dA_p, and A_pA_p. It has long been postulated that this elevated heterozygosity of allopolyploids may be responsible for their capacity to colonize habitats from which the diploids are excluded.

Soltis and Soltis (1989, 1991) have further studied these goatsbeards using restriction fragment analysis (RFLP: see Box D in Chapter 3) of nuclear ribosomal DNA (rDNA) and chloroplast DNA (cpDNA). The diploid species differ in both respects; moreover, polymorphic variants of rDNA are found in different populations of *T. porrifolius*. The tetraploid species have the mixtures of rDNA predicted by Ownbey's hypothesis of their origin. Populations of both *T. mirus* and *T. micellus* differ in which variant rDNA from the *porrifolius* parent they carry, implying that *each of these tetraploids arose more than once* (Figure 16.24B). Moreover, the cpDNA of *T. micellus* is identical to that of *T. dubius* in one region, but identical to that of *T. pratensis* elsewhere. Because the chloroplast DNA is transmitted only through the egg, the maternal parent of *T. micellus* must have been *T. dubius* in the one case and *T. pratensis* in the other. In several other groups of plants, DNA markers have similarly shown that each allopolyploid species has had multiple origins (Brochmann et al. 1992). Thus *polyploidy may generally be an exception to the principle that a biological species has a unique, monophyletic origin* (cf. Chapter 15).

The incidence of polyploidy   Botanists generally agree that allopolyploidy accounts for far more polyploid species than autopolyploidy, presumably because autopolyploids generally have low fertility. In many genera, hybridization has given rise to *polyploid series*, with various multiples of the basic chromosome number (as in *Chrysanthemum*, cited above), or *polyploid complexes*, in which lineages with different basic numbers have hybridized and become poly-

FIGURE 16.24   Newly arisen allotetraploid species of goatsbeards (*Tragopogon*). (A) The flower heads of the diploid species *T. porrifolius* (1), *T. dubius* (2), and *T. pratensis* (3), and of the fertile tetraploid species *T. mirus* (4, from 1 × 2) and *T. miscellus* (5, from 2 × 3). The unlabeled flower heads are the several diploid hybrids, which are mostly sterile. (B) Molecular evidence for the multiple origin of the hybrid species *T. mirus* (M_1, M_2) from *T. porrifolius* (P_1, P_2) and *T. dubius* (D). These fragments of a ribosomal RNA-encoding DNA sequence, the result of cleavage by a restriction enzyme, were sorted by size in a gel. The diploids D, P_1, and P_2 (*T. porrifolius* from two geographic populations) differ in profile because their DNA has been cleaved at different sites, owing to differences in their base pair sequences. Both populations of *T. mirus* (M_1 and M_2) share the fragment 11.2 from *T. dubius* and the fragments 5.9 and 5.3 from *T. porrifolius*, as predicted by the hypothesis of the hybrid origin of *T. mirus*. The two populations of *T. mirus* differ in the presence or absence of fragments 6.2 and 3.8, corresponding to the difference between *T. porrifolius* populations P_1 and P_2. Thus *T. mirus* has arisen at least twice. (A from Ownbey 1950; B after Soltis and Soltis 1991.)

(A)

(B)

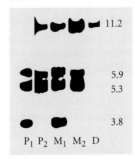

ploid. Figure 16.21B, for example, shows part of a phylogeny postulated for the genus *Clarkia*, in which taxa with gametic numbers 12, 17, and 18 have arisen from hybridization between species with smaller chromosome numbers.

In the fern genus *Ophioglossum*, the somatic number ranges from about 240 to 1260, the highest number known in plants. In many genera, polyploids are found at higher altitudes, or in more stressful environments, or at higher latitudes—especially in glaciated terrain—than their diploid relatives.

As many as 35 percent of flowering plant species have chromosome numbers that are multiples of the lowest number found in their genus (Stebbins 1950; Goldblatt 1979). However, many groups of plants have probably inherited their high chromosome numbers from an ancient polyploid ancestor. For instance, all living horsetails (Equisetaceae)—a group that dates from the Carboniferous—have $2n = 216$, and the lowest number in the fern genus *Ophioglossum* is $2n = 240$. Botanists have speculated that in these and many other cases, the ancestral basic number was 15 or fewer; if so, the ancestry of perhaps more than 75 percent of vascular plant species includes polyploidy (Grant 1971). This hypothesis has been supported by the large cells—a feature typical of polyploids—found in a number of early angiosperm fossils (Masterson 1994).

Establishment and fate of polyploid populations

The processes by which polyploid species become established are not fully understood. If a newly arisen tetraploid within a diploid population crosses at random, its reproductive success should be lower than that of diploids, because many of its offspring will be sterile triploids, formed by backcrossing with the surrounding diploids. The dynamics are formally the same as in the model of heterozygote inferiority, in which the fitnesses of A_1A_1, A_1A_2, and A_2A_2 are 1, $1 - s$, and 1. (Here, A_1A_1 and A_2A_2 represent the fully fertile diploid and tetraploid respectively, and A_1A_2 represents the almost sterile triploid.) Theoretically, several conditions might enable the new polyploid to increase and form a viable population (Fowler and Levin 1984; Thompson and Lumaret 1992; Rodríguez 1996): self-fertilization, higher fitness than the diploid, or niche separation from the diploid—i.e., occupation of a different habitat.

Most polyploid taxa differ from their diploid progenitors in habitat and distribution (Stebbins 1950; Lewis 1979b). In western Africa, for example, the diploid grass *Eragrostis cambessediana* is found in wet habitats, the tetraploid in drier habitats at the bases of dunes, and the hexaploid in the very dry, hot dunes. It is not clear how much of the difference in such cases is due to evolutionary changes in the polyploids since their origin, and how much is a direct consequence of polyploidy itself. Considerable evidence suggests that the latter is a very important factor (Lewis 1979b; Levin 1983b). Increases in ploidy have direct consequences such as larger cell size, higher water content, slower development, and often greater size and higher levels of some enzymes, hormones, and compounds that can protect the plant from herbivores and pathogens. Thus *many polyploids may be capable of occupying new ecological niches immediately upon their origin*, and this ability may both compensate for and reduce the incidence of reduced fitness caused by intercrossing with their progenitors.

Although polyploidy may confer new physiological and ecological capabilities, it does not, as far as we know, confer major new morphological characteristics such as differences in the structure of flowers or fruits. Thus *polyploidy does not cause the evolution of new genera or other higher taxa* (Stebbins 1950).

Does ploidy necessary increase, or can it decrease? There is considerable evidence that polyploids can give rise to populations with fewer chromosome complements, *but only soon after the polyploid has arisen* (Stebbins 1950; Grant 1971). In time, the *duplicate loci* derived from the diploid ancestors of an allopolyploid experience different mutations and *diverge*. Some diverge in function; for example, they may be differentially regulated and expressed in different tissues. In other cases, one of the duplicate loci loses its function, and the remaining locus functions as if diploid. Among suckers, a family of fishes derived from a tetraploid ancestor about 50 million years ago, studies of enzymes have shown that about half of the duplicate loci are no longer expressed (Ferris and Whitt 1979). Because one functional member of a pair of duplicate loci presumably suffices, mutations that reduce or abolish the function of the other member probably become fixed by genetic drift (Li 1980). Such changes would prevent the reversion of a tetraploid lineage to diploidy because the diploid would lack function at some loci.

Origin of Species by Hybridization

Hybridization sometimes gives rise to distinct species with the same ploidy as their parents. Among the great variety of recombinant offspring produced by F_1 hybrids between two species, certain genotypes may be fertile but reproductively isolated from the parent species, and may increase in frequency, forming a distinct population. G. Ledyard Stebbins (1950), a major figure in the Evolutionary Synthesis, detailed this hypothesis and reviewed examples; Verne Grant (1971) named it RECOMBINATIONAL SPECIATION.

Imagine that two species differ by two or more complementary genetic factors that greatly lower the fertility of the F_1 hybrid. These might be two independent translocations, such that parts of chromosomes *A* and *B* of species 1 are reciprocally translocated in species 2, and similarly for chromosomes C and D. As Figure 16.25 shows, only a minority of the gametes of the F_1 will carry balanced chromosome complements (without deficiencies or duplications), but an F_2 that is homozygous for one of these combinations will be fertile, although postzygotically isolated from other such homozygotes and from the parent species because the progeny of crosses to those will be largely sterile. Several researchers have produced such "hybrid species" of plants experimentally. For example, Grant (1966a) crossed *Gilia malior* and *G. mudocensis*; from the 99 percent sterile F_1, he obtained several fertile F_2 plants and reared inbred lines by self-fertilization. One of these combined the morphologi-

FIGURE 16.25 A model of the origin of diploid hybrid species by recombinational speciation. The upper three rows show four pairs of chromosomes in species P_1 and P_2 and their F_1 hybrid offspring. With reference to P_1, P_2 carries a reciprocal translocation between chromosomes A and B, and another between chromosomes C and D. These chromosomes are denoted by letters with subscripts (A_b, etc.). Most gametes of the F_1 hybrid are unbalanced (cf. Figure 17 in Chapter 10). For example, a gamete with chromosomes $ABCD_c$ lacks some of the genetic material of chromosome D. A small fraction of the F_1 gametes are balanced, lacking no genetic material. F_2 homozygotes for these balanced combinations have full gene complements and are likely to be fertile, but progeny of the backcross to either parent are largely sterile, as in the F_1. Because of allelic differences between P_1 and P_2, the fertile F_2 homozygotes are likely to have new combinations of genes and phenotypic characteristics. (After Grant 1971.)

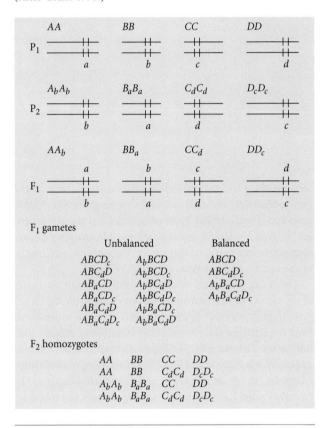

(Figure 16.27). *Helianthus annuus*, the most widespread species (and the ancestor of the cultivated sunflower), forms almost fully sterile hybrids with *H. petiolaris fallax*, but the diagnostic DNA markers of both species are present in three other distinct species (*H. anomalus*, *H. paradoxus*, *H. deserticola*) that have arisen by hybridization. The derivative species are fully fertile, have fewer allozyme polymorphisms than the parent species, and are genetically isolated from them by chromosomal barriers. Most interestingly, the derivative species have unique characteristics. All of them grow in much drier, sandier soil than either parent species; they have morphological features (e.g., lack of pubescence) possessed by neither parent; and *H. paradoxus* lacks a major class of chemical compounds possessed by both parents, and has a much later flowering season than either. This is one of several examples in which *hybridization, by generating diverse gene combinations on which selection can act, has been a source of novel morphological and ecological features.*

FIGURE 16.26 Two species of *Gilia* (A, B) and a fertile hybrid (C) derived from recombination by selective breeding. The hybrid is intermediate to the parent species in leaf shape and in the length of the stem between leaves. (From Grant 1966b.)

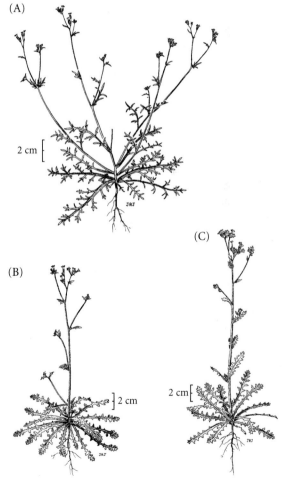

cal features of the two parent species and was intersterile with them (Figure 16.26).

Hybrid speciation seems to be rare in animals, although a few possible examples have been described (Bullini 1994), such as the minnow *Gila seminuda* in western North America (DeMarais et al. 1992). It may be a common mode of speciation in plants, however (Rieseberg and Wendel 1994; Abbott 1992). Diploid species of hybrid origin have been identified by morphological, chromosomal, and molecular characters. For example, Loren Rieseberg and coworkers (Rieseberg 1991; Rieseberg et al. 1990) have used both nuclear (rDNA) and chloroplast DNA markers to infer the phylogeny of one section of the sunflower genus, *Helianthus*

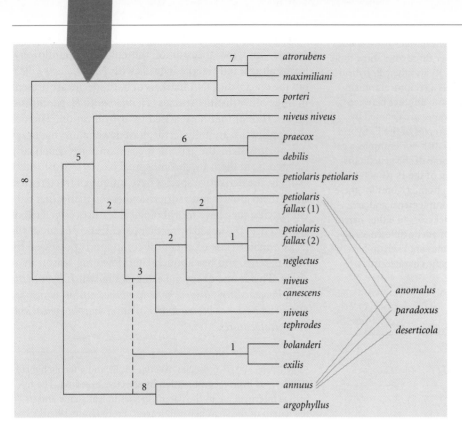

FIGURE 16.27 A phylogeny of part of the sunflower genus, *Helianthus*, based on sequences of chloroplast DNA and nuclear ribosomal DNA. The diploid species *H. anomalus*, *H. paradoxus*, and *H. deserticola* have arisen from hybrids between *H. annuus* and two subspecies of *H. petiolaris*. The numbers of synapomorphic base pair substitutions are shown along the branches of the phylogenetic tree. (After Rieseberg and Wendel 1993.)

Rates of Speciation

The "rate of speciation," like the rate of origin of higher taxa (see Chapter 25), can refer to two related concepts: the rapidity with which populations of a species evolve reproductive isolation, and the number of species, per ancestral species, that arise per unit of time (Figure 16.28). The latter is usually related to the growth in diversity of a clade. The rate of increase in the number of species in a monophyletic group, however, depends on both the rate of origin and the rate of extinction of species. Chapter 25 treats the effects of speciation rates on changes in diversity in the fossil record.

The time required for speciation—i.e., for the evolution of reproductive isolation—is highly variable. At the rapid end of the spectrum is instantaneous speciation by polyploidy. The time required for speciation by other modes can also be relatively short. The "big island" of Hawaii is less than 800,000 years old, but at least 24 species of *Drosophila* have apparently originated there. About 300 species of cichlids in Lake Victoria are thought to have evolved from one ancestor in about 200,000 years,[*] and several species are restricted to a small neighboring lake thought to be only 4000 years old (Fryer and Iles 1972). Several species of pupfish (*Cyprinodon*) have developed in the Death Valley region of

Nevada since the extensive lakes that existed there in the Pleistocene were reduced to isolated springs 20,000–30,000 years ago (Figure 16.29). Not only species, but also some genera, such as *Thalarctos* (polar bear) and *Microtus* (voles), are thought to have evolved during the Pleistocene—i.e., within the last 1.8 million years (Stanley 1979).

However, geological evidence, combined with estimates of divergence time based on molecular data (assuming an approximate molecular clock), suggests that the more usual time required for full reproductive isolation is about 3 million years. This estimate is based on organisms as different as *Drosophila* (Coyne and Orr 1989a) and snapping shrimps (*Alpheus*), in which sister species in the Caribbean and the eastern Pacific have achieved varying degrees of reproductive incompatibility since the Isthmus of Panama arose about 3.5 Mya (Knowlton et al. 1993). In some cases, populations remain reproductively compatible despite much longer isolation. It has probably been more than 20 million years since the opening of the North Atlantic separated Europe from eastern North America, yet American and European sycamores (*Platanus*, Figure 16.30) and plantains (*Plantago*) form fully fertile hybrids when crossed (Stebbins 1950; Stebbins and Day 1967), and the European and American forms of some birds, such as the treecreeper (*Certhia familiaris*) and the raven (*Corvus corax*), are so similar that taxonomists classify them as the same species. Paleontologists have described many cases in which organisms that are morphologically indistinguishable from modern species extend back in the fossil record 5 to 10 million years or more (Stanley 1979), although the fossil record seldom provides

---

[*]Recent evidence that the entire Lake Victoria basin may have been dry in the late Pleistocene has led Johnson et al. (1996) to suggest that this monophyletic group of species has evolved within the last 12,400 years. If this hypothesis is verified, the speciation rate of these fishes would be by far the most rapid known.

FIGURE 16.28 Two meanings of "rate of speciation." (A and B) After two populations are isolated by a geographic barrier, reduction of their *potential* capacity for gene exchange, represented by the horizontal connecting lines, evolves more rapidly (A) or more slowly (B). The rate of speciation—i.e., of attaining reproductive isolation—is higher in A. (C and D) The speciation rate is said to be higher in a clade such as C, in which the number of descendant species arising per original species is higher per unit of time. This may be due to a higher rate of evolution of reproductive isolation, as in A compared with B, but it may also be due to other factors, such as greater opportunity for geographic isolation. (E) The increase in number of species per unit of time is influenced not only by the rate of speciation, but also by the rate of extinction. The speciation rate in this clade is the same as in C, but the extinction rate is higher.

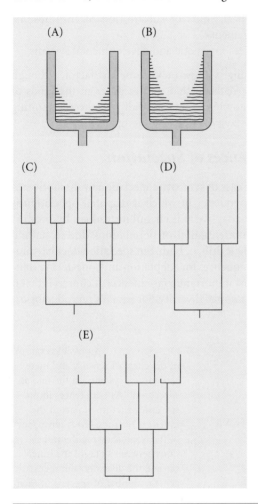

strong sexual selection and rapid divergence in sexually selected characters. In both groups, moreover, related species have different ecological specializations, and their environments harbor few other species that might act as competitors. These latter factors undoubtedly *permit* newly arisen species to persist and to diverge along one or another line of ecological specialization, resulting in adaptive radiation, but whether or not they hasten the evolution of reproductive isolation is not known.

In addition to ecological opportunities that allow an increase in the number of species, some intrinsic characteristics of organisms may affect rates of speciation. A possible example is illustrated by the frogs, in which the inner ear organ ranges from short to long. The shortest organ, which is sensitive to a narrow range of low frequencies, is found in primitive families of frogs, whereas the longest organ, sensitive to a far broader range of frequencies, is found in the more "advanced" frog families that have arisen more recently (Figure 16.31). The lineage with the longest ear organ, consisting of several families, is more diverse by far than all the other frogs. Ryan (1990) suggests that the greater acoustic range of the long ear organ, because it determines the range of male call frequencies that females can perceive, provides greater opportunities for sexual isolation and speciation. Likewise, about half of all bird species are oscines (songbirds), which have a vocal apparatus that enables them to produce highly complex songs that differ greatly among species; possibly this ability has facilitated speciation (Fitzpatrick 1988). For both frogs and birds, however, the pos-

FIGURE 16.29 Four species of desert pupfish that are thought to have evolved in the Death Valley region of California and Nevada within the last 30,000 years, when their local aquatic habitats became isolated. *Cyprinodon diabolis*, restricted to a single spring, is remarkable for its small size and lack of pelvic fins. (After Brown 1971b.)

Cyprinodon diabolis

Cyprinodon nevadensis

Cyprinodon salinus

Cyprinodon radiosus

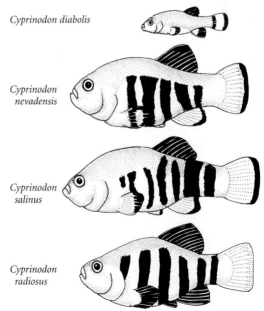

evidence on the evolution of characters that might affect reproductive isolation.

Some of the factors that promote rapid speciation seem fairly evident, but they are so often intercorrelated that it is difficult to determine their individual effects. For example, the explosive speciation of cichlids in the African Rift Valley lakes and of *Drosophila* in the Hawaiian islands appears to be associated with several factors that should promote speciation: low dispersal rates and abundant opportunities for isolation of populations formed by colonization; and

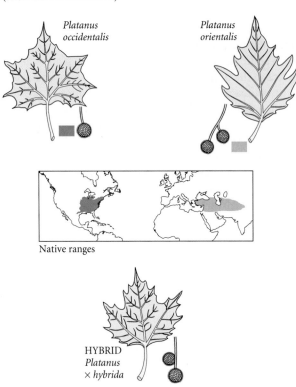

FIGURE 16.30 Speciation can be very slow in some cases, as in the American and Oriental plane trees (sycamores). Although these two forms have been geographically isolated for at least 20 million years, their hybrid, the London plane, is fully fertile. It is widely planted on city streets in temperate regions. (After Raven et al. 1992.)

sibility that ecological and other factors are responsible for the differences in diversity cannot be ruled out.

Although rigorous studies of the factors affecting speciation rate are few, these factors are generally thought to be:

1. Abundant opportunities for isolation of populations, afforded by topographic or other barriers
2. Low dispersal rates, reducing gene flow among populations
3. Ecological specialization, which may limit species to isolated patches of habitat and afford the opportunity for divergence
4. Strong sexual selection, which may cause divergence in mate recognition systems
5. Bottlenecks in population size, which may facilitate genetic peak shifts or substitution of chromosome rearrangements
6. Differences in genetic and developmental systems

Almost nothing is known about the last factor, although some have speculated that species differ in the kinds of epistatic gene systems that can readily be altered, resulting in pre- or postzygotic isolation (Templeton 1980).

Consequences of Speciation

Except in the case of sympatric speciation by disruptive selection, the frequency of which is disputed, speciation is probably not adaptive in itself, but instead is a by-product of genetic differentiation of populations due to natural selection, genetic drift, or both. But speciation does have important consequences for adaptation and long-term evolution. The most important consequence, of course, is that it is the sine qua non of diversity. For sexually reproducing or-

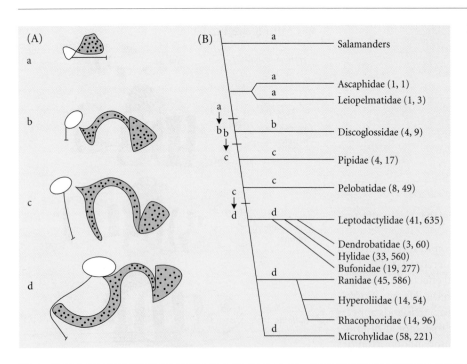

FIGURE 16.31 A possible example of the effect of sensory and mate recognition systems on the rate of speciation. (A) Differences in the size and structure of part of the inner ear of frogs. (B) A phylogeny of the frogs, showing the transitions between states of the inner ear and the diversity (number of genera, number of species) of each family. The clade with state d is by far more diverse than all the other frogs, which have more primitive states of this characteristic. (After Ryan 1990.)

ganisms, every branch in the great phylogenetic tree of life was a speciation event, in which populations became reproductively isolated and therefore capable of independent, divergent evolution, including, eventually, the acquisition of those differences that mark genera, families, and still higher taxa. Speciation, then, stands at the border between "microevolution"—the genetic changes within and among populations—and "macroevolution"—the evolution of the higher taxa in all their glorious diversity.

Some paleontologists have argued that speciation may be required for substantial evolution to occur at all. In proposing the hypothesis of *punctuated equilibrium*, Eldredge and Gould (1972) (see also Stanley 1979; Gould and Eldredge 1993) argued, from the observation that many fossil lineages change little over the course of millions of years (see Chapter 6), that broadly distributed, well-established species are perhaps incapable of evolving substantially due to internal constraints imposed by complex epistatic interactions of genes. They suggested, based on Mayr's (1954) proposal that peak shifts in small populations can rapidly give rise to morphologically divergent new species, that most evolutionary changes in morphology are triggered by and associated with speciation—i.e., with the divergence of highly localized, reproductively isolated populations. Upon spreading from their locus of origin, these new species would be abundant enough to be registered in the fossil record, thus "punctuating" otherwise static lineages with bursts of change (see Figure 13C in Chapter 6).

Examples of fossil lineages that do or do not conform to the punctuated equilibrium model are described in Chapter 6, and we will consider the matter further in Chapter 24. We shall here mention one line of evidence based not on the fos-sil record, but on extant species. If phenotypic divergence requires successive speciation events, then the amount of divergence between the most different species in a clade should be greater, the more speciation events separate them (Figure 16.32). Thus a species-rich clade should show greater phenotypic divergence than a species-poor clade of the same age, such as a species-poor sister group. If evolutionary change does not require speciation, then the amount of phenotypic divergence accrued between lineages should be independent of the species richness of a clade. This test was first applied by John Avise and Francisco Ayala (1976) and by Douglas and Avise (1982) to divergence in allozymes and morphology in two groups of fishes that differ in species diversity. They found no evidence for the punctuated equilibrium model, but their test has been criticized because they used two unrelated groups of fishes. More such studies would be useful to further test this hypothesis.

Population geneticists, who maintain that natural selection can readily modify the characteristics of populations, are generally dubious about the punctuationists' proposal that species cannot readily evolve unless they speciate (Charlesworth et al. 1982). The evidence from the genetic properties of contemporary populations seems to conflict with the *pattern*, frequently encountered in the fossil record, of long-term morphological stasis punctuated by rapid morphological change (Gould and Eldredge 1993). Futuyma (1987, 1989) proposed a simple hypothesis that may reconcile this conflict. A pivotally important feature of sexually reproducing species (in contrast to asexual forms) is recombination, which breaks down the combinations of genes that provide different adaptations. Suppose, for example, that because of the distribution of

FIGURE 16.32   The difference between the models of phyletic gradualism (G) and punctuated equilibrium (P) in the predicted relationship between the number of species and the range of morphological variation among species in a clade. Phyletic gradualism predicts no correlation (G1, G2) because phenotypic evolution does not depend on speciation. The punctuated equilibrium model holds that phenotypic evolution occurs chiefly when local populations become new species. Thus a clade with a high speciation rate (P2) would have more opportunities for phenotypic divergence than an equally old clade with a low speciation rate (P1).

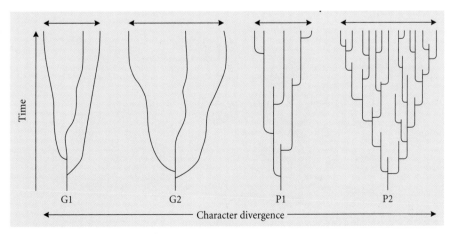

available food sizes, selection on the beak size of a bird species is disruptive, favoring small and large beaks. If beak size is an additive polygenic trait, a randomly mating population will not develop a bimodal distribution: because of recombination, most individuals will be near a single, perhaps intermediate, beak size (see Chapter 14). Thus phenotypes that are ideally adapted to different resources will persist only if there is a barrier to free interbreeding—either spatial segregation of populations or reproductive isolation.

Different local populations may respond rapidly to selection and diverge in, say, adaptation to different resources. But local populations are ephemeral: as climate and other ecological circumstances change, the distribution of habitats and resources changes, and new sites are colonized while old populations become extinct. Sooner or later, then, divergent populations will be brought into contact, and much of the divergence that has occurred will be lost by interbreeding (Figure 16.33). If these events happen frequently enough relative to the time scale of the fossil record, morphological divergence will be registered only as brief excursions or not at all, presenting the appearance of stasis. But if the divergent populations become reproductively isolated, they can retain their distinctive characters indefinitely, even when, in the course of following their habitats, they become sympatric. Thus speciation, while not accelerating evolution, provides morphological changes with enough permanence to be registered in the fossil record, and morphological change becomes associated with speciation. Moreover, because reproductive isolation confers a degree of permanence to the first divergent steps, the stage is set for still further divergence (Figure 16.33). Thus speciation can facilitate long-term directional changes in a lineage (anagenesis) by retaining, stepwise, the advances made in any one direction.

This hypothesis predicts, perhaps counterintuitively, that directional evolutionary change should be more pronounced during times of climatic stability (when the geographic isolation of populations lasts long enough for reproductive isolation to evolve) than during times of climatic change (when the distribution of habitats and populations changes, promoting gene flow). Paleontologists have described just such a pattern. Despite the great environmental changes during the Pleistocene, for example, early Pleistocene fossil beetles are identical to living species. Their geographic distributions have changed radically, however, and it is to these incessant changes in distribution, resulting in gene flow, that Coope (1979) attributes their morphological stability. Sheldon (1987, 1990), likewise, has found that gradual morphological evolution in trilobites was more pronounced in stable than in unstable environments.

Whether this proposed linkage between anagenesis and cladogenesis be true or not, the history of evolution is less a story of single lineages changing progressively over time than a story of evolutionary radiations, of diverse species springing from common ancestors, diverging, and in turn giving rise to new radiations. This is the history that has given rise to the great adaptive differences between orders of mammals and classes of vertebrates. As Ernst Mayr (1963, 621) said, "Speciation, the production of new gene complexes capable of ecological shifts, is the method by which evolution advances. Without speciation, there would be no diversification of the organic world, no adaptive radiation, and very little evolutionary progress. The species, then, is the keystone of evolution."

FIGURE 16.33 A model of how speciation might facilitate long–term evolutionary change in morphological and other phenotypic characters. Horizontal solid triangles signify geographic isolation of populations; open triangles, the breakdown of geographic isolation; vertical solid triangles, the evolution of reproductive isolation. (A) Geographically isolated populations diverge due to natural selection, but the divergence is lost by interbreeding if reproductive isolation has not evolved and the geographic barrier breaks down. (B) Divergence is permanent, and may continue to evolve after breakdown of the geographic barrier, if reproductive isolation has evolved. (C) Repetition of the process in B can enable still further departures from the ancestral character state to become permanent.

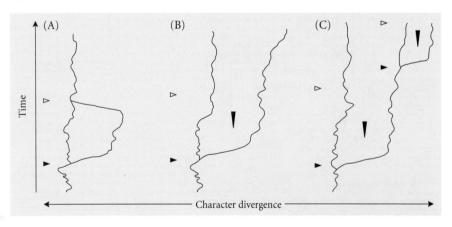

Summary

1. Because reproductive isolation between populations is usually based on multiple genetic differences, each of which contributes only partial isolation, a problem in understanding speciation is explaining how such discrete sets of genetic differences can arise, because recombination among partially isolated forms should oppose progress toward further isolation. A further problem is explaining how alleles that reduce the fitness of heterozygous hybrids, thus contributing to reproductive isolation, can increase in frequency despite seeming selection against them. The several possible modes of speciation are the ways in which these obstacles are circumvented or surmounted.

2. Probably the most common mode of speciation is allopatric speciation, in which gene flow between populations is reduced by topographic or habitat barriers, allowing genetic divergence by natural selection and/or genetic drift. Reproductive isolation evolves, at least in part, as a by-product of such divergence. The abundant evidence for allopatric speciation includes variation in reproductive compatibility among geographic populations of species, the parapatric distribution of many closely related species, the correspondence of genetic discontinuities with present or past topographic barriers, and the evolution of incipient reproductive isolation among laboratory populations.

3. In vicariant allopatric speciation, a widespread species becomes sundered by a topographic barrier, and one or both populations diverge from the ancestral state. In a simple model of the evolution of reproductive isolation, allele substitutions that do not reduce the fitness of heterozygotes occur at different loci in each population (e.g., A_2 replaces A_1 in one population, and B_2 replaces B_1 in the other). The different substitutions cause prezygotic incompatibility, such that $A_2A_2B_1B_1$ and $A_1A_1B_2B_2$ do not mate with each other, or the epistatic interaction of the substituted alleles (A_2, B_2) reduces the viability or fertility of the hybrid ($A_1A_2B_1B_2$). In a related model, successive allele substitutions in one population result in reproductive isolation from the other population, which is not altered.

4. The genetic changes that cause the evolution of reproductive isolation in allopatric populations may be caused by divergent ecological or sexual selection. In only a few cases have the effects of ecological selection been well documented. Abundant evidence suggests that sexual selection is an important cause of ethological (sexual) isolation in animals, but this evidence is mostly circumstantial. More research on the causes of evolution of reproductive isolation is needed.

5. Prezygotic isolation occurs mostly while populations are allopatric; whether or not it is often *reinforced* by natural selection to prevent interbreeding and production of unfit hybrids is uncertain. Alleles that enhance assortative mating are favored by selection only if they are associated (in linkage disequilibrium) with alleles that lower the viability or fertility of hybrid offspring. Recombination, by breaking down the association among these loci, opposes selection for assortative mating. In only a few cases is there strong evidence for reinforcement of prezygotic isolation—namely, stronger isolation between populations that are in contact with each other than between allopatric populations of the same species. Furthermore, natural selection ordinarily cannot favor reinforcement of postzygotic isolation (i.e., increased inviability or sterility of hybrids). Thus speciation is perhaps only rarely an adaptive process, but instead is usually a by-product of the processes (which, however, may be adaptive) that cause genetic divergence.

6. Peripatric speciation, or founder effect speciation, is a hypothetical form of allopatric speciation in which a colony, because of its founding by few individuals or later bottlenecks in population size, undergoes genetic drift, followed by natural selection for a new adaptive genetic configuration (a shift between adaptive peaks). Such a change at a few loci may alter the "genetic environment" for other loci, so that further genetic change follows, due to epistatic selection, and reproductive isolation occurs as a by-product. These changes need not be caused by any difference in the environment. The rapid evolution of highly divergent species in localized, isolated sites suggests that this *pattern* of speciation is common, but evidence for the *process* of peak shift in natural populations is equivocal. Because of limited evidence and theoretical difficulties, this hypothesis is controversial.

7. In theory, speciation can be caused by the divergence of spatially segregated populations connected by gene flow, as long as divergent selection is stronger than gene flow. This process, known as parapatric speciation, is hard to document, but a few examples are known.

8. Sympatric speciation is the origin of a biological barrier to gene exchange within an initially randomly mating population. Most models of sympatric speciation envision the evolution of assortative mating (sexual isolation) between two phenotypes that are favored by disruptive selection (intermediate phenotypes having low fitness). This is unlikely for the same reason that reinforcement is unlikely: recombination breaks down the association between genes for the disruptively selected phenotypes and genes for assortative mating. This is a lesser problem, however, if disruptive selection favors preference for two distinctly different resources or habitats, and mating occurs within the chosen habitat. Then assortative mating is a direct, rather than an indirect, consequence of disruptive selection. A few examples, especially in host-specific herbivorous insects, suggest that sympatric speciation can occur, as do some laboratory experiments. The prevalence of sympatric speciation is still arguable.

9. Speciation by polyploidy, a doubling or greater increase in the number of chromosome complements, is common in plants, especially by polyploidization of hybrids between genetically divergent populations. The polyploid is postzygotically isolated from its diploid ances-

tors because backcross progeny produce aneuploid gametes and have low fertility. Polyploidy is thus the one mode of speciation that occurs instantaneously, by a single genetic event in a single individual. Establishment of a polyploid population, however, probably requires ecological or spatial segregation from the diploid ancestors, because backcrossing to the diploids would reduce the polyploid's reproductive success. Polyploids often differ physiologically from their progenitors and are successful colonizers of different, often more stressful, environments. Polyploid species can have multiple origins. Some natural polyploid species have been recreated by deliberate breeding.

10. In plants, some genotypes of diploid hybrids are fertile and are reproductively isolated from the parent species, and so give rise to new species (recombinational speciation).

11. Factors that favor a high rate of speciation include (*a*) abundant topographic barriers that provide opportunities for allopatric divergence; (*b*) low dispersal rates; (*c*) strong sexual selection; (*d*) probably ecological specialization; and (*e*) perhaps bottlenecks in population size. In addition, persistence of newly formed species is favored by (*f*) ecological opportunity ("vacant niches"). Species may also differ in genetic properties (e.g., degree of epistasis) that affect the likelihood and rate of evolution of reproductive incompatibility, but these are poorly understood. The rate of speciation is highly variable. Aside from instantaneous speciation by polyploidy, the time required for speciation ranges from a few thousand to more than 20 million years; perhaps about 3 million years is average.

12. Speciation is the source of all the diversity of sexually reproducing organisms, and is the event responsible for every branch in phylogeny. Whether or not it contributes importantly to evolutionary change in morphological and other characters, as suggested by the hypothesis of punctuated equilibria, is controversial. It has also been suggested that speciation facilitates long-term progressive divergence not by accelerating morphological change, but by preventing such changes from being broken down by recombination when divergent populations eventually come into contact with ancestral phenotypes. Speciation may therefore be important not only for diversification, but also for long-term evolutionary "progress."

Major References

The most useful references on speciation also treat the nature of species, and are listed at the end of Chapter 15.

Problems and Discussion Topics

1. Why is it difficult to demonstrate that speciation has occurred parapatrically or sympatrically? If allopatric distributions of differentiated populations are taken as evidence for allopatric speciation, why should parapatrically or sympatrically distributed species not be evidence for parapatric or sympatric speciation?

2. Explain why many authors think that reinforcement of prezygotic isolation is unlikely to occur.

3. Coyne and Orr (1989a) found that sexual isolation is more pronounced between sympatric populations than between allopatric populations of the same apparent age, and took this finding as evidence for reinforcement of sexual isolation. It might be argued, though, that any pairs of sympatric populations that were not strongly sexually isolated would have merged, and would have been unavailable for study. Thus the degree of sexual isolation might be biased in sympatric compared with allopatric populations. How might one rule out this possible bias? (Read Coyne and Orr after suggesting an answer.)

4. Suppose two allopatric populations were exposed to the same ecological sources of selection. Is it possible that even without a peak shift, genetic changes caused by ecological selection could cause reproductive isolation to evolve? How might this happen? (Cf. Chapter 14.)

5. Genetic models of allopatric speciation show that reproductive isolation can evolve due to genetic changes in both populations, or in only one. Suggest an experimental protocol to determine which has occurred.

6. Suppose that full reproductive isolation between two populations has evolved. Can speciation in this case be reversed, so that the two forms merge into a single species? Under what conditions is this probable or improbable? Referring to the discussion of "parallel speciation" in sticklebacks, can a single biological species arise more than once, i.e., polyphyletically? How might this possibility depend on the nature of the reproductive barrier between such a species and its closest relative?

7. Sexual selection for elaborate male characteristics can be due to female preference, to competition among males, or both (see Chapter 20). We have discussed sexual selection by female preference as an agent of speciation. Might sexual selection by male competition also cause speciation? How might this occur?

8. It is often supposed that speciation by peak shift, if it occurs, should be faster than speciation by natural selection alone (speciation by gradual divergence). Postulate conditions under which speciation by peak shift, once initiated, could be slow, and conditions under which speciation by gradual divergence could be very rapid.

9. The heritability of a preference for different habitats or host plants might be high or low. How might this affect the likelihood of sympatric speciation by divergence in habitat or host preference?

10. Biological species of sexually reproducing organisms usually differ in morphological or other phenotypic traits. The same is often true of taxonomic species of asexual organisms such as bacteria and apomictic plants. What factors might cause discrete phenotypic "clusters" of organisms in each case?

PART IV

Character Evolution

One of the major goals of evolutionary biology is the development of a unified theory of evolutionary processes, a body of general principles that apply to all characteristics of all organisms. Mutation, recombination, gene flow and isolation, genetic drift, and the various forms of natural selection (Chapters 9–14) are the primary, most general ingredients of this theory. Slightly less general are the principles of cladogenesis and speciation (Chapters 5, 15, and 16) and the emerging principles of macroevolution (which we shall treat in Chapters 23 through 25). At the same time, the subject matter—indeed, the lifeblood—of evolutionary biology is the astonishing diversity of organisms and their characteristics. Evolutionary theory would be a pale, lifeless abstraction, perhaps pleasing to mathematicians but vacuous for biologists, if it could not be pressed into the service of explaining the features and diversity of living things. This is the raison d'être, the reason for being, of evolutionary biology, and indeed much of biology as a whole.

Much of evolutionary biology in the first half of the twentieth century focused on developing a unified theory, for those who forged the Evolutionary Synthesis of the 1930s and 1940s had to expunge erroneous hypotheses and establish the validity and sufficiency of the processes that form the core of modern evolutionary biology. This task having been accomplished, evolutionary biologists from the 1960s onward expanded their view and began to develop more detailed hypotheses to explain particular kinds of characteristics: Why do species vary in temperature tolerance, in longevity and reproductive rate, in whether or not mates form pair-bonds, in whether reproduction is sexual or asexual? How do species adapt to their interactions with other species, and how do such evolutionary changes affect the diversity and structure of ecological communities? What patterns of evolution do DNA sequences, chromosome numbers, and other aspects of the genome display, and what are the causes of these patterns? From such questions, there have developed fields of inquiry such as evolutionary physiology, evolutionary ecology, behavioral ecology, and molecular evolution.

The chapters in this section provide a glimpse—and no more—into some of these topics, showing how evolutionary theory is being developed to explain the diversity of various kinds of traits, as well as evolution at the molecular level and the evolution of interactions among species. At times we will be concerned with determining whether or not natural selection has played an important role, but in many instances, the circumstantial evidence for adaptive value is so great that we will be more concerned with developing general theories of adaptation to explain when natural selection will favor one feature rather than another. Fecundity, for example, is so obviously related to fitness that we ask not whether natural selection has favored high fecundity, but how it has done so. At the same time, we will bear in mind that a species' features are not solely a product of its particular selective milieu, but are also affected by its phylogenetic heritage.

Chapter 17 is devoted to questions about why such great diversity in form (morphology) and function (physiology) has evolved. From the many possible topics embraced by evolutionary morphology and physiology, I have chosen to emphasize patterns of adaptation to three factors—temperature, water availability, and body size—that have particularly wide-ranging consequences, and to highlight modern phylogenetic and genetic approaches to such problems as whether or not organisms are optimally adapted, how tolerance to environmental variation evolves, and why species have limited ecological and spatial distributions.

In Chapter 18, we shift from the abiotic features to the biotic features of an organism's environment: the other species with which it interacts. We ask how coevolution in interacting species has affected adaptation and diversification. Interactions between competing species, mutualists, predators and their prey, and parasites and their hosts may all evolve in diverse and sometimes surprising ways, some of which have implications for the human battle against infectious diseases.

Another important focus of contemporary evolutionary biology is the evolution of life histories, the subject of Chapter 19. Why some species have long life spans and others short, why some have many offspring and others few, why some achieve maturity rapidly and others slowly are questions that lie at the core of evolutionary theory, for these are the major components of fitness itself, the key concept in the theory of natural selection.

Almost every aspect of the ecology and evolution of animals is affected by their behavior. Chapter 20 emphasizes both the impact of behavior on animal evolution and theories that account for the evolution of some of the most important and curious of behaviors: those associated with mating and with social interactions. This chapter provides a foundation for the highly controversial topic (treated in Chapter 26) of the role of evolution of human behavior.

Some of the most curious and puzzling features of organisms are their diverse "genetic systems": variations among species in sex ratio, sex determination, level of inbreeding, recombination rates, and whether they reproduce sexually or asexually. Chapter 21 shows how the theory of evolutionary genetics can explain many of these features, the subject of some of the most active and inventive research in evolutionary biology.

In Chapter 22, we shift focus from phenotypic traits to the structure and composition of the genome itself, and explore the vigorous field of molecular evolution. Although we have already encountered uses of molecular techniques and have adopted a molecular perspective on many topics in previous chapters, the evolution of DNA sequences as such are the main subject here. In attempting to account for the different rates of DNA sequence evolution, the evolution of noncoding DNA, and the origin of gene structure—of gene families and of new genes and new functions—we achieve a deeper understanding of the foundations of evolution and of the living world.

CHAPTER 17

Form and Function

Among the first questions almost anyone is likely to ask about an organism is what uses its features serve. One need not be a biologist to ask why a cat has whiskers and slit-shaped pupils. And it requires only intellectual curiosity, not biological sophistication, to wonder how a plant manages to survive in a desert, or how a small bird can survive the bitter cold of a Canadian winter. These questions about form and function are part of the subject matter of morphology and physiology, two of the most venerable biological disciplines.

Morphology and physiology contribute to the study of evolution in two important ways. First, because many of the differences among organisms are morphological and physiological, it is by studying these features that we can begin to understand the reasons for biological diversity. We can ask why some mammals have slit-shaped and others circular pupils, why species of fish vary in shape, why some plants can grow only in marshes and others only in deserts. Second, morphological and physiological studies contribute to our understanding of the mechanisms by which organisms function. Understanding such mechanisms is critically important, for it can often tell us which modifications might be advantageous, what constraints might limit evolution, and thus what pathways of evolutionary change are more likely than others.

Morphology and Physiology

Morphology, the study of form and anatomy, includes several related disciplines (Wake 1992). FUNCTIONAL MORPHOLOGY is the study of the way that form or structure causes, permits, or constrains the function or performance of organisms. BIOMECHANICS uses the principles of engineering and physics to analyze form and function. ECOMORPHOLOGY studies the correlation of organisms' form and function with their environment. EVOLUTIONARY MORPHOLOGY explicitly aims to integrate development, ecology, biomechanics, and phylogenetic analysis to answer questions about the evolution of organismal complexity.

Physiology, the study of organismal functions, is closely related to both morphology and biochemistry, and includes COMPARATIVE PHYSIOLOGY (Prosser 1973; Hill and Wyse 1989), which contrasts physiological functions among taxonomic groups and among organisms that occupy different environments, and ECOLOGICAL PHYSIOLOGY or PHYSIOLOGICAL ECOLOGY (Schmidt-Nielsen 1990; Sibly and Calow 1986; Feder et al. 1987), which asks how physiological traits fit organisms for diverse ecological circumstances. Students of these fields have generally assumed that features are adaptive, and ask *how* organisms are adapted. The recently developed field of EVOLUTIONARY PHYSIOLOGY, instead of merely placing an evolutionary interpretation on physiological traits, uses evolutionary methods such as analyses of genetic variation, natural and artificial selection, and phylogeny to study physiological characters (Garland and Carter 1994; Sibly and Calow 1986).

Morphological and Physiological Adaptations

A useful framework for thinking about the evolution of any trait, such as a morphological or physiological characteristic, is the relationship among trait, performance, and fitness (Arnold 1983):

$$\text{morphology} \xrightarrow{r_{m,p}} \text{performance} \xrightarrow{r_{p,f}} \text{fitness}$$

where $r_{m,p}$ is the correlation between morphology and performance and $r_{p,f}$ is the correlation between performance and fitness. Physiologists often can estimate $r_{m,p}$, but estimating $r_{p,f}$ is very challenging. One must understand not only the *benefits* that a feature might provide, but also its *costs* and *constraints*. Biomechanics and physiology can provide insights into many constraints, such as those set by physics and by the materials of which organisms are made. Predicting the cost (in fitness) of a feature is often more difficult, because a character state may be advantageous in some contexts but disadvantageous in others. Costs are often referred to as TRADE-OFFS.

519

Functional morphology and ecological physiology, by assuming that features have been shaped to serve particular aspects of performance, have provided abundant insights into the adaptive value of many characteristics of organisms. A few examples follow.

Flight in Birds

Bird flight, analyzed by principles of engineering (Maynard Smith 1971; Vogel 1988), illustrates some of the trade-offs arising from simple physics, and shows clearly how we can account for some of the differences among species.

Flight requires that the wings provide LIFT (force opposing gravity) and THRUST (propulsive force). Forward movement is opposed by DRAG, due to friction between body and air and to turbulence. Bird wings, like those of airplanes, create lift by causing the air to flow faster over the curved upper surface than the flatter lower surface. Thrust is provided mostly by the long primary feathers at the end of the wing, which are tilted vertically forward during the downstroke, converting lift to thrust much as a propeller blade does. The "cost" of flight—the work needed to keep a bird aloft—increases with "wing loading," the weight of the body relative to the area of wing surface.

Using aerodynamic principles, much of the variation among bird species in the size and shape of the wings can be explained as adaptations for different modes of flight and for reducing its energetic cost (Pennycuick 1975). Reducing the wingbeat frequency, for instance, reduces the energetic cost of flight. Many large, heavy-bodied species that spend much of their time on the wing reduce their flight costs by low wing loading, provided by long wings that are broad at the tip, as in pelicans, eagles, and vultures (Figure 17.1). Many vultures and other broad-winged birds that stay aloft for long periods fly with the primary feathers separated by slots. Each feather acts like a miniature pointed wing, which reduces drag and adds lift. Hence such birds can stay airborne at a low air speed, and flap their wings only infrequently. Broader wing tips, however, cause more turbulence (drag). Drag is reduced by pointed wings, which are characteristic of many birds, such as falcons and swallows, that specialize in sustained, high-speed flight.

Long wings reduce energy costs, but they cannot be moved fast enough to provide high acceleration or maneuverability. Birds that forage in vegetation and must circumvent obstacles typically have relatively short, rounded wings that provide less lift than the long wings of falcons and other birds of the open air. Short-winged birds are capable of sustained flight only by energetically costly continuous flapping. Ground-living birds such as pheasants fly mostly to escape predators; they require high acceleration and maneuverability, and have short wings with broad tips. Many ground-living birds have high wing loading; and some, such as the great bustard (*Otis tarda*), have achieved the maximal weight that is theoretically possible for a flying bird (about 12 kg). Greater weight has evolved only in flightless groups such as ostriches and penguins.

FIGURE 17.1 Outlines of birds that differ in wing dimensions and behavior. (A) Turkey vulture, a large soaring bird; note the long, broad wings and widely separated primaries. (B) Falcon, a fast-flying aerial hunter with narrow, pointed wings. (C) Swallow, a fast-flying aerial insect eater. (D) Jay, a forest-dwelling bird with broad, rounded wings that allow it to maneuver through trees. (E) Bobwhite quail, a ground-dwelling bird that accelerates rapidly in short flights to escape predators. (A–C, E after Terres 1980; D after Burton 1990.)

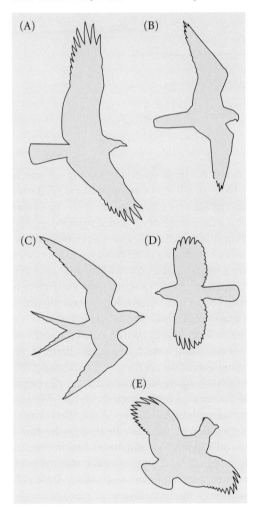

Animals in Hot Deserts

Organisms that live in hot deserts confront the dual problems of overheating and excessive loss of water. Moreover, evaporative cooling (by sweating, for example) is one of the major mechanisms by which body heat can be dissipated, but the dearth of water in a desert makes this a luxury that organisms can ill afford. Desert living organisms illustrate clearly both convergent evolution and the variety of adaptive "solutions" to an adaptive "problem" (Schmidt-Nielsen 1990; Hill and Wyse 1989).

Many desert-dwelling animals avoid excessive heat by staying in shade or foraging at night. Some enter torpor (estivation) during the hottest season. A few can tolerate high body temperatures during the day. Camels, for example, allow their body temperature to rise to over 40°C during the day, and let it drop to about 34° during the night. The increase in daytime body temperature reduces the loss of water that would be required to maintain a constant temperature by evaporative cooling, and reduces the rate at which heat is gained, because the rate at which heat flows from a source (such as air) to a sink (such as the body) is proportional to the difference in their temperatures. Insulation such as fur or feathers also reduces the rate at which the body gains heat from the environment.

Many desert animals dissipate heat by evaporative cooling. Sweating is an active process accomplished by sweat glands that secrete salts, with water following the concentration gradient. However, the resulting loss of salts may cause osmotic stress. Dogs and many other birds and mammals avoid this problem by panting: an increase in respiration rate that expels warm, moisture-laden air from the lungs. Panting can cool the brain, the most heat-sensitive organ. In gazelles, arterial blood flows through a network of capillaries that then rejoin the arteries that serve the brain (Figure 17.2). This capillary network lies in a venous sinus that comes from the nasal passages, where the blood has been cooled by panting. The arterial blood is therefore cooled, by loss of heat to the cooler venous blood, before it enters the brain—an example of a COUNTERCURRENT HEAT EXCHANGER. The nasal heat exchanger is not a specific adaptation to gazelles' hot environment, for it is characteristic of their family (Bovidae) as a whole, including cattle and sheep, which occupy diverse environments. Gazelles illustrate that the adaptations of desert organisms are seldom truly novel, but instead are modifications of structures and functions typical of the larger clades to which they belong.

Selection on desert organisms favors mechanisms that facilitate the acquisition of water and minimize its loss. For example, the skin may be relatively impermeable to water. Cutaneous water loss in a desert-dwelling lizard, the chuckwalla (*Sauromalus obesus*), is only one-fourth of that in the iguana (*Iguana iguana*), an inhabitant of wet tropical forests. Herbivorous lizards and birds excrete potassium and other salts that are ingested in their food not in urine, which

FIGURE 17.2 The countercurrent heat exchanger of a gazelle. Before reaching the brain, warm arterial blood passes in capillaries through a pool of venous blood that has been cooled in the nasal region. (After Taylor 1972.)

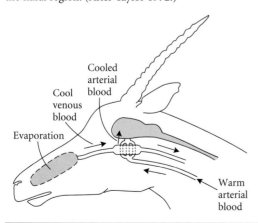

Cooled arterial blood

Cool venous blood

Evaporation

Warm arterial blood

FIGURE 17.3 In some lizards and birds, osmotic balance is maintained by salt glands that excrete concentrated salt solution that drains through the nostrils. (A) A Galápagos marine iguana (*Amblyrhynchus cristatus*). Traces of excreted salt are visible on the face. (B) Skull of a gull (*Larus*). The salt glands are located in depressions above the eyes. (A courtesy of A. F. Bennett; B from Schmidt-Nielsen 1990.)

(A)

(B)

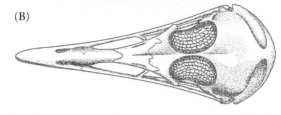

would entail a great loss of water, but by nasal glands that produce a highly concentrated salt solution expelled through the nostrils. These same nasal glands are used by marine reptiles and birds to excrete sodium chloride—a nice example of a structure that provides adaptation, or preadaptation, to quite different environments (Figure 17.3).

In insects, reptiles, and birds, nitrogenous wastes are excreted as insoluble uric acid. Rather little water is required for this process. Moreover, some desert insects (and others that subsist on dry diets such as wool or dry wood) can reabsorb almost all of this water through the wall of the rectum, depositing an almost dry mass of uric acid and fecal matter.

Mammals, in contrast, excrete nitrogenous wastes as water-soluble urea, filtered from the blood plasma by the kidneys, which reabsorb much of the water. Some desert mammals, such as kangaroo rats, economize on water by producing astonishingly concentrated urine, accomplished by modification of the length of the kidney tubules and their spatial relationships to the capillaries that receive the reabsorbed water.

Plants in Warm, Dry Environments

Plants generally cannot avoid insolation and high temperatures by behavioral means, although the shape and orientation of their leaves and stems may reduce heat load. Commonly, plants keep their tissues below lethal temperatures by dissipating heat through emission of infrared radiation, convection, and evaporative cooling. The latter is by far the most effective. Transpiration, the release of water vapor, cools the plant, at the cost of losing water. Moreover, photosynthesis (and, consequently, growth and reproduction) requires transpiration. The problem faced by a plant in a hot, dry environment, therefore, is how to photosynthesize and avoid overheating with minimal expenditure of water (Fitter and Hay 1981). Like the adaptations of animals, those of plants in dry environments illustrate trade-offs and constraints.

In terrestrial vascular plants, water is drawn up through the vascular tissues by transpiration, which occurs mostly through the stomata, openings on leaf surfaces between pairs of guard cells. The stomata also are the sites of entry for carbon dioxide, which then diffuses into the leaves' mesophyll cells, where it is fixed and used to synthesize glucose.

Two other fundamentals bear mention before we describe plants' adaptations to heat and water stress. One is the BOUNDARY LAYER, the layer of relatively undisturbed air cloaking a leaf or any other object. The boundary layer can be much hotter than the surrounding atmosphere, and it impedes the flow of CO_2 and water vapor. The thicker the boundary layer, the stronger the impedance. As wind speed increases, the thickness of the boundary layer decreases. Moving air thins the boundary layer more effectively if it passes over a narrow surface than a broad one, so a small or highly dissected leaf has a thinner boundary layer than a large, undivided leaf. Thus in the sun, a smaller leaf will have a lower surface temperature and a higher rate of CO_2 uptake—but also a higher rate of water loss.

The second basic principle, which is *fundamentally important to all of biology*, is *the relation of surface area to volume* (Figure 17.4). If L is a linear dimension (a length) of an object, then the surface area, A, is proportional to L^2, and the volume, V, is proportional to L^3. The relation of area to volume, then, is $A \propto V^{2/3}$.* Thus the surface/volume ratio of geometrically similar objects (i.e., with the same shape) declines as size increases. The mass of an organism (or a part, such as a leaf) is usually proportional to its volume. Per unit of mass, then, the flux of heat, water, or gases across the surface is lower for the larger of two organs or organisms of the same shape.

The diverse adaptations for life in dry environments are most pronounced in desert-dwelling plants (xerophytes), but also characterize plants that occupy other dry environments such as treetops (e.g., epiphytic bromeliads and orchids that grow on other plants' branches), as well as

---

*The symbol $\propto$ means "proportional to."

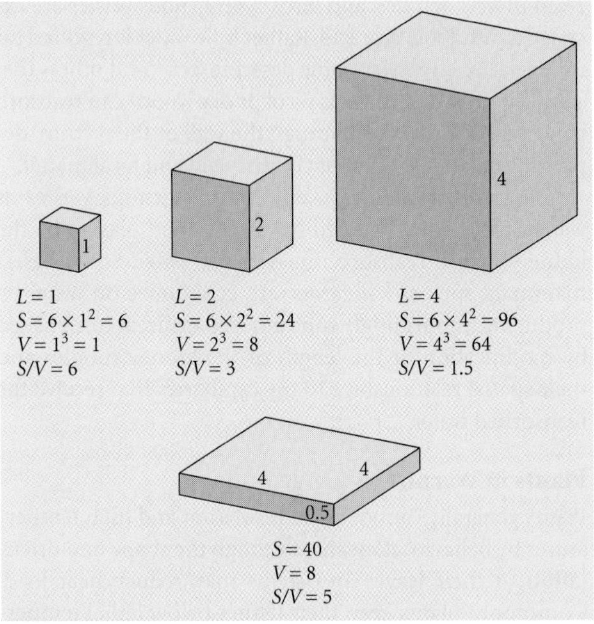

FIGURE 17.5 Root systems of desert plants. (A) Shallow roots extend far beyond the parts of a creosote bush (*Larrea*) seen above ground. (B) The deep taproot of a young mesquite tree (*Prosopis*). (A after Solbrig et al. 1977; B after Fisher et al. 1946.)

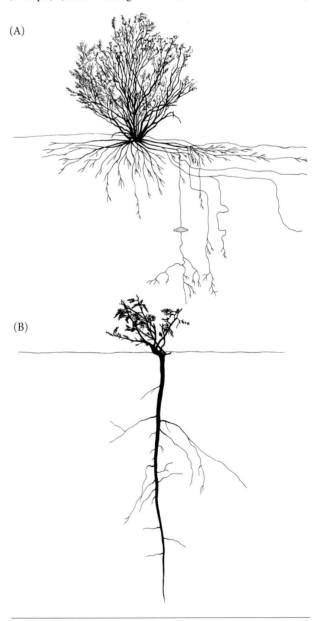

(A)

(B)

FIGURE 17.4 The effect of size on surface/volume (*S/V*) ratio is illustrated by three geometrically similar figures (cubes). As the linear dimension (*L*) increases, *S/V* decreases. The lower figure has the same volume (8) as the cube with *L* = 2, but has greater surface area due to its thinner shape.

$L = 1$
$S = 6 \times 1^2 = 6$
$V = 1^3 = 1$
$S/V = 6$

$L = 2$
$S = 6 \times 2^2 = 24$
$V = 2^3 = 8$
$S/V = 3$

$L = 4$
$S = 6 \times 4^2 = 96$
$V = 4^3 = 64$
$S/V = 1.5$

$S = 40$
$V = 8$
$S/V = 5$

cold environments (where soil water freezes and is therefore unavailable) and salt marshes (where plants lose water by osmosis to the hyperosmotic saline environment).

Many woody plants conserve water by dropping their leaves during the dry season. Adaptations for *acquiring* water include extensive root systems: broadly spreading, shallow systems that capture rainwater as it enters the soil, or long taproots that absorb groundwater deep below the surface (Figure 17.5). A few desert plants and epiphytes have specialized water-absorbing cells on the leaf surface.

Many desert plants have convergently evolved relatively thick, small or dissected leaves (Figure 17.6). The thickness of the leaf reduces the surface/volume ratio and therefore the rate of water loss; the small size or dissection into lobes or leaflets reduces the boundary layer and therefore the heat load. The reduction of surface/volume ratio is carried even further in the many leafless desert plants in which green stems are the site of photosynthesis (see Figure 3 in Chapter 4).

Other morphological adaptations to desert life include stomata sunk in pits below the leaf surface, an arrangement that reduces the flux of water more than the flux of CO_2; and dense coats of hairs (in many plants) or spines (in some cacti) that reduce thermal stress and transpiration both by reflecting solar radiation and by keeping the hot boundary layer away from the surface of the leaf (Figure 17.7). Some plants that lack such insulation, such as certain prickly pear cacti (*Opuntia*), can tolerate much higher tissue temperatures than most plants can.

Closing the stomata reduces water loss, but it reduces photosynthesis and growth, because CO_2 cannot enter the plant. This problem is ameliorated in the Crassulaceae (stonecrops) and several other groups that have independently evolved CRASSULACEAN ACID METABOLISM (CAM). They open the stomata during the night, when water loss by transpiration is lower, and admit CO_2, which is fixed by a spe-

FIGURE 17.7 This cholla cactus (*Opuntia*) in Arizona is densely coated with spines that reduce its heat load by reflecting sunlight. They also provide a formidable defense against browsing mammals. (Photograph by the author.)

cial enzyme (PEP carboxylase) and stored as organic acids that are metabolized during the day into CO_2 and thence into glucose. A related mechanism that reduces water loss (C_4 photosynthesis) has evolved in several other groups of plants.

Body Size

Probably no other single feature of an organism has as many implications as its size, so understanding the implications of size is essential for understanding the evolution of many morphological and physiological features (Peters 1983; McMahon and Bonner 1983; Calder 1984; Schmidt-Nielsen 1984). The range in size of organisms, as measured by mass, spans 10^{21} orders of magnitude, from the smallest bacteria to the blue whale.

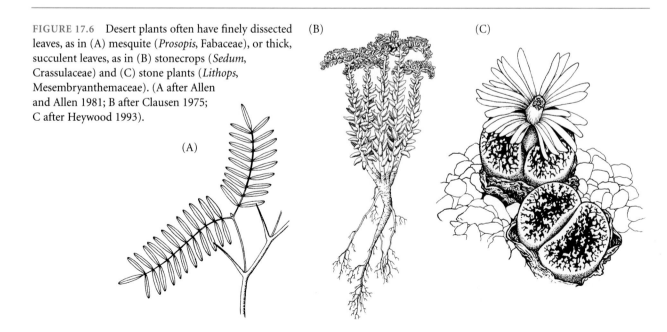

FIGURE 17.6 Desert plants often have finely dissected leaves, as in (A) mesquite (*Prosopis*, Fabaceae), or thick, succulent leaves, as in (B) stonecrops (*Sedum*, Crassulaceae) and (C) stone plants (*Lithops*, Mesembryanthemaceae). (A after Allen and Allen 1981; B after Clausen 1975; C after Heywood 1993).

(A)

(B)

(C)

Selection on Size

A vast number of factors can impose selection on size. For example:

Small organisms take less time to grow to maturity, so their generation time is shorter. Consequently, their rate of increase (i.e., fitness) is greater, all else being equal.

Small organisms require less food than large ones, and can inhabit smaller microhabitats.

Large plants can produce more seeds, large animals can carry more eggs, and large males often win contests with smaller males, so large size has reproductive advantages that may outweigh the reproductive advantages of rapid development (see Chapters 19 and 20).

Large animals, although they require more food, often can overpower, handle, or swallow a greater range of food items.

Predators may select for small or large size in prey species, depending on the predator's size and mode of feeding.

We now turn to several other implications of body size that have great significance for the study of evolution. Throughout, we will use body mass (weight) as a measure of size.

Allometry and Isometry

We are often interested in how one feature or measurement of an organism—say, y—changes as a function of another measurement—say, x. These measurements can refer to any characters: two dimensions of a leaf or a tooth, or the size of eggs relative to the number per clutch, or the wing area of a bird relative to its weight. Frequently, this relationship can be described by the **allometric equation**:

$$y = bx^a$$

If we divide both sides by x, and if $a = 1$, then $y/x = b$. That is, y changes as a constant proportion (b) of x, and "shape" stays constant as "size" is altered. This relationship is called ISOMETRY.

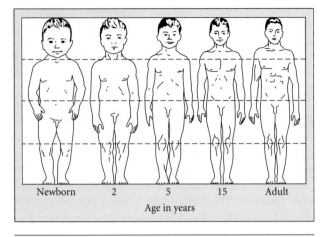

FIGURE 17.8   Allometric growth in humans. Individuals of different ages, drawn at equal heights, show proportionately less rapid growth of the head and more rapid growth of the legs than of the body as a whole. (After Sinclair 1969.)

Newborn   2   5   15   Adult

Age in years

Often, however, the several measurements do not change proportionately. For example, as humans grow in size, the head grows less rapidly, and the legs more rapidly, than the body (mass) as a whole (Figure 17.8). Such disproportionate changes with growth are called **allometric growth**. In such a relationship, the ALLOMETRIC COEFFICIENT (a in the above equation) is not equal to 1. If $a < 1$, y increases less rapidly than x; if $a > 1$, y increases more rapidly. Thus if x represents body mass, $a < 1$ in the case of human head growth, and $a > 1$ in the case of leg growth. If y and x are negatively correlated, $a < 0$.

It is often convenient to visualize these relationships by taking the logarithm (perhaps to base 10) of both sides of the equation, yielding

$$\log y = \log b + a \log x$$

which describes a straight line with slope a and intercept $\log b$. Figure 17.9 illustrates some such relationships for hypo-

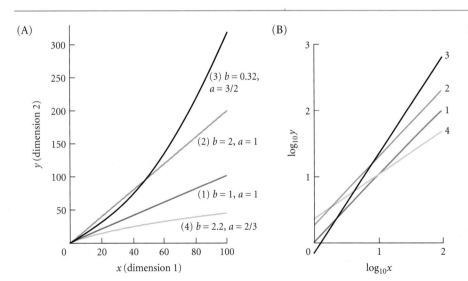

(A)

y (dimension 2)

(3) $b = 0.32$, $a = 3/2$

(2) $b = 2$, $a = 1$

(1) $b = 1$, $a = 1$

(4) $b = 2.2$, $a = 2/3$

x (dimension 1)

(B)

$\log_{10} y$

$\log_{10} x$

FIGURE 17.9   Hypothetical curves showing relationships between two measurements, y and x, according to the allometric equation $y = bx^a$. (A) Arithmetic plots. Curves 1 and 2 show isometric growth ($a = 1$), in which y is a constant multiple of x. Curves 3 and 4 show positive ($a > 1$) and negative ($a < 1$) allometry, respectively. Logarithmic plots of the same curves. Curves 1 and 2 have slope equal to 1. The slope is greater than 1 in curve 3, and less than 1 in curve 4.

thetical data. Figure 17.10 shows real data on the relationships between mass of the testes and body mass in primates.

Much of morphological evolution can be described in terms of the allometric relationships among the dimensions of an organism—i.e., its shape (Chapters 23, 24). Elongation of the digits accounts for the shape of a bat's wing, for example, and an increase in the length and thickness of the central toe relative to the side toes describes one of the major trends in horse evolution. Allometric relationships to body size are especially pervasive.

FIGURE 17.10   Log-log plot of testes mass in relation to body mass in some primates with polygamous (black circles) and monogamous (colored circles) mating systems. Testing for a correlation between testes mass and mating system requires correcting for body mass. Two approaches are illustrated. (A) Calculation of separate regression lines for the two sets of species. The regression for polygamous species has a higher intercept. (B) A single regression for all species yields the "predicted" testes mass for any body mass. The deviations between the observed and predicted values, shown by the vertical lines, can be analyzed to see if they are correlated with mating system, or perhaps with other variables. (After Harvey and Pagel 1991.)

(A)

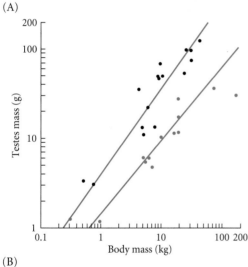

(B)

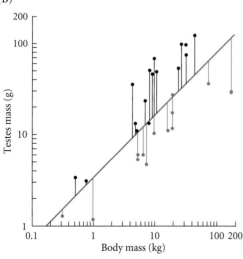

Allometry is also important in analyzing adaptation. For example, changes in body size are automatically accompanied by changes in all features that are correlated with body size, such as the weight of testes.

The comparative method uses correlations among taxa to infer the adaptive significance of various features by establishing whether or not they are correlated with differences in the species' environments or ways of life (see Chapter 12). It is often important to be sure that the size of a feature is not simply an effect of a species' body size, before concluding that it is an adaptation. That is, it is important to standardize by body size in making such comparisons. For instance, in order to test the hypothesis that selection favors greater sperm production, and hence greater testis weight, in polygamous than in monogamous primates (Chapter 12), it is necessary to show that the testes of polygamous species are larger than those of monogamous species *with the same body mass*. This is indeed the case (Figure 17.10A). In another approach, we can propose that body mass predicts, or "explains," much of the variation in testis weight, but not all. The *deviations* from the overall allometric relationship, then, require explanation. These differ between polygamous and monogamous species (Figure 17.10B). Thus, analyzing deviations from the relationship predicted by allometry provides evidence that natural selection has adjusted testis weight in relation to the species' breeding systems.

Allometry and Adaptation

Allometric relationships with body mass are often the *consequences of adaptation*. For example, structures that support an organism must change disproportionately in shape (or be made of stronger materials) as weight increases over a great range. Without a change in shape, the ratio of mass to cross-sectional area of a tree trunk would increase as the 3/2 power of its height, so resistance to compressional fracture would decrease, the larger a tree is. We would therefore expect that natural selection for resistance to fracture would favor a growth pattern in trees whereby the diameter of their trunks increases proportionately faster than their height ($a > 1$) (McMahon and Bonner 1983). The same principle of adaptation explains why the leg bones of heavy vertebrates such as elephants and ostriches are far thicker in relation to their length than the homologous bones in small animals (Figure 17.11).

Perhaps the most famous relationship between a physiological variable and body size is the "mouse to elephant" curve for basal metabolic rate in mammals (Figure 17.12), which shows that energy consumed (P) increases with body mass (m) approximately as $P = 70m^{0.75}$. However, for reasons still debated among physiologists, the MASS-SPE-CIFIC metabolic rate, P^*, the energy consumed per gram of body mass, is lower, the larger the animal: $P^* \propto m^{-0.25}$. Thus, the metabolism of a shrew or sparrow is far higher, per gram, than that of a lion or ostrich. Hence, to maintain a constant body temperature, small homeotherms must eat at a prodigious rate (especially when it is cold).

FIGURE 17.11 Left front leg, in lateral view, of (A) a domestic cat and (B) an Indian elephant, drawn to roughly equal heights. Note the much greater relative thickness of the elephant's bones, necessary to support its far greater weight. The radius and three of the digits of the elephant are not shown in this view. (After Yapp 1965.)

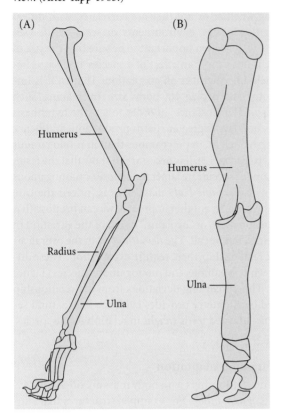

In a cold environment, large size is advantageous in birds and mammals because they lose heat more slowly, and thus require less food to maintain a constant body temperature. This is the traditional explanation of BERGMANN'S RULE, which states that birds and mammals in cold regions frequently are larger in body size than populations of the same species in warmer regions (Figure 17.13).

Body size and resource use Very small animals not only need less food, but can use smaller bits of it. Thus a seed beetle (Bruchidae) can complete its entire development within a single small seed, and a small hymenopterous wasp (Trichogrammatidae) accomplishes the same within a single moth egg. The optimal food items for a species often depend on its size, because size affects both the amount of food it needs and the size of the structures used for obtaining and manipulating food. For instance, large species of finches prefer larger seeds, which they can handle more efficiently with their larger bills than small species (Figure 17.14). Differences in body size are among the most common features that apparently enable species to coexist (see Chapter 4).

Constraints associated with body size Body size imposes constraints on the evolution of many other features, and may itself be constrained because of its many functional consequences. For example, in many fishes and other animals that carry eggs within the body until they are deposited, the space available within the body can limit fecundity. Function often demands that a structure, such as the skeleton, vary allometrically with body mass, so that it

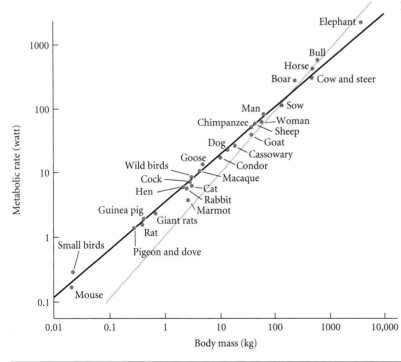

FIGURE 17.12 A log-log plot of the total metabolic rate of some mammals and birds in relation to body mass. The gray line represents a slope of 1, showing that smaller animals have higher metabolic rates, relative to body mass, than larger animals. (From Schmidt-Nielsen 1984.)

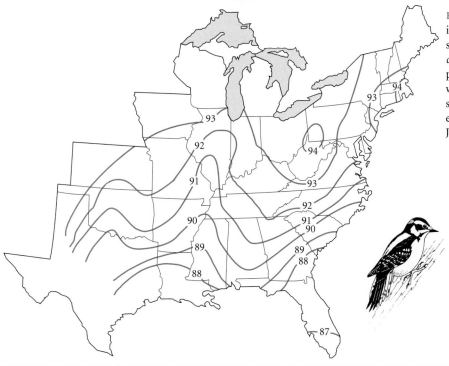

FIGURE 17.13 A north-south cline in wing length, a measure of body size, in the downy woodpecker (*Picoides pubescens*). Each line connects populations with the same mean wing length (in mm). The increase in size with latitude in this species exemplifies Bergmann's rule. (After James 1970.)

is not free to evolve independently in size and form. Because an organism must allocate energy and materials among growth, maintenance, and reproduction, the greater allocation to skeletal and other structures in larger organisms often comes at a cost in terms of reproductive rate. Functional relationships may well constrain body size itself in some groups of organisms. In insects, for example, oxygen and carbon dioxide diffuse through tubes

(tracheae) that ramify throughout the body. Many authors have hypothesized that the rate of diffusion limits insects to a relatively small size. In the same vein, it seems highly unlikely that an animal the size of a blue whale could ever support itself on land. Any of the many characteristics that are correlated with body size could, in principle, act as agents of natural selection that could affect the evolution of body size itself.

FIGURE 17.14 The relationship between seed value and seed size for three species of sparrows. The chipping sparrow is smallest, and the white-crowned sparrow largest. Pulliam measured the rate at which each species can husk seeds of different sizes. Seed value—the rate at which a bird can consume seed mass—increases with seed size at first, but then declines because larger seeds take longer to husk. Larger bird species, with larger bills, can efficiently husk a greater range of seed sizes, so

seed value peaks at a larger seed size for a larger bird. By measuring the daily energy requirement of each species, Pulliam calculated the range of seed sizes that could meet this requirement for each species; this range is shown by the horizontal lines ($\bar{v}_w$, $\bar{v}_j$, $\bar{v}_c$). Although all the sparrow species husk small seeds equally efficiently, the minimal seed size that can sustain a bird is lower for small birds because of their lower energy requirement. (After Pulliam 1985.)

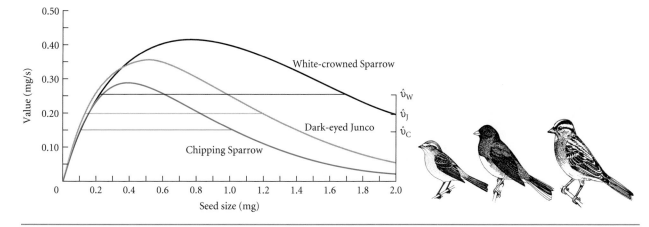

Adaptation and Constraint

The developing fields of evolutionary physiology and morphology (Feder et al. 1987; Sibly and Calow 1986; Wake 1992; Garland and Carter 1994) ask such questions as: Are the differences between closely related species adaptive? Are organisms optimally designed? What are the constraints on adaptation? Why and how do different traits evolve in concert? By what steps have complex traits evolved? Models of physiological function, studies of genetic variation and selection within populations, the comparative method, and phylogenetic analysis are among the methods used to address such questions.

Optimal Design and Constraints

Many models have been advanced to predict or explain various phenotypic characters, including morphology, life histories, reproductive systems, behavior, and others, on the assumption that natural selection has been able to *optimize* the feature, *within specified constraints*. Such **optimality models** have been quite controversial (e.g., Gould and Lewontin 1979; Rose et al. 1987; Maynard Smith 1978b; Williams 1992b). Unless, on the basis of biological knowledge, we can specify some constraints, we would have to predict that the optimally designed organism would be immortal, would have infinitely great fecundity, and would be able to live everywhere and do everything.

The most frequently invoked constraints assume a **principle of allocation**, whereby energy or some other limiting factor leads to a *trade-off* among two or more activities, functions, or structures. For example, a plant near the end of a growing season may have a fixed amount of energy, E, that can be allocated to forming the ovules (O) or the stamens (S) of its flowers. Thus $E = O + S$, and the question is what the optimal values of O and S might be in a given environment, subject to the constraint that E is fixed. There is a trade-off between O and S, expressed by the negative correlation between them (Figure 17.15). If we knew how the number of ovules and of stamens is related to reproductive success, we could estimate the effect on fitness of each possible combination of ovule mass and stamen mass (Figure 17.15B) and predict what attainable combination of O and S would be best, *given* the constraint that $E = O + S$. Other kinds of constraints may affect predictions of optimal design, such as the mechanical constraints that affect the evolution of birds' wings. Of course, there are several reasons why a feature may not be optimal at all (Chapter 12).

One of the most thoroughgoing hypotheses of optimal design is the principle of SYMMORPHOSIS (Taylor and Weibel 1981), which states that each structure should be just enough to meet functional needs, and no more. This means that no single component of a complex system, such as the respiratory system of a mammal, should be rate-limiting. For instance, the volume of the lungs should not exceed the capacity of the trachea to conduct air, and the trachea should conduct neither more nor less air than the lungs can accommodate. This hypothesis assumes ECONOMY OF DESIGN, because developing and maintaining biological structures is assumed to be costly.

Garland and Huey (1987) tested for symmorphosis in the respiratory systems of twelve species of Bovidae (cattle, sheep, etc.) and Viverridae (a family of carnivores that includes mongooses). They analyzed other authors' data on VO_2max (the maximal rate of oxygen consumption

---

FIGURE 17.15 Finding the optimal allocation between two structures or functions, such as mass of stamens (*s*) and ovules (*o*). (A) Assuming a fixed amount of energy or biomass that can be allocated, there is a linear trade-off between the mass of *s* and of *o* that the organism can develop. (B) From data on the reproductive success of various phenotypes that differ in *s* and *o*, a fitness landscape (as in Chapter 14) can be calculated. Each contour line connects possible combinations of *s* and *o* with the same fitness. More peripheral contours represent lower fitness. (C) Combining A and B, the point (*) on the trade-off line that is tangent to the highest contour is the optimal mixture of *s* and *o*, given that *s* + *o* cannot be further increased. (After Sibley and Calow 1986.)

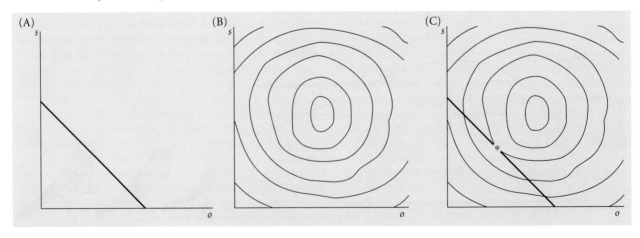

during exertion), DLO_2 (the diffusion capacity of the lungs), the density of mitochondria in muscles, and the density of capillaries in muscles. According to the principle of symmorphosis, these variables should all be strongly correlated with one another. However, after the researchers controlled for allometric relation to body mass, these variables proved to be uncorrelated. Thus if DLO_2 is not correlated with VO_2max, something else must limit VO_2max in at least some of the species—and the same holds for the density of mitochondria and capillaries. Therefore these structural features cannot all be so finely adjusted to one another that no one of them limits aerobic respiration. At least in this case, the hypothesis of symmorphosis is not supported. A likely reason is that structures often have more than one function. The lungs of mammals, for instance, serve not only for uptake of oxygen, but also for release of carbon dioxide and for vocalization functions not considered in an analysis of this kind.

Evolutionary loss of useless structures

The principle of economy of design also implies that organisms should not have nonfunctional organs. It certainly is true that *in general*, organs are vestigial or lacking in species in which, because of evolutionary changes in habits or habitat, these features no longer serve a function. The classic examples are parasites and cave-dwelling animals (Figure 17.16). Obligate cave dwellers, for example, typically have lost their eyes and pigmentation, in whole or in part.

Unused structures may be reduced and lost because it is costly to develop and maintain them, so that selection favors reduction; because they are disadvantageous for other reasons; because they are negatively correlated, by pleiotropy, with other structures; or because they are selectively neutral, and degenerative mutations are fixed by genetic drift. Probably all four causes operate in different instances, but we do not know their relative frequency (Fong et al. 1995). We described a study of cave-dwelling amphipod crustaceans in Chapter 14, in which the authors concluded that eyes degenerate as a pleiotropic effect of selection for longer antennae. The principle of allocation (a kind of pleiotropy) accounts for the loss or reduction of wings in many flightless insects, because in species that are polymorphic for wing development, individuals with reduced wings and wing muscles often have greater fecundity (Harrison 1980; Roff 1990). Some unused structures are probably disadvantageous for mechanical or other reasons, as hind legs probably were for the ancestors of whales and snakes. The role of genetic drift in the loss of neutral unused structures is not yet adequately understood. At the molecular level, however, it is well known that "extra" copies of genes that have arisen by duplication may become silenced and evolve into nonfunctional pseudogenes (see Chapter 22).

Experimental Analyses of Design

As an example of an analysis of optimal design, constraint, and adaptive differences among populations and species, we will describe some of the work by Ward Watt and Joel Kingsolver on temperature regulation in butterflies of the genus *Colias* (e.g., Watt 1968; Kingsolver 1983; Kingsolver and Watt 1984).

FIGURE 17.16 Reduction and loss of structures in parasites. (A) The dodder (*Cuscuta*), a flowering plant lacking leaves and chlorophyll, obtains nutriment via structures inserted into the stem of its host plant. (B) Female of a rhizocephalan barnacle, *Sacculina*, attached to a crab. The parasite ramifies throughout the crab's body from a saclike structure housing the ovaries. *Sacculina* lacks the appendages, calcareous plates, and other features of typical barnacles. (C) An ectoparasite of bats, the fly *Stylidia* (Nycteribiidae). These wingless, eyeless flies hold the head reflexed over the thorax. (A after Peterson and McKenny 1968; B from Brusca and Brusca 1990; C from Theodor 1967.)

(A)

(B)

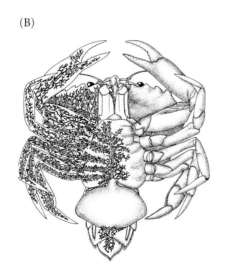

(C)

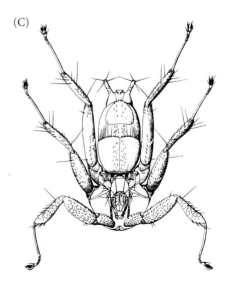

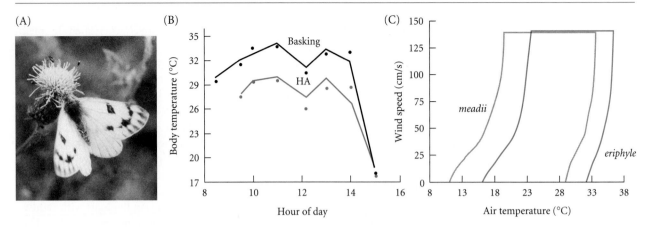

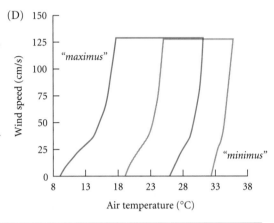

FIGURE 17.17   (A) A sulfur butterfly in basking posture, with wings held perpendicular to the sunlight. (B) Body temperatures measured in sulfur butterflies (*Colias meadii*) in basking posture (black circles) and heat avoidance posture (colored circles) at a montane site on July 27, 1980. The values predicted by a model incorporating air temperature, wind speed, and the morphological characteristics of the butterflies are shown by gray and colored lines for basking and heat avoidance temperatures respectively. (C) The model that successfully predicted temperature in *Colias* butterflies in part B is used here to calculate the combinations of wind speed and air temperature at which two species (*C. meadii* and *C. eriphyle*) would be expected to maintain a body temperature suitable for flight activity. The "envelopes" differ because *C. meadii* has thicker fur and darker wings. (D) Similar "flight space" diagrams calculated for two hypothetical species of *Colias*. "*C. minimus*" lacks fur and dark melanin pigment; "*C. maximus*" has entirely black wings and the longest fur plausible for a butterfly of this size. (Photograph courtesy of J. Kingsolver; B–D after Kingsolver 1983.)

Dark areas on the yellow wings of sulfur butterflies (*Colias*) absorb solar radiation (Figure 17.17A). The absorbed heat is conducted to the body, which bears an insulating coat of fur. These butterflies cannot fly until the body temperature (T_b) reaches 30°–40°C, but overheating ($T_b > 45°$) reduces fecundity. Flight is necessary for feeding, laying eggs, and finding mates (see Chapter 13). Especially at high elevations, daily flight activity time may be so limited by low temperatures that the butterflies succeed in laying fewer eggs than they could given more flight time. Thus fitness may be proportional to the time available for flight.

Using the physical principles of heat transfer, Kingsolver developed a quantitative model that predicts a butterfly's body temperature (T_b) based on the solar absorptivity of its wings, the insulating capacity of its fur, and the energy of solar radiation, temperature of the ground on which the insect sits while basking, air temperature, and wind speed. Using thermocouple probes to measure T_b, he determined how it varied with environmental conditions in four species of *Colias* that are found at different elevations and differ in wing pattern and fur thickness. The measured values were close to those predicted by the model (Figure 17.17B), which also predicted when butterflies should be flying during the day.

Kingsolver's model showed that T_b is most affected by wing absorptivity (determined by melanization of the base of the hindwing) and, to a lesser extent, by fur thickness, in

relation to air temperature and wind speed. *Colias philodice*, the species from the lowest elevation (where it is warmest), has the lowest wing absorptivity and fur thickness, and the high-altitude species *C. meadii* has the highest. Kingsolver's model shows that *C. meadii* should be able to achieve a body temperature sufficient for flight at a lower air temperature or higher wind speed than *C. philodice*, but that it would also overheat at a lower temperature (Figure 17.17C). Thus the morphological differences between the species appear to be adaptations to the species' different environments, but they also constrain each species from flying at certain air temperatures that are favorable for the other. Kingsolver used his model to calculate the air temperature/wind speed combinations at which hypothetical *Colias* species, with either entirely black or entirely yellow wings, could fly. These calculations (Figure 17.17D) show the limits of temperature to which *Colias* could adapt by evolution of its wing patterns.

This study of *Colias* butterflies illustrates (1) *the interaction of features* (wing pattern, fur thickness) *that affect a vital physiological function* (thermal balance), (2) *trade-offs in adaptation* (to low versus high temperatures), (3) *constraints* on the limits of adaptation (that could be achieved to different temperature regimes by evolution of these characters), and perhaps (4) the *factors limiting the ecological distribution of species* (to habitats with a temperature regime sufficient for reproduction).

The Evolution of Tolerance

Studies of genetic differences in physiology among species and populations, and of the physiological changes that individual organisms display when temperatures or other factors change, may help us to answer several important evolutionary questions: By what mechanisms can species adapt to extreme conditions? How do individual organisms tolerate environmental variation? Is there a limit to the range of conditions to which a species can adapt? Do limits of tolerance restrict the ecological and geographic distributions of species? What prevents a species from adapting further and extending its range indefinitely?

Responses to Stress

Ary Hoffmann and Peter Parsons (1991) define stress as an environmental factor that causes a change in a biological system that is potentially injurious and is likely to reduce fitness.

In animals, a series of responses occur sequentially in response to stresses (Slobodkin and Rapoport 1974; Levins 1968; Hoffmann and Parsons 1991): (1) changes in *behavior*, (2) *hormone-modulated* biochemical and physiological *changes*, (3) slower, longer-lasting changes in physiological functions (*acclimation*), (4) in some instances, *developmental changes* in morphology, and (5) at the population level, *genetic changes* due to differences among genotypes in survival and reproduction rates caused by the stress. If the responses of individual organisms (1–4) cannot fully compensate for the stress, fitness is reduced, and genetic (evolutionary) changes (5) may occur. These changes often will be alterations of one or more of the behavioral, hormonal, physiological, or developmental responses.

Some changes in physiological state develop over the course of days or weeks, and are similarly slow to reverse course if the environment returns to its previous condition. These changes, usually called physiological **acclimation,** alter an individual's tolerance of factors such as temperature and salinity. Figure 17.18 illustrates how in two fish species, both the upper and lower lethal temperatures differed when the fish were tested after being held at different temperatures long enough to become acclimated.

In many cases, responses to stresses or other environmental changes entail developmental changes in morphology (see Chapter 4). Some are reversible, such as seasonal changes in fat content and coloration. In other instances, different genetic programs are activated during development, giving rise to often irreversible "alternative adaptations" (West-Eberhard 1986). For instance, in many butterflies, individuals that become adults in spring have darker wings that those that mature in summer. Kingsolver (1995) tested the hypothesis that the developmental switch is adaptive by rearing western white butterflies (*Pontia occidentalis*) in the laboratory under simulated spring and summer conditions, so that they developed different pigmentation patterns. Both phenotypes were marked and released in a natural habitat in both spring and summer, and their survival estimated by the fraction subsequently recaptured. During the

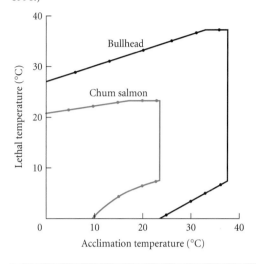

FIGURE 17.18 Acclimation to higher temperatures in two fish species increases both the upper and lower lethal temperatures (the temperature extremes at which they die). The chum salmon is a coldwater fish with a narrower temperature tolerance than the bullhead catfish, which normally experiences wide seasonal variation in temperature. Read upward from an acclimation temperature (say, 20°) to find first the lower, and then the upper, lethal temperature. (After Schmidt-Nielsen 1990.)

summer, butterflies with the pale "summer phenotype" survived longer than those with the dark "spring phenotype."

Adaptation to Varying Environments

Adaptation to varying environments, according to theory developed by Richard Levins (1968), depends on the *temporal pattern of environmental changes* and the *tolerance* of each phenotype that might arise by mutation and recombination (Futuyma and Moreno 1988; Hoffmann and Parsons 1991). The environmental variation (in, say, temperature) is fine-grained if an individual organism experiences several environmental states repeatedly, and coarse-grained if it experiences only one or another state during its lifetime. Each phenotype may have a *narrow* tolerance or a *broad* tolerance (Figure 17.19). Levins' model makes the following predictions:

Phenotype tolerance	Environment	Expected evolutionary outcome
Broad	Fine-grained	Intermediate "generalist" phenotype
Broad	Coarse-grained	Intermediate "generalist" phenotype
Narrow	Fine-grained	One or another "specialist" phenotype
Narrow	Coarse-grained	Mixture of "specialist" phenotypes*

*Polymorphism for specialized phenotypes, each better adapted to one or the other environmental state, is the "optimal" solution; however, if each phenotype is a distinct genotype, the polymorphism may or may not be stable (see Chapter 13).

FIGURE 17.19 (A) Fitness or some measure of performance of two hypothetical phenotypes in relation to an environmental variable such as temperature. Phenotype 2, a generalist, has a broader tolerance than phenotype 1, a specialist, but has lower fitness or performance in its optimal environment than the specialist, illustrating a cost of adaptation. (B) An illustration of the cost of adaptation. Maximal running speed at various temperatures was measured for males of a small parasitoid wasp (*Aphidius ervi*) from numerous sib families. The log (ln) of the breadth of the performance curve (see Figure 17.20) is plotted against the speed for individual wasps (black circles) and family means (colored circles). The negative relationship implies a cost of broad tolerance, as shown in A, and the relationship among family means suggests that it has a genetic basis. (B after Gilchrist 1996.)

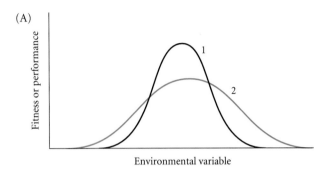

(A)

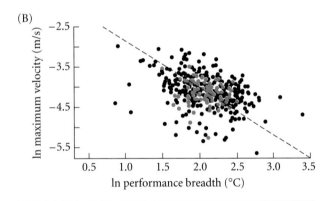

(B)

FIGURE 17.20 Possible evolutionary changes in thermal tolerance in response to selection for increased performance at high temperatures. The same principles apply to evolution of the norm of reaction of any character. (A) The performance curve of the ancestral phenotype. Its optimum temperature is *a*, but selection now favors increased performance at temperature *b*. (B) Increased breadth of the performance curve (gray line) may evolve if there is no trade-off between performance at higher and at lower temperatures—i.e., if there is no genetic correlation between them. (C) The entire curve may be shifted by selection if there is a trade-off—i.e., if a negative genetic correlation exists between performance at higher and at lower temperatures. (D) Selection for performance at high temperatures also increases performance at low temperatures (general stress resistance), but maximal performance may be reduced if a trade-off is assumed, as in Figure 17.19B. (After Huey and Kingsolver 1993.)

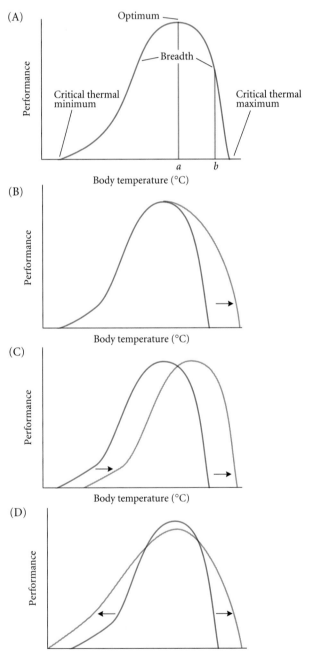

These predictions assume a *trade-off*, or a *cost of adaptation*, so that the broader the tolerance of a phenotype, the lower its fitness, even in the environment to which it is best adapted (Figure 17.19A). Therefore we should expect broadly tolerant genotypes to be replaced by narrow specialists if the environment is constant.

How might we expect the breadth of tolerance to evolve? Suppose Figure 17.20 represents the relationship between body temperature in an ectothermic animal and its fitness, or some measure of performance, such as running speed, that is correlated with its fitness (Figure 17.20A). If the climate becomes warmer, mean body temperature increases and the mean performance is reduced. If suitable genetic variation exists, the population will evolve increased performance at the new temperature. Performance at other temperatures will not be affected if there is no trade-off between physiological functions at different temperatures (Figure 17.20B). Alternatively, if performance at high and low temperatures is negatively correlated, tolerance of low temperatures may decline (Figure 17.20C). Another possibility is the evolution of GENERAL STRESS RESISTANCE (Hoffmann and Parsons 1991), whereby adaptation to temperatures at both extremes is achieved, but at the cost of lower maximal performance (Figure 17.20D). Several kinds of studies of physiological characteristics have begun to provide information on the evolution of tolerance.

Studies of acclimation We have described a study of two fish species in which individuals that were acclimated to high temperatures displayed high performance at high, but reduced performance at low, temperatures (see Figure 17.18). This common pattern, resembling that shown in Figure 17.20C, implies that the physiological properties of heat-acclimated and cold-acclimated individuals are different and perhaps incompatible. Indeed, we know that quite often, different forms (isozymes) of each of many enzymes are expressed, and function best, in animals acclimated to different temperatures (Hochachka and Somero 1984; Prosser 1986).

Comparative studies There is some evidence that species from relatively constant thermal environments have a lower capacity for thermal acclimation than species from more variable environments (Prosser 1986). For example, many amphibian species from the temperate zone are capable of substantial acclimation, but salamanders (*Bolitoglossa*) that inhabit constantly cold mountaintops in Central America have little such ability (Feder 1978). Probably the narrowest breadth of thermal tolerance known is found among the notothenioid fishes, which are mostly restricted to waters near Antarctica. Some of them occupy waters that vary only between −1.40° and −2.15°C throughout the year, and tolerate temperatures no higher than +6° (Eastman 1993).

Raymond Huey and his colleagues tested the sprint speeds of 19 species of iguanid lizards by chasing them down a runway at each of several temperatures (Huey and Kingsolver 1993). For each species, they determined the optimal temperature (at which speed was maximal) and the critical thermal maximum and minimum temperatures (at which the lizards could not run). They also obtained data on the body temperature at which each species usually is active in the field. Their analysis (Figure 17.21) showed that the optimal temperature for sprinting is correlated with the field body temperature, implying adaptation to different thermal environments. Perhaps most interesting is that the critical thermal maxima and minima are not correlated, suggesting that there is not necessarily a trade-off. Nor are performances at high and low temperatures positively correlated, as might be expected if there existed a generalized adaptation to thermal stress.

If, as the lizard study suggests, there is not necessarily a trade-off between physiological performance at high and low temperatures, why are cold-adapted species intolerant of high temperatures, and why are tropical salamanders incapable of temperature acclimation? We are not sure.

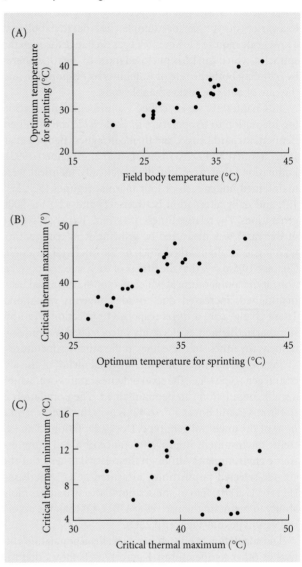

FIGURE 17.21 Correlations for 19 species of iguanid lizards between (A) optimum temperature for sprinting and field body temperature; (B) critical thermal maximum and optimum temperature for sprinting; and (C) critical thermal maximum and critical thermal minimum. Note that the last two variables are not correlated, suggesting that they can evolve independently. (After Huey and Kingsolver 1993.)

These species have no need for adaptations to temperatures they never experience. The loss of such adaptations during evolution, as in Antarctic fishes, might be due to unknown factors of natural selection, or simply to genetic drift.

Genetic studies Several studies have used artificial selection in order to determine if tolerances to different stresses are correlated, either positively or negatively (Hoffmann and Parsons 1991). For example, Ary Hoffmann and Peter Parsons (1989, 1991) selected three lines of *Drosophila melanogaster* by breeding from individuals that survived desiccation. After ten generations, these populations had evolved not only greater tolerance of desiccation, but also greater resistance to starvation, gamma radiation, ethanol vapor, and heat shock (although they did not become more resistant to ether, acetone, or cold). Hoffmann and Parsons interpreted their data as *evidence of a general increase in resistance to stress,* and they attributed it to a decrease in metabolic rate, which declined in the selected stocks. A lower metabolic rate can be expected to enhance resistance to a variety of stresses; for example, flies that metabolize at a lower rate require a lower rate of gas exchange through the spiracles, which can be kept closed more of the time, thereby reducing both the rate of water loss by evaporation and the rate of entry of toxic ethanol vapor.

In Chapter 10, we described experiments on selection for adaptation to temperature in *Escherichia coli* by Albert Bennett, Richard Lenski, and their colleagues (Bennett et al. 1992; Bennett and Lenski 1993; Lenski and Bennett 1993). Populations that were initially genetically identical were maintained under each of four thermal regimes (32°, 37°, 42°, and daily alternation between 32° and 42°) for 2000 generations. The relative fitness of a strain (its per capita rate of increase) was measured by growing it in competition with a genetically marked stock of the ancestral strain. In Chapter 10, we saw that the fitness of each selected population, at the temperature at which that population had been maintained, increased due to advantageous mutations. These changes are a *direct* response to selection. We now summarize Bennett et al.'s study of *correlated* responses— namely, the fitness of these lines at the other temperatures.

Figure 17.22A illustrates the "thermal niche" of the ancestral genotype—i.e., the span of temperatures over which a stable population can be maintained. The population is stable at 42°, but not at 43°, so 42° is near the *upper thermal limit* of the ancestral genotype. Does adaptation to an extreme environment improve tolerance of a still more extreme environment? Although the relative fitnesses of the six 42°-selected populations, measured at 42°, increased (Figure 17.22B), five of the six populations had no more ability to maintain themselves at 43° or 44° than the ancestral genotype did (Figure 17.22C).

Did adaptation to one temperature alter a population's fitness at other temperatures? Figure 17.22B shows that the 42°-selected stocks had greater fitness at and near 42° (the direct response to selection), but their fitness at lower temperatures did not differ from the ancestor's; there was no correlated response. The same was true of the 37°-selected populations, except that they suffered a reduction of fitness at 42°. This correlated response might suggest that there is a trade-off between adaptation to 37° and to 42°, but the population that became adapted to an even lower temperature, 32°, showed no such trade-off when tested at 42°. Finally, the population that became adapted to alternation between 32° and 42° had a higher relative fitness than the ancestor both at these temperatures and at 37°, suggesting that there was no "cost" of evolving broader temperature tolerance.

These experiments with *E. coli* provided no evidence for either a generalized adaptation to stress or a "cost" of adaptation to one temperature at the expense of fitness at other

FIGURE 17.22 Evolution of temperature tolerance in experimental populations of *Escherichia coli.* (A) The "thermal niche" of the ancestral population. Values of absolute fitness below zero indicate that the population could not persist in serial dilution cultures at that temperature. (B) Fitness at several temperatures of lines selected at 37°C (circles) and 42°C (triangles). A relative fitness of 1.00 equals that of the ancestral strain, with which these lines competed. The adaptation of the 37°- and 42°-selected populations to those temperatures was not accompanied by alteration of fitness at lower temperatures. (C) Absolute fitness of the ancestral strain (colored circles) and of lines selected at 42°C (triangles). The populations adapted to 42° could not persist at 44°, and only one of six could persist at 43°. (After Lenski and Bennett 1993.)

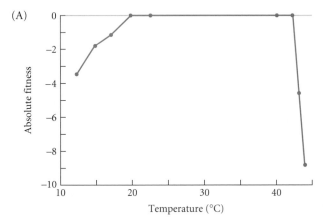

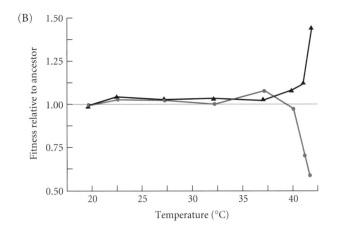

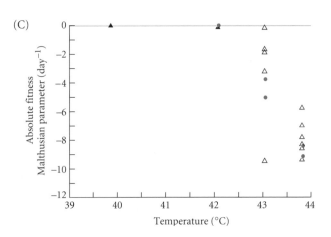

temperatures. Selection at a particular temperature resulted in adaptation that was specific to that temperature to an astonishing degree. In contrast, among laboratory populations of *Drosophila melanogaster* reared for 15 years at 18°, 25°, or 28°, the 28° populations best survived brief exposure to a 41° heat shock, and the 18° populations survived least well (Cavicchi et al. 1995). Thus a genetic correlation between fitness at different temperatures was found in *Drosophila*, but not in bacteria. It will be important to determine which is the most common pattern.

What Limits the Geographic Range of a Species?

From ecological studies, we know that some range limits are set by biotic factors, such as interspecific competition and predation, and some by abiotic factors, such as temperature and water availability (see Chapter 4). For instance, plants that typically are found in seemingly stressful environments such as salt marshes are limited on one side by intolerable stresses such as high salt levels; but their absence from less stressful habitats (e.g., those lacking salt) seems to be due to an inability to compete with other species, for they flourish if grown under these conditions without competition (Hoffmann and Parsons 1991).

As an ecological problem, the question of why a species has a restricted distribution, given its present physiological and other features, can be answered, even if with difficulty. But range limits pose an evolutionary problem that has not been solved. A species has adapted to the temperature, salt levels, or other conditions that prevail just short of the edge of its range. Why, then, can it not become adapted to the slightly more stressful conditions just beyond its present border, and extend its range slightly? And if it did so, why could it not then become adapted to still more demanding conditions, and so expand its geographic range (or its altitudinal or habitat distribution) indefinitely over the course of time? These questions pose starkly the problem of what limits the extent of adaptive evolution, and we do not know the answers. We will discuss several hypotheses, citing little evidence because little exists (Hoffmann and Blows 1994; Bradshaw 1991).

The simplest hypothesis is lack of genetic variation for tolerance of a physiological stress. In general, this may not be likely, because populations usually display genetic variation for tolerance of single stresses such as temperature (Hoffmann and Parsons 1991). Moreover, suitable mutations might be expected to arise eventually. In some cases, though, the physiological mechanism required for successful adaptation may be so particular and so complex that it is indeed unlikely to evolve. For instance, freezing tolerance in plants requires special biochemical mechanisms, and the upper latitudinal and altitudinal distributions of many plants are limited to frost-free regions. Some plant species lack the genetic variation necessary to survive on soils impregnated with toxic metals (Bradshaw 1991).

Successful colonization of sites beyond the current border may require numerous coincident adaptive changes, and hence an improbable concatenation of genetic variants for many characteristics. For instance, adaptation to the seasonal temperature regime beyond the species border might require the evolution not only of physiological tolerance of more extreme temperatures, but also of seasonal (phenological) timing of reproduction and growth, ability to feed on different prey species, and ability to compete with species better adapted to that climate.

A third hypothesis is that trade-offs exist between adaptation to conditions within and beyond the margin of the range. We have already reviewed some of the equivocal evidence for this familiar hypothesis. Trade-offs limit adaptation to a new environment only if, due to substantial gene flow between marginal and central populations, the genes continue to be subject to selection in the old environment as well.

A fourth hypothesis, advanced by Ernst Mayr (1963), is related to the previous one. Mayr proposed that gene flow from the main range of a species into marginal populations prevents them from further adaptation, by breaking down adaptive *combinations* of interacting genes. The demography of marginal populations may contribute to the importance of gene flow, for small marginal populations often become reduced in size or become extinct (Järvinen and Vaisenin 1984), and their ranges are recolonized from interior populations.

In a possible example of gene flow that prevents extension of the species' range, Stephen Stearns and Richard Sage (1980) found that mosquitofish (*Gambusia affinis*), both from a freshwater stream and from the brackish estuary into which the stream flows, had lower survival and growth in fresh than in brackish water. In contrast, *Gambusia* from a population that had been introduced into a freshwater lake in the Hawaiian Islands performed much better in fresh water. The authors suggested that gene flow from the estuarine population, a factor not affecting the Hawaiian population, may have hindered the adaptation of the stream population to fresh water.

If gene flow prevents frustrating adaptation to extralimital conditions, a marginal population might be better able to adapt and expand its range if it could not exchange genes with interior populations, due to physical isolation, parthenogenesis or self-fertilization, or the evolution of reproductive isolation, i.e., speciation. Perhaps, then, some species lineages have evolved broader ranges than we credit them with, simply because we call the adapted extralimital populations different species. Mayr (1963) drew attention to many instances in which a broadly distributed species has apparently given rise to localized satellite species that occupy different environments beyond its margins (Figure 17.23)—a pattern that conforms to, though it does not prove, the hypothesis. Parthenogenetic and polyploid populations often occupy more "stressful" habitats, including higher latitudes, than their sexual diploid progenitors, although this pattern has several possible explanations (Stebbins 1950; Glesener and Tilman 1978). But apomictic and polyploid taxa also have range limits, of course, so the effect of gene flow ultimately cannot be a sufficient explanation of geographic range limits.

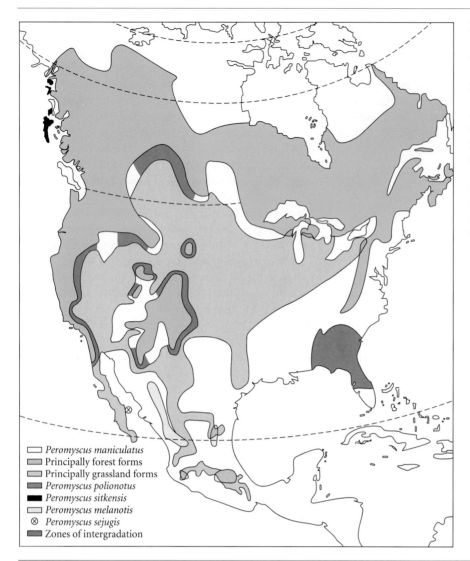

FIGURE 17.23 Phenotypically different forest and grassland forms of the deer mouse, *Peromyscus maniculatus*, are distributed across much of North America. *Peromyscus polionotus*, which has been derived from *P. maniculatus*, and three other species have smaller ranges to the south. Such patterns are consistent with, but do not prove, the hypothesis that interruption of gene flow by reproductive isolation enables populations to spread beyond the border of the species from which they arose. (After Blair 1950.)

Legend:
- ☐ *Peromyscus maniculatus*
- Principally forest forms
- Principally grassland forms
- *Peromyscus polionotus*
- *Peromyscus sitkensis*
- *Peromyscus melanotis*
- ⊗ *Peromyscus sejugis*
- Zones of intergradation

Summary

1. Morphological and physiological traits are often adaptive, and can sometimes be analyzed by assuming optimal design, within certain biophysical, biochemical, or other constraints.

2. Some physical principles, such as the relationship between surface area and volume (or body mass), have broad implications for much of an organism's morphology and physiology.

3. Body size (or mass) is an overridingly important feature with numerous implications for structure, function, and evolution. Differences in body mass are often associated with differences in the shape of certain structures (*allometry*). Body size is correlated with many other variables, including rate of heat exchange as well as generation time and therefore rate of population growth.

4. Many adaptations have costs; for example, they require energy or compromise other organismal functions. Constraints on the breadth of conditions to which a species' morphology can provide adaptation can be revealed by functional analyses based on biophysical principles.

5. The evolved responses of individual organisms to stressful changes in the environment include behavior (such as evasion of the stress), physiological acclimation, and developmental changes in morphology. Genetic changes may occur if these responses do not maintain high relative fitness.

6. Assuming tradeoffs, evolution in a fluctuating environment may result in a population that is monomorphic for either a generalized or a specialized genotype, or one that is polymorphic for specialized genotypes, depending on the ranges of tolerance of the genotypes and the frequency of the environmental change. There is also some evidence for "general stress tolerance," i.e., correlated resistance to a variety of stresses. The factors that limit a species' breadth of physiological tolerance are not fully understood.

7. We do not understand why species do not incrementally evolve tolerance of more and more extreme conditions, which would result in indefinite expansion of their geographic range. Insufficiency of appropriate genetic variation, inability to adapt to multiple environmental variables, and disruption of new adaptive gene complexes by gene flow are among the plausible reasons for the failure of species to become more broadly adapted and distributed.

Major References

Vogel, S. 1988. *Life's devices: The physical world of animals and plants.* Princeton University Press, Princeton, NJ. A well-written introduction to the biomechanical aspects of functional morphology.

Wainwright, P. C., and S. M. Reilly (editors). 1994. *Ecological morphology: Integrative organismal biology.* University of Chicago Press, Chicago. Contemporary approaches to morphological function and evolution.

Schmidt-Nielsen, K. 1990. *Animal physiology: Adaptations to environment.* Fourth edition. Cambridge University Press, Cambridge. A textbook by a leader of the field.

Fitter, A. H., and R. K. M. May. 1981. *Environmental physiology of plants.* Academic Press, London. A fairly technical treatment.

Hochachka, P. W., and G. N. Somero. 1984. *Biochemical adaptation.* Princeton University Press, Princeton, NJ. The standard work on this subject.

Sibly, R. M., and P. Calow. 1986. *Physiological ecology of animals: An evolutionary approach.* Blackwell Scientific, Oxford. A short introduction to evolutionary models as well as evidence on resource acquisition and allocation and their physiological consequences, emphasizing optimal models.

Garland, T., Jr., and P. A. Carter. 1994. Evolutionary physiology. *Annual Review of Physiology* 56: 579–621. A reference-rich review of contemporary approaches, especially those that employ genetic and phylogenetic methods.

Hoffmann, A. A., and P. A. Parsons. 1991. *Evolutionary genetics and environmental stress.* Oxford University Press, Oxford. A population biological approach to understanding adaptation to stressful environments.

Problems and Discussion Topics

1. Contrast the adaptations of birds and mammals to hot environments with those you would expect in species that inhabit cold environments. Referring to sources on environmental physiology (e.g., Schmidt-Nielsen 1990), do you find your expectations to be met?

2. What are the advantages and disadvantages of endothermy (metabolic maintenance of a constant body temperature) compared with ectothermy?

3. Why is there so much variation in shape among fish species? Among land plants?

4. If desert animals and plants could tolerate higher body temperatures, they would require less water for evaporative cooling. At a biochemical and cellular level, why is overheating deleterious or lethal? What characteristics would need to be modified for tolerance of higher temperatures to evolve? Have such modifications occurred in any organisms?

5. Leroi et al. (1994b) acclimated some *E. coli* bacteria to 32°C and others to 41.5°, and then placed them in competition with each other at each temperature. The competing groups were genetically identical except for a neutral marker allele. The 32°-acclimated bacteria proved to have higher fitness both at 32° and at 41.5°. Why is this a surprising result? What might account for it? What does this experiment say about the assumption that acclimation is beneficial?

6. The allometric equation ($y = bx^a$) may be used to describe the size of a feature (y) relative to body size (x) at different stages in the growth of an individual, or among different species at the same stage of development. If related species achieve different adult body sizes, can knowledge of the ontogenetic allometric equation be used to predict y, relative to x, among the adults of the different species? (See Gould 1966; Riska and Atchley 1985.)

7. Design a research program to determine why the altitudinal distribution of a plant species does not evolve to be greater.

8. Is there any evidence that genotypes with a greater than average ability to acclimate to high temperatures have a lower than average capacity to acclimate to low temperatures? What might you expect, and why?

CHAPTER 18

The Evolution of Interactions among Species

Many features of species are adaptations to interactions with other species. Adaptation to other species, moreover, has contributed to the diversification of some evolutionary lineages, and may well have had an important role in the increase of diversity throughout evolutionary time (see Chapter 25). The evolutionary responses of species to one another have many important ecological consequences, affecting, for example, the structure of communities.

Kinds of Interactions

From the point of view of individuals of any one species (a focal species), most of the other species with which they interact can be classified as

1. *Resources:* nutrition or habitat
2. *Competitors* for resources such as food, space, or habitat
3. *Enemies:* species for which the focal species is a consumable resource
4. *Commensals:* species that profit from but have no effect on the focal species

The effects of these four classes on the fitness of individuals of the focal species are, respectively, positive, negative, negative, and zero. Mutualistic symbionts provide resources to each other, and fall in the first category.

Although ecologists usually classify interactions by their effect on population growth in each of the interacting species, the evolutionary effect of an interaction depends on its effect on the fitness of individual organisms, not on populations, and these may be different. For example, the interaction between a predator population P and a prey population V may have a positive effect on V's population growth if the predator also feeds on a preferred (or more susceptible) species W that competes with V for food. In terms of population growth, P and V are beneficial to each other, but the interaction is certainly not beneficial to those members of population V that are eaten. Selection favors genotypes of population V that can avoid predation, even though the interaction between populations P and V might be classified as an indirect mutualism.

Two species may engage simultaneously in more than one kind of interaction. Moreover, the nature and strength of the interaction may vary among individuals and populations, depending on environmental conditions, genotype, age, and other factors (Thompson 1994). Seed-dispersing animals such as squirrels, for example, benefit some individual plants (seeds) by dispersing them, but harm others by eating them.

In the broad sense, predators (or enemies) include conventional predators, parasites, parasitoids (insects that kill their host only after they have completed their own development), and grazers (which consume only parts of their victims). MUTUALISMS are interactions in which the fitness of an *individual* of each species is enhanced, on average, by their interaction. Each species can be viewed as a resource for the other. Whether an interaction is mutualistic or antagonistic can be a delicate distinction, depending on how much resource one partner extracts from the other. Some mutualists are SYMBIOTIC, meaning that they are very intimately associated.

Coevolution

Ordinarily, the adaptation of a species to a physical feature such as temperature does not induce that physical feature to change. Likewise, a species may become adapted to another species without inducing any evolutionary change in the latter. For instance, some barnacle species live only on sea turtles and whales, but these vertebrates do not have any known adaptations to encourage or discourage barnacle growth. However, reciprocal evolution in the other species is a possibility, and it is this possibility that distinguishes selection in interspecific interactions from selection stemming from abiotic variables. *Reciprocal genetic change in*

539

FIGURE 18.1 Three kinds of coevolution. In each pair of graphs, the horizontal axis represents evolutionary time (t), and the vertical axis shows the state of a character (z) in one or more species of parasites (P, with character z_P, upper graphs) and in one or more species of hosts (H, with character z_H, lower graphs). These could also represent predator and prey species or mutualistically interacting species. The characters (z) mediate the effect of the interaction on individuals' fitness. (A) Specific coevolution, in which each of two species imposes selection on the other, as indicated by the arrows. (B) Guild coevolution, in which each of several P species interacts with each of several H species. It is assumed that a character in each species in a guild evolves similarly, although at different rates. (C) Escape-and-radiate coevolution. Several P species interact with several H species, one of which evolves a major new defense. P species previously associated with this host become extinct. The host lineage diversifies. Later, a P lineage, formerly associated with different hosts, adapts to the host clade and diversifies. The phylogenies of the P and H clades need not match (see Figure 18.2).

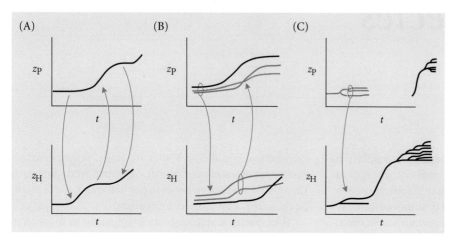

interacting species owing to natural selection imposed by each on the other is **coevolution** in the narrow sense.

Coevolution can take many forms, and includes several concepts (Figure 18.1; Table 18.1; see Futuyma and Slatkin 1983; Thompson 1982, 1989, 1994). In its simplest form, two species evolve in response to each other (SPECIFIC COEVOLUTION, Figure 18.1A). One scenario for specific coevolution is an "*arms race*" or "*escalation*" between enemies and their victims, whereby each change in one species is countered by change in the other. Darwin was the first to envision such directional coevolution when he envisioned predatory mammals such as wolves and ungulates such as deer evolving ever greater fleetness. GUILD COEVOLUTION, sometimes called *diffuse* or *multispecies coevolution* (Figure 18.1B), is conceptually similar, except that several or many species, such as a suite of parasites and a suite of host species, are involved. Some authors restrict "diffuse coevolu-

tion" to cases in which a species' response to selection by one interacting species is accompanied by genetically correlated changes in its interactions with other species (Hougen-Eitzman and Rausher 1994).

ESCAPE-AND-RADIATE COEVOLUTION (Figure 18.1C) is John Thompson's (1989) term for a scenario proposed by Paul Ehrlich and Peter Raven (1964) in a paper that stimulated much of the current research on coevolution of plants and herbivores. According to their hypothesis, a plant species, host to diverse herbivorous insects, evolves a new defense that rids the plant of most of its enemies. This evolutionary escape from most herbivores enables the plant species to diversify into numerous species that share the novel defense. At some later time, one or more species of herbivorous insects becomes adapted to one or more of these plants, and gives rise to clades of specialized species associated with various of the species in the diverse clade of

---

Table 18.1 Concepts of coevolution

A. Coevolution as a process of reciprocal adaptive response
   1. Specific coevolution: Coevolution of two (or a few) species
   2. Guild coevolution (diffuse, or multispecific, coevolution): Coevolution among sets of ecologically similar species
   3. Escape-and-radiate coevolution
   4. Cospeciation (induced by interaction)
B. Coevolution as a pattern, detected by phylogenetic analysis
   1. Cospeciation (coincident speciation)
   2. Parallel cladogenesis

plants. A similar hypothesis could apply to other groups of parasites (or mutualists) and their hosts.

COSPECIATION (Figure 18.2A) is the correlated speciation of two associated lineages. In some cases, which may be regarded as true coevolution, speciation in one lineage(e.g., a parasite) induces speciation in the other (e.g., a host). An example was described, in Chapter15, of speciation in a hymenopteran wasp that was apparently induced by the divergence of the symbiotic bacterium *Wohlbachia*. What we might term CONCORDANT SPECIATION may occur, however, due simply to the contemporaneous isolation and genetic divergence of each of two interacting species, without their interaction's playing any causal role.

The term *coevolution* has sometimes been used broadly to include *patterns of correlated evolution* among lineages, such as *parallel cladogenesis*, a pattern of matched (concordant) phylogenies of two associated lineages, such as a group of hosts and their parasites (Figure 18.2A). In such cases, parallel cladogenesis might or might not have been caused by true coevolution (reciprocal change) between the lineages.

Phylogenetic Perspectives on Species Associations

Although the fossil record seldom provides evidence about which species interacted in the past, phylogenetic studies of contemporary associations can cast light on how long interacting lineages have been associated, on whether they have speciated concordantly, on whether they have diversified in parallel or by "escape-and-radiate" coevolution, and in some cases, on whether they have undergone progressive adaptation to each other.

Contemporary associations, such as of parasites and hosts, may have arisen by one or both of two processes: *uninterrupted association with concordant divergence*, and *colonization*, or switching from one host to another (see Figure 18.2). The methods used for resolving these questions (e.g., Page 1990, 1993) are related to those used in historical biogeography to determine whether species on different land masses achieved their distribution by vicariance or colonization (see Chapter 8). We might expect that concordant divergence would be most common in groups of parasites or symbionts that are directly transferred from one individual host to another, perhaps by physical contact between parents and offspring, and that colonization, or host switching, would be more likely if parasites leave their hosts to search for new ones, or broadcast their offspring into the environment at large.

The rather few studies of the phylogenies of associated lineages have provided partial support for the expected effect of transmission mode on phylogenetic patterns. Nancy Moran and Paul Baumann (1994) contrasted two groups of bacterial associates of insects, both of which are vertically transmitted from mother to offspring through infected eggs or embryos. Bacteria of the genus *Buchnera* are ENDOSYMBIOTIC with (live within) aphids. The bacteria supply the essential amino acid tryptophan to their hosts, which suffer severely reduced growth and early death if they are treated with antibiotics. The phylogeny of these bacteria, inferred from DNA sequences, is completely concordant with that of their aphid hosts, inferred from morphological characters (Figure 18.3A). The simplest interpretation of this pattern is that the association between *Buchnera* and aphids dates from the origin of this insect family, that there has been little if any cross-infection between aphid lineages, and that the bacteria have diverged in concert with their hosts' divergence. Several aphid lineages have left fossils in 80-million-year-old amber, and the family is thought to have originated 200–250 million years ago. The high level of DNA sequence divergence among *Buchnera* lineages is consistent with such an ancient association. In fact, Moran and her associates used the fossil aphids to calibrate the rate of sequence divergence in

FIGURE 18.2 Congruent and incongruent phylogenies of hosts (black lines) and host-specific parasites or mutualists (colored lines). Each parasite lineage is specialized on the host with which it is closely associated in the diagrams. (A) Largely congruent phylogenies are due to several instances of concordant speciation, which may be due to the parasite-host interaction (cospeciation) or not. Host lineage B′ is free of parasites, perhaps due to the evolution of a new defense or to its invasion of a new geographic region, unaccompanied by its parasite. (B) Discordant phylogenies of interacting species. The parasite lineage, derived from one associated with an entirely different host, "colonized" the host clade and diversified after the host clade had diversified (see the diagram of "escape-and-radiate coevolution" in Figure 18.1). (After Mitter et al. 1991.)

(A)

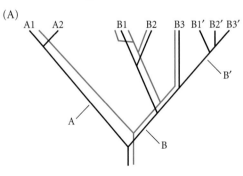

(B)

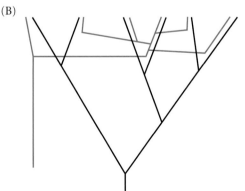

Buchnera, thus providing one of the few estimates of the rate of molecular evolution in bacteria (but one that was consistent with the few other estimates based on different methods).

Bacteria of the genus *Wolbachia* also are transmitted through the eggs of their hosts, but in contrast with *Buchnera,* they provide no benefit, and in fact often reduce host fitness (see Chapter 16). The phylogeny of *Wolbachia* strains shows little concordance with the phylogeny of their arthropod hosts (Figure 18.3B), implying that despite the usually vertical mode of transmission, cross-infection (host switching) among different lineages of insects and other arthropods has been rather frequent.

Coevolution of Enemies and Victims

We turn now to the *processes* of evolutionary change in interacting species, beginning with interactions between enemies and victims: predators and their prey, parasites and their hosts, herbivores and their host plants.

It is obvious that predators have adaptations for obtaining food, and that virtually all species have adaptations for lessening the likelihood that they will become food. However, *the coevolution of predator and prey could take several courses.* Considering just one species of predator and prey, coevolution might (1) continue indefinitely in an unending escalation or arms race (Dawkins and Krebs 1979); (2)

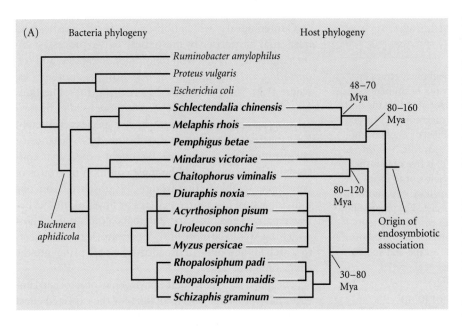

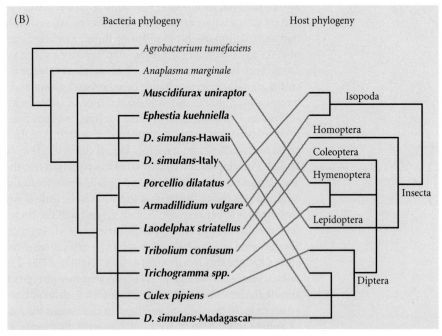

FIGURE 18.3 Phylogenies of some arthropods and their associated bacteria. In both diagrams, names of bacteria are in normal type and names of arthropod species are in boldfaced type. The bacterial phylogeny shows the relationship among the bacteria isolated from the species of arthropods indicated in boldface. (A) The phylogeny of bacteria included under the name *Buchnera aphidicola* is perfectly congruent with that of their aphid hosts. Several related bacteria were included as outgroups in this analysis. The estimated ages of the aphid lineages are based on fossils and/or biogeography. (B) The phylogeny of *Wolbachia* bacteria from various arthropods is not highly congruent with the phylogeny of their hosts, implying that the bacteria have been transferred among arthropod lineages. (After Moran and Baumann 1994.)

result in a stable genetic equilibrium, either monomorphic or polymorphic; (3) cause continual cycles (or irregular fluctuations) in the genetic composition of both species; or (4) lead to the extinction of one or both species.

All the models we consider assume evolution by individual, not group, selection. Hence enemies do not profit from sparing their victims' lives, except in the case of certain parasites that are transmitted only from one living host to another; such a parasite, if it is so virulent as to kill its host rapidly, may have low fitness because it cannot be transmitted. No such advantage of restraint applies to conventional predators or to many kinds of parasites. That a population of highly efficient predators might extinguish its prey and then be at risk of extinction does not mitigate the force of *individual selection* for greater efficiency in capturing prey.

Most models treat a single character in each species, both of which can be measured on the same scale, such as speed of both prey and predator, or shell hardness in a mollusc and crushing power in a predatory fish. Some authors have employed polygenic models in which new variation arises by mutation. Other have developed allele frequency models, such as "gene-for-gene" models based on the genetics of virulence and resistance that have been described for several pathogenic fungi and their plant hosts (Burdon 1987; Frank 1992).

Theory of Enemy–Victim Coevolution

Both allele-frequency and quantitative genetic models of enemy–victim coevolution have been developed. Both a character of the victim (prey) that provides resistance or defense and a character of the enemy (predator) that enables it to overcome the victim are inherited either as specified alleles at one or more loci or as quantitative, polygenic traits. In such models, the outcome of coevolution depends on whether or not these traits have costs. If they do not, the prey species evolves maximal defense (or resistance), and the predator species evolves maximal predatory proficiency. In principle, coevolution may then take the form of an "arms race," in which both characters evolve indefinitely, at rates that depend on the rate of origin of favorable mutations. It may be assumed, however, that greater resistance and greater predatory proficiency have costs, i.e., that they diminish fitness in ways other than by conferring resistance or proficiency only. (For example, they may be energetically costly.) In this case, both the mean resistance of the prey and the mean proficiency of the predator may fluctuate over time, and may or may not stabilize at lower than the maximal possible values. Two models illustrate these points.

Gene-for-gene models GENE-FOR-GENE INTERACTIONS were first described in cultivated flax (*Linum usitatissimum*) and flax rust (*Melampsora lini*), a basidiomycete fungus (Flor 1956). Similar systems have been described or inferred in several dozen other pairs of plants and fungi, as well as in cultivated wheat (*Triticum*) and one of its major pests, the Hessian fly (*Mayetiola destructor*). In each such system, the host has several loci at which a dominant allele (*R*) confers resistance to the parasite. At each of several corresponding loci in the parasite, a recessive allele (*v*) confers virulence—the ability to attack and damage a host with a particular *R* allele (Table 18.2). At the molecular level, it appears that plants with a particular *R* allele have a receptor that reacts to a product of the corresponding gene (with a *V* allele) in the parasite. Parasites homozygous for a recessive allele (*v*)

Table 18.2 Gene-for-gene interactions between a parasite and its host

ONE-LOCUS INTERACTION

PATHOGEN GENOTYPE	HOST GENOTYPE		
	RR	*Rr*	*rr*
VV	–[a]	–	+[b]
Vv	–	–	+
vv	+	+	+

TWO-LOCUS INTERACTION

PATHOGEN GENOTYPE	HOST GENOTYPE			
	R₁-R₂-	*R₁-r₂r₂*	*r₁r₁R₂-*	*r₁r₁r₂r₂*
$V_1\text{-}V_2\text{-}$	–	–	–	+
$V_1\text{-}v_2v_2$	–	–	+	+
$v_1v_1V_2\text{-}$	–	+	–	+
$v_1v_1v_2v_2$	+	+	+	+

Source: Frank (1992).

[a]– indicates that the host genotype is resistant to the pathogen genotype.

[b]+ indicates that the pathogen genotype can grow on a host of a given genotype (i.e., the pathogen is virulent and the host is susceptible).

do not produce this product, and so do not elicit a defensive reaction in the host, which they are therefore able to attack. Hosts homozygous for a recessive allele (r) are susceptible to both V- and vv parasites because they lack the receptor and thus lack a defensive response to the parasite.

If resistance (R) alleles are assumed to have costs, then any particular resistance allele (R_i) of the host will decline in frequency when the corresponding virulence allele (v_i) of the parasite has high frequency, because then R_i seldom confers resistance, and the energetic (or other) cost of resistance diminishes the genotype's fitness. Hence a different R allele, R_j, conferring resistance to the prevalent parasite genotype, increases in frequency. Selection then favors a corresponding allele, v_j, which enables the parasite to attack R_j-bearing hosts. Selection is thus frequency-dependent, with the fitness of each genotype in one species depending on the allele frequencies in the other (Burdon 1987, Frank 1992, Seger 1992). Computer simulations show that genotype frequencies can cycle or fluctuate irregularly (Figure 18.4). In accord with this model, studies have revealed substantial differences among populations of a wild Australian flax and its associated rust in the frequencies of several resistance and virulence genotypes (Jarosz and Burdon 1991). It is likely that the rust extinguishes local populations of flax, and that sites are recolonized by random samples of genotypes not adapted to the prevalent genotypes of the other species—thus complicating the pattern further.

FIGURE 18.4 A computer simulation of genetic changes at a resistance locus in a host (top) and a virulence locus in a parasite (bottom). The host is diploid and has three alleles; the parasite is haploid and has six alleles. Each parasite genotype can overcome the defenses of one of the six host genotypes (e.g., parasite P_1 can attack host H_1H_1). Both populations remain polymorphic, and fluctuate irregularly in genetic composition. (After Seger 1992.)

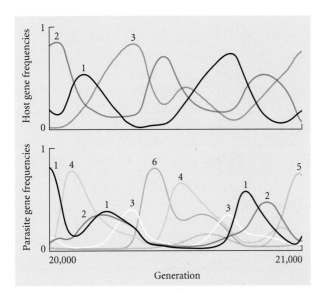

Predator-prey models with quantitative traits   In a quantitative genetic model (see Chapter 14) of predator–prey coevolution, Irma Saloniemi (1993) found that both the population densities and the mean values of the traits of a prey and its predator can cycle (see Figure 18.5). In some cases, the system is unstable: the evolving predator extinguishes its prey, and then itself becomes extinct. Increasing the levels of genetic variation in the prey's defensive feature and the predator's ability to attack tends to stabilize the interaction, yielding long-term coexistence and very slow evolution toward an equilibrium state for each character. Saloniemi assumed that the species' adaptations to each other have costs. Such costs, coupled with the effects of evolution on population densities, can cause evolution in counterintuitive directions; for example, *a prey species might evolve a lower level of defense* if its predator becomes so rare that the cost of defense becomes a more important selective factor than predation.

Costs of Adaptation

The costs of adaptations play an important role in the theory of coevolution. Such costs might take several forms. As is the case for any character, the necessary investment of energy or materials may be costly ("economic costs"), or the character, if sufficiently elaborated, may interfere with other functions for mechanical or biochemical reasons. Another possible cost is that adaptations to one predator (or prey) may be incompatible with adaptations to others (i.e., they are negatively genetically correlated). Several authors (Slobodkin 1974; Stenseth and Maynard Smith 1984; Rosenzweig et al. 1987) have suggested that because of such conflicts among adaptations, coevolution among multiple species should arrive at an equilibrium, or stasis. Thus the more predators a prey species must withstand, and the more prey species a predator relies on, the more stabilizing selection acts. Once an equilibrium is reached, directional selection, resulting in coevolutionary change, should occur only when the cast of interacting species changes (perhaps because of extinction) or when rare mutations occur that "break through" the constraints. This theory might account for some of the character stasis that is so evident in the fossil record (Chapter 6).

Empirical studies have provided evidence of costs in some cases, but not in others. For instance, plant defenses against herbivores include numerous chemical compounds (Rosenthal and Berenbaum 1992) that can account for up to 10 percent or more of a plant's energy budget. Such high levels of chemical defense are especially typical of slowly growing plant species, suggesting that they impose economic costs (Coley et al. 1985).

May Berenbaum and her colleagues grew each of numerous lines of wild parsnip (*Pastinaca sativa*), differing genetically in their level of toxic furanocoumarins, in both a greenhouse and a field (Berenbaum and Zangerl 1988). Outdoors, the more heavily defended lines suffered less attack from webworms, and matured more seeds. In the greenhouse, however, free from insect attack, the lines with

FIGURE 18.5 Computer simulation of coevolution between predator and prey due to a quantitative character (e.g., running speed) with low genetic variance in each species. (A) Changes in population density. Each point on the spiral plots the density of the predator against that of the prey at one point in time. The populations undergo decreasing oscillations and approach an equilibrium near the lower left. (B) Changes in the characters of the two species. Each point on the oscillating line plots the mean character of the predator against that of the prey at one point in time. The characters begin at the upper left, with a low value for the prey character (e.g., speed) and a high value for the predator character. The end of the simulated evolution is at the tip of the "tail" at the lower right. The prey character changes in a series of damped oscillations. The predator character steadily decreases, illustrating the counterintuitive conclusion that predators need not evolve steadily greater proficiency if, as assumed in the model, proficiency has a cost. (After Saloniemi 1993.)

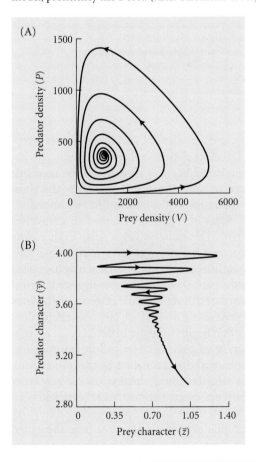

higher levels of furanocoumarins had lower seed production. Costs of this kind may explain why plants are not even more strongly defended than they are, and thus why they are still subject to insect attack. However, not all such studies of plants have yielded evidence of costs (Simms and Rausher 1989; Fritz and Simms 1992).

Costs due to negative genetic correlations in susceptibility to different predators are illustrated by the terpenoid compounds (cucurbitacins) of cucumber plants (*Cucumis sativus*): they enhance resistance to spider mites, but attract certain cucumber-feeding leaf beetles (Dacosta and Jones

1971). However, when David Maddox and Richard Root (1990) planted field plots with 18 genotypes of goldenrod (*Solidago altissima*) and repeatedly measured the density of 17 goldenrod-feeding insect species, they found that some species of insects tended to attack the same genotypes; that is, the genetic correlation of the plants' susceptibility to these insects was *positive*, implying that defensive characteristics are effective against a suite of enemies. Only a few of the 136 possible correlations in susceptibility to different insect species were negative.

On the other side of the coin, does adaptation of a predator to one prey species diminish its level of adaptation to other kinds of prey? If so, this not only could limit the rate of coevolution, but also could explain why animal species are specialized in diet, often to an extreme degree (Futuyma and Moreno 1988). Moreover, the theory that competition between species becomes reduced by the evolution of differences in resource use is based on the assumption of tradeoffs in proficiency (see below). Herbivorous insects, which frequently feed on only one or a few plant species, have been extensively studied in this respect (Jaenike 1990; Futuyma and Keese 1992). Heidi Appel and Michael Martin (1992), for example, measured the oxygen consumption of tobacco hornworms (*Manduca sexta*) fed diets that either contained or lacked nicotine, the toxic alkaloid in the insects' natural host. They found no evidence of metabolic (economic) cost. Several other investigators have measured genetic correlations in the growth rate (or other components of fitness) of an insect species on each of several plant species; in most cases, they have not found the negative correlations that would provide evidence of costs. The lesson of this research may be that economic *costs cannot be assumed a priori, but must be demonstrated.* On the other hand, several studies of vertebrate predators have found demonstrated trade-offs in handling different types of prey. For example, sticklebacks of the *Gasterosteus aculeatus* complex include benthic (bottom-dwelling) forms, with deep bodies and few gill rakers, and limnetic (open-water) forms, with slender bodies and long gill rakers (which are used to filter small prey items from the water). The former feed on benthic invertebrates and the latter on open-water plankton, and each feeds more efficiently on its typical food than on the other type's food (Figure 18.6; Schluter 1993).

Empirical Studies of Coevolution between Enemies and Victims

Brood parasitism Brood-parasitic birds and their hosts provide some of the most fully documented evidence of coevolution between "predators" and "prey" (Rothstein 1990). Certain species of cuckoos and a few other birds lay eggs only in the nests of certain other bird species. Depending on the species, the parasite nestling ejects the host's eggs or nestlings from the nest, stabs and kills them, or outcompetes them for food; usually, the host ends up rearing only the parasite. Adults of host species do not treat parasite nestlings any differently from their own young, but some host species do rec-

FIGURE 18.6   (A) The limnetic (L) and benthic (B) morphs of three-spined sticklebacks, which differ in body form, mouth size, and the number of gill rakers, the short projections on the inner edge of the gill arches shown at the left. (B) Success of prey capture by each morph in both benthic and limnetic habi-

tats. Each form was more successful, in terms of the volume of prey captured per strike, in its own habitat. Hybrids (H) had lower success than expected if they had equaled the average of their parents (gray line). (After Schluter 1993.)

(A)

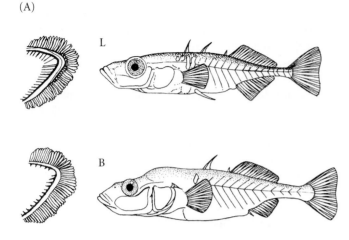

(B)

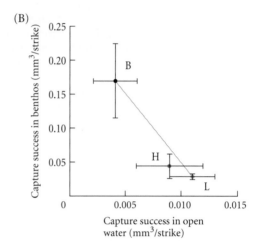

ognize parasite eggs, and either eject these or desert their nest and start a new nest and clutch.

The most striking counteradaptation among brood parasites is egg mimicry, which has been extensively studied by Nicholas Davies and colleagues (1989) in the European cuckoo (*Cuculus canorus*), the bird whose song is mimicked by cuckoo clocks. In any one region, the cuckoo population contains several apparently genetically different types that have different preferred hosts, and lay eggs closely resembling those of their preferred hosts (Color Plate 10). Some individuals lay nonmimetic eggs. Some host species accept cuckoo eggs, some frequently eject them, and others desert parasitized nests. By tracing the fate of artificial cuckoo eggs placed in the nests of various bird species, Davies and coworkers found that species that would be suitable cuckoo hosts discriminate against eggs unlike their own to a greater extent than do species that, due to nest site or feeding habits, would be unsuitable hosts. This finding suggests adaptation to parasitism. Surprisingly, among suitable host species, those that are rarely parasitized by cuckoos did not differ in discriminatory behavior from those commonly preferred as hosts. However, among the preferred hosts, those species whose eggs are mimicked by cuckoos rejected artificial eggs more often than those not mimicked. Moreover, populations of two species in Iceland, where cuckoos are absent, accepted artificial cuckoo eggs, whereas in Britain, where those species are favored hosts, they rejected such eggs. These studies, then, provide evidence of adaptation on the part of both the parasite (egg mimicry) and its hosts (rejection behavior).

Why is rejection of cuckoo eggs as likely among nonparasitized (but potentially suitable) species as among par-

asitized species, and why do some parasitized species accept cuckoo eggs? Davies et al. believe that there are LAGS in coevolution. Among the hosts of brood-parasitic cowbirds (*Molothrus ater*) in North America, some fail to discriminate because the range of the cowbird has expanded recently, so the hosts have not yet adapted to parasitism (Rothstein 1990). The same principle may explain why some species accept cuckoo eggs. Conversely, some suitable but nonparasitized hosts of the cuckoo may have been parasitized in the past. Davies et al. have suggested the following sequence of coevolutionary stages: (1) Cuckoos begin to parasitize a new host, which lacks discrimination. (2) Discrimination begins to evolve in the host. (3) An egg-mimetic genotype of cuckoo arises and increases in frequency. (4) Enhanced discrimination by the host evolves. (5) Due to this enhanced discrimination, cuckoo genotypes that use this host are selected against, and the cuckoo population switches to other hosts. The abandoned host retains its discriminatory ability for some time thereafter. Thus, cuckoos illustrate that the interaction of a species with several other species may change over time.

Fossil evidence   Although changes in the characters of individual lineages of prey and predators are difficult to trace in the fossil record, the evolution of defensive and "offensive" features in interacting clades often suggests that diffuse coevolution has occurred. For example, during the Mesozoic, molluscs became subject to new, highly effective predators, such as shell-crushing fishes and crustaceans that could either crush or rip shells. The diversity of shell form in bivalves and gastropods then increased, as various lineages evolved thicker shells, thicker margins of the shell aperture, or spines and other excrescences that could foil at least some predators (Figure 18.7).

Plants and herbivores

Plants are protected against herbivores by features such as spines, hairs, and especially defensive chemicals that are toxic or deter herbivores from feeding (Strong et al. 1984a; Rosenthal and Berenbaum 1992; Bernays and Chapman 1994; Thompson 1994). For example, most crucifers—mustards and their relatives (Brassicaceae)—bear glucosinolates, milkweeds (Asclepiadaceae) have cardenolides, and many willows (Salicaceae) bear compounds related to salicylic acid, the effective ingredient in aspirin. Some adapted insects, however, are not only tolerant of their host plants' compounds, but use them to recognize the host. For example, some crucifer-feeding insects, such as larvae of the cabbage white butterfly (*Pieris rapae*), are actually stimulated to feed by glucosinolates; in fact, some taste cells on the mouthparts of the larvae of *P. rapae* are specific glucosinolate receptors.

Although some authors have proposed that the herbivore-deterring features of plants evolved to serve other functions, some evidence implies that their main function is defense (Futuyma and Keese 1992). Some chemical defenses have been lost in plant lineages that have evolved a superior replacement, suggesting that related lineages retain the features because of their defensive function, not out of physiological necessity. The best example is found among Neotropical trees of the genus *Acacia*, which contain cyanide-releasing glycosides, except in certain species that support colonies of aggressive stinging ants (Rehr et al. 1973). Also, many compounds are deployed in highest concentration in the most "valuable" parts of plants, such as young leaves and seeds, the destruction of which would most diminish plant fitness (McKey 1979). In the same vein, plants that typically grow in environments low in light or nutrients, and which can therefore replace consumed tissues only slowly, generally invest a higher proportion of their energy in defensive compounds than species that grow in richer environments (Coley et al. 1985, Feeny 1976).

The adaptations of herbivores that have breached plants' chemical defenses include detoxifying enzymes. For example, many species in the carrot family (Apiaceae) contain toxic furanocoumarins such as xanthotoxin, and are fed on mostly by specialized insects such as larvae of the black swallowtail butterfly (*Papilio polyxenes*), which has higher activity of a detoxifying enzyme in the midgut, and can degrade xanthotoxin much more rapidly than the southern armyworm (*Spodoptera frugiperda*), a generalized feeder (Berenbaum 1983, Ivie et al. 1983).

Among host-specialized insects, related species often feed on related plants. For example, all members of the butterfly tribe Heliconiini (Color Plate 7) feed as larvae on passionvines (Passifloraceae), and most of the several thousand species of true fruit flies in the subfamily Tephritinae feed on the sunflower family (Asteraceae). Many such groups originated in the early Tertiary, and have evidently remained associated with the same groups of plants for more than 40 million years (Ehrlich and Raven 1964; Mitter and Farrell 1991). The phylogeny of the insects generally does not closely correspond with that of their host plants, which suggests that as new species evolve, they may switch to new host plants, but only if these are in the same family as their ancestors' hosts.

Based on an analysis of the host plant relationships of butterflies and the chemical features of their host plants, Ehrlich and Raven (1964), as we noted earlier, postulated a history of coevolution in which plants and herbivorous insects have enhanced each other's diversity (see Figure 18.1C). As this hypothesis would predict, the number of species of insects, within each of several families, that feed on a family of plants in Europe is correlated with the number of species in that plant family (Ward and Spalding 1993; Figure 18.8).

There is also some evidence that the evolution of novel defenses may have promoted plant diversification, as Ehrlich and Raven postulated. The hypothesis that a feature has contributed to diversification can be tested by comparing the diversity of sister taxa that differ in this respect. Because sister taxa are equally old, a difference in their diver-

FIGURE 18.7   Some features of molluscs that provide protection against predatory crustaceans and fishes. Spines in the bivalve *Hysteroconcha lupanaria* (A) and the gastropod *Lambis scorpio* (B) may provide protection against being swallowed by fishes, and may reduce the effectiveness of crushing predators. The very smooth shell of *Cypraea mauritiana* (C) deprives predators of purchase, and its narrow, toothed aperture (D) prevents predators from reaching the gastropod's body. (From Vermeij 1993; photographs by M. Graziose; courtesy of G. Vermeij.)

(A)

(B)   (C)   (D)

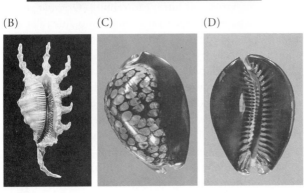

sity must be due to different *rates* of diversification. Comparing several pairs of sister taxa enables one to determine if a feature is consistently associated with a greater rate of diversification. Brian Farrell and colleagues (1991) applied this method to 16 plant taxa in which herbivore-deterring latex (as in milkweeds) or resin (as in pines) is delivered through special canals to sites of injury. In 14 of the 16 cases, the lineage possessing this feature had more species than its sister group (Figure 18.9). This is what we would expect if alleviation of herbivory were to enable lineages to diversify more rapidly.

Although many specialized herbivores display specific adaptations to certain groups of plants, many plant defenses appear to provide protection against a broad array of herbivores. Perhaps coevolution, on the part of most plants, has been more diffuse than specific. One of the few fairly clear examples of specific coevolution between plants and herbivores is afforded by *Heliconius* butterflies and their larval hosts, the Passifloraceae (Williams and Gilbert 1981, Gilbert 1983). The complex cyanogenic glycosides of these plants are thought to be a barrier to almost all herbivorous insects except *Heliconius*. Females lay eggs on tender young foliage, a sparse food for which larvae compete. Apparently to deter competition, the eggs of *Heliconius*, unlike those of related butterflies, are bright yellow. Females tend to avoid laying on foliage that already bears eggs of this color. Several species of *Passiflora* have independently evolved an adaptation to deter herbivory by *Heliconius*: the buds, stipules, or foliar nectar glands (depending on the species) are modified in form and color to resemble *Heliconius* eggs (Figure 18.10). Thus, this insect-plant interaction has affected the evolution of both the butterflies, which have evolved resistance to the plants' defensive compounds, and the plants, which have evolved ways of deterring egg laying by the butterflies.

FIGURE 18.8 The relationship between the number of species per plant family in Britain and the number of species of aphids that use plants in that family as hosts. The colored circles represent families that include trees. (After Ward and Spalding 1993.)

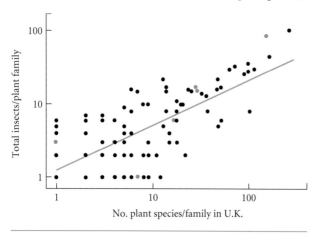

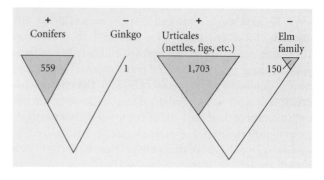

Infectious Disease and the Evolution of Parasite Virulence

The two greatest challenges that a parasite faces are moving itself or its progeny from one host to another (transmission) and overcoming the host's defenses. Some parasites are transmitted *vertically*, from a host parent to her offspring, as in the case of *Wohlbachia* bacteria, which are transmitted in insects' eggs. Other parasites are transmitted *horizontally* among hosts in a population, via the external environment (e.g., human rhinoviruses, cause of the common cold, are discharged by sneezing), via contact between hosts (e.g., venereal diseases such as the gonorrhea bacterium), or via VECTORS (e.g., the malaria-causing protozoan and the yellow fever virus, both transmitted by mosquitoes).

Some parasites are relatively *avirulent*, causing little reduction in the host's fitness. *Virulent* parasites reduce the host's survival or reproduction (e.g., some fungal parasites of plants and a variety of botflies and helminths that attack animals effectively castrate the host).

The level of virulence depends on the evolution of both host and parasite. Undoubtedly the best-documented example of coevolutionary changes in virulence is presented by myxoma virus and European rabbits (*Oryctolagus cuniculus*), which had been introduced into Australia and became rangeland pests (Fenner and Ratcliffe 1965; May and Anderson 1983b). The virus, from a South American rabbit, was introduced in 1950 to control the European rabbit. Following the introduction, wild rabbits were tested for resistance to a standard strain of virus, and virus samples from wild rabbits were tested for virulence in a standard laboratory strain of rabbits. Not surprisingly, *the rabbits evolved greater resistance* to the virus (Figure 18.11A). More interestingly, the virus evolved from an initial state of high virulence to a lower, intermediate virulence level; but although some almost avirulent strains were detected,

FIGURE 18.10 Mimicry of butterfly eggs by passionvines. (A) Two eggs of a *Heliconius* butterfly, placed on a tendril of a passionvine (*Passiflora*), the sole larval food. (B) Modified nectary glands at the base of the leaf petiole of *Passiflora auriculata* resemble *Heliconius* eggs in shape, size, and color. (C) Another of the many egg-mimicking structures in passionvines: modified tips of the stipules of *Passiflora cyanea*. (Photographs courtesy of L. E. Gilbert.)

(A)

(B)

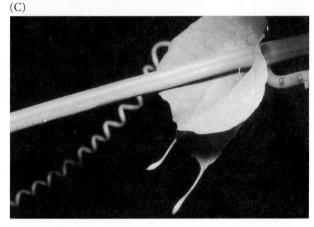

(C)

FIGURE 18.11 Coevolution in rabbits and myxoma virus after the virus was introduced into the rabbit population in Australia. (A) Mortality in field-collected rabbits challenged with a standard virus strain declined as the wild population experienced more epizootics (epidemics during which the virus killed many rabbits). That is, average resistance increased after successive episodes of strong natural selection. (B) Virus samples from the wild, tested on a standard rabbit stock, were graded from high (1) to low (5) virulence. Average virulence decreased over time, but stabilized at an intermediate level. (A after Fenner and Ratcliffe 1965; B after May and Anderson 1983a.)

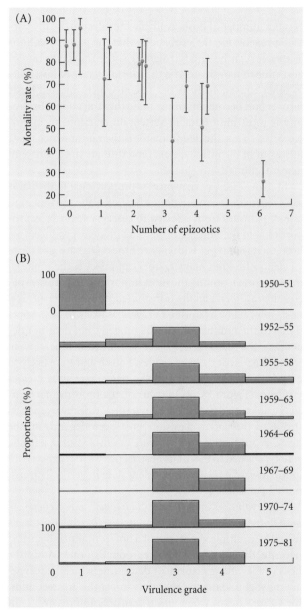

the virus population as a whole did not become avirulent (Figure 18.11B).

Theory of the Evolution of Virulence

Many biologists, including medical biologists, hold the naive belief that parasites generally evolve to be benign, because the survival of the species depends on that of the host population. However, *a parasite may evolve to become benign (or even beneficial), moderately virulent, or highly virulent,* depending on its biological features and ecological circumstances (May and Anderson 1983b, Anderson and May 1982, 1991, Bull 1994, Ewald 1994).

Consider the fitness of genotypes of a parasite such as the myxoma virus, which reproduces within an individual rabbit. Assume, at first, that a host is infected by only a single, asexually reproducing parasite, and so carries only one parasite genotype. Robert May and Roy Anderson (1983a) measure the fitness of a parasite genotype as its net reproductive rate, R_0, the number of new infections produced by an infected host:

$$R_0 = bN/(v + d + r)$$

N is the number of hosts available for infection by the parasite progeny, b is the probability that the progeny will infect each such host, v (virulence) is the mortality rate of hosts due to parasitism, d is the mortality rate of hosts due to other causes, and r is the rate at which infected hosts recover and become immune to further infection. (Immunity is most important in vertebrate reactions to microparasites.) Thus the denominator is the rate at which hosts move out of the infected class (and thus are not a source of new infections).

In many cases, v (virulence) is proportional to the parasite's reproductive rate *within* a host. If b, the transmission rate of parasites from infected to susceptible hosts, is proportional to the number of parasite progeny produced per host, then genotypes with higher fitness (R_0) are also more virulent. Although many factors can affect b, the mode of transmission is among the most important. In parasite species that are *horizontally* transmitted, the number of new hosts infected by a parasite's progeny is likely to be proportional to the number of parasite progeny, but does not depend on the reproductive success of the individual host in which the parent parasite resides. Thus b and v are likely to be positively correlated, and the parasite evolves greater virulence (Figure 18.12) In contrast, the progeny of a *vertically* transmitted parasite are "inherited" directly by the offspring of the parent parasite's host, so b depends on the hosts' reproductive success, and the fitness of a parasite genotype is tied to that of its host. Hence we may expect evolution toward a relatively avirulent state in vertically transmitted parasites.

The evolution of virulence, moreover, may depend on both individual selection and group selection. If a host is infected by multiple genotypes of a horizontally transmitted parasite, differences in their reproductive rate within the host (hence, their virulence) determine which genotypes are most abundant and therefore infect the greatest number of new hosts, when and if transmission occurs. Thus individual selection may favor higher virulence. However, the parasites within a single host may form a TRAIT GROUP, a term David Sloan Wilson (1980, 1983) has applied to temporary "populations" that differ in the mean of some trait. Because high average virulence of such a group may cause the death of the host—and of the parasite population—before much transmission can occur, *group selec-*

FIGURE 18.12 The theoretically expected relationship between the fitness of a parasite genotype and its virulence for (A) vertically transmitted and (B) horizontally transmitted parasite species. Virulence here means the probability of inducing the host's death within a given time, and fitness is measured by the number of new hosts infected per original host.

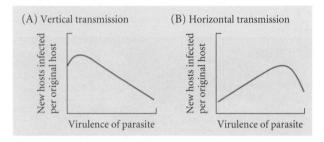

tion may tend to favor low virulence (Bremermann and Pickering 1983; Nowak and May 1994). The interplay between individual and group selection can have complicated consequences for the evolution of virulence (Nowak and May 1994).

Virulence in natural populations of parasites

Some parasites are more virulent in new hosts than in their normal hosts, but others greatly reduce survival or reproduction in their normal hosts (Holmes 1983, Dobson 1988). A study by Dieter Ebert (1994) supports the hypothesis that adaptation in parasites can lead to higher virulence. Spores of a microsporidian protozoan (*Pleistophora intestinalis*) are ingested by *Daphnia magna*, a filter-feeding planktonic crustacean; the parasite reproduces in the gut epithelium, and daughter spores are released with the host's feces. In experimental pairs of infected and uninfected *Daphnia*, the greater the number of parasites in the infected individual, the more likely was the infection of the other. Ebert experimentally infected *Daphnia* from widely separated geographic populations with parasites from several populations found near Oxford, England. The parasites produced more spores in *Daphnia* from their own or nearby populations than in those from distant populations, indicating higher adaptation to their local hosts (Figure 18.13A). Moreover, they failed to infect a high proportion of the distant *Daphnia* strains. The parasites caused greater mortality, and greater reduction of growth and reproduction, in their own or nearby host populations than in hosts from distant populations (Figure 18.13B), and mortality was correlated with the hosts' load of spores. Thus the transmission rate of this parasite is correlated with its reproductive rate within the host, which in turn is correlated with virulence. Its more virulent effect on sympatric than on allopatric host populations contradicts the naive hypothesis that parasites will evolve toward lower virulence.

Allen Herre (1993) found support for the hypothesis that vertically transmitted parasites should generally be less

FIGURE 18.13 The fitness of three strains of a microsporidian parasite and its effects on various populations of its host species, the water flea *Daphnia magna*. Each of three strains, indicated by different symbols, was tested in hosts from the strain's locality of origin (solid circles) and in hosts from localities at various distances away (open circles; the distance scale is logarithmic). (A) The number of parasite spores (log scale) produced per host is greatest when the parasite infects individuals from its source location, showing that parasites are best adapted to local host populations. (B) and (C) show that the mortality of the host and reduction in its reproductive success are greatest for sympatric or nearby populations, showing that the parasite is most virulent in the host population with which it has coevolved. (After Ebert 1994.)

(A)

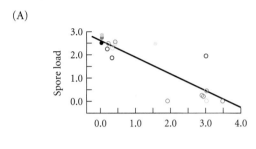

(B)

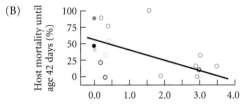

(C)

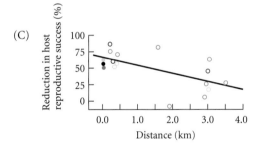

by single females (whether of species that typically have single or multiple foundresses). From the presence or absence of nematodes, he determined whether the single female had been infected, and counted the number of young wasps as a measure of her reproductive success. From theory, Herre expected nematodes associated with single-foundress wasp species to be relatively avirulent, since they are vertically transmitted from an adult wasp to her offspring. Parasites of multiple-foundress species, however, are often horizontally transmitted, so their reproductive success is decoupled from that of the wasp, and Herre expected them to be more virulent (i.e., to cause greater reduction of the host's reproductive success). As Figure 18.14 shows, this is exactly what Herre found.

Mutualisms

In mutualistic interactions among species, *individuals* of each species obtain some benefit, on average, that is thought to enhance their fitness relative to what it would be, absent the interaction. In SYMBIOTIC mutualisms, individuals usually are intimately associated for much or all of their lifetimes. Other mutualisms are very temporary associations between hosts and "visitors" (Thompson 1994). Many symbiotic mutualisms, especially between endosymbiotic prokaryotes and eukaryotic hosts, provide biochemical benefits to both species (Douglas 1994). Some mutualisms have promoted the evolution of extreme, highly specialized, adaptations. For example, flowers that are pollinated by long-tongued moths such as hawkmoths usually have a long, tubular corolla, are often white, and are fragrant at dusk or at night. Darwin predicted that in Madagascar, there must exist a hawkmoth with a 7-cm proboscis capable of pollinating the long-spurred white orchid *Angraecum arachnites*. More than a century

FIGURE 18.14 The average reproductive success of nematode-infected fig wasps, relative to that of uninfected wasps (set at 1.0). The reduction in reproductive success is greater in species that usually have multiple-foundress broods, and therefore provide greater opportunity for horizontal transmission of nematodes between unrelated individual hosts. Thus the virulence of the nematodes is greater in species with a greater opportunity for horizontal transmission. (After Herre 1993.)

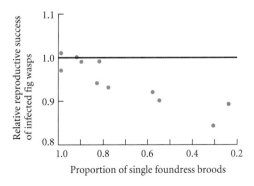

virulent than horizontally transmitted parasites in a study of the parasitic nematodes (*Parasitodiplogaster*) of fig wasps (Agaonidae). Female wasps enter the hollow inflorescence of a fig tree, pollinate the flowers, lay eggs, and die in situ. Their offspring develop and mate, and the females depart to repeat the cycle. Depending on the species, either a single female or (usually) several females (foundresses) enter the fig inflorescence. Each wasp species has its own species of parasitic nematode. Young nematodes, hatching from eggs laid in the fig inflorescence, crawl onto newly emerging female wasps, are carried to new inflorescences, and at some point enter the wasp, complete their development, and then emerge and lay eggs.

Herre determined, from the number of bodies of adult female wasps in fig inflorescences, which had been entered

FIGURE 18.15 The Madagascan orchid *Angraecum arachnites* bears nectar in an exceedingly long spur (left), and is pollinated by the long-tongued hawkmoth *Panogena lingens* (right). (Courtesy of L. A. Nilsson.)

FIGURE 18.16 Yucca moths and their relatives (Prodoxidae). (A) Yucca moths of the genus *Tegeticula* not only lay eggs in yucca flowers, but use specialized mouthparts to actively pollinate the flowers in which they oviposit. (B) Certain species of *Greya* oviposit in flowers, but pollinate only incidentally. (C) A phylogeny of major lineages of Prodoxidae, based on DNA sequence data. Lineages that oviposit (O) in flowers are shown by double lines, those that only pollinate (P) by colored branches, and those that do both by solid black branches. Pollination has evolved three times in this family. (Photograph courtesy of O. Pellmyr; B from Pellmyr and Thompson 1992; C after Pellmyr et al. 1996b.)

(A)

(B)

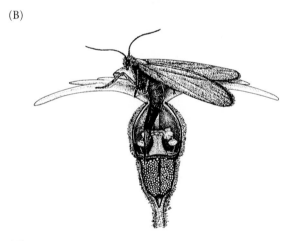

(C)

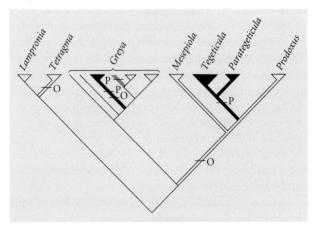

later, such hawkmoths were shown to be, indeed, the pollinators of this plant (Figure 18.15; Nilsson et al. 1985).

Origins of Mutualisms

In *The Origin of Species*, Darwin challenged his readers to find an instance of a species' having been modified solely for the benefit of another species, "for such could not have been produced through natural selection." No one has met Darwin's challenge. In mutualisms, each species provides a benefit to the other not because of any kind of altruism, but because one species exploits an intrinsic property of the other (e.g., aphids necessarily excrete sugar-rich honeydew, which is exploited by ants), or because individuals of one or both species profit from fostering the association. For example, some species of *Pseudomyrmex* ants inhabit certain species of acacia trees, which have specially modified thorns in which the ants reside and special protein-rich structures on which the ants feed. The ants vigorously sting herbivores and kill other plants that encroach on their tree, for it is in their interest to defend their home and food; likewise, the fitness of the acacia plant depends on the vigor of its ant colony, so it is advantageous to provide the ants with food and housing. The benefit each party provides to the other, at some cost in energy or reproduction, is the cost it pays for the benefit it receives.

Some mutualisms arise from parasitic relationships. For example, the nitrogen-fixing *Rhizobium* bacteria in the root nodules of legumes are related to the plant-pathogenic *Agrobacterium*, and may have evolved from a parasitic state (Young and Johnston 1989). Parasitism is also the origin of several remarkable cases in which insects both consume the seeds and are the sole pollinators of the flowers in which their larvae develop. These insects include the wasps that are obligate pollinators of figs and the yucca moths (*Tegeticula* and *Parategeticula*), which have a rather similar relationship to yuccas (*Yucca*, Agavaceae) (Figure 18.16A). The female moth inserts eggs into the flowers that she pollinates, and the larvae consume some of the many seeds that develop. The closest relatives of yucca moths feed on developing seeds, but do not generally pollinate (Figure 18.16C). However, one related moth genus (*Greya*) incidentally pollinates flowers as it lays eggs in the ovaries (Figure 18.16B), illustrating a transition from "parasitism" to mutualism (Pellmyr and Thompson 1992, Brown et al. 1994).

Conflict and Stability

Because individual mutualists profit from each other, we might expect one or both species to evolve greater exploitation. Yucca moths, for example, could lay so many eggs that all of the yucca's seeds would be consumed. Evolution of overexploitation could thus destabilize a mutualism and lead to the extinction of one or both species.

Two related questions thus arise. At a proximate level, what mechanisms prevent overexploitation, and do they consist of suppression of the symbiont by the host, or restraint on the part of the symbiont? Does coevolution involve increased exploitation by one species and countervailing measures by the other (an "arms race"), or do symbionts evolve restraint as an "evolutionarily stable strategy" (see Chapter 20)?

Selection will always favor protective mechanisms in one or both species to prevent overexploitation (Bull and Rice 1991). *Whether or not selection also favors restraint depends on the degree to which individuals' genetic self-interest depends on the fitness of their individual associates.* The principles that govern the evolution of virulence in parasites apply similarly to mutualisms. Thus, we should expect restrained exploitation by a mutualist, and even evolution of enhanced benefits to its associate, if the mutualist is vertically transmitted from host to host, or if its association with the host is lifelong (or at least extends through much of its lifetime). In contrast, mutualists that only temporarily visit individual hosts, and move horizontally from one to another, would not be expected to evolve restraint, but instead might evolve so as to extract maximal benefit from the hosts at minimal cost to themselves. Indeed, we may expect CHEATING—exploitation without reciprocation—to evolve frequently in such cases.

Most vertically transmitted endosymbionts do not proliferate excessively within their hosts. The *Buchnera* bacteria in aphids, for example, are restricted to special cells (mycetocytes) and divide at a decreasing rate as the host ages (Douglas 1994). This observation suggests restraint, but it is not known whether the rate of endosymbiont replication is controlled by the endosymbiont or by the host.

An example of how evolutionary stability can be achieved is provided by the interaction between yucca species and the moths that are their sole pollinators (Pellmyr and Huth 1994). Typically, a yucca moth lays only a few eggs in each flower, so that only a few of the many developing seeds are consumed by the larvae. The moth could lay more eggs per flower—indeed, she distributes eggs among many flowers—so why does she lay so few in each? The answer lies, in part, in the fact that the plant does not have

enough resources to mature all of its many (often 500–1500) flowers into fruits. Pellmyr and Huth hand-pollinated all the flowers on some plants, and found that only about 15 percent yielded mature seed-bearing fruits—the rest were aborted, and dropped from the plant. When the researchers counted the number of moth eggs in maturing and aborted fruits, they found that the fruits with the most eggs were most likely to be aborted (Figure 18.17). Thus the plant does not invest substantial resources in fruits that are likely to have few surviving seeds. Fruit abortion imposes strong selection on moths not to lay too many eggs in a flower, because the larvae in an aborted flower or fruit perish. Thus the moth has evolved restraint by individual selection and self-interest.

However, it is by no means certain that mutualisms generally are stable over evolutionary time. An interaction may be invaded by a third species, which may replace one of the original mutualists by competition. For example, endosymbiotic yeasts appear to have replaced *Buchnera* bacteria in an aphid tribe that, because it is phylogenetically nested among *Buchnera*-carrying aphid lineages, presumably evolved from a *Buchnera*-carrying ancestor (Moran and Baumann 1994). Or, a mutualism may cease to be a mutualism if one of the species evolves and stops providing benefits to the other—that is, if it becomes a "cheater" (Bronstein 1992). The evolution of cheating or of overexploitation might well drive both species to extinction, though this would be difficult to prove.

Cheating is common among pollinating bees, which frequently chew a hole through the base of a corolla near the nectaries, and thereby obtain nectar without entering the flower and contacting the sexual organs. Plants, too, cheat by not providing fitness-enhancing rewards to pollinators. Many orchids secrete no nectar; the extreme in deception is practiced by orchid species that release a mimic of a female insect's sex pheromone, and are visited by males that accomplish pollination while "copulating" with the flower (Color Plate 6).

FIGURE 18.17 The proportion of pollinated yucca flowers that were retained, rather than aborted, in relation to the number of moth eggs deposited in them. Flowers with fewer eggs were more likely to be retained until they developed into fruits. (After Pellmyr and Huth 1994.)

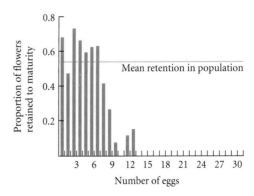

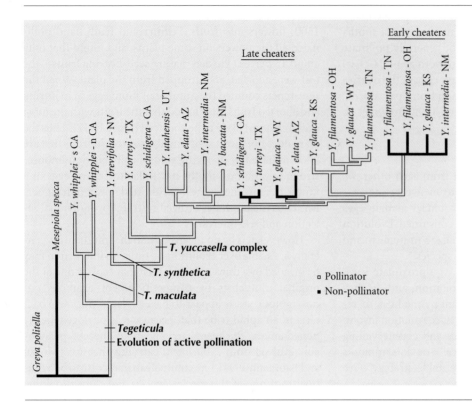

FIGURE 18.18 Phylogeny of the yucca moth genus *Tegeticula*, and two closely related genera. The *Tegeticula* populations in this phylogenetic analysis are labelled by the name of the host (*Yucca* species) from which they were collected. Three *Tegeticula* species are named; species limits in the *T. yuccasella* complex are uncertain. Most *Tegeticula* taxa actively pollinate (open branches in the phylogeny). Three lineages (solid branches), denoted "late cheaters" and "early cheaters," lay eggs in *Yucca* flowers, but do not pollinate. The phylogeny indicates that they have evolved from pollinating ancestors. (After Pellmyr et al. 1996a.)

A phylogenetic analysis of the yucca moths (*Tegeticula*) has shown that cheating evolved from mutualism at least once, and perhaps thrice (Figure 18.18; Pellmyr et al. 1996a). Each of several species of *Tegeticula* lays eggs in the ovaries of yucca flowers that have already been pollinated by a mutualistic species. The cheater species do not pollinate, and they lay so many eggs that the larvae consume most or all of the seeds. They circumvent the plant's abortion response to high numbers of eggs by laying their eggs after the critical period in which fruit abortion occurs.

Evolution of Competitive Interactions

Competition is frequent among species of plants, seed and nectar feeders, carnivores, and decomposers, although it is apparently less frequent among terrestrial herbivores (Schoener 1983; Connell 1983; Hairston 1989). It has long been postulated that competing species evolve to reduce competition. Such evolution could result in a greater difference in resource use, and in features associated with resource use, between sympatric than between allopatric populations of a pair of species, a pattern that Brown and Wilson (1956) termed **character displacement**.* We might also expect that if a species were isolated from competing species, it might expand its diet (or

use of other resources), and might display greater individual variation in trophic structures. This pattern has been termed **ecological release**. The highly influential ecologist G. Evelyn Hutchinson (1957) suggested that if three or more species evolve to minimize competition, they may become equally spaced apart in their use of resources, and that a character correlated with resource (such as bill size in birds adapted to eating different seeds) may then display a CONSTANT RATIO (or at least some minimal ratio) between adjacent species. From the few examples he was able to draw on, Hutchinson suggested that each such species might be, on average, about 1.3 times as large (in body length, beak length, etc.) as its next smaller competitor (Figure 18.19).

A Model of the Evolution of Competing Species

Of the several kinds of models of the coevolution of competing species that have been developed, possibly the most realistic are quantitative genetic models developed by Montgomery Slatkin (1980) and by Mark Taper and Ted Case (1985, 1992a,b). These models assume that individuals may compete both with conspecific individuals and with other species for a variable resource (such as seeds of different sizes; Figure 18.20A). A variable phenotypic character *z* (such as bill size of a seed-eating bird) determines the efficacy with which that phenotype consumes resources. Each phenotype has a "utilization curve" that describes its "optimal" resource and its "niche width," the variety of resources it can consume. The fitness of a phe-

---

*Brown and Wilson used this term for the *pattern*, whatever its cause might be. Another possible cause of greater difference between sympatric than allopatric populations is divergence to reduce the frequency of mating or hybridization between species (see Chapter 16).

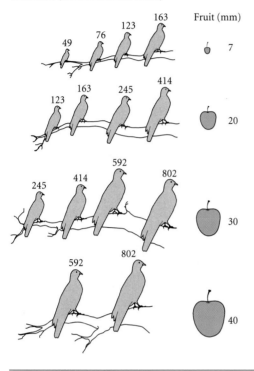

FIGURE 18.19 A schematic illustration of niche partitioning among eight species of fruit pigeons (*Ptilinopus* and *Ducula*) from lowland rain forests in New Guinea. The species differ in weight (shown in grams). Tree species with increasingly large fruits attract increasingly larger species of pigeons. Among the pigeons feeding on any one tree species, each pigeon species weighs about 1.5 times more than the next smaller species. The smaller species of pigeons feed from smaller, more peripheral branches. (After Diamond 1973.)

notype is proportional to an individual's average resource income. The phenotype (z) has additive genetic variance V_A, which might differ between the species.

The per capita resource income of a phenotype, and therefore its fitness, depends on (1) the efficacy with which it can utilize each kind of resource, (2) the abundance of each such resource, and (3) the extent to which it competes with other phenotypes (of both its own and other species), which diminish the resource supply. The distribution of abundance of various resources (the "K-curve" in Figure 20) is assumed to be bell-shaped; small and large seeds, for example, are assumed to be less abundant than medium-sized seeds. The resource distribution therefore selects against extreme values of z (and therefore on resource utilization); i.e., it imposes *stabilizing selection*. The intensity of competition between phenotypes is proportional to the overlap of their resource utilization curves; it therefore increases as a function of their breadth and decreases as a function of the distance between them (Figure 18.20B). All else being equal, a phenotype's fitness decreases in proportion to the intensity of competition it experiences, so those phenotypes have highest fitness that are spaced apart

most widely. Therefore, competition imposes *diversifying selection* on z and on resource utilization.

The results of this model, which is analyzed by computer simulation, include both intuitively obvious and surprising outcomes. When evolution in a *single* species, free of competitors, is simulated, the character mean $\bar{z}$ (e.g., mean beak size) evolves to an equilibrium that is positioned at the peak of the K-curve (Figure 18.20C). That is, the species evolves to utilize the most abundant resources, due to stabilizing selection imposed by the K-curve. However, the diversifying selection imposed by competition among phenotypes results in persistence of multiple phenotypes. Genetic variance in z, and therefore between-phenotype variation in resource use, will be greater, the greater the variety of resources. The broader the resource use of each phenotype, the less genetic variance there will be.

When two (or more) species compete in this model, the outcome may be *either divergence or convergence* of their character means ($\bar{z}$) and resource utilization. If each phenotype has broad resource use, selection for divergence between species is weak, because increasing the difference between phenotypes lowers competition only slightly. Then if the resource spectrum (the K-curve) is narrow, the stabilizing selection against small or large phenotypes causes the two species to *converge* toward the same mean ($\bar{z}$), and within-species genetic variation in resource use is also reduced or lost (Figure 18.20D). Consequently, competition between the species increases, and one of the species may eventually become extinct.

If, conversely, the K-curve is broad relative to each phenotype's niche width, selection due to scarcity of extreme (small or large) resources is weaker than selection due to competition, and initially similar species diverge until these opposing selection pressures are equal (Figure 18.20E). A pattern of character displacement may result from this divergence.

Because recombination among loci restricts the variance in the character that determines resource use (Chapter 14), two species may not fully utilize a broad spectrum of resources. In that case, one or more additional species, differing from the first two, may be able to invade the community. Both the invaders and the previous residents may then evolve further shifts that minimize competition. When divergence occurs, the species become spaced apart along the resource spectrum, but at narrower intervals than when only two species compete.

Evidence for Coevolution of Competing Species

Geographic variation Character displacement—greater difference between sympatric than allopatric populations—has provided some evidence for the coevolution of competing species. Such evidence must be carefully interpreted, to be sure that other factors are not responsible for the pattern (Grant 1972, 1975; Taper and Case 1992a).

In Chapter 9, we described an example of character divergence in the Galápagos ground finches *Geospiza fortis* and

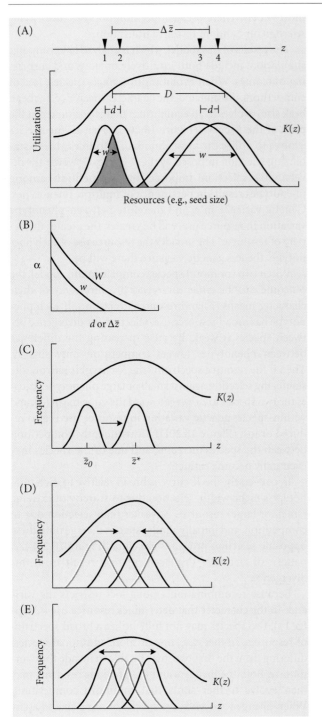

FIGURE 18.20 Formulation and results of a model of the evolution of competitors. (A) The model. Each phenotype z (e.g., beak size) has a bell-shaped curve portraying its utilization of resources (e.g., seeds of different sizes). Here, phenotypes 1 and 2 belong to one species, and 3 and 4 to another. The difference in mean phenotype between the species is $\Delta \bar{z}$, and the difference between the average resources used by the two species is D. The breadth of each phenotype's resource utilization curve (its within-phenotype niche width) is w, which is greater for phenotypes 3 and 4 than for 1 and 2. The degree of competition between two phenotypes i and j (α_{ij}) is proportional to the overlap of their utilization curves, shown as darkened regions for phenotypes 1 and 2 and for 2 and 3. The area of overlap is directly proportional to w and inversely proportional to d, the difference in the phenotypes' mean utilization curves. The abundance of different resources is shown by the curve $K(z)$, which specifies the equilibrium density that a population composed solely of each possible phenotype (z) would attain. (B) The competition coefficient between phenotypes, α, is a function of the difference between phenotypes within a species (d) or between species ($\Delta \bar{z}$). (C) Evolution of a solitary species, assuming heritability of the variation among phenotypes. (Here the utilization curve of the species as a whole is shown, not the variation among conspecific phenotypes.) The mean phenotype evolves to an equilibrium ($\bar{z}^*$) that matches the most abundant resources, thus maximizing population density. (D) Evolution of two species from initial positions (gray curves) to equilibrium positions (black curves) when resources have a narrow frequency distribution and the species are initially very different. Character convergence occurs. (E) The same, when resources have a broad frequency distribution and the species are initially similar. Character displacement occurs.

G. fuliginosa, which differ more in bill size where they coexist than they do where they occur singly. These species are often food-limited, differences in their bill size are correlated with the efficiency with which the birds process seeds that differ in size and hardness, and bill size is highly heritable (Grant 1986). Another well-documented example is provided by sticklebacks of the *Gasterosteus aculeatus* complex in northwestern North America (see Figure 18.6), where numerous lakes each have a benthic and an open-water form,

which differ in body shape, mouth morphology, and the number and length of the gill rakers. Other lakes have only a single form of stickleback, with intermediate morphology (Schluter and McPhail 1992). Several other groups of lake-dwelling fishes show similar patterns (Robinson and Wilson 1994).

Populations of birds and other species on islands, where species diversity is low, often occupy a greater variety of habitats and use more kinds of food than do the same or

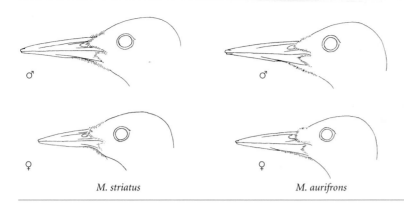

FIGURE 18.21 Character release. The difference in beak size between the sexes is greater in *Melanerpes striatus*, the only species of woodpecker on the island of Hispaniola, than in continental species such as *M. aurifrons*, which is sympatric with other species of woodpeckers. Beak size is correlated with differences in feeding behavior. (After Selander 1966.)

M. striatus   M. aurifrons

related species where competing species occur. For example, the sole finch species on Cocos Island in the Pacific Ocean has a much broader diet, and forages in more different ways, than do any of its relatives in the Galápagos Islands, where there are many more species (Chapter 20). Such cases illustrate ecological release. Morphological variation is usually not much greater in populations that are free from interspecific competition, which suggests that the broader niche of the population stems largely from expansion of the niche of individual phenotypes (Case 1981). However, sexual dimorphism in morphology and resource use is often more pronounced in bird populations on islands than on continents. For example, Robert Selander (1966) found that in the only species of woodpecker on the West Indian island of Hispaniola, sexual dimorphism in the length of the beak and tongue is about twice as great as in continental species that coexist with other woodpecker species (Figure 18.21), and the sexes differ in where and how they obtain food.

Community patterns Nonrandom patterns of differences among sympatric species in resource use, or in characters such as body size or trophic structures thought to reflect resource use, can provide evidence for the effects of competition (Simberloff and Boecklen 1981; Colwell and Winkler 1984; Dayan et al. 1989). For example, the three species of forest-dwelling bird-eating hawks (*Accipiter*) in the United States differ in body size and differ correspondingly in the size of the prey species they usually take. Similar differences in size characterize sympatric species of *Accipiter* throughout the world, and are greater than those expected by chance alone (Figure 18, Chapter 4).

Have such patterns come about through a history of evolutionary adjustment of species to one another, or through ecological exclusion or extinction of species that are too similar in resource use? A phylogenetic perspective can sometimes help to answer this question, as illustrated by studies of *Anolis* in the West Indies. These insectivorous lizards present striking, different patterns in the Lesser and Greater Antilles. In the Lesser Antilles, each island has either one or two species of anoles, which differ from island to island. Almost without exception, a "two-species island"

has a large species and a small species, whereas the species on "one-species islands" are of intermediate size. Jonathan Losos (1990c, 1992) found that all the small species form a single clade, as do the large species (with one exception) (Figure 18.22A). If small and large size have evolved from medium size only once (the most parsimonious interpretation), then evolutionary size adjustment has not occurred independently on each of the two-species islands. Rather, each such island has been colonized by a small species and by a large species.

The *Anolis* species of the Greater Antilles, such as Jamaica and Puerto Rico, have a very different history. Each of these large islands has anoles that largely correspond in habitat use and morphology (see Chapter 8 and Color Plate 9). Phylogenetic studies indicate that each island has a monophyletic group of species that have evolved independently of their ecological counterparts on other islands. Assuming that the common ancestor of two sister species belonging to different ecomorphs had an intermediate morphology and habitat use, Losos has drawn the remarkable inference that the sequence in which the several ecomorphs evolved was much the same on Jamaica and Puerto Rico (Figure 18.22B).

Evolution and Community Ecology

Many community ecologists hope to discover and explain regular patterns in the structure of communities. Examples such as the bird-eating hawks, Antillean anoles, and the correlation between the species diversity of plant families and of the insects that feed on them (see Figures 4.18, 18.22, 18.8) show that regularities in community structure exist. Ecological processes alone, such as competitive exclusion, may explain some patterns, as suggested by the sizes of anoles in the Lesser Antilles (see Figure 18.22A). However, many examples in this chapter show that evolutionary adjustments of species to one another, some fostering coexistence and others leading to extinction, have also played an important role. Moreover, the origin and extinction of species and higher taxa over the course of many millions of years have left their mark on the number, diversity, and interactions among species in contemporary communities, as we shall see in Chapter 25.

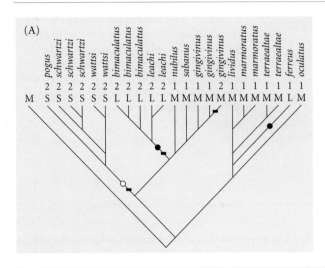

(A)

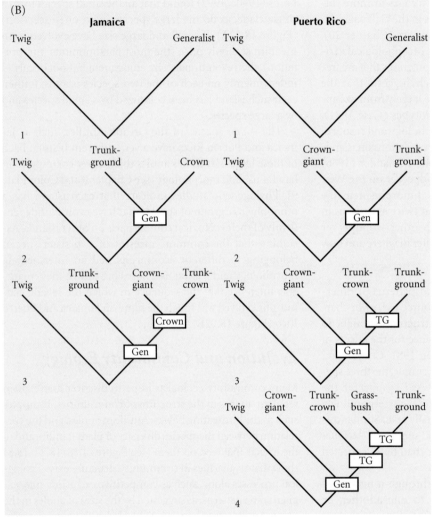

FIGURE 18.22 (A) Phylogeny of *Anolis* species and populations in the small islands of the Lesser Antilles. For each population, the number indicates the number of species on the island that it occupies, and the letter indicates body size (S = small, M = medium, L = large). The distribution of character states implies that the common ancestor was a medium-sized lizard on a one-species island. Body size has increased twice (solid circles) and decreased once (open circle). Two-species islands usually have a small and a large species, not due to independent character displacement on each island, but rather to colonization by members of small- and large-bodied clades. (B) The evolution of *Anolis* community structure in Jamaica and Puerto Rico. The species on each island form a monophyletic group, although similar ecomorphs, using corresponding habitats, occur on the two islands. The ecomorphs of the common ancestors have been reconstructed, using parsimony, from the phylogenies of the living species. From this reconstruction, the history of ecological divergence on each island can be inferred, and is shown in sequential stages from top to bottom. Symbols for the ecomorphs of common ancestors include Gen = generalist, TG = trunk-ground. A "crown-giant" is a large lizard that lives in the crowns of trees. (From Losos 1992.)

Summary

1. Interactions among species are a major source of natural selection on many characters, and have influenced the evolution of the numbers and phenotypic diversity of species. Often, adaptation of one species to another is not accompanied by reciprocal evolutionary change, but when it is, *coevolution* is said to occur.

2. Genetic, ecological, and phylogenetic studies all provide important insights into coevolution. Phylogenetic studies can provide information on the age of associations between species and on whether or not they have codiversified or acquired adaptations to each other. For example, the phylogenies of certain groups of host species and their associated symbionts or parasites are congru-

ent, implying a long history of association and cospeciation, whereas phylogenies are incongruent in other cases, implying more recent associations and/or shifts of symbionts between host lineages.

3. Coevolution need not lead to stable coexistence. Predator-prey and parasite-host systems can evolve to extinction, toward a stable genetic and ecological equilibrium, or toward indefinite fluctuations in genetic composition and population densities. Generally, prey species are expected to evolve improved defenses, and predators and parasites greater ability to capture and exploit prey; however, especially if these adaptations are costly, evolution may proceed in the opposite direction, at least for a while.

4. The evolution of virulence in parasites (including pathogenic microorganisms) depends on many factors. Often selection on parasites within hosts (individual selection) favors greater virulence, because as the parasite's reproductive rate within the host increases, the parasite extracts more resources from its host. However, early mortality of hosts that harbor excessively virulent groups of parasites may prevent their effective transmission; thus a kind of group selection can favor lower virulence. Evolution of greater virulence is expected if parasites are transmitted horizontally among hosts, rather than vertically from parent to offspring.

5. Mutualisms, in which individuals of each of two species obtain benefits from the other, are best viewed as reciprocal exploitation. Selection favors genotypes that provide benefits to the other species if this action yields benefits in return. Thus the conditions that favor avirulence in parasites, such as vertical transmission, can also favor the evolution of mutualism. Some mutualisms have evolved from parasitic relationships, and some have evolved into parasitic relationships when one species has evolved the ability to "cheat" by not reciprocating. The stability of mutualisms can be due to one species' preventing overexploitation by the other, or to the evolution of "restraint" when this is individually advantageous.

6. Evolutionary responses to competition among species may lead to divergence in resource use and sometimes in morphology (character displacement), or to convergence and competitive exclusion. Character displacement has been best documented empirically. If multiple species compete, character displacement can result in species' evolving evenly "spaced" differences in their use of resources. These differences may be reflected in morphological differences among the species. Much the same pattern, however, can arise by ecological processes of "species sorting"—i.e., extinction of excessively similar species. Phylogenetic analyses, coupled with ecological studies, suggest that both processes occur, and can give rise to repeatable patterns in community structure.

Major References

Futuyma, D. J., and M. Slatkin (editors.) 1983. *Coevolution.* Sinauer Associates, Sunderland, MA. Although somewhat out of date, the essays in this book, treating many aspects of coevolution, provide a foundation for reading the later literature. The same holds for the essays in M. H. Nitecki (ed.), *Coevolution* (University of Chicago Press, Chicago, 1983).

Thompson, J. N. 1994. *The coevolutionary process.* University of Chicago Press, Chicago. This and the same author's earlier book, *Interaction and coevolution* (Wiley, New York, 1982) provide extensive discussions of the evolution and ecology of many interactions, especially among plants and their herbivores and pollinators.

May, R. M., and R. M. Anderson. 1990. Parasite-host coevolution. *Parasitology* 100: s89–s101. This active area of research has not yet received a comprehensive book-length treatment. This article and several other references cited in this chapter (e.g., Ewald 1994; Anderson and May 1991; Bull 1994) provide an entry into the literature.

Taper, M. L., and T. J. Case. 1992. Coevolution among competitors. *Oxford Surveys in Evolutionary Biology* 8: 63–109. This review provides an introduction to the large literature on the subject.

Problems and Discussion Topics

1. The immune system of mammals confers protection against many microbial parasites. Is it an example of coevolution? Some parasitic protozoans and bacteria evade recognition by the immune system by rapidly altering their surface proteins. Does this exemplify coevolution? What criteria, in general, are required to demonstrate that features that mediate the interactions between species are consequences of coevolution?

2. Compare the likely course of coevolution between a specialized parasite and a host that either is or is not attacked by other species of parasites as well.

3. How might phylogenetic analyses of predators and prey, or of parasites and hosts, help to determine whether or not there has been a coevolutionary "arms race"?

4. The generation time of a tree species is likely to be 50 to 100 times longer than that of many species of herbivorous insects and parasitic fungi, so its potential rate of evolution should be slower. Explain why trees, or other organisms with long generation times, have not become extinct due to the potentially more rapid evolution of their natural enemies.

5. Design an experiment to determine whether greater virulence is more advantageous in a horizontally transmitted or a vertically transmitted parasite.

6. Some authors have suggested that selection by predators may have favored host specialization in herbivorous insects (e.g., Bernays and Graham 1988). How might this occur? Compare the pattern of niche differences among species that might diverge due to predation with the pattern that might evolve due to competition for resources.

7. Discuss the kinds of evidence that might best test Ehrlich and Raven's hypothesis of "escape-and-radiate" coevolution. (Note that their hypothesis includes several components, which may require different tests.)

8. How would the existence of genetic variation within and among populations of a parasite (for virulence) and a host (for resistance) affect the rate of coevolution of the species? Under what conditions would it reduce the rate of change?

9. Provide a hypothesis to account for the extremely long nectar spur of the orchid *Andraecum arachnites* (Figure 18.15) and the long proboscis of its pollinator. How would you test your hypothesis?

10. Which of the processes discussed in this chapter might tend to increase the number of species of organisms, and which might tend to decrease it?

CHAPTER 19

The Evolution of Life Histories

Natural selection consists of differences among genotypes in fitness, or reproductive success. The components of fitness, as described in Chapter 13, include survival, mating success, and number of offspring. We might, therefore, expect all organisms to have evolved long life spans and very high fecundity. However, these and other components of the life history vary enormously among species. Students of life history evolution seek to understand why there is such great diversity in traits that are so closely allied with fitness. A few illustrations of this diversity will show the scope of this field of study.

Life History Phenomena

Among the traits in which species differ are the number and size of their offspring, their age at first reproduction, and their life span.

Number and size of offspring Many bivalves, echinoderms, other marine invertebrates, and fishes such as sturgeon release thousands or millions of tiny eggs in each spawning. Elms and many other trees, similarly, may produce hundreds of thousands of small seeds per year, and orchids release up to 10^9 seeds that are hardly larger than the spores of fungi. In contrast, a coconut palm (*Cocos nucifera*) may produce a few dozen nuts per year, each weighing several kilograms. A blue whale (*Balaenoptera musculus*) gives birth to a single offspring that weighs as much as an adult elephant, and a kiwi (*Apteryx*) lays a single egg that weighs 25 percent as much as its mother (Figure 19.1).

Age distribution of reproduction A newly laid egg of *Drosophila melanogaster* may be a reproducing adult 10 days later, and a parthenogenetic aphid may carry an embryo even before she herself is born. In contrast, periodical cicadas (*Magicicada*) feed underground for 13 or 17 years before emerging to live as adults, which have a reproductive span of less than a month (Figure 19.2). Development to reproductive maturity in primates ranges from less than a year in some marmosets to 13 or more years in humans. Many species reproduce repeatedly (**iteroparity**), whereas others reproduce only once, and then die (**semelparity**).

Life span Sea anemones and corals are known to live for close to a century, and may have the physiological potential to live indefinitely. This may also be true of some plants that propagate vegetatively by rhizomes or other means. Great ages have been estimated for single clones of plants, such as huckleberry (*Gaylussacia brachycerium*), that may be more than 13,000 years old. However, most or all sexually reproducing organisms, if they escape predation or other extrinsic causes of death, undergo physiological changes (*senescence* or *aging*) that increase their likelihood of death as they grow older. Among animals, life spans range from about 10 days in some rotifers to more than 2 centuries in some molluscs, and individual trees may attain ages of 3200 (giant sequoia, *Sequoia gigantea*) or even 4600 years (bristlecone pine, *Pinus aristata*) (Nooden 1988).

Life history theory, a subject in evolutionary ecology, addresses the conditions that favor the evolution of variations in these DEMOGRAPHIC traits.

Major Life History Traits and Fitness

Life history traits affect the growth rates of populations. At first, we shall consider only the female's reproductive schedule and life span, deferring discussion of the male's. Three major books on the evolution of life histories (Stearns 1992; Roff 1992; Charlesworth 1994b) provide much of the basis for this chapter.

Differences among genotypes in life history traits (and in male reproductive success) determine differences in fitness, since fitness is measured by the per capita rate of increase of a genotype, r ($r = b - d$, where b and d are the per

FIGURE 19.1 An x-ray photograph of a kiwi (*Apteryx*), showing the relatively enormous egg, which weighs 25 percent of the female's body weight. (Photograph by permission of the Otorohanga Zoological Society.)

capita rates of birth and death). Under most circumstances, relative values of *r* predict the course of evolution by natural selection. Thus *evolutionary change in a demographic feature*, such as the reproductive rate at a certain age or the length of the reproductive lifetime, *constitutes change in a component of fitness*. Evolution of any other character, such as body size or the activity of an enzyme, occurs by natural selection only insofar as variation in that character affects one or more of the demographic traits we are considering here (or male reproductive success). The demographic traits are therefore "summary" features, under which are hidden virtually all of an organism's physiological and morphological characteristics.

Individual Selection and Group Selection

Why do codfish produce hundreds of thousands of eggs? Is it to compensate for their high mortality and thus to assure the survival of the species? Why do lemmings migrate and (according to unsubstantiated myth) drown themselves when they reach the sea? To avoid overpopulation, which could cause extinction of the species? Why do people senesce and die "of old age?" To make room for the vigorous new generation that will propagate the species?

If your answer was "yes" to any of these questions, which assume that the traits have evolved to benefit the species rather than individuals, either you have assumed that these characteristics did not evolve by Darwinian natural selection (selection among individuals), or you have not fully assimilated the meaning of natural selection.

Differences in fecundity and life span, because they are components of fitness, must have evolved at least partly by selection. In general, selection among populations, the only possible cause of evolution of a trait harmful to an individual but beneficial to the population or species, is a weaker force than selection among individuals (Chapter 12), and this must be especially true for life history traits, which manifestly affect the fitness of individuals.

In this light, the possibility of future extinction due to excessive population growth is irrelevant to, and cannot affect, the course of natural selection among individuals. A human genotype that could live and reproduce forever would have a higher *r* than one that dies at age 80 or so (all other things being equal), so we should expect evolution toward an indefinitely long life span, even if overpopulation should ensue. Likewise, we should expect fecundity to evolve to ever higher levels (all other things being equal). Instead of viewing species' fecundity as having evolved to balance mortality, *we should consider the level of mortality to be the ecological consequence of the level of fecundity*, since most populations are regulated by density-dependent factors such as limited food and other resources (Williams 1966).

The problem, therefore, is to understand what advantage low fecundity or short life span—or the genes that underlie them—might provide to individual organisms, rather than to whole populations or species.

FIGURE 19.2 Adult and nymph of a periodical cicada (*Magicicada septendecim*). (Courtesy of C. Simon.)

Life Tables and the Rate of Increase

We will distinguish life histories with *discrete, nonoverlapping generations* from those with *overlapping generations*. The first is exemplified by an annual insect in which all eggs hatch at about the same time in the spring, and all surviving adults reproduce and then die at about the same time in the fall. In contrast, a species with overlapping generations, such as humans, often includes several cohorts that were born at different times and reproduce repeatedly. In this case, the length of a generation (GENERATION TIME) is generally defined as the *average age of the mothers of newborn offspring*.

In what follows, *time*, in units such as days or years, is denoted as *t*. The *age* of an individual is *x*, using the same units as time. The *proportion of newly laid eggs or newborn offspring that survive* to age *x* is l_x. The average ("expected") *number of offspring* born to a female that is alive at age *x* is denoted m_x. The average *age at maturity*—i.e., at *first reproduction*—of a female is denoted *a*, and the average age at last reproduction as *z*. In many species, such as *Drosophila* and many mammals, m_x at first increases with age after reproduction commences and later declines. In contrast, m_x increases with age in many fishes, trees, and other organisms with INDETERMINATE GROWTH—i.e., growth in body size throughout life.

Values of l_x and m_x, for all ages *x*, can be measured for a population without reference to genetic variation, or a separate table could be prepared, in principle, for each genotype in a population. From these values, two measures of the per capita rate of increase can be calculated: R_0 and *r* (Table 19.1; also Box 19.A). R_0 is the average growth in population size per female per *generation*. The growth in population size per female per *unit of time* is *r*. Since we may be interested in explaining evolutionary changes in generation time, we will be more concerned with how alterations of life history components change *r* than with how they change R_0. Box 19.A describes how *r* can be calculated from a life table.

From the influence of l_x and m_x on *r*, several *important conclusions* can be derived.

1. All else being equal, increasing l_x—survival to any age *x*—up to and including the reproductive ages will increase *r*. If, as for human females, there is a postreproductive life span (when $m_x = 0$), changing the probability of survival to advanced postreproductive ages does not alter *r*. If, however, the reproductive period is extended into older ages, then increasing survival to these ages does increase *r*.

2. All else being equal, increasing m_x (fecundity at any age *x*) increases *r*. (This includes increasing m_x from 0 to a higher value in older age classes.)

3. Offspring produced at an early age increase fitness more, because they contribute more to population growth, than the same number of offspring produced at a later age—that is, they have greater "value" in terms of fitness. For instance, if the average fecundity is equal at ages 2 and 3 ($m_2 = m_3$), 2-year-old females contribute more to the future population size than do 3-year-olds. Consider the newborn females of a particular cohort. First, fewer will survive to age 3 than to age 2, so the 2-year-olds collectively will leave more offspring. Second, the offspring of the 2-year-olds, because they are born sooner, will themselves contribute offspring before the offspring of the 3-year-olds do so. Even if all 2-year-olds survive to age 3, the population would grow faster if only 2-year-olds reproduced than if only 3-year-olds reproduced. Thus, the contribution to population growth decreases as females age, unless this decrease is compensated by greater fecundity at later ages.

4. A corollary of the previous point is that the earlier in life reproduction begins, the greater the rate of increase, *r*, will be (all else being equal).

If, then, genotypes differ in their values of age-specific survival and fecundity, the characteristics that would maximize *r*, and therefore fitness, are (1) higher survival through the reproductive ages, (2) higher fecundity at each reproductive age, (3) higher fecundity especially early in life, (4) a longer reproductive life span, and (5) an earlier age of first reproduction. *These, then, are the life history characteristics that we would expect to evolve in all organisms*—based on our considerations so far.

The Evolution of Demographic Traits

General Principles

Like many theories of adaptation, life history theory can be applied by asking how we would expect a trait, such as age at first reproduction, to evolve under specified environmental circumstances. We might, for example, identify the average levels of mortality suffered by different age classes, the rate of growth in body size, and the correlation between body size and fecundity as variables that should be measured for a particular fish species in order to predict its "optimal" age at first reproduction—i.e., the age at first reproduction that, if it evolved, would maximize fitness. If we can specify a few generally important variables, we should be able to derive general quantitative predictions about expected patterns of life history. For instance, if forest trees generally suffer higher mortality early in life (as seedlings or saplings) than as adults, and if the pattern is reversed for small herbs, then we can predict that these two classes of plants will differ, on average, in age at first reproduction.

The theoretically *optimal life history traits* can be calculated by using either *genetic models* (Chapters 13, 14) or *phenotypic models*. Phenotypic models assume that genetic variation is available and simply calculate what life history would maximize fitness. In most instances, genetic and phenotypic models of life history evolution lead to similar conclusions (Charlesworth 1994b). We will use phenotypic models here.

Constraints

As noted above, knowing that traits evolve so as to maximize fitness, we might naively expect ever greater fecundity, ever longer life, and ever earlier maturation to evolve. That all organisms are nevertheless limited in these respects may be attributed to various *constraints*. Constraints on life history evolution may be classified in three overlapping categories, often termed phylogenetic, genetic, and physiological constraints.

Phylogenetic constraints

As is true of most characteristics of organisms, the history of evolution has bequeathed to each lineage certain features that limit variation in its life history characters (Harvey and Pagel 1991). Insects, for example, neither molt nor grow after attaining the adult (reproductive) stage, so a female's body cannot accommodate more eggs with age. Many insects, such as sphinx moths (Sphingidae), feed as adults, and so obtain energy and protein that enables them to form successive clutches of eggs; however, the adults of giant silkworm moths (Saturniidae) lack functional mouthparts, so their fecundity is limited by the amount of material stored when they fed as larvae. In most groups of birds, the number of eggs per clutch varies within and among species, but for unknown reasons, all species in the order Procellariiformes (albatrosses, petrels, and others) lay only a single egg.

Some lineages have properties that probably cannot rapidly evolve in response to different environmental selection pressures. For example, complex structures such as mouthparts seldom re-evolve once lost (see Chapter 24), so it is unlikely that adult silkworm moths could acquire functional mouthparts, and hence unlikely that they could evolve the capacity to lay multiple clutches of eggs. In other cases, a feature may be free to evolve toward its optimum, but the optimum is similar in related species because of their similar ecology (Harvey and Pagel 1991).

Genetic variation and genetic constraints

From the perspective of population genetics, constraints on the evolution of a life history trait could consist of either *a lack of genetic variation* or *genetic correlations* among traits owing to *pleiotropy* (see Chapter 14). With the exception of apparently invariant traits (such as clutch size in albatrosses), life history traits (such as age-specific fecundity and longevity) usually display polygenic, additive genetic variation (Roff 1992). In fact, the additive genetic variance of life history traits is apparently higher than that of other traits (Houle 1992). Stringent selection on these components of fitness is probably balanced by a high rate of origin of genetic variation by mutation at the very many loci that affect summary traits such as survival and fecundity (Houle et al. 1996; see Chapter 14).

Antagonistic pleiotropy, giving rise to a negative genetic correlation between traits, exists when genotypes manifest an inverse relationship between different components of fitness. This relationship may be due to physiological *trade-offs* (see below); for example, genotypes may display a negative relationship between fecundity and subsequent survival if they all acquire the same amount of energy or other resources but differ in the amount they allocate to reproduction versus their own maintenance. However, there may also be genetic variation in the amount of resource harvested. This would give rise to a *positive correlation* between reproduction and survival that could equal or exceed the *negative correlation* owing to genetic variation in allocation (Figure 19.3; van Noordwijk and deJong 1986; Bell and Koufopanou 1986; Houle 1991). The allocation trade-off may still constrain evolution, but the trade-off may be difficult to detect in this case.

Physiological constraints

The principal physiological constraints on the evolution of life history traits are the *trade-offs*, already noted, *in the allocation* of an individual's resources to different functions. Such trade-offs may be

FIGURE 19.3 Factors giving rise to positive or negative genetic correlations between life history traits such as survival (or growth) and reproduction. (A) Variation at loci such as *A* affects the amount of energy or other resources that an individual acquires from the environment. Such loci might affect features such as foraging behavior, digestive efficiency, or photosynthetic rate. The resources acquired are allocated in proportions *x* and $1 - x$ to functions such as growth or self-maintenance and to reproduction. Variation at loci such as *B* affects allocation. (B) Colored circles represent genotypes that differ in loci such as *A*; variation in resource acquisition causes a positive genetic correlation between life history traits. Variation at loci such as *B* (black circles) causes a negative genetic correlation between life history traits. The net, overall, genetic correlation depends on the relative magnitudes of variation in acquisition versus allocation.

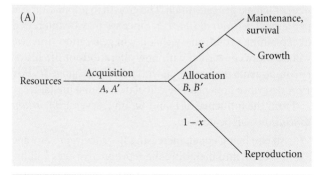

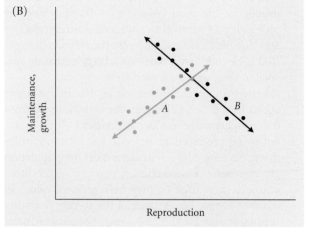

Life Tables and the Rate of Increase

In order to understand how r, the per capita rate of increase in number of a genotype or population, depends on age-specific survival (l_x) and fecundity (m_x), we will assume that a population growing at rate r achieves a *stable age distribution*—i.e., that the proportion of individuals in each age class x remains constant over time (see Box A in Chapter 4). This assumption implies that the number of individuals in each age class (e.g., newborns, $x = 0$) grows at the same rate, r, as does the population as a whole. When generations overlap and population growth is approximately continuous, the growth of the population over time is $dN/dt = rN$, which implies (using integral calculus) that N_t, the population size after t time units, will be

$$N_t = N_0 e^{rt}$$

where N_0 is the initial number (at $t = 0$) and $e = 2.718$, the base of natural logarithms. Thus the number of newborns (age $x = 0$) also grows at the rate e^{rt}—i.e., by a factor of e^r in each time interval. (Here and in all that follows, we count only females if the species has separate sexes, and assume that actual numbers are roughly doubled if the sex ratio is 1:1.)

Each female of age x contributes, on average, m_x offspring. The number of such females equals the number born x time units ago, multiplied by the fraction (l_x) that have survived to age x. If the population has been growing at a constant rate r and has a stable age distribution, then the number born at time $t - x$ (i.e., x time units in the past) is $n_{t-x} = Ke^{-rt}$, where K is the number being born at the present ($t = 0$). Thus the number of offspring produced at present by females of age x is $K = (Ke^{-rx})(l_x)(m_x)$. Dividing both sides by K yields $1 = e^{-rx}l_x m_x$. The total number of offspring produced by females of *all* age classes is then found by summing this expression over all ages (x) from the first (a) to the last (z) age of reproduction:

$$1 = e^{-ra}l_a m_a + e^{-r(a+1)}l_{a+1}m_{a+1} + \ldots e^{-rz}l_z m_z$$

which may be written

$$l = \sum_{x=a}^{x=z} e^{-rx}l_m m_x \tag{A.1}$$

Equation (A.1) provides the basis for calculating r, the per capita rate of increase of a population or genotype, if the age schedules of survival (l_x) and fecundity (m_x) have been measured. (There is no explicit solution for r, which must be found by iteration or "trial and error.") It also expresses the relative contribution of each age class to population growth; you may think of this as the relative likelihood that a newborn individual (the "1" individual on the left side of Equation A.1) is produced by a female of age x. Note that in the term $e^{-rx}l_x m_x$, the proportional contribution of progeny by females of age x, the product $e^{-rx}l_x$ must decline with age because (1) survival— l_x—necessarily declines with age, and (2) e^{-rx} is a negative exponential term, falling off as x increases. Therefore *the contribution to population growth must necessarily decrease as females age, except insofar as this drop-off is compensated by an increase in fecundity (m_x) with age*. In the example in Table 19.1, m_x does not increase with age enough to fully compensate for the decline of e^{-rx} and l_x, so there is a steady decline in the contribution of females to population growth as they age.

The average lifetime production of offspring by a newborn female is $R_0 = \sum l_x m_x$ (which equals 1.932 in Table 19.1). R_0 is the per capita rate of population growth per generation (see Chapter 13). If several age classes reproduce, generation time—the average age of mothers of newborn offspring—is approximately

$$T = \frac{\sum_{x=a}^{x=z} (x l_x m_x)}{\sum_{x=a}^{x=z} l_x m_x} \tag{A.2}$$

The relation between the per capita growth rate per unit of *time* (r) and per *generation* (R_0) is approximately $r \approx (\ln R_0)/T$. For the data in Table 19.1, $T = 2.981$, and $r \approx (\ln R_0/T) = 0.221$, slightly lower than the value of r calculated exactly from Equation (A.1). If a population is not growing (i.e., is stable), $r = 0$ and $R_0 = 1$.

In Equation (A.1), we note that the current contribution to population growth of a female of some age x is $e^{-rx}l_x m_x$. The average (expected) future contribution of such a female is the sum of the contributions she might make at all subsequent ages y ($y > x$) up to the maximum age (z). Charlesworth (1994b) has shown that the *sensitivity of fitness* to a small change in the probability of survival to age x is proportional to the expected future contribution to population growth, divided by the generation time, or

$$S_s(x) = \frac{\sum_{y=x}^{y=z} e^{-ry}l_y m_y}{T} \tag{A.3}$$

This function is constant up to the age of first reproduction, and declines thereafter due to the monotonic decline of the terms e^{-ry} and l_y with increasing y. Similarly, Charlesworth has shown that the sensitivity of fitness to a small change in fecundity at age x is given by

$$S_m(x) = \frac{e^{-rx}l_x}{T} \tag{A.4}$$

Because both these functions decline with age, a small increment in either probability of survival or fecundity adds less to r (fitness), the greater the age at which the increment is expressed. Table 19.1 includes calculations of $S_s(x)$ and $S_m(x)$ for a hypothetical data set.

manifested as negative *phenotypic correlations* between, say, fecundity and subsequent survival. If such a trade-off is to affect the course of evolution, however, it must have a genetic basis. Trade-offs cannot be assumed a priori; they must be documented by one of several methods (Reznick 1985):

1. Measuring *correlations between the means* of two or more traits in *different populations* or species. A weakness of this approach is that the correlations may be a consequence of other characters that differ among populations, and thus may not represent a causal relationship such as an allocation trade-off.

2. *Manipulating one trait and determining the effect on other traits.* For instance, many investigators have found that altering the number of eggs in birds' nests can affect the proportion of young that fledge and the subsequent survival and reproduction of both the fledglings and their parents. This approach assumes, but does not demonstrate, that a genetic change in clutch size would have similar effects.

3. Measuring *genetic correlations between traits within populations*, using the breeding methods of quantitative genetics (see Chapter 14). A negative correlation in this case is likely to indicate a real trade-off. However, a real allocation trade-off may be masked by a positive genetic correlation owing to genetic variation in the ability to acquire resources (see above), or simply to mutations that have a deleterious effect on both components of fitness. Moreover, since genetic correlations are altered when allele frequencies or linkage disequilibrium among loci change, they may not represent a long-term constraint (Clark 1987; Charlesworth 1994b).

4. *Imposing selection* (artificial or natural) on one trait and determining whether there are correlated changes in other traits. This approach has provided the most evidence for trade-offs (Reznick 1985; Stearns 1992).

The last two methods can provide evidence for a genetic, not merely a phenotypic, basis of a trade-off, so most researchers consider them best.

Evidence for a Cost of Reproduction

The assumption that reproduction is costly—that there exist trade-offs between reproductive effort at one age and components of fitness at other ages—underlies much of life history theory, and is supported by considerable evidence. We will describe only a few examples of such evidence here.

In *Drosophila* and many other insects, mating activity and egg production reduce the longevity of one or both sexes, and virgins live longer than nonvirgins (Fowler and Partridge 1989; Bell and Koufopanou 1986). Female blue tits (*Parus caeruleus*) whose egg clutches were halved or doubled by experimental manipulation showed increased and decreased survival, respectively, relative to control birds

with unmanipulated clutches (Figure 19.4; Nur 1984). A negative correlation between reproduction and growth was found among populations of guppies (*Poecilia reticulata*) that were reared under identical laboratory conditions, thus demonstrating that the correlation had a genetic basis (Reznick 1983). Experiments with many bird species have shown that if one adds eggs to a nest, the female is likely to lay fewer eggs the next year (Roff 1992). Law et al. (1979) grew individuals of each of many families of meadow grass (*Poa annua*) in a randomized array, and found that families that, on average, produced more inflorescences in their first season produced fewer in the second, and also achieved less vegetative growth (Figure 19.5).

The Theory of Life History Evolution

Two major considerations form the foundation for the theory of the evolution of life history traits: (1) the *sensitivity of fitness* to an alteration of the trait and (2) *trade-offs* among traits. Because reproduction early in life contributes more to the rate of population growth than an equal production of offspring later in life, the effect of a given magnitude of increase in fecundity or survival on fitness (*r*) depends on the age at which the change is expressed. Likewise, the effects of a given cost of reproduction on subsequent survival, growth, or reproduction can vary depending on the age at which the cost is incurred.

A useful theoretical approach (Charlesworth 1994b) is *sensitivity* analysis, i.e., calculating how greatly fitness

FIGURE 19.4 Survival of adult blue tits following breeding in relation to the size of the broods they produced. Survival through the spring was estimated by recaptures of marked individuals. The pattern for females (colored circles) suggests a cost of reproduction, although the pattern for males (black circles) does not. The vertical bars are measures of variation. (After Nur 1984.)

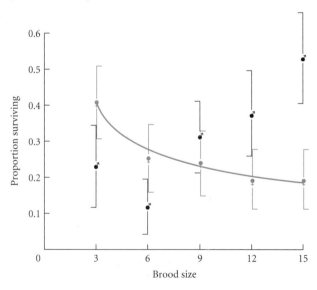

FIGURE 19.5 The number of inflorescences per plant in the grass *Poa annua* is lower in the second flowering season, the higher it was in the first season. Each point represents the mean of the individuals in a single family. Plants from two different habitats, indicated by black and colored circles, were grown together in the same plot. (After Law et al. 1979.)

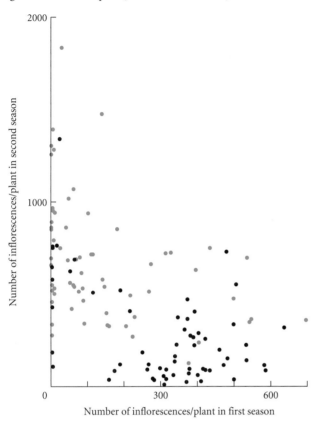

would be altered by a slight alteration of a demographic variable. Consider, for example, the hypothetical data in Table 19.1, and ask how much greater (or less) a genotype's fitness (r) would be, compared with r (0.228) for the hy-

pothetical genotype in the table, if any one of the values of l_x or m_x were altered slightly. For example, an increase in survival to the first age class (l_1) would obviously increase fitness. On the other hand, increasing survival from age 5 to age 6 (l_5) would not alter fitness at all, because this species does not reproduce beyond age 5 ($m_6 = 0$). A genotype that survives past the reproductive period contributes no more descendants to subsequent generations than one that does not, so it has no greater fitness. Therefore, *natural selection does not favor postreproductive survival.* (It may be advantageous, however, if postreproductive parents care for their offspring, as in humans.)

Because the contribution of a cohort to population growth declines with age, a slight increase in survival at an advanced age (large x) increases r less than an equal increase in survival at an early age (small x). Thus, *after reproduction begins, the selective advantage of survival declines with age.*

Likewise, *the selective advantage of a given increment in fecundity declines with age.* A simple reason is that cumulative mortality at each age lessens the likelihood that a female will be alive to reproduce. Thus a 5 percent increase in fecundity at age 5 will increase a genotype's r less than the same increase at age 4, because there will be fewer 5-year-olds than 4-year-olds. Another reason is that an offspring born to a young female contributes more to population growth than one born to an older female.

At any age, an individual has a certain amount of energy (or material resources such as protein) that can be allocated to reproduction or to its own maintenance (without which it cannot survive) and perhaps to its further growth. The fraction allocated to reproduction is the **reproductive effort** at age x. The decline with age in the effect of fecundity on fitness implies that it should be advantageous to reproduce as early in life as possible, and furthermore, to maximize reproductive effort at this age. Hence no energy would be reserved for subsequent growth, maintenance, or reproduction. A semelparous life history, with maximal re-

Table 19.1 **A hypothetical example of a life table**

AGE CLASS	NUMBER OF SURVIVORS	FRACTION OF SURVIVORS		AVERAGE FECUNDITY			SENSITIVITY OF r TO m_x	SENSITIVITY OF r TO l_x
x		l_x	m_x	$l_x m_x$	e^{-rx}	$e^{-rx}l_x m_x$	$S_m(x)$	$S_s(x)$
0	1000	1.000	0.00	0.00	1.000	0.000	0.335	0.334
1	750	0.750	0.00	0.000	0.796	0.000	0.200	0.334
2	600	0.600	1.20	0.720	0.634	0.456	0.128	0.182
3	480	0.480	1.40	0.672	0.505	0.339	0.081	0.068
4	360	0.360	1.03	0.396	0.402	0.159	0.049	0.018
5	180	0.180	0.96	0.144	0.320	0.046	0.019	0.018
6	100	0.100	0.00	0.000	0.255	0.000	0.011	—
Sums:				1.932		1.000		

Source: After Stearns (1992).

$r = 0.228$

$R_0 = 1.932$

Generation time (T) = 2.981

productive effort followed by death, would therefore appear to be most advantageous. Obviously, not all organisms have such a life history. We will now examine why life histories vary among species.

Life Span and Senescence

Most organisms in which germ cells are distinct from somatic tissues undergo physiological degeneration with age, a process called AGING or SENESCENCE. The evolutionary theory of why this occurs, treated in depth by Michael Rose (1991), begins with the principle that the selective advantage of an increased probability of survival, and therefore *the advantage of maintaining and repairing somatic tissues, declines with age.* This theory assumes that genes that affect physiological processes underlying survival or reproduction may be expressed only at certain ages. For instance, Huntington's disease, a degenerative disease of the human nervous system that leads to death, is caused by a dominant allele that is usually not expressed until the age of 30 to 40.

Two major theories of senescence build on this foundation. Peter Medawar (1952) proposed that deleterious mutations that affect later age classes accumulate in populations at a higher frequency than those that affect earlier age classes because selection against them is weak. This hypothesis predicts that genetic variation in fitness-related traits (such as those affecting survival) should be greater in late than in early age classes.

George Williams (1957) proposed the existence of antagonistic pleiotropy between the effects of genes early and later in life. An allele that is advantageous early but deleterious later in life has a selective advantage because of the greater contribution of earlier age classes to fitness. One class of genes that may have such effects are those that increase reproductive effort, reducing the energy and materials allocated to maintenance and repair. This hypothesis, like Medawar's, suggests that the physiological manifestations of senescence may be diverse, but in contrast to Medawar's mutation accumulation hypothesis, they need not be genetically variable. It predicts that there should be genetic trade-offs between components of fitness expressed early and later in life.

Evidence on the evolution of senescence

The antagonistic pleiotropy hypothesis advanced by Williams (1957) has been confirmed by experiments with *Drosophila melanogaster* (Rose and Charlesworth 1981; Luckinbill et al. 1984; Leroi et al. 1994a). In these experiments, some laboratory populations were selected for delayed reproduction by rearing only offspring of old parents, whereas other populations were propagated only from offspring of young parents. Periodically, samples from each population were scored for longevity and age-specific fecundity. The average longevity and the fecundity of old females increased in populations selected for reproduction late in life, but the fecundity of young females in these populations declined

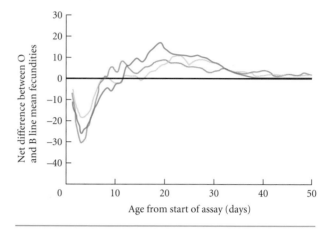

FIGURE 19.6   The difference in fecundity (eggs laid per female per day) between populations of *Drosophila melanogaster* selected for late reproduction (O) and for early reproduction (B). The fecundity of the B line is taken as a standard, and is set at zero. The three O populations, selected for late reproduction, have relatively high fecundity at later ages, but depressed fecundity at early ages. (After Rose 1984.)

(Figure 19.6). The negative relationship between early reproduction and both longevity and later reproduction has been attributed to the allocation of lipid reserves between these functions. Natural selection for increased early reproduction would therefore tend to cause a correlated decline in later reproduction and longevity. Because early reproduction is advantageous, we would therefore expect the evolution of a shorter reproductive life span, accompanied by senescence and early death. These experiments are among the most striking confirmations of evolutionary hypotheses that had been posed far in advance.

The evidence for Medawar's mutation accumulation hypothesis is mixed. It predicts that additive genetic variance in fitness components should be greater in older age classes. In *Drosophila*, the predicted relationship was found for mortality rate and male mating success, but not for female fecundity (Charlesworth 1994).

The Evolution of Age Schedules of Reproduction

Since early reproduction is correlated with lowered subsequent reproduction (as in the experiments by Rose and Luckinbill), and since there is always an advantage to reproducing earlier in life, we should expect organisms to reproduce at the *earliest possible age* and to reproduce only once. We must therefore ask why so many species are nevertheless iteroparous.

An iteroparous genotype can have a higher rate of increase than a semelparous genotype if adults have high enough survival rates to reproduce repeatedly and if older parents produce enough offspring to compensate for their lower value. This is especially true if the rate of population growth is low—i.e., if *r* is nearly zero (see Box 19.B)—as may be the case in a population that is regulated by density-dependent factors (such as competition for resources) that cause high juvenile mortality. Thus *iteroparity is likely to be advantageous if juvenile mortality is high, adult mortality is low, and the population growth rate is low.* Conversely, a semelparous life history is likely to evolve if the population growth rate is high (at least at times), juvenile survival is high, and adult survival is low (in which case the probability of surviving to reproduce a second time is low).

In an iteroparous life history, an individual of age x allocates some energy not to reproduction, but to self-maintenance and perhaps growth—and therefore to subsequent reproduction. In species with indeterminate growth, such as many plants and fishes, fecundity is often correlated with body mass. In such species, clearly, saving energy for growth is like an investment for future reproduction. As individuals age, however, the benefit of withholding energy from reproduction declines, both because larger bodies require more energy for maintenance, and because the intrinsic demographic disadvantages of reproducing late in life become greater. Therefore, we would expect, as a rule, that *species with indeterminate growth should be iteroparous,* but that *the proportion of energy or other resources devoted to reproduction should increase with age* (Williams 1966; Charlesworth 1994b).

Reproducing at an early age may increase the risk of death, decrease growth, or decrease subsequent fecundity so as to lower r compared with what it would be if reproduction were deferred. *The factors that favor delayed maturation* are therefore much the same as those that favor

iteroparity: *a high cost of reproduction and low adult mortality.* It has been suggested, for example (Gadgil and Bossert 1970), that some migratory salmon spawn only once, and then die, because of the great energetic effort required to swim upstream to their spawning sites—a cost of reproduction so great that it is unlikely that a fish could repeat the performance. Delayed maturation is advantageous in these species, enabling salmon to grow to a size at which they can reproduce prolifically at their only opportunity.

Evidence on age schedules of reproduction   The following examples, from among many, illustrate studies that have tested three of the predictions made by the theory of the evolution of the age schedule of reproduction.

Reproductive effort, in each reproductive episode, is expected to be lower in iteroparous than in semelparous organisms. Comparisons among species generally support this prediction (Roff 1992). For example, inflorescences make up a lower proportion of plant weight in perennial than in annual species of grasses (Wilson and Thompson 1989). Similar contrasts may be found within species. For instance, a colony of the short-lived marine colonial ascidian *Botryllus schlosseri* consists of modules, called zooids, each of which produces a variable number of new zooids by asexual budding, and then degenerates. After several cycles of asexual budding, sexually produced offspring develop in each mature zooid. Richard Grosberg (1988) found that there are two genetically determined types of colonies (Figure 19.7). Semelparous colonies reproduced sexually only once and

---

FIGURE 19.7   Polymorphism in the life history of the colonial ascidian *Botryllus schlosseri* illustrates trade-offs between components of fitness. (A) Cross-section of an iteroparous colony, showing young asexually produced zooids (small ovals) nestled among mature zooids (elongated structures). (B) Numbers of colonies, out of 200, that produced various numbers of clutches of offspring, plotted against age at first reproduction. The semelparous morph produces one clutch at an early age; iteroparous colonies begin reproducing later in life. (C) The fecundity of the semelparous morph is higher than that of the iteroparous colonies. (A courtesy of R. Grosberg; B and C after Grosberg 1988.)

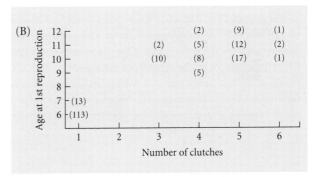

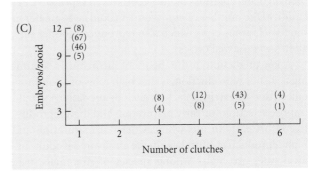

A Model of the Evolution of Semelparity and Iteroparity

If the chance of surviving beyond a single reproductive season is low (or if the act of reproduction greatly reduces the probability of survival), then a single reproductive episode (semelparity) is likely to be advantageous. But if the probability of survival is high, an iteroparous genotype may have a greater rate of increase (r) than a semelparous genotype. Eric Charnov and William Schaffer (1973) showed this by the following theory:

Suppose that an iteroparous genotype first reproduces at age a, has a probability l_a of surviving to this age, and has a constant probability of survival (P) and a constant fecundity (m_i) at each age thereafter. Its rate of increase, r, is found from Equation A.1 (in Box 19.A):

$$1 = l_a m_i(e^{-ra} + Pe^{-r(a+1)} + P^2e^{-r(a+2)} + \ldots)$$

which equals

$$1 = (l_a m_i e^{-ra})(1 - Pe^{-r})$$

A semelparous genotype with the same values of a and l_a, but with fecundity m_s, has a rate of increase r_s, found from Equation A.1 in Box 19.A:

$$1 = l_a m_s e^{-r_s a}$$

Since both equations equal 1, we may set them equal to each other and find the condition under which the genotypes have equal fitness, i.e., $r_s = r$. This condition is

$$\frac{m_i}{m_s} = 1 - Pe^{-r}$$

For example, if $r = 0.01$ and $P = 0.81$, then $1 - Pe^{-r} = 0.20$, and the genotypes have equal fitness if the fecundity of the semelparous genotype is five times that of the iteroparous genotype. If m_s is less than $5 \times m_i$, the iteroparous genotype has the higher rate of increase, and iteroparity will evolve. Figure B1 shows how the advantage of iteroparity increases, the greater the rate of adult survival (P) and the lower the rate of population growth.

This is an example of a simple, general optimality model. It specifies conditions, such as P and r, under which either an iteroparous or semelparous phenotype would be optimal, assuming that genetic variation for such life histories exists. For the sake of generality, the model makes simplifying assumptions—such as a constant survival probability that is the same for all adult age classes—that are unlikely to hold exactly in any real population. Thus the model is unlikely to be useful for making a quantitatively accurate prediction for any particular species. It does, however, specify the conditions that enable us to make *general* predictions: for example, species with high P are more likely to be iteroparous than those with low P.

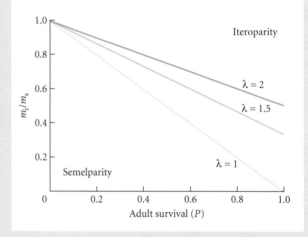

FIGURE 19.B1 Lines of equal fitness for semelparity and iteroparity. Combinations of m_i/m_s and adult survival (P) above a line favor iteroparity; those below a line favor semelparity. Conditions for the evolution of these life histories are given for three rates of population growth (λ, which is related to r). (After Roff 1992.)

---

then died; iteroparous colonies reproduced repeatedly. Fecundity (the number of embryos per zooid) was greater in the semelparous type, which also reproduced sooner (i.e., after fewer asexual cycles).

A key prediction of the theory of life histories, assuming a cost of reproduction, is that high rates of adult mortality (due to extrinsic factors such as predation) should impose selection for early maturation and high reproductive effort early in life, whereas delayed maturation and high reproductive effort later in life are favored if adult survival rates are high. Among species of mammals, fish, lizards and snakes, and other groups, species that have long life spans in nature also mature at a later age, as predicted (Figure 19.8; Promislow and Harvey 1991; Shine and Charnov 1992).

The relationship between age at maturity and mortality has been particularly well documented by David Reznick, John Endler, and colleagues, who have studied guppies (*Poecilia reticulata*) in Trinidad (e.g., Reznick and Endler 1982; Reznick et al. 1990). In some streams, this small fish is heavily preyed upon by the cichlid fish *Crenicichla alta*, which attacks large (mature) guppies. In other streams, the major predator is a cyprinodontid fish, *Rivulus hartii*, which feeds on small (juvenile) guppies. As

FIGURE 19.8 The relationship between the annual rate of mortality and the age at female sexual maturity in natural populations of 16 species of snakes (colored circles) and lizards (black circles). The average life span is proportional to the annual survival rate. Both groups of organisms, especially lizards, conform to the prediction that delayed onset of reproduction is most likely to evolve in species with low rates of adult mortality. (After Shine and Charnov 1992.)

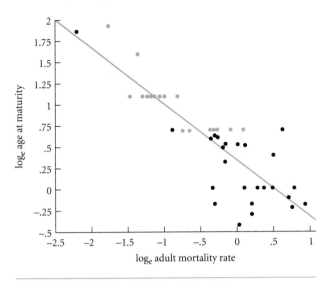

In 1976, the researchers introduced some guppies from a stream with *Crenicichla* into a stream that harbored *Rivulus*, but no guppies. Thus a guppy population that had experienced adult predation was now living in a stream in which juveniles suffered the greatest predation. Eleven years (30–60 generations) later, guppies from both the source and experimental populations were captured and reared in the laboratory, and their life history characteristics were measured. Several of these features had evolved, in the experimental population, toward the state characteristic of guppies from *Rivulus*-dominated sites: the fish matured later, at a larger size, and had smaller broods of larger offspring. This experiment in a natural population showed that natural selection can rapidly alter life history characteristics in the predicted direction.

The prediction that reproductive effort should increase with age in iteroparous species has been confirmed in studies of diverse species of animals and plants (Roff 1992). For example, Daniel Piñero and colleagues (1982) measured allocation to vegetative and reproductive structures, in relation to age, in the understory palm *Astrocaryum mexicanum* in southern Mexico. Fruits make up an increasing proportion of the plant's biomass, and roots a decreasing proportion, with age (Figure 19.9). Moreover, the trees reproduce more frequently as they age, accounting for much of the increase in reproductive effort.

Number and Size of Offspring

theory predicts, guppies from *Crenicichla*-dominated streams mature at a smaller size, reproduce more frequently, and have a higher reproductive effort (weight of embryos relative to weight of mother). They also produce the smallest, and greatest number of, offspring, a point that will be treated below.

In the preceding pages, we have discussed the factors that influence the optimal reproductive effort, but we have not discussed whether this effort should be directed toward producing many offspring or few. All else being equal, a genotype with higher fecundity has higher fitness than one with lower fecundity. If this is the case, why don't all or-

FIGURE 19.9 The distribution of (A) total biomass and (B) annual net production among various organs of the palm *Astrocaryum mexicanum* in Veracruz, Mexico. The proportional allocation to reproduction increases with age. (After Piñero et al. 1982.)

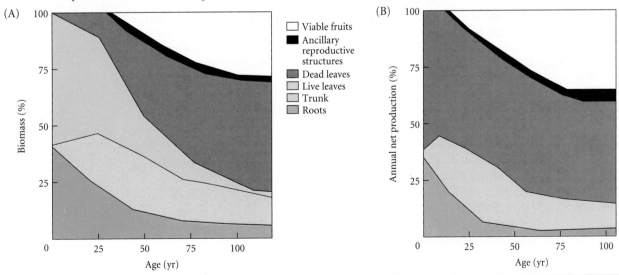

ganisms have huge numbers of offspring? The very low fecundity of species such as humans, albatrosses, and the California condor (*Gymnogyps californianus*, now facing extinction) requires explanation.

An early insight into this problem was provided by the British ecologist David Lack (1954), who found that starlings (*Sturnus vulgaris*) have the greatest number of surviving offspring if they lay no more than five or six eggs. The parents are unable to feed larger broods adequately, so the number of survivors from larger broods is less than the number from more modest clutches. Lack therefore proposed that a parent's fitness is maximized by laying an *optimal* clutch size, namely, that which yields the greatest number of surviving offspring.

A similar effect often arises from a *trade-off between the number and size of offspring* due to the finite amount of energy and materials that parents can allot to yolk, endosperm, or nourishment of embryos. A negative correlation between number and size of offspring (eggs, seeds, or live-born young) has been reported for comparisons among species in many groups, and for variation within many species of plants and animals (Roff 1992; Stearns 1992). For example, human twins have a lower birth weight, on average, than single-born infants. There is also a great deal of evidence that animals born or hatched at a larger size are superior in components of fitness such as survival and growth rate. The survival of juvenile side-blotched lizards (*Uta stansburiana*), for example, is correlated with size (Ferguson and Fox 1984). Barry Sinervo (1990) experimentally manipulated hatchling size in a related lizard, *Sceloporus occidentalis*, by removing yolk from eggs with a hypodermic needle. Smaller hatchlings ran more slowly, which probably would reduce their survival in the wild.

A modest clutch size can be optimal for many other reasons (Stearns 1992). We have already noted, for example, that the survival of parent birds may be reduced by increasing the size of their clutches.

The Evolution of the Rate of Increase

In most analyses of natural selection, the per capita rate of increase (*r*) of a genotype is the measure of its fitness. We might suppose, then, that the trend in all species should be to evolve higher rates of increase. Yet the potential rate of population growth has certainly evolved to a low level in humans, condors, and many other species. How can this apparent paradox be resolved?

The critical distinction is between the *intrinsic* rate of natural increase of a population (r_m) and the *instantaneous* rate of increase (*r*) (cf. Chapter 4). The intrinsic, or maximum potential, per capita rate of increase, r_m, is the rate that obtains when a population experiences no density-dependent effects on birth or death rates. The *actual, realized* rate of increase, *r*, is likely to be lower than r_m, especially if density-dependent factors reduce birth rates or increase death rates. In the simplest formulation, *r* declines linearly with *N*, the population density: $r = r_m - cN$, where the constant

$-c$ is the slope. Substituting this for *r* in the elementary equation for population growth, $dN/dt = rN$, we obtain

$$\frac{dN}{dt} = (r_m - cN)N$$

At equilibrium, when density dependence has reduced the population growth rate to zero, $dN/dt = 0 = (r_m - cN)N$, and the equilibrium value of *N* is $N = r_m/c$. This equilibrium density is often referred to as the "carrying capacity" of the population, denoted *K*. (A little algebra will show that substituting *K* for r_m/c in the above equation yields the logistic equation, introduced in Chapter 4.)

Figure 19.10 illustrates a hypothetical case in which two genotypes differ in both r_m and *c*. The growth rate of genotype *B* is less sensitive to density (*c* is smaller in absolute value), so $K = r_m/c$ is higher. At low densities, genotype *A* has the higher realized rate of increase (*r*), but at higher densities *B* has the higher *r*. As density (*N*) increases, *A* stops growing in number (it reaches its carrying capacity, K_A), but *B* continues to increase, finally reaching its carrying capacity at a high density. As individuals die, they are more frequently replaced by *B* than *A* individuals, since *B* has a higher *r* at high density, and so genotype *B* replaces *A*. Thus at high densities, the genotype with the higher *K* is fitter. In fact, *K* may be taken as a measure of the fitness of a genotype in a population that is near its equilibrium density (Stearns 1992; Charlesworth 1994b). In this example, genotype *B* will replace *A* if the high density persists long enough; but the population's *potential*, intrinsic rate of increase (r_m), which would be expressed if we grew a sample from this population at low density, will decline. Thus a genotype that has higher *r*, and therefore higher fitness, in a high-density environment may have a lower *intrinsic* rate of increase, r_m. Notice, moreover, that this simple analysis implies that under density-dependent conditions, *evolu-*

FIGURE 19.10 A model of density-dependent natural selection. The per capita rate of increase, *r*, declines for genotypes A and B as population density (*N*) increases. The intrinsic rate of increase (r_m), expressed at very low density, is lower for genotype B, but this genotype has a selective advantage at high density. (After Roughgarden 1971.)

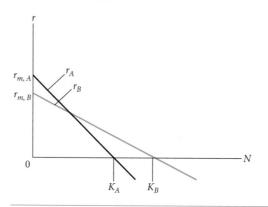

tion tends to increase population density, since genotypes that attain higher density (higher *K*) are fitter.(Population density may not be maximized, however, if selection is frequency-dependent: see Charlesworth 1994b.)

In the preceding pages, we saw that delayed maturation and low fecundity could evolve if trade-offs with subsequent survival and reproduction were sufficiently strong. Clearly the evolution of such features would reduce r_m, the potential rate of population increase. We have said little about the environments in which such traits might be advantageous, except to note in passing that this might be the case when adult mortality and the rate of population growth are low. For many, although not all, organisms, these conditions are likely to exist in populations that occupy relatively stable environments in which predation or competition for resources causes heavier mortality among juveniles than adults. The mortality of seedling trees in a mature forest, for example, is exceedingly high, but if a tree does survive beyond the sapling stage—perhaps because a treefall has opened a light gap—it is likely to have a long life.

When, because of population density or simple scarcity of resources, competition for resources is intense, juvenile survival can be enhanced by large size, which can confer a competitive advantage or simple resistance to starvation. Hence producing large eggs or offspring—and fewer of them—may be advantageous. Adult mortality may be low in these circumstances, but only if sufficient resources are allocated to survival rather than to reproduction, so a low level of reproductive effort may evolve. These and similar considerations have led many authors to suggest that traits associated with a low intrinsic rate of increase—delayed maturation, production of few, large offspring, a long life span—are likely to evolve in species that occupy stable, competitive, or resource-poor environments. These generalizations seem to hold widely. For example, some species of beetles, fish, and other animals that inhabit caves develop very slowly and produce large eggs at an extraordinarily low rate (Culver 1982). Reproductive effort in plant species that occupy disturbed or early successional environments is generally higher than in plants that typically grow in mature, "climax" communities (Figure 19.11; Harper 1977; Piñero et al. 1982).

In an influential early contribution to life history theory, the ecologists Robert MacArthur and Edward O. Wilson (1967) coined the terms **r-selection** and **K-selection** for the patterns of selection found in populations that are, respectively, relatively free of density-dependent restraint and subject to strong density dependence. They suggested that a genetic trade-off may exist between high r_m and high *K* (as in Figure 19.10), and specified some of the characteristics expected in "*r*-selected" versus "*K*-selected" species. Eric Pianka (1970) extended their treatment, proposing that *r*-selected organisms should be characterized by semelparity, rapid development, early reproduction, small body size, and a high potential rate of increase (r_m), whereas *K*-selected organisms should display the opposite traits. Some authors have criti-

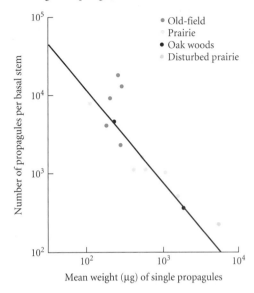

FIGURE 19.11 The relationship between number and weight of propagules (seeds) among species of goldenrods (*Solidago*) found in different habitats. Colonizing species, in old-field habitats, produce smaller seeds than species that grow in more stable prairie habitats, where intense competition is likely to favor larger offspring. (After Werner and Platt 1976.)

cized the concepts of *r*- and *K*-selection, and have urged that they be abandoned (Roff 1992; Stearns 1992; Charlesworth 1994b), because they do not take into account important elements of the life history theory we have discussed, and because many organisms do not fit either of the two proposed syndromes of traits. Nevertheless, many organisms do fit the predictions fairly well, so it is sometimes useful to use MacArthur and Wilson's terms, with the understanding that they do not begin to capture the full variety or causes of life histories (Boyce 1984).

Male Reproductive Success

Our discussion so far has been limited to the life history traits of females. The reproductive success of a male phenotype depends on its average survival to each reproductive age, its age of first reproduction, its age schedule of reproduction, and the survival of its offspring, just as for females. In addition, a male's reproductive success in each breeding season is affected by the number of mates he obtains and their average fecundity, for these determine the number of his offspring.

Much of the theory described for females applies to males as well. A substantial cost of reproduction, for example, may impose selection for delayed maturation and iteroparous reproduction. A special agent of selection that usually applies with greater force to males than to females is *competition for mates*. Competition among males for females is one form of *sexual selection*, which we explore in Chapter 20 (see also Chapter 16).

Effects of Sexual Selection on Life Histories

The mechanisms of male competition for females include *contests between males* and *mate choice by females* (Andersson 1994). Contests between males include behaviors such as physical combat (in some species of deer, for example) and defense of territories to which females come to mate and/or to nest (as in many birds). In sexual selection by female choice, females choose a mate from among a number of males, which usually engage in a display of some kind, as in frogs that aggregate and call. In either case, the male's behavior—whether displaying to females or confronting other males—is often very costly. Vocalizations by frogs, courtship displays by grouse, and battles among male deer and elephant seals require enormous expenditures of energy, and males often lose 20–30 percent of their body weight during a season of such activities (Figure 19.12; Andersson 1994, 238–240). Males that are larger or invest more energy in competition for females are usually more successful. The costs of reproduction and the advantages of large size explain why males of many species begin to breed at a later age than females, and why, among grouse and some other groups, delayed maturation of males is more frequent in species that are polygynous (each male potentially mates with more than one female) than in monogamous species, in which competition for mates is less intense.

Variant Male Life Histories

Many interesting variations in male life histories exist. One is **alternative mating tactics**, often assumed by males that differ in size or other morphological features. In some species, large males display and/or defend territories, whereas small males do not, but rather "sneak" about, intercepting females. In some instances, "sneaker" males have lower reproductive success than displaying males, so their behavior is probably not an adaptation. They are probably making the best of a bad situation, being unable to compete successfully. In other instances, however, the sneaking tactic appears to be an alternative adaptation, yielding the same fitness as the display tactic. A likely example is presented by the Pacific coho salmon (*Oncorhynchus kisutch*), which migrates from the sea to spawn in rivers (Gross 1984). Hooknose males are large and red and have hooked jaws and enlarged teeth, which they use in fights over females; jack males are smaller, resemble females, and do not fight. Hooknose males spend a year and a half in the sea, and jacks half a year. Hooknose males fight for positions near the sites to which females come to spawn, whereas jacks hide behind rocks near such sites. Apparently these behaviors are about equally successful, for when Gross estimated fitness from data on survival to breeding age and frequency of mating, he calculated the relative fitnesses of jack and hooknose males to be 0.95:1.00.

A fascinating life history variation found in some plants, annelid worms, fishes, and other organisms is *sex change* (**sequential hermaphroditism**), whereby an individual changes from female to male during its lifetime (**protogy-**

ny) or vice versa (**protandry**). This phenomenon is frequently associated with changes in reproductive success as an individual grows in size (Figure 19.13). If larger females produce more eggs, but the reproductive success of males does not increase commensurately with size, then it can be advantageous for an individual to switch from male to female as it grows. Conversely, if large size confers success in male competition for females, individuals may enjoy greater reproductive success if they become male after growing to a large size; protogyny is then advantageous. Anemone fishes (*Amphiprion*), for example, form monogamous pairs, so there is little competition among males for mates. Because female fecundity increases with size, it is understandable that these fishes are protandrous. In related polygamous fishes, however, there is strong competition among males for mates, and these species are protogynous (Warner 1984b).

FIGURE 19.12 Male competition for females requires large expenditures of energy. (A) Groups of male sage grouse display together on an arena (a lek). The one or few dominant males are likely to be chosen as mates by all the females that visit the lek. (B) The energy expended by displaying male sage grouse increases with the number of displays (struts). In this group, only the two most active males (circled points) were successful in mating. (A courtesy of R. H. Wiley; B after Vehrencamp et al. 1989.)

(A)

(B)

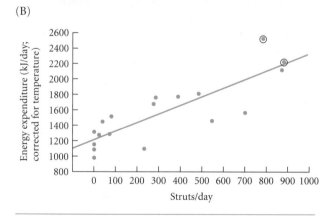

FIGURE 19.13 A model for the evolution of sex change in sequential hermaphrodites. (A) When reproductive success increases equally with body size in both sexes, there is no selection for sex change. (B) A switch from female to male (protogyny) is optimal if male reproductive success increases more steeply with size than female reproductive success. (C) The opposite relationship favors the evolution of protandry, in which males become females when they grow to a large size. (After Warner 1984b.)

(A)
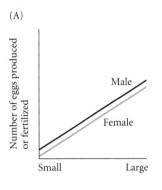

(B)
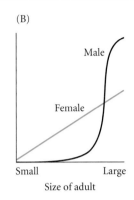

(C)
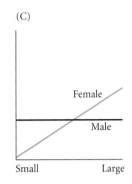

Robert Warner (1984a) studied the bluehead wrasse (*Thalassoma bifasciatum*), a fish species in which some individuals start life as "initial-phase" males and then become "terminal-phase" males, whereas others start as females and then become terminal-phase males (Figure 19.14). Initial-phase males resemble females and spawn in groups, whereas terminal-phase males are brightly colored and defend territories. Although the fecundity of females increases as they grow in size, it does not match the number of eggs a large terminal-phase male typically fertilizes. Warner (1984a) found that females become males at about the size at which terminal-phase males enjoy greater reproductive success.

The Evolution of Dispersal

In addition to the age-specific schedules of survival and reproduction, many other features of organisms may be considered aspects of their life histories. One of the most important is dispersal.

Dispersal may be defined as movement of organisms from their location at birth to other locations at which they reproduce. Dispersal is usually relatively undirected (except as it is influenced by factors such as prevailing winds or currents), in contrast to **migration**, which is cyclical movement between specific regions (as in migratory birds and salmon).

For dispersal to evolve (or not), there must be alternative alleles that promote or inhibit it, perhaps by affecting behavior or physical features. Natural selection against alleles that promote dispersal is almost inevitable, for several reasons:

1. Within a population, there is *automatic* selection against "dispersal alleles" because they remove themselves from the population, necessarily lowering their own frequency (Van Valen 1971). Suppose you propagated *Drosophila* in a cage from which wild-type

FIGURE 19.14 (A) A diagram of the two pathways by which terminal-phase males develop in the bluehead wrasse. (B) A terminal-phase male bluehead wrasse (top) and several females (bottom). Initial-phase males resemble females. (Photographs courtesy of R. Warner.)

(A)
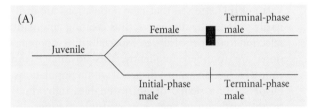

(B)
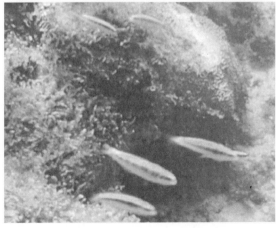

flies could escape by flight, but flightless flies with the *vestigial*-winged mutation could not; after a few generations, the *vestigial* allele would predominate.

2. As a result of selection, most populations are better adapted to local conditions than to the different environments that prevail elsewhere. A "stay-at-home" allele should confer higher fitness than a "dispersal allele," because it keeps its bearers in the environment to which they are adapted (Balkau and Feldman 1974).

3. The perils of dispersal are great, and the mortality of dispersers is often high.

Despite these disadvantages, virtually all species disperse, and many have elaborate adaptations for doing so. These include the diverse features of fruits and seeds that are clearly adaptations for carriage by animals, wind, or water (van der Pijl 1982); the features of the planktonic larvae of many molluscs, echinoderms, and other marine animals that keep them afloat (Thorson 1950); and hormonally mediated changes in the wing muscles of some insects that are correlated with a dispersal phase in the life cycle (Johnson 1969).

The advantages of dispersal, which can more than outweigh the disadvantages, are many. Among them are:

1. Dispersal makes colonization possible. Leigh Van Valen (1971) stressed that new populations are ordinarily founded by dispersers, thus creating automatic selection for dispersal alleles. Especially if new populations are founded frequently, this advantage can counter selection against dispersal within populations.

2. Dispersal in a sexual organism enables individuals to outbreed and to engage in multiple matings, increasing the number of lineages that inherit their genes.

3. An individual's birthplace is not necessarily the location in which it will have the highest expected fitness. For one thing, local environmental conditions change, and in many cases they are guaranteed to change for the worse: a site in the early stages of succession becomes untenable, as succession proceeds, for "pioneer" species. Likewise, the host of a parasite eventually dies. A perhaps even more ubiquitous advantage of dispersal arises from density regulation (Hamilton and May 1977). In a stable population, only $2/N$ of the N offspring that a pair produces in the course of a generation survive to reproduce. If dispersal, however hazardous, were to increase the probability of survival and reproductive success even slightly beyond this value, it would have a selective advantage.

There is considerable evidence that the advantages of dispersal include escape from deteriorating environments, including those in which the deterioration is caused by population growth. For example, Robert Denno and colleagues (e.g., Denno and Roderick 1992) have shown that in a small insect, the salt-marsh planthopper (*Prokelisia marginata*), females are genetically variable in their capacity to develop into either flightless or long-winged adults. Those competent to develop long wings do so in response to high densities or deterioration of the grasses on which they feed, and can colonize more nutritious plants. Males are always long-winged, presumably due to selection for the ability to find mates. A related species, *P. dolus*, suffers less diminution of reproduction and survival when crowded, and long-winged females develop at higher densities than in *P. marginata*.

Summary

1. The components of fitness (the per capita rate of increase of a genotype, r) are the age-specific values of survival and reproductive output. Natural selection on morphological and other phenotypic characters exists to the extent that they are correlated with these life history traits. Life history traits can be understood as the product of individual selection, not group selection.

2. Were it not for constraints, organisms would evolve indefinitely long potential life spans and indefinitely great fecundity, commencing early in life. Such life histories seldom evolve because of constraints, especially trade-offs such as those between reproduction and survival and between the number and size of offspring.

3. Because age classes differ in their contribution to population growth (r), the effect of changes in survival (l_x) or fecundity (m_x) on fitness depends on the age at which such changes are expressed. The effect of changes in m_x and (after reproduction begins) in l_x on fitness declines with age, so selection for reproduction and survival at advanced ages is weak. Ordinarily, there is no selection for postreproductive survival.

4. Consequently, senescence (physiological aging) evolves. Senescence may be a result of the negative pleiotropic effects on later age classes of genes that have advantageous effects on earlier age classes (a hypothesis supported by evidence), or of high frequencies of deleterious alleles that are expressed only late in life (a hypothesis with equivocal evidence).

5. If reproduction is very costly (in terms of growth or survival), repeated (iteroparous) and/or delayed reproduction may evolve, provided that reproductive success at later ages more than compensates for the loss of fitness incurred by not reproducing earlier. Otherwise a semelparous life history, in which all the organism's resources are allocated to a single reproductive effort early in life, is optimal. Iteroparity is especially likely to evolve if juvenile mortality is high relative to adult mortality and if population density is stable.

6. The optimal clutch size is often less than the maximal potential clutch size because of trade-offs between the number of offspring and their size or survival, or be-

tween the number of offspring and the survival or subsequent reproductive success of the parents.

7. Genotypes that have higher rates of increase (r) under conditions of high population density may have inferior rates of increase when density is low. Therefore, in a stable (high-density) population, life history traits may evolve that lower the intrinsic rate of increase (r_m), the rate of increase expressed at low density.

8. The evolution of age-specific levels of reproductive effort by males is governed by similar principles as in females. Delayed maturation may evolve if larger males are more successful in attracting or competing for mates. Similar principles explain phenomena such as sequential hermaphroditism (sex change with age) and alternative mating behaviors.

9. Dispersal has several advantages and disadvantages, and its evolution may be affected by both individual selection and group selection.

Major References

Three major treatments of the evolution of demographic features, in increasing order of technical and mathematical complexity, are:

Stearns, S. C. 1992. *The evolution of life histories.* Oxford University Press, Oxford.

Roff, D. A. 1992. *The evolution of life histories: Theory and analysis.* Chapman and Hall, New York.

Charlesworth, B. 1994. *Evolution in age-structured populations.* Second edition. Cambridge University Press, Cambridge.

Problems and Discussion Topics

1. Connell (1970) provided the following estimates, for a population of the barnacle *Balanus glandula*, of the *number* of survivors a_x from an initial cohort of 1 million larvae, over the course of 8 years, and the average number of offspring per surviving female:

Age (x) (years)	a_x	l_x (to be calculated)	m_x
0	1,000,000		0
1	62		4600
2	34		8700
3	20		11,600
4	16		12,700
5	11		12,700
6	7		12,700
7	2		12,700
8	2		12,700

(A) From the numbers of survivors (a_x), calculate values of l_x (set $l_0 = 1.0$). Then calculate the replacement rate

(R_0), the average generation time ($T = \sum x l_x m_x / R_0$), and the approximate value of r (from the relationship $r \approx (\ln R_0)/T$). (B) Calculate the value of r if the probability of survival from age 1 to age 2 were increased by 10 percent. (Note: The probability of survival from ages x to $x + 1$ equals a_{x+1}/a_x. Alteration of this probability at age 2 changes all subsequent values of l_x.) Do the same for a 10 percent increase in survival from age 5 to age 6. Do the same for a 10 percent increase in m_3 (only) and in m_6 (only). How does the change in r depend on the age at which l_x or m_x is altered?

2. Life history traits often have lower heritability than morphological traits (Mousseau and Roff 1987). How can this be, if their additive genetic variance (V_A) is generally equal or higher? How might one test the hypothesis that the V_A of life history traits is high because of mutation-selection balance at more loci than in morphological traits?

3. Female parasitoid wasps search for insect hosts in which to lay eggs, and they can often discriminate among individual hosts that are more or less suitable for their offspring. Behavioral ecologists have asked whether or not the wasps' willingness to lay eggs in less suitable hosts varies with age. On the basis of life history theory, what pattern of change would you predict? Does life history theory make any other predictions about animal behavior?

4. When the supply of food is very low, individual organisms frequently have very few offspring or do not reproduce at all. How would you tell whether this is an adaptive "strategy" or a physiological necessity? Review the evidence cited by Culver (1982) that cave-dwelling animal species have low fecundities and begin reproduction at advanced ages, and discuss whether these species have adaptive, "K-selected" life histories or are simply forced by scarcity of food to reproduce at low rates.

5. Many authors have argued that aging (senescence) is a property of somatic tissues, not of the germ line, and therefore that asexually reproducing organisms such as bacteria should not undergo senescence (Williams 1957; Buss 1987; Bell 1988; Rose 1991). Explain why this should be, and find out whether empirical studies support this hypothesis.

6. In some species of boobies, herons, and other birds, one nestling almost inevitably kills one or more of its siblings. This behavior can be interpreted as advantageous to the survivors, which get more food from their parents. Why, then, do the females lay "excess" eggs? Why does such "siblicide" not occur in all species? How would you expect clutch size to evolve if it is advantageous to parents to produce more offspring, but advantageous to each offspring to have fewer siblings? (See Parker and Mock 1987.)

7. Within many species of birds and mammals, clutch size is larger in populations at high latitudes than in populations at low latitudes. Species of lizards and snakes at high latitudes often have smaller clutches, and are more

frequently viviparous (bear live young rather than laying eggs), than low-latitude species (see references in Stearns 1992). What selective factors might be responsible for these patterns?

8. A likely advantage of iteroparity, not stressed in this chapter, is that if juvenile survival is extremely low, an iteroparous genotype may have an advantage because rare favorable conditions for the survival of its offspring may occur during its lifetime. (That is, the iteroparous genotype, unlike a semelparous genotype, doesn't put all its eggs in one basket.) Discuss how this principle may explain the high prevalence of iteroparity among forest trees, and the relatively high incidence of semelparity in roadside weeds.

9. Why is it that some plants have conspicuous adaptations for seed dispersal and others do not?

CHAPTER 20

The Evolution of Behavior

The study of the evolution of behavior is as old as the study of evolution itself. Darwin addressed many problems in behavior not only in *The Origin of Species*, but also in *The Descent of Man* (1871) and *The Expression of the Emotions in Man and Animals* (1872). From the 1930s through the 1950s, the modern study of behavior (ETHOLOGY) was initiated by Konrad Lorenz, Niko Tinbergen, and Karl von Frisch, who studied not only the mechanisms of behavior, but also its adaptive function and phylogenetic development. Since the 1970s, behavioral biologists have used explicit evolutionary models to understand how natural selection has shaped behavior. This approach is often labeled BEHAVIORAL ECOLOGY, and includes the study of social behavior, sometimes termed SOCIOBIOLOGY (Wilson 1975).

Behaviors as Phenotypic Traits

Some behavioral traits are affected by memory and learning, and so are very malleable within an individual's lifetime. This makes it difficult for some beginning students to understand how behavior can evolve. From the viewpoint of evolutionary biology, however, *behavioral traits are like any other class of characters.* They display both genetic and nongenetic variation, they differ among populations and species, and their evolution, including both homology and homoplasy, can be traced by phylogenetic methods. They are subject to evolution by natural selection, and indeed, much of the evolutionary study of behavior addresses the adaptive value of behaviors and the ways in which selection has shaped them.

Variation in Behavioral Traits within Species

For decades, scientists and nonscientists alike have argued about whether the behavior of humans and other animals is "instinctive" or "learned"—whether it is due to "nature" or "nurture." INSTINCTS have been defined as behaviors that appear in fully functional form the first time they are per-

formed (Alcock 1998); they are usually supposed to be genetically determined and "hard-wired." LEARNING is expressed as modification of an individual's behavior in response to specific experiences. An individual's behavior may also be modified by many other factors, such as age, physiological condition, and environmental conditions.

The dichotomy between instinct and learning or between nature and nurture is oversimplified. It is more useful to think of the magnitude of phenotypic variation (variance) and the proportions of its genetic and nongenetic ("environmental") components (see Chapters 9 and 14). Thus the expression of a behavioral trait may vary among genotypes, and within a genotype it may vary due to variation in previous experience (learning) and other factors that contribute to "environmental" (nongenetic) variance. Genotypes may vary in the extent to which they express different behaviors in response to learning or other factors. Evolutionary factors such as natural selection determine which genotypes—i.e., which reaction norms—exist in a population. A genetic component of variation within and among populations and species has been demonstrated for many behavioral traits (Ehrman and Parsons 1976; Mousseau and Roff 1987; Boake 1994; see Chapters 9, 14, 15).

Differences among Species

Many, if not most, differences in species-typical behaviors have a largely genetic basis (see Chapters 14 and 15). In many cases, there is little or no opportunity for learning: monarch butterflies, for example, migrate from northeastern North America to Mexico only once, in the company of equally naive consorts.

In some cases, however, traits spread through populations as individuals learn from others, so that the trait is CULTURALLY INHERITED. For example, some species of "higher" passerine birds (oscines) do not acquire species-typical song unless they hear it when young. Populations of white-crowned sparrows (*Zonotrichia leucophrys*) in western North America have somewhat different songs as a

FIGURE 20.1 Song dialects of white-crowned sparrows from two localities near San Francisco, California. The songs of three males from each locality are shown as sound spectrographs, in which frequency is displayed against time. (After Alcock 1998; based on sonograms by P. Marler.)

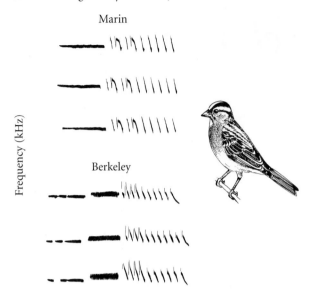

Marin

Berkeley

Frequency (kHz)

consequence of young birds' learning songs from their fathers and other adult neighbors (Figure 20.1).

At least some behaviors of all animal species can be modified by experience. But as anyone knows who has tried to teach a cat to retrieve or perform other acts that a dog would readily learn, *species differ greatly in which behaviors can be modified, and to what extent.* In western North America, male marsh wrens (*Cistothorus palustris*) commonly have a repertoire of more than twice as many song types, learned from other males, as marsh wrens in the eastern part of the continent. Young western wrens in captivity, exposed to tape recordings of 200 song types, learned about 100 of them, compared with about 40 in the case of eastern birds (Kroodsma and Canady 1985). Moreover, the song control nuclei in the brain are larger in western birds. In so-called "lower" passeriform birds (suboscines), such as tyrant flycatchers, learning apparently plays little role in the development of species-typical songs (Kroodsma 1984).

If genotypes differ in the degree to which a behavior can be modified by experience, the extent to which a behavior is "innate" or "learned" can evolve by natural selection. The extent to which any particular behavior of an animal species is learned frequently appears to be adaptive, conforming to the species' ecology and way of life (Johnston 1982; Kamil 1994). Royal terns (*Sterna maxima*) nest close together, and both their eggs and young may be easily displaced into other terns' nests. Herring gulls (*Larus argentatus*) nest farther apart, and their eggs are unlikely to be displaced, but the chicks can wander into other nests. Another gull, the kittiwake (*Rissa tridactyla*), nests on cliff ledges, and neither eggs nor chicks can move from one ledge to another (Figure 20.2). Royal terns learn to distin-

guish their own eggs; herring gulls can recognize their own chicks, but not their own eggs; and kittiwakes recognize their nest sites, but neither their eggs nor their chicks (Shettleworth 1984). If species are selected to learn only those specific tasks for which learning is advantageous, then learning ability will be highly dependent on the context in which the species lineage has evolved. Clark's nutcracker (*Nucifraga columbiana*), a relative of the jays, stores seeds in thousands of locations. Not surprisingly, this nutcracker has a far greater ability to learn and remember locations than do pigeons, which do not cache food (Kamil 1994).

Phylogenetic Studies of Behavior

Recent studies have used modern methods of phylogenetic analysis (see Chapter 5) to elucidate the evolutionary history of behavioral traits (Wenzel 1992). The preferred method is to estimate phylogenetic relationships from nonbehavioral data (such as morphological or molecular characters) and then "map" the behavioral character states on this phylogenetic tree, using the principle of parsimony.

FIGURE 20.2 (A) A pair of herring gulls at their nest on a sand dune. (B) Black-legged kittiwakes nest on cliff ledges. Unlike herring gulls, kittiwakes do not recognize their offspring, which cannot wander from one nest to another. (A by Tom Vezo; B courtesy of John Alcock.)

(A)

(B)

An example is Richard Prum's (1990) phylogenetic analysis of the extraordinary courtship displays of male manakins (Pipridae), small fruit-eating tropical birds. The brightly colored males perform these displays on LEKS: sites at which males aggregate, each performing on a small defended territory. Females visit the lek to mate, then rear the young by themselves. Some species have extraordinarily complex displays: The courtship of *Corapipo gutturalis,* for instance, includes a rapid, buzzing flight above the forest canopy, then a descent to a fallen log, accompanied by a complex call and snapping sounds produced by the wing feathers. On the log, the male jumps from one place to another, doing an about-face in midair, walks backward on the log while rapidly opening and closing his wings, and assumes various stationary postures such as holding his chin down against the log while holding his tail erect (Figure 20.3).

Prum inferred a phylogeny of manakins from data on the anatomy of the syrinx (vocal apparatus). He then distinguished 44 behavioral characters that occur in the courtship repertoire of one or more species, and mapped the behavioral characters onto the phylogenetic tree derived from morphology (Figure 20.4), thereby tracing the probable historical sequence in which behavioral elements have been added to or deleted from the courtship displays of various lineages. Behavioral elements have been added to the courtship display sequentially during the evolution of many of the lineages (e.g., in the ancestry of *Corapipo gutturalis* and *Pipra filicauda*), and in only a few cases have elements been lost from a formerly more complex repertoire. Thus there has been a trend toward increasingly complex courtship behavior, probably driven by sexual selection (see below).

Prum describes, further, how complex behavioral elements have often been derived by elaboration of more primitive elements and by adding novel elements to an ancestral display sequence. In the *Corapipo-Masius-Ilicura* clade, for example, the "about-face log jump" (character 31) has been combined with other movements, such as the chin-down posture (3) and a bill-pointing posture (1). Finally, Prum makes the interesting point that prominently displayed *plumage traits have generally evolved after the be-* *havior in which they are featured.* For example, two display elements that evolved in the common ancestor of a group of three species are combined into a single "wing-shiver twist" in one of these species, *P. filicauda,* in which the male swishes his tail feathers across the female's face. The central tail feathers of *P. filicauda* are lengthened and highly modified—a morphological change that followed the evolution of the behavioral elements in which the feathers are displayed. This example bears on the important question of how behavior and morphology evolve in relation to each other, which we will discuss below.

Studying the Adaptive Value of Behaviors

Much contemporary study of behavior consists of developing and testing predictive hypotheses about the adaptive value of behaviors in order to understand how they are selectively advantageous. Some general principles will serve as a useful prelude to our treatment of some of these hypotheses.

Teleological and anthropomorphic descriptions of behavior
In Chapter 12, we discussed the problem of teleology and the nonexistence of purpose in evolution. To an even greater extent than most other features, *behaviors appear purposive:* male birds sing "in order to" attract mates, and so forth. It is difficult, for biologists as for anyone else, to discuss animal behavior without using anthropomorphic terms that connote consciousness and purpose. Thus, you will encounter in the evolutionary literature phrases such as "knowledge of an opponent's strength affects an animal's decision whether or not to escalate a conflict," and references to the evolution of selfish behavior, deceit, cheating, and altruism.

Biologists, however, do not really ascribe consciousness, forethought, or purpose to the behavior of individual animals, much less to the evolutionary processes that have shaped their behavior. The anthropomorphic language we use is shorthand for much more cumbersome ways of describing behavior and the agents of natural selection that have

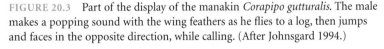

FIGURE 20.3   Part of the display of the manakin *Corapipo gutturalis.* The male makes a popping sound with the wing feathers as he flies to a log, then jumps and faces in the opposite direction, while calling. (After Johnsgard 1994.)

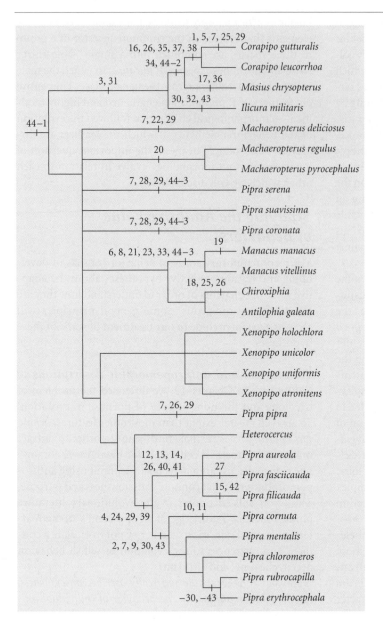

FIGURE 20.4 A phylogeny of part of the manakin family (Pipridae) showing evolutionary changes in 44 behavioral elements observed in the male courtship displays of one or more species. Minus signs indicate loss of an element from the display; all other numbers indicate gains (origins) of display elements. (After Prum 1990.)

molded it. The "decision" to escalate a conflict or not means, more or less, "which of the alternative physiological processes (escalating or not escalating) occurs or is predicted to occur." "Knowledge" of an opponent's strength means "processing by the brain of sensory stimuli emitted by the other animal that are correlated with its strength." Self-consciousness is not implied by the term "selfish behavior," which means "a behavioral act that, on average, causes higher reproductive fitness to the actor than alternative behaviors, which on average would increase the fitness of individuals other than the actor." The language used in the study of animal behavior uses words borrowed from everyday speech, but it is understood that they are analogies to human behaviors and do not imply conscious purpose. This kind of language runs the risk of being misunderstood, especially when biologists talk about "rape" in ducks, "infanticide" in monkeys, or "slavery" in ants, none

of which is intended to imply homology to the human behaviors denoted by these terms.

Levels of selection A second important point is that modern theory of the evolution of adaptive behavior looks for explanations almost exclusively in *natural selection among genes or individual organisms within populations*, rather than invoking selection among populations (group selection). Before the mid-1960s, it was not uncommon for biologists to explain behavior, especially social behavior, by invoking purported benefits to the population or species. Behaviors beneficial to the species, but not to the individual, would have to evolve by group selection, not individual selection (Williams 1966). Recognition that group selection is generally much weaker than individual selection (see Chapter 12) has led to the formulation of new models that account for ritualized displays and other behaviors in terms of selection within populations.

Methods of study Like the study of other kinds of adaptations (see Chapters 17 and 19), much modern evolutionary analysis of behavior consists of formulating and testing hypotheses about what kinds of character states might be expected to evolve under various conditions. The methods of *testing* these hypotheses include the following:

1. Comparing naturally occurring behavioral variants *within* species with respect to reproductive success or performance measures that are thought to be correlated with reproductive success.
2. Performing similar comparisons of artificially produced variants (e.g., observing responses of females to males with altered plumage or to computer-altered male songs).
3. Quantitatively comparing individuals' behaviors with the values predicted by quantitative theory.
4. Using the comparative method (see Chapters 5 and 17) to correlate the behavioral traits of different species with features of their environment.

Many hypotheses about the adaptive evolution of behavior are phenotypic models, which, like phenotypic models of life history evolution (see Chapter 19), assume that genetic variation has been plentiful enough to allow traits to evolve nearly to the *optimum* that natural selection dictates. The theoretical optimum is determined by the balance between the *benefit* and the *cost* of the behavior—between its advantages and disadvantages.

Optimization theory has been criticized (e.g., Gould and Lewontin 1979; Pierce and Ollason 1987) on the grounds (among others) that traits need not be optimal, or even adaptive (see Chapter 12). Defenders of optimization theory (e.g., Maynard Smith 1978b) argue that the theory does not really assume that all characteristics are optimal, but is simply a way of developing hypotheses. By comparing an organism's behavior against an "optimal design," it should be possible to identify not only the selective factors that shape behaviors, but also the constraints or evolutionary time lags that may result in suboptimal behaviors.

In the following pages, we describe a few of the many kinds of behavior that have been studied by evolutionary biologists. We begin with a simple example in which the source of selection is the external *ecological environment*, and proceed to cases in which the source of selection on an individual's behavior is *other members of the same species*.

Optimal Foraging Theory

OPTIMAL FORAGING THEORY is one of the most highly elaborated optimality theories. It treats questions such as which potential food items an animal may be expected to choose,

how its choices should be affected by risks of predation or competition, and when it should stop foraging in one patch and search for a more profitable one (Pyke 1984; Stephens and Krebs 1989).

We will describe only the simplest model of OPTIMAL DIET CHOICE, in which an individual predator is faced with two or more kinds of potential prey, the supply of which is not affected by competing individuals. The model assumes that the fitness of a forager increases linearly with its total intake of calories or some other measure of food value; that the time required to handle a food item detracts from the time available for searching; that prey are encountered sequentially and at random; and that the predator can assess the profitability and density of different prey types, perhaps by learning from experience, and adjust its behavior accordingly. It must be emphasized that this early model, described here only to illustrate the conceptual approach, has been largely superseded by newer, more elaborate models (Stephens and Krebs 1989).

For simplicity, let us assume that there are two types of prey (1 and 2) with different densities, so that they are encountered at rates λ_1 and λ_2 items per minute during T_s minutes of searching time; that these types yield E_1 and E_2 calories per item, respectively; and that once it captures a prey item, the predator must spend h_1 minutes handling an item of type 1 before it resumes searching, and h_2 minutes handling type 2. The prey items may be ranked in terms of profitability, measured as E/h, their caloric yield per unit of handling time. The model (Box 20.A) predicts that it should be optimal to eat only the more profitable prey type, if it is sufficiently abundant. The animal should eat the less profitable type only if the density of the more profitable type is insufficient to meet the predator's food requirement. Moreover, the less profitable prey should either be eaten whenever encountered or not at all (i.e., there should not be "partial preferences").

One of the first of many tests of this model was J. D. Goss-Custard's (1977) study of the redshank (*Tringa totanus*), a sandpiper that feeds on polychaete worms in mudflats at low tide. Goss-Custard recorded the frequency with which these birds ingested worms of different sizes, and independently estimated the worms' densities. The redshanks preferred larger, more profitable worms, and the higher the frequency with which they took large worms (and the higher their density), the less often they took small ones. However, the density of small worms was not correlated with the rate at which the birds took them—just as the model predicts. Goss-Custard concluded that the birds adaptively select the size of worms that maximizes the biomass ingested per unit of time. However, the birds did take some small worms when large worms were abundant, although at a lower rate. Thus they displayed a "partial preference," contrary to the prediction of the model. This has been generally true in studies of foraging in birds, bees, and other animals, and is one reason that more elaborate models have been developed.

BOX 20.A

A Model of Optimal Diet Choice

Prey items of types 1 and 2 yield a reward in calories of E_1 and E_2 per item, respectively, are encountered at rates λ_1 and λ_2, and require handling times h_1 and h_2 once captured (Charnov 1976). A nonselective predator (a generalist) foraging for time T_s will obtain $E = \lambda_1 T_s E_1 + \lambda_2 T_s E_2$ calories if it does not reject any prey, and the total time expended will be $T = (T_s) + (\lambda_1 T_s h_1 + \lambda_2 T_s h_2)$, where the terms in parentheses represent total search time and handling time respectively. Thus the average rate of intake will be

$$\frac{E}{T} = \frac{\lambda_1 E_1 + \lambda_2 E_2}{1 + \lambda_1 h_1 + \lambda_2 h_2}$$

Suppose prey type 1 is more profitable than type 2 in that it yields more calories per unit of handling time ($E_1/h_1 > E_2/h_2$). Then the optimal diet should be only type 1 if

$$\frac{\lambda_1 E_1}{1 + \lambda_1 h_1} > \frac{\lambda_1 E_1 + \lambda_2 E_2}{1 + \lambda_1 h_1 + \lambda_2 h_2}$$

i.e., if the average rate of caloric intake is greater for a specialist on type 1 than for a generalist. With a bit of algebra, this reduces to

$$\frac{\lambda_1 E_1}{1 + \lambda_1 h_1} > \frac{E_2}{h_2}$$

so specialization is favored if the specialist's average rate of intake is greater than the calories gained per unit of time spent handling a prey item of the less profitable type. Note that the abundance of the more profitable type (λ_1) affects whether or not a predator should specialize on it, whereas the abundance of the less profitable type does not (λ_2 has dropped out of the inequality).

Evolutionarily Stable Strategies

The effect of individuals' behavior on their fitness often depends on what other members of the population are doing. If members of a population compete for food, for instance, and a majority of the population is feeding on a superior food type, it will be to an individual's advantage to search for a less utilized food type, even if it has lower nutritional value; its greater quantity can compensate for its lower quality (thus maximizing E/T in the previous model of optimal foraging). As more and more individuals switch to this food, however, its profitability declines. In an optimal foraging model, an equilibrium is reached when the two food types yield equal benefits and the number of individuals adopting each behavior is such that their average fitness is equal. The population will display an "ideal free distribution" (Fretwell 1972) of the kind that can be observed on highways during rush hour, when so many drivers enter the fast lane that the average speed in all lanes becomes equally low.

In this and many other instances, the fitness associated with a particular trait is FREQUENCY-DEPENDENT: it depends on the traits of at least some other members of the population. The optimization theory for frequency-dependent traits uses the concept of an **evolutionarily stable strategy** (ESS). An ESS has been defined by John Maynard Smith (1982), one of the originators of the concept, as "*a strategy such that, if all the members of a population adopt it, then no mutant strategy could invade under the influence of natural selection*"—that is, it cannot be replaced by any other phenotype under the prevailing conditions. (If ecological conditions should change, however, they may determine a different optimal behavior, or ESS.) A strategy may be PURE, meaning that an individual always has the same phenotype, or MIXED, meaning that an individual's

phenotype varies over time, as is often the case with behavior. A mixed strategy is a description of an individual's variable phenotype, not a genetic polymorphism. (See Maynard Smith 1982 and Parker 1984 for more extensive introductions to ESS theory.)

ESS models frequently describe interactions between two individuals, each of which has one of two or more phenotypes (strategies). For each possible pairwise combination of strategies, an individual has a different *payoff*: the increment or decrement of fitness that it receives. The payoff depends not only on the individual's own phenotype, but also on that of the individual it interacts with. Among the set of hypothesized strategies, the ESS is the one that cannot be replaced by any other in the set under given environmental conditions.

One of the earliest applications of ESS theory was the analysis of conflicts between animals, such as males attempting to obtain a territory or a mate. In many species, aggressive encounters consist largely of "ritualized" displays (Figure 20.5), seldom escalating into physical conflicts that could cause injury or death. Although earlier authors assumed that ritualization had evolved for the good of the species, an ESS model shows that it can be *individually* advantageous not to escalate conflicts to that point.

Assume first that there are two possible pure strategies: "Hawk," which escalates either until it is injured or until the opponent retreats, and "Dove," which retreats as soon as its opponent escalates the conflict (Box 20.B). The winner of a conflict gains an increase in fitness of some amount V, on average, but if an individual is injured, its fitness is reduced by an amount C. The ESS model shows that Dove is never an ESS, because a Hawk genotype can always increase in a population of Doves. Hawk is an ESS if $V > C$, i.e., if the fitness gained on average in a contest with another Hawk is greater than the cost of injury. But

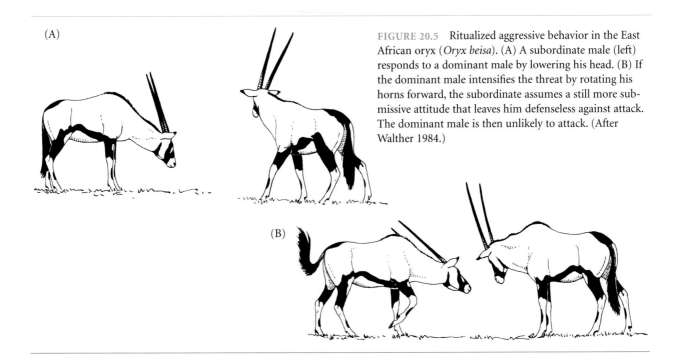

(A)

(B)

FIGURE 20.5 Ritualized aggressive behavior in the East African oryx (*Oryx beisa*). (A) A subordinate male (left) responds to a dominant male by lowering his head. (B) If the dominant male intensifies the threat by rotating his horns forward, the subordinate assumes a still more submissive attitude that leaves him defenseless against attack. The dominant male is then unlikely to attack. (After Walther 1984.)

what if $C > V$? Assume now that a mixed strategy, F, is possible, such that an individual will adopt Hawk behavior with probability P and Dove behavior with probability $1 - P$. F is an ESS if $P = V/C$. This equation suggests that the frequency or intensity of aggression (P) should be scaled to the value of the resource relative to the cost of injury.

Thus this model predicts that the optimal behavior is variable, and is contingent on conditions such as the value of the resource.

This model can be elaborated in many ways. For instance, contestants may vary in size and strength, introducing asymmetry into the likely outcome (Maynard

 BOX 20.B

ESS Analysis of Animal Conflict

Assume, following Maynard Smith (1982), that there are two strategies (phenotypes) in a population. Hawk (H) escalates until it is injured (with cost in fitness C) or until its opponent retreats (in which case Hawk gains the contested resource and receives a gain in fitness V). Dove (D) retreats when threatened by H, and in that context neither gains nor loses fitness. If two Doves meet, one gains the resource, so the average gain to an individual Dove in such encounters is $V/2$. A Hawk that encounters another Hawk wins with probability 1/2 and loses with probability 1/2, so its average "payoff" is $V/2 - C/2$. Thus we have a "payoff matrix:"

	Opponent	
Payoff to	**H**	**D**
H	$1/2(V - C)$	V
D	0	$V/2$

The fitness of a phenotype is w: if $w(x) > w(y)$ when phenotype y is rare, phenotype x is an ESS. If $E(x,y)$ is the payoff to x in a conflict with y, $E(x,x)$ is the payoff to x in a conflict with x, and so on, then for any two strategies I and J

$$w(I) = w_0 + (1 - p)E(I,I) + pE(I,J)$$

$$w(J) = w_0 + (1 - p)E(J,I) + pE(J,J)$$

where p and $1 - p$ are the frequencies of J and I in the population and w_0 is a "baseline" fitness. If J is a rare mutant and I is an ESS, $w(I) > w(J)$ by definition. From the expressions above, this will be true only if $E(I,I) > E(J,I)$ or if $E(I,I) = E(J,I)$ and $E(I,J) > E(J,J)$.

If Dove (D) were an ESS, I = D, J = H; but since $E(D,D) < E(H,D)$ (that is, $V/2 < V$), D is not an ESS. Hawk (H) is an ESS if $V > C$, because $E(H,H) = (V - C)/2$ is greater than $E(D,H) = 0$ if $V > C$. If, however, $V < C$, neither H nor D is an ESS.

Suppose, however, that $V < C$ and that there is a strategy F that entails playing H with probability P and D with probability $1 - P$. If genotypes vary in P, the fittest will be the one that plays each strategy often enough for its advantage to offset its disadvantage. That is, it must get an equal expected payoff from each randomly played strategy. From the payoff matrix, the expected payoffs from playing H and D with probabilities P and $1 - P$ are respectively $P(V - C)/2 + (1 - P)V$ and $P(0) + (1 - P)(V/2)$. Equating these and solving for P, $P = V/C$. Thus playing "Hawk" with probability V/C is an evolutionarily stable strategy (Maynard Smith 1982).

Smith 1982). In addition to Hawk and Dove (and mixed) strategies, we can postulate an Assessor strategy, in which the individual escalates the conflict if it judges the opponent to be smaller or weaker, but retreats if it judges the opponent to be larger or stronger. In most theoretical cases, the Assessor strategy is an ESS. In accord with these models, many animals do indeed react aggressively or not, depending on their opponent's size or correlated features. For example, male toads (*Bufo bufo*) clasp females for up to several days before their eggs are laid and fertilized. Mounted males are often aggressively displaced by larger males, but not by smaller ones. A male is unlikely to try to displace a mounted male that is larger than himself, or one that emits a deeper-pitched croak when touched—the pitch being correlated with body size (Davies and Halliday 1978).

The toads' croak is an "honest" ASSESSMENT SIGNAL: it imparts reliable information about the size of the toad, and hence its fighting ability or RESOURCE-HOLDING POTENTIAL. Other signals are CONVENTIONAL: they are correlated with dominance in contests, but need not be, a priori (as far as one can tell). Many male birds, for instance, have "badges," such as the black throat of the house sparrow (*Passer domesticus*), in which males with larger black bibs acquire better territories. In contrast to the calls of toads, the pitch of which is physically determined by body size, it is plausible that a mutant "deceptive" strategy could arise in house sparrows whereby males could develop larger bibs without having the strength or resource-holding potential that the badge is supposed to signify. In this case, *the conventional signal would be unstable* in evolutionary time, for it would no longer convey reliable information, and selection could then favor genotypes that ignored it altogether, in which case the signal, having lost its function, would be lost in subsequent evolution because is costly to produce. Thus, most existing signals of resource-holding potential are probably honest assessment signals (Grafen 1990; Johnstone and Norris 1993).

ESS models have been used to analyze many problems, not only in animal behavior but also in sex ratio and sex allocation, life histories, and other traits such as growth patterns in plants.

Sexual Selection

The Concept of Sexual Selection

Darwin devoted more than half of *The Descent of Man, and Selection in Relation to Sex* (1871) to one of his most important concepts, sexual selection. Sexual selection has been defined as "*selection that arises from differences in mating success (number of mates that bear or sire progeny over some standardized time interval)*" (S. J. Arnold 1994), and has been comprehensively reviewed by Andersson (1994).

Sexual selection was Darwin's solution to the problem of why conspicuous traits such as the bright colors, horns, and displays of males of many species have evolved. He proposed two forms of sexual selection: *contests between males* for access to females, (sometimes called "intrasexual selection") and *female choice* (or "preference") of some male phenotypes over others ("intersexual selection"). These remain the most commonly cited bases for sexual selection (Table 20.1).

We will treat sexual selection as a kind of natural selection, and for convenience use *ecological selection* to refer to forms of selection other than sexual selection. It is important to note that sexually dimorphic characters, including those associated with reproduction, can evolve for many reasons beside sexual selection, the critical feature of which is competition among individuals for reproductive success. Thus the ability of a male moth to find a female by responding to her pheromone has presumably evolved by ecological, not sexual, selection, since it is necessary for his

Table 20.1 **Mechanisms of competition for mates and characters likely to be favored**

MECHANISM	CHARACTERS FAVORED
1. Same-sex contests	Traits improving success in confrontation (e.g., large size, strength, weapons, threat signals); avoidance of contests with superior rivals
2. Mate preference by opposite sex	Attractive and stimulatory features; offering of food, territory, or other resources that improve mate's reproductive success
3. Scrambles	Early search and rapid location of mates; well-developed sensory and locomotory organs
4. Endurance rivalry	Ability to remain reproductively active during much of season
5. Sperm competition	Ability to displace rival sperm; production of abundant sperm; mate guarding or other ways of preventing rivals from copulating with mate

Source: After Andersson (1994).

reproductive success whether he must compete with other males or not. However, features that evolved by ecological selection are often elaborated or modified by sexual selection. The penis of damselflies, for example, may have originated by ecological selection to enable internal fertilization, but in some species it has spines that are used for removing other males' sperm from the female's reproductive tract, enhancing the likelihood that the last male to mate with her will father her offspring (Figure 20.6; Waage 1986).

Sexual selection exists because females produce few large gametes and males produce many small gametes (see Chapter 21). *This creates an automatic conflict* between the reproductive "strategies" of the sexes: a male can mate with many females, and often suffers little reduction in fitness if he should mate with an inappropriate female,

FIGURE 20.6 An elaborate mechanism for improving a male's likelihood of paternity: the penis of the black-winged damselfly *Calopteryx*. (A) The end of the penis includes a lateral horn at the top. (B) Close-up of the lateral horn, showing spinelike hairs and a clump of a rival's sperm. (A and B courtesy of J. Waage.)

(A)

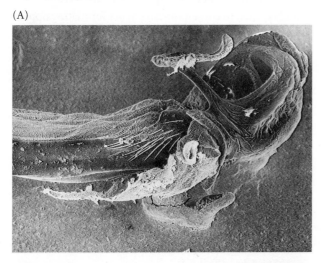

(B)

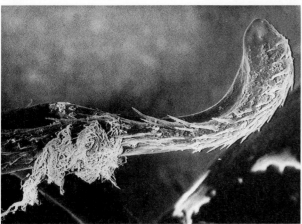

FIGURE 20.7 Variance in reproductive success of males versus females. (A) The male elephant seal defends a harem of smaller females against other males. (B) A plot of the number of offspring produced by each of 140 elephant seals over the course of several breeding seasons. Fewer than 10 percent of the males sired nearly all the pups, whereas most females weaned at least one offspring. (A by Burney LeBoeuf; B after Gould and Gould 1989, based on data of B. J. LeBoeuf and J. Reiter.)

(A)

(B)

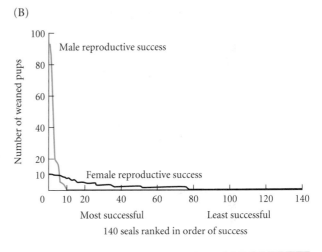

whereas all a female's eggs can be fertilized by a single male (usually), and her fitness can be significantly lowered by inappropriate matings (Trivers 1972). Commonly, then, *females are a limiting resource for males*, which compete for mates, but males are not a limiting resource for females. Moreover, females, because of the greater cost of error they incur, may be expected to resist mating unless presented with the "right" stimuli. Because a male is capable of multiple matings, *variation* (variance) *in mating success is generally greater among males than among females* ("Bateman's principle": Figure 20.7) and indeed is a measure of the intensity of sexual selection (Wade and Arnold 1980).

In some species, however, males are a limiting resource for females. This is illustrated by cases of so-called *sex role reversal* in some birds and fishes, in which the male cares for the young. Female phalaropes (Color Plate 11) can lay more eggs than a male can tend; they are more brightly colored than males and compete with one another for mates (Oring 1986). During mating, a male katydid (Tettigoniidae) provides a female not only with sperm, but also with a large, nutritious spermatophore, which the female eats. The nutrients in the spermatophore enable her to lay more eggs. The sex-role behavior of several species of these grasshoppers is situation-dependent (Gwynne 1990). When high-quality food is abundant, males can rapidly develop spermatophores, and they compete for access to females—the usual pattern in animals. When food is sparse or low in quality, however, females attempt to obtain spermatophores by mating with more males, and compete among themselves for males, which become more choosy in selecting mates.

Sexual Selection by Contests

In many animals, males engage in contests that determine which will gain access to females or to resources (such as territories) to which females are attracted. Males often compete through visual displays of bright colors or other ornaments, many of which make a male look larger. The males of many mammals possess weapons such as horns or tusks that can inflict injury (Figure 20.8).

The role of visual and vocal signals in male competition has been experimentally studied in many birds, including the red-winged blackbird (*Agelaius phoeniceus;* reviewed by Andersson 1994). The territorial display of the male blackbird includes several vocalizations and the exposure of a bright red shoulder patch, or epaulet. Males that were experimentally muted were likely to lose their territories to intruders; however, intruders were deterred if tape recordings of male song were played on a territory whose resident has been removed. More extensive song repertoires (a greater number of song types) were more effective in this respect. When the epaulets of territorial males were painted over to varying extents, those with the largest epaulets were avoided by potential trespassers. When mounted specimens with similarly altered epaulets were placed in occupied territories, they sustained attacks by the resident males in proportion to their epaulet size (Figure 20.9). The territorial males appear to respond to the epaulet as if it were a signal of an impending hostile takeover.

In sexual selection by male contest, directional selection for greater size, weaponry, or display features can, theoretically, cause *runaway sexual selection* of ever more extreme traits (West-Eberhard 1983). Such "escalation" can be limited by opposing ecological selection if the "cost" of larger size or weaponry becomes sufficiently great. Nevertheless, the equilibrium value of the trait is likely to be greater than if only ecological selection were operating. As Darwin noted, the muted coloration and lack of exaggerated display characters in both the females and nonbreeding males of many species implies that these features of breeding males are ecologically disadvantageous.

Paternity Insurance and Sperm Competition

Closely related to contests among males for mating opportunities are the numerous ways in which a male may reduce the likelihood that other males' sperm will fertilize his mates' eggs (Parker 1970; Thornhill and Alcock 1983; Birkhead and Hunter 1990; Birkhead and Møller 1992). Common tactics include male defense of females within his territory and mate guarding, as in many frogs, crustaceans, and insects, in which the male may remain mounted on the female for as long as she is producing fertilizable eggs. In some species of *Drosophila*, snakes, and other animals with

FIGURE 20.8 Some armaments of male mammals used in contests over mates: (A) The antlers of red deer (*Cervus elaphus*). (B) The extraordinary canine teeth of the babirusa (*Babyrousa babyrussa*), a tropical Asian pig. (A courtesy of T. Clutton-Brock; B after Wallace 1869.)

(A)

(B)

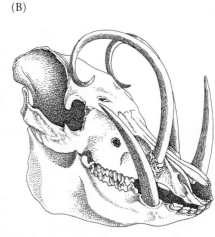

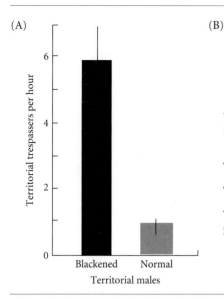

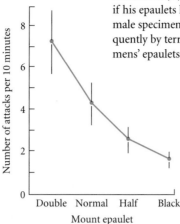

(A)

(B)

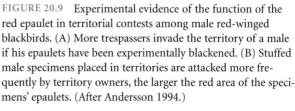

FIGURE 20.9 Experimental evidence of the function of the red epaulet in territorial contests among male red-winged blackbirds. (A) More trespassers invade the territory of a male if his epaulets have been experimentally blackened. (B) Stuffed male specimens placed in territories are attacked more frequently by territory owners, the larger the red area of the specimens' epaulets. (After Andersson 1994.)

internal fertilization, mating males deposit a "copulatory plug" in the female's vagina that prevents further mating, at least for a while.

This brings us to a subtle but probably very important form of sexual selection, SPERM COMPETITION, which occurs when the sperm of two or more males have the opportunity to fertilize a female's eggs (Parker 1970; Birkhead and Hunter 1990). In some such cases, a male enhances his chances of fertilization simply by *producing more sperm.* This explains why some polygamous species of primates have larger testes, relative to body size, than monogamous species (see Figure 24 in Chapter 12).

A common phenomenon in insects and some other animals with internal fertilization is SPERM PRECEDENCE, whereby most of a female's eggs are fertilized by the sperm of only one of the males with which she has mated—usually the last male. By mating female *Drosophila melanogaster* sequentially with wild males and with males of a genetically marked strain, and then scoring progeny for the genetic markers, Andrew Clark and colleagues (1994) showed that wild males vary genetically in sperm precedence—both in their ability to "displace" the sperm of earlier-mating males of the genetically marked stock, and in the resistance of their sperm to displacement by a later-mating genetically marked male. There was no correlation between displacement ability and resistance among the different wild lines, which moreover were differently correlated with different alleles at several accessory gland protein loci. These loci have diverged among *Drosophila* species at a high rate (Thomas and Singh 1992), perhaps due to runaway sexual selection or sperm precedence.

Sexual Selection by Mate Choice

The evolution of sexually selected traits (usually in the male) by mate choice (usually by the female) presents some of the most difficult and controversial problems in evolutionary biology (Kirkpatrick and Ryan 1991, Maynard Smith 1991, Andersson 1994).

Examples of female choice Females of many species mate preferentially with males that have larger, more intense, or more exaggerated characters such as color patterns, ornaments, vocalizations, or display behaviors. In Chapter 12, for instance, we described Malte Andersson's (1982) experiment with the long-tailed widowbird, in which males with experimentally shortened or lengthened tail feathers attracted fewer or more mates respectively. By presenting females of the Central American túngara frog (see Figure 9 in Chapter 16) with a choice between tape recordings of male calls with various components added or deleted, Michael Ryan (1985) showed that females prefer calls that include not only a "whine," but also a terminal "chuck" sound. Males do not always produce chucks, apparently because predatory bats (*Trachops cirrhosus*) are also more attracted to calls that include chucks. This is an example of a *conflict* between *ecological selection* and *sexual selection.* Female choice may also have costs, such as the time spent searching for acceptable mates (Andersson 1994).

A striking aspect of sexually selected traits is how extraordinarily numerous, and seemingly arbitrary, these traits may be, and how much they may vary among related species. Closely related species of hummingbirds, manakins, birds of paradise, cichlid fishes, and many other groups show astonishing differences in colors, ornaments, and displays that are thought to have evolved by sexual selection (cf. Figure 13 in Chapter 15 and Figure 3 in this chapter). Why do these traits—and female preference for them—exist?

The problem and some possible solutions The problem is, *why should females have a preference for exaggerated male traits,* especially for features that seem so arbitrary, and even dangerous for the males that bear them?

An early hypothesis was that females choose males with distinctive characters to avoid mating with other species and producing unfit hybrid progeny. However, we have seen that there is little evidence for this hypothesis (see Chapter 16).

Several other hypotheses for female preferences have been proposed, which have been the subject of much study and debate. These hypotheses can be divided into three categories: those relating to (1) *direct* and (2) *indirect benefits* to choosy females, and to (3) *sensory bias*, preference that is a by-product of a sensory system that has been selected in other contexts. The category of indirect benefits includes several distinct hypotheses.

Direct benefits of mate choice

The least controversial hypothesis applies to species in which the male provides a direct benefit to the female or her offspring, so that the female's choice of certain males over others may improve her immediate survival chances or reproductive success. Under these circumstances there is selection pressure on females to recognize males that are superior providers by some feature that is correlated with their ability to provide. Once this capacity has evolved in females, their mate choices select for males with the distinctive, correlated character. Among the benefits a male may provide are nutrition, a superior territory with resources for rearing offspring, or parental care.

For example, the coloration of male house finches (*Carpodacus mexicanus*) varies from orange to bright red. The variation is partly heritable and partly affected by the quality of the male's diet. Females prefer to mate with bright males in both the laboratory and the field. Their preference appears to be directly advantageous, because brighter males bring food to nestlings at a higher rate (Hill 1991).

Indirect benefits of mate choice: runaway sexual selection

The most difficult problem in accounting for the evolution of female preferences is presented by species in which the male provides no direct benefit to either the female or her offspring, but contributes only his genes. After mating, the male has no further association with the female or her offspring.

The first model to account for such preferences was proposed verbally by the pioneering population geneticist R. A. Fisher (1930) and developed mathematically by Peter O'Donald (1962), Russell Lande (1981), and Mark Kirkpatrick (1982b), who all used somewhat different models that arrived at broadly similar conclusions. These models describe what Fisher called a *runaway process*, in which the evolution of a male trait and a female preference, once initiated, becomes a *self-reinforcing* process of further, conceivably unending, change. The model provides an example of how, in theory, *the intrinsic dynamics of genetic change can cause nonadaptive, even maladaptive, evolution.*

Kirkpatrick's two-locus model may be framed, for simplicity, for a haploid organism (Figure 20.10A). Males of genotypes T_1 and T_2 have frequencies t_1 and t_2 respectively. T_2 has a more exaggerated trait, such as a longer tail, that increases its risk of predation so that its viability is $1 - s$ relative to T_1. The locus P affects female preference. Females of genotype P_1 (with frequency p_1) do not discriminate among males, but those of genotype P_2 (with frequency p_2) prefer T_2 males, perhaps because of some quirk of the visual system or the brain. It is assumed that alleles P_1 and P_2 do not affect survival or fecundity, and thus are selectively *neutral*. Although the expression of genes P and T is sex-limited, both sexes carry both genes and transmit them to offspring.

The essence of the runaway selection hypothesis is that since P_2 females prefer long-tailed (T_2) males, they will tend to have long-tailed sons that will have greater than average mating success because they too will be preferred by some females. The sisters of long-tailed males will inherit a preference (P_2) for long-tailed males, since they are the progeny of females with this preference. Linkage disequilibrium therefore develops between the T_2 and P_2 alleles. The male trait and the female preference thus become *genetically correlated*, so that any increase in the frequency (t_2) of the male trait is accompanied by an increase in the frequency of the preference (p_2) through *hitchhiking* (Figure 20.10A).

If allele P_2 increases to some modest frequency, possibly by genetic drift, or possibly because P_2-bearing females are less likely to mate with other species, T_2 males have a slight mating advantage because they are both acceptable to P_1 females and preferred by the still rare P_2 females. Thus t_2 increases slightly. Because of linkage discrimination between T_2 and P_2, P_2 also increases in frequency. *As P_2 increases, T_2 males have a still greater mating advantage because they are preferred by more females; thus the association of the P_2 allele with the increasing T_2 allele can increase P_2 still further.*

Lande's model, in which both the male trait and the female preference are polygenically inherited, is similar, except that new genetic variance due to mutation can allow for indefinite evolution of both characters. Under some theoretical conditions, the process "runs away" toward preference for ever longer tails (Figure 20.10B).

If female choice carries an ecological cost, the runaway process may lead to an equilibrium at which the male trait and the female preference are less extreme. Moreover, a cost of female preference can cause cyclical evolution, in which both preference and trait evolve first in one direction (toward, say, longer tails), then in the other, and then back again over the course of many generations (Iwasa and Pomiankowski 1995). If there exists genetic variation in female responses to each of several or many male traits, different traits or combinations of traits may evolve, depending on initial genetic conditions (Pomiankowski and Iwasa 1993). Thus, runaway sexual selection can follow different paths in different populations, so that *populations may diverge in mate choice and become reproductively isolated.* Thus sexual selection is a powerful potential cause of speciation (see Chapter 16). *Runaway selection of this kind could also explain the extraordinary variety of male ornaments* among different species of hummingbirds and many other groups.

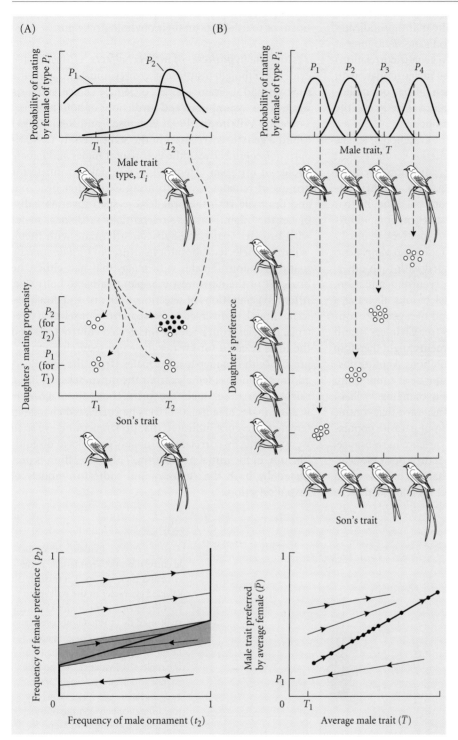

FIGURE 20.10 Models of sexual selection by female choice. (A) A model with two haploid loci. The different preferences of P_1 and P_2 females for T_1 versus T_2 males (top graph) cause a genetic correlation between preference and the male trait (middle graph), because P_2 females, mating preferentially with T_2 males, have P_2T_2 offspring (i.e., long-tailed sons and T_2-preferring daughters). Any change in the frequency (t_2) of P_2 thus causes a corresponding change in p_2. In the lower panel, each point in the space represents a possible population with some pair of frequencies p_2 and t_2, and the vectors show the direction of evolution. When p_2 is low, t_2 declines due to ecological selection, so p_2 declines through hitchhiking. When p_2 is high, sexual selection for T_2 males outweighs ecological selection, so t_2 increases, and p_2 also increases through hitchhiking. Along the solid line, allele frequencies are not changed by selection, but may change by genetic drift. (B) A polygenic model of the runaway process. Progeny of females with different mating preferences (top panel) tend to inherit their mother's preference and their father's trait, establishing a genetic correlation (middle panel). Changes in the average male trait ($\bar{T}$) depend on the average female preference ($\bar{P}$), which in turn evolves by hitchhiking (lower panel). One such evolutionary change is shown by the solid line, on which the dots show the value of $\bar{P}$ and $\bar{T}$ at various points in time. (Lower graph in A after Pomiankowski 1988.)

Indirect benefits of mate choice: indicators of genetic quality

Because females risk substantial loss of fitness by the reproductive failure that may follow from an unsuitable mating, an appealing hypothesis is that females should evolve to choose males with high genetic quality, so that their offspring will inherit "GOOD GENES" and so have a superior prospect of survival and reproduction. Any male trait that is correlated with genetic quality—i.e.,

any INDICATOR of "good genes"—could be used by females as a guide to advantageous matings, so selection would favor a genetic propensity in females to choose mates on this basis.

There are several kinds of indicator (good genes) models (Andersson 1994). In condition-dependent indicator models, for example, males with allele T can develop the indicator trait (e.g., long tail feathers) only if they are in good physiological condition owing to their also carrying allele *B* at another locus. The presence of the trait thus indicates to females that the male has the "good gene" *B*, as long at there is linkage disequilibrium (i.e., a genetic correlation) between alleles *T* and *B*. If P_2 females prefer to mate with *T*-bearing (and *B*-bearing) males, then linkage disequilibrium will develop between the preference allele P_2 and the trait allele *T*, just as in the runaway models.

Indicator models can account for the evolution of female preference for a male ornament *if fitness is heritable*—i.e., if there is additive genetic variance (V_A) in fitness (variation in fitness that is correlated between parents and their offspring) (Pomiankowski 1988). A theoretical difficulty that has stimulated much discussion is that at genetic equilibrium under natural selection, there is no additive variance in fitness (although there may be V_A for individual *components* of fitness that are negatively correlated with each other; see Chapters 14 and 19). However, recurrent mutation, summed over all loci, introduces deleterious alleles at a high rate (see Chapters 10 and 14) and may generate considerable additive variance for fitness (see Chapters 10 and 14). Males with well-developed ornaments might therefore be those that carry few deleterious mutations. Another possibility is that allele frequencies may not be at equilibrium because the environment is continually changing. William Hamilton and Marlene Zuk (1982) suggested that continual change in the genetic composition of parasite populations may maintain additive genetic variance in resistance, and therefore fitness, in populations of their hosts (see Figure 4, Chapter 18). They proposed that this process could result in selection of female preference for males with well-

developed ornaments, which indicate resistance to parasitism or to its effects on their physiological vigor.

Testing the hypotheses of indirect effects Considerable evidence, although insufficient to discriminate between the runaway and indicator models of female choice, supports some of the assumptions and predictions of both models.

Gerald Wilkinson (1993; Wilkinson and Reillo 1994) found a genetic correlation in the stalk-eyed fly *Cyrtodiopsis dalmanni* between the eye span of males and female choice of eye span (Figure 20.11A). Males compete for groups of females by confronting each other head-on, and the male with the broader eye span invariably displaces the other. Females preferentially settle near males with broader spans and subsequently mate with them. Wilkinson selected laboratory cultures of *C. dalmanni* by mating random females with males that had either the broadest or the narrowest eye span relative to body size. After ten generations of selection, both up- and down-selected lines had diverged in eye span from unselected controls. Females from both the unselected control lines and the lines selected for greater male eye span chose to aggregate with broader-headed males (from the up-selected lines), whereas females from the down-selected lines, selected for lesser male eye span, chose the narrower-headed males (Figure 20.11B). The correlated change in female preference indicated that preference and eye span were genetically correlated, probably by linkage disequilibrium, in the initial population. This correlation is predicted by both the runaway and indicator models of sexual selection.

(A)

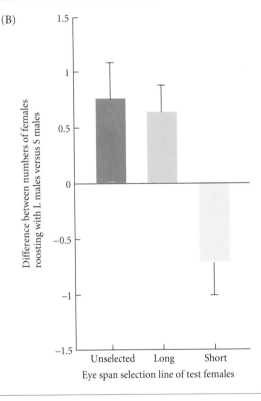

(B)

FIGURE 20.11 Evidence of a genetic correlation between female preference and male eye span, a sexually selected character in the stalk-eyed fly *Cyrtodiopsis dalmanni*. (A) A male stalk-eyed fly (*Cyrtodiopsis whitei*) with three females. The eyes are borne at the end of longer stalks in the male. (B) Evidence of a genetic correlation between female preference and male eye span, shown by the difference between the number of females roosting with L and S males, artificially selected for long (L) and short (S) eye stalks. "Unselected," "long," and "short" refer to females from unselected, L, and S populations, respectively. Females from the "S" population ("short") roosted more often with S than L males, providing evidence of a genetic correlation in preference. (A courtesy of G. S. Wilkinson; B after Wilkinson and Reillo 1994.)

Among the tests of Hamilton and Zuk's parasitism hypothesis is an extensive study by Anders Møller (1994) of the barn swallow (*Hirundo rustica*). The outermost tail feathers of this pair-bonding species are elongated in both sexes, but are about 15 percent longer in the male. Møller found that females prefer longer-tailed males, whether naturally so or experimentally altered. Longer-tailed males obtain mates earlier, have second broods more often, and so have more offspring than shorter-tailed males. The nestling offspring of naturally long-tailed males have fewer blood-feeding mites than the offspring of shorter-tailed males. The difference in mite load apparently has a genetic basis, because the correlation between the mite loads of parents and their offspring holds even for offspring that have been reared by foster parents (Figure 20.12). Møller also showed, by experimentally manipulating mite density, that mite infestation lowers

the weight of young birds, and so probably affects their fitness.

Alas, the swallow study and similar studies do not prove that the good genes hypothesis is the explanation for female choice (Kirkpatrick and Ryan 1991; Andersson 1994). Females may prefer mating with long-tailed, relatively mite-free males not because they have good genes, but because they are less likely to infect the female and her young. Avoidance of infection would be a *direct* benefit to the female, which is a noncontroversial advantage of mate choice, as we have seen. A second hypothesis is equally simple: The development of an individual's ornament is likely to be inhibited by any debilitating factor, including parasitism. Thus genetic variation in resistance to debilitation will be expressed by ornament size (and other traits). This would be true no matter when or why female choice evolved. Even if the species acquired its parasite after the female preference had evolved, there would be a correlation between the male phenotype preferred by females and genes that affect both the male's parasite resistance and the expression of his ornament. So far, no one has devised definitive ways of choosing between the runaway selection hypothesis and the indicator hypothesis.

The dual role of sexually selected traits

Anders Berglund and collaborators (1996) have suggested that the male ornaments preferred by females also serve to establish dominance in contests between males, and first evolved in this context. According to this hypothesis, females gain fitness by choosing highly ornamented males because their sons, inheriting the exaggerated ornament, will enjoy higher reproductive success due to their superiority in competition for mates. Because of the linkage disequilibrium that develops between female preference alleles and male trait alleles (as in the runaway process described earlier), alleles for female preference will be transmitted to more grandchildren than alleles for random mating. We noted earlier that a broad eye span in stalk-eyed flies both establishes dominance in male contests and is preferred by females. Berglund et al. list many other species in which male traits act as both "armaments" and "ornaments" and are favored by both "intrasexual" and "intersexual" selection. To date, this hypothesis has not been explicitly tested.

Pre-existing biases and sensory exploitation

Several authors have suggested that female preference for a male trait may evolve before the male trait does (reviewed by Ryan 1990, 1994; Basolo 1995)—that is, that females may have a pre-existing SENSORY BIAS toward the trait. The preference may have evolved because it was adaptive in another context, or simply because of the organization of the sensory system. Animals frequently show greater responses to SUPERNORMAL STIMULI that are outside the usual range of stimulus intensity (Tinbergen 1951). Moreover, some species display preferences for novel stimuli for reasons that are not understood. Nancy Tyler Burley (personal communication) found that females of two species of grassfinches preferred males fitted with artificial

FIGURE 20.12 Evidence that a genetically variable sexually selected male trait, tail length in the barn swallow, may be correlated with resistance to a parasitic mite. (A) In their parents' nest, the number of mites per nestling is negatively correlated with their father's tail length. (B) Cross-fostered offspring reared in other birds' nests show the same relationship to their father's tail length, suggesting that mite resistance is genetically based. (After Møller 1994.)

(A)

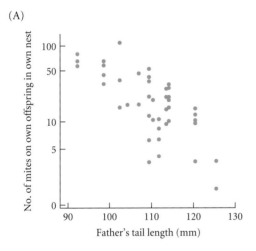

(B)

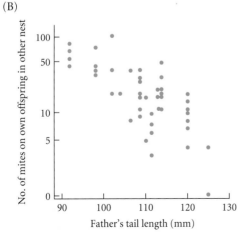

crests, even though none of the many species of grass-finches have crests (Figure 20.13).

In other cases, properties of the sensory system may explain preferences for stimuli beyond the normal range. We noted above that female túngara frogs (*Physalaemus pustulosus*) prefer male calls that include a low-frequency chuck. Part of the inner ear organ of the female is specifically tuned to these low frequencies (Ryan 1990). Using phylogenetic comparisons, Ryan and his collaborators (Ryan 1990; Ryan and Rand 1993a) found that the chuck is included only in the calls of *P. pustulosus* and its sister species. However, the inner ear of females of a more distantly related species, *P. coloradorum*, is also tuned to this low frequency. As predicted, when females of *P. coloradorum* were given a choice between a recording of the normal whinelike call of their own species and the same call to which the chuck of *P. pustulosus* had been appended, they preferred the latter. This finding supports the hypothesis that a sensory system conferring a preference for a low-frequency chuck evolved before the male signal itself.

Alexandra Basolo (1990, 1995) has found similar evidence for a visual signal in fishes of the genus *Xiphophorus*. Some species (swordtails) have a sword, a strikingly colored extension of the lower rays of the caudal fin. Other species of *Xiphophorus* (platyfish) lack this structure. Female swordtails prefer males with longer swords. In each of two swordless species of platyfish and in a swordless member of the sister genus, Basolo attached colored plastic swords, resembling those of a swordtail, to the tails of

some males, and colorless swords to control males. Females of these species, given a choice between a male of each type, generally chose the male with the colored plastic sword, even though the males of their own species lack this ornament.

If a pre-existing female bias is weak, it might simply initiate a process such as runaway sexual selection. If it is a strong, "open-ended" bias, then highly exaggerated male traits may evolve without any further evolution of female preference (Kirkpatrick and Ryan 1991). Future studies, which may well integrate neurobiology and evolutionary biology (Ryan 1994), will no doubt shed more light on this process.

Social Interactions and the Evolution of Cooperation

Because natural selection is based on *individual* advantage, "selfish" traits should increase in frequency if they are heritable. Thus cooperative interactions in which individuals apparently dispense benefits to others, often at a cost to themselves, seem antithetical to evolution by natural selection. Until the 1960s, biologists commonly assumed that cooperation had evolved because it increased the fitness of a population or species by reducing the risk of extinction or increasing the rate at which new populations were founded—that is, by group selection. The modern study of cooperation, for the most part rejecting group selection (Chapter 12), flows largely from two sources. One was the formulation by William Hamilton (1964) of a theory of "altruism" based on interactions among related individuals. The other was the realization, articulated most forcefully by George Williams (1966), that group selection is usually a weaker process than individual selection (Figure 17 in Chapter 12). A mutant "selfish" allele could easily replace the "altruistic" allele at a higher rate than populations proliferate or become extinct. Thus *a population of "altruists" is unstable; it can be invaded by a "selfish" or "cheater" allele that will increase to fixation.*

Theories of Cooperation and Altruism

Many seemingly altruistic traits are behaviors that seem disadvantageous to the individual. Four major classes of hypotheses have been advanced to explain such traits without relying on traditional group selection: MANIPULATION, INDIVIDUAL ADVANTAGE, RECIPROCATION, and KIN SELECTION.

Manipulation   A donor may dispense aid to a recipient because it is being manipulated by the recipient. This is obvious in many interactions between species, especially in parasitism. For instance, nestling brood-parasitic birds (see Chapter 18) successfully solicit care from the foster parents, which respond to the parasites' begging behavior as they do to their own offspring. Young parasitic whydahs (African finches) have markings on the lining of the mouth that mimic those of their hosts' young. Brood parasitism also occurs within species. Presumably the host cannot distinguish its own eggs from the interloper's, and is duped into performing behavior that reduces its own reproductive success.

FIGURE 20.13   Males of two species of grassfinches (zebra finch, *Taeniopygia guttata*, left, and long-tailed finch, *Poephila acuticauda*, right), fitted with feathers serving as artificial crests. Although none of the 120 species of grassfinches has a crest, females of these species preferred males with white artificial crests, compared to normal males. The experiment suggests that preferences for male features may exist before male features evolve, and may initiate the evolution of male traits by sexual selection. (Photograph by K. Klayman, courtesy of N. T. Burley.)

Individual advantage   In many cases, the donor is likely to receive a direct benefit from its cooperative behavior. Perhaps the simplest example is aggregating behavior, as in schools of fish. Countering the cost of such behavior—competition for food—is the benefit of protection from predation owing to safety in numbers. Hamilton (1971) referred to this as the principle of the SELFISH HERD, and pointed out that it will often be advantageous to each individual to be as close as possible to the center of the group, thus using other group members as shields against approaching predators. The effect of this behavior on the part of each individual is to increase the compactness and cohesion of the group. Among other possible advantages of aggregation, each individual may obtain more food by joining a cooperatively hunting group than by foraging alone (Wilson 1975).

Reciprocation   According to the hypothesis of RECIPROCAL ALTRUISM (Trivers 1971), it can be advantageous for individual A to help individual B if B will reciprocate in the future—as is common in human interactions. The theory presupposes repeated interactions between individuals, perhaps because they recognize each other, or because they are associated for some time, as is the case in many interspecific mutualisms. Reciprocal altruism is likely to be an unstable evolutionary strategy because a genotype that cheats by receiving, but not repaying, aid will have higher fitness than a helpful genotype. However, a "tit-for-tat" strategy of reciprocation, in which an individual first acts cooperatively, but thereafter does whatever the other individual does (i.e., help or refuse to help), is stable (is an ESS)as long as the number of encounters between individuals is variable and unpredictable (Axelrod and Hamilton 1981).[*]

Rather few examples of reciprocal altruism within species have been documented in nature. Vampire bats (*Desmodus rotundus*), which feed on mammalian blood, form roosting groups, in which some individuals feed regurgitated blood to other group members that have been unsuccessful in foraging. Blood donors feed not only related individuals, but also unrelated ones, as long as they are familiar members of the same group—and the recipients of blood reciprocate (Wilkinson 1988).

Interactions among Related Individuals

Basic principles   Possibly the most important explanation for altruism is the theory of **inclusive fitness**, a concept discussed by Fisher, Haldane, and several other authors, but developed explicitly in two seminal papers by William Hamilton (1964).

The fundamental principle of inclusive fitness is that the increase or decrease in frequency of an allele is affected not only by the allele's effect on the fitness of the individual bearing it (direct fitness), but also by its effect on the fitness of other individuals that carry copies of the same allele (indirect fitness). Usually these other individuals are the bearer's relatives, or kin, so selection based on inclusive fitness is often called KIN SELECTION (Maynard Smith 1964).[†]

The simplest example of a trait that has evolved by kin selection is parental care. Suppose that females with allele *A* care for their young and thus enhance their survival, whereas females lacking this allele do not. Although caring for offspring may reduce a female's lifetime fecundity or chance of survival (her individual fitness), *A* may nonetheless spread, because the enhanced survival of the copies of *A* inherited by the female's offspring can more than compensate for the reduction of her individual fitness. If *A* caused a female to dispense care to young individuals in the population at random, *it could not increase in frequency*, because the fitness of all genotypes in the population, whether they carried *A* or not, would be equally enhanced on average, so the only difference in fitness among genotypes would be the reduction in fitness (the cost) associated with caring for young. This cost would cause the frequency of *A* to decrease.

The *inclusive fitness* of a genotype is the average fitness individuals of that genotype would have in the absence of social interactions, augmented by certain fractions of the increase and/or decrease in fitness that these individuals cause both in themselves and other individuals by interacting with them. These fractions are the coefficients of relationship. Thus w_i, the inclusive fitness of genotype *i*, may be expressed as

$$w_i = a_i - c_{ii} + \sum r_{ij}b_{ij}$$

In this equation, a_i is the "base" fitness in the absence of interaction and c_{ii} is the deleterious effect of an individual's altruistic behavior on its own fitness. The final term is the increment in fitness (b_{ij}) of individual *j* due to the action of *i*, weighted by the coefficient of relationship between them (r_{ij}). The summation sign indicates that inclusive fitness takes into account all individuals to which *i* directs its ac-

---

[*]If the number of interactions is predictable, the cheater genotype has higher fitness. Suppose A acts first, and is helpful, and there are always four encounters, with actions performed by A, B, A, and finally B. If B refuses to help on encounter 4, helpful behavior by A on encounter 3—i.e., imitating B's helpful behavior in encounter 2—is disadvantageous, so the tit-for-tat strategy doesn't work.

[†]Even without knowledge of Mendelian genetics, Darwin, with his extraordinary insight, anticipated the principle of kin selection. In *The Origin of Species*, he referred to the "neuter" or sterile female workers of ants and other social insects as "a special difficulty, which at first appeared to me insuperable, and actually fatal to the whole theory" of natural selection, "for these neuters often differ widely in instinct and in structure from both the males and fertile females, and yet, from being sterile, they cannot propagate their kind." How, then, can their distinctive features have evolved by selection? Darwin noted that cattle breeders wish the flesh and fat to be marbled together; but since an animal with the desired condition has already been slaughtered, breeders succeed by propagating from the same stock, i.e., individuals related to that animal. Natural selection, likewise, can increase the frequency of a behavioral or morphological trait of sterile worker insects if relatives of individuals that express the trait enjoy enhanced reproductive success. Darwin noted, incidentally, that social insects provided evidence against Lamarck's theory of evolution by inheritance of acquired characteristics: "For peculiar habits, confined to the workers or sterile females, however long they might be followed, could not possibly affect the males and fertile females, which alone leave descendants. I am surprised that no one has hitherto advanced this demonstrative case of neuter insects, against the well-known doctrine of inherited habit, as advanced by Lamarck."

tion. Under most circumstances, a genotype with a higher inclusive fitness than another increases in frequency (Michod 1982).

"Hamilton's rule," which follows from the expression for inclusive fitness, states that *an altruistic trait can increase in frequency if the benefit (b) received by the donor's relatives, weighted by their relationship (r) to the donor, exceeds the cost (c) of the action to the donor's fitness.* That is, altruism spreads if

$$rb > c$$

The coefficient of relationship r usually refers to the fraction of the donor's genes that are identical by descent with any of the recipient's genes (Grafen 1991). For example, at an autosomal locus in a diploid species, half of a mother's gene copies are identical by descent with genes in one of her offspring, since the offspring inherits one of its mother's two gene copies; thus $r = 0.5$ for mother and offspring. For two full siblings, $r = 0.5$ also, because they have a probability of 0.25 of having inherited copies of the same gene from their mother, and the same from their father. (Box 20.C shows the calculation of r for various relationships in both diploid and haplodiploid species.) Thus, if a helpful act were almost sure to cost the donor its life, c, the cost in fitness, would be 1. Nevertheless, the loss of a copy of an "altruism gene" would be more than compensated if the donor were a parent that by so acting increased the number of her surviving offspring by more than two ($b > 2$, $r = 0.5$, $c = 1$ implies $rb > c$), and the same would be true if an individual's self-sacrifice saved more than two siblings from death. The more distantly related the beneficiaries are to the donor, the greater the benefit to them must be for an altruistic behavior to spread.

It may seem, and rightly so, that *from a genetic point of view, altruism that evolves by kin selection is not altruism at all:* it is behavior that is indirectly beneficial to the donor, because it enhances her or his inclusive fitness. Thus we will henceforth use the word *altruism* to mean "apparently altruistic" cooperative behavior.

There are two ways in which beneficial acts can be more commonly dispensed to relatives than to nonkin. Individuals may be able to distinguish kin from nonkin and behave differently toward them. Alternatively, the population may be structured so that interacting individuals are more likely to be related than unrelated. A nestling bird, for example, is more likely to be associated with siblings than with unrelated individuals. Prairie dogs are aggregated into groups (coteries), with members of a coterie more closely related to one another than to members of other coteries (see Chapter 11).

Levels of selection Distinctions among levels of selection (see Chapter 12) can become difficult in theories of the evolution of altruism (Sober 1984). Hamilton's rule and the definition of inclusive fitness are framed in terms of the inclusive fitness of individuals with one or another genotype, and so express evolution by individual selection. Others view kin selection as *genic selection* (Dawkins 1989). According to this view, an allele for altruism achieves higher fitness than an allele for selfishness by increasing the survival and replication of other copies of itself, carried by other bodies.

Still others view kin selection as a special form of group selection (Wade 1979a, 1980b; D. S. Wilson 1980), consisting of differences in the contributions of the groups to the size of the total population. Consider groups, such as families of young birds in nests, that exist only temporarily. Interactions occur within the groups, which then are broken down by mixing together in a single gene pool. If these groups differ in the frequency of an allele for altruism, *A*, such as one that influences a nestling bird to compete less intensely with its siblings for food brought by the parents, the survival of altruists will, on average, be lower than that of nonaltruists (*aa*) that compete more vigorously. *Thus within groups, the frequency of A declines.* But if groups (nests) containing altruists produce more surviving offspring than groups that lack altruists, in which all nestlings compete intensely and most die, then allele *A* may increase in the population as a whole (Figure 20.14). Altruistic behavior is always disadvantageous within groups, even groups of relatives (Williams and Williams 1957), but can increase in frequency because it increases the contribution of altruist-containing groups to the entire population.

Michael Wade (1980b) has demonstrated the effects of group structure in an experiment with the flour beetle *Tribolium castaneum*, in which larvae benefit by cannibalizing eggs (see Chapter 12). In some experimental populations, lar-

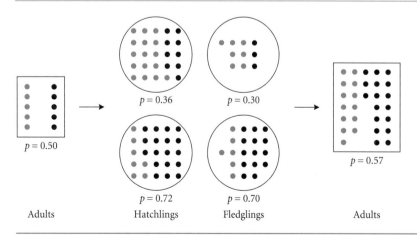

Adults
$p = 0.50$

Hatchlings
$p = 0.36$
$p = 0.72$

Fledglings
$p = 0.30$
$p = 0.70$

Adults
$p = 0.57$

FIGURE 20.14   Kin selection viewed as selection among groups, or "structured demes." A pool of adults (left) breaks into small groups, such as pairs of birds. The frequency of an allele *(p)* for cooperative behavior among nestmates (black circles) varies among families. Within each family group, the frequency of the allele declines due to its cost in individual fitness. However, when the young fledge and re-form a pool of adults, the frequency of the allele will have increased if the productivity of family groups with a high frequency of the allele is greater.

FIGURE 20.15 The difference in average rates of cannibalism between experimental populations of flour beetles (*Tribolium*) in which larvae were offered related eggs (full sibs, $r = 0.50$, or half-sibs, $r = 0.25$) and those in which they were offered unrelated eggs. Both the mating system and the degree of relationship affected the rapidity with which cannibalism declined due to genetic change. (After Wade 1980a.)

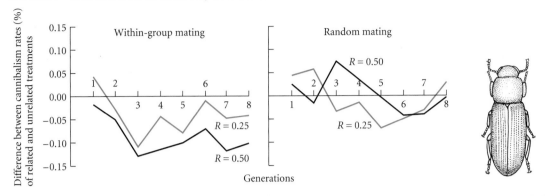

vae were provided with eggs drawn from the population at random, while in other populations, the eggs provided were either full sibs or half-sibs of the larvae. Each generation was bred from both the cannibals and those eggs that had survived cannibalism. For each of the treatments described above, some populations were mixed into randomly mating gene pools in each generation, from which colonists were drawn to found a new set of groups (thus genetic variance among groups was reduced by random mating). In other populations, groups were maintained separately from one generation to another. Over the course of eight generations, the rate of cannibalism declined in those populations in which larvae fed on their relatives (Figure 20.15), and the effect was more pronounced when the larvae were fed full-sib eggs ($r = 0.5$) than half-sib eggs ($r = 0.25$), as the theory of kin selection predicts. However, cannibalism declined only in those populations that were divided into inbreeding subpopulations. Wade interpreted these results to mean that altruism (i.e., refraining from cannibalism) evolved through the differential productivity of kin groups.

Evidence for Evolution by Kin Selection

We will now consider a few of the many phenomena in which kin selection is thought to have played an evolutionary role.

Kin recognition and cannibalism

Many animals behave differently toward individuals that are likely to be related to them than toward others (Hepper 1991). The larvae of several species of frogs and salamanders can develop facultatively into a morphologically distinctive cannibalistic form. One such species is the spadefoot toad *Scaphiopus bombifrons*, the tadpoles of which develop into detritus- and plant-feeding omnivores if they eat these materials early in life, or into carnivores, with large horny beaks and large mouth cavities, if they eat animal prey. In nature, their prey includes conspecific tadpoles. David Pfennig and collaborators (1993) found that omnivores associated more with their siblings than with nonrelatives, whereas carnivores did the opposite. Carnivores also ate

siblings less frequently than they ate unrelated individuals. In another study, Pfennig et al. (1994) found similar results in tiger salamanders (*Ambystoma tigrinum*), which also can develop into cannibals. The cannibalistic salamanders can distinguish different levels of relationship, because they not only ate siblings and cousins less frequently than nonrelatives, but also ate siblings less frequently than cousins!

Cooperative breeding

In at least 220 species of birds, young are reared not only by their parents, but by other individuals as well (Brown 1987, Emlen 1991, Stacey and Koenig 1990) In most species, these helpers are young birds that are helping their parents to rear their siblings. In the Florida scrub jay (*Aphelocoma coerulescens*), they appear to derive a twofold benefit. Young birds do not breed because the suitable habitat is fully occupied by territory holders. The helpers' best alternative is to gain the indirect fitness benefit, by kin selection, of rearing younger siblings. They also gain a direct benefit by extending the family's territory at the expense of neighboring birds, and then inheriting part of it (Woolfenden and Fitzpatrick 1990).

Hans-Ulrich Reyer (1990) conducted an 8-year study of the pied kingfisher (*Ceryle rudis*), an African fish-eating bird that excavates nesting burrows in sandy banks. Females incubate eggs and brood nestlings more than males do, which puts them at greater risk of death from predators or collapse of the burrow, so the sex ratio among breeding-age birds is about 1.6 males per female. Many nests are tended not only by the breeding pair, but also by one or more young, unmated males, which help to defend the nest and feed the young for one or two breeding seasons.

By marking each bird so that their relationships and behavior could be monitored, Reyer found that young male helpers fall into two categories. Primary helpers are immediately accepted by a mated pair, and invest as much effort in nest defense and feeding the young as the breeders do. Secondary helpers are frequently excluded by breeding males, are finally accepted by a breeding pair only after the eggs hatch, and are only about one-fourth as active as the breeders in feeding the young. In almost all cases, secondary helpers are unrelated to the breeders they help, but primary helpers are the sons, from a previous year, of the mated pair they join.

Reyer found that at sites where food is difficult to obtain, the number of surviving offspring is increased by the presence of either kind of helper (Figure 20.16). If a breeding male dies, a secondary helper is likely to become his widow's mate in the next breeding season, but primary helpers never become the mates of their widowed mothers. (Reyer suggests that there is active avoidance of incest.) Finally, the annual survival rate of breeders (57 percent) and primary helpers (54 percent) is lower than that of secondary helpers (74 percent), probably because they work harder and so have a greater risk of predation and physiological stress.

Reyer calculated the inclusive fitness of primary and secondary helpers in their first 2 years, which, together with other data, addresses the question of why a young male should adopt any one of four courses of action: (1) breed; (2) be a primary helper; (3) be a secondary helper; (4) be a "delayer"—i.e., none of the above. Table 20.2 shows the average inclusive fitness for each of these strategies, calculated as the product of (a) the probability that a male can adopt the strategy (e.g., breed); (b) the increment in the number of young successfully fledged as a consequence of the male's contribution (e.g., for a helper, the difference between the number of young fledged from nests that do and do not have helpers, divided by the number of helpers); and (c) the coefficient of relationship between the helper and the young birds referred to in b.

Breeding yields the highest fitness, but it is not an option for most young males, simply because almost all females have (older) mates due to the unbalanced sex ratio. Not helping is the worst strategy, because helpers obtain direct or indirect fitness benefits, or both. Inclusive fitness, consisting of both direct and indirect (via kinship) fitness components, is nearly equal for primary and secondary helpers

(Table 20.2). Primary helpers gain substantial indirect fitness because they rear their siblings or half-sibs. Secondary helpers accrue little indirect fitness, since they are generally unrelated to the young that they help, but their direct fitness is higher because they have higher survival rates and a much greater probability of mating with the female they helped. This case illustrates that an apparently altruistic behavior can be advantageous to the donor, either because it provides a direct benefit (increasing the chance of subsequent mating by secondary helpers) and/or an indirect benefit.

Social insects Of all the seeming altruism to be found in the animal kingdom, surely the most extreme is the behavior of the workers in social insects, which often forego reproduction entirely, sacrifice their lives to defend their colony against enemies, and devote all their energies to rearing other members of the colony.

Species are EUSOCIAL if they possess three traits (Wilson 1971): individuals cooperate in caring for the young; there is reproductive division of labor, with nearly or completely sterile workers helping more fecund individuals; and generations overlap, so that offspring assist their parents. Eusociality is known in one species of mammal, the naked mole-rat (*Heterocephalus glaber*) of Africa (Sherman et al. 1991), in all of the 2200 species of termites (order Isoptera), in many Hymenoptera, and in a few other insects.

Eusociality has evolved independently in at least ten lineages of the order Hymenoptera. The workers in all social Hymenoptera are female. The eusocial Hymenoptera include many species of vespoid wasps, all the ants (except some that have descended from eusocial ancestors and become brood parasites or slave-makers), and a small percentage of the species of bees, which display all gradations from solitary to eusocial (Wilson 1971, Spradbery 1973,

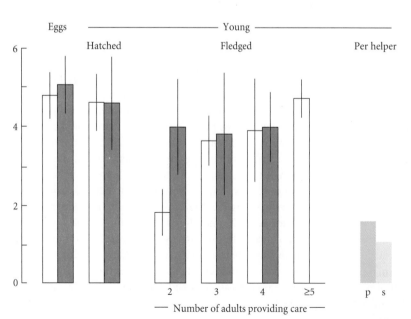

FIGURE 20.16 Survival of young pied kingfishers as a function of the number of adults providing care. The average numbers of eggs, hatchlings, and fledged young per nest are shown for two different localities (open and solid bars). The number of fledged young increases with the number of adults providing care; groups of more than two adults include male helpers. The number of surviving young per helper, over and above the number reared by parents without helpers, is shown at right for primary (p) and secondary (s) helpers. (After Reyer 1990.)

Table 20.2 Direct, indirect, and inclusive fitness values for primary and secondary helpers in pied kingfishers during their first 2 years of life

		GAIN IN FITNESS		
STATUS	YEAR	DIRECT	INDIRECT	INCLUSIVE
First-year breeder	1	0.96	0	0.96
	2	0.80	0	0.80
Total		1.76	0	**1.76**
Primary helper	1	0	0.45	0.45
	2	0.42	0.20	0.62
Total		0.42	0.65	**1.07**
Secondary helper	1	0.04	0	0.04
	2	0.87	0.01	0.87
Total		0.87	0.05	**0.92**
Delayer	1	0	0	0
	2	0.30	0	0.30
Total		0.30	0	**0.30**

Source: From Reyer (1990).

Michener 1974, Hölldobler and Wilson 1990, Ross and Matthews 1991). In most eusocial species, a colony is founded by a single queen or by several. When multiple queens (foundresses) participate in colony foundation, they are related to each other in many, but not all, species.

Kin selection is a leading hypothesis for the evolution of eusociality. Hamilton (1964) suggested that the haplo-diploidy of Hymenoptera predisposed them to become eusocial, because it affects the coefficients of relationship (see Box 20.C). Females are diploid and develop from fertilized eggs; males are haploid and develop from unfertilized eggs. Whereas r in "ordinary" diploid species is 0.5 both between parent and offspring and between full siblings, a female hymenopteran is more closely related to her sister ($r = 0.75$) than to her daughter ($r = 0.5$), and is less closely related to her brother ($r = 0.25$). In a colony with a single, singly mated queen, the queen's reproductive daughters (future queens) are the sisters of the (female) workers, so the inclusive fitness of a female may be greater if she acts as a worker, rearing reproductive sisters, than if she rears daughters—especially if, in order to rear daughters, she would have to establish her own nest, with a low probability of success. Thus if we apply Hamilton's rule that altruism can evolve if $rb > c$, and set the cost c at 1, rearing sisters instead of daughters is advantageous if $b > 1/r$, which yields $b > 2.0$ for diploid species, but only $b > 1.33$ ($1/0.75$) for haplodiploids. Thus the conditions for the evolution of helping relatives should be more easily satisfied in haplodiploids. (However, this theory does not explain why eusociality evolved in termites, in which both sexes are diploid and both act as workers.)

Kin selection is one of at least three hypotheses for the evolution of eusociality (Seger 1991). Another is parental manipulation, suggested by Richard Alexander (1974), who proposed that queens suppress reproduction by some of their daughters and induce them to rear their reproductive sisters and brothers. A third hypothesis, advocated by Mary Jane West-Eberhard (1978), is mutualism: workers may have greater average *personal* reproductive success by helping in a colony than by founding a new nest, because they may be either *cryptic reproductives* or *hopeful reproductives*. All three hypotheses, as Seger (1991) says, take the altruism out of altruism.

These hypotheses are not mutually exclusive, and there is some evidence in favor of each. For example, unmated workers lay haploid (male) eggs in many species, and so reproduce cryptically. Perhaps the most definitive test of kin selection versus parental manipulation has been based on an insightful theory concerning sex ratios proposed by Robert Trivers and Hope Hare (1976).

Within a colony of eusocial insects, there are genetic *conflicts of interest*. The queen is equally related ($r = 0.5$) to her reproductive daughters and sons. According to the prevailing theory of the evolution of sex ratio, to be discussed in Chapter 21, the queen's fitness would be maximized by investing equally in reproductive daughters and sons (as in diploid species). Workers, however, are more closely related to their reproductive sisters ($r = 0.75$) than to their brothers ($r = 0.25$), so if they controlled the sex ratio, they could maximize their inclusive fitness by an investment ratio of 3:1 in favor of females. (Workers could accomplish this by discriminating among larvae and providing more care to females.) Thus the sex ratio among reproductive offspring should differ depending on whether the workers "voluntarily" rear their siblings because this enhances their own inclusive fitness, or whether they are manipulated by their mother into helping. In multiple-queen colonies, the coefficient of relationship between workers and female offspring is lower than 0.75, because they are not necessarily sisters, so we would expect the sex ratio to be closer to 1:1 in such cases.

Calculating Coefficients of Relationship

Several measures of relatedness have been proposed (Grafen 1991). Figure 20.C1 illustrates the method of calculating the coefficient of relationship (*r*) that is most frequently used in studies of the evolution of social interactions (e.g., in Hamilton's rule). The coefficient of relationship is the average proportion of the genes of an individual A that are present, and identical by descent, in another individual B. Individuals A and B are the donor and recipient, respectively, of some interaction. They are indicated by solid symbols, with bold arrows indicating the pathways of inheritance by which A's genes may be traced to B. For example, the genes of a mother (A) "flow" directly to her offspring (B), but genes in two siblings A and B are traced from A to B via both their parents (see Full siblings, row 3). Thin lines indicate mates that do not contribute to common ancestry.

Diploid Species

Any autosomal gene has a probability of 0.5 of transmission to an offspring. One method (method 1) of calculating *r* is to examine each pathway of common descent, count the number of steps *L* in the pathway, calculate 0.5^L, and sum these values over pathways. For instance, two full siblings (row 3) share two pathways of common descent, through their mother and through their father, each of length *L* = 2. Thus $r = (0.5)^2 + (0.5)^2 = 0.5$. Alternatively (method 2), one can calculate the proportion of A's genes that are inherited from a parent, multiply by the probability that one such gene has also been inherited by B, and sum this product over A's parents. For instance, half of sibling A's genes are from her mother; a gene A inherited from her mother was also inherited by her sibling B with probability 0.5, so $(0.5 \times 0.5) = 0.25$; and similarly for genes from her father, so $r = 0.25 + 0.25 = 0.5$. For diploid species, the coefficients of relationship between relatives are symmetrical: either may be labeled A or B.

Haplodiploid Species

The haploid male transmits his single gene copy to his mate's daughters with a probability of 1, but transmits no genes to his mate's sons. This pattern creates asymmetries in relationship; for example, if A is mother and B is son, *r* = 0.5, but if A is son and B is mother, *r* = 1.0 (because the probability that the son's gene is carried by the mother is 1).

It is easiest to calculate *r* for haplodiploids by method 2. Consider (row 3) the relationship of a female, A, to her sister. Half of A's genes are from her mother, and a gene she inherited was also inherited by her sister, B, with a probability of 0.5. Half of A's genes were inherited from her father, and her sister, B, must have inherited the same genes with a probability of 1.0. Therefore, $r = (0.5)(0.5) + (0.5)(1.0) = 0.75$. However, the second term in this formula is zero for *r* between female A and her brother B, because B inherited no genes from A's father, so *r* = 0.25. Note that in haplodiploids, a female is more closely related to her full sister (*r* = 0.75) than to her daughter (*r* = 0.5).

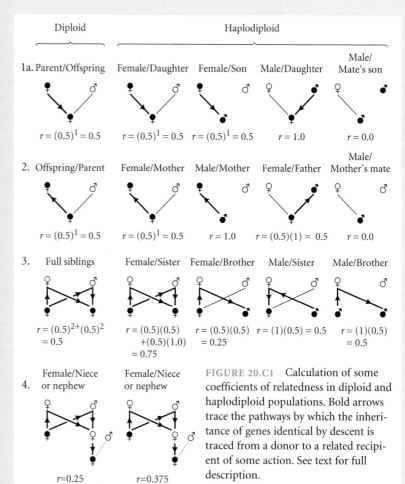

FIGURE 20.C1 Calculation of some coefficients of relatedness in diploid and haplodiploid populations. Bold arrows trace the pathways by which the inheritance of genes identical by descent is traced from a donor to a related recipient of some action. See text for full description.

Data on the sex ratio in 49 species of ants (Figure 20.17) are highly consistent with the theory's prediction if workers are in control: in single-queen species, the ratio centers around 3:1 female: male, but is closer to 1:1 in multiple-queen species. It is even closer to 1:1 in the few slave-making species for which data are available. In these cases, the workers that rear the slave-maker's brood are members of other species, captured by the slave-maker's own workers. The slaves have no genetic interest whatever in the captor's reproductive success, and so are not expected to alter the sex ratio by preferential treatment. By contrast, the nearly 3:1 female: male ratio in single-queen species provides evidence for worker control, and thus for the kin selection theory of altruism in eusocial insects.

The Role of Behavior in Evolution

Much of behavioral ecology asks how the agents of evolution have shaped various kinds of behavior. By shifting our viewpoint, we can ask how behavior has affected the course of organisms' evolution (Wcislo 1989 and Bateson 1988). For example, speciation is affected by the propensity for dispersal, which affects gene flow among populations; by mating behaviors, which may diverge and confer prezygotic isolation; and by the use of habitat, food, and other resources, which can affect the rate of genetic and phenotypic divergence and in some cases confer ecological isolation (see Chapter 16).

By virtue of their behavior, animals are not merely passive objects of external agents of selection (the "environment"), but instead actively determine, in part, the environment that impinges upon them (Lewontin 1983a, Plotkin 1988, Odling-Smee et al. 1996). Beavers literally construct their own niches by constructing dams, and spiders by building webs, but all animals, by their behavioral choices of habitat and food, affect selection on their morphological and physiological features.

Behavior can contribute to the evolutionary *stability* of other traits, *shielding* them from selection. By rapid behavioral responses to changes in their environment, individuals often avert physiological stresses that might cause mortality and therefore possible selection (see Chapter 17). By choosing the food or habitat to which it is already adapted, an animal maintains selection for those existing adaptations. This is likely to be an important cause of evolutionary conservatism in morphology and physiology. For instance, many species of dragonflies, diving beetles, and other groups of aquatic insects can be found in the arid American West, but all inhabit scattered pools or streams, and none has evolved any adaptations to terrestrial life, despite the scarcity of water. Rather few species became adapted to the greatly altered climatic conditions of the Pleistocene; most merely shifted their geographic ranges, following the spatial movement of their habitats (see Chapter 25). Moreover, behavioral selection of a habitat and morphological or physiological adaptations to that habitat can be self-reinforcing. To the degree that a population occupies only one of many available habitats, selection will improve fitness in that habitat at the expense of fitness in other habitats (assuming trade-offs in adaptation: cf. Chapter 17). This reinforces natural selection for choice of that same habitat—which maintains or strengthens stabilizing selection on other characters (Holt 1987).

On the other hand, the behavior of animals can allow or generate *directional selection* for change in other traits (Wcislo 1989), as sexual selection illustrates. Perhaps most importantly, *behavior can initiate shifts in ecological niche*, and so has probably been a key factor in the evolution of animal diversity. Ernst Mayr (1960, 1963) has strongly advanced this hypothesis, writing that "A shift into a new niche or adaptive zone is, almost without ex-

FIGURE 20.17   A plot of the number of species of ants that have different "investment ratios" in the production of males relative to reproductive females. The investment ratio is the proportion of biomass (number × weight) of males. Corresponding to the prediction from kin selection theory, multiple-queen species produce more males than single-queen species, and the few slave-making species examined produce still more, on average. Numbers are the number of species studied. (After Seger 1991.)

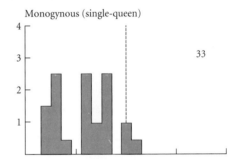

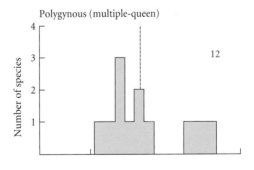

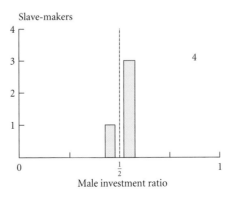

ception, initiated by a change in behavior. The other adaptations to the new niche, particularly the structural ones, are acquired secondarily. With habitat and food selection—behavioral phenomena—playing a major role in the shift into new adaptive zones, the importance of behavior in initiating new evolutionary events is self-evident" (Mayr 1963, 604).

The most obvious test of this hypothesis would be to map the distribution of the several states of a behavior, such as different food preferences, and of a morphological character associated with that behavior, such as the condition of the mouthparts, on a phylogenetic tree. Evidence for the critical role of behavior in niche shifts would be found if an apomorphic behavior appeared "earlier" (more basally) on the tree than the apomorphic structural feature that served the new function. Few such analyses have been performed, although Prum's (1990) study of manakin courtship provided evidence that certain components of the display were found to have evolved before the plumage characters that highlight them.

There is considerable circumstantial evidence that behavior initiates shifts of ecological niche. For instance, one of Darwin's finches, *Pinaroloxias inornata*, is the only member of this group of birds on Cocos Island, an isolated island about halfway between the Galápagos Islands and Central America that harbors only three other species of land birds. Although beak morphology is quite uniform among members of this species, individuals differ markedly in their diets and in their modes of foraging (Figure 20.18; Werner and Sherry 1987). Among the relatives of this finch that inhabit the Galápagos Islands, such

niche differences are associated with great differences in bill morphology (see Chapter 5).

Systematists have described many instances in which species have evolved apparently recent ecological shifts with little or no accompanying morphological change of the kind that might be expected on functional grounds. Darwin noted, in *The Origin of Species*, that a parasitic wasp differs from the many terrestrial members of its large family (Proctotrupidae) in that it forages underwater, laying its eggs in the eggs of dragonflies, yet "it exhibits no modification in structure in accordance with its abnormal habits." Several groups of birds, such as flycatchers and swallows, specialize in feeding on insects in midair, and are structurally modified for this habit. The same behavior is part of the repertoire of some woodpeckers, terns, and gulls that do not differ structurally from related species with different feeding habits (Figure 20.19).

Although behavior may play a key role in animal evolution, the issue requires more rigorous study, as do many related questions: How long is the evolutionary lag between a behavioral shift and structural or physiological adaptation to a new niche? What is the role of genetic change in behavior versus behavioral plasticity? And do differences among animal groups in the complexity or plasticity of their behavior account for differences in their diversity or their rate of evolution?

FIGURE 20.18 Difference in foraging behavior among five individual Cocos Island finches in a single tree. The rows represent the five birds; the columns, different behaviors; the intensity of shading is proportional to the proportion of "N" foraging actions in each category. Individuals differ in behavior, and thus in the food they obtain, despite their structural similarity. (After Werner and Sherry 1987.)

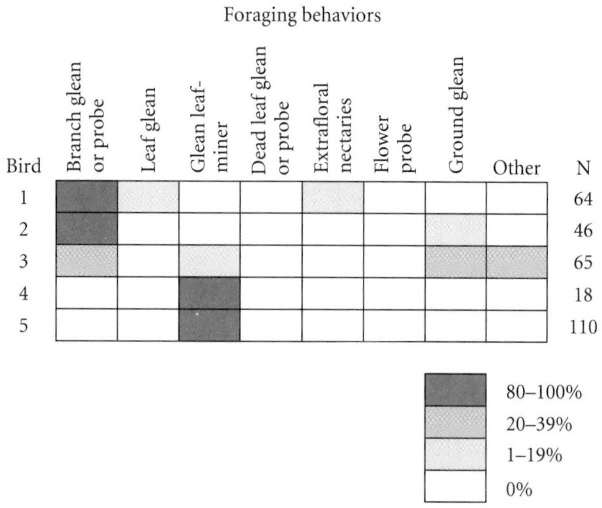

Foraging behaviors

Bird	Branch glean or probe	Leaf glean	Glean leaf-miner	Dead leaf glean or probe	Extrafloral nectaries	Flower probe	Ground glean	Other	N
1									64
2									46
3									65
4									18
5									110

- 80–100%
- 20–39%
- 1–19%
- 0%

FIGURE 20.19 Structurally similar, closely related birds that differ in feeding behavior. (A) The black tern (*Chlidonias niger*) feeds mostly on flying insects. When these are not available, it plucks fish from near the water's surface. (B) Most terns, such as the least tern (*Sterna antillarum*), dive into the water to catch fish. (A and B after the National Geographic Society 1987.)

(A)

(B)

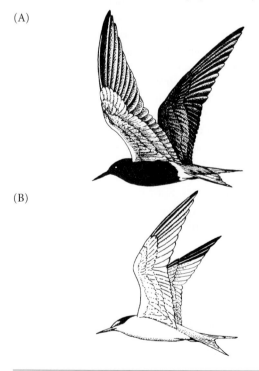

Summary

1. Behavioral traits, like other traits, display both genetic and nongenetic variation to different degrees. Nongenetic variation in behavior can include various forms of learning. Some learned behaviors are culturally rather than genetically inherited, giving rise to nongenetic differences among populations, but many behavioral differences among populations and species have a partly genetic basis. Species differ, often adaptively, in what they can learn.

2. Although behavioral biology uses some anthropomorphic terms, most behavioral biologists do not attribute consciousness to animals or impute purpose to their behavior.

3. Although some behaviors seem superficially to be directed toward the benefit of the population, evolutionary theory explains behavior on the basis of the advantages that genes for behavior confer on their individual bearers.

4. Many behaviors are analyzed by means of optimality theories, which postulate how behaviors should evolve in response to the balance between their benefits and costs. The theory of evolutionarily stable strategies (ESS) is a form of optimality theory applied to interactions among individuals, especially conflicts, in which the advantage of one or another behavior is frequency-dependent. If an ESS is fixed in a population, no alternative strategies can replace it. The ESS may be a "mixed strategy," such as escalating a conflict or retreating from it with some probability. ESS theory can account for phenomena such as ritualized displays in competitive interactions.

5. Sexual selection consists of differences in reproductive success caused by competition for mates. It stems largely from contests within one sex (usually male) for access to the other, and from choice among potential mates (usually by females). Sexual selection accounts for many male ornaments and display behaviors and for many cases of sexual dimorphism in size. Contests between males often favor such traits, as do mate guarding and other methods of ensuring a male's paternity.

6. Sexual selection by female choice accounts for many exaggerated male traits that are ecologically disadvantageous. Several hypotheses have been proposed to account for female choice. The runaway sexual selection hypothesis proposes that genetic correlation between a male trait and a female preference causes an ecologically maladaptive snowballing effect (positive feedback), so that both the trait and the preference evolve toward more extreme values. Indicator ("good genes") hypotheses, in contrast, hold that the preference is ecologically adaptive because the male trait indicates high genetic fitness. In some cases, sensory bias, due to the organization of the sensory system, may play a role in female preferences. Some data are consistent with each hypothesis, but do not unambiguously indicate which is correct.

7. Seemingly altruistic behaviors are directly advantageous to the donor; some represent manipulation of the donor by the recipient; some are maintained by reciprocity between individuals that repeatedly interact; and some are due in part to kin selection.

8. In kin selection, a donor's altruistic behavior affects not only its own fitness, but also that of relatives that share the donor's genes, so that enhancing their fitness increases the frequency of genes for the altruistic behavior. Altruism will not evolve unless benefits are directed more toward kin than nonkin, which requires kin recognition or spatial proximity of kin (nonrandom distribution).

9. Behavior can affect the evolution of animals' morphological and physiological features. Behaviors such as habitat selection can have a conservative effect by maintaining a consistent regime of selection on other characters. However, changes in behavior, such as choice of diet or habitat, can initiate shifts of ecological niche, altering selection on other characters and thus setting the pace of evolution.

Major References

Alcock, J. 1998. *Animal behavior: An evolutionary approach.* Sixth edition. Sinauer Associates, Sunderland, MA. An easily readable introduction to the study of behavior, emphasizing evolutionary, especially adaptive, explanations.

Krebs, J. R., and N. B. Davies. 1993. *An introduction to behavioural ecology.* Third edition. Blackwell Scientific, Oxford. A clearly written treatment, emphasizing adaptation, models and experimental tests of models.

Krebs, J. R., and N. B. Davies (editors). 1991. *Behavioural ecology: An evolutionary approach.* Third edition. Blackwell Scientific, Oxford. Essays by specialists, providing a somewhat advanced treatment of most topics in behavioral ecology. Several topics are covered only in earlier editions of this book.

Andersson, M. 1994. *Sexual selection.* Princeton University Press, Princeton, NJ. A comprehensive review of the topic, emphasizing experimental tests.

Wilson, E. O. 1971. *The insect societies.* Harvard University Press, Cambridge, MA. Despite its age, this is still a superb, comprehensive treatment of the biology of social insects.

Problems and Discussion Topics

1. Assume that for an individual bird, the risk of predation declines, the larger the flock it joins, but that its food intake also declines with flock size, due to competition. Can you devise a model that predicts how large the flock will be when it becomes advantageous to leave and join a smaller flock?

2. In langur monkeys and lions, both of which live in social groups, adults have been observed to kill very young members of the group ("infanticide"). Propose an adaptationist hypothesis to account for this behavior, and predict which group members, under what conditions, might be expected to practice it. See Parmigiani et al. (1992).

3. Many species of albatrosses and other seabirds nest in colonies on islands. The adults forage in flocks in nearby waters. Some ecologists who study seabirds turn to their favorite topic of conversation in a bar one evening, and one says, "I can imagine an albatross genotype that destroys the eggs or nestlings of its colonymates when the parents are away foraging, and leaves them dead without eating them. Do you think such a behavior would evolve?" One of his companions says, "Yes, because it would make more food available for the albatross and its offspring." Another says, "No, because it would threaten the survival of the population." The fourth says, "You're both wrong. I have a different explanation for why albatrosses don't kill others' chicks." What does the fourth ecologist say?

4. A male that stays with his mate and helps to rear their young may benefit by enhancing his offspring's survival. A male that leaves immediately after mating, in search of more females, may benefit by fathering more offspring. These alternatives (monogamy and polygamy) are both observed among mammals, birds, and many other taxa. Monogamy is more prevalent in birds than in mammals. (Monogamy refers here to formation of a pair-bond, and does not rule out "extra-pair copulation" by both sexes—which indeed occurs frequently in monogamous species of birds.) What factors would favor the evolution of either mating system? How would you test your hypothesis? Why are birds more frequently monogamous than mammals? (See Emlen and Oring 1977 and Clutton-Brock 1991.)

5. How can there be sexual selection in pair-bonding species with a 1:1 sex ratio, since every individual presumably obtains a mate?

6. According to the theory of genetic drift (see Chapter 11), all copies of a gene in a population can ultimately be traced back to a single ancestral gene copy. Thus every member of a population is related, even if very distantly, to all other members of the population. Does it follow, then, that altruistic behavior, dispensed to any and all members of the population, should evolve by kin selection? If not, why not?

7. Although the worker females in eusocial Hymenoptera do not mate, some of them can lay haploid eggs that develop into sons. In what ways would doing so affect the inclusive fitness of a worker? How would it affect the fitness of her mother, the queen? How would other workers be expected to react to these male larvae? Is there a genetic conflict of interest within a colony of social insects? What is known about egg laying by workers?

8. Males of the sailfin molly (*Poecilia latipinna*), a freshwater fish with internal fertilization, may either court females or attempt forced copulation without courting. Larger males more frequently court; smaller males more frequently attempt forced copulation. Variation in size is largely determined by alleles at a locus on the Y chromosome. How might you determine whether the propensity for one or the other behavior is a size-dependent adaptive "strategy" without an independent genetic basis, or is independently genetically determined? See Travis (1994) after you have designed your experiment. Discuss the broader implications of one or the other answer for the joint evolution of behavior and morphology.

9. Great controversy surrounds the field of human sociobiology, which attempts to explain many human behaviors (e.g., adoption, rape) as evolved adaptations (see Chapter 26). Discuss which of the methods of testing hypotheses about behavioral adaptation that we have described in this chapter can be successfully used in explaining such behaviors in humans. What are their strengths and weaknesses? (See Kitcher 1985 for a critique and Alcock 1998 for an approving description of this area of study.)

10. The field of experimental psychology includes studies of nonhuman animals as models for understanding perception, learning, and other aspects of human behavior. Discuss how an evolutionary perspective may contribute to this field, and to the implications that can be drawn about humans from animal models.

CHAPTER 21

The Evolution of Genetic Systems

The great diversity of organisms extends to their genetic systems. Organisms may be sexual or asexual, may be haploid, diploid, or polyploid, may have separate sexes or not, may have few or many chromosomes, may self-fertilize or outcross. These differences greatly affect patterns of genetic variation, and therefore the responses of populations to natural selection and other factors of evolution. Discovering why and how each of these characters evolved poses some of the most challenging problems in evolutionary biology, and is the subject of some of the most creative contemporary research on evolution.

Short-Term versus Long-Term Advantage

Evolution cannot occur without genetic variation, so a species lacking mutation and recombination would be destined for extinction, perhaps in short order. Thus these phenomena are necessary for the survival of species, and this fact has been cited for more than a century as the reason for their existence. Biology students have learned that sex is adaptive because it provides the variability required for adaptation ever since that great advocate of Darwinism, August Weismann, proposed this hypothesis in 1889. But by now, you the reader recognize that arguments that invoke benefits to the species rely on group-level or species-level selection, which is generally suspect because it is ordinarily a weak agent of evolution. That is, a feature that is advantageous to a population only in the long term, by increasing its likelihood of persistence, is likely to be replaced by a "selfish" gene or feature that benefits the individual organism, even if that feature should ultimately cause the extinction of the population. The question, then, is whether or not natural selection *within* populations can account for mutation, recombination, and other features of the genetic system. Does the evolution of these features depend on their long-term consequences for populations, or their short-term consequences for individual organisms and genes? We shall illustrate the problem with a short discussion of mutation, and then turn to the much more complicated topic of sex and recombination.

The Evolution of Mutation Rates

Two hypotheses have been proposed to explain rates of mutation: either the mutation rate has evolved to some *optimal* level; or it has evolved to the *minimal* possible level. The thermodynamics and physical chemistry of DNA replication imply that mutation rates should be much higher than they actually are, but many mutations are corrected by "proofreading" repair enzymes. Thus variation in the efficacy of these enzymes provides potential genetic variation in the genomic mutation rate. According to the hypothesis of optimal mutation rate, selection has favored somewhat inefficient repair enzymes. According to the hypothesis of minimal mutation rate, mutation exists only because repair is as complete as it can be, or because selection is not strong enough to favor investment of energy in a more efficient repair system. According to this hypothesis, *the process of mutation is not an adaptation.*

Group selection would favor an optimal (greater than zero) mutation rate, because genetically invariant species would become extinct, leaving only species that experience mutation. We do not know how fast this process would occur, because it would depend on the rate of change of environmental factors that could cause extinction: the faster the environment changes, the higher the mutation rate required to avert extinction (Lynch and Lande 1993).

Alternatively, we can ask how evolution *within* populations affects the mutation rate. This can be done by postulating a locus that affects the mutation rate of other genes. This locus might, for example, encode a DNA repair enzyme. For simplicity, let us assume that alleles at this locus do not differ in their *direct* effect on the fitness of their bearers—they do not directly affect survival or reproduction. They have only an *indirect* effect via the mutations they allow or prevent. Such a locus is often called a MODIFIER of mutation, and the action of natural selection on such a

605

locus is termed SECOND-ORDER SELECTION. Models of second-order selection on modifier loci are the most useful devices for understanding the evolution of genetic systems.

A mutation called *mutT* in *Escherichia coli* is one of several mutations that are known to increase the mutation rate throughout the genome. Two strains of *E. coli*, differing only in whether or not they carried the *mutT* allele, were placed in a chemostat by Cox and Gibson (1974). The *mutT* allele rose in frequency and approached fixation. This result is explicable if *mutT* caused an increase in both deleterious and beneficial mutations. Cell lineages with new deleterious mutations would be quickly eliminated, but genotypes with a beneficial mutation would increase. Because *E. coli* is mostly asexual, the *mutT* allele would be linked to the beneficial mutation it caused, and would increase along with it by *hitchhiking* (Figure 21.1). Thus *the fate of alleles that affect the mutation rate is determined by linkage disequilibrium with the mutations they cause* (or permit).

Consider now a population in which there is recombination (Leigh 1973). The great majority of mutations caused by a mutation-enhancing allele (*M*) are deleterious, but a few are beneficial. Recombination will separate *M*

from a beneficial mutation within a few generations, and so *M* will not hitchhike to high frequency. Usually, though, *M* will be associated with deleterious mutations, and will therefore decline in frequency for as long as it is linked to one or another of them. Segments of genetic material bearing both *M* and deleterious mutations are likely to be eliminated before recombination can free *M* from closely linked harmful partners. Therefore *natural selection within sexual populations will tend to eliminate any allele that increases mutation rates, and mutation rates should evolve toward the minimal achievable level.* This process is likely to be much faster than the rate at which populations or species become extinct for lack of genetic variation. Thus, although mutation is necessary for the long-term survival of a species, the mutation rate should evolve toward zero. Consequently, most evolutionary biologists believe that the existence of mutation is not an adaptation in most organisms, but rather an unavoidable effect of the physics and chemistry of DNA replication.

Sex and Recombination

"No area of evolutionary biology offers the curious investigator a more fascinating mixture of strange phenomena and deep intellectual puzzles than the evolution of sex and its consequences" (Stearns 1987, 13). Before explaining why this is so, it is best to define some terms.

SEX usually refers to the union (SYNGAMY) of two genomes, usually carried by gametes, followed at some later time by REDUCTION, ordinarily by the process of meiosis. A synonym for sex, thus defined, is AMPHIMIXIS. Sex often, but not always, involves OUTCROSSING, whereby the genomes of two individuals are partly or wholly combined.

Asexual reproduction may be carried out by VEGETATIVE PROPAGATION, in which an offspring arises from a group of cells, as in plants that spread by runners or stolons, or from a single cell, such as an unfertilized egg in many parthenogenetic plants and animals. The most common kind of parthenogenesis is APOMIXIS, in which meiosis is suppressed, and the offspring are genetically identical to their mother, except for whatever new mutations may have arisen in the cell lineage from which the egg arose.

Both sexual and asexual reproduction occur within certain species, some of which include both sexual and asexual genotypes.

Sexually reproducing species need not have distinct female and male sexes, which are defined by a difference in the size of their gametes. In many fungi, such as yeasts, and in many algae, such as *Chlamydomonas*, cells of the same size unite if they differ in MATING TYPE (+ or –). Such species are termed ISOGAMOUS. ANISOGAMOUS organisms have large (eggs) and small (sperm) gametes, defining female and male sexual functions. Species in which individuals are either female or male, such as willow trees and mammals, are termed DIOECIOUS or GONOCHORISTIC (in botany and zoology respectively). Those in which an individual can produce both kinds of gametes, such as roses and earthworms, are COSEXUAL or

FIGURE 21.1 Changes in the frequency of a mutation-enhancing allele (*MutT*), relative to the wild-type allele in a culture of *Escherichia coli*. The *MutT* strain carried a genetic marker (*Lac*–) at another locus that was easier to score, but which indicated the frequency of *MutT*-bearing cells. The frequency of *MutT* fluctuated, but generally increased, due to its association with new mutations. In a sexual population, the mutation-enhancing allele would not remain linked to beneficial mutations, and would not be expected to increase. (After Cox and Gibson 1974.)

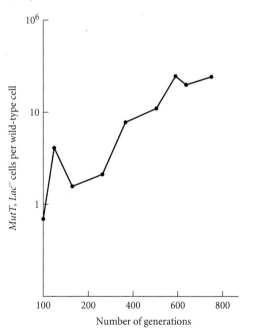

HERMAPHRODITIC. Flowering plants exhibit many other variations, such as the condition in species that include both cosexual and single-sex (e.g., female) individuals.

The Problem with Sex

The traditional explanation of the existence of recombination and sex is that it increases the rate of adaptive evolution of a species, either in a constant or a changing environment. This explanation poses two questions: (1) Is the premise that sex increases the rate of evolution true? (2) Even if it is true, can it really explain the origin and persistence of sex? Is this one of the few characteristics of organisms that owes its existence to group selection?

Even if they do increase evolutionary rates, recombination and sex have at least two *disadvantages*. One is that *recombination destroys adaptive combinations of genes*. Suppose that in a haploid organism, the combinations *AB* and *ab* have high fitness, perhaps because they are adapted to two different resources, whereas *Ab* and *aB* have lower fitness. In a constant environment (but a heterogeneous one, with two available resources), the mean fitness of the population will be lowered if recombination between *AB* and *ab* produces some *Ab* and *aB* progeny. More importantly, the fitness of an *AB* (or *ab*) individual will be lower if it engages in sex than if it reproduces apomictically (by mitosis), because fewer of its progeny will be viable.

We can model this situation (Figure 21.2) by postulating a third locus, *R*, that governs recombination (a recombination modifier). If *R* cells are asexual and *r* cells are sexual, then *r* will be selected against because it often will be associated with the unfit recombinant genotypes it has brought into existence, whereas *R* will increase in frequency because of its unbroken association with *Ab* and *ab*. Unless the environment should change quite soon, favoring *aB* and/or *Ab* over *AB* and *ab*, *R* will become fixed, yielding an asexual population.

The second disadvantage of sex is even more serious—so serious that it makes the existence of sexual reproduction one of the most difficult puzzles in biology. This disadvantage is the *cost of sex*, which is most easily understood in organisms such as typical sexual animals as the "cost of males."[*] In most such species, half of all offspring are male. However, all the offspring of an asexual female are female (because they inherit their mother's sex-determining genes). If sexual and asexual females have the same fecundity, then a sexual female will have only half as many grandchildren as an asexual female (Figure 21.3). Therefore,

---

[*]More generally, the cost of sex is a "cost of meiosis," which applies to all anisogamous species. The simplest way to think about the genetic cost of meiosis is to contrast two genotypes of females with equal fecundity, one of which, a sexual genotype, carries an allele for meiosis in heterozygous condition, while the other, an asexual genotype, does not. The allele for meiosis is transmitted to only half the eggs of the sexual female, whereas the allele for parthenogenesis is transmitted to all the eggs of the asexual female. Thus an allele for meiosis has only half the rate of increase. Bell (1982) provides a mathematical treatment of the cost of meiosis.

FIGURE 21.2 Selection against an allele (*r*) that promotes recombination, if allele combinations *Ab* and *aB* have lower fitnesses than *AB* and *ab*. Allele *R*, which suppresses recombination, increases in frequency because of its association with the favored genotypes *AB* and *ab*. The diagram pictures an organism, such as some algae, in which a diploid zygote undergoes meiosis, and the dominant part of the life cycle is haploid. The same principle holds for organisms in which the diploid phase of the life cycle is dominant.

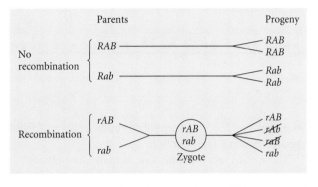

the rate of increase of an asexual genotype is approximately twice as great as that of a sexual genotype (all else being equal), and an asexual mutant should very rapidly be fixed.

In the *long term*, of course, the evolution of asexuality might doom a population to extinction, but in the *short term*, generation by generation, asexuality has a *twofold advantage* over sexuality. The problem, therefore, is to discover whether there are any *short-term advantages* of sex that can overcome its short-term disadvantages. This problem is the subject of books by Ghiselin (1974), Williams (1975), Maynard Smith (1978a), Bell (1982), Stearns (1987), and Michod and Levin (1988), and of incisive review papers by Charlesworth (1989) and Kondrashov (1993).

Can Recombination Rates Evolve?

There is abundant evidence that the rate of crossing over is genetically controlled and can evolve rapidly (Brooks 1988). For example, the frequency of crossing over between particular pairs of loci can be altered by artificial selection in *Drosophila melanogaster* without affecting the frequency of crossing over elsewhere on the chromosome. Asexual lineages are known in many otherwise sexual species of plants and animals, such as crustaceans and insects. Hampton Carson (1967), using artificial selection, was able to obtain parthenogenetic strains from a sexual population of *Drosophila mercatorum*. However, asexuality is unknown in some taxa, such as birds[†] and mammals, possibly because of unknown constraints that prevent its origin. One sexually reproducing clade of mites has probably arisen from a parthenogenetic ancestor (Norton et al. 1993), but other-

---

[†]Parthenogenetic strains of domestic turkeys have been produced by artificial selection, but no parthenogenetic birds are known in nature.

FIGURE 21.3 The advantages and disadvantages of an allele *S*, which codes for sexual reproduction, compared with an allele *s*, coding for asexual reproduction. Circles represent females, squares males. Each female produces four equally fit offspring, but the frequency of the *S* allele drops rapidly from 1/2 in the first generation to 1/5 by the third generation due to the production of males. However, the *S* allele would increase if the environment were to change so that the genotype *aabb* (for example) had much a higher survival rate than other genotypes.

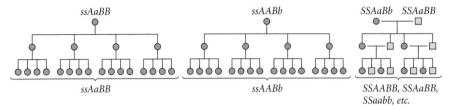

wise, there are few documented examples of reversal from asexual to sexual reproduction.

These observations imply that in many, though not all, groups of organisms, asexuality and/or lower recombination rates can evolve (Hebert 1987; Bierzychudek 1987). If so, the ubiquity of sex and recombination must generally be due to some advantage, whether in the short or long term.

Hypotheses for the Advantage of Sex and Recombination

As we examine some of the postulated advantages of recombination, bear in mind that any degree of advantage might account for the existence of *recombination* (e.g., crossing over), given that a species is sexual; but in order to account for the persistence of *sex* (rather than apomixis), its advantage must be great enough to offset its twofold cost.

Kondrashov (1993) distinguished hypotheses that propose some *immediate benefit* to the individual organism's progeny from hypotheses that invoke *variation and selection*.

The leading hypothesis for an *immediate benefit of recombination* is that *molecular recombination facilitates repair of damaged DNA* (Bernstein and Bernstein 1991). According to this model, breaks and other lesions in a DNA molecule can be repaired by copying from an intact sequence on a homologous chromosome. According to this hypothesis, the formation of new gene combinations is a by-product of the molecular mechanism of DNA repair, not the raison d'être of recombination or sex. Critics of this hypothesis (Maynard Smith 1988; Kondrashov 1993) argue that it fails to explain the elaborate mechanisms of meiosis and syngamy, and that DNA repair does not require these mechanisms; permanently diploid or polyploid apomicts, of which there are many, could also repair damage this way. Most evolutionary biologists would grant that the *origin* of recombination may have been due to its role in DNA repair, but believe that the evolution of meiosis and distinct mating types or sexes, and the *maintenance* of sex in most species, must be attributed to other causes involving variation and selection.

Variation and Selection

Recombination can have an effect on the frequency distribution of genotypes within a population, or among the progeny of a mating pair, only when there exists linkage disequilibrium among alleles at two or more loci—that is, when some combinations, such as *AB* and *ab*, are present in excess and others, such as *Ab* and *aB*, are deficient. Recombination decreases the frequency of excess combinations and increases that of deficient ones. Linkage disequilibrium can be caused by genetic drift or by epistatic selection (see Chapters 11 and 14). If "like" alleles (such as those that increase or decrease a trait) are nonrandomly associated, the variance of the trait will be greater than under linkage equilibrium; if "unlike" alleles are associated, it will be less. *The presumed advantage of recombination is that it reduces the frequency of nonoptimal combinations and can potentially increase the frequency of superior combinations.* Hypotheses for the advantage of recombination can be classified largely on the basis of whether they propose that linkage disequilibrium is caused by genetic drift or by epistatic selection, and whether suboptimal genotypes are present in excess because the environment has changed or because of deleterious mutations (Table 21.1).

Fixation of rare beneficial mutations   Suppose adaptation, perhaps to a newly changed aspect of the environment, depends on newly arising beneficial mutations *A* and *B*. These mutations arise in different lineages, and so are dissociated: the coefficient of linkage disequilibrium (*D*) between them is negative ($D < 0$) due to chance. In an asexual population, the favorable *AB* combination arises only when a second mutation (*B*) occurs in a growing lineage that has already experienced mutation *A* (or vice versa). In a sexual population, the *AB* combination can be formed more rapidly (Figure 21.4). [According to a related hypothesis, sex can be advantageous because it creates homozygotes for a beneficial mutation more rapidly than these can arise in an asexual population (Kirkpatrick and Jenkins 1989).] However, *this difference in evolutionary rate between asexual and sexual populations holds only for large populations.* In small populations, mutations are so few that the first (*A*) is likely to be fixed by selection before the second (*B*) arises, whether the population is asexual or sexual. Thus *recombination may or may not speed up adaptive evolution*, but in any case, this model describes a *long-term* advantage of

Table 21.1 Hypotheses for the advantage of sex

| CAUSE OF LINKAGE DISEQUILIBRIUM | CAUSE OF EXCESS OF NONOPTIMAL GENOTYPES | |
	CHANGE OF ENVIRONMENT	DELETERIOUS MUTATIONS
Genetic drift	Faster fixation of beneficial new mutations	Muller's ratchet
	Lottery model (offspring in diverse environments)	
	Sibling competition for heterogeneous resources	
Epistatic selection	Faster adaptation to new environment, based on pre-existing variation	Elimination of deleterious mutations
	Responsiveness to fluctuating environment (shifting optimum)	

After Kondrashov (1993)

sex—an advantage to the population—that is unlikely to counter the short-term disadvantage arising from the cost of meiosis.

FIGURE 21.4 Effects of recombination on the rate of evolution. *A*, *B*, and *C* are new mutations that are advantageous in concert. In asexual populations (1 and 3), combination *AB* (or *ABC*) is not formed until a second mutation, such as *B*, occurs in a lineage that already bears the first mutation, *A*. In a large sexual population, independent mutations can be assembled more rapidly by recombination, so adaptation is more rapidly achieved. In small populations, however (panels 3 and 4), the interval between the occurrence of favorable mutations is so long that a sexual population does not adapt more rapidly than an asexual population. (After Crow and Kimura 1965.)

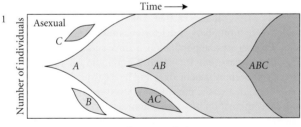

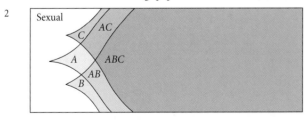

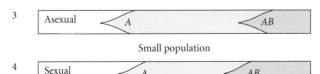

Environmental deterministic hypotheses In environmental deterministic hypotheses, selection acts not on new mutations, but on existing genetic variation, by reorganizing it into new combinations. The most likely circumstance in which this reorganization can provide an advantage to sex is when a polygenic character is subject to stabilizing selection, but the optimum fluctuates due to a fluctuating environment (Maynard Smith 1980). Let us assume that alleles *A*, *B*, *C*, *D* ... additively increase a trait such as body size, and alleles *a*, *b*, *c*, *d* ... decrease it. Stabilizing selection for intermediate size reduces the variance and creates negative linkage disequilibrium, so that combinations such as *AbCd* and *aBcD* are present in excess (see Chapter 14). If selection fluctuates so that larger size is favored, combinations such as *ABCD* may not exist in an asexual population, but they can arise rapidly in a sexual population. This can provide not only a long-term advantage to sex (a higher rate of adaptation of the population), but a short-term advantage as well, because sexual parents are likely to leave more surviving offspring than asexual parents.

Heterogeneous habitats Spatial, rather than temporal, variation in the environment can also provide a strong advantage to sex (Ghiselin 1974; Williams 1975; Bell 1982). Before explaining this hypothesis, we must point out that a genetically variable asexual population, consisting of various clonal genotypes, may actually utilize a variety of habitats or resources more effectively than a sexual population. Suppose an individual animal of a certain size can effectively feed on only a narrow range of particle sizes. The distribution of body size in a sexual population will be more or less normal due to recombination, so that some food types, such as the largest and smallest, will be underutilized (Figure 21.5). However, if different asexual genotypes span the range of body sizes, each genotype can increase in frequency to the level set by the abundance of its favored resource. Robert Vrijenhoek (1990, 1994) has illustrated this principle with his work on Mexican fishes (*Poeciliopsis*) in

FIGURE 21.5 The "frozen niche variation" model, showing how a group of asexual genotypes can utilize resources more fully than a sexual species. The horizontal axis represents an array of resources, such as food particles of different sizes, and the vertical axis represents the efficiency with which these resources are used. Each of two sexual species (S_1 and S_2) varies in a polygenic character, such as mouth size, that affects its efficiency. Because of recombination, genotypes with extreme character states are rare in both sexual species, so the populations of the sexual species have bell-shaped resource utilization curves, and fail to exploit some resources. In contrast, a population of asexual genotypes (A_1–A_5), each with a morphology suited to a part of the resource spectrum, can fully utilize the resources. Such a population can drive the sexual species to extinction. (After Vrijenhoek 1994.)

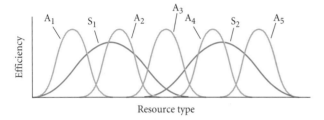

which females of hybrid origin lack recombination and transmit an intact haploid maternal genome to their progeny. The paternal genome is not transmitted during the female's meiosis. Maternal genomes ("hemiclones") differ in many characteristics; for example, one clone is mostly insectivorous, and another feeds on detritus. Populations that include combinations of several clones are more abundant, in aggregate, than sexual species, as we would expect if they can collectively use a wider range of resources. If a combination of asexual genotypes were to use the full spectrum of resources, it could drive a sexual population to extinction (Case and Taper 1986).

Heterogeneity of resources could, however, favor recombination in two related ways. First, suppose offspring disperse at random into small patches of habitat where competition occurs, and only one or a few of the individuals best adapted to conditions in a patch survive. Williams (1975) described this situation as a lottery, in which a sexual female has many "tickets" with different numbers (different offspring genotypes), but an asexual female has numerous copies of the same ticket. The probability is greater that the sexual female will have the "winning ticket." The second hypothesis assumes that even within an apparently homogeneous habitat patch, genotypes can differ in their use of limiting resources; for example, plant genotypes may vary in the ratios of various nutrients they require for growth. If *siblings compete*, then a patch of habitat can sustain more progeny from a sexual family than from an asexual family, because asexual siblings compete more intensely (Bell 1982; Price and Waser 1982).

Hypotheses based on deleterious mutations Several hypotheses postulate that the advantage of sex lies in ridding populations of deleterious mutations, which arise at a high rate (Chapter 10). Bear in mind that the *number* of deleterious mutations arising per generation (U) is the product of the mutation rate per locus (u), the number of loci (n), the number of copies of a locus per genome ($c = 1$ in haploids, 2 in diploids), and the size of the population (N): $U = Nunc$.

Herman Muller (1964), who won a Nobel Prize for his role in discovering the mutagenic effect of radiation, proposed a hypothesis that has been named Muller's ratchet. It assumes that back mutation from deleterious to wild-type alleles is so rare that it can be ignored. In an asexual population, deleterious mutations at various loci create a spectrum of genotypes carrying 0, 1, 2, ... m mutations. The zero class declines over time because its members experience new deleterious mutations. Moreover, the zero class may be lost by chance, despite its superior fitness, especially if the population is small. Thus all remaining genotypes have at least one deleterious mutation (Figure 21.6). Sooner or later, by the same process, the one-mutation class is lost, and all remaining individuals carry at least two mutations. The zero- and one-mutation classes cannot be regenerated in an asexual population, so the accidental loss of superior genotypes is an irreversible process—a ratchet. With each such loss, the ratchet advances, and the population consists of increasingly inferior genotypes. This process is likely to lower population size, which increases the rate at which the least mutation-laden genotypes are lost by genetic drift. Thus there may be an accelerated decline of fitness—a "*mutational meltdown*"—leading to extinction (Lynch et al. 1993). In contrast, *recombination in a sexual population reconstitutes the least mutation-laden classes of genotypes* by generating progeny with new combinations of favorable alleles.

FIGURE 21.6 Muller's ratchet. The frequency of individuals with different numbers of deleterious mutations (0–10) is shown for an asexual population at three successive times. The least loaded class (0 in top graph, 1 in middle graph) is lost over time, both by genetic drift and by its acquisition of new mutations. In a sexual population, class 0 can be reconstituted, since recombination between genomes in class 1 that bear different mutations can generate progeny with none. (After Maynard Smith 1988.)

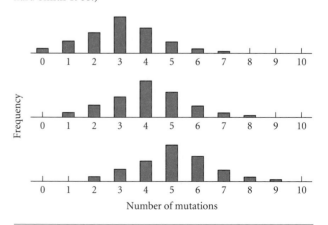

Muller's ratchet has a potentially strong effect only in rather small populations. Although Muller's ratchet may describe a long-term advantage for recombination, it is not clear that it can provide a short-term advantage.

Another hypothesis based on deleterious mutations, advocated by Alexey Kondrashov (1988, 1993), does not require genetic drift, and thus can apply to large populations. It assumes that deleterious mutations interact, so that each "genetic death" rids the population of more than one deleterious mutation, and fewer genetic deaths are required to improve the population's mean fitness than would be required if the mutations were purged individually (Figure 21.7). Selection in any population will create negative linkage disequilibrium among deleterious mutations at different loci: by eliminating genotypes with, say, both *a* and *b*, it leaves a residue of *Ab* and *aB* chromosomes (where uppercase and lowercase represent wild-type and deleterious alleles respectively). Thus the variance in fitness is reduced, so selection is less effective in further purging the population of deleterious genes. Recombination, however, breaks down linkage equilibrium and increases the variance. Thus it creates *ab* combinations anew, providing opportunities for selection to purge both *a* and *b* at once. With a sufficiently high mutation rate per genome, this process can provide a strong short-term advantage for a gene that promotes sex and recombination (Kondrashov 1982; Feldman et al. 1980).

Evaluating Hypotheses for the Advantage of Sex

What had seemed a difficult puzzle turns out to have many proposed solutions. We can describe only some of the evidence, which is consistent with several hypotheses, but does not prove any one of them to be the major explanation of sex.

The long-term advantage of sex If a parthenogenetic lineage were able to persist for many millions of years, as do many sexual lineages, then it should have diverged greatly from its sexual relatives, and should have given rise to a morphologically and ecologically diverse clade. The only asexual groups in which such diversity exists, betokening persistence since an ancient origin, are the rotifer order Bdelloida, which may be 10 My old (Judson and Normark 1996,) certain groups of mites (Norton et al. 1993), and prokaryotes [although recombination may be quite frequent in some groups of bacteria (Dykhuizen and Green 1991; Maynard Smith 1994)]. Because of their enormous populations and small genomes, bacteria might persist indefinitely without sex because they should be almost immune to Muller's ratchet (Bell 1988).

Most asexual lineages, however, have arisen recently from sexual ancestors, for they are very similar to closely related sexual species, and, indeed, retain structures that once had a sexual function. The dandelion *Taraxacum officinale*, for example, is completely apomictic, but is very similar to sexual species of *Taraxacum*, retaining nonfunctional stamens and the brightly colored petal-like structures that in its sexual relatives serve to attract cross-pollinating insects. The recent origin of most asexual forms indicates that asexual lineages survive for only a short time.

Although the evident recency of most parthenogenetic lineages indicates that sex confers a long-term advantage by lessening the risk of extinction, there is no evidence that extinction occurs frequently enough to prevent the replacement of sexual with apomictic genotypes within populations. Species in which both sexual and apomictic reproduction occurs pose a conspicuous challenge to this hypothesis, because sexual genotypes persist despite the presence of apomictic genotypes. So let us turn to other hypotheses.

Short-term advantages of sex Evidence on the short-term advantages of sex is fragmentary and inconclusive. There is no empirical evidence that more rapid fixation of new beneficial mutations in large sexual populations can

FIGURE 21.7 Kondrashov's model, in which deleterious mutations are eliminated more rapidly from sexual than from asexual populations. (A) Deleterious mutations have synergistic effects on fitness, so that fitness is greatly lowered in individuals with more than some threshold number (*T*). Therefore, mutations are eliminated by selection more rapidly if they are in the company of other mutations than if they are borne singly. (B, C) Before selection in any given generation, the frequency distribution of mutations per individual is broader in a sexual population, due to recombination, than in an asexual population. Because a greater fraction of the mutations are carried by individuals with more than *T* mutations in a sexual than in an asexual population, the frequency of each deleterious mutation is reduced more rapidly in the sexual population. (B, C after Maynard Smith 1988.)

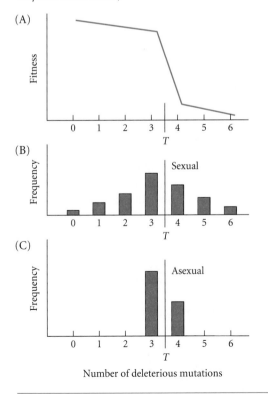

maintain sexual reproduction, and some doubt that these circumstances are common enough to account for the ubiquity of sex (Kondrashov 1993). The hypothesis that sex is advantageous under temporally fluctuating selection for change in the mean of a polygenic character is a theoretically most attractive explanation (Charlesworth 1989), but it has received little study. The response to selection for motility was more rapid in chromosomally monomorphic populations of *Drosophila* than in populations polymorphic for inversions that suppress crossing over (Carson 1959). Also, the rate of crossing over increased in populations of *Drosophila* selected for DDT resistance (Flexon and Rodell 1982).

Evidence that the advantage of sex is its role in purging populations of deleterious mutations consists only of evidence for the two major assumptions of Kondrashov's model: a high deleterious mutation rate and synergistic (epistatic) effects of such mutations. There is some evidence for both, especially a high genomic mutation rate (Kondrashov 1993). This model, though, remains very attractive because it is so general.

Evidence for the advantage of sex in heterogeneous environments includes the geographic distribution of related asexual and sexual taxa. In both plants and animals, asexual taxa are very often found at higher latitudes and/or altitudes, in physically harsher but biotically less complex environments, than their sexual relatives (D. Levin 1975; Bell 1982; Glesener and Tilman 1978; Bierzychudek 1987). Many authors have inferred that sex may be most advantageous where a diverse community of parasites, predators, and competitors, themselves undergoing genetic change, imposes selection on a species, favoring the continuing assembly of different gene combinations. However, the geographic pattern may have other explanations. For example, apomicts have greater colonizing ability because they do not have to find mates at low density (Stebbins 1950) and many of them are polyploid, which may provide greater physiological resistance to harsh environmental conditions (Bell 1982; Bierzychudek 1987; see Chapter 16).

Especially in plants, experiments have often shown that different genotypes perform best in each of many sites or microenvironments. In the same vein, the yield of biomass or seed in cereals (e.g., wheat) and other crops is often greater in plots sown with mixtures of genotypes than those sown with any single genotype. These observations suggest that genetically diverse siblings, by using somewhat different combinations of resources, should compete with each other less intensely than apomictically produced siblings. Some direct evidence on this point has been provided in a series of experiments by Janis Antonovics and his colleagues (summarized by Bierzychudek 1987). In one experiment, as described in Chapter 13, they found that individual grass plants had higher fitness if grown in competition with different genotypes than with the same genotype. In another experiment, focal plants were surrounded either by sexually produced siblings or by unrelated individuals. Attack by aphids caused high seedling mortality, and the survivorship of plants surrounded by genetically similar siblings was considerably lower than that of plants with genetically dis-

similar neighbors. Such experiments provide evidence for frequency-dependent selection that would favor genetically diverse offspring.

Haploidy and Diploidy

We turn now to some features of the genetic system that are consequences of, or associated with, sexual reproduction. Ironically, the evolution of some of these features is better understood than that of sex itself.

Why are so many eukaryotes diploid rather than haploid during most of their life cycle? Several investigators have shown, using mathematical models, that diploidy is advantageous because it masks partially or entirely deleterious mutations (Kondrashov and Crow 1991; Perrot et al. 1991; Jenkins and Kirkpatrick 1995). Thus an allele that extends the diploid phase of the life cycle can increase in frequency. However, this advantage holds only if there is recombination. Without recombination, diploids accumulate twice as many deleterious mutations as haploids (because they have twice as many genes), and this disadvantage outweighs the masking effect. But if recombination occurs, the allele determining diploidy can be dissociated from genomes that carry high numbers of mutations, and so is not reduced in frequency by association with them. This model (Jenkins and Kirkpatrick 1995) suggests that diploids should be sexual, but many parthenogenetic diploids do exist. However, it is unlikely that a diploid lineage would become haploid. Because a diploid population accumulates deleterious recessive alleles, almost any haploid genotype derived from a diploid population would have low fitness, and an allele causing haploidy would therefore be eliminated. Thus diploidy could evolve from haploidy, but not vice versa—an example of *irreversibility in evolution* (Bull and Charnov 1985).

The Evolution of Sexes

Why should there be distinct female and male sexes in many organisms? And why should there be only two sexes?

We will address the second question first. In many algae, fungi, and protists, individuals of two or more morphologically indistinguishable MATING TYPES can mate only with individuals of different mating types. According to a model developed by Yoh Iwasa and Akira Sasaki (1987), when the time available for mating is too short for cells to find a second partner if the first is unsuitable (i.e., is the same mating type), a rare mating type allele has greater fitness than a common one, because most potential partners will be suitable. Rare mating types increase in frequency, and many mating types will be equally abundant at equilibrium.

If, however, the time available for mating is long, so that each cell can encounter many potential partners until it finally finds one of a different mating type, then all cells with rare mating types eventually find partners, and only the most common type forms an unmated residue. Having lower average fitness than the others, the most common type declines in frequency. This process is repeated until there are two common types with equal frequencies, and others with lower frequencies. In contrast to the previous model, the rarer types do not have a frequency-dependent advantage in finding mates, so they may be lost by genetic drift, leaving only two equally abundant mating types ("sexes") in the population.

Phylogenetic evidence indicates that anisogamous organisms have evolved from isogamous ancestors. Explaining the origin of distinct sexes, then, requires an explanation for the evolution of large versus small gametes. Models by Charlesworth (1978), Hoekstra (1987), and other authors show that anisogamy will evolve if one genotype is favored because the large size of its gametes enhances the survival of the offspring, and another is favored because it can make many gametes. If alleles for different gamete sizes occur in a population that has two alleles at a mating type locus, an association will develop between alleles at the two loci (linkage disequilibrium), so that each type of gamete will tend to unite with the other, rather than with its own kind. Individuals that produce a third, intermediate, gamete size enjoy neither the advantage of size nor the advantage of numbers, and thus have lower fitness. Thus two types of gametes, of highly disparate size, are expected to evolve.

Sex Ratios, Sex Allocation, and Sex Determination

Among organisms with distinct female and male sexual functions (defined by gametes of different sizes), some are dioecious (have separate sexes in various sex ratios), and some are hermaphroditic, with either simultaneous or sequential female and male functions. The theory of sex allocation has been developed to explain this variation (Charnov 1982; Bull and Charnov 1988; Frank 1990). Some of this theory is complex and difficult. We will touch on only a few high points, beginning with sex ratio in dioecious species.

Sex Ratios in Randomly Mating Populations

We will define the SEX RATIO as the proportion of males, and must distinguish the sex ratio in a population (the POPULATION SEX RATIO) from that in the progeny of an individual female (an INDIVIDUAL SEX RATIO). Why is it that, in many species, both of these are approximately 0.5? In many groups of animals and some plants, the 1:1 segregation of sex-determining X and Y chromosomes in meiosis may constrain the possible variation in sex ratio. However, genes causing MEIOTIC DRIVE (see Chapter 12) of one or the other sex chromosomes are known in *Drosophila*, mosquitoes, lemmings, and some other species; individuals bearing such a "driven" chromosome produce an excess of X- or Y-bearing sperm, causing a biased sex ratio among their progeny. Moreover, female Hymenoptera can determine the sex ratio of their progeny by allowing each egg to be fertilized or not (haplodiploid sex determination: see Chapter 20); and in some species of turtles and other organisms,

sex is determined partly or entirely by environmental conditions. Thus the potential for evolution of altered sex ratios exists, at least in some species.

Why the sex ratio should have any particular value, such as 0.5, was a puzzle to Darwin, who did not see how it could evolve by individual selection. If all females have the same total number of progeny, why should a genotype with a sex ratio of 0.5 have an advantage over any other? The solution to the puzzle was provided by the great population geneticist R. A. Fisher (1930), who realized that *because every individual has both a mother and a father, females and males must contribute equally, on average, to the ancestry of subsequent generations, and must therefore have the same average fitness. Therefore, individuals that vary in individual sex ratio can differ in the number of their grandchildren (and later descendants), and thus differ in fitness if this is measured over two or more generations.*

In a large, randomly mating population, the fitness of a genotype that produces a given individual sex ratio depends on the population sex ratio, which in turn depends on the frequencies of the various sex ratio genotypes. That is, the evolution of the sex ratio is governed by *frequency-dependent selection, which will favor genotypes whose progeny are biased toward that sex that is in the minority in the population as a whole.*

For example, suppose the population sex ratio is 0.25 (1 male to 3 females) because the population consists of a genotype with this individual sex ratio. Suppose each female has 4 offspring. The average number of progeny per female is 4, but since every offspring also has a father, the average number of progeny sired by a male is 12 (since each male mates with 3 females on average). Thus a female has 4 grandchildren through each of her daughters and 12 through each of her sons, for a total of 24 grandchildren. Now suppose a rare genotype with an individual sex ratio of 0.50 (2 daughters and 2 sons) enters the population. Each such individual has $2 \times 4 = 8$ grandchildren through her daughters and $2 \times 12 = 24$ grandchildren through her sons, for a total of 32. Since this is a greater number of descendants than the mean number per individual of the prevalent genotype, any allele that causes a more male-biased individual sex ratio in this female-biased population will increase in frequency. The converse holds if the population sex ratio is male-biased: an allele for a female-biased individual sex ratio will increase. If, however, the population is monomorphic for a genotype that produces an even sex ratio (0.5), no genotype with a male-biased or female-biased sex ratio can spread, because if it were to increase in frequency, the population sex ratio would become skewed, providing an advantage to the prevalent unbiased genotype by the mechanism we have just reviewed (Figure 21.8). Therefore, a genotype that produces a sex ratio of 0.5 represents an *evolutionarily stable strategy* (ESS; see Chapter 20).* It can be proven that the fittest genotype has the maximal product $m \times f$, where m and f are (in the simplest case)

---

*Fisher's theory actually holds that the fittest genotype *allocates equal resources* to daughters and sons (see Charnov 1982).

FIGURE 21.8 Frequency-dependent selection on sex ratio. Mutants may arise in a population that vary in individual sex ratio ($\hat{r}$, the proportion of sons among a female's progeny). The fitness of each such mutant, based on its average number of grandchildren, depends on the sex ratio in the population (r). The average fitness of individuals in the population equals 2. The gray line shows the fitness of each possible mutant when the population sex ratio is 0.25—i.e., when 25 percent of the population is male. The fitness of mutants is then directly proportional to the proportion of sons among their offspring, and is greater than the mean fitness of the prevalent genotype if the mutant's sex ratio ($\hat{r}$) exceeds 0.25. Any such mutant will therefore increase in frequency. Conversely, if the population sex ratio is 0.75, the fitness of mutants, shown by the colored line, is inversely proportional to their individual sex ratio, and a mutant increases if $\hat{r}$ is less than 0.75. The lines cross at r = 0.50, which is the evolutionarily stable strategy, or ESS. (After Charnov 1982.)

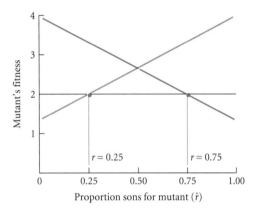

The Evolution of Sex Ratios in Structured Populations

In many species, mating occurs not randomly among members of a large population, but within small groups descended from one or a few founders. After one or a few generations, progeny emerge into the population at large, then colonize patches of habitat and repeat the cycle. In many species of parasitoid wasps, for example, the progeny of one or a few females emerge from a single host and almost immediately mate with each other; the daughters then disperse in search of new hosts. Such species often have sex ratios different from 0.5, the ratio predicted by Fisher's theory. Usually they have a preponderance of females.

The British evolutionary theorist and entomologist William Hamilton (1967) explained such "sex ratios" by what he termed LOCAL MATE COMPETITION. Whereas in a large population a female's sons compete for mates with many other females' sons, they compete only with one another in a local group founded by their mother (and compete largely with one another if the group was founded by just a few females). Hamilton reasoned that the founding females' genes would be propagated most prolifically by producing mostly daughters, with only enough sons to in-

FIGURE 21.9 Changes in sex ratio (proportion of males) in four experimental populations of platyfish, initiated with sex ratios of (A) 0.25 and (B) 0.78. The sex ratio evolved to nearly 0.5 within only two generations. Black and colored lines represent replicate populations. (After Basolo 1994.)

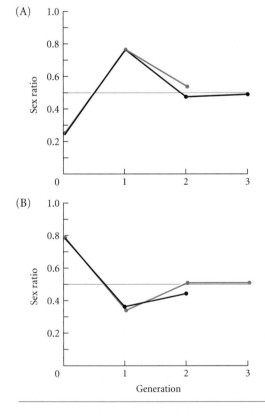

the numbers of male and female progeny (Charnov 1982). (Note that $m \times f$ is maximized when $m = f$.)

The theory can be tested only if there is genetic variation in individual sex ratio. Such variation exists in the platy *Xiphophorus maculatus*, which has three kinds of sex chromosomes, W, X, and Y. Female fish are XX, WX, or WY, and males are XY or YY. Of the six possible crosses, four yield a sex ratio of 0.5, but XX mated with YY yields all sons, and WX mated with XY yields 0.75 sons. Alexandra Basolo (1994) set up experimental populations in which each chromosome was marked with a different color pattern allele, so that genotype frequencies could be measured. Two populations were initiated with 75 percent females [the genotype frequencies were 0.15 WX, 0.60 XX (females), 0.15 YY, 0.10 XY (males)], and two were initiated with only 22 percent females [the genotype frequencies were 0.15 WX, 0.07 WY (females), 0.78 XY (males)]. Within only two generations, the sex ratio evolved nearly to 0.5 (Figure 21.9), due to changes in the frequencies of the three chromosome types that corresponded to theoretical predictions. The Y chromosome (borne only by males), for example, increased in frequency in the initially female-biased populations and decreased in the male-biased populations.

seminate all of them (additional sons would be redundant, since they all carry their mother's genes, including genes for sex ratio). Examining this problem further, David Sloan Wilson and Robert Colwell (1981; see also Colwell 1981; Frank 1986; Antolin 1993) showed that within each local group, even if founded by a single female, the sex ratio favored by individual selection is 0.5. However, *groups founded by female-biased genotypes contribute more individuals (and genes) to the population as a whole than groups founded by unbiased genotypes.* The difference among groups in production of females increases the frequency, in the population as a whole, of female-biasing alleles. This theory is identical in structure to the group selectionist interpretation of the evolution of "altruism" by kin selection, as illustrated in Figure 14 in Chapter 20. The greater the number of founders of a group, the more nearly even the optimal sex ratio will be.

Evidence supporting this theory was obtained by Allen Herre (1985, 1987), who measured the individual sex ratio in several species of fig wasps (Agaonidae), in which progeny develop and mate within a fig inflorescence before dispersing to lay eggs in other inflorescences. In these species, both local mate competition (Hamilton's theory) and inbreeding*are expected to favor the evolution of a female-biased sex ratio. Among species of fig wasps, the average level of inbreeding is higher, the lower the average number of foundresses (ovipositing females) per inflorescence; this number can be determined readily since foundresses die inside the inflorescence after oviposition. If, using the mechanism that all Hymenoptera share, foundresses can adjust the sex ratio of their offspring in response to the number of cofoundresses they encounter, we would expect that lower (more female-biased) sex ratios will be found, the lower the number of cofoundresses. Herre found that species differ as predicted: those that presumably experience greater inbreeding have lower sex ratios. Each species, moreover, adjusts its individual sex ratio to the number of cofoundresses (Figure 21.10).

A similar adaptive plasticity in sex ratio has been demonstrated in experiments on parasitoid wasps, such as *Nasonia vitripennis*, in which the progeny of one or more females develop in a fly pupa and mate with each other immediately upon emergence. The second wasp that lays eggs in a host can detect previous parasitization, so we would expect her to adjust her individual sex ratio to a higher value than that

---

*The expected effect of inbreeding on the evolution of sex ratio is a consequence of kin selection (see Chapters 12, 20) in these haplodiploid species, in which the relatedness of daughter to mother (the proportion of the daughter's genes that are identical by descent with those of the mother) increases with inbreeding, whereas the relatedness of son to mother does not (since sons do not have fathers, all of a male's genes are necessarily identical by descent with his mother's). Thus as inbreeding increases, a mother, by having a greater proportion of daughters, will transmit more genes to subsequent generations. This inbreeding effect does not apply to autosomal genes if both sexes are diploid.

FIGURE 21.10   Evidence for the theory of sex ratio evolution in fig wasps. (A) The sex ratio predicted by local mate competition theory as a function of the number of foundresses of a group. Each curve shows the expected sex ratio under a different level of inbreeding, measured by the proportion, $1/n$, of sib mating within the group (for example, the curve labeled 0.5 is expected if 1/2 the matings are between sibs). (B) Empirical data on sex ratio for three species of fig wasps. Females of each species increased the sex ratio of their progeny in response to the number of individual females laying eggs within a single fig inflorescence. The species depicted by the upper, middle, and lower curves have different proportions of sib mating: 1/1.99, 1/1.40, and 1/1.13 respectively. (After Herre 1985.)

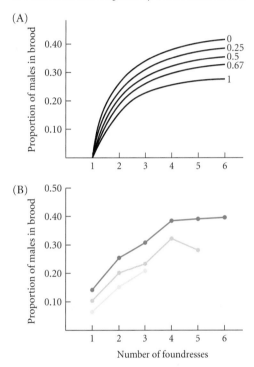

of the first female. John Werren (1980) calculated the theoretically optimal sex ratio of a second female, then measured the sex ratios of second females by exposing fly pupae successively to strains of *Nasonia* that were distinguishable by an eye color mutation. On the whole, his data fit the theoretical prediction very well (Figure 21.11).

Sex Allocation, Hermaphroditism, and Dioecy

Closely related to the evolution of sex ratio is the topic of sex allocation. What evolutionary factors affect how hermaphroditic species allocate resources to female and male functions? And why, in many taxa, are all resources allocated to one or the other—in other words, why do some taxa have separate sexes? We can treat these complex topics only superficially.

Some animals and plants switch sex function once during their lifetimes, from male to female (PROTANDRY) or vice versa (PROTOGYNY). Whether it is advantageous to begin as one sex or the other depends on how reproductive success

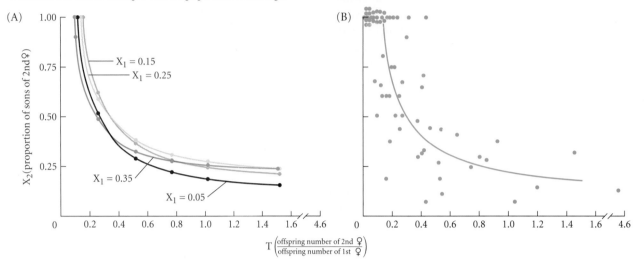

FIGURE 21.11 Adaptive adjustment of sex ratio by a parasitoid wasp. (A) The theoretically optimal sex ratio for the progeny of the second female to parasitize a host (X_2), as a function of the ratio of the number of her offspring to that of the first female (T), and as a function of the first female's sex ratio (X_1). (B) The points show the actual proportions of sons of second females in an experimental population. Although variation is high, the trend fits the curved line, which is the theoretical curve, as in A, based on the average sex ratio of first females' progeny. The fewer the progeny of the second female, the more of them are sons. This "strategy" is adaptive because these sons have the opportunity to mate with the many daughters of the first female. (After Werren 1980.)

as a male or as a female changes as a function of an individual's age or size (see Figure 13 in Chapter 19). If the gain in reproductive success via female function increases more steeply with these characters than the gain via male function, protandry is likely to be advantageous; if the reverse, then protogyny is expected.

One of many examples that conforms to the theory is the jack-in-the pulpit (*Arisaema triphyllum*), which can produce either male or female inflorescences at any point in its life, but is more likely to be female if it is large and/or occupies a resource-rich environment. The development of seeds and fruit imposes an additional cost on females, which therefore must allocate more resources to reproduction than males. Because larger plants can photosynthesize more and store more resources, it is evidently advantageous for sexual function to change with size (Figure 21.12).

The theory of whether it is advantageous to be hermaphroditic (cosexual) or dioecious depends in part on whether the species is a strict outbreeder or is capable of self-fertilization (selfing). Assume that in an outbreeding species, allocation of an individual's resources to one sex function (say, egg production) decreases allocation to the other sex function—i.e., that there is a trade-off. Fitness accrued through either function increases as allocation to that function increases, but it may not increase linearly (Figure 21.13A). If greater allocation to either sexual function yields diminishing returns, in terms of fitness, the relationship between fitness accrued through male function and that accrued through female function is convex: greater total fitness can be obtained by playing both sex roles than by playing exclusively one. Thus selection favors cosexuality. In cosexuals, the optimum allocation to female versus male function (e.g., to seed versus pollen production) is that combination that maximizes the product $m \times f$, where these represent the number of offspring produced via male and via female function respectively. (Note the correspondence to the optimal individual sex ratio in dioecious species, described above.)

If, conversely, reproductive success via either male or female function increases more than linearly with increasing

FIGURE 21.12 The size-frequency distribution of male and female jack-in-the-pulpit plants (*Arisaema triphyllum*). Plants switch from male to female as they grow in size. (After Policansky 1981.)

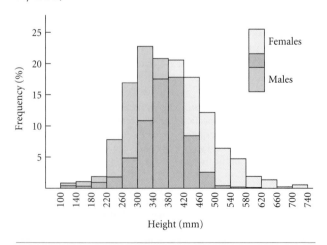

FIGURE 21.13 The theory of sex allocation. (A) Reproductive success of an individual as a function of the fraction of resources allocated to one sexual function (say, female) rather than the other. Increasing allocation to that sexual function may yield decelerating (i), linear (ii), or accelerating (iii) gains in reproductive success. (B) Reproductive success (RS) gained through female function, plotted against that gained through male function. Because resources are allocated between these functions, there is a trade-off between the reproductive success an individual achieves through either sex function. This trade-off is linear, and the sum of female RS and male RS equals 1.0 at each point on the trade-off curve, if RS is linearly related to allocation (curve ii in A). If the gain in RS is a decelerating function of allocation to one sex or the other (curve i in A), then the fitness of a hermaphrodite exceeds that of a unisexual individual (with RS = 1.0 for one sexual function and 0 for the other). If RS is an accelerating function of allocation (curve iii in A), then the trade-off curve is concave, and dioecy is stable—that is, a hermaphrodite's fitness is lower than that of either unisexual type. (After Charnov 1982 and Thomson and Brunet 1990.)

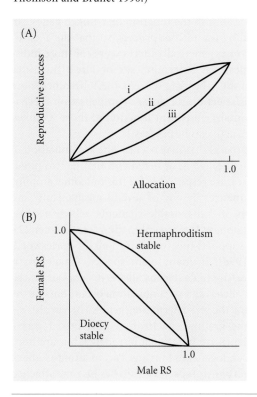

(A)

Reproductive success

i
ii
iii

1.0

Allocation

(B)

1.0

Hermaphroditism stable

Female RS

Dioecy stable

1.0

Male RS

allocation to that function, the relationship between fitness obtained through male and through female sex function is concave (Figure 21.13A), and hermaphrodites have lower fitness than either "pure" sex. Thus dioecy should evolve. One factor that may affect the shape of the trade-off curve (Figure 21.13B) is the cost of developing the structures required for the two sex functions. If male and female functions require different structures, a pure male or female pays the cost of developing only one set of structures, whereas a hermaphrodite pays both costs and thus is likely to have lower fitness.

The Evolution of Sex-Determining Mechanisms

Sex determination
The sex of an individual of a dioecious species can be determined in an astonishing variety of ways (Bull 1983). In ENVIRONMENTAL SEX DETERMINATION, an environmental factor, experienced early in development, determines sex. Temperature, for example, determines sex in crocodilians and in many turtles and lizards (Janzen and Pauktis 1991).

GENETIC SEX DETERMINATION comes in many forms. Most commonly, genes specifying sex are carried on one or both of two sex chromosomes. We are most familiar with species, such as mammals, *Drosophila*, and some dioecious plants, in which females are homogametic (XX) and males are heterogametic (XY), but the opposite is the case in birds and lepidopterans, and there are many taxa (for example, flies, amphibians, and lizards) in which some species have heterogametic males and others heterogametic females. In yet other species, the heterogametic sex is XO, meaning that it has a single X chromosome rather than an XY pair. Sex is determined by a small number of regulatory genes that activate male or female developmental pathways; in mammals, for example, a small region on the Y chromosome is responsible for maleness. Other than sex-determining genes, Y chromosomes usually bear relatively few functional genes, some of which may be homologous with loci on the X; the rest of a Y chromosome generally includes nonfunctional DNA. X and Y chromosomes recombine with each other only in short regions or not at all. In some species there is DOSAGE COMPENSATION, so that gene activity is equal in both sexes. In mammals, for example, one of the two X chromosomes of females is inactivated in somatic tissues, and in *Drosophila*, X-linked loci in the male have greater transcription rates.

The evolution of sex chromosomes
There are many examples of intermediate levels of genetic and structural differentiation between X and Y chromosomes, implying that they originated from a pair of homologous chromosomes much like a pair of autosomes. This observation provides the starting point for a theory of how and why sex chromosomes have evolved in many groups of organisms (Charlesworth 1991, Rice 1996).

The first steps in this model can be illustrated by some plants in which there exist cosexual and female individuals (gynodioecy; Figure 21.14A). Femaleness and maleness are determined by single, separate loci F and M, which affect the development of female and male sex organs. The superscripts f and s denote alleles for fertility and sterility. Suppose a recessive allele for male sterility (M^s), which transforms cosexuals into females, increases in frequency due to a reproductive advantage, as described in the section on sex allocation, resulting in a gynodioecious population. According to the theory of sex ratio, males will then be favored over cosexuals, since male function is more advantageous,

FIGURE 21.14 A model for the evolution of dioecy and sex chromosomes in plants. (A) In a population of cosexuals, some individuals are transformed by a male sterility mutation into females. The population is then polymorphic for cosexuals and females, a condition called gynodioecy. Other cosexuals are transformed into males by a female sterility mutation. If this allele increases in frequency so that cosexuals are eliminated, a dioecious population results. (B) A genetic model of the steps in A, showing the mutations for male sterility (M^s) and female sterility (F^s). The combinations M^sF^f and M^fF^s define proto-X and proto-Y chromosomes. (After Charlesworth 1991.)

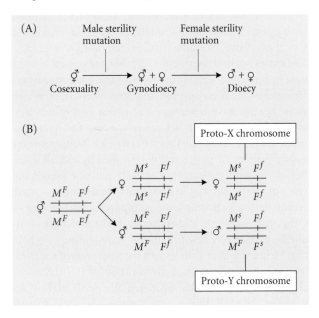

the greater the proportion of females in the population. Thus a dominant allele for female sterility (F^s) may increase in frequency, replacing the remaining cosexuals with males. Because individuals with both male sterility and female sterility alleles would have very low fitness, the evolution of dioecy from cosexuality is most likely to occur if the two loci are closely linked, so that the prevalent combinations of alleles are M^sF^f (a "proto-X" chromosome that determines a female when homozygous) and M^fF^s (a "proto-Y" chromosome that determines a male when heterozygous; Figure 21.14B).

At this stage, the chromosomes carry other functional, freely recombining genes besides sex-determining loci. Many such examples are known in animals and, especially, plants. However, any allele that is advantageous in one sex but disadvantageous in the other will spread if it is associated with a sex-determining allele. Suppose a mutation A' is advantageous to males but disadvantageous to females. If it is spread by recombination from M^fF^s (male-determining) chromosomes to M^sF^f (female-determining) chromosomes, it may not increase in frequency because of its deleterious effect on females. However, it can increase if it is restricted to M^fF^s chromosomes due to low recombination. An allele that reduces recombination in this region

will then have an advantage because it is associated with highly fit A' males. In this way, selection for reduced recombination between the sex-determining loci and other loci on the chromosome arises. (Genetic suppressors of recombination in specific chromosome regions have been well documented, as we noted above.) When recombination has been suppressed throughout much of the proto-X/proto-Y pair, the stage is set for several processes that lead to the genetic inactivation of most Y-linked genes and the accumulation of heterochromatin—the characteristics of "typical" Y chromosomes.

The first of these processes is Muller's ratchet. We have seen how, in an asexually reproducing species that lacks recombination, genomes with the fewest deleterious mutations may be progressively lost by genetic drift in small populations, resulting in an ever-mounting load of mutations. Less loaded genotypes cannot be reconstituted in the absence of recombination. Exactly the same principle applies to a non-recombining chromosome in a genome that otherwise engages in recombination. Thus mutations that reduce or abolish the function of Y-linked genes can become fixed by the ratchet process, so that the Y chromosome becomes genetically inert except for those indispensable genes that determine sex or affect functions important in the Y-bearing sex. (As the Y becomes more inert, there is selection for dosage compensation mechanisms to maintain equivalent function of those X-linked genes that are active in both sexes.)

Another process that may occur is the accumulation of repeated (satellite) DNA, or heterochromatin. Two molecular processes, are responsible for this outcome: amplification of sequences by any of several mechanisms, and accumulation of transposable elements, which become trapped in regions of low recombination (see Chapter 22).

Evidence supports much of this model (Charlesworth 1991; Rice 1996). In many plant species, femaleness is caused by recessive male sterility alleles, maleness is caused by dominant alleles, and recombination occurs outside the region carrying the sex-determining genes. There is also evidence for the gradual inactivation of Y-linked genes. Some such genes are functional in rodents, but are nonfunctional pseudogenes in humans. Part of an autosome of the *Drosophila obscura* species group has become attached (by translocation) to the Y chromosome of one member of this group, *Drosophila miranda*. This segment of the Y chromosome was probably added less than 5 million years ago, so it might be expected to be still deteriorating. As predicted, the density of transposable elements is far greater on the new segment of the Y than on any other chromosome, including the homologous chromosome in females. Moreover, many transposable elements have been inserted into genes on the new segment of the Y, and have reduced or abolished their function (Steinemann et al. 1993). In a rather complex experiment, William Rice (1996) showed that autosomal alleles that increase the fitness of one sex but reduce that of the other can increase in experimental populations of *Drosophila*, transforming the autosomes into proto-sex chromosomes.

The Evolution of Inbreeding and Outbreeding

Inbreeding (see Chapter 13) occurs when the probability that alleles in offspring will be identical by descent is greater than in a randomly mating population due to the union of gametes from related individuals or, in the extreme, from the same individual by self-fertilization (SELF-ING). Some of the most interesting variations in genetic systems are those caused by other biological characteristics that promote or prevent the extreme inbreeding that results from self-fertilization. Selfing can potentially occur in some hermaphroditic animals (such as certain snails) and in a great many plants, in which it has been most extensively studied. The evolution of selfing and avoidance of selfing has been reviewed by Jarne and Charlesworth (1993) and Uyenoyama et al. (1993).

Variation in Breeding Systems

Most cosexual plants can both self-pollinate and export pollen to the stigmas of other plants (OUTCROSSING). In many such species, some self-fertilization occurs. (The progeny of such species are referred to as "self" or "outcross" progeny, depending on their parentage.) The proportion of progeny, of a single plant or in the population as a whole, derived from selfing versus outcrossing can be estimated by several techniques, chiefly the use of genetic markers such as polymorphic allozymes. For instance, *Aa* progeny grown from the seeds of *AA* plants (the "ovule parents") must have been sired by other plants (the "pollen parents"). At least some *AA* progeny of *AA* plants may also be outcross progeny, and the proportion that are can be estimated from the allele frequencies in the population of plants from which pollen may have come. In practice, procedures for estimating outcrossing rates are more complex than we need go into here.

Plant species deviate from this "partial selfing" condition in many ways that can either increase or decrease the incidence of self-fertilization. Characteristics that *promote outcrossing* include:

Asynchronous male and female function. For example, pollen may be shed before or after the plants' stigmas are receptive.

Monoecy, i.e., separate female and male flowers on the same plant.

Dioecy, i.e., separate sexes.

Self-incompatibility. In gametophytic self-incompatibility, the growth of a pollen grain is inhibited if its allele (S_i) at the self-incompatibility locus matches either allele in the stigma (S_iS_j) on which it lands. Self-incompatibility is based on several proteins, in different plant families, that have been derived from proteins

with quite different functions (deNettancourt 1977; Uyenoyama 1995; Matton et al. 1994). Comparisons of the DNA sequences of self-incompatibility alleles within and among species (Ioerger et al. 1990; Clark 1993) have revealed immense variation in the encoded amino acid sequences—up to 40 percent—among alleles within species. Some alleles within a species are less closely related to each other than to genes from other species that diverged 27 to 36 million years ago (Figure 21.15). That a polymorphism should have been retained for such a long time is explicable only by balancing selection (Clark 1993). This selection is frequency-dependent, since more stigmas in a population will be receptive to pollen carrying rare *S* alleles than common ones (see Chapter 13).

HETEROSTYLY is a polymorphism, first studied by Darwin (1877a), for either two (distyly) or three (tristyly) forms of flowers, borne by different plants. In each morph, the style is situated at one of two (or three) levels and the anthers at a different level (Figure 21.16). There is considerable evidence that, as Darwin proposed, pollen is most effectively delivered between, rather than within, morphs, from the anthers of one plant to a stigma of the same height on another plant (Barrett 1992b).

FIGURE 21.15 A phylogeny, or genealogy, of some genes at the self-incompatibility locus in three species of Solanaceae: *Nicotiana alata* (nic), *Petunia inflata* (pet), and *Solanum chacoense* (sol). The tree is based on DNA sequences, and the branch lengths are proportional to the number of amino acid substitutions. Note that some alleles in *Nicotiana* (tobacco) are more closely related to alleles in the other two plant genera than to each other, indicating that multiple gene lineages have persisted in *Nicotiana* since before the several genera diverged 27 to 36 million years ago. (After Ioerger et al. 1990.)

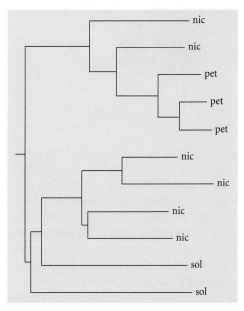

FIGURE 21.16 (A) A diagram of flower morphs in distylous and tristylous species. Ovals represent anthers; forks, stigmas. Compatible pollinations are shown by arrows; other pollinations usually produce little or no seed due to incompatibility. L, M, and S refer to long-, mid-, and short-styled morphs. (B) The long-, mid-, and short-styled morphs of the tristylous species *Eichhornia paniculata*. (A after Barrett 1992b; B courtesy of S. C. H. Barrett; drawings by E. Campolin.)

(A)

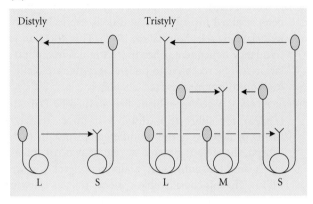

(B)

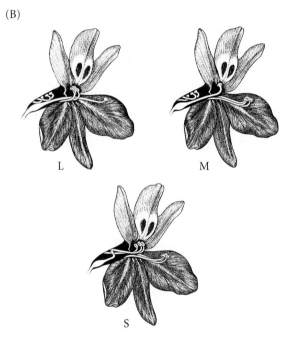

In contrast, many plants have evolved a strong tendency toward self-fertilization within flowers. In fact, a shift from cross-pollination to predominant selfing may have been more common than any other single evolutionary change in angiosperms (Stebbins 1974). Wheat (*Triticum*) is a familiar example of a normally self-fertilizing plant. Selfing species are, of course, self-compatible, and they commonly display a syndrome of floral characteristics that includes a reduction in the size of the flower and many of its parts, the reduction or loss of features such as scent and markings that attract pollinators, a reduction in the

amount of pollen produced, and alteration of the length of the pistil and stamens so that the anthers and stigma are adjacent (Figure 21.17; Wyatt 1988). In extreme cases, flowers are CLEISTOGAMOUS: they remain budlike and do not open at all.

Consequences of Inbreeding

To recapitulate briefly some points made in Chapter 11, selfing and other forms of inbreeding redistribute genes from the heterozygous to the homozygous state without altering allele frequencies. Homozygosity increases with each generation of inbreeding, resulting ultimately in a variety of completely homozygous genotypes (inbred lines). This process ordinarily increases the genetic variance of a character. As homozygosity increases, the opportunity for recombination between heterozygotes at two or more loci decreases; thus any initial linkage disequilibrium can become permanent.

The increasing homozygosity in an inbreeding population is usually accompanied by INBREEDING DEPRESSION, a reduction in the mean values of characteristics related to fitness. Whether inbreeding depression is caused by homozygosity for deleterious recessive alleles that are maintained in populations solely by recurrent mutation, or by homozygosity at overdominant loci, at which heterozygotes have highest fitness (see Chapter 13), has long been debated. Most population geneticists hold that recessive deleterious alleles are largely responsible, but the possibility of a few overdominant loci, which could have an important effect on mean fitness, cannot be ruled out (Uyenoyama et al. 1993).

When a population of outcrossing plants begins to inbreed, mean fitness declines at first. However, deleterious re-

FIGURE 21.17   An example of the evolution of selfing from outcrossing. (A) The mid-styled morph of *Eichhornia paniculata*, one of three morphs in an outcrossing population in Brazil (see Figure 21.16). The stigma is distant from the three short and three long stamens, and thus receives outcross rather than self pollen. (B) A modified mid-styled morph from a monomorphic, selfing population in Jamaica. Three of the anthers are situated close to the stigma, which thus receives mostly self pollen. The flower is smaller and less contrastively marked than outcrossing flowers, and is thus less conspicuous to pollinating insects. (Drawings by E. Campolin, courtesy of S. C. H. Barrett.)

(A)   (B)

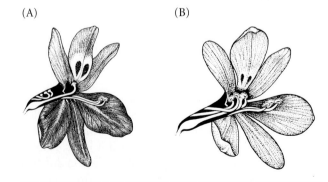

cessive alleles that had been masked in heterozygotes by dominant alleles now become exposed to and eliminated by selection, which may therefore bring an inbred population to a genetic equilibrium at which its mean fitness can equal or even exceed that of the initial outbreeding population (Lande and Schemske 1985).

Evidence for this theoretical course of evolution has been described by Spencer Barrett and Deborah Charlesworth (1991), who studied the tropical American aquatic plant *Eichhornia paniculata*, a tristylous species in which most populations have outcrossing rates exceeding 90 percent. Some populations, however, consist of only one morph, in which the flowers are reduced in size and are almost exclusively self-fertilizing (see Figure 21.17). For five generations, Barrett and Charlesworth self-pollinated plants from a highly outcrossing Brazilian population and from an almost exclusively selfing Jamaican population. They then cross-pollinated plants to produce outcross progeny of both the Brazilian and Jamaican lines. They retained seeds from all the generations and, at the end of the experiment, grew them under uniform conditions in a greenhouse. In the ordinarily outcrossing Brazilian lines, flower number declined as inbreeding proceeded (Figure 21.18A), but showed a dramatic increase in crosses between the inbred lines (heterosis, or "hybrid vigor"). In contrast, the naturally selfing Jamaican lines showed neither inbreeding depression nor heterosis when crossed with each other (Figure 21.18B). This result is exactly what would be expected if natural selection had purged the Jamaican population of deleterious alleles.

Advantages of Inbreeding and Outcrossing

Beginning with a partially selfing plant, in which individuals both self-fertilize and disperse pollen, let us consider what factors might favor evolution either toward exclusive selfing or toward exclusive outcrossing. R. A. Fisher (1941) first pointed out that *a partial selfer automatically has a strong selective advantage over an exclusive outcrosser*, because it can transmit genes in three ways: through its ovules, through its pollen by selfing, and through its pollen as a sire of outcross progeny. An exclusively outcrossing genotype, however, transmits genes only through its ovules and through its outcrossed pollen. On average, the ratio of genes transmitted by partial selfers and by outcrossers is 3:2, and partial selfing has a 50 percent advantage. This calculation assumes that so little pollen is used for selfing that it hardly affects the amount dispersed by outcrossing, that the cost (e.g., in energy) of producing self and outcross progeny is equal, and that all offspring have equal fitness.

Another advantage of partial selfing (or full selfing) is RE-PRODUCTIVE ASSURANCE: a plant is almost certain to produce some seeds by selfing, even if it is not pollinated due to scarcity of pollinators, low population density, or other adverse environmental conditions.

For exclusive outcrossing to evolve, its advantage must be greater than the 50 percent disadvantage described in

FIGURE 21.18   The mean number of flowers produced by *Eichhornia paniculata* plants from (A) Brazil and (B) Jamaica in the first outcross generation (O1), in five generations of selfing (S1–S5), and in outcrosses between selfed plants after five generations (O5). The naturally outbred Brazilian population displayed inbreeding depression (compare O1 through S5) and heterosis in outcrossed plants (O5), but the naturally inbred Jamaican population did not. (After Barrett and Charlesworth 1991.)

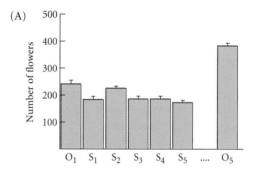

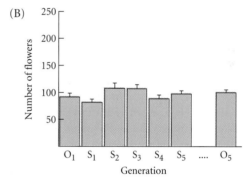

Fisher's model. The most important advantage is avoidance of inbreeding depression. An allele promoting outcrossing will increase if the fitness of outcross progeny is sufficiently greater than that of self progeny (Lloyd 1992). Similarly, dioecy may evolve as a mechanism to prevent production of unfit selfed progeny (Charlesworth and Charlesworth 1978).

Whatever factors might promote the evolution of self-fertilization would have to overcome two obstacles: inbreeding depression, and a reduction in the plant's genetic contribution to the population if increased selfing reduces the amount of outcross pollen it produces. Cleistogamous flowers, for example, sire no progeny by outcrossing.

One possible compensation for these disadvantages is that exclusive selfers may expend fewer resources on pollination. For example, exclusive selfers have small flowers and produce little pollen. Although this advantage may stabilize selfing after it has evolved, it seems unlikely to account for the origin of exclusive selfing, because a slight initial reduction of flower size seems unlikely to offset the advantages of outcrossing (Jarne and Charlesworth 1993).

Another possible advantage of selfing is that it preserves highly fit genotypes, by preventing the "outbreeding depression" that may result from crossing (Thornhill 1993). For example, Nikolas Waser and Mary Price (1989) crossed *Ipomopsis aggregata* plants separated by various geographic distances, planted the seeds under uniform conditions, and measured fitness-related characteristics in the progeny. The progeny had higher fitness if their parents were moderately distant than if their parents were situated either close together (which Waser and Price attributed to inbreeding depression), or far apart (outbreeding depression). Waser and Price postulated that distant populations have different coadapted gene combinations, perhaps adapted to different environmental conditions. (See Chapters 9 and 15 for examples of local coadaptation.) Recombination between such gene combinations, resulting in progeny with lower fitness, would be prevented by selfing. Whether or not this effect provides a sufficient explanation for the evolution of selfing is not known.

Counter to intuition, selfing might evolve because it produces homozygous genotypes (Holsinger 1991, Uyenoyama et al. 1993). Although most homozygous inbred lines are less fit than outbred genotypes, some few of them might have higher fitness. If such a highly fit gene combination includes an allele for selfing, it is perpetuated without recombination, and may replace the outcrossing genotypes. Selfing can evolve, according to this model, if inbreeding depression is caused by homozygosity for deleterious recessive alleles. If it is caused by homozygosity at loci with heterozygote advantage, the lower fitness of the homozygotes will cause the elimination of the selfing allele that they carry.

The most frequently cited advantage of selfing is *reproductive assurance*: even if selfers produce progeny of low fitness, they are guaranteed to reproduce, whereas reproduction by outcrossing depends on a sufficient supply of pollinators and a high enough population density of the plant species. There is abundant support for this hypothesis (Jarne and Charlesworth 1993; Wyatt 1988). The "selfing syndrome" described earlier is especially common in plants that grow in harsh environments, where insect visitation is low or unpredictable, and on islands, where populations are sparse for some time after colonization.

Of the ecological and genetic factors that may favor a change of breeding system, only the advantage of selfing due to reproductive assurance in pollinator-poor environments seems to have been well documented. Thus a subject rich in theory awaits further study, to which ecological, genetic, and phylogenetic approaches will all undoubtedly contribute.

Avoidance of Inbreeding in Animals

Whether or not adaptations to avoid inbreeding have evolved in dioecious animals is more difficult to determine than in plants, in which characteristics such as self-incom-

patibility can hardly be attributed to any other possible advantage. *The major factors that may reduce inbreeding in animals are dispersal* of one or both sexes so that close relatives are unlikely to mate; and *behavioral avoidance of matings with relatives* even when the opportunity is presented (reviewed by numerous authors in Thornhill 1993).

Dispersal before breeding can be advantageous for many reasons; whether or not it is an adaptation to avoid inbreeding is difficult to determine. However, there has been considerably research on "incest avoidance": avoiding matings with relatives. In most species capable of kin recognition, there is little evidence that matings between relatives occur any less often than would be expected by chance, except in some mammals and perhaps some birds.

In some species of vertebrates, close relatives, such as siblings or parents and their offspring, appear not to mate with each other. For example, chimpanzees (*Pan troglodytes*) consort with a favorite male, often their brother, until they reach reproductive age (at about 10 years of age). When they become reproductive, they change their affiliation, and consort and mate with other males (Pusey 1980). Although there are plenty of opportunities for mating between siblings and between mothers and sons, no such copulations were seen, despite extensive observations. In several species of rodents, sibling pairs have a much reduced likelihood of mating; for example, only 6 percent of sibling pairs of prairie voles (*Microtus ochrogaster*) bred, whereas 78 percent of nonsibling pairs did so (McGuire and Getz 1981). There is some evidence that mating in house mice occurs preferentially between individuals that differ at the major histocompatibility (MHC) loci, which also play an important role in the immune response (Potts et al. 1991).

Behavioral avoidance of inbreeding inevitably brings to mind the well-known "incest taboo," often codified in religious and civil law, in human societies (see Chapter 26). This is a highly controversial topic, with respect to both the actual incidence of inbreeding ("incest") and the interpretation of the social taboo. Societies vary as to which kin matings are prohibited (Ralls et al.1986). Moreover, the incidence of closely incestuous sexual activity is evidently far higher than society has generally been ready to recognize, especially in the form of sexual attention forced upon young women by fathers and uncles. Even if this activity is not consensual, it raises doubt about whether a strong, genetically based aversion to incest has evolved in our species (at least in males). The explanations offered by anthropologists to account for incest taboos include others besides the hypothesis that they have evolved to avoid inbreeding depression. For example, some anthropologists hold that outbreeding is a social device (not an evolved genetic trait) to establish coalitions between families or larger groups, in order to reap economic and other benefits of cooperation.

Summary

1. Features of genetic systems, such as rates of mutation and recombination, sexual versus asexual reproduction, sex ratio, and rates of inbreeding, differ among species and therefore must have evolved by changes in genes that modify these features. Although these characteristics (like all others) can affect the survival of species, we seek their explanations in natural selection and other processes *within* species.

2. Because most mutations that affect fitness are deleterious, alleles that increase mutation rates are generally selected against (except when closely linked to an advantageous allele). In general, therefore, we would expect mutation rates to evolve to the minimal achievable level, even if this should reduce genetic variation and increase the possibility of a species' extinction.

3. In a constant environment, alleles that decrease the recombination rate have a selective advantage because they lower the proportion of their bearers' offspring that have unfit recombinant genotypes. In sexual organisms with separate male and female functions, asexual reproduction has approximately a twofold advantage over sexual reproduction because only half of the offspring of sexuals (i.e., the females) contribute to population growth, whereas all of the (all-female) offspring of asexuals do so. Therefore, the prevalence of recombination and sex requires some explanation.

4. Among eukaryotes, asexual lineages have evolved many times, but almost all extant asexual lineages have arisen very recently, implying that asexual populations have a high extinction rate. Thus sex has a group-level advantage in the long term, but it is doubtful that this could prevent asexual reproduction from replacing sexual reproduction in the short term.

5. The major hypotheses for the short-term advantage of sex are: (*a*) Recombination may have evolved as a means of repairing damaged DNA. This hypothesis may explain the origin of recombination, but probably not the evolution of eukaryote sex. (*b*) The rate of fixation of combinations of advantageous mutations may be higher in a sexual population, but only if it is large. This is a long-term, and probably rather infrequent, advantage that does not solve the short-term problem. (*c*) Recombination enables the mean of a polygenic character to evolve to new optima in a fluctuating environment; alleles promoting sex and recombination are preserved by their association with the newly favored genotypes. There is some evidence supporting this hypothesis.(*d*) In a heterogeneous environment, a sexual parent's genetically diverse progeny may partition resources more efficiently than an asexual parent's. (*e*) In small asexual populations, genotypes with few deleterious mutations may be lost by genetic drift, so that the population steadily acquires a greater genetic load that may reduce its fitness (Muller's ratchet). In a sexual population, mutation-free genotypes can be reassembled by recombination. (*f*) Deleterious mutations that interact to reduce fitness can be more effectively purged by natural selection in sexual than in asexual populations, thus maintaining higher mean fitness.

6. The advantage of diploidy over haploidy is that it masks deleterious recessive mutations. Haploid populations could evolve to be diploid, but the reverse is unlikely to occur because the first such individuals would suffer low fitness due to exposed deleterious mutations.

7. Female and male sexes are defined by the difference in the size of their gametes. The prevalent theory holds that distinct sexes evolved because it is advantageous on the one hand to produce numerous gametes (sperm), and on the other hand to produce large gametes (eggs) because the zygotes' survival is enhanced by greater size.

8. In large, randomly mating populations, a genotype with a 1:1 individual sex ratio cannot ordinarily be replaced by a genotype with any other, because a genotype with a biased individual sex ratio produces fewer grandchildren, per capita, through the offspring belonging to the sex that is produced in excess.

9. When populations are characteristically subdivided into small groups whose offspring then colonize patches of habitat anew, a female-biased sex ratio can evolve because female-biased groups contribute a greater proportion of offspring to the population as a whole. This difference in productivity is a form of group selection.

10. The evolution of hermaphroditism versus dioecy depends on how reproductive success via female or male function is related to the allocation of an individual's energy or resources to each function. Dioecy is advantageous if the reproductive "payoff" from one or the other sexual function increases disproportionately with allocation to that function. Dioecy may also be advantageous because it prevents self-fertilization, which may result in offspring with low fitness (inbreeding depression).

11. A hypothesis, supported by some evidence, has been developed to explain how sex chromosomes evolve.

12. In some animals, and especially in plants, various lineages have evolved mechanisms that promote either outcrossing or self-fertilization. All other things being equal, a genotype that both self-fertilizes and outcrosses has a 50 percent advantage over an obligate outcrosser. However, outcrossing can be advantageous because it prevents inbreeding depression in an individual's progeny. Conversely, self-fertilization may evolve for several reasons: fewer resources need be expended on reproduction; reproduction is assured even if the population is sparse or pollinators are few; and an allele for selfing may increase if it is associated with advantageous homozygous genotypes.

13. In animals with separate sexes, inbreeding consists of mating among close relatives. In a few species, there is evidence that such matings are avoided.

Major References

The important references for the many topics treated in this chapter are too numerous to list here in full, and may be found throughout the text and in the bibliography. On the *evolution of sex and recombination*, some relatively recent summaries are the books by Maynard Smith (1978), Bell (1982), Stearns (1987), and Michod and Levin (1988) and articles by Charlesworth (1989) and Kondrashov (1993). The books cited also treat some of the related issues. On the *evolution of sex ratios and sex allocation*, see Charnov's (1982) book and articles by Bull and Charnov (1988) and Antolin (1993). On the *evolution of sex chromosomes*, see Charlesworth (1991) and Rice (1996), and on the *evolution of sex-determining mechanisms* in general, see Bull (1983). The *evolution of inbreeding and outcrossing* is treated by Jarne and Charlesworth (1993), Uyenoyama et al. (1993), and Holsinger (1996).

Problems and Discussion Topics

1. Is there a cost of sex in hermaphroditic species? Is there a cost of meiosis in isogamous species such as certain algae?

2. Populations of some species of fish, insects, and crustaceans consist of both sexually and obligately asexually reproducing individuals. Would you expect such populations to become entirely asexual or sexual? What factors might maintain both reproductive modes? How might studies of such species shed light on the factors that maintain sexual reproduction?

3. Some mites, scale insects, and gall midges display "paternal genome loss" (or *pseudoarrhenotoky*, a fine word for parlor games). Males develop from diploid (fertilized) eggs, but the chromosomes inherited from the father become heterochromatinized and nonfunctional early in development, so that males are functionally haploid. How might this peculiar genetic system have evolved? (See Bull 1983.)

4. *Genetic conflicts* occur within a genome, stemming from the "selfish gene" principle that any gene that bequeaths more copies to subsequent generations than a competing gene has a selective advantage, by definition. Consider a maternally inherited gene, such as those carried by mitochondria or (in most plants) chloroplasts. Explain why you might expect such genes often to cause male sterility (i.e., sterility of the male sexual organs), and find out whether this is in

fact the case. (See Gouyon and Couvet 1987; Frank 1989.)

5. The Y chromosome of mammals carries genes for growth factors that enhance the ability of a fetus to extract nourishment from its mother. The X chromosome carries genes that suppress these Y-linked genes. Can you devise a hypothesis to explain these observations, based on the principle of genetic conflict? (See Hurst et al. 1996.)

6. Whereas some species of plants grow as single stems, others grow by runners or stolons, so that the many stems of a single genetic individual may cover many acres. Assuming a trade-off between such vegetative spread and sexual reproduction, what circumstances may favor the evolution of vegetative propagation? How would you measure the fitness of genotypes that differ in their allocation to sexual reproduction versus vegetative spread?

7. Many parthenogenetic "species" of plants and animals are known to be genetically highly variable. Determine from the literature whether this genetic variation is due to mutation within asexual lineages or to multiple origins of asexual genotypes from a sexually reproducing ancestor.

8. In most Diptera (true flies), females are XX and males are XY. This is generally true of the housefly (*Musca domestica*), but great variation in the chromosomal constitution of both sexes occurs in natural populations of this species. One sample, for instance, included males with the combinations XX, XY, XO, YO, and YY, and females that were XX, XY, and XO (White 1978). (O represents absence of a chromosome.) Suggest hypotheses to explain this variation, and discuss how this species might be used to study the evolution of sex determination.

9. The number of chromosome pairs in the genome ranges from one (in a species of ant) to several hundred (in some butterflies and ferns). Within a single genus of butterflies (*Lysandra*), the haploid number varies from 24 to about 220 (White 1978). Is it likely that chromosome number evolves by natural selection? Is there any evidence for or against this hypothesis?

10. Suggest a hypothesis to account for the fact that crossing over does not occur in the males of *Drosophila* or most other Diptera, nor in the females of at least some Lepidoptera. (For other examples of variation among species in recombination rates, see White 1973.)

Molecular Evolution

During the last several decades, the spectacular growth of knowledge in molecular biology has transformed the entire field of biology, including evolutionary biology. *Molecular biology*, in this context, refers chiefly to the study of the structure of DNA, RNA, and proteins, and of processes such as replication, transcription, translation, and regulation of the levels of gene products. Evolutionary biology uses both information on molecular structures and processes and the techniques developed for molecular study.

Aims and Methods in the Study of Molecular Evolution

The study of molecular evolution uses methods such as PCR (polymerase chain reaction), DNA sequencing, and various techniques for isolating or localizing gene products such as RNA transcripts and proteins (see Box D in Chapter 3 for a sketch of some such techniques.) The information obtained by these methods is used in evolutionary studies in three overlapping ways:

1. *Molecular markers as tools.* Molecular markers can be used to analyze traditional problems in evolution. In such studies, the identity of the genes (or gene products) is not itself of primary interest; the researcher is concerned only with whether or not the genes are suitable markers for the problem at hand. Examples of such studies include enzyme electrophoresis and DNA haplotype analysis for estimating genetic variation within and among populations (see Chapter 9), for estimating population structure, gene flow, and breeding systems (see Chapters 11 and 21), and for describing genetic differences among species (see Chapter 15); genealogical analysis of haplotypes (gene trees) for inferring histories of population size, gene flow, and selection (see Chapters 9, 13 and 16); and DNA sequencing to obtain information on phylogenetic relationships among species and higher taxa (see Chap-

ters 5 and 15, inter alia). Avise (1994) has comprehensively reviewed the use of molecular markers in the study of organismal evolution.

2. *Molecular evidence on the evolution of phenotypes.* Some studies identify genes that affect particular biochemical, physiological, or morphological characters in order to obtain a deeper understanding of the processes responsible for phenotypic evolution. The identity of the genes is all-important in such studies. Such research programs are more difficult than those that use molecules only as markers, so information is only beginning to accrue. Examples of such studies include the identification of genes underlying polygenic variation (see Chapter 14) and comparative studies of genes that govern developmental processes (see Chapter 23).

3. *Study of the evolution of genes and genomes as such.* DNA sequences and the structure of the genome are features that, like behavior, morphology, or other phenotypic characters, are studied in their own right. What generalizations can we make about the mechanisms and evolutionary processes that have governed the rates and patterns of variation and change in genes and genomes? Impressive advances have been made in this area, the chief subject of this chapter. Some information and insight have come from phylogenetic and population-level studies of the kind described under the two preceding categories of molecular evolutionary research. Thus these categories are not entirely distinct.

Some studies of processes governing molecular evolution use experimental techniques such as transformation (incorporating DNA from one organism into the genome of another). To date, however, most studies of the evolution of genes and genomes have relied on analyzing patterns of variation within and among populations and species. In some investigations, a researcher obtains all or most of the data him- or herself. In others, the researcher analyzes data bases such as GenBank, in which sequence data from published studies are stored.

We have already described many of the methods by which such data are analyzed. One of the most important approaches is *phylogenetic analysis* (see Chapter 5), including methods such as maximum parsimony, neighbor-joining, and maximum likelihood, whereby the genealogy of sequences, or of the populations or species from which the sequences are sampled, is estimated. We noted in Chapter 5 that paleontological or geological data can sometimes be used to calibrate the rate at which sequences have evolved. We also noted that, using the principle of parsimony, ancestral sequences can often be estimated by reconstructing the probable ancestral state of each character (e.g., each nucleotide position).

The other major class of methods for analyzing sequence data are those of *population genetics*, in which data are frequently compared with expectations derived from mathematical models. These models include both classic models of processes such as genetic drift and natural selection and the newer COALESCENCE MODELS (see Chapter 11), by which evolutionary processes can be inferred from the genealogy, sequence differences, and levels of variation of DNA sequences. The reader should review the methods described in previous chapters (especially Chapters 5, 11, and 13) before reading the rest of this chapter.

Phylogenetic Insights

Histories of molecular evolutionary change, inferred by phylogenetic (genealogical) analyses, provide evidence bearing on many processes. We provide two illustrations here, and others below.

Resurrecting extinct genes   RETROPOSONS (retrotransposons) are transposable elements (see Chapter 3) that encode reverse transcriptase. RNA transcripts of a retroposon are reverse-transcribed into DNA copies, which become inserted into various sites in the genome. As in genes generally, transcription of the retroposon's DNA is initiated in an upstream promoter region. Mammalian genomes carry numerous copies of a retroposon known as *L1*. In mice (*Mus*), some copies of *L1* have an active promoter, denoted *A*, and others have an inactive promoter, *F*. (These latter copies rely for transcription on promoters of nearby genes.) The approximately 200-bp (base pair) sequence of the *F* promoter varies among copies of the element and among strains of mice, and it is thought to have been inactivated by various mutations over the course of about 6 My (million years).

Molecular biologist Nils Adey and colleagues (1994) set out to "resurrect" the functional ancestral promoter from which the inactive *F* sequences had evolved. Using a phylogenetic analysis of 30 different *F* promoter sequences, they used the principle of parsimony to construct a best estimate of the ancestral sequence. (For example, at nucleotide position 213, some sequences had G and others C, and the phylogenetic tree implied that C was probably ancestral; similar inferences were made for all other variable positions.) The investigators then synthesized the inferred an-

cestral sequence and attached it to a REPORTER GENE, one whose protein product would be detected if the promoter enabled the gene to be transcribed. These constructs were then inserted into mouse cells grown in tissue culture. The experiment was wildly successful: the levels of protein produced by genes with the synthetic promoter equaled those produced by control genes to which the naturally active *A* promoter was attached. This experiment supports the hypothesis that the inactive copies have been derived from a functional sequence, and demonstrates the validity of phylogenetic inference.

Inferring recombination and gene transfer   In the absence of recombination, all genes in the genome have the same history of transmission, and therefore identical genealogies. Therefore, if genealogies of different genes differ, but are well supported by data, recombination may be suspected. Daniel Dykhuizen and colleagues (Dykhuizen and Green 1991; Guttman and Dykhuizen 1994) found that the phylogenetic tree of strains of the bacterium *Escherichia coli* differed depending on which of three gene sequences was analyzed. They concluded that, contrary to previous inferences that *E. coli* is almost entirely asexual, exchange of genetic material must in fact be rather frequent.

Similarly, a VIROGENE (a gene derived by reverse transcription of a gene of a virus) in the genomes of several closely related species of cats is very similar in sequence to the same gene in baboons (Benveniste 1985), an observation utterly inconsistent with all other evidence on the relationships of these animals (see Figure 24 in Chapter 10). This is one of the few documented cases of **horizontal gene transfer** among different species of eukaryotes, i.e., movement of genes among organisms by means other than parent-to-offspring transmission. Although some authors have suggested that horizontal gene transfer may have been frequent enough in the history of plants and some other groups to account for character distributions usually ascribed to convergent evolution, no evidence yet supports a substantial role for horizontal gene transfer in eukaryote evolution (Kidwell 1993).

However, horizontal transfer seems to be rather common between some kinds of bacteria (Syvanen 1994). This phenomenon can have important medical consequences; for example, *Neisseria gonorrheae*, the bacterium that causes gonorrhea, seems to have acquired resistance to penicillin by horizontal transfer of genes from nonpathogenic species (Spratt et al. 1994).

Evolution of DNA Sequences

Beginning in the late 1970s, data on variation in DNA sequences became available, first glimpsed imperfectly through variation in restriction sites (RFLPs) and later in the form of full nucleotide sequences. These data have become a major focus for population geneticists who aim to discern which evolutionary processes have played important roles in sequence evolution. In particular, the neutralist-selectionist controversy, which had centered on allozymes, has shifted focus to DNA sequence variation. This debate concerns whether variation within populations and divergence among populations and species is attributable more to random genetic drift or to natural selection. We have already touched on this issue (see Chapter 11), but will expand on it here. Throughout this chapter, bear in mind the difference between a *mutation* (a change in a single gene copy) and a *substitution* (the partial or complete replacement of a nucleotide or longer sequence by another throughout an entire population or species).

Patterns of Variation in DNA Sequences

The first data on intraspecific variation in the complete sequence of a gene came from Martin Kreitman's (1983) study of the alcohol dehydrogenase (*Adh*) locus in *Drosophila melanogaster* (Chapter 9). Among 11 gene copies, *only one of 2721 sites displayed an amino acid-changing (nonsynonymous, or replacement) substitution, but 42 sites displayed synonymous substitutions. A much greater proportion of sites varied in the introns than in the exons.* A genealogy ("phylogeny") of the sequences provided evidence that recombination within this region had given rise to some of the variation among gene copies (Figure 22.1). More recent data on various protein-encoding genes have shown similar patterns.

Some of the earliest work on DNA divergence among species was performed by Allan Wilson, Wesley Brown, and their colleagues on mitochondrial DNA (mtDNA), using both restriction site variation and DNA sequencing (Wilson et al. 1977; Brown et al. 1979, 1982). The mitochondrial genes, which lack introns, are carried on a singular, circular "chromosome," are inherited in most species only through the egg, and ordinarily do not undergo recombination. Several patterns found in these early data have proved to be general, not only in mitochondrial but also in nuclear genes. When the proportion of base pairs that differ between pairs of species is plotted against the time since their common ancestor (estimated from the fossil record), the curve rises and then levels off (Figure 20 in Chapter 11). This is because over a sufficiently long time, multiple substitutions occur at the same sites, so that earlier substitutions are erased by later ones. *Hence the observed number of differences between two distantly related species will be less than the number of substitutions that have actually transpired.* For mtDNA of mammals, the relation between sequence divergence and time since common ancestry is linear for only 5–10 My; it is only in this portion of the curve that the number of nucleotide differences provides a direct estimate of the rate of sequence evolution. This rate, in mammalian mtDNA, is five to ten times greater than the average rate of sequence evolution in nuclear genes, possibly because the DNA polymerase that replicates mtDNA lacks the proofreading ability of the polymerase that replicates the nuclear genome, so that mtDNA has a higher mutation rate. However, mtDNA and nuclear DNA evolve at more nearly equal rates in some groups of invertebrates.

FIGURE 22.1 A schematic example of the history of mutational events as inferred from a gene tree. A through K represent variable base pairs in a longer DNA sequence. Mutations unique to single gene copies, such as *j* and *k*, must have occurred more recently than those shared by several copies, such as *a* and *b*. The change *c** alters an amino acid, so the protein product of copies 6 and 7 is a different electromorph than that of copies 1–5. The greater homogeneity of sequence of copies 6 and 7 than of any other pair reflects the recency of this electrophoretically detectable mutation. The variation among copies 1–4 (shown within the box) could be explained by postulating that either the pair of mutations *f* and *g* or the pair *h* and *i* occurred twice. It is more parsimonious, however, to postulate that each mutation occurred once, and that recombination occurred between sites G and H.

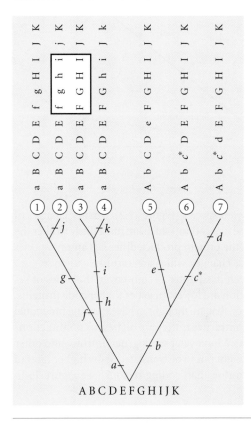

When Brown et al. compared the mtDNA sequences of several primate species, they found that *transitions* (substitutions of one purine for the other, or of one pyrimidine for the other) *greatly outnumbered transversions* (substitutions of a purine for a pyrimidine, or vice versa) in comparisons of recently diverged species, although the two types of substitutions were nearly equally frequent after greater divergence times. Thus transversional substitutions occur at a much lower rate, and take a longer time to accumulate. This observation, which has proved to be general for both mitochondrial and nuclear DNA, was a surprise, because twice as many transversions as transitions can occur among the possible nucleotide substitutions.

Another observation on mtDNA that has proved general is that *silent (synonymous) nucleotide substitutions accumulate faster than amino acid-replacing (nonsynonymous) sub-*

FIGURE 22.2 Divergence by synonymous (silent) and non-synonymous (replacement) substitutions in a mitochondrial gene of mammals, plotted against estimated time since common ancestry. Synonymous substitutions have occurred at about six times the rate of nonsynonymous substitutions. (After Brown et al. 1982.)

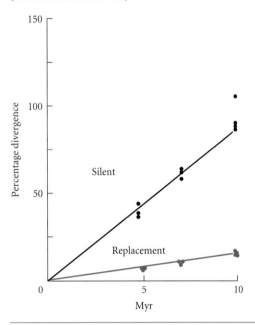

stitutions between species (Figure 22.2). That is, changes of the second base in a codon, which are most likely to alter the amino acid in the protein product, differ less among species than third-base changes, which are least likely to do so.

We have seen in Chapter 11, moreover, that rates of sequence evolution are lowest in genes that encode proteins, such as histones, that are thought to have exacting functional constraints; contrariwise, they are highest in nonfunctional pseudogenes. Moreover, pseudogenes, introns, and other sequences thought not to be translated show higher rates of sequence evolution than coding regions (see Figure 23 in Chapter 11).

Interpreting Variation in DNA Sequences: Theory

The *neutral theory of molecular evolution*, championed by Motoo Kimura (1983, 1991) and further elaborated by several other population geneticists, provides the "null hypothesis" against which data on molecular variation are compared. In other words, patterns of molecular variation are assumed to be explicable by genetic drift of selectively neutral mutations unless they depart significantly from those predicted by the neutral theory. We here recapitulate this theory (from Chapter 11) and describe how neutral patterns would be altered by natural selection.

If u_0 is the rate of selectively neutral mutation per gene per generation, the *substitution rate* (the rate of replacement in a population) is independent of population size, and is $k = u_0$ per generation. (The substitution rate per unit

of time is k/g, where g is the length of a generation.) Thus two populations or species that have been isolated for t generations are expected to differ by $2u_0t$ substitutions. Variation within a population at equilibrium, measured by the frequency of heterozygotes per nucleotide site, is $4N_eu_0$, where N_e is the effective population size. (The quantity $4N_eu_0$ is often denoted by the Greek letter theta, θ.) It has been shown that for neutral mutations, θ can be estimated either by S, the number of segregating (variable) nucleotide sites in a sample of gene copies (sequences), or by d, the average number of nucleotide differences between pairs of sequences in a sample (Watterson 1975; Tajima 1983). Note that if two classes of sequences (such as different genes, or introns versus exons, or different codon positions) differ in u_0, the one with the higher u_0 will display both higher intraspecific variation and higher interspecific divergence.

Natural selection can modify these patterns in three ways. PURIFYING SELECTION eliminates or reduces the frequency of deleterious mutations in a population. POSITIVE (directional) SELECTION fixes a sequence (a gene) that includes an advantageous mutation. BALANCING SELECTION, owing to factors such as heterozygote advantage, frequency-dependent selection, or variable selection (see Chapter 13), maintains variant sequences in a population. We shall now consider each of these in turn.

If the total mutation rate is u_T, but some mutations are deleterious and are eliminated by purifying selection, then a fraction f_0 of the mutations are neutral, giving $u_0 = u_Tf_0$ as the neutral mutation rate. Reducing f_0 reduces u_0, and therefore both the substitution rate (k) and the level of polymorphism (θ). *This reduction should be greater, the stronger the constraints on the function of the gene product.* Thus, according to the neutral theory, we might expect k and θ to be lower for functional genes than for pseudogenes, for exons than for introns, for second-base positions than for third-base positions in codons—as indeed they are. Recall also (from Chapters 11 and 13) that if a mutation is *slightly deleterious* and has a disadvantage s, its fate is largely determined by genetic drift if s is much less than $1/(2N_e)$, but is largely governed by selection otherwise. Therefore, *more mutations are effectively neutral if the effective population size (N_e) is low, and this increases the rate of substitution by genetic drift (k).* Likewise, because genetic drift is faster in smaller populations, lowering N_e reduces the level of variation within populations.

We now consider the *effects of selection on neutral variation at sites that are closely linked to the selected site. Positive directional selection* reduces variation at closely linked sites. Suppose an advantageous mutation occurs in a gene for which neutral variation exists in the population. Assume that there is little or no recombination in the vicinity of the advantageous mutation. If this mutation is fixed by selection, then all the copies of the gene in the population will be descended from the single copy in which the mutation occurred. The neutral variant sites linked to this mutation will also be fixed by hitchhiking. Thus all neutral variation in the gene is eliminated, swept away by a SELECTIVE SWEEP, and variation is only slowly reconstituted as new neutral mutations occur among the copies of the advantageous gene. Selective sweeps will eliminate variation over a longer segment of DNA in regions of the genome that have lower rates of recombination (Maynard Smith and Haigh 1974; Kaplan et al. 1989; Wiehe and Stephan 1993). The effects of a selective sweep are evident in gene genealogies. Consider two unlinked loci, one that has been evolving solely by genetic drift of neutral mutations (Figure 22.3A) and one that has experienced a selective sweep (Figure 22.3B). Compared with copies of the neutrally evolving gene, the copies of the gene that was fixed by selection are descended from a more recent common ancestor (the one in which the favorable mutation occurred); they have had less time to accumulate different neutral mutations, and so are more similar in sequence. A selective sweep resembles a bottleneck in population size in that it reduces variation and increases the genealogical relatedness among gene copies, but a bottleneck affects the entire genome, not just the portion surrounding an advantageous mutation.

The effects of *balancing selection* are opposite to those of positive directional selection. Again, assume that recombination is low in the vicinity of a nucleotide site at which two variants are maintained by selection in a polymorphic state. All the gene copies in the population are descended from two ancestral copies (bearing the original selectively advantageous alternative nucleotides), each of which was the progenitor of a lineage of genes that have accumulated neutral mutations in the vicinity of the selected site (Figure 22.3C). Thus, compared with a gene with solely neutral variation, a DNA sequence subjected to balancing selection displays elevated variation in the vicinity of the selected site (Strobeck 1983). In a genealogy of sequences sampled from the population, the common ancestor of all the sequences may be older than if they had been evolving solely by genetic drift, because selection has maintained two gene lineages longer (Figure 22.3C). In fact, the polymorphism may have been maintained by selection for so long that speciation has occurred in the interim. In this case, both lineages of genes may have been inherited by two (or more) species, and some gene copies in each species may be genealogically more closely related to genes in the other species than to other genes in the same species.

Purifying selection against deleterious mutations reduces neutral polymorphism at closely linked sites. Brian Charlesworth and colleagues (1993; Charlesworth 1994a), who have termed this effect **background selection**, have pointed out that when a copy of a deleterious mutation is eliminated from a population, selectively neutral mutations linked to it are eliminated as well (Figure 22.3D). Thus, for a region of DNA within which the recombination rate is low, the effective population size is reduced to the proportion of gametes that are free of deleterious mutations. Consequently, the level of heterozygosity for neutral mutations is

reduced. The reduction is greatest if the mutation rate is high, if the mutations are strongly deleterious, and if the recombination rate is very low.

Evidence on the Causes of Sequence Evolution

The neutral theory successfully predicted that variation and divergence would be greater, the lower the functional consequences of a mutation. The high rates of change at third-base positions, in noncoding sequences such as introns, and in pseudogenes are generally accepted as confirmations of the neutral theory.

Experiments have bolstered this conclusion. For example, Robert DuBose and Daniel Hartl (1991) found that amino acid-replacing (nonsynonymous) differences among nine species of bacteria in the gene encoding alkaline phosphatase were concentrated in certain regions of the gene: those corresponding to the protein's helices rather than its β sheets, and to the carboxyl rather than the amino ends of the helices. (These terms refer to aspects of the protein's three-dimensional structure.) The authors induced mutations in these various regions and measured their effects on the enzyme's activity. Mutations impaired enzyme function most if they occurred in beta sheets or amino ends. Thus these portions of the gene have evolved slowly because alterations of the corresponding portions of the protein impair its function.

The principle that high variation in a DNA sequence, within or among species, indicates functional unimportance, compared to relatively invariant nucleotide positions, is now routinely used by molecular biologists who compare sequences in different species to screen for functionally important sites. It exemplifies the usefulness of evolutionary biology for molecular studies.

Thus variation and divergence, once taken to indicate adaptive evolution, are now viewed as evidence of genetic drift. This change represents a major shift in evolutionary theory, which evolves as new theory and evidence accrue. But although the once heretical neutral theory seems to have triumphed, it contains the seeds of its own potential refutation, for it provides theoretical criteria for detecting selection. While few deny that much molecular variation is selectively neutral, evidence mounts that selection of molecular variants may be fairly common. The several lines of evidence are summarized by Kreitman and Akashi (1995) and Ballard and Kreitman (1995).

Experiments

A time-honored test for selection is to initiate experimental populations with different frequencies of genetic variants, and to see if changes in the frequencies are too consistent or rapid to be explained by genetic drift. In several such experiments, the frequencies of mitochondrial haplotypes in experimental populations of *Drosophila* have changed too rapidly to be accounted for by genetic drift (Ballard and Kreitman 1995). Because the mitochondrial genome is a single

FIGURE 22.3 Schematic genealogies of gene copies in a small population, illustrating the effects of three modes of selection on nucleotide diversity, compared to the neutral model. The branching shows the descent of gene copies from ancestral copies. The contemporary population is at the top of each diagram, and the ancestry of contemporary gene copies is marked by the boldfaced gene tree. Ovals represent selectively neutral mutations, each at a different nucleotide site. In each diagram, some gene lineages (thin branches) become extinct by chance. (A) Neutral mutations only. The 12 contemporary gene copies vary by 7 mutations. (B) Positive selection due to an advantageous mutation (marked by the asterisk). In a selective sweep, this gene lineage replaces all others. The contemporary gene copies are descended from a more recent common ancestor than in A, and so differ by fewer (5) neutral mutations. (C) Balancing selection. A mutation (marked by an asterisk) forms a balanced polymorphism for alleles A and A′, each representing a gene tree (portrayed by boldfaced dashed vs. solid branches). Both gene lineages are older than the lineage of the contemporary gene copies in panel A, so more mutational differences (13) among the gene copies have accumulated. (D) Background selection. Gene lineages become extinct not only by chance, but also because of deleterious mutations (marked by ×) to which they are linked, which eliminate some copies bearing linked neutral mutations, leaving 4 mutational differences among contemporary gene copies.

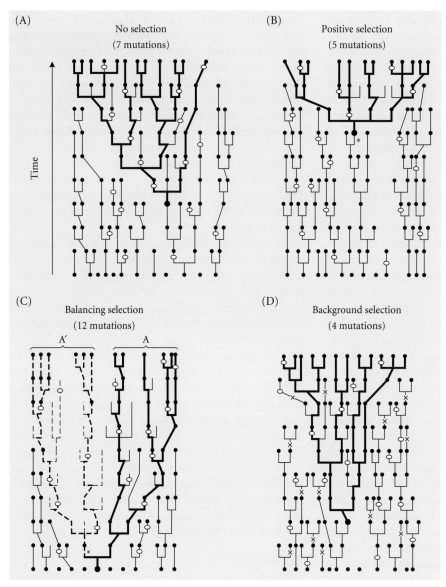

(A) No selection (7 mutations)

(B) Positive selection (5 mutations)

(C) Balancing selection (12 mutations)

(D) Background selection (4 mutations)

linkage group without recombination, it is impossible to say which of several variant sites is the target of selection. In an experiment with *Escherichia coli*, Daniel Dykhuizen and Daniel Hartl (1980) backcrossed six electrophoretic alleles at the 6-PGD (6-phosphogluconate dehydrogenase) locus into a common genetic background, in order to minimize the possibility that apparent selection at this locus might be due to unseen genes at neighboring loci. Pairs of genotypes, distinguished by a neutral genetic marker (see Chapter 11 and Figure 8 in Chapter 12), were placed together in chemostats. When gluconate, the substrate for this enzyme, was the limiting energy source, most pairs of alleles did not change in frequency: there was no evidence of selection, even though a selection coefficient of as little as 0.005 would have been detectable. Dykhuizen and Hartl then backcrossed these alleles into a strain that had a nonfunctional allele at another locus (*edd*) that provides an alternative pathway for metabolizing gluconate. Some 6-PGD alleles that in the first experiment were neutral proved under these circumstances to differ in fitness. Thus although these alleles may well be neutral under most circumstances, there exists the *potential* for selection if either the ecological environment (available resources) or the genetic "environment" (genetic constitution at other loci) changes.

Rates of sequence evolution If the neutral mutation rate (u_0) remains constant for a gene over time and among lineages, the rate of substitution should be constant (a molecular clock), varying no more than one would expect of a random (Poisson) process. (This rate can and does vary among kinds of sequences, since they differ in u_0 due to differences in intrinsic total mutation rate, u_T, or to functional constraints that affect the neutral rate.) We noted in Chapter 11 that despite some clocklike rates of sequence evolution, the rates of replacement substitution in many cases vary significantly more among lineages than the neutral theory would predict (Gillespie 1991; Langley and Fitch 1974). For example, replacement substitution rates have been lower in the hominoid primate lineage (humans and apes) than in other primates or other mammals (Goodman et al. 1971), and have been higher in rodents than in primates or artiodactyls (Seino et al. 1992; Easteal and Collet 1994; Figure 22.4). The higher rates in rodents have been ascribed to their shorter generation times (Wu and Li 1985), or to a higher mutation rate. However, rodents and primates have had the same rate of synonymous substitutions, *assumed* to be selectively neutral (Bulmer et al. 1991; Easteal and Collet 1994). This finding seems to imply that the total mutation rate (u_T) is the same, per unit of time (not per generation), in these lineages. If so, the higher rate of replacement substitution in the rodent lineage could be the consequence either of positive selection or of lower effective population size, which renders slightly deleterious mutations effectively neutral. These alternatives are difficult to distinguish.

The most unequivocal evidence for positive directional selection from comparisons among species is provided by cases in which the rate of replacement substitution has exceeded the rate of synonymous substitution, which is generally assumed to estimate the neutral rate. In Chapter 15, we noted one such instance: among species of abalone (*Haliotis*), genes encoding sperm lysin protein differ more by amino acid-replacing substitutions than by synonymous substitutions. The same observation has been made on the major histocompatibility (MHC) loci of humans

FIGURE 22.4 Diagrams of the number of nucleotide substitutions in 14 nuclear genes among marsupials (M), rodents (R), primates (P), and artiodactyls (A), plotted as phylogenetic diagrams, with branch lengths proportional to the number of substitutions per 100 base pairs. (A) Nonsynonymous substitutions have occurred at a higher rate in rodents (8.9) than in the lineage from the rodent/primate ancestor to modern primates (1.2 + 5.7 = 6.9). The rate has also been lower in primates (5.7) than in artiodactyls (7.5). (B) Synonymous substitutions have occurred no faster in the rodent lineage (28.0) than in the lineage leading to either primates (12.3 + 17.6 = 29.9) or artiodactyls (12.3 + 17.5 = 29.8). In both diagrams, marsupials are an outgroup, and the substitutions along this branch cannot be partitioned into those between the common ancestor of marsupials and placental mammals, and those between that ancestor and modern marsupials. (After Easteal and Collet 1994.)

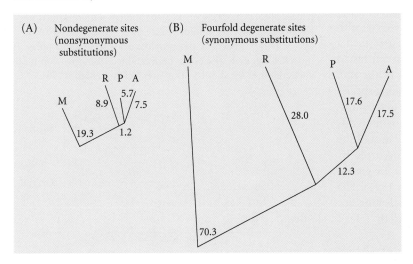

(A) Nondegenerate sites (nonsynonymous substitutions)

(B) Fourfold degenerate sites (synonymous substitutions)

and mice (Nei and Hughes 1991), which encode proteins that bind foreign peptides (antigens) and present them to the immune system.

Convergent evolution The independent evolution of similar features in unrelated but ecologically similar organisms provides strong evidence for adaptation (see Chapter 12). A most striking instance of convergence of protein sequences is that of lysozymes, bacteriolytic enzymes that aid in preventing infections. In ruminant artiodactyls (e.g., cow), colobine monkeys (e.g., langur) and the hoatzin *Opisthocomus hoazin* (a relative of the cuckoos), the lysozyme has acquired a new function: it is expressed in the digestive tract, and digests fermentative bacteria that break down plant material. The amino acid sequence of lysozyme in these groups has converged to a considerable extent (Kornegay et al. 1994).

Visual pigments in vertebrate retinas also illustrate convergent evolution (Yokoyama and Yokoyama 1996). These pigments consist of vitamin A aldehyde bound to a protein called opsin. Structural differences among opsins determine their maximal absorption of different wavelengths of light, thus determining color vision. Both humans and a characiform fish, the tetra *Astyanax fasciatus*, have a red-sensitive and a green-sensitive opsin. DNA sequencing showed that these proteins have evolved by independent gene duplications in the two lineages. Nevertheless, three of the amino acids that distinguish the red-sensitive and green-sensitive opsins are the same in the human and the fish. When these amino acids were substituted in vitro in cow rhodopsin (a different visual pigment), the maximal wavelength absorption was shifted toward green or red, showing that these amino acids play a critical functional role. This experiment shows how evolutionary analysis can direct researchers toward a deeper understanding of biochemical functions.

Sequence variation within and among species Several methods use patterns of sequence variation within species to test for deviations from the neutral model that might be caused by natural selection (Box 22.A). For example, the neutral theory predicts that nucleotide polymorphism within species should be correlated with divergence among species. In Chapter 11, we described a study by McDonald and Kreitman (1991), who contrasted synonymous and nonsynonymous variation in the *Adh* (alcohol dehydrogenase) locus within and among species in the *Drosophila melanogaster* group. Relative to the number of synonymous differences, more nonsynonymous differences among species were found than would be expected had the nonsynonymous changes occurred by genetic drift. Similar results have been found for several other genes in this species group (e.g., Eanes 1994). The most likely interpretation is that positive selection for adaptive amino acid substitutions has enhanced the rate of divergence among species.

Several researchers have looked for evidence of selective sweeps, which reduce levels of silent nucleotide variation in the vicinity of a site that has undergone directional selection. This effect should be most pronounced in regions of low recombination. For instance, in *Drosophila melanogaster*, the *cubitus-interruptus* locus (*ci*) is one of the few genes on the tiny fourth chromosome, which does not undergo recombination. Although the sequence of *ci* differs substantially between *D. melanogaster* and *D. simulans*, almost no sequence variation was found within either species. Likewise, sequence variation in loci located in chromosome regions that differ in recombination rate is positively correlated with the recombination rate (Aquadro et al. 1994). These observations may indicate that positive directional selection has occurred, but background selection can account for some of the correlation (Hudson and Kaplan 1995; Charlesworth 1996). Thus, it is uncertain how pronounced the effects of positive directional selection may prove to be.

Some evidence for balancing selection has been revealed by studying the genealogy of genes. A striking example, described in Chapter 21, has been found in the self-incompatibility alleles of solanaceous plants such as potato and petunia. These gene lineages are so old that many of the alleles in different genera of plants that diverged more than 30 Mya are more closely related to each other than to other alleles in their own species (see Figure 15 in Chapter 21). The strong frequency-dependent selection that is known to maintain polymorphism at this locus (see Chapters 13 and 21) explains how these gene lineages have persisted for so long. Similarly, allele lineages of certain of the major histocompatibility loci (MHC) are older than the divergence between humans and chimpanzees (Figure 22.5). Because variant MHC proteins may differ in specificity for different antigens, heterozygotes may have a broader spectrum of

FIGURE 22.5 The phylogenetic relationships among six alleles in humans (*H*) and four alleles in chimpanzees (*C*) at the major histocompatibility (MHC) loci A and B. Both species have loci A and B, which form monophyletic clusters, indicating that the two loci arose by gene duplication before speciation gave rise to the human and chimpanzee lineages. At both loci, each chimpanzee allele is more closely related (and has a more similar nucleotide sequence) to a human allele than to other chimpanzee alleles. This phylogenetic tree provides evidence that the polymorphism in each species has been retained from their common ancestor—i.e., for about 5 My. (After Nei and Hughes 1991.)

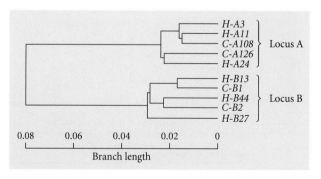

BOX 22.A

Detecting Natural Selection from DNA Sequence Data

Several methods have been devised to test for natural selection on DNA sequences by determining whether or not the pattern of nucleotide variation deviates significantly from the patterns predicted by the neutral theory. All of these tests are based on the principle that selection at one site will affect variation at closely linked sites, as described in the main text. These tests are statistically not very powerful, meaning that weak selection is likely to cause deviations from the neutral expectation that are too small to enable the neutral hypothesis to be rejected. The following brief descriptions omit mathematical and statistical aspects of the tests that are important in practice.

1. Tajima's (1989) test. The neutral theory specifies that the equilibrium level of nucleotide polymorphism in a DNA sequence is determined by $\theta = 4N_e u$–i.e., by the effective population size (N_e) and the neutral mutation rate (u) per nucleotide site. It can be shown that under the neutral model, θ can be estimated by either of two measurable quantities: S, the number of variable nucleotide sites in a sample of DNA sequences (gene copies), adjusted for the sample size, and k, the average number of nucleotide differences between pairs of sequences (pairs of gene copies). According to the neutral theory, k should equal S, since both are estimates of the same quantity, θ. However, k depends on the population frequencies of the variant nucleotides at each site, whereas S does not. Selection will not alter S appreciably, but by altering frequencies, it can cause k to deviate from the neutral value, θ. Frequencies of different neutral variants will be more nearly equal than they "should" be if they are linked to a polymorphic site that is maintained by balancing selection, and more uneven if they are linked to a site that has been fixed by directional selection. Thus the difference $d = k - S$ tests for a deviation from neutrality. Under the neutral hypothesis, $d = 0$. For balancing selection, $d > 0$, and for directional selection, $d < 0$. Tajima's formulation enables one to determine whether d is statistically significantly different from zero.

2. Fu and Li (1993) proposed a test that uses a genealogy (gene tree) of a sample of DNA sequences, estimated by phylogenetic analysis. The origins of individual mutations are mapped on the gene tree. For instance, a singleton mutation, unique to one gene copy, is termed "exterior," and is located at a tip of the tree. An "interior" mutation, in contrast, is shared by several gene copies, and thus must have originated deeper in the tree, on an interior branch. From coalescent theory (see Chapter 11), one can predict the distribution of interior versus exterior mutations under the neutral model. Fu and Li calculated a statistic, D, that is positive if there is an excess of interior mutations (as expected if balancing selection maintains polymorphism at a linked site), and is negative if there is an excess of exteri-

or, recently arisen mutations (as expected if a selective sweep has eliminated older mutations). This test and Tajima's test generally give similar results.

3. McDonald and Kreitman's (1991) test is described in Chapter 11.

4. Hudson, Kreitman, and Aguadé (1987) proposed a test that, like the McDonald-Kreitman test, contrasts intraspecific polymorphism with interspecific divergence. This HKA test, as it is called, contrasts two or more different loci, which may differ in their level of neutral polymorphism simply because they differ in functional constraint. The level of divergence between species is used, in effect, as a "control" that enables one to estimate the amount of neutral polymorphism expected, given differences in functional constraint. The data required are the number of nucleotide differences (D_1 and D_2) at the two loci between two closely related species, and the number of variable nucleotide sites (S_1 and S_2) in a sample of copies of each of the two loci from at least one of the species. According to the neutral theory, for each gene locus, the expected value of S is $S \propto \theta$, or $S \propto N_e u$, and the expected value of D is $D \propto T$, where T is the time since speciation. Loci 1 and 2 may have different values of θ (θ_1 and θ_2), since their neutral mutation rates may differ, but N_e is the same for both loci, and thus cancels out of the calculations. The unknowns θ_1, θ_2, and T are estimated from the observed quantities S_1, S_2, D_1, and D_2, using a set of simultaneous equations. In effect, θ_1 is estimated from both S_1 and D_1, so it is possible for S_1 to deviate from its predicted value (the estimate θ_1) if the level of intraspecific polymorphism and the level of divergence between species have not both been governed exclusively by neutral evolution (mutation and drift). If S is sufficiently greater than predicted by θ, then there is more polymorphism than expected from the level of divergence, and balancing selection at a closely linked site is likely; if S is sufficiently smaller, either directional selection or background selection has probably occurred.

Figure 22.A1 illustrates the application of the HKA test to three contiguous sequences, *Adh* (the alcohol dehydrogenase locus), *Adh-dup*, and a noncoding region upstream (5') of the *Adh* locus, in *Drosophila melanogaster* (Kreitman and Hudson 1991). The study employed 11 copies of each sequence from *D. melanogaster* and 1 copy from *D. simulans*. The authors estimated S and θ for intervals of 100 nucleotides, shifted site by site down the entire sequence. The observed level of nucleotide variation was less than expected in the *Adh-dup* region, suggesting that directional selection of an advantageous mutation may have reduced the level of closely linked variation. In contrast, variation was far higher than expected in exon 4 of the *Adh* gene. This region includes the one amino acid-changing polymorphism—at site 1490—in this gene, which, as indicated by several independent lines of evidence, is affected by balancing or positive selection (see Chapter 13). This example confirms that the HKA test can successfully detect the operation of natural selection at the molecular level.

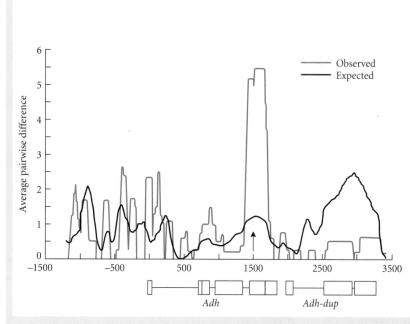

FIGURE 22.A1 The observed level of nucleotide polymorphism in the *Adh* and *Adh-dup* genes of *Drosophila melanogaster* is indicated by the colored line. The black line shows the variation expected under a neutral model, calculated according to the HKA method. The structure of the *Adh* and *Adh-dup* genes is shown below the graph, with boxes representing exons. An untranslated region, not shown, lies upstream of *Adh*. The vertical axis represents the average number of nucleotide differences, per 100 base pairs, between pairs of genes. The horizontal axis marks nucleotide positions, with the beginning of the *Adh* gene set at 0. Observed polymorphism is far greater than expected in the region of the amino acid polymorphism at site 1490 (vertical arrow), which is maintained by balancing selection. (After Kreitman and Hudson 1991.)

responses, and thus may have higher fitness than homozygotes (Nei and Hughes 1991).

Selection on silent substitutions

Our discussion of sequence evolution has assumed that selection does not act on silent variation in protein-encoding genes. However, evidence for such selection has mounted ever since the discovery of **codon usage bias** in the early 1980s (Grantham et al. 1980; Ikemura 1981). Certain of the synonymous codons for the same amino acid occur at significantly higher frequencies than others throughout certain genes—especially those that are highly expressed (i.e., that produce a lot of protein). For example, in highly expressed genes in yeast (*Saccharomyces cerevisiae*), 98 percent of the codons that specify glutamic acid use the codon CTT in the DNA, and only 2 percent use the synonymous codon CTC (Bulmer 1988). The bias is lower (67 versus 33 percent) in weakly expressed genes of this species. Codon usage bias is strong in microorganisms such as bacteria and yeast, weaker but nevertheless pronounced in *Drosophila*, and not at all evident in mammals (Shields et al. 1988).[*]

Evidently natural selection favors some synonymous codons over others. The observation that the degree of codon usage bias in a gene is correlated with its level of expression suggests that the favored codon is the one that matches the most abundant of the several transfer RNAs that transport a specific amino acid to the site of protein assembly on the ribosomes (Ikemura 1981; Gouy and Gautier 1982). Protein synthesis may thereby be more accurate, rapid, and efficient, which may be especially important for proteins that are needed in large quantities (Bulmer 1991).

Because selection for a favored codon is presumably quite weak, purifying selection against synonymous mutations can be effective only if the effective population size is large; in smaller populations, such mutations will be effectively neutral, so little codon usage bias should evolve. This may explain why codon usage bias is pronounced in microorganisms, which are thought to have huge effective populations, less pronounced in *Drosophila*, and weak or absent in mammals, which have much smaller populations (see Hartl et al. 1994).

If codon usage bias is the consequence of purifying selection, then the stronger the bias, the smaller the number of effectively neutral mutations. Thus according to the neutral theory, we should expect lower rates of evolutionary change by random fixation of synonymous substitutions in genes with high than with low codon usage bias. This is just what Paul Sharp and Wen-Hsiung Li (1987) found when they compared the sequences of 23 genes in the closely related bacteria *Escherichia coli* and *Salmonella typhimurium* (Figure 22.6).

---

[*]In mammalian genomes, large regions (*isochores*) differ in the proportion of G–C versus A–T base pairs. These differences affect both noncoding regions and synonymous codon frequencies in coding regions (Bernardi et al. 1985). They may be due to mutational bias, rather than to gene-specific selectivity.

FIGURE 22.6 A measure of the number of nucleotide substitutions, per synonymous site, between the bacteria *Escherichia coli* and *Salmonella typhimurium*, plotted against an index of codon usage bias for each of 23 loci. Fewer nucleotide substitutions in a gene indicate a lower rate of synonymous substitution. The lower rate of evolution in genes with high codon usage bias is evidence of purifying selection against "unfavored" codons. (After Sharp and Li 1987.)

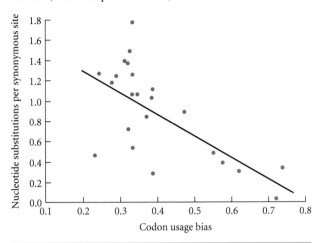

The evolution of nucleotide sequences can also be affected by selection for stability of RNA. Recall that ribosomal RNA (rRNA) has a secondary structure consisting of loops and stems (see Figure 5 in Chapter 7). The stems are formed by complementary sequences of Watson-Crick base pairs (C–G and A–U) that are rather distantly located in the molecule's linear sequence. Sequence comparisons among species have shown that although the secondary structure is conservative, stems vary in nucleotide sequence, but Watson-Crick pairing is generally preserved. These findings suggest that fixation of a mutation at one position, perhaps by genetic drift, is often followed by a *compensatory evolutionary change* in the complementary position, presumably due to selection for proper pairing that enhances the stability of the secondary structure. For example, nucleotides 184 and 199 in the 18s rRNA of Diptera are paired in one of the molecule's stems. These sites are occupied by the Watson-Crick pair of bases G and C in the ancestor of higher Diptera, but a mutation from C to U, causing failure to pair, was fixed in the ancestor of *Drosophila melanogaster* and *D. orena* (Figure 22.7). The nonpairing couplet G–U persists in *D. melanogaster*, but the compensatory mutation from G to A, establishing the Watson-Crick pair A–U, has been fixed in *D. orena*.

Like rRNA, the RNA transcripts (pre-mRNA) of protein-encoding genes also have stems and loops. Parts of these transcripts are encoded by introns, and are spliced out when the mature messenger RNA (mRNA) is formed. There is some evidence that the nucleotide sequence of introns is affected by selection, perhaps for stability of pre-mRNA. Stephen Schaeffer and Ellen Miller (1993) sequenced 99 *Adh* gene copies from natural populations of

Drosophila pseudoobscura. They found substantial linkage disequilibrium between polymorphic sites within two of the introns—that is, the bases at different sites were correlated. Having ruled out other explanations, Schaeffer and Miller postulated that these combinations of nucleotides are maintained by epistatic selection (see Chapter 14). David Kirby and Wolfgang Stephan (1995) used phylogenetic comparisons among species to show that these regions form stems in the secondary structure of the pre-mRNA, which probably increase its stability.

Gene Families and New Gene Functions

Gene Families

As we have remarked in Chapter 3 and elsewhere, the genome includes many **gene families**: sets of two or more loci with similar DNA sequences. The example cited in Figure 15 of Chapter 3 is the family of hemoglobins, consisting in humans of several genes in each of two (α and β) subfamilies. The members of a gene family can arise by **unequal crossing over** (Figure 21 in Chapter 10), which generates one chromatid with a lower, and one with a greater, number of copies. The greater the number of copies, the more likely unequal exchange becomes, because a copy on one chromosome can pair with any of the copies on another. A chromosome with a particular number of copies may be fixed by either genetic drift or natural selection (Ohno 1970; Ohta 1988). Repeated sequences (motifs) within genes can also arise by unequal crossing over and other

FIGURE 22.7 Phylogeny of three species of *Drosophila* and an outgroup, the tsetse fly (*Glossina morsitans*). At sites corresponding to positions 184 and 199 of the 18s rRNA of *D. melanogaster*, a shift from the Watson-Crick pair G–C has occurred in the ancestor of *D. melanogaster* and *D. orena* to the nonpairing couplet G–U. A second mutation has restored Watson-Crick pairing in *D. orena*. (After Hancock et al. 1988.)

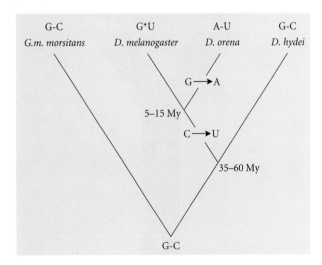

molecular processes such as "replication slippage" (Dover 1987). Direct evidence that unequal crossing over gives rise to gene duplication is provided by cases of individual humans who carry three, rather than the usual two, copies of the α-hemoglobin gene on one of their chromosomes, and by the anomaly known as Lepore hemoglobin, in which the N-terminal part of the δ-hemoglobin chain is fused to the C-terminal part of the β-chain owing to deletion of the DNA between these regions.

Duplicate loci may diverge in sequence, and sometimes in function, under the influence of mutation, genetic drift, and selection. (We will shortly see that this process can also be reversed.) If this occurs, the phylogenetic relationships among the related loci within a species can be traced, just as the phylogeny of genes in different species can. The corresponding genes in different species are referred to as **orthologous** loci, while the genes at different sites, in the same or different species, are **paralogous** (Figure 22.8). Using amino acid sequences, for example, Morris Goodman and colleagues (1982a,b) traced the history of vertebrate globins (Figure 22.9). In an ancestor of modern gnathostomes (jawed vertebrates), duplication of an ancestral globin gene gave rise to myoglobin (the single-chain globin in muscle) and hemoglobin nearly 500 Mya. A duplication of the hemoglobin locus gave rise to ancestral α- and β-globins, each of which, by further duplication, has given rise to different numbers of loci in different lineages. Note that the relationships of orthologous genes (e.g., the α-chains of *Xenopus* [clawed frog], chicken, and human) mirror the phylo-

FIGURE 22.8 If a gene duplication occurs in an ancestral lineage and is inherited by two descendant species (1,2), the corresponding genes (white *or* colored) are orthologous. Two noncorresponding genes (white *and* colored) are paralogous.

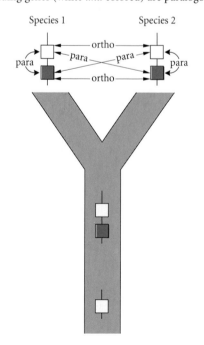

FIGURE 22.9 A partial phylogeny of globin genes, including those in a plant (leghemoglobin), an insect (the midge *Chironomus*), the jawless lamprey (agnathan vertebrate), and several jawed (gnathostome) vertebrates. Colored circles mark gene duplications; black circles, speciation events. The dates of the events are estimated from the fossil record. (After Hardison 1991; based on data of M. Goodman et al.)

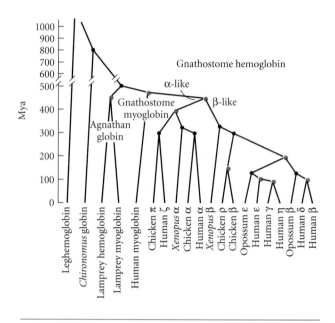

genetic relationships among the species that bear them, whereas a sample of paralogous genes (such as chicken Hb π, human δ, and *Xenopus* β) may not.

These duplicate globins have diverged in function (Hardison 1991). For example, the several members of the α and β subfamilies are expressed at different times in development. Among enzyme-encoding loci, the different but related enzymes that are expressed in different tissues or at different developmental stages are called ISOZYMES.

Using the principle of parsimony, one can estimate where in the phylogeny each character change has occurred (see Chapter 5). Goodman et al. (1982a,b; see also Ohta 1994) used this method to estimate when each amino acid substitution occurred in the history of the vertebrate globins. They concluded that substitutions accelerated for some time after gene duplication occurred, and postulated that the acceleration was evidence of *positive* selection of different mutations in duplicate genes for improvement of their new, divergent functions. In contrast, Kimura (1983) attributed the acceleration to random fixation of neutral mutations during the period when the genes were redundant, so that purifying selection was relaxed.

Not all duplicate genes acquire different functional roles. Many gene families, such as those encoding histones and the several ribosomal RNAs, are almost homogeneous in DNA sequence and collectively produce massive quantities of the same gene product. In contrast, one or more members of a gene family may acquire mutations that prevent transcription or translation, and therefore become "silent," nonfunctional *pseudogenes*. If a pseudogene persists in a population, it is selectively neutral, or nearly so, and further mutations may be fixed by genetic drift, so that its sequence diverges indefinitely. The hemoglobin gene family in humans includes three pseudogenes (ψ in Figure 15, Chapter 3), which display some sequence homology to both the exons and the introns of the functional globin genes. In addition to such "traditional" pseudogenes, there exist *processed pseudogenes*, which are thought to have arisen by reverse transcription of a mature mRNA transcript into DNA, which was then incorporated into the genome (Walsh 1985). For example, the nucleotide sequence of the $\psi\alpha3$ pseudogene of the mouse *Mus musculus* is homologous to that of the α-globin gene, except that the sequences corresponding to the introns are precisely lacking, exactly as expected if the pseudogene had arisen by reverse transcription. Many gene families include pseudogenes, and Walsh (1985) has postulated that processed pseudogenes may constitute about 9 percent of the total DNA of eukaryotes. Very old pseudogenes are presumably unrecognizable because sequence homology with functional genes is lost as mutations accumulate.

Evolution of Novel Functions

How do the members of a gene family acquire different sequences and functions? Two hypotheses have been advanced. The traditional hypothesis holds that a newly duplicated gene is redundant; its presence or absence is selectively nearly neutral, and it is free to accumulate mutations, which usually abolish its function (in which case it becomes a pseudogene), but occasionally provide a new, advantageous function. Austin Hughes (1994) calls this the MDN ("mutation during nonfunctionality") model. Hughes has proposed, as an alternative hypothesis, that duplicate genes arise from an ancestral gene that has two functions and then become specialized for one function or the other. In birds, for example, the same gene that encodes the enzyme argininosuccinate lyase also encodes one of the crystalline proteins of the eye lens.

In the traditional (MDN) model, the fate of a recently arisen duplicate locus is determined by a race between fixation of "null" mutations that silence the gene and render it nonfunctional (so that it becomes a pseudogene) and advantageous mutations that confer a new function. Tomoko Ohta (1988) and Bruce Walsh (1995) have modeled this process, and have concluded that fixation of advantageous mutations, and hence the evolution of a new functional gene rather than a pseudogene, is likely only if $4N_esr$ is high ($\gg 1$), where N_e is the effective population size, s is the selective advantage of an advantageous mutation, and r is the ratio of advantageous to null mutations. Thus, in populations of small or moderate size, most duplicate genes probably become pseudogenes. If, as in Hughes's model, an ancestral gene already has two functions, its duplicate descendants are presumably more likely to experience mutations that enhance one function or the other, and to avoid becoming pseudogenes.

In support of his hypothesis, Hughes noted that functional members of a gene family have sometimes been found to differ more by nonsynonymous than by synonymous nucleotide substitutions, as in the case of the olfactory receptor genes of a catfish, which may be specialized for binding different odorants. As we have seen, a higher rate of nonsynonymous than synonymous substitution is strong evidence of positive selection.

Concerted Evolution in Gene Families

When loci duplicated in an ancestral species are carried by two or more descendant species, we would expect the sequence difference between paralogous loci *within* each species to be at least as great as between orthologous loci in *different* species, since different mutations will be fixed by genetic drift and perhaps by selection. The globin gene family illustrates this principle, as do many others. However, an analysis of restriction sites in the gene family encoding rRNA in the clawed frog *Xenopus laevis* showed that the nucleotide sequence was homogeneous among the genes and differed from the sequence, equally homogeneous, of these genes in *Xenopus borealis* (Brown et al. 1972). Note that this homogeneity is among multiple loci within an individual organism, not merely among individuals of a species. This remarkable uniformity is now known to characterize many gene families (Dover et al. 1982). For example, about 400 copies of a sequence carrying the 18S rRNA and 28S rRNA genes, separated by an intergenic spacer, are distributed over five chromosomes in the human genome (Figure 22.10). Each (or most) of these units has a recognition site for the restriction enzyme *Hpa*I near the beginning of the intergenic spacer that is not found in the chimpanzee or other apes (Arnheim 1983). This restriction site must have arisen since the divergence of the human lineage. Either the same mutation has arisen and been fixed independently in all 400 genes—which is most unlikely—or the mutation has been spread from one gene to others by some homogenizing process.

Mechanisms of concerted evolution The ongoing evolution of nucleotide sequence homogeneity among the members of a gene family is termed **concerted evolution** (Zimmer et al. 1980). It can be caused by several molecular mechanisms, including unequal crossing over and gene conversion, followed by fixation of a homogenized array by genetic drift or selection (Figure 22.11).

Unequal crossing over in a gene family with different sequences can give rise to chromosomes in which all the copies have only one of the variant sequences. Such a chromosome may be fixed by genetic drift or natural selection, establishing a homogeneous gene family in the population. Different mutations may again accumulate among the copies, but the same process of homogenization may recur. Whether or not concerted evolution occurs therefore depends on the relative rates of the diversifying process of mutation and the homogenizing processes of unequal crossing over and genetic drift or selection.

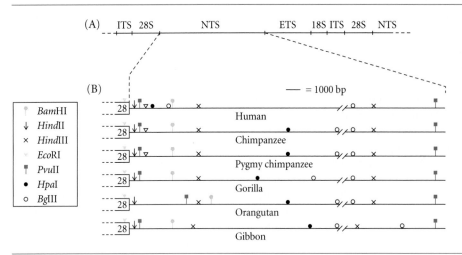

FIGURE 22.10 Restriction enzyme maps of the nontranscribed spacer (NTS) regions in the multiple rRNA genes of humans and apes. (A) The organization of one of the repeated units. (B) The sites recognized by seven restriction enzymes (named in the box at left). Each symbol indicates a site (a short sequence) cleaved by a specific restriction enzyme. Some such sites are unique to one or another species, but are present in most or all copies in the genome of that species. (After Arnheim 1983.)

Concerted evolution by gene conversion In Chapter 11, we described the phenomenon of *gene conversion*. In its simplest form, one allele in a heterozygote acquires the nucleotide sequence, in whole or in part, of its counterpart on the homologous chromosome, due to the operation of repair enzymes that copy a DNA sequence from one homologue to the other. Gene conversion can also occur between similar sequences (such as members of a gene family) on the same chromatid or on nonhomologous chromosomes. Experimental studies of gene conversion in yeast and other fungi have revealed instances of both UNBIASED conversion, in which either of two variant sequences is equally likely to be converted to the other, and BIASED conversion, in which one sequence is considerably more likely to convert the other (Nagylaki and Petes 1982).

Gene conversion, unlike unequal crossing over, does not alter the number of copies, or REPEATS, of a nucleotide sequence; but when it occurs among repeated sequences, each such event has a homogenizing effect. Population genetic models of gene conversion have shown that its con-

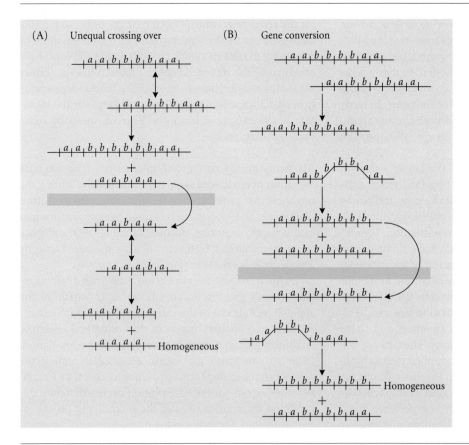

FIGURE 22.11 Mechanisms of concerted evolution of a gene family that initially includes two variant sequences, *a* and *b*. The shaded bar separates meiotic events in two generations. (A) Unequal crossing over generates a chromosome lacking some *b* sequences. If this chromosome increases in frequency, subsequent unequal crossovers between two such chromosomes may generate a homogeneous array. Many other possible histories of crossing over, coupled with changes in frequency, can result in uniformity. (B) Gene conversion events, in which two repeats on one chromosome are converted from *a* to *b* by copying from the elevated portion of the other chromosome. Unlike unequal crossing over, gene conversion does not alter the number of repeats. The homogeneous chromosome may increase in frequency by drift or selection and become fixed in the population, but repeated gene conversion events may retard this process. (After Arnheim 1983.)

tribution to concerted evolution is very complex (e.g., Nagylaki and Petes 1982; Ohta and Dover 1984), and that biased gene conversion can accomplish homogenization much more readily than unbiased conversion.

Probably both unequal crossing over and gene conversion contribute to concerted evolution. For example, the X and Y chromosomes of *Drosophila melanogaster* carry several hundred rRNA genes that are homogeneous within each of the sex chromosomes and fairly similar between them. The number of gene copies on the X chromosome varies considerably within a single natural population (Dover et al. 1982; Lyckegaard and Clark 1991). This variation in copy number is most readily explained by unequal crossing over, but gene conversion may explain the similarity of sequence between genes on the X and Y chromosomes, because these chromosomes seldom cross over with each other. Circumstantial evidence that biased gene conversion may be important in evolution has been found in a 20,000-member gene family that is homogeneous within, but different between, species of mice (*Mus*) that probably diverged less than 2 million years ago (Dover 1982). The mathematical models referred to earlier suggest that such a huge gene family would be unlikely to undergo concerted evolution so rapidly without conversion bias.

The evolutionary significance of concerted evolution

The most striking cases of concerted evolution are in gene families, such as those encoding histones and ribosomal RNA, in which large amounts of a gene product are required. Other gene families seem generally not to evolve in concert: within the globin family, for example, both the several functionally different globin genes and the pseudogenes differ in sequence (but see Hardison 1991). Although the intergenic spacers in rRNA genes differ among, and are homogeneous within, species in the *Drosophila melanogaster* group, the coding regions of these same genes do not differ among species, and so have not undergone concerted (or unconcerted) divergence (Dover et al. 1982). Such observations imply that natural selection may strongly influence whether or not concerted evolution occurs. In particular, uniformity versus diversity within gene families seems to depend on whether diverse, or uniform but plentiful, gene products are adaptive.

Transposable Elements

Barbara McClintock startled geneticists in 1948 when she found "genes" in maize (corn) that move among different sites in the genome, causing mutations in other genes as they do so. It appears now that all organisms have such genetic factors, now called *mobile* or *transposable elements* (TEs). As noted in Chapter 10, there are many kinds of TEs. Some transpose via direct DNA replication, but most appear to be retrotransposons that encode reverse transcriptase and are reverse-transcribed from an RNA copy to a DNA copy, which inserts itself into the genome. Each of about 50 families of TEs is dispersed throughout the genome of *Droso-*

phila melanogaster; they collectively constitute about 10 percent of the total DNA. Some families of TEs are ancient; for example, the type of transposable element that McClintock found in maize occurs also in other, distantly related, members of the grass family such as bamboo (Huttley et al. 1995).

Effects of Transposable Elements

In bacteria, some TEs carry genes that are advantageous to the host, such as genes for antibiotic resistance and for metabolism of novel substrates (Kleckner 1981; Condit and Levin 1990). In many such instances, these genes appear to be host genes that have been captured by TEs. However, most TEs do not carry genetic information for the host organism's phenotype, but *only information for their own replication*. Thus, they may be thought of as *genomic parasites* embedded in the genome of their host.

The *chief consequence* (not function) *of TEs for their hosts is mutation*, which they cause by inserting themselves into control regions of genes and altering regulation, or into coding regions and affecting (usually abolishing) gene function (Rubin 1983). Most of the morphological mutations used in *Drosophila* genetic research seem to be caused by insertion of TEs (Finnegan and Fawcett 1986). TEs have been used in experiments to increase mutation in polygenic characters (see Chapter 10). On average, TE-induced mutations lowered fitness in these experiments. TEs have characteristic DNA "signatures," but to date, DNA sequencing has yielded no evidence of adaptive TE-associated mutations in natural populations. In natural populations of *Drosophila*, most TEs occur at any one chromosome site at very low frequencies, as we would expect if the mutations they cause at these sites are deleterious.

As explained in Chapter 10, recombination between transposable element copies at different sites on a chromosome, or on different chromosomes, often causes *structural alterations* such as inversions, translocations, and deletions. (Such recombination is termed ECTOPIC EXCHANGE.) A sequence of host genes flanked by recombining TEs may be deleted from one location and inserted elsewhere in the genome (Figure 22.12). Some mutations at the *white* locus in *Drosophila melanogaster* are caused by inserts of such transported sequences, which may be as much as 14 kb long (Zachar and Bingham 1982). *Transposable elements may therefore play a major role in the evolution of the genome's architecture*, and may account for some of the differences among species in the location of genes on the chromosomes. However, *many, if not most, alterations of chromosome structure lower fitness*, either because genes are deleted or because heterozygotes for structural rearrangements segregate aneuploid gametes (see Chapters 13 and 15).

Retrotransposons, by producing reverse transcriptase, can enable reverse transcription of mRNA into cDNA, and are probably largely responsible for the *processed pseudogenes* that are abundant in the genomes of humans, mice, and other mammals (Walsh 1985). Whether or not an inserted

FIGURE 22.12 Alteration of genome structure by recombination between homologous transposable elements, represented by boxes with arrows showing their sequence polarity. The first recombination shown results in deletion of the B–C region from the chromosome. The deleted material may be lost or, as shown here, may be inserted elsewhere in the genome by another recombination event.

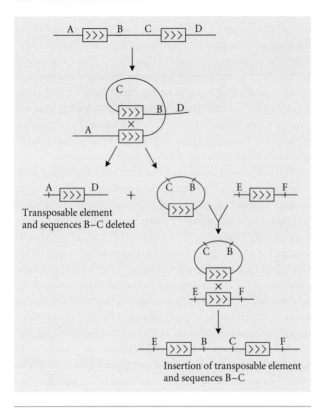

Transposable element and sequences B–C deleted

Insertion of transposable element and sequences B–C

pseudogene can generate additional copies of itself by further reverse transcription depends on the location of its promoter and the kind of RNA polymerase that transcribes the DNA sequence into RNA. Some genes, such as those encoding tRNAs, are transcribed by RNA polymerase III, which recognizes an internal promoter within the transcribed region. Reverse-transcribed copies of such a gene may proliferate, because they themselves can be transcribed. This is the case in the *Alu* sequence, a short DNA sequence with more than 500,000 copies distributed throughout the chromosomes in the human genome.

Evolutionary Dynamics of Transposable Elements

In general, transposable elements do not provide any adaptive service to the organism that bears them, and carry only the genetic information that enables them to proliferate within the genome. If a mutation enhanced the rate of transposition, so that more copies of an element were transmitted via the host's gametes to subsequent generations, the mutant element would increase in number and frequency relative to other TEs. This is a prime example of *selection not at the level of the individual organism, but at the level of the gene* (or "replicator": Dawkins 1989). Transposable elements do not exist because they serve the organism, but because they autonomously propagate themselves. They have been termed **selfish DNA** (Doolittle and Sapienza 1980; Orgel and Crick 1980), and may be viewed as parasites, much like viruses, of the genome in which they reside. Selfish DNA is different from functionless DNA that is replicated in the same proportion as functional genes; these sequences have been termed "ignorant DNA" (Dover 1980).

The capacity for the spread of TEs is spectacularly illustrated by the *P* element in *Drosophila melanogaster* (Kidwell 1983; Engels 1983). The effects of this transposable element depend on whether the cytoplasmic background of *P*-bearing chromosomes is compatible ("P" cytoplasm) or incompatible ("M" cytoplasm) with the element. In "P" cytoplasm, *P* elements have a low transposition rate and cause no detectable untoward effects. When a sperm bearing *P* elements fertilizes an "M" egg, however, the transposition rate is increased more than 20-fold in the offspring, which suffer *hybrid dysgenesis*, a syndrome that includes a high incidence of mutations and chromosomal aberrations, recombination in males (which ordinarily does not occur in *Drosophila*), and considerable sterility. Laboratory stocks derived from flies captured before the 1950s lack *P* elements and have "M" cytoplasm. At present, however, almost all members of the species, throughout much of the world, carry *P* elements. The low levels of sequence variation among *P* elements in *D. melanogaster*, the absence of *P* elements in species closely related to *D. melanogaster*, and the high sequence similarity between the *P* elements of *D. melanogaster* and those found in distantly related species such as *D. willistoni* indicate that *P* elements recently infected *D. melanogaster* from another species, and have spread throughout the world in a few decades.

Given such a high potential rate of increase, what controls the number of copies of a transposable element in the genome? Brian Charlesworth and Charles Langley (1989) have concluded that the number is controlled by natural selection, owing to the deleterious effects of TEs on the organisms that carry them. If insertion of a TE causes either a deleterious mutation or a deleterious chromosome rearrangement, selection will eliminate copies of the TE along with the deleterious alleles or rearrangements with which they are associated. Surprisingly, selection against deleterious TE-associated gene mutations seems to play only a minor role in regulating copy number. For example, recessive or partly recessive mutations should be more rapidly eliminated if sex-linked than if autosomal, because they are not sheltered by dominant alleles in males. Thus one would expect TEs to be reduced in frequency on X chromosomes compared with autosomes, but no such difference has been documented.

However, selection against chromosome rearrangements caused by ectopic exchange among TEs does seem to play a role in regulating copy number. Because ectopic exchange should be more frequent in regions of the genome with high recombination rates, TEs should be eliminated more rapidly, and thus have a lower frequency in such regions than in regions with little recombination. The best evidence for the hypothesis has come from studies of inversion polymorphism in *Drosophila melanogaster*. Recall that crossing over is effectively suppressed in heterozygotes for a chromosome inversion (Chapter 10), but not in chromosome homozygotes. Homozygotes for an inversion that has a lower frequency in a population than the "standard" chromosome are less frequent than "standard" homozygotes, so a rarer sequence undergoes crossing over less often than a more abundant sequence. The hypothesis of selection against TEs that cause ectopic exchange therefore predicts that the density of transposable elements should be higher in minority inversions (those with low frequency) than in the most frequent chromosome sequence. Exactly this pattern has been found in *Drosophila melanogaster* by two groups of investigators (Eanes et al. 1992; Sniegowski and Charlesworth 1994). The dynamics of transposable elements are an outstanding example of a *conflict between levels of selection*—between selection at the level of the gene and at the level of the organism.

Highly Repetitive DNA and Genome Size

The total amount of DNA per cell nucleus, often referred to as the C-VALUE, varies enormously among species. Some average differences exist among higher taxa (see Figure 13 in Chapter 3); for instance, frogs generally have less DNA than salamanders, and fungi less than plants. Even discounting the effects of polyploidy, closely related species can differ greatly in C-value (e.g., MacGregor 1982; Gold et al. 1992). For example, species in the salamander genus *Plethodon* differ threefold in C-value, in parallel with their differences in chromosome size, although, remarkably, not in chromosome number or shape. This variation is largely due to differences in the amount of highly repetitive DNA, which is evenly distributed among and within chromosomes.

Noncoding, highly repetitive DNA can account for more than 90 percent of eukaryote genomes (John and Miklos 1989). It is arranged in tandem arrays of repeated sequences, each 2 to more than 2000 base pairs in length. These sequences are referred to as *microsatellite, minisatellite,* and *satellite DNA,* depending on the length of the repeat. Most highly repetitive DNA has no obvious function for the organism.

The nucleotide sequence of the repeats in a tandem array of satellite DNA (to use the general term) varies due to mutation, and the number of repeats varies due to unequal crossing over and slippage replication (Figure 22.13). So much variation is generated that for human populations, minisatellite DNA provides each individual with a unique "DNA fingerprint" (Jeffreys et al. 1991). (Such "fingerprints" are increasingly used as evidence in criminal trials.) Computer simulations by Wolfgang Stephan (1989) and

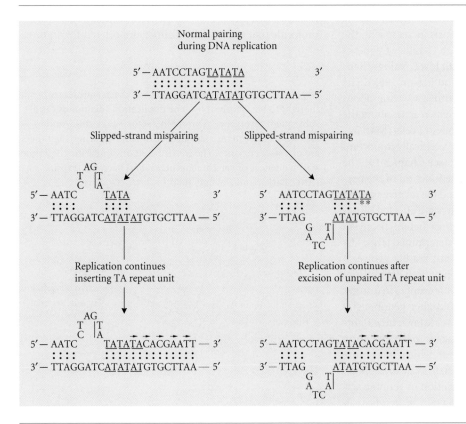

FIGURE 22.13 Generation of short duplications or deletions by replication slippage, due to mispairing between contiguous repeats (underlined). Dots indicate base pairing; small arrows, the direction of DNA replication. At left, slippage in the $3' \to 5'$ direction results in the insertion of one TA unit. At right, slippage in the $5' \to 3'$ direction, together with excision of the unpaired TA unit marked by asterisks, results in the deletion of one TA unit. (After Li and Graur 1991.)

others have shown that the joint action of unequal crossing over, slippage replication, and mutation can generate highly repetitive structures from any initial nucleotide sequence. Depending on the relative rates of these processes, the repeats may be long or short, and complex structures, such as short repeats nested within long repeats, can arise and be fixed by genetic drift. Some satellite DNA indeed has complex structure of this kind.

Highly repetitive sequences are more likely to build up in chromosome regions with low than with high recombination rates (Charlesworth et al. 1986; Walsh 1987; Charlesworth et al. 1994). Unequal crossing over is more frequent, the greater the number of repeats, because of the greater number of possible matches between sequences. If the recombination rate is high, chromosomes bearing only a single copy of the sequence may arise. Should a single-copy chromosome be fixed by genetic drift, the population is likely to remain in this state for a long time, since unequal crossing over is less likely to generate variation in copy number from a single copy than from multiple-copy chromosomes. If, on the other hand, the recombination rate is low, variants with a low copy number should arise less frequently, and a chromosome with multiple copies may remain intact long enough to be fixed by genetic drift. This model provides an explanation for the abundance of moderately and highly repetitive DNA in regions with low recombination, such as telomeres (chromosome tips), the vicinity of centromeres, and Y chromosomes.

Whether or not the nucleotide sequence of satellite DNA affects fitness is not known. However, the total *amount* of DNA does have biological effects, influencing both cell size and the rate of cell division. Both the time between mitotic cell divisions and the duration of meiosis increase with the amount of DNA, and species with high C-values frequently develop more slowly than those with low C-values (Bennett 1982; Rees et al. 1982; Sessions and Larson 1987). The salamander *Plethodon vehiculum*, with a C-value about twice that of the similar species *P. cinereus*, reaches about the same adult size, but with half the number of cells (MacGregor 1982). The rate of development is a critically important component of fitness and life history (see Chapter 19), the number of cells can strongly affect the function of the nervous system and many other organs, and the relative size and shape of many of an organism's features are affected by developmental rate (see the discussion of heterochrony in Chapter 23). If genome size, which is determined largely by the amount of satellite DNA, does not readily evolve, it could therefore constrain the evolution of both life histories and morphology; conversely, selection for changes in features such as the rate of development could alter genome size (Sessions and Larson 1987). It is not yet known how tightly coupled the evolution of genome size is to the evolution of life histories and morphology, nor which may more strongly affect the evolution of the other. It is safe to say, however, that because so much of the variation in genome size among species is due to noncoding DNA, differences in morphological "complexity" among taxa are not correlated with total DNA content (Cavalier-Smith 1985).

Evolution of Novel Genes and Proteins

The very first, unknown forms of life must have had very few organized nucleic acid sequences. Where, then, did all the many thousands of different functional genes in modern organisms come from?

We already have part of the answer: gene duplication has given rise to gene families, the members of which have diverged in sequence and function. The globin gene family described earlier, like many other gene families, is a modest example: although its members play different roles in different tissues and developmental stages, the major function of all the globins is oxygen transport. However, phylogenetic comparisons of amino acid and nucleotide sequences have revealed sequence similarity even among proteins with very different functions, which must have arisen by gene duplication in the very distant past. For example, many of the proteases of eukaryotes, including trypsin, chymotrypsin, elastase, carboxypeptidase, and phospholipase, are similar enough in sequence to suggest that they have arisen from a common ancestor (Barker and Dayhoff 1980). Moreover, several enzymes that play roles in blood clotting and in the breakdown of blood clots, such as thrombin, plasmin, and kallikrein, are related in sequence to the digestive proteases (Figure 22.14). Many other such examples have been described (Doolittle 1981).

New genes may also arise by *duplication of domains*. The DOMAINS of a protein are compact, continuous regions of the molecule that are spatially distinct and often differ in func-

---

FIGURE 22.14 An evolutionary tree of diverse proteases thought to have evolved by gene duplication (diamonds) and subsequent divergence. Thrombin, factor Xa, plasmin, and kallikrein are proteins associated with blood clotting; the others are digestive enzymes. The tree is based on similarities in amino acid sequences of vertebrate proteins and bacterial trypsin. (After Barker and Dayhoff 1980.)

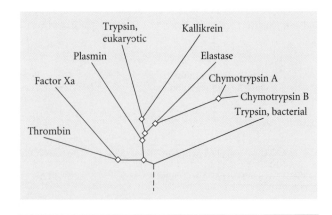

tion. For example, each of the three domains of ovomucoid, a protein in chicken egg white that inhibits trypsin, has a trypsin binding site. The three coding regions for these domains are similar in sequence and are separated by introns. However, each protein domain corresponds to two exons, so there is partial, but not complete, correspondence between exons and domains (Figure 22.15). This protein has undoubtedly arisen by two successive duplications of an ancestral gene with two exons (Li and Graur 1991).

A second part of the answer to our question is provided by increasing evidence that a single gene can serve two or more functions. In some cases, this is simply because the encoded product is versatile: we have already noted that the same protein serves as a crystallin in the eye lens of birds and as argininosuccinate lyase in other tissues. More intriguing are cases in which a single gene encodes two or more different polypeptides, produced by differential splicing of the primary transcript (pre-mRNA) into mRNA or by cleavage of the protein product into several functionally different polypeptides. The hormones corticotropin, melanotropin, and endorphin, for example, are all produced by cleavage of a primary protein product (Ohta 1988). In all such cases, it is easy to imagine that duplica-

tion of the gene, in whole or in part, could give rise to separate genes encoding different products (Jensen 1976; Hughes 1994)

A third answer to the question is that ancestral genes may have been joined to form modules in a new, larger gene. Soon after introns were discovered in eukaryote genes, the molecular biologist Walter Gilbert (1978) proposed that exons are ancestral genes that have been mixed and matched, like so many Lego pieces, into many combinations to form modern genes (see also Doolittle 1978; Darnell 1978). Evidence for this process, called **exon shuffling**, is provided by cases in which the exons of a gene correspond to different domains of the protein product, and by instances in which related proteins have different combinations of the same domains or modules. Gilbert (1978) suggested that introns facilitate exon shuffling (though this is clearly not their function).

Among the first described instances of exon shuffling were several of the proteins involved in the formation and dissolution of blood clots, mentioned earlier. These proteins consist of various mixtures of modules, or domains, among which are those we may call K, G, F, and C. Urokinase is an activator of the protein plasminogen, converting

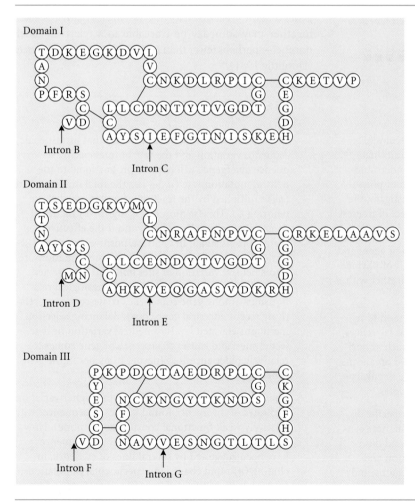

FIGURE 22.15 The three functional domains of the chicken ovomucoid protein, thought to have arisen by successive duplication of an ancestral gene encoding one such unit. The three domains are fairly similar in structure (e.g., the cysteine-cysteine bonds C–C) and in amino acids (denoted by letters) at corresponding positions. The coding region of the gene that encodes this protein has introns (D, F) that separate the sequences encoding the domains, but each domain corresponds to two exons that are also separated by introns (C, E, G) at corresponding points. (After Li and Graur 1991.)

An example of proteins that have evolved by exon shuffling. (A) Each of several proteins involved in blood clot formation and dissolution consists of a combination of the same modules (K, G, F, C) and a sequence (the cross-hatched bar) homologous to the protein-digesting enzyme trypsin. (B) The module-encoding part of the genes for two of the blood proteins, TPA and UK, and the genes that encode fibronect in (FN) and epidermal growth factor (EGFP). (After Li and Graur 1991.)

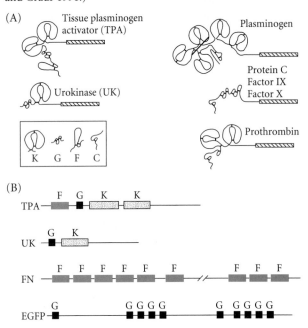

it into plasmin, which dissolves the fibrous protein (fibrin) in blood clots. Urokinase has both a G and a K module, each encoded by a separate exon (Figure 22.16). Epidermal growth factor, a protein with a very different function, as its name implies, consists of several G modules. Another activator of plasminogen is TPA (tissue plasminogen activator), which, unlike urokinase, activates plasminogen only when it is bound to fibrin. TPA has not only the K and G modules, but also an F module, the fibrin-binding moiety. Repeats of the F module constitute another protein, fibronectin, which promotes cell adhesion and also binds fibrin. Plasminogen itself, which is activated by TPA and urokinase, has an F module and several K modules. Finally, TPA and the other proteins associated with blood clotting include the sequence, mentioned earlier, that is homologous with digestive proteases such as trypsin, and which has protease activity. Thus TPA, the most complex of the lot, includes modules homologous to sequences in digestive proteases, epidermal growth factor, plasminogen, and fibronectin. The junctions among the modules correspond to the borders between the gene's exons and introns.

Although not all genes manifest such a nice correspondence between exons and functional modules of their protein products, so many sequence similarities have been described among modules of functionally quite different genes that the great diversity of genes in all organisms taken together may someday be traceable to a relatively small number—perhaps fewer than 1000—of ancestral sequences (Doolittle 1981).

Summary

1. Research in molecular evolution uses molecules (e.g., DNA sequences) as markers for solving traditional problems in organismal evolution; seeks understanding, at the molecular level, of the evolution of phenotypic features such as physiology; and investigates the evolution of genes and genomes as an object of interest in itself.

2. The processes and history of the evolution of genes and genomes are studied largely by the methods of phylogenetic inference and population genetics, together with information on molecular mechanisms.

3. Phylogenies (genealogies) of genes can be used to reconstruct (literally or figuratively) ancestral sequences, and can provide evidence of phenomena such as horizontal gene transfer among distantly related organisms. Horizontal gene transfer appears to be rare in eukaryotes, but may be fairly common in prokaryotes.

4. Studies of intraspecific variation and interspecific divergence in DNA sequences generally reveal more synonymous than nonsynonymous variation and divergence within coding regions, greater variation and divergence in noncoding than in coding regions, and more transitions than transversions.

5. The neutral theory (that most observed variation and divergence is due to genetic drift of selectively neutral or nearly neutral mutations) predicts that intraspecific sequence variation and the rate of interspecific sequence divergence will both be proportional to the neutral mutation rate (u_0)—i.e., the total mutation rate (u_T) multiplied by the fraction of mutations that are neutral (f_0). This fraction (f_0) is higher if functional constraints are weak or absent and if the effective population size is small. In contrast, positive directional selection, causing fixation of advantageous mutations, reduces neutral variation at sites linked to the selected site and causes closer-than-neutral genealogical relationships among gene copies (i.e., coalescence to a relatively recent ancestral gene copy). Balancing selection maintains elevated levels of neutral variation near selected sites, and causes coalescence of gene copies to relatively older ancestors (deeper branches in the gene genealogy).

6. The neutral theory is supported by the higher variation and rates of divergence found in sites or sequences with relatively weak functional constraints (or none). However, instances of selection on nucleotide sequences have been suggested by several lines of evidence, including (a) rapid change in experimental populations, (b) higher than expected variation in rates of sequence divergence among lineages, (c) greater rates of nonsynonymous than synonymous evolution, (d) convergent

evolution of functionally similar proteins, (*e*) differences in the ratio of synonymous to nonsynonymous substitutions among species compared with variation within species, (*f*) elevated variation in certain parts of genes, taken to indicate linkage to a balanced polymorphism, (*g*) reduced variation in regions of the genome with low recombination rates, which might indicate linkage to recently fixed advantageous mutations, and (*h*) "deep" gene genealogies, including polymorphisms that are older than the species in which they occur.

7. Codon usage bias appears to be due to selection, in part for rapidity of translation, which is enhanced by favoring codons that are recognized by the most abundant tRNAs. As predicted by population genetic theory, the rate of divergence is greater for genes with low than high codon usage bias, and the degree of bias is lower in species with low than high effective population sizes.

8. Selection may influence sequences in introns because complementary bases in the stems of the primary RNA transcript (pre-mRNA) stabilize its structure.

9. Gene families can evolve by unequal crossing over and fixation of the arrays of multiple genes thus produced. Members of gene families may become silent, nonfunctional pseudogenes, remain homogeneous in sequence and function, or diverge in sequence and function. Divergence in function may occur if redundant loci acquire different advantageous mutations, or if the descendants of an ancestral gene with multiple functions become differently specialized for those functions.

10. Members of some gene families undergo concerted evolution, remaining homogeneous in sequence despite divergent mutation, due to unequal crossing over or gene conversion (which may be unbiased or biased) followed by fixation of homogeneous arrays. Most homogeneous arrays are probably fixed by genetic drift, but natural selection may contribute to the homogeneity of gene families that produce a product required in large quantities. Gene families with functionally distinct members do not undergo concerted evolution.

11. Transposable elements produce copies of themselves, often by reverse transcription, that are inserted into the genome, where they frequently cause gene mutations. Recombination between TEs at different sites can cause alterations of chromosome structure, and the genes flanked by recombining TEs may be inserted at new sites in the genome. These effects are generally deleterious to the organism, and so TEs persist in spite of, rather than because of, natural selection at the organismal level. Selection among variant TEs (selection at the "gene" level) favors variants with high transposition rates. Although TEs collectively can constitute more than 10 percent of total DNA, they occupy any one site in the genome only infrequently. At least in *Drosophila*, the number of copies appears to be limited by natural selection at the organismal level, which eliminates TEs along with the deleterious chromosome alterations (and perhaps the gene mutations) that they cause. TEs are examples of "selfish DNA," and are an instance of conflict between selection at the level of the gene and the level of the organism.

12. Highly repetitive (satellite) DNA may constitute more than 90 percent of the genome. Changes in the number of repeats occur by unequal crossing over and other molecular mechanisms, which, together with genetic drift, can generate complex patterns, such as repeats within repeats. Population genetic theory also explains why satellite DNA is most prevalent in regions with low recombination. The amount, though probably not the nucleotide sequence, of satellite DNA can be affected by natural selection, and because DNA content affects the size of cells and the rate of cell division, the amount of repetitive DNA may influence the rates of organisms' growth and development, and possibly their life histories and morphology.

13. Over the course of evolutionary history, the number of functionally different genes has increased. Among the mechanisms that have engendered novel genes are (*a*) gene duplication and divergence, (*b*) duplication and further multiplication of domains or motifs within genes, (*c*) specialization, in duplicate loci, for the several functions served by the ancestral gene, and (*d*) exon shuffling, the union in various combinations of different ancestral functional domains (represented by exons) into larger genes. Introns are thought to have facilitated exon shuffling.

Major References

Li, W.-H. 1997. *Molecular evolution.* Sinauer Associates, Sunderland, MA. The most comprehensive textbook on the subject.

Golding, B. (editor). 1994. *Non-neutral evolution: Theories and molecular data.* Chapman and Hall, New York. Short essays, some empirical and others highly theoretical, by leading authors on the evidence for selection on DNA sequences.

Kimura, M. 1983. *The neutral theory of molecular evolution.* Cambridge University Press, Cambridge. Despite enormous advances in the study of molecular evolution since its publication, this exposition by the father and major advocate of the neutral theory is already among the classics in the literature of evolution.

Avise, J. C. 1994. *Molecular markers, natural history and evolution.* Chapman and Hall, London. A superbly written, comprehensive description of how molecular data can serve as tools in studying evolutionary processes, the evolutionary history of organisms, and conservation biology. Includes descriptions of molecular and analytical methods.

Problems and Discussion Topics

1. Explain how the level of polymorphism for synonymous base pairs may be affected by selection at a site to which they are linked.

2. If the effective size of the population of red-winged blackbirds in North America were 40 million, and the rate of neutral mutation were 10^{-8} per nucleotide, what proportion of nucleotide sites in an autosomal gene

would you expect to vary within a sample? What would be the proportion for a gene located on the X chromosome (of which males carry two, and females one)? What would the expected values be if the effective population size were only 50,000 instead? (Cf. the study of this species described in Chapter 11.)

3. Recall from Chapter 11 that inbreeding (due to self-fertilization or mating between relatives) alters the proportions of heterozygotes and homozygotes, but does not in itself alter allele frequencies. Nevertheless, partially self-fertilizing species of plants have much lower levels of allozyme variation than obligately outcrossing species (Hamrick and Godt 1990; Schoen and Brown 1991). List the several possible causes of this difference.

4. Recall from Chapter 11 that the rate of fixation of strictly neutral mutations, per generation, equals the mutation rate, and does not depend on population size. *Drosophila melanogaster* and *D. simulans* are more closely related to each other than to *D. yakuba*. Eanes et al. (1996) determined which of the nucleotide differences between the *G6pd* genes of *D. melanogaster* and *D. simulans* had arisen in each of these species by using *D. yakuba* as an outgroup. They found that more than twice as many synonymous substitutions have occurred in *D. melanogaster* as in *D. simulans*. However, more than twice as many amino acid-replacing substitutions have occurred in *D. simulans* as in *D. melanogaster*. There is independent evidence, from levels of molecular polymorphism, that the effective population size of *D. melanogaster* is much smaller than that of *D. simulans*. Explain the difference between these species in (*a*) synonymous and (*b*) nonsynonymous substitutions.

5. Relative to overall bias in codon usage in the genes of *Drosophila* species, one may designate any particular nucleotide change as a mutation to a "preferred" (i.e., common) or an "unpreferred" codon. For several genes, the ratio of preferred to unpreferred codons that have

become fixed in *Drosophila simulans* is 10:9, whereas this ratio is 8:46 for *D. melanogaster* (Akashi 1995; Eanes et al. 1996). Bearing in mind (as in question 4) that N_e has historically been larger for *D. simulans*, explain the difference in codon usage bias between these species.

6. Discuss ways in which studies of nucleotide variation, of the kind described in this chapter for *Drosophila*, might be used to locate the mutations responsible for harmful genetic disorders in humans.

7. The phenomenon of hybrid dysgenesis caused by *P* elements in *Drosophila* has suggested to some authors that transposable elements may cause speciation. What are the theoretical difficulties of this hypothesis? What evidence might be used to test it?

8. Large differences between populations in the amount of highly repetitive DNA might be expected to interfere with chromosome pairing in F1 hybrids, and thus to contribute to speciation by lowering hybrid fertility. Does evidence support this hypothesis? (See Flavell 1982; Rees et al. 1982.)

9. Is there any evidence that genome size ever decreases in evolution? Why might you expect it to do so?

10. In mammals and birds, extensive regions of the genome (isochores) are homogeneous, in that some regions have a high proportion of G–C base pairs, and others a high proportion of A–T pairs. A high proportion of the same type of pairs characterizes both the exons and introns of the genes in an isochore, as well as the regions in which the functional genes are embedded. What hypotheses might account for the existence of isochores? (See Bernardi et al. 1988; Wolfe et al. 1989.)

11. Why do most nuclear genes in eukaryotes have introns, and how did they originate? (See Cavalier-Smith 1991; Roger and Doolittle 1993.)

Macroevolution: Evolution above the Species Level

The most wonderful fact of evolution is that from a remote and simple ancestral form of life, such astonishing diversity should have evolved. That orchids and palms, butterflies and beetles, humans and pythons should have common ancestors—nay, that they should have the same ancestor as a bacterium—staggers the imagination. It seems incredible—and so it is, to those who cannot accept the reality of evolution. No principle of biology is as difficult for opponents of evolution to accept. And from the very beginning of evolutionary science—ever since *The Origin of Species*—no topic in evolution has been as controversial among biologists as the causes of the great changes that have marked the evolution of the higher taxa. We began our survey of evolution with a description of evolutionary history writ large (Part II); we examined the theory of evolutionary mechanisms that, at the level of populations and species, should account for evolutionary transformations (Part III); we then saw how this theory could account for a multitude of wondrous, puzzling adaptations (Part IV). Now the wheel has come full circle: we must ask whether the principles of evolutionary theory can account for the history with which we began.

We shall focus, in chapters 23 and 24, on the difficult questions posed by the evolution of major changes in morphology—the features that often characterize the higher taxa and reflect their physiological and ecological diversity. We must ask whether major changes ever arise by large "jumps" (*saltations*), or whether Darwin was correct in his belief that *Natura non facit saltum* ("nature does not make leaps"). That is, can we smoothly extrapolate from slight changes within species to the big differences that distinguish categories such as families and orders? Where do the novel features of the higher taxa come from, and how do their complex adaptations, composed of interacting elements, evolve? Can we explain very rapid evolution? Has there been enough time for known evolutionary mechanisms to account for the origin of diversity? What factors have limited or constrained the rates and directions of evolution? What have been the major trends in evolution? Is the history of life a history of progress?

As if these questions were not difficult enough, we turn in Chapter 25 to the full sweep of the diversity of life, and ask what factors have governed its history. Why have some groups diversified so much more prolifically than others? Why have some become extinct and others not? What has been the effect of the great mass extinctions on life's history?

These questions are mostly historical in nature. As in any historical science—geology, astronomy, human history—we can seldom demonstrate that an event had one cause rather than another in the same way that we can demonstrate the cause of, say, a physiological observation such as a low blood sugar level. We cannot do experiments to see whether the outcome would be different if we changed the hypothesized factor—we can't rerun the Cretaceous without a mass extinction. We cannot cross birds with dinosaurs to see what genes account for simple forelegs versus wings, and we cannot even identify or isolate for molecular analysis the genes responsible for this transformation (at least not yet).

Consequently, we can seldom demonstrate unequivocally the causes of a major evolutionary event. We can, however, propose that the fundamental physical, chemical, biochemical, genetic, and ecological principles that we document by studying living organisms have always operated in the same way, unless we have strong evidence to the contrary. Armed with this principle of uniformitarianism, we can ask whether current evolutionary and ecological theory is sufficient to explain the history of evolution. If historical events, inferred from paleontological and phylogenetic studies, are *compatible* with our theory of evolutionary mechanisms, and if they are *sufficiently* explicable by this theory, then we are justified in accepting the current theory until further evidence requires us to abandon it in favor of another. Our explanation is not *necessarily* the right one; but we may accept it as an adequate hypothesis until we are forced to abandon it.

An important question, then, is whether the history of evolution presents us with observations that call for some explanatory principle other than those embodied in contemporary evolutionary theory. Quite a few biologists have thought so, and some do still (e.g., Ho and Saunders 1984). But to have force, the critics must postulate alternative hypotheses, which we can then evaluate. For example, several paleontologists early in the twentieth century postulated that evolutionary trends are caused not by Darwinian selection, but by internal driving forces with their own momentum, which cause structures to evolve in a predetermined direction, even to the point at which they cause the species' extinction. The extinct "Irish elk," with enormous antlers, was cited as evidence for this theory (see Chapter 6). George Gaylord Simpson's (1944) counterargument included the following points: First, mutations do not show such directionality, so we know of no directional mechanism at this level. Second, in many living species with extremely elaborated structures, such as ungulates and beetles with enormous horns, the structures have a demonstrable function, as in sexual displays, and thus are likely to have a selective advantage. Third, some taxa with extreme structures survived for millions of years, so the structures per se obviously didn't cause their extinction (although given a change in environment, they might have). Finally, the fossil record shows few such unidirectional trends in any case: in the majority of cases, related lineages evolve in various directions. Thus the anti-Darwinians' hypothesis can be rejected because it makes predictions contrary to the data, and in fact is motivated by a supposed observation (straight-line evolution) that is generally false to begin with. Thus, although we cannot prove that trends in antler size or other features are caused by natural selection, this remains a more acceptable explanation, fully compatible with the data, than alternatives such as intrinsic drives.

In judging whether or not the theory of evolutionary mechanisms is sufficient to explain the history of evolution, we should ask what a "sufficient explanation" means. Explanations vary in depth, and usually the sufficiency of an explanation depends on precisely what the question is. An explanation that is sufficient for one person may not be for another who desires greater detail. In the 1940s, for example, geneticists purported to "explain" heredity by describing statistical patterns of inheritance of hypothetical molecules ("genes") and patterns of phenotypic effects (e.g., dominance, polygeny). The explanation went deeper when genes were found to encode enzymes and other proteins, and deeper yet when DNA was found to be the genetic material and its structure was elucidated; our explanation of heredity continues to grow as studies of gene action, gene regulation, and genetic control of development proceed. Thus explanations grow more comprehensive, often by becoming more reductionist—although it is ultimately necessary to integrate the particulars of a reductionist explanation into a unified whole.

In the same way, much of evolutionary theory is cast in terms of generalized, even abstract, concepts such as genes, mutation rates, and selection coefficients, with selection often cast in terms of ecological concepts such as competition. At this level of generality and abstraction, we cannot explain in detail why the climbing organs of grapevines are modified leaves but those of English ivy are aerial roots, or even why these plants should have evolved the climbing habit at all. We would need to know about the developmental biology of the plants' ancestors in order to judge whether their leaves or roots were more likely to vary genetically; we would need to know about the ecological circumstances these ancestors faced to understand why a vinelike rather than an erect growth form was advantageous. Likewise, if we were to ask why, at this time, the behaviorally complex, tool-using, landscape-changing, ecologically dominant animal species is a primate rather than a rodent or a lizard, the answer will not come from the generalized theories of population genetics or ecology. (That is, we could not predict, given only information on, say, Cretaceous animals, that a humanlike species would someday evolve.) The explanation instead will have something to do with the particular evolutionary histories of animal lineages, whereby the prerequisites for the evolution of human behavior became established in one lineage rather than the others.

The following chapters will present reasons to conclude that evolutionary theory is sufficient to explain the general features of evolutionary history, and that no evolutionary events discovered so far are incompatible with the theory. However, the general theory of evolutionary mechanisms is (*a*) incomplete and (*b*) neither capable of nor designed for predicting most individual evolutionary events (past or future). Like most scientific theories, evolutionary theory is incomplete in several respects, most conspicuously in that, like the theory of heredity, it lacks a sufficient body of principles for translating between genes and phenotypes.In other words, it requires a theory of developmental biology. Evolutionary theory can predict individual evolutionary events in principle, but seldom in practice, because to do so we would need an enormous amount of detailed genetic and ecological information in order to make corporeal the genetic variances and selection coefficients in the theory. These genetic properties and ecological circumstances, moreover, are uniquely different for every species as a consequence of its individual evolutionary history.

CHAPTER 23

Development and Evolution

Several kinds of information are necessary to account for both the *similarities* and the *differences* among organisms. Population genetics frames many of its explanations in terms of the operation of natural selection, genetic drift, and gene flow on genetic variation. But population genetics cannot, a priori, predict the source or form of natural selection, for which ecological or other information may be required. Likewise, population genetics does not include a theory of the *origin* of the genetically based phenotypic variation on which selection may act. Although it accepts that a species of fish may vary in the number of scales along its body, but not in the number of pairs of eyes, we require some other source of information to explain why.

Selection can have an effect only if there exists genetic variation, expressed in phenotypic characters. Some phenotypic traits, such as the structure and activity of some proteins, have a fairly direct, almost one-to-one, relationship to the genes that encode them; but even in the case of proteins, the relationship is frequently inexact because of post-translational modifications. For morphological features, the relationship between gene structure and phenotype is far more complex, involving both the influences of the environment and interactions within the developing organism. The processes that intervene during the development of an organism (its *ontogeny*) between primary gene action and the phenotypic trait are collectively called *epigenesis*. We should like, ideally, to understand developmental biology well enough to find "epigenetic rules," generalizations about the processes of development, that might help us to answer questions about the evolution of morphological and other phenotypic characters. Some such questions are:

1. Can developmental processes give rise to discontinuous new phenotypes, and have such phenotypes frequently contributed to evolution? Does development account for evolutionary transitions that appear to lack intermediate states?

2. Can development account for the origin of novel features?

3. Can developmental processes help to explain the evolution of complex features?

4. Does development bias or constrain the direction of evolution—i.e., the variety of possible modifications of a lineage?

5. Do constraints arising from developmental processes account for lack of evolutionary change—i.e., for stasis in the fossil record or for uniformity among related species?

6. Above all, how do developmental processes evolve?

Of all the major fields of biology that bear on evolutionary theory, developmental biology has been least incorporated into evolutionary science until very recently. Darwin declared in a letter to the botanist Asa Gray that "embryology is to me by far the strongest single class of facts in favor of a change of forms," and his followers, especially Ernst Haeckel, used embryological similarities as a major source of evidence for phylogenetic relationships (see below). However, when, early in the twentieth century, embryology made the transition from a purely descriptive field to an experimental science concerned with discovering the mechanisms of development, developmental and evolutionary studies diverged. During the Evolutionary Synthesis, evolutionary biologists focused on population genetics, and may not have known how to assimilate developmental data into their evolutionary theory. With a few exceptions—notably C. H. Waddington (1957)—experimental embryologists made few attempts to relate their data to evolutionary theory. Worse yet, the perspectives of embryologists, with their focus on mechanisms that operate within individual organisms, and of evolutionary theorists, who stress gradual gene frequency change at the population level, have often seemed incompatible. This tension still pervades some recent discussions of the role of development in evolution (cf. Goodwin et al. 1983; Ho and Saunders 1984). Only in about the last decade has developmental biology begun again to play a major role in evolutionary biology.

The emerging field of "evolutionary embryology" (Müller 1991), or "evolutionary developmental biology" (Raff 1996), seeks to determine how embryological processes have been modified in evolution and how developmental mechanisms have influenced the course of evolution. Developmental explanations are complementary, not mutually exclusive, to explanations in terms of molecular and population genetics. For example, an experimental embryologist may show that the form of a bone depends on the stresses that muscles impose on it during development. It would be wrong, though, to conclude that differences among species in the shape of this bone lack a genetic basis. Allele substitutions can alter the bone's shape via their effect on the intermediate processes (such as muscular insertions and stresses) that give it form. Furthermore, although developmental and genetic descriptions account for the PROXIMATE mechanisms that determine bone shape, they do not explain why alleles effecting one bone shape rather than another characterize the species. For this, we require an ULTIMATE explanation, such as natural selection or genetic drift.

Approaches to Studying Development and Evolution

On the basis of comparisons of the morphological features of different organisms, evolutionary biologists since the late nineteenth century have developed *descriptive models* of changes in development whereby an ancestral morphology might be transformed into a derived morphology. For instance, the greatly elongated digits that support the membrane of a bat's wing suggest a model in which the growth rate of the digits, relative to that of the rest of the body, increased during the origin of bats. Such formal models are not based on information about the mechanisms of development, but they may suggest mechanisms, such as elevated rates of cell division in the developing fingers of bats. Formal models of this kind still play an important role in describing patterns of evolutionary change in morphology.

Beginning early in the twentieth century, developmental biologists used the methods of experimental embryology to gain insights into the mechanisms of development, through manipulations such as removing (ablating) parts of embryos or transplanting organ rudiments between embryos. By such means, it was discovered, for example, that the developing dermis of vertebrate embryos, derived from mesoderm, induces the epidermis, which is derived from ectoderm, to develop scales, hairs, or feathers. In a few instances, a chemical basis for such processes was determined; for example, it was found that thyroxin, secreted by the thyroid gland, triggers metamorphosis of amphibian larvae to the adult form. Provided with such information, evolutionary biologists could sometimes gain insight into evolutionary changes in developmental pathways. For ex-

ample, some salamander species achieve reproductive maturity without undergoing metamorphic change in the rest of their morphology. As predicted, such salamanders produce little or no thyroxin.

Insights from the methods of classic experimental embryology, however, usually do not provide descriptions of developmental processes at the molecular level. In the past decade, some of the most exciting contemporary advances in all of biology have been made by using the techniques of molecular biology to reveal how genes affect development. Evolutionary biology has both contributed to these advances and profited from them, for we are beginning to learn how developmental processes evolve at a genetic level.

We shall describe the developmental bases of evolutionary change first in macroscopic, nonmolecular terms, and then in terms of contemporary molecular genetics.

Ontogeny and Phylogeny

Before *The Origin of Species*, biologists were already well aware that species are often more similar as embryos than as adults. Karl Ernst von Baer noted in 1828 that *the features common to a more inclusive taxon* (such as the subphylum Vertebrata) *often appear in development before the specific characters of lower-level taxa* (such as orders or families). This generalization is now known as VON BAER'S LAW. Probably the most widely known example is the similarity of many tetrapod vertebrate embryos, which display gill slits, a notochord, segmentation, and paddle-like limb buds before the features typical of their class or order become apparent (Figure 23.1). Another example is provided by the metatarsals of both birds and artiodactyls (e.g., cows), which are initiated as separate elements like those of other tetrapods, and only later become fused into the single bone characteristic of these taxa.

Darwin viewed embryological similarities as the most important source of evidence for the reality of evolution, and cited many examples in *The Origin of Species*. The relation between embryology and evolution was developed at length by one of Darwin's most enthusiastic supporters, the German biologist Ernst Haeckel. Haeckel coined the words *ontogeny* (the development of the individual organism) and *phylogeny* (the evolutionary history of species), and in 1866 issued his famous BIOGENETIC LAW: "Ontogeny recapitulates phylogeny." By this, Haeckel meant that in the course of its development, an individual successively passes through the adult forms of all its ancestors, from the very origin of the first cell to the present. Haeckel thus supposed that by studying embryology, one could read a species' phylogenetic history, and therefore infer directly phylogenetic relationships among organisms.

Repeated in biology textbooks ever since, Haeckel's "law" is one of the most famous maxims in biology. But by the end of the nineteenth century, it was already clear that the law rather seldom holds. The real development of organisms differs in several important ways from Haeckel's simple scheme (Gould 1977):

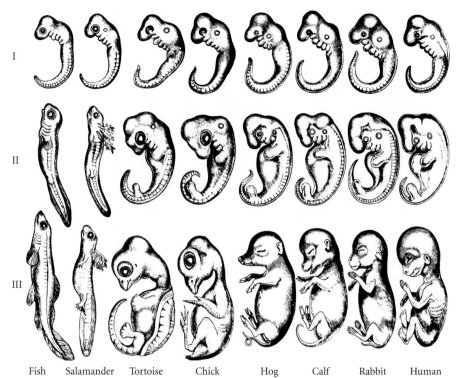

FIGURE 23.1 An illustration of von Baer's law: three stages in the development of several vertebrates. All the vertebrate classes share many common features early in development; many distinguishing features of the classes and orders appear later. (From Romanes 1901.)

Fish   Salamander   Tortoise   Chick   Hog   Calf   Rabbit   Human

1. Adult features of ancestors seldom appear as intermediate stages in the ontogeny of a descendant. von Baer's law is more often closer to the truth: different descendants of a common ancestor are likely to share early embryonic characters with one another and, presumably, with their ancestor. For example, although the embryos of mammals, "reptiles," and fishes all have gill pouches, in mammals and "reptiles" neither these features nor any others even fleetingly acquire the form typical of adult fishes.

2. Various features develop at different rates, relative to each other, in descendants than in their ancestors. Thus the ancestral morphology as a whole does not make a coordinated appearance during ontogeny. We shall shortly treat this subject under the topic of *heterochrony* (another word that Haeckel coined).

3. The preadult stages of any organism must grow and survive; to do so, they require stage-specific adaptations suitable for the particular environment of the species. Thus the horny beak of tadpoles, the innumerable adaptations of the larvae of insects and other invertebrates, and the deciduous egg tooth used by birds to break out of the eggshell are newly evolved features in these several lineages. In the extreme case, intermediate stages in the ontogeny of an ancestor may be lost completely from the ontogeny of a descendant; for example, direct-developing sea urchins such as *Heliocidaris erythrogramma* arrive at the adult morphology without passing through the planktonic larval stage that is typical of the ancestral ontogeny.

4. Haeckel's biogenetic law postulates that phylogenetically new features are added to the end of the ancestral ontogeny (TERMINAL ADDITION). The strongest blow to Haeckel's law was dealt by the discovery of *paedomorphic* species (from *paedos*, "child")—species in which the juvenile morphology of the ancestor is retained throughout life. In these cases, there has been terminal subtraction of stages of the ancestral ontogeny. The best-known examples are axolotls, members of the salamander genus *Ambystoma* in which metamorphosis does not occur, and larval features such as gills and the tail fin are retained throughout reproductive life (Figure 23.2). In the ancestral ontogeny, displayed by most species of *Ambystoma*, the aquatic larva metamorphoses into a terrestrial adult that lacks gills and the tail fin.

There are, to be sure, many cases in which certain features of an ancestor are recapitulated in the ontogeny of a descendent; for example, the metatarsals of a bird, as we saw above, at first develop separately (the ancestral condition) before becoming fused together. Still, the biogenetic law is honored more often in the breach than in the observance, and it is certainly not an infallible guide to phylogenetic history.

Developmental Principles of Evolutionary Change

What alterations of developmental pathways could account for evolutionary changes in structure? The findings of evolutionary embryology are usually discussed in phenomeno-

(A)

(B)

(C)

logical terms such as "change in rate of growth," and ultimately will require more reductionistic explanations than are presently available. Several developmental themes cast light on morphological evolution (Müller 1990, 1991; Raff 1996; G. P. Wagner 1996). These themes, derived mostly from nonmolecular studies of development and from comparisons among taxa, are (1) *individualization*, (2) *dissociation*, (3) *heterochrony*, (4) *heterotopy*, (5) *developmental interactions*, (6) *thresholds*, (7) *pattern formation*, and (8) *plasticity*.

Individualization and Dissociation

To a considerable degree, the body of an organism consists of *modules*—distinct units that have distinct genetic specifications, development patterns, locations, and interactions with other modules (Riedl 1978; Raff 1996). Modules are less distinct to the degree that their development is governed by pleiotropic effects of the same genes. In the extreme, several or many units of the body may have the same genetic specification and develop the same form. In such a case, they have no distinct individual identities, and may be considered aspects of a single character. This is often the case with uniform, repeated structures such as legs, teeth, and leaves (see Chapter 5). Such structures are termed SERIALLY HOMOLOGOUS if they are arrayed along the body axis, and HOMONYMOUS if they are not. In most "reptiles," for instance, including the synapsid ancestors of mammals, the teeth are uniform in shape (Figure 23.3A). In such cases, we cannot specify homology between individual teeth among different species.

One of the most important phenomena in morphological evolution is the acquisition of distinct identities by such units, a process called **individualization** (Wagner 1989b;

FIGURE 23.3   Acquisition and loss of individualization. (A) Most "reptiles," such as the extinct lizard *Kuehneosaurus*, have "homodont," uniform teeth that are not individualized. (B) Typical mammals, such as the elephant shrew, have "heterodont," differentiated teeth. (C) In the primitive Eocene whale *Prozeuglodon*, differentiation of the teeth has been reduced. (D) In modern toothed cetaceans, such as dolphins, the teeth are homodont, no longer individualized. (A after Romer 1966; B–D after Vaughan 1986.)

(A)

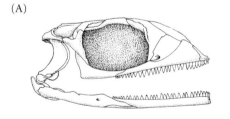

(B)

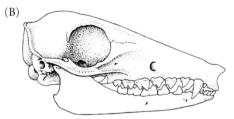

(C)

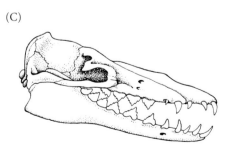

(D)

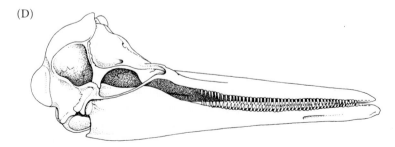

Müller and Wagner 1996). Individualization often entails **dissociation**: an evolutionary change in development whereby features acquire more independent genetic control, and thus can evolve more independently (Raff 1996). Individualization is therefore the basis of *mosaic evolution* (see Chapter 5). For instance, the teeth of mammals are individualized: they are differentiated into incisors, canines, premolars, and molars, and each tooth typically has a different shape (Figure 23.3B). Individual teeth, such as the first upper molar, are recognizably homologous among mammalian taxa. Individualization can not only be gained; it can also be lost. The distinction among most types of vertebrae has been lost in snakes, and toothed whales, descended from mammals with differentiated teeth, have uniform teeth (Figure 23.3C, D).

Günter Wagner (1996) has defined an individualized unit of phenotype as a complex of characters, or subunits, that serve a particular function and are tightly connected by pleiotropic effects of genes that have few effects on other such units. Each unit therefore develops relatively independently of others. Distinct units may evolve by PARCELLATION or by INTEGRATION (Figure 23.4), due in part to changes in the pattern of pleiotropic effects of genes. Wagner has proposed (see also Riedl 1978; Cheverud 1996) that natural selection may give rise to individualizing patterns of pleiotropy because genotypes with the right degree of integration of and independence among characters are most likely to survive episodes of directional selection. If, for example, selection on birds frequently favors changes in either bill size or leg length, but not both at the same time, genotypes in which these characters are independent can respond to selection, but those in which leg length covaries with bill size are selected against. Conversely, evolution of bill size may be more rapid if the upper and lower bill covary than if they are genetically uncoupled (see Chapter 14). Although there is some evidence that functionally related structures are genetically and developmentally correlated (see Chapter 14), this topic has been little explored, and the evolutionary theory of individualization has been little tested.

Heterochrony

HETEROCHRONY (Gould 1977, 1982c; McKinney and McNamara 1991) is broadly defined as *an evolutionary change in the timing or rate of developmental events*. Heterochrony has played an important role in studies of morphological evolution, for a vast variety of evolutionary changes can be described under this heading. However, several developmental mechanisms other than changes in timing can give rise to phenotypic changes that appear to be heterochronic (Raff 1996).

Heterochronic changes may be relatively *global*, affecting many characters simultaneously, or *local*, affecting one or a few characters. Global changes are illustrated by cases in which the time of development of most somatic features (those other than gonads and related reproductive structures) is altered relative to the time of maturation of the gonads (i.e., initiation of reproduction). The gonads of axolotls, for example, mature at about the same age and body size as those of their metamorphosing relatives, but development of the soma from the larval to the postmetamorphic condition is delayed—in fact, it does not occur. In local heterochronies, individual features of descendants become dissociated: they grow at different rates or are formed at different times than in their ancestors, relative to other features or to the body as a whole. Thus different features may show different heterochronic patterns. In fact, few examples of truly global heterochrony have been described. Even in axolotls, a shift from juvenile hemoglobin to adult hemoglobin occurs, just as in metamorphosing species of *Ambystoma* (Shaffer and Voss 1996).

Many evolutionary changes can be described as if local heterochronies had altered the *shape* of one or more characters. For example, an increased rate of elongation of the digits "accounts for" the shape of a bat's wing. Often, such differences can be described by simple mathematical models, an approach called *theoretical morphology*.

Theoretical morphology Assuming that shape differences among related organisms arise from changes in the relative rate of growth of various dimensions, D'Arcy Thompson (1917) suggested some simple mathematical transformations under which, by altering a minimal number of parameters, the shape of one species could be derived from another. Each landmark on an organism is assigned coordinates on a grid (for example, point 2c near the dorsal base of the pectoral fin in Figure 23.5). Then, for example, replacing the x coordinates with hyperbolas and the y coordinates with a system of circles centered near the head transforms the shape of a puffer fish into that of an ocean sunfish.

A more modern example is David Raup's (1962, 1966) theoretical analysis of snail morphology. A snail shell is ba-

FIGURE 23.4 Two ways in which a modular organization of individualized characters can evolve. In *parcellation*, pleiotropic effects of genes on different phenotypic units are reduced. In *integration*, initially independent characters are molded into modular complexes by the evolution of new pleiotropic effects. (After G. P. Wagner 1996.)

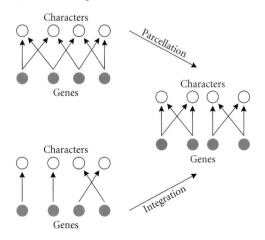

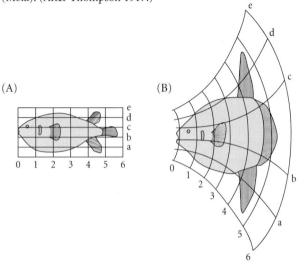

(A)   (B)

cause many structures change disproportionately with overall size.) The ALLOMETRIC COEFFICIENT, *a*, describes their relative growth rates. If $a = 1$, the two structures or dimensions increase at the same rate, and shape does not change. If *y* increases faster than *x*, as for human leg length relative to body size or weight, $a > 1$ (positive allometry); if it increases relatively slowly, as for human head size, $a < 1$ (negative allometry) (see Figure 9 in Chapter 17). These curvilinear relationships between *y* and *x* often appear more linear if transformed to logarithms, yielding the equation for a straight line, $\log y = \log b + a \log x$.

Allometric relationships can be measured in several ways (Cock 1966; Gould 1966). ONTOGENETIC ALLOMETRY expresses the relationship between *y* and *x* at different stages of the development of individual organisms. STATIC ALLOMETRY describes the relationship between *y* and *x* among different organisms at the same age or developmental stage. Static allometry can be described within or among species (intraspecific and interspecific allometry). In static allometry, variation in the relationship between *y* and *x* might have three sources:

1. The organisms all grew along the same allometric trajectory, but stopped at different points, reaching different values of both *y* and *x*.
2. The organisms had different allometric coefficients (*a*) during growth, attaining perhaps the same value of *x* but different values of *y* (compare curves 1 and 4 in Figure 9B in Chapter 17).
3. The organisms differed in *b*, but not in *a*, attaining different values of *y* at the same value of *x* (curves 1 and 2 in Figure 9B in Chapter 17).

Many allometric relationships are adaptive (see Chapter 17). For example, the amount of material absorbed through the wall of the intestine depends on its surface area, which increases as the square of body length, whereas the mass of tissue served by the organ increases as the cube. Function is maintained if the allometric relationship between intestine length (*y*) and body length (*x*) is 3/2, and approximately such a relationship holds within and among species.

sically a cone that expands as the shell grows at its aperture. The cone spirals around an axis as it grows (Figure 23.6). Raup refers to the cross-sectional outline of the aperture as the *generating curve*. He developed a computer model of growth with four variable parameters: the shape of the generating curve; the distance (*D*) between the generating curve and the axis about which it spirals; the rate at which the generating curve expands in size (*W*); and the distance that the generating curve moves down along the axis in one full revolution (*T*). For example, the shell will form a tight spiral if *D* is small, but an open spiral if *D* is large enough; it will form a high, narrow spire if the aperture grows slowly in size and moves rapidly down the axis; it will have a planispiral form (like a watchspring) if it does not move down the axis at all. The striking result of Raup's analysis is that virtually all real snails resemble one or another of the computer-generated forms. This analysis suggests that the diversity of gastropod shells may arise from alterations of just a few simple developmental "rules" of growth. However, we do not know whether the mechanisms of growth correspond to the simple variables in the model.

Allometry   The different rates of growth of the several dimensions of a snail's shell are an example of **allometric growth**, which refers to the *differential rate of growth of different measures of an organism* during its ontogeny. For example, during human growth, the head grows at a lower rate than the body as a whole, and the legs at a higher rate (see Figure 8 in Chapter 17). As described in Chapter 17, one way of describing such patterns is the allometric equation, (Huxley 1932)

$$y = bx^a$$

where *y* and *x* are two measurements such as the height and width of a tooth, or the size of the head and the body. (In many studies, *x* is a measure of body size, such as weight, be-

A model of allometry and heterochrony   Using an allometric framework, Pere Alberch and his colleagues (1979) developed a general model of heterochrony. Suppose that in an ancestral form, *x* represents body size, and *y* the size of some feature that begins to develop at age α and stops growing at age β. Heterochronic evolution may result in an "underdeveloped" state of character *y* in the adult descendant, resembling a juvenile state in its ancestor. Such a character state is **paedomorphic**. Alternatively, character *y* in the descendant may be "overdeveloped," or exaggerated, relative to its condition in the ancestor; such a state is **peramorphic**. Both paedomorphosis and peramorphosis can result from three kinds of evolutionary changes in development: alterations of the *rate* of development (k_y), or of the *duration* of development due to

FIGURE 23.6 A simple model of variation in the rate of growth in several dimensions accounts for variation in the form of gastropod shells. (A) The model specifies the shape of the aperture, or generating curve, the axis of coiling, the size ratio (*W*) of successive generating curves, the distance (*D*) of the generating curve from the axis, and the proportion (*T*) of the height of one generating curve that is covered by the succeeding generating curve in one full revolution. (B) Results of some computer simulations, showing the effects of varying the parameters *W*, *D*, and *T*. Most real snail shells closely resemble one or another of the simulated forms. (A after Raup 1962, 1966; B from Raup 1966; courtesy of D. Raup.)

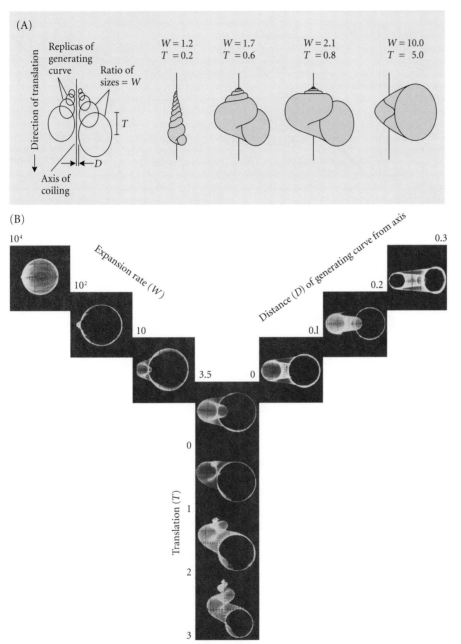

a change in the age at which growth of character *y begins* (onset) or *ends* (offset).

Figure 23.7 illustrates some effects of changes in the rate of growth or the time of offset. If the age of offset is extended from β to β + Δβ, character *y* is exaggerated, or peramorphic (Figure 23.7B). The evolutionary process resulting in a delayed offset of growth is called HYPERMORPHOSIS. During its ontogeny, the feature *y* follows the same growth trajectory as in the ancestor, so at some stage of development (namely, age β) the descendant has the same shape that its ancestor had as an adult. That is, the feature *recapitulates* the ancestral adult form, as in Haeckel's biogenetic law. Peramorphosis can also result from an *acceleration* of growth in *y* (i.e., an increase in k_y), a process that does not result in recapitulation (Figure 23.7D).

Conversely, truncation of development (cessation of growth at age β − Δβ) results in a paedomorphic character (Figure 23.7C). This process is called PROGENESIS. In con-

FIGURE 23.7 A diagrammatic representation of some forms of heterochrony, expressed as logarithmic plots of two structures or dimensions. The *x* axis might represent body size, the *y* axis, some character such as leg length. In each figure, the black line has a slope of 1 (isometry). Ontogenetic change in an ancestor, from age α to age β, is represented by the gray line. (A) The slope in the ancestor is > 1, so allometry is positive. (B) Extension of growth (hypermorphosis) to the colored circle in a descendant results in an exaggerated, peramorphic, structure *y*. (C) Development ceases at an earlier age (progenesis), leading to a juvenile structure at maturity (paedomorphosis). (D) Acceleration of development of *y* leads to peramorphosis. (E) A decrease in the rate of development of *y* relative to *x* (neoteny) results in a paedomorphic condition of *y*.

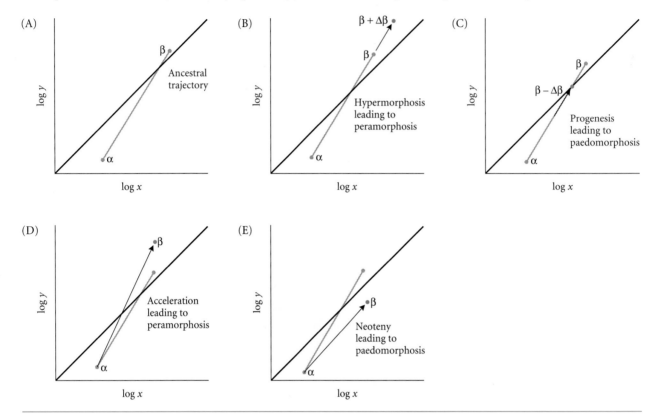

trast to Haeckel's law, the terminal stages of the ancestor's development are deleted in the descendant. Progenesis is usually associated with smaller adult body size than in the ancestor. Paedomorphosis also results if the descendant attains the same body size as the ancestor, but the growth rate (k_y) of character *y* is reduced (Figure 23.7E). This evolutionary process is called NEOTENY.

Although these changes are conceptually straightforward, it can be difficult to distinguish them in practice. For example, the bulbous cranium and short jaws of a human skull superficially resemble features of a juvenile chimpanzee, and thus have been cited as an example of neoteny. However, the bulbous human cranium is produced by prolonged brain growth (hypermorphosis), unaccompanied by comparable growth of the jaws (McKinney and McNamara 1991).

Examples of heterochrony At the genetic level, evolutionary shifts in the timing of gene expression represent one form of heterochrony. For example, a protein that appears to be instrumental in the formation of the internal skeleton of sea urchin larvae is expressed later in the developmental sequence of some species than in others (Wray and McClay 1989).

Allometric differences among species may be analyzed with reference to ontogenetic trajectories, or simply by means of static allometric comparisons among adults. For example, in the sea urchin *Rotuloidea*, of the Miocene, the skeleton is composed of small, simple ambulacral plates with slightly irregular edges (see Figure 13 in Chapter 6). In its descendants, the plates are larger, and the marginal irregularities form increasingly deep indentations, due to both an earlier onset and an accelerated rate of growth of some parts of the circumference, but not others. The allometric increase in the depth of the indentations is evident both in a comparison of the adult stages of ancestors and descendants and in the ontogeny of a descendant such as *Heliophora orbicularis*.

In the horse lineage from *Hyracotherium* to *Equus* (see Chapter 6), the length of the face and the depth of the jaw became peramorphic due to both hypermorphosis and acceleration (Figure 23.8A; Radinsky 1984). Because of the allometric relationship between these features and body size, they increased along with body size in the browsing forms during the Eocene and Oligocene (hypermorphosis). In the Miocene, *Merychippus* adapted to grazing on grasses, and the allometric relation of face length and jaw depth to body size became steeper (i.e., their growth became accelerated, in-

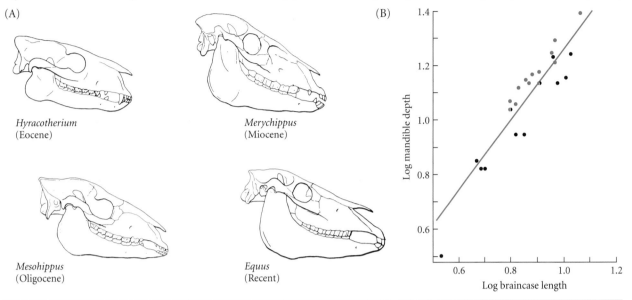

FIGURE 23.8 (A) Skulls of four equid taxa in the lineage from *Hyracotherium* to the modern horse, *Equus*. Note that the face is relatively longer and the mandible is relatively deeper in the more recent taxa, and that the posterior molars are shifted forward of the eye socket in *Merychippus* and *Equus*, which have higher-crowned teeth. All figures are drawn to the same size. (B) Mandible depth and braincase length (a measure of size) in various fossil horse taxa. The teeth are low-crowned in species represented by black circles, and high-crowned in those represented by colored circles. The latter species lie above the line, indicating an acceleration of development of mandible depth in high-crowned species. (A from Radinsky 1984; B after Radinsky 1984.)

creasing *a*) (Figure 23.8B). Thus among the later horses in the lineage, from *Merychippus* to *Equus*, the jaw is relatively much deeper than in the earlier forms, and the entire tooth row is situated anterior to the eye socket rather than extending beneath it, as in earlier horses (Figure 23.8A). These changes accommodate the bases of the molars, which extend deep into the jawbones. These teeth descend throughout life to compensate for abrasion of the grinding surfaces, an adaptation for grazing on silica-laden grasses.

Peramorphosis is very common in comparisons of species that differ in body size. When allometry is positive (*a* > 1), large species often have bizarrely elaborate structures. Antler size, for instance, increases disproportionately with body size in deer, and the largest species, the extinct "Irish elk" (*Megaloceros giganteus*), had truly huge antlers (Figure 31 in Chapter 6).

Many peramorphic features are not associated with evolutionary changes in body size, but are the result of accelerated development. For example, the related flying fishes, needlefishes, and halfbeaks all begin life with short jaws and reach similar body size, but one or both jaws in the halfbeaks and needlefishes undergo accelerated growth relative to the jaws of flying fishes (Figure 23.9).

Paedomorphosis is abundantly illustrated by salamanders, which provide many examples of both neoteny and progenesis. The paradigmatic example of *neoteny* is the axolotl (*Ambystoma mexicanum*), a member of the tiger salamander complex, in which many neotenic forms have evolved independently (see Figure 23.2). In the "normal" tiger salamander life cycle, the terrestrial adult loses the external gills and tail fin of the aquatic larval stage. The axolotl and other neotenic forms grow to normal adult size and reproduce while remaining aquatic, retaining gills and many other larval features. Some populations of *Ambystoma* are facultatively neotenic, capable of metamorphosis if their aquatic habitat dries up, whereas others are obligately neotenic. Metamorphosis is controlled in salamanders by the hypothalamus, which releases a hormone (TRH) that stimulates the pituitary gland to release another hormone (TSH), which in turn stimulates the thyroid gland to release thyroxin. Thyroxin acts on various tissues, stimulating metamorphic change. In *A. mexicanum*, this hormonal cascade does not occur. Crosses between neotenic and non-neotenic forms suggest that two genes determine whether or not metamorphosis takes place (Shaffer and Voss 1996). Injection of thyroxin induces metamorphosis in *A. mexicanum*, suggesting that in this species, neoteny is caused by failure of the hypothalamus to release TRH.

Genetic distances (see Chapter 9) among neotenic and non-neotenic *Ambystoma* populations are low, and neotenic adults closely resemble the larvae of their non-neotenic relatives (although they are much larger). Thus these neotenes appear to have evolved recently. Some other salamanders are derived from more ancient neotenic ancestors, and differ much more from their closest non-neotenic relatives (Figure 23.10). Permanently aquatic in habit, they have evolved in many respects since the initial neotenic event. In some such forms, such as the European cave-dwelling *Proteus anguinus*, injection of thyroxin does not induce metamorphosis, the tissues having lost their ability to respond to it.

Progenesis is illustrated by several lineages of dwarf salamanders, such as the tropical American genus *Thorius*

FIGURE 23.9   Differences in jaw length, attributable to acceleration of growth, among closely related families of fishes. (A) "Normal" jaws in a flying fish (Exocoetidae). (B) The lower jaw is lengthened in a halfbeak (Hemiramphidae). (C) Both jaws are lengthened in a needlefish (Belonidae). (From Jordan and Evermann 1973.)

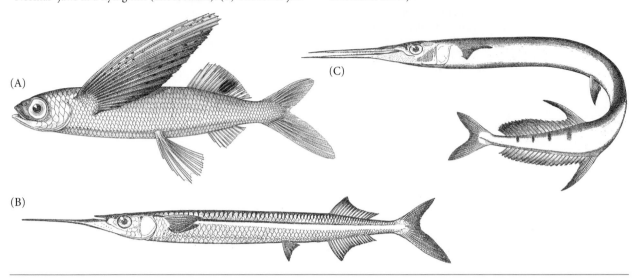

(A)

(B)

(C)

(family Plethodontidae), in which some species reach an adult body length of only 14 millimeters. Species of *Thorius* have many features characteristic of the juveniles of larger species of salamanders, as if their development had been abbreviated (Figure 23.11): they have fewer or no premaxillary teeth, and many of the bones of the skull are poorly ossified, remaining partly or entirely cartilaginous. In fact, some of the bones that typically develop last in salamanders may fail altogether to appear. James Hanken and his colleagues (Hanken 1984, 1985; Hanken and Wake 1993) have described other interesting features of *Thorius*. First, the eyes, nasal capsules, and brain of *Thorius* are proportionately larger than in other salamanders, having escaped great re-

duction in size, presumably because of natural selection. (These visual predators would suffer loss of visual acuity if the retina were too small.) Fitting large eyes and brain into a small head has resulted in major changes in the shape and organization of many of the skull bones compared with those of other salamanders. Second, Hanken has found that certain features vary much more within species of *Thorius* than in other salamanders. For example, the eight carpal (wrist) bones can be fused in as many as nine different combinations, all of which were found in a small sample from a single population (Figure 23.12). Hanken has suggested that because low body weight imposes less force on the legs, stabilizing selection on carpal structure is weak, allowing vari-

FIGURE 23.10   Some paedomorphic salamanders. (A) Adult *Eurycea neotenes*, with external gills. (B) A highly derived neotenic descendant of *Eurycea*, the blind cave salamander *Typhlomolge rathbuni*. (C) An unrelated species, *Amphiuma means*, that is highly divergent from other salamanders and is evidently derived from an ancient neotenic ancestor. The gills are retained in the adult, but are internal. (A after Conant 1958; B and C after Noble 1931.)

(A)   (B)   (C)

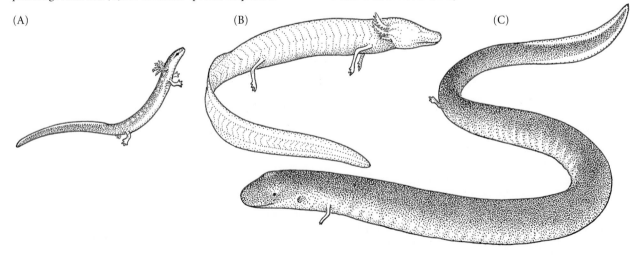

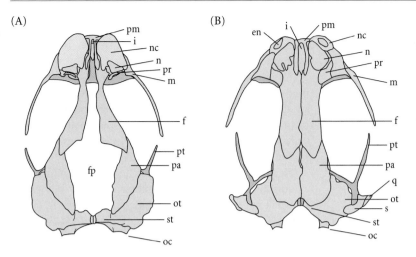

FIGURE 23.11 The skull of the dwarf salamander *Thorius narisovalis* (A), compared with that of a "typical" nonpaedomorphic relative, *Pseudoeurycea goebeli* (B). Among the modifications of the skull of *Thorius* is the open space (fp) between the paired frontal (f) and parietal (pa) bones, which fail to meet as they do in *Pseudoeurycea*. The other structures indicated are external nares (en), internasal (i), maxilla (m), nasal capsule (nc), occipital (oc), otic (ot), premaxillary (pm), prefrontal (pr), pterygoid (pt), stapes (s), and supratemporal (st). (After Hanken 1984.)

ation to persist. He speculated, as have previous authors who have made similar observations on small organisms, that genetic drift could fix different structural phenotypes in different populations. Should any such population give rise to a lineage of larger species, that lineage would be characterized by a major new structural organization, because of its history of relaxed selection and genetic drift. Thus both the high variability and the adaptive reorganization of features in miniaturized, progenetic species could be a source of evolutionary novelties (Hanken and Wake 1993).

Many authors have speculated that both progenesis and neoteny have been important in the origin of higher taxa (Garstang 1922; deBeer 1958; Gould 1977). For example, the byssal threads with which adult mussels and other sessile bivalves attach themselves to substrates are a feature only of the postmetamorphic juvenile stage in most other bivalves (Stanley 1972). Angiosperms differ from other plants in that both female and male gametophytes are greatly reduced,

consisting only of the embryo sac and pollen. They represent the extreme in a trend among vascular plants toward reduction of the gametophyte, which in the closest relatives of vascular plants (e.g., mosses) is the dominant phase of the life cycle. The reduction of the gametophyte has been described as paedomorphic (Takhtajan 1976; Guerrant 1988).

Heterotopy

HETEROTOPY is an evolutionary change in the topological position within an organism at which a phenotypic character is expressed. This term refers to changes in the cells or tissues in which gene activation or other developmental events occur. It does not ordinarily include changes in the relative positions of structures that can be described by changes in shape (such as the evolutionary shift, in some groups of fishes, of the position of the pelvic fins farther forward on the body). Studies of the distribution of gene products have revealed many heterotopic differences

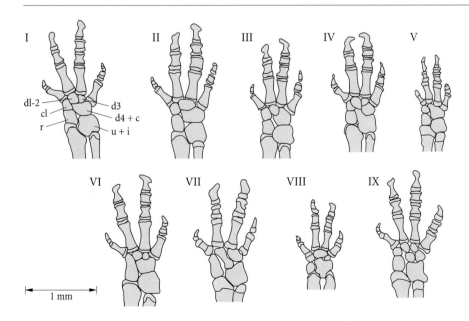

FIGURE 23.12 Variation in the number and arrangement of carpal (wrist) bones in dwarf salamanders (*Thorius*). Pattern I is the primitive arrangement, found both in *Thorius* and in related genera. The other patterns are common or rare variants found in various species of *Thorius*. (After Hanken 1985.)

FIGURE 23.13 Heterotopic expression of the protein Meso1 in two species of sea urchins at three early developmental stages. The protein is expressed in the cells shown in black. (After Wray and McClay 1989.)

Developmental stage

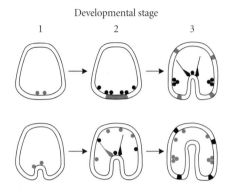

among species in sites of gene expression (Figure 23.13). For instance, certain species of deep-sea squids have organs on the body that house light-emitting bacteria. The light is diffracted through small lenses made up of the same two proteins that compose the lenses of the squids' eyes (Raff 1996). We shall later describe heterotopic differences in the expression of genes that play exceedingly important roles in development.

Heterotopic differences among species are very common in plants. For example, the major photosynthetic organs of most plants are the leaves, but this function is carried out in the superficial cells of the stem in cacti and many other plants that occupy dry environments. In many unrelated species of lianas (woody vines), roots grow along the aerial stem. In some lianas these serve as holdfasts, while in others they grow down to the soil from the canopy high above.

The bones of vertebrates provide many examples of heterotopy. For example, many phylogenetically new bones have arisen as sesamoids, bones that develop in tendons or other connective tissues subject to stress (Müller 1990). Many dinosaurs had ossified tendons in the tail, and the giant panda (*Ailuropoda melanoleuca*) is famous for having two extra "fingers," each consisting of a single sesamoid (Figure 23.14).

Changes in Tissue Interactions

During the course of embryonic development of many organisms, a group of cells differentiates into a specific tissue or structure in response to signals from other tissues or groups of cells. These epigenetic interactions can evolve. For example, the development of the lower jaw in vertebrates begins with the differentiation of its core structure, Meckel's cartilage, in response to inductive signals from other tissues. In amphibians, pharyngeal endoderm plays this role; in birds, cranial ectoderm; in mammals, the epithelium of the mandible (B. K. Hall 1983; Wagner 1989b). Changes may occur in the response to the inductive signal as well as in its source. For instance, the dermis of vertebrates induces the differentiation of epidermal structures. The embryonic epidermis of a lizard, when grafted onto the dermis of a

mouse embryo, develops scales in a pattern typical of mouse hairs (Dhouailly 1973). Thus, the epidermis can "read" the induction signal of a distantly related species, but it reacts by developing its own species-specific structures. This experiment, moreover, suggests that the signal from the dermis, and the capacity to respond to it, have not changed substantially in almost 300 million years, since the divergence of the lineages leading to lizards and mammals. *Some developmental mechanisms are therefore very conservative in evolution*—a point to which we shall return.

New kinds of interactions may be less important in the evolution of novel features than the context (time or place) in which interactions occur (Müller and Wagner 1991). For example, several groups of rodents carry food in cheek pouches located within the mouth, which develop by invagination of the oral epithelium. Some heteromyoid rodents, such as kangaroo rats and pocket gophers, have cheek pouches that open outside the mouth and are lined with fur. In these, the invagination during development is shifted slightly anteriorly, to the epithelium of the lip rather than the mouth cavity. It thereby acquires an external opening, and interacts with extraoral dermis, which performs its usual role of inducing the development of hair (Brylski and Hall 1988).

Thresholds

Complex dynamic systems often have nonlinear properties, so that a small change in one variable component can, by crossing a threshold, lead to discontinuous differences in other components. We saw some examples of such effects in development when we considered threshold traits, such as the number of toes in guinea pigs (see Chapters 3 and 14). Thresholds in developmental dynamics could have discontinuous effects in evolution. For example, reptilian scales

FIGURE 23.14 The wrist and part of the hand of the giant panda. The metacarpals of the five digits are numbered; the phalanges of these digits are not shown. The two extra "digits" are sesamoid bones that develop from ossifications in tendons, not by the developmental process that gives rise to true digits. (After Müller 1990.)

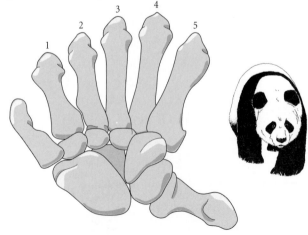

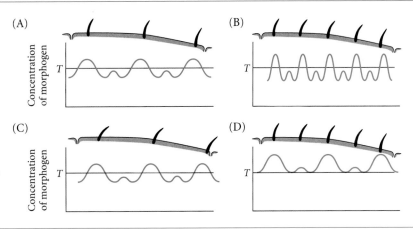

FIGURE 23.15 Hypothetical changes in an underlying prepattern (A) of a morphogen, which might be, for example, a material that induces bristle formation. If the concentration of morphogen exceeds a threshold, T, a bristle is formed at that site. Change in the kinetics of morphogen synthesis can change the spacing pattern (B) or the position (C) of the morphogen peaks. Increasing morphogen concentration can change the number of bristles (D).

and avian feathers form from papillae that develop as evaginations of epidermal epithelium around a dermal core, whereas mammalian hairs form from epithelial invaginations, which become the hair follicles. An epithelium is a sheet of cells attached to a relatively inelastic basement membrane. George Oster and Pere Alberch (1982) have described a mechanical model that shows how changes in the shape of groups of epithelial cells, owing to the contraction of intracellular filaments, can cause the basement membrane to buckle. Whether it buckles inward or outward, forming a follicle or a papilla, depends on slight differences in the mechanical properties of the cells and the membrane. This is not to say that a simple switch determines the many differences between feathers and hairs, only that invagination versus evagination might be a simple threshold effect.

Pattern Formation

PATTERN FORMATION may be defined as the process by which the spatial pattern of cellular differentiation is specified (Wolpert 1982). The spatial distributions of legs, scales, and petals are examples of patterns. Many models of pattern formation assume that a structure develops at a particular site on an embryo because chemical cues specifying its differentiation exceed a requisite threshold at that site. Such a chemical cue is termed a MORPHOGEN. Several models have followed the lead of a highly creative mathematician, Alan Turing (1952),* who showed that if two compounds that react to yield a chemical morphogen diffuse across an area (a "field"), very slight irregularities in their concentration can give rise to regularly spaced peaks

in the concentration of morphogen. The pattern depends on the size and shape of the field, the rates of diffusion of the precursors, and the kinetics of their reaction. Changes in the arrangement of the wave pattern or in the total concentration of the morphogen can then alter the distribution of peaks that exceed the threshold necessary for the structures to develop (Figure 23.15).

Applying an extension of Turing's model to pigment formation, J. D. Murray (1981) has predicted that if the field within which the reaction occurs is small, the morphogen wave pattern will be one-dimensional, giving rise to bands or stripes; above a critical field size, it should form a two-dimensional pattern of spots. Thus on a surface with a tapering cylindrical form, such as a leg or tail, spots may develop in the thick basal region, but only bands will develop near the tip—a pattern observed in many mammals (Figure 23.16).

FIGURE 23.16 An example of possible constraints on the phenotype arising from physical chemistry. Markings on the tails of mammals resemble theoretical patterns arising from a mathematical model of the kinetics of a biochemical reaction: (A) genet (*Genetta genetta*, a viverrid carnivore); (B) cheetah (*Acinonyx jubatus*); (C) jaguar (*Panthera onca*). (From Murray 1981.)

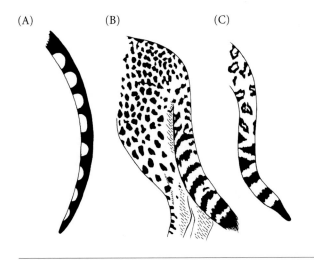

---

*Turing's name may be familiar in another context: he was the father of the concept of the modern computer. He is also credited with playing a crucial role in the defense of Britain during the Second World War by deciphering German codes. After the war, the government, instead of celebrating his contributions, denied his security clearance and subjected him to electric shock therapy when he was discovered to be homosexual. He committed suicide soon after writing his seminal paper on pattern formation. His life is the subject of a biography (A. Hodges, *Alan Turing: The Enigma*) and a play (H. Whitemore, *Breaking the Code*).

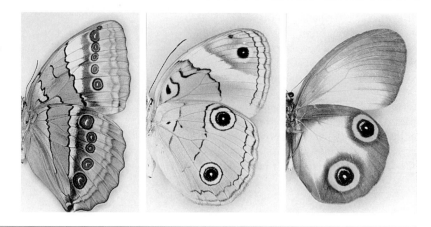

FIGURE 23.17 Patterns in the wings of nymphalid butterflies, showing variation in eyespots and other pattern elements. (From Nijhout 1991; courtesy of F. Nijhout.)

Murray's theory suggests that there should be no striped mammals with spotted tails.

Many morphological differences among organisms can be described as differences in pattern, such as the number of carpels or petals in a flower, of toes on the feet of mammals and "reptiles," and of veins in the wings of insects. It is possible that many such features evolve by relatively slight changes in chemical and developmental reactions.

A promising start in understanding the developmental basis of changes in pattern has been made in recent studies of the wing patterns of butterflies. These patterns are formed by the deposition of various pigments in the scales (modified hairs) that coat the wing. The wing develops from two sheets of cells, dorsal and ventral, that constitute the imaginal wing disc (see Chapter 3 on imaginal discs). The patterns on the dorsal and ventral sides often differ. The immense variety of color patterns among the 17,000 species of butterflies arises, in large part, from differences in the expression of a few widely shared pattern components, such as eyespots, crossbands, and rays of pigmentation paralleling the wing veins (Figure 23.17). Investigators such as Frederik Nijhout (1991) have shown that pattern elements such as eyespots are organized by specific groups of cells in the imaginal disc. If a certain group is ablated, for example, an eyespot fails to develop; if it is transplanted to another part of the wing disc, a new eyespot forms there.

Using DNA probes from genes known to control wing development in *Drosophila*, Sean Carroll and colleagues (1994) were able to determine the pattern of expression of the homologous genes in the developing wing discs of the butterfly *Precis coenia*. They focused attention especially on the development of the several conspicuous eyespots that are centered within certain of the "wing cells" (areas of the wing bounded by specific wing veins). Two genes are expressed in thin rays extending proximally to different, precise distances from the margin of the wing disc (Figure 23.18A,B). Their position, corresponding to the parallel rays and crossbands of pigmentation in many butterfly species, suggests that the expression of these and other genes establishes a *coordinate system* within each "wing cell," pro-

viding information to cells (in the usual sense) about their position relative to the wing veins and along the proximal/distal axis.

One gene, *Dll*, is expressed initially as a ray extending inward from the margin, along the middle of each "wing cell" (Figure 23.18C). As development proceeds, its expression is lost, except in the "wing cells" bounded by veins Cu_1 and Cu_2 and by M_1 and M_2. Here, *Dll* expression intensifies, and remains only as a spot in the center of the dorsal sheet of wing disc cells (Figure 23.18D). These are precisely the sites that have been shown by ablation and transfer experiments to be required for the dorsal eyespots to develop (Figure

FIGURE 23.18 Patterns of gene expression governing the development of color patterns in the hindwing of the butterfly *Precis coenia*. (A and B) Arrows indicate the expression of genes *wg* and *dpp* respectively, which are thought to establish a coordinate system that regulates the expression of other genes. (C) The gene *Dll* is first expressed in the distal portion of each "wing cell." Later in development (D), expression is limited to the spot indicated by the arrow, which corresponds to the pigmented eyespot in the mature wing (E). These genes also govern the polarity of wing development in *Drosophila*, where they have different spatial and temporal patterns of expression. (From Carroll et al. 1994; courtesy of S. Carroll)

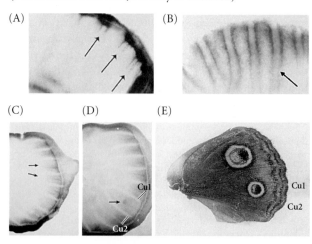

(A)   (B)

(C)   (D)   (E)

23.18E). Thus the pattern of pigment distribution in these parts of the wing is reflected in, and perhaps caused by, the pattern of expression of a gene that is known to organize wing development in *Drosophila*. Carroll et al. point out that if the expression of *Dll* in other "wing cells" were to follow the course it does in the Cu$_1$–Cu$_2$ and M$_1$–M$_2$ "cells," instead of dissipating, it would cause a row of eyespots parallel to the wing margin, as is displayed by many other species of butterflies.

Developmental Plasticity and Integration

A genetic alteration of phenotype can contribute to evolution only if it runs the gauntlet of natural selection: it must be at least effectively neutral, if not positively advantageous. Selection has not only external (ecological) sources, but also internal ones. For an organism to have any prospect of viability, its parts must function harmoniously. For instance, early in the history of horses (see Figure 23.8), a mutation that greatly increased the height of the molars so that their roots grew into the eye socket would surely have been lethal in any environment these animals were likely to encounter, since horses need functional eyes. Thus selection is imposed by the functional interrelationships among parts. This requirement must often impose limits on the rate and direction of evolution (see Chapter 24).

The need for functional integration would limit evolution less, however, if development were so organized that genetic change in one part automatically induced coordinated changes in functionally interacting parts. From a genetic point of view, this would be described as pleiotropy—but pleiotropy of a special, adaptively organized kind. Suppose, for example, that the length of each bone, muscle, blood vessel, and nerve in a leg were controlled by different genes—i.e., that there were no pleiotropy. Response to selection for longer legs would be extremely slow, for individuals with a longer than average femur would not usually have muscles and nerves of the appropriate length, and the leg could not function. But in fact, leg length responds readily to selection (cf. basset hounds vs. greyhounds), because the several organ systems grow harmoniously. They are developmentally integrated, and we can think of genes governing leg length as having pleiotropic effects on all these organs. Thus the existence of independent genetic variation in a large number of functionally related characters can retard the response to selection, whereas genetic correlations among such characters can enhance it (Wagner 1988; Cheverud 1988b; see Chapter 14).

We saw in Chapter 14 that in at least some instances, functionally related features do tend to be genetically correlated. Embryologists have also described many cases in which experimentally altering a feature induces compensatory changes in the growth of other features, thus tending to maintain function. Such changes, due not to genetic change but rather to the immediate reaction of the developmental system, fall under the heading of DEVELOPMENTAL PLASTICITY.

For example, Victor Twitty (1932) exchanged optic vesicles between embryos of the salamanders *Ambystoma maculatum* and *A. tigrinum*. The latter is a larger species, with larger eyes. In normal development, an optic nerve grows from the developing retina of the optic vesicle to the optic center on the opposite side of the midbrain. Extrinsic eye muscles, those that move the eyeball, develop from the head myotomes and insert on the developing eye. Twitty found that the transplanted eyes grew to the size normal for the donor species rather than the recipient. The optic centers of *A. maculatum* embryos with an implanted *A. tigrinum* eye became enlarged by cell proliferation; conversely, their size was reduced in *A. tigrinum* larvae bearing an implanted eye of *A. maculatum*. The extrinsic eye muscles of *A. maculatum* developed to a far greater size, consisting of more muscle fibers, in response to an implanted *A. tigrinum* eye. Thus the brain and muscles showed a developmental plasticity that would go a long way toward maintaining normal function.

Developmental Genetics and Evolution

Much of development is due to the expression (transcription and translation) of different genes in different cells. The consequent differences in enzymes and structural proteins determine the functional differences among cells, including the ways in which they interact with other cells or tissues and the ways in which they contribute to morphogenesis (the shaping of form). As we noted in Chapter 3, the expression of genes that produce cell-specific proteins (EFFECTOR GENES) is controlled by other genes (REGULATORY GENES), the expression of which is in turn regulated by yet other genes in a more or less hierarchical fashion. The study of development includes analyses of how the products of the effector genes build different tissues and organs, as well as analyses of the genetic pathways of development: which genes regulate which others. Research on genetic pathways is one of the most important topics in contemporary biology, and is yielding unprecedented insights into the mechanisms of development—and into the developmental basis of evolution.

Hox Genes and Other Regulatory Genes

An important body of knowledge has arisen from studies of the *homeotic genes* of *Drosophila*. Recall from Chapter 3 that mutations of these genes alter the identity of body segments. For example, a mutation of the *bithorax* gene (*bx*) transforms the third thoracic segment (T3), which normally bears halteres, into the form of the second thoracic segment (T2), which bears wings (see Figure 17 in Chapter 3). Another mutation of a gene in the same region, *Ubx*, transforms both T3 and A1 (the first abdominal segment) into T2-like segments. Deletion of this region has the same effect, so *bx* and *Ubx* are "loss of function" mutations. Because T2 is normal in *bx* and *Ubx* flies, but T3 and A1 are transformed, the morphology of T2 is the "default state," which itself does not require normal *bx* or *Ubx* genes to function, and which is acquired by T3 and A1 unless these genes function normally.

Drosophila has eight homeotic genes, which control the differentiation of all the body segments. These genes constitute two linked clusters in *Drosophila*, and one cluster in several other groups of insects. Remarkably, and for unknown reasons, the sequence of the genes along the chromosome matches the sequence along the body of the segments whose identity they control.

At a critical stage in early development, different combinations of homeotic genes are transcribed at different positions along the anterior-posterior (A-P) axis. This pattern of expression is hierarchically controlled by several "levels" of other genes (see Figure 18 in Chapter 3). In turn, the protein products of the homeotic genes directly or indirectly control the transcription of numerous other genes, including effector genes that build the features typical of a particular segment. The product of the *Ubx* gene, for example, is thought to regulate the expression of 85 to 170 "target" genes (Carroll 1995). It does so by binding to a four-base-pair-long regulatory sequence of the target gene. The critical DNA-binding region of the homeotic gene's protein product is encoded by a sequence of about 180 bases, called the HOMEOBOX. The homeobox is found in all the homeotic genes, although its sequence differs slightly from one to another. The homeotic genes that control A-P patterning are called HOX GENES. The identity of a prospective segment—whether it will have the features of T2 or T3, for example—is determined by the *combination* of Hox genes expressed in that region. By "selecting" target genes, the Hox genes switch development of a region of the body along one pathway or another.

Many genes other than Hox genes encode DNA-binding proteins and have regulatory functions. A few of these will figure in the ensuing discussion.

The Evolution of Hox Genes

The function of Hox genes was at first thought to be the control of segmentation in insects, but this idea proved to be wrong. When DNA sequences of *Drosophila* Hox genes were obtained, it became possible to search for similar sequences in other animals. *Hox genes have been found in all animal phyla examined, whether segmented or not.* All phyla, from Cnidaria (coelenterates) to Chordata, have multiple Hox genes, with very similar homeobox sequences, implying that a gene family, stemming from repeated duplications of an ancestral gene, has been inherited from the common ancestor of all metazoans, which lived more than 550 million years ago. Moreover, genes with homeobox sequences are carried by all eukaryotes, including fungi and plants.

Much attention has been focused on the Hox genes of vertebrates. Within the phylum Chordata, the sister group of the subphylum Vertebrata is the Cephalochordata. These animals, known as amphioxus (Figure 15 in Chapter 7), have a dorsal nerve cord, but lack paired sense organs or a brain. They have a single array of at least 10 different Hox genes (Figure 23.19; Holland and Garcia-Fernández 1996). In vertebrates, 13 different Hox genes have been found. Based on DNA sequences, many of these can be matched not only with the Hox genes of amphioxus, but also with particular Hox genes of insects, and the order of homologous Hox genes within an array is the same in insects and vertebrates! However, the entire array has been quadrupled in vertebrates, so that there are four "clusters" (*a* to *d*) of Hox genes (Figure 23.19). Any one Hox gene (say, number 9) may therefore have up to four *paralogous* representatives; each such group is known as a "paralogous group." In most of the 13 paralogous groups, a gene has been lost from one or more of the four clusters, but different genes have been lost in mammals and teleost fishes. Mice and humans have the same 38 Hox genes. The members of a paralogous group are not identical in either sequence or function, for their patterns of expression differ.

The Evolution of Regulatory Gene Function

Divergence among Hox genes and their functions in different lineages of animals is thought to underlie many dif-

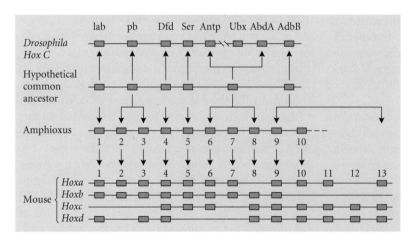

FIGURE 23.19 The Hox genes of *Drosophila*, amphioxus, and mouse, showing suggested evolutionary changes from a hypothesized Hox gene cluster in their common ancestor. *Drosophila* has two clusters of linked genes. Amphioxus has a single cluster. In vertebrates, duplication of the cluster has produced four paralogous clusters (*Hoxa–Hoxd*), with two to four members in each of the paralogous groups (1–13). (After Holland and Garcia-Fernández 1996.)

ferences in body plans, both among and within phyla. At present, some of these hypotheses are speculative, but others are well supported by evidence.

Hox genes do not encode structures. Rather, they are transcription-regulating factors with spatial and temporal patterns of expression that provide *positional information:* they evoke or repress activity in other genes according to the position of the cell in the developing body. Structures develop in each region in response to a "Hox code," the combination of Hox genes active in that region. For example, the combined expression of *Hoxb-1, Hoxa-1, Hoxa-3,* and *Hoxd-4* marks the region in which the first vertebra develops, and similar combinations mark other regions of the body axis. The genes that respond to a Hox code, and the structures they build, differ among organisms. For example, the Hox genes expressed in regions that mark part of the developing hindbrain of the mouse are homologous to the Hox genes expressed in the same segmental positions in amphioxus—but amphioxus develops only a simple nerve cord rather than a brain (Figure 23.20).

Evolution of Hox (and other regulatory) genes and their interactions with target genes might affect morphological evolution in several ways.

Changes in number   The single array of Hox genes found in most animals undoubtedly arose by duplication of an ancestral gene, followed by divergence in sequence and function (see Chapter 22 on the origin of new genes). The entire array has been duplicated in vertebrates. Increasing the number of potential Hox combinations may have set the stage for greater differentiation of body parts, and thus for the evolution of more complex body plans. It has been suggested, for example, that the duplication of Hox clusters near the time of the origin of the vertebrates may have permitted (although not directly caused) the evolution of greater morphological complexity than in amphioxus (Holland and Garcia-Fernández 1996).

Changes in spatial expression   Changes in the spatial expression of Hox genes may underlie morphological evolution. Such changes could be due to changes either in the "upstream" genes that regulate Hox gene expression or in the responses of the Hox genes to these upstream regulators. In the brine shrimp (*Artemia*), for instance, the eleven postgnathal/pregenital segments are identical and bear legs, whereas of the pregenital segments of insects, only the three thoracic segments bear legs. The Hox genes *Antp, Ubx,* and *abdA* are all expressed throughout the prospective thorax of the shrimp, but in insects, these genes have separate, overlapping domains of expression that distinguish thorax from abdomen (Averoff and Akam 1995; Figure 23.21). The shrimp and insects differ by heterotopic differences in gene expression.

Changes in gene interactions   *Changes in the interactions* between regulatory genes and their "downstream" target genes are probably one of the most important bases of morphological evolution (Carroll 1995; Raff 1996). In such cases, the mode of action and the spatial domain of expression of the Hox genes may remain much the same, but different genes come to be regulated by their products. In *Drosophila,* normal function of *Ubx* is required for segment T3 to develop halteres rather than wings, and loss-of-function mutations of *Ubx* transform T3 into a wing-bearing segment resembling T2. One might expect, therefore, that *Ubx* would not be expressed in the prospective T3 segment of four-winged insects such as butterflies—but in fact it is expressed there, just as it is in *Drosophila.* Therefore, the differences between a wing and a haltere, and indeed, between the other characteristic features of T2 and T3, are due to different responses by downstream genes that are regulated by *Ubx,* not to differences in *Ubx* itself.

This principle is also illustrated by one of the most dramatic results in experimental developmental biology. Anatomically, the compound eyes of insects (Figure 23.22) have almost nothing in common with the eyes of vertebrates (see Figure 20A in Chapter 5). They share only the universal property of photoreceptors: they contain visual pigments (opsins) that absorb light and transduce the energy into signals carried by neurons to the brain. The *eyeless* (*ey*) mutation in *Drosophila* causes partial or complete absence of the compound eyes. The *Small eye* (*Sey*) mutation causes failure of eye development in mice. Its homologue in humans is *Aniridia.* Remarkably, the sequences of

FIGURE 23.20   Homology between the neural tube of an amphioxus larva (A) and a vertebrate embryo (B), based on cellular organization. The posterior part of the vertebrate's developing brain is divided into segments (r1–r8). The anterior limits of expression of Hox genes 1 and 3 suggest that the vertebrate hindbrain evolved by elaboration of part of the neural tube, as seen in amphioxus. Abbreviations: d, diencephalon; mes, mesencephalon; n, notochord; n.t., neural tube; t, telencephalon. (After Holland and Garcia-Fernández 1996.)

(A)   (B)

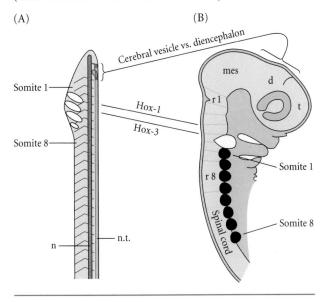

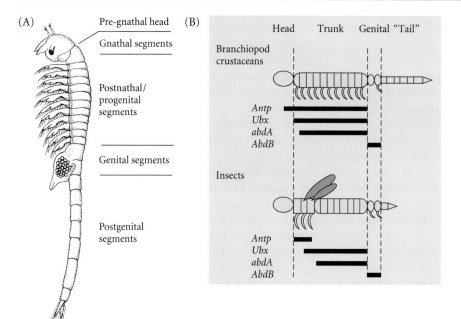

(A)
Pre-gnathal head
Gnathal segments
Postnathal/ progenital segments
Genital segments
Postgenital segments

(B)
Head   Trunk   Genital "Tail"

Branchiopod crustaceans

Antp
Ubx
abdA
AbdB

Insects

Antp
Ubx
abdA
AbdB

FIGURE 23.21 Differences in the domains of expression of Hox genes may be responsible for the evolution of segment differentiation in arthropods. (A) The brine shrimp *Artemia*, a crustacean in which the segments between the gnathal (mouthpart) segments and the genital segments all have the same form. (B) Differences between crustaceans and insects in the pattern of expression of Hox genes that are known to control segment differentiation in insects. The first three postgnathal segments of insects are differentiated into a distinct thorax, and differ strikingly from those of the crustacean in their Hox gene expression. (After Averoff and Akam 1995.)

the protein products of these genes are 94 percent identical between *Drosophila* and mammals, and the genes are normally expressed at similar sites in the developing nervous system and the embryonic eye primordia.

By clever molecular manipulations, Georg Halder and his colleagues (1995) were able to obtain, in *Drosophila*, ectopic expression of the normal form of the *Drosophila ey* gene—that is, they were able to activate the gene in parts of the developing fly where it is normally not expressed. Their experimental flies developed anatomically normal compound eyes on the legs, wings, and other structures (Figure 23.23A,B). This regulatory gene, therefore, acts as a switch that directly or indirectly turns on the many genes required for eye morphogenesis. Halder et al. estimate that there are more than 2500 such genes. Halder and his colleagues then used similar techniques to transfer the normal *Sey* gene of the mouse into the *Drosophila* genome and to obtain ectopic expression. Astonishingly,

the mouse gene also elicited anatomically normal ectopic compound eyes (Figure 23.23C). We infer, first, that this gene has evolved little in sequence over the more than 550 million years—perhaps more than a billion—since arthropods and chordates diverged from their last common ancestor. Second, it has retained its function—i.e., controlling the development of photoreceptors. Third, its target genes must have changed greatly over time, since the thousands of genes engaged in eye development must differ profoundly between taxa with such different eyes. Indeed, it is unlikely that the common ancestor of arthropods and chordates had *any* anatomical structures homologous to vertebrate structures such as lens, iris, and cornea, or to arthropod structures such as rhabdom and crystalline cone. The common ancestor probably had simple aggregations of light-sensitive cells with visual pigments, such as those seen in certain flatworms, annelids, and molluscs.

FIGURE 23.22 The compound eye of an insect, in sagittal section (A). It is composed of multiple units (ommatidia), one of which is shown enlarged (B). Each ommatidium receives light from only a fraction of the image registered by the entire eye. Compare with the vertebrate eye (Figure 20A in Chapter 5), in which the entire image is projected by the lens onto the retina. (After Chapman 1982.)

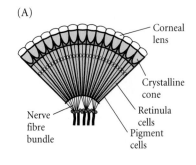

(A)
Corneal lens
Crystalline cone
Retinula cells
Pigment cells
Nerve fibre bundle

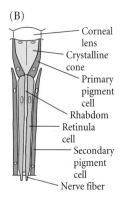

(B)
Corneal lens
Crystalline cone
Primary pigment cell
Rhabdom
Retinula cell
Secondary pigment cell
Nerve fiber

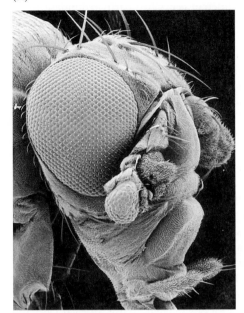

(A)

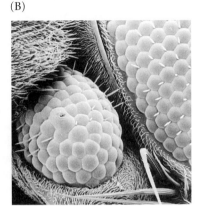

(B)

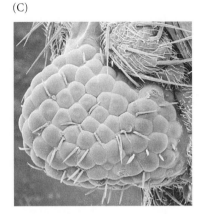

(C)

FIGURE 23.23 Ectopic eyes formed in *Drosophila* by gene transfer. (A) A small eye on the base of the antenna, formed by ectopic expression of the *Drosophila ey* gene. (B) A close-up of an ectopic eye showing the same morphology as the large, normal compound eye nearby. (C) An eye on a *Drosophila* leg, formed by ectopic expression of the mouse gene *Sey*. (From Halder et al. 1995; A and B courtesy of W. G. Gehring; C courtesy of Georg Halder)

The Problem of Homology

The concept of homology (see Chapter 5) is fundamental to all of comparative biology and systematics. The pectoral fin of a fish, the foreleg of a salamander, and the wing of a bird are in some sense "the same" structure, and are said to be homologous, despite profound differences in their structure, function, and mode of development. The evidence used to *hypothesize* homology is usually some combination of (1) similarity in position relative to other body structures; (2) similarity in at least some structural features; and (3) intermediate forms, either among species (fossil or extant) or during ontogeny.

Since Darwin, the *explanation* of homology has been inheritance of the structure, with more or less modification, from a common ancestor in which the structure first evolved. This explanation soon became part of the definition of homology (Donoghue 1992; Wagner 1989b). Thus for contemporary phylogenetic systematists, a character is homologous in two or more taxa only if it has been uniquely derived from their common ancestor; in other words, a homologous character is a synapomorphy. This is the PHYLOGENETIC or HISTORICAL HOMOLOGY CONCEPT (Patterson 1982; Rieppel 1992). The hypothesis that the character is homologous, based on the observational criteria listed above, is *evaluated* by the *congruence* of the character with a phylogeny of the taxa derived from other characters. Thus, according to the phylogenetic homology concept, a character possessed by two taxa is homoplasious rather than phylogenetically homologous if other characters imply a phylogeny in which the character originated twice, even if it is identical in form and (as far as we can tell) in its genetic and developmental basis (see Chapter 5).

However, the phylogenetic homology concept often conflicts with our expectation that homologous characters should have similar genetic and developmental bases, and that genetically and developmentally similar characters should be homologous. Such conflicts have led several authors to urge that we require not only a phylogenetic homology concept, but also a BIOLOGICAL HOMOLOGY CONCEPT (Roth 1988, 1991; G. P. Wagner 1989a, 1996). Several phenomena illustrate their argument.

A character that undergoes evolutionary reversal is considered by phylogenetic systematists not to be phylogenetically homologous between the taxa in which it is ancestral and the taxa in which it reappears after having been lost. For example, the second molar, present in most carnivores, was lost in the cat family (Felidae) in the Miocene. However, it has reappeared in the lynx genus (*Lynx*)—a clear case of evolutionary reversal. Phylogenetically, the presence of the tooth in *Lynx* and in nonfelid carnivores is homoplasious, not homologous. But developmentally and genetically, it may well be the "same" tooth: simple changes in the concentration of a morphogen at a particular site in the jaw could cause it to develop or not (Raff 1996).

Characters display structural and developmental similarity not only among species, but also among conspecific organisms and within individual organisms. For example, a developmental homology exists between many female and male genital structures in vertebrates, such as the penis and the clitoris, which develop from the same embryonic structure. Repeated structures on the same organism, such as legs, scales, and

Development and Evolution 669

leaves, likewise share a developmental program, just as homologous structures of related species do, but they cannot be termed phylogenetically homologous.

We would expect phylogenetically homologous characters in different taxa to have similar genetic and developmental bases, since they have evolved by modification of the developmental programs of a common ancestor. Often, this is the case. For instance, the same differences in Hox gene expression mark the boundary between cervical and thoracic vertebrae in birds and in mammals (Burke et al. 1995). However, both classic embryology and developmental genetics provide many examples in which *phylogenetically homologous structures have different developmental and genetic foundations*. We have seen that different tissues induce differentiation of Meckel's cartilage in the primordium of the lower jaw in different classes of vertebrates. Digits differentiate sequentially from back (postaxial) to front (preaxial) in all tetrapods except salamanders, in which the sequence is reversed. Eye lenses are made up of different crystallin proteins in different classes of vertebrates, and even in different groups of birds and mammals. Each such protein is closely related (by gene duplication) to one or another enzyme that performs a different function elsewhere in the body, including lactate dehydrogenases and heat-shock proteins (Wistow 1993; Raff 1996). In all these cases, the developed structure is phylogenetically homologous across taxa, but its genetic and developmental basis is not. The evolution of such features seems analogous to replacing the stones of a cathedral while retaining its form.

Conversely, *homologous genetic controls may underlie structures that are not phylogenetically homologous.* That is, morphogenetic signals need not be specific to the structure induced, and may be phylogenetically older than the anatomical characters that depend on them (Müller and Wagner 1996; see Color Plate 12). This fact was already known in the 1930s, from an experiment in which the oral mesoderm of a salamander induced transplanted frog epidermis to develop into frog mouthparts, and the converse transplant caused salamander balancer organs to develop on a frog embryo. The frog's mouthparts and the salamander's balancers develop from different gill arches and are not homologous, but both develop in response to a phylogenetically conserved mesodermal signal (see Figure 19 in Chapter 3). The induction of eyes in *Drosophila* by a gene that governs eye development in mice is a dramatic example of the same principle.

Thus, phylogenetic homology at the genetic level and at the morphological level may not correspond to each other. In some cases, the genetic and developmental underpinnings of a structure have shifted over time even while the structure has been retained, as if responsibility for its development had been transferred from one set of genes to another in different evolutionary lineages. Such a transfer could occur if different genes had partly redundant functions, so that one could replace another without altering development. This is likely to be the case with polygenic characters, in which, as we have seen, genetic change can occur without altering the phenotype (see Chapter 14). In other cases, at least some regulatory genes retain the same general function, such as providing positional information, over vast periods of evolutionary time, while various target genes gain or lose the ability to be regulated by them. Hence new genes may be "recruited" or "co-opted" into the developmental pathway regulated by a certain switch gene, and different structures may develop as a consequence.

Biologically, then, the homology of structures can change over time. The eyes of insects and vertebrates are not homologous, despite the homology of some of the genes that control their development. The digits of salamanders and frogs are homologous, despite differences in their development. The molars of lynxes and nonfelid carnivores are probably homologous in a genetic and developmental sense, but not in a phylogenetic sense. In each case, the homologous structure is an individualized part of the organism, a module that retains its individualized identity among taxa in a clade. *Biological homology* may be defined as "*the establishment and conservation of individualized structural units in organismal evolution*" (Müller and Wagner 1996). A major task for evolutionary biology in the future will be to understand how biologically homologous structural units are established and conserved, and how biological homology and phylogenetic homology are related to each other.

Implications of Development for Evolution

The neo-Darwinian theory of evolutionary mechanisms is cast largely in terms of the action of natural selection, genetic drift, and gene flow on genetic variation. Response to selection on a character depends on the existence of genetic variation in that character. Selection on or drift of one character induces correlated change in other characters if they are genetically correlated, as by pleiotropy. Two features can evolve independently only if they are correlated imperfectly or not at all—i.e., only if each feature exhibits some independent genetic variation. Of course, there cannot be any genetic variation in a character that the organism lacks (for example, leg number in snakes).

An important contribution of the developmental perspective is the capacity to describe, for various kinds of organisms, the conditions under which these genetic properties obtain. Just as ecological studies make concrete and particular the abstract and general concept of natural selection, developmental studies should provide a framework for understanding which characters will vary, how they will covary, and indeed, whether or not they may come into existence. Most developmental phenomena can be described in the language of genetics, but knowledge of the structure and development of an organism is essential for understanding why features vary and covary as they do. The deeper understanding that development provides has important implications for our perspective on several evolutionary topics. Among these are (1) constraints on evolution, (2) nonadaptive characters, and (3) the continuity and discontinuity of evolutionary change.

Constraints

Most evolutionary biologists have understandably been preoccupied with explaining the great variety of features that organisms exhibit. Only recently has attention turned to the question of why particular lineages of organisms are not even more variable and diverse than they are. It is difficult to say why there are no photosynthetic starfishes, no viviparous birds or turtles, no frogs that reproduce in the tadpole stage, no palm trees in Canada. Any science fiction writer or biologist can readily imagine creatures that have never existed (Figure 23.24). We can gain insight into evolution if we consider the kinds of constraints that are possible and attempt to study them. Maynard Smith et al. (1985) and Stearns (1986) provide overviews of this difficult and controversial topic.

A constraint may be *universal*, applying to all organisms, or *local*, meaning that it applies to certain clades. Local constraints are sometimes termed *phylogenetic constraints*, because the constraint exists on account of certain features of the organisms bequeathed to them by their particular phylogenetic ancestors. (Phylogeny in itself does not constrain; only the features do.)

A causal classification of constraints might include the following terms encountered in the evolutionary literature:

1. *Physical constraints*. Physical constraints include limitations set by physics and chemistry. It is unlikely, for example, that a mobile terrestrial animal with the mass of a blue whale could evolve, because known biological materials could not provide enough support. We may include under this heading certain "logical constraints," such as the necessary trade-off between the number and size of units (such as eggs) to which a fixed quantity of material can be allocated, and "fabricational constraints" arising from the properties of materials such as bone (Seilacher 1973). The specific effects of many physical constraints are phylogenetically local. In insects, for example, oxygen and carbon dioxide are exchanged chiefly by diffusion through narrow tubes (tracheae) ramifying throughout the body. The limits on diffusion rates are thought to set an upper limit on insect body size that does not apply to terrestrial vertebrates, with their very different respiratory system.

2. *Selective constraints*. A feature may not have evolved in a clade because it has always been disadvantageous. The disadvantage might be purely ecological, or it might arise because the feature interferes with the function of other features.

3. *Functional constraints*. Some features may not evolve because they would impair function. Functional constraints may be viewed as a class of selective constraints.

4. *Developmental constraints*. Maynard Smith et al. (1985) define a developmental constraint as "a bias on the production of variant phenotypes caused by the structure, character, composition, or dynamics of the developmental system." In other words, properties of the developmental system make some phenotypes unlikely to arise. The two most common manifestations of developmental constraints are (*a*) the absence or paucity of variation, including the absence of a developmental foundation for a proposed character that does not exist, and (*b*) strong correlations (negative or positive) among characters during development. Such manifestations do not in themselves tell us what the developmental constraints might be.

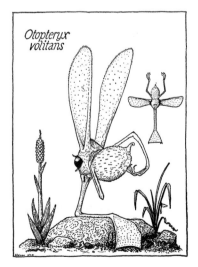

FIGURE 23.24 In a work both amusing and well informed by evolutionary principles, Harald Stümpke imagined an adaptive radiation of "Rhinogradentia" or "snouters," mammals with noses elaborated for diverse functions. *Otopteryx* flies backwards, using its ears as wings and its nose as a rudder. *Orchidiopsis* feeds on insects attracted to its petal-like nose and ears. Stümpke's fantasy illustrates some of the many conceivable phenotypes that have never evolved. (From Stümpke 1957.)

5. *Genetic constraints* likewise include (*a*) the paucity or absence of genetically based phenotypic variation and (*b*) genetic correlations among characters. Developmental constraints can be described in genetic terms, and vice versa.

Developmental and selective constraints merge in some instances, as when a developmentally correlated effect of a gene causes inviability, as does the recessive lethal mutation that increases digit number in guinea pigs (see Chapter 3).

Developmental and genetic constraints that limit the amount and pattern of variation on which natural selection can act might have several *consequences for evolution:*

1. Absence of features. Evolutionary lineages might fail to evolve features that we suppose might be adaptive, were they to appear.
2. Directional trends. Directional trends in the evolution of certain features are often found within a lineage. Development might cause a bias in the kinds of mutational variations on which selection can act, so that certain directions of evolution are more likely than others.
3. Parallel evolution. Parallelism, which is extremely common in evolution, could be due simply to similar selective regimes acting on different species. However, selection might also be acting on similarly biased patterns of variation in related species owing to shared developmental constraints (Mayr 1963, Wake 1991).
4. "Standardization" during evolution. In some phyletic lines, the morphological variation among species appears more exuberant early in their history than later (Gould 1989). Some authors have speculated that a factor contributing to this pattern might be the evolution of increasingly restrictive developmental constraints on the phenotypic variation available to selection (Riedl 1978; Gould 1989).
5. Low rates of evolution. Several factors might be responsible for low evolutionary rates; constraints on variation are at least a hypothetical possibility.
6. Embryological similarities. Because early developmental events may set the stage for later ones, alteration of an early process might disrupt later development, with disastrous effect. Thus early processes would be subject to stabilizing selection, and early developmental stages would display less evolutionary change than later ones. The result would be von Baer's law.

A developmental perspective on constraints From a genetic perspective, paucity of genetic variation in a character and genetic correlations among characters may sometimes constrain or influence the direction of evolution (Chapter 14). Developmental biology holds the promise of providing deeper insight into such constraints. The nonexistence of certain imaginable phenotypes seems to cry out for a developmental explanation. For instance, the genome of every holometabolous insect encodes instructions for both larval and adult characters. The tarsal (foot) structure of larval holometabolous insects (e.g., beetles, sawflies,

moths) is never expressed in adult insects, nor are the compound eyes or the wings of the adult ever found in larvae, either as a variant or as a species-typical character.* It is hard to imagine that these features could never have been advantageous to larvae, considering their enormous variety of lifestyles. That such features have not evolved does not prove that they cannot evolve, but it does suggest that research might uncover a developmental explanation.

Surely one of the most stringent constraints on the origin of variation must be the simple lack of a feature to vary. For example, dicotyledonous plants form secondary xylem, enabling the trunk of a tree to grow in diameter, but monocots such as palms and bananas have lost the capacity to form secondary xylem, and their stem diameter cannot increase (Maynard Smith et al. 1985). Likewise, the planktonic larval feeding stages have been lost in many lineages of echinoderms and other invertebrates, and have never been regained (Strathmann 1978). Although, as the lynx molar illustrates, a lost feature may be regained if the underlying developmental mechanism remains intact, the genes responsible for an unexpressed developmental pathway will ultimately degenerate into pseudogenes if they have no other function, so that the original structure is unlikely to be recovered (Raff 1996).

An example of a phylogenetically local constraint that casts light on *parallel evolution* is digit development in amphibians. The central and postaxial digits are the first to develop in frogs, as in most other tetrapods, whereas in salamanders digit development proceeds from preaxial to postaxial. Pere Alberch and Emily Gale (1985) treated the limb buds of salamander (*Ambystoma*) and frog (*Xenopus*) embryos with the mitotic inhibitor colchicine, causing some digits or phalanges to fail to develop. As they predicted, the missing elements were preaxial in the frog and postaxial in the salamander (Figure 23.25). Reduction in the number of digits or phalanges is a common, parallel, evolutionary trend in amphibians. Almost invariably, the preaxial elements have been lost in frogs, and the postaxial elements in salamanders, evolution having followed a pattern consistent with the developmental data.

Studies of the developmental biology of the tetrapod limb have led to several models suggesting that variation should follow predictable patterns (Stock and Bryant 1981; Hinchliffe and Johnson 1980). For example, limbs have been reduced or lost in many lizards, salamanders, and other tetrapods (Figure 23.26). The proximal-distal sequence of development in the limb implies that in the evolution of limb reduction, digits will always be lost before more proximal elements such as the humerus, and this appears to be the case: no instances are known of vertebrates that lack proximal but possess distal bones.

A phylogenetic perspective on constraints Faced with the observation that some phenotype has never evolved in a particular clade (as far as we can tell), we can

---

*As far as I know—D. J. F.

FIGURE 23.25 (A) Dorsal view of the left hind foot of the four-toed salamander, *Hemidactylium scutatum*. The four-toed condition is normal for this species. (B) Left hind foot of an axolotl, *Ambystoma mexicanum*, treated by an inhibitor of mitosis at the limb bud stage. (C) Right hind foot of the same axolotl, showing the normal five-toed condition. The experimentally treated foot is smaller than the control, lacks a postaxial toe and some phalanges, and resembles the condition in *Hemidactylium*. (From Alberch and Gale 1985; photographs courtesy of P. Alberch.)

(A)    (B)

(C)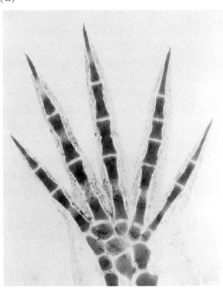

the Carboniferous, all tetrapods have had a maximum of five digits—except for a few ichthyosaurs (Figure 23.27). Maynard Smith and Sondhi's experiment might suggest that consistent bilateral asymmetry cannot evolve—but the most conspicuous feature of the male narwhal is its greatly elongated upper left incisor (see Figure 31 in Chapter 14), and some of our own internal organs are consistently asymmetrical. Some variations may be developmentally unlikely, but that does not necessarily mean that they are impossible.

Constraints, however, may have important evolutionary consequences even if they are not global or absolute. Although a constraint may ultimately be breakable, the evolution of a particular lineage may proceed along another line altogether if, in fact, the constraint is not broken when selection favors a new adaptation. That a lineage of ichthyosaurs evolved more than five digits does not imply that any

FIGURE 23.26 Limb reduction in lizards. (A) The Australian skink *Lerista distinguenda*, with 4 digits instead of the ancestral 5-toed state. (B) The hind limb of *Lerista picturata* has 1 or 2 digits, and the forelimb 2, 1, or none. (C–E) The skeletal structure of two species of South African *Chamaesaura* skinks with reduced limbs (C, E), compared with a species (D) with complete limb structure. (A, B courtesy of H. G. Cogger; C–E after Bellairs 1970.)

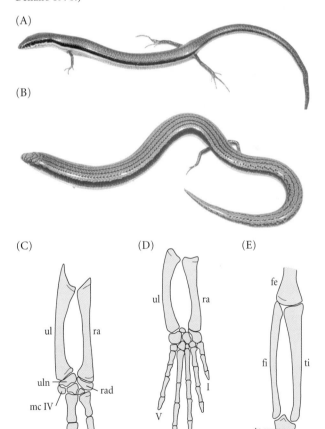

hypothesize either that selection did not favor the trait, or that it could not evolve because of developmental or genetic constraints. We should be equally cautious about assuming either alternative.

Features sometimes do evolve that we might suppose to be developmentally improbable. Knowing that frogs generally have not had teeth in the lower jaw since the Jurassic, we might have concluded that these are a developmentally "forbidden morphology," only to be proved wrong by *Amphignathodon*, which has precisely this trait (see Chapter 5). Since

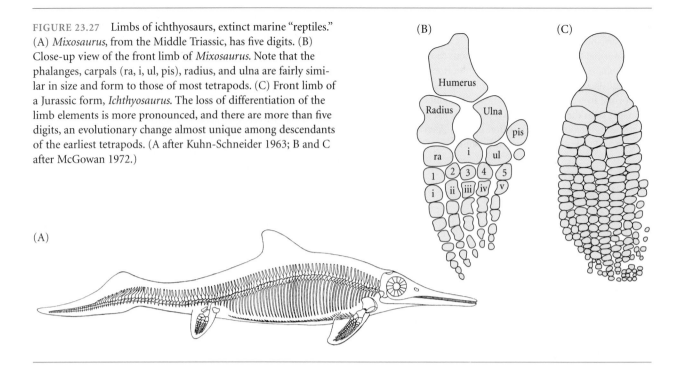

FIGURE 23.27 Limbs of ichthyosaurs, extinct marine "reptiles." (A) *Mixosaurus*, from the Middle Triassic, has five digits. (B) Close-up view of the front limb of *Mixosaurus*. Note that the phalanges, carpals (ra, i, ul, pis), radius, and ulna are fairly similar in size and form to those of most tetrapods. (C) Front limb of a Jurassic form, *Ichthyosaurus*. The loss of differentiation of the limb elements is more pronounced, and there are more than five digits, an evolutionary change almost unique among descendants of the earliest tetrapods. (A after Kuhn-Schneider 1963; B and C after McGowan 1972.)

tetrapod lineage could do so at any time. Indeed, special developmental circumstances may have made the ichthyosaurs' polydactyly possible. Perhaps because they served only as a scaffold for the flipperlike forelimbs, the digits of ichthyosaurs, and the phalanges that compose them, are more uniform than those in most tetrapods. They may have lost their individualized developmental programs, and so have come under a simpler control system allowing for variation in number, as is the case with many uniform serially repeated structures.* In tetrapods with individualized digits, the option of increased digit number may arise rarely or not at all. Thus in the giant panda (*Ailuropoda melanoleuca*), the two extra digits of the seven-digit hand, used for manipulating bamboo, are not true digits, but sesamoid bones (see Figure 23.14). The well-known principle that different lineages evolve different "solutions" to similar adaptive "problems," rather than developmentally equivalent organs (see Chapter 5), is doubtless a consequence, in part, of developmental constraints that vary among lineages and over time.

Nonadaptive Characters

If two or more features are developmentally correlated (or, to use the language of genetics, genetically correlated by pleiotropy), strong selection on one feature may induce nonadaptive (or even maladaptive) change in the more weakly selected features. A developmental perspective suggests that many features of organisms may fit this description. In paedomorphic salamanders such as *Thorius*, for example, the reduction or absence of some skull bones, especially those that develop last, appears to be a simple consequence of the truncated paedomorphic development of the organism, rather than an adaptation. Sexually dimorphic features provide many probable examples of nonadaptive correlated evolution. Nipples on men serve no evident function, and many female birds have muted or rudimentary versions of the color patterns or crests that are highly developed sexually selected features of males, but serve no known function in females. In many such instances, it is likely that the nonadaptive characters have been "permitted" by natural selection because they are not strongly disadvantageous. The presence of crests in female birds is understandable if crest development is genetically correlated between the sexes, but strong sexual dimorphism is expected if selection favors crests only in the males (Lande 1980).

Discontinuity of Evolutionary Change

Several of the developmental phenomena we have discussed raise the possibility that some features might at least occasionally evolve not by successive small steps, but by abrupt, discontinuous changes. *Thresholds*, by definition, are discontinuities; developmental *correlations*, as among allometrically related characters, can cause concerted evolution in functionally related features; and developmental *plasticity* may accommodate substantial change in a character so that it does not impair fitness. Changes in the site, timing, or level of activity of regulatory genes, such as those caused by homeotic mutations, have the potential to drastically and abruptly alter the phenotype.

---

*Similarly, toothed whales are among the few placental mammals with more than 11 teeth on either side of the jaw. Whale teeth, numbering up to 60 and varying greatly within and among species, are uniform, evidently lacking individualized developmental programs.

Most neo-Darwinian evolutionary biologists have followed Darwin in the belief that evolutionary change generally transpires gradually, by small successive changes in individual characters. However, this orthodoxy has frequently been challenged. We shall pursue this question in the next chapter.

Summary

1. An organism's features develop, first, as a consequence of temporally and spatially localized activity of genes that contribute importantly to cell and tissue differentiation, and second, as a consequence of epigenetic processes, such as tissue interactions, physical stresses, and hormonal signals, that affect morphogenesis. The evolution of any feature is a consequence of genetic change that directly or indirectly alters any of these processes. An important, though neglected, role for developmental biology in evolutionary studies is to specify the kinds of genetically based variation that can arise and be subject to natural selection or genetic drift.

2. Development is studied both at the molecular level of gene expression and at the epigenetic level of differentiation and morphogenesis. Both kinds of studies provide examples of rapid evolutionary change in developmental patterns as well as extraordinarily slow (conservative) change.

3. The developmental role of a gene may change during evolution; for example, a regulatory gene may come to govern the expression of a different battery of genes lower in the hierarchy of gene action. Conversely, the form of a phylogenetically homologous organ in different organisms may remain rather constant even though the genetic and developmental pathways that give rise to it undergo change. These possibilities pose a difficult question: What is the biological basis of phylogenetic homology?

4. Repeated (homonymous or serially homologous) structures that are uniform in an ancestor, or structures that are developmentally strongly coupled, can become individualized, or dissociated, by coming under the control of different genes. They can then evolve more or less independently (mosaic evolution). Conversely, independent characters can come under common developmental control, and then evolve coordinately.

5. Some evolutionary changes in ontogeny are terminal additions to or alterations of the development of a feature. When evolution proceeds by terminal addition, the early embryonic stages of different species are more similar than later stages (von Baer's law). In some such instances, a juvenile stage may resemble the adult form of an ancestor with respect to some features (Haeckel's biogenetic law). Some terminal alterations constitute truncation of the full course of development, so that the feature retains the juvenile form of an ancestor. This is one of the many ways in which the biogenetic law is violated.

6. From a developmental perspective, many evolutionary changes in morphology can be explained by changes in the rate or duration of growth of different features (heterochrony), changes in the timing and spatial location of gene expression, thresholds, pattern-forming mechanisms, tissue interactions, and developmental plasticity and integration.

7. Many differences among organisms are differences in shape, caused by differences in the rate or duration of growth of certain characters (or dimensions of a character) relative to others, or of the organism as a whole (heterochrony). Allometry, the differential rate of growth of an organism's features during development, can generate shape differences among species when heterochronic evolution occurs. Changes in either the rate or duration of the development of a feature can generate morphology that is either more exaggerated than the ancestor's adult form (peramorphosis) or less exaggerated, resembling a juvenile state of the ancestor (paedomorphosis).

8. Heterotopy, a change in the topological location of the development of a feature, presumably stems from changes in the location of gene expression, and can be a major source of evolutionary novelties.

9. A morphological feature often develops when some underlying process surpasses a threshold. Genetic changes in such processes can thereby give rise to discontinuous differences in morphology. Thresholds are important in pattern formation, the process that specifies the number and spatial distribution of homonymous or serially repeated structures. Alterations of pattern are common in evolution.

10. Some evolutionary changes are based on changes in the hierarchical genetic control of differentiation. Regulatory genes such as Hox genes may be altered in number and in the site and timing of their expression. More frequently, different target genes come under their control, so that the same regulatory genes come to regulate different structures. For this and other reasons, phylogenetically homologous structures and biologically homologous structures may differ.

11. Development imposes constraints on evolution, for it makes certain variations more likely to arise than others, providing selection with a restricted or biased range of variation on which to act. Many such constraints are phylogenetically "local," applying only to certain clades. Such constraints may affect the direction of evolution, by making some paths of evolution more likely than others. However, the importance of developmental constraints in evolution is unclear.

12. To some extent, an organism's development is compartmentalized, so that certain groups of tissues develop as integrated units. A consequence of this integration is

developmental plasticity, in that an alteration in one element may induce coordinated change in the development of interacting elements. This plasticity facilitates evolution in that the structure can respond as a whole to a genetic change in the development of any one part; thus in some instances, substantial alterations can occur without loss of function. These developmental correlations can constrain evolution in other instances, however, because the elements may not be able to change in response to divergent selection, or because certain genetic changes have deleterious pleiotropic effects.

Major References

Raff, R. A. 1996. *The shape of life: Genes, development, and the evolution of animal form.* University of Chicago Press, Chicago. A fairly comprehensive, informal survey of the subject by a leading researcher in the field.

McKinney, M. L., and K. J. McNamara. 1991. *Heterochrony: The evolution of ontogeny.* Plenum, New York. The authors treat this important subject in depth. Also see chapters in M. L. McKinney (editor), *Heterochrony in evolution: A multidisciplinary approach* (Plenum, New York, 1988).

Gould, S. J. 1977. *Ontogeny and phylogeny.* Harvard University Press, Cambridge, MA. Covers the biology, and especially the early history, of the subject.

Müller, G. B., and G. P. Wagner. 1991. Novelty in evolution: Restructuring the concept. *Annual Review of Ecology and Systematics* 22: 229–256. The authors attempt to join developmental biology and neo-Darwinian theory.

Problems and Discussion Topics

1. List ways in which information on embryology and development provides evidence for evolution.

2. Among species of mammals, the allometric coefficient relating the brain weight of adults to their body weight is about 0.66 in most orders. Within most species, the coefficient is about 0.25 at different stages during ontogeny (Lande 1979b; Harvey and Pagel 1991). Do the models of heterochrony in this chapter explain the difference?

3. If two allometrically related traits show a very strong correlation among species, does this provide evidence that the traits are developmentally constrained, and could not change independently? How else might a strong correlation be interpreted?

4. Under what ecological circumstances might you expect the evolution of paedomorphic or peramorphic lineages? Draw on the theory of life history evolution (see Chapter 19), especially theory on the optimal age of first reproduction. (See Gould 1977.)

5. How has evolutionary biology contributed to our understanding of the basic processes of development?

6. Can modern methods of studying gene expression be used to test hypotheses of evolution by heterochrony? For example, could they be used to test the hypothesis that the byssal threads of mussels are a juvenile character, retained paedomorphically by adults?

7. Similar serially homologous structures are often more variable if they are numerous than if they are few. For instance, the number of scutellar bristles is more constant (almost always four) than the number of abdominal bristles (averaging about ten) in *Drosophila*. What developmental mechanisms might account for this? (See Maynard Smith 1960 for a model.)

8. Formulate a hypothesis on a genetic-developmental mechanism by which ancestrally identical serially homologous structures might begin to acquire differences—i.e., individual identity.

9. A mutation of the *Ubx* gene of *Drosophila* transforms halteres (balancers) into wings. The halteres of Diptera are derived from (homologous to) hindwings. Is it likely that a single mutation of the *Ubx* gene originally caused the evolution of halteres? If not, what alternative evolutionary history do you propose?

10. If mutations such as those of the *Ubx* gene can drastically change morphology in a single step, why should most evolutionary biologists maintain that evolution has generally proceeded by successive small steps?

CHAPTER 24

Pattern and Process in Macroevolution

In this chapter, we examine what one of the architects of the Evolutionary Synthesis, Bernhard Rensch (1959), called "evolution above the species level." Another architect of the Synthesis, the paleontologist George Gaylord Simpson (1944), likewise addressed the "tempo and mode" of evolution in the long term. Have higher taxa evolved, as Darwin proposed, by successive slight changes, or by the sudden origin of drastically new features? Can—or must—each of the successive slight changes envisioned by Darwin be adaptive? Can evolutionary theory account for the rates at which evolution has occurred? How do complex, functionally integrated systems evolve? Are there trends in evolution, and how can we explain them?

Gradualism and Saltation

If, between even the most extremely different organisms, there existed a full panoply of all possible intermediate forms, each differing from similar forms ever so slightly, we would have little doubt that evolution is a history of very slight changes. Such is often the case when we examine differences among individuals in a population, among populations of a species, and among closely related species, such as those in the same genus. Moreover, quite different species are often connected by intermediate forms, so that it becomes arbitrary whether the complex is classified as two genera (or subfamilies, or families) or as one (see Chapter 5). Nonetheless, there exist many conspicuous gaps among phenotypically similar clusters of species. No living species bridge the gap between whales (Cetacea) and other mammals, nor between vascular plants and bryophytes (e.g., mosses). Most conspicuous are gaps such as those among the animal phyla, each of which has a body plan (or *Bauplan;* plural *Baupläne*) very different from those of most or all other phyla. Such gaps are the subject of probably the most enduring controversy in evolutionary biology.

Biologists can entertain three hypotheses to explain phenotypic gaps:

1. Distinct groups of organisms do not stem from common ancestors, but instead represent independent origins of life. We can dismiss this hypothesis, based on extensive evidence from molecular biology and cell structure for the common origin of all living things (see Chapter 5). Moreover, the gaps we wish to explain range in magnitude from truly impressive differences (e.g., among animal phyla), to markedly different variations on common morphological themes (e.g., among orders of mammals), to discrete differences in only a few features of otherwise similar organisms (e.g., the folding fangs of vipers compared with the fixed, erect teeth of other snakes: see Chapter 12). The number of discretely different "kinds" of organisms depends only on the magnitude of the differences we use to define them.

2. The gaps exist because of the origin and fixation of single mutations, or other drastic genetic changes, that radically alter the phenotype (Figure 24.1A). We consider this hypothesis of MACROMUTATION or SALTATION (from Latin *saltus,* "a jump") below.

3. Most differences among higher taxa have arisen by the successive fixation of slight phenotypic changes from the state of their common ancestors. If we had a complete fossil record of all the ancestral populations, it would present a gradual history of phenotypic change. We lack such a record in most instances because the fossil record is extremely sparse.

The more rapid a gradual change is, the less likely it is to be preserved in the very incomplete fossil record (Figure 24.1B). The hypothesis of PUNCTUATED EQUILIBRIUM (see Chapter 6) holds (among other things) that most evolutionary changes in morphology, although perhaps continuous in the sense of passing through many intermediate stages, have been so rapid that the fossil record presents the appearance of a discontinuous change. This proposition is very different from the saltation hypothesis, which holds

FIGURE 24.1 Two hypotheses accounting for phenotypic gaps in a fossil lineage. (A) Saltation, or macromutation. A single mutational change (asterisk), in a single individual, changes morphology discontinuously. Its descendants vary around the new morphology and replace the ancestral form. (B) The mean of a quantitative character changes gradually, but so rapidly that the intermediate stages may not be recovered from a coarse fossil record.

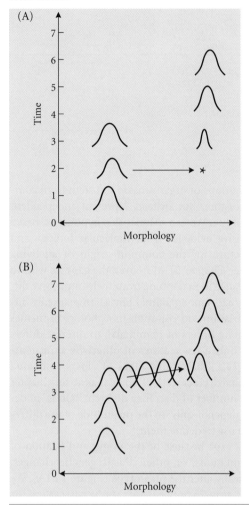

that intermediates never existed—that mutated individuals differed drastically from their parents.

If two very different contemporary organisms have diverged gradually from a common ancestor, neither that ancestor nor any of the intervening descendants need have been intermediate between the contemporary forms, because the modifications in each phyletic line may have been very different (Figure 24.2). The most recent common ancestor of whales and bats, for example, did not have a mixture of cetacean and chiropteran features, nor were its front legs intermediate between fins and wings. Instead, it probably had the shrewlike characteristics of the early generalized placental mammals (see Chapter 7).

If each of two diverging lineages (such as those leading to whales and bats) generated reproductively isolated species at each stage of divergence, and if each such species persisted to the present with little further change, the most

divergent species would today be connected by an array of intermediates, including the phenotype of their common ancestor (Figure 24.3A). We can observe many instances today of such "morphoclines," especially in clades that have recently experienced adaptive radiation. The family Bovidae, for example, includes sheep, goats, antelopes, cattle, and all manner of intermediates among these poorly defined groups. However, the fossil record is the only possible source of evidence for gradual evolution (1) if intervening species with intermediate morphology become extinct (Figure 24.3B); (2) if the intervening species themselves continue to evolve (Figure 24.3C); or (3) if speciation seldom occurs (Figure 24.3D). If the fossil record does not yield intermediate forms, it is extremely difficult to tell if we have simply not found them, or if they never existed.

In adjudicating between gradualism and saltationism, it is important to distinguish between the evolution of *taxa* and of their *characters*. Higher taxa often differ in many characters: for example, not only feathers, but a great many other features distinguish modern birds from dinosaurs. Gradualists hold that higher taxa did not arise discontinuously, with all their new features in place, but rather that many of their characters evolved independently and sequentially (*mosaic evolution:* see Chapter 5). Almost all contemporary biologists agree on this point. There is some disagreement, however, on whether each of the distinguishing *characters* of a higher taxon—the reduction and fusion of birds' tail vertebrae, for example—necessarily evolved by sequential slight changes, or discontinuously.

Arguments for Gradualism

Darwin was well aware of "sports"—what we now call mutations with large, discontinuous effects on one or more features—but held that most evolution is based on the slight individual differences that we observe within populations (which we now recognize as polygenic variation). This gradualist position was reaffirmed by neo-Darwinians such as Ronald Fisher (1930), and has been the

FIGURE 24.2 Two very different taxa may have evolved gradually from a common ancestor, but without any form ever having existed that was precisely intermediate between them.

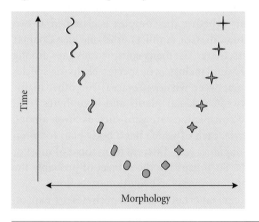

FIGURE 24.3 Conditions under which contemporaneous species do or do not provide evidence on whether a character diverged gradually. (A) Punctuational evolution, in which newly arisen species "freeze" intermediate phenotypes and survive to the present. (B) The same pattern, but most intermediate species become extinct. (C) A pattern similar to B, but in which some evolutionary change occurs within species. (D) Episodic, but almost entirely anagenetic, evolution of diverging lineages that seldom speciate. Evidence of gradual evolution is provided by living species in case A, but must be sought in the fossil record in the other cases.

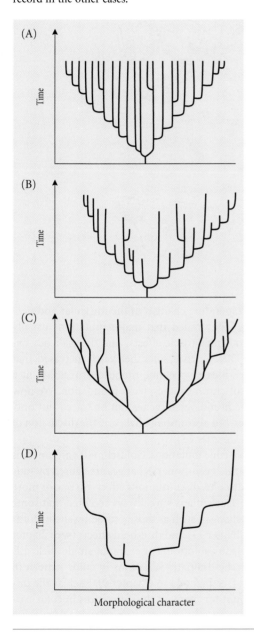

Intermediates Undoubtedly the most extensive evidence for gradualism comes from the extremely numerous intermediate phenotypes among living organisms and among fossils (see Chapters 5 and 6). Nevertheless, gaps do exist among well-marked taxa, and to extrapolate from examples of continuity to the conclusion that all gaps merely represent missing data may not be warranted.

Functional considerations One reason for Darwin's gradualism was his belief that organisms are so intricately and harmoniously constructed that a major alteration of any one feature would impair function, and so would decrease, rather than increase, the likelihood of survival. As one character evolves incrementally, however, natural selection can shape compensatory changes in functionally interacting characters. For example, the pygostyle (short, fused tail vertebrae) of a bird supports feathers used for steering and braking; if a mutant descendant of *Archaeopteryx* had had a pygostyle, but not the changes in muscles and nerves necessary to manipulate the tail feathers, it would have crashed (Maynard Smith 1958). Genetic studies show that the pleiotropic effects of major mutations seldom alter a suite of functionally interacting characters in an adaptively coordinated fashion.

Although this argument applies to many features, it does not necessarily apply to all. Whether function would be impaired by, say, a mutation from separate to fused petals—a distinguishing feature of several families and orders of plants—cannot be judged except by empirical investigation.

Fitness effects of mutations Mutations with discrete, large effects, such as those studied in *Drosophila*, usually have diverse pleiotropic effects; for example, Dobzhansky (1927) found that 10 of 12 arbitrarily chosen mutations of eye color or wing shape in *Drosophila* also affected the shape of the spermatheca (part of the female reproductive system). Not surprisingly, many of these "classic" mutations greatly lower viability or other components of fitness, and their frequency typically declines rapidly if they compete with "wild-type" alleles in experimental populations (Dobzhansky 1970). From this evidence, many authors have argued that the deleterious effects of major mutations usually outweigh their advantageous effects.

However, we cannot from this evidence conclude that *all* major mutations necessarily lower fitness. Moreover, mutations of small effect—those that contribute to polygenic variation—may be no more free of deleterious effects (Orr and Coyne 1993). Polygenic characters that have been artificially selected in laboratory populations generally revert toward the original phenotype when selection is relaxed. Both deleterious pleiotropic effects of alleles affecting the selected character and hitchhiking of deleterious alleles linked to the selected loci contribute to this effect (see Chapter 14).

Genetic differences The number and phenotypic effects of genes underlying phenotypic differences among

majority opinion ever since (reviewed by Maynard Smith 1983). The *chief arguments of contemporary gradualists* are based on (1) intermediates among both living and extinct species; (2) functional considerations; (3) fitness effects of mutations; and (4) the genetics of species differences.

higher taxa cannot ordinarily be determined because such organisms usually cannot be crossed. We are limited to information about those species that can be crossed and which yield some fertile hybrids. As we saw in Chapter 15, such evidence indicates that the various characters that distinguish a pair of species usually recombine in the F_2 or backcross generation. This evidence of multiple gene differences supports the hypothesis of mosaic evolution. The individual character differences also are frequently polygenic, conforming to the gradualist hypothesis. There are instances, however—especially in plants—in which one or two loci appear to account for much of the character difference. In a review of this topic, Allen Orr and Jerry Coyne (1993) concluded that, all in all, there is insufficient evidence for the traditional view that ecologically adaptive characters have usually evolved by selection of mutations of small rather than large effect.

Arguments for Saltation

An extreme saltation would be a radical transformation of many characteristics in a single mutational event. For example, the paleontologist Otto Schindewolf (1936, 1950), who believed that the differences among higher taxa have arisen discontinuously, declared that "the first bird [*Archaeopteryx*] hatched from a reptilian egg." A similar position was held by Richard Goldschmidt, who had made a major contribution to genetics by stressing that genes act by controlling the rates of biochemical and developmental reactions (Goldschmidt 1938). Despite his excellent work on the genetics of variation in moths, he argued in *The Material Basis of Evolution* (1940) that species and higher taxa arise not from the genetic variation that resides within species, but instead "in single evolutionary steps as completely new genetic systems." He denied gene mutations a role in evolution, believing instead that higher taxa arise by complete "repatterning" of the chromosomal material. These "systemic mutations" would give rise to bizarre creatures, most of which would have no chance of survival; but some few would be "hopeful monsters" adapted to new ways of life.

The examples Goldschmidt cited were actually far more modest than what he seemed to envision in his theory: most were pronounced transformations of individual organs. A mutation that transforms the halteres of *Drosophila* into more winglike structures resembled, he said, the rudimentary wings of a fly (*Termitoxenia*) that inhabits termite nests (Figure 24.4). The Manx breed of domestic cat, in which a mutation produces a stubby tail with fused vertebrae, is, said Goldschmidt, "just a monster. But a mutant of Archaeopteryx producing the same monstrosity was a hopeful monster because the resulting fanlike arrangement of the tail feathers was a great improvement in the mechanics of flying."

Prominent figures in the Evolutionary Synthesis (e.g., Mayr 1942; Simpson 1944) and later neo-Darwinians (e.g., Charlesworth 1982; Templeton 1982b) criticized Goldschmidt sharply. They argued that the vast majority of "monsters" would be hopeless—as we noted above, an *Archaeopteryx* with a suddenly shortened tail could not fly

FIGURE 24.4   An example offered by Goldschmidt as a possible case of saltational evolution. (A) The size and venation of the wing are greatly reduced in *Termitoxenia*, a fly that inhabits termite nests. (B) Wing size and venation are greatly reduced in *Drosophila* by a single mutation, *tetraptera*. Goldschmidt suggested that a similar mutation caused saltational evolution of the wing of *Termitoxenia*. (After Goldschmidt 1940.)

(A)

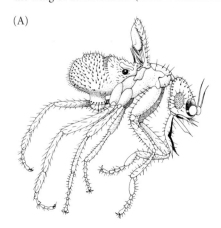

(B)

without corresponding changes in the muscular and nervous systems. They noted that major mutations usually have deleterious pleiotropic effects. And there already existed by the 1930s evidence, since then vastly increased, that Goldschmidt's idea of systemic mutations, residing in a repatterning of chromosomes, lacks foundation. We now know that the homology of genes can be traced not only among species, but also among phyla (see the discussion of Hox genes in Chapter 23).

Without systemic mutations, Goldschmidt's argument is merely that some evolutionary changes are caused by mutations with discrete, discontinuous effects on one or more features. Neo-Darwinians have always accepted that some such genetic changes, such as those caused by allelic dominance (see Chapter 13) and threshold effects (see Chapter 14), play a role in evolution. For example, single allele differences underlie dramatic variations in color pattern in many species (see Figure 4 in Chapter 9). Such allelic differences account for many of the differences in color pattern among geographic races of unpalatable *Heliconius* butterflies, each of which is a Müllerian mimic of another species (Color Plate 7). Phenotypes that deviated only slightly from one pattern toward another, lacking sufficient protective resemblance to the other species, presumably would suffer a disadvantage. Thus, it is likely that the evolution of one mimetic pattern to another was initiated by a mutation of large effect, followed by selection of alleles with smaller effects that "fine-tuned" the phenotype (Figure 24.5). Genetic analysis of the color patterns has confirmed this hypothesis. Crosses between the races of *Heliconius* showed that single alleles account for the major differences in color pattern, while "modifier" loci with small effects govern the finer differences (Turner 1977).

Some contemporary biologists hold that discontinuous changes in individual organs or organ systems may have contributed to the origin of novel or substantially altered features (e.g., Alberch 1980; Gould 1980; Stanley 1979). They rest their case chiefly on the developmental phenomena discussed in Chapter 23. Threshold effects in development cause continuous variation in underlying variables to be expressed discontinuously at the morphological level. Some functionally interdependent features are developmentally and genetically correlated, so that function might be retained despite discontinuous mutational change. Likewise, the compensatory effects of phenotypic plasticity during development might preserve function.

Perhaps the most dramatic single-gene differences yet described are those with strong heterochronic effects, such as the few allelic differences between metamorphosing and neotenic *Ambystoma* salamanders (see Chapter 23). But the allele for neoteny does not engender new organs; it merely acts as a switch that regulates how far various organs proceed along complex developmental pathways already established in these species' remote ancestors.

In summary, substantial differences in certain characters of some species have surely evolved by selection on single alleles with large effects. Some of the novel features (synapomorphies) of higher taxa may well have arisen this way, although we do not yet know that to be the case. But there is no evidence that single "macromutations" have been responsible for the multiple characters, and the substantial differences in morphological organization, that distinguish most higher taxa. The abundant evidence of mosaic evolution indicates that in this sense, evolution has been overwhelmingly gradual.

Selection and the Evolution of Novelty

No one doubts that many of the features that distinguish higher taxa, especially complex features, are adaptive. Yet ever since Darwin proposed that most of evolution is caused by the action of natural selection on small variations, critics have balked at the notion that each small step, from the slightest initial alteration to the full complexity of form displayed by later descendants, could have been guided by selection. Can all intermediate states possibly be adaptive? What functional advantage can there be, the critics ask, in an incompletely developed eye, or in a leg so slightly modified into a wing that it cannot serve for flight? This skepticism is one of the two chief reasons that anti-Darwinians have postulated saltation—the other being the phenotypic gaps among higher taxa that we have already discussed. We shall discuss this problem in two related contexts: How can we account for the incipient evolution of new features? And how can we account for the evolution of highly complex adaptations?

Incipient and Complex Features

Several hypotheses can account for the first stages of the evolution of features that, when fully developed, serve a clear adaptive function.

1. *Upon first appearance, the feature is sufficiently well developed* to provide an advantage. The major mutations in the wing patterns of mimetic butterflies, discussed above, provide examples. The pattern produced by the initial mutation, although imperfectly resembling an unpalatable model, is sufficiently similar to it to provide a selective advantage. It is then refined by additional substitution of alleles with small effects.

2. *The new feature* is not adaptive initially, but *is a developmental by-product* of other adaptive features. At

---

FIGURE 24.5   A model of the evolution of color pattern in Müllerian mimics such as certain butterflies. The colored curves represent degree of protection against predators for two unpalatable species that initially differ in color pattern. Selection favors convergence of the less abundant species B toward the pattern of the more abundant species A, because predators more often learn to avoid the more abundant species. The wavy black lines indicate the distribution of phenotypes within each species. Mutations that only slightly alter the phenotype of species B will reduce fitness, but a mutation of large effect that causes members of species B to acquire a phenotype within the range *x* to *y* will be selectively advantageous, so B initially evolves modest resemblance to A. After B has evolved to the left of phenotype *y*, allele substitutions with small effects that bring it closer to the peak for A will be advantageous. (After Charlesworth 1990.)

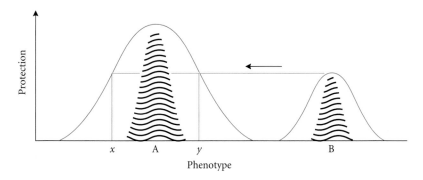

some later point, it becomes modified to serve an adaptive function. For instance, by excreting nitrogenous wastes as crystalline uric acid, insects lose less water than if they excreted ammonia or urea. Excreting uric acid is surely an adaptation, but the white color of uric acid is not. However, pierine butterflies such as the cabbage white butterfly (*Pieris rapae*) sequester uric acid in their wing scales, imparting to the wings a white color that plays a role in thermoregulation and probably in other functions.

3. *One of several ancestral functions of a feature becomes accentuated*, especially if another feature takes on its other original functions. In the synapsid ancestors of mammals, for example, the lower jaw, articulating with the skull by the articular and quadrate bones, received sound vibrations that were transmitted via the jaw joint to the inner ear. As the dentary-squamosal articulation of the jaw evolved, the articular and quadrate bones became smaller, looser, and functionally specialized for sound transmission, and now form two of the middle ear ossicles of mammals (see Chapter 6).

4. A *change in the function* of a feature *alters the selective regime*, leading to its modification. Already recognized by Darwin, this is one of the most important principles of macroevolution (Mayr 1960), and every group of organisms presents numerous examples. The wings of auks and several other aquatic birds are used in the same way in both air and water; in penguins, the wings have become entirely modified for underwater flight (Figure 24.6). The ability of an electric eel (*Electropho-*

rus electricus) to kill prey and defend itself by electric shock is an elaboration of the much weaker electric fields generated by other fishes in the same family (Gymnotidae, knifefishes), which use their electricity for orientation and communication in murky waters. The hairs, or setae, on the surface of insects' bodies universally serve functions that depend on registering changes in pressure, such as proprioception and detection of air currents, but they have been modified to serve many other functions as well (Figure 24.7). For example, the flattened, scalelike setae of moths and butterflies enable them to escape spiderwebs (by shedding scales and fluttering free) and bear the pigments that provide their wing patterns.

Intermediate steps in the evolution of complex features are generally advantageous. Most new features are modifications of pre-existing structures with altered functions, as we have just seen. Although in its later stages of modification a structure may be highly complex, with its function depending on the intricate coordination of many parts, it is a mistake to think that *only* such complexity can provide an adaptive advantage. Wings provide an example of this principle; so do eyes.

Vertebrates and cephalopods (squid, octopus) have remarkably similar eyes that are the epitome of complex organs. Darwin acknowledged, in *The Origin of Species:* "that the eye, with all its inimitable contrivances for adjusting the focus to different distances, for admitting different amounts of light, and for the correction of spherical and chromatic

---

FIGURE 24.6 Modification of a character in association with a change in its function. Both penguins and auks have evolved wings modified for underwater "flight" (stage C) from ancestors in which the wings are used for aerial flight (stage A). In both lineages, some species have partially modified wings, used for both functions (stage B). Modifications of the wing skeleton are shown at left: from bottom to top, an aerial gull, a flying auk, the flightless great auk, an extinct flightless auk, and a penguin. (After Storer 1960.)

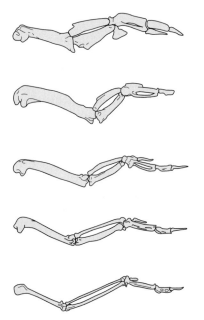

	Southern hemisphere Petrel — Penguin stock	Adaptive stage	Northern hemisphere Gull — Auk stock
	Penguins	Wings used for submarine flight only — Stage C	Great auk
	Diving petrels	Wings used for both submarine and aerial flight — Stage B	Razorbill
	Petrels	Wings used for aerial flight only — Stage A	Gulls

FIGURE 24.7 Hairs have been modified for many functions in insects. (A) End of the abdomen of a male danaid butterfly, showing extrusible hair pencils that waft sex pheromones toward females during courtship. (B) Close-up of scales on a butterfly's wing. (C) Fringes of long hairs on the legs of a diving beetle (Dytiscidae), used for swimming. See also Figure 1B in Chapter 1, showing modified hairs on a bee, used for collecting pollen. (A after Scott 1986; B after Nijhout 1991; C after White 1983.)

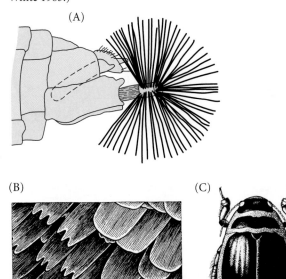

(A)

(B)

(C)

aberration, could have been formed by natural selection seems, I freely confess, absurd in the highest possible degree." But he then proceeded to supply examples of animals' eyes as evidence that "if numerous gradations from a perfect and complex eye to one very imperfect and simple, each grade being useful to its possessor, can be shown to exist; if further, the eye does vary ever so slightly, and the variations be inherited, which is certainly the case; and if any variation or modification in the organ be ever useful to an animal under changing conditions of life, then the difficulty of believing that a perfect and complex eye could be formed by natural selection, though insuperable by our imagination, can hardly be considered real."

Since Darwin's time, a great deal of information on the photoreceptive organs of various animals has been amassed (Salvini-Plawen and Mayr 1977; Osorio 1994). These organs are exceedingly diverse in structure and function, ranging from small groups of merely light-sensitive cells to the complex structures, capable of registering precise images, found in many arthropods, some molluscs, and vertebrates.

Many protists, such as dinoflagellates, have an "eyespot" consisting of an aggregation of visual pigment associated with the chloroplast or the base of the flagellum. These organisms can move in response to changes in light intensity. This photoreceptive structure resembles that of ciliated photosensitive cells that are widely distributed among the animal phyla.

In few instances has the phylogenetic sequence of photoreceptor evolution been clarified; in its place, we can recognize grades of complexity among unrelated animals that show the adaptive feasibility of each stage (Burggren and Bemis 1990). The simplest grade is a mere aggregation of a few or many photosensitive cells, found in some flatworms, rotifers, annelid worms, vertebrates (lamprey larvae), and others. The next grade is a simple epidermal cup lined with photic cells; this structure, which can provide some information on the direction of a light source through the differential illumination of different parts of the cup, has evolved independently in numerous lineages of flatworms, cnidarians, molluscs, polychaetes, cephalochordates, and others (Figure 24.8A). From this grade, there are numerous transition series to "pinhole eyes" and thence to "closed eyes," in which translucent cells or cell secretions (vitreous mass) act as a rudimentary lens (Figure 24.8B). Closed eyes, usually with some kind of lenslike structure, have evolved independently in cnidarians, snails, bivalves, polychaete worms, arthropods, and vertebrates. A closed eye with a lens enables the organism to more accurately determine the direction of incident light and to orient by it, to detect movement of objects, and, by the principle of the pinhole camera, to form at least elementary images. Image formation reaches its apogee in insects, in which each element (ommatidium) of the compound eye subtends a small angle of the field of view, enabling the many elements together to provide a detailed mosaic image; and in cephalopods and vertebrates, in which muscles move the lens or alter its shape in order to focus.

In Chapter 23, we described experimental results indicating that a homologous gene initiates eye development in both insects and vertebrates. Insect and vertebrate *eyes as such* are surely not homologous, since they differ so greatly in structure. It is likely, instead, that this gene primitively had the more general function of governing the differentiation of photosensitive cells, which are so widely distributed among animals that they doubtless predate the divergence of the animal phyla. In various phyla, other genes that organize the development of eye structures came under the control of this "master" gene. Salvini-Plawen and Mayr (1977) estimated that at least 15 lineages have independently evolved eyes with a distinct lens. The evolution of eyes is apparently not so improbable! Each of the many grades of photoreceptors, from the simplest to the most complex, serves an adaptive function. Simple epidermal photoreceptors and cups are most common in slowly moving or burrowing animals; highly elaborated structures are typical of more mobile animals. The mystery of how a simple eye could be adaptive is no great mystery after all.

Functional Integration and Complexity

The parts of an organism are often functionally interdependent, so the selective value of variation in any one character must often depend on the state of other features. For this reason, many biologists feel that evolutionary theory must be

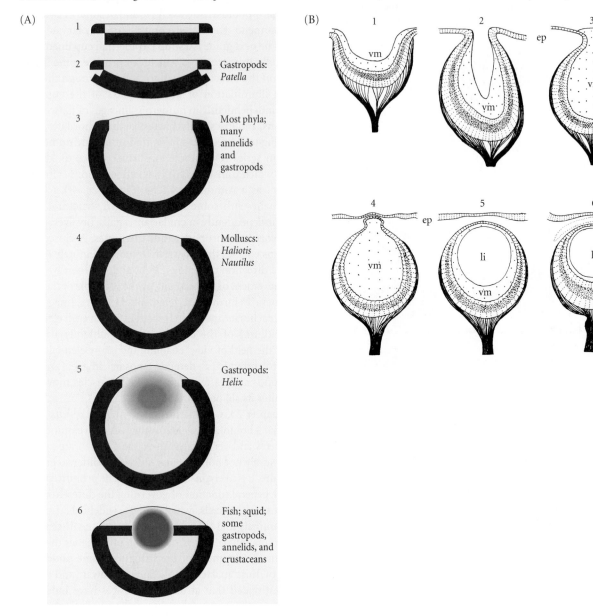

FIGURE 24.8 Intermediate stages in the evolution of complex eyes. (A) Schematic diagram of stages as seen in various animals: 1, a photosensitive epithelium sandwiched between a protective skin and a pigmented layer of cells; 2, a simple eye cup; 3, a deeper cup, providing more information on the direction of a light source; 4, gradual evolution toward a "pinhole" eye; 5, a refractive lens, resulting from increased protein concentra- tion; 6, a flat, pigmented iris surrounding the lens provides for better focusing. (B) Detailed diagrams of increasingly complex eyes among gastropods (snails): 1, eye cup of *Patella*; 2, deeper cup in *Pleurotomaria*; 3, pinhole eye of *Haliotis*; 4, closed eye of *Turbo*; 5, lens eye of *Murex*; 6, lens eye of *Nucella*. ep, epithelium; la, lacuna; li, lens; vm, vitreous protein mass. (A from Osorio 1994; B from Salvini-Plawen and Mayr 1977.)

(A)

1

2   Gastropods: *Patella*

3   Most phyla; many annelids and gastropods

4   Molluscs: *Haliotis* *Nautilus*

5   Gastropods: *Helix*

6   Fish; squid; some gastropods, annelids, and crustaceans

(B)

1   vm

2   ep   vm

3   ep   vm

4   ep   vm

5   li   vm

6   la   li

framed not only in terms of population genetics, with its focus on simple models of a few genes or characters, but also in terms of whole organisms (Wake and Roth 1989).

By its nature, complexity is difficult to understand and to describe with a few simple principles. Nonetheless, we must ask how it evolves, and what its evolutionary consequences may be.

Complex functional design   Advantageous changes in organismal design may not occur until *constraints* have

been relieved. Some constraints may reside in genetically controlled patterns of development, as we saw in Chapter 23. Other constraints are *functional*. The greater the number and degree of functional integration of interacting parts, the more stringent constraints on evolution are likely to be, and the rarer will be evolutionary "breakthroughs" to new organismal designs. Thus the evolution of a complex design often limits severely the variety of evolutionary pathways that a lineage can subsequently take. In most salamanders, for example, the hyobranchial bones and muscles that support the tongue also move the floor of the mouth, forcing air into the lungs. Many salamanders catch prey with their sticky tongues, but can extrude the tongue only slightly, because of a conflict between the respiratory and feeding functions of the hyobranchial apparatus (Wake and Roth 1989). Thus *structural coupling of different functions* can act as a constraint.

Key innovations frequently lift functional constraints. The term "key innovation" (or "key adaptation") is used differently by various authors. Usually it refers to an evolutionary change that (1) provides adaptations to one or more new ecological niches or adaptive zones and (2) enables a lineage to diversity greatly (Mayr 1960; Liem 1973). The continuously growing incisors of rodents (the largest order of mammals) are an example. Yet another criterion is (3) that the change set in motion the evolution of other, functionally interacting, features (Levinton 1988). The evolution of feathers of sufficient size for flight (as in *Archaeopteryx*) set the stage for the many other flight-related features of later birds. As we shall see (in Chapter 25), it can be very difficult to demonstrate that a new feature was the cause of a group's diversification (Cracraft 1990; Levinton 1988); moreover, many groups that have not radiated extensively have features that may deserve to be called key adaptations. Sheep, antelopes, and other ruminants are by far the most diverse of the three suborders of Artiodactyla; they use bacterial symbionts to digest refractory plant materials in a complex stomach that many biologists would surely call a key adaptation (see Chapter 22). But a similar adaptation is found in leaf-eating monkeys, which are not especially diverse, and in a South American leaf-eating bird, the hoatzin (*Opisthocomus hoatzin*), that is the sole member of its family. Thus it is perhaps best to define a key innovation as a character that provides entry into a substantially different ecological niche or adaptive zone, whether or not it increases the rate of diversification.

Two mechanisms by which functional constraints may be lifted are *functional divergence of redundant elements* and *decoupling of functions and structures.*

Functional divergence of redundant elements

Functional divergence is abundantly illustrated by the evolution of functionally different proteins due to the divergence of duplicated genes (see Chapter 22; Lauder and Liem 1989). A spectacular example is the evolution of the vertebrate globins (Figure 9 in Chapter 22), that originated by successive duplications of a gene that ancestrally encoded a myoglobin-like molecule. The duplication that first gave rise to distinct myoglobin and monomeric hemoglobin was followed by the divergence of the α- and β-globins that together comprise the hemoglobin of jawed vertebrates. Subsequent duplications of both the α- and β-chains occurred in various lineages. In some cases, these duplicates then became specialized for the different functions that the ancestral gene had served.

At a morphological level, repeated structures may acquire independent functions. Ancestrally similar appendages on each body segment diverged into various mouthparts, locomotory appendages, and genitalic structures in the ancestors of crustaceans and insects. The walking legs of trilobites, for example, carried gills and a modified base used for feeding; these three functions are decoupled in most other arthropods, in which they are carried out by different appendages or by different structures altogether (e.g., the tracheae of terrestrial arthropods).

Decoupling of functions

Two detailed studies of feeding mechanisms in vertebrates illustrate the evolutionary opportunities that may follow from decoupling of functions. The Labroidei, embracing more than 1700 species of cichlids, damselfishes, and wrasses, are distinguished from related percoid fishes by their feeding apparatus. Generalized percoids have, within the throat, upper and lower "pharyngeal jaws" that function chiefly to transport food into the esophagus. The "oral jaws" that surround the mouth have the tasks of collecting and manipulating food. The upper pair of pharyngeal jawbones in generalized percoids are disconnected from the base of the cranium, and the two separate lower pharyngeal jawbones lack muscular connections to the cranium. In the Labroidei, in contrast, the upper pharyngeal jaws are connected to and rotate on a cranial joint; the two lower elements are fused into a single massive bone; and a muscle originating on the cranium has shifted its insertion from a gill arch to the lower pharyngeal jaw, so that the upper and lower jaws can be forcefully appressed. Karel Liem (1973) has suggested that this versatile apparatus is a key innovation that, having assumed the function of manipulating food, has freed the oral jaws for collecting food in many ways generalized percoids are incapable of. Feeding modes are especially diverse in the Cichlidae (see Figure 30 in Chapter 5), which display an astonishing variety of feeding habits, matched by extraordinary variation in the teeth of the oral jaws.

In most families of salamanders, larvae are aquatic and adults are terrestrial. The larvae have gills supported by branchial arches (part of the hyobranchial skeleton), and feed by suction created by opening the mouth while lowering the throat with the hyobranchial skeleton and muscles. Adults breathe by forcing air into the lungs with the hyobranchial apparatus, which also controls the tongue. They catch arthropod prey by slight extrusion of the sticky tongue, accompanied by lunging forward and snapping their jaws. However, lungs have been lost in the family Plethodontidae, the largest family of salamanders. They breathe through the skin, as other salamanders do to a lesser degree. Some plethodontids have aquatic, gilled larvae, which is a plesiomorphic (ancestral) trait. Others, such as the primarily Neotropical bolitoglossines, are completely terrestrial, and develop directly into the adult form.

David Wake and his colleagues (Lombard and Wake 1977; Wake 1982) believe that the loss of lungs has relieved plethodontids of a functional constraint on the evolution of the tongue. Compared with other salamanders, plethodontids can project the tongue farther and with greater versatility, a capacity that reaches its apogee in the bolitoglossines, which can shoot their extraordinarily long tongues at a greater speed than any other movement known in vertebrates (Figure 24.9). The hyobranchial skeleton, no longer used for ventilating the lungs, has been modified into a set of long elements that can be greatly extended from a folded configuration by highly modified hyobranchial muscles (Figure 24.10). These modifications are most pronounced, and the capacity for tongue extension is greatest, in plethodontids with direct development, probably because the branchial bones no longer support larval gills. Freed from the compromises imposed on a hyobranchial apparatus that in ancestral salamanders serves multiple functions, the bolitoglossines have evolved an impressive new adaptation.

Acquisition of functional complexity   A common argument against Darwinian evolution is based on what is sometimes termed "irreducible complexity." This term refers to complex organismal features that cannot function except by the coordinated action of all their components. The vertebrate eye and complex molecular systems are frequently cited as examples. It is assumed that an organism's fitness would be reduced by elimination or alteration of any one of the interacting components. From this assumption, opponents of evolution argue that the functional complex must have required all of its components from the beginning. The entire functional complex could not have arisen in a single mutational step—a point on which Darwinians and anti-Darwinians agree. Anti-Darwinians, however, argue that such systems could therefore not have evolved at all. This is an entirely unjustified conclusion.

The anti-evolutionary argument ignores two points. First, intermediate stages in the evolution of complex systems can be shown to exist and to have adaptive value, as we have illustrated by the evolution of eyes and of multiple hemoglobins (see also the sections on gene duplication and exon shuffling in Chapter 22.) Second, a component of a functional complex that was initially merely *superior* often be-

FIGURE 24.9   A bolitoglossine salamander, *Hydromantes supramontis*, capturing prey with its extraordinarily long tongue. Tongue extension in these salamanders is among the most rapid movements recorded for vertebrates. (From Deban et al. 1997; courtesy of S. Deban.)

FIGURE 24.10   Modification of the hyobranchial apparatus for tongue extension in plethodontid salamanders. (A) Primitive condition of the apparatus, in dorsal view, in a plethodontid (*Eurycea*) with only slight tongue extension. When the tongue is retracted, joint 2, between elements bb and cbII, is elevated above the horizontal. (B) Same, when tongue is extended. Muscles are shown surrounding the hyobranchial elements, which are adducted toward the midline, causing tongue extension. Joint 2 is depressed toward the horizontal plane. (C) Modified hyobranchial apparatus, at rest, in a bolitoglossine salamander (*Bolitoglossa*). (D) Modified muscles and ligaments (l) lower joint 2 nearly to the horizontal in bolitoglossines, enabling greater extension of the tongue. (After Lombard and Wake 1977.)

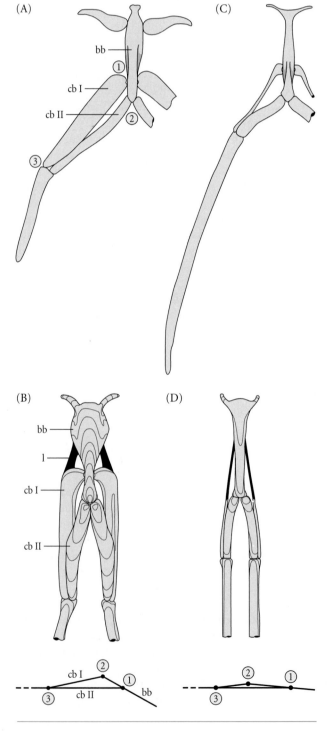

comes *essential*, as long as its function remains necessary. It becomes indispensable because other characters evolve to *become functionally integrated* with it. For instance, visual acuity of the kind afforded by the vertebrate eye is indispensable for eagles, since their way of hunting prey has been acquired and made possible only by such acuity—but this does not mean that the earliest vertebrates required precisely focused images. The high metabolic rate of adult and embryonic mammals depends on their multiple, functionally differentiated hemoglobins, and most modern mammals cannot survive without maintaining high metabolism. Lampreys, however, survive with a single, monomeric hemoglobin, and the embryos of live-bearing snakes and lizards develop without relying on specialized embryonic hemoglobins. Thus, eagles and mammals have *acquired* dependence on the elements of complex eyes and complex hemoglobins. Such dependence, indeed, is often lost, as we see in the many vertebrates with degenerate eyes and in certain Antarctic fishes that lack hemoglobin altogether (Eastman 1993).

Rates of Evolution

The great variation in rates of evolution (see Chapter 6), documented both by phylogenetic comparisons of living species and the fossil record, has stimulated some of the more vigorous arguments in evolutionary biology.

George Gaylord Simpson (1944, 1953), who pioneered the study of evolutionary rates, distinguished **phylogenetic rates** of evolution, by which he meant the rates at which single characters or complexes of characters evolve, from **taxonomic rates**, the rates at which taxa with different characteristics replace one another. Taxonomic rates are cruder indices of evolutionary rates than are rates based on quantitative analyses of individual features, but they can provide a more encompassing sense of the rates at which taxa evolve.

Taxonomic Rates

The tadpole shrimp *Triops cancriformis* (Figure 24.11) is indistinguishable from 180-million-year-old Triassic fossils that are given the same name. The oldest known living animal species, it is a "living fossil" that, at least in morphological features, has evolved amazingly slowly. The coelacanth, horseshoe crab, and ginkgo tree likewise have changed little since the Mesozoic or earlier (see Chapter 6). In contrast, numerous families of ammonoids (see Chapter 7) evolved within only 7 or 8 My from a few genera that survived the mass extinction at the end of the Permian (Stanley 1979).

Simpson proposed that the average rate of evolution in a group can be measured by the inverse of the average duration of a species (or genus). The *Hyracotherium*-to-*Equus* horse lineage, for example, is divided into eight genera that collectively span 45 My, for an average duration of 5.6 My per genus. If the lineage had changed so slowly as to be divided by paleontologists into only four genera, the average duration would be 11.2 My, reflecting a lower rate of evolution. Simpson compared the durations of genera of bivalves and of the mammalian order Carnivora (Figure 24.12), and found that they averaged 78 My and 6.5 My re-

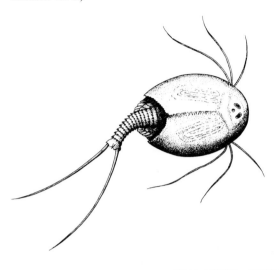

FIGURE 24.11 A "living fossil," the tadpole shrimp *Triops cancriformis*. This freshwater crustacean, found in temporary pools in arid regions of Eurasia and northern Africa, has undergone no evident morphological evolution since the Triassic. (From Kaestner 1970.)

FIGURE 24.12 George Gaylord Simpson, who integrated paleontology into the Evolutionary Synthesis, calculated curves showing the percentage of genera of (A) bivalves (pelecypods) and (B) carnivores surviving for various durations. Black lines represent extinct genera; colored lines, genera with living members. Although more recent calculations indicate considerably shorter survival spans for bivalves, their average is still far greater than that for mammal genera. (After Simpson 1944.)

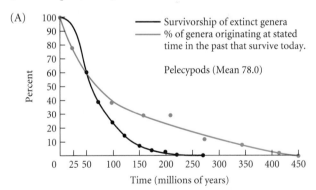

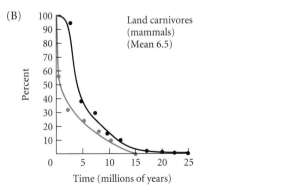

spectively. He concluded that bivalves evolve, on average, much more slowly than carnivorous mammals.

A potential problem with Simpson's analysis of genera is that a genus may be terminated either by true extinction or by pseudoextinction. If most seeming extinctions are pseudoextinctions, then the duration of genera is indeed an inverse measure of the rapidity of evolution. If most genera are soon terminated by real extinction, however, their short durations may indicate not rapid evolution, but bad luck.

Rates of Character Change

Measuring changes in individual characters or groups of characters provides a more rigorous approach to evolutionary rates than the turnover rates of genera or other taxa.

One of the important changes in the evolution of the equid lineage leading to the modern horse was an increase in the height of the cheek teeth (see Chapter 6). Simpson (1944) gives the mean tooth height of *Mesohippus* as 8.36 mm (with a standard deviation $s = 0.04$) and of *Merychippus* as 34.08 mm ($s = 2.01$). The two samples are separated in time by about 15 million years. The *average* rate of change is $(34.08 - 8.36)/15 = 1.71$ mm/My, or about 0.000005 mm per generation, assuming an average generation time of 3 years. Even though the tooth increased fourfold in height, and even though this is one of the faster rates of change documented in the fossil record, it is still an astonishingly low rate of change.[*]

As we saw in Chapter 6, fine sampling of a fossil sequence often shows fluctuations in character means as well as variable rates of change: long periods of little change are interspersed with short periods of substantial change. Thus over a long time span, the average rate of change is likely to be low, obscuring the rapid changes that occurred at certain times during the longer period. Philip Gingerich (1983) found just this pattern when he compiled, from the paleontological literature, rates of evolution of characters of many species, expressing these rates in units called DARWINS (1 darwin equals a change in the natural logarithm of the mean by a factor of 2.718 per million years). The greater the time interval between presumed ancestors and descendants, the lower the rates of change reported (see Figure 39 in Chapter 6). For samples separated by about 1 My, almost no rates exceeded 1 darwin. The low rates of evolution in the fossil record contrast strongly with those measured in species that have colonized new regions in historical times (such as the house sparrow, *Passer domesticus*, introduced into North America from Europe about a century ago). Comparisons of introduced and indigenous populations of such species show small but rapid changes in various features, at an average rate of about 400 darwins.

We would hardly expect a feature to sustain a rate of 400 darwins for a million years (i.e., to change more than a thousandfold). Even if genetic variation were not eroded by selection, most characters would be selectively disadvantageous if changed a hundredfold, or even tenfold, unless there were functionally compensatory changes in other features. Nevertheless, *the most striking feature of evolution in the fossil record is that it is, on average, very slow.* The major problem posed by evolution in the long term is *not* whether there has been enough time for it to occur; it is, instead, why it is usually so sluggish.

Genetic Approaches to Evolutionary Rates

Although it is not possible to study the genetics of most fossil species directly, models from quantitative genetics can be applied to the fossil record if we assume that estimates of variables such as heritability and mutation rate can be extrapolated to extinct populations from similar living species. Although we may be tempted to assume that a directional change in a feature was caused by natural selection, a random process such as genetic drift can also result in a trend in a population. A neutral allele that drifts to fixation from, say, $p = 0.50$ displays such a trend; an effectively neutral polygenic character will drift similarly (see Chapter 14). Several authors, using models for quantitative characters, have asked whether rates of change correspond to those we would expect from genetic drift alone, or whether they require explanation by natural selection (Lande 1976b; Turelli et al. 1988; Lynch 1990).

In the genetic drift model, the genetic variation in a character will arrive at an equilibrium at which variation arising by mutation (the mutational variance, V_m: see Chapter 14) balances the loss of genetic variation due to genetic drift. At this point, the rate of change in the character is limited by V_m, and does not depend on the effective population size—just as the long-term rate of substitution of neutral alleles at a single locus is independent of population size (see Chapter 11). A given rate of mutation can then support a certain maximal rate of evolution of the character by genetic drift. Therefore, if a rate of evolution calculated from paleontological data exceeds that implied by the mutation rate of the character, V_m, evolution has been too fast to be explained by genetic drift alone. If the rate is very much less than that permitted by V_m, it is slower than expected from genetic drift, and it may be inferred that stabilizing selection has acted on the character. The mutation rate, V_m, cannot be measured in extinct organisms, but it may be assumed to have been similar to the many estimates of V_m that have been made for morphological characters of living species[†] (see Chapters 10 and 14).

Michael Lynch (1990) used this mutation-drift equilibrium model to analyze data on skeletal features of seven groups of mammals. Some of the data were paleontological, such as tooth size in the well-known horse lineage. In other

---

[*]For several reasons, it is often more accurate to describe evolutionary rates in terms of the logarithms of the measurements. Using natural logs, tooth height in this example changed by $3.53 - 2.12 = 1.41$, or 0.095/My.

[†]To be exact, the quantity V_m/V_E is used in these analyses: the mutational variance scaled by the environmental variance of the character. This scaling permits comparisons among different characters that differ in their means.

cases, he compared living species, estimating their average rate of divergence based on fossil evidence of the age of their common ancestor; for instance, cranial dimensions of leopards, cheetahs, and ocelots were analyzed in this way. In almost all cases, the rate of change was much lower than the minimal rate expected if mutation and genetic drift were the only factors operating. For instance, horse teeth have evolved only 20 percent as fast, on average, and cat cranial dimensions only 10 percent as fast, as the minimal rate of expected neutral evolution. Lynch's analysis of human traits was particularly interesting. He assumed that human, chimpanzee, and gorilla arose almost simultaneously about 9 Mya, that this lineage diverged from the orangutan about 13 Mya, and that the major human "races" separated 150,000–75,000 years ago. Skeletal differences among the apes and human all are about what we might expect of characters evolving by genetic drift with a mutation rate (V_m) on the order of 10^{-7} to 10^{-6}. Since these values are far less than the lower end of the range of observed mutation rates, 10^{-4}, the low rates are probably best explained by stabilizing selection. However, the rate of divergence of human features from those of the apes is, on average, about ten times higher than the rate of divergence of the apes from one another: the rate of morphological evolution in the human lineage, although slow, nonetheless appears to have been accelerated. Skull differences among human "races" also have evolved relatively rapidly; on average, they would require that the mutation rate be 3.2×10^{-4} in order for mutation and drift to explain the data. This value is about thrice the lower boundary of observed mutation rates (10^{-4}), and thus are compatible with evolution by genetic drift (Figure 24.13). The changes in cranial capacity and other skeletal features

of humans have long been recognized to be among the most rapid long-term evolutionary changes known.

Do these analyses force us to conclude that when evolutionary change occurs, it is caused by mutation and genetic drift, rather than natural selection? This conclusion would go too far, for two reasons. First, an independent perspective, that of functional morphology, leads us to suspect that selection must have played a role in some of these cases: it is implausible that hypsodonty in horses was not an adaptation to a diet of grasses (see Chapter 6), or that the increase in cranial capacity in humans was not adaptive. Second, these average rates, as we emphasized earlier, may mask short episodes of rapid change driven by selection. In his analysis of mammals, Lynch found, as did Gingerich, that rates of evolution are lower if they are calculated over longer intervals. He suggested that characteristics may evolve for a while by neutral mutation and drift, but that they stop diverging when they encounter increasingly stringent stabilizing selection. Evolution by natural selection to a new adaptive optimum, followed by stabilizing selection around that optimum, could also account for low average rates. Before selection can be ruled out, it will be necessary to apply these models to the rapid changes over short intervals that are revealed in some high-resolution fossil records.

Punctuated Equilibrium Revisited

Rates of evolution, so low in many instances, can in some instances be so high that new taxa appear without fossil evidence of the intervening steps that most biologists believe existed. Simpson (1944), noting the sudden appearance of higher taxa such as orders of mammals, described this pattern as "quantum evolution," suggesting that it might mark

FIGURE 24.13 Mean rates of evolution of various morphological characters of mammals, based on both fossil material (*Merycochoerus, Brachycrus, Hyopsodus*, horses, Adapidae, *Homo erectus*, cave bear) and differences among extant taxa (Sciuridae, Felidae, *Homo sapiens*, and CGO = chimpanzee, gorilla, orangutan). The symbols Δ_{min} and Δ_{max} indicate minimal and maximal rates compatible with explanation by mutation and genetic drift. Note that rates are plotted on a log scale. (After Lynch 1990.)

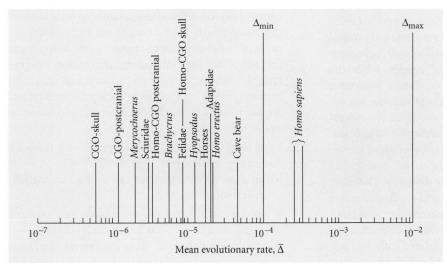

rapid evolution from one adaptive phenotype to another by way of intermediate stages that were less adaptive and therefore too brief to have left fossil traces. The botanist Verne Grant (1963) coined the term "quantum speciation" to describe the very rapid evolution of some species. And the leading authority on speciation, Ernst Mayr (1954), suggested that most species originate by rapid evolution in small colonies (peripatric speciation: see Chapter 16), and proposed that this process could account for many of the gaps in the fossil record.

Niles Eldredge and Stephen Jay Gould (1972) adopted Mayr's hypothesis as the theoretical explanation for the pattern of stasis, interspersed with the sudden appearance of new phenotypes, that they called punctuated equilibrium (see Chapter 6). Newly arisen phenotypes, they argued, represent new biological species that had spread into new areas where they left fossil remains, after originating elsewhere so rapidly and in so small an area that few or no fossils record their transformation. Moreover, Eldredge and Gould extended Mayr's hypothesis that strong epistatic relationships among genes prevent widespread, long-established species from evolving very much (see Chapter 16). According to their hypothesis, speciation in a localized population, in which coadaptation is disrupted by a founder event, is usually *necessary* for evolutionary change in morphology. The parent species from which the diverging colony is derived, not having experienced a genetic disruption, remains recalcitrant to natural selection and other forces of change. This hypothesis was proposed to explain the origin not of higher taxa (unlike Simpson's "quantum evolution"), but of closely related species, such as the trilobites that Eldredge had studied, which were similar except for the number of rows of eye lenses (see Figure 10 in Chapter 6). Eldredge and Gould's model grants that the morphological evolution of each new species may be caused by natural selection, and that the species may pass through intermediate stages (see Figure 12C in Chapter 6). The changes appear discontinuous because of their rapidity and geographic localization, but need not be due to discontinuous mutations.

Eldredge and Gould (1972) and Steven Stanley (1975, 1979) pointed out an important macroevolutionary implication of this model: If morphology evolves only during speciation (and only in the "daughter" species, while the "parent" species remains unchanged), then a long-term evolutionary trend over many millions of years in, say, the body size of horses cannot be the consequence of a succession of changes by natural selection acting on a single lineage. The trend must arise in a series of speciation events. One possibility is "directed speciation"—each new species deviates from its ancestor in the same direction. The other possibility, favored by these authors, is that a character in a newly arisen species may diverge from the parent species' character in either direction, but the rate of diversification of different descendant lineages is correlated with the character in question. For example, if large-bodied species either speciate more rapidly or become extinct less frequently

than small-bodied species, mean body size in the clade as a whole will display an increasing trend—even if each new species deviates from its parent equiprobably in either direction (see Figure 18 in Chapter 12). Stanley called such a correlation between a character state and the rate of speciation or extinction *species selection*.

The punctuational perspective *departs radically from traditional neo-Darwinian theory* on two points: namely, by proposing that (1) species generally cannot evolve new morphologies except when a small population becomes a new, reproductively isolated species; and (2) long-term trends in morphology are generally the consequence not of natural selection sorting genotypes within species, but of a higher-level sorting process among species. Traditional natural selection, in this view, may cause the divergence of each new species from its ancestor, but is not the principal agent of long-term evolutionary trends.

Critiques of the punctuational model Both the pattern and the hypothesis advanced by the punctuationists have been criticized by some paleontologists and population biologists (see, for example, Hoffman 1989; Levinton 1988; Charlesworth et al. 1982; Turner 1986; Stebbins and Ayala 1981; Maynard Smith 1983; and defenses of the model by Gould and Eldredge 1977, 1993; Gould 1982b; Stanley 1979). The chief criticisms have been:

1. There is abundant evidence, from geographic variation and adaptive changes observed within recent history, that populations can evolve in response to selection without speciating (see Chapters 9, 13, 14).

2. Stable lineages in the fossil record are not absolutely static in phenotype; they fluctuate, sometimes quite rapidly, around a constant long-term mean. Population geneticists interpret this pattern as evidence that the lineages are capable of change, but stabilizing selection keeps them near an optimum phenotype that may fluctuate, but does not change appreciably (Charlesworth 1984). In animals, habitat selection—the behavioral capacity to choose a suitable environment—may contribute importantly to this stability (Maynard Smith 1983). Even if the macroscopic aspects of the environment change, many animals can find much the same microenvironment they used before. (In deserts, for example, aquatic and marsh-inhabiting insects occupy patches of aquatic habitat and are no more desert-adapted than their relatives in wetter regions.)

3. Most fossil sequences are incomplete, with temporal resolution seldom less than 100,000 years, during which a polygenic character can evolve substantially (Turner 1986).

4. If populations are free to evolve by ordinary natural selection, then the process of selection within species can be very much faster than the rate of speciation and extinction of whole species. If environmental factors favor a progressive change in the features of indi-

viduals, the direction of evolution will be dominated by individual selection rather than by the weaker force of species selection. These arguments cast doubt on the role of species sorting (or species selection in the broad sense) as a mechanism of macroevolution.

Tests of the hypothesis The validity of the punctuated equilibria hypothesis cannot be tested merely by observing that fossil lineages shift rapidly from one long-lasting phenotype to another, because this pattern could equally well be explained by directional selection (for example, if the environment changed). The critical test is whether morphological change is ordinarily accompanied by bifurcation of the lineage—i.e., true speciation (Gould 1982c; Gould and Eldredge 1993). It can be difficult to tell whether variation in a fossil sample represents different species or not, but speciation may be implied if one form gives way to two distinct phenotypes that coexist in the same place and time, since reproductive isolation can be inferred from their distinctness while coexisting (see Chapter 15). Many punctuated fossil sequences do not show such evidence of speciation (Figure 24.14), but others do (Gould and Eldredge 1993). One of the best examples is provided by Alan Cheetham's (1986) analysis of Miocene to Pliocene bryozoans of the genus *Metrarhabdotos*. Several almost static lineages coexist with lineages that appear, from phylogenetic analysis, to be their descendants (see Figure 11 in Chapter 6). Almost no intermediates were found between ancestral species and their descendants, suggesting that the new species evolved within about 160,000 years, the average interval between samples. The new species evolved at a much higher rate than the rate of evolution within their ancestors, which was nearly zero. *Metrarhabdotos* provides perhaps the best evidence for punctuated equilibrium, but we do not yet know how typical this pattern is.

Directions, Trends, and Progress

Some historians have argued that Darwinian evolutionary theory owed its inspiration largely to the optimistic ideology of progress in nineteenth-century Western society. For many decades after the publication of *The Origin of Species*, much of educated society, although misunderstanding and doubting Darwin's theory of natural selection as the mechanism of evolution, accepted the historical reality of evolution—and viewed it as a cosmic history of progress. As humanity had been the highest earthly link in the pre-evolutionary Great Chain of Being, just below the angels, so humans were seen as the supreme achievement of the evolutionary process, and western Europeans as the pinnacle of human evolution. Even though Darwin distinguished himself from his contemporaries by denying the necessity of progress or improvement in evolution (Fisher 1986), virtually everyone else except for a few biologists viewed progress as an intrinsic, even defining, property of evolution. In this section, we shall examine the nature and

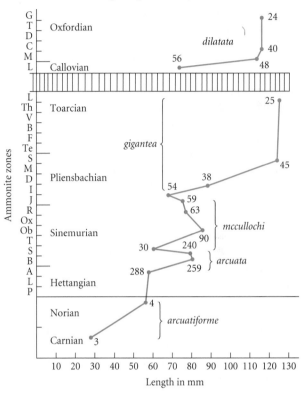

FIGURE 24.14 Punctuated change without biological speciation (bifurcation) in Jurassic oysters of the genus *Gryphaea*. The length of the left valve is plotted for samples (number of specimens indicated) in a stratigraphic series of zones, distinguished by different ammonite faunas, of about a million years' total duration. Sequential members of the same lineage are given different names (chronospecies). Compare with Figure 11 in Chapter 6, illustrating a case in which morphological evolution is associated with biological speciation. (After Hallam 1982.)

possible causes of trends in evolution, and ask whether the concept of evolutionary progress is meaningful.

A trend may be described objectively as a directional shift over time. "Progress," in contrast, implies betterment, which requires a value judgment of what "better" might mean. We will come to terms with evolutionary "progress" after discussing trends.

Trends: Kinds and Causes

We can distinguish UNIFORM TRENDS from NET TRENDS (or directions, or "progress") (Ayala 1988). A trend is uniform if a feature changes in the same direction in all evolving lineages at all times. If, in contrast, a feature evolves in both directions, but more frequently in one direction than in the other, so that the mean shifts in all lineages taken together, a net trend is the result. GLOBAL TRENDS characterize all of life. Phylogenetically LOCAL TRENDS may be discerned within individual clades, such as Mammalia or Hominidae. Global trends, if they exist, must emerge from the summation of local trends.

Trends may also be classified by their causes (McShea 1994; P. J. Wagner 1996).

Trends due to chance Due to genetic drift, a character will change *entirely at random,* by a *random walk* away from its initial state, just as an allele's frequency ends up at 0 or 1 under random genetic drift (Raup 1977; Raup and Gould 1974; Fisher 1986). In order to ascribe an apparent trend to a deterministic mechanism rather than to chance, it is necessary either to (1) study replicate histories (e.g., many phyletic lines) in order to show statistically that the trend is more consistent than expected from chance alone, or to (2) show that the trend is correlated with a hypothesized causal factor, such as a relevant aspect of the environment.

Passive trends due to boundaries If, in a group of initially similar lineages, a character state initially lies near a boundary imposed by some constraint, then there is more room for movement away from than toward the boundary, so variation increases and the mean shifts away from the boundary. Such a shift is called a PASSIVE TREND. However, the original (plesiomorphic) character state may be found in some lineages throughout the history of change (Figure 24.15A). The boundary could be due to history: for example, the earliest organisms were so simple that their descendants could only become more complex. It could also be due to constraints arising from developmental processes, natural selection, or both. The number of digits in tetrapod vertebrates, for instance, has decreased from five far more often than it has increased. Whether this is due to developmental or selective constraints is not clear.

Steven Stanley (1973) used the principle of passive trends to explain Cope's rule in mammals. Cope's rule states that body size tends to increase in evolution. Suppose that selection favors a variety of sizes in different species, so that variation within a clade increases over time. If the ancestor of the clade is closer to the lower than to the upper functional limit on size, the increase in variance will carry with it an increase in mean size. Stanley offered evidence that the generalized ancestors of many of the orders of mammals (e.g., various ungulate orders) were relatively small, and argued that generalist mammals are unlikely to be large, because large size often is accompanied by morphological features that enforce some degree of specialization. (Large mammals are unlikely to be able to both climb and burrow, for example). Generalists, Stanley suggested, are more likely than specialists to evolve the major new ways of life that frequently characterize new higher taxa. Therefore higher taxa are more likely to stem from small than from large ancestors.

Active trends due to constraints In an active trend, some process, such as natural selection, biases change in one direction. Over time, fewer lineages possess the ancestral character state, and the entire frequency distribution is shifted (Figure 24.15B).

Although it is difficult to be certain, it is likely that some evolutionary changes are *irreversible* (Bull and Charnov 1985). That is, boundaries, or constraints, may evolve that act as *ratchets*—mechanisms that make reversal unlikely. For example, highly differentiated sex chromosomes, such as the X and Y chromosomes of mammals, are unlikely to revert to the less differentiated ancestral state (see Chapter 21) because most of the genes that were originally on the Y chromosome (and were the same as those on the X) long ago became nonfunctional (Bull and Charnov 1985).

EPIGENETIC RATCHETS are evolved modifications of developmental patterns that prevent the regeneration of the ancestral condition (Levinton 1988). The thoracic seg-

FIGURE 24.15 Computer simulations of the diversification of a clade, showing (A) a passive trend and (B) an active (driven) trend in a feature such as a morphological character. The histograms at the top show the frequency distribution after 50 time units. (A) A character shift in either direction is equally likely, but the character cannot go beyond the boundary at left. The mean increases, but many lineages retain the original value. (B) The entire distribution is shifted by a bias in the direction of change, caused by a factor such as natural selection. (After McShea 1994.)

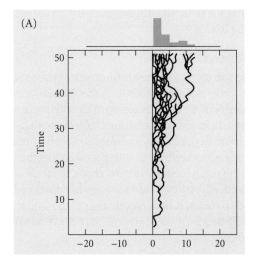

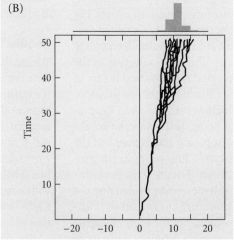

ments of insects, for example, have acquired individual developmental identities, and would be unlikely to regain the undifferentiated state postulated for ancestral arthropods (see Chapter 23). The most extreme epigenetic ratchets probably are represented by evolutionary losses of organs, which are unlikely to evolve again in the same form. SELECTIVE RATCHETS are features that are not easily dispensed with because of their role in functional integration. The notochord, for example, provides support in amphioxus, as it presumably did in ancestral chordates. It no longer serves that function in vertebrates, but it nonetheless appears during development, and is indispensable because it induces the development of the nervous system. The Austrian biologist Rupert Riedl (1978) used the term BURDEN to describe the "responsibility" that certain features bear in their developmental and functional roles. The greater a feature's burden, the less likely it is to become dismantled or reversed.

Epigenetic and selective ratchets may be especially associated with the evolution of ecological specialization. Specialization often entails extreme modifications, commonly including the loss of structures: baleen whales lack teeth and instead use combs of hairlike baleen for filter feeding; orchids have minute seeds that lack endosperm and rely on mycorrhizal fungi to nourish the seedlings. Such highly specialized organisms are unlikely to evolve the features of more generalized ancestors.

Active trends due to individual selection   Natural selection within populations is surely among the most common causes of trends. At least two broad categories of directional selection can be distinguished.

Selection for efficiency. In a relatively constant ecological environment, features may evolve toward an optimal state that maximizes functional "efficiency," subject to certain constraints. This explanation surely applies, for example, to those lineages of horses (see Chapter 6) that became progressively more adapted for running (by modification of the toes and metapodials) and for chewing grass (by modification of the cusp pattern and height of the cheek teeth).

Coevolution. In Chapters 18 and 21, we noted that evolution resulting from certain kinds of interactions, both within and between species, can take the form of a "runaway" process or "arms race." For example, female choice or male-male conflict can impose directional selection on display characters and weapons. Prey species are selected for improved defenses, and predators for more effective ways to overcome such defenses. The directional evolution of such characters is thought to be limited only by the "costs" that they ultimately incur.

Active trends due to lineage selection   As described above, the mean character state among lineages in a clade will shift if the rate of speciation or extinction of lineages depends on the character state (see Figure 18 in Chapter 12). Such species selection, or lineage selection, will occur only

if it is not opposed by more rapid evolution *within* species. Advocates of punctuated equilibrium argue that species selection is a major cause of evolutionary trends because, they suggest, species generally evolve little except during the speciation process, so sustained directional evolution within species does not occur. It is also possible that each species is free to evolve at any time, but that most species are so well adapted to their microhabitats or ways of life that they experience no individual selection for change, even while they differ in rates of speciation or extinction. For instance, Williams (1966) noted that if the extinct passenger pigeon had harbored a host-specific louse, the lice on the last surviving pigeon would have been as well adapted to their habitat as ever; the "global" event, the slaughter of the pigeons, would not have affected the relative fitnesses of their genotypes. Williams suggested that "running out of niche" may be a common cause of extinction that does not affect the course of natural selection within species.

A good example of a trend due to species selection is the increase in the ratio of nonplanktotrophic to planktotrophic species in several lineages of Cenozoic gastropods (Figure 24.16). Species that lack a planktonic larval dispersal stage

FIGURE 24.16   A trend caused by species selection. The vertical bars show the stratigraphic distributions of species of volutid snails. Although nonplanktotrophic species had shorter durations, they arose by speciation at a higher rate, so the ratio of nonplanktotrophic to planktotrophic species increased over time. (After Hansen 1980.)

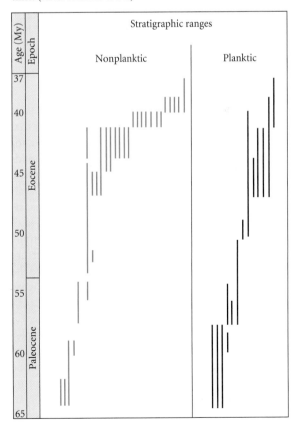

are more susceptible to extinction than planktonic species. However, they more than compensate by their higher rate of speciation, probably because their lower rate of dispersal reduces the rate of gene flow among populations (Hansen 1980; 1982; see also Jablonski and Lutz 1983; Jackson and McKinney 1990).

Active trends due to hitchhiking Many characters are strongly correlated among taxa, simply because closely related species are more similar, on the whole, than distantly related species. Therefore, if any one character causes one clade to become richer in species than other clades due to its effect on the rate of speciation or extinction, all the correlated characters will tend toward greater frequency. This process has been called SPECIES HITCHHIKING (Levinton 1988), by analogy with the hitchhiking of alleles that are correlated, by linkage disequilibrium, with an advantageous allele at another locus.

Trends due to species hitchhiking are undoubtedly very common. For example, clades of insects that eat green plants generally contain more species than their nonherbivorous sister groups, suggesting that the herbivorous habit somehow causes increased diversity (Mitter et al. 1988). One such group is the Lepidoptera (moths and butterflies), most of which have herbivorous larvae. In contrast to their sister group, the much less diverse Trichoptera (caddisflies), adult Lepidoptera have coiled, sucking mouthparts (rather than nonfunctional mouthparts) and scales (rather than hairs) on their wings. If herbivory is the cause of the greater diversity of the Lepidoptera, then coiled mouthparts and scales have become more prevalent among insects at large due to their association with the herbivorous habit.

Discriminating among Kinds of Trends: An Example

Paleobiologists have begun to analyze the several processes that can yield evolutionary trends, bearing in mind that several processes can act simultaneously. Peter Wagner, for example (1996), has studied trends in clades of gastropods (snails), encompassing about 370 species, that lived during the 100 My interval from the late Cambrian through the Silurian. He analyzed numerous ancestor-descendant transitions, inferred from phylogenetic analyses. Overall, later gastropods had, on average, a higher spire (or shell torque, ST) than earlier forms, higher inclination of the aperture (AI) relative to the axis of the shell, and a lower width of the sinus (SW)—in effect, the breadth of the aperture (Figure 24.17). Wagner applied the following tests for several possible causes of each trend.

An active trend would be discerned if, *within* the plesiomorphic range displayed by the earliest gastropods, the frequency distribution of character states shifted in the same direction as the overall trend. This would not rule out the possibility that the overall trend was also due in part to expansion of the variance away from a boundary near which the ancestors lie. For all three characters, Wagner found evidence for both a *passive* trend, owing to expansion of the variance, and for an *active* trend, especially in SW.

A biased *production* of derived character states in one direction, as might be expected if natural selection within species generally favored a similar character in several or many species, is implied if the transitions between ancestors and descendants in various lineages are disproportionately in the same direction. This pattern was displayed for the trend toward lower SW (Figure 24.18).

An active trend due to the evolution of a *constraint* (ratchet) may be implied if the direction of a transition between an ancestor and descendant is correlated with the ancestor's character state (for example, if an increase is more likely to occur in lineages that already have high values than in those with low values). Wagner found that increases in ST were more frequent in descendants of species with high ST than in those of species with low ST.

In species selection, the probability of speciation or extinction is *causally* affected by the state of a character. In species hitchhiking, the character state is merely *correlated* with the speciation or extinction rate, as must be true of all characters that differ between lineages that vary in their rates of speciation or extinction. To distinguish hitchhiking from species selection, Wagner performed computer simulations, mimicking the observed rates of speciation and extinction, to determine how the frequency distribution of each character would shift if it merely went along for the ride. The observed shift in one character, AI, was no greater than expected, according to the simulations, so the trend in this feature may be due to hitchhiking with other characters that did affect the rate of diversification. However, at certain times during the history of this clade, ST and SW were both correlated with heightened survival of lineages, above and beyond what would be expected according the simulations of hitchhiking. Lineages with narrow sinuses (low SW), for example, survived the mass extinction at the end of the Ordovician more frequently than expected by chance.

Thus, this analysis suggests that trends may have multiple causes. A trend may have both active and passive components, as was true of all the characters studied, and it may be driven by multiple dynamic factors. Both individual selection and species selection, for example, seem to have contributed to the trend in sinus width. Macroevolutionary patterns such as trends, therefore, may often have multiple explanations.

Are There Global Trends in Evolutionary History?

It is probably safe to say that no uniform trends can be discerned in the history of evolution. For every proposed trend, exceptions can be cited. Are there, nevertheless, *net* active trends, directions that have been taken by most evolving lineages at most times? Many such trends have

FIGURE 24.17 Trends in three characters of the shells of early Paleozoic gastropods. All the characters are angles, measured in radians. All show an increase in variance, demonstrating that the shift in the mean is partly attributable to a passive trend. The entire distribution of sinus width (C) has shifted, indicating an active trend in that character as well. The shift in aperture inclination (B) was shown to be partly attributable to species hitchhiking. The evolution of shell torque (A) apparently had several causes. (After P. Wagner 1996.)

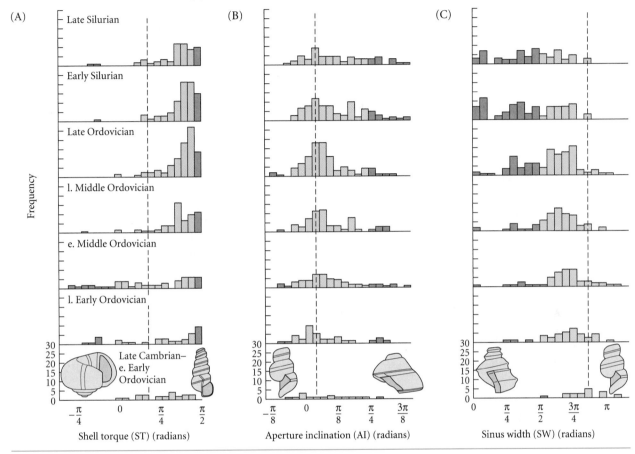

FIGURE 24.18 The trend toward lower sinus width in gastropod shells shown in Figure 24.17 is due partly to biased production of derived character states. The number of decreases was greater than the number of increases among ancestor-descendant comparisons. (After P. Wagner 1996.)

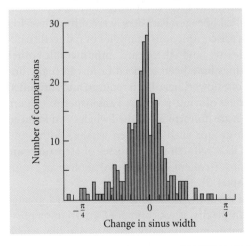

been postulated, but most of the evidence for these trends consists only of anecdotal examples, not analyses of the proportion of lineages that have evolved in one direction or the other. An exception is David Jablonski's (1996) analysis of Cope's rule, the generalization that body size tends to increase in evolution (LaBarbera 1986). Jablonski calculated changes in the mean size of the species in each of a number of clades of gastropod and bivalve molluscs of the Atlantic coastal plain of North America from the late Cretaceous, and found that an active trend toward decrease in size was as common as an active trend toward increase in size (Figure 24.19).

Complexity No one has devised a measure of complexity that can be used to compare very different organisms. In an attempt to capture our intuitive sense of complexity, Daniel McShea (1991) has argued that complexity of structure is proportional to the number of different kinds of parts of which an organism is composed, and to the irregularity of their arrangement (e.g., a simple, repeating pattern such as a checkerboard is not com-

FIGURE 24.19 A test of the generality of Cope's rule of phyletic size increases. Each point in A represents a single genus of late Cretaceous bivalves, and shows the change in both the maximum size and minimum size of species in that genus from the beginning to the end of the interval. Most points would be expected to lie in the upper right quadrant if there were an active trend toward increase in size. The data points (A) and percentages (B) show that there is no such trend. (After Jablonski 1996.)

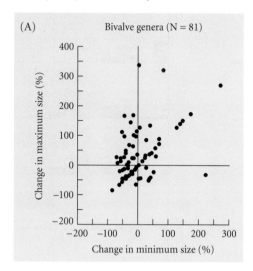

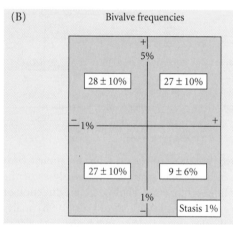

plex compared with, say, an abstract painting by Jackson Pollock).

Among the few features that we can imagine comparing among all forms of life are the DNA content, the number of different coding sequences (genes) in the DNA, and the number of cell types. In all these respects, bacteria, which among living organisms are closest to the presumed common ancestor of all life, are "simpler" than eukaryotes. But since the earliest forms of life must have had small genomes, the subsequent increase in these features may represent merely a passive, perhaps inevitable, expansion of variation and shift in the mean. Whether the increase has continued throughout the Phanerozoic is doubtful. Mammals have more DNA per cell than insects, but hu-

mans have only about 18 percent as much DNA as some salamanders, and only about 2 percent as much as the African lungfish (see Figure 13 in Chapter 3). However, much of the genome of eukaryotes consists of noncoding and/or repetitive DNA. The number of *different* coding sequences is known for very few species—certainly not enough for a phylogenetic analysis to determine whether or not it has generally increased in different lineages. Although the number of recognizably different cell types is greater in animals and plants than in protists and fungi, and may be greater in vertebrates than in most invertebrate phyla (Figure 24.20), we do not know if this feature has increased within any phyla, much less in the animal kingdom as a whole, since the phyla first appeared in the Precambrian.

At the level of multicellular organs, it is far easier to compare complexity within higher taxa than among all animals or all plants. We can, indeed, be confident that the complexity of certain structures has increased in many individual instances, as in the origin and elaboration of the wings of insects and of the teeth and brains of mammals. In many lineages of angiosperms, the flower has evolved from a regular state, with a large and variable number of similarly formed petals, stamens, and carpels (Figure 24.21A), to the condition seen in plants such as peas (Fabaceae), in which the flower parts are reduced in number, but are more differentiated in form, are partly fused to each other, and are more functionally integrated. In peas, a bilaterally symmetrical corolla restricts access to all insects except certain bees, and the stamens are situated so as to deposit pollen on a specific part of the insect's body (Figure 24.21B).

In many such examples, however, there has not been a sustained trend over time, but rather one or a few major episodes in which the feature has evolved from one plateau to another. Thus, the earliest mammals, such as morganucodonts (see Chapter 7), were as complex in their teeth and most other features as later mammals, just as the number of cell types and the complexity of the body plan in each animal phylum has not increased appreciably since the Cambrian. Moreover, complexity has decreased in many individual lineages. Parasites derived from free-living ancestors are conspicuous examples of reduced structural complexity, and there are innumerable other examples. Wings have been reduced or lost in many lineages of insects; flowers have been reduced and simplified in many lineages of wind-pollinated angiosperms (Figure 24.21C); teeth are uniform in toothed whales and reduced or lacking in groups such as baleen whales, sloths, and anteaters. In fact, reductions and losses of structures are among the most frequent events in morphological evolution (see Chapter 5).

There has certainly been a net trend toward greater structural complexity, if we compare Cambrian or post-Cambrian organisms with early Precambrian life, which included only prokaryotes; but there is little evidence that

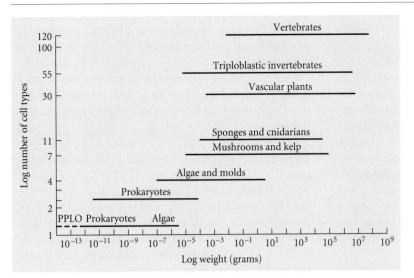

FIGURE 24.20 The number of recognizably different types of cells in various groups of organisms, plotted against the range of body weights. Note that both variables are plotted on a logarithmic scale. Although some groups, such as vertebrates, have more kinds of cells, there is little evidence that the number has increased within these groups over evolutionary time. (After Bonner 1988.)

it has been sustained since the Cambrian. In one of the few tests for net trends within individual higher taxa, McShea (1993) studied the complexity of the vertebral column in several lineages of mammals, and found about equal numbers of increases and decreases. Thus, although many individual instances of increased complexity can be cited,

we do not know if there has been a broad trend during the history of life.

Efficiency of design Innumerable examples exist of improvements in the design of features that serve a specific function. The mammal-like reptiles, for example, show

FIGURE 24.21 Two trends in the evolution of flower form. (A) The ancestral condition of angiosperm flowers, illustrated by the tulip tree *Liriodendron* (Magnoliaceae). This flower has multiple, separate, spirally arranged carpels surrounded by numerous spirally arranged stamens, which in turn are surrounded by radially arranged petals and sepals. (B) In many lineages, flowers have become more complex. The flowers of the pea family (Fabaceae, here illustrated by broom, *Sarcothamnus*) are bilaterally symmetrical, and the parts are fewer and functionally integrated. This flower has a single carpel, ten stamens arranged to contact a specific part of an insect's body, and five bilaterally arranged petals, the lower two of which are fused into a canoe-shaped structure that houses the sexual organs until they are released, as shown here, by a pollinator. (C) Another trend is reduction of the flower, illustrated by wind-pollinated plants such as willows (*Salix*, Salicaceae). The inflorescences contain many small female or male flowers, each consisting only of one or more united carpels or stamens. (A from Mitchell and Beal 1979; B from Scagel et al. 1966; C from Gleason 1952.)

(A)   (B)   (C)

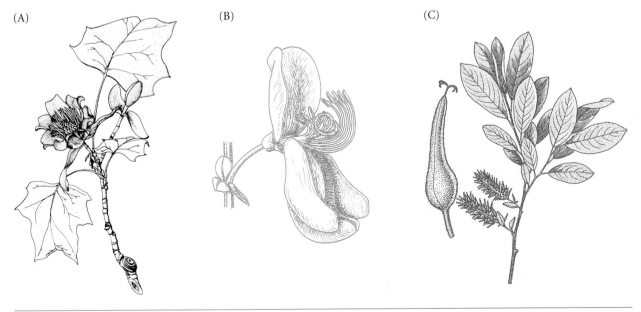

trends in feeding and locomotory structures associated with higher metabolism and activity levels, culminating in the typical body plan of mammals (see Chapter 6). Decreases in efficiency—in adaptation, really—might occasionally occur in small populations in which genetic drift overwhelms the force of selection, but most apparent decreases simply reflect adaptation to new ways of life and new standards of adaptive design. Aphids, for example, seem to be inefficient in assimilating carbohydrates from plant sap, but this is because their growth is limited by the sparser amino acids; they simply do not need the excess carbohydrates. There may well be a global trend toward greater efficiency (Ghiselin 1995). However, efficiency and effectiveness are difficult to measure, and must always be defined relative to the task set by the context—by the organism's environment and way of life, which differ from species to species.

Survival and reproduction

Directional selection among individual organisms consists of the replacement of less fit with more fit genotypes—namely, those with a higher replacement rate. But although the chief components of fitness are individual survival and reproductive rates, we saw in Chapter 19 that neither of these necessarily increases during evolution. Fecundity may evolve to be high, as in oysters and orchids, or low, as in albatrosses and humans; the maximal life span of individuals likewise may be lengthened, as in redwoods and elephants, or shortened, as in annual plants and aphids. Whether the intrinsic rate of natural increase increases or decreases during evolution depends on the ecological context, including population density, within which the species evolves.

Specialization

Evolution of increased specialization in habitat or resource use has occurred frequently, and might well be a net trend (Futuyma and Moreno 1988). In virtually every taxon, species with morphological and physiological adaptations to specialized habitats, resources, and ways of life have evolved from more generalized ancestors: the descendants of the earliest, generalized predatory mammals, for example, include giraffes, anteaters, and bats. In at least some cases, however, the reverse has occurred. For example, phylogenetic analysis has shown that some species of herbivorous insects with broad diets have evolved from more host-specific ancestors, and our own species is perhaps the greatest generalist of all time. The degree of ecological specialization is very difficult to compare among species, and no explicit analysis of trends in specialization have been performed.

Autonomy and homeostasis

Almost the converse of increasing specialization is the possibility that organisms tend to evolve homeostatic mechanisms that provide them with greater independence from the environment (autonomy). Such features enable organisms to function over a greater range of environmental conditions, offering opportunities for greater ecological generalization. Mecha-

nisms of autonomy include physiological homeostasis (such as the ability to acclimate to different temperatures), developmental plasticity, neural and behavioral complexity (including the capacity for learning), and social behavior. We have seen examples of both increases and decreases in such features (see Chapters 17 and 19), each of which must be viewed as adaptive under some circumstances, but not others.

Longevity of species

One might think that if anything should show a trend in the history of evolution, it would be the longevity not of individuals, but of species—the time they persist before succumbing to extinction. Surely species longevity would be a measure of increase in adaptedness. But with a moment's reflection, we can see that this need not be so. The consequence of selection is the adaptation of a population to the currently prevailing environment, not to future environments, so selection does not imbue a species with insurance against environmental change. The common trend toward increased specialization might even imply that lineages tend to evolve a shorter prospect of survival if specialized taxa are at greater risk of extinction than generalists.

Extinctions of taxa in the fossil record can be analyzed, like the deaths of individuals in a population, by plotting the fraction, of those "born" at a particular time, that survive over different time spans. If the probability of extinction is constant, the proportion surviving to time t declines exponentially; plotted logarithmically, the curve becomes a straight line. Two very different such analyses are presented in Figures 24.22 and 24.23. In Figure 24.22, the survivorship of marine animal families that became extinct during the Phanerozoic is plotted. Each curve begins with those families that were present during a geological stage, sets these at 100%, and traces the fraction still alive at each subsequent geological stage. In between mass extinctions (the steep dips that coincide for many curves), the declines are fairly linear, suggesting a fairly constant extinction rate. Moreover, the decline is generally at least as steep late in geological history as it is earlier, suggesting that the average longevity of lineages has not increased.

Figure 24.23 presents some of Leigh Van Valen's (1973) analyses of taxon survivorship. The abscissa represents not geological time, but the *age* of taxa when they became extinct—the interval between the origin and extinction of, say, a genus of ammonites. The rather straight curves suggest, again, an approximately constant probability of extinction, implying that older taxa are no more likely to survive another, say, 10 million years than are younger taxa. This pattern suggests that *as the evolution of a group proceeds, it becomes neither more nor less resistant to new changes in the environment.* Van Valen explained these patterns with his **Red Queen Hypothesis**, which states that the environment of a taxon is continually deteriorating because of the continual evolution of other species (competitors, predators, parasites), so that like the Red Queen

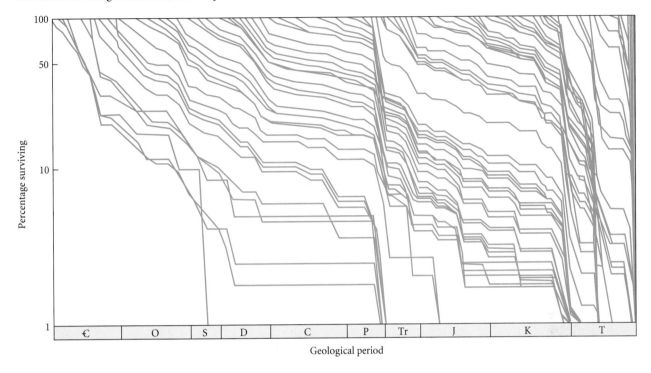

FIGURE 24.22 Survivorship of marine families over Phanerozoic time. Each curve traces the proportion of surviving families among those present during a certain geological stage (thus each stage begins with 100 percent). Steep portions of a curve show intervals of high extinction; shallow portions, intervals of low extinction rates. Note that except for times of mass extinction (such as the end of the Permian, Cretaceous, and Eocene), the slopes do not change appreciably between the early and late Phanerozoic. (After Flessa et al. 1986.)

in Lewis Carroll's *Through the Looking-Glass*, each species has to run (i.e., evolve) as fast as possible just to stay in the same place. There is always a roughly constant chance that it will fail to do so. The pattern, however, is consistent not only with this hypothesis of continual coevolution (see Chapter 18), but also with a continual assault by environmental changes of every kind, each carrying a risk of extinction.

Diversity Diversity—the number of lower-level taxa such as species that are included in a higher taxon—increases if the rate of origination exceeds the rate of extinction. Although many higher taxa have dwindled to extinction, and although several mass extinctions have temporarily reduced diversity, the total number of taxa (species, genera, families) on earth has increased fairly steadily since the beginning of the Mesozoic era (251 Mya). The causes of this increase are discussed in Chapter 25. Note that the trend toward greater diversity describes a collective property of life, and thus is qualitatively different from trends in the properties of individual organisms or species, such as body size or structural complexity.

The Question of Progress

Before Darwinian evolution became widely accepted, the Great Chain of Being, or *Scala Naturae*, was one of the most enduring and powerful of Western concepts (see Chapter 1). All living things were arranged on a linear scale surmounted by the human species—or European man, to be more exact. Darwinism did not erase this world view. Western society "temporalized" the chain: many of those who accepted evolution, including many biologists, and even Darwin himself in unguarded moments, interpreted the history of evolution as a grand progress upward through the "higher" animals toward the emergence of mankind. Many people who accept evolution still conceive it as a purposive, progressive process, culminating in the emergence of consciousness and intellect. "Man is the measure of all things," said the Roman philosopher Protagoras—a sentiment widely shared. Even some evolutionary biologists have taken some of the qualities we most prize in ourselves, such as intellect or empathy, as criteria of progress, and, quite ignoring the innumerable lineages that have not evolved at all in these directions, have seen in evolution a history of progress toward the emergence of humankind. (For some history of these ideas, see Nitecki 1988.)

In considering whether evolution embodies "progress," we should guard against this parochial anthropocentrism. If we were to recognize only one indisputable generalization about evolution, it would have to be the fact of divergence, of diversification. Were a rattlesnake, an orb-web spider, or

FIGURE 24.23 Taxonomic survivorship curves for (A) families and genera of Ammonoidea and (B) extinct species of Foraminifera. Each point represents the number of taxa that persisted in the fossil record for a given duration, irrespective of when the taxa originated during geological time. The approximate linearity of these semilogarithmic plots suggests that the probability of extinction is independent of the age of a taxon. (After Van Valen 1973.)

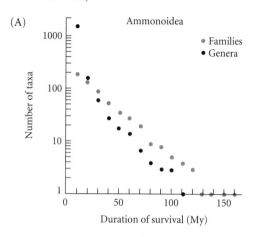

(A)

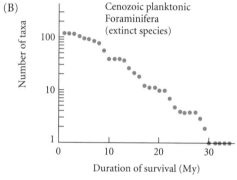

(B)

a gymnotid fish capable of conscious reflection, it doubtless would measure evolutionary progress by the elegance of an animal's venom delivery system, its web architecture, or its ability to communicate by electrical signals. The great majority of animal lineages, to say nothing of plants and fungi, show no evolutionary trend toward greater "intelligence" (however it might be defined and measured), which must be seen as a special adaptation appropriate to some ways of life, but not others. Nor must we suppose that humans, or any creature with our mental faculties, were des-

tined inevitably to evolve, any more than were ammonites or vampire bats. Natural selection and mutation utterly lack foresight and purpose. Every step in the long evolutionary history of humans—or vampire bats—was contingent on genetic and environmental events that could well have been otherwise; had dinosaurs not become extinct, for instance, the adaptive radiation of the mammals might never have occurred (Gould 1989). As far as we can tell, it is only by sheer accident that our ancestors escaped extinction—especially mass extinctions such as those that ended the Permian and Cretaceous periods. [Such considerations have led reflective evolutionary biologists to conclude both that humanity's existence has never been predetermined in any sense, and that the likelihood that any other "humanoid" species exists elsewhere in the near reaches of the universe—especially any with which we have the faintest hope of communicating—is negligible (Simpson 1964; Mayr 1988b).]

"Progress" usually implies directional movement toward a goal, which in turn implies foresight. None of the mechanisms of evolution has foresight, and the process of evolution cannot have a goal, any more than the universe can have the goal of achieving maximal entropy. But is it nevertheless possible that the history of evolution is one of progress in the sense of improvement, even if this is not a goal?

The problem is that improvement requires a criterion of measurement. We might take "fitness" as such a criterion, since natural selection is supposed generally to increase it. But fitness increases only relative to previously common genotypes, and only relative to the context within which selection proceeds. Within the context of grazing on grasses, for example, horses with higher-crowned teeth may well represent an improvement over those with lower crowns, but this change did not make them better than those species that browsed shrubs. In other words, we can provide many instances in which effectiveness at accomplishing some task has increased, but these tasks differ for each species.

It is difficult, if not impossible, to specify a universal criterion by which to measure improvement that is not laden with our human-centered values. Most, although not all, evolutionary biologists have therefore concluded that we cannot objectively find progress or improvement in evolutionary history (see Ruse 1993, 1996; Nitecki 1988).

Summary

1. Higher taxa arise not in single steps, by macromutational jumps (saltations), but by the acquisition, often sequentially, of genetically independent characters (mosaic evolution). Most such characters evolve gradually, through intermediate stages, but some characters evolve discontinuously due to gene substitutions with large effects and developmental phenomena such as thresholds.

2. New features commonly are advantageous even at their inception. Most frequently, they are derived by modification of pre-existing characters to serve new functions. Highly complex characters such as the vertebrate eye can often be shown to have evolved gradually, through functional intermediate stages.

3. Some changes in complex characters are "key innovations" that provide adaptation to substantially different

ecological niches or adaptive zones. A frequent precondition for the evolution of such features is the decoupling of their functional and structural integration with other organs, so that they become free to evolve independently. Another important source of evolutionary novelty is the functional and structural divergence of repeated (serially homologous) structures.

4. Rates of evolution can be measured either by the turnover (extinction and origination) of taxa within a clade or by the rate of change in individual characters. The latter approach shows that on average, evolutionary rates are so low that known processes such as mutation, genetic drift, natural selection, and speciation can readily account for them. The average rate of evolution of most characters is very low because long periods of little change ("stasis") are averaged with short periods of rapid evolution; but even the rapid shifts can readily be explained by known processes such as natural selection.

5. "Punctuated equilibrium" is a description of a pattern in the fossil record in which one long-stable (static) phenotype shifts rapidly to a related, but distinguishable, phenotype. "Punctuated equilibrium" is also the hypothesis that such shifts represent morphological changes that accompany, and are largely restricted to, speciation in small populations. This hypothesis has been criticized because responses to selection do not depend on speciation. However, we do not know which of several explanations for stasis are the most important.

6. A corollary of the hypothesis of punctuated equilibrium is that long-term trends in morphology may be the consequence of differences in rates of speciation and extinction among clades (species selection, *sensu lato*), rather than of natural selection within species. Other explanations for such trends include association, due to phylogenetic history, with characters that affect speciation or extinction rates ("species hitchhiking"); the origin of a character near a boundary, due to history or the limits of function; individual selection, originating in functional efficacy or in coevolution within and among species; and "irreversibility" due to selective and developmental constraints.

7. Probably no feature exhibits a trend common to all lineages. There may have been net trends in features such as genome size, body size, structural complexity, and degree of specialization, but it is not known whether the number of lineages displaying each such trend exceeds the number that do not. There exists no clear evidence for an increase in the longevity of species or higher taxa in geological time.

8. If "progress" implies a goal, then there can be no progress in evolution. Eschewing goal-directedness, we still cannot identify objective criteria by which the history of evolution can be shown to be one of "improvement." Characters improve in their capacity to serve certain functions, but these functions are specific to the ecological context of each species. By no *objective* standard can we say that amphibians are superior to fishes, or humans to frogs.

Major References

Rensch, B. 1959. *Evolution above the species level.* Columbia University Press, New York. A somewhat outdated but insightful treatment of many of the topics of this chapter, by a key contributor to the Evolutionary Synthesis.

Nitecki, M. M. (editor). 1990. *Evolutionary innovations.* University of Chicago Press, Chicago. A symposium on the topic.

Wake, D. B., and G. Roth (editors). 1989. *Complex organismal functions: Integration and evolution in vertebrates.* Wiley, Chichester, England. Proceedings of a workshop on the evolution of complex functional systems, using vertebrates as examples.

Levinton, J. S. 1988. *Genetics, paleontology, and macroevolution.* Cambridge University Press, Cambridge. Discusses many of the topics of this chapter.

Problems and Discussion Topics

1. Snapdragons, like other species of the family Scrophulariaceae, have bilaterally symmetrical flowers, derived from the radially symmetrical condition. A mutation in the *cycloidea* gene makes snapdragon flowers radially symmetrical. The *cycloidea* gene has been mapped and sequenced, and was found to be asymmetrically expressed in the developing flower, where it probably regulates other genes that contribute to flower development (Luo et al. 1996). Is it possible to determine whether this gene was the primary basis of the original evolution of bilateral symmetry in the Scrophulariaceae? How could one determine whether bilateral symmetry evolved continuously through intermediate stages or by a single discrete ("saltational") change at this locus?

2. Despite the close relationship between humans and chimpanzees, and despite the genetic variation in most of their morphological characteristics, there are no records of humans giving birth to babies with chimpanzee morphology. Would you expect this to occur, if humans evolved gradually from apelike ancestors? If not, why not?

3. Find and evaluate the evolutionary literature that describes how wings could have evolved gradually, through advantageous intermediate steps.

4. There is some evidence that duplication and divergence of Hox genes has been associated with the evolution of more complex body plans (see Chapter 23). What is the difference between the hypothesis that such changes in Hox genes *enabled* (or permitted) the evolution of complexity, and the hypothesis that they *caused* it? What data might provide evidence for each hypothesis?

5. Functional complexity often evolves as different features become "coadapted" to each other, so that high fitness comes to depend on their joint occurrence and interaction. Describe how this process might lead to "irre-

versible" evolution. Under what conditions might reversal occur, or complexity decrease? Provide examples.

6. Chapters 9–14 described examples of rapid evolution and provided evidence on genetic variation that might lead one to expect that evolution should often be rapid. This chapter, however, provides evidence that morphological evolution is often very slow. What hypotheses could account for low rates of morphological evolution? What evidence might favor each of them?

7. What is the theoretical basis for the hypothesis of punctuated equilibrium? How valid is the theory? What are the theoretical objections to punctuated equilibrium? (Review the discussion of speciation by peak shifts in Chapter 16. References cited in this chapter might also be consulted.)

8. What evidence would be required to support or refute the hypothesis that a trend was caused by species selection rather than individual selection?

9. List some possible criteria for complexity and for progress. Now compare lobsters, earthworms, honeybees, sheep, rattlesnakes, and humans. Can these organisms be objectively arrayed along one or more scales according to these criteria? (You may need to do background research on their morphology or other characteristics.)

10. Stephen Jay Gould (1989) and others have argued that the evolution of a self-conscious, intelligent species (i.e., humans) was historically contingent: it would not have occurred had any of a great many historical events been different. The philosopher Daniel Dennett (1995) and others have disagreed, arguing that convergent evolution is so common that if humans had not evolved, some other lineage would probably have given rise to a species with similar mental abilities. What do you think, and why? If Gould's position is right, what are its philosophical implications, if any?

CHAPTER 25

The Evolution of Biological Diversity

Few topics in biology are more challenging than biological diversity. Why are there more species of rodents than of primates, of flowering plants than of ferns? Why do tropical forests have many more species of plants, insects, and birds than temperate zone forests? How has the diversity of species changed over time, and why?

Answering such questions requires integrating information from ecology, systematics, paleontology, and geology with evolutionary theory.[*]

Diversity can be approached from the perspectives of ecology or of evolutionary history, which are not mutually exclusive (Ricklefs and Schluter 1993a). Ecologists focus primarily on factors that operate over short time scales to influence diversity within local habitats or regions (Diamond and Case 1986; Strong et al. 1984b; Huston 1994). On a scale of thousands of years, however, changes in climate may cause changes in the species composition of communities. And on a scale of millions of years, extinction and speciation, as well as climatic, plate tectonic, and other changes, alter the numbers of species and their geographic distributions. In explaining patterns of species diversity, then, ecologists must consider not only short-term ecological processes, but also long-term changes that determine the potential pool of species from which the members of a community might be drawn.

The long-term evolutionary approach, adopted chiefly by paleontologists and systematists, seeks to describe and explain patterns of change in diversity, and in the origination and extinction of taxa, on a scale of millions of years. This approach invokes not only ecological processes such as interspecific interactions, but also changes in climate, sea level, and the configuration of continents, and even collisions with extraterrestrial bodies, to explain major extinctions, evolutionary radiations, and variation in diversity.

*Chapters 4 (Ecology), 7 (A History of Life on Earth), and 8 (The Geography of Evolution) provide especially important background to this chapter.

Measuring Diversity

Two major approaches to studies of diversity can be distinguished by their methodology. Community ecologists generally go out into the field and sample taxa of interest (e.g., plants or birds or copepods) within a narrowly circumscribed area (perhaps a lake or a plot of several hectares of forest). On larger geographic scales, systematists and paleontologists often estimate the diversity in a region by compiling existing records, such as the publications or museum specimens that have been accumulated by many investigators, into faunal or floral lists of species.

Simple counts of species (referred to as SPECIES DIVERSITY or SPECIES RICHNESS) are often used to express diversity. At the level of the local community, most species are relatively unambiguously definable, but the count depends strongly on the intensity of sampling.

On a broad geographic scale (the scale usually treated by compiling faunal or floral lists), it is often ambiguous whether allopatric populations are different species or not (see Chapter 15), so species counts are sensitive to the practices of the systematists who have studied the group. Because higher taxa such as genera or families often provide a more complete fossil record than individual species, most paleontological studies of diversity employ counts of higher taxa. Counts of fossil taxa may be inaccurate if many of them are chronospecies (see Chapter 15), which may become "extinct" by pseudoextinction—i.e., by evolving into a form that is given a different name. Also, many higher taxa, especially extinct taxa, are paraphyletic groups defined by morphological distinctness from other taxa that arose from them (e.g., the Archosauria "should," but do not, include birds along with dinosaurs). Because strictly cladistic classifications do not permit paraphyletic groups (see Chapter 5), some cladists have attacked the practice of counting higher taxa (e.g., Smith and Patterson 1988). However, patterns of change in diversity are not greatly affected by including paraphyletic taxa (Sepkoski and Kendrick 1993).

703

Despite such pitfalls, the major patterns we shall discuss appear to be so pronounced as to be robust to these problems. There is no doubt, for example, that tropical forests generally have far more species of trees than high-latitude forests. The overall pattern of changes in diversity through the Phanerozoic has hardly been altered at all by numerous recent discoveries and changes in the taxonomy of fossil invertebrates and vertebrates, although the total number of known taxa has increased (Sepkoski 1993; Benton 1994).

Ecological Approaches to Contemporary Patterns of Diversity

Factors That Influence Diversity

The elementary model of island biogeography (see Chapter 4) suggests that the equilibrium number of species on an island (or in a specific local region) is determined by the rate of extinction relative to the rate of colonization by new species from a pool—perhaps in a nearby region—of potential colonizers. The species pool is determined by *long-term* historical factors: the origination, extinction, and movement of species in the past. The species from this pool that we find at a given locality are determined both by *local short-term processes* and by *"mesoscale" regional processes* that operate on a somewhat longer time scale (Holt 1993; Ricklefs and Schluter 1993b).

The *local processes* that can drive local populations to extinction (see Chapter 4) include (1) severe climatic events such as harsh winters and other density-independent catastrophes such as floods; (2) changes in the habitat; (3) predation, parasitism, and disease; (4) competition from other species; (5) extirpation of prey or other critical resources; and (6) random fluctuations in birth and death rates, which are especially likely to extinguish small populations. (In extreme cases, inbreeding depression may also hasten the extinction of small populations.) All of these factors are more important, the smaller the area occupied by the population; for example, extinction rates are higher on small than on large islands (Mayr 1965; MacArthur and Wilson 1967; Diamond and Case 1986).

Resource partitioning increases the likelihood that resource-limited species can coexist. This pattern implies some degree of specialization, which can increase susceptibility to extinction if the species' special resources fluctuate in abundance. For this reason, many authors, beginning with Alfred Russel Wallace (1878), have suggested that climatic stability and/or predictability, by ameliorating fluctuations in resources, should promote the coexistence of many species (Pianka 1966, 1983). However, evidence that climatic stability explains major patterns of community diversity is sparse (see Chapter 8). Competitive exclusion proceeds more slowly, the more similar species are in their competitive ability. Thus if the aggregate population of several very similar species is very large, one will ultimately dominate (by a random walk process analogous to genetic drift), but it may take a very long time for the process to run its course. This hypothesis might explain the high diversi-

ty of tree species in tropical forests (Hubbell and Foster 1983; but see Ashton 1969).

High local species diversity seems to be favored by intermediate levels of disturbance, because many species persist by using patches of habitat from which disturbance has removed superior competitors. Excessive disturbance, of course, reduces diversity.

Primary productivity, which in terrestrial environments is correlated with temperature and rainfall, determines the biomass of consumers (animals) and decomposers. High productivity might sustain high species diversity, because the population density of top carnivores and other rare species might be so low in unproductive environments that they would have a high rate of extinction. In general, however, species richness of both plants and animals appears to be greatest at *intermediate* productivity levels (Figure 25.1), for reasons that are not fully understood (Rosenzweig and Abramsky 1993; Huston 1994).

The persistence of species in a local community may depend on *mesoscale* processes in the larger surrounding region (Holt 1993; Pulliam 1988). Many species require different habitats in different seasons or at different stages of the life cycle. Because local populations become extinct, persistence of a species in a local region depends on recolonization from other populations. The importance of these factors has been made clear by the many studies that have shown a decline in numbers of species on islands or in patches of fragmented habitat where immigration has been so reduced that it is insufficient to maintain positive population growth or to compensate for extinction (Diamond 1984b).

Are Communities Saturated with Species?

Whether or not communities are typically in a state of ecological equilibrium is an important question that bears on how variation in diversity, in space and time, should be in-

FIGURE 25.1 Species diversity of plants at sites in Israel, Turkey, and Spain in relation to primary productivity. Productivity is correlated with rainfall. Diversity is greatest at intermediate productivity levels. (After Rosenzweig and Abramsky 1993.)

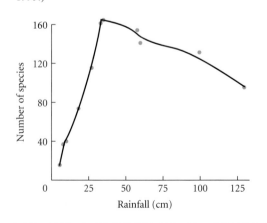

terpreted. An equilibrium community is one that cannot be invaded by more species. One possible reason for equilibrium is that all resources are fully utilized and partitioned among competing species. An equilibrium community can arise from processes of *assembly* and of *coevolution*. In the process of ASSEMBLY, an area is stocked by sequential invasions of species from a large regional pool; each invader persists only if it does not fully overlap in resource use with superior competitors. In the process of coevolution, species evolve adjustments in resource use so that they do not fully overlap, and new species, adapted to underutilized resources, may arise (see Chapter 18).

Several kinds of observations described in Chapters 4 and 8 suggest that communities may be near equilibrium: experimental evidence of interspecific competition in nature; patterns of geographic replacement of similar species (e.g., "checkerboard" patterns); resource partitioning among ecologically similar sympatric species; and the convergence of resource partitioning patterns among independently evolved communities on different islands or continents (Cody and Diamond 1975).

However, such patterns are not universal, suggesting that not all communities are at equilibrium.

The relationship of local to regional species richness
If species richness in a local community is limited by competition and other interspecific interactions within the community, it does not depend on the species richness of the larger surrounding region. If, however, interspecific interactions play a minor role, local species diversity depends on the relative rates of local extinction versus colonization of species from elsewhere in the region; the local species constitute a sample of the regional pool, and their number should be proportional to the number of species in the pool. Thus, "interactive" and "noninteractive" communities can be contrasted, and can be distinguished by plotting local species richness against regional species richness across a number of sites that vary in regional diversity (Figure 25.2A; Cornell 1993). For example, the parasite "communities" residing in different species of North American sunfishes (Figure 25.2B) conform to the interactive model, whereas the species diversity of cynipid gall wasps that attack oaks in California fit the noninteractive model (Figure 25.2C). If noninteractive community structures should prove to be common, communities could hold more species than they actually have.

Community convergence
Chapter 8 presented several striking examples of independently assembled communities that have similar organisms and are similar in number of species and in the way they partition resources (see also Chapter 18). In some cases, communities have developed similar structure even though, for historical reasons, they differ in overall diversity (Schluter and Ricklefs 1993). For example, the diversity of lizards in Australia is much higher than that in southern Africa, no matter what habitat type is examined. Nevertheless, wetlands have fewer species than

FIGURE 25.2 Distinguishing interactive from noninteractive communities. (A) Theoretical curves of species richness at local sites versus broad geographic regions. The total regional richness sets an upper boundary on local richness. Interactions among species set a ceiling on the richness at a local site. If interactions do not limit local richness, the number of species at a local site should increase steadily with the number in the broader region (proportional sampling). (B) A similar plot for parasites of sunfishes. "Regional richness" is the total number of parasite species recorded from a species of sunfish (named above the points) throughout the fish's range. The number of parasites found in local populations of a fish species varies, but appears to have a ceiling, as in the interactive model. (C) A similar plot, in which each point represents the number of species of cynipid gall wasps on seven species of oaks (as numbered), each sampled in several localities. These data conform to the noninteractive model. (A, C after Cornell 1985; B after Aho and Bush 1993.)

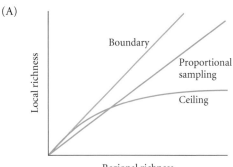

(A)

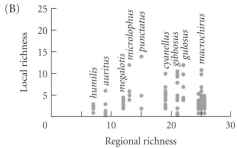

(B)

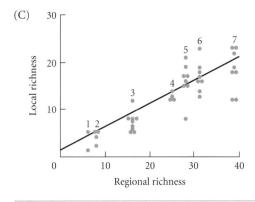

(C)

deserts in both regions (Figure 25.3A), suggesting that habitat differences have consistent effects on ecological or evolutionary processes that affect diversity. On the other hand, the numbers of bird species in habitats that differ in foliage height diversity (e.g., shrubland versus forest), al-

The Evolution of Biological Diversity 705

FIGURE 25.3 Evidence for and against convergence of community structure. (A) Due to differences in evolutionary history, Australia has more species of lizards than southern Africa, but the greater number of species in deserts than in wetlands in both regions indicates convergence of community structure. (B) Sites that differ in the structural diversity of vegetation differ in the number of breeding bird species. The relationship is similar in North America (shaded-color circles) and Australia (solid-color circles), indicating convergence, but is dissimilar in southern South America (black circles), indicating lack of convergence. (After Schluter and Ricklefs 1993.)

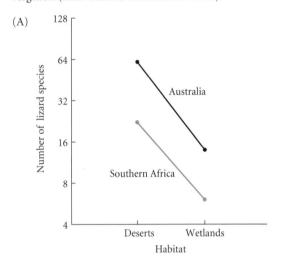

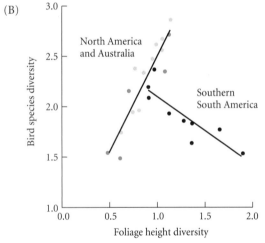

though exceedingly similar between Australia and North America, display a radically different relationship in southern South America (Figure 25.3B). Such a community might well be open to the addition of more species.

Effects of History on Contemporary Diversity Patterns

Classic biogeographers, especially those with a systematic or paleontological background, generally invoked long-term evolutionary history to explain global patterns of variation in species diversity (e.g., Darlington 1957; Fischer 1960). Although ecologists have generally attempted to explain such

patterns as effects of short-term ecological processes, long-term evolutionary events have greatly affected patterns of contemporary diversity (Ricklefs and Schluter 1993b). The species diversity of trees in the north temperate zone strikingly illustrates the effects of history (Latham and Ricklefs 1993). The three major regions occupied by the moist temperate forest biome are Europe, northeastern Asia, and eastern North America. Although these regions contain forested areas in a ratio of 1:1:1.3, the ratio of numbers of tree species is 1:6:2. These differences in species diversity are paralleled by the diversity at higher taxonomic levels. A greater proportion of taxa in temperate Asian forests belong to primarily tropical groups than in Europe or America. These differences are not correlated with contemporary patterns of climate.

For about the first 40 My of the Cenozoic, the earth was generally warmer than it is today. Forests were spread across northern America and Eurasia, and many genera were distributed more broadly than they are today. The temperate flora of North America was separated from the tropical American flora by a broad seaway, and that of Europe was disjunct from the African flora (Figure 25.4); but the northern Asian flora graded into the tropical flora, as it does today, from Siberia to the Malayan Peninsula. Thus, in Asia there was greater opportunity for tropical lineages to spread into and adapt to more temperate climates. Probably for this reason, northeastern Asia in the Tertiary had more genera of trees than either Europe or eastern North America.

In the late Tertiary and the Quaternary, global cooling culminated in the Pleistocene glaciations, which extended farther south in Europe and eastern North America than in Asia. These glaciations devastated the flora of America and especially of Europe, where southward migration was blocked by the Alps, the Mediterranean Sea, and deserts. The continuous corridor to the Asian tropics, however, provided refuge for the Asian flora. A far greater proportion of genera became extinct in Europe and North America than in Asia. Thus contemporary differences in diversity among these regions appear to have been caused by (1) a long history, extending through much of the 65 million years of the Cenozoic, of differences in opportunity for invasion, adaptation, and diversification, and (2) a recent history of differential extinction.

The History of Diversity and Its Causes

Most of the paleontological data used to study the history of diversity are counts of taxa inferred to have existed, originated, or become extinct during various time intervals throughout the Phanerozoic. Because the fossil record is incomplete, such data are often imprecise, and can sometimes be biased (i.e., misleading). The dating of fossils is often accurate to only about 5 to 6 My. Estimates of the diversity during different geological stages and periods may be biased

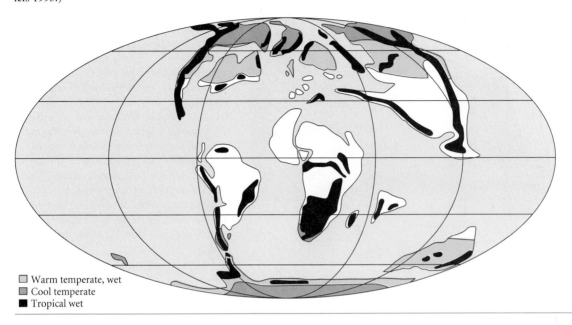

FIGURE 25.4   The distribution of the warm temperate (colored areas) and cool temperate (gray areas) biomes at the end of the Cretaceous. Note that these biomes extended much farther north in the Cretaceous than they do at present (see Figure 2 in Chapter 4 for contemporary biomes). White regions show land areas with other biomes. (After Latham and Ricklefs 1993.)

☐ Warm temperate, wet
☐ Cool temperate
■ Tropical wet

because they vary in duration, and because more recent geological times are represented by greater volumes and areas of fossiliferous rock (Raup 1972; Signor 1985). A special bias is created by fossil taxa that are still alive today. Because living taxa are far better known than those from previous times, they will have apparently longer durations and lower extinction rates than those known only from the fossil record. Thus diversity will seem to increase as we approach the present, a bias that David Raup (1972) called the "pull of the Recent." Paleobiologists have developed methods to *estimate* diversity from the apparent numbers. For instance, sampling bias may be estimated by calculating the proportion of "Lazarus taxa" in a sample—those that are "resurrected" after a long geological interval in which they were not found (Jablonski 1995). Such taxa provide an estimate of the proportion of taxa that were present during an interval, but are not actually documented by fossils from that time.

Taxonomic Diversity through the Phanerozoic

By far the most complete fossil record has been left by marine animals with hard parts (shells or skeletons). From the paleontological literature, John Sepkoski (1984, 1993; Sepkoski and Hulver 1985) has compiled data on the stratigraphic ranges of more than 4000 marine skeletonized families and 20,000 genera throughout the 543 million years of the Phanerozoic. His plot of the diversity of families (Figure 25.5A) shows a rapid increase in the early Cambrian, coinciding with the first appearance of most animal phyla (Box 25.A). After the Cambrian, *several substantial decreas-*

es (notably at or near the end of the Ordovician, Devonian, Permian, Triassic, and Cretaceous) interrupt an *overall pattern of increase followed by relative constancy in the Paleozoic,* and then a *fairly steady increase throughout the Mesozoic and Cenozoic, reaching an all-time maximum in the Pliocene and Pleistocene.* The number of marine invertebrate species is believed to have increased as well (Signor 1985), averaging perhaps 40,000 in the Paleozoic and Mesozoic and about 240,000 in the Cenozoic (Valentine et al. 1978). Individual phyla and classes (Figure 25.5 B–C) show a variety of "diversity profiles"—i.e., patterns of change in diversity.

Sepkoski (1984) has used statistical methods to distinguish three EVOLUTIONARY FAUNAS in the marine realm, the components of which overlap in time. Trilobites, inarticulate brachiopods, and several other groups make up the Cambrian fauna; a later Paleozoic fauna is dominated by articulate brachiopods, crinoids (sea lilies), and others; and a Modern fauna, composed largely of gastropod and bivalve molluscs, sea urchins, malacostracan crustaceans (shrimps, crabs), and bony fishes, dominates the Mesozoic and Cenozoic (Sepkoski and Miller 1985).

On land, the estimated species diversity of vascular plants (Figure 25.6A) increased only modestly from the Carboniferous to the mid-Cretaceous, after which it *steadily and dramatically increased as the angiosperms diversified* (Niklas et al. 1980). Insect diversity shows a *steady increase since the Permian* (Figure 25.6B), attributable, in large part, to the low rate at which insect families have become extinct (Labandeira and Sepkoski 1993). The pattern for terrestrial vertebrates is similar to that for plants (Benton 1990;

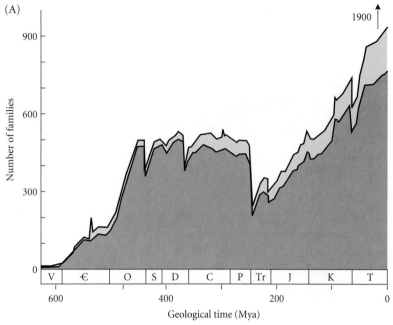

(A)

1900 ↑

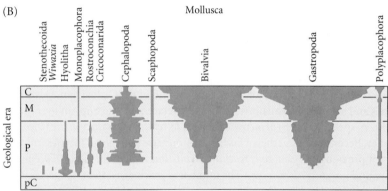

FIGURE 25.5 (A) Taxonomic diversity of marine animal families during the Phanerozoic. The lightly shaded area represents families that are rarely preserved; the solid area, families that have a more reliable fossil record. The arrow and the figure "1900" indicate the approximate number of marine animal families today, including those rarely preserved as fossils. Symbols along the *x*-axis represent geological periods. (V, Vendian; €, Cambrian; O, Ordovician; S, Silurian; D, Devonian; C, Carboniferous; P, Permian; Tr, Triassic; J, Jurassic; K, Cretaceous; T, Tertiary.) (B) A spindle diagram of changes in the number of known families in various classes of marine arthropods from the Precambrian (below) through the Cenozoic (above). The width of a spindle at any point is proportional to the number of families at that time (C) The same, for classes of Echinodermata. (A after Sepkoski 1984; B, C after Bambach 1985.)

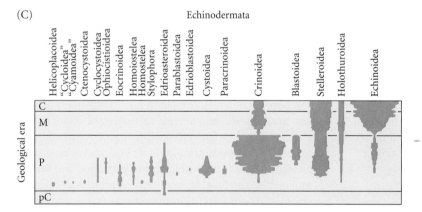

Figure 25.6C). Mammals and birds account for the great increase during the Tertiary.

The overall pattern of Phanerozoic diversity has been one of rapid increase in the early Paleozoic, a lower rate of increase during the middle and late Paleozoic, and a precipitous drop during the great mass extinction at the end of the Permian, followed by a fairly steady increase thereafter, from the Triassic to the late Cenozoic. Underlying this pattern is a history of origination and extinction of various taxa, some of which we

noted in Chapter 7. We shall now examine these patterns, and their possible causes, in more detail.

Patterns of Origination and Extinction

Null Hypotheses

A change in the number (*N*) of taxa is a consequence of the rates, per taxon, of origination of new taxa (*S*) and of ex-

FIGURE 25.6   Changes in the number of known (A) species of vascular land plants, (B) families of insects, and (C) families of nonmarine tetrapod vertebrates. (A, C after Benton 1990; B after Labandeira and Sepkoski 1993.)

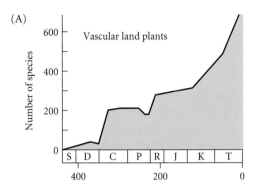

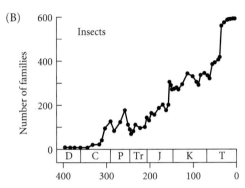

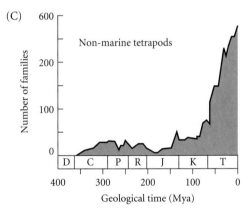

fluctuations in rates of origination and extinction. Several investigators have employed computer simulations to see how diversity might change according to this "null hypothesis" (Raup et al. 1973; Gould et al. 1977). The patterns of change in diversity observed in the fossil record often differ significantly from those exhibited by these "random clades" (Stanley et al. 1981). For example, Peter Ward and Philip Signor (1985), in an analysis of 83 families of Jurassic and Cretaceous ammonites, found that uniformity of diversity through time is not correlated with the average duration of survival of genera within families, whereas there is such a correlation in "random clades."

Origination and Extinction Rates through Time

Sepkoski's study of skeletonized marine taxa, which measured origination and extinction rates as the percentage of taxa in a geological interval that appeared in that interval for the first or last time, showed *higher rates of origin of new families in the Cambrian and Ordovician than at any subsequent time* (Sepkoski 1993: Figure 25.7). Raup and Sepkoski (1982), by calculating either the percentage or the number of families becoming extinct per million years, concluded that extinction rates seem to have generally declined since the Cambrian. However, this pattern of so-called "normal" or **background extinction** is punctuated by *five episodes of exceptionally high extinction rates*, the so-called **mass extinctions** at the end of the Ordovician, the late Devonian, the Permian/Triassic (P/Tr) boundary, the end of the Triassic, and the Cretaceous/Tertiary (K/T) boundary (Figure 25.8). Gilinsky and Good (1991) confirmed most of these conclusions, but could not statistically demonstrate an overall decline in background extinction probability.

FIGURE 25.7   Origination rates of marine animal families during the Phanerozoic. The vertical axis represents the percentage of families recorded from a geological stage that originated during that stage. The colored line represents an analysis performed in 1982; the black line, an analysis performed in 1992, based on more extensive data. (After Sepkoski 1993.)

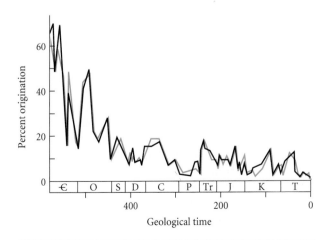

tinction (*E*). The net rate of increase (R) in the number of taxa per unit of time (e.g., per million years) is *S - E*. By analogy to simple population growth, the change (Δ*N*) per time interval (Δ*t*) may be written

$$\frac{\Delta N}{\Delta t} = N(S - E) \quad or \quad \frac{\Delta N}{\Delta t} = RN$$

A set of taxa (a genus with *N* species, say, or a class with *N* families) *may become extinct if R < 0, i.e., if the "per capita" extinction rate (E) rises* or *the origination rate (S) falls*, so that *S < E*.

Just as allele frequencies within a population may change by random genetic drift rather than natural selection, changes in a taxon's diversity may arise from random

An Enigma: The Evolution of Animal Body Plans

Disparity, the degree of anatomical difference among organisms, is greatest among higher taxa. In metazoan animals, it is most pronounced among classes, and especially among phyla. The disparity of animal life presents two knotty problems: first, accounting for its sudden origin, and second, determining whether or not it has increased over time.

The earliest recorded multicellular animal life is the Ediacaran fauna of the late Precambrian (Vendian), about 560 Mya, consisting largely of surface-dwelling, soft-bodied animals that may or may not be metazoans. Of the modern animal phyla in the late Vendian and the early Cambrian that have skeletonized body parts, all (except Bryozoa) appear in an interval of perhaps 30 My or less (Bowring et al. 1993). Moreover, this "Cambrian explosion" includes a great variety of organisms (such as *Hallucinogenia* and other peculiar members of the Burgess Shale fauna: see Chapter 7) that seem not to fit into any modern phyla. Classes of skeletonized marine animals also originated in a relatively short time (Figure 25.A1; Erwin et al. 1987). More of these classes became extinct in the early Paleozoic than originated later, so there were more in the early Paleozoic than since.

The well-preserved Ediacaran fauna lacks modern phyla (except for coelenterates), and fossilized burrows (trace fossils) of the kind made by many Cambrian organisms appear only in the late Vendian. Many authors therefore believe that the diverse body plans of animal phyla really evolved in a very short span of time.

Darwin and many other authors, however, have proposed that the apparent "explosion" of body plans is an artifact of preservation: that the fossil record is too poor to have recorded their earlier history. Gregory Wray and colleagues (1996) have argued for this position on the basis of molecular data. They applied the relative rate test (see Chapter 5) to DNA sequences of homologous genes from chordates, arthropods, and other phyla, using plants, fungi, and bacteria as outgroups. According to their analysis, the rate of sequence divergence among phyla has been fairly constant, so that the sequences can be used as a "molecular clock." They calibrated the "clock" by the earliest fossils of several classes within phyla, and used the calibration to estimate the times at which the phyla diverged. They concluded that chordates and echinoderms diverged about 1000 Mya (1 billion years ago), and that these deuterostomes diverged from arthropods and other protostomes about 1200 Mya—nearly twice as long ago as the beginning of the Cambrian period. Wray et al. argued that the morphological features, especially skeletons and shells, that characterize recognizable members of these lineages may well have evolved much later, in lineages that had become separate long before. Whether or not their conclusions will withstand challenges remains to be seen.

Even if the hard parts that contribute to the diverse body plans of different phyla evolved long after the lineages diverged, the body plans themselves evidently evolved very rapidly. Why and how this happened is one of the great mysteries of evolutionary biology (Valentine et al. 1991; Erwin 1993b). Both ecological and developmental hypotheses have been proposed. Some hold that diversification was stimulated by the increase in atmospheric oxygen, and probably in primary productivity, that occurred in the late Precambrian (Vermeij 1987). Others have suggested that vacant ecological niches, or "adaptive space," fostered diversification. "Adaptive space," however, was nearly as vacant after the end-Permian extinction as in the Precambrian, but no new classes or phyla appeared then (Erwin et al. 1987). Erwin (1993b) and Valentine et al. (1991) have suggested that the answer lies in the evolution of key innovations that established multicellularity and the organizational properties that control metazoan development. These properties include collagen and other molecules that enable cells to adhere to each other; epithelial sheets of cells and the basement membranes and other systems that regulate their behavior during development; and the hierarchies of gene action responsible for the differentiation of cell types and the positional information that governs this process (e.g., Hox genes: see Chapter 23). Erwin has suggested that the evolution of these developmental capacities provided the basis for diverse body plans, in so rare a watershed event that it has never since been equaled.

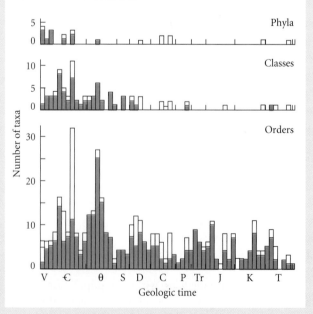

FIGURE 25.A1 The distribution of times of first appearance of animal phyla, classes, and orders in the fossil record. Most major body plans, represented by classes and especially phyla, arose early in the history of animal life. Open parts of bars represent poorly skeletonized taxa. (After Erwin et al. 1987.)

Why has it not been equaled? Many paleontologists have concluded that developmental processes in the earliest metazoans must have been malleable, and only later became canalized into relatively fixed pathways that differed from one lineage to another. Perhaps because of the evolution of increasing "burden" (Riedl 1978; see Chapter 23), the various phyla and classes acquired fundamental patterns of development that could not be greatly altered. If so, then the extinction of classes or phyla could not be compensated by the evolution of similarly disparate forms, and the total disparity of animal life (at the level of these *Baupläne*) would be irreversibly impoverished. Gould (1989) has maintained that in this sense, the history of life is *not* one of ever-increasing variety (disparity), but precisely the opposite: a steady decline from a Cambrian heyday of animal variety. However, morphological disparity among and within phyla is very difficult to measure, and the dispute surrounding Gould's proposal (Briggs et al. 1992; also *Science* 258:1816–1818, 1992) has not been resolved.

Correlated Rates of Speciation and Extinction

If a clade increases exponentially in diversity over the course of *t* million years, the net rate of increase in the number of taxa, *R*, can be estimated from the equation for exponential growth (see Chapter 4):

$$N_t = N_0 e^{Rt}$$

where N_0 and N_t, the numbers of taxa at the beginning and end of the interval, are estimated from the fossil record (N_0, representing the earliest species, is assumed to equal 1). Since $R = S - E$, the difference between the rate of origination and the rate of extinction, *S* can be calculated from estimates of *R* and *E*. From counts of the numbers of species and of extinctions at various times, Steven Stanley (1979) estimated the rates of increase, *R*, and extinction, *E*, of species in several clades during periods of rapid increase in diversity (Figure 25.9). The two rates are positively correlated, implying (since $R = S - E$) that taxa with high extinction rates also have high speciation rates: i.e., they have a high rate of TURNOVER. Thus, for example, both the rates of extinction and of speciation have been higher in mammals than in bivalves. Among families of bivalves and mammals that have increased rapidly in diversity during the Cenozoic, an average family of bivalves would take about 11 My to double in species number, whereas an average family of mammals would take only 3.15 My (Figure 25.10). Among the most rapidly radiating groups are murid rodents (mice and rats) and colubrid snakes, diverse groups that arose in the Miocene and have doubling times of about 1.98 My and 1.24 My respectively. Some taxa have, at times, increased even more rapidly. Following the end-Permian extinction, for example, one of the three known surviving genera of ammonites increased from perhaps 1–10 species to 134 species within 7–10 My (Stanley 1979). This rate implies that the average species speciated within 0.96 to 2.5 My.

These calculations lead to a very important conclusion: even in the most rapidly proliferating groups, a duration of about a million years per speciation event is more than sufficient to account for the evolution of great diversity.

The correlation between origination and extinction rates seems to hold for both animals and plants (Niklas et al. 1983). Stanley (1990) offers several possible reasons for this correlation.

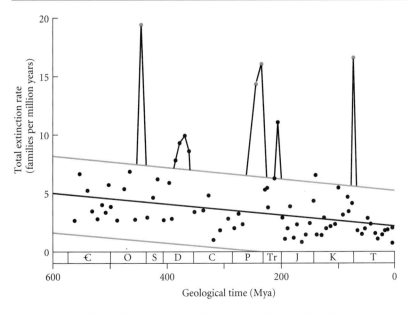

FIGURE 25.8 Extinction rates of marine animal families during the Phanerozoic. The solid black regression line, fitting points that represent fewer than eight extinctions per million years, suggests a decline in the "background" extinction rate. The colored points, representing several of the generally recognized mass extinctions, deviate significantly from the "background" cluster. (After Raup and Sepkoski 1982.)

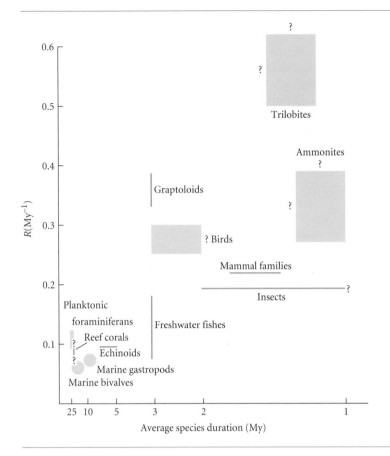

FIGURE 25.9   Rates of increase in the number of species, per million years (R), plotted against average species duration for various taxa. Taxa with shorter species durations (toward the right) have higher extinction rates, but their higher diversification rates (R) indicate that they also have higher speciation rates. For several groups, a range of values of R and/or species duration is given. Note that species duration decreases (and extinction rate increases) toward the right. (After Stanley 1979.)

1. Stenotopy versus eurytopy. Ecologically specialized species are likely to be more vulnerable than generalized species to changes in their environment (Jackson 1974). They may also be more likely to speciate because of their more patchy distribution (see Chapter 16), and newly formed species may be more likely to persist for a while by specializing on different resources and thus avoiding competition. Evidence on the relationship of extinction rate to specialization is presented below.

2. Population dynamics. Species with low or fluctuating population sizes are especially susceptible to extinction. Some authors believe that speciation is also

enhanced by small or fluctuating population sizes, although this hypothesis is controversial (see Chapter 16).

3. Dispersal and geographic range. Species with a high capacity for dispersal are expected to have low rates of speciation due to high rates of gene flow. They also may be expected to have broad geographic ranges and to escape local environmental changes, both of which would lower their risk of extinction. Although not all fossil taxa conform to this prediction, the extinction rates of Cretaceous and Paleocene gastropods are lower in taxa with high dispersal ability (Hansen 1980; Jablonski 1986b).

Extinction

Causes of Extinction

Extinction has been the fate of almost all the species that have ever lived, but little is known of its causes. Biologists agree that extinction is caused by failure to adapt to changes in the environment. Ecological studies of contemporary populations and species point to *habitat destruction* as by far the most frequent cause of extinction; a few examples of extinction due to introduced predators or diseases, and fewer still due to interspecific competition, are also known (Diamond 1984a; Lawton and May 1995).

If population growth becomes negative because of a persistent deterioration of the environment, *extinction is averted* only *if the species shifts its geographic range* by colonizing more favorable localities, *or if it adapts by genetic*

FIGURE 25.10   Patterns of increase in species diversity for an average mammal family and an average bivalve family over the course of 20 My. These idealized diagrams are based on calculations of R from paleontological data. (After Stanley 1979.)

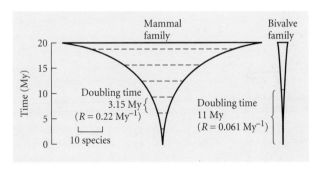

change that sufficiently improves fitness to reestablish positive population growth.

Massive long-distance emigration in response to failure of food resources is sometimes seen in certain birds and mammals, but on the whole, there is little evidence that species shift their geographic ranges in immediate response to changes such as deterioration of habitat or incursion of new predators or competitors. Shifts of geographic range are likely to occur only if the environmental change, while reducing the suitability of formerly occupied sites, makes other sites more suitable than they were before. The factor most likely to do this is change in climate, as occurred repeatedly during the Pleistocene.

The likelihood that populations will escape extinction by genetic change depends on how rapidly the environment (hence, the optimum phenotype) changes relative to the rate at which a character evolves. For characters under prolonged directional selection, the rate of evolution depends on the mutation rate and on the rate of population growth, because smaller populations experience fewer mutations. Thus an environmental change that reduces population size also reduces the chance of adapting to it (Lynch and Lande 1993). Because a change in one environmental factor, such as temperature, may bring about changes in other factors, such as the species composition of a community, the survival of a species might require evolutionary change in several or many features. All in all, biologists do not have enough genetic or ecological information to predict the likelihood that species will survive an environmental change, such as the global warming that many climatologists expect will occur rapidly over the course of the next century or so (Kareiva et al. 1993).

Mass Extinctions: When and Why?

The history of extinction is dominated by the five so-called mass extinctions listed above, when very high proportions of taxa became extinct (Table 25.1). The end-Permian extinction was the most drastic, eliminating about 51 percent of marine families, 82 percent of genera, and 80–90 percent of marine species (Erwin 1993a). On land, major changes in plant assemblages occurred, several orders of insects became extinct, and the dominant amphibians and therapsids were replaced by new groups of therapsids (including the ancestors of mammals) and diapsids (including the ancestors of dinosaurs). The second most severe mass extinction, in terms of the proportion of taxa affected, occurred at the end of the Ordovician. Less severe, though far more frequently discussed, was the "K/T event," or end-Cretaceous extinction, which had a major effect on marine communities, terrestrial plants, and vertebrates; the last of the dinosaurs became extinct at this time.

Many factors have been suggested as *possible causes* of these mass extinctions (see Elliott 1986; Donovan 1989; Erwin 1993a; and Jablonski 1995 for introductions to this literature). Many authors favor explanations based on changes in *climate and/or sea level*, due in part to changes in the configuration of land masses. Douglas Erwin (1993a), for example, points out that near the end of the Per-

mian, climates became less equable, volcanic activity increased, and sea level dropped (regression). Thus several factors may have acted together. *Regressions of sea level*, which reduce especially the area of shallow epicontinental seas, *accompanied each of the five mass extinctions*, and so have been widely considered a likely cause. However, not all regressions in earth history have been accompanied by heightened extinction rates (Jablonski 1991).

Discussion of mass extinctions took a major new turn when Walter Alvarez and his colleagues (1980) suggested that the *end-Cretaceous extinction* had been caused by the *impact of an extraterrestrial body*—an asteroid or large meteorite—that struck the earth with a force great enough to throw a pall of dust into the atmosphere that darkened the sky and lowered temperatures. Marine phytoplankton, incapable of normal photosynthesis under these conditions, would be expected to suffer (and indeed, their extinction rate was very high), and so then would the rest of the marine food chain. The Alvarez hypothesis is supported by *chemical and mineral anomalies at the K/T boundary* at many sites throughout the world, and by a large *crater*, buried under the coast of the Yucatán Peninsula of Mexico, that is thought to mark the impact (Sharpton et al. 1996). There is now broad agreement among paleontologists that an impact extinguished the last dinosaurs and many other forms of life (Sheehan and Fastovsky 1992; Marshall and Ward 1996). A much more controversial question is whether mass extinction events during the Phanerozoic occurred with such regular periodicity as to imply a common cause, perhaps due to some unknown astronomical cycle (Raup and Sepkoski 1984). Whether or not mass extinctions have been truly periodic has not been resolved. In any case, there is no direct evidence that any mass extinctions besides the K/T event were caused by an astronomical impact.

Table 25.1 **Mass extinctions and other major extinction events**[a]

EXTINCTION EVENT	AGE (× 10⁶ YEARS)	FAMILIES (%)[a]	GENERA (%)[a]	SPECIES (%)[a]
Late Eocene	35.4	—	15	35 ± 8
End-Cretaceous	65.0	16	47	76 ± 5
Early Late Cretaceous (Cenomanian)	90.4	—	26	53 ± 7
End-Jurassic	145.6	—	21	45 ± 7.5
Early Jurassic (Pliensbachian)	187.0	—	26	53 ± 7
End-Triassic	208.0	22	53	80 ± 4
End-Permian	245.0	51	82	95 ± 2
Late Devonian	367.0	22	57	83 ± 4
End-Ordovician	439.0	26	60	85 ± 3

Source: After Jablonski (1991, 1995).

[a]Percentages of families, genera, and species of skeletonized marine invertebrates that became extinct during the five major mass extinctions (uppercase) and several lesser extinction events (lowercase) during the Phanerozoic. Values for genera were calculated directly from fossil data; those for species are estimated from statistical analyses of the numbers of species per genus.

The Evolution of Biological Diversity

Selectivity of Extinction

Several researchers have attempted to determine whether mass extinctions were "selective," preferentially extinguishing taxa with certain characteristics rather than others. Most such studies have focused on the end-Cretaceous and end-Permian extinctions.

David Jablonski (1986a,b,c; 1995), in his analyses of late Cretaceous bivalves and gastropods, found that during times of background extinction, the duration of survival of taxa was greater if they had planktonic development (larvae dispersed by currents) or if they consisted of numerous species, especially if these had broad geographic ranges. In contrast, during the end-Cretaceous mass extinction, planktotrophic and nonplanktonic taxa had the same extinction rate, the geographic range of *species* did not influence their chance of survival, and the survival of genera was not influenced by their species richness, but broad geographic distribution of a *genus* enhanced its chance of survival. Both before and during the end-Permian extinction, however, survival of gastropods was *greater for species with wide geographic and ecological distributions*, and for *genera consisting of many species* (Erwin 1993a). Extinction appears to have been random with respect to other characteristics, such as mode of feeding.

During times characterized by *background extinction* rates, the most general characteristic of taxa with *long durations* (low extinction rates) was a *broad geographic distribution*, whereas narrowly distributed taxa tended to have brief geological durations (Boucot 1975). Broad distribution is often correlated with high dispersal ability and with high abundance (Brown 1995).

Extinction and the Problem of Adaptation

The final aspects of extinction that we will consider are the effects of *geological time* and the *age* of taxa. In this context, *age* refers to the evolutionary "life span" of a taxon. Thus a taxon that became extinct 100 My after its origin was "older" at that time than one that became extinct 20 My after its origin, regardless of when during geological time these events occurred.

As we saw above, Raup and Sepkoski (1982) described an *apparent decline in the background extinction rate of marine families over geological time* (see Figure 25.8). It is not yet known whether this decline is real; it may be due to artifacts such as the "pull of the Recent" (Pease 1992; Gilinsky and Good 1991). But if the decline were real, what would that mean? One possibility (Flessa and Jablonski 1985) is that the rate of extinction of families has declined simply because the number of species per family has increased; recall that high species richness of a taxon lowers its extinction rate. A more intriguing possibility is that during the history of life, the world has become populated with progressively better "adapted" organisms, more resistant to extinction.

Although this conclusion seems reasonable, *evolutionary theory does not necessarily predict any such pattern*, because natural selection cannot prepare species for changes in the environment. If the environmental changes that threaten extinction are numerous in kind, we should not ex-

pect much carryover of "extinction resistance" from one change to the next.

Higher taxa differ consistently in extinction rate (see Figure 12 in Chapter 24), possibly because of their characteristic differences in features such as dispersal ability or habitat (Ward and Signor 1983; Gilinsky 1988). Moreover, the Paleozoic evolutionary fauna that Sepkoski recognized had a higher extinction rate than the Modern evolutionary fauna (Figure 25.11; Van Valen 1984; Erwin et al. 1987). Thus the average longevity of families in the Mesozoic, most of which belong to Sepkoski's Modern fauna, was greater than in the Paleozoic. In this sense, the average rate of extinction has declined over the course of earth history.

Within higher taxa, the extinction rate of species does not change with the taxon's age—i.e., the time since its origin (see Figure 23 in Chapter 24). As noted in Chapter 24, the constancy of the risk of extinction could be explained either by Van Valen's Red Queen hypothesis (failure to adapt to coevolving competitors or other antagonistic species) or by failure to adapt to any of the multitude of other environmental changes that continually occur.

Tiers of Evolutionary Change

If the exquisite adaptations of species were insufficient to protect them from environmental changes during times of background extinction, we might expect these adaptations to be even less sufficient during periods of mass extinction, which surely were caused by more profound environmental changes. During the end-Cretaceous extinction, and perhaps others, the characteristics that were correlated with survival seem, as we have seen, to have differed from those during "normal" times. Jablonski (1986a) has suggested that different "macroevolutionary regimes" operate during "normal" times than during episodes of mass extinction, when normal evolutionary processes are disrupted and evolution is channeled in new, unpredictable directions by the loss of

FIGURE 25.11 Survivorship curves for families of skeletonized marine animals that originated in the early Paleozoic and in the Mesozoic, corresponding to the "Paleozoic fauna" and the "Modern fauna." Each curve shows the percentage of families that survived for different lengths of time. Groups that arose in the Mesozoic had lower extinction rates. (After Erwin et al. 1987.)

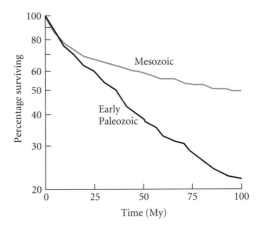

many lineages and the opportunity for other lineages to diversify (see below). Diverse taxa with superb adaptive qualities may succumb because they happen, by chance, not to have some critical feature that might have saved them from extinction. Evolutionary trends initiated in "normal" times may be cut off at an early stage; for example, the shell-drilling habit (as in modern oyster drills) evolved in a Triassic gastropod lineage, but was lost when this lineage became extinct in the late Triassic mass extinction (Fürsich and Jablonski 1984). Thus a new adaptation that might have led to a major adaptive radiation evolved at the wrong time, and did not originate again until 120 My later.

Stephen Jay Gould (1985) suggests that there have been "tiers" of evolutionary change, each of which must be understood in order to comprehend the full history of evolution. These tiers, he proposed, include genetic change *within* populations and species, *differential proliferation and extinction of species* ("species selection") during "normal" geological times, and the *shaping of the biota by mass extinctions*. Mass extinctions have extinguished diverse taxa, however well adapted they might have been, have ended evolutionary trends, and have reset the stage for new evolutionary radiations, initiating evolutionary histories that are largely decoupled from earlier ones.

Origination and Diversification

We turn now to the question of why increases in diversity have been greater in some lineages than in others and at some times than at others, and why diversity has tended to increase ever since the end-Permian extinction. Among the major factors that have fostered diversification are release from competition, the evolution of key adaptations, specialization, coevolution, and provinciality (Signor 1990, Benton 1990).

Release from Competition

Evolutionary biologists concluded long ago that lineages often diversify most rapidly when presented with *ecological opportunity* in the form of "vacant niches" not occupied by other species (Mayr 1942; Simpson 1944). Among living organisms, the most conspicuous examples of this phenomenon are the numerous adaptive radiations in which one or a few species that have colonized isolated islands or bodies of water have given rise to "species flocks." Examples include the several hundred species of amphipod crustaceans in Lake Baikal in Siberia; the more than 200 species of crickets, more than 700 species of drosophilid flies, and the diverse honeycreepers of the Hawaiian archipelago (Wagner and Funk 1995; see Chapter 4); and many "species flocks" of fishes, such as the spectacular radiation of cichlids in the African Great Lakes (Echelle and Kornfield 1984; Chapter 5). In many cases these species occupy diverse niches that in other regions are occupied by taxonomically disparate organisms, and many have evolved unusual new ways of life. For example, the Hawaiian Islands harbor a genus of damselflies that, uniquely for the order Odonata (damselflies and dragonflies), have terrestrial rather than aquatic larvae. These islands also

are home to a group of 15 species in the cosmopolitan moth genus *Eupithecia*. Unlike *Eupithecia* elsewhere, and unlike almost all other Lepidoptera, which are herbivorous, the larvae (inchworms) of the Hawaiian species are specialized for predation (Montgomery 1982; Figure 25.12).

The fossil record is replete with instances in which the reduction or extinction of one group of organisms is followed or accompanied by the proliferation of an ecologically similar group. The drop in "gymnosperm" diversity that accompanied the Cenozoic radiation of the angiosperms is one example (see Chapter 8); another, frequently cited, is the radiation of the mammals following the late Cretaceous extinction of the dinosaurs. Many examples, in fact, involve the replacement of one group by another after mass extinctions (Simpson 1944).

Several hypotheses can account for these patterns (Benton 1996; Sepkoski 1996). One taxon may have been more resistant than another to the environmental factors that caused the extinctions. The later group may have *caused* the extinction of the earlier group by competitive exclusion: this hypothesis is called **displacement**. Or, an incumbent earlier taxon may have *prevented* an ecologically similar taxon from diversifying. Extinction of the incumbent taxon would then vacate ecological "niche space," permitting the second taxon to radiate. This process has been called **incumbent replacement** (Rosenzweig and McCord 1991).

It is difficult to demonstrate that the decline and rise of two or more ecologically similar groups was caused by competitive displacement. For example, there is dispute about whether the post-Permian decline of brachiopods was caused by the diversification of ecologically similar bivalves after the Permian mass extinction, or simply reflects the greater decimation that the brachiopods had suffered (Gould and Calloway 1980; Sepkoski 1996). David Krause and his colleagues have provided plausible examples of competitive displacement from early North American mammals (Krause 1986; Maas et al. 1988). The multituber-

FIGURE 25.12 An inchworm (*Eupithecia*) in the Hawaiian Islands, holding a *Drosophila* that it has captured with its unusually long legs. Predatory behavior is extremely unusual in the order Lepidoptera. (Photograph by W. P. Mull, courtesy of W. P. Mull and S. L. Montgomery.)

culates were an early, nonplacental order of mammals, extending from the late Jurassic to the early Oligocene. In size, morphology, and probable feeding habits, they were very similar to rodents, and like them had bladelike incisors (see Figure 32 in Chapter 7). Rodents first appeared in North America in the late Paleocene, probably having invaded from Asia. Both the diversity (Figure 25.13) and abundance of multituberculates declined in rather close correlation with the diversification of rodents, which strongly suggests competitive displacement. The increase in rodent diversity was also accompanied by a decline of plesiadapiforms, a Paleocene group of mammals often classified as primitive primates, but probably similar to rodents in their habits.

Michael Rosenzweig and Robert McCord (1991) illustrated incumbent replacement with an analysis of turtles (Figure 25.14). The "stem group" of turtles, the amphichelydians, could not retract their heads and necks into their shells. Modern turtles consist of cryptodires, which retract their heads by bending the neck vertically into an S-shape within the shell, and pleurodires, which partially protect themselves by flexing the neck laterally under the shell margin. Cryptodires slowly, and at different times, replaced amphichelydians in Eurasia and in North America, where the replacement was accelerated during the K/T extinction. Pleurodires replaced amphichelydians in the early Tertiary in South America, and in the Pleistocene in Australia. Rosenzweig and McCord suggest that even though the two methods of flexing the neck provide better protection against predation, the cryptodires and pleurodires could not radiate until niches were opened by the extinction of amphichelydian species.

Paleontologists generally believe that a similar process of *replacement*, rather than displacement, accounts for the diversification of many lineages following mass extinctions. Commonly cited examples include the replacement of amphibians by reptiles as the dominant terrestrial vertebrates of the Triassic, and of dinosaurs by mammals in the Tertiary. In each case, the later group had arisen long before, persisting at low diversity but radiating rapidly only after the extinction of the earlier group. Thus, *mass extinctions may have been paramountly important in the history of life*, for the demise of incumbent groups may have enabled the flowering of others.

Key adaptations and adaptive space Suppose we postulate that a lineage has diversified greatly because of a *key adaptation* (see Chapter 24) or because it has invaded a new ADAPTIVE ZONE (often these events will be associated). If such an evolutionary change has occurred only in a single diverse lineage, it is hard to rule out other possible causes of diversification. But if the change has occurred convergently in many lineages, and if it is consistently associated with high diversity, we may conclude that the evolutionary novelty or new niche is causally associated with diversification. By comparing the diversity of each such clade with that of a sister group that retains the plesiomorphic (ancestral) condition, we can ascribe the difference in diversity to different *rates* of diversification, rather than different spans of time over which diversification has occurred, because sister groups, by definition, are equally old.

Charles Mitter, Brian Farrell, and colleagues (Mitter et al. 1988; Farrell et al. 1991) applied this method, called REPLICATED SISTER-GROUP COMPARISONS, to herbivorous insects and plants. The habit of feeding on the vegetative tissues of green plants has evolved at least 50 times in insects, usually from predatory or detritus-feeding ancestors. To test the hypothesis that entry into the herbivorous adaptive zone has promoted diversification, Mitter et al. surveyed the literature on insect phylogeny, and found 13 cases in which the nonherbivorous sister group of an herbivorous clade had been identified with fairly high confidence. In 11 of these cases, the herbivorous lineage had more species than its sister group, a statistically significant correlation. These researchers then tested Ehrlich and Raven's (1964) postulate that evolution of defenses against herbivores has promot-

FIGURE 25.13　The diversity of rodents in North America increased as that of the ecologically similar multituberculates declined, one of several lines of evidence suggesting competitive displacement. (After Krause 1986.)

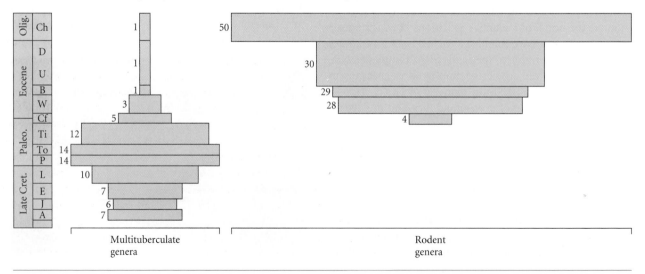

FIGURE 25.14 (A) Early turtles (amphichelydians), here represented by the reconstructed skeleton of the earliest known turtle (*Proganochelys quenstedti*, Upper Triassic), could not retract their heads for protection. (B) The Papuan snakeneck turtle (*Chelodina parkeri*), a pleurodiran turtle that flexes its neck laterally. (C) A softshell turtle (*Apalone mutica*), a cryptodiran turtle that retracts the head by flexing the neck vertically. The suborders Pleurodira and Cryptodira replaced the amphichelydians. (A courtesy of E. Gaffney, American Museum of Natural History; B and C courtesy of G. Zug, Division of Amphibians and Reptiles, National Museum of Natural History, Smithsonian Institution; photograph of *Apalone* by R. W. Barbour.)

(B)

(A)

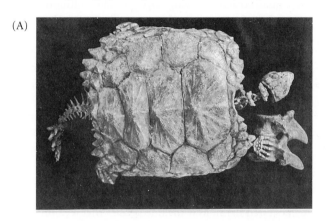

(C)

ed diversification in plants (see Figure 9 in Chapter 18). Canals that transport latex or resin, which deter herbivorous insects, have evolved independently in at least 40 lineages of plants, and in 16 of these cases, the sister group of a canal-bearing lineage has been identified by plant systematists. Mitter and his colleagues found that 13 of these canal-bearing taxa are richer in species than their sister groups that lack canals, and concluded that canals may be a key adaptation that has fostered diversification.

Although few fossil taxa have been analyzed by sister-group comparisons, many examples of prolific diversification appear to be attributable to key adaptations and occupation of new adaptive zones (Benton 1990, 1996; Bambach 1985). Among these groups are the Echinoidea (sea urchins), which primitively are more or less globular, spined, sluggish herbivores with a centrally located anus above, opposite a mouth surrounded by jaws (Figure 25.15A). Beginning in the Triassic, the subclass Euechi-

(A)

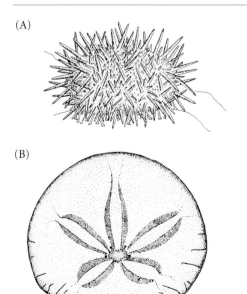

(B)

FIGURE 25.15 Echinoid diversity. (A) A sea urchin of the order Echinacea. (B) A sand dollar of the order Gnathostomata. (C) Diversification of the sand dollars (Gnathostomata) and heart urchins (Atelostomata) greatly increased the diversity of the echinoids. (A, B from Brusca and Brusca 1978; C after Bambach 1985.)

(C)

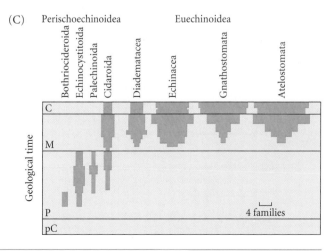

noidea began a major radiation (Figure 25.15C), attributed partly to the evolution of stronger jaws and a strong support for the jaw muscles, which enabled these urchins to use a greater variety of foods; some are even carnivorous, and others can bore into rocky substrates. An even greater diversification, however, accounting for half of the living echinoids, began in the Jurassic with the evolution of the flattened, bilaterally symmetrical heart urchins and sand dollars (Figure 25.15B). These urchins became specialized for feeding on fine particles of organic sediment, collected as the animal burrows either at considerable depth (heart urchins) or shallowly, in shifting sands (sand dollars). The key features allowing this major shift of habitat and diet include the flattened form, situation of the anus near the edge of the body, a variety of highly modified tube feet that can capture fine particles and transfer them to the mouth, and food grooves on the upper surface in which particles are carried to the edge and from there to the mouth (Bambach 1985; Simms 1990).

The *history of increase in marine animal diversity* through the Phanerozoic *can be explained*, in part, by *increases in the occupancy of ecological space* accomplished by evolutionary innovations such as those of the sand dollars. During the late pre-Cambrian and the early Cambrian, when the phyla and many classes originated, diversity increased very rapidly, but the variety of "guilds," or ecological lifestyles, was limited. Most animals were either EPIFAUNAL, rooted to or slowly creeping over the substrate, or shallowly INFAUNAL, burrowing only slightly below the surface. Few swam actively in open water. As the Paleozoic evolutionary fauna diversified in the Ordovician, organisms evolved that occupied a greater variety of "tiers," both above the substrate and below the surface (Ausich and Bottjer 1985). This increase in "tiering" is but one aspect of the *increase in the number of guilds of animals* occupying epifaunal, infaunal, and open-water habitats (Figure 25.16). Richard Bambach (1985) has argued that an increasingly diverse guild structure, owing largely to the evolution of key adaptations, has been the sine qua non of the evolution of diversity.

Specialization From studies of modern organisms, we know that much diversity resides in the great numbers of related species that differ from each other only subtly. Anoline lizards that forage in different microhabitats, hawks that feed on birds of different sizes: these and many other examples have appeared in earlier chapters of this book. Almost every lineage that has evolved a key adaptation allowing utilization of a new resource or habitat has given rise to species that utilize it in diverse specialized ways. Bats, having acquired flight, now include species that feed on fruit, nectar, insects, frogs, fishes, and blood.

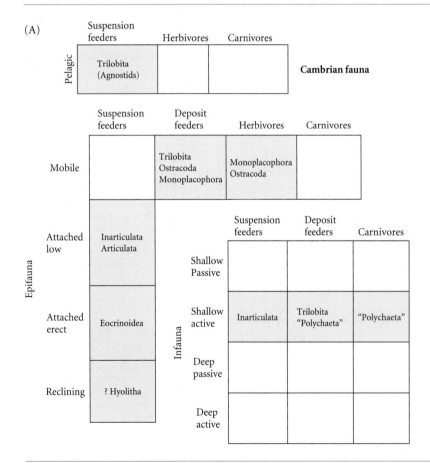

FIGURE 25.16 Modes of life common among marine animals in (A) the Cambrian, (B) Ordovician to Permian, and (C) Mesozoic and Cenozoic, showing the increase in guild diversity over time. The columns distinguish diet types, and the separate blocks distinguish pelagic, epifaunal, and infaunal animals. (After Bambach 1985.)

(B)

Middle and Upper Paleozoic fauna

Pelagic

	Suspension feeders	Herbivores	Carnivores
Pelagic	Conodontophorida Graptolithina ? Cricoconarida		Cephalopoda Placodermi Merostomata Chondrichthyes

Epifauna

	Suspension feeders	Deposit feeders	Herbivores	Carnivores
Mobile	Bivalvia	Agnatha Monoplacophora Gastropoda Ostracoda	Echinoidea Gastropoda Ostracoda Malacostraca Monoplacophora	Cephalopoda Malacostraca Stelleroidea Herostomata
Attached low	Articulata Edrioasteroida Bivalvia Inarticulata Anthozoa Stenolaemata Sclerospongia			
Attached erect	Crinoidea Anthozoa Stenolaemata Demospongia Blastoidea Cystoidea Hexactinellida			
Reclining	Articulata Hyolitha Anthozoa Stelleroidea Cricoconarida			

Infauna

	Suspension feeders	Deposit feeders	Carnivores
Shallow passive	Bivalvia Rostroconchia		
Shallow active	Bivalvia Inarticulata	Trilobita Conodontophorida Bivalvia Polychaeta	Merostomata Polychaeta
Deep passive			
Deep active		Bivalvia	

(C)

Mesozoic-Cenozoic fauna

Pelagic

	Suspension feeders	Herbivores	Carnivores
Pelagic	Malacostraca Gastropoda Mammalia	Osteichthyes Mammalia	Osteichthyes Chondrichthyes Mammalia Reptilia Cephalopoda

Epifauna

	Suspension feeders	Deposit feeders	Herbivores	Carnivores
Mobile	Bivalvia Crinoidea	Gastropodea Malacostraca	Gastropoda Polyplacophora Malacostraca Ostracoda Echinoidea	Gastropoda Malacostraca Echinoidea Stelleroidea Cephalopoda
Attached low	Bivalvia Articulata Anthozoa Cirripedia Gymnolaemata Stenolaemata Polychaeta			
Attached erect	Gymnolaemata Stenolaemata Anthozoa Hexactinellida Demospongia Calcarea			
Reclining	Gastropoda Bivalvia Stelleroidea Anthozoa			

Infauna

	Suspension feeders	Deposit feeders	Carnivores
Shallow passive	Bivalvia Echinoidea Gastropoda	Bivalvia	Bivalvia
Shallow active	Bivalvia Polychaeta Echinoidea	Bivalvia Echinoidea Holothuroidea Polychaeta	Gastropoda Malacostraca Polychaeta
Deep passive	Bivalvia		
Deep active	Bivalvia Polychaeta Malacostraca	Bivalvia Polychaeta	Polychaeta

Paleontologists who study terrestrial plants (Niklas et al. 1983) and animals (Benton 1990) have suggested that subdivision of niches has increased over time, because individual fossil deposits, which record local communities for the most part, contain more species in later than in earlier geological periods. The increase in species number seems greater than the increase in the variety of major growth forms or guilds, suggesting that the ever greater number of species coexisted by more finely partitioning similar resources. However, it is difficult to document subtle ecological differences among similar extinct species, and the extent to which local community diversity has increased due to greater resource partitioning within adaptive zones has not yet been established.

Coevolution Interactions among species are thought to promote the evolution of diversity in several ways (see Chapter 18). *Species serve as resources for other species,* so the diversification of one group can support the diversification of others. For instance, each of the more than 700 species of figs is the sole resource of a different species of pollinating fig wasp; the wasps, in turn, are parasitized by species-specific nematodes.

Predation can enhance the diversity of prey species by imposing selection for diverse mechanisms of escape and defense. We have already described how lineages of vascular plants with canals that carry herbivore-deterring chemicals have diversified at a greater rate than their unmodified sister taxa. During the Jurassic, Cretaceous, and early Cenozoic, many groups of sharks, bony fishes, decapod crustaceans (lobsters and crabs), and gastropods evolved the ability to feed on molluscs by crushing or drilling through their shells (Figure 25.17). Likewise, defenses against predation evolved in many lineages of molluscs that diversified rapidly in the Mesozoic and Cenozoic. Various gastropods evolved a thick shell wall, a long, narrow shell into which the animal could deeply withdraw, or a thickened or narrow aperture (see Figure 7 in Chapter 18). The new defenses of bivalves included spines or crenulations along the margin of the shell, and escape by rapidly burrowing through sediments.

Provinciality The degree to which the world's biota is partitioned among geographic regions is called PROVINCIALITY. A faunal province is a region containing high numbers of distinctive, localized taxa. The fauna and flora of the contemporary world are divided into more biogeographic realms, and provinces within realms (see Chapter 8) than ever before in the history of life (Valentine et al. 1978). A *trend from a cosmopolitan distribution of taxa to more localized distributions has persisted throughout much of the Mesozoic and Cenozoic, and is thought by many paleontologists to be one of the most important causes of the increase in global diversity* during this time (Signor 1990). Among marine animals, the number of faunal provinces dropped to an all-time low in the early Triassic, when most of the higher taxa that had survived the end-Permian mass extinction were so

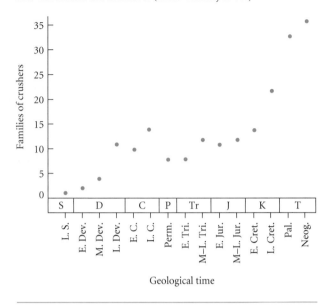

FIGURE 25.17 Increase in the number of families of marine animals specialized for predation by crushing or breaking shells. Eurypterid and crustacean arthropods, cephalopods, and vertebrates are included. (After Vermeij 1987.)

highly cosmopolitan that paleontologists recognize only a single, worldwide province. During the Jurassic and Cretaceous, and especially the Tertiary, marine animals were distributed among an increasing number of latitudinally arranged provinces in both the Atlantic and Pacific regions. Among terrestrial vertebrates, likewise, a distinct fauna developed on each major land mass during the later Mesozoic and the Cenozoic, and the broad latitudinal distributions typical of many dinosaurs and other Mesozoic groups gave way to the much narrower latitudinal ranges of today's vertebrates. The increasing Cretaceous and Cenozoic diversity of vascular plants, however, seems less attributable to increased provinciality than to evolutionary innovations that expanded plants' occupation of adaptive space (see above), and to Cenozoic mountain building, which increased the diversity of habitats and of plant community types (Niklas et al. 1980; Signor 1990).

Changes in the distribution of land masses due to plate tectonics are the fundamental cause of this trend. Once Pangaea began to break apart in the Triassic, land masses (including the continental shelves) slowly separated and approached their current distribution, arrayed almost from pole to pole along a wider latitudinal span than ever before in the earth's history. This deployment of the continents created two increasingly disjunct ocean systems, the Indian-Pacific and the Atlantic, and established a pattern of ocean circulation that created a stronger latitudinal temperature gradient than ever before. This temperature gradient has been especially pronounced since the end of the Eocene, about 38 Mya, and it is since then that the development of latitudinal chains of marine faunal provinces has been most pronounced (Valentine et al. 1978). It is also possible that diversity, especially of terrestrial species, increased due to simple fragmentation of land masses, which allowed for divergent evolution and prevented the interchange of species that, by competition or predation, might lower diversity.

Does Diversity Attain an Equilibrium?

Is there a *limit* on the diversity of life that the earth could sustain, and has this limit ever been reached? Has the diversity of life, now or at any time in the past, ever been at an *equilibrium*, and if so, what factors determine such an equilibrium?

The answer to the first question is that although a theoretical limit to diversity must exist (because of the finite input of energy to the earth), this limit almost surely has never been attained. Diversity from the Miocene to the present seems to be at an all-time high (see Figure 25.5), and there is no reason to suppose that it could not be higher (except for the fact, accidental in an evolutionary sense, that our own species has initiated the next mass extinction: see Box 25.B).

The second question is much more difficult, and paleontologists have not agreed on an answer. The most common model of diversity analogizes the number (N) of taxa (at whatever level) with the number of individuals in a population. If, by analogy with birth and death rates in a population, the per taxon rate of origination of new taxa (S) declines and the per capita rate of extinction (E) increases as the number of taxa grows, (i.e., if the rate of diversification is *diversity-dependent*), then the number of taxa will reach a stable equilibrium (K) (see Chapter 4),

There is some evidence for such equilibria. First, some biotas have displayed little growth in diversity over substantial spans of time, suggesting equilibrial conditions. Examples include post-Ordovician Paleozoic marine invertebrates and vascular plants over periods of 80–100 My in both the late Paleozoic and early Mesozoic (see Figures 25.5 and 25.6). Second, we have seen that the rate of increase of diversity in many taxa was high after mass extinctions and declined thereafter, a pattern that is most plausibly explained by increasing competition as ecological "niche space" was filled. Third, there is considerable evidence that the diversity and taxonomic composition of certain community types has been quite stable (Boucot 1978). For example, the diversity of ecologically equivalent species in Ordovician and in Devonian benthic communities, 70 My apart, was very similar, with ostracods dominating the supratidal zone, trilobites the mid-intertidal, and so on (Walker and Laporte 1970).

On the other hand, many taxa have diversified without causing evident diminution of others; many community types have by no means maintained stability of species number and niche partitioning; and both marine and terrestrial diversity since the late Mesozoic has increased spectacularly and rather steadily. How can this conflicting evidence, for equilibrium and against it, be reconciled?

A biota may remain near an equilibrium state for a few million years, but shift toward a different equilibrium when conditions change. At least three kinds of changes have al-

tered conditions for living organisms. First, changes in the physical environment of the earth alter the theoretical equilibrium. The increasing latitudinal stratification of climate that led to greater provinciality in the Cenozoic is a clear example.

Second, recovery from mass extinctions can take a long time. For example, it took bivalve and gastropod molluscs about 25 My to regain their late Cretaceous diversity level after the K/T extinction (Hansen 1988). If severe extinctions were frequent enough, equilibrium would never be attained. In other cases, the taxa that survive a mass extinction event attain a *new*, different *equilibrium level* of diversity, due to new patterns of competition and other interactions. There is no reason, for example, to expect the diversity of large herbivores and carnivores to be the same when they are mammals as when they are dinosaurs.

A third kind of event that creates *new equilibria* is the evolution of taxa that use *new resources or habitats*—i.e., expansion into new ecological niche space. Adaptations to new niche space account for much of the increase in the diversity of life.

This principle bears on the suggestion by Timothy Walker and James Valentine (1984) that ecosystems might have long-sustained *stable equilibrial numbers* of species, and *yet have many empty niches* that in principle could be occupied. That is, the equilibrium number of species might be considerably less than the maximal possible number. Walker and Valentine's model (Figure 25.18), inspired by MacArthur and Wilson's theory of island biogeography

FIGURE 25.18 Two models for the number of species at equilibrium over evolutionary time. The number of species originating per unit of time (S) increases, the more potential ancestors there are, but decreases at higher diversity levels, dropping to zero at the maximal number that resources can sustain (N_{max}). The number of extinctions per unit of time may be diversity-dependent (curve E) or diversity-independent (E'). Walker and Valentine's model assumes diversity-independent extinction, in which a fixed fraction of existing species become extinct per unit of time. This results in higher equilibrium diversity (N^*) than if extinction is diversity-dependent (K). (After Walker and Valentine 1984.)

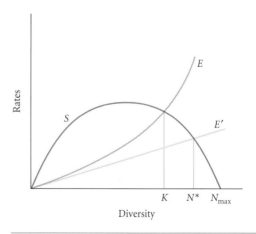

BOX 25.B

The Next Mass Extinction: It's Happening Now

None of the mass extinctions of the past was caused by the actions of a single species. But that is happening now. Several thousand years ago, humans initiated extinctions that now are accelerating to the point that within the next few centuries, the diversity of life will almost certainly plummet at a pace perhaps greater than ever before.

Beginning about 12,000 years ago, coincident with human migration from northeastern Asia to America, the Pleistocene fauna of large mammals and birds—horses, giant bison, camels, mammoths, sabertooths, ground sloths, giant condors, and many others—was almost entirely extinguished. Similar extinctions of large vertebrates occurred after humans first colonized other regions such as Madagascar, New Zealand, and Australia. These extinctions may have been caused by human hunting (Martin and Klein 1984).

The pace of extinction increased as Europeans colonized the world. Island faunas in particular suffered as settlers introduced animals such as goats, rats, pigs, and later mongooses, that destroyed vegetation and preyed on native birds and other animals. Since then, the human threat to the earth's biodiversity has accelerated steadily, due to two related factors: ever more powerful technology and exponential growth of the world's human population (see Figure 15 in Chapter 4), which will double to 11 billion by 2035. The per capita rate of population growth is generally greatest in the developing, chiefly tropical and subtropical, countries; but the per capita impact

on the world's environment is greatest in the most highly industrialized countries. An average American, for example, has perhaps 13 times the environmental impact of an average Brazilian, and 140 times that of an average Kenyan, because the United States is so profligate a consumer of resources (harvested throughout the world) and energy (the currency by which we can measure impacts ranging from strip mines and oil spills to insecticides and production of the "greenhouse gases" that cause global warming).

Some species are threatened by hunting, and others by human introduction of species into new regions. For example, introduction of the Nile perch (as a food fish) into Lake Victoria in Africa has extinguished about two-thirds of the 300 or more endemic cichlids there. But by far the greatest cause of extinction is the destruction of habitat. It is largely for this reason that 29 percent of the species of North American freshwater fishes and 20 percent of North American freshwater mussels are endangered or are already extinct, that more than 200 of the 20,000 plant species of the United States have become extinct, with more than 600 others likely to follow by the year 2000, and that about 10 percent of the world's bird species are considered endangered by the International Council for Bird Preservation.

The numbers of species likely to be lost are highest in tropical wet forests, which by 1989 had been reduced to half their prehistoric area, and are being destroyed at a rate of about 1.8 percent per year—a rate that is accelerating. As E. O. Wilson

FIGURE 25.B1 Regions of forest and scrub habitat in which exceptionally high numbers of terrestrial species are in danger of extinction due to human activity. These areas contain many endemic species, and are subject to severe habitat de-struction. Many biomes such as prairies, lakes, rivers, and coral reefs, are not shown here. (After Meffe and Carroll 1997.)

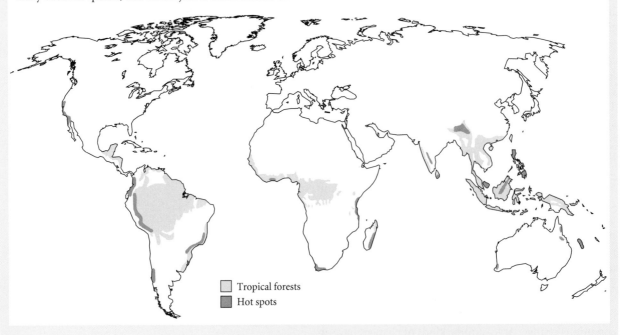

(1992) said, "in 1989 the surviving rain forests occupied an area about that of the contiguous forty-eight states of the United States, and they were being reduced by an amount equivalent to the size of Florida each year." With even rather conservative assumptions about the relationship between area and species diversity, several authors have estimated that perhaps 10 to 25 percent of tropical rain forest species—accounting for as much as 5–10 percent of the earth's species diversity—will become extinct in the next 30 years. To this toll must be added extinctions caused by the destruction of species-rich coral reefs, pollution of other marine habitats, and losses of habitat in many areas, such as Madagascar and the Cape Province of South Africa, that harbor unusually high numbers of endemic species (Figure 25.B1).

If mass extinctions have happened naturally in the past, without humans, why should we be so concerned about this one? Different people have different answers, ranging from utilitarian to aesthetic to spiritual. Some will point to the many thousands of species that are used by humans today, ranging from familiar foods to the innumerable species used by peoples throughout the world for fiber, herbal medicines, and spices; some will note that thousands of species support small industries that cater to tropical fish hobbyists, orchid fanciers, gardeners, and shell collectors; others will cite the economic value of ecotourism (the most important source of income for Costa Rica, for example) and the enormous popularity of bird-watching in some countries. Biologists will argue that among the invertebrates, plants, fungi, and microorganisms are thousands of species that may (as many already have) prove useful as pest control agents or as sources of medicinal compounds (such as taxol, a compound in yew trees that is used for treating cancer) or industrially valuable materials (for example, the polymerase chain reaction, the basis of DNA amplification and a key procedure in biotechnology, uses DNA polymerase from archaebacteria that inhabit hot springs). "Biological prospecting," searching for medically or industrially

useful compounds in plants and other organisms, already shows promise of developing into an important (and controversial) business.

These utilitarian concerns, however, are only part of the rationale for conserving diversity. Many people (including this author) cannot bear to think that future generations will be deprived of tigers, sea turtles, and macaws. And they share with millions of others, who care little for such taxonomic distinctions, a deep renewal of spirit in the presence of unspoiled nature. Still others, turning entirely away from a human-centered perspective that values nature only for its economic or aesthetic returns, feel that it is in some sense cosmically unjust to extinguish, forever, the species with which we share the earth.

Conservation is an exceedingly complicated topic; it requires not only a concern for other species, but compassion and understanding of the very real needs of people whose lives depend on clearing forests and making other uses of the environment. It requires that we understand not only ecology and other biological disciplines, but also global and local economics, politics, and social issues ranging from the status of women to the reactions of the world's peoples and their governments to what may seem like elitist, if not imperialistic, Western ideas. Anyone who undertakes work in conservation must deal with some of these complexities. But everyone can play a helpful role, however small. We can try to waste less; influence people about the need to reduce population growth (surely the most pressing problem of all) by means of birth control and economic development; support conservation organizations; patronize environment-conscious businesses; stay aware of current environmental issues; and communicate our concerns to elected officials at every level of government. Few actions of an enlightened citizen of the world can be more important.

For those seeking further information, E. O. Wilson's *The Diversity of Life* (1992) is a compelling introduction to the gravity of the crisis, and Meffe and Carroll's *Principles of Conservation Biology*, Second Edition (1997) is an outstanding textbook.

(see Chapter 4), assumes that the per species rate of extinction, E, is density-independent, but that the per species rate of origin of new species, S, declines as the number of species (N) approaches the maximum (N_{max}) that the environment can support. (In effect, this means that due to competition, newly arisen species cannot become established, and become extinct so quickly that they are not registered in the fossil record.) Thus $S = S_0(1 - N/N_{max})$, where S_0 is the highest possible rate of speciation. The resulting equation for the rate of change in the number of species is

$$\frac{\Delta N}{\Delta t} = \left(S - E\right)N = \left[S_0\left(1 - \frac{N}{N_{max}}\right) - E\right]N$$

which, when $\Delta N/\Delta t = 0$, yields the equilibrium number of species (N^*):

$$N^* = (1 - E/S_0)N_{max}$$

The proportion of niches that are occupied is $N^*/N_{max} = 1 - E/S_0$, so the proportion of "empty niches" is E/S_0. E can be estimated fairly easily from paleontological data, and S_0 can be estimated from the maximal rate of origin of new species observed in a taxon's history. In this way, Walker and Valentine estimated E/S_0 at 38–54 percent for various taxa. For certain bivalves, for example, they calculated E/S_0 as 0.07/0.194 (per My), or 36 percent "empty niches." This result implies that there has generally been "room" for more species of clams than actually exist, because of extinction.

As Valentine (1985a) pointed out, the maximal speciation rates (S_0) calculated from fossil material are generally far lower than they might be in theory, or have been in certain groups of living species, such as African lake cichlids. If speciation required 10^5 years, a single species could give rise to 1024 new species over the course of a million (10^6) years. Yet the speciation rate of ammonites, one of the most rapidly diversifying groups of all time, is estimated at only 1.35 species/species/My. Valentine suggests that the low rate of speciation is due to the difficulty populations have in crossing the "valleys" between the adaptive peaks (see Chapter 14) that might represent empty niches. A related possibility is that many new species are ecologically so similar to their "parent" or "sister" species that they rapidly become extinct due to competition. Empty niches may abound, but the new species are too "far away" from them to evolve readily to fill them. These "distant" empty niches, then, are effectively small adaptive zones that become occupied only when one lineage or another finally evolves the right "key adaptation." If this all sounds too metaphorical, recall the adaptive zone occupied by carnivorous Triassic snails that drilled through bivalve shells to attack their prey—a zone that remained empty for 120 My following the extinction of these gastropods, until another gastropod lineage evolved the same habit.

Thus, species diversity may be in equilibrium or not, depending on the time scale. On an ecological time scale of thousands of years, a regional biota may approach diversity equilibrium due to competition and other interactions among species. On a longer time scale (perhaps hundreds of thousands to several millions of years), and on a larger spatial scale, extinction and speciation determine an approximate longer-term equilibrium, which, however, is less than maximal because extinction continually creates empty niches. On a still longer time scale (tens of millions of years or more), diversity equilibria begin to vanish because of major changes in the environment, major extinction events, changes in provinciality, and the evolution of strikingly different kinds of organisms with ever more diverse ways of life.

Summary

1. Ecological studies of contemporary communities show that the species diversity in a locality or region depends on the pool of available species and on ecological processes, including interspecific interactions. Interactions such as competition appear to limit local diversity in some cases, but not others.

2. Large-scale patterns of diversity (e.g., differences among continents) have been strongly affected by a long-term history of speciation, adaptive diversification, extinction, and isolation from other biotas, as well as by the history of the region and its climate.

3. Analyses of levels and changes of diversity in the fossil record require estimation procedures to correct for the incompleteness of the record and the biases (systematic errors) that this often causes. Nevertheless, the record of skeletonized marine animals indicates a very rapid increase in diversity at the beginning of the Cambrian, a slower increase thereafter to an approximate equilibrium that lasted for almost two-thirds of the Paleozoic, a mass extinction at the end of the Permian, and an increase (with interruptions) since the beginning of the Mesozoic, accelerating in the Cenozoic. Terrestrial plants and vertebrates show a similar pattern, except that their diversity was relatively stable for much of the Mesozoic. The diversity of families and lower taxa was higher toward the end of the Cenozoic than ever before in the history of life, although it may very recently have been lowered slightly by Pleistocene extinctions.

4. The "background" rate of extinction (in between so-called mass extinctions) may have declined during the Phanerozoic, but this is uncertain. The rate of extinction of lower-level taxa does not change as a function of time since the origin of the higher taxon that includes them. Hence, the earth shows no clear tendency to become increasingly populated by extinction-resistant species. This is undoubtedly because new adaptations do not offer security against subsequent environmental changes.

5. Within many higher taxa, the rate of origination of new families has declined over time, perhaps because of increasing competition as "niche space" becomes filled.

6. Five major mass extinctions (at the ends of the Ordovician, Devonian, Permian, Triassic, and Cretaceous), as well as several less pronounced episodes of heightened extinction rates, have occurred. Their causes are generally unknown, although changes in sea level and in the disposition of land masses, which would affect climate, are among the leading hypotheses. There is strong evidence for an impact of an extraterrestrial body at the end of the Cretaceous, which evidently caused the extinction of many taxa, including the last dinosaurs.

7. Broad geographic distribution and ecological amplitude, rather than adaptation to "normal" conditions, enhanced the likelihood that taxa would survive mass extinctions. An important consequence of these events was the diversification of many of the surviving lineages, due in part to the extinction of other groups that had occupied similar adaptive zones.

8. The increase in diversity over time appears to have been caused mostly by adaptation to vacant or underutilized ecological niches ("adaptive space"), often as a consequence of the evolution of key adaptations, and by increasing provinciality (differentiation of the biota in different geographic regions) owing to the separation of land masses in the Mesozoic and Cenozoic and the consequent development of greater latitudinal variation in climate. Coevolution and the evolution of finer resource partitioning (specialization) have probably also contributed to the growth in diversity.

9. Newly diversifying groups have usually replaced competitors well after these became extinct, rather than displacing them. Although origination rates are diversity-dependent, declining as diversity increases, extinction rates seem seldom to depend on diversity. Diversity tends toward an equilibrium, although continual ex-

tinction may ensure the existence of "empty niches." Such an equilibrium, however, can change over geological time because of changes in the configuration of continents, climates, and other aspects of the environment, and because organisms evolve new ways of using habitats and resources. Diversity has almost certainly never attained its possible maximum.

10. The next mass extinction, at a perhaps unprecedented rate, has already begun.

Major References

Ricklefs, R. E., and D. Schluter (editors). 1993. *Species diversity in ecological communities*. University of Chicago Press, Chicago. Essays by many authorities on the causes of species diversity in contemporary communities; many adopt the new emphasis on the role of historical and large-scale effects. For an ecological perspective, see also M. A. Huston, *Biological diversity* (Cambridge University Press, Cambridge, 1994) and J. H. Brown, *Macroecology* (University of Chicago Press, Chicago, 1995).

Lawton, J. H., and R. M. May (editors). 1995. *Extinction rates*. Oxford University Press, Oxford. Analyses of extinction by ecologists and paleobiologists.

Signor, P. O. 1990. The geological history of diversity. *Annual Review of Ecology and Systematics* 21: 509–539. A succinct review.

Valentine, J. W. (editor). 1985. *Phanerozoic diversity patterns: Profiles in macroevolution*. Princeton University Press, Princeton, NJ. Analytical essays by leading paleontologists on many aspects of the subject.

Taylor, P. D., and G. P. Larwood (editors). 1990. *Major evolutionary radiations*. Clarendon Press, Oxford. Authors of this symposium review and analyze the diversification of many groups of organisms.

Problems and Discussion Topics

1. Distinguish between the rate of speciation in a higher taxon and its rate of diversification. What are the possible relationships between the present number of species in a taxon, its rate of speciation, and its rate of diversification?

2. What factors might account for differences among taxa in the number of extant species? Suggest methods for determining which factor might actually account for an observed difference.

3. Ehrlich and Raven (1964) suggested that coevolution with plants was a major cause of the great diversity of herbivorous insects, and Mitter et al. (1988) presented evidence that the evolution of herbivory was associated with increased rates of insect diversification (see Chapter 18). However, the fossil record suggests that increase in the number of insect families was not accelerated by the explosive diversification of flowering plants (Labandeira and Sepkoski 1993). Suggest some hypotheses to account for this apparent conflict, and ways to test them.

4. Gould (1989) argued that "disparity"—the diversity of body plans—has declined since the early Paleozoic. However, we have encountered many examples of new morphological adaptations that have evolved by modification of ancestral features, which would lead us to expect that morphological diversity should beget more morphological diversity. Discuss possible resolutions of this apparent contradiction.

5. How might you test the hypothesis that coevolution among marine predators and their prey affected their diversity during the Mesozoic "marine revolution"?

6. A factor that might contribute to increasing species numbers over time is the evolution of increased specialization in resource use, whereby more species coexist by finely partitioning resources. Discuss ways in which, using either fossil or extant organisms, one might test the hypothesis that a clade is composed of increasingly specialized species over the course of evolutionary time. Describe how you might test this hypothesis with a group of organisms with which you are familiar.

7. This chapter cites explanations of increasing diversity based on the notion that lineages evolve adaptations that enable them to fill vacant ecological niches or adaptive zones (sets of similar ecological niches). However, many ecologists hold that a niche cannot be identified except by observing the species that occupies it (cf. the definition of *niche*, Chapter 4). Can vacant niches be identified? Is a vacant niche or adaptive zone a meaningful or useful concept? Can you specify any niches or adaptive zones that are vacant at this time?

8. What effects is global warming likely to have on biological diversity? Consider changes in the climate of different geographic regions, in sea level, and in the geographic ranges of species, the likelihood that species will adapt to environmental changes, and possible causes of extinction. (See Kareiva et al. 1993 for some aspects of these problems.)

9. Debate the following proposition, citing examples to support your argument: It is essential for a professional evolutionary biologist to have detailed knowledge of the systematics and biology of at least one living or extinct higher taxon of organisms.

10. Debate the following proposition: In order to conserve biodiversity, the single most important thing we can do is reduce the growth rate of the human population.

CHAPTER 26

Human Evolution and Variation

H omo sum: *humani nil a me alienum puto.*
(I am human: nothing human is foreign
to me.)

Terence, about 1 B.C.

The mechanisms of coevolution, the causes of linkage disequilibrium, or the history of species diversity may be of absorbing interest to relatively few people, but human evolution evokes almost universal interest. This is the topic that ignites the creationists' fulminations against evolution, the topic that for others promises to solve the mystery of "the paragon of animals" and to provide insight into the potentialities and limitations of "human nature." Evolutionary biology indeed has much to say about the human condition, but so do anthropology and sociology, psychology and history, philosophy and the arts. People are too complex to be understood from the narrow perspective of biology or of any other single way of knowing.

Controversy and Objectivity

Most scientists believe that science generally approaches ever nearer to an objective description and understanding of its subject matter—objective both in the sense that one scientist's description or conclusion can be achieved independently by others, and in the sense that his or her conclusions are not dictated by emotion, desire, or a priori expectations. Objective conclusions, moving us closer to a description of reality or "truth," are achieved by the social process of science, which is one of competition—a kind of natural selection—among ideas and evidence, stemming in part from competition among scientists (Hull 1988). However, individual scientists are often far from objective.

Nowhere is this more true than in the area of human evolution and variation. Because of their social and philosophical implications, these subjects stir up more than the usual amount of passion in scientists, as in everyone else. Like everyone else, scientists are influenced by their times and circumstances. Some hold, or have held, highly progressive, liberal views, such as the population geneticist Theodosius Dobzhansky (1962) and the anthropologist Sherwood Washburn (1963), who celebrated human diversity and combated racism. Others have held opinions that have had unfortunate consequences. Many of scientists' early ideas on human races, for example, served to legitimize the racist beliefs of the societies in which those ideas held sway (Gould 1981). Freud and Jung have been accused of rationalizing sexism by developing theories that, reflecting the conventional wisdom of their times, assumed intrinsic differences between the sexes in aggressiveness, emotionality, and capacity for leadership. Some psychologists, sharing the prevailing view that a homosexual orientation is pathological, "proved" that homosexuals are neurotic and maladjusted. Later, controlled studies have revealed no differences in mental health between homosexuals and heterosexuals (Hooker 1957; Saghir and Robbins 1973).

Scientists who hold that there are biological differences between races or sexes are not necessarily, or even usually, racist or sexist in their social views. However, because the conclusions they voice can have such far-reaching consequences, it is important to scrutinize them carefully for valid evidence. Furthermore, in reading the literature on these subjects, it is critical to distinguish between statements about data, which may lay some claim to objectivity, and statements of the author's opinion, which embody the author's own values.[*] Values are neither true nor false; they are subjective or conventional, not statements of fact, and so cannot be verified or falsified by science. Science cannot tell us whether charitable acts, racist policies, or murder are good or evil; for this, we need a fount of values such as ethical philosophy—not science.

---

[*]In writing this chapter, I cannot assure the reader that it is untouched by my own values and opinions, which may be described as liberal. Readers must try to distinguish the evidence cited and the social implications I may impute to it.

Phylogenetic Relationships

"Light will be thrown on the origin of man and his history." These few words are Darwin's only reference to human evolution in the entire *Origin of Species*. Another 12 years would pass before he published on the subject, in *The Descent of Man, and Selection in Relation to Sex* (1871). But by 1859, when *The Origin of Species* was published, it was already obvious that if humans shared a common ancestry with other species, their closest relatives must be the great apes: the Asian orangutan (*Pongo pygmaeus*) and the African gorilla (*Gorilla gorilla*) and chimpanzees (*Pan troglodytes*, the common chimpanzee, and *P. paniscus*, the pygmy chimpanzee or bonobo, not yet described at that time) (Figure 26.1). Before *The Descent of Man* was published, Darwin's most enthusiastic supporters, Ernst Haeckel in Germany and Thomas Henry Huxley in England, described evidence for common ancestry with the apes. Darwin drew heavily on their work in *The Descent of Man*, in which he summarized masses of anatomical, embryological, physiological, and behavioral evidence for our common ancestry with the vertebrates, and with apes in particular.

A widely accepted estimate of the phylogeny of the major groups of living primates, based on anatomical and especially molecular sequence data (e.g., Bailey et al. 1991), is portrayed in Figure 26.2. The suborder Anthropoidea includes the New World monkeys and marmosets (Platyrrhini) and the Old World monkeys, apes, and humans (Catarrhini) (Table 26.1). The superfamily Hominoidea includes the gibbons and the great apes and humans. Traditionally, humans (*Homo*) have been placed in a separate family (Hominidae) from the orangutan, gorilla, and chimpanzees (family Pongidae). Thus, HOMINOID refers to the ape and human lineage, and HOMINID to the clade that includes humans, but excludes the apes.

The traditional assignment of humans to a separate family from the great apes might imply that *Homo* diverged from the apes before they diverged from each other. However, abundant molecular data, as well as cladistic analyses of morphological traits, show that human, chimpanzee, and gorilla form a monophyletic clade, sharing a more recent common ancestor with each other than with the orangutan (Goodman 1963; Sarich and Wilson 1967; Bailey et al. 1991; Shoshani et al. 1996).

(A)

(B)

FIGURE 26.1   Representative primates. (A) A prosimian, the lemur *Varecia variegatus*. (B) A New World monkey, the howler monkey *Alouatta villosa*. (C) An Old World monkey, the colobus monkey *Colobus polykomos*. (D) The gibbon *Hylobates lar* (Hylobatidae). (E) The orangutan, *Pongo pygmaeus*. (F) The gorilla, *Gorilla gorilla*. (G) The common chimpanzee, *Pan troglodytes*. (Photographs by the New York Zoological Society.)

(C)

(D)

Table 26.1 Two alternative classifications of the living primates

TRADITIONAL CLASSIFICATION	RECENTLY PROPOSED CLASSIFICATION
Order Primates	Order Primates
Suborder Prosimii	Suborder Prosimii
Infraorder Lemuriformes (lemurs, galagos, lorises)	Infraorder Lemuriformes (lemurs)
	Infraorder Lorisoformes (galagos, lorises)
Infraorder Tarsiiformes (tarsiers)	Suborder Tarsiiformes (tarsiers)
Suborder Anthropoidea	Suborder Anthropoidea
Infraorder Platyrrhini (New World primates)	Infraorder Platyrrhini (New World primates)
Infraorder Catarrhini	Infraorder Catarrhini
Superfamily Cercopithecoidea (Old World monkeys)	Superfamily Cercopithecoidea (Old World monkeys)
Superfamily Hominoidea	Superfamily Hominoidea
Family Hylobatidae (gibbons)	Family Hylobatidae (gibbons)
Family Pongidae (*Pongo, Gorilla, Pan*)	Family Hominidae
Family Hominidae (*Homo*)	Subfamily Ponginae (*Pongo*)
	Subfamily Homininae (*Gorilla, Pan, Homo*)

Source: Traditional classification after Fleagle (1988); recently proposed classification based on recent interpretations of the phylogenetic relationships (after Shoshani et al. 1996).

[a]Taxa of lower rank have been omitted except in Hominoidea. *Pongo* = orangutan, *Gorilla* = gorilla, *Pan* = chimpanzees, *Homo* = human.

(E)

(F)

(G)

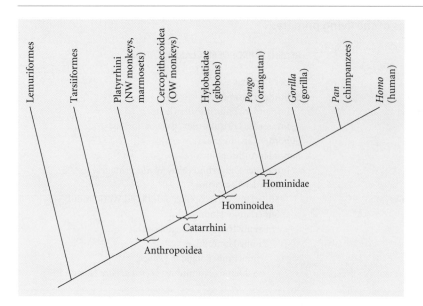

FIGURE 26.2 The phylogeny of the major groups of primates, based on morphological and molecular analyses. The names along the interior branches of the tree are the higher taxa to which the branches are assigned. Following recent suggestions, *Pongo*, *Gorilla*, *Pan*, and *Homo* are all assigned to the family Hominidae in this diagram.

The relationships among *Homo, Pan* (chimpanzee), and *Gorilla* have until recently been unresolved because of their close molecular similarity. Overall, the DNA sequences of these species differ by only about 1 percent, and even in a rapidly evolving pseudogene, they differ at less than 2 percent of nucleotide positions (Bailey et al. 1991; Table 26.2). However, both chromosome features (Borowik 1995) and many molecular studies have now provided evidence that humans and chimpanzees are sister groups, more closely related to each other than to the gorilla. In Chapter 5, we cited Sibley and Ahlquist's (1984) study, based on DNA-DNA hybridization, and the analysis of the sequence of the ψη-globin pseudogene by Miyamoto and colleagues (1987). Many other DNA sequence analyses have supported this conclusion. In a statistical analysis of 14 independent sets of DNA sequence data, MaryEllen Ruvolo (1997) found such strong support for the sister-group relationship of humans to chimpanzees that, she concluded, "the problem of hominoid phylogeny can be confidently considered solved." Most mol-

ecular analyses have indicated that *Homo* and *Pan* diverged about 4.6–5.0 million years ago (Takahata 1995). The gorilla lineage diverged from the *Homo-Pan* lineage about 0.3–2.8 million years earlier (Ruvolo 1997).

What are the implications of the apparent phylogeny of hominoids and the divergence of the human lineage about 5 Mya? Although the common ancestor of humans and chimps need not have closely resembled either, it surely had many of the anatomical features of chimpanzees and gorillas, since these retain so many features of their common ancestor. Thus the common ancestor of humans and chimps was probably a largely arboreal African ape, with an opposable big toe, long arms relative to its legs, luxuriant body hair, and an ape-sized brain. It may well have walked on its knuckles, as do African apes today, but surely not on its hind feet only. It lived in groups with complex social relationships among the members, and did not form extended monogamous pair-bonds. Many of these features have long been postulated, but a few have been newly suggested by the recent phylogenetic analyses. More significantly, the distinctive features of our species, especially our brain size, capacity for language, and phenomenally elaborate cognitive and intellectual abilities, have evolved very rapidly. The only direct evidence that might tell us the history of morphological changes must come from the fossil record.

The Fossil Record

Interpreting the Hominid Fossil Record

There is no fossil record of the gorilla or chimpanzee lineages, which is not surprising if the ancestors of these species lived, as these species do today, in moist tropical forests, which are unconducive to fossilization. The only fossils that illuminate the origin and evolution of the hominids are classified, in fact, in the Hominidae. The oldest informative material dates to the early Pliocene, about 4.4 million years ago, but new material has been discovered at so high a rate

Table 26.2 **Divergence between nucleotide sequences of the ψη-globin pseudogene among orangutan (*Pongo*), gorilla (*Gorilla*), common chimpanzee (*Pan*), and human (*Homo*)**[a]

	PONGO	GORILLA	PAN	HOMO
PONGO		3.39	3.42	3.30
GORILLA			1.82	1.69
PAN				1.56
HOMO				0.38

Source: Data from Bailey et al. (1991).

[a]The percentage of divergence between two human sequences is given in the lower right cell, and divergences between *Homo* and other species are calculated using the average of these two sequences. Values are not corrected for multiple substitutions.

in the last few years that older hominid fossils may yet be found.

In certain respects, the hominid fossil record is excellent, and in other ways it is dreadfully inadequate. It provides unequivocal evidence of general, more or less unidirectional trends in many characters, such as cranial capacity, a measure of brain size (Figure 26.3). The fossils document mosaic evolution, showing that some features began evolving toward the modern condition before others. They show beyond any doubt that modern humans evolved through many intermediate steps from ancestors that in most anatomical respects were apelike. However, specimens of fossil hominids are too few, and too widely separated in time and space, to answer fully our questions about the details of hominid history. Thus, paleoanthropologists hold differing opinions on how many distinct evolving lineages (evolutionary species) existed at various times, and which earlier populations were ancestral to which later ones. Useful summaries of the hominid fossil record can be found in Simons (1993), Grine (1993), Wood (1992), and Ciochon and Fleagle (1993).

Australopithecines

The earliest fossil hominids are called AUSTRALOPITHECINES (Figure 26.4), described from a few localities in South

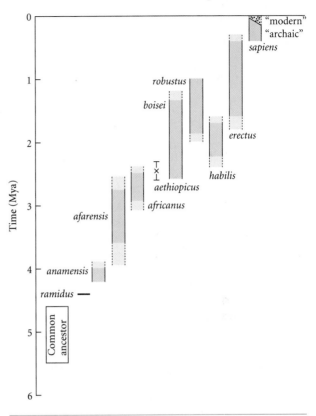

FIGURE 26.4 The approximate temporal distribution of hominid taxa in the fossil record. Uncertainty, indicated by dashed lines, is due to both imprecise dates of material and ambiguity in assignment of some specimens to one or another taxon. Fossils of the common ancestor of hominids and chimpanzees are not known. The × marks the date of *aethiopicus*.

FIGURE 26.3 Estimated body weights (A) and brain volumes (B) of hominid fossils, showing a steady, fairly gradual increase in brain volume, even though body size has not increased appreciably during the past 2 million years. The arrows indicate modern mean values. (After Jones et al. 1992.)

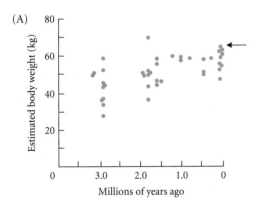

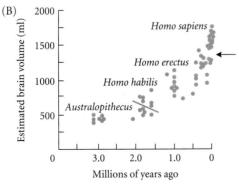

Africa and eastern Africa. The earliest known specimens, fragments from Ethiopia and Kenya, have been named *Ardipithecus ramidus* (dated at 4.4 Mya in the early Pliocene) and *Australopithecus anamensis* (dated at 4.2–3.9 Mya) (White et al. 1994; Leakey et al. 1995). The most extensive and informative early fossil material, from Tanzania and Ethiopia, has been named *Australopithecus afarensis* (Johanson and White 1979; Kimbel et al. 1994). It includes the famous "Lucy" (Figure 26.5), one of the most complete early hominid skeletons (about 40 percent of the skeleton has been recovered).

These forms, perhaps representing a single evolving lineage (species), have many "primitive" or ancestral features, indicating that they are not far removed from the common ancestor of humans and chimpanzees. These features include strong prognathism (lower face extending far beyond the eyes), a flat cranial base, relatively large canine teeth, long arms relative to the legs, a small brain (about 400 cc), and curved bones in the fingers and toes. These bones, similar to those of apes, imply that *afarensis* climbed trees. At the same time, the structure of the pelvis and hindlimb clearly shows that *anamensis* and *afarensis* were bipedal. Indeed, in 1978, Mary Leakey and coworkers found footprint traces of

FIGURE 26.5 Skeletal remains of *Australopithecus afarensis*. This famous specimen, nicknamed "Lucy," is unusually complete. Photograph © Science VU/Visuals Unlimited.

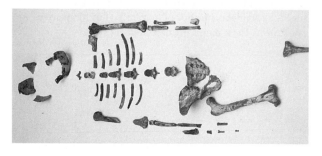

three individuals in rock formed from volcanic ash near an *afarensis* site in Tanzania. There is no evidence, anatomical or archaeological, that *afarensis* made stone or bone tools (Susman 1994).

Thus *afarensis*, nonhuman in most respects, had made the first critical step toward the human condition by walking upright on its hind legs. Another important feature of *afarensis* is its dentition: relative to Pliocene hominoids and modern pongids, the cheek teeth are larger and have thicker enamel, and the canines are reduced. In later hominids, the canines would become reduced still further.

Following *afarensis* in the late Pliocene of Africa, the number of hominid species, and the relationships among them, have not yet been resolved (Figure 26.6). There were at least two separate lineages, the "robust" and "gracile" australopithecines. The robust forms, often referred to as *Paranthropus*, had large molars and premolars and other features adapted for powerful chewing; they probably fed on tubers and hard plant materials (Figure 26.7D). *Paranthropus* fossils have been assigned to three species (*robustus, boisei, aethiopicus*), and extend from about 2.4 Mya to perhaps 1.8 or even 1.0 Mya. The robust australopithecines became extinct without having contributed to the ancestry of modern humans. Interestingly, *Paranthropus* appears to have made and used stone and bone tools (Susman 1988).

A key figure in hominid evolution is *Australopithecus africanus*, a "gracile" form from South Africa that has been dated only approximately, from perhaps 3.0 Mya to about 2.0 Mya. *A. africanus* differs from *A. afarensis* in only a few features, most notably a greater cranial capacity (averaging about 450 cc) (Figure 26.7C). Whether *africanus* is in the line of direct ancestry of modern humans, or became extinct without issue, is debated (Wood 1992). There is no clear evidence that *africanus* made tools.

Origin and Evolution of *Homo*

The earliest fossils referred to the genus *Homo*, from Tanzania, Ethiopia, and South Africa, range from about 2.4 to 1.6 Mya (latest Pliocene and early Pleistocene). They may include two species, but are generally all referred to *Homo habilis* (Wood 1992). *Homo habilis* is the epitome of a missing link no longer missing (Figure 26.7E). The oldest specimens are very similar to *Australopithecus africanus*, and the younger ones grade into the later form *Homo erectus*. Compared with *Australopithecus*, *Homo habilis* more nearly resembles modern humans in its greater cranial capacity (610 to nearly 800 cc), reduced prognathism, and shorter tooth row. Although the limbs retain apelike proportions that suggest an ability to climb, the structure of the leg and foot indicates that its bipedal locomotion was more nearly human than that of the australopithecines. *Homo habilis* is associated with stone tools (referred to as Olduwan technology), and with animal bones that bear cut marks and other signs of hominid activity (Potts 1988).

Homo habilis grades into later hominid fossils, from about 1.6 million to about 300,000 years (300 Ky) ago, referred to *Homo erectus*. *Homo erectus* grades into *Homo sapiens* later in its history. Most, though not all, authorities think that *habilis*, *erectus*, and *sapiens* are a single evolutionary lineage. In most respects, *erectus* from the middle Pleistocene onward has fairly modern human features: the skull is rounded, the face is steeper and less prognathous than in earlier forms, the teeth are smaller, and the cranial capacity is greater, averaging about 1000 cc and evidently increasing over time (Figure 26.7F). At least 1 Mya (perhaps as far back as 1.7 Mya), *erectus* spread from Africa into Asia, extending eastward to China and Java. Throughout its range, *erectus* is associated with stone tools, termed the Acheulian culture, that are more diverse and sophisticated than the Olduwan tools of *H. habilis*. The use of fire was widespread by half a million years ago.

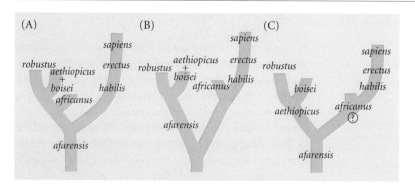

FIGURE 26.6 Three hypotheses of the phylogenetic relationships among fossil hominids, advocated by different authors. Hypothesis C is perhaps the most widely accepted, although with uncertainty about whether *Australopithecus africanus* is a separate lineage from *Homo habilis* or is its direct ancestor. (After Grine 1993 and Jones et al. 1992.)

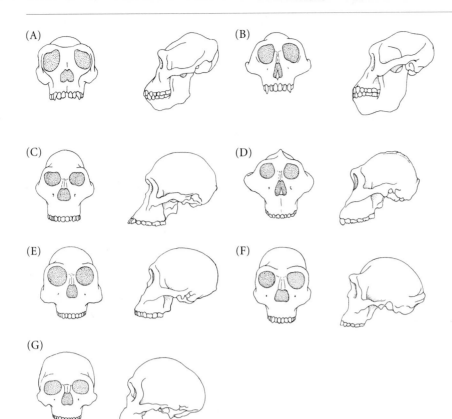

FIGURE 26.7 Frontal and lateral reconstructions of the skulls of a chimpanzee and some fossil hominids. (A) Chimpanzee. Note large canine teeth, projecting incisors, low forehead, prominent face and brow ridge. (B) *Australopithecus afarensis*. Some of the same features as in the chimpanzee are evident. (C) *Australopithecus africanus*. Note small canines, higher forehead. (D) *Australopithecus boisei*. The large molars, heavy cheekbones, and slight crest on the skull are elements of a complex of characters adapted for chewing tough foods such as roots. (E) *Homo habilis*. The face projects less, and the skull is more rounded, than in *A. africanus*. (F) *Homo erectus*. Note the still more vertical face and rounded forehead. (G) *Homo sapiens*, Neanderthal type. Rear of skull is more rounded than in *H. erectus*. (A, B after Jones et al. 1992; C–G after Howell 1978.)

Although there is morphological gradation between *Homo erectus* and *Homo sapiens*, it is possible that some populations of *sapiens* were reproductively isolated from populations that retained the morphology of *erectus*, and therefore constituted a separate biological species (not just a chronospecies, or morphologically altered descendant). Asian specimens referred to *sapiens* date back to 40,000 years. However, Javan specimens referred to *erectus* have been found in strata recently dated at 27,000–53,000 years (Swisher et al. 1996). This evidence, as well as genetic evidence to which we shall soon turn, suggests that multiple contemporaneous species of hominids have existed not only in the remote, but also in the recent, past.

Thus, throughout hominid evolution, different features evolved at different rates (mosaic evolution). On average, brain size (cranial capacity) increases throughout hominid history, although not at a constant rate, and there are progressive changes, from *afarensis* to *africanus* to *erectus* to *sapiens*, in many other features, such as the teeth, face, pelvis, hands, and feet. The very fuzziness of the taxonomic distinctions among the named forms attests to the mosaic and gradual evolution of hominid features. Although many issues remain unresolved, the most important point is fully documented: *modern humans evolved from an apelike ancestor*.

Homo sapiens

Most hominid fossils from about 0.3 Mya onward, as well as some African fossils about 0.4 My old, are referred to as *Homo sapiens*. Mean cranial capacity increased during the history of *sapiens*, from about 1175 cc at 0.2 Mya to its modern mean of 1400 cc. Middle Pleistocene specimens, called "archaic *sapiens*," differ in relatively minor respects from the "anatomically modern *sapiens*" that followed them in the late Pleistocene.

The best-known populations of archaic *Homo sapiens* are the Neanderthals[*] (*Homo sapiens neanderthalensis* or, according to some authors, *Homo neanderthalensis*). Neanderthals had dense bones, thick skulls, and projecting brows; contrary to the popular image of a stooping brute, they walked fully erect, had brains as large as or even larger (up to 1500 cc) than ours, had a fairly elaborate culture, including a variety of stone tools (Mousterian culture), and probably practiced ritualized burial of the dead. Their remains date from an earlier age (about 110 Kya) in the Middle East than in Europe (about 50 Kya).

"Modern *sapiens*," anatomically virtually indistinguishable from today's humans, appeared earlier in Africa (ca. 170 Kya) than elsewhere. It overlapped with Neanderthals in the Middle East for much of the Neanderthals' history, but abruptly replaced them in Europe about 40 Kya years ago. By 12,000 years ago, and possibly earlier, modern humans had spread from northeastern Asia across the Bering Land

---

[*]Or *Neandertals*, a frequently used modern spelling. The name means "Neander valley," the site in Germany where they were first found. *Thal* is the old, and *Tal* the modern, spelling of the German word for valley. In either spelling, it is pronounced "tal."

Bridge to northwestern North America, and thence rapidly throughout the Americas.

"Upper Paleolithic" culture began about 40 Kya. The earliest of several successive cultural "styles" in Europe, the Aurignacian, is marked by stone tools more varied and sophisticated than those of the Mousterian culture. Moreover, culture became more than utilitarian, for art, self-adornment, and possible mythical or religious beliefs are increasingly evident from about 35 Kya onward. Agriculture, which enormously increased population density and began the human transformation of the earth, is about 11,000 years old. There is, at least at present, no way of knowing which of these cultural advances were associated with genetic changes in the capacity for reason, imagination, and awareness, but they are not paralleled by any increase in brain size or other anatomical changes.

Causes of Hominid Evolution

What were the advantages of the changes that transpired between the apelike ancestor of *Australopithecus* and modern *Homo sapiens*? Why did hominids evolve erect posture and bipedal locomotion, changes in the structure of the feet and hands, reduced canine teeth, and above all, a larger brain, language, and a capacity for reason that seems to exceed any other species'?

Most of the methods we can use to demonstrate the adaptive value of the characteristics of other organisms (see Chapter 12) are difficult or impossible to apply to these questions. All modern humans are bipedal, for example, so we cannot assess the effects on fitness of variation in this character—and even if we could, such variation in the context of modern societies would tell us nothing about how it affected fitness in a prehominid ancestor, living in an entirely different ecological and social environment. The comparative method, which may provide insight into such issues as why some species have few and others many offspring, is of little or no use in testing hypotheses about the evolution of unique characteristics such as human intelligence. Consequently, most proposed explanations of hominid characters are highly speculative, and these hypotheses are very difficult to test. Some anthropologists, indeed, feel that there is little point in speculating on these questions. For our purposes, it suffices only to note a few of the hypotheses that have been advanced (Lovejoy 1981; Fedigan 1986).

Evidence on such hypotheses is indirect, consisting mostly of inferences from studies of other primates and from ethnography (the study of contemporary cultures), and material evidence—anatomy and artifacts—from paleoanthropology. Much of human evolution is presumed to have been influenced by the social milieu of early hominids, and the behavior of other primates and of contemporary hunter-gatherer cultures has been supposed, rightly or wrongly, to provide insight into what that social structure may have been like.

The erect posture and bipedal locomotion are the first major documented changes toward the human condition. The earliest australopithecines apparently lived in forests, judging from the animal and plant remains with which the specimens are associated. Thus the old idea that an erect posture enabled them to peer over savannah grasses to watch for predators is perhaps less plausible than the hypothesis that bipedalism freed the arms for carrying food back to the social unit, especially to an individual's mate and offspring. Food sharing occurs in chimpanzees, which have a complex social structure that includes matrilocal family groups and "friendships."

Chimpanzees make and use a variety of simple tools, such as stone and wooden hammers used to crack nuts and twigs fashioned to "fish" termites out of their nests. It is possible that the advantages of using a greater variety of tools selected for greater intelligence and brain size. However, many authors, beginning with Darwin, have emphasized that social interactions, such as learning how to provide parental care, forming cooperative liaisons with other group members, and competing for resources within and among groups, would place a selective premium on intelligence, learning, and communication. Some authors, drawing on the theory of reciprocal altruism (see Chapter 20), have proposed that the need to detect cheaters in social exchanges placed a premium on advanced cognitive abilities (Alexander 1979). Selection for greater intelligence, for whatever reasons, entailed selection for a larger brain. The effects of this change included the evolution of a larger birth canal in the female pelvis, and more importantly, a longer interval between birth and sexual maturity, characterized by prolonged growth, learning, and dependence on parents.[*]

The Origin of Modern Human Populations

The Transition from Archaic to Modern *Homo sapiens*

Homo erectus and archaic *Homo sapiens* were broadly distributed throughout Africa and Asia by about a million years ago. How are these ancient populations related to the different human populations of today? This has long been a controversial question.

Based on the morphology of fossil specimens, advocates of the *multiregional hypothesis* (Wolpoff 1989) hold that archaic *sapiens* populations in Africa, Europe, and Asia all evolved into modern *sapiens*, with gene flow spreading modern traits among the various populations (Figure 26.8A). According to this hypothesis, there should exist genetic differences among modern Africans, Europeans, and Asians that trace back to the genetic differences that developed among populations of *erectus* and archaic *sapiens*

---

[*]Many human features, such as skull shape and sparse body hair, have been interpreted as neotenic changes associated with the prolonged juvenile period of human development (Gould 1977). However, some authors argue that many of these features are actually hypermorphic (see Chapter 23) relative to their condition in African apes (Shea 1988; McKinney and McNamara 1991).

FIGURE 26.8 Two hypotheses on the origin of modern humans. (A) The multiregional hypothesis posits a single wave of expansion by *Homo erectus* from Africa to parts of Asia and Europe, and continuity of descent in each region to the present day. The horizontal dotted lines represent gene flow, by which derived characteristics of *H. sapiens* were spread throughout the populations. (B) The replacement hypothesis proposes that populations of archaic *H. sapiens*, derived from *H. erectus*, became extinct when modern *H. sapiens* expanded in a second wave of colonization. (C) The expected form of a gene genealogy according to the multiregional hypothesis. Haplotypes endemic to Africa (squares), Asia (circles), and Europe (triangles) should have a long history of divergence, and hence differ by many nucleotide substitutions. Solid symbols represent recent movement of genes among regions. (D) The gene genealogy expected under the replacement hypothesis is shallower, with fewer nucleotide differences among haplotypes because of their recent common ancestry, and regional clades of haplotypes are less distinct. In both C and D, haplotypes from Africa, the region of origin, vary more in nucleotide sequence and form more basal branches than those endemic to other regions.

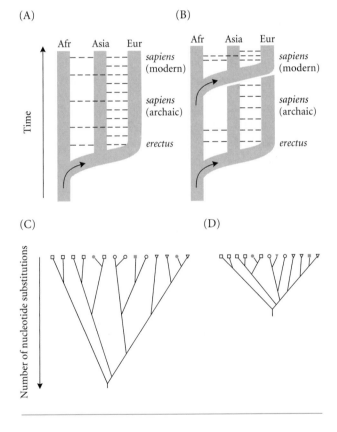

stantial extent (Figure 26.8B). That is, the modern *sapiens* that evolved from archaic *sapiens* in Africa was reproductively isolated from Eurasian populations of archaic *sapiens*—it was a distinct biological species. According to this hypothesis, most of the world's populations of archaic *sapiens* became extinct due to competition, and most genes in contemporary populations are descended from those carried by the population of modern *sapiens* that spread from Africa. The multiregional and replacement hypotheses make different predictions about the form of human gene trees (Figure 26.8C, D; see Chapter 22).

Among the first molecular studies supporting the replacement hypothesis were those carried out in Allan Wilson's laboratory at the University of California, Berkeley. These workers studied sequence diversity in mitochondrial DNA (Cann et al. 1987; Vigilant et al. 1991), and calibrated the rate of sequence evolution within the human species by the divergence between humans and chimpanzees. Their estimate of the gene tree (Figure 26.9) coalesced to a single ancestral gene copy between 166,000 and 249,000 years ago. The gene tree presented by the Berkeley group implied that

FIGURE 26.9 A gene tree of 135 mitochondrial DNA haplotypes, published in the first major molecular study of the origin of modern human populations. The tree is arranged in a circle to save space. A chimpanzee haplotype was used as an outgroup. The phylogenetically basal haplotypes in this diagram came from African populations, as expected if modern non-African populations differentiated after their ancestors migrated out of Africa. Later analyses of these data showed that this tree is not more parsimonious than others that fail to support an African origin. However, subsequent studies of other DNA sequences have provided support for an African origin of the human species. (After Vigilant et al. 1991.)

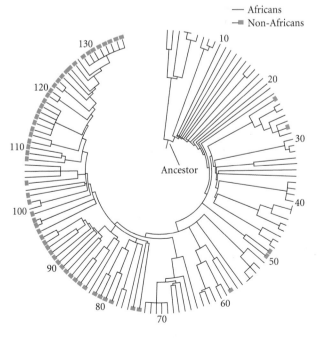

nearly a million years ago. Thus, for example, some genes in European populations today would be descended from those carried by European Neanderthals.

In contrast, the *replacement hypothesis*, often called the *out-of-Africa hypothesis* (Vigilant et al. 1991), holds that after *H. erectus* spread from Africa to Asia and Europe and evolved into archaic *H. sapiens*, modern *sapiens* evolved from archaic *sapiens* in Africa, spread throughout the world in a second expansion, and replaced the populations of archaic *sapiens* without interbreeding with them to any sub-

the ancestral sequence came from an African population. (The carrier of this ancestral, maternally inherited gene was dubbed "mitochondrial Eve," and the popular press mistakenly interpreted this study to mean that a single woman existed at that time. In a finite population of any size, however, all gene copies will be descended from a single copy.)

Although the analyses in these early papers have been criticized, some later studies of both mtDNA and nuclear sequences have arrived at similar conclusions (Ruvolo et al. 1993; Horai et al. 1995; Takahata 1995). Those studies have found that gene genealogies are rooted among African haplotypes, and have reported greater sequence variation among Africans than among non-Africans, as would be expected if non-Africans had undergone a bottleneck in population size when they were founded. In some DNA sequences, levels of divergence among copies from populations throughout the world are low enough to suggest that their common ancestor existed as little as 156,000 years ago (Goldstein et al. 1995). If, as in the multiregional hypothesis, contemporary human populations were descended from populations of archaic *Homo sapiens* (and of *H. erectus*) throughout both Africa and Eurasia, genetic drift should have been slower than in the replacement scenario, and the genes should have accumulated more mutational differences than are observed.

Some studies of DNA sequences, on the other hand, seem not to support the replacement hypothesis. For example, Rosalind Harding and her collaborators (1997), in a study of the β-hemoglobin gene, found that the level of sequence divergence among gene copies is greater in Asian than in African populations. In both regions, the number of mutational differences among haplotypes is great enough to imply coalescence to an ancestral gene copy more than 200,000 years ago—before the transition from archaic to anatomically modern *Homo sapiens* in the fossil record. This would suggest that archaic *sapiens* in Asia was not a reproductively isolated species, but instead contributed to the ancestry of contemporary human populations.

Thus at this time, genetic data yield conflicting evidence on the origin of modern *Homo sapiens*. The conflict might arise from differences among sequences (e.g., mitochondri-al compared to nuclear genes) in the extent to which natural selection has shaped contemporary variation (Hey 1997). For instance, background selection may have been more important in pruning the mitochondrial gene tree, resulting in more recent common ancestry of contemporary mitochondrial haplotypes (see Figure 3D in Chapter 22). The origin of modern humans provides an especially interesting opportunity for population genetics to reveal history.

Migrations

Allele frequencies at loci encoding enzymes and blood types have been used to infer relationships among human populations and movement among geographic regions. Genetic differences among populations parallel their linguistic differences to some extent (Figure 26.10), suggesting that both genes and languages have a common history of divergence in isolation (Cavalli-Sforza et al. 1994). Thus these genetic relationships provide insights into the history of modern human populations (Figure 26.11). Luca Cavalli-Sforza and colleagues (1994) suggest that modern humans spread first from Africa to southern Asia, and thence to Australia and northern Asia. From northern Asia, populations spread into Europe (at about the time the Neanderthal archaic *sapiens* disappeared), and into America via the Bering Land Bridge that connected Siberia and Alaska.

These events occurred long before the emergence of agriculture, at a time when human populations were sparse and subsisted by hunting and gathering. Agriculture arose 10,000–11,000 years ago, probably independently in central Africa, the Middle East, and Mexico through northern South America. It resulted in dramatic increases in population size, accompanied by movements of populations that have continued ever since. Some such movements can be inferred from clines in allele frequencies. From such genetic data, for example, it appears that from about 8000 to 5500 B.C., agriculture spread from the Middle East northward and westward into Europe by "demic diffusion" rather than by "cultural diffusion"—that is, by the movement of agriculturalist peoples rather than by cultural spread from one sedentary population to another (Sokal et al. 1991; Cavalli-Sforza et al. 1994).

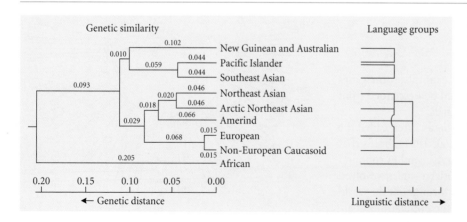

FIGURE 26.10 The average degree of genetic differentiation among groups of human populations, measured by allele frequency differences at numerous enzyme and blood group loci. More similar populations may be more closely related, although gene flow among populations may also contribute to their similarity. The diagram at right indicates which populations fall into the same language groups. (After Cavalli-Sforza et al. 1994.)

FIGURE 26.11 The possible major routes of expansion of modern human populations in the last 100,000 years or so, based largely on genetic data. These dates are very approximate, and may be revised in light of future research. (After Cavalli-Sforza et al. 1994.)

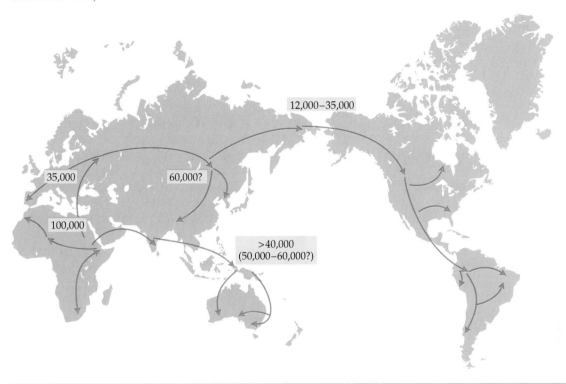

Racial and Ethnic Groups

Homo sapiens is a single biological species. There exist no biological barriers to interbreeding among human populations, although there frequently are cultural barriers—for example, there is assortative mating in Australia between people of Scottish and Irish origin. Nevertheless, cultural barriers and social taboos against interbreeding often break down. For instance, despite racist barriers in the United States, a blood group allele *Fy* that is fairly abundant in European populations, but virtually absent in Africa, was found to have a frequency of 0.11 in the black population of Detroit, Michigan, from which it was calculated that the admixture of genes from the white population has been 26 percent (Cavalli-Sforza and Bodmer 1971).

Physical characteristics such as skin color, hair texture, shape of the incisors, and stature vary geographically in humans, and have been used by various authors to define anywhere from 3 to more than 60 "races." As the variation in this estimate suggests, the number of races is arbitrary, depending only on the number of features studied and the degree of difference used to make distinctions. Each supposed racial group can be subdivided into an indefinite number of distinct populations. In Africa, for example, Congo pygmies are the shortest of humans, and the Masai are among the tallest.

As in other species, there is extensive genetic variation within individual human populations. For example, individuals are heterozygous at about 10–14 percent of loci screened by electrophoresis and blood typing (Lewontin 1972; Nei and Roychoudhury 1974), and many physical and behavioral traits also vary. Some alleles have higher frequencies in certain populations than in others; for instance, the allele for sickle-cell hemoglobin has a higher frequency in central Africa than elsewhere, and a mutation causing cystic fibrosis is most prevalent in northern Europe. However, the pattern of overall genetic variation among populations differs substantially from traditional racial divisions (Figure 26.12). Morphologically similar peoples are not necessarily genetically most similar overall; for example, Philippine and Malay Negritos share some morphological features with Africans, but are genetically as distinct from them as are morphologically less similar populations.

Genetic differences among human populations consist of allele frequency differences only, and the genetic "distances" (see Chapter 9) among major "racial groups" are less than those commonly found among subspecies in other mammals (Nei and Roychoudhury 1982; Lewontin 1972). At no known loci are "races" or other regional populations fixed for different alleles (Cavalli-Sforza et al. 1994). Although not all alleles occur in all populations, about 85 percent of the genetic variation in the human species is among individuals within populations, and only about 8 percent is among the major "races" (Nei and Roychoudhury 1982). Thus, "if everyone on earth became extinct except for

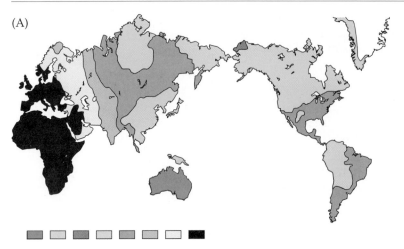

(A)

FIGURE 26.12 (A) A division of the world's human populations into eight classes of genetic similarity, based on overall difference and similarity at numerous enzyme and blood-group loci. The eight classes represented in the key are arrayed in order of increasing difference. (B) The geographic distribution of skin color, classified in eight grades of pigmentation intensity. Some considerable differences between the two maps are evident. (After Cavalli-Sforza et al. 1994.)

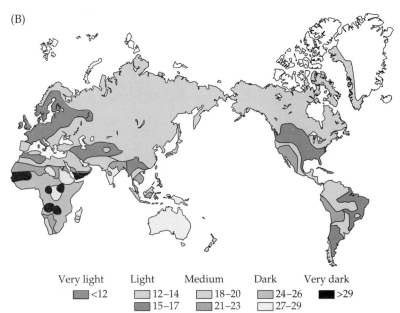

(B)

Very light	Light	Medium	Dark	Very dark
<12	12–14	18–20	24–26	>29
	15–17	21–23	27–29	

the Kikuyu of East Africa, about 85 percent of all human variability would still be present in the reconstituted species" (Lewontin et al. 1984).

Similarly complex patterns of geographic variation in other species have led some taxonomists to reject the division of species into subspecies or races. Among humans, likewise, the concept of distinct races is scientifically poorly justified. Moreover, in view of the overwhelming genetic similarity of human populations, there is little reason to presume, a priori, that a trait will vary substantially among so-called "races."

Genetic Variation in Human Populations

Much of the genetic variation in human populations may well be selectively nearly neutral; indeed, neutrality has been assumed in some of the analyses we have discussed. However, a great deal of human genetic variation, especially in genes that can cause disease or deficiency, is certainly

or probably subject to natural selection. Much of the huge field of human genetics concerns such traits.

Human genetics is inseparable from evolutionary genetics. Genes underlying hereditary disorders can be located by tracing the inheritance of a trait in family pedigrees and by its association with genetic markers (linkage mapping). Because humans cannot be experimentally crossed like fruit flies, such analyses depend on estimates of allele frequencies in populations and on the theory of linkage disequilibrium among loci (see Chapter 14), both of which are the subject matter of population genetics. Many of the statistical methods for determining inheritance, as well as the relative importance of genes and environment, have been developed by evolutionary geneticists. (See Weiss 1993 for an introduction to the methods and content of human genetics.)

Our knowledge of human genetics is growing explosively as geneticists use molecular methods to identify and sequence genes, identify gene products and their functions, and identify the mutations that may impair those functions. To facilitate the growth of such knowledge, geneticists throughout the world are contributing to the HUMAN GENOME PROJECT, an effort initiated in 1989 to sequence all 3 billion base pairs, including perhaps 100,000 functional genes, in the human genome (Cooper 1994).

Polymorphisms

The causes of a few human polymorphisms are known in part. The best understood are the several variant hemoglobins, such as those causing sickle-cell disease and thalassemia, that, despite their severely debilitating effects in homozygous condition, are maintained because heterozygotes have heightened resistance to malaria (see Chapters 13 and 22). The same is true of a mutation causing a deficiency of G6PD (glucose-6-phosphate dehydrogenase), which causes hemolytic anemia in 400 million people worldwide if they ingest certain foods or suffer certain infections, but which greatly reduces the risk of malaria in heterozygotes (Ruwande et al. 1995). In other cases, mutant alleles have high frequencies for unknown reasons, but are potentially subject to natural selection. For example, a deletion in the CRC-5 chemokine receptor gene, which has a frequency of about 0.10 in Caucasian populations, seems to confer resistance to the human immunodeficiency virus (HIV), the cause of AIDS, by blocking its entry into cells (Samson et al. 1996). However, the spread of HIV is much too recent to account for the high frequency of this allele. In yet other cases, such as the highly variable major histocompatibility gene complex (MHC), the high degree of sequence divergence among haplotypes implies that the polymorphism is very old and thus has been maintained by balancing selection, but the agents of selection are unknown (see Figure 8 in Chapter 22).

Rare Deleterious Alleles

About 5 percent of the human population displays genetic disabilities of one or another kind, and genetic defects account for about 20 percent of infant deaths and perhaps half of all miscarriages (McConkey 1993). Almost everyone carries at least one recessive allele somewhere among his or her genes that, if homozygous, would be life-threatening (see Chapter 9). Deleterious alleles occur at a great many loci. Most of them persist in populations at low frequencies, probably due to a balance between recurrent mutation and purifying selection (see Chapter 13). Mutations that exert their deleterious effects late in life, such as those that cause Huntington's disease, are subject to only weak selection, and so are expected to attain relatively high equilibrium frequencies (see Chapters 13 and 19). Although some genetic disabilities are simply inherited as single alleles with high penetrance (i.e., most individuals with the genotype in question express the phenotype), many display complex inheritance. For instance, Alzheimer's disease, entailing gradual loss of memory and cognitive abilities, can be caused by mutations at any of at least four loci, whose effects differ in age of onset and in their interaction with environmental risk factors such as smoking (Pericak-Vance and Haines 1995).

The molecular bases of genetic diseases are increasingly well understood. One of the first triumphs of molecular human genetics was the identification of the gene that, in a mutant state, causes cystic fibrosis, a lethal syndrome of pulmonary obstruction, insufficiency of pancreatic enzyme function, and other disorders. The identification and sequencing of this gene, aided by principles of linkage analysis derived partly from population genetics, enabled researchers to determine the normal function of its product, which serves as an ion channel in cell membranes. Recessive mutations cause cystic fibrosis in one out of every 2500 persons of northern European ancestry—i.e., an aggregate frequency of 0.02. More than 350 different mutations of the gene are thought to produce the pathology, but more than 70 percent of cases are attributed to a single three-base-pair deletion that deletes a single amino acid from the protein product. Studies of this gene in mice have raised the possibility that this mutation is advantageous in heterozygotes because it reduces loss of body fluids due to bacterial infections (Zielenski and Tsui 1995).

The explosive growth in our understanding of human genetics, like the fire brought to humankind in the Greek legend of Prometheus, can be used for both good and ill. Genetic markers already make it possible to screen people for hundreds of mutations that could impair either their children or themselves in later life. In some cases, such knowledge will provide opportunities for therapeutic interventions, and in many cases, it will help couples decide whether or not to have potentially affected children. But if therapy is not feasible, individuals may not wish to know that they should expect a gloomy genetic fate. Moreover, genetic information could easily be used to deny people employment, health or life insurance, or other needs. Techniques of genetic engineering, already developed and tested in animals, could be used for GENE THERAPY: inserting normal genes into somatic cell lines, or even germ cell lines, of individuals bearing defective genes, thus ameliorating their deficiencies (Lyon and Gorner 1995; Sokol and Gewirtz 1996). For economic and other practical reasons, it is unlikely that gene therapy will ever be practiced on a large enough scale to alter the course of human evolution, but even on a limited scale, genetic engineering and genetic knowledge raise many ethical questions that society has hardly begun to answer (Kitcher 1996).

Natural Selection and the Evolutionary Future

Is the human species continuing to evolve? To answer this question, we need to know how the two major agents of allele frequency change, genetic drift and natural selection, are acting now and are likely to act in the future.

Until the advent of agriculture, humans subsisted by hunting and gathering. Based on studies of the few contemporary populations with this mode of life, it is likely that population density was low and that most populations consisted of rather small groups whose travels were quite localized. Confined by topographic barriers that we today can surmount in hours, such populations would have diverged by genetic drift at nearly neutral loci. Much of the geographic variation in the human species is undoubtedly a consequence of this population structure. Even today, many agricultural populations are highly sedentary. Allele frequency differences have been discerned, for example, among villages along the shore of Lake Atitlán in Guatemala (Cavalli-Sforza and Bodmer 1971). However, the enormous growth in population sizes, especially since the Industrial Revolution, coupled with our vastly greater mobility, at least in industrialized societies, will make population differentiation by genetic drift much slower now and in the foreseeable future, and will blur previously existing differentiation.

The development of ever more complex culture and technology over the last 10,000 years or more has reduced mortality from many agents that must have afflicted our remote ancestors, and thus has reduced the intensity of natural selection on many characters. Modern technology, education, medicine, and especially improved sanitation have greatly reduced prereproductive mortality in industrialized nations, reducing the opportunity for natural selection. The opportunity for selection can be measured by the variance in prereproductive survival or in reproduction. James Crow (1966) calculated the opportunity for selection in three populations in Chile: a nomadic Andean tribe, an agricultural village, and an industrial town (Table 26.3). The prereproductive mortality rate was highest for the nomads, but the industrial society had the greatest variance in family size. Overall, the least "modern" of the three populations had the highest index of opportunity for selection. In the United States, variance in fertility now accounts for over 90 percent of the total opportunity for selection, so the traits most likely to evolve by selection are those correlated with fertility. It is not clear what traits these may be.

Many conditions, such as nearsightedness, that would have increased the risk of death for our hunter-gatherer ancestors can now be remedied by medical or technological means, such as eyeglasses. Phenylketonuria, caused by a recessive mutation, no longer need cause mental retardation and death; it is cured by prescribing a diet without the amino acid phenylalanine. However, purifying selection still acts on mutations that cause prereproductive mortality, such as those responsible for cystic fibrosis. And it is important to bear in mind that vast numbers of people throughout the world lack money, education, and access to health care, and so are subject to potential sources of natural selection that are greatly attenuated in affluent, industrialized societies.

Selection acts on at least some quantitative phenotypic characters in human populations. Birth weight, for instance, is subject to stabilizing selection (see Figure 20 in Chapter 14). In a study of Michigan residents, fertility appeared to be bimodally distributed with respect to IQ ("intelligence quotient") score, suggesting a pattern of disruptive (diversifying) selection (Table 26.4). However, there is little evidence of *directional* selection on physical, physiological, or mental traits in modern societies. Hence, we should not expect humans of the future to look much different from those of today. Perhaps the characteristics most likely to continue to evolve are those affecting resistance to infectious diseases. Antibiotics and medical care are unavailable to many millions of people, and hundreds of thousands die yearly from malaria, schistosomiasis, bacterial diarrhea, and other infections. The pathogens that cause malaria, tuberculosis, and other diseases have evolved resistance to antibiotics in many regions, posing a crisis in health care even in affluent nations (Neu 1992). New pathogens have emerged, such as hantavirus and, most devastatingly, the human immunodeficiency virus (HIV) that causes AIDS. Thus there still exists considerable scope for natural selection to act on genetic variation for resistance.

Human Behavior

No topics in evolutionary biology are more controversial than those concerning the evolution and genetics of human behavioral characters, including the cognitive abilities described as "intelligence." Humans are, to our knowledge, unique among species in their capacity for symbolic, syntactical language; they are enormously more capable of learning and transmission of information than any other species; and they are clearly extremely flexible in their behavior. How and why these capabilities evolved is a subject

---

Table 26.3 Demographic differences and indices of opportunity for selection in three contemporary Chilean populations

	INDUSTRIAL COASTAL TOWN	UPLAND PASTORAL VILLAGE	HIGHLAND NOMADIC SHEPHERDS
Mean number of children	4.3	5.9	6.1
Variance in number of children	8.5	7.5	6.4
Proportion surviving to adulthood	0.87	0.75	0.42
Selection index (mortality), I_m	0.15	0.33	1.38
Selection index (fertility), I_f	0.45	0.22	0.17
Index of total opportunity for selection, I	0.67	0.62	1.78

Source: From Crow (1966).

Table 26.4 Reproduction and relative fitness in relation to IQ[a]

IQ RANGE	SAMPLE SIZE	PERCENTAGE LEAVING NO OFFSPRING	AVERAGE NUMBER OF OFFSPRING	PER CAPITA RATE OF INCREASE (r)	AVERAGE GENERATION TIME (T) IN YEARS	RELATIVE FITNESS CALCULATED BY NUMBER OF OFFSPRING	RELATIVE FITNESS CALCULATED AS e^{rT}
≥120	82	13.41	2.598	+0.00889	29.42	1.000	1.0000
105–119	282	17.02	2.238	+0.00389	28.86	0.8614	0.8674
95–104	318	22.01	2.019	+0.00033	28.41	0.7771	0.7838
80–94	267	22.47	2.464	+0.00745	28.01	0.9484	0.9600
69–79	30	30.00	1.500	−0.01000	28.76	0.5774	0.5839

Source: From Bajema (1963).

[a]Individuals (non-immigrant whites) were scored for IQ at mean age 11.6 years in 1916–1917 in Kalamazoo, Michigan; survival and reproduction through age 45 were recorded. Calculations of the generation time and the per capita increase are based on the equations presented in Chapter 4. The relative fitness calculated from e^{rT} expresses the rate of population growth of a class relative to that of the ≥ 120 class, taking into account the generation time of the class relative to that of the whole population.

of great speculation, but little direct evidence. At the same time, the very fact that these abilities have evolved, and must therefore have a genetic foundation, implies to some researchers that some specific behaviors have a biological, genetic foundation and may, indeed, be constrained by our genes. In contrast, other researchers, pointing to our capacity to learn and to the immense variation in almost every human behavior, hold that genes have surrendered their sovereignty to culture—that human behavior is overwhelmingly a product of social conditioning and learning. The resolution of this debate—about the relative importance of "nature" and "nurture," as it is often phrased—may have far-reaching social and political implications (Lewontin et al. 1984; Degler 1991). This is an area in which the evaluation of evidence is sometimes colored by the social and ideological views of scientists and nonscientists alike.

Culture

Culture, a major topic in anthropology, is based on learning. Cultural traits, ranging from words to marriage rules to religious beliefs, can be transmitted vertically (from older to younger generations) or horizontally (as when we imitate our peers). Almost everyone agrees that the great majority of cultural differences among groups are consequences of cultural transmission rather than genetic differences, for pronounced cultural differences exist among geographically contiguous peoples that are genetically indistinguishable and interbreed. However, some population differences may be consequences of an interaction between culture and genetics. For example, many Asian people are genetically incapable of digesting milk products as adults, and these foods are traditionally not part of their diet—but whether the cultural practice of not drinking milk has influenced allele frequencies, or vice versa, is not known. Some authors have developed models of GENE-CULTURE COEVOLUTION (e.g., Boyd and Richerson 1985; Cavalli-Sforza and Feldman 1981; Durham 1991), in which genotypes may differ in their propensity to adopt a cultural trait, and culture in turn may affect fitness and thus alter allele frequencies.

Cultural "evolution" differs importantly from biological evolution in that a trait that an individual acquires during his or her lifetime can be transmitted in a way reminiscent of Lamarckian inheritance (which does not occur with genetic traits: see Chapter 3). Consequently, cultural change can occur far more rapidly than genetic evolution. Moreover, the selection guiding cultural change is selection of the traits themselves, not of the individuals practicing them: the automobile replaces the horse because of its perceived advantage, not because motorists have more children than equestrians. The advantage may be illusory, of course, as illustrated by cultural traits such as smoking. Nor do cultural changes follow simply from the desires or actions of a society's members; they may arise in complex ways from historical events and from economic and political forces that arise from a society's structure and govern its members' behavior.

The Evolution of Human Behavior

Arguments about genetic influences on human behavior concern two related but distinct problems. In some cases the question is whether behavioral *differences* among individuals or groups are based on genetic *differences* among them. In other cases, the question is whether or not behavior can be attributed to a genetic foundation held *in common* by virtually all human beings. The trait in question is sometimes described as an element of "human nature"—although this is a poorly defined concept. A universally held genotype may be postulated to explain either a supposedly universal human behavior such as avoidance of incest, or a variable phenotype such as behavioral differences between the sexes, which could be due to genes shared by females and males, but expressed differently. The two kinds of hypotheses must be tested by very different methods, because it is difficult or impossible to identify genes for invariant traits. Thus evidence on the issue of genetic universals is especially hard to evaluate. In both cases, an extreme alternative hypothesis is that environmental effects—culture and learning—fully account for the observed behavior. Extreme proponents of "environmentalism" hold that the newborn child's mind is a tabula rasa—a blank slate on which the child's experiences inscribe the guidelines of its cognitive abilities, emotional responses, and manifest behaviors. Most people do not subscribe to either extreme environmentalism or extreme genetic determinism, but grant that both play some role.

Biological Foundations of Human Behavior

Cognitive psychologists use the word MIND to denote the information-processing functions of the brain (Cosmides et al. 1992). This usage reflects the conviction, almost universal among cognitive psychologists and neurobiologists, that "mind" does not exist independently of the physical structure and activity of the brain, and that all mental activities such as cognition and emotion have a strictly material, physical foundation in the nervous system and sense organs. The existence of such a physical foundation is evident from the fact that emotional and cognitive processes of all kinds can be affected by physical injuries and other kinds of brain dysfunctions, and is further supported by many kinds of neurobiological and psychological research. Not only the sense organs, but also the brain centers responsible for processing sensory input and for generating responses to sexual and many other stimuli, are homologous among humans and other vertebrates.

Many brain functions and behaviors of humans are evident to at least some degree in other mammals, most strikingly in our closest relatives, the great apes. Most animals have some capacity for learning and memory, though they vary greatly in how much they learn and what they learn most readily (see Chapter 20). A rudimentary capacity for "culture," based on imitation of other individuals, is found in some species of social animals; in this way, for example, blue tits in England acquired the habit of pecking through the paper lids of milk bottles on people's doorsteps (Fisher and Hinde 1948). In primates, cultural "traditions" may vary from troop to troop. Japanese macaques (*Macaca fuscata*) developed a variety of cultural traditions spread by learning, such as separating wheat from sand by floating it on water. Chimpanzees likewise learn from their elders how to crack nuts with clubs and how to fashion twigs for extracting termites from their nests—illustrating further that tool making and tool use are not unique to humans (Figure 26.13). Considerable controversy surrounds the question of whether or not any nonhuman species has the capacity to learn or produce symbolic language. Captive chimpanzees, bonobos, and gorillas can learn and use sign language and other representational modes of communication, with an apparent command of both syntax and representational meaning (Jones et al. 1992).

More controversial is the interpretation of behavioral similarities between humans and other primates such as chimpanzees. For instance, male chimpanzees, like males of many other species, fight for access to females. Chimpanzees have intricate social relationships based on individual recognition. They share food preferentially with relatives, engage in apparent "reciprocal altruism," cooper-

FIGURE 26.13 Tool use in wild common chimpanzees (*Pan troglodytes*). A female chimpanzee in the Tai National Park, Ivory Coast, pounds a nut with a wooden club, as her one-year-old son watches. (Photograph courtesy of C. Boesch.)

ate in hunting prey, and may avoid incest (see Chapter 20; deWaal 1989; Goodall 1986). Some students of human behavior argue that such behaviors are homologous between humans and chimpanzees, and that the parallel behaviors in humans are derived from those in the common ancestor, and thus have a genetic foundation. Others caution that it is extremely difficult to determine whether human behaviors are evolutionarily homologous to those of other species, and argue that they probably have a cultural basis.

Human Nature

"Human nature" is a colloquial term expressing the notion that humans universally share an intrinsic disposition toward particular behaviors—a disposition residing in our genes. The suckling behavior of infants is an example; so too, according to some authors, are aggression, male dominance, heterosexuality, and many other traits. Historically, prevailing views about the extent to which human behaviors are genetically or culturally determined have shifted repeatedly (Degler 1991). From the 1940s through the 1960s, the cultural view prevailed. In the 1970s, publications such as E. O. Wilson's *Sociobiology* (1975) revived interest among some biologists, anthropologists, and psychologists in the evolutionary foundations of human behavior, an interest that has grown despite great controversy.

What may be "natural" to humans is difficult to say, and the term may well be meaningless. The human genotype clearly has a very broad *norm of reaction* (see Chapter 3) with respect to most behaviors: any of a great many phenotypes may develop, depending on the circumstances. The norm of reaction for growth form in white oaks (*Quercus alba*) ranges from spindly and short-branched if the tree grows in a forest to broadly spreading if it grows in the open; it would be absurd to consider either form unnatural. Similarly, if a person should develop to be brave or cowardly, heterosexual or homosexual, selfish or generous depending on early experience or social influence, no one of these phenotypes can be viewed as more "natural" than another. Some behaviors, such as wearing navel rings or composing symphonies, are rare or "abnormal" (i.e., very different from the average), but not "unnatural," for human nature is our behavioral norm of reaction, which includes everything that people do. It makes no sense to judge a behavior as moral or immoral, ethical or unethical, on the basis of whether or not it is "natural."

It is certainly true that our behavior is genetically determined, in the sense that our capacities reside in the organization of our nervous system, and would be very different if we had the brain of a chimpanzee or if we were six inches tall or lacked opposable thumbs. More interesting is the question of whether we possess genes for particular behaviors, or for neuropsychological mechanisms that generate specific responses to certain stimuli or environmental circumstances.

Sociobiology

Most researchers who seek an evolutionary basis for human behavior do so within a Darwinian, adaptationist framework. This approach, given impetus by evolutionary biologists such as E. O. Wilson (1975) and R. D. Alexander (1979), uses theories such as kin selection, reciprocal altruism, and sexual selection (see Chapter 20) to generate predictions about human behaviors and cognitive mechanisms. Advocates of this approach claim support for their hypotheses from cross-cultural data or from behavior within a single culture, such as their own. However, the adaptationist approach, to evolutionary problems in general and to human behavior in particular, has been criticized for relying too exclusively on natural selection, and for failing to sufficiently credit other explanations for traits, such as culture, genetic drift, genetic correlations, or lags in adaptation to present environments (see Chapters 12 and 20). Both advocates and critics of the adaptationist interpretation of human behavior, though, agree that any genetic foundations for human behavior must have evolved many thousands of years ago, especially during the Pleistocene, that technology and the social dynamics of modern societies are much too recent to have affected the genetic bases of behavior, and that our evolved behavioral propensities cannot be expected to be adaptive in the modern world. To take a simple example, even if a liking for sugar and fats was favored by selection in ancestral human (or earlier primate) populations, it is not beneficial to members of modern industrial societies.

The adaptationist approach, known as human sociobiology, Darwinian social science, or human behavioral ecology, has been employed to interpret a great many behavioral traits. For example, adaptationists often invoke selection to avoid inbreeding, and thus production of genetically deficient offspring, as an explanation of culturally ubiquitous prohibitions against marriage between very close relatives. The hypothesis of an innate aversion to sex with relatives seems to conflict with the rather high incidence of some such activity in practice (sexual molestation of young women by fathers and uncles is all too common). However, it may be supported by reports that unrelated people who have grown up *as if* they were siblings, such as those reared in Israeli kibbutzim, show little sexual or romantic interest in each other, despite their parents' encouragement (Shepher 1983).

As one more example of evolutionary interpretation of human behavior, we may consider sex-typical behaviors and mate preferences (Symons 1979; Barash 1982; Daly and Wilson 1983). The theory of sexual selection (see Chapter 20) holds that because of sex differences in parental investment in offspring, increasing the number of mates can increase a male's fitness more than a female's. Females are a "limiting resource" for which male animals compete, while the reverse is seldom the case. In some species, a female's reproductive success may be enhanced by choosing mates that provide superior resources (e.g., food, territory, defense) for her and her offspring. From these principles, some authors have predicted that we should expect men to be more competitive and aggressive than women, to be more sexually promiscuous, to defend their mates against the sexual attentions of other men (and thereby treat their mates as "property"), to be less nurturing toward children (since the certainty of paternity is lower than the certainty of maternity), and to be sexually most attracted to women who bear signs of high reproductive potential (i.e., young, slender, and unblemished). According to this reasoning, women are expected to be more nurturing to children (because each child represents a large reproductive investment), to be less competitive, and to be attracted to men who are most likely to provide resources for them and their children: men who are socially dominant, economically successful, and older (since social dominance and economic success are age-dependent).

Expectations such as these are borne out by data on many cultures—including our own. Therein lies the rub, for as critics argue, the expectations derived from evolutionary theory might equally be viewed as derived from our own experience, and the evolutionary theory itself might be as much a rationalization for familiar sex roles as an independent predictor of those roles. "Social constructionists" argue that the sex differences described are rooted in our culture rather than in our genes, that there exists great variation among cultures (e.g., in female traits preferred by men), and that even traits that are virtually uniform among cultures, such as male dominance, are cultural norms that stem from the simple fact that men are larger and stronger. It would not be surprising if some sex role differences in humans, which resemble those in some other primates, had an evolved, genetic foundation. However, it has not yet been possible to develop clear criteria for evaluating the validity of the sociobiological versus the social constructionist interpretations.

Evolutionary Psychology

Related to sociobiology is evolutionary psychology (Barkow et al. 1992), which is "psychology informed by the fact that the inherited architecture of the human mind is the product of the evolutionary process," and which expects "to find a functional mesh between adaptive problems and the structure of the mechanisms that evolved to solve them" (Cosmides et al. 1992). Everyone agrees, for example, that humans, unlike most or all other species, have a genetic capacity for language (although the particular language a person speaks is determined entirely by social learning, not by genes). The linguist Noam Chomsky (1986) goes further, arguing that all languages share certain fundamental syntactical properties ("universal grammar"), and that specialized modules, or "mental organs," in the mind must therefore exist and be responsible for the acquisition and expression of language. Whether or not this mental architecture (if it exists) evolved by natural selection is debated (Pinker and Bloom 1992).

Evolutionary psychologists believe that the mind includes many such modules, specialized for different kinds of cognitive tasks, much as the various organs of the body serve different adaptive functions. Like sociobiologists, they expect adaptationist reasoning, together with informed suppositions about the adaptive problems our Pleistocene ancestors faced, to generate predictions about adaptive psychological mechanisms. For instance, Leda Cosmides and John Tooby (1992) proposed that in social exchanges among unrelated individuals ("reciprocal altruism"), cheating is an ever-present threat, so there should have been selection to detect cheaters and withhold food or other goods from them. They therefore predicted that mental mechanisms for reasoning about social exchanges should have design features that are not activated in nonsocial contexts. They presented college students with problems that had the same logical form, but different content, and found that the students solved the problems more frequently if they described cheating than if they did not. However, we cannot rule out the plausible hypothesis that Americans learn from an early age to be especially sensitive to the possibility of cheating, rather than having a genetically programmed sensitivity.

Opponents of evolutionary psychology and sociobiology, in contrast, envision the mind as a relatively content-free mechanism that can be applied to solving any of a vast variety of problems, rather like a general-purpose computer that can be supplied with a virtually limitless variety of algorithms. Certainly the extraordinary capabilities of the human mind encourage this view, but research and debate in biology, psychology, and anthropology will continue for a long time to come.

Variable Behavioral Traits

Determining the genetic contribution to *variation* in traits is easier than showing a genetic basis for universal characters—but for human behaviors, this task is still very difficult. Humans, unlike fruit flies or corn, cannot be deliberately crossed or reared in different, experimentally altered environments. Family members typically share not only genes, but also environments: siblings, for example, typically experience similar parental treatment, schooling, and social and cultural milieux. These and other environmental factors indisputably cause variation in most behaviors, so in order to document a genetic component of variation, researchers must surmount this confounding of genetic and environmental sources of variation.

Most variable human behavioral traits are continuously rather than discretely distributed. They are usually described by the statistical methods of quantitative genetics, although molecular methods of mapping genes (QTL mapping) have also been employed (see Chapter 14). Recall that the variance in a phenotypic trait (such as IQ score, or "Intelligence Quotient") may have several components: V_G (genetic variance), V_E (environmental variance), $V_{G \times E}$ (variance due to genotype × environment interaction—i.e., different responses to an environment by different genotypes), and cov(G,E), the variation arising from correlations between an individual's genotype and environment. That is,

$$V_P = V_G + V_E + V_{G \times E} + \text{cov(G,E)}$$

Many studies seek to estimate the *heritability*, $h^2 = V_G/V_P$, of traits such as IQ by measuring correlations between relatives (see Chapter 14). In order to do so, it is necessary to estimate or, better, to eliminate the correlation between the environments experienced by those relatives [cov(G,E)]. $V_{G \times E}$ cannot be eliminated, nor can it be estimated except by replicating each genotype in a variety of representative environments. This can be done with plants and *Drosophila* (see Figure 13 in Chapter 9), but not with humans. If the variation that actually arises from G×E interactions is included in the estimate of V_G, heritability will be overestimated. Whether it is or not, failure to estimate $V_{G \times E}$ can result in serious biological misjudgments, for we will fail to recognize that genotypes that perform poorly in one set of environments may perform well in others. Moreover, if genotypes differ strongly in their responses to different environments, estimates of V_G and V_E, and therefore of heritability, may be greatly altered by changing the environmental setting. This is one reason why some population geneticists maintain that heritability measurements are not very informative and may even be misleading (Lewontin 1974a; Feldman and Lewontin 1975).

It is exceedingly important to bear in mind that a genetic basis for a character does *not* mean that the trait is fixed or unalterable (see Chapters 3, 9, and 14). Because heritability is a ratio, the same population may display high heritability of a trait if the environment is relatively uniform (V_E is low) or low heritability if the environment is more variable. *A heritability value, therefore, may hold only for the particular population and the particular environment in which it was measured, and cannot be reliably extrapolated to other populations.* By the same token, high heritability of variation *within* populations does not mean that differences *among* populations have a genetic basis. A character may display high heritability within a population, yet be altered dramatically by a change in the environment. Twin studies have suggested that the heritability of human height is 0.8 or more, yet in many industrialized nations, mean height has increased considerably within the twentieth century due to nutritional and other environmental factors, and Italian-Swiss immigrants to the United States gained an average of 4 cm in height within a single generation (Cavalli-Sforza and Bodmer 1971). Whether or not a characteristic can be substantially changed by the environment cannot be determined merely by showing that it has a genetic component.

The heritability of human traits is estimated by correlations among relatives such as siblings. Twins are favorite subjects, since monozygotic (MZ), so-called "identical," twins should be more similar than dizygotic (DZ, "nonidentical") twins if the variation has a genetic component. The genetic correlation between DZ twins should be no greater than between non-twin siblings. However, the genetic and environmental components of variation are confounded if siblings are reared together, and MZ twins may show higher concordance (similarity) than DZ twins because they are often treated more similarly as children. Only if twins or other siblings are *reared apart*, in *uncorrelated environments*, can the correlation between them serve as an estimate of heritability in the general population that they represent. For this reason, studies of people *adopted* as children are critically important. The genetic component of variation is estimated by correlations between twins or other siblings reared apart, or by adoptees' correlations with their biological parents. In a variant of this method, the correlations between parents and their adopted children are contrasted with the those parents' correlations with their biological (nonadopted) children, which are expected to be higher if the variation has a genetic component. There is a great risk that heritability will be overestimated in such studies, because adoption agencies often place children in homes that are similar (in factors such as religion and socioeconomic status) to those of their siblings. Some modern studies try to measure such environmental correlations and take them into account.

We shall briefly describe studies of variation in two characteristics that illustrate the methods by which genetic influences on human behavior are sought. Research on both traits—sexual orientation and IQ—has been highly controversial, both because some studies have been flawed and because these are socially and politically highly charged subjects.

Sexual Orientation

Students of human sexuality distinguish a person's biological SEX from his or her GENDER IDENTITY, or self-identification of one's sex. Transsexuals psychologically identify with the opposite sex, but almost all heterosexuals and homosexuals do not. GENDER ROLE refers to the "sex-typed" behaviors and personality traits that society designates as masculine or feminine. SEXUAL ORIENTATION refers to whether a person is more strongly sexually aroused by members of the opposite sex, the same sex, or both (Bailey and Zucker 1995). SEXUAL BEHAVIOR refers to an individual's engagement in sexual activity with members of the same or the opposite sex. Sexual orientation and behavior may differ; for instance, preadolescent boys in some cultures engage in sex with adult men as part of a ritualized entry into heterosexual adulthood; heterosexually oriented prisoners often engage in sex with each other; many homosexuals marry and have children.

Both heterosexual and homosexual behavior and, apparently, orientation occur in almost all human societies, among which attitudes range from strict prohibition to tolerance or encouragement (Churchill 1967). Sexual orientation in Western societies is usually scored on the "Kinsey scale," from 0 (entirely heterosexual) through intermediate scores (bisexual) to 6 (entirely homosexual). The scores are obtained by asking individuals about their orientation. Because gay men and lesbians in many societies suffer hostility and may therefore be reluctant to divulge their sexual orientation, the frequency of homosexual orientation is uncertain, although it appears to be much the same in different societies, regardless of the level of tolerance (Whitam and Mathy 1986). In the United States, the distribution of Kinsey scores is continuous, but most people lie near one extreme or the other. The proportion of predominantly homosexual men is probably about 2 to 4 percent, although estimates range up to 10 percent. Probably about 1 to 2 percent of women are predominantly lesbian (Bailey and Zucker 1995).

Many hypotheses have been proposed to explain homosexual orientation, many of them based on the conviction that it is pathological. Modern researchers tend to be less judgmental, viewing sexual orientation as part of the normal variation in human behavior. Psychological theories, based mostly on subjects in psychotherapy, have generally ascribed male homosexuality to dysfunctional relationships of the child with his parents, such as with an absent or emotionally distant father or an excessively close or dominant mother. Comprehensive studies of more representative samples of subjects have provided no unequivocal support for these hypotheses, and none at all for the common myth that people become gay through seduction by older homosexuals (Bell et al. 1981). It has become clear that sexual orientation begins to develop at a very young age, and that the best predictor of whether a child will be homosexual as an adult is nonconformity to gender role—childhood behaviors typically associated with the opposite sex (Bailey and Zucker 1995). Although one may choose one's sexual *behavior*, it is clear that gay men and lesbians do not choose their *orientation* any more than heterosexuals do. Moreover, sexual orientation appears not to be changeable by psychotherapy or other calculated attempts to alter it (Haldeman 1991).

Reports that homosexual and heterosexual men differ in their responses to hormones and in the size of three areas of the brain have been offered as evidence of a biological basis for sexual orientation, but these data are controversial (LeVay 1993; Hamer and Copeland 1994). Studies of twins and adopted children strongly suggest, but do not prove, a genetic influence. In one study, 52 percent of the MZ co-twins of gay men were homosexual or bisexual, compared with 22 percent of male DZ co-twins and 11 percent of adoptive brothers. Heritability was estimated as lying in the range of 0.31 to 0.74 (Bailey and Pillard 1991). However, because the twins were reared together, the role of common environment is uncertain, and the researchers were puzzled by the fact that a lower proportion of non-twin brothers than of DZ co-twins were reported to be gay. MZ co-twins, DZ co-twins, and adoptive sisters of lesbians showed levels of concordance similar to those in the study of men, but with nearly equal proportions of lesbian DZ sisters and non-twin sisters (Bailey et al. 1993). It is clear from these studies that not all the variation in sexual orientation has a genetic basis (since about half of the MZ twin pairs differed), but how much of it does have a genetic basis is unclear.

The strongest evidence for a genetic contribution to sexual orientation has been provided by Dean Hamer and his collaborators (Hamer et al. 1993; Hu et al. 1995; Hamer and Copeland 1994). When they interviewed a large sample of gay men, they found that a much higher proportion of their male relatives on the mother's side of the family than on the father's side were gay (Figure 26.14). This pattern suggested that a gene or genes on the X chromosome might contribute to sexual orientation, since men inherit their single X chromosome from their mothers.

To test this hypothesis, the researchers used the AFFECTED PAIRS METHOD (Weiss 1993) to search for X-linked genetic markers. This method tests the hypothesis that a trait shared by pairs of relatives is caused by an allele A' at an as yet unmapped locus. Suppose that a mother's X chromosome genotype were $AM_1/A'M_2$ and a father's genotype were AM_1/Y, where M_1 and M_2 are alleles at a marker locus. Fathers transmit Y chromosomes to their sons. If A and M are linked on the X chromosome and the recombination rate is r, then $1 - r$ of the gametes of the AA' mother will be non-recombinant, giving rise to affected (A'-bearing) offspring who also share M_2. If r is 0, then M marks the trait-causing locus itself, and all of the affected siblings will carry the marker allele. A strong association of a trait with a marker thus indicates that the trait is indeed affected by a gene and provides directions to its location.

Hamer et al. studied 40 pairs of gay brothers. Using 22 polymorphic DNA markers distributed over the length of the X chromosome, Hamer et al. found about 50 percent concordance for several markers in a region (Xq28) near one end of the chromosome (Figure 26.15). In a later study of 33 other pairs of gay brothers, they found similar results, and also showed that significantly less than 50 percent of heterosexual brothers of these gay men shared the Xq28 markers—just as predicted by the genetic hypothesis. They found, also, that pairs of lesbian sisters did not share X-linked markers any more often than expected by chance. They had expected this result, since homosexuality in men and in women is not associated in the same families.

These results strongly suggest that a polymorphic X-linked gene contributes to variation in sexual orientation in

FIGURE 26.14 A sample pedigree of the distribution of homosexual and heterosexual orientation. Squares represent males and circles females; solid-color squares represent homosexual, and shaded-color squares heterosexual, individuals. The homosexual man in generation III had male homosexual relatives in his mother's family, but not in his father's. Such a pattern, if consistent among families, suggests the possibility of a polymorphism for a gene on the X chromosome. (After Hamer et al. 1993.)

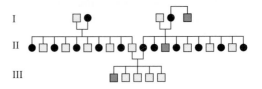

men, but that different genetic factors, if any, contribute to variation in women. Because the pairs of gay brothers were not all concordant for the Xq28 markers, other genetic and/or environmental factors must contribute to sexual orientation as well. Because Hamer's team used a carefully selected sample in order to maximize their chances of finding any X-linked genes that might exist, their study does not estimate the allele frequency in the population, and cannot indicate what fraction of gay men carry the hypothesized X-linked allele. Thus, although this study strongly suggests that some variation in sexual orientation has a genetic basis, it does not indicate what proportion of the variation in the population as a whole has a genetic basis.

The proportion of predominantly homosexual people in the population is higher than we would expect if alleles at one or more loci contribute importantly to homosexuality and if gay men and lesbians truly have had low reproductive rates. One possible explanation is that for much of human history, homosexuals have reproduced nearly as prolifically as heterosexuals, perhaps obeying the social demand to marry and rear children, even if they also sought sexual and emotional gratification in same-sex relationships. In ancient Greece, for instance, it was customary for adult men, married or not, to have close emotional and sexual relationships with younger men (see, for example, Plato's *Symposium*). Another possibility is that selection against alleles that increase the propensity for homosexual orientation is balanced by mutation at several or many loci, all with similar phenotypic effects. The genetic component would then be the sum of the effects of several or many individually rare alleles. A third possibility is that such alleles have advantageous pleiotropic effects and are maintained by balancing selection (see Chapter 13). In discussing human sociobiology, E. O. Wilson (1978) suggested that homosexuality may have evolved by kin selection, whereby homosexuals helped to rear their relatives' children, and that it is "a distinct beneficent behavior that evolved as an important element of early human social organization."

Almost all scientific knowledge that bears on people's lives can be used for good or ill. Some gay men and lesbians welcome evidence suggesting a genetic basis for sexual orientation, in the hope that heterosexual society will be more

FIGURE 26.15 A DNA profile illustrating concordance and discordance of an X-linked marker with sexual orientation. Gels are shown for seven pairs of gay brothers and for one or both parents of five of the pairs. Symbols are as in Figure 26.14. Four alleles (numerals 1, 2, 3, 6) can be distinguished by their mobility. "D" indicates cases in which the brothers' alleles are concordant by descent: the brothers share one of their mother's two alleles. "S" indicates concordance by state: the brothers share an allele, but the mother's allele is unknown. The symbol "—" indicates discordance: the brothers inherited different alleles from their mother. Such cases do not support the hypothesis that sexual orientation is affected by a gene linked to this marker. n indicates that the family is noninformative, because the brothers necessarily share the homozygous allele carried by their mother. (After Hamer et al. 1993.)

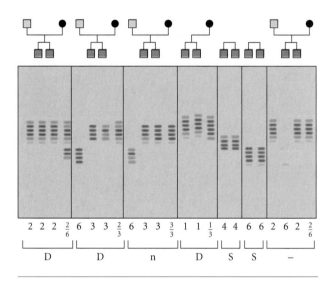

accepting of them, given evidence that a homosexual orientation is not freely chosen. Others fear that if "gay genes" are found, they will be used by people determined to view homosexuality as an illness for technological "cures" or to identify and abort fetuses that carry such genes. Their fears are understandable. It is only through education, and by insisting on the dignity and rights of all people, that we can guard against further abuses in the future.

Intelligence

To a greater extent than in almost any other context, the "nature versus nurture" debate has centered around variations in intelligence, or, more properly, in IQ ("intelligence quotient") score, since no one knows how to define intelligence other than by an IQ score. IQ tests are supposed to be "culture-free," but they have been strongly criticized as favoring white, middle-class individuals (see, for example, Kamin 1974; Gould 1981; Fischer et al. 1996). Moreover, IQ scores are often taken to mean that cognitive abilities are all correlated with one another and contribute to a single "general factor" of intelligence ("g"), whereas it is clear that abilities such as verbal comprehension, spatial visualization, and memory vary independently to some extent, and that some features, such as social and musical skills, are not correlated with IQ at all (Gould 1981; Fraser 1995).

IQ testing has a long, sordid history of social abuse (Gould 1981). Based on the belief that the low scores of some immigrant groups signaled genetically fixed inferiority, the United States adopted an Immigration Restriction Act in 1924 that excluded most eastern and southern Europeans, including untold numbers of Jews who would otherwise have escaped the approaching Holocaust. In the 1970s, Jensen (1973) renewed the debate by claiming that IQ is 80 percent heritable and that it therefore cannot be improved by education, so that programs of compensatory education are doomed to failure. Most recently, a similar inference was drawn in *The Bell Curve* (Herrnstein and Murray 1994), as was the inference that African-Americans, who on average score lower than European-Americans on IQ tests, have genetically lower intelligence. This book has been widely criticized for shoddy science and scholarship and for attempting to justify the social inequality among "races" in America (Fraser 1995; Fischer et al. 1996).

Even if the heritability of IQ, or of various cognitive abilities, were to prove to be very high, it need not be fixed. There is abundant evidence, in fact, that education and an enriched environment can substantially increase IQ score (Fraser 1995). For example, the mean IQ of black children adopted by white families in Minnesota equaled the national white average, about 15 points higher than the mean for the black population (Scarr and Weinberg 1976). The difference between the mean scores of American blacks and whites has declined in recent years.

Many of the earlier studies of the heritability of IQ were flawed and, in one important case, probably fraudulent (Hearnshaw 1979; Lewontin et al. 1984). Recent studies have corrected many of the flaws; for example, they assess the similarity of the family environment in which separated twins were reared. These studies include the Minnesota Study of Twins Reared Apart (Bouchard et al. 1990) and the Colorado Adoption Project (Plomin 1994; DeFries et al. 1994). The Minnesota study included 56 sets of monozygotic twins and triplets reared apart (MZA). If similarity due to common environment were negligible, then the correlation of such twin pairs would directly estimate the genetic variance, V_G, and therefore the broad-sense heritability, h_B^2, of IQ scores. These twin pairs, tested at an average age of 41 years, varied in the time spent together between birth and separation, and especially in time together between reunion and testing, but neither of these variables was correlated with the twins' degree of similarity in IQ. The heritability of IQ score was estimated as 0.70, and the heritability estimates of various personality traits, psychological interests, and social attitudes ranged from 0.34 to 0.50. Various features of the environments in which the twins had been reared appeared to be poorly correlated, and were considered not to have contributed to the correlation in IQ.

The Colorado Adoption Project is a long-term study in which tests of IQ and other traits are administered to adopted and nonadopted children at intervals from 1 year of age onward. Scores are compared for biological siblings reared together, for adoptive (nonbiological) "siblings" reared together (i.e., an adopted child reared with a non-adopted child), for children and their adoptive parents, and for biological parents and their adopted and nonadopted children. The variance (V_P) in a phenotype score such as IQ may be written as $V_P = V_G + V_c + V_e$, where V_G is the genetic variance, V_c is the variance due to a common, or shared, home environment, and V_e is the variance due to unshared environmental factors, unique to each individual. If V_P is set at 1, this expression may be rewritten as $1 = h^2 + c^2 + e^2$. Variances are estimated from correlations. For biological siblings reared together, who share both genes and home environment (Figure 26.16), the expected correlation is $r_n = h^2/2 + c^2$. For adoptive siblings reared together, the expected correlation is $r_a = c^2$. Thus c^2 is estimated directly from the latter equation, enabling h^2 to be estimated from the equation for r_n. There is some evidence that nongenetic maternal effects, owing to variation in the uterine environment, account for some of the similarity of twins and other siblings, so that heritability of IQ is commonly overestimated (Devlin et al. 1997).

The results of the Colorado study to date suggest that the heritability of IQ score is about 0.5, and is much the same at ages 1 through 9. The proportion of variance due to common environment (c^2) was estimated at 0.11–0.24. The investigators found that the genetic component of specific cognitive abilities such as verbal comprehension, spatial visualization, perceptual speed, and accuracy was correlated to a considerable degree with overall IQ score (i.e., with the hypothetical general factor, "g"), but that some uncorrelated genetic variation in these abilities emerged at age 7, implying that overall "intelligence" includes several components that are partly independent of each other.

Perhaps the most interesting result, in this and other studies, was evidence of a substantial genetic × environmental interaction (G × E) (Plomin and McClearn 1993; Plomin 1994). For instance, the researchers observed features of the homes and mothers' treatment of both adopted and nonadopted pairs of siblings when they were the same age, and also scored the mothers' descriptions of the family environment at those times. For both methods, the environmental scores of biological siblings were more highly correlated than those of adoptive siblings. The researchers hypothesized that genetically different traits in children may elicit different responses from parents and peers, and that children's different genotypes influence their perceptions of and responses to the same objective experiences. In other words, the individual may create his or her own environment, a process in which genetic differences in cognitive, personality, or physical traits may play a role. The environmental differences thus engendered may, in turn, affect the development of cognitive abilities and personality traits. This would mean that "nature" and "nurture" covary and interact, and cannot meaningfully be distinguished.

Probably the most incendiary question about IQ is whether genetic differences account for differences in aver-

FIGURE 26.16 A diagram of the expected influences of genetic and environmental factors on the correlation between siblings (P_1 and P_2) in behavioral or other characteristics. The genotypes (G_1, G_2) of two siblings are expected to be correlated at 0.5 if they are biological sibs, and at 0.0 if they are adopted, unrelated sibs. Their shared, common environment (C_1, C_2) is correlated at 1.0 by definition. The specific, unshared environmental experiences of each individual (SE_1, SE_2) are uncorrelated. In principle, the variance among individuals can then be partitioned into heritable genetic effects (h), common environmental effects (c), and specific environmental effects (e). (After DeFries et al. 1994.)

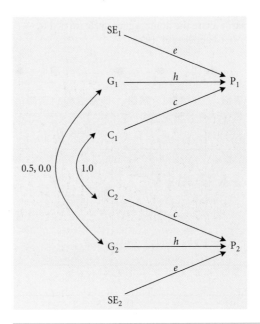

age IQ scores among "racial" or ethnic groups, such as the 15-point difference (one standard deviation) between the average scores of European-Americans and African-Americans. Even if the heritability of IQ is 0.5–0.7 *within* groups, the average difference *between* blacks and whites could be entirely due to their very different social, economic, and educational environments. Virtually all the evidence indicates that this is exactly the case (Nisbett 1995). In studies of black children adopted into white homes or reared in the same residential institutions with white children, the two "racial" groups had similar scores. One study correlated individuals' IQ scores with their degree of admixture of alleles for several blood groups that differ in frequency between European and African populations. The correlation between IQ and degree of European ancestry was nearly zero. Finally, the average IQ of German children fathered by white American soldiers during World War II was nearly identical to that of children with black American fathers. Thus several lines of evidence converge on the conclusion that these "races" do not differ genetically in IQ.

If there is, indeed, substantial heritability of IQ *within* populations, what are its practical, social implications? They may be few. Because a heritability estimate applies only to the population and environment in which that estimate has been made, the chief use of heritability values is in designing programs of artificial selection for desirable traits in domesticated plants and animals. At this time, thankfully, no one proposes to artificially select humans for IQ, or anything else. As we have seen, a rich social and educational environment—which we should hope to provide for all children anyway—can boost IQ scores even if heritability is substantial. A child's IQ score is a very poor predictor of his or her adult socioeconomic status (Gardner 1994), so we need not fear, despite some claims (Herrnstein and Murray 1994), that society will become stratified into an elite with genetically high IQ and an underclass with genetically low IQ. If anything, IQ studies may bear the encouraging message that children have diverse cognitive talents and diverse responses to their environment, which may lead to the diversity of interests and abilities that society depends on.

Evolution and Society

Evolution is central to modern biology and to the use of biology in modern society. Knowledge, principles, and methods that have been developed in the study of evolution are used to improve crops and domesticated animals, to combat agricultural pests and human pathogens, to study inherited diseases, to find useful natural products, and in many other applications. From a philosophical point of view, surely little can be more satisfying than to have attained an understanding of our origin and that of other living beings, and to assert, with Darwin, that "there is grandeur in this view of life," in which "from so simple a beginning endless forms most beautiful and most wonderful have been, and are being evolved."

But knowledge, and the power that comes with knowledge, can be used to our detriment as well as to our good. Herbert Spencer, a contemporary of Darwin, promulgated the philosophy of "Social Darwinism," the doctrine that human progress is the outcome of competition and struggle among individuals, races, and nations, just as "the survival of the fittest"—Spencer's term for natural selection—is the engine of evolutionary "progress." Spencer opposed state aid to the poor, state support for education, and regulation of business as barriers to competition and therefore to progress. Social Darwinism was immensely popular in early twentieth-century America, especially among business giants such as John D. Rockefeller (Hofstadter 1955). Moreover, it went hand in hand with the common belief that intelligence, criminality, and other traits were genetically fixed and could not be altered by the environment. These beliefs led to a flourishing eugenics movement in the first few decades of the century, which advocated encouraging "superior" people to have more children and discouraging or preventing "inferior" people from reproducing. Likewise, these beliefs were used to justify racism, colonialism, and domination of some peoples by others. "Aryanism," the ideology of Teutonic superiority that grew into the evil racism of the Nazis, was promulgated before *The Origin of Species* was published, but later justified itself by the "natural law" of natural selection.

The risk that evolution and genetics will be abused continues. The hypotheses and data of human sociobiology, evolutionary psychology, and human genetics develop largely out of scientists' quest for intellectual understanding, not out of any desire to justify social norms. But there are elements in the larger society who would use genetic data to discriminate in employment, who would use heritability to justify social and economic inequality, who would invoke sociobiological hypotheses to deny gay people equal rights and to keep women in the kitchen. Evolutionary scientists, like other scientists, do not set social policy, but they can and perhaps should bear responsibility for alerting the public to the uses to which their discoveries might be put. They can point out misunderstandings and illogical interpretations of evolutionary theory, such as the "naturalistic fallacy"—the belief that what is natural is good, and therefore provides moral guidance for human conduct. Social Darwinism was based on this fallacy. So are the beliefs that homosexuality is morally wrong because it does not lead to reproduction, and that women should be subservient to men because sex roles may have an evolutionary origin.

There exists no philosophical or scientific foundation for the belief that human behavior should conform to "natural laws." Natural "laws," like the "laws" of gravity or natural selection, are simply regularities in the mechanical processes of nature, not rules of conduct (Toulmin 1957; Collins 1959). Darwin's great defender Thomas Henry Huxley rightly said that "cosmic evolution may teach us how the good and evil tendencies of man have come about but, in itself, is incompetent to furnish any better reason why what we call good is preferable to what we call evil than what we had before." Further, he said, "the influence of the cosmic process [natural selection] on the evolution of society is the greater the more rudimentary its civilization. Social progress means a checking of the cosmic process at every step and the substitution for it of another, which may be called the ethical process ... Let us understand, once for all, that the ethical progress of society depends, not on imitating the cosmic process, still less in running away from it, but in combating it" (Paradis and Williams 1989). That is, evolution contains neither morality nor immorality, provides no philosophical basis for aesthetics or ethics, and provides no guidance for us as we devise our ethics, ideals, and aspirations.

But evolutionary biology, like other knowledge, can serve the cause of human freedom and dignity. Together with those of the other biological sciences, its applications in health science, food production, and environmental management can help to relieve disease and hunger. It helps us to appreciate both the unity and diversity of humankind. And as we reflect on our common origin with other forms of life, we may even come to feel at one with, and to care for, those "endless forms most beautiful and most wonderful."

Summary

1. Humans are most closely related to gorillas and chimpanzees, and almost certainly share their most recent common ancestor with the chimpanzees (the common chimpanzee and the bonobo). Overall, humans and chimpanzees are about 99 percent identical in DNA sequence, and probably diverged about 4.6–5 million years ago (Mya).

2. The oldest known fossil hominids date from the early Pliocene (4.4 Mya) of Africa. Early hominids had many apelike features, including adaptations for climbing trees, but were at least partly bipedal. Other human features evolved later, including a fairly gradual increase in brain size. Several hominid species inhabited Africa until about 1 Mya. One of them, *Homo habilis* (1.9–1.6 Mya), is the first known tool maker. Its apparent descendant, *H. erectus*, spread from Africa into Europe and Asia about 1 Mya, and evolved into "archaic" *Homo sapiens*.

3. Some, but not all, studies of DNA sequences suggest that "modern" *Homo sapiens* spread out of Africa in a second wave of dispersal between 0.4 and 0.15 Mya, and replaced populations of archaic *H. sapiens* without interbreeding with them. This dispersal pattern, if confirmed, would account for the high degree of genetic similarity among all modern human populations. Although there exist allele frequency differences among populations, most of the genetic variation in the human species can be found within any one region.

4. The genetic variation within human populations includes much that is probably neutral, some polymorphisms that are maintained by natural selection (such as sickle-cell and some other hemoglobin variants), and many deleterious alleles, at thousands of loci, that are probably maintained by the balance between mutation and purifying natural selection. These alleles account for many kinds of genetic diseases, which are being studied by both population genetic and molecular genetic methods. In contemporary populations, there is probably little directional selection for change in physical characteristics, but selection for resistance to infectious diseases may be substantial.

5. Learning and a rudimentary capacity for culture and possibly language exist in other primates. These capacities have evolved to a prodigious degree in humans. The capacity for language and culture has a genetic basis, but differences in language and culture are not based on genetic differences. Because cultural information is transmitted both within and across generations, cultural change can proceed at immensely greater rates than genetic evolution.

6. The evolutionary and genetic bases of human behavior are highly controversial. Many universal or very widely shared traits are considered by some researchers to be manifestations of adaptive traits that evolved in our Pleistocene ancestors. Such hypotheses cannot be tested by genetic methods. One approach to uniform traits (sociobiology) compares behaviors with predictions from kin selection, sexual selection, and other theories of adaptation. Correspondence to such predictions, as for some widely observed sex differences in behavior, may be interpreted as evidence of genetically based evolved behaviors, but can often be just as plausibly attributed to culture. Evolutionary psychology, which posits and tests for the existence of adaptive mental mechanisms, faces similar problems of interpretation.

7. Correlations between relatives may indicate a genetic component (heritability) in variable behavioral traits, but information is required on relatives that have been reared apart in order to avoid the confounding of genetic and environmental sources of variation. The existence of genes affecting a variable trait can also be indicated by linkage of the trait to genetic markers. This method has provided evidence suggesting that some variation in sexual orientation is based on one or more genes on the X chromosome.

8. IQ ("intelligence quotient") scores may have a heritability of 0.5 or greater, based on comparisons of adoptive and biological siblings. Different cognitive abilities may have partly independent genetic variation, and some variation in IQ may be due to genetic differences in the way children experience their environment. A genetic component does not mean that a trait is unalterable, and it is clear that IQ can be substantially increased by education or environmental enrichment. Heritable variation within groups does not mean that average differences among groups have any genetic basis, and there is considerable evidence that "racial" groups do not differ genetically in IQ.

9. In the past, evolutionary theory and genetics have been misinterpreted and misappropriated to lend support to eugenics, imperialism, Social Darwinism, sexism, racism, and other abuses of human rights and dignity. Contemporary theories and knowledge can also be abused. It is important for scientists to alert the public to possible abuses, and for citizens to be sufficiently informed on the principles of evolution and genetics to recognize false doctrines.

Major References

Jones, S., R. Martin, and D. Pilbeam (editors). 1992. *The Cambridge encyclopedia of human evolution*. Cambridge University Press, Cambridge. A comprehensive collection of short essays on the subjects of this chapter.

Ciochon, R. L., and J. G. Fleagle (editors). 1993. *The human evolution sourcebook*. Prentice-Hall, Englewood Cliffs, NJ. Reviews and analyses of the hominid fossil record and the origins of culture.

Cavalli-Sforza, L. L., P. Menozzi, and A. Piazza. 1994. *The history and geography of human genes*. Princeton University Press, Princeton, NJ. On patterns of genetic variation and their historical interpretation.

Mange, E. J., and A. P. Mange. 1993. *Basic human genetics.* Sinauer Associates, Sunderland, MA. An introduction to the methods and data of human genetics.

Wilson, E.O. 1975. *Sociobiology: The new synthesis.* Harvard University Press, Cambridge, MA. The last chapter in this overview of the evolution of social behavior treats human behavior. It caused considerable controversy.

Caplan, A. L. (editor). 1978. *The sociobiology debate: Readings on ethical and scientific issues.* Harper and Row, New York. An introduction to this controversial area. See also G. W. Barlow and J. Silverberg (editors), *Sociobiology: Beyond nature/nurture* (Westview Press, Boulder, CO, 1980) for more essays; E. O. Wilson, *On human nature* (Harvard University Press, Cambridge, MA, 1978) for an advocate's view of human sociobiology; and P. Kitcher, *Vaulting ambition* (MIT Press, Cambridge, MA, 1985) for a critique of human sociobiology. Evolutionary psychology is described by advocates in J. H. Barkow, L. Cosmides, and J. Tooby (editors), *The adapted mind* (Oxford University Press, New York, 1992).

Plomin, R., and G. E. McClearn (editors). 1993. *Nature, nurture, and psychology.* American Psychological Association, Washington, DC. Essays by behavior geneticists on IQ and other measures of behavior.

Gould, S. J. 1981. *The mismeasure of man.* W. W. Norton, New York. A history and critique of modern mental testing and its antecedents. For a vigorous critique of claims for genetic determination of human behavior, see R. C. Lewontin, S. Rose, and L. J. Kamin, *Not in our genes: Biology, ideology, and human nature* (Pantheon, New York, 1984).

Problems and Discussion Topics

1. Mated pairs of gibbons form long-lasting pair-bonds, chimpanzees form temporary pair-bonds, and female bonobos (pygmy chimpanzees) reinforce social bonds by sexual activity with each other. Discuss how, if at all, studies of nonhuman primates can shed light on the sexual and other social behaviors of human ancestors.

2. What evidence would support the hypothesis that genes from European Neanderthals and from Asian populations of archaic *Homo sapiens* persist in different human populations today?

3. Suppose that in a study of male siblings reared together by their biological parents, positive correlations (on a scale of 0 to 1) of 0.75 are found for monozygotic twins, 0.55 for dizygotic twins, and 0.45 for non-twin brothers in some aspect of personality, such as extroversion. What hypotheses can explain such data, and how would you test them?

4. Imagine that in a relatively isolated population of 100,000 people, a recessive allele that causes a severely disabling defect of the nervous system has a frequency of 0.05. Suppose that the normal and deleterious alleles have been sequenced, the deleterious allele can be detected in heterozygotes by DNA testing, and gene therapy has been developed, whereby the deleterious allele can be replaced by the normal allele in isolated ova that can be returned to a woman's uterus after test-tube fertilization. Discuss the possible methods and feasibility of eliminating the deleterious allele from the population, and the ethical questions each method would raise.

5. Based on your knowledge of evolutionary theory, describe (*a*) how resistance to antibiotics arises in disease-causing bacteria and viruses, and (*b*) what can be done to retard or prevent resistance.

6. What have evolutionary biologists found out about the origin of HIV (the virus that causes AIDS), and about the evolutionary changes it is undergoing in the human population? (See Leigh Brown and Holmes 1994.)

7. Suppose it should turn out that men are indeed biologically more prone to aggressiveness and social dominance than women, or less prone to nurture children. Should any social policies be affected by such discoveries? What might they be, and why?

8. For each of the following observations about Western culture, think of an evolutionary and a nonevolutionary (cultural) explanation, and discuss how they might be distinguished. (*a*) Young people are often more adventurous and less conservative than older people. (*b*) Most people would rather rear their own children than adopt. (*c*) Many people accept religious or political doctrines with little question. (*d*) Children (reputedly) don't like spinach.

9. Many conservatives have embraced evidence (even flawed evidence) that IQ is highly heritable, and have argued that expensive social programs designed to improve intellectual skills should be abandoned because they are doomed to fail. Liberals have generally been skeptical that IQ is highly heritable, perhaps because they oppose the conservatives' agenda. Compose a letter to a conservative and to a liberal member of Congress, explaining the implications of current evidence for such social policy decisions and offering a recommendation.

10. The 1986 predecessor of this book included this discussion topic, which you may wish to update: Read Shelley's poem *Ozymandias*, reflect on what has happened in the last few million years of evolutionary history and the last few thousand years of human history, and discuss the likely fate over the next 5000 years of such institutions as democracy, technology, the United States, and the Soviet Union.

Evolutionary Biology in the Future

Like many other sciences, evolutionary biology is progressing at an accelerating pace. In the last few decades, it has expanded in scope, has employed ever more sophisticated tools, and has inspired and attracted greater and greater numbers of young scientists. It has formed increasingly close ties with other disciplines, such as molecular biology, developmental biology, behavioral biology, physiology, and ecology. As every chapter in this book attests, many long-standing questions about evolution remain unanswered, and whole new areas of inquiry are being opened to exploration. What might the future hold?

In the next few decades, evolutionary studies will be affected by changes in research tools and methods, and by a focus on new subjects and questions. Traditional knowledge and training will remain important: the field will continue to require some researchers who are adept in mathematics, others who hold skills in molecular techniques, and still others who are deeply familiar with the biology, diversity, and systematics of individual groups of living or extinct organisms. Their work, however, will take advantage, as it has already, of methodological advances in the areas of molecular technology and information technology. Evolutionary biologists, like biologists in other fields, will find diverse uses for increasingly easy and automated methods of sequencing DNA, studying gene expression, inducing specific mutations, and transferring genes among organisms. Complete sequences of genomic DNA, already accomplished for yeast and bacteria such as *Escherichia coli* and *Haemophilus influenzae*, will become available for the genomes of other bacteria, *Drosophila*, the plant *Arabidopsis*, and humans. Evolutionary biologists will use these data as platforms for launching studies of other organisms. Advances in computer technology and information exchange will support increasingly sophisticated theoretical analyses, and huge data bases of DNA sequences and other information will provide ever broader and more detailed portraits of the history of life, the geographic deployment of biodiversity, and patterns of molecular and phenotypic evolution.

What might we accomplish with such tools? We can look forward to building a complete tree of life, a well-grounded phylogeny of all living species (*if* enough systematists, knowledgeable in the diversity of organisms, are trained and employed). We will then have not only the intellectually satisfying reward of knowing much more about the origins of diversity, but also a framework for interpreting patterns and inferring causes of molecular and phenotypic diversity. We can look forward, too, to compiling increasingly detailed information on the patterns and causes of genetic variation within species by coupling molecular descriptions with experiments in natural and laboratory populations.

As evolutionary molecular biology and evolutionary developmental biology grow, we will achieve an ever deeper understanding of the mechanisms by which biochemical pathways and anatomical and physiological features evolve, and the mechanisms by which new species originate. Progress in many areas, in fact, will come to be limited not by technological constraints, but by elementary biological information. Probably most of the world's species have not yet been described; we do not know the precise geographic distributions of most of the species that are known; we are ignorant of the feeding habits and life histories of most of the world's insects and other small organisms. These gaps in our knowledge, not technical limitations, will set the pace of our advances in understanding patterns of adaptation and biodiversity—an understanding that is fundamentally important in evolutionary ecology and conservation biology. The sister fields of evolution and ecology depend not only on molecularly adept researchers, but also on those who know organisms.

Evolutionary biology will also be transformed by expansion into new or understudied areas, such as the evolution of nervous systems, brain function, and mechanisms of behavior. Perhaps most importantly, researchers trained in evolutionary biology will increasingly apply their knowledge to practical societal concerns. Evolutionary geneticists will use models and molecular tools to explore the causes of human genetic diversity, especially as manifested in genetic diseases, and the nascent science of Darwinian medicine (Nesse and Williams 1994) will suggest hypotheses and experiments that will cast new light on the functions and malfunctions of the human body. Evolutionary ecologists will provide insights into the evolution of virulence and drug resistance by disease-causing organisms, will model epidemics, and will suggest not only technological, but also environmental interventions to slow the spread of infectious diseases.

The potential will soon exist, as never before, to explore the living world for natural products useful to humans. Penicillin, quinine, salicylic acid (the effective ingredient in aspirin) and taxol (a suppressant of certain cancers) are products of plants and fungi that have found medicinal uses; many hundreds of natural products are used as flavors, dyes, emulsifiers, and preservatives; biologists use snake venoms to study the function of the nervous system and Taq polymerase, from a hot-springs bacterium, to sequence DNA. The millions of living species produce many thousands of other potentially useful compounds. Not only can they be harvested directly, but in many cases, it will be possible to transfer the genes responsible for their biosynthesis from wild species into crops, or into laboratory cultures of bacteria that can then synthesize the compounds in quantity. For instance, insect-resistant tobacco plants have been engineered by transferring a gene for an insecticidal compound from the bacterium *Bacillus thuringiensis*. Evolutionary biologists can help to make the search for useful compounds and genes ("bioprospecting") far more efficient. Systematists can point to species that might harbor compounds similar to those in related species, but more potent. Evolutionary ecologists can study the adaptive raison d'être of various compounds, such as the chemical defenses of plants against herbivores and pathogens, and thus may point the way toward species that are likely to have desirable characteristics.

Evolutionary biology has been, and will continue to be, important in conservation and in environmental management. Degraded, polluted soils may be reclaimed by finding species or genotypes of plants adapted to these conditions. Microbial systematists and geneticists are searching for bacteria with the ability to digest crude oil in oil spills and to metabolize environmental toxins. Systematists are using phylogenetic information as a guide to finding the native lands from which pests such as the tiger mosquito have been inadvertently spread by humans, so that the ecology and natural enemies of such species can be studied. Conservation biologists are using evolutionary principles extensively in designing ways to prevent inbreeding depression and maximize genetic diversity in captive breeding programs for endangered species, in estimating the population sizes that natural populations must retain in order to remain viable, and in employing systematics and biogeography to locate areas of high biodiversity where conservation efforts are most urgently needed. Many evolutionary biologists have been leaders of the movement to preserve biodiversity—one of the most important challenges we face.

Both in basic research and applied research, then, evolutionary biology looks forward to an exciting, socially relevant, and intellectually rewarding future.

Elementary Statistics

Statistics play a very important role in evolutionary biology in two ways. First, statistical concepts and theory are important elements of theoretical models, especially in population genetics and quantitative genetics. Second, in evolutionary biology as in other sciences, statistical analysis is essential for evaluating the significance of data (such as whether or not two variables are really correlated). This appendix (some of which repeats material in Box C of Chapter 9) provides only the most elementary introduction to those elements of statistics needed to understand the material in the text.

Frequency Distributions

For a continuous variable such as body weight, some number of individuals in the population of interest will have each of many possible values. The word *population*, in statistical terminology, refers to all members of some class of objects, perhaps including past and future members as well as those in the present. For a biologist, a statistical "population" might be all the bacteria in a culture vessel at a certain time, or all members of one natural population of a species, or all members of a species. Ordinarily, we *estimate* the properties of the population from a *sample*. Thus the *sample mean* estimates the *population mean*, and similarly for other properties. The *frequency distribution* of the variable in the population—the proportion of individuals with each value of the variable—is estimated from the frequency distribution in the sample. Many, but not all, variables have a roughly *normal*, or bell-shaped, distribution (Figure C1 in Chapter 9). The mathematical characteristics of the normal distribution are often used both in mathematical modeling and in analyzing data on characteristics that are approximately normally distributed. Often, a frequency distribution that is not normally distributed on an original scale of measurement (such as grams) can be made approximately normal by a mathematical transformation, such as taking the logarithm of each measurement.

The "location" of a normal distribution is usually described by its **arithmetic mean**, commonly known as the *average*. If X_i is the value of the variable in the ith specimen (e.g., $X_3 = 4$ grams for mouse #3), the sample mean is

$$\bar{x} = \frac{\sum_i X_i}{n}$$

where n is the sample size and $\sum_i$ represents the sum from $i = 1$ to $i = n$.

For a variable with discrete, discontinuous values, such as the number of offspring per brood, there may be n_1 individuals with value X_1, n_2 with value X_2, and so on. Then if $n_i/n = f_i$, the *frequency* of individuals with value X_i, the mean may be calculated as

$$\bar{x} = \sum_i \left(f_i X_i \right)$$

A special case, important in genetics, is the binomial distribution, in which the probability (or proportion) of events of type i is p_i. Let there be two possible events, 1 (e.g., heads if a coin is tossed) and 0 (tails), with probabilities p and q respectively. Since $q = 1 - p$, the weighted sum of values in n trials (coin tosses) is $n[(p)(1) + (1 - p)(0)] = np$. Dividing by n, we find the mean of the probability distribution, $\bar{x} = p$.

For mathematical reasons, the most useful measure of the variation in a series of measurements is the **variance**, defined as the mean value of the squared deviations of the values from the arithmetic mean of the population or sample:

$$V = \frac{\left(X_1 - \bar{x}\right)^2 + \left(X_2 - \bar{x}\right)^2 + \ldots + \left(X_n - \bar{x}\right)^2}{n} = \frac{1}{n} \sum \left(X_i - \bar{x}\right)^2$$

Although the denominator is n for the variance of a population, $n - 1$ is used in calculating the variance of a sample, for statistical reasons.

For a discontinuous variable, for which $f_i = n_i/n$, the population variance may be written as

$$V = \sum_i f_i \left(X_i - \bar{x} \right)^2$$

Extending this to the binomial distribution, $f_0 = 1 - p$ and $f_1 = p$. Since the mean is $\bar{x} = p$, the variance is $V = (1 - p)(0 - p)^2 + (p)(1 - p)^2 = p(1 - p)$. This is the variance of the probability distribution of heads and tails, for example. In repeated sets of n tosses of a coin, the proportion of heads will vary from one set to another, and the smaller n is, the larger the variation will be. The variance, in repeated sets of n tosses, is

$$V = \frac{p(1-p)}{n}$$

This concept is important in genetics. If the proportion of A alleles in a population is p, repeated samples of $2n$ gene copies (from n diploid individuals) will vary in allele frequencies with variance $p(1 - p)/2n$.

Because it is expressed in squared units (gm^2, for example), the variance is less easily visualized than its square root, the **standard deviation**, S (often denoted s or σ). That is, $S = \sqrt{V}$. For a normal distribution, S constitutes a mathematically fixed fraction of the area under the curve (i.e., a fixed proportion of the population). For example, 68 percent of the population falls within one standard deviation on either side of the mean, and 99.7 percent falls within three standard deviations on either side (Figure C1 in Chapter 9). If S is large, 68 percent of the population will spread out farther from the mean than if S is smaller. The absolute magnitude of S and V may be related to the mean; for example, S would be greater for body weight in a sample of elephants than in an equally variable sample of mice. In order to compare variation in such cases, the *coefficient of variation* (CV), which standardizes the standard deviation by the mean, is sometimes used: $CV = 100(S/\bar{x})$.

Analysis of Variance

An important property of variances (but not of standard deviations) is that variances arising from different components of variation add up to the *total variance*. For instance, if a sample consists of several groups of individuals, the total variance equals the variance among the individuals within each group plus the variance among the means of the several groups (see Box C in Chapter 9 for a simple example.)

The groups may be defined by different conditions or treatments that could cause variation, such as growth under high or low light, or possession of different genotypes. If, for instance, we measure the weight of several plants of each of three genotypes, the total variance is the sum of a genetic component (the variance among the means of the three genotypes) and a nongenetic component (the variance among individuals of the same genotype).

There may be many components to the total variance. Table 1 offers some hypothetical weights of plants grown under four combinations of low and high levels of nitrogen and phosphorus. In case *a*, the total variance is the sum of (1) the average variance within each of the four treatments, (2) the variance among rows (nitrogen levels), and (3) the variance among columns (phosphorus levels). In case *b*, however, there is a fourth component of variance due to the *interaction* between the two fertilizers ("the whole is greater than the sum of its parts"). In statistical terms, the fertilizers do not interact in case *a*; they are said to have wholly *additive* effects. Additive effects and interactions are important in evolutionary genetics, in which we are frequently interested in partitioning the phenotypic variance into that arising from differences among genotypes, from environmental causes, and from interactions between the two. The statistical model used for such partitioning is the *analysis of variance*, which was first developed by R. A. Fisher, one of the founders of evolutionary genetics.

Correlations among Variables

Table 2 presents hypothetical data on the body length (X) and the tail length (Y) of five specimens. The mean total length, $\bar{z}$, is the sum of the separate means $\bar{x}$ and $\bar{y}$. The variance of the total length, V_Z, does not equal the sum of the variances V_X and V_Y, because X and Y are correlated. Rather, V_Z is found to be

$$V_Z = \frac{1}{n} \sum_i \left[(X_i + Y_i) - (\bar{x} + \bar{y}) \right]^2$$

$$= \frac{1}{n} \sum (X_i - \bar{x} + Y_i - \bar{y})^2$$

$$= \frac{1}{n} \left[\sum (X_i - \bar{x})^2 + \sum (Y_i - \bar{y})^2 + 2\sum (X_i - \bar{x})(Y_i - \bar{y}) \right]$$

$$= V_X + V_Y + 2\,\mathrm{cov}(X, Y)$$

Table 1 Hypothetical effects of fertilizers on plant weights

		(a) NO INTERACTION					(b) INTERACTION	
		PHOSPHORUS					PHOSPHORUS	
		Low	High				Low	High
NITROGEN	Low	9,11	14,16		NITROGEN	Low	9,11	14,16
	High	19,21	24,26			High	19,21	29,31

Increased phosphorus adds 5, on average

Increased nitrogen adds 10, on average

(High N, high P) − (low N, low P) = 15, on average

High phosphorus adds 5 if nitrogen is low, 10 if nitrogen is high

High nitrogen adds 10 if phosphorus is low, 15 if phosphorus is high

Table 2　Correlation and regression

SPECIMEN	TAIL LENGTH (Y)	BODY LENGTH (X)	TOTAL LENGTH (Z)
1	7	9	16
2	9	9	18
3	8	10	18
4	10	12	22
5	12	13	25
Σ	46	53	99
Mean	$\bar{y} = 9.2$	$\bar{x} = 10.6$	$\bar{z} = 19.8$
Variance	$V_Y = 2.96$	$V_X = 2.64$	$V_Z = 10.56$

Covariance　$\text{cov}(X,Y) = 4.96$
Correlation coefficient　$r_{XY} = 0.89$
Regression coefficient, Y on X　$b_{Y \cdot X} = 0.939$
Y-intercept　$a = -0.753$
Regression equation　$Y = -0.753 + 0.939X$

That is, the variance is inflated by twice the term $(1/n)$ $\sum (X_i - \bar{x})(Y_i - \bar{y})$, the *covariance*, which measures the dispersion of values around the joint mean, $\bar{x}, \bar{y}$.

The covariance is closely related to the **correlation coefficient** between X and Y,

$$r_{XY} = \frac{\dfrac{1}{n}\sum_i \left(X_i - \bar{x}\right)\left(Y_i - \bar{y}\right)}{S_X S_Y}$$

where S_X and S_Y are the standard deviations of X and Y respectively. The correlation coefficient measures the degree of association between the two variables. It ranges from $+1$ for variables that are perfectly positively correlated to -1 for variables that are perfectly negatively correlated. The higher the absolute value of r_{XY}, the less variation each variable displays independently of the other. In calculating a correlation coefficient, it is assumed that neither variable "causes" the other: neither can be specified as a dependent or independent variable. That is, body length does not cause tail length; both, however, may be consequences of other "causes," such as age or nutrition.

If, however, X is an independent variable and Y a dependent variable, so that X in some sense "causes" Y, we can calculate another measure of association between them, which is designed to *predict* Y from X. Suppose, for instance, that in Table 2, Y represents the mean tail length of each of five broods of offspring and X that of their parents. The tail length of offspring depends on that of their parents, because it depends on the different alleles that the parents transmit. The data in Table 2 are plotted in Figure 1. If, as in Figure 1, the points approximate a straight line, the predictive equation is the linear equation $Y = a + b_{Y \cdot X} X$, where $b_{Y \cdot X}$ is the **coefficient of regression** of Y on X, and is the slope of the line illustrated in the figure. Of all possible straight lines that can be drawn through the points, this is the one from which the sum of the squared deviations of the Y values is minimal, so a regression analysis of this kind is called a *least squares regression*. It is the line that best pre-

dicts Y (e.g., offspring tail length) if X (e.g., parent tail length) is known. The regression coefficient describing the relationship between the phenotypes of parents and their offspring is used to calculate the heritability of a trait (see Chapters 9 and 14).

The regression coefficient $b_{Y \cdot X}$ is related to the correlation coefficient

$$b_{Y \cdot X} = r_{XY}\frac{S_Y}{S_X}$$

but it does not equal the correlation coefficient unless X and Y have equal standard deviations. The regression coefficient can be calculated as

$$b_{X \cdot Y} = \frac{\sum_i \left[\left(X_i - \bar{x}\right)\left(Y_i - \bar{y}\right)\right]^2}{\sum_i \left(X_i - \bar{x}\right)^2}$$

FIGURE 1　Example of linear regression of a dependent variable Y on an independent variable X. The regression line has a slope b and Y-intercept a such that the sum of squared deviations (d) of the points from the line $\left(\sum_i d_i^2\right)$ is minimized.

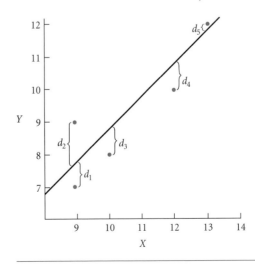

Hypothesis Testing

Throughout the natural and social sciences, statistics are used to determine whether or not the means, variances, regressions, or other descriptions of data lend support to hypotheses. Usually, statistical tests are used to judge whether the data support a *null hypothesis* or an *alternative hypothesis*. The alternative hypothesis is a prediction about the population derived from theory or prior observations, such as the expectation that highly ornamented male birds will obtain more mates than unadorned males, or that larger parents will have larger offspring. In statistical terms, then, the alternative hypothesis might be that the mean number of mates differs between ornamented and unadorned males, or that there is a positive regression of offspring size on parent size. The null hypothesis is that there is no difference, in the population from which the sample is taken, between the two kinds of males, or among the offspring of different parents. The null hypothesis specifies that the difference between the true means equals zero, or that the true regression coefficient equals zero.

Statistical tests are designed to determine whether, on the basis of data from a sample, the null hypothesis about the population can be confidently rejected. If so, we are justified in accepting an alternative hypothesis. Statistical testing is necessary because, although our *best estimate* of the true means or the true regression coefficient is provided by the means or the regression in our samples, the sample values can deviate from the true values by chance. For example, the means for samples of ornamented and unadorned males will probably differ by chance, even if the true mean of the population of ornamented males (i.e., all the ornamented males that exist) is identical to the true mean of the population of unadorned males. The question, then, is, how great a difference between the two sample means do we need in order to be confident that the population means truly differ? Or, how great does the regression coefficient have to be before we can reject the hypothesis that it equals zero in the population we have sampled?

Statistical tests are based on two principles. First, the more variable the population is, the more likely it is that the sample mean (or regression) will deviate by chance from the population mean (or regression). Second, the smaller the sample is, the greater this chance deviation is likely to be. Conversely, our sample estimates approach the true population values more closely, the larger the sample size and the lower the variation. Thus, as sample size increases, we become able to more confidently detect smaller differences between true population means. Still, even if we have huge samples, even if the population is not very variable, and even if the differences between sample means are large, there is still a nonzero probability that the sample means will differ even though the population means do not.

Statistical tests, by applying probability theory to our data—such as the differences between sample means, the variances of each sample, and the sample size—allow us to estimate the probability that the difference between sample means could have arisen by chance alone in the process of sampling from populations with identical means. (Similar statements apply to other quantities of interest, such as regression coefficients.) That is, the tests estimate the probability that by accepting an alternative hypothesis (that the population means truly differ), we are rejecting a true null hypothesis (that the population means do not differ). Expressions such as "$P < 0.001$," or "$P < 0.005$" mean that the probability is less than 0.001 or 0.005 that our observations (e.g., a difference between two sample means) could be due to chance alone, and that the null hypothesis could therefore be true. In most areas of study, $P < 0.05$ is the conventional criterion for saying that a result is "statistically significant," and for accepting the alternative hypothesis rather than the null hypothesis.

This book treats concepts more thoroughly than methods, and relates published claims of differences or relationships only when they were reported to be statistically significant.

Contending with Creationism

We cannot pretend to understand fully all the mechanisms of evolution, and there are many differences of opinion among evolutionary biologists concerning the relative importance of those mechanisms we do know about. However, the historical reality of evolution—the descent, with modification, of all organisms from common ancestors—has not been in question, among scientists, for well over a century. It is as much a scientific fact as the atomic constitution of matter or the revolution of the earth around the sun.

Nevertheless, many people do not believe in evolution. More than 40 percent of Americans believe that the human species was created directly by God, rather than evolving from a common ancestor with other primates (and from a much more remote common ancestor with all other species). This appendix provides a capsule summary of arguments for evolution and against special creation for readers who may find it necessary to deal with this issue. I have provided a more complete treatment of most of these points in *Science on Trial* (Futuyma 1995).

Creationists and Other Skeptics

In the United States and other Western countries, most disbelievers in evolution are Christians (or, more rarely, Jews) who find it to conflict with religious beliefs derived from their interpretation of the Bible, especially the first chapters of Genesis, which portray God's creation of the heavens, the earth, plants, animals, and humans within six "days." It should be understood that many Western religions understand these biblical descriptions to contain symbolic truths, not literal, scientific ones, and that many deeply religious people believe in evolution, viewing it as the natural mechanism by which God has enabled creation to proceed. For example, Pope John Paul II affirmed the validity of evolution in 1996, while emphasizing that there is no conflict between evolution and the Catholic Church's theological doctrines. However, disbelievers in evolution reject this position. Most of them believe in "special cre-

ation," the doctrine that each species, living and extinct, was created independently by God, essentially in its present form. Confronted with direct evidence of ongoing evolution (e.g., industrial melanism in moths, insecticide resistance in many species of insects), most creationists allow that slight changes can occur, and even that very similar species can arise from a common ancestor. However, they deny that higher taxa (genera, families, etc.) have evolved from common ancestors, and they most vigorously assert that the human species was specially created by God.

The beliefs of creationists vary considerably. The most extreme are those who interpret every statement in the Bible literally. They believe in a young universe and earth (less than 10,000 years old), a deluge that drowned the earth, and an ark in which Noah preserved a pair of every living species. They must therefore deny not only evolution, but also most of geology and physics (including radioactive dating and astronomical evidence of the great age of the universe). The fossil record, in their view, represents species that did not survive the deluge. Some creationists grant the antiquity of the earth and the rest of the universe, but otherwise share many or most of the biblical literalists' beliefs, including special creation. Still other nonbelievers in evolution, including a very few scientists, present supposedly rational arguments against evolution, and instead of specifically invoking the biblical account as an alternative, argue that the only possible explanation of biological phenomena is "intelligent design"—i.e., creation by an intelligent Creator.

Most of the efforts of activist creationists are devoted to suppressing the teaching of evolution in schools, or at least insisting on "equal time" for their views. In the United States, however, the Constitutional prohibition of state sponsorship of religion has been interpreted by the courts to mean that the biblical, or any other explicitly religious, version of the origin of life's diversity cannot be promulgated in public schools. Creationists have therefore adopted the camouflage of "scientific creationism," which consists of attacks on and supposed disproofs of evolution, and teaching that the complexity of living things can be explained only by intelligent design. God and the Bible are not named in the versions of creationists' writings intended for distribution in public schools, although they are prominent in versions intended for Christian schools. Bills requiring public schools to give "creation science" equal time have been introduced in state legislatures, but the U.S. Supreme Court in 1987 found one such bill unconstitutional because it "endorses religion by advocating the religious belief that a supernatural being created humankind," and concluded that the law was written "to restructure the science curriculum to conform with a particular religious viewpoint."

Refuting Creationist Arguments

Creationists explain the existence of diverse organisms and their characteristics as miracles: as the result of direct supernatural intervention. It is impossible to predict miracles or to do experiments on supernatural processes, so creationists do not do original research in support of their theory. (About the only quasi-exception to this statement was their claim to have found commingled human and dinosaur footprints in fossilized sediments in a riverbed in Texas, supposedly showing that these organisms were contemporaneous. Even most creationists now acknowledge that the "human" prints are a mixture of fraudulent carvings and natural depressions.) Thus "creation science," rather than providing positive evidence of creation, consists entirely of attempts to demonstrate the falsehood or inadequacy of evolutionary science, and to show that biological phenomena must, by default, be the products of intelligent design. Here are some of the most commonly encountered creationist arguments, together with capsule counterarguments. The main text should be consulted for more extensive support.

1. **Evolution is outside the realm of science because it cannot be observed.**

   Most of science depends not on direct observation, but on testing hypotheses against the predictions they make. The structure of DNA, for instance, is known from data that conformed to predictions of the Watson-Crick model, and has still not been directly observed. Besides, although we have not observed, and cannot expect to observe, the origin of new higher taxa, modest evolutionary changes of characters have been observed in many species (see Chapters 13 and 14), new species of plants have arisen within the last century (see Chapter 16), and natural allopolyploid species of plants have been "re-created" by hybridizing diploid species (see Chapter 16).

2. **Evolution cannot be proved.**

   Nothing in science is ever absolutely proved. "Facts" are hypotheses in which we can have very high confidence, because of massive evidence in their favor and the absence of contradictory evidence. Abundant evidence

from every area of biology and paleontology supports evolution, and there exists no contradictory evidence.

3. **Evolution is not a scientific hypothesis because it is not testable: no possible observations could refute it.**

   Many conceivable observations could refute or cast serious doubt on evolution, such as finding incontrovertibly mammalian fossils in incontrovertibly Precambrian rocks. In contrast, any puzzling quirk of nature could be attributed to the inscrutable will and infinite power of a supernatural intelligence, so creationism is untestable.

4. **The orderliness of the universe, including the order manifested in organisms' adaptations, is evidence of intelligent design, just like the orderliness of a machine designed for a particular function.**

   Order in nature, such as the structure of crystals, arises from natural causes, and is not evidence of intelligent design. The order displayed by the correspondence between organisms' structures and their functions is the consequence of natural selection acting on genetic variation, as has been observed in many experimental and natural populations (see Chapters 13 and 14).

5. **Evolution of greater complexity violates the second law of thermodynamics, which holds that entropy (disorder) increases.**

   The second law applies only to closed systems, such as the universe as a whole. Order and complexity can increase in local, open systems due to an influx of energy. This is evident in the development of individual organisms, in which biochemical reactions are powered by energy derived ultimately from the sun.

6. **It is almost infinitely improbable that even the simplest life could arise from nonliving matter. The probability of random assembly of a functional nucleotide sequence only 100 bases long is $1/4^{100}$, an exceedingly small number. And scientists have never synthesized life from nonliving matter.**

   It is true that a fully self-replicating system of nucleic acids and replicase enzymes has not yet arisen from simple organic constituents in the laboratory, but the history of scientific progress shows that it would be foolish and arrogant to assert that what science has not accomplished in a few decades cannot be accomplished. Critical steps in the probable origin of life, such as abiotic synthesis of purines, pyrimidines, and amino acids and self-replication of short RNAs, have been demonstrated in the laboratory (see Chapter 7). And there is no reason to think that the first self-replicating or polypeptide-encoding nucleic acids had to have had any particular sequence. If there are many possible sequences with such properties, the probability of their formation rises steeply. Moreover, the origin of life is an entirely different problem from the modification and diversification of life once it has arisen. Knowledge of the latter does not require knowledge of the former.

7. **Mutations are harmful, and do not give rise to complex new adaptive characteristics.**

Most mutations are indeed harmful, and are purged from populations by natural selection. Some, however, are beneficial, as shown in many experiments (see Chapter 10). Complex adaptations usually are based not on single mutations, but on combinations of mutations that jointly or successively increase in frequency due to natural selection.

8. **Natural selection merely eliminates unfit mutants, rather than creating new characters.**

"New" characters, in most cases, are modifications of pre-existing characters, altered in size, shape, developmental timing, or organization (see Chapters 5, 23, and 24). This is true at the molecular level as well (see Chapter 22). Natural selection "creates" such modifications by increasing the frequency of alleles at several or many loci so that combinations of alleles, initially improbable because of their rarity, become probable. Observations and experiments on both laboratory and natural populations have demonstrated the efficacy of natural selection (see Chapters 10, 13, and 14).

9. **Chance could not produce complex structures.**

This is true, but natural selection is a deterministic, not a random, process. The random processes of evolution, mutation and genetic drift, do not in themselves result in the evolution of complexity, as far as we know. Indeed, when natural selection is relaxed, complex structures, such as the eyes of cave-dwelling animals, slowly degenerate, due in part to fixation of neutral mutations by genetic drift.

10. **Complex adaptations such as wings, eyes, and biochemical pathways could not have evolved gradually because the first stages would not have been adaptive. The full complexity of such an adaptation is necessary, and this could not arise in a single step by evolution.**

The response to this common objection has two parts. First, many such features, such as hemoglobins and eyes, do show various stages of increasing complexity among different organisms (see Chapters 5, 22, and 24). "Half an eye"—an eye capable of discriminating light from dark, but incapable of forming a focused image—is indeed better than none. Second,

many structures have been modified for a new function after being elaborated to serve a different function (see Chapter 5).

11. **If an altered structure, such as the long neck of the giraffe, is advantageous, why don't all species have that structure?**

This naive question ignores the fact that different species and populations have different ecological niches and environments, for which different features are adaptive.

12. **If gradual evolution had occurred, there would be no phenotypic gaps among species, and classification would be impossible.**

Many disparate organisms are connected by intermediate species, and in such cases, classification into higher taxa is indeed rather arbitrary (see Chapter 5). In other cases, gaps exist because of the extinction of intermediate forms.

13. **The fossil record does not contain any transitional forms representing the origin of major new forms of life.**

This very common claim is flatly false, for there are many such intermediates (see Chapter 6). Creationists sometimes use rhetorical subterfuge in presenting this argument, such as defining *Archaeopteryx* as a bird because of its feathers, and then claiming that there are no known intermediates between reptiles and birds.

14. **The fossil record does not objectively represent a time series, because strata are ordered by their fossil contents, and then are assigned different times on the assumption that evolution has occurred.**

Even before *The Origin of Species* was published, geologists who did not believe in evolution recognized the temporal order of fossils that are characteristic of different periods, and named most of the geological periods. Since then, radioactive dating and other methods have established the absolute dates of geological strata.

15. **The similarities among organisms that biologists ascribe to common ancestry—i.e., to homology—are actually examples of a common design used by the Creator.**

Anything can be "explained" by the will of an omnipotent Creator, but we have no way to obtain information about the Creator, and no way to test this hypothesis. Certainly, many homologous features make no adaptive sense (e.g., gill slits in both fishes and mammalian embryos) and are hard to envision as the products of optimal, intelligent design. Evolutionary theory predicts that the degree of similarity between homologous features should generally decline with time since common ancestry, and this is generally the case (e.g., DNA sequences: see Chapter 22).

16. **Vestigial structures are not vestigial, but functional.**

According to creationist thought, an intelligent Creator must have had a purpose, or design, in each element of his creation. Thus all features of organisms must be functional. For this reason, the existence of adaptations is not a strong argument for evolution. However, nonfunctional and even maladaptive structures are expected if evolution is true, especially if a change in an organism's environment or way of life has rendered them superfluous or harmful. Organisms display many such features at both the morphological and molecular levels (see Chapters 5, 14, and 22).

17. **There are no fossil intermediates between apes and humans; australopithecines were merely apes. And there exists an unbridgeable gap between humans and all other animals in cognitive abilities.**

The array of fossil hominids shows numerous stages in the evolution of posture, hands and feet, teeth, facial structure, brain size, and other features. The mental abilities of humans are indeed developed to a far greater degree than in other species, but many of our mental faculties seem to be present in more rudimentary form in other primates and mammals (see Chapter 26). Both functional and nonfunctional DNA sequences are extremely similar between humans and African apes.

18. **Disagreements among evolutionary biologists show that Darwin was wrong. Even prominent evolutionists have abandoned the theory of natural selection, and the entire study of evolution is in disarray.**

Disagreements among scientists exist in every field of inquiry, and are in fact the fuel of scientific progress. They stimulate research, and are a sign of vitality. Creationists misunderstand or misinterpret evolutionary biologists who have argued (*a*) that the fossil record displays abrupt shifts rather than gradual change (punctuated equilibrium); (*b*) that many characteristics of species may not be adaptations; (*c*) that evolution may involve mutations with large effects as well as those with small effects; and (*d*) that natural selection does not explain certain major events and trends in the history of life. In fact, none of the evolutionary biologists who hold these positions deny the central proposition that adaptive characteristics evolve by the action of natural selection on random mutations. All these debates arise from different opinions on the relative frequency and importance of factors known to influence evolution: large-effect vs. small-effect mutations, genetic drift vs. natural selection, individual selection vs.

species selection, adaptation vs. constraint, and so forth (see Chapters 12, 23, and 24). These arguments about the relative importance of different processes do not at all undermine the strength of the evidence for the historical fact of evolution—i.e., descent, with modification, from common ancestors. On this point, there is no disagreement among evolutionary biologists.

Evidence for Evolution

In addition to answering creationist arguments, a person who has read this book should be able to cite many lines of evidence for evolution. These include the following:

1. The temporal order of appearance of taxa in the fossil record, from Precambrian prokaryotes to Cenozoic plants, vertebrates, and other organisms that increasingly resemble living forms, refutes any claim for a single creation event, or for a single catastrophe (such as a flood) as the cause of extinctions.

2. For well-preserved taxa, and even for some taxa with spotty fossil records, there is commonly a correspondence between the temporal order of appearance in the fossil record and the phylogenetic position of the taxa, as inferred from independent data. (For example, jawless vertebrates appear as fossils before jawed fishes, and these appear before tetrapods, which are cladistically more "advanced" based on molecular and morphological data from living representatives.)

3. Intermediates, or transitions from ancestral to derived forms, have been described from the fossil record. Examples range from slight changes within genera to transitions among orders and classes (see Chapter 6).

4. The geographic distributions of higher taxa are explicable in terms of geological events (such as continental drift), isolation, and migration from regions of origin (see Chapter 8). Ecological analogues in isolated regions (e.g., Australian marsupials and placental mammals) are not closely related, and are evidence of convergent evolution. The geographic distributions of taxa make no sense if attributed to an intelligent designer.

5. The distribution of character states among organisms is generally hierarchical—i.e., species with derived states of some characters are nested within more inclusive groups exhibiting derived states of other characters (for example, the amnion characterizes a subset of tetrapod vertebrates, and the tetrapod condition, a subset of jawed vertebrates.) This pattern is expected if different modifications occur in different descendants of common ancestors. Based on this supposition, we can infer phylogenies. Hypotheses of phylogenetic relationship based on one class of data (e.g., morphology) are usually supported in large part by

other classes of data (e.g., DNA sequences). Such "consilience," or correspondence, of different lines of evidence is one of the strongest tests of the validity of a scientific hypothesis.

6. Phylogenetic hypotheses are supported not only by functional characters (e.g., functional genes), but also by nonfunctional characters (e.g., pseudogenes) and selectively neutral characters (e.g., synonymous base pair substitutions).

7. Organisms have many features that make no sense in terms of functional design. These include pseudogenes, immense amounts of noncoding DNA, vestigial structures, and features that make a temporary appearance during embryonic development.

8. Many characters display transition series—i.e., slight stepwise differences—among related organisms (see Chapter 5). Different stages in the elaboration of complex structures (e.g., eyes) are evident among closely or distantly related organisms (see Chapter 24).

9. Molecular evolutionary studies provide evidence of the ways in which new genes, biochemical functions, and genetic regulators of development come into existence (see Chapters 22 and 23).

10. Especially at the molecular level, characters are shared by disparate organisms (e.g., homeobox genes are shared by all eukaryotes, even though they serve very different developmental functions in different taxa). The mitochondria and chloroplasts of eukaryotes share DNA sequence homology with bacteria, as predicted by the endosymbiotic hypothesis. All organisms share functionally arbitrary characteristics such as the DNA (or RNA) code and the incorporation of only L amino acids into proteins (rather than the D optical isomers). Such evidence points to the common origin of all organisms.

11. Those who postulate creation by a beneficent Creator may be hard put to explain phenomena such as the extinction of well over 90 percent of the species that have ever lived, coevolutionary "arms races" among antagonistic species (such as pathogens and their hosts), the nasty behavior of animals toward members of their own species (including so-called infanticide, siblicide, deception, and many other interactions predicted by selection theory), and "selfish genes" such as transposable elements (see Chapter 21).

12. Selection theory has anticipated and successfully predicted many phenomena, such as patterns of life history, behavior, and numerous aspects of genetic systems (see Chapters 19–21).

13. All the proposed causes of evolution (e.g., mutation, genetic drift, various forms of natural selection, the effects of isolation and gene flow) have been extensively documented by experiment and observation (see Chapters 10–14). Artificial selection has shown that the store of genetic variation within populations and the new variation arising by mutation provide the basis for extensive change in many characteristics (see Chapters 10 and 14).

14. Modest evolutionary changes in many features have been documented within the last century in a variety of organisms (see Chapters 13 and 14).

Glossary

Most of the terms in this glossary appear at several or many places in the text of this book. Terms that are used broadly in biology or are used in this book only near their definition in the text are not included.

adaptation A process of genetic change of a population, owing to natural selection, whereby the average state of a character becomes improved with reference to a specific function, or whereby a population is thought to have become better suited to some feature of its environment. Also, *an* adaptation: a feature that has become prevalent in a population because of a selective advantage owing to its provision of an improvement in some function. A complex concept; *see* Chapter 12.

adaptive peak That allele frequency, or combination of allele frequencies at two or more loci, at which the mean fitness of a population has a (local) maximum. Also, the mean phenotype (for one or more characters) that maximizes mean fitness.

adaptive radiation Evolutionary divergence of members of a single phylogenetic line into a variety of different adaptive forms; usually the taxa differ in the use of resources or habitats, and have diverged over a relatively short interval of geologic time. The term **evolutionary radiation** describes a pattern of rapid diversification without assuming that the differences are adaptive.

adaptive zone A set of similar **ecological niches** occupied by a group of (usually) related species, often constituting a higher taxon.

additive effect The magnitude of the effect of an allele on a character, measured as half the phenotypic difference between homozygotes for that allele compared to homozygotes for a different allele.

additive genetic variance That component of the **genetic variance** in a character that is attributable to additive effects of alleles.

allele One of the several forms of the same gene, presumably differing by mutation of the DNA sequence, and capable of segregating as a unit Mendelian factor. Alleles are usually recognized by their phenotypic effects; DNA sequence variants, that may differ at several or many sites, are usually called *haplotypes*.

allele frequency The proportion of gene copies in a population which are a given allele; i.e., the probability of finding this allele when a gene is taken randomly from the population; *see* **gene frequency**.

allometric growth Growth of a feature during ontogeny at a rate different from that of another feature with which it is compared.

allopatric Of a population or species, occupying a geographic region different from that of another population or species. *See* **parapatric**, **sympatric**.

allopolyploid A **polyploid** in which the several chromosome sets are derived from more than one species.

allozyme One of several forms of an enzyme coded for by different alleles at a locus. *See* **isozyme**.

altruism Conferral of a benefit on other individuals at an apparent cost to the benefactor. *See* Chapter 20.

anagenesis Evolution of a feature over an arbitrary period of time.

aneuploid Of a cell or organism, possessing too many or too few of one or more of the chromosomes, compared to other chromosomes.

apomixis Parthenogenetic reproduction in which an individual develops from one or more mitotically produced cells that have not experienced recombination or syngamy.

apomorphic A derived state of a **character**, with reference to another state. *See* **synapomorphy**.

aposematic Coloration or other features that advertise noxious properties; warning coloration.

artificial selection Selection by humans of a consciously chosen trait or combination of traits in a (usually captive) population; differing from natural selection in that the criterion of survival and reproduction is the trait chosen, rather than fitness as determined by the entire genotype.

assortative mating Nonrandom mating on the basis of phenotype; usually used for positive assortative mating, the propensity to mate with others of like phenotype.

autopolyploid A **polyploid** in which the several chromosome sets are derived from the same species.

autosome A chromosome other than a sex chromosome.

Bauplan A fundamental "body plan" or body structure of a group of organisms (such as a class or phylum), differing substantially from that of other groups. (German, "building plan," or blueprint; pronounced "bow′plon"; pl. *Baupläne*.)

benthic Inhabiting the bottom, or substrate, of a body of water.

biological species A population or group of populations within which genes are actually or potentially exchanged by interbreeding, and which are reproductively isolated from other such groups.

bottleneck A severe, temporary reduction in population size.

canalization The operation of internal factors during development that reduce the effect of perturbing influences, thereby constraining variation in the phenotype around one or more modes.

carrying capacity The population density that can be sustained by limiting resources.

category In taxonomy, one of the ranks of classification (e.g., genus, family). *See* **taxon**.

character A feature, or trait. **Character state**, one of the variant conditions of a character.

character displacement Usually refers to a pattern of geographic variation, in which a character differs more greatly between sympatric than between allopatric populations of two species; sometimes used for the evolutionary process of accentuation of differences between sympatric populations of two species, owing to reproductive or ecological interactions between them.

chronospecies A segment of an evolving lineage preserved in the fossil record that differs enough from earlier or later members of the lineage to be given a different binomial (name). Not equivalent to biological species.

clade The set of species descended from a particular ancestral species.

cladistic Pertaining to branching patterns; a cladistic classification classifies organisms on the basis of the historical sequences by which they have diverged from common ancestors.

cladogenesis Branching of lineages during phylogeny.

cladogram A branching diagram depicting relationships among taxa, i.e. an estimated history of the relative sequence in which they have evolved from common ancestors. Used by some authors to mean a branching diagram that displays the hierarchical distribution of derived character states among taxa.

cleistogamy Self-pollination within a flower that does not open.

cline A gradual change in an allele frequency or in the mean of a character over a geographic transect.

clone A lineage of individuals reproduced asexually, by mitotic division.

coadapted gene pool A population or set of populations in which prevalent genotypes are composed of alleles at two or more loci that confer high fitness in combination with each other, but not with alleles that are prevalent in other such populations.

coefficient of variation (C.V.) The standard deviation divided by the mean, multiplied by 100. C.V. $= 100 \times (s/\overline{x})$.

coevolution Strictly, the joint evolution of two (or more) ecologically interacting species, each of which evolves in response to selection imposed by the other. Sometimes used loosely to refer to evolution of one species caused by its interaction with another.

cohort Those members of a population that are of the same age.

commensalism An ecological relationship between species in which one is benefited but the other is little affected.

comparative method A procedure for inferring the adaptive function of a character, by correlating its states in various taxa with one or more variables, such as ecological factors hypothesized to affect its evolution.

competition An interaction between individuals of the same species or different species, whereby resources used by one are made unavailable to others.

competitive exclusion Extinction of a population due to competition with another species.

concerted evolution Maintenance of a homogeneous nucleotide sequence among the members of a gene family, which sequence evolves over time.

conspecific Belonging to the same species.

convergent evolution Evolution of similar features independently in different evolutionary lineages, usually from different antecedent features or by different developmental pathways.

correlation A statistical relationship that quantifies the degree to which two variables are associated (*see correlation coefficient* in Appendix I). For *phenotypic correlation*, *genetic correlation*, *environmental correlation* as applied to the relationship between two traits, *see* Chapter 14.

coupling Of a gamete or chromosome, bearing at two or more loci alleles that have been designated to be alike in some way (e.g., both wild type rather than mutant).

Creationism The doctrine that each species (or perhaps higher taxon) of organism was created separately in much its present form, by a supernatural creator.

deme A local population, usually a small, panmictic population.

demographic Referring to processes that change the size of a population, i.e., birth, death, dispersal.

density-dependent Affected by population density.

deterministic Causing a fixed outcome, given initial conditions. *See* **stochastic**.

diploid A cell or organism possessing two chromosome complements; *ploidy* thus refers to the number of chromosome complements (*see* **haploid**, **polyploid**).

directional selection Selection for a higher or lower value of a character than its current mean value.

dispersal In population biology, movement of individual organisms to different localities; in biogeography, extension of the geographic range of a species by movement of individuals.

disruptive selection Selection in favor of two or more modal phenotypes and against those intermediate between them. Equals *diversifying selection*.

divergence The evolution of increasing difference between lineages in one or more characters.

diversification Evolutionary increase of the number of species in a clade, usually accompanied by divergence in phenotypic characters.

dominance Of an allele, the extent to which it produces when heterozygous the same phenotype as when homozygous. Of a species, the extent to which it is numerically (or otherwise) predominant in a community.

ecological niche The range of combinations of all relevant environmental variables under which a species or population can persist; often more loosely used to describe the "role" of a species, or the resources it utilizes.

ecotype A genetically determined phenotype of a species that is found as a local variant associated with certain ecological conditions.

efficient cause Aristotle's term for the mechanical reason for an event.

endemic Of a species, restricted to a specified region or locality.

environment Usually, the complex of external physical, chemical, and biotic factors that may affect a population, an organism, or the expression of an organism's genes; more generally, anything external to the object of interest (e.g., a gene, an organism, a population) that may influence its function or activity. Thus, other genes within an organism may be part of a gene's environment, or other individuals in a population may be part of an organism's environment.

epigenetic Developmental; pertaining especially to interactions among developmental processes above the level of primary gene action.

epistasis An effect of the interaction between two or more gene loci on the phenotype or fitness, whereby their joint effect differs from the sum of the loci taken separately.

equilibrium An unchanging condition, as of population size or genetic composition. Also the value (of population size, allele frequency) at which this condition occurs. An equilibrium need not be stable. *See* **stability**, **unstable equilibrium**.

ESS *See* **evolutionarily stable strategy**.

essentialism The philosophical view that all members of a class of objects (such as a species) share certain invariant, unchanging properties that distinguish them from other classes. Also called "typological thinking."

ethological Behavioral.

evolution In a broad sense, the origin of entities possessing different states of one or more characteristics, and changes in their proportions over time. *Organic evolution*, or *biological evolution*, is a change over time of the proportions of individual organisms differing genetically in one or more traits. Such changes transpire by the origin and subsequent alteration of the frequencies of genotypes from generation to generation within populations, by the alterations of the proportions of genetically differentiated populations of a species, or by changes in the numbers of species with different characteristics, thereby altering the frequency of one or more traits within a higher taxon.

evolutionarily stable strategy (ESS) A phenotype such that, if almost all individuals in a population have that phenotype, no alternative phenotype can replace it or invade the population. The ESS may be a plastic phenotype, i.e. one that develops differently, or that can be altered within an individual's lifetime, under different conditions.

exon A part of an interrupted gene that is translated into a polypeptide.

fecundity The quantity of gametes, usually eggs, produced.

final cause Aristotle's term for a goal, attainment of which is the reason for being of an event or process.

fitness The success of an entity in reproducing; hence, the average contribution of an allele or genotype to the next generation or to succeeding generations. **Relative fitness** is the average contribution of an allele or genotype compared with that of another allele or genotype.

fixation Attainment of a frequency of 1 (i.e., 100 percent) by an allele in a population, which thereby becomes **monomorphic** for the allele.

founder effect The principle that the founders of a new colony carry only a fraction of the total genetic variation in the source population.

frequency In this book, usually used to mean *proportion*; e.g., the frequency of an allele is the proportion of gene copies belonging to that allelic state.

fugitive species One that occupies temporary environments or habitats and so does not persist for many generations at any one site.

function The way in which a character contributes to the fitness of an organism.

gene The functional unit of heredity. A complex concept; *See* Chapter 3.

gene family Two or more loci with similar nucleotide sequences that have been derived from a common ancestral sequence.

gene flow The incorporation of genes into the gene pool of one population from one or more other populations.

gene frequency Equals **allele frequency**.

gene pool The totality of the genes of a given sexual population.

gene tree A diagram representing the history by which gene copies have been derived from ancestral gene copies in previous generations.

genetic distance Any of several measures of the degree of genetic difference between populations, based on differences in allele frequencies.

genetic drift Random changes in the frequencies of two or more alleles or genotypes within a population.

genetic load Any reduction of the mean fitness of a population owing to the existence of genotypes with lower fitness than that of the most fit genotype.

genetic variance Variation in a trait within populations, as measured by the variance that is due to genetic differences among individuals.

genic selection A form of selection in which the single gene is the unit, such that the outcome is determined by fitness values assigned to different alleles. A complex and controversial concept (*see* Chapter 12). This term is restricted by some authors to cases in which an allele can increase due to its higher replication rate, even though it lowers the fitness of the organisms that bear it; other authors use it more generally, to mean differences in fitness between alleles, averaged over the variety of genotypes in which they occur. *See* **individual selection**, **kin selection**, **natural selection**.

genotype The set of genes possessed by an individual organism; often, its genetic composition at a specific locus or set of loci singled out for discussion.

grade A group of species that have evolved to the same stage in one or more characters, and typically constitute a **paraphyletic**, rather than a **monophyletic**, group.

gradualism The proposition that large changes in phenotypic characters have evolved through many slightly different intermediate states.

group selection The differential rate of origination or extinction of whole populations (or species, if the term is used broadly) on the basis of differences among them in one or more characteristics. *See* **interdemic selection**, **species selection**.

habitat selection The capacity of an organism (usually an animal) to choose a habitat in which to perform activities.

haploid A cell or organism possessing a single chromosome complement, hence a single gene copy at each locus.

haplotype A DNA sequence that differs from homologous sequences at one or more sites.

heritability The proportion of the **variance** among individuals in a trait, that is attributable to differences in genotype. For heritability in the narrow and broad senses, *see* Chapter 14.

heterochrony   An evolutionary change in phenotype based on an alteration of timing of developmental events.

heterokaryotype   A genome or individual that is heterozygous for a chromosomal rearrangement such as an inversion. A *homokaryote* is homozygous in this respect.

heterosis   Equivalent to *hybrid vigor*: the superiority in one or more characteristics (e.g., size, yield) of crossbred organisms compared with inbred organisms, as a result of differences in the genetic constitutions of the uniting parental gametes.

heterozygosity   The average heterozygosity in a population is the proportion of loci at which a randomly chosen individual is heterozygous, on average.

heterozygote   An individual organism that possesses different alleles at a locus.

heterozygous advantage   The manifestation of higher fitness by heterozygotes than by homozygotes at a specific locus.

hitchhiking   Change in the frequency of an allele due to linkage with a selected allele at another locus. *See also species hitchhiking*, Chapter 24.

homeostasis   Maintenance of an equilibrium state by some self-regulating capacity of an individual.

homeotic mutation   A mutation that causes a transformation of one structure into another of the organism's structures.

homology   Possession by two or more species of a trait derived, with or without modification, from their common ancestor. Also, **homologous chromosomes**, the members of a chromosome complement that bear the same genes.

homoplasy   Possession by two or more species of a similar or identical character state that has not been derived by both species from their common ancestor; embraces convergence, parallel evolution, and evolutionary reversal.

homozygote   An individual organism that has the same allele at each of its copies of a genetic locus.

hybrid   An individual formed by mating between unlike forms, usually genetically differentiated populations or species; occasionally in genetics, the offspring of a mating between phenotypically distinguishable genotypes of any kind.

hybrid zone   A region in which genetically distinct populations come into contact and produce at least some offspring of mixed ancestry.

hypermorphosis   An evolutionary increase of the duration of ontogenetic development, resulting in features that are exaggerated compared to those of the ancestor.

identical by descent   Two or more gene copies are identical by descent if they have been derived from a single gene copy in a specified common ancestor of the organisms that carry the copies.

inbreeding   Mating between relatives that occurs more frequently than if mates were chosen at random from a population.

inbreeding depression   Reduction, in inbred individuals, of the mean value of a character (usually one correlated with fitness).

inclusive fitness   The fitness of a gene or genotype measured by its effect on survival or reproduction both of the organism bearing it, and of the genes, identical by descent, borne by the organism's relatives.

individual selection   A form of natural selection consisting of non-random differences among different genotypes (or phenotypes) within a population in their contribution to subsequent generations. *See* **genic selection**, **natural selection**.

interaction   Strictly, the dependence of an outcome on a combination of causal factors, such that the outcome is not predictable from the average effects of the factors taken separately. **Genotype × environment interaction** (Chapter 14) is consequently variation in phenotype arising from the difference in the effect of environment on the expression of different genotypes. More loosely, an interplay between entities that affects one or more of them (as in interactions between species).

interdemic selection   Group selection of populations within a species.

intrinsic rate of natural increase   The potential per capita rate of increase of a population with a stable age distribution, whose growth is not depressed by the negative effects of density.

intron   A part of an interrupted gene that is not translated into a polypeptide.

inversion   A 180° reversal of the orientation of a part of a chromosome, relative to some standard chromosome.

isolating mechanism   A genetically determined difference between populations that restricts or prevents gene flow between them. The term does not include spatial segregation by extrinsic geographic or topographic barriers.

isozyme (isoenzyme)   One of several forms of an enzyme, produced by different loci in an individual organism's genome.

iterative evolution   The repeated evolution of similar phenotypic characteristics at different times during the history of a clade.

iteroparous   A life history in which individuals reproduce more than once.

karyotype   The chromosome complement of an individual.

key adaptation   An adaptation that provides the basis for using a new, substantially different habitat or resource; *see* Chapter 24.

kin selection   A form of selection whereby alleles differ in their rate of propagation by influencing the impact of their bearers on the reproductive success of individuals (kin) who carry the same alleles by common descent.

Lamarckism   The theory that evolution is caused by inheritance of character changes acquired during the life of an individual, due to its behavior or to environmental influences.

lineage   A series of ancestral and descendant populations, through time; usually refers to a single evolving species, but may include several species descended from a common ancestor.

lineage sorting   The process by which each of several descendant species acquires a single gene lineage, of the several gene lineages carried by a common ancestral species; hence, the derivation of a monophyletic gene tree, in each species, from the paraphyletic gene tree inherited from their common ancestor.

linkage   Occurrence of two loci on the same chromosome: they are functionally linked only if they are so close together that they do not segregate independently in meiosis.

linkage equilibrium and linkage disequilibrium   If two alleles at two or more loci are associated more frequently (or less frequently) than predicted by their individual frequencies, they are in linkage disequilibrium; if not, they are in linkage equilibrium.

locus A site on a chromosome occupied by a specific gene; more loosely, the gene itself, in all its allelic states.

logistic equation An equation describing the idealized growth of a population subject to a density-dependent limiting factor. As density increases, the rate of growth gradually declines, until population growth stops.

macroevolution A vague term for the evolution of great phenotypic changes, usually great enough to allocate the changed lineage and its descendants to a distinct genus or higher taxon.

maternal effect A nongenetic effect of the mother on the phenotype of the offspring, owing to factors such as cytoplasmic inheritance, transmission of disease from mother to offspring, or nutritional conditions.

mean Usually the arithmetic mean or average; the sum of n values, divided by n. The mean $\bar{x} = (x_1 + x_2 + \ldots + x_n)/n$.

meiotic drive Used broadly to denote a preponderance (>50 percent) of one allele among the gametes produced by a heterozygote; results in genic selection.

meristic trait A discretely varying, countable trait; e.g., number of vertebrae.

metapopulation A set of local populations, among which there may be gene flow and extinction and colonization.

microevolution A vague term for slight, short-term evolutionary changes within species.

migration Used in theoretical population genetics as a synonym for gene flow among populations; in other contexts, directed large-scale movement of organisms that does not necessarily result in gene flow.

mimicry Similarity of certain characters of two or more species, due to convergent evolution owing to an advantage conferred by resemblance. Common types include *Batesian* mimicry, in which a palatable *mimic* suffers lower predation due to its resemblance to an unpalatable *model*; and *Müellerian* mimicry, in which each of two or more unpalatable species enjoys reduced predation due to their similarity.

modifier gene A gene that is recognized by its alteration of the phenotypic expression of genes at one or more other loci.

monomorphic A population in which virtually all individuals have the same genotype at a locus. *Cf.* **polymorphism**.

monophyletic Of a taxon, consisting of species all of which are derived from a common ancestral taxon. In cladistic taxonomy, the term describes a taxon consisting of all the known species descended from a single ancestral species.

mosaic evolution Evolution of different characters within a lineage or clade at different rates, hence more or less independently of one another.

mutation An error in replication of a nucleotide sequence, or any other alteration of the genome that is not manifested as reciprocal recombination. A complex concept; *see* Chapter 10.

mutualism A symbiotic relation in which each of two species benefits by their interaction.

natural selection The differential survival and/or reproduction of classes of entities that differ in one or more characteristics; the difference in survival and/or reproduction is not due to chance, and it must have the potential consequence of altering the proportions of the different entities, to constitute natural selection. Thus natural selection is also definable as a partly or wholly deterministic difference in the contribution of different classes of entities to subsequent generations. Usually the differences are inherited. The entities may be alleles, genotypes or subsets of genotypes, populations, or in the broadest sense, species. A complex concept; *see* Chapter 12. *See also* **genic selection**, **individual selection**, **kin selection**, **group selection**.

negative feedback A dynamic relation whereby the product of a process inhibits the process that produces it, usually enhancing stability.

neo-Darwinism Usually used to describe the modern belief that natural selection, acting on randomly generated genetic variation, is a major but not the sole cause of evolution. Properly, the belief, advocated by a few biologists in the late nineteenth century, that natural selection is the sole mechanism of evolution.

neoteny Heterochronic evolution whereby development of some or all somatic features is retarded relative to sexual maturation, resulting in sexually mature individuals with juvenile features. *See* **paedomorphosis**, **progenesis**.

neutral alleles Alleles that do not differ measurably in their effect on fitness.

norm of reaction The set of phenotypic expressions of a genotype under different environmental conditions. *See* **phenotypic plasticity**.

normal distribution A bell-shaped frequency distribution of a variable. The expected distribution if many factors with independent, small effects determine the value of a variable; the basis for many statistical formulations; *see* Appendix I.

ontogeny The development of an individual organism, from fertilized zygote until death.

orthologous Refers to corresponding members of a multigene family in two or more species. *See* **paralogous**.

outgroup A taxon that diverged from a group of other taxa before they diverged from each other.

overdominance The expression by two alleles in heterozygous condition of a phenotypic value for some characteristic that lies outside the range of the two corresponding homozygotes.

paedomorphosis Possession in the adult stage of features typical of the juvenile stage of the organism's ancestor.

panmixia Random mating among members of a population.

parallel evolution The evolution of similar or identical features independently in related lineages, thought usually to be based on similar modifications of the same developmental pathways.

paralogous Refers to the relationship between two different members of a gene family, within a species or in a comparison of different species. *See* **orthologous**.

parapatric Populations that have contiguous but non-overlapping geographic distributions.

paraphyletic Refers to a specified taxon or portion of a phylogenetic tree or gene tree that is derived from a single ancestor, but does not include all the descendants of that ancestor.

parsimony Economy in the use of means to an end (*Webster's New Collegiate Dictionary*); the principle of accounting for observations by that hypothesis requiring the fewest or simplest assumptions that lack evidence; in systematics, the principle of invoking the minimal number of evolutionary changes to infer phylogenetic relationships.

parthenogenesis Virgin birth; development from an egg to which there has been no paternal contribution of genes.

PCR (polymerase chain reaction) A technique by which the number of copies of a DNA sequence is increased, by replication *in vitro*.

peak shift Change in allele frequencies within a population from one to another local maximum of mean fitness, by passage through states of lower mean fitness.

peramorphosis An exaggerated state of a character in a descendant population compared to the state in an ancestor, and corresponding to the state expected if ontogenetic growth of the ancestor's character had been prolonged.

peripatric Of populations, situated peripheral to most of the populations of a species; **peripatric speciation**, speciation by evolution of reproductive isolation in such populations.

phenetic Pertaining to phenotypic similarity, as in a phenetic classification.

phenotype The morphological, physiological, biochemical, behavioral, and other properties of an organism, manifested throughout its life; or any subset of such properties, especially those affected by a particular allele or other portion of the **genotype**.

phenotypic plasticity The capacity of an organism to develop any of several phenotypic states, depending on the environment; usually this capacity is supposed to be adaptive.

phylogeny The history of descent of a group of taxa such as species from their common ancestors, including the order of branching and sometimes absolute ages of divergence; also applied to the genealogy of genes derived from a common ancestral gene.

planktonic Living in open water; *see* **benthic**.

pleiotropy The phenotypic effect of a gene on more than one characteristic.

polygenic character A trait whose variation is based wholly or in part on allelic variation at more than a few loci.

polymorphism The existence within a population of two or more genotypes, the rarest of which exceeds some arbitrarily low frequency (say, 1 percent); more rarely, the existence of phenotypic variation within a population, whether or not genetically based.

polyphyletic Of a taxon, composed of members derived by evolution from ancestors in more than one ancestral taxon; hence, composed of members that do not share a unique common ancestor.

polyploid Possessing more than two entire chromosome complements.

polytopy Geographic variation in which each of one or more distinctive forms is found in each of several separate localities, between which other forms are distributed.

polytypy The existence of named geographic races or subspecies within a species.

population A group of conspecific organisms that occupy a more or less well defined geographic region and exhibit reproductive continuity from generation to generation; ecological and reproductive interactions are more frequent among these individuals than with members of other populations of the same species.

position effect A difference in the phenotypic expression of a gene caused by a change in its location on the chromosomes.

preadaptation Possession of the necessary properties to permit a shift into a new niche or habitat. A structure is preadapted for a new function if it can assume that function without evolutionary modification.

progenesis A decrease during evolution of the duration of ontogenetic development, resulting in retention of juvenile features in the sexually mature adult. *See* **neoteny**, **paedomorphosis**.

prezygotic Before union of the nuclei of uniting gametes; usually refers to events in the reproductive process that cause reproductive isolation.

postzygotic After union of the nuclei of uniting gametes; usually refers to inviability or sterility that confer reproductive isolation.

pseudogene A nonfunctional member of a gene family.

race A poorly defined term for a set of populations occupying a particular region that differ in one or more characteristics from populations elsewhere; equivalent to **subspecies**. In some writings, a distinctive phenotype, whether or not allopatric from others.

radiation *See* **adaptive radiation**.

recapitulation The ontogenetic passage of an organism's features through stages that resemble the adult features of its phylogenetic ancestors.

refugia Locations in which species have persisted while becoming extinct elsewhere.

regression In geology, withdrawal of sea from land, accompanying lowering of sea level; in statistics, calculation of a function that best predicts a dependent from an independent variable.

reinforcement Evolution of enhanced reproductive isolation between populations, due to natural selection for greater isolation.

relict A species that has been "left behind," for example, the last survivor of an otherwise extinct group. Sometimes, a species or population left in a locality after extinction throughout most of the region.

relictual The geographic distribution of a species or group that persists in localities that it occupied at an earlier time, but which is extinct over much of its former range.

replacement substitution A base pair substitution in DNA that results in an amino acid substitution in the protein product.

reproductive effort The proportion of energy or materials that an organism allocates to reproduction rather than to growth and maintenance.

repulsion Of a gamete or chromosome, bearing at two or more loci alleles that have been designated to be unlike in some way. *Cf.* **coupling**.

response to selection The change in the mean value of a character over one or more generations, due to selection.

restriction enzyme An enzyme that cuts double-stranded DNA at specific short nucleotide sequences. Variation in this sequence within a population results in variation in DNA sequence lengths after treatment with a restriction enzyme, or **restriction fragment length polymorphism** (RFLP).

reticulate evolution Union of different lineages of a clade by hybridization.

RFLP *See* **restriction enzyme**.

saltation A jump; a discontinuous mutational change in one or more phenotypic traits, usually of considerable magnitude.

scala naturae The "scale of nature," or chain of being: the pre-evolutionary concept that all living things were created in an orderly series of forms, from lower to higher.

selection Nonrandom differential survival or reproduction of classes of phenotypically different entities. *See* **natural selection**, **artificial selection**.

selection coefficient The difference between the mean relative fitness of individuals of a given genotype and those of a reference genotype.

semelparous A life history in which individuals (especially females) reproduce only once in their lifetime.

semispecies One of several groups of populations that are partially but not entirely isolated from each other by biological factors (isolating mechanisms).

serial homology A relationship among repeated, often differentiated, structures of a single organism, defined by their similarity of developmental origin; for example, the several legs and other appendages of an arthropod.

sex-linked A gene carried by one of the sex chromosomes; it may be expressed phenotypically in both sexes.

sexual reproduction Production of offspring whose genetic constitution is a mixture of that of two potentially genetically different gametes.

sexual selection Differential reproduction owing to variation in the ability to obtain mates.

sibling species Species that are difficult or impossible to distinguish by morphological characters.

silent substitution A nucleotide substitution in a DNA sequence that does not alter the amino acid sequence of the protein product.

sister taxa Two species or higher taxa derived from an immediate common ancestor, and are therefore each other's closest relatives.

speciation Evolution of reproductive isolation within an ancestral species, resulting in two or more descendant species.

species In the sense of biological species, the members of a group of populations that interbreed or potentially interbreed with each other under natural conditions; a complex concept (*see* Chapter 15). Also, a fundamental taxonomic category to which individual specimens are assigned, which often but not always corresponds to the biological species.

species selection A form of **group selection** in which species with different characteristics increase (by speciation) or decrease (by extinction) in number at different rates, because of a difference in their characteristics.

stability Often used to mean constancy; more often in this book, the propensity to return to a condition (a stable equilibrium) after displacement from that condition.

stabilizing selection Selection against phenotypes that deviate in either direction from an optimal value of a character.

standard deviation The square root of the **variance**.

stasis Absence of evolutionary change in one or more characters for some period of evolutionary time.

stochastic Random.

strata Layers of sedimentary rock that were deposited at different times.

subspecies A named geographic race; a set of populations of a species that share one or more distinctive features and occupy a different geographic area from other subspecies.

substitution The complete replacement of one allele by another within a population or species; the term **nucleotide substitution** usually means the complete replacement of one nucleotide pair by another within a lineage over evolutionary time. *Cf.* **fixation**.

supergene A group of two or more functionally related loci between which recombination is so reduced that they are usually inherited together as a single entity.

superspecies A group of semispecies.

symbiosis An intimate, usually physical, association between two or more species.

sympatric Of two species or populations, occupying the same geographic locality so that the opportunity to interbreed is presented.

synapomorphy A derived character state that is shared by two or more taxa, and is postulated to have evolved in their common ancestor.

synthetic theory The theory of evolutionary processes that emerged during the "evolutionary synthesis," and which emphasized the coaction of random mutation, selection, genetic drift, and gene flow.

taxon (pl. taxa) The named taxonomic unit (e.g., *Homo sapiens*, Hominidae, or Mammalia) to which individuals, or sets of species, are assigned. **Higher taxa** are those above the species level. *See* **category**.

teleology The belief that natural events and objects have purposes, and can be explained by their purposes.

territory An area or volume of habitat defended by an organism or a group of organisms against other individuals, usually of the same species; **territorial behavior**, the behavior by which the territory is defended.

transgression Incursion of sea upon land, owing to rise in sea level.

translocation The transfer of a segment of a chromosome to another, nonhomologous, chromosome; or the chromosome formed by the addition of such a segment.

transposable element A DNA sequence, copies of which become inserted into various sites on the chromosomes.

unstable equilibrium An unchanging state, to which a system (e.g., a population density or allele frequency) does not return if disturbed.

variance (σ^2, s^2, V) The average squared deviation of an observation from the arithmetic mean; hence, a measure of variation. $s^2 = [\Sigma(x_i - \bar{x})^2]/(n-1)$, where $\bar{x}$ is the mean and n the number of observations. *See* Appendix I.

vestigial Occurring in a rudimentary condition, as a result of evolutionary reduction from a more elaborated, functional character state in an ancestor.

viability Capacity for survival; often refers to the fraction of individuals surviving to a given age, and is contrasted with inviability due to deleterious genes.

vicariance Separation of a continuously distributed ancestral population or species into separate populations, due to the development of a topographic or ecological barrier.

wild type The allele, genotype, or phenotype that is most prevalent (if there is one) in wild populations; with reference to the wild-type allele, other alleles are often termed mutations.

zygote A single-celled individual formed by union of gametes. Occasionally used more loosely to refer to an offspring produced by sexual reproduction.

Literature Cited

Abbott, R. J. 1992. Plant invasions, interspecific hybridization and the evolution of new plant taxa. *Trends Ecol. Evol.* 7: 401–405. [16]

Adey, N. B., T. O. Tollefsbol, A. B. Sparks, M. H. Edgell, and C. A. Hutchison III. 1994. Molecular resurrection of an extinct ancestral promoter for mouse L1. *Proc. Natl. Acad. Sci. USA* 91: 1569–1573. [22]

Aho, J. M., and A. O. Bush. 1993. Community richness in parasites of some freshwater fishes from North America. In R. E. Ricklefs and D. Schluter (eds.), *Species Diversity in Ecological Communities*, pp. 185–193. University of Chicago Press, Chicago, IL. [25]

Akashi, H. 1995. Inferring weak selection from patterns of polymorphism and divergence at "silent" sites in *Drosophila* DNA. *Genetics* 139: 1067–1076. [22]

Alberch, P. 1980. Ontogenesis and morphological diversification. *Am. Zool.* 20: 653–667. [24]

Alberch, P., and E. A. Gale. 1985. A developmental analysis of an evolutionary trend: Digital reduction in amphibians. *Evolution* 39: 8–23. [23]

Alberch, P., S. J. Gould, G. F. Oster, and D. B. Wake. 1979. Size and shape in ontogeny and phylogeny. *Paleobiology* 5: 296–317. [23]

Alberts, B., D. Bray, J. Lewis, M. Raff, K. Roberts, and J. D. Watson. 1995. *Molecular Biology of the Cell.* Third edition. Garland, New York. [3]

Alcock, J. 1998. *Animal Behavior: An Evolutionary Approach.* Sixth edition. Sinauer Associates, Sunderland, MA. [20]

Alexander, R. D. 1968. Life cycle origins, speciation, and related phenomena in crickets. *Q. Rev. Biol.* 43: 1–41. [15]

Alexander, R. D. 1974. The evolution of social behavior. *Annu. Rev. Ecol. Syst.* 5: 325–383. [20]

Alexander, R. D. 1979. *Darwinism and Human Affairs.* University of Washington Press, Seattle. [26]

Alexander, R. D., and R. S. Bigelow. 1960. Allochronic speciation in field crickets, and a new species, *Acheta veletis. Evolution* 14: 334–346. [15]

Allard, R. W. 1960. *Principles of Plant Breeding.* Wiley, New York. [3]

Allen, K. C., and D. E. G. Briggs (eds.). 1989. *Evolution and the Fossil Record.* Smithsonian Institution Press, Washington, DC. [7]

Allen, M. K., and C. Yanofsky. 1963. A biochemical and genetic study of reversion with the A-gene A-protein system of *Escherichia coli* tryptophan synthetase. *Genetics* 48: 1065–1083. [10]

Allen, O. N., and E. K. Allen. 1981. *The Leguminosae.* University of Wisconsin Press, Madison. [17]

Alvarez, L. W., W. Alvarez, F. Asaro, and H. V. Michel. 1980. Extraterrestrial cause for the Cretaceous-Tertiary extinction. *Science* 208: 1095–1108. [25]

Anderson, E. 1936. Cytology in its relation to taxonomy. *Bot. Rev.* 3: 335–363. [10]

Anderson, E. 1948. Hybridization of the habitat. *Evolution* 2: 1–9. [15]

Anderson, E. 1949. *Introgressive Hybridization.* Wiley, New York. [15]

Anderson, R. M., and R. M. May. 1982. Coevolution of hosts and parasites. *Parasitology* 85: 411–426. [18]

Anderson, R. M., and R. M. May. 1991. *Infectious Diseases of Humans.* Oxford University Press, Oxford. [18]

Anderson, W. W. 1966. Genetic divergence in M. Vetukhiv's experimental populations of *Drosophila pseudoobscura.* 3. Divergence in body size. *Genet. Res.* 7: 255–266. [14]

Andersson, M. 1982. Female choice selects for extreme tail length in a widowbird. *Nature* 299: 818–820. [12]

Andersson, M. 1984. The evolution of eusociality. *Annu. Rev. Ecol. Syst.* 15: 165–189. [20]

Andersson, M. B. 1994. *Sexual Selection.* Princeton University Press, Princeton, NJ. [16]

Andrews, C. W. 1901. On the extinct birds of Patagonia. I. The skull and skeleton of *Phororhacos inflatus* Ameghino. *Trans. Zool. Soc. Lond.* 15: 55–86. [7]

Andrews, S. M., and T. S. Westoll. 1970. The postcranial skeleton of *Eusthenopteron foordi* Whiteaves. *Trans. R. Soc. Edinburgh* 68: 207–329. [6]

Antolin, M. F. 1993. Genetics of biased sex ratios in subdivided populations: Models, assumptions, and evidence. *Oxford Surv. Evol. Biol.* 9: 239–281. [21]

Antonovics, J. 1968. Evolution in closely adjacent plant populations. V. Evolution of self-fertility. *Heredity* 23: 219–238. [16]

Antonovics, J., A. D. Bradshaw, and R. G. Turner. 1971. Heavy metal tolerance in plants. *Adv. Ecol. Res.* 7: 1–85. [14]

Appel, H. M., and M. M. Martin. 1992. Significance of metabolic load in the evolution of host specificity of *Manduca sexta. Ecology* 73: 216–228. [18]

Aquadro, C. F., D. J. Begun, and E. C. Kindahl. 1994. Selection, recombination, and DNA polymorphism in *Drosophila.* In B. Golding (ed.), *Non-neutral Evolution: Theories and Molecular Data*, pp. 46–56. Chapman & Hall, New York. [22]

Armour, J. A. L., et al. 1996. Minisatellite diversity supports a recent African origin for modern humans. *Nature Genetics* 13: 154–160. [26]

Arnheim, N. 1983. Concerted evolution of multi-gene families. In M. Nei and R. K. Koehn (eds.), *Evolution of Genes and Proteins*, pp. 38–61. Sinauer Associates, Sunderland, MA. [22]

Arnold, A. P. 1994. Critical events in the development of bird song: What can neurobiology contribute to the study of the evolution of behavior? In L. A. Real (ed.), *Behavioral Mechanisms in Evolutionary Ecology*, pp. 219–237. University of Chicago Press, Chicago, IL. [20]

Arnold, M. L. 1992. Natural hybridization as an evolutionary process. *Annu. Rev. Ecol. Syst.* 23: 237–261. [15]

Arnold, M. L., and S. A. Hodges. 1995. Are natural hybrids fit or unfit relative to their parents? *Trends Ecol. Evol.* 10: 67–71. [15]

Arnold, S. J. 1981a. Behavioral variation in natural populations. I. Phenotypic, genetic, and environmental correlations between chemoreceptive responses to prey in the garter snake, *Thamnophis elegans. Evolution* 35: 489–509. [14]

Arnold, S. J. 1981b. Behavioral variation in natural populations. II. The inheritance of a feeding response in crosses between geographic races of the garter snake, *Thamnophis elegans. Evolution* 35: 510–515. [9]

Arnold, S. J. 1983. Morphology, performance and fitness. *Am. Zool.* 23: 347–361. [17]

Arnold, S. J. 1994. Is there a unifying concept of sexual selection that applies to both plants and animals? *Am. Nat.* 144: S1–S12. [20]

Ashton, P. S. 1969. Speciation among tropical forest trees: Some deductions in the light of recent research. *Biol. J. Linn. Soc.* 1: 155–196. [25]

Atwood, R. C., L. K. Schneider, and F. J. Ryan. 1951. Selective mechanisms in bacteria. *Cold Spring Harbor Symp. Quant. Biol.* 16: 345–355. [12]

Augspurger, C. K. 1984. Seedling survival of tropical tree species: Interactions of dispersal distances, light gaps, and pathogens. *Ecology* 65: 1705–1712. [4]

Ausich, W. I., and D. J. Bottjer. 1985. Phanerozoic tiering in suspension-feeding communities on soft substrata: Implications for diversity. In J. W. Valentine (ed.), *Phanerozoic Diversity Patterns*, pp. 255–274. Princeton University Press, Princeton, NJ. [25]

Austin, O. 1985. *Families of Birds*. Golden Press, New York. [5]

Averoff, M., and M. Akam. 1995. *Hox* genes and the diversification of insect and crustacean body plans. *Nature* 376: 420–423. [23]

Avise, J. C. 1989. Gene trees and organismal histories: A phylogenetic approach to population biology. *Evolution* 43: 1192–1208. [11]

Avise, J. C. 1994. *Molecular Markers, Natural History, and Evolution.* Chapman & Hall, New York. [22]

Avise, J. C., and F. J. Ayala. 1976. Genetic differentiation in speciose versus depauperate phylads: Evidence from the California minnows. *Evolution* 30: 46–58. [16]

Avise, J. C., and R. M. Ball, Jr. 1990. Principles of genealogical concordance in species concepts and biological taxonomy. *Oxford Surv. Evol. Biol.* 7: 45–67. [15]

Avise, J. C., J. Arnold, R. M. Ball, E. Bermingham, T. Lamb, J. E. Neigel, C. A. Reeb, and N. C. Saunders. 1987. Intraspecific phylogeography: The mitochondrial DNA bridge between population genetics and systematics. *Annu. Rev. Ecol. Syst.* 18: 489–522. [11]

Avise, J. C., R. M. Ball, and J. Arnold. 1988. Current versus historical population sizes in vertebrate species with high gene flow: A comparison based on mitochondrial DNA lineages and inbreeding theory for neutral mutations. *Mol. Biol. Evol.* 5: 331–344. [11]

Avise, J. C., B. C. Bowen, T. Lamb, A. B. Meylan, and E. Bermingham. 1992. Mitochondrial DNA evolution at a turtle's pace: Evidence for low genetic variability and reduced microevolutionary rate in Testudines. *Mol. Biol. Evol.* 9: 433–446. [11]

Axelrod, R., and W. D. Hamilton. 1981. The evolution of cooperation. *Science* 211: 1390–1396. [20]

Ayala, F. J. 1988. Can "progress" be defined as a biological concept? In M. H. Nitecki (ed.), *Evolutionary Progress*, pp. 75–96. University of Chicago Press, Chicago, IL. [24]

Ayala, F. J., M. L. Tracey, L. G. Barr, J. F. McDonald, and S. Pérez-Salas. 1974a. Genetic variation in natural populations of five *Drosophila* species and the hypothesis of selective neutrality of protein polymorphisms. *Genetics* 77: 343–384. [15]

Ayala, F. J., M. L. Tracey, D. Hedgecock, and R. C. Richmond. 1974b. Genetic differentiation during the speciation process in *Drosophila. Evolution* 28: 576–592. [15]

Bailey, J. M., and R. C. Pillard. 1991. A genetic study of male sexual orientation. *Arch. Gen. Psychiatry* 48: 1089–1096. [26]

Bailey, J. M., and K. J. Zucker. 1995. Childhood sex-typed behavior and sexual orientation: A conceptual analysis and quantitative review. *Dev. Psychol.* 31: 43–56. [26]

Bailey, J. M., R. C. Pillard, M. C. Neale, and Y. Agnei. 1993. Heritable factors influence sexual orientation in women. *Arch. Gen. Psychiatry* 50: 217–223. [26]

Bailey, W. J., D. H. A. Fitch, D. A. Tagle, J. Czelusniak, J. L. Slightom, and M. Goodman. 1991. Molecular evolution of the $\psi\eta$-globin gene locus: Gibbon phylogeny and the hominoid slowdown. *Mol. Biol. Evol.* 8: 155–184. [26]

Bajema, C. J. 1963. Estimation of the direction and intensity of natural selection in relation to human intelligence by means of the intrinsic rate of natural increase. *Eugenics Q.* 10: 175–187. [26]

Baker, A. J. 1980. Morphometric differentiation in New Zealand populations of the house sparrow (*Passer domesticus*). *Evolution* 34: 638–653. [14]

Baker, A. J., and A. Moeed. 1987. Rapid genetic differentiation and founder effect in colonizing populations of common mynas (*Acridotheres tristis*). *Evolution* 41: 525–538. [11]

Baldwin, G. G., and R. H. Robichaux. 1995. Historical biogeography and ecology of the Hawaiian silversword alliance (Asteraceae). In W. L. Wagner and V. L. Funk (eds.), *Hawaiian Biogeography: Evolution on a Hot Spot Archipelago*, pp. 259–287. Smithsonian Institution Press, Washington, DC. [5]

Balkau, B. J., and M. W. Feldman. 1974. Selection for migration modification. *Genetics* 74: 171–174. [19]

Ballard, J. W. O., and M. Kreitman. 1995. Is mitochondrial DNA strictly a neutral marker? *Trends Ecol. Evol.* 10: 485–488. [22]

Bambach, R. K. 1985. Classes and adaptive variety: The ecology of diversification in marine faunas through the Phanerozoic. In J. W. Valentine (ed.), *Phanerozoic Diversity Patterns: Profiles in Macroevolution*, pp. 191–253. Princeton University Press, Princeton, NJ. [25]

Barash, D. P. 1982. *Sociobiology and Behavior*. Elsevier, New York. [26]

Barker, W. C., and M. O. Dayhoff. 1980. Evolutionary and functional relationships of homologous physiological mechanisms. *BioScience* 30: 593–599. [22]

Barkow, J. B., L. Cosmides, and J. Tooby (eds.). 1992. *The Adapted Mind: Evolutionary Psychology and the Generation of Culture.* Oxford University Press, New York. [26]

Barlow, G. W., and J. Silverberg (eds.). 1980. *Sociobiology: Beyond Nature/Nurture.* Westview Press, Boulder, CO. [26]

Barnes, R. D. 1987. *Invertebrate Zoology.* Saunders College Publishing, Philadelphia, PA. [7]

Barraclough, T. G., P. H. Harvey, and S. Nee. 1995. Sexual selection and taxonomic diversity in passerine birds. *Proc. R. Soc. Lond.* B 259: 211–215. [16]

Barrett, S. C. H. 1989. Mating system evolution and speciation in heterostylous plants. In D. Otte and J. A. Endler (eds.), *Speciation and Its Consequences*, pp. 257–283. Sinauer Associates, Sunderland, MA. [15]

Barrett, S. C. H. (ed.). 1992a. *Evolution and Function of Heterostyly.* Springer-Verlag, Berlin. [21]

Barrett, S. C. H. 1992b. Heterostylous genetic polymorphisms: Model systems for evolutionary analysis. In S. C. H. Barrett (ed.), *Evolution and Function of Heterostyly*, pp. 1–29. Springer-Verlag, Berlin. [21]

Barrett, S. C. H. 1995. Mating-system evolution in flowering plants: Micro- and macroevolutionary approaches. *Acta Bot. Neerl.* 44: 385–402. [15]

Barrett, S. C. H., and D. Charlesworth. 1991. Effects of a change in the level of inbreeding on the genetic load. *Nature* 352: 522–524. [21]

Barton, N. H. 1989. Founder effect speciation. In D. Otte and J. A. Endler (eds.), *Speciation and Its Consequences*, pp. 229–256. Sinauer Associates, Sunderland, MA. [16]

Barton, N. H., and B. Charlesworth. 1984. Genetic revolutions, founder effects, and speciation. *Annu. Rev. Ecol. Syst.* 15: 133–164. [13]

Barton, N. H., and A. Clark. 1990. Population structure and process in evolution. In K. Wöhrmann and S. K. Jain (eds.), *Population Biology: Ecological and Evolutionary Viewpoints*, pp. 115–173. Springer-Verlag, Berlin. [13, 14]

Barton, N. H., and K. S. Gale. 1993. Genetic analysis of hybrid zones. In R. G. Harrison (ed.), *Hybrid Zones and the Evolutionary Process*, pp. 13–45. Oxford University Press, New York. [15]

Barton, N. H., and G. M. Hewitt. 1981a. The genetic basis of hybrid inviability in the grasshopper *Podisma pedestris*. *Heredity* 47: 367–383. [15]

Barton, N. H., and G. M. Hewitt. 1981b. Hybrid zones and speciation. In W. R. Atchley and D. S. Woodruff (eds.), *Evolution and Speciation: Essays in Honor of M. J. D. White*, pp. 109–145. Cambridge University Press, Cambridge. [15]

Barton, N. H., and G. M. Hewitt. 1985. Analysis of hybrid zones. *Annu. Rev. Ecol. Syst.* 16: 113–148. [15]

Barton, N. H., and S. Rouhani. 1987. The frequency of shifts between alternative equilibria. *J. Theor. Biol.* 125: 397–418. [14]

Barton, N. H., and M. Turelli. 1989. Evolutionary quantitative genetics: How little do we know? *Annu. Rev. Genet.* 23: 337–370. [14]

Basolo, A. L. 1990. Female preference pre-dates the evolution of the sword in swordtail fish. *Science* 250: 808–810. [20]

Basolo, A. L. 1994. The dynamics of Fisherian sex-ratio evolution: Theoretical and experimental investigations. *Am. Nat.* 144: 473–490. [21]

Basolo, A. L. 1995. Phylogenetic evidence for the role of a pre-existing bias in sexual selection. *Proc. R. Soc. Lond.* B 259: 307–311. [20]

Bateman, A. J. 1947. Contamination of seed crops. II. Wind pollination. *Heredity* 1: 235–246. [11]

Bateman, A. J. 1948. Intra-sexual selection in *Drosophila*. *Heredity* 2: 349–368. [20]

Bateson, P. 1988. The active role of behaviour in evolution. In M.-W. Ho and S. W. Fox (eds.), *Evolutionary Processes and Metaphors*, pp. 191–207. Wiley, Chichester. [20]

Baum, D. A. 1992. Phylogenetic species concepts. *Trends Ecol. Evol.* 7: 1–2. [15]

Baum, D. A., and K. L. Shaw. 1995. Genealogical perspectives on the species problem. In P. C. Hoch and A. G. Stephenson (eds.), *Experimental and Molecular Approaches to Plant Biosystematics*, pp. 289–303. Monographs in Systematic Botany. Missouri Botanical Garden, St. Louis, MO. [15]

Beaver, R. A. 1979. Host specificity of temperate and tropical animals. *Nature* 281: 139–141. [8]

Beeman, R. W., K. S. Friesen, and R. E. Denell. 1992. Maternal-effect selfish genes in flour beetles. *Science* 256: 89–92. [12]

Begon, M., J. L. Harper, and C. R. Townsend. 1990. *Ecology: Individuals, Populations, and Communities.* Blackwell Scientific, Boston. [4]

Behrensmeyer, A. K., J. D. Damuth, W. A. DiMichele, R. Potts, H.-D. Sues, and S. L. Wing (eds.). 1992. *Terrestrial Ecosystems Through Time: Evolutionary Paleoecology of Terrestrial Plants and Animals.* University of Chicago Press, Chicago, IL. [7]

Bell, A. P., M. S. Weinberg, and S. K. Hammersmith. 1981. *Sexual Preference: Its Development in Men and Women.* Indiana University Press, Bloomington. [26]

Bell, G. 1982. *The Masterpiece of Nature: The Evolution and Genetics of Sexuality.* University of California Press, Berkeley. [21]

Bell, G. 1988. *Sex and Death in Protozoa.* Cambridge University Press, Cambridge. [19]

Bell, G., and V. Koufopanou. 1986. The cost of reproduction. *Oxford Surv. Evol. Biol.* 3: 83–131. [19]

Bell, M. A., J. V. Baumgartner, and E. C. Olson. 1985. Patterns of temporal change in single morphological characters of a Miocene stickleback fish. *Paleobiology* 11: 258–271. [6]

Bellairs, A. 1970. *The Life of Reptiles.* Universe Books, New York. [8, 23]

Bengtson, S. (ed.). 1995. *Early Life on Earth.* Columbia University Press, New York. [7]

Bennett, A. F. 1991. The evolution of activity capacity. *J. Exp. Biol.* 160: 1–23. [17]

Bennett, A. F., and R. E. Lenski. 1993. Evolutionary adaptation to temperature. II. Thermal niches of experimental lines of *Escherichia coli*. *Evolution* 47: 1–12. [17]

Bennett, A. F., R. E. Lenski, and J. E. Mittler. 1992. Evolutionary adaptation to temperature. I. Fitness responses of *Escherichia coli* to changes in its thermal environment. *Evolution* 46: 16–30. [17]

Bennett, M. D. 1982. Nucleotypic basis of the spatial ordering of chromosomes in eukaryotes and the implication of the order for genome evolution and phenotypic variation. In G. A. Dover and R. B. Flavell (eds.), *Genome Evolution*, pp. 239–261. Academic Press, New York. [22]

Benson, L. 1982. *The Cacti of the United States and Canada.* Stanford University Press, Stanford, CA. [4]

Benton, M. J. 1983. Dinosaur success in the Triassic: A noncompetitive ecological model. *Q. Rev. Biol.* 58: 29–55. [5]

Benton, M. J. 1985. Classification and phylogeny of the diapsid reptiles. *Zool. J. Linn. Soc.* 84: 97–164. [5]

Benton, M. J. (ed.). 1988a. *The Phylogeny and Classification of the Tetrapods.* Clarendon Press, Oxford. [7]

Benton, M. J. 1988b. Relationships of the placental mammals: A comparison of phylogenetic trees. *Trends Ecol. Evol.* 3: 40–45. [7]

Benton, M. J. 1990. The causes of the diversification of life. In P. D. Taylor and G. P. Larwood (eds.), *Major Evolutionary Radiations*, pp. 409–430. Clarendon Press, Oxford. [25]

Benton, M. J. 1994. Palaeontological data, and identifying mass extinctions. *Trends Ecol. Evol.* 9: 181–185. [25]

Benton, M. J. 1996. On the nonprevalence of competitive replacement in the evolution of tetrapods. In D. Jablonski, D. H. Erwin, and J. Lipps (eds.), *Evolutionary Paleobiology*, pp. 185–210. University of Chicago Press, Chicago, IL. [25]

Benveniste, R. E. 1985. The contributions of retroviruses to the study of mammalian evolution. In R. I. MacIntyre (ed.), *Molecular Evolutionary Genetics*, pp. 359–417. Plenum, New York. [22]

Berenbaum, M. R. 1983. Coumarins and caterpillars: A case for coevolution. *Evolution* 39: 163–179. [18]

Berenbaum, M. R., and A. R. Zangerl. 1988. Stalemates in the coevolutionary arms race: Synthesis, synergisms, and sundry other sins. In K. C. Spencer (ed.), *Chemical Mediation of Coevolution*, pp. 113–132. Academic Press, San Diego, CA. [18]

Berenbaum, M. R., A. R. Zangerl, and J. K. Nitao. 1986. Constraints on chemical coevolution: Wild parsnip and the parsnip webworm. *Evolution* 40: 1215–1228. [18]

Berglund, A., A. Bisazza, and A. Pilastro. 1996. Armments and ornaments: An evolutionary explanation of traits of dual utility. *Biol. J. Linn. Soc.* 58: 385–399. [20]

Berglund, A., G. Rosenqvist, and I. Svensson. 1989. Reproductive success of females limited by males in two pipefish species. *Am. Nat.* 133: 506–516. [20]

Bermingham, E., and J. C. Avise. 1986. Molecular zoogeography of freshwater fishes in the southeastern United States. *Genetics* 113: 939–965. [11, 16]

Bernardi, G., D. Mouchiroud, C. Gautier, and G. Bernardi. 1988. Compositional patterns in vertebrate genomes: Conservation and change in evolution. *J. Mol. Evol.* 28: 7–18. [22]

Bernardi, G., B. Olofsson, J. Filipski, M. Zerial, J. Salinas, G. Cuny, M. Meunier-Rotival, and F. Rodier. 1985. The mosaic genome of warm-blooded vertebrates. *Science* 228: 953–958. [22]

Bernays, E. A., and R. F. Chapman. 1994. *Host-Plant Selection by Phytophagous Insects*. Chapman & Hall, New York. [18]

Bernays, E. A., and M. Graham. 1988. On the evolution of host specificity by phytophagous arthropods. *Ecology* 69: 886–892. [18]

Bernstein, C., and H. Bernstein. 1991. *Aging, Sex, and DNA Repair*. Academic Press, San Diego, CA. [21]

Berry, A. J., J. W. Ajioka, and M. Kreitman. 1991. Lack of polymorphism on the *Drosophila* fourth chromosome resulting from selection. *Genetics* 129: 1111–1117. [22]

Berthold, P., A. J. Heibig, G. Mohr, and U. Querner. 1992. Rapid microevolution of migratory behavior in a wild bird species. *Nature* 360: 668–670. [14]

Berven, K. A. 1987. The heritable basis of variation in larval developmental patterns within populations of the wood frog (*Rana sylvatica*). *Evolution* 41: 1088–1097. [14]

Berven, K. A., D. E. Gill, and S. J. Smith-Gill. 1979. Countergradient selection in the green frog, *Rana clamitans*. *Evolution* 33: 609–623. [9]

Bierzychudek, P. 1987. Patterns in plant parthenogenesis. In S. C. Stearns (ed.), *The Evolution of Sex and Its Consequences*, pp. 197–217. Birkhäuser Verlag, Basel. [21]

Birch, L. C., T. Dobzhansky, P. D. Elliott, and R. C. Lewontin. 1963. Relative fitness of geographic races of *Drosophila serrata*. *Evolution* 17: 72–83. [4]

Birkhead, T. R., and F. M. Hunter. 1990. Mechanisms of sperm competition. *Trends Ecol. Evol.* 5: 48–52. [20]

Birkhead, T. R., and A. P. Møller. 1992. *Sperm Competition in Birds: Evolutionary Causes and Consequences*. Academic Press, London. [20]

Bishop, J. A. 1981. A neo-Darwinian approach to resistance: Examples from mammals. In J. A. Bishop and L. M. Cook (eds.), *Genetic Consequences of Man Made Change*, pp. 37–51. Academic Press, London. [13]

Bishop, J. A., and L. M. Cook (eds.). 1981. *Genetic Consequences of Man Made Change*. Academic Press, London. [13, 14]

Björkman, O., and P. Holmgren. 1963. Adaptability of the photosynthetic apparatus to light intensity in ecotypes from exposed and shaded habitats. *Physiol. Plant* 16: 889–914. [9]

Blair, W. F. 1950. Ecological factors in speciation of *Peromyscus*. *Evolution* 9: 253–275. [17]

Blair, W. F. 1960. *The Rusty Lizard, a Population Study*. University of Texas Press, Austin. [11]

Blatchley, W. S., and C. W. Leng. 1916. *Rhynchophora or Weevils of North Eastern America*. The Nature Publishing Co., Indianapolis, IN. [4]

Boag, P. T. 1983. The heritability of external morphology in Darwin's ground finches (*Geospiza*) on Isla Daphne Major, Galápagos. *Evolution* 37: 877–894. [9]

Boake, C. R. B. (ed.). 1994. *Quantitative Genetic Studies of Behavioral Evolution*. University of Chicago Press, Chicago, IL. [14, 20]

Boardman, N. K., A.-W. Linnane, and R. M. Smillie (eds.). 1971. *Autonomy and Biogenesis of Mitochondria and Chloroplasts*. North-Holland, Amsterdam. [7]

Boardman, R. S., A. H. Cheetham, and A. J. Rowell (eds.). 1987. *Fossil Invertebrates*. Blackwell Scientific Publications, Palo Alto, CA. [7]

Bodmer, W., and M. Ashburner. 1984. Conservation and change in the DNA sequences coding for alcohol dehydrogenase in sibling species of *Drosophila*. *Nature* 309: 425–540. [10]

Bonaparte, J. F. 1978. El Mesozoico de America del Sur y sus tetrapodos. *Opera Lilloana* 26: 1–596. [7]

Bonnell, M. L., and R. K. Selander. 1974. Elephant seals: Genetic variation and near extinction. *Science* 184: 908–909. [11]

Bonner, J. T. 1988. *The Evolution of Complexity*. Princeton University Press, Princeton, NJ. [24]

Borowik, O. A. 1995. Coding chromosomal data for phylogenetic analysis: Phylogenetic resolution of the *Pan-Homo-Gorilla* trichotomy. *Syst. Biol.* 44: 563–570. [26]

Borror, D. J., D. M. DeLong, and C. A. Triplehorn. 1981. *An Introduction to the Study of Insects*. Saunders College Publishing, Philadelphia, PA. [5, 19]

Borst, P., and D. R. Greaves. 1987. Programmed gene rearrangements altering gene expression. *Science* 235: 658–667. [12]

Bouchard, T. J. Jr., D. T. Lykken, M. McGue, N. L. Segal, and A. Tellegen. 1990. Sources of human psychological differences: The Minnesota study of twins reared apart. *Science* 250: 223–228. [26]

Boucot, A. J. 1975. *Evolution and Extinction Rate Controls*. Elsevier, Amsterdam. [25]

Boucot, A. J. 1978. Community evolution and rates of cladogenesis. In M. K. Hecht, B. Wallace, and G. T. Prance (eds.), *Evolutionary Biology*, vol. 11: 545–655. Plenum, New York. [25]

Boureau, E. 1964. *Traité de paléobotanique*, vol. III. Masson, Paris. [7]

Bowler, P. J. 1989. *Evolution: The History of an Idea*. University of California Press, Berkeley. [2]

Bowring, S. A., J. P. Grotzinger, C. E. Isachsen, A. H. Knoll, S. M. Pelechaty, and P. Kolosov. 1993. Calibrating rates of early Cambrian evolution. *Science* 261: 1293–1298. [7]

Boyce, M. S. 1984. Restitution of *r*- and *K*-selection as a model of density-dependent natural selection. *Annu. Rev. Ecol. Syst.* 15: 427–447. [19]

Boyd, R., and P. J. Richerson. 1985. *Culture and the Evolutionary Process.* University of Chicago Press, Chicago, IL. [26]

Bradley, R. D., J. J. Bull, A. D. Johnson, and D. M. Hillis. 1993. Origin of a novel allele in a mammalian hybrid zone. *Proc. Natl. Acad. Sci. USA* 90: 8939–8941. [15]

Bradshaw, A. D. 1991. Genostasis and the limits to evolution. The Croonian Lecture, 1991. *Phil. Trans. R. Soc. Lond.* B 333: 289–305. [14, 17, 23]

Brandon, R. N. 1990. *Adaptation and Environment.* Princeton University Press, Princeton, NJ. [12]

Breed, M. D., and R. E. Page (eds.). 1989. *The Genetics of Social Evolution.* Westview Press, Boulder, CO. [20]

Bremermann, H. J., and J. Pickering. 1983. A game-theoretical model of parasite virulence. *J. Theor. Biol.* 100: 411–426. [18]

Breeuwer, J. A. J., and J. H. Werren. 1990. Microorganisms associated with chromosome destruction and reproductive isolation between two insect species. *Nature* 346: 558–560. [15]

Breeuwer, J. A. J., and J. H. Werren. 1993. Effect of genotype on cytoplasmic incompatibility between two species of *Nasonia. Heredity* 70: 428–436. [15]

Briggs, D. E. G., and P. R. Crowther (eds.). 1990. *Palaeobiology: A Synthesis.* Blackwell Scientific, Oxford. [6]

Briggs, D. E. G., R. A. Fortey, and M. A. Wills. 1992. Morphological disparity in the Cambrian. *Science* 256: 1670. [25]

Brink, A. S. 1956. Speculations on some advanced mammalian characteristics in the higher mammal-like reptiles. *Palaeontol. Afr.* 4: 77-96. [6]

Britten, R. J. 1986. Rates of DNA sequence evolution differ between taxonomic groups. *Science* 231: 1393–1398. [5, 11]

Britton-Davidian, J., H. Sonjaya, J. Catalan, and G. Cattaneo-Berrebi. 1990. Robertsonian heterozygosity in wild mice: Fertility and transmission rates in Rb (16.17) translocation heterozygotes. *Genetica* 80: 171–174. [15]

Broadhead, T. W., and J. A. Waters. 1980. *Echinoderms: Notes for a Short Course.* Studies in Geology 3. University of Tennessee Dept. of Geological Sciences, Knoxville. [7]

Brochmann, C., P. S. Soltis, and D. E. Soltis. 1992. Recurrent formation and polyphyly of Nordic polyploids in *Draba* (Brassicaceae). *Am. J. Bot.* 79: 673–688. [16]

Brodie, E. D. III. 1989. Genetic correlations between morphology and antipredator behaviour in natural populations of the garter snake *Thamnophis ordinoides. Nature* 342: 542–543. [14]

Brodie, E. D. III. 1992. Correlational selection for color pattern and antipredator behavior in the garter snake *Thamnophis ordinoides. Evolution* 46: 1284–1298. [14]

Brodie, E. D. III. 1993. Homogeneity of the genetic variance-covariance matrix for antipredator traits in two natural populations of the garter snake *Thamnophis ordinoides. Evolution* 47: 844–854. [14]

Brodie, E. D. III, and E. D. Brodie, Jr. 1990. Tetrodotoxin resistance in garter snakes: An evolutionary response of predators to dangerous prey. *Evolution* 40: 651–659. [14]

Bronstein, J. L. 1992. Seed predators as mutualists: Ecology and evolution of the fig/pollinator interaction. In E. Bernays (ed.), *Insect-Plant Interactions*, pp. 1–44. CRC Press, Boca Raton, FL. [18]

Brooks, D. R., and D. A. McLennan. 1991. *Phylogeny, Ecology, and Behavior.* University of Chicago Press, Chicago, IL. [5]

Brooks, L. D. 1988. The evolution of recombination rates. In R. E. Michod and B. R. Levin (eds.), *The Evolution of Sex*, pp. 87–105. Sinauer Associates, Sunderland, MA. [21]

Brown, D. D., P. C. Wensink, and E. Jordan. 1972. A comparison of the ribosomal DNAs of *Xenopus laevis* and *Xenopus mulleri:* The evolution of tandem genes. *J. Mol. Biol.* 63: 59–73. [22]

Brown, J. H. 1971a. Mammals on mountaintops: Nonequilibrium insular biogeography. *Am. Nat.* 105: 467–478. [8]

Brown, J. H. 1971b. The desert pupfish. *Sci. Am.* 225(5): 104–110. [16]

Brown, J. H. 1988. Species diversity. In A. A. Myers and P. S. Giller (eds.), *Analytical Biogeography*, pp. 57–89. Chapman & Hall, London. [8]

Brown, J. H. 1995. *Macroecology.* University of Chicago Press, Chicago, IL. [17]

Brown, J. H., and A. C. Gibson. 1983. *Biogeography.* Mosby, St. Louis, MO. [8]

Brown, J. L. 1987. *Helping and Communal Breeding in Birds.* Princeton University Press, Princeton, NJ. [20]

Brown, J. M., O. Pellmyr, J. N. Thompson, and R. G. Harrison. 1994. Mitochondrial DNA phylogeny of the Prodoxidae (Lepidoptera: Incurvarioidea) indicates rapid ecological diversification of yucca moths. *Ann. Entomol. Soc. Am.* 87: 795–801. [18]

Brown, W. L. Jr., and E. O. Wilson. 1956. Character displacement. *Syst. Zool.* 5: 49–64. [16]

Brown, W. M., J. M. George, and A. C. Wilson. 1979. Rapid evolution of animal mitochondrial DNA. *Proc. Natl. Acad. Sci. USA* 76: 1967–1971. [11, 22]

Brown, W. M., E. M. Prager, A. Wang, and A. C. Wilson. 1982. Mitochondrial DNA sequences of primates: Temp and mode of evolution. *J. Molec. Evol.* 18: 225–239. [22]

Brundin, L. Z. 1965. On the real nature of transantarctic relationships. *Evolution* 19: 496–505. [8]

Brundin, L. Z. 1988. Phylogenetic biogeography. In A. A. Myers and P. S. Giller (eds.), *Analytical Biogeography*, pp. 343–369. Chapman & Hall, London. [8]

Brusca, G. J., and R. C. Brusca. 1978. *A Naturalist's Seashore Guide.* Mad River Press, Eureka, CA. [25]

Brusca, R. C., and G. J. Brusca. 1990. *Invertebrates.* Sinauer Associates, Sunderland, MA. [5]

Bryant, E. H., and L. M. Meffert. 1988. Effect of an experimental bottleneck on morphological integration in the housefly. *Evolution* 42: 698–707. [14]

Bryant, E. H., and L. M. Meffert. 1990. Multivariate phenotypic differentiation among bottlenecked lines of the housefly. *Evolution* 44: 660–668. [14]

Bryant, E. H., S. A. McCommas, and L. J. Combs. 1986. The effect of an experimental bottleneck upon quantitative genetic variation in the housefly. *Genetics* 114: 1191–1211. [14]

Brylski, P., and B. K. Hall. 1988. Ontogeny of a macroevolutionary phenotype: The external cheek pouches of geomyoid rodents. *Evolution* 42: 391–395. [23]

Bull, J. J. 1983. *Evolution of Sex Determining Mechanisms.* Benjamin Cummings, Menlo Park, CA. [21]

Bull, J. J. 1994. Perspective: Virulence. *Evolution* 48: 1423–1437. [18]

Bull, J. J., and E. L. Charnov. 1985. On irreversible evolution. *Evolution* 39: 1149–1155. [21]

Bull, J. J., and E. L. Charnov. 1988. How fundamental are Fisherian sex ratios? *Oxford Surv. Evol. Biol.* 5: 97–135. [21]

Bull, J. J., and W. R. Rice. 1991. Distinguishing mechanisms for the evolution of cooperation. *J. Theor. Biol.* 149: 63–74. [18]

Bullini, L. 1994. Origin and evolution of animal hybrid species. *Trends Ecol. Evol.* 9: 422–426. [14]

Bulmer, M. 1988. Evolutionary aspects of protein synthesis. *Oxford Surv. Evol. Biol.* 5: 1–40. [22]

Bulmer, M. 1991. The selection-mutation-drift theory of synonymous codon usage. *Genetics* 129: 897–907. [22]

Bulmer, M., K. H. Wolfe, and P. M. Sharp. 1991. Synonymous nucleotide substitution rates in mammalian genes: Implications for the molecular clock and the relationship of mammalian orders. *Proc. Natl. Acad. Sci. USA* 88: 5974–5978. [22]

Burdon, J. J. 1987. *Diseases and Plant Population Biology.* Cambridge University Press, Cambridge. [18]

Bürger, R., G. P. Wagner, and F. Stettinger. 1989. How much heritable variation can be maintained in finite populations by mutation-selection balance? *Evolution* 43: 1748–1766. [14]

Burggren, W. W., and W. E. Bemis. 1990. Studying physiological evolution: Paradigms and pitfalls. In M. H. Nitecki (ed.), *Evolutionary Innovations*, pp. 191–228. University of Chicago Press, Chicago, IL. [24]

Buri, P. 1956. Gene frequency drift in small population of mutant *Drosophila. Evolution* 10: 367–402. [11]

Burian, R. M. 1988. Challenges to the evolutionary synthesis. *Evol. Biol.* 23: 247–270. [2]

Burke, A. C., C. E. Nelson, B. A. Morgan, and C. Tabin. 1995. *Hox* genes and the evolution of vertebrate axial morphology. *Development* 121: 333–346. [23]

Burley, N. 1986. Comparison of the band color preferences of two species of estrildid finches. *Anim. Behav.* 34: 1732–1741. [16]

Burton, R. 1990. *Bird Flight.* Facts on File, New York. [17]

Bush, G. L. 1969. Sympatric host race formation and speciation in frugivorous flies of the genus *Rhagoletis* (Diptera, Tephritidae). *Evolution* 23: 237–251. [16]

Bush, G. L. 1975. Modes of animal speciation. *Annu. Rev. Ecol. Syst.* 6: 334–364. [16]

Buss, L. S. 1987. *The Evolution of Individuality.* Princeton University Press, Princeton, NJ. [19]

Butlin, R. 1989. Reinforcement of premating isolation. In D. Otte and J. A. Endler (eds.), *Speciation and Its Consequences*, pp. 158–179. Sinauer Associates, Sunderland, MA. [16]

Butlin, R. K., and M. G. Ritchie. 1989. Genetic coupling in mate recognition systems: What is the evidence? *Biol. J. Linn. Soc.* 37: 237–246. [15]

Cabe, P. R. 1993. European starling (*Sturnus vulgaris*). In A. Poole and F. Gill (eds.), *The Birds of North America*, No. 48. Academy of Natural Sciences, Philadelphia. [8]

Cabot, E. L., A. W. Davis, N. A. Johnson, and C.-I. Wu. 1994. Genetics of reproductive isolation in the *Drosophila simulans* clade: Complex epistasis underlying hybrid male sterility. *Genetics* 137: 175–189. [15]

Caccone, A., and J. R. Powell. 1989. DNA divergence among hominoids. *Evolution* 43: 925–942. [5]

Cadle, J. E. 1985. The neotropical colubrid snake fauna (Serpentes: Colubridae): Lineage components and biogeography. *Syst. Zool.* 34: 1–20. [8]

Cain, A. J., and P. M. Sheppard. 1954. Natural selection in *Cepaea. Genetics* 39: 89–116. [13]

Cairns, J., J. Overbaugh, and S. Miller. 1988. The origin of mutants. *Nature* 335: 142–145. [10]

Calder, W. A. III. 1984. *Size, Function, and Life History.* Harvard University Press, Cambridge, MA. [17]

Camin, J., and R. Sokal. 1965. A method for deducing branching sequences in phylogeny. *Evolution* 19: 311–326. [5]

Campbell, N. A. 1993. *Biology.* Third edition. Benjamin Cummings, Redwood City, CA. [3]

Campos-Ortega, J. A., and Y. N. Jan. 1991. Genetic and molecular bases of neurogenesis in *Drosophila melanogaster. Annu. Rev. Neurobiol.* 14: 399–420. [14]

Cann, R. L., M. Stoneking, and A. C. Wilson. 1987. Mitochondrial DNA and human evolution. *Nature* 325: 31–36. [26]

Cano, R. J., H. N. Poinar, N. J. Pieniazek, A. Acra, and G. O. Poinar. 1993. Amplification and sequencing of DNA from a 120-135-million-year-old weevil. *Nature* 363: 536–538. [5]

Caplan, A. L. (ed.). 1978. *The Sociobiology Debate: Readings on Ethical and Scientific Issues.* Harper & Row, New York. [26]

Caple, G., R. P. Balda, and W. R. Willis. 1983. The physics of leaping animals and the evolution of pre-flight. *Am. Nat.* 121: 455–476. [6]

Cardé, R. T., and T. C. Baker. 1984. Sexual communication with pheromones. In W. J. Bell and R. T. Cardé (eds.), *Chemical Ecology of Insects*, pp. 355–383. Chapman & Hall, London. [15]

Carpenter, R. M., and L. Burnham. 1985. The geological record of insects. *Annu. Rev. Earth Planet. Sci.* 13: 294–314. [7]

Carroll, R. L. 1982. Early evolution of reptiles. *Annu. Rev. Ecol. Syst.* 13: 87–109. [5]

Carroll, R. L. 1988. *Vertebrate Paleontology and Evolution.* W. H. Freeman, New York. [6, 7]

Carroll, R. L., and D. Baird. 1972. Carboniferous stem-reptiles of the family Romeriidae. *Bull. Mus. Comp. Zool.* 143: 321–363. [6]

Carroll, S. B. 1995. Homeotic genes and the evolution of arthropods and chordates. *Nature* 376: 479–485. [23]

Carroll, S. B., J. Gates, D. N. Keys, S. W. Paddock, G. E. F. Panganiban, J. E. Selegue, and J. A. Williams. 1994. Pattern formation and eyespot determination in butterfly wings. *Science* 265: 109–114. [23]

Carson, H. L. 1959. Genetic conditions which promote or retard the formation of species. *Cold Spring Harbor Symp. Quant. Biol.* 24: 87–105. [21]

Carson, H. L. 1967. Selection for parthenogenesis in *Drosophila mercatorum. Genetics* 55: 157–171. [21]

Carson, H. L. 1975. The genetics of speciation at the diploid level. *Am. Nat.* 109: 83–92. [16]

Carson, H. L. 1982. Speciation as a major reorganization of polygenic balances. In C. Barigozzi (ed.), *Mechanisms of Speciation*, pp. 411–433. Alan R. Liss, New York. [16]

Carson, H. L. 1983. Chromosome sequences and interisland colonizations in the Hawaiian *Drosophila. Genetics* 103: 465–482. [8]

Carson, H. L., and B. A. Clague. 1995. Geology and biogeography of the Hawaiian Islands. In W. L. Wagner and V. A. Funk (eds.), *Hawaiian Biogeography*, pp. 14–29. Smithsonian Institution Press, Washington, DC. [8]

Carson, H. L., and K. Y. Kaneshiro. 1976. *Drosophila* of Hawaii: Systematics and ecological genetics. *Annu. Rev. Ecol. Syst.* 7: 311–345. [5, 16]

Carson, H. L., and A. R. Templeton. 1984. Genetic revolutions in relation to speciation phenomena: The founding of new populations. *Annu. Rev. Ecol. Syst.* 15: 97–131. [16]

Carson, H. L., D. E. Hardy, H. T. Spieth, and W. S. Stone. 1970. The evolutionary biology of the Hawaiian Drosophilidae. In M. K. Hecht and W. C. Steere (eds.), *Essays in Evolution and Genetics in Honor of Theodosius Dobzhansky*, pp. 437–543. Appleton-Century-Crofts, New York. [5, 15]

Case, T. J. 1981. Niche packing and coevolution in competition communities. *Proc. Natl. Acad. Sci. USA* 78: 5021–5025. [18]

Case, T. J., and M. L. Taper. 1986. On the coexistence and coevolution of asexual and sexual competitors. *Evolution* 40: 366–387. [21]

Castillo-Chávez, C., S. A. Levin, and F. Gould. 1988. Physiological and behavioral adaptation to varying environments: A mathematical model. *Evolution* 42: 986–994. [14]

Cavalier-Smith, T. 1985. Introduction: The evolutionary significance of genome size. In T. Cavalier-Smith (ed.), *The Evolution of Genome Size*, pp. 1–36. Wiley, New York. [22]

Literature Cited

Cavalier-Smith, T. 1987. Eukaryotes with no mitochondria. *Nature* 326: 332–333. [7]

Cavalier-Smith, T. 1991. Intron phylogeny: A new hypothesis. *Trends Genet.* 7: 145–148. [22]

Cavalli-Sforza, L. L., and W. F. Bodmer. 1971. *The Genetics of Human Populations*. W. H. Freeman, San Francisco, CA. [3]

Cavalli-Sforza, L. L., and M. W. Feldman. 1981. *Cultural Transmission and Evolution: A Quantitative Approach*. Princeton University Press, Princeton, NJ. [26]

Cavalli-Sforza, L. L., P. Menozzi, and A. Piazza. 1994. *The History and Geography of Human Genes*. Princeton University Press, Princeton, NJ. [26]

Cavener, D. R., and M. L. Clegg. 1981. Multigenic response to ethanol in *Drosophila melanogaster*. *Evolution* 35: 1–10. [14]

Cavicchi, S., D. Guerra, V. La Torre, and R. B. Huey. 1995. Chromosomal analysis of heat-shock tolerance in *Drosophila melanogaster* evolving at different temperatures in the laboratory. *Evolution* 49: 676–684. [17]

Chapman, R. F. 1982. *The Insects: Structure and Function*. Harvard University Press, Cambridge, MA. [23]

Charlesworth, B. 1978. The population genetics of anisogamy. *J. Theor. Biol.* 73: 347–357. [21]

Charlesworth, B. 1982. Hopeful monsters cannot fly. *Paleobiology* 8: 469–474. [24]

Charlesworth, B. 1984. Some quantitative methods for studying evolutionary patterns in single characters. *Paleobiology* 10: 308–318. [6, 24]

Charlesworth, B. 1989. The evolution of sex and recombination. *Trends Ecol. Evol.* 4: 264–267. [21]

Charlesworth, B. 1990. The evolutionary genetics of adaptation. In M. Nitecki (ed.), *Evolutionary Innovations*, pp. 47–70. University of Chicago Press, Chicago, IL. [24]

Charlesworth, B. 1991. The evolution of sex chromosomes. *Science* 251: 1030–1033. [21]

Charlesworth, B. 1994a. The effect of background selection against deleterious mutations on weakly selected, linked variants. *Genet. Res.* 63: 213–227. [22]

Charlesworth, B. 1994b. *Evolution in Age-Structured Populations*. Cambridge University Press, Cambridge. [19]

Charlesworth, B. 1996. Background selection and patterns of genetic diversity in *Drosophila melanogaster*. *Genet. Res.* 68: 131–149. [22]

Charlesworth, B., and D. Charlesworth. 1973. A study of linkage disequilibrium in populations of *Drosophila*. *Genetics* 73: 351–359. [9]

Charlesworth, B., and D. Charlesworth. 1978. A model for the evolution of dioecy and gynodioecy. *Am. Nat.* 112: 975–997. [21]

Charlesworth, B., and C. H. Langley. 1989. The population genetics of *Drosophila* transposable elements. *Annu. Rev. Genet.* 23: 251–287. [22]

Charlesworth, B., and S. Rouhani. 1988. The probability of peak shifts in a founder population. II. An additive polygenic trait. *Evolution* 42: 1129–1145. [14]

Charlesworth, B., R. Lande, and M. Slatkin. 1982. A neo-Darwinian commentary on macroevolution. *Evolution* 36: 474–498. [16]

Charlesworth, B., C. H. Langley, and W. Stephan. 1986. The evolution of restricted recombination and the accumulation of repeated DNA sequences. *Genetics* 112: 947–962. [22]

Charlesworth, B., M. T. Morgan, and D. Charlesworth. 1993. The effect of deleterious mutations on neutral molecular variation. *Genetics* 134: 1289–1303. [22]

Charlesworth, B., P. Sniegowski, and W. Stephan. 1994. The evolutionary dynamics of repetitive DNA in eukaryotes. *Nature* 371: 215–220. [22]

Charnov, E. L. 1976. Optimal foraging: Attack strategy of a mantid. *Am. Nat.* 110: 141–151. [20]

Charnov, E. L. 1982. *The Theory of Sex Allocation*. Princeton University Press, Princeton, NJ. [21]

Charnov, E. L., and W. M. Schaffer. 1973. Life history consequences of natural selection: Cole's result revisited. *Am. Nat.* 107: 791–793. [19]

Cheetham, A. H. 1986. Tempo of evolution in a Neogene bryozoan: Rates of morphologic change within and across species boundaries. *Paleobiology* 12: 190–202. [24]

Cheetham, A. H. 1987. Tempo of evolution in a Neogene bryozoan: Are trends in single morphological characters misleading? *Paleobiology* 13: 286–296. [6]

Chesser, R. K. 1983. Genetic variability within and among populations of the black-tailed prairie dog. *Evolution* 37: 320–331. [11]

Cheverud, J. 1982. Phenotypic, genetic, and environmental morphological integration in the cranium. *Evolution* 36: 499–516. [14]

Cheverud, J. M. 1988a. A comparison of genetic and phenotypic correlations. *Evolution* 42: 958–968. [14]

Cheverud, J. M. 1988b. The evolution of genetic correlation and developmental constraints. In G. de Jong (ed.), *Population Genetics and Evolution*, pp. 94–101. Springer-Verlag, Berlin. [14, 23]

Cheverud, J. M. 1996. Developmental integration and the evolution of pleiotropy. *Am. Zool.* 36: 44–50. [23]

Chomsky, N. 1986. *Knowledge of Language: Its Nature, Origin, and Use*. Praeger, New York. [26]

Chrispeels, M. J., and D. E. Sadava. 1994. *Plants, Genes, and Agriculture*. Jones & Bartlett, Boston, MA. [4]

Christiansen, B. 1977. Habitat preference among amylase genotypes in *Asellus aquaticus* (Isopoda, Crustacea). *Hereditas* 87: 21–26. [13]

Christiansen, F. B. 1984. The definition and measurement of fitness. In B. Shorrocks (ed.), *Evolutionary Ecology*, pp. 65–79. Blackwell Scientific, Oxford. [13]

Christiansen, F. B. 1990. Natural selection: Measures and modes. In K. Wöhrmann and S. K. Jain (eds.), *Population Biology: Ecological and Evolutionary Viewpoints*, pp. 27–81. Springer-Verlag, Berlin. [13]

Christiansen, F. B., O. Frydenberg, and V. Simonsen. 1977. Genetics of *Zoarces* populations. X. Selection component analysis of the *Est III* polymorphism using samples of successive cohorts. *Hereditas* 87: 129–150. [13]

Churchill, W. 1967. *Homosexual Behavior Among Males: A Cross-Cultural and Cross-Species Investigation*. Prentice-Hall, Englewood Cliffs, NJ. [26]

Cifelli, R. 1969. Radiation of Cenozoic planktonic Foraminifera. *Syst. Zool.* 18: 154–168. [6]

Ciochon, R. L., and J. G. Fleagle (eds.). 1993. *The Human Evolution Sourcebook*. Prentice-Hall, Englewood Cliffs, NJ. [26]

Clark, A. G. 1987. Senescence and the genetic correlation hang-up. *Am. Nat.* 129: 932–940. [19]

Clark, A. G. 1993. Evolutionary inferences from molecular characterization of self-incompatibility alleles. In N. Takahata and A. G. Clark (eds.), *Mechanisms of Molecular Evolution*, pp. 79–108. Sinauer Associates, Sunderland, MA. [21]

Clark, A. G., M. Aguadé, T. Prout, L. G. Harshman, and C. H. Langley. 1995. Variation in sperm displacement and its association with accessory gland protein loci in *Drosophila melanogaster*. *Genetics* 139: 189–201. [1, 20]

Clark, J., and R. L. Carroll. 1973. Romeriid reptiles from the Lower Permian. *Bull. Mus. Comp. Zool.* 147: 353–407. [6]

Clark, J. B., W. P. Maddison, and M. G. Kidwell. 1994. Phylogenetic analysis supports horizontal transfer of *P* transposable elements. *Mol. Biol. Evol.* 11: 40–50. [22]

Clarke, B. 1962. Balanced polymorphism and the diversity of sympatric species. *Syst. Assoc. Publ.* 4: 47–70. [13]

Clarke, P. H. 1974. The evolution of enzymes for the utilisation of novel substrates. In M. J. Carlile and J. J. Skehel (eds.), *Evolution in the Microbial World*, pp. 183–217. Cambridge University Press, Cambridge. [10]

Clarkson, E. N. K. 1993. *Invertebrate Palaeontology and Evolution.* Chapman & Hall, London. [7]

Clausen, J., D. D. Keck, and W. M. Hiesey. 1940. Experimental studies on the nature of species. I. Effect of varied environments on western North American plants. Carnegie Institution of Washington Publication no. 520: 1–452. [9]

Clausen, J., D. D. Keck, and W. M. Hiesey. 1947. Heredity of geographically and ecologically isolated races. *Am. Nat.* 81: 114–133. [9]

Clausen, R. T. 1975. *Sedum of North America North of the Mexican Plateau.* Cornell University Press, Ithaca, NY. [17]

Clayton, G. A., and A. Robertson. 1957. An experimental check on quantitative genetic theory. II. The long-term effects of selection. *J. Genet.* 55: 152–170. [14]

Clutton-Brock, T. 1991. *The Evolution of Parental Care.* Princeton University Press, Princeton, NJ. [20]

Coates, M. I., and J. A. Clack. 1990. Polydactyly in the earliest known tetrapod limbs. *Nature* 347: 66–69. [6]

Coates, M. I., and J. A. Clack. 1991. Fishlike gills and breathing in the earliest known tetrapod. *Nature* 352: 234–236. [6]

Cock, A. G. 1966. Genetical aspects of metrical growth and form in animals. *Q. Rev. Biol.* 41: 131–190. [23]

Coddington, J. A. 1988. Cladistic tests of adaptational hypotheses. *Cladistics* 4: 3–22. [12]

Cody, M. L., and J. M. Diamond (eds.). 1975. *Ecology and Evolution of Communities.* Harvard University Press, Cambridge, MA. [25]

Coe, M., and H. Beentje. 1991. *A Field Guide to the Acacias of Kenya.* Oxford University Press, Oxford and New York. [12]

Cogger, H. G. 1992. *Reptiles and Amphibians of Australia.* Cornell University Press, Ithaca, NY. [23]

Cohan, F. M. 1984. Can uniform selection retard random genetic divergence between isolated conspecific populations? *Evolution* 38: 495–504. [14]

Cohen, R. S., and L. Lauden (eds.). 1983. *Physics, Philosophy, and Psychoanalysis.* Reidel, Dordrecht. [8]

Colbert, E. H. 1955. *Evolution of the Verbebrates.* First edition. Wiley, New York. [6]

Colbert, E. H. 1980. Evolution of the Vertebrates. Third edition. Wiley, New York. [6]

Coley, P. D., J. P. Bryant, and F. S. Chapin III. 1985. Resource availability and plant antiherbivore defense. *Science* 230: 895–899. [18]

Collins, J. 1959. Darwin's impact on philosophy. *Thought* 34: 185–248. [1]

Colwell, R. K. 1981. Group selection is implicated in the evolution of female-biased sex ratios. *Nature* 290: 401–404. [21]

Colwell, R. K., and D. W. Winkler. 1984. A null model for null models in biogeography. In D. R. Strong, D. Simberloff, L. G. Abele, and A. B. Thistle (eds.), *Ecological Communities: Conceptual Issues and the Evidence*, pp. 344–359. Princeton University Press, Princeton, NJ. [18]

Conant, R. 1958. *A Field Guide to Reptiles and Amphibians.* Houghton Mifflin, Boston, MA. [8]

Condit, R., and B. R. Levin. 1990. The evolution of antibiotic resistance plasmids: The role of segregation, transposition and homologous recombination. *Am. Nat.* 135: 573–596. [22]

Connell, J. H. 1970. A predator-prey system in the marine intertidal region. I. *Balanus glandula* and several predatory species of *Thais. Ecol. Monogr.* 40: 49–78. [19]

Connell, J. H. 1971. On the role of natural enemies in preventing competitive exclusion in some marine animals and in rain forest trees. In P. J. den Boer and G. R. Gradwell (eds.), *Dynamics of Population*, pp. 298–310. Centre for Agricultural Publishing and Documentation, Wageningen, The Netherlands. [4]

Connell, J. H. 1983. On the prevalence and relative importance of interspecific competition: Evidence from field experiments. *Am. Nat.* 122: 661–696. [4]

Coope, G. R. 1979. Late Cenozoic fossil Coleoptera: Evolution, biogeography, and ecology. *Annu. Rev. Ecol. Syst.* 10: 249–267. [7, 16]

Cooper, J. D., R. H. Miller, and J. Patterson. 1990. *A Trip Through Time: Principles of Historical Geology.* Merrill, Columbus, OH. [6]

Cooper, N. G. (ed.). 1994. *The Human Genome Project: Deciphering the Blueprint of Heredity.* W. H. Freeman, San Francisco, CA. [26]

Cope, J. C. W., and P. W. Skelton (eds). 1985. *Evolutionary case histories from the fossil record.* Special Papers in Palaeontology 33: 1–203. [6]

Cornell, H. V. 1985. Species assemblages of cynipid gall wasps are not saturated. *Am. Nat.* 126: 565–569. [25]

Cornell, H. V. 1993. Unsaturated patterns in species assemblages: The role of regional processes in setting local species richness. In R. E. Ricklefs and D. Schluter (eds.), *Species Diversity in Ecological Communities*, pp. 243–252. University of Chicago Press, Chicago, IL. [25]

Cosmides, L., and J. Tooby. 1992. Cognitive adaptations for social exchange. In J. B. Barkow, L. Cosmides, and J. Tooby (eds.), *The Adapted Mind*, pp. 163–228. Oxford University Press, New York. [26]

Cosmides, L., J. Tooby, and J. B. Barkow. 1992. Introduction: Evolutionary psychology and conceptual integration. In J. B. Barkow, L. Cosmides, and J. Tooby (eds.), *The Adapted Mind*, pp. 3–15. Oxford University Press, New York. [26]

Costa, R., A. A. Peixoto, J. R. Thackeray, R. Dalgleish, and C. P. Kyriacou. 1991. Length polymorphism in the threonine-glycine-encoding repeat region of the *period* gene in *Drosophila. J. Mol. Evol.* 32: 238–246. [10]

Cowen, R. 1990. *History of Life.* Blackwell Scientific, London. [5]

Cowley, D. E., and W. R. Atchley. 1988. Quantitative genetics of *Drosophila melanogaster*. II. Heritabilities and genetic correlations between sexes for head and thorax traits. *Genetics* 119: 421–433. [23]

Cox, E. C., and T. C. Gibson. 1974. Selection for high mutation rates in chemostats. *Genetics* 77: 169–184. [21]

Coyne, J. A. 1974. The evolutionary origin of hybrid inviability. *Evolution* 28: 505–506. [16]

Coyne, J. A. 1976. Lack of genic similarity between two sibling species of *Drosophila* as revealed by varied techniques. *Genetics* 84: 593–607. [9]

Coyne, J. A. 1984a. Correlation between heterozygosity and rate of chromosome evolution in animals. *Am. Nat.* 123: 725–729. [16]

Coyne, J. A. 1984b. Genetic basis of male sterility in hybrids between two closely related species of *Drosophila. Proc. Natl. Acad. Sci. USA* 81: 4444–4447. [15]

Coyne, J. A. 1989. Genetics of sexual isolation between two sibling species, *Drosophila simulans* and *Drosophila mauritiana. Proc. Natl. Acad. Sci. USA* 86: 5464–5468. [15]

Coyne, J. A. 1992. Genetics and speciation. *Nature* 355: 511–515. [15]

Coyne, J. A. 1993. The genetics of an isolating mechanism between two sibling species of *Drosophila*. *Evolution* 47: 778–788. [15]

Coyne, J. A. 1994. Ernst Mayr and the origin of species. *Evolution* 48: 19–30. [15]

Coyne, J. A., and B. Charlesworth. 1986. Location of an X-linked factor causing sterility in male hybrids of *Drosophila simulans* and *D. mauritiana*. *Heredity* 57: 243–246. [15]

Coyne, J. A., and H. A. Orr. 1989a. Patterns of speciation in *Drosophila*. *Evolution* 43: 362–381. [16]

Coyne, J. A., and H. A. Orr. 1989b. Two rules of speciation. In D. Otte and J. A. Endler (eds.), *Speciation and Its Consequences*, pp. 180–207. Sinauer Associates, Sunderland, MA. [15]

Coyne, J. A., N. Barton, and M. Turelli. 1997. A critique of Wright's shifting balance theory of evolution. *Evolution* 51: 643–671. [14]

Coyne, J. A., I. A. Boussy, T. Prout, S. H. Bryant, J. S. Jones, and J. A. Moore. 1982. Long-distance migration of *Drosophila*. *Am. Nat.* 119: 589–595. [11]

Coyne, J. A., W. Meyers, A. P. Crittenden, and P. Sniegowski. 1993. The fertility effects of pericentric inversions in *Drosophila melanogaster*. *Genetics* 134: 487–496. [15]

Coyne, J. A., J. Rux, and J. R. David. 1991. Genetics of morphological differences and hybrid sterility between *Drosophila sechellia* and its relatives. *Genet. Res.* 57: 113–122. [15]

Cracraft, J. 1974. Phylogeny and evolution of the ratite birds. *Ibis* 116: 494–521. [8]

Cracraft, J. 1989. Speciation and its ontology: The empirical consequences of alternative species concepts for understanding patterns and processes of differentiation. In D. Otte and J. A. Endler (eds.), *Speciation and Its Consequences*, pp. 29–59. Sinauer Associates, Sunderland, MA. [15]

Cracraft, J. 1990. The origin of evolutionary novelties: Pattern and processes at different hierarchical levels. In M. H. Nitecki (ed.), *Evolutionary Innovations*, pp. 21–44. University of Chicago Press, Chicago, IL. [24]

Crawley, M. J. 1983. *Herbivory: The Dynamics of Animal-Plant Interactions*. University of California Press, Berkeley. [4]

Crespi, B. J. 1992. Eusociality in Australian gall thrips. *Nature* 359: 724–726. [20]

Crow, J. F. 1958. Some possibilities for measuring selection intensities in man. *Hum. Biol.* 30: 1–13. [26]

Crow, J. F. 1966. The quality of people: Human evolutionary changes. *BioScience* 16: 863-867. [26]

Crow, J. F. 1993. Mutation, mean fitness, and genetic load. *Oxford Surv. Evol. Biol.* 9: 3–42. [13]

Crow, J. F., and M. Kimura. 1965. Evolution in sexual and asexual populations. *Am. Nat.* 99: 439–450. [21]

Crow, J. F., and M. Kimura. 1970. *An Introduction to Population Genetics Theory*. Harper & Row, New York. [11]

Crow, J. F., W. R. Engels, and C. Denniston. 1990. Phase three of Wright's shifting-balance theory. *Evolution* 44: 233–243. [14]

Crumpacker, D. W., and J. S. Williams. 1973. Density, dispersion, and population structure in *Drosophila pseudoobscura*. *Ecol. Monogr.* 43: 499–538. [11]

Cruzan, M. B., and M. L. Arnold. 1993. Ecological and genetic associations in an *Iris* hybrid zone. *Evolution* 47: 1432–1445. [15]

CSIRO (Commonwealth Scientific and Industrial Research Organisation). 1991. *The Insects of Australia*. Second edition. Cornell University Press, Ithaca, NY. [5, 17]

Culver, D. C. 1982. *Cave Life: Evolution and Ecology*. Harvard University Press, Cambridge, MA. [19]

Curtsinger, J. W., P. M. Service, and T. Prout. 1994. Antagonistic pleiotropy, reversal of dominance, and genetic polymorphism. *Am. Nat.* 144: 210–228. [14]

Czelusniak, J., M. Goodman, D. Hewett-Emmett, M. L. Weiss, P. J. Venta, and R. E. Tashian. 1982. Phylogenetic origins and adaptive evolution of avian and mammalian haemoglobin genes. *Nature* 298: 297–300. [22]

Dacosta, C. P., and C. M. Jones. 1971. Cucumber beetle resistance and mite susceptibility controlled by the bitter gene in *Cucumis·sativus*. *Science* 172: 1145–1146. [18]

Daly, M., and M. Wilson. 1983. *Sex, Evolution, and Behavior*. Willard Grant Press, Boston, MA. [20, 26]

Darlington, C. D. 1939. *The Evolution of Genetic Systems*. Cambridge University Press, Cambridge. [16, 21]

Darlington, P. J. 1957. *Zoogeography: The Geographical Distribution of Animals*. Wiley, New York. [25]

Darnell, J. E. Jr. 1978. Implications of RNA-RNA splicing in evolution of eukaryotic cells. *Science* 202: 1257–1260. [22]

Darwin, C. 1859. *The Origin of Species by Means of Natural Selection, or the Preservation of Favored Races in the Struggle for Life*. Modern Library, New York. [1]

Darwin, C. 1871. *The Descent of Man, and Selection in Relation to Sex*. John Murray, London. (Reprinted 1981, Princeton University Press, Princeton, NJ.). [20, 26]

Darwin, C. 1872. *The Expression of the Emotions in Man and Animals* (1965 reprint). University of Chicago Press, Chicago, IL. [20]

Darwin, C. 1877a. *The Different Forms of Flowers on Plants of the Same Species*. John Murray, London. [21]

Darwin, C. 1877b. *The Various Contrivances by which Orchids are Fertilised by Insects*. (Reprinted 1984, University of Chicago Press, Chicago, IL.) [12]

Davidson, D. W., R. S. Inouye, and J. H. Brown. 1984. Granivory in a desert ecosystem: Experimental evidence for indirect facilitation of ants by rodents. *Ecology* 65: 1780–1786. [4]

Davies, N. B., and T. R. Halliday. 1978. Deep croaks and fighting assessment in toads *Bufo bufo*. *Nature* 275: 683–685. [20]

Davies, N. B., A. F. G. Bourke, and M. de L. Brooke. 1989. Cuckoos and parasitic ants: Interspecific brood parasitism as an evolutionary arms race. *Trends Ecol. Evol.* 4: 274–278. [18]

Davis, M. B. 1976. Pleistocene biogeography of temperate deciduous forests. *Geoscience and Man* 13: 13–26. [7]

Dawkins, R. 1986. *The Blind Watchmaker*. W. W. Norton, New York. [12]

Dawkins, R. 1989. *The Selfish Gene*. New edition. Oxford University Press, Oxford. [12, 20, 22]

Dawkins, R., and J. R. Krebs. 1979. Arms races between and within species. *Proc. R. Soc. London B* 205: 489–511. [18]

Dawley, R. M., and J. P. Bogart (eds.). 1989. *Evolution and Ecology of Unisexual Vertebrates*. New York State Museum, Albany. [20]

Dawson, P. S. 1970. Linkage and the elimination of deleterious mutant genes from experimental populations. *Genetica* 41: 147–169. [13]

Dayan, T., D. Simberloff, E. Tchernov, and Y. Yom-Tov. 1989. Inter- and intraspecific character displacement in mustelids. *Ecology* 70: 1526–1539. [18]

Dean, A. M., D. E. Dykhuizen, and D. L. Hartl. 1986. Fitness as a function of β-galactosidase activity in *Escherichia coli*. *Genet. Res.* 48: 1–8. [12]

DeBach, P. 1966. The competitive displacement and co-existence principles. *Annu. Rev. Entomol.* 11: 183–212. [4]

Deban, S. M., D. B. Wake, and G. Roth. 1997. Salamander with a ballistic tongue. *Nature* 389: 27–28. [24]

deBeer, G. R. 1958. *Embryos and Ancestors*. Clarendon Press, Oxford. [23]

DeBlase, A. F., and R. E. Martin. 1981. *A Manual of Mammalogy*. Wm. C. Brown, Dubuque, IA. [14]

DeFries, J. C., R. Plomin, and D. W. Fulkner. 1994. *Nature and Nurture During Middle Childhood.* Blackwell, Cambridge, MA. [26]

Degler, C. N. 1991. *In Search of Human Nature: The Decline and Revival of Darwinism in American Social Thought.* Oxford University Press, New York. [26]

DeMarais, B. D., T. E. Dowling, M. E. Douglas, W. L. Minckley, and P. C. Marsh. 1992. Origin of *Gila seminuda* (Teleostei: Cyprinidae) through introgressive hybridization: Implications for evolution and conservation. *Proc. Natl. Acad. Sci. USA* 89: 2747–2751. [16]

Dempster, E. R. 1955. Maintenance of genetic heterogeneity. *Cold Spring Harbor Symp. Quant. Biol.* 22: 25–31. [13]

de Nettancourt, D. 1977. *Incompatibility in Angiosperms.* Springer-Verlag, Berlin. [15, 21]

Dennett, D. C. 1995. *Darwin's Dangerous Idea: Evolution and the Meanings of Life.* Simon & Schuster, New York. [1, 24]

Denno, R. F., and M. S. McClure (eds.). 1983. *Variable Plants and Herbivores in Natural and Managed Systems.* Academic Press, New York. [4]

Denno, R. F., and G. K. Roderick. 1992. Density-related dispersal in planthoppers: Effects of interspecific crowding. *Ecology* 73: 1323–1334. [19]

DePomerai, D. 1990. *From Gene to Animal: An Introduction to the Molecular Biology of Animal Development.* Cambridge University Press, New York. [3]

de Queiroz, K., and M. J. Donoghue. 1988. Phylogenetic systematics and the species problem. *Cladistics* 4: 317–338. [15]

de Queiroz, K., and M. J. Donoghue. 1990. Phylogenetic systematics or Nelson's version of cladistics? *Cladistics* 6: 61–75. [15]

DeSalle, R. 1995. Molecular approaches to biogeographic analysis of Hawaiian Drosophilidae. In W. L. Wagner and V. A. Funk (eds.), *Hawaiian Biogeography,* pp. 72–89. Smithsonian Institution Press, Washington, DC. [8]

DeSalle, R., J. Gatesby, W. Wheeler, and D. Grimaldi. 1992. DNA sequences from a fossil termite in Oligo-Miocene amber and their phylogenetic implications. *Science* 257: 1933–1936. [5]

Desmond, A., and J. Moore. 1991. *Darwin.* Warner Books, New York. [2]

Devlin, B., M. Daniels, and K. Roeder. 1997. The heritability of IQ. *Nature* 388: 468–471. [26]

deWaal, F. B. M. 1989. Food sharing and reciprocal obligations among chimpanzees. *J. Hum. Evol.* 18: 433–459. [26]

Dhouailly, D. 1973. Dermo-epidermal interactions between birds and mammals: Differentiation of cutaneous appendages. *J. Embryol. Exp. Morphol.* 30: 587–603. [23]

Diamond, J. M. 1973. Distributional ecology of New Guinea birds. *Science* 179: 759–769. [18]

Diamond, J. M. 1975. Assembly of species communities. In M. L. Cody and J. M. Diamond (eds.), *Ecology and Evolution of Communities,* pp. 342–444. Harvard University Press, Cambridge, MA. [8]

Diamond, J. M. 1984a. Historic extinctions: A Rosetta Stone for understanding prehistoric extinctions. In P. S. Martin and R. G. Klein (eds.), *Quaternary Extinctions: A Prehistoric Revolution,* pp. 824–862. University of Arizona Press, Tucson. [25]

Diamond, J. M. 1984b. "Normal" extinctions of isolated populations. In M. Nitecki (ed.), *Extinctions,* pp. 191–246. University of Chicago Press, Chicago, IL. [25]

Diamond, J., and T. J. Case (eds.). 1986. *Community Ecology.* Harper & Row, New York. [25]

Diehl, S. R., and G. L. Bush. 1989. The role of habitat preference in adaptation and speciation. In D. Otte and J. A. Endler (eds.), *Speciation and Its Consequences,* pp. 345–365. Sinauer Associates, Sunderland, MA. [16]

Diver, C. 1929. Fossil records of Mendelian mutants. *Nature* 124: 183. [13]

Dobson, A. P. 1988. The population biology of parasite-induced changes in host behavior. *Q. Rev. Biol.* 63: 139–165. [18]

Dobzhansky, Th. 1927. Studies on manifold effects of certain genes in *Drosophila melanogaster. Zeitschr. Indukt Abstammungs-und Vererbungslehre* 43: 330–338. [24]

Dobzhansky, Th. 1935. A critique of the species concept in biology. *Phil. Sci.* 2: 344–355. [15]

Dobzhansky, Th. 1936. Studies on hybrid sterility. II. Localization of sterility factors in *Drosophila pseudoobscura* hybrids. *Genetics* 21: 113–135. [16]

Dobzhansky, Th. 1937. *Genetics and the Origin of Species.* Columbia University Press, New York. [2, 9]

Dobzhansky, Th. 1943. Genetics of natural populations. IX. Temporal changes in the composition of populations of *Drosophila pseudoobscura. Genetics* 28: 162–186. [12]

Dobzhansky, Th. 1948. Genetics of natural populations. XVIII. Experiments on chromosomes of *Drosophila pseudoobscura* from different geographic regions. *Genetics* 33: 588–602. [12]

Dobzhansky, Th. 1951. *Genetics and the Origin of Species.* Third edition. Columbia University Press, New York. [15]

Dobzhansky, Th. 1955. A review of some fundamental concepts and problems of population genetics. *Cold Spring Harbor Symp. Quant. Biol.* 20: 1–15. [14]

Dobzhansky, Th. 1956. What is an adaptive trait? *Am. Nat.* 90: 337–347. [14]

Dobzhansky, Th. 1962. *Mankind Evolving: The Evolution of the Human Species.* Yale University Press, New Haven, CT. [26]

Dobzhansky, Th. 1970. *Genetics of the Evolutionary Process.* Columbia University Press, New York. [13]

Dobzhansky, Th., and H. Levene. 1948. Genetics of natural populations. XVII. Proof of operation of natural selection in wild populations of *Drosophila pseudoobscura. Genetics* 33: 537–547. [13]

Dobzhansky, Th., and B. Spassky. 1967. Effects of selection and migration on geotactic and phototactic behavior of *Drosophila.* I. *Proc. R. Soc. Lond.* B 168: 27–47. [9]

Dobzhansky, Th., and B. Spassky. 1969. Artificial and natural selection for two behavioral traits in *Drosophila pseudoobscura. Proc. Natl. Acad. Sci. USA* 62: 75–80. [9]

Dobzhansky, Th., and S. Wright. 1943. Genetics of natural populations. X. Dispersion rates in *Drosophila pseudoobscura. Genetics* 28: 304–340. [11]

Dominey, W. J. 1984. Effects of sexual selection and life history on speciation: Species flocks in African cichlids and Hawaiian *Drosophila.* In A. A. Echelle and I. Kornfield (eds.), *Evolution of Fish Species Flocks,* pp. 231–249. University of Maine at Orono Press, Orono. [16]

Donoghue, M. J. 1992. Homology. In E. F. Keller and E. A. Lloyd (eds.), *Keywords in Evolutionary Biology,* pp. 170–179. Harvard University Press, Cambridge, MA. [23]

Donoghue, M. J. 1994. Progress and prospects in reconstructing plant phylogeny. *Ann. Missouri Bot. Gard.* 81: 405–418. [7]

Donoghue, M. J., J. A. Doyle, J. Gauthier, and A. G. Kluge. 1989. The importance of fossils in phylogeny reconstruction. *Annu. Rev. Ecol. Syst.* 20: 431–460. [6]

Donovan, S. K. (ed.). 1989. *Mass Extinctions: Processes and Evidence.* Columbia University Press, New York. [25]

Doolittle, W. F. 1978. Genes in pieces: Were they ever together? *Nature* 272: 581–582. [22]

Doolittle, W. F. 1981. Similar amino acid sequences: Chance or common ancestry? *Science* 214: 149–159. [22]

Doolittle, W. F., and C. Sapienza. 1980. Selfish genes, the pheno-typic paradigm and genomic evolution. *Nature* 284: 601–603. [22]

Douglas, A. E. 1994. *Symbiotic Interactions.* Oxford University Press, New York. [4, 18]

Douglas, M. E., and J. C. Avise. 1982. Speciation rates and mor-phological divergence in fishes: Tests of gradual versus rec-tangular modes of evolutionary change. *Evolution* 36: 224–232. [16]

Dover, G. A. 1980. Ignorant DNA? *Nature* 285: 618–620. [22]

Dover, G. A. 1982. Molecular drive: A cohesive mode of species evolution. *Nature* 299: 111–117. [22]

Dover, G. A. 1987. DNA turnover and the molecular clock. *J. Mol. Evol.* 26: 47–58. [22]

Dover, G. A., S. Brown, E. Coen, J. Dallas, T. Strachan, and M. Trick. 1982. The dynamics of genome evolution and spe-cies differentiation. In G. A. Dover and R. B. Flavell (eds.), *Genome Evolution*, pp. 343–372. Academic Press, New York. [22]

Doyle, J. A., and M. J. Donoghue. 1986. Seed plant phylogeny and the origin of angiosperms: An experimental cladistic analysis. *Bot. Rev.* 52: 321–431. [7, 25]

Drake, J. W. 1974. The role of mutation in microbial evolution. In M. J. Carlile and J. J. Skehel (eds.), *Evolution in the Mi-crobial World*, pp. 41–58. Cambridge University Press, Cambridge. [10]

Dressler, R. L. 1990. *The Orchids: Natural History and Classifica-tion.* Harvard University Press, Cambridge, MA. [12]

DuBose, R. F., and D. L. Hartl. 1991. Evolutionary and struc-tural constraints in the alkaline phosphatase of *Es-cherichia coli*. In R. K. Selander, A. G. Clark, and T. S. Whittam (eds.), *Evolution at the Molecular Level*, pp. 58–76. Sinauer Associates, Sunderland, MA. [22]

Durham, W. 1991. *Coevolution: Genes, Culture, and Human Di-versity.* Stanford University Press, Stanford, CA. [26]

Dykhuizen, D. E. 1990. Experimental studies of natural selec-tion in bacteria. *Annu. Rev. Ecol. Syst.* 21: 393–398. [10]

Dykhuizen, D. E., and L. Green. 1991. Recombination in *Es-cherichia coli* and the definition of biological species. *J. Bacteriol.* 173: 7257–7268. [15, 21]

Dykhuizen, D. E., and D. L. Hartl. 1980. Selective neutrality of 6PGD allozymes in *E. coli* and the effects of genetic back-ground. *Genetics* 96: 801–817. [22]

Eanes, W. F. 1994. Patterns of polymorphism and between spe-cies divergence in the enzymes of central metabolism. In B. Golding (ed.), *Non-neutral Evolution: Theories and Mol-ecular Data*, pp. 18–28. Chapman & Hall, New York. [22]

Eanes, W. F., M. Kirchner, J. Yoon, C. H. Biermann, I.-N. Wang, M. A. McCartney, and B. C. Verrelli. 1996. Historical selec-tion, amino acid polymorphism and lineage-specific diver-gence at the *G6PD* locus in *Drosophila melanogaster* and *D. simulans*. *Genetics* 144: 1027–1041. [22]

Eanes, W. F., C. Wesley, and B. Charlesworth. 1992. Accumula-tion of P elements in minority inversions in natural popu-lations of *Drosophila melanogaster*. *Genet. Res.* 59: 1–9. [22]

Easteal, S. 1990. The pattern of mammalian evolution and the relative rate of molecular evolution. *Genetics* 124: 165–171. [11]

Easteal, S. 1991. The relative rate of DNA evolution in primates. *Mol. Biol. Evol.* 8: 115–127. [11]

Easteal, S., and C. Collet. 1994. Consistent variation in amino-acid substitution rate, despite uniformity of mutation rate: Protein evolution in mammals is not neutral. *Mol. Biol. Evol.* 11: 643–647. [22]

Eastman, J. J. 1993. *Antarctic Fish Biology: Evolution in a Unique Environment.* Academic Press, San Diego, CA. [17]

Eberhard, W. G. 1986. *Sexual Selection and Animal Genitalia.* Harvard University Press, Cambridge, MA. [15, 20]

Ebert, D. 1994. Virulence and local adaptation of a horizontally transmitted parasite. *Science* 265: 1084–1086. [18]

Echelle, A. A., and I. Kornfield (eds.). 1984. *Evolution of Fish Species Flocks.* University of Maine at Orono Press, Orono. [5, 25]

Edwards, A., and L. Cavalli-Sforza. 1964. Reconstruction of evo-lutionary trees. In V. Heywood and J. McNeill (eds.), *Phe-netic and Phylogenetic Classification*, pp. 67–76. Publication no. 6. Systematics Association, New York. [5]

Ehrlich, P. R., and P. H. Raven. 1964. Butterflies and plants: A study in coevolution. *Evolution* 18: 586–608. [7]

Ehrlich, P. R., and P. H. Raven. 1969. Differentiation of popula-tions. *Science* 165: 1228–1232. [13]

Ehrlich, P. R., and J. Roughgarden. 1987. *The Science of Ecology.* MacMillan, New York. [4]

Ehrman, L. 1964. Genetic divergence in M. Vetukhiv's experi-mental populations of *Drosophila pseudoobscura*. I. Rudi-ments of sexual isolation. *Genet. Res.* 5: 150–157. [16]

Ehrman, L. 1965. Direct observation of sexual isolation be-tween allopatric and between sympatric strains of the different *Drosophila paulistorum* races. *Evolution* 19: 459–464. [15]

Ehrman, L., and P. A. Parsons. 1976. *The Genetics of Behavior.* Sinauer Associates, Sunderland, MA. [20]

Ehrman, L., and P. A. Parsons. 1981. *Behavior Genetics and Evo-lution.* McGraw-Hill, New York. [10]

Eicher, D. L. 1976. *Geologic Time.* Prentice-Hall, Englewood Cliffs, NJ. [6]

Eigen, M. 1971. Self-organization of matter and the evolution of biological macromolecules. *Naturwissenschaften* 58: 465–523. [7]

Eldredge, N. 1971. The allopatric model and phylogeny in Pale-ozoic invertebrates. *Evolution* 25: 156–167. [6]

Eldredge, N., and S. J. Gould. 1972. Punctuated equilibrium: An alternative to phyletic gradualism. In T. J. M. Schopf (ed.), *Models in Paleobiology*, pp. 82–115. Freeman, Cooper and Co., San Francisco. [6, 24]

Eltringham, H. 1916. On specific and mimetic relationships in the genus *Heliconius*. *Trans. Ent. Soc. Lond.* 1916: 101–155. [5]

Elliott, D. K. (ed.). 1986. *Dynamics of Extinctions.* Wiley, New York. [25]

Ellstrand, N., and J. Antonovics. 1984. Experimental studies of the evolutionary significance of sexual reproduction. I. A test of the frequency-dependent selection hypothesis. *Evo-lution* 38: 103–115. [13]

Emlen, S. T. 1991. Evolution of cooperative breeding in birds and mammals. In J. R. Krebs and N. B. Davies (eds.), *Be-havioural Ecology*, Third edition, pp. 301–337. Blackwell Scientific, Oxford. [20]

Emlen, S. T., and L. W. Oring. 1977. Ecology, sexual selection, and the evolution of mating systems. *Science* 197: 215–223. [20]

Emmons, L. H. 1990. *Neotropical Rain Forest Mammals: A Field Guide.* University of Chicago Press, Chicago, IL. [8]

Endler, J. A. 1973. Gene flow and population differentiation. *Science* 179: 243–250. [13]

Endler, J. A. 1977. *Geographic Variation, Speciation, and Clines.* Princeton University Press, Princeton, NJ. [15]

Endler, J. A. 1980. Natural selection on color patterns in *Poecilia reticulata*. *Evolution* 34: 76–91. [12]

Endler, J. A. 1982. Pleistocene forest refuges: Fact or fancy? In G. T. Prance (ed.), *Biological Diversification in the Tropics*, pp. 641–657. Columbia University Press, New York. [8]

Endler, J. A. 1983a. Natural and sexual selection on color patterns in poeciliid fishes. *Envir. Biol. Fishes* 9: 173–190. [20]

Endler, J. A. 1983b. Testing causal hypotheses in the study of geographic variation. In J. Felsenstein (ed.), *Numerical Taxonomy*, pp. 424–443. Springer-Verlag, Berlin. [8]

Endler, J. A. 1986. *Natural Selection in the Wild*. Princeton University Press, Princeton, NJ. [12]

Endress, P. K. 1994. *Diversity and Evolutionary Biology of Tropical Flowers*. Cambridge University Press, Cambridge. [11]

Engels, W. R. 1983. The P family of transposable elements in *Drosophila. Annu. Rev. Genet.* 17: 319–344. [10, 22]

Erwin, D. H. 1991. Metazoan phylogeny and the Cambrian radiation. *Trends Ecol. Evol.* 6: 131–134. [7, 25]

Erwin, D. H. 1993a. *The Great Paleozoic Crisis: Life and Death in the Early Permian*. Columbia University Press, New York. [7, 20]

Erwin, D. H. 1993b. The origin of metazoan development: A paleobiological perspective. *Biol. J. Linn. Soc.* 50: 255–274. [25]

Erwin, D. H., J. W. Valentine, and D. Jablonski. 1997. The origin of animal body plans. *Am. Sci.* 85: 126-137. [7]

Erwin, D. H., J. W. Valentine, and J. J. Sepkoski, Jr. 1987. A comparative study of diversification events: The early Paleozoic versus the Mesozoic. *Evolution* 41: 1177–1186. [25]

Ewald, P. W. 1994. *Evolution of Infectious Disease*. Oxford University Press, Oxford. [18]

Faegri, K., and L. van der Pijl. 1971. *The Principles of Pollination Ecology*. Pergamon, New York. [11]

Falconer, D. S. 1981. *Introduction to Quantitative Genetics*. Longman, London. [14]

Farrell, B. D., and C. Mitter. 1993. Phylogenetic determinants of insect/plant community diversity. In R. E. Ricklefs and D. Schluter (eds.), *Species Diversity in Ecological Communities*, pp. 253–266. University of Chicago Press, Chicago, IL. [8]

Farrell, B., D. Dussourd, and C. Mitter. 1991. Escalation of plant defenses: Do latex and resin canals spur plant diversification? *Am. Nat.* 138: 881–900. [18, 25]

Farris, J. 1973. On the use of the parsimony criterion for inferring evolutionary trees. *Syst. Zool.* 22: 250–256. [5]

Feder, J. L., C. A. Chilcote, and G. L. Bush. 1990. Geographic pattern of genetic differentiation between host-associated populations of *Rhagoletis pomonella* (Diptera: Tephritidae) in the eastern United States and Canada. *Evolution* 44: 570–594. [16]

Feder, J. L., T. A. Hunt, and G. L. Bush. 1993. The effects of climate, host plant phenology and host fidelity on the genetics of apple and hawthorn infesting races of *Rhagoletis pomonella. Entomol. Exp. Appl.* 69: 117–135. [16]

Feder, M. E. 1978. Environmental variability and thermal acclimation of metabolism in neotropical and temperate zone salamanders. *Physiol. Zool.* 51: 7–16. [17]

Feder, M. E., A. F. Bennett, W. W. Burggren, and R. B. Huey (eds.). 1987. *New Directions in Ecological Physiology*. Cambridge University Press, Cambridge. [17]

Fedigan, L. M. 1986. The changing role of women in models of human evolution. *Annu. Rev. Anthropol.* 15: 25–66. [26]

Feduccia, A. 1996. *The Origin and Evolution of Birds*. Yale University Press, New Haven, CT. [6]

Feeny, P. P. 1976. Plant apparency and chemical defense. In J. W. Wallace and R. L. Mansell (eds.), *Biochemical Interaction between Plants and Insects*, pp. 1–40. Plenum, New York. [18]

Feeny, P. P. 1992. The evolution of chemical ecology: Contributions from the study of herbivorous insects. In G. A. Rosenthal and M. R. Berenbaum (eds.), *Herbivores: Their Interactions with Secondary Plant Metabolites*, Second edition, pp. 1–44. Academic Press, San Diego, CA. [12]

Feldman, M. W., and R. C. Lewontin. 1975. The heritability hang-up. *Science* 190: 1163–1168. [26]

Feldman, M. W., F. B. Christiansen, and L. D. Brooks. 1980. Evolution of recombination in a constant environment. *Proc. Natl. Acad. Sci. USA* 77: 4838–4841. [21]

Felsenstein, J. 1965. The effect of linkage on directional selection. *Genetics* 52: 349–363. [14]

Felsenstein, J. 1976. The theoretical population genetics of variable selection and migration. *Annu. Rev. Genet.* 10: 253–280. [13]

Felsenstein, J. 1979. Excursions along the interface between disruptive and stabilizing selection. *Genetics* 93: 773–795. [14]

Felsenstein, J. 1981. Skepticism towards Santa Rosalia, or why are there so few kinds of animals? *Evolution* 35: 124–138. [16]

Felsenstein, J. 1983. Parsimony in systematics: Biological and statistical issues. *Annu. Rev. Ecol. Syst.* 14: 313–333. [5]

Felsenstein, J. 1985. Phylogenies and the comparative method. *Am. Nat.* 125: 1–15. [12]

Felt, F. E. 1917. Key to North American insect galls. Bulletin of the New York State Museum, Albany, NY, pp. 5–310. [4]

Fenner, F., and F. N. Ratcliffe. 1965. *Myxomatosis*. Cambridge University Press, Cambridge. [18]

Ferguson, G. W., and S. F. Fox. 1984. Annual variation of survival advantage of large juvenile side-blotched lizards (*Uta stansburiana*): Its causes and evolutionary significance. *Evolution* 38: 342–349. [19]

Ferris, S. D., and G. S. Whitt. 1979. Evolution of the differential regulation of duplicate genes after polyploidization. *J. Mol. Evol.* 12: 256–317. [16]

ffrench-Constant, R. H., R. T. Roush, D. Mortlock, and G. P. Dively. 1990. Isolation of dieldrin resistance from field populations of *Drosophila melanogaster* (Diptera: Drosophilidae). *J. Econ. Entomol.* 83: 1733–1737. [10]

Finnegan, D. J., and D. H. Fawcett. 1986. Transposable elements in *Drosophila melanogaster. Oxford Surv. Eukaryotic Genes* 3: 1–62. [22]

Fischer, A. G. 1960. Latitudinal variation in organic diversity. *Evolution* 14: 64–81. [25]

Fischer, C. S., M. Hout, M. S. Jankowski, S. R. Lucas, A. Swidler, and K. Voss. 1996. *Inequality by Design: Cracking the Bell Curve Myth*. Princeton University Press, Princeton, NJ. [26]

Fisher, C. E., J. L. Fults, and H. Hopp. 1946. Factors affecting the action of oils and water-soluble chemicals in mesquite eradication. *Ecol. Monogr.* 16: 109–126. [17]

Fisher, D. C. 1986. Progress in organismal design. In D. M. Raup and D. Jablonski (eds.), *Patterns and Processes in the History of Life*, pp. 99–117. Springer-Verlag, Berlin. [24]

Fisher, R. A. 1930. *The genetical theory of natural selection*. Clarendon Press, Oxford. [10, 20]

Fisher, R. A. 1941. Average excess and average effect of a gene substitution. *Ann. Eugenics* 11: 53–63. [21]

Fisher, R. A., and R. A. Hinde. 1948. The opening of milk bottles by birds. *British Birds* 42: 347–357. [20]

Fitch, J. E., and R. J. Lavenberg. 1968. *Deep Water Fishes of California*. University of California Press, Berkeley. [4]

Fitch, W. M., and E. Margoliash. 1970. The usefulness of amino acid and nucleotide sequences in evolutionary studies. *Evol. Biol.* 4: 67–109. [5]

Fitter, A. H., and R. K. M. Hay. 1981. *Environmental Physiology of Plants*. Academic Press, London. [17]

Fitzpatrick, J. W. 1988. Why so many passerine birds? A response to Raikow. *Syst. Zool.* 37: 71–76. [16]

Flavell, R. B. 1982. Sequence amplification, deletion, and rearrangement: Major sources of variation during species divergence. In G. A. Dover and R. B. Flavell (eds.), *Genome Evolution*, pp. 301–323. Academic Press, New York. [22]

Fleagle, J. G. 1988. *Primate Adaptation and Evolution*. Academic Press, San Diego, CA. [26]

Flessa, K. W., and D. Jablonski. 1985. Declining Phanerozoic background extinction rates: Effect of taxonomic structure? *Nature* 313: 216–218. [25]

Flessa, K. W. et al. 1986. Causes and consequences of extinction. In D. M. Raup and D. Jablonski (eds.), *Patterns and Processes in the History of Life*, pp. 235–257. Springer-Verlag, Berlin. [24]

Flexon, P. B., and C. F. Rodell. 1982. Genetic recombination and directional selection for DDT resistance in *Drosophila melanogaster. Nature* 298: 672–674. [21]

Flor, H. H. 1956. The complementary genic systems in flax and flax rust. *Adv. Genet.* 8: 29–54. [18]

Fong, D. W. 1989. Morphological evolution of the amphipod *Gammarus minus* in caves: Quantitative genetic analysis. *Am. Midl. Nat.* 121: 361–378. [14]

Fong, D. W., T. C. Kane, and J. C. Culver. 1995. Vestigialization and loss of nonfunctional characters. *Annu. Rev. Ecol. Syst.* 26: 249–268. [14]

Fontdevila, A. 1992. Genetic instability and rapid speciation: Are they coupled? *Genetica* 86: 247–258. [15]

Ford, E. B. 1971. *Ecological Genetics*. Chapman & Hall, London. [9]

Fortey, R. A., D. E. G. Briggs, and M. A. Wills. 1996. The Cambrian evolutionary "explosion": Decoupling cladogenesis from morphological disparity. *Biol. J. Linn. Soc.* 57: 13–33. [7]

Fowler, K., and L. Partridge. 1989. A cost of mating in female fruitflies. *Nature* 338: 760–761. [19]

Fowler, N. L., and D. A. Levin. 1984. Ecological constraints on the establishment of a novel polyploid in competition with its diploid progenitor. *Am. Nat.* 124: 703–711. [16]

Frank, S. A. 1986. Hierarchical selection theory and sex ratios. I. General solutions for structured populations. *Theor. Pop. Biol.* 29: 312–342. [21]

Frank, S. A. 1989. The evolutionary dynamics of cytoplasmic male-sterility. *Am. Nat.* 133: 345–376. [21]

Frank, S. A. 1990. Sex allocation theory for birds and mammals. *Annu. Rev. Ecol. Syst.* 21: 13–55. [21]

Frank, S. A. 1992. Models of plant-pathogen coevolution. *Trends Genet.* 8: 213–219. [18]

Fraser, S. (ed.). 1995. *The Bell Curve Wars: Race, Intelligence, and the Future of America*. BasicBooks, New York. [26]

Freed, L. A., S. Conant, and R. C. Fleischer. 1987. Evolutionary ecology and radiation of Hawaiian passerine birds. *Trends Ecol. Evol.* 2: 196–203. [5]

Fretwell, S. D. 1972. *Seasonal Environments*. Princeton University Press, Princeton, NJ. [20]

Fritz, R. S., and E. L. Simms (eds.). 1992. *Plant Resistance to Herbivores and Pathogens: Ecology, Evolution, and Genetics*. University of Chicago Press, Chicago, IL. [18]

Frost, S. W. 1959. *Insect Life and Insect Natural History*. Dover Books, New York. [4]

Fryer, G. 1959. Some aspects of evolution in Lake Nyasa. *Evolution* 13: 440–451. [5]

Fryer, G., and T. D. Iles. 1972. *The Cichlid Fishes of the Great Lakes of Africa*. T. F. H. Publications, Neptune City, NJ. [5, 25]

Fu, Y.-X., and W.-H. Li. 1993. Statistical tests of neutrality of mutations. *Genetics* 133: 693–700. [22]

Fürsich, F. T., and D. Jablonski. 1984. Lake Triassic naticid drillholes: Carnivorous gastropods gain a major adaptation but fail to radiate. *Science* 224: 78–80. [25]

Futuyma, D. J. 1983. Selective factors in the evolution of host choice by phytophagous insects. In S. Ahmad (ed.), *Herbivorous Insects: Host-Seeking Behavior and Mechanisms*, pp. 227–244. Academic Press, New York. [14]

Futuyma, D. J. 1987. On the role of species in anagenesis. *Am. Nat.* 130: 465–473. [16]

Futuyma, D. J. 1989. Macroevolutionary consequences of speciation. In D. Otte and J. A. Endler (eds.), *Speciation and Its Consequences*, pp. 557–578. Sinauer Associates, Sunderland, MA. [16]

Futuyma, D. J. 1990. Observations on the taxonomy and natural history of *Ophraella* Wilcox (Coleoptera: Chrysomelidae), with a description of a new species. *J. New York Entomol. Soc.* 98: 163–186. [5]

Futuyma, D. J. 1991. A new species of *Ophraella* Wilcox (Coleoptera: Chrysomelidae) from the southeastern United States. *J. New York Entomol. Soc.* 99: 643–653. [15]

Futuyma, D. J. 1995. *Science on Trial: The Case for Evolution*. Sinauer Associates, Sunderland, MA. [1, 6, 7]

Futuyma, D. J., and M. C. Keese. 1992. Evolution and coevolution of plants and phytophagous arthropods. In G. A. Rosenthal and M. A. Berenbaum (eds.), *Herbivores: Their Interactions with Secondary Plant Metabolites*, vol. 2, pp. 439–475. Academic Press, New York. [18]

Futuyma, D. J., and G. C. Mayer. 1980. Non-allopatric speciation in animals. *Syst. Zool.* 29: 254–271. [16]

Futuyma, D. J., and G. Moreno. 1988. The evolution of ecological specialization. *Annu. Rev. Ecol. Syst.* 19: 207–233. [17, 20, 24]

Futuyma, D. J., and S. C. Peterson. 1985. Genetic variation in the use of resources by insects. *Annu. Rev. Entomol.* 30: 217–238. [13]

Futuyma, D. J., and M. Slatkin (eds.). 1983. *Coevolution*. Sinauer Associates, Sunderland, MA. [18]

Futuyma, D. J., M. C. Keese, and D. J. Funk. 1995. Genetic constraints on macroevolution: The evolution of host affiliation in the leaf beetle genus *Ophraella. Evolution* 49: 797–809. [14, 18]

Gadgil, M., and W. H. Bossert. 1970. Life historical consequences of natural selection. *Am. Nat.* 104: 1–24. [19]

Gagné, R. J. 1989. *The Plant-Feeding Gall Midges of North America*. Cornell University Press, Ithaca, NY. [4]

Galiana, A., F. González-Canelas, and A. Moya. 1996. Postmating isolation analysis in founder-flush experimental populations of *Drosophila pseudoobscura. Evolution* 50: 941–944. [16]

Galiana, A., A. Moya, and F. J. Ayala. 1993. Founder-flush speciation in *Drosophila pseudoobscura*: A large-scale experiment. *Evolution* 47: 432–444. [16]

Gardner, H. 1994. Cracking open the IQ box. In S. Fraser (ed.), *The Bell Curve Wars*, pp. 23–35. BasicBooks, New York. [26]

Garland, T. Jr., and P. A. Carter. 1994. Evolutionary physiology. *Annu. Rev. Physiol.* 56: 579–621. [17]

Garland, T. Jr., and R. B. Huey. 1987. Testing symmorphosis: Does structure match functional requirements? *Evolution* 41: 1404–1409. [17]

Garstang, W. 1922. The theory of recapitulation: A critical restatement of the biogenetic law. *J. Linn. Soc. Zool.* 35: 81–101. [23]

Gavrilets, S., and A. Hastings. 1996. Founder-effect speciation: A theoretical reassessment. *Am. Nat.* 147: 466–491. [16]

Gee, H. 1990. Fossil fishes and fashion. *Nature* 348: 194–195. [6]

Georghiou, G. P. 1972. The evolution of resistance to pesticides. *Annu. Rev. Ecol. Syst.* 3: 133–168. [13]

Ghiselin, M. T. 1969. *The Triumph of the Darwinian Method*. University of California Press, Berkeley. [12]

Ghiselin, M. T. 1974. *The Economy of Nature and the Evolution of Sex*. University of California Press, Berkeley. [20]

Literature Cited

Ghiselin, M. T. 1995. Perspective: Darwin, progress, and economic principles. *Evolution* 49: 1029–1037. [24]

Gibson, R. M., and J. W. Bradbury. 1985. Sexual selection in lekking Sage Grouse: Phenotypic correlates of male mating success. *Behav. Ecol. Sociobiol.* 18: 117–123. [16]

Gilbert, L. E. 1983. Coevolution and mimicry. In D. J. Futuyma and M. Slatkin (eds.), *Coevolution*, pp. 263–281. Sinauer Associates, Sunderland, MA. [18]

Gilbert, S. F. 1997. *Developmental Biology*. Fifth edition. Sinauer Associates, Sunderland, MA. [3, 23]

Gilbert, W. 1978. Why genes in pieces? *Nature* 271: 501. [22]

Gilchrist, G. W. 1996. A quantitative genetic analysis of thermal sensitivity of the locomotor performance curve of *Aphidius ervi*. *Evolution* 50: 1560–1572. [17]

Gilinsky, N. L. 1988. Survivorship in the Bivalvia: Comparing living and extinct genera and families. *Paleobiology* 14: 370-386. [24]

Gilinsky, N. L., and J. J. Good. 1991. Probabilities of origination, persistence, and extinction of families of marine invertebrate life. *Paleobiology* 17: 145–166. [25]

Gill, D. E. 1989. Fruiting failure, pollinator inefficiency, and speciation in orchids. In D. Otte and J. A. Endler (eds.), *Speciation and Its Consequences*, pp. 458–481. Sinauer Associates, Sunderland, MA. [4]

Gill, F. B. 1995. *Ornithology*. Second edition. W. H. Freeman, New York. [12]

Gilles, A., and L. F. Randolph. 1951. Reduction in quadrivalent frequency in autotetraploid maize during a period of 10 years. *Am. J. Bot.* 38: 12–16. [16]

Gillespie, J. H. 1984. Pleiotropic overdominance and the maintenance of genetic variation in polygenic characters. *Genetics* 107: 321–330. [14]

Gillespie, J. H. 1991. *The Causes of Molecular Evolution*. Oxford University Press, New York. [11]

Gillette, D. D., and C. E. Ray. 1981. Glyptodonts of North America. *Smithsonian Contrib. Paleobiol.* 40: 1–255. [7]

Gimelfarb, A. 1989. Genotypic variation for a quantitative character maintained under stabilizing selection without mutations: Epistasis. *Genetics* 123: 217–227. [14]

Gingerich, P. D. 1983. Rates of evolution: Effects of time and temporal scaling. *Science* 222: 159–161. [6, 24]

Gingerich, P. D. 1990. Stratophenetics. In D. E. G. Briggs and P. R. Crowther (eds.), *Palaeobiology: A Synthesis*, pp. 437–442. Blackwell Scientific, Oxford. [6]

Gingerich, P. D., S. M. Raza, M. Arif, M. Anwar, and X. Zhou. 1994. New whale from the Eocene of Pakistan and the origin of cetacean swimming. *Nature* 368: 844–847. [7]

Gish, D. T. 1974. *Evolution: The Fossils Say No!* Creation-Life Publishers, San Diego, CA. [6]

Gleason, H. A. 1952. *The New Britton and Brown Illustrated Flora of the Northeastern United States and Adjacent Canada*. New York Botanical Garden, New York. [12, 24]

Glesener, R. R., and D. Tilman. 1978. Sexuality and the components of environmental uncertainty: Clues from geographic parthenogenesis in terrestrial animals. *Am. Nat.* 112: 659–673. [17, 21]

Glover, P. M., and B. P. Hames (eds.). 1989. *Genes and Embryos*. Oxford University Press, Oxford. [3]

Goddard, H. H. 1920. *Human Efficiency and Levels of Intelligence*. Princeton University Press, Princeton, NJ. [26]

Godfray, H. C. J. 1994. *Parasitoids: Behavioral and Evolutionary Ecology*. Princeton University Press, Princeton, NJ. [11]

Gold, J. R., C. J. Ragland, and J. B. Woolley. 1992. Evolution of genome size in North American fishes. In R. L. Mayden (ed.), *Systematics, Historical Ecology, and North American freshwater fishes*, pp. 534–550. Stanford University Press, Stanford, CA. [22]

Goldblatt, P. 1979. Polyploidy in angiosperms: Monocotyledons. In W. H. Lewis (ed.), *Polyploidy: Biological Relevance*, pp. 219–239. Plenum, New York. [16]

Goldblatt, P. (ed.). 1993. *Biological Relationships Between Africa and South America*. Yale University Press, New Haven, CT. [8]

Golding, B. (ed.). 1994. *Non-neutral Evolution: Theories and Molecular Data*. Chapman & Hall, New York. [22]

Goldschmidt, R. B. 1938. *Physiological Genetics*. McGraw-Hill, New York. [24]

Goldschmidt, R. B. 1940. *The Material Basis of Evolution*. Yale University Press, New Haven, CT. [24]

Goldstein, D. B., and K. E. Holsinger. 1992. Maintenance of polygenic variation in spatially structured populations: Roles for local mating and genetic redundancy. *Evolution* 46: 412–429. [14]

Goldstein, D. B., A. R. Linares, L. L. Cavalli-Sforza, and M. W. Feldman. 1995. Genetic absolute dating based on microsatellites and the origin of modern humans. *Proc. Natl. Acad. Sci. USA* 92: 6723–6727. [26]

Golenberg, E. M., and E. Nevo. 1987. Multilocus differentiation and population structure in a selfer, wild emmer wheat, *Triticum dicoccoides*. *Heredity* 58: 451–456. [11]

Golenberg, E. M., D. E. Giannasi, M. T. Clegg, C. J. Smiley, M. Durbin, D. Henderson, and G. Zurawski. 1990. Chloroplast DNA sequences from a Miocene *Magnolia* species. *Nature* 344: 656–658. [5]

Goodall, J. 1986. *The Chimpanzees of Gombe*. Harvard University Press, Cambridge, MA. [26]

Goodman, M. 1963. Man's place in the phylogeny of primates as reflected in serum proteins. In S. L. Washburn (ed.). *Classification and Human Evolution*, pp. 204–234. Aldine, Chicago, IL. [5]

Goodman, M., J. Barnabas, G. Matsuda, and G. W. Moor. 1971. Molecular evolution in the descent of man. *Nature* 233: 604–613. [22]

Goodman, M., A. E. Romero-Herrera, H. Dene, J. Czelusniak, and R. E. Tashian. 1982a. Amino acid sequence evidence on the phylogeny of primates and other eutherians. In M. Goodman (ed.), *Macromolecular Sequences in Systematic and Evolutionary Biology*, pp. 115–191. Plenum, New York. [5, 22]

Goodman, M., M. L. Weiss, and J. Czelusniak. 1982b. Molecular evolution above the species level: Branching patterns, rates, and mechanisms. *Syst. Zool.* 31: 376–399. [22]

Goodman, M., B. F. Koop, J. Czelusniak, D. H. A. Fitch, D. A. Tagle, and J. L. Slightom. 1989. Molecular phylogeny of the family of apes and humans. *Genome* 31: 316-335. [5]

Goodnight, C. J. 1988. Epistasis and the effects of founder events on the additive genetic variance. *Evolution* 42: 441–454. [14]

Goodwin, B. C., N. Holder, and C. C. Wylie (eds.). 1983. *Development and Evolution*. Cambridge University Press, Cambridge. [23]

Goodwin, D. 1986. *Crows of the World*. British Museum (Natural History), London. [15]

Goss-Custard, J. D. 1977. Predator responses and prey mortality in the redshank *Tringa totanus* (L.) and a preferred prey *Corophium volutator* (Pallas). *J. Anim. Ecol.* 46: 21–35. [20]

Gottlieb, L. D. 1974. Genetic confirmation of the origin of *Clarkia lingulata*. *Evolution* 28: 244–250. [16]

Gottlieb, L. D. 1984. Genetics and morphological evolution in plants. *Am. Nat.* 123: 681–709. [15]

Gould, J. L. and C. G. Gould. 1988. *The Honey Bee*. Scientific American Library, W. H. Freeman, New York. [1]

Gould, S. J. 1966. Allometry and size in ontogeny and phylogeny. *Biol. Rev.* 41: 587–680. [6, 23]

Gould, S. J. 1974. The origin and function of "bizarre" structures: Antler size and skull size in the "Irish elk," *Megaloceros giganteus*. *Evolution* 28: 191–220. [6, 23]

Gould, S. J. 1977. *Ontogeny and Phylogeny*. Harvard University Press, Cambridge, MA. [23]

Gould, S. J. 1980. Is a new and general theory of evolution emerging? *Paleobiology* 6: 119–130. [24]

Gould, S. J. 1981. *The Mismeasure of Man*. Norton, New York. [26]

Gould, S. J. 1982a. Change in developmental timing as a mechanism of macroevolution. In J. T. Bonner (ed.), *Evolution and Development*, pp. 333–346. Springer-Verlag, Berlin. [23]

Gould, S. J. 1982b. Darwinism and the expansion of evolutionary theory. *Science* 216: 380–387. [24]

Gould, S. J. 1982c. The meaning of punctuated equilibrium and its role in validating a hierarchical approach to macroevolution. In R. Milkman (ed.), *Perspectives on Evolution*, pp. 83–104. Sinauer, Sunderland, MA. [12]

Gould, S. J. 1985. The paradox of the first tier: An agenda for paleobiology. *Paleobiology* 11: 2–12. [25]

Gould, S. J. 1989. *Wonderful Life: The Burgess Shale and the Nature of History*. W. W. Norton, New York. [7, 23, 25]

Gould, S. J., and C. B. Calloway. 1980. Clams and brachiopods—ships that pass in the night. *Paleobiology* 6: 383–396. [25]

Gould, S. J., and N. Eldredge. 1977. Punctuated equilibria: The tempo and mode of evolution reconsidered. *Paleobiology* 3: 115–151. [6]

Gould, S. J., and N. Eldredge. 1993. Punctuated equilibrium comes of age. *Nature* 366: 223–227. [6]

Gould, S. J., and R. C. Lewontin. 1979. The spandrels of San Marco and the Panglossian paradigm. *Proc. R. Soc. Lond. B* 205: 581–598. [20]

Gould, S. J., and E. S. Vrba. 1982. Exaptation—a missing term in the science of form. *Paleobiology* 8: 4–15. [12]

Gould, S. J., D. M. Raup, J. J. Sepkoski Jr., T. J. M. Schopf, and D. S. Simberloff. 1977. The shape of evolution: A comparison of real and random clades. *Paleobiology* 3: 23–40. [25]

Gouy, M., and C. Gautier. 1982. Codon usage in bacteria: Correlation with gene expressivity. *Nucleic Acids Res.* 10: 7055–7084. [22]

Gouyon, P. H., and D. Couvet. 1987. A conflict between two sexes, females and hermaphrodites. In S. C. Stearns (ed.), *The Evolution of Sex and Its Consequences*, pp. 245–261. Birkhäuser-Verlag, Berlin. [21]

Grafen, A. 1990. Biological signals as handicaps. *J. Theor. Biol.* 144: 517–546. [20]

Grafen, A. 1991. Modelling in behavioural ecology. In J. R. Krebs and N. B. Davies (eds.), *Behavioural Ecology*, Third edition, pp. 5–31. Blackwell Scientific, Oxford. [20]

Grant, B. R. 1985. Selection on bill characters in a population of Darwin's finches: *Geospiza conirostris* on Isla Genovesa, Galápagos. *Evolution* 39: 523–532. [14]

Grant, B. R., and P. R. Grant. 1989. *Evolutionary Dynamics of a Natural Population: The Large Cactus Finch of the Galápagos*. University of Chicago Press, Chicago, IL. [14]

Grant, P. R. 1972. Convergent and divergent character displacement. *Biol. J. Linn. Soc.* 4: 39–68. [18]

Grant, P. R. 1975. The classical case of character displacement. *Evol. Biol.* 8: 237–337. [18]

Grant, P. R. 1986. *Ecology and Evolution of Darwin's Finches*. Princeton University Press, Princeton, NJ. [5, 14]

Grant, V. 1963. *The Origin of Adaptations*. Columbia University Press, New York. [24]

Grant, V. 1966a. The origin of a new species of *Gilia* in a hybridization experiment. *Genetics* 54: 1189–1199. [16]

Grant, V. 1966b. The selective origin of incompatibility barriers in the plant genus *Gilia*. *Am. Nat.* 100: 99–118. [16]

Grant, V. 1971. *Plant Speciation*. Columbia University Press, New York. [10, 15, 16]

Grant, V. 1993. Origin of floral isolation between ornithophilous and sphingophilous plant species. *Proc. Natl. Acad. Sci. USA* 90: 7729–7733. [16]

Grant, V., and K. A. Grant. 1965. *Flower Pollination in the Phlox Family*. Columbia University Press, New York. [16]

Grantham, R., C. Gautier, and M. Gouy. 1980. Codon frequencies in 119 individual genes confirm consistent choices of degenerate bases according to genome type. *Nucleic Acids Res.* 8: 1893–1912. [22]

Graur, D., and D. G. Higgins. 1994. Molecular evidence for the inclusion of cetaceans within the order Artiodactyla. *Mol. Biol. Evol.* 11: 357–364. [7]

Gray, M. W. 1989. Origin and evolution of mitochondrial DNA. *Annu. Rev. Cell. Biol.* 5: 25–50. [7]

Greene, E. 1989. A diet-induced developmental polymorphism in a caterpillar. *Science* 243: 643–646. [4]

Greenwood, P. J. 1974. The cichlid fishes of Lake Victoria, East Africa: The biology and evolution of a species flock. *Bull. Brit. Mus. Nat. Hist. (Zool.)* Suppl 6: 1–134. [16]

Greenwood, P. J., and P. H. Harvey. 1982. The natal and breeding dispersal of birds. *Annu. Rev. Ecol. Syst.* 13: 1–21. [11]

Greenwood, P. J., P. H. Harvey, and C. M. Perrins. 1978. Inbreeding and dispersal in the great tit. *Nature* 271: 52–54. [11]

Gregory, W. K. 1951. *Evolution Emerging*. MacMillan, New York. [6]

Griffiths, A. J. F., J. H. Miller, D. T. Suzuki, R. C. Lewontin, and W. M. Gelbart. 1993. *An Introduction to Genetic Analysis*. Fifth edition. W. H. Freeman, New York. [3]

Grimaldi, D. A. 1987. Phylogenetics and taxonomy of *Zygothrica* (Diptera: Drosophilidae). *Bull. Am. Mus. Nat. Hist.* 186: 103–268. [5]

Grine, F. E. 1993. Australopithecine taxonomy and phylogeny: Historical background and recent interpretations. In R. L. Ciochon and J. G. Fleagle (eds.), *The Human Evolution Source Book*, pp. 198–210. Prentice-Hall, Englewood Cliffs, NJ. [26]

Grosberg, R. K. 1988. Life-history variation within a population of the colonial ascidian *Botryllus schlosseri*. I. The genetic and environmental control of seasonal variation. *Evolution* 42: 900–920. [19]

Gross, M. 1984. Sunfish, salmon, and the evolution of alternative reproductive strategies and tactics in fishes. In G. W. Potts and R. J. Wootton (eds.), *Fish Reproduction: Strategies and Tactics*, pp. 55–75. Academic Press, London. [19]

Grotzinger, J. P., S. A. Bowring, B. Z. Saylor, and A. J. Kaufman. 1995. Biostratigraphic and geochronologic constraints on early animal evolution. *Science* 270: 598–604. [7]

Guerrant, E. O. Jr. 1988. Heterochrony in plants. The intersection of evolution and ontogeny. In M. L. McKinney (ed.), *Heterochrony in Evolution*, pp. 111–133. Plenum, New York. [23]

Gupta, A. P., and R. C. Lewontin. 1982. A study of reaction norms in natural populations of *Drosophila pseudoobscura*. *Evolution* 36: 934–948. [9]

Guttman, D. S., and D. E. Dykhuizen. 1994. Clonal divergence in *Escherichia coli* as a result of recombination, not mutation. *Science* 266: 1380–1383. [22]

Gwynne, D. T. 1990. Testing parental investment and the control of sexual selection in katydids: The operational sex ratio. *Am. Nat.* 136: 474–484. [20]

Literature Cited

Haffer, J. 1969. Speciation in Amazonian forest birds. *Science* 165: 131–137. [8]

Hairston, N. G. 1980. Evolution under interspecific competition: Field experiments in terrestrial salamanders. *Evolution* 34: 409–420. [4]

Hairston, N. G.. 1989. *Ecological Experiments: Purpose, Design, and Execution*. Cambridge University Press, Cambridge. [4, 18]

Hairston, N. G., F. E. Smith, and L. B. Slobodkin. 1960. Community structure, population control, and competition. *Am. Nat.* 94: 421–425. [4]

Haldane, J. B. S. 1922. Sex-ratio and unisexual hybrid sterility in animals. *J. Genet.* 12: 101–109. [15]

Haldane, J. B. S. 1932. *The Causes of Evolution*. Longmans, Green, New York. [13]

Haldane, J. B. S. 1949. Suggestions as to the quantitative measurement of rates of evolution. *Evolution* 3: 51–56. [6]

Haldane, J. B. S., and S. D. Jayakar. 1963. Polymorphism due to selection of varying direction. *J. Genet.* 58: 318–323. [13]

Haldeman, D. C. 1991. Sexual orientation conversion therapy for gay men and lesbians: A scientific examination. In J. C. Gonsiorek and J. D. Weinrich (eds.), *Homosexuality: Research Implications for Public Policy*, pp. 149–160. Sage, Newbury Park, CA. [26]

Halder, G., P. Callaerts, and W. J. Gehring. 1995. Induction of ectopic eyes by targeted expression of the *eyeless* gene in *Drosophila*. *Science* 267: 1788–1792. [23]

Hall, B. G. 1982. Evolution on a petri dish: The evolved β-galactosidase system as a model for studying acquisitive evolution in the laboratory. *Evol. Biol.* 15: 85–150. [10]

Hall, B. G. 1983. Evolution of new metabolic functions in laboratory organisms. In M. Nei and R. K. Koehn (eds.), *Evolution of Genes and Proteins*, pp. 234–257. Sinauer Associates, Sunderland, MA. [10]

Hall, B. G. 1988. Adaptive evolution that requires multiple spontaneous mutations. I. Mutations involving an insertion sequence. *Genetics* 120: 887–897. [10]

Hall, B. K. 1983. Epigenetic control in development and evolution. In B. C. Goodwin, N. Holder, and C. C. Wylie (eds.), *Development and Evolution*, pp. 353–379. Cambridge University Press, Cambridge. [23]

Hall, B. K. 1992. *Evolutionary Developmental Biology*. Chapman & Hall, London. [14]

Hall, J. C., and C. P. Kyriacou. 1990. Genetics of biological rhythms in *Drosophila*. *Adv. Insect Physiol.* 22: 221–298. [10]

Hallam, A. 1982. Patterns of speciation in Jurassic *Gryphaea*. *Paleobiology* 8: 354–366. [24]

Hamburgh, M. 1970. *Theories of Differentiation*. Elsevier, New York. [3]

Hamer, D. H., and P. Copeland. 1994. *The Science of Desire*. Simon & Schuster, New York. [26]

Hamer, D. H., S. Hu, V. L. Magnuson, N. Hu, and A. M. L. Pattatucci. 1993. A linkage between DNA markers on the X chromosome and male sexual orientation. *Science* 261: 321–327. [26]

Hamilton, W. D. 1964. The genetical evolution of social behavior, I and II. *J. Theor. Biol.* 7: 1–52. [20]

Hamilton, W. D. 1967. Extraordinary sex ratios. *Science* 156: 477–488. [21]

Hamilton, W. D. 1971. Geometry for the selfish herd. *J. Theor. Biol.* 31: 295–311. [20]

Hamilton, W. D., and R. M. May. 1977. Dispersal in stable habitats. *Nature* 269: 578–581. [19]

Hamilton, W. D., and M. Zuk. 1982. Heritable true fitness and bright birds: A role for parasites? *Science* 218: 384–387. [20]

Hamrick, J. L., and M. J. Godt. 1990. Allozyme diversity in plant species. In A. H. D. Brown, M. T. Clegg, A. L. Kahler, and B. S. Weir (eds.), *Plant Population Genetics, Breeding, and Genetic Resources*, pp. 43–63. Sinauer Associates, Sunderland, MA. [22]

Hancock, J. M., D. Tautz, and G. Dover. 1988. Evolution of the secondary structure and compensatory mutations of the ribosomal RNAs of *Drosophila melanogaster*. *Mol. Biol. Evol.* 5: 393–414. [22]

Hanken, J. 1984. Miniaturization and its effects on cranial morphology in plethodontid salamanders, genus *Thorius* (Amphibia: Plethodontidae). I. Osteological variation. *Biol. J. Linn. Soc.* 23: 55–75. [23]

Hanken, J. 1985. Morphological novelty in the limb skeleton accompanies miniaturization in salamanders. *Science* 229: 871–874. [23]

Hanken, J., and D. B. Wake. 1993. Miniaturization of body size: Organismal consequences and evolutionary significance. *Annu. Rev. Ecol. Syst.* 24: 501–519. [23]

Hansen, T. A. 1980. Influence of larval dispersal and geographic distribution on species longevity in neogastropods. *Paleobiology* 6: 193–207. [24, 25]

Hansen, T. A. 1982. Modes of larval development in Early Tertiary neogastropods. *Paleobiology* 8: 367–377. [24]

Hansen, T. A. 1988. Early Tertiary radiation of marine molluscs and the long-term effects of the Cretaceous-Tertiary extinction. *Paleobiology* 14: 37–51. [25]

Hanski, I., and M. E. Gilpin. 1991. Metapopulation dynamics: Brief history and conceptual domain. *Biol. J. Linn. Soc.* 42: 3-16. [11]

Hard, J. J., W. E. Bradshaw, and C. M. Holzapfel. 1992. Epistasis and the genetic divergence of photoperiodism between populations of the pitcher-plant mosquito, *Wyeomyia smithii*. *Genetics* 131: 389–396. [14]

Hard, J. J., W. E. Bradshaw, and C. M. Holzapfel. 1993. The genetic basis of photoperiodism and its evolutionary divergence among populations of the pitcher-plant mosquito, *Wyeomyia smithii*. *Am. Nat.* 142: 457–473. [14]

Harding, R. M., S. M. Fullerton, R. C. Griffiths, J. Bond, M. J. Cox, J. A. Schneider, D. S. Moulin, and J. B. Clegg. 1997. Archaic African and Asian lineages in the genetic ancestry of modern humans. *Am. J. Hum. Genet.* 60: 772-789. [26]

Hardison, R. C. 1991. Evolution of globin gene families. In R. K. Selander, A. G. Clark, and T. S. Whittam (eds.), *Evolution at the Molecular Level*, pp. 272–289. Sinauer Associates, Sunderland, MA. [22]

Harland, S. C. 1936. The genetical conception of the species. *Biol. Rev.* 11: 83–112. [15]

Harper, J. L. 1977. *The Population Biology of Plants*. Academic Press, London. [4, 19]

Harris, H. 1966. Enzyme polymorphisms in man. *Proc. R. Soc. Lond. B* 164: 298–310. [9]

Harrison, R. G. 1977. Parallel variation at an enzyme locus in sibling species of field crickets. *Nature* 266: 168–170. [13]

Harrison, R. G. 1979. Speciation in North American field crickets: Evidence from electrophoretic comparisons. *Evolution* 33: 1009–1023. [15]

Harrison, R. G. 1980. Dispersal polymorphisms in insects. *Annu. Rev. Ecol. Syst.* 11: 95–118. [17]

Harrison, R. G. 1990. Hybrid zones: Windows on evolutionary process. *Oxford Surv. Evol. Biol.*, 7: 69–128. Oxford University Press, New York. [15]

Harrison, R. G. (ed.). 1993. *Hybrid Zones and the Evolutionary Process*. Oxford University Press, New York. [15]

Hartl, D. L., and A. G. Clark. 1989. Second edition. *Principles of Population Genetics*. Sinauer Associates, Sunderland, MA. [9]

Hartl, D. L., and A. G. Clark. 1997. Third edition. *Principles of Population Genetics*. Sinauer Associates, Sunderland, MA [9, 14]

Hartl, D. L., E. N. Moriyama, and S. A. Sawyer. 1994. Selection intensity for codon bias. *Genetics* 138: 227–234. [22]

Harvey, P. H., and M. D. Pagel. 1991. *The Comparative Method in Evolutionary Biology*. Oxford University Press, Oxford. [12]

Hastings, J. R., and R. M. Turner. 1965. *The Changing Mile*. University of Arizona Press, Tucson, AZ. [4]

Haverschmidt, F. 1968. *Birds of Surinam*. Oliver & Boyd, London. [8]

Hayman, P., J. Marchant, and T. Prater. 1986. *Shorebirds: An Identification Guide to the Waders of the World*. Houghton Mifflin, Boston, MA. [5]

Hearnshaw, L. S. 1979. *Cyril Burt, Psychologist*. Hodder and Stoughton, London. [26]

Hebert, P. D. N. 1987. The comparative evidence. In S. C. Stearns (ed.), *The Evolution of Sex and Its Consequences*, pp. 175–195. Birkhäuser, Basel. [21, 22]

Hecht, M. K. 1952. Natural selection in the lizard genus *Aristelliger*. *Evolution* 6: 112–124. [14]

Hecht, M., J. H. Ostrom, G. Viohl, and P. Wellnhofer (eds.). 1985. *The Beginning of Birds*. Freunde des Jura-Museums Eichstätt, Willibaldsburg, Germany. [6]

Hedrick, P. W. 1986. Genetic polymorphisms in heterogeneous environments: A decade later. *Annu. Rev. Ecol. Syst.* 17: 535–566. [13]

Hedrick, P. W., M. E. Ginevan, and E. P. Ewing. 1976. Genetic polymorphism in heterogeneous environments. *Annu. Rev. Ecol. Syst.* 7: 1–32. [13]

Heimburger, M. 1959. Cytotaxonomic studies in the genus *Anemone*. *Can. J. Bot.* 37: 587–612. [16]

Heinemann, J. A. 1991. Genetics of gene transfer between species. *Trends Genet.* 7: 181–185. [10]

Held, L. I. J. 1991. Bristle patterning in *Drosophila*. *Bioessays* 13: 633–640. [14]

Hennig, W. 1966. *Phylogenetic Systematics*. University of Illinois Press, Urbana. [5]

Hepper, P. G. (ed.). 1991. *Kin Recognition*. Cambridge University Press, Cambridge. [20]

Herre, E. A. 1985. Sex ratio adjustment in fig wasps. *Science* 228: 896–898. [21]

Herre, E. A. 1987. Optimality, plasticity, and selective regime in fig wasp sex ratios. *Nature* 329: 627–629. [21]

Herre, E. A. 1993. Population structure and the evolution of virulence in nematode parasites of fig wasps. *Science* 259: 1442–1445. [18]

Herrnstein, R. J., and C. Murray. 1994. *The Bell Curve: Intelligence and Class Structure in American Life*. The Free Press, New York. [26]

Hersh, A. H. 1930. The facet-temperature relation in the Bar series in *Drosophila*. *J. Exp. Zool.* 57: 283–306. [3]

Hershkovitz, P. 1977. *Living New World Monkeys (Platyrrhini)*. University of Chicago Press, Chicago, IL. [8]

Heslop-Harrison, J. 1982. Pollen-stigma interaction and cross-incompatibility in the grasses. *Science* 215: 1358–1364. [15]

Hewitt, G. M. 1989. The subdivision of species by hybrid zones. In D. Otte and J. A. Endler (eds.), *Speciation and Its Consequences*, pp. 85–110. Sinauer Associates, Sunderland, MA. [15]

Hewitt, G. M. 1996. Some genetic consequences of ice ages, and their role in divergence and speciation. *Biol. J. Linn. Soc.* 58: 247–276. [15, 16]

Hey, J. 1997. Mitochondrial and nuclear genes present conflicting portraits of human origins. *Mol. Biol. Evol.* 14: 166-172. [26]

Hey, J., and R. M. Kliman. 1993. Population genetics and phylogenetics of DNA sequence variation at multiple loci within the *Drosophila melanogaster* species complex. *Mol. Biol. Evol.* 10: 804–822. [15]

Heywood, V. H. (ed.). 1993. *Flowering Plants of the World*. Oxford University Press, Oxford. [4, 17]

Hill, G. E. 1991. Plumage coloration is a sexually selected indicator of male quality. *Nature* 350: 337–339. [20]

Hill, R. W., and G. A. Wyse. 1989. *Animal Physiology*. Harper Collins, New York. [17]

Hill, W. G. 1982. Predictions of response to artificial selection from new mutations. *Genet. Res.* 40: 255–278. [14]

Hill, W. G., and A. Caballero. 1992. Artificial selection experiments. *Annu. Rev. Ecol. Syst.* 23: 287–310. [14]

Hillis, D. M. 1987. Molecular versus morphological approaches to systematics. *Annu. Rev. Ecol. Syst.* 18: 23–42. [5]

Hillis, D. M. 1988. Systematics of the *Rana pipiens* complex: Puzzle and paradigm. *Annu. Rev. Ecol. Syst.* 19: 39–63. [16]

Hillis, D. M., J. J. Bull, M. E. White, M. R. Badgett, and I. J. Molineaux. 1992. Experimental phylogenetics: Generation of a known phylogeny. *Science* 255: 589–592. [5]

Hillis, D. M., B. K. Mable, and C. Moritz. 1996a. Applications of molecular systematics: The state of the field and a look to the future. In D. M. Hillis, C. Moritz, and B. K. Mable (eds.), *Molecular Systematics*, second edition, pp. 515–543. Sinauer Associates, Sunderland, MA. [5]

Hillis, D. M., C. Moritz, and B. K. Mable (eds.). 1996b. *Molecular Systematics*. Second edition. Sinauer Associates, Sunderland, MA. [5]

Hinchliffe, J. R., and D. R. Johnson. 1980. *The Development of the Vertebrate Limb*. Clarendon Press, Oxford. [23]

Hinde, R. A. 1970. *Animal Behaviour, a Synthesis of Ethology and Comparative Psychology*. McGraw-Hill, New York. [16]

Ho, M.-W., and P. T. Saunders, (eds.). 1984. *Beyond Neo-Darwinism: An Introduction to the New Evolutionary Paradigm*. Academic Press, London. [23]

Hobbs, H. H., M. S. Brown, J. L. Goldstein, and D. W. Russell. 1986. Deletion of an exon-encoding cysteine-rich repeat of low-density lipoprotein receptor alters its binding specificity in a subject with familial hypercholesterolemia. *J. Biol. Chem.* 261: 13114–13120. [10]

Hochachka, P. W., and G. N. Somero. 1984. *Biochemical Adaptation*. Princeton University Press, Princeton, NJ. [17]

Hodges, S. A., and M. L. Arnold. 1994. Floral and ecological isolation between *Aquilegia formosa* and *Aquilegia pubescens*. *Proc. Natl. Acad. Sci. USA* 91: 2493–2496. [15]

Hoekstra, R. F. 1987. The evolution of sexes. In S. C. Stearns (ed.), *The Evolution of Sex and Its Consequences*, pp. 59–91. Birkhäuser Verlag, Basel. [21]

Hoelzel, A. R. (ed.). 1992. *Molecular Genetic Analysis of Populations: A Practical Approach*. IRL Press at Oxford University Press, Oxford. [3]

Hoffman, A. 1989. *Arguments on Evolution: A Paleontologist's Perspective*. Oxford University Press, New York. [6, 24]

Hoffmann, A. A., and M. W. Blows. 1994. Species borders: Ecological and evolutionary perspectives. *Trends Ecol. Evol.* 9: 223–227. [17]

Hoffman, A. A., and P. A. Parsons. 1989. Selection for increased dessication resistance in *Drosophila melanogaster*: Additive genetic control and correlated responses for other stresses. *Genetics* 122: 837–845. [17]

Hoffmann, A. A., and P. A. Parsons. 1991. *Evolutionary Genetics and Environmental Stress*. Oxford University Press, Oxford. [17]

Hofstadter, R. 1955. *Social Darwinism in American Thought*. Beacon Press, Boston, MA. [1]

Holland, P. W. H., and J. Garcia-Fernández. 1996. Hox genes and chordate evolution. *Dev. Biol.* 173: 382–395. [23]

Hölldobler, B., and E. O. Wilson. 1990. *The Ants.* Harvard University Press, Cambridge, MA. [12, 20]

Hollocher, H., and C.-I. Wu. 1996. The genetics of reproduction isolation in the *Drosophila simulans* clade: X vs. autosomal effects and male vs. female effects. *Genetics* 143: 1243–1255. [15]

Holmes, J. C. 1983. Evolutionary relationships between parasitic helminths and their hosts. In D. J. Futuyma and M. Slatkin (eds.), *Coevolution*, pp. 161–185. Sinauer Associates, Sunderland, MA. [18]

Holsinger, K. E. 1991. Inbreeding depression and the evolution of plant mating systems. *Trends Ecol. Evol.* 6: 307–308. [21]

Holsinger, K. E. 1996. Pollination biology and the evolution of mating systems in flowering plants. In M. K. Hecht, B. Wallace, and G. T. Prance (eds.), *Evolutionary Biology*, vol. 29, pp. 107–149. Plenum, New York. [21]

Holt, R. D. 1987. Population dynamics and evolutionary processes: The manifold roles of habitat selection. *Evol. Ecol.* 1: 337–347. [20]

Holt, R. D. 1993. Ecology at the mesoscale: The influence of regional processes on local communities. In R. E. Ricklefs and D. Schluter (eds.), *Species Diversity in Ecological Communities*, pp. 77–88. University of Chicago Press, Chicago, IL. [25]

Honeycutt, R. L., and R. M. Adkins. 1993. Higher level systematics of eutherian mammals: An assessment of molecular characters and phylogenetic hypotheses. *Annu. Rev. Ecol. Syst.* 24: 279–305. [7]

Hooker, E. 1957. The adjustment of the male overt homosexual. *J. Projective Techniques* 21: 18–31. [26]

Hopson, J. A. 1966. The origin of the mammalian middle ear. *Am. Zool.* 6: 437–450. [6]

Hopson, J. A., and A. W. Crompton. 1969. Origin of mammals. *Evol. Biol.* 3: 15–72. [6]

Horai, S., K. Hayasaka, R. Kondo, K. Tsugane, and N. Takahata. 1995. Recent African origin of humans revealed by complete sequences of hominoid mitochondrial DNAs. *Proc. Natl. Acad. Sci. USA* 92: 532–536. [26]

Hougen-Eitzman, D., and M. D. Rausher. 1994. Interactions between herbivorous insects and plant-insect coevolution. *Am. Nat.* 143: 677–697. [18]

Houle, D. 1989. The maintenance of polygenic variation in finite populations. *Evolution* 43: 1767–1780. [14]

Houle, D. 1991. Genetic covariance of fitness correlates: What genetic correlations are made of and why it matters. *Evolution* 45: 630–648. [19]

Houle, D. 1992. Comparing evolvability and variabiilty of quantitative traits. *Genetics* 130: 195–204. [19]

Houle, D. 1994. Adaptive distance and the genetic basis of heterosis. *Evolution* 48: 1410–1417. [13]

Houle, D., D. K. Hoffmaster, S. Assimacopoulos, and B. Charlesworth. 1992. The genomic mutation rate for fitness in *Drosophila*. *Nature* 359: 58–60. [10]

Houle, D., K. A. Hughes, D. K. Hoffmeister, J. Ihara, S. Assimacopoulos, D. Canada, and B. Charlesworth. 1994. The effects of spontaneous mutation on quantitative traits. I. Variances and covariances of life history traits. *Genetics* 138: 773–785. [10]

Houle, D., B. Morikawa, and M. Lynch. 1996. Comparing mutational variabilities. *Genetics* 143: 1467–1483. [19]

House, M. R. (ed.). 1979. *The Origin of Major Invertebrate Groups.* Academic Press, New York. [7]

Howard, D. J. 1993. Reinforcement: Origin, dynamics, and fate of an evolutionary hypothesis. In R. G. Harrison (ed.), *Hybrid Zones and the Evolutionary Process*, pp. 46–69. Oxford University Press, New York. [16]

Howard, D. J., and R. G. Harrison. 1984. Habitat segregation in ground crickets: The role of interspecific competition and habitat selection. *Ecology* 65: 69–76. [4]

Howe, C. J., T. J. Beanland, A. W. D. Larkum, and P. J. Lockhart. 1992. Plastid origins. *Trends Ecol. Evol.* 7: 378–383. [7]

Howell, F. C. 1978. Hominidae. In V. J. Maglio and H. B. S. Cooke (eds.), *Evolution of African Mammals*, pp. 154–248. Harvard University Press, Cambridge, MA. [26]

Hu, S., A. M. L. Pattatucci, C. Patterson, L. Li, D. W. Fulker, S. S. Cherny, L. Kruglyak, and D. H. Hamer. 1995. Linkage between sexual orientation and chromsome Xq28 in males but not females. *Nature Genetics* 11: 248–256. [26]

Hubbell, S. P., and R. B. Foster. 1983. Diversity of canopy trees in a neotropical forest and implications for conservation. In S. Sutton, T. C. Whitmore, and A. Chadwick (eds.), *Tropical Rain Forest: Ecology and Management*, pp. 25–41. Blackwell, Oxford. [25]

Hubbs, C. L., L. C. Hubbs, and R. E. Johnson. 1943. Hybridization in nature between species of catostomid fishes. *Contrib. Lab. Vert. Biol. Univ. Mich.* 22: 1–76. [15]

Hudson, R. R. 1990. Gene genealogies and the coalescent process. In D. J. Futuyma and J. Antonovics (eds.), *Oxford Surv. Evol. Biol.* 7: 1–44. Oxford University Press, New York. [11]

Hudson, R. R., and N. L. Kaplan. 1995. Deleterious background selection with recombination. *Genetics* 141: 1605–1617. [22]

Hudson, R. R., M. Kreitman, and M. Aguadé. 1987. A test of neutral molecular evolution based on nucleotide data. *Genetics* 116: 153–159. [22]

Huey, R. B. 1978. Latitudinal patterns of between-altitude faunal similarity: Mountains might be "higher" in the tropics. *Am. Nat.* 112: 225–229. [8]

Huey, R. B., and J. G. Kingsolver. 1993. Evolution of resistance to high temperature in ectotherms. *Am. Nat.* 142: S21–S46. [17]

Huffaker, C. B., and C. E. Kennett. 1969. Some aspects of assessing efficiency of natural enemies. *Can. Entomol.* 101: 425–447. [4]

Hughes, A. L. 1994. The evolution of functionally novel proteins after gene duplication. *Proc. R. Soc. Lond.* B 256: 119–124. [22]

Hull, D. L. 1973. *Darwin and His Critics.* Harvard University Press, Cambridge, MA. [1]

Hull, D. L. 1988. *Science as a Process.* University of Chicago Press, Chicago, IL. [26]

Hurd, L. E., and R. M. Eisenberg. 1975. Divergent selection for geotactic response and evolution of reproductive isolation in sympatric and allopatric populations of houseflies. *Am. Nat.* 109: 353–358. [16]

Hurst, L. D., A. Atlan, and B. D. Bengtsson. 1996. Genetic conflicts. *Q. Rev. Biol.* 71: 317–364. [21]

Huston, M. 1979. A general hypothesis of species diversity. *Am. Nat.* 113: 81–101. [8]

Huston, M. A. 1994. *Biological Diversity: The Coexistence of Species on Changing Landscapes.* Cambridge University Press, Cambridge. [25]

Hutchinson, E. W., and M. R. Rose. 1990. Quantitative genetic analysis of postponed aging in *Drosophila melanogaster*. In D. E. Harrison (ed.), *Genetic Effects on Aging* II, pp. 65–85. Telford Press, Caldwell, NJ. [19]

Hutchinson, G. E. 1957. Concluding remarks. *Cold Spring Harbor Symp. Quant. Biol.* 22: 415–427. [4]

Hutchinson, G. E. 1968. When are species necessary? In R. C. Lewontin (ed.), *Population Biology and Evolution*, pp. 177–186. Syracuse University Press, Syracuse, NY. [16]

Hutchinson, J. 1969. *Evolution and Phylogeny of Flowering Plants*. Academic Press, New York. [5]

Huttley, G. A., A. F. MacRae, and M. T. Clegg. 1995. Molecular evolution of the *AC/DS* transposable-element family in pearl millet and other grasses. *Genetics* 139: 1411–1419. [22]

Huxley, J. S. 1932. *Problems of Relative Growth*. MacVeagh, London. [23]

Ikemura, T. 1981. Correlation between the abundance of *Escherichia coli* transfer RNAs and the occurrence of the respective codons in its protein genes: A proposal for a synonymous codon choice that is optimal for the *E. coli* translational system. *J. Mol. Biol.* 151: 389–409. [22]

Imms, A. D. 1957. *A General Textbook of Entomology*. Methuen, London. [5]

Ioerger, T. R., A. G. Clark, and T.-H. Kao. 1990. Polymorphism at the self-incompatibility locus in Solanaceae predates speciation. *Proc. Natl. Acad. Sci. USA* 87: 9732–9735. [21]

Ivie, G. W., D. L. Bull, R. C. Beier, N. W. Pryor, and E. H. Vertli. 1983. Metabolic detoxification: Mechanisms of insect resistance to plant psoralens. *Science* 221: 374–376. [18]

Iwabe, N., K. Kuma, M. Hasegawa, S. Osawa, and T. Miyata. 1989. Evolutionary relationship of archaebacteria, eubacteria, and eukaryotes inferred from phylogenetic trees of duplicated genes. *Proc. Natl. Acad. Sci. USA* 86: 9355–9359. [7]

Iwabe, N., K. Kuma, H. Kishino, M. Hasegawa, and T. Miyata. 1991. Evolution of RNA polymerases and branching patterns of the three major groups of Archaebacteria. *J. Mol. Evol.* 32: 70–78. [7]

Iwasa, Y., and A. Pomiankowski. 1995. Continual change in mate preferences. *Nature* 377: 420–422. [20]

Iwasa, Y., and A. Sasaki. 1987. Evolution of the number of sexes. *Evolution* 41: 49–65. [21]

Jablonski, D. 1986a. Background and mass extinctions: The alternation of macroevolutionary regimes. *Science* 231: 129–133. [25]

Jablonski, D. 1986b. Causes and consequences of mass extinctions: A comparative approach. In D. K. Elliott (ed.), *Dynamics of Extinction*, pp. 183–229. Wiley, New York. [24]

Jablonski, D. 1986c. Evolutionary consequences of mass extinctions. In D. M. Raup and D. Jablonski (eds.), *Patterns and Processes in the History of Life*, pp. 313–329. Springer-Verlag, Berlin. [25]

Jablonski, D. 1991. Extinctions: A paleontological perspective. *Science* 253: 754–757. [25]

Jablonski, D. 1995. Extinctions in the fossil record. In J. H. Lawton and R. M. May (eds.), *Extinction Rates*, pp. 25–44. Oxford University Press, Oxford. [25]

Jablonski, D. 1996. Body size and macroevolution. In D. Jablonski, D. H. Erwin, and J. H. Lipps (eds.), *Evolutionary Paleobiology*, pp. 256–289. University of Chicago Press, Chicago, IL. [24]

Jablonski, D., and R. A. Lutz. 1983. Larval ecology of marine benthic invertebrates: Paleobiological implications. *Biol. Rev.* 58: 21–89. [24]

Jablonski, D., and D. M. Raup. 1995. Selectivity of end-Cretaceous marine bivalve extinctions. *Science* 268: 389–391. [25]

Jablonski, D., D. H. Erwin, and J. H. Lipps (eds.). 1996. *Evolutionary Paleobiology: Essays in Honor of James W. Valentine*. University of Chicago Press, Chicago, IL. [6]

Jablonski, D., S. J. Gould, and D. M. Raup. 1986. The nature of the fossil record: A biological perspective. In D. M. Raup and D. Jablonski (eds.), *Patterns and Processes in the History of Life*, pp. 7–22. Springer-Verlag, Berlin. [6]

Jackson, J. B. C. 1974. Biogeographic consequences of eurytopy and stenotopy among marine bivalves and their biogeographic significance. *Am. Nat.* 104: 541–560. [25]

Jackson, J. B. C., and A. H. Cheetham. 1990. Evolutionary significance of morphospecies: A test with cheilostome Bryozoa. *Science* 248: 579–583. [6]

Jackson, J. B. C., and F. K. McKinney. 1990. Ecological processes and progressive macroevolution of marine clonal benthos. In R. M. Ross and W. D. Allmon (eds.), *Causes of Evolution: A Paleontological Perspective*, pp. 173–209. University of Chicago Press, Chicago, IL. [24]

Jaeger, R. G. 1971. Moisture as a factor influencing the distribution of two species of terrestrial salamanders. *Oecologia* 6: 191–207. [4]

Jaenike, J. 1990. Host specialization in phytophagous insects. *Annu. Rev. Ecol. Syst.* 21: 243–274. [18]

Jain, S. K., and D. R. Marshall. 1967. Population studies on predominantly self-pollinating species. X. Variation in natural populations of *Avena fatua* and *A. barbata*. *Am. Nat.* 101: 19–33. [11]

James, F. C. 1970. Geographic variation in birds and its relationship to climate. *Ecology* 51: 365–390. [17]

Janzen, D. H. 1967. Why mountain passes are higher in the tropics. *Am. Nat.* 101: 233–249. [8]

Janzen, D. H. 1970. Herbivores and the number of tree species in tropical forests. *Am. Nat.* 104: 501–528. [4]

Janzen, D. H., and P. S. Martin. 1982. Neotropical anachronisms: The fruits the gomphotheres ate. *Science* 215: 19–27. [12]

Janzen, F. J., and G. L. Pauktis. 1991. Environmental sex determination in reptiles: Ecology, evolution, and experimental design. *Q. Rev. Biol.* 66: 149–179. [21]

Jarne, P., and D. Charlesworth. 1993. The evolution of the selfing rate in functionally hermaphroditic plants and animals. *Annu. Rev. Ecol. Syst.* 24: 441–466. [21]

Jarosz, A. M., and J. J. Burdon. 1991. Host-pathogen interactions in natural populations of *Linum marginale* and *Melampsora lini*: II. Local and regional variation in patterns of resistance and racial structure. *Evolution* 45: 1618–1627. [18]

Jarvik, E. 1955. The oldest tetrapods and their forerunners. *Sci. Monthly* 80: 141–154. [6]

Jarvik, E. 1980. *Basic Structure and Evolution of Vertebrates*. Academic Press, London. [6]

Järvinen, A., and R. A. Vaisanen. 1984. Reproduction of pied flycatchers (*Ficedula hypoleuca*) in good and bad breeding seasons in a northern marginal area. *Auk* 101: 439–450. [17]

Jeffreys, A. J., A. MacLeod, K. Tamaki, D. L. Neil, and D. G. Monkton. 1991. Minisatellite repeat coding as a digital approach to DNA typing. *Nature* 354: 204–209. [22]

Jenkins, C. D., and M. Kirkpatrick. 1995. Deleterious mutation and the evolution of life cycles. *Evolution* 49: 512–520. [21]

Jenkins, F. A. Jr., and F. R. Parrington. 1976. The postcranial skeletons of the Triassic mammals *Eozostrodon, Megazostrodon* and *Erythrotherium*. *Phil. Trans. R. Soc. Lond.* B 273: 387–431. [6]

Jensen, A. R. 1973. *Educability and Group Differences*. Harper & Row, New York. [26]

Jensen, R. A. 1976. Enzyme recruitment in evolution of new function. *Annu. Rev. Microbiol.* 30: 409–425. [22]

Jerison, H. J. 1973. *Evolution of the Brain and Intelligence*. Academic Press, New York. [23]

Jinks, J. L., J. M. Perkins, and H. S. Pooni. 1973. The incidence of epistasis in normal and extreme environments. *Heredity* 31: 263–269. [14]

Johanson, D. C., and T. D. White. 1979. A systematic assessment of early African hominids. *Science* 203: 321–330. [26]

John, B. 1981. Chromosome change and evolutionary change: A critique. In W. R. Atchley and D. Woodruff (eds.), *Evolution and Speciation*, pp. 23–51. Cambridge University Press, London. [15]

John, B., and G. L. G. Miklos. 1988. *The Eukaryotic Genome in Development and Evolution*. Allen and Urwin, London. [3, 22]

Johnsgard, P. A. 1983. *The Hummingbirds of North America*. Smithsonian Institution Press, Washington, DC. [5]

Johnsgard, P. A. 1994. *Arena Birds: Sexual Selection and Behavior*. Smithsonian Institution Press, Washington, DC. [20]

Johnson, C. G. 1969. *Migration and Dispersal of Insects by Flight*. Methuen, London. [19]

Johnson, T. C., et al. 1996. Late Pleistocene desiccation of Lake Victoria and rapid evolution of cichlid fishes. *Science* 273: 1091–1093. [16]

Johnston, R. F., and R. K. Selander. 1964. House sparrows: Rapid evolution of races in North America. *Science* 144: 548–550. [6, 9]

Johnston, R. F., and R. K. Selander. 1971. Evolution in the house sparrow. II. Adaptive differentiation in North American populations. *Evolution* 25: 1–28. [14]

Johnston, T. D. 1982. Selective costs and benefits in the evolution of learning. *Adv. Stud. Behav.* 12: 65–106. [20]

Johnstone, R. A., and K. Norris. 1993. Badges of status and the cost of aggression. *Behav. Ecol. Sociobiol.* 32: 127–134. [20]

Jones, D. F. 1924. The attainment of homozygosity in inbred strains of maize. *Genetics* 9: 405–418. [11]

Jones, R., and D. C. Culver. 1989. Evidence for selection on sensory structures in a cave population of *Gammarus minus* Say (Amphipoda). *Evolution* 43: 688–693. [14]

Jones, S., R. Martin, and D. Pilbeam (eds.). 1992. *The Cambridge Encyclopedia of Human Evolution*. Cambridge University Press, Cambridge. [26]

Jonsson, L. 1992. *Birds of Europe*. Princeton University Press, Princeton, NJ. [15]

Jordan, D. S., and B. W. Evermann. 1973. *The Shore Fishes of Hawaii*. Charles E. Tuttle, Rutland, VT. [23]

Jordan, K. 1896. On mechanical selection and other problems. *Novit. Zool.* 3: 426–525 (cited by Mayr 1963). [16]

Judson, O. P., and B. J. Normark. 1996. Ancient asexual scandals. *Trends Ecol. Evol.* 11: 41–46. [21]

Kaestner, A. 1970. *Invertebrate Zoology*. Volume III. Wiley, New York. [24]

Kamil, A. C. 1994. A synthetic approach to the study of animal intelligence. In L. A. Real (ed.), *Behavioral Mechanisms in Evolutionary Ecology*, pp. 11–45. University of Chicago Press, Chicago, IL. [20]

Kamin, L. J. 1974. *The Science and Politics of IQ*. Wiley, New York. [26]

Kaneshiro, K. Y. 1983. Sexual selection and the direction of evolution in the biosystematics of Hawaiian Drosophilidae. *Annu. Rev. Entomol.* 28: 161–178. [16]

Kaneshiro, K. Y., R. G. Gillespie, and H. L. Carson. 1995. Chromosomes and male genitalia of Hawaiian *Drosophila*. In W. L. Wagner and V. A. Funk (eds.), *Hawaiian Biogeography*, pp. 57–71. Smithsonian Institution Press, Washington, DC. [8]

Kaplan, W. L., R. R. Hudson, and C. H. Langley. 1989. The "hitch-hiking" effect revisited. *Genetics* 123: 887–899. [22]

Kareiva, P. M., J. G. Kingsolver, and R. B. Huey (eds.). 1993. *Biotic Interactions and Global Change*. Sinauer Associates, Sunderland, MA. [25]

Karn, M. N., and L. S. Penrose. 1951. Birth weight and gestation time in relation to maternal age, parity, and infant survival. *Ann. Eugenics* 16: 147–164. [14]

Kasper, A. E., H. N. Andrews, and W. H. Forbes. 1974. New fertile species of *Psilophyton* from the Devonian of Maine. *Am. J. Bot.* 61: 339–359. [7]

Katakura, H., and T. Hosogai. 1994. Performance of hybrid ladybird beetles (*Epilachna*) on the host plants of parental species. *Entomol. Exp. Appl.* 71: 81–85. [15]

Katakura, H., M. Shioi, and Y. Kira. 1989. Reproductive isolation by host specificity in a pair of phytophagous ladybird beetles. *Evolution* 43: 1045–1053. [15]

Kaufman, P. K., F. D. Enfield, and R. E. Comstock. 1977. Stabilizing selection for pupa weight in *Tribolium castaneum*. *Genetics* 87: 327–341. [14]

Kearsey, M. J., and B. W. Barnes. 1970. Variation for metrical characters in *Drosophila* populations. II. Natural selection. *Heredity* 25: 11–21. [14]

Kellogg, D. E. 1975. Character displacement in the radiolarian genus, *Eucyrtidium*. *Evolution* 29: 736–749.

Kellogg, D. E., and J. D. Hays. 1975. Microevolutionary patterns in Late Cenozoic Radiolaria. *Paleobiology* 1: 150–160. [6]

Kemp, T. S. 1982. *Mammal-like Reptiles and the Origin of Mammals*. Academic Press, London. [6]

Kermack, K. A., F. Mussett, and H. W. Rigney. 1973. The lower jaw of *Morganucodon*. *Zool. J. Linn. Soc.* 53: 87–175. [6]

Kermack, K. A., F. Mussett, and H. W. Rigney. 1981. The skull of *Morganucodon*. *Zool. J. Linn. Soc.* 71: 1–158. [6]

Kerster, H. W. 1964. Neighborhood size in the rusty lizard, *Sceloporus olivaceus*. *Evolution* 18: 445–457. [11]

Kettlewell, H. B. D. 1955. Selection experiments on industrial melanism in the Lepidoptera. *Heredity* 10: 287–301. [13]

Kidd, K. K., and L. L. Cavalli-Sforza. 1974. The role of genetic drift in the differentiation of Icelandic and Norwegian cattle. *Evolution* 28: 381–395. [11]

Kidston, R., and W. H. Lang. 1921. On Old Red Sandstone plants showing structure from the Rhynie chert bed, Aberdeenshire, Part IV. Restorations of the vascular cryptogams, and discussion of their bearing on the general morphology of Pteridophyta and the origin of the organization of land plants. *Trans. R. Soc. Edinburgh* 32: 477–487. [7]

Kidwell, M. G. 1983. Evolution of hybrid dysgenesis determinants in *Drosophila melanogaster*. *Proc. Natl. Acad. Sci. USA* 80: 1655–1659. [10, 22]

Kidwell, M. G. 1993. Lateral transfer in natural populations of eukaryotes. *Annu. Rev. Genet.* 27: 235–256. [22]

Kim, J. 1993. Improving the accuracy of phylogenetic estimation by combining different methods. *Syst. Biol.* 42: 331–340. [5]

Kimbel, W. H., D. C. Johanson, and Y. Rak. 1994. The first skull and other discoveries of *Australopithecus afarensis* at Hadar, Ethiopia. *Science* 368: 449–451. [26]

Kimura, M. 1955. Solution of a process of random genetic drift with a continuous model. *Proc. Natl. Acad. Sci. USA* 41: 144–150. [11]

Kimura, M. 1965. A stochastic model concerning the maintenance of genetic variability in quantitative characters. *Proc. Natl. Acad. Sci. USA* 54: 731–736. [14]

Kimura, M. 1968. Evolutionary rate at the molecular level. *Nature* 217: 624–626. [11]

Kimura, M. 1981. Possibility of extensive neutral evolution under stabilizing selection with special reference to nonrandom usage of synonymous codons. *Proc. Natl. Acad. Sci. USA* 78: 5773–5777. [14]

Kimura, M. 1983. *The Neutral Theory of Molecular Evolution*. Cambridge University Press, Cambridge. [22]

Kimura, M. 1991. Recent development of the neutral theory viewed from the Wrightian tradition of theoretical population genetics. *Proc. Natl. Acad. Sci. USA* 88: 5969–5973. [22]

King, J. L., and T. H. Jukes. 1969. Non-Darwinian evolution. *Science* 164: 788–798. [11]

King, M. 1993. *Species Evolution: The Role of Chromosome Change.* Cambridge University Press, Cambridge. [15]

Kingsolver, J. G. 1983. Thermoregulation and flight in *Colias* butterflies: Elevational patterns and mechanistic limitations. *Ecology* 64: 534–545. [17]

Kingsolver, J. G. 1995. Fitness consequences of seasonal polymorphism in western white butterflies. *Evolution* 49: 942–954. [17]

Kingsolver, J. G., and W. B. Watt. 1984. Mechanistic constraints and optimality models: Thermoregulatory strategies in *Colias* butterflies. *Ecology* 65: 1835–1839. [17]

Kirby, D. A., S. V. Muse, and W. Stephan. 1995. Maintenance of pre-mRNA secondary structure by epistatic selection. *Proc. Natl. Acad. Sci. USA* 92: 9047–9051. [22]

Kirkpatrick, M. 1982a. Quantum evolution and punctuated equilibrium in continuous genetic characters. *Am. Nat.* 119: 833–848. [16]

Kirkpatrick, M. 1982b. Sexual selection and the evolution of female choice. *Evolution* 36: 1–12. [20]

Kirkpatrick, M. 1987. Sexual selection and female choice in polygynous animals. *Annu. Rev. Ecol. Syst.* 18: 43–70. [16]

Kirkpatrick, M., and C. D. Jenkins. 1989. Genetic segregation and the maintenance of sexual reproduction. *Nature* 339: 300–301. [21]

Kirkpatrick, M., and M. J. Ryan. 1991. The evolution of mating preferences and the paradox of the lek. *Nature* 350: 33–38. [20]

Kitagawa, O. 1967. Genetic divergence in M. Vetukhiv's experimental populations of *Drosophila pseudoobscura*. IV. Relative viability. *Genet. Res.* 10: 303–312. [16]

Kitchell, J. A. 1990. The reciprocal interaction of organism and effective environment: Learning more about "and." In R. M. Ross and W. D. Allmon (eds.), *Causes of Evolution: A Paleontological Perspective,* pp. 151–169. University of Chicago Press, Chicago, IL. [18]

Kitcher, P. 1982. *Abusing Science: The Case Against Creationism.* MIT Press, Cambridge, MA. [1]

Kitcher, P. (Dec. 1984/Jan. 1985). Good science, bad science, dreadful science, and pseudoscience. *JCST* 168–173. [8]

Kitcher, P. 1985. *Vaulting Ambition.* MIT Press, Cambridge, MA. [20]

Kitcher, P. 1996. *The Lives to Come: The Genetic Revolution and Human Possibilities.* Simon & Schuster, New York. [26]

Klauber, L. M. 1972. *Rattlesnakes: Their Habits, Life Histories, and Influence on Mankind.* University of California Press, Berkeley. [8]

Kleckner, N. 1981. Transposable elements in prokaryotes. *Annu. Rev. Genet.* 15: 341–404. [22]

Kliman, R. M., and J. Hey. 1993. DNA sequence variation at the *period* locus within and among species of the *Drosophila melanogaster* complex. *Genetics* 133: 375–387. [15]

Knoll, A. H. 1986. Patterns of change in plant communities through geologic time. In J. Diamond and T. J. Case (eds.), *Community Ecology,* pp. 125–141. Harper & Row, New York. [7]

Knoll, A. H. 1992. The early evolution of eukaryotes: A geological perspective. *Science* 256: 622–627. [7]

Knowlton, N., L. A. Weigt, L. A. Solórzano, D. K. Mills, and E. Bermingham. 1993. Divergence in proteins, mitochondrial DNA, and reproductive compatibility across the Isthmus of Panama. *Science* 260: 1629–1632. [16]

Koehn, R. K., and J. J. Hilbish. 1987. The adaptive importance of genetic variation. *Am. Sci.* 75: 134–141. [13]

Kondrashov, A. S. 1986. Multilocus model of sympatric speciation. III. Computer simulations. *Theor. Pop. Biol.* 29: 1–15. [16]

Kondrashov, A. S. 1988. Deleterious mutations and the evolution of sexual reproduction. *Nature* 336: 435–440. [21]

Kondrashov, A. S. 1993. Classification of hypotheses on the advantage of amphimixis. *J. Hered.* 84: 372–387. [21]

Kondrashov, A. S., and J. F. Crow. 1991. Haploidy or diploidy: Which is better? *Nature* 351: 314–315. [21]

Kondrashov, A. S., and J. F. Crow. 1993. A molecular approach to estimating the human deleterious mutation rate. *Hum. Mutation* 2: 229–234. [10]

Koop, B. F., and M. Goodman. 1988. Evolutionary and developmental aspects of two hemoglobin β-chain genes (εM and βM) of opossum. *Proc. Natl. Acad. Sci. USA* 85: 3893–3897. [22]

Koop, B. F., D. A. Tagle, M. Goodman, and J. L. Slightom. 1989. A molecular view of primate phylogeny and important systematic and evolutionary questions. *Mol. Biol. Evol.* 6: 580–612. [26]

Kornegay, J. R., J. W. Schilling, and A. C. Wilson. 1994. Molecular adaptation of a leaf-eating bird: Stomach lysozyme of the hoatzin. *Mol. Biol. Evol.* 11: 921–928. [22]

Kornet, D. J. 1993. Permanent splits as speciation events: A formal reconstruction of the internodal species concept. *J. Theor. Biol.* 164: 407–435. [15]

Koufopanou, V., and G. Bell. 1991. Developmental mutants of *Volvox*: Does mutation recreate patterns of phylogenetic diversity? *Evolution* 45: 1806–1822. [10]

Krause, D. W. 1986. Competitive exclusion and taxonomic displacement in the fossil record: The case of rodents and multituberculates in North America. In K. M. Flanagan and J. A. Lillegraven (eds.), *Vertebrates, Phylogeny, and Philosophy,* pp. 95–117. Contrib. Geol. Special Paper, University of Wyoming, Laramie. [25]

Krebs, J. R., and N. B. Davies. 1987. *An Introduction to Behavioural Ecology.* Third edition. Blackwell Scientific, Oxford. [20]

Krebs, J. R., and N. B. Davies (eds.). 1991. *Behavioural Ecology: An Evolutionary Approach.* Blackwell Scientific, Oxford. [20]

Kreitman, M. 1983. Nucleotide polymorphism at the alcohol dehydrogenase locus of *Drosophila melanogaster. Nature* 304: 412–417. [9]

Kreitman, M. 1987. Molecular population genetics. *Oxford Surv. Evol. Biol.* 4: 38–60. [9]

Kreitman, M., and H. Akashi. 1995. Molecular evidence for natural selection. *Annu. Rev. Ecol. Syst.* 26: 403–422. [13, 22]

Kreitman, M., and R. R. Hudson. 1991. Inferring the evolutionary histories of the *Adh* and *Adh-dup* loci in *Drosophila melanogaster* from patterns of polymorphism and divergence. *Genetics* 127: 565–582. [22]

Krimbas, C. B. 1984. On adaptation, neo-Darwinian tautology, and population fitness. *Evol. Biol.* 16: 1–57. [12]

Kristensen, N. P. 1981. Phylogeny of insect orders. *Annu. Rev. Entomol.* 26: 135–157. [5]

Kristensen, N. P. 1991. Phylogeny of extant hexapods. In CSIRO, *The Insects of Australia,* Second edition, pp. 125–140. Cornell University Press, Ithaca, NY. [5, 7]

Kroodsma, D. E. 1984. Songs of the alder flycatcher (*Empidonax alnorum*) and willow flycatcher (*Empidonax traillii*) are innate. *Auk* 101: 13–24. [20]

Kroodsma, D. E., and R. A. Canady. 1985. Differences in repertoire size, singing behavior, and associated neuroanatomy among marsh wren populations have a genetic basis. *Auk* 102: 439–446. [20]

Kruckeberg, A. R. 1957. Variation in fertility of hybrids between isolated populations of the serpentine species, *Streptanthus glandulosus* Cook. *Evolution* 11: 185–211. [9]

Kuhn-Schneider, E. 1963. I sauri del Monte Giorgio. *Communicazione dell' Istituto di Paleontologia dell' Università di Zurigo* 20: 811–854. [23]

Kukalová-Peck, J. 1991. Fossil history and the evolution of hexapod structures. In CSIRO (ed.), *The Insects of Australia*, pp. 141–179. Cornell University Press, Ithaca, NY. [7]

Kyriacou, C. P. 1990. The molecular ethology of the *period* gene in *Drosophila*. *Behav. Genet.* 20: 191–211. [10]

Labandeira, C. C., and J. J. Sepkoski Jr. 1993. Insect diversity in the fossil record. *Science* 261: 310–315. [25]

Labandeira, C. C., B. Beale, and F. Hueber. 1988. Early insect diversification: Evidence from a lower Devonian bristletail from Québec. *Science* 242: 913–916. [6]

LaBarbera, M. 1986. The evolution and ecology of body size. In D. M. Raup and D. Jablonski (eds.), *Patterns and Processes in the History of Life*, pp. 69–98. Springer-Verlag, Berlin. [24]

Lack, D. 1947. *Darwin's Finches*. Cambridge University Press, Cambridge. [5]

Lack, D. 1954. *The Natural Regulation of Animal Numbers*. Oxford University Press, Oxford. [19]

Lai, C., R. F. Lyman, A. D. Long, C. H. Langley, and T. F. C. Mackay. 1994. Naturally occurring variation in bristle number and DNA polymorphisms at the *scabrous* locus of *Drosophila melanogaster*. *Science* 266: 1697–1702. [14]

Lake, J. A. 1990. Origin of the Metazoa. *Proc. Natl. Acad. Sci. USA* 87: 763–766. [7]

Laland, K. N., F. J. Odling-Smee, and M. W. Feldman. 1996. The evolutionary consequences of niche construction: A theoretical investigation using two-locus theory. *J. Evol. Biol.* 9: 293–316. [20]

Lande, R. 1976a. The maintenance of genetic variability by mutation in a polygenic character with linked loci. *Genet. Res.* 26: 221–235. [14, 23]

Lande, R. 1976b. Natural selection and random genetic drift in phenotypic evolution. *Evolution* 30: 314–334. [14, 24]

Lande, R. 1978. Evolutionary mechanisms of limb loss in tetrapods. *Evolution* 32: 73–92. [14]

Lande, R. 1979a. Effective deme sizes during long-term evolution estimated from rates of chromosome rearrangement. *Evolution* 33: 234–251. [10, 16]

Lande, R. 1979b. Quantitative genetic analysis of multivariate evolution, applied to brain: Body size allometry. *Evolution* 33: 402–416. [14, 23]

Lande, R. 1980. Sexual dimorphism, sexual selection and adaptation in polygenic characters. *Evolution* 34: 292–305. [14, 20]

Lande, R. 1981. Models of speciation by sexual selection on polygenic traits. *Proc. Natl. Acad. Sci. USA* 78: 3721–3725. [20]

Lande, R. 1982. Rapid origin of sexual isolation and character divergence in a cline. *Evolution* 36: 213–223. [16]

Lande, R. 1984. The expected fixation rate of chromosomal inversions. *Evolution* 38: 743–752. [16]

Lande, R., and S. J. Arnold. 1983. The measurement of selection on correlated characters. *Evolution* 37: 1210–1226. [14]

Lande, R., and D. W. Schemske. 1985. The evolution of self-fertilization and inbreeding depression. I. Genetic models. *Evolution* 39: 24–40. [21]

Langley, C. H., and W. M. Fitch. 1974. An examination of the constancy of the rate of molecular evolution. *J. Mol. Evol.* 3: 161–177. [5, 22]

Latham, R. E., and R. E. Ricklefs. 1993. Continental comparisons of temperate-zone tree species diversity. In R. E. Ricklefs and D. Schluter (eds.), *Species Diversity in Ecological Communities*, pp. 294–314. University of Chicago Press, Chicago, IL. [25]

Lauder, G. V. 1981. Form and function: Structural analysis in evolutionary morphology. *Paleobiology* 7: 430–442. [5]

Lauder, G. V., and K. F. Liem. 1989. The role of historical factors in the evolution of complex organismal design. In D. B. Wake and G. Roth (eds.), *Complex Organismal Functions: Integration and Evolution in Vertebrates*, pp. 63–78. Wiley, Chichester, England. [24]

Law, R., A. D. Bradshaw, and P. D. Putwain. 1979. The cost of reproduction in annual meadow grass. *Am. Nat.* 113: 3–16. [19]

Lawrence, P. A. 1992. *The Making of a Fly: The Genetics of Animal Design*. Blackwell, Cambridge. [3]

Lawton, J. H., and R. M. May (eds.). 1995. *Extinction Rates*. Oxford University Press, Oxford. [25]

Lazarus, D. 1983. Speciation in pelagic Protista and its study in the planktonic microfossil record: a review. *Paleobiology* 9: 327-340. [6]

Leakey, M. G., C. S. Felbel, I. McDougall, and A. Walker. 1995. New four-million-year-old hominid species from Kanapoi and Allia Bay, Kenya. *Nature* 376: 565–571. [26]

Lederberg, J., and E. M. Lederberg. 1952. Replica plating and indirect selection of bacterial mutants. *J. Bacteriol.* 63: 399–406. [10]

Lee, D. W., and J. H. Richards. 1991. Heteroblastic development in vines. In F. E. Putz and H. A. Mooney (eds.), *The Biology of Vines*, pp. 205–243. Cambridge University Press, Cambridge. [12]

Lee, Y.-H., T. Ota, and V. D. Vacquier. 1995. Positive selection is a general phenomenon in the evolution of abalone sperm lysin. *Mol. Biol. Evol.* 12: 231–238. [15]

Lees, D. R. 1981. Industrial melanism: Genetic adaptation of animals to air pollution. In J. A. Bishop and L. M. Cook (eds.), *Genetic Consequences of Man Made Change*, pp. 129–176. Academic Press, London. [13]

Leigh, E. G. Jr. 1973. The evolution of mutation rates. *Genetics* Suppl. 73: 1–18. [22]

Leigh Brown, A. J., and E. C. Holmes. 1994. Evolutionary biology of human immunodeficiency viruses. *Annu. Rev. Ecol. Syst.* 25: 127–165. [26]

Lennington, S., and K. Egid. 1985. Female discrimination of male odors correlated with male genotype at the T locus: A response to T-locus or H-2 locus variability? *Behav. Genet.* 15: 53–67. [12]

Lenski, R. E., and A. F. Bennett. 1993. Evolutionary responses of *Escherichia coli* to thermal stress. *Am. Nat.* 142: S47–S64. [17]

Lenski, R. E., and B. R. Levin. 1985. Constraints on the coevolution of bacteria and virulent phage: A model, some experiments, and predictions for natural communities. *Am. Nat.* 125: 585–602. [18]

Lerner, I. M. 1954. *Genetic Homeostasis*. Oliver & Boyd, Edinburgh. [11]

Leroi, A. M., A. F. Bennett, and R. E. Lenski. 1994a. Temperature acclimation and competitive fitness: An experimental test of the beneficial acclimation assumption. *Proc. Natl. Acad. Sci. USA* 91: 1917–1921. [19]

Leroi, A. M., A. K. Chippindale, and M. R. Rose. 1994b. Long-term laboratory evolution of a genetic life-history trade-off in *Drosophila melanogaster*. 1. The role of genotype-by-environment interaction. *Evolution* 48: 1244–1257. [17]

LeVay, S. 1993. *The Sexual Brain*. MIT Press, Cambridge, MA. [26]

Levene, H. 1953. Genetic equilibrium when more than one ecological niche is available. *Am. Nat.* 87: 331–333. [13]

Levin, B. R., and R. E. Lenski. 1983. Coevolution in bacteria and their viruses and plasmids. In D. J. Futuyma and M. Slatkin (eds.), *Coevolution*, pp. 99–127. Sinauer Associates, Sunderland, MA. [18]

Levin, B. R., and R. E. Lenski. 1985. Bacteria and phage: A model system for the study of the ecology and coevolution of hosts and parasites. In D. Rollinson and R. M. Anderson (eds.), *Ecology and Genetics of Host-Parasite Interactions*, pp. 227–242. Academic Press, London. [18]

Levin, D. A. 1975. Pest pressure and recombination systems in plants. *Am. Nat.* 109: 437–451. [21]

Levin, D. A. 1978. Genetic variation in annual *Phlox*: Self-compatible versus self-incompatible species. *Evolution* 32: 245–263. [11]

Levin, D. A. 1979. The nature of plant species. *Science* 204: 381–384. [15]

Levin, D. A. 1983a. Plant parentage: An alternate view of the breeding structure of populations. In C. E. King and P. S. Dawson (eds.), *Population Biology: Retrospect and Prospect*, pp. 171–188. Columbia University Press, New York. [10, 11]

Levin, D. A. 1983b. Polyploidy and novelty in flowering plants. *Am. Nat.* 122: 1–25. [10, 16]

Levin, D. A. 1984. Immigration in plants: An exercise in the subjunctive. In R. Dirzo and J. Sarukhán (eds.), *Perspectives on Plant Population Ecology*, pp. 242–260. Sinauer Associates, Sunderland, MA. [11]

Levin, D. A., and H. W. Kerster. 1974. Gene flow in seed plants. *Evol. Biol.* 17: 139–320. [11]

Levin, H. L. 1996. *The Earth through Time*. Fifth edition. Saunders College Publishing, Fort Worth, TX. [7]

Levins, R. 1968. *Evolution in Changing Environments*. Princeton University Press, Princeton, NJ. [17]

Levinton, J. S. 1988. *Genetics, Paleontology, and Macroevolution*. Cambridge University Press, Cambridge. [6]

Lewin, B. 1985. *Genes II*. Wiley, New York. [3]

Lewin, B. 1990. *Genes IV*. Oxford University Press, Oxford. [3]

Lewin, B. 1994. *Genes V*. Oxford University Press, Oxford. [3]

Lewis, H. 1962. Catastrophic selection as a factor in speciation. *Evolution* 16: 257–271. [16]

Lewis, H., and M. Lewis. 1955. The genus *Clarkia*. *Univ. Calif. Publ. Bot.* 20: 241–392. [16]

Lewis, W. H. (ed.). 1979a. *Polyploidy: Biological Relevance*. Plenum, New York. [16]

Lewis, W. H. 1979b. Polyploidy in species populations. In W. H. Lewis (ed.), *Polyploidy: Biological Relevance*, pp. 103–144. Plenum, New York. [16]

Lewontin, R. C. 1964. The interaction of selection and linkage. II. Optimum models. *Genetics* 50: 757–782. [12, 14]

Lewontin, R. C. 1972. The apportionment of human diversity. *Evol. Biol.* 6: 381-398. [26]

Lewontin, R. C. 1974a. The analysis of variance and the analysis of causes. *Am. J. Hum. Genet.* 26: 400–411. [14, 26]

Lewontin, R. C. 1974b. *The Genetic Basis of Evolutionary Change*. Columbia University Press, New York. [9]

Lewontin, R. C. 1983a. Gene, organism, and environment. In D. S. Bendall (ed.), *Evolution from Molecules to Men*, pp. 273–285. Cambridge University Press, Cambridge. [20]

Lewontin, R. C. 1983b. The organism as subject and object of evolution. *Scientia* 118: 63–82. [9]

Lewontin, R. C., and J. L. Hubby. 1966. A molecular approach to the study of genic heterozygosity in natural popula-

tions. II. Amount of variation and degree of heterozygosity in natural populations of *Drosophila pseudoobscura*. *Genetics* 54: 595–609. [9]

Lewontin, R. C., and M. J. D. White. 1960. Interaction between inversion polymorphisms of the two chromosome pairs in the grasshopper, *Moraba scurra*. *Evolution* 14: 116–129. [14]

Lewontin, R. C., S. Rose, and L. Kamin. 1984. *Not in Our Genes: Biology, Ideology, and Human Nature*. Pantheon, New York. [26]

Li, W.-H. 1980. Rate of gene silencing at duplicate loci: A theoretical study and interpretation of data from tetraploid fishes. *Genetics* 95: 237–258. [16]

Li, W.-H. 1997. *Molecular Evolution*. Sinauer Associates, Sunderland, MA. [22]

Li, W.-H., and D. Graur. 1991. *Fundamentals of Molecular Evolution*. Sinauer Associates, Sunderland, MA. [22]

Li, W.-H., M. Tanimura, and P. M. Sharp. 1987. An evaluation of the molecular clock hypothesis using mammalian DNA sequences. *J. Mol. Evol.* 25: 330–342. [11]

Liem, K. F. 1973. Evolutionary strategies and morphological innovations: Cichlid pharyngeal jaws. *Syst. Zool.* 22: 425–441. [24]

Liou, L. W., and T. D. Price. 1994. Speciation by reinforcement of premating isolation. *Evolution* 48: 1451–1459. [16]

Lipps, J. H., and P. W. Signor III. 1992. *Origin and Early Evolution of the Metazoa*. Plenum, New York. [7]

Lloyd, D. G. 1992. Evolutionarily stable strategies of reproduction in plants: Who benefits and how? In R. Wyatt (ed.), *Ecology and Evolution of Plant Reproduction*, pp. 137–168. Chapman & Hall, New York. [21]

Lloyd, J. E. 1966. Studies on the flash communication system in *Photinus* fireflies. *Misc. Publ. Mus. Zool. Univ. Mich.* 130: 1–195. [15]

Loeschcke, V. (ed.). 1987. *Genetic Constraints on Adaptive Evolution*. Springer-Verlag, Berlin. [14]

Löfstedt, C. 1990. Population variation and genetic control of pheromone communication systems in moths. *Entomol. Exp. Appl.* 54: 199–218. [15]

Löfstedt, C., W. M. Herrebout, and J.-W. Du. 1986. Evolution of the ermine moth pheromone tetradecyl acetate. *Nature* 323: 621–623. [16]

Lofsvold, D. 1986. Quantitative genetics of morphological differentiation in *Peromyscus*. I. Tests of the homogeneity of genetic covariance structure among species and subspecies. *Evolution* 40: 559–573. [14, 23]

Lombard, R. E., and D. B. Wake. 1977. Tongue evolution in the lungless salamanders, family Plethodontidae. II. Function and evolutionary diversity. *J. Morphol.* 153: 39–80. [24]

Long, A. D., S. L. Mullaney, L. A. Reid, J. D. Fry, C. H. Langley, and T. F. C. Mackay. 1995. High resolution mapping of genetic factors affecting abdominal bristle number in *Drosophila melanogaster*. *Genetics* 139: 1273–1291. [14]

Losos, J. B. 1990a. Ecomorphology, performance capability, and scaling of West Indian *Anolis* lizards: An evolutionary analysis. *Ecol. Monogr.* 60: 369–388. [8]

Losos, J. B. 1990b. The evolution of form and function: Morphology and locomotor performance in West Indian *Anolis* lizards. *Evolution* 44: 1189–1203. [8]

Losos, J. B. 1990c. A phylogenetic analysis of character displacement in Caribbean *Anolis* lizards. *Evolution* 44: 558–569. [18]

Losos, J. B. 1992. The evolution of convergent structure in Caribbean *Anolis* communities. *Syst. Biol.* 41: 403–420. [8, 18]

Lovejoy, C. O. 1981. The origins of man. *Science* 211: 341–350. [26]

Lucas, S. G. 1994. *Dinosaurs: The Textbook*. Wm. C. Brown, Dubuque, IA. [7]

Luckinbill, L. S., R. Arking, M. J. Clare, W. C. Cirocco, and S. A. Buck. 1984. Selection for delayed senescence in *Drosophila melanogaster*. *Evolution* 38: 996–1003. [19]

Lundberg, J. G. 1993. African-South American freshwater fish clades and continental drift: Problems with a paradigm. In P. Goldblatt (ed.), *Biological Relationships between Africa and South America*, pp. 156–199. Yale University Press, New Haven, CT. [8]

Luo, D., R. Carpenter, C. Vincent, L. Copsey, and E. Coen. 1996. Origin of floral asymmetry in *Antirrhinum*. *Nature* 383: 794-799. [24]

Luria, S., and M. Delbrück. 1943. Mutations of bacteria from virus sensitivity to virus resistance. *Genetics* 28: 491–511. [10]

Luria, S. A., S. J. Gould, and S. Singer. 1981. *A View of Life*. Benjamin Cummings, Menlo Park, CA. [8]

Lutz, F. E. 1918. *Field Book of Insects*. Knickerbocker, New York. [4]

Lyckegaard, E. M. S., and A. G. Clark. 1991. Evolution of ribosomal RNA gene copy number on the sex chromosomes of *Drosophila melanogaster*. *Mol. Biol. Evol.* 8: 458–474. [22]

Lynch, J. D. 1988. Refugia. In A. A. Myers and P. S. Giller (eds.), *Analytical Biogeography*, pp. 311–342. Chapman & Hall, London. [8]

Lynch, M. 1988. The rate of polygenic mutation. *Genet. Res.* 51: 137–148. [10]

Lynch, M. 1990. The rate of morphological evolution in mammals from the standpoint of the neutral expectation. *Am. Nat.* 136: 727–741. [24]

Lynch, M. 1994. Neutral models of phenotypic evolution. In L. A. Real (ed.), *Ecological Genetics*, pp. 86–108. Princeton University Press, Princeton, NJ. [14]

Lynch, M., and W. G. Hill. 1986. Phenotypic evolution by neutral mutation. *Evolution* 40: 915–935. [14]

Lynch, M., and R. Lande. 1993. Evolution and extinction in response to environmental change. In P. M. Kareiva, J. G. Kingsolver, and R. B. Huey (eds.), *Biotic Interactions and Global Change*, pp. 234–250. Sinauer Associates, Sunderland, MA. [21, 25]

Lynch, M., R. Bürger, D. Butcher, and W. Gabriel. 1993. The mutational meltdown in an asexual population. *J. Hered.* 84: 339–344. [21]

Lyon, J., and P. Gorner. 1995. *Altered Fates*. W. W. Norton, New York. [26]

Maas, M. C., D. W. Krause, and S. G. Strait. 1988. The decline and extinction of Plesiadapiformes (Mammalia: ?Primates) in North America: Displacement or replacement? *Paleobiology* 14: 410–431. [25]

MacArthur, R. H., and J. H. Connell. 1966. *The Biology of Populations*. Wiley, New York. [4]

MacArthur, R. H., and E. O. Wilson. 1967. *The Theory of Island Biogeography*. Princeton University Press, Princeton, NJ. [4]

MacFadden, B. J. 1986. Fossil horses from "Eohippus" (*Hyracotherium*) to *Equus*: Scaling, Cope's law, and the evolution of body size. *Paleobiology* 12: 355–369. [6]

MacFadden, B. J. 1988. Horses, the fossil record, and evolution: A current perspective. In M. K. Hecht, B. Wallace, and G. T. Prance (eds.), *Evolutionary Biology*, vol. 22, pp. 131–158. Plenum, New York. [6]

MacFadden, B. J. 1992. *Fossil Horses: Systematics, Paleobiology, and Evolution of the Family Equidae*. Cambridge University Press, New York. [6]

MacGregor, H. C. 1982. Big chromosomes and speciation amongst Amphibia. In G. A. Dover and R. B. Flavell (eds.), *Genome Evolution*, pp. 325–341. Academic Press, New York. [22]

Mackay, T. F. C., and C. H. Langley. 1990. Molecular and phenotypic variation in the *achaete-scute* region of *Drosophila melanogaster*. *Nature* 348: 64–66. [14]

Mackay, T. F. C., R. F. Lyman, and M. S. Jackson. 1992. Effects of P element insertions on quantitative traits in *Drosophila melanogaster*. *Genetics* 130: 315–322. [10]

Macnair, M. R. 1981. Tolerance of higher plants to toxic materials. In J. A. Bishop and L. M. Cook (eds.), *Genetic Consequences of Man Made Change*, pp. 177–207. Academic Press, New York. [13, 14]

Macnair, M. R., and P. Christie. 1983. Reproductive isolation as a pleiotropic effect of copper tolerance in *Mimulus guttatus*. *Heredity* 50: 295–302. [16]

Maddison, W. 1995. Phylogenetic histories within and among species. In P. C. Hoch and A. G. Stevenson (eds.), *Experimental and Molecular Approaches to Plant Biosystematics*, pp. 273–287. Monographs in Systematic Botany, 53. Missouri Botanical Garden, St. Louis. [11]

Maddison, W. P., and D. R. Maddison. 1992. *MacClade*, Version 3.0. Sinauer Associates, Sunderland, MA. [5]

Maddox, G. D., and R. B. Root. 1990. Structure of the selective encounter between goldenrod (*Solidago altissima*) and its diverse insect fauna. *Ecology* 71: 2115–2124. [18]

Mallet, J. 1993. Speciation, raciation, and color pattern evolution in *Heliconius* butterflies: Evidence from hybrid zones. In R. G. Harrison (ed.), *Hybrid Zones and the Evolutionary Process*, pp. 226–260. Oxford University Press, New York. [16]

Mallet, J., and N. Barton. 1989. Strong natural selection in a warning-color hybrid zone. *Evolution* 43: 421–431. [13]

Malmgren, B. A., W. A. Berggren, and G. P. Lohmann. 1983. Evidence for punctuated gradualism in the Late Neocene *Globorotalia tumida* lineage of planktonic Foraminifera. *Paleobiology* 9: 377–389. [6]

Mange, E. J., and A. P. Mange. 1993. *Basic Human Genetics*. Sinauer Associates, Sunderland, MA. [26]

Margulis, L. 1993. *Symbiosis in Cell Evolution*. Second edition. W. H. Freeman, San Francisco, CA. [7, 19]

Marshall, C. R., and P. D. Ward. 1996. Sudden and gradual molluscan extinctions in the latest Cretaceous of western European Tethys. *Science* 274: 1360–1363. [25]

Martin, A. P., G. J. P. Naylor, and S. R. Palumbi. 1992. Rates of mitochondrial DNA evolution in sharks are slow compared with mammals. *Nature* 357: 143–155. [11]

Martin, P. S., and R. G. Klein (eds.). 1984. *Quaternary Extinctions: A Prehistoric Revolution*. University of Arizona, Tucson. [8]

Martínez Wells, M., and C. S. Henry. 1992a. Behavioural responses of green lacewings (Neuroptera: Chrysopidae: *Chrysoperla*) to synthetic mating songs. *Anim. Behav.* 44: 641–652. [15]

Martínez Wells, M., and C. S. Henry. 1992b. The role of courtship songs in reproductive isolation among populations of green lacewings of the genus *Chrysoperla* (Neuroptera: Chrysopidae). *Evolution* 46: 31–43. [15]

Masterson, J. 1994. Stomatal size in fossil plants: Evidence for polyploidy in majority of angiosperms. *Science* 264: 421–424. [16]

Mather, K. 1949. *Biometrical Genetics: The Study of Continuous Variation*. Methuen, London. [3]

Mather, K. 1983. Response to selection. In M. Ashburner, H. L. Carson, and J. N. Thompson, Jr., (eds.), *The Genetics and*

Biology of Drosophila, pp. 155–221. Academic Press, New York. [14]

Mather, K., and B. J. Harrison. 1949. The manifold effect of selection. *Heredity* 3: 1–52; 131–162. [14]

Matthew, W. D., and W. Granger. 1917. The skeleton of *Diatryma*, a gigantic bird from the Lower Eocene of Wyoming. *Bull. Am. Mus. Nat. Hist.* 37: 307–326. [7]

Matton, D. P., N. Nass, A. G. Clark, and E. Newbigin. 1994. Self-incompatibility: How plants avoid illegitimate offspring. *Proc. Natl. Acad. Sci. USA* 91: 1992–1997. [21]

May, R. M. 1981. Population biology of parasitic infections. In K. J. Warren and E. F. Purcell (eds.), *The Current Status and Future of Parasitology*, pp. 208–235. Josiah Macy Jr. Foundation, New York. [4]

May, R. M. 1990. How many species? *Phil. Trans. R. Soc. Lond.* B 330: 293–304. [5]

May, R. M., and R. M. Anderson. 1983a. Epidemiology and genetics in the coevolution of parasites and hosts. *Proc. R. Soc. Lond.* B 219: 281–313. [18]

May, R. M., and R. M. Anderson. 1983b. Parasite-host coevolution. In D. J. Futuyma and M. Slatkin (eds.), *Coevolution*, pp. 186–206. Sinauer Associates, Sunderland, MA. [18]

Maynard Smith, J. 1958. *The Theory of Evolution.* Penguin Books, Baltimore, MD. [24]

Maynard Smith, J. 1960. Continuous, quantized and modal variation. *Proc. R. Soc. Lond.* B 152: 397–409. [23]

Maynard Smith, J. 1964. Group selection and kin selection. *Nature* 201: 1145–1147. [20]

Maynard Smith, J. 1966. Sympatric speciation. *Am. Nat.* 100: 637–650. [16]

Maynard Smith, J. 1971. *Mathematical Ideas in Biology.* Cambridge University Press, Cambridge. [17]

Maynard Smith, J. 1976. Group selection. *Q. Rev. Biol.* 51: 277–283. [12]

Maynard Smith, J. 1978a. *The Evolution of Sex.* Cambridge University Press, Cambridge. [21]

Maynard Smith, J. 1978b. Optimization theory in evolution. *Annu. Rev. Ecol. Syst.* 9: 31–56. [17, 20]

Maynard Smith, J. 1980. Selection for recombination in a polygenic model. *Genet. Res.* 35: 269–277. [21]

Maynard Smith, J. 1982. *Evolution and the Theory of Games.* Cambridge University Press, Cambridge. [20]

Maynard Smith, J. 1983. The genetics of stasis and punctuation. *Annu. Rev. Genet.* 17: 11–25. [24]

Maynard Smith, J. 1988. The evolution of recombination. In R. E. Michod and B. R. Levin (eds.), *The Evolution of Sex*, pp. 106–125. Sinauer Associates, Sunderland, MA. [21]

Maynard Smith, J. 1991. Theories of sexual selection. *Trends Ecol. Evol.* 6: 146–151. [20]

Maynard Smith, J. 1994. Estimating the minimum rate of genetic transformation in bacteria. *J. Evol. Biol.* 7: 525–534. [21]

Maynard Smith, J., and J. Haigh. 1974. The hitch-hiking effects of a favourable gene. *Genet. Res.* 23: 23–35. [22]

Maynard Smith, J., and R. Hoekstra. 1980. Polymorphism in a varied environment: How robust are the models? *Genet. Res.* 35: 45–57. [13]

Maynard Smith, J., and K. Sondhi. 1960. The genetics of a pattern. *Genetics* 35: 1039–1050. [14]

Maynard Smith, J., and E. Szathmáry. 1995. *The Major Transitions in Evolution.* W. H. Freeman, San Francisco, CA. [7]

Maynard Smith, J., et al. 1985. Developmental constraints and evolution. *Q. Rev. Biol.* 60: 265–287. [23]

Mayr, E. 1942. *Systematics and the Origin of Species.* Columbia University Press, New York. [15]

Mayr, E. 1947. Ecological factors in speciation. *Evolution* 1: 263–288. [16]

Mayr, E. 1954. Change of genetic environment and evolution. In J. Huxley, A. C. Hardy, and E. B. Ford (eds.), *Evolution as a Process*, pp. 157–180. Allen and Unwin, London. [16]

Mayr, E. 1960. The emergence of evolutionary novelties. In S. Tax (ed.), *The Evolution of Life*, pp. 157–180. University of Chicago Press, Chicago, IL. [24]

Mayr, E. 1963. *Animal Species and Evolution.* Harvard University Press, Cambridge, MA. [15]

Mayr, E. 1965. Avifauna: Turnover on islands. *Science* 150: 1587–1588. [25]

Mayr, E. 1982a. *The Growth of Biological Thought: Diversity, Evolution, and Inheritance.* Harvard University Press, Cambridge, MA. [2]

Mayr, E. 1982b. Processes of speciation in animals. In C. Barigozzi (ed.), *Mechanisms of Speciation*, pp. 1–19. Alan R. Liss, New York. [16]

Mayr, E. 1983. How to carry out the adaptationist program? *Am. Nat.* 121: 324–334. [14]

Mayr, E. 1988a. Cause and effect in biology. In E. Mayr (ed.), *Toward a New Philosophy of Biology*, pp. 24–37. Harvard University Press, Cambridge, MA. [22]

Mayr, E. 1988b. The probability of extraterrestrial intelligent life. In E. Mayr (ed.), *Toward a New Philosophy of Biology*, pp. 67–74. Harvard University Press, Cambridge, MA. [24]

Mayr, E. 1988c. *Toward a New Philosophy of Biology.* Harvard University Press, Cambridge, MA. [1]

Mayr, E. 1992. A local flora and the biological species concept. *Am. J. Bot.* 79: 222–238. [15]

Mayr, E. 1997. *This Is Biology.* Harvard University Press, Cambridge, MA.

Mayr, E., and P. D. Ashlock. 1990. *Principles of Systematic Zoology.* McGraw-Hill, New York. [5]

Mayr, E., and W. B. Provine (eds.). 1980. *The Evolutionary Synthesis: Perspectives on the Unification of Biology.* Harvard University Press, Cambridge, MA. [2]

Mayr, E., and C. Vaurie. 1948. Evolution in the family Dicruridae (birds). *Evolution* 3: 238–265. [9]

McCauley, D. E. 1993. Evolution in metapopulations with frequent local extinction and recolonization. In D. J. Futuyma and J. Antonovics (eds.), *Oxford Surv. Evol. Biol.* 9: 109–134. Oxford University Press, Oxford. [11]

McClintock, B. 1948. Mutable loci in maize. *Carnegie Institution of Washington Year Book* 47: 155–169. [22]

McCommas, S. A., and E. H. Bryant. 1990. Loss of electrophoretic variation in serially bottlenecked populations. *Heredity* 64: 315–321. [11]

McConkey, E. H. 1993. *Human Genetics: The Molecular Revolution.* Jones & Bartlett, Boston, MA. [26]

McDonald, J. H., and M. Kreitman. 1991. Adaptive protein evolution at the *Adh* locus in *Drosophila*. *Nature* 351: 652–654. [11]

McGowan, C. 1972. The distinction between latipinnate and longipinnate ichthyosaurs. *Roy. Ont. Mus., Life Sci. Occ. Papers* 20: 1–8. [23]

McGrew, W. C. 1992. *Chimpanzee Material Culture.* Cambridge University Press, Cambridge. [26]

McGuire, M. R., and L. L. Getz. 1981. Incest taboo between sibling *Microtus ochrogaster*. *J. Mammal.* 62: 213–215. [21]

McKaye, K. R., T. Kocher, P. Reinthal, R. Harrison, and I. Kornfield. 1984. Genetic evidence for allopatric and sympatric differentiation among color morphs of a Lake Malawi cichlid fish. *Evolution* 38: 215–219. [16]

McKenzie, J. A., and G. M. Clarke. 1988. Diazinon resistance, fluctuating asymmetry and fitness in the Australian sheep blowfly, *Lucilia cuprina*. *Genetics* 120: 213–220. [14]

McKey, D. 1979. The distribution of secondary compounds within plants. In G. A. Rosenthal and D. H. Janzen (eds.),

Herbivores: Their Interaction with Secondary Plant Metabolites, pp. 55–133. Academic Press, New York. [18]

McKinney, M. L., and K. J. McNamara. 1991. *Heterochrony: The Evolution of Ontogeny*. Plenum, New York. [23, 26]

McKitrick, M. C., and R. M. Zink. 1988. Species concepts in ornithology. *Condor* 90: 1–14. [15]

McMahon, T. A., and J. J. Bonner. 1983. *On Size and Life*. Scientific American Books, W. H. Freeman, New York. [17]

McNamara, K. J. 1988. Heterochrony and the evolution of echinoids. In C. R. C. Paul and A. B. Smith (eds.), *Echinoderm Phylogeny and Evolutionary Biology*, pp. 149–163. Clarendon Press, Oxford. [6]

McNeilly, T. 1968. Evolution in closely adjacent plant populations. III. *Agrostis tenuis* on a small copper mine. *Heredity* 23: 199–208. [13]

McNeilly, T., and J. Antonovics. 1968. Evolution in closely adjacent plant populations. IV. Barriers to gene flow. *Heredity* 23: 205–218. [16]

McShea, D. W. 1991. Complexity and evolution: What everybody knows. *Biol. Phil.* 6: 303–324. [24]

McShea, D. W. 1993. Evolutionary change in the morphological complexity of the mammalian vertebral column. *Evolution* 47: 730–740. [24]

McShea, D. W. 1994. Mechanisms of large-scale evolutionary trends. *Evolution* 48: 1747–1763. [24]

Medawar, P. B. 1952. *An Unsolved Problem of Biology*. H. K. Lewis, London. [19]

Meeuse, B. J. D. 1961. *The Story of Pollination*. Ronald Press, NY. [24]

Meffe, G. K., and C. R. Carroll. 1997. *Principles of Conservation Biology*. Second edition. Sinauer Associates, Sunderland, MA. [1, 24, 25]

Meffert, L. M., and E. H. Bryant. 1991. Mating propensity and courtship behavior in serially bottlenecked lines of the housefly. *Evolution* 45: 293–306. [16]

Meise, W. 1928. Die Verbreitung der Aaskrähe (Formenkreis *Corvus corone* L.). *J. Ornithol.* 76: 1–203. [15]

Menken, S. B. J. 1987. Is the extremely low heterozygosity level in *Yponomeuta rorellus* caused by bottlenecks? *Evolution* 41: 630–637. [16]

Meyer, A., and S. I. Dolven. 1992. Molecules, fossils, and the origin of tetrapods. *J. Mol. Evol.* 35: 102–113. [6]

Meyer, A., and A. C. Wilson. 1990. Origin of tetrapods inferred from their mitochondrial DNA affiliation to lungfish. *J. Mol. Evol.* 31: 359–364. [6]

Meyer, A., T. D. Kocher, P. Basasibwaki, and A. C. Wilson. 1990. Monophyletic origin of Lake Victoria cichlid fishes suggested by mitochondrial DNA sequences. *Nature* 347: 550–553. [15]

Michener, C. D. 1974. *The Social Behavior of the Bees: A Comparative Study*. Harvard University Press, Cambridge, MA. [20]

Michener, C. D., and D. A. Grimaldi. 1988. A *Trigona* from Late Cretaceous amber of New Jersey. *Am. Mus. Novitates* 2917: 1–10. [7]

Michener, C. D., and R. R. Sokal. 1957. A quantitative approach to a problem in classification. *Evolution* 11: 130–162. [5]

Michod, R. E. 1982. The theory of kin selection. *Annu. Rev. Ecol. Syst.* 13: 23–55. [20]

Michod, R. E., and B. R. Levin (eds.). 1988. *The Evolution of Sex: An Examination of Current Ideas*. Sinauer Associates, Sunderland, MA. [21]

Millais, J. G. 1897. *British Deer and Their Horns*. Henry Sotheran and Co., London.

Miller, R. B. 1981. Hawkmoths and the geographic patterns of floral variation in *Aquilegia caerulea*. *Evolution* 35: 763–774. [16]

Miller, S. L. 1953. Production of amino acids under possible primitive earth conditions. *Science* 117: 528–529. [7]

Mitchell, R. S., and E. D. Beal. 1979. Contributions to a Flora of New York State II: Magnoliaceae through Ceratophyllaceae. New York State Museum Bulletin 435, Albany. [24]

Mitchell-Olds, T. 1995. The molecular basis of quantitative genetic variation in natural populations. *Trends Ecol. Evol.* 10: 324–328. [14]

Mitter, C., and B. Farrell. 1991. Macroevolutionary aspects of insect-plant relationships. In E. Bernays (ed.), *Insect-Plant Interactions* III, pp. 35–78. CRC Press, Boca Raton, FL. [18]

Mitter, C., B. Farrell, and D. J. Futuyma. 1991. Phylogenetic studies of insect-plant interactions: Insights into the genesis of diversity. *Trends Ecol. Evol.* 6: 290–293. [18]

Mitter, C., B. Farrell, and B. Wiegmann. 1988. The phylogenetic study of adaptive zones: Has phytophagy promoted insect diversification? *Am. Nat.* 132: 107–128. [25]

Mittler, J. E., and R. Lenski. 1992. Experimental evidence for an alternative to directed mutation in the *bgl* operon. *Nature* 356: 446–448. [10]

Mitton, J. B., and M. C. Grant. 1984. Associations among protein heterozygosity, growth rate, and developmental homeostasis. *Annu. Rev. Ecol. Syst.* 15: 479–499. [13]

Miyamoto, M. M., and M. Goodman. 1986. Biomolecular systematics of eutherian mammals: Phylogenetic patterns and classification. *Syst. Zool.* 35: 230–240. [7]

Miyamoto, M. M., B. F. Koop, J. L. Slightom, M. Goodman, and M. R. Tennant. 1988. Molecular systematics of higher primates: Genealogical relations and classification. *Proc. Natl. Acad. Sci. USA* 85: 7627–7631. [5]

Miyamoto, M. M., J. L. Slightom, and M. Goodman. 1987. Phylogenetic relations of humans and African apes from DNA sequences in the ψη-globin region. *Science* 238: 369–373. [5]

Møller, A. P. 1994. *Sexual Selection and the Barn Swallow*. Oxford University Press, Oxford. [20]

Montgomery, S. L. 1982. Biogeography of the moth genus *Eupithecia* in Oceania and the evolution of ambush predation in Hawaiian caterpillars (Lepidoptera: Geometridae). *Entomologia Generalis* 8: 27–34. [25]

Moore, J. A. 1957. An embryologist's view of the species concept. In E. Mayr (ed.), *The Species Problem*, pp. 325–338. American Association for the Advancement of Science, Washington, DC. [16]

Moore, J. A. 1961. A cellular basis for genetic isolation. In W. F. Blair (ed.), *Vertebrate Speciation*, pp. 62–68. University of Texas Press, Austin. [15]

Moore, W. S., and J. T. Price. 1993. Nature of selection in the northern flicker hybrid zone and its implications for speciation theory. In R. G. Harrison (ed.), *Hybrid Zones and the Evolutionary Process*, pp. 196–225. Oxford University Press, New York. [9]

Moran, N. A., and P. Baumann. 1994. Phylogenetics of cytoplasmically inherited microorganisms of arthropods. *Trends Ecol. Evol.* 9: 15–20. [18]

Moran, P., and I. Kornfield. 1993. Retention of an ancestral polymorphism in the mbuna species flock (Teleostei: Cichlidae) of Lake Malawi. *Mol. Biol. Evol.* 10: 1015–1029. [15]

Morgan, T. H. 1903. *Evolution and adaptation*. MacMillan, NY. [2]

Moritz, C., C. J. Schneider, and D. B. Wake. 1992. Evolutionary relationships within the *Ensatina eschscholtzii* complex confirm the ring species interpretation. *Syst. Biol.* 41: 273–291. [15]

Morton, W. F., J. F. Crow, and H. J. Muller. 1956. An estimate of the mutational damage in man from data on consanguineous marriages. *Proc. Natl. Acad. Sci. USA* 42: 855–863. [9]

Mousseau, T. A., and H. Dingle. 1991. Maternal effects in insect life histories. *Annu. Rev. Entomol.* 36: 511–534. [14]

Mousseau, T. A., and D. A. Roff. 1987. Natural selection and the heritability of fitness components. *Heredity* 59: 181–197. [14, 19]

Moyle, P. B. 1993. *Fish: An Enthusiast's Guide.* University of California Press, Berkeley. [4]

Moyle, P. B., and J. J. Cech Jr. 1982. *Fishes: An Introduction to Ichthyology.* Prentice-Hall, Englewood Cliffs, NJ. [4, 6]

Moy-Thomas, J. A., and R. S. Miles. 1971. *Palaeozoic fishes.* Second edition. Chapman & Hall, London. [6]

Mukai, T., S. I. Chigusa, L. E. Mettler, and J. F. Crow. 1972. Mutation rate and dominance of genes affecting viability in *Drosophila melanogaster. Genetics* 72: 335–355. [10]

Müller, G. B. 1989. Ancestral patterns in bird limb development: A new look at Hampé's experiment. *J. Evol. Biol.* 2: 31–47. [5]

Müller, G. B. 1990. Developmental mechanisms at the origin of morphological novelty: A side-effect hypothesis. In M. H. Nitecki (ed.), *Evolutionary Innovations*, pp. 99–130. University of Chicago Press, Chicago, IL. [23]

Müller, G. B. 1991. Experimental strategies in experimental embryology. *Am. Zool.* 31: 605–615. [23]

Müller, G. B., and G. P. Wagner. 1991. Novelty in evolution: Restructuring the concept. *Annu. Rev. Ecol. Syst.* 23: 229–256. [23]

Müller, G. B., and G. P. Wagner. 1996. Homology, *Hox* genes, and developmental integration. *Am. Zool.* 36: 4–13. [23]

Muller, H. J. 1939. Reversibility in evolution considered from the standpoint of genetics. *Biol. Rev.* 14: 261–280. [16]

Muller, H. J. 1940. Bearing of the *Drosophila* work on systematics. In J. S. Huxley (ed.), *The New Systematics*, pp. 185–268. Clarendon Press, Oxford. [15, 16]

Muller, H. J. 1942. Isolating mechnisms, evolution, and temperature. *Biol. Symp.* 6: 71–125. [15]

Muller, H. J. 1964. The relation of recombination to mutational advance. *Mutat. Res.* 1: 2–9. [21]

Munger, J. C., and J. H. Brown. 1981. Competition in desert rodents: An experiment with semipermeable exclosures. *Science* 211: 510–512. [4]

Müntzing, A. 1930. Über Chromosomenvermehrung in Galeopsis-Kreuzungen und ihre phylogenetische Bedeutung. *Hereditas* 14: 153–172. [16]

Murray, J. D. 1981. A pre-pattern formation mechanism for animal coat markings. *J. Theor. Biol.* 88: 161–199. [23]

Murray, J., E. Murray, M. S. Johnson, and B. Clarke. 1988. The extinction of *Partula* on Moorea. *Pacific Sci.* 42: 150–153. [8]

Myers, A. A., and P. S. Giller (eds.). 1988. *Analytical Biogeography.* Chapman & Hall, London. [8]

Nagylaki, T., and T. D. Petes. 1982. Intrachromsomal gene conversion and the maintenance of sequence homogeneity among repeated genes. *Genetics* 100: 315–337. [22]

Nason, J. D., N. C. Ellstrand, and M. L. Arnold. 1992. Patterns of hybridization and introgression in populations of oaks, manzanitas, and irises. *Am. J. Bot.* 79: 101–111. [15]

National Geographic Society. 1987. *Field Guide to the Birds of North America.* Second edition. National Geographic Society, Washington, D.C. [15, 20]

Naveira, H. F. 1992. Location of X-linked polygenic effects causing sterility in male hybrids of *Drosophila simulans* and *D. mauritiana. Heredity* 67: 57–72. [15]

Neel, J. V. 1983. Frequency of spontaneous and induced "point" mutations in higher eukaryotes. *J. Hered.* 74: 2–15. [10]

Nei, M. 1987. *Molecular Evolutionary Genetics.* Columbia University Press, New York. [9]

Nei, M., and A. L. Hughes. 1991. Polymorphism and evolution of the major histocompatibility complex loci in mammals. In R. K. Selander, A. G. Clark, and T. S. Whittam (eds.), *Evolution at the Molecular Level*, pp. 222–247. Sinauer Associates, Sunderland, MA. [22]

Nei, M., and A. K. Roychoudhury. 1982. Genetic relationship and evolution of human races. *Evol. Biol.* 14: 1–59. [26]

Nei, M., T. Maruyama, and R. Chakraborty. 1975. The bottleneck effect and genetic variability in populations. *Evolution* 29: 1–10. [11]

Nei, M., T. Maruyama, and C.-I. Wu. 1983. Models of evolution of reproductive isolation. *Genetics* 103: 557–579. [16]

Neigel, J. E., and J. C. Avise. 1986. Phylogenetic relationship of mitochondrial DNA under various demographic models of speciation. In E. Nevo and S. Karlin (eds.), *Evolutionary Processes and Theory*, pp. 515–534. Academic Press, London. [15]

Nelson, G., and N. Platnick. 1981. *Systematics and Biogeography: Cladistics and Vicariance.* Columbia University Press, New York. [8]

Nestmann, E. R., and R. F. Hill. 1973. Population genetics in continuously growing mutator cultures of *Escherichia coli. Genetics* 73: 41–44. [12]

Neu, H. C. 1992. The crisis in antibiotic resistance. *Science* 257: 1064–1073. [26]

Nevo, E. 1991. Evolutionary theory and processes of active speciation and adaptive radiation in subterranean mole rats, *Spalax ehrenbergi* superspecies, in Israel. *Evol. Biol.* 25: 1–125. [15]

Nijhout, H. F. 1991. *The Development and Evolution of Butterfly Wing Patterns.* Smithsonian Institution Press, Washington, DC. [5, 23]

Niklas, K. J., B. H. Tiffney, and A. H. Knoll. 1980. Apparent changes in the diversity of fossil plants. *Evol. Biol.* 12: 1–89. [25]

Niklas, K. J., B. H. Tiffney, and A. H. Knoll. 1983. Patterns in vascular land plant diversification. *Nature* 303: 614–616. [25]

Nilsson, L. A. 1983. Processes of isolation and introgressive interplay between *Platanthera bifolia* (L.) Rich. and *P. chlorantha* (Custer) Reichb. (Orchidaceae). *Bot. J. Linn. Soc.* 87: 325–350. [15]

Nilsson, L. A. 1992. Orchid pollination biology. *Trends Ecol. Evol.* 7: 255–259. [12]

Nilsson, L. A., L. Jonsson, L. Ralison, and E. Randrianjohany. 1985. Monophily and pollination mechanisms in *Angraecum arachnites* Schltr. (Orchidaceae) in a guild of long-tongued hawkmoths (Sphingidae). *Biol. J. Linn. Soc.* 26: 1–19. [18]

Nisbett, R. 1995. Race, IQ, and scientism. In S. Fraser (ed.), *The Bell Curve Wars*, pp. 36–57. BasicBooks, New York. [26]

Nitecki, M. H. (ed.). 1988. *Evolutionary Progress.* University of Chicago Press, Chicago, IL. [12, 24]

Nitecki, M. H. (ed.). 1990. *Evolutionary Innovations.* University of Chicago Press, Chicago, IL. [24]

Nixon, K. C., and Q. D. Wheeler. 1990. An amplification of the phylogenetic species concept. *Cladistics* 6: 211–233. [15]

Noble, G. K. 1931. *The Biology of the Amphibia.* McGraw-Hill, New York. [5]

Noodén, L. D. 1988. Whole plant senescence. In L. D. Noodén and A. C. Leopold (eds.), *Senescence and Aging in Plants*, pp. 391–439. Academic Press, San Diego, CA. [19]

Norton, R. A., J. B. Kethley, D. E. Johnston, and B. M. O'Connor. 1993. Phylogenetic perspectives in genetic systems and reproductive modes of mites. In D. Wrensch and M.

Ebbert (eds.), *Evolution and Diversity of Sex Ratio in Insects and Mites*, pp. 8–99. Chapman & Hall, London. [21]

Novacek, M. J., A. R. Wyss, and M. C. McKenna. 1988. The major groups of eutherian mammals. In M. J. Benton (ed.), *The Phylogeny and Classification of the Tetrapods*, vol. 2, pp. 31–71. Clarendon Press, Oxford. [7]

Nowak, R. M. 1991. *Walker's Mammals of the World*. Johns Hopkins University Press, Baltimore, MD. [23]

Nowak, M. A., and R. M. May. 1994. Superinfection and the evolution of parasite virulence. *Proc. R. Soc. Lond.* B 255: 81–89. [18]

Nur, N. 1984. The consequences of brood size for breeding blue tits. I. Adult survival, weight change and cost of reproduction. *J. Anim. Ecol.* 53: 479–496. [19]

Oakeshott, J. G., J. B. Gibson, P. R. Anderson, W. R. Knib, D. G. Anderson, and G. K. Chambers. 1982. Alcohol dehydrogenase and glycerol-3-phosphate dehydrogenase clines in *Drosophila melanogaster* on different continents. *Evolution* 36: 86–96. [9]

Odling-Smee, F. J., K. N. Laland, and M. W. Feldman. 1996. Niche construction. *Am. Nat.* 147: 641–648. [20]

O'Donald, P. 1962. The theory of sexual selection. *Heredity* 17: 541–552. [20]

Ohno, S. 1970. *Evolution by Gene Duplication*. Springer-Verlag, New York. [22]

Ohno, S., C. Stenius, L. Christian, and G. Schipmann. 1969. *De novo* mutation-like events observed at the 6PGD locus of the Japanese quail, and the principle of polymorphism breeding more polymorphisms. *Biochem. Genet.* 3: 417–428. [10]

Ohta, T. 1988. Multigene and supergene families. *Oxford Surv. Evol. Biol.* 5: 41–65. [22]

Ohta, T. 1994. Further examples of evolution by gene duplication revealed through DNA sequence comparisons. *Genetics* 138: 1331–1337. [22]

Ohta, T., and G. A. Dover. 1984. The cohesive population genetics of molecular drive. *Genetics* 108: 501–521. [22]

Olson, E. C., and R. L. Miller. 1958. *Morphological Integration*. University of Chicago Press, Chicago, IL. [14]

Orgel, L. E. 1973. *The Origins of Life: Molecular and Natural Selection*. Wiley, New York. [22]

Orgel, L. E. 1994. The origin of life on earth. *Sci. Am.* October: 77–91. [7]

Orgel, L. E., and F. H. C. Crick. 1980. Selfish DNA: The ultimate parasite. *Nature* 284: 604–606. [22]

Orians, G. H., and R. T. Paine. 1983. Convergent evolution at the community level. In D. J. Futuyma and M. Slatkin (eds.), *Coevolution*, pp. 431–458. Sinauer Associates, Sunderland, MA. [8]

Oring, L. W. 1986. Avian polyandry. In R. F. Johnson (ed.), *Current Ornithology*, pp. 309–351. Plenum, New York. [20]

Orr, H. A. 1989. Genetics of sterility in hybrids between two subspecies of *Drosophila*. *Evolution* 43: 180–189. [15]

Orr, H. A., and J. Coyne. 1993. The genetics of adaptation: A reassessment. *Am. Nat.* 140: 725–742. [24]

Ortí, G., and A. Meyer. 1997. The radiation of characiform fishes and the limits of resolution of mitochondrial ribosomal DNA sequences. *Syst. Biol.* 46: 75–100.

Osborn, H. F., and C. C. Mook. 1921. *Camarasaurus, Amphicoelias* and other sauropods of Cope. *Mem. Am. Mus. Nat. Hist.*, n.s., 3(3): 245–387. [7]

Osorio, D. 1994. Eye evolution: Darwin's shudder stilled. *Trends Ecol. Evol.* 9: 241–242. [24]

Oster, G., and P. Alberch. 1982. Evolution and the bifurcation of developmental programs. *Evolution* 36: 444–459. [23]

Oster, G. F., and E. O. Wilson. 1978. *Caste and ecology in the social insects*. Princeton University Press, Princeton, NJ. [12]

Ostrom, J. H. 1976. *Archaeopteryx* and the origin of birds. *Biol. J. Linn. Soc.* 8: 91–182. [6, 12]

Otte, D., and J. A. Endler (eds.). *Speciation and Its Consequences*. Sinauer Associates, Sunderland, MA. [15]

Owen, D. 1980. *Camouflage and Mimicry*. University of Chicago Press, Chicago, IL. [4]

Ownbey, M. 1950. Natural hybridization and amphiploidy in the genus *Tragopogon*. *Am. J. Bot.* 37: 489–499. [16]

Page, R. D. M. 1988. Quantitative cladistic biogeography: Constructing and comparing area cladograms. *Syst. Zool.* 37: 254–270. [8]

Page, R. D. M. 1990. Temporal congruence and cladistic analysis of biogeography and cospeciation. *Syst. Zool.* 39: 205–226. [18]

Page, R. D. M. 1993. Genes, organisms, and areas: The problem of multiple lineages. *Syst. Biol.* 42: 77–84. [18]

Page, R. D. M. 1994. Maps between trees and cladistic analyses of historical associations among genes, organisms, and areas. *Syst. Biol.* 43: 58–77. [8]

Palmer, A. R. 1982. Predation and parallel evolution: Recurrent parietal plate reduction in balanomorph barnacles. *Paleobiology* 8: 31–44. [6]

Pamilo, P., and M. Nei. 1988. Relationships between gene trees and species trees. *Mol. Biol. Evol.* 5: 568–583. [11, 15]

Paradis, J., and G. C. Williams. 1989. *Evolution and Ethics: T. H. Huxley's Evolution & Ethics with New Essays on its Victorian and Sociobiological Context*. Princeton University Press, Princeton, NJ. [26]

Parker, G. A. 1970. Sperm competition and its evolutionary consequences in the insects. *Biol. Rev.* 45: 525–567. [20]

Parker, G. A. 1984. Evolutionarily stable strategies. In J. R. Krebs and N. B. Davies (eds.), *Behavioural Ecology*, Second edition, pp. 30–61. Blackwell Scientific, Oxford. [20]

Parker, G. A., and D. Mock. 1987. Parent-offspring conflict over clutch size. *Evol. Ecol.* 1: 161–174. [19]

Parker, S. P. (ed.). 1982. *Synopsis and Classification of Living Organisms*. McGraw-Hill, New York. [7]

Parmigiani, S., F. vom Saal, and B. Svare (eds.). 1992. *Infanticide and Parental Care*. Harwood Academic Press, London. [20]

Parsley, R. L. 1988. Feeding and respiratory strategies in Stylophora. In C. R. C. Paul and A. B. Smith (eds.), *Echinoderm Phylogeny and Evolutionary Biology*, pp. 347–361. Clarendon Press, Oxford. [7]

Paterson, H. E. H. 1982. Perspectives on speciation by reinforcement. *S. Afr. J. Sci.* 78: 53–57. [16]

Paterson, H. E. H. 1985. The recognition concept of species. In E. S. Vrba (ed.), *Species and Speciation*, pp. 21–29. Transvaal Museum Monograph No. 4, Pretoria, South Africa. [15]

Patterson, C. 1982. Morphological characters and homology. In K. A. Joysey and A. E. Friday (eds.), *Problems of Phylogenetic Reconstruction*, pp. 21–74. Academic Press, London. [23]

Patterson, C., D. M. Williams, and C. J. Humphries. 1993. Congruence between molecular and morphological phylogenies. *Annu. Rev. Ecol. Syst.* 24: 153–188. [5]

Patton, J. L. 1969. Chromosome evolution in the pocket mouse, *Perognathus goldmani* Osgood. *Evolution* 23: 645–662. [10]

Patton, J. L. 1972. Patterns of geographic variation in karyotype in the pocket gopher, *Thomomys bottae* (Eydoux and Gervais). *Evolution* 20: 574–586. [11]

Patton, J. L., and S. Y. Yang. 1977. Genetic variation in *Thomomys bottae* pocket gophers: Macrogeographic patterns. *Evolution* 31: 697–720. [11]

Paul, G. 1988. *Predatory Dinosaurs of the World*. Simon and Schuster, New York. [6]

Pease, C. M. 1992. On the declining extinction and origination rates of modern taxa. *Paleobiology* 18: 89–92. [25]

Pellmyr, O. 1986. Three pollination morphs in *Cimicifuga simplex*; incipient speciation due to inferiority in competition. *Oecologia* 68: 304–307. [16]

Pellmyr, O., and C. J. Huth. 1994. Evolutionary stability of mutualism between yuccas and yucca moths. *Nature* 372: 257–260. [18]

Pellmyr, O., and J. N. Thompson. 1992. Multiple occurrences of mutualism in the yucca moth lineage. *Proc. Natl. Acad. Sci. USA* 89: 2927–2929. [18]

Pellmyr, O., J. Leebens-Mack, and C. J. Huth. 1996a. Non-mutualistic yucca moths and their evolutionary consequences. *Nature* 380: 155–156. [18]

Pellmyr, O., J. N. Thompson, J. M. Brown, and R. G. Harrison. 1996b. Evolution of pollination and mutualism in the yucca moth lineage. *Am. Nat.* 148: 827–847. [18]

Pennycuick, C. J. 1975. Mechanics of flight. In D. S. Farner, J. R. King, and K. C. Parkes (eds.), *Avian Biology*, pp. 1–75. Academic Press, New York. [17]

Pericak-Vance, M. A., and J. L. Haines. 1995. Genetic susceptibility to Alzheimer Disease. *Trends Genet.* 11: 504–508. [26]

Perrot, V., S. Richerd, and M. Valeró. 1991. Transition from haploidy to diploidy. *Nature* 351: 315–317. [21]

Peters, R. H. 1983. *The Ecological Implications of Body Size*. Cambridge University Press, Cambridge. [17]

Peterson, R. T., and M. McKenny. 1968. *A Field Guide to the Wildflowers*. Houghton Mifflin, New York. [17]

Petit, C., and L. Ehrman. 1969. Sexual selection in *Drosophila*. *Evol. Biol.* 3: 177–223. [13]

Pfennig, D. W., H. K. Reeve, and P. W. Sherman. 1993. Kin recognition and cannibalism in spadefoot toad tadpoles. *Anim. Behav.* 46: 87–94. [20]

Pfennig, D. W., P. W. Sherman, and J. P. Collins. 1994. Kin recognition and cannibalism in polyphenic salamanders. *Behav. Ecol.* 5: 225–232. [20]

Pianka, E. R. 1966. Latitudinal gradients in species diversity: A review of concepts. *Am. Nat.* 100: 33–46. [25]

Pianka, E. R. 1970. On *r*- and *K*-selection. *Am. Nat.* 104: 592–597. [19]

Pianka, E. R. 1983. *Evolutionary Ecology*. Harper & Row, New York. [19, 25]

Pianka, E. R. 1986. *Ecology and Natural History of Desert Lizards: Analysis of the Ecological Niche and Community Structure*. Princeton University Press, Princeton, NJ. [8]

Pielou, E. C. 1991. *After the Ice Age: The Return of Life to Glaciated North America*. University of Chicago Press, Chicago, IL. [7]

Pierce, G. J., and J. G. Ollason. 1987. Eight reasons why optimal foraging theory is a complete waste of time. *Oikos* 49: 111–118. [20]

Pimentel, D., et al. 1992. Environmental and economic costs of pesticide use. *Bioscience* 42: 750–760. [1]

Pimm, S. L. 1982. *Food Webs*. Chapman & Hall, London. [4]

Piñero, D., J. Sarukhán, and P. Alberdi. 1982. The costs of reproduction in a tropical palm, *Astrocaryum mexicanum*. *J. Ecol.* 70: 473–481. [19]

Pinker, S., and P. Bloom. 1992. Natural language and natural selection. In J. B. Barkow, L. Cosmides, and J. Tooby (eds.), *The Adapted Mind*, pp. 451–493. Oxford University Press, New York. [26]

Plomin, R. 1994. *Genetics and Experience: The Interplay Between Nature and Nurture*. Sage Publications, Thousand Oaks, CA. [26]

Plomin, R., and G. McClearn (eds.). 1993. *Nature, Nurture & Psychology*. American Psychological Association, Washington, DC. [26]

Plotkin, H. C. 1988. Learning and evolution. In H. C. Plotkin (ed.), *The Role of Behavior in Evolution*, pp. 133–164. MIT Press, Cambridge, MA. [20]

Policansky, D. 1981. Sex choice and the size-advantage model in jack-of-the-pulpit (*Arisaema triphyllum*). *Proc. Natl. Acad. Sci. USA* 78: 1306–1308. [21]

Pomiankowski, A. 1988. The evolution of female mate preferences for male genetic quality. *Oxford Surv. Evol. Biol.* 5: 136–184. [16]

Pomiankowski, A., and Y. Iwasa. 1993. Evolution of multiple sexual preferences by Fisher's runaway process of sexual selection. *Proc. R. Soc. Lond.* B 253: 173–181. [20]

Popham, E. J. 1942. Further experimental studies on the selective action of predators. *Proc. Zool. Soc. Lond.* 112: 105–117. [13]

Porter, K. R. 1972. *Herpetology*. W. B. Saunders, Philadelphia, PA. [12]

Potts, R. 1988. *Early Hominid Activity at Olduvai*. Aldine de Gruyter, New York. [26]

Potts, W. K., C. J. Manning, and E. K. Wakeland. 1991. Mating patterns in seminatural populations of mice influenced by MHC genotype. *Nature* 352: 619–621. [21]

Pough, R. H. 1951. *Audubon Water Bird Guide*. Doubleday, New York. [9]

Press, F., and R. Siever. 1982. *Earth*. W. H. Freeman, San Francisco, CA. [6]

Price, M. V., and N. M. Waser. 1982. Population structure, frequency-dependent selection, and the maintenance of sexual reproduction. *Evolution* 36: 35–43. [21]

Price, P. W. 1980. *Evolutionary Biology of Parasites*. Princeton University Press, Princeton, NJ. [4]

Price, T., M. Turelli, and M. Slatkin. 1993. Peak shifts produced by correlated responses to selection. *Evolution* 47: 280–290. [16]

Proctor, N. S., and P. J. Lynch. 1993. *Manual of Ornithology: Avian Structure and Function*. Yale University Press, New Haven, CT. [4]

Promislow, D. E. L., and P. H. Harvey. 1991. Mortality rates and the evolution of mammalian life histories. *Acta Oecologica* 220: 417–437. [19]

Prosser, C. L. (ed.). 1973. *Comparative Animal Physiology*. Saunders, Philadelphia, PA. [17]

Prossser, C. L. 1986. *Adaptational Biology: Molecules to Organisms*. Wiley, New York. [17]

Prosser, C. L., and F. A. Brown Jr. 1961. *Comparative Animal Physiology*. Saunders, Philadelphia, PA. [4]

Provine, W. B. 1971. *The Origins of Theoretical Population Genetics*. University of Chicago Press, Chicago, IL. [2]

Provine, W. B. 1986. *Sewall Wright and Evolutionary Biology*. University of Chicago Press, Chicago, IL. [14]

Pruett-Jones, S. G., and M. A. Pruett-Jones. 1990. Sexual selection through female choice in Lawes' Parotia, a lek-mating bird of paradise. *Evolution* 44: 486–501. [16]

Prum, R. O. 1990. Phylogenetic analysis of the evolution of display behavior in the neotropical manakins (Aves: Pipridae). *Ethology* 84: 202–231. [20]

Pulliam, H. R. 1985. Foraging efficiency, resource partitioning, and the coexistence of sparrow species. *Ecology* 66: 1829–1836. [17]

Pulliam, H. R. 1988. Sources, sinks, and population regulation. *Am. Nat.* 132: 652–661. [25]

Purnell, M. A., and D. H. von Bitter. 1992. Blade-shaped conodont elements functioned as cutting teeth. *Nature* 359: 629–631. [7]

Purves, W. K., G. H. Orians, H. C. Heller, and D. Sadava. 1998. *Life: The Science of Biology*. Fifth edition. Sinauer Associates, Sunderland, MA. [3, 5]

Pusey, A. E. 1980. Inbreeding avoidance in chimpanzees. *Anim. Behav.* 28: 543–552. [21]

Pyke, G. H. 1984. Foraging theory: A critical review. *Annu. Rev. Ecol. Syst.* 15: 523–575. [20]

Radinsky, L. B. 1966. The adaptive radiation of the phenacodontid condylarths and the origin of the Perissodactyla. *Evolution* 20: 408–417. [6]

Radinsky, L. B. 1984. Ontogeny and phylogeny in horse skull evolution. *Evolution* 38: 1–15. [6, 23]

Raff, R. A. 1996. *The Shape of Life: Genes, Development, and the Evolution of Animal Form*. University of Chicago Press, Chicago, IL. [23]

Raff, R. A., and T. C. Kaufman. 1983. *Embryos, Genes, and Evolution: The Developmental-Genetic Basis of Evolutionary Change*. Macmillan, New York. [3]

Ralls, K., P. H. Harvey, and A. M. Lyles. 1986. Inbreeding in natural populations of birds and mammals. In M. E. Soulé (ed.), *Conservation Biology: The Science of Scarcity and Diversity*, pp. 35–56. Sinauer Associates, Sunderland, MA. [21]

Rand, D. M., and R. G. Harrison. 1989. Ecological genetics of a mosaic hybrid zone: Mitochondrial, nuclear, and reproductive differentiation of crickets by soil type. *Evolution* 43: 432–449. [15]

Raup, D. M. 1962. Computer as aid in describing form in gastropod shells. *Science* 138: 150–152. [23]

Raup, D. M. 1966. Geometric analysis of shell coiling: General problems. *J. Paleontol.* 40: 1178–1190. [23]

Raup, D. M. 1972. Taxonomic diversity during the Phanerozoic. *Science* 177: 1065–1071. [25]

Raup, D. M. 1977. Stochastic models in evolutionary paleontology. In A. Hallam (ed.), *Patterns of Evolution*, pp. 59–78. Elsevier, Amsterdam. [24]

Raup, D. M., and S. J. Gould. 1974. Stochastic simulation and the evolution of morphology—towards a nomothetic paleontology. *Syst. Zool.* 23: 305–322. [24]

Raup, D. M., and J. J. Sepkoski Jr. 1982. Mass extinctions in the marine fossil record. *Science* 215: 1501–1503. [25]

Raup, D. M., and J. J. Sepkoski Jr. 1984. Periodicity of extinctions in the geologic past. *Proc. Natl. Acad. Sci. USA* 81: 801–805. [25]

Raup, D. M., S. J. Gould, T. J. M. Schopf, and D. S. Simberloff. 1973. Stochastic models of phylogeny and the evolution of diversity. *J. Geol.* 8: 525–542. [25]

Rausher, M. D. 1984. The evolution of habitat preference in subdivided populations. *Evolution* 38: 596–608. [13]

Rausher, M. D. 1993. The evolution of habitat preference: Avoidance and adaptation. In K. C. Kim and B. A. McPheron (eds.), *Evolution of Insect Pests: Patterns of Variation*, pp. 259–283. Wiley, New York. [14, 20]

Raven, P. H. 1976. Systematics and plant population biology. *Syst. Bot.* 1: 284–316. [15]

Raven, P. H., R. F. Evert, and S. E. Eichorn. 1992. *Biology of Plants*. Fifth edition. Worth Publishing Co., New York. [16]

Raymond, M., A. Callaghan, P. Fort, and N. Pasteur. 1991. Worldwide migration of amplified insecticide resistance genes in mosquitoes. *Nature* 350: 151–153. [10]

Reed, S. C., and E. W. Reed. 1950. Natural selection in laboratory populations of *Drosophila*. *Evolution* 4: 34–42. [12]

Rees, H., G. Jenkins, A. G. Seal, and J. Hutchinson. 1982. Assays of the phenotypic effects of changes in DNA amounts. In G. A. Dover and R. Flavell (eds.), *Genome Evolution*, pp. 287–297. Academic Press, London. [22]

Reeve, H. K., and P. W. Sherman. 1993. Adaptation and the goals of evolutionary research. *Q. Rev. Biol.* 68: 1–32. [12, 14]

Regan, C. T. 1914. Fishes. British Antarctic Expedition, 1910. *Zoology* 1: 1–54. [17]

Rehr, S. S., P. P. Feeny, and D. H. Janzen. 1973. Chemical defenses in Central American non-ant acacias. *J. Anim. Ecol.* 42: 405–416. [18]

Remington, C. L. 1968. Suture-zones of hybrid interaction between recently joined biotas. *Evol. Biol.* 2: 321–428. [15]

Rendel, J. M. 1967. *Canalisation and Gene Control*. Logos Press, London. [14]

Rensch, B. 1959. *Evolution Above the Species Level*. Columbia University Press, New York. [24]

Reyer, H.-U. 1990. Pied kingfishers: Ecological causes and reproductive consequences of cooperative breeding. In P. B. Stacey and W. D. Koenig (eds.), *Cooperative Breeding in Birds*, pp. 527–557. Cambridge University Press, Cambridge. [20]

Reznick, D. 1983. The structure of guppy life histories: The tradeoff between growth and reproduction. *Ecology* 64: 862–873. [19]

Reznick, D. 1985. Cost of reproduction: An evaluation of the empirical evidence. *Oikos* 44: 257–267. [19]

Reznick, D., and J. A. Endler. 1982. The impact of predation on life history evolution in Trinidadian guppies (*Poecilia reticulata*). *Evolution* 36: 160–177. [19]

Reznick, D., H. Bryga, and J. A. Endler. 1990. Experimentally induced life-history evolution in a natural population. *Nature* 346: 357–359. [19]

Rice, W. R. 1996. Evolution of the Y sex chromosome in animals. *BioScience* 46: 331–343. [21]

Rice, W. R., and E. E. Hostert. 1993. Laboratory experiments on speciation: What have we learned in forty years? *Evolution* 47: 1637–1653. [16]

Rice, W. R., and G. W. Salt. 1990. The evolution of reproductive isolation as a correlated character under sympatric conditions: Experimental evidence. *Evolution* 44: 1140–1152. [16]

Ricklefs, R. E. 1989. Speciation and diversity: Integration of local and regional processes. In D. Otte and J. A. Endler (eds.), *Speciation and Its Consequences*, pp. 599–622. Sinauer Associates, Sunderland, MA. [8]

Ricklefs, R. E. 1990. *Ecology*. Third edition. W. H. Freeman, New York. [17]

Ricklefs, R. E., and D. Schluter (eds.). 1993a. *Species Diversity in Ecological Communities*. University of Chicago Press, Chicago, IL. [25]

Ricklefs, R. E., and D. Schluter. 1993b. Species diversity: Regional and historical influences. In R. E. Ricklefs and D. Schluter (eds.), *Species Diversity in Ecological Communities*, pp. 350–363. University of Chicago Press, Chicago, IL. [25]

Ridley, M. 1983. *The Explanation of Organic Diversity: The Comparative Method and Adaptations for Mating*. Oxford University Press, Oxford. [12]

Riedl, H., and B. A. Croft. 1978. The effects of photoperiod and effective temperatures on the seasonal phenology of the codling moth (Lepidoptera: Tortricidae). *Can. Entomol.* 110: 455–470. [14]

Riedl, R. 1978. *Order in Living Organisms: A Systems Analysis of Evolution*. Wiley, New York. [6]

Rieppel, O. 1992. Homology and logical fallacy. *J. Evol. Biol.* 5: 701–715. [23]

Rieseberg, L. H. 1991. Homoploid reticulate evolution in *Helianthus*: Evidence from ribosomal genes. *Am. J. Bot.* 78: 1218–1237. [16]

Rieseberg, L. H., and J. F. Wendel. 1993. Introgression and its consequences in plants. In R. G. Harrison (ed.), *Hybrid Zones and the Evolutionary Process*, pp. 70–109. Oxford University Press, New York. [16]

Rieseberg, L. H., R. Carter, and S. Zona. 1990. Molecular tests of the hypothesized hybrid origin of two diploid *Helianthus* species (Asteraceae). *Evolution* 44: 1498–1511. [16]

Riggs, E. S. 1934. A new marsupial saber-tooth from the Pliocene of Argentina and its relationships to other South American predaceous marsupials. *Trans. Am. Phil. Soc.*, n.s., 24: 1–32. [5]

Riley, M. A., S. R. Kaplan, and M. Veuille. 1992. Nucleotide polymorphism at the xanthine dehydrogenase locus in *Drosophila pseudoobscura*. *Mol. Biol. Evol.* 9: 56–69. [9]

Ringo, J. M., D. Wood, R. Rockwell, and H. Dowse. 1985. An experiment testing two hypotheses of speciation. *Am. Nat.* 126: 642–661. [16]

Riska, B., and W. R. Atchley. 1985. Genetics of growth predicts patterns of brain-size evolution. *Science* 229: 668–671. [17]

R'Kha, S., P. Capy, and J. R. David. 1991. Host-plant specialization in the *Drosophila melanogaster* species complex: A physiological, behavioral, and genetical analysis. *Proc. Natl. Acad. Sci. USA* 88: 1835–1839. [15]

Roach, D. A., and R. D. Wulff. 1987. Maternal effects in plants. *Annu. Rev. Ecol. Syst.* 18: 209–236. [14]

Robertson, A. 1960. A theory of limits in artificial selection. *Proc. R. Soc. Lond.* B 153: 234–249. [14]

Robinson, B. W., and D. J. Wilson. 1994. Character release and displacement in fishes: A neglected literature. *Am. Nat.* 144: 596–627. [18]

Rodríguez, D. J. 1996. A model for the establishment of polyploidy in plants. *Am. Nat.* 147: 33–46. [16]

Roelofs, W. L., T. J. Glover, X.-H. Tang, I. Sreng, P. S. Robbins, E. E. Eckenrode, C. Löfstedt, B. S. Hansson, and B. O. Bengtsson. 1987. Sex pheromone production and perception in European corn borer moths is determined by both autosomal and sex-linked genes. *Proc. Natl. Acad. Sci. USA* 84: 7585–7589. [15]

Roff, D. A. 1990. The evolution of flightlessness in insects. *Ecol. Monogr.* 60: 389–421. [17]

Roff, D. A. 1992. *The Evolution of Life Histories: Theory and Analysis.* Chapman & Hall, New York. [19]

Roger, A. J., and W. F. Doolittle. 1993. Why introns-in-pieces? *Nature* 364: 289–290. [22]

Rohlf, F. J., and G. D. Schnell. 1971. An investigation of the isolation-by-distance model. *Am. Nat.* 105: 295–324. [11]

Romanes, G. J. 1901. *Darwin and After Darwin.* Open Court Publishing, London. [23]

Romer, A. S. 1956. *Osteology of the Reptiles.* University of Chicago Press, Chicago, IL. [5]

Romer, A. S. 1966. *Vertebrate Paleontology.* University of Chicago Press, Chicago, IL. [5, 23]

Romer, A. S., and T. S. Parsons. 1986. *The Vertebrate Body.* Saunders College Publishing, Philadelphia, PA. [6]

Roose, M. L., and L. D. Gottlieb. 1976. Genetic and biochemical consequences of polyploidy in *Tragopogon*. *Evolution* 30: 818–830. [16]

Rose, M. R. 1982. Antagonistic pleiotropy, dominance, and genetic variation. *Heredity* 48: 63–78. [14]

Rose, M. R. 1984. Laboratory evolution of postponed senescence in *Drosophila melanogaster*. *Evolution* 38: 1004–1010. [19]

Rose, M. R. 1991. *The Evolutionary Biology of Aging.* Oxford University Press, Oxford. [19]

Rose, M. R., and B. Charlesworth. 1981. Genetics of life history in *Drosophila melanogaster*. II. Exploratory selection experiments. *Genetics* 97: 187–196. [19]

Rose, M. R., P. M. Service, and E. W. Hutchinson. 1987. Three approaches to trade-offs in life history evolution. In V. Loeschcke (ed.), *Genetic Constraints on Adaptive Evolution*, pp. 91–105. Springer-Verlag, Berlin. [17]

Rosen, D. E., P. L. Forey, B. G. Gardiner, and C. Patterson. 1981. Lungfishes, tetrapods, paleontology, and plesiomorphy. *Bull. Am. Mus. Nat. Hist.* 167: 159–276. [6]

Rosenberg, A. 1985. *The Structure of Biological Science.* Cambridge University Press, Cambridge. [22]

Rosenthal, G. A., and M. A. Berenbaum (eds.). 1992. *Herbivores: Their Interactions with Secondary Plant Metabolites.* Second edition. Academic Press, San Diego, CA. [4, 18]

Rosenzweig, M. L. 1975. On continental steady states of species diversity. In M. L. Cody and J. M. Diamond (eds.), *Ecology and Evolution of Communities*, pp. 121–140. Harvard University Press, Cambridge, MA. [8]

Rosenzweig, M. L., and Z. Abramsky. 1993. How are diversity and productivity related? In R. E. Ricklefs and D. Schluter (eds.), *Species Diversity in Ecological Communities*, pp. 52–65. University of Chicago Press, Chicago, IL. [8]

Rosenzweig, M. L., and R. D. McCord. 1991. Incumbent replacement: Evidence for long-term evolutionary progress. *Paleobiology* 17: 202–213. [25]

Rosenzweig, M. L., J. S. Brown, and T. L. Vincent. 1987. Red Queens and ESS: The coevolution of evolutionary rates. *Evol. Ecol.* 1: 59–94. [18]

Ross, H. H., C. A. Ross, and J. R. P. Ross. 1982. *A Textbook of Entomology.* Fourth edition. Wiley, New York. [5]

Ross, K. G., and R. W. Matthews (eds.). 1991. *The Social Biology of Wasps.* Cornell University Press, Ithaca, NY. [20]

Roth, V. L. 1988. The biological basis of homology. In C. J. Humphries (ed.), *Ontogeny and Systematics*, pp. 1–26. British Museum (Natural History), London. [23]

Roth, V. L. 1991. Homology and hierarchies: Problems solved and unresolved. *J. Evol. Biol.* 4: 167–194. [23]

Rothstein, S. I. 1990. A model system for coevolution: Avian brood parasitism. *Annu. Rev. Ecol. Syst.* 21: 481–508. [18]

Roughgarden, J. 1971. Density-dependent natural selection. *Ecology* 52: 453–468. [19]

Roughgarden, J. 1995. *Anolis Lizards of the Caribbean: Ecology, Evolution, and Plate Tectonics.* Oxford University Press, New York. [8]

Rouhani, S., and N. H. Barton. 1987. Speciation and the "shifting balance" in a continuous population. *Theor. Pop. Biol.* 31: 465–492. [14]

Roush, R. T., and J. A. McKenzie. 1987. Ecological genetics of insecticide and acaricide resistance. *Annu. Rev. Entomol.* 32: 361–380. [13]

Rowe, F. W. E., and J. Gates. 1995. Echinodermata. In A. Wells (ed.), *Zoological Catalogue of Australia*, vol. 33. CSIRO Australia, Melbourne. [7]

Ruben, J. 1991. Reptilian physiology and the flight capacity of *Archaeopteryx*. *Evolution* 45: 1–17. [6]

Rubin, G. M. 1983. Dispersed repetitive DNAs in *Drosophila*. In J. A. Shapiro (ed.), *Mobile Genetic Elements*, pp. 329–361. Academic Press, New York. [22]

Ruse, M. 1979. *The Darwinian Revolution.* University of Chicago Press, Chicago, IL. [12]

Ruse, M. 1993. Evolution and progress. *Trends Ecol. Evol.* 8: 55–59. [24]

Ruse, M. 1996. *Monad to Man: The Concept of Progress in Evolutionary Biology.* Harvard University Press, Cambridge, MA. [24]

Ruvolo, M. 1997. Molecular phylogeny of the hominoids: Inferences from multiple independent DNA sequence data sets. *Mol. Biol. Evol.* 14: 248–265. [26]

Ruvolo, M., S. Zehr, M. von Dornum, D. Pan, B. Chang, and J. Lin. 1993. Mitochondrial COII sequences and modern human origins. *Mol. Biol. Evol.* 10: 1115–1135. [26]

Ruwande, C., et al. 1995. Natural selection of hemi- and heterozygotes for G6PD deficiency in Africa by resistance to severe malaria. *Nature* 376: 246–249. [26]

Ryan, M. J. 1985. *The Túngara Frog: A Study in Sexual Selection and Communication.* University of Chicago Press, Chicago, IL. [20]

Ryan, M. J. 1990. Sexual selection, sensory systems, and sensory exploitation. *Oxford Surv. Evol. Biol.* 7: 157–195. [16]

Ryan, M. J. 1994. Mechanisms underlying sexual selection. In L. A. Real (ed.), *Behavioral Mechanisms in Evolutionary Ecology*, pp. 190–215. University of Chicago Press, Chicago, IL. [20]

Ryan, M. J., and A. S. Rand. 1993a. Sexual selection and signal evolution—the ghost of biases past. *Phil. Trans. R. Soc. Lond.* B 340: 187–195. [16, 20]

Ryan, M. L., and A. S. Rand. 1993b. Species recognition and sexual selection as a unitary problem in animal communication. *Evolution* 47: 647–657. [16]

Sage, R. D., W. R. Atchley, and E. Capanna. 1993. House mice as models in systematic biology. *Syst. Biol.* 42: 523–561. [15]

Sage, R. D., D. Heyneman, K.-C. Lim, and A. C. Wilson. 1986. Wormy mice in a hybrid zone. *Nature* 324: 60–63. [15]

Saghir, M. T., and E. Robbins. 1973. *Male and Female Homosexuality.* Williams and Wilkins, Baltimore, MD. [26]

Saint-Seine, P. de. 1955. Sauropterygia. In J. Piveteau (ed.), *Traité de paleontologie*, vol. 5, pp. 420–428. Masson, Paris. [7]

Saitou, N., and M. Nei. 1987. The neighbor-joining method: A new method for reconstructing phylogenetic trees. *Mol. Biol. Evol.* 4: 406–425. [5]

Saloniemi, I. 1993. A coevolutionary predator-prey model with quantitative characters. *Am. Nat.* 141: 880–896. [18]

Salvini-Plawen, L. V., and E. Mayr. 1977. On the evolution of photoreceptors and eyes. *Evol. Biol.* 10: 207–263. [24]

Samson, M., et al. 1996. Resistance to HIV-1 infection in Caucasian individuals bearing mutant alleles of the CCR-5 chemokine receptor gene. *Nature* 382: 722–725. [26]

Sanderson, M. J., and M. J. Donoghue. 1989. Patterns of variation in levels of homoplasy. *Evolution* 43: 1781–1795. [5]

Sanderson, N. 1989. Can gene flow prevent reinforcement? *Evolution* 43: 1223–1235. [16]

Sansom, J. J., M. P. Smith, H. A. Armstrong, and M. M. Smith. 1992. Presence of the earliest vertebrate hard tissue in condonts. *Science* 256: 1308–1311. [7]

Sarich, V. M., and A. C. Wilson. 1967. Immunological time scale for hominid evolution. *Science* 179: 1144–1147. [5]

Scagel, R. F., R. J. Bandoni, G. E. Rouse, W. B. Schofield, J. R. Stein, and T. M. C. Taylor. 1966. *An Evolutionary Survey of the Plant Kingdom.* Wadsworth, Belmont, CA. [24]

Scarr, S., and R. A. Weinberg. 1976. IQ test performance of black children adopted by white families. *Am. Psychol.* 31: 726–739. [26]

Schaeffer, B. 1952. Rates of evolution in coelacanth and subholostean fishes. *Evolution* 10: 201–212. [6]

Schaeffer, B. 1956. Evolution in the subholostean fishes. *Evolution* 10: 201–212. [6]

Schaeffer, S. W., and E. L. Miller. 1993. Estimates of linkage disequilibrium and the recombination parameter determined from segregating nucleotide sites in the alcohol dehydrogenase region of *Drosophila pseudoobscura*. *Genetics* 135: 541–552. [22]

Scharloo, W. 1987. Constraints in selective response. In V. Loeschcke (ed.), *Genetic Constraints on Adaptive Evolution*, pp. 125–149. Springer-Velag, Berlin. [14]

Scharloo, W. 1991. Canalization: Genetic and developmental aspects. *Annu. Rev. Ecol. Syst.* 22: 65–93. [14]

Schemske, D. W. 1983. Breeding system and habitat effects on fitness components in three neotropical *Costus* (Zingiberaceae). *Evolution* 37: 523–539. [11]

Schindewolf, O. H. 1936. *Paläontologie, Entwicklungslehre und Genetik.* Borntraeger, Berlin. [24]

Schindewolf, O. H. 1950. *Grundfrage der Paläontologie.* Schweitzerbart, Jena, Germany. [24]

Schlegel, M. 1994. Molecular phylogeny of eukaryotes. *Trends Ecol. Evol.* 9: 330–335. [7]

Schliewen, U. K., D. Tautz, and S. Pääbo. 1994. Sympatric speciation suggested by monophyly of crater lake cichlids. *Nature* 368: 629–632. [16]

Schluter, D. 1993. Adaptive radiation in sticklebacks: Size, shape, and habitat use efficiency. *Ecology* 74: 699–709. [18]

Schluter, D., and P. R. Grant. 1984. Determinants of morphological patterns in communities of Darwin's finches. *Am. Nat.* 123: 175–196. [8]

Schluter, D., and J. D. McPhail. 1992. Ecological character displacement and speciation in sticklebacks. *Am. Nat.* 140: 85–108. [18]

Schluter, D., and L. M. Nagel. 1995. Parallel speciation by natural selection. *Am. Nat.* 146: 292–301. [16]

Schluter, D., and R. E. Ricklefs. 1993. Convergence and the regional component of species diversity. In R. E. Ricklefs and D. Schluter (eds.), *Species Diversity in Ecological Communities*, pp. 230–240. University of Chicago Press, Chicago, IL. [25]

Schmalhausen, I. I. 1949. *Factors of Evolution.* Blakiston, Philadelphia, PA. [14]

Schmidt, G. D., and L. S. Roberts. 1981. *Foundations of Parasitology.* C. V. Mosby, St. Louis, MO. [4]

Schmidt-Nielsen, K. 1984. *Scaling: Why Is Animal Size so Important?* Cambridge University Press, Cambridge. [17]

Schmidt-Nielsen, K. 1990. *Animal Physiology: Adaptation and Environment.* Fourth edition. Cambridge University Press, Cambridge. [17]

Schoen, D. J., and A. H. D. Brown. 1991. Intraspecific variation in population gene diversity and effective population size correlated with the mating system in plants. *Proc. Natl. Acad. Sci. USA* 88: 4494–4497. [22]

Schoener, T. W. 1970. Size patterns of West Indian *Anolis* lizards. II. Correlations with the sizes of particular sympatric species: Displacement and convergence. *Am. Nat.* 104: 155–174. [8]

Schoener, T. W. 1983. Field experiments on interspecific competition. *Am. Nat.* 122: 240–285. [4, 18]

Schoener, T. W. 1984. Size differences among sympatric, bird-eating hawks: A worldwide survey. In D. R. Strong, Jr., D. Simberloff, L. G. Abele, and A. B. Thistle (eds.), *Ecological Communities: Conceptual Issues and the Evidence*, pp. 254–281. Princeton University Press, Princeton, NJ. [4]

Schoener, T. W. 1988. Ecological interactions. In A. A. Myers and P. S. Giller (eds.), *Analytical Biogeography*, pp. 255–297. Chapman & Hall, London. [8]

Schwartz, R., and M. Dayhoff. 1978. Origins of prokaryotes, eukaryotes, mitochondria, and chloroplasts. *Science* 199: 395–403. [7]

Scotese, C. R., and W. W. Sager (eds.). 1989. *Mesozoic and Cenozoic Plate Reconstructions.* Elsevier, New York. [7]

Scott, J. A. 1986. *The Butterflies of North America.* Stanford University Press, Stanford, CA. [24]

Searle, J. B. 1993. Chromosomal hybrid zones in eutherian mammals. In R. G. Harrison (ed.), *Hybrid Zones and the Evolutionary Process*, pp. 309–353. Oxford University Press, New York. [16]

Seehausen, O., J. J. M. van Alphen, and F. Witte. 1997. Cichlid fish diversity threatened by eutrophication that curbs sexual selection. *Science* 277: 1808–1811. [16]

Seger, J. 1991. Cooperation and conflict in social insects. In J. R. Krebs and N. W. Davies (eds.), *Behavioural Ecology*, Third edition, pp. 338–373. Blackwell Scientific, Oxford. [20]

Seger, J. 1992. Evolution of exploiter-victim relationships. In M. J. Crawley (ed.), *Natural Enemies: The Population Biology of Predators, Parasites, and Disease*, pp. 3–25. Blackwell Scientific, Oxford. [18]

Seilacher, A. 1973. Fabricational noise in adaptive morphology. *Syst. Zool.* 22: 451–465. [23]

Seino, S., G. I. Bell, and W.-H. Li. 1992. Sequences of primate insulin genes support the hypothesis of a slower rate of molecular evolution in humans and apes than in monkeys. *Mol. Biol. Evol.* 9: 193–203. [22]

Selander, R. K. 1966. Sexual dimorphism and differential niche utilization in birds. *Condor* 68: 113–151. [18]

Selander, R. K. 1970. Behavior and genetic variation in natural populations. *Am. Zool.* 10: 53–66. [11]

Selander, R. K. 1976. Genic variation in natural populations. In F. J. Ayala (ed.), *Molecular Evolution*, pp. 21–45. Sinauer Associates, Sunderland, MA. [9]

Selander, R. K., A. G. Clark, and T. S. Whittam (eds.). 1991. *Evolution at the Molecular Level*. Sinauer Associates, Sunderland, MA. [22]

Selander, R. K., W. G. Hunt, and S. Y. Yang. 1969. Protein polymorphism and genic heterozygosity in two European subspecies of the house mouse. *Evolution* 23: 379–390. [10]

Sepkoski, J. J. Jr. 1984. A kinetic model of Phanerozoic taxonomic diversity. III. Post-Paleozoic families and mass extinctions. *Paleobiology* 10: 246–267. [25]

Sepkoski, J. J. Jr. 1993. Ten years in the library: New data confirm paleontological patterns. *Paleobiology* 19: 43–51. [25]

Sepkoski, J. J. Jr. 1996. Competition in macroevolution: The double wedge revisited. In D. Jablonski, D. H. Erwin, and J. Lipps (eds.), *Evolutionary Paleobiology*, pp. 211–255. University of Chicago Press, Chicago, IL. [25]

Sepkoski, J. J. Jr., and M. L. Hulver. 1985. An atlas of Phanerozoic clade diversity patterns. In J. W. Valentine (ed.), *Phanerozoic Diversity Patterns: Profiles in Macroevolution*, pp. 11–39. Princeton University Press, Princeton, NJ. [25]

Sepkoski, J. J. Jr. and D. C. Kendrick. 1993. Numerical experiments with model monophyletic and paraphyletic taxa. *Paleobiology* 19: 168–184. [25]

Sepkoski, J. J. Jr., and A. I. Miller. 1985. Evolutionary faunas and the distribution of Paleozoic marine communities in time and space. In J. W. Valentine (ed.), *Phanerozoic Diversity Patterns: Profiles in Macroevolution*, pp. 153–190. Princeton University Press, Princeton, NJ. [25]

Sessions, S. K., and A. Larson. 1987. Developmental correlates of genome size in plethodontid salamanders and their implications for genome evolution. *Evolution* 41: 1239–1251. [22]

Shaffer, H. B., and S. R. Voss. 1996. Phylogenetic and mechanistic analysis of a developmentally integrated character complex: Alternate life history modes in ambystomatid salamanders. *Am. Zool.* 36: 24–35. [23]

Shapiro, A. M., and A. H. Porter. 1989. The lock-and-key hypothesis: Evolutionary and biosystematic interpretation of insect genitalia. *Annu. Rev. Entomol.* 34: 231–245. [15]

Sharp, P. M., and W.-H. Li. 1987. The rate of synonymous substitution in enterobacterial genes is inversely related to codon usage bias. *Mol. Biol. Evol.* 4: 222–230. [22]

Sharpton, V. L., L. M. Marin, J. L. Carney, S. Lee, G. Ryder, B. C. Schuraytz, P. Sikora, and P. D. Spudis. 1996. A model of the Chicxulub impact basin based on evaluation of geophysical data, well logs, and drill core samples. *Geol. Soc. Am. Special Papers* 307: 55–74. [25]

Shaw, D. D., A. D. Marchant, N. Contreras, M. L. Arnold, F. Groeters, and B. C. Kohlmann. 1993. Genomic and environmental determinants in a narrow hybrid zone: Cause or coincidence? In R. G. Harrison (ed.), *Hybrid Zones and the Evolutionary Process*, pp. 165–195. Oxford University Press, Oxford. [15]

Shaw, K. L. 1995. Biogeographic patterns of two independent Hawaiian cricket radiations (*Laupala* and *Prognathogryllus*). In W. L. Wagner and V. A. Funk (eds.), *Hawaiian Biogeography*, pp. 39–56. Smithsonian Institution Press, Washington, DC. [8]

Shea, B. T. 1988. Heterochrony in primates. In M. L. McKinney (ed.), *Heterochrony in Evolution: A Multidisciplinary Approach*, pp. 237–266. Plenum, New York. [26]

Shear, W. A., and J. Kukalová-Peck. 1990. The ecology of Paleozoic terrestrial arthropods: The fossil evidence. *Can. J. Zool.* 68: 1807–1834. [7]

Sheehan, P. M., and D. E. Fastovsky. 1992. Major extinctions of land-dwelling vertebrates at the Cretaceous-Tertiary boundary, eastern Montana. *Geology* 20: 555–560. [25]

Sheldon, P. R. 1987. Parallel gradualistic evolution of Ordovician trilobites. *Nature* 330: 561–563. [6, 16]

Sheldon, P. R. 1990. Shaking up evolutionary patterns. *Nature* 345: 772. [16]

Shepher, J. 1983. *Incest: A Biosocial View*. Academic Press, New York. [26]

Sherman, P. W., J. U. M. Jarvis, and R. D. Alexander (eds.). 1991. *The Biology of the Naked Mole-rat*. Princeton University Press, Princeton, NJ. [20]

Shettleworth, S. J. 1984. Learning and behavioural ecology. In J. R. Krebs and N. B. Davies (eds.), *Behavioural Ecology*, Second edition, pp. 170–194. Blackwell Scientific, Oxford. [20]

Shields, D. C., P. M. Sharp, D. G. Higgins, and F. Wright. 1988. "Silent" sites in *Drosophila* genes are not neutral: Evidence of selection among synonymous codons. *Mol. Biol. Evol.* 5: 704–716. [22]

Shine, R., and E. L. Charnov. 1992. Patterns of survival, growth, and maturation in snakes and lizards. *Am. Nat.* 139: 1257–1269. [19]

Short, L. L. 1965. Hybridization in the flickers (*Colaptes*) of North America. *Bull. Am. Mus. Nat. Hist.* 129: 307–428. [9]

Shoshani, J. 1986. Mammalian phylogeny: Comparison of morphological and molecular results. *Mol. Biol. Evol.* 3: 222–242. [7]

Shoshani, J., C. P. Groves, E. L. Simons, and G. F. Gunnell. 1996. Primate phylogeny: Morphological and molecular results. *Mol. Phyl. Evol.* 5: 102–154. [5]

Shrimpton, A. E., and A. Robertson. 1988. The isolation of polygenic factors controlling bristle score in *Drosophila melanogaster*. I. Association of third chromosome sternopleural bristle effects to chromosome sections. *Genetics* 118: 437–443. [14]

Shubin, N., D. B. Wake, and A. J. Crawford. 1995. Morphological variation in the limbs of *Taricha granulosa* (Caudata: Salamandridae): Evolutionary and phylogenetic implications. *Evolution* 49: 874–884. [10]

Sibley, C. G. 1950. Species formation in the red-eyed towhees of Mexico. *Univ. Calif. Publ. Zool.* 50: 109–193. [15]

Sibley, C. G. 1954. Hybridization in the red-eyed towhees of Mexico. *Evolution* 8: 252–290. [15]

Sibley, C. G. 1957. The evolutionary and taxonomic significance of sexual dimorphism and hybridization in birds. *Condor* 49: 166–189. [16]

Sibley, C. G., and J. E. Ahlquist. 1984. The phylogeny of the hominoid primates as indicated by DNA-DNA hybridization. *J. Mol. Evol.* 20: 2-15. [26]

Sibley, C. G., and J. E. Ahlquist. 1987. DNA hybridization evidence of hominoid phylogeny: Results from an expanded data set. *J. Mol. Evol.* 26: 99–121. [5]

Sibley, C. G., and J. E. Ahlquist. 1990. *Phylogeny and Classification of Birds: A Study in Molecular Evolution.* Yale University Press, New Haven, CT. [5, 8, 16]

Sibly, R. M., and P. Calow. 1986. *Physiological Ecology of Animals: An Evolutionary Approach.* Blackwell Scientific, Oxford. [17]

Signor, P. W. III. 1985. Real and apparent trends in species richness through time. In J. W. Valentine (ed.), *Phanerozoic Diversity Patterns: Profiles in Macroevolution*, pp. 129–150. Princeton University Press, Princeton, NJ. [25]

Signor, P. W. III. 1990. The geological history of diversity. *Annu. Rev. Ecol. Syst.* 21: 509–539. [25]

Simberloff, D., and W. Boecklen. 1981. Santa Rosalia reconsidered: Size ratios and competition. *Evolution* 35: 1206–1228. [18]

Simmons, M. J., and J. F. Crow. 1977. Mutations affecting fitness in *Drosophila* populations. *Annu. Rev. Genet.* 11: 49–78. [13]

Simms, E. L., and M. D. Rausher. 1989. The evolution of resistance to herbivory in *Ipomoea purpurea*. II. Natural selection by insects and costs of resistance. *Evolution* 43: 573–585. [18]

Simms, M. J. 1990. The radiation of post-Palaeozoic echinoderms. In P. D. Taylor and G. P. Larwood (eds.), *Major Evolutionary Radiations*, pp. 287–304. Clarendon Press, Oxford. [25]

Simons, E. L. 1979. The early relatives of man. In G. Isaac and R. E. F Leakey (eds.), *Human Ancestors*, pp. 22–42. W. H Freeman, San Francisco, CA. [7]

Simons, E. L. 1993. Human origins. *Science* 245: 1343–1350. [26]

Simpson, G. G. 1944. *Tempo and Mode in Evolution.* Columbia University Press, New York. [6]

Simpson, G. G. 1951. *Horses: The Story of the Horse Family in the Modern World and through Sixty Million Years of History.* Oxford University Press, New York. [6]

Simpson, G. G. 1953. *The Major Features of Evolution.* Columbia University Press, New York. [6]

Simpson, G. G. 1961. *Principles of Animal Taxonomy.* Columbia University Press, New York. [6, 15]

Simpson, G. G. 1964. *This View of Life: The World of an Evolutionist.* Harcourt, Brace and World, New York. [22]

Sinclair, D. 1969. *Human Growth After Birth.* Oxford University Press, Oxford. [23]

Sinervo, B. 1990. The evolution of maternal investment in lizards: An experimental and comparative analysis of egg size and its effect on offspring performance. *Evolution* 44: 279-294. [19]

Singer, M. C., C. D. Thomas, and C. Parmesan. 1993. Rapid human-induced evolution of insect-host associations. *Nature* 366: 681–683. [14]

Sinnott, E. W., L. C. Dunn, and T. Dobzhansky. 1958. *Principles of genetics.* Fifth edition. McGraw-Hill, New York. [3]

Sites, J. W. J., and C. Moritz. 1987. Chromosomal evolution and speciation revisited. *Syst. Zool.* 36: 153–174. [15]

Skibinski, D. D. F., and R. D. Ward. 1982. Correlations between heterozygosity and evolutionary rate of proteins. *Nature* 298: 490–492. [11]

Skutch, A. F. 1973. *The Life of the Hummingbird.* Crown Publishers, New York. [15]

Slatkin, M. 1977. Gene flow and genetic drift in a species subject to frequent local extinctions. *Theor. Pop. Biol.* 12: 253–262. [11]

Slatkin, M. 1979. Frequency- and density-dependent selection on a quantitative character. *Genetics* 93: 755–771. [14]

Slatkin, M. 1980. Ecological character displacement. *Ecology* 61: 163–177. [18]

Slatkin, M. 1983. Genetic background. In D. J. Futuyma and M. Slatkin (eds.), *Coevolution*, pp. 14–32. Sinauer Associates, Sunderland, MA. [14]

Slatkin, M. 1985a. Gene flow in natural populations. *Annu. Rev. Ecol. Syst.* 16: 393–430. [11]

Slatkin, M. 1985b. Rare alleles as indicators of gene flow. *Evolution* 39: 53–65. [11]

Slatkin, M. 1996. In defense of founder-flush speciation. *Am. Nat.* 147: 493–505. [16]

Slatkin, M., and W. P. Maddison. 1989. A cladistic measure of gene flow inferred from the phylogenies of alleles. *Genetics* 123: 603–613. [11]

Slobodkin, L. B. 1974. Prudent predation does not require group selection. *Am. Nat.* 108: 665–678. [18]

Slobodkin, L. B., and A. Rapoport. 1974. An optimal strategy of evolution. *Q. Rev. Biol.* 49: 181–200. [4, 17]

Slobodkin, L. B., F. E. Smith, and N. G. Hairston. 1967. Regulation in terrestrial ecosystems, and the implied balance of nature. *Am. Nat.* 101: 104–124. [4]

Smith, A. B., and C. Patterson. 1988. The influence of taxonomic method on the perception of patterns of evolution. *Evol. Biol.* 23: 127–216. [25]

Smith, A. B., B. Lafay, and R. Christen. 1992. Comparative variation of morphological and molecular evolution through geologic time: 28S ribosomal RNA versus morphology in echinoids. *Phil. Trans. R. Soc. Lond.* B 338: 365–382. [5, 11]

Smith, A. G., and J. C. Briden. 1977. *Mesozoic and Cenozoic Paleocontinental Maps.* Cambridge University Press, Cambridge, MA. [7]

Smith, R. L. (ed.). 1984. *Sperm Competition and the Evolution of Animal Mating Systems.* Academic Press, Orlando, FL. [20]

Smith, T. B. 1993. Disruptive selection and the genetic basis of bill size polymorphism in the African finch *Pyrenestes*. *Nature* 363: 618–620. [13]

Smocovitis, V. B. 1996. *Unifying Biology: The Evolutionary Synthesis and Evolutionary Biology.* Princeton University Press, Princeton, NJ. [2]

Sniegowski, P. D., and B. Charlesworth. 1994. Transposable element numbers in cosmopolitan inversions from a natural population of *Drosophila melanogaster*. *Genetics* 137: 815–827. [22]

Sniegowski, P. D., and R. E. Lenski. 1995. Mutation and adaptation: The directed mutation controversy in evolutionary perspective. *Annu. Rev. Ecol. Syst.* 26: 553–578. [10]

Snodgrass, R. E. 1935. *Principles of Insect Morphology.* McGraw-Hill, New York. [5, 6]

Sober, E. 1984. *The Nature of Selection: Evolutionary Theory in Philosophical Focus.* MIT Press, Cambridge, MA. [12, 20]

Sober, E. 1988. *Reconstructing the Past: Parsimony, Evolution, and Inference.* MIT Press, Cambridge, MA. [5]

Sober, E. 1993. *Philosophy of Biology.* Westview Press, Boulder, CO. [1]

Sober, E., and R. Lewontin. 1982. Artifact, cause, and genic selection. *Phil. Sci.* 47: 157–180. [12]

Sokal, R. R., N. L. Oden, and C. Wilson. 1991. Genetic evidence for the spread of agriculture in Europe by demic diffusion. *Nature* 351: 143–145. [26]

Literature Cited

Sokol, D. L., and A. M. Gewirtz. 1996. Gene therapy: Basic concepts and recent advances. *Critical Reviews in Eukaryotic Gene Expression* 6: 29–57. [26]

Solbrig, O. T., M. A. Barbour, J. Cross, G. Goldstein, C. H. Lowe, J. Morello, and T. W. Yang. 1977. The strategies and community patterns of desert plants. In G. H. Orians and O. T. Solbrig (eds.), *Convergent Evolution in Warm Deserts*, pp. 67–106. Dowden, Hutchinson and Ross, Stroudsburg, PA. [17]

Soltis, D. E., and P. S. Soltis. 1989. Allopolyploid speciation in *Tragopogon*: Insights from chloroplast DNA. *Am. J. Bot.* 76: 1119–1124. [16]

Soltis, P. S., and D. E. Soltis. 1991. Multiple origins of the allotetraploid *Tragopogon mirus* (Compositae): rDNA evidence. *Syst. Bot.* 16: 407–413. [16]

Soulé, M. 1966. Trends in the insular radiation of a lizard. *Am. Nat.* 100: 47–64. [16]

Spassky, B., N. Spassky, H. Levene, and T. Dobzhansky. 1958. Release of genetic variability through recombination. I. *Drosophila pseudoobscura*. *Genetics* 43: 844–867. [10]

Spencer, H. G., B. H. McArdle, and D. M. Lambert. 1986. A theoretical investigation of speciation by reinforcement. *Am. Nat.* 128: 241–262. [16]

Spiegelman, S. 1970. Extracellular evolution of replicating molecules. In F. O. Schmitt (ed.), *The Neuro Sciences: A Second Study Program*, pp. 927–945. Rockefeller University Press, New York. [7]

Spiess, E. B. 1968. Low frequency advantage in mating of *Drosophila pseudoobscura* karyotypes. *Am. Nat.* 102: 363–379. [13]

Spradbery, J. P. 1973. *Wasps: An Account of the Biology and Natural History of Solitary and Social Wasps.* University of Washington Press, Seattle. [20]

Spratt, B. G., N. E. Smith, J. Zhou, M. O'Rourke, and E. Feil. 1994. The population genetics of the pathogenic *Neisseria*. In S. Baumberg, J. P. W. Young, E. M. N. Wellington, and J. R. Saunders (eds.), *Population Genetics of Bacteria*, pp. 143–160. Cambridge University Press, Cambridge. [22]

Srb, A. M., R. D. Owen, and R. S. Edgar. 1965. *General Genetics.* W. H. Freeman, San Francisco, CA. [10]

Stace, C. A. 1989. Hybridization and the plant species. In K. M. Urbanska (ed.), *Differentiation Patterns in Higher Plants*, pp. 115–127. Academic Press, New York. [10]

Stacey, P. B., and W. D. Koenig (eds.). 1990. *Cooperative Breeding in Birds.* Cambridge University Press, Cambridge. [20]

Stanley, S. M. 1972. Functional morphology and evolution of byssally attached bivalve molluscs. *J. Paleontol.* 46: 165–212. [23]

Stanley, S. M. 1973. An explanation for Cope's rule. *Evolution* 27: 1–26. [24]

Stanley, S. M. 1975. A theory of evolution above the species level. *Proc. Natl. Acad. Sci. USA* 72: 646–650. [6, 12]

Stanley, S. M. 1979. *Macroevolution: Pattern and Process.* W. H. Freeman, San Francisco, CA. [6, 24, 25]

Stanley, S. M. 1990. The general correlation between rate of speciation and rate of extinction: Fortuitous causal linkages. In R. M. Ross and W. D. Allmon (eds.), *Causes of Evolution: A Paleontological Perspective*, pp. 103–127. University of Chicago Press, Chicago, IL. [25]

Stanley, S. M. 1993. *Earth and Life Through Time.* Second edition. W. H. Freeman, New York. [6, 7]

Stanley, S. M., P. W. Signor III, S. Lidgard, and A. F. Karr. 1981. Natural clades differ from "random" clades: Simulations and analyses. *Paleobiology* 7: 115–127. [25]

Stearn, C. W., and R. L. Carroll. 1989. *Paleontology: The Record of Life.* Wiley, New York. [5, 7]

Stearns, S. C. 1986. Natural selection and fitness, adaptation and constraint. In D. M. Raup and D. Jablonski (eds.), *Patterns and Processes in the History of Life*, pp. 23–44. Springer-Verlag, Berlin. [23]

Stearns, S. C. (ed.). 1987. *The Evolution of Sex and Its Consequences.* Birkhäuser-Verlag, Basel. [21]

Stearns, S. C. 1992. *The Evolution of Life Histories.* Oxford University Press, Oxford. [19]

Stearns, S. C., and R. D. Sage. 1980. Maladaptation in a marginal population of the mosquito fish, *Gambusia affinis*. *Evolution* 34: 65–75. [17]

Stebbins, G. L. 1950. *Variation and Evolution in Plants.* Columbia University Press, New York. [10]

Stebbins, G. L. 1957. Self-fertilization and population variability in the higher plants. *Am. Nat.* 41: 337–354. [15]

Stebbins, G. L. 1971. *Chromosomal Evolution in Higher Plants.* Addison-Wesley, Reading, MA. [16]

Stebbins, G. L. 1974. *Flowering Plants: Evolution Above the Species Level.* Belknap Press of Harvard University Press, Cambridge, MA. [3]

Stebbins, G. L. 1982. Plant speciation. In C. Barigozzi (ed.), *Mechanisms of Speciation*, pp. 21–39. Alan R. Liss, New York. [15]

Stebbins, G. L., and F. J. Ayala. 1981. Is a new evolutionary synthesis necessary? *Science* 213: 967–971. [24]

Stebbins, G. L., and K. Daly. 1961. Changes in the variation pattern of a hybrid population of *Helianthus* over an eight-year period. *Evolution* 15: 60–71. [16]

Stebbins, G. L., and A. Day. 1967. Cytogenetic evidence for long continued stability in the genus *Plantago*. *Evolution* 21: 409–428. [16]

Stebbins, G. L., E. B. Matzke, and C. Epling. 1947. Hybridization in a population of *Quercus marilandica* and *Quercus ilicifolia*. *Evolution* 1: 79–88. [15]

Stebbins, R. C. 1949. Speciation in salamanders of the plethodontid genus *Ensatina*. *Univ. Calif. Pub. Zool.* 48: 377–526. [15]

Stebbins, R. C. 1954. *Amphibians and Reptiles of Western North America.* McGraw-Hill, New York. [9, 15, 17]

Stehli, F. G., and S. D. Webb (eds.). 1985. *The Great American Biotic Interchange.* Plenum Press, New York. [8]

Steinemann, M., S. Steinemann, and F. Lottspeich. 1993. How Y chromosomes become inert. *Proc. Natl. Acad. Sci. USA* 90: 5737–5741. [21]

Stenseth, N. C., and J. Maynard Smith. 1987. Coevolution in ecosystems: Red Queen evolution or stasis? *Evolution* 38: 870–880. [18]

Stent, G. S., and R. Calendar. 1978. *Molecular Genetics.* W. H. Freeman, San Francisco, CA. [10]

Stephan, W. 1989. Tandem-repetitive noncoding DNA: Forms and forces. *Mol. Biol. Evol.* 6: 198–212. [22]

Stephens, D. W., and J. R. Krebs. 1989. *Foraging Theory.* Princeton University Press, Princeton, NJ. [20]

Stern, C. 1973. *Principles of Human Genetics.* W. H. Freeman, San Francisco, CA. [11]

Stevens, G. C. 1989. The latitudinal gradient in geographic range: How so many species coexist in the tropics. *Am. Nat.* 113: 240–256. [8]

Stewart, W. N. 1983. *Paleobotany and the Evolution of Plants.* Cambridge University Press, Cambridge. [7]

Stewart, W. N. 1993. *Paleobotany and the Evolution of Plants.* Second edition. Cambridge University Press, Cambridge. [7]

Stock, C. 1925. Cenozoic gravigrade edentates of western North America with special reference to the Pleistocene Megalonychinae and Mylodontidae of Rancho La Brea. Carnegie Institution of Washington Publication no. 331: 1–206. [7]

Stock, G. B., and S. V. Bryant. 1981. Studies of digit regeneration and their implications for theories of development and evolution of vertebrate limbs. *J. Exp. Zool.* 216: 423–433. [23]

Storer, R. W. 1960. Evolution in the diving birds. *Proc. XII Int. Ornithol. Cong.* 694–707. [24]

Strathmann, R. R. 1978. The evolution and loss of larval feeding stages of marine invertebrates. *Evolution* 32: 894–906. [23]

Strauss, S. Y. 1994. Levels of herbivory and parasitism in host hybrid zones. *Trends Ecol. Evol.* 9: 209–214. [15]

Strickberger, M. W. 1968. *Genetics.* Macmillan, New York. [3]

Strobeck, C. 1983. Expected linkage disequilibrium for a neutral locus linked to a chromosomal arrangement. *Genetics* 103: 545–555. [22]

Strong, D. R., Jr., J. H. Lawton, and T. R. E. Southwood. 1984a. *Insects on Plants: Community Patterns and Mechanisms.* Blackwell, Oxford. [4, 18, 25]

Strong, D. R., Jr., D. Simberloff, L. G. Abele, and A. B. Thistle (eds.). 1984b. *Ecological Communities: Conceptual Issues and the Evidence.* Princeton University Press, Princeton, NJ. [25]

Stümpke, H. 1957. *Bau und Leben der Rhinogradentia* (English translation: *The Snouters).* Doubleday, New York. (Reprinted 1981, University of Chicago Press). [23]

Sturmbauer, C., and A. Meyer. 1992. Genetic divergence, speciation, and morphological stasis in a lineage of African cichlid fishes. *Nature* 358: 578–581. [15]

Suomalainen, E., A. Saura, and J. Lokki. 1987. *Cytology and Evolution in Parthenogenesis.* CRC Press, Boca Raton, FL. [21]

Susman, R. L. 1988. Hand of *Paranthropus robustus* from Member 1, Swartkrans: Fossil evidence for tool behavior. *Science* 240: 781–784. [26]

Susman, R. L. 1994. Fossil evidence for early hominid tool use. *Science* 265: 1570–1573. [26]

Suzuki, D. T., A. J. F. Griffiths, J. H. Miller, and R. C. Lewontin. 1989. *An Introduction to Genetic Analysis.* W. H. Freeman, San Francisco, CA. [3]

Swanson, W. J., and V. D. Vacquier. 1995. Extraordinary divergence and positive Darwinian selection in a fusagenic protein coating the acrosomal process of abalone spermatozoa. *Proc. Natl. Acad. Sci. USA* 92: 4957–4961. [15]

Swisher, C. C. III, et al. 1996. Latest *Homo erectus* of Java: Potential contemporaneity with *Homo sapiens* in southeast Asia. *Science* 274: 1870–1874. [26]

Swofford, D. L., G. J. Olsen, P. J. Waddell, and D. M. Hillis. 1996. Phylogenetic inference. In D. M. Hillis, C. Moritz, and B. K. Mable, (eds.), *Molecular Systematics,* second edition, pp. 407–514. Sinauer Associates, Sunderland, MA. [5]

Symons, D. 1979. *The Evolution of Human Sexuality.* Oxford University Press, Oxford. [26]

Syvanen, M. 1984. The evolutionary implications of mobile genetic elements. *Annu. Rev. Genet.* 18: 271–293. [22]

Syvanen, M. 1994. Horizontal gene transfer: Evidence and possible consequences. *Annu. Rev. Genet.* 28: 237–261. [22]

Szathmáry, E. 1989. The integration of the earliest genetic information. *Trends Ecol. Evol.* 4: 200–204. [7]

Szathmáry, E. 1993. Coding coenzyme handles: A hypothesis for the origin of the genetic code. *Proc. Natl. Acad. Sci. USA* 90: 9916–9920. [7]

Szymura, J. M. 1993. Analysis of hybrid zones with *Bombina.* In R. G. Harrison (ed.), *Hybrid Zones and the Evolutionary Process,* pp. 261–289. Oxford University Press, New York. [15]

Tajima, F. 1983. Evolutionary relationship of DNA sequences in finite populations. *Genetics* 105: 437–460. [22]

Tajima, F. 1989. Statistical method for testing the neutral mutation hypothesis by DNA polymorphism. *Genetics* 123: 585–595. [22]

Takahata, N. 1989. Gene genealogy in three related populations: Consistency probability between gene and population trees. *Genetics* 122: 957–966. [11]

Takahata, N. 1995. A genetic perspective on the origin and history of humans. *Annu. Rev. Ecol. Syst.* 26: 343–372. [26]

Takhtajan, A. 1976. Neoteny and the origin of flowering plants. In C. B. Beck (ed.), *Origin and Early Evolution of Angiosperms,* pp. 207–219. Columbia University Press, New York. [23]

Tan, C. C. 1946. Mosaic dominance in the inheritance of color patterns in the lady-bird beetle, *Harmonia axyridis. Genetics* 31: 195–210. [9]

Taper, M. L., and T. J. Case. 1985. Quantitative genetic models for the coevolution of character displacement. *Ecology* 66: 355–371. [18]

Taper, M. L., and T. J. Case. 1992a. Coevolution among competitors. *Oxford Surv. Evol. Biol.* 8: 63–109. [18]

Taper, M. L., and T. J. Case. 1992b. Models of character displacement and the theoretical robustness of taxon cycles. *Evolution* 46: 317–333. [18]

Tauber, C. A., and M. J. Tauber. 1989. Sympatric speciation in insects: Perception and perspective. In D. Otte and J. A. Endler (eds.), *Speciation and Its Consequences,* pp. 307–344. Sinauer Associates, Sunderland, MA. [16]

Tauber, M. J., C. A. Tauber, and S. Masaki. 1986. *Seasonal Adaptations of Insects.* Oxford University Press, New York. [4]

Taylor, C. B. 1996. More arresting developments: S RNases and interspecific incompatibility. *Plant Cell* 8: 939–941. [15]

Taylor, C. R. 1972. The desert gazelle: a paradox resolved. In G. M. O. Maloiy (ed.), *Comparative physiology of desert animals,* pp. 215–227. Academic Press, London. [17]

Taylor, C. R., and E. R. Weibel. 1981. Design of the mammalian respiratory system. I. Problem and strategy. *Respir. Physiol.* 44: 1–10. [17]

Taylor, D. W., and L. J. Hickey (eds.). 1996. *Flowering Plant Origin, Evolution, and Phylogeny.* Chapman & Hall, New York. [5]

Taylor, G. E. J., L. E. Pitelka, and M. J. Clegg (eds.). 1991. *Ecological Genetics and Air Pollution.* Springer-Verlag, Berlin. [14]

Taylor, P. D., and G. P. Larwood (eds.). 1990. *Major Evolutionary Radiations.* Clarendon Press, Oxford. [25]

Templeton, A. R. 1977. Analysis of head shape differences between two interfertile species of Hawaiian *Drosophila. Evolution* 31: 630–642. [15]

Templeton, A. R. 1980. The theory of speciation via the founder principle. *Genetics* 94: 1011–1038. [16]

Templeton, A. R. 1982a. Genetic architectures of speciation. In C. Barigozzi (ed.), *Mechanisms of Speciation,* pp. 105–121. Alan R. Liss, New York. [16]

Templeton, A. R. 1982b. Why read Goldschmidt? *Paleobiology* 8: 474–481. [24]

Templeton, A. R. 1989. The meaning of species and speciation: A genetic perspective. In D. Otte and J. A. Endler (eds.), *Speciation and Its Consequences,* pp. 3–27. Sinauer Associates, Sunderland, MA. [15]

Templeton, A. R. 1996. Experimental evidence for the genetic-transilience model of speciation. *Evolution* 50: 909–915. [16]

Terres, J. K. 1980. *The Audubon Encyclopedia of North American Birds.* Alfred A. Knopf, New York. [17]

Thaler, D. S. 1994. The evolution of genetic intelligence. *Science* 264: 224–225. [10]

Theodor, O. 1967. *An Illustrated Catalogue of the Rothschild Collection of Nycteribiidae (Diptera) in the British Museum*

(Natural History). British Museum (Natural History), London. [17]

Thewissen, J. G. M., and S. T. Hussain. 1993. Origin of underwater hearing in whales. *Nature* 361: 444–445. [7]

Thewissen, J. G. M., S. T. Hussain, and M. Arif. 1994. Fossil evidence for the origin of aquatic locomotion in archaeocete whales. *Science* 263: 210–212. [7]

Thomas, S., and R. S. Singh. 1992. A comprehensive study of genetic variation in natural populations of *Drosophila melanogaster*. VII. Varying rates of genic divergence as revealed by two-dimensional electrophoresis. *Mol. Biol. Evol.* 9: 507–525. [20]

Thompson, D. W. 1917. *On Growth and Form*. Cambridge University Press, Cambridge. [23]

Thompson, J. N. 1982. *Interaction and Coevolution*. Wiley, New York. [18]

Thompson, J. N. 1988. Evolutionary ecology of the relationship between oviposition preference and performance of offspring in phytophagous insects. *Entomol. Exp. Appl.* 47: 3–14. [14]

Thompson, J. N. 1989. Concepts of coevolution. *Trends Ecol. Evol.* 4: 179–183. [18]

Thompson, J. N. 1994. *The Coevolutionary Process*. University of Chicago Press, Chicago, IL. [18]

Thompson, J. N., and R. Lumaret. 1992. The evolutionary dynamics of polyploid plants: Origins, establishment and persistence. *Trends Ecol. Evol.* 7: 302–307. [16]

Thomson, J. D., and J. Brunet. 1990. Hypotheses for the evolution of dioecy in seed plants. *Trends Ecol. Evol.* 5: 11–16. [21]

Thorne, B. L., and J. M. Carpenter. 1992. Phylogeny of the Dictyoptera. *Syst. Entomol.* 17: 253–268. [5]

Thornhill, N. W. (ed.). 1993. *The Natural History of Inbreeding and Outbreeding*. University of Chicago Press, Chicago, IL. [19]

Thornhill, R., and J. Alcock. 1983. *The Evolution of Insect Mating Systems*. Harvard University Press, Cambridge, MA. [20]

Thorson, G. 1950. Reproduction and larval ecology of marine bottom invertebrates. *Biol. Rev.* 25: 1–45. [19]

Tilley, S. G., P. A. Verrell, and S. J. Arnold. 1990. Correspondence between sexual isolation and allozyme differentiation: A test in the salamander *Desmognathus ochrophaeus*. *Proc. Natl. Acad. Sci. USA* 87: 2715–2719. [9]

Tilman, D. 1988. *Plant Strategies and the Dynamics and Structure of Plant Communities*. Princeton University Press, Princeton, NJ. [4]

Tinbergen, N. 1951. *The Study of Instinct*. Oxford University Press, Oxford. [20]

Tizard, I. R. 1988. *Immunology: An Introduction*. Saunders, Philadelphia, PA. [12]

Toulmin, S. 1957. Contemporary scientific mythologies. In S. Toulmin, R. W. Hepburn, and A. MacIntyre (eds.), *Metaphysical Beliefs*, pp. 50–65. SCM Press, London. [26]

Travis, J. 1989. The role of optimizing selection in natural populations. *Annu. Rev. Ecol. Syst.* 20: 279–296. [14]

Travis, J. 1994. Size-dependent behavioral variation and its genetic control within and among populations. In C. R. B. Boake (ed.), *Quantitative Genetic Studies of Behavioral Evolution*, pp. 165–187. University of Chicago Press, Chicago, IL. [20]

Trivers, R. L. 1971. The evolution of reciprocal altruism. *Q. Rev. Biol.* 46: 35–57. [20]

Trivers, R. L. 1972. Parental investment and sexual selection. In B. Campbell (ed.), *Sexual Selection and the Descent of Man*, pp. 136–179. Heinemann, London. [20]

Trivers, R. L., and H. Hare. 1976. Haplodiploidy and the evolution of the social insects. *Science* 191: 249–263. [20]

Turelli, M. 1984. Heritable genetic variation via mutation-selection balance: Lerch's zeta meets the abdominal bristle. *Theor. Pop. Biol.* 25: 138–193. [14]

Turelli, M. 1988. Phenotypic evolution, constant covariances, and the maintenance of additive variance. *Evolution* 42: 1342–1347. [14, 23]

Turelli, M., and H. A. Orr. 1995. The dominance theory of Haldane's rule. *Genetics* 140: 389–402. [15]

Turelli, M., J. H. Gillespie, and R. Lande. 1988. Rate tests for selection on quantitative characters during macroevolution and microevolution. *Evolution* 42: 1085–1089. [24]

Turing, A. M. 1952. The chemical basis of morphogenesis. *Phil. Trans. R. Soc. Lond.* B 237: 3772. [23]

Turner, J. R. G. 1971. Two thousand generations of hybridisation in a *Heliconius* butterfly. *Evolution* 25: 471–482. [15]

Turner, J. R. G. 1977. Butterfly mimicry—the genetical evolution of an adaptation. *Evol. Biol.* 10: 163–206. [24]

Turner, J. R. G. 1981. Adaptation and evolution in *Heliconius*: A defense of NeoDarwinism. *Annu. Rev. Ecol. Syst.* 12: 99–121. [16]

Turner, J. R. G. 1986. The genetics of adaptive radiation: A neo-Darwinian theory of punctuational evolution. In D. M. Raup and D. Jablonski (eds.), *Patterns and Processes in the History of Life*, pp. 183–207. Springer-Verlag, Berlin. [24]

Twitty, V. C. 1932. Influence of the eye on the growth of its associated structures, studied by means of heteroplastic transplantation. *J. Exp. Zool.* 61: 333–374. [23]

Ursprung, H., K. D. Smith, W. H. Sofer, and D. T. Sullivan. 1968. Assay systems for the study of gene function. *Science* 160: 1075–1081. [3]

Uyenoyama, M. K. 1995. A generalized least-squares estimate for the origin of sporophytic self-incompatibility. *Genetics* 139: 975–992. [21]

Uyenoyama, M. K., K. E. Holsinger, and D. M. Waller. 1993. Ecological and genetic factors directing the evolution of self-fertilization. *Oxford Surv. Evol. Biol.* 9: 327–381. [21]

Val, F. C. 1977. Genetic analysis of the morphological differences between two interfertile species of Hawaiian *Drosophila*. *Evolution* 31: 611–629. [15]

Valentine, J. W. 1985a. Biotic diversity and clade diversity. In J. W. Valentine (ed.), *Phanerozoic Diversity Patterns: Profiles in Macroevolution*, pp. 419–424. Princeton University Press, Princeton, NJ. [25]

Valentine, J. W. (ed.). 1985b. *Phanerozoic Diversity Patterns: Profiles in Macroevolution*. Princeton University Press, Princeton, NJ. [25]

Valentine, J. W. 1990. The macroevolution of clade shape. In R. M. Ross and W. D. Allmon (eds.), *Causes of Evolution: A Paleontological Perspective*, pp. 128–150. University of Chicago Press, Chicago, IL. [25]

Valentine, J. W., S. M. Awramik, P. W. Signor III, and P. M. Sadler. 1991. The biological explosion at the Precambrian-Cambrian boundary. *Evol. Biol.* 25: 279–356. [7, 25]

Valentine, J. W., T. C. Foin, and D. Peart. 1978. A provincial model of Phanerozoic marine diversity. *Paleobiology* 4: 55–66. [25]

van der Pijl, L. 1982. *Principles of Dispersal in Higher Plants*. Springer-Verlag, Berlin. [19]

van Noordwijk, A. J., and G. deJong. 1986. Acquisition and allocation of resources: Their influence on variation in life history tactics. *Am. Nat.* 128: 137–142. [19]

van Noordwijk, A. J., and W. Scharloo. 1981. Inbreeding in an island population of the great tit. *Evolution* 35: 674–688. [11]

Van Tyne, J., and A. J. Berger. 1959. *Fundamentals of Ornithology*. Wiley, New York. [5, 8]

Van Valen, L. 1971. Group selection and the evolution of dispersal. *Evolution* 25: 591–598. [19]

Van Valen, L. 1973. A new evolutionary law. *Evol. Theory* 1: 1–30. [24]

Van Valen, L. 1976. Ecological species, multispecies, and oaks. *Taxon* 25: 233–239. [15]

Van Valen, L. 1984. A resetting of Phanerozoic community evolution. *Nature* 307: 50–52. [25]

Varley, G. C. 1949. Population changes in German forest pests. *J. Anim. Ecol.* 18: 117–122. [4]

Varley, G. C., and G. R. Gradwell. 1968. Population models for the winter moth. *Symp. R. Entomol. Soc. Lond.* 9: 132–142. [4]

Vaughan, T. A. 1986. *Mammalogy*. Third edition. Saunders College Publishing, Philadelphia, PA. [23]

Vehrencamp, S. L., J. W. Bradbury, and R. M. Gibson. 1989. The energetic cost of display in male sage grouse. *Anim. Behav.* 38: 885–896. [19]

Vermeij, G. J. 1987. *Evolution and Escalation: An Ecological History of Life*. Princeton University Press, Princeton, NJ. [18, 25]

Vermeij, G. J. 1993. *A Natural History of Shells*. Princeton University Press, Princeton, NJ. [18]

Via, S. 1994. The evolution of phenotypic plasticity: What do we really know? In L. A. Real (ed.), *Ecological Genetics*, pp. 35–57. Princeton University Press, Princeton, NJ. [14]

Via, S., and R. Lande. 1985. Genotype-environment interaction and the evolution of phenotypic plasticity. *Evolution* 39: 505–522. [14]

Via, S., R. Gomulkiewicz, G. de Jong, S. M. Scheiner, C. D. Schlichting, and P. H. Van Tienderen. 1995. Adaptive phenotypic plasticity: Consensus and controversy. *Trends Ecol. Evol.* 10: 212–217. [14]

Vigilant, L., M. Stoneking, H. Harpending, K. Hawkes, and A. C. Wilson. 1991. African populations and the evolution of human mitochondrial DNA. *Science* 253: 1503–1507. [26]

Voelker, R. A., H. E. Schaffer, and T. Mukai. 1980. Spontaneous allozyme mutations in *Drosophila melanogaster*: Rate of occurrence and nature of the mutants. *Genetics* 94: 961–968. [11]

Vogel, S. 1988. *Life's Devices: The Physical World of Animals and Plants*. Princeton University Press, Princeton, NJ. [17]

Vouidibio, J., P. Capy, D. Defaye, E. Pla, J. Sandrin, A. Csink, and J. R. David. 1989. Short-range genetic structure of *Drosophila melanogaster* populations in an Afrotropical urban area and its significance. *Proc. Natl. Acad. Sci. USA* 86: 8442–8446. [13]

Vrba, E. S. 1980. Evolution, species and fossils: How does life evolve? *S. Afr. J. Sci.* 76: 61–84. [12]

Vrba, E. S. 1984. What is species selection? *Syst. Zool.* 33: 318–328. [12]

Vrba, E. S., and N. Eldredge. 1984. Individuals, hierarchies and processes: Towards a more complete evolutionary theory. *Paleobiology* 10: 146–171. [12]

Vrijenhoek, R. C. 1990. Genetic diversity and the ecology of asexual populations. In K. Wöhrmann and S. Jain (eds.), *Population Biology: Ecological and Evolutionary Viewpoints*, pp. 175–197. Springer Verlag, Berlin. [21]

Vrijenhoek, R. C. 1994. Unisexual fish: Model systems for studying ecology and evolution. *Annu. Rev. Ecol. Syst.* 25: 71–96. [20, 21]

Waage, J. K. 1979. Reproductive character displacement in *Calopteryx* (Odonata: Calopterygidae). *Evolution* 33: 104–116. [16]

Waage, J. K. 1986. Evidence for widespread sperm displacement ability among Zygoptera (Odonata) and the means for predicting its presence. *Biol. J. Linn. Soc.* 28: 285–300. [20]

Waddington, C. H. 1953. Genetic assimilation of an acquired character. *Evolution* 7: 118–126. [14]

Waddington, C. H. 1957. *The Strategy of the Genes*. Allen and Unwin, London. [23]

Wade, M. J. 1977. An experimental study of group selection. *Evolution* 31: 134–153. [12]

Wade, M. J. 1979a. The evolution of social interactions by family selection. *Am. Nat.* 113: 399–417. [20]

Wade, M. J. 1979b. The primary characteristics of *Tribolium* populations group selected for increased and decreased population size. *Evolution* 33: 749–764. [12]

Wade, M. J. 1980a. An experimental study of kin selection. *Evolution* 34: 844–855. [20]

Wade, M. J. 1980b. Kin selection: Its components. *Science* 210: 665–667. [20]

Wade, M. J. 1992. Sewall Wright: Gene interaction and the Shifting Balance Theory. *Oxford Surv. Evol. Biol.* 8: 35–62. Oxford University Press, New York. [14]

Wade, M. J., and S. J. Arnold. 1980. The intensity of sexual selection in relation to male sexual behavior, female choice, and sperm precedence. *Anim. Behav.* 28: 446–461. [20]

Wagner, A., G. P. Wagner, and P. Similion. 1994. Epistasis can facilitate the evolution of reproductive isolation by peak shifts: A two-locus two-allele model. *Genetics* 138: 533–545. [16]

Wagner, G. P. 1988. The influence of variation and of developmental constraints on the rate of multivariate phenotypic evolution. *J. Evol. Biol.* 1: 45–66. [14, 23]

Wagner, G. P. 1989a. The biological homology concept. *Annu. Rev. Ecol. Syst.* 20: 51–69. [5, 23]

Wagner, G. P. 1989b. The origin of morphological characters and the biological basis of homology. *Evolution* 43: 1157–1171. [23]

Wagner, G. P. 1996. Homologues, natural kinds and the evolution of modularity. *Am. Zool.* 36: 36–43. [23]

Wagner, M. 1868. *Die Darwin'sche Theorie und das Migrationsgesetz der Organismen*. Dunker und Humbolt, Leipzig (cited by Mayr 1963). [16]

Wagner, P. J. 1996. Contrasting the underlying patterns of active trends in morphologic evolution. *Evolution* 50: 990–1007. [24]

Wagner, R. P., B. H. Judd, B. G. Saunders, and R. H. Richardson. 1980. *Introduction to Modern Genetics*. Wiley, New York. [3]

Wagner, W. L., and V. A. Funk (eds.). 1995. *Hawaiian Biogeography: Evolution on a Hotspot Archipelago*. Smithsonian Institution Press, Washington, DC. [8]

Wagner, W. L., D. R. Herbst, and S. H. Sohmer. 1990. *Manual of the Flowering Plants of Hawaii*. University of Hawai'i Press and Bishop Museum Press, Honolulu, HI. [5]

Wainwright, P. C., and S. M. Reilly (eds.). 1994. *Ecological Morphology: Integrative Organismal Biology*. University of Chicago Press, Chicago, IL. [17]

Wake, D. B. 1982. Functional and developmental constraints and opportunities in the evolution of feeding systems in urodeles. In D. Mossakowski and G. Roth (eds.), *Environmental Adaptation and Evolution*, pp. 51–66. G. Fischer, Stuttgart. [24]

Wake, D. B. 1991. Homoplasy: The result of natural selection, or evidence of shared design limitations? *Am. Nat.* 138: 543–567. [14, 23]

Wake, D. B., and G. Roth (eds.). 1989. *Complex Organismal Functions: Integration and Evolution in Vertebrates*. Wiley, Chichester, England. [24]

Wake, D. B., K. P. Yanev, and M. M. Frelow. 1989. Sympatry and hybridization in a "ring species": The plethodontid salamander *Ensatina eschscholtzii*. In D. Otte and J. A. Endler (eds.), *Speciation and Its Consequences*, pp. 134–157. Sinauer Associates, Sunderland, MA. [15]

Wake, M. 1992. Morphology, the study of form and function, in modern evolutionary biology. *Oxford Surv. Evol. Biol* 8: 278–346. [6]

Walker, K. R., and L. F. Laporte. 1970. Congruent fossil communities from Ordovician and Devonian carbonates of New York. *J. Paleontol.* 44: 928–944. [25]

Walker, T. D., and J. W. Valentine. 1984. Equilibrium models of evolutionary species diversity and the number of empty niches. *Am. Nat.* 124: 887–899. [25]

Wallace, A. R. 1869. *The Malay Archipelago*. MacMillan & Co., London. [20]

Wallace, A. R. 1878. *Tropical Nature and Other Essays*. MacMillan, New York. [25]

Wallace, B. 1968. *Topics in population genetics*. W. W. Norton, New York. [12]

Wallace, B., and M. Vetukhiv. 1955. Adaptive organization of the gene pools of *Drosophila* populations. *Cold Spring Harbor Symp. Quant. Biol.* 20: 303–310. [14]

Walsh, J. B. 1982. Rate of accumulation of reproductive isolation by chromosome arrangements. *Am. Nat.* 120: 510–532. [16]

Walsh, J. B. 1985. How many processed pseudogenes are accumulated in a gene family? *Genetics* 110: 345–364. [10, 22]

Walsh, J. B. 1987. Persistence of tandem arrays: Implications for satellite and simple-sequence DNAs. *Genetics* 115: 553–567. [22]

Walsh, J. B. 1995. How often do duplicated genes evolve new functions? *Genetics* 139: 421–428. [22]

Walther, F. R. 1984. *Communication and Expression in Hoofed Mammals*. Indiana University Press, Bloomington. [20]

Ward, L. K., and D. F. Spalding. 1993. Phytophagous British insects and mites and their food-plant families: Total numbers and polyphagy. *Biol. J. Linn. Soc.* 49: 257–276. [18]

Ward, P. D., and P. W. Signor III. 1983. Evolutionary tempo in Jurassic and Cretaceous ammonites. *Paleobiology* 9: 183–198. [25]

Ward, P. D., and P. W. Signor III. 1985. Evolutionary patterns of Jurassic and Cretaceous ammonites: An analysis of clade shape. In J. W. Valentine (ed.), *Phanerozoic Diversity Patterns*, pp. 399–418. Princeton University Press, Princeton, NJ. [25]

Warner, R. R. 1984a. Deferred reproduction as a response to sexual selection in a coral reef fish: A test of the life historical consequences. *Evolution* 38: 148–162. [19]

Warner, R. R. 1984b. Mating behavior and hermaphroditism in coral reef fishes. *Am. Sci.* 72: 128–136. [19]

Warner, R. R., and E. T. Schultz. 1992. Sexual selection and male characteristics in the bluehead wrasse, *Thalassoma bifasciatum*: Mating site acquisition, mating site defense, and female choice. *Evolution* 46: 1421–1442. [16]

Waser, N. M., and M. V. Price. 1989. Optimal outcrossing in *Ipomopsis aggregata*: Seed set and offspring fitness. *Evolution* 43: 1097–1109. [21]

Washburn, S. L. 1963. The study of race. *Am. Anthropol.* 65: 521–531. [26]

Watt, W. B. 1968. Adaptive significance of pigment polymorphism in *Colias* butterflies. I. Variation of melanin pigment in relation to thermoregulation. *Evolution* 22: 437–458. [17]

Watt, W. B. 1977. Adaptation at specific loci. I. Natural selection on phosphoglucose isomerase of *Colias* butterflies: Biochemical and population aspects. *Genetics* 87: 177–184. [13]

Watt, W. B. 1983. Adaptation at specific loci. II. Demographic and biochemical elements in the maintenance of the *Colias* PGI polymorphism. *Genetics* 103: 691–724. [13]

Watt, W. B., P. A. Carter, and S. M. Blower. 1985. Adaptation at specific loci. IV. Differential mating success among glycolytic allozyme genotypes of *Colias* butterflies. *Genetics* 109: 157–175. [13]

Watt, W. B., R. C. Cassin, and M. B. Swan. 1983. Adaptation at specific loci. III. Field behavior and survivorship differences among *Colias* PGI genotypes are predictable from *in vitro* biochemistry. *Genetics* 103: 725–739. [13]

Watterson, G. A. 1975. On the number of segregating sites in genetical models without recombination. *Theor. Pop. Biol.* 7: 256–276. [22]

Wcislo, W. T. 1989. Behavioral environments and evolutionary change. *Annu. Rev. Ecol. Syst.* 20: 137–169. [20]

Weber, K. E. 1992. How small are the smallest selectable domains of form? *Genetics* 130: 345–353. [14]

Weber, K. E., and L. T. Diggins. 1990. Increased selection response in larger populations. II. Selection for ethanol vapor resistance in *Drosophila melanogaster* at two population sizes. *Genetics* 125: 585–597. [14]

Weichert, C. K. 1958. *Anatomy of the Chordates*. McGraw-Hill, New York. [23]

Weis, A. E., W. G. Abrahamson, and M. C. Andersen. 1992. Variable selection on *Eurosta's* gall size. I. The extent and nature of variation in phenotypic selection. *Evolution* 46: 1674–1697. [14]

Weishampel, D. B., P. Dodson, and H. Osmólska. 1990. *The Dinosauria*. University of California Press, Berkeley. [7]

Weismann, A. 1889. *Essays upon Heredity and Kindred Biological Problems*. Clarendon Press, Oxford. [21]

Weiss, K. M. 1993. *Genetic Variation and Human Disease*. Cambridge University Press, Cambridge. [26]

Wenzel, J. W. 1992. Behavioral homology and phylogeny. *Annu. Rev. Ecol. Syst.* 23: 361–381. [20]

Werner, E. E., and D. J. Hall. 1979. Foraging efficiency and habitat switching in competing sunfishes. *Ecology* 60: 256–264. [4]

Werner, P. A. 1979. Competition and coexistence of similar species. In O. T. Solbrig, S. Jain, G. B. Johnson, and P. H. Raven (eds.), *Topics in Plant Population Biology*, pp. 287–310. Columbia University Press, New York. [19]

Werner, P. A., and W. J. Platt. 1976. Ecological relationships of co-occurring goldenrods (*Solidago*: Compositae). *Am. Nat.* 110: 959–971. [19]

Werner, T. K., and T. W. Sherry. 1987. Behavioral feeding specialization in *Pinaroloxias inornata*, the "Darwin's Finch" of Cocos Island, Costa Rica. *Proc. Natl. Acad. Sci. USA* 84: 5506–5510. [20]

Werren, J. H. 1980. Sex ratio adaptations to local mate competition in a parasitic wasp. *Science* 208: 1157–1159. [21]

Werren, J. H., U. Nur, and C.-I. Wu. 1988. Selfish genetic elements. *Trends Ecol. Evol.* 3: 297–302. [12]

West-Eberhard, M. J. 1978. Temporary queens in *Metapolybia* wasps: Nonreproductive helpers without altruism? *Science* 200: 441–443. [20]

West-Eberhard, M. J. 1979. Sexual selection, social competition, and evolution. *Proc. Am. Phil. Soc.* 123: 222–234. [16]

West-Eberhard, M. J. 1983. Sexual selection, social competition, and speciation. *Q. Rev. Biol.* 58: 155–183. [20]

West-Eberhard, M. J. 1986. Alternative adaptations, speciation, and phylogeny (a review). *Proc. Natl. Acad. Sci. USA* 83: 1388–1392. [15, 17]

West-Eberhard, M. J. 1989. Phenotypic plasticity and the origins of diversity. *Annu. Rev. Ecol. Syst.* 20: 249–278. [14]

West-Eberhard, M. J. 1992. Adaptation: Current usages. In E. F. Keller and E. A. Lloyd (eds.), *Keywords in Evolutionary Biology*, pp. 13–18. Harvard University Press, Cambridge, MA. [12]

Westoby, M., M. R. Leishman, and J. M. Lord. 1995. On misinterpreting the "phylogenetic correlation." *J. Ecol.* 83: 531–534. [12]

Westoll, T. S. 1949. On the evolution of the Dipnoi. In G. L. Jepsen, G. G. Simpson, and E. Mayr (eds.), *Genetics, paleontology, and evolution*, pp. 121–184. Princeton University Press, Princeton, NJ. [6]

Wheeler, D. A., C. P. Kyriacou, M. L. Greenacre, Q. Yu, J. E. Rutila, M. Rosbash, and J. C. Hall. 1991. Molecular transfer of a species-specific behavior from *Drosophila simulans* to *Drosophila melanogaster*. *Science* 251: 1082–1085. [15]

Wheeler, W. M. 1910. *Ants: Their Structure, Development and Behavior*. Columbia University Press, New York. [12]

Whitam, F. L., and R. M. Mathy. 1986. *Male Homosexuals in Four Societies*. Praeger, New York. [26]

White, M. J. D. 1973. *Animal Cytology and Evolution*. Cambridge University Press, Cambridge. [21]

White, M. J. D. 1978. *Modes of Speciation*. W. H. Freeman, San Francisco, CA. [13, 21]

White, R. E. 1983. *A Field Guide to the Beetles of North America*. Houghton Mifflin, Boston, MA. [4, 24]

White, T. D., G. Suwa, and B. Asfaw. 1994. *Australopithecus ramidus*, a new species of early hominid from Aramis, Ethiopia. *Nature* 371: 306-312. [26]

Whitlock, M. C., P. C. Phillips, F. B. G. Moore, and S. J. Tonsor. 1995. Multiple fitness peaks and epistasis. *Annu. Rev. Ecol. Syst.* 26: 601–629. [14]

Whitlock, M. C., P. C. Phillips, and M. J. Wade. 1993. Gene interaction affects the additive genetic variance in subdivided populations with migration and extinction. *Evolution* 47: 1758–1769. [14]

Whittemore, A. T., and B. A. Schaal. 1991. Interspecific gene flow in oaks. *Proc. Natl. Acad. Sci. USA* 88: 2540–2544. [15]

Whittington, H. B. 1985. *The Burgess Shale*. Yale University Press, New Haven, CT. [7]

Wiehe, T. H. E., and W. Stephan. 1993. Analysis of a genetic hitchhiking model and its application to DNA polymorphism data from *Drosophila melanogaster*. *Mol. Biol. Evol.* 10: 842–854. [22]

Wijsman, E. M. 1984. The effect of mutagenesis on competitive ability in *Drosophila*. *Evolution* 38: 571–581. [10]

Wiley, E. O. 1978. The evolutionary species concept reconsidered. *Syst. Zool.* 27: 17–26. [15]

Wilkinson, G. S. 1988. Reciprocal altruism in bats and other mammals. *Ethol. Sociobiol.* 9: 85–100. [20]

Wilkinson, G. S. 1993. Artificial sexual selection alters allometry in the stalk-eyed fly *Cyrtodiopsis dalmanni* (Diptera: Diopsidae). *Genet. Res.* 62: 213–222. [20]

Wilkinson, G. S., and P. R. Reillo. 1994. Female choice response to artificial selection on an exaggerated male trait in a stalk-eyed fly. *Proc. R. Soc. Lond.* B 255: 1–6. [20]

Wilkinson, G. S., K. Fowler, and L. Partridge. 1990. Resistance of genetic correlation structure to directional selection in *Drosophila melanogaster*. *Evolution* 44: 1990–2003. [14]

Williams, E. E. 1972. The origin of faunas: Evolution of lizard congeners in a complex island fauna—a trial analysis. *Evol. Biol.* 6: 47–89. [8, 18]

Williams, E. E. 1983. Ecomorphs, faunas, island size, and diverse end points in island radiations of *Anolis*. In R. B. Huey, E. R. Pianka, and T. W. Schoener (eds.), *Lizard Ecology: Stud-*

ies of a Model Organism, pp. 326–370. Harvard University Press, Cambridge, MA. [8]

Williams, E. E. 1989. Old problems and new opportunities in West Indian biogeography. In C. A. Woods (ed.), *Biogeography of the West Indies: Past, Present, and Future*, pp. 1–46. Sandhill Crane Press, Gainesville, FL. [8]

Williams, G. C. 1957. Pleiotropy, natural selection, and the evolution of senescence. *Evolution* 11: 398–411. [19]

Williams, G. C. 1966. *Adaptation and Natural Selection*. Princeton University Press, Princeton, NJ. [12]

Williams, G. C. 1975. *Sex and Evolution*. Princeton University Press, Princeton, NJ. [21]

Williams, G. C. 1988. Huxley's evolution and ethics in sociobiological perspective. *Zygon* 223: 383–407. [1]

Williams, G. C. 1992a. *Gaia*, nature worship and biocentric fallacies. *Q. Rev. Biol.* 67: 479–486. [14]

Williams, G. C. 1992b. *Natural Selection: Domains, Levels, and Challenges*. Oxford University Press, New York. [12, 17]

Williams, G. C., and D. C. Williams. 1957. Natural selection of individually harmful social adaptations among sibs with special reference to social insects. *Evolution* 11: 32–39. [20]

Williams, J. A., and S. B. Carroll. 1993. The origin, patterning, and evolution of insect appendages. *BioEssays* 15: 567–577. [23]

Williams, K. S., and L. E. Gilbert. 1981. Insects as selective agents on plant vegetative morphology: Egg mimicry reduces egg laying by butterflies. *Science* 212: 467–469. [18]

Willis, J. H., J. A. Coyne, and M. Kirkpatrick. 1991. Can one predict the evolution of quantitative characters without genetics? *Evolution* 45: 441–444. [14]

Williston, S. W. 1925. *The Osteology of the Reptiles*. Harvard University Press, Cambridge, MA. [7]

Wills, M. A., D. E. G. Briggs, and R. A. Fortey. 1994. Disparity as an evolutionary index: A comparison of Cambrian and Recent arthropods. *Paleobiology* 20: 93–130. [7]

Wilson, A. C., S. S. Carlson, and T. J. White. 1977. Biochemical evolution. *Annu. Rev. Biochem.* 46: 573–639. [5, 22]

Wilson, A. M., and K. Thompson. 1989. A comparative study of reproductive allocation in 40 British grasses. *Funct. Ecol.* 3: 297–302. [19]

Wilson, D. S. 1980. *The Natural Selection of Populations and Communities*. Benjamin Cummings, Menlo Park, CA. [18]

Wilson, D. S. 1983. The group selection controversy: History and current status. *Annu. Rev. Ecol. Syst.* 14: 159–187. [18]

Wilson, D. S., and R. K. Colwell. 1981. The evolution of sex ratio in structured demes. *Evolution* 35: 882–897. [21]

Wilson, D. S., and M. Turelli. 1986. Stable underdominance and the invasion of empty niches. *Am. Nat.* 127: 835–850. [13]

Wilson, E. O. 1971. *The Insect Societies*. Harvard University Press, Cambridge, MA. [12, 20]

Wilson, E. O. 1975. *Sociobiology: The New Synthesis*. Harvard University Press, Cambridge, MA. [20, 26]

Wilson, E. O. 1978. *On Human Nature*. Harvard University Press, Cambridge, MA. [26]

Wilson, E. O. 1992. *The Diversity of Life*. Harvard University Press, Cambridge, MA. [1, 5, 25]

Wilson, E. O., and W. L. Brown. 1953. The subspecies concept and its taxonomic applications. *Syst. Zool.* 2: 97–111. [5]

Wilson, E. O., F. M. Carpenter, and W. L. Brown Jr. 1967. The first Mesozoic ants. *Science* 157: 1038–1040. [6]

Wistow, G. 1993. Lens crystallins: Gene recruitment and evolutionary dynamism. *Trends Biochem. Sci.* 18: 301–306. [23]

Witte, F., T. Goldschmidt, J. Wanink, M. V. Oijen, K. Goudswaard, E. Witte-Mass, and N. Bouton. 1992. The destruction of an endemic species flock: Quantitative data on the decline of the haplochromine cichlids of Lake Victoria. *Env. Biol. Fishes* 34: 1–28. [4]

Woese, C. R. 1987. Bacterial evolution. *Microbiol. Rev.* 51(2): 221–271. [7]

Wolfe, K. H., P. M. Sharp, and W.-H. Li. 1989. Mutation rates differ among regions of the mammalian genome. *Nature* 337: 283–285. [10, 22]

Wolpert, L. 1982. Pattern formation and change. In J. T. Bonner (ed.), *Evolution and Development*, pp. 169–188. Springer-Verlag, Berlin. [23]

Wolpoff, N. 1989. Multiregional evolution: The fossil alternative to Eden. In P. Mellars and C. Stringer (eds.), *The Human Revolution*, pp. 62–108. Princeton University Press, Princeton, NJ. [26]

Wood, B. 1992. Origin and evolution of the genus *Homo. Nature* 355: 783–790. [26]

Wood, R. J. 1981. Insecticide resistance: Genes and mechanisms. *In* J. A. Bishop and L. M. Cook (eds.), *Genetic consequences of man made change*, pp. 53–96. Academic Press, London. [13]

Wood, R. J., and J. A. Bishop. 1981. Insecticide resistance: Populations and evolution. In J. A. Bishop and L. M. Cook (eds.), *Genetic Consequences of Man Made Change*, pp. 97–127. Academic Press, London. [13]

Wood, T. K., and S. I. Guttman. 1982. Ecological and behavioral basis of reproductive isolation in the sympatric *Enchenopa binotata* complex (Homoptera: Membracidae). *Evolution* 36: 233–242. [16]

Wood, T. K., and M. C. Keese. 1990. Host-plant-induced assortative mating in *Enchenopa* treehoppers. *Evolution* 44: 619–628. [16]

Woodburne, M. O., and B. J. MacFadden. 1982. A reappraisal of the systematics, biogeography, and evolution of fossil horses. *Paleobiology* 8: 315–327. [6]

Woodruff, R. C., B. G. Slatko, and J. N. Thompson. 1983. Factors affecting mutation rate in natural populations. In M. Ashburner, H. L. Carson, and J. N. Thompson (eds.), *Genetics and Biology of* Drosophila, pp. 39–124. Academic Press, New York. [10]

Woolfenden, G. E., and J. W. Fitzpatrick. 1990. Florida scrub jays: A synopsis after 18 years of study. In P. B. Stacey and W. D. Koenig (eds.), *Cooperative Breeding in Birds*, pp. 239–266. Cambridge University Press, Cambridge. [20]

Wootton, R. J. 1990. Major insect radiations. In P. D. Taylor and G. P. Larwood (eds.), *Major Evolutionary Radiations*, pp. 187–208. Clarendon Press, Oxford. [7]

Wray, G. A., and D. R. McClay. 1989. Molecular heterochronies and heterotopies in early echinoid development. *Evolution* 43: 803–813. [23]

Wray, G. A., J. S. Levinton, and L. H. Shapiro. 1996. Molecular evidence for deep pre-Cambrian divergences among metazoan phyla. *Science* 274: 568–573. [25]

Wright, S. 1931. Evolution in Mendelian populations. *Genetics* 16: 97–159. [11]

Wright, S. 1935. The analysis of variance and the correlations between relatives with respect to deviations from an optimum. *J. Genet.* 30: 243–256. [14]

Wright, S. 1941. On the probability of fixation of reciprocal translocations. *Am. Nat.* 75: 513–522. [16]

Wright, S. 1968–1978. *Evolution and the Genetics of Populations.* 4 vols. University of Chicago Press, Chicago, IL. [3, 12]

Wright, S. 1982. Character change, speciation, and the higher taxa. *Evolution* 36: 427–443. [14]

Wu, C.-I., and A. W. Davis. 1993. Evolution of postmating reproductive isolation: The composite nature of Haldane's rule and its genetic basis. *Am. Nat.* 142: 187–212. [15]

Wu, C.-I., and W. H. Li. 1985. Evidence for higher rates of nucleotide substitution in rodents than in man. *Proc. Natl. Acad. Sci. USA* 82: 1741–1745. [5, 11, 22]

Wu, C.-I., and M. F. Palopoli. 1994. Genetics of postmating reproductive isolation in animals. *Annu. Rev. Genet.* 28: 283–308. [15]

Wu, C.-I., N. A. Johnson, and M. F. Palopoli. 1996. Haldane's rule and its legacy: Why are there so many sterile males? *Trends Ecol. Evol.* 11: 281–284. [15]

Wu, C.-I., D. E. Perez, A. W. Davis, N. A. Johnson, E. L. Cabot, M. F. Palopoli, and M.-L. Wu. 1993. Molecular genetic studies of postmating reproductive isolation in *Drosophila*. In N. Takahata and A. G. Clark (eds.), *Mechanisms of Molecular Evolution*, pp. 199–212. Sinauer Associates, Sunderland, MA. [15]

Wyatt, R. 1988. Phylogenetic aspects of the evolution of self-pollination. In L. D. Gottlieb and S. K. Jain (eds.), *Plant Evolutionary Biology*, pp. 109–131. Chapman & Hall, London. [21]

Wynne-Edwards, V. C. 1962. *Animal Dispersion in Relation to Social Behaviour.* Oliver & Boyd, Edinburgh. [12]

Yapp, W. B. 1965. *Vertebrates: Their Structure and Life.* Oxford University Press, New York. [17]

Yokoyama, S., and R. Yokoyama. 1996. Adaptive evolution of photoreceptors and visual pigments in vertebrates. *Annu. Rev. Ecol. Syst.* 27: 543–567. [22]

Yoo, B. H. 1980. Long-term selection for a quantitative character in large replicate populations of *Drosophila melanogaster*. I. Response to selection. II. Lethals and visible mutants with large effects. *Genet. Res.* 35: I. 1–17, II. 19–31. [14]

Young, J. P. W., and A. W. B. Johnston. 1989. The evolution of specificity in the legume-rhizobium symbiosis. *Trends Ecol. Evol.* 4: 341–349. [18]

Zachar, Z., and P. M. Bingham. 1982. Regulation of *white* locus expression: The structure of mutant alleles of the *white* locus of *Drosophila melanogaster*. *Cell* 30: 529–541. [22]

Zielenski, J., and L.-C. Tsui. 1995. Cystic fibrosis: Genotypic and phenotypic variations. *Annu. Rev. Genet.* 29: 777–807. [26]

Zimmer, E. A., S. L. Martin, S. M. Beverly, Y. W. Kan, and A. C. Wilson. 1980. Rapid duplication and loss of genes coding for the α chains of hemoglobin. *Proc. Natl. Acad. Sci. USA* 77: 2158–2162. [22]

Zouros, E. 1981. The chromosomal basis of sexual isolation in two sibling species of *Drosophila*: *D. arizonensis* and *D. mojavensis. Genetics* 97: 703–718. [15]

Zouros, E. 1987. On the relation between heterozygosity and heterosis: An evaluation of the evidence from marine mollusks. In M. C. Rattazi, J. G. Scandalios, and G. S. Whitt (eds.), *Isozymes: Current Topics in Biological and Medical Research*, pp. 255–270. Alan R. Liss, New York. [13]

Zouros, E., K. Lofdahl, and P. A. Martin. 1988. Male hybrid sterility in *Drosophila*: Interactions between autosomes and sex chromosomes in crosses of *D. mojavensis* and *D. arizonensis. Evolution* 42: 1321–1331. [15]

Zuckerkandl, E., and L. Pauling. 1962. Molecular disease, evolution, and genetic heterogeneity. In M. Kasha and B. Pullman (eds.), *Horizons in Biochemistry*, pp. 189–225. Academic Press, New York. [11]

Zuckerkandl, E., and L. Pauling. 1965. Evolutionary divergence and convergence of proteins. In V. Bryson and H. J. Vogel (eds.), *Evolving Genes and Proteins*, pp. 97–166. Academic Press, New York. [5]

Zug, G. R. 1993. *Herpetology.* Academic Press, New York. [25]

Zuk, M., R. Thornhill, J. D. Ligon, K. Johnson, S. Austad, S. H. Ligon, N. W. Thornhill, and C. Costin. 1990. The role of male ornaments and courtship behavior in female mate choice of red jungle fowl. *Am. Nat.* 136: 459–473. [16]

Index

A

Abiotic features, 59–60
Absolute fitness, 367
Acacia, adaptation in, 357–358
Acclimation, 531, 533
Acorn weevil, 67
Acrocentric chromosomes, 31, 288
Activation, by transcription factor, 47–48
Active trends
  due to constraints, 692–693
  due to hitchhiking, 694
  due to individual selection, 693
  due to lineage selection, 693–694
Actualism, 18
Adaptation and Natural Selection, 351
Adaptations, 4, 205
  allometry and, 525–527
  constraint and, 528–530
  costs of in coevolution, 544–545
  creationism and, 760
  definitions of, 227, 354–355
  evolution of, 28
  examples of, 337–341
  explanations of, 341–343
  methods for recognizing, 356–360
  misconceptions regarding, 360–362
  morphological and physiological, 519–522
  nonadaptive traits and, 355–356
  problem of and extinction, 714
  to temperature, 278–280
  to varying environments, 531–535
Adaptationist interpretation, of human behavior, 743
Adaptive divergence, 91, 92
Adaptive evolution, 24
  directional selection and, 375, 377–381
  genetic correlation and, 435
Adaptive geographic variation, 259
Adaptive landscapes, 392
  multiple loci in, 402–409
  for quantitative characters, 428
Adaptive peaks, 392, 407
Adaptive radiation, 100, 117

of grasses, 191
of mammals, 191–196
Adaptive significance, 358
Adaptive surface, 402–403
Adaptive topographies. *See* Adaptive landscapes
Adaptive valley, 392
Adaptive value, of behavior, 581–583
Adaptive zone, 716. *See also* Ecological niches
Additive effects, 39, 247, 399, 756
Additive genetic variance, 413–414, 416
Additive inheritance, 413
Additive model, 399–400
Adhesion, 54
Apomorphy, 95
Aerobic respiration, 169
Affected pairs method, 746
African longclaw, 65
Age of Fishes. *See* Devonian
Age of Reptiles, 182, 187–190
Age schedules of reproduction, evolution of, 568–571
Agelaius phoeniceus. See Red-winged blackbird
Age-specific survivorship, 72–73
Aggressive mimicry, 82
Aging. *See* Senescence
Agriculture
  evolutionary biology and, 6–7
  human populations and, 736, 739–740
Agrostis tenuis. See Grass
Aligning genetic sequences, 98
Alle alle. See Dovekie
Allele differences, evolution and, 680–681
Allele frequencies, 234–236, 245. *See also* Gene flow; Genetic drift; Selection
  changes in due to genetic drift, 302
  directional selection and, 375, 377–381
  effect of inbreeding on, 308
  estimating gene flow using, 318
  fitness and, 368–369
  random fluctuations in, 300–301
Allele frequency models, 397
Allele frequency variance, 300–301, 303

Allele substitution, models of, 486
Alleles, 26, 33. *See also* Genes
  defined, 231
  frequencies of. *See* Allele frequencies
  rare deleterious, in humans, 739
Allelopathic chemicals, 75
Allen's rule, 258, 356
Allochronic isolation, 501
Allochthonous taxa, 214
Allometric coefficient, 524, 656
Allometric equation, 524
Allometric growth, 524
Allometry, 156, 524–525, 656–658
  adaptation and, 525–527
Allopatric populations, 135
  biological species concept and, 456
  defined, 450
Allopatric speciation, 482, 483
  defined, 482
  ecological selection and, 488–489
  evidence for, 484–485
  genetic models, 485–487
  kinds of, 482–483
  natural selection for reproductive isolation and, 491–493
  peripatric speciation, 493–498
  role of genetic drift, 487–488
  role of natural selection, 488
  sexual selection and, 489–491
Allopatric speciation alternatives
  parapatric speciation, 498–499
  sympatric speciation, 499–504
Allopolyploids, 287, 505
Allotetraploid, meiosis in, 504–505
Allozygous, 307
Allozymes, 242
Alpine tundra, 63
Alternative equilibria, 390–393
Alternative hypothesis, 758
Alternative mating tactics, 574
Altruism
  in social insects, 598–599, 601
  theories of, 594–595
Altruistic trait, 351. *See also* Kin selection
Ambystoma spp. *See* Salamanders
Amino acid sequences, use in phylogenetic studies, 119
Ammodramus maritimus. See Seaside sparrow

Ammonoids, 185
Amniotes
  evolution of, 145–146
  Mesozoic, 187
  phylogeny of, 96
Amphibians
  Carboniferous, 182
  evolution of, 143–146
Amphimixis, 34, 606
Amphioxus, 176–178
Anaerobic prokaryotes, 169
Anagenesis, 85, 89, 447
  vs. cladogenesis, 513–514
Analysis of variance, 756
Anaphase, 31–32
Anaximander, 17
Ancestors, 280
Ancestral character state, 94
Aneuploid chromosome complement, 286
Aneuploid gamete, 287
Angiosperms
  Cenozoic, 191
  diversification of, 707
  evolution of and relation to insects' evolution, 186–187
  life cycle of, 35
Angler, 64
Animal body plans, evolution of, 710–711
Animal conflict, evolutionarily stable strategy analysis of, 585
Animal phyla, Cambrian explosion of, 172–176, 710
Animals. *See also specific species*
  adaptation to deserts, 520–521
  Devonian, 179–180
  genetic variation at allozyme loci in, 243
  heritability in, 252
  hybrid speciation in, 509
  inbreeding in, 311–312, 622
  polyploidy among, 504–505
Anisogamous species, 34, 606
Anisota senatoria. See Moths
Anoles. *See Anolis* spp.
Anolis spp.
  community patterns of coevolution in, 557–558
  systematics and ecology of, 221–222
Antagonistic pleiotropy, 438, 564, 568

Antagonistic selection, 385
Anthropomorphic descriptions of behavior, 581–582
Anti-Darwinian theories, 23
Antirrhinum majus. See Snapdragon
Ants, adaptations in, 340–341
Apes, phylogenetic relationship with humans, 103–105. *See also* Hominoid; Primates
Apis mellifera. See Honeybee
Apomixis, 34, 606
 role in plant speciation, 504
Aposematic pattern, 82
Apple maggot fly, sympatric speciation in, 501–502
Aquatic ecosystems, 63–65
Aquatic life, 63–65. *See also* Marine life
 Carboniferous and Permian, 182
 Cenozoic, 191
Archaean era, 130, 166, 169
Archaebacteria, 169–170, 171–172
 rRNA structure in, 170
Archaeopteryx, 189. *See also* Dinosaurs
 skeleton, 152
Archaeopteryx lithographica, 153–154
Archaic seed plants, 181
Archosauromorph diapsids, 187–190
Ardipithecus ramidus, 731
Area cladogram, 207
Argument from design, 342
Arisaema triphyllum. See Jack-in-the-pulpit
Aristotle, 5, 17
Arithmetic mean, 250, 755
Armadillo, 65
Aroids, adaptation in, 337, 338
Arrowhead, phenotypic plasticity in, 70
Artemisia carruthii. See Big sagebrush
Artemisia tridentata, 81
Artenkreis, defined, 450
Arthropods, Mesozoic, 185–186
Artificial selection, responses to, 251–253, 418–421
Ascidian, age schedules of reproduction in, 569–570
Asellus aquaticus. See Isopod
Asexual reproduction, 34
 role in plant speciation, 504
Assembly, 705
Assessment signal, 586
Associative dominance, 385
Assortive mating, 500
Astrocaryum mexicanum. See Palm
Asynchronous male and female function, outcrossing and, 619
Atta laevigata, 340–341
Australian grasshopper, adaptive landscapes in, 404
Australian lungfish, 143
Australopithecines, fossil record, 731–732
Australopithecus afarensis, 731–732, 733
Australopithecus africanus, 732, 733
Australopithecus anamensis, 731
Australopithecus boisei, skull, 733

Autapomorphy, 95
Author, of species name, 90
Autochthonous taxa, 214
Autonomy, homeostasis and, 698
Autopolyploids, 287, 505
Autoradiography, 50
Autosomes, 33
Autotetraploids, meiosis in, 504–505
Autozygous, 307
Avena fatua. See Wild oats
Average, arithmetic, 250, 755
Average fitness, 367
Average heterozygosity, 243

B
Back mutation, 272
Backcross, 33
Background extinction, 709
Background selection, 629, 630
Bacteria. *See also Archaebacteria; Buchnera* spp.; *Escherichia coli;* Eubacteria; *Wohlbachia*
 mutations for biochemical abilities in, 280–281
 temperature adaptation in, 278–280
Balance of nature, adaptation and, 361–362
Balance school, of natural selection, 384
Balancing selection, 628–629, 630, 632
Barn swallow, indirect effects in, 593
Barnacle, phylogeny of, 158
Basal lineage, 113
Base pairs, 43
Bateman's principle, 587
Batesian mimicry, 82
Baupläne, 172
Bees, evolution of and role as pollinators, 186–187
Behavior. *See also* Human behavior
 adaptive value of, 581–583
 as phenotypic trait, 579–581
 phylogenetic studies of, 580–581
 role in evolution, 601–602
Behavioral ecology. *See* Evolutionary behavior
Behavioral effects of mutations, 276
Behavioral traits within species, variation in, 579
The Bell Curve, 747
Benthic biota, 63–64
Bergmann's rule, 258, 259, 356, 526–527
Bering land bridge, 190, 197, 733–734, 736
Biarmosuchus skull, 149
Biased gene conversion, 638
Bible
 creationism and, 759–760
 evolution and, 9
Big sagebrush, 81
 adaptation in, 357
Binomial distribution, 755–756
Binomial nomenclature, 87
Biochemical abilities, mutations for, 280–281
Biogenetic law, 652–653, 657
Biogeographic analyses, historical, 209–214

Biogeographic evidence, for evolution, 201–203
Biogeographic provinces, 65
Biogeographical realms, 64–65, 204
Biogeography. *See also* Vicariance biogeography
 defined, 201
 historical, 207–208
 role of paleontology, 201
 role of systematics, 201
Biogeography, ecological approaches
 community convergence, 220–222
 competition effects on distribution, 219–220
 gradients in species diversity, 222–225
 interspecific interactions, 218–219
 island biogeography, 216–217
Biological evolution, 4. *See also* Evolution
 Darwin and, 5
Biological homology, 669–670
 defined, 670
Biological species concept, 27, 263, 449–452. *See also* Isolating mechanisms; Reproductive isolation
 defined, 448
 limitations of, 453–456
 usefulness of, 453
Biology, relation to evolutionary biology, 1
Biomass, 64
Biomechanics, 519
Biomes
 at end of Cretaceous, 707
 terrestrial, 60–63
Biometric approach, to genetic variation, 248
Biometrical school of evolution, 24
Bioprospecting, 754
Biotas, regional, 214–216
Biotic adaptations, 351–352
Biotic features, 59
Birds
 adaptations for flight, 520
 brood parasitism among, 545–546
 Cenozoic, 191
 character gradation in, 113–114
 convergent evolution of giant, 192
 cooperative breeding in, 597–598
 evolution of, 152–154
 evolutionary radiation in, 117
 eye sagittal section, 357
 living ratite families, 206
 Mesozoic, 189
 pollination adaptations in flowers, 111–112
Biston betularia. See Peppered moth
Bivalents, 505
Black-bellied seedcracker, multiple-niche polymorphism in, 387–388
Blackcap, evolution of migration in, 426
Black-tailed prairie dog, estimating gene flow in, 318–319
Blattodea, 102
Blending inheritance, 232

Bluehead wrasse, sequential hermaphroditism in, 575
Body size, 523–527
Bombina spp. *See* Fire-bellied toad
Bonnet, Charles, 18
Bony fishes, 178
 evolution of, 143–144
 respiratory and digestive systems of, 5
 three grades of, 157
Bootstrapping, 99
Botryllus schlosseri. See Ascidian
Bottlenecks
 effects on genetic drift, 392
 effects on population size, 304
Boundaries, passive trends due to, 692
Boundary layer, 522
BP. *See* Base pairs
Breeding systems, variation in, 619–620
Bridgewater Treatises, 19
Brood parasitism, 545–546
Brown rat, evolution of warfarin resistance in, 378–380
Bryozoans, speciation in, 136–138
BSC. *See* Biological species concept
Buchnera spp.
 endosymbiotic species, 541–542
 mutualism in, 553
Burden, 156, 693
Burgess shale, 172
 representative animals, 174
Butterflies. *See also* Lepidoptera
 evolution of color pattern in, 664–665, 681
 evolution of food sources in, 426
 optimal design and constraint in, 529–530

C
Cacti, American, 62
Calibrated rates, 121–122
Calopteryx aequabilis. See Damselflies
CAM. *See* Crassulacean acid metabolism
Camarosaurus, 189
Cambrian archaeocyathid, 173
Cambrian fauna, 707
Cambrian revolution, 172–176, 710
Canalization, 441–442
Cannibalism, kin recognition and, 597
Captorhinida, 145–147
Captorhinomorphs, 187
Carboniferous, terrestrial and aquatic life, 180–183
Carrying capacity, 75
Catasetum saccatum. See Orchids
Catastrophic speciation, 497
Category, taxonomic, 87
Causation, problem of, 353
Cave organisms, vestigial features in, 423–424
cDNA. *See* Complementary DNA
Cell death, 54
Cell differentiation, control of, 49, 52–54
Cell division, 31–-37
 effects of timing on, 54
Cell populations, developmental activities of, 54–55

Cenozoic, 130, 190–191
  aquatic life, 191
  mammals, 191–196
  Pleistocene events, 197–199
  terrestrial life, 191
Census size, 302
Centromere, 31
Cepaea nemoralis. See Land snails
Certhia spp. *See* Treecreepers
Ceryle rudis. See Pied kingfisher
Cetacea, reconstructions, 196
Chambers, Robert, 19
Chance, 4
  natural selection and, 349–350
  trends due to, 692
Chaparral, 62
Characiform fishes, 213–215. *See also* Fishes
  phylogeny, 215
Character displacement, 260, 554
  coevolution of competing species and, 555
Character evolution
  estimating the history of, 107
  rates of, 111–113
Character release, 557
Character scoring, in cladistic analysis, 97
Character states, 93
  distribution of as evidence for evolution, 762–763
  homology and, 109
Characters, 93. *See also* Character displacement; Character evolution; Character states; Metric characters
  adaptive value of, 108
  defining in phylogenetic analysis, 98
  form and function, 115–116
  geographically variable, 259–262
  homology and, 109
  rate of change, 688
Cheating, 553
Checkerboard distribution, 218
Checkerspot butterflies, evolution of food sources in, 426
Chemical compounds, evolutionary biology and, 7
Chetverikov, Sergei, 24, 239
Chiasmata, 32
Chimeric genes, 52
Chimpanzee. *See also* Hominoid; Primates
  classification of, 103–105, 107
  evolution of, 728–730
  molecular clocks and, 120–121
  skull, 733
Chloroplasts, 31, 33
  in eukaryotes, 170
Christian thought, on creation, 17–18. *See also* Bible
Chromatids, 31
Chromosome differences, postzygotic isolation and, 473–475
Chromosome inversions, 288–289
Chromosome number, changes in, 289–290
Chromosome rearrangements, 288–292
  speciation by, 496–497
Chromosomes
  meiosis, 31–33
  Mendelian genetics, 33–37

variant patterns in eukaryotes, 34, 36
Chronospecies, 132, 451
CI. *See* Consistency index
Cicada, 67
Cichlid fishes. *See also* Fishes
  decline in species of, 78, 80
  evolutionary radiation in, 117, 119
  genetic differences among, 468, 469
  rate of speciation in, 510–511
  sexual selection in, 490
  sympatric speciation in, 502–503
Circular overlap, 455–456
Clades, 95, 132, 207. *See also* Monophyletic groups
  evolutionary radiation and, 117–118
  vs. grade, 156–157
  model of distribution of, 207
Cladistic analyses, 105
  limitations of, 97, 100
Cladistic variance school, of vicariance biogeography, 208–209
Cladistics, 94–97, 100
Cladists, 95
Cladogenesis, 85, 89, 447
  vs. anagenesis, 513–514
Cladograms, 95
  amniotes to mammals, 148
Classification, evolution and, 88–89, 91–92. *See also* Cladistics; Phenetics; Systematics; Taxonomic practice
Cleaner wrasse, mimicry in, 82
Cleistogamous flowers, 308, 620
Climate, species diversity and, 222–225
Climate changes, Pleistocene, 197–198
Climate patterns, global, 59–60
Climate regime, 206–207
Climax species, 66–67
Clines, 258–259, 371
  hybrid zones and, 464–468
  selection, gene flow, and, 382–383
Clone, 51
Cloning DNA, 51
Coadapted alleles, 385
Coadapted gene pool, 406
Coalescence models, 626
Coalescent theory, 327, 468–469
  of genetic drift, 298–299
Coarse-grained variation, 386–387
Codon usage bias, 634
Codons, 45
Coefficient of linkage disequilibrium, 401
Coefficient of regression, 757
Coefficient of relationship, 313
  calculating, 600
Coefficient of selection, 368
Coefficient of variation, 159, 756
Coelacanth, 143
  change in rate of evolution of, 158
Coevolution, 220, 539–541, 693, 705, 720. *See also* Mutualism
  of competing species, 554–555
  computer simulation of, 545

of enemies and victims, 542–548
Cohesion species concept, defined, 448
Colaptes auratus. See Northern flicker
Coleoptera, 102
Colias spp. *See* Sulfur butterflies
Collembola, 102–103
Colonization, dispersal and, 576
Commensals, 539
Common ancestor, 95
Common ancestral population, 4
Common descent, 21
Common phylogenetic pattern, 209
Common-garden method, 234
Communities, 65–68
  distinguishing interactive from noninteractive, 705
  species saturation and, 704–706
Community convergence, 220–222, 705–706
Community ecology, 59
  evolution and, 557
Community patterns, in coevolution of competing species, 557
Comparative method, 111
  of recognizing adaptations, 359–360
Comparative physiology, 519
Compartmentation, 166, 167
Competence, 53
Competing species, evolution and coevolution of, 554–555
Competition, 75–76
  effects on distribution, 219–220
  interspecific, 218–219
Competitive exclusion principle, 75–76, 461
Competitive interactions, evolution of, 554–557
Competitors, 539
Complementary DNA (cDNA), 52. *See also* Deoxyribonucleic acid
Complexity, 695–697
  functional integration and, 683–687
  as method for recognizing adaptation, 356
Components of fitness, 368–370
Computer models, 229
Concerted evolution, 291n
  evolutionary significance of, 639
  by gene conversion, 638–639
  in gene families, 637–639
  mechanisms of, 638
Concordant speciation, 541
Condylarths, 194–195
Conflicting selection pressures, 345
Congenital differences, 233
Coniferous forest, 63
Conodonts, 174–175
Consensus trees, 99
Conservation biology, 753
  evolutionary biology and, 7
Conservation management, evolutionary biology and, 7
Conservative characters, 112, 159
Consilience, 763
Consistency indices, 99
Conspecific populations, 262, 263
Conspecific species, 448–449

Constant ratio, 554
Constitutive transcription, 38
Constraints
  active trends and, 692–693
  adaptation and, 528–530
  developmental perspective on, 672–673
  on evolution, 671–674, 684–685
  on life history evolution, 563–566
  on loci, 321
  phylogenetic perspective on, 672–674
Contact zones, 484–485
Contests, sexual selection by, 588–589
Continuity of evolutionary change, 115
Continuous variation, 24, 41, 42
Conventional signal, 586
Convergence. *See* Convergent evolution
Convergent evolution, 65, 91–92, 110–111, 123, 220–221, 259, 632
  examples in animals, 65
Cooperation
  evolution of, 594–599, 601
  theories of, 594–595
Cooperative breeding, 597–598
Cope's rule, 157, 695–696
  passive trends and, 692
Corapipo gutturalis. See Manakins
Corn, gene flow in, 317–318
Correlated evolution, 398
  of quantitative traits. *See under* Quantitative traits
Correlated response, 410
Correlated selection, 428–429
  defined, 428
Correlation, 359–360, 674
Correlation coefficient, 429, 757
Corvus spp. *See* Crows
  as distinguished by morphology, 448
  narrow hybrid zone of, 454
Coryanthes speciosa. See Orchids
Cosexual species, 34, 606–607
Cospeciation, 541
Countercurrent heat exchanger, 521
Countergradient variation, 259
Coupling gametes, 400
Covariance, 416, 757
Crabs, differences among species of in physiological homeostasis, 70
Crassulacean acid metabolism, 523
Creation science, 759–760
Creationism
  creationists, 759–762
  refuting creationist arguments, 760–762
Cretaceous, 182, 185
Crickets
  allele frequency in, 371
  phylogeny of on Hawaiian Islands, 212
Crocodilians, 188
Cross-fertilization, 36
Crossing over, 32. *See also* Recombination; Unequal crossing over
Crossing technique, 241
Crown group, 95

Crows
  as distinguished by morphology, 448
  narrow hybrid zone of, 454
Culex pipiens. See Mosquito
Cultural evolution, vs. biological evolution, 741
Culturally inherited trait, 579
Culture, genetics and, 741
Cuvier, Georges, vs. Lamarck, 19
CV. *See* Coefficient of variation
C-value, 641
Cycad, 186
Cynodonts, 152
Cynognathus, 148, jaw, 149
Cynomys ludovicianus. See Black-tailed prairie dog
Cyprinodon spp. *See* Desert pupfish
Cyrtodiopsis dalmanni. See Stalk-eyed fly

D

Damselflies, reproductive character displacement in, 493
Darwin (unit of measurement), 159, 688
Darwin, Charles
  on adaptations, 338, 354
  on behavior and shifts of ecological niche, 602
  on coevolution, 540
  on common descent, 11, 89, 122, 337
  on competition, 76
  on embryology and evolution, 651, 652
  on evolution of behavior, 576
  on evolution of body plans, 710
  evolutionary theory after, 23–24. *See also* Evolutionary synthesis
  evolutionary theory of, 4–6, 21–23
  evolutionary thought prior to, 17–18
  on fossil record, 129
  on functional modification, 682–683
  on gradualism, 678–679
  on human evolution, 7–8, 728
  hypothetico-deductive method and, 359
  on importance of biogeography to, 201–203
  life of, 19–21
  on morality and evolution, 362
  on natural selection, 11, 21–22, 342
  on origin of mutualisms, 552
  on perfection and adaptation, 361
  phylogenetic tree (illus), 88–89
  on principle of kin selection, 595n
  on sex ratio, 613
  on sexual selection, 586
  on speciation, 481
  on species concept, 448
Darwin's finches. *See under Geospiza*
Darwinian evolutionary theory, progress and, 691
Darwinian medicine, 754
Darwinian social science, 743
Datura spp. *See* Jimsonweeds

De Vries, Hugo, 24
  mutation and, 268
Deaths, nonselective vs. selective, 368
Deer, diploid chromosome complements, 292
Deer mouse, geographic range of, 536
Deinonychus, 188–189
Deleterious alleles, persistence in natural populations, 381
Deleterious mutations, advantage of sex and, 610–611
Deletions, genetic, 47, 290–292
Demes, 300
Demographic traits, 561
  evolution of, 563–566
Denaturation, annealing, and hybridization of DNA, 50–51
Density, population, 71
Density-dependent limiting factors, 73–75
Density-dependent natural selection, 572
Density-independent limiting factors, 73–75
Deoxyribonucleic acid (DNA), 43, 231. *See also* Amino acids; Deoxyribonucleic acid sequences; Molecular biology
  double helix, 44
  genome size and highly repetitive, 641–642
  range of contents, 46
  variation at level of, 244–245
Deoxyribonucleic acid polymerase enzymes, 43
Deoxyribonucleic acid replication, 43
Deoxyribonucleic acid sequence evolution, 626–627
  causes of sequence evolution, 629–632, 634–635
  detecting natural selection from DNA sequences, 633–634
  interpreting variation in DNA sequences, 628–629
  patterns of variation in DNA sequences, 627–628
Deoxyribonucleic acid sequences
  on sequencing gel, 52
  replacement hypothesis and, 735–736
  use in phylogenetic studies, 119–120
Derived character state, 94
The Descent of Man, and Selection in Relation to Sex, 21, 123, 728
  evolution of behavior in, 579
  sexual selection in, 586
Descent with modification, 4, 19
  Darwin on, 11, 89, 122, 137
Desert, 62
Desert plants
  adaptations in, 522, 523
  convergent growth in, 62
Desert pupfish, rate of speciation in, 510–511
Design
  efficiency of, 697–698
  as method for recognizing adaptation, 356–357
Desmognathus ochrophaeus. See Dusky salamanders
Development. *See also* Ontogeny

evolution and, 651–652, 670–675
  hierarchical control of gene expression in, 48–49
  implications for evolution, 670–675
  mechanisms of, 49, 52–55
Developmental biology, role in evolutionary theory, 651
Developmental constraints, on evolution, 671
Developmental genetics, evolution and, 665–669
Developmental noise, 248
Developmental plasticity, integration and, 665
Developmental principles of evolutionary change, 653–654
  changes in tissue interactions, 662
  developmental plasticity and integration, 665
  heterochrony, 655–661
  heterotypy, 661–662
  individualization and dissociation, 654–655
  pattern formation, 663–665
  thresholds, 662–663
Developmental switches, 70
Devonian, marine life in, 176–178
Diapause, 70
Diapsids, 187
Dicots, flowers of, 110
Dicrocoelium dendriticum. See Fluke
Differential extinction, 352
Differential speciation, 352
Differentiation, 48–49
  cellular, 49, 52–54
  defined, 49
Diffuse coevolution, 540
Dinosaurs, 188
  evolution of, 152–154
Dioecious species, 34, 606
Dioecy, 615–617
  outcrossing and, 619
Diploid cell, 33
Diploid chromosome complements, 292
Diploid generation, eukaryotic reproduction and, 34
Diploid species, calculating coefficients of relationship in, 600
Diploids, 286
Diploidy, 612
Diptera, 102
Direct pleiotropy, 55
Directed mutation, 285
Directed speciation, 690
Directional mutation, 156
Directional selection, 156, 366, 693. *See also* Selection
  behavior and, 601–602
  measuring, 422
  responses to, 418
  at two loci, 398–399
Directional trends, evolution and, 672
Discrete generations, 71
Discrete variations, 24
Disjunct distributions, 205
  explanations for, 208–209
Disparity, 172, 710
Dispersal
  correlated rates of speciation and extinction and, 712
  defined, 575

evolution of, 575–576
Dispersal hypothesis, 208–209
Displacement, 715–716
Disruptive selection, 366, 500
  responses to, 421–422
Dissociation, individualization and, 654–655
Distance, geographic identity and, 260
Distributions
  disjunct, 208–209
  effects of competition on, 219–220
  patterns of, 203–205
Divergence, 207
  temporal course of, 461–462
Diversification, origination and, 715–721, 723–724
Diversity, 699. *See also* Species diversity
  ecological approaches to, 704–706
  equilibrium and, 721, 723–724
  factors that influence, 704
  history of, 706–708
  measuring, 703–704
The Diversity of Life, 723
Diversity patterns, effects of history on, 706
Diving petrel, 65
DNA. *See* Deoxyribonucleic acid
DNA-DNA hybridization method, 119
Dobzhansky, Theodosius, 4, 24–25, 26, 240–241, 252–253, 727
  on adaptive mutations, 282
  on biological species concept, 450
  on coadapted gene pool, 406
  on dependence of fitness on environment, 370–371
  Drosophila studies in hybrid sterility, 471
  Drosophila studies in natural selection, 344
  on fitness effects of mutations, 679
  genetic recombination studies of *Drosophila*, 284–285
  on isolating mechanisms, 457
  on natural selection, 384–385
  on reproductive isolation, 491
  on speciation, 481
Dollo's law, 99, 156
Domains, 642–643
Dominance, 37, 39, 238–239, 399. *See also* Competitive exclusion principle; Competitive interactions; Sexual selection
  effect on genotype, 414
Dominance variance, 413–416
Dominant allele, 39
Dosage compensation, 617
Dovekie, 65
Downy woodpecker, as example of Bergmann's rule, 527
Drag, 520
Drongos, geographic variation in, 257, 259
Drosophila genetics, 24
Drosophila melanogaster
  acclimation in, 533–534
  artificial selection in, 419–421
  differentiation in, 48–49
  DNA sequencing in, 244

Index

evolution of senescence in, 568
fitnesses of two-locus genotypes, 404–405
genetic drift in, 304–305
genetic map, 37, 40
geographic variation in,
253–255, 256
mutation rates in, 274, 275
polymorphism in, 239–241
polytene chromosomes, 46
QTL mapping in, 410–411
real reaction norms in, 41
sexual selection in, 345
speciation in, 469–470
spread of transposable elements
in, 640–641
sympatric speciation in, 503
Drosophila pseudoobscura
artificial selection in, 252–253
estimating fitnesses in, 370–371
estimating gene flow in, 317
heterozygosity and, 385
inversions in, 288–289
lethal alleles in, 240–241
polymorphism and heterozygosity in, 242–243
selection for phototaxis and
geotaxis, 254
Drosophila serrata, fecundity
and age-specific survivorship in, 73
Drosophila spp. *See also specific
species*
allele frequencies in, 260
biogeographical analysis in
Hawaiian Islands, 210–212
dispersal hypothesis and, 208
evolutionary radiation in, 117
genetic recombination in,
284–286
genome, 45
homeotic genes of, 665–666,
667–669
hybrid sterility in, 471–473
model of hierarchical control of
development in, 53
natural selection experiments
in, 344–345
norm of reaction in, 39
peripatric speciation in, 496, 497
phenotypic differences among
(illus), 470–471
prezygotic and postzygotic isolation in, 462
rate of speciation, 510-511
real reaction norms in, 41
replacement and synonymous
substitutions in, 323
role in early genetic laboratory
studies, 24–25
sequence variation in, 632
sexual isolation in, 477
threshold trait evolution in,
441–442
Drosophila willistoni, divergence
in, 461
Dry forest, 61
Duplications, of genetic material,
290–292
Dusky salamanders
progenesis in, 659–661
sexual isolation in, 260–261
Dwarf willow, 63

E

Ear apparatus, evolution of, 150
Earth's atmosphere, pattern of cir-
culation, 60
Echinoderms
diversity in, 717
extinct and living classes, 177
Ecological approaches, to diversi-
ty, 704–706
Ecological biogeography, 201
Ecological equilibrium, 216
Ecological genetics, 13
Ecological isolation, 500–501
Ecological isolation and
adaptation, role in plant
speciation, 504
Ecological niches, 68–69
Ecological physiology, 519
Ecological release, 554
Ecological selection, 586–587
vs. sexual selection, 589
speciation and, 488–489
Ecological species concept,
defined, 448
Ecological succession, 66–67
Ecology, 13. *See also* Biogeogra-
phy, ecological approaches
diversity and, 703
etymology, 59
Ecomorphology, 519
Ecomorphs, 221
Economy of design, 528
Ecosystems, 59, 361–362
aquatic, 63–65
Ecotypes, 254, 259
defined, 450
Ectopic exchange, 639
Edentates, 192–194
Ediacaran fauna, 172
Effect hypothesis, 353
Effective population size, 302–304
haplotype diversity and,
327–329
Effectively neutral allele, 321
Effector genes, 665
Efficient causes, 5
Egg mimicry, 546,, 549
Eichhornia paniculata
evolution of self-fertilization in,
458–459
outcrossing in, 621
Elaphe obsoleta. *See* Rat snake
Electromorph, 242
Electrophoresis, 242–244
genetic drift studies and, 307
Electrophoretic gel, 242
Embryological similarities, evi-
dence for evolution, 122
Embryology. *See* Development
Embryonic induction, 53
Empedocles, 17
Enchenopa binotata. *See* Treehop-
pers
Endemism, 201
level of, 203
Endosymbionts, 170
Endosymbiotic species, 541–542
Enemies, 539
Enemy-victim coevolution,
542–548
Ensatina eschscholtzii. *See*
Salamanders
Environment
adaptation to, 531–535

dependence of fitness on,
370–371
genotype and, 39
phenotypic variation and, 233
selection in variable, 387
Environmental coefficient, 429
Environmental deterministic
hypotheses, 609
Environmental effects, 26
Environmental factors, mutation
effects on fitness and, 278
Environmental management, 754
evolutionary biology and, 7
Environmental sex determination,
617
Environmental variance, 248, 274
responses to, 69–71
Environmentalism, 741–742
Environments, 59
Epicontinental sea, 135–136
Epifaunal animals, 718
Epigenesis, 49, 651
Epigenetic ratchets, 692–693
Epistasis, 42, 399–400
genetic variance and, 442–444
Epistatic allele, 42
Epistatic interactions, 473
Epistatic variance, 415–416, 443
Equidae. *See also* Horse lineage
evolution of, 159–160
evolutionary relationships of,
139–142
Equilibrium, diversity and, 721,
723–724
Equilibrium community, 705
Equilibrium frequency, 316
Equilibrium hypotheses, of species diversity gradients,
223–225
Equilibrium species diversity, on
islands, 217
Equus, 139–142. *See also* Horse
lineage
Erythrina flabelliformis, adaptation in, 112
Escape-and-radiate coevolution,
540–541
Escherichia coli
adaptation and temperature in,
278–280
adaptive mutation experiments,
282–283
mutations for biochemical abilities in, 280–281
natural selection experiments
in, 343–344
rRNA structure, 170
second-order selection in, 606
selection on silent substitution
in, 634–635
thermal acclimation in, 534–535
ESS. *See* Evolutionarily stable
strategy
*Essay on the Principle of Popula-
tion*, 20
Essence, Plato's 6
Essentialism, 6, 17
Essentialist notion of species, 448
Establishment, of populations,
206
Estimation, of mutation rates,
271–273
Estivation, 62
Ethics
evolution and, 8–9
natural selection and, 362

Ethnic groups, 737–738
Ethological isolation, 457–458
Ethology, 579
Eubacteria, 169–170, 171–172
rRNA structure in, 170
Eucyrtidium spp. *See* Radiolarians
Eukaryotes, 170–172
cell division in, 31–32
genetic variation in, 283–285
phylogeny of, 173
rRNA structure in, 170
variant patterns of reproduction
and meiosis in, 34, 36
Eukaryotic gene, diagram, 44
Euphydryas editha. *See* Checkerspot butterflies
Euplectes progne. *See* Long-tailed
widowbird
European starling, range expansion of, 204
Eurosta solidaginis. *See* Goldenrod
gall fly
Eurycea spp., paedomorphism in,
659–660
Eurytopy, 712
Eusocial species, 598–599, 601
Eustatic events, 207
Eusthenopteron, 144
Eutherian mammals, 146, 190,
192–196
evolutionary radiation in, 117
Evolution. *See also* Coevolution;
Concerted evolution; Con-
vergent evolution; Development; Developmental
principles of evolutionary
change; Fossil record, evolution in; Genetic drift; Molecular evolution
after Darwin, 23–24. *See also*
Evolutionary synthesis
of age schedules of reproduction, 568–571
biogeographic evidence for,
201–203
classification and, 88–89, 91–92
community ecology and, 557
of competitive interactions,
554–557
of cooperation, 594–599, 601
defined, 4
of demographic traits, 563–566
of dispersal, 575–576
effect of genetic correlation on,
433–437
etymology, 4
evidence for, 122–123, 762–763
explanations of history of,
648–649
as fact vs. theory, 11–12
genetic constraints on, 436–437
geography of. *See* Biogeography
gradualism in, 113–115
of humans, 728–734, 741–742
of inbreeding and outbreeding,
619–622
iterative, 157
of kin selection, 597–599, 601
mechanisms of, 28
molecular insights into, 29
of mutation rates, 605–606
of natural selection, 365
natural selection and future,
739–740
nonadaptive, 297
in nonliving systems, 166

of novel genes and proteins, 637, 642–644
of novelty, 681–687
of parasite virulence, 548–551
of quantitative characters, 418–422
of quantitative traits, 425–428
rate of increase of, 572–573
rates of, 687–691
relation to biology, 3–4
relation to phylogenetic inference, 100
role of behavior in, 601–602
of semelparity and iteroparity, 570
of senescence, 568
of sex chromosomes, 617–619
of sex ratios in structured populations, 614–615
of sex-determining mechanisms, 617–619
of sexes, 612–613
society and, 749
three models of, 139
of tolerance, 531–535
Evolution Above the Species Level, 25
Evolution rates, 157–162
Evolution: The Modern Synthesis, 25
Evolutionarily stable strategy
analysis of animal conflict, 585
of behavior, 584–586
sex ratios and, 613–614
Evolutionary behavior, 14, 579
Evolutionary biology
contemporary studies, 28–29
future of, 753–754
importance of, 6–8
relation to other biological disciplines 1, 13
structure of, 12–14
themes of, 3
Evolutionary change
discontinuity of, 674–675
speciation and, 514
tiers of, 714–715
Evolutionary change principles
character evolution rates, 111–113
character form and function, 115–116
clades and evolutionary radiation, 117–118
gradualism of evolution, 113–115
homologous features, 108–110
homoplasy, 110–111
phylogenetic analysis and evolutionary trends, 116–117
Evolutionary developmental biology, 13, 652, 753
Evolutionary ecology, 13, 14, 753
Evolutionary embryology, 652
Evolutionary faunas, 707
Evolutionary genetics, 13, 14
Evolutionary history, 648–649
diversity and, 703
patterns in, 29
trends in, 694–695
Evolutionary molecular biology, 753
Evolutionary morphology, 14, 519
Evolutionary perspective, 4–6
Evolutionary physiology, 14, 519
Evolutionary psychology, 744
Evolutionary radiation, 100, 165

clades and, 117–118
Evolutionary rates, 687–691
correcting for size, 159
genetic approaches, 688–689
vs. time interval, 162
Evolutionary reversals, 110
Evolutionary species, 450–452
defined, 448
Evolutionary synthesis, 24–28
major tenets, 26–28
mutation and, 268
Evolutionary theory
Darwin's, 21–23
Mendelism and, 232–233
prior to nineteenth century, 17–18
vs. transformational, 22
Evolutionary trends, 116–117, 155–157
Evolvability, 417
Exaptations, 355
Exon shuffling, 643–644
Exons, 44, 270–271
comparison of DNA sequence, 272
Expected distribution, 99
Exploitation, 75
The Expression of Emotions in Man and Animals, 21
evolution of behavior, 579
Extinct genes, resurrecting, 626
Extinct organisms, phylogenetic analysis and, 105–107
Extinction
causes of, 712–713
correlated rates of and speciation, 711–712
in fossil record, 698
gene flow and, 316–317
island biogeography and, 216–217
mass. *See* Mass extinction
patterns of origination and, 708–709, 711–712
problem of adaptation and, 714
selectivity of, 714
Extranuclear genome, 31
Eye
sagittal section of bird's, 357
evolution of, 682–683, 684
example of convergent evolution, 111

F

F$_2$ breakdown, 406
Facts
defined,
vs. hypotheses, 9–11
Fecundity, fitness and, 368–370
Female choice, in mate selection, 345, 589–594
Female mating success, 369–370
Females, evolution of sexes, 612–613
Fern, life cycle, 35
Fig wasps
sex ratio evolution in, 615
virulence in, 551
Figs, mutualism with fig wasps, 81
Final causes, 5
Finches. *See under Geospiza*
Fine-grained variation, 386–387
Fire-bellied toad, hybrid zones and, 465–466, 467
Fisher, R. A., 24–25, 298
on adaptive evolution, 407
on analysis of variance, 756

on gradualism, 678–679
on mass selection, 409
on runaway sexual selection, 590
on selfing and outcrossing, 621
on sex ratio, 613
Fishes. *See also* Bony fishes; Characiform fishes; Cichlid fishes; Jawless fishes; Sticklebacks
biogeographical analysis in Central America, 209–210
Cenozoic, 191
Mesozoic, 185
Fissions, 474
chromosomal, 289
Fitness, 398–402
adaptive landscapes and, 402–409
competition and, 554–555
components of, 368–370
defined, 349, 366–367
dependence of fitness on environment, 370–371
effects of mutations on, 278–281
F$_2$ breakdown and, 406
inbreeding and, 620–621
indicator models and, 592
life history traits and, 561–563
as measure of progress, 700
modes of selection, 366
mutation effects on, 679
pictorial representation of genotypes, 398
relative fitness and rate of change, 367–368
selection models and, 376
species interaction and, 539
symbiotic mutualisms and, 551
theory of life history evolution and, 566–568
Fitness surface, 429
Fixation index, 314
Fixation probability, selection and genetic drift and, 393
Fixation time, 301
Fixed allele, 299
Flanking repeats, model of origin of, 270
Flour beetles
cannibalism in, 596–597
directional selection in, 378–379
evolution of population size in, 346–348
genetic differences in fitness in, 443
stabilizing selection on, 422
Flowering plants, gradualism in, 115
Flowers
cleistogamous, 308, 620
pollination adaptations in, 111–112
trends in form, 696–697
Fluke, life cycle, 79, 80
Focal group, 97
Food chain. *See* Trophic level
Food source, evolution of in checkerspot butterflies, 426
Foraminifera, character change in, 133–134
The Formation of Vegetable Mould, Through the Action of Worms, 21
Fossil DNA, phylogenetic analysis and, 105–107

Fossil evidence, for enemy-victim coevolution, 546–547
Fossil record
angiosperms in, 186
Australopithecines in, 731–732
creationism and, 759–760, 761, 762
Devonian plants in, 179
dispersal hypothesis and, 208
explanation for gaps in, 677–679
extinction in, 698
history of diversity and, 706–708
hominid, 730–732, 734
Homo, 732–734
mammals in the Cenozoic, 191–192
phylogeny and, 154–155
reliability of, 129, 131–132
taxonomy in, 132
temporal distribution of insects in (illus), 183
temporal distribution of vascular plants in, 180
value in biogeographic history, 208
Fossil record, evolution in, 132–133, 688
of animal body plans, 710–711
of early animals, 173–174
history of, 27, 127
punctuated equilibrium and, 137
of speciation, 135–137
of species character changes 133–135
of tetrapod vertebrates, 143–154
of transitions among genera, 138–142
trends, 155–157
Fossil sample, speciation and, 691
Fossils, rock formation and, 127
Founder effect speciation. *See* Peripatric speciation
Founder effects, 304
Frameshift mutations, 269
Frequency, 69, 755
Frequency distribution, 409, 755–756
Frequency-dependent fitness, 584
Frequency-dependent selection, 389–390, 613–614
Freshwater ecosystems, 63–64
Freud, Sigmund, 5
Frog-eating bat, sexual selection and, 345–346
Frogs
Cenozoic, 191
rate of speciation, 511–512
Frozen niche variation model, 610
Fugitive species, 67
Full-sib families, comparisons of variance among, 417–418
Function, of feature, 350
Functional complexity, acquisition of, 686–687
Functional constraints, on evolution, 671
Functional divergence, of redundant elements, 685
Functional integration, complexity and, 683–687
Functional modification, 682–683
Functional morphology, 519
Functional response, 78
Functions, decoupling of, 685–686
Fusions, 474
chromosomal, 289

 Index

G

Gaia hypothesis, 362
Galápagos Islands, Darwin's trip to, 20
Gall midge, 81
Galton, Francis, 24
Gambusia affinis. See Mosquitofish
Gamete frequencies, 376
Gametes, 32
Gametic incompatibility, 459–460
Gametic selection, 369–370
Gametophyte, 34
Gap dynamics, 67
Gap genes, 49
Garter snake, genetic correlation in, 429, 431–433
Gasterosteus aculeatus. See Sticklebacks
Gasterosteus spp., coevolution of competing species in, 556
Gastropods, trends in, 694–695
Gender identity, 745
Gender role, 745
Gene amplification, 47
Gene conversion, 269. *See also* Recombination
  concerted evolution by, 638–639
Gene copy, defined, 231
Gene exchange
  restricted, 454–456
  as species criterion, 451
Gene expression, 47–48
  hierarchical control of, 48–49
Gene families, 46, 291, 635–637
  concerted evolution in, 637–639
  evolution of new function, 637
Gene flow, 237, 292, 314–315, 319–320, 371, 438
  effects on allele frequencies, 382
  estimates of, 317–319
  extinction and recolonization, 316–317
  genetic drift and, 316
  inbreeding, mutation, and, 315
  limits on geographic range of species and, 535
  models of, 315–316
  selection and, 381–383
  shifting balance theory and, 409
Gene flow barriers, 457–460
Gene frequency, 234. *See also* Allele frequencies
Gene function, evolution of new, 637
Gene interactions, 399–400, 438
  change in, 667–669
Gene mapping, 39
Gene migration. *See* Gene flow
Gene mutations, 267–268. *See also* Mutations
  defined, 267
  effects on fitness, 278–281
  phenotypic effects, 276–278
  point mutations and, 268–271
  as random process, 281–283
  rates of, 271–276
Gene pools, 27
Gene therapy, 739
Gene transfer, inferring, 626
Gene trees, 328, 330, 332, 468–469
  history of mutational events, 627
  of human species, 735–736
  population structure and, 327–333

species trees and, 332–333
Genealogical species concept, 452
Gene-culture coevolution, 741
Gene-environmental factor/phenotype mapping function, 440–442
Gene-for-gene interactions, 543–544
Genera, evolutionary transitions among, 138–142
General stress resistance, 532–533
Generalist, 66
Generation length, substitution rates and, 326
Generation time, 563
Genes. *See also* Alleles; Gene families; Gene flow; Gene mutations; Gene trees; Genetics; Horizontal gene transfer
  defined, 44, 231
  evolution of, 625, 642–644
Genet, 34, 71
Genetic analyses, methods in molecular, 50–52
Genetic approaches, to evolutionary rates, 688–689
Genetic assimilation, 441
Genetic basis
  of prezygotic isolation, 476–477
  of reproductive isolation, 477–478
  of species differences, 468–470
Genetic change, in trait under directional selection, 399
Genetic code, 45
  origin of, 168
Genetic constraints
  genetic variation and, 564
  on evolution, 672
Genetic correlation, 429–431
  defined, 428
  effect on evolution, 433–437
  epistatic component of genetic variance, 442–444
  evolution of quantitative characters by genetic drift, 437–438
  examples of, 431–433
  genetic variation in quantitative characters, 438–439
  genotype-environment interactions, 439–440
  threshold traits, 440–442
Genetic demography, 373
Genetic differences, underlying phenotypic differences, 679–680
Genetic disabilities, in humans, 739
Genetic distance, 256–257. *See also* Molecular clocks; Phylogenetic inference
Genetic drift, 26, 297–300, 365. *See also* Divergence
  concerted evolution and, 637
  evolution by, 300–304
  evolution of quantitative characters by, 437–438
  gene flow and, 316
  inbreeding and, 314
  interaction with selection, 392–393
  linkage disequilibrium and, 401
  model of, 688
  modern effects of, 739–740
  nonadaptive traits and, 355–356
  in real populations, 304–307

relative to genetic variation, 301–302
  sequence evolution and, 629
  shifting balance theory and, 409
  speciation by, 486, 487–488
  stabilizing selection and, 406
Genetic dynamics, in hybrid zones, 465–467
Genetic engineering, 52
Genetic map, 37
Genetic markers, 244
Genetic material, functions of, 43–48
Genetic models, of allopatric speciation, 485–487
Genetic polymorphism, 239
Genetic quality, mate choice as indicator, 591–592
Genetic sex determination, 617
Genetic variance, 248, 409, 413
  dependence on allele frequency, 416
  effect of inbreeding on additive, 309
  epistatic component of, 442–444
  estimating, 417–418
Genetic variation, 231
  absence of, 436
  external sources of, 292–294
  genetic constraints and, 564
  maintenance of in quantitative characters, 438–439
  principles of in populations, 234–239, 738–740. *See also* Genetic variation in natural populations
  recombination in, 285–286
  release of, 283–285
  in threshold traits, 253
Genetic variation in natural populations, 239–241
  estimating proportion of polymorphic loci, 242–244
  multiple loci, 245–247
  variation at the DNA level, 244–245
  variation in quantitative traits, 247–249, 251–253
Genetics. *See also* Population genetics; Quantitative genetics
  developmental, 665–669
  Mendelian, 33–37
  of postzygotic isolation, 471–476
Genetics and the Origin of Species, 24
Genic selection, 353–354, 596
Genome size, highly repetitive DNA and, 641–642
Genomes. *See also* Molecular data
  evolution of, 625
  organization of, 45–47
Genotype, 26. *See also* Phenotype
  defined, 231
  dominance, 37, 39
  environment and, 39
  fitness of, 366–367
  multiple loci, 41–43
  phenotypic variation and, 233
  threshold traits, 43
Genotype and phenotype, relation at single locus, 41
Genotype frequencies, 234–235, 245–246. *See also* Inbreeding
  deviations from expected, 372

Genotype-environment interactions, 248–249, 439–440
Geographic distribution, 123
  factors affecting, 205–207
Geographic identity, distance and, 260
Geographic isolation, defined, 450
Geographic patterns, as evidence for allopatric speciation, 484
Geographic range
  correlated rates of speciation and extinction and, 712
  limits on, 535–536
Geographic variation, 253–256
  coevolution of competing species and, 555–557
  patterns of, 257–259
Geographic variation in status, 454–455
Geographically variable characters, 259–262
Geography and level mode, of speciation, 482
Geological time
  extinction and, 714
  measuring, 128–129
Geological time scale, 129, 130
Geology
  fossil record, 129, 131–132
  geological time measurement, 128–129
  geological time scale, 129, 130
  plate tectonics, 127–128
  rock formation, 127
  sedimentary rocks, 129
Geospiza difficilis, ecological selection in, 488–489
Geospiza fortis
  bill depth in, 78
  heritability and, 248–249, 251–252
  selection in, 422–423
Geospiza spp.
  bill depth in, 78
  character displacement in, 260–261
  coevolution of competing species in, 555–556
  competition and distribution among, 219–220
  definitions of adaptation and, 355
  evolutionary radiation in, 117–118
GEPM. *See* Gene-environmental factor/phenotype mapping function
Germ cells, 32
Gibbons, classification of, 104. *See also* Hominoid; Primates
Gilia spp., hybrid speciation in, 508–509
Ginkgo, 186
Glacial episodes, Pleistocene, 197–198
Global climate patterns, 59–60
Global trends, 691
  in evolutionary history, 694–699
Globin gene family, 636
Globins, functional divergence in evolution of, 685
Gloger's rule, 259
Goatsbeards, speciation by polyploidy in, 506–507
Goldenrod gall fly, selection on size in, 424–425
Goldschmidt, Richard, 24, 25

on saltation, 680
Gondwanaland, 180, 182–183, 191
Gonochoristic species, 34, 606
Good genes, 591
Gorilla. See also Hominoid;
　Primates
　classification of, 103–105, 107
　evolution of, 728–730
　molecular clocks and, 120–121
Gradation, of characters, 113
Grade, 156–157
Gradual speciation, 482
Gradualism, 113–115
　arguments for, 678–680
　saltation vs., 677–681
Gradualness, 21–22
Grass, copper tolerance in, 383
Great chain of being, 17, 699
　Lamarck and, 18–19
Great tit, inbreeding in, 311–312
Ground finches, bill depth in, 78.
　See also under Geospiza
Ground plan, 95
Group selection, 347, 350–352, 562
　cooperation and, 594
　evolution of behavior and,
　　582–583
　kin selection and, 596
Gryllus spp. *See* Crickets
Guild coevolution, 540
Gulper, 64
Guppies
　conflicting selection pressures
　　in, 345–347
　relation between age at maturity
　　and mortality in, 570–571
Gymnosperms, 181
　Mesozoic, 185
Gynodioecy, 617–619
Gypsy moth, sex races of, 261–262

H

Habitat, 67
Habitat destruction, extinction
　and, 712
Habitat isolation, 457
Habitat selection, 69, 387
Hadley cells, 60
Haeckel, Ernst, 728
　on embryology and evolution,
　　651, 652
Haldane, J. B. S., 24–25, 159, 168,
　460
Haldane's rule, 460, 472
Half-sibs, comparisons of vari-
　ance among, 418
Hamilton's rule, 596
Haplodiploid species, calculating
　coefficients of relationship
　in, 600
Haploid cell, 33
Haploid generation, eukaryotic
　reproduction and, 34
Haploidy, 612
Haplospore, 34
Haplotype diversity, effective pop-
　ulation size and, 327–329
Haplotypes
　defined, 231
　lineage of, 452
Haptodus skull, 149
Hard selection, 386
Hardy, G. H., 236
Hardy-Weinberg distribution,
　derivation of, 235

Hardy-Weinberg equilibrium,
　236–238
Hardy-Weinberg genotype fre-
　quencies, 236, 238
Hardy-Weinberg principle,
　235–236, 300
　significance of, 236–238
Hares, fluctuation in abundance
　of, 78–79
Harmonia axyridis. See Ladybird
　beetle
Hawaiian honeycreeper
　bill evolution in, 113
　evolutionary radiation in, 117
Hawaiian Islands, *Drosophila* in,
　210–212
Hawaiian silversword alliance,
　evolutionary radiation in,
　117–118, 120
Hawks, coexisting species, 77
Health sciences, evolutionary biol-
　ogy and, 6
Heavy metal tolerance, evolution
　of in monkeyflowers,
　427–428
Helianthus annuus. See Sunflower
Hemiptera, 102
Hennig, Willi, 94
HENNIG86, 97
Herbivores, 77
　enemy-victim coevolution in,
　　547–548
Herbivory, 79–80
Hereditary similarity, 4
Hereditary variation, 11, 26
Heritability, 248, 409, 413,
　416–417
　of behavioral traits, 744–745
　estimating, 417–418
　of intelligence, 747–749
　theoretical equilibrium level of,
　　437–438
Heritability in the narrow sense,
　416–417
Hermaphroditic species, 34, 607
Hermaphroditism, 615–617
Heterochromatic regions, 45
Heterochrony, 138–139, 653
　allometry, 656–658
　defined, 655
　examples of, 658–661
　model of allometry and,
　　656–658
　theoretical morphology,
　　655–657
Heterogametic sex, 36, 460
Heterogeneous habitats, 609–610
Heterokaryotypes, 288
Heterosis, 384
Heterostyly, 247, 619
Heterotopy, 661–662
Heterozygosity
　decrease in due to genetic drift,
　　301–302
　decrease in due to inbreeding,
　　308
　genetic drift and, 304–305
　steady-state level, 323
Heterozygote, 33
Heterozygote advantage, fitness
　and, 384–385
Heterozygote disadvantage,
　391–392
Heterozygote superiority, 438
Hexapods, phylogeny of, 100–103
Hierarchical classification, 87
Hierarchical origin of life, 122

Higher taxa, 12, 87. *See also* Taxo-
　nomic practice
　species and, 263
Highly repetitive sequences, 45
Hindlimb, as homologous struc-
　ture, 109
Hirundo rustica. See Barn swallow
Histoire Naturelle, 18
Historical biogeography, 201,
　207–208. *See also* Dispersal;
　Vicariance biogeography
Historical contingency, 5–6
Historical homology concept, 669
History of evolution. *See* Evolu-
　tionary history
Hitchhiking, 343
　active trends due to, 694
　artificial selection and, 420
　nonadaptive traits and, 356
Hitchhiking effect, 350
HKA test, 633
Holotype, 90
Homeobox, 48, 666
Homeostasis, 69
　autonomy and, 698
Homeotic genes, 665–666
Homeotic mutations, 48, 276
Homeotic selector genes, 49
Hominid, 12, 728. *See also Homo*
　evolution of, 734
　fossil record, 730–732
Hominoid, 728. *See also* Primates
　histories of change of character,
　　108
　phylogenetic relationships,
　　728–730
Homo
　classification of, 103–105, 107
　evolution of, 728–730
　molecular clocks and, 120–121
　origin and evolution of,
　　732–734
Homo erectus, 732–733, 735
Homo habilis, 732–733
Homo neanderthalensis, 733
Homo sapiens, 12, 193, 732–734,
　735. *See also* Human behav-
　ior; Human behavioral
　traits; Human nature;
　Human populations;
　Humans
　transition from archaic to
　　modern, 734–738
Homogametic sex, 36, 460
Homokaryotypes, 288
Homologous, 89
Homologous chromosomes, 32
Homologous features, derived
　from common ancestors,
　108–110
Homologue, 32
Homology
　as evidence for evolution, 122
　problem of, 669–670
Homonymous structures, 654
Homoplasious character, 110
Homoplasy, 94, 99, 100, 110–111,
　220
Homozygosity, inbreeding and,
　620
Homozygote, 33
Honeybee, 4
Horizontal gene transfer, 292–294,
　626
Horned lizards, as species distin-
　guished by morphology, 463
Horse lineage. *See also* Equidae

evolutionary transitions in,
　139–142
　heterochrony in, 658–659
　measuring rates of evolution in,
　　159–161
　mutation-drift equilibrium
　　model in, 688–689
　rate of character change in, 688
　taxonomic rate of evolution in,
　　687–688
Hosts and parasites, adaptations
　in, 339–340
House mice
　allele frequencies in natural
　　populations of, 306
　introgressive hybridization in,
　　292–293
　selfish genetic elements in,
　　347–348
House sparrows, evolution of,
　160–161
Houseflies
　additive genetic variance in,
　　443
　genetic drift in, 305
　peripatric speciation in,
　　497–498
Housekeeping genes, 47
Hox genes, 665–666
　evolution of, 666–669
Human behavior, 740–744
　biological foundations of, 742
　evolution of, 741–742
Human behavioral ecology, 743
Human behavioral traits, 744–745
　intelligence, 747–749
　sexual orientation, 745–747
Human evolution, 728–730,
　732–734. *See also under*
　Human behavior
　Darwin on, 7–8
Human genome project, 739
Human hemoglobin
　genic vs. individual selection,
　　353
　heterozygote advantage and,
　　384–385
　pleiotropic effects of amino
　　acid substitution on, 55
Human nature, 742–743
Human population of the
　world, 74
Human populations
　genetic variation in, 738–740
　migrations of, 736–737
　origin of, 734–738
　racial and ethnic groups,
　　737–738
Human sociobiology, 743
Human species. *See* Hominid;
　Homo sapiens
Humans
　allometric growth in, 524
　classification of, 103–105, 107
　evolution of, 728–734
　modern mass extinction and,
　　722–723
　phylogenetic relationship with
　　apes, 103–105
　rates of substitution in, 324–326
　stabilizing selection for birth
　　weight, 424
Hummingbirds
　bill evolution in, 113
　secondary sexual characteristics,
　　459
Hutchinson, G. Evelyn, 68

Index

on constant ratio, 554
on speciation, 481
Hutton, James, 18
Huxley, Julian, 9, 25
Hybrid, defined, 450
Hybrid dysgenesis, 275
Hybrid inviability, 460
Hybrid speciation, polyploidy
    and, 504–510
Hybrid sterility, 460. *See also* Iso-
    lating mechanisms; Repro-
    ductive isolation
Hybrid vigor, 313, 384
Hybrid zone, 257, 464–465
    defined, 450
    fate of, 468
    genetic dynamics in, 465–467
    narrow, 454
Hybridization, 292–293, 464–468
    defined, 464
    speciation by, 508–510
Hybrids, speciation via, 482
Hypermorphosis, 657
Hypotheses, 9–11
    defined, 10
Hypothesis testing, 758
Hypothetico-deductive method,
    359
Hyracotherium, 139–140

I

Ichthyostegids, 144–145
Idealistic morphology, 108–109
Identical by descent, gene copies,
    307
Igneous rock, 127
Iguanidae, distribution of, 206
Immunological distance method,
    119
In situ hybridization, 52
Inbreeding, 307–309. *See also*
    Selfing
    advantages of, 621–622
    in animals, 622
    consequences of, 308–309,
        620–621
    evolution of, 619–622
    genotype frequencies, 307
    mutation, gene flow, and, 315
    in natural populations, 309–314
    pedigrees, 307
    relation to genetic drift, 314
Inbreeding coefficient, 307
    calculating from pedigrees, 313
    increase in with inbreeding, 308
Inbreeding depression, 309, 312,
        314, 384–385, 620
Incest taboos, 622
Inclusive fitness, 595–597
Incomplete dominance, 37, 39
Incumbent replacement, 715–716
Independence, in statistics, 360
Indeterminate growth, 563
Index fossils, 129
Indicator models, of good genes,
    591–592
Indirect effects, testing hypotheses
    of, 592–593
Individual advantage, altruism
    and, 595
Individual selection, 347, 352, 562
    active trends due to, 693
    evolution of behavior and,
        582–583
    kin selection and, 596
Individual sex ratio, 613

Individualization, dissociation
    and, 654–655
Individualized characters, 109
Individuals
    evolution and, 4
    interactions among related,
        595–597
Individuation, 117
Induction, genetic specificity of, 53
Infaunal animals, 718
Infectious disease, evolution of
    parasite virulence and,
        548–551
Ingroup, 97
Inherent drive, 155–156. *See also*
    Preadaptation
Inheritance of acquired character-
    istics, 18–19, 232
Insecticide resistance, evolution
    of, 378–380
Insectivorous Plants, 21
Insects
    adaptations in, 340–341
    Carboniferous, 181–182
    Cenozoic, 191
    diversification of, 707
    Mesozoic, 186
    phylogeny of orders of, 100–103
Insertion sequences, 270
Insertion/deletion polymor-
    phisms, 245
Instantaneous rate of increase, 71
Instincts, defined, 579
Integration, 655
    developmental plasticity and,
        665
Intelligence, human
    genetic basis of, 747–749
    heritability of, 747–749
    reproduction and fitness in
        relation to, 740–741
Intelligent design, 759, 760
Intensity of selection, 422
Interaction variance, 416
Interbreeding, as species criteria,
    451
Interdemic selection, 347, 409
Interference, 75
Intermediate forms, evidence for
    evolution, 123
Intermediate phenotypes, 679
International Codes of Botanical
    Nomenclature and of Zoo-
    logical Nomenclature, 90
Internodal species concept, de-
    fined, 448
Interspecific competition, 75. *See
    also* Competitive interaction
    islands and, 218–219
Intragenic recombination, 269
Intraspecific competition, 75
Intrinsic rate of natural increase,
    72
Introgression, 466
    defined, 450
Introgressive hybridization, 292
Introns, 44, 270–271, 325
Inverse frequency-dependent
    selection, 389–390
Inversion polymorphisms, 288
Inversions, 473
    chromosomal, 288
Investment ratios, 601
IQ. *See* Intelligence, human
Iris spp., polyploid complex and,
    287

Irish elk, 156
Irreducible complexity, 686
Irreversibility, 156
Island biogeography, 216–217,
    721–722
Island models, 315
Islands
    interspecific competition and,
        218–219
    seed dispersal and, 202
Isogamous species, 34, 606
Isogenic genetic background, 411
Isolating mechanisms, 457. *See
    also* Reproductive isolation
Isolation by distance models,
    315–316
Isometry, 524–525
Isopod, multiple-niche polymor-
    phism in, 388–389
Isozymes, 242n, 636
Iterative evolution, 157
Iteroparity, 561, 568–571
Iteroparous reproduction, 72

J

Jack-in-the-pulpit, sex allocation
    in, 616
Jawless fishes, 177–178. *See also*
    Fishes
Jenkin, Fleeming, 232
Jimsonweeds, reciprocal transloc-
    cations in, 464
Jump dispersal, 206, 208
    in tropical American snakes, 210
Jurassic, 182, 185

K

Kangaroo rats, interspecific com-
    petition among, 76
Karyotype alterations. *See also*
    Chromosomes
    chromosome rearrangements,
        288–292
    polyploidy, 286–288
Key adaptations, 716–717
Key beds, 129
Key innovations, 685
Keyacris scurra. *See* Australian
    grasshopper
Kin recognition, cannibalism and,
    597
Kin selection, 361, 595–597
    evolution by, 597–599, 601
Kondrashov's model, 611, 612
Kropotkin, Peter, 9
K-selection, 573

L

Ladybird beetle
    geographic variation in, 253,
        255
    variation in pattern due to poly-
        morphic locus, 240
Lagosuchus, 188
Lags, in coevolution, 546
Lamarck, Jean Baptiste de, 18–19,
    232
    vs. Darwin, 21–22
Land mass distribution
    biogeographic analysis and,
        209–210
    Carboniferous, 180
    Cenozoic, 190–191
    Mesozoic, 182, 184–185

Pleistocene, 197
Land snails
    genetic demography in, 373
    geographic variation in,
        371–372
Latimeria chalumnae. *See*
    Coelacanth
Latitude, species diversity and,
    223–225
Latitudinal gradient, 223
Laurasia, 182–183, 213
Leaf beetles
    diagnosis of new species of, 464
    DNA sequences in, 52
Leaf gall, 67
Leaf-cutter ant, 340–341
Learning, defined, 579
Least squares regression, 757
Leks, 581
Leopard frog, allopatric speciation
    in, 484–485
Lepidoptera, 102–103. *See also*
    Butterflies
    parallel evolution of wing pig-
        ment patterns in, 112
Lepomis punctatus. *See* Sunfishes
Lepomis spp., niche partitioning
    in, 77
Lethal allele, 240–241
Level of selection, 347
Life, 520
    defining, 166
    emergence of, 166–168
    time before its emergence,
        165–166
Life histories
    effect of sexual selection on,
        573–574
    male, 574–575
Life history evolution, 566–568
    age schedules of reproduction,
        568–571
    evolution of rate of increase,
        572–573
    life span and senescence, 568
    number and size of offspring,
        571–572
Life history phenomena, 561
Life history traits, fitness and,
    561–563
Life span, 561
    senescence and, 568
Life tables, 563, 567
    rate of increase and, 565
Lineage selection, active trends
    due to, 693–694
Lineage sorting, 469
Linkage disequilibrium, 245–247,
    285, 400–402, 608. *See also*
    Linkage equilibrium
    artificial selection and, 420
    decay of, 246
    genetic correlation and,
        429–431
    in hybrid zones, 466
    inbreeding and, 309
Linkage equilibrium, 245, 247, 285
Linked loci, 36
Linnaean classification system,
    87–88. *See also* Taxonomic
    practice
Linnaeus, Carolus, 17, 18, 87
    on primate classification, 103
Lipid membranes, formation of,
    167–168
Local mate competition, 614

Index

Local populations, 71
Local processes, diversity and, 704
Local trends, 691
Location, of normal distribution, 755
Loci, multiple, 41–43, 245–247
Locus, 33
    defined, 231
Logistic equation, 75
Log-log plot, 526
Long-tailed widowbird, sexual selection in, 345–346
Lungfishes, 213
    change in rate of evolution in, 158
    phylogenetic relationships in, 215
    respiratory and digestive systems in, 5
Lycopsids, 179
Lymantria dispar. See Gypsy moth
Lynx, fluctuation in abundance, 78–79
Lysenko, Trofim, 232n

M

MacClade, 97
Macroevolution, 25, 28, 447
    defined, 85
    vs. microevolution, 477–478
Macromutation. *See* Saltation
Macronyx croceus. See African longclaw
The Major Features of Evolution, 25
Malaria, heterozygosity and, 385
Male mating success, 369–370
Male reproductive success, 573–575
    natural selection and, 345–346
Males
    evolution of sexes, 612–613
    sexual selection by contests in, 588–589
Malthus, 20
Mammals. *See also specific species*
    adaptive radiation of, 191–196
    cladogram, 148
    convergent evolution in, 190, 221
    evolution of, 146–151
    origin of, 151–152
    respiratory and digestive systems of, 5
    skeletal ground plan of, 147
Manakins, phylogenetic study of behavior in, 581, 582
Manipulation, altruism and, 594
Map unit, 37
Marginal overdominance, 387
Marine ecosystems, 64–65
Marine life. *See also* Aquatic life
    Mesozoic, 185
    Silurian and Devonian, 176–178
Marine reptiles, extinct, 187
Mark-recapture method, of estimating gene flow, 317
Marsupials, 190, 192
    evolutionary radiation in, 117
Marx, Karl, 5, 9
Mass extinction, 165, 709, 713. *See also* Extinction
    Cretaceous, 185
    modern, 722–723
    Ordovician, 176
    Triassic, 185
Mass selection, 409

Mass-specific metabolic rate, 525
Mate choice, sexual selection by, 589–594
The Material Basis of Evolution, 680
Maternal chromosomes, 33
Maternal effect gene, 49
Maternal effects, phenotypic variation and, 233–234
Mathematical models, 229
    of population growth, 72
Mating type, 606, 612
Matorral, 62
Maximum likelihood method, for phylogenetic trees, 98
Mayr, Ernst, 25
    on biological species concept, 263, 450
    on Darwin's theory of evolution, 21–22
    on ecological niche and behavior, 601–602
    on limits on geographic range of species, 535
    on molecular evolution, 320
    on peripatric speciation, 483, 487, 493–494
    on punctuated equilibrium and, 136, 137
    on reproductive isolation, 491
    on speciation, 481–482, 514
    on species concept, 448–449
    on teleonomic processes, 342
McDonald-Kreitman test, 633
Meadowlark, 65
Mean, 69, 250–251
Mean fitness, plots of, 379
Mechanical isolation, 459
Mechanistic explanations, evolution and, 5
Mediterranean vegetation, 62
Megaloceros giganteus. See Irish elk
Meiosis, 31–33
    chromosome segregation in, 287
    variant patterns in eukaryotes, 34, 36
    variations in, 36
Meiotic drive, 33, 237, 613
Mendel, Gregor, 22–24, 232
Mendelian genetics, 33–37, 232–233
    mutationist theories and, 24
Mendelian ratios, 38–39
Meristic character, 247, 250
Mesohippus, 140–141. *See also* Horse lineage
Mesozoic, 130, 182–185
    marine life, 185
    terrestrial plants and arthropods, 185–186
    terrestrial vertebrates, 187–190
Mesozoic marine revolution, 185
Messenger RNA (mRNA), 44. *See also* Ribonucleic acid
    genetic code, 45
    transcripts, 48–49
Metacentric chromosome, 31, 288
    fission of, 290
Metamorphic rock, 127
Metaphase, 31–32, 505
Metapopulation, 300
Metatheria, 146
Metazoan phylogeny, 175–176
Methanococcus vannielli, rRNA structure in, 170
Metric characters, 156, 247, 250
Metric trait, variation in, 251
Microcoryphia, 102, 154

Microevolution, 447
    vs. macroevolution, 477–478
Microhabitats, 67–68
Microorganisms, natural selection in, 343–344
Microsatellite DNA, 641
Midges, vicariance biogeography of, 212–213
Migration, 70, 575
    Darwin on, 202
    evolution of in blackcaps, 426
    human, 736–737
Milkweed, 62
Milkweed beetle, estimates of gene flow in, 317–318
Millidarwin, 159
Mimic, 82
Mimicry, 82
Mimulus cardinalis, adaptation in, 112
Mimulus guttatus. See Monkeyflower
Mind, defined, 742
Minimum age, 208
Minisatellite DNA, 641
Mitochondria, 31, 33
    in eukaryotes, 170
Mitochondrial DNA (mtDNA), 327. *See also* Deoxyribonucleic acid
    sequence evolution in, 627–628
Mitosis, 31, 32
    morphogenesis and, 54
Mobile elements. *See* Transposable elements
Model, 82, 228–229, 357
Moderately repetitive sequences, 46
Modern fauna, 707
Modern *sapiens*, 733–734
Modern synthesis. *See* Evolutionary synthesis
Modifier alleles, 430
Modifier locus, 605–606
Molecular biology, 625
Molecular clocks, 120–122, 210, 320, 322
Molecular data, role in phylogenetic analysis, 118–122
Molecular differences, among related species, 468
Molecular evolution, 14
    aims and methods of, 625–626
    constancy of rate of, 122
    of DNA sequences, 626–634
    neutral theory of, 320–327
    rates of and neutral theory, 323–327
Molecular genetic analyses, methods in, 50–52
Molecular markers, 625
Molecular variation, 375
Monet, Jean Baptiste Pierre Antoine de. *See* Lamarck
Monkeyflower
    adaptation in, 112
    evolution of heavy metal tolerance in, 427–428
Monocots, flowers of, 110
Monoecy, outcrossing and, 619
Monomeric compounds, 166
Monomorphic locus, 239
Monophyletic groups, 91, 92
Monstera tenuis, 338
Morality
    natural selection and, 362

natural world and, 9
Morgan, Thomas Hunt, 24
    on mutations, 268
Morganucodon, 149–151
Morphogen, 54, 663
Morphogenesis, 49
    mechanisms of, 54
Morphological adaptations, 519–522
Morphological integration, 435–436
Morphology, 519
    evolution of in house sparrows, 428
    as species criteria, 448–449, 463
    theoretical, 655–656
Mosaic development, 52
Mosaic evolution, 112, 114, 145
    individualization and, 655
Mosaic hybrid zone, 467–468
Mosquito, insecticide resistance in due to mutation, 277–278
Mosquitofish, limits on geographic range of, 535
Moths, 67
    fluctuations in density of, 74
mRNA. *See* Messenger RNA
mtDNA. *See* Mitochondrial DNA
Muller, Herman
    on biological species concept, 450
    on speciation, 481
Muller's ratchet, 610–611, 618
Müllerian mimicry, 82, 391
Multiple loci, 41–43, 245–247
Multiple loci evolution
    adaptive landscapes and, 402–409
    directional selection and, 398–399
    gene interactions and, 399–400
    linkage disequilibrium and, 400–402
    principles of, 397–398
Multiple stable equilibria, 390
Multiple-niche polymorphism, 386–387
Multiregional hypothesis, 734–735
Multispecies coevolution, 540
Muntiacus spp. *See* Deer
Mus musculus. See House mice
Mus spp. *See* House mice
Musca domestica
    additive genetic variance in, 443
    genetic drift in, 305
    reproductive divergence in, 497–498
Mussel, allele frequency in, 383
Mutation during nonfunctionality model, 637
Mutation rates
    evolution of, 605–606
    evolutionary implications of, 272–276
Mutational variance, 275, 418
Mutation-drift equilibrium model, 688–689
Mutationist theories, 24
Mutations, 4, 26. *See also* Frameshift mutations; Gene mutations; Homeotic mutations; Neutral mutations; Point mutations
    artificial selection and, 421
    for biochemical abilities, 280–281

concerted evolution and, 637
creationism and, 761
defined, 231, 267
directional, 156
effects on fitness, 278–281, 679
in evolutionary synthesis theory, 24
fixation of rare beneficial, 608–609
genetic drift and, 437–438
Hardy-Weinberg principle and, 237
history of concept of, 268
inbreeding, gene flow, and, 315
limits of, 276–278
linkage disequilibrium and, 401
molecular evolution and, 320
neutral, 327
phenotypic effects of, 276–278
as random process, 281–283
rates of. *See* Mutation rates
transposable elements and, 639
Mutation-selection balance model, 439
Mutualism, 80–81, 361, 539, 551–554. *See also* Coevolution
evolution of eusociality and, 599
origins of, 552–553
vs. parasitism, 81
Mutualists, phylogenies of, 541
Mycorrhiza, 67
Myriapod, 102
Mytilus edulis. See Mussel

N

Nasonia spp., symbiont-induced incompatability in, 476
Natural populations, 27, 297, 306. *See also* Populations
inbreeding in, 309–314
persistence of deleterious alleles in, 381
selection in, 422–428
virulence of parasites in, 551–552
Natural products, evolutionary biology and, 7
Natural selection, 4–6, 26. *See also* Selection
creationism and, 760–761
Darwin on, 11, 21–22, 342
definition of, 349–350
detecting from DNA sequence data, 633–634
directional, 156
dispersal and, 575–576
evolution by, 365
evolutionary future and, 739–740
in evolutionary synthesis theory, 24
experimental studies of, 343–349
fitness and, 366–371
functional studies of, 373–375
Hardy-Weinberg principle and, 237
intensity of, 424–425
levels of, 350–354
measuring on quantitative characters, 422
methods of studying, 371–375
misconceptions regarding, 360–362

neutral theory of molecular evolution and, 320, 323–327
in nonliving systems, 166
rejection of theory of, 23–24
for reproductive isolation, 491–493
speciation by, 486, 488
strength of, 393–395
Natural Theology 19, 342
Naturalistic fallacy, 9, 362
Neanderthals, 733
Nei's index of genetic distance, 260, 261
Nei's index of genetic identity, 260
Neighborhood, 315
Neighbor-joining method, for phylogenetic trees, 98
Neoceratodus forsteri, 143
Neo-Lamarckian theories, 23, 155, 232n
Neoteny, 658–659
Neotropical biogeographic realm, taxa endemic to, 205
Net reproductive rate, 71
Net trends, 691, 696–697
Neutral alleles, 300
Neutral mutation rate, 321
Neutral mutations, 327, 628, 630
Neutral theory of molecular evolution, 273, 628–629
Neutralist-selectionist debate, 320, 375, 395
Newt, mutational limits in, 276–277
Newton, Isaac, 5, 18
Niche partitioning, 77, 555
Niches, 259–260
Nicotiana longiflora. See Tobacco
Nile perch, 80
Node, 95, 121
Nomenclature, taxonomic practice and, 90–91
Nonadaptive characters, 674
Nonadaptive evolution, 297
Nonadaptive traits, 355–356
Nonadditive genetic variance, 415–416
Nondisjunction, 289
Nonequilibrium hypothesis, 216
Nonselective deaths, 368
Norm of reaction, 39, 248
Normal distribution, 250, 760
Northern flicker, hybrid zone, 257–258
Nuclear genome, 31
Null hypothesis, 9–10, 297, 708–709, 758
Numerical response, 78
Numerical taxonomy. *See* Phenetics

O

Oecophylla smaragdina. See Weaver ants
Offspring
number and size of and optimal reproductive effort, 571–572
species differences in number and size of, 561
Ontogenetic allometry, 656
Ontogeny, 4, 55, 652–653. *See also* Development
Operon, 47
Ophraella spp. *See* Leaf beetles
Opposing selective factors, 385
Optimal design, 528–530

Optimal diet choice, model of, 583–584
Optimal foraging theory, 583
Optimality models, 528–529
Optimization theory, 583
Orangutan. *See also* Hominoid; Primates
classification of, 103–105, 107
evolution of, 728–739
molecular clocks and, 120–121
Orchids, 337–339
adaptation in, 337–338
mechanical isolation in, 459
Ordovician, 176
Organic adaptations, 351–352
Organic evolution. *See* Biological evolution
Organic molecules, 166–167
Organic progression, Lamarck's theory of, 19
Organisms
commonalities among, 3
selection of, 350–352
The Origin of Species, 4, 7, 8, 11, 21, 23, 76, 122, 337
behavior and shifts of ecological niche in, 602
biogeographic facts in, 202
classification in, 88
embryology and evolution in, 652
evolution of behavior in, 579
fossil record in, 129
functional modification in, 682–683
human evolution in, 728
kin selection in, 595n
Origination, 217
diversification and, 715–721, 723–724
extinction and patterns of, 708–709, 711–712
Ornithischia, 188, 190
Orthogenetic theories, 23–24
Orthogenetic trends, 139
Orthologous genes (orthologous loci), 98, 636
Oscillographs, of song types, 458
Osteolepiformes, evolution of, 143–145
Outbreeding, evolution of, 619–622. *See also* Sexual reproduction
Outbreeding depression, 621–622
Outcrossing, 36, 309–311, 606, 619
advantages of, 621–622
Outgroups, 97
Out-of-Africa hypothesis, 735
Overdominance, 384, 438
Owen, Richard, 108–109

P

Paedomorphic character state, 656–657
Paedomorphic species, 654
Paedomorphosis, 659–660
Pair-rule genes, 49
Palaeodictyoptera, reconstruction of a, 181
Paleobiology, 12
new directions in, 14
Paleontology, role in biogeography, 201
Paleozoic era, 130, 172–183
Paleozoic fauna, 707
Paley, William, 19, 342

Palm, age and allocation to reproduction in, 571
Pan. See also Hominoid; Primates; Chimpanzee
classification of, 103–105, 107
evolution of, 728–730
molecular clocks and, 120–121
skull, 733
Panaxia dominula. See Scarlet tiger moth
Pangaea, 182–183
Pangolin, 65
Panmictic population, 307
Paracentric inversion, 288, 473
Parallel cladogenesis, 541
Parallel evolution, 110–111, 672
Parallel speciation, 488–489
Parallel trends, 117, 156–157
Parallelism, 110
Paralogous genes (paralogous loci), 98, 171, 636
Parapatric populations, 257
defined, 450
Parapatric speciation, 482, 483, 498–499
Paraphyletic group, 91
Parasaurolophus, 190. *See also* Dinosaurs
Parasite virulence, evolution of and infectious disease, 548–551
Parasites, 77
coevolution of, 543–544
role in coevolution, 539–541
virulence in natural populations of, 551–552
Parasites and hosts, adaptations in, 339–340
Parasitic relationships, mutualisms arising from, 552
Parasitism, 78–79
vs. mutualism, 81
Parasitoid wasps, sex ratio evolution in, 615–616
Parasitoids, 77
Paratypes, 90
Parcellation, 655
Parental manipulation, evolution of eusociality and, 599
Parent-offspring regressions, 417
Parsimony, 580–581
phylogenetic inference and, 96–97
Parthenogenesis, 34
Parus major. See Great tit
Passer domesticus. See House sparrow
Passive trends, 692
Paternal chromosomes, 33
Pattern formation, 663–665
defined, 663
PAUP, 97
PCR. *See* Polymerase chain reaction
Peak, in adaptive landscapes, 403
Peak shift, 392–394, 409, 483. *See also* Founder effects
speciation by, 486–487. *See also* Peripatric speciation
Pedigrees, 312
calculation of inbreeding coefficients in, 313
Pelecanoides magellani. See Diving petrel
Peppered moth, 372
adaptation in, 358

natural selection in, 394
Per capita rate of increase, 71
Peramorphic character state, 656–657
Peramorphosis, 659–660
Pericentric inversion, 288, 473, 474–475
Period of oscillation, 69
Periodic selection, 343–344
Peripatric speciation, 483, 486–487, 493–498. *See also* Speciation
Perissodactyla, 195–196
Permian, terrestrial and aquatic life, 182–183
Permian mass extinction, 182
Perognathus goldmani. See Pocket mouse
Peromyscus maniculatus. See Deer mouse
Peromyscus spp., peripatric speciation in, 495–496
Petroica multicolor. See Robin
Phacops rana. See Trilobites
Phanerozoic, 172
    extinction rates of marine animal families, 711
    origination rates in, 709
    taxonomic diversity in, 707–708
Phasmatodea, 102
Phenetics, 92–94
Phenogram, 93
Phenotype, 26. *See also* Genotype
    defined, 231
    definition of natural selection and, 349
    development of, 651–652
    dominance and, 37, 39
    molecular evidence and evolution of, 625
    multiple loci and, 41–43
    threshold traits and, 43
Phenotype and genotype, relation at single locus, 41
Phenotypic character, effect of inbreeding, 308
Phenotypic correlation, 429
Phenotypic differences among species, 470–471
Phenotypic effects
    estimating mutation rates using, 272–274
    of mutations, 276–278
Phenotypic gaps
    creationism and, 761
    explanations for, 677–678
Phenotypic models, of optimal life history traits, 563–566
Phenotypic plasticity, 70, 439–440
Phenotypic traits, behaviors as, 579–581
Phenotypic variance, 409
    components of, 413–414, 416–417
    recombination and, 285
    sources of, 233–234
Philosophie Zoologique, 18
Philosophy, evolutionary thought and, 8
Phlox spp., self-incompatibility in, 311
Photoperiod, evolution in codling moths, 426
Photoreceptors, evolution of, 682–683, 684
Phrynosoma spp. *See* Horned lizards

Phyletic gradualism, 513
    model of, 137, 139
Phyletic rate. *See* Phylogenetic rate
PHYLIP, 97
Phylogenetic analysis, 98–99
    of apes and humans, 103–105
    of behavior, 580–581
    evolutionary trends and, 116–117
    extinct organisms and, 105–107
    of insects, 100–103
    molecular data in, 118–122
    of sequence data, 626
Phylogenetic constraints
    on evolution, 671
    on life history evolution, 564
Phylogenetic correlation, problem of, 360
Phylogenetic homology concept, 669
Phylogenetic hypotheses
    evaluating, 105
    as evidence for evolution, 763
Phylogenetic inference, parsimony and, 96–97
Phylogenetic perspectives, on species associations, 541–542
Phylogenetic rate, of evolution, 158–159, 687–688
Phylogenetic relationships
    hominoid, 728–730
    among mammals in Cenozoic, 191–193
Phylogenetic school, of vicariance biogeography, 208–209
Phylogenetic species concepts, 452–453
    defined, 448
Phylogenetic Systematics, 94
Phylogenetic tree, 19, 88–89, 92–93
    creating, 97
    evaluating, 99
    methods for estimating, 98–99
Phylogeny, 85, 652–653. *See also* Phylogenetic analysis
    dispersal hypothesis and, 208–209
    fossil record and, 154–155
    hypothetical, 94
    to test neutral theory of molecular evolution, 324
Physalaemus pustulosus. See Túngara frogs
Physical constraints, on evolution, 671
Physiological acclimation, 70
Physiological adaptations, 519–522
Physiological constraints, on life history evolution, 564, 566
Physiological ecology, 59, 519
Physiology, 519
Phytosaurs, 188
Picoides pubescens. See Downy woodpecker
Pied kingfisher, cooperative breeding in, 597–598, 599
Pitcher-plant mosquito, role of epistasis in evolution of, 443–444
Plankton, 63
Plants. *See also specific species*
    adaptation to deserts, 521–523
    biological species concept and, 454
    character variation in, 116

Devonian, 178–179
    enemy-victim coevolution in, 547–548
    evolution of dioecy and sex chromosomes in, 618
    genetic variation at allozyme loci in, 243
    heterotopy in, 662
    life cycles of, 35
    Mesozoic, 185–186
    polyploidy in, 287
    selfing and outcrossing in, 309–311
    short-term advantages of sex in, 612
    speciation in, 504–510
Plasmids, 51, 293–294
Plasticity, 674
Plate tectonics, 127–128
Plates, 127
Plato, 6
    essentialism and, 17
Pleiotropy, 55, 234
    artificial selection and, 420
    genetic correlation and, 429–431
    nonadaptive traits and, 356
Pleistocene
    events in, 197–199
    extinctions, 214–215
    speciation in and latitudinal diversity gradients, 224–225
Plesiadapsis, 194
Plesiomorphy, 95
Plethodon spp. *See* Salamanders
Pocket gopher, estimating gene flow in, 318–319
Pocket mouse, geographic variation in karyotype of, 291
Poecilia reticulata. See Guppies
Point mutations, 268–271
Poisson distribution, 282, 324
Polacanthus, 190
Polarity, of evolution, 97
Pollen, gene flow and, 317
Pollination mechanisms, in orchids, 337–338
Polygenes, 409
Polygenic, 247
Polygenic characteristics, 41
Polygenic trait, 27
Polygenic variation, 398
Polymerase chain reaction, 7, 120
    general protocol, 51
Polymorphic locus, 239–240
    estimating the proportion of, 242–244
Polymorphism, 247, 320, 323, 531n, 739
    maintained by selection, 384–390
    trade-offs between fitness components and, 569
Polyphyletic taxon, 92
Polyploid, defined, 504
Polyploid complex, 287
Polyploid organisms, 33
Polyploidy, 286–288, 482, 504–508
    incidence of, 507–508
    speciation by, 506–507
Polytypic species, defined, 450
Pongo. See also Hominoid; Primates; Orangutan
    classification of, 103–105, 107
    evolution of, 728–730
    molecular clocks and, 120–121
Population cages, 344

Population dynamics, correlated rates of speciation and extinction and, 712
Population ecology, 59
Population genetic mode, of speciation, 482
Population genetics, 12–13
    sequence data analysis and, 626
Population growth, mathematical models, 72
Population mean, 755
Population sex ratio, 613
Population size. *See also* Effective population size
    coalescence time in relation to, 328
    effect on selection efficacy, 392
    evolution of in flour beetles, 346–348
Population structure, 297
    gene trees and, 327–333
    history of, 330–332
    spatial patterns, 298
Populational speciation, 22
Populations, 4. *See also* Effective population size; Genetic drift; Human populations; Natural populations; Variation among populations
    common ancestral, 4
    descent of gene copies in, 298
    divergence of, 4, 407–408
    evolution and, 4
    factors limiting growth, 73–75
    genetic variation in, 234–249, 251–253
    geographic distribution of, 257
    growth of, 71–73
    in Hardy-Weinberg principle, 237
    natural selection and, 371–372
    response to perturbations, 372
    statistical, 755
    structure of, 71
Position, homology and, 109
Position effects, 288
Positional information, 53–54
Positive assortative mating, 457
Positive frequency-dependent selection, 389, 390–391
Positive selection, 628–629, 630
Posttranscriptional control, 48
Postzygotic barriers, 457, 460
Postzygotic isolation
    chromosome differences and, 473–475
    genetics of, 471–476
Potentilla glandulosa. See Sticky cinquefoil
Prairie, 62
Preadaptation, 155–156, 355
Preaptation, 355n
Precambrian, 169–172
Precipitation, 59–63
Predation, 77–78, 720
Predator, 77
Predator-prey interaction, 78
Predator-prey models of coevolution, 544
Predictability, 69
Prediction, 359
Prezygotic barriers, 457–460
Prezygotic isolation
    genetic basis of, 476–477
    reinforcement of, 491–492
Primary hybrid zone, 464–465
Primary kingdoms, 171

 Index

Primary productivity, 704
Primates. See also specific species
  adaptive radiation of, 192–193
  classification of, 91, 92, 103–105, 729
  phylogeny of, 106
  relation of testes mass to body mass in, 359–360, 525
Primer sequences, 51
Primitive character state. See Ancestral character state
Principle of allocation, 528
Principles of Conservation Biology, 723
Priority, nomenclature and, 90
Probability, 37
Probability distribution, 300, 303
Probainognathus, 148–149
  skull, 149
Probe, 51
Proboscidea, skulls, 195
Processed pseudogenes, 271, 637, 639–640
Procynosuchus, 148, skull, 149
Progenesis, 659–661
Progress, evolution and question of, 361, 699–700
Prokaryotes, 169–170
  cell division in, 31
Promoter region, 47
Prophase, 31–32, 36
Protandry, 574, 615–616
Protein electrophoresis, 50, 119
  to estimate polymorphic loci, 242–244
Protein enzymes, evolution of, 168
Proteins, evolution of novel, 642–644. See also Molecular biology
Proterozoic, 130, 169, 171–172
Protogyny, 574, 615–616
Prototheria, 146
Provinces, 204
Provinciality, 720–721
Proximate mechanisms, 652
PSC. See Phylogenetic species concepts
Pseudocopulation, 337
Pseudoextinction, 132
Pseudogenes, 46, 104, 325, 637
Psychology, evolutionary, 744
Pterosaurs, 188
Punctuated equilibrium, 137, 139, 447, 495, 513, 677–678, 689–691
  critiques of model, 690–691
Punctuated gradualism, 137, 139
Punctuational evolution, 679
Purifying selection, 377, 628–629, 630
Pyrenestes ostrinus. See Black-bellied seedcracker

Q
QTLs. See Quantitative trait loci
Quadrivalents, 505
Quantitative characters. See also Metric characters
  evolution of, 418–422, 437–438
  maintenance of genetic variation in, 438–439
  natural selection on, 419, 422
Quantitative genetics, 397

components of phenotypic variation, 413–414, 416–417
  estimating genetic variance and heritability, 417–418
  genetic dissection of quantitative traits, 409, 410–411
  nonadditive genetic variance, 415–416
  response to selection, 409, 412–413
Quantitative trait loci, 409
Quantitative traits
  adaptive landscapes and, 406–409
  genetic dissection of, 409, 410–411
  mapping, 410, 471
  predator-prey models, 544
  rapid evolution of, 425–428
  variation in, 247–249, 251–253
Quantitative traits, correlated evolution of
  correlated selection, 428–429
  genetic constraints on, 436–437
  genetic correlation and, 429–436
Quantum evolution, 689–690. See also Divergence
Quantum speciation, 690
Quasispecies, 167

R
Race, defined, 450
Racial groups, 737–738
  IQ and, 747–749
Racism, scientific, 747–748, 749
Radiation, early evolution of animals and rapid, 174
Radioactive decay, 128
Radiolarians, speciation in, 135, 136
Ramets, 34, 71
Rana pipiens. See Leopard frog
Random clades, 709
Random genetic drift, 26, 228, 237, 297. See also Genetic drift
  computer simulations, 301
Random walk, 300
Randomly mating populations, sex ratios in, 613–614
Randomness, mutation and, 281–283
Range, 69
Range expansion, 205, 208
  in tropical American snakes, 210
Rassenkreis, defined, 450
Rat snake, geographic subspecies of, 257–258
Rate of change, relative fitness and, 367–368
Rate of gene flow, 315–316
Rate of increase
  evolution of, 572–573
  life tables and 563, 565
Ratite birds, living families of, 206
Rattlesnakes, eastern and western diamondbacks, 215–216
Rattus norvegicus. See Brown rat
Reaction norms, evolution of, 440
Real populations, genetic drift in, 304–307
Real reaction norms, 41
Realized heritability, 418
Rearrangement, genetic, 47
Recessive allele, 39

Recessive mutations, accumulation of, 275
Reciprocal altruism, 595
  hominid evolution and, 734
Reciprocal exchange, 292
Reciprocal translocation, 289–290, 473, 474
Recognition species concept, defined, 448
Recolonization, gene flow and, 316–317
Recombination, 26, 32
  diploidy and, 612
  effects on rate of evolution, 609–610
  erosion of variation by, 285–286
  inferring, 626
  sequence changes arising from, 269
  sex and, 606–612
  as source of variation, 283–286
Recombination rates, evolution of, 607–608
Recombination repeats, 271
Recombinational speciation, 508–510
Red Queen hypothesis, 698–699, 714
Reduction, 606
Reductional division, 32
Reductionism, selfish gene and, 353–354
Redundant elements, functional divergence of, 685
Red-winged blackbird, neutral mutation rates in, 327–329
Refuges, 197–198
Refuges of fish, evidence for genetic divergence in, 485
Refugia. See Refuges
Regional biotas, composition of, 214–216
Regression, 136
Regression equation, 409
Regulative development, 52–53
Regulatory gene function, evolution of, 666–669
Regulatory genes, 665–666
Reinforcement of prezygotic isolation, 491–492
Related individuals, interactions among, 595–597
Relational pleiotropy, 55
Relationship, calculating coefficients of, 600
Relative fitness, rate of change and, 367–368
Relative rate test, 121, 324
Relaxed selection, 419
Religion, evolution and, 5, 8–9, 17–18, 759–760
Rensch, Bernhard, 25, 677
Repeats, 638
Repetitive sequences, 45
Replacement hypothesis, 735–736
Replacement substitution, 245, 321
Replica plating, 282, 284
Replicated sister-group comparisons, 716
Reporter gene, 626
Reproduction. See also Asexual reproduction; Fecundity; Inbreeding; Reproductive isolation; Selfing; Sexual reproduction

age distribution of, 561
  costs of, 566
  evolution of age schedules of, 568–571
  modern evolution and, 740
  survival and, 698
  variant patterns in eukaryotes, 34, 36
Reproductive assurance, 621–622
Reproductive barriers, 457
  genetics of postzygotic isolation, 471–476
  genetics of prezygotic isolation, 476–477
  genetics of reproductive isolation, 477–478
Reproductive character displacement, 491–492
Reproductive effort, 567–568
Reproductive isolation, 27. See also Inbreeding; Isolating mechanisms; Selfing; Speciation
  defined, 450
  genetics of, 477–478
  natural selection for, 491–493
Reproductive success, 367
  male, 573–575
  variance in males vs. females, 587–588
Reptiles, extinct marine, 187. See also Amniotes
The Republic, 6
Repulsion gametes, 400
Resource, 67, 539
Resource isolation, 457
Resource partitioning, 704
Resource use, body size and, 526
Resource-holding potential, 586
Response times, 69
Response to selection, 349, 409
Restriction enzymes, 50
Restriction fragment-length polymorphism, 50, 244
Restriction sites, use in phylogenetic studies, 119
Retention indices, 99
Retroelements, 270
Retrogenes, 270–271
Retroposons, 270, 626
Retrotransposons, 639–640
Retroviruses, 270
Reverse transcriptase, 52
Reverse transcription, 270
Revision, of species, 90
RFLP. See Restriction fragment-length polymorphism
Rhagoletis pomonella. See Apple maggot fly
Rhamphorhynchus, 189
Rhyniopsid, 179
Ribonucleic acid (RNA), 43, 231
  catalytic properties of, 166–167
Ribonucleic acid polymerase enzyme, 44
Ribonucleic acid replication, scenario for origin of, 167
Ribosomal RNA (rRNA), 44. See also Ribonucleic acid
Ring species, 455–456
  defined, 450
RNA. See Ribonucleic acid
Robertsonian rearrangements, 474–475
Robin, peripatric speciation in, 494

Rock formation, 127
Rodents
  adaptive radiation of, 194
  competitive displacement in, 716
  rates of substitution in, 324–325
Rooting, 97
Rotulid echinoids, evolutionary transitions in, 135–140
Rousseau, Jean Jacques, 18
rRNA. *See* Ribosomal RNA
r-selection, 573
Runaway sexual selection, 490, 590
Rusty lizards, estimating gene flow in, 317

S

Sabertooth condition, convergent evolution of, 111
Saber-toothed blenny, mimicry in, 82
Saccharomyces cerevisiae. See Yeast
Sagittaria sagittifolia. See Arrowhead
Salamanders
  decoupling of function in, 685–686
  developmental plasticity in, 665
  interspecific competition among, 75, 76
  paedomorphosis in, 659–660
  progenesis in, 659–661
  reduction of competition in, 77
  speciation in, 455–456
Salix reticulata. See Dwarf willow
Salmonella typhimurium, selection on silent substitution in, 634–635
Saltation, 24, 113, 647
  arguments for, 680–681
  vs. gradualism, 677–681
Salvia henryi, adaptation in, 112
Sample, 398
  defined, 755
Sample mean, 755
Sampling error, 238
  allele frequencies and, 297–298
Sandpipers, bill variation in, 113–114
Sarcopterygii, evolution of, 143–145
Satellite DNA, 45, 641–642. *See also* Deoxyribonucleic acid
Saurischia, 188–189. *See also* Dinosaurs
Savannah, 61–62, 191
Saxifraga cernua, disjunct distribution of, 203
Scala Naturae. See Great chain of being
Scarlet tiger moth, allele frequency in, 234–235, 238
Sceloporus olivaceus. See Rusty lizard
Schindewolf, Otto, 113
  on saltation, 680
Schmalhausen, I. L., 441
Science, objectivity and, 727
Science on Trial, 759
Scientific creationism, 759
Scientific method, 9-11
Sea lettuce, life cycle of, 35
Sea urchins, relationships among genera, 105, 107
Search image, 390

Seaside sparrow, restriction fragment-length polymorphism, 50
Secondary contacts, 464
Secondary hybrid zone, 464–465
Secondary succession, 66
Second-order selection, 606
Sedimentary rock, 127, 129
Seed dispersal, 61–63
  gene flow and, 317
Seed predator, 77
Segment polarity genes, 49
Segregation distortion, 33, 237
Selection. *See also* Artificial selection; Directional selection; Frequency-dependent selection; Group selection; Individual selection; Sexual selection
  antagonistic, 385
  evolution of novelty and, 681–687
  gene flow and, 381–383
  interaction with genetic drift, 392–393
  levels of, 596–597
  modes of, 366, 367
  in natural populations, 422–428
  opportunity for, 740
  polymorphism maintained by, 384–390
  recombination and, 608–611
  variable, 385–389
Selection differential, 409
Selection gradients, 422
Selection in Relation to Sex, 586
Selection models
  directional selection, 375, 377–381
  models with constant fitness, 376
  persistence of deleterious alleles in natural populations, 381
  selection and gene flow, 381–383
Selection plateau, 419
Selection response, of quantitative character, 409, 412–413
Selection theory, as evidence for evolution, 763
Selective advantage, 368
Selective constraints, on evolution, 671
Selective deaths, 368
Selective ratchets, 693
Selective sweep, 629
Selectivity, of extinction, 714
Self-fertilization, 36, 308–311
  as barrier to gene flow, 458–459
  role in plant speciation, 504
Self-incompatibility, 310–311
  outcrossing and, 619
Selfing, 309–311. *See also* Inbreeding
  evolution from outcrossing, 619–620
Selfish DNA, 641
Selfish gene, reductionism and, 353–354
Selfish genetic elements, 347–349
Selfish herd, 595
Semelparity, 561
  evolution of, 570
Semispecies, 454
  defined, 450
Senescence, life span and, 568
Sensitivity analysis, 566–567

Sensory bias, 593–594
Sequence evolution. *See* DNA sequence evolution
Sequencing DNA, 51–52
Sequential hermaphroditism, 34, 574–575
Serially homologous structures, 654
Sex
  defined, 606
  vs. gender identity, 745
  hypotheses for advantage of, 608–611
  long-term advantages of, 611
  recombination and, 606–612
  short-term advantages of, 611–612
Sex allocation, 615–617
Sex chromosomes, 33
  evolution of, 617–619
Sex determination, 617
Sex pheromones, 458
Sex ratios
  defined, 613
  evolution of in structured populations, 614–615
  in randomly mating populations, 613–614
Sex-determination, evolution of mechanisms of, 617–619
Sexes, evolution of, 612–613
Sexual behavior, 745
Sexual differentiation, in eukaryotes, 34
Sexual isolation, 260. *See also* Reproductive isolation
Sexual orientation, 745–747
  variation in, 746–747
Sexual reproduction, 34
Sexual selection, 345–346, 349. *See also* Divergence
  concept of, 586–588
  by contests, 588
  effect on life histories, 573–574
  by mate choice, 589–594
  paternity insurance and sperm competition, 588–589
  speciation and, 489–491
Sexually selected traits, dual role of, 593
Shared derived characters, 94
Shifting balance theory, 408–409
Sibling species, 449
  defined, 450
Silent substitution, 321
  selection on, 634–635
Silurian, marine life in, 176–178
Silurian brachiopod, 173
Simpson, George Gaylord, 24, 25, 26, 142, 677
  on quantum evolution, 689–690
  on rates of evolution, 687–688
Single cell morphogenesis, 54
Siphonaptera, 102
Sister chromatids, 31
Sister groups, 95
Sister species, 450
Sister taxa, 95
Snail shell morphology, theoretical, 655–656, 657
Snakes
  adaptations in skulls of, 339, 340
  biogeographic analysis of tropical American, 210–211
  Cenozoic, 191
  fossil record of, 187

genetic correlation in, 429, 431–433
Snapdragon, flower, 4
Social Darwinism, 9, 749
Social insects
  adaptations in, 340–341
  altruism in, 598–599, 601
Social interactions, and the evolution of cooperation, 594–599, 601
Sociobiology, 579, 743
Sociobiology, 742
Soft inheritance, 232
Soft selection, 386
Somatic cells, 31
Sorting, 4
Spacers, 45
Spatial variation, 386, 609–610
Special creation, 759
Specialist, 66
Specialization
  diversity and, 718, 720
  as net trend, 698
Speciation, 27, 135–137, 207, 447. *See also* Allopatric speciation; Cospeciation; Founder effects; Origination
  by chromosome rearrangement, 496–497
  consequences of, 512–514
  correlated rates of and extinction, 711–712
  defined, 132
  ecological selection and, 488–489
  by genetic drift, 486, 487–488
  by hybridization, 508–510
  modes of, 481–482
  by natural selection, 486, 488
  by peak shift, 486–487. *See also* Peripatric speciation
  by polyploidy, 506–507
  rates of, 510–512
  sexual selection and, 489–491, 590
Speciation event, 447
Species, 27. *See also* Geographic distribution; Speciation; Taxonomic practice
  character change within, 133–135
  coevolution of competing, 554–555
  defined, 132, 263
  differences among, 460–463
  differences in behavior among, 579–580
  estimating mutation rates from comparisons among, 273
  higher taxa and, 263
  hybridization and, 292
  limits on geographic range of, 535–536
  Linnaean classification system of, 87–88
  models for number of at equilibrium over evolutionary time, 721
  naming, 90–91
  number of, 87–88
  rates of increase, 712
  role in communities, 65–66
  variation within and among, 323, 632, 634
Species associations, phylogenetic perspectives, 541–542
Species classification, 330–332

Species concepts, 447–448
biological species concept, 449–452
history of, 448–449
limitations of biological species concept, 453
phylogenetic species concept, 452–453
usefulness of biological species concept, 453
Species diagnosis, 453, 462–463
Species differences, genetic basis of, 468–470
Species diversity, 703
gradients in, 222–225
Species hitchhiking, 694
Species interactions, 539
competition, 75–76
herbivory, 79–80
mimicry, 82
mutualism, 80–81
parasitism, 78–79
predation, 77–78
trophic level coexistence, 76–77
Species longevity, 698–699
Species niches, 68–69
Species richness, 703
relationship of local to regional, 705
Species selection, 352–353, 690, 693
vs. species hitchhiking, 694
Species trees, 468–469
gene trees and, 332–333
Specific coevolution, 540
Specific mate recognition system, 457, 490
Speciose, vs. species-rich, 64n
Spencer, Herbert, on Social Darwinism, 749
Sperm competition, 588–589
Sperm precedence, 589
Sphecomyrna freyi, 155
Sphenodon punctatus. See Tuatara
Sphenopsid, 179
Spontaneous mutation rates, 273–274
Sporophyte, 34
Spurge, 62
Stabilizing selection, 366, 400, 688
evidence of, 424–425
responses to, 421–422
Stable equilibrium, 345, 381
Stalk-eyed fly,
indirect effects in, 592
sexual selection in, 592
Standard deviation, 159, 250–251, 756
measuring evolution using, 160
Static allometry, 656
Statistically significant differences, 360
Statistics
analysis of variance, 756
correlation among variables, 756–757
frequency distributions, A755–756
hypothesis testing, 758
Steady state, 323
Stebbins, G. Ledyard, 25, 26
on speciation by hybridization, 508
Stem group, 95
Stenotopy, 712
Stepped cline, 382, 465, 466, 467

Stepping-stone models, 315
Sticklebacks. *See also* Fishes
character change in, 134–135
coevolution of competing species in, 556
parallel speciation in, 488–489
predatory adaptation in 545, 546
Sticky cinquefoil, geographic variation in, 254–256
Strata, 129
chronostratigraphic correlation of, 129
Stratigraphic column, 129
Stress
defined, 531
responses to, 531
Stromatolites, 169
Structural habitats, 221–222
Structure, homology and, 109
Structured populations, evolution of sex ratios in, 614–615
Sturnella magna. See Meadowlark
Sturnus vulgaris. See European starling
Suboptimal design, as evidence for evolution, 123
Subspecies, 257
defined, 450
Substitution, neutral theory of molecular evolution and, 321, 324–325
Succession, ecological, 66–67
Sulfur butterflies
natural selection in, 373–375
optimal design and constraint in, 529–530
Sunbird, bill evolution, 113
Sunfishes
gene tree, 331
mtDNA haplotypes in, 329–330
niche partitioning in, 77
Sunflower, hybrid speciation in, 509–510
Supernormal stimulus, 490, 593–594
Superposition, 129
Superspecies, 454
defined, 450
Survival, reproduction and, 698
Survivorship, 698–700
Survivorship curves, 714
Sycamores, rate of speciation, 512
Sylvia atricapilla. See Blackcap
Symbiont-induced incompatibility, 475–476
Symbiosis, 80–81
Symbiotic mutualism, 81, 551
Symbiotic mutualists, 539
Symmorphosis, 528–529
Sympatric hybridization, 454–455
Sympatric populations, 135, 257
defined, 450
Sympatric speciation, 482, 483, 499–504
defined, 499
Symplesiomorphy, 95
Synapomorphy, 95
Synapsids, 189
Synapsis, 32
Synchronic populations, 135
Syngamy, 606
Synonym, in nomenclature, 90
Synonymous codons, 45
Synonymous substitution, 245, 323–326

Synthetic theory, 26
Systema Naturae, 17, 87
Systematics, 12, 85, 87. *See also* Evolutionary change principles; Taxonomic practice
new directions in, 14
role in biogeography, 201
Systematics and the Origin of Species, 25, 449

T

T7 bacteriophages, phylogeny of, 105, 107
Tadpole shrimp, taxonomic rate of evolution in, 687
Tajima's test, 633
Taricha granulosa. See Newt
Taxa, 12
Taxon, 87
Taxon age, extinction and, 714
Taxon selection, 352
Taxonomic diversity, through the Phanerozoic, 707–708
Taxonomic frequency rate, of evolution, 157–158
Taxonomic practice, nomenclature and, 90–91
Taxonomic rates, of evolution, 687–688
Taxonomy, in fossil record, 132
Tectonic events, 207
Teleological descriptions of behavior, 581–582
Teleological statements, 342
Teleonomic processes, defined, 342
Teleost fishes, 191
Telophase, 31–32
Temperate broadleaf forest, 62
Temperate grassland, 62
Temperature, 59–63
Temperature adaptation, 278–280
Tempo and Mode in Evolution, 25
Temporal fenestrae, 147
Temporal fluctuation, 386
Temporal isolation, 457
Temporal patterns, 372
Tension zone, 465, 467, 474
Terminal addition, 653
Terminal taxa, 95
Termites, phylogenetic relationship with roaches and mantids, 106, 108
Terrestrial biomes, 60–63
Terrestrial life
Carboniferous and Permian, 180–183
Cenozoic, 191
Devonian, 178–180
Territories, 75
TEs. *See* Transposable elements
Tetraopes tetraophthalmus. See Milkweed beetle
Tetraploids, 33, 287
Tetrapod vertebrate classes
amniotes to mammals, 146–151
amphibians to amniotes, 145–146
dinosaurs to birds, 142–154
origin of mammals, 151–152
Sarcopterygii to amphibians, 143–145
Thalassoma bifasciatum. See Bluehead wrasse
Thamnophis elegans. See Garter snake

Thamnophis ordinoides. See Garter snake
Theist position, evolution and, 9
Theory, defined, 11
Thermal acclimation, 531–533
Thermocline, 63–64
Thomomys bottae. See Pocket gopher
Thorius narisovalis. See Salamanders
Three-point test cross, 39
Threshold traits, 43, 253
evolution of, 440–442
model of, 43
Thresholds, 662–663, 674
Thrinaxodon, 148
skull and skeleton, 149
Thrust, 520
Thysanura, 102
Tibicen sayi. See Cicada
Time-averaged sample, 131
Tissue interactions, changes in, 662
Tobacco, inheritance of continuously varying trait in, 42
Tolerance, evolution of, 531–535
Total variance, 756
Trace fossils, 172
Tracheophyta, 178–179
Trachops cirrhosus. See Frog-eating bat
Trade-offs, 424, 519, 564, 566
between number and size of offspring, 572
theory of life history evolution and, 566–568
Tragopogon spp. *See* Goatsbeards
Trait group, 550
Transcription complex, 47
Transcription factors, 47
Transcription of DNA, 44
Transcriptional control
genetic, 47
model of, 48
Transfection, 52
Transfer RNA (tRNA), 44. *See also* Ribonucleic acid
Transformational theory, 21, 232
vs. variational, 22
Transformations, of morphology, 655–656
Transgression, 136
Transilience, 483
Transition
in point mutation, 268
sequence evolution and, 627
Translation, of genetic code, 45
Translocations, chromosomal, 288–289
Transposable elements, 52, 269–271, 639–641
evolutionary dynamics of, 640–641
Transposons, 270
Transversions
in point mutation, 268
sequence evolution and, 627
Tree of life, 171
Treecreepers, as sibling species, 449
Tree-fungus beetle, 67
Treehoppers, sympatric speciation in, 501–502
Trends. *See also* Evolutionary trends
active, 692–694
discriminating among, 694

due to chance, 692
global trends in evolutionary history, 694–699
kinds of, 691
passive, 692
Triassic, 182, 184
Tribolium castaneum. See Flour beetles
Triceratops, 189–190. *See also* Dinosaurs
Trichoptera, 102–103
Trickle gene flow, 316
Trilobites, 174
    character change in, 134
    peripatric speciation in, 494
    speciation in *Phacops rana*, 135–137
Trimerophytes, 179
Triops cancriformis. See Tadpole shrimp
Triploids, 286
Triticum dicoccoides. See Wild wheat
tRNA. *See* Transfer RNA
Trophic level, coexistence of species at, 76–77
Tropical environments, species diversity and, 222–225
Tropical scrub forest, 61–62
Tropical seasonal forest, 61
Tropical wet forest, 61
Truncation selection, 251, 253, 412, 418
Tuatara, 204
Tundra, 63
Túngara frogs, sexual selection in, 345–346, 490
Turnover, 711
Turtles, replacement in, 716–717
Type specimen, 90
Typological notion of species, 448
Tyrannosaurus, 188–189. *See also* Dinosaurs

U

Ulva. See Sea Lettuce
Unbiased conversion, 638
Underdominance, 391
Unequal crossing over, 290–291, 292, 635–636
concerted evolution and, 637, 639
Ungulates, adaptive radiation of, 194–196
Uniform traits, genetic variation in, 253
Uniform trends, 691
Uniformitarianism, 18
Unreduced gametes, 286
Unresolved taxonomic relationships, in phylogenetic trees, 99
Unstable equilibrium, 392
Useless structures, evolutionary loss of, 529

V

Valley, in adaptive landscapes, 403
Variable selection, 385–389, 438
Variables, correlations among, 756–757
Variance, 69, 159, 248, 250–251
    additivity of, 251
    allele frequency and, 300–301, 303
    analysis of, 756
    statistical, 755–756
Variance selection, 422
Variation, 4. *See also* Genetic variation; Geographic variation; Phenotypic variation
    in breeding systems, 619–620
    coarse-grained vs. fine-grained, 386–387
    Darwin on, 6
    at the DNA level, 244–245, 375, 627–629
    estimating components of, 248–249, 251
    polygenic, 398
    in quantitative traits, 247–249, 251–253
    recombination and, 283–286, 608–611
    within and among species, 323
Variation among populations
    genetic distance, 256–257
    geographic, 253–256
    geographic variation patterns, 257–259
geographically variable characters, 259–262
Variation and Evolution in Plants, 25
The Variation of Plants and Animals Under Domestication, 21
Variational theory of evolutionary change, 21, 232
Variety, defined, 450
The Various Contrivances by Which Orchids are Fertilised by Insects, 21, 338
Vectors, 548
Vegetative propagation, 34, 606
Vertebrates. *See also* Cambrian revolution; *specific species*; Tetrapod vertebrate classes
    extinct Paleozoic classes, 178
    first known, 177–178
    Mesozoic, 187–190
    phylogeny of, 91, 106
Vestiges of the Natural History of Creation, 19, 20
Vestigial characters, 122–123
Viability selection, 370
Vicariance biogeography, 208–209
    of midges, 212–213
Vicariance hypothesis, 208–209
Vicariant speciation, 483
Victims, coevolution with enemies, 542–548
Viola spp. *See* Violets
Violets, leaf forms, 293
Virogenes, 293–294, 626
Virulence
    in natural populations of parasites, 551–552
    theory of evolution of, 549–551
Vitalism, 282n
von Baer, Karl Ernst, 652
von Baer's law, 652–653

W

Waddington, Conrad, on canalization, 441–442
Wallace, Alfred Russel, 20, 22, 232, 704
    on biogeographic provinces, 65
    importance of biogeography to, 201–203
on natural selection, 337
Water hyacinths, evolution of self-fertilization in, 458–459
Weaver ants, 341
Weinberg, W., 236
Weismann, August, 23, 232
    on adaptiveness of sex, 605
Whipnose, 64
White-crowned sparrows, song dialects as culturally inherited traits in, 579–580
Wild oat, self-fertilization in, 309–310
Wild wheat, outcrossing in, 310–311
Wild-type *Drosophila*, 42
Wohlbachia, as endosymbiotic species, 542
Wright, Sewall, 24–25, 370
    on adaptive landscapes, 392, 402
    definition of selection, 349
    on estimating gene flow, 317
    on fixation index, 314
    on inbreeding theory, 307
    on population genetics, 297
    on shifting balance theory, 408
    on threshold traits, 43
Wyeomyia smithii. See Pitcher-plant mosquito

X

Xenodontine snake, 210–211

Y

Yangchuanosaurus shangyouensis, 131
Yeast, rRNA structure in, 170
Yucca moths, mutualism in, 552–554

Z

Zacryptocerus varians, 340–341
Zygote, 33
Zygothrica, heads of male flies, 115
Zygotic selection, 368–369

About the Book

Editor: Andrew D. Sinauer
Project Editors: Kerry L. Falvey, Carol J. Wigg
Copy Editor: Norma Roche
Production Manager: Christopher Small
Book Production: Maggie Haddad
Art: Precision Graphics, Nancy Haver, Abigail Rorer
Book Design: Jean Hammond
Cover Design: MBDesign
Book Manufacturer: Courier Westford, Inc.
Cover Manufacturer: Henry N. Sawyer Company, Inc.

 Index